extra-tropical South African Orchids. London (William Wesley) 1893-1913, 3 vols. Oct. (*Icon. orchid. austro-afric.*). → **Abbreviation of short-title for use in taxonomic publications**

→ **pagination of parts and their dates of publication**

vol.	part	*plates*	dates	pages
1	1	*1-50*	15 Aug 1893	[i]-vi, [1, note], l.p., [2 p. ind.]
	2	*51-100*	20 Aug 1896	[i-iii], l.p., [3 p. ind.], app. [2]p., preface and t.p. to vol. 1 [viii]p.
2	1	*1-100*	Apr-Jun 1911	[i-vi], l.p. [3 p. ind.]; page iv: Apr 1911, Nat. Nov. Nov 1911, BR inscr. 15 Jun 1911.
3	1	*1-100*	1913	[i-v], l.p.; page v: 12 Feb 1913, Nat. Nov. Dec 1913.

→ **Special notes about this book**   **Libraries where copies seen by authors**

The plates are accompanied by one or two pages letterpress. *Copies*: BR, HH. – 36 plates of vol. 3 were first published by H. Bolus in his "Orchids of the Cape Peninsula," Trans. South Afr. Philosoph. Soc. 5(1): 75-200. 1888. The 300 (partly) coloured plates are by H. Bolus and F. Bolus. F. Bolus edited the third volume; preface by H. M. L. Bolus. – Portrait as frontispiece to vol. 2.
Ref.: BM 1: 193, 6: 106; Kew 1: 291; NI 197. → **References concerning this book**
Tyrrell-Glynn and Levyns, Flora africana 7. 1968.

birth/death dates

Bolus, Harriet Margaret Louisa (née **Kensit**) (1877-1970), South African botanist, daughter-in-law of Harry Bolus. (*L. Bolus*). → **Proposed citation form as author**

HERBARIUM and TYPES: BOL, other material BM, GRA, NBG, PRE, SAM.
Ref.: IH 2: 82. → **References to collections**
Tölken, Index herb. austro-afr. 1971.

→ **References in standard biographical and bibliographical works**
BIBLIOGRAPHY and BIOGRAPHY: Barnhart 1: 214; BL 1: 28, 54; BM 6: 106; Bossert p. 44; Kew 1: 291; Langman p. 391.   **Articles in books, journals etc.** ←
Anon., Flowering plants South Africa 23. 1943 (portr.).
Herre, Kakteen und andere Sukkulenten 12(1): 1-3. 1961 (portr.).
E.G.H.O., Forum botanicum 8: 68-69. 1970 (Died at Cape Town on 5 Apr 1970. Harry Bolus was her uncle and father-in-law).
Levyns and Jessop, J. S. Afr. Bot. 36(4): 319-330. 1970 (bibl.).
Herre, Kakt. Sukk. 21: 139, 170, also in The genera of the Mesembryanthemaceae 46-47. 1971 (portr.).
Glen, Aloe 13(3): 84, 87. 1975 (portr.).

→ **Generic and journal names based on names of person**
EPONYMY: *Bolusanthemum* Schwantes (1928). – *Note*: For other eponyms based on the name Bolus, cf. supra, sub Harry Bolus.

→ **Book number**   **short-title**
626. *Notes on Mesembryanthemum and allied genera.* Cape Town 1928-1958. Oct. (*Notes Mesembryanthemum*). → **Abbreviation of short-title**
Part I: 1 Jul 1928 (reprinted from articles appearing in S. A. Garden & Country Life August-December, 1927), reprint 1928 (156 pp.).

Part II:

→ **Data on place of publication, dates of whole or parts, etc.**

pages	dates	pages	dates	pages	dates
1-16	9 Nov 1928	147-160	22 Nov 1929	309-336	29 Jan 1932
17-32	21 Dec 1928	161-176	20 Feb 1930	337-356	24 Jun 1932
33-48	24 Jan 1929	177-192	9 Mai 1930	357-376	6 Dec 1932
49-64	12 Apr 1929	193-208	15 Aug 1930	377-396	19 Mai 1933
65-80	3 Mai 1929	209-224	12 Nov 1930	397-416	16 Oct 1933
81-94	6 Jun 1929	225-244	12 Feb 1931	417-436	26 Jan 1934
95-110	4 Jul 1929	245-268	1 Mai 1931	437-452	23 Mai 1934
111-129	16 Aug 1929	269-292	3 Jul 1931	453-472	17 Aug 1934
131-146	4 Nov 1929	293-308	24 Sep 1931	473-508	11 Feb 1935

TAXONOMIC LITERATURE

Volume IV: P-Sak

Regnum vegetabile, a series of publications for the use of plant taxonomists published under the auspices of the International Association for Plant Taxonomy, edited by Frans A. Stafleu.

Volume 110

Frans A. Stafleu and Richard S. Cowan

Taxonomic literature

A selective guide to botanical publications and collections with dates, commentaries and types

Volume IV: P-Sak

Second edition

Bohn, Scheltema & Holkema, Utrecht/Antwerpen
dr. W. Junk b.v., Publishers, The Hague/Boston
1983

© 1983, Frans A. Stafleu, Utrecht, The Netherlands
No part of this book may be reproduced by film, microfilm or any other means without written permission from the publisher

Complete work ISBN 90 313 0224 4; Volume 4 ISBN 90 313 0549 9

D/1983/3407/22

Library of Congres Cataloging in Publication Data
(Revised for volume 4)

Stafleu, Frans Antonie.
  Taxonomic literature.

  (Regnum vegetabile; v.   )
  Includes bibliographical references and indexes.
  1. Botany–Classification–Bibliography. 2. Botany–Catalogs and collections. 3. Botanists–Biography–Bibliography. 4. Plant collectors–Biography–Bibliography. 5. Botany–Nomenclature. 6. Botanical literature. I. Cowan, Richard S., joint author. II. Title. III. Series.
QK96.R4   vol. 110   581′.012s   77-468327
ISBN 90-313-02244 (set) [016.581′012]

LIBRARY

Contents

Introduction	VI
Abbreviations	VIII
Taxonomic literature and collections P-Sak	I
Index to titles	1083
Index to names	1159

Introduction

The fourth volume of TL-2 represents a further step in the evolution of the technology used to produce this work. Thanks to the imaginative approach of the publishers to our work we are now preparing the text with modern word processing equipment so that we produce a manuscript to work with at the same time as a direct instruction to a CPT 6000 composer in the printing shop which produces the type for an entire volume in five or six hours.

The first volume was produced from ordinary manuscript by means of type setting, in lead, on a Monotype. The second volume was also produced from typed manuscript but by means of an early phototype composer, Monophoto 400/8, which was essentially working in the same way as the Monotype. Volume three, still produced with a typewriter by us, was processed by the printer by means of Lasercomp 3000.

The fourth volume therefore is typeset in fact by ourselves with phototype being produced at the printers on the basis of our coded instructions on magnetic disks. TL-2, as published so far, therefore illustrates the to many of us extraordinary and rapid evolution of typesetting and printing during the past decade. Our making the fullest use of that development was made possible by the technical manager of Bohn, Scheltema and Holkema, Mr. D.G. den Besten, whose dedicated input and imaginative approach to our problems has been invaluable. In the technical execution he was aided by Rudo Romijnders, of the printers, Koninklijke Drukkerij Van de Garde B.V. Apart from the growing technological help we are pleased to report that the support given by the botanical community to our work is still growing. This support comes directly from the users (scientists as well as librarians) and from institutions and foundations.

In the United States we have the benefit of the liberal support by the National Science Foundation and of the New York Botanical Garden as the administrator of the N.S.F. grant. This support is now secure until the end of the project on 31 December 1987. In the Netherlands we enjoyed continued support by the Netherlands Foundation for pure research (Z.W.O.). This Foundation support from both sides of the Atlantic Ocean is to us the best proof that 'Taxonomic Literature' is well received by the world community of botanists.

The librarians at our main libraries continue to give us all the help we ask for and we realize that we are asking a great deal of them. We therefore once more express our gratitude to: S. Baker (Academy of Natural Sciences, Philadelphia), P. v.d. Berg and J. v.d. Kaa (Institute of systematic Botany, Utrecht), J. Brenan (New York Botanical Garden), H.V. Burdet and Mme Dubugnon (Conservatoire et Jardin botaniques de Genève), R. Clarysse and Mme Dumont (Jardin national de Botanique, Meise nr. Bruxelles), A. Fusonie (National Agricultural Library, Beltsville), I. Haesler (Botanische Staatssammlungen, München), G. Kaye (Farlow Herbarium of Harvard University), R. Kiger (Hunt Library, Pittsburgh), C. Lange (Missouri Botanical Garden), L. Linus

(New York Botanical Garden), C.R. Long (New York Botanical Garden), J. Marquart (Smithsonian Institution), W. Olsen (National Agricultural Library, Beltsville), Ruth Schallert (Smithsonian Institution), E. Schwabe, K. Stehr and (for volume 5), P. Hirsch (Botanisches Museum, Berlin-Dahlem), M.E. Thomas (Gray-Arnold Libraries of Harvard University). In addition to professional librarians who give liberally of their time and knowledge, there are many other contributors who help by finding all manner of obscure data – historians, bibliographers, taxonomists, students, and friends. Among the scores who have helped us, we mention the following with much gratitude: T. Ahti (Helsinki), O. Almborn (Lund), T.M. Barkley (Manhattan, Kan.), S. Bluhan (Washington), C.D.K. Cook (Zürich), M.R. Crosby (St. Louis, Mo.), W.A. Deiss (Washington), M.F. Denton (Seattle, Wash.), S. Eichhorn (Washington), J.J. Engel (Chicago), D. Ernet (Graz), J. Ewan (New Orleans, La.), E. Farr (Washington), G.F. Follmann (Kassel), R. Grether (Mexico City), W. Greuter (Berlin), P. Holberton (Washington), P. Isoviita, J. Jalas, T. Kopponen, I. Kukkonen (Helsinki), H.W. Lack (Berlin), K. Mägdefrau (München), J. Mears (Philadelphia), D.H. Pfister (Cambridge, Mass.), R.W. Pohl (Ames, Iowa), R.K. Rabeler (East Lansing, Mich.), D.P. Rogers (Urbana-Champaign, Ill.), E.D. Rudolph (Columbus, Ohio), G. Sayre (Cambridge, Mass.), R. Schmidt (Berkeley, Calif.), W.T. Stearn (London), R.L. Stuckey (Columbus, Ohio), J.H. Thomas (Stanford, Calif.), R.A. Townsend (Sydney), E.G. Voss (Ann Arbor, Mich.), G. Wagenitz (Göttingen). To our deep regret we lost in C. Steinberg of the Istituto botanico, Firenze, and S.J. Lipschitz, Komarov Botanical Institute, Leningrad two of our most dedicated collaborators; their death was a blow to us as well as to their many friends and colleagues and to plant taxonomy in general.

With our documentation becoming more complete the relative number of smaller publications increases. Many of these are independently paginated reprints or preprints from papers published in regular journals. In the case of preprints we supply abbreviated titles because these are independent publications having priority over the journal publication. For reprints we do not provide abbreviated titles because such texts should be cited from the journals.

We should like to point out once more that we deal with the taxonomic literature (in a wide sense) between 1753 and 1939. Publications by authors who published both before and after 1939 are treated only in a regular fashion if dating from before 1940. However, we may add additional information on important works of such authors published outside our period in notes or under 'Composite works'. Also, in order to make full use of the information presented by us it is good to scan the paragraphs 'Biography and bibliography' and 'Biofile' for references to bibliographies of the relevant authors. The paragraphs now called 'Biofile' contain references to publications on the authors and their works which are not referred to by means of abbreviations; the latter appear under 'Biography and Bibliography'.

References to manuscripts and correspondence are frequently given under 'Herbarium and types'. These references are incidental; we can alas not make an effort to give extensive coverage to archival material in view of the restricted time available. With the publication of this fourth volume we are still on schedule. We still hope to reach the end of the alphabet before 1988, volume 5 in the second half of 1985, and volume 6 in the second half of 1987. The seventh volume will contain indexes, errata and additions as well as, perhaps, an overflow from the last part of the alphabet. The text for this volume, as far as our part is involved, should be ready by the end of 1987. We realize that much is still to be done in this short span of time but with the heartwarming support from our colleagues, librarians and publishers we hope to able to reach our goals without undue delay.

Utrecht, Washington
July 1983

Additional Abbreviations

This list contains abbreviations not listed in volumes one to three. The abbreviations of herbaria are those used in the Index herbariorum vol. 1, ed. 7, 1981.

B-G	Bibliothek Deutsche Gartenbau-Gesellschaft, Technische Universität, Berlin (W).
B-TU	Bibliothek Technische Universität Berlin (W).
Biol.-Dokum.	Biologie Dokumentation der deutschen biologischen Zeitschriftenliteratur 1796-1965.
Christiansen	CHRISTIANSEN, W. & W., Botanisches Schrifttum Schleswig-Holsteins, 1936.
Ewan ed. 1	EWAN, J. Rocky Mountain naturalists, Univ. Denver Press, Denver, Colo. 1950.
Ewan ed. 2	EWAN, J. and N.D., Biographical dictionary of Rocky Mountain naturalists, Regnum vegetabile 107, Utrecht/Antwerpen 1981.
H-UB	Helsinki University Library, Helsinki, Finland.
NSUB	Niedersächsische Universitätsbibliothek, Göttingen, B.R.D.
SIA	Smithsonian Institution Archives, Washington, D.C., U.S.A.

Pabst, Carl (1825/6-1863), German (Saxonian) botanist from Halle; collected plants for Van Houtte at the Cape Verde island Mayo (Jun 1846) and subsequently on the Brazilian island of Santa Catharina; settled as a colonist in Itajahi in the Donna Franziska settlement of the Hamburger Colonisations-Gesellschaft. (*C. Pabst*).

HERBARIUM and TYPES: B (extant), GOET, HAL.

BIBLIOGRAPHY and BIOGRAPHY: Rehder 5: 644; Urban-Berl. p. 309, 322, 380.

BIOFILE: Anon., Bot. Zeit. 10: 64. 1852 (Brazilian colls.); Österr. bot. W. 2: 44. 1852. Schlechtendal, D.F.L. von, Bot. Zeit. 21: 328. 1863 (d., 37 yrs. old). Werner, K., Wiss. Z. Univ. Halle, Math.-Nat. 4(4): 777. 1955. Braz. coll. at HAL).

Pabst, Gustav (x-1911), German (Thuringian) botanist, artist and surveyor at Gera. (*G. Pabst*).

HERBARIUM and TYPES: HB.

NOTE: Not to be confused with the Brazilian botanist Guido Frederico João Pabst (1914-1980); see e.g. Orchid Rev. 88: 158. 1980, Amer. Orch. Soc. Bull. 49: 881. 1980; Kakt. Orchid. Rundschau 4: 55. 1980; Orch. Dig. 44: 136. 1980; Taxon 29: 702. 1980, 30: 559-561. 1981; Bradea 3: 65-76. 1980 (bibl.); Orquideologica 14(2), in memoriam, 1981; Brittonia 32: 329. 1980; J. Bromeliad Soc. 31: 220-221. 1981.

BIBLIOGRAPHY and BIOGRAPHY: BM 4: 1405; DTS 1: 216; Kelly p. 165; Kew 4: 208; LS 19695; NI 1427, 1479, 1771; Rehder 5: 644; SBC p. 128 (Pab.); TL-2/6526; Tucker 1: 534.

BIOFILE: Anon., Hedwigia 52: (82). 1912 (d. Aug 1911); Bradea 3(10): 65-76. 1980 (portr.).

COMPOSITE WORKS: Editor of Köhler's *Medizinal-Pflanzen* (TL-2/3806).

EPONYMY: *Pabstia* Garay (1973) is named for the botanist Guido F. Pabst (1914-1980), director of the Herbarium Bradeanum, Rio de Janeiro, Brazil, see note above.

7175. *Cryptogamen-Flora* enthaltend die Abbildung und Beschreibung der vorzüglichsten Cryptogamen Deutschlands und der angrenzenden Länder. Die Flechten und Pilze mit circa 900 Abbildungen in Farben- und Schwarzdruck auf 39 Tafeln und mit 19 in den Text gedruckten Holzschnitten. I. Theil: Flechten von O. Müller und G. Pabst. II. Theil: Pilze. Herausgegeben von G. Pabst. Gera (C.B. Griesbach's Verlag) 1875-1876. Qu. (*Crypt.-Fl.*).
Co-author (part 1): Walther Otto Müller (1833-1887); see TL-2/6526.
Preface material: 1876, p. [i*, iii*] [i]-iv.
1 (*Flechten*): Apr 1874 (Bot. Zeit. ann. 6 Mar 1874; Hedwigia Apr 1874; publ. Bot. Zeit. 1 Mai 1874), v-xxviii, [xxix], *pl. 1-12* (uncol. liths.).
2 (*Pilze*): 1875 (p. [v]: Jul 1875; Flora 21 Sep 1875; Hedwigia Sep 1875), published in parts, parts 1/2: p. 1-16, *pl. 1-6*, p. [i, v], [1]-98, *pl. 1-7, 7a, 8-21, 21a, 22-23* (col. liths.).
Copies: B, B-S, BR, G, MICH, NY, Stevenson, USDA. – For part *3* (Moose, *1*:) Lebermoose: see below. – Asher (Cat. 14, no. 135, 1974) offered a complete copy at D.fl. 950.= (US $ 400.=).
Ref.: R., Flora 57: 336. 21 Jul 1874 (rev. *Flechten*), 58: 430-431. 21 Sep 1875 (rev. 1, 2); Österr. bot. Z. 24: 126. 1874, 25: 368-369. 1875.

7176. *Die Lebermoose* mit circa 500 Abbildungen in Farben- und in Schwarzdruck. Herausgegeben von G. Pabst. Mit Zeichnungen von W.O. Müller und G. Pabst. Gera (C.B. Griesbach's Verlag) 1877. Qu. (*Lebermoose*).
Artist: Walther Otto Müller (1833-1887).
Publ.: 1877 (Hedwigia Jan 1878), p. [i, iii], *1 pl.*, [1, expl. pl.], [1]-36, *pl. 1-8* (partly col. liths.). *Copies*: B, B-S, MO, USDA. – Also issued as part 3(1) of the *Cryptogamen-Flora*.

– The B copy has a different t.p.: "*Die Moose* mit Abbildungen in Farben- und Schwarzdruck herausgegeben von G. Pabst. Mit Zeichnungen von W.O. Müller und G. Pabst. i. Abtheilung: *Lebermoose.*" Gera (C.B. Griesbach's Verlag) 1877.

Paccard, Ernesto (*fl.* 1905), Uruguayan botanist. (*Paccard*).

HERBARIUM and TYPES: Unknown.

BIBLIOGRAPHY and BIOGRAPHY: BL 1: 259, 311; Tucker 1: 534.

7177. *Lista de algunas plantas medicinales* de las Repúblicas Oriental y Argentina ... Montevideo (Talleres A. Barreiro y Ramos ...) 1905. Oct. (*Lista pl. med.*).
Publ.: 1905, p. [1]-77, *23 pl.* (col.). *Copy*: USDA.

Pacher, David (1816-1902), Austrian clergyman and botanist; ord. 29 Jul 1840; in various clergical ranks in Carinthia 1841-1902, at Leoben ober Gmünd 1840-1843, Glödnitz and St. Lorenzen 1843-1845, at Sagritz 1845-1851, at Tröppolach 1851-1861, at Tiffen 1861-1875, from 1875-1902 dean at Obervellach. (*Pacher*).

HERBARIUM and TYPES: KL; further material BASSA, IBF, REG, WRSL. – Some letters at G.

BIBLIOGRAPHY and BIOGRAPHY: AG 6(1): 355, 12(2): 618; Barnhart 3: 40; BFM nos. 862, 863; BM 4: 1496; CSP 4: 732, 12: 553, 17: 667; DTS 1: 216, 6(4): 32, 182; Hegi 4(1): 384, 6(2): 1087; IH 1 (ed. 6): 362, (ed. 7): 339, 2: (in press); Kew 4: 209; Morren ed. 10, p. 34; Rehder 5: 644.

BIOFILE: Anon., Bot. Jahrb. 32 (Beibl. 72): 4. 1903; Bot. Not. 1903: 34; Flora 26: 404. 1843; Österr. bot. Z. 52: 291. 1902 (d. 28 Mai 1902).
Fürnrohr, A.E., Flora 31: 565. 1848 (pl. at REG).
Hausmann, F. v., Fl. Tirol. 1164. 1851.
Janchen, E., Cat. fl. austr. 1: 31, 34. 1956.
Jabornegg, M. von, Carinthia 92: 93-98. 1902 (b. 5 Sep 1816, d. 29 Mai 1902).
Kneucker, A., Allg. Bot. Z. 8: 172. 1902.

7178. *Flora von Kärnten* von Dechant David Pacher und Markus Freiherr von Jabornegg. Herausgegeben vom naturhistorischen Landesmuseum von Kärnten. I. Theil Gefässpflanzen ... Klagenfurt (Druck von Ferdinand v. Kleinmayer) 1881-1894, 3 parts, suppl. Oct. (*Fl. Kärnten*).
Authorship : David Pacher ("Dechant" = dean) was author of all parts except the
. *Verzeichniss* of vernacular names in part 3 which was by Gustav Adolf Zwanziger.
   Markus (Freiherr) Jabornegg von Gamsenegg und Möderndorf (1837-1910) was collaborator but is not cited as author on the alternative t.p.'s.
Alt. title: Systematische Aufzählung der in Kärnten wildwachsenden Gefässpflanzen bearbeitet von Dechant David Pacher. – Also published (and mostly to be cited from) as "Hefte" of the Jahrb. Naturhist. Landes-Mus. Kärnten ("*Jahrb.*"); the publication in the journal is often earlier than the reprint.
1 (Akotyl., Monokotyl.): late 1880 or Jan-Feb 1881 (in *journal*: Nat. Nov. Feb(2) 1881; Bot. Zeit. 25 Mai 1881; *reprint* 26 Mai 1882, id. Nat. Nov. Mai(1) 1882; BC Apr 1882; Bot. Zeit. 26 Mai 1882; t.p. reprint: 1882), p. [ii-iii], [1]-257, [258, err.], [i]-viii (index). – *Jahrb.* Heft 14: i-[iii], [1]-257, [258] 1880.
2 (Dikotyl. Conif.-Hypopit.): Jan-Apr 1884 (Bot. Zeit. 27 Jun 1884; Nat. Nov. Mai(1) 1884), p. [ii-iii], [1]-353, [i]-iii, err., add., [iv]-xvi index. – *Jahrb.* Hefte 15: [i], [1]-192. 1882, 16: [i], [1]-161. 1884.
3 (Dikotyl. dialypet.; Verz.): 1887 (Nat. Nov. Feb 1888; Bot. Centralbl. 30 Jan-3 Feb 1888), p. [ii-iii], [1]-420, [i]-xvii, (index, add.), [i]-xxix. – "Verzeichniss der in Kärnten volksthümlichen deutschen Pflanzennamen" von G.A. Zwanziger. – *Jahrb.* Hefte 17: [47]-216. 1885, 18: [81]-284. 1886, 19: [i], [1]-83. 1888.
Nachträge: Nov-Dec 1894 [ÖbZ Nov-Dec; Bot. Zeit. 16 Jan 1895; Nat. Nov. Jan 1895;

Bot. Centralbl. 9 Jan 1895), p. [i], [1]-235, [236, err., add.]. – *Jahrb.* Hefte 22: 25-160. 1893, 23: [85]-184. 1894.
Copies (reprint): BR, FI, G, GJO, MO, NY. – (Journal issue: B (p.p.)).
Note: we are grateful to D. Ernet, Graz, for information on this publication.
Ref.: Freyn, J.F., Bot. Centralbl. 7: 75-76. 1881, 18: 239-241. 1884, 36: 173-174. 1888 (rev.).
Wiesbaur, Österr. bot. Z. 38: 65-67. 1888.

Packard, Alpheus Spring, (1839-1905), American geologist, zoologist and botanist; MD Harvard 1864; professor of zoology and geology at Brown University 1878-1905; secretary of the U.S. Entomological Commission (to investigate the Rocky Mountain locust) 1876; Dr. med. Bowdoin 1864, LL.D. id. 1891. (*Packard*).

HERBARIUM and TYPES: Unknown. Entomological colls. at Mus. comp. Zool. Harvard. Correspondence and photograph at Smithsonian Archives.

BIBLIOGRAPHY and BIOGRAPHY: Barnhart 3: 40; BM 4: 1496-1497, 8: 969; CSP 4: 733, 8: 549-550, 10: 978-979, 12: 553-554, 17: 668-670; De Toni 1: xcvi; Ewan, ed. 1: 276, ed. 2: 166; Herder p. 392; Kew 4: 210; Langman p. 560; Lenley p. 319. ME 1: 217, 3: 633 (publications); Nickles p. 807-808; NW p. 54; PH p. 509; Quenstedt p. 325; Rehder 5: 644; SO 1248C; Tucker 1: 534.

BIOFILE: Allibone, S.A., Crit. dict. lit. 1480. 1870.
Anon., Appleton Cycl. Amer. biogr. 4: 619-620. 1888 (biogr. of father (b. 23 Dec 1798, d. 13 Jul 1884 and son A.S.P. b. 19 Feb 1839)); Canad. Entom. 37: 111-112. 1905 (obit.); Entom. News 16(4): 97-98, *pl. 6*, 1905 (obit., portr.).
Benjamin, M., Harper's Weekly 34: 925-926. 1890 (Agassiz's pupils).
Carpenter, M.M., Amer. midl. Natural. 33: 76. 1945 (refs. to biogr.).
Cockerell, T.D.A., Biogr. Mem. Natl. Acad. Sci. 9: 181-236. 1920 main biogr., portr., bibl., b. 19 Feb 1839, d. 14 Feb 1905).
Dexter, R.W., Bull. Brooklyn Entom. Soc. 52: 57-66, 101-112. 1957.
Essig, E.O., Hist. entom. 727-729. 1931 (biogr., bibl., portr., biogr., refs.).
Geiser, S.W., Natural. Frontier 18, 210, 233, 235, 237. 1948.
Gilbert, P., Comp. biogr. lit. deceased entom. 285. 1977. (q.v. for further biogr. refs.).
Glass, B. et al., Forerunners of Darwin 79, 84. 1968.
Henshaw, S., Bull. Div. Entom. USDA 16: 1-44. 1887 (bibl.) (biogr., entom. bibl. of 339 nos.; b. 19 Feb 1839).
Holland, W.J., Entom. mon. Mag. 41: 140-141. 1905 (obit.).
Howard, L.O., Hist. appl. entom. (Smiths. misc. coll. 84) 14-19, 81-83, 93, 102, 169, 172, 177, 207, 473, 537, *pl. 2*. 1930.
Kingsley, J.S., Pop. Sci. Mon. 33: 260-267. 1888 (biogr.).
Kingsley, J.S. & C. Barus, Science ser. 2. 21: 401-406. 1905 (obit.).
Kuznezov, N.Y., Rev. russe Entom. 5: 189-190. 1905.
Mallis, E., Amer. Entom. 296-302. 1971 (portr.).
Mead, A.D., Pop. Sci. Mon. 67: 43-48. 1905 (biogr., portr.).
Osborn, H., Fragm. entom. hist. 1: 1, 29, 34, 35, 52, 99, 100, 147, 148, 188, 205, 211, 238-240, 244, 314, *pl. 10*.
Packard, A.S., The Labrador coast, a journal of two summer cruises ... New York 1891, vi, 513 p.
Poggendorff, J.C. Biogr.-Lit. Handw.-Buch 3: 997-998. 1898, 4(2): 1108. 1904 (bibl.), 5(2): 933. 1926.
Russell, F.W., Mount Auburn biographies 1953, p. 123-124.
Smith, J.B., Psyche, Cambr. 12: 33-35. 1905 (portr.).
Weiss, H.B., Pioneer cent. Amer. entom. 241-244. 1936.
W.F.K., Entomologist 38: 143-144. 1905.
Woodward, A.S., Proc. Linn. Soc. London 1905: 45-46. (obit.).

COMPOSITE WORKS: *American Naturalist*, editor vols. 1-20, 1868-1886.

7179. *The Labrador Coast* a journal of two summer cruises to that region. With notes on

its early discovery, on the Eskimo, on its physical geography, geology and natural history, ... New York (N.D.C. Hodges, publisher, ...), London (Kegan Paul, French, Trübner & Co.) 1891. Oct. (*Labrador Coast*).
Publ.: 1891, frontisp., p. [1]-7, 1-513, *5 pl.*, 5 maps. *Copies*: MICH, MO, NY. – "Catalogue of the plants, compiled by John Macoun" on p. 451-474.

Packe, Charles (1826-1896), British naturalist and mountaineer of independent means in Leicestershire; B.A. Oxford 1849; explorer of the Alps, Pyrenees and the Sierra Nevada. (*Packe*).

HERBARIUM and TYPES: CGE; further material in E-GL and MANCH.

BIBLIOGRAPHY and BIOGRAPHY: Barnhart 3: 40; BB p. 234; CSP 8: 550; Desmond p. 475; DTS 1: 216; IH 2: (in press); Jackson p. 278; Rehder 5: 644.

BIOFILE: Anon., J. Bot. 35: 415-416. 1897 (herb. to CGE); Proc. Linn. Soc., Lond. 1896/7: 66 (obit., d. 16 Jul 1896).
Hedge, I.C. & Lamond, J.C., Index coll. Edinburgh herb. 116. 1970.
Horwood, A.R. and C.W.F. Noel, Fl. Leicestershire ccxix. 1933.
Kent, D.H., Brit. herb. 71. 1957.
W.P.H.S., The alpine Journal 18: 236-242. 1897 (on P. and the Pyrenees).

7180. *A guide to the Pyrenees.* Especially intended for the use of mountaineers. London (Longman, Green, Longman, Roberts & Green) 1862. (*Guide Pyrenees*).
Ed. 1: 1862, p. [i]-xiv, [xv], [1]-130, frontisp., 3 maps, 1 diagr. *Copy*: Natl. Library Scotland (inf. J. Edmondson). – Main botanical items on p. 4, 54-55, 57, 81, 83; table climate and vegetation on p. 104.
Ed. 2: 1867, n.v.

Paczoski (Patschosky, Patschotsky), **Joseph Conradovich** (1864-1941) Ukranian-born Polish botanist; assistant at the Dept. Botany, Univ. Kiew 1888-1894, id. at the botanical garden St. Petersburg 1894-1895, at Dublany 1895-1897; director of the Natural History Museum, Cherson 1897-1920 and professor at the Polytechnic ib. 1918-1922; director of the Steppe Reserve at Ascania Nova 1922-1923; repatriated to Poland 1923; director National Park Bialowieza 1923-1928; professor of phytosociology at the University of Poznan 1925-1931; Dr. phil. h.c. Poznan 1932; from 1931 active as a private scientist; early phytosociologist and ecologist. (*Pacz.*).

HERBARIUM and TYPES: KRAM, other original material at LE; further material at C, E, POZ. – Some letters at G.

BIBLIOGRAPHY and BIOGRAPHY: Bossert p. 297; CSP 17: 671; Hirsch p. 222; Kew 4: 210; LS 19708; Rehder 1: 366, 443.

BIOFILE: Alechin, W.W., Bot. Not. 1924: 192-193 (P was the first to have a clear concept of phytosociology as of 1891).
Anon., Chron. bot. 7(7): 353. 1943 (d. 14 Feb 1942).
Hedge, I.C. & Lamond, J.M., Index coll. Edinb. herb. 116. 1970.
Kusnezov, N.J., Bot. Jahrb. 15 (Lit.): 69, 83-85, 87-88. 1892.
Lipschitz, S., Florae URSS fontes p. 222 [index]. 1975.Lipski, V.I., Imp. St. Petersb. bot. Zad. 3: 386-392. 1915 (biogr., bibl., portr.).
Maycock, P.F., Ecology 48: 1031-1034, 1967 (biogr., bibl.).
Pax, F., Bibl. schles. Bot. 139. 1929.
Radde, G., Grundz. Pfl.-Verbr. Kaukasusländ. 9, 10, 17. 1899 (Veg. Erde 3).
Szafer, W., Regn. veg. 71: 386, 387. 1970.
Szymkiewicz, D., Bibl. fl. polsk. 15, 110-111. 1925.
Verdoorn, F., Chron. bot. 1: 240. 1935 (retirement; chair merged with general botany; 70th birthday 9 Dec 1934), 2: 255. 1936, 4: 445. 1938.
Voronow, A.G., Biol. Moskovsk. Obsh. Ispyt. Prir. 69: 130-134. 1964. (100th anniversary birth).

Wodziczko, A., Acta Soc. bot. Polon. 9 (suppl.): i-xiii. 1932 (tribute, bibl., b. 8 Dec 1864, pioneer of phytosociology).

NOTE: Paczoski's fundamental "Social life of plants" (Życie gromadne róslin) originally published in Wszechswiat 15, nos. 26, 27, 28, Warsaw 1896, was republished in Krakow, 1930 as Bibljoteka Botaniczna vol. 2, p. [1]-40. (*Copy*: U).

7181. Osnovniya cherti razvitiya flori yugo-zapadnoi Rossii ... *Grundzüge der Entwickelung der Flora in Südwest-Russland*. Kherson (Parovaya tipo-litografia nasl. O.D. Khodushinoi) 1910. Oct. (*Grundz. Entw. Fl. S.W.-Russl.*).
Publ.: 1910 (p. xxxiv: 19 Nov 1909; printer's mark on p. [iii*] 9, Feb 1910), p. [i*-iii*], [i]-xxxiv, [1]-430, map. *Copies*: BR, PH.

7182. *Lasy Białowiczy* (Die Waldtypen von Białowicża). Poznań (Nakladem panstwowes rady ochrony przyrody ...) 1930. (*Lasy Białowieczy*).
Publ.: 1930, p. [i-iv], [1]-175, *pl. 1-6*, map. *Copies*: FI, GB, GOET, H, Illinois Univ. – Państwowa rada och rony przyrody. Monografje naukowe nr. 1.

Page, William Bridgewater (1790-1871), British nurseryman at Hammersmith and Southampton. (*Page*).

HERBARIUM and TYPES: Unknown.

BIBLIOGRAPHY and BIOGRAPHY: Barnhart 3: 41; BB p. 234; BM 4: 1498; Desmond p. 475; Henrey 2: 357; Jackson p. 415; Kew 4: 211; PR 6892; Rehder 1: 53; SO 809a; TL-2/see J. Kennedy; Tucker 1: 534.

BIOFILE: Anon., Gard. Mag. 11: 60-61. 1835 (description of P's botanic garden and nursery).
Britten, J., J. Bot. 54: 241. 1916 (Kennedy was the real author).
Johnson, G.W., A history of English gardening 301. 1829 ("His father-in-law, Mr. Kennedy of the Hammersmith Nursery, is considered to be the author" of the [*Prodromus*]).

EPONYMY: *Pagella* Schoenland (1921) is named for Mary M. Page (1867-1925), botanical artist at the Bolus Herbarium, Cape Town.

7183. Page's *Prodromus*; a general nomenclature of all the plants, indigenous and exotic, cultivated in the Southampton Botanic Gardens: arranged, alphabetically, as they are considered hardy or tender to the climate of Britain, under their different characters of trees, shrub's, herbaceous, &c. The generic and specific names, after the Linnaean system; with the English names, propagation, soil, height, time of flowering, native country, &c. &c. Also, occasional hints for their cultivation. An appendix, containing selected lists of annuals; all the choicest kinds of fruit, now in general cultivation, with their characters, &c. and a short tract on the sexual system, from the Philosophia botanica of Linnaeus. By William Bridgewater Page, London (John Murray, ...) 1818. Qu. (*Prodromus*).
Publ.: 1818 (p. [3]: Mar 1818, p. [1]-279. *Copies*: MO, NY(2), USDA. – By tradition John Kennedy (1759-1842) is considered to be the author of the nomenclatural part of this work. – Tucker (1: 534) and Henrey (2: 357) mention an issue dated 1817 (not seen by us), giving 279 [281] p. for both.
Ref.: Anon., Flora 5: 509-510. 28 Aug 1822.

Paget, Sir James (1814-1899), British surgeon and botanist, baronet 1871, LL.D. Cambridge 1874. (*Paget*).

HERBARIUM and TYPES: NWH (1000) and K.

BIBLIOGRAPHY and BIOGRAPHY: Barnhart 3: 41; BB p. 234; BL 2: 260, 698; BM 4: 1499; CSP 4: 739, 8: 553, 10: 981, 12: 554, 17: 674 (q.v. for further biogr. refs. in the medical literature); Desmond p. 475; DNB suppl. 3: 240-242; IH 1 (ed. 7): 339, 2: (in press); Jackson p. 262, 493; Kew 4: 211; LS 19717; PH 1; Rehder 1: 398, 528.

BIOFILE: Allan, M., The Hookers of Kew 93, 221. 1967.
Anon., Bot. Centralbl. 81: 352. 1900; Bot. Not. 1900: 47; J. Bot. 38: 62. 1900, 24: 96. (herb. to NWH); Leopoldina 36: 52-53. 1900; Proc. Linn. Soc. London 112: 79-80. 1900 (obit., d. 30 Dec 1899); (J.H.), Proc. Roy. Soc. Lond. 75: 136-140. 1905 (obit.).
Babington, C.C., Mem. journ. bot. corr. lxxxix. 1897.
Bloomfield, E.N., Entom. mon. Mag. 36: 89. 1900.
Bridson, G.D.R. et al., Nat. hist. mss. res. Brit. Isl. 271. 45. 1980.
Huxley, L., Life letters J.D. Hooker 2: 558 [index]. 1918.
Jackson, B.D., Bull. misc. Inf. Kew 1901: 50 (pl. at K).
Kent, D.H., Brit. herb. 71. 1957.
Kneucker, A., Allg. bot. Z. 6: 76. 1900.
Paget, S., Encycl. brit. ed. 20. 20: 451-452. 1911 (see also index vol. p. 589); Memoirs and letters of Sir James Paget, 1901, ix, 438 p. (portr., biogr., letters) (p. 25-28 reprinted J. Bot. 42: 298-299. 1904).

EPONYMY: *Pagetia* F. v. Mueller (1866).

7184. *Sketch of the natural history of Yarmouth* and its neighbourhood, containing catalogues of the species of animals, birds, reptiles, fish, insects, and plants, at present known ... London (published by Longman, Rees, & Co., ...) 1834. (*Sketch nat. hist. Yarmouth*).
Co-author (for zoology): Charles John Paget (1814-1899), brother of J. Paget.
Publ.: Oct-Dec 1834 (p. xxxii: 11 Oct 1834), p. [i*], [i]-xxxii, [1]-88. *Copies*: E (inf. J. Edmondson), K, PH. – The PH copy has a different imprint: "Yarmouth: printed and published by F. Skill, Quay; sold in London by Longman, Rees and Co., ...".

Paglia, Enrico (1834-1889), Italian botanist at Mantua. (*Paglia*).

HERBARIUM and TYPES: FI.

BIBLIOGRAPHY and BIOGRAPHY: BL 2: 368, 698; Bossert p. 297; CSP 4: 740, 8: 553, 10: 981, 12: 554, 17: 675; Jackson p. 319; LS 19718; Quenstedt p. 325; Saccardo 1: 120, 2: 80.

BIOFILE: Strambio, G., R. Ist. Lomb., Rendic. ser. 2. 22: 112. 1889 (d.).

7185. *Saggio di studi naturali sul territorio mantovano* ... Mantova (V. Guastalla Tipografo-Editore) 1879. Oct. (*Sagg. stud. nat. mantov.*).
Publ.: 1879 (p. iv: Apr 1879). p. [i*-iv*], i-iv, [1]-[508], [i]-ix. *Copy*: Univ. of Illinois (inf. D.P. Rogers).

Pahnsch, Gerhard (x-1880), Esthonian botanist. (*Pahnsch*).

HERBARIUM and TYPES: Unknown.

BIBLIOGRAPHY and BIOGRAPHY: Barnhart 3: 41; CSP 12: 554; Herder p. 200; Rehder 1: 364.

7186. *Beitrag zur Flora Ehstlands* .... Dorpat (Verlag der Dorpater Naturforscher-Gesellschaft) 1881. (*Beitr. Fl. Ehstlands*).
Publ.: 18 Apr-30 Jun 1881 (censor 18 Apr 1881; Bot. Zeit. 29 Jul 1881; Nat. Nov. Jul(1). 1881), p. [i-ii], [1]-51. *Copy*: H. – Reprinted from Arch. Nat., Liv., Ehst.-Kurl. (Dorpat) ser. 2. 9(3): 237-287. 1884.

Paiche, Philippe (1842-1911), Swiss botanist at Genève; studied Potentilla, Hieracium and Rosa. (*Paiche*).

HERBARIUM and TYPES: G (via herb. Boissier); some further material at BR and E. – Letters at G.

BIBLIOGRAPHY and BIOGRAPHY: AG 6(1): 251, 12(2): 732; Barnhart 3: 41; DTS 6(4): 182; Hegi 4(2): 960; IH 2: (in press); Rehder 2: 291.

BIOFILE: Beauverd, G., Bull. Soc. bot. Genève ser. 2. 3: 262-264. 1911 (obit., bibl.).
Briquet, J., Bull. Soc. bot. Suisse 50a: 359-360. 1940 (bibl., biogr., b. 21 Dec 1842, d. 23 Aug 1911; epon.).
Hedge, I.C. & Lamond, J.M., Index coll. Edinb. herb. 116. 1970.
Miège & Wuest, Saussurea 6: 132. 1975.
Paiche, Ph., Réempoisonnage des plantes d'herbiers, 2 p., Genève 1901, Bull. Herb. Boiss. ser. 2. 1: 330-331. 28 Feb 1901.

7187. *Observations sur quelques espèces critiques du genre Hieracium* ... Genève (Imprimerie Romet, ...) 1894. Oct.
Publ.: Dec 1894 (t.p.; journal issue Dec 1894; Nat. Nov. Apr 1895), p. [i], [1]-33. *Copy*: BR. – Reprinted and to be cited from Bull. Trav. Soc. bot. Genève 7: [199]-231. 1894. – AG 12(2): 732: (translated:) "His *Observ*. ... contains many erroneous records ...".

Paige, Edward Winslow (1844-1918), American lawyer and botanist at New York; A.B. Union Coll. 1864; LL.D. Hobart Coll. 1887. (*Paige*).

HERBARIUM and TYPES: Unknown.

BIBLIOGRAPHY and BIOGRAPHY: Barnhart 3: 42; BL 1: 202, 311; ME 3: 494.

7188. *Catalogue of the flowering plants of Schenectady County* ... Albany (van Benthuysen's Steam Printing House) 1864. Qu. (*Cat. fl. pl. Schenectady Co.*).
Publ.: 1864 (p. [3]: Dec 1864), p. [1]-48. *Copy*: USDA.

Pailleux, [Nicolas] **Auguste** (1812-1898), French botanist; owner and director of an embroidery firm at Saint Pierre-lez-Calars ca. 1840-1871; in retirement at Crosnes nr. Villeneuve-Saint-Georges dedicating himself to horticulture developing new cultivars of vegetables; introduced the chinese artichoke ("Crosne"). (*Pailleux*).

HERBARIUM and TYPES: Unknown.

BIBLIOGRAPHY and BIOGRAPHY: BL 1: 35, 311; BM 4: 1499; Bossert p. 298; Bret. p. 1060; CSP 17: 676; Kew 4: 211; MW p. 375; Plesch p. 353; Rehder 5: 645; Tucker 1: 534-535.

BIOFILE: Bourguignon, L., Rev. hort. 1898: 176 f., reprinted in Potager curieux ed. 3, 1899, p. vi-ix (b. 10 Sep 1812).
Merrill, E.D., Contr. US. natl. Herb. 30(1): 232. 1947.

7189. *Le Soya* sa composition chimique ses variétés, sa culture et ses usages .... Paris (Librairie agricole de la Maison rustique) 1881. Oct. (*Soya*).
Publ.: 1881, p. [1]-127. *Copy*: FI. – Reprinted and to be cited from Bull. Soc. Acclim. ser. 3. 7: 414-471. Sep 1880, 538-596 early 1881.

7190. *Le potager d'un curieux* histoire, culture et usages de 100 plantes comestibles peu connues ou inconnues ... Paris (Librairie agricole de la maison rustique ...) 1885. Oct. (*Potag. cur.*).
Co-author: Désiré George Jean Marie Bois (1856-1946).
Ed. 1: 1884, published in Bull. Soc. nat. Accclimation ser. 4. 1: 44-75. Jan, 131-158. Feb, 259-288. Mar, 363-391. Apr, 465-492. Jun, 570-598. Jul, 653-678. Aug, 728-747. Sep, 824-847. Oct, 896-913. Nov, 945-962. Dec 1884, (journal, with orig. covers, at USDA). – Based upon Pailleux's own horticultural experiments at Crosnes.
Reprint: 1885 (t.p.; BSbF 13 Nov 1885), p. [1]-294, [1, err.]. *Copies*: FI, G.

Ed. 2: 1892 (BSbF 27 Mai 1892; Nat. Nov. Jul(1) 1892), p. [i]-xii, [1]-589. *Copy*: MO. – "*Le potager d'un curieux* ... 200 plantes ... deuxième édition entièrement refaite ... Paris (id.) 1892. Oct.

Ed. 3: 1899 (Nat. Nov. Mar(2) 1899), p. [i]-xvi, [1]-678. *Copy*: USDA. – "*Le potager d'un curieux* ... 250 plantes ... troisième édition entièrement refaite ..." Paris (id.) 1899. Oct. – Rev. P. Hariot, Bull. Soc. bot. France 46: 61-63. 1899.

Ed. [4]: 1927-1937. The following work may be considered "comme la quatrième édition ...": D. Bois (Désiré G.J.M. Bois, 1856-1946, see TL-2/1: 254), *Les plantes alimentaires chez tous les peuples* et à travers les ages. Histoire, utilisation, culture, phanérogames, léguminières ... Paris (Paul Chevalier ...) 1927-1937, 4 vols. (in series Encyclopédie biologique). *Copy*: USDA.

1: 1927 (printing finished 8 Mai 1927), p. [1]-593, [1, colo.].
2: 1928 (id. 25 Nov 1928), p. [1]-637, [638, err.], [1, colo.].
3: 1934 (id. 24 Dec 1934), p. [i-iii], [1]-289, [1, colo.].
4: 1937 (id. 15 Apr 1937), p. [i-iii], [1]-600, [1, err.], [1, colo.].

Paillot, Justin (1829-1891), French botanist; high school teacher at Nans; from 1863-1885 pharmacist at Besançon; from 1886 at Rougemont. (*Paill.*).

HERBARIUM and TYPES: Acquired by the botanical laboratory of the Faculty of Sciences ces, Besançon, through A. Magnin, see Bornet, 1893. Other material at AUT, BR, C, DBN, E, G, MANCH, P, PC, W. – *Exsiccatae*: (1) *Flora Sequaniae exsiccata ou herbier de la Flore du Franche-Comté*, see below. Sets at BR, W.
(2) *Flora cryptogama Sequaniae exsiccata*, fasc. 1-7, nos. 1-328, 1873-1897. Sets at G, PC, STR, (see Sayre 1969).
Some letters at G.

BIBLIOGRAPHY and BIOGRAPHY: Barnhart 3: 42; BL 2: 149, 156, 699; BM 4: 1499; CSP 8: 553, 12: 554, 17: 677; GR p. 342; IH 2: (in press); Jackson p. 283; Kew 4: 211; Morren ed. 10, p. 60; PR 523; TL-2/518; Tucker 1: 535.

BIOFILE: Anon., Bull. Soc. bot. France 15 (bibl.): 240. 1869, 22 (bibl.): 111. 1875 (on exsicc.), 26 (bibl.): 95-96. 1879, 38: 404. 1891 (d.), 38 (bibl.): 192. 1892 (herb. for sale).
Beck, C., Bot. Centralbl. 34: 150. 1888 (exs.).
Bornet, Ed., Bull. Soc. bot. France 40 (bibl.): 64. 1893 (herb. to Besançon).
Candolle, Alph. de, Phytographie 438. 1880 (coll.).
Hedge, I.C. & J.M. Lamond, Index coll. Edinb. herb. 116. 1970 (pl. E).
Lamy, D., Occas. Pap. Farlow Herb. 16: 12, 126. 1981 (corr. with F.F.G. Renauld).
Quincy, C., Bull. Soc. Sci. nat. Saône-et-Loire 37: 47. 1911.
Sayre, G., Mem. N.Y. Bot. Gard. 19: 40. 1969. (on exsicc.).

COMPOSITE WORKS: *Billotia*, ou notes de botanique, publiées par Vital Bavoux, Albert Guichard, Paul Guichard et Justin Paillot, vol. 1, 1864-1869. Text accompanying the *Flora exsiccata de C. Billot*.

7191. *Flora Sequaniae exsiccata* ou herbier de la flore de Franche-Comté ... Besançon (Imprimerie de Dodivers ...)1872. Oct.
Co-author: Xavier Vendrely (1837-1908).
Publ.: 1872 (t.p.), p. [1]-36. Text accompanying exsiccatae; to be cited from Mém. Émul. Soc. Doubs ser. 4. 6: 75-100. 1872 and 7: 514-521. 1873.
Continuation: published in Mém. Soc. Émul. Doubs ser. 4. 10: 477-494. 1876, ser. 5. 5: 12-69. *2 pl*. 1881 (BSbF rd 9 Dec 1881), 6: 108-115. Jan-Mar 1890, ser. 6. 10: 353-362. 1896, ser. 7. 4: 380-388. 1900, 7: 291-309. 1903, 10: 152-156. 1906. We have seen no reprints with independent pagination (but see BM 4: 1499). – The total series consisted of parts 1-3 [3 by Paillot alone], 4 and 7 by Paillot & Vendrely, 5 by Paillot, Vendrely, C. Flagey and F. Renauld and parts 8-11 by Vendrely. Part 6 is lacking; its place is taken by Renauld, F. et al., Mém. Soc. Émul. Doubs ser. 5. 7: 162-200. 1883 (BSbF Nov-Dec 1883: reprint pagination 113-151).

Paine, John Alsop (1840-1912), American botanist; M.A. Hamilton College 1862; Dr. phil. id. 1874; employed by the board of regents of the State of New York in connection with the flora of the State in 1862-1867; teaching at various colleges and editing journals from 1868; in Eastern Palestine for the Palestine Exploration Society 1872-1873; curator at the Metropolitan Museum of Art, New York, from 1889. (*Paine*).

HERBARIUM and TYPES: Material at GH, K, MICH, NYS, US.– Some letters at G.

BIBLIOGRAPHY and BIOGRAPHY: Backer p. 658; Barnhart 3: 42; BL 1: 202, 311; BM 4: 1499; CSP 8: 553, 12: 555; DAB (by W.F. Albright); Hortus 3: 1201 ("Paine"); Jackson p. 364; Kew 4: 212; Lenley p. 465; ME 2: 726; Pennell p. 615; PH 73, 164; PR 6895, 10613; Rehder 1: 314; Tucker 1: 535; Zander ed. 10, p. 699, ed. 11, p. 797.

BIOFILE: Allibone, S.A., Crit. dict. Engl. lit. 1483. 1870.
Anon., Hedwigia 53: (95). 1912 (d. 24 Jul 1912); Nat. Nov. 34: 513. 1912 (d.); Natl. Cycl. Amer. biogr. 13: 456-457. 1906 (biogr., portr., b. 14 Jan 1840); Torreya 12: 200. 1912 (d.).
Gray, A., Amer. J. Sci. ser. 2. 41: 130-132. Jan 1836 (rev.).
Jackson, B.D., Bull. misc. Inf. Kew 1901: 50 (Syrian pl. K).
McKeen Cattell, J., Amer. men. sci. 242. 1906.

7192. *Catalogue of plants found in Oneida County* and vicinity ... [From the report of the Regents of the University of the State New York, presented March 22, 1865]. Qu. (*Cat. pl. Oneida Co.*].
Publ.: Oct-Dec 1865 (p. 140: Oct 1865, Amer. J. Sci. Jan 1866), p. [1]-140. *Copies*: FH, G, MO, NY, US. – Reprinted from the Rep. Regents Univ. New York [presented 22 Mar] 1865: [53]-205 (*copy*: USDA). – (Ann. Rep. State Cabinet Nat. Hist. New York no. 11). See also J.V. Haberer, Rhodora 7: 92-97, 106-110. 1905.
Ref.: Lubrecht, H., Early Amer. bot. works 1967, no. 108.

Painter, William Hunt (1835-1910), British clergyman (entered Church of England 1861) and botanist at Barbon, High Wycombe, Edgbaston, Derby and Bristol; rector of Stinchley 1894-1901; died in Shrewbury. (*Painter*).

HERBARIUM and TYPES: ABS; duplicates at ABD, BIRM, BM, DBN, DBY, GL, K, LIV, NOT, OXF.

BIBLIOGRAPHY and BIOGRAPHY: Barnhart 3: 42; BB p. 234; BL 2: 234, 266, 698; BM 4: 1499; Clokie p. 220; CSP 10: 982, 17: 678; Desmond p. 475; IH 2: (in press); Kew 4: 212; Rehder 1: 404; Tucker 1: 535.

BIOFILE: Anon., Bot. Not. 1911: 28 (d. 12 Oct 1910); J. Bot. 27: 160. 1889 (ann. Contr.), 35: 366. 1897 (suppl.), 40: 128. 1902 (suppl.), 47: 451. 1909 (herb. to ABS), 48: 296. 1910 (d.); Hedwigia 51(1): 88). 1911 (d. 12 Oct 1910).
Bagnall, J.E., J. Bot. 27: 318-320. Oct 1889 (rev. Contr. Fl. Derbyshire).
Bridson, G.D.R. et al., Nat. hist. mss. res. Brit. Isl. 84.21. 1980.
Britten, J., J. Bot. 49: 125-126. 1911 (portr.; obit., b. 16 Jul 1835).
Druce, G.C., Bot. Exch. Club Brit. Isl. Rep. 1910: 533-534. 1911 (obit.).
Edees, E.S., Fl. Staffordsh. 20-21. 1972.
Freeman, R.B., Brit. nat. hist. books 271. 1980.
Horwood, A.R. & Noel, C.W.F.N., Fl. Leicestershire ccxxxvi-ccxxvii. 1933.
Jackson, B.D., Bull. misc. Inf. Kew 1901: 50 (*Rubus* at K).
Kent, D.H., Brit. herb. 71. 1957 (herb. ABS; dupl. BIRM, BM, DBY, GL, K, NOT, OXF).
Murray, G., Hist. coll. BM(NH)1: 172. 1904 (1730 Brit. pl. at BM).
Painter, W.H., Naturalist 1899: 241-272 (list of Derbishire mosses).

7193. *A contribution to the Flora of Derbyshire.* Being an account of the flowering plants, ferns, and characeae found in the county ... London (George Bell & Sons, ...), Derby (E. Clulow, Jun.) 1889. Oct. (*Contr. Fl. Derbyshire*).

Publ.: Aug-Sep 1889 (pref.: Jul 1889; J. Bot. Oct 1889; Nat. Nov. Oct(2) 1889; Bot. Zeit. 28 Feb 1890), p. [i-v], map, [1]-156. *Copy*: MO.
Re-issue: 1891 (fide Nat. Nov. Feb(1) 1892, no. 1188).
Supplement: Notes supplementary to the Flora of Derbyshire, *in*: Naturalist 1899: 177-208, 241-272. 1902; 5-12. 1 Jan 1902. – A reprint with unchanged pagination, with additional t.p., introd. and index was published in Leeds (Chorley and Pickersgill) 1902 (see Bot. Centralbl. 15 Apr 1902 and J. Bot. Mar 1902).

Palacký, Johann Baptist [Jan] (1830-1908), Bohemian politician, geographer, botanist and ichthyologist; Dr. phil. Praha 1850; lecturer in geography, later also in botany at Praha 1856-1866; 1878-1885; extraordinary professor 1885-1902; travelled widely mainly in Europe. (*Palacký*).

HERBARIUM and TYPES: PR, PRC; some further material at W. – Some letters at G.

BIBLIOGRAPHY and BIOGRAPHY: Barnhart 3: 42; BJI 1: 43, 2: 129; BL 1: 97, 311; BM 4: 1500; Bossert p. 298; CSP 8: 554, 10: 982-983, 12: 555, 17: 679-680; Futák-Domin p. 454; Herder p. 167; Hortus 3: 1201 ("Palacký"); IH 2: (in press); Kew 4: 213; Langman p. 561-562; Maiwald p. 244, 251, 252; Morren ed. 10, p. 33; MW p. 375; Rehder 5: 645; TL-2/6111; Tucker 1: 535.

BIOFILE: Anon., Bonplandia 9: 12. 1861; Bot. Centralbl. 108: 32, 1908 (d.); Bot. Jahrb. 43 (Beibl. 98): 57. 1909; Nat. Nov. 24: 658. 1902 (retirement); Österr. bot. Z. 40: 140. 1890 (algae at W), 58: 223. 1908 (d.).
Degen, A., Magy. bot. Lap. 7: 275. 1908 (d.).
Domin, K., Bull. Acad. Géogr. bot. 17: 249, 333. 1908 (obit., b. 30 oct 1830, d. 22/23 Feb 1908).
Iltis, H., Life of Mendel ed. 2. 21. 1966.
Kneucker, A., Allg. bot. Z. 8: 208. 1902 (member Acad. int. Géogr. bot.), 14: 214. 1926.
Merrill, E.D., Contr. U.S. natl. Herb. 30(1): 233. 1917, Enum. Phil. pl. 4: 214. 1926.
Poggendorff, J.C., Biogr. lit. Handw.-Buch 3: 1000. 1898, 4(2): 1112-1113. 1904, 6(3): 1939. 1938 ((b. 10 Oct 1830, d. 22 Feb 1908, bibl).
Urban, I., Symb. ant. 1: 122. 1898.
Wittrock, V.B., Acta Horti Berg. 3(2): 4. 1903.
–ský, Österr. bot. Z. 39: 2-4. 1889 (biogr., portr.).

EPONYMY: *Palackya* L. Crié (1889) is very likely named for Palacký, but no derivation is given by Crié.

7194. *Pflanzengeografische Studien* ... I. Erläuterungen zu Hooker et Bentham Genera plantarum ... Prag (Druck von Dr. Ed. Grégr. -Selbstverlag der Gesellschaft) 1864. Qu. (*Pfl.-geogr. Stud.*).
Orig.: Apr-Jul 1864 (p. 67: 24 Mar 1864; t.p. reprint 1864; Flora 31 Aug 1864), p. [1]-67. *Copy*: G. – Preprinted from Abh. k. böhm. Ges. Wiss. ser. 5. 13(3). 1864 (1865).
Continuation(1): 1883, p.[1]-84; published as Abh. k. böhm. Ges. Wiss. ser. 6. 12(2). 1883.
Continuation (2): 1884 (t.p. preprint), p. [1]-50. *Copy*: BR. – Preprinted from id. ser. 6. 12(11). 1884 (1885).
Ref.: Engler, A., Bot. Jahrb. 5 (Lit.): 9. 1883.
Kornhuber, A., Österr. bot. Z. 37: 178-179. 1887.

7195. *Catalogus plantarum madagascariensium* ... Pragae [Praha](sumptibus auctoris) 1906-1907, 5 fasc. Oct. (*Cat. pl. madagasc.*).
1: Mai-Jun 1906 (ÖbZ Mai-Jun 1906), p. [1]-55.
2: Jan-Mar 1907 (Nat. Nov. Mai (2) 1907; ÖbZ 1,Apr 1907; AbZ 15 Apr 1907), p. [i], [1]-38.
3: Jan-Jun 1907 ("ultimus"; AbZ 15 Jul 1907; Nat. Nov. Sep (1, 2) 1907; ÖbZ Jun-Jul 1907), p. [i-ii], [1]-89, [1, 2, err., epilogue].
4: Oct-Dec 1906 (t.p.; ÖbZ Oct-Dec 1906), p. [i], [1]-60.
5: Jan-Aug 1907 (AbZ 15 Oct 1907; Nat. Nov. Sep (1, 2) 1907), p. [i-ii], [1]-57.
Copies: BR, G, MO, NY, USDA.
Ref.: Palacký, J.B., Allg. bot. Z. 13: 173-174. 1907.

7196. *Filices madagascarienses* ... Prague [Praha] (sumptibus auctoris ...) 1906. Oct. (*Fil. madagasc.*).
Publ.: Jan-Jun 1906 (t.p.; p. [3]: 28 Dec 1905; AbZ 15 Jul 1906; Nat. Nov. Mai (2) 1907; ÖbZ Mar-Mai 1907), p. [1]-32. *Copies*: BR, MO, NY, USDA.

Palanza, Alfonso (1851-1899), Italian botanist and plant collector; sometime private teacher at Napoli; from 1894 high school teacher at the Liceo di Bitonto. (*Palanza*).

HERBARIUM and TYPES: A general and a local (Bari) herbarium were acquired by the province of Bari for its technical institute. We do not know whether this collection was later transferred to BI. – Some material at FI (inf. C. Steinberg). – Some letters at G.

BIBLIOGRAPHY and BIOGRAPHY: BL 2: 385, 698; Bossert p. 298; CSP 17: 681; Hortus 3: 1201 ("Palanza"); Saccardo 2: 80.

BIOFILE: Anon., Nat. Nov. 22: 65. 1900 (d. 26 Jul 1899).
Geremicca, M., Bull. Orto bot. Napol. 3: 72. 1913 (portr., b. 6 Sep 1851).
Jatta, A., Bull. Soc. bot. ital. 1899 (7/8): 159-160 (obit.).

7197. *Flora della terra di Bari* per Alfonso Palanza ... Trani (Tipografia dell' editore V. Vecchi) 1900. Qu. (*Fl. Bari*).
Published by: Antonio Jatta (1852-1912).
Publ.: Jan-Jun 1900 (BSbF rd 27 Jul 1900; Bot. Centralbl. 12 Dec 1900), p. [i], [1]-90. *Copies*: FI, G. – Reprinted from *La terra di Bari* 3: 155-244. 1900. For preliminary publications see Nuov. Giorn. bot. Ital. 4: 277-288. Jul 1897, Bull. Soc. bot. Ital. 1898: 150-158. Jul 1898, 195-202. Nov 1898.
Ref.: Solla, R.F., Bot. Centralbl. 85: 143-145. 1901.

Palassou, Pierre Bernard (Palasso) (1745-1830), French clergyman; naturalist and explorer of the Pyrenees. (*Palassou*).

HERBARIUM and TYPES: MPU.

BIBLIOGRAPHY and BIOGRAPHY: Barnhart 3: 43; BM 4: 1503; Colmeiro 1: cxci; Dryander 3: 142; IH 2: (in press); PR ed. 1: 7704; Quenstedt p. 326; Rehder 1: 535; Tucker 1: 535.

BIOFILE: Lacroix, A., Figures de savants 4: 43. 1938.
Lapeyrouse, P., Hist. abr. pl. Pyren. xvi, xxviii. 1813.
Zittel, K.A. v., Hist. geol. palaeontol. 102-103. 1901.

7198. *Suite* des *mémoires* pour servir à *l'histoire naturelle* des *Pyrénées*, et des pays adjacens ... à Pau (de l'imprimerie de A. Vignancour, ...) 1819. Oct. (*Suite mém. hist. nat. Pyrénées*).
Publ.: 1819, p. [i]-xxiii, [1]-428, [1-2, table, err.]. *Copy*: PH. – Original Mém. hist. nat. Pyrénées, 1815, not seen by us.

7199. *Supplément aux mémoires pour servir à l'histoire naturelle des Pyrénées*, et des pays adjacens; suivis de recherches relatives aux anciens camps de la Novempopulanie ... à Pau (De l'imprimerie de A. Vignancour, ...) 1821. Oct. (*Suppl. mém. hist. nat. Pyrénées*).
Publ.: 1821, p. [1]-205, [1-2, table, err.]. *Copy*: PH. – Novem-populanie (Novempopulania), province of ancient Gallia covering the present Gascogne, Béarn, and Comminges.

7200. *Observations* pour servir à *l'histoire naturelle* et civile *de la Vallée d'Aspe*, d'une partie de la Basse-Navarre et des pays circonvoisins, avec les preuves de l'exactitude de plusieurs faits relatifs aux Pyrénées ... à Pau (de l'imprimerie de Vignancour, ...) 1828. Oct. (*Observ. hist. nat. Vallée d'Aspe*).
Publ.: 1828, p. [1]-201, 102 [for 202], [1-2, tabl.], [1, err.]. *Copy*: PH.

Paláu y Verdéra, Antonio (x-1793), Spanish botanist at the Madrid botanical garden. (*Paláu*).

HERBARIUM and TYPES: Unknown.

BIBLIOGRAPHY and BIOGRAPHY: Barnhart 3: 43; BM 4: 1503, 8: 972; Colmeiro 1: cxci, penins. p. 172-173; Dryander 5: 99, 100; Herder p. 94; Kew 4: 213; Langman p. 562; PR (alph.), 5426, 6858; Rehder 5: 645; SO 16, 17, 471, 587, 595a, 718a, 2617; TL-2/4760; Tucker 1: 535; Zander ed. 10, p. 699, ed. 11, p. 797.

BIOFILE: Munoz Mediña, D.J., Hist. desarr. bot. España 11, 12. 1969.
Willkomm, M., Grundz. Pflanzenverbr. Iber. Halbinsel 6. 1896 (Veg. Erde 1).

COMPOSITE WORKS: With Casimiro Gomez Ortega: *Curso elemental de botánica*, Madrid, 1785 (SO 587), ed. 2, Madrid 1795, 2 vols.

EPONYMY: *Palaua* Cavanilles (1785); *Palaua* Ruiz et Pavon (1794); *Palava* A.L. de Jussieu (1789, *orth. var.*); *Palava* Persoon (1806, *orth. var.*); *Palavia* Poiret (1816, *orth. var.*); *Palavia* Schreber (1791, *orth. var.*).

7201. *Explicacion de la filosofia, y fundamentos botanicos* de Linneo, con la que se aclaran y entienden facilmente las instituciones botanicas de Tournefort. Su autor Don Antonio Paláu y Verdéra, ... parte theoretica. Con privilegio. Madrid (por Don Antonio de Sancha) Año de 1778. Oct. (*Expl. filos. fund. bot.*).
Publ.: 1778, p. [i-xvi], 1-312, *pl. 1-9* (uncol. copp.). *Copy*: USDA. – See SO 471.

7202. *Parte práctica de botánica* del caballero Cárlos Linneo, que comprehende las clases, órdenes, géneros especies y variedades de las plantas, con sus caracteres genéricos y especificos, sinónimos mas selectos, nombres triviales, lugares donde nacen, y propriedades. Traducida del Latin en Castellano é ilustrada por Don Antonio Paláu y Verdéra, ... De orden superior. Madrid (en la Imprenta real) 1784-1788), 8 vols. & atlas. Oct. (*Parte práct. bot.*).
1: 1784, portr., p. [i*-xii*], i-lvi, 1-796, *5 pl.* (uncol. copp.).
2: 1785, p. [i, iii], 1-918, *5 pl.* (id.).
3: 1785, p. [i-ii], 1-801.
4: 1786, p. [i, iii], 1-914.
5: 1786, p. [i, iii], 1-788.
6: 1787, p. [i, iii], 1-925.
7: 1787, p. [i, iii], 1-927.
8: 1788, p. [i*], [i]-clxxviii, [1]-782, *2 pl.* (copp.).
Copies: HU (no. 670), NY, USDA. – See SO 16. Based on J.J. Reichard, *Systema plantarum*, 1779-1780, 4 parts, q.v.
Ref.: McVaugh, R., Contr. Univ. Mich. Herb. 11(3): 127. 1977.

7203. *Sistema de los vegetables* ó resumen de la parte prática de bótanica del Caballero Cárlos Linneo, que comprehende las clases, órdenos, generos y especies de las plantas, con algunas de sus variedades. Por Don Antonio Paláu de Verdéra, ... Madrid (en la Imprenta Real) 1788. Oct. (*Sist. Veg.*).
Publ.: Dec 1788 (p. xii: 26 Nov 1788), p. [i]-xii, 1-713. *Copies*: HU (no. 693), MO, USDA (2). – See SO 595a.

Paley, Frederick Apthorp (1815-1888), British Greek scholar and botanist; BA Cantab. 1838; LL.D. Aberdeen 1883. (*Paley*).

HERBARIUM and TYPES: Unknown.

BIBLIOGRAPHY and BIOGRAPHY: Barnhart 3: 43; BB p. 235; BL 2: 249, 260, 698; BM 4: 1503; CSP 8: 555; Desmond p. 470; DNB 43: 99-101; Jackson p. 195, 251, 258, 589; Kew 4: 214; Rehder 1: 400, 401.

BIOFILE: Anon., Encycl. brit. ed. 11. 20: 628. 1911.
Druce, G.C., Fl. Northamptonshire cxxi. 1930.
Freeman, R.B., Brit. nat. hist. books 271. 1980.
Gillow, J., Dict. Engl. catholics, 4: 234. 1887 (fide BB).
Glass, B. et al., Forerunners of Darwin 226, 251, 259, 277. 1968.

EPONYMY: *Paleya* Cassini (1826) is dedicated to William Paley (1743-1805), English philosopher.

7204. *A list of* four hundred *wild flowering plants*; being a contribution to the flora of Peterburough, with an introduction ... Peterburough (Robert Gardner. George Bell, ...) 1860. (*List wild fl. pl.*).
Publ.: 1860, p. [i]-xxviii, [1-26]. *Copy*: Natl. Libr. Scotland (inf. J. Edmondson).

Palézieux, Philippe de (1871-1957), Swiss botanist at Genève; student of L. Radlkofer; Dr. sci. nat. münchen 1899; from 1900-1939 as "Privatgelehrter" in Germany, often at Berlin-Dahlem, and traveling widely; returned to Genève 1939; curator of the Herbier Boissier 1943-1952. (*Paléz.*).

HERBARIUM and TYPES: G (1957 to Université de Genève; 1963 to Conservatoire); further material at A, C, MPU. – Letters at G.

BIBLIOGRAPHY and BIOGRAPHY: BFM no. 1273; BL 2: 571, 698; BM 8: 973; Bossert p. 298; IH 1 (ed. 1): 36, 2: (in press); Langman p. 562; MW p. 375; Rehder 2: 635.

BIOFILE: Baehni, C., Candollea 16: 207-210. 1958 (obit., b. 12 Feb 1871, d. 25 Nov 1957, bibl.).
Retz, B. de, Soc. franç. éch. pl. vasc. fasc. 9: 7. 1960.
Verdoorn, F., ed., Chron. bot. 7: 441. 1943.

7205. *Anatomisch-systematische Untersuchung des Blattes der Melastomaceen* mit Ausschluss der Triben: Microlicieen, Tibouchineen, Miconieen. Inaugural-Dissertation zur Erlangung der Doctorwürde bei der hohen philosophischen Facultät der k. bayer. Ludwig-Maximilians-Universität zu München vorgelegt am 24 Januar 1899 ... Genève (Imprimerie Romet, ...) 1899. Oct. (*Anat. syst. Unters. Blatt. Melast.*).
Publ.: 24 Jan 1899, p. [1]-83, [1, cont.], *pl. 1-3* (uncol.). *Copies*: G, U, US. – Also issued as Appendice no. 5, Bull. Herb. Boiss. 7: 1-32. Aug, 33-34. Sep, 65-83. Oct 1899.
Ref.: Roth, E., Bot. Centralbl. 87: 348-350. 1901 (rev.).

Palhinha, Ruy Telles (1871-1957), Azores-born Portugese botanist; high school teacher at Lisbon, later at the Botanical Institute of the University of Lisbon (director 1921-1941). (*Palhinha*).

HERBARIUM and TYPES: LISU; other material at C and L.

BIBLIOGRAPHY and BIOGRAPHY: Barnhart 3: 43; BFM 1529; BL 2: 464, 699; IH 2: (in press); Kew 4: 214; TL-2/1259.

BIOFILE: Anon., Broteria 3: 310. 1903; Taxon 7: 57. 1958 (obit.), 16: 55. 1967 (rev. of his posthumously publ. Cat. pl. vasc. Açores, 1966).
Gager, C.S., Brooklyn Bot. Gard. Rec. 27(3): 320. 1938 (dir. bot. gard. Lisboa 1921-[1941]).
Gonçalves da Cunha, A., Bol. Soc. brot. 32: vi-xx. 1958 (obit., bibl., portr., b. 4 Jan 1871, d. 13 Nov 1957).
Hansen, A., Bot. bibl. Azores 6. 1970 (publ. 1941-1966).
Quintanilha, A., Bol. Soc. port. Ci. nat. 7-12. 1972 (obit., portr.).
Tavares, C.N., Naturalia 7: 260-264. 1957/58 (b. 4 Jan 1871, d. 13 Nov 1957).

COMPOSITE WORKS: Editor of A.X. P. Coutinho, *Flora de Portugal*, ed. 2, 1939, TL-2/1259.

EPONYMY: *Palhinhaea* Franco et Vasconellos (1967).

Palibin, Ivan Vladimirovich (1872-1949), Russian botanist; Dr. phil. 1934; in various ways associated with the St. Petersburg botanical garden, later Komarov Botanical Institute, from 1895; from 1906-1910 at the Jeneve University; from 1910-1915 at the St. Petersburg botanical garden; from 1915-1922 at Batoem; 1923-1929 head botanist, 1929-1932 director of the Botanical Museum and Garden at Leningrad; from 1932-1949 head of the Palaeobotanical Department at Leningrad; organized a palaeobotanical cabinet at the Petroleum Institute 1931. (*Palib.*).

HERBARIUM and TYPES: LE.

BIBLIOGRAPHY and BIOGRAPHY: Backer p. 418; Barnhart 3: 43; BJI 2: 129; BM 4: 1504, 8: 973; Bret. 1042, 1095; CSP 17: 682; Hirsch p. 223; Hortus 3: 1201 ("Palib."); Kew 4: 214; Kleppa p. 173, 232, 292; MW p. 375-376, suppl. p. 275; TL-2/3857; Tucker 1: 535; Zander ed. 11, p. 797.

BIOFILE: Anon., Österr. bot. Z. 61: 455. 1911. (app. St. Petersb.).
Asmous, V.C., Chron. bot. 11(2): 101-106. 1947.
Borodin, J., Trav. Mus. bot. Acad. Sci. St. Petersb. 4: 87-88. 1908.
Krischtofovitch, A.V. and A. N., Bot. Zhurn. 1950(6): 684-688. (portr.).
Lipschitz, S., Bull. Soc. Natural. Moscou, biol. 79(4): 151. 1974, Florae URSS fontes 222 [index]. 1975.
Lipsky, V.I., Imp. St. Petersb. bot. Zad 3: 380-386. 1915 (portr., bibl.).
Shetler, S.G., Komarov Bot. Inst. 232 [index]. 1967.
VanLandingham, S., Cat. diat. 6: 3576. 1978.
Verdoorn, F., Chron. bot. 1: 273. 1935, 3: 254. 1937.

COMPOSITE WORKS: (1) *in* N.I. Kusnetzov, *Flora caucasica critica*:
(a) *Convolvulaceae*, 4(2): 1-32. 1912, 33-61. 1913.
(b) *Polemoniaceae*, 4(2): 61-66. 1913.
(c) *Cistaceae*, 3(9): 117-158. 1909.
(2) Contributed to V.L. Komarov et al., *Flora URSS*, vols. 3, 7, 9, 13, 14.

NOTE: We are grateful to S. Lipschitz, Leningrad, for information on Palibin (in litt. 7 Feb 1980).

EPONYMY: *Palibinia* E.P. Korovin (1931). *Note*: The etymology of *Palibiniopteris* V.D. Prinada (1956) could not be determined.

7206. *Revisio generis Enkianthus* Lour. ... St. Petersburg 1897. Oct.
Publ.: Jan-Mai 1897 (Bot. Centralbl. 30 Jun 1897), p. [1]-18. *Copy*: BR. – Published, and to be cited as Scripta bot. Hort. Univ. Petrop. 15: 1-18. 1897.
Ref.: Palibin, I., Bot. Centralbl. 73: 271-272. 1898 (rev.).

7207. *Conspectus florae Koreae* ... Petropoli [St. Petersburg] 3 parts, 1899-1901. Oct. (*Consp. fl. Koreae*).
1: 1899 (t.p. fascicle 1898, but p. [ii*] 1899, p. [2]: Jan 1898), p. [1]-127, [128, err.], *pl. 1-4*. Issued as Acta Horti Petrop. 17: 1-128, *pl. 1-4*. 1898 (1899).
2: 1900, p. [1]-52, reprinted from id. 18(2): 147-198. 1900.
3: 1901, [i, ii] (gen. t.p. and dates), p. [1]-70, reprinted from id. 19: 101-151. 1901. – General t.p., issued with part 3 dated 1899-1901.
Copies: BR, G, US.
Ref.: Chodat, R., Bull. Herb. Boissier ser. 1. 7: 496. Jun 1899 (rev.).

7208. *Résultats botaniques du voyage à l'Océan glacial sur le bateau brise-glace "Ermak"* pendant l'été de l'année 1901 ... Petersburg 1903-1906. Oct. (*Résult. bot. voy. Ermak*).
Publ.: 1903-1906, p. [i], [1]-128. *Copy*: G.

7209. *Contributions à l'histoire de la flore de la Transcaucasie occidentale* ... Genève (Imprimerie Romet, ...) 1908. Oct. (*Contr. fl. Transcaucas. occid.*).
Publ.: Jun 1908, p. [1]-14, *pl. 6*. *Copy*: G. – Reprinted and to be cited from Bull. Herb. Boissier ser. 2. 8(7): [445]-458. 1908.

Palisot de Beauvois, Ambroise Marie François Joseph (1752-1820), French traveller and botanist; from 1786-1787 in Oware and Bénin; 1788-1791 on Haiti; 1791-1798 in the United States (briefly in Haiti again 1793); back in Paris as teacher at the Athenée des Étrangers; member Institut de France 1806; councillor Université de Paris 1815. (*P. Beauv.*).

HERBARIUM and TYPES: G. – Delessert acquired the original herbarium, including the types of the *Flore d'Oware et de Bénin*, in 1820. An important set of duplicates is in the Jussieu herbarium at Paris (P-JU). For other duplicates see Index herbariorum (under Beauvois). For notes on the collections in the Delessert herbarium, see Hochreutiner (1898), Niles (1925) and Stafleu (1971). United States collections (1791-1798) partly destroyed, some material at B, DS, G, GH, PC, P-JU. – Some of the material at G have labels "Reliquiae Palisotianae" with a note indicating that these are plants from Oware and Bénin, meant to be described in vol. 3 of the *Flore*, identified by Berlin and intercalated in the Delessert herbarium in 1897. – Some *letters* at G.

BIBLIOGRAPHY and BIOGRAPHY: Ainsworth p. 312, 345; Backer p. 418-419; Barnhart 3: 43; BM 4: 1504; CSP 4: 743, 12: 555; Dryander 3: 443-444; DU 220; Frank 3(Anh.): 73-74; GR p. 70; Herder p. 469; IH 2: 63; Jackson p. 132, 351; Kew 4: 215; Langman p. 562; Lasègue p. 70-72, 578; LS 19719-19722; MD p. 196-198; ME 3: 634 [index]; MW p. 73; NAF ser. 2. 9: 125; NI 1480-1481, suppl. p. 53; PR 6896-6901 (ed. 1: 7705-7711); Rehder 5: 645; RS p. 132; SBC p. 128 (P. Beauv.); SK 4: ccv; Stevenson p. 1253; SY p. 19, 25, 28, 31, 58; TL-1/673, 939-942; TL-2/1293, 1409-1410, 1414, 2988, 4430, 5775, J. Clarion; Tucker 1: 535; Urban-Berl. p. 309; Zander ed. 10, p. 700, ed. 11, p. 798.

SUBJECT INDEX TO BIOFILE *Agrostology*: Niles (1925), Stafleu (1971), Stieber (1979). *Bibliography*: Merrill (1936), Niles (1925), Percheron (1837), Pritzel (1843), Stafleu (1964), Steenis-Krusemann (1962), Urban (1902); see Biogr. & Bibl. section above
Biographical accounts: Depping (1822, 1865), Fisquet (1862), Jussieu (1805), L.Hn. (1884), O. (1824), Poiret (1808), Rose (1850). *Bryology*: Crum (1952), Sayre (1977), Schuster (1966), Steere (1977).
Collections: Candolle (1880), Exell (1944), Hochreutiner (1898), Mears (1978), Merrill (1936), Sayre (1977), Stafleu (1963, 1971), Steinberg (1977); Lasègue, Urban (Berlin).
Correspondence: Amoureux (1822), Darlington (1843; with Baldwin). *Death notices*: Anon. (1820), Cuvier (1824).
Entomology: Papavero(1971), Percheron (1837).
Grasses: see *Agrostology*.
Mycology: Lütjeharms (1936), Medikus (1790); Ainsworth, L.S.
Nomenclature: Kuntze (1898).
North American stay: Camus (1967), Ewan (1967), Leroy (1967).
Obituaries: Cuvier (1824), Silvestre (1820), Thiébaut (1821).
Portraits: Milner (1906), Wittrock (1905).

BIOFILE: Amoureux, Mém. Soc. Linn. Paris 1: 79-720. 1822 (letters from Santo Domingo and Philadelphia).
Anon., Flora 3: 622. 1820 (d.).
Barnhart, J.H., Proc. Amer. philos. Soc. 76: 920. 1930 (dates Flore d'Oware).
Beer, G. de, The sciences were never at war 179, 186. 1960.
Camus, A. *in* J.F. Leroy, ed., Les botanistes français en Amérique du Nord avant 1880, Paris 1967, p. 109-111.
Candolle, Alph. de, Phytographie 438. 1880 (coll. in herb. Delessert).
Crum, H., Bull. Torrey bot. Club 79: 407-409. 1952 (on *Prodrome*).
Cuvier, G., Mém. Inst. de France 4: cccviii-cccxlvi. 1824 (obit., b. 28 Oct 1755, d. 21 Jan 1820; date of birth probably not correct).
Darlington, W., Reliq. baldw. 160. 1843.
Depping, G.B., *in* Michaud, biogr. univ. 32: 412-417. 1822, ed. 2: 32: 14-17. ca. 1865.
Exell, A.W., Cat. vasc. pl. S. Tome 7. 1944.
Ewan, J., *in* J.F. Leroy, ed., Les botanistes français en Amérique du Nord avant 1850, Paris 1947, p. 23, 37.

Ewan J. et al., Short hist. bot. U.S. 38, 90. 1969.
Fisquet, H., *in* Didot, Nouv. biogr. gén. 39: 86-88. 1862.
Gilbert, P., Comp. biogr. lit. deceased entom. 286. 1977.
Hochreutiner, B.E.G., Ann. Cons. Jard. bot. Genève 2: 79-101. 1898 (Reliquiae palisotianae).
Jessen, K.F.W., Bot. Gegenw. Vorz. 376, 404, 469, 471. 1864.
Jussieu, A.L. de, Décade philos. 10 nivôse, an xii, p. 1-8. 1804, reprinted in *Fl. Oware* 1: viii-xi. 1805.
Kuntze, O., Rev. gen. pl. 3(2): 154. 1898.
Lacepède et al., P.V. Séanc. Acad.-Sci., Paris 1: 532-534. 1910.
Lacroix, E., Figures de savants 4: 278-296. (n.v.).
Leclair, E., Arch. du Nord ser. 2: 278-296. (n.v.).
Leroy, J.F. et al., Les botanistes français en Amérique du Nord avant 1850, Paris 1947, p. 354 [index, many refs.]. 1957.
Letouzey, R. & C. Ntepe, Taxon 30: 794-499. 1981 (on *Culcasia scandens* P. Beauv.).
L.Hn., Dict. sci. méd. 72: 7-8. 1884.
Lütjeharms, W.J., Gesch. Mykol. 120, 175, 247, 259. 1936.
Mears, J.A., Proc. Amer. philos. Soc. 122: 170. 1978 (no P. Beauv. specimens located in Muhlenberg herb. at PH).
Medikus, F.K., Lettre à Méthérie 1790, 16 p. (see TL-2/5775, dispute on origin of fungi).
Merrill, E.D., Proc. Amer. philos. Soc. 76: 899-914. 1936 (bibl., collections), Chron. bot. 10: 280-286. 1946.
Milner, J.D., Cat. portr. Kew 89. 1906.
Niles, C.D., Contr. U.S. natl. Herb. 24: 135-214. 1925 (see also Jennings, O.E., Bryologist 29: 76-77. 1926) (bibl. study *Agrost.*; intr. by A. Chase).
Nissen, C., Zool. Buchill. 307. 1969.
O., Dict. Sci. méd., Biogr. méd. 6: 350-352. 1824.
Papavero, N., Essays hist. neotrop. dipterol. 1: 19-20, 190. 1971.
Percheron, A., Bibl. entom. 1: 305. 1837.
Poiret, J.L.M., *in* Lamarck, Encycl. méth. 8: 744-746. 1808.
Pritzel, G.A., Linnaea 19: 459. 1843.
Rose, H.J., New gen. biogr. dict. 10: 451-452. 1850.
Sayre, G.,Bryologist 80: 515. 1977 (bryoph. coll.).
Schuster, R.M., Hep. Anthoc. N. Amer. 1: 85. 1966 (*Carpolepidium* to replace *Plagiochila*).
Silvestre, Mém. Agr. Soc. r. centr. Agr. 51: 58-84. 1820 (obit.).
Stafleu, F.A., *in* C.L. Héritier, Sertum angl. fasc. ed.1963, p. xxxi; *in* Jussieu, A.L., Gen. pl. facsimile ed. 1964, p. xlii (on Jussieu's ref. to Palisot's *Dissert. mss.*); *in* Labillardière, J.J.H. de, Nov. Holl. pl. spec., facs. ed. xxxvi. 1966; Naturaliste canad. 98: 553-560. 1971 (*herb. Agrost.*).
Steenis-Kruseman, M.L. van, Fl. males. Bull. 19: 1143-1144. 1964.
Steere, W.C., Bot. Rev. 43: 293, 340. 1977 (contr. to N. Amer. muscology).
Steinberg, C.H., Webbia 32: 33. 1977 (material from Oware also in FI).
Stieber, M.T., Huntia 3(2): 124. 1979 (Chase manuscript notes on P.), 4(1): 84. 1981.
Taton, R., ed., Science 19-th cent. 386. 1965.
Thiébaut-de Berneaud, A., Éloge historique de A.M.F.J. Palisot de Beauvois, Paris 1821, 81 p. (biogr., portr., b. 27 Jul 1752) (see also Mém. Soc. Linn. Paris 1: 26-27. 1822).
Urban, I., Symb. ant. 3: 96-98. 1902.
Vallot, J., Bull. Soc. bot. France 29: 185. 1882.
Wittrock, V.B., Acta Horti Berg. 3(3): 109. 1905 (on portr.).

NAME: The *Dictionnaire de biographie française* uses the orthography *Pallisat de Beauvois*. We follow the usual orthography, also used by Palisot de Beauvois on the title pages of his own publications.

COMPOSITE WORKS: Contributed many articles to the *Dictionnaire des sciences naturelles*, vols. 1-6, 1804-1817 and the Nouveau dictionnaire d'histoire naturelle, 36 vols., 1816-1819.

HANDWRITING: Webbia 32: 34. 1977, Candollea 32: 201-202. 1977.

EPONYMY: *Belvisia* Desvaux (1814); *Palisota* H.G.L. Reichenbach ex Endlicher (1836, nom. cons.).

7210. *A scientific and descriptive catalogue of Peale's Museum*, by C.W Peale, ... and A.H.F.J. Beauvois, ... Philadelphia (printed by Samuel H. Smith, ...) 1797. (*Cat. Peale's Mus.*).
Senior author: Charles Willson Peale (1741-1827).
Publ.: 1797, p. [i]-xii, [1]-44. *Copy*: PH. − For Peale's Museum see e.g. ME 2: 57-58. − Mainly dealing with Quadrupeds.

7211. *Napoléone impériale*. Napoleonaea imperialis. Premier genre d'un nouvel ordre des plantes. Les Napoléonées. [Paris (De l'imprimerie de Fournier fils) 1804]. Fol. (*Napoléone*).
Publ.: between 8 Oct and 24 Dec 1804, 1 p., *1 pl*. − Copies in Bibliothèque nationale (Paris, Département des Estampes) and at the Bibliothèque centrale du Muséum d'Histoire naturelle (Paris; no. 63.736). For an exhaustive study see Heine (1967).
Ref.: Palisot de Beauvois, A.M.F.J., Décade philosophique 44: 198-205. 1805 (full version of paper delivered at Acad.).
Anon., Allg. teut. Gart.-Mag. 2: 223-225. 1805 (German translation).
Fischer, F.E., Mém. Soc. Natur. Moscou 1: 92-93. 1806, ed. 2: 65-66. 1811.
Heine, H., Adansonia ser. 2. 7(2): 115-140. 1967.

7212. *Flore d'Oware et de Bénin, en Afrique*, par A.M.F.J. Palisot-Beauvois, ... à Paris (De l'imprimerie de Fain [vol. 1 only: jeune] et Compagnie) 1804-1807 [1803-1820]. Fol. (*Fl. Oware*).
Publ.: The folio edition, with the above title, was published in 20 parts between Jan 1805 and Oct 1820. However, the first part was published twice: once in quarto (2 Oct 1803) and once in folio. The quarto edition was not continued.
Quarto ed.: 2 Oct 1803, p. [i*-v*], [i]-v, [1]-10, *pl. 1-6*. *Copies*: Bibliothèque nationale, Paris; Bibl. centr. Muséum d'hist. nat. (P). See Heine (1967) for full details. The text corresponds with p. 1-8 of the folio edition; Heine discusses nomenclatural consequences of his discovery of this forerunner of the *Flore*. − Title-page: "Flore d'Oware et de Bénin, première livraison, composée de six planches, savoir: ..." à Paris (chez Bleuet, libraire, place de l'École, no. 45) Vendémiaire an xii."
Folio ed.: 2 vols., title and imprint as in heading. − *Copies*: B, BR, G, MICH, MO, NY, Teyler, US, USDA; IDC 5277. For details on dates see TL-1/940. Some copies (e.g. B) have only *pl. 1-30* col., the rest uncol.
1: [*iii*]-*xii*, [*1*]-*100*, *pl. 1-60* (colour-printed stipple-engr. finished by hand; artists: B. Mirbel (*1-18*), Sophie de Luigné, Jean Gabriel Prêtre).
2: [*1*]-*95*, *pl. 61-120* (id.; by J.G. Prêtre).
Facsimile ed.: announced, but not yet published by J. Cramer (1974).

vol.	part	pages	plates	dates
1	1	[i*-v*], [i]-v, 1-80 [iii]-xii, [1]-8	*1-6*	2 Oct 1803 quarto late Jan 1805 folio
	2	9-16	*7-12*	late Jan 1805
	3	17-32	*13-18*	20 Mai 1805
	4	33-40	*19-24*	Jun 1805
	5	41-48	*25-30*	late Oct 1805
	6	49-60	*31-36*	Mai 1806
	7	61-72	*37-42*	late Sep 1806
	8	73-80	*43-48*	late Oct 1806
	9	81-88	*49-54*	10 Nov 1806
	10	89-100	*55-60*	23 Feb 1807
2	11	[1]-12	*61-66*	18 Apr 1808
	12	13-24	*67-72*	6 Aug 1810
	13	25-32	*73-78*	1810 or 1811, before 26 Aug
	14	33-44	*79-84*	26 Aug 1816

vol.	part	pages	*plates*	dates
2	15	45-52	*85-90*	26 Aug 1816
	16	53-	*91-96*	21 Feb 1818
	17	-72	*97-102*	20 Apr 1818
	18	73-	*103-108*	24 Mai 1819
	19	-84	*109-114*	19 Jul 1819
	20	85-95	*115-120*	16 Oct 1820

The genus described as *Napoleona* (vol. 2, p. 29, *t. 78*) was previously published in a separate publication by Palisot de Beauvois (between 6 Oct and 24 Dec 1804) as *Napoleonaea*.

The first five parts (*pl. 1-30*) seem to have been published with coloured plates only; the publisher's announcements mentioned issues with col. and uncol. plates. The folio edition was printed in 200 copies on vellum paper. – Palisot de Beauvois' original herbarium at Geneva contains the types of the *Flore d'Oware et de Bénin*. – Ten unpublished plates illustrating algae at P (Ms 202).

Ref.: Anon. [announcement of publication with many details in] Journal typographique [for] 25 pluviose 13 (14 Feb 1805).

Barnhart, J.H., [in Merrill, E.D.], Proc. Amer. philos. Soc. 76: 914-920. 1936.

Heine, H., Adansonia ser. 2. 7: 115-140. 1967.

Jussieu, A.L. de, Décade philosophique 43(10): 1-8. 1804 [review of quarto edition of part 1].

Marshall, H.S., Kew Bull. 6: 43-49. 1951.

Merrill, E.D., Proc. Amer. philos. Soc. 76: 914-920. 1936.

Palisot de Beauvois, A.M.F.J., Napoléone impériale. Napoleonaea imperialis. Paris [1804]; full text: Décade philosophique 44: 198-205. 1805.

Steenis-Kruseman, M.J. van, Fl. males. Bull. 19: 1143-1144. 1964 (dates presentation parts to Académie Sci.).

Steenis-Kruseman, M.J. van and W.T. Stearn, Flora malesiana ser. 1. 4(5): ccv. 1954.

7213. *Prodrome des cinquième et sixième familles de l'Aethéogamie. Les mousses. Les lycopodes*. Par A.M.F.J. Palisot-Beauvois, ... Paris (de l'imprimerie de Fournier fils) 1805. Qu. (*Prodr. aethéogam.*).

Publ.: 10 Jan 1805 (SY p. 19, 25), p. [i*], [i]-ii, [1]-114. *Copies*: BR, FH(3), G, H, MO-Steere, PH, USDA; IDC 5003. One of the FH copies is on much heavier paper than the other copies: obviously there was a special issue on velin type paper. Part of the text was originally published in *Mag. Encycl.* 9e Année, 5(19): 289-330 [=1-40 reprint], 21 Feb 1804 (sic), a further part in 5: 471-483, early Mar 1804. The completed work was presented to the Académie des Sciences on 1 Sep 1806; Sayre's dates are based on the Journal typographique and a review in Mag. Encycl. 9e Année, 5(22). Crum (1952) supplies a list of generic names and binary combinations which must be dated from 1804.

Engl. transl.: Konig and Sims, Ann. Bot. 2: 218-251. 1806.

Note: "Aetheogamie" stands for "Cryptogams", Palisot was of the opinion that there was nothing "cryptic" about the reproduction of these groups; it was at most "unusual" (*aethes, insolitae*).

7214. *Nouvelles observations sur la fructification des mousses et des lycopodes*. Lues à la Classe des Sciences physiques et mathématiques, le 22 avril 1811, par M. Palisot de Beauvois, ... [Paris (Imprimerie de Mme Veuve Courcier) 1811]. Qu. (*Nouv. observ. mousses*).

Publ.: Aug-Sep 1811 (Acad. 9 Sep 1811), p. [1]-32, *1 pl.* (uncol.). *Copies*: BR, G, MO, NY(2). – Reprinted from J. Phys. 73: 89-109, *pl. 1*. August 1811.

7215. *Essai d'une nouvelle agrostographie*; ou nouveaux genres des Graminées; avec figures représentant les caractères de tous les genres ... Imprimerie de Fain. Paris (chez l'auteur, ...) 1812. Oct. and Qu. (*Ess. Agrostogr.*).

Octavo issue: Dec 1812 (see Stafleu 1971), p. [i*-xii*], [i]-lxxiv, chart, [1]-182, [183-184, err.]. *Copies*: BR, MICH, NY, PH, U, US, USDA; IDC 7. – 500 copies issued.

Quarto issue: Dec 1812, pagination as above. *Copies*: FI, K (inf. C.E. Hubbard), MO. – 100 copies issued.
Atlas: Dec 1812, p. [1]-16, *pl. 1-25* (uncol. copp.). *Copies*: BR, PH, US, USDA; IDC 7. – The plates are by Jean Gabriel Prêtre, engraved by Canu.
Errata: some copies of the octavo edition have two extra items of errata, referring to p. 176 and 179. In addition some copies (e.g. at K and MO) carry a *supplément à l'errata* which must be of later date (see Niles p. 207; text reprinted in Stafleu (1971)).
Notes: part of the manuscript is at P (MS 241); Palisot's own copy, with notes, is at USDA. For a contemporary review see Desvaux, J. Bot. Agr. [Desvaux] 1: 63-77. 1813. – The facsimile edition "Zug 1967" mentioned in TL-1, was never published. The introduction was published by Stafleu (1971), who lists the herbaria used by Palisot for the *Essai*. – For a continuation of the *Agrostographie* see Lestiboudois, *Essai Cypér.*, TL-2/4430, in part based on a Beauvois manuscript.

7216. *Lettre de M. Palisot*, Bon de Beauvois, à M. Delamétherie. Paris [Imprimerie de Mme Ve Courcier)] 1814. Qu.
Publ.: Jul-31 Dec 1814 (letter dated 15 Jul 1814; journal issue for Jul 1814), p. [1]-11. *Copies*: FH, H, PH. – Reprinted from J. Phys. 76: 5-15. Jul 1814. To be cited from journal. – Preliminary publication on the subject treated in 1822 in his *Muscologie* (see below).

7217. *Mémoire sur les Lemna*, ou lentilles d'eau, sur leur fructification et sur la germination de leurs graines ... [Paris (Veuve Courcier impr.) 1816]. Qu.
Publ.: Feb 1816 (in journal), p. [1]-15, *pl. 1* (uncol. copp. by Jean Gabriel Prêtre). *Copies*: FI, G. – Reprinted from J. Phys. 82: 101-115, *pl. 1*. Feb 1816. To be cited from journal.

7218. *Muscologie ou traité sur les mousses*, ... Paris (de l'imprimerie d'Hautel) 1822 [1823], text Oct., plates Qu. (*Muscol.*).
Publ.: 7-31 Dec 1822 or 1823 (MD), p. [1]-88, *pl.1-11* (with text, uncol. copp., Jean Gabriel Prêtre). *Copies*: FI, H, MO-Steere, NY.; IDC 5064. – A separate with independent pagination reprinted from the Mémoires de la Société Linnéenne de Paris 1: 388-472, *pl. 1-11*. 1822 [or 1823]. The separate and the original publication in the Mémoires must be assumed to have been issued at the same time, probably even as late as Dec 1823 [fide W.D. Margadant].

Palla, Eduard (1864-1922), Moravia-born Austrian botanist; Dr. phil. Wien 1887; from 1888 at the University of Graz; habil. 1891; titular professor and "Adjunct" 1901; extra-ordinary professor 1909; ordinary professor 1913. (*Palla*).

HERBARIUM and TYPES: GZU; some duplicates of Java collections at W.

BIBLIOGRAPHY and BIOGRAPHY: Backer p. 658; Barnhart 3: 43; BJI 1: 43; BM 8: 973; CSP 17: 684-685; De Toni 4: xliv; DTS 1: 216, 6(4): 183; Futàk-Domin p. 454; Kew 4: 215; Klásterský p. 140; Langman p. 562; LS 19723- 19725; MW p. 376; Nordstedt p. 25; Quenstedt p. 326; RS p. 132 (author of *Schoenoplectus*); SK 1: 398, 8: lxxiii; TL-2/2421; Tucker 1: 535; Zander ed. 10, p. 699, ed. 11, p. 797.

BIOFILE: Anon., Biogr. Jahrb. 4: 365; Bot. Centralbl. 48: 239. 1891 (habil.), 84: 208. 1900 (to Buitenzorg), 86: 79. 1901 (adjunct Graz), 111: 576. 1909 (extraord. prof.); Bot. Archiv 3: 4-7. 1922; Bot. Not. 1922: 287 (d.); Nat. Nov. 23: 178. 1901 (adjunct Graz), 31: 554. 1909 (e.o. prof. Graz); Österr. bot. Z. 51: 107. 1901 (adjunct Graz), 59: 407 (e.o. prof.) 1909, 71: 152. 1922 (d. 5 Mai 1922).
Fritsch, K., Ber. deut. bot. Ges. 40: (86)-(89). 1923 (bibl., b. 3 Sep 1864; d. 7 Apr 1922).
Janchen, E., Cat. fl. austr. 1: 30. 1956.
Kneucker, A., Allg. bot. Z. 2: 199. 1896, 3: 14. 1897, 7: 80. 1901, 15: 184. 1909, 28/29: 56 (112). 1925 (d.).
Linsbauer, K., Grazer Tagespost 8 Apr 1922 (fide ÖbZ 71: 136. 1922).
Merrill, E.D., B.P. Bishop Mus. Bull. 144: 146. 1937; Contr. U.S. natl. Herb. 30: 233. 1947.
Palla, E., Allg. bot. Z. 13: 49. 1907.

COMPOSITE WORKS: (1) Editor of *Atlas der Alpenflora*, ed. 2, see under A. Hartinger, TL-2/2421.
(2) *Cyperaceae*, in W.D.J. Koch, *Syn. fl. deut. schweiz. Fl.* ed. 3. 3: 2515-2550. Jan 1905, 2551-2681. Sep 1905, TL-2/3804.
(3) *Cyperaceae*, in A. Usteri, Fl. São Paulo 154-162. 1911.

7219. *Zur Systematik der Gattung Eriophorum* [Bot. Zeit. 54, 1896]. Oct.
Publ.: 16 Aug 1896, p. [i, cover-t.p.], [141]-157, [158], *pl. 5* (uncol. lith. auct.). *Copy*: G. – Reprinted and to be cited from Bot. Zeit. 54: [141]-157, [158]. 1896.

7220. *Die Gattungen der mitteleuropäischen Scirpoideen*, Karlsruhe (Verlag von J.J. Reiff) 1900. Oct.
Publ.: Oct, Dec 1900 (in journal), p. [1]-10. *Copies*: FI, G. – Reprinted and to be cited from Allg. bot. Z. 1900: 199-201, Oct, 213-217. Dec 1900.

Pallas, Johann Dietrich (*fl.* 1758), German botanist. (*J. Pall.*).

HERBARIUM and TYPES: Unknown.

BIBLIOGRAPHY and BIOGRAPHY: Barnhart 3: 43; Dryander 3: 279; PR 6902 (ed. 1: 7713).

BIOFILE: Kirschleger, F., Fl. Alsace 2: xxxviii.

7221. Dissertatio medica inauguralis de *Chrysosplenio* quam adsistente divina gratia ex decreto gratiosae Facultatis medicae pro licentia privilegia et honores doctorales legitime consequendi solenni eruditorum examini submittit Johannes Dietericus Pallas tabernensis-alsata d. xxviii aprilis mdcclviii h.l.q.c. Argentorati [Strasbourg] (Typis Jo. Henrici Heitzii ...) [1758]. Qu. (*De Chrysospl.*).
Publ.: 28 Apr 1758, p [i-vi], 1-20. *Copy*: WU (inf. W. Gutermann).

Pallas, Peter [Pyotr] **Simon** (1741-1811), German (Berlin) botanist and geographer; explorer of Russia and Siberia, pioneer naturalist of Northern Asia; Dr. med. Leiden 1759; 1761-1766 working in Holland and England; to Russia 1767; from 1768-1774 in the Ural Mts. and Western Siberia; 1774-1793 in St. Petersburg; 1793-1794 in S. Russia and the Crimea; settled later in the Crimea 1795-1810; retired to Berlin 1810-1811. (*Pall.*).

HERBARIUM and TYPES: Pallas' main personal herbarium was sold during his lifetime to A.B. Lambert (Miller 1970). At the Lambert sale one part went to Robert Brown (now at BM), another part to William Robertson (now also at BM) whereas a third part was acquired by Babington probably for the Cambridge University herbarium (communication frm Mrs. H.P. Miller). There is indeed a set of Pallas plants at CGE but the acquisition books mention nothing with respect to their provenance. In addition, however, there are about 660 specimens (species) of Russian and Siberian plants, collected by Pallas, in the Willdenow herbarium at B (Urban-Berl.). These specimens were retained by Pallas after the sale of his herbarium to Lambert and were left to Willdenow. Further duplicates in BM, BR, C, CAS, CGE, FI, G, GJO, H, HAL, LD, LE (important set!), LINN, LIV, LZ, M, MANCH, MO, MW, OXF, S, SBT, UPS, W. – Some *letters* at G.

NOTE: (1) Pallas took part in the expedition which was sent out by the St. Petersburg academy, at the instigation of Catherine the Great, to make astronomical observations in connection with the conjunction of the Sun and Venus. Itinerary: Departure from St. Petersburg 21 Jun 1768, Moscow, Vladimir, Mourom, Arzamas, Kazan, Simbirsk (for the winter), March 1769 to Samara, Orenburg, Gourier, Caspian sea, back to Orenburg, Oufa (winter), 16 Mai 1770 over the Ural to Jekaterinenburg, Tchèliabinsk, Tobolsk, (Dec. 1771), Altai mountains, Omsk, Kolyvan, Tomsk,Krasnojarsk (7 March 1772 on to), Irkoutsk, Lake Baikal, Oudinsk, Sélinghinsk, Kiakhta, Amour River, Krasnojarsk (winter), Tara, Jaïtskoï-Gorodsk, Astrakhan, Tzaritzin, St. Petersburg, 30 July 1774. This journey is described in Pallas' *Reise*. From 1774-1793 Pallas resided mainly in

St. Petersburg working e.g. on his *Flora rossica*. From 1793 to 1794 he made an official journey to Southern Russia and the Crimea. Leaving St. Petersburg on 1 February 1793, Pallas, accompanied by the artist Geissler, travelled to Tzaritzin, Sarepta and Astrakan. On 26 August 1793 he left Astrakhan for the Caucasus (Georgievsk, Pod-Kouma, Constantinogorsk, Bechtau Mountain), spent the winter in the Crimea and returned to St. Petersburg via Perekop, Kherson, and Nicolaïev. Apart from his botanical publications Pallas is known for his publications on the peoples of Mongolia. Pallas was in contact with many of his fellow botanists in W. Europe, such as Joseph Banks.

NOTE: (2) A.B. Lambert acquired the Pallas herbarium in 1808 at a London auction. A copy of the catalogue of the auction, describing the collection, is at PH: "A catalogue of a valuable collection of antiquities, and natural history; consisting of ... together with the superb herbarium of those celebrated botanists Pallas & Noezens; ... which will be sold by auction by Messrs. King and Lochee, ... on Monday, April 25, 1809, and three following days", ... printed by J. Barker, ... [1808], p. [1]-31.

BIBLIOGRAPHY and BIOGRAPHY: ADB 25: 81-98; AG 1: 214, 5(1): 169, 174, 178, 215, 6(2): 875; Backer p. 419; Barnhart 3: 43; BM 4: 1504-1505, 8: 973; Bossert p. 298; Bret. p. 311; Clokie p. 220; CSP 4: 743-744; Dawson p. 644-646, 950; De Toni 1: xcvi; Dryander 3: 126, 172, 174, 557; DSB 10: 283-284 (q.v. for modern Russian studies); DU 221-223; Frank 3 (Anh.): 74; GF p. 70; GR p. 36, *pl.* [*16*], cat. p. 69; Hegi 4(1): 155; Henrey 2: 36; Herder p. 469; Hortus 3: 1201 ("Pall."); HU nos. 662, 672, (2(2): 427-428, 436-439); IH 1 (ed. 7): 339, 2: (in press), Jackson p. 7, 113, 124, 326, 443; JW 4: 400; Kew 4: 215; Lasègue p. 578 [index]; LS 19726; MW p. 376, suppl. p. 275; NI 1482-1484, see also 1: 224, 2: 722, suppl. p. 54; Plesch p. 353-354; PR 2154, 6903-6908 (ed. 1: 7714-7722); Quenstedt p. 326-327; Reher 5: 645-646; RS p. 133; Sotheby 578, 579; TL-1/414, 942a-946, and see index; TL-2/2049, 2203, 2285 and see indexes; TR p. 211-214, nos. 1049-1062; Tucker 1: 535-536; Urban-Berl. p. 380, 412, 415; Zander ed. 10, p. 699, ed. 11, p. 797.

SUBJECT INDEX TO BIOFILE:
Bibliography: Barlett (1949), Bourdeille (1892), Jackson (1900), Junk (1910-1913), Lipschitz (1975), Martini (1803), MacAtee (1949), Nissen (1969), Percheron (1837), Poggendorff (1863); see also section Biography and Bibliography above.
Biographical accounts: Anon. (Encycl. brit. 1911), Borodin (1908), Cuvier (1845), Jourdan (1824), König (1805), Köppen (1907), Marakuew (1877), Nehring (1890), Ratzel (1887), Rudolphi (1812) (important!), Smith (1839).
Collections: Candolle (1880), Kukkonen (1971), Lambert (1811), Lindemann (1886), Miller, J.S., Murray (1904), Rudolphi (1812), Rudolphi (1812), Stansfield (1935), Werner (1955).
Correspondence: Bridson (1980), Fries (1908-1917, with Linnaeus), Hall (1830), Hulth (1922-1943), Lefebvre (1976, v. Marum), Stieber (1981), Urness (1967, Pennant).
Exploration: Coats (1969), Herder (1888), Masterson (1948), Merrill (1954), Pusanov (1974).
Nomenclature: Cockerell (1926), Dandy (1967), Kuntze (1892), Lada-Mocarski (1969).
Obituary: Desvaux (1813).
Palaeontology: Andrews (1980), Zaunick (1925), Zittel (1901).
Portraits: Dörfler (1907), Stafleu (1973), Wittrock (1903, 1905).
Tombstone: Anon. (1852, 1853, 1855).
Zoology: Zaunick (1923).

BIOFILE: Andrews, H.N., The fossil hunters 361. 1980.
Anon., Bot. Zeit., Regensburg 3: 368. 14 Dec 1804 (P. on his Crimean farm near Achmetschet); Bot. Zeit. 10: 924-925. 1852 (tombstone), 11: 319. 1853 (id.); Bull. Soc. bot. France 2 (bibl.): 343. 1855 (tombstone); Flora 18, Int. Bl. 1: 24-26. 1835 (on *Fl. ross.* incl. plates 2(1); Encycl. brit. ed. 11. 20: 637-638. 1911.
Asmous, V.C., Chron. bot. 7(1): 14. 1942 (centenary birth).
Bartlett, H.H., 55 rare books 1949, no. 53 (*Fl. ross.*).
Borodin, J., Trav. Mus. bot. Acad. Sci. St. Petersb. 4: 88-89. 1908.
Bourdeille de Montresor, Bull. Soc. Imp. Natural. Moscou ser. 2. 7: 434-436. 1892 (bibl.)

Burckhardt, R.W., The spirit of system 50-51. 1977.
Candolle, Alph. de, Phytographie 438. 1880 (coll.).
Candolle, A.P. de, Mém. souvenirs 88-89. 1862.
Carpenter, N.M., Amer. Midl. Natural. 33(1): 77. 1945.
Coats, A.P., The plant hunters 52-56. 1969.
Cockerell, T.D.A., Torreya 26: 67-69. 1926 (Lamarck's names in Voy. 8, 1794).
Cuvier, Dict. Sci. nat. 61: 198-200. 1845.
Dandy, J.E., Index of generic names of vascular plants 1753-1774. Utrecht 17. 1967.
Desvaux, N.A., J. Bot. appl., Paris 1: 239-240. 1813.
Dörfler, I., Botaniker-Porträts no. 37. 1907 (portr.) (text R. Kräusel).
Ewan, J., Proc. Amer. philos. Soc. 103: 814. 1959; *in* F. Pursh, Fl. Amer. sept. facs. repr. 1979, intr. p. 23.
Fries, Th. M., Bref Skrifv. Linné ser. 1. 2: 349. 1908, 5: 185, 223. 1911, 6: 65, 194. 1912, 7: 127. 1917.
Fussel, G.E., More old Engl. farming books 106. 1948.
Gilbert, P., Comp. biogr. lit. deceased entom. 286. 1977.
Hall, H.C. van, Epist. Linnaei 1830 (letters P. to Linnaeus; TL-2/2285).
Herder, F, v., Bot. Jahrb. 9: 432, 442-443. 1888 (itin.)
Hulth, J.M., Bref Skrifv. Linné ser. 1. 8: 187, 188, 196. 1922, ser. 2. 1: 427 [index], 2: 296 [index]. 1943.
Jackson, B.D., J. Bot. 38: 189. 1900 (*Fl. ross.* at K); Proc. Linn. Soc. 134 (suppl.): 19. 1922 (pl. in LINN).
Jessen, K.F.W., Bot. Gegenw. Vorz. 276, 300, 345, 389, 470. 1864.
Jourdan, A.J.L., Dict. sci. méd., Biogr. méd. 6: 353-356. 1824 (bibl.).
Junk, W., Rara 45-46. 1910-1913.
Keppen, F.P., Uchenye trudy P.S. Pallasa, St. Petersburg 1895, 54 p. (n.v.).
König, C.D.E. & Sims, J., Ann. Bot. 1(2): 396-397. 1805.
Köppen, F.T., Bibliotheca Zool.-ross. 1:170-171. 1907 (references to Pallas biographies in Russian).
Kukkonen, I., Herb. Christ. Steven 76. 1971 (material at H).
Kuntze, O., Rev. gen. pl. 1: cxxxviii. 1892 (*Fl. ross.*).
Lada-Mocarski, V., Bibl. books Alaska 120-122. 1969 (on *Neue Nordische Beyträge*).
Lambert, A.B., Trans. Linn. Soc. 10: 256-265. 1811 (account of herbarium) (see also 11: 419).
Lefebvre, E. & J.G. de Bruyn, Martinus van Marum 6: 111, 183. 1976.
Lindemann, E. v., Bull. Soc. Imp. Natural. Moscou 61: 51-52. 1886 (coll. at MW).
Lipschitz, S , Florae URSS fontes 222 [index]. 1975.
Lysaght, A.P., Joseph Banks in Newfoundland 39, 94, 251, 332, 448. 1971.
McAtee, W.L., Auk 66: 210-211. 1949 (on *Reise*).
Mägdefrau, K., Gesch. Bot. 287. 1973.
Magnus, P., Verh. bot. Ver. Brandenburg 29: 170-171. 1888 (p.p. erroneous).
Marakuew, W., P.S. Pallas, sein Leben ... Moskou 1877 (fide Jackson, p. 7).
Martini, Bot. Zeit. (Regensburg) 2(5): 79-80. 14 Mai 1803 (ser. *Sp. astragal.*).
Masterson, James R., Bering's successors 1745-1780, contributions of Peter Simon Pallas to the history of Russian exploration toward Alaska, Univ. of Seattle Press, Seattle 1948, iv, 96 p. (Pacific Northwest Quart. 38(1, 2)).
Merrill, E.D., Chron. bot. 14: 208, 211. 1954 (bot. Cook's Voy.).
Miller, J.S., Taxon 19: 534-535. 1970 (sale Lambert herb., dispersal of Pallas' personal herb.).
Murray, G., Hist. Coll. BM(NH) 1: 172. 1904.
Nehring, A., Naturw. Wochenschr. 5: 243-245. 1890 (brief biogr.).
Nissen, C., Zool. Buchill. 307-309. 1969.
Nordenskiöld, E., Hist. biol. 260, 262, 353. 1949.
Palmer, T.S., Condor 30(5): 290. 1928 (ornithol. eponymy).
Percheron, A., Bibl. entom. 1: 305-306. 1837.
Poggendorff. J.C., Biogr.-lit. Handw.-Buch 2: 348. 1863.
Poiret, J.L.M., *in* Lamarck, Encycl. méth., Bot. 8: 744-746. 1808.
Pusanov, I.I., Byull. Mosk. Obshch. Isphyt. Prir., Biol. 79: 22-40. 1974 (Pallas in the Crimea 1794-1810).
Radde, G., Grundz. Pflanzenverbr. Kaukausländ. 2. 1899 (Veg. Erde 3).

Rajkov, B.E., Peter-Simon Pallas, Ocherki po istorii evel. idey v Rossii do Darwina, Moskva (Leningrad l947, 197 p.).
Ratzel, F., Allg. deut. Biogr. 25: 81-98. 1887 (good concise biogr., b. 22 Sep 1741, d. 8 Sep 1811).
Rudolphi, K.A., Beytr. Anthropol. allg. Naturgesch., Berlin 1-78. 1812 (biogr., portr., bibl., important original source).
Schlechtendal, D.F.L. von, Bot. Zeit. 13: 375-376. 1855 (memorial erected Berlin 1854).
Sherborn, J. Soc. Bibl. nat. Hist. 1: 142. 1938 (*Fl. ross.*).
Smith, C.H., Dogs, *in* Jardine, W., Naturalist's library (Mammalia 9), 25: 17-76. 1839 (mainly based on Cuvier's Éloge; some new material; bibl. incomplete).
Smith, E.C., Nature 148: 334-335. (brief biogr.).
Sontzov, A.A., Drevnyaya i nov. Rossia 1(3): 279-289. 1876 (Pallas in the Crimea).
Stafleu, F.A., *in* Daniels, G.S. & Stafleu, F.A., Taxon 22: 438. 1973 (portr.); Taxon 12: 75. 1963 (*Fl. ross.*).
Stansfield, H., Handb. guide herb. coll. Liverpool 41, 48-49, *pl. 7*. 1935 (coll., portr.).
Stearn, W.T., J. Arnold Arb. 22: 225-230. 1941 (on *Fl. ross.*).
Stieber, M.T. et al., Huntia 4(1): 84. 1981(arch. mat. HU).
Stresemann, E., Beitr. Vogelk., Leipzig 5: 207-209. 1957 ("Pallas" letzte Tage").
Svetovidov, A.N., Arch. nat. Hist. 10(1): 45-64. 1981 (on Zoographia rosso-asiatica; biogr. details).
Swainson, W., Taxidermy (Cabinet Encycl.) 282-287. 1840 (biogr. note).
Taton, R. ed., Science in the 19th century 360, 439-460, 560. 1965.
Thieneman, A., Zool. Anz. 36: 417-419. 1911 (on P's ideas on taxonomic relationships and use of a phylogenetic tree).
Urness, Carol, A naturalist in Russia. Letters from Peter Simon Pallas to Thomas Pennant. Minneapolis 1967 (rev. by J. Stannard, Science 159; 26 Jan 1968).
Van de Velde, A.J.J., Biol. Jaarb. Dodonaea 22: 227-242. 1955 (bibl. notices).
Werner, K., Wiss. Z. Univ. Halle, Math.-Nat. 4(4): 777. 1955 (pl. HAL).
Wittrock, V.B., Acta Horti Berg. 3(2): 175. 1903, 3(3): 189, xx. *pl. 31*, lxxx. 1905 (portr.).
Zaunick, R., Pallasia 3: 1-37. 1925 (P. as founder of knowledge on palaearctic vertebrates; portr., partial bibl.; important critical and well-documented study with notes on P's zoological activities).
Zittel, K.A. von, Hist. geol. palaeontol. 553. 1901.

COMPOSITE WORKS: Editor of S.G. Gmelin ... *Reise durch Russland*, vol. 4, 1784 (with a life of Gmelin by Pallas) see TL-2/2049.

HANDWRITING: *in* S.V. Lipschitz & I.T. Vasilczenko, Central Herbarium of the USSR 115, 116. 1968.

EPONYMY: (genera): *Neopallasia* P. Poljakov (1955); *Pallasa* Cothenius (1790, *orth. var.* of *Pallasia* Linnaeus f.); *Pallasia* Scopoli (1777); Pallasia Linnaeus f. (1782); *Pallasia* J.F. Klotzsch (1853); *Pallassia* Houttuyn (1775). *Note:* The etymology of *Pallasia* L'Héritier ex W. Aiton (1789) could not be determined; (journal:) *Pallasia*, Eine Zeitschrift für Wirbeltierkunde vornehmlich des paläarktischen Faunengebietes in Verbindung mit namhaften Zoologen des In– und Auslandes. Dresden. Vol. 1-3. 1923-1925. (Volume 1 is entitled *Zoologica palaearctica*, Eine Zeitschrift ... Auslandes).

7222. P.S. Pallas ... *Elenchus zoophytorum* sistens generum adumbrationes generaliores et specierum cognitarum succinctas descriptiones cum selectis auctorum synonymis ... Hagae-Comitum ['s Gravenhage; The Hague] (apud Petrum van Cleef) 1766. Oct. (*Elench. zooph.*).
Publ.: 1766 (pref. Idib. Mart. 1766), p. [i]-xvi, [17]-451. *Copy*: NY.
Dutch translation (1): 1768, 1 vol., p. [i]-xxxvi, [37]-50, [1]-654, *pl. 1-14* (uncol. copp. P. Boddaert). *Copy*: U. – *Lyst der plant-dieren*, bevattende de algemeene schetsen der geslachten en korte beschryvingen der bekende zoorten met de bygevoegde naamen der schryveren, ... vertaald en met aanmerkingen en afbeeldingen voorzien door P. Boddaert [Pieter Boddaert 1730-1796] ... te Utrecht (by A. van Paddenburgh en J. van Schoonhoven) 1768. Oct.

Dutch translation (2): 1768, 2 vols. *Copies*: BR, U. – "*Natuurlijke historie der plant-dieren* bevattende ... P. Boddaert ... Amsterdam (by J.B. Elwe) 1768. Oct.
1: [i]-xxxvi, [37]-50, [1]-274, *pl. 1-8* (hand-col. copp., P. Boddaert).
2: [i], *275-654, pl. 9-14.* (id.).
Dutch translation (3): 1798, 2 vols. *Copy*: BR. – "Natuurlijke historie der plant-dieren bevattende de algemeene schetzen der geslachten en korte beschryvingen der bekende zoorten, met de bygevoegde naamen der schryveren in het Latyn beschreeven door P.S. Pallas, ... vertaald, met aanmerkingen en afbeeldingen voorzien door P. Boddaert [Pieter Boddaert, of Vlissingen] ... te Amsterdam (by J.B. Elwe) 1798. Oct., 2 parts.
1: [i]-xxxvi, [37]-50, [1]-274, *pl. 1-8* (col. copp.).
2: [i], 275-654, *pl. 9-14.*
German translation: 1787, p. [1]-344, *pl. 1-12* (uncol. copp.). *Copy*: MICH. – P.S. Pallas ... Charakteristik der Thierpflanzen, ... Nürnberg (verlegt von der Raspischen Buchhandlung) 1787, translated by Christian Friedrich Wilkens and edited posthumously by Johann Friedrich Wilhelm Herbst. – See also Johann Samuel Schröter, *Namen Register* über die von dem seel. Herrn Inspector Wilkens übersetzte Charakteristik der Thierpflanzen ... Nürnberg (in der Raspischen Buchhandlung) 1798. (*Copy* at GOET; fide Zaunick 1925).

7223. P.S. Pallas ... *Miscellanea zoologica* quibus novae imprimis atque obscurae animalium species describuntur et observationibus iconibusque illustrantur ... Hagae Comitum ['s Gravenhage; The Hague] (apud Petrum van Cleef) 1766. Qu. (Misc. zool.).
Orig. ed.: 1 Nov-31 Dec 1766 (p. 6: 15 cal. Nov 1766), p. [3]-6, [vii]-xii, [1]-224, *pl. 1-14* (uncol. copp. J.J. Bijlaart). *Copy*: US.
Re-issue: 1778, p. [1]-6, [vii]-xii, [1]-224, *pl. 1-14. Copy*: US.
Note: An extended version of the text was incorporated in P.S. Pallas, *Spicilegia zoologica*, Berlin 1767-1780, not treated here.

7224. P.S. Pallas ... *Reise durch verschiedene Provinzen des russischen Reichs* ... St. Petersburg (gedruckt bey der Kayserlichen Academie der Wissenschaften) 1771-1776, 3 vols. Qu. *Reise russ. Reich.*).
Publ.: 1771-1776, in 3 "Theilen" with vols. 2 and 3 each subdivided in 2 "Büchern" (parts). *Copies*: MO, PH, Teyler; IDC 5016. – Pallas' generic names in this work are dealt with by Dandy (1967). The two copies seen by us show differences in the numbers of plates; the list below is that of the Teyler copy. – See also the analysis given by Nissen, Zool. Buchill. no. 3075.
1: 1771 (p.[xii]: 28 Apr 1770), p. [i*-iv*], i-vi, [1]-504, *pl.* (non botanical): *1-6, 6bis, 7, 8a, 8b, 9, 10a, 10b, 11*; (botanical): *A-I, K-L*, [in copy MO] (all uncol. copp.). Botany on p. 479-504.
2(1): 1773 (p. [vi]: 19 Apr 1772), p. [i-vi], [1]-368.
2(2): 1773, p. [i-ii], [369]-744, 2 maps, *pl.* (non botanical): *1/2/5* (in one), *3/4, 5 bis* (botanical!), *6-14, A/B, C/D, E/F, G/H/I,* (botanical): *K-T, T*, U, W-Z.* – Botany on p. 733-744 (in "Anhang").
3(1): 1776 (p. [i-xxii]: 10 Feb 1776 o.s.), p. [i-xxii], [1]-454, *pl. 1/2, 3-7,* map (copp.; non botanical).
3(2): 1776, p. [455]-760, [i-xxxii], *pl.* (non botanical): *8, 9,* 2 maps; (botanical): *A-G, G*, H, I, K-Z, Aa-Gg, Gg*, Hh, Ii, Kk-Nn* (copp.). – Botany on p. 710-760 (in "Anhang").
Shortened version: 1776-1778, 3 vols. "P.S. Pallas ... Reise ... in einem ausführlichem Auszüge ..." Frankfurt & Leipzig (bey Johann Georg Fleischer). Oct. *Copies*: MO, NY.
1: 1776, p. [i-viii], [1]-384, 1-52, *pl. A, B, C, C*, C**, D-O (uncol. copp.).*
2: 1776, p. [i-vi], [1]-464, 1-51, [52-56], *pl. i, ii, A/B, C/D, E/I, K-T, T*, U-Z.*
3: 1778, frontisp., p. [i-x], 1-488, 1-80, [1-24], *pl.* A-Z, Aa-Nn [*41*, not *51 pl.* see note on p. [24]].
Second ed.: 1801, vol. 1 only, p. [i*-xii*], [1]-504, [505-510], *pl. 1-4, 6* (for 5), *6, 7* (large), *7* (small), *8, 8a, 9, 9a, 9b, 11, 11, A-I, K-L. Copy*: PH (orig. covers, unopened). – Zweyte Auflage, St. Petersburg (gedruckt bey der Kayserlichen Akademie der Wissenschaften) 1801. Qu.

French translation (1): 1788-1793, 5 vols. and atlas. Paris. Amended. – "*Voyages de M.P.S. Pallas, en differentes provinces de l'Empire de Russie, et dans l'Asie septentrionale*"; traduits de l'allemand, par M. Gauthier de la Peyronie ... À Paris ([1: chez Lagrange...] Chez Maradan,...) 1788-l793. Qu. *Copies*: BR, FI, G, MICH, PH, U.
1: 1788, p. [i*-iii*], [i]-xxxii, [1]-773, [774-776, err.], plants on p. 741-773.
2: 1789, p. [i-iv], [1]-550, [1, err.], plants on p. 544-550.
3: 1793, p. [i-iii], [1]-491, [492], plants on p. 482-491.
4: 1793, p. [i-iii], [1]-722, [1-2, err.], plants on p. 680-772.
5: 1793, p. [i-iii], [1]-559, [560], "additions et éclaircissements" from the journals by "Gmelin neveu" et al. from p. 383, plants on p. 495-516.
6: 1793, atlas, p. [i-iii], map, *pl.*[vol. 1:] *2-3, 4/5, 6-29*; [vol. 2]: *2-4, 5/6, 7/8, 9-17*; [vol. 3]: *1-7, 8/9, 10/11, 12-20, 21/22*; [vol. 4]: *1/2, 3-8, 9/10, 11-14, 15/16, 17-19, 20/21, 22-23, 24/25, 26-31, 32/33, 34-35*; [vol. 5]: *1, 2/3, 4/5, 6/7, 8-21* (uncol. copp.).
French translation (2): 1794 (an II de la République), 8 vols. and atlas. *Copies*: G, MO (atlas wanting), NY, PH. – "*Voyages du professeur Pallas*, dans plusieurs provinces de l'empire de Russie et dans l'Asie septentrionale"; traduits de l'allemand par le C. [citoyen] Gauthier de la Peyronie. Nouvelle édition, revue et enrichie de notes par les cc. [citoyens] Lamarck, ... Langlès, ... Paris (chez Maradan, ...). Oct.
Botanical zoological notes by Jean Baptiste Antoine Pierre Monnet de Lamarck (1744-1829); for comments on his botanical notes see Cockerell (1926).
1: [iii]-xl, [1]-422.
2: [i, iii], [1]-490.
3: [i, iii], [1]-492.
4: [i, iii], [1]-499.
5: [1]-448.
6: [1]-455.
7: [1]-448.
8: [i*-iii*], [i]-viii, [1]-463, contains natural history descriptions and Lamarck's "notes et observations".
9, Atlas: p. [i], *pl. 1-108 (41-108* botanical).
Russian translation: mentioned by Trautvetter (TR).

NOTE: We regret that we have not been able to see more copies of the various editions and translations of the *Reise*. Our data are incomplete and do not cover the variation found in the various copies mentioned in the literature but not seen by us.

7225. Sammlungen historischer Nachrichten über *die Mongolischen Völkerschaften* ... St. Petersburg (gedruckt bey der Kayserlichen Academie der Wissenschaften) 1776-1801. Qu. (*Mongol. Völkersch.*).
1: 1776, p. [i]-xiv, [1]-232, *pl. 1-7* (uncol. copp.), charts. *2*: 1801, p. [i*], [i]-x, [1]-438, [1-2, err], *pl. 1-22* (id.). *Copies*: Teyler, U.

7226. *Enumeratio plantarum* quae in *horti* viri illustris atque excell. Dni. Procopi à *Demidof*, ... Mosquae vigent, recensente P.S. Pallas, ... Petropoli [St. Petersburg] (typis Acad. Imder. [sic]. Scientiarum) 1781. Oct. (*Enum. hort. Demidof*).
Publ.: late Dec 1781 (t.p.; p. xxix: 10 Dec 1781), frontisp., p. [i]-xxxi, [1]-163, *pl. 1-2* (col. copp.). *Copies*: HU, USDA; IDC 843. – Title also in Russian. For a bibliogr. analysis see HU 662.
Note: A new catalogue was published, anonymously, in 1786, see PR ed. 1: 2437 and Rehder 1: 41, "Enumeratio plantarum ordine alphabetico undique collectarum ex quatuor plagis mundi, adjecta botanicorum characterum descriptione, quae in horto Procopii a Demidow ... Mosquae vigent". Mosquae, 1786, xiv, 469 p.

7227. *Flora rossica* seu stirpium Imperii rossici par Europam et Asiam indigenarum descriptiones et icones. Jussu et auspiciis Catharinae II. augustae edidit P.S. Pallas ... Petropoli [St. Petersburg] (e Typographia imperiali J.J. Weitbrecht) 1784-1788[-1831], 2 vols. Fol. †. (*Fl. ross.*).
1(1): Oct-Dec 1784 (p. viii: mid. Oct 1784; Gent. Mag. rd. Aug 1785, p. [i*-v*], i-viii, [1]-80, *pl. 1-50, 8b* (col. copp. Karl Friedrich Knappe), p. [i*] is an engraved and coloured title page.

1(2): probably Jan-Jun 1789 (t.p. 1788; p. [i], [1]-114, *pl. 51-100* (id.).
   Copies (vol. 1): BR, G, H, HU, MICH, MO, NY, U, USDA; IDC 5089. – See HU 672 for a bibliographical analysis and a list of plants illustrated on *pl. 1-100, 8b*. See Jackson (1900) for different states. Junk (Cat. 215, no. 259, 1979) offered a copy at Hfl. 15.000 (US $ 7.500).
2(1): 1815 (date on cover copy NY), possibly published only in 1831, see Jackson (1900), cover, *pl. 1-24*, (K copy: *–25*), [i, index], plates hand-coloured copp. *Copy*: NY. – For a description of the copy at K see Jackson. – Cover title: "*Flora rossica* ... P.S. Pallas. Tomi ii, Pars i. Berolini, C.G. Schoene. mdcccxv." PR cites a copy with 1815 from B-S. The anonymous note from St. Petersburg in Flora (1835) speaks simply of the "von Pallas ohne Text hinterlassenen 25 Tafeln (Tab. 101-125 [sic])."
Annonce: 1782, p. [1]-4. *Copy*: G. – "Annonce d'un ouvrage botanique sur les arbres, arbustes et plantes de l'Empire de Russie, qui sera publié par ordre et sous les auspices de sa majesté imperiale. [p. 4:] S. Petersburg ce 28 juillet v. St. 1782."
Reprint: text (only): 1789-1790, 2 parts (vol. 1, parts 1 & 2). *Copies*: G, NY, USDA. – "*Flora rossica* seu ... Pallas" ... Francofurti [Frankfurt a. M.] et Lipsiae [Leipzig] (apud Joannem Georgium Fleischer).
   1: Apr 1789 (Leipzig Easter fair), p. [i]-xxii, [1]-191.
   2: Apr 1790 (id.), p. [i], [1]-229.

7228. P.S. Pallas ... *Bemerkungen auf einer Reise in die südlichen Statthalterschaften des russischen Reichs* in den Jahren 1793 und 1794 ... Leipzig (bey Gottfried Martini) 1799-1801, 2 vols. Qu. (*Reise südl. Statthaltersch. russ. Reich.*).
   1: 1799, p. [i]-xxxii, [1]-516, *pl. 1-25* (col. copp.), *vign. 1-14*.
   2: 1801 (ALZ 21 Dec 1801, p. [i]-xxiv, [1]-525, [526-527, err.], *pl. 1-27* (id.), 3 maps, *vign. 1-14*. *Copy*: Teyler.
English translation: 1802-1803. *Copies*: MICH, PH. – Travels through the southern provinces of the Russian empire, in the years 1793 and 1794 ... London (printed by A. Strahan, ...) 1802-1803, 2 vols. Qu. (*Trav. s. prov. Russ. emp.*). – Re-issued 1812, 4 vols. (n.v.).
   1: 1802, p. [i]-xxiii, [xxiv], [1]-552, *pl. 1-25*, maps 1-3.
   2: 1803, p. [i]-xxx, [xxxi, err.], [1]-523, *pl. 1-27*.
French translation(1): 2 vols. Paris, 21 Mar 1802 (JT), n.v., also Paris 1805, 2 vols. (see Nissen, Zool. Buchill. no. 3066).
Id. (*2*): 1811, 4 vols. and atlas. *Copies*: MO (atlas wanting). "*Second voyage de Pallas*, ou entrepris dans les gouvernemens méridionaux de l'empire de Russie, pendant les années 1793 et 1794 ... traduit de l'allemand par MM de la Boulaye, ... et Tonnelier, ...". À Paris (Chez Guillaume, ... Déterville, ...) 1811. Oct.
   1: [i]-xxviii, [1]-372, *pl. 1-3* ("vignettes").
   2: [1]-383, *pl. 5-14* [*pl. 4* not in MO].
   3: [i,iii], [1]-387, *pl. 15-21*.
   4: [i, iii], [1]-376, *pl. 22-28*.
   Atlas: *55 pl.*
   Other issues: 2 vols. in quarto with atlas, "papier vélin". The octavo edition was also issued on laid as well as wove paper.

7229. *Species astragalorum* descriptae et iconibus coloratis illustratae a P.S. Pallas Eq. ... cum appendice. Lipsiae [Leipzig] (sumptibus Godofredi Martini) 1800[-1803]. Fol. (*Sp. astragal.*).
Publ.: 1800-1803 (in 13 parts, the last of which appeared in Feb 1803. The first four parts contained 8 p. of text each and 26 plates in all (GGA 4 Apr 1801); further details wanting); p. [i]-viii, [1]-124, *pl. 1-20, 20a-d, 21-43, 43b, 44-58, 58b, 59-60, 60b, 61-69, 70a, 70b, 71-91* (in all *99 pl.*; col. engr. by C.G.H. Geissler), *pl. 57* may have a pasted cancellans slip "Caespitosus"; original drawings in the Sächsische Landesbibliothek, Dresden (still extant?). *Copies*: Barneby, H, MO, NY, Teyler, USDA; IDC 5267. – For a review see Martini (1803); for a collation see Collins in Sotheby 580.

7230. *Illustrationes plantarum* imperfecte vel nondum cognitarum, cum centuria iconum, recensente Petro Simone Pallas ... Lipsiae [Leipzig] (sumptibus Godofredi Martini) 1803 [-1806]. Fol. (*Ill. pl.*).

Publ.: 1803-1806, in parts (further information wanting), p. [i-iii], [1]-68, *pl. 1-59* (col. copp. of drawings by C.G.H. Geissler; DU speaks of etchings, coloured by hand). *Copies*: BR, FI, MO, NY, PH, Teyler.

Palm, Björn Torvald (1887-1956), Swedish botanist and traveller; Dr. phil. Stockholm, lecturer at the University of Stockholm 1915-1918 and 1932-1939, assistant for plant pathology at Buitenzorg, Java, 1916, director at a plant-breeding station for tobacco at Deli, Sumatra, 1920, on plantations in Guatemala 1926-1929; professor at the University of Illinois, Urbana, 1930-1932; expert of plant-pathology at Stockholm 1945-1947; travelled in Madagascar 1912-1914, in Dutch East Indies 1915-1926, in Central America and the West Indies 1926-1930. (*Palm*).

HERBARIUM and TYPES: Types unknown; some other material at S.

BIBLIOGRAPHY and BIOGRAPHY: GR p. 497; KR p. 568-569; LS suppl. 2082-20840.

BIOFILE: Arwidsson, T. & Cedergren, G., Generalregister till Svensk botanisk tidskrift 1-20: 26-27. 1932.
Fries, R., A short history of botany in Sweden: 46, 97. 1950.
Lindman, S., Sv. män o. kvinn. 6: 8-9. 1949 (portr.).
Palm, B., Tropikerna var mitt liv, Stockholm 1943. 318 p. (autobiography 1912-1918); portr., 64 figs., 2 maps.
Palm, B., Farväl till tropikerna, Stockholm 1944. 309 p. (autobiography 1919-1930); portr., 56 figs., 1 map.
Peterson, B., Generalregister till Svensk botanisk tidskrift 21-40: 29. 1949.

EPONYMY: *Palmomyces* Maire (1926).

NOTE: The entry for B.T. Palm was submitted by O. Almborn.

7230a. *De embryologia generum Asteris et Solidaginis*. Zur Embryologie der Gattungen Aster und Solidago. Stockholm (Isaac Marcus Boktryckeri-Aktiebolag). Oct. (*Embryol. Aster. et Solidag.*).
Publ.: Oct-Dec 1914, p. [1]-18, 29 figs. in text. *Copy* LD. Issued as Acta Horti Berg. 5(4).

7230b. Studien über Konstruktionstypen und Entwicklungswege *des Embryosackes der Angiospermen*. Inauguraldissertation mit Genehmigung der mathematisch-naturwissenschaftlichen Fakultät der Hochschule zu Stockholm zur öffentlichen Beurteilung vorgelegt von Björn Palm. Die Verteidigung wird am 29 Mai um 10 Uhr vormittags im Hörsaal 3 stattfinden. Stockholm (Svenska tryckeriaktiebolaget) 1915. Oct. (*Stud. Embryosack. Angiosperm.*).
Publ.: 10 Mai 1915, p. [i-iv], [1]-257, 53 figs. in text. *Copy*: LD.
Ref.: Juel, O., Bot. Centralbl. 131: 146-147. 1916.

7230c. *Svenska Taphrinaarter*. Uppsala (Almqvist & Wiksells Boktryckeri-A.-B. [1917]. Oct. (*Svenska Taphrimaarter*).
Publ.: printed 21 Mai 1917 (ms. presented 10 Jan 1917), p. [1]-41, 9 figs. in text. *Copy*: LD. – Issued as Ark. Bot. 15(4) (1917) 1918. N.B.: Vol. 15, fasc. 1-2[=nos. 1-12] was issued 7 Oct 1918. Reprints of nos. 1-4 were probably distributed in 1917.

Palmberg, Johannes Olai (1640-1691), Swedish physician, clergyman and botanist; studied medicine and botany at Turku from 1663, "med. et phys. lector" at Strängnäs (near Stockholm) from 1672; parish priest at Turinge (Södermanland) from 1688. (*Palmberg*).

HERBARIUM and TYPES: Unknown.

BIBLIOGRAPHY and BIOGRAPHY: KR p. 569-570; PR 6910.

BIOFILE: Afzelius, A. (ed.), Egenhändiga Anteckningar af Carl Linneus om sig sielf 7.

1823 ("I had got several books, which I read day and night, ... as Arvidh Månson, Rydaholm, Örta-Book, Tillands, Flora Aboensis, Palmberg, Serta Florea Svecana (poor guides!)". (Transl. from the Swedish).
Anon., Biographiskt lexicon övfer namnkunnige svenska män 11: 16. 1845.
Collin, I., Sveriges bibliografi. 1600-talet 1: 684. 1944.
Djurberg, V., Hygiea 62: 357-405. 1900 and 68: 145-159. 1906. (Comprehensive biography with references).
Eriksson, G., Bot. hist. Sverige 91, 97, 139-141 (ill.). 1969.
Fries, E., Botaniska utflygter 3: 51-52. 1864.
Holmström, J.A., Bot. Not. 1849: 109-112. (Serta Florea Svecana: "In spite of all shortcomings, as well in figures as in descriptions, this work was for a long time almost the only guide to the Swedish flora"). (Transl. from the Swedish).
Lindman, S., Sv. män o. kvinn. 6: 14. 1949.
Linnaeus, C. Flora Svecica: XI-XII. 1745 (Serta Florea Svecana: "Descriptiones imperfectae, figurae rudes, vires, compilatae").

NOTE: This entry was submitted by O. Almborn.

7230d. *Serta Florea Svecana* eller Den Swänske Örtekrantz Gudhi den aldrahögste til ähra; dem Naturälskarom til lust och behagh. Den Studerande Ungdomen i Strängnäs til en lijten påminnelse och underwijsning ... Strängnäs (Z. Asp.) 1683. Oct. (*Serta fl. svec.*).
Ed. 1: printed 2 Jun 1683 (sec. t.p.), p. [1]-123, register [1-11]. 123 figs. without text. *Copies*: Kungl. Bibl. (Royal Library, Stockholm); LD (univ. libr.), SBT.
Ed. 2: Serta Florea Svecana Eller Swenske Örtekranz Aff Johane Palmberg ... Uhti Strangnääs Sammanflätad. Cum gratia et Privilegio. Åhr Efter Christi Börd 1684. Tryckt aff Zach. Asp Capit. Bookt. (= printer to the Cathedral Chapter). Strängnäs 1684. Oct. – *Publ.* c. Jul 1684 (Privilegium by King Charles IX, dated 7 Mai 1684; preface dated 28 Mai 1684), p. [i-xvi], [1]-416, register [1-44], 130 figs. in text. *Copies*: Almborn (plates partly handcolored), LD.
Ed. 3: Serta Florea Svecana, Eller: Swenske Örte-Krantz, Af Johanne Palmberg ... Sammanflätad. Men nu å nyo oplagd och förbättrad samt vid slutet försedd med en wacker Tilökning af många härliga och hälsosamma Örter. Af J.B. Steinmejer. Stockholm (Joh. L. Horns Boktr.) 1733. Oct. – *Editor*: J.B. Steinmejer (x-1738), headmaster at the German School at Stockholm. – *Publ.*: 1733 (preface dated 88 Mar 1732), p. [i-vii], [1]-282, register [1-62], appendix (by J.B. Steinmejer) [1]-118, register of appendix (by J.B.S.) [1-10]. No figs. Reissue: 1738 (unchanged, with new t.p.). *Copies*: Almborn, KVA.

Palmer, Charles Mervin (1900-x), American botanist; MS Pennsylvania State Coll. 1925; subsequently assistant, instructor and professor of botany at Butler University, Indianapolis. (*C. Palmer*).

HERBARIUM and TYPES: BUT. – Some manuscript material in Smithsonian Archives.

BIBLIOGRAPHY and BIOGRAPHY: Barnhart 3: 44; Bossert p. 298; CSP 13: 238, 17: 687; Hirsch p. 223; NAF ser. 2. 6: 71; PH 617.

BIOFILE: Deam, C.C., Fl. Indiana 1118. 1910.
Verdoorn, F., ed., Chron. bot. 2: 321. 1936, 3: 284. 1937.

EPONYMY: *Palmeriamonas* B.V. Skvortzov (1968).

7231. *Algae of Marion County*, Indiana a description of thirty-two forms ... Indianapolis [Butler University Botanical Studies volume ii paper no. 1]. 1931. Oct.
Publ.: Feb 1931 (cover), p. [1]-24. *Copy*: NY. – Issued to and to be cited as Butler Univ. bot. Stud. 2(1), 1931.

Palmer, Edward (1831-1911), English-born American surgeon and botanist; professional collector of botanical, archeological and ethnological materials; emigrated to the U.S. 1849; from then on travelling and collecting mainly in the American West and in Mexico. (*Ed. Palmer*).

HERBARIUM and TYPES: US (major collections), other important holdings at GH, MO, NY; further material in many herbaria (see e.g. Knobloch 1979). Extensive details on itineraries, location of main sets and references to lists of identification and related taxonomic papers based on Palmer are found in McVaugh (1956). – Main repositories of letters and other archival material: Smithsonian Archives and Harvard University. Palmer is estimated to have collected some 100.000 specimens. For exsiccatae (algae) see Sayre (1969).

BIBLIOGRAPHY and BIOGRAPHY: Barnhart 3: 44; BB p. 235; Bl 1: 162, 311; Bossert p. 298; CSP 10: 984; Desmond p. 476; DSB 10: 285-286; Ewan, ed. 1: 40, 276-277, ed. 2: 166-167; IH 1 (ed. 6): 362, (ed. 7): 339, 2: (in press); Kew 4: 216; Langman p. 562-563; Lenley p. 320, 465; ME 3: 159, 163; Pennell p. 615; PH 164; Rehder 5: 646.

BIOFILE: Anon., Entom. News 22: 239. 1911.
Barnhart, J.H., New York Bot. Gard. 32: 40-41. 1911 (b. 12 Jan 1831, d. 10 Apr 1911).
Beaty, J.J., Pacific Discovery 17: 10-15. 1964 (portr., popular account with many ills.); Plants in his pack, New York 1964, 182 p. (biogr.).
Blake, S.F., Madroño 16(1): 1-4. 1961 (Palmer's visit to Guadeloupe Isl., Mexico).
Bye, R.A., Jr., in D.D. Fowler, ed., Great Basin cultural ecology, Reno, Nevada 1972, p. 87-104 (ethnobotany Paiute Indians and Ed. Palmer); Econ. Bot. 33(2): 135-162. 1980 (Palmer's first major Mex. coll., 1878, San Luis Potosi).
Cantelow, E.D. & H.C., Leafl. w. Bot. 8: 96-97. 1957 (eponyms given by A. Eastwood).
Coates, R.A., Great amer. natural. 48-53. 1974.
Geiser, S.W., Men of science in Texas 167. 1959.
Hedge, I.C. & Lamond, J.M., Index coll. Edinb. herb. 116. 1970.
Hitchcock, A.S. & A. Chase, Man. grasses U.S. 988. 1951 (epon.).
Jackson, B.D., Bull. misc. Inf. Kew 1901: 50, 51 (pl. K).
Jaeger, E.C., Source-book biol. names terms ed. 3. 318. 1966.
Knobloch, I.W., Pl. coll. N. Mexico 52. 1979 (b. 12 Jan 1831, d. 10 Apr 1911; the erroneous birth year 1821, often quoted, goes back to a typographic error in Safford(1911)).
McVaugh, R. & Kearney, T.H., Amer. midl. Natural. 29(3): 768-778. 1943 (P's coll. in Arizona 1869, 1876, 1877, itin.).
McVaugh, R., Brittonia 5(1): 64. 1943 (P on La Plata exped. 1853-1855); Edward Palmer, plant explorer of the American West. Norman 1956, xvii, 430 p. (main biography and bibliography; very detailed information on itineraries and collections); Contr. Univ. Mich. Herb. 9: 288-289. 1972.
Murray, G., Hist. coll. BM(NH)1: 172. 1904 (3.196 spec. at BM).
Palmer, T.S., Condor 30: 290-291. 1928 (ornithol. eponymy).
Rodgers, A.D., Amer. bot. 1873-1892. 334 [index]. 1944; John Torrey 298-300. 1942.
Rose, J.N., Contr. U.S. Natl. Herb. 1: 63-90, 91-116, 117-127, 129-134, 293-392. 1890-1895 (identification of pl. coll. 1889-1892).
Safford, W.E., Amer. Fern J. 1(6): 143-147. 1911; Pop. Sci. Mon. 78: 341-354. 1911 (b. 12 Jan 1821, err. for 1831, biogr., portr.). Bot. Gaz. 52(1): 61-63. 1911 (obit., portr.).
Sayre, G., Mem. New York Bot. Gard. 19(1): 91-92. 1969 (on *Algae floridanae* and *Algae bahamenses*); Mem. New York Bot. Gard. 19(3): 376-377. 1975.
Stieber, M.T. et al., Huntia 4(1): 84. 1981 (arch. mat. HU).
Thomas, J.H., Huntia 3: 16, 17. 1969, 1979 (portr.).
Wiggins, I., Fl. Baja Calif. 42. 1980.

HANDWRITING: McVaugh, R., Edward Palmer 1956, *pl.* opposite p. 142-143.

EPONYMY: *Malperia* S. Watson (1889); *Palmerella* A. Gray (1876); *Palmerocassia* Britton (1930).

Palmer, Elmore (1839-1910/1911), American botanist and physician at Dexter, Mich. (*Elm. Palmer*).

HERBARIUM and TYPES: Unknown.

BIBLIOGRAPHY and BIOGRAPHY: Barnhart 3: 44.

BIOFILE: Voss, E.G., Contr. Univ. Mich. Herb. 13: 85-86. 1978.

7232. *Catalogue of phaenogamous and acrogenous plants* found growing wild in the State of Michigan, compiled by Elmore Palmer, M.D., Dexter, Washtenaw County, Michigan ... (1877). Qu. (*Cat. phaenog. pl.*).
Publ.: 25 Mai 1877 (date on pamphlet; Bot. Gaz. Sep 1877), p. [1]-16. *Copies*: MICH(2), NY, US, USDA, E.G. Voss.

Palmer, Ernest Jesse (1875-1962), English-born American botanist and forester; came to the U.S. in 1878; collector for the Missouri Botanical Garden and the Arnold Arboretum; on the staff of the Arnold Arboretum 1921-1948; student of *Crataegus*. (*E.J. Palmer*).

HERBARIUM and TYPES: Main sets at A and MO; duplicates in many herbaria, e.g. B (extant), DUR, HPH, MOR, RSDR, SMS, UMO. – Manuscript material in Smithsonian Archives.

BIBLIOGRAPHY and BIOGRAPHY: Barnhart 3: 44; BJI 2: 129; BL 1: 164, 165, 188, 193, 194, 216, 311; Bossert p. 299; Desmond p. 476; Ewan ed. 2, p. 167; Hortus 3: 1201 ("Palmer"); IH 1 (ed. 6): 362, (ed. 7): 339, 2: (in press); Kew 4: 216; Langman p. 563; Lenley p. 321; MW suppl. p. 275, 292; NAF ser. 2. 5: 251; Pennell p. 615; PH 209; Roon p. 85; Tucker 1: 536; Urban-Berl. p. 380; Zander ed. 10, p. 699, ed. 11, p. 797.

BIOFILE: Allen, C.K., Garden J. 12: 135, 139. 1962 (obit.); portr.; Dr. sci h.c. Univ. Missouri to be conferred upon him in June 1962 (P. died Feb 1962)).
Bernstein, E.B., Journal of the Arnold Arboretum index vols. 1-10. 48. 1973.
Deam, C.C., Fl. Indiana 1118. 1910.
Hedge, I.C. & Lamond, J.M., Index coll. Edinb. herb. 116. 1970.
Kobuski, C.E., J. Arnold Arb. 43(4): 351-358. 1962 (obit., portr.; bibl. by Lazella Schwarten and Elizabeth M. Palmer); Rhodora 64: 194-200. 1962 (obit.; portr.; b. 8 Apr 1875, d. 25 Feb 1962).
Rouleau, E., Rhodora, Index to volumes 1-50. 375. 1953.
Sargent, C.S., J. Arnold Arb. 6: 1-128. 1925.
Sutton, S.B., Charles Sprague Sargent 286, 289, 295, 297-298, 325, 329. 1970.
Verdoorn, F., ed., Chron. bot. 3: 295. 1937, 7: 373. 1943.

COMPOSITE WORKS: *Crataegus*, in M.L. Fernald, Gray's *Manual of Botany*, ed. 8, p. 767-801. l950; *id.*, in Gleason, H.A., *The New Britton and Brown* 2: 338-375. 1952. – See also his *Synopsis of North American Crataegi* edited by Sargent (1925).

7233. *An annotated catalogue of the flowering plants of Missouri* ... [Ann. Mo. Bot. Gard. vol. 22, 1935].
Co-author: Julian Alfred Steyermark (1909-x).
Publ.: 30 Sep 1935, reprinted with special cover and original pagination (*copies*: MO, NY, USDA). Ann. Missouri Bot. Gard. 22: 375-756, *pl. 15-21.* 1935.
Ferns and fern allies: ib. 22: 105-122. 1932.
Additions: ib. 25: 775-794. 1938 (reprint with special cover and original pagination: USDA).

Palmer, Frédéric T. (*fl.* 1867), French horticulturist (*F. Palmer*).

HERBARIUM and TYPES: Unknown.

BIBLIOGRAPHY and BIOGRAPHY: Barnhart 3: 44; Hortus 3: 1201; Rehder 5: 646; Tucker 1: 536, 782.

7234. *Culture des cactées* suivie d'une description des principales espèces et variétés ... Paris (Librairie centrale d'agriculture et de jardinage ... Auguste Goin, éditeur) [1867]. 18-mo. (*Cult. cact.*).
Publ.: 1867 (Tucker 1: 782; Flora 18 Jan 1868 "Neuigkeit"), p. [1]-216. *Copies*: BR, G, NY.

Palmér, Johan Ernst (1863-1946), Swedish farmer and botanist at Tånga (Bohuslän), later at Ör (Dalsland). (*J.E. Palmér*).

HERBARIUM and TYPES: GB; further material at LD and A, C, E, H, L, U.

BIBLIOGRAPHY and BIOGRAPHY: Barnhart 3: 44; BL 2: 521, 699; KR p. 570 (b. 5 Jan 1863).

BIOFILE: Almborn, O., in lit. 29 Feb 1980 (d. 6 Apr 1946).
Hedge, I.C. & Lamond, J.M., Index coll. Edinb. herb. 116. 1970.
Nilsson-Leissner, G., Bot. Not. 1933: 64.

7235. *Förteckning över Göteborgs och Bohus Läns fanerogamer* och kärlkryptogamer ... [Uddevalla] (Eget Förlag) [1927]. Oct. (*Förteckn. Göteb. fanerog.*).
Publ.: 1927 (p. 6: preface dated Jan 1927), p. [1]-146. *Copies*: G, LD, NY. Cover title: "Bohusläns Flora Förteckning ... ". – Superseded by H. Fries, Göteborgs och Bohus läns fanerogamer och ormbumkar, 1945, ed. 2, 1971. According to Fries Palmér's modest flora "contains many misleading or wrong statements" (inf. O. Almborn).

Palmer, Johann Ludwig (1784-1836), German physician at Marbach; Dr. med. Tübingen 1817.(*J.L. Palmer*).

HERBARIUM and TYPES: Unknown.

BIBLIOGRAPHY and BIOGRAPHY: Herder p. 117; PR 3373, 6911; Rehder 1: 169.

7236. *Dissertatio inauguralis de plantarum exhalationibus*. Commentatio in concertatione civium Academiae tubingensis vi. novembris mdcccxvi. Praemio a rege Würtembergiae constituto ab ordine medicorum ornata. Quam consentiente eodem ordine praeside F.G. Gmelin ... pro gradu doctoris medicinae publice defendet die [ ] Apr mdcccxvii. Auctor Ioannes Ludovicus Palmer winnendensis. Tubingae (typis Ludovici Friderici Fues) [1817]. Oct. (*Diss. pl. exhal.*).
Publ.: Apr 1817, p. [i], [1]-45, [46]. *Copy*: G.

Palmer, Julius Auboineau (1840-1899), American mycologist. (*J.A. Palmer*).

HERBARIUM and TYPES: Unknown.

BIBLIOGRAPHY and BIOGRAPHY: Barnhart 3: 44; BM 4: 1506; Kelly p. 165; LS 19729-19735; Stevenson p. 1253.

BIOFILE: Rogers, D.P., Brief hist. mycol. N. America ed. 2. 71, 72, 74. 1981.

EPONYMY: *Palmeria* Greville (1865) honors John Linton Palmer, British Royal Navy surgeon; *Palmeria* F. v. Mueller (1864) was named for Sir John Frederick Palmer (1804-1871) English-born Australian politician.

7237. *Mushrooms of America*, edible and poisonous. Edited by Julius A. Palmer, Jr. Boston (published by L. Prang & Co.) 1885. (*Mushrooms Amer.*).
Publ.: 1885 (TBC Jun 1885; Nat. Nov. Aug(1) 1885; Bot. Zeit. 25 Sep 1885; Bot. Gaz. Nov 1885; J. Bot. Jun 1886, p. [1]-4, *pl. 1-12*, (col.). *Copies*: FH, NY, Stevenson, USDA.

7238. *About mushrooms* a guide to the study of esculent and poisonous fungi ... Boston (Lee & Shepard, publishers ...) 1894. (*Mushrooms*).
Publ.: 1894 (p. xiv: 1 Jan 1894; Bot. Zeit. 1 Apr 1895), p. [i]-xiv, [xv-xvi], *pl. 1-13* (uncol.), [1]-100. *Copies*: MICH, MO, NY, Stevenson.

Palmgren, Alvar (1880-1960), Finnish botanist; Dr. phil. Helsinki 1914; lecturer in botany at the College of technology, Helsinki 1916; professor of botany at Helsinki University 1928-1950. (*Palmgr.*).

HERBARIUM and TYPES: H (much *Carex* & *Hieracium*), other material at C, GB, LD, NY and S. – *Exsiccatae*: *Carices fulvellae Fries*, nos. 1-60, 1910. Sets at B, C, H and LD (total number of sets prepared: 9), for an index see Medd. Soc. Fauna Fl. fenn. 44: 219-222. 1918, Sv. bot. Tidskr. 12: 407-409. 1918, Luonnon Ystävä 22: 69-70. 1918. – Some letters at G.

BIBLIOGRAPHY and BIOGRAPHY: Barnhart 3: 45; BJI 2: 129; BL 2: 68, 69, 699; BM 8: 974; Bossert p. 299; Collander p. 392-398 (main bibl., biogr. refs.), Collander hist. p. 156 [index]; CSP 17: 689; Hirsch p. 223; IH 1 (ed. 1): 40, (ed. 2: 59, (ed. 3): 72, (ed. 4): 77, (ed. 6): 362, (ed. 7): 339, 2: (in press); Kew 4: 217; KR p. 570; LS 37618; Saelan p. 389.

BIOFILE: Anon., Bot. Not. 1902: 158 (stipend); Memor. Soc. Fauna Fl. fenn. 34: 140-141, 154-155. 1959, 37: 261-262, 273-274. 1962; Österr. bot. Z. 76: 88. 1927 (prof. bot.).
Haartman, L. v., Memor. Soc. Fauna Fl. fenn. 38: 46-63. 1963 (obit., b. 28 Apr 1880, d. 30 Nov 1960, brief bibl.).
Kneucker, A., Allg. bot. Z. 33: 64 (288). 1928 (prof. bot.).
Luther, H., Soc. Sci. fenn. Årsb. 46C(1): 1-20. 1969 (biogr., bibl., portr.).
Verdoorn, F., ed., Chron. bot. 1: 297. 1935, 2: 117, 118. 1936, 5: 134, 136, 308. 1939.
Wulff, E.V., Intr. hist. plant geogr. 37, 38, 46, 131, 132, 136. 1943.

7239. *Bidrag till kännedom om Ålands vegetation och flora* ... Helsingfors 1910 [-1911], 2 parts.
1: soon after 1 Mar 1910 (see p. 53), p. [1]-53, *pl. 1-12*.
2: soon after 31 Dec 1910 (see p. 16), p. [1]-16, *pl. 1-3*.
Copies: H, U. – Issued and to be cited as Acta Soc. Fauna Fl. fenn. 34(1) and 34(4). Completed journal volume issued 3 Feb-2 Mar 1912. Inf. P. Isoviita.

7240. *Hippophaës rhamnoides* auf Åland ... Helsingfors 1912. Oct.
Publ.: *9 Dec 1912 – 16 Jan 1913, in two issues. Copies*: H.
First issue: as Acta Soc. Fauna Fl. fenn. 36(3): [1]-188, *pl. 1-10*, map.
Thesis issue: p. [3]-188, *pl. 1-10*, map. "Hippophaës ... wird mit Genehmigung der philosophischen Fakultät der kaiserl. Alexander-Universität in Helsingfors am 16. Januar 1913 ... vorgelegt." Helsingfors 1912. [Note by O. Almborn: This is a normal phrase in Danish, Finnish or Swedish at that time. It means that the thesis should be available not only to the opponents but also to the general public about 3 weeks before the actual day on which it is publicly defended].

7241. *Studier öfver löfängsområdena på Åland*. Ett bidrag till kännedomen om vegetationen och floran på torr och på frisk kalkhaltig grund ... Helsingfors 1915-1917, 3 parts. Oct.
Publ.: in Acta Soc. Fauna Fl. fenn. 42(1), in three parts which (at least parts 2 and 3) were also distributed separately with reprint covers. *Copy*: H. – To be cited from journal (complete volume issued ca. 13 Mai 1917):
1: on or shortly after 13 Mai 1915, p. [i-iii], [1]-169, *pl. 1-8*.
2: on or shortly after Jul 1915, p. [171]-474.
3: on or circa 13 Mai 1917, p. [i-iii], general t.p. and ded.], [475]- 633, [634], *pl. 1-8*, 2 maps.
Information supplied by P. Isoviita. Part 3 was translated in German, see below: *Über Artenzahl und Areal*.

7242. *Die Entfernung als pflanzengeographischer Faktor* ... Helsingfors 1921. Oct.
Publ.: soon after 10 Jun 1921 (see p. 97), p. [i-iii], [1]-113, map. *Copy*: H. – Issued and to

be cited as Acta Soc. Fauna Fl. fenn. 49(1). 1921. Complete journal volume published 1 Nov 1921. Inf. P. Isoviita.

7243. *Über Artenzahl und Areal* sowie über die Konstitution der Vegetation. Eine vegetationsstatistische Untersuchung ... Helsingfors 1922. Oct. (*Artenz. Areal.*).
Publ.: Aug-Dec 1922 (Bot. Jahrb. 1 Jul 1923), p. [i]-vii, [1]-135, [136], tables 1-8, map.
Copies: B, G, H, NY. − Issued as Acta forest. fenn. 22(1): 1-138, tables 1-8, map. 1922.
− A translation of part 3 of Studier öfver löfängsområdena på Åland, see above.

7244. *Zur Kenntnis des Florencharakters des Nadelwaldes.* Eine pflanzengeographische Studie aus dem Gebiete Ålands ... Helsingfors 1922. Oct. †.
Publ.: 10 Sep-31 Dec 1922 (see p. 5), p. [1]-114, [115], map. *Copy*: H. − Reprinted and to be cited from Acta forest. fenn. 22(2): 1-115. 1922. Copies exist with title-pages with and without journal identification. Inf. P. Isoviita.

7245. *Die Artenzahl als pflanzengeographischer Charakter* sowie der Zufall und die säkulare Landhebung als pflanzengeografische Faktoren/Ein pflanzengeographischer Entwurf, basiert auf Material aus dem äländischen Schärenarchipel. Helsinki 1925. Oct.
Publ.: 8 Mai-3 Oct 1925 (p. 142: 8 Mai 1925; copy pres. to Soc. F.F. f. 3 Oct 1925), p. [1]-142, 2 maps. *Copies*: B, G, H, NY. − Reprinted and to be cited from Acta bot. fenn. 1(1): 1-142, 2 maps 1925. Also issued as Fennia 46(2), 1925. Inf. P. Isoviita.

7246. *Die Einwanderungswege der Flora* nach den Ålandsinseln ... I ... Helsingforsiae 1927. Oct.
Publ.: after 13 Mai 1927 (mss. submitted 3 Feb 1927; inf. P. Isoviita), p. [1]-198, [199, err.], maps 1-56 (2/page), 57. *Copies*: B, G, H, NY(2). − Reprinted and to be cited from Acta bot. fenn. 2(1): 1-199. 1927 (rev. W. Wangerin, Bot. Centralbl. ser. 2. 13: 36-37. 1928).

Palmstruch, Johan Wilhelm (1770-1811), Swedish soldier [ryttmästare, cavalry capt.)] and botanical artist. (*Palmstr.*).

ORIGINAL DRAWINGS: The drawings for *Svensk botanik* are in the library of the Royal Swedish Academy of Sciences at Stockholm (Andersson 1849; KR). Palmstruch, who was the artist of the undertaking, is not known to have had a herbarium.

BIBLIOGRAPHY and BIOGRAPHY: Barnhart 3: 45; BL 2: 514, 699; BM 4: 1507; Bossert p. 299; GR p. 480; Hegi 4(2): 733; Kew 4: 217; KR p. 98-100, 570 (b. 3 Mar 1770, d. 30 Aug 1811); NI 2223, see also 1: 229, 3: 73 (2349n); PR alph., 775, 10707, 11827; Rehder 1: 352; SY p. 98 [index]; TL-1/947; TL-2/517, see G.J. Billberg; see C.A.M. Lindman.

BIOFILE: Andersson, N.J., Bot. Not. 1849: 52-54 (early history).
Anon., Bot. Not. 1845: 104 (entry Biografiskt. lexikon 10(3)).
Borén, P.G., Bot. Not. 1920: 63-64 (years of publ. *Sv. bot.*).
Cronlund, E., Svenska män och kvinnor 6: 32-33. 1949 (portr.).
Fries, R.E., Short history botany Sweden 41. 1950.
Landin, B.-O., Biblis 1977-1978: 133. 1978.
Roemer, J.J., Coll. bot. 267-268. 1809 ("ratio operis" Sv. bot.).
Scheutz, N.J.W., Bot. Not. 1863: 71.
Wikström, J.E., Conspectus 254-259. 1831.
Wittrock, V.B., Acta Horti Berg. 3(2): 65. 1903, 3(3): 48. 1905.

EPONYMY: *Palmstruckia* (sic) Sonder (1860); *Palmstruckia* (sic) Retzius (1810).

7247. *Svensk botanik*, utgifven af J.W. Palmstruch och C.W. Venus. Första Bandet. Stockholm (tryckt hos C. Delén och J.G. Forsgren, 1802. Oct. (*Sv. bot.*).
Publ.: In all 11 vols., originally in monthly fascicles of 6 plates. Later three fascicles were issued each quarter; twelve fascicles make a volume. *Copies*: G, H, LD, NY, PH; IDC 5631.

1: 1802-1803, p. [i, iii], *pl. 1-72* with text (col. copp.), [i]-x, [i], [1]-12, [1, err.].
  Co-editor: Carl Wilhelm Venus (1770-1851).
  Author of text: Conrad Quensel (1767-1806), not on t.p.
  Imprint: See above.
  Ed. 2: 1803.
  Ed. 3: 1815, p. [i, iii], *pl. 1-72*, [i-xv], edited by Gustav Johann Bilberg (1772-1844), texts by Quensel except for those signed "S-Z": Olof Peter Swartz (1760-1818).
2: 1803-1804, p. [i], *pl. 73-144* (id.), [i]-xxii, [xxiii], [i], [1]-14, [1, err.].
  Editor: J.W. Palmstruch. *Author of text*: C. Quensel, except for 8 texts by O. Swartz.
  Imprint: Stockholm, tryckt hos C. Delén, 1803.
3: 1804-1805, p. [i], *pl. 145-216* (id.), [i]-xiv, [1]-11, [12].
  Editor: J.W. Palmstruch. *Author of text*: C. Quensel. *Imprint*: Stockholm, tryckt hos Hendrik A. Nordström, *1804*.
4: 1805-1806, p. [i], pl. 217-288 (id.), [i]-ix, [x], [1]-9.
  Editor: J.W.Palmstruch. *Author of text*: C. Quensel; p. [i]-ix signed by O. Swartz.
  Imprint: Stockholm, tryckt hos Carl Delén, 1805.
5: 1807-1808, p. [i], *pl. 289-360* (id.), [i]-[x], [1]-11.
  Editor: J.W.Palmstruch. *Author of text*: O. Swartz. *Imprint*: Stockholm, tryckt hos Carl Delén, 1807.
6: 1809-1811, p. [i], *pl. 361-432* (id.), [i]-[viii], [1]-8, [i]-iv.
  Editor: J.W. Palmstruch. *Author of text*: O. Swartz. *Imprint*: Stockholm, tryckt hos Carl Delén, 1809.
7: 1812-1814, p. [i], *pl. 433-504*, [1]-8, ind.
  Editor: Gustav Johan Billberg (1772-1844). *Author of text*: O. Swartz. *Imprint*: Stockholm, tryckt hos Olof Grahn, 1812.
8: 1816-1819, p. [i], *pl. 505-576*, [1]-8, ind.
  Editor: G.J. Billberg. *Authors of text*: fasc. 1-6 (*pl. 505-540*) by O. Swartz; 7-12 (*pl. 541-476*) by G.J. Billberg, text for *pl. 506* by E. Fries. *Imprint*: Stockholm, tryckt hos Carl Delén, 1819.
9: 1821-1825, p. [i], *pl. 577-684*, [1]-10, ind.
  Editor: fasc. 1-3 (*pl. 577-594*) G.J. Billberg, 4-12: Göran Wahlenberg (1780-1851). *Authors of text*: editors. – "Svensk botanik utgifven af Kongl. Vetenkaps Akademien ... Texten tryckt hos Palmblad & C. i Upsala 1823-1825."
10: 1826-1829, p. [i], pl. 649-720, [1-7].
  Editor and author: G. Wahlenberg. *Imprint*: as vol. 9, 1826-1829. A handwritten note in the NY copy states that Anderson told the writer that the text of plates 675-720 was by P.F. Wahlberg. However, this does not agree with Krok's statement that Wahlberg took over from *pl. 739* (vol. 11) onward.
11: 1830-1843, p. [i], *pl. 721-774*, [1-24, index].
  Editors [*and authors*]: G. Wahlenberg fasc. 1-3 (*pl. 721-738*); fasc. 4-9 (*pl. 739-774*): Pehr Frederick Wahlberg (1800-1877). – "Svensk botanik ... Texten tryckt hos P.A. Nordstedt & Söner i Stockholm 1843". Actually fasc. 1-3 were printed in Uppsala.

vol.	fasc.	plates	dates
1	1-8	*1-48*	1802
	9-12	*49-72*	1803
2	1-8	*73-120*	1803
	9-12	*121-144*	1804, post 27 Mai
3	1-6	*145-180*	1804
	7-12	*181-216*	1805, post 14 Jun
4	1-5	*217-246*	1805
	6-12	*247-288*	1806
5	1-7	*289-330*	1807
	8-12	*331-360*	1808
6	1-3	*361-378*	1809
	4-8	*379-408*	1810
	9-12	*409-432*	1811
7	1-4	*433-456*	1812
	5-7	*457-474*	1813

vol.	fasc.	plates	dates
	8-9	475-486	1814
	10-12	487-504	1815
8	1-3	505-522	1816
	4-5	523-540	1817
	7-8	541-558	1818
	10-12	559-576	1819
9	1-3	577-594	1821
	4-6	595-612	1823
	7-9	613-630	1824
	10-12	631-648	1825
10	1-3	649-666	1826
	4-6	667-684	1827
	7-9	685-702	1828
	10-12	703-720	1829
11	1-3	721-738	1830
	4-6	739-756	1836
	7-9	757-774	1838
	Index	p. [1]-24	1843

Parts 10-12 of volume 11 remained unpublished. Andersson (1849) reports, on the basis of a letter of the Royal Swedish Academy of Science, that the academy stopped publication of Svensk botanik because of the high costs. The Academy had requested the government for funds to have typical Swedish plants painted, the pictures to be deposited at the Academy for possible later use. Some of these plates were used for Thedenius, *Svensk Skolbotanik* (1852-1854) and for C.A.M. Lindman, *Bilder ur Nordens Flora* (1901-1905, 3 vols., suppl. 1926) (see KR p. 99, note 1) ed. 2. 3 vols. 1917-1926; ed. 3. 3 vols. 1921-1926), and *Nordens Flora*, Stockholm 1964, 3 vols., revised by M. Fries; Danish ed. *Billeder of Nordens Flora*, København 1965, 3 vols. (We are grateful to O. Almborn and B. Peterson for this information). *Artists*: J.W.Palmstruch (1-6), C.W. Venus (1); from 7 onward: Acharius, A.J. Agrelius, Laestadius (Lars Levi Laerstadt), Swartz, P.F. Wahlberg. See Junk, Cat. 214 (no. 300a) 1979 for a recently offered copy (Hfl. 18.000; US $ 9.000).

Palouzier, Émile (*fl.* 1891), French botanist and pharmacist at the "École supérieure de Pharmacie", Montpellier. (*Palouz.*).

HERBARIUM and TYPES: Unknown.

BIBLIOGRAPHY and BIOGRAPHY: BM 4: 1507.

7248. *Essai d'une monographie des fougères françaises* ... Montpellier (Typographie et Lithographie Charles Boehm ...) 1891. Oct. (*Essai monogr. foug. franç.*).
Publ.: 1891 (Nat. Nov. Nov(2) 1891; Bot. Zeit. 29 Jan 1892; Bot. Centralbl. 10 Dec 1891), p. [1]-103. *Copies*: G, MO.
Ref.: Roth, E., Bot. Centralbl. 50: 205-206. 1892.

Palun, Maurice (1777-1860), French botanist at Avignon. (*Palun*).

HERBARIUM and TYPES: AV.

BIBLIOGRAPHY and BIOGRAPHY: Barnhart 3: 45; BL 2: 209, 699; BM 4: 1507; IH 1(ed. 6): 362; (ed. 7): 339; Jackson p. 280; Kew 4: 217; Rehder 1: 413; Tucker 1: 536.

7249. *Catalogue des plantes phanérogames* qui croissent spontanément dans le territoire d'*Avignon* et dans les lieux circonvoisins rédigé par Maurice Palun ... Avignon (de l'Imprimerie de Fr.: Seguin Ainé) 1867. Oct. (*Cat. pl. phan. Avignon*).
Publ.: 1867, p. [i-iv], [1]-189. *Copies*: G, NY, USDA.

35

Palustre (*fl.* 1840), French physician and botanist. (*Palustre*).

HERBARIUM and TYPES: Unknown.

BIOFILE: PR 6912-6913 (ed. 1: 7726-7727); Rehder 1: 261.

7250. *Études de botanique,* ou classification des végétaux d'après les méthodes de MM de Jussieu et de Candolle. Avec deux nomenclatures botaniques, distinctes par leur désinences, l'une pour les familles et l'autre pour les sous-familles. Par Palustre, ... Poitiers (de l'Imprimerie de F.-A. Saurin, ...) 1840. Broadsheet. (*Étud. bot.*).
Publ.: 1840, p. [1]-34, [1-2]. *Copy*: G.
246

Pammel, Louis Hermann (1862-1931), American botanist and conservationist; M.S. Univ. Wisc. 1889; Dr. phil. St. Louis 1896; assistant to W.G. Farlow 1885-1886; assistant Shaw school of Botany, St. Louis, 1886-1889; from 1889 at Ames, Iowa, as professor of botany; Dr. Sci. Univ. Wisc. 1925. (*Pammel*).

HERBARIUM and TYPES: ISC; further material in many herbaria, e.g. A, B, C, CM, CS, DBN, E, F, FLAS, GH, K, MIN, MO, NY, OKL, OKU, POM, WELC, WTU. – Some letters at G.

BIBLIOGRAPHY and BIOGRAPHY: Barnhart 3: 45; BJI 2: 129; BL 1: 159, 178, 180, 191, 192, 222, 311, 2: 9, 699; BM 4: 1507; Bossert p. 299; CSP 17: 691-692; Ewan ed. 1: 277, ed. 2: 167-168; GR p. 236-237; Hirsch p. 223; IH 2: (in press); Kelly p. 165; Kew 4: 217-218; Langman p. 563; Lenley p. 321; LS 19740-19784, 37619-37641, suppl. 20844-20858; MW suppl. p. 275; NW p. 54; Pennell p. 615; PH 12; Rehder 5: 646; Stevenson p. 1253; TL-2/see F. Lamson-Scribner; Tucker 1: 536; Urban-Berl. p. 280, 380.

BIOFILE: Allison, E.M., Univ. Colo. Stud. 6: 69. 1908.
Anon., Amer. men Sci. 243. 1906; Bot. Centralbl. 38: 784. 1889 (app. Iowa); Österr. bot. Z. 39: 316. 1889 (id.).
Cratly, R.I., Dict. Amer. Biogr. 14: 197-198. 1934 (biogr., b. 19 Apr 1862, d. 23 Mar 1931).
Erwin, A.T., Science ser. 2. 73: 604-605. 1931 (obit.).
Gilman, J.C., Phytopathology 22: 669-674. 1932 (obit., portr., bibl.).
Graham, E.H., Ann. Carnegie Mus. Pittsburgh 26: 21. 1937.
Hedge, I.C. & Lamond, J.M., Index coll. Edinb. herb. 116. 1970.
Humphrey, H.B., Makers N. Amer. bot. 195-196. 1961.
Jackson, B.D., Bull. misc. Inf. Kew 1901: 50 (pl. K).
McVaugh, R., Edward Palmer 103. 1956.
Murrill, W.A., Hist. found. bot. Florida 19. 1945.
Osborn, H., Fragm. entom. hist. 339. 1937.
Pammel, L.H., Proc. Iowa Acad. Sci. 10: 57-68. 1903, 20: 133-149, map, 1913 (itin. Uinta Mts., Utah), 27: 51-73. 1920, 31: 45-68. 1924 (Century of botany in Iowa).
Rickett, H.W., Index Bull. Torrey bot. Club 76. 1955.
Rodgers, A.D, Liberty Hyde Bailey 501 [index]. 1949.
Shimek, B., Proc. Iowa Acad. Sci. 38: 55-56. 1931 (obit., followed by a detailed bibl. by C.M. King on p. 56-68).
Shope, P.F., Mycologia 21: 295. 1929.
Stieber, M.T. et al., Huntia 4(1): 84. 1981 (arch. mat. HU).

THESIS: *Anatomical characters of the seeds of Leguminosae,* chiefly genera of Gray's Manual, cover, p. [89]-263, [264-273], *pl. 7-35.* 29 Jun 1899, reprinted from Trans. Acad. Sci. St. Louis (*copies*: BR, MO, NY).

7251. *Weeds of southwestern Wisconsin* and southwestern Minnesota, ... Saint Paul, Minn. (the Pioneer Press Company) 1887. Qu. (*Weeds Wisc. Minn.*).
Publ.: 1887, p. [1]-20. *Copy*: NY.

7252. *Woody plants of Western Wisconsin* [Proc. Iowa Acad. Sci. 1892]. Oct.
Publ.: 1892, p. [1]-11. *Copy*: NY. – Reprinted from Proc. Iowa Acad. Sci. 1(2): 76-80. 1892. To be cited from journal.

7253. *Notes on the grasses and forage plants of Iowa*, Nebraska, and Colorado ... Washington (Government Printing Office) 1897. Oct. (*Notes forage pl. Iowa*).
Publ.: Sep-Dec 1897 (p.3: 10 Aug 1897; Nat. Nov. Feb(2) 1898), p. [1]-47. *Copies*: BR, MO, NY; IDC 7598. – Bulletin no. 9 U.S. Department of Agriculture. Division of Agrostology.

7254. *The Thistles of Iowa* with notes on a few other species ... Des Moines (B. Murphy, State Printer) 1901. Oct. (*Thistles Iowa*).
Publ.: 1901, p. [1]-26, *pl. 12-31*. *Copies*: FI, G. – Contr. Bot. Dept. Iowa State Coll. Agr. 19, 1901; reprinted from Proc. Iowa Acad. Sci. 8: 214-239. 1901.

7255. Iowa Geological Survey Bulletin no. 1. *The grasses of Iowa* ... Des Moines, Iowa 1901-1904 [1905], 2 parts. Oct. (*Grass. Iowa*).
1: Jan 1901, p. [i-vii], [1]-525, *pl.1-3*. *Authors*: L.H. Pammel; Julius Buel Weems (1865-1930); Frank Lawson-Scribner (1851-1938). – Also published (in part) as Bulletin 54, Iowa Agr. Coll. Exp. Station Ames, Iowa, p. [69]-344.
2: 1 Apr 1905 (see p. [ix]; t.p. 1904), p. [i]-xiii, [1]-436, *pl. A*. – *Authors*: L.H. Pammel; Elmer Drew Merrill (1876-1956), Carleton Roy Ball (1873-1958); Frank Lawson-Scribner; Cornelius Lott Shear (1865-1956); J. Smith; Thomas Henry Kearney (1874-1956). Iowa Geological Survey supplementary Report 1903, Des Moines, Iowa, 1904.
Copies: B, BR, G, MICH, NY, US, USDA.

7256. *A manual of poisonous plants* chiefly of eastern North America, with brief notes on economic and medicinal plants, and numerous illustrations ... Cedar Rapids, Iowa (The Torch Press) 1910-1911, 2 parts. (*Man. poison. pl.*).
1: 1910 (p. iv: 1 Jun 1909, frontisp., p. [i]-vi, [1]-150, *5 pl*. – Also reissued with part 2 and dated 1911.
2: 1911 (Nat. Nov. Oct(2) 1911), frontisp., p. [i]-v, [153]-977, *12 pl*.
Copies: FH, MO, NY, USDA; IDC 7599.
Consolidated issue: 1911, frontisp., p. [i]-viii, [ix, expl. pl.], [xi, err.], [xiii, h.t.], [1]-977, *17 pl*. *Copies*: MO, US.
Ref.: Mansfield, W., Torreya 21: 264-269. 10 Nov 1912.

7257. *The weed flora of Iowa* by L.H. Pammel and others. Des Moines (published for Iowa Geological Survey) 1913. Oct. (*Weed fl. Iowa*).
Collaborators ed. 1: Charlotte Maria King (1864-1937), John Nathan Martin (1875-1964), Jules Cool Cunningham (1879-1948), Ada Hayden (1884-1950), Harriette Susan Kellogg (1880-1916).
Ed. 1: 1913 (p. xiii: 18 Dec 1912; TBC 29 Mai 1914), p. [i]-xiii, err. slip, [1]-912. *Copies*: BR, FI, G, MICH, MO, NY; IDC 7600. – Iowa Geol. Survey Bull. 4.
Ed. 2: 1926 (p. vi: 9 Feb 1925, p. iii: 1 Mai 1925), p. [i]-vi, [vii], [1]-715. *Copies*: MO, NY. – Revised edition ... by L.H. Pammel and Charlotte M. King [Charlotte Maria King, 1864-1937] with the collaboration of J.N. Martin, Joseph Charles Gilman (1890-1966), J.C. Cunningham, Ada Hayden, Fred Dunway Butcher (1897-x), Donald Rockwell Porter (1900-x), Ralph Rudolph Rothacker (1894-1960) ... Des Moines (published for the Iowa Geological Survey by the State of Iowa) 1926. Oct.

7258. *Honey plants of Iowa* L.H. Pammel, ... and Charlotte M. King with the collaboration of Ada Hayden, ... J.N. Martin, ... Frank P. Sipe, ... Wm.S. Cook, ... Edna C. Pammel, ... Clarissa Clark, ... L.E. Yocum, ... L.A. Kenoyer, ... O.W. Park, ... Charles A. Hoffman, ... R.I. Cratty, and C.C. Lounsberry, ... Des Moines (published for the Iowa Geological Survey by the State of Iowa) 1930. Oct. (*Honey pl. Iowa*).
Publ.: 1930, frontisp., p. [i-ii], [1]-1192. *Copies*: BR, MICH, MO.
Collaborators: Charlotte Marie King (1864-1937), Ada Hayden (1884-1950), John Nathan Martin (1875-1964), Frank Perry Sipe (1888-?; M.Sc. Iowa 1923), William

Schlei Cook (1898-1979), Edna Caroline Pammel (1888-1933), Clarissa Clark (Grad. Iowa State Univ. 1912; deceased), Loewson Edwin Yocum (1890-?; instr. bot. ib. 1916-1919), Leslie Alva Kenoyer (1883-1964), Oscar Wallace Park 1889-1954), Charter Andrew Frederick Hoffman (1903-x), Robert Irvin Cratty (1853-1940), C. Claude Lounsberry (1872-1949).
(For information on Pammel's collaborators we are indebted to Becky Seim Owings and Richard W. Pohl of Iowa State University Ames, Iowa and to Mr. Lounsberry, Iowa Secr. Agric.).

Pampanini, Renato (1875-1949), Italian botanist; studied at Fribourg, Lausanne and Genève (with R.H. Chodat); Dr. phil. Fribourg 1902; from 1904-1933 at the Istituto botanico di Firenze, from 1912 as assistant; professor of botany at Cagliari 1933-1948, ultimately at Torino. (*Pamp.*).

HERBARIUM and TYPES: FI. – Other material at A, BM, E, G, GH, K, OXF, P, PAD, PI, SS, ULT, W. – For exsiccatae see A. Fiori et al., *Flora italica exsiccata*. – Some letters at G.

BIBLIOGRAPHY and BIOGRAPHY: Barnhart 3: 45; BFM nos. 1102, 1540; BJI 2: 129; BL 1: 32, 42, 43, 311, 2: 384, 390, 415, 417, 474, 699; BM 8: 975; Clokie p. 220; DSB 6(4): 183; Hirsch p. 223; Hortus 3: 1201 ("Pamp."); IH 1(ed. 6): 362, (ed. 7): 339, 2: (in press); Kew 4: 218-219; Langman p. 563; LS suppl. 20860-20862; MW p. 377, suppl. 275; Saccardo 2: 80; TL-2/6715; see also A. Fiori; Tucker 1: 536; Zander ed. 10, p. 699, ed. 11, p. 797.

BIOFILE: Anon., Hedwigia 71: (108). 1931 (app. Cagliari); Österr. bot. Z. 79: 380. 1930 (app. Cagliari).
Gager, C.S., Brooklyn Bot. Gard. Rec. 27(3): 269. 1938 (dir. bot. gard. Cagliari).
Hedge, I.C. & Lamond, J.M., Index coll. Edinb. herb. 116. 1970.
Lusina, G., Annali di Bot. 25(1, 2): 151, 165. 1958.
Martinoli, G., Lavori Ist. bot. Univ. Cagliari ser. 2. 1: 3-4. 1949/50 (obit., portr.).
Merrill, E.D., Enum. Philipp. pl. 4: 214. 1926; B.P. Bishop Mus. Bull. 144: 146. 1937 (Polynes. bibl.), Contr. U.S. Natl. Herb. 30(1): 233. 1947 (Pacific bibl.).
Negri, G., Nuovo Giorn. bot. Ital. 65: 771-805. 1958 (Memorial lecture, biogr., bibl., portr., b. 20 Oct 1875, d. 19 Jul 1949).
Pampanini, R., Fl. Rep. San Marino 23-25. 1930 (itin., coll.); Publicazioni di R. Pampanini 1903-1930, Cagliari 1930, 12 p.
Steinberg, C.H., Webbia 34: 62. 1979 (P left his herb. to FI).
Verdoorn, F., ed., Chron. bot. 1: 32, 40, 69, 196. 1935, 2: 28, 30, 34, 35. 1936, 3: 27, 188. 1937, 7: 40. 1943.
Zangheri, P., Renato Pampanini 1875-1949, Forli, 1950 (ex Archivio botanica 25(3/4): 269-276. 1950 (obit., b. 20 Oct 1875, d. 19 Jul 1949, discussion of publications; evaluation).

COMPOSITE WORKS: With A. Béguinot, G. Negri et al., *Stato conosc. veg. Italia* 1909, see TL-2/6715.

POSTHUMOUS PUBLICATION: *La flora del Cadore* catalogo sistematico delle piante vascolari. Publicato a cura della Magnifica comunità di Cadore dai prof.ri G. Negri e P. Zangheri. Forli, Tipografia Valbonesi, 1958, portr., map, [i]-xxviii, [1]-897. *Copy*: FI.

7259. *Essai sur la géographie botanique des Alpes* et en particulier des Alpes sud-orientales. Thèse présentée à la Faculté des sciences de l'Université de Fribourg (Suisse) pour obtenir le grade de docteur ès-sciences naturelles ... Fribourg (Imprimerie Fragnière frères) 1903. Oct. (*Essai geogr. bot. Alpes*).
Publ.: 1903 (Nat. Nov. Nov (1, 2) 1903), p. [i],[1]-215, *pl. 1-10*. *Copies*: FI, G, NY. – Also issued as Mém. Soc. frib. Sci. nat. 3(1), 1903. – Preliminary publication, with R. Chodat, in Le Globe, Genève 41; 63-132. 1902.
Ref.: Diels, L., Bot. Jahrb. 33(Lit.): 40-41. 8 Dec 1903.

7260. *Un manipolo di piante nuove* ... Estratto del Nuovo Giornale botanico italiano (nuova serie) vol. xiv, n. 4, Ottobre 1907.
Publ.: Oct 1907 (in journal), p. [1]-16. *Copy*: FI. – Reprinted and to be cited from Nuov. Giorn. bot. ital. ser. 2. 14(4), 591-606. Oct 1907.

7261. *Astralagus alopecuroides* Linneo (em. Pampanini) ... Firenze (Stabilimento Pellas ...) 1907. Oct.
Publ.: 31 Oct 1907 (in journal; Nat. Nov. Mai(1) 1908), p. [i-ii], [1]-155, *pl. 8-13*. *Copies*: BR, G. – Reprinted and to be cited from Nuov. Giorn. bot. Ital. ser. 2. 14(3): 327-481, *pl. 8-13*. Jul 1907 (publ. 31 Oct 1907). – For a review see Acta Hort. Univ. Jurjev. 12: 70-71. 1911.

7262. *Le piante vascolari raccolte dal Rev. P.C. Silvestri nell' Hu-peh durante gli anni 1904-1907 (e negli anni 1909, 1910)*. Firenze (Stabilimento Pellas ...) [*1910*-] *1911*. Oct.
Publ.: 1910-1911 (in journal; repr. issued Mai-Jun 1911), p. [3]-315, *pl. 1-7*. *Copies*: B, FI, G, NY. – Reprinted and to be cited from Nuovo Giorn bot. Ital. ser. 2. 17(2): 223-298. Apr, (3): 391-432. Jun, (4): 669-735. Aug 1910, 18(1): 80-143. Feb, (2): 161-223. Apr 1911.

7263. Societa italiana per lo studio della Libia. R. Pampanini. *Plantae tripolitanae* ab auctore anno 1913 lectae et repertorium florae vascularis Tripolitaniae. Firenze (Stabilimento Pellas ...) 1914. Oct. (*Pl. tripol.*).
Publ.: Nov-Dec 1914 (p. xiv: 5 Oct 1914; reviewed by Mattei in 1914; Nat. Nov. Jan (1, 2) 1915), p. [i]-xiv, map, [1]-334, [1, expl. pl.], *pl. 1-9*. *Copies*: B, FI, G, NY, USDA. – Published in series *La Missione Franchetti in Tripolitania* (Il Gebèl), Appendice 1. – Reviews cited in an advertisement by the "Società ... per lo studio della Libia" in the NY copy are dated 1914 (Mattei), Jan 1915 (Battandier et Trabut) and Jan 1915 (G. Barratte); the Society wrote to NY on 22 Mar 1915 "recentemente publicato").

7264. *Le Magnolie* ... Firenze (Tipografia M. Ricci ...) 1916. Oct.
Publ.: 1 Jun 1915-15 Dec 1916 (in journal), p. [1]-92, *pl. 1-5*. *Copies*: FI, G, USDA. – Reprinted and to be cited from Bull. Soc. Tosc. Orticult. 40: 127-134. Jun, 151-154. Jul, 170-173. 1 Aug, 181-184. 1 Sep, 199-202. 1 Oct, 213-218. 1 Nov, 229-234. 15 Dec 1915; 41: 6-8, *pl. 1-2*. 1 Jan, 23-26, *pl. 3-4*. 15 Feb, 40-45. 15 Mar, 58-62, *pl. 8*. 15 Apr, 77-78. 15 Mai, 101-107. 15 Jul, 122-125. 15 Aug, 135-141. 15 Sep, 151-157. 15 Oct, 167-173. 15 Nov, 183-189. 15 Dec 1916.

7265. *Piante di Bengasi e del suo territorio* raccolte dal Rev. P.D. Vito Zanon ... [Nuovo Giornale botanico italiano] 1916-1917. Oct.
Collector: Vito Zanon, see Pampanini, Prodr. fl. Cirenaica xxiii-xxiv. 1930.
1: Apr 1916, p. [1]-36. *Copy*: G. – Reprinted and to be cited from Nuovo Giorn. bot. ital. 23(2): 260-294. 1916.
2: Jul 1917, p. [1]-59. *Copy*: G. – Id. 24(3): 213-271. 1917.

7266. *Piante nuove della Republicca di San Marino*. San Marino (Tipografia Reffi & della Bolda) 1917. (*Piante San Marino*).
Publ.: 1917, p. [1]-5. *Copy*: FI. – Reprinted from Museum, Boll. Republ. San Marino 1(2): 140-144. 1917 (journal n.v.) continued ib. 4: 116, 118. 1920.

7267. *L'Esplorazione botanica dell' isola di Rodi* dal 1761 al 1922. [Firenze 1923].
Publ.: Nov-Dec 1923 (in journal), p. [1]-29, *4 pl.*, map. *Copy*: FI. – Reprinted and to be cited from L'Universo, Firenze 4(11-12): [859]-871, 955-971. 1923.

7268. *Il viaggio del botanico fiorentino Pier Antonio Micheli a Verona ed al Monte Baldo* nell' autunno del 1736 ... Verona (Tipografia Operaia-Vicolo Regina d'Ungheria) 1925. (*Viagg. Micheli*).
Publ.: 1925, p. [1]-43. *Copy*: FI. – Reprinted from Madonna Verona 38(69-72) 1925 (journal n.v.).

7269. *L'esplorazione botanica del Dodecaneso* dal 1787 al 1924 [Nuovo Giornale botanico italiano 1926].
Publ.: 1926, p. [1]-19. *Copy*: FI. – Reprinted and to be cited from Nuovo Giorn. bot. ital. ser. 2. 33: 20-38. 1926.

7270. *Prodromo della flora Cirenaica*. Forli (Tipografia Valbonesi) 1930. Oct. (*Prodr. fl. ciren.*).
Publ.: 1930 (t.p.), p. [i*-ii*], [i]-xxxviii, [1]-577, *pl. 1-6. Copies*: B, NY, USDA. – Also issued with a cover dated 1931, together with A. Maugini, Contr. prati Cirenaica on p. 579-665. *Copies*: FI, G.
Addenda: Archivio botanico 12(1): 17-53. 1936, 12(2): 176-180. 1936 (see below 1936) and Rend. Sem. Fac. Sci. r. Univ. Cagliari 8: 53-79. 1938.

7271. *Flora della Repubblica di San Marino*. San Marino (arti grafiche Sammarinesi di Filippo della Balda) 1930. Oct. (*Fl. San Marino*).
Publ.: 1930, p. [i-iii], [1]-228. *Copies*: FI, G, NY, USDA.

7272. Speditione italiana de Filippi nell' Himàlaya, Caracorùm et Turchestàn Cinese (1913-1914) – Serie ii. Resultati geologici e geografici. vol. 10; publicati sotto la direzione di Giotto Dainelli. R. Pampanini – D. Vinciguerra. *La flora del Caracorùm*. Bologna (Nicola Zanichelli editore) [1930]. Qu. (*Fl. Caracorùm*).
Publ.: 1930 (Nat. Nov. Aug 1930), p. [1]-290, [1, cont.], *pl. 1-17. Copies*: G, NY. – Botanical part of *Raccolte di Piante e di animali* (animals by Dr. Vinciguerra). Copies with *Raccolte* cover at FI, MICH and USDA, p. [i]-vii, [1]-314, [1], *pl. 1-8*.

7273. *Aggiunte e correzioni al "Prodromo della flora cirenaica"*. Forlì (Tipografia Valbonesi) 1936 (a. xiv). Oct.
1: Mar 1936, p. [1]-38. *Copy*: G. – Reprinted and to be cited from Archivio botanico 12(1): 17-53. Mar 1936.
2: Jun 1936, p. [1]-7, *pl. 4. Copy*: G. – Id. 12(2): 176-180. 1936.
See also above under *Prodr. fl. ciren.*

7274. *Contributo alla conoscenza della flora del Cadore* (Alpi orientali). Forli (S.A.C. Stab. Tip. Valbonesi) 1939 (a. xvii). Oct.
Publ.: Jun 1939 (in journal; p. 12: Apr 1939), p. [1]-12. *Copy*: NY. – Reprinted and to be cited from Archivio bot. 15(2): 90-99. Jun 1939. – The "xvii" refers to the Mussolini era.

Pamplin, William (1806-1899), British botanist; from 1839 botanical bookseller and publisher in Soho; retired to Llanderfel, North Wales 1863. (*Pamplin*).

HERBARIUM and TYPES: Llandderfel herb. at OXF (with corr. acquired by G.C. Druce); some ferns in AK, E and K. Further letters at CGE, K, UCNW (main collection) and National Library of Wales. – Pamplin published sets of exsiccata such as "Flora Dalmatica" and "Vicinity of Adelaide 1846" not provided with collector's names. – Correspondence with D.F.L. von Schlechtendal at HAL; further letters at G.

BIBLIOGRAPHY and BIOGRAPHY: Barnhart 3: 45; BB p. 235; BM 8: 975; Clokie p. 220-221; CSP 4: 748, 17: 692; Desmond p. 477; Henrey 2: 154; Jackson p. 248, 258, 589; Kew 4: 219; NI 917, 927, 2341; PR ed. 1: 7728; Rehder 1: 397, 400; TL-1/74; Urban-Berl. p. 322.

BIOFILE: Allen, D.E., The Victorian fern craze 24, 35. 1969; Soc. Bibl. nat. Hist. Newsletter 3: 9. 1979 (letters at CBG).
Anon., Bonplandia 1: 22. 1853; Bot. Centralbl. 80: 447. 1899 (d.); Bot. Not. 1900: 47; Bot. Zeit. 1: 408 (bookseller). 1843, 22: 48. 1864 (retires from book trade; moves to Llandderfel; business continued by Dulau & Co.); Gard. Chron. 1899: 59; Hedwigia 39: (40). 1900 (d. 9 Aug 1899); J. Bot. 37: 496. 1899; Österr. bot. Z. 50: 30. 1900 (d. 9 Aug 1899); Proc. Linn. Soc. London 112: 80-81. 1900 (b. 5 Aug 1806); Termt. Közlon. 32: 695. 1900.

Bridson, G.D.R. et al., Nat. hist. mss. res. Brit. Isl. 427 [index]. 1980.
Dolezal, H., Friedrich Welwitsch 41, 97, 125-126. 1974.
Dony, J.G., Hertfordshire 15. 1967.
Druce, G.C., Fl. Berkshire clxx-clxxi. 1897; Fl. Buckinghamshire xcviii. 1926; Fl. Oxfordsh. 395. 1896; J. Bot. 50: 76. 1912 (herb.).
Foggitt, G., Rep. Bot. Soc. Exch. Cl. Brit. Isl. 1932: 285 (herb. acquired by Druce).
Freeman, R.B., Brit. nat. hist. books 271. 1980(see no. 2909).
Hedge, I.C. & Lamond, J.A., Index coll. Edinb. herb. 116. 1970.
Kent, D.H., Brit. herb. 71. 1957; Hist. fl. Middlesex 21. 1975.
Kneucker, A., Allg. bot. Z. 6: 16. 1900.
Loudon, J.C. & W.J. Hooker, Gard. Mag. 15: 303-304. 1839 (succeeds Hunneman as bookseller).
Pearsall, W.H., Fl. Surrey 50. 1931.
Phillips, R.W., J. Bot. 37: 521-524. 1899 (portr.).
Stieber, M.T. et al., Huntia 4(1): 84. 1981 (arch. mat. HU).
Thompson, P., Essex Natural. 19: 76. 1920.
Trimen, H. & Thistleton Dyer, Fl. Middlesex 399. 1869.

COMPOSITE WORKS: Publisher of *The Phytologist*, a botanical journal, edited by A. Irvine [new series] vols. 1-6, 1855-1863.

7275. *A botanical tour in the highlands of Perthshire* by W.P. and A.I. ... London (William Pamplin, ...) 1857. Qu.
Authors: William Pamplin and Alexander Irvine (1793-1873), the latter editor of the *Phytologist*.
Publ.: 1857, p. [i]-viii, [1]-76. *Copy*: G. – Reprinted from the Phytologist 1: 417-424, 446-458, 475-486. 1856.

Pampuch, Albert (*fl.* 1840), German botanist, high school teacher at Tremessen (Trzemessno, now Poland) near Bromberg. (*Pampuch*).

HERBARIUM and TYPES: Unknown.

BIBLIOGRAPHY and BIOGRAPHY: BM 4: 1507; PR 6914 (ed. 1: 7729); Rehder 1: 380.

BIOFILE: Pfuhl, F., Z. naturw. Abt. deut. Ges. Kunst Wiss. Posen 14(11), 1907 (on a second botanical publication by Pampuch).
Szymkiewicz, D., Bibl. fl. Polsk. 112. 1925.

7276. *Flora tremesnensis*, oder systematische Aufstellung der in der Umgegend von Trzemessno bis jetzt entdeckten wildwachsenden Pflanzen, so wie auch vieler veredelten und exotischen, welche des Nutzens oder der Schönheit und Seltenheit wegen als Feld-, Garten– und Treibhausgewächse gehegt werden, als Wegweiser bei Anlegung von Herbarien für seine Schüler entworfen ... Trzemessno [Tremessen] (Druck und Verlag von Gustav Olawski) 1840. Oct. (*Fl. tremesn.*).
Publ.: 1840, p. [i-iii], [1]-70. *Copy*: W.

Pancher, Jean Armand Isidore (1814-1877), French gardener, botanist and collector ("jardinier colonial") in Tahiti (1849-1857) and in New Caledonia 1857-1869, in France 1869-1874; again to New Caledonia, for J. Linden 1874-1877. (*Pancher*).

HERBARIUM and TYPES: P, PC; duplicates at A, B, BM, BR, C, DBN, E, FI, H, K, L, LA, LE, MEL, MO, W.

BIBLIOGRAPHY and BIOGRAPHY: Backer p. 658; Barnhart 3: 46; IH 1(ed. 6): 362, (ed 7): (in press), 2: (in press); Kew 4: 219; Rehder 2: 648; Urban-Berl. p. 309, 380.

BIOFILE: Anon., Bull. Soc. bot. France 21 (bibl.): 240. 1875, 24 (bibl.): 95. 1877 (c.r.); 253. 1877; J.Bot. 15: 288. 1877 (d.).
Bescherelle, E., Ann. Sci. nat. Bot. ser. 5. 18: 184. 1873 (Florule bryologique de la Nouvelle Calédonie, in part based on Pancher coll.).

Brongniart, Ad. & A. Gris, Ann. Sci. nat. Bot. ser. 5. 13: 345- 346. 1871 (*Dacrydium Pancheri*).
Carrière, E.A., Revue hort. 49: 285-286. 1877 (d.).
Guillaumin, A., Les fleurs des jardins xl. 1929.
Hedge, I.C. & J.A. Lamond, Index coll. Edinb. herb. 116. 1970.
Jackson, B.D., Bull. misc. Inf. Kew 1901: 51 (440 spec. at K).
Merrill, E.D., Contr. U.S. natl. Herb. 30(1): 233. 1947.
Murray, G., Hist. coll. BM(NH) 1: 173. 1904 (880 spec. at BM).
Nylander, W., Ann. Sci. nat. Bot. ser. 4. 15: 37. 1861 (New Caled. lichens in part coll. by P.).
Pancher, J.A.I., Adansonia 10: 372-373. 1873 (descr. *Aralia tenuifolia*); Ill. hort. 28: 24-27. 1881 (notes on New Caled. from his corr.).
Sagot, J. Soc. centr. Hort. France 1879: 515-534 (reprint 20 p.) (obit., b. Jan 1814, d. Apr 1877).
Schlechter, R., Bot. Jahrb. 36: 6. 1905.
Steinberg, C.H., Webbia 32(1): 33. 1977.

NAME: Pancher was the junior author of the paper by "MM Vieillard et Panchet" in Bull. Soc. bot. France 3: 160-161. 1856.

EPONYMY: *Pancheria* A. Brongniart et Gris (1862); *Panchezia* [sic] Montrouzier (1860).

Pančic, Josef (Giuseppe Pancio; Josif Panchic) (1814-1888), Croatian botanist and physician; M.D. Budapest 1842; regional physician in Serbia 1846-1853; high school teacher (natural history) at the Belgrad Lyceum and director of the Belgrad botanical garden.(*Pančic*).

HERBARIUM and TYPES: Largest set definitely known to us at BP; other material at FI, G, GE, GOET, K, LE, LY, MANCH, MW, PAD, W and WRSL. Adamovic (1909) mentions a "Herb. Pančic" which, according to L. Glišic(in J.Bélic 1976, p. 43) is now kept at the Botanical Institute of the University of Belgrad. To our regret we failed to make contact with the present authorities. – Some letters at G.

BIBLIOGRAPHY and BIOGRAPHY: AG 2(1): 479, 4: 527, 5(1): 481; Backer p. 658; Barnhart 3: 46; BJI 1: 43; BM 4: 1508; Bossert p. 299; CSP 4: 748, 8: 559, 10: 986, 12: 556, 17: 693; Hegi 5(2): 1196; Herder p. 470; Hortus 3: 1201 ("Panč."); IH 2: (in press); Jackson p. 229, 314; Kanitz no. 184; Kew 4: 220; LS 37643; Morren ed. 10, p. 101; NI (alph.); PR 6917-6918; Quenstedt p. 327; Rehder 5: 647; Tucker 1:536; Zander ed. 10, p. 699, ed. 11, p. 797 (b. 6 Mai 1814, d. 8 Mar 1888).

BIOFILE: Adamovic, L., Veg.-Verh. Balkanländ. 1, 3-6, 20, 1909 (Veg. Erde 11).
Anon., Ann. Bot. 2: 422-423. 1889 (bibl.); Bonplandia 8: 170. 1860; Bot. Centralbl. 34: 159. 1888 (d. 8 Mar 1888), 38: 542. 1889 (succeeded by St. Jakšic); Bot. Gaz. 13: 136. 1888; Bot. Jahrb. 9 (Beibl. 22): 1. 1888; Bot. Not. 1889: 29 (d.); Bot. Zeit. 46: 305. 1888 (d.); Nat. Nov. 10: 121. 1880 (d.); Österr. bot. Z. 38: 147. 1888 (d.); Verh. bot. Ver. Brandenburg 29: xlvii. 1888(d.).
Beck, G., Bot. Centralbl. 33: 378. 1888, 34: 150. 1888 (666 pl. at W); Veg.-Verh. Illyr. Länd. 7, 21, 24, 39-40. 1901 (Veg. Erde 4).
Belic, J. et al., Josif Pančic, Beograd 1976, 439 p. (major work on Pančic with e.g. an "essai de biographie" by Jurišic on p. 411-428, an extensive bibliography on p. 429-432, and refs. to studies on P. on p. 433-436).
Bolle, C., Garden & Forest 1: 135. 1888 (d.).
Braun, H., Österr. bot. Z. 38: 257-262, 310-314. 1888 (obit., bibl., b. 6 Mai 1814, d. 8 Mar 1888).
Candolle, Alph. de, Phytographie 438. 1880 (pl. at BP and W).
Jackson, W.B., Bull. Misc. Inf. Kew 1901: 51. (pl. at K).
Jankovic, M.M., Glasnik bot. Zav. Bašte Univ. Beogr. 2: 5-8. 1967 (150 yrs. anniv. birth).
Jankovic, M.M. & Ž.V. Vasič, Glasnik Prirod. Muz. Beogr. 19: 3-24. 1964 (id., portr., sec. refs., bibl.).

Jowantschewits, W., Gartenwelt 7: 394-395. 1903 (biogr.).
Kanitz, A., Magy. Növén. Lap. 12: 40-42. 1888 (obit., bibl.).
Lindemann, E. v., Bull. Soc. imp. Natur. Moscou 61: 52. 1886.
Mayer, E. & Diklic, N., Nomenclator pančicanus, *in* Pančic-Sammelb., Serb. Akad. Wiss. Künste, nat.-math. Kl. 3-26, 1967 (bibl., list of taxa described by P. and eponymy).
Steinberg, C.H., Webbia 34: 61. 1979 (pl. at FI). Žujovic, J.M., Le docteur Josif Pančic, Belgrad 1889, 13 p., repr. from Ann. géol. Pénins. balkan. vol. 1, 1889 (b. 5 Apr 1844, d. 25 Feb 1888, bibl.).

COMPOSITE WORKS: see R. de Visiani for de Visiani & Pančic, *Plantae serbicae*, 1862-1870.

POSTAGE STAMPS: Yugoslavia 200 d. (1965) yv. $1035^1 = 2$.

EPONYMY: *Pancicia* Visiani ex Schlechtendal (1859).

7277. *Flora agri belgradensis* methodo analytica digesta ... Beogradu [Belgrad] (u Držampariji) 1865. Oct. (*Fl. agri belgr.*).
Ed. 1: 1865 (Flora 8 Nov 1865 "soeben" p. [i*, iii*], [i]-x, [1]-295. *Copies*: G. – Alternative title: "Flora u okolini beogradskoj ...".
Ed. 2: 1878, p. [i*, iii*], [i]-ix, [1]-472, [1-2]. *Copies*: G. – Flora agri belgradensis recusa.
Ed. 3: 1882, p. [i]-xvi, [1]-520. *Copy*: G. – Editio tertia.
Ed. 4: 1885, p. [i]-xvi, [1]-518. *Copy*: B. – Editio quarta.
Ed. 5: 1888, (p.x: Apr 1888), p. [i]-xxiii, [1]-535. *Copy*: B. – Editio quinta.
Ed. 6: 1892, p. [i]-xxiv, [1]-535. *Copy*: G. – Editio sexta. Obituary of Pančic on p. v-xiii (by Ž.J. Jurišic).
Ref.: Kanitz, A., Österr. bot. Z. 16: 219-220. 1 Jul 1866.

7278. *Šumsko drvece i Šiblje u Srbiji* ... Beogradu [Belgrad] (u Državioj Štamiariji) 1871. Oct.
Publ.: 1871, p. [i], [1]-184. *Copy*: G. – See Belic, J. et al., p. 431, no. 21 for journal publ.

7279. Flora Kneževine Srbije ... *Flora principatus Serbiae.* u Beogradu [Belgrad] (... Državio Štamiariji) 1874. Oct. (*Fl. Serbiae*).
Publ.: 1874, p. [i*], [i]-xxxiv, [1]-798, err. [1-6]. *Copies*: B, G.
Additamenta [Dodatak]: 1884 (Bot. Jahrb. 24 Oct 1884), p. [ii-iii], [1]-253, [1-2, index]. *Copies*: B, G. – "Dodatak fl. Knež. Srbije".
Reprint: 1976, Beograd (Državna štamparija). *Copy*: MO.

7280. *Elenchus plantarum vascularium* quas aestate a. 1873 in Crna gora [Montenegro] legit Dr. Jos. Pancic. Belgradi [in typographia status) 1875. Oct. (*Elench. pl. vasc.*).
Publ.: 1875, p. [i]-vii, [1]-106. *Copies*: B, G, H, NY, USDA.

7281. *Eine neue Conifere in den östlichen Alpen* ... Belgrad (in der fürstl.-serbischen Staatsdruckerei) 1876. (*Neu. Conif. Alp.*).
Publ.: 1876, p. [1]-8. *Copies*: B, G. – (Discovery of *Picea omorica*).
Ref.: Reichenbach, H.G., Bot. Zeit. 35: 121-122. 23 Feb 1877.

7282. *Elementa ad floram principatus Bulgariae* ... Beograd 1883. Oct. (*Elem. fl. Bulg.*).
Publ.: 1883 (Bot. Jahrb. 6 Mai 1884; Bot. Centralbl. 4-8 Mai 1884), p. [i], [1]-71. *Copies*: B, G. – Alternative title: Gráda za floru kneževine Bugarske.
Nova elementa ad floram principatus Bulgariae/Nova gráda za floru kneževine Bugarske: 1886 (Bot. Jahrb. 2 Nov 1886), p. [i], [1]-43. *Copy*: B.
Ref.: Uechtritz, M.F.S. v., Bot. Jahrb. 8(Lit.): 44-45. 1886.

7283. *Der Kirschlorbeer* im Süd-Osten von Serbien ... Belgrad [Beograd] (Königlich-Serbische Staatsdruckerei) 1887. (*Kirschlorbeer*).
Publ.: 1887 (ÖbZ 1 Dec 1887), p. [1]-8. *Copy*: B.

7284. *Regius hortus botanicus belgradensis* 1887. Belgradi (In Tipographia Status) 1888. Qu. (*Reg. hort. bot. belgr.*).

Publ.: Mar 1888 (p. 31: Mar 1888; published shortly after P's death 25 Feb 1888, fide Adamovic (1909) p. 5, cover t.p., p. [1]-31. *Copy*: G. – Alternative title on p. [1]: *Enumeratio plantarum vascularium florae Serbiae*. – A note by Kanitz in the Boissier copy says: "c'est une énumération des plantes de Serbie sensu lat., mais pas critique. Le pauvre Pančic l'a terminée sur son lit de mort mais n'a pu le revoir." Page 31 carries the name of Pančic as well as of Ž.J. Jurišic and J. Bornmüller.

Pandé, S.K. (1899-1960), Indian bryologist; D. Sc. Lucknow 1936; from 1899-1959, demonstrator, professor of botany at Lucknow and chairman of the Department of Botany of Saugar University. (*Pandé*).

HERBARIUM and TYPES: LWU.

BIBLIOGRAPHY and BIOGRAPHY: IH 2: (in press); MW suppl. 275; SBC p. 128.

BIOFILE: Anon., Rev. bryol. lichénol. 7: 130, 131. 1934, 8: 126. 1935, 9: 165. 1936, 20: 226. 1951, 24: 162. 1955, 25: 191, 390, 404. 1956 (bibl.), 31: 114. 1962 (d.).
Sayre, G., Bryologist 80: 515. 1977.
Schuster, R.M., Hepat. Anthoc. N. Amer. 1: 85. 1966 (bibl.).
Udar, R., J. Indian bot. Soc. 40: 292-293. 1961 (obit., portr.); Bull. Bot. Soc., Univ. Saugar 13(1, 2): 1-5. 1963 (obit., bibl., portr., b. 16 Feb 1899, d. 25 Nov 1960); Rev. bryol. lichénol. 33: 287-390. 1964 (bibl., portr.).
Verdoorn, F., ed., Chron. bot.1: 360. 1935, 2: 210, 392. 1936, 3: 183. 1937, 6: 300. 1941.

COMMEMORATION VOLUME: Bull. Bot. Soc., Univ. Saugar 13(1, 2). 1961, publ. 1963.

Panizzi-Savio, Francesco (1817-1893), Italian pharmacist at San Remo; author on the flora of San Remo. (*Panizzi*).

HERBARIUM and TYPES: material at FI, G, GE, TO. – Some letters at G.

BIBLIOGRAPHY and BIOGRAPHY: AG 3: 391; Barnhart 3: 46; Bossert p. 300; CSP 17: 695; IH 2: (in press); LS 19790a-b; Morren ed. 10, p. 87; NI 1: 253, no. 1486; Plesch p. 355; Rehder 1: 428; Saccardo 1: 121, 2: 80; Sotheby p. 583.

BIOFILE: Anon., Bonplandia 1: 23. 1853, 6: 49. 1858 (Leopoldina cogn.: Risso); Flora 40: 264. 1857.
Burnat, É., Bull. Soc. bot. France 30: cxxv. 1883 (bibl.).
Cavillier, F., Boissiera 5: 69. 1941.

7285. *Flora fotografata* delle piante più pregevoli e peregrine di Sanremo e sue adiacenze per Francesco Panizzi. Fotografie de Pietro Guidi. Sanremo 1870.(*Fl. fotogr. Sanremo*).
Publ.: 1870-1874. *Copy*: FI (inf. C. Steinberg). This copy, which is possibly incomplete, consists of 102 photographs in 10 fascicles without printed pages but with captions mentioning latin name, locality and flowering time. The author foresaw the publication of 200 photographs. G.K. (Bot. Zeit.31: 358-360. 6 Jun 1873) provides a list of the photographs so far published. A selection of 40 hand-coloured photographs "Flore de Marseille à Gènes" is mentioned in Sotheby 583. A prospectus dated Sanremo 1874, at FI, mentions no number of fascicles (decades) anticipated. NI(1486) mentions 153 fasc., plus 9 plates on Oleaceae, 20 on Hymenomycetes and 3 suppl.

Pansch, Adolf (1841-1887), German (Eutin/Lübeck) physician and botanist; at Kiel from 1864; participated in the second German North Polar Expedition 1869-1870 under K. Koldewey. (*Pansch*).

HERBARIUM and TYPES: Material at B, L and TU, and possibly also in BREM or KIEL.

BIBLIOGRAPHY and BIOGRAPHY: Barnhart 3: 46; BM 4: 1509; Christiansen p. 32, 35, 318; IH 2: (in press); Jackson p. 296; Kew 4: 222; Rehder 1: 309; TL-2/866; Urban-Berl. p. 267, 309, 380.

BIOFILE: Anon., Flora 52: 393. 1869 (participates in polar exp.), 54: 221-222. 1871 (report on bot. Greenland).
Fischer-Benzon, R.J.D., v., *in* Prahl, P., Fl. Schlesw.-Holst. 2: 45-46. 1890.
Kusnezow, N.J., Bot. Centralbl. 68: 258. 1896 (coll. TU).
Volkens, G.L.A., Verh. bot. Ver. Brandenburg 51: (85). 1910.

COMPOSITE WORKS: *Klima und Pflanzenleben auf Ostgrönland*, in vol. 1 of Hartlaub, G.& M. Lindeman, ed., *Die zweite Deutsche Nordpolarfahrt* in ... 1869 und 1870, Leipzig 3 Jun 1873 (announced by J. Bot. Mai 1873 and Bot. Zeit. 4 Apr 1873), see TL-2/866; p. [1]-96, *pl. 1*.

Pantanelli, Dante (1844-1913), Italian palaeontologist and geologist; Dr. math. phys. Pisa 1865; from 1882-1913 professor of mineralogy and geology at the University of Modena; student of Meneghini. (*D. Pantan.*).

HERBARIUM and TYPES: Unknown.

BIBLIOGRAPHY and BIOGRAPHY: Barnhart 3: 47; CSP 8: 560, 10: 987, 12: 556, 17: 696-697; De Toni 1: xc, 2: xcvi; Quenstedt p. 327; Saccardo 2: 80-81.

BIOFILE: Anon., Nat. Nov. 35: 620. 1913.
Poggendorff, J.C., Biogr.-lit. Handw.-Buch 3: 1003. 1898, 4(2): 1117. 1904 (bibl.).
Tassi, F., Bull. Lab. Orto bot. 7(1-4): 41. 1905 (b. 4 Jan 1844).

7286. Reale Accademia dei Lincei (anno cclxxix 1881-82). *Note microlitologiche sopra i calcari* ... Roma (con tipi del Salviucci) 1882. Qu. (*Not. microlit. calc.*).
Publ.: 1882, p. [1]-20, *pl. 1-2* (auct., uncol. lith.). *Copy*: NY.

Pantanelli, Enrico Francesco, (1881-1951), Italian mycologist; son of Dante Pantanelli; Dr. sci. Modena 1902; with Pfeffer at Leipzig 1902-1903; with E. Schulze at Zürich 1903-1904; with L. Kny at Berlin 1904; at the Istituto di Botanica, Roma, 1904-1906; at the Roma Stazione di Patologie vegetale 1906-1913; at Naples 1913-1914; with the Italian Department of Agriculture at Roma 1914-1920; at the Bari Stazione de Agric. sperimentale 1920-1947. (*E. Pantan.*).

HERBARIUM and TYPES: Material at ROPV.

BIBLIOGRAPHY and BIOGRAPHY: Barnhart 3: 47; CSP 17: 697; IH 2: (in press); LS 19794-19800, 37645-37655, suppl. 20871-20883; Saccardo 2: 81.

BIOFILE: Anon., Boll. (Contr.) Ist. bot. Catania 1: 183-184. 1955 (obit.); Bot. Centralbl. 99: 560. 1905 (to Roma); Jahrb. wiss. Bot. 56: 823. 1915 (Schüler Pfeffers); Österr. bot. Z. 55: 412. 1905 (lecturer Roma).
Ciferri, R., Taxon 1: 126. 1952; Ber. deut. bot. Ges. 68a: 251-253. 1955 (biogr., portr.).
Rivera, V., Ann. di Bot. 23(3): 555-574. 1951 (obit., bibl., portr., b. 18 Aug 1881, d. 5 Dec 1951).
Tassi, F., Bull. Lab. Orto bot. 7(1-4): 48. 1905 (b. 18 Aug 1881).
Verdoorn, F., Chron. bot. 3: 188. 1937.

7287. *Anatomia fisiologica delle Zygophyllaceae* (con 4 tavole) in Modena (coi tipi di G.T. Vincenzi e nipoti ...) 1900. Oct.
Publ.: 1900, p. [1]-93, *4 pl. Copy*: FI. – Reprinted from Atti Soc. Natural. Mat. Modena ser. 4. 2: 93-181. 1900.

Pantel, C. (*fl.* 1885), French botanist and geographer (*Pantel*).

HERBARIUM and TYPES: Some material at MPU.

BIBLIOGRAPHY and BIOGRAPHY: BL 2: 112, 699; BM 4: 1509.

BIOFILE: Granel de Solignac, L. et al., Naturalia monsp. Bot. 26: 27. 1976.

7288. *Formation et aspect du relief actuel des Cévennes* avec la liste des plantes qui croissent dans ce pays ... Paris (Librairie Henri Aniéré et A. Broussois ...) 1885. (*Form. relief Cévennes*).
Publ.: 1885 (Nat. Nov. Dec(2) 1885; J. Bot. Jan 1886, Bot. Zeit. 26 Feb 1886), p. [1]-77, [1, err.]. *Copies*: E, P (Bibl. centr.).

Pantling, Robert (1856-1910), British botanist; trained at Kew as a gardener; assistant Cinchona Dept., Bengal, 1879; curator Royal Botanic Gardens, Calcutta Dec 1879; at the Cinchona Dept. 1880; from 1897 as deputy superintendent. (*Pantl.*).

HERBARIUM and TYPES: Important sets at CAL and K; other material at B, BR, FI, G, GH, L, LE, P, UPS, W (material of Pantling and King).

BIBLIOGRAPHY and BIOGRAPHY: Barnhart 3: 47; BB p. 235; BM 4: 1509; Desmond p. 477; Hortus 3: 1201 ("Pantl."); IH 1 (ed. 7): 339, 2: (in press); Kew 4: 224; Morren ed. 10, p. 144; NI 256, 1051, 1052 (alph.); SK 1: 400; TL-2/3660; Tucker 1: 537; Urban-Berl. p. 380.

BIOFILE: Anon., J. Kew Guild 2: 491-492. 1910 (portr.); Trans. bot. Soc. Edinburgh 24: 101-103. 1911 (err. "Pant*h*ing").

COMPOSITE WORKS: see G. King, for Pantling and King, *The Orchids of the Sikkim Himalaya*, 1898 (see e.g. J. Bot. 36: 498-499. 1898 and TL-2/3660).

EPONYMY: *Pantlingia* Prain (1896).

7289. Some new *Orchids from Sikkim* ... Calcutta (printed at the Baptist Mission Press) 1895. Oct.
Senior author: George King (1840-1909).
1. 1895, p. [1]-16, reprinted and to be cited from J. Asiat. Soc. Bengal 64(2): 329-344. 1895 (vol. t.p. 1896).
2. 1896, p. [17]-38, id. 65(2): 118-134 (also 107). 1896 (vol. t.p. 1897). *Copy*: FI. – Preliminary publ. for TL-2/3660, Orch. Sikkim-Himalaya, q.v.

Pantocsek, Josef (1846-1916), Hungarian botanist and provincial physican; district physician at Tavarnok; from 1896 director of the hospital and "wirklicher Sanitätsrat" at Bratislava (Pozsony); botanical explorer of Montenegro. (*Pant.*).

HERBARIUM and TYPES: BP (for the present state of the diatom coll. see Krenner (1980)); further material at C, G, GOET, LE, PAD (Montenegro), W. – Some letters at G.

BIBLIOGRAPHY and BIOGRAPHY: AG 5(2): 719, 12(3): 51; Backer p. 658; Barnhart 3: 47; BM 4: 1509; Bossert p. 300; CSP 8: 560, 12: 556, 17: 698; De Toni 1: xcvi, 2: xc, cxxviii, 4: xliv; DTS 1: 216; Futák-Domin p. 455; IH 1 (ed. 7): 339; Herder p. 205; Jackson p. 314; Kew 4: 224; Morren ed. 10, p. 41; Quenstedt p. 327; Rehder 5: 648; Saccardo 1: 121; Tucker 1: 537; Zander ed. 10, p. 699, ed. 11, p. 797.

BIOFILE: Anon., Bot. Centralbl. 61: 271. 1896 (Sanitätsrath); Bot. Not. 1916: 282; Magy. bot. Lap. 15: 312; 1916; Hedwigia 63, Beibl. 1: 97. 1921; Österr. bot. Z. 22: 137; 1872 (journey Montenegro), 46: 239 (Sanitätsrath), 342, (dir. hosp. Bratislava). 1896.
Beck, G., Bot. Centralbl. 34: 150. 1888 (pl. W), 50: 317-318. 1892 (controversy Beck/P.); Veg. Verh. Illyr. Länd. 16, 21, 40. 1901 (Veg. Erde 4).
Candolle, Alph. de, Phytographie 439. 1880 (coll.).
Degen, A., Magy. bot. Lap. 15: 213-223. 1916 (obit., bibl., portr., b. 15 Oct 1846, d. 4 Sep 1916; herbarium donated to the Ärzte- und Naturf. Verein, Pressburg (Bratislava); library and diatom coll. still in the hands of the family).
Grisebach, A.H.R., Bot. Zeit. 30: 131. 1872 (will collect in Montenegro).
Kneucker, A., Allg. bot. Z. 2: 108, 172. 1896, 23: 32. 1917 (d.).

Krenner, J.A., Studia botanica hungarica 14: 9-28. 1980 (diatom coll. of 5000 slides at BP; 4000 destroyed in World-War II; 912 preparations now restored; list of taxa).
Koster, J., Taxon 8: 556. 1969.
Meister, F., Mikrokosmos 19167 (7/8): [1 p.] (obit., portr.).
Pax, F., Grundz. Pflanzenverbr. Karpathen 1: 29, 39, 43. 1898 (Veg. Erde 10).
Simonkai, L., Enum. fl. transsilv. xxvi. 1886.
Szymkiewicz, D., Bibl. Fl. Polsk. 112. 1925.
VanLandingham, S., Cat. diat. 6: 3576-3577, 7: 4213-4214. 1978.
Verseghy, K., Feddes Repert. 68(1): 125. 1963.
Wornhardt, W.W., Int. direct. diat. 35. 1968.

EPONYMY: *Pantocsekia* Grisebach ex Pantocsek (1873, *nom. rej.*); *Pantocsekia* Grunow ex Pantocsek (1886, *nom. cons.*).

7290. *Adnotationes ad floram et faunam Hercegovinae*, Crnagorae [Montenegro] et Dalmatiae ... Posonii [Pozsony, Bratislava, Pressburg] (typis C.F. Wigand) 1874. Oct. (*Adnot. fl. faun. Herceg.*).
Publ.: 1874 (ÖbZ 1 Mai 1874; Bot. Zeit. 5 Jun 1874; BSbF Oct 1874), p. [i, iii], [1]-143, [144]. *Copy*: G. – Reprinted from Verh. Ver. Naturk. Pressburg ser. 2. 2. 1974. Dedicated to A. Grisebach who participated in the editorial preparation. – See also ÖbZ 23: 4-6, 79-81, 265-268. 1873.

7291. *Beiträge zur Kenntniss der fossilen Bacillarien Ungarns* .... Nagy-Tapolcsány (Buchdruckerei von Julius Platzko) 1886-1893. Oct. (*Beitr. foss. Bacill. Ung.*).
Ed. 1: 1886-1893, in 3 vols., circa 30 copies printed. *Copies*: B (vol.1), FH (vol.1), Landesbibl. Graz (vols. 1, 2); PH (vols. 1-3).
1: (Marine Bacillarien): 1886, p. [1]-74, [75, index; 76, err.], *pl. 1-30* (uncol., auct.) with text.
2: (Brackwasser-Bacillarien): Mai-Jul 1889 (Bot. Centralbl. 28 Aug 1889; Bot. Z. 27 Sep 1889; Nat. Nov. Sep(1) 1889; ÖbZ 1 Oct 1889), p. [1]-123, *pl. 1-30*.
3: (Süsswasser-Bacillarien): 1893 (Hedwigia rd. Mai 1893; Nat. Nov. Apr(1) 1893; ÖbZ 1 Mai 1893), *pl. 1-42*, plates with text.
Ed. 2: 1903-1905, 2. verbesserte Auflage, 3 Teile. Berlin NW 5. Verlag von W. Junk. *Copies*: FH, MICH, NY; Landesbibliothek Graz (vol. 3).
1: 1903, p. [1]-76, [1, index], *pl. 1-30* (id.).
2: 1903 [?], p. [1]-122, [1, index], *pl. 1-30* (id.).
3: plates 1903, text 1905 (ÖbZ Oct 1905), p. [1]-118, *pl. 1-42* (id.). Imprint: Pozsony (Buchdruckerei C.F. Wigand) 1905.
Ref.: Grunow, A., Bot. Centralbl. 34: 174-176. 1888 (rev. vol. 1).

7292. *Resultate der wissenschaftlichen Erforschung des Balatonsees*. Mit Unterstützung der hohen Kön. Ung. Ministerien für Ackerbau und für Cultus und Unterricht herausgegeben von der Balatonsee-Commission der Ung. [sic] Geographischen Gesellschaft. Zweiter Band. Die Biologie des Balatonsees. Zweiter Theil die Flora. 1. Sektion Anhang. *Die Bacillarien des Balatonsees* von Dr. Josef Pantocsek ... Wien (Commissionsverlag von Ed. Hölzel) 1902. Oct. (*Bacill. Balatonsees*).
Publ.: 1902 (p. 4: 30 Mai 1901; Nat. Nov. Feb(1) 1903, p. [i-ii], [1]-112, [1-2, cont., err.], *pl. 1-17* (uncol.) w.t. *Copies*: G, NY. – Also published in Hungarian "A Balaton kovamoszatai" 1901.
Ref.: Gutwinski, R., Bot. Centralbl. 95: 69-71. 1904.

7293. *Beschreibung neuer Bacillarien* (novarum Bacillariarum descriptio) ... Pozsony [Bratislava, Pressburg] (Druck von C.F. Wigand), Berlin (Verlag R. Friedländer U.S.) 1909. *Beschr. neu. Bacill.*).
Publ.: Jun-Jul 1909 (Nat. Nov. Aug(2) 1909, Bot. Centralbl. 28 Sep 1909), p. [1]-13, [14], *2 tables. Copy*: PH.

7294. A lutillai ragpalában elöforduló *Bacillariák vagy Kovamoszatok* leirás. (Bacillarien des Klebschiefers von Lutilla) ...Pozsony [Bratislava, Pressburg] [Wigand K.F. Könyvnyomdája] 1913. (*Bacill. vagy Kovamoszatok*).

Publ.: 1913 (Bot. Centralbl. 23 Sep 1913), p. [1]-14, *2 pl.* with text. *Copy*: PH.
Ref.: Matouschek, F., Hedwigia 54: (139). 1914.

Panţu, Zacharia C. (Pantzu) (1866-1934), Roumanian botanist; at the Bucarest botanical institute 1890 (as naturalist), 1891-1934 (as curator). (*Panţu*).

HERBARIUM and TYPES: BUC; further material at C, CL, E. - Some letters at G.

BIBLIOGRAPHY and BIOGRAPHY: Barnhart 3: 47; BJI 2: 130; BM 4: 1509, 8: 976; Bossert p. 300; IH 2: (in press); Kew 4: 224; Tucker 1: 537.

BIOFILE: Anon., Hedwigia 75: (71). 1935 (d.), Österr. bot. Z. 83: 240. 1934 (d. 19 Mar 1934).
Hedge, I.C. & Lamond, J.M., Index coll. Edinb. herb. 116.1970.
Pax, F., Grundz. Pfl.-Verbr. Karpathen 2: 279. 1908 (Veg. Erde 10).
Pop, E., Bull. Jard. Mus. bot. Univ. Cluj 14: 85-94. 1934 (biogr., bibl., portr., b. 31 Jul 1866).
Verdoorn, F., Chron. bot. 1: 244. 1935 (portr.; d.).

7295. *Contribuţiuni la flora Ceahlaului.* I. Regiunea alpina şi subalpina. Beiträge zur Flora des Ceahlau i. Alpine und subalpine Region de Zach. C. Panţu şi A. Procopianu-Procovici ... Bucuresci [Bucarest] (Institutul de arte grafice şi editură Minerva ...) 1901. Oct. †. (*Contr. fl. Ceahlau.*).
Co-author: Aurel Procopianu-Procovici.*Publ.*: Sep 1901 (in journal; p. viii: Aug 1899; Nat. Nov. Jan(2) 1902; Bot. Zeit. 1 Apr 1902; J. Bot. Mar 1902), p. [i]-viii, 1-44, [1, err. on p. 3 of cover]. *Copies*: G, USDA. - Bull. Erbar. Inst. bot. Bucuresti 1: 80-131. 1901, continued 2: 81-103. 1902. - Romanian and German; diagnoses in Latin.

7296. Academia Română. *Contribuţiune la flora Bucegilor* de Zach. C. Panţu ... Bucureşti (Instit. de Arte grafice "Carol Göbl" ...) 1907. Oct. (*Contr. fl. Buceg.*).
Publ.: 1907 (AbZ 15 Sep 1907), p. [1]-32, *pl. 1-2. Copy*: G. - Issued as a separate memoir in Anal. Acad. Roman. 29. Mem. ştiinţ. 9: 281-312. 1907.

7297. Academia Română. *Contribuţiuni la flora Bucureştilor şi a imprejurimilor sale* ... Bucureşti [Bucarest] (Instit. de arte grafice Carol Göbl, ...) 1908-1912, 4 parts. (*Contr. fl. Bucureşt.*).
Publ.: in four parts which were separate Memoirs of the Analele Academiei Române, Memoriile secţ. ştiintifice. *Copies*: G, USDA.
1: 1908 (p. 8: 21 Apr 1908; AbZ 15 Oct 1908, p. [1]-96. Anal. 31, Mem. 1.
2: 1909 (p. 2: 14 Mai 1909; AbZ 15 Dec 1909), p. [1]-96. Anal. 32, Mem. 1.
3: 1910 (p. 2: 4 Mar 1910; AbZ 15 Mai 1910; Nat. Nov. Mar(1) 1911), p. [1]-94, [95, ind.]. Anal. 32, Mem. 3: 134-226. 1910.
4: 1912 (p. 3: 24 Apr 1912), p. [1]-164. Anal. 34, Mem. 21: 436-598. 1912.
Ref.: Panţu, Z.C., Bot. Jahrb. 46(Lit.): 3-4. 1911 (summary parts 2, 3).

7298. *Contribuţiuni nouă flora ceahlăului* ... [Bucureşti (Librăriile Socec & Comp., ...)] 1911. Oct.
Publ.: Dec 1911 (copy G signed by author 11 Dec 1911), p. [1]-54, [1, ind.]. *Copy*: G. - Reprinted and to be cited from Anal. Acad. Române ser. 2. 33: 293-347. 1911.

7299. *Orchidaceele din România.* Studiu monografic cu 50 tabele ... Bucureşti (Librăriile Socec & Comp., ...) 1915. Oct. (*Orchid. Român.*).
Publ.: 1915 (p. xii: 15 Apr 1915), p. [i]-xii, [1]-228, [229, 231-232], *pl. 1-50. Copies*: FI, G, PH.

Panzer, Georg Wolfgang Franz (1755-1829), German (Oberpfalz) physician, botanist and entomologist; Dr. med. Erlangen 1777; travelled in C. Europe 1778-1780; from 1780 regional physician and lecturer at the Collegium medicum, Nürnberg. (*Panz.*).

HERBARIUM and TYPES: At WBM (13.000) fide Leiblein, Flora 17: 269-270 (1834) but probably no longer extant; the directors of the various Bavarian herbaria informed us that the herbarium can not be located. The head librarian of the University of Würzburg, Dr. H. Thurn, could not find any indication that the herbarium was at Würzburg before 1945; however, some herbarium material at Würzburg has been saved but has not yet been sorted (inf. O.H. Volk). Extant Panzer material at: B-Willdenow (240, orig.); further material at BP (herb. Martius), BR, C, E, IBF, LE (material from Dauria), OXF. - Spiess (1891) refers to a "reichhaltiges, über 400 Foliobände umfassendes Herbarium ...". This may be the herbarium given to WBM; if so it was destroyed during World-War II (see IH 1 (ed. 6): 298). - Some letters at G.

NOTE: It is not quite certain that the Panzer who collected in Dauria is identical with G.W.F. Panzer. We have no information yet on the dates of the collections.

BIBLIOGRAPHY and BIOGRAPHY: ADB 25: 134; Backer p. 421; Barnhart 3: 47; BM 4: 1510; Bossert p. 300; Clokie p. 221; CSP 4: 750; Dryander 3: 85, 317; Frank 3 (Anh.): 74; Herder p. 147, 297; Hortus 3: 1201 ("Panz."); Kew 4: 224; Langman p. 563; Lasègue p. 337; MW p. 377; NI (alph.) see also 1: 99, 2: 1290, suppl. p. 54, 1486 p.; PR 5430, 5438, 6921-6924 (ed. 1: 7734-7737); Ratzeburg p. 475 (footnote); Rehder 2: 146; RS p. 133; SO 121, 577, 577a; TL-2/1130, 5563; Urban-Berl. p. 380, 415.

BIOFILE: Anon., Bonplandia 6: 219. 1858; Encycl. brit. ed. 11. 14: 369. 1911; Flora 12: 400. 1829 (obit.), 14(2) Int. Bl. 5: 17. 1832 (herb. for sale), 17: 269-270. 1834 (herb. to WM); Neuer Nekrolog der Deutschen 7: 530-533. 1829 (biogr., bibl.).
Bridson, G.D.R. et al., Nat. hist. mss. res. Brit. Isl. 229, 236. 1980.
Brockhaus, Conversations-Lexikon ed. 13. 12: 662. 1885.
Carpenter, M.M., Amer. Midl. Natural. 33(1): 77. 1945.
Dawson, W.R., Smith papers 72. 1934.
Eisinger, F., Int. entom. Z. 13(12): 89-92. 1919 (bibl.).
Gilbert, P., Comp. biogr. lit. deceased entom. 287. 1977.
Hedge, I.C. & Lamond, J.M., Index coll. Edinb. herb. 116. 1970.
Herder, F. v., Bot. Jahrb. 9: 432, 444. 1888 (plants from Dauria in herb. Fischer (at LE)).
Heufler, L. v., Flora 26: 590. 1843 (coll. IBF).
Jessen, G.F.W., Bot. Gegenw. Vorz. 405. 1864.
Leiblein, V., Flora 17: 269-270. 1834 (herb. to Würzburg).
Nissen, C., Zool. Buchill. 310. 1969 (b. 31 Mai 1735, d. 28 Jun 1829; zool. bibl.).
Schultes, J.A., Flora 5: 133-134. 1822 (describes Panzer herbarium).
Spiess, E., Abh. naturf. Ges. Nürnberg 8: 196-199. 1891 (bibl.).
Sturm, J., Faunus 1: 51. 1832.
Swainson, W., Taxidermy 287. 1840 (entom. publ.).
Stafleu, F.A., Linnaeus and Linnaeans 174, 176. 1971.
Z., Dict. Sci. méd., Biogr. méd. 6: 359. 1824 (bibl.).

COMPOSITE WORKS: (1) see G.F.; Christmann, *Vollst. Pflanzensyst.* 1777-1788, 14 vols., TL-2/1130, Panzer author of vols. 8-14.(2) Johann Martyns *Abbildung und Beschreibung seltener Gewächse*: neu übersetzt ... von G.W.F. Panzer, 1797, TL-2/5563.

NAME: Son of Georg Wolfgang Panzer (1729-1804), bibliographer at Nürnberg (translated M. Catesby, *Piscium, serpentum*, ... nec non plantarum quarundam *imagines* ... Nürnberg 1777.).

EPONYMY: *Panzera* Willdenow (1799); *Panzeria* J.F. Gmelin (1791); *Panzeria* Moench (1794). *Note: Panzera* Cothenius (1790) is very likely named for Panzer, but no etymology is given.

7300. *Observationum botanicarum specimen* auctore G.W.F. Panzero med. dr. ... Norimbergae [Nürnberg] et Lipsiae [Leipzig] (in officina libraria Ad. Gottl. Schneideriana) 1781. Oct. (*Observ. bot. spec.*).
Publ.: 1781 (p. viii: 1 Mai 1781; p. [i-viii], [1]-56.
Copies: B, MO, NY.

7301. *Beytrag zur Geschichte des ostindischen Brodbaums* mit einer systematischen Beschreibung desselben aus den ältern sowohl als neuern Nachrichten und Beschreibungen zusammengetragen von D. Georg Wolffgang Franz Panzer, ... Nürnberg (bey Gabriel Nicolaus Raspe) 1783. Oct. (*Beytr. Gesch. Brodbaums*).
Publ.: 1783 (p. [4]: 18 Sep 1783, p. [1]-45, *pl. 76*. *Copies*: FI, NY. – Reprinted from G.W.F. Panzer *in* G.F. Christmann und G.W.F. Panzer, *Vollst. Pflanzensyst.* 10: 337-381, *pl. 76*. 1783.

7302. Viro annis meritisque summe venerabili Georgio Wolfgango Panzero ... parenti suo optimo quinquagesimum muneris sacri annum pia mente gratulatur simulque quaedam de *D. Joanne Georgio Volcamero* ... additis duabus ad illum epistolis Hermann Boerhaave et Ios. Pitt. Tournefort antea nondum impressis exposit D. Georg. Wolfgang. Francisc. Panzer ... Norimbergiae [Nürnberg] d. vi. Januar. 1802. Qu. (*De Volcamero*).
Publ.: 6 Januar 1802, p. [i]-xv. *Copy*: G. – On Johann Georg Volckamer (1616-1693).

7303. *Ideen zu einer künftigen Revision der Gattungen der Gräser* ... München 1813. Qu. (*Id Publ.*: 1813 (t.p. preprint), p. [1]-62, *pl. 8-13* (uncol. copp.). *Copies*: MO, NY, US. – Preprinted from Denkschr. k. Akad. Wiss. München 4: [253]-312, *8-13*, 1813, publ. 1814. See also ICBN, nom. cons. no. 286.

Paoletti, Giulio (1865-1941), Italian botanist; 1890-1897 assistant at the Padova botanical garden; professor of natural history at the technical college of Melfi 1897. (*Paol.*).

HERBARIUM and TYPES: PAD.

BIBLIOGRAPHY and BIOGRAPHY: Barnhart 3: 47; BFM see no. 1512; BL 2: 333, 334, 408, 699; BM 4: 1510; CSP 17: 698; De Toni 1: xcvi, 2: xc; DTS 1: 217 xxii; Hortus 3: 1201 ("Paol."); IH 2: (in press); Kew 4: 224; LS 19805-19807, 23438a; Rehder 5: 648; Saccardo 1: 121; TL-1/365; TL-2/1781-1782; Tucker 1: 537; Zander ed. 10, p. 699, ed. 11, p. 749, 797.

BIOFILE: Anon., Bot. Centralbl. 41: 368. 1890 (asst. Padua), 70: 336. 1897 (Melfi); Bot. Jahrb. 23(Beibl. 57): 61. 1897; Nat. Nov. 19: 330. 1897 (Melfi); Österr. bot. Z. 40: 214. 1890 (Padua).
Kneucker, A., Allg. bot. Z. 3: 136. 1897.

COMPOSITE WORKS: (1) Contributed *Tuberaceae, Elaphomycetaceae, Onygenaceae, Edogonaceae* to P.A. Saccardo, *Sylloge fungorum* 8, 1889.
(2) With A. Fiori, q.v., *Iconographia florae italicae* 1895-1904, TL-2/1781.
(3) With A. Fiori, q.v., *Flora analitica d'Italia*, vols. 1-4, 1896-1908, TL-2/1782.
(4) With P.A. Saccardo, *Mycetes malacenses*, 1888.
(5) With G.B. de Toni & G.S Bullo, *Alcune notizie sul Lago d'Arquà-Petrarca*, Atti r. Ist. Ven. ser. 2. 3: 1-65, *pl. 5*. 1892, journal pagination (1149)-1213. Also issued separately.

7304. *Saggio di una monografia del genere Eutypa tra i pirenomiceti* ... [Venezia (presso la segreteria dell' Istituto ...) 1892]. Oct.
Publ.: 1892, p. [1]-68, *pl. 1*. – Published with double pagination in Atti r. Ist. venet. ser. 7. 3: 1373-1440. 1892.

7305. *Contribuzione alla flora del Bacino di Primiero* (Trentino) ... Padova (Stabilimento Prosperini) 1892. (*Contr. fl. Bacino*).
Publ.: Jan 1892 (p. 28: Nov 1891; Nat. Nov. Apr(1) 1893; p. 28: Nov 1891), p. [1]-28. *Copy*: WU. – Preprinted from Atti Soc. Ven.-Trent. Sci. nat. Padova ser. 2. 1: 3-28. 1893. See also Bull. Soc. Ven.-Trent. 5(3):132-134. 1893 on controversial listings.

7306. *Le Primule italiane* ... Padova (R. Stabilimento Prosperini) 1894. Oct.
Publ.: 1894 (p. 15: Apr 1894), p. [3]-15. *Copy*: BR. – Reprinted and to be cited from Bull. Soc. Ven.-Trent. Sci. nat. 5(4): 173-183. 1894.

Paolucci, Luigi (1849-1935), Italian zoologist and botanist; professor of natural history at the Ancona technical School. (*Paolucci*).

HERBARIUM and TYPES: Ancona; other material at FI and Urbino (inf. Steinberg). – Some letters at G.

BIBLIOGRAPHY and BIOGRAPHY: Barnhart 3: 47; BL 2: 371, 372, 699; BM 4: 1510; CSP 8: 560, 10: 988, 17: 699; Jackson p. 317; Kew 4: 224; LS 19809; Morren ed. 10, p. 81; Rehder 1: 428, 429; Tucker 1: 537.

7307. *Primo elenco delle piante* più caratteristiche *dei Monti Sibillini* Vellore(m. 2477. s.m.). Priore (m. 2334 s.m.) Sibilla (m.2213 s.m.) ... Ancona (Tipografia del Commercio) 1879. (*Primo elenc. piante Monti Sibill.*).
Publ.: 1879 (Bot. Centralbl. 26 Jan-6 Feb 1880. Nat. Nov. Feb(2) 1880; Bot. Zeit. 19 Mar 1880), p. [i-ii], [1]-46. *Copy*: FI.

7308. *Flora marchigiana* ossia revisione sistematica e descrittiva delle piante fanerogame spontanee finora raccolte nella regione delle Marche oltre quelle pià estesamente coltivate e che talora inselvatichiscono ad uso specialmente degli agricoltori, periti-agronomi, farmacisti, medici, veterinari ecc. ... Pesaro (premiato Stab. Tipo-Lit. Federici) 1890 [1891]. Oct. (*Fl. marchig.*).
Publ.: 1891 (cover; t.p. 1890; atlas 1891; Nat. Nov. Mai(2) 1891; J. Bot. Jun 1891; Bot. Zeit. 31 Jul 1891), p. [iii]-xxv, 1-656.
Atlas: 1891, p. [i-iii], *pl. 1-45* with text, [1-5, index]. *Copies*: B, G, NY.
Preliminary publ.: Ancona, 1884, n.v. "Flora marchigiana ossia elenco ... Introduzione". 32 p. (see Bot. Centralbl. 3-7 Nov 1884; Bot. Jahrb. 31 Jul 1885; Bot. Zeit. 26 Dec 1884; Nat. Nov. Nov(1) 1884).
Addenda: Malpighia 9: 125-135. 1895; Nuovo Giorn. bot. Ital. ser. 2. 7: 96-114. 1900. For further literature see BL 2: 371.

7309. Nuovi materiali e richerche critiche sulle *piante fossili terziarie* dei Gessi di Ancona ... Ancona (A. Gustavo Morelli, ...) 1896. Oct. (*Piante foss. terz.*).
Publ.: Jun-Sep 1896 (p. xix: Jun 1896; Bot. Zeit. 1 Oct 1896; ÖbZ Sep 1896), p. [i]-xix, [1]-158, *pl. 1-24* (uncol.). *Copy*: USGS.

7310. *I funghi mangerecci della regione Marchigiana* col raffronto delle specie velenose affini ... Ancona (Stab. Tip. Mengarelli) 1901. Oct. (*Fung. mang.*).
Publ.: 1901, p. [1]-22, *6 pl. Copy*: FI (Bibl. nazionale).

Pape, Georg Karl von (1834-1868), German (Prussian) botanist and magistrate at Lüneburg. (*Pape*).

HERBARIUM and TYPES: Material at GOET and HAN.

BIBLIOGRAPHY and BIOGRAPHY: AG 6(2): 913; Barnhart 3: 47; CSP 8: 560, 12: 556; Hegi 4(3): 1518; IH 2: (in press).

BIOFILE: Pape, G.C. von, Jahresber. [Abh.] naturw. Ver. Bremen 1: [85]-120. 1868 (Verzeichniss der in der Umgegend von Stade beobachteten Gefässpflanzen; reprint with orig. pag. at B).

Papenfuss, George Frederick [Frederik] (1903-1981), South African-born American algologist; came to U.S. 1926, naturalized 1945; Dr. phil. Johns Hopkins 1933; assistant and lecturer Johns Hopkins 1929-1934; at Univ. of Cape Town 1935-1939; at University of Lund 1939-1940; at Univ. of Hawaii 1940-1942, with Univ. of California, Berkeley, from 1942-1971, from 1953-1971 as full professor. (*Papenfuss*).
HERBARIUM and TYPES: UC; other material at GRA, J and LD. – Manuscript material and letters at UC and Smithsonian Archives.

BIBLIOGRAPHY and BIOGRAPHY: IH 1 (ed. 1): 9, (ed. 2): 22, (ed. 3): 25, (ed. 4): 26, (ed.

5): 18, (ed. 6): 25, 362, (ed. 7): 26, 339, 2: (in press); Kew 4: 225; Lenley p. 321; MW suppl. p. 275; Roon p. 85.

BIOFILE: Anon., The Berkeley Gazette 11 Dec 1981 (d.); Dept. Bot. UC, death notice (d. 8 Dec 1981), Forum bot. 9(10): 85-86. 1971, 12(3): 5-6. 1974; Who's who in America ed. 38. 2: 2375. 1975.
Chiang, Y.M., Univ. Calif. Publ. Bot. 58: 82. 1970 (bibl. Cryptonemiac.).
Ewan, J. et al., Short Hist. Bot., U.S. 76, 81. 1969.
Gunn, M. & L.E. Codd, Bot. expl. S. Afr. 270. 1981 (portr., epon., coll.).
Koster, J., Taxon 18: 556. 1969.
Levyns, M.R.B., Insnar'd with flowers 133. 1977.
Merrill, E.D., Contr. natl. Herb. U.S. 30(1): 233-234. 1947 (Pacific bibl.).
Papenfuss, G.F, Curriculum vitae, mimeographed, Berkeley 1939 (incl. bibliography); Israel J. Bot. 17: 108. 1968. − [G.F. Papenfuss died 8 December 1981 at Berkeley, Calif., note by F.A.S.].
Prescott, G.W., Contr. bibl. antarct. subantarct. alg. 292-293. 1979.
Rickett, H.W., Index Bull. Torrey bot. Club 76. 1955.
Verdoorn, F., Chron. bot. 2: 327. 1936, 4: 179. 1938, 5: 267, 292, 295. 1939, 6: 21, 300. 1941, 7: 355. 1943.

EPONYMY: *Papenfussia* H. Kylin (1938); *Papenfussiella* H. Kylin (1940); *Papenfussiomonas* T.V. Desikachary (1972).

NOTE: Useful modern bibliographies, including lists of a number of P's own publications, published by Papenfuss are: *A history, catalogue, and bibliography of Red Sea Benthic algae* (Israel J. Bot. 17(1-2): 1-118. 1968) and *Catalogue and bibliography of antarctic and subantarctic benthic marine algae* (Antarctic Research Series 1: 1-76. Mai 1964).

Papp, Constantin (1896-1972), Roumanian botanist; Dr. phil. Iaşi 1926; in various functions at Iaşi University 1920-1964. (*Papp*).

HERBARIUM and TYPES: IASI (bryol. herb); further material at B, C, G and I (phan.). − Some letters at G.

BIBLIOGRAPHY and BIOGRAPHY: Barnhart 3: 47; BJI 2: 130; Hirsch p. 224; IH 1 (ed. 1): 43, (ed. 2): 61, (ed. 3): 75, (ed. 4): 81, (ed. 5): 81; Kew 4: 225; MW suppl. p. 275; Roon p. 85; SBC p. 128.

BIOFILE: Anon., Rev. bryol. lichénol. 5(1). 1932.
Mihai, Gh., Anal. ştiinţ. Univ. Iaşi, Biol. 19(2): 479-481. 1973 (obit., portr.).
Sayre, G., Bryologist 80: 515. 1977.
Stefureac, T.I., Stud. şi cercet. Biol., Bot. 25(5): 467-472. 1973 (obit., portr., b. 1 Jan 1896, d. 17 Aug 1972), Rev. bryol. lichénol. 39: 663-667. 1974 (bibl.).
Verdoorn, F., Chron. bot. 3: 228. 1937.

NOTE: bryological publications listed and reviewed in Rev. bryol. 53: 29. 1926; n.s. 1: 64, 72. 1928; 1: 113. 1929; 3: 152. 1930; 4: 210. 1932; 9: 158. 1936; 10: 101, 107, 161, 165. 1935; 11: 126. 1939; 18: 190. 1949; 20: 309. 1951; 32: 319. 1965; 34: 958. 1967; 36: 376. 1969; 38: 628-629. 1972.

7311. *Monographie der europäischen Arten der Gattung Melica* L. ... Leipzig (Verlag von Max Weg) 1932. Oct.
Publ.: 1 Dec 1932 (in journal; Nat. Nov. Mar 1933), reprinted with special cover but original pagination from Bot. Jahrb. 65(2/3): [275]-348. 1932.

7312. *Monographie der asiatischen Arten der Gattung Melica* L. ... Bucureşti (Monitorul oficial ... 1937.) Oct.
Publ.: 1937, p. [1]-81, *pl. 1-10. Copy*: US. − Published and to be cited as Acad. Rom. Mem. ştiint. ser. 3. 12(9). 1937.

Pappe, Karl [Carl] **Wilhelm Ludwig** (1803-1862), German-born (Hamburg) South African botanist; Dr. med. Leipzig 1827; to South Africa 1835 as practicing physician; colonial botanist from 1848. (*Pappe*).

HERBARIUM and TYPES: SAM; further material at B, BM, CAL, DBN, E, FI, FR, HBG, K, L (alg.), M (lich.), MO, NBG, P, S, SAM, W. – Letters at Kew.

BIBLIOGRAPHY and BIOGRAPHY: Barnhart 3: 47; BB p. 235; BM 4: 1510-1511; Frank 3(Anh.): 74; Herder p. 187; Hortus 3: 1201; IF p. 723; IH 1(ed. 6): 362, (ed. 7): 339, 2: (in press); Jackson p. 347, 589; Kew 4: 225; PR 6925-6930 (ed. 1: 7738-7739); Rehder 5: 648-649; Tucker 1: 537; Urban-Berl. p. 309, 380; Zander ed. 10, p. 699, ed. 11, p. 797.

BIOFILE: Anderson, F. & D. Geary-Cooke, Veld & Flora 61(2): 12-14. 1975 (portr.).
Anon., Bonplandia 6: 49, 324. 1858 (cognomen Leopoldina: Thunberg), 6: 324. 1858 (Cape botanist), 7: 68. 1859; Bot. Zeit. 1: 303. 1843 (letter from Cape), 16: 272. 1858 (Cape Botanist); Flora 40: 624. 1857, 41: 593. 1858 (Cape botanist); Gard. Chron. 24 Jan 1863; Gartenflora 12: 399. 1863 (d.); J. Bot. 1: 62. 1863.
Bullock, A.A., Bibl. S. Afr. bot. 82. 1978.
Burrows, Hist. med. S. Afr. 1958 (n.v.).
Candolle, Alph. de, Phytographie 439. 1880 (coll.).
Geary-Cooke, D. & F. Anderson, Veld & Flora 61(2): 12-14. 1975 (portr.).
Gunn, M. & L.E. Codd, Bot. expl. S. Afr. 71, 270-272. 1981 (biogr., coll., itin., portr.; major source).
Harvey, W.H. & Sonder, O.W., Flora Capensis 1: xi. 1894.
Hedge, I.C. & Lamond, J.M., Index coll. Edinb. herb. 116. 1970.
Jackson, B.D., Bull. misc. Inf. Kew 1901: 51. ("large set" at K).
Jessen, G.F.W., Bot. Gegenw. Vorz. 472. 1864.
MacOwan, P., Trans. S. Afr. philos. Soc. 4: li. 1887.
Müller, R.H.W., & Zaunick, R., Friedr. Traug. Kützing 290-291. 1960.
Newton, L.M., Phycol. Bull. 1: 17. 1952.
Pearson, H.H.W., Bull. misc. Inf. Kew 1910: 373.
Phillips, E.P.,S. Afr. J.Sci. 27: 41, 46, 47. 1930.
Steinberg, C.H., Webbia 32: 33. 1977.
Tölken, H.R., Ind. herb. austro-afr. 9, 48. 1971 (coll. SAM).
Tyrrell-Glynn, W. & Levyns, M.L., Flora africana 53. 1963 (bibl.).

HANDWRITING: Gunn, M. & L.E. Codd, Bot. expl. S. Afr. 271. 1981.

EPONYMY: *Pappea* Ecklon et Zeyher (1834-1835); *Pappea* Sonder (1862).

7313. *Enumerationis plantarum phaenogamarum lipsiensium specimen.* Dissertatio inauguralis botanica quam gratiosi medicorum ordinis auctoritate. In Academia lipsiensi pro summis in medicina atque chirurgia honoribus rite capessendis illustris ictorum ordinis concessu in auditorio juridico die xx. novembris anni mdcccxxvii publice defendet Carolus Guilielmus Ludovicus Pappe ... Lipsiae [Leipzig] (impressit Hirschfeld) [1827]. Oct. (*Enum. pl. lips. spec.*).
Publ.: 20 Nov 1827, p. [iii]-xx, [1]-42, [43, theses]. *Copies*: B, G, NY.

7314. *Synopsis plantarum phaenogamarum agro lipsiensi indigenarum* ... Lipsiae [Leipzig] (sumptibus Leopoldi Vossii) 1828. Oct. (*Syn. pl. agr. lips.*).
Publ.: (p. x: 23 Oct 1827; Linnaea Oct 1828 rev.), p. [i]-xx, [1]-85. *Copies*: G, NY, USDA. – Pages [i]-xx, [1]-42 are a re-issue of the *Enum pl. lips. spec.* (see above).
Ref.: Schlechtendal, D.F.L. von, Linnaea 4 (Litt.): 169-170. 1828.

7315. *Florae capensis medicae prodromus*, or an enumeration of South African indigenous plants, used as remedies by the colonists of the Cape of Good Hope ... Cape Town (A.S. Robertson, ...) 1850. Oct. (in fours). (*Fl. cap. med. prodr.*).
Ed. 1: 1850 (p. vii: 22 Oct 1850), p. [i]-vii, [1, index], [1]-32. Copies: NY, USDA. – Preceded by "A list of South African Plants used as remedies ..." 1847.

Ed. 2: Jan-Mar 1857(p. vi: 10 Oct 1856), p. [i]-vi, [1]-54. *Copy*: BR. – "Second edition: with corrections ..." Cape Town (W. Brittain, ...) 1857. Oct (in fours).
Ed. 3: 1868. (n.v.).
Ref.: Hooker, W.J., J. Bot. Kew Gard. Misc. 9: 125-127. Apr 1857.
Schlechtendal, D.F.L. von, Bot. Zeit. 10: 827. 1852, 15: 484-486. 10 Jul 1857. (rev. ed 2).

7316. *Silva capensis*, or a description of South African forest-trees, and arborescent shrubs, used for technical and oeconomical purposes by the colonists of the Cape of Good Hope ... Cape Town (printed by van de Sandt de Villiers & Co.) 1854. Oct. (in fours). (*Silva cap.*).
Ed. 1: 27-31 Dec 1854 (t.p.; p. [2]: 27 Dec 1854), p. [i-iii], [1]-53. *Copies*: G, MO, NY, USDA; IDC 428. – The cover has a different imprint: Cape Town (L. Taats) Leipzig (K.F. Koehler) 1854.
Ed. 2: 1862 (p. v: 12 Jun 1862), p. [i-v], [1]-59, [1, index]. *Copies*: NY, IDC 5146. – "Second, revised and enlarged edition ..." London (Ward & Co., ...) 1862. Oct. (in fours).
Ref.: Schlechtendal, D.F.L. von, Bot. Zeit. 13: 725-726. 12 Oct 1855.

7317. *Synopsis filicum Africae australis*; or, an enumeration of the South African ferns hitherto known ... Cape Town (W. Brittain, ...) 1858. Oct. (in fours). (*Syn. fil. Afr. austr.*).
Co-author: Rawson William Rawson (1812-1899).
Publ.: Jan-Jul 1858 (Bonplandia 1 Sep 1858), p. [i]-viii, [1]-57. *Copies*: BR, G, MO, NY(2); IDC 7240. – The NY copies have a different imprint: Cape Town (Saul Solomon and Co., ...) 1858. Oct. (in fours).
Ref.: Anon., Bonplandia 6: 321-322. 1 Sep 1858.

Pâque, Égide (Pâques) (1850-1918), Belgian clergyman (Jesuit priest) and botanist; ord. 7 Sep 1886; from 1888-1892 at Charleroi; from 1892-1906 id. at the Collège Notre-Dame de la Paix at Namur; from 1906-1910 at Antwerp; ultimately at Bruxelles. (*Pâque*).

HERBARIUM and TYPES: BR. – Some letters at G.

BIBLIOGRAPHY and BIOGRAPHY: Barnhart 3: 47; BL 1: 25, 311, 2: 33, 35, 43, 699; BM 4: 1511; CSP 17: 700; De Toni 1: xcvi; GR p. 698; IH 2: (in press); Kew 4: 226; LS 19811-19812a, 37668-37670, suppl. 20943-20947; Morren ed. 10, p. 46; Nordstedt p. 25; Rehder 1: 73; Tucker 1: 537.

BIOFILE: Anon., Bull. Soc. Bot. Belg. 54: 103-104. 1919 (obit., b. 8 Nov 1850, d. 18 Mar 1918).
Evens, F.M.J.G., Gesch. algol. Belgie 186. 1944 ("Egmont Pâque").
De Wildeman, É. de, Bull. Soc. Bot. Belg. 54: 103-104. 1921.

7318. *Catalogue des plantes plus ou moins rares* observées dans les environs *de Turnhout* par E. Pâques, S.J. Gand [Gent] (Imprimerie C. Annoot-Braeckman) 1880. Duod.
Publ.: 1880 (BSbF rd. 12 Nov 1880) p. [1]-23. *Copy*: KNAW. – Reprinted from Bull. Soc. bot. France 19(1): 7-25. 1880. To be cited from journal. – Further notes by Pâque, l.c. 21(2): 22-28. 1882, 22(1): 29-43. 1883 and by J. Adriaensen & P. Haeck, Bot. Jahrb. Dodonaea 4: 240-250. 1892.

7319. *De Vlaamsche volksnamen der planten* van België, Fransch-Vlaanderen en Zuid-Nederland met aanduiding der toepassingen en der genezende eigenschappen der planten ... Namen (Ad. Wesmael-Charlier, ...) 1896. Oct. (*Vlaamsche volksnam. pl.*).
Publ.: Jun-Jul 1896 (p. 11: 8 Dec 1895; ÖbZ Jul 1896; Bot. Zeit. 1 Aug 1896; Nat. Nov. Aug(1) 1896), p. [1]-569. *Copies*: BR, MO.
Bijvoegsel: 1913 (t.p. 1912, cover 1913), p. [1]-156. *Copy*: BR.

7320. *Guide de l'herborisateur en Belgique* (plantes phanérogames et cryptogames spontanées

ou fréquemment cultivées) ... nouvelle édition entièrement remaniée et complétée ... Namur (Librairie classique de Wesmael-Charlier, ...) 1900. Oct. (*Guide herbor. Belgique*).
Publ.: 1900 (Bot. Zeit. 16 Sep 1900; Nat. Nov. Jul(1) 1900), p. [i-iv], [1]-117. *Copies*: BR, G.

7321. *Flore* analytique et descriptive des provinces *de Namur et Luxembourg* (plantes indigènes et cultivées) accompagnée d'une carte botanique, des étymologies des noms, des propriétés des plantes, etc. ... Namur (Librairie classique et scientifique de Ad. Wesmael-Charlier, ...) 1902. Oct. (*Fl. Namur Luxemb.*).
Publ.: 1902 (J. Bot. Oct 1902; Nat. Nov. Mar(1) 1903), p. [i]-xxxii, map, [1]-594, [1]. *Copies*: BR, NY.

7322. Botanique-Série v – Bas– et Moyen-Congo. Notes botaniques sur la région du Bas– et Moyen-Congo fascicule 1. *Plantes principales de la région de Kisantu* leur nom indigène, leur nom scientifique, leurs usages ...Bruxelles 1910. Qu. (*Pl. Kisantu*).
Co-author: Justin Gillet (1866-1943).
Publ.: Jan-Mai 1910 (BSbF séance 27 Mai 1910; Bot. Centralbl. 17 Jan 1911), p. [i]-ix, [1]-120, *pl. 1-19*. *Copies*: BR, MICH, MO, NY.

Pardé, Léon Gabriel Charles (1865-1943), French dendrologist; trained at the École nationale forestière (1888); teacher and later director of the Arboretum national des Barres nr. Nogent-sur-Vernisson (Loiret). (*Pardé*).

HERBARIUM and TYPES: Unknown. – Some letters at G.

BIBLIOGRAPHY and BIOGRAPHY: Barnhart 3: 48; Hortus 3: 1201 ("Pardé"); Kew 4: 226-227; Langman p. 565; MW p. 377-378, suppl. p. 2575; NI 1487-1488, see also 1: 254; PFC 3(2): xii; Rehder 5: 649; Tucker 1: 537-538.

BIOFILE: Cavillier, F., Boissiera 5: 69. 1941.
Guinier, Ph., Rev. hort., Paris 115: 408-410. 1944 (obit., portr.), Rev. Bot. appl. 24: 294. 1944 (obit., d. 14 Jul 1943).
Pardé, L., Les feuilles, Paris, 1941, ed. 2. 1952.
Verdoorn, F., ed., Chron. bot. 2: 127. 1936, 3: 109. 1937.

7323. *Arboretum national des Barres*. Énumération des végétaux ligneux indigènes et exotiques qui y sont cultivés avec l'indication de leurs pays d'origine, de leurs synonymes, des caractères qui permettent de distinguer facilement les espèces le plus souvent confondues et la discussion de leurs exigences et de leurs qualités comme essences forestières dans nos pays. Travail exécuté sous la direction de Mr. L. Daubrée ... par L. Pardé ... Paris (Librairie des sciences naturelles Paul Klincksieck ...) 1906. Oct. (*Arbor. Barres*).
Publ.: text and atlas volumes Sep-Dec 1906 (p. 11: Sep 1906; Nat. Nov. Feb(1) 1907). *Copies*: BR, G, MO, NY, PH, USDA.
Texte: [1]-397, [1, cont.].
Atlas: p. [1]-20, *pl. 1-94, 34 bis* (photos), 22 charts.
Ref.: Fliche, P., Bull. Soc. bot. France 54(7): 568-570. 1907 (rev.).

7324. *Iconographie des conifères* fructifiant en France ... Paris (Librairie des sciences naturelles Paul Klincksieck, Leon Lhomme, succr. ...) [1912-1924]. †. (*Iconogr. conif.*).
Publ.: Unbound plates in dated covers. *Copies*: BR, FI, MO, USDA.
Livr. 1: Nov 1912, *5 col. pl., 5 uncol. pl.*, 5 captions, numbered *21, 39, 41, 112, 114*.
Livr. 2: 2 Feb 1913, id., numbered *16, 18, 19, 42, 71*.
Livr. 3: Jun 1913, *6 col., 5 uncol. pl.*, 5 captions, numbered *17, 22, 46, 86, 86 bis, 89*.
Livr. 4: Nov 1913, *5 col. pl., 5 uncol. pl.*, 5 captions, numbered *29, 30, 43, 100, 134*.
Livr. 5: Mai 1914, *6 col. pl., 5 uncol. pl.*, 5 captions, numbered *26, 44, 61, 98, 98 bis, 132*.
Livr. 6: Jul 1914, *5 col. pl., 5 uncol. pl.*, 5 captions, numbered *20, 49, 69, 73, 99*.
Prospectus: 1911 (with *2 pl.*).
The original plans called for 28 livraisons, a total of 150 col. plates and for 140 photographs. The plates are by Mme Guillo-Kastner.

7325. *Les Conifères* ... Paris (La Maison rustique ...) [1937]. Oct. (*Conifères*).
Ed. 1: 1937 (p. 7: Feb 1937; BR rd 8 Dec 1937; Nat. Nov. Dec 1937), p. [1]-294, *61 pl.* (photos in text). *Copies*: BR, G, USDA.
Reprint (1): Oct-Dec 1946 (p. 7: Aug 1946, p. 294: Oct-Dec 1946), p. [1]-294, *61 pl.* (id.). *Copies*: FI, USDA.
Reprint (2): 1955 (p. 7: Feb 1955), p. [1]-294, *61 pl.* (id.).
Copy: NY.

Pardo de Tavera, Trinidad Herménégilde José (1857-1925), Philippine physician, anthropologist, philologist and botanist; studied medicine at Paris; professor of anatomy at the Universidad de Santo Tomás; later heavily involved in local politics as well as representing the Philippines abroad. (*Pardo*).

HERBARIUM and TYPES: Unknown.

BIBLIOGRAPHY and BIOGRAPHY: Barnhart 3: 48; BL 1: 122, 311; Kew 4: 227; Rehder 3: 91; Tucker 1: 538.

BIOFILE: Anon., Encicl. Univ. ilustr. 41: 1447-1448.
Merrill, E.D., Enum. Phil. pl. 4: 214. 1926.

7326. *Plantas medicinales de Filipinas* ... Madrid (Bernardo Rico ....) 1892. Oct. (*Pl. med. Filip.*).
Publ.: Mai-Aug 1892 (p. [5]: 13 Apr 1892; Nat. Nov. Sep(1) 1892), p. [1]-339, [1]. *Copy*: NY.

7327. *The medicinal plants of the Philippines* by T.H. Pardo de Tavera .... translated and revised by Jerome B. Thomas, Jr., ... Philadelphia (P. Blakiston's Son & Co. ...) 1901. Oct. (*Med. pl. Philipp.*).
Publ.: 1901, p. [i]-xvi, [17]-269. *Copies*: NY, PH, USDA.

Parfitt, Edward (1820-1893), British gardener, entomologist, mycologist and lichenologist; gardener in Norfolk and from 1848-1860 in Exeter; librarian of the Devon and Exeter Institution 1861-1893. (*Parfitt*).

HERBARIUM and TYPES: TOR (orig. herbarium and 12 vols. of 1530 unpublished drawings of Devon fungi); British lichens and letters at BM.

BIBLIOGRAPHY and BIOGRAPHY: Barnhart 3: 48; BB p. 236; BM 4: 1511; CSP 4: 756, 8: 561-562, 10: 989-990, 17: 704; Desmond p. 477-478; De Toni 1: xcvi, 4: xliv; DNB 43: 205; GR p. 408; IH 1 (ed. 7): 339, 2: (in press); Kew 4: 227; Krempelh. 1: 306; LS 19817.

BIOFILE: Anon., Entom. mon. Mag. 29: 73. 1893 (b. 17 Oct 1820, d. 15 Jan 1893, "1892"); Leopoldina 29 (17/18): 159. Sep 1893 (d.); Zool. Anz. 16: 297. 1893 (d.).
Boase, G.C., Notes and Queries ser. 8. 4: 262-263. 1893 (contains text of a letter by Parfitt giving details about his life and work).
Britten, J., J. Bot. 31: 160. 1893.
Freeman, R.B., Brit. nat. hist. books 272. 1980.
Hawksworth, D.L. & M.R.D. Seaward, Lichenology Brit. Isl. 24, 129-130, 198. 1977.
Kent, D.H., Brit. herb. 71. 1957.
Martin, W.K. & G.T. Fraser, Fl. Devon 774-775. 1939.

Paris, [Jean] Édouard Gabriel [Narcisse] (1827-1911), French soldier and bryologist; collected in France and Algeria; retired from the military as général de brigade 1889; commander of the Légion d'Honneur. (*Par.*).

HERBARIUM and TYPES: REN; other material at B, BM, BP, BR, C, F, FI, JE, K, L, LD, MO, MW, NY, PC, S, STU, W. – 223 letters from P. to Brotherus are in the Helsinki University Library. – *Exsiccatae*: *Iter boreale africanum* (2 cent., see Buchinger, Flora 50:

80. 1867). – Some letters at G; correspondence with V.F. Brotherus (223 letters) at H (Univ. Libr.).

BIBLIOGRAPHY and BIOGRAPHY: Barnhart 3: 48; BM 4: 1521; Bossert p. 300; CSP 17: 705; GR p. 342; Herder p. 266; IH 2: (in press); Kelly p. 166; Kew 4: 228; Lenley p. 321; LS 19821-19822; MW p. 378; NAF ser. 2. 9: 125; PR (alph.); SBC p. 128; Tucker 1: 538; Urban-Berl. p. 310.

BIOFILE: A.M.S., Bryologist 15(6): 97-98. 1912.
Anon., Bot. Gaz. 17: 267; 1892 (request collaboration *Index bryologicus*); Bot. Not. 1911: 190 (b. 8 Nov 1827, d. 30 Apr 1911); Bull. Soc. bot. France 58: 261. 1911 (d.).
Britton, E.G., Bull. Torrey bot. Club 19: 273. 1892 (request collaboration *Index bryologicus*).
Cosson, É., Comp. fl. atl. 1: 75-76. 1881.
Cosson & Durieu, Exped. Sci. Algér. Bot. 2: xi. 1868.
Husnot, T., Rev. bryol. 26: 39. 1899 (prix Montagne), 34: 79. 1907 (prix Milne-Edwards), 35: 16. 1908 (prix Desmazières), 38: 93-95, 120. 1911 (obit., herb. to REN).
Koponen, T., *in* G.C.S. Clarke, ed., Bryophyte Systematics 164. 1979 (corr. with Brotherus).
Lamy, D., Occas. Pap. Farlow Herb. 16: 121, 126. 1981 (corr. F.F.G. Renauld).
Lindemann, E. v., Bull. Soc. imp. Natural. Moscou 61(1): 52. 1886.
Maire, R., Progr. conn. bot. Algérie 110, 175. 1931.
Merrill, E.D., Contr. U.S. natl. Herb. 30: 236. 1947 (Pacific bibl.).
Murray, G., Hist. coll. BM(NH) 1: 172. 1904 (N. Afr. pl.).
Paris, J.E.G.N., Bull. Soc. bot. France 14: 197-225, 268-290. 1867 (travel report Sahara), 17 (bibl.): 144. 1870 (exch. pl. Algeria), 39: 53-56. 1892 (letter on the projected *Index bryol.*), 41: 30. 1894 (id.), 45: 151-157. 1898 (id.); Rev. bryol. 25: 41-42. 1898 (inf. suppl. Index), 27: 99-100. 1900 (exsicc. for sale); 38: 48. 1911 (*Florule bryol. Guinée*).
Sayre, G., Bryologist 80: 515. 1977.
Steinberg, C., Webbia 34: 61. 1979 (pl. FI).

EPONYMY: *Parisia* V. F. Brotherus. (1906).

7328. *Index bryologicus* sive enumeratio muscorum hucusque cognitorum adjunctis synonyma distributioneque geographica locupletissimis quem conscripsit E.G. Paris ... (ex Actis Societatis Linnaeanae Burdigalensis). Parisiis (apud Paul Klincksieck ...) 1894-1898. Oct. (*Index bryol.*).
Publ.: in 17 parts in the Act. Soc. Linn. Bordeaux, reprinted in 5 parts with above t.p. Paris undertook the compilation in 1864 at the suggestion of W.P. Schimper. His military career, however, withheld him from making progress until his retirement in 1889; p. [i*-iii*], [i]-vi, [1]-1379, [1380, err.]. *Copies* reprint: B, BR, FI, G, H, MO, NY, WU. – We are grateful to Mrs. S.W. Greene for the major part of the contents of the following table:

Reprint part	pages reprint	date (Nat. Nov.)	Actes ser.	vol.	part	Actes pages	Actes dates	BM dates rd.
1	[i]-vi, [1]-324	Nov 1894	5.	6.	1	15-62	Mai 1894	26 Mai 1894
					2	63-126	Jul 1894	18 Jul 1894
					3	127-324	Nov 1894	8 Dec 1894
2	325-644	Oct 1895		9.	1	1-64	Mai 1895	9 Jul 1895
					2	65-128	Jun 1895	3 Aug 1895
					3	129-192	Jul 1895	10 Aug 1895
					4	193-256	Aug 1895	1 Jan 1896
					5	257-320	Sep 1895	1 Jan 1896
3	645-964	Dec 1896			6	321-384	Oct 1895	1 Jan 1896
				10.	1	1-48	Mai 1896	25 Jun 1896
					2	49-112	Jul 1896	25 Jul 1896

part	pages	date	Actes		pages	dates	BM rd.
3		Dec 1896	10.	3	113-176	Aug 1896	7 Nov 1896
				4	177-240	Nov 1896	23 Jan 1896
	[p. 949-964 preprinted]			5	241-256	Apr 1897	24 Apr 1897
4	965-1284	Feb 1898	6.	1. 1	1-160	Sep 1897	30 Oct 1897
				2	161-320	1898	26 Mar 1898
5	1285-1380	Sep 1898		3	321-416	Mai 1898	24 Mai 1898

Supplementum primum: Mai-Jun 1900 (BSbF rd. 27 Jul 1900; Bot. Zeit. 16 Sep 1900; ÖbZ Mai-Jun 1900; Nat. Nov. Jul(1) 1900; Bot. Gaz. 15 Oct 1900; Rev. bryol. Dec 1900), p. [i-iv], [1]-334 ["234"]. *Copies*: BR, G, H, MO, NY(3), WU. – "Index bryologicus ... E.P. Paris. Officier de l'instruction publique. Mémoires de l'Herbier Boissier. Supplementum primum". Genève et Bale (Georg & Cie., ...) Lyon (même maison) 1900. Qu.

Reviews and notices of reprint:

1: Nat. Nov., Nov(2) 1894; Gepp., A., J. Bot. 33: 26-29. Jan 1898; Rev. bryol. 21(6): Dec 1894; Bot Zeit. 1 Feb 1895; Hedwigia 34 (hepat.): 25-26. 15 Feb 1895; Bot. Centralbl. 60: 402. 18 Dec 1894; J.Bot., Morot 8: lxiii. 16 Nov 1894; Hedwigia 34: (25). Jan-Feb 1895.

2: Nat. Nov. Oct(2) 1895; Gepp., A., J. Bot. 34: 143. Mar 1896 (dates part 2 "Dec 1895"); Hedwigia 35: (56). Mar-Apr 1896.

3: Nat. Nov. Jan(1) 1897; Gepp., A., J. Bot. 35: 63-64. Feb 1897 (dates part 3 "Dec 1896"); Hedwigia 36: (28). Jan-Feb 1897.

4: Nat. Nov. Mar(1) 1898; Rev. bryol. 25: 55. Jun 1898; Bot. Gaz. 28 Jul 1898; Hedwigia 37: (76). 9 Apr 1898.

5: Nat. nov. Sep(2) 1898; Gepp., A., J. Bot. 36: 462-463. Nov 1898; Rev. bryol. 26: 22. Feb 1899.

Note: An anonymous note in the NY copy cites as dates for the reprint *1*: 21 Nov 1894, *2*: 30 Nov 1895, *3*: 18 Dec 1896, *4*: 28 Feb 1898, *5*: 20 Aug 1898.

Editio secunda: 1903-1906, in 5 vols. (27 fasc.; dates given in vols.). "*Index bryologicus sive enumeratio muscorum ad diem ultimam anni 1900 cognitorum adjunctis synonymia distributioneque geographica locupletissimis ... editio secunda ...*" Paris (Librairie scientifique A. Hermann ...). Oct.

1: [i-iii], [1]-384, *2*: [i-ii], [1]-375, *3*: [i-ii], [1]-400, *4*: [i-ii], [1]-368, *5*: [i-ii], [1]-160, 15 tables, 1 map. *Copies*: FI, H, MO, NY, U, USDA; IDC 1143.

vol.	fasc.	pages	dates
1	1	1-64	1 Dec 1903
	2	65-128	2 Jan 1904
	3	129-192	5 Feb 1904
	4	193-256	2 Mar 1904
	5	257-320	1 Apr 1904
	6	321-384	1 Mai 1904
2	1[7]	1-64	10 Jun 1904
	2[8]	65-128	8 Jul 1904
	3[9]	129-192	8 Aug 1904
	4[10]	193-256	1 Sep 1904
	5[11]	257-320	29 Sep 1904
	6[12]	321-375	1 Nov 1904
3	13	1-72	10 Dec 1904
	14	73-136	10 Mai 1905
	15	137-200	3 Jun 1905
	16	201-264	20 Jun 1905
	17	265-328	15 Jul 1905

vol.	fasc.	pages	dates
3	18	329-400	17 Aug 1905
4	19	1-56	1 Sep 1905
	20	57-120	21 Sep 1905
	21	121-184	10 Oct 1905
	22	185-248	29 Oct 1905
[4]	23	249-312	25 Nov 1905
	24	313-368	10 Dec 1905
5	25	1-72	6 Jan 1906
	26	73-136	5 Feb 1906
	27	137-160	4 Apr 1906

Notes on publication: Rev. bryol. 30: 110-111. 1903, 33: 88. 1906, Bull. Soc. bot. France 51: 96. 1904.

7329. *Musci japonici* a R.P. Faurie anno 1900 lecti (Extrait du Bulletin de l'Herbier Boissier ...) 1902. Oct.
Collector: Père Urbain Jean Faurie (1847-1915).
Publ.: p. [1]-16. 4 Nov 1902, p. [17]-22. 5 Dec 1902. *Copies*: G, U. – Reprinted and to be cited from Bull. Herb. Boissier ser. 2. 2: 918-933. 4 Nov, 988-993. 5 Dec 1902.

7330. *Florule bryologique de la Guinée française* ... Paris (Librairie scientifique A. Hermann ...) 1908. Oct. (*Fl. bryol. Guinée*).
Publ.: 1 Dec 1908 (date of issue stamped on cover of Mémoire; Rev. bryol. Apr 1909; Nat. Nov. Jan(1) 1909), p. [1]-66. *Copies*: BR, G, H. – Published as Mém. Soc. bot. France 14, 1908, belonging to Bull. Soc. bot. France 55, 1908. The G copy has a special cover with the above title and imprint; the BR copy is the regular issue of Mém. 14.

7331. *Collatio nominum brotherianorum* et indicis bryologici ... Parisiis (Libraria J.-B. Baillière et filii ...) [1909] Oct. (*Coll. nom. broth.*).
Publ.: 1910 (p. [v]: Jun 1909; BSbF séance 4 Oct 1910; Nat. Nov. Feb(1) 1911; Rev. bryol. Jan-Feb 1911; FH rd. Jul 1911), p. [i-v], [1]-37. *Copies*: FH, G, H, MO, U. – Collation of the names accepted by V.F. Brotherus in his treatment of the *Bryales* in EP, Nat. Pfl.-Fam. 1901-1909 with the *Index bryol.*

7332. *Liste des mousses et hépatiques* offertes en échange par M. le général Paris [Dinard, Ille-et-Vilaine 1906-1910], 4 nos. (*Liste mouss. hépat.*).
1: Jun-Jul 1906, p. [1]-8 (Hedwigia 45: (205). 18 Aug 1906).
2: Nov-Dec 1907, p. [1]-7 (Hedwigia 47: (100). 3 Jan 1908).
3: Jan 1909, p. [1]-8 (Hedwigia 48: (197). 10 Mai 1909).
4: Nov 1910, p. [1]-8.
Copies: BR, H, PH.

Parish, Samuel Bonsall (1838-1928), American botanist; B.A. New York Univ. 1858; collected extensively in Southern California; settled as fruit grower in the San Bernardino Valley, Calif. 1872; from 1920 hon. curator of the herbarium and lecturer in Californian botany at Stanford University. (*Parish*).

HERBARIUM and TYPES: DS (over 30.000; rd. 1917), further important sets at ASUF, CM, NA, UC and US; other material at A, B, BM, BR, C, CART, CAS, E, F, GH, GOET, ILL, JE, JEPS, K, KIEL, L, LCU, LE, LIV, MANCH, MICH, MIN, MO, MSC, NMC, NY, NYS, P, PH, POM, RSA, TEX, UTC, W, WRSL, WS, WTU. Library acquired by POM. – Parish offered sets of Southern California Plants for sale as *Select plants of Southern California*, see e.g. a 1 p. pamphlet [1895] with this title in G. – Some letters at G.

BIBLIOGRAPHY and BIOGRAPHY: Barnhart 3: 48; BL 1: 166, 169, 312; Bossert p. 301; CSP

17: 705; Ewan ed. 1: 29, 126, 205, 218; GR p. 237; Hirsch p. 224; Hortus 3: 1201 ("Parish"); IH 1 (ed. 7): 339; Kew 4: 232; Langman p. 565; Lenley p. 321; Morren ed. 10, p. 117; NW p. 54; Rehder 5: 649; Tucker 1: 538-539; Urban-Berl. p. 381.

BIOFILE: Anon., Bot. Soc. Amer. Publ. 99: 26-28. 1929 (b. 13 Jan 1838, d. 5 Jun 1928); Madroño 1: 202, 270-271. 1929 (d.); Torreya 28: 88. l928.
Cantelow, E.D. & H.C., Leafl. w. Bot. 8: 97. 1957.
Ewan, J. et al., Leafl. w. Bot. 7: 56, 73. 1953.
Hedge, I.C. & Lamond, J.M., Index coll. Edinb. herb. 116. 1970.
Jepson, W.L., Science ser. 2. 69: 63. 1929; Univ. Calif. Publ. Bot. 16(12): 427-444, *pl. 32*. 1932 (portr., bibl.; main biogr.).
Murray, G., Hist. coll. BM(NH) 1: 172. 1904.
Parish, S.B., Madroño 1: 71-75. 1922 (address annual dinner Cal. bot. Soc. 1914, portr.).
Rodgers, A.D., Amer. bot. 1873-1892, p. 334 (index). 1944.
Stieber, M.T., Huntia 4(1): 84. 1981 (arch. mat. HU).
Sutton, S.B., Charles Sprague Sargent 113. 1970.
Thomas, J.H., Contr. Dudley Herb. 5(6): 159. 1961; Huntia 3: 35-37. 1969 (1979).

EPONYMY: *Parishella* A.Gray (1882) is also dedicated to his brother W.F. Parish (-). *Note*: *Parishia* J.D.Hooker (1860) commemorates Rev. Charles S.P. Parish (1822-1897) a collector of plants in Burma and the Andaman Isl.

7333. *Plants of Southern California* collected in the Counties of San Bernardino, San Diego and Los Angeles, by S.B. & W.F. Parish, San Bernardino [Oquawka, Ill., 1882]. Oct. (*Pl. S. Calif.*).
Publ.: Jan 1882 (Jan 1882 on separate sheet "to botanists" accompanying the pamphlet), p. [i], [1]-8. *Copies*: G, NY, PH, USDA.
Supplementary list: undated, p. [1-4]. *Copies*: NY, USDA.

7334. *A catalogue of plants collected in the Salton Sink* by S.B. Parish ... Washington, D.C. (published by the Carnegie Institution of Washington) 1913. (*Cat. pl. Salton Sink*).
Publ.: 1913, p. [i], [1]-11, 1 map. *Copies*: MO, NY. – "Printed in advance from "*The Salton Sea*: a study ..." Publ. 193. Carnegie Institution of Washington, 1913" (statement on cover reprint). The preprint has independent pagination and corresponds with p. 103-114 of *The Salton Sea*. – The complete article by Parish, including plant ecology, occupies p. 85-114: "Plant ecology and floristics of Salton Sink".

7335. *An enumeration of the Pteridophytes and Spermatophytes of the San Bernardino Mountains*, California ... (Reprinted from the Plant World ...) 1917. Oct.
Publ.: Plant World 20(6): 63-178. Jun, (7): 208-223. Jul, (8): 245-259. Aug 1917. – Reprints with special covers and original pagination at FI, MO, and NY.

7336. *The immigrant plants of Southern California* (reprinted from the Bulletin of the Southern California Academy of Sciences ...) 1920. Oct.
Publ.: Oct 1920, p. [1]-30. *Copy*: NY. – Reprinted with special t.p. [1], and to be cited from Bull. S. Calif. Acad. Sci. 19(4): 3-30. 1920.

Parisot, Charles Louis (1820-1890), French botanist and pharmacist at Belfort. (*Parisot*).

HERBARIUM and TYPES: material at FI, L, P.

BIBLIOGRAPHY and BIOGRAPHY: BL 2: 130, 699; PR 6931; Rehder 1: 411.

BIOFILE: Parisot, L., Bull. Soc. bot. France 5: 535. 1858 (trip Kaiserstuhl).
Steinberg, C.H., Webbia 32: 33. 1977.

7337. *Notice sur la flore des environs de Belfort* ... Besançon (Imprimerie de Dodivers et Ce, ...) 1858. Oct. (*Not. fl. Belfort*).

Publ.: 1858 (BSbF Dec 1858), p. [i, iii], [1]-108. *Copies*: G, NY. – Preprinted or reprinted from Mém. Soc. Émul. Doubs 3: 57-164. 1858 (1859, ?) (copy journal publ.: G).
Ref.: Anon., Bull. Soc. bot. France 5: 567-570. 1858.

Parke, Mary (1908-x), British phycologist at the Marine Biological Association, Plymouth, until March 1973. (*Parke*).

HERBARIUM and TYPES: Type cultures with the culture collection of algae & protozoa, Cambridge, England. – Some manuscript material in Smithsonian Archives.

BIBLIOGRAPHY and BIOGRAPHY: Bossert p. 301; IH 2: (in press); Kew 4: 232; Roon p. 85.

BIOFILE: Anon., Brit. phycol. J. 13: 1-2. 1978 (portr.; tribute on 70th birthday).
Parke, M., personal comm. to senior author, 1957 (on suppl. cult.).

EPONYMY: *Parkia* R. Brown (1826) honors Mungo Park (1771-1806) British surgeon and traveller in Africa. The etymology of *Parka* Fleming ex H. Miller (1857) could not be determined.

7338. *A contribution to knowledge of the Mesogloiaceae* and associated families ... Liverpool (The University Press of Liverpool) 1933. (*Contr. Mesogloiac.*).
Publ.: Mai 1933 (forwarding slip in BR copy), p. [1]-43, errata slip, *pl. 1-11*. *Copies*: BR, NY.

Parker, Richard Neville (1884-1958), British botanist in the Indian Forest Service 1905-1939, forest botanist Dehra Dun 1922-1932, conservator of forests Punjab, 1932. (*R. Parker*).

HERBARIUM and TYPES: DD; important collection also at K; other material at A, B, BOL, CAL, DPU, G, GH, MO, NBG, NY, PH, PRE. – Some letters at G.

BIBLIOGRAPHY and BIOGRAPHY: Barnhart 3: 50; Desmond p. 479; Hortus 3: 1201 ("R. Parker"); IH 1 (ed. 7): 339; Kew 4: 233; MW p. 378; TL-2/1607; Zander ed. 10, p. 699, ed. 11, p. 797.

BIOFILE: Gunn, M. & L.E. Codd, Bot. expl. S. Afr. 272. 1981.
Raizada, M.B., Indian Forester 85(4): 270. 1959 (obit., portr., b. 14 Dec 1884, d. 12 Apr 1958).
Stewart, R.R., Pakistan J. Forestry 17: 353. 1967.
Verdoorn, F., ed., Chron. bot. 6: 300. 1941.

COMPOSITE WORKS: Co-author vol. 3 of Duthie, *Fl. Gangetic plain* 1915-1920, TL-2/1607.

EPONYMY: *Parkerella* A. Funk (1976) was named for A.K. Parker (1922-1974), Canadian forest pathologist. *Parkeria* W.J Hooker (1825) and the derived genus *Parkeroidea* B. Renault (1901) honor C.S. Parker (x-1869) British botanist and plant collector. The etymology of *Parkerella* E. Munier-Chalmas ex L. Morellet et J. Morellet (1922) could not be determined.

7339. *A forest flora for the Punjab* with Hazara and Delhi ... Lahore (printed by the Superintendent, Government Printing, Punjab) 1918. Qu. (*Forest fl. Punjab*).
Ed. 1: 1918 (p. [578]: printers mark 7 March 1918), p. [i*], [i]-xxxv, [1]-577, [578].
Copies: G, US; IDC 5345.
Ed. 2: 1924 (p. [592]: 15 Mai 1924; MO rd. 7 Jul 1925, p. [i*-ii*], [i]-xxxv, [1]-591, [592]. *Copies*: MO, NY.
Reprint: of 1918 ed.: 1973. *Copies*: BR, MICH. – M/S. Bishen Singh Mahendra Pal Singh ... Dehra Dun and M/S. Periodical Experts ... Delhi.

7340. *Common Indian trees* and how to know them ... Delhi (Manager of Publications) 1933. (*Common Ind. trees*).

Ed. 1: 1933 (p. 46: 16 Oct 1933), p. [i-iii], [1]-46, *pl. 1-40* (*uncol.*).
Reprints: 1940, 1948, idem. Printed at the Survey of India Offices, Dehra Dun. *Copies*: MO (1940), USDA (1948). *Cover: Forty trees Common in India*, Forest Research Institute, Dehra Dun.

Parkhurst, Howard Elmore (1848-1916), American botanist. (*Parkhurst*).

HERBARIUM and TYPES: Some Jamaican material at NA.

BIBLIOGRAPHY and BIOGRAPHY: Barnhart 3: 50; IH 2: (in press); Tucker 1: 539.

7341. *Trees, shrubs and vines of the northeastern United States, their characteristic landscape features fully described for identification by the non-botanical reader; together with an account of the principal foreign hardy trees, shrubs and vines cultivated in our country, and found in Central Park, New York City ...* New York (Charles Scribner's Sons) 1903. (*Trees n.-e. U.S.*).
Publ.: 1903 (J. NYBG Sep 1903; TBC 5 Oct 1903), p. [i]-viii, [ix], 1-451, *pl. 1-15, 1-10, 1-9, 1-4, 1, [1-3]*. *Copies*: NY, USDA.

Parkinson, Charles Edward (1890-1945), Scottish botanist; assistant conservator of forests, Andaman Isl. 1912; India Forest Service, Burma, 1916; Forest Botanist Burma 1925; Forest Research Institute, Dehra Dun 1929. (*C. Parkinson*).

HERBARIUM and TYPES: material at A, CAL, DD, E, K, NY, S.

BIBLIOGRAPHY and BIOGRAPHY: BL 1: 93, 312; BM 8: 983; Desmond p. 479; IH 1 (ed. 7): 339; 2: (in press); Kew 4: 235.

BIOFILE: Anon., Empire For. J. 24(1): 5. 1945 (d.).
Bridson, G.D.R. et al., Nat. hist. mss. res. Brit. Isl. 269-279. 1980.
Hedge, I.C. & Lamond, J.M., Index coll. Edinb. herb. 117. 1970.
Verdoorn, F., ed., Chron. bot. 2: 180, 208. 1936, 3: 180. 1937, 5: 292. 1939.

7342. *A forest flora of the Andaman Islands* an account of the trees, shrubs and principal climbers of the Islands ... Simla (Superintendent, Government Central Press) 1923. Oct. (*Forest fl. Andaman Isl.*).
Publ.: 1923 (p. 325: 23 Apr 1923), p. [i*, t.p.], errata slip, frontisp. with [1] p. text, [iii*, cont.], [i]-v foreword, [i]-v preface, [i]-xiii introd., [1, list lit.], 1-325, *pl. 1-6*. *Copies*: MO, USDA.
Reprint: 1972, pagination as 1923 original. *Copies*: B, BR, G, NY. – M/s Bishen Singh Mahendra Pal Singh ... Dehra Dun.

Parkinson, James (1755-1824), English surgeon, apothecary and palaeontologist at Hoxton, East London. (*Js. Parkinson*).

COLLECTIONS: Unknown.

BIBLIOGRAPHY and BIOGRAPHY: Andrews p. 311; Barnhart 3: 50; BB p. 236; BM 4: 1523; CSP 4: 760; Desmond p. 479; DNB 43: 314-315; HU 764; Jackson p. 176; PR 6932 (ed. 1: 7747); Quenstedt p. 328.

BIOFILE: Allen, D.E., The naturalist in Britain 68-69, 71. 1976.
Allibone, S.A., Crit. dict. Engl. lit. 1508. 1870 (bibl.).
Andrews, H.N., The fossil hunters 53-55. 1980.
Anon., Encycl. brit. ed. 11. 20: 832. 1911.
Bridson, G.D.R. et al., Nat. hist. mss. res. Brit. Isl. 72.2. 1980.
Thackray, J.C., J. Soc. Bibl. Nat. Hist. 7(4): 451-466. 1976 (major treatise of his *Organic remains*).
Ward, Ann. Rep. U.S. Geol. Survey 5: 401-402. 1885.
Zittel, K.A. v., Hist. Geol. Palaeontol. 127, 118, 129, 130, 386, 389, 405. 1901.

NOTE: For a major treatise (bibliography, dates, notes) of the *Organic remains of a former world* 1804-1811 (3 vols.; botany in vol. 1) see Thackray (1976). In view of Thackray's extensive treatment we refrain from treating this work in any detail.

7343. *Organic remains of a former world.* An examination of the mineralized remains of the vegetables and animals of the antediluvian world; generally termed extraneous fossils ... The first volume; containing the vegetable kingdom. London (printed by C. Whittingham, ... and published by J. Robson, ...) 1804. Qu. (*Organ. remains*).
Ed. 1: Jul 1804 (Thackray (1976)), p. [i]-xii, 1-471, [473-479], 11 copper engr. (except for frontisp. handcol.).
Further editions: ed. 2 (early 1811), ed. 3 (late 1819 or 1820), ed. 4 (1833), see Thackray (1976).

Parkinson, John (1567-1650), British apothecary and herbalist; King's herbarist; had a garden in Long Acre, London. (*Jn. Parkinson*).

HERBARIUM and TYPES: Unknown.

BIBLIOGRAPHY and BIOGRAPHY: AG 3: 21; Ainsworth p. 40, 309; Barnhart 3: 50; BB p. 236; Blunt p. 72-73, 102; BM 4: 1523; Bossert p. 301; Clokie p. 221; Desmond p. 479 (q.v. for secondary refs.); De Toni 1: xcvi; DNB 43: 35; Dryander 3: 60, 604; GR p. 409; Herder p. 83, 363; Jackson p. 28; Kew 4: 235; NI 1489-1490, 3: 54; Plesch p. 351; PR 6933-6934, ed. 1: 7748-7749; Rehder 1: 285, 530; Sotheby 584, 585; TL-2/4909; Tucker 1: 539-540.
BIOFILE: Anon., J. Bot. 41: 254. 1903 (reprint Parad. terr. announced), 42: 32, 214, 320. 1904 (published).
Arber, A., Herbals 32 [index]. 1953.
Druce, G.C., Fl. Buckinghamshire lxix-lxx. 1926; Fl. Northamptonshire xlv-xlvi. 1930.
Ewan, J. & N., John Banister 482 [index]. 1970.
Green, J.R., Hist. bot. 48-53. 1914.
Gunther, R.T., Early Brit. botanists 265-271. 1922.
Hadfield, M., Pioneers in gardening 19, 25-27. 1951.
Hawksworth, D.L. & M.R.D. Seaward, Lichenology Brit. Isl. 2, 130. 1977.
Kent, D.H., Hist. fl. Middlesex 13. 1975.
Lange, C., Missouri Bot. Gard. Bull. 58(1): 4-10. 1970.
Martin, W.K. & G.T. Fraser, Fl. Devon 769. 1939.
Pearsall, W.H., Fl. Surrey 44. 1931.
Pulteney, R., Sketches 1: 138-154. 1790.
Raven, C.E., English natural. 248-273. 1947.
Riddelsdell, Fl. Gloucestershire cix. 1948.
Trimen, H. & W.T. Thistleton Dyer, Fl. Middlesex 372. 1869.
White, J.W., Fl. Bristol 52-53. 1872.
Wittrock, V.B., Acta Horti Berg. 3(2): 96. 1903 (on portr.).
Wolley-Dod, A.H., Fl. Sussex xxxvi. 1970.

EPONYMY: *Parkinsonia* L. (1753), *Parkinsona* Cothenius (1790).

NOTE: Author of *Paradisi in sole paradisus terrestris* (1629) and *Theatrum botanicum* (1640), falling outside our period.

Parkinson, Sydney (1745-1771), British woollen draper; employed by Joseph Banks to work on the latter's collections, 1767; sent out with Cook on the Endeavour; died at Batavia. (*S. Parkinson*).

DRAWINGS: Sydney Parkinson was draughtsman on Cook's first voyage in the Endeavour (1768-1771). He does not seem to have collected himself (for the collections see Banks and Solander) but he made a number of drawings which are at BM (19 volumes of water-colour drawings of plants and animals).

BIBLIOGRAPHY and BIOGRAPHY: Barnhart 3: 50; BB p. 237; BM 4: 1523, 8: 983; Dawson p. 342, 661, 662, 669; Desmond p. 480; DNB 43: 317-318; Dryander 3: 185; DSB 10: 323-324 (by Lysaght); Hortus 3: 1201 ("Parkins."); HR; Jackson p. 223; Kew 4: 235; NI 74, 1123, see also 1: 113, 237 and suppl. p. 80; PR 6935 (ed. 1: 7750); Rehder 3: 94, 283; RS p. 133; SK 1: 401, 4: xciv; 5: cccix; TL-1/33, 948; Tucker 1: 540; Zander ed. 10,p. 699, ed. 11, p. 797.

BIOFILE: Allibone, S.A., Crit. dict. Engl. lit. 1508. 1870.
Anderson, A.W., Gard. Chron. 135: 24-25 (portr.). 1954.
Anon., Encycl. brit. ed. 11. 10: 734. 1911.
Beaglehole, J.C., ed., The journals of Captain James Cook, i, the voyage of the Endeavour, Cambridge 1955. (see index p. 674).
Beddie, M.K., Bibliogr. Cook, ed. 2, 1970, nos. 4601-4614, see also p. 869.
Bridson, G.D.R. et al., Nat. hist. mss. res. Brit. Isl. 427 [index]. 1980.
Cameron, H.C., Sir Joseph Banks 16, 19, 39, 42, 43. 1952.
Colenso, W., Trans. New Zealand Inst. 10: 108-134. 1877 (biogr., extr. from journal).
Dandy, J.E., Index generic names 1753-1774, p. 17-18. 1967 (Accepts P's binomials; the hyphenation "was surely a printing accident, the real intention being clear").
Dixson, W., J. Proc. r. Austral. hist. Soc. 7: 100-101. 1921.
Lemmon, K., Golden age plant hunters 226 [index]. 1968.
Lysaght, A.M., Joseph Banks in Newfoundland and Labrador, 1766, London 1971, p. 102-103, 478 [index]; see also Emu 56: 129-130. 1956 (zool., distr. records).
Maiden, J.H., Sir Joseph Banks 8, 36, 62-64. 1909.
Merrill, E.D., Contr. U.S. natl. Herb. 30(1): 236. 1947 (Pacific bibl.), Chron. bot. 14: 326-363, 375. 1954 (the botany of Cook's voyages; notes on Parkinson's botanical names).
Parkinson, S., Journal voyage South Sea v-xxiii. 1773.
Sawyer, F.C., J. Soc. Bibl. Nat. Hist. 2: 190-193. 1950 (drawings Cook's Voy.).
Smith, B., European vision and the South Pacific 1768-1850, p. 3-4, 10-13, 15-17, 19, 21, 24, 28-32, 34, 37-38, 51, 77, 79, 88-89, 94, 116, 121, 127, 128, 133. 1960.
Smith, E., Life Joseph Banks 344 [index]. 1911.
Stafleu, F.A., Linnaeus and the Linnaeans 228, 229. 1971.
Stearn, W.T., *in* R. Brown, *Prodr.* facs. ed. 1960, preface p. 10; Endeavour 27: 3-10. 1968 (botanical results Endeavour voyage); Nat. Hist. Mus. S. Kensington 19, 280, 282, 295, 326. 1981.
St. John, H., Biol. J. Linn. Soc. 4(4): 305-310. 1973 (validly publ. sci. names in *Die Pflanzen der Insel Outahitée*); Taxon 12: 203. 1963 (rejects P's binomials because they are hyphenated; see Dandy 1967).
Willson, E.J., James Lee and the Vineyard Nursery, London 1961 (P. taught natural history painting to Lee's daughter Ann).
Woodward, B.B., Hist. Coll. BM(NH) 1: 27, 44. 1904 (drawings).

POSTAGE STAMPS: Australia 5 c. (1970) yv. 410.

7344. *A journal of a voyage to the South Seas*, in His Majesty's ship, the Endeavour. Faithfully transcribed from the papers of the late Sydney Parkinson, draughtsman to Joseph Banks, Esq., on his late expedition with Dr. Solander, round the world; embellished with views and designs, delineated by the author, and engraved by capital artists. London (printed for S. Parkinson, sold by Richardson, Urquhart, Evans, Hooper, Murray, Leacroft, Riley) 1773. Qu. (*J. voy. South Seas*).
Editor and author of biographical sketch: Stanfield Parkinson.
Ed. 1: Jul 1773 (rd. by Gentleman's Mag.), portr., p. [i]-xxiii, [1]-212, [1-2], *pl. 1-27*. *Copies*: NY, Teyler. – Parkinson mentions a number of new generic and specific names, mostly quite casually, with little or no descriptive material and clearly without the intention of publishing new scientific names. In a few cases it will be possible to argue that a name is validly published, but Merrill (1954) suggests to "reject most or all of them". It should be pointed out, however, that the ICBN accepts Parkinson names as validly published in two cases (nos. 1946 and 3848), a view supported by Dandy (1967).
Ed. 2: 1784, (p. ii: 1 Jun 1784), frontisp. portr. auth., [i*], [i]-xxiii, [1]-353, [1-2, err.],

pl. 1-27 (incl. frontisp.; col. copp. auct.). – "A journal ... Banks, Bart., in Bart., in his expedition with Dr. Solander round the world; and embellished with twenty-nine views and designs ... to which is now added remarks on the preface, by the late John Fothergill, and an appendix ..." London (printed for Charles Dilly, ... and James Phillips, ...) 1784. Qu. *Copies*: Teyler; IDC 5632. – Plants Otaheite on p. 37-50.
French: 1797, *1*: [1]-12, [i]-lvi, [1]-344, *4 pl.*, 2: [1]-309. *Copy*: FI. – "Voyage autour du monde sur le vaisseau ... l'Endeavour, par Sidney Parkinson, dessinateur attaché à M. Banks; Précédé ... Ouvrage traduit de l'anglais par le C. Henri. Paris (Imprimerie de Guillaume) 1797.
Plants of use for food, medicine &c. in Otaheite, in *Journal* p. 37-50. German translation in Der Naturforscher, Halle, 4: 220-258. 1774 (Die Pflanzen der Insel Outahitée ... mit Anmerkungen erläutert von Z. [Halle 1774). For notes on scientific names in this translation see St. John (1973).

Parks, Harris Bradley (1879-?), American botanist, insect ecologist and agriculturist; B.S. Blackburn, Col. 1900; teacher at Sitka, Alaska 1905-1911, at Palmer College 1912-1917; entomologist at Texas A. and M. 1918-1920; agriculturist id. from 1922. (*Har. B. Parks*).

HERBARIUM and TYPES: ASTC, TAES.

BIBLIOGRAPHY and BIOGRAPHY: Barnhart 3: 50; BL 1: 215, 216, 312; IH 1 (ed. 1): 27, 94; Kew 4: 235; TL-2/1236.

BIOFILE: Anon., Amer. men sci., ed. 5. 858. 1933, ed. 6: 1085. 1938 (b. 10 Jun 1879).
Osborn, H., Fragm. entom. hist. 274. 1937.
Verdoorn, F., ed., Chron. bot. 3: 362. 1937.

COMPOSITE WORKS: with V.L. Cory, *Cat. fl. Texas* 1938, see TL-2/1236.

7345. *Valuable plants native to Texas* ... (Texas Agricultural Experiment Station, Bulletin no. 551). August 1937 (1938). (*Val. pl. Texas*).
Publ.: 18 Jan 1938 (see BL 1: 215), p. [1]-173. *Copies*: BR, MO, NY. – Blake: "The "value" of most of the species included is hypothetical".

Parks, Harold Ernest (1880-1967), American mycologist and plant collector; in the U.S. army in China and the Philippines 1898-1899; salesman at Tacoma 1900-1910, at Santa Cruz, Calif. 1910-1914; post office clerk San José, Calif. 1914-1921, from 1921-1928 collector for UC; associate curator UC 1928-1950, hon. curator 1950-1967. (*H.E. Parks*).

HERBARIUM and TYPES: UC; further material at A, AMES, BISH, BPI, C, E, F, G, GB, K, NSW, NY, URM, W, WELC.

BIBLIOGRAPHY and BIOGRAPHY: Barnhart 3: 50; BJI 2: 130; Ewan (ed. 2): 168; IH 1 (ed. 6): 362, (ed. 7): 339, 2: (in press); Kelly p. 166; Kew 4: 235.

BIOFILE: Bonar, L., Madroño 20: 373-377. 1970 (obit., portr., b. 5 Aug 1880, d. 5 Mar 1967, list of published records of P's colls.).
Brotherus, V.F., Univ. Calif. Publ. Bot. 12(3): 45-48. 1924 (Tahitian mosses coll. by W.A. Setchell and H.E. Parks).
Hedge, I.C. & J.M. Lamond, Index coll. Edinb. herb. 117. 1970.
Hultén, E., Bot. Not. 1940: 337. (Alaskan coll.).
Parks, H.E., Univ. Calif. Publ. Bot. 12(4): 49-59. 14 Mai 1926 (Tahitian fungi coll. by W.A. Setchell and H.E. Parks).
Setchell, W.A., Univ. Calif. Publ. Bot. 12(5): 61-142, *pl. 7-22.* 1926. (Tahitian algae id.), 12(6): 143-240, *pl. 13-36.* 1926 (Tahitian spermatophytes id.).
Smith, A.C., Flora vitiensis nova 1: 57. 1979 (Fiji coll.).
Stieber, M.T. et al., Huntia 4(1): 84. 1981 (arch. mat. HU).
Verdoorn, F., ed., Chron. bot. 2: 299. 1936.

EPONYMY: *Parksia* E. K. Cash (1945).

Parlatore, Filippo (1816-1877), Italian botanist; Dr. med. Palermo 1837; prosector and professor of anatomy at the University of Palermo; from 1842-1877 professor of botany and director of the botanic garden in Florence; founder of the herbarium of the botanical museum at Florence. (*Parl.*).

HERBARIUM and TYPES: FI; further material at AWH, BP, BR, CM, E, G, GE, GOET, K, LE, LY, MW, NA, NAP, OXF, P, PA, PAL, PI, PR, W. – Parlatore's manuscripts and library are at the city library of Palermo. His letters to A. Massalongo are at the Biblioteca communale di Verona. Other letters at G.

BIBLIOGRAPHY and BIOGRAPHY: AG 2(1): 246; Backer p. 658; Barnhart 3: 51; BJI 1: 42; BL 2: 158, 335-336, 394, 699; BM 4: 1523-1524; Bossert p. 301; Cesati p. cix-cxviii; Clokie p. 221; CSP 4: p. 760-761, 8: 564, 10: 992, 12: 558, 17: 513; DTS 1: 218; Frank 3(Anh.): 74; Hegi 1: 256; Herder p. 470; Hortus 3: 1201 ("Parl."); IH 1(ed. 6): 362, (ed. 7): 339, 2: (in press); Jackson p. 7, 129, 132, 145, 314, 315, 343, 432; JW 1: 452; Kew 4: 235-237; KR p. 570-571; Langman p. 566; Lasègue p. 112, 297, 343; Moebius p. 139; Morren ed. 2, p. 26; MW p. 378; NI 1491; PFC 1: xlix; PR 6937-6944 (ed. 1: 7752-7768); Rehder 5: 650; RS p. 133; Saccardo 1: 122, 2: 81, cron. p. xxviii; SO 3501; TL-1/949-950, see Webb; TL-2/999, see T. Caruel, A. Mori; Tucker 1: 540; Zander ed. 10, p. 699, ed. 11, p. 797-798.

BIOFILE: Anon., Bonplandia 1: 33, 42, 46, 93, 143, 168, 228. 1853, 2: 67, 105, 239. 1854, 3: 15, 79, 261, 343. 1855, 4: 151. 1855, 8: 62. 1861, 10: 216, 217, 260. 1862; Bot. Not. 1851: 127 (in Sweden), 1878: 44 (d.); Bot. Zeit. 9: 463, 887. 1851 (arctic trip), 10: russian decoration), 11: 198. 1853 (knighthood), 28: 576. 1870. (prof. bot. at Pharmac. School); 35: 678. 1877 (d.); Bull. Soc. bot. France 12 (bibl.): 238. 1865, 24 (bibl.): 189. 1877, 30: cxxv-cxxvi; Encycl. 11. 7: ed. 257. 1911; Flora 36: 31, 288. 1883 (knighthood), 50: 476. 1867 (corr. memb. München), 60: 448 ("844"). 1877(d.); Gard. Chron. ser. 2. 1: 599. 1874 (portr.); Österr. bot. Z. 20: 189. 1870, 27: 390. 1877 (d.).

Beck, G., Bot. Centralbl. 34: 150. 1888 (pl. W.).
Béguinot, A., Archivio bot. 4(1): 72-77. 1928 (3 letters).
Britten, J., J. Bot. 15: 320. 1877.
Burci, E. et al., Nuovo Giorn. bot. ital. ser. 2. 34(5): 967-999. 1928 (alla memoria di F.P., portr.; main address by N. Passerini).
Burnat, E., Bull. Soc. bot. France 30: cxxv-cxxvi. 1883 (b. 8 Aug 1816, d. 9 Sep 1877).
Cavillier, F.G., Boissiera 5: 69-70. 1941.
Cesati, V. de, Mém. Mat. Fis. Soc. ital. Sci. ser. 3. 3: cix-cxviii. 1879.
Giacomini, V., Regn. veg. 71: 8, 90, 98. l970.
Haynald, L., Parlatore Fülöp, Emlékbeszéd. Kolozsvárt 1878. (46 p., repr. from Magy. Növen. Lap. 2: 97-140. 1879, also Magy. Tudom. Akad. Evkön. 16(4), 42 p. On.), German: Denkrede auf Philipp Parlatore, (Leipzig 2) Budapest 1879, (Lit. Ber. Ungarn 2: 1-63. 1879).
Hedge, I.C. & Lamond, J.M., Index coll. Edinb. herb. 117. 1970.
Hooker, J.D., in W.J. Hooker, London J. Bot. 1: 205. 1842, 6: 54. 1847.Huxley, L., Life letters J.D. Hooker 1: 419. 1918.
Jackson, B.D., Bull. misc. Inf. Kew 1901: 51 (pl. K).
Jessen, K.F.W., Bot. Gegenw. Vorz. 372. 1864.
Lindemann, E. v., Bull. Soc. imp. Natural. Moscou 61: 53. 1886.
Massalongo, C., Nuovo Giorn. bot. ital. ser. 2. 34: 1333-1342. 1928 (corr. with A. Massalongo).
Moggi, G., OPTIMA Second Meeting, Firenze, 23-29 May 1977, Abstracts, (1 p.). 1977; Natural. Siciliano 2: 97-108; Webbia 34(1): 51-57. 1979 (general account of floristic work, further biogr., refs.).
Mussa, E., Nuovo Giorn. bot. ital. ser. 2. 34: 1068-1077. 1928 (corr. with V. Gioberti, Italian nationalist).
Negri, G., Nuov. Giorn. bot. ital. n.s. 34: 972-999. 1927 (portr., main biogr. account).

Parlatore, F., in W. Hooker, J. Bot. Kew Gard. Misc. 1: 370-371. 1849 (letter to Webb 1, 4: 56-59.
Seemann, B., Bonplandia 1:33, 66, 93, 143. 1853.
Stafleu, F.A., The great Prodromus 30. 1966.
Steinberg, C.H., OPTIMA, second meeting, Firenze 23-29 May 1977, Abstr. (1) p. 1977; Webbia 32: 11, 33. 1977, 34(1): 59-62. 1979 (P and medit. coll. FI).
Tassi, F., Bull. Lab. Orto Bot. 7: 6, 33-34. 1905.
Wittrock, V.B., Acta Horti Berg. 3(3): 179, *pl. 136*. 1905 (portr.).

COMPOSITE WORKS: (1) A.P. de Candolle, *Prodr.*: (a) *Coniferae*, 16(2): 361-521, 685. med. Jul 1868, (b) *Gnetaceae*, 6(2): 347-360, 685. med. Jul 1868.
(2) with P.B. Webb, q.v., *Florula aethiopico-aegyptica* (1851).

MEMORIAL PUBLICATION: Nuovo Giornale botanico italiano 34(5): 967-999. 1928, papers by E. Burci, A. Pucci, N. Passerini and G. Negri.

HANDWRITING: Candollea 32: 204. 1977; Monogr. Biol. Canar. 4: 31. 1973 (fig. 5).

EPONYMY: Fondazione Filippo Parlatore per lo studio della flora e della vegetatione italiana (see Webbia 27: 583-589. 1972). *Parlatoria* Boissier (1842); *Parlatoria* Meisner (1843); *Parlatorea Barbosa Rodrigues* (*1877*); *Parlatoria* Boissier (1842).

7346. *Rariorum plantarum* et haud cognitarum *in Sicilia sponte provenientium* ... auctore Philippo Parlatore. Panormi [Palermo] (ex Typographeo diarii literarii) 1838-1840, 2 fasc. Oct. (*Rar. pl. Sicilia*).
Fasciculus 1: 1838, p. [1]-16, *1 pl.* (PR: *2 pl.*), first published in Giorn. Sci. Lett. Arti Sicilia 1837 and 1838 (n.v.).
Fasciculus 2: 1840, p. [3]-21, *pl.* [? cited but not seen by PR].
Copies: FI, G. – Information incomplete.

7347. *Flora panormitana* sive plantarum prope Panormum sponte nascentium enumeratio auctore Philippo Parlatore ... Panormi [Palermo] (ex Typographeo Petri Pensante) 1839. † Oct. (*Fl. panorm.*).
1(1): *1839*, p. [i]-xxxi, [*1*], [*1*]-*64*.
1(2): *1839*, p. *65-128*.
Copies: FI, G, NY. – Planned to consist of 3 volumes and an atlas.

7348. *Observations sur quelques plantes d'Italie*, ... Paris (imprimé chez Paul Renouard, ...) 1841. Oct.
Publ.: Mai *1841*, p. [*1*]-*12*. *Copy*: G. – Reprinted from Ann. Sci. nat. 15: 294-303. Mai 1841. To be cited from journal.

7349. *In nonnullas Filaginis Evacisque* species ex naturali Compositarum familia *observationes* ... [Pisa 1841]. Oct.
Publ.: 1841 (p. 8: 12 Mar 1841), p. [1]-8. *Copies*: F, G. – Reprinted from Giorn. Tosc. Sci. med. fis. e nat. 1(2): 179-184. 1841.

7350. *Sulla botanica* in Italia e sulla necessità di formare un erbario generale in Firenze (discorso ai botanici radunati nel terzo congresso italiano ... Parizi [Paris] (coi torchi della signora de Lacombe, ...) 1841. Oct.
Publ.: 1841, p. [1]-19, [20]. *Copies*: FI, G. – For a follow up see Parlatore, Giorn. bot. ital. 1(3): 18-32. 1846: *Sullo stato attuale dell'erbario centrale italiano*, ... (also reprinted s.l.n.d., p. [1]-11. *Copies*: FAS, FI, G.

7351. *Plantae novae* vel minus notae opusculis diversis olim descriptae generibus quibusdam speciebusque novis adjectis iterum recognitae ... Parisiis (Gide, editori) 1842. Oct. (*Pl. nov.*).
Publ.: 2 Apr 1842 (BF), p. [1]-87, [88, index]. *Copies*: BR, FI, G, NY, USDA; IDC 8319.
Ref.: Schlechtendal. D.F.L. von, Bot. Zeit. 1: 447-453. 30 Jun 1843.

7352. Come possa considerarsi *la botanica* nello stato attuale delle scienze naturali prolusione letta in occasione dell'apertura della cattedra di botanica e di fisiologia vegetale nell'I e R. museo di fisico e storia naturale di Firenze il 1.; dicembre 1842 ... Firenze (nella Tipografia Piatti) 1842. Oct. (*La botanica*).
Publ.: Dec 1842, p. [1]-35. *Copies*: FI, G, NY.

7353. *Notizia sulla Pachira alba* della famiglia delle Bombacee letta nella tornata ordinaria del 5 marzo 1843 della I. e R. Accademia de'Georgofili ... Firenze (per la Società tipografica) 1843. Oct. (*Not. Pachira alba*).
Publ.: 1 Apr 1843 (date journal; sent from Florence to PAW 4 Apr 1843), p. [1]-7. *Copies*: F, G. – Reprinted from Gazzetta toscana delle scienze medico-fisiche anno primo n. 4, 1 Aprile (n.v.).

7354. *Lezioni di botanica comparata* ... Firenze (per la Società tipografica) 1843. Oct. (*Lezioni bot. comp.*).
Publ.: 1843, p. [1]-238. *Copies*: FI, G, NY.

7355. *Giornale botanico italiano* compilato per cura della sezione botanica dei congressi scientifici italiani da Filippo Parlatore ... Firenze (per la Società tipografica) 1844-1852. Oct.
Organization of the journal: This is one of the more complicated botanical journals in its structure. The confusion engendered by the structural complication is exacerbated by the fact that a complete set of the journal is not known to us for certain; the FH copy may be complete. The following table summarizes what we have been able to understand and infer from the several partial runs we have studied. The journal was divided into two "years" (anno, anni), although publication took place over several years (1844 to 1852), with each "year" subdivided into two volumes. Each volume was issued in fascicles but some of these have not been seen and are indicated by enclosing fascicle numerals in brackets. Added to all this, the contents appeared in three parts, each with independent pagination: Parte prima (original research reports), Parte seconda (reviews and announcements of botanical works), and Parte terza (news and notes).
Dates: The dates given below are not very meaningful but in some instances may be helpful. The covers often bear the year of publication, as the title pages do, but the two dates don't always agree. Some of the earlier fascicles have covers with month/year dates but since these usually indicate six-month periods, they are of limited value. Some of the dates from pages in the text are also given below when they seem to affect the interpretation of the title-page date.
Copies: E, FI, FH, G, NY, USDA. The FH copy may lack some pages but is certainly the most complete we have seen; it is bound without fascicle covers.

Tom.	Fasc.	parte prima (orig. works)	parte seconda bot. lit.	parte terza (news//notes)	t.p. dates	other dates
1	[1-3]	1-184	1-28	1-24	1844	Jan-Jun 1844
	4	185-236	29-44	25-32	1844	Apr 1844 (cover); 15 Sep 1843
	5//6	236-238	45-60	33-64	1845	Mai-Jun 1844 (cover); 10 Mai 1844 (p. 236)
	7//8	1-96	81-116	41-48*	1845	11 Mai 1845 (p. 57)
2		97-395, 2 p. index	117-123		1844	1845 (cover); Jul-Dec 1844 (cover)
1	1//2	3-118	1-59	1-48	1846	
	3//4	117-236	1-8	49-68	1846	
	5//6	237-354		69-84	1847	

Tom.	fasc.	parte prima	seconda	terza	t.p.	other
2	7//8	1-164		85-92	1847	
	[9]	[165-220]		[93-116]	[1849]	Feb 1849 (p. 210)
	10-12	221-294	9-41	117-140	1852 (no t.p.).	

Ref.: Schlechtendal, D.F.L. von, Bot. Zeit. 3: 725-728, 744-747, 765-767. 1845.

7356. *Monografia delle Fumariée* presentata alla sezione botanica del quarto congresso degli scienzati italiani in Padova nel settembre del 1842 ... Firenze (per la Sociéta tipografica) 1844. Qu. (*Monogr. Fumar.*).
Publ.: Mar-Jun (ded. 11 Feb 1844; PAW rd. 27 Jun 1844; Flora rd. Jun 1844; Bot. Zeit. 27 Dec 1844), p. [i]-x, [1]-110, *1 pl.* (*uncol., by T. Heldreich*). *Copies*: FI, G(2), L, MO, NY, USDA, WU. – Reprinted or preprinted with independent pagination from Parlatore, F., Giorn. bot. ital. 1: 50-105, 124-178. *1 pl.* 1844. – For a continuation see Nicotra, *Fumar. ital.* 1897, TL-2/6795.

7357. *Sullo spirito delle scienze naturali* nel secolo passato e nel presente prolusione letta per l'apertura del corso di botanica il 4 dicembre 1843 ... Firenze (per la Società tipografica) 1844. Qu. (*Spirito sci. nat.*).
Publ.: 1844, p. [1]-21. *Copies*: FAS, FI.

7358. *Maria Antonia* novello genere della famiglia delle Leguminose ... Firenze (Coi Torchi della Società Tipografica) 1844. Qu. (*Maria Antonia*).
Publ.: 19 Dec 1844 (p. 7; Flora rd. Feb 1845), p. [1]-8, *1 pl.* (uncol.). *Copies*: FAS, FI, MO, NY, USDA, WU. – The new generic name is consistently written in two words and is therefore not validly published in this publication (species: *Maria Antonia orientalis*). – Issued with Giorn. bot. ital. 1(1)(2).
Ref.: Schlechtendal, D.F.L., von, Bot. Zeit. 3: 365-366. 30 Mai 1845.

7359. *Descrizione di due nove specie di piante orientali* del professore Filippo Parlatore s.l.n.d. [Firenze, 1844].
Publ.: Mai-Jun 1844, p. [1]-3. *Copies*: FAS, FI. – Reprinted and to be cited from Giorn. bot. ital. 1: 307-309. Mai-Jun 1844.

7360. *Flora palermitana* ossia descrizione delle piante che crescono spontanee nella valle di Palermo ... Firenze (per la Società tipografica) 1845. Qu. †. (*Fl. palerm.*).
Publ.: 1845, (p. [vi]: 1 Mar 1845), p. [i]-xxii, [1]-442. *Copies*: FI, US, USDA. – Only vol. 1 (parts 1, 2) were published: part 1: 1-218, part 2: 219-442. – Preprinted or reprinted from Giorn. bot. ital. 1(1)(2): 69-208, 284-375, 2(1)(1): 60-173, 326-350, 2(2) fasc. 7/8: 83-166. 1845.

7361. *Flora italiana*, ossia descrizione delle piante che crescono spontanee o vegetano come tali in Italia e nelle isole ad essa aggiacienti; disposta secondo il metodo naturale ... Firenze (Tipografia Le Monnier) 1848-1896, 10 vols. Oct. (*Fl. ital.*).
Authors of vols. 6-10: Théodore (Teodoro) Caruel (1830-1898) (all parts except those listed below), Enrico Tanfani (1848-1892, 7(1), 8(1), 9(2) pp.), Giovanni Arcangeli (1840-1921, (7(2)), Antonio Mori (1847-1902, (8: Plumb.)), Lodovico Caldesi (1822-1884, (8: Prim.)), Parlatore (9(1): Viol., Dros. Elat.).
1: 1848[-1850], p. [i-viii], [1]-568 (cover fasicolo 1 dated 29 Apr 1848, that of fasc. 2-5: 1850.
2: 1852[-1857], p. [1]-638 (cover parte 1ª: 1852, seconda: 1857).
3: 1858[-1860], p. [1]-690 (cover parte prima 10 Jul 1858, seconda 1 Mai 1860).
4: 1867 [1868-1869], p. [1]-623 (cover parte prima 1868, seconda 1869).
5: 1873 [1873-1876], p. [1]-671 (cover parte prima dated 1873, seconda 1875).
6: s.d. [1884-1886], p. [1]-971 (cover parte prima dated Sep 1884, seconda Aug 1885, tertia Jun 1886).

7: s.d. [1887-1893], p. [1]-300 (cover parte prima dated Mar 1887, seconda Apr 1893).
8: s.d. [1888-1889], p. [1]-773 (cover parte prima dated Jul 1888, seconda Mar 1889, tertia Oct 1889).
9: s.d. [1890-1893], p. [1]-1085 (cover parte prima dated Mar 1890, seconda Feb 1892, tertia 1893).
10: 1894, p. [1]-234. Cover dated Apr 1894.
Index: 1896, p. [1]-31. Cover dated Apr 1896.
Copies: BR, FI, G, MO, NY, USDA; IDC 5059.

vol.	part	pages	dates	vol.	part	pages	dates
1	1	1-96	29 Apr 1848	6	1	1-336	Sep 1884
	2	97-568	Oct-Nov 1850		2	337-656	Aug 1885
2	1	1-220	Mai-Jul 1852		3	657-971	Jun 1886
	2	221-638	late 1857	7	1	1-256	Mar 1887
3	1	1-160	10 Jul 1858		2	257-300	Apr 1893
	2	161-690	1 Mai 1860	8	1	1-176	Jul 1888
4	1	1-288	mid 1868		2	177-560	Mar 1889
	2	289-623	late 1869		3	561-773	Oct 1889
5	1	1-320	Apr 1873	9	1	1-232	Mar 1890
	2	321-671	late 1875		2	232-624	Feb 1892
					3	625-1085	Feb 1893
				10		1-234	Apr 1894
				Index		1-31	Apr 1896

The contents and dates of the fascicles can be established from the copies in original covers at G, MO and USDA. These data are confirmed by the sources mentioned in TL-1/950.
Imprints: 1-3 as above, 4-9(2): Firenze (Tip. dei successori Le Monnier), 9(3) – index: Firenze (Stabilimento tipografico fiorentino).
Title vols. 6-10, ind.: "Filippo Parlatore, *Flora italiana* continuata da Teodoro Caruel ...".
Prospectus: 17 Feb 1848, see Bot. Zeit. 6: 359. 5 Mai 1848.
Ref.: Anon., Bull. Soc. bot. France 4: 967-968. 1858, 6: 165-166. 1859, 7: 524-527. 1861, 15 Bibl.: 201. 1868.
  Ascherson, P., Bot. Zeit. 28: 201-205. 1870 (rev. vol. 4).
  Caruel, T., Bull. Soc. bot. France 36: 257-273. 1889.
  Briquet, J., Prodr. Fl. Corse 1: xlix. 1910.
  Cavillier, F., Boissiera 5: 69-70. 1941.
  Drude, O., Bot. Jahrb. 6(Lit.): 73. 1885.
  Koehne, E., Bot. Zeit. 44: 341-343. 1886 (rev. 6(1)).

7362. *Colpo d'occhio sulla vegetatione d'Italia* del prof. Filippo Parlatore [Firenze (Tip. di M. Cecchi) 1850].
Publ.: 1850, p. [1]-8. *Copies*: FAS, FI. – Reprinted from Giorn. ital. Sci. Med. Il Progresso 1(24) & 2(1), (n.v.).

7363. *Viaggio alla catena del Monte Bianco* e al Gran San Bernardo eseguito nell' agosto del 1849 ... Firenze (Tipografia Le Monnier) 1850. Qu. (*Viagg. Monte Bianco*).
Publ.: 1850 (p. xi: 1 Mar 1850), p. [i]-xi, [1]-216, [1, index], 1 chart, [1, err.]. *Copies*: FI, G, NY.
Ref.: S-l, Bot. Zeit. 9: 833-834. 21 Nov 1851 (rev.).

7364. *Mémoire sur le Papyrus des anciens* et sur le Papyrus de Sicile, ... Paris (Imprimerie impériale) 1853. Qu. (*Mém. Papyrus*).
Publ.: 1853, p. [i-iii], [1]-34, *pl. 1-2* (uncol.). *Copies*: BR, FI, G, USDA. – Preprinted from Mém. Acad. Sci., Paris, Sav. étr. 12: 469-502. 1854. Preliminary publ. in C.R. Acad. Sci. 35: 211-217. 1852.
Ref.: Anon., Bonplandia 3: 9-10. 1855 (rev.).

7365. *Viaggio per le parti settentrionali di Europa* fatto nell'anno 1851 ... parte prima narrazione del viaggio con una geografica. Firenze (Tipografia Le Monnier) 1854. Oct. (*Viagg. parti sett. Eur.*).
Publ.: 1854 (J. Bot. Kew Gard. Misc. Oct. 1854), p. [i]-viii, [1]-392, map. *Copies*: FI, PH.

7366. *Nuovi generi e nuove specie di piante monocotiledoni* ... Firenze (Tipografia Le Monnier) 1854. Oct. (*Nuov. gen. sp. monocot.*).
Publ.: 1854, p. [1]-60, [61, index, err.]. *Copies*: FI, G, MO, NY, USDA; IDC 6145.

7367. *Due nuovi generi di piante monocotiledoni* ... [Firenze (Le Monnier) 1858]. Oct.(*Due nuov. gen. monocot.*).
Publ.: Mai 1858 (p. 2), p. [1]-8. *Copies*: FI, G.

7368. *Coniferas novas nullas* descripsit Philippus Parlatore [Florentiae (ex Typis Le Monnier) 1863]. (*Conif. nov.*).
Publ.: Jan 1863 (p. 4), p. [1]-4. *Copies*: FI, G.

7369. *Considérations sur la méthode naturelle en botanique* ... Florence (Imprimerie de Félix Le Monnier) 1863. Oct. (*Consid. méth. nat.*).
Publ.: 1863 (p. 3: 20 Feb 1863; BSbF rd. 10 Jul 1863), p. [1]-73. *Copies*: BR, FAS(2), FI, G(2), MO.
Ref.: Anon., Flora 46: 561-566. 22 Dec 1863.
Fournier, E., Bull. Soc. bot. France 10: 171-173. 1863.

7370. *Enumeratio seminum* in horto botanico regii musei florentini physices ac naturalis historiae anno mdccclxiii collectorum quae cum aliis seminibus sunt commutanda [Firenze 1864]. Oct. (*Enum. semin.*).
Publ.: 1864 p. 24: Feb 1864), p. [1]-28. *Copy*: G. – Contains descriptions of new taxa on p. 25-28. We have in addition seen the seed lists published by Parlatore for 1860, 1862 and 1863 which also contain descriptive material (copies at FI).

7371. *Studi organografici sui fiori e sui frutti delle Conifere* ... Firenze (con i tipi di M. Cellini e c. ...) 1864. (*Stud. organogr. conif.*).
Publ.: 1864, p. [1]-29, *3 pl. Copy*: FI, G. Preprinted from Ann. r. Mus. Stor. nat. Firenze 1: 155-181. 1866.

7372. *Le specie dei Cotoni* ... Firenze (Stamperia reale) 1866. Qu.(text), Fol. (atlas). (*Sp. Cotoni*).
Publ.: 26 Feb 1866 (date from letter offering a copy to Alph. de Candolle by Commissione reale per la coltivatione del Cotone p. [6]: 27 Nov 1865; Flora "Neuigkeit" 28 Mai 1866; BSbF 13 Jun 1866). *Text*: [1]-62, [63-64], *Tavole*: [i, iii], *pl. 1-6* (liths. by A. Menici, *1-5 col.*). *Copies*: BR, FAS, FI, G, MO, NY.
Ref.: Seemann, B., J. Bot. 4: 268-271. 1866.

7373. *Les collections botaniques* du musée royal de physique et d'histoire naturelle *de Florence* au printemps de mdccclxxiv ... Florence (Imprimerie Successeurs Le Monnier) 1874. Oct. (*Coll. bot. Florence*).
Publ.: 1874 (J. Bot. Jul 1874), p. [1]-163, *pl. 1-17* (uncol.). *Copies*: BR, FAS, FI, G, MO, U, US.

7374. *Todaroa* novum Umbelliferarum genus olim descriptum et iterum recognitum a Philippo Parlatore. Panormi [Palermo] (in aedibus Francisci Lao) 1876. Qu. (*Todaroa*).
Publ.: 1876 (Bot. Zeit. 2 Mar 1877), p. [1-10], *1 pl.* (uncol. lith. A. Mattolini). *Copies*: BR, FAS, FI, MO. – *Todaroa* was first published by Parlatore *in* P.B. Webb et S. Berthelot, Hist. nat. Iles Canaries 3(2.2): 155. Jan 1843, see ING 3: 1765. 1979.

7375. *Études sur la géographie botanique de l'Italie* ... Paris (Librairie J.B. Baillière et Fils ...) 1878. Oct. (*Étud. géogr. bot. Italie*).
Publ.: 1878 (p. 7: 10 Sep 1877), p. [1]-76. *Copies*: BR, FI, G, NY. – Posthumously

published by P. de Tchihatchef, appended to A. Grisebach, La végétation du globe, vol. 2, 1878, translated by P. de Tchihatchef.

7376. *Tavole per una "Anatomia delle piante aquatiche'* opera rimasta incompiuta di Filippo Parlatore. Firenze (coi tipi dei successori Le Monnier) 1881. Qu. (*Tav. Anat. piante aquat.*).
Editor: Théodore Caruel (1830-1898).
Publ.: Mar-Apr 1881 (p. 3: Feb 1881; BSbF rd. 22 Apr 1881; Bot. Zeit. 24 Jun 1881), p. [1]-24, *pl. 1-9* (uncol. liths. A. Mattolini). *Copy*: G. – Pubbl. [no. 8] Ist. Studi sup. Firenze, Sci. fis. nat.

Parmentier, Antoine Augustin (1737-1813), French economic botanist and military pharmacist, especially active in promoting the potato and, later, the sugar beet in France; "apothicaire aide-major" at the Invalides 1766; member of the Conseil de Santé 1782-1813; first army pharmacist 1800; inspector general of the health service 1805-1813. (*A. Parm.*).

HERBARIUM and TYPES: Unknown.

BIBLIOGRAPHY and BIOGRAPHY: Backer p. 425; Barnhart 3: 51; BM 4: 1524; Bossert p. 301; CSP 4: 762-763, 12: 558; Dryander 3: 430, 559, 622, 626 (as "Jean" P.); Frank 3(Anh.): 74; Henrey 2: 613, 615; 3: 99 (no. 1201); Jackson p. 194; Kew 4: 237; Langman p. 566; Plesch p. 357; PR 6956, 7855 (ed. 1: 7770-7784); Rehder 5: 650; Sotheby 587; TL-1/290; TL-2/see Déterville; Tucker 1: 540; Zander ed. 10, p. 699, ed. 11, p. 798 (b. 17 Apr 1737, d. 17 Dec 1813).

BIOFILE: Dehérani, P.P. et al., Bull. Soc. natl. Acclim. 35: 263-273. 1888 (dedication statue at Neuilly; addresses).
Hallynck, P., Grand Larousse encycl. 8: 198. 1963.
Jourdan, A.J.L., Dict. sci. méd., Biogr. méd. 6: 366-369. 1824.
Poggendorff, J.C., Biogr. lit. Handw.-Buch 2: 362-364. 1858 (bibl., b. 17 Apr 1737, d. 17 Dec 1813).
Silvestre, A.F., Mém. Agric., Soc. r. centr. Agric. 46: 38-59. 1815, also as reprint, 25 p.
Stieber, M.T. et al., Huntia 4(1): 84. 1981 (arch. mat. HU-Adanson).
Weemaels, F., Natuur Wereld 31(4): 181-183. 1969.
Wittrock, V.B., Acta Horti Berg. 3(3): 109. 1905.

COMPOSITE WORKS: (1) Co-editor of *Annales de Chimie et de Physique* 33-88, 1799-1813. (2) Contributed to Déterville, *Nouv. dict. hist. nat.*, ed. 1, 1802-1804, ed. 2, 1816-1819.

POSTAGE STAMPS: France 12 f. (1956) yv. 1081.

EPONYMY: *Parmentaria* Fee (1824); *Parmentariomyces* Ciferri et Tomasselli (1953); *Parmentiera* A.P. de Candolle (1838); *Parmentiera* Rafinesque (1840); *Parmentieria* Trevisan (*1860, orth. var.* of *Parmentaria* Fée).

7377. *Examen chymique des pommes de terre*. Dans lequel on traite des parties constituantes du Bled ... à Paris (Chez Didot, le jeune, ...) 1773. Duod. (*Exam. pommes de terre*).
Publ.: 1773 ("approbation" 10 Apr 1773, p. [i-xxiv, [1]-248, [1-3, approb., 4 err.]. *Copy*: MO. – See also his *Ouvrage économique sur les pommes de terre, le froment et le riz*, Paris (Monory) 1774, xxiv, 248 p. (*copy*: MO).

7378. *Traité de la chataigne*, ... À Bastia; et se trouve à Paris (chez Monory, ...) 1780. Oct. (*Traité chataigne*).
Publ.: 1780, p. [i]-xxviii, [1]-160.

Parmentier, Joseph Julien Ghislain (1755-1852), Belgian landscape gardener, botanist; maire and burgomaster of Enghien. (*J. Parm.*).

HERBARIUM and TYPES: Unknown.

BIBLIOGRAPHY and BIOGRAPHY: Barnhart 3: 51; BM 4: 1524; Jackson p. 419 (sub A.A.P.); PR 6957 (ed. 1: 7785); Rehder 1: 49; Tucker 1: 540.

BIOFILE: Crépin, F., Biogr. natl. Belgique 16: 649-650. 1901 (further biogr. refs.).
Gager, C.S., Brooklyn Bot. Gard. Rec. 15: 12-13. 1926.
Steenis-Kruseman, M.J. van, *in* Forbes, R.J., ed., Martinus van Marum 3: 150-152. 1971.

HANDWRITING: Candollea 32: 205-206. 1977.

7379. *Catalogue des plantes cultivées dans les jardins de M.ʳ Joseph Parmentier*, maire à Enghien, département de Jemappe, à l'époque du 1.ᵉʳ janvier 1808. À Bruxelles (de l'Imprimerie d'André Leduc, ...) [1808]. Oct. (in fours). (*Cat. pl. Parm.*).
Publ.: 1808, p. [i-iv], [1]-84. *Copy*: BR.

7380. *Catalogue des plantes cultivées dans le jardin de M.r Joseph Parmentier*, ... à l'époque du 1.er Mai 1812. Bruxelles (Imprimerie de M.-E. Rampelbergh, ...) [1812]. Oct. (*Cat. pl. Parm.*).
Publ.: 1812, p. [ii-v], [1]-33. *Copy*: BR.
Suppl.: announced for Jan 1813 (on p. [ii] of orig.), n.v.

7381. *Catalogue des arbres et plantes, cultivés dans les jardins de M.r Joseph Parmentier*, ... à l'époque du 1ᵉʳ Mai 1818 ... Bruxelles (chez F. Demanet, ... G. Choppinet, ...) 1818. Oct. (*Cat. arbr. Parm.*).
Publ.: 1818, p. [i-vi], [1]-88, [1-2, err.]. *Copy*: BR.

7381a. *Exposé succinct de produits* du règne végétal et animal dans le canton *d'Enghien*, province de Hainaut, ... Bruxelles 1819. Oct. (*Expos. prod. Enghien*).
Publ.: 1819, p. [1]-176, [1, err.]. *Copies*: BR, KNAW.

Parmentier, Paul Evariste (1860-?), French botanist; Dr. phil. Besançon 1898; from 1901 professor of botany at the University of Besançon; director of the Station agronomique de Franche-Comté. (*P. Parm.*).

HERBARIUM and TYPES: Unknown. (Some material collected by "P. Parmentier" in CGE and P). – Some letters at G.

BIBLIOGRAPHY and BIOGRAPHY: Barnhart 3: 51; BJI 1: 43; BL 2: 113, 699; CSP 17: 713; Hirsch p. 225; IF p. 723; IH 2: (in press); Kew 4: 237; Langman p. 566; LS 19832, 20997; MW p. 378; Rehder 1: 253, 5: 650; Tucker 1: 540.

BIOFILE: Anon., Bull. Acad. Géogr. bot. 9: 12. 1900, 11: 9-10. 1902, 13: 9. 1904, 19: 8. 1909; Hedwigia 39: (71). 1900(prix Thore); Nat. Nov. 24: 69. 1902 (app. Besançon).

7382. *Contribution à l'étude du genre Pulmonaria* ... Besançon (Imprimerie Dodivers et Cie, ...) 1891. Oct. (*Contr. étud. Pulmonaria*).
Publ.: Aug-Oct 1891 (p. 4: Dec 1890; Bot. Centralbl. 27 Nov 1891;ÖbZ Dec 1891; Nat. Nov. (1) 1891, Bot. Zeit. 26 Feb 1892), p. [1]-24. *Copies*: G, MO. – Preprinted from Mém. Soc. Émul. Doubs 6: 185-206. 1892.

7383. *Histologie comparée des Ébénacées* dans ses rapports avec la morphologie et l'histoire généalogique de ces plantes ... Paris (G. Masson, éditeur ...) 1892. Oct. (*Histol. Ébén.*).
Publ.: 1892 (p. 2: Apr 1892; Nat. Nov. Jan(1) 1893; Bot. Zeit. 1 Jun 1893, p. [i-iii], [1]-155, *pl. 1-4* (uncol., auct.). *Copies*: B, BR, G, MO, NY. – Issued as Ann. Univ. Lyon 6 (fasc. 2) 1892. – The MO copy has a cover with the imprint Paris (Librairie J.-B. Baillière et fils ...), Lyon (A. Rey, ...) 1892.
Ref.: Roth, E., Bot. Jahrb. 17 (Lit.): 32-33. 29 Sep 1893.

7383a. *Flore nouvelle de la chaine jurassique* & de la Haute-Saone à l'usage du botaniste herborisant ... Autun (Imprimerie Dejussieu père et fils) 1895. Oct. (*Fl. nouv. chaine juras.*).

Publ.: 1895 (p. 3: Nov 1894; Bot. Centralbl. 27 Dec 1895; Nat. Nov. Mar(2) 1896), p. [i-iv], [1]-307. *Copies*: B, BR, G, NY. – Bull. Soc. Hist. nat. Autun 7: 125-431. 1895.
Ref.: Malinvaud, E., Bull. Soc. bot. France 42: 704. 1895 (rev.).

7384. *Recherches sur les Épilobes de France* ... [Extrait de la Revue générale de Botanique ... 1896]. Oct.
Publ.: 1896, p. [1]-28, *pl. 1-3*. *Copy*: G. – Reprinted and to be cited from Rev. gén. Bot. 8: 23-50, *pl. 1-3*. 1896.

7385. *Recherches anatomiques et taxinomiques sur les Onothéracées et les Haloragacées* [Paris (G. Masson, ...) 1897]. Oct. (*Rech. Onothér.*)
Publ.: Jun-Aug 1897 (copy reprint rd. at BR 15 Aug 1897; Nat. Nov. Jun(2) 1897), p. [65]-149, *pl. 1-6* (uncol. liths auct.) *Copies*: B, BR, G. – Preprinted with special cover but original pagination from Ann. Sci. nat., Bot. ser. 8. 3(3-6): [65]-149, *pl. 1-6*. Sep 1897 (cover of issue: "publié en septembre 1897"), issue rd. at USDA 3 Nov 1897).

7386. *Recherches anatomiques et taxinomiques sur les Rosiers* ... [Paris (Masson et Ce. ...) 1898]. Oct.
Publ.: Aug 1898 (date on cover journal; BR rd. 7 Oct 1898. Nat. Nov. Sep (1) 1898), reprinted with special cover but original pagination from Ann. Sci. nat., Bot. ser. 8. 6(1-3): [1]-175, *pl. 1-8*. Aug 1898. *Copies*: B, BR, MICH.

7387. *Leçons de botanique* appliquée à l'horticulture et notions d'horticulture pratique ... Paris (Vigot frères, éditeurs ...) 1924. Oct. (*Leçons bot.*).
Publ.: 1924 (p. xii: Jun 1923), p. [i]-xii, [1]-392. *Copy*: USDA.

Parnell, Richard (1810-1882), British ink manufacturer, ichthyologist and botanist, born at Bramford Speke, Devon; studied medicine at Edinburgh, Paris and London; died in Edinburgh; co-founder (1836) of the Edinburgh botanical Society. (*Parn.*).

HERBARIUM and TYPES: E (e.g. first set of grasses; see Cleghorn (1886) p. 7); further material at CGE, K, LINN, LIV.

BIBLIOGRAPHY and BIOGRAPHY: Barnhart 3: 51; BB p. 237; BM 4: 1524; CSP 4: 763-764, 17: 714; Desmond p. 480; Hortus 3: 1201 ("Parn."); IH 2: (in press); Jackson p. 239, 246; Kew 4: 237; NI 1492; PR 6958-6959 (ed. 1: 7786).

BIOFILE: Allibone, S.A., Crit. dict. Engl. lit. 1510. 1870.
Anon., Bot. Centralbl. 13: 72. 1883 (d.); Bot. Not. 1882: 30 (d.).
Bridson, G.D.R. et al., Nat. hist. mss. res. Brit. isl. 143.21, 143.22. 1980.
Britten, J., J. Bot. 21: 30-31, 382. 1883.
Cleghorn, H., Trans. Bot. Soc. Edinburgh 16: 6-8. 1886 (obit., d. 28 Oct 1882).
Freeman, R.B., Brit. nat. hist. books 273. 1980.
Hedge, I.C. & Lamond, J.M., Index coll. Edinb. herb. 117. 1970.
Jackson, B.D., Bull. misc. Inf. Kew 1901: 51 (pl. K).
Kent, D.H., Brit. herb. 72. 1957.
Trail, J.W.H., Scott. Natural. 7: 43. 1884 (obit.).

7388. *The grasses of Scotland* ... Edinburgh (William Blackwood and Sons), London (id.) 1842. Oct. (*Grass. Scotland*).
Publ.: Oct-Dec 1842 (p. vii: 26 Sep 1842), p. [i]-xxi, [1], [1]-152, err. slip, *pl. 1-66* (uncol. auct.). *Copies*: BR, US. – Also included in *Grass. Britain*.

7389. *The grasses of Britain* ... Edinburgh (William Blackwood and sons), London (id.) [1842]-1845. Oct. (*Grass. Britain*).
Publ.: a combination of the *Grass. Scotland* with additional *Grasses of Britain*, 1845 (p. vi: 1 Mar 1845; Bot. Zeit. 29 Aug 1845), p. [i]-xxvii, [1842 part:] [i]-xxi, [i], [1]-152, *pl. 1-66*, [1845 part continued:] [153]-310, [1, err.], *pl. 67-142* (uncol. liths. auct.). *Copies*: BR, G, MO, NY.

Parodi, Domingo (1823-1890), Italian-born pharmacist; to Montevideo 1833; pharmacist ib. 1843; in Paraguay as physician and pharmacist 1853-1867; returned to Montevideo 1869, active as pharmacist until 1879; in Buenos Aires from 1878, Dr. pharm. ib. 1881, lecturer in pharmacy at the Faculty of medical sciences from 1884. (*D. Parodi*).

HERBARIUM and TYPES: Material at BAF and K.

BIBLIOGRAPHY and BIOGRAPHY: Barnhart 3: 51; BL 1: 253, 312; BM 4: 1524; CSP 12: 558, 17: 714; IH 2: (in press); Jackson p. 376; Kew 4: 237; KR p. 506 sub Munck af Rosenschöld); Rehder 5: 650; Tucker 1: 540.

BIOFILE: Anon., Anal. Soc. Ci. Argent. 29: 44-45. 1890; Bol. Acad. nac. Ci. Cordoba 19(2): 324-325. 1913 (bibl.); Bot. Not. 1909: 60, 82 (on plagiarism as reported by Hassler; note by Malme).
Hassler, E., Bull. Herb. Boissier ser. 2. 8: 979-980, 985-986. 1908 (1909) (on plagiarism, see note below).
Jackson, B.D., Bull. misc. Inf. Kew 1901: 51 (171 spec. at K).
Malme, G.O., Sv. bot. Tidskr. 3(1): (1)-(4). 1909 (on P's use of Munck af Rosenschöld's notes and material).

EPONYMY: *Neoparodia* Petrak et Ciferri (1932); *Ophioparodia* Petrak et Ciferri (1932); *Parodia* Spegazzini (1923); *Parodiella* Spegazzini (1880); *Parodiellina* Hennings ex Arnaud (1918); *Parodiellina* Viégas (1944); *Parodiellinopsis* Hanford (1946); *Parodiodia* A.C. Batista (1960); *Parodiopsis* Maublanc (1915); *Pseudoparodia* Theissen et H. Sydow (1917); *Pseudoparodiella* F.L. Stevens (1927).

NOTE: See Hassler (1908) and Malme (1909) on Parodi's plagiarisms. They demonstrate that D. Parodi acquired herbarium specimens and manuscript notes of Eberhard Munck af Rosenschöld (1811-1868) after the latter had been murdered in Paraguay. The main conclusions are:
(1) that the *Plantas usuales* del Paraguay were copied from C.F.P. von Martius, *Syst. mat. med. bras.* (TL-2/5541) and *Pflanzennamen in der Tupisprache* (TL-2/5548), supplemented in the second edition by material taken from Hieronymus, *Pl. diaph. fl. argent.* 1882 (TL-2/2751).
(2) that the *Contribuciones* are copies of the notes left by Munck with the abbreviated name of the latter ("Ros.") changed to Parodi.
(3) that Domingo Parodi "était dépourvu des connaissances botaniques les plus élémentaires".
The plants at BAF and K are probably all collections made by Munck. – Parodi, in Anal. Soc. Ci. Argentina 4: 81. 1877 states that "Everardo Munk ... desapareció ... durante veinte años ... y quedando ignorada hasta su tumba!" The anonymous author of D. Parodi's obituary in the Anal. Soc. Ci. Argent. (1890) states that Parodi "ha dejudo obra que no morirá" and of "su espíritu recto y sereno..." "Tan cierto es que la honorabilidad y el saber se sobreponen à todas las acciones humanas!"

7390. *Notas sobre algunas plantas usuales del Paraguay* de Corrientes y de Missiones ... Buenos Aires (Imprenta de Pablo E. Coni, ...) 1877. Oct.
Ed.1: Aug 1877-Mar 1878 (in journal), p. [1]-61. *Copies*: FI, MO, USDA. – Reprinted and to be cited from Anal. Soc. Ci. Argentina 4: 80-86. Aug; 124-135, Sep; 12-217. Oct; 243-251, Dec 1877, 5: 33-45. Jan-Mar 1878.
Ed. 2: 1886, [i]-xxvii, [1]-123. *Copy*: MICH.
Note (1): See Hassler (1908) and BL (1:253) for a note on the origin of the data listed by Parodi. The treatise is a plagiarism of Martius, *Syst. mat. med. bras.* and Martius, *Pflanzen-Namen in der Tupi-Sprache*. Ed. 2 contains in addition elements from Hieronymus, *Plantae diaphoricae*.
Note (2): See Bull. Soc. bot. France 25(bibl.): 171. 1879 for a "Flora de la republica Argentina y Paraguay", Buenos Aires (Pablo E. Coni) 1877, not seen by us, containing the *Notas* and fasc. 1-2 of the *Contribuciones* (see below).

7391. *Contribuciones à la flora del Paraguay* ... Buenos Aires (Imprento de Pablo E. Coni, ...) 1877-1879, 4 fasc. Oct. (*Contr. fl. Paraguay*).
Fasc. 1 (Convolvuláceas): 1877 (J. Bot. Jan 1878), p. [1]-32, [1, ind.].
Fasc. 2 (Urtic.-Nyctag.): Jan-Mar 1878, p. [35]-64, [1, ind.]. Also published in Anal. Soc. Ci. Argentina 5: 87-96, 152-162, 206-214. Jan-Mar 1878.
Fasc. 3 (Laur.-Piperac.): 1878 (Nat. Nov. Mar(2) 1879, p. [65]-106, [1, ind.]. Also published in Anal. Soc. Ci. Argentina 5: 260-277, 315-320. Jan-Mar 1878, 6: 37-47, 90-96. Apr-Jun 1878.
Fasc. 4 (Mirtáceas): Jan-Jun 1879 (J. Bot. Aug 1879), p. [109]-159, [160, ind.]. Also published in Anal. Soc. Ci. Argentina 7: 61-74, 114-122, 173-188, 213-224. Jan-Jun 1879.
Copies: FI(2,3), G(1), MO(1-4), NY(1-4), USDA(4). – See note above on plagiarism.

7392. *Apuntes sobre la familia de las Nictagíneas* ... Buenos Aires (Imprenta de Pablo E. Coni, ...) 1882.
Publ.: Dec 1882 (in journal; Nat. Nov. Aug(2) 1883; Bot. Centralbl. 3-7 Sep 1883); Bot. Zeit. 28 Sep 1883), p. [1]-18. *Copy*: USDA. – Reprinted, and to be cited, from Anal. Soc. Ci. Argentina 14: 255-270. Dec 1882.

7393. *Contribuciones à la flora del Paraguay* ... Familia de las *Amarantáceas*. Buenos Aires (Imprenta de "La Nacion" ...) 1892. (*Contr. fl. Paraguay, Amarant.*).
Publ.: 1892, p. [1]-183, [1]-6. Copy: USDA. – The Amarantaceae are new and occupy p. 3-19; p. 20-183 are the same as the Contribuciones of 1877-1879, p. 1-160.

Parodi, Lorenzo Raimundo (1895-1966), Argentinian botanist (esp. agrostologist); studied with L. Hauman at the Univ. of Buenos Aires, Ir. agron. ib. 1918; professor of botany at the Faculties of Agronomy in Buenos Aires (1926-1966) and La Plata (1922-1947) and at the Museo de La Plata (1934-1947). (*Parodi*).

HERBARIUM and TYPES: BAA (with his private library); other material in B, BAF, DPU, F, GB, GH, K, L, LP, NY, P, S, US, W. – Some correspondence in Smithsonian Archives and G.

NOTE: Not known to have been related to Domingo Parodi.

BIBLIOGRAPHY and BIOGRAPHY: Barnhart 3: 51; BJI 2: 130; BL 1: 233, 236, 312; Bossert p. 301; Hortus 3: 1201; IH 1 (ed. 4): 38, (ed. 6): 362, (ed. 7): 339, 2: (in press); Kew 4: 238-240; Langman p. 566; Lenley p. 321; Morren ed. 10, p. 142; Roon p. 85; TL-2/see A. Lorenz; Zander ed. 10, p. 699, ed. 11, p. 798.

BIOFILE: Anon., Bol. Soc. arg. Bot. 11: 67. 1966 (d.).
Boelcke, O., Bol. Soc. arg. Bot. 12: 1-6. 1968 (obit., bibl. by A. Burkart, portr., b. 23 Jan 1895, d. 21 Apr 1966), 17: 350-354. 1968.
Braun-Blanquet, J., Vegetatio 20: 394-395. 1970 (obit.).
Burkart, A. y Cabrera, A.L., Darwiniana 5: 9-19. 1941 (portr., bibl.).
Burkart, A., Physis 26: 141-144. 1966, Taxon 16: 522-533. 1967 (bibl.); Bol. Soc. argent. Bot. 13: 350-354. 1971.
Garaventa H., A., Revista univ., Chile 52 (Anales 30): 167-175. 1967 (obit., bibl. Chilean publ.).
Hauman, L. & A. Castellanos, Physis 5: 274, 283. 1922 (bibl.).
Moldenke, H.N., Plant Life 2: 75. 1946 (1948).
Padilla, F.J., Kurtziana 3: 241-245. 1966 (obit., portr.).
Parodi, L.R., Bol. Soc. arg. Bot. 9: 1-68. 1961 (50 years bot. Argentina).
Steenis, C.G.G.J. van, Fl. males. Bull. 21: 1379. 1966, 23: 1675. 1969.
Stieber, M.T., Huntia 3(2): 124. 1979 (notes, mss. A. Chase).
Verdoorn, F., ed., Chron. bot. 2:27, 28, 35, 66, 68, 132. 1936, 3: 53, 54, 55, 327, 331, 336. 1937, 4: 563. 1938, 5: 176, 244, 306. 1939, 6: 21, 36, 45, 172. 1940, 7: 19, 43, 44, 91, 102, 235, 263, 351. 1943.

COMPOSITE WORKS: Editor (permanent director) of the *Revista Argentina de Agronomía*, 1934-1962.

MEMORIAL PUBLICATION: Bol. Soc. arg. Bot. 12, 1968, *Suma agrostológica* editada en memoria de L.R.P. por O. Boelcke, A. Burkart y A.L. Cabrera.

EPONYMY (genera): *Lorenzochloa* J.R. Reeder et C.G. Reeder (1969); *Parodianthus* N.S. Tronosco (1941); *Parodiella* J.R. Reeder et C.G. Reeder (1968); *Parodiodendron* Hunziker (1969); *Parodiodoxa* O.E. Schulz (1929); (journal): *Parodiana*. Buenos Aires. Vol. 1-x. 1981-x.

7394. *Chlorideas* de la Republica Argentina ... Buenos Aires (imprenta y casa editoria "Coni" ...) 1919. Oct.
Publ.: Dec 1919 (in journal), p. [1]-107. *Copy*: MO. – Reprinted and to be cited from Revista Fac. Agron. Vet., Buenos Aires 2: 233-335. 1919.

7395. *Gramíneas bonaerenses* clave para la determinación de los géneros ... Extracto de la Revista des Centro Estudiantes de Agronomía y Veterinaria, Universidad de Buenos Aires no 120, pág. 17 y sig. y no 121, pág. 118 y sig., julio y agosto de 1925. Buenos Aires (Imprenta de la Universidad) 1925. Oct. (*Gram. bonaer.*).
Ed. 1: not seen; issued as *Clave para la determinacion de los generos de Gramineas ... de Buenos Aires, 1916.*
Ed. 2: Jul-Aug 1925, p. [1]-73. *Copy*: B. – Reprinted from Revista Centro Est. Agron. Vet., Buenos Aires 120: 17-42. Mai-Jun, 121: 111-155. Jul-Aug 1925.
Ed. 3: Oct 1939 (p. 98: 30 Sep 1939; see also ed. 5, p. 7), p. [1]-98. *Copies*: MICH, NY (dedicated by author 8 Nov 1939), USDA.
Ed. 4: 27-31 Dec 1946 or early 1947 (p. 112: printing finished 27 Dec 1946; t.p. 1946), p.[1]-112. *Copy*: NY.
Ed. 5: 20 Oct-31 Dec 1958 (p. [143] printing finished 20 Oct 1958), p. [1]-142, [143]. *Copies*: FAS, FI, NY. – Buenos Aires (ACME Agency, ...).
Ed. 5, reprint: Dec 1964 (p. [143] printing finished 30 Nov 1964), p. [1]-142, [143]. *Copy*: MICH.

7396. *Revisión de las gramíneas argentinas del género "Diplachne"* ... Buenos Aires (Imprenta de la Universidad) 1927. Oct.
Publ.: Oct 1927 (in journal), p. [1]-24, [25, index]. *Copy*: MICH. – Reprinted and to be cited from Revista Fac. Agron. Vet., Buenos Aires 6(2): 21-43. Oct 1927.

7397. *Revisión de las gramíneas argentinas del género "Sporobolus"* ... Buenos Aires (Imprenta de la Universidad) 1928. Oct.
Publ.: 30 Jul 1928 (date stamped on cover of reprint), p. [i], [115]-167, [168, index], *2 pl. Copy*: MICH. – Reprinted from Revista Fac. Agron. Vet., Buenos Aires 6(2): 115-168. 1928.

7398. *Ensayo fitogeográfico sobre el partido de Pergamino* estudio de la pradera pampeana en el Norte de la provincia de Buenos Aires (Imprenta de la Universidad) 1930. Oct.
Publ.: Oct 1930 (in journal, Nat. Nov. Feb 1931), p. [i], [65]-271, *pl. 1-16*. *Copies*: B, G, NY. – Reprinted with special t.p. and original pagination from Rev. Fac. Agron. Vet., Buenos Aires 7:[65]-271, *pl. 1-16*. 1930. To be cited from journal.

7399. *Contribución al estudio* de las gramíneas *del género Paspalum* de la flora Uruguaya ... Buenos Aires (Imprenta y Casa Editora "Coni" ...) 1937. Oct.
Publ.: 21 Dec 1937 (p. 250), p. [i], [211]-250, *1 pl. Copy*: U. – Reprinted and to be cited from Revista Museo La Plata ser. 2. 1 (Bot.): 211-250. 1937.

7400. *Estudio crítico* de las Gramineas austral.-americanas del genero *Agropyron* ... Buenos Aires (Imprenta y Casa Editora "Coni" ...) 1940. Oct.
Publ.: 27 Mai 1940 (p. 63), p. [i], [1]-63, *pl. 1-12. Copy*: U. – Reprinted and to be cited from Revista Museo La Plata ser. 2. 3(Bot.): 1-63. 1940.

Parrique, Géraud ("Frère Gasilien") (1851-1907), French school teacher, clergyman, lichenologist and bryologist at Sorbiers (Loire) (*Parrique*).

HERBARIUM and TYPES: Material at E, H, PC and W. – The original herbarium was owned by Bouly de Lesdain at Dunkerque [Duinkerken], France, but destroyed in 1940. After Frère Gasilien changed his name back to Parrique, he formed a new herbarium (Gyelnik 1939). Its location is not known to us.

BIBLIOGRAPHY and BIOGRAPHY: Barnhart 3: 51; GR cat. p. 69; IH 2: (in press).

BIOFILE: Anon., Bull. Soc. bot. France 54: 128. (will issue a series of *Parmelia exsicc.*), 752 (obit.). 1907.
Camus, F., Bull. Soc. bot. France 54: 752. 1907 (obit. giving many words but no information).
Gyelnik, V.K., Ann. Hist.-nat. Mus. natl. Hung., Bot. 32: 204. 1939 (on herb.).
Hedge, I.C. & Lamond, J.M., Index coll. Edinb. herb. 117. 1970.
Husnot, F., Rev. bryol. 31: 52. 1904 (lichens for sale).
Laundon, J.R., Lichenologist 11: 15. 1979 (herb. destr. 1940).
Roux, C., Ann. Soc. bot. Lyon 35: 145, 148. 1911 (bibl.).

7401. *Cladonies de la flore de France* ... Bordeaux (Ancienne Imprimerie J. Durand, L. Delbrel & Cie., successeurs ...) 1905. Oct.
Publ.: 1905, p. [1]-76. *Copy*: FH. – Reprinted from Actes Soc. Linn. Bordeaux 59: [45]-124. 1904 (1905). To be cited from journal.

Parrot, Johann Jacob Friedrich Wilhelm (1792-1841), German-born (Baden) physician and botanist; studied at Dorpat (1809-1811); to the Crimea and the Caucasus 1811-1812; military physician with the Russian army 1815; professor of medicine at Dorpat 1821; visited the Pyrenees in 1824, and climbed Ararat 1829; to the North Cape 1837. (*Parrot*).

HERBARIUM and TYPES: Unknown. Some material in H, LE and MW.

BIBLIOGRAPHY and BIOGRAPHY: Backer p. 426; Barnhart 3: 51; BM 4: 1525; CSP 4: 764-766; Herder p. 56; IH 2: (in press); Kew 4: 240; Lasègue p. 341, 399, 422, 429, 503; Moebius p. 378; PR ed. 1: 7787; Rehder 5: 650; Tucker 1: 540.

BIOFILE: Anon., Bot. Zeit. 10: 726. 1852.
Embacher, F., Lexikon Reisen 225. 1882.
Kukkonen, I., Herb. Christian Steven 77. 1971.
Lindemann, E. von, Bull. Soc. imp. Natural. Moscou 61: 53. 1886.
Poggendorff, J.C., Biogr.-lit. Handw.-Buch 2: 367. 1863 (b. 14 Oct 1792, d. 15 Jan 1840).
Pritzel, G.A., Linnaea 19(3): 459. 1846.

COMPOSITE WORKS: with M. von Engelhardt, *Reise in die Krym und den Kaukasus*, 2 vols., 1815.

EPONYMY: *Parrotia* C.A. Meyer (1831) is named for this author according to Backer, as well as the derived name *Parrotiopsis* (Niedenzu) Schneider (1905). *Parrotia* Walpers (1849) is an *orth. var.* of *Barrotia* Gaudichaud-Beaupré (1841).

7402. *Reise zum Ararat von Dr. Friedrich Parrot* ... unternommen in Begleitung des Herren ... Wassili Fedorov, ...Maximil. Behaghel von Adlerskron, ... Julius Hehn und Karl Schiemann ... Berlin (In der Haude und Spenersche Buchhandlung ...) 1834, 2 vols. (*Reise Ararat*).
1: 18-23 Aug 1834 (Hinrichs), map, p. [i-iv], [1]-262, [1, err.], *4 pl.*, botany on p. 180-188.
2: 18-23 Aug 1834 (id.), p. [i, iii], [1]-198, [i, err.,], *3 pl.*, no botany. *Copies*: G, H (Univ. Libr.), HH, L (Univ. Libr.).
English: 1846, p. [i]-xi, [12]-389. PH. – *Journey to Ararat* ... translated by W.D. Cooley. New York (Harper & brothers, ...) 1846.

Parry, Charles Christopher (1823-1890), English-born American botanist and explorer; to the United States 1832; Dr. med. Columbia 1846; with the Mexican Boundary Survey 1849-1861; exploring the Rocky Mountains 1861-1867; with the Pacific Railroad Survey 1867-1869; with the U.S. Dept. of Agriculture 1869-1872; from then on free-lance explorer in Utah, Nevada, California and Mexico; "king of Colorado Botany" (J.D. Hooker). (*Parry*).

HERBARIUM and TYPES: ISC (ca. 18.000; formerly at Davenport Academy of Natural Sciences); other important set at US; further material at BCL, BM, BRU, C, CM, DS, E, F, FH, FI, G, GH, IA, K, LE, M (lich.), MICH, MO, NA, NEB, NY, OXF, P, P-DU, PH, QPH, SD, UC, US (first set Mexic. Boundary Survey), YU. The herbarium left by Parry at his death (deposited in the Davenport Acad.), was arranged for sale by Mrs. E.R. Parry (Bot. Gaz. 16: 297). 1891; see also E.C. Parry 1891, for a catalogue) and contained 18.000 specimens. It was acquired by Iowa Agricultural College (US $ 5000.=). The herbarium at Davenport (Bot. Gaz. 15: 68. 1890) was the same as that acquired by ISC, and must have included a major early collection (erroneously said to have numbered 30.000 specimens) deposited at Davenport in the course of 1878 (Bull. Torrey bot. Club 6: 280. 1878, see also Parry (1878)). – Parry published only a fraction of the novelties discovered by him; his collections were worked up by J. Torrey (duplicates of his coll. of Mexican Boundary Survey 1849-1852 are therefore at NY), G. Engelmann, Asa Gray, E.L. Greene and S. Watson. – We are grateful to Mark L. Gabel (Dept. of Botany, Iowa State Univ.), Donald G. Herold (Davenport Public Museum, Iowa), Lyman Benson and Joseph Ewan for their help in establishing the history of the herbarium. – Letters by Parry are e.g. at G, ISC, MO, ND, and in the Smithsonian Archives.

BIBLIOGRAPHY and BIOGRAPHY: Backer p. 426; Barnhart 3: 52; BB p. 237; Bossert p. 301; Clokie p. 221; CSP 4: 767, 8: 565, 10: 994, 12: 358, 17: 717; DAB 14: 261-262 (by W.L. Jepson, F.;"best available biography" Ewan); Desmond p. 480; Ewan (ed. 1): see below, (ed. 2): 168-169; Herder p. 225, 300, 318; Hortus 3: 1201 ("Parry"); IH 1 (ed. 6): 362, (ed. 7): 339; 2: (in press); Jackson p. 589; Kew 4: 240-241; Langman p. 566-567; Lenley p. 465-466; LS 19835-19836; ME 1: 213, 217-218, 3: 635, 737; Morren ed. 10, p. 121; NW p. 54; Pennell p. 616; PH 364; PR 2682; Rehder 5: 650-651; TL-1/1314; TL-2/see E.M. Durand, W.H. Emory, E. Hall; Tucker 1: 13, 540-541; Urban-Berl. p. 381; Zander ed. 10, p. 699, ed. 11, p. 798.

BIOFILE: Anon., Bot. Centralbl. 1/2: 736. 1880 (to California and Oregon with George Engelmann), 44: 96. 1890(d.), 57: 236. 1894 (herb. ISC); Bot. Gaz. 15: 66-68. 1890 (b. 28 Aug 1823, d. 20 Feb 1890); Bot. Not. 1891: 46 (d.); Bull. Torrey Bot. Club 21: 82. 1894 (herb. acquired by ISC); Davenport Morning Democrat 20 Feb 1890 (well-informed original account; copy in Davenport Public Library); Nat. Nov. 12: 219. 1890 (d.); ÖbZ 40: 214. 1890 (d.).
Barnhart, J.H., J. New York Bot. Gard. 28: 134. 1927 (b. 28 Aug 1823, d. 20 Feb 1890).
Blankinship, J.W., Montana Agric. Coll. Sci. Studies Bot. 1: 8. 1904.
Britton, N.L., Bull. Torrey bot. Club 17: 74-75. 1890.
Candolle, Alph. de, Phytographie 439. 1880 (herb. Davenport, dupl. G; HH).
Clarke, J.M., James Hall of Albany 274, 275. 1923.
Daniel, T.F., Taxon 30(1): 48. 1981 (Mexican Boundary Survey material directly to Torrey).
Engelmann, G., West. Amer. Sci. 7: 271. 1891 (letter to Parry).
Ewan, J., Trail & Timberline 268, 1941, Rocky Mt. natural. 34-44, 124, 125, 126, 278-279, 354. 1950 (extensive details; bibl., biogr., portr.), ed. 2: 168-169; *in* Century Progr. nat. Sci. 57. 1955.
Ewan, J. et al., Leafl. W. Bot. 7(3):72. 1953, Short hist. bot. U.S. 10, 43, 47. 1969.
Geiser, S.W., Field and Laboratory 27: 168-169. 1959.
Graham, E.H., Ann. Carnegie Mus. 26: 18. 1937.
Gray, J.L., Letters A. Gray 468, 628, 636. 1893.
Hedge, I.C. & Lamond, J.M., Index coll. Edinb. herb. 117. 1970.
Hemsley, W.B., Biol. Centr. Amer. 4: 131. 1887.
House, H.D., N.J. State Mus. Bull. 328: 14. 1942.

Humphrey, H.B., Makers N. Amer. Bot. 196-198. 1961.
Jackson, B.D., Bull. misc. Inf. Kew 1901: 51. (pl. K).
Kelly, H.A., Some American Medical botanists 180-186. 1914.
Knobloch, I.W., Plant coll. N. Mexico 52. 1979.
Knowlton, C.C., Phil. Soc. Washington 12: 497-500. 1895 (obit.).
Lemmon, J.G., Pacific Rural Press 39: 385. 12 Apr 1890.
Leon, N., Bibl. bot.-mexic. 357. 1895.
McKelvey, S.D., Bot. Expl. Trans. Mississippi West 1290-1850, p. 1031-1038. 1955.
McVaugh, R., Edward Palmer 427 [index]. 1956.
Milner, J.D., Cat. portr. Kew 89. 1906.
Murray, G., Hist. coll. BM(NH) 1: 172. 1904 (pl. BM).
Orcutt, C.R., West Amer. Sci. 7: 1-5. 1890 (obit., portr.).
Pammel, L.H., Proc. Iowa Acad. Sci. 31: 49-50. 1924.
Parish, S.B., Plant World 12(7): (1-7, repr.). Jul 1909.
Parry, C.C., Proc. Davenport Acad. Sci. 2: 279-282. 1878 (descr. of his herb.).
Parry, E.C., Catalogue of the herbarium of the late Dr. Charles C. Parry of Davenport, Iowa, 82 p. 1891.
Preston, C.H. & Parry, Mrs. C.C., Biographical sketch of Charles Christopher Parry, Proc. Davenport Acad. Sci. 6: 35-52. 1897 (portr., bibl. by Mrs. Parry; "deficient in ... information of particular value to the naturalist" Ewan).
Reveal, L., in A. Cronquist et al., Intermountain Fl. 1: 57. 1972.
Rickett, H.W., Index Bull. Torrey bot. Club 76. 1955.
Rodgers, A.D., Amer. bot. 1873-1892, p. 334 (index) 1944; Bernard Eduard Fernow 32, 83, 96. 1951; John Torrey 347 (index). 1942.
Sayre, G., Mem. New York Bot. Gard. 19(3): 377. 1975.
Schofield, E.K., Brittonia 30: 404. 1978 (material Mexican Boundary Survey, 1850-1852, at NY).
Sutton, S.B., Charles Sprague Sargent 48, 85-88, 113, 121. 1970.
Thomas, J.H., Contr. Dudley Herb. 5(6): 58, 60. 1961 (portr.), Huntia 3: 14-15. 1969.
Urban, I., Symb. ant. 1: 138. 1898, 3: 98. 1902 (colls. S. Domingo).
White, C.A., Ann. Iowa ser. 3. 7: 413-424. 1906 (biogr., portr., bibl. by Mrs. Parry).
Wiggins, I., Fl. Baja Calif. 42. 1980.
Wittrock, V.B., Acta Horti Berg. 3(2): 180. 1903.

COMPOSITE WORKS: *Report on the United States and Mexican Boundary Survey* [dir. W.H. Emory], *Botany of the Boundary*, Washington 1859, introduction by C.C. Parry (see TL-1/1314 and TL-2 under Torrey, J.).

NOTE: Parry contributed many articles to newspapers, which are often still of interest but difficult of access; they are listed by Mrs. Parry (see C.H. Preston(1897) and C.A. White (1907)), e.g. the *San Francisco Bulletin* (set in California State Library), the *Chicago Evening Journal*, the Davenport journal *Davenport Gazette*, and *Davenport Daily Democrat*.

EPONYMY: *Neoparrya* Mathias (1929); *Parryella* Torrey et A. Gray ex A. Gray (1868).

7403. *Physiographical sketch* of that portion of the Rocky Mountain Range, at the head waters of South Clear Creek and east of Middle Park: with an enumeration of the plants collected in this district in the summer months of 1861 ... [From the American Journal of Science and Arts, vol. xxxiii, 1862]. Oct.
Publ.: 1862, p. [1]-22. *Copy*: G. – Reprinted and to be cited from Amer. J. Sci. 33: 231-243, 404-411. 1862. Continued ib. 34: 249-261, 330-341. 1862.

7404. *Botanical observations in Western Wyoming*, with notices of rare plants and descriptions of new species collected on the route of the North-western Wyoming Expedition under Captain W.A. Jones ... by Dr. C.C. Parry. Salem (Printed at the Salem Press) 1874. Oct.
Publ.: Apr 1874, p. [1]-25. *Copies*: FH, MO. – Reprinted and to be cited from the American Naturalist 8: 9-14. Jan 1874, 102-108. Feb 1874, 175-180. Mar 1874, 211-215. Apr 1874. – See also Ewan, ed. 1, p. 126.

7405. *Arctostaphylos*, Adans. Notes on the United States Pacific Coast species, from recent observation of living plants, including a new species from Lower California ... [From Proceedings Davenport Academy of Natural Sciences, vol. iv]. 1884. Oct.
Publ.: Jan-Feb 1884 (Bot. Gaz. Mar 1884). Reprinted without change of pagination from Proc. Davenport Acad. Sci. 4: 31-37. 1884.

7406. *Chorizanthe* by C.C. Parry [From Proceedings Davenport Academy of Natural Sciences, vol. iv.] Davenport, Iowa (Glass & Hoover, printers, Davenport) 1884. Oct.
Publ.: Apr1884 (NY copy rd 7 Mai 1884), p. [i, iii], 45-63. *Copies*: FH, NY. – Reprinted with special title page and original pagination from Proc. Davenport Acad. Sci. 4: 45-63. 1884. To be cited from journal.

7407. *Ceanothus*, L. A synoptical list, comprising thirty-three species, with notes and descriptions ... Reprinted from vol. V. Proceedings of the Davenport Academy of Sciences ... 1889. Oct.
Publ.: 9 Feb 1889, title reprint on recto of p. 162, 162-176. *Copy*: G. – Reprinted and to be cited from Proc. Davenport Acad. Sci. 5: 162-176. 1889.

7408. *Catalogue of the herbarium of the late Dr. Charles C. Parry* of Davenport, Iowa. Oquawka, Ills. (printed by H.N. Patterson) [1891]. Oct. (*Cat. herb. Parry*).
Authorship: Preface signed by Mrs. E.R. Parry.
Publ.: 1891 (pref. Jul 1891; Bot. Centralbl. 28 Oct 1891), p. [1]-82. *Copies*: FH, MO, NY.

Parry, William Edward (1790-1855), British explorer; made three expeditions to the Arctic (north-west passage), 1819-20, 1821-23, 1824-25; DCL Oxford 1829; knighted 1829; rear admiral 1852. (*W. Parry*).

HERBARIUM and TYPES: BM (first voyage) and K (second-fourth voyages); other material at BRISTM, C, DBN, E, HAMU, LY, MANCH, NA, NY (mosses), OXF, PH, S (collections made by Parry as well as by his associates W.H. Hooper, A. Fisher and R. McCormick.

BIBLIOGRAPHY and BIOGRAPHY: AG 1: 208; Backer p. 426; Barnhart 3: 52; BB p. 237; BM 4: 1525-1526; Clokie p. 221-222; CSP 4: 767, 12: 558; Desmond p. 481; DNB 43: 392-393; IH 2: (in press); Kew 4: 241; Lasègue p. 84, 324, 326, 373, 502; PR 1234, 6960; Rehder 1: 308; TL-1/163, 164, 951-953; TL-2/831-832, see R. McCormick.

BIOFILE: Allan, Mea, The Hookers of Kew 112. 1967.
Allibone, S.A., Crit. dict. Engl. lit. 1515. 1870 (bibl.).
Anon., Bot. Zeit. 13: 568. 1855 (brief obit.); Flora 38: 528. 1855 (obit.); Hertha 12: frontisp. 1828 (portr.).
Beechey, F.W., J.R. Geogr. Soc. 26: clxxxii-clxxxv. 1856 (obit.).
Bridson, G.D.R. et al., Nat. hist. mss. res. Brit. Isl. 75.49, 75.53. 1980.
Britton, N.L., J. New York Bot. Gard. 8: 31. 1907 (Parry mosses at NY).
Embacher, F., Lexik. Reisen 225-226. 1882.
Hedge, I.C. & Lamond, J.M., Index coll. Edinb. herb. 117. 1970.
Huxley, L., Life letters J.D. Hooker 1: 15, 166. 1912.
Jackson, B.D., Bull. misc. Inf. Kew 1901: 51 (pl. K).
Meyer, F.G. & Elsasser, S., Taxon 22: 375-404, 386. 1973 (material in NA).
Murray, G., Hist. coll. BM(NH) 1: 172. 1904 (pl. BM).
Nissen, C., Zool. Buchill. 311. 1969.
Parry, E., Memoirs of rear-admiral Sir W. Edward Parry, New York 1858, xii, 341 p. (by his son; biogr., portr.).
Poggendorff, J.C., Biogr.-lit. Handw.-Buch 2: 367-368. 1863 (b. 19 Dec 1790, d. 8 Jul 1855; bibl.).
Polunin, N., Bot. Canad. Eastern Arctic 1: 11-12, 20. 1940, J. Bot. 80: 81-94. 1942 (studies on Parry plants at Bristol and Durham); Rhodora 54: 45-55 (id. at MANCH), 213-216 (id. at Roy. Geogr. Soc.), 271-285 (id. at OXF). 1952; Canad. Field Natural. 64: 45-51. 1950 (id. at Scott Polar Research Inst. Cambridge).

Rodgers, A.D., William Starling Sullivant 175, 178. 1940.
Sayre, G., Bryologist 80: 515. 1977; Mem. New York Bot. Gard. 19(3): 378. 1975.Steere, W.C., Bryologist 74: 430. 1971 (bryoph. first voyage, 1819-20, not traced).

EPONYMY: *Parrya* R. Brown (1823); *Parryodes* S.M.H. Jafri (1957); *Parryopsis* Botschantzev (1955).

7409. *Journal of a voyage for the discovery of a North-West passage* from the Atlantic to the Pacific; performed in the years 1819-'20, in His Majesty's Ships Hecla and Griper ... with an Appendix containing the scientific and other observations ...The second edition. London (John Murray, ...) 1821. Qu. (*J. voy. N.-W. pass.*).
Ed. 1: Mai 1821 (P & W p. 435; Gent. Mag. Jun 1821), n.v. – A supplement to the Appendix was published in Jan 1824, with a contribution by R. Brown, for which see TL-1/164, TL-2/832 and below.
Ed. 2: 1821 (plates dated 21 Mar 1821) p. iii: Mai 1821, map, p. [i*-viii*], [i]-xxix, [xxxi], [1]-310 [1], *19 pl.* & maps, p. [i]-clxxix [= appendix]. Differs only in title page and in having a few corrections made in the text. *Copies*: MO, NY.
Dutch: 1822 (p. viii: 1 Sep 1822), p. [i]-xii, [1]-332, [334], 2 maps, 2 uncol. maps [no bot. app.]. *Copy*: Kon. Bibl. 's Gravenhage. – Reis ter ontdekking van eene noordwestelijke door vaart, uit de Atlantische in de Stille Zee, gedaan in de jaren 1819 en 1820 door de schepen The Hecla en The Griper, onder het bevel van W.E. Parry ... Amsterdam (Johannes van der Hey) 1822. Oct.

7410. *A supplement to the Appendix of Captain Parry's Voyage* for the discovery of a north-west passage, in the years 1819-20. Containing an account of the subjects of natural history. London (John Murray, ...) 1824. Qu. (*Suppl. App. Parry's Voy.*).
Publ.: 1 Jan-5 Feb 1824 (see TL-2/832), p. [i-viii], [clxxxi]-cccx, *pl. A-D* (plants), *1-2* (animals), by Franz Bauer. *Copies*: FI, MO, NY. – *The botany*, by Robert Brown (1773-1858), on p. cclix-cccx, was preprinted as *Chloris melvilliana* in 1823. For further details see nos. 831 and 832. The abbreviated reference to the botanical section is *J. voy. n.-w. passage, Bot.*, but reference should always be to the earlier *Chlor. melvill.*

7411. *Journal of a second voyage for the discovery of a North-West passage* from the Atlantic to the Pacific; performed in the years 1821-22-23, in his majesty's ships Fury and Hecla, under the orders of captain William Edward Parry, ... London (John Murray, ...) 1824. Qu. (*J. sec. voy.*).
Publ.: Mar 1824 (P & W; p. [iii]: Mar 1824; Gent. Mag. Apr 1824), frontisp., p. [i*-viii*], [i]-xxx, [xxxi], [1]-571, [572, err.], *37 pl.* & maps (uncol.). *Copies*: MO, NY.

7412. *Appendix to Captain Parry's Journal of a second voyage* for the discovery of a north-west passage from the Atlantic to the Pacific, performed in His Majesty's ships Fury and Hecla, in the years 1821-22-23 ...London (John Murray, ...) 1825. Qu. (*App. Parry J. sec. voy.*).
Publ.: 1825 (plates dated 1 Feb 1825; P & W Jan 1827), p. [i-iii], [1]-432, *2 pl. Copies*: MO, NY. – Botanical appendix by Professor [W.J.] Hooker on p. [381]-430, *2 pl.*

7413. *Journal of a third voyage for the discovery of a North-West Passage* from the Atlantic to the Pacific; performed in the years 1824-25 in his Majesty's ships Hecla and Fury, under the orders of Captain William Edward Parry, ... London (John Murray, ...) 1826. Qu. (*J. third voy.*).
Publ.: Aug 1826 (see TL-1), *1 pl.* p. [i]-xxvii, [xxviii], [1]-186, [1, note], appendix p. [1]-150, *4 [2?] pl.* (uncol.). *Copies*: MO, NY. *Botanical Appendix* by Professor [W.J.] Hooker, in app. p. [121]-131.

7414. *Narrative of an attempt to reach the North Pole*, in boats fitted for the purpose, and attached to his Majesty's ship Hecla, in the year mdcccxxvii, under the command of Captain William Edward Parry, ...London (John Murray, ...) 1828. Qu. (*Narr. attempt North Pole*).
Publ.: 1828 (p. xxii: 24 Mar 1827; plates dated Jan 1828, p. [iii]: Jan 1828, p. [i]-xxii, [1], [1]-229, *5 pl.*, map. *Copies*: MO, NY. – *Botanical appendix* by Professor [W.J.] Hooker on p. [207]-222.

Parsons, Frances Theodora (formerly Mrs. William Starr Dana; née Smith) (1861-1952), American popular writer on botany. (*F. Parsons*).

HERBARIUM and TYPES: Unknown.

BIBLIOGRAPHY and BIOGRAPHY: Barnhart 3: 52; Ewan (ed. 1): 169; Lenley p. 321.

BIOFILE: Clute, W.N., Fern Bull. 7: 74-75. 1899; 10: 20-21. 1902 (portr.).
Coulter, J.M., Bot. Gaz. 27: 484. 1899.
Ewan, J. et al., Short hist. bot. U.S. 99. 1969.
Tilton, G.H., Fern Lover's Comp. 212, 214. 1922 (n.v.).

EPONYMY: *Parsonsia* P. Browne (1756, *nom. rej.*) and *Parsonsia* R. Brown (1810, *nom. cons.*) were named in memory of James Parsons (1705-1770) an English medical doctor who studied seeds microscopically.

7415. *How to know the ferns* a guide to the names, haunts and habits of our common ferns .... Illustrated by Marion Satterlee and Alice Josephine Smith ... New York (Charles Scribner's Sons) 1899. (*How to know ferns*).
Publ.: Mai 1899 (p. vii: 6 Mar 1899; Bot. Centralbl. 25 Mai 1899; BG: 22 Jun 1899; Fern Bull. Jul 1899; Bot. Zeit. 16 Oct 1899; Nat. Nov. Mai (1) 1899), frontisp. p. [i]-xiv, [xv], 1-215, *pl. 1-42*. *Copies*: MO, US.
Ed. 2: 1899 (Hedwigia 10 Aug 1899); front., p. [i]-xiv, [xv], 1-215, *pl.1-42*. *Copy*: NY. – Second edition. New York (id.) 1899.

Parsons, Mary Elizabeth (1859-1947), American popular writer on wild plants, born in Chicago; to California as a young girl, first at San Rafael and later at Kentfield. (*M. Parsons*).

HERBARIUM and TYPES: Unknown.

BIBLIOGRAPHY and BIOGRAPHY: Kew 4: 242; Rehder 1: 327.

BIOFILE: Ewan, J. et al., Leafl. W. Bot. 7(3): 56. 1953.
Howell, J.T., *in* M.E. Parsons, Wild fl. Calif. 1955 ed., preface.
Parsons, M.E., Amer. Fern J. 51: 74. 1961 (copy letter to Maxon).

7416. *The wild flowers of California* their names, haunts and habits by Mary Elizabeth Parsons illustrated by Margaret Warriner Buck. San Francisco (William Doxey, ...) 1897. Oct. (*Wild fl. Calif.*).
Ed. 1: 1897 (p. xii: 15 Oct 1897; Bot. Centralbl. 14 Apr 1898 Nat. Nov. Apr(2) 1898), p. [i]-xlviii, [1]-410 (209 pl. in pagination). *Copies*: E, MO, UC, USDA.
Sixth Thousand: 1904 (J. NYBG Nov 1905), p. [i]-xlviii, [1]-411. *Copy*: USDA. San Francisco (Payot, Upham & Company).
Eighth Thousand: 1907 (preface Nov 1906), p. [i]-cvi, [1]-417. – Plates newly made following the destruction of the previous plates in the April 1906 earthquake. (Inf. R. Schmid).
Twelfth Thousand: 1912, p. [i*], [i]-cvi, [1]-417. *Copy*: USDA. San Francisco (Cunningham, Curtiss & Welch).
Sixteenth Thousand: 1916, p. [i*], [i]-cvi, [1]-417. *Copy*: MO.
Twentieth Thousand: 1921 (copy US signed by author 4 Dec 1921), p. [i*], [i]-cvi, [1]-417. *Copy*: US.
Twenty-seventh Thousand: 1955, col. frontisp. + 5 other col. pl., p. [i*], [i]-cvi, [1]-423. *Copies*: BR, UC. – California Academy of Sciences, San Francisco 1955, with an apppendix by J.T. Howell on p. 418-423, updating nomenclature and giving some biographical details. Illustrations by Margaret Warriner Buck (incl. the col. ills. here published for the first time).

Parsons, Thomas Henry (*fl.* 1926), British gardener, trained at Kew; from 1913 curator of the Peradeniya botanical Garden. (*T. Parsons*).

HERBARIUM and TYPES: Unknown.

BIBLIOGRAPHY and BIOGRAPHY: Barnhart 3: 52; Kew 4: 242.

BIOFILE: Anon., Bull misc. Inf. Kew 1913: 417 (app. Peradeniya).
Gager, C.S., Brooklyn Bot. Gard. Rec. 27: 183. 1938 (dir. Peradeniya from 1914).

7417. *An alphabetical list of plants in the Royal Botanic Gardens. Peradeniya, Ceylon* ... Colombo (Printed by H. Ross Cottle, ...) 1926. Oct. (*Alph. list pl. Peradeniya*).
Publ.: 1926 (J. NYBG Sep 1926), p. [i-iii], [1]-165. *Copy*: USDA.

Parvela, August Armas (1885-1953), Finnish botanist. (*Parvela*).

HERBARIUM and TYPES: H, OULU.

BIBLIOGRAPHY and BIOGRAPHY: Barnhart 3: 52; BL 2: 79, 81, 90, 91, 699; Collander p. 401-402 (bibl.).

BIOFILE: Kalela, A., Luonnon Ystävä 49: 42. 1945.
Metsävainio, K., Luonnon Tutkija 57: 83-84. 1953 (obit., portr., b. 23 Jan 1885).
Verdoorn, F., ed., Chron. bot. 2: 117. 1936.

7418. *Oulaisten pitäjän kasvisto* ... Helsinki 1921.
Publ.: 1921 (submitted 5 Feb 1921), p. [1]-78. *Copy*: USDA. – Published and to be cited as Acta Soc. Fauna Fl. fenn. 49(3).

Pascal, Diego Baldassare (1768-1812), Italian botanist of French origin; professor of botany and director of the botanical garden at Parma 1795-1802. (*Pascal*).

HERBARIUM and TYPES: GDOR.

BIBLIOGRAPHY and BIOGRAPHY: Rehder 1: 61; Saccardo 1: 122, cron. p. xxviii.

EPONYMY: *Pascalia* C.G. Ortega (1797).

Pascher, Adolf (1881-1945), Bohemia-born German botanist; studied at Praha with Beck von Mannagetta; Dr. phil. Praha 1905; habil. Praha 1909; assistant at the Praha botanical institute 1904- 1912, extra-ordinary professor of pharmaceutical botany from 1912, of systematic botany (ordinary) from 1927, director of the botanical institute and garden 1933-1945. (*Pasch.*).

HERBARIUM and TYPES: According to J. Gerloff (in litt.) Pascher had no herbarium. – Some letters and portrait at G.

BIBLIOGRAPHY and BIOGRAPHY: Barnhart 3: 52; BFM 2481; BJI 2: 130; BM 4: 1527, 8: 986; Bossert p. 302 (err. Pasher); Christiansen nos. 123, 154, 155; Futák-Domin p. 456; Hirsch p. 225; Kew 4: 243; LS suppl. 21006-21008; Maiwald p. 230, 237; Moebius p. 67, 69, 73, 83, 352; MW p. 378-379; Nordstedt suppl. p. 12; Stevenson p. 1253; TL-2/see indexes.

BIOFILE: Anon., Bot. Centralbl. 110: 320. 1909 (habil. Praha), 120: 448. 1912 (prof. bot. Praha); Bot. Zeit. 67: 200. 1909 (habil. Praha); Mycol. Centralbl. 1: 352. 1912 (prof. German University); Hedwigia 49: (51). 1909 (habil. Praha), 53: (96). 1912 (prof. bot. Praha), 53: (235). 1913 (id., corr.), 62: (147). 1921 (dir. bot. gard. Praha), 74: (141). 1934 (id., definitive); Nat. Nov. 31: 217. 1909 (habil.), 34: 512. 1912 (prof. syst. bot. Praha); Öster. bot. Z. 62: 351. 1912 (app. Praha), 68: 344. 1919 (dir. bot. gard. Praha), 77(4). 1928 (ord. prof.), 78(1): 96. 1929 (corr.), 82: 360. 1933 (reappointed dir. bot. gard. Praha).
Cupp, E.E., Mar. plankton diat. W. Coast N. Amer. 219. 1943.
Ewan, J. et al., Short hist. bot. U.S. 79. 1969.

Fritsch, F.E., Nature 162: 287-288. 1948 (obit.).
Hartmann, M. & J. Buder, eds., Arch. Protistenk. 98(3/4): i-ii, portr. 1953 (tribute, preceding his autobiography).
Junk, W., Rara 2: 166. 1926-1936.
Koster, J.T., Taxon 18: 556. 1969.
Pascher, A., Beih. Bot. Centralbl. 48(2): 317-332. Oct 1931 (review of his general ideas on algal classification); Arch. Protistenk. 98(3/4): iii-xxxii. 1953 (detailed autobiography and bibliography).
Pohl, F., Ber. deut. bot. Ges. 68a: 117-20. 1956; Generalreg. Beitr. Bot. Centralbl. 1-17, p. 13, 1931, 18-50, p. 12. 1942.
Verdoorn, F., ed., Biologia 1: 12. 1947 (death by suicide during last weeks of World War II); Chron. bot. 1: 31, 40, 112, 123, 138. 1935, 2: 34, 168. 1936, 4: 454. 1938, 6: 143, 162, 300. 1941, 7: 352. 1943.

COMPOSITE WORKS: (1) *Heterokonten*, x, ii, 1092 p., 1939, *in* Rabenhorst, L., *Krypt.-Fl. Deutschland* vol. 11.
(2) Editor *Bot. Centralbl.* (from 1924 with Uhlworm, from 1930 alone), vols. 40 (1924)-45 (1929) and 46 (1930)-62 (1944).
(3) Botanical co-editor *Archiv für Protistenkunde* 38 (1918)-97 (1945).

EPONYMY: *Pascherella* W. Conrad (1926); Pascheriella O. Korshikov (1928); *Pascherina* P.C. Silva(1959); *Pascherinema* G. De Toni (1936).

7419. *Uebersicht über die Arten der Gattung Gagea* [Separatabdruck... "Lotos" 1904. Nr. 5]. Oct.
Publ.: 1904 (p. 23: Jul 1904; Nat. Nov. Apr (1) 1905), p. [1]-23. *Copy*: B. – Reprinted and to be cited from S.B. Lotos, Prag 52(5): 109-131, 1904, preliminary notice ib. 51: 105-107. 1903 (1904).
Critical note: see A. Terracciano, *Per la priorità delle mie Gagearum novarum diagnoses*, Palermo 1904, p. [1]-7. *Copy*: G. – Reprinted from Boll. Soc. Orticola di Palermo 2(4), 31 Dec 1904.

7420. *Studien über die Schwärmer einiger Süsswasseralgen*. Stuttgart (E. Schweizerbartsche Verlagsbuchhandlung (E. Nägele)) 1907. Qu. (*Stud. Schwärm. Süssw.-Alg.*).
Publ.: Dec 1907 (Nat. Nov. Jan(2) 1908; Bot. Centralbl. 10 Mar 1908; ÖbZ Mar 1908), p. [ii-iii], [1]-115, [116, cont.], *pl. 1-8* (charts). *Copies*: B, BR, G, MO. – Issued as Bibliotheca botanica Heft 67, 1907.
Ref.: Heering, W.C.A., Bot. Centralbl. 108: 414-416. 1908.
Senn, G., Bot. Zeit. 6(2): 104-107. 16 Mar 1908.

7421. *Chrysomonaden aus dem Hirschberger Grossteiche*, Untersuchungen über die Flora des Hirschberger Grossteiches I. Teil ... Leipzig (Verlag von Dr. Werner Klinkhardt) 1910. Qu. (*Chrysomon. Hirschb. Grossteich.*).
Publ.: Oct-Nov 1910 (ÖbZ Nov-Dec 1910; Nat. Nov. Nov(2) 1910), p. [1]-66, *pl. 1-3* (p.p. col.). *Copy*: NY. – Issued as Band I, Monographien und Abhandlungen zur Internationalen Revue der gesamten Hydrobiologie und Hydrographie.

7422. *Flagellaten und Rhizopoden* in ihren gegenseitigen Beziehungen/Versuch einer Ableitung der Rhizopoden von Adolf Pascher ... Jena (Verlag von Gustav Fischer) 1917. Oct.
Publ.: 1917, p. [i-iii], [1]-88. *Copies*: BR, FH, NY. – Reprinted and to be cited from Arch. Protistenk. 38(1): 1-88. 1917.

7423. *Die Süsswasser-Flora* Deutschlands, Österreichs und der Schweiz bearbeitet von ... G. Beck von Mannagetta und Lerchenau ..., O. Borge..., J. Brunntaler ..., W. Heering ..., R. Kolkwitz ..., E. Lemmermann ..., J. Lütkemüller ..., W. Mönkemeyer ..., W. Migula ..., M. v. Minden ..., A. Pascher ..., V. Schiffner ..., A.J. Schilling ..., H. v. Schönfeld ..., C. Warnstorf ..., F.N. Wille ..., A. Zahlbruckner ... herausgegeben von Prof. Dr. A. Pascher (Prag) Heft 1 [-7, 9-12, 14-25] ... Jena (Verlag von Gustav Fischer) 1913-1936. Oct. (*Süsswasserflora*).

Authors: Beck von Mannagetta und Lerchenau, Günther (1856-1932) (no volumes issued specifically under his name).
Borge, Oscar Fredrik Andersson (1862-1938), 9.
Brunnthaler, Josef (1871-1914), 5.
Czurda, Viktor (1897-x), 9 (ed. 2).
Geitler, G.L. (1899-x), 12.
Glück, Christian Maximilian Hugo (1869-1940), 15.
Heering, Wilhelm Christian August (1876-1916), 6, 7.
Hustedt, Friedrich Carl (1886-1968), 10 (ed. 2).
Kolkwitz, Richard (1873-1956), [13].
Lemmermann, Ernst Johann (1867-1915), 1, 2, 5.
Lütkemüller, Johannes (1850-1913), [8].
Migula, Emil Friedrich August Walther (1863-1938), 11.
Minden, Max D. von (1871-?),
Mönkemeyer, Wilhelm (1862-1938), 14, (eds. 1-2)
Pascher, Adolf (1881-1945), 1, 2, 4, 5, 9, 11, 12.
Paul, Hermann Paul Gustav (1876-1964), 14 (ed. 2).
Schiffner, Victor Felix (1862-1944), 14 (eds. 1, 2).
Schiller, Josef (1877-1960), 11.
Schilling, August Jakob (1865-?), 3.
Schönfeld, H. von (*fl.* 1913), 10.
Warnsdorf, Carl Friedrich (1837-1921), 14.
Zahlbrückner, Alexander (1860-1938), [13].
Publ.: In Hefte; the series remained incomplete: 8 and 13 were never published. *Copies*: B, BR, C, G, MO, NY, U; IDC 1072.

Heft 1: Jan 1914 (BR rd 6 Feb 1914; J. Bot. Sep 1914; Nat. Nov. Feb(1) 1914), p. [i]-iv, [1]-138. – Flagellatae 1, Allgemeiner Teil (Pascher: 1-29), Pantostomatinae (Lemmermann: 30-51), Protomastiginae (id.: 52-121), Distomatinae (id.: 122-133).
Heft 2: Jul 1913 (Bot. Centralbl. 12 Aug 1913; BR rd 2 Sep 1913; ÖbZ Sep 1913; Nat. Nov. Jan(1) 1914; J. Bot. Dec 1913), p. [i]-iv, [1]-192, – Flagellatae 2 (Pascher: 1-114, Lemmermann: 115-185). 115-185). (Pascher's part also published as a separate, 1913. *Copy*: B).
Heft 3: Jul-Aug 1913 (BR rd 2 Sep 1913; ÖbZ Jul-Aug 1913; J. Bot. Dec 1913), p. [i]-iv, [1]-66. Dinoflagellatae (A.J. Schilling). – Hefte 3, 9, and 10 were already announced by ÖbZ Mai 1913.
Heft 4: Jan-Mar 1927 (p. iv: Dec 1926; MO rd. 28 Mai 1927; Nat. Nov. Apr 1927), p. [i]-vi, [1]-506.
Heft 5: Nov-Dec 1915 (p. iv: Oct 1915; BR rd 5 Jan 1915; Nat. Nov. Jan 1916), p. [i]-iv, [1]-250. – Chlorophyceae 2 (Pascher: 1-20, Lemmermann: 21-51, Brunnthaler: 52-205, Pascher: 206-236).
Heft 6: 1914 (p. iv: Dec 1921; ÖbZ Mai 1914; J. Bot. Sep 1914), p. [i]-iv, [1]-250. – Chlorophyceae 3 (Heering: 1-244).
Heft 7: Nov-Dec 1921 (p. iv: Oct 1912; Nat. Nov. Jan 1922), p. [i]-iv, [1]-103. – Chlorophyceae 4 (Heering: 3-99).
Heft 8: (planned: Desmidiaceae von J. Lütkemüller & R. Grönblad) not published.
Heft 9: Jul 1913 (p. iv: Dec 1912; Bot. Centralbl. 12 Aug 1913; BR rd. 2 Sep 1913; J. Bot. Dec 1913), p. [i]-iv, [1]-51. – Zygnemales (Pascher: 1-11, Borge: 12-48).
Heft 10: Jul-Aug 1913 (p. iv: Dec 1912; BR rd. 2 Sep 1913; J. Bot. 1913), p. [i]-iv, [1]-187. – Diatomeae (H. v. Schönfeldt).
Heft 11: 1 Jun-15 Jul 1925 (p. iv: Mai 1925; BR rd 18 Jul 1925; Nat. Nov. Aug 1925), p. [i]-iv, [1]-250. – Heterokontae (Pascher: 1-118), Phaeophyta (id.: 119-133), Rhodophyta (id. & J. Schiller: 134-206), Charophyta (Migula: 207-243).
Heft 12: Jul-Oct 1925 (p. iv: Mai 1925; BR rd. 20 Oct 1925; ÖbZ Jul-Oct 1925; Nat. Nov. Jan 1926), p. [i]-viii, [1]-481. Cyanophyceae (Geitler: 1-450), Cyanochloridinae (id. & Pascher: 451-463).
Heft 13: planned: Schizomycetes, R. Kolkwitz; Fungi, A. von Minden; Lichenes, A. Zahlbruckner), not published, but announced Mycol. Centralbl. 3(1): 43. 1913.
Heft 14: Jan 1914 (BR rd. 20 Jan 1914; NYBG Mar 1914; ÖbZ Jan-Feb 1914), p. [i]-iv, [1]-222. – Bryophyta (Warnstorf: 3-38 [Sphagnales], Mönkemeyer: 39-168 [Bryales], Schiffner: 169-224 [Hepat.]).

Heft 15: Mai-Jun 1936 (p. iv: Mar 1936; Nat. Nov. Aug 1936; MO copy rd. 1 Aug 1936, BR copy rd. 24 Jun 1936), p. [i]-xx, [1]-486. – Pteridophyten & Phanerogamen (H. Glück). – Review: E. Janchen, Österr. bot. Z. 85: 316-317. 1936.

Ed. 1. Arrangement by taxa.

Taxon	Author(s)	Heft	Pagination	Figures	Date
Bacillariales	Schönfeldt, H. v.	10	[1]2-173	*1-379*	1913
Bryales	Mönkemeyer, W.	14	[39]40-168	*A-H, 1-62*	1914
Bryophyta	Pascher, A.	14	[1]2		1914
Charophyta	Migula, W.	11	208-243	*I-XIV*	1925
Chlorophyceae-I	Pascher, A.	4	[1]2-19	*1-19*	1927
Chlorophyceae-II	Pascher, A.	5	[1]2-20		1915
Chlorophyceae(pp)	Pascher, A.	5	[206]207-229	*1-34*	1915
Chlorophyceae-III	Heering, W.	6	1-244	*1-384*	1914
Chlorophyceae-IV	Heering, W.	7	[3]4-99	*1-94*	1921
Cyanochloridinae	Geitler, L. & Pascher, A.	12	[451]452-463	*1-14*	1925
Cyanophyceae	Geitler, L.	12	[1]2-450	*1-560*	1925
Dinoflagellatae	Schilling, A. J.	3	[1]2-64	*1-69*	1913
Distomatinae	Lemmermann, E.	1	[122]123-133	*228-252*	1914
Eugleninae	Lemmermann, E.	2	[115]116-185	*181-398*	1913
Flagellatae-I	Pascher, A.	1	[1]2-29	*1-15*	1914
Flagellatae-I	Pascher, A.	2	[1]2-114	*1-180*	1913
Hepaticae	Schiffner, V.	14	[169]170-214	*1-158*	1914
Heterokontae	Pascher, A.	11	[1]2-118	*1-96*	1925
Pantostomatinae	Lemmermann, E.	1	[30]31-51	*16-67*	1914
Phaeophyta	Pascher, A.	11	[119]120-133	*1-7*	1925
Phanerogamen	Glück, H.	15	21-463	*8-258*	1936
Protococcales	Brunnthaler, J.	5	[52]53-205	*34-330*	1915
Protomastiginae	Lemmermann, E.	1	[52]53-121	*68-227*	1914
Pteridophyta	Glück, H.	15	[v]vi-xx-21	*1-7*	1936
Rhodophyta	Pascher, A. & Schiller, J.	11	[134]135-206	*1-94*	1925
Sphagnales	Warnstorf, K.	14	[3]4-38	*1-16*	1914
Tetrasporales	Lemmermann, E.	5	[21]22-51	*1-33*	1915
Volvocales	Pascher, A.	4	[20]21-498	*20-451*	1927
Zygnemales	Borge, O.	9	[12]13-48	*1-79*	1913
Zygnemales	Pascher, A.	9	[1]2-11	*A-K*	1913

Ed. 1. Arrangement by volume.

Heft	Pagination	Figures	Date	Authors	Taxon
1	[1]2-29	*1-15*	1914	Pascher, A.	Flagellatae-I
1	[30]31-51	*16-67*	1914	Lemmermann, E.	Pantostomatinae
	[52]53-121	*68-227*			Protomastiginae
	[122]123-133	*228-252*			Distomatinae
1	[134]135-138		1914	Anon.	Index
2	[1]2-114	*1-180*	1913	Pascher, A.	Flagellatae II
2	[115]116-185	*181-398*	1913	Lemmermann, E.	Eugleninae
2	[186]187-192		1913	Anon.	Index
3	[1]2-64	*1-69*	1913	Schilling, A. J.	Dinoflagellatae
3	[65]66		1913	Anon.	Index
4	[1]2-19	*1-19*	1927	Pascher, A.	Chlorophyceen-I General Part
4	[20]21-498	*20-451*			Volvocales
4	[499]500-506		1927	Anon.	Index
5	[1]2-20		1915	Pascher, A.	Chlorophyceae-II General

Heft	Pagination	Figures	Date	Authors	Taxon
5	[21]22-51	1-33	1915	Lemmermann, E.	Tetrasporales
5	[52]53-205	34-330	1915	Brunnthaler, J.	Protococcales
5	[206]207-229	1-34	1915	Pascher, A.	Chlorophyceae (pp)
5	[230]231-236	35-39	1915		Supplement
5	[237]238-250		1915	Anon.	Index
6	1-244	1-384	1914	Heering, W.	Chlorophyceae-III
6	[245]246-250		1914	Anon.	Index
7	[1]2		1921	Pascher, A.	Introd.
7	[3]4-99	1-94	1921	Heering, W.	Chlorophyceae-IV
7	[100]101-103		1921	Anon.	Index
9	[1]2-11	A-K	1913	Pascher, A.	Zygnemales-General
9	[12]13-48	1-79	1913	Borge, O.	Zygnemales
9	[49]50-51		1913	Anon.	Index
10	[1]2-173	1-379	1913	Schönfeldt, H. v.	Bacillariales
10	[174]175-187		1913	Anon.	Index
11	[1]2-118	1-96	1925	Pascher, A.	Heterokontae
11	[119]120-133	1-7	1925	Pascher, A.	Phaeophyta
11	[134]135-206	1-94	1925	Pascher, A. & J. Schiller	Rhodophyta
11	[207]208-243	I-XIV	1925	Migula, W.	Charophyta
11	[244]245-250		1925	Anon.	Index
12	[1]2-450	1-560	1925	Geitler, L.	Cyanophyceae
12	[451]452-463	1-14	1925	Geitler, L. & A. Pascher	Cyanochloridinae
12	[464]465-481		1925	Anon.	Index
14	[1]2		1914	Pascher, A.	Bryophyta-Gen.
14	[3]4-38	1-16	1914	Warnstorf, K.	Sphagnales
14	[39]40-168	A-H, 1-62	1914	Mönkemeyer, W.	Bryales
14	[169]170-214	1-158	1914	Schiffner, V.	Hepaticae
15	[v]vi-21	1-7	1936	Glück, H.	Pteridophyta
15	21-463	8-258	1936	Glück, H.	Phanerogamen
15	[464]465-486		1936	Anon.	Index

SECOND EDITION

Heft 9: Jun-Sep 1932 (p. iv: Apr 1932; Hedwigia Oct 1932; MO rd. 12 Oct 1932; Nat. Nov. Jan 1933), p. [i]-iv, [1], [1]-232. – Zygnemales (Czurda).

Heft 10: Nov-Dec 1930 (p. iv: Oct 1930; Nat. Nov. Feb 1931; MO rd. 28 Feb 1931), p. [i]-viii, [1]-466. – Diatomeae (Hustedt).

Heft 14: Mar-Mai 1931 (p. iv: Feb 1931; MO rd.25 Jun 1931; Hedwigia Aug 1931; Nat. Nov. Sep 1931), p. [i]-viii, [1]-252. – Bryophyta (Paul: 1-46) [Sphagn.], Mönkemeyer: 47-197 [Bryol.], Schiffner: 198-243 [Hepat.]).

Ed. 2. Arrangement by taxa.

Taxon	Author	Heft	Pagination	Figures	Date
Bacillariophyta-Centrales	Hustedt	10	81-119	37-94	1930
Bacillariophyta-General	Hustedt	10	[1]2-81	1-36	1930
Bacillariophyta-Pennales	Hustedt, F.	10	119-449	95	1930
Bryales-General	Mönkemeyer, W.	14	[47]48-57	A-G	1931
Bryales-System.	Mönkemeyer, W.	14	57-197	1-76	1931
Dicotyledonae	Glück, H.	15	170-463	84-258	1931
Gymnospermae	Glück, H.	15	21-22		1931
Hepaticae	Schiffner, V.	14	[198]199-243	1-158	1931
Monocotyledonae	Glück, H.	15	22-170	8-83	1931

Taxon	Author	Heft	Pagination	Figures	Date
Pteridophyta	Glück, H.	15	[1]2-21	*1-7*	1931
Sphagnales-General	Paul, H.	14	[1]2-7	*1-3*	1931
Sphagnales-Systematics	Paul, H.	14	8-46	*4-23*	1931
Zygnemales	Czurda, V.	9	[1]2-53	*1-33*	1932
Zygnemales	Czurda, V.	9	53-210	*34-226*	1932

Ed. 2. Arrangement by volume

Heft	Pagination	Figures	Date	Author	Taxon
9	[1]2-53	*1-33*	1932	Czurda, V.	Zygnemales-Gen.
	53-210	*34-226*	1932		Zygnemales
9	211-232		1932	Anon.	Indices
10	[1]2-81	*1-36*	1930	Hustedt, F.	Bacillariophyta-General
10	81-119	*37-94*	1930		Bacillariophyta-Centrales
	119-449	*95-875*	1930		Bacillariophyta-Pennales
10	450-466		1930	Anon.	Index
14	[1]2-7	*1-3*	1931	Paul, H.	Sphagnales-Gen.
	8-46	*4-23*	1931		Sphagnales-Syst.
14	[47]48-57	*A-G*	1931	Mönkemeyer, W.	Bryales-General
14	57-197	*1-76*	1931		Bryales-Syst.
14	[198]199-243	*1-158*	1931	Schiffner, V.	Hepaticae
14	[244]245-252			Anon.	Index
15	[1]2-21	*1-7*	1931	Glück, H.	Pteridophyta
	21-22		1931		Gymnospermae
	22-170	*8-83*	1931		Monocotyledonae
	170-463	*84-258*	1931		Dicotyledonae

NEW EDITION: *Süsswasserflora von Mitteleuropa* announced 1978, editors H. Ettl, J. Gerloff, H. Heynig. First volumes published: 3: *Xanthophyceae* 1, by Hanu Ettl, xiv, 530 p., 1978; 4: *Xanthophyceae* 2, by Alfred Rieth, xiv, 147 p., 1980; 20: *Schizomycetes* by J. Häusler, 1982, x, 588 p., 23: *Pteridophyta* and *Anthophyta*. 1. Theil: *Lycopodiaceae bis Orchidaceae*, by S.J. Caspar and H.K. Krausch (rev. H. Manitz, Philippia 4: 428-429. 1981), 403 p., *109 pl.*, 1980 (see Taxon 30: 552-553. 1981 (new edition of Heft 15); 24: idem, (2) *Saururaceae* bis *Asteraceae* by S.J. Caspar and H.K. Krausch, 1981. In all 24 volumes planned. (Gustav Fischer Verlag, Stuttgart).
Ref.: Engler, A., Bot. Jahrb. 51 (Lit.): 42. 1914, 52 (Lit.): 20. 1915, 54 (Lit.): 37-38. 1916.
West, G.S., J. Bot. 51: 362-363. 1913, 52: 254. 1914.

7424. *Heterokonten* von Prof. Dr. A. Pascher mit einem Beitrag von Prof. Dr. W. Vischer, Basel ... Leipzig (Akademische Verlagsgesellschaft M.B.H.) [1937] 1939. Oct. (*Heterokonten*).
Publ.: 1937-1939, in six parts, p. [i]-ii, [iii]-x, [1]-1092. *Copy*: U. – Alternative title: Dr. L. Rabenhorst's *Kryptogamen-Flora* von Deutschland, Österreich und der Schweiz, zweite, vollständig neu bearbeitete Auflage elfter Band ..." q.v.
1: 1937 (Nat. Nov. Jun 1937), p. [i]-ii, [1]-160.
2: 1937 (Nat. Nov. Sep-Oct 1937), p. 161-320.
3: 1938 (Nat. Nov. Mar 1938), p. 321-480.
4: 1938 (Nat. Nov. Mar 1938), p. 481-640.
5: 1938 (Nat. Nov. Feb 1939 as of 1938), p. 641-832.
6: 1939 (Nat. Nov. Mar-Apr 1939 as of 1939), p. 833-1092, [iii-x].

Pasquale, Fortunato (1856-1917), Italian botanist and agronomist at Napoli; son of Giuseppe Antonio Pasquale; connected with the R. Istituto Tecnico di Napoli. (*F. Pasq.*).

HERBARIUM and TYPES: NAP; some mycological material at FI and PAD. – Some letters at G.

BIBLIOGRAPHY and BIOGRAPHY: Barnhart 3: 53; BL 2: 336-343, 699; CSP 10: 996, 17: 722-723; Jackson p. 204; Kew 4: 243; LS 19845, suppl. 21009; Morren ed. 10, p. 85; Rehder 5: 651; Saccardo 1: 123; Tucker 3: 542.

BIOFILE: Cavara, F., Bull. Orto bot. Univ. Napoli 5: 319-320. 1918 (obit., portr., b. 6 Aug 1856, d. 28 Feb 1917).
Ciferri, R., Taxon 1: 126. 1952.

7425. *Atlante di piante medicinali* ... da servire pel compendio di botanico dei professori V. Tenore e G.A. Pasquale. Napoli (Stabilimento tipografico Lanciano e c. ...) 1880. (*Atlante piante med.*).
Publ.: Jan-Feb 1880 (Bot. Zeit. 19 Mar 1880), p. [i], *pl. 1-57*. *Copy*: FI. – To accompany V. Tenore & G.A. Pasquale, *Comp. bot.* 1847, ed. 3, 1870, see under V. Tenore.

Pasquale, Giuseppe Antonio (1820-1893), Italian botanist, Dr. med. Napoli 1842; assistant to Tenore at Napoli 1840-1848; practicing physician 1848-1860; from 1860 professor of botany at the R. Collegio medico; succeeded Cesati as professor of botany and director of the botanical garden of the University 1883; disabled from 1884. (*Pasq.*).

HERBARIUM and TYPES: NAP. – Other material at AWH, C, FI, K, P. – Some letters at G.

BIBLIOGRAPHY and BIOGRAPHY: Barnhart 3: 53; BL 1: 34, 312, 2: 338, 385-386, 699-770; BM 4: 1527; Bossert p. 302; CSP 4: 770, 8: 567-568, 10: 996, 12: 559, 17: 723; De Toni 1: xcvi, 4: lv; GR p. 536; Herder p. 117, 132, 432; Hortus 3: 1201 ("Pasq."); Jackson p. 73, 89, 100, 104, 109, 435; Kew 4: 243-244; Krempelh. 3: 204; Langman p. 243-244; LS 19846-19851; Morren ed. 2, p. 26, ed. 10, p. 85; PR 6965-6970 (ed. 1: 7794); Rehder 5: 651-652; Saccardo 1: 123 (q.v. for further biogr. refs.), 2: 82, cron. p. xxviii; TL-2/4504; Tucker 1: 542; Urban-Berl. p. 310.

BIOFILE: Anon., Bot. Centralbl. 16: 255. 1883 (app. prof. bot. Napoli), 53: 367. 1893 (d.), 58: 78. 1894 (further biogr. refs.); Bot. Jahrb. 17 (Beibl. 40): 32. 1893; Bot. Not. 1894: 47; Bot. Zeit. 42: 59, 160. 1846 (app. Napoli), 51(2): 108, 188. 1893; J. Bot., Morot 7: 96, 1903 (d.); J. Bot. 31: 160. 1892; Nature 47: 421. 1893; Nat. Nov. 5: 236. 1883 (app. Napoli), 15: 123. 1893 (d.), Österr. bot. Z. 34: 38. 1884, 43: 192. 1893.
Arcangeli, G., Bull. Soc. bot. Ital. 1893 (4): 210-211 (obit.).
Balsamo, F., Bull. Orto bot. Napoli 3: 52-53. 1913.
Carpenter, M.M., Amer. Midl. Natural. 33(1): 77. 1945.
Ciferri, R., Taxon 1: 126. 1952.
Gager, C.S., Brooklyn Bot. Gard. Rec. 27(3): 276. 1938.
Gilbert, P., Comp. biogr. lit. deceased entom. 289. 1977.
Jackson, B.D., Bull. misc. Inf. Kew 1901: 51. (pl. K).
Lusina, G., Annali di Bot 25: 166. 1958 (bibl. Lazio).
Paladino, G., Atti Accad. Pontan. 23: 1-16. 1893 (obit., bibl., b. 30 Oct 1820, d. 23 Feb 1893).
Pasquale, G.A., Opera e titoli del Cav. Prof. Giuseppe Antonio Pasquale. s.l.n.d. 14 p. (*copy*: G) (bibl.).
Sayre, G., Bryologist 80: 515. 1977.
Wittrock, V.B., Acta Horti Berg. 3(2): 164, *pl. 32*. 1903 (portr.); 3(3): 179, 1905.

COMPOSITE WORKS: (1) V. Tenore & G.A. Pasquale, *Compendio di botanica*, 1847 (for this and further eds. see sub Tenore).
(2) V. Tenore & G.A. Pasquale, *Atlante di botanica popolare*, 3 vols., 1872-1886 (id.).

7426. *Flora medica della Provincia di Napoli* ossia descrizione delle piante medicinali che nascono spontaneamente nel perimetro della provincia, con la indicazione de' luoghi in cui vegetano in essa ed in altre località del regno; l'epoca della fioritura; i nomi vernacoli; e gli usi igienici, terapeutici, ed economici. Da servire di guida ai giovani medici e farmacisti, non meno che a coloro i quali volessero provvedersene per uso delle proprie famiglie... Napoli (Dai Tipi di Azzolino e Compagno...) 1841. Oct. (*Fl. med. Napoli*).
Co-author: Giulio Avellino.
Publ.: 1841, p. [i]-vi, [7]-200. *Copies*: FI, G.

7427. *Relazione sullo stato* fisico-economico-agrario *della prima Calabria ulteriore* memoria... Napoli (Tipografia nel R. Albergo de' Poveri) 1863. Qu. (Relaz. stato Calabria).
Publ.: 1863, [1]-432, 1 map, *1 pl. Copy*: MO. – Atti Ist. Sci. nat. Napoli, vol. 11.

7428. *Catalogo del real orto botanico di Napoli* con prefazione, note e carta topografica... Napoli (Stabilimento tipografico Ghio) 1878. Qu. (*Cat. ort. bot. Napoli*).
Publ.: Jan-Mar 1867 (p. xviii: Nov 1866; BSbF rd. 23 Feb-26 Apr 1867), p. [i]-xxxi, [1]-114, plan garden. *Copies*: FI, G, MO, USDA.

7429. *Proposta d'un nuovo genere di Leguminosi* fondato sulla Trigonella coerulea... Napoli (Stamperia del Fibreno...) 1866.
Publ.: Nov 1866, p. [1]-6. *Copy*: FI. – Reprinted and to be cited from Rendic. r. Accad. Sci. Napoli 5(11): 408-411. Nov 1866.

7430. *Flora vesuviana* o catalogo ragionato delle piante del Vesuvio confrontate con quelle dell'isola di Capri e di altri luoghi circostanti... [Napoli 1869]. Qu. (*Fl. vesuv.*).
Publ.: Aug 1869 (mss. submitted 3 Oct 1868; Flora "neu" 10 Sep 1869; Bot. Zeit. "neu" 24 Sep 1869), p. [1]-142. *Copies*: FI, G. MO. – Also issued as Atti r. Accad. Sci. fis. mat., Napoli, 4(6). 1869. Copy presented at a meeting at Catania, Aug 1869, fide Flora 53: 141. 28 Apr 1870.
Ref.: Ascherson, P.F.A., Bot. Zeit. 28: 418-419. 1 Jul 1870.

7431. *Di un viaggio botanico al Gargano*... Napoli (Stamperia del Fibreno...) 1872. Qu. (*Viagg. bot. Gargano*).
Co-author: Gaetano Licopoli (1833-1897).
Publ.: 1872 (ms. submitted 13 Apr 1872), p. [i-ii], [1]-31. Copies: FI, G. – Also issued in Atti R. Acad. Sci. fis. mat., Napoli (session 6 of 13 Apr 1872) vol. 5(18), 1872. – Also treated as TL-2/4504.

7432. *Notizie botaniche* relative alle province meridionali d'Italia memoria... Napoli (Tipografia dell'Accademia reale delle Scienze...) 1881. Qu. (*Not. bot.*).
Publ.: Nov-Dec 1881 (t.p. 1881; p. 12: 2 Nov 1881; Bot. Centralbl. 24-28 Apr 1882), p. [i], [1]-12, *1 pl.* (no. 9). *Copies*: G, USDA. – Preprinted from Atti r. Accad. Sci. fis. mat., Napoli, 9(9). Sep 1882. – Was preceded by paper with the same title in R.C. Accad. Napoli 15, Dec 1876 (reprinted p. [1]-4. *Copy*: FI) and 20: 218-221. Dec 1878 (publ. 1879) (reprinted p. [1]-4. *Copy*: FI).

7433. *Cenni sulla flora di Assab*... Napoli (Tipografia della reale Accademia delle Scienze...) 1885. Qu. (*Cenni fl. Assab*).
Publ.: 1885, p. [1]-12. *Copy*: FI. – Issued as a memoir in Atti r. Accad. Sci. Napoli ser. 2. 1(12): 1-12. 1881 (in complete series).

Passerini, Giovanni (1816-1893), Italian botanist; Dr. med. Parma 1836; at Milano 1843-1844; from 1844-1849 and 1853-1893 professor of botany and director of the botanical garden at Parma. (*Pass.*).

HERBARIUM and TYPES: PARMA and PI (flora of Tuscany); other material at B, E, FI, H (fungi), LUCCA, M (lich.), MI (crypt.), MW, PAD, PI, WRSL. – Passerini contributed to the *Erbario crittogamico italiano* and to Roumeguère, *Fungi selecti exsiccati*. – Correspondence with D.F.L. von Schlechtendal at HAL. – Some further letters at G.

BIBLIOGRAPHY and BIOGRAPHY: AG 3: 201; Barnhart 3: 53; BL 2: 333, 353, 700; BM 4: 1528; Bossert p. 302; CSP 4: 771-772, 8: 568, 10: 996-997, 12: 560, 17: 724; DTS 1: 218; Frank 3 (Anh.): 74; GR p. 520, cat. p. 69; Hawksworth p. 185; Herder p. 179, 432; IH 1 (ed. 6): 362, (ed. 7): 339, 2: (in press); Jackson p. 317, 320; JW 1: 452; Kelly p. 166; Kew 4: 244; LS 19855-19892; Morren ed. 2, p. 27, ed. 10, p. 86; NAF 7: 1081; PR 1646, 6974-6976, 10614 (ed. 1: 7800); Rehder 5: 652; Saccardo 1: 123, 2: 82, cron. p. xxix; Stevenson p. 1253; TL-2/1075; Tucker 1: 542; Urban-Berl. p. 280; Zander ed. 10, p. 699, ed. 11, p. 732, 798 (b. 16 Jun 1816, d. 17 Apr 1893).

BIOFILE: Anon., Bonplandia 6: 196. 1858; Bot. Centralbl. 49: 191. 1892 (retirement), 52: 426. 1892 (replacement), 54: xxiv, 191, 1893 (d.); Bot. Not. 1894: 47 (d.); Bot. Zeit. 51: 188. 1893 (d.); Bull. Soc. entom. ital. 25: 318. 1893; Nat. Nov. 14: 101. 1892 (retirement), 15: 197. 1893 (d.), Ver. bot. Ver. Brand. 35: lxxviii. 1894 (d.).
Arcangeli, G., Bull. Soc. bot. ital. 1893(7): 379-380.
Carpenter, M.M., Amer. Midl. Natural. 33(1): 77. 1945.
Cavillier, F.G., Boissiera 5: 70. 1941.
Chiarugi, A., Nuovo Giorn. bot. Ital. ser. 2. 57: 641. 1950 (herb. Napoleone Passerini also at PI).
De Toni, G.B., Boll. Ist. bot. Univ. Parmense 1892-93: 5-11. 1893 (obit., portr., b. 16 Jun 1816, d. 17 Apr 1893; Nuova Notarisia 6: 5-16. 143-145. 1895 (bibl.; inauguration bust).
Gager, C.S., Brooklyn Bot. Gard. Rec. 27(3): 280. 1938 (dir. bot. gard.).
Gilbert, P., Comp. biogr. lit. deceased entom. 289. 1977.
Hedge, I.C., & Lamond, J.M., Index coll. Edinb. herb. 117. 1970.
Howard, L.O., Hist. applied entom. 248, *pl. 5.* 1930 (portr.) (Smiths. misc. Coll. 84).
Magnus, P., Hedwigia 32: 154-156. 1893; Verh. bot. Ver. Brandenburg 35: xxvi-xxvii. 1894 (obit.).
Wittrock, V.B., Acta Horti Berg. 3(3): 179. 1905.

COMPOSITE WORKS: Cesati, V., Gibelli, G. and Passerini, G., *Compendio della flora italiana,* 1863-1886, 2 vols. The information given in TL-2/1075 has to be amended as follows: The first fascicle received its license for printing on 19 Aug 1867. Gibelli was responsible for the later part of the work which runs to fasc. 41, 1897, and ends with *pl. 129.* Oreste Mattirolo published *plates 130-137* together with the *Index generale* in 1901 (preface Jul 1901). The text ends indeed with p. 906, but the plates run until *137.* A complete copy is at FI (inf. C. Steinberg).

EPONYMY: *Passeriniella* Barlese (1892); *Passerinula* P.A. Saccardo (1875).

7434. *Flora Italiae superioris* methodo analytica. Thalamiflorae praemissa synopsi familiarum phanerogamiae ... Mediolano [Milano] (apud Sanctum Bravetta) 1844. Oct. (*Fl. Ital. sup.*).
Publ.: 1844, p. [i]-viii, [1]-134, [1, err.]. *Copies*: FI, G, NY. – "Part I", only part published.

7435. *Flora dei contorni di Parma* esposta in tavole analitiche con alquante nozioni generali intorno alle piante un dizionario esplicativo de' termini tecnici e une lista di nomi volgari e i rispondenti latini di G. Passerini ... Parma (tipografia Carmignani) 1852. (*Fl. Parma*).
Publ.: 1852 (preface dated 28 Apr 1852), p. [i]-xlviii, [1]-408. *Copy*: FI.
Facsimile ed.: 1976, p. [ii*-iii*, new t.p.], [i]-xlviii, [1]-408, [1, colophon dated 1976]. *Copy*: BR. – Bibliotheca botanica collana divetta da Pietro Zangheri, Arnoldo Forni Editore, 300 copies.
Addenda: see C. Avetta & V. Casoni, Malpighia 11: 209-224. 1897, and C. Avetta, Malpighia 12: 164. 1898.

7436. *Mazzetto di fiori* per la festa dell'8 gennajo 1855 formato con alcune piante nuove o poco conosciute de r. Orto botanico. Parma (Tipografia reale) 1855. Qu. (*Mazz. fior.*).
Publ.: 1855, p. [1]-11, *1 pl. Copy*: FI.

7437. *Primo elenco di funghi parmensi* ... Genova (coi tipi del R.I. de'Sordo-muti) 1867.
Publ.: 1867 (preface 14 Mar 1867; Flora 30 Jun 1868), p. [3]-46. *Copy*: WU (inf. W. Gutermann). – Reprinted from Commentario critt. ital. 2(3): 435-476. 1867.
Continuation: II: Nuovo Giorn. bot. Ital. 4: 48-84. 31 Jan 1872.
 III: ib. 9: 235-267. 10 Jul 1877.
 IV: Atti Soc. critt. ital. 2: 20-47.
 V: Nuovo Giorn. bot. ital. 13(4): 267-283. 1 Oct 1884.

Passy, Antoine François (1792-1873), French botanist, geologist and politician; studied in Brussels; with the Napoleonic Elba army 1813-1815; from 1815 in France; prefect of the dépt. de l'Eure 1830; later in various administrative positions with the ministry of the interior; from 1848 in retirement at Gisors. (*Passy*).

HERBARIUM and TYPES: BR.

BIBLIOGRAPHY and BIOGRAPHY: Barnhart 3: 53; BL 2: 37, 700; BM 4: 1528; CSP 4: 772, 8: 568, 12: 560, 17: 726; Herder p. 171; Kelly p. 166; LS 19899-19900; PR 2199; Quenstedt p. 329-330; Rehder 5: 652; Tucker 1: 542.

BIOFILE: Anon., Bull. Soc. bot. France 20 (C.R.): 226. 1873.
Cosson, E., Bull. Soc. bot. France 21 (C.R.): 131-145. 1874, Notice biographique sur M. Antoine-François Passy, Inst. de France 1874, 27 p. (main biogr; b. 23 Apr 1792, d. 8 Oct 1873.
Crépin, F., Guide bot. Belg. 228. 1878.
Drouyn de Lhuys, Bull. mens. Soc. Acclim. ser. 3. 2: 346-357. 1875 (extensive obit.), reprinted Paris 1875, 20 p.
Du Mortier, B.C., Bull. Soc. Bot. Belg. 12: 257. 1873 (obit.).
Poggendorff, J.C., Biogr.-lit. Handw.-Buch 3: 1007. 1898 (bibl., b. 23 Apr 1792, d. 10 Oct 1873).

7438. *Florul. bruxellensis* seu catalogus plantarum circa Bruxellas sponte nascentium auctoribus A. Dekin et A.F. Passy. Bruxellis (ex Tipis Weissenbruch) 1814. Oct. (*Fl. brux.*).
Co-author: Adrien Dekin (x-1823).
Publ.: 1814 (p. iv: 15 Mai 1814), p. [i]-x, [11]-72. *Copies*: BR, FI, G, KNAW, MO.

Paterson, William (1755-1810), British naturalist, traveller and administrator; collected in South Africa 1777-1779; to India 1781-1785, to Australia 1791, collected for Banks; Lieut.-Governor New South Wales 1800-1810. (*W. Paterson*).

HERBARIUM and TYPES: BM; further material at CGE, E and G. – See Desmond p. 482 for mss.

BIBLIOGRAPHY and BIOGRAPHY: AG 3: 532; Backer p. 427; Barnhart 3: 54; BB p. 238; BM 4: 1529; Dawson p. 654-656; Desmond p. 482; DNB 44: 26; Herder p. 56; HR; HU 706; Jackson p. 349; Kew 4: 246; Lasègue p. 278-279, 446; NI 1: 236, 3: 9; Plesch p. 357-358; PR ed. 1: 7803; Rehder 5: 652; Sotheby 590; Tucker 1: 543.

BIOFILE: Baines, J.A., Vict. Natural. 90: 248-249. 1973.
Bridson, G.D.R. et al., Nat. hist. mss. res. Brit. Isl. 255.22. 1980.
Britten, J., J. Linn. Soc. Bot. 45: 45-46. 1920 (Cape coll.).
Brown, R., Prodr. 303. 1810 (eponymy).
Bullock, A.A., Bibl. S. Afr. bot. 82. 1978.
Cottage Gardener 8: 328-329, 351, 364-365, 378-379. 1851; 9: 3. 1851.
Dyer, R.A., S. Afr. biol. Soc. Pamphl. 14: 44-62. 1949.
Forbes, V.S., Pioneer travellers of South Africa 81-92. 1965 (S.Afr. journey; b. 10 Aug 1755).
Gunn, M. & L.E. Codd, Bot. Expl. S. Afr. 70, 273-275. 1981 (extensive inf. on itin. S. Afr.; portr., secondary lit.).
Hedge, I.C. & J.M. Lamond, Index coll. Edinb. herb. 117. 1970 (ferns Norfolk Isl. E.?).
Hooker, J.D., Fl. Tasm. 1(3): cxxiv. 1859.

Hutchinson, J., Botanist in S. Africa 620-623. 1966.
Maiden, J.H., J. Proc. R. Soc. N.S.W. 42: 116. 1908; Sir Joseph Banks 104, 197. 1909.
Miller, H.S., Taxon 19: 535. 1970 (CGE material via Lambert).
Norton, J., Proc. Linn. Soc. N.S.W. 25: 768. 1900.
Poggendorff, J.C., Biogr.-lit. Handw.-Buch 2: 374. 1863 (b. 17 Aug 1755, d. 21 Jul 1810).
Thunberg, C.P., Travels 4: 271. 1796.
White, A. & Sloane, B.L., The Stapelieae 1, 19, 156. 1933. 1: 20 [index], 86. 1937 (Expl. Orange River).

EPONYMY: *Patersonia* R. Brown (1807, *nom. cons.*). Note: The etymology of *Pattersonia* J.F. Gmelin (1792) and its orthographic variants *Patersonia* Poiret (1816) and *Pathersonia* Poiret (1816) have not been established, but possibly may also commemorate him. *Neopatersonia* Schönland (1912) was named for Mrs. T.V. Paterson, a botanical collector in South Africa. Toponyms: Paterson River (N.S.W.), town of Paterson (Tasmania).

7439. *A narrative of four journeys* into the country of the Hottentots and Caffraria. In the years one thousand seven hundred and seventy-seven , eight, and nine ... London (printed for J. Johnson, ...) 1789. Qu. (*Narr. journ. Hottent.*).
Publ.: 1789, p. [i]-xii, [1]-171, map, i-iii, [iv], *16 pl.* (col. copp., dated 30 Mar 1789). *Copy*: NY.
German transl.: 1790 (p. viii: 5 Sep 1789), p. [i-xiv], [1]-170, [1, binder, err.], *pl. 1-16* (uncol. copp.). *Copy*: BR. – "Wilhelm Patterson's [sic] *Reisen in das Land der Hottentotten und der Kaffern während der Jahre* 1777, 1778 und 1779. Aus dem Englischen übersetzt und mit Anmerkungen begleitet von Johann Reinhold Forster [1729-1798], ..." Berlin (bei Christian Friedrich Voss und Sohn) 1790. Oct. (*Reis. Land Hottent.*).
Other eds.: English (1790) and French (by B. de Laborde 1790) n.v.

Patouillard, Narcisse Théophile (1854-1926), French mycologist; studied at the École de Pharmacie, Paris; lived in Poligny 1881-1884 and Fontenay-sous-Bois 1884-1885; from 1885 in Paris; from 1898-1926 living at Neuilly-sur-Seine; assistant at PC from 1901. (*Pat.*).

HERBARIUM and TYPES: FH (50.000 specimens; bought 1927); types in part also at PC; other material at AUT, B, BR, CUP, PAD, PAV. See Pfister (1977) for an annotated index of fungi described by Patouillard and for notes on his private herbarium. PC has 3000 plates from Patouillard's note books (see Heim 1971; some plates publ. as suppl. to Rev. mycol.). Other plates of fungi used by P. are at FH. – *Exsiccatae*: with E. Doassans, *Champignons figurés et déséchés* (with atlas) 2 vols., Paris 1880-1883. Contributed also to T. Vestergren, *Micromycetes rariores selecti.* – *Letters and portrait at G.*

BIBLIOGRAPHY and BIOGRAPHY: Ainsworth p. 74, 226, 266, 284, 320, 345; Barnhart 3: 54; BM 4: 1529; Bossert p. 302; CSP 12: 560-561, 17: 731-732; GR p. 292, cat. p. 69; Hawksworth p. 185, 197; IH 1 (ed. 6): 362, 2: (in press); Kelly p. 166-170, 255; Kew 4: 247; KR p. 571; Lenley p. 322; LS 20016-20143, 37693-37714, suppl. 21031-21076; Morren ed. 10, p. 68; MW p. 379; NAF 9: 445; Rehder 5: 652; SBC p. 129; Stevenson p. 1253; TL-1/49, 955; TL-2/354, 640, 1604, 2393, 4112, see Lagerheim, N.G. v.; Tucker 1: 543; Urban-Berl. p. 280.

BIOFILE: Ainsworth, G.C., Dict. fungi ed. 6, p. 427. 1971.
Anon., Bull. Soc. bot. France 55: 144 (Officer instr. publ.), 367 (death of Doassans). 1908; J. Bot. 63: 144; 1926; Hedwigia 41: (100). 1902 (prix Thore); J. Mycol. 2. 1905. frontisp. portr.).
Arnaud, G., Rev. Bot. appl. Agric. colon. 6: 333-335. 1926 (obit., b. 2 Jul 1854, d. 31 Mar 1926).
Bary, A. de, Bot. Zeit. 38: 831. 1880 (on fasc. 1 of *Champignons figurés et désséchés*).
Donk, M.A., Introductory note to facs. repr. Patouillard, *Essai tax. Hymenomyc.*, Amsterdam 1963.
Heim, R., Ann. Crypt. exot. 1: 25-36. 1928 (portr., obit.); Rev. Mycol. 36(2): i-xv. 1971 (dedic., evaluation, portr.).

Laundon, J.R., Lichenologist 11: 15. 1979 (coll. FH).
Lloyd, C.G., Syn. Polystictus, Cincinnati, Ohio, 1910 (frontisp. portr. of P.).
Maire, R., Progr. Conn. bot. Algérie 181, 182, 190. 1931.
Mangin, N., Bull. Soc. mycol. France 43: 8-23. 1927 (portr., bibl. of 241 nos., main biogr.), reprinted in Coll. mycol. pap. vol. 1. 1978.
Merrill, E.D., Contr. natl. Herb. U.S. 30: 237-238. 1947.
Pfister, D.H., Annotated index to fungi described by N. Patouillard, Baltimore, Md., 1977 (contr. Reed Herb. 25), vi, 211 p.
Rogers, D.P., Brief hist. mycol. N. Amer. ed. 2. 43, 56. 1981.
Sayre, G., Bryologist 88: 515. 1977 (bryophytes at PC).
Singer, R., Agaricales ed. 2. 1962, p. 822, 831, *pl. 2* (portr. with caption).
Talbot, P.H.B., Taxon 17: 620-628. 1968 (Fossilized pre-patouillardian taxonomy?).
Taton, R., ed., Science in the 19th century 383. 1965.
Urban, I., Symb. ant. 3: 10. 1902, 5: 10. 1904.
Vogelenzang, L. ed., Patouillard, Collected mycological papers, 3 vols., Amsterdam 1978 (the contents (in vol. 1) constitute the most complete bibliography so far of P's writings; contains also a reprint of Mangin 1927; not reproduced e.g. *Tabulae analyticae fungorum*, and *Essai taxonomique*).
Wittrock, V.B., Acta Horti Berg. 3(3): 110, *pl. 143*. 1905 (portr.).

COMPOSITE WORKS: (1) with N.L. Lagerheim (q.v.) *Sirobasidium* (1892), TL-2/4112. (2) Duss, A., *Énum. champ. Guadeloupe* 1903, 94 p.,(all nomenclature by P.), TL-2/1604.

INDEX TO FUNGI DESCRIBED BY PATOUILLARD: See Pfister (1977).

HANDWRITING: Candollea 32: 379-380. 1977.

EPONYMY: *Patouillardea* C. Roumeguère (1885); *Patouillardiella* C. Roumeguère (1890, *orth. var.* of *Patouillardiella* Spegazzini); *Patouillardina* Bresadola ex J. Rick (1906); *Patouillardina* Arnaud (1917); *Patoullardiella* Spegazzini (1889).

7440. *Tabulae analyticae fungorum*. Descriptions & analyses microscopiques des champignons nouveaux, rares ou critiques ... Paris ([fasc. 1: Poligny (Jules Gindre, ...), subseq. fascicles:] Librairie C. Klincksieck ...) 1883-1889, 2 series, 7 fasc. Qu. †. (*Tab. anal. fung.*).

Série	fascicule	nos.	pages	*plates*	dates
1	1	1-100	[1]-40	*[1-32]*	Jan-Apr 1883
	2	101-200	[41]-85	*[33-64]*	Oct-Dec 1883
	3	201-300	[87]-134	*[65-96]*	Jan-Jun 1884
	4	301-400	[137]-180	*[97-128]*	Jan-Mar 1885
	5	401-500	[181]-232	*[129-160]*	Jan-Mai 1886
2	6(1)	501-527	[1]-	*[161-168]*	Jan-Aug 1886
	(2)	528-550	-	*[169-176]*	1886
	(3)	551-575	-30	*[177-184]*	1886
	(4)	576-605	31-42	*[185-192]*	Jan-Aug 1887 (t.p. 1886)
	7	606-700	[i], 43-75	*[193-224]*	Jan-Apr 1889

Ser. 1: p. [i-iii], [1]-232, *pl. 1-160* (col. liths.).
Ser. 2: p. [1]-75, *pl. 161-224* (id.).
Copies: B, BR, FH, FI, G, NY, Stevenson, USDA.

7441. École supérieure de Pharmacie de Paris. Année 1883-1884. No. 4. *Des Hyménomycètes* au point de vue de leur structure et de leur classification. Thèse pour l'obtention du diplome de pharmacien de 1e classe présentée et soutenue le 18 mars 1884 par Narcisse Patouillard ... Lille (Imprimerie L. Danel ...) 1884. Qu. (*Hyménomycètes*).
Publ.: 18 Mar 1854, p. [1]-51, *pl. 1-4*, (uncol. liths.). *Copies:* FH, G, Stevenson.
Reprinted and continued: in Journal de Micrographie 8: 38-44, 101-108, 158-166, 221-227,

266-273, 338-342, *4 pl.*, [continuation:] 385-390, 436-443, 447-485, 532-540, 579-586, 619-627; 9: 19-27, 70-78, 117-121. 1885.
Coll. mycol. pap.: Thesis (no. 22): 1: 157-208, reprint and cont. (nos. 23, 24): 1: 209-328. 1978.

7442. *Les hyménomycètes d'Europe.* Anatomie générale et classification des champignons supérieurs ... Paris (Librairie Paul Klincksieck ...) 1887. Oct. (*Hyménomyc. Eur.*).
Publ.: Jan-Mar 1887 (22 Apr 1887 rd by BSbF; before 15 Mai, date rev. Boudier; Bot. Zeit. 29 Apr 1887; Nat. Nov. Mai (1, 2) 1887; Grevillea Jun 1887), p. [i-xi], [1], [1]-166, [1], *pl. 1-4* (uncol. liths.) with text. *Copies*: B, BR, FH, G, MO, NY, PH, Stevenson. *Cover*: Matériaux pour l'histoire des champignons vol. 1. – The book has priority over Saccardo, Syll. fung. 5, issued 28 Mai 1887.
Coll. mycol. pap.: no. 36, *in* 1: 362-541. 1978.
Ref.: Boudier, E., Bull. Soc. bot. France 34 (Bibl.): 16-18. 1887 (went to press before 15 Mai, see p. 46) (see also ib. (C.R.) 150).
Fischer, Ed., Bot. Zeit. 45: 484-485. 29 Jul 1887.

7443. *Contributions à l'étude des champignons extra-européens* ... [Société Mycologique de France 1887]. Oct.
Publ.: 1 Oct 1887 (cover journal), p. [i], [1]-13, *pl. 9-11* (partly col. liths.). *Copy*: NY. – Reprinted, and to be cited from, Bull. Soc. mycol. France 3(2): 119-131, *pl. 9-11.* 1887.
Coll. mycol. pap.: no. 38, *in* 1: 544-559. 1978.

7444. *Champignons de la Nouvelle Calédonie* ... [Poligny (G. Cottez, imprimeur)] 1888. Oct.
Publ.: Jan-Apr 1888 (in journal; 3(3) not precisely dated, 4(1): 1 Jun 1888), p.[1]-11, *pl. 17*. *Copy*: FH. – Reprinted, and to be cited from Bull. Soc. mycol. France 3(3) 168-178, *pl. 17.* 1888 (for 1887).
Coll. mycol. pap.: no. 39, *in* 1: 560-571. 1978.

7445. *Champignons du Vénézuela* et principalement de la région du Haut-Orénoque par MM. N. Patouillard & A. Gaillard ... [Lons-le-Saunier 1888].
Co-author and collector: Albert Gaillard (1858-1903).
Part [1]: 1 Jun 1888 (in journal 4(1)), p. [1]-40, *pl. 6-13*. *Copy*: FH(2). – Reprinted and to be cited from Bull. Soc. mycol. France 4(1): 7-46, *pl. 6-13.* 1888.
Part 2: 1889, p. [1]-39, *pl. 18-19*. *Copy*: FH. – Reprinted and to be cited from ib. 4: 92-129, *pl. 18-20*. 1889. – The reprint has a special t.p.: "Champignons ... Haut-Orénoque récoltés en 1887 par M.A. Gaillard ..." Lons-le-Saunier (Imprimerie et lithographie Lucien Declume ...). 1889.
Coll. mycol. pap.: no. 45, *in* 1: 591-638, no. 47, *in*: 1: 642-682. 1978.

7446. *Sur quelques espèces de Meliola* nouvelles ou peu connues ... [Revue mycologique, juillet 1888]. Oct.
Publ.: Jul 1888, p. [1]-8, *pl. 69*. *Copy*: FH. – Reprinted, and to be cited from Rev. mycol. 10: 134-141. 1888.
Coll. mycol. pap.: no. 56, *in* 1: 715-723. 1978.

7447. *Fragments mycologiques* ... [Extrait du Journal de Botanique ... Paris, J. Mersch ... 1888-1890]. Oct.
Publ.: 8 publications in J. de Bot., Morot, to be cited from the journal. Reprinted (type shifted), p.[1]-70. *Copies*: FH, NY. – Original publication as follows:
Coll. mycol. pap.: nos. 49, 50, 51, 52, 54, 61, 62, 63, 64, 65, 72, 73, *in* 1: 688-706, 710-711, 752-779, 818-837. 1978.

Fragment	J. de Bot.	dates
1	2: 49-53, *pl. 1*	16 Feb 1888
2	2: 146-151	1 Mai 1888
3	2: 216-218	1 Jul 1888
4	2: 267-270	16 Aug 1888

Fragment	J. de Bot.	dates
5	2: 406-407	16 Nov 1888
6	3: 23-27	16 Jan 1889
	3: 33-37, *pl. 1*	1 Feb 1889
7	3: 165-168	16 Mai 1889
	3: 256-259	1 Aug 1889
8	3: 335-343	16 Oct 1889
9	not published	
10	4: 197-200	16 Mai 1890
11	4: 253-358	16 Jul 1890

7448. *Sur quelques champignons extra-européens* ... Lons-le-Saunier (Imprimerie et lithographie Lucien Declume ...) 1889. Oct.
Publ.: 1 Mar 1889 (cover journal 4(3)), p. [1-4]. *Copies*: FH, NY. – Reprinted, and to be cited from Bull. Soc. mycol. France 4(3): 71-73. 1889 (for 1888).
Coll. mycol. pap.: no. 46, *in* 1: 639-641. 1978.

7449. *Note sur trois espèces mal connues d'hyménomycètes* ... Paris (au siège de la Société [mycologique de France]). 1889. Oct.
Publ.: 1 Mai 1889 (cover journal, 5(2)), p. [1]-6. *Copies*: FH, NY. – Reprinted and to be cited from Bull. Soc. mycol. France 5(2): 30-33. 1889.
Coll. mycol. pap.: no. 57 (1: 724-727. 1978).

7450. *Le genre Ganoderma* ... Lons-le-Saunier (Imprimerie et lithographie Lucien Declume ...) 1889. Oct.
Publ.: 1 Mai 1889 (cover journal, 5(2)), p. [1]-19, *pl. 10-11*, (col.). *Copies*: FH, NY. – Reprinted and to be cited from Bull. Soc. mycol. France 5(2): 64-80, *pl. 10-11*. 1889.
Coll. mycol. pap.: no. 58, *in* 1: 728-746. 1978.

7451. *Sur la place du genre Favolus* dans la classification ... Lons-le-Saunier (Imprimerie et lithographie Lucien Declume ...) 1890. Oct.
Publ.: 1 Apr 1890 (cover journal 6(1)), p. [1]-5. *Copy*: FH. – Reprinted, and to be cited from Bull. Soc. mycol. France 6(1): xix-xxi. 1890.
Coll. mycol. pap.: no. 67, *in* 2: 785-787. 1978.

7452. *Dussiella* nouveau genre d'Hypocréacées ... Lons-le-Saunier (Imprimerie et lithographie Lucien Declume ...) 1890. Oct.
Publ.: 15 Jul 1890 (p. 5: 10 Apr 1890; cover journal, 6(2)), p. [1]-5. *Copy*: FH. – Reprinted, and to be cited from Bull. Soc. mycol. France 6(2): 107-109. 1890.
Coll. mycol. pap.: no. 68, *in* 2: 788-790. 1978.

7453. *Le genre Podaxon pl. xvii* ...Lons-le-Saunier (*Imprimerie et lithographie Lucien Declume ...*) *1890*. Oct.
Publ.: 31 Dec 1890 (p. 11: 11 Sep 1890; cover journal, 6(4), p. [1]-11, *pl. 17*. *Copies*: FH, NY. – Reprinted, and to be cited from Bull. Soc. mycol. France 6(4): 159-167, *pl. 17*. 1890.
Coll. mycol. pap.: no. 70, *in* 2: 793-802. 1890.

7454. *Quelques champignons de la Chine* récoltés par M. l'abbé Delavay ... [Revue mycologique, no 47. – Juillet 1890] Oct.
Collector: Abbé Pierre Jean Marie Delavay (1834-1895).
Publ.: Jul 1890, p. [1]-4, *pl. 117bis*. *Copies*: FH, G. – Reprinted, and to be cited from Rev. mycol. 2: 133-136. 1890.
Coll. mycol. pap.: no. 75, *in* 2: 840-844. 1890.

7455. *Champignons de l'Équateur* par MM. N. Patouillard & G. de Lagerheim ... [1-3:] Lons-le-Saunier (Imprimerie et lithographie Lucien Declume) 1891-1895, 5 parts.
Co-author: Nils Gustav Lagerheim (1860-1926); Urédinés., Ustilaginés, Peronosporés and Chytridinés exclusively by L.

Publ.: five papers ("pugilli") published (1-3, 5) in *Bull. Soc. mycol. France* and (4) *Bull. Herb. Boissier* and to be cited from the journal publication.

Pugillus	reprint pag.	journal pag.	dates	Nat. Nov.
1	[1]-29, *pl. 11-12*	7(3): 158-184	30 Sep 1891	Oct(2) 1892
2	[1]-30, *pl. 11-12*	8(3): 113-140	31 Jul 1892	
3	[1]-42, *pl. 8-10*	9(2): 124-164	31 Mar 1893	
		9(3): 145-165	31 Jul 1893	Jan(1, 2) 1895
4	[1]-22, *pl. 2*	3(2): 53-74	Feb 1895	Mar(2) 1895
5	[i], [1]-22	11(4): 205-234	31 Dec 1895	Mar(2) 1896

Copies: FH (with type-written index to genera; reprints of parts 1-4; part 5: journal publ.), G. – See also KR p. 374, no. 119.
Coll. mycol. pap.: nos. 79 (2: 861-889), 89 (2: 969-998), 96 (2: 1023-1067), 109 (2: 1109-1132), 113 (2: 1147-1176).

7456. *Remarques sur l'organisation de quelques champignons exotiques* ... Lons-le-Saunier (Imprimerie et lithographie Lucien Declume ...) 1891. Oct.
Publ.: 31 Mar 1891 (cover journal, 7(1)), p. [1]-10, *pl. 4* (uncol.). *Copies*: FH, NY. – Reprinted, and to be cited from Bull. Soc. mycol. France 7(1): 42-49. 1891.
Coll. mycol. pap.: no. 76 (2: 845-853. 1978).

7457. *Polyporus bambusinus* nouveau polypore conidifère ... Lons-le-Saunier (Imprimerie et lithographie Lucien Declume) 1891. Oct.
Publ.: 30 Jun 1891 (cover journal, 7(2)), p. [1]-5. *Copies*: FH, NY. – Reprinted, and to be cited from Bull. Soc. mycol. France 7(2): 101-103. 1891.
Coll. mycol. pap.: no. 77 (2: 854-856. 1978).

7458. *Podaxon squamosus* nov. sp. ... Lons-le-Saunier (Imprimerie et lithographie Lucien Declume ...) 1891. Oct.
Publ.: 31 Dec 1891 (cover journal, 7(4)), p. [1-2], *pl. 13*. *Copies*: FH, NY. – Reprinted, and to be cited from Bull. Soc. mycol. France 7(4): 210. 1891.
Coll. mycol. pap.: no. 80 (2: 890-891. 1891).

7459. *Quelques espèces nouvelles de champignons extra-européens.* [Extrait de la Revue mycologique, no. 51. – Juillet 1891]. Oct.
Publ.: Jul 1891, p. [1]-3. *Copies*: FH, NY. – Reprinted, and to be cited from Rev. mycol. 13: 135-138. 1891.
Coll. mycol. pap.: no. 84 (2: 914-917. 1978).

7460. Exploration scientifique de la Tunisie. (*Énumération des champignons observés en Tunisie* ... Paris (Imprimerie nationale) 1892. (*Énum. champ. Tunisie*).
Publ.: 1892, p. [i*-iii*], [i]-iv, [1]-19. *Copies*: G, NY.*Coll. mycol. pap.*: no. 85 (2: 918-941. 1978).

7461. Exploration scientifique de la Tunisie. *Illustration des espèces nouvelles rares ou critiques de champignons de la Tunisie* par N. Patouillard. Planches i-v dessinées d'après nature par M.N. Patouillard. Paris (Imprimerie nationale) 1892-1895, 2 fasc. Fol. (*Ill. champ. Tunisie*; alternative, see no. 640: *Explor. sci. Tunisie, Ill. champ.*).
Part 1: Jan-Apr 1892 (Nat. Nov. Mai 1892), cover, *pl. 1-2* with text. *Cover title*: "Exploration ... *Illustrations de la partie botanique.* Espèces nouvelles de Champignons ..." Paris (id.) 1892.*Part 2*: 1895 (Nat. Nov. Mar 1896), *pl. 3-5* with text, and t.p. as in heading.
Copies: MICH, MO, Stevenson. – See TL-2/640 for further details.
Coll. mycol. pap.: no. 86 (2: 942-952. 1978).

7462. *Phlyctospora maculata* nouveau gastéromycète de la Chine occidentale ... Lons-le-Saunier (Imprimerie et Lithographie Lucien Declume) 1892. Oct.

Publ.: 30 Nov 1892 (cover journal, 8(4)), p. [1]-4. *Copy*: FH. – Reprinted, and to be cited from Bull. Soc. mycol. France 8(4): 189-190. 1892.
Coll. mycol. pap.: no. 90 (2: 999-1000. 1978).

7463. *Septobasidium* nouveau genre d'hyménomycètes heterobasidiés ... [(Extrait du Journal de Botanique. Numéro du 16 Février 1892)]. Oct.
Publ.: 16 Feb 1892, p. [1]-4. *Copies*: FH, NY. – Reprinted, and to be cited from J. Bot. Morot, 6: 61-64. 1892.
Coll. mycol. pap.: no. 91 (2: 1001-1004. 1978).

7464. *Quelques champignons asiatiques nouveaux ou peu connus* ... Genève (Imprimerie Romet, ...) 1893. Oct.
Publ.: Mai 1893, p. [1]-4 (imprint: see cover). *Copies*: FH, G, U. – Reprinted, and to be cited from Bull. Herb. Boiss. 1(5): 300-303. 1893.
Coll. mycol. pap.: no. 94 (2: 1014-1017. 1978).

7465. *Le genre Skepperia Berk.* ... Lons-le-Saunier (Imprimerie et lithographie Lucien Declume) 1893. Oct.
Publ.: 10 Jan 1893 (cover journal, 9(1)), p. [1]-6. *Copy*: FH. – Reprinted, and to be cited from Bull. Soc. mycol. France 9(1): 1-4, *pl. 1*. 1893.
Coll. mycol. pap.: no. 95 (2: 1018-1022. 1978).

7466. *Quelques champignons du Thibet* ... [Paris (J. Mersch, ...) 1893]. Oct.*Publ.*: 16 Sep 1893, p. [1-2]. *Copy*: FH. – Reprinted without identification from J. Bot., Morot, 7: 343-344. 1893.
Coll. mycol. pap.: no. 99 (2: 1076-1077. 1978).

7467. *Le genre Phlebophora Lév.* ... Lons-le-Saunier (Imprimerie et lithographie Lucien Declume) 1894. Oct.
Publ.: 15 Jan 1894 (cover journal, 10(1)), p. [1]-4. *Copies*: FH, NY. – Reprinted, and to be cited from Bull. Soc. mycol. France 10(1): 55-56. 1894.
Coll. mycol. pap.: no. 101 (2: 1080-1081. 1978).

7468. *Espèces critiques d'hyménomycètes* ... Lons-le-Saunier (Imprimerie et lithographie Lucien Declume ...) 1894. Oct.
Publ.: 30 Jun 1894 (p. 9: 8 Mar; cover journal 10(3)), p. [1]-9, *pl. 3* (uncol.). *Copies*: FH, NY. – Reprinted, and to be cited from Bull. Soc. mycol. France 10(2): 75-81. 1894.
Coll. mycol. pap.: no. 102 (2: 1082-1089. 1978).

7469. *Asterodon* nouveau genre de la famille des Hydnacées ... Lons-le-Saunier (Imprimerie & Lithographie Lucien Declume ...) 1894. Oct.
Publ.: 15 Nov 1894 (p. 4: 14 Jun 1894; cover journal, 10(4)), p. [1]-4., *pl. 5* (col.). *Copies*: FH, NY. – Reprinted, and to be cited from Bull. Soc. mycol. France 10(3): 129-130. 1894.
Coll. mycol. pap.: no. 103 (2: 1090-1092. 1894).

7470. *Le genre Lopharia Kalch.* ... Lons-le-Saunier (Imprimerie et Lithographie Lucien Declume ...) 1895. Oct.
Publ.: 31 Mar 1895 (cover journal, 11(1)), p. [1]-5, *pl. 1* (uncol.). *Copies*: FH, NY. – Reprinted and to be cited from Bull. Soc. mycol. France 11(1): 13-15. 1895.
Coll. mycol. pap.: no. 110 (2: 1133-1136. 1978).

7471. *Quelques espèces nouvelles de champignons africains* ... Lons-le-Saunier (Imprimerie & Lithographie Lucien Declume ...) 1895. Oct.
Publ.: 31 Mai 1895 (cover journal, 11(2)), p. [1]-6, *pl. 11* (uncol.). *Copy*: NY. – Reprinted, and to be cited from Bull. Soc. mycol. France 11(2): 85-88. 1895.
Coll. mycol. pap.: no. 111 (2: 1137-1141. 1895).

7472. *Énumération des champignons récoltés par les rr. pp. Farges et Soulié dans le Thibet oriental & le Su-Tchuen* ... Lons-le-Saunier (Imprimerie et Lithographie Lucien Declume ...) 1895. Oct.

Collectors: Paul Guillaume Farges (1814-1912); Jean André Soulié (1858-1905).
Publ.: 30 Sep 1895 (cover journal, 11(3)), p. [1]-6, *pl. 13*. *Copy*: FH. – Reprinted, and to be cited from Bull. Soc. mycol. France 11(3): 196-199. 1895.
Coll. mycol. pap.: no. 112 (2: 1142-1146. 1978).

7473. *Liste des champignons récoltés en Basse-Californie* par M. Diguet ... [Paris (J. Mersch ...) 1896]. Oct.
Co-author: Paul Auguste Hariot (1854-1917).*Publ.*: 1 Aug 1896 (in journal), p. [1]-3, *pl. 2*. *Copy*: FH. – Reprinted, and to be cited from J. Bot., Morot 10: 250-252. 1896.
Coll. mycol. pap.: no. 120 (2: 1201-1204. 1978).

7474. *Cyclostomella* nouveau genre d'Hémihystériés ... [Extrait du Bulletin de l'Herbier Boissier ... 1896]. Oct.
Publ.: Sep 1896 (in journal), p. [1]-2. *Copies*: FH, G, NY. – Reprinted, and to be cited from Bull. Herb. Boissier 4(9): 655-656. 1896.
Coll. mycol. pap.: no. 116 (2: 1182-1183. 1978).

7475. *Exploration scientifique de la Tunisie. Catalogue raisonné des plantes cellulaires de la Tunisie* ... Paris (Imprimerie nationale) 1897. Oct. (*Cat. pl. cell. Tunisie*).
Publ.: Jan-Apr 1897 (Bot. Zeit. 16 Jun 1897; Nat. Nov. Mai(1) 1897; Bot. Centralbl. 26 Mai 1897; Hedwigia Mai-Jun 1897), p. [i*-iii*], [i]-xxiv, [1]-158. *Copies*: FH, MICH, MO, NY, Stevenson. – *Fungi* by Patouillard on p. 20-135; *Mousses* by Émile Bescherelle (1828-1903) on p. [1]-13; *Characées* by Jean François Gustave Barratte (1857-1920) on p. 13-16; *Algues* by Camille François Sauvageau (1861-1936) on p. 16-19; *Lichens* by Auguste Maria Hue (1840-1917) on p. [136]-151.
Additions: See next item.
Coll. mycol. pap.: no. 122 (2: 1209-1326, fungi only, 1978).

7476. *Additions au catalogue des champignons de la Tunisie* ... Lons-le-Saunier (Imprimerie et Lithographie Lucien Declume) 1897. Oct.
Publ.: 30 Oct 1897 (cover journal, 13(4)), p. [1]-22, *pl. 13*. *Copies*: FH, NY. – Reprinted, and to be cited from Bull. Soc. mycol. France 13(4): 197-216, *pl. 13*. 1897.
Coll. mycol. pap.: no. 125 (2: 1354-1374. 1978).
Additions(2): *in* C.R. Congr. Soc. Sav. Paris 1908: 242-257. 1909; Coll. mycol. pap. 3: 1880-1895 (Nat. Nov. Dec(1) 1909).

7477. *Quelques champignons nouveaux récoltés au Mexique par Paul Méry* [i.e. Maury] ... Lons-le-Saunier (Imprimerie et Lithographie Lucien Declume ...) 1898. Oct.
Collector: Paul Jean Baptiste Maury (1858-1893).
Publ.: 31 Apr 1898 (in journal, 14(2)), p. [1]-7, *pl. 7*. *Copies*: FH(2), NY. – Reprinted, and to be cited from Bull. Soc. mycol. France 14: 53-57. 1898. – The second FH copy is accompanied by 56 original drawings and photographs.
Coll. mycol. pap.: no. 129 (2: 1402-1407. 1978).

7478. *Champignons nouveaux* ou peu connus ... Lons-le-Saunier (Imprimerie et lithographie Lucien Declume ...) 1898. Oct.
Publ.: 31 Aug 1898 (in journal, 14(3)), p. [1]-10. *Copy*: NY. – Reprinted, and to be cited from Bull. Soc. mycol. France 14(3): 149-156. 1898.
Coll. mycol. pap.: no. 130 (2: 1408-1415. 1978).

7479. *Quelques champignons de Java* ... Lons-le-Saunier (Imprimerie et lithographie Lucien Declume) 1898. Oct.
Publ.: 1898 (Nat. Nov. Jan(1) 1899), p. [1]-19. *Copy*: FH. – Reprinted, and to be cited from Bull. Soc. mycol. France 14(4): 182-198. 1898.
Coll.mycol. pap.: no. 131 (2: 1416-1432. 1898).

7480. *Champignons du Nord de l'Afrique* ... Lons-le-Saunier (Imprimerie et lithographie Lucien Declume ...) 1899. Oct.
Publ.: 31 Jan 1899 (cover journal, 15(1)), p. [1]-8, *pl. 4*. *Copy*: NY. – Reprinted, and to be cited from Bull. Soc. mycol. France 15(1): 54-59, *pl. 4*. 1899.
Coll. mycol. pap.: no. 132 (2: 1433-1439. 1978).

7481. *Champignons de la Guadeloupe* ... Lons-le-Saunier (Imprimerie et lithographie de Lucien Declume ...) 1899. Oct.
Collector: [Père] Antoine Duss (1840-1924), orig. herb. at B. (partly destroyed).
Publ.: 31 Jul 1899 (in journal, 15(3)), p. [1]-21, *pl. 9-10*. *Copy*: NY. – Reprinted, and to be cited from Bull. Soc. mycol. France 15(3): 191-209, *pl. 9-10*. 1899. – See Urban (1902).
Coll. mycol. pap.: no. 133 (2: 1440-1460. 1978).

7482. Description d'une nouvelle espèce d'Auriculariacés *Septobasidium langloisii* .... Lons-le-Saunier (Imprimerie et lithographie Lucien Declume) 1900. Oct.
Publ.: 31 Jan 1900 (cover journal, 16(1)), p. [1]-4. *Copies*: FH, NY. – Reprinted, and to be cited from Bull. Soc. mycol. France 16(1): 54-55. 1900.
Coll. mycol. pap.: no. 134 (2: 1461-1462. 1978).

7483. Université de Paris. École supérieure de Pharmacie Année 1900-1901. No. 2. *Essai taxonomique sur les familles et les genres des Hyménomycètes.* Thèse pour l'obtention du diplôme de docteur de l''Université de Paris (Pharmacie). Présentée et soutenue le [ ] 1900 par N. Patouillard ... Lons-le-Saunier (Imprimerie et Lithographie Lucien Declume) 1900. Oct. (*Essai tax. Hyménomyc.*).
Publ.: 1900 (Bot. Centralbl. 13 Feb 1901; Nat. Nov. Mar(1) 1901), p. [i-ii], [1]-184. *Copies*: MO, Stevenson.
Trade issue: 1900, p. [i-ii], [1]-184. *Copies*: BR, NY, PH, USDA.
Facsimile reprint: of thesis issue: 1963, p. [i-ii], [1]-184. *Copy*: FAS. – Réimpression A. Asher & Co. [Amsterdam] 1963. Preface by M.A. Donk.

7484. *Champignons de la Guadeloupe*, recueillis par le R.P. Duss, (2ᵉ Série) ... Paris (Au siège de la Société [mycologique de France]...) 1901. Oct.
Collector: Antoine Duss (1840-1924).
Publ.: 28 Feb 1901 (cover journal,; Nat. Nov. Jul(2) 1901 p. [1]-14, *pl. 7* (uncol.). *Copy*: NY. – Reprinted, and to be cited from Bull. Soc. mycol. France 16(4): 175-188. 1901.
Coll. mycol. pap.: 139 (2: 1490-1504. 1978).

7485. *Champignons Algéro-Tunisiens* nouveaux ou peu connus ...Paris (au siège de la Société [mycologique de France] ...) 1901. Oct.
Publ.: 31 Aug 1901 (cover journal; Nat. Nov. Feb(1) 1902), p. [1]-7, *pl. 6-7* (col.). *Copy*: FH. – Reprinted, and to be cited from Bul. Soc. mycol. France 17: 182-188. 1901.
Coll. mycol. pap.: no. 141 (2: 1508-1516. 1978).
Suite (1): 25 Jan 1902, Bull. Soc. mycol. France 18: 47-53. 1902.
Suite (2): 1904, p. [1]-4, *pl. 5* (col.). *Copies*: FH, NY. – Reprinted, and to be cited from Bull. Soc. mycol. France 20(2): 51-54. 1904.
Coll. mycol. pap.: no. 158 (2: 1708-1712. 1978).

7486. *Champignons de la Guadeloupe*, recueillis par le R.P. Duss, (3ᵉ Série) ... Paris (au Siège de la Société [de la Société mycologique de France]) 1902. Oct.
Publ.: 15 Mai 1902 (cover journal; Hedwigia 5 Aug 1902; Nat. Nov. Dec(1) 1902, p. [1]-16. *Copy*: NY. – Reprinted, and to be cited from Bull. Soc. mycol. France 18(2): 171-186. 1902.
Coll. mycol. pap.: no. 145 (2: 1532-1547. 1978).

7487. *Note sur trois champignons des Antilles* ... Extrait des "Annales mycologici" (vol. 1, no. 3, 1903). Oct.
Publ.: 31 Mai 1903 (in journal), p. [1]-4. *Copies*: FH, NY. – Reprinted, and to be cited from Annales mycologici 1(3): 216-219. 1903.
Coll. mycol. pap.: no. 148 (2: 1558-1561. 1978.).

7488. *Rollandina*, nouveau genre de Gymnoascés ... Paris (au Siège de la Société [mycologique de France] ...) 1905. Oct.
Publ.: 5 Apr 1905 (in journal, 21(2)), p. [1]-3, *pl. 5*. *Copy*: FH. – Reprinted, and to be cited from Bull. Soc. mycol. France 21(2): 81-83. 1905.
Coll. mycol. pap.: no. 161 (3: 1721-1725. 1978).

7489. *Fungorum novorum decas prima* ... Paris (au Siège de la Société [mycologique de France] ...) 1905. Oct.
Co-author: Paul Auguste Hariot (1845-1917).
Publ.: 5 Apr 1905 (cover journal, 21(2)), p. [1]-3. *Copy*: FH. – Reprinted, and to be cited from Bull. Soc. mycol. France 21(2): 84-86. 1905.
Coll. mycol. pap.: no. 162 (3: 1726-1728. 1978).

7490. *Champignons recueillis par M. Seurat*, dans la Polynésie française, ... Paris (au Siège de la Société [mycologique de France] ...) 1906. Oct.
Collector: Léon Gaston Seurat (1872-?).
Publ.: 28 Feb 1906 (in journal; FH rd. 13 Apr 1906), p. [1]-18, *pl. 1-2*. *Copy*: FH. – Reprinted, and to be cited from Bull. Soc. mycol. France 22(1): 45-62. 1906.
Coll. mycol. pap.: no. 164 (3: 1735-1754. 1978).

7491. *Le Ratia*, nouveau genre de la série des Cauloglossum ... Paris (au Siège de la Société [mycologique de France] ...) 1907. Oct.
Eponymous collector: Auguste-Joseph Le Rat (1872-1910).
Publ.: 30 Apr 1907 (FH rd. 13 Jun 1907; cover journal 23(1)), p. [1]-3. *Copies*: FH, NY. – Reprinted, and to be cited from Bull. Soc. mycol. France 23(1): 50-52. 1907. – The two term generic name *Le Ratia* is not validly published in this publication because it is not hyphenated.
Coll. mycol. pap.: no. 173 (3: 1781-1784. 1978).

7492. *Note sur trois espèces d'Hydnangium* de la flore du Jura ... Paris (au siège de la Société [mycologique de France] ...) 1910. Oct.
Publ.: 5 Jul 1910 (FH rd. 8 Nov 1910; cover journal 26(2)), p. [1]-5. *Copy*: FH. – Reprinted, and to be cited from Bull. Soc. mycol. France 26(2): 199-204. 1910.
Coll. mycol. pap.: no. 196 (3: 1922-1927. 1978).

7493. *Liste des champignons recueillis à San Thomé* comuniqués [sic] par m. de Seabra ... Lisbonne [Lisboa] (Oficinas Gráficas da Biblioteca Nacional) 1922. Oct.
Publ.: 1922. Reprint cover with above imprint, p. [35]-39. *Copy*: FH. – Reprinted and to be cited from Bull. Soc. Port. Sci. nat. 9: 35-39. 1922.
Coll. mycol. papers: not included.

7494. *Contribution à l'étude des champignons de Madagascar* ... Tananarive (Imprimerie moderne de l'Emyrne G. Pitot & Cie.) 1928. Qu. (*Contr. champ. Madagascar*).
Publ.: 1928 (journal fasc. so dated; Nat. Nov. Jun 1929), p. [1]-49, *pl. 1-2*. *Copies*: BR, FH, G, NY, USDA. – Issued as Mém. Acad. Malgache fasc. 6, 1928.
Coll. mycol. pap.: no. 261 (3: 2315-2363. 1928).

7495. *Collected mycological papers* recueil des essais mycologiques chronologically arranged and edited by L. Vogelenzang ... Amsterdam (A. Asher & Co. B.V.) 3 vols. Oct. (*Coll. mycol. pap.*).
Publ.: Dec 1978, 3 vols., ISBN 90-6123-327-5, with (in vol. 1) a preface by editor, a list of contents in chronological order with bibliographical details (p. iii-xxi) a reprint of L.N. Mangin, Notice nécrologique (with portr. and bibl.), Bull. Soc. mycol. France 43: 8-23. 1927. The volumes show the original pagination of the papers and have a continuous pagination of their own. Lacking from this reprint: *Essai tax. Hyménomyc.* (1900), *Tab. anal. fung.* (1883-1889) and *Liste champ. San Thomé* (1922). – *Copies*: CBS, FAS, L.
1: [i]-xxi, [1], [1]-784.
2: [i-ii], 785-1693.
3: [i-ii], 1695-2400.
Ref.: Stafleu, F.A., Taxon 28: 294. 1979 (rev.).

Patout, Marie Rose (*fl.* 1864), French herbalist and midwife at Toulon (*Patout*).

HERBARIUM and TYPES: Unknown.

BIBLIOGRAPHY and BIOGRAPHY: BL 2: 208, 700; Jackson p. 291; Rehder 5: 652; Tucker 1: 543.

7496. *Abrégé des plantes médicinales* croissant dans les environs de Toulon ... Toulon (imprimerie et lithographie d'E. Aurel, ...) 1864. Oct. (*Abr. pl. méd.*).
Publ.: 1864 (p. 8: 4 Jul 1864), p. [1]-85. *Copy*: G. – We thought that one midwife/botanist would not be out of place in TL-2; the book is a simple enumeration of medicinal plants growing in the wild around Toulon.

Patraw, Pauline Mead (*fl.* 1936), American botanist (*Patraw*).

HERBARIUM and TYPES: Unknown.

BIBLIOGRAPHY and BIOGRAPHY: BL 1: 164, 217, 312; GR p. 237; Kew 4: 247.

COMPOSITE WORKS: with Clifford Presnall, *Plants of Zion National Park*, Zion-Bryce Museum Bulletin no. 1, Jun 1937, 69 p., *15 pl.* (Madroño 4: 198, by H.L. Mason).

7497. Preliminary *check-list of plants of Grand Canyon National Park* (United States Department of the Interior National Park Service Grand Canyon National Park Technical Bulletin no.6) [1932]. (*Checkl. pl. Grand Canyon*).
Preliminary list: Feb 1932, mimeographed, p. [i, iii], [1]-47, [1]-10. *Copy*: NY.
Check-list: Jun 1936, p. [i*], i-ix, 1-75. *Copy*: NY. "Check-list of plants of Grand Canyon National Park" Natural History Bulletin no. 6, Grand Canyon Natural History Association.

Patrick, Ruth Myrtle (1907-x), American diatomologist at the Philadelphia Academy of Natural Sciences; married Charles Hodge 1931; Ph. D. Univ. of Virginia 1934, at the Academy of Sciences from 1937 in various positions, curator of the Limnology Dept. from 1947; chairman of this department 1947-1973, chairman board of trustees Academy 1973-1976, from 1976 honorary chairman; Tyler award 1975; chairman Section of Population Biology, Evolution and Ecology of the National Academy of Sciences. (*R. Patrick*).

HERBARIUM and TYPES: PH; other material at F, FH, and G. – Correspondence and manuscript material at Philadelphia (PH), G and Smithsonian Archives.

BIBLIOGRAPHY and BIOGRAPHY: Barnhart 3: 54; Bossert p. 302; IH 1 (ed. 1): 76, (ed. 2): 96, 97, (ed. 3): 122, (ed. 4): 135, (ed. 5): 143, (ed. 6): 213; Kew 4: 247; Langman p. 570; PH 209, 456, 618, 939; Roon p. 86.

BIOFILE: Anon., Biogr. News 1: 334. 1974, 2: 407. 1975; Taxon 29: 306. 1980; Who's who of American women ed. 11. 633. 1979 (full details).
Emberlin, Diane, Contributions of women, Science 106-123. 1977 (portr.; "She speaks for the rivers").
Ewan, J. et al., Short hist. bot. U.S. 79. 1969.
Fryxell, G.A., Beih. Nova Hedwigia 35: 363. 1975.
Holden, C., Science 188: 997-999. 1975 (Tyler award, portr.).
Koster, J.T., Taxon 18: 556. 1969.
Mandel, B., Biogr. News 2: 407. 1975.
Patrick, R., Biographical data for Dr. Ruth Patrick, Philadelphia 1978, 10 p. (curr. vit.; bibl.).
Pennell, F.W., Bartonia 22: 30. 1943.
VanLandingham, Cat. diat. 6: 3577-3578, 7: 4214. 1978 (bibl.).
Verdoorn, F., ed., Chron. bot. 4: 452. 1938, 5: 234. 1939.
Wornardt, W.W., Int. direct. diat. 35, 110-115. 1968 (bibl.).

COMPOSITE WORKS: with Charles W. Reimer, *The diatoms of the United States* (exclusive of Alaska and Hawaii), vol. 1, Mai 1966 (Monogr. 13, Acad. Nat. Sci., Philadelphia).

THESIS: *A taxonomic and distributional study of some diatoms from Siam and the Federated Malay States*, Proc. Acad. nat. Sci. Philadelphia 88: 367-470, *pl. 11*, 20 Oct 1936 (reprints exist with original pagination and special cover).

Patrick, William (*fl.* 1831), Scottish clergyman and botanist. (*W. Patrick*).
HERBARIUM and TYPES: Unknown.

BIBLIOGRAPHY and BIOGRAPHY: Barnhart 3: 54; BB p. 238; BL 2: 306, 700; BM 4: 1529; CSP 8: 571; Jackson p. 255, 257; Kew 4: 247; PR 6982 (ed. 1: 7804); Rehder 5: xx; Tucker 1: 543.

BIOFILE: Freeman, R.B., Brit. nat. hist. books 274. 1980.

7498. *A popular description of the indigenous plants of Lanarkshire*, with an introduction to botany, and a glossary of botanical terms ... Edinburgh (Daniel Lizars), Glasgow (W.E. M'Phun and A. Lottimer), Hamilton (James Thomson) 1831. Duod. (in sixes). (*Indig. pl. Lanarkshire*).
Ed. 1: 1831, map, p. [i]-xxxiv, [1]-399. *Copies*: NY, USDA.
Ed. 2: 1832. *Copy*: HH. – "*Popular* ... terms ... second edition. Edinburgh (id.), Glasgow (id.), Hamilton (id.). 1832.

Patterson, Flora (née Wambaugh) (1847-1928), American mycologist; married Edward Patterson 12 Aug 1869; A.M. Wesleyan Coll., Cincinatti 1882; id. Univ. Iowa 1895; at Radcliffe Coll., Harvard 1892-1895; pathologist, later mycologist, in the U.S. Dept. of Agriculture 1896-1923; in retirement in Brooklyn, N.Y. (*F. Patterson*).

HERBARIUM and TYPES: Material at BPI, NY and PAV. – See Stevenson (1971) and Patterson, F., USDA Bur. Plant Ind., Bull. 8: 1-31. 1902 and USDA Agric. Circ. 195: 1-50. 1922 for sets of reference collections sent to US agricultural experiment stations and other mycological herbaria.

BIBLIOGRAPHY and BIOGRAPHY: Barnhart 3: 55; BM 4: 1529; Bossert p. 302; CSP 17: 734; Ewan ed. 1: 77, 279, ed. 2: 170; Hirsch p. 226; Kelly p. 170; LS 37715-37177, suppl. 21077-21080; MW p. 379; NW p. 54; Rehder 4: 489; Stevenson p. 1253.

BIOFILE: Anon., Science ser. 2. 67: 189. 1928 (d. 5 Feb 1928); Torreya 28: 40. 1928; Who was who in Amer. 1: 942. 1968 (b. 15 Sep 1847, d. 6 Feb 1928).
Charles, V.K., Mycologia 21(1): 1-4. 1929 (portr., bibl.).
Galloway, F.W., Phytopathology 18(11): 877-879. 1928 (obit., bibl., d. 5 Feb 1928).
Pammel, L.H., Proc. Iowa Acad. Sci. 31: 58-59. 1924.
Rickett, H.W., Index Bull. Torrey bot. Club 76. 1955.
Rogers, D.P., Brief hist. mycol. N. Amer. ed. 2. 31, 32. 1981.
Shope, P.F., Mycologia 21: 296. 1929 (coll. Colorado 1916-1919; material to Lloyd; now BPI).Stevenson, J., Beih. Nova Hedw. 36: 260-264. 1971; Taxon 4: 183. 1935.

EPONYMY: The etymology of *Pattersonia* J.F. Gmelin (1792) and its orthographic variants *Patersonia* Poiret (1816) and *Pathersonia* Poiret (1816) has not been established, but possibly may commemorate William Paterson (1755-1810), q.v.

7499. *A collection of economic and other fungi* prepared for distribution ... Washington (Government Printing Office) 1902. Oct. (*Coll. econ. fung.*).
Publ.: 1902, p. [1]-31. *Copies*: G, U. – Bureau of Plant Industry, Bull. 8.

7500. *Mushrooms and other common fungi* ... Bulletin of the U.S. Department of Agriculture no. 175... Washington, D.C. (Government Printing Office) 1915. Oct. (*Mushrooms*).
Co-author: Vera Katharine Charles (1877-1954).
Publ.: 29 Apr 1915, p. [1]-64, *pl. 1-38*. *Copies*: MO, NY.

7501. *A list of fungi* (*Ustilaginales* and *Uredinales*) prepared for exchange ... Washington, D.C. (Government Printing Office) 1922. Oct. (*List fungi, Ustil., Ured.*).

Co-authors: William Webster Diehl (1891-?), Edith Katherine Cash (1890-x).
Publ.: Feb 1922, p. [1]-50. *Copy*: G.

Patterson, Harry Norton (1853-1919), American printer, botanist and explorer, editor proprietor of the Oquawka [Ill.]. Spectator 1884-1908 (*H. Patt.*).

HERBARIUM and TYPES: F (30.000 acquired for F by C.F. Millspaugh); further material in many herbaria e.g. C, CART, CM, DS, E, G, GH, HBG, ILL, ISC, L, LY, MANCH, MICH, MO, MSC, NY, NYS, OXF, QK, RPM, RSA, US, VT, WELC, WRSL. – Issued sets of exsiccatae: *Flora of Colorado* (seen: nos. 167-304, 1892), accompanied by printed lists. Manuscript material in Smithsonian Archives.

BIBLIOGRAPHY and BIOGRAPHY: Barnhart 3: 55; BL 1: 174, 175, 312; Bossert p. 302; Clokie p. 22; Ewan ed. 1: 61, 73-78, 80, 154, 279 (portr., biogr., bibl., b. 15 Feb 1853, d. 23 Mai 1919), ed. 2: 170; Kew 4: 247-248; Langman p. 570; Morren ed. 10, p. 119; NW p. 54; Pennell p. 616; Rehder 5: 652; Tucker 1: 145, 543.

BIOFILE: Allison, E.M., Univ. Colo. Stud. 6: 69. 1909 (coll.).
Coulter, J.M., Bot. Gaz. 30: 144. 1900 (herb., 30.000 sheets, acquired by F).
Ewan, J., Trail and Timberline 293: 62-63. 1943 (portr., chronology, further references to – mainly local and rare – eulogies and obituaries).
Ewan, J. et al., Short hist. bot. U.S. 45. 1969.
Hedge, I.C. & Lamond, J.D., Index coll. Edinb. herb. 117. 1970.
Hitchcock, A.S. & A. Chase, Man. Grass. U.S. 988. 1951.
Kibbe, A.L., ed., Harry Norton Patterson. Afield with plant lovers and collectors, botanical correspondence ... 1870-1919, 565 p., 1953.
Sargent, C.S., Silva 4: 24. 1891.

COMPOSITE WORKS: With S. Watson, q.v., *Catalogue of plants* collected in the years 1871, 1872, and 1873, Washington 1874.

NOTE: Patterson printed the early sets of the Gray Herbarium Index. ("Gray Cat.").

EPONYMY: Mt. Sniktau, Colorado, commemorates Patterson's brother E.H.N. Patterson as well as Patterson himself. "Sniktau" was the pseudonym used by P's brother as publisher of the Georgetown Miner, Colo. See Ewan (1943).

7502. *A list of plants collected in the vicinity of Oquawka*, Henderson County, Ills., ... Oquawka Spectator Print 1874. (*List pl. Oquawka*).
Publ.: Jan 1874 (p. i: Jan 1874), p. [i], [1]-18. *Copies*: MO, NY, USDA. – Cover title only.

7503. *Catalogue of the phaenogamous and vascular cryptogamous plants of Illinois* native and introduced. Oquawka, Ill. ("Spectator" Print) 1876. (*Cat. pl. Illinois*).
Publ.: Mar 1876 (p. [v]: 1 Mar 1876), p, [i-v], [1]-54. *Copies*: MICH, NY, USDA.

7504. *Check-list of North American Gamopetalae*, from Gray's Synoptical flora ... Oquawka, Ill. (H.N. Patterson, printer) s.d. [1880]. (*Checkl. N. Amer. Gamopet.*).
Publ.: 1880 (Bot. Gaz. Dec 1880), p. [i], [1]-43. *Copy*: MICH, NY, US, USDA. *Check-list of North American Gamopetalae after Compositae*. From Gray's Synoptical flora and later publications ... Oquawka, Ill. (id.). s.d., p. [1]-12. *Copies*: MICH, MO, NY. – Also 1880? We have not been able to establish a date for this pamphlet, which, as the 1880 list, is a simple enumeration of names.

7505. *Check list of North American Polypetalae*. From Watson's Bibliographical index and later publications ... [H.N. Patterson, Publisher, Oquawka, Ills.] Id. (*Check list N. Amer. Polypet.*).
Publ.: 1881 (Bot. Gaz. Apr 1881), p. [1]-20. *Copies*: MICH, MO, NY. – Errata slip with MO copy. A simple enumeration of names. – E. Wycoff, Bibl. contr. Lloyd Libr. 9: 396 dates this pamphlet at 1874.

7506. *Check-list of North American plants*, including Mexican species which approach the U.S. Boundary. Oquawka, Ills. (H.N. Patterson, publisher) 1887. Oct. (*Checkl. N. Amer. pl.*).
Publ.: Jan 1887 (TBC Feb 1887; US copy(1) handwritten note: Mar 1887), p. [3]-151.
Copies: MO, NY, PH, US(2), USDA. – Errata slip with US (first copy).
Ref.: Anon., Bull. Torrey bot. Club 14: 36. 1887.

7507. *Patterson's numbered check-list of North American plants* north of Mexico ... Oquawka, Ill. (H.N. Patterson, publisher) January, 1892. (*Numb. checkl. N. Amer. pl.*).
Publ.: in 4 issues, differing only in paper and binding, all Jan 1892.
Issue: "Style A. – Price $ 1.50" Herbarium list, p. [i-iii], p. [1]-158. *Copies*: MO, PH.
Issue: "Style B. – Price $ 1.75": Jan 1892, p. [i-iii], [1]-158. *Copies*: NY, US, USDA. – Herbarium list, printed on one side.
Issue: "Style C. – Price 75c.": Jan 1892, p. [i-iii], [1]-158. *Copies*: US, USDA. – Mailing list, light paper, paper covers.
Issue: "Style D. – Price $ 1.00": Jan 1892, p. [i-ii], [1]-158. *Copies*: NY, US. – As issue C, but printed on one side.*Collaborators*: William Trelease (1857-1945): Epilobium; Charles Frederick Millspaugh (1854-1923): Euphorbia; Michael Schuck Bebb (1833-1895): Salix; George Vasey (1822-1893): Gramineae; George Edward Davenport (1833-1907): Filices and Ophioglossaceae.
General advice: Sereno Watson (1826-1892).
Number of species: 10.706 (12.794 taxa).
Ref.: Britton, N.L., Bull. Torrey bot. Club 19: 101. 5 Mar 1892.

Patterson, Paul Morrison (1902-x), American bryologist; A.M. Univ. N. Carolina 1927; asst. prof. botany Davidson Coll. 1927-1928; assoc. prof. S. Carolina Univ. 1930-1933; Dr. phil. Hopkins Univ. 1933; from 1934-1966 prof. biology Hollins College, Va. (*P. Patt.*).

HERBARIUM and TYPES: NY (bryophytes). – Some material in Smithsonian Archives.

BIBLIOGRAPHY and BIOGRAPHY: Barnhart 3: 55; Bossert p. 303; GR p. 237; IH (ed. 6): 362, 2: (in press); LS 21081-21082; SBC p. 129.

BIOFILE: Anon., Rev. bryol. n.s. 4: 50. 1931, Rev. bryol. lichénol. 8: 126. 1935, 19: 242. 1950 (president Amer. bryol. Soc.); 19: 250. 1950, 34: 394. 1966.
Ewan, J. et al., Short hist. bot. U.S. 95. 1969.
Rogers, D.P., Brief hist. mycol. N. Amer. 31-32. 1977.
Schuster, R.M., Hepat. Anthoc. N. Amer. 1: 86. 1966 (bibl.).
Steere, W.C., Garden J. 1968 (Jul/Aug): 116 (Bryoph. herb. Virginia, 8000 specimens, to NY in 1966).
Verdoorn, F., ed., Chron. bot. 2: 404. 1936.

Pattison, Samuel Rowles (1809-1901), British palaeobotanist of Launceston, Cornwall. (*Pattison*).

HERBARIUM and TYPES: Unknown.

BIBLIOGRAPHY and BIOGRAPHY: Barnhart 3: 55; BM 4: 1530; CSP 8: 571, 10: 1002, 17: 735; Desmond p. 482; Jackson p. 177; Quenstedt p. 330.

BIOFILE: Allibone, S.A., Crit. dict. Engl. litt. 1527. 1870 (bibl.).
Anon., Geol. Mag. ser. 2. Dec 4. 9: 48. 1902 (obit., d. 27 Nov 1901); Science 65: 109. 1901 (obit.).
Boase, G.C. and W.P. Courtney, Bibliotheca Cornubiensis 430-431. 1874 (fide Desmond).
Bridson, G.D.R. et al., Nat. hist. mss. res. Brit. Isl. 240.117. 1980.
H.B.W., Quart. J. Geol. Soc. Lond. 58: lxii. 1902 (obit.).

7508. *Chapters on fossil botany* ... London (Simpkin, Marshall & Co.) 1849. Duod. (in sixes). (*Chapt. foss. bot.*).
Publ.: 1849, p. [iii]-xii, [13]-223. *Copies*: NY, PH.

Patze, Carl August (1808-1892), German (Berlin) botanist and pharmacist; from 1830-1836 in pharmacies in Berlin and Charlottenburg (1829-1830 in Breslau); from 1837-1874 owner of a pharmacy in Königsberg i. Pr. (*Patze*).

HERBARIUM and TYPES: Patze had a considerable private herbarium (ca. 7.000 sheets); it went to the old Königsberg herbarium which was destroyed in World War II. Some material coming from Patze is still at C.

BIBLIOGRAPHY and BIOGRAPHY: AG 4: 224; Barnhart 3: 55; BM 4: 1530; CSP 17: 735; Herder p. 470; IH 2: (in press); Morren ed. 10, p. 14; PR 6983; Rehder 1: 383; TL-2/see E.H.F. Meyer.

BIOFILE: Abromeit, J., Schr. phys. ökon. Ges. Königsberg 33: 131-133. 1892 (b. 24 Sep 1808).
Candolle, Alph. de, Phytographie 439. 1880 (types at KBG).
Caspary, R., Lebensbeschr. Ost– und Westpreuss. Bot., Festschr. 50j. Best. Preuss. bot. Ver. 254-255. 1912 (bibl.).

EPONYMY: *Patzea* R. Caspary (1872).

7509. *Flora der Provinz Preussen* ... Königsberg (Verlag der Gebrüder Bornträger) 1850. Oct. (*Fl. Preuss.*).
Co-authors: Ernst Heinrich Friedrich Meyer (1791-1858), Ludwig Elkan (1815-1850).
Publ.: in 3 parts (see pref. p. iv). *Copies*: B, BR, G, NY, REG; IDC 5633.
1: [1]-176. Apr 1848 (Hinrichs 9-12 Aug, Flora 9 Jun 1848).
2: 177-383. Feb 1849 (Hinrichs 23-24 Apr, Flora Apr 1849).
3: [i]-xl, 385-599. Mar 1850 (Hinrichs 29 Mai-1 Jun, Flora Apr 1850).
Ref.: Fürnrohr, A.E., Flora 31: 648-653. 28 Oct 1848, 32: 377-379. 28 Jun 1849, 33: 603-607. 14 Oct 1850.
Tournay, R., Taxon 12: 36. 1963.
Schlechtendal, D.F.L. von, Bot. Zeit. 6: 743-744. 20 Oct 1848, 7: 591-592. 10 Aug 1849.

Patzelt, Joseph Eduard (*fl.* 1842), Austrian botanist and physician in Wien, later in Bucarest. (*Patzelt*).

HERBARIUM and TYPES: Some material in DBN, E, K and PR.

BIBLIOGRAPHY and BIOGRAPHY: Barnhart 3: 55; BM 4: 1530; Kláštreský p. 140; PR 6984 (ed. 1: 7806); Rehder 1: 440.

BIOFILE: Hedge, I.C. & Lamond, J.M., Index coll. Edinb. herb. 117. 1970.
Neilreich, A., Verh. zool.-bot. Ges. Wien 5 (Abh.): 57. 1855.

7510. *Wildwachsende Thalamifloren der Umgebungen Wien's.* Inaugural-Dissertation von Joseph Eduard Patzelt, ... Wien (gedruckt bei Carl Heberreuter) 1842. (*Thalamifl. Wien*).
Publ.: 1842, p. [i]-viii, [9]-90, [1, index]. *Copy*: NY.

Pau [y Español], **Carlos** (1857-1937), Spanish botanist and pharmacist at Segorbe. (*Pau*).

HERBARIUM and TYPES: MA; further material in AK, B (extant; Herb. hisp. 1893-1894), BR, E-GL, FI, G, GB, GOET, K, L, LIVU, LRMANCH, PH, PI, W.

BIBLIOGRAPHY and BIOGRAPHY: AG 5(1): 734; BL 1: 46, 312, 2: 700 [index]; BM 4: 1530;

Bossert p. 303; CSP 17: 735-736; Hortus 3: 1201 (Pau); IH 1 (ed. 7: 339; Kew 4: 248; Rehder 5: 653; Tucker 1: 543; Urban-Berl. p. 384; Zander ed. 10, p. 699-700, ed. 11, p. 798.

BIOFILE: Allorge, P., Rev. bryol. lichén. 7: 304. 1935 (on P's musci from Marocco).
Anon., Bull. Acad. Géogr. bot. 9: 16. 1900; Encicl. Univ. illustr. 42: 928-929 (portr., b. 10 Mai 1857).
Bellot, E., Anal. r. Acad. Farm. 8(1): 1-33. 1942 (biogr., add. bibl.).
Bolós, A., Coll. bot. 4(2): 203-205. 1954 (add. bibl.) (incl. titles listed by Bellot).
Carrasco, M.A., Contribución a la obra taxonómica de Carlos Pau. Trab. Dep. Bot. Univ. Madrid 7: 35-37. 1975 (add. bibl.), 8: 1-331. 1975 (indexes to taxa).
Fernández-Galiano, E., Anal. r. Acad. Farm. 25: 235-236. 1959 (add. bibl.).
Font Quer, P. y J. Cuatrecasas, eds., Cavanillesia 8(8): 115-132. 1937 (main bibl.).
Hedge, I.C. & Lamond, J.M., Index coll. Edinb. herb. 117. 1970.
Verdoorn, F., ed., Chron. bot. 4: 72. 1938 (d.).
Willkomm, M., Grundz. Pfl.-Verbr. Iber. Halbinsel 18, 26. 1896 (Veg. Erde 1).

MEMORIAL PUBLICATION: Cavanillesia 8(8), 1937 (portr., bibl., b. 10 Mai 1857, d. 9 Mai 1937).

HANDWRITING: Rec. Auckland Inst. Mus. 12: 118. 1975; Candollea 32: 381-382. 1977.

EPONYMY: *Paua* Caballero (1916). *Note*: *Pauahia* F.L. Stevens (1925) honors the Hawaiian Princess Bernice Pauhai; *Pauella* Ramamurthy et Sebastine (1967) and *Pauia* D.B. Deb et R. Dutta (1965) were named in honor of H. Santapau, Director of the Botanical Survey of India.

7511. *Notas bótanicas à la flora española* ... Madrid (Escuela Tipográfia del Hospicio ...) 1887. Qu. (*Not. bot. fl. españ.*).
Publ.: 6 fasc., 1887-1895. *Copies*: BR, G, NY.
 1: 1887, (p. 6: Jul 1887; Nat. Nov. Sep (1, 2). 1888, p. [1]-40.
 2: 1888, p. [1]-40.
 3: 1889, p. [1]-40.
 4: 1891, p. [1]-52. Reprinted from Semanario farmaceutico. French translation by O. Debeaux, Rev. bot. 9: 681-700. 1892.
 5: 1892, p. [1]-28.
 6: 1895 (p. [52: 16 Jun 1895), p. [1]-115. – *Imprint*: Segorbe (Imprenta de Romaní y Suay ...) 1895.
Ref.: Willkomm, M., Bot. Centralbl. 35: 238. 1888 (critical rev. part 1).

7512. *Gazapos botánicos* gazados en las obras del Senor Colmeiro que es director del Jardin botánico de Madrid, ... Segorbe (Imprente y libreria de Federico Romaní y Suay) 1891. Oct. (*Gazapos bot.*).
Publ.: 1891, p. [1]-72, [1, 4 notes and err.]. *Copy*: G. – Critical appraisal of the botanical work of Miguel Colmeiro y Penido (1816-1901): "Hablar mal del que obra mal, hupo, es hacerle justicia" Forner (on t.p.).

7513. *Herborizaciones por Valldigna, Jálisa y Sierra Mariola* en los meses de Abril, Mayo y Junio de 1896. [Anal. Soc. Esp. Hist. Nat.]. 1899. Oct.
Publ.: 1899, p. [1]-42. *Copies*: G, MO. – Anal. Soc. Esp. Hist. nat. 27: 411-452. 1899.

7514. *Plantas de Persia y de Mesopotamia* recogidas por D. Fernando Martínez de la Escalera ... Madrid 1918. Oct. (*Pl. Persia Mesop.*).
Publ.: *10 Jun 1918, p.[3]-48, pl. 1-5* (no. 2 col.). *Copy*: BR. – Issued as Trab. Mus. nac. Ci. nat., Bot. 14, 1918.

7515. *Nueva contribución al estudio de la flora de Granada* pel Dr. Carles Pau ... Barcelona (Museu de Ciències Naturales) 1922. Qu. (*Nueva contr. fl. Granada*).
Publ.: Sep 1922, p. [1]-74, *pl. 1-10* (uncol.). *Copy*: BR. – Published as vol. 1(1) of Mem. Mus. Ci. nat. Barcelona, Bot.

7516. Contribución a la flora española *plantas de Alméria* ... Barcelona (Museo de Ciencias naturales) 1925. Qu. (*Pl. Alméria*).
Publ.: 15 Mai 1925, p. [1]-34, *pl. 1-4* (uncol.). *Copies*: BR, NY. – Mem. Mus. Ci. nat. Barcelona, Bot. 1(3), 1925.

Paul, Hermann Karl Gustav (1876-1964), German bryologist; student of A. Engler; Dr. phil. Berlin 1902; assistant at the Kön. bayerischen Moorversuchsstation Bernau, Bayern, 1902-ca. 1943. (*H. Paul*).

HERBARIUM and TYPES: M ("Paul's Sammlungen gingen, insgesammt wohl komplett, in vielen Lieferungen geschenkweise zu" Hertel 1980); further material in B, E, H, L, W.

BIBLIOGRAPHY and BIOGRAPHY: Barnhart 3: 56; BJI 2: 131; Christiansen no. 112; Futák-Domin p. 457; GR p. 119; Hirsch p. 226; LS suppl. 21088-21091; SBC p. 129; Stevenson p. 1253; TL-2/406, 1711, see H. Marzell; Urban-Berl. p. 280, 310.

BIOFILE: Anon., Bot. Jahrb. 31 (Beibl. 70): 30. 1902.
Familler, I., Denkschr. bay. bot. Ges. Regensburg 11: 20. 1911 (bibl.).
Hedge, I.C. & Lamond, J.M., Index coll. Edinb. herb. 117. 1970.
Hertel, H., Mitt. Bot. München 16: 415. 1980.
Iltis, H., Verh. naturf. Ver. Brünn 50: 312. 1912.
Isoviita, P., Ann. bot. fenn. 3: 260. 1966.
Mägdefrau, K., Hoppea 37: 147, 157-158. 1978 (portr.).
Paul, H., Lebenslauf, *in* his Beitr. Biol. Laubmoosrhiz. p. [2]. 1902 (b. 6 Aug 1876); Feddes Rep. 58: 10-11. 1955 (relations with Th. Herzog).
Pax, F., Bibl. schles. Bot. 113, 43. 1929.
Poelt, J., Ber. bayer. bot. Ges. 37: 69-76. 1964 (obit., b. 6 Aug 1876, d. 22 Jan 1964, bibl.).
Sayre, G., Bryologist 80: 515. 1977.
Schindler, L., Landw. Jahrb. Bayern 27: 65-72. 1950.
Verdoorn, F., ed., Chron. bot. 1: 442 [index]. 1935, 2: 159, 160, 166. 1936, 3: 139. 1937, 4: 445. 1938, 5: 113. 1939.

COMPOSITE WORKS: (1) EP, *Die natürlichen Pflanzenfamilien* ed. 2, *Sphagnaceae*, 10: 105-125. Mai-Jun 1924.
(2) A. Pascher, *Süssw.-Fl. Mitteleur.* ed. 2, *Sphagnales, in* 14:[1]-46. 1931 (see e.g. Rev. bryol. lichén. 4: 95. 31 Dec 1931).

EPONYMY: *Paulia* Fée (1836) is dedicated to Fée's late son Paul; *Paulia* C.G. Lloyd (1916) is named for J.T. Paul, an Australian collector of fungi; *Paulina* Krempelhuber (1869) is an *orth. var.* of *Pulina* Adanson (1763); *Paulomyces* F.W. Sommer (1954) and *Paulophyton* E. Dolianiti (1954) honor Paulo Erichsen de Oliveira, Brazilian paleobotanist.

7517. *Beiträge zur Biologie der Laubmoosrhizoiden* ... Inaugural-Dissertation zur Erlangung der Doktorwürde von der philosophischen Fakultät der Friedrich-Wilhelms-Universität zu Berlin genehmigt und mit den beigefügten Thesen öffentlich zu verteidigen am 12. November 1902 ... Leipzig (Wilhelm Engelmann) 1902. Oct. (*Beitr. Biol. Laubm.-Rhiz.*).
Publ.: 12 Nov 1902, p. [1]-48, [1, 2, vita]. *Copies*: G, NY.
Regular issue: in A. Engler, Bot. Jahrb. 32(2): 231-274. 24 Apr 1903.

7518. *Ueber Hypnum turgescens* T. Jensen eine systematisch-geographische Studie von Dr. H. Paul, München [Bryol. Zeitschrift Bd. 1, Heft 10. 1918]. Oct. †.
Publ.: Sep 1918, p. 145-160. *Copies*: HBG, M. – Reprinted and to be cited from Bryol. Z. 1(10): 145-160. 1918. – The *Bryologische Zeitschrift* was discontinued after Heft 10 of volume 1. The last part of Paul's paper was still to follow. After the discontinuation of the Bryol. Z. Paul's paper was published again and then completely in Kryptogamische Forschungen, München, 6: 408- 419. Jul 1924. (Reprinted with ad. pagination 1-12). *Copy*: H). The original publication contains on p. 155 the publication of *Loeskypnum* nov. gen. (from Loeske and Hypnum; see eponymy under Loeske. Information K. Mägdefrau).*Bryologische Zeitschrift*: vol. 1, Heft 1-10, 1916-1918 (all published), founded

and edited by Leopold Loeske; Heft 1: 1-16. Jul 1916, 2/3: 17-48. Aug 1916, 4: 49-64. Sep 1916, 5: 65-80. Oct. 1916, 6: 81-96. Jan 1917, 7: 97-112. Jan 1917, 8: 113-128. 15 Jul-3 Oct 1918, 9: 129-144. 15 Jul-3 Oct 1918, 10: 145-160. 15 Jul-3 Oct 1918. *Copies*: B(1-7), DUIS(1-10, ex. F. Koppe), JE (1-10), Mägdefrau(1-7), NICH (1-7), NY (2-5). (Inf. K. Mägdefrau, T. Koponen & P. Isoviita.).
Ref.: Koponen, T. & P. Isoviita [paper on Loeske's *Bryol. Z.*, in press].

Paulet, Jean Jacques (1740-1826), French physician and mycologist. Studied medicine at Montpellier; Dr. med. ib. 1764; practicing physician in Paris 1764-1802; in retirement at Fontainebleau. (*Paul*).

HERBARIUM and TYPES: Unknown.

BIBLIOGRAPHY and BIOGRAPHY: Ainsworth p. 184, 226, 279-280, 308, 312, 346; Backer p. 428; Barnhart 3: 56; BM 4: 1530-1531; Dryander 3: 225, 348, 552; Hawksworth p. 185; Herder p. 244; Jackson p. 22, 162; Kelly p. 170; Kew 4: 250; LS 20159-20166; NI 1496; Plesch p. 359; PR 5262, 6985-6989, 9203 (ed. 1: 7807-7811); Rehder 5: 653; Stevenson p. 1253; TL-1/676, 957-958; TL-2/909, 4215, 4470; Tucker 1: 543.

BIOFILE: Anon., Cat. gén. livr. impr. Bibl. natl. 131: 594-598. 1935 (bibl.).
Dawson, W.R., Smith papers 72. 1954.
Demoulin, V. et al., Taxon 30(1): 58-59. 1981 (maintains that *Traité champ.* was not published in 1793).
Donk, M.A., Persoonia 1: 176-179. 1960, 2: 201-202. 1962, Beih. Nova Hedwigia 5: 10-11. 1962 (notes on Paulet's nomenclature).
Killermann, S., Z. Pilzk. 10: 34-35. 1931 (b. 26 Apr 1740, d. 4 Aug 1826).
Laplanche, M.C., Dict. iconogr., champ. sup. 1894 (correlates Paulet's names of hymenomycetes with the Friesian names.), TL-2/4215 (rev. Britzelmayr, Bot. Centralbl. 61: 195-196. 1895).
Lütjeharms, W.J., Gesch. Mykol. 10-12, 247, 259. 1936.
O., Dict. sci. méd., Biogr. méd. 6: 378-380. 1824.
Raab, H., Schweiz., Z. Pilzk. 48: 15, 16, 55. 1970-1973.
Rivier, G., Bull. Soc. franç. Hist. Méd. 22: 216-218. 1928.(on Paulet as "praticien parisien"; b. 27 Apr 1740, d. Oct 1826).
Terson, Bull. Soc. mycol. France 44(4): 377-378. 1928 (on Paulet manuscripts; taken from G. Rivier).
Uellner, W., Fungorum libri bibliothecae Joachim Schliemann nos. 1317-1319. 1976.
Wittrock, V.B., Acta Hort. Berg. 3(3): 110. 1905.

EPONYMY: *Pauletia* Cavanilles (1799).

7519. *Traité des champignons*, ouvrage dans lequel, on trouve, après l'histoire analytique & chronologique des découvertes & des travaux sur ces plantes, suivie de leur synonimie botanique & des tables nécessaires, la description détaillée, les qualités, les effets les differens usages non-seulement des champignons proprement dits, mais des truffes, des agarics, des morilles, & des autres productions de cette nature, avec une suite d'expériences tentées sur les animaux, l'examen des principes pernicieux de certaines espèces, & les moyens de prévenir leurs effets ou d'y remédier; le tout enrichi de plus de deux cents planches où ils sont représentées avec leurs couleurs, & en général leurs grandeurs naturelles, & distribués suivant une nouvelle méthode ... À Paris (De l'Imprimerie nationale exécutive du Louvre) [1790-]1793. 2 vols. Qu. (*Traité champ.*).
1: 1790, (p. xxxvi, 15 Jun 1790), p. [i*-v*], [i]-xxxvi, [1]-629. *Copies*: B, Paris (Bibl. natl.). With imprint: " A Paris, de l'Imprimerie royale, mdcclxxxx". The Paris copy has a second t.p., dated 1793, immediately following the first, 1790. All other copies seen by us are as above and as follows:
1: 1793, p. [i*, iii*], [i]-xxxvi, [1]-629.
2: 1793, p. [i*, iii*], [i]-viii, [1]-476, index [1-11], [1, err.].
Copies: BR, FH(2), G, LG, NY, Paris (Bibl. natl., second copy), Stevenson, USDA; IDC 5634. The second FH copy has bound in the prospectus of 1808 (see below).
Plates: published between 1808 and 1835 as *Iconographie des champignons*, in fascicles of

which the details are not yet known. According to Nissen there were 42 fasc. of which nos. 31-42 came out posthumously. The maximum number of plates is said to be *232* on *217* leaves; however, we have been able to find only 223 numbers. The Stevenson copy (which lacks t.p. and introduction) has the following plates:

1: portr. Paulet, *1-17*, *17bis*, *18-19*, *20bis*, *21-22*, *22bis*, *23-24*, *24bis*, *25-35*, *35bis*, *36-39*, *69bis*, *70-100*, *100bis* (col. copp. Fossier del.) [Lacking 97bis].

2: 101-129, 129bis, 130, 132-135, 135bis, 136-139, 139bis, 140, 140bis, 141-148, 148ter, 149-156, 156bis, 156bis, 157-168, 168bis, 169-183, 183bis, 184-190, 190bis, 191-195, 197-204 (id.). [Lacking 112bis, 122bis, 166bis].

Plates: re-issue with new text: see J.H. Léveillé, *Iconographie des champignons de Paulet*, Paris 1855, q.v. (TL-2/4470).

Note on date: V. Demoulin et al. (1981) hold that the text was not published in 1790 and 1793, but in 1808. The evidence contained in the 1808 prospectus that no copies at all of the text were distributed in 1793 is not convincing. The circumstance that there are two issues (with different imprints), of vol. 1, one dated 1790 and one 1793 could point at the distribution of at least a few copies. It is important to note Demoulin et al.'s reference to the difference in nomenclature in the text and on the plates. However, the fact that Paulet, in 1808, no longer accepted the 1790-1793 names has no bearing on their nomenclatural standing if the dates of 1790-1793 are accepted as that of effective publication of the text. Dr. Demoulin informs us that most names in dispute occur in the second volume (1793). The question remains unresolved and depends on the interpretation of the sentence in the 1808 prospectus quoted by Demoulin.

7520. *De la mycétologie*, ou traité historique, graphique, culinaire, et médical des Champignons [Paris circa 1812]. (*Mycétologie*).
Publ.: The only copy known to us is that mentioned by Donk (1960), which is apparently a re-issue of the old sheets with a new title-page [1835]. This re-issue may be different for each copy because sample plates are added. For further details see Donk (1960).
Ref.: Donk, M.A., Persoonia 1: 176-179. 1960.

7521. *Prospectus du traité* historique, graphique, culinaire, et médical *des champignons*, 2 volumes in 4;; de l'imprimerie royale. Par M. Paulet, docteur en médicine. Prix, 18 fr. broché, 24 fr. relié, y compris le Prospectus, sans les figures coloriées de cet ouvrage, qui seront vendues avec, ou séparément, et qui paraîtront par livraison de plusieurs planches, de mois en mois. Les deux premièrs sont en vente. On commence par les champignons de bonne qualité. Se trouve à Paris, chez M.me Huzard, ... et au dépôt général, Tessier, ... 1808. Qu. (*Prosp. Traité champ.*).
Publ.: 1808, p. [i], [1]-88. *Copies*: FH, LG. – The prospectus accompanies a re-issue of the 1793 sheets (not seen by us) together with the first two fascicles of the plates. – See p. 42 for statement on delay in publication *Traité champ*.
Other issue: s.d. p. [1]-52. *Copy*: MICH. – "Prospectus du traité des champignons, imprimé par ordre du gouvernement, à l'Imprimerie royale. 2 volumes in 4; et 232 planches (bis et doubles comprises), de champignons gravés et coloriés d'après nature. Prix 230 francs. Par Jean-Jaques Paulet, ... s'adresser à M. Hoüel Paulet, chez MM Mallet frères, ...". The reference to the "Imprimerie royale" refers to the original 1790 issue: the number of plates mentioned (232) would refer to the completed work (ca. 1835).

7522. *Examen d'un ouvrage* qui a pour titre: Illustrationes Theophrasti, in usum botanicorum praecipue peregrinantium. Autore Joh. Stackhouse, ... À Melun (de l'imprimerie de Lefèvre-Compigny;) et se trouve à Paris (chez M. de Huzard, ...) 1816. Oct. (*Exam. ouvr.*).
Author of pamphlet discussed: John Stackhouse (1742-1819).
Publ.: 1816 (p. 61: 1 Mai 1816), p. [i], [1]-61, err. slip. *Copy*: G.

7523. *Flore et faune de Virgile*, ou histoire naturelle des plantes et des animaux (reptiles, insectes) les plus intéressans à connaître et donc ce poete a fait mention. À Paris (chez Madame Huzard, ...) 1824. Oct. (*Fl. faune Virgile*).
Publ.: 1824, front., p. [i*-vi*], [i]-xix, [1]-159, *pl. 1-3* (uncol.). *Copy*: G.

Paulin, Alphons (1853-1942), Austrian botanist in Slovenia; director of the botanical garden of the University of Ljubljana (Laibach) 1886-1931; Dr. phil. Graz 1880. (*Paulin*).

HERBARIUM and TYPES: GJO; further material at BM, E, GB, GZU, LE. *Exsiccatae: Flora exsiccata Carniolica*, nos. 1-1000, 1901-1907, see below under *Beitr. Veg. Verh. Krains* and Allg. bot. Z. 7: 143. 1901. – See Wraber (1966) for contents and history and continuation of the series of exsiccatae, with critical notes on centuries 19 and 20.

BIBLIOGRAPHY and BIOGRAPHY: BFM nos. 1135-1137; BM 4: 1531; CSP 17: 739; Hirsch p. 226; IH 1 (ed. 7): 339, 2: (in press).

BIOFILE: Hedge, I.C. & Lamond, J.M., Index coll. Edinb. herb. 117. 1970.
Wraber, T., Acad. Sci. Art. sloven., Class. 4, Hist. nat., diss. 9(3): 127-163. 1966 (on his *Flora exsiccata carniolica* with index to generic names for all printed schedae), also separately as *Paulinova* "Flora exsiccata carniolica" xix. in xx. centurija, Ljubljana 1966, p. [1]-40. *Copy*: WU. (Also: handwriting, portr., brief biogr., b. 14 Sep 1853, d. 1 Dec 1942).

7524. Schedae ad floram exsiccatam Carniolicam. *Beiträge zur Kenntniss der Vegetationsverhältnisse Krains* ... Laibach [Ljubljana] (1-2: Otto Fischer; 3: im Selbstverlage ...) 1901-1904, 3 Hefte. Oct. †. (*Beitr. Veg.-Verh. Krains*).
1: Jun 1901 (p. viii: Mai 1901; Nat. Nov. Jul(2) 1901; ÖbZ Jul-Aug 1901; Bot. Zeit. 16 Sep 1901; Allg. bot. Z. 15 Sep 1901), p. [ii]-viii, [1]-104. – Alt. t.p.: *Schedae ad floram exsiccatam Carniolicam* auctore A. Paulin. I. Centuria i. et ii. Labaci [Laibach] O. Fischer 1901.
2: 1902 (ÖbZ Dec 1902-Jan 1903; Bot. Centralbl. 4 Aug 1903; Bot. Zeit. 16 Mai 1903), p. [ii-iii], 105-214. – Id.: "..II. Centuria iii. ed. iv. ...".
3: Jan-Mai 1904 (ÖbZ Mai-Jun 1904; Nat. Nov. Jul(1) 1904; Centralbl. 4 Jul 1904; G copy rd. 7 Oct 1904; Bot. Zeit. 16 Jul 1904; Allg. bot. Z. 15 Sep 1904), p. [ii-iii], 215-308. – Id.: "... III. Centuria v. ed. vi. Labaci 1904." – The pages [305]-308 contain an index to part 3.
4: 1905 (ÖbZ Aug/Sep 1906), p. [i], 305-340. The pages 305-308 contain new text "... IV. Centuria vii. et viii. Labaci. 1905.
5: 1907 (n.v., fide Wraber), p. [i], 341-378, V. Centuria ix. et x.
For the continuation, parts 6-10, cent. 11-20, 1926-1936 see Wraber: the labels were no longer printed but simply handwritten. *Copies*: B (1-3), FI (1-2), G(1-3), NY (1-3), WU (1-4). – For reviews of the exsiccatae, cent. 1-6, see F. Matouschek, Bot. Centralbl. 89: 350-351. 1902, 92: 344-345. 1903, 96: 15. 1904.

7525. *Die Farne Krains* ... Laibach [Ljubljana] (Buchdruckerei von Ig. v. Kleinmayr & Fed. Bamberg) 1906. Qu. (*Farne Krains*).
Publ.: 1906, p. [1]-44. *Copies*: FI, M. – Reprinted or preprinted from Jahresber. k.k.I. Staatsgymnasiums in Laibach 1906 (journal publ. n.v.). – Krain, or Carniola, Kranj., is now part of Slovenia, Yugoslavia.

7526. *Übersicht der in Krain bisher nachgewiesenen Formen aus der Gattung Alchemilla L.* ... Laibach [Ljubljana] (Buchdruckerei Ig. v. Kleinmayr & Fed. Bamberg) 1907. Qu. (*Übers. Krain Alchemilla*).
Publ.: 1907, p. [1]-19. *Copy*: FI. – Reprinted or preprinted from Jahresber. k.k.I. Staatsgymnasiums Laibach 1907 (journal publ. n.v.).

Paull, Charles Leslie Fairbanks (*fl.* 1904), American botanist, AM Brown Univ. 1898, assistant Kansas State College 1902-1904; from 1904 with the USDA Bureau of Plant Industry. (*Paull*).

HERBARIUM and TYPES: Some material at CM.

BIBLIOGRAPHY and BIOGRAPHY: Barnhart 3: 57; NW p. 54.

7527. *A key to the Spring flora of Riley* and Pottawatomie Counties in flower and fruit, March to June, inclusive. By Leslie F. Paull, A.M., ... Manhattan, Kansas (Kansas State Agricultural College, published by the Botanical Department) 1903. (*Key spring fl. Riley*).
Publ.: 1903 (p. [2]: 7 Mai 1903), p. [1, err.], [3, t.p.]-95. *Copies*: US, USDA.

Paulli [Pauli], **Johan** (1732-1804), Danish botanist and physician in Copenhagen. (*Paulli*).

HERBARIUM and TYPES: Unknown.

BIBLIOGRAPHY and BIOGRAPHY: Barnhart 3: 56; BM 4: 1531; Dryander 3: 557; Frank 3(Anh.): 74; PR 6991 (ed. 1: 7817); Rehder 3: 66.

BIOFILE: Christensen, C., Danske bot. hist. 1: 105, 110. 1924, 2: 52-53. 1924 (bibl., b. 1732, d. 28 Feb 1804).
Warming, E., Bot. Tidsskr. 12: 58-59. 1880.

EPONYMY: *Paullia* Cothenius (1790, *orth. var.*) and *Paullinia* Linnaeus (1753) were named for Simon Paulli (1603-1680), botanical teacher and physician to the king of Denmark.

7528. *Dansk oeconomisk Urte-Bog*, hvori endeel vilde Vaexter og Urter beskrives, og deres Nytte vises, sammendraget af Johan Paulli. Kiøbenhavn (Trykt hos Thomas Larsen Borup, ...) 1761. Oct. (*Dansk oecon. Urte-Bog*).
Publ.: 1761, p. [i-xvi], [1]-510. *Copy*: USDA. – See Christensen, 1924, p. 53, for the relation of this book to J.B. Ehrhart, Oecon. Pflanzenhist. 1753-1762, TL-2/1647.

Paulsen, Ove Vilhelm (1874-1947), Danish botanist; M.Sc. Copenhagen 1897; Dr. phil. ib. 1911; in the Danish West Indies 1895-1896, on the Pamir expedition 1898-1899; from 1900-1918 curator at the Botanical Museum, Copenhagen; professor of botany at the Danish pharmaceutical college 1920-1944. (*Paulsen*).

HERBARIUM and TYPES: C. – Further material at F, KIEL, LE, NY, S, WRSL. – Some letters at G.

BIBLIOGRAPHY and BIOGRAPHY: Barnhart 3: 57; BL 2: 53, 700; BM 8: 989; CSP 17: 740; Ewan ed. 2: 170; Hirsch p. 226; Kew 4: 250-251; Kleppa p. 6, 11, 58; LS 20171; MW p. 379; PH 450; TL-2/see C. Ostenfeld; Tucker 1: 543.

BIOFILE: Anon., Bot. Centralbl. 83: 176. 1900 (return from C. Asia); Nat. Nov. 20: 214. 1898 (to Pamir, C. Asia), 22: 107. 1920 (prof. bot. Copenh.); Österr. bot. Z. 67: 404. 1918 (inspector Bot. Mus.), 69: 192. 1920 (prof. bot.).
Christensen, C., Danske bot. litt. 1880-1911: 202-204. 1913 (portr., bibl., b. 22 Mar 1874); Danske bot. hist. 1: 760, 766, 773, 810, 812, 825, 828, 831, 835. 1926, Danske bot. litt. 1912-1939: 80-82 (portr., bibl.).
Ewan, J. et al., Short hist. bot. U.S. 130. 1969.
Fedtschenko, B.A., Bot. Zhurn. 30: 39-40. 1945.
Fox Maule, A., Københavns Universitet 1479-1979, 13 Bot.: 223- 224. 1979.
Hansen, A., Dansk bot. Ark. 21: 69. 1963 (q.v. for further biogr. refr.); Bocagiana 51: 10. 1980 (coll. Madeira).
Jepson, W.L., Madroño 1: 12-18. 1916 (on Int. phytogeogr. Excursion, 1913).
Jespersen, P., Naturh. Tidende Copenhagen, 11: 42-44. 1947 (obit., portr., b. 22 Mar 1874, d. 29 Apr 1947).
Jessen, K., Bot. Tidsskr. 48: 135-140. 1947 (obit., portr.).
Kiaerskou, H., Bot. Tidsskr. 23: 42-43. 1900.
Kneucker, A., Allg. Bot. Z. 6: 11. 1900 (return from Pamir), 24/25: 32. 1920.
Kornerup, T., Overs. medd. Grønland 1876-1912, p. 96. 1913.
Millspaugh, C.F., Publ. Field Mus. 68, Bot. 1: 458. 1902.
Rydberg, P.A., Augustana Coll. Libr. Publ. 5: 49. 1907.
Steemann Nielsen, E., Naturens Verden 31: 33-37. 1947 (d. 29 Apr 1947, obit., portr.).

Urban, I., Symb. ant. 1: 14. 1898, 3: 99. 1902.
VanLandingham, S.L., Cat. diat. 7: 4214. 1978.
Wainio, E.A., Bot. Tidsskr. 26(2): 241. 1904 (on his Pamir lichens).
Verdoorn, F., ed., Chron. bot. 1: 361. 1935, 2: 174, 188. 1936, 3: 43, 96. 1937, 4: 266. 1938, 6: 300. 1941.

EPONYMY: *Paulsenella* Chatton (1920); *Paulseniella* Briquet (1907).

7529. *Om Vegetationen paa de dansk-vestindiske Øer* ... Kjøbenhavn (Det Nordiske Forlag ...) 1898. Oct. (*Veg. dansk-vestind. Øer*).
Senior author: Frederik Christian Emil Børgesen (1866-1956).
Publ.: Jan-Mar 1898 (Bot. Centralbl. 6 Apr 1898; Nat. Nov. Apr(2) 1898; Flora 23 Apr 1898), p. [i-iv], [1]-113, [114, cont.], *pl. 1-11* with text. – Preprinted from Bot. Tidsskr. 22: [1]-114, *pl. 1-11. Copies*: LD, NY. – See also Urban (1898), p. 14-15.
Ref.: Pedersen, M., Bot. Centralbl. 74: 143-149. 1898 (rev.).

7530. *Traek af Vegetationen i Transkaspiens Lavland* ... Forsvaret finder sted torsdag den 14 December 1911 ... Kjøbenhavn (Gyldendalske Boghandel. Nordisk Forlag ...) 1911. Oct.(*Traek veg. Transkasp.*).
Publ.: shortly before 14 Dec 1911 (date defense thesis for Dr. phil.), p. [i-iii], [1]-238, map, [1, theses]. *Copy*: NY. – Preprinted from Bot. Tidsskr. 32(1): [1]-239, map. Jun 1912. – Thesis rd. by Nat. Nov. Mai(1) 1912 & ÖbZ Mai 1912.
English translation: 1912 (ÖbZ Jun-Jul 1912), p. [i-vii], [1]-279, map. *Copies*: LD, MICH, PH. – *Studies on the vegetation of the Transcaspian lowlands*, Copenhagen (Gyldendalske Bokhandel – Nordisk Forlag) 1912, published in the series *The second Danish Pamir expedition* conducted by O. Olufsen. See also O. Paulsen, Bot. Centralbl. 122: 20-21. 1913.
Ref.: Diels, L., Bot. Jahrb. 49 (Lit.): 13. 1913 (rev. Engl. transl.).

7531. The second Danish Pamir expedition conducted by O. Olufsen ... *Studies in the vegetation of Pamir* ... Copenhagen (Gyldendalske Boghandel. Nordisk Forlag) 1920. Oct. (*Stud. veg. Pamir*).
Publ.: 1920, (ÖbZ Nov-Dec 1920), p. [i-ix], [1]-132, map. *Copies*: B, LD, MICH, NY. – See also Nature 107: 270-274. 1921 (extract).
Ref.: Diels, L., Bot. Jahrb. 57 (Lit.): 20-21. 1922.

Paulson, Robert (1857-1935), British school teacher, mycologist and lichenologist in London. (*Paulson*).

HERBARIUM and TYPES: BM (600); further material in LIV.

BIBLIOGRAPHY and BIOGRAPHY: Arct. Bibl. 2: 13176-13179; Barnhart 3: 57; BL 2: 240, 700; Bossert p. 303; CSP 17: 740; Desmond p. 483; GR p. 378, cat. p. 69; IH 2: (in press); Kelly p. 170, 255; Kew 4: 251; LS 20172-20173, 37722-37723, suppl. 21093-21115; MW p. 379.

BIOFILE: Anon., J. Queckett micr. Club ser. 3. 1(4): 180. 1935 (obit.).
Bisby, G.R., Trans. Brit. mycol. Soc., Fifty-year index 18. 1952.
D.J.S., Essex Natural. 25: 62-63, *pl. 2*. 1935 (obit., portr.).
Hawksworth, D.L. & M.R.D. Seaward, Lichenol. Brit. Isles 29, 130-131, 198. 1977.
Kent, D.H., Brit. herb. 72. 1957.
Laundon, J.R., Lichenologist 11: 15. 1979.
Lister, G., J. Bot. 73: 232. 1935.
Ramsbottom, J., Trans. Brit. mycol. Soc. 30: 5, 13. 1948.
Thompson, P., Proc. Linn. Soc. London 147: 186. 1935 (obit., d. 1 Mar 1935).

Pauquy, Charles Louis Constant (1800-1854), French botanist and physician; studied at Paris; practicing physician at Amiens. (*Pauquy*).

HERBARIUM and TYPES: Unknown.

BIBLIOGRAPHY and BIOGRAPHY: Barnhart 3: 57; BM 4: 1532; CSP 4: 782; Kew 4: 251; PR 7000-7001 (ed. 1: 7828-7829); Rehder 5: 654; RS p. 133.

7532. *De la Belladone*, considérée sous ses rapports botanique, chimique, pharmaceutique, pharmacologique et thérapeutique, etc.; thèse présentée et soutenue à la Faculté de Médecine de Paris, le 2 Avril 1825, ... à Paris (de l'imprimerie de Didot le Jeune, ...) 1825. Qu. (*Belladone*).
Publ.: on or shortly after 2 Apr 1825, p. [1]-61, *1 pl. Copy*: FI.

7533. *Statistique botanique ou flore du département de la Somme*, et des environs de Paris, description de toutes les plantes qui y croissent spontanément distribuées suivant la méthode naturelle d'une part et le système de Linnée de l'autre, avec l'indication de leur lieu natal, de leur durée, de la couleur de leurs fleurs et de leur emploi dans la médecine, les arts, et l'économie domestique, par Ch. Pauquy, ... Amiens (Imprimerie de R. Machart, ...) 1831. Oct. (in fours). (*Statist. bot. Somme*).
Publ.: 1831, p. [i*-v*], [i]-xi, [1-2], [1]-635. *Copies*: G, NY.
Re-issue: 1834, p. [i*-v*]. [i]-xi, [1-2], [1]-635. *Copies*: B,NY. – Paris(J.B. Baillière et Delloye), Amiens (Caron-Vitet et Allo-Poiré). 1834. Oct. (in fours).

Pauwels, Jan Lodewijk Huibrecht (1899-x), Belgian linguist; Dr. phil. Leuven 1923; professor of Dutch linguistics ib. from 1942. (*Pauwels*).

HERBARIUM and TYPES: Unknown.

BIBLIOGRAPHY and BIOGRAPHY: BL 2: 427, 700.

BIOFILE: Anon., Grote Winkler Prins Encycl. 15: 189. 1972 (bibl.).

7534. *Enkele bloemnamen in de Zuidnederlandsche dialecten* door Dr. J.L. Pauwels met de medewerking van Dr. L. Grootaers ... s'Gravenhage (Martinus Nijhoff) 1933. Oct. (*Bloemnam. Zuidnederl. dial.*).
Co-author: Ludovic Grootaers (1885-1956).
Publ.: 1933 (Nat. Nov. Apr 1934), p. [i]-viii, [1]-321, 7 maps. *Copy*: BR.

Pavillard, Jules (1868-1961), French diatomologist, protistologist and phytogeographer; Lic. sci. nat. Bordeaux 1890; Dr. sci. Paris 1905; high school teacher at Montpellier 1895-1912; maître de conférences (1908), professeur adjoint (1909), professeur de cryptogamie et de cytologie (1924) and from 1927-1937, professor of botany and director of the botanical garden at the University of Montpellier, succeeding Ch. Flahault. (*Jul. Pavillard*).

HERBARIUM and TYPES: Some material at STR.

BIBLIOGRAPHY and BIOGRAPHY: Barnhart 3: 57; BJI 2: 131; BL 2: 164, 700; BM 4: 1533, 8: 990; Kelly p. 170; Kew 4: 251; LS 20178, suppl. 21121-21124.

BIOFILE: Anon., Bot. Centralbl. 107: 496. 1908 (app. Montpellier), 111: 576. 1909 (prof. adj.); Bot. Zeit. 67: 352. 1909 (id.); Hedwigia 68: (48). 1928; Österrr. bo.t Z. 77: 80. 1928 (dir. bot. gard. MPU).
Bourelly, P., Bull. Soc. phycol. France 8: 8. 1962 (obit.).
Buvat, R., Rev. gén. Bot. 66: 225-231. 1958 (obit., bibl., portr. of Jean Pavillard (son of Jules), b. 25 Aug 1915, d. 11 Aug 1956).
Cupp, E.E., Mar. plankton diatoms west coast N. Amer. 219. 1943.
Emberger, L., Bull. Soc. bot. France 110: 53-58. 1963 (obit., portr., bibl., b. 7 Apr 1868, d. 5 Nov 1961).
Tregouboff, G., Bull. Mus. océanogr. Monaco 1247: 1-16. 10 Oct 1962 (biogr., bibl., portr.).VanLandingham, S.L., Cat. Diat. 6: 3578, 7: 4214. 1978.
Verdoorn, F., ed., Chron. bot. 1: 30, 31. 1935, 2: 28, 29, 30. 1936, 3: 102. 1937, 4: 181. 1938, 5: 138, 392. 1939.

COMPOSITE WORKS: (1) *Les diatomées planctoniques* dans l'Atlantique subtropical, *in* Travaux cryptogamiques dédiés à Louis Mangin. Paris, 1931, p. 289.
(2) *Bacillariales*. Report on the Danish oceanographical expeditions 1908-1910 to the Mediterranean and adjacent seas, 2, biol. J. 4, 72 p.

EPONYMY: *Pavillardia* Kofoid et Swezy (1920); *Pavillardinium* G. De Toni (1936).

NOTE: Not to be confused with his son, Jean Pavillard (1915-1958), see Rev. gen. Bot. 66: 225-231. 1959; GR p. 343.

7535. *Myxomycètes des environs de Montpellier* ...Paris (au siège de la Société [mycologique de France]) 1903. Oct.
Co-author: Joannnes Joseph Lagarde (1866-?).
Publ.: 1903, cover (with title and imprint), [1]-25. *Copy*: G. – Reprinted and to be cited from Bull. Soc. mycol. France 19(2): 81-105. 1903.

7536. Thèses présentées à la Faculté des Sciences de Paris pour obtenir le grade de docteur ès sciences naturelles par Jules Pavillard ... 1re thèse. – *Recherches sur la flore pélagique* (Phytoplankton) *de l'Étang de Thau*. 2me thèse. – propositions ... soutenues le [in ink: 5] juin 1905 devant la commission d'examen ... Montellier (Imprimerie Gust. Firmin, Montane et Sicardi ...) 1905. Oct. (*Rech. fl. pélag. Étang Thau*).
Thesis issue: 5 Jun 1905, p. [1]-116, [1, theses], 1 map, *pl. 1-3*. *Copy*: MO.
Trade issue: Jun-Jul 1905 (Nat. Nov. Aug(1) 1905), p. [1]-116, chart, map, *pl. 1-3*. *Copy*: FH. – Trav. Inst. bot. Univ. Montpellier, série mixte, mém. no. 2.
Ref.: Hariot, P., Bot. Centralbl. 99: 378-379. 1905 (rev.).

7537. *Remarques sur la nomenclature phytogéographique* 26 juillet 1919 [Montpellier]. Oct. (*Rem. nomencl. phytogéogr.*).
Publ.: 26 Jul 1919 (cover), p. [1]-27. *Copy*: NY.

Pavlov, Nikolai Vasilievich (1893-1971), Russian botanist at the Academy of Sciences of Kazakhstan at Alma Ata; specialist on Allium, Astragalus and Artemisia. (*Pavlov*).

HERBARIUM and TYPES: MW (first set Kazakhstan and Mongolian coll.), LE (first set Kamchatka 1935 coll.); further material at A, B (extant "Montes Karatau ex Talas-Alatau 1928, 1931), E, MO (Caucasus, Gorki, Moskou) and NY.

BIBLIOGRAPHY and BIOGRAPHY: Barnhart 3: 57; BFM no. 1555; Hirsch p. 227; Hortus 3: 1201 (Pavlov); Kew 4: 252; MW p. 379-380, suppl. p. 276; Roon p. 86; TL-2/3857.

BIOFILE: Asmous, V.C., Chron. bot. 11(2): 106. 1947.
Bagdasarova, T.V., & Gubanov, I.A., Novosti Sist. Vyssh. Rast. 12: 335-355. 1975 (hist. of types of P. in MW).
Bykov, B.A., Izv. Akad. Nauk. Kazakh. SSR., Biol. 7(2): 74-76. 1969.
Goloskokov, V.P., Izv. Akad., Nauk. Kaz. SSR., Biol. 5: 18-21. 1974 (obit.).
Goloskokov, V.P., & Filatova, W.S., Bot. Zhurn. 54: 795-798. 1969 (75th birthday, portr., biogr. notice, bibl.).
Gubanov, I.A. and Kultiasov, I.M., in Akademiya Nauk Kzakhskoi SSR, Inst. Bot., Flora ... Kazakhstana 6-21. 1975 (fide Kew Record); bibl. by B.A. Bykov on p. 3-5.
Hedge, I.C. & Lamond, J.M., Index coll. Edinb. Herb. 117. 1970.
Kultiasov, I.M., & Gubanov, I.A., Byull. Mosk. Obshsh. Ispyt. Prir., Biol. 77: 149-155. 1972 (obit., portr.).
Lipschitz, S., Florae URSS fontes 222 [index]. 1975.
Pavlov, N.V., Personal communication to senior author 1957 (on coll.).
Verdoorn, F., ed., Chron. bot. 5: 89. 1939.

COMPOSITE WORKS: Komarov, Fl. URSS: (1)*Polygonaceae* (*Atraphaxis, Calligonum*), 5: 501-594. 22 Jun 1936.
(2) *Cruciferae* (*Stroganovia*), 8: 524-535. 20 Mai 1939.
(3) *Boraginaceae* (*Rochelia*), 19: 548-565. 5 Feb 1953.

HANDWRITING: V.N. Pavlov et al., Gerb. Moskovsk. Univ. 141. 1978.

EPONYMY: The derivation of *Pavlova* R. Butcher (1952) was not given.

7538. *Flora central'nogo Kazakhstana* ... Izdanie Narodnogo Komissariata Zemledelnya KASSR. 1928. Oct. (*Fl. centr. Kazakh.*).
Part 1: 1928 (p. 12: 28 Feb 1928), p. [1]-178, [1], [i]-xi, [xii].*Copy*: USDA.
Part 2: 1935 (p. 6: 15 Dec 1934), p. [1]-549, [550]. *Copy*: G. – Moskva 1935. Oct.*Part 3*: 1938 (p. [2]: 21 Mar 1938), p. [1]-428, [429, cont.]. *Copy*: G. – Moskva 1938.

Pavon y Jiménez, José Antonio (1754-1844), Spanish botanist; with H. Ruiz in South America 1778-1788. (*Pav.*).

HERBARIUM and TYPES: MA; duplicates: B (350 + types saved, extant), B-W, (109 extant), BC, BM, BR, CGE, FI, G, H, K, MPU, MW, NY, OXF, P, US, W. – The important collection at FI (4994 specimens) was bought by Webb from Pavon; it contains in addition to plants from Peru and Chile collected by Ruiz & Pavon, collections by Moçino & Sessé from Mexico (marked "N.E." New England), Cuba, Puerto Rico; by Tafulla from Quito and Guayaquil, and by Née from the Philippines. See Pichi-Sermolli (1949) and Steinberg (1977). – BM has Ruiz & Pavon manuscripts, as well as lists of plants sent by Pavon to A.B. Lambert. FI has a set of 28 plates of species of *Laurus* given by Pavon to Webb. – Letters at G.

BIBLIOGRAPHY and BIOGRAPHY: AG 4: 865; Backer p. 429; Barnhart 3: 57; Clokie p. 222; Colmeiro 1: cxcii, Bot. Penins. p. 181; CSP 4: 782, 8: 574; DSB 15: 470-471 (b. 22 Apr 1754, d. 1840, bibl.); Henrey 2: 36; Hortus 3: 1201 (Pav.); IH 1 (ed. 6): 362, (ed. 7): 339, 2: (in press); Lasègue p. 579 [index]; Lenley p. 466; NI 1: 158, suppl. p. 58, see also nos. 1698-1699; PR 7002, 7894,7896-7897 (ed. 1: 7830); Rehder 5: 654; TL-1/1123-1125, see p. 523; TL-2/3143, see D. Bouchet, J. Dombey, J.F. Jacquin, A.B. Lambert, M.E. Moricand; Tucker 1: 544; Urban-Berl. p. 381, 415; Zander ed. 10, p. 700, ed. 11, p. 798, 809.

NOTE: for a more detailed list of references to the South American explorations by Ruiz and Pavon, incl. subsequent publications, see under H. Ruiz Lopez.

BIOFILE: Acosta-Solis, M., Inst. Ecuat. Ci. nat. Contr. 65: 18. 1968.
Anon., Proc. Linn. Soc. 1: 208-209. 1844 (1849).
Alvarez Lopez, Anal. Inst. bot. Cavanilles 12: 5-112. 1954.
Barreiro, R.P.A.J., Relaciones del viaje ... Hipólito Ruiz, Madrid 1931 (n.v.).
Bellot, F., Lagascalia 7: 51-53. 1977 (commemor.).
Bernardi, L., Mus. Genève 174: 8-15. 1977 (portr., brief account exped.).
Bridson, G.D.R. et al., Nat. hist. mss. res. Brit. Isl. 229.415. 1980.
Candolle, Alph. de, Phytographie 439, 445. 1880 (coll.).
Coats, A.M., The plant hunters 338, 363-367, 373. 1969.
Dawson, W.R., Smith papers 72. 1934.
Granel de Solignac, L. & Bertrand, L., Naturalia monsp., bot., 18: 273. 1967 (coll. APU).
Herrera, F.L., Estudios sobre la flora del Departemento del Cuzco Lima 1930, p. 8.
Hiepko, P., Private communication to authors 25 Jul 1980.
Howard, J.E., Examination of Pavon's collection of Peruvian bark, contained in the British Museum, 1853, vi, 47, iv. p.; originally published Pharm. J. Trans. 11: 489-498. 1 Mai, 557-564. 1 Jun 1852, 12: 11-15. 1 Jul, 58-62. 1 Sep, 125-129. 1 Sep, 173-180. 1 Oct, 230-235. 1 Nov 1852, 12: 338-342. 1 Jan 1853.
Jackson, B.D., Bull. misc. Inf. Kew 1901: 51 (part of herbarium at K).
Jessen, K.F.W., Bot. Gegenw. Vorz. 466. 1864.
Kühnel, J., Thadd. Haenke 43, 74, 75, 76, 153, 159, 160, 178. 1960.
Kukkonen I., Herbarium of Christian Steven 78. 1971.
Lack, H.W., Willdenowia 9: 177-198. 1979 (Ruiz & Pavon specimens extant at B and B-Willd.).
Lindemann, E.V., Bull. Soc. Imp. Natur. Moscou 61: 53. 1886 (one specimen).

Moldenke, H.N., Plant life 2: 75. 1946 (1948).
Munoz Mediña, D.J., Historia del desarrollo de la botanica en España 15-16. 1969.
Murray, G., Hist. coll. BM(NH) 1:173. 1904.
Papavero, N., Ess. hist. neotrop. dipterol. 1: 35, 111. 1971.
Pichi-Sermolli, R.E.G., Nuovo Giorn. bot. ital. 56(4): 699-701. 1949 (FI Pavon coll.).
Reiche, K., Grundz. Pfl.-Verbr. Chile 5-6. 1907 (Veg. Erde 8).
Ron Alvarez, Mª Eugenia, Aportación al conocimiento de la histografia del botánico da Josá Antonio Pavon y Jiménez. Anales de la Real Academia de Farmacia [Madrid] 36(4): 599-631. 1970 (important recent study on Pavon, based on original documentation and providing many new details).
Ruiz, H., Relación del Viaje ... Madrid 1931, ed. 2, Madrid 1952, English version by R. Dahlgren, Field Mus. Bot. 21: 1-372. 1969.
Stafleu, F.A., *in* C.L. L'Héritier, Sertum angl. facs. ed. 1963, p. xx; *in* Jussieu, A.L., Gen. pl., facs. ed. xlii. 1964; Linnaeus and the Linnaeans 289. 1971.
Stearn, W.T., Nat. Hist. Mus. S. Kensington 285. 1981 (coll.).
Steele, A.R., Flowers for the king 371-372 [index]. 1964 (Important study; many details!).
Steinberg, C., Webbia 32(1): 8-9. 1977. (coll. FI).
Turrill, W.B., Bull. misc. Inf. Kew 1920: 58.
Urban, I., Symb. ant. 3: 99. 1902.
Weberbauer, A., Pfl.-Welt peruan. Anden 2-4. 1911. (Veg. Erde 12).
Woodward, B.B., Hist. coll. BM (NH) 1: 45 (mss. at BM).

COMPOSITE WORKS: See under H. Ruiz Lopez for the various "Ruiz et Pavon" publications.

HANDWRITING: Candollea 32: 383-384. 1977; Lady Smith, Mem. corr. J.E. Smith 1: facsim. 1832; see also Ron Alvarez (1970).

EPONYMY: *Pavona* Cothenius (1790, *orth. var.* of *Pavonia* Cavanilles); *Pavonia* Cavanilles (1787, *nom. cons.*); *Pavonia* Ruiz (1794); *Pseudopavonia* Hassler (1909).

7539. *Disertacion botanica* sobre los generos Tovaria, Actinophyllum, Araucaria y Salmia, con la reunion de algunos que Linneo publicó como distintos, por don Joseph Pavón, professor de historia natural, y socio de varias academias. s.l.n.d. [Madrid 1791]. Qu.
Publ.: 1797 (in journal), p. [1]-14. *Copies*: FI (given by Pavon to R. Desfontaines; later acquired by P.B. Webb), GOET. – Reprinted and to be cited from Mem. r. Acad. méd. Madrid 1: 191-204. 1797 (journal publ. in Natl. Libr. Medicine, U.S.A.).

7540. *Illustrations of the Nueva Quinologia of Pavon,* with coloured plates by W. Fitch, and observation on the barks described.
Editor and in part author: John Eliot Howard (1807-1883).
Publ.: 1859-1862, p. [i-vi], [i]-xvi, *30 pl.*
See also J.E. Howard, Examination of Pavon's collections of Peruvian barks contained in the British Museum, London 1853, vi, 47. iv p. (reprinted from Pharmaceutical Journal, June 1852), and H. Karsten, Flora 43: 600-608. 1860.

Pawlowski, Bogumił (1898-1971), Polish botanist; Dr. phil. Cracóv 1922; at the Botanical Institute of the Polish Academy of Sciences, Cracóv; specialist on the floras of Poland and the Balkan peninsula. (*Pawl.*).

HERBARIUM and TYPES: KRA and KRAM; duplicates at BUT, C, GB, GH, M, US. – Letters at G.

BIBLIOGRAPHY and BIOGRAPHY: BFM no. 1480, 1481; BJI 2: 131; Bossert p. 303; Futák-Domin p. 458-459; Hirsch p. 227; Hortus 3: 1201 (Pawl.); IH 1 (ed. 7): 339; Kew 4: 252-253; Roon p. 86.

BIOFILE: Anon., Ochr. Přzyr. 38: portr. 1973 (fide Kew Record); Preslia 44: 278. 1972 (obit.).

Asmous, V.C., Chron. bot. 11(2): 106. 1947.
Braun-Blanquet, J., Vegetatio 25: 281-282. 1972 (obit.).
Futák, J., Biológia, Bratislava 27: 307-310. 1972 (obit., portr., b. 25 Nov 1898, brief bibl.).
Jasiewicz, A., Folia biol. 20: 319-320. 1972 (obit., portr., d. 27 Jul 1971 through an accident on a field trip on Mount Olympus).
Pawlowski, B., Flora tatrorum plantae vasculares, tomus 1, 1956.
Pax, F., Bibl. schles. bot. 24. 1929.
Poirion, L., Riviera sci. 59: 17. 1972 (obit.).
Szafer, W., Concise hist. bot. Cracow 162 [index]. 1969; Fragm. flor. geobot. 16(1): 17-23. 1970 (portr., biogr., preceded by a bibliography on p. 4-16.).
Szymkiewicz, D., Bibl. fl. Polsk. 112. 1925.
Verdoorn, F., ed., Chron. bot. 1: 239, 365. 1925, 2: 254, 395. 1936, 3: 222. 1937, 5: 21, 391, 392, 393.
Zarzycki, K., Vegetatio 25: 278-281. 1972 (obit., portr.).
Zarzycki, K. et al., Acta Soc. bot. Pol. 42: 3-20. 1973 (obit., portr., list of new taxa and new combinations).

COMPOSITE WORKS: with W. Szafer and St. Kulczyński, *Rośliny polskie* Lwow, Warszawa 1924, 736 p., see W. Szafer.

7541. *Studien über mitteleuropäische Delphinien* aus der sogenannten Sektion Elatopsis ... Cracovie (Imprimerie de l'Université) 1934, 5 parts. Oct.

1-2: 1934, cover t.p., p. [29]-44, [67]-81, *pl. 3*. *Copy*: U. – Reprinted and to be cited from Bull. Acad. Polon. Sci. Lettr., Sci. mat. nat. ser. B. 1: 29-34, 67-81, *pl. 3*. 1934.
3-5: 1934, cover t.p., p. [93]-106, chart, [149]-164, 2 charts, *pl. 5*, [165]-181. *Copy*: U. – Reprinted from id. 1: 93-106, 149-164, 165-181. 1934.

Pax, Ferdinand Albin (1858-1942), German botanist (born in Bohemia); Dr. phil. Breslau 1882; assistant with A. Engler at Kiel 1883-1884; with id. to Breslau 1884 as lecturer, from 1889-1893 again with Engler at Berlin; from 1893-1925 professor of botany and director of the botanical garden at Breslau; one of the most productive of Engler's collaborators. (*Pax*).

HERBARIUM and TYPES: Many types of Pax's work for the Engler publications were in the pre-1943 collections of B; other types are at WRSL; relatively large sets of material collected by Pax himself are at BP and WRSL; other material at CR, KIEL, L, and PR. – *Exsiccatae*: with G.H.E.W. Hieronymus : *Herbarium cecidiologicum*, 1892-1907, see Hieronymus. – Letters and portrait at G.

BIBLIOGRAPHY and BIOGRAPHY: AG 4: 247, 5(1): 624, 7: 408, 12(1): 202, 12(3): 231; Backer p. 429; BFM nos. 263, 1888, 2382; BJI 1: 44(bibl.), 2: 131-132 (bibl.); BL 2: 35, 312; BM 4: 1534, 8: 991; Bossert p. 303; CSP 10: 1008, 12: 563, 17: 746-747; DTS 1: 218-219, 6(4): 183; Futák-Domin p. 459-460; GR p. 456; Herder p. 334, 336; Hirsch p. 227; Hortus 3: 1201 (Pax); IH 1 (ed. 7): 339; Kew 4: 253-254; Klášterský p. 141; Kleppa p. 291; Langman p. 571; LS 20181-20182, 37735, suppl. 21126; Maiwald p. 238, 242, 252, 253; Morren ed. 10, p. 14; MW p. 380, suppl. p. 276; Rehder 5: 654; TL-2/1592, 1711-1713 and see index vol. 2; Tucker 1: 544; Urban-Berl. p. 381; Zander ed. 10, p. 700, ed. 11, p. 798; Zep.-Tim. p. 90.

BIOFILE: Anon., Bot. Centralbl. 25: 387. 1886 (habil. Breslau), 40: 239. (Custos Berlin), 54: 127. 1893 (prof. Breslau); Bot. Jahrb. 11 (Beibl. 25): M. 1889 (to B as "Custos"), 17: (Beibl. 40): 33. 1893 (to BRSL as prof. bot.); Bot. Centralbl. 54: 127. 1893 (prof. bot. Breslau); Bot. Zeit. 44: 207. 1886 (habil. Breslau), 47: 784. 1889 (custos Berlin), 51: 138. 1893 (prof. bot. Breslau)); Hedwigia 51: (243). 1911 (Geheimrat); Mycol. Centralbl. 3: 205. 1913 (rector Breslau); Nat. Nov. 8(6): 80. 1886 (habil. Breslau), 11: 369. 1889 (Custos Berlin), 15: 159. 1893 (prof. bot. Breslau), 38: 64. 1916. Österr. bot. Z. 39: 456. 1889 (Berlin), 43: 192. 1893 (succeeds Prantl at Breslau).
Borza, Al., Bul. Grad. Bot. Muz. bot. Univ. Cluj 25: 133-142. 1945 (biogr., portr.).
Bullock, A.A., Bibl. S. Afr. bot. 82-83. 1978.

Gager, C.S., Brooklyn Bot. Gard. Rec. 27: 221. 1938 (dir. bot. gard. Breslau 1893-1926).
Hoffmann, K., Revist. Sudam. Bot. 7: 22-23. 1942 (portr.).
Kräusel, R., Ber. deut. bot. Ges. 68(a): 93-100. 1955 (bibl.).
Mägdefrau, K., Gesch. Bot. 158. 1973.
Pax, F., Beitr. Kenntn. Ovulum p. [42]. 1882. (Lebenslauf, b. 26 Jul 1858, born Königinhof, Böhmen [now Dvur Králové nad Labem], worked 4 semesters with F. Cohn); Bibl. schles. Bot. [see index]. 1929.
Schultes, R.E., Regn. Veg. 71: 279-281. 1970, Bot. Rev. 36: 261-262. 1970 (portr.).
Szymkiewicz, D., Bibl. fl. polsk. 113. 1925.
Verdoorn, F., ed., Chron. bot. 1: 130, 135. 1935, 2: 360. 1936, 3: 127. 1937, 4: 277. 1938, 5: 100. 1939, 7(7): 353. 1943 (d. 1 Mar 1942).
Wittrock, V.B, Acta Horti Berg. 3(2): 138. *pl. 27*. 1903 (portr.).

COMPOSITE WORKS: (1) EP, *Die natürlichen Pflanzenfam.*: *Aceraceae*, 3(5): 263-272. Mai 1893.
Aizoaceae, 3(1.B): 33-48. Apr 1889, 49-51. Mai 1889.
Amaryllidaceae, 2(5): 97-124. Sep 1887.
Buxaceae, 3(5): 130-135. Jun 1892.
Callitrichaceae, 3(5): 120-123. Jun 1891.
Capparidaceae, 3(2): 209-237. Mar 1891, 276. Mai 1891.
Caryophyllaceae, 3(1.B): 61-94. Mai 1889.
Cyperaceae, 2(2): 98-126. Jan 1888 (see K. Schumann, Bot. Centralbl. 38: 859-861. 1889.
Dioscoreaceae, 2(5): 130-137. Sep 1887.
Empetraceae, 3(5): 123-127. Jun 1891.
Euphorbiaceae, 3(5): 1-48. Mai 1890, 49-96. Jun 1890, 97-119. Jun 1891, 456-458. Apr 1896.
Haemodoraceae, 2(5): 92-96. Jun 1887.
Hernandiaceae, 3(2): 126-129. Mar 1889.
Hippocastanaceae, 3(5): 273-276. 2 Apr 1895.
Iridaceae, 2(5): 137-144. Sep 1887, 145-160. Mar 1888.
Lauraceae, 3(2): 106-126. Mar 1889.
Monimiaceae, 3(2): 94-96. Jul 1888, 97-105. Mar 1889.
Moringaceae, 3(2): 242-244. Mai 1891.
Myrsinaceae, 4(1): 84-96. Dec 1889, 97. Jun 1890.
Pittosporaceae, 3(2.A):106-114. Mar 1891.
Plumbaginaceae, 4(1): 116-125. Jun 1890.
Portulacaceae, 3(1.B): 51-60. Mai 1889.
Primulaceae, 4(1): 98-116. Jun 1890.
Salicaceae, 3(1): 29-37. Jan 1888.
Stackhousiaceae, 3(5): 231-233. Mai 1893.
Staphyleaeae, 3(5): 258-262. Mai 1893.
Taccaceae, 2(5): 127-130. Sep 1887.
Tovariaceae, 3(2): 207-208. Mar 1891.
Velloziaceae, 2(5): 125-127. Sep 1887.
"Die Gramineen und Cyperaceen" by Hackel und Pax, were also issued as a reprint, 130 p., Leipzig 1887, (Nat. Nov. Jan (2) 1888), n.v., probably simultaneously with vol. 2.2. Jan 1888.
(2) Engler, A., *Pflanzenreich*:
Aceraceae, iv, 163 (Heft 8): 1-89. 7 Jan 1902. (rev.: K. Schumann, Bot. Jahrb. 90: 14-15. 1902.
Euphorbiaceae, iv, 147, in 10 Hefte 1910-1924, see under Engler, no. 1713.
Primulaceae, (with R. Knuth), iv, 237 (Heft 22): 1-386. 14 Nov 1905. (Rev.: C. Mez, Bot. Centralbl. 10: 299-301. 1906).
(3) Durand, Th. & H. Pittier, *Primit. fl. costaric.*, contr. *Euphorbiaceae* to vol. 2(5), see TL-2(1592).
(4) EP, *Die natürlichen Pflanzenfamilien*,ed. 2:
Aizoaceae,16c: [179]-233, 587. Jan-Apr 1934.
Amaryllidaceae, with K. Hoffmann, 15a: 391-430. Apr 1930.
Bretschneideraceae, 17b: [699]-700. 1936.
Callitrichaceae, with id., 19c: 236-240. 1931.
Capparidaceae, with id., 17b: [146]-223. 1936.

Caryophyllaceae,with id., 16c: [275]-364, 587. Jan-Apr 1934.
Dysphaniaceae, with id., 16c: [272]-274. Jan-Apr 1934.
Euphorbiaceae, with id., 19c: 11-233. 1931.
Haemodoraceae, 15a: 386-390. Apr 1930.
Moringaceae, 17b: [693]-698. 1936.
Portulacaceae, with K. Hoffmann, 16c: [234]-262, 587. Jan-Apr 1934.
Taccaceae, 15a: 434-437. Apr 1930.
Tovariaceae, with K. Hoffmann, 17b: [224]-226. 1936.
Velloziaceae, 15a:431-434. Apr 1930.
(5)*Prantl's Lehrbuch der Botanik* herausgegeben und neu bearbeitet von Dr. Ferdinand Pax, eds. *9-14*, 1894-1916. See under Prantl.
(6) with H. Winkler, *Vegetationsbilder aus den Südkarpathen*, in G. Karsten und H. Schenck, Vegetationsbilder, 15. Reihe, Heft 8. *Taf. 43-48*, Jena 1924 (6 pl. and 13 p. text).
(7) collaborator, J. Mildbraed, *Wiss. Erg. Deut. Zentral-Afrika-Exp.* 1907-1908, Band 2, Botanik, 1914, see TL-2/6008.

HANDWRITING: Candollea 32: 385-386. 1977.

EPONYMY: *Neopaxia* Ö. Nilsson (1966); *Paxia* Gilg (1891); *Paxia* Ö. Nilsson (1966); *Paxina* O. Kuntze (1891); *Paxiodendron* Engler (1895); *Paxiuscula* W.G. Herter (1941).

PROMOVENDI: F. Fedde, W. Grosser, K. Schott, H. Winkler, W. Limpricht, R. Knuth, Rich. Schulz, R. Kirchner, Z. v. Szabo, A. v. Lingelsheim, V. Engler, R. Kräusel, E. Lindemann, K. Reitter, F. Brieger, K. Meyer (see Kräusel 1955).

7542. *Beitrag zur Kenntniss des Ovulums* von Primula elatior Jacq. und officinalis Jacq. Inaugural-Dissertation welche nebst beigefügten Thesen mit Genehmigung der Philosophischen Facultät der Universität Breslau zur Erlangung der Doctorwürde Freitag, den 17. März 1882, mittags 12 Uhr im Musiksaale der Universität ... öffentlich vertheidigen wird Ferdinand Pax. Breslau (Breslauer Genossenschafts-Buchdruckerei, E. Gen.) 1882. Oct. (*Beitr. Kenntn. Ovul.*).
Publ.: 17 Mar 1882 (Nat. Nov. Mai(2) 1882; Flora rd. 1-21 Jul 1882; Bot. Zeit. 30 Jun 1882, p. [1]-41, [42, vita], [1, theses].*Copies*: NY, WU.
Ref.: Köhne, E., Bot. Centralbl. 10(22): 316-318. 29 Mai-2 Jun 1882.

7543. *Monographie der Gattung Acer* von Dr. Ferd. Pax [three papers in Bot. Jahrb. 1885-1889; Berlin]. Oct.
1: Bot. Jahrb. 6(4): 287-375, *pl. 5*. 9 Jun 1885.
2: Bot. Jahrb. 7(2): 177-263. 31 Dec 1885. *Nachträge und Ergänzungen* 1889 (Nat. Nov. Jul(1) 1889), Bot. Jahrb. 11: [72]-83. 2 Apr 1889.
Reprints with original pagination (except for a special t.p. [i] for the Nachträge): G, NY.

7544. *Beiträge zur Morphologie und Systematik der Cyperaceen*. Habilitationsschrift, welche nebst beigefügten Thesen mit Zustimmung der Philosophischen Fakultät der Königlichen Universität Breslau zur Erlangung der venia docendi Donnerstag, den 11. März 1886, mittags 12 Uhr im Musiksaale der Universität gegen die Herren Dr. Th. Schube, Dr. O. Mütter öffentlich vertheidigen wird Dr. phil. Ferdinand Pax. Leipzig (Wilhelm Engelman) 1886. (*Beitr. Morph. Syst. Cyper.*).
Publ.: on or shortly before 11 Mar 1886 (Centralbl. 15-19 Mar 1886; Nat. Nov. Apr(1) 1886), p. [i], [1]-32, *pl. 2*. *Copies*: G, H, MO. – Preprinted from Bot. Jahrb. 7: 287-318. *pl. 2*. 28 Mai 1886.
Ref.: Roth, E., Bot. Centralbl. 26: 253-256. 1886 (rev.).

7545. *Monographische Übersicht über die Arten der Gattung Primula* ... Leipzig (Verlag von Wilhelm Engelmann) 1888. Oct. (*Monogr. Primula*).
Publ.: 9 Oct 1888 (Nat. Nov. Nov(1) 1888; Bot. Zeit. 28 Dec 1888), p. [1]-169. *Copies*: B, G, L, MICH, MO, NY. – Reprinted and to be cited from Bot. Jahrb. 10: 75-192. 6 Jul 1888, 193-241. 9 Oct 1888.
Ref.: Dewar, D., J. Bot. 27: 59-61. Feb 1889.
   Franchet, A., Bull. Soc. bot. France (bibl.): 35-37. 1889.

7546. *Allgemeine Morphologie der Pflanzen* mit besonderer Berücksichtigung der Blüthenmorphologie ... Stuttgart (Verlag von Ferdinand Enke) 1890. Oct. (*Allg. Morph. Pfl.*).
Publ.: Apr-Mai 1890 (p. vi: Easter 1890; Nat. Nov. Mai(2) 1890; Bot. Zeit. 30 Mai 1890, Flora 5 Jul 1890; J. Bot. Sep 1890, ÖbZ Mai 1890, p. [i]-x, [1]-404. *Copies*: B, G, MO, NY, US.
Ref.: Goebel, K., Flora 73: 274-275. 1890.
Jost, L., Bot. Zeit. 48: 541-543. 1890.
Niedenzu, F.J., Bot. Jahrb. 12 (Lit.): 25-26. 1890.

7547. *Führer durch den königlichen botanischen Garten* der Universität *Breslau* ... mit einem Plane des Gartens. Breslau (Ferdinand Hirt, ...) 1895. Oct. (*Führ. bot. Gart. Breslau*).
Ed. 1: 1895 (Nat. Nov. Mai (1) 1895; Bot. Zeit. 16 Mai 1895), map, p. [1]-63. *Copy*: G.
Ed. 2: 1903 (Nat. Nov. Jul(2) 1903), p. [1]-63, map. *Copy*: MO.

7548. *Über die Gliederung der Karpathenflora* ... [Jahresber. Schles. Ges. vaterl. Kult. 1896]. Oct.
Publ.: 1896, p. [1]-12. *Copy*: G. – Reprinted and to be cited from Jahresber. Schles. Ges. vaterl. Kult. 1896(2b): 15-26.

7549. *Grundzüge der Pflanzenverbreitung in den Karpathen* ... Leipzig (Verlag von Wilhelm Engelmann) 1898-1908, 2 vols. (*Grundz. Pfl.-Verbr. Karpath.*).
1: 1898 (p. vi: Dec 1897, Nat. Nov. Mai(2) 1898; Bot. Centralbl. 15 Jun 1898; ÖbZ 1 Jul 1898), p. [i]-viii, [1]-269, [270, err.]. – Veg. Erde vol. 2.
2: 1908 (p. vi: Jun 1908, Nat. Nov. Dec(1) 1908; ÖbZ Nov 1908), p. [1]-viii, [1]-321, [322, err.]. – Veg. Erde vol. 10.
Copies: NY, US. – See also TL-2/1712.
Facsimile ed.: Jun 1974, Vaduz (A.R. Gantner Verlag K.-G. [J. Cramer, Lehre], *1*: ISBN 3-7682-0986-5, *2*: ISBN 3-7682-0998-9. *Copy*: FAS.
Ref.: Diels, L., Bot. Zeit. 67(2): 69-70. 1909.
Gilg, E., Bot. Jahrb. 26 (Lit.): 31 Jan 1899.
Gehrman, K., ib. 42 (Lit.): 51-52. 29 Dec 1908.
Höck (Luckenwalde), Bot. Centralbl. 75: 243-247. 1898.
Solms, H., Bot. Zeit. 56: 299-300. 1898.
Wangerin, W., Bot. Centralbl. 111: 591-598. 1909 (rev.).

7550. *Schlesiens Pflanzenwelt* eine pflanzengeographische Schilderung der Provinz ... Jena (Verlag von Gustav Fischer) 1915. Oct. (*Schles. Pflanzenw.*).
Publ.: *1915* (*Nat. Nov. Dec* (*1, 2*) *1915; ÖbZ Aug-Nov 1915*), *p.* [*i*]*-vi,* [*1*]*-313, map*. *Copies*: B, BR, FI, G, NY.
Ref.: Diels, L., Bot. Jahrb. 54 (Lit.): 26. 1926.

7551. Beiträge zur polnischen Landeskunde, Reihe A, Band i, *Pflanzengeographie von Polen* (Kongress-Polen) ... Berlin (Dietrich Reimer (Ernst Vohsen)) 1918. Oct. (*Pfl.-Geogr. Polen*).
Publ.: 1918 (p. [vii]: 1 Apr 1918; ÖbZ 1 Dec 1918), p. [ii]-xvi, [1]-148, *pl. 1-8*. *Copy*: G. – Preliminary publications: Die Pflanzengeogr. Gliederung Polens, Z. Ges. Erdk., Berlin 1917(5): 280-284; Die Pflanzenwelt Polens, *in* Handbuch von Polen 179-212. 1918.

7552. *Pflanzengeographie von Rumänien* ... Halle (Druck von Erhardt Karras, ...) 1919. Qu.
Publ.: 1920 (mss. presented 5 Mai 1919; Nat. Nov. Oct (1, 2) 1920), p. [i-iv], [1]-262, *pl. 5-12*. *Copies*: B, MO. – Reprinted and to be cited from Nova Acta Leop. 105(2): 81-342. 1919.

Paxton, [Sir] **Joseph** (1803-1865), British horticulturist, editor, landscape gardener, railroad promotor and speculator, builder of glass structures (Crystal Palace), architect, civil engineer, and politician; knighted 1851. (*Paxt.*).

HERBARIUM and TYPES: Paxton is not known to have made herbarium material.

BIBLIOGRAPHY and BIOGRAPHY: Backer p. 429; Barnhart 3: 58; BB p. 239; BM 4: 1534; Bossert p. 303; Bret. p. 550; Desmond p. 483 (q.v. for further refs.); DNB 44: 103-104; Frank 3 (Anh.): 74-75; GF p. 65, 85; Herder p. 88, 363; Hortus 3: 1201 ("Paxt."); Jackson p. 12, 408, 471, 472; Kew 4: 254; Langman p. 571; Lasègue p. 529; MW p. 381, suppl. p. 276-277; NI 1498-1499, 2351, see also 1: p. 129-131, suppl. p. 54; Plesch p. 238; PR 5369, 7003-7005 (ed. 1: 7832-7834); Rehder 5: 654-655; Sotheby 591, 592; TL-1/959; TL-24659, 4661; Tucker 1: 544; Zander ed. 10, p. 710, ed. 11, p. 781, 798.

BIOFILE: Allibone, S.A., Crit. dict. Engl. lit. 1530-1531. 1870 (bibl.).
Anon., Amer. J. Sci. 40: 140. 1865; 41: 264. 1866; Bonplandia 1: 67. 1853, 2: 7, 296. 1854 (member of parliament); Bot. Zeit. 10: 238. 1852, 23: 211. 1865; Flora 48: 350, 445, 1865; Gard. Chron. 1924: 120; Gartenflora 1: 66. 1852 (knighthood), 14: 221-222. 1865 (obit.).
Allan, M., The Hookers of Kew 95, 184. 1967.
Allen, D.E., The Victorian fern craze 36, 42. 1969.
Bentham, G., Proc. Linn. Soc. 1865/6: lxxxi-lxxxiii (obit.).
Bridson, G.D.R. et al., Nat. hist. mss. res. Brit. Isl. 138: 36, 143.21. 1980.
Chadwick, G.F., The works of Sir Joseph Paxton, London 1961 (major study of P as a gardener, builder of glass structures, and architect), (bibl.).
Fletcher, H.H., Story R. Hort. Soc. 1804-1968, p. 552 [index]. 1969.
Gloag, J., Mr. Loudon's England 21, 47, 57, 69, 196, 206. 1970.
Hadfield, M., Pioneers in gardening 170-182. 1951.
Heeps, A.P., Morris Arbor. Bull. 19(1): 8-13. 1968.
Jessen, K.F.W., Bot. Gegenw. Vorz. 392, 394. 1864.
Lemmon, K., Golden age plant hunters 226 [index]. 1968.
Markham, V.R., Paxton and the bachelor Duke, London 1935 (main biography; see also Times Lit. Suppl. 11 Apr 1935 and Observer 28 Apr 1935); b. 3 Aug 1803, d. 8 Jun 1865).
Pearson, C.E., Agriculture 58(3): 132-135. 1951 ("father ... British glasshouse industry", portr.).
Reinikka, M.A., A history of the Orchid 164-168. s.d. (portr.).
Seemann, B., J. Bot. 3: 231-232. 1865.
Simmonds, A., J. R. Hort. Soc. 59: 477-481. 1934 ("gardener-architect"); Hort. who was who 47-52. 1948.
Stearn, W.T., J. Soc. Bibl. nat. Hist. 3: 103. 1955 (on the Magazine of Botany); Taxon 14: 293-298. 1965 (Kew and the Crystal Palace); see also Fl. males. Bull. 12: 489-490. 1956.
Wright, R., Gardener's tribute p. 42-43, 49. 1949.

EPONYMY: *Paxtonia* Lindley (1838).

7553. *The horticultural register and general magazine* ... London (published by Baldwin and Cradock, ...) [1831]-1836. 5 vols. Oct. (*Hort. reg.*).
Authorship: vol. 1: Paxton with Joseph Harrison (x-c. 1855), vols. 2 & 3: Paxton, vols. 4 and 5: James Main (c. 1775-1846).
Publ.: dated monthly issues. *Copy*: NY.
1: 1 Jul 1831-1 Dec 1832, nos. 1-18, p. [i*-iv*], [i]-v, [6]-860.
2: 1 Jan 1833-1 Dec 1833, nos. 19-30, p. [i*], [i-iv], [1]-572.
3: 1 Jan 1834-1 Dec 1834, nos. 31-42, p.[i*-v*], [i]-ii, [3]-529, *1 pl.*
4: 1 Jan 1835-1 Dec 1835, nos. 43-54, p. [i-iv], [1]-487, *1 pl.*
5: 1 Jan 1836-1 Dec 1836, nos. 55-66, p. [i-vi], [1]-476, *1 pl.*

7554. *Paxton's Magazine of botany*, and register of flowering plants ... London (Orr and Smith, Paternoster Row) [1833] 1834-1849, 16 vols. Qu. (*Paxton's Mag. Bot.*).
Publ.: in monthly parts of which 12 made up a volume. "The parts were issued near the first of each month, each has a roman numeral giving the number of the part at the foot of its first page, while garden operations for the month of issue are described on its last pages. Hence the contents and date of each part can be deduced with little

trouble"(W.T. Stearn). Vol. 12 was published e.g. from Feb 1845-Jan 1846, vol. 13, Feb 1846-Jan 1847.

The series contains 723 coloured plates hors texte (copper engravings and lithographs) as well as woodcuts in the text. The artists were F.W. Smith (engravings), S. Holden (lithographs) and C.J. Flemin and O. Jewitt (woodcuts). The numbers of plates quoted below are those of the NY copy. The copies do not always have all the plates: we are not certain that 723 is the exact number of plates issued.
Copies: B, BR, FI, G, MO, NY; IDC 5300.

1: 1834 (t.p. date; p. viii: 20 Dec 1834), p. [i]-xii, [1]-278, *42 pl.*
2: 1836 (p. viii: 20 Dec 1835), p. [i]-xii, [1]-270, [1-5, index], *43 pl.*
3: 1837 (p. viii: 20 Dec 1836), p. [i]-vii, [1-4, index], [1]-280, *43 pl.*
4: 1838 (p. viii: 20 Dec 1837), p. [i]-xii, 1-279, *43 pl.*
5: 1838 (t.p. date; p. viii: 20 Dec 1838), p. [i]-xii, [1]-280, *50 pl.*
6: 1839 (p. viii: 20 Dec 1839), p. [i]-xii, [1]-280, *48 pl.*
7: 1840 (p. viii: 20 Dec 1840), p. [i]-xii, [1]-280, *48 pl.*
8: 1841 (p. viii: 30 Dec 1841), p. [i]-xii, [1]-280, *46 pl.*
9: 1842 (p. viii: 20 Dec 1842), p. [i]-xii, [1]-279, *48 pl.*
10: 1843 (p. viii: 20 Dec 1843), p. [i]-xii, [1]-280, *48 pl.*
11: 1844 (p. viii: 20 Dec 1844), p. [i]-xii, [1]-280, *48 pl.*
12: 1846 (.........), p. [i]-xii, [1]-280, *48 pl.*
13: 1847 (p. viii: 19 Dec 1846), p. [i]-xii, [1]-280, *48 pl.*
14: 1848 (p. viii: 20 Dec 1847), p. [i]-xii, [1]-280, *48 pl.*
15: 1849 (p. [vii]: 23 Dec 1848), p. [i]-xi, [1]-314, *48 pl.*
16: 1849 (p. 3: 12 Jan 1849), p. [i]-xii, [1]-376, *24 pl.*
Reissue vol. 1: 1841. – See Stearn (1956) for precise contents of fascicles of vols. 12 and 13. – Part 1 of vol. 1 came out in Feb 1834, no. 2 Mar 1934, etc. – A set with 723 plates (717 of flowers, 6 of garden-design) was offered by Junk (Cat. 214, no. 238, 1979) at Hfl. 9.800.=.

7555. *A pocket botanical dictionary*, comprising the names, history, and culture of all plants known in Britain; with a full explanation of technical terms ... London (J. Andrews, ...) 1840. Oct. (in fours). (*Pocket bot. dict.*).
Co-author: John Lindley (1799-1865).
Ed. [*1*]: 1840 (p. ix: Jul 1840), p. [i]-xii, [1], 1-354. *Copy*: NY.
Ed. [*2*]: 1849, p. [i]-xii, [1], 1-339, 1-72 [suppl.], [1]. *Copies*: NY, USDA. "A new edition with a supplement containing all the new plants since its appearance." London (Bradbury & Evans, ...) 1849. Oct.
Re-issue of ed. [2]: 1853, p. [i]-xii, [1], 1-339, suppl. 1-72, [1]. *Copy*: FI. – "A new edition ... 1853". Oct.
New edition: 1868, (p. iv: Mai 1868), p. [i]-xii, [1]-623. *Copies*: MO, USDA. – "*Paxton's botanical dictionary* comprising the names, history, and culture of all plants in Britain; with a full explanation of technical terms. New edition including all the new plants up to the present year revised and corrected by Samuel Heteman [fl. 1840-1868], secretary for nearly forty years to the late Sir Joseph Paxton." London (Bradbury, Evans, & Co, ...) 1868. Oct.

7556. *Paxton's flower garden* ... London (Bradbury and Evans, ...) [1850]-1853. Qu. 3 vols. (*Paxton's Fl. Gard.*).
Co-author: John Lindley (1799-1865).
Publ.: in parts 1850-1853, see TL-2/4659. *Copies*: BR, G, MO. – (Junk, Cat. 206, no. 239, 1977 offered a copy at Hfl. 1800.=).
1: [i]-iv, [1]-194, *pl. 1-36* (col. by L. Constans).
2: [i-ii], [1]-186, *pl. 37-72.*
3: [i-ii], [1]-178, *pl.73-100.*
Revised ed.: 1882-1884, 3 vols. *Copies*: AMD, NY, US. – Revised by Thomas Baines [1822-1875], London, Paris & New York (Cassel & Company, Limited).
1: 1882, p. [i]-iv, [1]-195, *pl. 1-36*. *2*: 1883, [i-iii], [1]-183, *pl. 37-72*. *3*: 1884, p. [i-iii], [1]-179, *pl. 72-108.*
Ref.: Garay, L.A., Taxon 18: 711-712. 1969 (information also in TL-2/4659).

Payer, Jean-Baptiste (1818-1860), French botanist; Dr. sci. nat. Paris 1840, Dr. med. Paris 1852; teacher at the École normale, and, from 1852, at the Faculté de Médecine, Paris; in politics and administration 1848-1851. (*Payer*).

HERBARIUM and TYPES: Mosses at PC (Sayre 1977). – Portrait at G.

BIBLIOGRAPHY and BIOGRAPHY: Barnhart 3: 58; BFM no. 2428; BM 4: 1534-1535, 8: 991; CSP 4: 789-790; Frank 3 (Anh.): 75; GR p. 292-293, cat. p. 69; Herder p. 117, 238; Hortus 3: 1201 (Payer); Jackson p. 42, 148, 156; Kelly p. 170; Kew 4: 254; Krempelh. 2: 197; LS 20190; Moebius p. 153; NI 1500; PR 22, 7008-7015 (ed. 1: 7841-7842); Rehder 5: 656; SO 646; TL-2/27, 1736; Tucker 1: 544.

BIOFILE: Anon., Bonplandia 1: 34, 86, 103, 155, 241. 1853, 3: 94, 180. 1855 (memb. Acad. Sci.), 9: 14. 1861 (death notice, err. "47" years old), see also Table art. orig. 1854-1893: 182-183 (list publ. in BSbF); Flora 44: 96. 1861; Österr. bot. Z. 11: 101. 1861 (d., err. "47" years old).
Beck, G., Bot. Centralbl. 34: 150. 1888 ("Plantae spitzbergenses" at W; status not clear).
Constantin, J., Ann. Sci. nat., Bot. ser. 10. 16: cxxiv (portr.).
Fürnrohr, A.E., Flora 36: 32, 80. 1852 (app. Fac. méd.).
Junk, W., Rara 2: 175. 1926-1936.
Mägdefrau, K., Gesch. Bot. 165, 280. 1973.
Sachs, J. v., Hist. botanique 201, 572. 1892.
Sayre, G., Bryologist 80: 515. 1977.
Schlechtendal, D.F.L. von, Bot. Zeit. 18: 395-396. 1860.
Stafleu, F.A., *in* Lawrence, G.H.M., ed., Adanson 1: 245. 1963 (also: p. 103 by J.P. Nicolas).
Taton, R., ed., Science in the 19th century 377. 1965.
Wittrock, V.B., Acta Horti Berg. 3(3): 110. 1905.

COMPOSITE WORKS: Adanson, M., *Familles naturelles des plantes* ...deuxième édition ... 1847 [1864], edited and supplemented by Payer, see TL-2/27. – *Other issue*: "Histoire de la botanique et plan des familles naturelles des plantes de Michel Adanson, deuxième édition, préparée par l'auteur, publiée sur ses manuscrits par MM. Alexandre Adanson et J. Payer. Paris[,] Victor Masson et fils ... imprimé en mdccclvii publié en 1864", cover, p. [i*-iii*], [i]-v, [vii], [1]-300, *1 pl.* (uncol. lith.). *Copy*: G. (*Review*: D.F.L. von Schlechtendal, Bot. Zeit. 24: 148-149. 1866).

EPONYMY: *Payera* Baillon (1878, *nom. cons.*); *Payeria* Baillon (1860, *nom. rej.*).

7557. Académie de Paris, Faculté des Sciences. Thèse pour le doctorat ès-sciences naturelles soutenue le mercredi, 3 juin 1840, par Jean Baptiste Payer, avocat, licencié ès-sciences naturelles, ... Zoologie : de la forme animale ... Phytologie: *Essai sur la nervation des feuilles dans les plantes dicotylées*. Géologie ... [Paris 1840]. Qu. (*Essai nerv. feuill. dicot.*).
Publ.: 3 Jun 1840, p. [1]-28, [1]-12, *1 pl.*, [13]-14. *Copy*: G.

7558. École de pharmacie de Strasbourg. *Des classifications et des méthodes en histoire naturelle*. Thèse soutenue le lundi, 5 février 1844 ... Paris (Imprimerie de Lacour et Maistrasse, ...) 1844. Qu. (*Classif. méth. hist. nat.*).
Publ.: 5 Feb 1844 (date on which thesis was defended; Flora rd. Apr 1844), p. [i, iii], [1]-36. *Copies*: FI, L, NY.
Ref.: Schlechtendal, D.F.L. von, Bot. Zeit. 3: 665-666. 20 Sep 1844 (rev.).
K-r, Flora 27: 380-383. 21 Jun 1844 (rev.).

7559. *Familles naturelles des plantes* avec des figures sur bois représentant les caractères des genres par J. Payer ... faisant suite à la seconde édition des Familles naturelles d'Adanson. *Algues et Champignons*. Paris (Victor Masson ...) 1848. Qu. †. (*Fam. pl., alg. champ.*).
Publ.: 1848 (cover with above text and date), text starts with p. 15, Thallophytes), p. [15]-112. *Copy*: USDA.

Completed issue: 1849, p. [i*-iii*], [i]-iv, [1]-222, [1, cont.]. *Copy*: MO. – "*Familles naturelles des plantes* par J. Payer ... faisant suite à la seconde édition des Familles naturelles d'Adanson. Paris[,] Victor Masson ... 1849". This is the first issue 1849 of the *Bot. crypt.* (see below) published in 1850.

7560. *Botanique cryptogamique* ou histoire des familles naturelles des plantes inférieures par J. Payer ... avec 1105 gravures sur bois représentant les principaux caractères des genres. Paris (Victor Masson ...) 1850. Qu. (*Bot. crypt.*).
Ed. 1: 1850 (but p. [15]-112 issued 1848, and the entire work issued in 1849 as *Fam. nat. pl.*, see above; Bot. Zeit. 10 Jan 1851), p. [i*-iii*], [i]-vi, [1]-222, [1, cont.]. *Copies*: Almborn, B, FH, FI, G, LD, MO, NY, Stevenson, US, USDA.
Ed. 2: 1868 (p. [i]: 31 Jan 1868; BSbF after 15 Mai 1868; Flora 30 Jul 1868), p. [i*-iii*], [i]-viii, [1]-234. *Copies*: B, BR, FH, FI, MO, NY (226 p.), Stevenson (226 p.), USDA.
– The copies with 226 p. lack the index but have an extra unnumbered [1] p. table of contents. – "Deuxième édition, revue et annotée par H. Baillon [Henri Ernest Baillon (1827-1895)] ..." Paris (F. Savy, ...) 1868. Qu.

7561. Faculté de Médecine de Paris. No. 269. Thèse pour le doctorat en médecine présentée et soutenue le 27 août 1852, par Jean-Baptiste Payer ... *De la famille des Malvacées* ... Paris (Rignoux, imprimeur de la Faculté de Médecine, ...) 1852. Qu. (*Fam. Malv.*).
Publ.: 27 Aug 1852 (t.p.; p. 39: 13 Aug 1852), p. [1]-39. *Copies*: NY (2).

7562. *Traité d'organogénie comparée de la fleur* ... Paris (Librairie de Victor Masson ...) [1854]-1857. Qu. (*Traité organogén. fl.*).
Publ.: 8 Apr 1854 (livr. 1)-1857 (complete, Bot. Zeit. 14 Aug 1857), in 15 parts of 48 p. and *10 pl.* each, p. [i]-viii, [1]-748, (1, err., add.), atlas: [i]-viii, *pl.1-154* (by Payer and Faguet). *Copies*: BR, FI, G, LD, MO, NY; IDC 7667. – Publication was originally on a monthly basis but this schedule was abandoned. The book is an often overlooked but extremely rich source of systematic and morphogenetic information.
Facsimile ed.: 1966, Lehre (J. Cramer), ISBN 3-7682-0346-8, [i*-iii*], followed by facsimile. Historiae naturalis classica 47. *Copies*: B, FAS, NY. – See Taxon 15: 330. 1966, Bot. Jahrb. 88 (Lit.): 37-38. 1968.

7563. *Éléments de botanique* ... Première partie organographie ... Paris (Victor Masson ... Langlois et Leclerq) 1857. Duod. (in sixes). †. (*Élém. bot.*).
Publ.: 1857, p. [i]-xii, [1]-276. *Copies*: B, BR, FI, G, MO, NY.
Ref.: Schlechtendal, D.F.L. von, Bot. Zeit. 16: 86-87. 26 Mar 1858 (rev.).
Z., Flora 41: 21-26. 14 Jan 1858 (rev.).

7564. *Leçons sur les familles naturelles des plantes* faites à la Faculté des sciences de Paris par J.-B. Payer ... ouvrage continué par H. Baillon ... Première partie. Paris (Librairie de G. Masson ...) [1860-]1872. †. (*Leçons fam. nat.*).
Co-author: Henri Ernest Baillon (1827-1895).
Publ.: in 11 parts between 1860 and 1872, p. [i]-vi, [1], [1]-408. *Copies*: G, MO, USDA.

Payne, Frederick William (1852-1927), British diatomist; schoolmaster at the Brighton Grammar School 1873-1876; at the City of London School, 1876-1919; BA London 1871. (*Payne*).

HERBARIUM and TYPES: BM (6000 slides).

BIBLIOGRAPHY and BIOGRAPHY: Barnhart 3: 58; BB p. 239; BM 8:991; Desmond p. 484; IH 1 (ed. 6): 362 (err. "G.W."), 2: (in press); Kew 4: 254-255.

BIOFILE: Fryxell, G.A., Beih. Nova Hedw. 35: 363. 1975.
Kent, D.H., Brit. Herb. 72. 1957.
Koster, J.Th., Taxon 18: 556. 1969.
Newton, L.M., Phycol. Bull. 1: 17. 1952.
Rendle, A.B., J. Bot. 65. 177-178, 208. 1927. (bibl.).
Wornardt, W.W., Int. direct. diat. 35. 1968.

7565. Diatomaceae. *Liostephania* and its allies. Illustrated by seventy-seven original figures ... London (Wheldon & Wesley. Ltd., ...) 1922. (*Liostephania*).
Publ.: Dec 1922 (preface: Dec 1922; Nat. Nov. Mai, Jul 1923; FH rd. 5 Jun 1924), p. [1]-30, *pl. 1-4*, with text. *Copies*: FH, PH, US.

Payot, Vénance (1826-1902), French (Savoie) commercial naturalist and mountain guide; local administrator and botanist at Chamonix; botanical explorer of the Mont Blanc region. (*Payot*).

HERBARIUM and TYPES: G (deposited by the Musée d'Annecy, 1969), material at B, BM, BR, E, FI, G, H, LG, MANCH, OXF, PC. For exsiccatae see Sayre (1975). Genève (G) has "Vénance Payot, *Herbier des Alpes*" with printed labels; the complete text of these labels, brought together in a 38 leaf quarto volume is also at G. – Portrait at G.

BIBLIOGRAPHY and BIOGRAPHY: AG 12 (3): 147; Barnhart 3: 58; BL 2: 158-159, 700; BM 4: 1535; Clokie p. 222; CSP 4: 790-791, 8: 575-576, 10: 1008, 12: 563, 17: 748; De Toni 1: xcvi; GR p. 343; Herder p. 433; IF p. 723; IH (2): (in press); Jackson p. 343; Kelly p. 170; Kew 4: 255; Morren ed. 10, p. 61; PR 7016-7018; Rehder 5: 656; Saccardo 1: 124, 2: 82, cron. xxix; SBC p. 129; Tucker 1: 544; Urban-Berl. p. 310.

BIOFILE: Anon., Bot. Not. 1903: 34; Bull. Soc. bot. France, Table art. orig. 1854-1893: 183; Hedwigia 42: (66). 1903 (d.); Nat. Nov. 25: 224. 1903 (d.).
Bonnot, E.J., Bull. Soc. bot. France 108: 80-110. 1962.
Bureau, É., Bull. Soc. bot. France 49: 106. 1902.
Charpin, A., Saussurea 1: 17-21. 1971, Musées de Genève 121: 16-18. 1972 (portr., biogr. note, b. 25 Jun 1826, d. 13 Mar 1902).
Harmand, J.H.A.J., Bull. Soc. bot. France 49: 168-169. 1902 (obit., b. 25 Jun 1826, d. 13 Mar 1902).
Hedge, I.C. & Lamond, J.M., Index coll. Edinb. herb. 117. 1970.
Miège & Wuest, Saussurea 6: 134. 1975 (deposit Payot herbarium at G (1969) by Musée d'Annecy).
Sayre, G., Mem. New York Bot. Gard. 19(3): 378. 1975; Bryologist 80: 515. 1977.

HANDWRITING: Musées de Genève 121: 18. 1972, Candollea 32: 387-388. 1977.

7566. *Guide du botaniste au jardin de la Mer de Glace* ou catalogue des plantes qui croissent à cette limite de la végétation, accompagné d'un aperçu d'une notice sur l'avancement des deux principaux glaciers de la vallée ... Genève (Imprimerie Ch. Gruaz, ...) 1854. (*Guide Mer de Glace*).
Publ.: 1854, p. [1]-13, [14-15]. *Copies*: FI, G. – The "Jardin de la Mer de Glace" (Jardin du Courtil) is a rocky island in the Talèfre glacier (BL 2: 158). – See also Martins, C., Bull. Soc. bot. France 12: 144-162. 1865 and further references given by BL.

7567. *Catalogue phytostatique* de plantes cryptogames cellulaires ou guide du lichénologue au Mont-blanc et sur les montagnes, entre les vallées de Sixt, Diozaz, Servoz, Chamounix, Bérard, Valorsine, Trient, Champé, Essert, Ferret, Entrèves, Allée-blanche, Chapui et Mont-Joie, comprises dans un rayon de 200 kilomètres ... Lausanne (Imprimerie Blanchard) 1860. Oct. (*Cat. phytostat.*).
Publ.: 1860 (mss. submitted 7 Mar 1860), p. [1]-32. *Copies*: BR, G(2), NY. Preprinted from Bull. Soc. vaud. Sci. nat. 6 (no. 47): 421-450. 1861. The cryptogams mentioned in this *Catalogue* were offered for sale by Payot at 15 fr/century. The title of the journal article was "*Flore de Chamounix* ... Famille des Lichens", "par V. Payot" sic.

7568. *Catalogue des fougères, prêles et lycopodiacées des environs du Mont-Blanc* ou énumération détaillée des plantes acotylédones vasculaires qui naissent dans les vallées de Sixt, Servos, Diozaz, Bérard, Valorsine, Treint, Champé, Essert, Ferret, Allée-Blanche, Chapiu [sic], Mont-Joie, comprises dans un rayon de 200 kilomèt. autour de celle de Chamounix suivi d'un catalogue des mousses et des lichens des mêmes localités ... Paris (Joël Cherbuliez, ...) Genève (id.) 1860. Oct. (*Cat. foug. Mont-Blanc*).
Publ.: 1860, p.[i]-xi, [13]-70, map. *Copies*: G(2), MO, NY.

7569. *Énumération des mousses* nouvelles, rares et peu connues des environs *du Mont-Blanc*... [Lausanne 1865]. Oct.
Publ.: 1865 (mss. submitted 5 Apr 1865), p. [1]-8. *Copies*: G, NY. – Reprinted and to be cited from Bull. Soc. vaud. Sci. nat. 53: 354-361. 1865.

7570. *Guide du botaniste* ou catalogue des plantes rares de la Suisse française rangées par localités (pour 19 courses) ... 1878. (*Guide bot. Suisse franç.*).
Publ.: 1878 (p. [3]: 29 Apr 1878; rev. 28 Aug-1 Sep 1878), p. [1]-110, [1-2, cont., err.]. *Copy*: G.

7571. Deuxième partie *Plantes cryptogames* vasculaires et cellulaires *Florule du Mont-Blanc* ou guide du botaniste et du touriste sur les Alpes pennines excursions phytologiques (fougères ferns) ... Genève (Henri Trembley ...) 1881. Qu. (*Pl. crypt., Fl. Mont-Blanc*).
Publ.: 1881, p. [i*-iii*], [i]-ii, [1]-22. *Copies*: BR, G(2), NY(2). For the phanerogamic part see below, no. 7572; for the second part of the cryptogams (bryophytes) see no. 7573.– The ferns were elaborated with the help of "Mr. le Dr. Wild de Meran", this was Julius Milde (1824-1872) who resided at Merano for some time for health reasons. Nat. Nov. Sep(2) 1886. mentions a Paris 1886 issue.

7572. *Florule du Mont-Blanc*. Guide du botaniste et du touriste dans les Alpes pennines ... *Phanérogames.* Paris (Librairie Sandoz et Thuillier ...), Neuchatel (Librairie Jules Sandoz), Genève (Librairie Desrogis). s.d. [1882]. Oct. (*Fl. Mont-Blanc, Phan.*).
Publ.: Jan-Feb 1882 (Nat. Nov. Mar 1882; Bot. Centralbl. 3-7 Apr 1882; Bot. Zeit. 28 Apr 1882), p. [i-ii], [1]-291. *Copies*: BR, G, NY.
Ref.: Saint-Lager, J.B., Ann. Soc. bot. Lyon 9(2): 391-396. 1882.

7573. *Florule bryologique* ou guide du botaniste au Mont-Blanc 2me partie des cryptogames ou Muscinées des Alpes pennines ... Genève (Henri Trembley ...) 1886. Qu. (*Fl. bryol. Mont-Blanc*).
Publ.: 1 Jan-12 Feb 1886 (Bot. Centralbl. 31 Jan-4 Feb 1887; BSbF rd. 12 Feb 1886; Bot. Zeit. 28 Mai 1886; Nat. Nov. Mai(2) 1886; Flora 21 Feb 1887; Hedwigia Jan-Feb 1887), p. [i*-iii*], [i]-iii, [1]-78. *Copies*: BR, G, LD, NY(2).
Ref.: Zukal, H., Österr. bot. Z. 37: 108-109. 1887 (rev.).

7574. *Lichens recueillis sur le Massif du Mont-Blanc* par M. Vénance Payot et determinées par M. l'Abbé Harmand ... Paris (Librairies-imprimeries réunies ...) 1901. Oct.
Author: Julien Herbert Auguste Jules Harmand (1844-1915), addendum to TL-2/2:54-55.
Publ.: 1901, cover (with above title and imprint), p. [65]-91. *Copy*: G. – Reprinted and to be cited from Bull. Soc. bot. France 48: [65]-91. 1901.

Payrau, Vincent (*fl.* 1900), French botanist and pharmacist, Dr. pharm., Paris, 1900, resident pharmacist of the Paris hospitals. (*Payrau*).

HERBARIUM and TYPES: Unknown.

BIBLIOGRAPHY and BIOGRAPHY: BM 4: 1535; Kew 4: 255; Rehder 3: 690.

7575. *Recherches sur les Strophanthus* ... Paris (Société d'éditions scientifiques ...) 1900. Oct. (*Rech. Strophanthus*).
Publ.: 1900 (Nat. Nov. Nov(2) 1900), p. [1]-176, *pl. 1-10*. *Copies* MICH, PC. – We have not seen the thesis issue of this publication.
Ref.: Roth, E., Bot. Centralbl. 87: 420-421. 1901 (rev., mentions *11 pl.*).

Payson, Edwin Blake (1893-1927), American botanist; Dr. phil. Univ. Washington 1921; BA Univ. Wyoming 1917; professor of botany Univ. Wyoming 1921-1927. (*Payson*).

HERBARIUM and TYPES: RM; further material at DBN, E, F, GH, MIN, MO, MSC, NY, POM, US. For itinerary and collections see Ewan (1951).

BIBLIOGRAPHY and BIOGRAPHY: Barnhart 3: 59; BJI 2: 132; Bossert p. 303; Ewan ed. 1: 255, 279-280, 300. 1950 (with itin. by L.E.B. Payson), ed. 2: 170-171; Hirsch p. 227; Hortus 3: 1201 (Payson); IH 1 (ed. 7): 339; Kew 4: 255; Langman p. 572; Lenley p. 322; NAF 28B(2): 350; NW p. 54; PH 209; RS p. 133.

BIOFILE: Anon., Ann. bot. Soc. 96: 8. 1928; Laramie (Wyo.) Republican 16 Mai 1927 (fide Ewan); Torreya 27: 60. 1927.
Hedge, I.C. & Lamond, J.M., Index coll. Edinb. herb. 117. 1970.

7576. *The North American species of Aquilegia* ... Washington (Government Printing Office) 1918. Oct.
Publ.: 14 Oct 1918, p. [i]-vii, 133-157, ix, *pl. 8-14*. *Copies*: U, US. – Reprinted and to be cited from Contr. U.S. natl. Herb. 20(4): 133-157, *pl. 8-14*. 1918.

7577. *A monograph of the genus Lesquerella* ... [St. Louis 1922]. Oct.
Publ.: 14 Jan 1922 (according to footnote on p. [103]; the reprint cover states: reprinted Apr 1921), reprinted with special cover but with original pagination, and to be cited from Ann. Missouri bot. Gard. 8: 103-236. 1921 (1922). *Copies*: reprint G, ·MO, NY.

7578. *A monographic study of Thelypodium* and its immediate allies [St. Louis 1923]. Oct.
Publ.: 16 Feb 1923 (see footnote p. 233; the cover has Sep 1922), reprinted and to be cited from Ann. Missouri Bot. Gard. 9: 233-324. 1922 (1923). *Copies* reprint: BR, G, MO, NY.

7579. *A monograph of the section Oreocarya of Cryptantha* [St. Louis 1927]. Oct.
Publ.: 8 Oct 1927 (footnote on p. 211; date on cover of reprint: Sep 1927). Reprinted and to be cited from Ann. Missouri Bot. Gard. 14: 211-358. *pl. 25-30*. 1927. *Copies* reprint: G, MO, NY.

Peacock, John T. (*fl.* 1878], British country gentleman of Sudbury House, Hammersmith; collector of succulent plants. (*Peacock*).

HERBARIUM and TYPES: Unknown.

BIBLIOGRAPHY and BIOGRAPHY: Desmond p. 484; Tucker 1: 544.

BIOFILE: Anon., Bull. misc. Inf. Kew 1897: 232 (living plants from Peacock's collections were purchased by Kew in 1889).

7580. *List of succulent plants* in the collection of Mr. J.T. Peacock at Sudbury House, the Octagon House at Kew, the Alexandra Palace conservatories, and the Royal Botanical Gardens, Regent's Park. London (Judd and Co., ...) 1878. (*List succ. pl.*).
Publ.: 1878, *plate*, p. [i], [1]-16. *Copy*: USDA.

Pearl, Raymond (1879-1940), American botanist, entomologist, student of human populations, teacher and editor; Ph.D. Univ. Mich. 1902; instructor in zoology Univ. Michigan 1902-1905, in London 1905-1906, at Univ. Pennsylvania 1906-1907; head dept. biol. Maine Agric. Exp. Station 1907-1918; professor of biometry and vital statistics Johns Hopkins, Baltimore 1918-1925; director Institute of Biological Research ib. 1925-1930; professor of biology ib. 1930-1940; D.Sc. Dartmouth 1919. (*Pearl*).

HERBARIUM and TYPES: Unknown. Some manuscripts in Smithsonian Archives.

BIBLIOGRAPHY and BIOGRAPHY: Barnhart 3: 60; CSP 17: 752; Hirsch p. 227; Kew 4: 257; Lenley p. 322; PH 450, 939.

BIOFILE: Anon., Genetics 26(1): frontisp., portr. on p. i. 1941 (member editorial board of Genetics 1916-1940; b. 3 Jun 1879, d. 17 Nov 1940); J. Wash. Acad. Sci. 31(3): 127. 1941 (obit.); New York Times 18 Nov 1940.
Gilbert, P., Comp. biogr. lit. deceased entom. 291. 1977.

Greenwood, Major, Nature 147: 140. 1941; Ecology 22: 408. 1941.
Henderson, L.J., Amer. Philos. Soc. Yearb. 1940: 431-433. 1941 (obit.).
Jennings, H.S., Nat. Acad. Sci. Biogr. Mem. 22: 295-347. 1943 (full biogr. and bibl., portr.).
Miner, J.R. & J. Berkson 52: Sci. Mon. 52: 192-194. 1941 (b. 3 Jun 1879, d. 17 Nov 1940, portr., obit.).
Park, T., Dict. Amer. biogr. 22(suppl. 2) 521-522. 1958 (biogr., bibl.).
Reed, L.J., Science ser. 2. 92: 595-597. 1940 (obit.).

COMPOSITE WORKS: Founder and editor until his death of *The Quarterly Review of Biology and human Biology* (vol. 1, 1926-x).

7581. *Variation and differentiation in Ceratophyllum* ... Washington, D.C; (Published by the Carnegie Institution of Washington) February, 1907. Oct. (*Var. diff. Ceratophyllum*).
Co-authors ("with the assistance of"): Olive M. Pepper and Florence J. Hagle.
Publ.: Feb 1907, p. [1]-136, *2 charts*. *Copies*: BR, MO. – Carnegie Institution of Washington, Pub. no. 58.

Pearsall, William Harrison (1860-1936), British schoolmaster and botanist at Dalton-in-Furness until 1925; retired to Kent; secretary of the Botanical Society of the British Isles 1931-1936. (*Pearsall*).

HERBARIUM and TYPES: CGE, LIV, OXF, SLBI.

BIBLIOGRAPHY and BIOGRAPHY: Barnhart 3: 60; BJI 2: 132; BL 2: 252, 267, 701; Clokie p. 222; Desmond p. 485; Hirsch p. 227; IH 2: (in press); Kew 4: 257-258.

BIOFILE: Allen, D.E., The naturalist in Britain 263. 1976.
Bridson, G.D.R. et al., Nat. hist. mss. res. Brit. Isl. 427 [index]. 1980.
Dallmann, A.A., ed., North West Natural. 11(4): 375-376. 1936 (obit.).
Hall, P.M., J. Bot. 74: 352-353. 1936 (brief bibl.), Bot. Soc. Exch. Cl. Brit. Isl. Rep. 11(3): 211-212. 1936 (obit., b. 6 Jun 1860, d. 12 Aug 1936)
Hawksworth, D.L. & M.R.D. Seaward, Lichenology Brit. Isl. 131-132. 1977 (bibl.).
Kent, D.H., Brit. herb. 72. 1957 (coll.).
Martin, W.K. and G.T. Fraser, Fl. Devon 779. 1939.
Verdoorn, F., ed., Chron. bot. 1 161, 168, 175. 1955, 2: 193. 1936, 3: 153, 154, 169. 1937, 4: 181, 273, 534. 1938, 5: 168, 426. 1939, 6: 22, 260, 300. 1941.
Woodhead, T.W., Naturalist, Yorks., Mar 1923: 113, 118.

COMPOSITE WORKS: (1) edited Salmon, C.E., *Fl.Surrey*, 1931.
(2) Edited *Bot. Soc. Exch. Club Brit. Isles, Report*, 1931-1935.
(3) With his son, William Harold Pearsall, *Phytoplankton of the British Lakes*, J. Linn. Soc. Bot. 43: 55-73. 1925.

NOTE: For P's son William Harold Pearsall (1891-1964), plant ecologist, sometime editor of *J. Ecol.*, professor of botany at Sheffield (1938-1944) and University College, London (1944-1957) see e.g. Desmond p. 485, GR p. 409 and Biogr. Mem. Fellows R. Soc. 17: 511-540. 1971.

Pearson, Arthur Anselm (1874-1954), British businessman and mycologist at Cleckheaton; authority on agarics; treasurer of the British mycological Society 1919-1946, president 1931, 1952. (*A. Pearson*).

HERBARIUM and TYPES: E and K.

BIBLIOGRAPHY and BIOGRAPHY: Barnhart 3: 60; BL 2: 269, 700; BM 8: 992; Desmond p. 485; IH 2: (in press); Kelly p. 170; Kew 4: 258; LS suppl. 21134-21135.

BIOFILE: Anon., J. Bot. 56: 280; 1918; The Times 16 Mar 1954 (fide Desmond); Trans. Brit. mycol. Soc. 36: frontisp. 1953 (portr.).

Bisby, G.R., Trans. Brit. mycol. Soc., fifty-year index 618. 1952.
Blackwell, E.M., Naturalist 877: 66. 1961.
Bridson, G.D.R. et al., Nat. hist. mss. res. Brit. Isl. 255.1, 255.230, 269.282. 1980.
Hedge, I.C. & Lamond, J.M., Index coll. Edinb. herb. 117. 1970.
Heim, R., Rev. de Mycol. 19(1) suppl. 130-131. 1954.
Le Gal, M., Bull. Soc. mycol. France 73(1): 13-17. 1957 (portr., obit.).
Orton, P.D., Trans. Brit. mycol. Soc. 37: 321-323. 1954 (obit., portr. in group, b. 12 Apr 1874, d. 13 Mar 1954).
Ramsbottom, J., Trans. Brit. Mycol. Soc. 30: 6, 8, 9, 10, 13, 14. 1948.
Wakefield, E.M., Nature 173: 709. 1954 (obit.), Proc. Linn. Soc. London 165(2): 218-219. 1955 (obit.).

7582. *List of the fungi of Epping Forest* ... Stratford (Essex Museum, ...) 1938. (*List fung. Epping Forest*).
Publ.: 1938, p. [1]-24. *Copy*: USDA. – See Essex Natural. 26: 123-129. 1938 for additional details.
Ref.: Ramsbottom, J., J. Bot. 76: 298. 1938.

Pearson, Henry Harold Welch (1870-1916), English botanist; M.A. Cambridge 1900, D. Sc. ib. 1907; assistant curator Cambridge herbarium 1898-1899; assistant for India at Kew 1899-1903; from 1903-1916 Harry Bolus professor of botany at the South African College, Cape Town; founder of the Kirstenbosch National Botanic Gardens of South Africa. (*H. Pearson*).

HERBARIUM and TYPES: Important sets at BOL, PRE, and SAM; further material at A, BM, CGE, COI, GRA, K, KMG, MO, NH, NY, STE, WAG.

BIBLIOGRAPHY and BIOGRAPHY: BB p. 240; BJI 2: 132; BL 1: 29, 56, 237, 312; BM 4: 1537, 8: 993; Bossert p. 304; CSP 17: 753; Desmond p. 485-486; Hortus 3: 1201 (H. Pearson); IH 2: (in press); Kew 4: 258-259; Langman p. 573; MW p. 381; Rehder 5: 656; TL-2/2448, 7055; Tucker 1: 545; Urban-Berl. p. 381; Zander ed. 10, p. 700, ed. 11, p. 798.

BIOFILE: Anon., Bot. Centralbl. 75: 223. 1898 (asst. Cambridge), Bot. Jahrb. 26 (Beibl. 61): 4. 1899 (asst. Cambridge); 32 (Beibl. 72): 5. 1903 (appointment Cape Town); Bot. Not. 1917: 141 (d.); Bull. misc. Inf. Kew 1916: 277-281 (obit., bibl.); Gard. Chron. 60: 238, 250. 1916 (obit., portr.); Magy. Bot. Lap. 15: 312. 1916 (d.); Nat. Nov. 25: 271. 1903 (prof. bot. Cape Town); Österr. bot. Z. 66: 408. 1916 (d.).
Bews, J.W., S. Afr. J. Sci. 13: 352-355. 1917 (obit., portr.).
Britten, J., J. Bot. 54: 375. 1916, 55: 62. 1917 ("The military occupation of South Africa by the Union Forces has afforded an opportunity for the preservation of Welwitschia ...").
Bullock, A.A., Bibl. S. Afr. bot. 83. 1978.
Evans, I.B.P., Trans. R. Soc. S. Afr. 7: 139-145. 1919 (obit.).
Exell, A.W., Bol. Soc. Brot. ser. 2.26: 218. 1952 (pl. BM, K).
Gager, C.S., Brooklyn Bot. Gard. Rec. 27(3): 327. 1938 (dir. bot. gard. Kirstenb. 1913-1916).
Gossweiler, J., Bot. Soc. Brot. ser. 2. 13: 301. 1939.
Gunn, M. & L.E. Codd, Bot. expl. S. Afr. 96, 275-276. 1981 (biogr., coll., portr.).
Huxley, L., Life letters J.D. Hooker 2: 24. 1918.
Kneucker, A., Allg. bot. Z. 23: 32. 1917 (death), 4: 68. 1898 (appointed Cambridge), 9: 76. 1903 (appointed Cape Town).
Moldenke, H.N., Plant Life 2: 75. 1946 (1948).
Levyns, M.H.B., Dict. S. Afr. biogr. 2: 536. 1972.
Nelmes & Cuthbertson, Curtis's Bot. Ma. Dedic. 1827-1927, p. 346-347. 1932. (portr.).
Pearson, H.H.W., J. Linn. Soc., Bot. 35: 375-390, *pl. 9*. 1902 (on Dischidia); Bull. misc. Inf. Kew 1910: 372-380 (on natl. bot. gard. S. Afr.).
Phillips, E.P., S. Afr. J. Sci. 27: 55, 60, 65, 69, 71.
Pole Evans, I.B., Trans. R. Soc. S. Afr. 3: 139-145. 1918.
Seward, A.C., Nature 98: 211. 1916 (obit.), Ann. Bolus Herb. 2: 131-147. 1917 (obit.,

portr., bibl., b. 28 Jan 1870, d. 3 Nov 1916), Ann. Bot. 31: i-xviii. 1917 (id.), Proc. Linn. Soc. London 129: 57-60. 1917 (obit.); Proc. R. Soc. ser. B, 89: lx-lxvii. 1917.
Smith, Mathilda, J. Kew Guild 1917: 377-378.
White, A., and B.L. Sloane, The Stapelieae, 36, 90, 1933, ed. 2, 106, 132, 461, 520. 1937 (portr.).

COMPOSITE WORKS: (1) *Verbenaceae, in* Harvey & Sonder, *Fl. cap.* 5(1): 180-224. Jun 1902, 225-226. Mai 1910; *Welwitschiaceae, in* id. 5(2, suppl.):1-3. 1933; *Gnetaceae, in* id. 6(2): 328-333. Nov. 1917; *Thymelaeaceae, in* id. 6(1, 2): 212-255. Dec 1910.
(2) Editor, *Annals of the Bolus Herbarium* 1-2(3), 1914-1917.
(3) Pearson's *On the collections of dried plants obtained in South-West Africa by the Percy Sladen Memorial Expeditions, 1908-1911,* were published in *Ann. S. Afr. Mus.* 9: 1-19. 28 Feb 1911, 21-90. 30 Mai 1912, 129-192. 30 Oct 1913, 193-272. (Apr 1915, 355-418. 30 Jun 1917.

DEDICATION: Botanical Magazine, vol. 140. 1914.

EPONYMY: *Pearsonia* Dümmer (1912); *Pearsoniella* Fritsch et Rich (1924).

7583. *Gnetales* ... Cambridge (At the University Press) 1929. Oct. (*Gnetales*).
Editor: Albert Charles Seward (1863-1941), assisted by Mary Gladys Thoday (x-1943).
Publ.: 1929 (p. vi: 16 Mai 1929; Nat. Nov. Apr 1930 portr. author), p. [i]-vi, [1], [1]-194, *pl. 1-3. Copies*: BR, FI, G, L, MO, NY, U.
Ref.: Rendle, A.B., J. Bot. 68: 284-285. Sep 1930.

Pearson, William Henry (1849-1923), British businessman (yarn agent) and hepaticologist at Eccles nr. Manchester; from 1902 at Manchester; one of the founders of the British Bryological Society; M.Sc. Manchester 1911. (*Pearson*).

HERBARIUM and TYPES: BM (9000 Brit. [and some exotic] Hepat.; acquired 1902; many types, esp. of *Hepat. Brit. Isl.*) and MANCH (exot.); further material in DBN, E, GL, H, K, L, OXF, PC, W, WELC. – *Exsiccatae*: (1) with Benjamin Carrington (1827-1893), q.v.: *Hepaticae britannicae exsiccatae*, fasc. 1-4 nos. 1-290. 1878-1890. Sets: see under Carrington. –
(2) *Canadian hepaticae,* collected and distributed by John Macoun, named by W.H. Pearson, 116 nos., 1891, (see note by E.G. Britton in Bull. Torrey bot. Club 19: 97. 1892), sets at BM, and FH, see IH 2: 487. – Further sets of British hepatics were issued after 1890; the contents of these sets varied; they do not constitute regular exsiccatae; see e.g. Rev. bryol. 30: 60. 1903, 35: 84. 1908, 37: 28. 1910, 38: 96. 1911.)
Some letters at G.

BIBLIOGRAPHY and BIOGRAPHY: Barnhart 3: 61; BB p. 240; BM 4: 1537, 8: 993; Clokie p. 222; CSP 10: 1011, 17: 754; Desmond p. 486; IH 1 (ed.6): 362, (ed. 7): 339, 2: (in press); Kew 4: 259-260; Kleppa p. 147, 148, 149, 150,.154); Lenley p. 322; NI 1501; SBC p. 129.

BIOFILE: Anon., Bot. Not. 923: 383; Brit. Bryol. Soc., Rep. 1: 39-40. 1923; Bryologist 26: 41-42. 1923; Lancashire Naturalist 2: 182. 1908 (portr.), 15: 197-198. 1923 (obit.); Rev. bryol. 30: 36. 1903 (Hepat. herb. acquired by BM).
Beck, G., Bot. Centralbl. 19(34): 253. 1884 (Hep. at W.).
Bridson, G.D.R. et al., Nat. hist. mss. res. Brit. Isl. 42.3. 1980.
Britten, J., J. Bot. 40: 399. 1902 (Hepat. herb. acquired); 61: 184. 1923 (died 19 Apr 1923).
Broome, H.C., Bryologist 27: 96-101. 1925 (main bibl.).
Bullock, A.A., Bibl. S. Afr. bot. 83. 1978.
Chamberlain, E.B., Bryologist 27: 12-14. 1924 (portr.).
Hedge, I.C. & J.M. Lamond, Index coll. Edinb. herb. 117. 1970 (coll. at E).
H.E.S., Rucksack Club J. 5: 209-211. 1924.
Jackson, B.D., Bull. misc. Inf. Kew 1901: 51 (pl. K), Proc. Linn. Soc. London 135: 45. 1923 (obit.).
Murray, G., Hist. coll. BM(NH) 173. 1904 (herb.).

Pearson, W.H., J. Linn. Soc., Bot. 46: 13-44. *pl. 2-3.* 1922 (Hepat. New Caled.).
Sayre, G., Bryologist 80: 502-521. 1977; Mem. New York Bot. Gard. 19(2): 236. 1971.
Schuster, R.M., Hepat. Anthoc. N. Amer. 1: 86. 1966.
Watson, W., J. Bot. 61: 194-197. 1923.
Weiss, F.E., Mem. Proc. Manchester Lit. Philos. Soc. 68: iii-iv. 1924.

NOTE: Pearson published many papers, e.g. in J. Bot. and Irish naturalist, describing a single new species. Reprints with independent pagination were prepared by Pearson but we refrain from listing them here. The papers should all be cited from the journals. A good set of these independent reprints is at FH; others are at NY.

7584. *Hepaticae natalenses* a clarissima domina Helena Bertelsen missae ... Christiania [Oslo] (i Commission hos Jacob Dybwad/A.W. Brøggers Boktrykkeri) 1886. Oct. *(Hepat. natal.).*
Collector: Helena Bertelsen (no further details known to us).
Publ.: Mai-Jul 1886 (p. 19: printed 12 Apr 1886; Nat. Nov. Jul(2) 1886; Bot. Zeit. 24 Sep 1886; Rev. Bryol. Oct 1886), p. [1]-19, [20], *pl. 1-12* (uncol.). *Copies*: FH, G(2), H, NY. − Christiania Vid.-Selsk. Forh. 1886(3).

7585. *Hepaticae knysnanae* sive hepaticarum in regione capensi "Knysna" Africae australis a fabro ferrario Hans Iversen lectarum recensio ... Christiana (in commission by Jacob Dybwad/printed by A.W. Brøggers) 1887. Oct. *(Hepat. knysn.).*
Collector: "ab Hans Iversen lectarum ..." at Knysna, Cape of Good Hope. We have no particulars on this collector, except that he was a Norwegian blacksmith at Knysna ("fabro ferrario") who corresponded with F.C. Kiaer.
Publ.: Aug-Sep 1887 (printed 30 Jul 1887; Bot. Zeit. 28 Oct 1887; Nat. Nov. Oct(2) 1887; Rev. bryol. Dec 1887), p. [1]-15, [16], *pl. 1-6. Copies*: FH, G(2), H, NY. − Christiania Vid.-Selsk. Forh. 1887(9).

7586. Geological and natural history survey of Canada ... *List of Canadian Hepaticae* by Wm. Hy. Pearson. Montreal (Wm. Foster Brown & Co.) 1890. Oct. *(List Canad. Hepat.).*
Publ.: Nov 1890 (US rd. 3 Dec 1890; BSbF 9 Jan 1891; Rev. bryol. Apr 1891; Nat. Nov. Jan(1) 1891), p. [i-iii], [1]-31, *pl. 1-12* (uncol., auct.). *Copies*: B, BR, FH, MO, MO-Steere, NY(2), U-V; US(2), USDA. − Based on collections made by John Macoun (1832-1920) and Coe Finch Austin (1831-1880), all in P's own herbarium.

7587. *Frullaniae madagascarienses* praecipue e collectionibus Borgeni ... Christiania (i Commission hos Jacob Dybwad/A.W. Brøggers boktrykkeri) 1891. Oct. *(Frull. madag.).*
Collector: M. [G?] Borgen, collected in Madagascar 1869-1877.
Publ.: 13 Feb-15 Mar 1891 (p. 9: printed 13 Feb 1891; Nat. Nov. Mar(2) 1891; BC 22 Apr 1891), p. [1]-9, *pl. 1-4* (uncol., auct.). *Copies*: FH, G. − Christiania Vid.-Selsk. Forh. 1890(2).

7588. *Lejeuneae madagascarienses* ... Christiana (i Commission hos Jacob Dybwad/A.W. Brøggers bogtrykkeri) 1892. *(Lejeun. madag.).*
Publ.: Sep-Oct 1892 (mss. submitted 22 Apr 1892; printed 19 Aug 1892; Nat. Nov. Nov(2) 1892), p. [1]-9, *pl. 1-2* (uncol., auct.). *Copies*: FH, G, NY. − Christiania Vid.-Selsk. Forh. 1892(8).

7589. *Hepaticae madagascarienses* notes on a collection made by Rev. M. Borgen, Rev. Borchgrevink and Rev. Dahle, 1877-82 ... Christiania (i Commission hos Jacob Dybwad/A.W. Brøggers bogtrykkeri) 1893. Oct. *(Hepat. madag.).*
Collectors: M.[G?] Borgen (fl. 1887); Borchgrevink; Dahle; three missionaries on whom we have no details.
Publ.: 17 Feb-31 Mar 1893 (printed 17 Feb 1893; Nat. Nov. Apr(1) 1893), p. [1]-11, *1 pl.* (uncol., auct.). *Copies*: FH, G, NY. − Christiania Vid.-Selsk. Forh. 1892(14), publ. 1893.

7590. *The Hepaticae of the British Isles* being figures and descriptions of all known British species ... London (Lovell Reeve & Co., Limited ...) 1902, 2 vols. Oct. *(Hepat. Brit. Isl.).*

1 (text): Aug 1899-Sep 1902 (in 29 parts, see below), p. [i]-vi, [1], [1]-520.
2 (plates): Aug 1899-Sep 1902 (id., 8 plates per part), p. [i]-vii, *pl. 1-228* (col., author).
Copies: BR (*col. pl.*), FH (orig. covers; *col. pl.*), FI, MO(2), NY, U-V; IDC 5417.

part	pages	plates	dates	part	pages	plates	dates
1	1-32	*1-8*	Jun-Jul 1899	15	273-288	*113-120*	Nov 1900
2	33-48	*9-16*	Jul-Sep 1899	16	289-304	*121-128*	1901
3	49-64	*17-24*	Sep-Dec 1899	17	305-320	*129-136*	1901
4	65-80	*25-32*	Sep-Dec 1899	18	321-336	*137-144*	1901
5	81-96	*33-40*	Sep-Dec 1899	19	337-352	*145-152*	1901
6	97-128	*41-48*	1900	20	353-368	*153-160*	1901
7	129-144	*49-56*	1900	21	369-384	*161-168*	1901
8	145-160	*57-64*	1900	22	385-400	*169-176*	1901
9	161-176	*65-72*	1900	23	401-416	*177-184*	1901
10	177-208	*73-80*	1900	24	417-432	*185-192*	1901
11	209-224	*81-88*	1900	25	433-448	*193-200*	1901
12	225-240	*89-96*	1900	26	449-464	*201-208*	Dec 1901
13	241-256	*97-104*	Sep 1900	27	465-480	*209-216*	Jan-Apr 1902
14	257-272	*105-112*	Oct-Nov 1900	28//29	481-520	*217-228*	Apr-Jul 1902

Copies were issued with coloured and with uncoloured plates drawn by the author.
– The months of publication given above are inferred on the basis of notes in the literature and on receipt-dates at FH; the years are given on the original covers. The announcement called for monthly parts.
Ref.: Farmer, J.B., Nature 66: 385-386. 1902.
Waddell, C.H., Irish Naturalist 11: 226-228. 1902.

7591. The Manchester Museum. Museum Handbooks. *Catalogue of Hepaticae (Anacrogynae) in the Manchester Museum* arranged according to Stephani's "Species hepaticarum" by Wm. Hy. Pearson, A.L.S. Manchester (Sherratt & Hughes, ...) London (Dulau & Co., ...) 1910. (*Cat. Hepat. Manchester Mus.*).
Publ.: 1910, p. [1]-31. *Copies*: BR, PH, US, USDA.

Pease, Arthur Stanley (1881-1964), American classicist and botanist; A.B. Univ. Illinois 1902; Ph. D. Harvard 1906; taught Latin and Greek at Harvard (1906-1909), Univ. Illinois (1909-1924), Amherst Coll. (1924-1926), Harvard (1932-1950); president Amherst College (1927-1932); active plant collector, made trip around the world 1926-1927. (*Pease*).

HERBARIUM and TYPES: GH (over 37.000 nos.); other material at A, C, DBN, DPU, E, F, MT, NEBC, US and VT. – Manuscript material in Smithsonian Archives.

BIBLIOGRAPHY and BIOGRAPHY: Barnhart 3: 61; BL 1: 197, 312; Bossert p. 304; IH 1(ed. 6): 362, (ed. 7): 339, 2: (in press); Kew 4: 260; NW p. 55; Pennell p. 616; SK 5: cccx; Tucker 1: 545.

BIOFILE: Arditti, J. et al., Orchid biol. 1: 31, 44, 132, 154, 159, 174. 1977.
Bean, R.C., Rhodora 66: 3-5. 1964 (obit., portr., b. 22 Sep 1881, d. 7 Jan 1964).
Fernald, M.L., et al., Amherst Grad. Quart. 64: 262-271. 1927 (portr., b. 22 Sep 1881).
Hedge, I.C. & J.M. Lamond, Index coll. Edinb. herb. 117. 1970.
Pease, A.S., Flora of Northern New Hampshire Cambridge, Mass., 1964 (see Taxon 14: 204. 1965), p. [i]-v, 1-278.
Verdoorn, F., ed., Chron. bot. 1: 306, 375. 1935, 4: 94, 572. 1938, 5: 277. 1939.

7592. *Vascular flora of Coös county*, New Hampshire ... Boston (printed for the [Boston] Society [of natural History] July, 1924. Oct.
Publ.: Jul 1924 (t.p. reprint), p. [39]-388, *pl. 5-11* with text. *Copies*: BR, G, MO, NY, US, USDA. – Reprinted (with special cover) and to be cited from Proc. Boston Soc. nat. Hist. 37(3): 39-388, *pl. 5-11*. Jul 1924.
Ref.: Bartlett, H.H., Bot. Gaz. 78(4): 466-467. Dec 1924 (rev.).

Peattie, Donald Culross (1898-1964), American botanist; studied at Univ. Chicago (1915-1917; no degree). (*Peattie*).

HERBARIUM and TYPES: material at CHARL, F, GH, NCU, NY. – Manuscript material in Smithsonian Archives.

BIBLIOGRAPHY and BIOGRAPHY: Barnhart 3: 61; BL 1: 177, 205, 312; BM 8: 993; Kew 4: 260; Langman p. 573, 616; Lenley p. 323; Pennell p. 616; PH 209.

BIOFILE: Copeland, D.C., Madroño 3(8): 369-370. 1936 (rev. of *Green Laurels*).
Deam, C.C., Fl. Indiana 1118. 1910.
Joseph, R.V., et al., ed., Biogr. Index 2: 755. 1953, 3: 739. 1956, 4: 649. 1960 , 7: 739. 1968 (further biogr. refs.).
Peattie, D.C., Cargoes and harvests, New York and London, 1926, vii, 311 p.; Green Laurels, the lives and achievements of the great naturalists, New York 1936, 368 p.; A natural history of trees of Eastern and Central North America, 1950.

7593. *Flora of the Indiana dunes* a handbook of the flowering plants and ferns of the Lake Michigan coast of Indiana and of the Calumet district ... Chicago, U.S.A. (published by Field Museum of Natural History) 1930. Oct. (*Fl. Indiana dunes*).
Publ.: 1930 (MO copy rd. 30 Jun 1930), p. [1]-432, map. *Copies*: BR, G, MICH, MO, NY, US.
Further notes: Buhl, C.A., Amer. midland Natural. 16: 248-253. 1935.

Peck, Charles Horton (1833-1917), American botanist; A.B. Union Coll. (Schenectady, N.Y.) 1859, M.A. ib. 1862, D. Sc. hon. ib. 1908; high school teacher at Sandlake Collegiate Inst. 1852-1855 and 1859-1862; id. at Albany 1862; Botanist at the State Cabinet of natural History of New York 1867-1883; State Botanist of New York 1883-1915. (*Peck*).

HERBARIUM and TYPES: NYS; other material at B, BUF, CUP, K, M (lich.), MASS, MICH, MSC, NY (impt.), PUR, QK, WRSL. – Peck correspondence is e.g. at MICH (via H.A. Kelly) and NY. See House (1923) for an index to Peck's note books at NYS. Other manuscript material in Smithsonian Archives and in G.

BIBLIOGRAPHY and BIOGRAPHY: Ainsworth p. 229, 318, 346; BL 1: 200, 202, 312; BM 4: 1538; Bossert p. 304; CSP 8: 579, 10: 1012, 12: 564, 17: 758; DAB (by W.R. Maxon); Frank (Anh.): 75; GR p. 194, cat. p. 69; Hawksworth p. 185; Hortus 3: 1201 (C.H. Peck); IH 1 (ed. 7): 339; Kelly p. 170-172, 256; Kew 4: 260; Krempelh. 3: 160; Lenley p. 323, 466; LS 20208-20304, 23448, 37748-37758, suppl. 21112-21150; ME 1: 218, 3: 636; Morren, ed. 10, p. 126; NAF 7: 1081-1082, 9: 446-447 (bibl.); Pennell p. 616; PH 26, 132, 164, 906; Rehder 5: 657; Stevenson p. 1253; TL-2/see Howe, E.C., Kelly, H.A.; Tucker 1: 545; Urban-Berl. p. 280.

BIOFILE: Ainsworth, G.C., Dict. fungi ed. 6. 428. 1971.
Anon., Albany Evening J. 12 Jul 1917 (fide Maxon); Bot. Not. 1917: iii (d.); J. Mycol. 8: portr. opp. p. 3. 1902 (portr.); Hedwigia 60: (92). 1918 (d. 11 Jul 1917); J. New York Bot. Gard. 9: 4. 1908 (material at NY); Mycologist 9: 317. 1917; Nat. Nov. 37: 134. 1915 (retirement); Torreya 17: 170. 1917.
Atkinson, G.F., Bot. Gaz. 65: 103-108. 1918 (obit., portr., b. 30 Mar 1833, d. 11 Jul 1917).
Bessey, C.E., Science ser. 2. 40: 48. 1914 (on Peck's reports).
Britton, N.L., ed., J. New York Bot. Gard. 18: 93 (d.), 231-232. 1917 (obit.).
Burnham, S.H., Mycologia 11: 32-39. 1919 (early years).
Clarkes, J.M., Charles Hall of Albany 385, 391. 1923.
Ewan, J. et al., Short hist. bot. U.S. 170. 1969.
Gilbertson, R.L., Mycologia 54: 460-465. 1962 (index to Peck taxa 1909-1915).
Haines, J.H., McIlvainea 3(2): 3-10. 1978 (portr.).
Harsha, D.A., Noted living Albanians 325-330. 1891.
Hesler, L.R., Biogr. Sketches N. Amer. Mycol. 1975 (mss.).

House, H.D., Bibl. bot. N.Y. State 9-17, 20, 22, 24, 40-53, 77, 97, 109, 111, 139, 159. 1941, 180, 181, 188. 1942.; Index to the notebooks of Dr. Charles Morton Peck. 1923 (typewritten; copies at BPI, FH, MICH, MO, NY, NYS,), Index to 32 notebooks (1868-1913); (at NYS); further notes by House in Mycologia 15: 192-193. 1923.
Humphrey, H.B., Makers N. Amer. Bot. 198-200. 1961.
Lloyd, C.G., Mycol. notes 38: 509-511. 1912 (portr.).
Maxon, W.R., Dict. Amer. biogr. 14: 372-373. 1974 (biogr.).
Murrill, W.A., Hist. found. bot. Florida 28. 1945.
Overholts, L.O., New York State Mus. Bull. 205/206: 67-166, *pl. 1-23*. 1 Jul 1919 (Species of *Poria* descr. by Peck).
Peterson, R.H., *in* L. Vogelenzang, ed., C.H. Peck, Annual Reports of the State Botanist 1888. 1972, vol. 1, p. 3-25 (portr., b. 30 Mar 1833).
Rickett, H.W., Index Bull. Torrey bot. Club 77. 1955.
Rogers, D.P., Brief hist. mycol. N. America 12-13. 1977, ed. 2.85 [index]. 1981.
Sayre, G., Bryologist 80: 515. 1977 (musci at NYS, NY).
Seaver, F.J., North American Cup-fungi 352, 362-363. 1942 (partial bibl.).
Steere, W.C., Bot. Rev. 43: 293. 1977.
Verdoorn, F., ed., Chron. bot. 6: 355. 1941.
Wittrock, C.B., Acta Horti Berg. 3(3): 200. 1905.
Wood, R.D., Monogr. Charac. 1: 833. 1965 (publ. on Chara).

HANDWRITING: J. Mycol. 8: plate opp. p. 3. 1902.

INDEXES taxa described by Peck; New York State Mus., Bull. 131: 51-190. 1909 and by Gilbertson, R.L., Mycologia 54: 460-465. 1962; for indexes to Peck's original note books at NYS see House (1923) and Petersen (1980).

EPONYMY: *Chapeckia* M.E. Barr (1978); *Neopeckia* P.A. Saccardo (1883); *Peckia* Clinton (1878); *Peckiella* (P.A. Saccardo) P.A. Saccardo (1891); *Peckifungus* O. Kuntze (1891). *Note*: *Peckia* Vellozo (1825) is probably dedicated to the memory of W.D. Peck (1763-1822), q.v.; *Peckichara* L. Grambast (1957) and *Peckisphaera* L. Grambast (1963) are named for Raymond E. Peck (1904-x), an American geologist at the University of Missouri.

7594. (B.) *List of mosses of the State of New York* ... [Albany (C. Wendell, printer) 1866.] Oct.
Publ.: 2 Apr 1866, p. [42]-70 in Nineteenth Annual Report of the Regents of the University of the State of New York on the condition of the State Cabinet of Natural History, Albany 1866. *Copies*: MO-Steere.

7595. Twenty-second *annual report* of the regents of the University of the State of New York, on the condition of the State Cabinet of natural history and the historical and antiquarian collection annexed thereto ... Albany (The Argus Company, Printers) 1869. Oct. Contains: *Report of the Botanist.* (*Ann. Rep. State Bot.*).
Publ.: The reports of the *State Botanist* over the years 1868-1912 (C.H. Peck), contains much valuable, especially mycological information. The bibliography of these reports is involved and is described in great detail by Petersen (1980). We saw sets at MICH and NY and partial sets at MO, PH and USDA. In view of the intricacy of the matter we follow here Petersen's account.
Facsimile ed.: Annual Reports of the State Botanist. 1868-1912, edited by L. Vogelenzang, Rijksherbarium, Leiden, volume 1 (1868-1977) [and following, not yet published Fall 1980], ISBN 90-70153-12-2 (series), 1980, Boerhaave Press, p.o. Box 1051, 2302 BB Leiden, Holland. With an introduction by R.H. Petersen (portr., biogr. note, extensive bibl. of Reports), p. [i-iv], [1]-25, facsimile of reports 1868-1877. *Copy*: FAS. – The reports for 1866, 1867 and 1913 will not be published in the reprint edition.
List of reports for 1866-1913.
Every report came out in various issues, all duly listed by Petersen (1980). We list them here mentioning the earliest issue in first instance. The categories of issues are:
a. A.D. or S.D. – Assembly, c.q. Senate Document, as presented to the New York State legislature.

b. Ann. Rep. – Annual Report of the New York State Museum [Nat. Hist.].
c. Bull. – Bulletin of idem.
d. Reprints [i.e. separates] or preprints of the Report of the Botanist from one of the above.

The date of publication listed after the year for which the report was made always refers to the earliest issue, listed as no. 1.

1866: publ.: 1867.
   1: Ann. Rep. 20(ed. 1): 403-410. 1868.
   2: A.D. 239, vol. 12, 90th session 1867, p. 403-410.
   3: Ann. Rep. 20 (revised ed.): 159-166. 1868.
   4: Ann. Rep. 20 (revised ed.): 159-166. 1870.
1867: publ.: 1870.
   1: S.D. 92, vol. 7, 91th session, p. 19-20. 1868.
   2: Ann. Rep. 21: [23]-24. 1871.
1868: publ. Mai-Dec 1869.
   1: Ann. Rep. 22: [25]-106. 1869.   2: S.D. 87, vol. 6, 92nd session, p. 25-106.
   3: reprint from 1, same pagination. 1869.
1869: publ. 23 Mar 1872.
   1: Preprint from Ann. Rep., p. [27]-135, *pl. 1-6* and [*7-12*].
   2: A.D. 133, 93rd. session, see Petersen (1980).
   3: Ann. Rep. 23, app. C.: [27]-135, *pl. 1-6* and [*7-12*].
   Publication delayed by fire (Atkinson 1918, p. 105; TBC rev. Apr 1872).
1870: publ. Jan 1872.
   1: Ann. Rep. 24: [41]-108, *pl. 1-4*. 1872.
   2: S.D. 68, vol.4, 94th session, p. 41-108, *pl. 1-4*.
   Presented by Peck as for "1869", actually, however, for 1870.
1871: Sep 1873.
   1: Preprint from Ann. Rep. 25: [57]-122, *pl. 1-2*, [*3-4*].
   2: Ann. Rep. 25: [57]-123. 1873.
   3: S.D. 83, vol. 4, 95th session, same pagination.
1872: Apr 1874.
   1: Preprint from Ann. Rep. 26: [35]-91.
   2: Ann. Rep. 26: [35]-91. 8 Mai 1874.
   3: S.D. 109, vol. 5, 96th session, same pagination.
1873: Dec 1875.
   1: Ann. Rep. 27: [73]-116, *pl. 1-2*, [1-2].
   2: S.D. 102, vol. 6, 97th session "17 Apr 1874" same pagination.
   3: Reprint from Ann. Rep., same pagination.
1874: Dec 1876 (or 3 Jun 1877, fide House).
   1: Preprint from Ann. Rep. 28: [31]-88, *pl. 1-2*, [1-2].
   2: S.D. 71, vol. 6, 98th session, "30 Mar 1875", same pagination.
   3: Ann. Rep. 28: [31]-88, *pl. 1-2*, [1-2]. 1875.
   4: Ann. Rep. id. but. "State Museum edition".
1875: probably Nov 1878.
   1: Preprint from Ann. Rep. 29: [29]-82, *pl. 1-2*, [1-2].
   2: Ann. Rep. 29: [29]-85, *pl. 1-2*, [1-2].
   3: S.D. 64, vol. 5, "18 Apr 1876", "1877", same pagination.
1876: Sep 1878.
   1: Preprint from Ann. Rep. 30: [23]-78, *pl. 1-2*, [1-3].
   2: Ann. Rep. 30: [23]-78, *pl. 1-2*, [1-3].
   3: Ann. Rep. id. but. "State Museum Edition", cover 1879.
   4: S.D. 63, vol. 5, "13 Apr 1877, same pagination.
1877: late 1878.
   1: S.D. 42, vol. 2, 101st session p. [19]-60. 1878.
   2: Ann. Rep. 31: 19-60. 1879.
   3: Reprint of 2, same pagination.
1878: early Mai 1880.
   1: A.D. 89, vol. 6, 102nd session, p. [17]-72, *pl. 1-2*.
   2: Ann. Rep. 32: 17-72, *pl. 1-2*, [1-2].

1879: Jun 1883.
  1: A.D. 120, vol. 8, 103 rd. sess. (distr. with doc. 104th session), p. 11-49, *pl. 1-2*.
  2: Ann. Rep. 33: [11]-49, *pl. 1-2*, [1-2]. 1880 (1883).
  3: Reprint of 2, same pagination. 1884.
1880: early Jun 1883.
  1: A.D. 127, vol. 8, 104th sesion, p. [24]-58, *pl. 1-4*. 1881 (1883).
  2: Ann. Rep. 34: [24]-58, *pl. 1-4*, [1-4]. 1881 (1883).
  3: Reprint of 2, pagination unchanged, 1884.
1881: Dec 1884.
  1: S.D. 38, vol. 8, 105th session, p. [125]-164. 1882 (1884).
  2: Ann. Rep. 35: [125]-164. 1884.
  3: Reprint of 2, same pagination. 1884.
1882: Dec 1884.
  1: S.D. 53, vol. 6, 106th session, p. 29-49. 1883 (1884).
  2: Ann. Rep. 30: [29]-49. 1883 (1884).
  3: Reprint of 2, same pagination. 1884.
1883: sections A-B: Dec 1884, C-F: Mai 1887.
  1: S.D. 60, vol. 1(2), 107th session, p. 63-68. 1884.
  2: Ann. Rep. 37: 63-68. 1884.
  3: Reprint of 2, 1884, same pagination.
  4: Bull. 1(2): 5-66, *pl. 1-2*. 1887 (all sections).
1884: Oct 1885.
  1: Ann. Rep. 38: [77]-138, *pl. 1-3*, [1-3] 1885.
  2: A.D. 23, vol. 2, 108th session, p. 77-138. 1885.
1885: Sep 1887.
  1: Ann. Rep. 39: 30-73, *pl. 1-2*, [1]. 1886 (1887).
  2: A.D. 104, vol. 9, 109th session, p. 30-73, *pl. 1-2*. 1887.
1886: Dec 1887.
  1: A.D. 115, vol. 10, 110th session, p. 39-75.
  2: Ann. Rep. 40: [39]-75. 1887.
  3: Reprint of 2, same pagination, 1920.
1887: Dec 1888.
  1: S.D. 19, vol. 4, 111th session, p. 51-122, *fig. 1-4*. 1888.
  2: Ann. Rep. 41: 51-122, *fig. 1-4*. 1888.
  3: Reprint of 2, same pagination. 1888.
1888: early Dec 1889.
  1: S.D. 65, vol. 11, 112th session, p. 101-144, *pl.1-2*. 1889.
  2: Ann. Rep. 42: 101-144, *pl.1-2*. 1889.
  3: Reprint as S.D. no. 67, with ref. to S.D. 65, p. 5-48, *pl. 1-2*.
1889: Oct 1890.
  1: Ann. Rep. 43: 51-97, *pl. 1-4*. 1890.
  2: S.D. 51, vol. 4, 113th session, p. 51-97, *pl. 1-4*. 1890.
  3: Reprint as S.D. no. 61, p. 1-54, *pl.1-4*. 1890.
1890: Dec 1891.
  1: Preprint as S.D. no. 77 but not published in that series, p. 1-75, *pl. 1-4*, [1-4]. 1891.
  2: S.D. 75, vol. 7, 114th session, p. 117-187, *pl. 1-4*. 1891.
  3: Ann. Rep. 44: 117-187, *pl. 1-4*. 1892.
  4: Reprint of 3, same pagination. 1892.
1891: early Dec 1893.
  1: Preprint, as S.D. no. 66 but not published in the series, p. 1-42. 1893.
  2: S.D. 64, vol. 10, 115th session, p. 65-102. 1892.
  3: Ann. Rep. 45: 65-102. 1892.
1892: early Apr 1894.
  1: Ann. Rep. 46: 85-149. 1893.
  2: S.D. 39, vol. 6, 116th session.
  3: Reprint, as S.D. no. 41, not published in series, 1-69. 1893.
1893: Dec 1894.
  1: Preprint, p. [1]-48, from Ann. Rep. 46: 131-174. 1894.
  2: Ann. Rep. 46: 131-174. 1894.
  3: S.D. 87, vol. 11, 117th sessions, p. 131-174. 1894.

1894: Apr 1897 (or 1896?).
  1: Preprint from Ann. Rep., p. [1]-241, *pl. A, 1-43* (copy: PH).
  2: Ann. Rep. 48(1): 103-337. 1895.
  3: As 2, "second edition", p. [1]-241, *pl. A, 1-43*. 1897.
  4: S.D. 67, vol. 15, 118th session, p. 101-337. 1895 (1897?).
    Note: Both Bot. Gaz. 23: 384-385. 1 Mai 1897 and TBC 24: 316. 29 Jun 1897 date this report from "1896" in accordance with the t.p. date of the preprint.
1895: Nov 1897.
  1: Preprint from Ann. Rep., p. [1]-69, *pl. 44-49*. 1897.
  2: Ann. Rep. 49(1): 17-83, *pl. 44-49*. 1895 (1897).
  3: S.D. 55, vol. 8, 119th session, p. 19-83, *pl 44-49*. 1896 (1897).
1896: Mai 1898.
  1: Ann. Rep. 50(1): 77-159. 1898.
  2: S.D. 48, vol. 12, 120th session, p. 77-159. 1898.
  3: Reprint of 1, same pagination, dated 1897 (reprint rev. TBC 12 Mai 1898).
1897: probably Dec 1898.
  1: Ann. Rep. 5(1), app. 2: 267-321, *pl. A-B, 50-56*. 1899 (see Petersen for date).
  2: S.D. 48, vol. 9, 121st session, p. 267-321, *pl. A, B, 50-56*. 1899.
  3: Reprint of 1, same pagination. 1899.
1898: Oct 1899.
  1: Bull. 25(Bot. 3): 619-688, *pl. 57-61*. Oct 1899 (date on fly leaf, covers: 1900.
  2: S.D. 60, vol. 13, 122nd sessions, p. 619-688, *pl. 57-61*. 1899.
  3: Ann. Rep. 52(1): 619-688, *pl. 57-61*. 1900.
    Note: also re-issued in Mem. N.Y. State Mus. 4(3), Nov 1900 (re-issue of relevant parts of 48-52. Reports).
1899: probably Apr 1901.
  1: Ann. Rep. 53(1): 823-867, +865-867 index, + *pl. A-D*. 1901.
  2: S.D. 46, vol. 18, 123rd session, p. 823-867, *pl. A-D*. 1900 (1902).
  3: Reprint from 1, same pagination, dated 1900 (but Torreya: 22 Mai 1901, recently distributed; TBC 18 Apr 1901, referring either to 1 or 3 or both).
1900: Feb 1902.
  1: Preprint from Ann. Rep., p. 131-199, *pl. E-I*, 69-76. 1902.
  2: Ann. Rep. 54(1): app. 2: 131-199, *pl. E-I*, 69-76. 1902.
  3: S.D. 46, vol. 14, app. 2, 124th session, p. 131-199, *pl. E-I*, 69-76. 1901 (t.p.).
1901: Nov 1902.
  1: Bull. 54 (Bot. 5): 931-984, *pl. K-L*, 77-81. 1902.
  2: S.D. 38, vol. 21, 125th session, p. 931-984, *pl. K-L*, 77-81. 1903.
  3: Ann. Rep. 55(1), app.: 931-984, *pl. K-L*, 77-81. 1903.
1902: Mai 1903.
  1: Bull. 67 (Bot. 6): 3-194, *pl. M-N*, 82-84. 1903.
  2: S.D. 42, vol. 15, 126th sess., p. 3-194, *pl. M-N*, 82-84. 1904.
1903: 1904.
  1: Bull. 75 (Bot. 7): 3-68, *pl. o*, 84-86. 1904.
  2: S.D. 40, vol. 23, 127th sess., p. 3-68, *pl. o*, 84-86. 1904.
  3: Ann. Rep. 57(2), app. 6: 3-68, *pl. o*, 84-86. 1905.
1904: Jul 1905.
  1: Bull. 94 (Bot. 8): 5-58, *pl. P-R*, 87-93. 1905.
  2: S.D. 12, vol. 7, app. 6, 128th sess., p. 5-58, *pl. P-R*, 87-93. 1906.
  3: Ann. Rep. 58(4), app. 6: 5-58, *pl. P-R*, 87-93. 1906.
1905: Aug 1906.
  1: Bull. 105 (Bot. 9): 5-106, *pl. S-T*, 94-103. 1906.
  2: A.D. 66, vol. 21, app. 6, 129th sess., p. 5-106, *pl. S-T*, 94-103.
  3: Ann. Rep. 59(2), app. 6: 5-106, *pl. S-T*, 94-103.
1906: Jul 1907.
  1: Bull. 116 (Bot. 10): 5-117, *pl. 104-109*. 1907 (also as N.Y. St. Educ. Dept. Bull. 404. Jul 1907).
  2: A.D. 68, vol. 24, app. 4, 130th sess., p. 5-117, *pl 104-109*. 1907.
  3: Ann. Rep. 60(1), app. 4: 5-117, *pl. 104-109*. 1908.
1907: 15 Aug 1908.
  1: Bull. 122: 5-175, *pl. 110-114*. 1908 (also as N.Y.St. Educ. Dept. Bull. 429, 15 Aug 1908).

2: A.D. 31, vol. 19, app. 3, 131st sess., p. 5-175, *pl. 110-174.*
3: Ann. Rep. 61(2), app. 3: 5-175, *pl. 110-114.* 1908.
1908: 1 Jul 1909.
1: Bull 131: 5-202, *pl. U-V*, 115-116. 1909 (also N.Y. St. Educ. Dept. Bull. 450, 1 Jul 1909. – Contains on p. 59-190 a list of taxa described by Peck.
2: A.D. 64, vol. 35, app. 5, 132nd session, p. 5-202, *pl. U-V*, 115-116. 1909.
3: Ann. Rep. 62(2), app. 5: 5-202, *pl. U-V*, 115-116. 1909.
1909: 1 Mai 1910.
1: Bull. 139: 5-114, *pl. ii-iii, 117-120, W-Z.* 1910 (also as N.Y. St. Educ. Dept. Bull. 47. 1 Mai 1910).
2: A.D. 45, vol. 23, app. 4, 133rd sess., p. 5-114, *pl. ii-iii, 117-120.*
3: Ann. Rep. 63(2), app. 4: 5-114, *pl. ii-iii, 117-120. 1910:* 15 Mai 1911.
1: Bull. 150: 5-100, *pl. iv, vi, 121-123.* 1911 (also as N.Y. St. Educ. Dept. Bull. 495. 15 Mai 1911.
2: S.D. 55, vol. 27, app. 4, 134th sess., p. 5-100, *pl. iv, vi, 121-123.* 1911.
3: Ann. Rep. 64(2), app. 4: 5-100, *pl. iv, vi, 121-123.* 1912.
1911: 1 Mar 1912.
1: Bull. 157: 5-139, *pl. vii-viii, 124-130.* 1912 (also as N.Y. St. Educ. Dept. Bull. 514, 1 Mar 1912).
2: A.D. 34, vol. 23, app. 4, 135th sess., p. 5-139, *pl.vii-viii, 124-130.* 1912.
3: Ann. Rep. 65(2), app. 4: 5-139, *pl. vii-viii, 124-130.* 1913.
1912: Sep 1913.
1: Bull. 167: 5-137, *pl. ix-x, 131-132.* 1913 (also as N.Y. St. Educ. Dept. Bull. 550, 1 Sep 1913).
2: A.D. 43, vol. 29, app. 4, 136th sess., p. 5-137, *pl. ix-x, 131- 132.* 1914.
3: Ann. Rep. 66(2), app. 4: 5-137, *pl. ix-x, 131-132.* 1914.
Note: according to Mycologia 6: 40. 1914 this report was published 28 Nov 1913.
1913: 1 Jun 1915.
1: Bull. 176, 1915 (N.Y. St. Educ. Dept. Bull. 592, 1 Jun 1915 (publication of some Peck names by his successor H.D. House).
2: A.D. 68, vol. 23, app. 4, 137th session, p. 5-77. 1914 (1915).
3: Ann. Rep. 67, app. 4, 1915.

7596. *United States species of Lycoperdon* ... (read before the Albany Institute Feb 4th, 1879], [1879]. (*U.S. Lycoperdon*).
Publ.: Feb-Mar 1879 (read 4 Feb 1879; TBC Apr 1897, BF Jun 1879; Nat. Nov. Sep(1) 1879; Bot. Zeit. 19 Sep 1879), p. [1]-34. *Copies*: MO, NY, USDA. – Preprinted or reprinted from Trans. Albany Inst. 9: 286-318. 1879 (issue for Feb).

7597. *Contributions to the botany of the State of New York* ... Albany (Charles van Benthuysen & Sons) 1887. Oct.
Publ.: Mai 1887, p. [1]-66, *pl. 1*. *Copy*: G. – Published and to be cited as Bull. N.Y. State Mus. nat. Hist. 1(2): 1-66, *1 pl.* Mai 1887.

7598. *Boleti of the United States* ... Albany (University of the State of New York) 1889. Qu.
Publ.: Sep 1889, p. [71]-166. 1889, issued as and to be cited from Bull. New York State Museum 2(8): 71-166. Sep 1889. *Copies*: G, NY, Stevenson, USDA.
Ref.: Anon., Bot. Gaz. 15: 22. Jan 1890.

7599. *Mushrooms and their use* ... Cambridge, Mass. (Cambridge Botanical Supply Company) 1897. Oct. (*Mushrooms*).
Publ.: Mai 1897, p. [ii, err.], [1]-80. *Copies*: NY, USDA. – Reprinted from the Cultivator and Country Gentleman, of Albany, N.Y.,31 Mai-20 Sep 1894 (n.v.).
Ref.: Arthur, J.C., Bot. Gaz. 24: 64. 31 July 1897.

7600. *Plants of North Elba* Essex County, N.Y. ... Albany (University of the State of New York) 1899. Oct.Publ.: Jun 1889, p. [65]-266, map, issued as and to be cited from Bull. New York State Mus. 6(28): [65]-266. 1899. *Copies*: B, G, MO, NY, Stevenson, E.G. Voss.

7601. *Report of the State Botanist on edible fungi of New York* 1895-1899 ... Albany (University of the State of New York) 1900. Oct.
Publ.: Nov 1900, p. [129]-234, *pl. 44-68* (col., auct.). Reprinted, and to be cited from, Mem. N.Y. State Mus. 4(3): [129]-234, *pl. 44-48*. 1900. *Copies*: NY, Stevenson, USDA.
Re-issue, with changes and additions, of papers on this subject in the 48th-52nd Reports of the State Botanist.
Ref.: Davis, B.M., Bot. Gaz. 31: 358-359. 1901.

7602. *The fungi of Alaska* ... [from Harriman Alaska Expedition, Washington Academy of Sciences], [1904]. (*Fung. Alaska*).
Co-authors: Pier Andrea Saccardo (1845-1920), William Trelease (1857-1945).
Publ.: 1904 (BG 24 Jun 1904; Bot. Centralbl. 25 Mai 1904), p. [11]-64, *pl. 2-7* (partly col.). *Copies*: Stevenson, USDA. – Alaska: giving the Results of the Harriman Alaska Expedition ... 5 (Cryptogamic botany): [11]-64, *pl. 2-7* (for full series see BM 5: 2269).
Copies: FH, G, NY, US; IDC 6391.
Facsimile ed.: 1965, Amsterdam (A. Asher & Co.), 2 vols., ISBN 90-6123-212-0, as above with extra t.p.'s dated 1965 preceding material as listed above. *Copies*: BR, FAS, PH.

Peck, Franz Gustav Magnus (1817-1892), German (Niederlausitz) botanist at Görlitz; studied law at Berlin; from 1839-1843 law court official (Kammergerichtsreferendar, "Auskultator") at Görlitz; 1843-1845 id. Berlin; from 1845 judge at various stations, ultimately president of the district court at Schweidnitz 1867-1883; from 1884-1892 in retirement at Görlitz. (*F. Peck*).

HERBARIUM and TYPES: GLM.

BIBLIOGRAPHY and BIOGRAPHY: CSP 17: 758; Futàk-Domin p. 461; GR p. 119; Morren ed. 10, p. 16; Rehder 5: 657.

BIOFILE: Anon., Bot. Centralbl. 53: 271. 1893 (d.); Österr. bot. Z. 43: 112. 1893 (d.).
Ascherson, P., Ber. deut. bot. Ges. 11: (32)-(34). 1893 (obit., b. 1 Mar 1817, d. 21 Dec 1892). (See also p. (222)).
Pax, F., Bibl. schles. Bot. 41, 147. 1929.

7603. *Die Flora der Umgegend von Schweidnitz*. Ein Beitrag zur Pflanzenkunde der Provinz Schlesien von F. Peck, Kreisgerichts-Direktor in Schweidnitz [Görlitz 1871].
Orig.: 1871, p. [16]-56. *Copy*: B. – Reprinted and to be cited from Abh. naturf. Ges. Görlitz 14: [16]-56. 1871.
Nachtrag [1]: 1875, p. [56]-68. *Copy*: B. – Id. Abh. naturf. Ges. Görlitz 15: [56]-68. 1875.
Nachtrag 2: 1887, p. [93]-96. *Copy*: B. – Id. Abh. naturf. Ges. Görlitz 19 [93]-96. 1887.

Peck, Morton Eaton (1871-1959), American botanist; B.A. Cornell College, Iowa, 1895; at Willamette University, Salem, Oregon 1908-1941, ultimately as chairman of the Department of Biology; botanical explorer of British Honduras and, for over fifty years, of Oregon; D. Sc. Cornell 1940, LL.D. Willamette 1955. (*M. Peck*).

HERBARIUM and TYPES: WILLU (35.000) (on permanent loan at OSC), other material e.g. at B, BHC, DS, F, GH, ISC, K, L, MICH, MO, NY, OSC, PH, POM, WTU. – Field note books in Smithsonian Archives.

BIBLIOGRAPHY and BIOGRAPHY: Barnhart 3: 62; BL 1: 180, 210, 312; Bossert p. 304; Hortus 3: 1201 ("Peck"); IH 1 (ed. 1): 84, (ed. 2): 106, (ed. 3): 136, (ed. 4): 151, (ed. 5): 162, (ed. 6): 362, (ed. 7): 339, 2: (in press); Kew 4: 260; NAF 28B: 350, ser. 2. 5: 251; NW p. 55; Pennell p. 616; PH 209.

BIOFILE: Anon., Capitol, J., Salem, Ore, 4 Dec 1959; J. New York Bot. Gard. 21: 180. 1920 (Peck herb. material rd.; The Oregonian 25 Mar 1941 (retirement), 9 Jan 1955 (portr.); The Oregon Statesman 25 March 1941 (retirement; portr.); 11 Jun 1941 (LL.D. h.c.), 4/5 and 6 Dec 1959 (d. 4 Dec 1959); Willamette Alumnus 1(4): 4. 1944

(portr.), 9(2): 2. 1953, 9(3). 1953. 6(2) [sic!]: 8. 1958. 11(1): 4-5. 1963; Willamette Collegian 71(11): 6. 1959 (portr.).
Chambers, K.L., Taxon 26: 160-161. 1977 (on Peck and the Peck herbarium).
Constance, L., Madroño 11(1): 22-24. 1951 (80th birthday, portr., tribute), Taxon 8(6): 165-167. 1960 (portr.).
Gilkey, H.M., Bull. Torrey bot. Club 87(4): 280-282. 1960 (obit., portr., bibl., b. 12 Mar 1871).
Peck, M.E., Amer. J. Bot. 12: 33-49, 69-81 ("Preliminary sketch of the plant regions of Oregon").
Reveal, J., *in* A. Cronquist et al., Intemountain Fl. 1: 67. 1972 (portr.).
Stieber, M.T. et al., Huntia 4(1): 84. 1981 (arch. mat. HU).
Thomas, J.H., Huntia 3: 22, 23. 1969 (portr.).
Verdoorn, F., ed., Chron. bot. 3: 310. 1937.

EPONYMY: *Peck herbarium*, at WILLU, see e.g. Chambers (1977).

NOTE: A rich collection of newspaper cuttings and other, usually ephemeral, printed documents on Peck is at NY.

7604. *A manual of the higher plants of Oregon* ... Portland, Oregon (Binfords & Mort, ...) [1941]. Oct. (*Man. pl. Oregon*).
Ed. 1: 9 Mai 1941 (fide Oregon Stateman, Salem, Ore., 9 Mai 1941; see also Sunday Oregonian 1 Jun 1941, cuttings at NY; J. NYBG Sep 1941), p. [1]-866, [1, 2. err.]. *Copies*: BR, G, MO, NY. – *New combinations*: Madroño 6: 13-15. 17 Jan 1941, 133-137. 15 Oct 1941; *preliminary publ.*: Proc. Biol. Soc. Washington 47: 185-188. 2 Oct 1934, Torreya 32: 147-153. Nov-Dec 1932. See also Leafl. West. Bot. 7: 177-192. 10 Dec 1954.
Ed. 1961, frontisp., p. [1]-936. *Copies*: B, NY. Oregon (id.) [1961].

Peck, William Dandridge (1763-1822), American botanist and entomologist; B.A. Harvard 1782; first professor of natural history at Harvard 1805-1822. (*W. Peck*).

HERBARIUM and TYPES: The only herbarium specimens made by Peck definitely known to us are the – relatively few – specimens seen by FAS in the Thunberg herbarium at UPS; and a small number at PH. – The specimens at UPS stem from the White Mountains in New Hampshire. – Diary 1805-1808 in Widener Archives, Harvard University.

BIBLIOGRAPHY and BIOGRAPHY: Backer p. 429; Barnhart 3: 62; Bossert p. 304 (b. 8 Mai 1763; d. 3 Oct 1822); CSP 4: 798; DAB (by L.O. Howard); IH 2: (in press); Jackson p. 448; ME 1: 218; Rehder 5: 657; Tucker 1: 545.

BIOFILE: Bridson, G.D.R. et al., Nat. hist. mss. res. Brit. Isl. 255: 125. 1980.
Darlington, W., Reliq. Baldwin. 39, 41. 1843.
Dawson, W.R., Smith papers 73. 1954 (letters to J.E. Smith).
Essig, E.O., Hist. entomology 729-732. 1931 (biogr., b. 8 Mai 1763, d. 8 Oct 1822).
Ewan, J., Regnum veg. 71: 43. 1970 (visit Bonpland); *in* F. Pursh, Fl. Amer. sept., facs. repr. 1979, p. 23-24. 1979.
Ewan, J. et al., Short hist. bot. U.S. 6, 7, 36. 1969.
Gager, C.S., Brooklyn Bot. Gard. Rec. 27(3): 369. 1938 (first dir. bot. gard. Harvard, 1807-1822).
Gilbert, P., Comp. biogr. lit. deceased entom. 291. 1977 (q.v. for further biogr. refs.).
Graustein, J.E., Thomas Nuttall 475 [index]. 1967.
Hedrick, U.P., Hist. hortic. America 230, 422. 1950.
Howard, L.O., Smithson. misc. Coll. 84: 11, 30, 72. 1930.
Kelly, H.A. & W.L. Burrage, Amer. med. biogr. 900-901. 1920.
Mallis, A., Amer. entom. 13-16. 1971.
Mears, J.A., Proc. Amer. philos. Soc. 122: 166. 1978 (visit Pursh, see Pursh 1814, p. xv).
Osborn, H., Fragm. entom. hist. 7, 11, 13, 26, 93, 98, 238, 239, *pl.* 7. 1937.
Sutton, S.B., Charles Sprague Sargent 42. 1970.
Weiss, H.B., Pioneer cent. Amer. entom. 50-52. 1936.

EPONYMY: *Peckia* Vellozo (1825) is dedicated to his memory according to Backer. However, Vellozo states, "In memoriam D. Peck, Botanici Gallii." and Backer's attribution must be incorrect.

7605. *A catalogue of American and foreign plants* cultivated in the botanic garden, Cambridge, Massachusetts ... Cambridge (printed by Hilliard and Metcalf, ...) Qu. (in twos). 1818. (*Cat. Amer. pl.*).
Publ.: 1818, p. [i]-iv, [1]-60. *Copies*: MO, NY. – Also issued as an appendix to Rep. J. Mass. Hort. Soc. 5(1). 1818.

Peckolt, Theodor (1822-1912), German (Niederlausitz) botanist; trained as a pharmacist in Triebel, 1837-1841; settled in Brazil 1847; practicing pharmacist in Rio de Janeiro 1848; exploring the province of Rio de Janeiro 1848-1850; pharmacist at Cantagallo 1851-1867; id. again in Rio de Janeiro from 1868 ("pharmacia e drugeria Peckolt"). (*Peckolt*).

HERBARIUM and TYPES: BR (main set); further material at B, KIEL, MB, R, W, WRSL.

BIBLIOGRAPHY and BIOGRAPHY: Barnhart 3: 62; BL 1: 239, 240, 312; BM 4: 1538; CSP 8: 580, 10: 1012, 12: 564-565; Frank 3 (Anh.): 75; Herder p. 410; IH 2: (in press); Jackson p. 373, 374; Kew 4: 260; KR p. 571; Morren ed. 2, p. 34, ed. 10, p. 139; PR 7022; Rehder 5: 657; Tucker 1: 545; Urban-Berl. p. 381.

BIOFILE: Anon., Apoth. Zeit. 27: 761. 1912; Biogr. Jahrb. 18: 50*; Bot. Centralbl. 122: 32. 1913 (d.); Bot. Not. 1913: 95 (d.); Nat. Nov. 34: 588. 1912.
Bridson, G.D.R. et al., Nat. hist. mss. res. Brit. Isl. 263.1. 1980.
Carpenter, H.C., Jornal do Comercio, Rio de Janeiro, 20 Oct 1912 (fide Ihering).
Hein, W.H., *in* W.H. Hein & H.D. Schwarz, Deut. Apoth. Biogr. 2: 485-486 (b. 13 Jul 1822, d. 21 Sep 1912).
Hoehne, C., Jard. bot. São Paulo 145. 1941 (portr.).
Ihering, H. von, Rev. Mus. Paulista 9: 55-84. 1914 (obit., bibl., b. 13 Jul 1822, d. 21 Sep 1912).
Junk, W., Rara 2: 183. 1926-1936.
Kneucker, A., Allg. bot. Z. 19: 32. 1913 (death, err. Perkolt).
Reber, B., Gallerie hervorragender Therapeuten und Pharmakognosten, Genève 1897, p. 125-129 (portr., bibl.).
Siedler, P., Ber. deut. pharm. Ges. 24: 81-97. 1914 (obit., bibl.).
Tavares, J.S., Broteria, Bot. 14(1): 59-64, *pl. 1*.1916 (portr.).
Urban, I., Fl. bras. 1(1): 74-75. 1906 (itin., coll.).

EPONYMY: *Peckoltia* Fournier (1885).

NOTE: Peckolt published a series of papers entitled *Brasilianische Nutz– und Heilpflanzen*, in Pharm. Rundschau 11(1893)-15(1897) and its successors Pharm. Arch. 1(1898)-4(1901), and Pharm. Rev. 20: 1902 ("Medicinal plants in Brazil"; for details see BL 1: 239. We did not succeed in finding a complete copy of Peckolt's *Historia das plantas medicinaes e uteis do Brazil*, 8 fasc., 1888-1914; for details see BL 1: 240.

7606. *Historia das plantas alimentares e de gozo do Brazil* contendo generalidades sobre a agricultura brasileira, a cultura, uso e composição chimica de cada uma dellas ... Rio de Janeiro (em casa dos editores proprietarios Eduardo & Henrique Laemmert ...) 1871-1884. Oct. (*Hist. pl. alim. Brasil*).
1: 1871, p.[i]-xvi, [1]-142.
2: 1874, p. [1]-102.
3: 1878 (p. vii: Aug 1877), p. [i-vii], [1]-175. – Other title "Monographia do milho e da mandioca ...".
4: 1882 (p. [v]: 8 Oct 1882), p. [i-vii], [1]-199, [200, index].
5: 1884, p. [i-vii], [1]-167.
Copies: USDA (2).

7607. *Volksbenennungen der brasilianischen Pflanzen* und Produkte derselben in brasilianischer (portugiesischer) und von der Tupisprache adoptierten Namen ... Milwaukee (Pharmaceutical Review Publishing Co.) 1907. (*Volksbenenn. bras. Pfl.*).
Publ.: 1907 (Bot. Zeit. 16 Aug 1909; Nat. Nov. Apr(2) 1909), p. [ii-iii], p. [1]-252. *Copy*: MO. – Consolidated reprint of a series of papers in Pharm. Arch. 1(1898)-6(1903) and Pharm. Rev. 23(1905)-24(1906) for details see BL 1: 240. – Pharmaceutical Science Series, Monographs. No. 15.

Pedicino, Nicola Antonio (1839-1883), Italian diatomologist; teacher at a technical college at Napoli; from 1872-1877 at the agricultural college and director of its botanical garden at Portici nr. Napoli; from 1877-1883 professor of botany and director of the botanical garden at Roma, succeeding De Notaris. (*Pedicino*).

HERBARIUM and TYPES: RO; further material at AWH, FI, PAD, PAL, PI, POR, W, WAG. – Letters and portrait at G.

BIBLIOGRAPHY and BIOGRAPHY: Barnhart 3: 63; BM 4: 1538; Bossert p. 304; Cesati p. 58, nos. 308-309; CSP 6: 739, 8: 580, 10: 1012, 12: 565; De Toni 1: xcvi, 2: xci; Herder p. 3, 117; IH 1(ed. 6): 362, (ed. 7): 339, 2: (in press); Kew 4: 261; Morren ed. 2, p. 29; NI 3: 53; PR 7023; Saccardo 1: 124 (b. 12 Jul 1839), 2: 82.

BIOFILE: Anon., Bot. Centralbl. 15: 255. 20-24 Aug 1883 (death); Bot. Gaz. 9: 14. 1884; Bot. Not. 1884: 39 (d. 2 Aug 1883); Bot. Zeit. 41: 614. 1883 (d. 2 Aug 1883); Bull. Soc. bot. France 25 (bibl.): 42. 1878, 30 (bibl.): 143. 1883 (d.); Revue bot. 2(15): 157. 1883 (b. 12 Jul 1839, d. 2 Aug 1883).
Balsamo, F., Bull. Orto bot. Napoli 3: 55. 1913.
Beck, G., *in* Fritsch, K. et al., Bot. Anst. Wiens 75. 1894 (pl. at W).
Bureau, E., Bull. Soc. bot. France 30: 257 (bibl.:) 143. 1884 (d.).
Gager, C.S., Brooklyn Bot. Gard. Rec. 27(3): 285, 287. 1938 (director bot. gard. Napoli, Roma).
Kanitz, A., Magy. növen. Lap. 7: 122. 1883 (obit., bibl.).
VanLandingham, S.L., Cat. Diat. 6: 3578, 7: 4215. 1978.

7608. *Pochi studi sulle diatomee* viventi presso alcune terme dell'isola d'Ischia ... Napoli (Stamperia del Fibreno ...) 1867. Qu. (*Pochi stud. diatom.*).
Publ.: 1867 (BSbF rd. Mai-Nov 1867), p. [i-iv], [1]-18, [1, expl. pl.], *pl. 1-2* (uncol., by M. Colaneri and Pedicino). *Copies*: G, NY. – Reprinted from Atti r. Accad. Sci. fis. mat., Napoli, 3(20); see also Rendic. Accad. Sci. Napoli 6: 70-72. 1867.
Ref.: L.R., Hedwigia 9(12): 188-190. Dec 1870.

Peekel, Gerhard (1876-1949), German missionary and plant collector; ord. 1902 Hiltrup; from 1904 in the Bismarck Archipelago; from 1908 at New Mecklenburg (New Ireland). (*Peekel*).

HERBARIUM and TYPES: B (partly destroyed), BO (orig. after 1939); further material at BPI, L and Z.

BIBLIOGRAPHY and BIOGRAPHY: IH 2: (in press); Kew 4: 262; SK 1: 402 (itin., coll.), 4: 106, 8: lxxiv; Urban-Berl. p. 381.

BIOFILE: Diels, L., Notizbl. Berl. 10: 273-285. 1930 (Plantae peekelianae papuanae; for further publ. on Peekel's coll. see SK 1: 402).
Pilger, R., Bot. Jahrb. 74: 653. 1949 (d. 19 Feb 1949).
Sleumer, H., Taxon 9: 90. 1960.
Steenis, C.G.G.J. van, Fl. males. Bull. 44: 194, 5: 124. 1949 (d. 19 Feb 1949).
Streit & Dindinger, Bibl. missionum 21: 458-460. 1955 (fide SK).

EPONYMY: *Peekelia* Harms (1920); *Peekeliodendron* Sleumer (1937); *Peekeliopanax* Harms (1926).

7609. *Illustrierte Flora des Bismarck-Archipels* für Naturfreunde, manuscript vols. 1-12, 13 (Nachtrag 1), 14 (Nachtrag 2), c. 1300 p. text, 940 drawings, microphotocopy of manuscript.
Publ.: Not effectively published by the microfiche (IDC 8096); for the importance of this manuscript see Sleumer 1960.

Pée-Laby, E. (*fl.* 1891), French algologist; Dr. Sci; assistant at the botany department of the Faculté des Sciences, Toulouse (*Pée-Laby*).

HERBARIUM and TYPES: Unknown.

BIBLIOGRAPHY and BIOGRAPHY: CSP 17: 761; GR p. 343; Kelly p. 172; Kew 4: 261; LS 20309-20311, suppl. 21154-21155; Nordstedt suppl. p. 12; Rehder 5: 657.

BIOFILE: Astre, G., Bull. Soc. Hist. Nat. Toulouse 101: 187. 1966.
Comère, J., Bull. Soc. bot. France 46: 169. 1899.

7610. *Flore analytique et descriptive des cryptogames cellulaires* (mousses, hépatiques, champignons, lichens, algues) *des environs de Toulouse*. Tableaux dichotomiques pour la détermination facile des espèces ...Toulouse (Édouard Privat, ...) 1896. Oct. (*Fl. crypt. cell. Toulouse*).
Publ.: 1896 (Bot. Zeit. 1 Jan 1897; Nat. Nov. Dec (2) 1896), p. [1]-262. *Copy*: BR.

Pehersdorfer, Anna (1849-1925), Austrian botanist, teacher at a girl's college in Steyr. (*Pehersd.*).

HERBARIUM and TYPES: Unknown; some material at G.

BIBLIOGRAPHY and BIOGRAPHY: Barnhart 3: 63; CSP 17: 762; GR p. 456-457 (b. 22 Jul 1849, d. 3 Apr 1925); LS 20375, 37784; Rehder 1: 71, 5: 657.

BIOFILE: Janchen, E., Cat. fl. austr. 1: 23. 1956 (bibl.).

7611. *Botanische Terminologie* alphabetisch geordnet. Handbuch zur Auffindung aller in der Botanik vorkommenden lateinischen Kunstausdrücke und solcher deutschen, welche einer Erklärung bedürfen ... Steyr (Druck von W. Fiedler Strezek, Wien) 1897. Oct. (*Bot. Termin.*).
Ed. 1: Feb 1897 (preface: 28 Jan 1897; ÖbZ Nov 1897; Nat. Nov. Dec (2) 1897, Mar (2) 1898), p. [1]-58. *Copy*: WU. – "Ein gutgemeintes Büchlein" (ÖbZ).
Ed. 2: 1901, p. [i]-iv, [1]-104. *Copy*: USDA. – Stuttgart (Verlag von K.G. Lutz) 1901. Oct.
Ref.: Kneucker, A., Allg. bot. Z. 5: 149. 1899. 8: 34. 1902.
Oltmanns, F., Bot. Zeit. 60(2): 125. 16 Apr 1902. (rev. ed. 2).

7612. *Die Flechten des Bezirkes Steyr* in Oberösterreich von Anna Pehersdorfer Fachlehrerin in Steyr ... Steyr (Im Selbstverlage der Verfasserin ...) 1908. Oct. (*Flecht. Steyr*).
Publ.: 1908 (Nat. Nov. Jul(2) 1908; ÖbZ for Aug-Sep 1908), p. [1]-32, [33-40, figs. by Sepp Urban on p. 33-38]. *Copy*: O. Almborn.

7613. *Kleine Auslese der interessantesten Pflanzen aus der Flora von Steyr*, welche dieselbe charakterisieren ... Steyr (Selbstverlag ... Druck von Emil Haas & Cie.). 1907. Oct. (*Kl. Ausl. Steyr*).
Publ.: 1907, p. [1]-21. *Copy*: WU. – Reprinted and to be cited from Alpenbote, 1907.

P'ei Chien (1903-x), Chinese botanist; AB Stanford 1927; AM ib. 1928, studied in U.S. 1925-1930, later at the Biological Laboratory of the Science Society of China, Nanking. (*P'ei*).

HERBARIUM and TYPES: NSM.

BIBLIOGRAPHY and BIOGRAPHY: Barnhart 3: 63 (orig. inf.); Hortus 3: 1201 (P'ei); Kew 4: 263; MW p. 381-382, suppl. p. 277; Zander ed. 10, p. 700, ed. 11, p. 798.

BIOFILE: Moldenke, H.N., Plant Life 2: 75. 1946 (1948).

7614. *The Verbenaceae of China* ... Shanghai (The Science Society of China ...) 1932. (*Verben. China*).
Publ.: 1932, p. [i-v], 1-193, *pl. 1-33. Copies*: USDA. Published as Mem. Sci. Soc. China 1(3): 1-193, *pl. 1-33*.
Preliminary and add. publ.: Contr. Biol. Lab. Sci. Soc. China Bot. Ser. 6(5): 35-38. 1931 (phytogeogr.), 7: 205-213, *pl. 1*. 1932 (additional), Sinensia 2: 65-77. 1932.

Peirce, George James (1868-1954), Manila-born (of American parents) American botanist; Dr. phil. Leipzig 1894 (with Pfeffer); instructor in botany, Bloomington, Ind. 1895; plant physiologist at Stanford from 1897-1934; all-round botanist; studied the effect of air pollution on plants. (*Peirce*).

HERBARIUM and TYPES: Some material at NY.

BIBLIOGRAPHY and BIOGRAPHY: Barnhart 3: 63; Bossert p. 305 (b. 13 Mar 1868, d. 15 Oct 1954); CSP 17: 763; GR p. 238; Kelly p. 172, 256; Kew 4: 263; LS 20377-20382; Rehder 2: 447; Tucker 1: 545.

BIOFILE: Anon., Bot. Centralbl. 63: 222. 1895 (app. Indiana), 70: 399. 1897 (app. Stanford); Bot. Jahrb. 21 (Beibl. 53): 61. 1896; Hedwigia 74: (142). 1934 (retirement from Stanford); Jahrb. Wiss. Bot. 56: 823. 1915 (publ. of his Leipzig period with Pfeffer); Madroño 10: frontisp. portr. 1950; Nat. cycl. Amer. biogr. 41: 479. 1956; Nat. Nov. 17: 358. 1895 (app. Bloomington), 19: 330. 1897 (app. Stanford); Österr. bot. Z. 60: 287. 1910 (app. Stanford); Rev. bryol. 34: 127. 1907.
Blinks, L.R., Plant Physiology 30(3): 295-296. 1955 (obit., d. 15 Oct 1954); Ber. deut. bot. Ges. 68a: 320-321. 1955.
Ewan, J. et al., Leafl. W. Bot. 7: 73, 76. 1953; Short hist. bot. U.S. 48, 95. 1969.
Kneucker, A., Allg. bot. Z. 1: 200. 1894.
Murrill, W.A., Hist. found. bot. Florida 40. 1945.
Pierce, G.J., A textbook of plant-physiology, New York 1903; The physiology of plants. New York 1926, 363 p. (GH cat.).
Schuster, R.M., Hepat. Anthoc. N. Amer. 1: 87. 1966.
Smith, G.M., Science 121: 349-350. 1955 (obit., 13 Mar 1868, d. 15 Oct 1954).
Stieber, M.T., Huntia 4(1): 84. 1981 (arch. mat. HU).
Verdoorn, F., ed., Chron. bot. 2: 361. 1936, 3: 23. 1937, 4: 86, 168. 1938.
Wittrock, V.B., Acta Horti Berg. 3(3): 200. *pl. 149* 1905 (portr.).

Peirson, Frank Warrington (1865-1951), American botanist and plant collector, from 1902 realtor at Altadena, Calif.; AB Haverford 1889, AM ib. 1890; plant collector in the Pacific States. (*Peirson*).

HERBARIUM and TYPES: RSA (12-14.000); other material at A, BH, CAS, DS, GH, JEPS, NEBC, NO, UT, WELC.

BIBLIOGRAPHY and BIOGRAPHY: Barnhart 3: 63; BL 1: 169, 312; Ewan ed. 2: 172; IH (ed. 6): 362, (ed. 7): 339, 2: (in press); Lenley p. 324.

BIOFILE: Anon., Taxon 1: 63. 1952 (herb. at RSA).
Cantelow, E.D. & H.C., Leafl. W. Bot. 8: 97. 1957 (biogr. notice, b. 11 Dec 1865, d. 1 Mai 1951).
Ewan, J. et al., Leafl. W. Bot. 7(3): 94. 1953.
Munz, P.A., El-Aliso 2(4): 339-340. 1952 (obit., portr., account of collecting, bibl.).
Reveal, J., *in* A. Cronquist et al., Intermountain flora 1: 71. 1972.
Thomas, J.H., Huntia 3: 40. 1969.

7615. *Plants of Rock Creek Lake Basin* Inyo County, California/ a check list ... Los Angeles, California (printed by Belmont Adult Evening School Print Shop) 1938. (*Pl. Rock Creek Lake Basin*).
Publ.: 1938, p. [i-iv], 1-16. *Copy*: USDA. – A four-page list of addenda was published in 1942.

Pellanda, Giuseppe (*fl.* 1904), Italian botanist. (*Pellanda*).

HERBARIUM and TYPES: Material in C, E, GB, GE, NAP, PI, and WU.

BIBLIOGRAPHY and BIOGRAPHY: BL 2: 379, 701.

BIOFILE: Hedge, I.C. & J.M. Lamond, Index coll. Edinb. herb. 118. 1970.

7616. *La flore estiva dei Monti d'Oropa* ... Biella (Tipografia G. Testa) 1904, [1906]. Oct. (*Fl. est. Mt. d'Oropa*).
Publ.: 1904 or 1906 (see BL 2: 379; Bot. Centralbl. 25 Oct 1904, (probably announcement; cover 1906, Bot. Centralbl. 17 Jul 1906 (with precise details), Nat. Nov. Mai (2). 1906), p. [i]-vii, [1]-682. *Copy*: FI.

Pellegrin, François (1881-1965), French botanist; from 1912 at the Museum d'Histoire naturelle, Paris; Dr. Sci. Paris 1908; Croix de Guerre 1914-1918. (*Pellegr.*).

HERBARIUM and TYPES: Pellegrin did not collect in any great quantity and did not travel in the tropics. His types are mainly at P. – Letters and portrait at G.

BIBLIOGRAPHY and BIOGRAPHY: Barnhart 3: 64; BJI 2: 132; BL 1: 36, 43, 313, 2: 160, 701; BM 8: 994; Hortus 3: 1201 (Pellegr.); IH 2: (in press); Kew 4: 263-264; LS suppl. 21184; MW p. 382; NI 1008; Roon p. 86; TL-2/4275; Tucker 1: 540.

BIOFILE: Anon., Taxon 14: 206. 1965; Hedwigia 70: (74). 1930 (sous-directeur at P.).
Aubréville, A., Adansonia ser. 2 5(3): 285-287. 1965 (obit., portr., b. 25 Sep 1881, d. 9 Apr 1965).
Bullock, A.A., Bibl. S. Afr. bot. 83. 1978.
Gagnepain, F., *in* H. Lecomte, Fl. Indoch. vol. prél. 45. 1945 (*pl. 12*).
Gaussen, H., Monde Pl. 60: 6. 1965 (obit.).
Halle, N., Adansonia ser. 2 5(3): 289-296 (main bibl.).
Léandri, J., Bull. Soc. bot. France 12: 182-184. 1965 (obit., portr.); Taxon 14(8): 249-250. 1965 (portr.); Aetfat Bull. 17: 11-12. 1966.
Pellegrin, F., Notice sur les titres et les travaux scientifiques de François Pellegrin, Paris 1931, 52 p. (bibl.).
Verdoorn, F., ed., Chron. bot. 2: 34, 35, 131, 134. 1936, 3: 27, 112, 115, 350. 1937, 4: 55, 87. 1938, 6: 300. 1941.

COMPOSITE WORKS: (1) Lecomte, H., *Fl. gén. Indochine*:
Dichapetalaceae 1(7): 796-801. Aug 1911; 1(suppl.) 6: 728-730. Dec 1948.
Gesneraceae: 4(5): 487-565. Aug 1930.
Lentibulariaceae 4(5): 467-487. Aug 1930.
Meliaceae 1(7): 723-795. Aug 1911, 1(suppl.) 5: 683-700. Dec 1946, 1(suppl.) 7: 701-728. Dec 1948.
Orobanchaceae 4(4): 461-464. Jan 1927, 4(5): 465-466. Aug 1930.
Plumbaginaceae 3(6): 748-753. Feb 1930.

EPONYMY: *Pellegrinia* Sleumer (1935); *Pellegriniodendron* Léonard (1955); *Aubregrinea* Heine (1960), also honors A. Aubréville (1897-x).

7617. Thèses présentées à la Facultés des sciences de Paris pour obtenir le grade de docteur ès sciences naturelles par M. François Pellegrin. 1re Thèse. – *Recherches anatomiques sur la classification des Genêts et des Cytises* ... soutenues le [mss. or stamped: 24] juin 1908, ... Paris (Masson et Cie, ...) 1908. Oct. (*Rech. anat. Genêts*).

Publ.: 24 Jun 1908, p. [i-iii], [129]-319 [320], *pl. 1*, [1, thèses]. *Copies*: B, NY. – Reprinted or preprinted from Ann. Sci. nat., Bot. ser. 9. 7: [129]-319 [320].

7618. *La flore du Mayombe* d'après les récoltes de M. Georges Le Testu, ... Caen 1924-1928. 1938 Qu. (*Fl. Mayombe*).
Collector: Georges Marie Patrice Charles Le Testu (1877-1967).
1: 1 Aug 1924, p. [1]-125, [126, expl. pl.], *pl. 4-11*. – Mém. Soc. Linn. Normandie 26(2), 1924.
2: 5 Dec 1928, p. [1]-83, [1, expl. pl.], *pl. 1-6*, map. – Ib. Ser. 2, Bot. 1(3). 1928.
3: 15 Dec 1938, p. [1]-114, [1, expl. pl.], *pl. 1-8*, [1]-viii index. – Ib. 1(4) 1938 plus [i-iv], general t.p. etc. for consolidated reprint 1924-1938.
Copies: B, BR, G, MO, NY, US. – The precise dates of publication are given on the covers of the journal issues.

Pelletan, Jules (1833-1892), French microscopist and diatomist. (*Pelletan*).

HERBARIUM and TYPES: Unknown.

BIBLIOGRAPHY and BIOGRAPHY: Barnhart 3: 64; BM 4: 1539; CSP 17: 769-770; De Toni 1: xcvi, 2: xci-xcii (bibl.); Jackson p. 220; Kew 4: 264; Morren ed. 10, p. 58; MW p. 382; Rehder 1: 36; TL-2/2228.

BIOFILE: VanLandingham, S.L., Cat. Diat. 6: 3578, 7: 4215. 1978.

COMPOSITE WORKS: (1) *Journal de Micrographie* 1-6(5), Mai 1877-Mai 1892.
(2) editor of Habirshaw, *Cat. diatomées*, Paris 1879 (n.v.). We have not been able to find this book, announced by Nat. Nov. 1879. However, Habirshaw's *Bibliographie des diatomées* was published by Pelletan in J. Micrographie 3(1): 368-370, 410-414, 453-455, 497-500. 1879, 4: 40-42, 98-104, 1880, 5: 146-148. 1881 (285 nos.). This *Bibliographie* is almost certainly identical with the "Cat. diatomées". (Correction and addition to TL-2/2228 French ed.).

7619. *Le microscope* son emploi et ses applications ... Paris (G. Masson, éditeur ...) 1876. Oct. (*Microscope*).
Publ.: 1876 (Bot. Zeit. 20 Apr 1877), p. [i]-viii, [1]-772, *4 pl. Copies*: FI-BN, PH. – Botany on p. 311-589.
Rev.: Anon., Bull. Soc. bot. France 23 (bibl.): 95. 1876.

7620. *Les Diatomées* histoire naturelle, préparation, classification & description des principales espèces ... Paris (Journal de Micrographie ...) 1888-1889. Oct. (*Diatomées*).
Co-authors: Introduction by Julien Marc Deby (1826-1895); exposé classification: Paul Charles Michel Petit (1834-1912); list french diat. Maurice Peragallo (1853-?).
1: 1888 (J. Bot. Jun 1888; Bot. Centralbl. 15-19 Oct 1888; Nat. Nov. Apr 1888 ann.), p. [i]-xiii, [xiv-xv], [1]-322, *pl. 1-5* (uncol. liths.) with text.
2(1): 1889 (Bot. Zeit. 31 Mai 1889), p. [i]-viii, [1]-210, *pl. 6-7* (id.).
2(2): 1890 (t.p. 1889; Bot. Zeit. 26 Dec 1900, p. [i, h.t.], 211-364, [1, expl. pl.], *pl. 9-10*.
Copies: B, FH(3), NY, US.
Re-issue: 1891 (t.p.), 2 vols. *Copies*: BR, FI, G, MO, PH. – Les diatomées ... espèces ... Introduction, par J. Deby ... Liste des Diatomées françaises, par H. Peragallo ..." Paris (Librairie J.B. Baillière et fils ...) 1891. Oct.
1: [i]-xiii, [xiv-xv], [1]-322.
2: [i]-viii, [1]-364.

Pelletier, [Pierre] **Joseph** (1788-1842), French botanist and pharmacist at Paris; Dr. Sci. Paris 1812, active at the École de pharmacie at Paris from 1814. (*Pellet.*).

HERBARIUM and TYPES: Unknown.

BIBLIOGRAPHY and BIOGRAPHY: Backer p. 431, 658; Barnhart 3: 64 ("Pierre Joseph"); BM 4: 1539; Bossert p. 305; CSP 4: 806-810; Frank (Anh.): 75; GR p. 365 (b. 22 Mar

1785, sic); Krempelh. 1: 580; Lenley p. 321; LS 20394; PR 7028 (b. 22 Mar 1788, d. 20 Jul 1842), (ed. 1: 7854-7855); Rehder 3: 494, 5: 658; Tucker 1: 546; Zander ed. 10, p. 700, ed. 11, p. 798 (b. 22 Mar 1788, d. 20 Jul 1842).

BIOFILE: Poggendorff, J.C., Biogr.-lit. Handw.-Buch 2: 392-393. 1863 (d. 19 Jul 1842; bibl.; "Josephe").
Wittrock, V.B., Acti Horti Berg. 3(3): 110. 1905.

EPONYMY: *Pelleteria* Poiret (1825, *orth. var.* of *Pelletiera* A.F.C. Saint-Hilaire); *Pelletiera* A.F.C. Saint-Hilaire (1822). The derivation of *Pelletieria* Seward (1913) could not be determined.

7621. Essai sur la nature des substances connues sous le nom de *Gommes résines*. Thèse soutenue devant la Faculté des Sciences de l'Université impériale 1e [ink: 22] aout 1812, par J. Pelletier, pharmacien, ... Paris 1812. Qu. (*Gomm. résin.*).
Publ.: 22 Aug 1812, p. [1]-32. *Copy*: G.

Pelloe, Emily Harriet (Mrs. Theodore Pelloe, née Sundercombe) (1878-1941), Australian botanist and artist, living mainly in Western Australia from 1901. (*Pelloe*).

HERBARIUM and TYPES: Unknown.

BIBLIOGRAPHY and BIOGRAPHY: HR; Kew 4: 264.

BIOFILE: Anon., The West Australian 16 Apr 1941 (p. 1, 6) (fide HR).
Karg, A.T., Taxon 23: 616. 1974.
Pescott, E.E., Vict. Natural. 58: 15. 1941 (obit.).

7622. *Wild flowers of Western Australia* ... Melbourne (C.J. DeGaris Publishing House) 1921. (*Wild fl. W. Australia*).
Publ.: 1921, p. [1]-124, *8 pl. Copies*: NY, US.
Ref.: Anon., Vict. Natural. 38(12): 134. Apr 1922 (rev.).

7623. *West Australian Orchids* ... Perth, Western Australia (published by Emily H. Pelloe) 1930. (*W. Austral. Orchid.*).
Publ.: 1930, p. [i]-iv, [1]-77, frontisp., *pl. 1-3* (col.). *Copies*: MO, NY.

Pelourde, Fernand (1884-1916), French palaeobotanist; Dr. phil. Paris 1907; "préparateur" at the Muséum d'Histoire naturelle, Paris (*Pelourde*).

HERBARIUM and TYPES: Palaeobotanical types at Mus. Hist. nat., Paris.

BIBLIOGRAPHY and BIOGRAPHY: Barnhart 3: 65 (b. 7 Jul 1884, d. 16 Feb 1916).

BIOFILE: Mangin, L., Bull. Mus. Hist. nat. Paris 1916(2): 66-68 (obit., b. 7 Jul 1884, d. 16 Feb 1916, bibl.).

EPONYMY: *Pelourdea* A.C. Seward (1917).

7624. *Paléontologie végétale* cryptogames cellulaires et cryptogames vasculaires ... Paris (Octave Doin et fils, éditeurs ...) 1914. (*Paléontol. vég.*).
Preface: Charles René Zeiller (1847-1915).
Publ.: 1914 (Bot. Centralbl. 27 Jan 1914, ann., 12 Mai 1914), p. [i]-xxviii, [1]-360. *Copy*: FI. – Encycl. sci. bibl. paléont.
Ref.: Zeiller, R., Bot. Centralbl. 125: 428-429. 28 Apr 1914 (dates book at 1913).

Pena, Pierre (*fl.* 1520-1600), French botanist and physician; collaborated with L'Obel at Montpellier (1565) and in England 1566-1572; from then onward practicing physician in France, ultimately as chief physician of Henri III. (*Pena*).

HERBARIUM and TYPES: Unknown. Legré (1899) states that Pena and L'Obel took their herbaria with them to London in 1566, but no trace of plant collections made by them is known.

BIBLIOGRAPHY and BIOGRAPHY: Barnhart 3: 65; BB p. 240; BM 4: 1541; Desmond p. 487; Dryander 3: 57; DTS 1: xxii; Frank 3(Anh.): 75; HA 1: 351-353; Herder p. 83; HU 183; Jackson p. 26; JW 5: 248; Kew 4: 264; Moebius p. 32; NI 1502, see also 1: 61; Oudemans p. 221-222; PR 7029 (ed. 1: 658); Rehder 5: 658; Saccardo 2:83; TL-2/4308, 4906,4907; see L'Obel; Tucker 1: 546.

BIOFILE: Briquet, J., Ber. Schweiz bot. Ges. 50a: 362-364. 1940 (bibl., biogr., epon.).
Cavillier, F., Boissiera 5: 70. 1941.
Haller, A. de, Hist. stirp. Helv. 1: xiii. 1768.
Kent, D.H., Hist. Fl. Middlesex 12. 1975.
Legré, L., La botanique en Provence au xvième siècle, Pierre Pena et Mathias de Lobel, Marseille 1899 (see TL-2/4308) ("... doit être considéré un chef d'oeuvre de sagacité" Briquet 1940; Indigénat en Provence du Styrax officinal. Pierre Pena et Fabri de Peirese. Marseille 1901, 23 p. (n.v., see J. Bot. 40: 116. 1902), see also Bull. Soc. bot. France 44: xi-xlvii. 1897, 46 (bibl.): 544-545. 1899.
Magnin, A., Bull. Soc. bot. Lyon 31: 18. 1906.
Martins, C., Le jardin des plantes de Montpellier 12. 1854.
O., Dict. Sci. med., Biogr. méd. 6: 387. 1824.
Planchon, J.E., Rondelet et ses disciples, Montpellier 1866, p. 18, 34-37.
Wright, R., Gardener's tribute 230 f., 237 s.d.

EPONYMY: *Penaea* Linnaeus (1753).

7625. *Stirpium adversaria nova* perfacilis vestigatio luculentaque accessio ad priscorum, presertim Dioscorides & recentiorum materiam medicam. Quinus praediem accedit altera pars. Conjectaneorum de plantis appendix. De succis medicatis et metallicis sectio. Antiquae et novatae medicinae lectiorum remediorum thesaurus opulentissimus, de succedaneis libellus continentur ... [Londini (Thomas Purfoot) 1570] 1571. Fol. (*Stirp. advers. nov.*).
Co-author: Matthias L'Obel (share of authors not established).
Publ.: Jan 1571 (see p. 442), p. [i-xxi], 1-442. *Copy*: USDA.
Subsequent editions: see e.g. L'Obel, PR 7029, Dryander 3: 57, Legré (1899), p. 33-38 and Louis (1958). The 1576 and 1605 editions were reissues of the sheets printed in 1570 by Thomas Purfoot. The altera pars in the 1605 ed. is by L'Obel.
Ref.: Corbett, M., Arch. nat. Hist. 10(1): 111-117. 1981 (on emblematic t.p.).

Peña, Rafael (*fl.* 1901), Bolivian botanist. (*R. Peña*).

HERBARIUM and TYPES: Unknown.

BIBLIOGRAPHY and BIOGRAPHY: BL 1: 237, 312; BM 4: 1541; Kew 4: 264.

7626. *Flora cruceña*. Apuntes ... Sucre (Imp. "Bolivár" de M. Pizarro ...) 1901. Qu. (*Fl. cruc.*).
Ed. [*1*]: 1901, p. [i, iii, v], [1]-287. *Copy*: US.
Ed. [*2*]: 1944, p. [i*-iii*], i-ix, 3-475. *Copy*: NY. – La Paz, Bolivia (Editorial del Estado ...).

Penard, Eugène (1855-1954), Swiss protozoologist, botanist and author of children's books; bank employee at Genève 1872-1881; studied at Edinburgh, Heidelberg and Genève; Dr. Sci. Genève 1888; in Colorado 1891; from 1892-1898 private tutor in Germany and Russia; from 1898 private scientist in Genève; one of the founders of the Société botanique de Genève. (*Penard*).

HERBARIUM and TYPES: G; further material at NY, W. – Portrait and letters at G.

BIBLIOGRAPHY and BIOGRAPHY: Barnhart 3: 65; BM 4: 1541, 8: 995; Bossert p. 305; CSP 17: 74-775; DSB; Ewan ed. 1: 89, 282 (portr., Colorado trip), ed. 2: 172; IH 2: (in press); LS 21226; TL-2/6474.

BIOFILE: Allison, E.M., Univ. Colorado Studies 6: 69-70. 1909.
Britton, N.L. & A.M. Vail, Bull. Herb. Boissier 3: 197-221. Mai 1895 (Enum. pl. Colorado coll. by P.; Contr. Herb. Columbia Coll. no. 75; has introduction by Penard with Colorado itin.; visit 1891, not 1892 as mentioned in the title of this paper).
Burdet, H.M., Saussurea 8: 93-108. 1977 (on P's hypothesis on the origin of the autochtonous alpine flora of 1908; portr., b. 16 Sep 1855, d. 5 Jan 1954; brief biogr.).
Corliss, J., Trans. Amer. microsc. Soc. 97(4): 431-432. 1978 (portr.).
Dottrens, E., Rev. Suisse Zool. 61(1): 3-8. 1954 (portr., obit., bibl.).
Engeln, O.D. von, Proc. Geol. Soc. Amer. 1958: 169-172. 1959.
Ewan, J., Trail and Timberline 352: 84-87. 1948 (portr. and see Ewan ed. 1 and ed. 2).
Guyénot, E., Arch. d. Sci. 7: 33-37. 1954 (obit.).
Miège, J., Musées de Genève 160: 15. 1975 (portr.).
Mueller Arg., J. Bull. Herb. Boissier 3(5): 199-201. Mai 1895 (plants collected by P., see TL-2/6474).
Nissen, C., Zool. Buchill. 312. 1969.
Weibel, R., Trav. Soc. bot. Genève 3: 41. 1956 (obit.; co-founder Soc. bot. Genève).
Wunderlich, E., Geogr. Anz. 29: 297-306. 1928.

EPONYMY: *Penardia* J. Cash (1904).

7627. Contributions à l'étude des Dino-flagellés. Recherches sur le *Ceratium macroceros* avec observations sur le Ceratium cornutum ... Genève (Imprimerie L.-É. Privat, ...) 1888. Qu. (*Ceratium macroceros*).
Publ.: Mai 1888 (authorization to print 1 Mai 1888; Bot. Centralbl. 34: 337. 11-15 Jun 1888; Bot. Zeit. 29 Jun 1888), p. [1]-43, [1, note], *pl. 1-3* (no. 1 col.; auct.). *Copies*: FH(2), G.
Ref.: Möbius, M., Bot. Centralbl. 37: 131-132. 30 Jan 1889.
Hieronymus, G.H.E.W., Bot. Zeit. 47: 93-97. 8 Feb 1889 (critical rev.).

7628. *Les Péridiniacées du Léman* ... Genève (Imprimerie P. Dubois, ...) 1891. Oct. (*Péridin. Léman*).
Publ.: 1891, p. [1]-63, *pl. 1-5* (uncol.). *Copy*: G. – Reprinted from Bull. Trav. Soc. bot. Genève 6: 1-63, *pl. 1-5*. 1891.

7628a. *Encore la Chlamydomyxa* ... [Bull. Boiss. ser. 2. vol. 5. 1905]. Oct.
Publ.: 31 Mai 1905 (in journal), p. [1]-10. *Copy*: U. – Issued in Bull. Herb. Boissier ser. 2. 5(6): 527-526. 31 Mai 1905.

Penfold, [Mrs.] **Jane Wallas** (*fl.* 1820-1850), British botanical artist and naturalist; lived in Madeira. (*Penfold*).

HERBARIUM and TYPES: Unknown.

BIBLIOGRAPHY and BIOGRAPHY: Barnhart 3: 65; BB p. 240; BM 4: 1542; Desmond p. 487; Herder p. 218; Jackson p. 352; Kew 4: 266; NI 1503, see also 1: 122; PR 7030 (ed. 1: 7858); Rehder 1: 497.

BIOFILE: Britten, J., J. Bot. 57: 97-99. 1919, Times Literary Suppl. 6 Mar 1919.

7629. *Madeira flowers, fruits, and ferns*, a selection of the botanical productions of that island, foreign and indigenous, drawn and coloured from nature, ... London (printed and published by Reeve, Brothers, ...) 1845. Qu. (*Madeira fl.*).
Publ.: Feb-Jul 1845 (p. [v]: Feb 1845; Bot. Zeit. 1 Aug 1845), p. [i-x], *pl. 1-20* (col., auct.). *Copies*: MO, NY. – A curiosity. See Britten (1919) and Collins (Sotheby 594) for further details and for a note on the compendium volume by the author's daughter, Augusta J. Robley, *Selection of Madeira flowers*, fol., 1845 (n.v.). The botanical text is by William Lewes Pugh Garnons (1791-1863) (see Desmond p. 245).

Penhallow, Davis Pearce (1854-1910), American actuo– and palaeobotanist; B.S. Amherst Agr. Coll. 1873; professor of botany and chemistry at the College of Agriculture, Sapporo (Japan) 1876-1880; at Harvard 1880-1882; at the Houghton Experimental Farm, N.Y., 1882-1883; from 1883-1910 professor of botany at McGill University, Montreal; D.Sc. Montreal 1904. (*Penh.*).

HERBARIUM and TYPES: Unknown. – Manuscripts and correspondence in Smithsonian Archives.

BIBLIOGRAPHY and BIOGRAPHY: Andrews p. 312; Barnhart 3: 66; BJI 1: 44; BL 1: 133, 144, 312; BM 4: 1542, 8: 996; Bossert p. 305; CSP 10: 1021, 17: 778-779; De Toni 4: xliv; Hortus 3: 1201 ("Penh."); Kew 4: 266; Langman p. 574; Lenley p. 324; LS 20398-20404; MW p. 382; Quenstedt p. 333; Rehder 1: 38, 80, 5: 658; Tucker 1: 546-547.

BIOFILE: Andrews, H.N., The fossil hunters 229. 1980.
Anon., Bot. Centralbl. 116: 32. 1911 (d. 20 Oct 1910); Bot. Not. 1911: 28 (d. 20 Oct 1910); J. New York Bot. Gard. 11: 260-261. 1910 (b. 25 Mai 1854, d. 26 Oct 1910); Nat. Nov. 32: 544; 1910 (d.); Österr. bot. Z. 61: 47. 1911 (d.); Proc. Trans. Roy. Soc. Canada ser. 3. 5: vii-x. 1912 (obit., portr.); Torreya 10: 260. 1910.
Barlow, A.E., Bull. Geol. Soc. America 22(1): 15-19. 1911 (obit., bibl.).
Deane, W., Rhodora 13: 1-4, 56. 1911.
Derick, C.M., Canad. Rec. Sci. 9: 387-390. 1915 (obit., portr., b. 25 Mai 1854, d. 20 Oct 1910).
Jeffrey, E.C., Bot. Gaz. 51: 142-144. 1911 (obit., portr.).
Kneucker, A., Allg. bot. Z. 17: 32. 1911 (death at sea).
Penhallow, D.P., Trans. Roy. Soc. Canada 5. (sect. 4): 45-61. 1887, review of Canadian botany), ser. 2. 3 (sect. 4): 16, 21, 50-51. 1897 (bibl.).
Rodgers, A.D., Amer. bot. 1873-1892, p. 73, 145, 174-175, 198, 285, 313. 1944.
Stieber, M.T., Huntia 4(1): 84. 1981 (arch. mat. HU).

COMPOSITE WORKS: Editor, *Canadian Record of Science* 1888-1890; associate editor *American Naturalist* 1897-1907; editor of palaeobotany, *Botanisches Centralblatt* 1902-1907.

EPONYMY: *Penhallowia* O. Kuntze (1903).

7630. *The botanical collector's guide*: a manual for students and collectors; containing directions for the collection and preservation of plants and the formation of a herbarium ... Montreal (E.M. Renouf, ...) 1891. Duod. (*Bot. coll. guide*).
Publ.: Sep-Oct 1891 (preface: Aug 1891; Bot. Gaz. Nov 1891), p. [1]-125. *Copy*: NY.

7631. *A manual of the North American Gymnosperms* exclusive of the Cycadales but together with certain exotic species ... Boston, U.S.A. (Ginn & Company, publishers the Athenaeum Press) 1907. (*Man. N. Amer. Gymnosp.*).
Publ.: 1907 (Bot. Centralbl. 28 Jan 1908; Nat. Nov. Jun(1) 1908, p. [i]-viii, [1]-374, [1], *pl. 1-54* (2 p.), *55* (photos). *Copies*: G, MICH, MO, NY, USDA.
Ref.: Chrysler, M.A., Bot. Centralbl. 107: 318. 24 Mar 1908 (rev.).

7632. *Report on the Tertiary plants of British Columbia* collected by Lawrence M. Lambe in 1906 together with a discussion of previously recorded tertiary floras ... Ottawa (Government Printing Bureau) 1908. Qu. (*Tert. pl. Brit. Columb.*).
Collector: Lawrence Morris Lambe (1863-1919).
Publ.: 1908 (p. [3]: 3 Feb 1908), p. [1]-167. *Copies*: NY. – Canada Department of Mines, Geological Survey Branch [publ.] no. 1013.

Penland, Charles William Theodore (1899-x), American botanist; and ecologist; professor of botany at Colorado College (Boulder) from 1922. (*Penland*).

HERBARIUM and TYPES: Material at AMES, COCO, GH, M (lich.), MASS, US.

BIBLIOGRAPHY and BIOGRAPHY: Barnhart 3: 66; BJI 2: 132; Bossert p. 305; Ewan ed. 1: 282, ed. 2: 173; Hirsch p. 228; IH 2: (in press).

BIOFILE: Acosta-Solis, M., Contr. Inst. Ecuat. Ci. nat. 65: 63-64. 1968 (in Ecuador Mar-Jul 1939).
Moldenke, H.N., Plant Life 2: 75. 1946 (1948).
Penland, C.W.T., Letter to J. Ewan, Nov 1980 ("I am still alive and kicking although I will be 81 years old in less than a month" in response to our request for information).

7633. Colorado College Publication. *The alpine vegetation of the southern Rockies* and the Ecuadorean Andes ... Colorado Springs, Colorado May, 1941. Oct.
Publ.: Mai 1941, p. [1]-30. *Copy*: NY. – Colo. Coll. Publ. Studies 32, 1941.

Pennant, Thomas (1726-1798). British zoologist and antiquary; sheriff of Flintshire; gentleman-naturalist; correspondent of Linnaeus. (*Pennant*).

HERBARIUM and TYPES: all natural history coll., incl. fossils, at BM, Letters at LINN, Chester Public Library, Fitzwilliam Museum, Cambridge, and Queen's College Oxford. – Forty-four of the letters making up Gilbert White's *The natural history and antiquities of Selborne* were written to Pennant. – The manuscript of the *Flora indica* (Outlines Globe 4) is at A. A collection of [animal] fossils is at BM. – Some letters at G.

BIBLIOGRAPHY and BIOGRAPHY: BB p. 240-241 (b. 14 Jun 1726, d. 16 Dec 1798); BM 8: 996-997; Bossert p. 305; Dawson p. 660-664; Desmond p. 487; DNB 44: 320-323; DSB 10: 509-510 (bibl.); Henrey 2: 26, 157, 158; Kew 4: 267; ME 3: 350; NI 1: 118; Quenstedt p. 333; Rehder 5: 658; SK 4: xcv; TL-1/687, 960; TL-2/4522, see Kalm, P.; Tucker 1: 547.

BIOFILE: Allen, D.E., The naturalist in Britain 24, 36, 53, 67, 95, 215. 1976; Newsletter Soc. Bibl. nat. Hist. 13: 11-12. 1982 (calendar papers).
Allen, E.G., The history of American ornithology before Audubon, 1951 (Trans. Amer. philos. Soc. ser. 2. 41(3)), see p. 589 [index].
Allibone, S.A., Crit. dict. Engl. lit. 1553-1554. 1870 (bibl.).
Archer, H.R., Huntia 1: 204-209. 1964 (P's copy of Kalm's *Travels*).
Berkeley, E. & D.S., Dr. Alexander Garden 377 [index]. 1969; Dr. John Mitchell 96-97. 1974.
Blunt, W., The compleat naturalist 154, 180. 1971.
Bridson, G.D.R. et al., Nat. hist. mss. res. Brit. Isl. 428 [index]. 1980.
Cameron, H.C., Sir Joseph Banks 5, 16, 58. 1952.
Cuvier, G., Biogr. univ. 33: 315-318. 1823.
Dawson, W.R., Smith papers 73. 1934.
Fletcher, H.R., Royal Bot. Garden Edinburgh 94. 1970.
Freeman, R.B., Brit. nat. hist. books 275-276. 1980.
Fries, Th.M., Bref Skrifv. Linné 1(5): 203. 1911, 1(6): 194, 204. 1912.
Gilbert, P., Comp. biogr. lit. deceased entom. 292. 1977.
Gunther, R.W.T., Early Science in Oxford 11: 131-132. 1937 (fide DSB).
Hawksworth, D.L. & M.R.D. Seaward, Lichenology Brit. Isles 132. 1977.
Hulth, J.M., Bref Skrifv. Linné 2(1): 377, 378, 387, 388, 401. 1916.
Lada-Mocarski, V., Bibl. books Alaska 144. 1969 (on "Arctic zoology").
McAtee, W.L., J. Soc. Bibl. nat. Hist. 4(2): 100-124. 1963 (N. Amer. birds).
Matheson, C., Ann. Sci. 10(3): 258-271. 1954 (portr., corr. with Morris brothers).
Merrill, E.D., J. Arnold Arb. 29: 186-192. 1918 (on the manuscript of Pennant's *Flora indica* published in his *Outlines of the Globe* 4: 237-317. 1800; list of the new binomials).
Moore, S., Fl. Cheshire lxxxvi. 1899.
Nissen, C., Zool. Buchill. 312. 1969 (b. 14 Jan 1726, d. 16 Dec 1798; zool. bibl.).
O., Dict. sci. méd. 6: 387-389. 1824.
Pennant, T., The literary life of the late Thomas Pennant, Esq. by himself. London 1793 (fide DSB).
R.B.N., Geol. Mag. Dec 5. 10(4): 192. 1913 (on P's collection of fossils at BM).
Rose, H.J., New general biographical dictionary 11: 23-25. 1850.
Smith, E., Life Joseph Banks 344 [index]. 1911.
Smith, P., Mem. corr. J.E. Smith 1: 300, 448-449. 1832.
Swainson, W., Taxidermy 288-291. 1840.

Urness, C., A naturalist in Russia 187 [index]. 1967 (Urness also contr. entry in DSB).
Zittel, K.A. von, Hist. geol. paeontol. 114. 1901.

EPONYMY: *Pennanta* Cothenius (1790, *orth. var.*); *Pennantia* J.R. Forster et J.G.A. Forster (1776).

7634. *The view of the Malayan Isles*, New Holland, and the Spicy Islands ... vol. iv. London (... sold by John White, ...) 1800. Oct. (*View Malay. Isl.*).
Publ.: 1800. *Copy*: E. – Alternative title: Outlines of the Globe vol. 14, *Flora indica* on p. 237-317. For a detailed discussion see Merrill (1948).

Pennell, Francis Whittier (1886-1952), American botanist; travelled and collected widely in the Americas; president of the Philadelphia botanical Club 1942-1952; Ph.D. Univ. Pennsylvania 1913; associate curator at NY 1914-1921; from 1921 Curator of botany at PH; eminent authority on Scrophulariaceae, botanical historian and bibliographer. (*Pennell*).

HERBARIUM and TYPES: PH (main part); Scrophulariaceae 1911-1914 at PENN; Philadelphia local 1914-1921 and Colombian coll. made with E.P. Killip and T.E. Hazen (4500) at NY. – Duplicates in many herbaria e.g. A, AMES, B, BOG, CHARL, CM, COL, DS, F, GH, ILL, K, MO, NO, NY, NYS, PENN, RM, SPAWH, TEX, US (large set). – Archival and manuscript material in G, NY, PH and Smithsonian Archives.

BIBLIOGRAPHY and BIOGRAPHY: Barnhart 3: 66; BJI 2: 133; BL 1: 211, 212, 312; BM 8: 997; Bossert p. 305; Ewan ed. 1: 282, ed. 2: 173; Hirsch p. 228; Hortus 3: 1201 ("Penn."); IH 1 (ed. 1): 76, (ed. 2): 88, (ed. 3): 111, (ed. 6): 362, (ed. 7): 339, 2: (in press); Kew 4: 267; Langman p. 575; Lenley p. 324; MW p. 382, suppl. p. 278; NI suppl. p. 12, 56; NW p. 55; Pennell p. 616; PH p. 511 [index]; TL-2/781; Tucker 1: 547; Urban-Berl.p. 381; Zander ed. 10, p. 700; ed. 11, p. 798 (b. 4 Aug 1886, d. 2 Feb 1952).

BIOFILE: Anon., J. New York Bot. Gard. 25: 96. 1924; New York Times 5 Feb 1952, p. 29.
Benner, W.M., Bartonia 26: 16-17. 1952.
Camp, W.H., Taxon 1: 83. 1952.
Deam, C.C., Fl. Indiana 1118. 1910 (coll.).
Dix, W.L., Bull. Torrey bot. Club 80: 95-96. 1953 (obit.).
Ewan, J., Southw. Louisiana J. 7: 40-42. 1967.
Ewan, J. et al., Short hist. bot. U.S. 17, 150. 1969.
Graham, E.H., Ann. Carnegie Mus. 26: 21. 1937.
Henrey, M.G. et al., Bartonia 26: 1-15. 1952 (portr.; addresses delivered at memorial meeting 28 Feb 1952; detailed bibl. by V.T. Phillips and J. Ewan).
Herrera, F.L., Est. fl. dept. Cuzco. 21-22. 1930.
Knobloch, I.W., Pl. coll. N. Mexico 53. 1979 (b. 4 Aug 1886, d. 3 Feb 1952; coll.).
McVaugh, R., Contr. Univ. Mich. Herb. 9: 289-290. 1972 (Mexican itin.).
Moldenke, H.N., Plant Life 2: 75. 1946 (1948).
Murrill, W.A., Hist. found. bot. Florida 19. 1945.
Pennell, F.W., Bartonia 22: 28-29. 1943.
Rickett, H.W., Index Bull. Torrey bot. Club 77. 1955.
Steenis, C.G.G.J. van, Fl. males. Bull. 10: 353. 1953 (d.).
Steere, W.C., Bot. Rev. 43: 293. 1977.
Verdoorn, F., ed., Chron. bot. 1: 299, 312, 335, 389. 1935, 2: 316, 353, 364. 1936, 3: 263. 1937, 4: 278, 435, 447. 1938, 5: 110, 173, 234, 278, 295. 1939, 6: 86, 87, 172, 263, 7: 83, 237. 1943.

COMPOSITE WORKS: (1) Editor of *Bartonia* 1924-1952.
(2) NAF, *Fabaceae – Eisenhardtia* 24(1): 34-40. 25 Apr 1919.
(3) *In* H.B. Davis, *Life and work of Cyrus Guernsey Pringle*, 1937: Indices of states, places and railroad routes of Mexico.

EPONYMY: *Pennellia* Nieuwland (1918); Pennellianthus F.S. Crosswhite (1970).

7635. *Flora of the Conowingo barrens* of southeastern Pennsylvania [Proc. Acad. nat. Sci. Philadelphia 1910] 1911. Qu.
Publ.: 13 Jan 1911 (on special cover of reprint). p. 541-584. *Copy*: NY. – Reprinted and to be cited from Proc. Acad. nat. Sci. Philadelphia for October 1910: 541-584, publ. 13 Jan 1911.
Further notes: 11 Feb 1913, p. 520-539. *Copy*: NY. – Reprinted and to be cited from Proc. Acad. nat. Sci. Philadelphia for December 1912: 520-530. 11 Feb 1913.

7636. *Scrophulariaceae of the southeastern United States* ... New York 1920.
Publ.: 11 Mar 1920 (t.p. special cover reprint). Reprint, issued as Contr. N.Y. Bot. Gard. 221, from Proc. Acad. nat Sci. Philadelphia 1919: 224-291. 11 Mar 1920.

7637. *Scrophulariaceae of the Central Rocky Mountain States* ... Washington (Government Printing Office) 1920. Oct.
Publ.: 29 Apr 1920 (stamped copy at US) *in* Contr. U.S. natl. Herb. 20(9): [i]-v, 313-381, vii-viii. 1920; to be cited from journal.

7638. *The Scrophulariaceae of eastern temperate North America* ... Philadelphia [The Academy of Natural Sciences of Philadelphia, ...] 1935. (*Scroph. e. N. Amer.*).
Publ.: 27 Nov 1935 (p. [ii], Nat. Nov. Mai 1936), p. [i]-xiv, [1], 1-650. *Copies*: B, G, MO(2), NY, US. – Monograph no. 1. Academy of Natural Sciences of Philadelphia.
– *Preliminary publ.*: Torreya 19: 107-119, 143-152, 161-171, 205-216, 1919, 235-242. 1920.
Ref: Gleason, H.A., Torreya 36: 41-45. 27 Apr 1936.
Keck, D.D., Madroño 3(6): 253-255. 1936.

7639. *The Scrophulariaceae of the western Himalayas* ... Philadelphia [The Academy of Natural Sciences of Philadelphia ...] 1943. (*Scroph. w. Himal.*).
Publ.: 10 Sep 1943 (p. [ii], MO rd 25 Sep 1943), p. [i]-vii, [1], [1]-163, *pl. 1-25* (uncol.). *Copies*: B, FI, G, MO, NY, US. – Monograph no. 5 Acad. nat. Sci. Philadelphia.

Pennier de Longchamp, Pierre Barthélemy (*fl.* 1766), French mycologist and physician at Avignon (*Pennier*).

HERBARIUM and TYPES: Unknown.

BIBLIOGRAPHY and BIOGRAPHY: LS 20405; PR 7031 (ed. 1: 7859).

BIOFILE: Lütjeharms, W.J., Gesch. Mykol. 247. 1936.

7640. *Dissertation physico-médicale, sur les truffes* et sur les champignons ... à Avignon (Chez Roberty & Guilhermont, ...) 1766. Duod. (*Diss. Truffes*).
Publ.: 1766, p. [i]-vi, 1-59. *Copy*: Stevenson.

Pennington, Leigh H [Humboldt] (1877-1929), American phytopathologist and mycologist; B.A. Univ. Mich. 1907, Dr. phil. 1909; from 1910 at Syracuse University as professor of botany, from 1914-1929 as head of the dept. of forest botany. (*Pennington*).

HERBARIUM and TYPES: MICH and SYR; other material at NY.

BIBLIOGRAPHY and BIOGRAPHY: Barnhart 3: 66; Hirsch p. 228; IH 2: (in press); Kelly p. 173; LS 37799-37800, suppl. 21229-21235; Pennell p. 616; PH 186, 567; TL-2/781. Pennington, L.H. 1157
BIOFILE: Anon., Bot. Soc. America, Proc. 24: 15-17. 1930 (obit., b. 26 Oct 1877, d. 23 Apr 1929; Phytopathology 20(5): 467. 1930 (b. 26 Oct 1877, d. 24 Apr 1929); Science ser. 2. 69. 619. 1929 (d. 3 Apr 1929); Torreya 30: 115. 1930.
Murrill, W.A., Hist. found. bot. Florida 36. 1945.
Spaulding, P., Phytopathology 22(11): 873-877. 1932 (obit., portr., bibl., d. 24 Apr 1929).

COMPOSITE WORKS: (1) *Marasmius, in Fl. N. Amer.* 9(4): 250-286. 30 Apr 1915.
(2) *Coprinus, in* Kauffman, *Agaric. Michigan* 1919, see TL-2/3548.

NAME: Pennington himself held that "H" was his full middle name (Barnhart).

Penzig, Albert Julius Otto [later: Alberto Giulio Ottone] (1856-1929), German (Silesian) botanist, from 1878 in Italy, Italian citizen 1882; Dr. phil. Breslau 1877; at Pavia with Garovaglio 1878-1879; with Saccardo at Padua 1879-1883; at the Stazione agronomica, Modena 1883-1886; professor of botany and director of the botanical garden, Genova 1887 (Jan)-1929; in Java Nov 1896-Apr 1897. (*Penz.*).

HERBARIUM and TYPES: GE (partly destroyed); other material at B, BO, DBN, DOMO, G, GE, GH, L, MOD, PAD, PAV, PR, ROPV, W, Z. – Letters and portrait at G.

BIBLIOGRAPHY and BIOGRAPHY: Ainsworth p. 231, 322, 346; Barnhart 3: 66; BFM no. 2383, 3243; BJI 2: 44, 133; BL 1: 35, 113, 312, 2: 333, 336, 357, 362, 363, 368, 480, 701; BM 4: 1545; Bossert p 305; CSP 10: 1022-1023, 12: 566, 17: 781; De Toni 2: xcii; GR p. 436-437, cat. p. 69; Hawksworth p. 185, 197; Herder p. 117, 167, 244, 327, 329; Hirsch p. 228; IH 1(ed. 6): 362, (ed. 7): 339, 2: (in press); Jackson p. 89, 108, 320, 493; Kelly p. 173; Kew 4: 269-270; LS 20407-20426, suppl. 21239; Moebius p. 305; Morren ed. 10, p. 85; NI 1504-1505, see also 1: 20, 150, 244, 2: 179, suppl. p. 13; Plesch p. 359; Rehder 5: 659; Saccardo 1: 124, cron. p. xxix; SK 1: 403 (itin., portr.), 4: cxxiv, 8: lxxiv; Stevenson p. 1254; TL-1/169; TL-2/861; Tucker 1: 547-548; Urban-Berl. p. 310, 382.

BIOFILE: Anon., Biogr. Jahrb. 11: 363 (b. 25 Mar 1856, d. 6 Mar 1929); Bot. Centralbl. 1/2: 96. 1880; 29: 63. 1887 (app. Genova); 48: 366. 1891 (return from Ethiopia), 70: 399. 1897 (app. member Ist. Veneto); Bot. Jahrb. 8(Beibl. 18): 1. 1887; Bot. Zeit. 38: 190. 1880 (asst. Padua), 45: 62. 1887 (app. Genova); J. Bot. 69: 88. 1931; Nat. Nov. 5: 43. 1883 (app. Modena), 9: 40. 1887 (app. Genova), 18: 532. 1896 (to Buitenzorg, Singapore, Ceylon); Nuova Notarisia 1890: 138 (app. Genova).; Österr. bot. Zeit. 33: 101, 1883 (app. Modena), 37: 72. 1887 (app. Genova), 47: 343. 1897 (return from Java).
Béguinot, A., Archivio bot. 5(1): 108-109. 1929, 6:(1): 60-88. 1930 (biogr., bibl., more careful than other bibliographies), Ber. deut. Ges. 47: (96)-(102). 1929 (obit.).
Burnat, E., Bull. Soc. bot. France 30: cxxvi. 1883.
Cavillier, F., Boissiera 5: 70-71. 1941.
Ciferri, R., Taxon 1: 126, 127. 1952 (coll. in Italian herb.).
Kneucker, A., Allg. bot. Z. 3: 152. 1897 (plants from Java at GE).
Laundon, J.R., Lichenologist 11: 15. 1979 (cites lichens from PAD and G).
Polacci, G., Atti Ist. bot. G. Brioso Pavia ser. 4. 4: i-x. 1933 (obit., portr., bibl.).
Savelli, R., Nuovo Giorn. bot. ital. ser. 2. 37(4): 756-788. 1930 (biogr., portr., bibl., b. 25 Mar 1856, d. 6 Mar 1929).
Sirks, M.J., Ind. natuurond. 200. 1915.
Verdoorn, F., ed., Chron. bot. 1: 37. 1935, 5: 299. 1939.
White, A. & B.L. Sloane, The Stapeliae 32. 136. 1933, ed. 2. 121, 133, 857, 981. 1937.

COMPOSITE WORKS: (1) Edited P. Bubani, *Flora pyrenaea* 1897-1901, 4 vols., see TL-2/861, further information: *vol. 1*: publ. Sep-Oct 1897 (Bot. Centralbl. 3 Nov 1897; ÖbZ Nov 1897; rev. by L. Diels, Bot. Centralbl. 74: 283-284. 1898), *vol. 2*: publ. Dec 1899 (Bot. Centralbl. 14 Feb 1900), *vol. 3*: publ. Mar 1901 (Bot. Centralbl. 27 Mar 1901, 24 Apr 1901), *vol. 4*: publ. Jan-Feb 1902 (ÖbZ Feb-Mar 1902).

THESIS: *Untersuchungen über Drosophyllum lusitanicum Lk.*, Breslau 1877, 46 p. (Bot. Zeit. 19 Oct 1877).

EPONYMY: *Penzigia* P.A. Saccardo (1888); *Penzigiella* M. Fleischer (1906); *Penzigina* O. Kuntze (1891).

7641. *Il monte generoso* schizzo geografica botanica dal dott. Otto Penzig. Pavia (Tipografia dei fratelli Fusi) 1879. Oct.

Publ.: 7 Apr 1879 (in journal; Flora 11 Jun 1879; ÖbZ 1 Aug 1879; Nat. Nov. Aug(1) 1879), p. [3]-23. *Copy*: G. – Reprinted and to be cited from Nuov. Giorn. bot. ital. 11: 129-147. 7 Apr 1879.

7642. *Funghi agrumicoli.* Contribuzione allo studio dei funghi parassiti degli agrumi ...Padova (Tip. del Seminario, Lit. P. Fracanzani) 1882. Oct. (*Fung. agrum.*).
Publ.: Jul 1882 (t.p.; Nat. Nov. Nov(1) 1882; Bot. Zeit. 29 Dec 1882), p. [i], [1]-124, *pl.* [in all 136] *1121-1224, 1144A-D, 1148A-D, 1184A-D, 1188A-D, 1192A-D, 1196A-D, 1200A-D, 1204A-D*. *Copies*: FH, FI, G, NY, Stevenson. – Plates crudely coloured lithographs; the numbering links up with Saccardo, P.A., Fungi italici; the A-D numbers are extra plates not included in Saccardo. Text reprinted from Michelia 2: 385-508. 1882.
Seconda contribuzione allo studio dei funghi agrumicoli: Apr-Mai 1884 (p. 2: Jan 1884; rd. 23 Mar 1884; Bot. Zeit. 27 Jun 1884), in series *Note micologiche*, p. [1]-28, *5 pl.* (uncol.). *Copies*: FH, FI, NY. – Reprinted from Atti r. Ist. Ven. Sci. ser. 6. 2: 665-692. 1884. Imprint reprint: Venezia (Tipografia di G. Antonelli) 1884.
Ref: Penzig, O., Bot. Centralbl. 14: 80-81. 16-20 Apr 1883 ("Autoreferat").

7643. Note micologiche del professore O. Penzig. *Funghi della Mortola* (con 2 tavole). Venezia (Tipografia di G. Antonelli) 1884. Oct. (*Funghi Mortola*).
Publ.: Apr-Mai 1884 (rd 23 Mar 1884; Bot. Zeit. 29 Jun 1884; Nat. Nov. Jun(2) 1884; Hedwigia Jul 1884), p. [1]-25, [26], *pl. 4-5*. *Copies*: FH, FI, NY. – Reprinted from Atti r. Ist. ven. Sci. ser. 6. 2: 639-663. 1884. – Continuation from ib. ser. 6. 2: 595-597. 1884 (read Feb 1884) (no separately paged reprint known to us).

7644. Note micologiche del professore O. Penzig. *Appunti sulla flora micologia del Monte Generoso.* Venezia (Tipografia di G. Antonelli) 1884. Oct. (*Appunti fl. micol. Mt. Generoso*).
Publ.: 1884 (ÖbZ 1 Sep 1884), p. [1]-21. *Copies*: FH, FI, NY. – Reprinted from Atti r. Ist. ven. Sci. ser. 6. 2: 577-597. 1884.

7645. *Studi botanici sugli agrumi* e sulle piante affini ... Roma (Tipografia eredi Botta) 1887. Oct. (with atlas, Fol.). (*Stud. bot. agrum.*).
Text: 1887 (p. vi: Mar 1887; Bot. Zeit. 30 Sep 1887; Nat. Nov. Sep(1), Oct(2) 1887; J. Bot. Nov 1887), p. [i]-vi, [1]-590. *Copies*: BR, FH, FI, G, MO, NY. Published as *Annali di Agricoltura*, 116, 1887 (separate t.p.).
Atlante [*Atlas*]: 1887 (id.), p. [i], *pl. 1-58* with text (uncol., auct.). *Copies*: BR, FI, G, MO, NY, Stevenson. Published as *Annali di Agricoltura*, 116, 1887 (separate t.p.).
Ref.: Duchartre, P., Bull. Soc. bot. France 34 (bibl.): 97-100. Jul-Dec 1887.
Engler, A. Bot. Jahrb. 9(Lit.): 28-29. 30 Dec 1887.
Karsten, G., Bot. Zeit. 46: 227. 6 Apr 1888.

7646. *Pflanzen-Teratologie* systematisch geordnet von Dr. O. Penzig ... Genua (Druck von Angelo Ciminago) 1890. 2 vols. Oct. (*Pfl.-Teratol.*).
Ed. 1: 2 vols., 1890-1894. *Copies*: FI, H, M, PH, U, UC, WU.
1: Oct-Sep 1890 (ÖbZ Oct 1890, Nat. Nov. Nov(1) 1890; Bot. Zeit. 26 Dec 1890; J. Bot. Jan 1891), p. [iii]-xx, [1]-540.
2: Jul-Aug 1894 (p. vii: Jun 1894; Bot. Zeit. 1 Sep 1894; ÖbZ Jul 1894), p. [i]-vii, [1]-594.
Ed. 2: 1921-1922, 3 vols. Berlin (Verlag von Gebrüder Borntraeger, ...) Oct. *Copies*: B, BR, FI, UC; IDC 7383.
1: 1921, p. [i]-xvi, [1-2], [i]-xi, [1]-283 (p. 1-160 Nat. Nov. Feb 1921, ÖbZ: 161-283, Mai-Jun 1921).
2: 1921, p. [i-ii], [1]-548 (p. 1-160 Nat. Nov.: Feb 1921; ÖbZ: 161-310. Mai-Jun 1921, 320-548. Jan 1922).
3: 1922, p. [i-ii], [1]-624 (Nat. Nov.: p. 1-160. Mar 1922, p. 161-320. Mai 1922, p. 320-480. Jul 1922, p. 480-624. Aug/Sep 1922).
Supplément (in French), 164 p. 1923 fide ÖbZ 72: 367. 1 Oct 1923.
Ref: Magnus, P., Bot. Zeit. 49: 28-29. 1891, 52(2): 280-281. 1894.
Rendle, A.B., J. Bot. 60: 31. Dec 1922 (rev. vols. 1-3).

7647. *Considérations générales sur les anomalies des Orchidées* ... Cherbourg (Imprimerie Émile Le Maout) 1894. Oct.
Publ.: 1894, p. [1]-28. *Copy*: G. – Reprinted and to be cited from Mém. Soc. natl. Sci. nat. Cherbourg 29: 79-104. 1894.

7648. *Diagnoses fungorum novorum* in insula Java collectorum. Series prima [-tertia]. Genova (Tipografia di Angelo Ciminago ...) 1897 [-1902]. Oct.
Co-author: Pier Andrea Saccardo (1845-1920).
1: 1897, p. [1]-27 [28]. Reprinted and to be cited from Malpighia 11 (9-10): 387-409. 1897.
2: 1898 (p. 490 journal: 20 Feb 1898; Nat. Nov. Apr 1898), p. [1]-42. Ib. 11(11-12): 491-530. 1898.
3: 1902 (t.p. reprint; Nat. Nov. Jan(1) 1902, Bot. Centralbl. 29 Jul 1902), p. [i], [1]-60. [1]. Ib. 15(7-9): 201-260. 1902.
Copies: BR (no. 3 only), FH, FI, G, NY (2 sets), USDA (no. 3 only).

7649. Res ligusticae xxvii. *Florae ligusticae synopsis* ... Genova (tipografia r. Istituto Sordo-Muti) 1897. Oct.
Orig. ed.: 28 Oct-6 Nov 1897 (in journal; Bot. Zeit. 16 Jun 1898; Nat. Nov. Apr(1) 1898), p. [1]-11. *Copies*: G, MO. – Reprinted and to be cited from Ann. Mus. civ. Storia nat. Genova ser. 2. 18(38): [423]-531 [=p. 3-111]. 28 Oct-6 Nov 1897.
Suppl.: Oct-Dec 1925, p. [1]-30. *Copy*: G. – Reprinted from Archivio bot. 1(3): 187-204, 1(4): 246-259. 1925.

7650. *Flora popolare ligure*. Primo contributo allo studio dei nomi volgari delle piante in Liguria. Genova (tipografia di Angelo Ciminago ...) 1897. (*Fl. pop. ligure*).
Publ.: 1897 (Nat. Nov. Mai(1) 1898; Bot. Centralbl. 18 Mai 1898; Bot. Zeit. 16 Jun 1898), p. [1]-101, [102]. *Copies*: G(2), NY. Reprinted from Atti Soc. ligust. Sci. nat. geogr. 8(3): 249-308, 8(4): 341-379. 1897.

7651. *Amallospora* nuove genere di Tuberculariee ... Genova (Tipografia di Angelo Ciminago ...) 1898. Oct.
Publ.: 1898 (p. 460 of journal dated 1 Nov 1897), p. [1]-6, [7], *pl. 10* (uncol.). *Copies*: G, NY. Reprinted, and to be cited from Malpighia 11 (11-12): 461-464, *pl. 10*. 1897, publ. 1898.

7652. *Die Myxomyceten der Flora von Buitenzorg* ... Leiden (Buchhandlung und Druckerei vormals E.J. Brill) 1898. Oct. (*Myxomyc. Fl. Buitenzorg*).
Publ.: 1898 (p. 7: Jan 1898; Bot. Zeit. 1 Aug 1898; Bot. Centralbl. 10 Aug 1898; Nat. Nov. Jul(1, 2), Aug(1) 1898), p. [2]-83. *Copies*: BR, FH, G, MO, NY, PH, Stevenson. – Alt. t.p.: "*Flore de Buitenzorg ... 2me partie. Myxomycètes ...*".

7653. *Note sul genere Mycosyrinx*. Genova (tipografia di Angelo Ciminago ...) 1900. Oct.
Publ.: 1900 (t.p. reprint), p. [1]-13, *pl. 19-20* (uncol.). *Copies*: BR, NY. Reprinted and to be cited from Malpighia 13: 522-532, *pl. 19-20*. 1899, publ. 1900.

7654. *Flore* coloriée de poche *du littoral méditerranéen* de Gênes à Barcelone y compris la Corse. 139 planches coloriées et 5 planches noires représentant 144 espèces ... Paris (Librairie des Sciences naturelles Paul Klincksieck ...) 1902. 16-mo. (*Fl. litt. médit.*).
Artists: Mme B. Hérincq (*fl.* 1900), Charles Émile Cuisin, & Jobin.
Publ.: Feb-Mar 1902 (p. vii: Feb 1902; ÖbZ Feb-Mar 1902; Nat. Nov. Apr(2) 1902; Bot. Centralbl. 20 Aug 1902), p. [i]-vii, *pl. 1-144* (2 pl., drawings by Cuisin, Mme Hérincq, Jobin, 139 col., 5 uncol.) with text 1[verso of vii]-144, 146-161. *copies*: B, BR, FI, G, MO, NY. – Bibliothèque de poche du naturaliste, xiii.

7655. *Icones fungorum javanicorum* ... Leiden (Buchhandlung und Druckerei vormals E.J. Brill) 1904. Oct. (*Icon. fung. jav.*).
Co-author: Pier Andrea Saccardo (1845-1920).
Text: Jan-Apr 1904 (Bot. Zeit. 1 Jan 1904 ann., 1 Jun 1904 publ.; Bot. Centralbl. 3 Mai 1904; ÖbZ Mar-Apr 1904; Nat. Nov. Mai(2) 1904), p. [i-vi], [1]-124.

Plates: Jan-Apr 1904 (id.), p. [i], *pl. 1-80* (uncol.; photozincotypes).
Copies: B, BR, MO, NY, Stevenson, USDA; IDC 6293. A prospectus by Brill, dated Aug 1903, promised the book for the end of the year 1903; Bot. Zeit. announced it on 1 Jan 1904, but gave the precise details only on 1 Jun 1904; Nat. Nov. idem in Dec(1) 1903 and Mai(2) 1904 respectively. Some copies have partly coloured plates.
Refs.: Fischer, Ed., Bot. Zeit. 62(2): 356. 1 Dec 1904.

7656. *Flora popolare italiana* raccolta dei nomi dialettali delle principali piante indigene e coltivate in Italia ... Genova (A cura dell'autore Orto botanica della R.a. Università) 1924. Oct. (*Fl. pop. ital.*).
1: 1902 (Nat. Nov. Mai 1925; MO rd. 14 Mai 1925; BR rd. 4 Jul 1925; ÖbZ Jun-Sep 1925), p. [i]-xv, [1]-541.
2: 1924 (id.), p. [1]-615.
Copies: BR, G, MO, USDA. – See BL 2: 336 for a note on the coverage of the various dialects; see also Merlo, C., L'Italia dialettale 1: 273-275. 1925.
Reprint (facsimile): 1972, Bologna Edagricole [Edizione Agricole] (printing finished Aug 1972), 2 vols. *Copy*: B.
1: [i]-xvi, facs. [i]-xv, [1]-541.
2: [i-iv], facs. [1]-615.

Peola, Paolo (1869-?), Italian palaeobotanist; teacher at the Genova Technical Institute. (*Peola*).

HERBARIUM and TYPES: Unknown.

BIBLIOGRAPHY and BIOGRAPHY: Andrews p. 312; Barnhart 3: 66; Bossert p. 305; CSP 17: 782; Hirsch p. 228; Saccardo 1: 124.

7657. *Flora fossile braidese* ... Bra [Cuneo, Piedmont] (Tipografia Stefano Racca ...) 1895. (*Fl. foss. braid.*).
Publ.: 1895 (Bot. Centralbl. 22 Jan 1896; Bot. Zeit. 16 Feb 1896), p. [1]-128, [1-6, index; 1, cont.]. *Copy*: USGS.

7658. *Flora del Langhiano torinese* ... Bologna (tipografia Gamberini e Parmeggiani) 1899. Oct.
Publ.: 31 Dec 1899 (date on cover reprint), p. [1]-16. *Copy*: FI. – Reprinted and to be cited from Riv. ital. paleont. 5(4): 95-108. 1899.

Pepoon, Herman Silas (1860-1941), American botanist; head instructor in botany and agriculture, Lake View High School, Chicago, Ill. (*Pepoon*).

HERBARIUM and TYPES: ILLS; further material at F, ILL, ISM, MICH, MSC.

BIBLIOGRAPHY and BIOGRAPHY: Barnhart 3: 67; BL 1: 175, 312; IH 2: (in press); Pennell p. 616.

BIOFILE: Deam, C.C., Fl. Indiana 1118. 1910.
Verdoorn, F., ed., Chron. bot. 1: 319. 1935, 7: 233, 278. 1943.

7659. *Studies of plant life* a series of exercises for the study of plants ... Boston, U.S.A. (D.C. Heath & Co., ...) 1900. Oct. (*Stud. pl. life*).
Co-authors: Walter Reynoldo Mitchell (1861-?), Fred Baldwin Maxwell (1862-1907).
Publ.: Jul-Dec 1900 (p.v.: Jun 1900, Bot. Centralbl. 23 Jan 1901; TBC 18 Apr 1901), p. [i]-xii, [1]-95. *Copy*: NY.

7660. *An annotated flora of the Chicago area* with maps and many illustrations from photographs of topographic and plant features ... Chicago, Illinois 1927. (*Annot. fl. Chicago*).
Publ.: 1927 (BR rd. 4 Oct 1927; p. 137, 1 Mai 1927, frontisp.), p. [i]-xxii, [1]-554. *Copies*: B, BR, MICH, NY, US, USDA.
Supplement: C.A. Buhl, Bull. Chicago Acad. Sci. 5(2): 5-12. 31 Dec 1934.
Ref.: Gleason, H.A., Torreya 28: 6-9. 23 Feb 1928.

Peragallo, Hippolyte (1851-?), French diatomologist. (*H. Perag.*).

HERBARIUM and TYPES: AWH and FH; further material at PC, PH, TALE. For *exsiccatae* (*Diatomées du monde entier, Diatomées de France*) see Tempère. – Letters at G.

BIBLIOGRAPHY and BIOGRAPHY: Barnhart 3: 67; BM 4: 1546, 8: 998; CSP 17: 783; De Toni 2: xcii-cxiii, 4: xliv; IH 1 (ed. 6): 362, (ed. 7): 339, 2: (in press); Kew 4: 271; MW p. 383; NI 1506; Saccardo 1: 124-125; TL-1/1291.

BIOFILE: Anon., Bot. Centralbl. 113: 192. 1910. (prix Montagne).
Cavillier, F., Boissiera 5: 71. 1941.
Frison, Ed., Henri Ferdinand van Heurck 47, 58. 1959.
Fryxell, G.A., Beih. Nova Hedw. 35: 363. 1975.
Hanna, G.D., J. Palaeontol. 4(3): 296-297. 1930 (dates publ. Tempère & Peragallo, *Diat. monde* ed. 2, see under Tempère).
Koster, J.W., Taxon 18: 556. 1969.
Merrill, E.D., B.P. Bishop Mus. Bull. 144: 149. 1937; Contr. U.S. natl. Herb. 30: 239. 1947.
Wornardt, W.W., Int. direct. diat. 35, 36. 1968.
VanLandingham, S.L., Cat. diat. 6: 3578-3579, 7: 4215. 1978 (bibl.).

COMPOSITE WORKS: (1) With J.A. Tempère, q.v., *Diatomées du monde entier*, & *Diatomées collection J. Tempère & H. Peragallo*.

EPONYMY: *Peragallia* Schütt (1895).

7661. *Diatomées du midi de la France* notions sommaires sur les diatomées leur récolte, leur préparation, leur examen suivies d'une liste des diatomées récoltées dans la Provence, le Bas Languedoc, la vallée de la Garonne, et les Pyrénées ... Toulouse (Imprimerie Durand, Fillous et Lagarde ...) 1884. Oct. (*Diatom. midi France*).
Publ.: Sep-Dec 1884 (t.p. 1884; Nat. Nov. Mai(1) 1885; Bot. Zeit. 26 Jun 1885), p. [1]-88. *Copies*: BR, G, US; IDC 6389. – Repr. from Bull. Soc. Hist. nat. Toulouse 18: 189-272. Sep-Dec 1884.

7662. Diatomées du midi de la France. *Diatomées de la Baie de Villefranche* (Alpes-maritimes) ... Toulouse (Imprimerie Durand, Fillous et Lagarde, ...) 1888. Oct. (*Diatom. Baie Villefranche*).
Publ.: 1888 (Nat. Nov. Dec(1) 1888), p. [1]-100, *pl. 1-6* (uncol., auct.). *Copies*: BR, US. – Reprinted from Bull. Soc. Hist. nat. Toulouse 22: 13-100, *pl. 1-6*. 1888. – Reprint cover at BR has imprint Paris (Librairie J.-B. Baillière et fils ...) 1888.
Ref.: Möbius, M., Bot. Centralbl. 41: 48-51. 1890 (ext. review).

7663. *Monographie du genre Pleurosigma* et des genres alliés ... (extrait du Diatomiste 1890-1891) à Paris (Chez M.J. Tempère ...) [1891]. Qu. (*Monogr. Pleurosigma*).
Publ.: Mar 1891 (p. 1-16, *pl. 1-5*); Jun 1891 (p. 17-35, *pl. 6-10*, p.[i-iii], [1]-35, *pl. 1-10* with text (uncol.), [i-iii], index]. *Copies*: BR, NY, Teyler. – Issued as an independent publication (with special pagination) with *le Diatomiste* 1(4, 5) see Diatomiste 1: ii (table).
Ref.: Pantocsek, J., Bot. Centralbl., Beih. 2(3): 161-163. 1892 (rev.).

7664. *Diatomées marines de France* et des districts maritimes voisins par MM. H. et M. Peragallo ... Texte. Éditées par M.J. Tempère micrographe-éditeur, à Grez-sur-Loing (S.-et-M.) 1897-1908, text and atlas. Oct. (*Diatom. mar. France*).
Co-author: Maurice Peragallo (1853-?).
Texte: 1897-1908, p. [i*-v*], [i]-xii, tables [i], [1]-48, main text: [i]-iii, [1]-491, [492-493, err.], published as follows:
 Preface material: 1908.
 Tables (Tableaux synoptique & systématique): 1908 ([i], [1]-48).
 Première partie – Rhaphidées: in parts of 16 p.: 1897-1899 (p. iii: 1 Mar 1897; Nat. Nov. Jun(1) 1899 compl.), p. [i]-iii, [1]-236.

Deuxième partie – Pseudo-Raphidées: in parts of 16 p.: 1899-Apr 1901, p. 237-364.
Troisième partie – Anaraphidées: in parts of 16 p.: Mai 1901-1908, p. 365-491.
Atlas: 1897-1908, p. [i-ix], *pl.* (with text) *1-118, 119*, 119-124, 124a, 125-137* (uncol. or *(30-36, 121, 122, 131, 137)* col., auct.) (title pages sometimes placed all in front [they were issued simultaneously in 1908] or those for parts 2 and 3 precede *pl. 51* and *90*). The issue of the plates was irregular, and took place independently of the text, including provisional issues. The definitive sequence as given in the preface material is:
Copies: FH, G, NY, US; IDC 6391.

1-8	Feb 1897	*81-88*	Jan 1901
9-16	Mai 1897	*89-96*	Apr 1901
17-24	Oct 1897	*97-104*	Feb 1902
25-32	Jan 1898	*105-110, 112, 113*	Oct 1902
33-40	Apr 1898	*113, 114-119*	Apr 1908
41-48	Nov 1898	*120-124, 136, 137*	Aug 1907
49-56	Jan 1899	*124a-127*	Feb 1904
57-64	Sep 1899	*128-137*	Aug 1904
73-80	Apr 1900		

Facsimile ed.: 1965, Amsterdam (A. Asher & Co.), 2 vols., ISBN 90-6123-212-0, as above with extra t.p.'s dated 1965 preceding material as listed above. *Copies*: BR, FAS, PH.
Note: A publisher's announcement of Mai 1901 states that publication took place in fascicles of 4 plates with text and 16 pages of text; this is confirmed by notices in the contemporary literature.
Ref.: Junk, Rara 2: 207. 1926-1936.

Peragallo, Maurice (1853-?), French diatomologist; brother of H. Peragallo. (*M. Perag.*).

HERBARIUM and TYPES: See H. Peragallo.

BIBLIOGRAPHY and BIOGRAPHY: Barnhart 3: 67; BM 4: 1546; Kew 4: 271; TL-2/2675, 2677.

BIOFILE: Cavillier, F., Boissierra 5: 71. 1941.
VanLandingham, S.L., Cat. diat. 6: 3578-3579, 7: 4215. 1978.

7665. *Catalogue général des Diatomées* par Maurice Peragallo. Clermont-Ferrand 5. Cours Sablon (1897), 2 vols. (*Cat. gén. Diatom.*).
1: 1897 (t.p.), p. [i*], i-xviii, 1-471, [1, err. in copy B; 1-12 add. in copy NY].
2: 1903 (?, fide VanLandingham), p. 472-973.
Copies: B, NY. – BR has a prospectus consisting of p. iii-vi, 1-14. – Lithographed manuscript, printed on one side; effectively published. See Junk, Rara p. 158 for a copy with 2 p. addenda, and Asher, Cat. 14, no. 141. (1974) for a copy dated 1900 with 11 p. add.

7666. *Deuxième expédition antarctique française* (1908-1910) commandée par le Dr. Jean Charcot. Sciences naturelles; documents scientifiques. Botanique *Diatomées* d'eau douce et Diatomées d'eau salée par le commandant Maurice Peragallo ... Paris (Masson et Cie, ...) 1921. Qu. (*Deux. exped. antarct. franç., Diatom.*).
Publ.: 1921, Nat. Nov. Mar 1921, p. [i-iv], [1]-98, *pl. 2-6* (uncol., auct.). *Copy*: FH.

Pérard, Alexandre Jules César (1834-1887), French botanist; bank-clerk, later high school teacher at Montluçon. (*Pérard*).

HERBARIUM and TYPES: AUT; further material at BR, G, MW. – A "Collection de Menthes" was taken from the original herbarium, and sold by P's daughter to G (via Briquet) in Mai 1892.

BIBLIOGRAPHY and BIOGRAPHY: Backer p. 658 (err.); Barnhart 3: 67; BL 2: 117, 134, 701;

CSP 8: 587-588, 10: 1024, 17: 784; GR p. 344; Hortus 3: 1201 (Pérard); IF p. 723; Jackson p. 287; Kelly p. 173; Kew 4: 271; Krempelh. 3: 13, 69; LS 20428-20429; Rehder 1: 414; TL-2/5985; Tucker 1: 548; Zander ed. 10, p. 700, ed. 11, p. 799.

BIOFILE: Anon., Bot. Zeit. 46: 708. 1888 (library for sale); Bull. Soc. bot. France 34(1): 415. 1887 (d.); Bull. Soc. Hist. nat. Autun 24(2): 180-182. 1911 (on herbarium); Rev. bryol. 14: 80. 1887.
Berthier, B., Bull.Soc. Hist. nat. Autun 24 (proc.-verb.): 100-103. 1911 (Pérard's herb. given to society by Moriot; brief biogr. note and bibl.; b. 12 Jun 1834, d. 17 Jun 1887).
Klincksieck, P., Bot. Zeit. 46: 708. 1888 (sale library).
Lindemann, E. v., Bull. Soc. imp. Natural. Moscou 61(1): 54. 1886 (130 specimens at MW).
Miège & Wuest, Saussurea 6: 131, 1975 (*Mentha* at G).
Moriot, J., Notice sur Alexandre-Jules-César Pérard, 8 p., Bizeneuille, 16 Oct 1888 (biogr., bibl., b. 12 Jun 1834, d. 15 Jun 1887) (pamphlet bound with the copy at G of the FI Bourbonnais part 1).

COMPOSITE WORKS: with A. Migout, *Excursion botanique dans les montagnes du Bourbonnais*, 1881, see TL-2/5985.

7667. *Catalogue raisonné des plantes* croissant naturellement ou soumises à la grande culture dans l'arrondissement *de Montluçon* (Allier) avec une notice sur la distribution des végétaux dans cette contrée, des observations sur les fougères, un essai de classification des menthes de la flore française et une étude anatomique de l'Agropyrum caesiam Presl ... Pari (F. Savy, ...) 1869-1871. Oct.
Publ.: in Bull. Soc. bot. France 16, 17 and 18 and to be cited from journal.
Reprint: (with above title): 1871, p. [i-vii], [1]-248, *pl. 3*. *Copies*: B, G, NY, USDA.
Journal publ.: Bull. Soc. bot. France:
16: 154-164, 177-194, 241-242, 255-269, 303-310, 346-354, 394-403, 448-450. 1869.
17: 28-38, 62-73, 96-104, 114-122, 132-142, 156-163, 198-207, 331-353, 353-358, 358-363, 363-384. 1870.
18: 272-283, 318-330, 382-389, 433-436, 436-439, *pl. 3*. 1871.
Supplément: 1878 (p. 29: 3 Jan 1878; Nat. Nov. Apr(2) 1879, p. [i], [1]-29, *pl. 3* (col. lith auct.), [only in B copy, cf. main publ.]. *Copies*: B, G(2). "Supplément du catalogue raisonné des plantes de l'arrondissement de Montluçon (Allier) avec une liste de quelques Menthes nouvelles ou peu connues ..." Montluçon (Prot., ...), Paris (Coccoz, ...) 1878.

7668. *Classification du genre Mentha* ... Moulins (Imprimerie de C. Desroziers). Oct.
Publ.: 1876 (or 1877?) (mss. submitted 28 Jul 1876), p. [1]-60, [1]. *Copies*: G, MO. – Reprinted and to be cited from Bull. Soc. Émul. Allier (séance 28 Jul 1876).

7669. *Revue monographique du genre Mentha* ... no. 1. Montluçon (Imprimerie Prot (breveté)), ... 1878. Oct. †. (*Rev. monogr. Mentha*).
Publ.: 15 Feb-15 Apr 1878 (p. 2: 15 Feb 1878; Nat. Nov. Apr(2) 1879), p. [i], [1]-27. *Copies*: G(2), MO.

7670. *Flore du Bourbonnais* comprenant le département de l'Allier et une partie des départements du Cher, de la Creuse, du Puy-de-Dôme et de la Nièvre ... Montluçon (Ch. Moulin, ...) 1884-1886. Oct †. (*Fl. Bourb.*).
1: 1884 (p. 2: 1 Sep 1884; Nat. Nov. 1 Sep 1885; Bot. Centralbl. 14-18 Sep 1885), p. [i], [1]-112.
2: 1886 (Nat. Nov. Jun(3) 1887; Bot. Zeit. 29 Jul 1887), p. [i], [1]-48.
Copy: G.

Pereboom, Cornelis (*fl.* 1788), Dutch physician and botanist at Amsterdam. (*C. Pereb.*).

HERBARIUM and TYPES: Unknown.

BIBLIOGRAPHY and BIOGRAPHY: BM 4: 1546; Herder p. 91; PR 7033, 9499 (ed. 1: 7866); Rehder 1: 254; Tucker 1: 548.

7671. *Systema characterum plantarum*, seu dictionarium rerum botanicum: filio conscriptum, et ab ipso figuris illustratum: autore Cornelio Pereboom ... Lugduni Batavorum [Leiden] (prostat apud Samuelem & Joannem Luchtmans bibliopolas.) 1788. Qu. (*Syst. charact. pl.*).
Publ.: 1788, p. [i-viii], 1-311, [312, index]. *Copies*: G, NY, USDA.

Pereboom, Nicolaas Ewoud (*fl.* 1787), Dutch botanist, son of Cornelis Pereboom. (*N. Pereb.*).

HERBARIUM and TYPES: Unknown.

BIBLIOGRAPHY and BIOGRAPHY: BM 4: 1546; Dryander 3: 27; NI 1507; PR 7034 (ed. 1: 7867); Rehder 1: 70; SO 721a.

7672. *Materia vegetabilis*, systemati plantarum, praesertim philosophiae botanicae, inserviens: characteribus, quoscunque ill. Linnaeus indicavit, delineatis ... Lugduni Batavorum [Leiden] (apud S. & J. Luchtmans) 1787-1788, 3 Dec. Qu. (*Mat. veg.*).
1: 1787, p. [i]-viii, 9-32, *pl. 1-10* (uncol., copp., auct.).
2: 1787, p. [i-iii], 33-52, *pl. 11-20* (id.).
3: 1788, p. [i-iii], 53-74, *pl. 21-30* (id.).
Copies: B(1-3), G(1-3), MO(2, 3), NY(1-3), USDA(1-3).

Pereira, Huascar (x-1926), Brazilian botanist. (*H. Pereira*).

HERBARIUM and TYPES: Unknown.

BIBLIOGRAPHY and BIOGRAPHY: BL 1: 242, 312.

EPONYMY: *Pereiria* Lindley (1838) honors Jonathan Pereira (1804-1853), English pharmacologist.

7673. Pequeña contribução para um *diccionario das plantas uteis* do estado de S. Paulo (indigenas e aclimadas) ... São Paulo (Typographia Brasil de Rothschild & Co., ...) 1929. (*Dicc. pl. uteis*).
Publ.: 1929 (p. 5: 8 Mar 1926), p. [1]-779. *Copy*: USDA.

Pérez Arbeláez, Enrique (1896-1972), Colombian botanist; Dr. phil. Burgos 1918; Dr. theol. ib. 1926; Dr. Sci. München 1928. (*Perez Arb.*).

HERBARIUM and TYPES: COL; further material at CAS, MA, MEDEL, NY, PH, US. – Archival material in Smithsonian Archives.

BIBLIOGRAPHY and BIOGRAPHY: Barnhart 3: 68; BL 1: 249, 250, 312; Bossert p. 306 (b. 17 Mar 1896); IH 1 (ed. 7): 339, 2: (in press); Kew 4: 273; MW 383; NI suppl. p. 7; TL-2/see J.C.B. Mutis y Bosio.

BIOFILE: Arango, M., Orquideologia 7(1): 53, 55-56. 1972 (obit., portr.).
Perry, Rev. Acad. Colomb. C. 14: 69. 1972 (1973) (obit., portr.).
Pinto, P., Caldasia 11: 6-7. 1975 (obit., b. 1 Mar 1896, d. 22 Jan 1972).
Verdoorn, F., ed., Chron. bot. 6: 172, 457, 458. 1941.

7674. *Die natürliche Gruppe der Davalliaceen* (Sm.) Kfs. Unter Berücksichtigung der Anatomie und Entwicklungsgeschichte ihres Sporophyten ... Jena (Verlag von Gustav Fischer) 1928. Oct. (*Nat. Gruppe Davall.*).
Publ.: 1928, p. [1]-96. *Copies*: G, NY. – Issued as Bot. Abh., Goebel 14, 1928.

7675. *Plantas útiles de Colombia* tomo i generalidades, criptogamas, gimnospermas y

monocotiledoneas. Bogota (Imprenta Nacional) 1935. Oct. †. *(Pl. útil. Colombia).*
Publ.: 1936 (cover; t.p.: 1935), p. [i], [1]-172. *Copies*: MO, US.
Ed [*2*]: 1947 (G copy rd. 11 Dec ; colo. 1 Nov 1946), p. [i*-iii*], [i]-iv, frontisp. [1]-537, [1, colo.], *pl. 1-80. Copies*: G, NY. – Large part of the stock destroyed by fire. – "Plantas útiles de Colombia ensayo de botánica colombiana aplicada" Bogota (... Imprenta Nacional) 1947. Oct.
Ed. 3: 1956(p. 833: printing finished Mai 1956), p. [1-2], front., [3]-831, [832], [1, colo. Mai 1956], *pl. 1-44. Copies*: BR, G, T. Cleef. – Tercera redaccion muy corregida ... Madrid (Sucesores de Rivadeneyra (S.A.) ..., Bogotá (... Camacho Roldan ...) 1956. Oct. – On p. 6 a bibliography of the author.

7676. *Plantas medicinales más usadas en Bogotá* [suplemento al Boletin de Agricultura, Bogota] 1934. Oct. *(Pl. med.).*
Publ.: Mar 1934, p. [1]-112. *Copy*: NY. – Issued as supplement to the Bot. Agric., Bogota, 32, 1934.

7677. *Plantas medicinales y venenosas de Colombia* estudio botanico, etnico, farmaceutico, veterinario y forense. Bogota (Editorial Cromos) 1937. Oct. *(Pl. med. venen. Colombia).*
Publ.: 1937, p. [1]-295. *Copy*: MO.

Pérez Lara, José María (1841-1918), Spanish botanist, studied at Sevilla; B.A. 1862; expelled from the University 1863; henceforth living on his ample private means at Jerez. *(Pérez Lara).*

HERBARIUM and TYPES: MAF; some further material at W.

BIBLIOGRAPHY and BIOGRAPHY: Barnhart 3: 68; BL 2: 486, 701; BM 4: 1547; Colmeiro 1: cxciii; CSP 10: 1025, 17: 788; IH 1 (ed. 7): 339; Rehder 1: 422; Tucker 1: 548-549.

BIOFILE: Galiano, E.F., Anal. Inst. bot. Cavanilles 32: 693-698. 1975 (portr.).
Willkomm, M., Bot. Centralbl. 31: 14-16. 1887 (rev. Fl. gadit. 1), 38: 796. 1889 (2), 39: 231-232. 1889 (3).

HANDWRITING: Anal. Inst bot. Cavanilles 32: 695. 1975.

EPONYMY: *Perezlaria* T. Delevoryas et R.E. Gould (1971) is named for Ing. J. Pérez Larios, a present-day geologist in Mexico; *Perezia* Lagasca (1811), and likely also the derived generic name *Pereziopsis* J.M. Coulter (1895), are named for Laurentio Perez, a 15th century pharmacist and traveller from Toledo.

7678. *Florula gaditana* seu recensio celer omnium plantarum in provincia gaditana hucusque notarum ... Madrid (Imprenta de Fortanet, ...) 1886-1896, 5 parts. Oct.
1: 30 Sep 1886 (in journal; Nat. Nov. Jul(1) 1887; Bot. Centralbl. 4-8 Jul 1887), p. [1]-127. – Reprinted and to be cited from Anal. Soc. Esp. Hist. nat. 15(2): [349]-475. 1886. Journal issue with double pagination.
2: 30 Sep 1887 (in journal; Bot. Centralbl. 17 Jun 1889), p. [129]-228. – Ib. 16(2): [273]-372. 1887.
3: 1889 (Bot. Centralbl. 20 Aug 1889; Nat. Nov. Sep(2) 1889; Bot. Jahrb. 25 Apr 1890), p. 230-337. – Ib. 18: [35]-143. 1889.
4(1): 30 Apr 1891 (in journal), p. [339]-410. – Ib. 20: [23]-94. 1891 (see also p. 2-3).
4(2): 1892 (in journal), p. [411]-500. – Ib. 21: [191]-280. 1892.
5(1): 15 Mai 1896 (in journal), p. [501]-557. – Ib. 24: [279]-335. 1896.
5(2): 15 Nov 1896 (in journal), p. [559]-608. – Ib. 25: [173]-222. 1896.
5(3): 31 Jul 1898 (in journal), p. [611]-682. – Ib. 27: [21]-92. 1898.
Addenda: Mem. Soc. Esp. Hist. nat. 2(1): [5]-62. 1903. – Later add.: Font Quer, P., Bol. Soc. Esp. Hist. nat. 27: 39-64. 1927.
Copies: G. – Journal publ.: BR, US.

Perini, Carlo (1817-1883), Italian physician and botanist; Dr. med. Padua; practicing at Triest. *(Perini).*

HERBARIUM and TYPES: sets of Veneto-trentine plants with those of his brother Agostino in FI; other material at BASSA, BR, PAD and PI. *Exsiccatae: Flora tridentina exsiccata* [1854], 300 specimens, n.v., for details see DTS 1: 222.

BIBLIOGRAPHY and BIOGRAPHY: Barnhart 3: 68; BL 2: 336, 701; BM 4: 1548; Bossert p. 306; DTS 1: 221; Herder p. 179; Jackson p. 269, 317; NI 1508, see also 1: 249; PR 7039; Rehder 1: 424, 427; Saccardo 1: 125, 2: 83.
Perini, C. 1240
BIOFILE: Fischer, E., Gutenberg Jahrb. 1933: 211 (no. 85).
Hausmann, F. v., Fl. Tirol 1164-1165. 1854.
Heufler, L. v., Flora 26: 596. 1843.
Parlatore, F., Giorn. bot. ital. anno 1, 1(3): 622. 1845.

7679. *Flora dell'Italia settentrionale e del Tirolo meridionale rappresentata colla fisiotipia.* (*Fl. Italia sett.*).
Co-author: Agostino Perini (1802-1878), brother of Carlo P.
Publ.: in fascicles of *10 pl.* each from 1854-1865; *386 pl.* in three centuries; the fourth century remained incomplete (see Fischer 1933). The plates are nature prints. NI cites a copy with 386 plates in the Bibl. civica, Trento, noting that other copies have mostly only 300 pl.
Ref.: Fischer 85.
Anon., Flora 38: 656. 7 Nov 1855 (cent. 1).
Schlechtendal, D.F.L. von, Bot. Zeit. 13: 38-39. 12 Jan 1855 (cent. 1), 14: 623-624. 29 Aug 1856.

Perkins, Mrs. E.E. (*fl.* 1830-1840), British botanist at Chelsea; "professor of botanical flower painting" (t.p. Elem. bot.). (*E. Perkins*).

HERBARIUM and TYPES: Unknown.

BIBLIOGRAPHY and BIOGRAPHY: Barnhart 3: 69; BB p. 242; Desmond p. 490; Herder p. 95; Jackson p. 39; NI 619; PR ed. 1: 7873.

BIOFILE: Allibone, S.A., Crit. dict. Engl. lit. 1563. 1870.

7680. *The elements of botany*, with illustrations ... London (Thomas Hurst, ...) 1837. Oct. (*Elem. bot.*).
Publ.: 1837 (dedications: 15 Feb 1837), p. [i-ii], *pl. 17* as frontisp., [iii]-xxxiii, [xxxiv], [1]-268, *pl. 1-16, 18* (uncol., auct.). *Copy*: USDA. – PR (ed. 1) mentions also an issue with coloured plates.

Perkins, George Henry (1844-1933), American botanist, zoologist and geologist; Dr. phil. at Yale 1869, from 1869-1933 at the University of Vermont in various positions, incl. 56 years as curator of the University Museum; vice-president of the University 1907-1931; State geologist of Vermont 1898-1933; LL.D. Univ. Vermont 1911, Litt. D. Knox College 1912. (*G. Perkins*).

HERBARIUM and TYPES: VT. – Archival material in Smithsonian Archives.

BIBLIOGRAPHY and BIOGRAPHY: Andrews p. 312; Barnhart 3: 69; BM 4: 1548; Bossert p. 306; CSP 8: 591, 10: 1028, 17: 794; Hirsch p. 129; IH 2: (in press); Lenley p. 324; PH 567; Quenstedt p. 334; Rehder 5: 661; Tucker 1: 549.

BIOFILE: Anon., in Appleton's Cycl. Amer. Biogr. 4: 729. 1888; N.Y. World Telegram 13 Sep 1933 (at NY); Science ser. 2. 78: 259. 1933 (d. 12 Sep 1933); Vermont bot. bird. clubs, joint Bull. 6: 3-4. 1920 (portr., brief biogr., b. 25 Sep 1944), 16: frontisp. + text. 1934 (portr.).
Day, M., Rhodora 3: 260. 1901.
Fairchild, H.L., Proc. Geol. Soc. America 1933: 235-241, *pl. 7*. 1934 (obit., bibl., portr., b. 25 Sep 1844, d. 12 Sep 1933).
Osborn, H., Fragm. entom. hist. 69. 1937.

7681. *A general catalogue of the flora of Vermont* ... Montpelier, Vt. (Freeman Steam Printing House and Bindery) 1882. Oct. (*Gen. cat. fl. Vermont*).
Publ.: 1882, (LC rd. 17 Oct 1882), front., p. [1]-49. *Copies*: NY(2), US.

7682. From the tenth report of State board of agriculture. *Catalogue of the flora of Vermont*, including phaenogamous and vascular cryptogamous plants growing without cultivation ...Burlington (Free Press Association) 1888. Qu. (*Cat. fl. Vermont*).
Publ.: Sep 1888 (LC rd. 4 Oct 1888, TBC 26 Nov 1888; Nat. Nov. Nov(2) 1888), p. [1]-74. *Copies*: MO, US, USDA. – Also published in Tenth Vermont agric. Rep. 1887-1888: 231-302. 1888. – Preceded by *Gen. cat. fl. Vermont* 1882 (see above) and by a series of papers in *Archives of Science*: 1(2): 65-71. Jan 1871 (*Notes on the flora of Vermont*), and, as *Catalogue of the flowering plants of Vermont*, in 1(5): 161-166. Oct 1872, 1(6): 181-190. Jan-Apr 1873, 1(7): 215-218. Jan 1874, 1(8): 231-234. Apr 1874, 1(9): 252-253. Jul 1874.

Perkins, Janet Russell (1853-1933), American botanist; studied at the Univ. of Wisconsin 1867-1871; private teacher in Hildesheim and studying music and languages at Paris 1871-1874; from 1875-1895 private teacher at Chicago and (for three years) travelling in Europe California, Açores and Hawaii; to Berlin, studying botany e.g. with Engler, Gilg, and Kny 1895-1898; at Heidelberg 1898; again at Berlin as voluntary associate of Engler from 1898; Dr. phil. Heidelberg 1900. (*Perkins*).

HERBARIUM and TYPES: Original material and types at B (mostly destroyed); other material at A, F, G, GH, K. – Letters and portrait at G.

BIBLIOGRAPHY and BIOGRAPHY: Barnhart 3: 69; BJI 1: 44-45, 2: 133; BM 4: 1548, 8: 999; Bossert p. 306; CSP 17: 794; Hortus 3: 1201 (Perkins); IH 2: (in press); Kew 4: 275; Langman p. 579; MW p. 383; Plesch p. 359; SK 4: cxvi; cxxxii; TL-2/1713, see C.C. Mez; Rehder 2: 211, 222; Tucker 1: 549; Urban-Berl. p. 282.

BIOFILE: Anon., Hedwigia 74: (139). 1934 (d. 14 Jul 1933), Science ser. 2. 78: 87. 1933 (d. 7 Jul 1933).
Bullock, A.A., Bibl. S. Afr. bot. 84. 1978.
Merrill, E.D., Enum. Philipp. pl. 4: 215. 1926.
Perkins, J.R., "Lebensbericht" on p. [1] of her thesis, Monogr. Gatt. Mollinedia, Leipzig 1900 (b. 20 Mai 1853).
Stieber, M.T., Huntia 3(2): 124. 1979 (mss. notes A. Chase on P.).
Weberbauer, A., Pflanzenwelt Peruan. Anden 33. 1911 (Veg. Erde 12).

COMPOSITE WORKS: (1) Engler, *Pflanzenreich*, *Monimiaceae* iv. 101 (Heft 4): 1-122. 21 Jun 1901 (with E. Gilg); supplement iv. 101 (Heft 49): 1-67. 10 Oct 1911. – See also *Nachtrag*, in Bot. Jahrb. 31: 743-748. 29 Aug 1902 and reviews by K. Schumann; Bot. Centralbl. 89: 574-575. 1902 and G.G.P. Leeke, Bot. Centralbl. 120: 235-236. 1912.
(2) Id. *Styracaceae* iv. 241 (Heft 30): 1-111. 3 Sep 1907; preliminary publ. in Bot. Jahrb. 31: 478-488. 1902, 43: 214-217. 1909; rev. by C. Mez, Bot. Centralbl. 107: 77. 1908.
(3) Collaborator, J. Mildbraed, *Botanik*, Wiss. Erg. deut. Z.-Afr. Exped. 2, 1914.

7683. *Monographie der Gattung Mollinedia.* Inaugural-Dissertation zur Erlangung der Doctorwürde der hohen Naturwissenschaftlich-matematischen Fakultät der Ruprecht-Karls-Universität zu Heidelberg ... Leipzig (Wilhelm Engelmann) 1900. Oct. (*Monogr. Mollinedia*).
Thesis issue: 23 Feb 1900 (in journal; Nat. Nov. Mar (2) 1900), p. [3]-52, [1, Lebensbericht], *pl. 9-10* (uncol., J. Pohl). *Copies*: BR, NY(2), US.
Regular publication: in Bot. Jahrb. 27(5): 636-683. 23 Feb 1900.

7684. *Fragmenta florae Philippinae* contributions to the flora of the Philippine Islands ... Leipzig (Gebrüder Borntraeger), Paris (Paul Klincksieck), London (Williams & Norgate) 1904-1905. Oct. (*Fragm. fl. Philipp.*).
Co-authors: O. Beccari, A. Brand, C. de Candolle, E.B. Copeland, P. Graebner, C. Mez, R. Pilger, L. Radlkofer, W. Ruhland, R. Schlechter, K.O. von Seemen, O. Warburg.

Fasc. 1: 12 Mar 1904 (p. iv: Feb 1904; see cover fasc. 2; ÖbZ Mar-Apr 1904; Nat. Nov. Mai(1) 1904; Torreya 13 Mai 1904; Bot. Zeit. 16 Mai 1904; J. NYBG Mai 1904), p. [i]-iv, [1]-66.
Fasc. 2: 30 Jun 1904 (cover p. 1; US rd. 29 Aug 1904; TBC 4 Oct 1904; Nat. Nov. Aug(1) 1904), cover with title, p. [67]-152, *pl. 1-3* (uncol.).
Fasc. 3: 20 Feb 1905 (cover p. 1; US rd. 7 Apr 1905; Bot. Centralbl. 11 Apr 1905; TBC 27 Mai 1905; Nat. Nov. Mar(2) 1905), cover id., p. [153]-212, *pl. 4.*
Copies: B, BR, G, MO, US, USDA.
Ref.: Perkins, J., Bot. Jahrb. 34 (Lit.): 1-2. 17 Jun 1904 preliminary summary.
Mez, C., Bot. Centralbl. 69-71, 346-347. 1905.

7685. *The Leguminosae of Porto Rico* ... Washington (Government Printing Office) 1907. Oct.
Publ.: 10 Jun 1907 (p. ii), Contr. U.S. natl. Herb. 10(4): [i]-[v], 133-220, vii-ix. To be cited from journal. (*Copies*: BR, G, MO, NY, U, US). – Prepared at Berlin 1901/1902 by Miss Perkins as scientific aid in USDA.

7686. *Beiträge zur Flora von Bolivia* ... [Leipzig (Wilhelm Engelmann) 1912]. Oct. (*Beitr. Fl. Bolivia*).
Publ.: 27 Aug 1912 (in journal; cover reprint dated 1912; Nat. Nov. Mar(1) 1914), p. [i, recto of p. 170 with reprint title], [170]-233. *Copies*: BR, G. – Partly preprinted with special cover and preprint title from Bot. Jahrb. 49(1): [170]-176. 27 Aug 1912, 177-233. 14 Jan 1913.

7687. *Übersicht über die Gattungen der Monimiaceae* sowie Zusammenstellung der Abbildungen und der Literatur über die Arten dieser Familie bis zum Jahre 1925 ... Leipzig (Verlag von Wilhelm Engelmann) 1925. Oct. (*Übers. Gatt. Monim.*).
Publ.: Sep 1925 (ÖbZ Jun-Sep 1925; copy at MO stamped (by Perkins?) 28 Sep 1925, rd. 9 Dec 1925), p. [i-ii], [1]-55, [56, ind.]. *Copies*: B, G, MO, NY, U. – Earlier papers on Monimiaceae by this author: (1) in Pflanzenreich, see above, (2) in Bot. Jahrb. 31: 743-748. 29 Aug 1902 (Nachtrag to treatment Pflanzenreich), 25: 547-577. 1898, 27: 636-682. 1900, 28: 660-705. 1901, 45: 422-425. 1911, 52: 191-218. 1915, 55 (Beibl. 118): 1-3. 1916.
Ref.: Krause, K., Bot. Jahrb. 60: 71. 1 Apr 1926.

7688. *Übersicht über die Gattungen der Styracaceae* sowie Zusammenstellung der Abbildungen und der Literatur über die Arten dieser Familie bis zum Jahre 1928 ... Leipzig (Verlag von Wilhelm Engelmann) 1928. Oct. (*Übers. Gatt. Styrac.*).
Publ.: 1928, p. [1]-35, [36]. *Copy*: U.

Perktold, Josef Anton (1804-1870), Austrian cryptogamist and clergyman; entered the Praemonstratenser Convent of Wilten 1825 (baptismal name Anton); ordained 1829; clergyman at Tarrenz 1829-1830, Ellbögen nr. Innsbruck 1830-1840, Tulfes 1834-1836, Vill 1836-1842, Igels 1842-1852, ultimately again at Wilten as canon of its Convent. (*Perktold*).

HERBARIUM and TYPES: IBF; duplicates (lich.) at M.

BIBLIOGRAPHY and BIOGRAPHY: CSP 4: 829, 17: 794; DTS 1: 222-223 (b. 14 Jul 1804; d. 27 Oct 1870; bibl.); GR p. 437; GR, cat. p. 69; IH 2: (in press); Kew 4: 275; Krempelh. 1: 492, 538, 591, 3: 105; LS 20444-20448; PR (alph.) (ed. 1: 7874); SBC p. 129.

BIOFILE: Anon., Bot. Not. 1871: 36 (d. 29 Oct 1870); Bot. Zeit. 28: 744. 1870 (d. 27 Oct 1870); Flora 33: 44. 1870, 53: 448. 1870 (d.); Österr. bot. Z. 21: 48. 1871 (d.).
Dalla Torre, K.W. v., Z. Ferdinandeum, Innsbruck ser. 3. 35: 211-291. 1891 (biogr., bibl., identif. of P's cryptogams by author, E. Hampe (Hepat.), L. v. Heufler & O. Sendtner (Musci), F. Arnold (Lich.). (b. 14 Jul 1804, d. 29 Oct 1870).
Hausmann, F., Fl. Tirol. 1165. 1854.
Heufler, L.v., Flora 26: 593. 1843 ("treibt ...mit wahrem Feuereifer Botanik ...").
Sayre, G., Bryologist 80: 515. 1977.

Schlechtendal, D.F.L., Bot. Zeit. 1: 202-203. 24 Mar 1843 (rev. of his Umbilicarien von Tirol, *in* Neue Z. Ferdinand 8: 54-67. 1842, see DTS 1: 222 (1)).

Perleb, Karl [Carl] Julius (1794-1845), German botanist and physician; professor of natural history at the university of Freiburg in Breisgau 1818-1845; director of the botanical garden ibid. 1828-1845. (*Perleb*).

HERBARIUM and TYPES: FB (fide Bot. Zeit. 4: 136. 1846; further details lacking). – Letters at G.

BIBLIOGRAPHY and BIOGRAPHY: ADB 25: 379-380; Barnhart 3: 69; BM 4: 1548; Frank 3 (Anh.): 75 (b. 20 Jun 1794); GR p. 119-120 (d. 11, not 8, Jun 1845); Hegi 5(2): 1093; Herder p. 470; Kew 4: 275; Krempelh. 1: 110, 2: 9; MD p. 198-200; PR 1467, 7040-7042 (ed. 1: 7876-7878); Rehder 1: 46, 86, 260; TL-2/985; Tucker 1: 549.

BIOFILE: Anon., Bot. Not. 1845: 220 (d); Flora 28: 432. 1845 (d. 8 Jun 1845, left herb. to Univ. Freiburg), 29: 48. 1846 (endowment); Bot. Zeit. 3: 504. 1845 (d. 11 Jun 1845), 4: 136. 1846 (d. 8 Jun 1845, but see GR); Flora 22: 528. 1839 (Hofrath).
Candolle, A.P. de, Mém. souvenirs 150. 1862.
Gager, C.S., Brooklyn Bot. Gard. Rec. 27: 226. 1938 (dir. bot. gard Freiburg).
Hasskarl, J.K., Allg. Sach. Namenreg. Flora 172. 1851.
Jessen, K.F.W., Bot. Gegenw. Vorz. 421. 1884.
Oehlkers, F., Ber. deut. bot. Ges. 68: (2). 1955.
Pritzel, G.A., Linnaea 19: 459. 1847 (d. 8 Jun 1845).
Wunschmann, G., Allg. deut. Biogr. 25: 379-380. 1887 (biogr., b. 20 Jun 1794, d. 8 Jun 1845).

COMPOSITE WORKS: Translated Candolle, A.P. de, *Essai propr. méd. pl.* ed. 2, into German, *Vers. Arzneikr. Pfl.* Aarau 1818, see TL-2/985.

EPONYMY: *Perlebia* C.F.P. Martius (1828, fide Margadant, 1968); *Perlebia* A.P. de Candolle (1829).

7689. *Conspectus methodi plantarum naturalis* in usum auditorum typis exscribi curavit Dr. C.J. Perleb ... Friburgi Brisgoviae [Freiburg i. Breisgau] (excud. Fr. Xav. Wangler) 1822. Qu. (*Consp. meth. pl. nat.*).
Publ.: 1822, p. [1]-42. *Copy*: BM. – See MD p. 199 for a collation and notes.

7690. *Lehrbuch der Naturgeschichte des Pflanzenreichs* ... Freiburg im Breisgau (Druck und Verlag von Friedrich Wagner) 1826. (*Lehrb. Naturgesch. Pflanzenr.*).
Ed. [*1*]: Sep-Oct 1826 (Linnaea Oct-Dec 1926; Flora 28 Oct 1828), p. [i]-xii, [xiii-xiv], [1]-422, [423-426]. *Copy*: NY. – See MD p. 199-200 for a collation and notes. – Special issue of the botanical part of his *Lehrbuch der Naturgeschichte* 1, 1826 (*Copy*: G).
Ed. 2: 1852 ("ist erschienen", Bot. Zeit. 13 Sep 1852).
Ref.: Anon., Flora 9: 618, 640, 671-672. 1826.

7691. *Natalitia augustissimi atque potentissimi principis Ludovici Guilielmi Augusti* ... Academiae Alberto-Ludovicianae Rectoris magnificentissimi ad diem ix Februarii rite pieque celebrando Senatus academici nomine indicit D. Carolus Perleb ... Disseritur simul *de horto botanico friburgensi.* Friburgi brisigavorum [Freiburg i. Breisgau] (typis Friderici Wagner) 1829. Qu. (*Hort. bot. frib.*).
Publ.: 9 Feb 1829, p. [i]-xxviii, [1]-38, map. *Copies*: G, NY, REG.
Ref.: Anon., Lit.-Ber. Flora 1: 46-47. Jan-Feb 1831.

7692. *Clavis classium,* ordinum et familiarum atque index generum regni vegetabilis. Diagnostische Uebersichtstafeln des natürlichen Pflanzensystems. Nebst vollständigem Gattungsregister. Von Dr. C.J. Perleb ... Freiburg im Br. (bei Adolph Emmerling) 1838. Qu. (*Clav. class.*).
Publ.: Jan-Mar 1838 (p. v: 20 Jun 1837; Flora 5 Feb-28 Apr 1838, p. [i]-viii, [1]-94.

Copies: G, MO, NY, USDA, WU. – Important source of publication of names of higher taxa.
Ref.: Anon., Lit. Ber. Flora 9: 186-189. 1839.

Pernhoffer, Gustav, Edler von Bärnkron (ca. 1830-1899), Austrian amateur botanist; friend of A. Kerner v. Marilaun. (*Pernh.*).

HERBARIUM and TYPES: Material from Austria at AK, B, C, GZU, LE, S and W. – Pernhoffer contributed to the *Flora exsiccata austro-hungarica* (see A. Kerner) and issued *Hieracia seckauensia exsiccata* (fasc. 2, 1895 in GOET) fasc. 1-2, 1894-1895, 104 nos. (see also Pernhoffer 1896).

BIBLIOGRAPHY and BIOGRAPHY: AG 12(1): 31; Barnhart 3: 69; CSP 17: 796; DTS 1: 223, 6(4): 184 (d. 17 Mai 1899); Hegi 5(4): 2467; IH 2: (in press); Rehder 1: 441, 444; Urban-Berl. p. 382.

BIOFILE: Anon., Bot. Centralbl. 79: 111. 1899 (d.); Bot. Jahrb. 27 (Beibl. 64): 19. 1900; Bot. Not. 1900: 47; Flora 28: 432. 1845 (d. 8 Jun 1845), 29: 48. 1846 (herb. to FB); Österr. bot. Z. 49: 218. 1899 (d.).
Fritsch, C., Verh. zool.-bot. Ges. Wien 49(6): 311-312. 1899 (obit., d. 17 Mai 1899, 69 yrs. old).
Goulding, J.H., Rec. Auckland Inst. Mus. 12: 112. 1975.
Janchen, E., Cat. fl. Austr. 1: 30. 1956.
Kneucker, A. v., Allg. bot. Z. 5: 136. 1899 (17 Mai 1899).

7693. *Verzeichniss der in der Umgebung von Seckau* in Ober-Steiermark *wachsenden Phanerogamen* und Gefässkryptogamen, einschliesslich der wichtigern cultivirten Arten ... [Verh. Zool. bot. Ges. Wien 1896]. Qu.
Publ.: 16 Dec 1896 (in journal, fide cover Heft 9; mss. submitted 30 Sep 1896; Nat. Nov. Nov(1) 1897), p. [1]-43. *Copy*: G. – Reprinted and to be cited from Verh. Zool.-bot. Ges. Wien 46(9): 384-425. 1896.
Preliminary publication on *Hieracium*: Die Hieracien der Umgebung von Seckau in Ober-Steiermark, in Österr. bot. Z. 44: 315-319. Aug, 362-365. Sep, 430-434. Oct, 477-479. Dec 1894, 46: 36-38. Jan, 74-79. Feb, 112-119. Mar, 154-158. Apr, 196-197. Mai, 236-238. Jun, 268-270. Jul 1896. – Text accompanying his *Hieracia seckauensia exsiccata* of which sets were deposited (p. 270) in W, PR, PRC, GR, BP, B, GOET, M, Oborny, S.

Pernitzsch, Heinrich (*fl.* 1825), German (Saxonian) forester and botanist at Heidelbach (Sachsen); "königlicher Sächs. Oberförster". (*Pernitzsch*).

HERBARIUM and TYPES: Unknown.

BIBLIOGRAPHY and BIOGRAPHY: BM 4: 1549; Herder p. 470; MD p. 200; PR 7043 (ed. 1: 7880); Rehder 5: 662; Tucker 1: 549.

7694. *Flora von Deutschlands Wäldern* mit besonderer Rücksicht auf praktische Forstwissenschaft ... Leipzig (in der Baumgärtnerschen Buchhandlung) 1825. Oct. (*Fl. Deutschl. Wäld.*).
Publ.: 1825 (p. vi: Mar 1825), p.[i]-vi, [vii], [1]-332, [1, err.]. *Copies*: B, FI, MO, NY. – For a collation and notes see MD p. 200.

Perreymond, Jean Honoré (1794-1843), French botanist, school inspector and music composer at Fréjus. (*Perreymond*).

HERBARIUM and TYPES: Material at B, G, P, PC; Burnat (1883) was not able to trace the original herbarium.

BIOFILE: AG 1: 170, 6(2): 509; Barnhart 3: 70; BL 2: 208, 701; BM 4: 1550; IH 2: (in press); Kew 4: 276; Lasègue p. 102, 296; PR 7046 (ed. 1: 7884); Rehder 1: 408; Urban-Berl. p. 382.

BIOFILE: Burnat, E., Bull. Soc. bot. France 30: cxxvi. 1893 (b. 13 Jan 1794, d. 18 Jul 1843).
Cavillier, F.G., Boissiera 5: 71-72. 1941.
Reynier, A., Rev. hort. Bouches-du-Rhône 1896: 32-34. (two letters to Achintre).

EPONYMY: *Perreymondia* Barnéoud (1845).

7695. *Plantes phanérogames* qui croissent aux environs *de Fréjus*, avec leur habitat et l'époque de leur fleuraison. Paris (F.-G. Levrault, ...), Fréjus (Aragon père, ...) 1833. Oct. (in fours). (*Pl. phan. Fréjus*).
Publ.: Sep-Dec 1833 (t.p.; p. [vii]: 1 Sep 1833; Arch. Bot. 23 Dec 1833), p. [i-vii], [1]-90, [1-2, err.]. *Copies*: G, NY, USDA.

Perrier, Alfred (1809-1866), French botanist; Dr. med. Paris 1835; from 1836 practicing physician and amateur naturalist at Caen. (*A. Perrier*).

HERBARIUM and TYPES: CN.

BIBLIOGRAPHY and BIOGRAPHY: Barnhart 3: 70; CSP 4: 832, 8: 592; Kew 4: 276.

BIOFILE: Chevalier, A., Monde des plantes ser. 2. 6: 31-32, 52-53. 1897 (herb. at CN).
Fournier, E., Bull. Soc. bot. France 9: 545-546. 1863, 11 (bibl.): 23-24. 1864.
Letacq, A.L., Invent. pl. crypt. vasc. Orne 250-251. 1896 (bibl.) (several publ. together with Pierre Modeste Duhamel, 1813-1890); Bull. Soc. Amis Sci. nat. Rouen ser. 5. 42: 110-111. 1907 (biogr. sketch, bibl., b. 30 Sep 1809, d. 22 Sep 1866).
Morière, P.G., Bull. Soc. Linn. Normandie ser. 2. 2: 161-171. 1868 (biogr., portr.).

Perrier de la Bâthie, Eugène Pierre [baron] de (1825-1916), French landowner, agronomist and botanist at Conflans (Albertville), Savoie; professor of agriculture at Albertville 1875-1900. (*E. Perrier*).

HERBARIUM and TYPES: G(25.000; rd. 1921); further material at AUT, B, BR, CGE, CN, FI, L, MANCH, P, PC. – Letters and portr. at G.

BIBLIOGRAPHY and BIOGRAPHY: AG 6(1): 356; (b. 9 Jun 1825), 12(3): 355; Barnhart 3: 70; BL 2: 198, 199, 701; Frank 3 (Anh.): 75; Hortus 3: 1201; IH 1 (ed. 7): 339, 2: (in press); Kew 4: 276; LS 20466-20470, 37818; Morren ed. 10, p. 58; Rehder 1: 411; Saccardo 2: 83; TL-2/see J. Offner; Tucker 1: 549-600; Urban-Berl. p. 382.

BIOFILE: Anon., Bot. Not. 1917: 141 (d.); Österr. bot. Z. 67: 48. 1918 (d.).
Beauverd, G., Bull. Soc. Bot. Genève ser. 2. 8(7-9): 353-355. 1916 (obit., bibl.).
Correvon, H., Bull. Soc. Hort. Genève 6(6): 104-106. 1916 (obit.).
Kneucker, A., Allg. bot. Z. 23: 48. 15 Mar 1919 (d. 31 Mai 1916).
Miège, J. & J. Wuest, Saussurea 6: 133. 1975.
Moldenke, H.N., Plant Life 2: 75. 1946 (1948).
Offner, J., Bull. Soc. Hist. nat. Savoie ser. 2. 18: 103-111 (biogr., bibl.), repr. 9 p. Chambéry 1918).
Steinberg, C.H., Webbia 32: 33. 1977.
Weibel, R., Musées de Genève 65: 11-14. 1966 (biogr., portr., b. 9 Jun 1825, d. 31 Mai 1916).

HANDWRITING: *Candollea* 32: 389-390. 1977.

7696. *Indication de quelques plantes nouvelles*, rares ou critiques, observées en Savoie, spécialement dans les provinces Savoie-propre, Haute-Savoie et Tarentaise, suivie d'une revue de la section Thylacites du genre Gentiana ... Chambéry (Imprimerie Bachet) 1855. Oct. (*Indic. pl. nouv.*).
Co-author: André Songeon (1826-1905).
Publ.: 1855 (BSbF: rd. at session 26 Dec 1856), p. [1]-46. *Copies*: G(2). – Above title taken from cover. Reprinted from Ann. Soc. Hist. nat. Savoie 1854: 153-198. 1855.
Ref.: Anon., Bull. Soc. bot. France 4: 722-725. 1857.

7697. *Notes sur des plantes nouvelles ou peu connues de la Savoie* . . . . [Haguenau (imprimerie et lithographie de V. Edler) 1859] Oct.
Senior author: André Songeon (1826-1905).
1: Dec 1859 (t.p. Annotations), p. [1]-15. *Copies*: G(2). – Reprinted from Billot, P.C., Annotations à la flore de France et d'Allemagne 181-193. Dec 1859 (TL-2/518).
2: 1866, *in* Billotia 1: 72-81. 1866.
Other publication: Notes sur quelques plantes nouvelles ou intéressantes de la Savoie et des pays voisins, Bull. Herb. Boiss. ser. 1. 2: 425-437 [1-13]. Jun 1894. *Copy*: U.

7698. *Aperçu sur la distribution des espèces végétales dans les Alpes de la Savoie* [Extrait du Bulletin de la Société botanique de France 1863]. Oct.
Co-author: André Songeon (1826-1905).
Publ.: Sep-Dec 1863 (session juillet-août 1863), p. [1]-12. *Copy*: U. – Reprinted and to be cited from Bull. Soc. bot. France 10: 675-686. 1863.

7699. Excursions en Tarentaise *Guide du botaniste* . . . Moutiers (F. Ducloz, . . .) 1894. Oct. (*Guide bot.*).
Publ.: 1894, p. [3]-86 [87]. *Copies*: G. – Reprinted from Laissus, En Savoie, La Tarentaise, Moutiers, 1894.

7700. *Nouvelles observations sur les Tulipes de la Savoie.* [Bull. Herb. Boiss. ser. 2, tome 5(5)] 1905. Oct.
Publ.: 30 Apr 1905, p. [1]-3. *Copy*: G. – Reprinted from Bull. Herb. Boiss. ser. 2. 5(5): 507-509. 1905. – See also Bull. Ass. Protect. Pl. 6: 11-14. 1888 (also in Rev. hort. 36, Sep 1887).

7701. *Catalogue raisonné des plantes vasculaires de Savoie* Départements de la Savoie et de la Haute-Savoie Plateau du Mont-Cenis . . . Paris (Librairie des sciences naturelles Paul Klincksieck Léon Lhomme . . .) 1917-1928. Oct., 2 vols. (*Cat. rais. pl. Savoie*).
Editor and author of preface and supplement: Jules Offner (1873-1957).
1: 1917 (p. ix: 5 Jul 1916), p. [i]-xlv, [1], [1]-433, 1 map.
2: 1928 (Nat. Nov. Jun 1928), p. [i-iii], [1]-415. *Imprint*: Le Carriol (Lot) (Ancienne Librairie Klincksieck . . .).
Copies: B, BR, USDA. – Also published as Mém. Acad. Sci. Belles-Lettr. Arts Savoie ser. 5. vol. 4 (xlv, 433 p., map), 1917 and 5(415 p.) 1928.
Additions: Offner, J., C.R. Ass. franç. Avanc. Sci. 57: 297-299. 1933; Thommen, É. & A. Becherer, Bull. Soc. bot. Genève ser. 2. 33: 109-130. 1942.

Perrier de la Bâthie, [Joseph Marie] **Henri** [Alfred] (1873-1958), French botanist; nephew of E. Perrier; soldier in Indochina (Tonkin), from 1896-1933 living in Madagascar (1914-1918 in the French army in Europe), later in Paris. (*H. Perrier*).

HERBARIUM and TYPES: P, PC; duplicates e.g. in B, C, G, ILL, K, L, MO, NH, NY, TAN. – Letters at G.

BIBLIOGRAPHY and BIOGRAPHY: Barnhart 3: 70; BJI 2: 133; BM 8: 1000; Kew 4: 276-277; Plesch p. 359; Tucker 1: 550; Urban-Berl. p. 382; Zander ed. 10, p. 700, ed. 11, p. 799.

BIOFILE: Humbert, H., C.R. Séances Acad. Sci. 247(22): 1921-1923. 1958 (obit., d. 2 Oct 1958); J. Agric. trop. Bot. appl. 5(1-3): 863-867. 1958 (obit., b. 11 Aug 1873, d. 2/3 Oct 1958, brief bibl.); Not. Syst. 16: 1-5. 1960 (portr.); C.R. AETFAT 4: 133, 137, 138. 1962.
Lacroix, A., Not. hist. quatre bot. 21-39. 1938 (biogr., sketch (superficial), work on Madagascar); Figures de Savants 3: 140. 1938.
Leandri, J., AETFAT Bull. 9: 52-55. 1959; Taxon 9: 1-3. 1962 (portr.).
Leroy, J.F., Science and nature 31: 2. 1959 (hommage by quoting from his Vég. malg.).
Verdoorn, F., Chron. bot. 1: 353. 1935, 2: 125, 131, 132, 229, 385. 1936, 3: 112, 199. 1937.
Wulff, E.V., Intr. hist. plant geogr. 112, 114. 1943.

COMPOSITE WORKS: (1) Founder and editor, with René Viguier (1880-1931) and Henri Chermezon (1885-1939) of *Archives de botanique, Bulletin mensuel*, Caen, vols. 1-4, 1927-1930.
(2) Contributed several treatments to H. Humbert, *Fl. Madagascar*, e.g. 140. *Flacourtiacées*, [1]-131. 1946, 142. *Turnéracées*, [1]-13, 1950, 143. *Passifloracées*, p. [1]-49. 1945, 151. *Combretacées*, p. [1]-83. 1954, 152. *Myrtacées*, p. [1]-80. 1953, 153. *Mélastomacées*, p. [1]-326. 1951, 161. *Myrsinacées*, p. [1]-148. 1953, 165. *Ébénacées*, p. [1]-137. 1952, 166. *Oléacées*, p. [1]-[89]. 1952, 178. *Bignoniacées*, p. [1]-91. 1938, 181. *Lentibulariacées*, p. [1]-22. 1955. 49. *Orchidées*, p. [1]-846. 1939-1941.
(3) With Raymond-Hamet, q.v., *Nouv. Contr. Crassulac. malgach.* 1914.

MEMORIAL VOLUME: *Travaux botaniques dédiées à la mémoire de Henri Perrier de la Bâthie 1873-1958.* Notulae systematicae 16(1-2), Paris, Oct 1960.

EPONYMY: *Bathiea* Drake del Castillo (1902); *Bathiea* R. Schlechter (1918); *Bathiorhamnus* R. Capuron (1966); *Neobathiea* R. Schlechter (1925); *Perriera* Courchet (1905); *Perrieranthus* Hochreutiner (1916); *Perrierastrum* Guillaumin (1931); *Perrierbambus* A. Camus (1924); *Perrieriella* R. Schlechter (1925); *Perrierodendron* Cavaco (1951); *Perrierophytum* Hochreutiner (1916).

7702. Annales du Musée colonial de Marseille ... *La végétation malgache* ... Marseille (Musée colonial ...), Paris (Librairie Challamel ...) 1921. (*Vég. malg.*).
Publ.: 1921 (t.p.: 1921; Nat. Nov. Apr 1922; ÖbZ Jan-Apr 1922), p. [i-iii], [1]-268, *3 pl.*, 1 map. *Copies*: B, BR. – Issued as Ann. Mus. Colon. Marseille ser. 3. 9, 1921.

7703. *Les Lomatophyllum et les Aloe de Madagascar* ... Caen (Imprimerie E. Lanier, ...) 1926. Qu.
Publ.: 1926, p [1]-58, [1, expl. pl.], *pl. 1-8*, 1 map. *Copy*: G. – Reprinted and to be cited from Mém. Soc. Linn. Normandie ser. 2, Bot. 1(1): 1-58. 1926.

7704. Colonie de Madagascar et dépendances. *Catalogue des plantes de Madagascar* publié par l'Academie malgache ... Tananarive (Imprimerie G. Pitot & Cie ...) 1930-1940. †. (*Cat. pl. Madag.*).
Publ.: Perrier contributed the texts of 34 families to this series. No editor is given; the publication remained unfinished. The dates given on the title-pages are often earlier than the printer's dates on the last page of each treatment. These later dates should be followed. Each fascicle is a separate publication; there is no serial numbering. *Copies*: BR, MO, PH, US.

Family	date t.p.	printer's date	pagination	author
Orchid.	Oct 1930	Apr 1931	[1]-60	H. Perrier
Anon.	Oct 1931	Feb 1932	[1]-11	L. Diels
Asclep.	Oct 1931	Dec 1931	[1]-24	P. Choux
Chlaen.	Oct 1931	Feb 1932	[1]-9, [11]	H. Perrier
Cyperac.	Feb 1931	Mai 1931	[1]-48	H. Chermezon
Dioscor.	Oct 1931	Feb 1932	[1]-12	H. Perrier
Menisp.	Oct 1931	Feb 1932	[1]-9	L. Diels
Pterid.	Jul 1931	Feb 1932	[1]-72	C. Christensen
Sapind.	Oct 1931	Mar 1932	[1]-14, [15]	P. Choux
Scrophul.	Oct 1931	Feb 1932	[1]-14, [15]	H. Perrier
Loranth.	Apr 1932	Sep 1932	[1]-13, [15]	H. Lecomte
Polygal.	Dec 1932	Aug 1933	[1]-9, [11]	H. Perrier
Sapot.	Apr 1932	Sep 1932	[1]-9, [11]	H. Lecomte
Palmae	Jul 1933	Jan 1934	[1]-26, [27]	H. Jumelle
Balsam.	Oct 1934	Jun 1935	[1]-18, [19]	H. Perrier
Chenop.	Feb 1934	Mar 1934	[1]-8	J. Leandri
Eric., Vacc.	Oct 1934	Mai 1935	[1]-12, [13, 15]	H. Perrier
Hydr. + 6 fam.	Oct 1934	Jun 1935	[1]-18, [19]	H. Perrier
Melast.	Feb 1934	Apr 1934	[1]-35	H. Perrier

Family	date t.p.	printer's date	pagination	author
Pod., Hydrost.	Feb 1934	Apr 1936	[1]-11	H. Perrier
Thymel.	Feb 1934	Mar 1934	[1]-9, [11]	J. Leandri
Xyr. + 14 fam.	Oct 1934	Jun 1935	[1]-22, [23]	H. Perrier
Euphorb.	Jan 1935	Mai 1935	[1]-51	J. Leandri
Acanth.	1939	1939	[1]-32	R. Benoist
Bignon.	1939	-	[1]-21	H. Perrier
Commel.	1939	-	[3]-15	H. Perrier
Liliac.	1939	-	[3]-29	H. Perrier
Pass. + 3 fam.	Jan 1940	-	[3]-27, [28]	H. Perrier
Theac., Ochn.	Mai 1940	-	[1]-19	H. Perrier

7705. *Les Mélastomacées de Madagascar* ... Toulouse (Imp. Henri Basuyau & Cie., ...) 1932. Oct. (*Mélast. Madag.*).
Publ.: 1932, p. [i-iv], [1]-292, *pl. 1-10*. *Copies*: B, MICH, NY. – Issued as Mém. Acad. malg. 12, 1932.

7706. *Les plantes introduites à Madagascar* listes des plantes cultivées, rudérales, messicoles ou naturalisées croissant dans l'île, suivie d'un aperçu sur les plantes autochtones devenues anthropophiles ... Toulouse (Imp. Henri Basuyau & Cie., ...) 1933. Oct. (*Pl. introd. Madag.*).
Publ.: 1933, p. [1]-80. *Copy*: PH. – Reprinted with changes and corrections from Rev. Bot. appl. Agr. col. 12: nos. 121, 122, 123, 124, 125, 126, 127, 128, 129, 130 and 131. 1932-1933.

7707. *Biogéographie des plantes de Madagascar* ... Paris (Société d'Éditions géographiques ...) 1936. Oct. (*Biogéogr. Madag.*).
Publ.: 13 Mar 1936 (letter in copy BR; BR rd. 17 Mar 1936; p. 156: printed Feb 1936; Nat. Nov. Jul 1936), p. [i, iii], [1]-156, *pl. 1-40*. *Copies*: BR, G, PH.

7708. *Les Bignoniacées de la région malgache* (Madagascar, Mascareignes, Seychelles et Comores) ... Marseille (Faculté des Sciences ... Musée coloniale ...) 1938. Oct. (*Bignon. malg.*).
Publ.: 1938, p. [1]-101, *pl. 1-8*. *Copy*: G. – Issued as Ann. Mus. colon. Marseille ser. 5, 6(1). 1938.

Perrin, Ida Southwell (née Robins; Mrs. Henry Perrin), (1860-?), British popular writer on botany and botanical artist. (*Perrin*).

HERBARIUM and TYPES: Unknown.

BIBLIOGRAPHY and BIOGRAPHY: BL 2: 215, 701; BM 8: 1000; Kew 4: 277; NI 1509; Tucker 1: 778-779.

7709. *British flowering plants* illustrated by three hundred full-page coloured plates reduced from drawings by Mr. Henry Perrin ... London (Bernard Quaritch, ...) 1914, 4 vols. (*Brit. fl. pl.*).
Introduction and descriptive notes by: George Edward Simmonds Boulger (1853-1922).
1: 1914, p. [i]-xlv, *pl. 1-66*, with text.
2: 1914, p. [i]-viii, *pl. 67-146*, id.
3: 1914, p. [i]-ix, *pl. 147-223*, id.
4: 1914 (Bot. Centralbl. 21 Jul 1914), p. [i]-viii, *pl. 224-300*, id. "To the scientific botanist the book hardly appeals ...". (Britten). Edition of 1000 numbered copies. – *Copies*: BR, NY.
Ref.: Britten, J., J. Bot. 52: 131-136. 1914.
New and revised edition in one volume: 1939, n.v.

Perrine, Henry (1797-1840), American botanist; introduced many tropical plants into the southern U.S.; consul at Campeachy, Yucatan, 1827-1831, 1832-1837, at Key West (Indian Key) from 1937. (*Perrine*).

HERBARIUM and TYPES: Most collections and many manuscripts destroyed when he was murdered by a band of Spanish Indians during a raid on Indian Key (Millspaugh, mss.); some material at NY.

BIBLIOGRAPHY and BIOGRAPHY: Barnhart 3: 70 (unpubl. inf.); Hortus 3: 1201 (Perrine); Kew 4: 277; Langman p. 579; Lenley p. 466; ME 2: 576; Zander ed. 10, p. 700, ed. 11, p. 799 (b. 5 Apr 1797, d. 7 Aug 1840).

BIOFILE: Anon., Army & Navy Chronicle 1840: 157-158 (the massacre at Indian Key; fide Millspaugh).
Bickel, M., *in* E. Slosson, Pioneer Amer. gard. 84-85. 1951 (fide Ewan 1957).
Darlington, W., Reliq. Baldwiniae 219. 1843.
Ewan, J., *in* J.F. Leroy, ed., Bot. franç Amér. nord 38. 1957; Southw. Louisiana J. 7: 21. 1967; *in* R.W. Long & O. Lakela, Fl. trop. Florida 4. 1971.
Ewan, J. et al., Short hist. bot. U.S. 7. 1969.
Millspaugh, C.F., Perrine, Dr. Henry, typescript, 5 p., dated 1904, at NY.
Murrill, W.A., Hist. found. bot. Florida 12. 1945.
Perrine, H.F., A true story of some eventful years ... Hutchinson Press, Buffalo, New York (303 p., fide Millspaugh).

7710. 25th Congres, 2d session.[Rep. no. 564] Ho. of Reps. Dr. Henry Perrine – *Tropical plants* ... In the House of Representatives, March 8, 1838 ... [Thomas Allen, print.]. Oct.
Publ.: 8 Mar 1838, p.[1]-99. *Copy*: NY. – See Langman, p. 579 for further details.

7711. 25th Congress, 2d session. [Senate] [300]. In Senate of the United States. March 12, 1838 ... *Report: The Committee on Agriculture*, ... [Blair & Rives, printers].
Publ.: 12 Mar 1838, p. [1]-142, *pl. 1-24*. *Copy*: NY. – Contains five papers by Perrine, mainly dealing with tropical plants introduced or fit to be introduced into South Florida; see also Langman p. 579.

Perrot, Émile Constant (1867-1951), French pharmacist, anatomist and botanist; Dr. sci. Paris 1899; succeeded G. Planchon as professor of materia medica at the École de Pharmacie, Paris, 1899, (aggrégé; prof. tit. 1902)-1937; Dr. h.c. Univ. Utrecht 1945. (*Perrot*).

HERBARIUM and TYPES: Unknown. – Letters and portrait at G.

BIBLIOGRAPHY and BIOGRAPHY: Barnhart 3: 70; BFM no. 2874; BJI 1: 45, 2: 133; BL 1: 37, 312, 2: 7, 105, 701; BM 4: 1551; CSP 17: 806; Hirsch p. 229; Kelly p. 173; Kew 4: 278; Langman p. 579-580; LS 20478, 37820, suppl. 21267; MW p. 383; Plesch p. 360; TL-2/579; Zander ed. 10, p. 700, ed. 11, p. 733, 799.

BIOFILE: Anon., Amer. Druggist 63: 22-23. 1922 (portr.); Bot. Centralbl. 95: 528. 1904 (memb. Pharm. Soc. Gr. Brit.); Bull. Soc. bot. France 55 (bibl.): 755. 1908, 56: 80. 1909 (chevalier Légion d'Honneur); Rev. Bot. appl. Agric. trop. 31: 648-658. 1951 (bibl.).
Chevalier, Aug., Rev. Bot. appl. Agric. trop. 31: 562-563. 1951.
Mascré, M., Mém. Soc. bot. France 1952: 181-187 (obit., portr.).
Paris, R., Ann. pharm. franç. 10: 711-727. 1952 (obit., portr., bibl., b. 14 Aug 1867, d. 16 Sep 1951).
Perrot, É., Notice sur les titres et travaux scientifiques de M. Émile Perrot. Lons-le-Saunier 1902, 65 p. (bibl. up to 1901), id. 1928, 51 p. (bibl. up to 1928), La culture des plantes médicinales. Paris 1947, viii, 382 p.
Raymond-Hamet, Rev. int. Bot. appl. Agric. trop. 31: 657-758. 1951 (on P's work on glycosides and alkaloids).
Verdoorn, F., ed., Chron. bot. 1: 14, 37. 1935, 2: 54, 128. 1936, 3: 36, 38, 110, 118. 1937, 4: 83, 118. 1938, 5: 299, 302, 1939.
Weitz, R., Rev. int. Bot. appl. Agric. trop. 31: 648-656. 1951 (biogr., bibl.).

EPONYMY: *Perrotia* Baudier (1901); *Perrotiella* Naumov (1916).

NOTE: We have no special information on the Württemberg pharmacist and botanist August Perrot, born 30 Nov 1869, who obtained the degree of Dr. phil. at Erlangen on 8 Mar 1897 on a thesis *Kernfrage und Sexualität bei Basidiomyceten*, Stuttgart (K. Hofbuchdruckerei zu Gutenberg (Carl Grüninger)) 1897. Oct., p. [1]-37, [38-39]. *Copy*: G. – Vita on p. 5.

7712. *Contribution à l'étude histologique des Lauracées.* Thèse pour l'obtention du diplôme de pharmacien de 1re classe présentée & soutenue le [ ]juillet 1891 ... Lons-le-Saunier (Imprimerie et lithographie, Lucien Declume ...)] 1891. Oct. (*Contr. histol. Laurac.*).
Publ.: Jul 1891 (Nat. Nov. Oct(2) 1891), p. [1]-62. *Copy*: G. – École supérieure de pharmacie de Paris année 1890-1891, no. 6.
Ref.: Roth, E., Bot. Jahrb. 16(Lit.): 6-7. 1892.

7713. Série A n° 320 n° d'ordre 979. Thèses présentées à la Faculté des sciences de Paris pour obtenir le grade de docteur ès sciences naturelles par M.E. Perrot, ... 1re thèse. – *Anatomie comparée des Gentianacées* ... soutenues le [ ] février 1899 ... Paris (Masson et c^{ie}, ...) 1899. Oct. (*Anat. comp. Gentian.*).
Publ.: Feb 1899, p. [i, iii], [105]-294, [1, 3], *pl. 1-9* (uncol., auct.). *Copy*: G. – Reprinted from Ann. Sci. nat., Bot. ser. 8. 7(2-6): 105-294, *pl. 1-9*. Feb 1899.
Ref.: Belzung, E., Bull. Soc. bot. France 46 (bibl.): 50-52. 1899 (rev.).
Roth, E., Bot. Centralbl. 83: 246-250. 1900 (extensive summary).

7714. *Les algues marines* utiles et en particulier les algues alimentaires d'Extrême-Orient ... [Paris (Masson et Cie, éditeurs ...) 1911]. Qu.
Publ.: 1911 (Nat. Nov. Sep(1) 1911), p. [1]-101, *pl. 1-10*. *Copies*: NY, PH. – Published in and to be cited from Ann. Inst. océanogr. 3(1): [1]-101, *pl. 1-10*. 1911.

7715. *Matières premières d'origine végétale* ... Paris (Librairie le François ...) 1924. Oct. (*Mat. orig. vég.*).
Publ.: 1924, p. [1]-43. *Copy*: G.

7716. *Plantes médicinales de France* ... Paris [1:] (Office national des Matières premières végétales ...), Saint-Aubin-d'Écrosville (Établiss^{ts} du Dr Auzoux ...) 1928-1943, 4 vols. Oct (*Pl. méd. France*).
1: 1928, p. [1]-64, *pl. 1-48* with text.
2: Mar 1934, p. [1]-72, *pl. 49-96, pl. A1-H1* id. – Paris (Centre de Documentation technique et économique sur les plantes médicinales ...).
3: Mai 1938, p. [1]-41, [42-50], *pl. 97-144, pl. A2-H2* (id.).
4: 1943 (dépot légal 15 Oct 1943), p. [1]-47, [48], *pl. 145-200, pl. A3-H3* (id.).
Copy: BR. – Earlier parts reprinted p.p. from Revue des Agriculteurs de France (n.v.).

7717. *Sur les productions végétales indigènes ou cultivées de l'Afrique occidentale française* (Sahara, Soudan, Nigérien, Haute Volta, Guinée). Rapport sur la mission ... Octobre 1927-Janvier 1928 ... Lons-le-Saunier (Imprimerie L. Declume) 1929. (*Prod. vég. Afr. occid.*).
Publ.: Jun 1929 (t.p.), p. [i]-viii, [1]-468, *pl. 1-23*, 2 maps. *Copy*: BR. – Trav. Off. natl. Mat. prem. veg., Notice 31.

Perrottet, George (Georges Guerrard) **Samuel** (1793-1870), Swiss-born French gardener, sericiculturist, agronomical botanist, traveller and director of tropical cultures; in the "Rhône" around the world 1819-1821; Sénégal 1824-1829; in Pondichéry 1834-1839 and, ultimately, as government botanist from 1843-1870. (*Perrottet*).

HERBARIUM and TYPES: P and PC; a considerable world-wide collection also at G; duplicates in many herbaria. *Plantae Senegambiae* (senegalenses) and *Plantae Pondicerianae*, irregular sets of herbarium specimens, were offered for sale (e.g. Bot. Zeit. 13: 183. 1855, 14: 599. 1856, 15: 175, 312. 1857; Bonplandia 6: 342. 1858, 10: 31. 1862; Flora 38: 112. 1855. – Letters and portrait at G.

BIBLIOGRAPHY and BIOGRAPHY: Backer p. 436; Barnhart 3: 71; BL 1: 53, 312; BM 4: 1551; Clokie p. 223; CSP 4: 835, 12: 568; Frank (Anh.): 75; Hortus 3: 1201 (Perrottet); Herder p. 217; IH 1 (ed. 6): 362, (ed. 7): 339 (set in DS); 2: (in press); Jackson p. 210; Kew 4: 298; Lasègue p. 89-94, 579 (index); MW suppl. p. 278; PR 3647, 7047-7052, 7601 (ed. 1: 7887-7892); Rehder 5: 662-663; SK 1: 404-405 (portr., coll., itin.), 5: cccx; TL-1/462; TL-22211, 6244, see also Delessert; Tucker 1: 550; Urban-Berl. p. 272, 310,

BIOFILE: Abonnenc, E. et al., Bibl. Guyane franç. 1: 210. 1957.
Anon., Biogr. gén. 39: 653-654. 1862. (bibl.); Bot. Zeit. 28: 408. 1870; Bull. Soc. bot. France 2: 560. 1855 (exsicc. for sale); 17(bibl.): 45. 1870, 17(c.r.): 112. 1870, 29: 186. 1882 (on his collecting in Sénégal); Bull. Soc. Linn., Paris 2: 18. 1824 (leaves France for Guadeloupe and Sénégal); Cat. gén. livr. impr. bibl. natl., Paris 134: 271-272. 1835; Flora 16: 272. 1833 (price sericiculture), 20: 239. 1837, 23: 720. 1840, 24: 575-576. 1841 (in Nilgherry Mts.), 53: 174. 1868 (d.).
Beck, G., Bot. Centralbl. 34: 150. 1888 (material at W).
Buchinger, Bot. Zeit. 13: 183. 1855 (Pondichéry plants for sale).
Burkill, I.H., Chapt. hist. bot. India 20, 55, 89, 156. 1965.
Candolle, Alph. de, Phytographie 439. 1880 (main set at G, considerable set also at W).
Candolle, A.P. de, Mém. souvenirs 373, 410. 1862.
Chavannes, E.L., Notices historiques sur MM. Samuel Perrottet et Louis Agassiz, ... extrait de la Feuille du Canton de Vaud, 18. année, cahiers no. 223 et 224. Lausanne (Imprimerie des Frères Blanchard) 1831, p. [1]-20, *copy*: G.
Hasskarl, J.V., Allg. Sach– Namenreg. Flora 172. 1851.
Hedge, I.C. & J.M. Lamond, Index coll. Edinb. herb. 118. 1970.
Jessen, K.F.W., Bot. Gegenw. Vorz. 469, 472. 1884.
Kukkonen, I., Herb. Chr. Steven 78. 1971 (material at H).
L.Hn., Dict. Sci. méd. 75. 517. 1887.
Lindemann, E. v., Bull. Soc. imp. Natural. Moscou 61(1): 54. 1886 (material at MW).
Masy, M., Verh. schweiz. nat. Ges. 90(1): 7-8. 1908.
Moreau, F., Rev. bryol. 21: 135. 1952 (Indian lich.).
Murray, G., Hist. coll. BM(NH)173. 1904 (coll. at BM include mosses).
Perrottet, G.S., Das Ausland 6: 285 ff. 1833, 7: 115 ff. 1834 ("Wanderungen in Senegambien"; series of 23 and 13 papers, see list in CSP 5: 568; *in* Bajot et Poiré, Ann. mar. col. Paris 27 (ser. 3, non off. tome 1): 897-972. 1842 (fide Urban 1902; Rapport mission Inde, Bourbon, Martinique, Guadeloupe).
Robinson, C.B., Philipp. J. Sci. 3: 303-306. 1908 (P. and the Philippines).
Sayre, G., Mem. New York Bot. Gard. 19: 378-379. 1975 (crypt. coll.).
Steinberg, C.H., Webbia 32: 33. 1977 (material at FI)
Urban, I., Symb. ant. 3(1): 99-100. 1902 (on his W. Ind. coll.).
Vallot, Bull. Soc. bot. France 29: 186. 1882 (coll. in Sénégal).
White, A., & B.L. Sloane, The Stapelieae ed. 2. 95, 96, 242, 264. 1937.
Wittrock, V.B., Acta Horti Berg. 3(3): 172-173. 1905.

COMPOSITE WORKS: (1) with Guillaumin and Richard, A., *Florae senegambiae tentamen* (1830-1833), see TL-2/2211.
(2) See Richard, A., for *Monographie des Orchidées recueillies dans les Nils-Gherries* (Indes-orientales) par M. Perrottet, ... Paris 1841, iii, 36 p., *12 pl.*

EPONYMY: *Perrotetia* A.P. de Candolle (1825); *Perrottetia Kunth (1824)*.

7718. *Catalogue raisonné des plantes introduites dans les colonies françaises de Bourbon et de Cayenne* et de celles rapportés vivantes des mers d'Asie et de la Guyane au jardin du roi à Paris ... Paris (de l'imprimerie de Lebel, ...) 1824. Oct. (*Cat. pl. intr. colon.*).
Publ.: Mai-Dec 1824 (t.p. preprint; journal issue for Mai 1924, see Robinson 1908), p. [i-ii], [1]-63. *Copies*: FI, G, NY. – Preprinted from Mém. Soc. Linn. Paris 3(1): 89-128 for Mai 1824, 1824, publ. as a whole only after Feb 1826, 3(2): 129-151. "1824" publ. 14 Jun 1826 (BF; see MD p. 182).
Ref.: Robinson, C.B. Philipp. J. Sci. 3: 303-306. 1908.

7719. *Art de l'indigotier* ou traité des indigofères tinctoriaux et de la fabrication de

l'indigo, suivi d'une notice sur le Wrightia tinctoria et sur les moyens d'extraire de ses feuilles le principe colorant qu'elles contiennent; ... Paris (Librairie de Mme Ve Bouchard-Huzard, ...) 1842. Oct. (*Art indigotier*).
Publ.: 1842, p. [i]-vii, [1]-219, chart. *Copy*: MO.

7720. *Catalogue des plantes du jardin botanique et d'acclimatation du gouvernement à Pondichéry.* Pondichéry (É.-V. Géruzet, ...) 1867. (*Cat. jard. bot. Pondichéry*).
Publ.: 1867 (BSbB rd. 28 Dec 1867-11 Dec 1868, p. [i]-ix, [x-xi], [1]-51. *Copy*: NY.
Ref.: Anon., Bull. Soc. bot. France 15(bibl.): 17. 1868.

Perroud, Louis François (1833-1889), French botanist; Dr. med. Paris 1858; connected with the Faculté de Médecine and hospital physician at Lyon from 1860-1889. (*Perroud*).

HERBARIUM and TYPES: LY.

BIBLIOGRAPHY and BIOGRAPHY: Barnhart 3: 71; BL 2: 118, 181, 701; Bossert p. 306; CSP 10: 1033, 12: 568, 17: 808; Jackson p. 278; Morren ed. 10, p. 64; Rehder 5: 663; Saccardo 2: 83; Tucker 1: 550.

BIOFILE: Beauvisage, Ann. Bot. 3: 485. 1890 (bibl.).
Boullu, Abbé, Bull. Soc. bot. France 36(1): 182-183. 1889 (letter Boullu to Malinvaud on death P., d. 26 Feb 1889).
Cavillier, F.G., Boissiera 5: 72. 1941 (b. 26 Feb 1833, d. 26 Feb 1889).
Holden, W., Bibl. contr. Lloyd Libr. 4: 175. 1911.
Magnin, A., Bull. Soc. bot. Lyon 32: 53-54, 136. 1907, 35: 47. 1910.
Saint-Lager, J.B., Ann. Soc. bot. Lyon 17: 291-298. 1890 (1891), bibl.; herb. left to Facultés de science et de médecine, Lyon; now LY).

Perry, Lily May (1895-x), Canadian-born American botanist; Dr. phil. Wash. Univ., St. Louis 1932; at Gray Herbarium 1925-1932, University of Georgia 1932-1934, Sweet Briar College 1934-1935; later at Arnold Arboretum (*L.M. Perry*).

HERBARIUM and TYPES: GH; other material at ACAD, K, L, MO, MTJB.

BIBLIOGRAPHY and BIOGRAPHY: BJI 2: 133; BL 1: 141, 312; Bossert p. 307 (b. 5 Jan 1895); Hortus 3: 1201 (L.M. Perry); IH 1 (ed. 1): 46, (ed. 2): 63, (ed. 3): 41, (ed. 4): 44, (ed. 5): 37; Kew 4: 278; Langman p. 580; NW p. 55; Roon p. 87; SK 1: cxlvii, 4: cxix, cxxxi, cxlvi, clii; Zander ed. 10, p. 700-701, ed. 11, p. 799 (b. 5 Jan 1895).

BIOFILE: Anon., J. Arnold Arb. 56(1): frontisp., [i]. 1975 (portr., tribute 80th birthday).
Jacobs, M., ed., Fl. males. Bull. 29: 2531. 1976 (80th birthday), 32: 3189. 1979 (second retirement).
Moldenke, H.N., Plant Life 2: 75. 1946 (1948).
Steenis, C.G.G.J. van, Fl. males. Bull. 13: 551. 1957 (to Europe), 26: 1999. 1972 (Dr. h.c. Arcadia Coll. 1971).
Steenis-Kruseman, M.J.van, Blumea 25(1): 41. 1979.
Verdoorn, F., ed., Chron. bot. 2: 329, 404. 1936, 3: 293, 294. 1937.

HANDWRITING: *Flora malesiana* 1: cli. 1950.

EPONYMY: *Perrya* Kitton (1874) commemorates Kitton's friend, Captain Perry, of Liverpool, England.

DEDICATION: J. Arnold Arb. 56(1). Jan 1975 (on the occasion of her 80th birthday; portr.).

7721. *Washington University doctoral dissertations. A revision of the North American species of Verbena* a dissertation presented to the Board of graduate studies in partial fulfillment of the requirements for the degree of doctor of philosophy, June 1932 by Lily May Perry

... Publications of Washington University Saint Louis [1933]. Oct. (*Rev. N. Amer. Verbena*).
Publ.: 10 Jul 1933 (p. 239), p. 239-362, *pl. 13-15*. *Copies*: G, MO, NY, USDA. – Reprinted with special cover (with above title) from Ann. Missouri bot. Gard. 20(2): 239-362, *pl. 13-15*. 1933.

7722. *A tentative revision of Alchemilla § Lachemilla* ... Cambridge, Mass. (published by the Gray Herbarium of Harvard University) 1929. Oct. (*Tent. rev. Lachemilla*).
Publ.: 18 Jul 1929 (copies so stamped), p. [1]-57. *Copies*: G, MO, NY. – Issued as Contr. Gray Herb. 84. 1929.

Perry, William Groves (1796-1863), British bookseller and botanist, curator of the Warwickshire Museum at Warwick. (*W.G. Perry*).

HERBARIUM and TYPES: WAR; further material at OXF.

BIBLIOGRAPHY AND BIOGRAPHY: Barnhart 3: 71; BB p. 242; BM 4: 1551, Clokie p. 223; Desmond p. 490; IH (ed. 7): 339; Jackson p. 261; Kew 4: 279; PR 7053; Rehder 1: 397.

BIOFILE: Amphlet, J. and C. Rea, Bot. Worcestershire xxii. 1909 (fide Desmond).
Bagnall, J.E., Fl. Warwickshire 494-497, 500. 1891.
Cadbury, D.A., Hawkes, J.G. and R.C. Readett, A computer-mapped flora. A study of the flora of Warwickshire 52-54. 1971 (herb.).
Irvine, A., ed., Phytologist 6: 605. 1863 (d. 25 Mar 1863; obit. notice).
Kent, D.H., Brit. herb. 72. 1957 (herb. at WAR).

7723. *A select list of plants found in Warwickshire*, by W.G. Perry, of the Museum, Leamington. [*in*: W. Dugdale, Warwickshire; being a concise topographical description of the different towns and villages in the County of Warwick, ... Coventry (printed by and for John Aston, ...) 1817]. Oct. (*Select. list. pl. Warwicks.*).
Publ.: 1817 (advertisement: 5 Jul 1817), l.c. p. 591-594. *Copy*: LC.

7723a. *Plantae varvicenses selectae*; or botanist's guide through the county of Warwick ... Warwick (printed by W. Perry; ...) 1820. Oct. (in fours). (*Pl. varvic. sel.*).
Publ.: 1820, p. [i-iv], [1]-120. *Copy*: HH.

Persoon, Christiaan Hendrik (1761/1762-1836), South African botanist (Dutch citizen), especially mycologist, who was educated (high school) at Lingen; studied theology at Halle 1783-1786, and medicine and natural sciences at Göttingen 1787-1802; Dr. phil. Leopoldina 1799; lived in Paris 1802-1836; left his herbarium and library to the Dutch government after having exchanged his earlier herbarium in 1828 for an annual pension with the same government. (*Pers.*).

HERBARIUM and TYPES: L; some further material at BM, E, G, GOET, H (lich. in herb. Acharius), LD, LG, NY, PC, STR. – Letters to Persoon at the University Library of Leiden; some letters by Persoon at G and HU.

BIBLIOGRAPHY and BIOGRAPHY: Aa 15: 210; AG 2(2): 62, 4: 628; Ainsworth p. 312, 313, 346 [index]; Backer p. 658 (err.); Barnhart 3: 72; BM 4: 1552, 8: 1000; Bossert p. 307; Bret. p. 275; CSP 4: 837-838; Dawson p. 667; Dryander 3: 90,225, 342-343, 444; DSB 10: 530-532 (by M.A. Donk); Frank (Anh.): 75; GR p. 704, *pl. [32]*, cat. p. 69; Hawksworth p. 185, 197; Herder p. 470; Hortus 3: 1201 (Pers.); IH 1 (ed. 6): 362, (ed. 7): 339, 2: (in press); Jackson p. 114, 162, 228; JW 1: 447, 3: 365, 4: 401, 5: 248; Kelly 173-174, 256; Kew 4: 279; KR p. 571; Krempelh. 2: 50-51; Lasègue p. 579 [index]; LS 20485-20501; Moebius p. 74, 97, 98, 141; MW p. 383; NAF 1: 167; NI 1511-1514, see also 1: 144, suppl. 39, 54; PH 438; PR 4187, 5430, 7054-7064, 9295 (ed. 1: 7893-7903); Rehder 5: 663; RS p. 133; SBC p. 129; SO 603; Stevenson p. 1254; TL-1/396, 515, 714 (ed. 15), 961-969, 1156; TL-2/322-323, 554, 793, 1409, 1464, 1968, 2885, 2956, 3387, 4215, 4455-4456, 4500, see indexes; Tucker 1: 550; Zander ed. 10, p. 701, ed. 11, p. 799.

BIOFILE: Anon., Bonplandia 8: 221-222. 1860 (on Humboldt, Link and Persoon at Göttingen); Flora 12: 185-186. 1829 (herb. bought by Netherlands), 19: 752. 21 Dec 1836. (d. Nov 1836).
Ainsworth, G.C., Nature 193: 22-23. 1962; Dict. fungi ed. 6, p. 435. 1971 (statement importance).
Baines, J.A., Vict. Natural. 90: 249. 1973 (erroneous).
Barnhart, J.H., J. New York Bot. Gard. 23: 165. 1922 (in many respects erroneous), Bartonia 16: 32. 1934.
Benjamin, C.R., Proc. Iowa Acad. Sci. 60: 92-94. 1953.
Bessey, C.E., Amer. Natural. 24(288): 1196. 1890 (letter of Persoon to [probably] Sowerby dated 2 Mai 1801, sending him a copy of *Syn. meth. fung.*).
Boivin, B., Brittonia 14: 327-331. 1962 (Persoon and the subspecies; correct use of subspecies in botany originated with Persoon in 1805 (*Syn. pl.*).
Bridson, G.D.R. et al., Nat. hist. mss. res. Brit. Isl. 428 [index]. 1980.
Candolle, A.P. de, Mém. souvenirs 141. 1862.
Chater, A.O. and R.K. Brummitt, Taxon 15: 143-149. 1966 (subspecies in the works of P.).
Chiovenda, E., Bull. Soc. bot. ital. 1923 (8-9): 122-126. 1923 (on Raddi and Persoon; correspondence).
Clausen, R.T., Rhodora 43: 157-167. 1961 (introduction "subspecies" by Persoon, 1805).
Darlington, W., Reliq. Baldwin. 171, 172, 173, 335. 1843.
Dawson, W.R., Smith papers 73-74. 1934.
Dennis, R.W.G., Kew Bull. 1952(3): 301-302 (fungi err. assigned by P. to the Discomycetes).
Donk, M.A., C.H. Persoon (unpubl. text address 1966 on restoration grave 3 p., 21 Jun 1966, copy FAS).
Durand, E.J., J. of Mycol. 13: 141-142. 1907 (relation mycol. writings T. Holmskjold to Persoon's *Comm. fung. clavaef.*).
Egeling, G., Österr. bot. Z. 11: 377-378. 1881.
Fée, A., in Parlatore, Ph., Giorn. bot. ital. anno 2, 1(3): 69-76, 1846; translated into French by Mme R. Rousseau *in* Bull. Soc. Bot. Belg. 30: 50-60. 1891; translated into Dutch by C.E. Destrée; Ned. kruidk. Arch. ser. 2. 6: 366-377. 1895 (biography in the romantic tradition and not quite reliable. Fée introduced the Hottentot notion for Persoon's mother; account of transaction sale first herbarium in 1828; personal account of P's later years).
Fitzpatrick, H.M., Mycologia 35: 256-257. 1943, 36: 177-187. 1944 (bibl. study of *Icon. pict. sp. fung.*).
Forbes, R.J., ed., Martinus van Marum 1: 392. 1969, 2: 289 (v. Marum visits P;), 290. 292, 307. 1970, 3: 137. 1971, 5: frontisp. portr. 1974.
Franken, J.L.M., Ann. Univ. Stellenbosch 15(B)(4): 1-100. 1937 (basic biographical study, mainly of P's life until 1800; much source material on family; English summary by R. de Zeeuw, Mycologia 31: 369-370. 1939); J.S. Afr. Bot. 4: 127-128. *pl. 39*. 1938 (on Persoon's death, grave at Le Père Lachaise, will and estate).
Greuter, W., Candollea 23: 90-96. 1968 (use of subspecies).
Gunn, M. & R.E. Codd, Bot. expl. S. Afr. 279. 1981.
Hasskarl, J.K., All. Sach- Namenreg. Flora 172-173. 1851.
Hasslow, O.J., Bot. Not. 1933: 610-614.
Hedge, I.C. & J.M. Lamond, Index coll. Edinb. herb. 118. 1970 (fungi in herb. M.C. Cooke).
Hesler, L.R., Biogr. sketches N. Amer. mycol., mss. 1975.
Hugo, C.E., J. Bot. Soc. S. Afr. 51: 48-51. 1965 (proposal to restore grave Persoon at Père Lachaise), 52: 13-16. 1966 (grave restored).
Jacobs, M., Fl. males. Bull. 31: 2972. 1979 (ref. to Petersen 1979; announces biogr. of Persoon by W.F.B. Jühlich).
Jessen, K.F.W., Bot. Gegenw. Vorz. 373, 405, 408, 461. 1864.
Junghuhn, F., Flucht nach Afrika 308-309. 1834 (describes visit to Persoon Oct 1834) (*in* M.C.P. Schmidt, Fr. Junghuhn, Biogr. Beitr. 1834).
Junk, W., Rara 2: 171-172. 1926-1936.
Kalkman, C. & P. Smit, ed., Blumea 25(1): 21, 83, 86, 96. 1979.

Killerman, S., Z. Pilzk. 4(=9) (6): 92-96.1925 (biogr.) 5(=10) (3/4): 50-57. 1925 (updates nomenclature of many Persoon species of fungi).
Laplanche, M.C., Dict. iconogr. champ. sup. 1894 (TL-2: 4215) (concordance nomencl.).
Laundon, G.F., Taxon 17: 179-180. 1968 (Persoon's Uredo names); Lichenologist 11: 16. 1979.
Lawrence, G.H.M., Huntia 1: 164. 1964 (letters to Persoon at HU in corr. C.H. Mertens).
Lefebvre, E. & J.G. de Bruyn, Martinus van Marum 6: 128, 285, 390. 1976.
Lek, H.A.A. v.d., Meded. Rijksherbarium 18: 1-12. 1913 (types *Polyporus*).
Le Turquier & Levieux, Concord. fig. crypt. Dillen (1820 TL-2/4455; concordance nomenclature Dillen et al. with Persoon), Concord. Persoon (1826, TL-2/4456, id. Persoon/Fries, DC, Bulliard).
Lindemann, E. v., Bull. Soc. imp. Natural. Moscou 61(1): 54. 1886 (incorrect).
Lloyd, C.G., Mycol. Notes 16: 158-160. 1904 (based on Fée; some mistakes corrected 1924); 17:173-176. 1904; 27: 345. 1907; 35: 464-473. 1910 (Polypores of Persoon's herb.), 73: 1301-1303. 1924 (translation of L. Verwoerd; the ancestors of Christiaan Hendrik Persoon).
Lütjeharms, W.J., Gesch. Mykol. 260 [index]. 1976; Nieuwe Rotterd. Courant 15 Nov 1936 (accepts 1 Jan 1763 as date of birth); Vakblad Biol. 18(2): 42-44. 1936; Forum bot. 4(9): 1-2. 1966.
Martin, G.W., Taxon 9: 1-3. 1960.
Mears, J.A., Proc. Amer. philos. Soc. 122(3): 170. 1978 (Muhlenberg sent fungi to Persoon; on offer by Persoon to sell his herbarium to Muhlenberg).
Murrill, W.A., Hist. found. bot. Florida 27. 1945 (partly incorrect).
Palmer, J.T., Nova Hedwigia 15: 152. 1968 (bibl. Gasteromyc.).
Peltereau, E., Bull. Soc. mycol. France 44: 57-62. 1928 (on Holmskjold and Persoon).
Persoon, C.H., Mém. Soc. Linn. Paris 3: 79, 421-424. 1825 (on making herbarium specimens of fungi).
Petersen, R.H., Mycotaxon 1: 149-188. 1975; Kew Bull. 31: 685-698. 1977.
Raab, H., Schweiz. Z. Pilzk. 59: 153-154, 158. 1971 (brief note).
Radt, C., J. Agric. trop. Bot. appl. 8: 542-543. 1961 (1962) (translation of Ainsworth 1962; brief note on bicentenary birth).
Ramsbottom, J., J. Bot. 53: 277-279. 1915 (on *Obs. mycol.*, letter P. to Sowerby; note by J. Britton on letter P. to R. Brown (Mai 1825); Trans. brit. mycol. Soc. 18: 187-188. 1933 (Engl. index to *Syn. meth. fung.*); Proc. Linn. Soc. London 146: 10-21. 1934 (relations and corr. J.E. Smith and P.; Persoon specimens at BM; pre-publ. sheets of *Syn. meth. fung.*; F.M.L.S. 16 Jul 1799; offer in 1818 to sell his herb. to Smith).
Reynolds, G.W., Aloes S. Africa 93. 1950 (on P's Aloes).
Rogers, D.P., Mycologia 36: 530. 1944 (dates publ. *Abh. essb. Schwämme* and *Mycol. eur.* 1); Brief hist. mycol. N. Amer. ed. 2: 6, 8, 10. 1981.
Roumeguére, C., Revue mycol. 11: 17-34. 1889 (corr. between J.B. Mougeot and P.).
Rousseau, M., Bull. Soc. Bot. Belg. 30(2): 50-60. 1891.
Sachs, J. v., Hist. botanique 219. 1892.
Sayre, G., Bryologist 80: 515. 1977 (bryoph. also at H).
Schmid, G., Z. Pilzk. 17(2): 54-60. 1933 (P's contribution to G.F. Hoffmann, *Abbild. Schwämme* (1790-1793), TL-2/2885).
Seaver, F.J., N. Amer. cup-fungi 363. 1942.
Singer, R., Taxon 9: 35-37. 1960 (Persoon, Syn., 1801, starting point?).
Stafleu, F.A., Linnaeus and Linnaeans 251. 1971.
Steenis-Krusemann, M.J. van, Blumea 25: 38. 1979.
Svrček, M., Česká Mykologie 15(2): 124-126. 1961 (bicent. birth).
Taton, R., ed., Science in the 19th cent. 382. 1965.
Verduyn de Boer, J.H., Die Huisgenoot 22(814): 27 (portr.), 85; 2(815): 39, 43. 1937 (letter by J.B. Hoffmann to P's father dated 27 Dec 1766 congratulating him on the birthday of Christiaan; accepted birth date 31 Dec 1761), Nieuwe Rotterd. Courant 2 Dec 1936.
Verwoerd, L., Contr. Lab. Phytopath. Stellenbosch 1-6. 1924, see also Lloyd 1924.
Verdoorn, F., ed., Chron. bot. 1: 336. 1935, 2: 21. 1936, 6: 238. 1941.
Z., Dict. sci. méd. 6: 396. 1824 (bibl.).

COMPOSITE WORKS: (1) *Commentatio de fungis clavaeformibus*, in Th. Holmskjold, *Coryphaei clavarias Ramariasque* 131-236. 1797, TL-2/2956; for separate issue see below, no. 7725.
(2) with L. de Brondeau, *Descr. champ. nouv.* (1824), TL-2/793.
(3) contributed to N.A. Desvaux, *J. Bot.* (108), TL-2/1409.
(4) *Fungi*, in C. Gaudichaud, *Voy. Uranie* (1826-1830), TL-2/1968.
(5) contributed many new taxa to G.F. Hoffmann, *Abbild. Schwämme* (1790-1793), TL-2/2885.
(6) Schaeffer, J.C., *Fung. Bav. Palat. nasc.* ed. nova, 4 vols., 1800, was edited by Persoon with an added commentary in vol. 4. See under Schaeffer.
(7) *Lichen xanthostigma*, in Acharius, E., Lichenogr. univ. 403. 1810.

EPONYMY: (genera): *Persoonia* A. Michaux (1803); *Persoonia* J.E. Smith (1798, *nom. cons.*); *Persoonia* Willdenow (1799); *Persooniana* Britzelmayr (1897); *Persooniella* H. Sydow (1922); (journal): *Persoonia*; a mycological journal. Leiden. Vol. 1-x. 1959-x.

NOTE: The date of birth of Persoon has not been established with certainty. It is likely that he was born on 31 Dec 1761, but Franken (1937), who consulted much original material, accepts 31 Dec 1762. Lütjeharms (1936) is also of the opinion that 31 Dec 1762 is probable but does not even exclude 1 Jan 1763. Verduyn de Boer (1937) and Donk (DSB 10: 530), however, accept 31 Dec 1761. Verwoerd (in Lloyd 1924), who was the first author to accept this date, deduced it from the Government Gazette (Cape Archives) of Friday 10 Mar 1837 ("died, at Paris, on the 16th of November, 1836, age 74 years, 10 months, and 16 days, Christiaan Hendrik Persoon, ...") signed by his sister, J.M. Storm, born Persoon). Persoon's mother was not a "Hottentot", a misunderstanding by Fée of the notion "Afrikaner" often repeated in the literature. Actually she was a Cape-born woman, Elsie [Elizabeth] Wilhelmina Groeneveld, who had a Cape-Dutch father (Johannes G.) and mother (Margareta Hatting). Persoon's father, Christiaan Daniel, was born in Usedom, Pommerania, Germany (orig. Persohn), Persoon's mother died May 1763.

The Gymnasium (then still called Seminarium) at Lingen, Germany where Persoon received his secondary education was the same as that where F.A.W. Miquel went to school in 1828-1829.

HANDWRITING: Lloyd, C.G., Mycol. Notes 35: 465. 1910; Bartonia 28: *pl.* 7. 1957; Candollea 32: 391-392. 1977; TCHAF voorjaar 1979, Leiden, p. 47.

CONCORDANCES: Many publications correlate Persoonian nomenclature with later conventions, see e.g. Killermann (1925), Laplanche (1894), Laundon (1968), Le Turquier (1826), Lloyd (1910), Schmid (1933).

7724. *Observationes mycologicae.* Seu descriptiones tam novorum, quam notabilium fungorum exhibitae a C.H. Persoon ... Lipsiae [Leizig] ([1]: apud Petrum Phillippum [sic] Wolf; [2:] Lipsiae et Lucernae [Luzern] (sumptibus Gessneri, Usterii et Wolfii) 1796-1799 [-1800], 2 parts. Oct. (*Observ. mycol.*).
1: 1796 (GGA: 21 Mai 1796; P. Usteri, Ann. Bot. 20: 115-130. 1796), p. [i-iv], [1]-115, [116, err.], *pl. 1-6* (handcol. copp., Besemann). – An amended version of a paper with the same title published in Usteri, Ann. Bot. 15: 1-39, *3 pl.* 1795 (not using the same type). A notice by Persoon on the *Obs. mycol.* vol. 1, 1796, with additions and corrigenda appears in Usteri, P., Ann. Bot. 21: 115-130. 1796.
2: Jun 1800 (t.p. 1799; p. vi: Apr 1799; Jun 1799 not yet out, fide letter to J.E. Smith; copies sent to Sowerby and to Smith on 19 Jun 1800; GGA 18 Oct 1800; second t.p. [ii*] present in MO copy, dated 1800: "*Animadversiones et dilucidationes circa varias fungorum species ...*"), p. [ii*], [i]-xii, [1]-106, [1, err.], *pl. 1-6* (id., by Besemann).
Copies: BR, MO, Stevenson; IDC 5095.
Studies: Killermann, (1925, p. 50-51), Laplanche (1894, p. 467-468), Ramsbottom and Britten (1915); TL-1/961. We are grateful to D.P. Rogers for various notes on the Observ. mycol., including a note on the probability of the 1800 date for vol. 2.
Facsimile ed.: 1967, New York (Johnson Reprint Corporation), East Ardsley, Wakefield, Yorkshire (S.R. Publishers Ltd.). Extra t.p. and colophon on p. [i*-ii*], followed by facsimile.
Copy: B.

7725. C.H. Persoonii *Commentatio de fungis clavaeformibus* sistens specierum hus usque notarum descriptiones cum differentiis specificis nec non auctorum synonymis ... Lipsiae [Leipzig] (apud Petrum Philippum Wolf) 1797. Oct. (*Comm. fung. clav.*).
Publ.: 1797 (rev. P. Usteri, Ann. Bot. 22: 104. 1797), p. [i-vi], [1]-124, [1, err.], *pl. 1-4* (col. copp., auct.). *Copies*: G, MO; IDC 6294. Published simultaneously in Holmskjold, T., *Coryph. Clav. Ramar.* 131-239[-240], *pl. 1-4*. 1797 TL-2/2956). See also TL-1/963. For concordance see Laplanche (1894, p. 456-457); for updated nomenclature e.g. Killermann (1925, p. 57); see also Durand (1907) and Peltereau (1928).

7726. Caroli a Linné equitis *Systema vegetabilium* secundum classes ordines genera species cum characteribus et differentiis. *Editio decima quinta* quae ipsa est recognitionis ab. Io. Andrea Murray institutae tertia procurata a C.H. Persoon ... Gottingae (typis et impensis Io. Christ. Dieterich) 1797. Oct. (*Syst. veg. ed. 15*).
Publ.: 1 Mai-9 Jun 1797 (p. iv: 20 Apr 1797; GGA 10 Jun 1797), p. [i]-xvi, [1]-1026, [1-19, indexes]. *Copies*: FAS, FI, G, MO, US(2), USDA. – See also TL-1/964, KR p. 428 and SO 603; for Murray eds. 13 and 14 see sub Murray. For the full series of Linnaean texts edited as *Systema vegetabilium* see Linnaeus, TL-2/4709.
French editions: See Jolyclerc, TL-2/3387, *Syst. sex. vég.*, originally published 1798. We have now seen a copy of the 1803 edition, "A Paris, Arthur Bertrand, Librairie, ... An xi-1803" at NY, p. [i-vi], [1]-789, [790, avis].

7727. *Tentamen dispositionis methodicae fungorum* in classes, ordines genera et familias. Cum supplemento adjecto ... Lipsiae [Leipzig] (apud Petrum Philippum Wolf) 1797. Oct. (*Tent. disp. meth. fung.*).
Orig. ed.: 14 Oct-31 Dec 1797 (p. iv: 14 Oct 1796; GGA 15 Feb 1798, p. [i]-iv, [1]-76, *pl. 1-4* (uncol. copp.). *Copies*: G, Stevenson; IDC 6295. The text on p. 1-48 is a second impression (from original type) of Persoon, *Dispositio methodica fungorum*, in Roemer, J.J., Neu. Mag. Bot. 1: 81-128. 1794; p. 49-76 constitute a *Supplementum* with new matter. For the use of the category subspecies in this work see Chater & Brummitt (1966); see also the note by Barnhart in NAF 9: 448.
Lisbon ed.: 1800. *Copy*: K (n.v.). Published from new type with alterations.
Note: The full paper by Persoon from which the *Dispositio* was taken, was entitled "*Neuer Versuch einer systematischen Eintheilung der Schwämme*", in Roemer, ib. p. 63-128.

7728. *Icones et descriptiones fungorum* minus cognitorum ... Lipsiae [Leipzig] (Bibliopolii Breitkopf-Haerteliani impensis) [1798-1800], 2 fasc. Qu. (in twos). (*Icon. descr. fung.*).
1: 1798 (GGA 19 Jan 1799), p. [i-vi], [1]-26, *pl. 1-7* (col. copp. Besemann).
2: 1800 (letter Persoon to Banks 18 Apr 1800, fide Dawson, "has dedicated the second part ... to Banks."), p. [i], [27]-60, *pl. 8-14* (id. but unsigned).
Copies: FI-BN, MO, NY; IDC 6296. – *Studies*: Killermann (1925), p. 51-52; Laplanche (1894) p. 458-460.

7729. *Commentarius* D. Iac. Christ. Schaefferi ... fungorum Bavariae indigenorum *icones pictas* differentiis specificis, synonymis et observationibus selectis illustrans. Auctore D.C.H. Persoon ... Erlangae (apud Ioan. Iac. Palm.) 1800. Qu. (*Comm. Schaeff. icon. pict.*).
Publ.: 1800 (p. [xxii]: Mar 1800; JT 16 Mar 1801), p. [i-xxii], [1]-130, [1-7, ind., 8 err.]. *Copies*: BR, FI, NY(2), USDA(2); IDC 6297. – For the new edition of Schaeffer, *Fung. Bav. Palat. nasc.* (4 vols., 1800) edited by Persoon, see J.C. Schaeffer.

7730. *Synopsis methodica fungorum*. Sistens enumerationem omnium huc usque detectarum specierum, cum brevibus descriptionibus nec non synonymis et observationibus selectis. Auctore D.C.H. Persoon. ... Gottingae [Göttingen] (apud Henricum Dieterich) 1801, 2 parts. Oct. (*Syn. meth. fung.*).
Publ.: 31 Dec 1801, conventional date under the International Code of Botanical Nomenclature, Art. 13.1(e) until the Sydney Congress (1981) as starting point for the nomenclature of Uredinales, Ustilaginales, and Gasteromycetes. The book has been under consideration as starting point for all fungi; for a discussion see e.g. Singer (1960), Martin (1960), and Petersen (1979). – The Congress at Sydney (1981) ac-

cepted proposal C to Article 13 (see Taxon 30: 106. 1981) whereby the starting point for nomenclature of fungi becomes Linnaeus, Sp. pl. (1753) with the provision that "names in the Uredinales, Ustilaginales, and Gasteromycetes adopted by Persoon (Syn. math. fung., 31 Dec 1801) ... are not affected by, and take priority over, homonymous or synonymous names published earlier".

1: [i]-xxx, [1]-240, *pl. 1-5* (uncol. copp.).
2: [i-ii], [241]-706, [1-2, index].
Copies: B, BR, FI, G, H, MO, NY, Stevenson; IDC 5023.
Index: 1808, *Index botanicus* sistens omnes fungorum species in D.C.H. *Persoonii Synopsi methodica fungorum* enumeratas una cum varietatibus et synonymis, confectus a D.G.H.L. [Georg Herman Lühnemann (also: Lünemann) 1780-1830]. Gottingae [Göttingen] (apud Henricum Dieterich) 1808. Oct. (*Ind. bot. Pers. Syn. meth. fung.*), p. [1]-36. *Copies*: B, BR, G.
First reprint: 1819, by T.I.M. Forster (see Trans. Brit. myol. Soc. 18: 187-188. 1933). - London : reprinted for T. and G. Underwood ... 1819.
Second reprint: 1952, in facsimile ed. *Syn. meth. fung.*, see below.
Facsimile ed.: 1952, New York, N.Y. (Johnson Reprint Corporation), with extra 1952 t.p., facsimile of text as above, with the index by G.H. Lühnemann.
Studies: Le Turquier & Levier (1826, concordance with DC), Killermann (1925, 1252, updated nom.), Lütjeharms (1936, p. 216-235, methodology), Singer (1960, starting point); Bessey (1890, copy sent to Sowerby 2 Mai 1801); Ramsbottom (1933, p. 13-16, on pre-publ. sheets sent to Smith, unpublished and containing several mistakes corrected in effectively publ. issue).

7731. *Icones pictae specierum rariorum fungorum* in Synopsi methodica descriptarum a C.H. Persoon. Fasciculus primus [-quartus]. Figures coloriées des espèces rares des champignons décrits dans l'ouvrage intitulé: Synopsis methodica fungorum ... à Paris et à Strasbourg (chez Amand Koenig, libraire) 1803-1808, 4 parts. Qu. (*Icon. pict. sp. fung.*).
Publ.: in four fascicles. Two issues: on "papier grand-raisin" and on "papier vélin", both with unsigned coloured copper engravings. Text in Latin and French. For details on dates see TL-1.
1: 13 Jun 1803, p. [i*], [i]-ii, [1]-14, [1, "table"], *pl. 1-6* (handcol. copp.).
2: 20 Jul 1804, p. [i, iii], 15-28, *pl. 7-12* (id.), [1, "table"].
3: 15 Apr 1805, p. [i], 29-44, *pl. 13-18* (id.), [1, "table"].
4: Jul-Sep 1808, p. [i, iii], 45-64, *pl. 19-24* (id.), [1, "table"].
Copies: B, G, MO (completed with photocopies of some of the missing pages, e.g. the "tables", thus now being made up as an "ideal copy" by H.M. Fitzpatrick (see his 1944 publ. for this and other copies thus completed), NY (id.), PH, Stevenson (1-3); IDC 5034.
Studies: Fitzpatrick (1944, bibliographic study), Killermann (1925, p. 52-53, nomenclature), Laplanche (1894, p. 461-462, concordance).

7732. *Synopsis plantarum,* seu enchiridium botanicum complectens enumerationem systematicam specierum hucusque cognitorum ... Pars prima [2(1): partis secundae sectio prima, 2(2): pars secunda] Parisiis lutetiorum [Paris] (apud Carol. Frid. Cramerum) et Tubingae (apud J.G. Cottam) 1805-1807, 2 vols. in 3. 16-mo. (*Syn. pl.*).
1: 1 Apr-15 Jun 1805 (p. xii: 12 Mar 1805; Gazette natl. 15 Jun 1805; Bot. Zeit. Regensb. 31 Jan 1806), p. [i]-xii, [1]-546.
2(1): Nov 1806, p. [i], [1]-272, t.p. imprint only "Tubingae, apud J.G. Cottam 1806", in USDA copy (Bot. Zeit. Regensb. 5: 321. 15 Nov 1806, "277" p.).
2(2): Sep 1807 (JGLF), p. [i-iv], 273-657, [1,-2, err.].
Copies: B, BR, FI, G, H, MO, NY, U, US(3), USDA; IDC 429.
Issues: on heavy and on light weight paper, publ. simultaneously.
The book is in 16-mo and printed in characters called "parisienne". The motto "in parvo copia" is particularly appropriate. The number of genera treated is 2.300, with circa 20.000 species. Many of the new taxa are based on specimens in the Jussieu herbarium at P.
Studies: Chater and Brummitt (1966; subgeneric names), Brizicky (1969, id.), Clausen (1941, id.), Boivin (1962, id.), Bentham (1881, on part L.C. Richard: New taxa often by L.C. Richard, but only to be attributed to him when his name is expressly in-

dicated), M. Breistoffer and M. Kerguélen drew our attention to the Gazette natl. 1805 reference.

Index: Lichtenstein, A.G.G., Index alphabeticus generum botanicorum quotquot a Willdenovio in speciebus plantarum et a Persoonio in synopsis plantarum recensentur. Helmstadt 1814. Oct.
Enlarged edition: see below, Species plantarum (1817-1821).
Index hort. Petrop.: 1816, *Index plantarum horti imperatoriae medico-chirurgicae Academiae*, quas secundum Synopsis Persoonii in systematicum ordinem redegit Jason Petrow, Doctor M. atque Botanices & Pharmacologiae Prof. P.O. Petropoli [St. Petersburg] (In Typographia Imperatoria) 1816, p. [1]-216. *Copy*: H (Univ. libr.). – We have no further information on Jason Petrow.

7733. *Species plantarum*, seu enchiridium botanicum, complectens enumerationem systematicam specierum hucusque cognitarum ... Petropoli [St. Petersburg] (Typis Caesar. Academiae Scientiarum] 1817-1821, 6 vols. Oct. (*Sp. pl.*).
1: 1817, p. [i]-vii, [1]-882.
2: 1819, p. [i], [1]-477
3: 1819, p. [i], [1]-464.
4: 1821, p. [i], [1]-455, [456], [i]-xii.
5: 1821, p. [i], [1]-436.
6: 1822, p. [i], [1]-287, [i]-viii.
Copies: H(1-6), HU(1-6), NY(1-3). Actually a slightly amended reprint of the *Synopsis plantarum*.

7734. *Traité sur les champignons comestibles*, contenant l'indication des espèces nuisibles précédé d'une introduction à l'histoire des champignons avec quatre planches coloriées ... Paris (chez Belin-Leprieur, libraire, ...) 1818. Oct. (*Traité champ. comest.*).
Orig. issue: 1818 (t.p.), p. [1]-10, [1]-276, [1-2, err.], [1, expl. pl.], *pl. 1-4* (handcol. copp.). *Copies*: FI, NY, Stevenson; IDC 6298.
Re-issue: 1819 (t.p.), p. [1]-10, *pl. 1-4*, (col.), [1, expl. pl.], [1]-276, [1-2, err.]. *Copies*: BR, MO, Stevenson.
Studies: Rogers (1944; relative dates with respect to Schweinitz, *Syn. fung. Carol. sup.* and Persoon, *Mycol. eur.*; Donk (unpubl. ms.) stressed the nomenclatural and taxonomic importance of the introduction.
German translations: see below, 1822, *Abh. essb. Schwämme*.

7735. *Abhandlung über die essbaren Schwämme*. Mit Angabe der schädlichen Arten und einer Einleitung in die Geschichte der Schwämme von C.H. Persoon, ... aus dem Französischen übersetzt und mit einigen Anmerkungen begleitet von J.H. Dierbach. Mit vier Kupfertafeln. Heidelberg (Neue Akademische Buchhandlung von Karl Grous) 1822. Oct. (*Abh. essb. Schwämme*).
Translation and *notes* by: Johann Heinrich Dierbach (1788-1845).
Publ.: Jul-Dec 1821 (t.p. 1822; Hinrich's Verzeichnis Jul-Dec 1821, sic, see Rogers 1944), p. [i]-xii, [1]-180, *pl. 1-4* (uncol. copp.). *Copies*: BR, MO, NY, Stevenson. – See Killermann (1925, p. 54 for nomencl.).
Ref.: Anon., Flora 396-397. 7 Jul 1823 (rev.).

7736. *Mycologia europaea* seu completa omnium fungorum in variis europaeae regionibus detectorum enumeratio, methodo naturali disposita; descriptione succincta, synonymia selecta et observationibus criticis additis. Elaborata a C.H. Persoon. Sectio prima [-tertia] ... Erlangae (impensibus Joanni Jacobi Palmii) 1822-1828. Oct. (*Mycol. eur.*).
1: 1 Jan-14 Apr 1822 (Flora 14 Apr 1822 "ist ... zu beziehen"; Beck 15 Jul 1822; Hinrich's Verzeichnis Jan-Jun 1822), p. [i], [1]-356, [1-2, admonitio], *pl. 1-12* (handcol. copp., Iac. Sturm sc.; captions to plates never publ.).
2: Jan-Jul 1825 (Flora 8(2); Beil. 1: 4. Jul-Oct 1825; Bull. Soc. Linn. Paris Sep 1825; Linnaea Jan 1826), p. [i], [1]-214, [1, err.], *pl. 13-22* (id., Poiteau pinx., I. Sturm sc.). Imprint "impensis Joannis Jacobi Palmii".
3: 1828 (Linnaea Lit.-Ber. 1828(4): 178-179. Oct-Dec 1828), p. [ii-iii], [1]-282, [1, err.], *pl. 23-30* (id., P. Duménil pinxit, I. Sturm sc.). Imprint: "in libraria Palmii". Alternative t.p.: "Monographia agaricorum ...".

Copies: B, BR, CBS, FI, G, KNAW, MO, NY, Stevenson, USDA; IDC 6299.
Studies: Laplanche (1894, p. 463-466, concordance), Killermann (1925, p. 54-57, nomenclature), Rogers (1944, p. 530).
Index: Chater and Brummitt (1966, p. 147, subsp.), D.P. R[ogers], A.M. R[ogers], & E.V. S[eeler], Index botanicus sistens omnes fungorum species in C.H. Persoonii Mycologia europaea enumeratas una cum synonymis iconumque explicatione. Cantabrigiae [Cambridge, Mass.] (apud Bibliothecum Farlowianam Universitatis Harvardianae) 1942, 37 p. *Friesian synonymy* (Agarics), see E. Fries, Linnaea 5: 689-731. 1830.

7737. *Enumeratio systematica specierum plantarum medicinalium*, e synopsi plantarum C.H. Persoon desumta. Roterodami [Rotterdam] (apud M. Wijt) 1829. Duod. (in sixes). (*Enum. pl. med.*).
Editor: "G.J.M." Gerrit Jan Mulder (1802-1880) in 1829 lecturer at the Rotterdam medical college.
Publ.: 14-19 Sep 1829 (Ned. Staatscour.; p. vi: Aug 1829), p. [1]-vi, [1]-84. *Copy*: U. - Abstracted from Persoon's *Syn. pl.*, see above.

Persson, Nathan Petter Herman (1893-1978), Swedish physician and bryologist; Dr. phil. h.c. Stockholm 1945; in various functions (1947-1969 as curator) at the Naturhistoriska Riksmuseum, Stockholm. (*Perss.*).

HERBARIUM and TYPES: S; further material at BM, C, GB, L, LD. - Some letters at G.

BIBLIOGRAPHY and BIOGRAPHY: Barnhart 3: 72; BL 2: 519, 701; BM 8: 1001; Bossert p. 307 (b. 1893); GR p. 497; IH 1 (ed. 1): 91, (ed. 2): 114, (ed. 3): 146, (ed. 4): 163, (ed. 5): 175, (ed. 6): 260; Kew 4: 279-280; Kleppa p. 146, 147, 148; KR p. 571; MW suppl. p. 278; NAF ser. 2. 3: 41; SBC p. 129.

BIOFILE: Anon., Rev. bryol. lichénol. 10-15 Dec 1937 (trip to Azores).
Crundwell, A.C. & J. Nyholm, J. Bryol. 11: 375-376. 1980 (obit.).
Fries, R.E., Short hist. bot. Sweden 52. 1950.
Hansen, A., Bocagiana 51: 10. 1980 (Madeira coll.).
Koponen, T., Luonnen Tutkija 83: 20-21. 1979 (obit., portr., 25 Jul 1893, d. 14 Aug 1978).
Lindman, S., Sv. män och kvinnor 6: 72-73. 1949 (portr.).
Müller, K., Rev. bryol. lichénol. 22: 1-2. 1953 (portr.).
Persson, N.P.H., Bot. Not. 1915: 96.
Sayre, G., Bryologist 80: 515. 1977.
Schuster, R.M., Hepat. Anthoc. N. Amer. 1: 87. 1966 (bibl.).
Verdoorn, F., ed., Chron. bot. 3: 237, 1937, 4: 182, 187, 479. 1938, 6: 300, 428. 1941.

EPONYMY: *Perssonia* M. Bizot (1969); *Perssoniella* Herzog (1952).

NOTE: Journal publications reviewed in Rev. bryol. 4: 210. 1932, 11: 121. 1939, 17: 75. 1948, 18: 94, 190. 1949, 20: 234, 305, 309, 310. 1951, 21: 301. 1952, 24: 29, 162. 1955, 30: 288, 295. 1961, 32: 306. 1963, 34: 945, 947. 1967, 36: 765, 786. 1970; see also Bot. Not. 1911: 235, 1912: 223, 1929: 229, 1932: 81.

7738. *Bladmossfloran i sydvästra Jämtland* och angränsande delar af Härjedalen. Uppsala (Almqvist & Wiksells Boktryckeri-A.-B.). 1915. Oct.
Publ.: Mar 1915 (printed 19 Feb 1915; Bot. Not. 1 Apr 1915, Nat. Nov. Mai(1, 2) 1915), p. [1]-70. *Copies*: LD, U. - Issued and to be cited as Ark. Bot. 14(3), 1915.

Perty, Joseph Anton Maximilian (1804-1884), German botanist and zoologist; Dr. med. Landshut 1826; Dr. phil. Erlangen 1828; habil. München 1831; at München until 1833; from 1833-1876 professor of zoology at Bern. (*Perty*).

HERBARIUM and TYPES: BERN (rd. 1902).

BIBLIOGRAPHY and BIOGRAPHY: Barnhart 3: 72; BM 4: 1553; CSP 4: 840, 8: 596-597, 10: 1035, 17: 810; De Toni 1: xcvii, 2: xcii; Frank 3(Anh.): 75; Herder p. 117, 253; Kew 4: 280; LS 20503-20504; Moebius p. 70, 72, 82; Nordstedt p. 25; PR 7065-7067; Rehder 2: 279.

BIOFILE: Anon., Berl. [Deut.] entom. Z. 28: 407. 1884 (d.); Psyche, Cambridge 4: 236. 1890 (d. 8 Aug 1884); Wiener entom. Z. 3: 224. 1884 (d.).
Bach-Gelpke, M., Samml. bern. Biogr. 1: 323-328. 1884.
Fischer, Ed., Verh. schweiz. naturf. Ges. 1914(2): 18-19, 23-24, 27.
Gilbert, P., Comp. biogr. lit. deceased entom. 294. 1977.
Hofman, A.W., Ber. deut. chem. Ges. 17(9): 1211. 1884, S.B. Bayer bot. Ges. 15: 170-174. 1886 (b. 17 Sep 1804).
Papavero, N., Essays hist. neotrop. dipterol. 1: 110-111, 193. 1971.
Perty, J.A.M., Erinnerungen aus dem Leben eines Natur- und Seelenforschers des 19. Jahrhunderts, Leipzig, Heidelberg 1879, viii, 486 p. (autobiogr., portr., bibl. 277-482).
Rytz, W., Mitt. naturf. Ges. Bern 5: 25. 1922 (herb. at BERN).

EPONYMY: *Pertya* C.H. Schultz-Bip. (1862).

7739. *Zur Kenntniss kleinster Lebensformen* nach Bau, Funktionen, Systematik, mit Specialverzeichniss der in der Schweiz beobachteten. Von Dr. Maximilian Perty, ... Bern (Verlag von Jent & Reinert. (Platzfirma: Jent und Gassmann)). 1852. Qu. (*Kenntn. kleinst. Lebensf.*).
Publ.: 1852 (p. vi: Mar 1852, but dedication dated "März 1855" (dedication added later?), p. [i-iv], [1]-3, dedic. to G.R. [i.e. C.G.] Ehrenberg, [iii]-viii, [1]-228, *pl. 1-17* (col. liths. auct.). *Copies*: G, NY.

Pescott, Edward Edgar (1872-1954), Australian botanist, pomologist and historian; principal of a school of horticulture at Burley, Victoria 1909-1916; from 1917-1937 government pomologist and seed-tester in the victorian Department of Agriculture. (*Pescott*).

HERBARIUM and TYPES: Material coll. ca. 1900 in MEL.

BIBLIOGRAPHY and BIOGRAPHY: Barnhart 3: 72; Bossert p. 307 (b. 11 Dec 1872; d. 31 Jul 1954); HR; Kew 4: 280.

BIOFILE: Hyam, G.N., Vict. Natural. 71: 166-168. 1954 (obit.).
Willis, J.H., Vict. Natural. 66: 103-104, 126. 1949.
Young, L., Vict. Natural. 71: 87. 1954 (obit.).

7740. *A census of the genus Acacia in Australia* ... [s.l.n.d., 1914]. (*Census Acacia*).
Publ.: 1914, p. 1-24. *Copy*: USDA. – No. 7741 has not been used.

7742. *The orchids of Victoria* ... Melbourne (The Horticultural Press Pty, Ltd.) 1928. (*Orchid. Victoria*).
Publ.: 1928, p. [1]-92, [93], [1, 2, suppl.], *1 col. pl.* in text, *14 uncol. photos. Copy*: G. – Preceded by a series of 9 papers with the same title in the Victorian Naturalist 1926-1927.

Pestalozzi, Anton (*fl.* 1890), Swiss botanist; assistant to the botanical museum at Zürich 1896. (*Pestal.*).

HERBARIUM and TYPES: Unknown. – Some letters at G.

BIBLIOGRAPHY and BIOGRAPHY: BM 4: 1553; Kew 4: 281; Tucker 1: 551.

BIOFILE: Anon., Nat. Nov. 18: 601. 1896 (assistant at Zürich, 19: 156. 1896 (id.)).
Kneucker, A., Allg. bot. Z. 3: 36. 1897 (appointed at Zürich).

EPONYMY: *Pestalotia* De Notaris (1841) is dedicated to Fortunato Pestalozza, an Italian medical doctor and botanist of the mid-19th Century. The following names are orthographic variants or derived from *Pestalotia* De Notaris: *Pestalopezia* F.J. Seaver (1942); *Pestalosphaeria* M.E. Barr (1975); *Pestalotiopsis* R.L. Steyaart (1949); *Pestalozzia* L. Crié (1878); *Pestalozziella* P.A. Saccardo et J.B. Ellis ex P.A. Saccardo (1882); *Pestalozzina* (P.A. Saccardo) P.A. Saccardo (1895) and *Pestalozzites* E.W. Berry (1917). *Pestallozia* Endlicher (1850, *orth. var.*) and *Pestalozzia* Zollinger et Moritzi (1846) are named for J.A. Pestalozzi (1746-1827), Swiss pedagogist.

7743. Mitteilungen aus dem Botanischen Museum der Universität Zürich, vii. Beiträge zur Kenntniss der Afrikanischen Flora. (Neue Folge). ...ix. *Die Gattung Boscia* Lam. ... Genève (Imprimerie Romet, 26, ...) 1898. Oct. (*Boscia*).
Thesis issue: Jun-Sep 1898 (in journal: Jun 1898; Nat. Nov. Nov(2) 1898), p. [i], [1]-152, *pl. 1-14* (uncol.) with text. *Copies*: BR, G, MO.
Regular issue: Jun-Sep 1898 (signatures so dated; ÖbZ Nov 1898), p. [1]-152, *pl.1-14* (uncol.) with text. *Copies*: BR, NY, U. – Issued as Appendix iii to Bull. Herb. Boissier 6, 1898.

Petagna, Luigi (1779-1832), Italian botanist and entomologist at Napoli; from 1812 professor of zoology, from 1813 also director of the Museo zoologico Napoli of the University of Napoli. (*L. Petagna*).

HERBARIUM and TYPES: Unknown. – Some letters at G.

BIBLIOGRAPHY and BIOGRAPHY: BM 4: 1553; CSP 4: 843; Rehder 5: 663; Tucker 1: 551.

BIOFILE: Costa, O.G., Proc. entom. Soc. London 4: xviii. 1847.
Gilbert, P., Comp. biogr. lit. deceased entom. 294. 1977.
Vulpes, B., Atti Ist. Incoragg. Sci. nat. Napoli 5: 287-310. 1834 (obit., b. 27 Aug 1779, d. 29 Mar 1832).

Petagna, Vincenzo (1734-1810), Italian botanist, physician and entomologist; professor of botany and director of the Monteoliveto botanical garden at Napoli. (*Petagna*).

HERBARIUM and TYPES: At the agricultural college of Portici (POR). – Some letters at G.

BIBLIOGRAPHY and BIOGRAPHY: Barnhart 3: 72; Bossert p. 307; Dryander 3: 23, 42, 149; Kew 4: 281; LS 20506-20507; PR 7070-7071 (ed. 1: 7906); Rehder 5: 663; Saccardo 1: 126, 2: 84, cron. p. xxix; SO 700c, 758a, 1230a; Tucker 1: 551; Zander ed. 10, p. 701, ed. 11, p. 799 (b. 17 Jan 1734, d. 6 Oct 1810).

BIOFILE: Anon., (Vulpes, D.B.?), Atti Ist. Incoragg. Sci. nat. Napoli 2: 340-342. 1818 (obit., b. 17 Jan 1734, d. 6 Oct 1810).
Balsamo, F., Bull. Orto bot. Napoli 3: 45. 1913 (biogr. sketch, portr., epon., b. 17 Jan 1730,d. 6 Oct 1810).
Comes, O., Atti congr. bot. int. Genova 1892: 124-126. 1893 (herb. at Portici). (see also Bot. Centralbl. 56: 137-138. 1893).
Costa, O.G., Proc. entom. Soc. London 4: xvii-xviii. 1847.
Gilbert, P., Comp. biogr. lit. deceased entom. 294. 1977.
Swainson, W., Taxidermy 292. 1840.
Wittrock, V.B., Acta Horti Berg. 3(2): 164. 1903, 3(3): 180. 1905.

EPONYMY: *Petagnaea* T. Caruel (1894); *Petagnana* J.F. Gmelin (1792); *Petagnia* Gussone (1827); *Petagnia* Rafinesque (1814).

7744. Vincentii Petagnae in regio neapolitano lyceo botanices professoris *Institutiones botanicae* ... Neapoli [Napoli] ([1:] Typis Josephi Mariae Porcelli ... [2-5:] Typis Petri Perger ...) 1785-1787, 5 vols. Oct. (*Inst. bot.*).
1: (De Philosophia botanica): 1785, p. [i]-xv, 1-285, [286, approb.], *pl. 1-10* (uncol. copp., Lomanto del.).

2: (De plantis in specie): 1787, p. [i-iv], 1-576.
3(id.):1787, p. [i], 577-1194.
4 (id.): 1787, p. [i], 1195-1766.
5 (id.): 1787, p. [i], 1767-2142, [1-48, index].
Copies: FI, G, USDA.

Petch, Thomas (1870-1948), British botanist (plant pathologist and mycologist) and schoolteacher at King's Lynn and Leyton; at the Royal Botanic Gardens, Peradeniya, Ceylon (Sri Lanka) 1905-1925; director Tea research institute of Ceylon 1925-1928. (*Petch*).

HERBARIUM and TYPES: K (fungi); other material at A, BM (slides), CGE, NY, PDA. – P's library, which included that of his father-in-law C. Plowright, was sold by Dawson's in 1949.

BIBLIOGRAPHY and BIOGRAPHY: Ainsworth p. 230, 332, 346; Barnhart 3: 72; CSP 17: 813; Desmond p. 491; IH 2: (in press); Kelly p. 174, 256; Kew 4: 281; Lenley p. 325; LS 20508-20521, 37822, suppl. 21281-21372 (extensive bibl.!); MW p. 384, suppl. p. 278; NAF 1: 167, 7: 1083; PH 567; Stevenson p. 1254; Tucker 1: 551.

BIOFILE: Ainsworth, G.C., Dict. fungi, ed. 6, 436. 1971; Trans. Brit. mycol. Soc. 67(1): 179-181. 1976 (centenary tribute; portr., b. 11 Mar 1870, d. 24 Dec 1948; influenced by C. Plowright).
Bisby, G.R., Trans. Brit. mycol. Soc., Fifty year Index 18-19. 1952.
Blackwell, E.M., Naturalist 877: 65. 1961 (biogr. sketch; reminescences J. Ramsbottom).
Bridson, G.D.R. et al., Nat. hist. mss. res. Brit. Isl. 229.396, 269.232. 1980.
Brooks, F.T., Nature 163: 202-203. 1949 (obit., d. 24 Dec 1949).
Doidge, E., Bothalia 5: 1015. 1950.
Grainger, J., Naturalist 829: 50. 1949 (obit., d. 24 Dec 1948).
Hawksworth, D.L. & M.R.D. Seaward, Lichenol. Brit. Isl. 132. 1977.
Hedge, I.C. & J.M. Lamond, Index coll. Edinb. herb. 118. 1970 ("T. Petch" contributed to Sydow and Petrak's series of exsiccatae).
Holttum, R., Taxon 19: 710. 1970.
J.F.R., Trans. Hull. sci. field Natural. Cl. 3(2): 182-183. 1904 (obit., portr.).
Laundon, J.R., Lichenologist 11: 16. 1979 (herb., slides).
Merrill, E.D., B.P. Bishop Mus. Bull. 144: 149. 1937. (Polynes.); Contr. U.S. natl. Herb. 30(1): 240. 1947.
Palmer, J.T., Nova Hedwigia 15: 152. 1968 (bibl. gasteromyc.).
Petch, T. and G.R. Bisby, The fungi of Ceylon, Colombo, Mar 1950, v, 11 p., map. (list of refs. contains a bibl. of T's writings on Ceylon fungi).
Ramsbottom, J., J. Bot. 61: 201-203. 1923 (review of *Diseases Tea-Bush*, 1923, and *Fungus diseases of crops*, 1922); Trans. Brit. mycol. Soc. 30: 7, 11, 13. 1948.
Stearn, W.T., Nat. Hist. Mus. S. Kensington 306. 1981.

COMPOSITE WORKS: (1) Editor *Annals r. Bot. Gard. Peradeniya* vols. 5-9, 1911-1925.
(2) with G.R. Bisby, *The fungi of Ceylon*, Colombo, Mar 1950, p. [i-v], map, [1]-111. *Copy*: BR.

EPONYMY: *Petchia* Livera (1926).

NOTE: The series of papers *Studies in entomogenous fungi* was published in Trans. Brit. Mycol. Soc. and Ann. R.B.G. Peradeniya:
1: in Trans. 7(1, 2): 89-132. 15 Jul 1921, 7(3): 133-167, *pl. 3-5*. 13 Dec 1921.
2: in Ann. 7(3): 167-278, *pl. 2-5*. Oct 1921.
3: in Trans.9(1, 2): 108-128, *pl. 2*. 29 Sep 1923.
4: in Trans. 10(1): 28-44. *pl. 1*. 26 Sep 1924.
5: in Trans. 10(2): 45-80. *pl. 2-3*. 26 Sep 1924.
6: in Trans. 10(3): 152-182. 15 Mai 1924.
7: in Trans. 10(3): 183-189, 190-201. 15 Mai 1925.
8: in Trans. 16(1): 55-75. 1931.

9: in Trans. 11(1/2): 50-66. *pl. 1.* 26 Aug 1926.
10: in Trans. 11(3/4): 251-266. Dec 1926.
Notes in entomogenous fungi were published in Trans. Brit. Mycol. Soc. *16*: 55-75. 4 Sep 1931, 209-245. 2 Mai 1932; *18*: 48-75. 16 Aug 1933; *19*: 161-194. 19 Feb 1935.

Pételot, [Paul] **Alfred** (1885-post 1940), French botanist; assistant at the Faculté des sciences, Nancy, 1908-1918; teacher at the Hanoi high school (lycée) 1918-1922; director of the entomological station of Cho-gangh 1922-1924; professor of botany at the Indochina University from 1924. (*Pételot*).

HERBARIUM and TYPES: P, PC; duplicates in A, B, BM, C, CAS, DS, E, GH, L, LU, MO, NCY (partly orig.), NY, UC, US.

BIBLIOGRAPHY and BIOGRAPHY: IH 1(ed. 6): 362, (ed. 7): 339, 2: (in press); Kew 4: 281; MW p. 384, suppl. p. 279; TL-2/5220.

BIOFILE: Gagnepain, F., Fl. gén. Indo-Chine tome prél. 45. 1944.
Hedge, I.C. & J.M. Lamond, Index coll. Edinb. herb. 118. 1970.
Henry, R., Rev. bryol. ser. 2. 1(1): 41. 1928 (mosses).
Moldenke, H.N., Plant Life 2: 76. 1946 (1948).
Tixier, R., Rev. bryol. lichén. 34: 127. 1966 (mosses).
Verdoorn, F., ed., Chron. bot. 1: 123. 1935, 2: 131. 1936, 3: 112. 1937, 4: 87. 1938, 6: 300. 1941, 7: 230. 1943.

COMPOSITE WORKS: With M. Magalon, *Élém. bot. indoch.* 1929, TL-2/5220.

EPONYMY: *Petelotia* Gagnepain (1928); *Petelotia* Patouillard (1924); *Petelotiella* Gagnepain (1929).

Peter, [Gustav] **Albert** (1853-1937), German (Prussian) botanist; Dr. phil. Königsberg 1874; private tutor at München 1874-1884 and working with Nägeli on Hieracium; habil. München 1884; "Custos" at the botanical garden, München 1884-1888; professor of botany and director of the botanical garden Göttingen 1888-1923; in retirement working at a Flora von Deutsch-Ostafrika. (*Peter*).

HERBARIUM and TYPES: GOET; collections from Africa 1913-1918, 1925-1926, however, at B (extant); other material at BERN, BM, BR, C, CGE, CORD, E, G, H, L, MANCH, S, US, W, WRSL. *Exsiccatae: Hieracia Naegelianae exsiccatae,* with C. von Naegeli, cent., nos. 1-400, München 1884-1886 (sets e.g. at BERN, BM, BP, CGE, CORD, G, GOET, H, K, L, LE, PRC, S, W; sets 1-3 publ. Sep 1884), set 1 published Oct 1884; for a description see Bot. Centralbl. 19: 185. 1884 and Bot. Zeit. 42: 527-528. 1884 (1-3) and 44: 660-661. 1886 (cent. 4). This series contains e.g. material supplied by G.J. Mendel. – Some letters at G.

BIBLIOGRAPHY and BIOGRAPHY: AG 12(1): 153; Barnhart 3: 72; BFM no. 392; BJI 2: 45; BL 1: 57, 312; BM 4: 1553; Bossert p. 307; CSP 10: 1036, 12: 569, 17: 813; De Toni 1: xcvii; DTS 1: 223; Hegi 6(2): 1214; Herder p. 304; Hortus 3: 1201 (Peter); IF suppl. 3: 216; IH 1(ed. 6): 362, (ed. 7): 339, 2: (in press); Jackson p. 88, 90; Kew 4: 282; Kleppa p. 243; LS 37861; Morren ed. 10, p. 18; Rehder 5: 663; TL-2/6618, see K.A.O. Hoffmann, G.J. Mendel, C.W. Nägeli; Tucker 1: 551; Urban-Berl. p. 382; Zander ed. 10, p. 701, ed. 11, p. 799.

BIOFILE: Anon., Bot. Centralbl. 18: 288. 1884 (habil. München), 34: 159. 1888 (to Göttingen), 41: 336. 1890 (member Kgl. Ges. Wiss., Göttingen), 114: 496. 1910 (Geheimrat); Bot. Zeit. 36: 798. 1878 (replaces A. Engler as curator at München), 42: 463. 1884 (habil. München), 46: 273. 1888 (app. Göttingen), 68: 208. 1910 (Geh. Regierungsdrat); Flora 61: 544. 1878 (app. München); Gard. Chron. ser. 3. 102: 385. 1937 (obit.); Hedwigia 27: 180. 1888 (app. Göttingen), 50: (193). 1910 (Geheimrat), 78: (74). 1938 (d. 4 Oct 1937); Mycol. Centralbl. 2(6): 319. 1912 (jubilee), 4(6): 319. 1914 (plans to return to Germany), 5(6): 307. 1915 (stranded in Africa because of

war); Nachr. Ges. Wiss. Göttingen, Jahresber. 1937/38: 6-9. 1938 (obit.); Nat. Nov. 10: 149. 1888 (succeeds Solms-Laubach at Göttingen); Österr. bot. Z. 29: 34. 1879 (appointment "Custos"), 38: 218. 1888 (app. Göttingen), 42: 95. 1891 (announcement; 100 nos. planned); 87: 80. 1938 (d. 4 Oct 1937).
Beck, G., Bot. Centralbl. 34: 150. 1888 (Hierac. naegel. exsicc. at W).
Gager, C.S., Brooklyn Bot. Gard. Rec. 27(3): 227. 1938 (dir. bot. gard. Göttingen).
Engler, A., ed., Bot. Jahrb. 9 (Beibl. 22): 1. 1888 (accepts Göttingen "nachdem einige andere Botaniker ... abgelehnt hatten").
Focke, W.O., Bot. Zeit. 43: 442-447. 10 Jul 1885 (rev. *Hierac. Mitteleurop.*).
Hedge, I.C. & J.M. Lamond, Index coll. Edinb. herb. 118. 1970
Hiepko, P., Willdenowia 8(2): 393. 1978 (duplicates extant in B).
Iltis, H., Life of Mendel ed. 2. 182, 197, 202. 1966.
Peter, G.A., Bot. Jahrb. 5: 203-238. 4 Mar 1884, 239-286. 16 Mai 1884, 448-496. 5 Sep 1884, 6: 111-136. 30 Dec 1884 (his first treatise on Hieracium sect. Pilos.); Bot. Zeit. 42: 527-528. 15 Aug 1884 (cent. 1-3), 44: 660-667. 24 Sep 1886 (cent. 4).
Pilger, R., Mitt. Bot. Gart. Mus. Berlin-Dahlem 1(1): 5, 17, 18. 1953 (herb. at B for the greater part extant).
Schmucker, Th., Ber. deut. bot. Ges. 56: (203)-(213). 1938 (obit., portr., bibl., b. 21 Aug 1853, d. 4 Oct 1937).
Seeland, H., Mitt. Herm. Roemer-Mus. Hildesheim 40: 43, 44, 46. 1936 (on Peter and A. Schlauter).
Szymkiewicz, D., Bibl. Fl. Polsk. 113. 1925.
Verdoorn, F., ed., Chron. bot. 4: 266, 271. 1938 (portr.).

COMPOSITE WORKS: (1) C. von Naegeli und A. Peter, *Die Hieracien Mittel-Europas* (1885), see von Naegeli.
(2) *Hieracium*, in Potonié, Ill. Fl. N. Mitt. Deutschl. ed. 3, p. 449-465. 1887 (Bot. Centralbl. 18-22 Apr., reprint).
(3) EP, *Die natürlichen Pflanzenfamilien*, ed. 1: (a) *Convolvulaceae*, in 4(3A): 1-40. Dec 1891; Nachtr. p. 375-377. Dec 1897.
(b) *Polemoniaceae*, in 4(3A): 41-48. Dec 1891, 49-53. Jun 1893; p. 377. Dec 1897.
(c) *Hydrophyllaceae*, in 4(3A): 54-71. Jun 1893; Nachtr. p. 377. Dec 1897.
(d) *Hieracium*, in 4(5): 375-387. 22 Mai 1894.
(4) Editor, *Botanische Zeitung* 67-68, 1909-1910.

EPONYMY: *Peterodendron* Sleumer (1936). *Note: Peteria* A. Gray (1852) was dedicated to Dr. Robert Peter (1805-1894), English-born American botanist; *Peteria* Rafinesque (1820) is an *orth. var.* of *Petesia* P. Browne (1756).

BOTANISCHE WANDTAFELN, coloured plates (71 × 90.5 cm) with accompanying text, Cassel 1892-1914. Part of the original drawings are still at GOET (inf. G. Wagenitz).

Plates	dates Bot. Zeit. [ÖbZ], [B.C.]	Nat. Nov.
1-2	24 Jun 1892	Jul(1) 1892
3-5	1 Mar 1893	Feb(1) 1893 rev. B.Z. 1 Jun 1897
6-7	[ÖbZ Feb 1893]	Mai(1) 1893
[6-11]	[ÖbZ Mai 1894]	
8-9		Dec(2) 1893
10-11		Jan(2) 1894
12-14		Sep(2) 1894
13	16 Sep 1894	
15-16	16 Feb 1895	Nov(1) 1894
18, 21	16 Dec 1895	Aug(2) 1895
19, 22	[Bot. Centralbl. 27 Feb 1895]	
23-30		Jan(2) 1901
31-40	[Bot. Centralbl. 11 Sep 1901]	Sep(1) 1901
41-50	1 Jun 1902	Jan(2) 1902
51-55		Mai(1) 1911
56-60		Aug(1) 1912

Plates	Nat. Nov.
61-65	Oct(1) 1913, Sep(2) 1913
66-70	Apr(1) 1914
71-75	Dec(1, 2) 1914

Ref.: Engler, A., Bot. Jahrb. 16 (Lit.): 9-10. 1892 (rev.) (100 plates planned).
Kienitz-Gerloff, J.H.E.F., Bot. Zeit. 51(2): 171-172. 1893.

7745. *Ueber spontane und künstliche Gartenbastarde der Gattung Hieracium* sect. *Piloselloidea.* Habilitationsschrift, der philosophischen Facultät der Ludwig-Maximilians-Universität zu München vorgelegt von Dr. A. Peter. Leipzig [Wilhelm Engelmann] 1884. Oct. (*Gartenbast. Hieracium*).
Publ.: Mar-Dec 1884, p. [i], [203]-286, [h.t. on verso of p. 448] [448]-496, [111]-136.
Copy: G. – Partly preprinted from Bot. Jahrb. 5(2): 203-238, 4 Mar 1884, 5(3): 239-286. 6 Mai 1884, 5(5): 448-496. 5 Sep 1884, 6(2): 111-136. 30 Dec 1884.
Ref.: Beck, G., Österr. bot. Z. 35: 104. 1885.

7746. *Flora von Südhannover* nebst den angrenzenden Gebieten, umfassend: das südhannoversche Berg– und Hügelland, das Eichsfeld, das nördliche Hessen mit dem Reinhardswalde u. dem Meissner, das Harzgebirge nebst Vorland, das nordwestliche Thüringen und deren nächste Grenzgebiete ... Göttingen (Vandenhoeck & Ruprecht) 1901, 2 vols. Oct. (in fours). (*Fl. Südhannover*).
1: Mar-Mai 1901 (p. xiv: Mar 1901; Nat. Nov. Jun(1) 1901; ÖbZ Mai-Jun 1901; Bot. Centralbl. 5 Jun 1901; Bot. Zeit. 16 Jun 1901), p. [ii]-xvi, [1]-323.
2: Mar-Mai 1901 (Bot. Jahrb. 19 Nov 1901; Bot. Zeit. 16 Jun 1901; Nat. Nov. Jun(1) 1901; ÖbZ Mai-Jun 1901; Bot. Centralbl. 5 Jun 1901; Allg. bot. Z. 15 Jul 1901). Bestimmungstabellen, p. B1-B137, map. *Copies*: BR, G.
Ref.: Appel, F.C.L.O., Bot. Centralbl. 87: 352-353. 1901.
Buchenau, F.G.P., Bot. Zeit. 59(2): 202-204. 1 Jul 1901.
Kneucker, A., Allg. bot. Z. 7: 136. 15 Jul 1901.
Matouschek, F., Bot. Centralbl. 89: 345-346. 1902.

7747. *Wasserpflanzen und Sumpfgewächse in Deutsch-Ostafrika* ... Berlin (Weidmannsche Buchhandlung) 1928. Oct. (*Wasserpfl. Deut.-Ostafr.*).
Publ.: 1928 (mss. submitted 23 Mar 1928; Nat. Nov. Oct 1928), p. [1]-129, [130], *pl. 1-19*. *Copies*: BR, MO, NY, USDA. – Abh. Ges. Wiss. Göttingen, math.-phys. Kl., N.F. 13(2). 1928.
Ref.: Engler, A., Bot. Jahrb. 62: 52-53. 1 Feb 1929.

7748. *Flora von Deutsch-Ostafrika* Zusammenstellung der in Deutsch-Ostafrika beobachteten farnartigen Gewächse und Blüthenpflanzen mit Literatur-Nachweisen, Angabe der Verbreitung auf der Erde und Bestimmungstabellen. Von Dr. Albert Peter (†) ... unter Mitwirkung von Frl. Dr. Ingeborg Haeckel ... und Dr. h.c. Georg Kükenthal ... Dahlem bei Berlin (Selbstverlag des Herausgebers, ...) 1929-1938. Oct.
Publ.: in Repertorium specierum novarum regni vegetabilis .. [ed. F. Fedde], Beihefte, Band 40(1-2). (*Copies*: BR, U, US).

Vol.	Lieferung	pages	*plates*	dates
1	1	1-144	*1-10*	1 Feb 1929
	2	145-208	*11-20*	15 Oct 1930
	3	209-336	*21-40*	10 Apr 1931
	4	337-384	*41-78*	31 Mar 1936
	5	385-416	*79-91*	10 Mai 1937
	6	417-540	-	30 Jun 1938
2	1	1-144	*1-20*	31 Jan 1932
	2	145-224	*21-40*	1 Nov 1932
	3	225-272	*41-44*	15 Apr 1938

Plantarum novarum, p. [1]-142, ib. 40(1) Anhang, [1]-112. 1 Feb 1929; p. 113-128. 31 Mar 1936; p. 137-142. 30 Jun 1938; p. 1-16. 31 Jan 1932; p. 17-32. 1 Nov 1932; p. 33-36. 15 Apr 1938.

Péterfi, Márton [Martin] (1875-1922), Hungarian (Transylvanian) botanist, especially bryologist; school teacher at Deva, from 1906 at Cluj; curator of the National Transilvanian Museum from 1908. (*Péterfi*).

HERBARIUM and TYPES: CL; further important set at BP; duplicates at B, BR, C, E, G, H, L, MANCH. For *exsiccatae, Bryotheca regni Hungariae exsiccatae*, see I. Györffi (additional information: *1*: nos. 1-50. 1916, *2-3*: nos. 51-150. 1919, with *Schedae* published in Bot. Muzeumi Füzetek 1: 10-73. 1915 and 3: 43-74. 1919). Manuscripts also at CL.

BIBLIOGRAPHY and BIOGRAPHY: Barnhart 3: 73; BM 8: 1001; CSP 17: 814; Futák-Domin p. 465; Moebius p. 438; SBC p. 129; TL-2/see Györffi, I., Urban-Berl. p. 310.

BIOFILE: Anon., Bot. Not. 1922: 287 (d. 30 Jan 1922); Hedwigia 64: (69). 1923 (d.); Österr. bot. Z. 71: 72. 1922 (d.); Rev. bryol. 31: 17. 1904, 34: 53. 1907.
Borza, A., Bull. Soc. Stiinte Cluj 1: 597-603. 1923 (portr., bibl.).
Györffy, I., Bot. Közl. 20: 117-128. 1924 (obit., bibl.).
Hedge, I.C., & J.M. Lamond, Index coll. Edinb. Herb. 118. 1970.
Kneucker, A., Allg. bot. Z. 17: 32. 1911 ("Adjunkt" at the Museum society of Siebenbürgen).
Pax, F., Grundz. Pflanzenverbr. Karpathen 2: 275, 278, 280. 1908 (Veg. Erde 10).
Sayre, G., Bryologist 80: 515. 1977.
Stefureac, T.I., Contr. Bot., Cluj 1976: 1-7 (315-321) (portr.).
Szymkiewicz, D., Bibl. Fl. Polsk. 55. 1925.
Verdoorn, F., ed., Chron. bot. 1: 245. 1935.

EPONYMY: *Peterfiella* Gerloff (1940) is named for Stefan Peterfi (1906-x), Romanian phycologist.

Petermann, Wilhelm Ludwig (1806-1855), German (Saxonian) botanist; habil. Leipzig 1835; suceeded Kunze as professor of botany in 1851. (*Peterm.*).

HERBARIUM and TYPES: LZ (destroyed); duplicates at B (mainly destroyed); a few specimens in MW and U. The herbarium was originally donated to the "Realschule" at Leipzig where it was combined with that of Otto Delitsch.

BIBLIOGRAPHY and BIOGRAPHY: AG 5: 18; Backer p. 659; Barnhart 3: 73; BM 4: 1554-1555; CSP 4: 844; Frank (Anh.): 75-76; Futák-Domin p. 465-466; GR p. 36, cat. p. 69; Herder p. 470; Hortus 3: 1201 (Peterm.); Jackson p. 116, 293, 300; Kew 4: 282; MW p. 384; PR 5432, 7073-7080 (ed. 1: 7907-7915); Rehder 5: 663; SO 26, add. 850ab, ac; Tucker 1: 551; Zander ed. 10, p. 701, ed. 11, p. 799 (b. 3 Nov 1806, d. 27 Jan 1855).

BIOFILE: Anon. [Seemann, B. or H.G. Reichenbach]. Bonplandia 3: 103-104. 1855 (obit., b. 3 Nov 1806, d. 27 Jan 1855); "... gräcisirende Terminologie ..."; "... nöthigten ihn wenig günstige äussere Verhältnisse mehr zu schreiben, als er sonst gethan hätte ..."; ... "zahlreiche "neue" Arten"); Bot. Not. 1858: 13; Bot. Zeit. 9: 824. 1851 (app. Leipzig), 13: 144. 1855 (d.), 28: 616. 1870 (herb. to "Realschule" at Leipzig); Bull. Soc. bot. France 2: 68. 1855; Flora 34: 770. 1851 (app. Leipzig), 38: 95, 368. 1855 (d. 27 Jan 1855); Österr. bot. W. 1: 404. 1851 (appointed successor to Kunze at Leipzig).
Candolle, Alph. de, Phytographie 439. 1880 (herb. LZ).
Hendrych, R., Novit. Bot., Praha 1966: 29-37. 1967 (list of subgenera in *Deutschl. Fl.*).
Lindemann, E. v., Bull. Soc. imp. Natural. Moscou 61: 54. 1886 (some material at MW).
Reichenbach, H.G., Bot. Zeit. 13: 183-184. 1855 (obit.).
Seemann, B., Bonplandia 1: 240. 1853, 3: 103-104. 1855.
Stafleu, F.A., Regn. veg. 71: 327, 339. 1970 (Miquel on P.).

EPONYMY: *Petermannia* Klotzsch (1854); *Note*: *Petermannia* H.G.L. Reichenbach (1841), *nom. rej.*) is very likely named for W.L. Petermann, but no etymology is given. *Petermannia* F. v. Mueller (1860, *nom. cons.*) honors A.H. Petermann (1822-1878), German geographer.

NOTE: Not to be confused with August H. Petermann (1822-1878), geographer and cartographer from 1854-1879 at the cartographical institute of Justus Perthes at Gotha; founder of [Petermann's] Mitteilungen aus Justus Perthes' Geographischer Anstalt.

7749. *De flore gramineo* adiectis graminum circa Lipsiam tam sponte nascentium quam in agris cultorum descriptionibus genericis. Dissertatio botanica quam ordinum Lipsiensium amplissimi philosophorum consensu et illustris ictorum concessu pro licentia de cathedra academia docendi rite obtinenda ... die xiv. m. jan. mdcccxxxv publice defendet auctor Gulielm. Ludov. Petermann Doct. philos. ... Lipsia [Leipzig] (typis Staritzii, typogr. acad.) [1835]. Oct. (*Fl. gramin.*).
Thesis issue (Habilitationsschrift): 14 Jan 1835 (t.p.), p. [i], [1]-80, *1 pl.*["*12A*"], 1 chart, [81, theses]. *Copies*: B, MO, NY(2).
Trade issue: 18-24 Jan 1835 (Hinrichs), p. [i-iv], [1]-80, *1 pl.*, 1 chart. *Copy*: KNAW. – "De flore gramineo ... dissertatio auctore Dr. Guil. Ludov. Petermann ..." Lipsiae (apud Ambros. Barth) 1835.
Ref.: Anon., Flora 5: 58-64. 14 Mai, 65-68. 14 Jun 1835.

7750. *Handbuch der Gewächskunde* zum Gebrauch bei Vorlesungen so wie zum Selbststudium ... Leipzig (Verlag von Johann Ambrosius Barth) 1836. Oct. (*Handb. Gewächsk.*).
Publ.: 7-13 Feb 1836 (p. x: 14 Jan 1836; Hinrichs 7-13 Feb 1836; Lit. Ber. Flora 7 Mai 1836), p. [i]-xxvi, [xxvii], [1]-690, [691, symbols]. *Copies*: B, NY.
Ref.: Winkler, E., Repert. ges. deut. Lit. (E.G. Gersdorf) 8: 261-262. 1836 (rev.).

7751. *Flora lipsiensis excursoria*, exhibens plantas phanerogamas circa Lipsiam tam sponte nascentes, quam in agris cultas, simul cum arboribus et fructibus pomerii lipsiensis, ... Lipsiae [Leizig] (sumptibus Ioannis Ambrosii Barth) 1838. Duod. (in sixes) (*Fl. lips. excurs.*).
Publ.: 1838 (p. iv: 14 Jan 1838; Flora 28 Apr 1838, p. [i]-x, [1], [1]-707, map. *Copies*: G, NY, USDA.

7752. *Das Pflanzenreich* in vollständigen Beschreibungen aller wichtigsten Gewächse dargestellt, nach dem natürlichem Systeme geordnet und durch naturgetreue Abbildungen erläutert ... Leipzig (Verlag von Eduard Eisenach) 1845. Qu. (*Pflanzenreich*).
Text: Aug 1838-1845, in 50 parts, p. [i], [1]-1010, [1-2, err.].
Atlas: 1838-1845, in 50 parts, p. [i], *pl. 1-282* (lithographs signed I.S.P.; two issues: col. and uncol.).*Copies*: B (uncol.), MO (col.), Wheldon and Wesley (Cat. 131, no. 97, 1974) cites a copy with an 1838 t.p. with different subtitle.
Other issue: 1845, as part Pflanzenreich in "*Die Naturgeschichte* in getreuen Abbildungen und mit ausführlicher Beschreibung derselben". Leipzig (id.) 1845. *Copy*: B. – Pagination as in regular issue; t.p. constitutes only difference.
Ed. 2 (first re-issue): 1847, p. [i], [1]-1010, [1-2, err.], atlas p. [i-iii]*pl. 1-282*. *Copy*: BR. – Leipzig (Verlag von Eduard Eisenach) 1847. Qu.
Ed. 2 (re-issue): 1857, p. [i], [1]-1010, [1-2, err.], *pl. 1-282* (id.). *Copies*: NY (uncol.), US. – Leipzig (Verlag von Julius Werner) 1857. Qu.

7753. *In Codicem botanicum linnaeanum index alphabeticus* generum, specierum ac synonymorum omnium completissimus composuit atque edidit Dr. Guil. Ludov. Petermann ... Lipsiae [Leipzig] (sumptum fecit Otto Wigand) 1840. Qu. (*Cod. linn. index*).
Publ.: 1840, p. [i]-iv, [1]-202. *Copies*: MO, US, USDA. *Index* to H.E. Richter, Caroli Linnaei opera ... *Codex botanicus Linnaeanus* ... 1835, ed. 2. 1840.

7754. *Flora des Bienitz* und seiner Umgebungen ... Leipzig (Friedrich Fleischer) 1841. (*Fl. Bienitz*).
Publ.: 1841, p. [i]-xviii, [1]-171, 1 map. *Copy*: FI.

7755. *Taschenbuch der Botanik* ... Leipzig (Friedrich Volckmar) 1842. Oct. (*Taschenb. Bot.*).
Publ.: 1842, (Bot. Not. Sep 1842; Flora rd. before 28 Dec 1842), p. [i], [1]-484, [1, err.], *pl. 1-12* (uncol.). *Copies*: B, NY.

7756. *Analytischer Pflanzenschlüssel* für botanische Excursionen in der Umgegend von Leipzig ... Leipzig (Carl Heinrich Reclam sen.) 1816. Oct. (*Anal. Pfl.-Schlüss.*).
Publ.: Jun-Jul 1846 ("soeben" Flora 14 Jul 1846; Bot. Zeit. 4 Sep 1846), p. [i*], [i]-clxvi, [1], [1]-592. *Copies*: B, BR, NY.
Neue Ausgabe: Mar-Apr 1879 (Nat. Nov. Apr(2) 1879; Bot. Zeit. 25 Apr 1879), n.v.
Ref.: Schlechtendal, D.F.L. v., Bot. Zeit. 4: 615-616. 4 Sep 1846.

7757. *Deutschlands Flora* mit Abbildungen sämmtlicher Gattungen auf 100 Tafeln ... Leipzig (Georg Wigand's Verlag) 1849. Qu. (*Deutschl. Fl.*).
Publ.: 1846-1849 (in twelve parts), p. [i], [1]-668, [1, err.], [1, order pl.], *atlas*: [i], *pl. 1-100* (uncol. liths.). *Copies*: BR, FI, G, MO, NY, PH (parts 1-5, 1846-1847, p. 1-248, *pl. 1-40*). – Junk (Rara 1: 16) mentions also copies with hand-coloured liths. – The PH copy has some parts of each figure coloured.
Ref.: Schlechtendal, D.F.L. v., Bot. Zeit. 4: 762-763. 30 Oct 1846 (Lief. 1).

Peters, Karl [Carl] (1865-1925), German (Prussian) botanist and gardener at Berlin and Berlin-Dahlem; from 1886-1925, subsequently as Reviergehilfe, Öbergärtner (1894), Garteninspektor (1907), and Oberinspektor (1913). (*K. Peters*).

HERBARIUM and TYPES: If any, at B.

BIBLIOGRAPHY and BIOGRAPHY: ADB 25: 479-483; Barnhart 3: 73; Kew 4: 282; Zep.-Tim. p. 91.

BIOFILE: Anon., Ber. deut. bot. Ges. 43: (2). 1926 (d. 3 Jun 1925); Hedwigia 66: (55). 1926; Möllers deut. Gärtn.-Zeit. 18: 538. 1903 (portr.).
Engler, A., Notizbl. Bot. Gart. Mus. Berlin 9: 443-447. 1925 (obit., b. 10 Apr 1865, d. 3 Jun 1925; account of P's role in shaping the Dahlem garden).
G.L., Gartenflora 74: 289. 1925 (obit., portr.).
Kneucker, A. Allg. bot. Z. 28/29: 56. 1925 (d. 3 Jun 1925).
Wittrock, V.B., Acta Horti Berg. 3(3): 147. 1905.

HANDWRITING: Möllers deut. Gärtn.-Zeit. 18: 538. 1903.

7758. *Führer zu einem Rundgang durch die Freiland-Anlagen des* königl. *Botanischen Gartens* ... Dahlem-Steglitz bei Berlin 1908. Oct. (*Führ. Bot. Gart.*).
Preface: Heinrich Gustav Adolf Engler (1844-1930).
Publ.: 1908 (p. viii: Mai 1908), p. [i]-viii, [1]-48, map. *Copies*: G, UC.
Other edition: 1919 (ÖbZ 1 Oct 1919), p. [1]-100, map. *Copy*: B. – By K. Peters, A. Engler & P. Gräbner [Karl Otto Robert Peter Paul Gräbner 1871-1933]. Berlin (Selbstverlag des Botanischen Gartens).

Peters, Wilhelm Carl Hartwig (1815-1883), German (Schleswig) zoologist, physician and traveller; 1842-1848 in S. and E. Africa and India; professor of medicine Berlin 1851; professor of zoology ibid. 1856. (*W. Peters*).

HERBARIUM and TYPES: B, partly extant; duplicates at A, BR, CGE, EA, K, LE.

BIBLIOGRAPHY and BIOGRAPHY: ADB 25: 489-493; Backer p. 437; Barnhart 3: 73; BL 1: 47, 312; BM 4: 1555; CSP 4: 847-851, 6: 741-742, 8: 600-603, 10: 1039-1043, 17: 817; Herder p. 56; Kew 4: 282; Langman p. 580; Lasègue p. 579; NI 1515; PR 7081; Quenstedt p. 335; Rehder 1: 492; SK 4: ccv; TL-1/970; TL-2/3984; Tucker 1: 551; Urban-Berl. p. 382.

BIOFILE: Anon., Bonplandia 8: 330. 1860; Bot. Zeit. 1: 439-440. (note on possible death),

590-591 (arrived safely in Luanda, Angola). 1843; Flora 26: 245. 1843. (to Africa on 5000 Thaler); Peterm. Mitt. 30: 104. 1884 (b. 22 Apr 1815, d. 21 Apr 1883); Proc. Linn. Soc. London 1882/83: 47-48. 1883 (obit.); Psyche 4: 59. 1883; Science 1: 438. 1883.
Ascherson, P., Bot. Zeit. 41: 368. 1 Jun 1883 (b. 2 Apr 1815, d. 20 Apr 1883); S.B. Ges. naturf. Fr. Berlin 1883: 67.
Becker, K. & E. Schumacher, S.B. Ges. naturf. Fr. Berlin ser. 2. 13(2): 145. 1973 (index to P's papers in this journal).
Bridson, G.D.R. et al., Nat. hist. mss. res. Brit. Isl. 229.32. 1980.
Dolezal, H., Friedrich Welwitsch 77, 95, 152, 189, 206, 214. 1974.
Dunning, J.W., Proc. entom. Soc. London 1883: xlvii.
Embacher, F., Lexik. Reisen 229. 1882.
Exell, A.W. & F.A. Hayes, Kirkia 6(1): 99. 1967 (coll.).
Exell, A.W. et al., Bol. Soc. Broter. ser. 2. 26: 218. 1952.
Gilbert, P., Comp. biogr. lit. deceased entom. 294. 1977.
Gomes e Sousa, A. de F., Expl. Flora Moçambique, Docum. trim. 57-58. 1949.
Hilgendorf, F., Allg. deut. Biogr. 25: 489-493. 1887 (extensive details, bibl.).
Moldenke, H.N., Plant Life 2: 76. 1946 (1948) (epon.).
Nissen, C., Zool. Buchill. 315. 1969 (b. 22 Apr 1815, d. 21 Jun 1883).
Türckheim v., Berl. entom. Z. 27: ii. 1883.
Wood, R.D., Monogr. Charac. 1: 833. 1965 (note on *Characeae*, in Peters, *Naturw. Reise* by A. Braun and possibly by K. Müller, p. 566).

EPONYMY: *Petersia* Klotzsch (1816); *Petersia* Welwitsch ex Bentham (1865) and its derived genus *Petersianthus* Merrill (1916).

7759. *Naturwissenschaftliche Reise nach Mossambique* auf Befehl seiner Majestäts des Königs Friedrich Wilhelm IV in den Jahren 1842 bis 1848 ausgeführt von Wilhelm C.H. Peters. Botanik ... Berlin (Druck und Verlag von Georg Reimer) [1861-]1862-1864, 2 parts. Qu. (*Naturw. Reise Mossambique*).
Vol. 6, Botanik, in two parts [1861-]1862-1864. (Vols. 1, 3-5, zoological, vol. 2 not published).
1: late 1861 (see preface 6(2); however, the first 34 sheets presented to the K. Akad. Wiss., Berlin Apr 1858), p. [i]-iv, [1]-304, *pl. 1-48* (uncol.; J.D.L. Franz Wagner del.).
2: 1864 (BSbF Apr 1864; "erschienen" Bot. Zeit. 4 Mar 1864), p. [i*], [i]-xxii, [305]-584, *pl. 48ª-60* (id.).
Copies: B, BR, FI, G, H, MO, NY, PH.
Authors: Nils Johann Andersson (1821-1880); Johann Otto Boeckeler (1803-1899); Carl August Bolle (1821-1909); Alexander Carl Heinrich Braun, (1805-1877); Christian August Friedrich Garcke (1819-1904); Justus Karl Hasskarl (1811-1894); Friedrich Wilhelm Klatt (1825-1897); Johann Friedrich Klotzsch (1805-1860); Carl Sigismund Kunth (1788-1850); Karl Müller (1818-1899); Heinrich Gustav Reichenbach (1824-1889); Joachim Steetz (1804-1862).
Ref.: Hiern, W.P., J. Bot. 18: 264. 1880 (vol. 1 post-dated).

Petersen, Hans (1836-1927), German (Schleswig) botanist and high school teacher at Sonderburg on Alsen (now Als) from 1862-1894; from 1895-1927 living in retirement at Schwesing (Husum). (*H. Petersen*).

HERBARIUM and TYPES: Left to the Arbeitsgemeinschaft für Floristik, Schleswig-Holstein; present location not known (see Christiansen 1927).

BIBLIOGRAPHY and BIOGRAPHY: BL 2: 54, 702; BM 4: 1556; Christiansen p. 318 [index]; CSP 17: 818.

BIOFILE: Christiansen, W., Die Heimat 37(9): 219. 1927 (obit., b. 20 Jan 1836, d. 12 Feb 1927).

7760. *Beitrag zur Flora von Alsen* ... Beilage zum Programm des königl. Realprogymnasiums zu Sonderburg. 1891. Nr. 290. Oct. (*Beitr. Fl. Alsen*).

Publ.: 1891 (Nat. Nov. Jul(1) 1891; Bot. Zeit. 29 Apr 1892), p. [1]-50. *Copy*: MO. – Names and distr. only. – Alsen, now [Danish:] Als, island in the Little Belt SE of Jutland, chief town Sonderburg, now Sønderborg.
Ref.: Knuth, P.E.O.W., Bot. Centralbl. 47: 212. 19 Aug 1891 (rev.).

Petersen, Henning Eiler [Ejler] (1877-1946), Danish botanist, mainly mycologist and algologist; M.Sc. København 1902; Dr. phil. ib. 1914; lecturer at the Polyteknisk Laereanstalt 1916-1930; lecturer ("docent") at University of Copenhagen 1929-1946. (*H.E. Petersen*).

HERBARIUM and TYPES: C.

BIBLIOGRAPHY and BIOGRAPHY: Barnhart 3: 73; BM 4: 1556; Christiansen no. 477; Hirsch p. 229; IH 2: (in press); Kelly p. 174; Kew 4: 282-283; LS 20537-20538, 37869-37871; NAF 2(1): 72; Stevenson p. 1254.

BIOFILE: Anon., Generalreg. Bot. Not. 112. 1939.
Christensen, C., Danske bot. Litt. 1880-1911, p. 238-239. 1913 (bibl., portr.); 1912-1939, p. 111-114. 1940 (portr., bibl.).
Christiansen, W., Heimat 37: 219. 1927 (n.v.).
Gram, K. Bot. Tidskr. 48: 122-126. 1946 (obit., portr., b. 22 Aug 1877, d. 22 or 23 Mai 1946).
Hansen, A., Dansk bot. Ark. 21(1): 69. 1963 (further biogr. refs.).
Kornerup, T., Overs. Medd. Grønland 1876-1912: 96. 1913.
Koster, J.Th., Taxon 18: 556. 1969.
Lind, J., Danish fungi herbarium Rostrup 36. 1913.
Verdoorn, F., Biologia 1: 7. 1949 (d.).
1464a
EPONYMY: *Petersenia* Sparrow (1934).

7761. *Danske arter af slaegten Ceramium* (Roth) Lyngbye ... København (Bianco Lunos bogtrykkeri) 1908. Qu.
Publ.: 1908 (Nat. Nov. Jul(1) 1909), p. [1]-58, *pl. 1-7*. *Copies*: BR, C, H, LD, NY. – Reprinted and to be cited from Kgl. Danske Vid. Selsk. Skr. ser. 7. Naturw. Afd. 5(2): [39]-83, summary in French [84]-96, *pl. 1-7*.

7761a. *Inledende studier over polymorphien hos Anthriscus silvestris* (L.)Hoffm. ... Kj°5benhavn (Vilhelm Priors kgl. Hofboghandel) 1914. Qu.
Publ.: Nov-Dec 1914 (t.p. thesis dated 1914; accepted by the faculty 5 Nov 1914 to be publicly defended), p. [i-vii], [1]-140, [141, theses], *pl. 1-18*. *Copies*:C, LD. – Preprinted from Dansk bot. Ark. 1(6): [1]-150, [150-152], *pl. 1-18*. 15 Jul 1915. – Inf. O. Almborn.

Petersen, Johannes Boye (1887-1961), Danish algologist; assistant at the Polyteknisk Laereanstalt 1916-1929, lecturer ib. 1929-1951; (cryptogamic "amanuensis") at the Copenhagen botanic garden 1920-1951; professor of botany at Copenhagen University 1951-1958. (*J.B. Petersen*).

HERBARIUM and TYPES: C; diatoms also at S.

BIBLIOGRAPHY and BIOGRAPHY: Barnhart 3: 74; BJI 2: 134; Hirsch p. 230; IH 2: (in press); Kew 4: 283; Kleppa p. 92.

BIOFILE: Christensen, C., Danske bot. litt. 1912-1939, p. 163-164. 1940 (bibl., portr., biogr. details).
Christensen, T., Bot. Tidsskr. 58: 131-133. 1962 (obit., portr., b. 5 Feb 1887, d. 7 Nov 1961).
Fox Maule, A., Københavns Universitet 1479-1979, 13(Bot.): 240. 1979.
Fryxell, G.A., Beih. Nova Hedw. 35: 363. 1975 (diatom coll.).
Hansen, A., Dansk. bot. Arkiv 21: 70. 1963 (further biogr. refs.).

Hansen, J.B., Phycologia 3: 45-49. 1963 (obit., portr., bibl.).
Koster, J.Th., Taxon 18: 556. 1969.
Nygaard, G., Revue algol. ser. 2. 6(2): 83-86. 1962 (obit., portr., bibl.).
VanLandingham, S.L., Cat. diat. 6: 3579-3580, 7: 4216. 1978.

COMPOSITE WORKS: (1) *Marine Cyanophyceae from Easter Island*, in C. Skottsberg, *Nat. Hist. Juan Fernandez* 2(Bot.): 461-463. 1926.
(2) *The fresh-water Cyanophyceae of Iceland*, 1923, in *The Botany of Iceland* 2(2, 7): [249]-324. 1923.
(3) *The aërial algae of Iceland*, 1928, in *The Botany of Iceland* 2(2, 8): [325]-447. 1928.

7762. *Studier over danske aërofile alger* ... København (Hovedkommissionaer: Andr. Fred. Høst & Søn, ... Bianco Lunos Bogtrykkeri) 1915. Qu. (*Stud. Dansk. aërof. alg.*).
Publ.: 1915 (cover and t.p. 1915; journal as a whole; Nat. Nov. Jan (1, 2) 1916), p. [1]-111, [112, cont.], [1, h.t.], *pl. 1-4* with text (uncol.). *Copies*: NY. – Preprinted or reprinted from Kgl. Danske Vidensk. Selsk. Skr. ser. 7, nat. math. afd. 12(7): [269]-379, [380], [1], *pl. 1-4*.
Ref.: Diels, L., Bot. Jahrb. 54 (Lit.): 63. 12 Mar 1917.

7763. *Studies on the biology and taxonomy of soil algae* ... København (C.A. Reitzels Forlag ...) 1935. Oct. (*Stud. soil alg.*).
Publ.: shortly before 27 Jun 1935 (date defense thesis), p. [i-vii], [1]-183. *Copy*: H. – Preprinted or reprinted from Dansk bot. Ark. 8(9): [1]-180. 1935.

Petersen, Karl (*fl.* 1929), German botanist. (*K. Petersen*).

HERBARIUM and TYPES: Unknown.

BIBLIOGRAPHY and BIOGRAPHY: Barnhart 3: 74; BFM nos. 86, 88, 89; Christiansen p. 318 [index]; Kew 4: 283.

7764. *Flora von Lübeck und Umgebung* ... [Lübeck 1929], 2 parts. Oct. (*Fl. Lübeck*).
1: 1929 (p. 3: Sep 1929), p. [1]-102.
2: 1929(?), p. [1]-211, [212, pasted-on err. slip], 1 map.
Copy: B. – Part 1 published in Mitt. Geogr. Ges. u. Nat. Mus. Lübeck ser. 2. 33. 1929.

Petersen, Niels Frederick (1877-1940), Danish-born American botanist; grew up in Plainview, Nebraska; M.S. Univ. Nebraska 1911, instructor of botany at Baton Rouge 1909-1913; in Panama 1914; at Univ. of Chicago 1915-1916; at Nevada Experiment Station 1917-1920; from 1921 at Nebraska Experiment Station; later settling again at Plainview, Nebr. (*N. Petersen*).

HERBARIUM and TYPES: Material from Panama at G and US, from South Dakota, Nebraska and Florida at NY, from Nevada at NESH.

BIBLIOGRAPHY and BIOGRAPHY: Barnhart 3: 74; BL 1: 195, 312; Bossert p. 307 (b. 22 Apr 1877; d. 28 Oct 1840); Hirsch p. 230; IH 2: (in press); Tucker 1: 551.

BIOFILE: Dwyer, J.D., Taxon 22: 563. 1973 (Panama coll. in Pittier's herb. at G).
Reifschneider, O., Biogr. Nevada Bot. 1844-1963, p. 104-105. 1964 (portr.).

7765. *Flora of Nebraska. A list of the conifers and flowering plants of the state with keys for their determination* ... [Lincoln, Nebraska (published by the author) 1912]. Oct. (*Fl. Nebraska*).
Ed. 1: 1912 (p. [iii]: Sep 1911, copyrighted 1912, TBC 9 Sep 1912; Nat. Nov. Oct(2) 1912), p. [i-iii], [1]-217. *Copies*: G, MO, NY, USDA.
Ed. 2: s.d. (PH rd. 11 Nov 1915), p. [i-iii], [1]-217. *Copies*: BR, PH. – "Second edition. Plainview, Nebraska published by the author".
Ed. 3: 1923 (p. [iii]: 1922), p. [i-iii], [1]-220. *Copies*: NY, US, USDA. – "Third edition" [Lincoln, Nebr.] (published by the author) [s.d.]. Oct.

Ref.: Stieber, M.T., Bull. Hunt Inst. bot. Docum. Fall/Winter 1980: 4-5 (on *Fl. Nebraska*).

Petersen, Otto Georg (1847-1937), Danish botanist, Dr. phil. Kjøbenhavn 1882; with the Botanical Museum 1878-1882, from 1882-1893 in various functions at the University and the Botanical Garden; from 1893-1902 lecturer, from 1902-1919 professor of botany at the Copenhagen Agricultural College. (*O. Petersen*).

HERBARIUM and TYPES: C.

BIBLIOGRAPHY and BIOGRAPHY: Backer p. 658; Barnhart 3: 74; BJI 2: 45; BL 2: 49, 56, 702; BM 4: 1556; Bossert p. 307 (b. 26 Mar 1847; d. 17 Jun 1937); CSP 10: 1043-1044, 12: 570, 17: 819; Herder p. 300; Hortus 3: 1201 (Petersen); Kew 4: 284; Langman p. 580; LS 37871; Morren ed. 10, p. 47, 48; Zander ed. 10, p. 701, ed. 11, p. 799.

BIOFILE: Anon., Bot. Centralbl. 16(47): 256. 1883 (lecturing pharm. bot.), 54: 352. 1893 (app. lecturer Agric. coll. Copenhagen); Bot. Jahrb. 17 (Beibl. 40): 33. 1893 (prof. bot. Agric. coll., succeeding J. Lange); Bot. Not. 1893: 181 (app. Agric. coll.); Generalreg. Bot. Not. 1839-1938, p. 112; Nat. Nov. 15: 237. 1893 (app. Agric. coll.).
Christensen, C., Danske bot. litt. 1880-1911: 54-57. 1913 (bibl., portr.), Danske bot. hist. 1: 672-673, 832-833, 881, 2: 475-480. 1926 (bibl.), Bot. Tidsskr. 44: 239-241. 1937 (obit., portr. b. 26 Mar 1847, d. 16 Jun 1937), Danske bot. litt. 1912-1939, p. 3. 1940 (refs. to obituaries).
Christiansen, W. & W., Bot. Schriftt. Schlesw.-Holst. 318 [index]. 1936.
Fox Maule, A., Københavns Univ. 1479-1979. 13. Bot. p. 211-212. 1979.
Lind, J., Danish fungi herb. Rostrup 36. 1913.
Paulsen, O.G., Nat. Verden 21: 419-421. 1937 (b. 21 Mar 1847, d. 17 Jun 1937; obit., portr.); Overs. k. Danske Vid. Selsk. Virksomh. 1937-1938: 17-65. 1938 (obit., portr.).
Ravn, F.K. et al., K. Veterin. Landbo højskoles Festskrift 1908, p. 451-453, 520.
Schumann, K., Bot. Centralbl. 43: 154-155. 1890 (rev. Lief. 21 of EP, Nat. Pflanzenfam., Musac.-Marantac.).
Urban, I., Symb. ant. 1: 122-123. 1898, 5: 10. 1907; Fl. bras. 1(1): 192-193. 1906 (biogr. details).
Verdoorn, F., ed., Chron. bot. 3: 11. 1937, 4: 72, 266, 271. 1938 (portr.).
Warming, E., Bot. Tidsskr. 12: 231. 1881 (bibl.), Dansk biogr. Lex. 13: 57-58. 1899 (biogr.).

COMPOSITE WORKS: (1) *in* Warming, *Symbolae: Marantaceae*, 33: 328-336. 1889 (publ. 8 Mai 1890, fide Fox Maule), *Zingiberaceae*, 33: 327. 1889 (8 Mai 1890), *Cannaceae*, 33: 328. 1889 (8 Mai 1890), *Musaceae* 33: 327. 1889 (8 Mai 1890).
(2) E.P., *Die natürlichen Pflanzenfamilien*, ed. 1:
  (a) *Cannaceae*, 2(6): 30-32. Oct 1888.
  (b) *Halorrhagidaceae*, 3(7): 226-237. 21 Nov 1893.
  (c) *Musaceae*, 2(6): 1-10. Oct 1888.
  (d) *Trigoniaceae*, 3(4): 309-311. Jul 1896.
  (e) *Vochysiaceae*, 3(4): 312-319. Jul 1896.
  (f) *Zingiberaceae*, 2(6): 10-30. Oct 1888.
(3) Martius, *Flora brasiliensis*:
  (a) *Cannaceae* 3(3) (fasc. 107): 63-80. 1 Jan 1890.
  (b) *Marantaceae*, 3(3): (fasc. 107): 81-172. 1 Jan 1890.
  (c) *Musaceae*, 3(3) (fasc. 107): 1-29. 1 Jan 1890.
  (d) *Zingiberaceae* 3(3) (fasc. 107): 29-62. 1 Jan 1890. (rev. P.H.W. Taubert, Bot. Centralbl. 42(2): 59. 9 Apr 1890.

HANDWRITING: Christensen, C., Danske bot. hist 1: 674. 1926.

7766. *Træer og buske* diagnoser til Dansk frilands-trævækst ... Kjøbenhavn og Kristiania [Oslo] (Gyldendalske Boghandel. Nordisk Forlag). 1916. Oct. (*Træer buske*).
Publ.: Mai-Jun 1916 (p. vi: Apr 1916; Nat. Nov. Jul(1, 2) 1916), p. [i*-iv*], [i]-ix, [1]-517. *Copies*: C, UC.

Petersen, [Lorents Christian] **Severin** (1840-1929), Danish mycologist, schoolmaster at Slotsbjergby near Slagelse, later at Sorø. (*S. Petersen*).

HERBARIUM and TYPES: C.

BIBLIOGRAPHY and BIOGRAPHY: CSP 17: 819; Kelly p. 174; LS 20539-20541, 37872, suppl. 21391-21392.

BIOFILE: Buchwald, N.F., Nat. Verden 1929: 241-244 (obit., portr.; b. 17 Mai 1840, d. 2 Mar 1929).
Christensen, C., Danske bot. litt. 1880-1911: 147. 1913 (b. 17 Mai 1840); Danske bot. hist. 1: 595, 596 (portr.), 881, 2: 463-464. 1926 (bibl.), Danske bot. litt. 1912-1939, p. 3. 1940 (b. 17 Mai 1840, d. 2 Mar 1929; biogr. refs.).
Lind, J., Danish fungi herb. Rostrup 37. 1913.
Rosenvinge, K., Bot. Tidsskr. 40: 445-448. 1929 (obit., portr.).
Wittrock, V.B., Acta Horti Berg. 3(3): 89. 1905.

7767. *Danske Agaricaceer Systematisk fremstilling af bladsvampe, iagttagne i Danmark* ... Kjøbenhavn (G.E.C. Gad) 1907-1911. Oct. (*Dan. Agaric.*).
Publ.: 1907-1911, in two parts, p. [i, iii], [1]-460. *Copies*: FH, H, LD, MICH, Stevenson.
1: 1907 (p. 4: Jan 1907; cover 1907; Bot. Centralbl. 28 Jan 1908; Nat. Nov. Mai(1) 1908, p. [i, iii], [1]-208.
2: 1911 (cover 1911; Nat. Nov. Dec(1) 1911; Bot. Centralbl. 9 Apr 1912; Mycol. Centralbl. 23 Apr 1912), p. [i], 209-460.

Pethybridge, George Herbert (1871-1948), British botanist (phytopathologist); Ph. D. Göttingen 1899; with Department of Agriculture and Royal College of Science at Dublin from 1900, Economic Botanist 1908-1923; mycologist and asst. director of the Plant Pathology Lab., Harpenden, 1923-1936; O.B.E. 1937. (*Pethybr.*).

HERBARIUM and TYPES: Some fungi at E (herb. M. Wilson). – Portr. at G.

BIBLIOGRAPHY and BIOGRAPHY: Barnhart 3: 75; BM 8: 1003; Desmond p. 491; Hawksworth p. 185; IH 2: (in press); Kelly p. 174; Kew 4: 284; LS 37873-37877, suppl. 21401-21434; Stevenson p. 1254.

BIOFILE: A.E.M., Trans. brit. mycol. Soc. 33 (1/2): 161-165. 1950 (obit., portr., bibl., b. 1 Oct 1871, d. 23 Mai 1948).
Bisby, R., Trans. brit. mycol. Soc., Fifty-year Index 19. 1952.
Cotton, A.D., Proc. Linn. Soc., London 160(2): 186-187. 1948 (obit.).
G.S., J. hort. Sci. 24(2): 69-71. 1948 (obit.).
Hedge, I.C. & J.M. Lamond, Index coll. Edinb. herb. 118. 1970 (coll.).
Kneucker, A., Allg. bot. Z. 16: 64. 1910 ("Island"= Ireland).
Large, E.C., Ann. applied Biol. 36(3): 414-417. 1949 (obit.).
Mitchell, M.E., Bibl. Irish lichenol. 52. 1971.
Moore, W.C., Nature 161: 1002. 1948 (obit.).
Praeger, R.L., Bot. Ireland 78. 1934; Some Irish naturalists 143. 1949.
Ramsbottom, J., Trans. Brit. mycol. Soc. 30: 9, 13. 1948.
Verdoorn, F., ed., Chron. bot. 1: 31, 165. 1935, 2: 28, 33. 1936, 3: 161. 1937, 6: 301. 1941.
1496
COMPOSITE WORKS: (1) With R.L. Praeger, *Vegetation of the District lying South of Dublin*, Proc. R. Irish Acad. 25B(6): 124-180, *pl. 7-12*. 2 Dec 1905.
(2) Assistant editor *Journal of Pomology and horticultural Science* 1936-1948 (vol. 23, 1948, dedicated to him).
(3) With J. Adams, *A census catalogue of Irish fungi*, Proc. R. Irish Acad. 28(A): 120-166. 8 Jun 1910. [John Adams, 1872-1950].

Petif [de la Gautrois], **Johann Friedrich Carl Ludwig Corentin** (1764-1845), German (Württemberg) botanist; Dr. med. Stuttgart; private tutor in Graubünden

1787-1789; municipal physician at Tuttlingen 1789-1802; director of a madder factory at Mussbach bei Neustadt (Pfalz) 1802-ca. 1834; from 1834-1839 at Haardt (ib.); from 1839-1845 with his daughter at Zeist (Netherlands). (*Petif*).

HERBARIUM and TYPES: U; some further material at FI. – Some letters written by Petif are at U.

BIBLIOGRAPHY and BIOGRAPHY: Barnhart 3: 75; BM 4: 1557; JW 3: 365; PR 7082 (ed. 1: 7918); Rehder 1: 378, 532; Tucker 1: 551.

BIOFILE: Anon., Alg. Konst. Letterbode 1847(1): 191 (herbarium for sale).
Martens, G. v. & C.A. Kemmler, Fl. Württemberg ed. 2. 780. 1865.
Schultz Bipont., C.H., Pollichia 14: 12-15. 1856 (biogr. details obtained from P's daughter Eleonore, various dates not quite correct).
Steinberg, C.H., Webbia 32: 33. 1977 (material at FI).
Wilde, J., Pfälzisches Museum 48(1/2): 1-5, 1931 (biogr., b. 20 Feb 1764, Petif as a botanist), 49(1/2): 1 p. 1932 (on his herbarium at U; d. 16 Dec 1845).

7768. *Enumeratio plantarum* in ditione *florae palatinatu sponte crescentium.* Post Pollichium, Kochium & Zizium denuo recensuit auxitque C. Petif, Med. Dr. Pars phanerogamica. Biponti [Zweibrucken] (typis Georgii Ritteri) 1830. Oct. (in fours). (*Enum. pl. fl. palat.*).
Publ.: 1830, p. [i]-viii, [1]-96, [1, err.]. *Copies*: G, USDA.

Petit, Emil Charles Nicolai [Nicolai Emil Charles] (1817-1893), Danish botanist and physician; one of the founders of "Det Naturhistoriske Selskab" (later Botanisk Forening); schoolteacher from 1843-1848; military physician 1841-1852; practicing physician 1848-1852; id. in Fredensborg 1852-1873; living at Copenhagen from 1878. (*Petit, E.*).

HERBARIUM and TYPES: C (large herbarium left to Bot. Forening); some further material at B, GE, MANCH, PC. – Christensen (1925) mentions a "Samling af Danmarks officinelle Planter. 1.-15. Hefte. – Kbh. Fol. " which he too had not seen.

BIBLIOGRAPHY and BIOGRAPHY: Barnhart 3: 75; BL 2: 54, 140, 702;; Christiansen p. 318 [index]; CSP 10: 1044, 17: 821; Kew 4: 284-285; Morren ed. 10, p. 48; PFC 1: xlix, 3(1): x.

BIOFILE: Anon., Bot. Tidsskr. 19: xix: 2894 (left foreign part of herb. to C, Danish and moss herb. to Botanisk Forening; left also a fund for botanical travelling).
Christensen, C., Danske bot. litt. 1880-1911: 6-7. 1913 (bibl., portr.); Danske bot. hist. 1: 429-431, 590-591, 882 (portr.), 2: 261-262. 1925.
Fischer-Benzon, R.D.J. von, *in* P. Prahl, Krit. Fl. Schlesw.-Holst. 2: 46. 1890.
Warming, E., Bot. Tidsskr. 12: 181-182. 1881 (bibl.); 19: 244-254. 1895 (obit., portr., brief bibl., b. 2 Feb 1817, d. 7 Nov 1893).

COMPOSITE WORKS: See A.S. Örstedt, *Reg. mar.*, TL-2/7014, Petit respondens.

Petit, Paul Charles Mirbel (1834-1913), French pharmacist and algologist, esp. diatomologist. (*P. Petit*).

HERBARIUM and TYPES: PC. – *Exsiccatae*: see J. Tempère for *Les diatomées de France* ... 1871-1891 at AWH, PC, PH, S.

BIBLIOGRAPHY and BIOGRAPHY: Barnhart 3: 75; BM 4: 1557, 8: 1003; CSP 8: 605, 10: 1044-1045, 12: 571, 17: 823; De Toni 1: xcvii, 2: xxviii,xciii-xciv (bibl.), cxiv, cxxix, 4: xliv; Jackson p. 289, 350, 403; Kew 4: 285; LS 20545-20546a, suppl. 21440; Morren p. 56; MW p. 384;TL-1/49; TL-2/354, 2390, 3116, 5779; Tucker 1: 551.

BIOFILE: Anon., Table art. orig. 1854-1893, Bull. Soc. bot. France 186-187 (bibl.).
Deby, J., Bibl. microsc. microgr. 3, Diat.: 44. 1882 (bibl.).

De Toni, G.B., Nuova Notarisia 25: 78-91. 1941 (obit., bibl., d. 27 Nov 1913).
Fryxell, G.A., Beih. Nova Hedw. 35: 363. 1975.
Koster, J.Th., Taxon 18: 556. 1969.
Urban, I., Symb. ant. 1: 123. 1898.
VanLandingham, S., Cat. diat. 6: 3580, 7: 4216-4217. 1978.

COMPOSITE WORKS: (1) Diatomacées, *in* P.A. Hariot, *Miss. sci. Cap Horn, Bot.* p. 111-140, *pl. 10*. Jan-Feb 1889, see TL-2/2390 (reprint listed by Nat. Nov. Mar(2) 1889).
(2) With A.M. Hue, *Exped. antarct., bot.*, lichens, see TL-2/3116.
(3) Co-author, J.A. Battandier, *Fl. Algérie* (1888-1897), see TL-2/354.

EPONYMY: *Petitia* M. Peragallo (1909); *Note: Petitia* J. Gay (1832) is named for a French botanist, Félix Petit; *Petitia* N.J. Jacquin (1760) and *Petita* Cothenus (1790, *orth. var.* of *Petitia* N.J. Jacquin) are probably named for François Petit (1664-1741), French surgeon.

7769. *Catalogue des Diatomées de l'Ile Campbell* et de la Nouvelle-Zélande par Paul Petit ... précédé d'une étude géologique des abords de l'Ile Campbell et de la Nouvelle-Zélande par Léon Périer ... Paris (Librairie de Al. Coccoz ...) 1877. Oct. (*Cat. Diatom. Ile Campbell*).
Co-author: Léon Périer (1835-?).
Publ.: 1877 (p. 40: Aug 1877; BSbF 22 Mar 1878; Hedwigia Mar 1878), p. [1]-40, *pl. 4-5* with text (uncol. liths., auct.). *Copies*: BR, FH, G(2), NY, PH; IDC 6392. – Reprinted from Fonds de la Mer 3: 164-198. 1877.
Ref.: Anon. Bull. Soc. bot. France 25 (bibl.): 53. 1878 (rev.); Hedwigia 17: 124-126, 129-131. 1878.

7770. *Liste des Diatomées et des Desmidiées* observées dans les environs de Paris précédée d'un essai de classification des diatomées ... Paris (Chez Alex. Coccoz, libraire ..) 1877. Oct. (*Liste Diatom. Desmid.*).
Publ.: preprint Mai 1877 (Petit deposited a copy of the combined reprint with BSbF 25 Mai 1877; Hedwigia Mai 1877), p. [1]-32, *2 pl. Copies*: BR, NY. – In part preprinted from Bull. Soc. bot. France 23(4): 372-383. 5 Mar-7 Mai 1877 and 24(1): 3-8. 18 Jun-14 Sep 1877, 34-36: Feb(?) 1877.

7771. *Spirogyra des environs de Paris* ... avec xii planches. Paris (Jacques Lechevalier ...) 1880. Oct. (*Spirogyra Paris*).
Publ.: Mai-Jun 1880 (p. 2: 1 Mai 1880; Bot. Centralbl. 28 Jun-2 Jul 1880; Nat. Nov. Jul(2) 1880; J. Bot. Aug 1880), p. [i-iii], [1]-37, [38-39], *pl. 1-12* (uncol., auct.). *Copies*: B, BR, NY.
Preliminary publ.: Bull. Soc. bot. France 21: 38-42, *pl. 1*. Mai 1874. (also reprinted with original pagination, *copy*: BR and with independent pagination, p. [1]-5, *copy*: G).
Ref.: Richter, Bot. Centralbl. 3/4 (51/52):1601-1602. 20-31 Dec 1880 (rev.).
N.L., Bot. Zeit. 38: 608. 27 Aug 1880, 640. 10 Sep 1880.

7772. *Révision des diatomées* de l'herbier des algues *de la Guadeloupe* et de la Guyane de messieurs Mazé & Schramm 1870-1877. [Nuova Notarisia 1898]. Oct.
Collection: Hippolyte Pierre Mazé (1818-1892).
Publ.: Jan 1898 (in journal), p. [1]-13, *pl. 7* (uncol. lith., auct.). *Copy*: G. – Reprinted from Nuova Notarisia 9: 1-13. 1898.

7773. *Diatomées rares* ou peu connues *des cotes françaises* de la Manche et de l'Océan atlantique [1898: Ass. franç. Avancem. Sci. 1898]; [1899] Paris (Secretariat de l'Association ...) s.d. Oct.
1898 (Congrès de Nantes, Aug 1898), p. [1]-8. *Copy*: G. – Reprinted and to be cited from [C.R.] Ass. franç. Avancem. Sci. 1898(2): 375-382, publ. 1898 or 1899.
1899 (Congrès de Boulogne-sur-Mer 1899), p. [1]-7. *Copy*: G. – Reprinted from id. 1899(2): 437-443, publ. 1899 or 1900.

7774. *Catalogue des diatomées provenant de Madagascar* ... Paris (secrétariat de l'Association [française pour l'Avancement des Sciences] ...) 1903. Oct.

Publ.: 1903 (fide Bot. Centralbl. 25 Mai 1904), p. [1]-10, *1 pl.* with text (uncol. lith. auct.). *Copies*: G, NY. – Reprinted and to be cited from [C.R.] Ass. franç. Avancem. Sci. (Montauban 1902): 590-599. 1903.
Ref.: Hariot, P., Bot. Centralbl. 95: 541. 1904.(rev.).

7775. *Diatomées recoltées en Cochinchine* par Monsieur D. Bois, ... déterminées par Paul Petit ... Padova (Tipografia del Seminario) 1904. Oct.
Publ.: Oct 1904 (in journal), p. [1]-8, *pl. 1* (uncol. lith., auct.). *Copy*: G. – Reprinted and to be cited from from Nuova Notarisia 15: 161-168. *pl. 1.* 1904.

Petitmengin, Marcel Georges Charles (1881-1908), French botanist; curator of botany (préparateur) at the Pharmaceutical college of Nancy; travelled in Greece with R. Maire (1906). (*Petitm.*).

HERBARIUM and TYPES: NCY (fide GR); further material at A, G, IBF, K, LE, MPU (herb. J. Vichet), NH. – Some letters and portr. at G.

BIBLIOGRAPHY and BIOGRAPHY: BFM no. 527; BL 2: 160, 174, 378, 702; GR p. 344; Hortus 3: 1201 (Petitm.); Kew 4: 285; LS 20547; MW p. 384-385; TL-2/5268; Zander ed. 11, p. 799.

BIOFILE: Anon., Bot. Not. 1908: 294 (d.); Bull. Soc. bot. France 53: 288. 1906 (to Greece), 55 (Bibl.): 587-588. 1908 (obit.); Hedwigia 48: (141). 1909 (d.); Österr. bot. Z. 59: 39. 1909 (d. 18 Oct 1908).
Bonati, G., Bull. Acad. int. Géogr. bot. 17: xiii, xiv-xvi. 1908 (obit., bibl.).
Camus, F., Bull. Soc. bot. France 55: 587-588. 1908 (obit.).
Kneucker, A., Allg. bot. Z. 15: 16. 1909 (d. 18 Oct 1908).
Werner, R.G., Bull. Acad. Soc. lorrain. Sci. 6(2): 115. 1966 (b. 3 Jan 1881, d. 9 Oct 1908).

COMPOSITE WORKS: With R.C.J.E. Maire, *Étud. pl. vasc. Grèce*, 1908, TL-2/5268.

EPONYMY: *Petitmenginia* Bonati (1911) ("Petimenginia") is very likely named for Petitmengin.

7776. *Contributions à l'étude des Primulacées sino-japonaises* (Bull. Herb. Boissier). 1907. Oct.
Publ.: 31 Mai 1907, p. [i], [1]-14. *Copy*: G. – Reprinted and to be cited from Bull. Herb. Boissier ser. 2. 7(6): [521]-534. 1907. Preliminary publications in Bull. Soc. Sci. Nancy Jan 1907 (n.v.) and Monde des Plantes 44: 14-15. 1 Mar 1907.

7777. *Flore analytique de poche de la Lorraine* et des contrées limitrophes par Julien Godfrin ... et Marcel Petitmengin ... Paris (A. Maloine, éditeur ...) 1909. 18-mo. (*Fl. Lorraine*).
Senior author: Julien Godfrin (1850-1913).
Text: Jan-Mar 1909 (t.p. 1909; p. viii: 15 Dec 1907; Bot. Centralbl. 24 Apr 1908, ann., 27 Apr 1909, reviewed), p. [i*], [i]-viii, [1]-239. *Copy*: B.
Atlas: 1913 (p. viii: 1 Feb 1913), p. [i]-viii, 1-229. (1608 figs.) *Copy*: BR.
Ref.: Cuisinier-Reclus, M., Bot. Centralbl. 110: 443-444. 27 Apr 1909.
 Malinvaud, E., Bull. Soc. bot. France 56: 398-399. 25 Jun 1909.

Petit-Radel, Philippe (1749-1815), French botanist. (*Petit-Radel*).

HERBARIUM and TYPES: Unknown.

BIBLIOGRAPHY and BIOGRAPHY: BM 4: 1537; PR 6, 7086; Rehder 1: 199; TL-2/1293.

BIOFILE: Anon., Flora Beilage 1: 40. 1822.
Candolle, A.P. de, Mém. souvenirs 148. 1862.

COMPOSITE WORKS: Collaborator, Cuvier, F., *Dict. Sci. nat.* see TL-2/1293.

Petiver, James (1658-1718), British botanist, entomologist and apothecary to the Charterhouse; demonstrator at Chelsea (1709), owner of an important natural history cabinet. (*Petiver*).

HERBARIUM and TYPES: BM-SL; further specimens in BM, FI, K, LE, MEL, OXF and P-JU. Petiver's collections were bought by Sir Hans Sloane in 1718. The herbaria number 106 *Horti sicci*. For a full account see Dandy.

MANUSCRIPTS and DRAWINGS: BM possesses (Banksian ms 88) "73 watercolour drawings of plants made at the Cape from living specimens for Dr. Martin Dolneus and given by him to J. Petiver; 67 of these were used by Petiver in the preparation of the plates for his "Gazophylacii Naturae ... decas nona" and bear reference to those plates in his handwriting" (Woodward). BM also possesses Petiver's annotated copy of Rumpf's Herbarium amboinense. See also Edwards (1968).

BIBLIOGRAPHY and BIOGRAPHY: Backer p. 437-438; Barnhart 3: 75; BB p. 242; BM 4: 1557-1558 (extensive list contents *Opera*); Bret. p. 33; Clokie p. 223-224 (extensive inf.); Desmond p. 491-492; DNB 45: 85-86; Dryander 3: 13, 17, 75, 95-97, 130, 134, 176, 180, 183, 187, 209, 225, 446, 447, 464, 465; DU 224; Frank 3 (Anh.): 76; GR p. 409; HA 2: 24-26; HE p. 45; Hegi 3: 583; Herder p. 22, 29, 77; Henrey 1: 286 [index], 2: 733 [index], 3: 100-102 (nos. 1208-1225) (bibl., main treatment); Jackson p. 591 [index; many titles!]; Kew 4: 285; KR 1: 23, 28, 471, 524, 2: 506; Lasègue p. 10, 12, 324, 434, 496; Lenley p. 325; LS 20548-20551; ME 3: 385; MW p. 385; NI 1516-1522, suppl. p. 54; PR 1251, 7087-7089 (ed. 1: 7925-7942 (!)); Quenstedt p. 335; Rehder 5: 664-665; SA 2: 591; Saccardo 1: 126, 2: 84; SK 4: lxxix, lxxxi, lxxxviii; Sotheby 595; TL-1/971; TL-2/see Bradley, R., Breyne, J., Buddle, A., Micheli, P.A.; Tucker 1: 551-552.

BIOFILE: Allen, D.E., The Naturalist in Britain 289 [index]. 1976.
Allibone, S.A., Crit. dict. Engl. lit. 1571. 1870.
Apperson, G.L., Bygone London life 96-104. 1903.
Bridson, G.D.R. et al., Nat. hist. mss. res. Brit. Isl. 428 [index]. 1980.
Briquet, J. Bull. Soc. bot. Suisse 50a: 369-370. 1940 (eponymy; author of first catalogue of plants found on the mountains about Geneva, 1709.
Coats, A.M., The plant hunters 88, 144-145, 203. 1969.
Copeman, W.S.C., Worshipful Soc. Apothecaries London 76-77. 1968.
Daggett, F.S., Auk 24: 448-449. 1907 (special copy of Gazophylacium).
Dandy, J.E., The Sloane herbarium 175-182. 1958 (coll.).
Darlington, W., Memorials 305. 1849.
Dolezal, H., Friedrich Welwitsch 143-144. 1974.
Drewitt, F.D., Apothec. Garden 105 [index]. 1922.
Edwards, P.I., J.S. Afr. Bot. 34: 243-253. 1968, J. Soc. bibl. nat. Hist. 8(4): 335-336. 1978 (on Petiver ills. of S. Afr. pl. in the Gazophylacium).
Ewan, J., Regn. veg. 71: 22, 25, 27-29, 50, 54. 1970 (P. and American botany).
Ewan, J. & N., John Banister 482 [index]. 1970 (many details, biogr., colls.).
Freeman, R.B., Brit. nat. hist. books 277. 1980.
Fries, Th.M., Bref Skrifv. Linné 1(5)" 260. 1911.
Gilbert, P., Comp. biogr. lit. deceased entom. 295. 1977 (further biogr. refs.).
Hawksworth, D.L. & M.R.D. Seaward, Lichenology Brit. Isl. 1568-1975, p. 4, 5, 133, 198. 1977.
Herder, F. v., Bot. Centralbl. 55: 259, 260. 1893 (Petiver material at LE).
Hindle, Brook, Pursuit sci. revol. Amer. 15, 16. 1956.
Howard, R.A., Bot. J. Linn. soc. 79: 71-74. 1979.
Horwood, A.R., Fl. Leicestershire clxxxvii. 1933.
Hulth, J.M., Bref Skrifv. Linné 1(8): 29, 33. 1922.
Jessen, K.F.W., Bot. Gegenw. Vorz. 240, 276, 284, 381. 1884.
Jourdan, Dict. Sci. méd., Biogr. méd. 6: 405-406. 1824.
Karsten, M.C., The old Company's Garden 71, 77, 81-87, 96. 1951.
Kent, D.H., Brit. herb. 72. 1957; Hist. fl. Middlesex 14. 1975.
Lisney, A.A., Bibl. Brit. Lepidopt. 1609-1799, p. 42-44. 1960 (bibl., portr.).
Löwegren, I., Naturaliekabinett Sverige 58, 89. 1952.

Murray, G., Hist. coll. BM(NH) 173. 1904 (coll.).
Nissen, C., Zool. Buchill. 315-316. 1969.
Pax, F., Bibl. schles. Bot. 2. 1929.
Pearsall, W.H., Fl. Surrey 45-46. 1931.
Percheron, A., Bibl. entom. 1: 314-315. 1837.
Pulteney, R., Sketches 2: 31-43. 1790.
Reynolds, G.W., Aloes S. Africa 80. 1950 (identif. Aloes in Gazophylacium).
Richardson, R., Extr. lit. sci. corr. 48, 50, 73, 109, 110. 1835.
Saccardo, P.A., J. Bot. 37: 227. 1899 (on his exsiccatae).
Saccardo, P.A. & A. Béguinot, Bull. Soc. bot. ital. 1901: 243-251. (P. as inventor of plantae exsiccatae).
Salmon, C.E., Fl. Surrey 45-46. 1931.
Séguier, J.F., Bibl. bot. 137-139. 1740.
Smith, J.E., in Rees, Cyclop. vol. 27 (alph.), Sel. corr. Linnaeus 2: 161-170. 1821.
Stafleu, F.A., Intr. Jussieu Gen. Pl., facs. repr. xxxi. 1964 (specimens in P-JU); Linnaeus and the Linnaeans 233. 1971.
Stearns, R.P., Proc. Amer. antiq. Soc. 62: 243-265. 1952 (1913) (James P., promotor of natural science).
Trimen, H. & W.T.T. Dyer, Fl. Middlesex 377-386. 1869.
Wilson, E.H., Plant hunting 1: 9, 2: 107. 1927.
Wolley-Dod, A.H., Fl. Sussex xxxvii-xxxviii. 1937 (on Journal to Hastings; Sloane ms no. 3340 at BM, transcribed in Phytologist 1862: 114.).
Woodward, B.B., Hist. coll. BM(NH) 28, 45. 1904 (mss., copy Rumphius).

HANDWRITING: J.E. Smith, Sel. corr. Linnaeus 2, 1821, (facs. pl.); Dandy, J.E., The Sloane herb. facs. no. 83; Clokie p. 279.

EPONYMY: *Petivera* Cothenius (1790, *orth. var.*); *Petiveria* Linnaeus (1753).

PUBLICATIONS: We refer to Henrey (1: 186 (index), 2: (index), 3: 100-102, bibl. nos. 1208-1225) for detailed treatment of Petiver's works. Dandy (1958) supplies many additional details in his account of the Petiver collections in the Sloane Herbarium. – The post-Linnaean issue of the original stock of folio's and separate issues (*Opera historiam naturalem spectantia* (1764, 1767) have no standing in botanical nomenclature and have to be regarded as pre-linnaean. An important critical bibliography of many Petiver items which are also of importance for botany is given by Lisney (1960).

Petkoff [Petkov], **Stefan Pavlikianoff** (1866-1956), Bulgarian botanist; habil. Sofia 1901; from 1906 professor of botany and director of the botanical garden of Sofia, ord. prof. 1911-1936. (*Petkoff*).

HERBARIUM and TYPES: SO (inf. S. Kozucharov).

BIBLIOGRAPHY and BIOGRAPHY: BJI 2: 134; Bossert p. 308 (b. 5 Jun 1866, d. 5 Feb 1951); Kew 4: 285-286; LS suppl. 21442; Nordstedt suppl. p. 12-13.

BIOFILE: Adamovíc, L., Vegetationsverh. Balkanländer 9, 20. 1909 (Veg. Erde 11).
Anon., Bot. Centralbl. 87: 160. 1901 (habil.), 102: 576. 1906 (app. Sofia), 117: 320. 1911 (ord. prof.); Hedwigia 40: (175). 1901, 46: (89). 1907, 52: (83). 1912; Nat. Nov. 33: 526. 1911 (ord. prof.); Österr. bot. Z. 60: 455. 1911 (ord. prof.), 85: 320. 1936 (retirement).
Kneucker, A., Allg. bot. Z. 13: 20. 1907, 17: 170. 1911.
Koster, J.Th., Taxon 18: 556. 1969.
Petkoff, St., Bull. Soc. Bulgarie 8: iii-xviii. 1939 (50 yrs of botany at Sofia).
Stojanov, N., Bull. Inst. bot. Sofia 2: 7-12. 1951 (obit., b. 17 Jun 1866, d. 8 Feb 1951).
Wood, R.D., Monogr. Charac. 1: 833. 1965 (bibl. Chara).

EPONYMY: *Petkovia* Stefanoff (1936).

FESTSCHRIFT: *Bulletin de la Société botanique de Bulgarie* 7, 1936, "Ce volume fait partie d'un

Recueil dédié à M. le Prof. St. Petkoff pour son 70-e anniversaire"; portr., biogr., bibl. on p. v-viii.

7778. *La flore aquatique et algologique de la Macédoine* du S.-O. avec 4 planches, 85 figures, une photographie, une carte géographique et un résumé en langue française. Philippopoli [Plovdiv, Bulgaria] (imprimerie Chr.G. Danoff) 1910. Oct. (*Fl. aquat. alg. Macéd.*).
Publ.: Jun 1910 (P. dated a copy sent to De Wildeman at BR 21 Jun 1910; ÖbZ Jun 1910), p. [i], [1]-189, *pl. 1-4*, [*5-6*]. *Copies*: G, US. – Text in Bulgarian. – Number of copies printed: 500.

Petrak, Franz (1886-1973), Austrian mycologist, born in Mährisch-Weisskirchen (later Hranice); studied in Vienna from 1906-1910 with R. von Wettstein; Dr. sci. ib. 1913; high school teacher in Wien until 1916; in the Austrian Army 1916-1918; from 1918-1938; living as a private scientist in his home town selling exsiccatae; at the Naturh. Museum, Wien 1938-1951; travelled in the United States Apr 1950-Feb 1951. (*Petr.*).

HERBARIUM and TYPES: W (main parts of mycological herbarium acquired in 1967 and 1974; small part of herbarium bought 1936), M (*Pilzherbarium* Petrak, 2. Satz, 5000, rd. 1959), PR (phanerogam herb. including main part of *Cirsium* coll.). *Exsiccatae* (list incomplete):
Mycotheca generalis, nos. 1-2100, 41 Lieferungen, 1928-1950. Sets: B (nos. 101-2100 extant), BPI, E, FH (compl.).
Flora Bohemiae et Moraviae exsiccata; nos. 1-2575 (series 1: fungi; ser. 2, Abt. 1, fungi; 2: lichens; 3: musci; 4: algae), B (extant), BM, BR, BRNO, C, DBN, E, FH, FI, GH, GZU, IBI, K, LAU, LD, LPS, M, NY, PAD, PR, PRC, RMS, S, SI, TLA, URM, W, WVA.
Cirsiotheca universa (fasc. 1-20, 1910-1927, 200 nos.; continued?). Sets at B, BM, C, G, K, LAU, PR, W.
Menthotheca universa. Sets at B, C, LAU.
Fungi albanici et bosniae exs., nos. 1-200. Set at FH.
Fungi polonici exs., nos. 1-650 (fasc. 1-26), set at FH.
Mycotheca carpathica, nos. 1-475 (fasc. 1-19), set at FH.
Fungi eichleriani (*Mycotheca eichleriana*) (ex herb. C.A. Eichler), nos. 1-298, 1908-1915 (Lief. 1, nos. 1-25, Sep 1908, see Allg. bot. Z. 14: 164. 1908; Lief. 14-15, nos 266-300, sic). Oct 1912). Sets at BR, DBN, W.
Letters at G.

BIBLIOGRAPHY and BIOGRAPHY: AG 12(1): 203; Ainsworth p. 282; BM 8: 1003; Bossert p. 308; GR p. 437, cat. p. 69; Hawksworth p. 185; Hortus 3: 1201 (Petrak); IH 1 (ed. 1): 103, (ed. 4): 182, (ed. 5): 198, (ed. 6): 362, (ed. 7): 339, 2: (in press); Kelly p. 174-175, 256; Kew 4: 286-287; Kláštersky p. 141; Langman p. 580; LS 37878, suppl. 21443-21490; MW p. 385, suppl. p. 279; NW p. 55; TL-2/2571, see G. Moesz.

BIOFILE: Anon., Hedwigia 63: (150-151), 1922 (sets up the Allg. mykolog. Tauschverein); Taxon 23: 224. 1974; Preslia 48(4): 305. 1976 (brief biogr., b. 9 Oct 1886, d. 9 Oct 1973).
Arx, J.A. von, Persoonia 9(1): 95-96. 1976.
Doidge, E., Bothalia 5: 1015-1016. 1950 (publ. on S. Afr. fungi).
Futák, J. & K. Domin, Bibl. fl. ČSR 466. 1960 (From 1907; err. sub E.R. Petrak).
Hedge, I.C. & J.M. Lamond, Index coll. Edinb. herb. 118. 1970.
Hertel, H., Mitt. Bot. München 16: 416. 1980 (*Pilzherbarium* 2. Satz, rd. 1959 at M).
Hiratsuka, N., Trans. mycol. Soc. Japan 15(3): vii-xii. 1974 (obit., portr.).
Janchen, E., Richard Wettstein 189. 1933 (ÖbZ vol. 82) (with von Wettstein 1906-1910; Cat. fl. austr. 1: 16. 1956.
Kneucker, A., Allg. bot. Z. 14: 164. 1908 (P. has bought herb. C.A. Eichler [Teplitz, Bohemia], distr. fungi as Mycotheca eichleriana and offers rest of herb. for sale).
Kohlmeyer, *in* Potztal, E., Willdenowia 2: 785. 1961 (Mycoth. gen.; extant).
Lohwag, K., Festschr. Franz Petrak vi-x. 1957 (Sydowia ser. 2, Beih. 1), (tribute, portr., b. 9 Oct 1886).
Pax, F., Bibl. schles. Bot. 14, 19, 90, 101, 117. 1929.

Pilát, A., Česká Mykol. 10(4): 255-256. 1956 (70th birthday, tribute, portr., bibl.), 21(1): 48-49. 1967 (80th birthday, portr.), 28(1): 60-61. 1974 (obit., portr.).

Rechinger, K.H., Sydowia 26: xix-xxviii. 1974 (portr.; extensive biography with many original details; important also for the history of the collections of W between 1938 and 1948).

Riedl, H., Sydowia 26: xxix-xxxii. 1974 (P. as a mycologist); Taxon 24: 348. 1975 (W rd. greater part of his herb.), Ann. Naturh. Mus. Wien 81: 661-664. Feb 1978 (portr.).

Sayre, G., Mem. New York Bot. Gard. 19(1): 40. 1969, 19(2): 236. 1971 (exsicc.).

Stafleu, F.A., Taxon 19: 282-283. 1970 (Petrak's lists and supplement, Kew 1969).

Stevenson, J., Beih. Nova Hedw. 36: 265-271 (detailed descr. of American material in *Mycotheca generalis*).

Verdoorn, F., ed., Chron. bot. 2: 34, 109, 374. 1936, 3: 27, 93. 1937.

GENERAL INDEX: G.J. Samuels, *An annotated index the mycological writings of Franz Petrak*, New Zeal. Dept. Sci. Ind. Res. Bull. 230, volume 1, A-B, Wellington New Zealand, 1981. (Contains a biogr. introduction, a list of herbaria holding exsicc., a bibliography and the index to taxa treated by Petrak, letters A-B).

NOTE: Petrak was a profilic writer. His enormous output is divided over nearly five hundred separate publications. Much of his mycological work was published in the *Annales mycologici* and its successor *Sydowia*. According to Riedl (1978) Petrak described nearly 400 new genera and thousands of new species of fungi. His private mycological herbarium consisted ultimately of nearly 100.000 specimens. "In seinen Sammlungen wie seinen Publikationen waren Petrak's Dimensionen fast zu gross für unsere nach raschen Erledigungen drängende Zeit; zu gross, auch für ihn selbst zu einer umfassenden Bewältigung."

COMPOSITE WORKS: (1) Collaborator Hegi, *Ill. fl. Mitteleur.* ed. 2, 3(1).

(2) published sets 33-47 of the *Cryptogamae exsiccatae*, edited by W, see TL-2/387; accompanied by *Schedae*, in Ann. Naturhist. Mus. Wien 51: 347-373. 1941, 52: 279-300. 1942, 56: 417-433. 1948, 58: 31-43. 1951; 61: 60-73. 1951 (parts with Asiatic taxa; cited from MW suppl.).

(3) editor of *Sydowia* 1-25, 1947-1972.

HANDWRITING: Trans. mycol. Soc. Japan 15(3): viii. 1974.

FESTSCHRIFT: Sydowia, ser. 2, Beih. 1, *Festschrift für Franz Petrak*, zu seinem 70. Geburtstage am 9 Oct 1956 herausgegeben von ... K. Lohwag (Wien), Verlag von Ferdinand Berger, Horn, N.-Ö., Austria, 1957.

EPONYMY: *Franzpetrakia* Thirumalachar & Pavgi (1957); *Petrakia* P. Sydow (1913); *Petrakiella* H. Sydow (1924); *Petrakina* Ciferri (1932); *Petrakiopeltis* Batista (1958); *Petrakiopsis* C.V. Subramanian et K.R.C. Reddy (1968); *Petrakomyces* C.V. Subramanian et K. Ramakrishnan (1953); *Pseudpetrakia* M.B. Ellis (1917).

7779. *Der Formenkreis des Cirsium eriophorum* (L.) Scop. in Europa ... Stuttgart (E. Schweizerbart'sche Verlagsbuchhandlung Nägele & Dr. Sproesser) Qu. (*Formenkr. Cirsium eriophorum*).

Publ.: 1912 (Nat. Nov. Dec(1) 1912; ÖbZ Oct-Nov 1912, p. [ii-iv], [1]-92, *pl. 1-6*, map. *Copies*: B, BR, G, NY. – Bibliotheca botanica Heft 78.

7780. *Aufzählung der von G. Woronoff im Jahre 1910 in Adzarien und Russisch-Lazistan gesammelten Cirsien* .... Tiflis (Kozlovskago ...) 1912. Oct.

Publ.: 1912, p. [1]-16. *Copy*: G. – Reprinted and to be cited from Trav. Jard. bot. Tiflis 12(1): 32-46. 1912.

7781. *Ueber einige Cirsium aus dem Kaukasus* ... Tiflis (Kozlovskago ...) 1912. Oct.

Publ.: 1912, p. [i], [1]-31. *Copy*: G. – Reprinted from Trav. Jard. bot. Tiflis 12(1): 1-31. 1912.

Neue Beiträge zur Kenntniss der Cirsien der Kaukasus: 1912, p. [i], [1]-14. *Copy*: G. – Reprinted from Moniteur du jardin botanique de Tiflis 24, 1912.

7782. *Beiträge zur Pilzflora von Mähren* und Oesterr.-Schlesien ... [Berlin, Ann. mycol.] 1914-1917. 4 parts.
Publ.: in Ann. mycol. and to be cited from journal.
1: Oct 1914 (Nat. Nov. Dec(1, 2). 1914), Ann. mycol. 12: 471-479. 1914.
2: 25 Mar 1915 (Nat. Nov. Mai (1, 2). 1915), Ann. mycol. 13(1): 44-51. 1915.
3: 28 Mai 1916 (Nat. Nov. Jun (1, 2). 1916), Ann. mycol. 14(3/4): 159-176. 1916.
4: 28 Feb 1917 (Nat. Nov. Apr (1, 2). 1917), Ann. mycol. 14(6): 440-484. 1917.

7783. *Die nordamerikanischen Arten der Gattung Cirsium* ... [Beih. Bot. Centralbl.] 1917. Oct.
Publ.: 5 Oct 1917 (Nat. Nov. Nov-Dec 1917), published in Beih. Bot. Centralbl. 35(2): 223-567. 5 Oct 1917. (Reprint with special cover but orig. pagination: G). – See also: *Die mexikanischen und zentralamerikanischen Arten der Gattung Cirsium*, Beih. Bot. Centralbl. 27(2): 207-255, *pl. 1-2.* 22 Nov 1910. "Nach eigener Aussage hatte [Petrak] eine umfassende Monographie der Gattung geplant und im Manuskript weitgehend fertiggestellt. Es war ihm jedoch nicht gelungen, die Typen einiger, von Japanischen Autoren aufgestellter Arten zur Ansicht zu erhalten, daher vernichtete er eines Tages sein Manuskript. Die Im Beiheft ... erschienene ausführliche Bearbeitung ... ist das einzige erhalten gebliebene grössere Bruchstück dieser Monographie" (Rechinger 1974). This must refer to the above mentioned publications on American *Cirsium* in Beih. Bot. Centralbl.

7784. *Mykologische Notizen* i-xiv [Annales mycologici] 1919-1941. Oct.
Publ.: A series of papers published and to be cited from *Annales mycologici*. Reprints were made; those seen by us all show original pagination.

No.	Ann. mycol.	date	Nat. Nov.
1	17: 59-100	10 Jun 1919	Jun 1920
2	19: 17-128	31 Jul 1921	Sep 1921
3	19: 176-223	31 Oct 1921	Jan 1922
4	20: 300-345	30 Dec 1922	Mar 1923
5	21: 1-69	28 Feb 1923	Apr 1923
6	21: 182-335	28 Jul 1923	Aug 1923
7	22: 1-182	20 Jun 1924	Jul 1924
8	23: 1-143	20 Aug 1925	Oct 1925
9	25: 193-343	10 Apr 1927	Mai 1927
10	27: 324-410	30 Dec 1929	Feb 1930
11	29: 339-397	30 Nov 1931	Jan-Feb 1932
12	32: 317-447	25 Sep 1934	-
13	38: 181-267	20 Jul 1940	-
14	39: 251-349	31 Dec 1941	-

7785. *Die Gattungen der Pyrenomyzeten, Sphaeropsideen, und Melanconieen* von F. Petrak und H. Sydow. 1. Theil: Die phaeosporen Sphaeropsideen und die Gattung Macrophoma. Dahlem bei Berlin (Verlag des Repertoriums, ...) [1926-] 1927. Oct. †. (*Gatt. Pyrenomyz.*).
Co-author: Hans Sydow (1879-1946) (private herb. destroyed 1943).
Publ.: as *Beiheft* 42 of F. Fedde, Repertorium specierum novarum regni vegetabilis, in 3 parts 1926-1927. *Copies*: B, BR, MO, Stevenson, U.
T.p.: 5 Sep 1927, p. [i-ii], (dates on p. ii and on covers of parts).
1. Lieferung: 30 Sep 1926, p. [1]-160.
2. Lieferung: 25 Jan 1927, p. 161-320.
3. Lieferung: 5 Sep 1927, p. 321-551.
Facsimile ed.: 27 Sep 1974, p. [i-ii], [1]-551. *Copy*: FAS. Reprint by Otto Koeltz Antiquariat, Koenigstein-Ts./B.R.D. 1974, ISBN 3-87429-071-9. – See Taxon 24: 169. 1975.
Preliminary papers: Petrak, F. & Sydow, H., *Kritisch-systematische Original-Untersuchungen*

über Pyrenomyzeten ..., in *Ann.-mycol.*: (1), 21: [349]-384. 15 Oct 1923; (2), 22: [318]-386. 15 Nov 1924; (3), 23: [209]-294. 31 Dec 1925; (4), 27: [87]-115. 25 Feb 1929.
Copies: (reprint with original pagination): MICH. – Another important paper for the understanding of P's system of *Fungi imperfecti* is his criticism of von Höhnel's modification of Saccardo's system, Petrak, Ann. mycol. 23: 1-11. 1925.

7786. *Verzeichniss der neuen Arten*, Varietäten, Formen, Namen und wichtigsten Synonyme [of Fungi], [Just's Botanischer Jahresbericht] 1930-1944. [*Petrak's lists*].
Publ.: Lists of new names of fungi known as "Petrak's lists", preceeding the present *Index of fungi*. The list numbers are those appearing in the reprint, see below *Index of fungi*. Saccardo's *Sylloge fungorum* (1882-1931) covers the literature up to 1919.

List (for)	Just's Botanischer Jahresbericht	date
1 (1920)	48(2): 184-256	1930
2 (1921)	49(2): 267-336	1931
3 (1922/28)	56(2): 241-697	8 Apr 1937 (p. 481-497: 5 Jul)
4 (1929)	57(2): 592-632	5 Apr 1938
5 (1930)	58(1): 447-570	1938
6 (1931)	60(1): 449-514	1939
7 (1932/35)	63(2): 805-1056	5 Sep 1944

Reprint and continuation: 1950-1953 as: Index of Fungi; 1920 [1921, ...]. List of new species and varieties of fungi, new combinations and new names published 1920 [1921, ...] by F. Petrak ... The Commonwealth Mycological Institute, Kew, Surrey (*Copy*: NY):

1	for 1920 (Just 48(2): 184-256)	1953
2	for 1921 (Just 49(2): 267-336)	1953
3	for 1922-1928 (Just 56(2): 241-697)	1953
4	for 1929 (Just 57(2): 592-632)	1953
5	for 1930 (Just 58(1): 447-570)	1952
6	for 1931 (Just 60(1): 449-514)	1952
7	not reprinted	
8	for 1936-1939, published only as *Index of fungi 1936-1939* ... by F. Petrak, Kew 1950, [1]-117.	

Cumulative index: 1956-1957, 2 parts. – *Index of fungi. Petrak's lists* 1920-1939. *Cumulative index*:
1(generic names; spec. epith.): 1956, p. [1]-168.
2: (host index ...): 1957, p. [169]-307.
Supplement: 1969, p. [1]-236. *Index of Fungi, a supplement to Petrak's lists 1920-1939*, Commonwealth Mycological Institute, Kew, Surrey, England, 1969 (see Taxon 19: 282-283. 1970).

Petrescu, Constantin C. (1879-1936), Roumanian botanist at Iaşi, published on the flora of the Dobragea (Dobruja) and Moldavia; with the Botanical laboratory, Iaşi, from 1911 (preparator), 1914 (assistant), 1920 (head assistant) until 1936. (*Petrescu*).

HERBARIUM and TYPES: I.

BIBLIOGRAPHY and BIOGRAPHY: Hirsch p. 230; IH 2: (in press); LS suppl. 21491.

BIOFILE: Răvăruţ, M., Bul. Grad. Muzeul. bot. Cluj. 17: 88-89. 1937 (obit., bibl., b. 5 Mar 1879, d. 25 Oct 1936).
Verdoorn, F., ed., Chron. bot. 4(2): 172. 1938.

Petri, Friedrich (1837-1896), German (Berlin) botanist; Dr. phil. Berlin 1863; pupil of A. Braun; from 1864-1896 teacher at the high school (Realschule) of Luisenstadt (Berlin; 1865 ordentlicher Lehrer, 1875 Oberlehrer, 1884 Professor); in his later years especially active in chemical technology and hygienics. (*F. Petri*).

HERBARIUM and TYPES: Unknown.

BIBLIOGRAPHY and BIOGRAPHY: Barnhart 3: 76; BM 4: 1558; CSP 12: 571, 17: 825; Jackson p. 108; Kew 4 : 287; PR 7092; Rehder 1: 252.

BIOFILE: Beyer, R., Verh. bot. Ver. Brandenburg 39: lv-lviii. 1897 (obit., b. 26 Mai 1837, d. 28 Nov 1896) (see also p. cxxii).
Petri, F., Vita, *in* his *De gen. Armeriae* p. [42]. 1863 (in his "theses": on p. [3] e.g.: "Dispositio foliorum optimam rationem nobis praebet ad inflorescentiam plantae cognoscendam".

EPONYMY: *Petriella* Curzi (1930); *Petriellidium* D. Malloch (1970).

NOTE: Not to be confused with the originator of the "Petri dish", the German bacteriologist R.J. Petri (1852-1921), who published e.g. *Das Mikroskop*, Berlin 1896, 249 p. – For the "development" of the Petri dish see R.C. Benedict, Torreya 29: 9-12. 9 Mar 1929.

7787. *De genere Armeriae*. Dissertatio inauguralis botanica quam consensu et auctoritate amplissimi philosophorum ordinis in alma litterarum universitate Friderica Guilelma ad summos in philosophia honores rite capessendos die xxiv. m. januarii a. mdccclxiii. H.L.Q.S. Publice defendet auctor Fridericus Petri berolinensis ... Berolini (typis expressit Gustavus Schade) [1863] Oct. (*Gen. Armeriae*).
Publ.: 24 Jan 1863 (t.p.; BSbF session Jan 1863), p. [i-iv], [1]-41, [42, vita; 43, theses].
Copies: B, G, MO, NY. – Dissertation defended under Alexander Braun.
Ref.: Fournier, E., Bull. Soc. bot. France 10: 240-241. 1863.

Petri, Lionello (1875-1946), Italian botanist. (*Petri*).

HERBARIUM and TYPES: Mycological type specimens at FI, FIPF and ROPV.

BIBLIOGRAPHY and BIOGRAPHY: Hawksworth p. 185; CSP 17: 825; Hirsch p. 230; IH 2: (in press); Kelly p. 175; LS 20553-20566, 37879-37992, suppl. 21496-21571 (bibl.); Saccardo 2: 84; Stevenson p. 1254.

BIOFILE: Ciferri, R., Taxon 1(8): 127. 1952.
Petri, L., Ann. mycol. 2: 412-438. 1904 (on *Tulostoma*).
Verdoorn, F., ed., Chron. bot. 3: 26, 37, 192. 1937, 4: 415. 1938, 5: 243. 1939.

7788. *Società botanica italiana. Flora italica cryptogama* pars I: Fungi *Gasterales* Secotiaceae, Lycoperdaceae, Sclerodermataceae auctore L. Petri. Fascicolo n. 5 ... Rocca S. Casciano (Stabilimento Tipografico Cappelli) 1909. Oct. (*Fl. ital. crypt., Gasterales*).
Publ.: 1 Mar 1909, (t.p.; Nat. Nov. Jun(2) 1909), p. [i], [1]-139, [140, err.]. *Copies*: FH, FI, MICH, NY, Stevenson.

Petri, Donald (1846-1925), Scottish botanist; M.A. Aberdeen 1867; to Australia 1868, teacher at Scotch College, Melbourne; inspector of schools Otago prov. govt., New Zealand 1874; chief inspector of schools, Auckland 1894. (*Petrie*).

HERBARIUM and TYPES: WELT (over 40.000); duplicates at BM, C, CHR, E, G, K, KIEL, MEL, MO, NA, NH, NY, WELC, Z. – Some letters at G.

BIBLIOGRAPHY and BIOGRAPHY: Backer p. 438; Barnhart 3: 76; BJI 1: 44; BL 1: 81, 83, 85, 101, 312; BM 4: 1558; CSP 10: 1045-1046, 12: 571, 17: 826-827; Desmond p. 492; Hortus 3: 1201 (Petrie); IH 1(ed. 6): 362, (ed. 7): 339, 2: (in press); Kew 4: 287; Rehder 5: 665.

BIOFILE: Anon., Hedwigia 69: (95). 1929 (d.).
Cheeseman, T.F., Manual New Zealand fl. xxxii. 1906.
Cockayne, L., Trans. Proc. New Zealand Inst. 56: vii-ix. 1926 (obit., portr., b. 7 Sep 1846, d. 1 Sep 1925).

Hamlin, B.G., Rec. Dominion Museum 3(2): 89-99. 1958 (itin., bot. expeditions, bibl.).
Hedge, I.C. & J.M. Lamond, Index coll. Edinb. herb. 118. 1970.
Jackson, B.D., Bull. misc. Inf. Kew 1901: 52 (pl. K).

COMPOSITE WORKS: *Gramina of the subantarctic islands of New Zealand*, in C. Chilton, ed., *The subantarctic islands of New Zealand* 2: 472-481. 1909 (art. 20).

EPONYMY: *Petriella* Zotov (1943).

7789. *List of the flowering plants indigenous to Otago* [Trans. Proc. New Zealand Inst.] 1896. Oct.
Publ.: Jun 1896 (in journal, Nat. Nov. Feb(2) 1897). Published *in* Trans. Proc. New Zeal. Inst. 28: 540-591. Jun 1896. – We have seen no reprint with independent pagination.
Supplement: Jun 1897, *in* Trans. Proc. New Zealand Inst. 29: 421-422. 1897.

Petrov, Vsevolod Alexeevič (1896-x), Russian botanist. (*Petrov*).

HERBARIUM and TYPES: LE.

BIBLIOGRAPHY and BIOGRAPHY: BM 8: 1004; IF suppl. 3: 216; Kew 4: 288; MW p. 385.

BIOFILE: Lipschitz, S.J., Fl. URSS fontes 222 [index]. 1975.
Verdoorn, F., ed., Chron. bot. 3: 256. 1937.

HANDWRITING: S.V. Lipschitz & I.T. Vasilczenko, Central Herb. USSR 116. 1968.

EPONYMY: *Petrovanella* Kylin (1956) is named for J. Petrová, a worker in Rhodophyceae.

7790. *Flora Iakutiae* ... fasc. i Pteridophyta – Poaceae ... Leningrad (Akademia Nauk SSSR) 1930. Oct. †. (*Fl. Iakut.*).
Publ.: Sep 1930 (p. xii: 1 Aug 1930; p. [ii*]: Sep 1830), p. [i*-ii*], [i]-xii, [1]-221. *Copies*: B, G, H, NY, US, USDA. – Edited by V.L. Komarov; only part issued.

Petrovič, Sava (1839-1889), Serbian botanist; educated at Belgrad and Paris (Faculté de médecine), M.D. 1868; physician to King Milan 1873; "Sanitätsoberst" (military physician) at Belgrad. (*Petrovič*).

HERBARIUM and TYPES: Material at B, BP, E, G, LY, W. – Some letters at G.

BIBLIOGRAPHY and BIOGRAPHY: AG 6(2): 762; Barnhart 3: 76; BM 4: 1559; Hortus 3: 1201 (Petrovič); IH 2: (in press); Kew 4: 288; Morren ed. 10, p. 101; Rehder 5: 665; Urban-Berl. p. 382; Zander ed. 10, p. 701, ed. 11, p. 799-800.

BIOFILE: Adamovic, L., Vegetationsverh. Balkanländ. 6, 20. 1909 (Veg. Erde 11).
Anon., Bot. Centralbl. 37: 256. 1889; Bot. Jahrb. 10 (Beibl. 23): 1. 1889; Bot. Not. 1890: 45; Österr. bot. Z. 39: 124. 1889 (d. 1 Feb 1889).
Beck, G. v., Bot. Centralbl. 34 (17/18): 150. 1888 (coll. at W); in K. Fritsch, Bot. Anst. Wiens 75. 1894 (pl. W).
Hedge, I.C. & J.M. Lamond, Index coll. Edinb. herb. 118. 1970.
Nedich, Ann. Bot. 3: 485-486. 1890 (bibl.).
Tucakov, J., Acta Hist. Med. Pharm. Vet. 1: 162-169. 1961 (on his "medicinal plants of Serbia (Lekovito bilje u Srbiji) of 1883, first work of its kind in the Serbian language; portr.).

7791. *Flora agri nyssani* ... Beograd (... Državna Štamnarija) 1882. Oct. (*Fl. agr. nyss.*).
Orig.: 1882 (p. iv: 1 Mar 1882), p. [i*], [i]-xxxii, [1]-950. *Copies*: B, G. Title also in Serbian: *Flora okoline Niša*.
Additamenta: 16 Jan 1886 (t.p. 1885; p. 281: Apr 1885; BR, G and NY copies signed by author 20 and 16 Jan 1886), p. [i]-vi, [1]-281. *Copies*: B, BR, G, NY. – "Additamenta

ad floram agri Nyssani ..." Beograd (id.) 1885. Oct., also "Dodatak flori okoline Niša".

Petrowsky [Petrowski, Petrovsky], **Andrei** [Andreas] **Stanislavovic** (1832-1882), Russian botanist; high school teacher at Jaroslaw. (*Petrowsky*).

HERBARIUM and TYPES: MW; duplicates B, CGE, E, LE, U. – According to a note by De Bary (Bot. Zeit. 26: 447. 1868), Petrowsky published a "Herbarium der Flora von Jaroslaw", cent. 1-2).

BIBLIOGRAPHY and BIOGRAPHY: Barnhart 3: 76; CSP 10: 1046; IH 2: (in press); LS 2: 20573a-b; MW p. 385; NAF 11(1): 91; TR 1067-1074 (bibl.); Urban-Berl. p. 382.

BIOFILE: Anon., Nat. Nov. 4: 120. 1882 (d. 25 Mar 1882).
Candolle, Alph. de, Phytographie 439. 1880 (herb. at Soc. imp. Natural. Moscou).
Hedge, I.C. & J.M. Lamond, Index coll. Edinb. herb. 118. 1970.
Herder, F. v., Bot. Centralbl. 58: 387. 1894 (pl. at LE are dupl. from the "Soc. imp. Natural. Moscou").
Lindemann, E. v., Bull. Soc. Natural. Moscou 61(1): 55. 1886.
Lipschitz, S., Florae URSS fontes 222 [index]. 1975.

NOTE: We have not seen his *Flora of the Jaroslaw district* (in Russian), published Moscow 1880, fasc. 1, iv, 77 p. (see Bot. Centralbl. 5(10): 310. 5-10 Mar 1881, rev. ib. 7(27): 16. 1881 by von Herder; Nat. Nov. Mar(1) 1881).

Petry, Arthur (1858-?), German botanist; high school teacher at Nordhausen. (*Petry*).

HERBARIUM and TYPES: Unknown.

BIBLIOGRAPHY and BIOGRAPHY: Barnhart 3: 76; BL 2: 134; BM 4: 1559; CSP 17: 830.

BIOFILE: Drude, O., Hercyn. Florenbezirk 10, 11, 16. 1902 (Veg. Erde 6).
Kellner, K., Beitr. Heimatk. Nordhausen 5: 23-43. 1980. (n.v.).

7792. *Die Vegetationsverhältnisse des Kyffhäuser Gebirges*. Inaugural-Dissertation der philosophischen Facultät der vereinigten Friedrichs-Universität Halle-Wittenberg ... Halle a. S. (Verlag von Tausch & Grosse) 1889. Qu. (*Veg.-Verh. Kyffhäus. Geb.*).
Publ.: Jun-Jul l889 (J. Bot. Sep 1889; Bot. Zeit. 26 Jul 1889; Flora 20 Dec 1889; ÖbZ 1 Aug 1889), p. [i], [1]-55, [56]. *Copy*: MO. Vita on p. [56].
Ref.: Geisenhagen, Flora 73: 203-204. 22 Apr 1890.
Reiche, C., Bot. Jahrb. 11 (Lit.): 58-59. 13 Sep 1889.
Rosen, Bot. Zeit. 48: 230-232. 11 Apr 1890.
Roth, E., Bot. Centralbl. 41(1): 23-24. 4 Jan 1890.

Petter, Franz (1798-1853), Austrian botanist; from 1823 high school teacher (German language) at Spalato (Central Dalmatia). (*Petter*).

HERBARIUM and TYPES: large set at W (*Plantae Dalmaticae*, 1344 specimens); further material at AAU, B (extant), BP, BR, CGE, CN, DBN, FI, G, GB, GOET, GJO, H, IBF, JE, KIEL, L, LE, LZ, M, MW, P, PAD, REG, S, WAG. – Some letters at G.

NOTE: Not to be confused with Franz Petters (1784-1866), see below. PR combined the two authors.

BIBLIOGRAPHY and BIOGRAPHY: AG 2(2): 375, 5(3): 50, 6(2): 269; Backer p. 438; Barnhart 3: 76; BM 4: 1559; CSP 4: 862; Hegi 4(3): 1148, 5(2): 1368; IH 2: (in press); Kanitz 149; Kew 4: 288; Lasègue p. 332, 402, 503; PR 7097 (ed. 1): 7959; Rehder 1: 447; Saccardo 1: 126 (b. 4 Feb 1798), 2: 84; SO add. 842bf; Urban-Berl. p. 382. – *Note*: References to "Versuch einer Geschichte der amerikanischen Agave, 1817" should be attributed to Franz Petters (1784-1866).

BIOFILE: AG 2(2): 375; Anon., Bot. Not. 1855: 159 (d. 7 Jul 1853); Bot. Zeit. 11: 662. 1853 (d.); Flora 12(1). Beil. 2: 24-34. 1829 (exsicc. Triest), 15(1) Int. Bl. 1-16. 1832 (exsicc. Dalmat.), 16(1). Int. Bl. 19-23. 1833 (id.), 26: 134, 257-263. 1843 (Bot. Bericht aus Dalmatien), 28: 251-254, 1845 (id.), 29: 464. 1846 (id.), 32: 573-576. 1849 (id.); 36: 600. 1853 (d. 7 Jul 1853); Österr. bot. W. 1: 4. 1851.
Beck, G. v., Bot. Centralbl. 34 (17/18): 150. 1888 (Pl. Dalm. at W); Veg.-Verh. Illyr. Länd. 6. 1901 (Veg. Erde 4); in Fritsch, K., Bot. Anst. Wiens 75. 1894 (pl. W).
Candolle, Alph. de, Phytographie 439. 1880 (coll.).
Kukkonen, I., Herb. Chr. Steven 79. 1971.
Lindemann, E. von, Bull. Soc. Natural. Moscou 61: 55. 1886.
Müller, R.H.W. & R. Zaunick, Friedr. Traugott Kützing 121, 125, 126, 128, 130, 291. 1960.
Petter, F., Flora 16, Int.-Bl. 2: 19-23. 1833. (exsicc.).
Schlechtendal, D.F.L. von, Bot. Zeit. 11: 662. 16 Sep 1853 (d. 7 Jul 1853; sic not 1858 as sometimes reported).

EPONYMY: *Petteria* K.B. Presl (1845, *Nom. cons.*). *Note*: *Pettera* H.G.L. Reichenbach (1841, *nom. rej.*) is likely named for this author but no etymology is given.

7793. *Botanischer Wegweiser in der Gegend von Spalato* in Dalmatien. Ein alphabetisches Verzeichniss der vom Verfasser in Dalmatien und insbesondere in der Gegend von Spalato gefundenen, wildwachsenden Pflanzen nebst Angabe ihrer Fundörter, Blütezeit, Ausdauer, gebräuchlichsten Synonymen, und der Klasse und Ordnung, welche sie im Linnéschen Sexual-System einnehmen. Mit einem Vorbericht. Ein botanisches Taschenbuch in Coupons-Form ... Zara (Gedruckt und im Verlage bei Battara). 1832. (in fours). (*Bot. Wegweis. Spalato*).
Publ.: 1832 ("eben ... erschienen" Flora 7 Dec 1832), p. [1]-32, 1-144. *Copy*: B.
Ref.: Anon., Flora 15: 720. 1832, 16: 47. 1833; Lit.-Ber. Flora 2: 280-284. Dec 1832 (rev.).

Petters, Franz (1784-1866), Bohemian botanist and clergyman; ordin. 1807; in various clergical positions at Schönwalde and Kratzau (Kragau); correspondent of Opiz. (*Petters*).

HERBARIUM and TYPES: PR.

BIBLIOGRAPHY and BIOGRAPHY: AG 2(2): 375; PR 7097 (err. sub F. Petter).

BIOFILE: Maiwald, P.V., Jahres-Ber. Stifts-Obergymn. Braunau 1901, p. 14 (on P. and Opiz; b. 16 Oct 1784, d. 4 Nov 1866).
Opiz, P.M., Flora 7(2), Beil. 2: 141. 1824 ("Ein genauer Pflanzenforscher").

7794. *Versuch einer Geschichte der amerikanischen Agave* besonders der gegenwärtig im Schlossgarten zu Friedland blühenden mit einer Einleitung über die Verbreitung einiger anderer interessanter Gewächse ... Zum Besten des Armeninstituts herausgegeben und zu haben zu Friedland in Böhmen, bey Rudolph Ledsebe, gräfl. Clam-Gablassischen Obergärtner 1817. (*Vers. Gesch. amer. Agave*).
Publ.: 1817, p. [1]-54. *Copy*: HU.

Petterson, Bror Johan (1895-x), Finnish botanist, curator at the Botanical Museum, University of Helsinki 1959-1965; travelled in the Canary Islands 1947, 1949, 1951 and Sudan & Egypt 1962. (*Petterson*).

HERBARIUM and TYPES: H; early collections destroyed during World War II.

BIBLIOGRAPHY and BIOGRAPHY: Barnhart 3: 76; BL 2: 65, 85, 456-702; Collander p. 413-414.

BIOFILE: Jalas, J., Luonnon Tutkija 69: 189-190. 1965 (portr., biogr. sketch).

7795. *Botaniska anteckningar från dyröya och några angränsande öar* vid Norska västkusten ... [Acta Societatis pro Fauna et Flora fennica, 62, n:o 5]. Helsingforsiae 1939. Oct.
Publ.: between 7 Oct 1939 an 6 Apr 1810 (Mem. Soc. Fauna Fl. fenn. 16: 77. 1940), cover t.p., [1]-36. *Copy*: H. – Inf. P. Isoviita.

Petunnikov, Alexej Nikolaievič (1842-1919), Russian amateur botanist at Moscow. (*Petunn.*).

HERBARIUM and TYPES: Some material at C, IBF, L and LE. – Contributed to *Herbarium florae Rossicae* (1898-x), edited by S.I. Korshinsky. – Some letters at G.

BIBLIOGRAPHY and BIOGRAPHY: Barnhart 3: 77; BM 8: 1004; Bossert p. 308 (b. 19 Nov 1902); CSP 4: 863, 8: 608, 12: 572, 17: 833; Herder p. 200; Kew 4: 288; MW p. 385; Rehder 5: 665; TL-2/5908; TR 1075; Tucker 1: 552.

BIOFILE: Anon., Österr. bot. Z. 72: 124. 1923 (d.).
Lipschitz, S.J., Bjull. Mosk. Obšč. Isp. Prir. Biol. 73(3): 5-23. 1968 (biogr., portr., bibl.);
Flora URSS fontes 222 [index]. 1975.

NOTE: Petunnikov's main works were published in Russian, such as his "Critical review of the flora of Moscow"; they are listed by Lipschitz (1975).

COMPOSITE WORKS: The third continuation of C.A. v. Meyer, *Fl. Tambov*, TL-2/5908, was published by P. in Bull. Soc. Natural. Moscou 1865(3): 121-146.

7796. *Die Potentillen Centralrusslands* (Acta Horti Petropolitani, vol. xiv, no. 1) 1895. Oct.
Publ.: 1895 (Nat. Nov. Apr(1) 1896; Allg. bot. Z. 15 Mar 1896), p. [1]-52, *pl. 1-11*. *Copy*: BR. – Published and to be cited as Acta Horti Petrop. 14(1): 1-52, *pl. 1-11*. 1895.

7797. *Svod botanicheskikh terminov* ... Moskva (Tipo-litegrafiya vysoch-aishe utverzhlepnago ...) 1898. Oct. (*Svod Bot. termin.*).
Ed. 1: Jan-Aug 1898 (Nat. Nov. Sep(1) 1898; Allg. bot. Z. 15 Sep 1898; Bot. Zeit. 16 Nov 1898), p. [i]-x, [xi], [1]-133. *Copy*: B.
Ed. 2: 1912 (Nat. Nov. Aug(1) 1913), p. [i]-xi, [1]-161. *Copy*: B.

Petzholdt, Georg Paul Alexander (1810-1889), German (Saxonian) botanist, published on palaeobotany, agriculture and peat research; originally physician at Dresden; from 1846 professor of agronomy and technology at Dorpat (Tartu); later in retirement at Freiburg i. Br. (*Petzh.*).

HERBARIUM and TYPES: TU. – Some letters at G.

BIBLIOGRAPHY and BIOGRAPHY: Andrews p. 312; Barnhart 3: 77; BM 4: 1559; CSP 4: 863, 8:608, 11: 1, 12: 572-573; Herder p. 36, 56, 352, 393; Jackson p. 180; PR ed. 1: 7960-7963; Quenstedt p. 336; TR 1076.

BIOFILE: Anon., Flora 29: 656. 1846 (accepts app. Dorpat as prof. of agricultural chemistry "mit dem Prädicate Hofrath"); Österr. bot. W. 2: 54. 1852.
Kusnezow, Bot. Centralbl. 68: 258. 1896 (pl. at TW).
Poggendorff, J.C., Biogr.-litt. Handwörterbuch 2: 420-421. 1863, 3: 1031. 1898 (b. 29 Jan 1810, d. 23 Apr 1889; bibl., biogr. refs.).

EPONYMY: *Petzholdtia* Unger (1842).

7798. *De Balano et Calamosyringe. Additamenta ad Saxoniae palaeologiam duo* ... Dresdae [Dresden] et Lipsiae [Leipzig] (prostat in libraria Arnoldia) 1841. Oct. (*De Balano*).
Publ.: Dec 1841 (p. 4: Cal. Nov 1841), p. [1]-34, [1, expl. pl.], *pl. 1-2* (uncol. lith.). *Copy*: PH.

7799. *Ueber Calamiten und Steinkohlenbildung* ... Dresden und Leipzig (in der Arnoldischen Buchhandlung) 1841. Oct.
Publ.: 1841 (Geol. Soc. London rd. 24 Nov 1841), p. [i]-vi, [1]-68, *pl. 1-8* (uncol., 1-6 liths., 7-8 copp.). *Copy*: NY.

7800. *Zur Naturgeschichte der Torfmoore* (zweiter Beitrag) ... Dorpat (Druyck von Heinrich Laakmann) 1862. Oct.
Publ.: 1862, p. [1]-32. *Copy*: MO. – Reprinted and to be cited from Arch. Naturk. Liv-, Est- und Kurlands 3: 75-104. 1862. *Erster Beitrag*: ib. : 1-28. 1861 ("Chemische Untersuchung des Torflagers von Awandus, ...").

Petzi, Franz von Sales (1851-1928), German (Bavarian) botanist; studied mathematics and physics at München; high school teacher at the Regensburg Gymnasium; editor and curator of the herbarium of the Regensburgische botanische Gesellschaft. (*Petzi*).

HERBARIUM and TYPES: Regensburg; dupl. E.

BIOFILE: Blümml, Allg. bot. Z. 5: 101-102. 1899 (rev. of his Floristische Notizen, Denkschr. bot. Ges. Regensburg 7: 109-126. 1898.
Hedge, I.C. & J.M. Lamond, Index coll. Edinb. herb. 118. 1970.
Pongratz, L., Acta Albertina Ratisbonensia 25: 111-112. 1963 (portr., bibl., b. 3 Apr 1851, d. 11 Sep 1928).

Petzold, [Carl] **Eduard Adolph** (1815-1891), German (Prussian) botanist and gardener; trained in practice at Matzdorf (1835-1838); and Neuenhof (1838-1840); at the forestry school of Eisenach 1840-1844; Hofgärtner at Ettersburg (1844-1848), and Weimar (1848-1852); from 1852-1881 director of the garden at Muskau (W. Schlesien); in retirement at Blasewitz (Sachsen) 1882-1891. (*Petzold*).

HERBARIUM and TYPES: Unknown. – *Correspondence* with D.F.L. von Schlechtendal in HAL.

BIBLIOGRAPHY and BIOGRAPHY: AG 4: 213 (err. Karl Friedrich Adolf; b. 19 Jan 1815), 6(2): 135; Backer p. 438 (err. K.Fr. A. Petzold); Barnhart 3: 77; BM 4: 1560; Bossert p. 308 (b. 14 Jan 1815, d. 10 Aug 1891); Herder p. 159, 363; Jackson p. 443; Kew 4: 289; Morren ed. 2, p. 8, ed. 10, p. 10; MW p. 385; PR 7098; Ratzeburg p. 393-399 (b. 14 Jan 1815, autobiogr.); Rehder 1: 208, 5: 665; Tucker 1: 552.

BIOFILE: Gresky, W., Eduard Petzold, der Geisteserbe des Fürsten Pückler als Hofgärtner in Ettersburg und Weimar, Erfurt, Verlag von Kurt Stenger, 1940. Sonderschr. Akad. gemeinnütz. Wiss. Erfurt Heft 13. (portrait, main biogr.).
Hampel, C., Gartenflora 40: 469-471. 1891 (d. 10 Aug 1891, portr., obit.).
Petzold, C.E.A., *in* Ratzeburg, J.T.Z., Forstwiss. Lex. 393-399. 1872 (autobiogr.; "Carl Eduard Adolph").
Wittrock, V.B., Acta Horti Berg. 3(2): 138. 1903 (Eduard Carl Adolph), 3(3): 147. 1905.

7801. *Arboretum muscaviense*. Über die Entstehung und Anlage des Arboretum Sr. königlichen Hoheit des Prinzen Friedrich der Niederlande zu Muskau nebst einem beschreibenden Verzeichniss der sämmtlichen, in demselben cultivirten Holzarten. Ein Beitrag zur Dendrologie der deutschen Gärten bearbeitet von E. Petzold ... und G. Kirchner, ... mit einem colorirten Plane des Arboretum zu Muskau. Gotha (in Commission bei W. Opetz) 1864. Oct. (*Arbor. muscav.*).
Co-author: G. Kirchner (x-1885).
Publ.: Mai 1864 (p. 5: Apr 1863; USDA has letter Petzold, 29 Mai 1864, accompanying a courtesy copy of the book; BSbF rev. 11 Nov 1864), p. [i]-vi, [1], [1]-830, chart. *Copies*: B, MO, NY, USDA. *Note*: The Muskau Arboretum was owned by Frederik Willem Karel van Oranje (1797-1881) and his wife Louise von Preussen. Frederik was a younger son of King Willem I of the Netherlands.
Ref.: Regel, E., Gartenflora 13: 350. Nov 1864 (rev.).
Schlechtendal, D.F.L. von, Bot. Zeit. 22: 231-232. 22 Jul 1864 (rev.).

Petzold, [Karl] Wilhelm (1848-1897), German (Saxonian) botanist and geologist; Dr. phil. Halle 1876; in the campaign in France 1870-1871; high school teacher at Neu-Brandenburg 1874-1877, id. at Weisenburg i. E. (Wissembourg) 1877-1879, id. at Braunschweig 1880-1897. (*W. Petzold*).

HERBARIUM and TYPES: Unknown.

BIBLIOGRAPHY and BIOGRAPHY: Barnhart 3: 77; BL 2: 128, 702; BM 4: 1560; CSP 17: 833; Jackson p. 312; Rehder 5: 665.

BIOFILE: Anon., Allg. bot. Z. 4: 16. 1898; Jahresber. Ver. Naturw. Braunschweig 10: 244-245 (obit., b. 18 Feb 1848, d. 23 Jul 1897).
Zimmermann, P., Biogr. Jahrb. 2: 386-387. 1898 (biogr., b. 9 Feb 1848, d. 24 Jul 1897; refers to a biogr. and bibl. in Jahresber. Städt. Oberrealschule Braunschweig, Ostern 1898, p. 21-24).
Lühmann, H., Jahresber. Ver. Nat.-Wiss. Braunschweig 10: 244-245. 1897 (obit., b. 18 Feb 1848, d. 23 Jul 1897).

THESIS: *Ueber die Vertheilung des Gerbstoffes in den Zweigen und Blättern unserer Holzgewächse.* Halle 1878, 30 p.

7802. *Verzeichniss der in der Umgegend von Weissenburg* im Elsass *wildwachsenden* und häufiger cultivirten *Gefässpflanzen* von Dr. W. Petzold. Beilage zum Programm des Gymnasiums zu Weissenburg. Weissenburg [Wissembourg] (Druck von F.C. Wentzel. - 2059) 1879. Qu. (*Verz. Weissenb. Gefässpfl.*).
Publ.: 1879 (Bot. Centralbl. 1-12 Mar 1880; Bot. Zeit. 12 Mar 1880; Nat. Nov. Apr(1) 1880), p. [i], [1]-45. *Copy*: MO. – "Beilage" to the Weissenburg Gymnasium program no. 428. 1879.

7802. *Verzeichniss der in der Umgegend von Weissenburg* im Elsass *wildwachsenden* und häufiger cultivirten *Gefässpflanzen* von Dr. W. Petzold. Beilage zum Programm des Gymnasiums zu Weissenburg. Weissenburg [Wissembourg] (Druck von F.C. Wentzel. - 2059) 1879. Qu. (*Verz. Weissenb. Gefässpfl.*).
Publ.: 1879 (Bot. Centralbl. 1-12 Mar 1880; Bot. Zeit. 12 Mar 1880; Nat. Nov. Apr(1) 1880), p. [i], [1]-45. *Copy*: MO. – "Beilage" to the Weissenburg Gymnasium program no. 428. 1879.

7803. *Die Bedeutung des Griechischen für das Verständnis der Pflanzennamen* ... wissenschaftliche Beilage zu dem Jahresberichte der Städtischen Realschule zu Braunschweig, Ostern 1886. Braunschweig (Hof-Buchdruckerei von Julius Krampe) 1886. Progr. Nr. 632. Qu. (*Bedeut. Griech. Pfl.-Namen*).
Publ.: Apr-Mai 1886(Bot. Zeit. 30 Jul 1886; Nat. Nov. Aug(1) 1886), p. [1]-38. *Copy*: GOET(Univ.-Bibl.). – Inf. G. Wagenitz.

Peyl, Josef (*fl.* 1863), Bohemian garden director at Kačina near Kuttenberg (Kutná Hora); published mainly on fungi. (*Peyl*).

HERBARIUM and TYPES: PR (phan.), PRM (fungi), WRSL (lich.); Maiwald states that Peyl's "Kryptogamensammlung" is at the Karolinenthaler [Praze-Karlíně] high school.

BIBLIOGRAPHY and BIOGRAPHY: Barnhart 3: 77; CSP 4: 864, 12: 573; Futák-Domin p. 468; IH 2: (in press); Jackson p. 163; Klášterský p. 142; LS 20578-20581; Maiwald p. 120, 141-142, 204, 216, 231; Rehder 3: 194; Stevenson p. 1254.

EPONYMY: *Peylia* Opiz ex Peyl (1857).

Peyre, B.L. (*fl.* 1823), French botanist, army surgeon with the 18th "Régiment d'Infanterie de Ligne". (*Peyre*).

HERBARIUM and TYPES: Unknown.

BIBLIOGRAPHY and BIOGRAPHY: BM 4: 1560; Herder p. 141; PR 7099 (ed. 1: 7968); Rehder 1: 407.

7804. *Méthode analytique-comparative de botanique*, appliquée aux genres de plantes phanérogames qui composent la flore française ... Paris (Chez Ferra jeune, libraire, ...) 1823. Fol. (*Meth. anal.-comp. bot.*).
Publ.: 1823, p. [i]-xvi, chart, [1]-62 (p. 49 fold-out). *Copies*: G, NY.

Peyritsch, Johann Joseph (1835-1889), Austrian botanist, especially plant teratologist; Dr. med. Wien 1864; naval physician at Pola 1864-1866; from 1866-1871 connected with the Vienna general hospital; from 1871-1878 at the Natural History museum ("2. custos"); habil. Univ. Wien 1874; professor of botany at Innsbruck (successor to A. Kerner) 1878-1889; especially interested in plant teratology and Laboulbeniales. (*Peyr.*).

HERBARIUM and TYPES: IB; diary 1880-1888 (Tirol, Vorarlberg) in mss. also at IB. See Magnus (1894) for a list of P's fungi at IB. Other material in M (via Radlkofer). Correspondence with D.F.L. von Schlechtendal at HAL; further letters at G.

NOTE: Peyritsch was born 20 Oct 1835 at Völkermarkt, Kärnten; on the same day and in the same region was born Hubert Leitgeb (Portendorf, Kärnten). P. died 14 Mar 1889 at Gries nr. Bozen, Leitgeb died 5 Apr 1888 at Graz.

BIBLIOGRAPHY and BIOGRAPHY: AG 6(1): 861; Ainsworth p. 237, 346; Barnhart 3: 77; Blunt p. 248; BM 4: 1560; Bossert p. 308; CSP 4: 864-865, 8: 609, 11: 1-2, 17: 834; DTS 1: 223-224; Frank (Anh.): 76; GF p. 70; Hawksworth p. 185; Herder p. 470; Hortus 3: 1201 (Peyr.); Jackson p. 93, 104, 105, 134, 170, 373; Kew 4: 289; Langman p. 581; LS 20583-20587; Moebius p. 104; Morren ed. 2, p. 14, ed. 10, p. 37; NI 1523, also 1: 192; PR 4843, 7100, 10019, 10615; Rehder 5: 665; Stevenson p. 1254 (err.); TL-1/1389; TL-2/3507, 3894, see Fenzl, E.; Tucker 1: 552.

BIOFILE: Anon., Bot. Centralbl. 37: 407. 1889; Bot. Jahrb. 11 (Beibl. 24): 1. 1889; Bot. Not. 1890: 45; Bot. Zeit. 37: 64. 1879 (app. Innsbruck), 47: 242. 1889 (d. 14 Mar 1889); Bull. Soc. bot. France 25 (bibl.): 190. 1879 (app. Innsbruck); Flora 54: 288 (curator W), 400 (id.) 1871, 56: 144. 1873 (habil. W), 61: 544. 1878 (app. Innsbruck); Leopoldina 25: 56. 1889 (bibl.); Nat. Nov. 1: 157. 1879 (app. Innsbruck); 11: 122. 1889 (d. 14 Mar 1889); Österr. bot. Z. 11: 371. 1861 (alpine journeys), 21: 289. 1871 (curator W), 28: 346. 1878 (app. Innsbruck), 39: 160. 1889 (d. 14 Mar 1889).Bayley Balfour, J., Ann. Bot. 3: 486-487. 1890 (bibl.).
Beck, G. v., Bot. Centralbl. 33: 379. 1888, 34: 87. 1888; *in* Fritsch, K. et al., Bot. Anst. Wiens 64. 1894 (curator W).
Dalla Torre, K.W. von, Ber. naturw.-med. Ver. Innsbruck 1890/91: 10-91. Oct 1891 (posthumous publ. of floristic notes).
Gager, C.S., Brooklyn Bot. Gard. Rec. 27(3): 160. 1938 (dir. bot. gard. Innsbruck 1878-1889).
Heinricher, Ber. deut. bot. Ges. 7: (12)-(20). 1889 (obit., bibl., b. 20 Oct 1835, d. 14 Mar 1889), see also p. (180).
Kanitz, A., Magy. növen Lap. 13:90-92. 1890 (obit., bibl.).
Kronfeld, M., Bot. Centralbl. 40: 133-135, 171-174, 204-205. 1889 (bibl.).
Magnus, P., Ber. naturw.-med.-Ver. Innsbruck 21: 25-73, *pl. 1*. 1894 (on fungi collected by P. at IB). (reprint, Innsbruck 1894, p. (1)-49, *pl. 1*).
Urban, I., Fl. bras. 1(1): 193-194. 1906 (biogr. data; b. 20 Oct 1835, d. 14 Mar 1889, bibl., biogr. refs.).
Verdoorn, F., ed., Chron. bot. 1: 18. 1935.

COMPOSITE WORKS: (1) Editor, with Theodor Kotschy, q.v., of *Plantae tinneanae*, 1867, see TL-2/3894.
(2) Co-author, with H. Wawra, q.v., of *Sertum benguelense*, 1860. (see TL-1/1389).
(3) *Erythroxylaceae*, *in* Warming, E., *Symb. bras.* 27: 156-157. 10 Mai 1883 (date Fox Maule).

(4) Martius, *Fl. bras.*:
(a) *Hippocrateaceae*, 11(1): 125-164, *pl. 42-49*. 1 Feb 1878 (fasc. 75).
(b) *Erythroxylaceae*, 12(1): 125-180, *pl. 23-32*. 1 Dec 1878 (fasc. 81).

EPONYMY: *Peyritschia* E. Fournier (1881); *Peyritschiella* R. Thaxter (1890).

7805. *Über einige Pilze aus der Familie der Laboulbenien* ... [aus dem lxiv. Bande der Sitzb. der k. Akad. der Wissensch. ... Wien (Druck der k.k. Hof– und Staatsdruckerei)] 1871. Oct.
Publ.: Nov-Dec 1871 (or early 1872?; Bot. Zeit. 12 Jul 1872) (mss. submitted 2 Nov 1871, "Nov.-Heft" 1871 of journal), p. [1]- 18, *pl. 1-2* (col.). *Copy*: G. – Reprinted or preprinted from S.B. k. Akad. Wiss., Wien 64(1): 441-458, *pl. 1-2*. 1871 (or 1872). References in literature: ÖbZ 1 Jul 1872, Bot. Zeit. 12 Jul 1872. This might indicate that publication took place later than Nov-Dec 1871.

7806. *Beiträge zur Kenntniss der Laboulbenien* ... [aus dem lxviii. Bande der Sitzb. der k. Akad. der Wissensch. ...] 1873. Oct.
Publ.: Nov-Dec 1873 (submitted 23 Oct 1873; in "Oct.-Heft" of Sitzungsberichte; Bot. Zeit. 19 Jun 1874), p. [1]-28, *pl. 1-3* (uncol. liths., Liepoldt). *Copy*: G. – Reprinted or preprinted from S.B. k. Akad. Wiss. 68(1): 227-245, *pl. 1-3*. 1873 (or 1874).
Note: The third main paper by P. on this group was *Ueber Vorkommen und Biologie von Laboulbeniaceen*, in id. 72(1): 377-385. 1875. (No reprint seen by us).

7807. *Zur Synonymie einiger Hippocratea-Arten* ... [aus dem lxx. Bande der Sitzb. der k. Akad. der Wissensch. I. Abth. Oct.-Heft. Jahrg. 1874, Wien (k.k. Hof– u. Staatsdruckerei)] 1874. Oct.
Publ.: 22 Oct-31 Dec 1874, p. [1]-23. *Copy*: G. – Reprinted or preprinted from S.B. Akad. Wiss., Wien, 70(Abth. I): 401-423. Oct 1874 (publ.Oct-Dec 1874).

7808. *Aroideae maximilianae*. Die auf der Reise Sr. Majestät des Kaisers Maximilian I. nach Brasilien gesammelten Arongewächse nach handschriftlichen Aufzeichnungen von H. Schott beschrieben von Dr. J. Peyritsch ... Wien (Druck und Verlag von Carl Gerold's Sohn) 1879. Fol. (*Aroid. maximil.*).
Original descriptions: Heinrich Wilhelm Schott (1794-1865).
Editors: subsequently: Heinrich Wawra, Ritter von Fernsee (1831-1887), Carl Georg Theodor Kotschy (1813-1866), Siegfried Reisseck (1819-1871) Eduard Fenzl (1808-1879) and, ultimately: Peyritsch.
Artists: Wenzel Liepoldt (1841-1901); (title:) and Joseph Selleny (1824-1875). – Originals of plates at W.
Publ.: Oct 1879 (p. [vi]: Jun 1879; Nat. Nov. Oct(2) 1879; Bot. Zeit. 5 Dec 1879; Engler reviewed the book on 10 Nov 1879), p. [i-viii], frontisp., [1]-53, *pl. 1-42* (chromoliths. W. Liepoldt). *Copies*: B, BR, FI, G, MO, NY, U; IDC 1033. – Schott worked only with the living plants brought back by the expedition and grown in Schönbrunn. Wawra relates that the herbarium specimens of aroids of the expedition were lost (by Schott). Most of the new species of the expedition were previously published by Schott in the Österr. bot. J. and in Bonplandia (see under Schott).
Ref.: Brown, N.E., J. Bot. 18: 59-60. Feb 1880.
Engler, A., Bot. Zeit. 37: 853-856. 26 Dec 1879 (rev. dated 10 Nov 1879), Bot. Jahrb. 1: 50-52. 30 Apr 1880 (important assessment by Schott's successor as arologist).
Wawra, H., Österr. bot. Z. 29: 400-402. 1879.

Peyronel, Beniamino (1890-x), Italian mycologist at the Istituto botanico dell'Università, Torina (*Peyronel*).

HERBARIUM and TYPES: ROPV.– Some letters at G.

NOTE: published as late as 1971: Inform. bot. Ital. 2(3): 163-167. 1970 (1971: on conservation of nature).

BIBLIOGRAPHY and BIOGRAPHY: Barnhart 3: 77; BL 2: 200, 703; Hirsch p. 231; IH 1(ed.

1): 94, (ed. 2): 19, (ed. 3): 152, (ed. 4): 170, (ed. 5): 184, (ed. 6): 273, 2: (in press); Kelly p. 175; Kew 4: 289; LS 21584-21627 (bibl.); Stevenson p. 1254.

BIOFILE: Ciferri, R., Taxon 1(8): 127. 1952 (types at ROPV).
Verdoorn, F., ed., Chron. bot. 3: 189. 1937, 4: 273. 1938.

EPONYMY: *Peyronelia* R. Ciferri et R.G. Fragoso (1927); *Peyronellula* Malan (1953); *Note*: *Peyronelina* P.J. Fisher, J. Webster et D.F. Kane (1927) and *Peyronellaea* G. Goidanich (1952) were probably also named for Peyronel.

7809. *Primo elenco di funghi di Val San Martino* o Valle della Germanasca. Contributo alla flora micologica delle Valli Valdesi del Piemonte ... Torino (Libreria Fratelli Bocca ...) 1926. Qu. (*Prim. elenc. fung. San Martino*).
Publ.: 1916, p. [i], [1]-58. *Copy*: Stevenson. – Issued as Mem. R. Accad. Sci. Torino (anno 1915-1916) ser. 2, vol. 66.

Pfeffer, Wilhelm Friedrich Philipp (1845-1920), German (Hesse-Nassau) plant physiologist, pharmacist, chemist and bryologist; Dr. phil. Göttingen 1865; assistant pharmacist in Chur 1865-1869; with Pringsheim in Berlin 1869-1870; with Sachs at Würzburg 1870-1871; habil. Marburg 1871, "dozent" ib. 1871-1873; professor of botany Bonn 1873-1877, at Basel 1877, at Tübingen 1878-1887; ultimately director of the botanical garden and of the plant physiology laboratory at Leipzig 1887-1920; leading plant physiologist of his era. (*Pfeff.*).

HERBARIUM and TYPES: Mosses in B and H. – Pfeffer's private library was bought by the Institute for Agricultural and Biological Sciences, University of Okayama, Japan. – Some letters at G.

BIBLIOGRAPHY and BIOGRAPHY: Ainsworth p. 102, 283; Barnhart 3: 78; Bossert p. 308; Collander hist. p. 45, 46, 48, 49, 56; CSP 8: 610-611, 11: 4, 12: 573, 17: 838-839; DSB 10: 574-578 (bibl.); DTS 5: lii; Frank 3 (Anh.): 76; Herder p. 117; IH 2: (in press); Jackson p. 70, 78, 81, 83, 149, 344; Kew 4: 290; Kleppa p. 9; LS 20592-20595, 37902; Moebius p. 454 [index]; Morren ed. 2, p. 6, ed. 10, p. 21; PR 7102, 10616; Rehder 5: 666; Saccardo 1: 126-127; SBC p. 129; TL-2/see M. Mioshi; Urban-Berl. p. 310.

BIOFILE: Andrews, F.M., Plant Physiology 4: 285-288. 1929 (obit., portr.).
Anon., Bot. Centralbl. 31: 191. 1887 (to Leipzig), 36: 384. 1888 (call to München declined), 37: 29. 1889 (Geh. Hofrath), 41: 31. 1890 (memb. Kön. Akad. Wiss., Berlin), 60: 287. 1894 (assumes, with Strasburger, editorship Jahrb. wiss. Bot.), 64: 447. 1895 (Maximilian-Orden, Bayern), 75: 63. 1898 (Dr. h.c. Cambr.), 81: 415. 1900 (corr. mem. Acad. Sci., Paris), 92: 592. 1903 (assoc. Natl. Acad. Sci., U.S.A.), 95: 464. 1904 (Otto Vahlbruch Preis), 108: 80. 1908 ("Pour le mérite"); Bot. Jahrb. 9 (Beibl. 20): 1. 1887; Bot. Not. 1920: 50 (death); Bot. Zeit. 29: 269. 1871 (habil. Marburg), 35: 231. 1877 (call to Basel), 45: 439. 1887 (call to Leipzig); Bull. Soc. bot. France 24 (bibl.): 95. 1877 (to Basel replacing Schwendener); Flora 54: 205. 1871 (habil. Marburg), 60: 320. 1877 (call to Basel), 61: 432. 1878 (call to Tübingen); Nat. Nov. 1: 47. 1879 (to Tübingen), 2: 142. 1880 (corr. Bay. Akad. wiss.), 9: 192. 1887 (to Leipzig as successor of Schenk; de Bary had first refusal), 12: 35. 1890 (corr. Akad. Wiss. Berlin; Engler elected member at the same session), 19: 217. 1897 (member Amer. Acad. Arts Sci.), 22: 176. 1900 (Acad. Paris), 25: 407. 1903 (Natl. Acad. Sci.); Österr. bot. Z. 37: 335. 1887 (app. Leipzig), 46: 79. 1896 (Maximilian-Orden, Bayern), 48: 78. 1898 (foreign member Roy.-Soc.), 48: 279. 1898 (Dr. h.c. Cambridge, England), 69: 88. 1920 (d. 21 Jan 1920).
Bergdolt, E., Karl von Goebel 261 [index]. 1941 (letters).
Bünning, E., Wilhelm Pfeffer, Apotheker, Chemiker, Botaniker, 1845-1920. Stuttgart 1975, 166 p. (biogr., bibl., portr., biogr. notices students; main treatment).
Campbell, D.H., Proc. Amer. Acad. Arts Sci. 57: 499-502. 1922 (obit.).
Cohen, E., Naturwissenschaften 3(10): 118-120. 1915 (on P. and physical chemistry).
Cranner, B.H., Naturen 44: 321-327. 1920 (portr.), Overs. vid. Selsk. Møter 1920: 73-79. 1921 (obit.).

Czapek, F., Naturwissenschaften 3(10): 120-121. 1915 (on P. and biophysics); Almanach Akad. Wiss. Wien 1920: 167-173.
De Toni, G.B., Nuova Notarisia 32: 70. 1921 (obit., not.).
Dörfler, I., Botaniker-Porträts no. 13. 1906 (portr.).
Fitting, H., Ber. deut. bot. Ges. 38: (30)-(63). 1920 (obit., bibl., portr.), Deut. biogr. Jahrb. 2: 578-582, 756-757. 1928 (biogr., bibl., biogr., refs., b. 9 Mar 1845, d. 31 Jan 1920); Decheniana 105/106: 8-9, pl. 2. 1952 (portr.; stay at Bonn).
Gager, C.S., Brooklyn Bot. Gard. Rec. 27(3): 233. 1938 (1887-1920 dir. bot. gard. Leipzig).
G.P.J., Science ser. 2. 51: 291-292. 1930 (obit.).
Goebel, K., Jahrb. bay. Akad. Wiss. 1920: 24-25. 1921 (obit.; "Goldene Äpfel in silbernen Schalen").
Haberlandt, G., Naturwissenschaften 3(10): 115-118. 1915 (biogr., tribute, b. 9 Mar 1845).
Harvey-Gibson, R.J., Outlines hist. bot. 69, 170, 175, 185, 205, 206, 214, 215, 220, 241, 243. 1914.
Jost, L., Naturwissenschaften 3(10): 129-131. 1915 (on P. and plant physiological methodology).
Kneucker, A., Allg. bot. Z. 4: 136. 1898 (Dr. h.c. Cambridge, England), 9: 140. 1903 (member Natl. Acad. Sci.) 24/25: 48. 1923 (death).
Kniep, H., Naturwissenschaften 3(10): 124-129. 1915 (on P. and the "Reizphysiologie").
Lehmann, E., Schwäbische Apotheker 183-184, 210. 1951 (early training as pharmacist).
Mägdefrau, K., Gesch. Bot. 312. [index]. 1973; Acta hist. leop. 9: 101. 1975 (portr.).
Newcombe, F.C., Bot. Gaz. 71: 152-154. 1921 (obit., portr., "there are few fields in plant physiology that have not been extended by his researches".).
Nordenskiöld, E., Hist. biol. 576, 578, 584, 595. 1949.
Ostwald, W., Chemiker-Zeit. 44(21): 145. 1920 (obit.).
Overbeek, J., Decheniana 105/106: 31. 1952 (at Landw. Hochsch. Bonn; pharmacognosy).
Pfeffer, W., Pflanzenphysiologie. Ein Handbuch ... 1881; ed. 2. 1, 1897, 2, 1904. (famous plant physiology handbook; for contemp. evaluation see Volkers, Bot. Jahrb. 26 (Lit.): 7-9. 1898, 34 (Lit.): 9-10. 1904.); Engl.: The physiology of plants, 2 vols., 1900-1906 (transl. by A.J. Stewart) French; Physiologie végétale, 2 vols., 1906-1912 (transl. J. Friedel).
Poggendorff, J.C., Biogr.-lit. Handw.-Buch 3: 1033-1034. 1898, 4(2): 1152. 1904, 5(2): 966. 1926, 6(3): 1998. 1938 (bibl.).
Pringsheim, E.G., Julius Sachs 299 [index]. 1932.
Pringsheim, H. & E.G., Ber. deut. chem. Ges. 53a: 36-39. 1920 (obit.).
Reinke, J., Mein Tagewerk 58, 60, 64, 125, 212. 1925.
Rodgers, A.D., Liberty Hyde Bailey 156-157, 160. 1949.
Ruhland, W., Ber. Verh. sächs. Akad. Wiss. Leipzig, Math.-phys. Kl. 107-124. 1923 (obit.).
Sayre, G., Bryologist 80: 515. 1977 (mosses at H).
Taton, R., ed., Science in the 19th cent. 610 [index]. 1965.
Veer, P.H.W.A.M. de, Leven en werk van Hugo de Vries. Groningen 69, 70. 1969.
Verdoorn, F., ed., Chron. bot. 2: 164. 1936.
V.H.B., Nature 105: 302. 1920 (obit.).
Wittrock, V.B., Acta Horti Berg. 3(2): 138. *pl. 28.* 1903; 3(3): xxix, 147. *pl. 94.* 1905.

COMPOSITE WORKS: (1) Editor, *Jahrb. wiss. Bot.* 27-59, 1895-1920 (until Mai 1912 with E. Strasburger).
(2) Editor, Kölreuter, J.G., *Vorläuf. Nachr. Geschl. Pfl.*, ed. Ostwald's Klassiker 41, 1893, see TL-2/3820.

FESTSCHRIFT (1) *Jahrb. wiss. Bot.* 56: 1-832. 1915 (contains, in addition to the scientific contributions, a portrait, dedication, bibliography 1865-1915 (p. 793-804), and a list of Pfeffer's students (p. 805-832).
(2) Die Naturwissenschaften 3(10): 115-140. 1915 (papers by G. Haberlandt et al. on Pfeffer, bibl., portr.).

HANDWRITING: Bünning, E., Wilhelm Pfeffer 30, 113. 1975.

7810. *Bryogeographische Studien* aus den rhätischen Alpen [Zürich 1869]. Qu. (*Bryogeogr. Stud.*).
Publ.: 1869 (p. [3]: Mar 1869; Bot. Zeit. 17 Dec 1869; BSbF Apr-Mai 1870 as of "1869"; Flora 22 Feb 1870), p. [1]-142, [1, cont.]. *Copies*: FH, G, H, MO-Steere, NY, U-V. – Neue Denkschr. allg. Schweiz. Ges. Naturw. 24(5). 1871.
Ref.: H.S., Bot. Zeit. 27: 859-860. 17 Dec 1869.
Juratzka, J., Hedwigia 8: 180-182. Dec 1869.

7811. *Die Entwickelung des Keimes der Gattung Selaginella* ... Bonn (Bei Adolph Marcus) 1871. Oct. (*Entw. Keim. Selaginella*).
Publ.: 1871 (J. Bot. Aug 1872), p. [i-v], [1]-80, *pl. 1-6* (uncol., auct.). *Copy*: BR. – Published in Bot. Abh. Morph. Physiol., ed. J. Hanstein, vol. 1, 1871.
Ref.: M'Nab, W.R., J. Bot. 10: 254-255. 1872.

7812. *Die Oelkörper der Lebermoose* [Flora, Regensburg, vol. 57, nos. 1-3] 1874. Oct.
Publ.: Jan 1874, p. [1]-25, *pl. 1. Copy*: H. – Reprinted and to be cited from Flora 57: 2-6. 1 Jan 1874, 17-27. 11 Jan 1874, 33-43, *pl. 1*. 21 Jan 1874.

Pfeiffer, Hans Heinrich (1896-x), German biosystematic botanist; Dr. phil. Washington, D.C.; specialist on Cyperaceae; ultimately at Bremen. (*H. Pfeiff.*).

HERBARIUM and TYPES: Types in so far as extant at BREM (personal comm. 1958). Letters and portrait at G.

BIBLIOGRAPHY and BIOGRAPHY: Barnhart 3: 78 (b. "1890", err.); BFM no. 2530; Bossert p. 308 (b. 20 Jan 1896); Hirsch p. 231; IH 1 (ed. 5): 25; Kew 4: 290; Moebius p. 221; MW p. 385-386; Roon p. 87.

BIOFILE: Bullock, A.A., Bibl. S. Afr. bot. 84. 1978.
Merrill, E.D., B.P. Bishop Mus. Bull. 144: 149. 1937 (Polynes. bibl.), Contr. natl. Herb. U.S. 30: 240. 1947 (Pacific Isl. bibl.).
Pfeiffer, H.H., Rev. Sudamer. Bot. 10(1): 3-6. 1951 (proposes Lophopyxidaceae, fam. nov.).
Poggendorff, J.C., Biogr.-lit. Handw.-Buch 6(3): 2000. 1938 (b. 10 Jan 1896, bibl.).
Pohl, F., Generalreg. Beih. Centralbl. 1: 13. 1931, 2: 12. 1942.
Verdoorn, F., ed., Chron. bot. 6: 301. 1941, 7: 191. 1943.

7813. *Revision der Gattung Ficinia* Schrad. von H. Pfeiffer, ... Druck und Verlag E. Gerst, Bremen. [1921]. Oct. (*Revis. Ficinia*).
Publ.: 28 Feb 1921 (fide copies so marked by author at NY and US; Nat. Nov. Apr 1922; ÖbZ Mai-Jun 1922), p. [1]-63. *Copies*: NY, US.

Pfeiffer, Johan Philip (1888-1947), Dutch chemical engineer, botanist and forester; Dr. phil. 1917; wood technologist at Semarang, Java 1912-1920; manager of an oil factory at Kediri, Java, 1920-1922; lecturer and professor at Delft technical College 1922-1929; from 1929 with the Batavian Oil Company at Amsterdam. (*J. Pfeiff.*).

HERBARIUM and TYPES: DELFT (Suriname); BO and L (Indon.).

BIBLIOGRAPHY and BIOGRAPHY: Andrews p. 312; Barnhart 3: 79; BL 2: 258, 312; BM 8: 1005; IH 2: (in press); Langman p. 581; SK 1: 405.

BIOFILE: Iterson, G. van, Jr., Vakbl. Biol. 28(2): 25-27. 1948 (obit., b. 24 Dec 1888).
Steenis, C.G.G.J. van, Fl. males. Bull. 3: 56. 1948 (d. 18 Nov 1947).

7814. *De houtsoorten van Suriname* ... Amsterdam (Uitgave van de Kon. Ver. Koloniaal Instituut, ... Druk de Bussy) 1926-1927, 2 vols. (*Houtsoort. Suriname*).
1: 1926 (p. [5]: Nov 1925; BR copy rd. 9 Sep 1926), p. [i-iii], [1]-505.

2: 1927 (p. [5]: Feb 1927), p. [i-iii], [1]-244.
Atlas: cover, *pl. 1-24* (12 lvs) (uncol. phot.) and *pl. 1-18* (id.).
Copies: B, BR, NY, U. – Kon. Ver. Koloniaal Instituut Amsterdam, Meded. 22, Afd. Handelsmuseum no. 6.

Pfeiffer, Louis [later: Ludwig] **Karl Georg** (1805-1877), German (Hesse-Nassau) botanist and malacologist; Dr. med. Marburg 1825; practicing physician at Kassel 1826-1833; military physician in Poland 1831; from 1833 active as Privatgelehrter; travelled to Cuba 1838-1839; Dr. phil. h.c. Marburg 1875. (*Pfeiff.*).

HERBARIUM and TYPES: KASSEL (no longer extant except for some ferns and mosses (see also Laubinger 1905)); some specimens at LE, MW and WRSL. – Dr. G. Follmann informs us that at least 90% of the Pfeiffer material at KASSEL was destroyed during World War II. – Correspondence with D.F.L. von Schlechtendal at HAL; other letters at G.

BIBLIOGRAPHY and BIOGRAPHY: ADB 25: 643-646 (by E. Wunschmann); Backer p. 659; Barnhart 3: 79; BFM no. 2345; BM 4: 1562-1563; Bossert p. 308; CSP 4: 872-873, 8: 611-612, 11: 4-5, 12: 574, 17: 840; Frank 3 (Anh.): 76; GF p. 70; GR p. 120; Hegi 5(2): 696; Herder p. 88, 187, 329; Hortus 3: 1201 (Pfeiff.); Jackson p. 15, 125, 126; Kelly p. 175; Kew 4: 290-291; Langman p. 581; Lasègue p. 455; LS 20601-20605; Morren ed. 2, p. 6; MW p. 386; NAF ser. 2. 4: 44; NI 1524, 1: 216, 3: 55; Nordstedt p. 25-26; PR 7103-7109, 10617, (ed. 1: 7974-7979); Rehder 1: 77, 5: 666; SBC p. 129; TL-1/972-976, 1149; TL-2/1044, see F. Otto; Zander ed. 10, p. 701, ed. 11, p. 800.

BIOFILE: Anon., Ber. Ver. Naturk. Cassel 24/25, 1878; Bonplandia 3: 33. 1855 (Leop. cogn. *Bradley*); 6:49. 1858 ("Ludwig Georg Carl", cogn. Leopoldina: Bradley); Bot. Not. 1878: 44 (b. 4 Jul 1805, d. 2 Oct 1877); Brockhaus' Conversations-Lexikon ed. 13. 12: 887. 1885 ("Louis Georg Karl"); Bull. Soc. bot. France 24 (bibl.): 237. 1878 (obit.); Hessische Morgenzeitung 1877, no. 8274, reprinted in Leopoldina 14(1-2): 7-9. 1878; J. Bot. 16: 32. 1878 (death).
Fischer, Th., [advertising pamphlet, Cassel Aug 1874, 2 p. for P's Nomenclator botanicus. Copy at U].
Hesse, R., *in* I. Schnack, ed., Lebensbilder aus Kurhessen und Waldeck 1830-1930, 6: 249-253. 1958 (biogr., brief bibl., sources; "Carl Georg Ludwig" and "Louis").
H.K., Monatschrift für Kakteenkunde 1(10): 135. 1892 (biogr. notice, portr.), see also Kakt. Orchid. Rundschau 4: 62. 1976.
Kanitz, A., Magy. novén. Lap. 1(11): 195-196. 1877 (obit.).
Lauburger, C., Abh. Ber. Ver. Naturk. Cassel 49: 81-102. 1905 (on Pfeiffer's moss herbarium at KASSEL).
Lindemann, E. von, Bull. Soc. Natural. Moscou 61(1): 55. 1886.
Pfeiffer, A.L., Die Familie Pfeiffer 61-70. 1886 ("Louis").
Pfeiffer, L., Kakt. Orchid. Rundschau 4(2): 26-29. 1979 (facsimile of 4 p. from Beschr. Syn. Cact.).
Sayre, G., Bryologist 80: 515. 1977.
Schmid, G., Chamisso als Naturforscher 171. 1942.
Stafleu, F.A., Regn. veg. 71: 301, 339. 1970 (relations Pfeiffer/Miquel).
Stearn, W.T., Cactus J. 8: 39-46. 1939 (on *Abb. Beschr. Cact.* q.v.; biogr. sketch; portr.).
Wittrock, V.B., Acta Horti Berg. 3(3): 147. 1905.
Zuchold, E.A., Jahresber. naturf. Ver. Halle 5: 46. 1853.

COMPOSITE WORKS: with J.H. Cassebeer, (but Pfeiffer main author): *Uebers. Kurhessen Pfl.* 1844, see TL-2/1044. The book was received at Regensburg in Mar 1844; and p. x is dated 24 Feb 1844, which would put the date of publication at 24-28 Feb or Mar 1844; Schlechtendal reviewed it for Bot. Zeit. only on 9 Mai 1845 (Bot. Zeit. 3: 326-327). A further copy is at FI.

NAME: We have not been able to establish the order of the Christian names given to Pfeiffer when he was baptized. He himself signed himself Louis but also Ludwig Pfeiffer in print and (in so far as we know) in correspondence. The Louis was latinized to

Ludovicus in Latin book titles. The names appear as Carl Georg Ludwig (in necrology in Hessische Morgenzeitung, hence quite possibly authentic), Karl Georg Ludwig, Ludwig Georg Karl and Ludwig Karl Georg (Karl also as Carl). We have no indication how Pfeiffer was baptized but follow his own apparent preference for "Louis", at any rate in his earlier publications (at least until 1855). A.L. Pfeiffer (1886) also lists our author as Karl Georg Louis (Inf. G. Follmann).

EPONYMY: *Pfeiffera* Salm-Dyck (1845); *Pfeifferago* O. Kuntze (1891).

7815. *Enumeratio diagnostica Cactearum* hucusque cognitarum. Auctore Ludovico Pfeiffer, ... Berolini (sumptibus Ludovici Oehmigke) 1837. Oct. (*Enum. diagn. Cact.*).
Publ.: 19-25 Feb 1837 (see TL-1/972), p. [i]-viii, [1]-192. *Copies*: B, FI, G, MO, NY, US, USDA.
Ref.: Anon., Lit. Ber. Flora 7: 21-24. 28 Feb 1837.

7816. *Beschreibung und Synonymik der in deutschen Gärten lebend vorkommenden Cacteen.* Nebst einer Uebersicht der grösseren Sammlungen und einem Anhange über die Kultur der Cactuspflanzen. Von Dr. Louis Pfeiffer in Kassel. Berlin (Verlag von Ludwig Oehmigke) 1837. Oct. (*Beschr. Synon. Cact.*).
Publ.: 19-25 Feb 1837 (see TL-1/973), p. [i]-vi ["iv"], [1]-231, [232, err.]. *Copies*: MO, NY, USDA.
Ref.: Anon., Lit. Ber. Flora 7: 21-24. 28 Feb 1837, 75-76. 28 Mai 1837.

7817. *Abbildungen und Beschreibung blühender Cacteen.* Von Dr. L. Pfeiffer, ... [vol. 1 only: und Fr. Otto, ...]. Figures des Cactées en fleur peintes et lithographiées d'après nature. Avec un texte explicatif... Cassel (Verlag [2: Druck und Verlag] von Theodor Fischer). [1838-] 1843-1850, 2 vols. Qu. (*Abbild. Beschr. Cact.*).
Co-author (vol. 1 only): Christoph Friedrich Otto (1783-1856).
Publ.: in twelve parts of five plates each with accompanying unpaginated text; 1: [i-iii], *pl. 1-30*, 2: [i-iii], *pl. 1-30*. *Copies*: BR (fully col.), G (fully col.), NY (partly col.); IDC 5635.

vol.	part	plates	dates	vol.	part	plates	dates
1	1	*1-5*	Oct 1838	2	1	*1-5*	Jan-Feb 1845
	2	*6-10*	Aug 1839		2	*6-10*	Oct 1846
	3	*11-15*	Sep 1839		3	*11-15*	Nov 1846
	4	*16-20*	Sep 1840		4	*16-20*	Oct 1847
	5	*21-25*	Mar 1842		5	*21-25*	Feb-Mar 1848
	6	*26-30*	Jul 1843		6	*26-30*	Jan-Feb 1850

The dates are based on the receipt of the parts by Hinrichs at Leipzig. The plates are lithographs by F.J.U. Prestele. The book was published in two editions: half-coloured (flowers only), and fully coloured on wove paper. – A copy with fully coloured plates was offered by Junk (Cat. 214, no. 243. 1979) at Hfl. 3.800.=.
Ref.: Junk, Rara 68. 1900-1913.
Stearn, W.T., Cactus 8: 39-46. 1939 (*General commentary* and list of plates with current names).

7818. *Einige Worte über die subalpine Flora des Meissners.* Dem Herrn Geheimen Hofrath Dr. Rich. Harnier zu Cassel am Jubiläumstage seiner den 24. September 1794 rühmlichst erlangten Doctorwürde in dankbarer Ergebenheit gewidmet von Dr. Louis Pfeiffer. [Kassel (Druck der Hotop'schen Officin) 1844]. Oct. (*Subalp. Fl. Meissn.*).
Publ.: 24 Sep 1844 (see t.p.), p. [1]-16. *Copy*: HH.

7819. *Flora von Niederhessen und Münden.* Beschreibung aller im Gebiete wildwachsenden und im Grossen angebaute Pflanzen. Mit Rücksicht auf Schulgebrauch und Selbststudium bearbeitet von Dr. Louis Pfeiffer, ... Kassel (Druck und Verlag von Theodor Fischer) 1847-1855, 2 vols. Duod. (in sixes). (*Fl. Niederhessen*).
1: Mai-Jul 1847 (p. xiv: Mar 1847; Hinrichs 21-24 Jul 1847; Bot. Zeit. 23 Jul 1847), p. [i]-l, [1]-428.

2: Jul 1855 (Hinrichs 2-3 Aug 1855; BSbF post 25 Aug 1855; Bot. Zeit. 12 Oct 1855), p. [i]-xiii, [1]-252.
Copies: B, G, NY.
Ref.: Schlechtendal, D.F.L. von, Bot. Zeit. 5: 535-537. 1847.

7820. *Vollständige Synonymik* der bis zum Ende des Jahres 1858 publicirten botanischen Gattungen, Untergattungen und Abtheilungen. Zugleich systematische Uebersicht des ganzen Gewächsreiches mit den neueren Bereicherungen und Berichtigungen nach Endlicher's Schema zusammengestellt von Dr. Ludwig Pfeiffer ... Kassel (Verlag von Theodor Fischer) 1870-1871. Oct. (*Vollst. Synon.*).
Publ.: 1870-1871, in two parts. – Second t.p.: "*Synonymia botanica* locupletissima generum, sectionem vel subgenerum ad finem anni 1858 promulgatorum in forma conspectus systematici totius regni vegetabilis schemati Endlicheriano adaptati ...". *Copies*: B, BR, FI, G, H, MO(2), NY, Stevenson.
Erste Hälfte: Oct 1870-Jan 1871 (Bot. Zeit. 13 Jan 1831 just publ.; ÖbZ 1 Mar 1871 "so-eben", Flora 2 Mai 1871 ("neu"), p. [ii*-iii*], [i]-viii, [1]-380.
Zweite Hälfte: Mar-Aug 1871 (not yet out fide adv. ÖbZ 1 Mar 1871; Reviewed Bot. Zeit. 1 Sep 1871, ÖbZ 1 Feb 1872; cover of combined volume dated 1871), p. 381-674, 364a-b (err.).
Note: 364a-b and 671-674 issued conjugate, see MO copy in original cover.
Erstes Supplement: Aug[?] 1874 (p. viii: Jan 1874; J. Bot. Nov 1874), p. [ii]-viii, [1]-45.
Copies: BR, G, MO, NY, Stevenson, U, USDA. – "Synonymik des botanischen Klassen-, Familien-, Gattungs- und Sektionsnamen. Erstes Supplement zu dem 1870 erschienenem Werke ..." Cassel (Verlag von Theodor Fischer) 1874; title on p. [ii]: "Synonymiae botanicae 1870 editae supplementum primum ...". Actually a supplement to both the *Vollst. Synon.* as well as to the *Nomencl. bot.* 1, with e.g. a listing of variants and errors contained in Steudel, *Nomenclator*, ed. 2, and D.N.F. Dietrich, *Syn. pl.* TL-2/1464. The BR and USDA copies, in original covers, have on p. [i*-ii*] the advertisement for the *Nomencl. bot.* of Aug 1874, by Theodor Fischer (see above Fischer 1874). This might point at publication in Aug 1874.
Neue Ausgabe: 1887, p. [ii*-iii*], [i]-viii, [1]-674, suppl.: [ii]-viii, [1]-45. *Copies*: US, USDA.
Ref.: Ascherson, P., Bot. Zeit. 586-588. 1 Sep 1871.

7821. *Nomenclator botanicus.* Nominum ad finem anni 1858 publici juris factorum, classes, ordines, tribus, familias, divisiones, genera, subgenera vel sectiones designantium enumeratio alphabetica. Adjectis auctoribus, temporibus, locis systematicis apud varios, notis literaris atque etymologicis et synonymis. Conscripsit Ludovicus Pfeiffer, ... Cassellis [Kassel] (sumptibus Theodori Fischeri) [1871-] 1873-1874, 2 vols. (4 parts). Qu. (*Nomencl. bot.*).
Publ.: In parts of 80 pages, the precise dates of which are not yet known to us. See below for details in so far as known to us. *Copies*: BR, FAS, FI, G, H, NY, U, US.
1(1): 1871-1873 (reviewed as a whole by Die Natur, 26 Jun 1873), p. [i-vi], [1]-808, p. [i]: t.p.; p. [iii-vi]: praefatio, Vorwort and publisher's note requesting subscriptions.
1(2): 1874-1875 (t.p. 1874), p. [i-vi], p. 809-1876, [1-2, err.], p. [i]: t.p., [iii-vi] list of subscribers.
2(1): 1874 (t.p.), p. [i], [1]-760.
2(2): 1874-1875 (t.p. 1874), p. [i], 761-1698.
Both volumes complete: Bot. Zeit. 27 Aug 1875, Flora 1 Sep 1875.

Vol. 1

Part	pages	Bot. Zeit.	other refs.	part	Bot. Zeit.
1	1-80	8 Dec 1871	J. Bot. Feb 1872; Flora 9 Dec 1871	19	16 Jan 1874
2	81-161	16 Feb 1872	J. Bot. Mar 1872	20	20 Mar 1874
3		10 Mai 1872	Flora 1 Mar 1872	21	29 Mai 1874
4		14 Jun 1872		22	26 Jun 1874

Part	Bot. Zeit.	Part	Bot. Zeit.
5	12 Jun 1872	23	24 Jul 1874
6	12 Jul 1872	24	21 Aug 1874
7	30 Aug 1872	25	27 Nov 1874
8	27 Sep 1872	26	27 Dec 1874
9	15 Nov 1872	27	12 Mar 1875
10		28/29	11 Jun 1875
11			Schluss
12	24 Jan 1873		
13	21 Feb 1873		
14	9 Mai 1873		
15	20 Mai 1873		
16	3 Oct 1873		
17	24 Oct 1873		
18	12 Nov 1873		

Vol. 2

Part	Bot. Zeit.	Part	Bot. Zeit.
1	12 Jul 1872	19	20 Mar 1874
2	30 Aug 1872	20/21	29 Mai 1874
3	27 Sep 1872	22	26 Jun 1874
4	15 Nov 1872	23	24 Jul 1874
5		24	21 Aug 1874
6	24 Jan 1873	25	27 Nov 1874
7	24 Jan 1873	26	25 Dec 1874
8	21 Feb 1873	27	12 Mar 1875
9	9 Mai 1873		Schluss
10	30 Mai 1873		
11	27 Jun 1873		
12	1 Aug 1873		
13/14	3 Oct 1873		
15	24 Oct 1873		
16	12 Dec 1873		
17	16 Jan 1874		
18	27 Feb 1874		

We have seen only one copy of the original preface and Vorwort (BR), and have also encountered a reprint of the Latin part (praefatio) in one other copy. All other copies lack this preface which is of especial interest because Pfeiffer often indicates *type-species* for generic names which constitute in many cases the first selection of a lectotype. In this "praefatio" he states: "species plantarum in libro meo omnio negliguntur, excepta indicatione illarum, quae typum generis novi aut novo modo circumscripti vel sectionis offerunt." German version ("Vorwort"): "Einzelne Arten sind nur dann genannt, wenn Sie als Typus neuer Gattungen oder Abtheilungen dienen." [Single species are mentioned only when they serve as type of new genera or sections]. Hence Pfeiffer is the first reference book to indicate type species in the sense of the International Code of Nomenclature in a more or less systematic way. In case Pfeiffer's choice can be shown to have been erroneous or based on a misinterpretation of the protologue it can of course be overruled as provided for by ICBN 8.1. – The book is an extremely important index of all names of taxa above the rank of species published up till 1858; it not only provides references to first uses, but also to later usage which often provides an opportunity to establish the date and place of valid publication of names so far cited with a reference to a non-valid publication.

The D.P. Rogers, G, NY and US copies show the division of the volumes 1 and 2 after p. 808 and 760 respectively, the U copy has divisions following p. 928 and 840.

Ref.: Anon., J. Bot. 15: 159. 1877 (continuation planned by Uhlworm); Bot. Zeit. 53(2): 359. 1899 (sold at reduced price).

Eichler, A.W., J. Bot. 12: 379-380. Dec 1874 (work completed).
Pfeiffer, L., Bot. Zeit. 6: 825-834.1848 (sample treatment and announcement of plans).

Pfeiffer, Norma Etta (1889-x), American botanist; Ph.D. Univ. Chicago 1913; taught at the University of North Dakota from 1913-1923, at the University of Wisconsin 1923-1932; at the Boyce Thompson Institute, Yonkers, N.Y. 1924-1954; in retirement living in Dallas, Texas. (*N. Pfeiff.*).

HERBARIUM and TYPES: Unknown.

BIBLIOGRAPHY and BIOGRAPHY: Barnhart 3: 79; Kew 4: 291; MW p. 386; NW p. 55.

BIOFILE: Bullock, A.A., Bibl. S. Afr. bot. 84. 1978.
Cattell, J. et al., ed., Amer. men Sci. ed 7. 1387. 1944; ed. 10, Phys.-Biol. L-R: 3167. 1961.
Verdoorn, F., ed., Chron. bot. 5: 276. 1939.

NOTE: when writing this entry, summer 1980, acquaintances of N.E. Pfeiffer at Grand Forks informed us that Dr. Pfeiffer was still happily alive; her adress in 1981 was in Dallas, Texas.

7822. The University of Chicago. *Morphology of Thismia americana*. A dissertation submitted to the Faculty of the Ogden graduate school of science in candidacy for the degree of doctor of philosophy (Department of botany) ... Reprinted from the Botanical Gazette, vol. lvii, no. 2 Chicago 1914. Oct.
Publ.: Feb 1914 (in journal), p. [i, recto of [122]], [122]-135, *pl. 7-11. Copy*: U. – Reprinted and to be cited from Bot. Gaz. 57(2): [122]-135. *pl. 7-11. Copy*: U.

7823. *Monograph of the Isoetaceae* ... (Reprinted from Annals of the Missouri Botanical Garden 9: 79-232. April 1922). Oct.
Publ.: 27 Nov 1922 (footnote p. 79; cover reprint: Apr 1922; Fern J. 3 Oct 1923). Reprinted with original pagination and special cover, and to be cited from Ann. Missouri Bot. Gard. 9: 79-232, *pl. 12-19.* 1922. *Copies* reprint: B, G, MO, NY.
Ref.: Palmer, T.C., Fern J. 13(3): 89-92. 3 Oct 1923.
Fuchs, H.P., Proc. Kon. Ned. Akad. Wet., C., 1981 (in press).

Pfeil, Friedrich Wilhelm Leopold (1783-1859), German (Prussian) forestry botanist; in active service during the Napoleonic wars 1813-1815; administrator of the property of the princess of Carolath in Silezia 1815; director of the Berlin Forestry College 1821; founder and director of the forestry school of Neustadt-Eberswalde, 1830-1859. (*Pfeil*).

HERBARIUM and TYPES: Unknown.

BIBLIOGRAPHY and BIOGRAPHY: Barnhart 3: 79; Futák-Domin p. 468; GR p. 120; KR 1: 430. 457, 602; LS 2: 20606; Moebius p. 395, 398; Ratzeburg p. 399-416 (important source, b. 28 Mar 1783, d. 7 Sep 1859); Rehder 5: 666-667; Tucker 1: 553, 779.

BIOFILE: Anon., Österr. bot. Z. 9: 370. 1859(d.).
Grunert, J.T., Forstliche Blätter 1: 1-60. 1861, 11: 207-218. 1866 (extensive biogr.; d. 4 Sep 1859, correct).
Hess, R., Lebensbilder hervorragender Forstmänner 269-274. 1885 (biogr., bibl.); Allg. deut. Biogr. 25: 648-655. 1887 (biogr., b. 28 Mar 1783; d. 4 Sep 1859).
Schlechtendal, D.F.L. von, Bot. Zeit. 17: 344. 1859 (d. "Anfang" September (1859), b. 1783), 18: 320. 1860 (d. 4 Sep 1858 sic).

COMPOSITE WORKS: *Kritische Blätter für Forst– und Jagdwissenschaft* 1-42(1), 1822-1859. Contains many contributions by Pfeil himself as well as autobiographical papers 27(1): 135. 1849, 33(1): 186. 1853, 41(2): 98. 1859, and obit. by Marcard in 42(2): 1. 1860, bibl. 45(2): 179. 1863; (n.v.).

7824. *Die deutsche Holzzucht.* Begründet auf die Eigenthümlichkeit der Forsthölzer und ihr Verhalten zu den verschiedenen Standorte. Letztes Werk von Dr. W. Pfeil, ... Leipzig (Baumgärtner's Buchhandlung) 1860. Oct. (*Deut. Holzzucht*).
Editor: Staatsanwalt Pfeil, son of F.W.L. Pfeil.
Publ.: 1860 (p. vi: late Nov 1859), p. [i]-viii, [1]-551. *Copy*: UC.
Ref.: Schlechtendal, D.F.L. von, Bot. Zeit. 18: 320. 21 Sep 1860.

Pfitzer, Ernst Hugo Heinrich (1846-1906), German (Prussian) botanist; Dr. phil. Königsberg 1867; habil. Bonn 1868; from 1868-1872 at Bonn; from 1872-1906 professor of botany and director of the botanical garden at Heidelberg; outstanding orchidologist. (*Pfitz.*).

HERBARIUM and TYPES: HEID. – P's once famous collection of living orchids was neglected after his death and finally died off during World War II. Letters and portrait at G.

BIBLIOGRAPHY and BIOGRAPHY: Backer p. 659; Barnhart 3: 79; BJI 1: 45; BM 4: 1563; Bossert p. 309; CSP 8: 612-613, 11: 5, 12: 574, 17: 842; De Toni 1: xcix, 2: xcv; Frank 3 (Anh.): 76-77; Herder p. 157, 288; Hortus 3: 1201 (Pfitz.); Jackson p. 81, 99, 109, 157, 511; Langman p. 581; LS 20608-20609, 37909-37911; Moebius p. 454 [index]; Morren ed. 2, p. 11, ed. 10, p. 25; MW p. 386; NI 1526; Nordstedt p. 26; PR 7110-7111; Rehder 5: 667; RS p. 133; TL-2/1713, see F. Noll; Tucker 1: 553; Zander ed. 10, p. 701, ed. 11, p. 800.

BIOFILE: Ames, Amer. Orch. Soc. Bull. 2(4): 56-58. 1934 (biogr., portr.).
Anon., Bot. Centralbl. 58: 288. 1894 (Geheimer Hofrath), 79: 78. 1899 (corr. memb. Akad. Wiss. Berlin), 98: 400. 1905 (Soc. roy. Bot. Belg.), 102: 656. 1906 (d.); Bot. Not. 1907: 20; Bot. Zeit. 26: 912. 1868 (habil. Bonn), 27: 784. 1869 (asst. Bonn), 30: 594. 1872 (to Heidelberg), 64 (2): 384. 16 Dec 1906 (d. 2 Dec 1906); Bull. Soc. bot. France 18 (bibl.): 190. 1871 (called to Petersburg; did not go); 19 (bibl.): 97. Nov 1872 (just appointed prof. bot. Heidelberg); Flora 52: 61 (habil. Bonn), 524 (asst. Bonn). 1869; Hedwigia 33: (98). 1894 (Geheimrath), 46: (89). 1807 (d.); Mitt. deut. dendr. Ges. 1917: 248 (portr., P. vice pres. 1892-1906); Nat. Nov. 28: 653. 1906 (d. 2 Dec err.! 1906); Österr. bot. Z. 20: 61. 1870 (assistant Bonn), 22: 340. 1872 (app. Heidelberg), 56: 491. 1906 (d. 30 Nov err. 1906).
Arditti, J., ed., Orch. biol. 1: 22, 97, 114. 1977.
Bergdolt, E., Karl von Goebel 39, 74. 1941.
Britten, J., J. Bot. 45: 40. 1907 (obit.).
Cogniaux, A., Bull. Soc. Bot. Belg. 43(3): 369-375. 1907 (obit.), repr. Gand. 1907, 7 p.
Dammer, U., Orchis 10: 73-74. 1907 (portr.).
Fitting, H., Decheniana 105/106: 7-8. 1952 (at Bonn).
Gager, C.S., Brooklyn Bot. Gard. Rec. 27(3): 229. 1938.
Gilg, E., Bot. Jahrb. 40 (Beibl. 90): 57-58. 1907 (obit.).
Hemsley, W.B., Bull. misc. Inf. Kew 1907: 68 (obit.).
Hoehne, C., Jard. bot. São Paulo 158-159. 1941 (portr.).
Kneucker, A., Allg. bot. Z. 12: 208. 1906 (d.).
Merrill, E.D., Enum. Philipp. pl. 4: 216. 1926.
Möbius, M., Ber. deut. bot. Ges. 26a: (33)-(47). 1908 (portr., bibl.); (Möbius was a pupil of Pfitzer; obit., b. 26 Mar 1846, d. 3 Dec 1906).
Pringsheim, E.G., Julius Sachs 124, 264. 1932.
Reinikka, M.A., A history of the orchid 255-257. 1972 (sketch; portr.).
Reinke, J., Mein Tagewerk 47, 122. 1925.
Senghas, E., Die Orchidee, Hamburg 17: 238-244. 1966 (on the collection of living orchids at HEID; notes on Pfitzer on p. 241-243; portr. The Pfitzer collection died off during World War II).
Tischler, G., Ernst Pfitzer, Gedächtnissrede gehalten am 21. Dezember 1906 ... Heidelberg 1907, sep., 29 p. and *in* Verh. nat. med. Ver. Heidelberg ser. 2. 8(3): 397-425. 1906 (obit., bibl., portr.).
VanLandingham, S.L., Cat. diat. 6: 3580, 7: 4217. 1978.
Verdoorn, F., ed., Chron. bot. 1: 154. 1935, 7: 175. 1943.

Wittmack, L., Gartenflora 56: 450-456. 1907 (obit., portr., b. 26 Mar 1846, d. 3 Dec 1906).
Wittrock, V.B., Acta Horti Berg. 3(2): 138, *pl. 28.* 1902 (portr.), 3(3): 147-148. 1905.

COMPOSITE WORKS: (1) Die Bacillariaceen (Diatomaceen) *in* Schenk's *Handbuch der Botanik* 2(28): 410-445. Jan-Mar 1882 (rev. Klebs, Bot. Zeit. 40: 909-910. 22 Dec 1882; Bot. Zeit. 31 Mar 1882.
(2) EP, *Die natürlichen Pflanzenfamilien, Orchidaceae,* 2(6): 53-96. 16 Oct 1888 (not Nov), 97-144. 5 Nov 1888, 145-192. 20 Dec 1888 (not Jan), p. 193-220. 9 Feb 1889 (not Mar). Suppl. in Nachtr. 2-4: 97-113. Aug 1897, Ergänzungsheft 1: 12-16. Apr 1906; reprint with special cover and orig. pagination "*Die Orchideen* dargestellt von E. Pfitzer ..." Leipzig 1888. *Copy*: BR.
(3) Engler, *Pflanzenreich*:
(a) *Orchidaceae, Pleonandreae* iv. 50 (Heft 12): 1-132. 27 Mar 1903 (rev. K. Schumann, Bot. Centralbl. 93: 118-119. 11 Aug 1903).
(b) with F. Kraenzlin, *Coelogyninae,* iv. 50 (iiB7) (Heft 32): 1-169. 26 Nov 1907 (Bot. Zeit. 16 Jan 1908).

7825. *Ueber die Schutzscheide der deutschen Equisetaceen.* Abhandlung geschrieben behufs Erlangung der Würde eines Doctors bei der philosophischen Facultät der Königl. Albertus-Universität zu Königsberg i. Pr. von Ernst H.N. Pfitzer. Oeffentlich vertheidigt den 9. April 1867 Vormittags 12 Uhr ... Königsberg (Druck der Universitäts-Buch– und Steindruckerei von E.J. Dalkowski) [1867]. Oct. (*Schutzscheide deut. Equiset.*).
Thesis: 9 Apr 1867, p. [1]-31, [32]. *Copies*: LC, Niedersachs. Staats– und Univ. Bibl. Göttingen.
Extended commercial ed. 1867, p. [i], [1]-66, *pl. 18-20.* Preprinted from Jahrb. wiss. Bot. 6: [297]-362, *pl. 18-20.* 1868. The thesis issue consists of part of this text.
Ref.: R., Bot. Zeit. 25: 350-351. 1 Nov 1867 (thesis issue).

7826. *Untersuchungen über Bau und Entwicklung der Bacillariaceen* (Diatomaceen) ... Bonn (bei Adolph Marcus) 1871. Oct. (*Unters. Bacill.*).
Publ.: Jan-Mai 1871 (Flora 23 Jun 1871; ÖbZ 1 Sep 1871), p. [i]-vi, [1]-189, *pl. 1-6* (col.; auct.). *Copies*: B,BR, FH, G. – Published as Heft 2 of J. Hanstein, Bot. Abh. Morph. Physiol., 1872.
Ref.: Anon., Bull. Soc. bot. France 19 (bibl.): 213-214. 1873.
Flora 54: 206, 432, 456-464. 1871 Hedwigia 10: 176. Nov 1871.

7827. *Beobachtungen über Bau und Entwicklung epiphytischer Orchideen* ... Separat-Abdruck aus den Verhandlungen des Heidelberger Naturhistorisch-Medizinischen Vereins ... Heidelberg (Verlag von Carl Winter's Universitätsbuchhandlung) [1877]. Oct.
Publ.: 1877 (Bot. Zeit. 27 Apr 1877), p. [1]-10. *Copy*: FI. – Reprinted and to be cited from Verh. Heidelb. Nat.-Med. Ver. ser. 2. 1: 493-502. 1877.

7828. *Der botanische Garten der Universität Heidelberg.* Ein Führer für dessen Besucher ... Heidelberg (Carl Winter's Universitätsbuchhandlung) 1880. Oct. (*Bot. Gart. Heidelberg*).
Ed. 1: Jan –Mai 1880 (Nat. Nov. Jun(1) 1880; Flora 21 Aug 1880; Bot. Zeit. 18 Jun & 9 Jul 1880), p. [i, iii], [1]-50, map. *Copies*: G, NY.
Ed. 2: Sep 1898 (rd. Sep 1898 by Radlkofer, Nat. Nov. Nov(1) 1898, Bot. Zeit. 1 Dec 1898), p. [i]-[iv], [1]-48. *Copy*: M (inf. I. Haesler). – 2. Auflage ... Heidelberg (id.) 1899 [sic].

7829. *Grundzüge einer vergleichenden Morphologie der Orchideen* ... Heidelberg (Carl Winter's Universitätsbuchhandlung) 1882. Qu. (*Grundz. Morph. Orchid.*).
Publ.: 1-9 Dec 1881 (Nat. Nov. Dec(2) 1881; soeben, Bot. Zeit. 9 & 30 Dec 1881; Flora 1 Jan 1882; Bot. Centralbl. 17-21 Apr 1882), frontisp. (col. lith. J.N. Fitch), [i]-iv, [1]-194, *pl. 1-3* (uncol. lith.), plates and figures by W.H. Fitch, J.N. Fitch and E. Pfitzer. *Copies*: B, BR, G, MO.
Ref.: Dalmer, Bot. Centralbl. 10: 86-89. 17-21 Apr 1882.
J.W., Österr. bot. Z. 32: 59-62. 1882 (extensive review).
Kränzlin, F., Bot. Jahrb. 3: 199-204. 9 Jun 1882 (extensive review).
Zacharias, E., Bot. Zeit. 40: 282-283. 28 Apr 1882.

7830. *Morphologische Studien über die Orchideenblüthe.* ... Heidelberg (Carl Winter's Universitätsbuchhandlung) 1886. Oct. (*Morph. Stud. Orchideenbl.*).
Publ.: Jan-Jul 1886 (Nat. Nov. Aug(3) 1886; Bot. Zeit. 24 Sep 1886; J. Bot. Nov 1886), p. [i-ii], [1]-139. *Copies*: B, G. – Heidelberg, Nat.-Med. Ver., Festschrift, Naturhist. Theil.
Ref.: Pax, F., Bot. Jahrb. 8: 77-78. 31 Dec 1886.

7831. *Entwurf einer natürlichen Anordnung der Orchideen* ... Heidelberg (Carl Winter's Universitätsbuchhandlung) 1887. Oct. (*Entwurf Anordn. Orch.*).
Publ.: 1887 (Bot. Zeit. 20 & 27 Mai 1887; Nat. Nov. Jun(1) 1887), p. [i-iii], [1]-108. *Copies*: B, FI, G, GRON, MO, NY, PH, USDA.
Note: Goebel, always ready not to hide his sarcasm under a bushel, writes to Sachs on 9 Dec 1887 "wenn ... die Lektüre von Pfitzer's Orchideenbuch [= *Entw.*] mich nicht im hohem Grade stumpfsinnig gemacht [hätte] ..." and earlier (to Sachs 27 Dec 1882): "... indem er seine Morphologie der Orchideen geschrieben und damit den Beweiss geliefert hat dass ..., wenn eine interessante Pflanzengruppe und ein Kopf zusammenstossen, und es hohl klingt, die Hohlheit nicht immer in der Pflanzengruppe zu liegen braucht" (see Bergdolt 1941). Koehne (1887): Die vorliegende Arbeit wird unzweifelhaft ... die botanische Systematik ... wo die Beschäftigung mit ihr kaum - noch für wissenschaftlich galt, wieder zu Ehren zu bringen ..."
Ref.: Koehne, E., Bot. Zeit. 45: 857-861. 1887.

7832. *Beiträge zur Systematik der Orchideen* von E. Pfitzer. [Bot. Jahrb. 19, 1894]. Oct.
Publ.: 13 Apr 1894 (in journal), p. [1]-42. *Copy*: FI. – Reprinted and to be cited from Bot. Jahrb. 19: [1]-42. 13 Apr 1894. A criticism of O. Kuntze, *Rev. gen.* (1891).

7833. *Übersicht des natürlichen Systems der Pflanzen.* Zum Gebrauch in Vorlesungen für Anfänger ... Heidelberg (Carl Winter's Universitätsbuchhandlung) 1894. Oct. (*Übers. Syst. Pfl.*).
Ed. [*1*]: Jun 1894 (Nat. Nov. Jun(1) 1894; Bot. Zeit. 1 Jul 1894; Bot. Centralbl. 24 Jul 1894; Flora 29 Oct 1894; J. Bot. Oct. 1894), p. [i]-iv, [1]-36 (one-sided). *Copies*: B, G, NY.
Ed. 2: 1902 (Nat. Nov. Jun(1) 1902; ÖbZ Jun 1902; Allg. bot. Z. 15-20 Sep 1902; J. NYBG Oct 1902), p. [i-iv], [1]-40 (one-sided). *Copy*: NY.
Ref.: Möbius, M., Bot. Centralbl. 61: 60-61. 1895.

Pflaum, Fritz (*fl.* 1897), German (Bavarian) botanist; Dr. phil. München 1897. (*Pflaum*).

HERBARIUM and TYPES: Unknown.

BIBLIOGRAPHY and BIOGRAPHY: Barnhart 3: 79; Rehder 5: 667; Tucker 1: 553.

7834. *Anatomisch systematische Untersuchung des Blattes der Melastomaceen* aus den Triben: Microlicieen und Tibouchineen. Inaugural-Dissertation zur Erlangung der Doctorwürde bei der hohen philosophischen Facultät der k. bayr. Ludwigs-Maximilians-Universität zu München. Vorgelegt am 28. April 1897 von Fritz Pflaum aus München. München (Druck von H. Kutzner, ...) 1897. Oct. (*Anat.-syst. Unters. Blatt. Melast.*).
Publ.: 28 Apr 1897 (Nat. Nov. Jan(2) 1898; Flora rd. 27 Oct 1897), p. [i-iv], [1]-91, *pl. 1-2*. *Copies*: G, USDA.

Pfuhl, Fritz C.A. (1853-1913), German (Pommeranian/Prussian) botanist and high school teacher at the Marien Gymnasium at Posen (now Poznan, Poland). (*Pfuhl*).

HERBARIUM and TYPES: Unknown; posssibly at POZ. – Some letters at G.

BIBLIOGRAPHY and BIOGRAPHY: Kew 4: 292; LS 20613-20621, 37912; Rehder 5: 668.

BIOFILE: Anon., Biogr. Jahrb. 18: 114* (b. 20 Mai 1853, d. 16 Jul 1913); Bot. Not. 1913: 298 (d.); Nat. Nov. 26: 253. 1904 (member Leopoldina); Verh. bot. Ver. Brandenburg 55: (77). 1913 (d.).

Kneucker, A., Allg. bot. Z. 16: 75-76. 15 Mai 1910 (rev. Pflanzengarten); 19: 160. 1913 (d.).
Könnemann, W., Aus dem Posener Lande 8(9): 385-390. 1913 (n.v., fide Allg. bot. Z. 19: 174. 1913).
Pax, F., Bibl. schles. Bot. 22. 1929.
Szymkiewicz, D., Bibl. fl. Polsk. 30, 114-115. 1925 (bibl.).

NOTE: Published *Der Pflanzengarten*, 1910, n.v. (Nat. Nov. Mar(1) 1910, AbZ 15 Mai 1910).

Pfund, Johann Daniel Christian (1813-1876), German botanist; born in Hamburg; practicing physician in Prague, from 1842-1845 also custos at PR; later physician in Alexandria, Egypt; collected in the Nile valley; died on a trip to Nubia, Kordofan and Darfur 1874-1876. (*Pfund*).

HERBARIUM and TYPES: B (mainly destroyed), HBG and PR; further material at C, CAIH, K, MO, NA, ROST. – Some letters at G.

BIBLIOGRAPHY and BIOGRAPHY: Barnhart 3: 80; BM 4: 1563; Bossert p. 309 (b. 8 Nov 1813, d. 21 Aug 1876); CSP 4: 880; Futák-Domin p. 468 (bibl.); Jackson p. 146; Kew 4: 292; Klásterský p. 142; Lasègue p. 333; Maiwald p. 281 [index]; PR 632, 4544, 6375, 7113 (ed. 1: 7980); Rehder 5: 668; TL-2/434, 439, 6241; Urban-Berl. p. 382.

BIOFILE: Anon., Peterm. [geogr.] Mitt. 25: 159. 1879.
Ascherson, P., Mitt. geogr. Ges. Hamburg 1878/79: 124-132. 1880 (notes on his Reisebriefe).
Embacher, F., Lexik. Reisen 231. 1882.
Friederichsen, L., Mitt. geogr. Ges. Hamburg 1876-77: 121-305. 1878 (P's Reisebriefe aus Kordofan und Darfur).
Gürke, M., Bot. Jahrb. 14: 287. 1891 (on his 1874-1876 trip).
Jackson, B.D., Bull. misc. Inf. Kew 1901: 52 (err. Pfundt; 758 sp. at K).
Pfund, J.D.C., Bot. Reisebr. aus Kordofan und Darfur. Hamburg 1878; also in Mitt. geogr. Ges. Hamburg 1876/7: 121-305. 1880 (see also Ascherson, 1880, above).
Wickens, G.E., Kew Bull. 24: 191-216. 1970 (biogr., bibl., biogr., refs., itin., b. 8 Nov 1813, d. Aug 1876).

COMPOSITE WORKS: *Skizzen zur Organographie und Physiologie der Classe der Schwämme*... Prag 1844, x, 67 p., a translation of J.F.C. Montagne, TL-2/6241, (rev. by K.M. Bot. Zeit. 2: 862. 13 Dec 1844). *Plantes cellulaires*, in Sagra, *Hist. Cuba* p. 239-291, published separately as *Esquisse organographique et physiologique sur la classe des champignons*, Paris 1841.

7835. *Monographiae generis Verbasci prodromus*. Deutschlands Bärtlinge oder Wollkräuter (Königskerzen), mit besonderer Berücksichtigung der böhmischen Arten. In ökonomisch-technischer und medizinischer Hinsicht bearbeitet von F. von Berchtold, in botanischer von Johannes Pfund ... Prag (Gedruckt bei Th. Thabor; ...) 1840. Oct. (*Monogr. Verbasci prodr.*).
Co-author: Friedrich Berchtold (1781-1876), see TL-2/434, 439.
Publ.: 1840 (p. 4: Sep 1840), p. [1]-80, *1 pl. Copies*: FI, MO. – Preprinted from Berchtold, Oekon.-techn. fl. Böhm. 3(1): 352- 424. 1841.

7836. *Publications de l'état-major général égyptien. Rapport* fait à s.e. le général Stone Pacha chef de l'état-major général égyptien *sur les spécimens botaniques* conservés au cabinet botanique de l'état-major au Caire colligés pendant les expéditions égyptiennes au *Kordofan* et au Darfour ... en 1875 et 1876 par le Dr. Pfund, naturaliste, déterminés et classés par le chevalier docteur J.H. Zarb. Le Caire [Cairo] (Imprimérie de l'état-major général) 1879. Qu. (*Rapp. spec. bot. Kordofan*).
Main author: J.H. Zarb.
Publ.: 6 Feb 1879 (p. 7: Jan 1879; USDA dedic. copy signed 6 Feb 1879), p. [1]-40.
Copy: USDA.

Phares, David Lewis (1817-1892), American botanist, active in Mississippi. (*Phares*).

HERBARIUM and TYPES: MISSA (grasses).

BIBLIOGRAPHY and BIOGRAPHY: Barnhart 3: 80; BL 1: 193, 312; CSP 17: 846; IH 2: (in press); Pennell p. 616.

7837. *Synopsis of the medical flora of the state of Mississippi* ... [First Annual Report of the Mississippi State Board of Health, December, 1877. Jackson Miss. (Power & Barksdale, State Printers) 1877]. Oct.
Publ.: Dec 1877, published in Ann. Rep. Miss. State Board Health 1: [141]-180. Dec 1877. Copy of journal with Mississippi Geological Survey.

Phelps, Almira (previously Lincoln, née Hart or Heart) (1793-1884), American popular writer on botany and teacher at the Troy Female Seminary; from 1838-1841 in charge of a seminary in West Chester, Pa., from 1841 id. of the Patapsco Female Institute, Ellicott City, Md.; married Samuel Lincoln 1817, John Phelps 1831. (*A. Phelps*).

HERBARIUM and TYPES: Unknown.

NOTE: Almira Heart [Hart] was born 15 Jul 1793 in Berlin, Conn.; she married (1817) Simeon (?) Samuel Lincoln, who died in 1823; in 1823 she married John Phelps, who died 1847. Mrs. Phelps herself died at Baltimore, Md., 15 Jul 1884. In her dedication on 20 Apr 1829 to her mother, in the *Familiar lectures*, she spelled her mother's name Mrs. Lydia Heart. The spelling Hart occurs in later sources.

BIBLIOGRAPHY and BIOGRAPHY: Barnhart 2: see Lincoln, A., 3: 80. BM 3: 1116; Bossert p. 309 (b. 15 Jul 1793; d. 15 Jul 1884); Ewan ed. 1: 50; Jackson p. 44; ME 1: 219, 3: 402-403; PR 5211 (ed. 1: 5906); Rehder 5: 669; Tucker 1: 553.

BIOFILE: Allibone, S.A., Crit. dict. Engl. lit. 1575. 1870.
Anon., Appleton's Cycl. Amer. Biogr. 4: 750-751. 1888 (d. 15 Jul 1884; biogr.); Bot. Gaz. 9: 135-136. 1884 (obit., d. 14 Jul 1884); Gard. Mon., Meehan 26: 253. 1884 (obit., d. 14 Jul 1884); Natl. Cycl. Amer. Biogr. 11: 359. 1909 (biogr., b. 15 Jul 1793, d. 15 Jul 1884).
Ewan, J. et al., Short hist. bot. U.S. 39, 115. 1969.
Voss, E.G., Contr. Univ. Mich. Herb. 13: 68. 1978.

EPONYMY: *Phelpsiella* Maguire (1958) is dedicated jointly to the ornithologists, natural historians and explorers in Venezuela, Dr. William H. Phelps, (1875-1965), Mr. William H. Phelps, Jr. (-), and the latter's wife, Kathleen Deery Phelps (-).

7838. *Familiar lectures on botany.* Including practical and elementary botany, with generic and specific descriptions of the most common native and foreign plants, and a vocabulary of botanical terms. For the use of higher schools and academies. By Mrs. Almira H. Lincoln, vice-principal of Troy Female seminary. Hartford (published by H. and F.J. Huntington; G. & C. & H. Carvill, New York; Richardson & Lord, Boston) 1829. Duod. (in sixes). (*Famil. lect. bot.*).
Ed. 1: Jul 1829 (date of legal deposit 4 Jul 1829, dedic. 20 Apr 1829), p. [i]-x, [11]-335, [336, err.], [1-4, index], *pl. 1-13*. *Copies*: US, USDA. – For further information on this and subsequent editions see Rehder 1: 8.
Ed. 2: 1831, p. [i]-xii, [13]-428, [1, err.], *pl. 1-13, 13 [14]*. *Copy*: HU. – Second edition: Hartford (id.) 1831.
Ed. 3: 1832, p. [i]-xii, [13]-440. *Copies*: UC, USDA.
Ed. 4: 1835, p. [i]-x, [xii], [13]-307, [1]-190, [1, err.], *pl. 1-7*. *Copies*: MO, USDA.
Ed. 5: 1836, p. [1-4, adverts.], [5]-246, [1]-186, *pl. 1-8* (in pagination). *Copies*: HU, LC, PH. – Familiar lectures on botany, practical, elementary and physiological; with an appendix, ... fifth edition, ... Hartford (published by F.J. Huntington). 1836. Duod.

Ed. 5: *New York Issue*: 1837, idem. *Copies*: HU, MARY. – New York (published by F.J. Huntington & Co. ...) 1837.
Ed. 7: 1838, p. [1]-246, [1]-186, *pl. 1-8*. *Copies*: LC, UC, USDA. – New York (published by F.J. Huntington & Co. ...).
Ed. 10: 1840, p. [1]-246, [1]-186, *pl. 1-8*. *Copy*: U. – New York (id.).
Ed. 12: 1841, p. [i-x], 11-246, [1]-186, *pl. 1-8*. *Copy*: USDA.
Ed. 13: 1841, p. [5]-241, [1]-186, *pl. 1-8* (in text). *Copy*: HU.
Ed. 17: 1842, p. [5]-246, [247], *pl. 1-8* w.t. [27]-186. *Copy*: LC (possibly incomplete).
New edition: 1845, p. [5]-246, [247-248], *pl. 1-8*, [27]-220. *Copies*: LC, USDA. – Reissued 1847, 1848, 1850.
New edition: 1852, p. [1]-297, [21]-208, *pl. 1-8*, w.t. *Copies*: LC, US, USDA. – Reissued 1854, 1855, 1858.
New edition: 1860, front., p. [1]-312, [315-316], *pl. 1-8* w.t., [21]-208. *Copy*: LC.
New edition: 1864, front., p. [1]-514, *pl. 1-8*. *Copy*: PH.

Phelps, William (1776-1856), British clergyman, topographer and botanist; B.A. Oxford 1797, rector at Oxcombe 1851. (*W. Phelps*).

HERBARIUM and TYPES: Unknown.

BIBLIOGRAPHY and BIOGRAPHY: Barnhart 3: 80; BB p. 243 (d. 17 Aug 1856); BM 4: 1564; Desmond p. 492 (d. 17 Aug 1856); DNB 45: 150; Herder p. 235; Jackson p. 233; Kew 4: 292; PR 7115 (ed. 1: 7981); Rehder 5: 669.

BIOFILE: Bridson, G.D.R. et al., Nat. his. mss. res. Brit. Isl. 224-238. 1980.

7839. *Calendarium botanicum*, or a botanical calendar: exhibiting, at one view, the generic and specific name, the class, order, and habitat of all the British plants, from the class monandria monogynia, to polygamia dioecia, inclusive. Arranged according to their time of flowering under each month of the year ... London (printed by Harding and Wright, ... for Lackington, Allen and Co., ...) 1810. Oct. (in fours). (*Calend. bot.*).
Publ.: 1810 (p. iii: 1 Sep 1809), p. [i]-vi, [1-4, expl. pl.], [1-3], err. slip, [1, h.t.], [1]-186, *pl. 1-5* with text (uncol. copp.).*Copies*: MO, NY.

Phelsum, Murk [Mark, Murck] **van** (*fl.* 1769, d. ca. 1780), Dutch naturalist; Dr. med. Franeker 1755; practicing physician at Sneek, Friesland, from 1764. (*Phelsum*).

HERBARIUM and TYPES: GRO (1200) (see Moll, 1918, and JW 5: 248), all without date and locality.

BIBLIOGRAPHY and BIOGRAPHY: Dryander 3: 67, 345; IH 2: (in press); Jackson p. 32; JW 4: 401, 5: 248; NNBW 10: 721; PR 7116 (ed. 1: 7982); Rehder 5: 669.

BIOFILE: Moll, J.W., Het herbarium van den Hortus Botanicus der Rijks-Universiteit te Groningen 23. 1918 (descr. herb. at GRO).
Wumkes, G.A., Stads- en Dorpskroniek van Friesland 1: 196, 272, 316. 1930 (sale library; "Murck" van Phelsum).

7840. *Explicatio partis IV Phytographiae* Leonardi Pluc'neti M.D. curante M. van Phelsum. Harlingae [Harlingen] (typis Volk. van der Plaats Junioris) 1769. Qu. (*Explic. Phytogr. Pluc'n.*).
Publ.: 1769, p. [i]-xii, [1]-35, [1-2]. *Copies*: FI, M, WU.

Philibert, Henri (1822-1901), French bryologist; educated at the École normale supérieure, Agr. phil. 1846; Dr. lit. 1865; high school teacher at Avignon, Pau, Grenoble, Montpellier, Chaumont, Angoulême 1843-1867; lecturer in philosophy at the Faculté des lettres, Aix-en-Provence 1867-1882; hon. professor, in retirement, from 1883. (*H. Philib.*).

HERBARIUM and TYPES: AUT. – Further material at B, H, MANCH, OXF, PC. – Letters to F. Renauld at PC (Lamy 1981); further letters (and portrait) at G.

BIBLIOGRAPHY and BIOGRAPHY: Barnhart 3: 80; BM 4: 1565; Bossert p. 309 (b. 15 Nov 1822, d. 14 Mai 1901); CSP 4: 881, 8: 614, 12: 574-575, 17: 847; DTS 1: 224 (err. sub J.C. Philibert); IH 1(ed. 6): 362, (ed. 7): 339, 2: (in press); KR p. 573; Lenley p. 326 (mss.); Morren ed. 10, p. 58; SBC p. 129; Urban-Berl. p. 310.

BIOFILE: Anon., Bot. Centralbl. 87: 431. 1901 (d.); Bot. Gaz. 32(1): 74. 1901; Bot. Not. 1901: 94; Hedwigia 40: (176). 1901 (herb. to AUT); Nat. Nov. 23: 520. 1901 (d.); Rev. bryol. 28: 108. 1901 (herb. to AUT); Torreya 1(6): 72. 1901.
Bischler, H. & S. Jovet-Ast, Rev. bryol. lichénol. 39(1): 45. 1973 (Corsican material).
Burnat, É., Bull. Soc. bot. France 30: cxxvi. 1883.
Cavillier, F., Boissiera 5: 72. 1941 (b. 15 Nov 1882, d. 14 Mai 1901).
Gillot, F.X., Proc. Verb. Soc. Hist. nat. Autun 14(2): 129-141. 1901 (biogr., portr., b. 15 Nov 1822, d. 14 Mai 1901, bibl., herb.).
Husnot, T., Rev. bryol. 28(4): 82-83. 1901 (d. 14 Mai 1901).
Kneucker, A., Allg. bot. Z. 7: 176. 1901.
J.M.H., Bryologist 4: 46. 1901.
Lamy, D., Occ. Pap. Farlow Herb. 16: 126. 1981 (corr. F.F.G. Renauld).
Magnin, A., Arch. Fl. jurass. 2(20): 94. 1901 (obit.); Bull. Soc. bot. Lyon 32: 138. 1907.
Quincy, C., Bull. Soc. Sci. nat. Saône-et-Loire 37: 38, 46-47. 1911.
Sayre, G., Bryologist 80: 515. 1977 (coll.).
Thériot, I., Rev. bryol. 28(5): 108. 1901, ser. 2. 4: 143. 1932 (herb. at AUT).

COMPOSITE WORKS: Co-editor, *Revue bryologique* 1878-1901.

EPONYMY: *Philibertiella* Cardot (1914).

Philibert, J.C. (*fl.* 1800), French botanist. (*J. Philib.*).

HERBARIUM and TYPES: Unknown.

NOTE: According to Quérard, 3: 108, the name Philibert is a pseudonym for a "M. Legendre", ancien Conseiller au Parlement; "des fautes graves obligèrent ce personnage à quitter le corps auquel il appartenait, et ... il se cache sous le nom de Philibert, qui était peut-être l'un de ses prénoms; ...".

BIBLIOGRAPHY and BIOGRAPHY: BM 4: 1565; Herder p. 95; Jackson p. 11; Kelly p. 175; Kew 4 : 293; NI 1527; Plesch p. 360; PR 7117-7121 (ed. 1: 7984-7988); Quérard 7: 123-124; Rehder 5: 669; SO 769; Sotheby 596, 597.

EPONYMY: *Philibertella* Vail (1897); *Philibertia* Kunth (1819); *Philibertia* K.M. Schumann (1895).

7841. *Introduction à l'étude de la botanique*, ouvrage orné de dix planches coloriées, contenant: un discours sur l'accord des sciences naturelles; un traité complet et comparé des organes des plantes et des functions de ces organes à toutes les époques de leur vie, ... avec des tables qui donnent à cet ouvrage la commodité d'un dictionnaire. Par J.C. Philibert ... Paris (de l'Imprimerie de Digeon, ...) An VIII [1799]. Oct., 3 vols. (*Intr. bot.*).
1: 1799, p. [i-iii], [1]-454, [1-2, err.].
2: 1799, p. [i-iii], [1]-658, [1-4].
3: 1799, p. [i-iii], [1]-524, *pl.* [*1-10*] (col. copp., unsigned).
Copies: BR, G, H, NY.
Re-issue: 21 Dec 1801 (JT), 3 vols., as above. Paris (Delalain) an X. (n.v.).

7842. *Exercices de botanique*, à l'usage des commerçans. Ouvrage élémentaire, orné de 157 planches, rédigé par J.C Philibert ... A Paris (De l'imprimerie de Crapelet. Chez Bossange, Masson et Besson). 1801, 2 vols. (*Exerc. bot.*).

Ed. 1: 1801, 2 vols. *Copies*: G, MO, NY.
1: [i, iii], [1]-234, *pl. 1-70* (col. copp.).
2: [i, iii], 235-438, *pl. 71-153, 89bis, 113bis, 118bis, 137bis*. Some plates after drawings by R. Turpin.
Ed. 2: 1806, 2 vols. *Copy*: USDA (vol. 2).
1: n.v.
2: [i-iv], 235-438, *pl. 71-153, 89bis, 113bis, 118bis, 137bis* (id.).

7843. *Dictionnaire universel de botanique*, contenant l'explication détaillée de tous les termes français et latins de botanique et de physique végétale ... Paris (Chez Merlin, ...) an XIII-1804. Oct. (*Dict. univ. bot.*).
Publ.: 3 vols. announced as available together on 26 Nov 1804 (JT).
1 (A-E): [i]-ix, [1]-601, [602, err.].
2 (F-O): p. [i-iii], [1]-551, [1, err.].
3 (P-Z): p. [i-iii], [551, sic]-1134, [1, err.], [1-10], *pl. 1-5* (uncol. copp.).
Copies: G, NY.

7844. *Dictionnaire abrégé de botanique*, faisant suite au exercices de botanique à l'usage des commerçans; par J.C. Philibert. Orné de 24 planches, contenant 236 figures. À Paris (Chez Bossange, Masson et Besson) An XI-1803. Oct. (*Dict. abr. bot.*).
Orig. issue: 19 Jul 1803 (JT), p. [i]-vi, [viii], [1]-180, *24 pl.* (col. copp.). *Copies*: Kon. Bibl. (The Hague), NY. – 500 copies (Oct., also 30 copies qu. on papier vélin) with col. or uncol. pl.
Reissue: 1806, p. [i]-vii, [1]-186, *24 pl.* (col. copp.). *Copy*: USDA.
Ed. 2: 1834, [i]-vii, [1]-186,. *Copy* offered by J. Meesters, Cat. 20, no. 90. 1979 (n.v.) (large paper).

Philip, Robert Harris (1852-1912), British diatomist at Hull. (*Philip*).

HERBARIUM and TYPES: Philip's own collections, as well as those by G. Norman catalogued by Philip, were in the Hull museum, but burnt in World War II. (Newton 1952).

BIBLIOGRAPHY and BIOGRAPHY: BB p. 243; Desmond p. 492 (b. 1852); Kew 4: 293; LS suppl. 21644; Nordstedt suppl. p. 13; TL-2/6065.

BIOFILE: Anon., The Naturalist 670: 327. 1912 (coll. and books to the Hull Museum).
Newton, L.C., Phycol. Bull. 1: 17. 1952.
Robinson, J.F., Trans. Hull. Sci. Field Natural. Club 4(4): 219-220. 1912 (obit., d. 13 Apr 1912, b. 1851).
T.S., The Naturalist 664: 150-151. 1912 (obit., d. 15 Apr 1912, portr.).
VanLandingham, S.L., Cat. Diat. 7: 4217. 1978.

COMPOSITE WORKS: see F.W. Mills, *The diatomaceae of the Hull district* 1901, TL-2/6065.

Philippar, François Haken [Aken] (1802-1849), Austrian-born botanist (of French parents); professor at the national agricultural college of Grignon, 1830, and at the "École normale primaire" at Versailles, 1832; founder of the "Jardin des Plantes de Versailles". (*Philippar*).

HERBARIUM and TYPES: Philippar presented a set of Swiss and Savoie plants to the Société Linnéenne de Paris on 7 Apr 1825; it is now at the Société Versaillaise des Sciences Naturelles, see Rouet, J.M. (1973).

BIBLIOGRAPHY and BIOGRAPHY: Barnhart 3: 80; CSP 4: 882, 12: 575; Jackson p. 167, 422; Kew 4: 293-294; PR 7122 (ed. 1: 7989-7993); Rehder 5: 669; SI A 192; Tucker 1: 554; SIA 192.

BIOFILE: Anon., Bot. Zeit. 7: 792. 1849 (d. med. Jun), 10: 23. 1852 (d. Jun 1849 fide J. Débats 24 Jun 1849); Bull. Linn., *in* Mém. Soc. Linn. Paris 1825 (2): 16. 1826.

Robinet, ["M."], Mém. Agric. Écon. rur. domest. 96: 190-202. 1855 (obit., b. 21 Jan 1802, d. Jun 1849).
Rouet, J.M., Rev. Féd. franç. Soc. Sci. nat. ser. 3. 12(51): 4-10. 1973 (on herb. at Versailles).

7845. *Catalogue des végétaux ligneux et des végétaux herbacés cultivés dans le jardin d'étude de l'Institut royale agronomique de Grignon* ... *année 1837*. Paris (Imprimerie de Mme Huzard (née Vallat la Chapelle), ...) 1837. Oct. (*Cat. vég. lign.*).
Publ.: 1837, p. [i-iii], 1-134, chart. *Copy*: NY.

7846. *Catalogue méthodologique des végétaux cultivés dans le jardin des plantes de la ville de Versailles*, précédé d'une notice historique sur les jardins royaux et sur les jardins particuliers de Versailles, sur les hommes qui, dans la botanique et dans l'horticulture, ont rendu des services à la ville, et du programme du cours public; avec plan & facsimile ... Versailles (Imprimerie de Montalant-Bougleux, ...) 1843. Qu. (*Cat. vég. jard. Versailles*).
Publ.: 1843 (Bot. Zeit. 27 Dec 1844), p. [1]-284, 4 charts and facsimile. *Copy*: NY.

Philippe, Xavier (earlier: Philippe Xavier Camus) (1802-1866), French naturalist, municipal librarian and plant salesman living at Bagnères-de-Bigorre. (*Philippe*).

HERBARIUM and TYPES: Bryophytes in PC, H and W. – Loret (1883) states that P's Pyrenean herbarium of 2000-3000 specimens was at the time at the Seminary of Oloron. Beck (1888) mentions 521 *Cryptogamae pyrenaicae* (227 lichens) at W. See also Sayre (1975) for further details. Hertel (1980) mentions lichens at M. Further material in BORD, G, MANCH and Soc. Versaill. Sci. nat.

NOTE: The M. in M. Philippe on the title-page of the *Fl. Pyren.* stands for "Monsieur". According to Ascherson and Graebner (AG 3: 616, 781) there was also a Philippe of Saint-Mandrier, near Toulon. This Philippe was director of the "jardin botanique de la marine impériale à Saint-Mandrier", see e.g. BSbF 7: 44. 1860. He published a list of plants of the Dépt. Var in BSbF 12: 191. 1865.

BIBLIOGRAPHY and BIOGRAPHY: AG 3: 656 (err.), 781 (identity); Barnhart 3: 80; BL 2: 113, 702; BM 4: 1565; Colmeiro 1: cxciii; Frank 3 (Anh.): 77; GR p. 344; Hortus 3: 1201 (Philippe); Jackson p. 278; Kew 4: 294; Krempelh. 1: 269; Plesch 361; PR 7123; Rehder 1: 412; Tucker 1: 554.

BIOFILE: Anon., Bull. Soc. bot. France 13 (bibl.): 47. 1866.
Beck, G., Bot. Centralbl. 34: 150. 1888 (crypt. at W).
Hertel, H., Mitt. Bot. München 16: 416. 1980.
Loret, H., Bull. Soc. bot. France 30: 50-57. 1883 (see also Freyn, Bot. Centralbl. 15: 398. 1883; a severely critical appraisal of the *Fl. Pyren.* which is said to have been copied from Grenier's Fl. France; gives details on herbarium).
Pée-Laby, E., Bull. Soc. Ramond 31: 42-49. 1896.
Sayre, G., Mem. New York Bot. Gard. 19(3): 379-380. 1975 (coll.); Bryologist 80: 515. 1977.

7847. *Flore des Pyrénées* par M. Philippe ... Bagnères-de-Bigorre (chez P. Plassot, ...) 1859, 2 vols. Oct. (*Fl. Pyren.*).
1: Mai 1859 (BSbF presented 29 Apr 1859; avis; vol. 2 to appear in June), p. [i, iii, v], [1]-605, [606, note on sale plants].
2: Jun-Dec 1859 (BSbF presented 27 Jan 1860), p. [i, iii], [1]-505.
Copies: BR, G, H, MO, NY, USDA. – For a critical notice see Loret (1883); see also BL 2: 113.
Ref.: Anon., Bull. Soc. bot. France 6: 240-241. 1859 (rev. vol. 1), 7: 288. 1860 (post Sep).

Philippi, Federico (baptized Friedrich Heinrich Eunom) (1838-1910), Chilean botanist and entomologist of German origin (born in Napoli), son of R.A. Philippi, whom he succeeded in 1874 as professor of botany at Santiago; from 1897 also director of the Museo nacional (de Chile). (*F. Philippi*).

HERBARIUM and TYPES: main collections at SGO; other material at BA, BM, BP, F, G, K, L, LE, LZ, MA, P, W. – Some letters at G.

BIBLIOGRAPHY and BIOGRAPHY: Barnhart 3: 80; BL 1: 128, 246, 247, 248, 312; BM 4: 1565; CSP 8: 614, 17: 848-849; Frank 3 (Anh.): 77; Herder p. 225; Hortus 3: 1201 (Phil. f.); Kew 4: 294; LS 20626, 37914; Morren ed. 10, p. 141; Rehder 5: 669; Tucker 1: 554; Urban-Berl. p. 382.

BIOFILE: Anon., Ber. deut. bot. Ges. 27: (144). 1910; Bot. Not. 1910: 176 (d.); Bot. Zeit. 65: 288. 1907 (app. ord. prof. bot. Santiago); Nat. Nov. 19: 601. 1897 (succeeds his father as dir. Mus. nac. Chile), 29: 482. 1907 (retirement from professorship), 32: 289. 1910 (d.).
Ball, J., J. Bot. 24: 65-67. 1886 (letter from F. Ph. on Atacama desert).
Gotschlich, B., Bol. Mus. nac. Hist. nat. Chile 1: 39-80. 1910 (biogr., portr., bibl.), 2: 264-298. 1910 (summary by M. Moore).
Kneucker, A., Allg. bot. Z. 16: 144. 1910.
Jackson, B.D., Bull. misc. Inf. Kew 1901: 52 (pl. K).
Lizer y Trelles, C.A., Curso entom. 1: 24. 1947.
Merrill, E.D., Bull. B.P. Bishop Mus. 144: 150. 1937, Contr., U.S. natl. Herb. 30: 240-241. 1947.
Neubert, Deut. Gart. Mag. 45: 184-190. 1892 (on F. Ph.'s journey to Antofagasta and Tarapacá as reported in R.A.Ph.'s *Verz.* of 1891).
Philippi, F. et R.A., Anal. Univ. Santiago 27(3): 289-352. 1865 (Excursion botánica en Valdivia, travelogue by F. Ph., descriptions of new plant taxa by R.A. Ph.).
Porter, C.E., Rev. chil. Hist. nat. Valparaiso 7: 105-107. 1903 (portr., brief bibl.), 14: 19-23. 1910 (obit.), 43: 10-15. 1939 (zool.); Anal. Soc. ci. Argentina 69: 147-149. 1910 (obit., portr., brief bibl.); Bull. Acad. int. Géogr. bot. 19: 97. 1910 (d. 16 Jan 1910).
Reiche, K., Grundz. Pfl.-Verbr. Chile 38. 1907 (Veg. Erde 8) (bibl.); Bol. Mus. nac. Chile 3(1): 235-241. 1911 (obits. from newspapers).
Sayre, G., Mem. New York Bot. Gard. 19(3): 380. 1975 (coll.).
Schaufuss, C., ed., Entom. Rundschau 27: 46. 1910 (obit.).
Turrill, W.B., Bull. misc. Inf. Kew 1920: 63.

COMPOSITE WORKS: See C. Reiche, *Flora de Chile*.

7848. *Catalogus plantarum vascularium chilensium* adhuc descriptarum ... Santiago de Chile (Imprenta Nacional, ...) 1881. Qu. (*Cat. pl. vasc. chil.*).
Publ.: Nov-Dec 1881 (p. v: 20 Oct 1881; Bot. Centralbl. 10-14 Apr 1882; Bot. Zeit. 31 Mar 1882; Nat. Nov. Mai(1) 1882), p. [i]-viii, [5]-377, [378, err]. *Copies*: B, FI, G, MICH, MO, NY, PH, US. – Reprinted from Anal. Univ. Santiago 59: 49-422. 1881.
Ref.: Engler, A., Bot. Jahrb. 3: 255. 18 Aug 1882.
Freyn, J.F., Bot. Centralbl. 10: 367. 1882 (rev.).

7849. *Memoria i catalogo de las plantas cultivadas en el Jardin botanico* hasta el 1.; de mayo de 1884. Santiago de Chile (Imprenta nacional, ...) 1884. Qu. (*Mem. pl. cult.*).
Publ.: 1884 (p. [3]: 29 Mai 1884; Nat. Nov. Sep(2) 1885; Bot. Zeit. 25 Dec 1885), p. [1]-83, map. *Copies*: MO, NY.

Philippi, Rudolph [Rudolf; later called Rodolfo Amando] **Amandus** (1808-1904), German (Prussian) botanist, studied in Berlin, Dr. phil. 1830, teacher at the technical school at Kassel 1835-1838, 1840-1851; in Italy 1830-1832, 1838-1840; emigrated to Chile 1851; at first as farmer and teacher at Valdivia; professor of botany and zoology at Santiago from 1853-1874, until 1897 director of the Museo nacional (de Chile). (*Phil.*).

HERBARIUM and TYPES: Collections before 1846: GOET; Chilean material and types: SGO; further material in many herbaria. *Plantae chilenses*, irregular exsiccatae, distributed by Hohenacker (see e.g. Bonplandia 2: 268. 1854, 6: 342. 1858). The largest set of Chilean material went to B (now mainly destroyed), other significant sets are at BAF, G, LE, P, SI, and W; lichens e.g. at M. – Correspondence with D.F.L. von Schlechtendal at HAL; letters and portrait at G.

BIBLIOGRAPHY and BIOGRAPHY: AG 2(2): 319; Backer p. 659; BJI 1: 45; BL 1: 246, 247, 248, 312; BM 4: 1565, 8: 1006; Bossert p. 309; CSP 4: 882-887, 8: 614-616, 11: 8-9, 17: 849-850; Frank 3 (Anh.): 77; Herder p. 56, 293; Hortus 3: 1201 (Phil.); IF p. 723-724; IH 1 (ed. 6): 362, (ed. 7): 339; Jackson p. 48, 119, 120, 371, 374, 403; Kew 4: 294-295; Lasègue p. 405, 503; Lenley p. 326; LS 20627, 37915-37918; Morren ed. 2, p. 35, ed. 10, p. 141; NAF ser. 2. 2: 167; NI 2: 622; PR 7124; PH 98, 567; Quenstedt p. 336 (b. 14 Sep 1808, d. 23 Jul 1904); Rehder 5: 669-670; RS p. 133; Saccardo 1: 127, 2: 84; TL-1/449, 977; TL-2/2178; Tucker 1: 554-555, 782; Urban-Berl. p. 280, 310, 382; Zander ed. 10, p. 701, ed. 11, p. 800.

BIOFILE: Anon., Ber. deut. bot. Ges. 22: 343. 1904 (d.); Bot. Centralbl. 73: 192. 1898 (Philippi, 90 yrs old, resigns directorship mus. Santiago); Bot. Jahrb. 37 (Beibl. 84): 3. 1906; Biogr. Jahrb. 10: 83*-84*. 1907; Bot. Not. 1905: 56 (d.); Bot. Zeit. 1: 222. 1843 (pl. from Chile to B), 12: 367. 1854 (letter on Atacama trip), 23: 36. 1865 (trip to Juan Fernandez); Bull. Soc. Bot. Belg. 42: 88-89. 1904 (obit.); Nat. Nov. 11: 26. 1889 (80th birthday), 19: 601. 1897 (retirement), 22: 224. 1900 (70 years dr. phil.), 26: 513. 1904 (d.); Österr. bot. Z. 39: 158. 1889 (plants rd.), 48: 119. 1898 (retirement), 54: 311. 1904 (d.); Torreya 4: 128. 1904.
Arana, D.B., El doctor don Rodolfo Amando Philippi, Santiago de Chile 1904, vii, 248 p.
Ascherson, P., Fl. Prov. Brandenburg 1: 10. 1864.
Bässler, M., Wiss. Z. Humboldt-Univ. Berlin, math. nat. 19: 299. 1970 (pl. at BHU).
Baur, E., Jahrb. Gesch. Oberdeut. Reichst. 16: 241. 1970 (P. material distr. by Hohenacker).
Beck, G., Bot. Centralbl. 33: 314. 1888;34: 150. 1888.
Burkart, A., Darwiniana 2: 140. 1928 (coll. SI).
Candolle, Alph. de, Phytographie 439. 1880 (coll.).
D.R., Bot. Centralbl. 37: 29-31. 1889 (80th birthday), see also Bull. Torrey bot. Club 16: 78. 1889.
Drathen, R., Revista Univ., Chile 37(1): 5-6. 1952 (brief tribute).
Embacher, F., Lex. Reisen 231. 1882.
Fennell, Abh. Ber. Ver. Naturk. Cassel 53: 170-176. 1913 (obit.).
Follmann, G., Philippia 1(1): 3-8. 1970 (tribute; list of biogr. refs. and bibl.).
Fürstenberg, P., Dr. Rudolph Amandus Philippi. Sein Leben und seine Werke. Santiago de Chile 1906, 39 p. (portr., bibl.). Reprinted from Verh. deut. wiss. Ver. Santiago 5, 1906.
Gilbert, P., Comp. biogr. lit. deceased entom. 296. 1977 (further biogr. ref.).
Gotschlich, B., Biografia del Dr. Rodulfo Amando Philippi (1808-1904). Santiago, Chile, 1904, vi, 185 p. (bibl. of 450 nos., portr., itin.).
Hantzsch, V., Biogr. Jahrb. 9: 186-191. 1906 (biogr., b. 14 Sep 1808 at Charlottenburg, d. 23 Jul 1904 at Santiago de Chile).
Hedge, I.C. & J.M. Lamond, Index coll. Edinb. herb. 118. 1970.
Herzog, Th., Pfl.-Welt Boliv. Anden 19. 1923 (Veg. Erde 15).
Hicken, C.M., Anal. Soc. ci. Argent. 58: 145-151. 1904 (obit., portr., brief bibl.).
Hohenacker, R.F., Bonplandia 2: 268. 1854, 10: 32. 1862 (offers *Pl. chil.* for sale); Bot. Zeit. 11: 678-679. 1853, 12: 743-744. 1854 (id.).
Jackson, B.D., Bull. misc. Inf. Kew 1901: 52. (*pl.* K).
Jessen, K.F.W., Bot. Gegenw. Vorw. 417, 465, 468. 1864.
Kneucker, A., Allg. bot. Z. 4: 52. 1898 (retired from directorship of Mus. nacional at 90).
Kurtz, H., Philippia 4: 97-107. 1979 (outline exped. Atacama desert 1853-1854).
Leveillé, A.A.H., Bull. Acad. int. Géogr. bot. 11: 82-88. 1902 (bibl., portr.).
Looser, G., Revista Argent. Agron. 7(2): 144-150. 1940 (dates publ. *Pl. nuev. chil.*).
Losch, P., Abh. Ber. Ver. Naturk. Cassel 49: 143-161. 1905 (obit.; Philippi left Kassel for political reasons on 17 Dec 1850 and resigned early 1851, stayed with C.L. Koch in Braunschweig before emigrating, to Chile on 20 Jul 1851, arr. 4 Dec 1851; letters).
Merrill, E.D., Bull. B.P. Bishop Museum 144: 150. 1937; Contr. U.S. natl. Herb. 30(1): 241. 1947 (bibl.).
Muñoz Pizarro, C., Las especias de plantas descritas por R.A. Philippi, 1960 (see below *General study*).

Murray, G., Hist. coll. BM(NH) 1: 173. 1904 (564 pl. at BM).
Nissen, C., Zool. Buchill. 316. 1919.
Ochsenius, C., Nachrichtsbl. deut. Malakol. Ges. 21: 1-4. 1889; Leopoldina 42: 16-20, 39-40, 53-56, 59-66. 1906 (biogr., b. 14 Sep 1808)
Papavero, Essays hist. neotr. dipterol. 1: 160. 1971, 2: 275-280. 1975.
Parodi, L.R., Bot. Soc. Arg. Bot. 9: 13. 1961.
Perez Moreau, R.A., Bibl. geobot. patagon. 29-30, 44-45, 68-69, 86, 88, 94, 95, 96, 100, 103, 104, 106. 1965.
Philippi, R.A., Bot. Zeit. 19: 377-384, 385-390. 1861; Spanish translation by G. Looser, La Farmacía Chiléna 1935 (3-5), 24 p. (Bot. excursion to Aconcagua); Export, Organ des Centralvereins für Handelsgeogr. 20: 402-45. 1898 (autobiogr., portr.); Bull. Acad. Géogr. bot. 13: 321-323. 1904 (autobiogr. notes).
Philippi, F. et R.A., Anal. Univ. Santiago 27(3): 289-352. 1865 (Excursion botánica en Valdivia, travelogue by F. Ph., descr. of new plant taxa by R.A. Ph.).
Poggendorff, J.C., Biogr.-lit. Handw.-Buch 2: 433. 1863, 3: 1034. 1898, 5(2): 969. 1926.
Porter, C.E., Anal. Soc. ci. Argent. 70: 284-286. 1910.
Reiche, K., Ber. deut. bot. Ges. 22: (68)-(83). 1904 (extensive and authoritative biogr., bibl., b. 14 Sep 1808, d. 23 Jul 1904); Grundz. Pfl.-Verbr. Chile 17-21, 38-39, 372. 1907 (Veg. Erde 8).
Sayre, G., Mem. New York Bot. Gard. 19(3): 380. 1975 (coll.).
Schofield, E.K., Brittonia 30: 404. 1978 (coll. at NY).
Stafleu, F.A., Miquel-Schlechtendal Corr., Regn. veg. 71: 309, 329. 1970 (error: Philippi's plants were distributed by Hohenacker, but Philippi was never employed as traveller for the Esslinger Reiseverein).
Steinberg, C.H., Webbia 32(1): 33. 1977 (material at FI).
Stieber, M.T., Huntia 3(2): 124. 1979 (notes A. Chase on Ph.).
Turrill, W.B., Bull. misc. Inf. Kew 1920: 62.
Verdoorn, F., Pl. pl. sci. Latin America xxvi. 1945.
Verdoorn, F., ed., Chron. bot. 3: 83. 1937, 5: 115. 1939, 7: 348. 1943.
Wittrock, V.B., Acta Horti Berg. 3(2): 100. 1903, 3(3): 84. 1905.
Zittel, K.A. von, Hist. geol. palaeont. 243, 401, 530. 1901.

GENERAL STUDY OF PH'S TAXA: C. Muñoz Pizarro, Las especies de plantas descritas por R.A. Philippi en el siglo xix. Estudio crítico en la identificación de sus tipos nomenclaturales. l960, Ediciones de la Universidad de Chile, 189 p.

COMPOSITE WORKS: Founder and editor of the *Anales del Museo nacional*, Chile, *vols. 1-16*, 1892-1904.

HANDWRITING: Wiss. Z. Humboldt-Univ. Berlin, math.-nat. 19: 298. 1970; D.B. Arana, El doctor don Rodolfo Amando Philippi, 1904, frontisp.

EPONYMY (genera): *Philippia* Klotzsch (1834); *Philippiamra* O. Kuntze (1891); *Philippicereus* Backeberg (1942); *Philippiella* Spegazzini (1897); *Philippimalva* O. Kuntze (1891). *Note*: *Philippiella* P.C. Silva (1959) and *Portphillippia* P.C. Silva (1970) are named for Port Phillip, Australia; *Philippinaea* Schlechter et Ames (1920) is named for the Philippine Islands; *Philippiregis* Ciferri et Tomaselli (1953) honors R. Philippo (1763-1817); *Philippodendrum* Poiteau (1837) and *Philippodendron* Endlicher (1840, *orth. var.*) honor Louis Philippe (1773-1850), King of France (1830-1848) and called the "Citizen King"; (journal): *Philippia*, Abhandlungen und Berichte aus dem Naturkundemuseum im Ottoneum zu Kassel, Germany. Vol. 1-x. 1970-x.

TRAVEL REPORTS: Ph. published several travel reports in the Botanische Zeitung:
(1) *Botanische Reise nach der Provinz Valdivia*, 16: 257-262. 27 Aug, 265-290. 3 Sep, 273-279. 10 Sep 1858; extract in Gartenflora 8: 21-26. 1859.
(2) *Excursion nach dem Ranco-See in der Provinz Valdivia*, 18: 305-311. 14 Sep, 313-318. 21 Sep 1860. Extract in Gartenflora 10: 295-298. 1861.
(3) *Botanische Excursion in die Provinz Aconcagua*, 19: 377-383. 20 Dec, 385-390. 27 Dec 1861. *Spanish translation* by Gualterio Looser, Revista Chil. Hist. Geogr. 74: 700-710.

1933, reprinted Santiago (Imprenta Universitaria ...) 1934, p. [i, t.p. on recto of p. 700], [700]-710. *Copies*: G, U.

POSTAGE STAMPS: Chile, $ 3.50, publ. 28 Dec 1978.

7850. *Reise durch die Wueste Atacama* auf Befehl der chilenischen Regierung im Sommer 1853-1854 unternommen und beschrieben von Doctor Rudolph Amandus Philippi ... Halle (Eduard Anton) 1860. Qu. (*Reise Atacama*).
Publ.: 1860 (Bonplandia 15 Dec 1860; Bot. Zeit. 11 Jan 1861; ÖbZ 1 Jul 1861), p. [i]-ix, [x], [1]-192, *Florula atacamensis*: [1]-62, map, *pl. 1-12* (general), *1-2* (petref.), *1-7* (zool.), *1-6* (bot., monochr. liths. auct.). *Copies*: BR, G, MO, NY, PH, Teyler; IDC 7124. – The *Florula* was also published separately: see below. – The references on p. [57]-59 of the *Florula* are to p. 181-230 which stand for 7-56. Apparently the pagination was meant to start with 175 (= 1? Florula).
Spanish: 1860, p. [i]-viii, [1]-236, map, *pl. 1-12, 1-2, 1-7, 1-6* (see above). *Copies*: MICH, PH. – *Viage al desierto de Atacama* hecho el orden del gobierno de Chile en el Verano 1853-1854 ... " Halle en Sajonia (Libreria de Eduardo Anton) 1860. Qu.
Ref.: Anon., Flora 44: 135-137. 7 Mar 1861 (rev.).
Schlechtendal, D.F.L. von, Bot. Zeit. 19: 14-16. 11 Jan 1861. (rev.).

7851. *Florula atacamensis* seu enumeratio plantarum, quas in itinere per desertum atacamense observavit Dr. R. Philippi ... cum tab. vi. Halis Saxonum [Halle] (Sumptibus Eduardi Anton) 1860. Qu. (*Fl. atacam.*).
Publ.: 1860, p. [i], [1]-62, *pl. 1-6* (uncol. lith. auct.). *Copies*: G, NY.

7852. *Descripción de las nuevas plantas* incorporadas últimamente en el herbario chileno ... Santiago (Imprenta nacional, ...) 1872. Qu. (*Descr. nuev. pl.*).
Publ.: 1872, p. [1]-88. *Copies*: B, G.
Ref.: G.K., Bot. Zeit. 31: 654. 10 Oct 1873.

7853. *Botánica. Sobre las especies chilenas del jenero Polyachyrus* ... Santiago de Chile (Imprenta nacional, ...) 1886. Oct.
Publ.: 1886, p. [1]-15, *1 pl. Copy*: NY. – Reprinted and to be cited from Anal. Univ. Chile 69, 1886.

7854. *Die tertiären und quartären Versteinerungen Chiles* ... Leipzig (F.A. Brockhaus) 1887. Qu. (*Tert. quart. Verstein. Chil.*).
Publ.: 1887, p. [i-iii], [1]-266, *pl. 1-58* (uncol.). *Copy*: US.

7855. *Verzeichniss der von D. Francisco Vidal Gormaz an den Küsten des nördlichen Chiles gesammelten Gefässpflanzen* ... Santiago (In Commision bei R. Friedländer & Sohn, Berlin) 1890.
Collector: Francisco Vidal Gormaz (*fl.*1860-1880).
Publ.: 1890, p. [1]-5. *Copy*: BR. – Reprinted and to be cited from Verh. deut. wiss. Ver. Santiago 2(2): 106-108. 1890.

7856. *Verzeichniss der von Friedrich Philippi auf der Hochebene der Provinzen Antofagasta und Tarapacá gesammelten Pflanzen* aufgestellt von Dr. R.A. Philippi. Leipzig (F.A. Brockhaus) 1891. Qu. (*Verz. Antofagasta Pfl.*).
Publ.: Sep-Oct 1891 (Nat. Nov. Oct(2) 1891; Bot. Zeit. 29 Dec 1891; Bot. Centralbl. 9 Jan 1892; ÖbZ Nov 1891), p. [i]-viii, [1]-94, [94bis], [95]-96, *pl. 1-2* (uncol.). *Copies*: B, BM, NY, USDA; IDC 5637. – Also published as Annales del Museo nacional de Chile, sect. 2. Bot. vol. 8, 1891, with the title *Catalogus praevius plantarum in itinere ad Tarapaca a Friderico Philippi lectarum. Copies*: BR, NY, US.
Ref.: Gunckel, T.H., Revista Univ. Chile 24(1): 13. 1939.
Taubert, P.H.W., Bot. Jahrb. 15 (Lit.): 112-113. 1892; Bot. Centralbl. 51: 170-172. 1892.
Kolb, M. et al., eds., *in*Neubert, Deut. Garten-Mag. 45: 184-190. 1892 (rev.).

7857. *Plantas nuevas Chilenas* ... Santiago de Chile (Imprenta Cervantes ...) 1892-1896. Oct.

Publ.: a series of papers published in the Anales de la Universidad, Santiago, vols. 81-94, 1892-1896. Reprinted with original pagination and special covers. To be cited from journal. See Looser (1940). The tome nos. correspond with those of C. Gay, *Hist. fis. pol. Chile, Bot.*

Annales	Pagination	dates	tome pl. nuev.
81	[65]-86	Mai 1892	1
	177-195	Jun 1892	
	329-347	Jul 1892	
	489-498	Aug 1892	
	[761]-775	Sep 1892	
82	[5]-24	Nov 1892	
	[305]-325	Jan 1893	
	[725]-740	Feb 1893	
	[895]-915	Mar 1893	
	[1095]-1106	Apr 1893	
84	[1]-32	Mai 1893	2
	[265]-289	Jun 1893	
	[433]-444	Jul 1893	
	[619]-634	Aug 1893	
	[743]-762	Sep 1893	
	[975]-983, *1 pl.*	Oct 1893	
85	[5]-18	Nov 1893	
	[167]-195	Dec 1893	
	[299]-324	Jan 1894	
	[491]-514	Feb 1894	3
	[699]-749	Mar 1894	
	[813]-844	Apr 1894	
87	[5]-24	Mai 1894	
	[81]-112	Jun 1894	
	[299]-330	Jul 1894	
	[399]-436	Aug 1894	4
	[585]-624	Sep 1894	
	[677]-713	Oct 1894	
88	[5]-38	Nov 1894	
	[245]-292	Dec 1894	
90	[5]-44	Jan 1895	
	[187]-230	Feb 1895	
	[337]-358	Mar 1895	
	[511]-566	Apr 1895	
	[607]-625	Mai 1895	5
	[759]-772	Jun 1895	
91	[5]-47	Jul 1895	
	[105]-160	Aug 1895	
	[243]-275	Sep 1895	
	[415]-432	Oct 1895	
	[487]-526	Nov 1895	
	[607]-635	Dec 1895	6
93	[5]-24, *pl. 1-5*	Jan 1896	
	[143]-166, *pl. 6*	Feb 1896	
	[261]-278	Mar 1896	
	[343]-352	Apr 1896	
	[475]-506	Mai 1896	
	[711]-735	Jun 1896	
94	[5]-34	Jun 1896	
	[155]-179	Aug 1896	
	[341]-362	Sep 1896	

Copies: B, BR, G, MO, USDA.

7858. *Botanische Excursion in das Araukanerland* ... Kassel (Druck von L. Döll) 1896. Oct.
Publ.: Aug 1896 (err. p. [2]: Jul 1896; Nat. Nov. Sep(1) 1896), p. [i-ii], [1]-31, [1-2, err.]. *Copies*: MO, NY. – Reprinted and to be cited from Abh. Ber. Ver. Naturk. Kassel 41: 1-31. 1896.

Phillips, Edwin Percy (1884-1967), South African botanist; Dr. sci. Univ. Cape 1915; with the South African Museum at Capetown 1917-1918; from 1918-1944 at the National Herbarium, Pretoria (chief (1939-1944). (*E. Phillips*).

HERBARIUM and TYPES: PRE and SAM (earlier coll.); duplicates A, BOL, E, G, GRA, K, L, MO, PRE, SAM, WELC, Z. –Collector' registers at PRE. – Letters at G.

BIBLIOGRAPHY and BIOGRAPHY: Barnhart 3: 81; BJI 2: 134; BL 1: 23, 29, 54, 92, 313; BM 8: 1007; Bossert p. 309 (b. 18 Feb 1884, d. 12 Apr 1967); Hortus 3: 1201 (E.P. Phillips); Kew 4: 297-298; LS 37919; NI 3: 74; SIA/HI 107; TL-2/2448; Zander ed. 10, p. 701, ed. 11, p. 800.

BIOFILE: Anon., J.S., Afr. biol. Soc. 7/8: 87. 1968 (obit.); S. Afr. Tydskr. Wet. 63(8): 332. 1967; The Public Servant 1967 (Mai): 7 (n.v.).
Bullock, A.A., Bibl. S. Afr. bot. 84-89. 1978 (main bibl.).
Goldblatt, P., Taxon 29: 690. 1980 (coll. MO).
Guillarmod, A.J., Musées de Genève 38: 12. 1963, Fl. Lesotho 54, 64. 1971 (on Fl. Leribe Plateau).
Gunn, M.D., Bothalia 10(1): 1-3. 1969 (obit., portr., b. 18 Feb 1884, d. 12 Apr 1967).
Gunn, M.D. & L.E. Codd, Bot. expl. S. Afr. 219, 280. 1981 (coll., portr., biogr.).
Hedge, I.C. & J.M. Lamond, Index coll. Edinb. herb. 118. 1970.
Karsten, M.C., The old Company's Garden 159-162. 1951.
Leistner, O.A., Taxon 16: 572-573. 1967 (b. 18 Feb 1884, d. 12 Apr 1967).
Levijns, M.R.B., Insnar'd with flowers 15. 1977.
Phillips, E.P., S. Afr. J. Sci. 27: 55, 65, 68, 69. 1930.
Tölken, H.R., Index herbariorum austro-africanorum 49. 1971 (coll.).
Verdoorn, F., ed., Chron. bot. 2: 2, 263, 266, 329. 1936, 3: 2, 232, 233. 1937, 5: 119, 510. 1939, 6: 301. 1941.
White, A. & B.L. Sloane, The Stapelieae 205. 1933, ed. 2, 20 [index]. 1937 (portr., p. 135.).

COMPOSITE WORKS: (1) *Flora capensis*, *Proteaceae* (with J. Hutchinson and O. Stapf), *in* 5(1, 3): 502-640. Jan 1912, 5(1, 4): 641-718. Jun 1912.
(2) *The flowering plants of South Africa*, editor vols. 1-24.

DEDICATION: *The flowering plants of Africa*, vol. 25, ed. R.A. Dyer, frontisp. portr. Phillips.

EPONYMY: *Susanna* E.P. Phillips, 1950, commemorates Mrs. Susan Phillips, née Kriel, second wife to E.P. Phillips.

7859. *A contribution to the flora of the Leribe Plateau and environs: with a discussion on the relationships of the floras of Basutoland, the Kalahari, and the South-Eastern Regions* ... [Annals of the South African Museum vol. xvi(1). 1917]. Oct.
Publ.: 21 Jun 1917 (as testified by publisher, see MO copy), p. 1-379, *pl. 1-7*. *Copies*: MO, NY. – Ann. S. Afr. Mus. 16(1): 1-379, *pl. 1-7*. To be cited from journal. Doctor's thesis.

7860. Botanical survey of South Africa. Memoir no. 9. *A preliminary list of the known poisonous plants found in South Africa* ... Pretoria (The Government Printing and Stationery Office) 1926. (*List poison. pl. S. Afr.*).
Publ.: early 1926 (printing finished 3 Oct 1925, 1000 copies), p. [1]-30, *pl. 1-20* (mostly col.). *Copies*: MO, USDA.

7861. *The genera of South African flowering plants* ... Cape Town (Cape Times limited, government printers) 1926. Oct. (*Gen. S. Afr. fl. pl.*).

Ed. 1: Nov 1926 (p. [iii]: 11 Jan 1926; MO copy rd. 28 Mai 1927; J. Bot. Aug 1927; personal copy C.A. Smith at PRE; Nov 1926, inf. O. Leistner), p. [i-vii], [1]-702. *Copies*: B, G, MO, NY, US, USDA. – Botanical Survey of South Africa, Memoir 10.
Ed. 2: 1951 (p. [vii]: 21 Sep 1950; published by 29 Nov 1951, fide O. Leistner), p. [i-xi], [1]-923. *Copies*: B, BR, FI, MO, US.
Ref.: Phillips, E.P., S. Afr. J. Sci. 30: 220-221. 1933 (additions).
Rendle, A.B., J. Bot. 65: 232-233. Aug 1927.

7862. *A brief sketch of the development of botanical science in South Africa* and the contribution of South Africa to botany ... Presidential address to Section C of the South African Association for the Advancement of Science, delivered July 9th. 1930. Oct.
Publ.: Nov 1930 (in journal), p. [i], [1]-44. *Copy*: G. – To be cited from S. Afr. J. Sci. 27: 39-80. Nov 1930.

7863. South African agricultural series – volume 6. *An introduction to the study of the South African grasses* with notes on their structure, distribution, cultivation, etc. (with one hundred and twenty-one plates and text figures) ... South Africa (Central News Agency, limited) 1931. Oct. (*Intr. S. Afr. grass.*).
Publ.: Jan-Mar 1931 (Nat. Nov. Apr 1931; J. Bot. Jun 1931), p. [1]-224, *pl. 1-26, 26a, 27-121*. *Copies*: B, BR, G, MO, NY, US, USDA. – South African agricultural series, volume 6.
Ref.: Anon., J. Bot. 69: 173-174. 1931.

7864. *The weeds of South Africa* ... Pretoria (printed ...by the Government Printer) 1938. Oct. (*Weeds S. Afr.*).
Publ.: 1938, p. [1]-229. *Copies*: MO, US. – Bull. 195, Union of S. Africa, Dept. Agr., Div. Bot. ser. no. 41.

Phillips, Henry (1779-1840), British botanist and horticultural writer; banker at Worthing, later living in London and Brighton. (*H. Phillips*).

HERBARIUM and TYPES: Unknown.

BIBLIOGRAPHY and BIOGRAPHY: Barnhart 3: 81; BB p. 243; BM 4: 1568; Desmond p. 493 (b. 1779, d. 8 Mar 1840); DNB 45: 201-202 ("1775-1838"); Jackson p. 195, 214, 407, 592; Kew 4: 299; Langman p. 582; Plesch p. 361; PR 7125-7129 (ed. 1: 7995-7998); Rehder 5: 670; Sotheby 598; Tucker 1: 555.

BIOFILE: Allibone, S.A., Crit. dict. Engl. lit. 1583. 1870.
Chadwick, G.F., The works of Sir Joseph Paxton 93-94. 1961.
Coats, A.M., Garden Hist. Soc. Newsletter 14: 2-4. 1971.
Fussell, G.E., Gard. Chron. 1950(2): 178. (on W. Herbert, E. Kent and P.).
Johnson, G.W., Hist. Engl. gardening 304-305. 1829 (bibl.).

7865. *Pomarium britannicum*: an historical and botanical account of fruits, known in Great Britain, ... London (printed for the author, and sold by T. and J. Allman, ...) 1820. Oct. (*Pom. brit.*).
Ed. 1: 1820, p. [i*-x*], [i]-vii, [viii, err.], [1]-378, *pl. 1-3* (col.). *Copies*: G, MO, NY.
Ed. 2: 1821 (Flora 7 Apr 1822), p. [i*-viii*], [i]-vii, [viii], *pl. 1-3* (col.), [1]-378. *Copies*: NY, UC, USDA. – London (Henry Colburn and Co., ...) 1821. Oct.
Ed. 3: 1823, p. [i*-iii*], [i]-ix, *pl. 1-3*, [1]-372. *Copy*: Univ. Illinois (inf. D.P. Rogers). – Third edition, considerably enlarged and improved.London (printed for Henry Colburn and Co.) 1823.
Ed. 3 (*other issue*): 1827, p. [i*-iii*], [i]-ix, *pl. 1-3*, [1]-372. *Copy*: USDA. – London (Henry Colburn, ...) 1827. Oct.

7866. *History of cultivated vegetables*: comprising their botanical, medicinal, edible, and chemical qualities; their natural history; and relation to art, science, and commerce ... London (Henry Colburn and Co. ...) 1822, 2 vols. Oct. (*Hist. cult. veg.*).
Ed. 1: Feb 1822 (see Flora 5: 352. 1822).

1: [i*, iii*], [i]-vii, [1-4], [1]-383, (p. [ii]: 24 Dec 1821).
2: [i, iii], [1]-430.
Ed. *2*: 1822. *Copy*: MO.
1: [i*, iii*], [i]-vii, [1]-383.
2: [i, iii*], [1]-480.

7867. *Sylva florifera*: the shrubbery historically and botanically treated; with observations on the formation of ornamental plantations, and picturesque scenery ... London (Printed for Longman, Hurst, Rees, Orme and Brown, ...) 1823, 2 vols. Oct. (*Sylva fl.*).
Publ.: 1823. *Copies*: B, G, MO, NY, USDA.
1: [i]-vi, [vii], [1]-336.
2: [i-iii], [1]-333.

7868. *Flora historica*: or the three seasons of the British parterre historically and botanically treated; with observations on planting, to secure a regular succession of flowers, from the commencement of spring to the end of autumn. To which are added, the most approved methods of cultivating bulbous and other plants, as practised by the most celebrated florists of England, Holland, and France ... London (printed for E. Lloyd and Son, ... and Archibald Constable, and Co. Edinburgh) 1824, 2 vols. Oct (*Fl. hist.*).
Ed. *1*: 1824. *Copies*: MO, USDA.
1: [i]-li, [1]-354.
2: [v]-xii, [1]-464.
Ed. *2*: 1829. *Copies*: MO, NY. – "The second edition revised ..." London (E. Lloyd and Son, ...) 1829, 2 vols. Duod.
1: [i]-xlvii, [1]-333.
2: [i]-viii, [1]-423.

7869. *Floral emblems* ... London (printed for Saunders and Otley, ...) 1825. Oct. (*Fl. emblems*).
Publ.: 1825 (p. iii: 28 Mai 1825, engr. t.p. and frontisp. (1 leaf)), p. [i]-xvi, *16 pl.* (col. printed, finished by hand, auct.), [1]-350. *Copy*: NY. – The Plesch copy had 352 p. and *20 pl.*

Phillips, John (1800-1874), British geologist and palaeobotanist; curator of the York Museum 1824-1840; professor of geology, Trinity Coll., Dublin 1844-1853; keeper Ashmolean museum, Oxford 1854-1870. (*J. Phillips*).

COLLECTIONS AND TYPES: Oxford University Museum and BM (see J.M. Edmonds 1977 and Bridson 1980).

BIBLIOGRAPHY and BIOGRAPHY: Andrews p. 312; Barnhart 3: 81; BB p. 243 (b. 25 Dec 1800, d. 24 Apr 1874); BM 4: 1568, 8: 1008; CSP 4: 888-890, 8: 617-618, 11: 9, 17: 853; De Toni 2: xcv; DNB 45: 207-208; Jackson p. 182; Kew 4: 297-299; Quenstedt p. 336-337; Rehder 1: 393; TL-2/896.

BIOFILE: Anon., Geol. Mag. dec. 1. 7: 301-306. 1870 (biogr., portr.), dec.2.1(5): 240. 1874 (b. 25 Dec 1800, d. 24 Apr 1874) repr. Amer. J. Sci. ser. 3. 7: 608. 1874.
Bridson, G.D.R. et al., Nat. hist. mss. res. Brit. Isl 428 [index]. 1980.
Davis, J.W., Proc. Yorks. Geol. Polytechn. Soc. ser. 2. 8(1): 3-20. 1883 (obit.); Hist. Yorks. Geol. Polytechn. Soc. 1837-1887, p. 119-135. 1889 (Proc. ser. 2, 10, 1889).
Edmonds, J.M., J. Soc. Bibl. nat. Hist. 8(2): 169-175. 1977 (on location of P's type specimens; many taxa were based on colls. by others; deals with the legend of John Phillips's lost fossil collection).
Evans, J., Quart. J. Geol. Soc. London 31: xxxvii-xliii. 1875 (obit.).
Freeman, R.B., Brit. nat. hist. books 1980, no. 2986.
Huxley, L., Life letters J.D. Hooker 1: 514, 2: 208, 512, 514, 515, 520. 1918.
Nissen, C., Zool. Buchill. 317. 1969.
Poggendorff, J.C., Biogr.-lit. Handw.-Buch 2: 436-437. 1863, 3: 1035. 1898 (bibl.).
Sheppard, T., Proc. Yorks. Geol. Soc. 22: 153-187. *pl. 10*. 1933 (biogr., portr., bibl.).

Sherborn, C.D., Where is the ... collection 107. 1940 ("In 1908 coll. was worked over by Lindsall Richardson").
Woodward, H.B., Hist. Geol. Soc. London 333 [index], portr. facing p. 112. 1907.
Woodhead, T.W., Naturalist Mar 1923: 100 (portr.).
Zittel, K.A. von, Hist. geol. palaeontol. 436, 554 [index]. 1901.

COMPOSITE WORKS: (1) Contributed to W. Buckland, *Geol. mineral.* ed. 3, 1858, see TL-2/896.
(2) with John William Salter, *Palaeobotanical appendix*, Great Brit. Geol. Survey Mem. 2(1): 331-386, *pl. 1-30.* 1848.

EPONYMY: *Phillipsia* K.B. Presl (1838).

7870. *Illustrations of the geology of Yorkshire*; or a description of the strata and organic remains of the Yorkshire coast: accompanied by a geological map, sections, and plates of the fossil plants and animals ... York (printed for the author, ...) 1829. Qu. (*Ill. geol. Yorkshire*).
Ed. 1: 1-10 Aug 1829 (Geol. Soc. London, rd. 11 Aug 1829), p. [i]-xvi, [1]-192, *pl. [o], 1-[9], 1-16. Copy*: Univ. Illinois (inf. D.P. Rogers).
Ed. 2: 1835-1836. Copies: Edinburgh Univ. Libr., PH, USGS.
Part 1: 1835 front., map, p. [i]-xii, [1]-184, *pl. 1-2 [3-9], 1-14,* [1, note].
Part 2: 1836, p. [i]-xx, [1]-253, *pl. 1-25.*
Ed. 3: 1875, edited by R. Etheridge (n.v.).

7871. *Figures and descriptions of the Palaeozoic fossils of Cornwall, Devon, and West Somerset*; observed in the course of the ordnance geological survey of that district ... London (Longman, Brown, Green & Longmans) 1841. (*Fig. foss. Cornwall*).
Publ.: 1841, p. [i]-xii, [1]-231, *60 pl. Copies*: Edinburgh Univ. Libr., UC (inf. J. Edmondson and R. Schmid).

Phillips, Reginald William (1854-1926), British botanist; B.A. Cambr. 1884; Dr. sci. 1898; head Dept. Biol. Bangor 1884; professor of botany ib. until 1922. (*R.W. Phillips*).

HERBARIUM and TYPES: UCNW.

BIBLIOGRAPHY and BIOGRAPHY: Barnhart 3: 82; BB p. 243; BL 2: 314, 315, 702; CSP 17: 583; Desmond p. 493; De Toni 4: xlv; Kew 4: 300; Rehder 5: 670.

BIOFILE: A.G., Proc. Linn. Soc. London 140: 128-129. 1928 (obit., b. 15 Oct 1854, d. 2 Dec 1926).
Anon., Nature 119: 92-93. 1927 (obit.); J. Bot. 65: 58. 1927.
Bridson, G.D.R. et al., Nat. hist. mss. res. Brit. Isl. 5.33, 20.17. 1980.
Dickinson, Phycol. Bull. 1: 13. 1952 (alg. herb. Bangor).
Druce, G.C., Bot. Soc. Exch. Cl. Brit. Isl. Rep. 8(1): 101. 1927 (d. 2 Dec 1926).
J.L.W., North Western Natural. 2(3): 185-187. 1927 (obit., portr., brief bibl.).
Williams, J.L. & Rendle, A.B., J. Bot. 65: 80-83. 1927 (bibl.).

COMPOSITE WORKS: (1) *Algae*, in *Encycl. brit.* ed. 11. 1: 585-598. 1910 (also in ed. 10, 1902).
(2) *The flora of Breconshire*, in T.R. Phillips *The Breconshire border* (publ. by D.J. Morgan, Talgarth 1926), p. 84-88 and appendix 1, p. i-xxiv (n.v.); see J. Bot. 63: 316. Nov 1926 (see also J. Bot. 30: 354-355. Dec 1892).

Phillips, Robert Albert (1866-1945), Irish botanist in Cork. (*R.A. Phillips*).

HERBARIUM and TYPES: DBN.

BIBLIOGRAPHY and BIOGRAPHY: Desmond p. 493.

BIOFILE: A.W.S., Irish Natural. J. 8: 391-394. 1946 (portr.).

Colgan, N., Fl. County Dublin xxx. 1904.
Praeger, G.L., Proc. R. Irish Acad. 7: cxxxvi. 1901 (bibl.), Botanist in Ireland 78. 1934; Some Irish natural. 143-144. 1949.

Phillips, Walter Sargeant (1905-1975), American botanist; AB Oberlin 1929, Dr. phil. Chicago 1935; taught English at Taiku, Shansi, 1929-1931; at University of Chicago 1932, University of Miami 1935, University of Arizona 1940, from 1947-1972 as head of the Department of Botany and curator of the herbarium; student of the Polypodiaceae. (*W.S. Phillips*).

HERBARIUM and TYPES: ARIZ; further material at BUS, MO, NYS, UC, US.

BIBLIOGRAPHY and BIOGRAPHY: Barnhart 3: 82; Bossert p. 309 (b. 20 Oct 1905); IH 1 (ed. 1): 96, (ed. 2): 121, (ed. 3): 156, (ed. 4): 174, (ed. 5): 188, 2: (in press); PH 209.

BIOFILE: Cowan, R.S., Taxon 24: 553. 1975 (d. 7 Apr 1975).Ewan, J. et al., Short hist. bot. U.S. 121. 1969.Mason, C.T., Amer. Fern J. 65(3): 70. 1975 (obit., b. 20 Oct 1905, d. 7 Apr 1975, list pterid. papers).
Stieber, M.T. et al., Huntia 4(1): 85. 1981 (arch. mat. HU).

Phillips, William (1822-1905), British botanist, mycologist and antiquary at Shrewsbury. (*W. Phillips*).

HERBARIUM and TYPES: acquired by BM (with drawings of fungi), now at K. – Duplicates in B, BP, C, CUP, DBN, E, NYS, PAD, S, W. *Exsiccatae: Elvellacei britannici*, fasc. 1-4, nos. 1-201, 1874-1881, sets in above-mentioned herbaria. Phillips contributed specimens to M.C. Cooke, *Fungi britannici* and to Plowright, *Sphaeriacei britannici*, q.v.

BIBLIOGRAPHY and BIOGRAPHY: Barnhart 3: 82; BB p. 243; BL 2: 263, 702; BM 4: 1569; CSP 8: 618, 11: 10, 12: 575, 17: 853-854; Desmond p. 494 (further biogr. refs.); De Toni 1: xcix; DNB suppl. 2, 3: 115-116; GR p. 278, cat. p. 69; Hawksworth p. 185; Jackson p. 259; Kelly p. 175-176; Kew 4: 300-301; LS 20628-20682, 37921-37924 (bibl.); Morren ed. 10, p. 77; Rehder 5: 670; Stevenson p. 1254; Urban-Berl. p. 280.

BIOFILE: Anon., Ann. Scot. nat. Hist. 57: 57-58. 1906 (obit.); Bot. Centralbl. 19: 253. 1884 (specimens at BM); Bot. Not. 1906: 45 (b. 4 Mai 1822, d. 22 Oct 1905); J. Bot. 13: 63. 1875 (*Elvellacei*, fasc. of 50 species available), 43: 336. 1905 (d. 22 Oct 1905).
Bridson, G.D.R. et al., Nat. hist. mss. res. Brit. Isl. 229.399. 1980.
Britten, J., ed., J. Bot. 43: 361-362. 1905 (portr.), 44: 184. 1906 (herb. and drawings of fungi acquired by BM).
Freeman, R.B., Brit. nat. hist. books 2987-2990. 1980.
Hamlin, R.T., Index Grevillea 160-161. 1978.
Hawksworth, D.L. & M.R.D. Seaward, Lichenol. Brit. Isl. 133. 1977 (bibl.).
Hedge, I.C. & J.M. Lamond, Index coll. Edinb. herb. 118. 1970 (specimens at E).
Jackson, B.D., Bull. misc. Inf. Kew 1901: 52 (coll. K), Proc. Linn. Soc. 118: 44-45. 1906 (obit.).
Kent, D.H., Brit. herb. 72. 1957.
Murray, G., Hist. coll. BM(NH) 1: 173. 1904 (exsicc.).
Palmer, J.T., Nova Hedwigia 15: 152. 1968.
Plowright, C.B., Gard. Chron. 1905(2): 331-332 (portr., obit.).
Praeger, R.L., Proc. R. Irish Acad. 7: cxxxvii. 1901.
Seaver, F.J., N. Amer. cup fungi 363. 1942.
Winter, Bot. Centralbl. 8: 91. 1881 (*Elvell. brit.* fasc. 4, publ. 1881).

EPONYMY: *Phillipsiella* M.C. Cooke (1878). *Note: Phillipsia* M.J. Berkeley (1881) is also very likely named for this author; *Phillipsia* Rolfe ex J.G. Baker (1895) is named for Mrs. E. Lort-Phillips, a plant collector in Somaliland; *Phillipsiella* Lemmermann (1899) honors Frederick W. Phillips, English protozoologist of the 19th Century.

7872. *British lichens*: how to study them ... Birmingham (Herald Printing Office, ...) 1880. Oct.

Publ.: 1880 (Grevillea Mar 1881; Nat. Nov. Oct(1) 1881 as of 1881), p. [1]-17, *pl. 5*.
Copy: G. – Reprinted and to be cited from the Midland Naturalist 3: 125-128, 167-172, 196-199, 241-245. 1880.

7873. *A manual of the British Discomycetes* with descriptions of all the species of fungi hitherto found in Britain,included in the family and illustrations of the genera ... London (Kegan Paul, French & Co., ...) 1887. Oct. (*Man. Brit. Discomyc.*).
Ed. 1: late 1887, or Jan-Feb 1888 (p. vii: Aug 1887; Grevillea Mar 1888; Nat. Nov. Jun(1) 1888; Bot. Centralbl. 16-20 Jan 1888; Bot. Zeit. 24 Feb 1888; Bot. Gaz. Apr 1888), p. [i]-xii, [1]-462, *pl. 1-12* (uncol., auct.). *Copies*: B, BR, FH, G, MO, NY, Stevenson. – "The international scientific series" vol. 61.
Ed. 2: 1893, p. [i]-xii, [1]-462, *pl. 1-12* (id.). *Copies*: B, FH, NY, Stevenson, USDA. – Second edition" London (Kegan Paul, French Trübner & Co., Ltd.) 1893. Unchanged reprint.
Ref.: Peck, C.H., Bot. Gaz. 13: 101-102. 1888.
Wettstein, Bot. Centralbl. 34: 197. 14-18 Mai 1888.

Phinney, Arthur John (1850-1942), American botanist; MD Pulte Med. Coll. 1877; in the Indiana Geological Survey 1881-1886; with the U.S. Geological Survey from 1890; practicing physician at Indianapolis. (*Phinney*).

HERBARIUM and TYPES: Unknown.

BIBLIOGRAPHY and BIOGRAPHY: Barnhart 3: 82; BL 1: 177, 313; CSP 17: 854; Dawson p. 669; Rehder 5: xvii; Tucker 1: 556.

BIOFILE: Deam, C.C., Flora of Indiana 1118. 1910 (no location of specimens known). Phinney, A.J., Ann. Rep. Indiana Dept. Geol. nat. Hist. 12: 196-236. 1883 (USDA copy rd. 12 Oct 1883); reprints have original pagination.

Phipps, Constantine John, second baron Mulgrave (1744-1792), British captain in the Royal Navy; arctic explorer. (*Phipps*).

HERBARIUM and TYPES: Plants collected on the voyage towards the North Pole in the RACECOURSE and CARCASS, 1773, with Israel Lyons, described by D.C. Solander (in *Voy. North Pole* p. 200-204) at BM.

BIBLIOGRAPHY and BIOGRAPHY: AG 2(1): 156; Backer p. 398; Barnhart 3: 82; BM 4 1570; Desmond p. 494; GR p. 409; Henrey 2: 256-257; Herder p. 56; Krempelh. 1: 185 (err. Phillips); Lasègue p. 395, 502.

BIOFILE: Bridson, G.D.R. et al., Nat. hist. mss. res. Brit. Isl. 428 [index] 1980 (mss.).
Dawson, W.R., Bull. BM(NH), Hist. 3(3): 88. 1965 (letter to Banks).
Hitchcock, A.S. & A. Chase, Man. Grass. U.S. 988. 1951 (epon.).
Hooker, W.J., Fl. bor. amer. 2: 238. 1840 (*Phippsia*).
Lemmon, K., Golden Age Plant Hunters 18, 19. 1968.
Lysaght, J. Soc. Bibl. nat. Hist. 4: 208-209. 1964 (corr. with J. Banks).
Nissen, C., Zool. Buchill. 317. 1969.

EPONYMY: *Phippsia* (Trinius) R. Brown (1823), ING 3: 1317, fide Hitchcock & Chase (1951).

7874. *A voyage to the North Pole* undertaken by his majesty's command 1773 ... London (printed by W. Bowyer and J. Nichols, for J. Nourse ...) 1774. (*Voy. North Pole*).
Publ.: Feb 1774 (Gent. Mag.; however, a copy of the book was presented to the King only on 5 Aug 1774, fide letter to J. Banks 6 Aug 1774), p. [i]-viii, 1-253, [1, binder], *11 pl.*, 3 maps. *Copies*: Edinburgh Univ. Libr. (inf. J. Edmondson), MICH. – For other English editions see BM 4: 1570.
Abstract: 1808, *in* J. Pinkerton. A general collection of the best and most interesting voyages and travels ... 1: 538-594. 1808. *Copy*: UC (inf. G. Schmid).

French: 1775, p. [i]-xii, 1-259, *pl. 1-12. Copy*: FI. – *Voyage au pole boréal*, fait en 1773, ... à Paris (chez Saillant & Nyon, ... Pissot, ...) 1775.

Phoebus, Philipp (1804-1880), German botanist; Dr. med. chir. Berlin 1827; from 1828 practicing physician at Berlin; 1831 prosector at the Charité; 1832 lecturer at the University of Berlin; practicing in the Harz 1835-1843, professor of medicine and pharmacology at Giessen from 1843-1865; Dr. phil. h.c. Giessen 1849; Geh. Medizinalrath 1865. (*Phoebus*).

HERBARIUM and TYPES: Unknown. The herbarium (6500 "species") was offered for sale by Phoebus himself in 1868 and was again on the market in 1880 and 1903. Correspondence with D.F.L. von Schlechtendal at HAL; portr. at G.

BIBLIOGRAPHY and BIOGRAPHY: Ainsworth p. 73, 74, 314, 315, 346; BM 4: 1570-1571; Bossert p. 309; CSP 4: 896, 8: 620; Frank 3 (Anh.): 77 (d. 1 Jul 1880); Herder p. 117; Kelly p. 176; Kew 302; LS 20692-20694; NI alph. (d. 1 Jul 1880); PR 1091, 7130-7131 (ed. 1: 7999); Ratzeburg p. 416-417; Rehder 5: 670; Stevenson p. 1254; TL-2/708; Tucker 1: 556.

BIOFILE: Anon., Bonplandia 1: 52, 223. 1853, 3: 33. 1855, 6(2): 49-50. 1858 (Leopoldina cogn.: Morgagni), 10: 318. 1862; Bot. Centralbl. 1/2: 766. 1880 (d. 1 Jul 1880); Bot. Zeit. 1: 664. 1843(app. Giessen), 7: 182. 1849 (Dr. phil. h.c. Giessen), 38: 510, 592. 1880 (b. 27 Mar 1804, d. 1 Jul 1880; herb. for sale), 61(2) 176. 1903 (id.); Flora 26: 571. 1843 (to Giessen); Österr. bot. Z. 18: 105. 1868 (herb. for sale), 30: 302. 1880 (d. 1 Jul 1880).
Müller, R.H.W. & R. Zaunick, Friedr. Traug. Kützing 249, 291. 1960.
Poggendorff, J.C., Biogr.-lit. Handw.-Buch 2: 439-440. 1863 (b. 27 Mar 1804, bibl.).
Wittrock, V.B., Acta Horti Berg. 3(3): 148. 1905.
Zuchold, E.A., Jahresb. nat. Ver. Halle 5: 46. 1853.

7875. *Deutschlands kryptogamische Giftgewächse* in Abbildungen und Beschreibungen ... Berlin (Bei August Hirschwald) 1838. Qu. (*Deutschl. krypt. Giftgew.*).
Publ.: Jun-Dec 1838 (p. ix: Jun 1838; Flora 7 Jan 1839), p. [ii]-xii, [1]-114, *pl. 1-9* (col. copp. auct.). *Copies*: B, FH, FI, Stevenson. – Issued as vol. 2 of Brandes, J.F. von et al. *Abb. Beschr. Deutschl. Giftgew.*, see TL-2/708 q.v.

7876. *Die Delondre-Bouchardatschen China-Rinden* ... Giessen (J. Rickersche Buchhandlung) 1864. Oct. (*Delondre-Bouch. China-Rind.*).
Publ.: 1864 (Flora 1-16 Jun 1864),p. [i-iv], [1]-75, table. *Copy*: L.
Ref.: A.W., Bot. Zeit. 23: 39-41. 3 Feb 1865.

Pia, Julius [von] (1887-1943), Austrian palaeobotanist; Dr. phil. Wien 1911; in various functions at the Natural History Museum ib. from 1913 (Volontär), 1916 (assistant), 1928 (Kustos I. Klasse). (*Pia*).

COLLECTIONS: W. – Portrait at G.

BIBLIOGRAPHY and BIOGRAPHY: Andrews p. 312; Barnhart 3: 83; BM 8: 1009; Bossert p. 310; GR p. 457; Hirsch p. 231; Kew 4: 302; TL-2/2789.

BIOFILE: Andrews, H.N., The fossil hunters 318-319. 1980.
Anon., Hedwigia 73(1): (84). 1933 (corr. member Akad. Wien).
Kneucker, A., Allg. bot. Z. 33: 64. 1928 (appointed extraord. professor at Vienna University).
Trauth, F., Ann. naturh. Mus. Wien 55: 19-49. 1947 (obit., b. 28 Jul 1887, d. 2 Jan 1943; bibl. portr.).
Verdoorn, F., Chron. bot. 1: 32. 1935, 2: 33. 1936, 3: 66. 1937, 4: 92. 1938, 5: 118. 1939, 6: 301. 1941.
Wood, R.D., Monogr. Charac. 1: 833. 1965.

COMPOSITE WORKS: *Thallophyta, in* H. Hirmer, *Handb. Paläobot.* 1927, TL-2/2789.

EPONYMY: *Piaea* R. Florin (1929); *Pianella* R. Radoičić (1962); *Pseudopiaea* H. Tryoff (1966). *Note*: *Piaella* A. Fucini (1936) is likely also named for this author, but there is no etymology given.

7877. *Die Siphonieae verticillatae* vom Karbon bis zur Kreide ... Wien (Verlag der zool.-botan. Gesellschaft) 1920. Oct. (*Siphon. vertic.*).
Publ.: 1920, p. [1]-263, *pl. 1-8* (uncol., auct.). *Copy*: G. – Issued as Abh. Zool.-bot. Ges., Wien 11, 1920.

7878. *Pflanzen als Gesteinsbildner* ... Berlin (Verlag von Gebrüder Borntraeger ...) 1926. Oct. (*Pfl. Gesteinsbild.*).
Publ.: 1926 (ÖbZ Feb/Mai 1926; J. Bot. Mai 1927), p. [i]-viii, [1]-355. *Copies*: BR, G, MO, USDA.
Ref.: Anon., J. Bot. 65: 90-91. 1927.

Picard, Casimir (1806-1841), French botanist and physician; studied medicine at Amiens; practicing physician at Abbeville 1828-1841; "administrateur" of the museum of Abbeville; founder of the Société Linnéenne du Nord de la France, 1838. (*Picard*).

HERBARIUM and TYPES: Unknown.

BIBLIOGRAPHY and BIOGRAPHY: Barnhart 3: 83; BM 4: 1571; CSP 4: 897-898; Jackson p. 144; Kew 4: 303; PR 7132-7133 (ed. 1: 8000-8001).

BIOFILE: Anon., Actes Soc. Linn. Bordeaux 12(38): 32-33. 1841 (d. 31 Mar 1831, bibl.). Morgand, T., Mém. Soc. roy. Émul. Abbeville 1841-1843: 449-456 (obit., b. 16 Dec 1806, d. 13 Mar 1841).

EPONYMY: *Picardaea* Urban (1903) honors Louis Picarda, French collector of plants in Haiti; *Picardenia* Steudel (1841) is an *orth. var.* of *Picradenia* W.J. Hooker (1833).

7879. *Observations botaniques sur le genre Sonchus*, ... Boulogne (Imprimerie de Le Roy-Mabille ...) s.d. [1835?]. Oct.
Publ.: 1835 (fide Jackson), p. [1]-16, *1 pl.* (uncol. lith.). *Copy*: G.

7880. Histoire naturelle. *Étude sur les Géraniées* qui croissent spontanément dans les départements de la Somme et du Pas-de-Calais, ... Boulogne (Imprimerie de Le Roy-Mabille) 1838. Oct. (in fours).
Publ.: 1838 (t.p.; Flora 7 Jan 1839), p. [1]-46. *Copy*: G. – Reprinted and to be cited from Mém. Soc. Agric. Boulogne-sur-Mer 1: 95-138. 1838. (copy reprint with orig. pag.: NY).

Picard, François (1879-1939), French entomologist and mycologist; professor of zoology at the École nationale d'Agriculture, Montpellier 1909-1921; at the Paris Laboratoire d'évolution des êtres organisés 1921-1939. (*F. Picard*).

HERBARIUM and TYPES: Unknown.

BIBLIOGRAPHY and BIOGRAPHY: Barnhart 3: 83; Hirsch p. 231; Kelly p. 176; Kew 4: 303; LS 37928, suppl. 21658-21663.

BIOFILE: Berland, F., Ann. Soc. entom. France 108: 173-181. 1940 (obit., entom., bibl.).

7881. *Contribution à l'étude des Laboulbéniacées d'Europe* et du nord de l'Afrique (avec planches) ... Paris (au siège de la Société [mycologique de France] ...) 1913. Oct.
Publ.: 20 Dec 1913 (in journal; USDA rd. 8 Apr 1914), cover, p. [1]-69, *pl. 29-32* (uncol.). *Copy*: USDA. – Reprinted and to be cited from Bull. Soc. mycol. France 29(4): 503-571. 1913.

Picbauer, Richard (1856-1955), Moravian mycologist and floristic botanist; teacher at Chvalkovice nr. Olmütz (1908), high school teacher at Olmütz until 1914; in military service 1914-1919; high school teacher at Brünn (Brno) 1919; Dr. phil. 1925 Brünn; from 1923 connected with the agricultural research station Brünn (Brno); (*Picb.*).

HERBARIUM and TYPES: BRNU, BRNM (phytopath. herb.); further material at BP, CUP, E, M (lich.), PRM, W. − Contributed to F. Petrak, *Flora Bohemiae et Moraviae-exsiccata*.

BIBLIOGRAPHY and BIOGRAPHY: Barnhart 3: 83; Bossert p. 310 (b. 2 Feb 1886, d. 30 Aug 1955); Futák-Domin p. 469; GR p. 671; Hirsch p. 231; IH 1 (ed. 6): 362, (ed. 7): 339, 2: (in press); Klášterský p. 142-143 (b. 2 Feb 1886, d. 30 Aug 1955); LS 20697a, 37931-37934, suppl. 21664-21679; Stevenson p. 1254.

BIOFILE: Anon., Preslia 9: 94. 1930.
Bandyš, Česká Mykol. 11(1): 56-60. 1957 (obit., portr., bibl., b. 2 Feb 1886, d. 30 Aug 1955).
Hedge, I.C. & J.M. Lamond, Index coll. Edinb. herb. 118. 1970.
Hertel, H., Mitt. Bot. München 16: 417. 1980 (lich. at M, Moravian, 1905-1919).
Petrak, F., Sydowia 11: 18-26. 1957 (obit., b. 2 Feb 1886, d. 30 Aug 1955, portr., bibl.).
Verdoorn, F., ed., Chron. bot. 1: 110. 1935, 2: 107. 1936.

Piccioli, Antonio (1741-1842), Italian botanist; curator of the Firenze botanical garden. (*A. Piccioli*).

HERBARIUM and TYPES: Unknown. − Letters at G.

BIBLIOGRAPHY and BIOGRAPHY: BM 4: 1571-1572; GF p. 70; Jackson p. 433; Kew 4: 303; NI 1527n; Plesch p. 361; PR 7134 (ed. 1: 8002-8005); Rehder 5: 671; Saccardo 1: 127 (b. 22 Apr 1794, d. 10 Jun 1842), 2: 84, cron. xxix; Sotheby 599; Tucker 1: 556.

7882. *Pomona toscana* che contiene una breve descrizione di tutti i frutti che si coltivano nel suolo toscano per servire alla collezione in gesso dei medesimi ... Firenze 1820. Qu. (*Pom. tosc.*).
Publ.: 1820, p. [i], [1-29]. *Copy*: G.

7883. *Catalogus plantarum horti botanici* musei imperialis et regalis *florentini* ... Florentiae [Firenze] (ex typographia Aloisii Pezzati) 1829. Oct. (*Cat. pl. hort. florent.*).
Publ.: 1829 (p. 4: Feb 1829), p. [1]-54. *Copy*: FI.

7884. *L'antotrofia* ossia la coltivazione de'fiori ... Firenze (per V. Batelli e figli) 1834. Oct. (*Antotrofia*).
Publ.: 1834, in two parts. *Copy*: FI-BN. − *1*: [1]-424, *pl. 1-36*, *2*: [425]-798, [1, err.], *pl. 37-72*.

7885. *Catalogo delle piante del giardino botanico* annessa all' T. e. R. Museo di fisica e storia naturale *di Firenze* l'anno 1841. Firenze (tipografia Galletti) 1841. Oct. (*Cat. giard. bot. Firenze*).
Publ.: 1841, p. [1]-40. *Copy*: FI.

Piccioli, Elvira (fl. 1932), Italian bryologist and high school teacher at Vittoria Veneto. (*E. Piccioli*).

HERBARIUM and TYPES: Unknown.

BIBLIOGRAPHY and BIOGRAPHY: Kew p. 303.

7886. *Les espèces européennes du genre Orthotrichum* thèse présenté à la Faculté des Sciences de l'Université de Neuchâtel pour l'obtention du grade de docteur ... Florence (Imprimerie "Le Cénacle") 1932. Oct. (*Esp. eur. Orthotrichum*).

Publ.: 1932 (p. [3]: 10 Oct 1931), p. [1]-128. *Copies*: G, PH (rd. 17 Apr 1933). – Issued as Trav. Inst. bot. Univ. Neuchâtel ser. 2, 1.

Piccioli, Giuseppe (*fl.* 1783-1818), Italian botanist; curator of a botanical garden at Firenze. (*G. Piccioli*).

HERBARIUM and TYPES: Unknown. – Letters at G.

BIBLIOGRAPHY and BIOGRAPHY: Jackson p. 434; PR 7135 (ed. 1: 8006); Rehder 5: 671; Saccardo 1: 127.

7887. *Hortus panciaticus* ossia delle piante esotiche e 'dei fiori esistenti nel giardino della villa detta la loggia presso a Firenze di proprieta' dell' illustriss. sig. Marchese Niccolò Panciatichi direttore della reale Accademia dei Georgofili descritto da Giuseppe Picciuoli (sic) custode del predetto giardino e socio aggiunto della suddetta real Accademia dedicato al merito distinto dello stesso signor Marchese. In Firenze (nella stamperia della Rovere da S. Mar. Magg. ...) 1783. Qu. (*Hort. panciat.*).
Publ.: 1783, 32 p., *1 col. pl. Copy*: HH (not analysed).

7888. *Catalogus plantarum horti botanici* musei imperialis et regalis *florentini*. Florentiae [Firenze] (ex Typographia Guilelmi [sic] Piatti) 1818. Duod. (in sixes). (*Cat. pl. hort. florent.*).
Publ.: 1818 (p. [iv]: id. Feb 1818), p. [i-iv], [1]-50. *Copies*: FI, G, NY, USDA.
Auctuarium: 1824, p. [1]-8. *Copy*: FI. – *Auctuarium ad catalogum plantarum horti botanici* Musei imperialis et regalis *Florentini* anno 1824.

Piccioli, Ludovico (1867-?), Italian botanist and forester; forestry inspector in the province of Siena; habil. Univ. Siena 1905; director of the experimental arboreta of Vallombrosa 1913-1922; later professor of silviculture at the Facoltà Agraria e Forestale della R. Università, Firenze. (*L. Piccioli*).

HERBARIUM and TYPES: Unknown.

BIBLIOGRAPHY and BIOGRAPHY: Barnhart 3: 83; BL 2: 336, 401, 702; BM 4: 1572; CSP 17: 869; Hirsch p. 231; Kew 4: 303; Langman p. 582; LS 21681; Rehder 5: 671; Saccardo 1: 127, 2: 84; Tucker 1: 556.
1877
BIOFILE: Anon., Nat. Nov. 27: 539. 1905 (habil. Siena).
Gager, C.S., Brooklyn Bot. Gard. Rec. 27(3): 272. 1938.
Tassi, F., La botanica nel Senese 47. 1905 (Bull. Lab. Orto bot. Siena), (b. 23 Jun 1867).
Verdoorn, F., ed., Chron. bot. 3, 188. 1937 (70th birthday), 5: 108. 1939 (retirement).

7889. *Guida alle escursioni botaniche* nei dintorni di Vallombrosa con chiavi analitiche per determinare i nomi delle piante che vi crescono ... Firenze (Tipografia dell'arte della Stampa ...) 1888. Oct. (*Guida escurs. bot.*).
Publ.: 1888 (p. 6: 27 Mar 1888; Nat. Nov. Jan(2) 1889; J. Bot. Apr 1889; Bot. Zeit. 26 Apr 1889), p. [1]-297. *Copies*: B, G.
Additions: Solla, R.F., Bull. Soc. bot. Ital. 1893: 52-60, 124-128, 197-207, 276-285, 381-393. 1893.

7890. *Le piante legnose italiane*. In Firenze (Pei tipi di Salvadore Landi ...) 1890-1903, 5 fasc. Oct. (*Piante legn. ital.*).
1: 1890 (cover date; Bot. Zeit. 29 Mai 1891; J. Bot. Mai 1891), p. [i]-vii, [1]-129, *1 pl.*
2: 1891, p. [131]-310.
3: 1894 (Bot. Zeit. 16 Jul 1895; Nat. Nov. Mar(2) 1895), p. 311-434.
4: 1896 (Bot. Zeit. 1 Dec 1896; Nat. Nov. Sep(2) 1896), p. 435-690.
5: Jun 1903 (G copy rd. via Friedländer Jun 1903; Nat. Nov. Jul(1) 1903), p. 691-991.
Copies: G (orig. cover), MO (fasc. 1).

7891. *Raccolta di vocaboli botanici e forestali italiani e tedeschi* ... Firenze (pei tipi di S. Landi ...) 1890. Oct. (*Racc. vocab. bot.*).
Publ.: 1890, p. [1]-100. *Copy*: FI.

7892. *La coltura dei Salici* con 46 figure. Firenze (Tipografia di Salvadore Landi ...) 1896. Oct. (*Colt. Salici*).
Publ.: 1896, p. [i]-viii, [1]-249. *Copy*: FI.

7893. *Monografia del Castagno* suoi caratteri varietà, coltivazione, prodotti e nemici ... Firenze (Tipografia di Salvatore Landi ...) 1902. Oct. (*Monogr. Castagno*).
Publ.: 1902 (p. ix: 31 Dec 1901; Bot. Centralbl. 24 Feb 1903), p. [i]-xi, [1]-178. *Copy*: G.

7894. *Gli arboreti sperimentali di Vallombrosa*. Firenze (Tipografia di M. Ricci ...) 1917. Oct. (*Arbor. Vallombrosa*).
Publ.: 1917, p. [1]-95. *Copy* FI.

7895. *Monografia del Carpino* ... Firenze (Tipografia di Mariano Ricci ...) 1924. Oct. (*Monogr. Carpino*).
Co-author: Floriano Speranzini.
Publ.: 1924, p. [1]-51. *Copy*: FI. – Reprinted from Ann. r. Ist. Sup. forest. naz. 9: 231-280. 1924.

Picco, Vittorio (latinized: *Picus*, gen. *Pici*) (*fl.* 1788), Italian botanist and physician at Torino. (*Picco*).

HERBARIUM and TYPES: Unknown.

BIBLIOGRAPHY and BIOGRAPHY: Ainsworth p. 32, 312; Barnhart 3: 83; BM 4: 1574 (Picus); Dryander 3: 92, 444, 534, 553 (Picus); Jackson p. 162; NAF ser. 2. 1: 32; PR 2047, 7136 (ed. 1: 8007); Saccardo 1: 127.

BIOFILE: Lütjeharms, W.J., Gesch. Mykol. 198, 247. 1936.

7896. Victorii Pici ... *Melethemata inauguralia*. Augustae Taurinorum [Torino] (excudebat Ioan. Mich. Briolus ...) s.d. [1788]. Oct. (*Meleth. bot.*).
Publ.: 1788, p. [i-iii], [1]-283, [284, err.], *pl. 1-2* (uncol. copp.). *Copies*: NY, Stevenson. – Contains: *De physica de fungorum generatione* (p. 1-103), *Ex materia medica de fungis* (p. 105-167), ..., *Ex theoria de symptomatibus quae fungorum venenatorum esum consequi solent* (p. 237-264), ... – The NY and Stevenson copies are bound with Josephi Antonii Dardana ... *In agaricum campestrem veneno in patria infamem* Acta ad amicissimum et amantissimum Victoricum Picum m.d. Augustae Taurinorum (id.) 1788, p. [1]-32, p. 32 dated 13 Nov 1787. (PR 2047; Saccardo 1: 60, 2: 39; 1743-1796).

Piccone, Antonio (1844-1901), Italian botanist; high school teacher at the Liceo Cristoforo Colombo in Genova. (*Picc.*).

HERBARIUM and TYPES: VER; other material at B, BP, CN, E, FI, GE, L, MW, PAD, PI, TO, VEN, W, WRSL. – Contributed to the *Erbario crittogamico italiano*, ser. 3. – Letters and portrait at G.

BIBLIOGRAPHY and BIOGRAPHY: Barnhart 3: 83; BJI 1: 45; BM 4: 1572; Bossert p. 310; Cesati, Saggio p. 59-60, 311-313; CSP 11: 16, 17: 869-870; De Toni 1: xcix-c, 2: xcv, 4: xlv; GR p. 536-537; IH 2: (in press); Jackson p. 217; Kew 4: 303; LS 20698-20702; Morren ed. 10, p. 84; MW p. 386; Nordstedt p. 26; PR 7137; Rehder 5: 671; Saccardo 1: 127; SBC p. 129 (Picc.); Stevenson p. 1254; Tucker 1: 556; Urban-Berl. p. 383.

BIOFILE: Anon., Bot. Centralbl. 86: 432. 1901 (d.), 98, 48. 1905 (herb. acquired by VER); Bot. Jahrb. 30 (Beibl. 68): 53. 1901, 31: (Beibl. 69): 39. 1901; Bot. Not. 1902: 94 (d. 21 Mai 1901); Nat. Nov. 23: 344. 1901 (d.), 27: 143. 1905 (herb. acquired by VER); Nuova Notarisia 12: 154. 1901 (b. 11 Sep 1844, d. 21 Mai 1901); Österr. bot. Z. 51: 319. 1901 (d. 21 Mai 1901).

Burnat, É., Bull. Soc. bot. France 30: cxxvi-cxxvii. 1883 (list publ. on Alpes mar.).
Cavillier, F., Boissiera 5: 72-73. 1941 (b. 11 Sep 1844, d. 21 Mai 1901).
De Toni, G.B., Della vita e delle opere di Antonio Piccone. Roma, 1902, i, 19 p., reprinted from Ann. R. Ist. bot. Roma 9(3): 167-185. 1902 (bibl., obit.).
Hedge, I.C. & J.M. Lamond, Index coll. Edinb. herb. 118. 1970.
Kneucker, A., Allg. bot. Z. 7: 144. 1901.
Koster, J. Th., Taxon 18: 556. 1969.
Lindemann, E. v., Bull. Soc. imp. Natural. Moscou 61(1): 53. 1886 (material at MW).
Merrill, E.D., Enum. Philipp. pl. 4: 216.1926; B.P. Bishop Mus. Bull. 144: 150. 1937; Contr. U.S. natl. Herb. 30(1): 241-242. 1947.
Papenfuss, G.F., Israel J. Bot. 17: 108. 1968 (bibl. Red. Sea. alg.).
Penzig, O., Cenni sulla vita e sulle opere di Antonio Piccone. Genova 1901, 11 p., (repr. from Atti Soc. Ligust. Sci. nat. Geogr. vol. 12), also Malpighia 15(2/3): 92-100. 1901. (obit., bibl.).
Sayre, G., Bryologist 80: 515. 1977 (musci at VER).
Sommier, S., Bull. Soc. ital. 1901: 200-201 (obit.).

COMPOSITE WORKS: *Alghe dell'isola del Giglio*, by A. Piccone & G.B. De Toni, p. [1]-10 (copy of reprint: USDA), reprinted from S. Sommier, L'isola del Giglio e la sua flora, Torino (C. Clausen), 1900.

EPONYMY: *Picconiella* De Toni f. (1936) very likely honors this author. *Picconia* Alph. de Candolle (1844) was named for J.B. Picconi who wrote about olives.

7897. *Elenco dei muschi di Liguria* ... Genova (Co' tipi del R.I. de' Sordo-Muti) 1863.
Orig.: 1863, p. [1]-50. *Copy*: H. – Reprinted and to be cited from Comment. Soc. Critt. ital., Genova 1(4): 240-287. 1863.
Suppl.: Jul 1876, Nuovo Giorn. bot. ital. 8: 368-377. 1876.
Ref.: Milde, F., Hedwigia 2(14): 122. 1863.

7898. Reale accademia dei Lincei ... *Catalogo delle alghe raccolte durante le crociere del cutter Violante e specialmente in alcune piccole isole mediterranee* ... Roma (coi tipi del Salviucci) 1879. Qu. (*Cat. alg. Violante*).
Publ.: 1879 (mss. submitted 6 Apr 1879; Nat. Nov. Feb(2) 1880; Bot. Zeit. 19 Mar 1880; Bot. Jahrb. 30 Apr 1880), p. [1]-19. *Copies*: FI, NY. – Reale Accad. dei Lincei, ser. 3, Mem. Cl. sci. fis. vol. 4.

7899. *Resultati algologici delle crociere del Violante* comandato dal capitano-armatore Enrico d'Albertis per A. Piccone. Genova (Tipografia del r. Istituto Sordo-Muti) 1883. Oct. (*Result. algol. Violante*).
Publ.: 24 Oct 1883 (see colophon; Bot. Zeit. 28 Dec 1883; Nat. Nov. Nov(2) 1883), p. [1]-39, [40, coloph.]. *Copies*: FH, MO, NY. – Reprinted from Ann. Mus. Civ. Storia nat. Genova 20: [106]-142. 24 Oct 1883 (date of part; volume publ. 1884).

7900. *Prime linee per una geografica algologica marina* ... Genova (Tipografia di Gaetano Schenone ...) 1883. Oct. (*Prim. linee geogr. algol. mar.*).
Publ.: 21 Oct 1883(see p. 55; Bot. Zeit. 28 Dec 1883; Bot. Centralbl. 26-30 Nov 1883; Nat. Nov. Nov(2) 1883), p. [1]-55. *Copies*: BR, MO. – Reprinted from Cronaca del r. Liceo Cristoforo Colombo 1882/1883 (journal n.v.).
Ref.: Engler, A., Bot. Jahrb. 5: 34-35. 4 Mar 1884.
Penzig, O., Bot. Centralbl. 16: 289-293. 1883 (rev.).

7901. *Crociera del Corsaro alle Isole Madera e Canarie del capitano Enrico d'Albertis. Alghe* per Antonio Piccone. Genova (Tipografia del r. Istituto Sordo-Muti) 1884. Oct. (*Crociera Corsaro, alg.*).
Publ.: Jul-Aug 1884 (p. 6: 8 Jun 1884; Bot. Zeit. 31 Oct 1884; Nat. Nov. Sep(1) 1884), p. [3]-60, *1 pl.* (col. lith.,R. Gestro). *Copies*: FH, FI, MO, NY. – Penzig (1901) cites this also from Atti Mus. civ. Genova 1883.
Ref.: Penzig, O., Bot. Centralbl. 21(7): 193-194. 9-13 Feb 1885.

7902. *Spigolature per la ficologia ligustica* [Estratto dal Nuovo Giornale botanico italiano ...) 1885. Oct.
Publ.: 1 Jul 1885 (in journal; Bot. Zeit. 27 Nov 1885; Nat. Nov. Aug(2) 1885), p. [189]-200. – Reprinted with special imprint but with orginal pagination and hence to be cited from Nuov. Giorn. bot. ital. 17(3): [189]-200. 1 Jul 1885.

7903. *Alghe del viaggio di circumnavigazione della Vettor Pisani* per Antonio Piccone. Genova (Tipografia del R. Istituto Sordo-Muti) 1886. Oct. (*Alg. viagg. Vettor Pisani*).
Publ.: Mar 1886 (in journal; Bot. Zeit. 24 Jun 1887; Nat. Nov. Jul(1) 1886), p. [1]-97, *pl. 1-2* (uncol. liths.). *Copies*: BR, FH, FI, NY. – Reprinted from Giorn. Soc. Letture Convers. sci. Genova, Mar 1886 (fide De Toni 1: xcix).
Preliminary notice: Nuov. Giorn. bot. ital. 17: 185-188. 1885.

7904. *Saggio di studi intorno alla distribuzione geografica delle alghe d'acqua dolce e terrestri* ... Genova (Tipografia di Angelo Cimanago ...) 1886. Oct.
Publ.: 1886 (Bot. Centralbl. 6-10 Dec 1886; Nat. Nov. Jul(1) 1886), p. [1]-49. *Copy*: W. – Reprinted from Giorn. Soc. Letture Convers. Sci. 1886(5) (n.v.).

7905. *Nuove alghe del viaggio di circumnavigazione della "Vettor Pisani"* Memoria di Antonio Piccone. Roma (Tipografia della r. Accademia dei Lincei) 1889. Qu.
Publ.: 1889 (mss. submitted 3 Feb 1889; Nat. Nov. Oct(2) 1889), p. [1]-57 (alt.: [i-ii], 9-63). *Copy*: NY. – Reprinted and to be cited from R. Accad. Lincei, ser. 4, Mem. Cl. Sci. fis. 6: 9-63. 1889.
Notizie preliminari: Nuove Giorn. bot. ital. 17: 185-188. 1 Jul 1885.

7906. *Elenco delle alghe crociera del Corsaro alle Baleari* per A. Piccone. Genova (Tipografia del r. Istituto Sordo-Muti) 1889. Oct. (*Elenc. alg. Corsaro*).
Publ.: Feb-Mar 1889, (p. 6: 6 Jan 1889; Nat. Nov. Apr(2) 1889; Bot. Zeit. 26 Jul 1889; Bot. Centralbl. 4 Sep 1889), p. [1]-22. *Copies*: BR, NY.

7907. *Manipolo di alghe del Mar Rossa* ... Roma (Tipografia della r. Accademia dei Lincei) 1889. Qu.
Publ.: Mar-Sep 1889 (mss. submitted 3 Feb 1889; Nat. Nov. Oct(2) 1889), p. [1]-68 (alt. [i-ii], [63]-78). *Copy*: NY. – Reprinted and to be cited from R. Accad. Lincei, Mem. Cl. Sci. fis. ser. 4. 6: 63-78.

7908. *Alcune specie di alghe del Mar di Sargasso* ... Roma (Tipografia della r. Accademia dei Lincei) 1889. Qu.
Publ.: Mar-Sep 1889 (mss. submitted 3 Feb 1889; Nat. Nov. Oct(2) 1889), p. [1]-11 (alt: [i-ii], [78]-86). *Copies*: NY. – Reprinted and to be cited from R. Accad. Lincei, Mem. Cl. Sci. fis. ser. 4. 6: [78]-86.

7909. *Nuove contribuzioni alla flora marina del Mar Rosso* ... [Genova (Tipografia Ciminago) 1901]. Oct.
Publ.: Jan-Jul 1901 (p. [20]; p. 2: 3 Jun 1890; Bot. Centralbl. 31 Jul 1901), p. [1]-18, [20, date]. *Copies*: BR, USDA. Reprinted and to be cited from Atti Soc. lig. Sci. nat. 11: 251-268. 1900 (publ. 1901), also published in Malpighia 14: 493-510. 1901.

Pichi-Sermolli, Rodolfo Emilio Giuseppe (1912-x), Italian botanist (pteridologist); Laureato Sci. nat. Firenze 1935; at the Herbarium Universitati Florentinae 1935-1950; from 1957-1959 at Sassari; later at Genova and Perugia. (*Pic.-Ser.*).

HERBARIUM and TYPES: FI, FT and GE but fern collections also in private herbarium; further material at A, BI, BR, MO, NY. – Letters at G and in I.A.P.T. archives; portrait at G.

BIBLIOGRAPHY and BIOGRAPHY: BFM no. 1405, 2319; BL 2: 338, 401, 703; Bossert p. 310 (b. 24 Feb 1912); GR p. 537; IF suppl. 4: 335-336; IH (ed. 1): 34, (ed. 2): 51, 111, (ed. 3): 62, 143, (ed. 5): 69, (ed. 7): 339; Kew 4: 303-305; Langman p. 582; Lenley p. 327; MW p. 280; NI 3: 26, 70; PFC 3(2): xvi; Roon p. 87; TL-1/1128.

BIOFILE: Lewalle, J., Rev. Univ. off. Bujumbara 1: 34-36. 1967 (n.v.).
Holman, J.H. & C. Jermy, Int. direct. pteridol. 1937, no. 258.
Pichi-Sermolli, R.E.G., Curriculum vitae, Firenze 18 Novembre 1955, 91 p. (bibl., itin.); Webbia 19: 887-888. 1965 (report on the *Adumbratio florae Aethiopicae*).
Stafleu, F.A., Regn. veg. 5: 2. 1954 (portr. in group).
Steenis, C.G.G.J. van, Fl. males. Bull. 3: 56. 1948 (at K), 14: 624. 1959.
Verdoorn, F., ed., Chron. bot. 4: 61. 1938, 6: 301. 1941.

COMPOSITE WORKS: (1) with A. Messeri et al., *Bulletino bibliografico della Botanica italiana* 7(1932)-9(3)(1938).
(2) *Index filicum, supplementum quartum*, Jan 1965, see TL-2/1128.

7910. *Ricerche botaniche nella regione del Lago Tana* e nel Semièn ... Roma (Reale Accademia d'Italia) 1838-xvi. Oct. (*Ric. bot. Lago Tana*).
Publ.: 1938, p. [1]-31. *Copy*: FI. – Reprinted from *Missione di studio al Lago Tana*, vol. 1.

Pichon, Thomas (*fl.* 1811), French botanist. (*T. Pichon*).

HERBARIUM and TYPES: Unknown.

BIBLIOGRAPHY and BIOGRAPHY: BM 4: 1572; Jackson p. 419; PR 7138 (ed. 1: 8009); Rehder 1: 57.

EPONYMY: The etymology of *Pichonia* Pierre (1890) was not given, but likely honors Pichon.

7911. *Catalogue* raisonné ou tableau analytique et descriptif *des plantes cultivées à l'école de botanique* du Muséum imperial maritime *du Port de Brest*, classées suivant le systême sexuel de Linné, corrigé par Joh. Frid. Gmelin, avec concordance des familles naturelles de Jussieu; précédé d'un abrégé du systême du premier de ces auteurs et de notes explicatives des différentes parties des plantes, et des abbréviations employées dans l'ouvrage; ... à Brest (de l'imprimerie impériale de la Marine) 1811. Oct. (in fours). (*Cat. pl. école bot. Brest*).
Co-author: Broca, "pharmacien entretenu de la Marine ...".
Publ.: 1811 (p. xiv: 21 Apr 1809), p. [i*-vi*], [i]-xiv, [1]-18, [1]-571, [572-573]. *Copies*: G, NY.

Pickard, Joseph Fry (1876-1943), British draper, tea merchant and botanist in Yorkshire (Leeds). (*Pickard*).

HERBARIUM and TYPES: BM and LES; further material at LIV.

BIBLIOGRAPHY and BIOGRAPHY: BL 2: 226, 278, 702; Desmond p. 494; IH 2: (in press); Kew 4: 305.

BIOFILE: Dallman, A.A., North Western Natural. 20(3/4): 285-287. 1945 (obit., portr., b. 3 Apr 1876, d. 18 Feb 1943).
Kent, D.H., Brit. herb. 73. 1957.
Sledge, W.A., Bot. Soc. Exch. Cl. Brit. Isl. Rep. 12(6): 649-650. 1946 (obit.).

COMPOSITE WORKS: Assisted A. Wilson with his *Fl. Westmorland* 1938 q.v.

Pickel, Bento José (1890-1963), German-born Brazilian botanist, explorer and plant collector; sometime professor at the agricultural college at Tapera (Pernambuco) and at São Paulo. (*Pickel*).

HERBARIUM and TYPES: IPA; other material at B, BH, BM, CAS, CM, DPU, DS, EAN, F, GH, IAC, IPA, LAM, LCU, MICH, MO, MTJB, NY, PH, POM, R, S, SP, U, UC, URM, US.

BIBLIOGRAPHY and BIOGRAPHY: Barnhart 3: 83; BL 1: 242, 313; IH 1(ed. 6): 362, (ed. 7): 339, 2: (in press); Kew 4: 305; PH 565; SIA 227.

BIOFILE: Anon., Österr. bot. Z. 80: 364. 1931 (Brazilian plants for sale); Studia entom. 6: 585. 1963 (obit.).
Pickel, B.J., Bol. Mus. nac. Rio de Janeiro 13: 63-132. 1938 (on his herb. coll. at Tapera, Pernambuco).
Stafleu, F.A., Taxon 12: 271. 1963 (b. 28 Jul 1906, erroneous [b. 1890] Markelsheim, Germany; d. 4 Apr 1963).
Verdoorn, F., ed., Chron. bot. 2: 88. 1936, 6: 172, 464. 1941, 7: 456. 1943.

Pickering, Charles (1805-1878), American botanist, zoologist and anthropologist; M.D. Harvard 1826; from 1827-1838 at Philadelphia; from 1838-1842 with Wilkes and Brackenridge on the U.S. Exploring Expedition; in Egypt, India, Arabia and E. Africa 1843-1844; from 1844 residing in Boston. (*Pickering*).

HERBARIUM and TYPES: General collection: PH; material from the U.S. Expl. Exp. at US; further material at DBN, DWC, K and OXF. – Some letters at G and HU.

BIBLIOGRAPHY and BIOGRAPHY: Barnhart 3: 84 (b. 10 Nov 1805; d. 17 Mar 1878); BM 4: 1572; Bossert p. 310; Clokie p. 224; CSP 4: 899, 8: 622; DAB (by F. Estelle Wells); Ewan ed. 1: 168, 283, 337, ed. 2: 174; Jackson p. 4, 222; Kew 4: 305-306; Langman p. 582; Lenley p. 466; ME 1: 219; MW p. 386, 638; NI 3: 12; Pennell p. 616; PH p. 513; PR 7139; Rehder 5: 672; SIA 7058; SK 1: 406-407; TL-2/see T. Nuttall; Tucker 1: 556.

BIOFILE: Allibone, S.A., Crit. dict. Engl. lit. 1589. 1870 (b. 10 Nov 1805, bibl.).
Anon., Appleton's Cycl. Amer. biogr. 5: 4. 1888 (biogr.).
Barnhart, J.H., Corr. Schweinitz and Torrey, Mem. Torrey bot. Club 16: 298. 1921; Bartonia 16: 32. 1934.
Bartlett, H.H., Proc. Amer. philos. Soc. 82(5): 646-650. 1940 (on his publ. of the Wilkes results).
Bouvé, T.T., Hist. sketch Boston Soc. nat. Hist. (Anniv. Mem. Boston Soc. nat. Hist. 1880) 189-192. 1880 (biogr., d. 17 Mar 1878).
Brewer, N.H., *in* S. Watson, Bot. Calif. 2: 555. 1880 (Calif. coll.).
Collins, F.S., Rhodora 14: 57-68. 1912 (general inf. on publ. U.S. Expl. Exp.).
Diehl, W.W., Mycologia 13: 38-41. 1921 (The fungi of the Wilkes exp.).
Ellery, H. & C.P. Bowditch, The Pickering genealogy, 3 vols. 1897 (n.v., fide DAB).
Embacher, F., Lex. Reisen 231-232. 1882.
E.S.K., Dict. Amer. biogr. 14: 562. 1934 (biogr., biogr. refs., b. 10 Nov 1805, d. 17 Mar 1878).
Ewan, J., Rocky Mt. Natural. 168, 283, 337. 1950; *in* Cent. progr. nat. sci. 57. 1955.
Ewan, J. et al., Short hist. bot. U.S. 115, 116, 117. 1969.
Geiser, S.W., Natural. Frontier 16. 1948.
Gordon, Bartonia 22: 8. 1943 (material at DWC).
Graustein, J.E., Thomas Nuttall 475 [index, many refs.]. 1962.
Gray, A., Proc. Amer. Acad. Sci. 13: 441-444. 1878 (obit.); Sci. Papers 2: 406-410. 1889.
Gray, J.L., Letters Asa Gray 337, 347. 1893.
Harshberger, J.W., Bot. Philadelphia 190-193. 1899.
Haskell, D.C., Bull. New York Publ. Library 44: 93-112. 1940, 45: 69-89, 509-532, 823-858. 1941, 46: 103-150. 1942 (general bibl. of publications U.S. Expl. Exped.).
Hedrick, U.P., Hist. hortic. America 418-419. 1950.
Hitchcock, A.S. & A. Chase, Man. grasses U.S. 988. 1951 (epon.).
Hughes, K.W., Bibl. Oreg. bot. 15-16. 1940.
Jackson, B.D., Bull. misc. Inf. Kew 1901: 52 (pl. K).
Kelly, H.A., Some Amer. med. bot. 151-153. 1929.
Lawrence, G.H.M., Huntia 1: 164. 1964 (some letters at HU).
Merrill, E.D., Contr. U.S. natl. Herb. 30(1): 242. 1947.
Pennell, F.W., Bartonia 21: 53. 1942.
Piper, C.V., Contr. U.S. natl. Herb. 11: 15. 1906.
Quisumbing, E., Philipp. geogr. J. 8: 25. 1964.

Robinson, J., Bull. Essex Inst. 12: 91-92. 1880 (work on botany Essex County).
Rodgers, A.D., John Torrey 87, 129, 178-182, 200, 291. 1942.
Ruschenberger, W.S.W., Proc. Acad. nat. Sci. Philadelphia 30: 166-170. 1878 (1879) (obit., review of publ.).
Scudder, H.C., Psyche, Cambr. 6: 57-60, 121-124, 137-141, 169-172, 185-187, 297-298, 345-346, 357-358. 1891 (correspondence between T.W. Harris, Th. Say and Pickering).
Smith, E., Bartonia 28: 11. 1949.
Thomas, J.H., Huntia: 3: 13. 1969.
Tyler, D.B., The Wilkes Expedition 432 [index, many refs.]. 1968.
Weiss, H.B., Pioneer cent. Amer. entom. 223-224. 1936.

EPONYMY: *Pickeringia* Nuttall (1834, *nom. rej.*); *Pickeringia* Nuttall ex Torrey et A. Gray (1840, *nom. cons.*).

7912. *United States Exploring Expedition* during the years 1838, 1839, 1840, 1841, 1842 under the Command of Charles Wilkes, U.S.N. vol. xv. [i.e. xix]. *The geographical distribution of animals and plants*, ... Boston (Little, Brown & Co.), London (John Murray, ...) 1854. Qu. (in twos), 2 parts. (*U.S. Expl. exped., Geogr. distr. pl.*).
Part 1, first issue (unofficial): 3 Mar 1854 (see Haskell p. 848). Number of copies printed of unofficial issue: 500; (official issue never published) volume number '15' erroneous since this had been given to Botany; should have been 19. See Haskell (1940) for many further details. This part remained incomplete; it contains the "Chronological observations on introduced animals and plants".
Part 1, second issue (unofficial): 1863 (Geol. Soc. London rd. Jul-Sep 1863), p. [i, iii], [1]-168, index [1-44]. *Copies*: MO, NY. – Boston (Gould and Lincoln, ...), London (Trübner and Company). 1863. Qu. (in twos). Half-title: "History of the introduction of domestic animals and plants".
Part 1, third issue (unofficial): 1864, p. [i, iii], [1]-168, index [1-44]. *Copy*: G. – "*The geographical distribution of animals and plants* ... a new issue". London (Trübner & Co., ...), Boston (Little, Brown, & Co.) 1864. Qu. (in twos).
Part 2 (unofficial): Mai 1876 (pref. 10 Jan 1876; pres. copy Mai 1876), p. [i], [1]-524, 4 maps. *Copies*: G, MO. "Part ii. Plants in their wild state ..." Salem, Mass. (Naturalist's Agency) 1876. Qu. (in twos).
Ref.: Haskell, D.C., Bull. New York Publ. Libr. 44: 847-849. 1950 (bibl. details).

7913. *Chronological history of plants*: man's record of his own existence illustrated through their names, uses and companionship ... Boston (Little, Brown and Company) 1879. Qu. (*Chron. hist. pl.*).
Publ.: 1-10 Jun 1879 (p. [v]: 1 Mai 1879; "just published" letter Mrs Pickering to Dr. Norris 11 Jun 1879, Tulane Univ.; KNAW rd Jul-Sep 1979; Bot. Zeit. 26 Sep 1879; Nat. Nov. Jul(1) 1879; Bot. Centralbl. 12-23 Jan 1880), portr. Pickering, [i]-xvi, [1]-1222. *Copies*: B, FI, G, MO, NY, USDA. – Contains in preface obituaries of P. by J.H. Morison (in part taken from Unitarian review Apr 1878), by Asa Gray reprinted from Proc. Amer. Acad. Arts Sci. 13: 441-444. 1878, and by W.W. Ruschenberger from Proc. Acad. nat. Sci. Philadelphia 30: 166-170. 1878 (1879). – We are grateful to J. Ewan for drawing our attention to Mrs Pickering's letter of 11 Jun 1879.
A facsimile ed. was announced, but not yet published, by Koeltz (1974).
Ref.: F.A.F., Bot. Zeit. 37: 576-578. 5 Sep 1879.

Pickett, Fermen Layton (1881-1940), American botanist; Dr. phil. Indiana Univ. 1914; at the Department of Botany at Washington State College from 1914, head of the Department from 1918, dean of the graduate school from 1930. (*Pickett*).

HERBARIUM and TYPES: Material at GH, NY.

BIBLIOGRAPHY and BIOGRAPHY: Barnhart 3: 84; Hirsch p. 232; IH 2: (in press); Lenley p. 327; NW p. 55; PH 567.

BIOFILE: Anon., Fern J. 30: 138-139. 1941; Proc. Indiana Acad. Sci. 50: 10. 1941 (obit.,

b. 10 Jan 1881, d. 27 Jun 1940); Rev. Bryol. 47: 76. 1920, 52: 30. 1925, n.s. 1(4): 216. 1928; Science ser. 2. 92: 7. 1940 (d.); Torreya 40: 140. 1904 (d. 26 Jun 1948).
Rickett, H.W., Index Bull. Torrey bot. Club 78. 1955.
Steere, W.C., Bot. Rev. 43: 293. 1977.
Stieber, M.T. et al., Huntia 4(1): 85. 1981 (arch. mat. HU).
Verdoorn, F., ed., Chron. bot. 3: 323. 1937, 6: 427. 1941.

Picquenard, Charles-Armand (1872-1940), French botanist; Dr. med. Paris 1900; pupil of Nylander at Paris; settled at Quimper 1900 as practicing physician, ardent amateur naturalist and boy scout leader. (*Picq.*).

HERBARIUM and TYPES: REN. – *Exsiccatae*: *Lichens du Finistère* (30 nos., 1904), sets at DUKE, M, NTM, NY, PC (see Lynge 1919). – Letters at G.

BIBLIOGRAPHY and BIOGRAPHY: Andrews p. 313; Barnhart 3: 84; BL 2: 135, 148, 166, 702; CSP 17: 880; GR p. 293 (b. 14 Mar 1872, d. 3 Jan 1940), cat. p. 69; IH 2: (in press); Kelly p. 176; LS 20724-20740; Rehder 1: 418; SBC p. 129; Tucker 1: 556-557.

BIOFILE: Abbayes, H. des, Bull. Soc. sci. Bretagne 17: 5-20. 1940 (obit., bibl., portr., b. 14 Mar 1872, d. 3 Jan 1940).
Anon., Bull. Acad. Géogr. bot. 12: 113. 1903, 13: 14. 1904, 19: 8. 1909; Rev. bryol. lichénol. 8: 131. 1935, 9: 169. 1936.
Laundon, J.R., Lichenologist 11(1): 16. 1979 (coll.).
Lynge, B., Nyt Mag. Naturv. 56: 421-422 ("1917") 1919 (list of lichens distributed in *Lichens du Finistère*).
Picquenard, C.-A., Bull. Soc. Sci. nat. Ouest France 2: 125-129, 138-146, 228-254, 342-355. 1893, 3: 116-128, 129-147, 179-200, 305-318. 1894 (Cat. pl. vasc. Finistère).
Sayre, G., Mem. New York Bot. Gard. 19(3): 380. 1975 (coll.).
Wood, R.D., Monogr. Charac. 1: 833. 1965.

7914. *Catalogue des plantes vasculaires spontanées du département d'Ille-et-Vilaine* ... Nantes (Secrétariat au Muséum d'histoire naturelle) [1897]. Oct.
Publ.: 31 Mar 1897 (journal fasc. so dated), p. [29]-128. *Copy*: G. – Reprinted with special cover but original pagination from Bull. Soc. Sci. nat. Ouest France 7(1): 29-128. 1897. *Copy*: G. – See BL 2: 166 for references to later publ. on Ille-et-Vilaine.

7915. Faculté de médecine de Paris. Année 1900. Thèse pour le doctorat en médecine presentée et soutenue le jeudi 18 janvier 1900, ... *La végétation de la Bretagne* étudiée dans ses rapports avec l'atmosphère et avec le sol ... Paris (Georges Carré et C. Naud, ...) 1900. (*Vég. Bretagne*).
Publ.: 18 Jan 1900 (Bot. Centralbl. 21 Mar 1900), p. [1]-64, *1 pl.* (uncol.). *Copy*: G. – Bibliography of P's main papers on p. 3-4.
Ref.: Flahault, C., Bull. Soc. bot. France 47(bibl.): 371-372. 1900 (rev.).

7916. *Lichens du Finistère* ... Le Mans (Institut international de bibliographie de Paris) 1904. Oct. (*Lich. Finistère*).
Publ.: Jan-Jul 1904 (Nat. Nov. Aug(2) 1904), p. [1]-72 (alt.: [i], 1-48, 109-132). *Copies*: FH, G, O. Almborn. – Reprinted with special cover and double pagination from Bull. Acad. int. Géogr. bot. 13: 48, 109-132. 1904.
Ref.: Hue, A.M., Bot. Centralbl. 98: 657-658. 1905.

7917. *Études sur les collections botaniques des frères Crouan* ... Concarneau (Laboratoire maritime de Concarneau) 1911-1912. 4 parts. Oct.
1 (Characées): Trav. sci. Lab. Zool. Phys. marit. Concarneau 3(4): 1-8. 1911. – To be cited from journal.
2 (Fucoideae): ib. 3(6): [1]-44. 1911. – id.
3 (Guerinea): ib. 4(3): [1]-5, *pl. 1-2*, Apr 1912. – id. *Copies*: NY (2, 3).
4 (Florideae): ib. 4(4): [1]-105. 1912.
Copies: BR (1-4), MO (1-2, 4), NY (2-4).

Piddington, Henry (1797-1858), British meteorologist and botanist; commander in the mercantile marine; from circa 1800 curator of the Museum of economic geology and coroner at Calcutta. (*Piddington*).

HERBARIUM and TYPES: Unknown.

NOTE: Piddington introduced (1848) the term and concept "cyclone" for disturbances (winds) with a whirling or circular course.

BIBLIOGRAPHY and BIOGRAPHY: Backer p. 444; Barnhart 3: 84; BB p. 243; BM 4: 1574; CSP 4: 904-905; Desmond p. 494; DNB 45: 256-257; Herder p. 211; Kew 4: 306; Jackson p. 383; PR 7140 (ed. 1: 8015); Rehder 5: 672; SO 842bg; Tucker 1: 557.

BIOFILE: Allibone, S.A., Crit. dict. Engl. lit. 1592. 1870.
Anon., *in* Balfour, E., ed., Cycl. India 3: 210. 1885.
Stewart, R.R., Taxon 31(1): 60. 1982.

EPONYMY: *Piddingtonia* A.P. de Candolle (1839).

7918. *An English index to the plants of India* ... Calcutta (printed at the Baptist Mission Press, ...) 1832. Oct. (*Index pl. India*).
Publ.: 1832, p. [i]-viii, [1]-235. *Copies*: G, NY, USDA.

7919. *A tabular view of the generic characters in Roxburgh's Flora indica* ... Calcutta (printed at the Baptist Mission Press, ...) 1836. Oct. (*Tab. gen. char.*).
Publ.: 1836 (pref. 30 Oct 1836), p. [i]-xi, [xii], [1]-156. *Copy*: NY. – Issued *in* Journal of the Asiatic Society, useful tables part iii, 1836 (xii, 156 p.); appendix to volume for 1836.

Piepenbring, Georg Heinrich (1763-1806), German (Kurhessen) pharmacist and botanist at Meinberg (Lippe), Karlshafen (Hessen); later professor of chemistry and pharmacy at Marburg; from 1805 at Rinteln. (*Piepenbr.*).

HERBARIUM and TYPES: Unknown.

BIBLIOGRAPHY and BIOGRAPHY: BM 4: 1574; PR 7141 (ed. 1: 8016); Rehder 5: 672; Tucker 1: 557.

BIOFILE: Dilg, P. & Ilg, W., Deut. Apoth. Zeit. 118: 251. 1978 (member Regensb. bot. Ges.).
Poggendorff, J.C., Biogr.-lit. Handw.-Buch 2: 447-448. 1863 (bibl., b. 5 Jan 1763, d. 6 Jan 1806).

Pieper, Gustav Robert (*fl.* 1908), German botanist. (*G. Piep.*).

HERBARIUM and TYPES: Unknown.

BIBLIOGRAPHY and BIOGRAPHY: Barnhart 3: 84; Christiansen p. 482, 635, 6873; CSP 17: 884; GR p. 120.

7920. *Systematische Übersicht der Phanerogamen* ... Leipzig (Verlag von Quelle & Meyer) 1908. Oct. (*Syst. Übers. Phan.*).
Publ.: Apr-Jun 1908 (Nat. Nov. Jul(2) 1908; Bot. Centralbl. 25 Aug 1908), p. [1]-36. *Copy*: BR.
Ref.: Anon., Öster. bot. Z. 59: 36. 1908.

Pieper, Philipp Anton (1798-1851), German (Prussian) botanist and practicing physician at Paderborn. (*Piep.*).

HERBARIUM and TYPES: MSTR.

BIBLIOGRAPHY and BIOGRAPHY: Barnhart 3: 84; BM 4: 1574; Herder p. 118; IH 2: (in press); Kew 4: 306; PR 7142-7143 (ed. 1: 8017-8018); Rehder 1: 185; SBC 129.

BIOFILE: Anon., Bot. Zeit. 9: 390. 1851 (d.); Flora 34: 336. 1851 (d. 15 Apr 1851).
Sayre, G., Bryologist 80: 515. 1977 (bryoph. at MSTR).
Poggendorff, J.C., Biogr.-lit. Handw.-Buch 2: 448. 1863 (b. 1798, d. 15 Apr 1851).

Pierce, Newton Barris (1856-1916), American phytopathologist for the U.S. government for the Pacific coast region 1889-1906; later private collector and breeder of rare plants at Santa Ana, Calif. (*N. Pierce*).

HERBARIUM and TYPES: BPI.

BIBLIOGRAPHY and BIOGRAPHY: BM 4: 1575; CSP 11: 20, 17: 885; Hortus 3: 1201 (N.B. Pierce); IH 2: (in press); Kelly p. 176-177; LS 20744-20753a, 37946-37948; PH 150; Rehder 5: 672; Stevenson p. 1254.

BIOFILE: Anon., Science ser. 2. 44: 814. 1916 (d. 13 Oct 1916).
Galloway, B.T., Phytopathology 7: 143-144. 1917 (obit.).
Pierce, N.B., The California Vine disease, Washington 1892, 222 p., 27 pl.
Rodgers, A.D., Liberty Hyde Bailey 138, 145, 151, 282, 338-339. 1949.

EPONYMY: *Piercea* P. Miller (1759) honors Hugh Percy "Piercy" (1715-1786), 1st Duke of Northumberland.

Pieri, Michele Trivoli (x-1834), Italian botanist and physician at Corfù. (*Pieri*).

HERBARIUM and TYPES: Unknown.

BIOFILE: BM 4: 1575; Kew 4: 306; PR 7144-7145 (ed. 1: 8019-8020); Rehder 1: 448; Saccardo 1: 128; Tucker 1: 557.

BIOFILE: Britten, J., J. Bot. 31: 355-356. 1893 (on the 1834/5 *Fl. Corcirese* of Mazziari, err. attributed to Pieri).
Saccardo, P.A., J. Bot. 32: 373. 1894 (*Flora Corcirese*, 1834-1835, was written by Mazziari).

NOTE: The *Flora corcirese* anonymously published in Ionios Antologi 1(2): 424-469, (3): 668-703, (4): 940-961. 1834, 2(1): 180-227. 1835 was attributed to Pieri e.g. by B.D. Jackson in *Index kewensis*. However, P.A. Saccardo (1894) has shown that this publication was written by Alessandro Domenico Mazziari (x-1857). See also ING 3: 1479 sub *Raddia* Mazziari (1834) (see also Britten 1893).

7921. *Della Corcirese flora. Centuria prima* del Dr. Michele T. Pieri. Corfu (Nella Tipografia Nazionale, ...) 1808. (*Corcir. fl., cent. 1*).
Cent. 1: 1808, p. [1]-175. *Copy*: FI. – Corcira = Corfu.
Cent. 1-3: 1814, p. [i-viii], [1]-141, [143-144]. *Copies*: FI, G. – *Della Corcirese flora, centuria prima, seconda, e terza*, ossia storia di piante trecento, appartenenti al suolo dell' isola di Corfu', ... Corfu' (nella Stamperia del Governo) 1814. Qu. (*Corcir. fl., cent. 1-3*).

7922. *Flora corcirensis centuriae prima et secunda sive enumeratio ducentarum plantarum quas in Insula Corcirae invenit Michael Trivoli Pieri* ... Corcirae [Corfu] (typis publicis) 1824. Qu. (*Fl. corcir.*).
Publ.: 1824, p. [i]-ix, [1]-85 (copy FI), [i]-v, [1]-78 (copy G).

Pierre, Jean Baptiste Louis (1833-1905), French botanist, born on Réunion; studied in Paris 1851, in 1855 at Strasbourg; from 1861-1865 at the Calcutta botanical garden; from 1865-1877 director at the Saigon botanical garden, exploring Cambodja, Cochinchina and S. Siam; back to France 1877, living near or in Paris. (*Pierre*).

HERBARIUM and TYPES: P, PC; further material from Indochina at A, B, BO, BPI, CAIH, E, F, G, GH, HK, K, L, MO, MPU, NY, U, US. – Paris received many plants from Pierre during his life-time (e.g. 9000 in 1878); his private herbarium and mss. notes were officially acquired in 1906. – Letters at G.

BIBLIOGRAPHY and BIOGRAPHY: Backer p. 444; Barnhart 3: 85; BJI 1: 45; BM 4: 1575; Bossert p. 310 (b. 23 Oct 1833, d. 30 Oct 1905); CSP 17: 886-887; Frank 3(Anh.): 77; GF p. 70; Hortus 3: 1201 (Pierre); IH 1 (ed. 6): 362, (ed. 7): 339, 2: (in press); Jackson p. 510; Kew 4: 307; Langman p. 583; Morren ed. 2, p. 39, ed. 10, p. 61, 146; MW p. 386; NI 1527, see also 1: 150; Rehder 5: 672-673; SK 8: lxxv; TL-1/978; Tucker 1: 557; Urban-Berl. p. 383; Zander ed. 10, p. 701, ed. 11, p. 800.

BIOFILE: Anon., Ber. deut. bot. Ges. 23: (90). 1905 (d.); Bot. Centralbl. 1-2: 415. 1880 (gold medal); Bot. Not. 1906: 45 (d. 30 Oct 1905); Bull. misc. Inf. Kew 1906: 121-122 (obit., cites autobiogr. notes; "incomplete studies"; left for India 1861 from Réunion); Bull. Soc. Bot. Belg. 42(2): 212. 1905 (d.); Bull. Soc. bot. France 52: 489. 1905 (d.); Nat. Nov. 28: 83. 1906 (d.).
Chevalier, A., J.-B.-Louis Pierre 1833-1905. Paris 1906, 15 p. (portr., bibl.), reprinted from L'agriculture pratique des pays chauds.
Gager, C.S., Brooklyn Bot. Gard. Rec. 27(3): 265. 1938 (dir. bot. gard. Saigon).
Gagnepain, F., Nouv. Arch. Mus. Hist. nat. Paris sér. 4. 8: xix-xxxi. 1906 (portr., bibl.); Bull. Soc. bot. France 53: 54-59. *pl. 1.* 1906 (biogr., bibl., b. 23 Oct 1883, d. 30 Oct 1905, portr.); in Lecomte, H., Fl. gén. Indochine, tome prél. 45-46. 1944.
Hedge, I.C. & J.M. Lamond, Index coll. Edinb. herb. 119. 1970.
Jackson, B.D., Bull. misc. Inf. Kew 1901: 52 (pl. K).
Leandri, J., Adansonia 3: 207-220. 1963 (biogr., portr., b. 23 Oct 1833, d. 30 Oct 1905).
Merrill, E.D., Enum. Philipp. pl. 4: 216. 1926.
Moldenke, H.N., Plant Life 2: 76. 1946 (1948) (eponymy).
Steenis-Kruseman, M.J. van, Blumea 25: 43. 1979 (coll.).

COMPOSITE WORKS: *Sapotaceae*, with I. Urban, *in Symb. antill.* 5(1): 95-176. 20 Mai 1904.

EPONYMY: *Pierranthus* Bonati (1912); *Pierrea* Hance (1877, nom. rej.); *Pierrea* F. Hance (1891, nom. cons.); *Pierreodendron* Engler (1907); *Pierreodendron* A. Chevalier (1917); *Pierrina* Engler (1909).

7923. *Flore forestière de la Cochinchine* ... Paris (Octave Doin, éditeur ...) [1880-1907], 4 volumes in 26 parts. (*Fl. forest. Cochinch.*).
1: [i], *pl. 1-100* (uncol. liths. by L. Pierre and E. Delpy), with text; *2: pl. 101-200; 3: pl. 201-300; 4: pl. 301-400.* – Copies may be bound in different ways; the book came in fascicles and we have seen no volume titles.
Publ.: in 26 parts, the dates of publication of parts 4-25 are given in the work itself (vol. 1, vol. 5). The announcements in *Nat. nov., J. Bot., Bot. Zeit., Bot. Jahrb.* and *Bot. Centralbl.* confirm these dates. The 400 lithographs are of drawings by E. Delpy.

fasc.	plates	dates	fasc.	plates	dates
1	*1-16*	[Nov. 1880]	14	*209-224*	1 Aug 1889
2	*17-32*	[Feb 1881]	15	*225-240*	1 Mar 1890
3	*33-48*	[Jul 1881]	16	*241-256*	1 Oct 1891
4	*49-64*	15 Mai 1882	17	*257-272*	1 Jun 1892
5	*65-83*	15 Feb 1883	18	*273-288*	1 Dec 1893
6	*84-96*	1 Nov 1883	19	*289-304*	1 Jul 1894
7	*97-112*	1 Jul 1885	20	*305-320*	1 Jul 1895
8	*113-128*	25 Jan 1887	21	*321-336*	1 Jul 1897
9	*129-144*	1 Jan 1888	22	*337-352*	1 Jul 1897
10	*145-160*	1 Feb 1888	23	*353-368*	1 Jul 1897
11	*161-176*	1 Mai 1888	24	*369-384*	1 Sep 1898
12	*177-194*	1 Dec 1888	25	*385-400*	15 Apr 1899
13	*195-208*	1 Feb 1889	26	Preface, indexes	15 Mar 1907

Copies: BR, G, HH, L, MICH, MO, NY; IDC 5638. A few copies were printed on "papier de Chine" and sold at 40 fr./fasc. (regular price 25 fr./fasc) (see BF 1892, no. 2202). The copies at G and NY have a typewritten list on which it is stated that fasc. 1 was published on 15 Apr 1879, 2 on 15 Feb 1880, and 3 on 1 Mar 1881. A facsimile reprint was announced, but not published, by Asher (1974).
Ref.: De Bary, A., Bot. Zeit. 42: 573-574. 1884 (fasc. 1-6).

7923a. *Notes botaniques*. Sapotacées ... Paris (Librairie des sciences Paul Klincksieck, ...) [1890-1891], 2 fasc. Oct. †. (*Not. bot.*).
1: 30 Dec 1890, p. [1]-36.
2: 5 Jan 1891, p. [37]-68, text ends abruptly; not continued.
Copies: BR, G, L, MO, NY, U, US. – Proofs of p.69-83 are at P. Pierre left an album with 147 unpublished plates of Sapotaceae drawn by E. Delpy, now at P. The planned monograph of this family never appeared. See also his *Genres nouveaux de Sapotacées*, Bull. Soc. bot. France 38: 50-51. 1891 (nomina nuda).

Pierrot, Philogène (1835-1896), French botanist, printer and newspaper publisher at Montmédy near Sedan; sometime burgomaster of the city and county-counsel. (*Pierrot*).

HERBARIUM and TYPES: Some material at BR.

BIBLIOGRAPHY and BIOGRAPHY: Barnhart 3: 85 (d. 16 Nov 1896); BL 2: 181, 702; IH 2: (in press); Jackson p. 287; TL-2/see J. Cardot.

BIOFILE: Anon., Bull. Soc. Bot. Belg. 35(2): 47. 1896 (1897), 36(2): 149. 1897 (1898). Werner, K.G., Bull. Acad. Soc. Lorraines Sci. 6(2): 115. 1966 (b. 21 Sep 1835, d. 16 Nov 1895).

7924. *Notice sur les plantes vénéneuses* dangereuses ou suspectes de l'arrondissement de Montmédy ... Montmédy (chez l'auteur et chez les libraires ...) [1868]. Duod. (in sixes) (*Not. pl. vénén.*).
Publ.: 1868, p. [i, iii], [1]-103. *Copy*: BR ("deuxième tirage"). – We did not see the first printing.

7925. *Additions à la flore de la Meuse* ... Mém. Soc. philom. Verdun 8: 296-327. 1877.
Publ.: 1877, p. 296-327. We have seen no copy with independent pagination; the copy at NY is an unchanged reprint without identification. To be cited from journal.

7926. *Liste des plantes vasculaires observés dans l'arrondissement de Montmédy* (Meuse). (*Liste pl. vasc. Montmédy*).
Co-author: Jules Cardot (1860-1934).
Publ.: Jan-Feb 1882 (BSbF 24 Feb 1882; Bot. Centralbl. 29 Mai-2 Jun 1882; Nat. Nov. Jun(1) 1882; Bot. Zeit. 30 Jun 1882). (n.v.).

7927. *Catalogue des plantes vasculaires* de l'arrondissement *de Montmédy* avec indication de leurs stations, propriétés et usages divers ... Montmédy (Imprimerie de Ph. Pierrot-Caumont) 1891[-1906] Oct. (*Cat. pl. vasc. Montmédy*).
Co-authors: Jules Cardot (1860-1934) and A. Vuillaume (*fl.* 1891).
Publ.: in parts in Soc. Amat. Natur. Nord Meuse, Mém., from 1898: Soc. Natur. Archéol. Nord Meuse, Mém. Sc. Nat. (see below) between 1891 and 1903, reprinted with continuous pagination 1906 (p. vii: Mar 1906; cover: 1906; Nat. Nov. Jan(1) 1907), p. [i*-iii*], [i]-vii, [1]-532. *Copies*: BR, G. – R. Tournay (mss. BR, 23 Nov 1962) has provided the following concordance:

Reprint pages	Journal vol.	pages	dates	
[i*-iii*]	-	-	1906	not in journal
[i]-iii	3	46-48	1891	preface 1891
v-vii	-	-	1906	not in journal

Repr. p.	vol.	pages	dates	
[1]-32	3	49-80	1891	SANNM, Mém.
36-64	4	17-49	1892	
64-112	5	48-96	1893	
112-160	6	39-87	1894	
160-208	7	50-98	1895	
208-256	8	50-98	1896	
256-320	9	50-114	1897	
320-352	10	36-68	1898	SANNM, Mém. Sci. nat.
352-400	12	69-117	1900	
400-449	13	10-59	1901	
449-465	14	1-17	1902	
465-502	15	1-38	1903	
503-523	16	18-38	1904	add., err.
525-532	-	-	1906	table, not in journal

Pieters, Adrian John (1866-1940), American botanist (plant physiologist and morphologist); Ph.D. Univ. Mich. 1915; assistant botanist USDA, Washington 1895-1900, botanist ib. 1900-1906; seed grower 1906-1910; instructor in botany Univ. Michigan 1912-1915, agronomist Bureau Plant Industry, USDA 1915-1926, principal agronomist 1926-1935; in charge of Lespedeza investigations 1935-1940; one of the founders of the Bureau of Plant Industry 1901; "father of Lespedeza". (*Pieters*).

HERBARIUM and TYPES: US; other material at CUP, MICH, and MSC.

BIBLIOGRAPHY and BIOGRAPHY: Barnhart 3: 85; BL 1: 208, 313; BM 4: 1575; Bossert p. 310 (b. 18 Nov 1866, d. 25 Apr 1940); CSP 17: 888; Hirsch p. 232; IH 2: (in press); Kelly p. 177; Kew 4: 308; LS 20761, 21720-21722; MW p. 386; Nordstedt suppl. p. 13; NW p. 55; Pennell p. 616; Rehder 5: 673; Tucker 1: 557.

BIOFILE: Anon., Bot. Jahrb. 27(Beibl. 64): 21. 1900 (app. USDA), 78: 32. 1899.
Cattell, J.M., ed., Amer. Men Sci. 253. 1906, ed. 2. 372. 1910, ed. 3. 544. 1921, ed. 4. 775. 1927, ed. 5. 885. 1933, ed. 6. 1118. 1938 (b. 18 Nov 1866).
Donelly, W.A. et al., The University of Michigan, an encyclopaedic survey 1448. 1956.
Kephart, L.W., Science ser. 2. 91: 610-611. 1940 (obit.).
Verdoorn, F., ed., Chron. bot. 2: 21, 308. 1936, 3: 276. 1937, 4: 593. 1938, 6: 20, 213. 1941, 7: 188. 1943.
Voss, E.G., Contr. Univ. Mich. herb. 13: 73. 1978.
Wood, R.D., Monogr. Charac. 1: 834. 1965.

7928. *The plants of Lake St. Clair* ... with a map. Results of a biological examination of Lake St Clair undertaken for the State Board of Fish Commisioners in the summer of 1893 under the supervision of J.E. Reighard. Lansing (Robert Smith & Co., ...) 1894. Oct. (*Pl. Lake St. Clair*).
Publ.: 1894 (p. 10: 3 Apr 1894; USDA rd. 25 Jun 1894), p. [1]-10, [11, 12], map. *Copies*: MO(3), NY, US, USDA. – Bulletin of the Michigan Fish Commission no. 2.

7929. U.S. Commission of fish and fisheries ... Contribution to the Biology of the Great Lakes. *The plants of Western Lake Erie*, with observations on their distribution, ... extracted from U.S. Fish Commission Bulletin for 1901. Pages 57 to 79. Plates 11 to 20. Washington (Government Printing Office) 1901. Oct. (*Pl. W. Lake Erie*).
Publ.: 1901 (Bot. Centralbl. 23 Dec 1901), p. [i], 57-79, *pl. 11-20* (photos). *Copy*: USDA.

7930. *The little book of Lespedeza* ... published by the author, Washington, D.C. 1934. (*Lespedeza*).
Publ.: 1934, p. [1]-94. *Copy*: USDA.

Pietsch, Friedrich Maximilian (1856-?), German (Saxonian) botanist; Dr. phil. Halle 1893; high school teacher at Gera. (*Pietsch*).

HERBARIUM and TYPES: Unknown.

BIBLIOGRAPHY and BIOGRAPHY: Barnhart 3: 86; BJI 1: 46; BM 4: 1576.

7931. *Die Vegetationsverhältnisse* der Phanerogamen-Flora *von Gera.* Inaugural-Dissertation zur Erlangung der Doctorwürde bei der Philosophischen Facultät der Vereinigten Friedrichs-Universität Halle-Wittenberg vorgelegt von Friedrich Maximilian Pietsch aus Chemnitz. Halle (Druck von Erhardt Karras) 1893. Oct. (*Veg.-Verh. Gera*).
Publ.: 1893 (Nat. Nov. Jul(2) 1893; Bot. Zeit. 1 Sep 1893; Bot. Centralbl. 9 Aug 1893), p. [1]-64, [1, vita]. *Copy*: MO.
Ref.: Roth, E., Bot. Centralbl. 55: 336-338. 1893, Bot. Jahrb. 18(Lit.): 36. 1894.

Pijl, Leendert van der (1903-x), Dutch botanist and flower-biologist; studied at Amsterdam University 1923-1927; high school teacher at Arnhem 1926-1927; to the Netherlands East Indies 1927; high school teacher at Bandung 1927; Dr. phil. Amsterdam 1933; in Japanese captivity during World War II; rector at Bandung Gymnasium 1947; professor of botany at Bandung University 1948-1954; retired to the Hague, Netherlands; professor of botany at Nijmegen 1965-1975. (*v.d. Pijl*).

HERBARIUM and TYPES: material at BO and K.

BIBLIOGRAPHY and BIOGRAPHY: Barnhart 3: 86; BFM 2621, 2658, 2681; BL 1: 114, 313; IH 2: (in press); Kew 4: 309; Moebius p. 360; SK 1: 419-420.

BIOFILE: Arditti, J., ed. et al., Orchid biology 1: 297 [index]. 1977.
Jacobs, M., Fl. males. Bull. 28: 2309, 2457, 2511-2513. 1975 (tribute; b. 21 Dec 1903 at Utrecht; biogr., bibl. notes, portr. after p. 2534).
Linskens, H.F., *in* Brantjes, N.B.M. & H.F.Linskens, Pollination and dispersal. Symposium 14 December 1973, offered to Prof. Dr. L. van der Pijl at his 70th birthday. 1973, University of Nijmegen, Dept. of Botany, 4-7 (tribute, portr.).
Rollins, R.C., Taxon 5: 2. 1956 (in group photogr.).
Steenis, C.C.G.J. van, Fl. males. Bull. 9: 285. 1952 (on leave), 11: 404. 1955 (retirement from Bandung).
Verdoorn, F., ed., Chron. bot. 1: 223, 228. 1935, 2: 240. 1936, 7: 95. 1943.
Westhoff, V., Farewell to Professor van der Pijl, Nijmegen 1973, 3 p.

COMPOSITE WORKS: Main books on flower biology: *Principles of dispersal in higher plants*, Berlin, 1969, ed. 2. 1972; with C.H. Dodson, *Orchid flowers*, Coral Gables, Florida, 1967; with K. Faegri: *Principles of pollination ecology*, Pergamon Press, 1966.

7932. *Über die Polyembryonie bei Eugenia.* Academisch proefschrift ter verkrijging van den graad van doctor in de wis- en natuurkunde aan de Universiteit van Amsterdam ... in het openbaar te verdedigen in de aula der Universiteit op Woensdag 4 October 1933, 's namiddags te $3^{1}/_{2}$ uur door Leendert van der Pijl geboren te Utrecht. Amsterdam (N.V. Drukkerij en Uitgeverij J.H. de Bussy) 1933. Oct. (*Polyembryon. Eugenia*).
Publ.: 4 Oct 1933, p. [1]-12, 113-187, [1-2, theses]. *Copy*: U. – Major part preprinted from Rec. Trav. bot. néerl. 31: 113-187. 1934.

Pike, Nicolas (1815-1905), American soldier, herpetologist and algologist; U.S. consul at Porto, Portugal 1856-1866 and Mauritius (1866-1876). (*Pike*).

HERBARIUM and TYPES: NY; further material at BM and TCD (through Harvey).

BIBLIOGRAPHY and BIOGRAPHY: Barnhart 3: 86; BM 8: 1011; CSP 8: 626, 17: 890; De Toni 1: c; IH 2: (in press); Kew 4: 310; SIA 7220; Tucker 1: 557.

BIOFILE: Anon., Science ser. 2. 21: 677. 1905 (obit., d. 11 Apr 1905); The New York Times 13 Apr 1905; Torreya 5: 74. 1905 (d. 11 Apr 1905).
Howe, M.A., J. New York Bot. Gard. 5: 86-87, 123. 1904 (algal herb. to NY), 9: 126. 1908, 11: 178. 1908.

Koster, J.Th., Taxon 18: 556. 1969 (coll.).
Palmer, T.S., Condor 30: 291-292. 1928 (on ornithological work).
Pike, N., Subtropical rambles in the land of the Aphanapteryx. Personal experiences, adventures and wanderings in and around the island of Mauritius. New York 1873, i, xviii, 511 p., *11 pl.*, maps.
Stiles, H.R., History of Brooklyn, N.Y. 2: 1327-1328. 1884 (fide ME 1: 219).

EPONYMY: Pike's Peak (Colorado) was named after his relative Zebulon Montgomery Pike (1779-1813) (see e.g. S.W. Geiser, Field and Laboratory 27: 173. 1959, and J. Ewan, Rocky Mt. Natural. 283. 1950). *Pikea*, Harvey (1853), ING 3: 1340.

Pilar, Georg [Gjuro] (1847-1893), Slavonian/Croatian palaeobotanist; professor at Zagreb (Agram) University. (*Pilar*).

COLLECTIONS: Unknown.

BIBLIOGRAPHY and BIOGRAPHY: Barnhart 3: 86; CSP 8: 626, 11: 22; Herder p. 471; Kelly p. 256; Quenstedt p. 339 (d. 19 Mai 1893).

BIOFILE: Tietze, E., Verh. k.k. geol. Reichsanst. 8: 187-188. 1893 (obit., d. 19 Mai 1893).
Zittel, K.A. von, Hist. geol. palaeontol. 238, 283. 1901.

7933. *Flora fossilis susedana* (Susedska fosilna flora – Flore fossile de Sused). Descriptio plantarum fossilium quae in lapicidinis ad Nedelja, Sused, Dolje etc. in vicinitate civitatis Zagrabiensis hucusque repertae sunt .. Zagrabiae [Zagreb, Agram] (apud Leopoldum Hartman ...) 1883. Qu. (*Fl. foss. sused.*).
Publ.: late 1883 or Jan-Feb 1884 (t.p. 1883; Bot. Zeit. 28 Mar 1884; Bot. Centralbl. 31 Mar-4 Apr 1884; LC rd. 29 Mai 1884), p. [i]-viii, [1]-163, [164, err], *pl. 1-15* (uncol.).
Copies: BR, USGS. Issued as Djela Jugoslavenske Akademiji, knjiga 4, 1883.
Ref.: Staub, M., Bot. Centralbl. 22: 172-175. 1885.

Pilát, Albert (1903-1974), Czechoslovakian mycologist; Dr. sci. Praha 1926 (with J. Velenovský); from 1930 at the National Museum, Praha, from 1948 as head of the botanical department, from 1965 idem of the mycological department; řad práce (labor medal) ČSSR 1969; corr. memb. Czechosl. Acad. Sci. 1959. (*Pilát*).

HERBARIUM and TYPES: PR (phan.), PRM (fungi); duplicates in e.g.; B, BM, DAO, E, FH, GH, K, MICH, OTB, PC, UPS, US, W. – *Exsiccatae: Fungi carpatici lignicoli*, nos. 1-265.

BIBLIOGRAPHY and BIOGRAPHY: BM 8: 1011; Bossert p. 310 (b. 2 Nov 1903); Futák-Domin p. 469-470; Hirsch p. 232; IH 1 (ed. 1): 78, (ed. 2): 99, 100, (ed. 3): 126, (ed. 4): 139, (ed. 5): 137, (ed. 6): 362 (err. "I.A. Pilát"), (ed. 7): 340, 2: (in press); Kew 4: 310-311; Klášterský p. 143; LS 21733-21843 (bibl. 1924-1930); MW p. 386-387, suppl. 280; Roon p. 88; Stevenson p. 1254; TL-2/see Kavina, K.

BIOFILE: Anon., Österr. bot. Z. 82: 275. 1933 (*Fungi carpatici* in preparation).
Cejp, K., Česká Mykol. 17: 169-173. *pl. 15*. 1963 (60th birthday, portr., bibl. 1954-1963).
Deyl, M., Preslia 35: 330-336. 1963 (tribute on 60th birthday, portr., bibl. 1923-1963), 45: 366-368. 1973 (70th birthday, bibl. 1963-1973).
Guzmán, G., Bol. Soc. mexic. mycol. 8: 138-139. 1974 (obit.).
Haller, R., Schweiz. Z. Pilzk. 32(1): 6. 1954 (50th birthday).
Hedge, I.C. & J.M. Lamond, Index coll. Edinb. herb. 119. 1970.
Herink, J., Česká Mykol. 27: 193-200. 1973 (70th birthday, bibl. 1968-1973; further biogr. refs.).
Herink, J. et Svrček, M., Čeksá Mykol. 7(4): 145-162. 1953 (portr., bibl. (1923-1953).
H.J., Westfäl. Pilzbriefe 9(8): 136. 1973 (1974) (obit.).
Kotlaba, F. & Z. Pouzar, Trans. brit. mycol. Soc. 65(1): 163-165. 1975 (obit., portr., b. 2 Nov 1903, d. 29 Mai 1974); Taxon 24: 399-400. 1975.

Moser, M., Z. Pilzk. 40 (3/4): 238-239. 1974 (obit., portr.; published in all 540 papers).
Neuhäuslova-Novotna, Z., Preslia 45(4): 366-368. 1973 (bibl.).
Palmer, J.T., Nova Hedwigia 15: 152. 1968 (Gasteromyc. in Fl. ČSR).
Pouzar, Z., & Svrček, M., J. nat. Mus., nat. Hist., Praha 144: 105-107. 1977 (fide Excerpta bot.).
Reid, D.A., Bull. Brit. mycol. Soc. 8(2): 86-87. 1974 (obit.).
Singer, R., Mycologia 76(3): 445-447. 1975 (obit.).
Stafleu, F.A., Taxon 23: 688. 1974 (b. 2 Nov 1903; d. 29 Mai 1974).
Svrček, M., Česká Mykol. 22: 241-246, *pl. 13.* 1968 (65th birthday, portr., bibl. 1964-1968); Živa 22(1): 21. 1974 (70th birthday, tribute, portr.), Friesia 21(1): 75-76. 1975 (obit., portr.).
Verdoorn, F., ed., Chron. bot. 1: 31, 113. 1935, 2: 31, 34, 109. 1936, 3: 27, 93. 1937, 4: 187, 190. 1938, 5: 309. 1939, 6: 301. 1941.
Wojewoda, W., Wiadomosci bot. 19(1): 3-5. 1975 (obit., portr.).
Zerova, M. Ja., & I.P. Vasser, Ukraj. bot. Zhurn. 30(4): 532-533. 1972 (tribute, portr.).

FULL BIBLIOGRAPHY: Česká Mycol. 7: 152-161. 1953 (1923-1953), 17: 170-173. 1963 (1954-1963), 22: 245-246. 1968 (1964-1968), 27: 197-200. 1973 (1968-1973).

COMPOSITE WORKS: (1) Founder and editor-in-chief of *Česká Mykologie.* (2) *Gasteromycetes*, in *Flora ČSR*, ser. B., vol. 1, 1958, 862 p.

EPONYMY: *Pilatia* Velenovský (1934). – For further eponyms (species names), see Kotlaba & Pouzar (1975).

7934. *Monographia Cyphellacearum Čechosloveniae* ... Praha (Přírodovědecká Fakulta ...) 1925, 2 parts. Oct. (*Monogr. Cyphell. Čech.*).
1: 1925, p. [1]-52. *2*: 1925, p. [1]-92, *pl. 1-5.* *Copies*: FH, G, MICH. – Publ. Fac. Sci. Charles 28. 29. 1925.

7935. *Atlas des champignons de l'Europe* ... Praha (Chez les éditeurs ...) 1934-1942, 4 vols. (*Atlas champ. Eur.*).
Co-editor: Karel Kavina (1890-1948).
1: *Amanita* by Rudolf Veselý, in 5 fascicles:
  1: Jan 1934, p. [1]-16, *pl. 1-8.*
  2: Mai 1934, p. [17]-32, *pl. 9-16.*
  3: Jun 1934, p. 33-48, *pl. 17-24.*
  4/5: Sep-Nov 1934, p. 49-80, *pl. 25-40.*
2: *Pleurotus* by Albert Pilát, t.p. 1935, in 11 facicles:
  6: Jan 1935, p. 1-16, *pl. 1-8.*
  7/8: Feb-Mar 1935, p. 17-48, *pl. 9-24.*
  9/10: Apr-Mai 1935, p. 49-96, *pl. 24-40.*
  11/13: Oct 1935, p. 97-144, *pl. 41–*
  14/15: Dec 1935, p. 145-176.
  16: *Jan 1936, p. 177-193, pl. –80.*
3: *Polyporaceae*, t.p. 1936, in 48 fascicles (17-73):
  1: Feb 1936, p. 1-16, *pl. 1-8.*
  2/3: Mar-Apr 1936, p. 17-48, *pl. 9-24.*
  4: Jun 1936, p. 49-64, *pl. 25-32.*
  5/6: Nov-Dec 1936, p. 65-96, *pl. 33-48.*
  7/8: Feb-Mar 1937, p. 97-112, *pl. 49-72.*
  9/10: Mai-Jun 1937, p. 113-128, *pl. 73-96.*
  11/12: Sep-Oct 1937, p. 129-144, *pl. 97-120.*
  13/14: Nov-Dec 1937, p. 145-176, *pl. 121-136.*
  15/16: Jan-Feb 1938, p. 177-208, *pl. 137-152.*
  17: Mai 1938, p. 209-224, *pl. 153-160.*
  18: Nov 1938, p. 225-240, *pl. 161-168.*
  19: Mar 1939, p. 241-256, *pl. 169-176.*
  20/21: Sep-Oct 1939, p. 257-288, *pl. 177-192.*
  22/25: Jan-Mai 1940, p. 289-336, *pl. 193-232.*

26/27: Jan-Feb 1941, p. 337-360, *pl. 233-248*.
28/34: Jun-Oct 1941, p. 361-472, *pl. 249-304*.
35/41: Jan-Mai 1942, *pl. 305-374*.
42/48: Jun-Oct 1942, p. 473-624.
4: *Omphalia* [and *Delicatula*] by Karel Cejp (1936-1938), p. [1]-152, *pl. 1-56*, in 8 fascicles:
1: Mai 1926, p. 1-16, *pl. 1-8*.
2/3: Sep-Oct 1936, p. 17-48, *pl. 9-24*.
4: Jan 1937, p. 49-64, *pl. 25-32*.
5: Apr 1937, p. 65-80, *pl. 33-40*.
6/7: Mar-Apr 1938, p. 81-128, *pl. 41-48*.
8: Jun 1938, p. 129-152, *pl. 49-56*.
Copies: B, BR, G, NY, Stevenson.

Pilger, Robert Knuds Friedrich (1876-1953), German botanist; Dr. phil. Berlin (with Engler) 1898; collected in Brazil 1898-1900; assistant at Berlin-Dahlem 1900; Custos 1908; habil. Berlin 1908; titular professor 1913, vice-director 1921, director 1945-1950. (*Pilg.*).

HERBARIUM and TYPES: B (partly destroyed; Matto Grosso material extant); further material at BM, NY, P, POM, RB, VT. – Letters and portrait at G.

BIBLIOGRAPHY and BIOGRAPHY: Backer p. 444-445; BFM no. 2430; BJI 1: 46, 2: 135-136; BL 1: 242, 313 ("Knuds"); BM 4: 1576, 8: 1011; Bossert p. 310; CSP 17: 891; Hirsch p. 232; Hortus 3: 1201 (Pilg.); IH 1 (ed. 1): 10, (ed. 6): 362, (ed. 7): 340, 2: (in press); Kew 4: 311; KR p. 574; Langman p. 583; Lenley p. 327; LS suppl. 21845; Moebius p. 138, 140, 142, 144, 146; MW p. 387, suppl. p. 280-281; NW p. 55; PH 209; SIA 226 (and HI 107); SK 4: cxix (portr.), cli; TL-2/1711, 1713, 4558, see Luetzelburg; Tucker 1: 557; Urban-Berl. p. 267, 383; Zander ed. 10, p. 701-702, ed. 11, p. 800; Zep.-Tim. p. 93 (b. 3 Jul 1876, d. 9 Jan 1953; "Knuds").

BIOFILE: Anon., Bot. Centralbl. 105: 672. 1907 (lecturer Berlin), 107: 80. 1908 (asst. B), 107: 656. 1908 (habil. Berlin), 110: 272. 1909 (Custos, B); Bot. Jahrb. 40 (Beibl. 92): 47. 1908; Nat. Nov. 29: 600. 1907 (lecturer Berlin, Charlottenburg Techn. Hochschule), 30: 361 (habil. Univ. Berlin), 616 (first assist. B). 1908, 31: 127. 1909 Custos); Hedwigia 39: (41). 1900 (returned from S. Amer.), 47: (157). 1918 (lecturer), 48: (197). 1909 (Custos), 53: (235). 1913 (professor); Österr. bot. Z. 58: 47. 1908 (lecturer), 59: 127. 1909 (Custos); Rodriguesia 16/17: 213. 1934.
Becker, K. & E. Schumacher, S.B., Ges. naturf. Freunde Berlin ser. 2. 13(2): 145. 1973 (bibl. S.B.).
Bullock, A.A., Bibl. S. Afr. bot. 90. 1978.
Eckardt, T., Willdenowia 4: 174. 1966.
Hubbard, C.E., Kew Bull. 1953(1): 162 (obit.).
Kneucker, A., Allg. bot. Z. 14: 16. 1908, 15: 48. 1909 (Custos Berlin-Dahlem).
Melchior, H., Taxon 2: 19-21. 1953; Bot. Jahrb. 76(3): 385-409. 1954 (portr., bibl.); Ber. deut. bot. Ges. 68(a): 293-297. 1955 (portr.).
Merrill, E.D., B.P. Bishop Mus. Bull. 144: 151. 1937.
Parodi, L.R., Rev. argent. Agron. 20(2): 107-114. 1953 (portr., bibl.).
Pfeiffer, H.H., Revista sudam. Bot. 10(6): 195. 1954 (obit.).
Pilger, R., Mitt. Bot. Gart. Mus. Berlin-Dahlem 1(1): 14, 16, 19. 1953 (duplicates of Matto Grosso material extant).
Prance, G., Acta amaz. 1(1): 49. 1971.
Steenis, C.G.G.J. van, Fl. males. Bull. 10: 354. 1953 (retirement).
Stieber, M.T., Huntia 3(2): 124. 1979 (notes Agnes Chase).
Urban, I., Fl. bras. 1(1): 75-76. 1906 (itin. Braz.).
Verdoorn, F., ed., Chron. bot. 1: 31, 32, 130. 1935, 2: 142. 1936, 3: 121. 1937, 4: 110. 1938, 6: 301. 1941, 7: 287, 288. 1943.
Weberbauer, A., Pfl.-Welt Peruan. Anden 33. 1911 (Veg. Erde 12).

COMPOSITE WORKS: (1) Engler, A., *Pflanzenreich*, *Taxaceae*, 18(iv, 5): 1-124. 8 Dec 1903; *Plantaginaceae*, 102 (iv, 269): 1-466. 22 Jun 1937.

(2) With C. Mez, *Gramineae, in* Perkins, J., *Fragm. fl. Philipp.* fasc. 2: 137-140, 145-150. 1904.
(3) *Arthrostylidium, in* Urban, I., Symb. ant. 2: 336. 20 Oct 1900, 337-343. 1 Oct 1901; *Arundinella, Cenchrus, Aristida, Leptochloa, Phragmites, Eragrostis, in* id. 4: 80-106. 1903; *Gramineae, in* id. 5: 288-289. 1907, id. 6: 1-4. 1909; *Juniperus,* id. 7: 478-481. 15 Aug 1913.
(4) *Gramineae, in* Engler, A., *Pflanzenw. Afrik.* 2(1): 114-192. 1908.
(5) Editor, *Ergänzungsheft* 2, Nachtr. 3, zu Engler und Prantl, *Natürliche Pflanzenfam.* 1908, 379 p.; id., with K. Krause, 3, Nachtr. 4, 1915, 381 p.
(6) *Angiospermen, in Handwörterbuch Naturwissenschaften* 1: 365-425. 1912 (ed. 2. 2: 60-121. 1932, revised by E. Janchen).
(7) *Pflanzengeographie, in* Francé, *Das Leben der Pflanze* 6: 117-243. 1913.
(8) *Die Algen.* Dritte Abteilung. *Die Meeresalgen, in* Lindau, G., *Kryptogamenflora für Anfänger* 4(3), 1916, vi, 29, 125 p., *11 pl.,* see under Lindau. – Pilger also acted as editor of the *Kryptogamenflora* after Lindau's death in 1923.
(9) EP, *Natürliche Pflanzenfamilien,* ed. 2, *Caryocaraceae,* 21: 90-93. 1925, *Bixaceae* 21: 313-315. 1925, *Cochlospermaceae,* 21: 316-320. 1925, *Cycadaceae,* 13: 44-81. 1926,*Ginkgoaceae,* 13: 98-108. 1926, *Coniferae* 13: 121-165, 199-402. 1926, *Mayacaceae,* 15a: 33-34. 1930; *Thurniaceae,* 15a: 58. 1930, *Rapateaceae,* 15a: 59-64. 1930; *Philydraceae,* 15a: 190-191. 1930; *Santalaceae,* 16b: 52-91, 339. 1936; *Gramineae, Panicoideae,* 14e: 1-208. 1941; *Gramineae* II, 14d: i-vi, 1-225. 1956.
(10) with H. Melchior, *Gymnospermae, in Syll. Pfl. Fam.* ed. 12, 1: 312-344. 1954.
(11) collaborator, G.W.J. Mildbraed, Wiss. Erg. deut. Zentr.-Afr. Exped., 2, *Botanik,* 1914, see TL-2/6008.

EPONYMY: *Pilgeria* W. Schmidle (1901), *Pilgeriella* P.C. Hennings (1900), *Pilgerochloa* Eig (1929), *Pilgerodendron* Florin (1930); see ING 3: 1340-1341 and Zep.-Tim. p. 93.

POSTHUMOUS PUBLICATION: *Das System der Gramineae* unter Ausschluss der Bambusoideae, Bot. Jahrb. 76(3): 281-384. Oct 1954, edited and completed by Eva Potztal.

7936. *Das System der Blütenpflanzen* mit Ausschluss der Gymnospermen ... Leipzig (G.J. Göschen'sche Verlagshandlung) 1908. Oct. (*Syst. Bl.-Pfl.*).
Ed. 1: Jul 1908 (Nat. Nov. Jul(2) 1908; ÖbZ Aug-Sep 1908; Bot. Zeit. 1 Dec 1908, 1 Jan 1909; NYBG Apr 1909; Allg. bot. Z. 15 Sep 1908), p. [1]-140. *Copies*: B, G, NY. – Sammlung Göschen 393.
Ed. 2: 1919 (Nat. Nov. Jan 1920) n.v.

7937. *Die Stämme des Pflanzenreiches* ... Leipzig (G.J. Göschen'sche Verlagsbuchhandlung) 1910. Oct. (*Stämme Pflanzenr.*).
Ed. 1: Apr-Mai 1910 (Nat. Nov. Mai(1) 1910; ÖbZ Jun 1910; Bot. Zeit. 16 Jul 1910; Allg. bot. Z. 15 Jul 1910), p. [1]-146. *Copies*: B, BR, NY. – Sammlung Göschen 485.
Ed. 2: 1921(?,t.p. but BR rd. 8 Jul 1922; Nat. Nov. Jun 1922; ÖbZ Mar-Sep 1922), p. [1]-119. *Copy*: BR. – Zweite, umgearbeitete Auflage. Berlin und Leipzig (Vereinigung wissenschaftlicher Verleger Walter de Gruyter & Co.) 1921. Oct.

7938. *Die Gattung Plantago in Zentral– und Südamerika* (Bot. Jahrb. 1928).
Publ.: 1 Jul 1928, p. [1]-112, *pl. 1-5. Copy*: B. – Reprinted and to be cited from Bot. Jahrb. 62: [1]-112, *pl. 1-5.* 1928.

Pillans, Neville Stuart (1884-1964), South African botanist; studied at Cambridge; at the Bolus Herbarium 1918-1953. (*Pillans*).

HERBARIUM and TYPES: BOL and NBG; further material at BH, BM, J, K, MO, PRE, SAM, STE, U.

BIBLIOGRAPHY and BIOGRAPHY: Barnhart 3: 86; Bossert p. 311 (b. 2 Mai 1884, d. 23 Mar 1904); Hortus 3: 1201 (Pillans); IH 1 (ed. 7): 340; Kew 4: 311; Kew-abbr. 168 (Pill.).

BIOFILE: Anon., South Afr. Forum bot. 2(5): 1-2. 1964 (d. 23 Mar 1964).
Bridson, G.D.R. et al., Nat. hist. mss. res. Brit. Isl. 269.285. 1980.

Bullock, A.A., Bibl. S. Afr. bot. 90. 1978.
Gunn, M.D. & L.E. Codd, Bot. expl. S. Afr. 281-282. 1981 (biogr. portr., coll.).
Hutchinson, J., Bot. S. Afr. 53-54. 1946 (portr.).
Oliver, E.G.K., J. Bot. Soc. S. Afr. 51: 42. 1965.
Phillips, E.P., S. Afr. J. Sci. 27: 65. 1930.
White, A. & B.L. Sloane, The Stapeliae 205 [index]. 1933, ed. 2. 20 [index]. 1937.

EPONYMY: *Pillansia* Bolus (1914).

Piller, Mathias (1733-1788), Austrian botanist and clergyman (S.J.); professor of natural history at the Theresianum in Vienna; later idem in Buda. (*Piller*).

HERBARIUM and TYPES: Unknown.

BIBLIOGRAPHY and BIOGRAPHY: AG 12(2): 474; Barnhart 3: 87; BM 4: 1577; Dawson p. 672; Herder p. 56; Kanitz 43-44, 64; Kew 4: p. 312; Plesch p. 362; PR 7149 (ed. 1: 8025); Rehder 5: 673; Sotheby 600; Tucker 1: 557; WU 22: 293-294.

BIOFILE: Kanitz, A., Bonplandia 10: 313. 1862.
Nissen, C., Zool. Buchill. 319. 1969.
Szinnyei, J., Magyar irók 10: 1163-1164. 1905 (b. 25 Apr 1733, bibl., further biogr. refs.).

7939. *Iter per Poseganam Sclavoniae* [Slavoniae] provinciam mensibus Junio, et Julio mdcclxxxii susceptum ... Budae [Budapest] (Typis regiae universitatis) 1783. Qu. (*Iter Poseg. Sclavon.*).
Co-author: Ludwig Mitterpacher von Mittenburg (1734-1814) (see Kanitz, 1862, p. 313).
Publ.: 1783, p. [1]-147, *pl. 1-16*, (uncol. copp. J. Kibler). *Copies*: NY, PH; IDC 2457. – Collation: A-S⁴T². *Pl. 2-9* are sometimes cited as coloured.

Pilling, Friedrich Oscar (1824-1897), German botanist and pomologist; high school teacher at Altenburg. (*Pilling*).

HERBARIUM and TYPES: Unknown.

BIBLIOGRAPHY and BIOGRAPHY: Barnhart 3: 87; TL-2/6527.

BIOFILE: Anon., Biogr. Jahrb. 4: 58* (d. 23 Nov 1897); Bot. Not. 1898: 45 (d.); Deut. bot. Monatschrift 16(1): 16. 1898 (d.); Nat. Nov. 19: 576. 1897.

7940. *Deutsche Flora* ... Gera (Verlag von Theodor Hofmann) 1894. Atlas and text. Oct. (*Deut. Fl.*).
Textbeilage: 1893-1894 (p. vii Ende Januar 1894), p. [i]-vii, [viii], [1]-264. *Copy*: B – (Erste Hälfte, p. 1-128: Bot. Zeit. 16 Dec 1893; Zweite Hälfte, [i]-[viii], 129-264. 1894).
Atlas: see Müller and Pilling, *Deutsche Schul-Flora* 1891-1893, see TL-2/6527 (Walter Otto Müller, 1833-1887).

Pilous, Zdněk (1912-x), Czechoslavak bryologist, Privatgelehrter at Hostinné. (*Pilous*).

HERBARIUM and TYPES: PR; in part still private; further material at BRNU and OP (5000 bryoph. especially Hepat.). *Exsiccatae*: *Musci čechoslovenici exsiccati*, fasc. 1-27, nos. 1-1350, 1945-1961, sets at B, FH, G, H, K, NY, PRC.
Bryotheca čechoslovenica, fasc. 1-3, nos. 1-90, 1937-1939, set at FH.
Bryotheca romanica, dec. 1-2, nos. 1-20, 1938, set at FH. *Sphagna čechoslovenica exsiccata*, fasc. 1-6, nos. 1-300, 1946-1950, sets at FH, H, NY, PRC.
Letters at G.

BIBLIOGRAPHY and BIOGRAPHY: Bossert p. 311; GR p. 671; IH 2: (in press); Kew 4: 312; Roon p. 88; SBC p. 129.

BIOFILE: Anon., Rev. bryol. lichénol. 18: 186. 1949, 24: 170. 1955, 25: 202-203, 398-399. 1956, 30: 295. 1961, 32: 320. 1963, 34: 395. 1966, 36: 377. 1969, 36: 787, 792. 1970. 38: 623. 1972 (bibl. not.).
Pilous, Z., Bryophyta, Sphagnidae (Flora ČSSR). Praha Mai-Jun 1972, 416 p.
Sayre, G., Mem. New York Bot. Gard. 19(2): 236-237. 1971 (exs.); Bryologist 80: 515. 1977.
Váňa, J., Preslia 44: 279-285. 1972 (tribute 60th birthday: portr., bibl. 1931-1972), Taxon 29: 675. 1980 (moss herb. private; hepatics in OP).

Pinatzis, Leonidas (1891-1964), Greek field-botanist, born at Kios (Gemlik, Asia minor), came to Greece as a refugee in 1922; energetic botanical explorer of Greece. (*Pinatzis*).

HERBARIUM and TYPES: Private (19.800; see Goulandris); duplicates at G, P, K and W.

BIOFILE: Goulandris, N., Ann. Musei Goulandris 4: 19-26. 1978.

Pinchot, Gifford (1865-1946), American botanist, forester, conservationist and politician, M.A. Yale 1901, M.A. Princeton 1904; studied forestry in France, Germany, Switzerland, Austria; consulting forester 1892-1898; forester USDA 1898-1910, governor of Pennsylvania 1923-1927, 1931-1935. (*Pinchot*).

HERBARIUM and TYPES: Unknown. Manuscripts and note books and diaries at the Library of Congress (3345 boxes).

BIBLIOGRAPHY and BIOGRAPHY: Barnhart 3: 87; BM 4: 1577; Bossert p. 311 (b. 11 Aug 1865, d. 4 Oct 1946); CSP 17: 900; Ewan ed. 2: 176; Kew 4: 312; Langman p. 584; Lenley p. 327; LS 37954-37955; NW p. 55; PH p. 513 [index]; Rehder 5: 673; SIA 74, 192, 204, 208, 7230; Tucker 1: 557-558.

BIOFILE: Anon., Bot. Centralbl. 75: 159. 1898; Bot. Jahrb. 26 (Beibl. 61): 4. 1899 (appointment Cornell); List of the more important publications of Gifford Pinchot (3 p., anon. brochure at MO); The New York Times 6 Oct 1946 (fide Verdoorn 1947).
Guinier, Ph., *in* J.F. Leroy, Bot. franç. Amér. nord 301. 1957 (french influence on P.).
Kneucker, A., Allg. bot. Z. 4: 152. 1898 (appointment Cornell).
McKeen Cattell, J., ed., Amer. Men Sci. 253. 1906, ed. 2. 372. 1910, ed. 3. 545. 1921, ed. 4. 776. 1927, ed. 5. 885. 1933, ed. 6. 1119. 1938 (b. 11 Aug 1865).
Nelson McGeary, Forester-politician, Princeton Univ. Press 1960, 481 p. (fide Ewan).
Palmer, T.S., Auk 65: 493-494. 1948.
Pinchot, G., Breaking new ground. 1947, 522 p. (autobiography).
Pinkett H.T., North Carolina hist. rev. 34: 346-357. 1957 (C. Pinchot at Biltmore).
Rodgers, A.D., Liberty Hyde Bailey 199, 202, 207-209, 359, 363, 380. 1949; Bernard Eduard Fernow 618 [index]. 1951.
Rosenberger, H.T., Cosmos Club Bull. 14(5): 2-6. 1961 (biogr., portr., bibl.).
Sutton, G.B., Charles Sprague Sargent 378-379 [index]. 1970.
Verdoorn, F., Biologia 1:4. 1947 (d.; err. "Clifford").
Verdoorn, F., ed., Chron. bot. 2: 296. 1936, 6: 18. 1941, 7: 35. 1943.

EPONYMY: *Geogr.*: Mount Pinchot, Glacier National Park, Montana; Gifford Pinchot National Forest, State of Washington.

7941. *Timber trees and forests of North Carolina* ... Winston (M.I. & J.C. Stewart) 1897 [i.e. 1898; 1897 date cancelled on t.p.]. Oct. (*Timb. trees N. Carolina*).
Co-author: William Willard Ashe (1872-1932).
Publ.: 20 Mai 1898 (date stamped on t.p.; Bot. Gaz. 17 Sep 1898), p. [1]-227, *pl. 1-23*.
Copies: MO, NY. – Bull. 6, North Carolina Geol. Surv., Raleigh.

Pinoy, Pierre Ernest (1873-1948), French physician and mycologist; Dr. med. Paris 1899, Dr. sci. Paris 1907; professor of microbiology and cryptogamy at the faculty of medicine of Algeria. (*Pinoy*).

HERBARIUM and TYPES: Unknown.

BIBLIOGRAPHY and BIOGRAPHY: Barnhart 3: 88; CSP 17: 903; Kelly p. 177; LS 37957-37960, 41696, suppl. 21853-21864; Stevenson p. 1254.

BIOFILE: Anon., Bot. Centralbl. 110: 160. 1909 (prix Montagne 1908).
Locquin, M., Bull. Soc. mycol. France 65(3/4): 93-96. 1949 (obit., portr., b. 3 Feb 1873, bibl.).
Maire, R., Progr. connaiss. bot. Algérie 174, 186. 1931.

EPONYMY: *Pinoyella* Castellani et Chalm. (1919).

Pinto, Joaquim de Almeida (x-1870), Brazilian botanist and pharmacist. (*J. de A. Pinto*).

HERBARIUM and TYPES: Unknown.

NAME: Also alphabetized as Almeida Pinto, J. de.

BIBLIOGRAPHY and BIOGRAPHY: Barnhart 3: 44; BL 1: 238, 283; Jackson p. 372; Kew 4: 313.

EPONYMY: *Pintoa* C. Gay (1846) is dedicated to General Antonio Pinto, an ex-president of Chile.

7942. *Diccionario de botanica brasileira* ou compendio dos vegetaes do Brasil, tanto indigenas como acclimados revista por una commissão da sociedade vellosiana, e approvada pela faculdade de medicina da corte. Contendo: una descripçao scientifica de cada familia a que pertencem, e outra vulgar ao alcance de qualquer intelligencia, seu emprego e differentes denominações nas diversas provincias do imperio, as propriedades medicas e venenosas, sua utilidade nas artes, industrias, economia domestica e na veterinaria coordinada redigido em grande parte sobre os manuscriptos do Dr Arruda Camara por Joaquim de Almeida Pinto ... e mandado imprimir por seu irmão o bacharel Zeferino d'Almeida Pinto. Rio de Janeiro (Typographia-Perseverança ...) 1873. Qu. (*Dicc. bot. bras.*).
Manuscript: Manoel Arruda da Camara (1752-1810).
Publ.: 1873 (p. vi: 2 Apr 1873; BSbF Dec 1873), p. [i]-xix, [1]-433, [1, err.], *15 pl. Copy*: USDA.

Pinto da Silva, Antonio Rodrigo (1912-x), Portuguese agronomic engineer and botanist; at the Estação agronómica nacional, scientific officer from 1937 (at Bélem-Lisboa 1937-1941, Sacavém 1942-1964, at Oeiras 1964-x), lately as investigador-coordenador; member (and chairman) of the Special Committee for Pteridophytes and Phanerogams, later Committee for Spermatophyta 1950-1969). (*Pinto da Silva*).

HERBARIUM and TYPES: LISE; some isotypes in COI, LISI, LISU, PO; further material at INA and US;

BIBLIOGRAPHY and BIOGRAPHY: BL 2: 464, 465, 467, 468, 471, 473, 716; Bossert p. 311 (b. 13 Mar 1912); IH 1 (ed. 1): 84, (ed. 2): 106, (ed. 3): 134, (ed. 4): 149, (ed. 5): 160, (ed. 6): 195, 2: (in press); Kew 4: 313; Roon p. 88; Zander ed. 10, p. 702, ed. 11, p. 800 (b. 13 Dec 1912).

BIOFILE: Anon., Grande encicl. luso-brasileira 28: 772. 1954 (biogr.).

Pio, Giovanni Batista (*fl.* 1813), Italian botanist. (*Pio*).

HERBARIUM and TYPES: Unknown.

BIBLIOGRAPHY and BIOGRAPHY: BM 4: 1578; Hortus 3: 1201 (Pio); Kew 4: 314; PR 7151 (ed. 1: 8028); Saccardo 1: 128; Zander ed. 10, p. 702, ed. 11, p. 800.

7943. *De Viola* specimen botanico-medicum Joannis Baptistae Pio a Mango e Montenotte praefectura ad medicinae lauream in schola medica taurinensis academiae anno mdcccxiii, mense maii, die xxv, hora v pomeridiana. In aedibus Academiae Taurinensis typis Vincentii Bianco [Torino 1813]. Qu. (*Viola*).
Publ.: 25 Mai 1813, p. [i-ii], [1]-52, *pl. 1-3* (Amati et Tela, uncol. copp.). *Copies*: G, HU, REG.

Piper, Charles Vancouver (1867-1926), American botanist; M.Sc. Univ. Washington 1892; professor of botany at Pullman, Washington 1893; M.Sc. Harvard 1900; from 1903 in Washington, D.C., with the Bureau of Plant Industry (forage crop investigations); D.Sc. Kansas Agric. College 1921. (*Piper*).

HERBARIUM and TYPES: Main collections and most types at US; private herbarium at WS (acquired 1926); further material e.g. in A, B, BUF, C, DBN, E, F, GH, L, MIN, MSC, NY, PH, PNH, POM, S, US, WCW, WS, WSP (fungi), WTU. Manuscripts, books, separates and notes at Washington State University Library (see NW p. 24-27). *Exsiccatae*: *Musci occidentali-americani* 1902, set at NY. – Letters at G, PH.

BIOFILE: Backer p. 659; Barnhart 3: 88; BJI 2, p. 136; BL 1, p. 219-220, 313; BM 4: 1578, 8: 1012; Bossert p. 311 (*b.* 16 Jun 1867, *d.* 11 Feb 1926); CSP 17: 905; (Ewan ed. 1: 159, 284, ed. 2: 176; Hortus 3: 1201 (Piper); IH 1 (ed. 6): 363, (ed. 7): 340; Kew 4: 314; Langman p. 584; Lenley p. 327-328; LS 20784, 37961, suppl. 21872-21873; MW p. 387, suppl. p. 281; NW p. 24-27, 55; Pennell p. 616; PH 209, 450, 939; Rehder 5: 674; RS p. 133; SIA 220, 221, 223, 225, 7262; SK 1: 408; Stevenson p. 1254; Tucker 1: 558, 779; Urban-Berl. p. 383; Zander ed. 10, p. 702, ed. 11, p. 800.

BIOFILE: Anon., Evening Star, Washington 12 Feb 1926 (DAB); J. New York Bot. Gard. 1: 39. 1900, 9: 4. 1908; Madroño 1: 278. 1929 (d. 11 Feb 1926); The official Record, USDA, 17 Feb 1926, p. 4.
Beattie, R.K., Proc. Biol. Soc. Washington 41: 61-66. 1928 (bibl.).
Croat, T.B., Flora Barro Colorado Isl. 1978.
Ewan, J. et al., Short hist. bot. U.S. 11. 1969.
Garby, L., Dict. Amer. biogr. 14: 632-633. 1934 (biogr., b. 16 Jun 1867, d. 11 Feb 1926).
Hedge, I.C. & J.M. Lamond, Index coll. Edinb. herb. 119. 1970.
Hitchcock, A.S., Proc. Amer. Acad. Arts Sci. 62: 275-276. 1928 (P. introduced the Sudan grass as forage plant).
Hughes, K.W., Contr. bibl. Oregon 23. 1940.
Hultén, E., Bot. Not. 1940: 320 (Alaskan coll.).
Humphrey, H.B., Makers of North American Botany 200-202. 1961.
Jones, G.N., Leafl. west. Bot. 1: 198. 1936 (P. and the young Naturalists' Society Washington State).
Oakley, R.A., Bull. Green Sect. U.S. Golf Ass. 6(3): 54-57. 1926 (portr.; golf turf investigations).
Osborn, H., Fragm. entom. hist. 201. 1937.
Pieters, A.J., Science ser. 2. 63: 248-249. 1926 (obit.).
Piper, C.V., Contr. U.S. natl. Herb. 11: 20. 1906.
Rickett, H.W., Index Bull. Torrey bot. Club 78. 1955.
Riggs, G.B., Washington hist. Quart. 20(3): 165, 166, 167. 1929.
Rodgers, A.D., Liberty Hyde Bailey 285, 347. 1949.
Sargent, C.S., Silva 9: 145. 1896.
Steere, W.C., Bot. Rev. 43: 293. 1977.
Stieber, M.T., et al., Huntia 4(1): 85. 1981 (mss. mat. HU).
Thomas, J.H., Huntia 3: 20. 969.Vinall, H.N., J. Amer. Soc. Agron. 18: 295-300. 1926 (obit., bibl., b. 16 Jun 1867, d. 11 Feb 1926).

EPONYMY: *Piperia* Rydberg (1901).

7944. *The flora of the Palouse region.* Containing descriptions of all the Spermatophytes and Pteridophytes known to grow wild in the area within thirty-five kilometers of Pullman, Washington ... Pullman, Wash. (Allen Bros., printers) 1901 (*Fl. Palouse reg.*).

Junior-author: Rolla Kent Beattie (1875-1960).
Publ.: 14 Mai 1901, p. [i*], [i]-viii, [1]-208. *Copies*: NY, US, USDA. – Piper sent a copy to N.L. Britton on 17 Mai 1901 (letter in NY copy).
Ref.: Davis, B.M., Bot. Gaz. 32(1): 62-63. 24 Jul 1901.

7945. *The flora of Mount Rainier*, *in*: Mazama, a record of Mountaineering in Pacific Northwest. 2(2): 93-117. Apr 1901, 2(4): 270-271. Dec 1905.
Publ.: Apr 1901 and Dec 1905, as above. No separately paginated reprint known to us.

7946. *Flora of the State of Washingon* ... Washington (Government Printing Office) 1906. Oct. (*Fl. Washington*).
Publ.: 8 Oct 1906, p. [1]-637, map, *pl. 1-22* (photos). *Copies*: B, BR, G, MO, NY, US, USDA(2); IDC 7601. – Contr. U.S. natl. Herb. 11, 1906.
Ref.: Engler, A., Bot. Jahrb. 38(Lit.): 66-67. 1907.
Solms, H., Bot. Zeit. 65(2): 121. 1907.
Trelease, W., Bot. Centralbl. 104: 184-187. 1907.

7947. *North American species of Festuca* ... Washington (Government Printing Office). 1906. Oct.
Publ.: 30 Mar 1906, p. [i-vi], 1-48, vii-ix, *pl. 1-15* (uncol.). *Copies*: BR, G, MO, NY, US, USDA. – To be cited from Contr. U.S. natl. Herb. 10(1): [i-vi], 1-48, vii-ix, *pl. 1-15*. 1906. – Supplementary notes: ib. 16: 197-199. 1913.

7948. *Agricultural varieties of the Cowpea* and immediately related species ... Washington (Government Printing Office) 1912. Oct. (*Agric. var. Cowpea*).
Publ.: 29 Feb 1912, p. [1]-160, *pl. 1-12* (photos). *Copy*: US.

7949. *Flora of Southeastern Washington* and adjacent Idaho ... Lancaster, Pa. (The New Era Printing Company) 1914. Oct. (*Fl. s.e. Washington*).
Junior author: Rolla Kent Beattie (1875-1960).
Publ.: 22 Jan 1914, map, p. [i]-xi, 1-296. *Copies*: G, NY, US, USDA.
Facsimile ed.: 1936, p. [i]-xi, 1-296. *Copy*: USDA. Lithoprinted O.S.C. Cooperative Association, Corvallis, Oregon.

7950. *Flora of the northwest coast* including the area west of the summit of the Cascade mountains forty-ninth parallel south to the Calapooia mountains on the south border of Lane county, Oregon ... Lancaster, Pa. (Press of the New Era Printing Company) 1915. Oct. (*Fl. n.w. coast*).
Junior author: Rolla Kent Beattie (1875-1960).
Publ.: 10 Nov 1915 (see p. [ii] and Torreya 15 Jun 1916), p. [i]-xiii, [1]-418. *Copies*: G, MO, NY, US, USDA; IDC 7602.
Ref.: Rydberg, P.A., Torreya 16: 143-146, 187. 1916.

7951. *A study of Allocarya* ... Washington (Government Printing Office) 1920. Oct.
Publ.: 13 Mar 1920 (U.S. copy so stamped: sic, not 29 Mar 1920 as printed in volume). Published as Contr. U.S. natl. Herb. 22(2): [i]-v, 79-113, vii. *Copies*: MO, US.

7952. *The Soy bean* ... New York, London (McGraw-Hill Book Company, Inc.) 1923. Oct. (*Soy bean*).
Co-author: William Joseph Morse (1884-1959).
Publ.: 1923 (ÖbZ Jan-Jun 1923), p. [i]-xv, 1-329. *Copies*: MO, USDA.
Other ed.: 1943, p. [i]-xv, 1-329. *Copy*: USDA. – New York (Peter Smith). Oct.

7953. *The American species of Canavalia and Wenderothia* ... Washington (Government Printing Office) 1925. Oct.
Publ.: 27 Apr 1925, to be cited from Contr. U.S. natl. Herb. 20(14): [i]-v, 555-588, vii-viii. 1925. *Copies*: MO, US.

7954. *Studies in the American Phaseolineae* ... Washington (Government Printing Office) 1926. Oct.

Publ.: 12 Jun 1926, to be cited from Contr. U.S. natl. Herb. 22(9): [i]-v, 663-701, vii-viii, *pl. 64. Copies*: MO, US.

Piper, Richard Upton (1818-1897), American physician and botanist, MD Dartmouth 1840; practicing physician at Boston and Chicago. (*R. Piper*).

HERBARIUM and TYPES: Unknown.

BIBLIOGRAPHY and BIOGRAPHY: Barnhart 3: 88 (b. 3 Apr 1818); Lenley p. 466; ME 3: 473; Rehder 1: 304; Tucker 1: 558.

BIOFILE: Allibone, S.A., Crit. dict. Engl. lit. 1601. 1870 (bibl.).
Anon., Appleton's cycl. Amer. biogr. 5: 30. 1888 (b. 3 Apr 1818, probably correct).
Kelly, H.A. & W.L. Burrage, Amer. med. biogr. 917. 1920; Dict. Amer. med. biogr. 970. 1928 (b. 3 Apr 1886, d. Aug 1897).

7955. *The trees of America* ... Entered, according to the act of Congress, in the year 1855, by R.U. Piper in the Clerk's Office of the District Court of the District of Massachusetts [Boston 1855]. (*Trees Amer.*).
1: 1855, p. [i-iv], [1]-16, *3 pl.* (uncol. liths.).
2: 1855, p. 17-32, *4 pl.*
3: 1858, p. 33-48, *3 pl.* "Boston, A. Williams & Co. ... 1858".
4: 1858, p. 49-64, *3 pl.* "Boston, A. Williams & Co. ... A. O. Moore ... New York, 1858".
Copies: NY, USDA.
Ref.: Anon., North Amer. Rev. 85: 178-204. Jul 1857 (rev. Nos. 1, 2).

Piquet, John (1825-1912), British pharmaceutical chemist and botanist at St. Hélier, Jersey. (*Piquet*).

HERBARIUM and TYPES: JSY; other material at CGE, MANCH, OXF (Druce); a collection of seaweeds was made by him for Van Heurck (AWH); see Kent (1957) and Perrédès (1912) for references to other special herbaria.

BIBLIOGRAPHY and BIOGRAPHY: Barnhart 3: 89; BB p. 244 (b. 16 Mar 1825, d. 5 Sep 1912); Clokie p. 221; Desmond p. 496.

BIOFILE: Druce, G.C., Bot. Exch. Cl. Soc. Brit. Isl. 3(3): 205-207. 1913 (obit., d. 5 Sep 1912), 6(1): 177. 1921 (herb. bought).
Kent, D.H., Brit. herb. 73. 1957.
Perrédès, P.É.F., J. Bot. 50: 371-374. 1912 (portr.).

EPONYMY: *Piquetia* H. Hallier (1921) and *Piquetia* N.E. Brown (1925) are likely named for Piquet.

Piré, Louis Alexandre Henri Joseph (1827-1887), Belgian botanist; high school teacher at the Bruxelles Athenaeum 1854-1883; in retirement at Spa (1884-1887). (*Piré*).

HERBARIUM and TYPES: BR; other material at BM, FI, K, L, PC. *Exsiccatae*: *Les mousses de la Belgique*, fasc. 1-2, 1870-1871, sets at B, BM, BR, K, LG. Sayre found no evidence of the existence of fasc. 2 (see Delogne, Fl. crypt. Belg. 1: 308. 1883). – Bull. Soc. Bot. Belg. 7: 401. 1868 mentions the receipt of a *Bryotheca belgica*. – Letters at G.

BIBLIOGRAPHY and BIOGRAPHY: Backer p. 659; Barnhart 3: 89; BL 2: 36, 703; BM 4: 1578; Bossert p. 311 (b. 4 Mar 1827, d. 6 Jul 1887); CSP 4: 918, 8: 628, 11: 25, 12: 578, 17: 905; Herder p. 433; Jackson p. 193, 272; Kew 4: 314; Morren ed. 10, p. 44; PR 7154; Rehder 5: 674; SBC p. 129; SO 2746b; TL-2/6541; Tucker 1: 558; Urban-Berl. p. 310.

BIOFILE: Anon., Bull. Soc. bot. France 34 (bibl.): 143. 1887 (d. 16 Jul 1887); Nat. Nov. 9: 216. 1887; Rev. bryol. 14: 80, 95. 1887.

Crépin, F., Ann. Bot. 1: 410-411, 1887 (bibl.); Bull. Soc. Bot. Belg. 29: 7-16. 1890 (biogr., bibl.); Guide bot. Belg. 475-476. 1878; Biogr. natl. Belgique 17: 556-558. 1903 (b. 3 Mar 1827, d. 16 Jul 1887).
Evens, F.M.J.C., Gesch. algol. België 186. 1944.
Jackson, B.D., Bull. misc. Inf. Kew 1901: 53 (exs.).
Sayre, G., Mem. New York Bot. Gard. 19(2): 238. 1971 (exs.); Bryologist 80: 515. 1977.
Seyn, E. de, Dict. biogr. sci. lettr. arts Belgique 2: 823. 1936.

COMPOSITE WORKS: Bulletin de la Société Royale Linnéenne de Bruxelles 1872-1887.

EPONYMY: *Pirea* J. Cardot (1894); *Pireella* J. Cardot (1913). *Note*: The source of *Pirea* T. Durand (1888) was not given.

7956. *Flore analytique du centre de la Belgique* ... Bruxelles (Comptoir universel d'imprimerie et de librairie, Victor Devaux et Cie., ...) 1866. Oct. (*Fl. centr. Belgique*).
Co-author: [Pierre] Félix Muller (1818-1896).
Publ.: Jul-Dec 1866, (p. viii: 11 Jun 1866), p. [i]-x, [1-2, err.], [1]-299, [300]. *Copies*: BR, G, NY. – Also treated as TL-2/6541.

7957. *Les sphaignes de la flore de Belgique* ... Bruxelles (Gustave Mayolez, ...) 1868. Oct.
Publ.: Jan-Jun 1868 (séance 1 Dec 1867; BSbF 7(1) publ. Aug 1868), p. [1]-21, *1 pl.*
Copy: H (inf. P. Isoviita). Reprinted and to be cited from Bull. Soc. Belg. 6(3): 323-339. 1868.

7958. *Recherches bryologiques.* [1, 2:] Revue de quelques genres de mousses pleurocarpes ... Gand [Gent] (Imprimerie C. Annoot-Braeckman, ...) 1868-1871, 5 fasc. Oct.
Publ.: a series of five papers published in the Bull. Soc. Bot. Belg. (BSBB) 1868-1871, to be cited from journal.
1: Aug 1868 (Cogniaux rd. his copy on 29 Nov 1868; BR), p. [1]-18, BSBB 7(1): 70-83. 1868.
2: Nov 1868 (Cogniaux id. 29 Nov 1868), p. [1]-27, BSBB 7(1): 181-203. 1868.
3(1): Aug 1869, p. [1]-31, BSBB 8: 109-135. 1869. – Subtitle: *Revue des mousses acrocarpes de la flore belge.*
3(2): 15 Mar 1870, p. [33]-76, BSBB 8: 406-449. 1870. – Subtitle: as 3(1).
4: Nov 1871, p. [1]-25, *pl. 1-2* (col. liths. A. Piré), BSBB 10: 86-106. 1871. – *Nouvelles recherches bryologiques.*
Copies: BR, G, H(1, 2).
Ref.: Anon., Bull. Soc. bot. France 18 (Bibl.): 164. Nov 1871.
Geheeb, A., Hedwigia 11: 76-77. Mai 1872, Flora 55: 253-254. 1 Jun 1872; Bot. Zeit. 30: 333-335. 10 Mai 1872 (fasc. 4).

7959. *Tableau des familles végétales* avec l'énumération des plantes les plus utiles ... première partie. Monopétales calyciflores ... Bruxelles (Imprimerie Félix Callewaert Père ...) 1876. Oct. †.
Publ.: 1876 (t.p. reprint), p. [i], [1]-60. *Copy*: G. – Originally published and to be cited from Bull. Soc. roy. linn. Bruxelles; 4(3/4): 46-75, (8): 179-189, (10): 203-221, 247-252. 1875 (all fascicle covers dated 1875).

7960. *Analyse des familles et des genres de la flore Bruxelloise* ... Bruxelles (Imprimerie Félix Callewaert père, ...) 1879. Oct. (*Anal. fl. Bruxell.*).
Ed. 1: 1879 (t.p. reprint; Bot. Zeit. 13 Aug 1880; Nat. Nov. Jun(2) 1880), p. [1]-39, *1 pl.*
Copy: BR. – Reprinted or preprinted from Bull. Soc. Linn. Bruxelles 8: 27-65. 1879.
Ed. 2: Apr-Jun 1883 (p. 6: 1 Mar 1883; Bot. Zeit. 31 Aug 1883; Bot. Centralbl. 23-27 Jul 1883; Nat. Nov. Jul(1) 1883), p. [1]-62, [63]. *Copy*: BR. – 2e édition. Bruxelles (Office de publicité A.-N. Lebègue & Cie, ...) s.d.

7961. *Les muscinées des environs de Spa* ... [Bull. Soc. Bot. Belg. 24(1) 1885]. Oct.
Co-author: Jules Cardot (1860-1934), Piré's son-in-law.
Publ.: Dec 1885, p. [1]-29, [30]. *Copies*: BR, G. – Reprinted and to be cited from Bull. Soc. Bot. Belg. 24(1): 326-350. 1885 (p. 350 dated 13 Dec 1885).

7962. *Les végétaux inférieurs* ... Bruxelles (Librairie classique de A.-N. Lebègue et Cie, ...) [1882]. Oct. (*Vég. inf.*).
Publ.: Jun-Nov 1882 (p. viii: 1 Mai 1882; Nat. Nov. Dec(1) 1882), p. [i]-viii, [9]-114. *Copy*: BR. – "Collection nationale".

7963. *Spicilège de la flore bryologique des environs de Montreux-Clarens* ... Bruxelles (Jardin botanique de l'État) 1882. Oct.
Publ.: 1882, p. [1]-14. *Copies*: BR, G. – Reprinted from Bull. Soc. Bot. Belg. 21(2): [1]-14. 1882.

Pirona, Giulio Andrea (1822-1895), Italian botanist; Dr. med. Padova 1846; from 1851-1887 lecturer at the Ginnasio liceale and Udine medical college as well as practicing physician at Udine, from 1887 with the title "professore effetivo". (*Pirona*).

HERBARIUM and TYPES: MFU; further material at E, FI and H. – The original collections at MFU are partly destroyed.

BIBLIOGRAPHY and BIOGRAPHY: AG 6(2): 422; Barnhart 3: 89; BL 2: 355-356, 703; BM 4: 1578-1579; Bossert p. 311; CSP 4: 919, 8: 628, 11: 25, 17: 906; Hegi 4(3): 1269; Herder p. 175; Kew 4: 315; Morren ed. 10, p. 88; PR 7156; Quenstedt p. 340; Rehder 1: 74, 427; Saccardo 1: 128-129, 2: 85, 155-156 (b. 20 Nov 1822, d. 28 Dec 1895), cron. p. xxix; Tucker 1: 558.

BIOFILE: Anon., Atti r. Ist. Veneto 61(1): 215-222. 1902 (obit., bibl., b. 20 Nov 1822, d. 28 Dec 1895).
Hedge, I.C. & J.M. Lamond, Index coll. Edinb. herb. 119. 1970.

COMPOSITE WORKS: *Vocabulario botanico friulano*, in Gian Jacopo Pirona (1789-1870), *Vocabulario friulano*. Venezia 1869, p. 481-526. (PR 7155); first published in Atti solenna distr. premi 1862 ginnasio Udine, p. 3-81, Udine 1862 (see BL 2: 355-356).

7964. *Florae forojuliensis syllabus* ... Utini [Udine] (Typis Literalis Vendrame) 1855. Oct. (*Fl. forojul. syll.*).
Publ.: Oct 1855 (p. 8: med Aug 1855; ÖbZ 20 Dec 1855; Flora rd. Oct 1855), p. [1]-170. *Copies*: G, NY. – On p. 170: "Del programma dell I.R. Ginnasio-Liceale di Udine pel 1855." The *Forum Julii* is identical with Friuli [German: Friaul] (Udine and Goritzia, now NE Italy and NW Yougoslavia), which was Austrian from 1797-1865, to Italy 1866-1918 (Western part only), to Italy 1918-1945 (as a whole), to Yugoslavia 1945 (Eastern part only). – For further details see letter Pirona in Saccardo 2: 155-156.*Ref.*: Fürnrohr, A.E., Flora 38: 654-656. 7 Nov 1855.
Schlechtendal, D.F.L. von, Bot. Zeit. 13: 830-831. 23 Nov 1855.

Pirotta, Pietro Romualdo (1853-1936), Italian botanist (educated in Pavia); director of the botanic garden at Modena 1880-1883, professor of botany at the University of Roma 1883 (extraord.), 1888. (ord.) and director of the botanic garden until 1928. (*Pirotta*).

HERBARIUM and TYPES: RO; other material at BP, C, CAG, F, FI, MOD, MPU, PAD, WRSL. – Letters and portrait at G.

BIBLIOGRAPHY and BIOGRAPHY: Barnhart 3: 89; BJI 1: 46, 2: 136; BL 1: 33, 313, 2: 352, 358, 361, 703; BM 4: 1579; Bossert p. 311 (b. 7 Feb 1853, d. 3 Aug 1936); Cesati p. 314; CSP 11: 25-26, 12:578, 17: 907; DTS 1: 226; GR p. 537; Herder p. 118; Hirsch p. 233; Hortus 3: 1201; IF p. 724, suppl. 1: 84; IH 1 (ed. 6): 363, (ed. 7): 340, 2: (in press); Kew 4: 316; Langman p. 585; LS 20787-20819a, 23445; Morren ed. 10, p. 81; MW p. 387; Rehder 5:674; Tucker 1: 558; Zander ed. 10, p. 702, ed. 11, p. 800 (b. 7 Feb 1853, d. 3 Aug 1936).

BIOFILE: Anon., Amer. J. Sci. ser. 5. 34: 408. 1937 (d. 4 Aug 1936); Bot. Centralbl. 16: 255; 1883, 34: 383. 1888; Bot. Zeit. 39: 20. 1881 (app. Modena), 41: 874. 1883 (app.

Roma); Nat. Nov. 5: 216. 1883 (to Roma); Österr. bot. Z. 31: 67. 1881 (app. Modena), 38: 254. 1888 (app. Roma), 85: 320. 1936 (d. 8 Aug 1936).
Carano, E., Ann. Bot., Roma 18(1): 1-6. 1928 (75th birthday), 21(2): 384-412. 1937 (reprint 29 p.), (bibl., portr., extensive biogr.).
Ciferri, R., Taxon 1(8): 126. 1952.
De Toni, G.B., Malpighia 20: 279. 1906 (on Pirotta at Modena).
Gager, C.S., Brooklyn Bot. Gard. Rec. 27: 275 (Modena 1880-1883), 287 (Roma 1884-1928, dir. bot. gard.). 1938.
Lusina, G., Annali bot. 25(1/2): 167. 1956 (bibl. of papers on flora of Rome).
Verdoorn, F., Chron. bot. 3: 191-192. 1937 (obit., portr., b. 7 Feb 1852, d. 3 Aug 1936).

COMPOSITE WORKS: (1) *Malpighia*, vols. 1-16, 1886-1902, editor.
(2) With Terracciano, A. & U. Brizi, *Flora della provincia di Roma*, in E. Abbate, *Guida della Provincia di Roma* 171-225. 1890 (pub. before 15 Oct 1890).
(3) Founder of the *Annali di Botanica*, Roma, editor vols. 1-16, 1903-1926; with E. Carano 17-18, 1927-1928.
(4) With F. Cortesi, Relazione sulle piante raccolte nel Karakoram, *in* F. de Filippi, La spedizione nel Karakoram e nell' Imalaia occidentale 1909, Bologna, (Nicola Zanichelli) 1912, p. [93]-110 (Bot. Centralbl. 17 Sep 1912) also as app. D. in English version: Karakoram and Western Himalaya 1909, New York (E.P. Dutton and Company, ...) 1912, p. [455]-469. *Copies*: PH.
(5) Gymnospermae, Pteridophyta, *in* L.A. di Savoia, *Il Ruwenzori* 475-476, 477-483. 1909.

EPONYMY: *Pirottaea* P.A. Saccardo (1878); *Pirottantha* Spegazzini (1917).

7965. *Funghi parassiti dei vitigni* .... Milano (Coi tipi di G. Bernardoni) 1877. Oct.
Publ.: 1877 (Bot. Zeit. 26 Oct 1877; Hedwigia Dec 1877), p. [3]-96, *pl. 10-13* (uncol. liths.). *Copy*: Stevenson. – Reprinted from Arch. trienn. Lab. bot. critt. Pavia 2/3: 129-215. 1877.

7966. *Flora del Modenese e del Reggiano* ... Modena (Tipografia di G.T. Vincenzi e nipoti) 1882. Oct. (*Fl. Moden.*).
Co-author: Giuseppe Gibelli (1831-1898), see TL-2/1, following 2009.
Publ.: 1882 (Nat. Nov. Dec(2) 1882; Bot. Centralbl. 29 Jan-2 Feb 1883), p. [1]-196. *Copies*: FI, NY. – Reprinted from Atti Soc. Natural. Modena ser. 3. 1: 29-220. 1882.
Suppl.: 1884 (Bot. Centralbl, 11-15 Aug 1884; Nat. Nov. Jan(2) 1885), p [1]-30. *Copies*: FI, NY. – Reprinted from id. ser. 3. 3: 1-30. 1884.

7967. *Contribuzione all'anatomia comparata della foglia I. Oleaceae* ... Roma (Tipografia della r. Accademia dei Lincei ...) 1885. Qu. (*Contr. anat. foglia, Oleac.*).
Publ.: 1885 (t.p. reprint), p. [1]-28, *pl. 2. Copy*: U. – Preprinted or reprinted from Annu. Ist. bot. Roma 2: 22-47. 1886.

7968. *Flora romana* ... Roma (Tipografia Enrico Voghera ...) 1900. Qu. †. (*Fl. Rom.*).
Co-author: Emilio Chiovenda (1871-1940).
Publ.: in 2 fasc. as Annuario R. Ist. Bot. Roma 10(1 and 2), 1900-1901.
1: 20-31 Dec 1900 (p. viii: 1 Nov 1899; p. 144: printing finished 20 Dec 1900; J. Bot. Morot Feb 1901; Nat. Nov. Aug(1) 1900), p. [i]-viii, [1]-144.
2: shortly after 15 Aug 1901 (p. 304: printing finished 15 Aug 1901; J. Bot. Morot Feb 1902), p. 145-304.
Copies: MO, NY. – Remained incomplete. Treats botanical history until Feliciani Scarpellini (b. 1762). Another botanical bibliography of the area is G. Lusina, Annali Bot. 25: 127-178. 1956.
Ref.: Kneucker, A., Allg. bot. Z. 8: 102. 15 Mai 1902.
Solla, Bot. Centralbl. 86: 422-427. 1901.

7969. *Flora della colonia Eritrea* ... Roma (Tipografia Enrico Voghera ...) 1903. Qu. (*Fl. Eritrea*).
Co-authors: Riccarda Almagia (1, 2), Béatrice Armari (2), Eva Boselli (2), Ernesta di Capua (2), Emilio Chiovenda (1, 2, 3), Fabrizio Cortesi (3).

Publ.: in 3 parts in Annuario R. Ist. Bot. Roma vol. 8 (1, 2, 3), 1903-1908:
1: Jan 1903 (p. 128: printing finished 31 Dec 1902; ÖbZ Jun 1903; Nat. Nov. Jul(1) 1903), p. [1]-128, *pl. 1-12* (uncol. liths. Chiovenda).
2: Mai 1904 (p. 264: printing finished 30 Apr 1904, p. 129-264; Bot. Centralbl. 14 Jun 1904), p. 129-264.
3: Jan 1908 (t.p.; p. 464: printing finished 31 Dec 1907; Nat. Nov. Mai(1) 1908), p. 265-464.
Copies: BR, FI, MO, NY – Fascicle 4, prepared by E. Chiovenda to complete the work was lost (see BL 2: 33; Chiovenda, Atti Cong. Studi Colon. 3: 50. 1931).

Piscicelli, Maurizio (1871-?), Italian explorer of northeastern tropical Africa; with the Duchess of Aosta in Somalia 1911. (*Piscicelli*).

HERBARIUM and TYPES: Some material at RO.

BIBLIOGRAPHY and BIOGRAPHY: SK 1: 408.

BIOFILE: Schubert, B.G. & G. Troupin, Taxon 4: 94-96. 1955 (see also Bull. Soc. Bot. Belg. 85: 5-8. 1952).

7970. *Nella regione dei Laghi Equatoriali* (sotto gli auspici della Reali Società geographica). Napoli (Libreria Luigi Pierro) [1913]. (*Reg. Laghi equat.*).
Publ.: on or shortly after 25 Nov 1913 (colophon: printing finished 25 Nov 1913), p. [i]-vii, [ix], [1]-479, [481, ind., 483 colo., 485-486 expl. map], map. *Copies*: FI, LC. – The new taxa described and illustrated in this work (based on material collected by Hélène, Duchess of Aosta) were previously published by L. Buscalioni and R. Buschler, Bot. Jahrb. 49: 457-512. 28 Mar 1913, 513-515. 17 Jun 1913. See Schubert and Troupin (1955) for further details.

Piso, Willem (ca. 1611-1678), Dutch physician, pharmacist and botanist; M.D. Caen 1633; practicing physician at Amsterdam; in Brazil as physician of the Dutch settlement (1633-1644) with Johan Maurits van Nassau. (*Piso*).

HERBARIUM and TYPES: Part of the Marcgrave herbarium is at C; see under G. Marcgrave.

NOTE: See G. Marcgrave (TL-2/3: 286-287) for further information; see NI 1533 for further information on the *Historia naturalis Brasiliae*. 1648.

BIBLIOGRAPHY and BIOGRAPHY: Aa 15: 322-333; Backer p. 447; Barnhart 2: 89; BM 3: 1579-1580; Dryander 3: 462; DSB 10: 621-622 (bibl., sec. refs.); Frank 3(Anh.) p. 77; Herder p. 56, 140; JW 3: 365, 4: 401-402, 5: 249; Kew 4: 317; Lasègue p. 343, 474, 505; Lenley p. 328; NI 1533, see also 1: 78, 82, 83; NNBW 9: 805-806; PR 972, 5911, 7157-7158, ed. 1: 8036-8037; Rehder 5: 674; TL-2/5546; see G. Marcgrave; Tucker 1: 558.

BIOFILE: Andel, M.A. van, Opusc. sel. neerl. arte med. 14: i-xxxviii. 1937 (see also ib. 10, 1931 for an account of de Bondt, who edited the second ed. of the Hist. nat. bras.).
Andrade-Lima, D. et al., Bot. Tidsskr. 71: 121-160. 1977 (see also sub Marcgrave).
Barnhart, J.H., J. New York Bot. Gard. 30: 154. 1929.
Bartlett, H.H., 55 Rare books 1949, no. 31.
Ewan, J. & N., John Banister 482 [index]. 1970; Regn. veg. 71: 19, 21, 50. 1970.
Fries, T.M., Bref Skrifv. Linné 1(5): 67. 1911.
Hulth, J.M., Bref Skrifv. Linné 2(1): 34. 1916.
Hulth, J.M. & A.H. Uggla, Bref Skrifv. Linné 2(2): 83, 84, 161, 162. 1943.
J., Dict. Sci. méd., Biogr. méd. 6: 425-426. 1824.
Jessen, K.F.W., Bot. Gegenw. Vorz. 269, 272. 1884.
Junk, W., Rara 1: 50. 1900-1913.
Papavero, N., Essays hist. neotrop. dipterol. 1: 3, 13. 1971.
Pickel, B.J., Revista fl. medic. 16(6): 211-280. 1949 (Piso & Marcgrave in Brazilian botany).

Poiret, J.L.M., Encycl. méth., Bot. 8: 748. 1808.
Urban, I., Fl. bras. 1(1): 76-77. 1906 (q.v. for further secondary literature).

EPONYMY: *Pisoa* Cothenius (1790, *orth. var.* of *Pisonia* Linnaeus); *Pisonia* Linnaeus (1753) and the derived genus *Pisoniella* (Heimerl) Standley (1911).

Pitard, Charles-Joseph Marie (Pitard-Briau) (1873-1927), French botanist; Dr. sci. nat. Bordeaux 1899, pharmacien supérieur 1901; from 1897-1902 at the Faculté des Sciences, Bordeaux; from 1902 curator, from 1904 professor of natural sciences at the École de médecine, Tours; travelled in the Canary Islands (1904/05, 1905/06) and North Africa. (*Pit.*).

HERBARIUM and TYPES: Pitard left his herbarium to John Briquet at G with the exception of the special herbaria of Morocco and the Canary Islands (2028 specimens) which were deposited at P and PC. Briquet deposited the Pitard herbarium at G (34.592 nos.). Other Canary Islands and Moroccan material collected by Pitard was later also acquired by G. Lichens at M. Duplicates in many herbaria. *Exsiccatae: Plantae canarienses* [*exsiccatae*] 1905-1906, 3 series, nos. 1-851. (Enumerated in Pitard, Iles Canaries 467-477. 1908), sets at e.g. BP, FI, G, GRO, L, MO, NSW, P, PAL, PC. Irregular exsiccatae were distributed with labels *Plantae tunetanae* and *Plantes du Maroc*. – Letters and portrait at G.

BIBLIOGRAPHY and BIOGRAPHY: Barnhart 3: 89 (err. "Pitar"); BL 1: 46, 61, 87, 313, 2: 152, 703; BM 4: 1580; CSP 17: 910; GR p. 293-294 (b. 30 Oct 1873, d. 29 Dec 1927), cat. p. 69; Hawksworth p. 185; Hortus 3: 1201; IH 1 (ed. 6): 363 (material at GRO), (ed. 7): 340, 2: (in press); Kelly p. 177; Kew 4: 317; Langman p. 585; LS suppl. 21882-21883; MW p. 387; Nordstedt suppl. p. 13; Rehder 2: 532, 5: 675; SBC p. 129; Tucker 1: 387, 558, 778, 779; TL-2/4275, 4634, see G. Negri; Urban-Berl. p. 310, 383; Zander ed. 10, p. 702, ed. 11, p. 800.

BIOFILE: Anon., Bull. Soc. bot. France 53: 96. 1906 (offer exsicc.), 54: 48. 1907 (id.).
Briquet, J. & F. Cavillier, Charles-Joseph Pitard 1873-1927. Notice biographique, précédée d'un hommage à la mémoire de C.J. Pitard par Raoul Mercier, Genève, 1930, 39 p., reprinted from Candollea 4: 202-240. 1930 (portr., main biogr., bibl.), see also J. Bot. 68: 288. 1930.
Briquet, J., Bull. Soc. bot. Suisse 50a: 373, 383. 1940 (biogr., bibl., epon., biogr. refs., b. 30 Oct 1873, d. 29 Dec 1927).
Gagnepain, F., *in* H. Lecomte, Fl. gén. Indochine tome prél. 46. 1944 (portr., *pl. 13*).
Hedge, I.C. & J.M. Lamond, Index coll. Edinb. herb. 1970 (Moroccan and Tunesian material).
Jovet-Ast, S. et H. Bischler, Rev. bryol. lichénol. 8: 3. 1972.
Lecomte, H., Bull. Mus. Hist. nat. 34: 125-128. 1928 (obit.), Charles-Joseph Pitard, Tours 1928, 6 p. (b. 30 Oct 1873, d. 29 Dec 1927; obit., portr.).
Maire, R., Progr. Connaiss. bot. Algérie 154-155, 177, 178, 190. 1931.
Miège, J. & Wuest, J., Saussurea 6: 133. 1975.
Sayre, G., Mem. New York Bot. Gard. 19(3): 380-381. 1975 (coll.).
Steenis-Kruseman, M.J. van, Blumea 25: 45. 1979.
Steinberg, C.H., Monogr. biol. Canar. 4: 33, fig. 40. 1973 (exsicc. FI; label).

COMPOSITE WORKS: Contributed to H. Lecomte, *Fl. gén. Indochine* (TL-2/4275):
Guttiferae, 1(4): 292-330. Mar 1910.
Ternstroemiaceae, 1(4): 330-352. Mar 1910.
Stachyuraceae, 1(4): 352-353. Mar 1910.
Ilicaceae, 1(8): 850-863. Jan 1912.
Celastraceae, 1(8): 863-894. Jan 1912.
Hippocrateaceae, 1(8): 895-907. Jan 1912.
Rhamnaceae, 1(8): 908-934. Jan 1912.
Rubiaceae, 3(1): 20-144. Dec 1922, 3(2): 145-288. Aug 1923, 3(3): 289-432. Feb 1924, 3(4): 433-442. Mai 1924.
Myrsinaceae, 3(6): 765-808. Feb 1930, 3(7): 809-877. Dec 1930.
Apocynaceae, 3(8): 1087-1182. Mar 1933, 3(9): 1123-1262. Jun 1933.

HANDWRITING: Monogr. biol. Canar. 4: 33, *fig. 40*. 1973.

THESIS: Recherches sur l'évolution et la valeur anatomique et taxonomique du péricycle des Angiospermes. Thèse de la Faculté de médecine et de pharmacie de Bordeaux. Mém. Soc. Sci. phys. nat. Bordeaux ser. 6, 1, 1901 (185 p.); see e.g. Gidon, Bot. Centralbl. 90: 21-22. 1902.

EPONYMY: *Pitardia* Battandier ex Pitard 1918 (fide Briquet 1940), ING 3: 1348.

7971. *Contribution à l'étude des Muscinées des îles Canaries* [Coulommiers (Imp. Paul Brodard) 1907]. Oct. *Contr. Musc. Canaries*).
Collaborators: Louis Corbière (1850-1941), Giovanni Negri (1877-1960).
Publ.: Jun 1907 (cover p. 3), p. [1]-44. *Copies*: BR, G. – Issued as Mém. Soc. Bot. France no. 7.
Ref.: Anon., Rev. bryol. 34: 124. Dec 1907.

7972. *Les Iles Canaries*. Flore de l'Archipel ... Paris (Librairie des sciences naturelles Paul Klincksieck ...) [1909]. Oct. (*Iles Canaries*).
Co-author: L. Proust (1878-1959); collaborator for Musci: Giovanni Negri (1877-1960), Hepaticae: Louis Corbière (1850-1941).
Publ.: Jan-Mar 1909 (p. 2: Mar 1908; Bot. Centralbl. 6 Apr 1909; Nat. Nov. Apr(2) 1909; Bot. Zeit. 16 Aug 1909; J. Bot. Mai 1909), p. [i-iii], [1]-502, [1, cont.], *pl. 1-19* (photos). *Copies*: B, BR, G, MO, NY, U; IDC 7404.
Facsimile ed.: 1973, Koenigstein-Ts./ B.R.D. (Otto Koeltz Antiquariat), ISBN 3-87429-050-6, p. [i-iii], [1]-502, [1, cont.], *pl. 1-19*. *Copy*: FAS.
For *Musci* (p. 415-463) see also Pitard (1907), above; for *Lichenes* see Pitard 1911 (below).
Supplementary publ.: Lindinger, L., Abh. Gebiet Auslandsk. Hamb. Univ. 21: 135-350. 1926 (*Beiträge zur Kenntniss von Vegetation und Flora der kanarischen Inseln*), see TL-2/4634.
Ref.: Britten, J., J. Bot. 47: 188-190. 1909.
Gagnepain, F., Bull. Soc. bot. France 56 (bibl.): 75-77. 1909.
Offner, J., Bot. Centralbl. 113: 185-187. 1910.
Stafleu, F.A., Taxon 22: 492-493. 1973.

7973. *Contribution à l'étude des lichens des îles Canaries* ... [Coulommiers (Paul Brodard)] 1911. Oct. (*Contr. lich. Canaries*).
Co-author: Julien Herbert Auguste Jules Harmand (1844-1915).
Publ.: Oct 1911 (p. 1 cover and p. 72), p. [1]-72. *Copies*: O. Almborn, FH. – Issued as Mém. Soc. Bot. France 22 (suppl. to Bull. Soc. bot. France 58). Addendum: fungi by Vouaux.

7974. *Exploration scientifique du Maroc* organisée par la Société de Géographie de Paris. Premier fascicule. *Botanique* (1912) par M.J. Pitard. Paris (Masson et Cie., éditeurs de la Société de Géographie ...) 1913. Qu. (*Explor. sci. Maroc, Bot.*).
Publ.: 1913 (p. viii: Feb 1913; ÖbZ Mai-Jun 1914; Nat. Nov. Jun 1914), p. [i]-xxix, [1]-187, *pl. 1-9* (photos). *Copy*: G.

7975. *Contribution à l'étude de la flore du Maroc*. Paris (Éditions scientifiques E. le Moult ...) 1931. Oct. (*Contr. fl. Maroc*).
Publ.: 1931 (Nat. Nov. Jun 1932), p. [1]-80. *Copy*: G.

Pitcher, Zina (1797-1872), American physician and botanist, MD Middlebury College, Vermont 1822; in service on the frontier 1822-1836; later in Michigan; 1837-1852 regent and professor at the University of Michigan; 1840-1843 mayor of Detroit. (*Pitcher*).

HERBARIUM and TYPES: NA; other material at MICH, NY, PH and WIS. – Letters at G.

BIBLIOGRAPHY and BIOGRAPHY: AG 5(3): 68; Backer p. 448; Barnhart 3: 89; Bossert p. 312 (b. 12 Apr 1797, d. 5 Apr 1872); IH 1 (ed. 6): 363, 2: (in press); Lenley p. 466; ME 1: 219-220, 2: 261, 637; Pennell p. 616; PH 567; Zander ed. 10, p. 702.

BIOFILE: Barnhart, J.H., Mem. Torrey Bot. Club 16: 298. 1921.
Geiser, S.W., Field & Laboratory 4(2): 51. 1936; Natural. Frontier 280. 1948.
Kelly, H.A., Cycl. Amer. med. biogr. 2: 274, 276. 1912; Some Amer. med. bot. 145-150. 1914 (portr.).
Kelly, H.A. & W.L. Burrage, Amer. med. biogr. 917-918. 1920 (b. 12 Apr 1797).
Meyer, F.G. and S. Elsasser, Taxon 22: 382-383. 1973.
Voss, E.G., Univ. Mich. herb. 13: 12-14, 16, 18, 20, 25, 27, 58. 1978 (portr.).

HANDWRITING: Taxon 22: 382. 1973.

EPONYMY: *Pitcheria* Nuttall (1834).

Pittier, Henry François (Pittier de Fabrega) (1857-1950), Swiss botanist and civil engineer, D. Sc. Lausanne 1885; Ph.D. Jena; at Lausanne 1882-1887; to Costa Rica in 1887; with USDA 1905-1919 exploring Central America, Colombia and Venezuela; from 1919-1950 in Venezuela; naturalized U.S. citizen. (*Pitt.*).

HERBARIUM and TYPES: BR, CR, US and VEN; duplicates at A, AMES, B (extant), BH, BM, C, F, G, GE, GH, H, K, L, LAU, LE, M, MO, NY, PC, PH, S, W, Z. – Pittier sent his Costa Rican plants (1887-1904) to Th. Durand at BR. Complete sets of *Plantae costaricenses exsiccatae* (circa 4200 nos.) were deposited at BR and CR (see Sayre 1975). Pittier had collaborated with Th. Durand already in his Swiss period (1877-1879, at Château-d'Oex). -Letters at G, PH, US (SIA), portrait at G.

BIBLIOGRAPHY and BIOGRAPHY: Barnhart 3: 90; BJI 2: 136; BL 1: 146, 147, 260, 313, 2: 589, 703; BM 4: 1580; Bossert p. 312 (b. 13 Aug 1857, d. 27 Jan 1950); CSP 11: 28-29, 12: 579, 17: 912; Hirsch p. 233; Hortus 3: 1201 (Pitt.); IF suppl. 4: 336; IH 1 (ed. 6): 363, (ed. 7): 340, 2: (in press); Kelly p. 177; Kew 4: 317-319; Langman p. 585-586; Lenley p. 328; PH 939; Plesch p. 362; Rehder 4: 107; SIA 221, 223, 226, 237 (incl. HI 107); TL-2/1590, 1592, see indexes; Tucker 1: 558, 779; Urban-Berl. p. 289, 310, 383.

BIOFILE: Anon., Bryologist 25: 60. 1922 (subscription to Venezuelan exsiccatae).
Bally,W., Verh. Schweiz. naturf. Ges. 131: 363-369. 1951 (portr., biogr., b. 13 Aug 1857).
Bernardi, L., Musées de Genève 201: 19-22. 1980 (portr., Mosé Bertoni & Pittier; d. 27 Jan 1950).
Braun, F.G. et al., Henry F. Pittier, Centenario de su nacimiento. San José, agosto de 1957, 55 p. (portr., bibl., bibliographical notes, partly reprinted from other journals, by F.G. Braun, J. Valerio, J.A. Echeverría, P.C. Standley, T. Lasser, A. Chase, L.C. Bolaños, O.C. Saborio).
Chase, A., J. Washington Acad. Sci. 40(7); 25 Jul 1950 (1 p.).
Chevalier, A., Rev. int. Bot. appl. Agric. trop. 31: 351-352. 1951 (obit.).
Croat, T.B., Flora of Barro Colorado Island 49. 1978 (coll. in Panama 1911, 1914/15).
Durand, Th., Bull. Soc. Bot. Belg. 27(2): 173-178. 1888, 29(4): 47. 1890, and see *Prim. fl. costaric.*, below (see also TL-2/1592).
Dwyer, J.D., Taxon 22: 557-576. 1973 (coll., itin., types for Panama).
Echeverría, J.A., Enrique Pittier, San José, Costa Rica 1950, 16 p.
Fuchs, H.P., Proc. Royal Neth. Acad. Sci. C(84)2: 167. 1981 (*Isoëtes*).
Gómez, L.D., Brenesia 14-15: 361-393. 1978 (fragments of letters by C. Wercklé to P.).
Herter, W., Rev. Sudamer. Bot. 10: 31-32, *pl.2*. 1951 (portr.).
Jackson, B.D., Bull. misc. Inf. Kew 1901: 53 (pl. K).
Jahn, A., Bol. Soc. venez. Ci. nat. 4(30): 1-43. 1937 (biogr., bibl. 1878-1937).
Lasser, T., Agriculture in the Americas 6(4): 183-184. 1946 (portr.); Bol. Soc. venezol. Ci. nat. 13(76) 1-5. 1951.
Marquez, M.O., Henry François Pittier, Universidad Central de Venezuela, Caracas 1968, serie bibliografico, 29 p. (bibl., portr.; further references to biographical notes).
Massart, J., Bull. Soc. R. belge Géogr. 36: 73-74. 1912 (relation with Th. Durand).
Murray, G., Hist. coll. BM(NH) 1: 175. 1904.
Naville, R., La Tribune de Genève 14 Feb 1950 (obit.).

Pittier, H., La evolución de las ciencias naturales y las exporaciones botánicas en Venezuela. Caracas 1920, 28 p., suppl. 14 to "Cultura Venezolana".
Phelps, W.H., Geogr. J. Rev. 41: 341-342. 1951.
Sayre, G., Mem. New York Bot. Gard. 19: 381. 1975.
Standley, P.C. et al., Ceiba 1: 129-141. 1950.
Steyermark, J.A., O. Huber, Flora del Avila 16, 931-932. 1978.
Stieber, M.T., Huntia 3(2): 124. 1979 (ms. notes A. Chase).
Tamayo, F., Bol. Soc. venezol. Ci. nat. 8(52): 5-13. 1942 (personal recollections).
Verdoorn, F., ed., Chron. bot. 2: 366. 1936, 3: 16, 300, 326. 1937, 4: 184, 185, 460. 1938, 5: 176, 280, 511. 1939, 6: 172, 186, 257, 258, 463. 1941, 7: 94, 132, 357. 1943.

COMPOSITE WORKS: (1) with Th. Durand, *Cat. fl. Vaud.*, 1882-1885, see TL-2/1590.
(2) with T. Lasser, L. Schnee, Z. Luces de Febres and V. Badillo: *Catalogo de la flora venezolana*, 1: [1]-423. 1945, 2: [1]-577. Sep 1947. *Copies*: BR, US.

EPONYMY: (*genera*): *Pittiera* Cogniaux (1891); *Pittierella* Schlechter (1906); *Pittierothamnus* Steyermark (1962). (journal): *Pittiera*. Publicacion del herbario de la Facultad de Ciencias Forestales de la Universidad de Los Andes, Merida, Venezuela. Vol. 1-x. 1967-x. (locality): Pittier National Park, Venezuela.

7976. *The flora of the Pays d'Enhaut* (Switzerland) a botanical account by Prof. H. Pittier. Chateau d'Oex 1885. (*Fl. Pays d'Enhaut*).
Publ.: 1885, p. [1]-16. *Copy*: BR – Preliminary publ.: Notice botanique sur les Alpes des Pays d'Enhaut, 1884, 8 p. (*copy* BR).

7977. *Viaje de exploracion al Valle del Rio Grande de Térraba* ... San José de Costa Rica (Tip. Nacional) 1891. Oct. (*Viaje Rio Grande Térraba*).
Publ.: 1891, p. [1]-138, [1, err; 1, cont.], map. *Copy*: MO. – Also in Anal. Ist. fis.-geogr. nac. 3, 1890.

7978. Instituto fisico-geografico nacional. *Primitiae florae costaricensis* par H. Pittier ... tome ii.-fascicule 1[-7] [tome iii, fascicule 1-2] San José de Costa Rica, A.C. 1898[-1907]. Oct. (*Prim. fl. costaric.*).
Vol. 1: see T.A. Durand, TL-2/1592.
Vol. 2: edited by H. Pittier alone. Contributions by various authors. *Copies*: BR, NY, PH, U, US.
1: 1898 (Bot. Centralbl. 7 Dec 1898; Bot. Gaz. 22 Dec 1898; Nat. Nov. Jul(2) 1899), p. [1]-126. – By J. Donnell Smith (*Polypetalae*). – Repr. from Ann. Inst. fis.-geogr. Nac. vol. 8).
2: 1898 (cover; TBC 17 Aug 1899. Bot. Zeit. 1 Oct 1899), p. [129]-216, [1, err.]. – By J. Donnell Smith (*Gymnopetalae*).
3: 1899 (cover; Nat. Nov. Apr(2), 1900), p. [217]-294, [295, err.]. – By Casimir de Candolle (*Piperaceae*).
4: 1900 (cover, Nat. Nov. Mai(1) 1900; Bot. Centralbl. 10 Oct 1900), p. [297]-317, [319, index]. By G. Lindau (*Acanthaceae*).
5: 1900 (cover, Nat. Nov. Aug(1) 1900; Bot. Zeit. 1 Oct 1900), p. [321]-337, [339]. By F. Pax (*Euphorbiaceae*).
6: 1900 (cover, Nat. Nov. Sep(1, 2) 1900; Bot. Zeit. 1 Oct 1900), p. [311]-365, [367]. By A. Engler (*Araceae*).
7: 1900 (cover, Nat. Nov. Jan(1) 1901), p. [373]-405 (index by H. Pittier).
Vol. 3: edited by H. Pittier.
1: Jan-Feb 1901 (Bot. Centralbl. 27 Mar 1901; Nat. Nov. Apr(1) 1901), p. [1]-62, [1-7]. *Copies*: BR, USDA. – Tome III-fascicule 1. – Filices, ... Rhizocarpaceae. Auctore H. Christ. San José de Costa Rica, A.C. 1901. Instituto de fisico-geografico nacional. Treatment continued from vol. 1(3), 1896, Filices by E. Bommer & H. Christ.
2: In Bulletin Herb. Boissier; Filices ctd. by H. Christ.
 (1): 31 Aug 1904, p. [1]-16. Bull. ser. 2. 4: 936-951.
 (2): 30 Sep 1904, p. 17-32, ib. 4: 957-972.
 (3): 31 Oct 1904, p. 33-48, ib. 4: 1089-1104.

(4): 30 Dec 1904, p. 49-63, ib. 5: 1-16.
(5): 28 Feb 1905, p. [65]-77, ib. 5: 248-260.
Vol. 2, folio edition: 1907, p. [77]-223. – *Copy*: BR. – This folio issue, with the text in two columns, without imprint, was issued in San José, 1907, according to a statement in Bot. Jahrb. 41(Lit.): 53. 1908. We do not know why the pagination starts with p. [77]. The contents of this seemingly unique issue are:
Polypetalae, pars, J. D. Smith, p. [81]-150.
Piperaceae, C. de Candolle, p. [151]-180.
Acanthaceae, G. Lindau, p. 181-189.
Euphorbiaceae, F. Pax, p. 191-198.
Araceae, A. Engler, p. 199-209.
Indexes, H. Pittier, p. 211-223.
These contents agree with those of fasc. 1 and 3-7 of the octavo edition, including also the *préface* dated 28 Nov 1897.

7979. *Ensayo sobre las plantas usuales de Costa Rica* ... Washington, D.C. (H.L. & J.B. McQueen, Inc.) 1908. (*Ensayo pl. usual. Costa Rica*).
Publ.: 5 Nov 1908 (p. vii: Sep 1908; Nat. Nov. Dec(2) 1908; Pittier signed his dedic. copy to Durand (BR) 5 Nov 1908), p. [i]-xi, [1]-176, *pl. 1-31*. *Copies*: B, BR, MO, NY, US, USDA(2).

7980. *New or noteworthy plants from Colombia and Central America* ... [8 papers published in Contr. U.S. natl. Herb. 1909-1922].
1: 22 Jan 1909, *in* 12(5): [i]-v, 171-181, *pl. 18-19*.
2: 11 Jun 1910, *in* 13(4): [i]-iv, 93-132, vii, *pl. 17-20*.
3: 5 Jan 1912, *in* 13(12): [i]-vi, 431-466, vii-viii, *pl. 78-96*.
4: 16 Apr 1914, *in* 18(2): [i]-vii, 69-86, ix-x, *pl. 42-56*.
5: 3 Mar 1916, *in* 18(4): [i]-vii, 143-171, *pl. 57-80*.
6: 15 Sep 1917, *in* 18(6): [i]-vii, 225-259, ix-x, *pl. 106*.
7: 18 Jun 1918, *in* 20(3): [i]-vii, 95-132, ix-x, *pl. 7*.
8: 9 Jan 1922, *in* 20(12): [i]-vi, 453-490, *pl. 27-30*, vii-viii.
Copies: MO, USDA.

7981. *Clave analítica de las familias de plantas fanerógamas de Venezuela* y partes adyacentes de la América tropical ... Caracas (Litografia del Comercio) 1917. Oct. (*Clav. fam. pl. Venez.*).
Ed. 1: 1917 (p. viii: 10 Sep 1917), p. [i]-viii, [1]-108. *Copies*: MICH, NY, USDA.

7982. *La evolución de las ciencias naturales* y las exploraciones botanicas *en Venezuela* ... Caracas (Tip. Cultura venezolana) 1920. (*Evol. ci. nat. Venez.*).
Publ.: 1920, p. [1]-28. *Copy*: MO. – Supplement to *Cultura venezolana* no. 14.

7983. *Esbozo da las formaciones vegetales de Venezuela* con una breve reseña de los productos naturales y agrícolas ... (complemento explicativo del mapa ecológico del mismo autor.) Caracas (Litografia Comercio) 1920. Qu. (*Esboz. form. veg. Venez.*).
Publ.: 1920 (p. [5]: 23 Oct 1920; PH copy rd. 11 Jan 1921), p. [1]-44. *Copy*: PH.

7984. Contribuciones para la flora de Venezuela. *Arboles y arbustos de Venezuela* ... Caracas (Tipografia americana) 1921-1929, 10 decades. Oct. (*Arb. arbust. Venez.*).
1: Jan 1921, p. [1]-19, reprinted from Bol. Comerc. Ind. 13: 417-433. 1921.
2/3: 1923, p. [21]-43, reprinted from id. 34, 1923.
4/5: 1925, p. [45]-73, reprinted from Bol. cient. tecn. Mus. Comerc. Venez. 1: 1926.
6/8: Aug-Sep 1927, p. [75]-103, reprinted from Bol. Minist. RR. EE. 1927 (8/9).
9/10: Dec 1929, p. [105]-132, reprinted from id. nos. 1929 (8-12).
Copies: MICH, MO, USDA.

7985. *Exploraciones botanicas y otras en la Cuenca de Maracaibo* ... Caracas (Tipografia mercantil) 1923. Oct. (*Explor. Cuenca de Maracaibo*).
Publ.: 1923, p. [1]-100, *13 pl*. *Copies*: G, MO, NY, PH, US. Reprinted from Bol. Comerc. Ind. 4, 1923.

7986. *Clave analítica de las familias de plantas superiores de la América tropical* ... Caracas (Lit. y Tip. del Comercio) 1926. (*Clav. fam. pl. Amér. trop.*).
Ed. 1: 1926 (p. viii: 1 Sep 1926; Nat. Nov. Feb 1927), p. [i]-viii, [1]-130. *Copies*: MICH, MO (photocopy), USDA.
Ed. 2: 1939 (p. vii: Jul 1939; Nat. Nov. Jul-Aug 1940), p. [i]-vii, [1]-94. *Copies*: MO, USDA. – Caracas (Tipografia la Nacion) 1939.

7987. *Manual de las plantas usuales de Venezuela* ... Caracas (Litografía del Comercio) 1926. (*Man. pl. usual. Venez.*).
Publ.: 1926 (Nat. Nov. Nov 1926), p. [i]-xvi, [1, ind.], [1, ills.], [1]-458, *pl. 1-42*. *Copies*: B, BR, G, MO, NY, US, USDA.
Suplemento: 1939 (US dedic. copy to "Killip" dated 13 Sep 1939), p. [i]-viii, [1, ind.], [1]-129, err. slip. *Copies*: BR, G, MICH, MO, NY, US, USDA. – Caracas (Editorial Elite) 1939.

7988. Contribuciones a la dendrologia de Venezuela. I. *Arboles y arbustos del orden de las leguminosas* ... Caracas (Tipografia americana) 1927. (*Arb. legum.*).
1 (Mimosaceas): Dec 1927, p. [1]-82. – Reprinted from Bol. Minist. RR. EE. [Relaciones exteriores]. 1927 (10-12): [385]-464. – Also in Trab. Mus. com. Venezuela 2: [31]-112. Dec 1927.
2 (Cesalpiniaceas): Mar 1928, p. [83]-148. – Reprinted from id. 1928(1-3): [87]-152. – Also in Trab. Mus. com. Venezuela 3: [113]-178. Mar 1928.
3 (Papilionaceas): Aug 1928, p. [149]-229. – Reprinted from id. 1928(4-7): [327]-409. – Also in Trab. Mus. com. Venezuela 4: [179]-259. Aug 1928.
Copies: G, MICH, NY, US, USDA.

7989. *Classificacion natural de las plantas* con especial mencion de las familias mas importantes de la flora de Venezuela y de las especies de interes economico ... Caracas (Tipografía americana) 1932. (*Classif. nat. pl.*).
Publ.: 15 Nov 1932 (copy sent to USDA by Pittier), p. [1]-140. *Copies*: MO, USDA.

7990. *Los musgos de Venezuela* ... Caracas (Lit. y Tip. del Comercio) 1937. Oct. (*Musg. Venez.*).
Publ.: 1937, p. [1]-38. *Copies*: MO-Steere, U. – Reprinted from Bol. Soc. venezuel. Ci. nat. 27. Aug-Oct. 1936 (n.v.).

7991. *Lista provisional de las gramíneas señaladas en Venezuela* hasta 1936, con notas acerca de su valor nutritivo, etc., ... Caracas (Cooperativa de artes graficas) 1937.
Publ.: Jan-Mar 1937 (U rd. 10 Apr 1937; PH rd. 28 Mai 1937), p. [1]-77, [1, index]. *Copies*: PH, U. – Minist. Agr., Bol. Técn. 1, 1937.

7992. *Genera plantarum venezuelensium* clave analítica de los generos plantas hasta hoy conocidas en Venezuela ... Caracas (Tipografia americana) 1939.(*Gen. pl. venez.*).
Publ.: Dec 1939 (p. 9: Nov 1939; USDA rd. 14 Mar 1940), p. [1]-354. *Copies*: FI, G, MO, US, USDA.

Pittoni, Josef Claudius, Ritter von Dannenfeldt (1797-1878), Austrian botanist and administrator ("Truchsess," sheriff). (*Pittoni*).

HERBARIUM and TYPES: W (20.000); other material at B (extant), BASSA, BM, BP, BR, CGE, CN, DBN, E, FI, G, GJO, IBF, K, MANCH, MW, OXF, P, REG, VEN.

BIBLIOGRAPHY and BIOGRAPHY: AG 4: 615, 12(2): 53; Backer p. 448; Barnhart 3: 90; Bossert p. 312 (b. 4 Jul 1797, d. 2 Apr 1878); Clokie p. 225; CSP 4: 922, 8: 630, 11: 29; DTS 1: 226; Hegi 4(2): 552; Herder p. 77; IH 2: (in press); Saccardo 2: 85.

BIOFILE: Anon., Bull. Soc. bot. France 13 (Bibl.): 95. 1866. 25 (bibl.): 44. 1878; Österr. bot. Z. 19: 292. 1869 (herb. sold to W), 28: 179. 1878 (died 2 Apr 1878, Görz).
Beck, G., Bot. Centralbl. 33: 314. 1888, 34: 150. 1888 (herb.); *in* K. Fritsch et al., Bot. Anst. Wiens 63, 75. 1892 (coll. W).

Dolezal, H., Friedrich Welwitsch 30, 34, 66, 93, 226-227. 1974.
Hedge, I.C. & J.M. Lamond, Index coll. Edinb. herb. 119. 1970.
Kanitz, A., Magy. növen. Lap. 2: 80. 1878 (obit., d. 2 Apr 1878).
Lack, H.W., Willdenowia 10: 81. 1980 (material at B).
Lindemann, E., Bull. Soc. imp. Natural. Moscou 61: 55. 1886 (sold herb. to W because of an unfortunate railroad speculation).
Murray, G., Hist. coll. BM(NH) 1: 174. 1904 (coll. BM).
Steinberg, C.H., Webbia 34: 61. 1979 (coll. FI).

EPONYMY: The genera *Pittonia* P. Miller (1754) and *Pittoniotis* Grisebach (1858), as well as the journal *Pittonia*; a series of papers relating to botany and botanists (Berkeley, California. Vol. 1-5. 1887-1905) honor Joseph Pitton Tournefort, *q.v.*

Pizarro, João Joaquim (*fl.* 1872-1887), Brazilian physician and botanist; Dr. med. Rio de Janeiro 1872; later professor of natural history at the Rio de Janeiro medical faculty. (*Pizarro*).

HERBARIUM and TYPES: R. – Further material was at B.

BIBLIOGRAPHY and BIOGRAPHY: Barnhart 3: 90; CSP 11: 29; Morren ed. 10, p. 151; Urban-Berl. p. 383.

BIOFILE: Sampaio, A.J. de, Arch. Mus. Rio de Janeiro 22: 47. 1919 (coll. R).
Urban, I., Fl. bras. 1(1): 77. 1906.

7993. [*Dissertação Solanaceas brasileiras*] Facultade de Medicina do Rio de Janeiro. Theses de habilitação ao lugar de lente oppositor da secção de sciencias accessorias por João Joaquim Pizarro ... Rio de Janeiro (Typographia – Perseverança ...) 1872. Qu. (*Diss. Solan. bras.*).
Publ.: 1872, p. [i-x], [2]-96. *Copy*: BR.

Planchon, [François] Gustave (1833-1900), French botanist and pharmacist; Dr. med. Montpellier 1859; prof. agr. méd. Montpellier 1860, prof. bot. Lausanne 1862; Dr. sci. Paris 1864; Pharm. 1. classe 1864; prof. at École de Pharmacie, Montpellier 1864-1866; prof. École supérieure de Pharmacie de Paris 1866; prof. id. 1886-1900; brother of J.E. Planchon. (*G. Planch.*).

HERBARIUM and TYPES: Some material in NA. – Letters at G.

BIBLIOGRAPHY and BIOGRAPHY: Barnhart 3: 90 (b. 29 Oct 1833, d. 13 Apr 1900); BL 2: 166, 703; BM 4: 1581, 8: 1014; Bossert p. 312; CSP 4: 930, 8: 631, 11: 29-30, 17: 916-917; Herder p. 175; Jackson p. 128, 203, 288; Kew 4: 319; Morren ed. 2, p. 21, ed. 10: p. 55; PR 7160-7165; Rehder 5: 675-676; Saccardo 1: 129; Tucker 1: 559.

BIOFILE: Anon., Bot. Centralbl. 52: 450. 1892; Bot. Zeit. 25: 56. 1867 (app. Paris); Nat. Nov. 22: 269. 1900 (d. 23 Apr 1900); Amer. Druggist Dec 1922: 22 (portr.); Bull. Soc. bot. France, Table art. orig. 187-188; Nature 61: 618. 1900.
Bourquelot, Bull. Acad. méd. Paris 43: 487-490. 1900 (obit.).
Bridson, G.D.R. et al., Nat. hist. mss. res. Brit. Isl. 263.2. 1980.
Guignard, L., Bull. Soc. bot. France 47: 147-152. 1900 (obit., b. 29 Oct 1833).
Moissan et al., Titres et travaux scientifiques de G. Planchon, Paris 1900, 37 p. (portr., appointments, bibl., tributes); J. Pharm. 11: 405-427. 1900.
Planchon, G., Notice sur les titres et travaux scientifiques de G. de Planchon, Paris, 1876, 15 p. (bibl., appointments, membership).
Reber, B., ed., Gallerie Therap. Pharmakogn. 273-279.1897 (bibl., portr.).
Seynes, J. de, Bull. Soc. bot. France 47: 129-130. 1900 (obit., brief bibl.).
Urban, I., Symb. ant. 1: 123. 1898.
Wittrock, V.B., Acta Horti Berg. 3(3): 111. 1905.

7994. *Des Globulaires au point de vue botanique et médical* ... Montpellier (Typographie de Boehm, ...) Oct. (*Globulaires*).
Publ.: 1859 (BSbF 29 Apr 1859), p. [i]-vi, [7]-59, chart. *Copies*: G, MO, NY.
Ref.: Anon., Bull. Soc. bot. France 6: 246-248. 1859.

7995. Faculté de médecine de Montpellier. Concours pour l'agrégation ... *Les principes de la méthode naturelle* appliqués comparativement à la classification des végétaux et des animaux. Thèse qui sera soutenue publiquement, le 31 Mai 1860 ... Montpellier (Typographie de Boehm & fils, ...) 1860. Oct. (*Princ. méth. nat.*).
Publ.: Mai 1860 (BSbF rd. 22 Jun 1860), p. [1]-112. *Copies*: G, MICH, MO, NY.

7996. Thèses présentées à la Faculté des sciences de Paris pour obtenir le grade de docteur ès sciences naturelles ... 1re thèse: Geologie ... 2me thèse: botanique. – *Des modifications de la flore de Montpellier depuis le xvie siècle jusqu'à nos jours* ... soutenues le [4, mss] avril 1864 ... Paris (F. Savy, ...) 1864. Qu. (*Modif. fl. Montpellier*).
Thesis issue: Apr 1864, p. [i-viii], [3]-57. *Copy*: MO (possibly incomplete). – The text of the geology thesis is not included, see below *Étud. tufs*....
Regular issue: 1864 (BSbF Dec 1864), p. [1]-57, [59, 2me thèse]. *Copies*: BR, G. – Paris (F. Savy, ...), Montpellier (Boehm et fils, ...) 1864. Qu.

7997. *Étude des tufs de Montpellier au point géologique et paléontologique* ... Paris (Savy, ...), Montpellier (Boehm et fils, ..) 1864. Qu. (*Étud. tufs Montpellier*).
Publ.: 1864, p. [i-vii], [1]-73, *pl. 1-3* (1-2 col. Planchon; 3, uncol. J. de Seynes). *Copy*: G.
– Text of the geological thesis defended on 4 Apr 1864, see above).

7998. *Des Quinquinas* ... Paris (Savy, ...) Montpellier (Typographie de Boehm & fils, ...) 1864. Oct. (*Quinquinas*).
Publ.: 1864 (BSbF rev. Dec 1864), p. [3]-150. *Copies*: B, BR, FI, G, MO, NY.
Ref.: Schlechtendal, D.F.L. v., Bot. Zeit. 24: 46. 1866.

7999. *Matériaux pour la flore médicale de Montpellier* et des Cévennes d'après Lobel ... Montpellier (Boehm & fils, imprimeurs-éditeurs ...) 1868. Qu. (*Mat. fl. méd. Montpellier*).
Publ.: 1868 (BSbF rev. Dec 1868), p. [1]-44. *Copies*: G, USDA. – Reprinted from Montpellier médical.

8000. *Traité pratique de la détermination des drogues simples* d'origines végétale ... Paris (Librairie F. Savy, ...) 1875. Oct. 2 vols. (*Traité drog. simpl.*).
1: Oct 1874, (t.p. 1875, but reviewed Bot. Zeit. 13 Nov 1874), p.[i]-vii, [1]-664.
2: Jan-Feb 1875 (Bot. Zeit. 12 Mar 1875), p. [i, iii], [1]-535, [536, err.].
Copy: MO.
Ref.: Bary, A., Bot. Zeit. 32: 747. 13 Nov 1874 (rev. part 1).

8001. *Sur les charactères et l'origine botanique du Jaborandie* [Paris (Imprimerie Arnous de Rivière et c^e, ...) 1875]. Oct.
Publ.: 1875, p. [1]-11. *Copy*: G. – Reprinted and to be cited from J. Pharm. Chimie 21: 295-305. 1875.

8002. *Étude sur les produits de la famille des Sapotacées* ... Montpellier (Imprimerie centrale du Midi (Hamelin frères)) 1888. 16-mo (*Étud. prod. Sapotac.*).
Publ.: 1888: p. [i]-ix, [11]-121. *Copy*: US.

8003. *Les drogues simples d'origine végétale* Paris (Octave Doin, ...) [1894-]1895. Oct., 2 vols. (*Drogues simpl. vég.*).
Co-author: Eugène Baptiste Collin (1845-1919).
1: Nov-Dec 1894 or early 1895 (p. ii: 20 Oct 1894; t.p. 1895; Bot. Zeit. 1 Mai 1895; Nat. Nov. Nov(2) 1894), p. [i*-iii*], [i]-ii, [1]-805, [1, err.].
2: Dec 1895 or early 1896 (Nat. Nov. Dec (1, 2) 1895; t.p. 1896), p. [i-iii], [1]-988.
Copies: BR, FI, USDA.

Planchon, Jules Émile (1823-1888), French botanist; Dr. sci. nat. Montpellier 1844;

assistant to William Jackson Hooker at Kew 1844-1848; professor of botany at the horticultural institute, Gent, 1849-1851; at the École de Pharmacie, Nancy, 1851-1853; assistant professor at Montpellier succeeding Dunal (Faculté des Sciences) 1853-1857), professor at id. as well as at the École de Pharmacie 1857-1881; at the Faculté de Médecine and director of the botanical garden, Montpellier 1881-1888; discovered Phylloxera 1868. (*Planch.*).

HERBARIUM and TYPES: Partly at MPU and partly at NA (in herb. Martindale); other material at FI and L. – Some letters at G and K; letters to D.F.L. von Schlechtendal at HAL; portrait at G.

BIBLIOGRAPHY and BIOGRAPHY: Backer p. 448; Barnhart 3: 90; BB p. 245 (b. 21 Mar 1823, d. 1 Apr 1888); BJI 1: 46; BL 1: 249, 313; BM 4: 1581-1582, 5: 2138; Bossert p. 312; Bret. p. 933; CSP 4: 930-932, 8: 631, 11: 30, 12: 579, 17: 917; Desmond p. 497; Frank 3(Anh.): 77; Herder p. 471; Hortus 3: 1201; IH 1 (ed. 6): 363, 2: (in press); Jackson p. 5, 91, 133, 147, 223, 434, 505; Kelly p. 177; Kew 4: 319-321; Langman p. 586-587; LS 20828-20836; Moebius p. 417; Morren, ed. 2, p. 19, ed. 10, p. 66; MW p. 388-389, suppl. p. 281; NI 1257; see also 1: 94, 95; Plesch p. 363; PR 603, 5335, 7166-7172 (ed. 1: 8042-8043), 9503; Quenstedt p. 340; Rehder 5: 676-677; RS p. 133; Saccardo 2: 85; Sotheby 602; TL-1/667, 688, 979; TL-2 /978, 999, 2348, 4334, 4621, 4622, 4625, 6949, 6950, 7018, see J.J. Lindley, G.H. Mettenius, W. Nylander; Tucker 1: 559-560; Urban-Berl. p. 383; Zander ed. 10, p. 702, ed. 11, p. 800.

BIOFILE: Allan, M., The Hookers of Kew 141, 187. 1967.
Anon., Bonplandia 1: vi [index]. 1853, 2: 8, 122. 1854, 3: 63, 228, 279. 1855, 5: 330. 1857 (app. Montpellier); Bot. Centralbl. 34: 95. 1888 (d. 1 Apr 1888, Montpellier); Bot. Jahrb. 9 (Beibl. 22: 1. 1888; Bot. Not. 1889; 29 (d.); Bot. Zeit. 9: 799. 1851 (app. Nancy), 11: 359. 1853 (to Montpellier), 46: 291. 1888 (d. 1 Apr 1888); Bull. Soc. bot. France, Table art. orig. 188-189; Flora 30: 30. 1867 (Silver medal Sorbonne); Gard. Chron. 1895(1): 461; Nat. Nov. 10: 149. 1888 (d.); Gartenflora 3: 39. 1854 (to Montpellier), 7: 96. 1858 (app. prof. Montpellier); Österr. bot. W. 2: 28. 1852 (to Nancy), 38: 183. 1888 (d.); Proc. Linn. Soc. London 1087/88: 95-96. 1890 (obit., b. 23 Mar 1823, d. 1 Apr 1888); Psyche, Cambr. 5: 156. 1889; Rev. hort. belg. étr. 14: 151. 1888.
Bentham, G., Fl. austral. 1: 8*. 1863.
Blanc, H., Bull. Soc. vaud. Sci. nat. 24: 218-219. 1888.
Bullock, A.A., Bibl. S. Afr. bot. 91. 1978.
Candolle, A.P. de, Mém. souv. 553. 1862.
Dehérain, P.P., Ann. agron. 40: 221-229. 1888 (obit.).
Flahault, C., L'oeuvre de J.-É. Planchon, Montpellier 1889, xxii p. (biogr., bibl, epon., taxa, b. 21 Mar 1823, d. 1 Apr 1888, further biogr. refs.) (Reprinted from Mém. Acad. Sci. Lettr. Montpellier, Sci. 13, 1889); Univ. Montpellier, Inst. Bot. 54-56. 1890 (further biogr. refs.).
Foex, G., J.É. Planchon, Montpellier (Ch. Boehm) s.d., 8 p.
Fürnrohr, A.E., Flora 34: 770. 1851 (app. Nancy), 40: 735. 1857 (app. Montpellier).
Granel de Solignac, L. & L. Bertrand, Les herbiers de l'institut botanique de Montpellier 275, 280. 1867 (herb. at MPU), Naturalia monspel. fasc. 18.
Hall, N., Botanists of the Eucalypts 103-104. 1978.
Jadin, F., Rev. gén. Bot. 28: 1-10. 1916.
Huxley, L., Life letters J.D. Hooker 1: 29, 175, 236, 328, 423. 1918.
Maiden, J.H., R. Soc. NSW 44: 152-153. *pl. 13.* 1910 (portr.).
Mandon, L., Ann. Soc. Hort. Hist. nat. Hérault ser. 2. 42: 202-206. 1910 (biogr.).
Merrill, E.D., Enum. Philipp. flow. pl. 4: 217. 1926; B.P. Bishop Mus. Bull. 144: 151. 1937; Contr. U.S. natl. Herb. 30: 242-243. 1947.
Morot, L., J. de Bot. 2: 140, 220-228. 1888 (obit., bibl.).
Morrow, D.W., Jr., Agric. Hist. 34(2): 71-76. 1960 (on P's visit to the US, 1873, to study phylloxera resistant grapevines).
Planchon, J.E., Installation de ... comme professeur de botanique ... Fac. méd. Montpellier, Montpellier 1881, 20 p. (ex. Montpellier médical Apr 1881).
Planchon, L., Ann. Bot. 2: 423-428. 1889 (bibl.).

Reber, B., Gallerie Therap. Pharmakogn. 387-388. 1897 (biogr. sketch).
Roumeguère, C., ed., Rev. mycol. 10: 169-170. 1888 (obit.).
Sabatier, A., Ann. Soc. Hort. Hist. nat. Hérault ser. 2. 20: 77-104. 1888 (analysis works), also repr. 32 p. Montpellier 1888; 50: 202-206. 1910 (portr., biogr. sketch).
Sabatier, A., et al., Inauguration du monument élévé à la mémoire de J.-É. Planchon le 9 décembre 1894. Montpellier 1895, 75 p. (portr., biogr., tributes).
Sahut, F. et al., Ann. Soc. Hort. Hist. nat. Hérault ser. 2. 20: 23-27. 1888.
Schmid, G., Goethe u.d. Naturw. 277. 1940 (on Pavonia (Goethea) strictiflora).
Seemann, B., Bonplandia 1: 23, 33, 43, 228. 1853.
Stafleu, F.A., The great prodromus 30. 1966.
Steinberg, C.H., Webbia 32: 33. 1977 (material at FI).
Taton, R., ed., Science in the 19th century 368. 1965.
Verdoorn, F., ed., Chron. bot. 3: 24, 67. 1937 (see also indexes vols. 1 and 2).
Wright, R., Gard. tribute 231. s.d.

COMPOSITE WORKS: (1) Co-editor, *Flore des Serres*, 1849-1881.
(2) Co-author, *Prelud. fl. columb.* 1853, with J.J. Linden, TL-2/4621.
(3) Co-editor, *Pescatorea*, 1854-1860, see J.J. Linden, TL-2/4622.
(4) Co-author, *Centralamer. Lobeliac.* 1857, with A.S. Örsted, see TL-2/7018.
(5) Co-author, *Pl. columb.* 1863, with J.J. Linden, TL-2/4625.
(6) Triana, J. & J.E. Planchon, *Prodromus florae nova-granatensis* 2 vols., 1862-1873, see under Triana. Planchon was co-author of vol. 1, *Phanerogamia*.
.(7) DC, *Prodr.*, *Ulmaceae* 17: 151-210. 16 Oct 1873 (see also his *Sur les Ulmacées*, Ann. Sci. nat. Bot. ser. 3. 10: 244-341. 1848).
(8) Alph. & C. de Candolle, *Monographiae phanerogamarum*, 5(2) (*Ampelideae*): [i-v], [305]-654, [655]. Oct. 1887 (rev. Freyn, Bot. Centralbl. 36: 204-206. 1888).

HANDWRITING: A. Sabatier et al., Inauguration ... 1895, plate facing p. 74. (see above bibl./biogr.).

EPONYMY: *Planchonella* Pierre (1890, *nom. cons.*); *Planchonella* Van Tieghem (1904); *Planchonia* Blume 1851-1852; ING 3: 1355. (fide Flahault 1889).

8004. *Mémoire sur les* développements et les charactères des vrais et des faux *arilles*, suivi de considérations sur les ovules de quelques Véroniques et de l'Avicennia ... Montpellier (Typographie et lithographie de Boehm, ...) 1844. (*Mém. arilles*).
Publ.: 1844, p. [i, iii], [1]-53, *pl. 1-3* (uncol. liths. auct.). *Copies*: FI, G, WU.

8005. *Description de deux genres nouveaux de la famille des Euphorbiacées* [London Journal of Botany 4, 1845]. Oct.
Publ.: Sep (or Aug-Oct) 1845, p. [1]-4. *Copy*: FI. – Reprinted and to be cited from W.J. Hooker, London J. Bot. 4: 471-474. 1845.

8006. *Sur les affinités des genres Henslowia*, Wall. (Crypteronia? Blume Quilamum? Blanco) *Raleighia*, Gardn. et *Alzatea* Ruiz et Pav. ... [London Journal of Botany 4. 1845]. Oct.
Publ.: Oct 1845, p. [1]-4. *Copy*: FI. – Reprinted and to be cited from W.J. Hooker, Lond. J. Bot. 4: 519-521. 1845.

8007. *Description d'un genre nouveau, voisin du Cliftonia*, avec des observations sur les affinités des Saurauya, des Sarracenia et du Stachyurus ... [London Journal of Botany 5, 1846]. Oct.
Publ.: Mai 1846, p. [1]-7. *Copy*: FI. – Reprinted and to be cited from W.J. Hooker, London J. Bot. 5: 250-256. 1846; also in Ann. Sci. nat., Bot. ser. 3. 6: 123-129. 1846. (genus *Purdiaea* Planch.).

8008. *Revue de la famille des Simaroubées*, ... [London Journal of Botany 5, 1846]. Oct.
Publ.: Nov 1846, p. 9-33. *Copy*: FI. – Reprinted and to be cited from W.J. Hooker, Lond. J. Bot. 5: 560-584. 1846.

8009. *La Victoria regia*, au point de vue horticole et botanique, avec des observations sur

la structure et les affinités des Nymphéacées, par J.E. Planchon ... et (pour la culture), par L. van Houtte, ... Gand [Gent] (Belgique) 1850-1851. Qu.
Co-author: Louis Benoît Van Houtte (1810-1876).
Publ.: p. [1]-52, *pl. 1-6* (uncol.), 7 (col. liths. Planchon). *Copies*: BR, FI, NY. – Reprinted from Flore des Serres 6(7): 193-224. 1850, p. 37-42 in 6(8): 249-254. 1850, p. 43-47 in 7(1): 25-29. 1851, p. 48-52 in 7(2): 49-53. 1851. – To be cited from journal. – See also J.G.C. Lehmann, TL-2/4334, for a note on Planchon's later publications on Nymphaea and the Nymphaeaceae (Revue hort. ser. 4. 2: 62-68. Feb 1853 (p. 61 dated 16 Feb 1853) and Ann. Sci. nat. Bot. ser. 3. 19: 17-63. 15 Mai 1853.

8010. *Description d'un genre nouveau du groupe des Thismiées*, ... [Annales des Sciences naturelles, ser. 3. 68. 1852]. Oct.
Publ.: 1852, p. [1]-3. *Copy*: FI. – Reprinted and to be cited from Ann. Sci. nat., Bot. ser. 3. 18: 319-320. 1852. (Genus *Stenomeris* Planch.).

8011. *Énumération succincte des espèces de la famille des Nymphéacées*. [Extrait de la Revue horticole, no. du 16 février 1853]. Oct.
Publ.: 16 Feb 1853, p. [1]-7. *Copy*: FI. – Reprinted and to be cited from Revue horticole ser. 4. 2: 62-68. 1853.

8012. *Études sur les Nymphéacées*, ... [Extrait des Annales des Sciences naturelles, tome xix (cahier no. 1) 1853]. Oct.
Publ.: Jan-Feb 1853, p. [1]-47. *Copy*: FI. – Reprinted and to be cited from Ann. Sci. nat., Bot. ser. 3. 19(1): 17-63. 1853. – See also A.E. Fürnrohr, Flora 36: 571-577. 2 Sep 1853.

8013. *Hortus donatensis*. Catalogue des plantes cultivées dans les serres de S. Ex. le prince A. de Démidoff à San Donato, près Florence ... Paris (Imprimerie de W. Remquet et Cie. ...) 1854-1858. Fol. (*Hort. donat.*).
Publ.: 1858 (BSbF Jul-Dec 1858), p. [i*], [i]-xxxiv, [1], [1]-225, *6 pl.* (1 view of garden, 5 chromoliths. of plants). *Copies*: BR, FI, G, NY, USDA.
Ref.: Steffatschek, A., Bonplandia 10: 234. 1862.

8014. *Mémoire sur la famille des Guttifères* ... Paris (Victor Masson et fils, ...) [1860-]1862. Oct.
Co-author: Jerónimo Triana (1826-1890).
Publ.: 1862 (BSbF rd. 14 Nov 1862), p. [i]-iv, [1]-336, *pl. 15-16* (uncol.lith. Planchon). *Copies*: BR, G, NY, US. – Consolidated reprint of a series of articles in Ann. Sci. nat. Bot. (to be cited from Journal): ser. 4.
See also below Planchon & Triana, *Sur la famille des Guttifères*, 1861. (also reprinted, p. [1]-15. *Copy*: G.

vol.	pages	dates
13	306-376, *pl. 15-16*	1860
14	226-367, *pl. 15-18*	1860
15(4)	240-319	1861
16	263-308, *pl. 17-18*	1862

8015. *Sur la famille Guttifères* ... [Paris (Imprimerie de L. Martinet, ...) 1861]. Oct.
Co-author: Jerónimo Triana (1826-1890).
Publ.: Mar-early Mai 1861, p. [1]-15. *Copy*: G. – Reprinted and to be cited from Bull. Soc. bot. France 8: 26-28. Mar 1861, 66-73, 96-100. Apr-early Mai 1861.

8016. *La vraie nature de la fleur des Euphorbes* expliquée par un nouveau genre d'Euphorbiacées ... [Paris (Imprimerie de L. Martinet) 1861]. Oct.
Publ.: 8 Mar-8 Apr 1861 (in journal), p. [1]-4. *Copy*: G. – Reprinted and to be cited from Bull. Soc. bot. France 8: 29-32. 1861.

8017. *Pierre Richer de Belleval* fondateur du Jardin des plantes de Montpellier. Discours

prononcé à la séance solennelle de rentrée des Facultés et de l'École supérieure de pharmacie le 15 Novembre 1869 ... Montpellier (Jean Martel Aîné, ...) 1869. Oct. *(Belleval)*.
Publ.: 15 Nov-31 Dec 1869, frontisp., facsim. handwriting, p. [1]-72. *Copies*: MO, NY.

8018. *Des limites naturelles des flores* et en particulier de la florule locale de Montpellier communication verbale faite devant le congrès scientifique de France le 8 décembre 1868 ... Montpellier (Jean Martel Aîné, ...) 1871. Oct. *(Limites nat. fl.)*.
Publ.: 1871, p. [1]-7. *Copy*: G. "Extrait des actes du Congrès scientifique, xxxve session".

8019. *Les vignes sauvages* des États-Unis de l'Amérique [Montpellier (Imprimerie centrale du Midi ...) 1874]. Oct.*Publ.*: Apr 1874 (in BSbF), p.[1]-7. *Copy*: G. – Reprinted and to be cited from Bull. Soc. bot. France 21: 107-111. Apr 1874, also issued in Bull. Soc. Hort. Hist. nat. Hérault 1874(2).

8020. *Le morcellement de l'espèce en botanique et le jordanisme* ... [Revue des deux Mondes 15 septembre 1874]. Oct.
Publ.: 15 Sep 1874, p. [1]-28. *Copy*: G. – Reprinted and to be cited from Revue des deux Mondes 15 Sep 1874 (5: 389-416.) – of importance for P's ideas on the species concept and the ideas of Alexis Jordan.

8021. *Les vignes américaines* leur culture, leur résistance au Phylloxera et leur avenir en Europe ... Montpellier (C. Coulet, ...) Paris (Adrien Delahaye, ...) 1875. 18-mo *(Vignes amér.)*.
Publ.: late 1875 (Bot. Zeit. 3 Dec 1871), p. [i]-xiv, [15]-240. *Copies*: B, FI, G, MO.
Preliminary reports: Le Phylloxera et les vignes américaines, Montpellier, 1873, 24 p.; Revue des deux Mondes ser. 3. 1: 544-566. 1 Feb 1874, 914-943. 15 Feb 1874. For many further details on Planchon and the American grapevines see Morrow (1960).

8022. *The Eucalyptus globulus*, from a botanic, economic and medical point of view, embracing its introduction, culture and uses ... Washington (Government printing office) 1875. Oct. *(Eucalyptus globulus)*.
Publ.: 1875 (p. 6: Mai 1875), frontisp., p. [1]-20. *Copy*: USDA. – Translated from Revue des deux Mondes 45(ser. 3. 7): 149-174. 1875.

8023. La botanique à Montpellier études historiques, notes et documents. *L'Herbier de Chirac* improprement dit de Magnol ... Montpellier (Typographie et lithographie Boehm et fils ...) 1884. Oct. *(Herb. Chirac)*.
Publ.: 1884, p. [1]-39, 5 p. facs. mss. *Copies*: FI, G.

8024. *Les vignes des tropiques du genre Ampelocissus*, considéré au point de vue pratique [Lyon (Imp. Waltener et Cie, ...) 1884-1885]. Oct.
Publ.: Dec 1884-Mar 1885 (in journal), p. [1]-31. *Copy*: G. – Reprinted and to be cited from La Vigne américaine 8(12): 370-381. Dec 1884, 9(1): 24-32. Jan 1885, 9(2): 44-51. Feb 1885, 9(3): 93-96. Mar 1885.
Planchon also revised and annotated the second edition of Louis Bazille's french version of Bush fils et Meissner, *Catalogue* illustré et descriptif *des vignes américaines*, Montpellier, Paris, 1885, 233 p.

Planchon, Louis David (1858-1915), French botanist at Montpellier; Dr. méd. Montpellier 1883; Pharm. sup. 1891, Dr. sci. 1900; from 1882-1894 at the Faculté de médecine; aggr. pharm. Paris 1894; associated with the École Supérieure de pharmacie, Montpellier 1883-1915; from 1901 as professor of matera medica; son of J.E. Planchon. *(L. Planch.)*.

HERBARIUM and TYPES: MPU.

BIBLIOGRAPHY and BIOGRAPHY: AG 12(2): 354; Barnhart 3: 90 (b. 3 Jul 1858, d. 8 Sep 1915); BJI 1: 46; BL 2: 163, 166, 703 (sic item on p. 166 correctly by Gustave Planchon); CSP 17: 917; Langman, p. 587; LS 20836a-20843; Morren ed. 10, p. 66; MW p. 389, suppl. p. 281; NI 1: 254; Plesch p. 363; Rehder 5: 677; Tucker 1: 560.

BIOFILE: Anon., Bull. Soc. bot. France 62: 214-215. 1916; Nat. Nov. 23: 280. 1901 (prof. mat. med. Montp.).
Granel de Solignac, L. et al., Natural. monsp. Bot. 26: 27. 1976 (pl. in herb. J. de Vichet).
Jadin, M.F., Rev. gén. Bot. 28: 1-10. 1916 (obit., bibl., b. 3 Jul 1858, d. 8 Sep 1915).
Mandon, L., Ann. Soc. Hort. Hist. nat. Hérault 50: 272, 279. 1910 (portr.).
Reber, B., Gallerie Therap. Pharmakogn. 388. 18987 (biogr. sketch).

8025. *Les champignons* comestibles et vénéneux de la région *de Montpellier* et des Cévennes aux points de vue économique et médical ... Montpellier (Imprimerie centrale du Midi (Hamelin frères) 1883. Oct. (*Champ. Montpellier*).
Publ.: 1883 (BSbF 14 Dec 1883; Nat. Nov. Jan(2) 1884; Bot. Zeit. 25 Jan 1884; Bot. Centralbl. 10-14 Dec 1883, 21-25 Apr 1884), p. [i]-x, [11]-223. *Copies*: FH(2), FI, MICH, Stevenson.

8026. *Étude sur les produits de la famille des Sapotées* .... Montpellier (Imprimerie centrale du Midi (Hamelin frères)) 1888. Oct. (*Étude Sapot.*).
Publ.: Jan-Jun 1888 (BSbF 27 Jul 1888; Morot 16 Aug 1888; Nat. Nov. Mar(1) 1889; Bot. Zeit. 25 Oct 1889), p. [i]-ix, [11]-121. *Copies*: B, FI, G, MO, NY, USDA; IDC 7135. – Thesis to become pharmacist.
Ref.: Morot, L., J. Bot. 2(16): 105-107. 1888.

8027. *Les Aristoloches.* Étude de matière médicale ... Montpellier (Imprimerie centrale du Midi (Hamelin frères)) 1891. Oct. (*Aristoloches*).
Publ.: Jan-Jun 1891 (BSbF 24 Jul 1891; Morot 16 Nov 1891; Nat. Nov. Sep(2) 1891; Bot. Zeit. 29 Jan 1892), p. [i]-viii, [9]-266. *Copies*: B, BR, FI, G, NY, USDA. – Thesis for "pharmacien supérieur".
Ref.: Roth, E., Bot. Centralbl. 3: 543-545. 1892 (rev.).
Morot, L., J. Bot. 5(22): ci-cii. 16 Nov 1891 (rev.).

8028. École supérieure de pharmacie de Paris. Thèse présentée au concours d'agrégation du 1er mai 1894 ... *Produits* fournis à la matière médicale par la famille *des Apocynées* ... Montpellier (Imprimerie centrale du Midi (Hamelin frères) 1894. Qu. (*Prodr. Apoc.*).
Thesis issue: Mai 1894, p. [i]-viii, [9]-364, *1 pl.* (uncol., B. Herincq). *Copy*: G. – "Thèse d'agrégation" (habil.).
Trade issue: Mai-Jun 1894 (BSbF 27 Jul 1894; Nat. Nov. Jul(1) 1894; Bot. Zeit. rev. 16 Aug 1894), p. [i]-viii, [9]-364, *1 pl.* (id.). *Copies*: B, BR, MO, NY.
Ref.: Hariot, P., Bull. Soc. bot. France 42(bibl.): 706-708. 1895.

Planellas Giralt, José (1850-1886), Spanish botanist; studied at Santiago; Dr. sci. nat. Madrid 1879; until 1883 professor of natural history at Barcelona; later director of the botanical garden and professor of botany at Habana, Cuba (1883-1886). (*Planellas*).

HERBARIUM and TYPES: Unknown.

BIBLIOGRAPHY and BIOGRAPHY: AG 5(2): 400; Barnhart 3: 90; BL 2: 503, 703; Colmeiro 1: cxciii; Jackson p. 340; Kew 4: 321; Morren ed. 10, p. 51; PR 7173; Rehder 5: 677; Tucker 1: 560.

BIOFILE: Conde, J.A., Hist. bot. Cuba 96, 97. 1958.
Urban, I., Symb. ant. 3: 101. 1902.

8029. *Ensayo de una flora fanerogámica gallega* ampliada con indicaciones acerca los usos médicos de las especies que se describen ... Santiago (Imprenta y litografia de D. Juan Rey Romero) 1852. Qu. (*Ensayo fl. gallega*).
Publ.: 1852, p. [1]-452. *Copies*: G, NY.
Ref.: W-m [M. Wilkomm], Bot. Zeit. 13: 191-197, 207-211. 1855.

Planer, Johann Jacob (1743-1789), German (Saxonian) botanist and physician; Dr. med. Erfurt 1778; prosector (1773), later professor (extraord. 1779, ord. 1783) of botany and chemistry at the University of Erfurt. (*Planer*).

PLANER

HERBARIUM and TYPES: Unknown.

BIBLIOGRAPHY and BIOGRAPHY: AG 4: 546 (b. 25 Jul 1743, d. 10 Dec 1789); Barnhart 3: 90; BM 4: 1582; Dryander 3: 160; Frank 3(Anh.): 77; Herder p. 88, 187, 244; Jackson p. 10; Kew 4: 321; LS 20845-20847; PR 1447, 5411, 7174-7176 (ed. 1: 8044-8046); Rehder 5: 677; SO 307, 308, 317, 2195; TL-2/6858.

BIOFILE: Hulth, J.M., Bref Skrifv. Linné 2(1): 141. 1916.
Jourdan, A.J.L., Dict. Sci. méd., Biogr. méd. 6: 430-431. 1824.
Manitz, H., Wiss. Z. Fr. Schiller Univ. Jena 24: 499. 1975.
Meusel, J.G., Lexik. teut. Schriftst. 10: 444-445. 1840.
Poggendorff, J.C., Biogr.-lit. Handw.-Buch 2: 464. 1863 (b. 25 Jul 1743, d. 10 Dec 1789).
Rothmaler, W., Mitt. thür. bot. Ver. ser. 2. 41: 58-59. 1933.

EPONYMY: *Planera* J.F. Gmelin (1791) and *Planera* Giseke (1792) are probably named for Planer but neither author gives etymology. ING 3: 1355.

8030. Karl von Linné, ... *Gattungen der Pflanzen* und ihre natürliche [sic] Merkmale, nach der Anzahl, Gestalt, Lage und Verhältniss aller Blumentheile. Nach der sechsten Ausgabe und der ersten und zweyten Mantisse übersetzt von Johann Jacob Planer, ... Gotha (bey Karl Wilhelm Ettinger) 1775, 2 vols. (*Gatt. Pfl.*).
Publ.: 1775.
1: p. [i-xviii], chart, [xix-xxiv], [1]-504.
2: p. [i], [505]-1100.
Copies: MO, U, USDA. – Essentially a translation of Linnaeus, Gen. pl. ed. 6 (1764) and Mantissa 1 and 2 (1767, 1771). – KR p. 431, SO 307.
Nachtrag: 1785, p. [1]-104, [i-iv]. – "Nachtrag zur sechsten Ausgabe der Gattungen..." Gotha (id.). Oct.

8031. Caroli a Linné ... *Termini botanici* classium methodi sexualis generumque plantarum characteres compendiosi. Recudi cum interpretatione germanica definitionum terminorum curavit Paulus Dietericus Giseke ... Editioni huic alteri accesserunt fragmenta ordinum naturalium Linnaei, nomina germanica Planeri generum, gallica et anglica terminorum, et indices. Hamburgi (Sumptibus B. Chr. Heroldi Viduae excudit Car. Wilh. Meyn) 1787. Oct. (*Term. bot.*).
Publ.: 1787 (p. xii: 4 Jan 1787), p. [i-xx], [1]-396. *Copies*: MO, US. A second edition of the Giseke edition (1781; SO 2194) of the *Termini botanici* with the addition of generic names taken from Planer (see KR p. 453-454, SO 2195).

8032. *Index plantarum, quas in agro Erfurtensi sponte provenientes* olim D. Joh. Philipp Nonne deinde D. Joh. Jacob Planer collegerunt. Gothae (apud Carol. Guilelm. Ettingerum). 1788. Oct. (*Index pl. agr. erfurt.*).
Publ.: 1 Jan-5 Feb 1788 (Manitz, 1975), p. [i-iv], [1]-284, [1, err.]. *Copies*: B, G, GOET, HU, NY, WU. – A follow up of J.P. Nonne, Fl. terr. erford. 1763, see TL-2/6857 and 6858.

8033. *Indici plantarum erffurtensium* [sic] fungos et plantas quasdam nuper collectas addit et praelectiones de historia naturali materiae medicae iii. ante idus octobris institutendas indicat Joh. Jac. Planer ... Erfordiae (Typis Nonnianis) [1788]. Oct. (*Indici pl. erffurt.*).
Publ.: Oct-Dec 1788 (p. iv; "iii ante idus octobris" is 13 Oct), p. [i-iv], [1]-44. *Copies*: BR, NY(2).

Platen, Paul Louis (1876-?), German palaeobotanist; student of J.P. Felix at Leipzig. (*Platen*).

HERBARIUM and TYPES: Unknown.

BIBLIOGRAPHY and BIOGRAPHY: Andrews p. 313; Barnhart 3: 91; BM 4: 1583; Ewan ed. 2: 176; LS 37970, suppl. 21908; Tucker 1: 560.

8034. *Untersuchungen fossiler Hölzer* aus dem Westen der Vereinigten Staaten von Nordamerika. Inaugural-Dissertation zur Erlangung der Doktorwürde bei der Philosophischen Fakultät der Universität Leipzig ... Leipzig (Druck von Gressner & Schramm) 1907. Oct. (*Unters. foss. Hölz.*).
Publ. 1907 (approved 16 Jul 1907; Bot. Zeit. 26 Jan 1900; Bot. Centralbl. 12 Jan 1909), p. [i]-xvi, [1]-155, [156, vita], *pl. 1-3* with text. *Copy*: USGS. – Also published in Sitzungsb. naturf. Ges. Leipzig 34: 1-164, *pl. 1-3.* 1907 (1909).
Ref.: Salfeld, H., Bot. Zeit. 67(2): 136-137. 16 Mai 1909.

Plawski, Alexander (*fl.* 1830), Polish botanist. (*Plawski*).

HERBARIUM and TYPES: Unknown.

BIBLIOGRAPHY and BIOGRAPHY: BM 4: 1584; PR 7182.

BIOFILE: Szymkiewicz, D., Bibl. Fl. Polsk. 18. 1925.

8035. *Słownik wyrazów botanicznych*, przez Alexandra Plawskiego. Wilno (Józef Zawadzki, własnym nakladem) 1830. Oct. (in fours). (*Słow. wyraz. bot.*).
Publ.: 1830, p. [i]-viii, [1]-296. *Copy*: NY.

Playfair, George Israel (1871-1922), Australian algologist; honorary curator of freshwater algae at the National Herbarium, Sydney, Australia. (*Playf.*).

HERBARIUM and TYPES: NSW; diatoms at PH. – Letters at G.

BIBLIOGRAPHY and BIOGRAPHY: Barnhart 3: 92; IH 1 (ed. 7): 340; LS 21918; NAF ser. 2. 6: 72; Nordstedt suppl. p. 13.

BIOFILE: Anon., Proc. Linn. Sci. N.S.W. 48(1): vi-vii. 1923 (obit., d. 8 Oct 1922; pastor at various N.S.W. churches 1900-1912, 1912-1918 Govt. Research Scholar at Sydney Univ.).
Fryxell, G.A., Beih. Nova Hedwigia 35: 363. 1975.
Koster, J.Th., Taxon 18: 556. 1969.
VanLandingham, S.L., Cat. Diat. 6: 3580-3581, 7: 4217. 1978.

8036. *Some new or less known Desmids* found in New South Wales ... [From Proceedings of the Linnean Society of New South Wales ... 1907]. Oct.
Publ.: 20 Jun 1907 (t.p. reprint so dated), p. [i, recto of 160], 160-201, *pl. 2-5. Copy*: G. – Reprinted and to be cited from Proc. Linn. Soc. N.S.W. 32(1): 160-201. 1907.

8037. *Some Sydney Desmids* .... [From the Proceedings of the Linnean Society of New South Wales] ... 1908. Oct.
Publ.: 20 Nov 1908 (reprint cover so dated), cover, p. [603]-628, *pl. 11-13. Copy*: G. – Reprinted and to be cited from Proc. Linn. Soc. N.S.W. 33(3): 623-628, *pl. 11-13.* 1908.

8038. *Polymorphism and life history in the Desmidiaceae* ... [From the Proceedings of the Linnean Society of New South Wales ... 1910]. Oct.
Publ.: 17 Sep 1910 (reprint cover so dated), cover, p. [459]-495, *pl. 11-14. Copy*: G. – Reprinted and to be cited from Proc. Linn. Soc. N.S.W. 35(2): 459-495, *pl. 11-14.* 1910.

8039. *Growth, development and life-history in the Desmidiaceae* ... Sydney (W.E. Smith Limited, ...) 1912. Oct.
Publ.: 1 Mai 1912 (t.p. reprint so dated), cover-t.p., p. [278]-298. *Copy*: G. – Reprinted and to be cited from Australas. Ass. Adv. Sci. 13: 278-298. 1912.

8040. *Plankton of the Sydney water-supply* ... [From the Poceedings of the Linnean Society of New South Wales ... 1912]. Oct.

Publ.: 19 Mar 1913 (t.p. reprint so dated), p. [1, t.p. reprint recto of p. 512], 512-552, *pl. 53-57. Copy*: G. – Reprinted and to be cited from Proc. Linn. Soc. N.S.W. 37(3): 512-552, *pl. 53-57.* 1913.

8041. *Contributions to a knowledge of the biology of the Richmond River* ... [From the Proceedings of the Linnean Society of New South Wales 1914]. Oct.
Publ.: 17 Jul 1914 (cover reprint so dated), cover-t.p., p. [93]-151, *pl. 2-8. Copy*: G. – Reprinted and to be cited from Proc. Linn. Soc. N.S.W. 39(1): 93-151, *pl. 2-8.* 1914.

8042. *The genus Trachelomonas* ... [From the Proceedings of the Linnean Society of New South Wales 1915]. Oct.
Publ.: 16 Jun 1915 (cover reprint so dated), cover t.p., p. [1]-41, *pl. 1-5. Copy*: G. – Reprinted and to be cited from Proc. Linn. Soc. N.S.W. 40(1): 1-41, *pl. 1-5.* 1915.

8043. *Freshwater algae of the Lismore district*: with an appendix on the algal fungi and schizomycetes. [From the Proceedings of the Linnean Society of New South Wales ... 1915]. Oct.
Publ.: 15 Sep 1915 (reprint t.p. so dated),p. [i, recto of 310], 310-362, *pl. 41-46. Copy*: G. – Reprinted and to be cited from Proc. Linn. Soc. N.S.W. 40(2): 310-362. *pl. 41-46.* 1915.

8044. *Oocystis and Eremosphaera* [from the Proceedings of the Linnean Society of New South Wales ... 1916]. Oct.
Publ.: 14 Jun 1916 (reprint so dated), cover t.p. (dated), p. [107]-147, *pl. 7-9. Copy*: G. – Reprinted and to be cited from Proc. Linn. Soc. N.S.W. 41(1): 107-147, *pl. 7-9.* 1916.

8045. *Australian freshwater phytoplankton. Protococcoideae* ... [From the Proceedings of the Linnean Society of New South Wales 1917]. Oct.
Publ.: 4 Apr 1917 (cover-t.p. so dated), cover-t.p., p. [823]-852, *pl. 56-59. Copy*: G. – Reprinted and to be cited from Proc. Linn. Soc. N.S.W. 41(4): 823-852, *pl. 56-59.* 1917.

8046. *Census of New South Wales fresh-water algae* ... Sydney (W.A. Gullick, ...) 1917. Oct. (*Census NSW fresh-wat. alg.*).
Publ.: 1917, p. [i], [219]-263. *Copies*: G, MO, USDA. – Issued in: J.H. Maiden & E. Betche, *A census of New South Wales Plants*, supplement 1. – Fresh water algae.

8047. *New and rare freshwater algae* ... [From the Proceedings of the Linnean Society of New South Wales 1918]. Oct.
Publ.: 18 Dec 1918 (cover-t.p. so dated), cover-t.p., p. [497]-543, *pl. 54-58. Copy*: G. – Reprinted and to be cited from Proc. Linn. Soc. N.S.W. 43(3): 497-543, *pl. 54-58.* 1918.

8048. *Peridineae of New South Wales* ... [From the Proceedings of the Linnean Society of New South Wales 1919 [1920]. Oct.
Publ.: 15 Mar 1920 (cover-t.p. so dated), p. [793]-818, *pl. 41-43. Copy*: G. – Reprinted and to be cited from Proc. Linn. Soc. N.S.W. 44(4): 793-818, *pl. 41-43.* 1920.

Plaz, Anton Wilhelm (1708-1784), German (Saxonian) botanist; Dr. med. Halle 1728; extraord. professor of botany Leipzig 1733, ord. id. 1749; later professor of physiology and therapeutics; dean of the medical faculty from 1773. (*Plaz*).

HERBARIUM and TYPES: Unknown.

BIBLIOGRAPHY and BIOGRAPHY: Barnhart 3: 92 (b. 2 Jan 1708); BM 4: 1584; Dryander 3: 250, 380, 381, 383, 385, 386, 397, 427, 564, 610; Herder p. 471; Kew 4: 323; LS 20862; PR 7183-7194 (b. 1706) (ed. 1: 8052-8064); Rehder 5: 678; TL-1/980.

BIOFILE: Anon., Comm. hebd. Sci. nat. med., Leipzig, 26: 546-555. 1784 (obit.). Jessen, K.F.W., Bot. Gegenw. Vorz. 327. 1864.

Jourdan, A.J.L., Dict. sci. méd., Biogr. méd. 6: 439-442. 1824.
Meusel, J.G., Lex. teut. Schriftst. 10: 453-456. 1810 (bibl.).
Rothmaler, W., Rep. spec. nov. 53(1): 36. 1944.
Zuchold, E.A., Jahresb. naturf. Ver. Halle 5: 46. 1853.

EPONYMY: *Plazia* Ruiz et Pavon (1794) commemorates Juan Plaza, a Spanish botanist and medical doctor.

8049. *De plantarum plethora* disserit et ad adveniendam orationem aditialem, qua professoris physiologiae ordinarii munus d. x. oct. mdccliv. hora ix. in auditorio philosophorum auspicabitur, invitat D. Antonius Guilielmus Plaz, ... [Lipsiae [Leipzig] (ex officina Langenhemiana) 1754]. Qu. (*Pl. pleth.*).
Publ.: 10 Oct 1754 (t.p.), p. [i]-xxxiii. *Copy*: NY.

8050. Ordinis medicorum in Academia lipsica h.t. procancellarius d. Andonius Guilielmus Plaz ... panegyrin medicam ad d. xviii Febr. mdcclxiii indicit *de Saccharo* nonnulla praefatus. Qu. (*De Saccharo*).
Publ.: 18 Feb 1763 (NZGS 14 Apr 1763), p. [i]-xvi. *Copy*: Edinburgh Univ. Libr.

8051. Ordinis medicorum in Academia lipsica h.t. procancellarius d. Antonius Guilielmus Plaz ... panegyrin medicam ad d. vi. julii a.r.g. mdcclxiv. Indicit *de plantarum sub diverso coelo nascentium cultura* praefatus [Leipzig 1764] Qu. (*De pl. cult.*).
Publ.: 6 Jul 1764, p. [i]-xii. Copy: Edinburgh Univ. Libr.

Plée, Auguste (1787-1825), French traveller for the Paris Muséum d'Histoire naturelle; from 1820-1825 in the West Indies. (*A. Plée*).

HERBARIUM and TYPES: P, PC; other material at B, G and L. – Original drawings at P, ms 65-73, 75.

BIBLIOGRAPHY and BIOGRAPHY: Barnhart 3: 92; Herder p. 175; Kew 4: 324; Lasègue p. 315, 492, 505; NI 1534, see also index; PR 7195-7196; Rehder 1: 94, 407, 5: xxi; Sotheby 603; TL-2/984, 986, 989; Tucker 1: 560; Urban-Berl. p. 383.

BIOFILE: Anon., Belg. hort. 17: 244, 256. 1867.
Baillon, H., Dict. bot. 3: 612. 1891.
Didot, Nouv. biogr. gén. 40: 463-464. 1862.
Guillaumin, A., Les fleurs des jardins xxxix. 1928.
Leland, A., Carnegie Inst. Washington Publ. 392(1): 274. 1932 (letter).
Leroy, F., ed., et al., Bot. franç. Amér. nord 7, 14, 28, 38 [Ewan], 193-201, [M. Raymond], 235. 1957.
Papavero, N., Essays hist. neotrop. dipterol. 1: 127-128, 192. 1971.
Pittier, H., Man. pl. usual. Venezuela 3. 1926.
Raymond, M., *in* J.F. Leroy, ed., Bot. franç. Amer. nord 193-201. 1957 (Plée and the Amer. flora; with notes on the "affaire Plée"; details on W. Ind. and Venez. explorations).
Rouse, P., Antiques 96: 763. 1969 (n.v.).
Urban, I., Symb. ant. 3: 101. 1902 (itin., coll.).
Verdoorn, F., ed., Chron. bot. 3: 366. 1937.

8052. *Herborisations artificielles* aux environs de Paris ou recueil de toutes les plantes qui y croissent naturellement, dessinées et gravées d'après nature, de grandeur naturelle, ... par Auguste Plée neveu, et François Plée, fils, ... à Paris (Chez les auteurs ...) 1811. Oct. (*Herbor. artif.*).
Co-author: François Plée.
Publ.: 1811 (p. [ii]: 20 Jul 1811), p. [i-ii], [1]-12, *100 unnumbered plates* (uncol. copp. hand-coloured by authors). *Copies*: G (*40 pl.*), NY (*100 pl.*). – The New York copy has the original cover of parts 1-10 ("Herborisations"). The contents are given om the wrappers but these are incomplete. Fascicles 1-3 contained 10 plates each, and were published by Auguste and François Plée; parts 4-10 contained 5 plates each and were

published by François Plée alone. The covers of the 11-17th parts are missing in the NY copy. With the three "double" parts 1-3, the 17 parts contained *100 plates* in all (ideal copy; NI mentions 104 pl., PR 99). The covers mention three issues: (1) with anatomical details in colour (NY); in full colour and in full colour on "papier grand raisin vélin". They were said to appear each month (from August 1811 onward).
Re-issue: PR mentions a new publication of the first two parts (20 pl. plus initial text), Paris (Lequien) 1830, not seen by us.

8053. *Le jeune botaniste*, ou entretiens d'un père avec son fils sur la botanique et la physiologie végétale. Ouvrage contenant, en abrégé, les principes de la physique végétale, ... à Paris (Chez Ferra aîné, ... et chez Lebel et Guibel, ..) 1812, 2 vols. 1812. Duod. (*Jeune bot.*).
1: 1812, frontisp., p. [i]-xvi, [1]-337, *pl. 1-8* (uncol. copp.).
2: 1812, n.v.
Copy: NY (vol. 1).

Plée, François (*fl.* 1844), French botanical artist and author. (*F. Plée*).

HERBARIUM and TYPES: Unknown.

BIBLIOGRAPHY and BIOGRAPHY: Blunt p. 228; BM 4: 1584; GF p. 70; Herder p. 175; Jackson p. 10, 274; Kew 4: 324; Lasègue p. 554; NI 1535, see also index; Plesch p. 363; PR 7197-7198 (ed. 1: 8067); Rehder 5: 678; Sotheby 604; TL-1/229; TL-2/989, 1093; Tucker 1: 560.

8054. *Types de chaque famille* et des principaux genres des plantes croissant spontanément en France exposition détaillée et complète de leurs caractères et de l'embryologie ... Paris (L'auteur, ... J.B. Baillière et fils, ...), Londres (Hippolyte Baillière, ...), New York (Baillière brothers, ...), Madrid (C. Bailly-Baillère, ...) 1844-1864. Qu. in twos. (*Types fam.*).
1: 1844-1860 (t.p.; livr. 1: Bot. Zeit. 27 Dec 1844), p. [1]-60, *pl. 1-75* with text (*2-75* handcol. or colour-printed copp. auct.; no. *1* plain).
2: 1844-1864 (t.p.), p. [i-iii], *pl. 76-142* ["*64*"], *143-160* (id.), [1-16, indexes].
Copies: B, BR, FI, G, MO, NY, USDA. – The book was published in fascicles each consisting of a single plate with text. Pritzel (ed. 1) mentions 8 p. introduction.
Ref.: Schlechtendal, D.F.L. von, Bot. Zeit. 3: 310-311. 2 Mai 1845 (rev. fasc. 1-6).

8055. *Glossologie botanique*ou vocabulaire donnant la définition des mots techniques usités dans l'enseignement; appendice indispensable des livres élémentaires et des traités de botanique par F. Plée, ... à Paris (Chez J.B. Baillière, ...), Londres (Chez H. Baillière, ...), New York (Chez H. Baillière, ...), Madrid (Chez Ch. Bailly-Baillière, ..) 1854. Duod. (in sixes). (*Gloss. bot.*).
Publ.: 1854 (p. 8: 15 Apr 1854), p. [1]-72. *Copies*: FI, NY.

Pleijel, Carl Gerhard Wilhelm (1866-1937), Swedish pharmacist and botanist at Alvesta, Småland (1911-1920), pharmacist at Stockholm 1920-1936. (*Pleijel*).

HERBARIUM and TYPES: S; other material at GB, LD, UPS.

BIBLIOGRAPHY and BIOGRAPHY: Barnhart 3: 92; BL 2: 515, 703; Bossert p. 312 (d. 23 Aug 1937); Kleppa 228; KR p. 574 (b. 3 Dec 1866).

BIOFILE: Anon., Generalreg. Bot. Not. 1839-1938, p. 66, 1939.
Lindman, S., Svenska män och kvinnor 6: 140. 1949.

Plenck, Joseph Jacob von [Plenk] (1738-1807), Austrian botanist, military physician, teacher at the military medical academy of Vienna. (*Plenck*).

HERBARIUM and TYPES: Unknown.

BIBLIOGRAPHY and BIOGRAPHY: AG 7: 240; Barnhart 3: 92; Blunt p. 158; BM 4: 1585; Bossert p. 312 (b. 28 Nov 1738, d. 24 Aug 1807); DU 225; GF p. 70; Hortus 3: 1201; Herder p. 91, 118, 132, 410; Jackson p. 200; Kew 4: 324; Langman p. 587; Lasègue p. 540; LS 20864; NI 1536; Plesch p. 364 (collation); PR 7199-7204 (ed. 1: 8069-8074); Rehder 5: 678; SO 724a, 759; Sotheby 605; TL-1/981; Tucker 1: 560, 779; Zander ed. 10, p. 702, ed. 11, p. 801 (b. 28 Nov 1738, d. 24 Aug 1807).

BIOFILE: Anon., Portraiten-Gallerie berühmter Aertzte und Naturforscher des Österreichischen Kaiserthumes *pl. 25.* 1838 (portr.).
Neilreich, A., Verh. zool. bot. Ges. Wien 5(Abh.): 34, 44, 48. 1855.
Poggendorff, J.C., Biogr.-lit. Handw.-Buch 2: 472. 1858.
Verdoorn, F., ed., Chron. bot. 4: 446. 1938.
Wittrock, V.B., Acta Horti Berg. 3(2): 84. 1903.

EPONYMY: *Austroplenckia* Lundell (1939); *Plenckia* Lundell (1939); *Plenckia* Reisseck (1861, nom. cons.). *Note*: Rafinesque does not give a derivation for *Plenckia* Rafinesque (1814, nom. rej.).

8056. Josephi Jacobi Plenck, ... *Bromatologia* seu doctrina de esculentis et potulentis. Viennae [Wien] (apud Rudolphum Graeffer) 1784. Oct. (*Bromatologia*).
Publ.: 1784, frontisp., p. [1]-428, [1-12, index]. *Copy*: NY (latin ed.). – Bromatologia: the science of food; p. 19-176 on edible plants. BM mentions latin and german editions with this same t.p.

8057. Josephi Jacobi Plenck ... *Icones plantarum medicinalium* secundum systema Linnaei digestarum, cum enumeratione virium et usus medici, chirurgici atque diaetetici ... Centuria. i[-viii]. Viennae [Wien] (apud Rudolphum Graeffer et Soc. [cent. vi: apud A. Blumauer, cent. vii sumptibus Librariae camesinianae ...] 1788 [-1812]. Fol. (*Icon. pl. med.*).
Publ.: 1788-1812, in parts. Volume 8 consisted only of two parts due to the death of the author (edited by Joseph Lorentz Kerndl). The book contains 758 hand-coloured engravings. 7 grosse[n] kostspielig aufgelegten Foliobände[n] [...], allein der Text ist ohne Werth und die Abbildungen sind meist Copien aus anderen Werken" (Neilreich). *Copies*: GOET (vols. 1-7), HU (vol. 1); Teyler, WU (1-7). The analysis below is of the copy at GOET (in original covers), submitted by G. Wagenitz, and on the copy at WU (inf. Gutermann).
Text, cent. 1: 1788, p. [i-xii], [1]-62, [63-64].
  cent. 2: 1789, p. [1]-80, [1-2, index].
  cent. 3: 1790, p. [1]-74, [1-2].
  cent. 4: 1791, p. [1]-93, [1-2].
  cent. 5: 1792, p. [1]-84, [1-2].
  cent. 6: 1794, p. [1]-65, [1-2]; 2 title-pages.
  cent. 7: 1803, p. [i]-iv, 1-58.
  cent. 8: 1812, not in copy GOET.
Plates, cent. 1, 1788.
  fasc. 1: pl. 1-25, "Hiezu kommt" (with this fascicle belongs:) t.p., dedic., pref., index cent. 1 (=[i-xii], text A-D.
  fasc. 2: pl. 26-50, id. text E-H.
  fasc. 3: pl. 51-75, id. text I-M.
  fasc. 4: pl. 76-100, id. text N-S.
Cent. 4: 1789.
  fasc. 5: pl. 101-125, id. text A-E.
  fasc. 6: pl. 126-150, id. text F-M.
  fasc. 7: pl. 151-175, id. text N-Q.
  fasc. 8: pl. 176-200, id. text R-X, "Erinnerung".
Cent. 3: 1790.
  fasc. 9: pl. 201-225, id. text. A-G.
  fasc. 10: pl. 226-250, id. text H-M.
  fasc. 11: pl. 251-275, id. text N-Q.
  fasc. 12: pl. 276-300, id. text R-U.

Cent. 4: 1791.
  fasc. 13: *pl. 301-325*, id. text A-H.
  fasc. 14: *pl. 326-350*, id. text I-O.
  fasc. 15: *pl. 351-375*, id. text P-S.
  fasc. 16: *pl. 376-400*, id. text T-Aa.
Cent. 5: 1792-1793, "Centuria vi. et ultima".
  fasc. 17: 1792, *pl. 401-425*, id. text A-K.
  fasc. 18: 1792, *pl. 426-450*, id. text L-P.
  fasc. 19: 1793, *pl. 451-475*, id. text Q-T.
  fasc. 20: 1793, *pl. 476-500*, id. text U-Z.
Cent. 6: 1794-1795.
  fasc. 21: 1794, *pl. 501-525*, id. text A-D.
  fasc. 22: 1794, *pl. 526-550*, id. text E-S.
  fasc. 23: 1795, *pl. 551-575*, id. text H-I.
  fasc. 24: 1794 [sic], *pl. 576-600*, no mention of text on cover.
Cent. 7: 1803-1805.
  fasc. 1: 1803, *pl. 601-625*, id. text A-D.
  fasc. 2: 1804, *pl. 626-650*, id. text E-G.
  fasc. 3: 1805, *pl. 651-675*, id. text H-L.
  fasc. 4: 1805, *pl. 676-700*, id. text M-P.
Cent. 8: 1812 (Teyler copy), *pl. 701-758*.
Ref.: DU 225, GF p. 70, NI 1536, PR 7201.
Neilreich, A., Verh. zool.-bot. Ver. Wien 5: 34. 1855.

8058. Josephi Jacobi Plenk [sic] ... *Elementa terminologiae botanicae* ac systematis sexualis plantarum. Viennae [Wien] (apud A. Blumauer) 1796. Oct. (*Elem. termin. bot.*).
Publ.: 1796, p. [1]-169, [170-172]. *Copies*: G, NY, USDA.
German: 1798, p. [1]-168. *Copy*: B. – "Joseph Jakob Plenk's ... *Anfangsgründe der botanischen Terminologie* und des Geschlechtssystems der Pflanzen". Wien (Bey Christ. Fridr. Wappler). Oct. 1798.
Spanish: 1802, p. [3]-222, [1, err.], [p. [1], h.t.? lacking in Hunt and FI copies?]. *Copies*: FI, HU. – *Elementos de la nomenclatura botánica*, y sistema sexual de las plantas, ... traducidos del latin al español para el uso de los discípulos de los Reales Colégios de Cirurgía Medica, por el fisico D. Juan Francisco Bahí [1775-1841], ... Barcelona (por la Compañia de Jordi, Roca, y Gaspár) 1802. Qu.

Plitt, Charles Christian (1869-1933), American lichenologist and bryologist; Ph.G. (pharmacy) Maryland College Pharmacy 1891; Dr. sci. 1921 Int. Acad. Sci., Baltimore; teacher at various public schools in Baltimore, later at Baltimore City College; ultimately professor of botany at the University of Maryland. (*Plitt*).

HERBARIUM and TYPES: BPI (17.000); US (flowering plants, ferns, lichens); other material at B, BM, BP, CM, DBN, H, LD, M, MICH, PH, and in the private herbarium of Clyde Reed (material rd. from Maryland Academy of Sciences, incl. notebooks). – See under H.E. Hasse for *Lichenes exsiccati ex herb. Dr. H.E. Hasse relicti*, distributed by Plitt.

BIBLIOGRAPHY and BIOGRAPHY: Barnhart 3: 92; Bossert p. 312 (b. 6 Mai 1869, d. 13 Oct 1933); Ewan ed. 2: 176; GR p. 194, cat. p. 69; Hawksworth p. 185; Hirsch p. 234; IH 1 (ed. 3): 24, (ed. 6): 363, 2: (in press); Kelly p. 177, 256; LS 37975, suppl. 21927-21935; SBC p. 129; TL-2/see Hasse, H.E.; Urban-Berl. p. 289.

BIOFILE: Anon., Bryologist 37: 93-95. 1935 (portr., bibl.); Rev. bryol. lichénol. 34: 377. 1966 (Plitt lichen herbarium at US); Science ser. 2. 78: 378. 1933 (d.).
Boufford, D.E., Taxon 29: 676. 1980 (coll. at CM).
Ewan, J. et al., Short hist. bot. U.S. 95. 1969.
Fessenden, G.R. & A.R. Goldberg, Wild Flower 30(4): 81-90. 1954 (extracts from journal of C.C. Plitt; b. 6 Mai 1869, d. 13 Oct 1933; (portr.)), 31: 5-13, 35-39, 48, 53-59. 1955.
Laundon, J.R., Lichenologist 11: 16. 1979 (herb. at US).
Stevenson, J.A., Taxon 4: 184. 1955 (herb. at BPI).

Plitzka, Alfred (1861-?), Austrian-Silesian/Moravian botanist; high school teacher at Neutitschein (Nový Jičín) 1894-1903. (*Plitzka*).

HERBARIUM and TYPES: PR.

BIBLIOGRAPHY and BIOGRAPHY: CSP 17: 926; Future-Domin p. 471; IH 2: (in press); Kláštersky p. 144.

8059. *Einiges über die Gymnospermen.* Ein Blick auf die Gymnospermen Linné's nebst eingehender Besprechung der gegenwärtig gleichbenannten Pflanzengruppe unter besonderer Berücksichtigung der bereits entschiedenen und noch schwebenden Streitfragen ... Neutitschein (Im Selbstverlage des Verfassers ...) 1896. Oct. (*Gymnospermen*).
Publ.: Jun-Aug 1896 (ÖbZ 1 Sep 1896; Nat. Nov. Sep(2) 1896), p. [1]-55, *1 pl. Copy:* B. – Reprinted from Jahresber. mähr. Landes-Oberrealschule in Neutitschein 1896: 1-55.

Plowright, Charles Bagge (1849-1910); British botanist and physician; MD Durham 1890; medical officer of health, Freebridge Lynn Rural District Council; from 1892-1894 Hunterian professor of comparative anatomy and physiology at the London Royal College of Surgeons. (*Plowr.*).

HERBARIUM and TYPES: BIRM; fungi also at K; *exsiccatae: Sphaeriacei britannici* 1873-1878, fasc. 1-3, nos. 1-300; exsiccatae and other material at B (extant), BIRM, C, CGE, CUP, DBN, E, FI, H, K, MANCH, MICH, MW, NYS, PAD, S, W. – Contributed also to Phillips, *Elvellacei britannici* and to M.C. Cooke, *Fungi britannici* (on labels often as "C.B.P."), and to C. Roumeguère, *Fungi selecti exsiccati.* – Letters and manuscripts at BM, G, K and PH.

BIBLIOGRAPHY and BIOGRAPHY: Ainsworth p. 302, 321, 331; Barnhart 3: 93; BB p. 246 (b. 3 Apr 1849, d. 24 Apr 1910); BM 4: 1586-1587; Bossert p. 312; CSP 8: 635, 11: 35-36, 12: 580, 17: 927-928; Desmond p. 498; GR p. 279, cat. p. 69; Hawksworth p. 185; IH 2: (in press); Kelly p. 178; Kew 4: 325; Lenley p. 466; LS 20871-20994, 37976-37985 (bibl.); Morren ed. 2, p. 23, ed. 10, p. 77; NI 3: 27; NAF 7: 1083; PH 26; Rehder 5: 678; Stevenson p. 1254; Urban-Berl. p. 280.

BIOFILE: Ainsworth, G.C., Dict. fungi ed. 6. 470. 1971.
Anon., Bot. Jahrb. 14, Beibl. 32: 70. 1892; Bot. Not. 1910: 176 (d.); Gard. Chron. 1910: 286. 30 Apr 1910 (obit.); Mycologia 2: 198. 1910; Nat. Nov. 32: 289. 1910 (d.); Nature 83: 287. 1910 (obit.); Österr. bot. Z. 41: 396. 1891 (app. London); Trans. Brit. mycol. Soc. 3: 231-232. 1910 (obit., b. 3 Apr 1849, d. 24 Apr 1919).
Bisby, G.R., Trans. Brit. mycol. Soc. Index p. 19. 1952.
Bridson, G.D.R. et al., Nat. hist. mss. res. Brit. Isl. 229.462. 1980.
Ewan, J. et al., Leafl. W. Bot. 7(3): 72. 1953.
Hanlin, R.T., Index Grevillea 161. 1978.
Hedge, I.C. & J.M. Lamond, Index coll. Edinb. herb. 119. 1970.
Jackson, B.D., Bull. misc. Inf. Kew 1901: 53 (fungi at K).
Murray, G., Hist. coll. BM(NH) 1: 174. 1904 (exsicc.).
Palmer, J.T., Nova Hedwigia 15: 153. 1968 (bibl. Gasteromyc.).
Ramsbottom, J., Trans. Brit. mycol. Soc. 30: 1, 2, 13. 1948.
Rickett, H.W., Index Bull. Torrey bot. Club 78. 1955.
Rogers, D.P., Brief hist. mycol. N. Amer. ed. 2. 16. 1981.
Saccardo, P.A., Syll. fung. 2: 635-639. 1883.
T.P., Trans. Norfolk Norw. Natural. Soc. 9(2): 275-282. 1911 (obit.).
Wittrock, V.B., Acta Horti Berg. 3(2): 96, *pl. 35.* 1903 (portr.).

EPONYMY: *Plowrightia* P.A. Saccardo (1883); *Plowrightiella* (P.A. Saccardo) Trotter (1926).

8060. *Monograph of the British Uredineae and Ustilagineae* with an account of their biology including the methods of observing the germination of their spores and their experimental culture ... London (Kegan Paul, Trench & Co., ...) 1889. Oct. (*Monogr. Brit. Ured.*).

Publ.: 1-15 Jan 1889 (L.M. Underwood copy at NY signed "January 1889"; Nat. Nov. Jan(2) 1889; ÖbZ 1 Mar 1889; Bot. Centralbl. 5 Mar 1889; Bot. Zeit. 26 Apr 1889; Bot. Gaz. Mai 1889; J. Bot. Apr 1889), p. [i]-vii, [1, cont.], [1]-347, *pl. 1-8* (uncol. liths. auct.). *Copies*: B, BR, FH, G, MO, NY, Stevenson, USDA.
Ref.: M.C.C. Grevillea 17: 62-64. Mar 1889 (rev.).
Möbius, M., Bot. Centralbl. 40: 138-140. 29 Oct 1889 (rev.).
Phillips, W., J. Bot. 27: 156-157. 1889 (rev.).
Poirault, G., J. Bot. Morot 3(15): lxix-lxx. 1889 (rev.).

Plues, Margaret (c. 1840-c. 1903), British botanist; entered a roman catholic convent at Weybridge in middle-age, ultimately mother superior; in her youth a popular writer on botany. (*Plues*).

HERBARIUM and TYPES: Unknown; some letters at BM.

BIBLIOGRAPHY and BIOGRAPHY: BB p. 2216; BM 4: 1587; Desmond p. 498; Jackson p. 236, 239, 240, 241, 244; Kelly p. 178; Kew 4: 325; NI 1537-1539; PR 7211.

BIOFILE: Allen, D.E., The victorian fern craze 56. 1969.
Allibone, S.A., Crit. dict. Engl. lit. 1610. 1870.
Freeman, R.B., Brit. nat. hist. books 282. 1980 (bibl.).
Hawksworth, D.L. & M.R.D. Seaward, Lichenol. Brit. Isl. 1568-1975 p. 16-17, 134. 1977.

EPONYMY: *Pluesia* Nieuwland (1916).

8061. *A selection of the eatable funguses of Great Britain* edited by Robert Hogg, ... and George W. Johnson, ... and illustrated by W.G. Smith, ... London (Journal of Horticulture & Cottage Gardener Office, ...) [1866]. (*Select. fung. Gr. Brit.*).
Editors: Robert Hogg (1818-1897), George William Johnson (1802-1886).
Publ.: 1866, p. [i*-iv*], [i]-vii, *pl. 1-9, 11-21, 23-24* (handcol. liths. by W.G. Smith, with text). *Copy*: USDA.

8062. *British ferns*: an introduction to the study of the ferns, lycopods, and equiseta indigenous to the British Isles, with chapters on the structure, propagation, cultivation, diseases, uses, preservation, and distribution of ferns ... London (L. Reeve and Co., ...) [1866]. Oct. (*Brit. ferns*).
Publ.: 1866, p. [i]-x, [1]-281, [1, h.t.], *pl. 1-16* (col. liths. W. Fitch). *Copies*: BR, NY, US.
Reissue: 1914, London (plates uncol., see Nat. Nov. Mai(1) 1914).

8063. *British grasses*: an introduction to the study of the Gramineae of Great Britain and Ireland ... London (L. Reeve and Co., ...) [1867]. Oct. (*Brit. grass.*).
Publ.: 1867 (p. vi; Mai 1867; Flora 8 Oct 1867; BSbF Dec 1868), p. [i]-vi, [vii-viii], [1]-307, *pl. 1-16* (col. liths. W. Fitch). *Copies*: BR, G, MO, NY. – The copy at G has a different imprint: "London: Reeve & Co., ... 1867". The other copies, with undated t.p. may be of a reissue (e.g. London 1914, see Bot. Centralbl. 21 Jul 1914).

Plukenet, Leonard (1642-1706), British botanist and physician, probably educated at Westminster School; practicing physician in London, later Queen's botanist to Mary II (wife of Willem III) and superintendent of Hampton Court. (*Pluk.*).

HERBARIUM and TYPES: Plukenet bequeathed his herbarium to his wife, afterwards it came in the hands of Dr. Moore, Bishop of Norwich; Sir Hans Sloane bought them from Dr. Moore in 1710, and they now constitute H.S. 83, 84, 85, 86, 87-105 of his herbarium (BM). Dandy (1958) gives a detailed description of these volumes, which are "models of neatness". The collectors are seldom indicated but can "be ascertained by reference to Plukenet's published works for which the specimens are typical, being in many cases the actual plants from which the figures in the Phytographia were taken". The importance of Plukenet's herbarium is evident from this circumstance; many later authors, including Linnaeus based species upon Plukenet's plates. A disturbing circumstance is that

Plukenet's labels (as was the case with Petiver's) were unattached to the specimens when received by Sloane. For all further details, including a list of collectors, see Dandy. – Other herbaria claiming the possession of some Plukenet material: L, LE and OXF (possibly mostly specimens from P's garden). – Plukenet's own copies of his books, with ms notes, are at BM.

BIBLIOGRAPHY and BIOGRAPHY: AG 1: 38; Backer p. 452; Barnhart 3: 93; BB p. 246 (b. Dec 1641, d. 6 Jul 1706); BM 4: 1587; Bossert p. 313; Bret. p. 33; Clokie p. 225 (specimens at OXF); Desmond p. 498 (b. 1642); De Toni 1: c; DNB 45: 432; Dryander 3: 66-67; GR p. 410; HE p. 96; Herder p. 84; Jackson p. 31; Kew 4: 325; Lasègue p. 12, 322, 324, 434; LS 29997-20999; MW p. 389; NI 1540-1543, see also 1: 103; Plesch p. 364; PR 3348, 7116, 7212, 9172 (ed. 1: 8086); Rehder 1: 228; SA 2: 594; SO 681C; Sotheby 606; SK 4: lxxx, lxxxi; TL-1/982; TL-2/see Buddle; Tucker 1: 561.

BIOFILE: Allen, D.E., The naturalist in Britain 10, 19. 1976.
Allibone, S.A., Crit. dict. Engl. lit. 1610. 1870.
Boulger, G.S., J. Bot. 38: 336-338. 1900 (on a copy of Ray, Cat. pl. Angl., 1760, with mss notes by P.).
Bretschneider, E., J. Bot., 21: 213. 1883 (with note by B.D. Jackson on dates of *Phytographia*).
Bridson, G.D.R. et al., Nat. hist. mss. res. Brit. Isl. 429 [index]. 1980 (mss).
Burkill, I.H., Chapt. Hist. Bot. India 9-10. 1965.
Dandy, J.E., The Sloane herbarium 36-41, 183-188. 1958.
Darlington, W., Memorials 308, 325. 1849.
Dony, J.G., Fl. Hertfordshire 11. 1967.
Ewan, J., Regn. veg. 71: 19, 23, 25-29, 32, 50. 1970.
Ewan, J. & N., John Banister 483. 1970; also J.Ewan *in* F. Pursh, Fl. Amer. sept. facs. repr. 1979, introd. p. 24. 1979.
Ewan, J. et al., Short hist. bot. U.S. 30, 31. 1969.
Freeman, R.B., Brit. nat. hist. books 282-283. 1980 (bibl.).
Fries, Th.M., Bref Skrifv. Linné 1(3): 67. 1909, 1(6): 189. 1912.
Giseke, P.D., Index linn. Leonh. Pluk. opera bot., p. i-x. 1779.
Guédès, M., Arch. nat. Hist. 19(1): 67-76. 1981 (impt. analysis of printed books).
Gunn, M. & L.E. Codd, Bot. expl. S. Afr. 37-38. 1981.
Hawksworth, D.L. & M.R.D. Seaward, Lichenol. Brit. Isl. 1568-1975, p. 5, 134, 198. 1977.
Herder, F. v., Bot. Centralbl. 55: 259. 1893 (some Plukenet material at LE).
Howard, R.A., Bot. J. Linn. Soc. 79: 71-85. 1979 (West-Indian records Plukenet and Petiver; updated nomenclature).
Hulth, J.M., Bref Skrifv. Linné 1(8): 29, 33. 1922, 2(1): 10, 11, 309, 2(2): 37, 38, 74, 96. 1943.
Jackson, B.D., J. Bot. 20: 338-342. 1882, 32: 247-248. 1894 (biographical details on the "Queen's botanist").
Jourdan, A.L.J., Dict. sci. méd., Biogr. méd. 6: 454-455. 1824.
Kent, D.H., Brit. herb. 73. 1957, Hist. Fl. Middlesex 15. 1975.
King, G., J. Bot. 37: 454-463. 1899 (P. on Indian pl.).
Kuntze, O., Rev. gen. pl. 1: cxxxviii. 1891.
Milner, J.D., Cat. portr. Kew 96. 1906.
Pearsall, W.H., Fl. Surrey 45. 1931.
Pulteney, R., Sketches 2: 18-29. 1790.
Raven, C.E., John Ray, ed. 2. 64, 231-232, 245, 301. 1950.
Reynolds, G.W., Aloes S. Africa 78. 1950.
Römer, J.J., Neu. Mag. Bot. 1: 329-331. 1794 (on mss).
Smith, J.E., *in* Rees, Cycl. vol. 27, alph.
Stafleu, F.A., Intr. Jussieu Gen. Pl. xlii. 1964; *in* fasc. ed. 1971 of N.J. Jacquin, Sel. stirp. amer. (1763).
Trimen, H. & W.T.T. Dyer, Fl. Middlesex 374-376. 1869.
White, A. et al., Succ. Euphorb. 1: 36-37. 1946.
White, A. & B.L. Sloane, The Stapelieae 14, 15. 1933, ed. 2. 77. 1937.
Wilson, E.H., Plant hunting 1: 33, 2: 107. 1927.

NOTE: For a discussion of contemporary opinion on Plukenet and on his relations with Ray, Petiver and Sloane, relations which were not always of the friendliest, see Dandy. Plukenet and Petiver are known for their criticism of colleagues expressed on labels of herbarium specimens sent abroad (e.g. to Tournefort and Vaillant). Linnaeus expressed himself favorably on Plukenet (Crit. Bot. no. 283): "Plukenetia is a plant with flowers of unique structure, even as Plukenet was unique among botanists" and "Plukenet prefers plants to any riches: he grudges nothing if he can only draw pictures of the rarities with which he has fallen more deeply in love than almost anyone else" (transl. Hort.).

GENERAL INDEX: P.D. Giseke, *Index linnaeanus in Leonhardi Plukenetii*, ... *opera botanica* ... Hamburg 1779, see below (no. 8065), preceded by M. van Phelsum, Explicatio Phytographiae L. Plukentii. Harlingae [Harlingen, Netherlands] 1769.

HANDWRITING: Dandy, J.E., The Sloane Herbarium, London 1938, facsimile no. 6; Clokie p. 279.

EPONYMY: *Plukeneta* Cothenius (1790, orth. var.); *Plukenetia* Linnaeus (1753), ING 3: 1375.

8064. Leonardi Plukenetii *Phytographia* sive illustriorum & miniis cognitarum icones tabulis aeneis summâ diligentiâ elaboratae; quarum unaquaeque titulis descriptoriis ex notis suis propriis & characteristicis desumptis insignita; ab aliis ejusdem sortis facile discriminatur. Pars prior ... Londini (sumptibus autoris) 1691. Qu. (*Phytographia*).
Pars prior: 1691, p. [i, engr. t.p., iii engr. ded., v engr. pref.], *pl. 1-72* (all *454 pl.* (2715 figs.) are copper engr. of drawings by J. Collins).
Pars altera: 1691, p. [i, iii], *pl. 73-113,* [first] appendix [i], *pl. 114-117,* altera appendix [i], *pl. 118-120.*
Pars tertia: 1692, p. [i-vii], *pl. 121-239,* [third] appendix *pl. 240-250.*
[*Pars quarta*]: 1694, *pl. 251-328,* [1-6], belongs to *Almagestum.*
Almagestum: 1696, p. [i-v], 1-402, [1, 2, abbrev.], plates publ. 1694 see pars quarta.
"Almagestum botanicum sive Phytographiae Pluc'netianae onomasticon methodo syntheticâ digestum exhibens stirpium exoticarum, rariorum, novarumque nomina, quae descriptionis locum supplere possunt ... adjiciuntur & aliquot novarum plantarum icones in gratiam phylophytosophorum in lucem nùnc editae. Trahit sua quemquè voluptas ... Londini (sumptibus autoris) 1696. Qu.
Almagesti ... mantissa: 1700, p. [i-v], 1-191, [1, app.], [1-28, index], *pl. 329-350.* "Almagesti botanici mantissa. Plantarum novissimè detectarum ultrà millenarium numerum complectens. Cui, tanquam pedi jam stantis Columneae plus ultra inscribere fas est. Cum indice otius operis ad calcem adjecto ... Londini (sumptibus autoris) 1700.
Amaltheum: 1705, p. [i-v], 1-216, [1-2, app.], [1-7, index], [1, err.], *pl. 351-454.* "Amaltheum botanicum. (i.e.)stirpium indicarum alterum copiae cornu millenas ad minimum & bis centum diversas species novas & indictas nominatim comprehendens; quarum sexcenae & insuper, selectis iconibus, aeneisque tabulis in gratiam phytosophorum exquisite & summo artificio illustrantur ... Londini 1705".
Copies: BR, NY, USDA.
Reprint[*1*]: [*1*]: 1720. Opera omnia botanica, in sex tomos divisa; viz. i, ii, iii, Phytographia ... Londini (Apud Guil. & Joan. Innys, ...) 1720, p. [i-vii], followed by original volumes. *Copies*: NY, USDA; IDC 6154.
Reprint [*2*]: 1769, secundò excusum, ediderunt T. Davies, ... 1769. [London]. *Copies*: US, USDA. – This post-linnaean reprint is not a new publication and does not count for botanical nomenclature.
Note: For an up-to-date summary of the knowldge of the bibliography of Plukenet's works see M. Guédès (1981). – Our analysis is that of the NY copy; for all further details however, see Guédès.
Ref.: Tenzel, F.B.R., Nomenclator systematicus in Leonhardi Plukenetii phytographiam, Erlangen, 1820, 106 p. (fide Rehder 1: 78).

8065. *Index linnaeanus in Leonhardi Plukenetii*, M.D. Opera botanica olim in privatos usus conscriptus, nunc vero in aliorum etiam commodum editus. Accessere variae in vitam et opera Plukenetii observationes partim ex ipsius msto auctore Paulo Dieterico Giseke, ... Index linnaeanus in Joannis Jacobi Dillenii Historiam muscorum ob similem usum ad-

ditus est auctore eodem ... prostat Hamburgi apud auctorem et Carolo Ernesto Bohn commissum. Typis Caroli Wilhelmi Meyn. 1779. Qu. (*Index linn. Pluk.*).
Author: Paul Dietrich Giseke (1741-1796).
Publ.: 1779, p. [i]-x, [1]-46. *Copy*: MO.

Plumier, Charles (1646-1704), French missionary, explorer and botanist; travelled in the West Indies 1689, 1693 and 1695. (*Plum.*).

HERBARIUM: P(local, of little importance). The types of Plumier's American plants are his original drawings. Further elucidation is provided, in part, by Surian's herbarium (see sub Surian); in several cases the Surian plants may be the actual types of Plumier species.

MANUSCRIPTS and DRAWINGS: The Plumier manuscripts and drawings often referred to by Adanson and Jussieu are the ca. 6000 original drawings made by Plumier which are now kept at the Muséum d'Histoire naturelle. This Plumier collection of drawings was originally at the Bibliothèque nationale which incorporates the holdings of the old Cabinet des Estampes du Roi. Since 22 Oct 1834 they have been at the *Bibliothèque centrale* of the Paris Muséum and constitute ms. 1-37, 1335. – The Groningen (University library) set of 508 copies was made by Claude Aubriet for Boerhaave in 1733, and used by Burman for the preparation of the *Plantarum americanarum fasciculus* (1-10) (262 plates published). The set was seen by Linnaeus during his stay in Holland. – Another set of copies (312) was acquired by Lord Bute and is now at BM (through Banks, Banksian ms. 1-5). An uncoloured set of copies of Plumier's drawings together with transcripts of descriptions is at K; a further set is at OXF. Clokie (1964) states that Vaillant "employed a draughtsman to copy Plumier's drawings after his death, and sold them to his friends". The correspondence between Vaillant and Sherard at OXF reveals that Sherard also bought such copies.

BIBLIOGRAPHY and BIOGRAPHY: AG 1: 100; Backer p. 452-453; Barnhart 3: 93; Blunt p. 128, 263; BM 4: 1587; Bossert p. 313 (b. 20 Apr 1646, d. 20 Nov 1704); Clokie p. 225-226; Dryander 3: 34, 186, 220; DSB 11: 47-48 (by P. Jovet and J.C. Mallet); Frank 3(Anh.): 77-78; GF p. 70; GR p. 345; Hegi 6(2): 1098; Herder p. 225, 271; HU 407, 554, see also 2(1): clxx; IH 1(ed. 6): 363, 2: (in press); Jackson p. 354, 359; Kew 4: 325; Langman p. 587; Lasègue p. 21, 37, 313, 315, 487, 505; LS 21001; Moebius p. 129; NI 1544-1548, see also 1: 100, 185, 3: 55; Plesch p. 365; PR 7213-7217 (ed. 1: 8087-8091); Rehder 5: 679; SO 635g; Sotheby 607, 608; TL-1/983, see J.D. Surian; TL-2/see Burman,J.; Tucker 1: 561.

BIOFILE: Andreas, C.H., In en om een botanische tuin. Hortus groningianus 1626-1966, p. 67. 1976 (copies Plumier drawings at GRO).
Bridson, G.D.R. et al., Nat. hist. mss. res. Brit. Isl. 429 [index]. 1980 (mss.).
Candolle, A. de, Phytographie 440. 1880 (drawings P.).
Coats, A.M., The plant hunters 331, 332, 359. 1969.
Crowley, W.R., Jr., Morton Arbor. Quart. 12(2): 25. 1976 (epon.).
Ewan, J., Regn. veg. 71: 21, 22, 32, 50. 1970 (links with American botany).
Fournier, P., Voy. déc. Sci. mission. natural. franç. 1: 53-59, 62, 63, 94. 1932 (West Indian work, portr., bibl., list of manuscripts at P).
Fries, Th.M., Bref Skrifv. Linné 1(2): 74, 200. 1908; 1(5): 74, 207. 1911, 1(6): 135. 1912, 1(8): 99. 1922.
Garidel, Hist. pl. Aix xxxii-xxxiv. 1715 (early inf. on life and publ.).
Gillis, W.T. & W.T. Stearn, Taxon 23: 185-191. 1974 (citation letters Burman to L. on *Pl. amer.*).
Guillaumin, A., Les fleurs des Jardins xxxv, *pl. 11*. 1929 (portr.).
Haegeman, J., Tuberous Begonias 7-17, 208, 209 (on Plumier's Begonias with ills of original drawings and manuscripts, b. 20 Apr 1646, d. 20 Nov 1704, portr.).
Hulth, J.M. & A.H. Uggla, Bref Skrifv. Linné 2(1): 5, 323, 324. 1916, 2(2): 36-145. 1943.J.D., *in* Swainson, Taxidermy 293-295. 1840.
Jessen, K.F.W., Bot. Gegenw. Vorz. 270, 281. 1864.
Keefe, R., Biologist 48(3-4): 49-50. 1966.

Lacroix, A., Figures de savants 3: 15, 53, 4: 104, 159, 163, 223. 1938.
Mägdefrau, K., Gesch. Bot. 48. 1973.
Magnin, A., Bull. Soc. bot. Lyon 31: 24. 1906, 32: 107. 1907 (P. at Lyon).
Marquis, Dict. Sci. méd., Biogr. méd. 6: 455-456. 1824 (bibl.).
Maurer, M., Begonian 45: 228-231. 1978 (portr.).
Michaud, Biogr. univ. ed. 2. 33: 536-539.
Nicolas, J.P., *in* G.H.M. Lawrence, ed., Adanson 1: 83-84, Stafleu, F.A., ib. p. 165-175, Margadant, W.D., ib. p. 364. 1963.
Percheron, A., Bibl. entom. 319-320. 1837.
Planchon, J.E. & J. Triana, Ann. Sci. nat., Bot. ser. 4. 13: 335-336. 1860 (on L's use of Plumier drawings; Plumier's *Clusia*).
Poiret, J.L.M., Encycl. méth., Bot. 8: 748-749. 1808.
Polhill, R.M. & Stearn, W.T., Taxon 25: 323-325. 1976 (Linnaeus' notes on Plumier drawings).
Questel, A., Fl. Guadeloupe 56. 1951.
Schumann, K., Monatschr. Kakt. 14: 77-79, 94-96. 1904 (on P's cacti; brief general info.).
Stafleu, F.A. et al., *in* L'Héritier, Sert. angl. facs. ed. 1963, p. lxxxvi, xcvi, 28.
Stafleu, F.A. *in* N.J. Jacquin, Obs. pl. amer. 1763, facsimile ed. 1971, p. F22-23.
Stahl, A., Est. Fl. Puerto Rico ed. 2. 1: 35-38, 49. 1936.
Steele, A.R., Flowers for the King 15, 18, 59, 146, 268n. 1964.
Urban, I., Symb. ant. 1: 123-130. 1898 (annotated bibl.), 3: 101-103. 1902 (itin., biogr. refs.; Surian plants are his originals).
Wagner, R., Repert. sp. nov., Beih. 111: 46-48. 1939 (on 1713 reissue of *Descr. pl. Amér.*).
Whitmore, P.J.S., Modern Lang. Rev. 54: 400-401. 1959. (Plumier as craftsman and botanist).
Wittrock, V.B., Acta Horti Berg. 3(2): 116. 1903, 3(3): 111. 1905.
Woodward, B.B., *in* Murray, G., Hist. coll. BM(NH) 1: 28, 45. 1904 (312 orig. drawings).
Wright, J., Garden, London, 105(2): 76-77. 1979 (portr.).

GENERAL COMMENTARY AND KEY TO PLUMIER'S TAXA: Urban, I., Rep. Spec. nov. Beih. 5: 1-196. 1920 (Plumiers Leben und Schriften nebst einem Schlüssel zu seinen Blütenpflanzen). Urban based his work on his studies of 1657 of the original drawings, mentioned above, namely those in the mss *Botanicon americanum* (8 vols., 1219 pl.), *Botanographia americana* (3 vols., 248 pl.) and *Antillarum insularum natur. icones* (1 vol., 190 pl.). – See also Fournier, Voyages et découvertes des missionaires naturalistes français 1: 58-59. 1932, PR 7213-7217.

HANDWRITING: Haegeman, J., Tuberous Begonias 1979, p. 10-15 (figs. 2-7); reproductions from the *Botanicon americanum* (1697) at P.

EPONYMY: *Plumeria* Linnaeus (1753) and the derived *Plumeriopsis* Rusby et Woodson (1937); *Plumiera* Adanson (1763); *Plumieria* Scopoli (1777); see ING 3: 1375.

8066. *Description des plantes de l'Amérique*. Avec leurs figures ...à Paris (L'imprimerie royale) 1693. Fol. (*Descr. pl. Amér.*).
Orig. ed.: 1693, p. [i-viii], [1]-94, [95-103, index, coloph.], *pl. 1-108* (uncol. copp.).
Copies: BR, NY, U, US, USDA; IDC 938. – Plates *1-50*, dealing with ferns, occur also in the *Traité des fougères*; *pl. 51-108* occur as *165-222* in *Filic. amér.*
Re-issue (of first edition with t.p. dated:) 1713, p. [i-viii], [1]-94, [95-103], *pl. 1-108*.
Copies: HU, MO, W. – Substitute (or misprinted original?) t.p. only: Paris (id.). 1713. See Wagner (1939). Stevenson (HU 389) considers the "mdccxiii" as a misprint which was corrected later. See HU for further bibliographical details. NI and Wagner consider the 1713 issue a "Neudruck".
Ref.: W. & W., Cat. 137, no. 74 (800).

8067. *Nova plantarum americanarum genera*, ... Parisiis (apud Joannem Boudot, ...) 1703. Qu. (*Nov. pl. amer.*).
Publ.: 1703 (p. [iv]: 27 Apr 1703), p. [i-viii], 1-52, [1-3, index], *Catalogus plantarum*

americanarum: 1-21, [22, err.], *pl. 1-40* (uncol. copp.). *Copies*: HU, NY, US; IDC 1116.
– For details see HU 407 (e.g. a3 as cancel).

8068. *Traité des fougères de l'Amerique* ... À Paris (de l'imprimerie royale) 1705. Fol. (*Traité foug. Amér.*).
Publ.: 1705, p. [i*, iii*], [i]-xxxvi, [1]-146, [1-5, table], [1-4, index], [1, coloph.], *pl. 1- 170, A, B* (uncol. copp.). *Copies*: BR, E (Univ. Libr.), NY, U; IDC 7245. – Contains 50 *pl.* from *Descr. pl. Amér.* (see above no. 8066); *164* plates occur also as *1-164* in *Filic. amer.*
Ref.: Fée, A.L.A., Hist. foug. Antill. 1866 (Mém. foug. 11; TL-2/1757), q.v. for a review of Plumier's ferns).
Jenman, G.S., Plumier's American ferns, iv, 7 p., Georgetown, Demerara 1889 (see BM 2: 931).

8069. *Plantarum americanarum fasciculus primus* [*-decimus*], continens plantas, quas olim Carolus Plumierius, botanicorum princeps detexit, eruitque, atque in insulis Antillis ipse depinxit. Has primum in lucem edidit, concinnis descriptionibus, & observationibus, aeneisque tabulis illustravit Joannes Burmannus, ... sumptibus auctoris, prostant Amstelaedami in Horto medico, atque apud viduam & filium S. Schouten, & Lugd. Batav. [Leiden] apud Gerard. Potvliet & Theodor. Haak. mdcclv [1775]. Fol. (*Pl. amer.*).
Editor and author of descriptions: Johannes Burman (1707-1779).
Publ.: in 10 parts, 1755-1760, the dates of which can be established (except that of no. 10) from the correspondence between Burman and Linnaeus. *Copies*: HU (with portr. Burman), MO, NY, U, US, USDA(2); IDC 1002.

Fasc.	pages	*plates*	published
1	[i-viii], 1-16	*1-25, 25**	27 Dec 1755
2	[i], 17-37	*26-50*	8 Feb 1756
3	[i], [39]-64	*51-75*	4 Jun 1756
4	[65]-87	*76-100*	26 Oct 1756
5	[i], [89]-116, [116 bis]	*101-125*	8 Feb 1757
6	[i], [117]-142	*126-150*	22 Aug 1757
7	[i], [143]-168	*151-175*	28 Mar 1758
8	[i], [169]-194	*176-201*	20 Jun 1758
9	[i], [195]-220	*202-226*	20 Mar 1759
10	[i], [221]-262, [1-4]	*227-262*	post 15 Feb 1760

All drawings are by Plumier, see above for details. See Stevenson for a complete collation and bibliographical details (HU 554), and Urban (1920) for an extensive discussion and for modern nomenclature.
Facsimile ed.: announced, but not yet published, by Asher (1974).
Ref.: HE p. 47, HU 554, NI 1547, SA 2: 594-595, TL-1/983.
Urban, I., Symb. ant. 1: 128-130. 1898; Rep. Spec. nov. Beih. 5: 26-100. 1920.

Plummer, Sarah Allen (1836-1923), American botanist and plant collector, active in California; married John Gill Lemmon in 1880. (*Plummer*).

HERBARIUM and TYPES: See J.G. Lemmon.

BIBLIOGRAPHY and BIOGRAPHY: Backer p. 453; Barnhart 2: 367, 3: 93; IH 2: (in press); Tl-2/see Lemmon, S.A.

BIOFILE: Brewer, W.H., *in* S. Watson, Botany Calif. 2: 558. 1880.
Thomas, J.H., Huntia 3: 32. 1969 (1979).

EPONYMY: *Plummera* A. Gray (1882).

Pluskal (Pluska-Moravičanský), **Franisček Sal.** (1811-1901), Moravian botanist; practicing physician at Lomnitz 1839-1857; later in Welechrad. (*Pluskal*).

HERBARIUM and TYPES: BRNU.

BIBLIOGRAPHY and BIOGRAPHY: BM 4: 1588; CSP 4: 949; Futák-Domin p. 471-472; Herder p. 77; Klášterský p. 144 (q.v. for further bibl. and biogr. details); LS 21002-21003; PR 7219-7220 (ed. 1: 8092); Rehder 5: 679.

BIOFILE: Anon., Österr. Bot. W. 2(7). 1852, Österr. bot. Z. 12: 400. 1862.
Hrabětová-Uhrová, A., Sborník Klubu přírodov. Brně 31: 13-19. 1959 (fide Klašterský).
Oborny, A., Verh. naturf. Ver. Brünn 21: 14, 16. 1882.
Pluskal, F.S., Verh. zool.-bot. Ver. 6: 363-372. 1856 (Zur Geschichte der Pflanzenkunde in Mähren).

8070. *Neue Methode die Pflanzen* auf eine höchst einfache Art gut und schnell für das Herbar *zu trocknen* ... Brünn (Gedruckt bei Franz Gastl) 1849. Oct. (*Neu. Meth. Pfl. trockn.*).
Publ.: 1849 (Flora rd. Aug-Sep 1849), p. [1]-40. *Copy*: Graz Landesbibliothek.
Ref.: Schlechtendal, D.F.L., Bot. Zeit. 10: 74-75. 1852.

Pluszczewski, Emile (1855-?), French pharmacist and botanist; curator at the Paris École supérieure de Pharmacie; Pharm. prem. cl. 1885. (*Plusz.*).

HERBARIUM and TYPES: Unknown.

BIOFILE: Rehder 2: 93.

8071. École supérieure de pharmacie de Paris. Année 1885 no. 1 *Étude de la famille des Pipéracées* au point de vue de la morphologie & de l'anatomie comparée. Thèse pour l'obtention du diplôme de Pharmacien de première classe presentée et soutenue le jeudi 19 mars 1885 ... Paris (Librairie Ollier-Henry, ...) 1885. Qu. (*Étud. Piper.*).
Thesis issue: 19 Mar 1885, p. [1]-77, [1, cont.], *pl. 1-8* (partly col. liths. P. Appel). *Copy*: G. – t.p.: b. 25 Mar 1855.
Trade issue: 1885, p. [1]-77, [1, cont.], *pl. 1-8* (id.). *Copy*: L.

Pobedimova, Evgeniia Georgievna (1898-1973), Russian botanist at the Komarov Botanical Institute; travelled in Central Asia (1932), Mongolia (1930, 1931), Caucasia (1946, 1949), Altai (1953) and Crimea (1947, 1949). (*Pobed.*).

HERBARIUM and TYPES: LE; other material e.g. at C and NY.

BIBLIOGRAPHY and BIOGRAPHY: BFM no. 179; GR p. 565; Hortus 3: 1201 (*Pobed.*); MW p. *389, suppl. p. 281*; Roon p. *88*; TL-2/*3857*.

BIOFILE: Pobedimova, E.G., personal communication to senior author 1957, incl. mss. bibl. 1936-1957 (b. 12 Nov 1898).

COMPOSITE WORKS: *Fl. USSR*:
Balsaminaceae, 14: 624-634. 9 Mar 1949.
Thymelaeaceae, 15: 481-515. 12 Dec 1949.
Apocynaceae, 18: 645-662. 16 Oct 1952.
Asclepiadaceae, 18: 663-718. 16 Oct 1952.
Labiatae (*Salvia, Schradera*), 21: 244-374. 13 Sep 1954.
Rubiaceae (*Asperula*) (et al.), 23: 193-285. 12 Dec 1958.
Rubiaceae (*Microphysa, Galium*), 23: 286-381. 12 Dec 1958.
Compositae (Gen. 1525-1527), 26: 147-184. 31 Oct 1961.
Compositae (*Cotula*), 26: 418-421. 31 Oct 1961.
Compositae (*Mutisineae*), 28: 589-598. 24 Jun 1963.

Pobéguin, Charles Henri Oliver (1856-1951), French botanist and colonial administrator in French Africa (Ivory Coast, Comores, Guinea, Equatorial Africa, Congo) 1892-1923. (*Pobég.*).

HERBARIUM and TYPES: P, PC; other material at A, BR, H, K, MO.

BIBLIOGRAPHY and BIOGRAPHY: Barnhart 3: 93; BL 1: 37, 313; BM 4: 1588; CSP 17: 930; IH 1 (ed. 7): 340 (err. Pobéquin); Kew 4: 326; Rehder 1: 491; Tucker 1: 561.

BIOFILE: Chevalier, A., ed., Rev. int. Bot. appl. Agric. colon. 31: 461-463. 1951 (obit., colls., b. 25 Feb 1856, d. 25 Jun 1951).
Hepper, F.N. & F. Neate, Pl. coll. W. Africa 65. 1971 (coll.).
Sayre, G., Mem. New York Bot. Gard. 19: 381. 1951 (coll.).

EPONYMY: *Pobéguinea* Jaques-Félix (1950).

8072. Côte occidentale d'Afrique. *Essai sur la flore de la Guinée française* produits forestiers, agricoles et industriels ... Paris (Augustin Challamel, ...) 1906. Oct. (*Essai fl. Guinée franç.*).
Publ.: 1906 (Nat. Nov. Sep(2) 1906; Bot. Centralbl. 9 Oct 1906), p. [1]-392, *pl. 1-80, 27bis* (photos). *Copies*: B, BR, FI, G, USDA.

8073. Bibliothèque d'agriculture coloniale. Les plantes médicinales de la Guinée ... Paris (Augustin Challamel, ...) 1912. Oct. (*Pl. méd. Guinée*).
Publ.: 1912, p. [1]-85. *Copy*: USDA.

Pocock, Mary Agard (1886-1977), South African phycologist; B.S. botany London University 1908 (honors degree ibid. 1921), high school teacher at Cheltenham 1909-1912, at Wijnberg (Cape) Girl's High School 1913-1917; worked with A.C. Seward at Cambridge 1919-1921; temporary lecturer at various South African universities and colleges from 1923 onward; turned to algology 1928 (Ph.D. 1932 Univ. Cape Town), D.Sc. h.c. Rhodes Univ. 1967. (*Pocock*).

HERBARIUM and TYPES: GRA; duplicates at A, B, BM, BOL, COI, KMG, PRE, RUH, SAM, STE.

BIBLIOGRAPHY and BIOGRAPHY: Bossert p. 313; IH 1 (ed. 6): 363, (ed. 7): 340; 2: (in press); Kew 4: 326; Roon, p. 88.

BIOFILE: Exell, A.W. & G.A. Hayes, Kirkia 6(1): 99. 1967 (Zambesian coll. at PRE).
Exell, A.W. et al., Bol. Soc. Broter. ser. 2. 26: 218. 1952.
Guillarmod, A.J., The Bluestocking 31: 40. 1970 (n.v.), Forum bot. 15(1): 1-2. Jan 1977 (90th birthday), Phycologia 17(4): 440-445. 1978 (portr., bibl., b. 31 Dec 1886, d. 10 Jul 1977).
Gunn, M.D. & L.E. Codd, Bot. expl. S. Afr. 283. 1981 (biogr. sketch, portr., coll.).
Koster, J.Th., Taxon 18: 556. 1969 (coll.).
Levyns, M.R.B., Insnar'd with flowers 139, 140. 1977.
Riddelsdell, H.J., Fl. Glouc. cli. 1948.
White, F., C.R. AETFAT 4 : 199. 1962 (N. Rhodes. coll.).

NOTE: The second given name, Agard, derives from the maternal grandmother Ann Elizabeth Agard; the likeness to the name of her phycological predecessors Agardh was incidental.

EPONYMY: *Pocockiella* G.F. Papenfuss (1943). *Note*: *Pocockia* Seringe ex A.P. de Candolle (1825) was probably named for Rev. Richard Pococke (1704-1765), English-born bishop, who earlier in life travelled and collected plants in Egypt and Arabia. *Pocockia* J.K. Lentin et G.L. Williams (1973) and *Pocockipites* D.C. Bharadwaj (1974) were named for S.A.J. Pocock (1928-x) Canadian geologist and palynologist.

Podpěra, Josef (1878-1954), Czechoslovak (Bohemian) botanist; studied at the Charles University, Praha 1897-1903;Dr. phil. Praha 1903; teacher of classical languages at various colleges 1903-1914, from 1914-1920 prisoner of war in Russia and the

Soviet Union, part of the time at the botanical institute of Ufa (n. Tomsk); from 1921 professor of botany at Brno; member Acad. Sci. ČSR 1953. (*Podp.*).

HERBARIUM and TYPES: BRNM; important material also at BRNU and PR; other material at BP, C, CERN, E, G, GB, H, LAU, LD, MANCH, MZ, NI, OLM, OXF, P, PC; Siberian bryoph. at TK). – *Exsiccatae: Bryophyta exsiccata reipublicae čechosloveniae*, see Sayre (1971). – *Flora exsiccata reipublicae Bohemicae Slovenicae*, 1925-1938, cent. *1-13* (sets ann. ÖbZ 75-88, 1926-1939). – Contributed to Petrak, *Flora bohemiae et moraviae exsicc.* ser. 2. – Letters and portrait at G.

BIBLIOGRAPHY and BIOGRAPHY: AG 7: 28; Backer p. 659; Barnhart 3: 93; BFM no. 692; BJI 2: 137; BM 4: 1589, 8: 1018; Bossert p. 313; Clokie p. 226; CSP 17: 935; Futák-Domin p. 472-476; GR p. 672; Hortus 3: 1201 (*Podp.*); IH 1 (ed. 1): 14, (ed. 6): 363, (ed. 7): 340, 2: (in press); Kew 4: 327; Klášterský p. 144; LS 37986, suppl. 21943-21945; Maiwald p. 234, 237, 244, 253; MW p. 389, suppl. 281; SBC p. 129; TL-2/4941; Tucker 1: 561; Zander ed. 11, p. 801.

BIOFILE: Anon., Allg. bot. Z. 9: 196. 1903 (high school teacher at Olmütz); 14: 164. 1908 (leaves Olmütz; teacher at high school (Brno)); Bot. Centralbl. 93: 448. 1903 (app. Olmütz), 108: 400. 1908 (app. Brünn); Österr. bot. Z. 53: 431. 1903 (app. Olmütz), 58: 455. 1908 (app. Brünn), 70: 152. 1921 (returns from Siberia; Siberian bryophyte colls. at TK); Preslia 9: 95. 1930.
Čermák, St., Česká Mykol. 8(2): 49-51. 1954 (obit., portr., b. 7 Nov 1878, d. 18 Jan 1954).
Gager, C.S., Brooklyn Bot. Gard. Rec. 27: 187. 1938 (dir. bot. gard. Brno from 1921).
Hedge, I.C. & J.M. Lamond, Index coll. Edinb. herb. 119. 1978.
Hendrych, R. Preslia 50: 281-285. 1978 (b. 8 Nov 1878, d. 18 Jan 1954).
Janko, J., [Russia in the work of Academician Josef Podpěra] Děj. Ved. Techn. 7: 164-166. 1974, n.v., fide Kew Record and Curr. Work. Hist. med. 84: 156. 1975.
Jedlička, J., Rev. bryol. lichénol. 24: 152-154. 1955 (portr., bibl.).
Krist, V., Čas. vlast. spolk. musejn. Olomouci 51: 186-189. 1938 (portr., biogr.).
Němec, B., Věstník česk. Akad. Věd. 63: 116-120. 1954 (obit., b. 7 Jan 1878, d. 19 Jan 1954; dates corect?).
Pax, F.,,Bibl. schles. Bot. 14, 19, 20, 101, 117. 1929.
Pospisil, V., Vědy přír. 64: 233-236. 1979 (fide Exc. bot.; 100th anniversary of birth).
Sayre, G., Mem. New York Bot. Gard. 19: 238-239. 1971; Bryologist 80(3): 515. 1977 (byoph. coll.).
Smarda, J., Preslia 26: 315-328. 1954 (obit., portr., bibl. by M. Smejkal; b. 7 Nov 1878, d. 18 Jan 1954).
Váňa, J., Taxon 29: 674. 1980 (coll. PR).
Verdoorn, F., ed., Chron. bot. 1: 31, 32, 109. 1935, 2: 28, 35, 36, 106. 1936, 3: 75, 89. 1937, 5: 113, 135, 311. 1939, 6: 301. 1941.

NOTE: P's most sizeable bryological publications appeared after 1939 (our closing date): the series *Bryum generis monographiae prodromus*, 16 parts, 1942-1973 (posthumous parts edited by Zdenek Pilous) and the *Conspectus muscorum europaeorum* (p. [1]-697, [699, approved for printing 13 Nov 1954], Praha (Práce Česk. Akad. Věd.) 1954. Oct., see also B. Fott, Taxon 4: 71-72. 1955 and S. Jovet-Ast, Rev. bryol. lichénol. 25: 402. 1956.

8074. *Bryologische Beiträge aus Südböhmen* ... Prag [Praha] (Verlag der königl.-böhmischen Gesellschaft der Wissenschaften ...) 1899. Oct.
Publ.: 1899 (Nat. Nov. Mar(2) 1900; Allg. Bot. Z. 15 Mai 1900), p. [1]-28. *Copy*: NY. – Issued as S.B. k. böhm. Ges. Wiss., math. nat. Cl. 1899(46): 1-27.

8075. *Monografické studie o českých druzích rodu Bryum* ... V Praze [Praha] (Nakladem České Akademie ...) 1901. Oct. (*Monogr. Bryum*).
Publ.: 1901 (Nat. Nov. Nov(1) 1901; Hedwigia 20 Oct 1901), p. [1]-85, *pl. 1-3*. *Copies*: B, FH, H, NY. – Issued as Rozpravy České Akad. 10(2, 2): 1-85. 1901.

8076. *Výsledky bryologického výzkuma Moravy* za rok 1903-04 ... v Prostějově [Prossnitz] (Nákladem Klubu přirodovědeckého ...) 1904. Oct. (*Výsl. bryol. výzk. Moravy*).

1, 1903-1904: 10 Aug 1904 (p. 30), p. [1]-30. Preprinted from Věstn. Klubu přirod. Prostějově 7: 3-30. 1905.
2, 1904-1905: 1905, ibid. vol. 8, 1906 (n.v.) (ÖbZ Sep 1905, 33 p.).
3, 1905-1906: Oct-Dec 1906 (mss. subm. Oct 1906, ÖbZ Mar-Apr 1907), p. [1]-82, [83]. Issued as Zprávy Kommisse pro přirod. prozkoumáni Moravi, Brno, vol. 2.
4, 1906-1907: Oct-Dec 1907 (mss. submitted 2 Oct 1907; reprint t.p. 1907), p. [1]-82, [83]. Issued as id. vol. 4.
5, 1907-1908: Nov-Dec 1908 (mss. 10 Oct 1908; reprint t.p. 1908), p. [1]-41, [43]. Issued as id. vol. 5.
6, 1909-1912: 1913 (p. 4: 1 Oct 1912; reprint t.p. 1913; ÖbZ 10 Oct 1913), p. [1]-49, *pl*. Reprinted from Čas. Moravsk. Mus. Zemsk. 13(1): 32-54, 13(2): 233-257. 1913. Index muscorum cites orig. pagination. Autoreferat ÖbZ 63: 435. 1913.
7, 1913-1922: 1923 (p. 2: 15 Oct 1922; t.p. reprint 1923), p. [1]-29. Reprinted from Sborníku Klubu přirod. Brně 5: 1-29. 1923.
Copy: H. – Information P. Isoviita. We have not seen the publications in the original journals.

8077. *Doplňky ku květeně moravské* ... v Brně (Tiskem Mor. Akc. Knihtiskárny ...) 1914. Oct.
Publ.: 1914, p. [1]-63. *Copy*: G. – Reprinted from Čas. Moravsk. Mus. Zemsk. 11(2): 238-253, 12(2): 265-280, 14(1): 49-65, 14(2): 414-428. To be cited from journal.

8078. *Ad bryophytorum cisuralensium cognitionem additamentum* ... Brno [Brünn] (Přirodo vedědecká fakulta) 1921. Qu.
Publ.: 1921 (p. [3]: 15 Jan 1921; Nat. Nov. Jul, Sep 1922), p. [1]-42. *Copies*: BR, FH, G, H, NY. – Publ. Fac. Sci. Univ. Masaryk 5: 1-42. 1921.

8079. *Plantae moravicae* novae vel minus cognitae ... Brno (Přirodovědecká Fakulta ...) [1922]. Oct. (*Pl. morav.*).
Publ.: 1922 (p. 3: Jan 1922; Nat. Nov. Sep 1922), p. [1]-35, *pl. 1-3*. *Copy*: G. – Publ. Fac. Sci. Univ. Masaryk 12, 1922.

8080. *Bryi generis sectionis Erythrocarpa* Kindb. *species europaeae* revisio critica ... Praha (Nakladem Vlastním ...) 1923. Oct.
Publ.: 1923 (NY rd. 1 Jun 1923), p. [1]-9. *Copies*: H, NY. – Reprinted and to be cited from Preslia 2: 81-89. 1923.

8081. *Květena Moravy* ve vzatazích systematických a geobotanických ... Brno (Práce přirod. Společn.) 1924-1928. 3 parts. Oct.
1: 1924, p. [1]-226, map. Reprinted and to be cited from Práce Morav. přirod. společn. 1(10): 393-618. 1924.
2: 1925, p. [1]-512, map. Reprinted id. 2(10): 271-782. 1925.
3: 1928, p. [1]-359. Reprinted id. 5(5): 57-415. 1928.*Copy*: B.

8082. *Schedae ad floram exsiccatam reipublicae bohemicae slovenicae* editam ab instituto botanico universitatis Brno, Moraviae (Č.S.R.) Centuria 1(-13) ... Brno [Brünn] (Tiskem Akciové moravské knihtiskárny v Brně ...) 1925-[1938]. (*Sched. fl. exs. bohem.*).
Publ.: 1925, preprints from Sborn. Klubu Prírod. Brno, vols. 8-20. *Copy*: G.
1: 1925 (date preprint), p. [1]-20. Sborn. 8: 135-153. 1926.
2: 1926 (id.), p. [1]-20. Sborn. 9: 132-151. 1927.
3: 1928 (id.), p. [1]-19. Sborn. 10: 104-122. 1928.
4: 1929 (id.), p. [1]-24. Sborn. 11: 155-173. 1929.
5: 1930 (id.), p. [1]-24. Sborn. 12: 111-134. 1930.
6: 1931 (id.), p. [1]-28. Sborn. 13: 93-118. 1931.
7: 1932 (id.), p. [1]-24. Sborn. 14: 179-201. 1931.
8: 1933 (id., Nat. Nov. Mar 1933), p. [1]-26. Sborn. 15: 123-149. 1933.
9: 1934 (id., Nat. Nov. Jun 1934), p. [1]-26. Sborn. 16: 153-178. 1934.
10: 1935 (id., Nat. Nov. Mai 1935), p. [1]-23. Sborn. 17: 122-144. 1935.
11: 1936 (id., Nat. Nov. Jul 1936), p. [1]-23. Sborn. 18: 123-145. 1936.
12: 1937 (id., Nat. Nov. Jun 1937), p. [1]-22. Sborn. 19: 90-111. 1937.
13: 1938 (id., p. [1]-20. Sborn. 20: 102-121. 1938.

8083. *Klíč k úplné květeně republiky Československé.* Jako druhé vydáni Polivkova klíče k úplné květeně zemí koruni české ... v Olomouci (Nakladatel – R. Promberger ...) 1928. Oct. (*Klíč květ. Československ.*).
Co-author: Karel Domin (1882-1953).
Publ.: 1928 (p. 6: 15 kvetna 1928), p. [1]-126, [1, h.t.], [1]-1088. *Copies*: G, USDA. The subtitle states that this is a second edition of F. Polívka, Klíč k úplné květně zemí koruni české, Olmouc 1912 (110 & 864 p.).. (Key to the complete flora of ...).

8084. *Musci insulae rossicae prope Vladivostok* ad bryophytorum orientis extremi cognitionem additamentum ...Brno [Brünn] (Vlastním nákladem vydává Přirodovědecká Fakulta ...) 1929. (*Musc. ins. ross. Vladiv.*).
Publ.: 1929 (p. 4: 15 Mai 1929; Rev. bryol. 14 Aug 1930; Hedwigia Aug 1930), p. [1]-40. *Copies*: BR, FH, H, NY. – Issued as Publ. Fac. Sci. Univ. Masaryk 116: 1-40. 1929.

Poech, Josef [Alois] (1816-1846), Bohemian botanist, born at Snedovice (Wegstädl) nr Praha, studied medicine, assistant to the botanical cabinet of Praha University; associated with Opiz. (*Poech*).

HERBARIUM and TYPES: PR (main set), REG (see Flora 1846), and W (musci; Austrian plants). *Exsiccatae: Musci bohemici*, nos. 1-100, Prag 1846 (see Sayre 1971).

BIBLIOGRAPHY and BIOGRAPHY: Barnhart 3: 94 (b. 22 Mai 1816, d. 20 Jan 1846); BM 4: 1590; CSP 4: 949; Hortus 3: 1201 (Poech); Jackson p. 314; Kew 4: 327; Klášterský p. 145; Maiwald p. 69, 119, 144, 145, 189, 256; PR 7221 (ed. 1: 8093); Rehder 5: 679; SBC p. 129; Tucker 1: 561; Zander ed. 10, p. 702, ed. 11, p. 801.

BIOFILE: Anon., Bot. Zeit. 4: 255. 1846 (d.); Flora 27: 256. 1844 (asst. Praha), 29: 94-95. 1846 (herb. to REG).
Beck, G., Bot. Centralbl. 34: 150. 1888.
Candolle, Alph. de, Phytographie 440. 1880 (at W).
Matouschek, F., Verh. zool. bot. Ges. Wien 50: 373-381. 1900 (on the *Musci bohemici*).
Pfund, J., Flora 29: 94-95. 1846 (obit., d. 20 Jan 1846, of tuberculosis; herbarium left to Regensb. bot. Ges.).
Pritzel, G.A., Linnaea 19: 459. 1846 (d. 20 Jan 1846).
Sayre, G., Mem. New York Bot. Gard. 19: 239. 1971 (on *Musci bohemici*); Bryologist 80: 515. 1977.

EPONYMY: *Poechia* P.M. Opiz (1852). Although the author does not give the etymology for *Poechia* Endlicher (1848), the name likely honors Poech.

8085. *Enumeratio plantarum hucusque cognitarum insulae Cypri* ... Vindobonae [Wien] (Typis Caroli Ueberreuter) 1842. Oct. (*Enum. pl. Cypr.*).
Publ.: 1842 (p. 6: 23 Jul 1842), p. [1]-42. *Copies*: FI, G, MO, NY.
Ref.: Fürnrohr, A.E., Flora 27: 240 (rd. Mar 1844), 453-454. 1844.

Poederlé, Eugène Joseph Charles Gilain Hubert d'Olmen baron de (1742-1813), Belgian landed gentleman, botanist, sylviculturist and agronomist living in Bruxelles and on his property at Saintes. (*Poederlé*).

HERBARIUM and TYPES: Unknown.

BIBLIOGRAPHY and BIOGRAPHY: Barnhart 3: 28 (sub Olmen); BM 4: 1590; Dryander 3: 423; Herder p. 381; Hortus 3: 1201 (Poederlé); PR 7222 (b. 20 Sep 1742; d. 17 Aug 1813), (ed. 1: 8094); Rehder 3: 106; Tucker 1: 561-562.
Poederlé, E. 2356
BIOFILE: Crépin, F., Guide bot. Belg. 224, 227. 1878; Biogr. natl. Belg. 17: 844-847. 1903 (d. 18 Aug 1813).
Morren, Ch., J. Agric. prat. 3: xxvii (n.v.).

8086. *Manuel de l'arboriste et du forestier belgiques*, par M. de Poederlé, l'aîné. Ouvrage extrait des meilleurs auteurs anciens & modernes, & soutenu d'observations recueillis dans différens voyages, & d'essais faits dans sa patrie ... à Bruxelles (chez J.L. de Boubers, ...) 1772. Oct. (*Man. arboriste belg.*).
Ed. 1: Mai-Dec 1772 (approb. 10 Apr 1772), p. [i*, iii*], [i]-x, [1-8], [1]-404, [1, approb.]. *Copy*: NY.
Suppl.: Oct-Dec 1779 (approb. 28 Sep 1779), p. [i*, iii*], [i]-iii, [iv-viii], [1]-358. *Copies*: NY. – "Supplément au Manuel ..." à Bruxelles (chez Emmanuel Flon, ...) 1779. Oct.
Reissue: 1774 (fide Tucker:) p. [x], x, 404 , copy HH, n.v.
Ed. 2: 1778, in 2 vols. "Seconde édition, augmentée de plusieurs articles ..." à Bruxelles (chez Emmanuel Flon, ...).
 1: [i*-iii*], [i]-v, [7]-375, [1-4].
 2: [i*, iii*], [1]-446, [1-4].
 Copy: UC (inf. R. Schmid).
Ed. 3: 1792 "Troisième édition, augmentée de plusieurs articles ..." à Bruxelles (chez Emmanuel Flon, ...). *1*: [i*, iii*], [i]-iv, [5]-284, [1-4]. *2*: [i, iii], [1]-342, [1-4].*Copy*: NY.

Pöll, Josef (1874-1940), Austrian hieraciologist; high school teacher at Bludenz and Innsbruck; later professor at the Innsbruck teacher's college. (*Pöll*).

HERBARIUM and TYPES: IBF. – Letters at G.

BIBLIOGRAPHY and BIOGRAPHY: AG 12(3): 490; BFM no. 837; BM 4: 1590; DTS 6(4): 185; Hegi 6(2): 1337; NI 2: 142, 204.

BIOFILE: Janchen, E., Cat. fl aust. 1: 45, 49. 1956 (bibl.).
Murr, J., Neue Übers. Farn. Blütenpfl. Vorarlberg Lichtenstein 1-2: xx. 1923.

COMPOSITE WORKS: *Hieracium* II, in Reichenbach, L. et H.G., *Ic. fl. Germ. Helv.* 19[2] auctoribus J. Murr, K.H. Zahn, J. Pöll, parts 1-3, 1904-1912; see under Reichenbach.

Poellnitz, Karl von (1896-1945), German agriculturist and botanist; studied at Leipzig University; specialist in succulent plants; owner of an estate at Oberlödla nr Altenburg, Thüringen. (*Poelln.*).

HERBARIUM and TYPES: B.

BIBLIOGRAPHY and BIOGRAPHY: Barnhart 3: 94; BJI 2: 137; Hortus 3: 1201 (Poelln.); Kew p. 327-328; Langman p. 588; MW p. 389-390, suppl. p. 281-282; NW p. 55; Zander ed. 10, p. 702, ed. 11, p. 801.

BIOFILE: Bullock, A.A., Bibl. S. Afr. bot. 124-125. 1978 (bibl.).
Legrand, D., Revista sudamer. Bot. 10(1): 26. 1951 (obit., b. 4 Mai 1896, d. 15 Feb 1945); Comunic. bot. Mus. Hist. nat. Montevideo 2(25): 1-9. 1952 (on P's publ. on Portulaca).
Merrill, E.D., B.P. Bishop Mus. Bull. 144: 151. 1937, Contr. US natl. Herb. 30: 243. 1947 (Polynes. bibl.).

EPONYMY: *Poellnitzia* Uitewaal (1940); × *Poellneria* G.D. Rowley (1973, *Poellnitzia* Uitewaal × *Gasteria* Duval).

8087. *Versuch einer Monographie der Gattung Portulaca* L. [Sonderabdruck aus Fedde, Repertorium ....1934]. Oct.
Publ.: 31 Dec 1934 (p. [i, t.p. of reprint on recto of p. 240])p. 240-320. *Copy*: B. – Reprinted and to be cited from Repert. Sp. nov., Fedde, 37: 240-320. 31 Dec 1934.

8088. *Zur Kenntniss der Gattung Echeveria* DC ... Dahlem bei Berlin (Verlag des Repertoriums, ...) 1936. Oct.

Publ.: 31 Jan 1936 (t.p.), p. 193-270. *Copies*: B, NY. – Reprinted with special cover but original pagination from Rep. Sp. nov., Fedde, 39: 193-270. 31 Jan 1936. To be cited from journal. Nat. Nov. cites this publication only in Dec 1937.

Poeppig, Eduard Friedrich (1798-1868), German botanist, zoologist and explorer; studied natural sciences (Dr. phil.) at Leipzig; travelled in Cuba (1822-1824), Pennsylvania (1824-1826), Chile (1827-1829), Peru and Brazil (1829-1832); professor of philosophy, zoology and botany at Leipzig (extraord., 1833, ord. 1846), director of the zoological institute 1834; curator of the natural history collections and director of the botanical garden. (*Poepp.*).

HERBARIUM and TYPES: W; duplicates in many herbaria, e.g. B, BM, BR, C, CGE, FI, G, GOET, H, HAL, HBG, JE, K, KIEL, L, LAU, LE, LZ, M, MO, NY, OXF, P, PC, PRC, US, WRSL. – Poeppig's own herbarium is at W. The total number of specimens brought home by Poeppig from his travels (1822-1824 Cuba, 1824-1826 Pennsylvania, 1827-1829 Chile, 1829-1832 Peru, Amazonas) was 17.000. The plants from Cuba, Pennsylvania and Chile were distributed by G. Kunze (Leipzig): an important set went to B (mainly destroyed), other sets are at LE and W. The main set of the Peruvian and Amazonas collections went to W with Poeppig's own herbarium. – Poeppig specimens from S. Africa came from his herbarium and were probably collected by Wilhelm Gueinzius (1814-1874); Poeppig himself never visited S. Africa. Poeppig's original drawings are also at W. – Correspondence with D.F.L. von Schlechtendal at HAL.

BIBLIOGRAPHY and BIOGRAPHY: ADB 26: 421-427; AG 3: 593; Barnhart 3: 94; BM 4: 1590; Bossert p. 313 (b. 16 Jul 1798, d. 4 Sep 1868); Clokie p. 226; CSP 4: 986-987, 8: 646, 12: 581, 17: 936-937; DTS 6(4): 16-17; Frank 3(Anh.): 78; Herder p. 56, 225; Hortus 3: 1201 (Poepp.); Jackson p. 374, 377; Kew 4: 328; KR p. 579; Langman p. 588; Lasègue p. 579 [index]; Maiwald p. 90, 105; ME 2: 367; NAF ser. 2. 2: 168; NI 1549; PR 7223-7225 (ed. 1: 8096-8098); Rehder 5: 679; SK 4: ccv-ccvi; Stevenson p. 1254; TL-1/984-985; TL-2/4037; Tucker 1: 562; Urban-Berl. p. 289, 310, 383; Zander ed. 10, p. 702, ed. 11, p. 801.

BIOFILE: Anon., Appleton's Cycl. Amer. biogr. 5: 47. 1888; Bonplandia 1: 203. 1853, 2: 7. 1854, 6: 50. 1858 (cogn. Leopoldina Hernandez II); Bot. Zeit. 4: 352. 1846 (said to have sold his collection to Boissier; obviously only a set), 4: 608. 1846 (collections acquired by W), 5: 120. 1847 (prof. zool. Leipzig), 10: 551. 1852 (decoration), 11: 128. 1853 (gold medal), 26: 672. 1868 (d. 9 Oct 1868); Bull. Soc. bot. France 15 (bibl.): 190. 1868 (death); Flora 7: 606. 1824, 16: 14. 1833, 30: 147. 1847 (prof. zool.), 34: 336. 1851 (temp. dir. bot. gard.), 36: 31, 190. 1852 (decor.); J. Bot. 7: 31. 1869 (death); Österr. bot. W. 1: 186. 1851 (temporary director of Leipzig botanical garden); Österr. bot. Z. 18: 367. 1868 (death).
Barros Araña, D., Anal. Univ. Chile 33: 136-139. 1869 (obit.).
Beck, G., Bot. Centralbl. 33: 314. 1888 (coll.), 34: 31. 1888 (drawings at W).
Candolle, Alph. de, Phytographie 440. 1880 (coll.).
Coats, A.M., Plant hunters 370-371. 1969 (general).
Embacher, F., Lexikon Reisen 235. 1882 (general).
Gilbert, P., Comp. biogr. lit. entom. 300. 1977.
Gunn, M.D. & L.E. Codd, Bot. Expl. S. Afr. 283. 1981 ("Poeppig specimens from S. Afr. were probably collected by Gueinzius).
Helm, J. Kulturpfl. 14: 353. 1966 (P. and Pritzel).
Herder, F. v., Bot. Centralbl. 55: 292. 1893 (material at LE).
Herrera, F.L., Est. Fl. Dept. Cuzco 9. 1930 (brief note).
Hoehne, C., Jard. bot. São Paulo 161 1941 (portr.).
Jackson, B.D., Bull. misc. Inf. Kew 1901: 53 (pl. K).
Jacobi, A., J. für Ornithol. 76(2): 436-440. 1928 (on P as ornithologist).
Jessen, K.F.W., Bot. Gegenw. Vorz. 347, 429, 467-468. 1864.
Keller, C., *in* E.F. Poeppig, Tropenvegetation und Tropenmenschen, Ostwalds Klassiker 249: 11-32. 1965 (bibl., biogr., portr., reprint of two lectures held in 1833 at Leipzig), see also Mattick, F., Bot. Jahrb. 85 (Lit.): 17-18. 1966.
Kühnel, J., Thad. Haenke 62, 63, 85, 144. 1960.

Kunze, G., Syn. pl. crypt. 1834 (Linnaea 9: 1-111. 1834), see TL-2/4037 (synopsis of P's cryptogams).
Looser, G., Revista Universitaria 15(3): 180-188. 1930 (portr., general).
Mägdefrau, K., Gesch. Bot. 67, 104, 267. 1973 (general).
Meyer, F.K. & H. Manitz, Jenaer Reden und Schriften 1974: 90 (material at JE).
Moldenke, H.N., Plant life 2: 76. 1946 (1948) (epon.).
Morren, Ed., Belg. hort. 1869: 14.
Murray, G., Hist. coll. BM(NH) 1: 174. 1904 (pl. BM).
Nissen, C., Zool. Buchill. 321-322. 1969.
Papavero, N., Essays hist. neotrop. dipterol. 2: 300. 1975.
Poeppig, E.F., Bot. Not. 1845: 91-98 (Swedish extract from *Reise* by C.J. Hartmann).
Poggendorff, J.C., Biogr.-lit. Handw.-Buch 2: 479. 1863.
Prance, G., Acta amaz. 1(1): 52. 1971 (coll.).
Reiche, K., Grundz. Pfl.-Verbr. Chile 10-11, 41. 1907 (Veg. Erde 8).
Sayre, G., Mem. New York Bot. Gard. 19: 381-382. 1975 (coll.).
Schlechtendal, D.F.L. von, Bot. Zeit. 3: 454-455. 3 Jul 1845 (rev. *Nova genera*).
Stafleu, F.A., Taxon 18: 321-323. 1969 (on *Nova genera*: brief evaluation); *in* Daniels, G.S. and F.A. Stafleu, Taxon 22(4): 444. 1973 (portr.).
Steenis-Kruseman, M.J. van, Fl. males. Bull. 15: 740. 1960 (dates *Nova gen. sp. pl.*); Blumea 25: 37. 1979 (pl. L).
Steinberg, C.H., Webbia 32: 33. 1977 (material at FI).
Stieber, M.T., Huntia 3(2): 124. 1979 (notes A. Chase).
Turrill, W.B., Bull. misc. Inf. Kew 1920: 60. (Chile).
Urban, I., Bot. Jahrb. 21, Beibl. 53: 1-27. (repr.: 3-29). 1896 (main biogr., bibl.), Symb. ant. 1: 130-131. 1898, 3: 103. 1902, Fl. bras. 1(1): 77-78. 1906 (b. 16 Jul 1798, d. 4 Sep 1868).
Weber, W., J. Bromeliad Soc. 31: 249-254. 1981 (Early essay P. on "epiphytism"; portr.).
Weberbauer, A., Pfl.-Welt peruan. And. 7-8, 35. 1911 (Veg. Erde 12).
Werner, K., Wiss. Z. Univ. Halle, Math.-Nat. 4(4): 777. 1955 (pl. HAL).
Wettstein, R. v., Österr. bot. Z. 49: 96. 1899 (pl. PRC).
Wittrock, V.B., Acta Hort Berg. 3(2): 138-139. 1903 (on portr.).
Zittel, K.A. von, Hist. geol. palaeontol. 290. 1901.

COMPOSITE WORKS: (1) Notes on the Cinchona trees of Huanuco, in Northern Peru, *in* H. Karsten, *Notes on the medicinal Cinchona barks of New Granada* p. 57-75. 1861; translated from *Reise* 2: 217-223, 257-264.
(2) Illustrirte Naturgeschichte des Thierreichs, 3 vols. (Leipzig 1847-1848 (main zoological publication).

EPONYMY: *Poeppigia* K.B. Presl (1830).

8089. *Fragmentum synopseos plantarum phanerogamum* ab auctore annis mdcccxxvii ad mdcccxxix in Chile lectarum. Dissertatio botanica qua ad audiendam orationem quam muneris professoris philosophiae extraordinarii in Universitate Lipsiensi rite suscipienda causa die xviii octobris mdcccxxxiii hora nona matutina in auditorio juris consultorum illustris ordinis consensu habiturus est invitat Eduardus Poeppig phil. dr. ... Lipsiae [Leipzig] (typis Elberti) [1833]. Oct. (*Fragm. Syn. Pl.*)
Publ.: 18 Oct 1833 (date of oration), p. [i, iii], [1]-30. *Copies*: G, MO, NY, USDA; – The title page shows (1) that Poeppig's degree was "Dr. phil.", (2) that his professorship was in "philosophy", a term standing for natural sciences (his actual charges were mainly zoological). The *Fragmentum* was published as a "Habilitationsschrift" at the occasion of the formal acceptance of his professorship by means of a public lecture "de civium ordinum inferiorum ad terras remotiores perpetua migratione, ...".
Ref.: Anon., Flora 16: 719-720. 1833.

8090. *Reise in Chile, Peru* und auf dem Amazonenstrome während der Jahre 1827-1832 ... Leipzig (Friedrich Fleischer, J.C. Hindrichssche Buchhandlung) [1834-]1835-1836, 2 vols. Qu. (*Reise Chile*).
1: 14-16 Dec 1834 (see TL-1; t.p. 1835), p. [i]-xviii, [1]-466, *1 map*.
2: 25-31 Jan 1836 (see TL-1), p. [i]-viii, [1]-464. *Atlas: pl. 1-16*, in folio.

Copies: B (regular issue), MICH ("Wohlfeile Ausgabe 1837. Ohne Atlas", "cheap edition" without the atlas).
Facsimile ed.: 1960, in one volume: [i]-xii, [1]-466, [i]-v, [1]-466, *pl.* [*1-16*], map, [1-2, music]. *Copies*: G, MO, NY, U, J. Wurdack. – Stuttgart (F.A. Brockhaus Komm.-Gesch. GMBH. Abt. Antiquarium) 1960.
Extracts: (1) Swedish, by C.J. Hartmann, Bot. Not. 1845: 91-98.
(2) German, by W. Drascher, *Im Schatten der Cordillera*, Stuttgart 1927.
(3) German, by H. Petersen, *Pampayaco*, Potsdam 1936.
(4) German, by H. Butze, *Über die Anden zum Amazonas*. Leipzig 1953, 1956.
Translations: (5) Danish, by Frederik Schaldemose, Eduard Pöppig's *Reise i Chili og Peru* ... Kjøbenhavn (... C. Steens Forlag ...) 1842, p. [1]-482. *Copy*: MO.
(6) Spanish, by C. Keller, *Un testigo en lu alborada de Chile*, Santiago 1960.
(items 2-5 n.v., fide C. Keller 1965).

8091. *Nova genera ac species plantarum* quas in regno chilensi peruviano et in terra amazonica annis mdcccxxvii ad mdcccxxxii legit Eduardus Poeppig et cum Stephano Endlicher descripsit iconibusque illustravit ... Lipsiae [Leipzig] (sumptibus Friderici Hofmeister) 1835-1845, 3 vols. Fol. (*Nov. gen. sp. pl.*).
Co-author (vols. 1 and 2): Stephan Ladislaus Endlicher (1804-1849).
1: 1835-1836, p. [i*-iv*], [i]-iv, [1]-62, *pl. 1-100* (copp. J. Zehner (1-3) and E.F. Poeppig; some copies coloured).
2: 1836-1838, p. [i], [1]-74, *pl. 101-200* (copp. E.F. Poeppig).
3: 1840-1845, p. [i]-iv, [1]-91, *pl. 201-300* (id.) (no. *260* as *160*).
Copies: BR (uncol.), G (uncol.), MO (col.), NY (col.), U; IDC 2264. – Published in 30 parts as follows:

vol.	part	pages	plates	dates
1	1	1-4	*1-10*	17-23 Mai 1835
	2-3	5-20	*11-30*	18-24 Oct 1835
	4-6	21-30	*31-60*	31 Jan-6 Feb 1836
	7-10	31-62 + iv	*61-100*	22-28 Mai 1836
2	1-2	1-12	*101-120*	20-26 Nov 1836
	3	13-16	*121-130*	Jan-Oct 1837
	4	17-20	*131-140*	Jan-Oct 1837
	5-8	21-60	*141-180*	Jan-Sep 1838
	9-10	61-74	*181-200*	16-21 Dec 1838
3	1-2	1-16	*201-220*	5-11 Jul 1840
	3-4	17-32	*221-240*	15-21 Aug 1841
	5-6	33-52	*241-260*	8-11 Mar 1843
	7-10	53-91, ind.	*261-300*	23-25 Jan 1845

The dates are those of receipt by Hinrichs at Leipzig (except those of parts 2(3-8)).
Facsimile ed.: 1968 [ii*-iv**], vol. 1: [i*-iii*], [i]-iv, [1]-62, *pl. 1-100*, vol. 2: [i], [1]-74, *pl.101-200*, vol. 3: [i, t.p.], [1 of text, with index p. iv on verso, p. [iii] of index with text p. 2 on verso], 3-91, *pl. 201-203*, in a single volume. *Copies*: B, FAS. – "Reprint 1968 Verlag von J. Cramer 3301 Lehre ...", Historiae naturalis classica 1865. See Taxon 18: 321-323. 1969. – ISBN 3-7682-0549-5.
Ref.: Schlechtendal, D.F.L. von, Bot. Zeit. 3: 454-455. 1845.

Poetsch, [Pötsch], **Ignaz Sigismund** (1823-1884), Bohemian-born Austrian botanist and physician; Dr. med. Wien 1849; until 1852 working in various hospitals in Wien; 1852-1854 practicing in Gaming; from 1854-1877 physician (Stifts– und Convictsarzt) at Kremsmünster; from 1875-1884 in retirement at Randegg. (*Poetsch*).

HERBARIUM and TYPES: Naturhist. Sammlungen des Obergymnasiums der Benediktiner at Seitenstetten; other material also in Stift Kremsmünster (pp. in Herbar Stieglitz). The herbarium of Schiedermayr at LI also contains Poetsch material (information Dr. F. Speta, Linz). Some lichens at M and WRSL. *Exsiccatae*: *Cladoniae austriaceae*, fasc. 1-2, nos. 1-325, 1873. Sets at L, W. – Letters an portrait at G.

BIBLIOGRAPHY and BIOGRAPHY: Barnhart 3: 94 (b. 29 Oct 1823, d. 24 Apr 1884); BM 4: 1590; Bossert p. 313 (d. 24 Apr 1884); CSP 4: 993-994, 8: 649, 12: 581; De Toni 4: xlv; Frank 3 (Anh.): 78; GR p. 437, cat. p. 69; Herder p. 192; Kew 4: 328; LS 21014-21024; Morren ed. 10, p. 32; Nordstedt p. 26; SBC p. 129; Stevenson p. 1254.

BIOFILE: Anon., Bonplandia 4: 398. 1856, 6: 193. 1858; Bot. Centralbl. 18: 223. 1884 (d. 23 Apr 1884 at Randegg); Bot. Gaz. 9: 149. 1884 (d.); Flora 67: 274. 1884 (d.); Nat. Nov. 6: 104. 1884 (d.); Österr. bot. Z. 12: 401. 1862 (at the time "Dr. d. Med., in Kremsmünster in Oberösterreich."); Österr. bot. Z. 24: 99. 1874 (on Cladoniae Austriaceae; Fortschrittsmedaille Weltausstellung Wien 1873), 31: 1-4. 1881 (portr.), 34: 186. 1884 (d. 23 Apr 1884); Rev. bot. Bull. mens. Soc. Franç. Bot. 3: 287. 1885 (death; b. 20 Oct 1823).
Familler, I., Denkschr. bayer. bot. Ges. Regensburg 11: 13, 20. 1911.
Kanitz, A., Magy. növen. Lap. 8: 63. 1884 (b. 20 Oct 1823, d. 23 Apr 1884).
Sayre, G., Mem. New York Bot. Gard. 19: 154. 1969 (exs.); Bryologist 80: 515. 1977 (coll. at W, bryophytes only?).
Wittrock, V.B., Acta Horti Berg. 3(2): 84. 1903, 3(3): 69. 1905 (on portr.).

EPONYMY: *Poetschia* Körber (1861).

8092. *Beitrag zur Flechtenkunde Niederösterreichs.* (Aus den Schriften des zoologisch-botanischen Vereins in Wien, 1857). Oct.
Publ.: 1857, p. [1]-8. *Copy*: G. – Reprinted and to be cited from Verh. zool. bot. Ges. Wien 7: 27-34. 1857.

8093. *Beitrag zur Kenntniss der Laubmoose und Flechten* von Randegg in Niederösterreich. (Aus den Schriften des zoologisch-botanischen Vereins in Wien, 1857). Oct.
Publ.: 1857, p. [1]-6. *Copy*: G. – Reprinted and to be cited from Verh. zool.-bot. Ges. Wien 7: 211-216. 1857.

8094. *Dritter Beitrag zur Kryptogamenkunde Oberösterreichs...* (Aus den Verhandlungen der k.k. zoologisch-botanischen Gesellschaft in Wien... besonders abgedruckt) [1858]. Oct.
Publ.: 1858 (after 3 Apr; date presentation mss.), p. [1]-8. *Copy*: G. – Reprinted and to be cited from Verh. zool.-bot. Ges. Wien 8: 277-284. 1858.

8095. *Systematische Aufzählung* der im Erzherzogthume Oesterreich ob der Enns bisher beobachteten *samenlosen Pflanzen* (*Kryptogamen*) ... Wien (Im Inlande besorgt durch W. Braumüller, ... für das Ausland in Commission bei F.A. Brockhaus in Leipzig) 1872. Oct. (*Syst. Aufz. Krypt.*).
Co-author: Karl B. Schiedermayr (1818-1895).
Publ.: Jun 1872 (Flora rd. 1873; Bot. Zeit. 16 Mai 1873 ann. 4 Jul 1873 publ.; ÖbZ 1 Jul 1873; Hedwigia Aug 1873), p. [i]-xlvii, [xlviii], [1]-384. *Copies*: B, FH(2), G, H, MO, Stevenson, USDA.
Nachträge: Jul-Aug 1894 (ÖbZ 1 Sep 1894; Nat. Nov. Oct(1) 1894; Bot. Zeit. 16 Nov 1894; Bot. Centralbl. 30 Oct 1894), p. [1]-216. *Copies*: FH, Stevenson, USDA. "*Nachträge zur systematischen Aufzählung* ... unter Mitwirkung der Herren Moriz Heeg und Dr. Siegfried Stockmayer bearbeitet von Dr. C.B. Schiedermayr ..." Wien (Druck von M. Schinkay ...). 1894. Oct.
Ref.: Geheeb, A., Rev. Bryol. 4(6): 92-93. 1877.
Sauter, A., Flora 56: 425-431. 21 Sep 1873 (rev.).

Poeverlein, Hermann (1874-1957), German administrator and botanist; Dr. jur. Erlangen 1898; in the judiciary and administration at Ludwigshafen (1904, assistant judge), Kemnath (1916) and Speyer (1919, Regierungsrat; 1932 Regierungsdirektor); from 1934 in semi-retirement at Augsburg; from 1941 retired, from 1945/6 at Karxheimzell and later at Ludwigshafen; specialist on Uredinales. (*Poeverl.*).

HERBARIUM and TYPES: M (main collections, inf. H. Merxmüller); private herbarium destroyed (Hepp); further material at B, BERN, BP, C, E, MICH, W. – *Exsiccatae*: *Flora exsiccata bavarica*, fasc. 1-6, 1898-1902, nos. 1-500; with notes and schedae published in the Denkschr. bot. Ges. Regensburg n.s. 1: 150-152, 164-172. 1900, 215-217, 237-241. 1901 and 1 (Beil. 2): 1-67. 1898, 2 (Beil. 2): 19-56, (Beil. 3): 1-74. 1904, 3 (Beil. 1): 1-70. 1905. Contributed also to Sydow, *Mycotheca germanica. Exsiccatae*: *Flora exsiccata rhenana*, fasc. 1, 1909, nos. 1-100, accompanied by schedae, see below. *Uredineen Süddeutschlands* (with E. Eichhorn), cent. 1-5. – Lettters and portrait at G.

BIBLIOGRAPHY and BIOGRAPHY: AG 12(1): 226; Barnhart 3: 94; BFM no. 635, 636; BJI 2: 137; BM 8: 1018; CSP 17: 937; DTS 6(4): 185-186; Futák-Domin p. 467-477; Hegi 7: 560; Kew 4: 328; LS suppl. 21986-21994; Rehder 5: 679; TL-2/183, 200.

BIOFILE: Anon., Österr. bot. Z. 67: 112. 1918 (app. Kemnath as "Bezirksamtmann").
Hedge, I.C. & J.M. Lamond, Index coll. Edinburgh herb. 119. 1970.
Hepp, E., Ber. bayer. bot. Ges. 31 (Nachtr.): xli-xlii. 1957; herb. and libr. destroyed at Augsburg during World War II).
Killermann, S., Denkschr. Regensb. bot. Ges. 21: v, xiii, 1940.
Kneucker, A., Allg. bot. Z. 8: 192. 1902 (to Ludwigshafen as assistant judge).
Künkele, T., Pfälzer Heimat 5(8): 103. 1954 (80th birthday; herb. destroyed in 1944).
Löhr, O., Pfälzer Heimat 8: 73-75. 1957 (obit., portr., bibl.); Pollichia ser. 3. 4: 208. 1957 (obit.).
Müllerot, M., Hoppea 34(2): 280. 1976 (publ. at REG).
Pongratz, L., Acta albert. ratisb. 25: 120-121. 1963 (portr.).
Verdoorn, F., ed., Chron. bot. 1: 244. 1935, 6: 263. 1941.

8096. *Die bayerischen Arten, Formen und Bastarde der Gattung Potentilla* [Separat-Abdruck aus "Denkschriften der kgl. botanischen Gesellschaft in Regensburg ...] 1898 [1899]. Oct. (*Bayer. Potentilla*).
Publ.: Mar-Apr 1899 (Flora 20 Mai 1899; Allg. bot. Z. 15 Mai 1899; J. Bot. Mai 1899, Nat. Nov. Jul 1899), p. [1]-121, [122]. *Copy*: G. – Reprinted and to be cited from Denkschr. k. bot. Ges. Regensburg 7: [147]-[268]. Dec 1898 or Jan-Apr 1899.

8097. *Die* seit Prantl's "Exkursionsflora für das Königreich Bayern (1. Auflage. Stuttgart 1884) erschienene *Literatur über Bayerns Phanerogamen– und Gefässkryptogamenflora* ... Regensburg (Verlag der Gesellschaft) 1889. Oct. (*Lit. Bayer. Phan.-Fl.*).
Publ.: 1899 (Allg. bot. Z. 15 Mai 1899; Nat. Nov. Aug(1) 1899), p. [1]-27. *Copy*: G. Issued as 1. Beilage, Denkschr. bot. Ges. Regensburg vol. 7.

8098. *Vorarbeiten zu einer Flora Bayerns.* Die bayerischen Arten, Formen und Bastarde der Gattung Alectorolophus ... Sonderabdruck aus Bd. X der Berichte der Bayer. Botan. Gesellschaft (Druck von Val. Höfling) 1905. Oct.
Publ.: 1905, p. [1]-24. *Copy*: G – Reprinted and to be cited from Ber. Bay. bot. Ges. 10: 1-24. 1905.

8099. *Die Rhinantheen Niederbayerns* [Sonderabdruck aus dem achtzehnten Jahresbericht des " Naturwissenschaftlichen Vereins Landshut") 1907.
Publ.: 1907 (Allg. bot Z. 15 Jun 1908), p. [1]-33. *Copies*: FI, G. – Reprinted and to be cited from Jahresber. Naturw. Ver. Landshut 18: 1-33. 1907.

8100. *Flora exsiccata rhenana* fasciculus 1. Nr. 1-100 ... herausgegeben von Dr. Hermann Poeverlein, Dr. Walter Voigtlaender-Tetzner und Friedrich Zimmermann. Karlsruhe (Buchdruckerei J.J. Reiff) 1909. Oct. (*Fl. exs. rhen.*).
Publ.: 20-25 Jul 1908 and Mar-Apr 1909 (p. iv: Mar 1909; Bot. Centralbl. 27 Apr 1909), p. [i-iv], [1]-16, 20-25 Jul 1908, [17]-28, Mar 1909. *Copy*: MO. – See also Allg. bot. Z. 14: 1-15. 1908.
Ref.: Anon., Herbarium 8: 59. 1909, 53: 29. 1920.

Poggenburg, Justus Ferdinand (1840-1893), German-born (Hannover) American botanist; at his death business manager of the "New Yorker Staats Zeitung". (*Poggenb.*).

HERBARIUM and TYPES: NY.

BIBLIOGRAPHY and BIOGRAPHY: Barnhart 3: 94; IH 2: (in press); Kew 4: 328; Pennell p. 616; PH 132; TL-2/775; Zander ed. 10, p. 702, ed. 11, p. 801.

COMPOSITE WORKS: see N.L. Britton, *Prelim. Cat.*, TL-2/775.

Pohl, Johann [Baptist] Emanuel (1782-1834), Bohemian-born Austrian botanist and traveller; Dr. med. Praha 1808; army physician 1809-1811; at the garden of J.M. de Canal (Freiherr von Hochberg) 1811-1817; travelled in Brazil with Spix, Martius et al. 1817-1821; afterwards curator at the Vienna Natural History Museum and the Vienna Brasilian Museum. (*Pohl*).

HERBARIUM and TYPES: Herbarium at W. Most types at W, others at LE and M. Duplicates in BM, BR, F, G, HBG (pterid.), LE, M, MICH, P (pterid.), PR, PRC, US. – A set of 12 nature prints, Pragae 1804, is mentioned by Weitenweber (1853). – Correspondence with D.F.L. von Schlechtendal at HAL; further letters and portrait at G.

BIBLIOGRAPHY and BIOGRAPHY: ADB 26: 369-370; Barnhart 3: 94 (b. 22 Feb 1782); Blunt p. 247-248; BM 4: 1591; Clokie p. 226-227; CSP 4: 960, 8: 640; DU 226; Futák-Domin p. 78, 478; GF p. 70; Herder p. 56, 225; Hortus 3: 1201 (Pohl); IH 1 (ed. 6): 363, (ed. 7): 340, 2: (in press); Jackson p. 206, 264, 371, 416; Kew 4: 329; Klástersky p. 145 (states that Pohl's private herb. was sold to AMD); Langman p. 588; Lasègue p. 332, 477, 505, 526; Maiwald p. 96-98, 106-107, 281 [index]; Moebius p. 310; NI 1550-1551, see also 1: 189, 191; Plesch p. 366; PR 7229-7332 (ed. 1: 8106-8109); Quenstedt p. 341; Rehder 5: 679; RS p. 134; SK 4: ccvi; Sotheby 609; TL-1/986-987; TL-2/see J.C. Mikan; Tucker 1: 562; WU 23: 28-31; Urban-Berl. p. 384; Zander ed. 10, p. 702, ed. 11, p. 801.

BIOFILE: Anon., Appleton's cycl. Amer. biogr. 5: 48. 1888; Bonplandia 8: 218. 1860 (coll. "im brasiliensischen Kabinet ..."); Flora 17: 384. 1834 (d. 23 Mai 1834).
Baer, W. & H.W. Lack, Pflanzen auf Porzellan 122. 1979.
Candolle, Alph. de, Phytographie 440. 1880 (coll. W; also LE, M).
Candolle, A.P. de, Mém. souvenirs 377, 379. 1862.
Domin, K., Věra přír. 22: 120-122. 1943 (on *Tentamen*).
Hertel, H., Mitt. Bot. München 16: 418. 1980 (lich. M).
Jessen, K.F.W., Bot. Gegenw. Vorz. 467. 1864.
Maiwald, P.V., Jahres-Ber. Stifts-Obergymn. Braunau 6. 1901 (b. 22 Feb 1782, d. 22 Mai 1834; friend of Opiz; on *Tentamen*).
Martius, C.F.P. von, Flora 20(2) Beibl.: 36-38. 1837.
Moldenke, H.N., Plant life 2: 76. 1946 (1948) (epon.).
Murray, G., Hist. coll. BM(NH) 1: 174. 1904 (pl. at BM through R. Brown).
Nissen, C., Zool. Buchill. 322. 1969.
Pandler, A., Mitt. nordböhm. Excurs.-Clubs 10: 158-159. 1887 (on *Tentamen*).
Papavero, N., Ess. hist. neotrop. dipterol. 1: 61, 62-65, 95, 118. 1971 (itin. Brazil).
Poggendorff, J.C., Biogr.-lit. Handw.-Buch 2: 484. 1863 (b. 22 Feb 1782, d. 22 Mai 1834, bibl.).
Pritzel, G.A., Linnaea 19: 459. 1846 (d. 23 Mai 1834).
Skalický, V., Nov. bot. Int. bot. Univ. Carol. Prag 1967: 37-44 (on *Tentamen*).
Schmid, G., Goethe u.d. Naturw. 560. 1940.
Stieber, M.T., Huntia 3(2): 124. 1979 (notes Agn. Chase on P.).
Urban, I., Fl. bras. 1(1): 78-82. 1906 (itin. Brazil).
Weitenweber, W.R., Lotos 3: 25-28. 1853 (biogr.; mentions a set of 12 "Pflanzenabdrücke" publ. in 1804, distributed among friends: *Adumbrationes plantarum* [no copy seen by us].
Wettstein, R. v., Österr. bot. Z. 49: 96. 1899 (pl. at PRC).
Wiltshear, F.G., J. Bot. 50: 171-174. 1912 (on *Tentamen*).

EPONYMY: *Pohlana* C.F.P. von Martius et C.G.D. Nees (1823). *Note*: the source of *Pohlia* J. Hedwig (1801) and the derived *Pseudopohlia* R.S. Williams (1917) was not given, but may be named for Johann Pohl. *Pohliella* Engler (1926) honors Joseph Pohl (see sub Julius Pohl).

8101. *Tentamen florae bohemicae.* Versuch einer Flora Böhmens ... Prag (Gedruckt bey Gottlieb Haase, ...) 1809-1814, 2 parts. Oct. (*Tent. fl. bohem.*).
1: Sep (prob.) 1809 (autograph presentation in copy BM; p. xx: Jan 1809), p. [i]-xxxii, [1]-302, [1, err.], *1 pl.* (uncol. or col. copp.). *Copies*: BM, KNAW (uncol.), MO (col.).
1, reissue: 1810, p. [i]-xxxii, [1]-302, [1, err.], *1 pl. Copies*: B, G, US; IDC 5640. – Prag (bei C.W. Enders und Compagnie) 1810.
2: 1814, p. [i]-vi, [1]-234. *Copies*: G, KNAW.
2, reissue: 1815, p. [i]-vi, [1]-234. *Copies*: B, BM, K, MO, US; IDC 5640. – Prag (bei C.W. Enders) 1815.
The references to Mikan's *Icones* are to an unpublished iconography of the Bohemian flora now at the library of Charles University, Praha (Skalický 1967). – Wiltshear (1912) lists the Pohl names at that time not yet listed by Index kewensis.

8102. *Des Freiherrn von Hochberg botanischer Garten zu Hlubosch* 1812. Geordnet von Johann Emmanuel Pohl, ... Prag [Praha] (gedruckt bei Franz Gerzabek, ...) [1812]. Oct. (in fours) (*Hochberg bot. gart.*).
Publ.: 1812, p. [i-xvi], [1]-58. *Copy*: M.

8103. *Plantarum Brasiliae icones et descriptiones* hactenus ineditae. Jussu et auspiciis Francisci primi, ... auctore Joanne Emanuele Pohl, ... Vindobonae [Wien], [1826-] 1827-1831 [-1833], 2 vols. Fol. (*Pl. bras. icon. descr.*).
1: 1826-1828, p. [i]-xvi, [1]-135, [136, index], *pl. 1-100* (*56* as "*57*"). (col. or uncol. liths. by Sandler).
2: 1828-1833, p. [i-iii], [1]-152, *pl. 101-200* (uncol. or col. liths.).
Copies: BR, G, MO, NY (*pl. 1-175* uncol., *176-200* col.), U, USDA (*pl. col.*); IDC 933.

vol.	part	pages	plates	dates
1	1	1-36	*1-25*	Aug 1826
	2	37-60	*26-50*	Mai-Oct 1827
	3	61-92	*51-75*	1827
	4	91-136, title, dedication, preface, list of subscribers	*76-100*	Mai-Dec 1828
2	1	1-40	*101-125*	1828 v. Jan 1829
	2	41-80(?)	*126-150*	Sep 1830
	3	81(?)-116(?)	*151(?)-175*	1830 or 1831
	4	116(?)-153	*176(?)-200*	Jan-Sep 1833

From the prospectus of the work published in Vienna on 1 Mai 1826 (in Lambert correspondence at K) it is clear that the ordinary issue with uncoloured plates was in small folio and available somewhat earlier than the large paper issue with the hand-coloured plates. The latter was delivered only on order. Both issues are on wove paper. – The 200 lithographs are of drawings by Wilhelm Sandler (only 2 prints signed). The lithographs were drawn with a pen and resemble engravings.
Ref.: Lack, H.W., *Zandera 1: 3-7. 1982.*

8104. *Reise im Innern von Brasilien.* Auf allerhöchsten Befehl seiner Majestät des Kaisers von Österreich, Franz des Ersten, in den Jahren 1817-1821 unternommen und herausgegeben von Johann Emanuel Pohl, ... Wien 1832-1837, 2 vols. Qu. (*Reise Brasil.*).
1: Oct-Dec 1832 (p. xiv: 4 Oct 1832), p. [i]-xxx, [xxxi, h.t.], [1]-448.
2: 1837 (Hinrichs rd. 21-27 Jun 1838), p. [i]-xii, [1, h.t.], [1]-641.
Atlas: *9 plates*: 7 uncol. copper engr. (views landscapes), 1 col. drawing (insects), 1 lith. (geogn.). Oblong; cover: *Atlas zur Beschreibung der Reise* in Brasilien von Dr. Johann Emanuel Pohl. Erster Theil, Wien 1832, portfolio, drawings by Thom. Endler (portfolio with cover MO copy).
Copies: MO, NY. – The section on *Die vorzüglich lästige Insekten* of vol. 1 was reprinted separately, Wien 1832, 20 p., *1 pl.*, that on geognosy id. as *Beiträge zur Gebirgskunde Brasiliens*, Wien 1832 (n.v.).
Portugese transl.: 1951, 2 vols. (400, 471 p., ill.), *Viagem no interior do Brasil*, ... Rio de Janeiro (Ministerio de Educação e Saúde, ...) 1951 (n.v.).

Pohl, Johann Ehrenfried (1746-1800), German (Saxonian) physician and botanist; professor of botany at Leipzig; from 1788 royal physician at Dresden. (*J.E. Pohl*).

HERBARIUM and TYPES: Unknown.

BIBLIOGRAPHY and BIOGRAPHY: Barnhart 3: 94 (b. 12 Sep 1746, d. 25 Oct 1800); BM 4: 1591; Dryander 3: 384, 415; Herder p. 118, 363; PR 7227-7228, ed. 1: 8103-8105.

8105. *Animadversiones in structuram ac figuram foliorum in plantis.* Amplissimi philosophorum ordinis auctoritate a.d. xxi decembr. [1771] publico eruditorum examini submittit M. Johannes Ehrenfried Pohl lipsiensis medicinae baccalaureus respondente Nathanael Godofredo Leske Muskavia lusato medicinae cultore. Lipsiae [Leipzig] (ex officina Langenhemia) [1771]. Qu. (*Animadv. struct. fig. fol.*).
Publ.: 21 Dec 1771, p. [i-iv], [1]-32. *Copy*: G. *Reprint*: *in* P. Usteri, Delect. opusc. bot. 1: 145-194. 1790.

8106. *De soli differentia in cultura plantarum* attendenda disserit et ad benivole audiendam orationem qua professionem botanices extraordinariam clementissime sibi demandatum a.d. xxii. dec. mdcclxxiii auspicabitur humanissime invitat Ioannes Ehrenfried Pohl ... [Leipzig (typ. Langenheim) 1773]. Qu. (*De soli diff.*).
Publ.: 22 Dec 1773, p. [i]-xviii. *Copy*: G.

Pohl, Julius (1861-1939), German botanist; professor of pharmacology at the German University of Prag (1895). (*Jul. Pohl*).

HERBARIUM and TYPES: Unknown.

NOTE: The J. Pohl who illustrated so many works by Adolf Engler and his school was Joseph Pohl (1864-1939), who joined Engler in Breslau for *Die natürlichen Pflanzenfamilien* and went with him to Berlin where he remained active at the Botanical Museum until l938. See H. Harms, Notizbl. Berlin 15: 142-144. 1940 [Eponymy: *Pohliella* Engler].

BIBLIOGRAPHY and BIOGRAPHY: Barnhart 3: 94; BM 4: 1591; CSP 4: 938-939; Kew 4: 329.

BIOFILE: Anon., Bot. Centralbl. 64: 32. 1895 (prof. pharmacol. Praha).
Kneucker, A., Allg. bot. Z. 1: 184. 1895 (appointed at Praha).

8107. *Botanische Mitteilung über Hydrastis canadensis* ... Stuttgart (Verlag von Erwin Nägele) 1894. Qu. (*Bot. Mitt. Hydrastis*).
Publ.: 1894 (Bot. Zeit. 16 Jul 1894; Nat. Nov. Mai(2) 1894), p. [i], [1]-12, *pl. 1-4* with text. *Copy*: NY. – Bibliotheca botanica Heft 29.

Pohle, Richard Richardowich (1869-1926), German-Latvian botanist, born in Riga of German parents; in military service at Berlin 1889/90; studied at Dresden Technical University 1894-1899, from 1895-1899 also assistant at the Dresden Botanical Institute; travelled in N. Europe 1898-1899; at University of Rostock 1900-1901; Dr. phil. ib. 1901; until 1914 at the Botanic Garden St. Petersburg; later professor of geography at the Technical University of Braunschweig. (*Pohle*).

HERBARIUM and TYPES: LE; further material at B, C, E, and H.

BIBLIOGRAPHY and BIOGRAPHY: Barnhart 2: 94 (b. 5/18 Aug 1869, d. 3 Aug 1926); BJI 2: 137; Hortus 3: 1201 (Pohle); IH 2: (in press); MW p. 390; Urban-Berl. p. 384.

BIOFILE: Anon., Verh. bot. Ver. Brandenburg 68: 279. 1926 (d. 4 Aug 1926).
Hedge, I.C. & J.M. Lamond, Index coll. Edinb. herb. 119. 1970 (mat. at E).
Isačenko, B., Bull. Jard. bot. URSS (Leningrad) 26: 86-88. 1927 (obit., portr., b. 5/18 Aug 1869).
Lipschitz, S., Florae URSS fontes 223 [index]. 1975.

Lipski, V.I., S. Petersb. bot. zad. 3: 404-406. 1915 (biogr., bibl., portr.).
Palmgren, A., Mem. Soc. Fauna Fl. fenn. 3: 98-99. 1927 (obit.).
Sayre, G., Bryologist 80: 515. 1977 (coll.).
Wulff, E.V., Intr. hist. plant geogr. 40, 46, 201. 1943.

COMPOSITE WORKS: *Vegetationsbilder aus Nordrussland,* in G. Karsten & H. Schenk, *Vegetationsbilder,* ser. 5, Heft 3-5. Jena (G. Fischer) 1907. Qu. *16 pl.* with text.

8108. *Pflanzengeographische Studien über die Halbinsel Kanin und das angrenzende Waldgebiet.* Bericht zweier Reisen, ausgeführt in den Sommern 1898/99 in der nordrussischen Waldregion (Gebiet der Flüsse: Pinega, Jula, Kuloi, Mesen und auf der Halbinsel Kanin). Inaugural-Dissertation zur Erlangung der Doctorwürde der hohen philosophischen Facultät der Landes-Universität Rostock ... [s.l.; 1901]. Oct. (*Pfl.-geogr. Stud. Kanin*).
Publ.: Jun 1901 (preface Jun 1901; Nat. Nov. Jul(1) 1903), p. [i-vii], [1]-112. *Copy:* US.

8109. *Materialy dlya poznaniya rastitel'nosti severnoy Rossii.* I. K flore mkhov severnoy Rossii. Petrograd 1915.
Publ.: 1915, p. [i]-viii, [1]-148, *pl. 1-10. Copy:* H (without plates), reprinted and to be cited from Trudy Imp. Bot. Sada Petra Velikago (Acta horti petrop.) 33(1): [i*-ii*], [i]-viii, [1]-148, *pl. 1-10.* – Inf. P. Isoviita.

8110. *Drabae asiaticae* Systematik und Geographie nord- und mittelasiatischer Draben ... Dahlem bei Berlin (Verlag des Repertoriums, ...) 1925. Oct. (*Drab. asiat.*).
Publ.: 30 Nov 1925, p. [i-iv], [1]-225. *Copies:* G, H, U. – Repert. sp. nov. Beih. 32.
Ref.: Krause, K., Bot. Jahrb. 60 (Lit.): 69. 1926.

Poirault, [Marie Henri] Georges (1858-1936), French botanist; son of J.P.F. Poirault; on oceanographic voyage of Talisman, 1883; with A.S. Famintzin, St. Petersburg 1886-1888; Dr. sci. Paris 1894; at Mus. Hist. nat. Paris, 1888-1889; director of the Villa Thuret at Antibes 1899-1936. (*G. Poirault*).

HERBARIUM and TYPES: PC. – Portrait and letters at G.

BIBLIOGRAPHY and BIOGRAPHY: Ainsworth p. 123, 322, 346; Barnhart 3: 95 (b. 3 Sep 1895, d. 10 Feb 1936); BM 4: 1591; CSP 12: 581, 637, 17: 944-945; GR p. 346; IF p. 724; IH 2: (in press); Kew 4: 329; Langman p. 589; LS 37991-37991a, suppl. 21996.

BIOFILE: Anon., Bull. Soc. bot. France 46: 80. 1899 (app. Villa Thuret); Nat. Nov. 21: 377. 1899 (app. Villa Thuret).
Cavillier, F.G., Boissiera 5: 73. 1941 (b. 3 Sep 1858, d. 10 Feb 1936).
Chevalier, A., Rev. Bot. appl. Agric. trop. 16: 403-413. 1936 (obit., b. 3 Sep 1858, d. 10 Feb 1936; bibl.).
Chouard, P., Rev. hort. 108: 143-144. 1936 (on the unpublished work of Poirault).
Gager, C.S., Brooklyn Bot. Gard. Rec. 27(3): 194. 1938.
Sauvageau, C., Rev. hort. 108: 141-142. 1936 (obit., portr.).
Verdoorn, F., ed., Chron. bot. 3: 103, 104. 1937, 4: 52, 53. 1938.

COMPOSITE WORKS: Translated O. Drude, *Handbuch der Pflanzengeographie* (1890) into French. *Manuel géographie botanique* (1897).

8111. *Recherches d'histoire végétale.* Développement tissus dans les organes végétatifs des cryptogames vasculaires ... St. Pétersbourg (Commissionnaires de l'Académie impériale des sciences ...) 1890. Qu.
Publ.: shortly after 1890 (mss. submitted 28 Apr 1887, imprimatur Mai 1890), p. [1-111], [1]-26, *pl. 1-5* (uncol., lith. auct.). *Copy:* G. – Mém. Acad. imp. Sci. St. Petersb. ser. 7. 37(14). 1890.

8112. Ministère de l'agriculture ... *Hortus thuretianus antipolitanus.* Catalogue des plantes cultivées au Jardin de la Villa Thuret à Antibes ... Cannes (Imprimerie Robaudy) 1933. Duod. (*Hort. thuret.*).

Publ.: 1933, p. [i], [1]-204. *Copies*: G, MO.
Ref.: Stehlé, H., Bull. Soc. bot. France 114(9): 429-440. Dec 1968 (revision of the genus *Mesembryanthemum* as treated in the Hort. thuret.).

Poirault, Jules Pierre François (1830-1907), French pharmacist and botanist; professor at the École de Médecine and director of the botanic garden at Poitiers. (*Poirault*).

HERBARIUM and TYPES: Unknown.

BIBLIOGRAPHY and BIOGRAPHY: Barnhart 3: 95; BL 2: 211, 703; Jackson p. 292; Kew 4: 330; LS 21038-21040; Rehder 1: 414.

BIOFILE: Anon., Bot. Centralbl. 105: 560. 1907 (d. Aug 1907), Bull. Acad. Géogr. bot. 16: 209. 1907 (d. Aug 1907); Bull. Soc. bot. Deux-Sèvres 18: *pl. 6.* 1907 (portr., b. 28 Sep 1830); Nat. Nov. 30: 71. 1908 (d.).
Kneucker, A., Allg. bot. Z. 13: 196. 1907.
Mangin, L., Bull. Soc. bot. France 54: 497. 1907 (d.).
Souché, B., Fl. Haut Poitou xvi. 1901.
Treub, M., Bull. Acad. Géogr. bot. 12: 49. 1903, 13: 14. 1904.

8113. *Plantes vasculaires du département de la Vienne.* Thèse présentée et soutenue à l'école supérieure de pharmacie de Paris le [] mai 1874 pour obtenir le diplôme de pharmacien de 1re classe ... Poitiers (Typographie de Henri Oudin, ...) 1874. Oct. (in fours). (*Pl. vasc. Vienne*).
Publ.: Mai 1874 (t.p.), p. [i], [1]-15, [1]-127. *Copy*: G. *Supplément*: 1883.
Commercial issue: 1875 (fide Bot. Zeit. 20 Apr 1877), Poitiers, Oudin, 127 p. (n.v.).
Ref.: Anon., Bull. Soc. bot. France 23 (Bibl.): 54-55. 1876.

Poiret, Jean Louis Marie (1755-1834), French clergyman and botanist; travelled in North Africa 1785-1786. (*Poir.*).

HERBARIUM and TYPES: P, PC (through Moquin-Tandon and Cosson); other material at BR, FI (herb. Desfontaines), H, P-LA (types), UPS (herb. Thunberg). Some letters at HU.

BIBLIOGRAPHY and BIOGRAPHY: AG 1: 214, 1 (ed. 2): 333, 5(1): 776; Backer p. 455; Barnhart 3: 95; BM 4: 1591; Bossert p. 313; Bret. p. 125; CSP 4: 963; DU 227; Frank 3 (Anh.): 78; Herder p. 471; Hortus 3: 1201 (Poir.); IF p. 724; IH 1 (ed. 6): 363, (ed. 7): 340, 2: (in press); Kew 4: 329-330; Langman p. 589; Lasègue p. 59, 316, 321, 441, 504; Lenley p. 466; MW p. 390; NI 349, 1552; Plesch p. 366; PR 1679, 5004-5005, 7233-7235 (ed. 1: 8113-8116); Rehder 5: 680; RS p. 111-112; SBC p. 129; SO 815; SK 4: xcvi; Stevenson p. 1254; TL-1/267, 315, 643-644, 988-989; TL-2/1091, 1293, 1409-1410, 1597, 4136-4137, see C.H.B.A. Moquin-Tandon; Tucker 1: 562; Zander ed. 10, p. 702, ed. 1, p. 801.

BIOFILE: Anon., Biogr. univ. Suppl. 77: 342-344. 1845; Bonplandia 1: 216. 1853.
Candolle, Alph. de, Phytographie 440. 1880 (herb.).
Candolle, A.P. de, Mém. souvenirs 58, 78. 1862.
Cardew, F., J. Roy. Hort. Soc. 78: 293-294. 1953 (on *Leçons de flore*, vellum copy, Lindley Library).
Cosson, E., Comp. fl. atl. 1: 10. 1881.
Cosson, E. & M.C. Durieu, Exp. sci. Alg. Bot. 2: xvii. 1868.
Cuvier, G., Dict. Sci. nat. portraits: [alph.] 1845 (portr.).
Fournier, P., Voy. déc. sci. mission. natural. franç. 1: 88, 90, 92, 98. 1932.
Guillaumin, A., Les fleurs des jardins xxxvii, *pl. 12.* 1929 (portr.).
Hall, N., Bot. Eucalypts 104. 1978.
Hitchcock, A.S. & A. Chase, Man. Grass. U.S. 988. 1951 (epon.).
Jessen, K.F.W., Bot. Gegenw. Vorz. 417, 471. 1864.
Kukkonen, I., Herb. Christian Steven 80. 1971 (mat. at H).
Lawrence, G.H.M., Huntia 1: 164. 1964 (some letters at HU).

Levot, Nouv. Biogr. gén. 39: 566-567. 1862.
Maire, R., Progr. conn. bot. Algérie *pl. 1.* 1931 (portr.).
Poiret, J.L.M., Encycl. méth., Bot. 8: 749-750. 1808 (autobiogr., account of N. Afr. journey).
Sayre, G., Bryologist 80: 515. 1977.
Stafleu, F.A., *in* G.S. Daniels and F.A Stafleu, Taxon 22: 590. 1973 (portr.).
Stearn, W.T., Roy. Hort. Soc. Exhib. manuscr. books 1954, p. 40 (describes the vellum copy in Lindley Library).
Steinberg, C.H., Mon. biol. Canar. 4: 31, fig. 6. 1973 (plants in herb. Desfontaines at FI); Webbia 32(1): 11. 1977.
Stieber, M.T., Huntia 3(2): 124. 1979 (notes Agn. Chase); 4(1): 85. 1981 (arch. mat. HU).
Wittrock, V.B., Acta Horti Berg. 3(2): 116. 1903.

COMPOSITE WORKS: (1) See Lamarck, (*Encycl.* (TL-2/4136), vols. 3 (from P onward), 4-8, suppl. 1-5) and *Tabl. encycl.* (TL-2/4137), vols. 5(2) and tome 3.
(2) See F.P. Chaumeton, Fl. méd., TL-2/1091.

HANDWRITING: Monogr. biol. Canar. 4: fig. 6. 1973; Webbia 32(1): 14. 1977.

EPONYMY: *Poiretia* Cavanilles (1797, *nom. rej.*), *Poiretia* J.F. Gmelin (1791, *nom. rej.*), *Poiretia* J.E. Smith (1808), *Poiretia* Ventenat (1807, *nom. cons.*); ING 3: 1383-1384.

8114. *Voyage en Barbarie*, ou lettres écrites de l'ancienne Numidie pendant les années 1785 & 1786, sur la réligion, les coutumes & les moeurs des maures & des arabes-bédouins; avec un essai sur l'histoire naturelle de ce pays ...à Paris (chez J.B.F. Née de la Rochelle, ...) 1789. Oct. 2 vols. (*Voy. Barbarie*).
1: 1789 (approb. 16 Apr 1788, registr. 20 Mai 1788), p. [i*, iii*], [i]-xxiv, [1]-363, [364, err.].
2: 1789 (p. [318]: 16 Apr 1788), p. [i, iii], [1]-315, [316-319].
Copies: FI, G, NY; IDC 5570.
German transl.: 1789, *Reise in die Barbarey* ... Strasbourg, 2 vols. Oct. (n.v.).

8115. *Leçons de flore*. Cours complet de botanique explication de tous les systèmes, introduction à l'étude des plantes par J.L.M. Poiret continuateur du dictionnaire de botanique de l'Encyclopédie méthodique suivi d'une iconographie végétale en cinquante-six planches coloriées offrant près de mille objets par P.J.F. Turpin. Ouvrage entièrement neuf, tome premier [-troisième]. Paris (C.L.F. Panckoucke, éditeur ...) 1819-1820, 3 vols. Qu. (*Leçons fl.*).
Author of vol. 3: Pierre Jean François Turpin (1775-1840).
Original quarto issue (25 copies): as above, vols. 1-2 by Poiret, vol. 3 by Turpin, together in 17 monthly parts the contents and dates of which are imperfectly known.
Copy: NY. – See also DU 227.
1: 1819, p. [i*-iii*], [i]-vi, [1]-250, lettre [1-4].
2: 1820, p. [1]-159, [1-3].
3: 1820, p. [i*-v*], [i]-ii, [9]-184, [1, err.], *pl. 1-56, 2bis, 4bis, suite 4bis, 36bis, 43bis, 44bis, 48bis, 56bis* (col. copp. Turpin), chart. Alternative t.p. [v*]: "*Essai d'une iconographie* élémentaire et philosophique *des végétaux* avec une texte explicatif par P.J.F. Turpin" (*Essai iconogr. vég.*).
Original octavo issue: 1819-1820, same letter press as for quarto edition but type shifted, "Livraisons" indicated.
Prospectus: 1819, p. [1]-8.
1: 1819, p. [i*, iii*], [i]-viii, [1]-278, letter [1-5].
2: 1820, p. [1]-174, [1-3], [1-4, adv.].
3: 1820, p. [i*-v*], [i]-ii, [9]-199, [200, err.], *pl.*: as in quarto issue, 2 uncol. charts.
Copies: B, BR, G, PH.
Other issues of orig. ed. (n.v.): folio (10 copies): same plates and pagination as quarto; Plesch (p. 366) reports a folio copy; see also Sotheby 613 (Sold 3.800). Two [?] copies (PR) were printed on vellum (one copy sold at Christie, 12 Nov 1975; other vellum copies are at the Österr. Nationalbibliothek and the Lindley Library (Stearn 1954, Cardew 1953).

Flore médicale issues: the *Leçons de flore* (vols. 1-2) and the *Essai d'une iconographie* (vol. 3 in above issues) were also published as part of the series *Flore médicale* edited by François Pierre Chaumeton (see TL-2/1091). Our information on the various issues is still incomplete because the complexity was the same as that of the *Flore médicale*. The information available on 1 Jun 1980 was the following:

Fl. méd. issue, Quarto: 1819-1820, vols. "septième et dernier", 7(1) and 7(2), by Poiret, vol. [7(3)] published as "*Essai d'une iconographie* ..." by P.J.F. Turpin. Title pages may have been shifted. The "ideal" composition is probably:

7(1): 1819, p. [i*-iii*], [1]-3, [i]-vi, [1]-200 as " *Flore médicale partie élémentaire* par J.L.M. Poiret ... tome septième et dernier. Paris (C.L.F. Panckoucke), éditeur ...).

7(2): 1820, p. [1]-159.

7(3): 1820, p. [i], [1]-184, [1, err.], *pl. 1-56, 2bis, 4bis, suite 4bis, 36bis, 43bis, 44bis, 48bis, 56bis*, (as above), chart. with two t.p.'s: [1]: "*Flore médicale* ... tome septième ..." and p. [3]: "*Essai d'une iconographie* ... Turpin".

Copy: NY.

Fl. méd. issue, Octavo: 1819-1820, idem:

7(1): 1819, p. [i*-xii*], [i]-viii, [1]-278, [1-5], text t.p. as in quarto issue: "tome septième et dernier".

7(2): 1820, p. [1]-174, [1-3], text t.p.: "tome septième – iie partie".

7(3): 1820, p. [i*-v*], [i]-ii, [9]-199, [200, err.], *pl.* as in quarto; "tome septième" and *Essai* title-pages.

Copies: B, BR, MICH, MO, NY, USDA. – "Livraisons" indicated in text.

Ed. 2 of text *Leçons de flore* (vols. 1 and 2 combined in a single volume, octavo edition entitled:) *Édition classique*: 1823, p. [i*-ix*], [i]-viii, [1]-367. *Copy*: MO. – We have not seen the accompanying volume for this edition of the Turpin *Essai* (vol. 3 in orig.).

8116. *Histoire philosophique, littéraire, économique des plantes de l'Europe* ... à Paris (chez Ladrange et Verdière, ...) 1825-1829, 7 vols. and atlas. Oct. (*Hist. philos. pl. Eur.*).

1: 4 Jun 1824, p. [i-iii], [i]-xxiv, [1]-494; atlas [1]-8, *pl. 1-16*.
2: 28 Dec 1824, p. [i-iii], [1]-532; atlas [1]-6, *pl. 16bis, 17-31*.
3: 19 Jul 1826, p. [i-iii], [1]-500; atlas [1]-6, *pl. 32-47*.
4: 24 Feb 1829, p. [i-iv], [1]-500; atlas [1]-8, *pl. 48-60*.
5: 19 Dec 1827, p. [i-iii], [1]-516; atlas [1]-7, *pl. 61-79*.
6: 12 Dec 1829, p. [i-iii], [1]-516; atlas [1]-7, *pl. 80-95*.
7: 12 Dec 1829, p. [i-iii], [1]-511; atlas [text n.v.], *pl. 96-127*.

Atlas: in 7 separate parts (of which no. 7 double) of *16 pl.* each (see above); col. liths. by A. Poiret, fils.

Copies: B, BR, MO, NY, USDA; IDC 5643.

Issues: regular (papier fin des Vosges) and on "papier satiné". *Advertising pamphlet*: [1824], p. [(1)]-(4) announcement, [1]-35 premier discours (from vol 1). *Copy*: G. – Five volumes anticipated; two issues: on "papier fin des Vosges" and on "papier satiné".

Poisson, Henri [-Louis] (1877-1963), French botanist; in Madagascar 1916-1954 as veterinary surgeon and inspector (1921-1934) and amateur botanist; Dr. sci. nat. Paris 1912. (*Poiss.*).

HERBARIUM and TYPES: P, PC, TAN. – Letters to Malinvaud at P.

BIBLIOGRAPHY and BIOGRAPHY: Barnhart 3: 95; BL 1: 98, 313; Hortus 3: 1201 (Poiss.); IH 2: (in press); Kew 4: 330; Langman p. 589; MW p. 390; NI 3: 7; Tucker 1: 562.

BIOFILE: Anon., Bull. Acad. malgache ser. 2. 41: 69-72. 1963 (obit., portr., b. 24 Aug 1877, d. 8 Mai 1963).
Aymonin, G.G., C.R. Congr. Soc. Sav. 102 (Limoges) (3): 57. 1977 (letters to Malinvaud).
Boiteau, P., Mém. Acad. malgache 43: 17-73. 1969 (extensive bibl.).
Bouriquet, P., Mém. Acad. malgache 43: 135-137. 1969 (mycol.).
Decary, R., Mém. Acad. malgache 43: 9-16. 1969 (biogr., b. 24 Aug 1877, d. 8 Mai 1963).

Humbert, H., C.R. AETFAT 4: 139. 1962.
Soucadaux, A., Mém. Acad. malgache 43: 131-133. 1969.
Verdoorn, F., ed., Chron. bot. 1: 36. 1935, 2: 219, 393. 1936.

COMMEMORATIVE VOLUME: Mém. Acad. malgache 43, 1969.

8117. *Recherches sur la flore méridionale de Madagascar* ... Paris (Augustin Challamel, ...) 1912. Oct. (*Rech. fl. mérid. Madagascar*).
Commercial issue: Jun 1912 (Nat. Nov. Jul(2) 1912; ÖbZ Jun-Jul 1912), p. [i-iii], [1]-230, *pl. 1-6, 6bis, 7-16* (photos). *Copies*: G, MO, MICH, USDA.
Thesis issue: 5 Jun 1912, p. [i-iii], [1]-230, [1, deuxième thèse], *pl. 1-6, 6bis, 7-16* . *Copies*: B, BR. – "Thèses présentées à la faculté des sciences de Paris pour obtenir le grade de docteur es sciences naturelles, ... 1re thèse. – *Recherches* ... soutenues le [in ink: 5 juin] 1912, ... Paris (Gauthier-Villars, ...) 1912. Oct." This thesis issue was possibly issued more or less at the same time as the commercial one.
Ref.: Lutz, L., Bull. Soc. bot. France 59: 797-798. 1912 (rev.).

Poisson, Jules (1833-1919), French botanist; at the Paris Muséum d'Histoire naturelle from 1843 (sic "élève jardinier") until 1909, from 1864 as "préparateur" (botany), from 1873 aide-naturaliste. (*J. Poiss.*).

HERBARIUM and TYPES: Unknown. – Letters at G.

BIBLIOGRAPHY and BIOGRAPHY: Barnhart 3: 95; BL 1: 46; BM 4: 1591; CSP 11: 41, 12: 581, 17: 946; Herder p. 471; Jackson p. 127, 402; Kew 4: 330; Langman p. 589; LS 21041-21052; Morren ed. 2, p. 20, ed. 10, p. 54; Rehder 5: 680; TL-2/253; Tucker 1: 562; Zander ed. 11, p. 801 (b. 29 Apr 1833, d. Sep 1919).

BIOFILE: Anon., Bull. Soc. bot. France 20 (Bibl.): 103. 1873 (promoted to aide naturaliste Paris Muséum); 53: 288 (gold medal), 544 (Légion d'honneur). 1906.
Lecomte, H., Bull. Mus. Hist. nat. Paris 25: 539-545. 1919 (obit., bibl., b. 29 Aug 1833, d. 29 Sep 1919).
Merrill, E.D., B.P. Bishop Mus. Bull. 144: 151-152. 1937 (Polynes. bibl.), Contr. U.S. natl. Herb. 30(1): 243. 1947 (Pacif. bibl.).

COMPOSITE WORKS: Contibuted to H. Baillon, *Dict. bot.* 1876-1892, 4 vols., see TL-2/253 (items signed "P.").

EPONYMY: *Poissonia* Baillon (1870). *Note*: *Poissoniella* Pierre (1890) is very likely named for Poisson.

8118. *Recherches sur les Casuarina et en particulier sur ceux de la Nouvelle-Calédonie* ... Paris (Typographie Lahure ..) 1876. Qu. (*Rech. Casuarina*).
Publ.: 1876 (BSbF post 10 Apr 1877), p. [1]-56, *pl. 4-7* (Faguet, col liths.). *Copies*: G, L. – Reprinted or preprinted from Nouv. Arch. Mus. Hist. nat., Paris 10: 59-111, *pl. 4-7*. (Copy reprint with journal pagination: USDA).
Ref.: Anon., Bull. Soc. bot. France 23 (Bibl.): 233-234. 1877.

8119. *Étude sur le nouveau genre Hennecartia de la famille des Monimiacées* ... Paris (Société d'Imprimerie et Librairie administratives et classiques Paul Dupont ...) 1885. Qu. (*Hennecartia*).
Publ.: 1885 (post 23 Jan in journal; BSbF 24 Jul 1885), p. [i], [1]-6, [7,expl. pl.], *1 pl*. (partly col. lith.). *Copy*: USDA. – Reprinted from Bull. Soc. bot. France 32-40. 1885, see J.J. Swart, ING 2: 802. Eponymous collector: Jules Hennecart (1797-1888), see IH 2: 269.
Ref.: Engler, A., Bot. Jahrb. 7: 93. 28 Mai 1886.

8120. *Sur trois espèces cactiformes d'Euphorbes* de la côte occidentale d'Afrique ... (Extrait du Bulletin du Muséum d'histoire naturelle 1902, no. 1, p. 60). Oct.
Publ.: Mar 1902 (p. 3), p. [1]-3. *Copy*: U. – Reprinted and to be cited from Bull. Mus. Hist. nat. 1902(1): 60-62. Mar 1902.

Poiteau, Pierre Antoine (1766-1854), French botanist and botanical artist; gardener at the Jardin des Plantes, Paris 1790; from 1796-1801 at Haiti; subsequently at the botanical garden of the École de médecine and (1855) Chef des Pépinières (Versailles); from 1819-1822 in French Guiana; 1820 professor at an agricultural college in Fromont; ultimately active as editor for the Société d'Horticulture. (*Poit.*).

HERBARIUM and TYPES: P, PC; important set also at G; further material at: B (Willd.), BR, E, F, FI, G, GH, H, K, L, LE, MO, NA, NEU, NY, PH, W. – For mss. *Florule de Saint-Domingue* see Bureau (1897), p. 26. – Portrait at G.

BIBLIOGRAPHY and BIOGRAPHY: Backer p. 659; Barnhart 3: 95; Blunt p. 180, 181, 219; BM 4: 1591-1592; CSP 4: 969-970, 12: 581, 17: 946; DU 228; Frank 3 (Anh.): 78; GF p. 71; Herder p. 471; Hortus 3: 1201; IH 2: (in press); Jackson p. 128, 421; Kew 4: 321; Langman 589; Lasègue p. 222-223, 266-268, 579 [index]; LS 21042-21043; MW p. 390; NI 1553-1554, see also index 2: 204 and 3: 80; PR 2467, 7236-7238, 7641 (ed. 1: 8118-8123); Rehder 5: 680-681; RS p. 134; Stevenson p. 1254; TL-1/523 [index]; TL-2/990, 1349, 1548, 1752, 3141-3142, 4069, 4071, 4952, 4959, 6745, 7065; Tucker 1: 562; Urban-Berl. p. 310, 384, 415; Zander ed. 10, p. 702-703, ed. 11, p. 801 (b. 23 Mar 1766, d. 27 Feb 1854).

Poiteau, P.A. 2454
BIOFILE: Anon., Gartenflora 3(9): 307-308. 1854 (d.).
Baer, W. & H.W. Lack, Pflanzen auf Porzellan 62. 1979.
Bureau, É., Nouv. Arch. Mus. ser. 3. 9: 1-94. 1897 (portr., bibl., main biography).
Candolle, Alph. de, Phytographie 440. 1880 (coll.).
Candolle, A.P. de, Mém. souvenirs 512. 1862.
Decaisne, J., Gartenflora 3: 307-308. 1854 (obit., from Flora des serres); Revue hort. ser. 4. 3: 115-120. 1854 (obit.).
Ewan, J., in J.F. Leroy, ed., Bot. franç. Amér. nord 38. 1957.
Guillaumin, A., Les fleurs des Jardins xxxviii, *pl. 13*. 1929 (portr.).
Hedge, I.C. & J.M. Lamond, Index coll. Edinb. herb. 119. 1970 (material at E).
Herder, F. v., Bot. Centralbl. 55: 292. 1893 (material at LE).
Jessen, K.F.W., Bot. Gegenw. Vorz. 262, 404, 467. 1864.
Kukkonen, I., Herb. Christ. Steven 80. 1971.
Lacroix, A., Figures de Savants 3: 132, 157.
Liron d'Airoles, J. de, Revue hort. 39: 456-458. 1867 (biogr. notice).
Meyer, F.G. & S. Elsasser, Taxon 22: 386. 1973 (material at NA).
Poiret, J.L.M., Encycl. méth., Bot. 8: 750-753. 1808 (itin.).
Robinet, M., Mém. Soc. imp. centr. Agr. 1856: 142-173. (q.v. for Poiteau's association with the Almanach du Bon Jardinier).
Sagot, P., Ann. Sci. nat. Bot. ser. 6. 10: 368, 374. 1880.
Sargent, C.S., Silva 2: 75. 1891 (Alexandre Poiteau 1766-1850), almost certainly incorrect; not a different person).
Steinberg, C.H., Webbia 32: 35. 1977 (material at FI).
Stieber, M.T., Huntia 3(2): 124. 1979 (mss. notes Agnes Chase).
Urban, I., Symb. ant. 1: 93. 1898, 3: 10, 103-106. 1902 (b. 23 Mar 1766, d. 27 Feb 1854, itin. Haiti, biogr. refs., coll.).

COMPOSITE WORKS: (1) Duhamel du Monceau, H.L., *Traité arbr. fruit.* ed. 2, 1806-1835, 6 vols, see TL-2/1548.
(2) Editor, *Revue horticole* sér. 1-4, 1832-1856.
(3) With A. Risso, (q.v.): *Hist. nat. Orangers* 1818-1820, ed. 2. 1872.
(4) Illustrated J.L.A. Loiseleur-Deslongchamps, *Fl.-gén. France* 1828 [1829], see TL-2/4959.

8121. *Monographie du genre Hyptis* de la famille des Labiées et qui a des rapports d'une part avec le Basilic, le Plectranthus et de l'autre avec la Cataire [Paris 1806]. Qu.
Publ.: 1806, p. [1]-19, *5 pl. Copy*: FI. –Reprinted and to be cited from Ann. Mus. Hist. nat. 7: 459-477, *pl. 27-31*. 1806.

8122. *Flora parisiensis* secundum systema sexuale disposita, et plantarum circa Lutetiam

sponte nascentium descriptiones, icones, characteres tum genericos, tum specificos, synonymiam, selectam, nomina vernacula, et usum cum locis natalibus exhibens; auctoribus A. Poiteau et P. Turpin. Tomus primus. Lutetiae Parisiorum (sumptibus F. Schoell, ...) 1808 [1813]. 8 parts. Fol. †. (*Fl. paris.*).
Co-author: Pierre Jean François Turpin (1775-1840).
Publ.: in 8 parts, 1808-1813, each containing 6 plates, p. [i-viii], [1]-36, *pl. 1-5, 7-10, 12-15, 17-20, 22-26, 28-32, 34-35, 37-41, 45-53, 59, 61, 68, 74-75* (col. copp. auct.). *Copies*: G (2 copies, first as above, dated 1808, col. pl.; second dated 1813, with uncol. pl.), NY (col.), PH (uncol.), USDA (1813, uncol.); IDC 5644. – The original covers carried the title in french: "*Flore parisienne*, contenant la description des plantes qui croissent naturellement aux environs de Paris; ..." Paris (from fasc. 2: Chez F. Schoell, ... et chez les auteurs ...). 1808-1809 (fasc. 7; cover fasc. 8 n.v.). – PR ed. 1 mentions issues with uncoloured plates, with coloured plates and 12 copies with coloured plates on "grand colombier vélin". – The copy at PH has p. "45-48" instead of 33-36, on smaller size paper.

livraison	plates	dates
1	1, 2, 5, 8, 22, 23	18 Jan 1808
2	4, 10, 13, 30, 48, 51	Jan 1808
3	3, 9, 14, 31, 39, 45	Feb 1808
4	19, 24, 29, 46, 49, 75	Mai 1808
5	15, 26, 34, 41, 50, 74	Jul 1808
6	18, 35, 47, 52, 59, 68	Jun 1809 [v. Dec? 1808]
7	7, 17, 28, 32, 38, 53	30 Oct 1809
8	12, 20, 25, 37, 40, 61	Feb 1813

Another 100 drawings at the Paris Muséum (ms. 113-114) were intended for this book.

8123. *Jardin botanique de l'École de médecine de Paris*, description abrégée des plantes qui y sont cultivées; ... Paris (Chez Méguignon-Marvis, Libraire, ...) 1816. Duod. (*Jard. bot. École méd.*).
Publ.: 1816, p. [i]-xiii, (xv-xvi), [1]-278, chart. *Copy*: NY.

Poivre, Pierre (1719-1786), French plant collector and administrator, travelling and collecting in China (1739-1745), Indochina, the Moluccas, the Philippines and Madagascar, for the French East India Company from 1748-1757, in France as agriculturist 1757-1767; from 1767-1773 "intendant" of Réunion and Île de France; from 1773 again in France (*Poivre*).

HERBARIUM and TYPES: P (also in P-JU). – Manuscripts at P and in Angers and Lyon. – Portrait at G.

BIBLIOGRAPHY and BIOGRAPHY: Backer p. 455; Barnhart 2: 95 (b. 17 Aug 1719, d. 6 Jan 1786); Bossert p. 313; Bret. p. 117-118; Lasègue p. 56, 316, 449, 504; PR ed. 1: 8125-8126, see also p. 359; SK 1: 409-410 (portr.), 4: xcvi, 8: lxxvi (itin., sec. refs.); TL-2/1: see P. Commerson.

BIOFILE: Abonnenc, E. et al., Bibl. Guyane franç. 1: 213. 1957.
Beer, G. de, The sciences were never at war 20-22. 1960.
Fournier, P., Voy. décr. sci. mission. natural. franç. 1: 88-90, 92, 2: 50. 1932.
Gager, C.S., Brooklyn Bot. Gard. Rec. 27(3): 303. 1938.
Jessen, K.F.W., Bot. Gegenw. Vorz. 472. 1864.
Lacroix, A., Figures de savants 3: 253 [index], 4: 33. 1938.
Magnin, A., Bull. Soc. bot. Lyon 31: 34, 47. 1906, 32: 111. 1907, 35: 21, 67. 1910.
Malleret, L., Proc. R. Soc. Mauritius 3(1): 117-130. 1968 (introd. Chinese pl. in Mauritius), Pierre Poivre, Paris, 1974 (Publ. École franç. Extr.-Orient 92), 723 p., 4 pl. (main biogr; portr., very detailed bibl. of secondary references).
Nicolas, J.P., *in* G.H.M. Lawrence, ed., Adanson 1: 23, 26. 1962.

Stafleu, F.A., *in* L'Heritier, Sert. angl., facs. ed. 1963, p. lix, lxxv, 8, Linnaeus and the Linnaeans 275. 1971.
Steenis, C.G.G.J. van, Fl. males. Bull. 23: 1675. 1969.
Verdoorn, F., ed., Chron. bot. 1: 21. 1935, 2: 471. 1936.

EPONYMY: *Poivrea* Commerson ex L.M.A.A. Du Petit-Thouars (1811).

OEUVRES COMPLÈTES: 1797, *Oeuvres complètes* de P. Poivre précédé de sa vie [par P.S. Dupont de Nemours], Paris (Fuchs) 1797. Oct., iv, 310 p. (see Abonnenc 1957).

Pokorny, Alois (1826-1886), Bohemian-born Austrian botanist; studied at the University of Wien 1844-1848; worked at the natural history museum (W) 1848-1849; from 1849-1886 high school teacher in Vienna, until 1864 at the academic Gymnasium, then director of the Realgymnasium of Wien-II; Dr. phil. Göttingen 1855; id. and habil. Wien 1857; lectured phytogeography at the University 1857-1868; developed nature printing with von Ettingshausen. (*Pokorny*).

HERBARIUM and TYPES: W (rd. 1887 by the Institute for Plant Physiology), contains also the herbarium of his brother Franz Pokorny (1809-1873); further material at IB/IBF, WRSL and WU (888 lichens) (inf. Gutermann). – Correspondence with D.F.L. von Schlechtendal at HAL.

BIBLIOGRAPHY and BIOGRAPHY: AG 4: 203-204; Barnhart 3: 95; Bossert p. 313; CSP 4: 970-971,8: 641, 11: 41, 17: 946; DTS 1: 226, xxii; Fischer nos. 75, 78; Futák-Domin p. 479; GR p. 457-458; Herder p. 191, 192; Jackson p. 4, 50, 56, 57, 86, 263, 267; Kanitz 283; Kew 4: 331; Klášterský p. 146; LS 21044-21052; Maiwald p. 162, 204, 237; Morren ed. 10, p. 30; MW p. 390; NI 1551, see also 1: 248-250; Nordstedt p. 26; PR 2756, 7239-7241; Rehder 5: 681; Saccardo p. 129; TL-2/1723, 1729; Tucker 1: 562-563; WU 23: 39-42.

BIOFILE: Anon., Bonplandia 4: 134, 195-196, 206-208 (extensive report on Physiotyp. pl. austr.) 222-223, 283, 369, 372, 397. 1856 (demonstration nature prints), 6: 289, 245, 380. 1858, 8: 204. 1860; Bot. Centralbl. 32: 27. 1887 (herb. to WU, "eine der grössten Privatsammlungen Oesterreichs); Bot. Zeit. 21: 248. 1863; Österr. bot. W. 2: 69. 1852 (promoted to "wirklichen Gymnasiallehrer"); Österr. bot. Z. 9: 234, 1859, (trip to Hungary), 25: 105. 1875, 37: 72. (death by apoplexy), 335 (herb. to WU). 1887; Termész. közl. 20: 472. 1888.
Beck, G., Bot. Centralbl. 34: 150. 1888 ("Crypt. austriac." at W).
Burgerstein, A., Österr. bot. Z. 37: 77-80. 1887 (b. 23 Mai 1826, d. 29 Dec 1886); Verh. zool.-bot. Ges. Wien 37: 673-678. 1887 (obit., bibl., herb.).
Gilbert, P., Comp. biogr. lit. entom. 300. 1977.
Kanitz, A., Magy. növen. Lap. 11: 94-95. 1887 (obit., b. 23 Mai 1826, d. 29 Dec 1886).
Neilreich, A., Verh. zool-bot. Ver. Wien 5: 61. 1855.
Oborny, A., Verh. naturf. Ver. Brünn 21(2): 14, 15. 1882.
Pluskal, F.S., Verh. zool.-bot. Ver. Wien 6: 369. 1856.
Skofitz, A., Österr. bot. Z. 13: 209-217. 1863 (portr., bibl.).
Verdoorn, F., ed., Chron. bot. 2: 22. 1936.
Voss, W., Versuch einer Geschichte der Botanik in Krain 1754-1883, 1: 57. 1884.
Wittrock, V.B., Acta Horti Berg. 3(2): 84. 1903.
Zittel, K.A. von, Hist. geol. palaeontl. 372. 1901.

COMPOSITE WORKS: (1) With Constantin Freiherr von Ettingshausen, q.v.:
Physiotyp. pl. austr., [1855]-1856, 6 vols., TL-2/1723, reissue and ser. 2. 1873, TL-2/1729.
– A set of the very rare 1856 original issue of the text and plates (*500*) is at USDA:
Text: Jan 1856, p. [i*-ii*], [i]-lxvii, [1]-276, *pl. 1-30*.
Plates: *1*: 1856, p. [i-x], *pl. 1-100* (details see TL-2/1723, 1729).
2: 1856, p. [i-iv], *pl. 101-200*.
3: 1856, p. [i-iv], *pl. 201-300*.
4: 1856, p. [i-iv], *pl. 301-400*.
5: 1856, p. [i-xi], *pl. 401-500*; the work was completed by 10 Apr 1856 when it was presented as a whole to the Akad. Wiss. (Bonplandia 4: 195. 1856).

(2) Pokorny, published an *"Illustrierte Naturgeschichte des Pflanzenreiches. Für die unteren Classen der Mittelschulen,"* as part of an *Illustrirte Naturgeschichte der drei Reiche*, ed. 1, which went through at least 22 editions (ed. 22, Leipzig 1903, by K. Fritsch). Italian translation: *Storia illustrata del regno vegetabile*, translated by T. Caruel, ed. 1, ed. 6, 1896.

EPONYMY: *Pokornya* Montrouzier (1860).

8124. *Die Vegetationsverhältnisse von Iglau*. Ein Beitrag zur Pflanzengeographie des böhmisch-mährischen Gebirges ... Wien. (In Commission bei W. Braumüller, ...) 1852. Oct. (*Veg.-Verh. Iglau*).
Publ.: 1852 (p. vi: 12 Feb 1852; Österr. bot. W. 31 Mar 1853; Bot. Zeit. 30 Sep 1853), p. [i]-viii, [1]-64, map, 6 charts. *Copies*: B, G, NY.
Ref.: S., Österr. bot. W. 3: 101-102. 31 Mar 1853.

8125. *Vorarbeiten zur Kryptogamenflora von Unter-Oesterreich*. I. Revision der Literatur. Nebst einer systematischen Aufzählung sämmtlicher in der vorhandenen Literatur angeführten Kryptogamen aus Unter-Oesterreich ... Wien (Gedruckt bei Karl Ueberreuter) 1854. Qu. (*Vorarb. Krypt.-Fl. U.-Oesterr.*).
Publ.: 1854, p. [1]-136. *Copy*: MO. – Reprinted or preprinted from Verh. zool.-bot. Ver., Wien 4: 35-168. 1854; the journal was received only in 1855 by *Flora* (38: 240. 21 Apr 1855) and BSbF (post 6 Apr).

8126. *Über die Nervation der Pflanzenblätter* mit besonderer Berücksichtigung der Österreichischen Cupuliferen ... Wien (aus der kaiserlich-königlichen Hof– und Staatsdruckerei) 1858. Qu. (*Nerv. Pfl.-Blätt.*).
Publ.: 1858, p. [i], [1]-32 (59 figs. in "Naturselbstdruck"). *Copy*: G. – Reprinted from Progr. k.k. akad. Gymn. Wien for 1858.

8127. *Beitrag zur Flora des ungarischen Tieflandes* [Wien 1860]. Qu.
Publ.: 1860 (mss. submitted 4 Apr 1860), p. [1]-8 and [283]-290. *Copy*: NY. – Reprinted, and to be cited from Verh. zool.-bot. Ver., Wien 10: 283-290. 1860.

8128. *Ueber das Wandern der Pflanzen*. Zwei Vorträge gehalten im Vereine zur Verbreitung naturwissenschaftlicher Kenntnisse in Wien am 13. u 20. Jänner 1862 ... Wien (In Commission bei Carl Gerold's Sohn) 1863. Oct. (*Wandern Pfl.*).
Publ.: 1863, p. [1]-53. *Copy*: G.

8129. *Plantae lignosae imperii austriaci*. *Österreichs Holzpflanzen* Eine auf genaue Berücksichtigung der Merkmale der Laubblätter gegründete floristische Bearbeitung aller im Österreichischen Kaiserstaate wild wachsenden oder häufig cultivirten Bäume, Sträucher und Halbsträucher ... Wien (Druck und Verlag der k.k. Hof– und Staatsdruckerei) 1864. Qu. (*Österr. Holzpfl.*).
Publ.: Jan 1864 (p. viii: 15 Nov 1863; ÖbZ 1 Feb 1864; Flora rd. 1-26 Feb 1864; Bot. Zeit. 14: 192. 14 Mar 1864; publ. Jan 1864), p. [i]-viii, [1, h.t.], ix-xxviii, [1, h.t.], [1]-524, [1, h.t.], *pl. 1-80* (*1-79* nature prints of 1640 leaves; *80* chart). *Copies*: B, BR, G, MO, NY. – Skofitz (ÖbZ Jul 1863) speaks about "das eben fertig gewordene Werk"; however, there is no indication that actual publication took place before 1 Jan 1864. – See Fischer no. 78.
Ref.: Seemann, B, J. Bot. 3: 29-31 Jan 1865.
Anon., Österr. bot. Z. 14: 163-165. 1 Mai 1864 (rev.), Flora 47: 359-364. 1864 (rev.).
Schlechtendal, D.F.L. von, Bot. Zeit. 22: 247-248. 1864 (rev.).
Regel, E., Gartenflora 15: 124-125. 1866.

8130. *Ueber den Ursprung der Alpenpflanzen* ... Wien (Verlag des Vereines [zur Verbreitung naturwissenschaftlicher Kenntnisse] Wien 1868. Oct. (*Urspr. Alpenpfl.*).
Publ.: 1868, p. [1]-26. *Copy*: Fl. – For a later publication with the same title see CSP 8: 641.

Polakowsky, Helmuth (1847-1917), German botanist; until 1880 in Berlin. (*Polak.*).

HERBARIUM and TYPES: material from Costa Rica at B (mainly destroyed), BM, C,

GOET, L, QK, W (301), WRSL (lich.). Bryophyte types were in herb. Müller (B, destroyed).

BIBLIOGRAPHY and BIOGRAPHY: Barnhart 3: 96; BL 1: 147, 313; CSP 11: 41, 12: 582, 17: 947-948; GR p. 121; Herder p. 225; Hortus 3: 1201 (Polak.); Kew 4: 322; LS 21054-21054a; Morren ed. 10, p. 7; Rehder 5: 681; SBC p. 129; Tucker 1: 563; Urban-Berl. p. 268, 280, 289, 310, 384.

BIOFILE: Anon., Hedwigia 59: (141), 1917 (d. 23 Jan 1917).
Beck, G., Bot. Centralbl. 34: 150. 1888 (material at W).
Hemsley, J.B., Biol. centr. amer. 4: 137. 1887.
Murray, G., Hist. coll. BM(NH) 1: 174. 1904 (pl. BM).
Sayre, G., Bryologist 80: 515. 1977.

EPONYMY: *Polakowskia* Pittier (1910).

8131. *La flora de Costa Rica*. Contribución al estudio de la fitogeografía centro-americana por el Dr. H. Polakowsky. Traducido del alemán por Manuel Carazo Peralta y anotado por H. Pittier. San José de Costa Rica (Tip. Nacional) 1890. (*Fl. Costa Rica*).
Annotations: Henri François Pittier (1857-1950).
Translator: Manuel Carazo Peralta.
Publ.: 1890 (Bot. Centralbl. 27 Mai 1891; Nat. Nov. Jun(2) 1891; MO copy signed by author 9 Nov 1891; TBC 10 Feb 1892; BSbF 12 Jun 1891),p. [1]-76, [1, err.]. *Copies*: BR, G, MO, NY. – Reprinted from Anal. Inst. fis.-geogr. nac. 2: 177-201. 1890. – Original text: *Die Pflanzenwelt von Costa-Rica*, Jahresber. Ver. Erdk. Dresden 16 (Wiss.): 25-124. Dec 1879 (map). See also Polakowsky, Linnaea 41: 545-598. 1877 (enum. pl.), Verh. bot. Ver. Brandenburg 19: 58-78. 20 Oct 1877 (id.).
Ref.: Schiffner, V., Bot. Centralbl. 52: 413-414. 1892.

Pole-Evans, Illtyd (Iltyd) **Buller** (1879-1968), British-born mycologist; M.A. Cantab.; South Africa 1905; mycologist to Transvaal Govt. 1905-1911; Chief Div. Mycology & Plant Pathology, Dept. Agr. S. Afr. 1911-1918; Director Botanical Survey S. Afr. 1918-1939; L.L.D. hon. Witwatersrand 1933. (*Pole-Evans*).

HERBARIUM and TYPES: PRE and PREM; other material at A, B, BOL, E, EA, GRA, K, L, MO, S, SRGH, US. – Letters at G.

BIBLIOGRAPHY and BIOGRAPHY: Bossert p. 314 ("Illtyd"); Desmond, p. 499 ("Iltyd", b. 3 Sep 1879, d. 16 Oct 1968); Hortus 3: 1201 (Pole-Evans); IH (ed. 7): 340; LS suppl. 22002-22011; Stevenson p. 1254; Urban-Berl. p. 275, 298.

BIOFILE: Anon., Bull. misc. Inf. Kew 1933: 304 (hon. degree Witwatersrand), 1939: 669 (retirement); The Public Servant Oct 1939 (n.v.).
Bullock, A.A., Bibl. S. Afr. Bot. 38-39, 91. 1978.
Guillarmod, A.J., Fl. Lesotho 54. 1971.
Gunn, M.D., Bothalia 10(2): 130-135. 1971 (portr., b. 3 Sep 1879, d. 16 Oct 1968).
Gunn, M.D. & L.E. Codd, Bot. expl. S. Afr. 196, 283-285. 1981 (detailed info).
Hedge, I.C. & J.M. Lamond, Index coll. Edinb. herb. 119. 1970 (material at E).
Levyns, M.R.B., Insnar'd with flowers 22. 1977.
Phillips, E.P., S. Afr. J. Sci. 27:65, 69, 70. 1930.
Pole-Evans, I.B., Fl. pl. S. Afr. 20: front. portr. 1940 (portr.) (dedication volume).
Verdoorn, F., ed., Chron. bot. 1: 250. 1935, 2: 27, 34, 375. 1936, 3: 27, 28, 155, 230, 232. 1937, 5: 295, 553. 1939.
White, A. et al., Succ. Euphorb. 1: 54-55. 1941.
White, A. & B.L. Sloane, The Stapelieae 205 [index]. 1933, ed. 2, p. 20 [index]. 1937.

COMPOSITE WORKS: Founder and editor of *Bothalia* 1921-1939.

EPONYMY: *Polevansia* B. de Winter (1966), ING 3: 1384.

8132. *The flowering plants of South Africa.* A magazine containing hand-coloured figures with descriptions of the flowering plants indigenous to South Africa ... London (L. Reeve & Co., Ltd., ...), South Africa (The Speciality Press of South Africa, Ltd. ... Johannesburg ... Capetown) 1921-1942, vols. 1-25. (*Fl. pl. S. Afr.*).
Publ.: volumes 1-19 were edited by Pole-Evans, 20-24 by Edwin Percy Phillips, 25-36 by Robert Allen Dyer (1900-x), 37-42 by Leslie Edward W. Codd (1908-x), 43-x by Donald Joseph Boomer Killick (1926-x). Authors of texts indicated separately. *Copies*: BR, MICH, PH (original covers with dates). The volumes came out in fascicles of 10 plates each with dated covers. Plates coloured.

1: 1921 (covers: Nov 1920-Aug 1921), p. [i-vi], *pl. 1-40* w.t., [i, index].
2: 1922 (covers Jan-Oct 1922), p. [i-iii], *pl. 41-80* (id.), [i].
3: 1923 (covers: Jan-Oct 1923), p. [i-iii], *pl. 81-120* (id.), [i].
4: 1924 (covers: Jan-Oct 1924), p. [i-iii], *pl. 121-160* (id.), [i].
5: 1925 (covers: Jan-Oct 1925), p. [i-iii], *pl. 161-200* (id.), [i].
6: 1926 (covers: Jan-Oct 1926), p. [i], *pl. 201-240* (id.), [i].
7: 1927 (covers: Jan-Oct 1927), p. [i-iii], *pl. 241-280* (id.), [i]. *Imprint* Ashford, Kent (L. Reeve & Co., ...), South Africa (id.).
8: 1928 (covers: Jan-Oct 1928), p. [i], *pl. 281-320* (id.), [i].
9: 1929 (p. [iii]: Oct 1929), p. [i-iii], *pl. 321-360* (id.), [i].
10: 1930 (covers: Jan-Oct 1930), p. [i-iii], *pl. 361-400* (id.), [i].
11: 1931 (covers: Jan-Oct 1931), p. [i-iii], *pl. 401-440* (id.), [i].
12: 1932 (covers: Jan-Oct 1932), p. [i-iii], *pl. 441-480* (id.), [i].
13: 1933 (covers: Jan-Oct 1933), p. [i-iii], *pl. 481-520* (id.), [i].
14: 1934 (covers: Jan-Oct 1934), p. [i-iii], *pl. 521-560*, [i].*Imprint*: Ashford Kent (L. Reeve & Co., Ltd., ...), South Africa (J.L. van Schaik Ltd. ... Pretoria).
15: 1935 (covers Jan-Oct 1935), p. [i-iii], *pl. 561-600*, [i].
16: 1936 (covers: Jan-Oct 1936), p. [i-iii], *pl. 601-604*, [i].
17: 1937 (covers: Jan-Oct 1937), p. [i-iii], *pl. 641-680*, [i].
18: 1938 (covers: Jan-Oct 1938), p. [i-iii], *pl.681-720*, [i].
19: 1939 (covers: Jan-Oct 1939, but preface note Nov 1939), p. [i-iii], *pl. 721-760*, [i].
20: 1940 (covers: Jan-Oct 1940), p. [i-ii], portr. Pole-Evans, [iii], *pl. 761-800*, [i].
21: 1941 (covers: Jan-Oct 1941), p. [i-ii], portr. Schönland, [iii], *pl. 801-840*, [i].
22: 1942 (covers: Jan-Oct 1942), p. [i-ii], portr. R. Marloth, [iii], *pl. 841-880*, [i].
23: 1943 (covers: Jan-Oct 1943), p. [i-ii], portr. H. Bolus, [iii], *pl. 881-920*, [i].
24: 1944 (covers: Jan-Oct 1944), p. [i-ii], portr. F.Z. v.d. Merwe, [iii-v], *pl. 921-960*, p. [1]-24.

Polgár, Sándor (1876-1944), Hungarian lichenologist; Dr. phil. Budapest 1913. (*Polg.*).

HERBARIUM and TYPES: DE (12.000); further material at BHU, BP, GB and GH. – Letters at G.

BIBLIOGRAPHY and BIOGRAPHY: Backer p. 659; GR p. 660 (b. 13 Sep 1876, d. Jul 1944, Auschwitz); Hirsch p. 234; Kew 4: 322; SBC p. 129.

BIOFILE: Bässler, M., Wiss. Z. Humboldt Univ. Berlin, Math.-Nat. 19(2/3): 299. 1970 (pl. BHU).
Sayre, G., Bryologist 80: 515. 1977.
Verseghy, K., Feddes Repert. 68(1): 125. 1963.

Polívka, Frantisek (1860-1923), Czech botanist; high school teacher at Olomouc (Olmütz). (*Polívka*).

HERBARIUM and TYPES: Unknown; the large Polívka herbarium at PR is of Jaroslav Polívka (1893-x).

BIBLIOGRAPHY and BIOGRAPHY: Barnhart 3: 96; Futák-Domin p. 483; Klášterský p. 147 (q.v. for further biogr. refs.).

8133. *Názorná květena* zemí koruny české obsahující též čelnější rostlini cizozemské, pěstvonané u nás pro užitek a okrasu ... Olomouc (Nákladem knihkupectvi R. Promberga) 1900-1904, 4 vols. Oct. (*Názorná květ.*).
1: 1904 (p. v: zacatkem prosince 1903), p. [i-v], [1]-476.
2: 1900, p. [i], [1]-682.
3: 1901, p. [i], [1]-620.
4: 1902, p. [i], [1]-712.
Copies: G, MICH.

Pollacci, Gino (1872-1963), Italian botanist; assistant at the Botanical Garden, Pavia (1896); habil. ib. 1903; lecturer at the botanical institute Pavia 1910-1913, at the cryptogamic laboratory ib. 1913-1915; in the Italian army 1915-1919; interim-director bot. inst. and garden 1919-1920; extraordinary professor of botany at Sassari 1920-1921; id. at Siena (also dir. bot. gard.) 1921-1927; from 1927-1942 director of the botanical institute and cryptogamic laboratory Pavia (*Pollacci*).

HERBARIUM and TYPES: PAV; further material at BRL, K, L, MIPU, O, PAD, PC, ROPV, S SIENA, W. – *Exsiccatae*: *Miceti patogeni* (fasc. 1-10, nos. 1-100), sets at FH, K, PC; *Fungi longobardiae exsiccatae* (no further details known to us). – Letters at G.

BIBLIOGRAPHY and BIOGRAPHY: Barnhart 3: 96; BL 2: 363, 369, 703; BM 8: 1019; Bossert p. 314; CSP 17: 952; Kew 4: 334; LS 21069-21075, suppl. 22016-22028; NI 1: 259; Rehder 5: 681; Saccardo 1: 130; Stevenson p. 1254; TL-2/6635.

BIOFILE: Anon., Bot. Centralbl. 67: 256. 1896 (appointed at Pavia), 93: 80. 1903 (habil.); Bot. Jahrb. 2 (Beibl. 55): 12. 1896 (appointed at Pavia); Nat. Nov. 18: 442. 1886 (asst. Pavia), 25: 520. 1903 (habil. Pavia); Österr. bot. Z. 69: 224. 1920 (succeeds Briosi at Pavia), 71: 71. 1922 (app. Siena).
Arditti, J. et al., ed., Orchid biol. 146, 151, 160, 164, 174. 1977.
Cavillier, F.G., Boissiera 5: 73. 1941.
Ciferri, R., Taxon 1: 127-128. 1952; Atti Ist. bot. Univ. Lab. critt. Pavia ser. 5. 21: 25-38. 1964 (obit., portr., bibl., b. 23 Mai 1872, d. 21 Oct 1963).
Ferro, G., Arch. biogeogr. ital. 39(3): 152-164. 1963 (obit., portr., bibl.).
Gager, C.S., Brooklyn Bot. Gard. Rec. 27(3): 280, 288. 1938.
Verdoorn, F., ed., 1: 199, 200. 1935,2: 372. 1936, 3: 187, 190. 1937, 4: 90. 1938, 5: 262. 1939.

EPONYMY: *Pollaccia* Baldacci et Ciferri (1937).

8134. *Micologia ligustica* ... Genova (Tipografia di Angelo Ciminago ...) 1897. Oct.
Publ.: 1897 (p. 6: Mai 1896; Nat. Nov. Jul(1, 2) 1897), p. [1]-112. *Copy*: FH. – Reprinted from Atti Soc. ligust. Sci. nat. 7(4): 283-350. 1896 and 8(1): 94-134. 1897. To be cited from journal.

8135. *Botanica farmaceutica* ... Milano (Casa editrice Dottor Francesco Vallardi) 1939. Oct. (*Bot. farm.*).
Co-author: [Siro] Luigi Maffei (1879-?).
Ed. 1: 1939, p. [i]-x, [1]-579. *Copies*: FI, USDA.
Ed. 2: 1944, p. [i]-xiii, [1]-621. *1 pl.* (col.). *Copy*: Natl. Libr. Med., U.S.A. – 2; edizione riveduta et aggiornata.
Ed. 3: 1949 (printed Apr 1949), p. [i]-xv, [1]-642, [643, date printing]. *Copy*: USDA.

Pollard, Charles Louis (1872-1945), American botanist; AM Colombia Univ. 1894; curator of plants U.S. natl. Mus. 1895-1903. (*Pollard*).

HERBARIUM and TYPES: US; duplicates at A, CU, E, F, GH, K, MIN, MISSA, MO, MSC, NY, POM. – Confusion is possible with material collected by Henry Minter Pollard (1886-1973). – Letters at G, NY, PH, US.

BIBLIOGRAPHY and BIOGRAPHY: Barnhart 3: 96 (b. 29 Mar 1872, d. 16 Aug 1945); BM 4:

1593; Bossert p. 314; CSP 17: 953-954; Hortus 3: 1201 (Pollard); IH 2: (in press); Langman p. 589; Lenley p. 331-332; LS 21076; NW p. 55; Pennell p. 616; PH p. 513; Rehder 5: 681; SIA 189, 201, 221, 222, 242, 422; TL-2/781.

BIOFILE: Anon., Bot. Centralbl. 78: 288. 1899 (sec. Wash. bot. Club).
Ewan, J., Southw. Louisiana J. 7: 33, 35. 1967.
McVaugh, R., Edward Palmer 112, 186, 346. 1956.
Moldenke, H.N., J. New York Bot. Gard. 42: 38. 1941 (material at NY).
Murrill, W.A., Hist. found. bot. Florida 19. 1945.
Rickett, H.W., Index Bull. Torrey bot. Club 79. 1955.
Urban, I., Symb. ant. 1: 131. 1898, 5: 10. 1904.

COMPOSITE WORKS: *North American Flora, Calycanthaceae, in* 22(3): 237-238. 12 Jun 1908.

8136. *The families of flowering plants* ... Supplement to the Plant World vols. iii, iv, and v. 1900-1902. Washington, D.C. (The Plant World Company) [1900-1902]. (*Fam. fl. pl.*). *Publ.*: 1900-1902, p. [i]-vii, [1]-253. *Copies*: NY, US, USDA. – The issues of the Plant World are given in parentheses in the following table.

Pages	dates	pages	dates
1-6	Jan 1900 (3, 1)	85-91	Dec 1900 (3, 12)
7-14	Feb 1900 (3, 2)	93-99	Jan 1901 (4, 1)
15-22	Mar 1900 (3, 3)	189-195	Jan 1902 (5, 1)
23-28	Apr 1900 (3, 4)	197-203	Jan 1902 (5, 2)
29-35	Mai 1900 (3, 5)	205-210	Jan 1902 (5, 3)
37-43	Jun 1900 (3, 6)	211-218	Jan 1902 (5, 4)
45-51	Jul 1900 (3, 7)	219-225	Jan 1902 (5, 6)
53-59	Aug 1900 (3, 8)	227-234	Jan 1902 (5, 7)
61-67	Sep 1900 (3, 9)	235-240	Jan 1902 (5, 8)
69-76	Oct 1900 (3, 10)	241-246	Jan 1902 (5, 10)
77-84	Nov 1900 (3, 11)	247-253	Jan 1902 (5, 11)

Pollexfen, John Hutton (1813-1899), British physician and algologist; MD Edinburgh 1835, B.A. Cantab. 1843; ordained 1844, from then on clergyman at Colchester, East Wetton and from 1874-1899 vicar at the parish of Middleton Tyas. (*Pollexf.*).

HERBARIUM and TYPES: BM; further material at CGE, E and U; fossil algae at Woodwardian Museum, Cambridge.

BIBLIOGRAPHY and BIOGRAPHY: Barnhart 3: 96; BB p. 246 ("recte Pollexsen"); Desmond p. 499.

BIOFILE: Anon., Bot. Not. 1900: 47 (death).
Batters, E.A.L., J. Bot. 37: 438-439. 1899 (obit., d. 5 Jun 1899)
Dickinson, C.I., Phycol. Bull. 1: 13. 1952
Hedge, I.C. & J.M. Lamond, Index coll. Edinb. herb. 119. 1970 (algae at E).

EPONYMY: *Pollexfenia* Harvey (1844).

Pollich, Johann Adam (1740-1780), German botanist and physician at Kaiserslautern; Dr. med. Strassbourg 1763; from 1764 devoting himself solely to natural history. (*Poll.*).

HERBARIUM and TYPES: Unknown.

BIBLIOGRAPHY and BIOGRAPHY: ADB 26: 393; AG 2(1): 317; Backer p. 456; Barnhart 3: 96; BM 4: 1593; Dryander 3: 156, 546; Frank 3(Anh.): 78; GR p. 37, cat. p. 69; Hawksworth p. 185; Hegi 4(1): 224; Herder p. 187; Hortus 3: 1201 (Pollich); Kelly p. 178-179; Kew 4: 334; Krempelh. 1: 481; LS 21078; PR 7247 (ed. 1: 8129); Rehder 1:

374; SBC p. 129; SO add. 675a; Zander ed. 10, p. 703, ed. 11, p. 801 (b. 1 Jan 1740, d. 24 Nov 1780).

BIOFILE: J., Dict. Sci. med., Biogr. méd. 6: 469. 1824.
Blättner, Dr. Johann Pollich und seine Zeit, Mitt. Saarpfälz. Ver. Naturk. Natursch. Pollichia ser. 2. 8. 1940.
Jessen, K.F.W., Bot. Gegenw. Vorz. 371. 1864.
Jung-Stilling, Rhein. Beitr. Gelehrsamkeit 1: 397-413. 1780 (n.v.), repr. Jahresber. Pollichia 22: 1-11. 1866 (b. 1 Jan 1740, d. 24 Dec 1780).
Kirschleger, F., Fl. Alsace 2: xliii-xliv. 1857.
Meusel, J.G., Lex. teut. Schriftst. 10: 495-496. 1810 (b. 1 Jan 1740, d. 24 Feb 1780).
Schultz-Bipontinus, C.H., Jahresber. Pollichia 22: 11-18. 1866 (add. biogr. details).

EPONYMY: (genera): *Polichia* Schrank (1781, *nom. rej.*); *Pollicha* Cothenius (1790, *orth. var.* of *Polichia* Schrank); *Pollichia* Medikus (1784, *nom. rej.*); *Pollichia* Willdenow (1787, *orth. var.* of *Polichia* Schrank); No etymology is given for *Pollichia* W. Aiton (1789, *nom. cons.*); (journal): *Pollichia*, ein naturwissenschaflicher Verein der Rheinpfalz (title varies). Dürkheim. Vol. 1-x. 1843-x.

8137. Johannis Adami Pollich ... *Historia plantarum in Palatinatu electorali* sponte nascentium incepta, secundum systema sexuale digesta ... Mannhemii [Mannheim] (apud Christ. Frid. Schwan ...) 1776-1777, 3 vols. Oct. (*Hist. pl. Palat.*).
1: 1776 (p. x: 28 Nov 1774), p. [i]-xxxii, [1]-454, *2 pl.* (uncol. copp. Egb. Verhelst).
2: 1777, p. [1]-664, *1 pl.*
3: 1777, p. [1]-320, *1 pl.*, [1-16, index].
Copies: B, BR, G, L, MICH, MO, NY, US; IDC 7133.

Pollini, Ciro (1782-1833), Italian botanist and physician; curator of the Verona botanical garden; teacher at the Verona Lyceum. (*Pollini*).

HERBARIUM and TYPES: VER; further material at C, GE, IBF, L, PAV. – Letters at G.

BIBLIOGRAPHY and BIOGRAPHY: AG 2(1): 42, 5(2): 207, 12(3): 365; Backer p. 456; Barnhart 3: 96 (b. 27 Jan 1782, d. 1 Feb 1833); BL 2: 413, 416, 703; BM 4: 1593; Bossert p. 314; Cesati p. 315-317; CSP 4: 976-977; De Toni 1: c; DTS 1: 227; GR p. 520, cat. p. 69; Hegi 5(2): 1448; Herder p. 471; Hortus 3: 1201 (Pollini); Jackson p. 323, 324, 438; Kew 4: 334-335; KR 1: 506, 549, 612; LS 21079-21083; PR 7248-7253 (ed. 1: 8130-8144); Rehder 5: 682; RS p. 134; Saccardo 1: 130-131 (b. 27 Jan 1782, d. 1 Feb 1833), 2: 85-86, cron. p. xxix; SBC p. 129; SO 798; Tucker 1: 563; Urban-Berl. p. 384.

BIOFILE: Anon., Bonplandia 4: 48. 1856 (Massalongo chooses cogn. Leop. Pollini); Flora 3: 633. 1820, 16: 288. 1833 (death).
Candolle, Alph. de, Phytographie 440. 1880 (coll. VER).
Candolle, A.P. de, Mém. souvenirs 145. 1862.
Hausmann, F. v., Fl. Tirol 1184. 1854.
Hoppe, D.H., Flora 3(1): Beil. 1/2: 21-22. 1820, 6(2). Beil. 3: 99. 1823 (bibl.).
Laundon, J.R., Lichenologist 11: 16. 1979 (lich. VER).
Sayre, G., Bryologist 80: 515. 1977.

EPONYMY: *Pollinia* K.P.J. Sprengel (1815, *nom. rej.*); *Pollinia* Trinius (1833) and the derived names *Pollinidium* Stapf ex Haines (1924) and *Polliniopsis* Hayata (1918).

8138. *Elementi di botanica* compilati da Ciro Pollini ... Verona (dalla Tipografia Moroni) 1810-1811, 2 vols. Oct. (*Elem. bot.*).
1: late Dec 1810 (t.p. 1810, p. [4]: 23 Dec 1810), p. [1]-392, *pl. 1-11* (uncol. copp., auct.).
2: 1811, p. [1]-530, *pl. 1-9 [10]* (id.).
Copies: G, NY.

8139. *Catalogo delle piante dell'orto botanico veronese* per l'anno 1814 con un cenno di varie

piante nuove. In Verona (dalla Tipografia Mainardi) 1814. Oct. (*Cat. piante orto veron.*).
Publ.: 1814, p. [1]-34. *Copy*: FI.

8140. *Saggio di osservazioni* e di sperienze sulla vegetatione degli *alberi* ... Verona (Tipografia Bisesti) 1815. Oct. (*Sagg. osserv. alb.*).
Publ.: 1815 (p. iv: 20 Mar 1815), p. [i-iv], [1]-160, [1-2, list publ. Pollini], [1, colo.], [1, err.]. *Copies*: FI, MO.

8141. *Horti et provinciae veronensis plantae* novae vel minus cognitae quas descriptionibus et observationibus exornavit Cyrus Pollinius. Fasciculus primus ... Ticini [Pavia] (ex Tipographia haered. Petri Galeatti) 1816. Qu. †. (*Hort. veron. pl.*).
Publ.: 1816, p. [1]-39, [40], *1 pl.* (uncol. copp. C. Pollini/V. Freddi). *Copies*: G, NY. – Reprinted from Giorn. Fis. Chim. Stor. nat. Pavia 9: 21-35, 94-101, 174-187. 1816.

8142. *Viaggio al Lago di Garda* e al Monte Baldo in cui si ragiona della cose naturali di quei luoghi aggiuntovi un cenno sulle curiosità del Bolca e degli altri monti Veronesi ... in Verona (dalla Tipografia Mainardi) 1816. Oct. (*Viagg. Lago di Garda*).
Publ.: 1816 (but cover p. 4 dated 1817), p. [1]-152, *1 pl. Copies*: FH, FI, G. – *Commentary*: Cenomio Euganeo, *Osservazioni intorno al Viaggio al Lago di Garda e al Monte Baldo del dottor Ciro Pollini*, s.l. 1817, p. [1]-76, [1, err.] (*copies*: FH, G) followed by "*Risposto di Eleuterio Benacense* alle Osservazioni di Cenomio Euganeo intorno al Viaggio al Lago di Garda e al Monte Baldo del dottor Ciro Pollini. Timepoli 1817, p. [1]-76, [1, err.). *Copy*: G.

8143. *Sulle alghe* viventi nelle terme Euganee con un indice delle piante rinvenute sui colli Euganei e un'appendice sopra alcune alghe della provincia Veronese. Lettera del sig. Ciro Pollini al sig. conte Francesco Rizzo Patarolo. Milano (coi Tipi di Giovanni Pirotta) 1817. (*Sulle alg.*).
Publ.: 1817, p. [1]-24, *1 pl.* (uncol.). *Copies*: FH, FI, NY. – Reprinted from Biblioteca italiana 7: 432 seq. (n.v.).

8144. *Sopra la teoria della riproduzione vegetale* del sig. Gallesio aggiuntevi alcune osservazioni fisiologiche. Lettera del sig. dottore Ciro Pollini al sig. conte Francesco Rizzo Patarolo a Venezia. Milano (presso Giuseppe Maspero in Santa Margherita) 1818. Oct. (*Teor. riprod. veg.*).
Publ.: 1818 (p. 5: 15 Aug 1816), p.[1]-24. *Copies*: FI, NY.

8145. *Flora veronensis* quam in prodromum florae Italiae septentrionalis ... Veronae (typis et expensis Societatis typographicae) 1822-1824, 3 vols. Oct. (*Fl. veron.*).
1: Nov-Dec 1822 (t.p.; p. xxxv: 5 Mai 1821, ded. 9 Nov 1922), p. [i*-iv*], [i]-xxxv, [1]-535, *pl. 1-2* (uncol. copp., auct.).
2: Nov-Dec 1822 (p. iii: 5 Nov 1822), p. [i-iii], [1]-754, *pl. 1-6* (id.).
3: Dec 1824 (fide Mém. Soc. Linn. Paris 4: (51). 1825/26.), p. [1]-898, *pl. 1-4* (id., Ronzani del.).
Copies: B, BR, FH, FI, G, MO, US, USDA.
Ref.: Moretti, G., Intorno alla Flora veronensis ... Milano, 1822, 36 p., TL-2/6301.
V., Flora veronensis ... [a review in "Antologia, fascicolo no. 56"], 9 p. *Copy*: G.

Polscher, W. (1831-1861), German (Prussian) high school teacher at Duisburg. (*Polscher*).

HERBARIUM and TYPES: Unknown.

BIBLIOGRAPHY and BIOGRAPHY: BM 4: 1593; PR 7255 (b. 21 Nov 1831, d. 27 Apr 1861).

8146. *Anleitung zur Bestimmung der* in der Umgegend von Duisburg wachsenden *Gräser* und Verzeichniss der daselbst vorkommenden Cruciferen, Umbelliferen, Compositen, Labiaten, Juncaceen und Cyperaceen. Aus dem Nachlasse des verstorbenen Reallehrers W. Polscher ... Duisburg (Gedruckt bei Joh. Erwich) [1861]. Oct. (*Anleit. Best. Gräs.*).
Publ.: Sep-Dec 1861 (p. iv: 27 Aug 1861), p. [i]-iv, [1]-28. *Copy*: MO. – "Abhandlung

zum Programm des königl. Gymnasiums und der Realschule zu Duisburg. Herbst 1861.

Polunin, Nicholas Vladimir (1909-x), British botanist, ecologist and conservationist; D.Sc. Oxon.; demonstrator and lecturer in botany Oxford 1938-1947; McDonald professor of botany, McGill Univ. 1947-1952; director Baghdad University Herbarium 1955-1958; guest prof. bot. Geneva 1959-1961; professor of botany Univ. of Ife, Nigeria 1962-1966. (*Polunin*).

HERBARIUM and TYPES: private; duplicates e.g. in BR, BUH, CAN, DS, E, FH, GH, K, MICH, MO, NY, OXF, US. – Letters at G.

BIBLIOGRAPHY and BIOGRAPHY: Barnhart 3: 96; BFM no. 1598, 1628; BL 1: 136, 137, 145, 313, 2: 222, 457, 460, 478, 703; Bossert p. 314 (b. 26 Jun 1909); Clokie p. 227; GR p. 410; Hortus 3: 1201 (Polun.); Kleppa p. 191, 268, 289, 294, 301; Kew 4: 335-336; Langman p. 590; PH 209.

BIOFILE: Anon., Nature 160: 600. 1947 (app. McGill).
Blakelock, R.A. & E.R. Guest, Fl. Iraq 1: 115. 1966.
Hedge, I.C. & J.M. Lamond, Index coll. Edinb. herb. 119. 1970.
Kay, E., ed., Dict. int. biogr. 11(2): 1325. 1975 (b. 26 Jun 1909).
Maeijer, E.A. de, ed., Who is who in Europe ed. 3. 2429. 1972.
Mains, E.B., *in* W.A. Donelly et al., The University of Michigan 1454. 1956.
Polunin, N.V; Bull Torrey bot. Club 77: 214-221. 1950, Taxon 2(2): 25-26. 1953 (specific and trivial decapitalization); Circumpolar arctic flora, Oxford 1959 (see Taxon 8: 275. 1959; Bot. Jahrb. 79: 11-13. 1960).
Sayre, G., Bryologist 80: 515. 1977.
Schuster, R.M., Hepat. Anthoc. N. Amer. 1: 88. 1966.
Verdoorn, F., ed., Chron. bot. 2: 190. 1936, 3: 293, 350. 1937, 4: 274, 279, 455, 564. 1938, 5: 117, 174, 288, 295, 302, 307, 508. 1939, 6: 22, 166, 167, 301. 1941, 7: 133, 232, 235. 1943.

8147. *Botany of the Canadian Eastern Arctic* ... [Ottawa (Canada Department of Mines and Resources) ...] 1940-1948. Oct. (*Bot. Canad. E. Arctic*).
1: 1940 (p. iii: 1 Jun 1938), p. [i]-vi, [1]-498, *pl. 1-8*; Natl. Mus. Canada, Bull. 92.
2: 1947 (p. iv: 20 Nov 1946), p. [i]-v, [1]-573, *pl. 1-18* (uncol.); Bull. 97.
3: 1948 (p. [iii]: 9 Mai 1947), p. [i]-vii, err. slip, [1]-304, *pl. 1-105*; Bull. 104.
Copies: MICH, NY, US, E.G. Voss.

Pomata y Gisbert, Eladio (*fl.* 1880), Spanish botanist. (*Pomata*).

HERBARIUM and TYPES: Unknown.

BIBLIOGRAPHY and BIOGRAPHY: Barnhart 3: 96; BM 4: 1594; Colmeiro 1: cxciv; CSP 11: 45; GR p. 763; Herder p. 177; Rehder 1: 422.

8148. *Catálogo de plantas* recolectadas al estado espontáneo en la provincia *de Toledo* ... [Anales de Historia natural XI] 1882. Oct.
Publ.: 1882, p. [1]-66, *3 pl.* (uncol.). *Copies*: BR, USDA. – Reprinted, and to be cited from Anal. Soc. Esp. Hist. nat. 11: 241-306. 1882.
Appendix: 2 Mai 1883, p. 67-86, reprinted from id. 12: 221-240. 1883.

Pomel, Auguste Nicolas (1821-1898), French botanist and geologist; deported to Algeria 1852; in various administrative functions 1856-1870; in various political posts in Algeria and Paris 1870-1880; in the "Service" of the geological mapping of Algeria 1882, director 1885-1895. (*Pomel*).

HERBARIUM and TYPES: AL; further material at LY and MPU.

BIBLIOGRAPHY and BIOGRAPHY: Andrews p. 313; Backer p. 459 (epon.); Barnhart 3: 96

(b. 20 Sep 1821, d. 2 Aug 1898); BM 4: 1594; CSP 4: 978-979, 8: 643, 11: 45, 17: 956-957; Hortus 3: 1201 (Pomel); IH 1 (ed. 6): 363, (ed. 7): 340, 2: (in press); Jackson p. 184, 348; Kew 4: 337; Morren ed. 10, p. 112; PR 7257; Quenstedt p. 342 (b. 21 Sep 1821); Rehder 5: 682; Tucker 1: 563; Zander ed. 10, p. 703, ed. 11, p. 801.

BIOFILE: Anon., Bot. Not. 1899: 48 (d.); Nat. Nov. 20: 475. 1898 (d.); Österr. bot. Z. 48: 471. 1898 (d.).
Battandier, J.A., Bull. Soc. bot. France 45: 205-208. 1898, 46: 281. 1899.
Cosson, E., Comp. fl. atl. 1: 77. 1881.
Cosson, E. & M.C. Durieu, Exp. sci. Alg. Bot. 2: xli. 1868.
Ficheur, E., Bull. Soc. géol. France ser. 3. 27: 191-223. 1899 (obit., bibl.).
Granel de Solignac, L. & L. Bertrand, Naturalia monsp. 18: 284. 1967 (coll. MPU).
Kneucker, A., Allg. bot. Z. 5: 16. 1899 (death).
Maire, R., Progr. conn. bot. Alger 125-132, *pl.* 7. 1931 (portr., discussion priority of Pomel names over those of Cosson).
Nissen, C., Zool. Buchill. 322-323. 1969 (b. 21 Sep 1821, d. 2 Aug 1898).
Poggendorff, J.C., Biogr.-lit. Handw.-Buch 3: 1057. 1898 (b. 20 Sep 1821), 4(2): 1183. 1904 (d. 20 Aug 1898).
Taton, R. et al., Science in the 19th century 341. 1965.
Ward, Ann. Report U.S. Geol. Surv. 5: 420. 1885.
Wolff, C., C.R. Acad. Sci. Paris 127: 1056-1057. 1898 (obit., b. 20 Sep 1821, d. 2 Aug 1898).
Zittel, K.A. von, Hist. geol. palaeontol. 214, 387, 420. 1901.

EPONYMY: *Pomelia* Durando ex Pomel (1860).

8149. *Matériaux pour la flore atlantique* ... [Caen (Imprimerie Dedebant et Alexandre, ...) 1860]. (*Mat. fl. atl.*).
Publ.: 1 Mar 1860 (BSbF), p. [1]-16. *Copies*: FI, G. ("une feuille très rare, Battandier, 1899).
Ref.: Fournier, E., Bull. Soc. bot. France 9: 673. 1864.

8150. *Nouveaux matériaux pour la flore atlantique* ... Paris (Savy, ...), Alger (Juillet St.-Lager, ...) 1874. Qu. (*Nouv. mat. fl. atl.*).
Part 1: 1874 (p. iii: 10 Mar 1874; J. Bot. Jun 1875), p. [i*, iii*], [i]-iii, [1]-260. Reprinted from Bull. Soc. Climatol. Alger 11, 1874.
Part 2: 1875 (cover dated 1875; BSbF r. 24 Mar 1876; Bot. Zeit. 13 Oct 1876), p. 257 [sic]-399. Reprinted from ib. 13, 1876.
Copies: B, BR, G, NY, USDA (part 1); IDC 5306.
Ref.: Anon., Bull. Soc. bot. France 21 (Bibl.): 211-214. 1875.

8151. Seconde Thèse. *Contributions à la classification méthodique des crucifères* [Alger (Typographie Adolphe Jourdan) 1882]. Qu. (*Contr. classif. crucif.*).
Publ.: Nov-Dec 1882 (or perhaps early 1883) (p. 22: Vu et approuvée: le 11 novembre 1882), p. [i], [1]-22, [24, expl. pl.], *1 pl. Copy*: WU (inf. W. Gutermann). – Thesis Paris, in order to be authorized to accept an official appointment in Algeria.

Ponce de Leon y Aimé, Antonio (1887-1961), Cuban botanist; Dr. Sci. Habana 1906; high school teacher until 1934, then professor of botany at Habana University; founder of the Sociedad Cubána de botánica (1945). (*A. Ponce de Leon*).

HERBARIUM and TYPES: HAJB; further material at SV.

BIBLIOGRAPHY and BIOGRAPHY: Barnhart 3: 97; Bossert p. 314 (b. 23 Jan 1887, d. 22 Feb 1961); IH 1 (ed. 2): 57, (ed.2): 69, (ed. 3): 74; Kew 4: 3381; Langman p. 590; Roon p. 89.

BIOFILE: Alain, Brother, Taxon 11: 64-65. 1962 (portr.).
Stafleu, F.A., Taxon 10: 126. 1961 (b. 23 Jan 1887, d. 22 Feb 1961).

COMPOSITE WORKS: Editor, *Revista de la Sociedad Cubana de Botánica*.

Ponce de Leon, José (*fl.* 1814), Spanish botanist. (*Ponce de Leon*).

HERBARIUM and TYPES: Unknown.

BIBLIOGRAPHY and BIOGRAPHY: Colmeiro penins. no. 116.

8152. *Sistema floro-sexual de botanica* por D. José Ponce de Leon. Granada (Imprente de D. Nicolas Moreno). 1814. Oct. (*Sist. fl.-sex. bot.*).
Publ.: 1814, p. [i-xvi], [1], 3-34, [1]-412. *Copy*: NY.

Poneropoulos, Eustathios (*fl.* 1880), Greek botanist. (*Ponerop.*).

HERBARIUM and TYPES: Some material from Greece at P (rd. 1875). – Letters at G.

BIBLIOGRAPHY and BIOGRAPHY: Barnhart 3: 97; IH 2: (in press); Jackson p. 488; Rehder 5: 682.

8153. *Stoicheia botanikes* ... en Athenais (ek tou typographeiou tes Philokalias) 1880. Oct. (*Stoich. bot.*).
Publ.: 1880 (preface 17 Aug 1879; BSbF Apr-Jul 1880; Bot. Zeit. 5 Nov 1880; Nat. Nov. Nov(2) 1880), p. [i]-xiii, [1]-432. *Copy*: MO.*Ref.*: Anon., Bull. Soc. bot. France 27 (bibl.): 60-61. 1880.

Pontarlier, Nicolas Charles (1812-1889), French botanist; studied at the Lycée Saint-Louis, Paris; at École norm. sup. 1831; régent Collège des Palmiers 1833, chargé de cours, Coll. r. Bourbon-Vendée 1839; régent Col. Vannes; at Lycée de Napoleon-Vendée, La Roche-sur-Yon 1848-1878. (*Pontarl.*).

HERBARIUM and TYPES: some material at AK, E and G. – According to Souché (1901) P's herbarium is at the library of Roche-sur-Yon, with a second set at the "Lycée" of that same town.

BIBLIOGRAPHY and BIOGRAPHY: Barnhart 3: 98; BL 2: 210, 703; GR p. 346 (b. 12 Feb 1812, d. 20 Apr 1889); Rehder 5: 682; Tucker 1: 564.

BIOFILE: Anon., Bull. Soc. bot. Deux Sèvres 23: 248, *pl. 2.* 1912 (portr., b. 12 Feb 1812, d. 20 Apr 1889).
Goulding, J.H., Rec. Auckland Inst. Mus. 12: 114. 1975.
Hedge, I.C. & J.M. Lamond, Index coll. Edinb. herb. 119. 1970.
Souché, B., Flore du Haut Poitou xvi. 1901 (herb.).

NOTE: Published, with Henri Nicolas Marichal (1812-1886), *Catalogue des plantes vasculaires et spontanées du Département de la Vendée* in Rev. Sci. nat. Ouest. 4: 37-64. Jan-Mar 1894, 107-136. Apr-Dec 1894 and 5: 26-45. Jan-Mar 1895, 102-124. Apr-Dec 1895. We saw no reprint with independent pagination; however, a reprint in one volume of 100 p. is mentioned by E. Malinvaud, Bull. Soc. bot. France 43: 649. 1896.

Pontén, Johan (Jon) Peter (1776-1857), Swedish botanist, physician and clergyman; Dr. phil. Greifswald 1800, ordained 1801, parish priest at Hultsjö from 1826, at Korsberga from 1837. (*Pontén*).

HERBARIUM and TYPES: UPS (see Löwegren 1952).

BIBLIOGRAPHY and BIOGRAPHY: Barnhart 3: 98; BM 4: 1595; CSP 4: 984; Herder p. 141; KR p. 574-575 (b. 2 Mai 1779 (sic, misprint for 1776), d. 23 Sep 1857); MW p. 391; PR 7267, 9292; Rehder 1: 33.

BIOFILE: Lindman, S., Svenska män och kvinnor 6: 150. 1949 (portr.).
Löwegren, Y., Naturaliekab. Sverige 361. 1952 (herb. of 4000 phanerogams at UPS).
Wittrock, V.B., Acta Horti Berg. 3(3): 48. 1905 (b. 2 Mai 1776, d. 23 Sep 1857).

COMPOSITE WORKS: Defended C.P. Thunberg *Dissertatio botanica de Hydrocotyle* 1798, (also in Thunberg, Diss. Acad. 2: 410-418. 1800), see under Thunberg.

8154. Dissertatio philosophico-botanica *de serie vegetabili*, quam venia ampliss. facult. philos. gryph. ... pro gradu philosophico proposuit auctor Johannes Petr. Pontén, smolandia-suecus, in aud. maj. die xviii sept mdccc. h.a.m.s. Gryphiae [Greifswald] (litteris I.H. Eckhardt, ...) [1800]. Qu. (*Ser. veg.*).
Publ.: 18 Sep 1800, p. [i-ii], [1]-14. *Copy*: Lund University Library (inf. O. Almborn).

Pool, Raymond John (1882-1967), American botanist; Dr. phil. Univ. Nebraska 1913; professor of botany and chairman, Department of Botany, University of Nebraska 1915-1948. (*Pool*).

HERBARIUM and TYPES: NEB; other material at MO, NY, WTU.

BIBLIOGRAPHY and BIOGRAPHY: Barnhart 3: 98; BJI 2: 138; BL 1: 195, 313; Bossert p. 314 (b. 23 Apr 1884, d. 2 Feb 1967); Ewan ed. 1: 284, ed. 2: 177 (Ph.D. 1908); Hirsch p. 235; Kelly p. 179; Kew 4: 339; LS 38012, suppl. 22046; Pennell p. 616; PH 209.

BIOFILE: Anon., Hedwigia 57: (151). 1916 (app. Univ. Nebraska succeeding C.E. Bessey); Plant Science Bull. 13(1): 8. 1967 (d. 2 Feb 1967).
Ewan, J. et al., Short hist. bot. U.S. 125. 1969.

8155. *The Erysiphaceae of Nebraska* [Lincoln, Nebraska, 1910]. Oct.
Publ.: Jan 1910, p. [1]-26. *Copy*: USDA. − Reprinted with double pagination from Univ. Nebraska 10(1): 59-84. 1910.

8156. *Handbook of Nebraska trees* a guide to the native and most important introduced species ... [The University of Nebraska] 1919.
Ed. 1: Mar 1919, p. [1]-171, *73 pl. Copies*: MICH, USDA. − Nebraska Conservation and Soil Survey, Bulletin 7.
Ed. 2: Mai 1929, p. [1]-179, *77 pl. Copies*: MICH, USDA.

8157. *Flowers and flowering plants* an introduction to the nature and work of flowers and the classification of flowering plants ... New York, London (McGraw-Hill Book Company, Inc.) 1929. (*Fl. pl.*).
Ed. 1: 1929 (p. xii: Oct 1929), p. [i]-xx, 1-378. *Copies*: MO, USDA.
Ed. 2: 1941 (p. ix: Apr 1941), p. [i]-xxiii, 1-428. *Copy*: USDA.
Ref.: Rendle, A.B., J. Bot. 68: 223-224. Jul 1930 (cites 1930 as date of imprint).

Pop, Emil (1897-1974), Roumanian botanist; Dr. phil. Cluj 1928; habil. ib. 1932; connected with the chair of systematic botany, Cluj from 1920 (assistant 1922, professor 1939). (*Pop*).

HERBARIUM and TYPES: CL; duplicates at E. − Letters at G.

BIBLIOGRAPHY and BIOGRAPHY: BJI 2: 138; Bossert p. 315; Futák-Domin p. 484; Hirsch p. 235; IH 2: (in press); Kew 4: 340.

BIOFILE: Boşcaiau, N. & V. Soran, Studii si Cercetari de biol. 26: 213-215. 1974 (obit., d. 14 Jul 1974, portr.).
Hedge, I.C. & J.M. Lamond, Index coll. Edinb. herb. 119. 1970.
Peterfi, S., Revue roum. Biol., ser. Bot. 12(2-3): 99-123. 1967 (tribute 70th birthday, portr., b. 13 Apr 1897; bibl. 157 items by S. Færcaş).
Resmerita, I., Natura, Roumania, 24(4): 88-90. 1972 (tribute, portr.).
Sælægeanu, N., Natura, Bucuresti 19(2): 87-90. 1967 (70th birthday; b. 13 Apr 1897; portr.).
Stefureac, T.I., Rev. bryol. lichénol. 42(4): 1017-1018. 1976.
Verdoorn, F., ed., Chron. bot. 1: 245, 246. 1935, 2: 262. 1936, 3: 42, 228. 1937, 5: 509. 1939, 6: 301. 1941, 7: 225. 1943.

COMPOSITE WORKS: Scientific co-ordinator of part of the *Flora republicii socialiste România*, vols. 1-13. 1952-1976.

8158. *Flora pliocenicæ dela Borsec.* Die pliozäne Flora von Borsec (Ostkarpathen). Cluj (Tipografia nationalæ s.a.) 1936. (*Fl. plioc. Borsec*).
Publ.: Mai 1936, p. [i]-iv, [1]-189, *pl. 1-22* (uncol.). *Copies*: BR, G. – Contr. bot. Cluj 2(8), Mai 1936.
Ref.: Hofmann, E., Österr. bot. Z. 86: 235-236. 1937.

Pope, Clara Maria (née Leigh) (*fl.* 1760s-1838), British flower painter, wife of the actor Alexander Pope (Francis Wheatley). (*C. Pope*).

HERBARIUM and TYPES: Unknown. Drawings of Paeonia at BM.

BIBLIOGRAPHY and BIOGRAPHY: BB p. 247; Blunt p. 4, 208, 211-212, *pl. 37*; BM 4: 1596 (drawings); Desmond p. 500; DNB 46; DU 85; NI 122, see also 1: 436-438; TL-2/1283-1284.

BIOFILE: Bridson, G.D.R. et al., Nat. hist. mss. res. Brit. Isl. 229.404. 1980.
Britten, J., J. Bot. 56: 126-127. 1918.
Burleigh, A.B., J. Roy. Hort. Soc. 58(2): 325. 1933.
Woodward, B.B., *in* G. Murray, Hist. coll. BM(NH) 1: 45. 1904 (11 water colours of Paeonia at BM).

COMPOSITE WORKS: (1) Curtis, S., *Monogr. Camellia* 1819, TL-2/1283, (2) *Beauties Flora* 1820, TL2/1284.

Pope, Willis Thomas (1873-1961), American botanist and horticulturist; M.S. Univ. Calif., D. Sci. Univ. Hawaii; horticulturist at USDA Hawaii Experiment Station. (*Pope*).

HERBARIUM and TYPES: Unknown.

BIBLIOGRAPHY and BIOGRAPHY: Barnhart 3: 99; BL 1: 110, 313; Kew 4: 340; Lenley p. 333.

BIOFILE: Anon., Hawaii. Forester 4(12): 370-371. 1907 (appointed acting dean Coll. agric. Hawaiian terr.; teacher at Doylestown nr Philadelphia 1900-1902; id. at Honolulu Normal School 1902-1907).
Merrill, E.D., B.P. Bishop Mus. Bull. 144: 152. 1937, Contr. natl. Herb. U.S. 30(1): 244. 1947.

8159. *Manual of wayside plants of Hawaii* including illustrations, descriptions, habits, uses and methods of control of such plants as have a wild nature of growth, exclusive of ferns ... Honolulu, Hawaii (published by Advertiser Publishing Co., Ltd.) 1929. Oct. (*Man. wayside pl. Hawaii*).
Publ.: 1929, p. [1]-289. *Copies*: R.S. Cowan, MO, NY, USDA. – *160 pl.* in text.
Reprint: 1968, p. [i-iv], [1]-289. *Copy*: US. – Rutland, Vermont & Tokyo, Japan (Charles E. Tuttle Co., ...) [1968].

Popenoe, Dorothy Kate (née Hughes) (1899-1932), British-born botanist; assistant at Kew 1918-1923 as collaborator of O. Stapf; to USDA Washington 1923; married Frederick Wilson Popenoe (17 Nov 1923). (*D. Popenoe*).

HERBARIUM and TYPES: US; dupl. at F and K; types of early work (as D. Hughes) at K.

BIBLIOGRAPHY and BIOGRAPHY: Barnhart 3: 99; Bossert p. 315; Desmond p. 500; IH 2: (in press); Lenley p. 333.

BIOFILE: Adamic, L., The house in Antigua 167-171, 224-228. 1937.

Anon., Bull. misc. Inf. Kew 1933: 304 (d. 31 Dec 1932); J. Bot. 71: 320. 1933 (d. 31 Dec 1932 at Guatemala City).

Popenoe, Frederick Wilson (called Wilson Popenoe) (1892-1975), American botanist and explorer; studied at Pomona College; agricultural explorer for USDA from 1913-1923; with the United Fruit Company from 1923-1957; from 1944 as director of the Escuela Agricola Panamericana, Honduras. (*Popenoe*).

HERBARIUM and TYPES: US; duplicates at BH, G, GH, K, L, LZ (destr.), NA, W.

BIBLIOGRAPHY and BIOGRAPHY: Barnhart 3: 99; BL 1: 148, 250, 313; Bossert p. 315 (b. 9 Mar 1892); Kew 4: 340; Langman p. 591; Lenley p. 333; MW p. 391; PH 59; Tucker 1: 564.

BIOFILE: Acosta-Solis, M., Natural Viaj. Ci. Ecuador 51-52, 123. 1968.
Adamic, L., The house in Antigua, 1937, x, 300 p., *18 pl.*, (on P. and his house in Guatemala).
Ewan, J. et al., Short hist. bot. U.S. 23. 1969.
Frost, M.J. & B.I. Judd, Econ. bot. 24(4): 471-478. 1970 (1971) (biogr., portr.; description of career).
Smiley, N., Bull. Fairchild trop. Gard. 39(4): 11-12. 1975 (obit., portr., b. 9 Mar 1892, d. 20 Jun 1975).
Stieber, M.T. et al., Huntia 4(1): 85. 1981 (arch. mat. HU).
Verdoorn, F., ed., Chron. bot. 2: 324. 1936, 5: 57. 1939, 7: 16, 120, 124, 217, 229, 336. 1943.

8160. *Manual of tropical and subtropical fruits* excluding the banana, coconut, pineapple, citrus fruits, olive and fig ... New York (The Macmillan Company) 1920. Oct. (*Man. trop. fruits*).
Publ.: Jul 1920 (p. [ii]), p. [i]-xv, [1]-474, *pl. 1-24*. *Copies*: B, BR, FI, G, MO, NY, US.

Poplu, Mme **M.C.** (*fl.* 1873), French botanist. (*Poplu*).

HERBARIUM and TYPES: Unknown.

BIBLIOGRAPHY and BIOGRAPHY: BL 2: 136, 704.

8161. *Flore des Rives de la Touque* et des Falaises de Trouville ... Pont-l'Évêque (imprimerie C. Delahais, ...) 1873. Oct. (*Fl. Rives Touque*).
Publ.: 1873 (p. vii: 17 Feb 1873), p. [i]-vii, [1]-98, *3 pl.* (uncol.). *Copy*: USDA.

Popov, Mikhail Grigorievič (1893-1955), Russian botanist and explorer of the flora of the Asiatic part of the Soviet Union. (*Popov*).

HERBARIUM and TYPES: VLA, WIR and LE; further material at C, E, MO, NY, PKM, RNMUT.

BIBLIOGRAPHY and BIOGRAPHY: Hortus 3: 1201; IH 1 (ed. 6): 363, (ed. 7): 340, 2: (in press); Kew 4: 340-341; Lenley p. 333; MW p. 391, suppl. p. 283-285; TL-2/3857; Zander ed. 10, p. 703, ed. 11, p. 801.

BIOFILE: Dobročaeva, D.M., Ukr. bot. Zhurn. 34(2): 205-211. 1977.
Govorukhina, V.A., Izv. Acad. nauk. Turkmen. SSR, Biol. Nauk. 3: 54-60. 1976. (portr.).
Hedge, I.C. &. J.M. Lamond, Index coll. Edinb. herb. 120. 1970.
Lipschitz, S.J., Bot. Zhurn. 41: 736-769. 1956 (obit., photogr., bibl., b. 18 Apr 1893, d. 18 Dec 1955), Florae URSS fontes 223. 1975.
Popov, M.G., Izbrannye sočineniia [opera selecta], Aschabad, Acad. Sci. TRSS, 1958, 488 p.
Rubtzov, N.I., Izv. Akad. Nauk. Kaz. SSR, Biol. 5: 15-18. 1974.
Shetler, S.G., Komarov bot. Inst. 59, 154. 1967.

COMPOSITE WORKS: (1) Contributed to Komarov, *Fl. URSS*:
(a) *Papaveracae*, 7: 573-717. 8 Jul 1937.
(b) *Leguminosae, Astralagus* (with others), 12: 1-873. 27 Sep 1946.
(c) *Boraginaceae* p.p., 19: 565-691. 5 Feb 1953.
(d) *Orobanchaceae-Mannag.*, 23: 115-116. 12 Dec 1958.
(2) *Flora srednej Sibiri* 1-2, 1957-1959, see Taxon 6: 241. 1957, 9: 30. 1960.

HANDWRITING: Lipschitz, S.J. & T.I. Vasilczenko, Central Herb. USSR 116. 1968.

EPONYMY: *Popoviocodonia* Federov (1957).

Popovici, Alexandru P. (1866-?), Rumanian botanist, professor of botany at the University of Jassi; pupil of W. Pfeffer; Dr. phil. Bonn 1893. (*Popovici*).

HERBARIUM and TYPES: I.

BIBLIOGRAPHY and BIOGRAPHY: Barnhart 3: 99; CSP 17: 967; Hirsch p. 235; IH 2: (in press); Kelly p. 179; LS 21113-21115, 38016, suppl. 22084.

BIOFILE: Anon., Jahrb. wiss. Bot. 56: 824. 1915.

8162. *Contribution à la flore cryptogamique de la Roumanie* ... Jassy (Imprimerie "Dacia" P. Iliescu & D. Grossu) 1902. (*Contr. fl. crypt. Roum.*).
Publ.: Apr 1902 (p. 14: 29 Dec 1901; journal Apr 1902), p. [1]-14. *Copy*: FH. – Reprinted from Anal. Sci. Univ. Jassy 2(1): 31-44. 1902.

8163. *Contribution à la flore mycologique de la Roumanie* ... Jassy (Imprimerie "Dacia" P. Iliescu & D. Grossu) 1903. Oct. (*Contr. fl. mycol. Roum.*).
Publ.: Oct 1903 (in journal), p. [1]-13. *Copy*: FH. – Reprinted from Ann. Sci. Univ. Jassy 3: 199-211. 1903.

8164. *Contribution à l'étude de la flore mycologique du Mont Ciahlæŭ* ... Jassy (Imprimerie "Dacia" P. Iliescu & D. Grossu) 1903. Oct. (*Contr. fl. mycol. Mt. Ciahlæŭ*).
Publ.: Sep-Dec 1903 (p. 65), p. [1]-65, [66, expl.], [1, err.]. *Copy*: FH.

Popp, Bonifaz (x-1892), German (Bavarian) botanist, Benedictine monk and high school teacher at Scheyern, Bayern. (*Popp*).

HERBARIUM and TYPES: Unknown.

BIBLIOGRAPHY and BIOGRAPHY: Barnhart 3: 99; Rehder 1: 388.

BIOFILE: Familler, I., Denkschr. bay. bot. Ges. Regensburg 11: 13, 20. 1911 (d. 16 Oct 1892).

8165. *Flora von Scheyern*. Programm der vollständigen Lateinschule im erzbischöflichen Knabenseminare zu Scheyern ... Pfaffenhofen a/Ilm (Buchdruckerei F.X. Herzogs Wwe). [1889-]1891. (*Fl. Scheyern*).
1: 1889 (?), p. [i]-xii, [1]-79, [80].
2: 1890 (?), p. [i], [81]-166.
3: 1891, p. [167]-217, map.
Copies: B, REG.

Porcher, Francis Peyre (1825-1895), American botanist; M.D. Univ. S. Carolina 1847; studied in Paris and Italy 1847-1849; practicing physician in Charleston 1849-1895. (*Porcher*).

HERBARIUM and TYPES: CHARL; duplicates at DBN, E and E-GL.

BIBLIOGRAPHY and BIOGRAPHY: Barnhart 3: 100; BL 1: 161, 213, 313; CSP 4: 987, 17:

969; IH 1 (ed. 6): 363, (ed. 7): 340; 2: (in press); Jackson p. 361; Kelly p. 179; Kew 4: 341; Lenley p. 466; LS 21119-21123; ME 1: 220, 3: 639; PH 364, 567; PR 7272; Rehder 5: 683; Tucker 1: 564.

BIOFILE: Allibone, S.A., Crit. dict. Engl. lit. 1640. 1870 (bibl.).
Anon., Appleton's Cycl. Amer. Biogr. 5: 70. 1888; Bot. Centralbl. 65: 320. 1896 (d. 20 Nov 1895); Bot. Jahrb. 21 (Beibl. 54): 31. 1896 (d.); Garden and Forest 10: 302. 1897 (d.); Österr. bot. Z. 2: 46. 159. 1896 (d. 20 Nov 1895).
Childs, A.R., Dict. Amer. Biogr. 15: 79-80. 1935 (biogr., b. 14 Dec 1825, d. 19 Nov 1895; biogr. refs.).
Ewan, J., Southw. Louisiana J. 7: 27, 29. 1967.
Ewan, J. et al., Short hist. bot. U.S. 9, 148. 1969.
Gee, W., Bull. Univ. S. Carolina 72: 46-48. 1918 (biogr., portr., bibl., b. 14 Dec 1825).
Hedge, I.C. & J.M. Lamond, Index coll. Edinb. herb. 120. 1970.
Porcher, W.P., *in* Kelly, H.A. & W.L. Burrage, Amer. med. biogr. 922-923. 1920, Cycl. Amer. med. biogr. 2: 277-288.
Kneucker, A., Allg. bot. Z. 2: 52. 1896.

8166. *A medico-botanical catalogue* of the plants and ferns *of St. John's*, Berkly, South Carolina. An inaugural thesis submitted to the dean and faculty of the medical college of the state of South-Carolina for the degree of M.D., ... Charleston (printed by Burges and James) 1847. Duod. (in sixes). (*Med.-bot. cat. St. John's*).
Publ.: 1847, p. [1]-55. *Copy*: NY. – Also as S.J. Med. Pharm. 2: 255-286, 397-417. 1847 (fide ME). See also Report on the indigenous medicinal plants of South Carolina, Trans. Amer. med. Ass. 2: 677-822. 1849.

8167. *The medicinal*, poisonous, and dietetic *properties, of the cryptogamic plants of the United States* ... New York (Baker, Godwin & Co., printers, ...) 1854. Oct.
Publ.: 1854, p. [1]-126. *Copies*: FI, MO, NY. – Reprinted and to be cited from Trans. Amer. med. Ass. 7: 167-284. 1854.

8168. *Resources of the Southern fields* and forests, medical, economical, and agricultural. Being also a medical botany of the confederate states; with practical information on the useful properties of the trees, plants and shrubs ... Charleston (Steam-power press of Evans & Cogswell, ...) 1863. Oct. (*Resour. S. fields*).
Ed. 1: 1863, p. [i]-xxv, [1]-601. *Copies*: MICH, MO, US, USDA.
Ed. 2: 1869, p. [i]-xv, [1]-733. *Copies*: USDA(2); IDC 7604. – Charleston (Walker, Evans & Cogswell, ...) 1869. Oct.

Porcius, Florian (1816-1906), Transylvanian botanist at Ó-Radna (Rodna); high school teacher in Zagra and Naszód; from 1849 in the civil service in various functions, from 1861-1877 at Naszód; in retirement at Ó-Radna. (*Porcius*).

HERBARIUM and TYPES: CL; other material at BP, E and LE. – The original herbarium also contains material collected by G. Linhart, V. Borbas, L. v. Simonkai, A. Kanitz, E. Hackel, A. Zimmeter, C. Nägeli, Peter, F. Pax, A. Degen, M. Gandoger. – Letters and portrait at G.

BIBLIOGRAPHY and BIOGRAPHY: AG 2(1): 495; Barnhart 3: 100 (b. 16 Aug 1816; err. d. 30 Mai 1907); BM 8: 1022; CSP 8: 646, 12: 584, 17: 969; Herder p. 192; IH 1 (ed. 6): 363, 2: (in press); Kew 4: 341; Morren ed. 10, p. 40; Rehder 5: 683; Tucker 1: 564.

BIOFILE: Anon., Bot. Not. 1907: 20 (d. 20 Mai 1906); Herbarium 9: 70. 1909 (herb. for sale; 7000 species).
Borza, A., Natura, Bucuresti, 18(6): 21-26. 1966 (on P's bot. terminology).
Ghişa, E., Comunicări de Botanicæ, a vii, consf. naţ geobot. 1970: 17-25. 1971 (explorer of Transilvania, portr., b. 16 Aug 1816, biogr., bibl.).
Hedge, I.C. & J.M. Lamond, Index coll. Edinb. herb. 120. 1970.
Kneucker, A., Allg. bot. Z. 12: 205. 1906 (d. 30 Mai 1906).
Pax, F., Grundz. Pfl.-Verbr. Karpathen 1: 25, 50, 55. 1898 (Veg. Erde 2(1)).

Prodán, G., Magy. Bot. Lap. 6: 204-212. 1907 (obit., portr., b. 16 Aug 1816, d. 30 Mai 1906).
Simonkai, L., Enum. fl. transilv. xxvi. 1886.

8169. *Enumeratio plantarum phanerogamicarum districtus quondam naszódiensis* ... Claudiopoli [Cluj, Klausenburg] (typ. Nic. K. Papp) 1878. Oct. (*Enum. pl. phan. naszód.*).
Publ.: 1878 (Nat. Nov. Feb(2) 1879; Flora rd. 21-31 Jan 1879), p. [i-iv], [1]-64. *Copies*: BR, FI, G, (given by A. Kanitz to E. Burnat in Mar 1879), NY, REG. – See also his Diagnosele plantelorŭ fanerogame şi criptogame vasculare, Anal. Acad. Rom. ser. 2. 14: 11-360. 1893.

8170. Academia romana. *Flora din fostulŭ districtŭ Romanescŭ alŭ Năsĕudului* in Transilvania ... Bucurescĭ (tipografia Academieĭ române ...) 1885. Qu. (*Fl. naseud.*).
Publ.: 1885, p. [i], [1]-140. *Copy*: WU (inf. W. Gutermann). Also issued as Anal. Acad. Rom. ser. 2. 7(2): 1-140. 1885.

Porsch, Otto (1875-1959), Austrian botanist (flower biologist); student of R. Wettstein; Dr. phil. Wien 1901; at the Botanical Institute, Graz 1900-1903 (with Haberlandt); 1903-1911 at the Botanical Institute of Vienna Univ.; habil. Wien 1906; from 1911-1919 at the University of Czernowitz (Cernæuţi); from 1919 at the Hochschule für Bodenkultur, Wien (from 1938 full professor); in Java Jan-Jun 1914. (*Porsch*).

HERBARIUM and TYPES: Some material at CERN.

BIBLIOGRAPHY and BIOGRAPHY: Barnhart 3: 100; BFM no. 2662; BJI 2: 138; BM 4: 1598; DTS 6(4): 186; Hirsch p. 236; Kew 4 341-342; Langman p. 592; Moebius p. 145, 360; SK 1: 412; Zander ed. 10, p. 703, ed. 11, p. 801.

BIOFILE: Anon., Biologia generalis 11: i-iv. 1935 (tribute 60th birthday, portr.); Bot. Centralbl. 85: 191. 1901 (app. Graz), 102: 192. 1906 (habil. Wien); Mycol. Centralbl. 1: 162. 1912 (app. prof. bot. Czernowitz); Nat. Nov. 23: 155. 1901 (asst. Graz), 28: 480. 1906 (habil. Wien), 34: 259 (lecturer and dir. bot. gard. Czernowitz), 411 (director inst. bot. ib.) 1912; Österr. bot. Z. 51: 38. 1901 (asst. Graz), 53: 175. 1903 (asst. Wien), 59: 127. 1909 (hon. lect. veterinary coll. Wien), 62: 103 (to Czernowitz), 247: (prof. bot., dir. bot. gard. Czernowitz) 1912; 68: 292. 1919 (Hochschule Bodenkultur Wien), 69: 272. 1920 (prof. bot. ib.), 87: 248. 1938 (had been suspended because of "völkerischer Gesinnung" (pro Nazi), now reinstated as prof. bot. Wien).
Cufodontis, G., Archivio bot. 9(3/4): 179. 1933 (on biol. exped. to Costa Rica in 1930).
Engler, A., Bot. Jahrb. 30 (Beibl. 68): 54. 1901 (ass. Graz), 32 (Beibl. 72): 5. 1903 (id. Wien).
Gager, C.S., Brooklyn Bot. Gard. Rec. 27(3): 322. 1938 (dir. bot. gard. Czernowitz 1912-1918).
Hoehne, F.C., Jard. bot. São Paulo 163. 1941 (portr.).
Janchen, E. Richard Wettstein, Österr. bot. Z. 82: 179-180. 1933 (biogr. diagnosis, b. 12 Sep 1875).
Kneucker, A., Allg. bot. Z. 7: 40. 1901 (asst. Graz), 9: 108. 1903 (id. Wien, Univ.), 15: 64. 1909 (temp. at Wien, veterinary coll.), 18: 136. 1912 (prof. at CERN).
Steenis, C.G.G.J. van, Flora males. Bull. 14: 619. 1959 (d. 2 Jan 1959).

COMPOSITE WORKS: (1) Collaborator, C.K. Schneider, *Illustriertes Handwörterbuch der Botanik*, Leipzig 1905, see Schneider.
(2) *Methodik der Blütenbiologie*, in Abderhaldens Handb. biol. Arbeitsmeth. 11(1, 4): 395-514. 1922 [Lief. 81] (see Engler, A., Bot. Jahrb. 58 (lit.): 49-50. 1923.

8171. *Die österreichischen Galeopsisarten* der Untergattung Tetrahit Reichb. Versuch eines natürlichen Systems auf neuer Grundlage ... Wien (Alfred Hölder ...) 1903. Oct. (*Österr. Galeopsis.*).
Publ.: 10 Mar 1903 (t.p.; Nat. Nov. Mai(1) 1903; Allg. bot. Z. 15 Jul 1903), p. [i-ii], [1]-

125, [126], *pl. 1-3 (1-2* col.). *Copies*: B, BR, G, NY, REG. – Published as Abh. zool.-bot. Ges. Wien 2(2), 1903.
Ref.: Ulbrich, E., Bot. Jahrb. 34 (Lit.): 69-70. 1905 (rev.).
Kneucker, A., Allg. bot. Z. 9: 105. 15 Mai 1903 (rev.).

8172. *Der Spaltöffnungsapparat im Lichte der Phylogenie.* Ein Beitrag zur "phylogenetischen Pflanzenhistologie" ... Jena (Verlag von Gustav Fischer) 1905. Oct. (*Spaltöffnungsapp. Phylog.*).
Publ.: Sep-Oct 1905 (p. x: Aug 1905; ÖbZ Oct 1905; AbZ 15 Dec 1905), p. [i]-xv, [xvi, err.], [1]-196, *pl. 1-4* (uncol.). *Copy*: U-V.

8173. *Versuch einer phylogenetischen Erklarung des Embryosackes und der doppelten Befruchtung der Angiospermen* ... Jena (Verlag Gustav Fischer) 1907. Oct. (*Vers. Embryosack. Angiosp.*).
Publ.: Oct-Dec 1907 (address given on 16 Sep 1907), p. [i]-v, [1]-49. *Copy*: FI.

Porsild, Alf Erling (1901-1977), Danish botanist, naturalized Canadian from 1939; son of M.P. Porsild; assistant botanist at the Danish biological station, Greenland 1922-1925; botanist in charge of surveys, Canadian Dept. of the Interior 1926-1935; at National Herbarium of Canada, Ottawa from 1936-1966; Dr. phil. Kjøbenhavn 1955 id. h.c. Acadia 1967, Waterloo 1973. (*Porsild*).

HERBARIUM and TYPES: CAN. – Duplicates in many herbaria, e.g. ALA, H, M. – Letters in G.

BIBLIOGRAPHY and BIOGRAPHY: Barnhart 3: 100 (b. 17 Jan 1901); BFM no. 1599; BJI 2: 138; BL 1: 132, 138, 156, 157, 313; BM 4: 100; Bossert p. 315; Hortus 3: 1201 (Porsild); IH 1 (ed. 1): 71, (ed. 2): 91, (ed. 3): 115, (ed. 4): 126, (ed. 5): 134, (ed. 6): 199, 363, (ed. 7): 340, 2: (in press); Kew 4: 342-343; Kleppa p. 217; MW suppl. p. 285; Roon p. 89; SIA 223, 227, 7176 (incl. HI 107).

BIOFILE: Anon., Canad. bot. Ass. Bull. 11(1): 10. 1978 (d.).
Böcher, T.W., Tidsskr. 73(1):62-63. 1979 (obit., portr., b. 17 Jan 1901; d. 14 Nov 1977).
Cowan, R.S., Taxon 28: 76. 1979.
Christensen, C., Danske bot. Litt. 1912-1939, p. 203-205. 1940 (portr., bibl.).
Ewan, J. et al., Short hist. bot. U.S. 24. 1969.
Hansen, A., Dansk bot. Ark. 21(1): 59. 1963.
Hultén, E., Bot. Not. 1940: 334 (Alaskan coll.).
Jörgensen, C.A. et al., Biol. Skr. Danske Vid. Sellsk. 9(4): 6, 7, 167-168. 1958.
Kornerup, T., Overs. Med. Grønland 1876-1912: 105. 1913.
Löve, A., Taxon 24: 163-164. 1975 (rev. Rocky Mt. wild fl.), Arctic Alp. Res. 12: 649-651. 1978 (obit.).
Morton, J.K., Canad. bot. Ass. Bull. 11(2): 27-32. 1978 (obit., portr., bibl., d. 13 Nov 1977).
Polunin, N. Bot. Canad. E. Arctic 1: 20. 1940.
Porsild, A.E., Rocky Mountain wild flowers, Ottawa 1974, 454 p., 258 coll. ills.
Porsild, A.E. & W.J. Cody, Vasc. pl. cont. Northw. Terr. Canada vi. l980 (portr., biogr. sketch).
Scotter, S.W., Canad. geogr. J. 97: 12-19. 1978 (on his reindeer-grazing investigations 1927-1928).
Solandt, O.M., Canad. geogr. J. 72: 182. 1966 (Massey medal 1966).
Soper, J.H. &,W.J. Cody, Can. Field-Natural. 92: 299-304. 1978 (bibl.).
Stieber, M.T. et al., Huntia 4(1): 85. 1981 (corr. C.R. Ball at HU).
Verdoorn, F., ed., Chron. bot. 6: 428. 1941.

EPONYMY: *Porsildia* A. Löve & D. Löve. 1976.

Porsild, Morton Pedersen (1872-1956), Danish botanist; mag. sci. Kjøbenhavn 1900; director of the Danish Arctic biological station at Disko from 1906 (1905)-1946. (*M. Porsild*).

HERBARIUM and TYPES: C; further material in CAN, F, GB, MO, NY, QK, S, US.

NAME: Earlier papers signed Morton Pedersen.

BIBLIOGRAPHY and BIOGRAPHY: Barnhart 3: 100 (b. 1 Sep 1872, d. 30 Apr 1956); BJI 2: 139; BL 1: 155, 157, 313, 2: 139, 408, 704; BM 8: 1022; Bossert p. 315; Hirsch p. 236; IH 2: (in press); Kew 4: 343; Kleppa p. 147; KR p. 576; Lenley p. 333; NW p. 55; SBC p. 129; SIA 7183; Tucker 1: 565; Urban-Berl. p. 384.

BIOFILE: Anon., Bot. Centralbl. 98: 560. 1905 (endowment Disko station); Bryologist 59(2): 160. 1956 (d. 30 Apr 1956); Hedwigia 45(2): (77). 1906 (Disko established).
Christensen, C., Danske Bot. Litt. 1880-1911: 165-166. 1913 (portr., bibl.), id. 1912-1939, p. 43-44. 1940 (portr., bibl.), Danske bot. Hist. 1: 760, 767, 822, 824-827, 849, 850, 852. 1926.
Ekblaw, E., Amer. Mus. J. 18: 581-589. 1918 (description of Arctic station at Godhavn, Disko).
Ewan, J. et al., Leafl. W. Bot. 7(3): 97. 1953.
Gager, C.S., Brooklyn Bot. Gard. Rec. 27(3): 191. 1938 (dir. bot. gard. Godhavn, Disko from 1906).
Gröntved, J., Bot. Tidsskr. 53: 117-119. 1956 (obit., b. 1 Sep 1872, d. 30 Apr 1956, portr.).
Hansen, A., Dansk bot. Ark. 21(1): 70. 1963.
Jörgensen, C.A. et al., Biol. Skr. Dan. Vid. Selsk. 9(4): 5, 6, 168. 1958.
Müller, D., Naturh. Tid. 20: 57-59. 1956 (obit.).
Sayre, G., Bryologist 80: 515. 1977.
Schuster, R.M., Hepat. Anthoc. N. Amer. 1: 88. 1966.
Verdoorn, F., ed., Chron. bot. 2: 335. 1936, 3: 173. 1937, 5 [see index]. 1939, 7: 188. 1943.
Wittrock, V.B., Acta Horti Berg. 3(3): 89. 1905.

8174. *List of vascular plants* from the South Coast *of the Nugsuaq peninsula* in West Greenland ... Kjøbenhavn (Bianco Lunos Bogtrykkeri) 1910. Oct. (*List pl. Nugsuaq penins.*).
Publ.: 1910 (p. 248: printed 20 Mai 1910), p. [i-ii], [239]-248. *Copies*: B, C, H, NY, WU. – Reprinted with original pagination and special cover from Meddel. Grønl. 47: [239]-248. 1910.

8175. *The flora of Disko Island* and the adjacent coast of West Greenland from 66;-71;N. Lat. with remarks on phyteogeography, ecology, flowering, fructification and hibernation ...Kjøbenhavn (Bianco Lunos Bogtrykkeri) 1926. Oct. (*Fl. Disko Isl.*).
Collaborator: Alf Erling Porsild (1901-1977).
Publ.: 1926 (err. "1920" corrected in ink to 1926, cover Medd.: 1926; Nat. Nov. Sep 1926; p. [156]: Oct 1925), p. [1]-155, [156]. *Copies*: B, BR, G, NY, US. – From Medd. Grønland 58; Arb. Danske Arkt. St. Disko 11(1).
Ref.: Kneucker, A., Allg. bot. Z. 33: 55. 1928.

8176. *Stray contributions to the flora of Greenland* I-V ... København (Bianco Lunos bogtrykkeri ...) 1930. Oct. (*Contr. fl. Greenland*).
I-V: 1930, p. [1]-44. Medd. Grønland 77; Arb. Danske Arkt. St. Disko 13.
VI-XII: 1935, p. [1]-94, *pl. 1-5.* – Medd. Grønland 93(3); Arb. Danske Arkt. St. Disko 15.
XIII-XVII: 1946, p. [1]-39, (p. 29: 27 Aug 1946). Medd. Grønland 134(4), Arb. Danske Arkt. St. Disko 16.
Copy: US.

8177. *Alien plants and apophytes of Greenland* ... København (C.A. Reitzels Forlag ...) 1932. Oct. (*Alien pl. Greenland*).
Publ.: 1932, p. [1]-85, *pl. 1-2* (maps). *Copies*: H, MO, US. – Medd. Grønland 92(1), 1932; Arb. Danske Arkt. St. Disko 14.

Porta, Pietro (1832-1923), Italian botanist and clergyman; in Vallaria 1856; later priest at Bolone, Val Vestino, Trento; explored S. Tirol, Spain, the Balearic Is. with R. Huter and G. Rigo; ultimately living in Cologna di Creto. (*Porta*).

HERBARIUM and TYPES: Original herbarium at Trento, Italy (Liceo Arcivescovile and Museo di Storia naturale); further material, often collected with G. Rigo, in many herbaria, e.g. B (extant), BHU, BM, BR, C, CGE, DBN, E, F, FI, G, GB, GE, IBF, JE, KIEL, K, L, LE, LY, M, MANCH, MPU, NAP, OXF, P, STU, TO, US, W, WU. – Letters and portrait at G.

BIBLIOGRAPHY and BIOGRAPHY: AG 3: 488, 12(2): 99, 12(3): 88 (d. 1 Jun 1923); Barnhart 3: 100; BL 2: 343, 704; BM 4: 1598; Clokie p. 227; CSP 11: 49, 17: 972; DTS 1: 229, 6(4): 186; Hegi 6(1): 96; Kew 4: 344; Rehder 5: 683; Saccardo 1: 130-131, 2: 87 (b. 5 Nov 1852), cron. p. xxix; Urban-Berl. p. 384; Zander ed. 11, p. 801.

BIOFILE: Anon., J. Bot. 29: 127. 1891 (Spanish coll.); Österr. bot. Z. 41: 187. 1891 (coll.), 46: 302. 1896 (distr. Porta-Rigo 1895 coll.), 48: 238. 1898 (further colls.; distr. by Rupert Huter).
Beck, G., Bot. Centralbl. 34: 149. 1888 (coll. at W).
Hedge, I.C. & J.M. Lamond, Index coll. Edinb. herb. 120. 1970.
Huter, R., Bot. Centralbl. 21: 185. 1885 (on Porta-Rigo-Huter coll.); Österr. bot. Z. 53-58, 1903-1908 (series of paper on plants collected by Huter, Porta and Rigo).
Porta, P., Nuovo Giorn. bot. Ital. 19: 276-344. 1888 (see also M. Willkomm, Bot. Centralbl. 36: 364-365. 18 Dec 1888), (also as reprint with special cover but orig. pagination, e.g. at G).
Willkomm, M., Grundz. Pfl.-Verbr. Iber. Halbinsel 21, 26. 1896 (Veg. Erde).

EPONYMY: The etymology of *Portaea* Tenore (1846) and *Portea* Pfeifer (1874, *orth. var.*) could not be determined. *Portalesia* F.J.F. Meyen (1834) honors Don Diego Portales, a Chilean politician of the 19th century.

8178. *Vegetabilia in itinere iberico austro-meridionali lecta* ... Rovereto (tipografia Giorgio Grigoletti) 1892. Oct. (*Veg. itin. iber.*).
Publ.: 1892, p. [i], [1]-74. *Copies*: G, WU. – Reprinted from Atti Accad. Agiati 9, 1891 (1892). (journal n.v.).

Portenschlag-Ledermayer, Franz [Edler] von (1772-1822), Austrian botanist; studied law, abandoned his career as a lawyer to practice botany; botanical explorer of the Austrian empire. (*Portenschl.*).

HERBARIUM and TYPES: W (11.700 nos.), other material at B (extant), FI, G, GJO, GZU (p.p. orig. herb.), H, MW, PAD, PR. – Letters at G.

BIBLIOGRAPHY and BIOGRAPHY: Backer p. 460; Barnhart 3: 100 (b. 13 Feb 1772, d. 7 Nov 1822); BM 4: 1598; Bossert p. 315; Frank 3 (Anh.): 78; IH 1 (ed. 7): 340; Hegi 6(1): 118; Herder p. 192; Hortus 3: 1201 (Portenschl.); Kanitz 110; Kew 4: 344; Lasègue p. 331, 332, 402, 502; PR 7275 (ed. 1: 8168); Saccardo 2: 87; Tucker 1: 565; Zander ed. 10, p. 703, ed. 11, p. 801.

BIOFILE: Anon., Flora 6: 286. 1823 (herb. to W), 15(2): 402, 403, 404. 1832 (herb.).
Beck, G., Bot. Centralbl. 33: 251. 1888, 34: 150. 1888 (herb. at W via Trattinick); *in* K. Fritsch et al. Bot. Anst. Wiens 56, 75. 1894 (herb. W); Veg.-Verh. Illyr. Länd. 5, 40. 1901 (Veg. Erde 4).
Candolle, Alph. de, Phytographie 440. 1888 (coll.).
Heufler, L. v., Flora 26: 595. 1843 (identif. herb. Joseph v. Giovanelli).
Kanitz, A., Bonplandia 10: 365. 1862 (b. 13 Feb 1772, d. 7 Nov 1822).
Kukkonen, I., Herb. Christian Steven 80. 1971 (material at H).
Lindemann, E. von, Bull. Soc. imp. Natural. Moscou 61(1): 56. 1886 (material at MW).
Neilreich, A., Verh. zool.-bot. Ver. Wien 5: 40. 1855.
Pax, F., Grundz. Pfl.-Verbr. Karpathen 1: 14. 1898 (Veg. Erde 2(1)).
Welden v., Flora 5(2) Beil.6: 81-82. 1823 (obit.).

EPONYMY: *Portenschlagia* R. de Visiani (1850); *Portenschlagiella* T.G. Tutin (1967). *Note*:The etymology of Portenschlagia Trattinick (1818) could not be ascertained but may honor this author.

8179. *Enumeratio plantarum in Dalmatia* lectarum a Francisco de Portenschlag-Ledermayer ... Zum Andenken des Verewigten von seinen Freunden. Wien (Im Verlage der Franz Härter'schen Buchhandlung) 1824. Oct. (*Enum. pl. Dalmatia*).
Authorship: The text was written by Leopold Trattinick (1764-1849), on behalf of some of Portenschlag's friends. The plates are based on plants collected by P and (except for no. 1, by author) drawn by Rochel (see p. 16). The text contains also a brief biography of P. by Trattinick.
Publ.: 1824, p. [1]-16, *pl. 1-12* (uncol.; Rochel del.). *Copies*: BR, FI, G, NY; IDC 679.
Ref.: Meyer, E., Gött. gel. Anz. 1825(196): 1959-1960. 7 Dec 1825 (rev.).

Porter, Carlos Emilio (1868-1942), Chilean naturalist; director of the zoology dept. of the Museo de Historia natural Valparaiso 1897-1910; natural history teacher at various colleges in Valparaiso and Santiago, e.g. at the Santiago military academy 1912-1918 and at the Instituto agronómico de Chile 1901-1927; director of the zoological museum of the Instituto 1914-1927, head of the invertebrate section of the Museo nacional 1912-1923, id. entomology section 1924-1927; professor of animal parasitology at the Faculty of Agronomy 1919-1939. (*C.E. Porter*).

HERBARIUM and TYPES: Some fungi at NY. – Letters at G.

BIBLIOGRAPHY and BIOGRAPHY: Barnhart 3: 100; BL 1: 246, 313; BM 4: 1598, 8: 1022; CSP 17: 972-973; Lenley p. 333.

BIOFILE: Anguita, B.F., Revista chil. hist. nat. 25: xi-xxiv. 1921 (tribute 25 yrs Revista; portr.).
Anon., Bot. Centralbl. 107: 256. 1908 (app. Santiago); Bull. Acad. Geogr. bot. 9: 12-13. 1900, 11: 10. 1902, 13: 9. 1904, 18: 8. 1909; J. econ. Entom. 36: 247. 1913 (d.); Le Monde des Plantes 14: 1. 1912 (on Naturalistas americanos); Nat. Nov. 30: 235. 1908 (app. Santiago); Österr. bot. Z. 58: 175. 1908 (app. Santiago); Physis 9(32): 155-156. 1928; Rev. Acad. colomb. Ci. exact. fis. nat. 3: 195. 1939 (obit.).
Birabén, M., Revista Mus. La Plata 1943: 133-135. 1944 (obit., portr.).
Carpenter, M.M., Amer. Midl. Natural. 33(1): 81. 1945.
Gigoux, E.E., Bol. Mus. nac. Hist. nat., Chile, 20: 107-109. 1942 (obit.).
Gilbert, P. Comp. biogr. lit. deceased entom. 301. 1977.
Howard, L.O., Smiths. misc. Coll. 84: 435-437. 1930 (entom. work).
Kneucker, A., Allg. bot. Z. 14: 68. 1908 (app. Santiago).
L.T., Revista Soc. entom. Argent. 11: 485-486. 1943 (obit.).
Larrain, A.F., Revista Entom. Rio de Janeiro 14(1-2): 321-324. 1943 (obit., portr.).
Looser, G., Rev. Argent. Agron. 10: 77-80. 1943 (obit., portr.).
Porter, C.E., Bibliografia del Prof. Cárlos E. Porter, Santiago de Chile 1913, 16 p. (bibl.); Hoja de servicios y actuacion scientifica del Prof. Dr. Carlos E. Porter, ... Santiago de Chile 1924, 7 p. (curr. vitae, publ.).
Reiche, K., Grundz. Pfl.-Verbr. Chile 41. 1907 (Veg. Erde 8).
Verdoorn, F., ed., Chron. bot. 7: 114, 351. 1943.
Zuniga, F.R., Revista chil. Hist. nat. 45: 7-9. 1943 (obit., d. 13 Dec 1942).

COMPOSITE WORKS: Editor *Revista Chilena de Historia natural*, Valparaiso, vols. 1-44, 1897-1942.

EPONYMY: *Bryoporteria* Thériot (1933); *Neoporteria* N.L. Britton et J.N. Rose (1922); *Porterula* Spegazzini (1920). *Note*: *Porterandia* Ridley (1940) commemorates George Porter(*fl.* 1800-1830), a plant collector in Penang; *Porteresia* Tateoka (1965) was dedicated to Prof. Roland Portères(1906-1974), French botanist and agronomist; *Porteria* W.J. Hooker (1851) was named in memory of Sir Robert Ker Porter (1777-1842) British Consul-General in Caracas, Venezuela (1826-1841); *Porterinema* Waern (1952) was named for Hobart C. Porter, an American botanist who specialized in the study of algae and vascular cryptogams and who first translated Strasburger's *Textbook of Botany* into English.

Porter, Cedric Lambert (1905-x), American botanist; Dr. phil. Univ. Washington 1937; instructor in botany Univ. Wyoming 1929-1932, professor of botany ib. 1932-1968. (*C.L. Porter*).

HERBARIUM and TYPES: RM; duplicates at DAO, G, GH, MICH, MIN, MO, MSC, MT, NY, OKLA, RM, TEX, UPS, WS, WTU.

BIBLIOGRAPHY and BIOGRAPHY: Barnhart 3: 101; Bossert p. 315 (b. 15 Jan 1905); Ewan (ed. 1): 284, (ed. 2): 177; Hortus 3: 1201 (C.L. Porter); IH 1 (ed. 1): 53, (ed. 2): 69, (ed. 3): 86, (ed. 4): 93, (ed. 5): 96, (ed. 7): 340; Kew 4: 344; Langman p. 592; LS 22000-22006; PH 33, 209; Roon p. 89; SIA 7231.

BIOFILE: Anon., Rev. bryol. lichénol. 5: 55. 1933, 8: 124. 1935, 9: 158, 162. 1936 (bryol. publ.).
Moldenke, H.N., List coll. Verbenac. suppl. 20. 1947.
Porter, C.L., Taxonomy of flowering plants, San Francisco, London 1959, xii, 452 p. (see Kramer, K.W., Taxon 9: 58. 1960).
Schuster, R.M., Hepat. Anthoc. N. Amer. 1: 88. 1966.
Stieber, M.T. et al., Huntia 4(1): 85. 1981 (arch. mat. HU).
Verdoorn, F., ed., Chron. bot. 2: 358, 364, 404. 1936, 3: 325, 326, 368. 1937.

Porter, Lilian E. (née Baker) (1885-1973), Irish lichenologist, daughter of John Gilbert Baker (1834-1920); M.Sc. Liverpool 1911; lecturer Bangor 1913-1916; demonstrator biol. Cork 1916-1919; examiner Nat. Univ. Ireland 1921-1945. (*L. Porter*).

HERBARIUM and TYPES: material at CRK, DBN, and GALW.

BIBLIOGRAPHY and BIOGRAPHY: GR p. 279 (b. 21 Sep 1885; orig. info., portr.), *pl.* [*34*]; LS 22114-22116; Roon p. 89.

BIOFILE: Anon., Rev. bryol. lichénol. 21: 195. 1952, 25: 410. 1956.
Cullinane, J.P., Irish Natural. J. 17(2): 48. 1971 (material at CRK).
Hawksworth, D.L. & M.R.D. Seaward, Lichenology Brit. Isl. 33, 198. 1977.
Mitchell, M.E., Bibl. Irish lichenol. 53-54. 1971 (bibl.); in litt. 8 Mai 1980 (d. 7 Mar 1973).

COMPOSITE WORKS: published a supplement to M.C. Knowles, *The Lichens of Ireland*, 1929 (TL-2/2: no. 3772) *in*: Proc. R. Irish Acad. 51 (B, 22): [345]-386. Mai 1948 (*copy*: NY).

NAME: A further important paper is *On the attachment organs of the common corticulous Ramalinae*, Proc. r. Irish Acad. 34(B2): 17-32, *pl. 2-4*. 7 Sep 1917 (inf. and copy: O. Almborn).

Porter, Thomas Conrad (1822-1901), American botanist; Dr. Div. Rutgers 1865; presbyterian clergyman, ordained 1843 or 1844, practicing 1846-1849; from 1849-1860 teaching natural sciences at Marshall and Franklin College, Mercersburg/Lancaster; from 1866-1897 professor of botany and zoology at Lafayette College, Pa. and minister at Easton, Pa. 1877-1884,; German scholar and poet. (*Porter*).

HERBARIUM and TYPES: PH; duplicates e.g. in BUF, CM, DWC, E, F, FMC(!), FI, GH, K, MASS, MICH, MO, MVSC (!), NA, NY, P, PAC, PC, PUR, RM, US, VT, W, WAB, WELC. – Porter's original herbarium was damaged – but for the greater part saved – from a fire in Pardee Hall at Lafayette College, Pa., in 1897. The herbarium was deposited by Lafayette College at PH in 1914. -Letters at G, GH, NY and PH.

BIBLIOGRAPHY and BIOGRAPHY: Barnhart 3: 101 (b. 22 Jun 1822, d. 27 Apr 1901); BL 1: 210, 313; Blankinship p. 22; BM 4: 1598-1599; Bossert p. 315; CSP 12: 584, 17: 974-975; DAB 15: 104-105 (B.W. Kunkel); further biogr. refs.);Ewan (ed. 2): 177; Herder p. 225;

Hortus 3: 1201 (T.C. Porter); IH 1 (ed. 6): 363; Jackson p. 363; Kew 4: 344; Langman p. 592; Lenley p. 334; LS 21134; ME 1: 220, 3: 639; Morren ed. 10, p. 129; NW p. 55; Pennell p. 616; PH p. 514; Rehder 5: 683; SIA 220, 233, 7183, 7231, 7256; TL-2/no. 775; Tucker 1: 13, 85, 565; Zander ed. 10, p. 703, ed. 11, p. 801.

BIOFILE: Allibone, S.A., Crit. dict. Engl. lit. 1647. 1870.
Anon., Bot. Centralbl. 87: 272. 1901 (d.); Bot. Gaz. 25: 71-72. 1898 (herb. damaged by fire), 31: 448. 1901 (d.); Bot. Jahrb. 31 (Beibl. 70): 28. 1902 (d. 27 Apr 1901); Bot. Not. 1902: 94; Meehan's Mon. 3: 79. 1893; Torreya 1: 60. 1901 (d.); 15: 19. 1915 (herb. deposited at PH).
Britton, N.L., Bull. Torrey bot. Cl. 28: 369-373. 1901.
Britton, N.L. et al., Addresses delivered at a celebration in honor of prof. Thos. Conrad Porter ... at Lafayette College, Easton, Pa. 1898, 48 p. (portr.), also published in Daily Free Press, Easton Pa., 20 Oct 1897 (copy at NY; Britton on progr. syst. bot. N. Amer., W.B. Scott on id. geology, J.M. Crawford on "Dr. Porter and Finnish literature; introd. E.D. Warfield).
Ewan, J., Trail and Timberline 305: 55-57. 1944 (portr., bibl., sources); Rocky Mt. Natural. 67-72, 80, 128, 160, 189, 263, 277, 284-285. 1950 (portr., extensive info.).
Ewan, J. et al., Short hist. bot. U.S. 92. 1969.
Groff, Bartonia 23: 48. 1945 (material at FMC).
Harshberger, J., Bot. Philadelphia 236-243. 1899 (portr., bibl.).
Heller, A.A., Plant World 4: 130-131. 1901 (obit., portr.).
Hitchcock, A.S. & A. Chase, Man. Grass. US 988. 1951 (epon.).
Kneucker, A., Allg. bot. Z. 7: 160. 1901 (death).
Meehan, T., Rhodora 3: 191-193. 1901 (obit., b. 22 Jan 1822, d. 27 Apr 1901; several details incorrect).
Pennell, F.W., Bartonia 22: 12. 1943 (herb.).
Porter, T.C., The Pennsylvania-German in the field of the natural sciences, p. 1-15, reprinted from Proc. Pennsylvania-German Society vol. 6, 1896.
Reveal, J., *in* A. Cronquist et al., Intermountain flora 1: 56. 1972.
Rickett, H.W., Index Bull. Torrey Club 79. 1955.
Rodgers, A.D., W.S. Sullivant 224, 282. 1940; John Torrey 268, 290. 1942; Amer. bot. 335 [index]. 1944.
Sargent, C.S., Silva 4: 28. 1892.
Shope, Mycologia 21: 295. 1929 (mycol. coll. Colo).
Voss, E.G., Botanical beachcombers and explorers, Contr. Univ. Mich. Herb. 13: 59, 74. 1978.
Voss, E.G. & A.A. Reznicek, Taxon 30(4): 866. 1981 (coll. also in F.J. Hermann herb. at MICH).
Wittrock, V.B., Acta Horti Berg. 3(3): 200. 1905.

COMPOSITE WORKS: (1) Contributed *Solidago, Aster* and *Mentha* to E.L. Rand & J.H. Redfield, *Flora of Mount Desert Island, Maine*, 1894.
(2) Contributed to *Catalogue of plants collected in Nevada*, Utah, Colorado, New Mexico, and Arizona ... by J.R. Rothrock et al., *U.S. Geogr. geol. surv. one hundr. merid.*, 6 (Bot.), chapter 4, 1878.
(3) *Catalogue of plants*, in F.V. Hayden, *Prel. rep. U.S. geol. surv. Montana* 477-498, 1872.

EPONYMY: *Porteranthus* N.L. Britton ex Small (1894); *Porterella* Torrey (1872).

8180. *Synopsis of the flora of Colorado,* ... Washington (Government Printing Office) 1874. Oct. (*Syn. fl. Colorado*).
Junior author: John Merle Coulter (1851-1928).
Publ.: 20 Mar 1874 (t.p.; TBC Apr 1874), p. [i-ix], [1]-180. *Copies*: B, BR , G, MICH, MO, NY, US(2), USDA; IDC 7606. – Department of the interior, U.S. Geological and geographical survey of the territories. F.V. Hayden, ... in charge. Miscellaneous Publications, no. 4.
Ref.: Eaton, D.C., Amer. J. Sci. 7: 520-522. Mai 1874. (advance repr. Apr 1874, 2 p.). Many descriptions shown to have been taken from Sereno Watson.

8181. *A list of the Carices of Pennsylvania* [Proc. Acad. nat. Sci. Philadelphia 1887].
Publ.: Jul 1887 (mss. submitted 22 Mar 1887), p. [1]-13. *Copy*: US. – Reprinted and to be cited from Proc. Acad. nat. Sci. Philadelphia 1887: 60-72: 14 Jun 1887, 73-80. 5 Jul 1887.

8182. *Rare plants of Southeastern Pennsylvania* [Easton, Pa., March 1900]. Oct. (*Rare pl. s.e. Pennsylvania*).
Publ.: Mar 1900 (p. 8: so dated), p. [1]-8. *Copy*: USDA. – (Tucker 1: 565 lists this also as an independent publication).

8183. *Flora of Pennsylvania* ... edited with the addition of analytical keys by John Kunkel Small, ... Boston (Ginn & Company, ...) 1903. Oct. (*Fl. Pennsylvania*).
Editor: John Kunkel Small (1869-1938) (provided keys); Porter had provided for the publication of his manuscript in his will.
Publ.: 15 Aug 1903 (see Torreya 3: 128. 22 Aug 1903; TBC 3 Sep 1903; date of preface is always date of publ. with Small), map, p. [i]-xv, [1]-362. *Copies*: MO, NY, PH; IDC 7605.

8184. *Catalogue of the bryophyta* (hepatics, anthocerotes and mosses) *and pteridophyta* (ferns and fern-allies) *found in Pennsylvania* ... Boston (Ginn & Company, ...) 1904. Oct. (*Cat. bryoph. pterid. Pennsylvania*).
Editor: John Kunkel Small (1869-1938).
Publ.: 6-27 Feb 1904 (note in FH copy; notes in LC copy; copyright 6 Feb 1904; copies rd. LC 27 Feb 1904), p. [i, iii], [1]-66. *Copies*: FH, MO, NY, U-V, USDA;

Porterfield, Willard Merritt, Jr. (1893-1966), American ethnobotanist; M.A. Franklin and Marshall College, Lancaster, Pa. 1915; teaching missionary in China; professor of biology St. John's University, Shanghai around 1933-1936; later connected with the Univ. Vermont Agric. Exp. Station, the New York Botanical Garden and the Soil Conservation Service of USA; with the C.I.A. during World War I and until 1963; Research consultant George Washington University 1963-1966. (*Porterf.*).

HERBARIUM and TYPES: Some material from Panama at NY.

BIBLIOGRAPHY and BIOGRAPHY: Barnhart 3: 101; Hirsch p. 236; IH 2: (in press); Kew 4: 345; Lenley p. 334; MW p. 392, suppl. p. 235.

BIOFILE: Anon., New York Times 27 Mar 1966, p. 86 (obit., d. 24 Mar 1966).
Rickett, H.W., Index Bull. Torey bot. Club 79. 1955.

8185. *Wayside plants and weeds of Shanghai* ... Shanghai (Kelly & Walsh, Limited) Hongkong. Singapore 1933. (*Wayside pl. Shanghai*).
Publ.: Oct-Dec 1933 (p. vii: 1 Oct 1933), p. [i*], [i]-xxx, [1]-232. *Copy*: USDA.

Poscharsky, Gustav Adolf (1832-1915), German gardener, botanist and plant collector; royal Saxonian garden inspector at the Dresden botanical garden 1866; ultimately living at Magyar Ovár (Ungarisch Altenberg). (*Posch.*).

HERBARIUM and TYPES: Unknown; plants collected by him at B (extant), E, KIEL, L, MANCH, PR and WB.

BIBLIOGRAPHY and BIOGRAPHY: AG 1: 60; Barnhart 3: 102; CSP 17: 976; Klášterský p. 148; LS 21142; Rehder 1: 448.

BIOFILE: Anon., Flora 49: 170. 1866 (app. Dresden: inspector; so far: "Obergehilfe").
Beck, G., Veg.-Verh. Illyr. Länd. 13, 40. 1901 (Veg. Erde 6).
Degen, A., Magy. bot. Lap. 14: 112. 1915 (d.).
Hedge, I.C. & J.M. Lamond, Index coll. Edinb. herb. 120. 1970.
Kneucker, A., Allg. bot. Z. 21: 140. 1916 (death reported in list of deaths of 1915.

NOTE: Not to be confused with Oskar Poscharsky, b. 21 Jul 1856, d. 7 Feb 1914, gardener at Dresden, later director of the tree nurseries of Laubegast (see e.g. Flora, Dresden 28/29: 59-60. 1915).

8186. *Beiträge zur Flora von Croatien und Dalmatien.* Eine Festschrift zur siebzigsten Stiftungs-Feier der Genossenschaft "Flora" ... Dresden (Buchdruckerei von Arthur Schönfeld) 1896. Oct. (*Beitr. Fl. Croat.*).
Publ.: Sep-Oct 1896 (Nat. Nov. Nov(1) 1896), p. [1]-64. *Copy*: G.

Poselger, Heinrich (1818-1883), German chemist and botanist; collector and student of succulent plants, collected plants on and near the U.S. and Mexican boundary 1849-1851; living on private means in Berlin, growing a rich collection of cacti and succulents. (*Poselg.*).

HERBARIUM and TYPES: Unknown; cacti from Boundary Survey were sent to D. Dietrich at Jena.

BIBLIOGRAPHY and BIOGRAPHY: Barnhart 3: 102; CSP 4: 992; Hortus 3: 1202 (Poselg.); Langman p. 593; Rehder 5: 684; Urban-Berl. p. 384; Zander ed. 10, p. 703, ed. 11, p. 801.

BIOFILE: Anon., Bot. Centralbl. 16: 191. 1883 (d. 4 Oct 1883).
Geiser, S.W., Field & Laboratory 4: 51-52. 1936, 27: 176-177. 1959 (we follow the data as given in the more detailed 1959 publication; b. 25 Dec 1818, d. 4 Oct 1883; bibl.); Natural. frontier 280. 1948.
Knobloch, J.W., Plant coll. N. Mexico 53. 1979 (based on Geiser 1936).
Meyer, R., Monatsschr. Kakteenk. 29(9): 97-100. 1919 (on his private collection of cacti).

Pospichal, Eduard (1838-1905), Bohemian-born Austrian botanist, sometime teacher at Litomyšl and, longtime at Triest. (*Pospichal*).

HERBARIUM and TYPES: TSM(Dalmatian herb.), further material at PR and TRM. – Letters at G.

BIBLIOGRAPHY and BIOGRAPHY: AG 12(2): 81; BFM no. 1067; BL 2: 357, 704; BM 4: 1601; CSP 12: 584; DTS 6(4): 186; Futák-Domin p. 484; Hegi 5(2): 1252, 6(2): 1252; IH 1 (ed. 7): 340, 2: (in press); Kew 4: 346; Klášterský p. 148 (b. 13 Jun 1838); Rehder 5: 684; Saccardo 2: 87, cron. p. xxx; Tucker 1: 565.

BIOFILE: Anon., Bot. Jahrb. 37 (Beibl. 84): 8. 1906 (d. Mai 1905); Bot. Not. 1906: 45 (d.); Nat. Nov. 27: 404. 1905 (d.); Österr. bot. Z. 55: 211. 1905 (d.).
Kneucker, A., Allg. bot. Z. 11: 112. 1905 (d.).

8187. *Flora des Flussgebietes der Cidlina und Mrdlina* ... Prag [Praha] (Druck von Dr. Ed. Gregr....) 1881. Oct. (*Fl. Cidlina*).
Publ.: Jul-Aug 1881 (Bot. Centralbl. 12-16 Sep 1881; Nat. Nov. Sep(2) 1881; Bot. Zeit. 30 Sep 1881), p. [i], [1]-103. *Copies*: B, G, PH. – Published as Arch. naturw. Landesdurchf. Böhmen 4(5). 1881.
Ref.: Freyn, J.F., Bot. Centralbl. 9: 302-305. 1882 (rev.).

8187a. *Flora des oesterreichischen Küstenlandes* ... Leipzig und Wien (Franz Deuticke) 1897-1899, 2 vols. Oct. (*Fl. oesterr. Küstenl.*).
1: Mar 1897 (p. v: Nov 1896; Allg. bot. Z. 15 Jun 1897; Bot. Centralbl. 11 Mar 1897; Bot. Zeit. 1 Mar 1897; ÖbZ 1 Apr 1897), p. [i]-xliii, [1]-574, [1, 2], charts 1-14.
2(1): Aug 1898 (Bot. Centralbl. 24 Aug 1898; Allg. bot. Z. 15 Sep 1898; Nat. Nov. Aug(1) 1898; ÖbZ 1 Oct 1898), p. [i], [1]-528, map.
2(2): Oct 1899 (p. 946: Jun 1899; ÖbZ Oct 1899; Bot. Centralbl. 15 Nov 1899; Nat. Nov. Nov(2) 1899; Allg. bot. Z. 15 Dec 1899; Bot. Zeit. 1 Jan 1900; Flora rd. 12 Apr 1900), p. [i], 529-946, *charts 15-25*.

Copies: B, BR, FI, G, MO, NY. – For an updated inventory see A. Cohrs, Repert. Spec. nov. 56: 66-143. 1954.
Ref.: Kneucker, A., Allg. bot. Z. 3: 86, 101. 1897, 4: 1898, 5: 196-197. 1899.
Ascherson, P., Bot. Zeit. 55(2): 305-316. 16 Oct, 321-329, 1 Nov 1897 (rev. vol. 1).
Drude, O., Bot. Zeit. 55(2): 276-277. 16 Sep 1897 (rev. vol. 1), 58(2): 99-101. 1 Apr 1900 (vol. 2).
Ross, H., Flora 86: 232-233. 20 Mai 1899 (vols. 1, 2(1)).

Pospíšil, Valentin (1912-x), Czechoslovak (Moravian) botanist (bryologist, geobotanist) at the Moravian museum at Brno. (*Pospíšil*).

HERBARIUM and TYPES: BRNM and OLM, see Klášterský.

BIBLIOGRAPHY and BIOGRAPHY: BFM no. 781; Bossert p. 315 (b. 26 Mar 1912); Futák-Domin p. 484; Klášterský p. 148 (b. 26 Mar 1912).

BIOFILE: Anon., Bibl. bot. čechosl. 1959/60 (publ. 1967), 1961-1962 (publ. 1968); Rev. bryol. lichénol. 25: 204. 1956, 32: 320. 1963, 36: 378. 1968, 36: 788. 1969, 39: 509. 1973, 41: 518. 1975 (notes on publ.).
Smejkal, M. & J. Vicherek, Preslia 44: 379-382. 1972 (portr., bibl., 60th birthday).

Post, Ernst Jacob Lennart von (1884-1951), Swedish phytogeographer and geologist, one of the founders of pollen analysis and; Fil. lic. Uppsala 1904; State geologist 1910 (1914); Dr. h.c. Stockholm 1927; professor of geology at Stockholm 1929. (*L. Post*).

HERBARIUM and TYPES: None. – Letters at G.

BIBLIOGRAPHY and BIOGRAPHY: Barnhart 3: 102 (b. 16 Jun 1884); BM 4: 1602, 8: 1024; Bossert p. 316 (d. 11 Jun 1951); Kew 4: 347; Kleppa p. 249; KR p. 576-577 (bibl. up to 1921); Nordstedt suppl. p. 13.

BIOFILE: Caldenius, C., Sv. män och kvinnor 6: 166-167. 1949 (portr.).
Ewan, J. et al., Short hist. bot. U.S. 12. 1969.
Fries, M., Rev. Palaeobot. Palyn. 4: 9-13. 1967 (on the epoch-making lecture of 1916).
Fries, R.E., Short hist. bot. Sweden 75-76. 1950.
Grönwall, K.A., Sv. uppslagsbok ed. 2. 23: 171-172. 1952 (portr.).
Lundqvist, G., Sv. geogr. årsb. 27: 96-106 (obit., portr.).
Manten, A.A., Rev. Palaeobot. Palyn. 1: 11-22. 1967 (founder modern palynology).
Selling, O.H., Sv. bot. Tidskr. 45(1): 275-296. 1951 (obit., portr., bibl., b. 16 Jun 1884, d. 11 Jan 1951).
Straka, H., Umschau Wiss. Techn. 1966 (13): 426 (50th anniversary of pollenanalysis).
Verdoorn, F., ed., Chron. bot. 2: 28, 29, 273. 1936, 3: 25, 42. 1937, 4: 102, 121, 122. 1938, 5: 512. 1939, 6: 231. 1941.

FESTSCHRIFT: Geologiska förenings i Stockholm Förhandlingar 66: 3. 1944 (portr.).

NOTE: For information on E.J.L. von Post we are grateful to O. Almborn, Lund.

8188. *Norrländiska torfmosse studier* ... Stockholm (Kungl. Boktryckeriet. P. A. Norstedt & Söner) 2 parts, 1906-1910. Oct. (*Norrl. torfmossestud.*).
1: Mai 1906 (copy LD rd 24 Mai 1906), p. [199]-308, 3 maps. *Copy*: LD (inf. O. Almborn). – Issued as Geol. Fören. Stockholm Förh. 28(4), 1906 and as Medd. Ups. Univ. Mineral.-Geol. Inst. 30.
2: Apr 1910, ib. 52(1): 63-90. 1930.

8189. *Pflanzen-physiognomische Studien auf Torfmoosen* in Närke ... Excursion A7 ... [Stockholm (P.A. Norstedt & Söner) 1910]. Oct. (*Pfl.-physiogn. Stud. Torfm.*).
Co-author: Johan Rutger Sernander (1866-1944).
Publ.: 1910, p. [i], [1]-48, maps 1-5. *Copy*: U. – Publ. no. 14, Geologorum conventus 11, 1910, Suecia.

8189a. *Skogsträdpollen i sydsvenska torvmosselagerföljder* ... Saertryk av Forh. ved 16. skand. naturforskermøte 1916 [1918]. Oct.
Publ.: 1918, p. [433]-465. *Copy*: LD (inf. O. Almborn). – Reprinted and to be cited from Forh. ved scand. Naturf. 16. Møte 1916: [433]-465. 1918. – The classical paper in which von Post introduced "pollen statistics" (pollen analysis).
Reprint: with introduction and translation by B. Davis, K. Faegri and Johs. Iversen, *in* Pollen et Spores 9(3): [375]-401. Dec 1967.

8189b. *Einige Aufgaben der regionalen Moorforschung* ... Stockhom (Kungl. Boktryckeriet. P.A. Norstedt & Söner) 1926. Oct. (*Aufg. region. Moorforsch.*).
Publ.: Mar(?) 1926, p. [1]-41. *Copy*: LD (inf. O. Almborn). – Sv. geol. Unders. ser. C, Avh. o. upps. 337, Årsbok 19(4), 1925.

8189c. *Pollenstatistika perspectiv på jordens klimathistoria* ... Stockholm 1944. Qu.
Publ.: 1944, published in Ymer 64(2): [79]-113. 1944 (reprinted with original publication). – Honorary lecture delivered on 24 Apr 1944 to the Swedish Anthropological and Geographical Society, on the occasion of the receipt by von Post of the Vega Medal.
English translation: *The prospect for pollen analysis* in the study of the earth's climatic history, New Phytologist 45(2): [193]-217. Dec 1946.

Post, George Edward (1838-1909), American physician, missionary and botanist; educated at New York Free Academy (City College), Univ. of New York medical dept., and at Union Theological Seminary; Dr. med. New York 1860; professor of surgery in the Syrian protestant College, Beirut and surgeon to the Johanniter Hospital, Beirut 1868-1909. (*Post*).

HERBARIUM and TYPES: BEI; further material at AAR, B, BEI, BM, C, E, F, FI, G, K, MICH, MO, NA, OXF, US, W, Z. – Letters and portrait at G.

BIBLIOGRAPHY and BIOGRAPHY: Barnhart 3: 102; BFM no. 1551; BJI 1: 46; BM 4: 1602, 8: 1024; Clokie p. 227; DAB (W.L. Wright, Jr.); IH 1 (ed. 6): 363, 2: (in press); Kew 4: 347; Morren ed. 10, p. 148; Rehder 5: 684; Tucker 1: 436, 565; Urban-Berl. p. 310; Zander ed. 10, p. 703, ed. 11, p. 802.

BIOFILE: Anon., Bot. Not. 1910: 16 (d. 29 Sep 1909); Nat. Nov. 31(21), 1909 (d.); New York Evening Post 8 Oct 1909 (fide Kelly); New York Observer 7 Oct 1909 (fide Kelly).
Beck, G., Bot. Centralbl. 19: 253. 1884.
Blatter, E., Rec. Bot. Surv. India 8(5): 485. 1933.
Burdet, H., Candollea 32: 359-369. 1977.
D.W., *in* Kelly, H.A., Cycl. Amer. med. biogr. 2: 280-281. 1920 (b. 17 Dec 1838, d. 29 Sep 1909).
Davis, P.H. & J.R. Edmondson, Notes R.B.G. Edinburgh 37(2): 278. 1979 (bibl. Fl. Turkey).
Hedge, I.C. & J.M. Lamond, Index coll. Edinb. herb. 120. 1970.
Jackson, B.D., Bull. misc. Inf. Kew 30, 53. 1901 (pl. K).
Kelly, H.A., Some Amer. med. bot. 192-203. 1914 (repr. 1929) (portr.).
Kneucker, A., Allg. bot. Z. 16: 164. 1910 (death).
Murray, G., Hist. BM(NH) 1: 174. 1904.

HANDWRITING: Candollea 32: 395-396. 1977.

EPONYMY: *Postia* E.P. Boissier et E. Blanche (1875).

8190. *Plantae postianae* fasciculus 1(-10) ... Lausanne (Imprimerie Georges Bridel & Cie.) 1890-1900, 10 parts. Oct. (*Pl. post.*).
Co-author fasc. 8-9: Eugène John Benjamin Autran (1855-1912).
1: 1890 (US rd. 24 Jan 1891), p. [1]-14.
2: Feb 1891 (t.p.), p. [1]-23.

3: Feb 1892 (t.p.), p. [1]-19, [1-2].
4: Mai 1892 (t.p.), p. [1]-12.
5: Jan 1893 (id.), p. [1]-17, repr. from Bull. Herb. Boissier 1(1): 15-32. 1893.
6: Aug 1893 (in journal), p. [1]-19, repr. from id. 1(8): [393]-411. 1893.
7: Apr 1895 (id.), p. [1]-18, repr. from id. 3(4): 150-167. 1895.
8: Sep 1897 (id.), p. [1]-7, repr. from id. 5(9): 755-761. 1897.
9: Mar 1899 (id.), p. [1]-16, repr. from id. 7(3): 146-161. 1899.
10: 15 Aug 1900 (id.), p. [1]-14, repr. from Mém. Herb. Boissier 18: 89-102. 1900.
Copies: B, BR, FI, G, NY, U, US.

8191. *Flora of Syria, Palestine, and Sinai*, from the Taurus to Ras Muhammad, and from the Mediterranean sea to the Syrian Desert ... Beirut, Syria (Syrian Portestant College) [1896]. Oct. (in fours). (*Fl. Syria*).
Ed. 1: 1896 (printing commenced 1883 according to p. [1], but actual publication took place in 1896, see J. Bot. 35: 59-60. 1891; Nat. Nov. Oct(2) 1896, ÖbZ 1 Oct 1896), map, p. [i], [1]-919. *Copies*: BR, G, MICH, MO, NY, US, USDA. – A fully arabic *Flora of Syria and Egypt*, vol. 1, Ranunc.-Cornac., was issued at Beirut, Mai 1884. *Copy* at G.
Ed. 2: 1932-1933, 2 vols., *copies*: BR, FI, MO, NY. – Flora of Syria, Palestine and Sinai/A handbook of the flowering plants and ferns, native and naturalized from the Taurus to Ras Muhammad and from the Mediterranean sea to the Syrian Desert ... second edition, extensively revised and enlarged by John Edward Dinsmore [1862-1951] ... American Press, Beirut. – American University of Beirut, Publ. Fac. Arts Sci., Nat. Sci., Nat. Sci. Ser. I. – Oct.*1*: 1932 (p. vi: 1 Mar 1932; p. 639 has dates of printing; Nat. Nov. Jan 1933), p. [i*, iii*], [i]-xxxxiii, [xxxxiv], [1]-639. – *2*: 1933 (p. 928 has dates of printing; Nat. Nov. Jul 1934), p. [i]-xviii, [1]-928, 5 maps.
Note: J.E. Dinsmore had published in 1911 *Die Pflanzen Palästinas* auf Grund eigener Sammlung und der Flora Posts und Boissiers Verzeichniss, with Arabian names by G. Dalman, Leipzig 1911, 122 p.

Post, Hampus Adolf von (1822-1911), Swedish botanist, geologist, mycologist, phytogeographer and phytosociologist; Prof. h.c. 1875; Dr. phil. h.c. Uppsala 1877, agricultural and soil chemist at Ultuna Lantbruksinstitut 1848-1851 and 1869-1892; director of a glass factory at Reijmyre 1852-1869. (*H. v. Post*).

HERBARIUM and TYPES: S (fungi; rd. 1911), including his unpublished drawings and descriptions. – Letters at G.

BIBLIOGRAPHY and BIOGRAPHY: Backer p. 659; Barnhart 3: 102; BL 2: 546, 557, 704; BM 4: 1602; Collander hist. p. 72, 75, 77; CSP 4: 993, 8: 649, 11: 51, 17: 979; KR p. 577-578 (b. 15 Dec 1822, d. 16 Aug 1911); Morren ed. 10, p. 107; NI 2: 204; PR p. 7277; Quenstedt p. 343; Rehder 5: 684; TL-2/1894.

BIOFILE: Anon., Generalreg. Bot. Not. 66-67. 1939; Hedwigia 52: (83). 1912 (d. 14 Aug 1911); Mycologia 4: 157. 1912 (herb. drawings to S).
Fries, R.E., Short hist. bot. Sweden 70, 136. 1950.
Grönwall, K.A., Sv. uppslagsbok ed. 2. 23: 170-171. 1952 (portr.).
Hesselman, H., Levnadsteckningar K. Sv. Vet.-Akad. Ledam. 7: 117-178. 1942 (printed 1943) (portr., bibl.).
Osvald, H., *in* R.E. Fries, Short hist. bot. Sweden 70, 136-137. 1950 (see also Fries on p. 70).
Poggendorff, J.C., Biogr.-lit. Handw.-Buch 3: 1061. 1898 (b. 15 Dec 1822, biogr.).
Post, L. von, Sv. män o. kvinnor 6: 164-165. 1949 (portr., bibl.).
Romell, L., Mycologia 4: 103. 1912 (obit.).
Sernander, R., Geol. Fören. Förh. 34(1): 139-177. 1912 (obit., portr., bibl., b. 15 Dec 1822, d. 16 Aug 1911); Sv. bot. Tidskr. 6: 318-325. 1912 (port., bibl.).
Wittrock, V.B., Acta Horti Berg. 3(2): 66-67, *pl. 5.* 1902, 3(3): 49, *pl. 110.* 1905. 2699
COMMEMORATIVE COIN: K. Sv. Vetenskapakademien skådepenning 1942.

EPONYMY: *Postia* E.M. Fries (1874) according to Donk, Persoonia 1: 273. 1961.

8192. *Försök till en systematisk uppställning af vextställena* i mellersta Sverige ... Stockholm (Adolf Bonnier) 1862. Oct. (*Förs. uppställn. vextställena*).
Publ.: 1862 (p. [3]: Apr 1862), p. [1]-41, [42-43]. *Copies*: LD, NY. – The paper was written in 1857 and intended for Botaniska Notiser. This journal, however, was not issued in 1859-1862 and the paper was printed separately. It is a pioneerwork in plant sociology (O. Almborn).

Post, Tom [Tomas] **Erik von** (1858-1912), Swedish botanist; son of Hampus von Post; director of the Uppsala seed testing station from 1887 and of the Esperanza field station for medicinal herbs near Landskrona from 1907. (*T. v. Post*).

HERBARIUM and TYPES: Types of O. Kuntze and T. v. Post in Kuntze's herbarium at NY. There is no material from T. v. Post at either LD, S, or UPS.

BIBLIOGRAPHY and BIOGRAPHY: Andrews p. 313; Barnhart 3: 102; BM 4: 1602; CSP 17: 980; Kew 4: 347; KR p. 578 (b. 8 Mar 1858, d. 30 Apr 1912); Langman p. 593; RS p. 134; TL-1/991; TL-2/4030-4031, see Kuntze, C.E.O.; Tucker 1: 565; Zander ed. 10, p. 703, ed. 11, p. 802.

BIOFILE: Anon., Bot. Centralbl. 119: 624. 1912 (d.); Bot. Not. 1912: 141-142 (b. 8 Mar 1858, d. 30 Apr 1912); Nat. Nov. 34: 461. 1912 (d.); Österr. bot. Z. 62: 287. 1912 (d.).
Lindman, S., Sv. män o. kvinnor 6: 165-166. 1919.

COMPOSITE WORKS: With Otto Kuntze (q.v.): *Nomenkl. Revis. höher. Pfl.-Gr.* 1900, see TL-2/4030 (contains T. v. Post, Wissenschaftliche Korrekturen und Ergänzungen zum Gesamt-Register II-IV von Engler's *Natürlichen Pflanzenfamilien*, Allg. bot. Z. 6: 150-164, 179-191. 1900).

8193. *Lexicon generum phanerogamarum* in de ab anno mdccxxxvii cum nomenclatura legitima internationali simul scientifica auctore Tom von Post. Opus revisum et auctum ab Otto Kuntze. Upsaliae [Uppsala] (Typis Wretmanianis) 1902. Oct. (*Lex. gen. phan., Prosp.*).
Co-author: Carl Ernst Otto Kuntze (1843-1907); Kuntze's name is mentioned as co-author for the ultimate work; this prospectus is signed (p. 10) by T. von Post.
Publ.: 1902, p. [i-ii], [1]-10. *Copies*: G, NY. – A prospectus announcing the *Lexicon* (see below) with recommendations by other botanists and a brief text "Quelques considérations sur la nomenclature botanique". – Further pamphlets announcing the publication of the Lexikon are at G. Among them is a "provisorischer Prospectus" by the ultimate publisher (Deutsche Verlags-Anstalt, Stuttgart) with requests for subventions, sample treatments and an "Entwurf" of the "Vorwort" as well as a pamphlet "Anfangs Juli 1903 erschien: ...." Actual publication of the completed work, however, took place 20-30 Nov 1903.
Other Issue: 1902, p. [i]-vi, [vii], [i*-ii*], [1]-10. *Copy*: LD. Title on p. [i]: Lexicon ... internationali auctore Tom van Post ...", p. [iii]-vi have a different "Entwurf" preface from [i*-ii*]. Inf. O. Almborn.
Ref.: Britten, J., J. Bot. 41: 222-223. 1903 (proofs of introduction).

8194. *Lexikon generum phanerogamarum* inde ab anno mdccxxxvii cum nomenclatura legitima internationali et systemate inter recentia medio autore Tom von Post. Opus revisum et auctum ab Otto Kuntze. Stuttgart (Deutsche Verlags-Anstalt) 1903. Oct. (*Lex. gen. phan.*).
First issue: 20-30 Nov 1903 (t.p. 1903; p. iv: 15 Oct 1903; copies distributed in Stockholm 14 Dec 1903 (KR), first review (A. Voss) dated 30 Nov 1903 stating "jetzt erschienen", Kew copy rd. Dec 1903), p. [i]-xlvii, [xlviii], [1]-714, 1, 3, 5, err.]. *Copies*: LD, US (Hitchcock). – The copy at LD differs in minor respects from the Hitchcock copy at US. O. Almborn informs us that it may be a "proofsheet issue".
Second issue: Dec 1903 (t.p. 1904; Bot. Zeit. 1 Jan 1904; AbZ 15 Jan 1904; Nat. Nov. Jan(1) 1904), p. [i]-xlvii, [xlviii, list of publ. O. Kuntze], [1]-714, [1, 3, 5, err.]. *Copies*: B, BR, FI, LD, MO, U, US.

Post and Kuntze accepted the year 1737 as the starting point date for botanical nomenclature. Many names cited in the *Lexicon* are therefore invalidly published or were validly published by later authors among them Post and Kuntze in the Lexicon itself. The book is based on Kuntze's *Codex brevis maturus nomenclaturae botanicae* (Lexicon p. ix-xli) which was rejected by the Vienna Congress of 1905. Barnhart (1904) in his review, speaking of Kuntze:"It is certainly unfortunate that he should regard himself as an infallible referee upon all points in dispute, and hurl anathemas at all who refuse to acknowledge his authority, characterizing their propositions as "dishonest", inexecutable", [sic] "false" and "lawless".

Ref.: Barnhart, J.H., Torreya 4: 42-44. 12 Mar 1904.
Britten, J., J. Bot. 42: 58-59. 1904 (see also p. 187.).
Brizicki, G.K., Taxon 18: 659. 1969 (on subgenera).
Jacobusch, E., Deut. bot. Monatschr. 1903(11/12) [reprint 1 p., dated Weihnachten 1903].
Mez, C., Bot. Centralbl. 95: 346-347. 5 Apr 1904.
Nathorst, A.G., Bot. Not. 1904: 63-64.
Nordstedt, C.F.O., ed., Bot. Not. 1904: 38-43 (*fiat justicia, pereat mundus*).
Voss, A., Der deutsche Gartenrat 13 Dec 1903 (rev.).

Postel, Emil A.W. (*fl.* 1856), German botanist, cantor and teacher at Parchwitz, Schlesien. (*Postel*).

HERBARIUM and TYPES: Unknown.

BIBLIOGRAPHY and BIOGRAPHY: BM 4: 1602; Jackson p. 62, 326 (mentions Russian edition, 1875, of his *Führer*); Kew 4: 347; PR 7278-7279; Rehder 5: 684.

BIOFILE: Pax, F., Bibl. schles. Bot. 41, 66. 1929.

8195. *Der Führer in die Pflanzenwelt.* Hülfsbuch zur Auffindung und Bestimmung der in Deutschland wildwachsenden Pflanzen ... Langensalza (Schulbuchhandlung d. Th. L.B.) 1856. Oct. (*Führ. Pflanzenw.*).
Ed. 1: 1856, p. [1]-752. *Copy*: G.
Ed. 2: 1858, p. [1]-752, *5 col. pl. Copies*: B, NY.
Ed. 4: 1866, p. [1]-791, col. frontisp. *Copy*: NY.
Ed. 5: 1871, p. [1]-850, *pl. 1-14*.
Ed. 7: 1876, p. [1]-866. *Copies*: NY, G. Wagenitz. – Reissued 1879.
Ed. 8: Feb 1881 (Bot. Centralbl. 5: 308. 7-11 Mar 1881; Nat. Nov. Mar(1) 1881; Bot. Zeit. 29 Apr 1889), p. [1]-866. *Copies*: NY.
Ed. 9: 1894 (Bot. Centralbl. 3 Oct 1894, rev. 63: 80. 1895), p. [1]-816. Reissued 1903, see Holden, W., Bibl. Contr. Lloyd Libr. 5: 234. 1912.

8196. *Vademecum für Freunde der Pflanzenwelt.* Taschenbuch zum Gebrauche bei botanischen Excursionen behufs der möglichst leichten Bestimmung aller in diesem Gebiete wildwachsenden oder häufig angebauten Gefäss-Pflanzen ... Langensalza (Schulbuchhandlung d. Th. L.B.) 1860. Oct. (*Vadem. Pflanzenw.*).
Publ.: 1860, p. [i]-viii, [1]-735, [736]. *Copies*: B, NY.
Russian translation: see Jackson p. 339.

Postels, Alexander Philipou (1801-1871), Esthonian-born geologist who collected plants on the voyage of the corvette *Seniavin* around the world. (*Postels*).

HERBARIUM and TYPES: LE. – The collections made during the voyage of circumnavigation of the *Seniavin* (1826-1829) are all at LE.

BIBLIOGRAPHY and BIOGRAPHY: Barnhart 3: 102 (b. 24 Aug 1801, d. 28 Jun 1871); BM 4: 1602; Bret. p. 322-323; CSP 4: 993; De Toni 1: c; Herder p. 231, 253; Jackson p. 155; Kew 4: 347; Lasègue p. 380; Moebius p. 88; MW p. 392; NI 226, 1557; PR 5687, 7280 (ed. 1: 8172); SK 8: lxxvi; TR 1091.

BIOFILE: Anon., Rusk biogr. slov. 1905: 626-628 (n.v.).
Borodin, I., Trav. Mus. bot. Acad. Sci. St. Petersb. 4: 94, 204. 1908 (coll.).
Ewan, J. et al., Short hist. bot. US 74. 1969.
Hultén, E., Fl. Kamtchatka 1: 9. 1927; Bot. Not. 1940: 300.
Jessen, K.F.W., Bot. Gegenw. Vorz. 465. 1864.
Lada-Mocarski, V., Bibl. books Alaska 392-394. 1969 (bibl. descr. of Ill. alg.).
Merrill, E.D., Brittonia 1(1): 4. 1931.
Nozikow, N., Russian voyages round the world, London, xx, 165 p., ca. 1946.
Schmid, G., Chamisso als Naturforscher 171 (nos 245, 246). 1942.
Shetler, S.G., The Komarov bot. institute 32. 1967.

EPONYMY: (genus): *Postelsia* F.J. Ruprecht (1852) very likely honors him, although no information is given by Ruprecht as to the derivation of the name; (journal): *Postelsia*; the yearbook of the Minnesota seaside station. St. Paul, Minnesota. Vol. 1-2, 1901-1906.

NOTE: The Corvette *Seniavin* under the command of Feodor Petrowitsch Luetke (1797-1882) left Kronstadt on 1 Sep 1828 together with the corvette Moller. The voyage went via Portsmouth, Teneriffa, Rio de Janeiro, Cape Horn and Valparaiso to the N.W. American Russian territories (later: Alaska), calling on Sitka Island, the Alaskan peninsula itself and continuing to Kamchatka. From Kamchatka a trip was made (end 1827) to the Caroline Islands, the Bonin group (Peel Island) and Bering Strait. The return voyage went from Petropavlovsk (Kamchatka) to Manilla, and the Cape of Good Hope, arriving at Kronstadt 6 Sep 1829. The naturalists on the expedition were C.H. Mertens and F.H. von Kittlitz q.v.; Alexander Postels was painter but collected algae also. The illustrations of the *Ill.-alg.* are lithographs of drawings made by Postels from living specimens during the expedition. Kittlitz published *Vier und zwanzig Vegetations-Ansichten* von Küstenländern und Inseln des Stillen Oceans, Siegen 1844-1846 (TL-2/3690); furthermore *Denkwürdigkeiten einer Reise nach dem russischen Amerika* ... Gotha 1858 (TL-2/3691). Lütke, the commander, published *Voyage autour du monde* Paris 1836.

8197. *Illustrationes algarum* in itinere circa orbem jussu imperatoris Nicolai I. Atque auspiciis navarchi Friderici Lütke annis 1826, 1827, 1828 et 1829 celoce Seniavin exsecuto in Oceano pacifico, inprimis septemtrionali ad littora rossica asiatico-americana collectarum. Auctoribus prof. Alexandro Postels et doct. Francisco Ruprecht. Petropoli [St. Petersburg] (Typis Eduardi Pratz) 1840. Broadsheet. (*Ill. alg.*).
Co-author: Franz Joseph Ruprecht (1814-1870).
Publ.: Nov-Dec 1840 (index dated 30 Oct 1840), p. [Russian:] [i*-vi*], [i]-vi, [1]-28, [1-2, index], [Latin:] [*-iv*], [i]-iv, [1]-22, [1-2, index], *pl.* [*o*], *1-40*, uncol liths. Postels.
Copy: NY; IDC 5283. – We have seen no copy with coloured plates; copies with coloured plates were e.g. offered by L. Voss, Flora 27: 160. 1844.
Facsimile reprint: 1963, p. [i**-ii**, t.p. reprint], facsimile of Latin part only: [*iv*], [i]-iv, [1]-22, [1-2, index], *pl.* [*o*], *1-40* (id.). *Copies*: B, FAS, G, MO. – Reduced to 23 × 33 cm. Weinheim J. Cramer) 1963, Historiae naturalis classica 29.
Ref.: Kützing, T., Bot. Zeit. 1: 494-497. 1843.

Posthumus, Oene (1898-1945), Dutch botanist (pteridologist, palaeobotanist, geneticist); Dr. phil. Groningen 1924; to Indonesia 1925; in Bogor and on Djambi Exped. 1925; geneticist at the Java Sugar Experiment Station, Pasuruan 1926-1939; from 1939 director of the general agricultural experiment station, Buitenzorg; murdered by extremists Nov 1945. (*Posth.*).

HERBARIUM and TYPES: main collections at BO; material collected during his Pasuruan period, in part collected with C.A. Backer, at PAS; other material at BRI, G, GRO, L (important set), P, SING. (inf. C.G.G.J. van Steenis). – Letters at G.

BIBLIOGRAPHY and BIOGRAPHY: Andrews p. 313-314; Backer p. 659; Barnhart 3: 102; BJI 2: 139; BL 1: 112, 313; BM 8: 1024; Hirsch p. 237; Hortus 3: 1202 (Posth.); IF supp. 3: 213-216, supp. 4: 336; IH 1 (ed. 1): 96, (ed. 2): 120, 2: (in press); Kew 4: 347; MW suppl. p. 285; SK 1: cxxiii, 412-414 (itin., portr.), 4: cxliv, cxlv, cxlix, cliv, clv, clvi; TL-2/228.

BIOFILE: Donk, M.A., Bull. Bot. Gard. Buitenzorg ser. 3. 18(2): 171-179. 1949 (P. as a pteridologist; bibl. portr.).
Merrill, E.D., Contr. natl. Herb. U.S. 30: 244. 1947.
Steenis, C.G.G.J. van, Fl. males. Bull. 1: 31. 1947 (kidnapped and killed), 3: 56. 1948 (Posthumus fern garden laid out at Bogor).
Verdoorn, F., ed., et al., Chron. bot. 1: 224, 229. 1935, 2: 242, 244, 245. 1936, 3: 203, 209. 1937, 4: 392, 541, 559. 1938, 5: 366, 524. 1939, 6: 141, 459. 1941.

COMPOSITE WORKS: Contributed to *Flora of Suriname* (ed. A.A. Pulle): *Pteridophyta*, 1 (suppl.): [1]-196. Oct-Dec 1928, see below, *The ferns of Surinam*.
(2) with C.A. Backer, *Varenfl. Java* 1939, TL-2/228; review e.g.: Smith, A.C., J. New York Bot. Gard. 40: 289. Dec 1939.
(3) *Inversicatenales*, in W. Jongmans, ed., *Fossilium, catalogus* part 12, p. [1]-56, 27 Jan 1926.

HANDWRITING: Flora malesiana ser. 1. 1: cli. 1950.

8198. *On some principles of stelar morphology.* Proefschrift ter verkrijging van den graad van doctor in de wis- en natuurkunde aan de Rijksuniversiteit te Groningen, ... te verdedigen op Dinsdag 3 Juni 1924, des namiddags te 4 ure, door Oene Posthumus, geboren te Farnsum ... Amsterdam (Drukkerij en Uitgeverij J.H. de Bussy) 1924. Oct. (*Princ. stelar morph.*).
Publ.: 3 Jun 1924, p. [i-vii], [111]-295, [296, contents], loose, theses: [1-3]. *Copy*: U. – Main part from Rec. trav. bot. néerl. 21: 111-296. 1924.

8199. *The ferns of Surinam* and of French and British Guiana ... Uitgegeven voor rekening van den schrijver. Malang, Java (N.V. Jahn's Drukkerij) 1928. Oct. (*Ferns Surinam*).
Original issue: 1928 (p. v: Sep 1928; BR rd. 6 Mai 1929), p. [i-vi], [1]-196. *Copies*: B, BR, G, MO, NY, U(2), US, USDA; IDC 7246.
Other issue: as vol. 1, supplement, of A.A. Pulle, ed., Flora of Suriname.

8200. *Ferns of Bawean* ... [Amsterdam (Kon. Akad. Wet.)] 1929. [i.e. 1930].
Publ.: early 1930 (submitted 21 Dec 1929), p. [1]-9. *Copy*: BR. – Reprinted from Proc. R. Acad. Sci., Amsterdam 32(10): 1362-1369. 1929 [i.e. 1930]; to be cited from journal.

8201. *On the ferns of Sumba* (Lesser Sunda Islands) ... [Amsterdam (Kon. Akad. Wet.)] 1930. Oct.
Publ.: late 1930 or early 1931 (submitted 25 Oct 1930), p. [1]-5. *Copy*: BR. – Reprinted from Proc. R. Acad. Sci., Amsterdam 33(8): 872-875. 1930.

8202. *Catalogue of the fossil remains, described as fern stems and petioles* ... Malang, Java (N.V. Jahn's Drukkerij) 1931. Oct. (*Cat. foss. fern stems*).
Publ.: 1931, p. [i-iii], [1]-234. *Copies*: G, U(2).
Ref.: Krause, K., Bot. Jahrb. 66: 45. 1934.

8203. *Malayan fern studies* I[II] ... Amsterdam (Uitgave van de N.V. Noord-Hollandsche Uitgevers-Maatschappij) 1937-1938, 2 parts. (*Malay. fern stud.*).
1: 28 Oct 1937 (inf. KNAW), p. [1]-67.
2: 11 Aug 1938 (id.), p. [1]-34, [35]. *Copies*: BR, G, MO, U. – Issued as Verh. Kon. Akad. Wet. Amsterdam, afd. Natuurk. 36(5) and 37(5).
3: The ferns of the Lesser Sunda Islands, Ann. Bot. Gard. Buitenzorg vol. hors série: 35-113. 1944.

Potanin, Grigorii Nikolajevič (1835-1920), Russian botanist and explorer of Eastern Siberia, China, Tibet and Mongolia. (*Potanin*).

HERBARIUM and TYPES: LE; other material at B, BR, DBN, E, FI, H, K, L, S, TK.

BIBLIOGRAPHY and BIOGRAPHY: Backer p. 461; Barnhart 3: 102; BM 4: 1602; Bret. p.

1007; CSP 17: 980; IH 1 (ed. 2): 88, (ed. 6): 363, (ed. 7): 340; 2: (in press); Kelly p. 179; MW p. 392 (bibl.), suppl. p. 285-286; TL-2/5735; Urban-Berl. p. 310, 384.

BIOFILE: Anon., Bot. Centralbl. 56: 288. 1893 (P. ends exploration because of death of his wife); Österr. bot. Z. 72: 124. 1923 (d.).
Blanc, É., Bull. Soc. bot. France 43: 59-64. 1896.
Borodin, I., Trav. Mus. bot. Acad. St. Pétersbourg 4: 94-95. 1908 (list of collectors and collections, flora sibirica).
Coats, A.M., Plant hunters 110, 114, 115-116, 135. 1969.
Embacher, F., Lex. Reisen 236. 1882.
Hedge, I.C. & J.M. Lamond, Index coll. Edinb. herb. 120. 1970.
Herder, F. v., Bot. Jahrb. 9: 432, 445. 1888; Bot. Centralbl. 58: 392. 11 Jun 1894 (coll. LE).
Sutton, S.B., Charles Sprague Sargent 200, 244. 1975.

HANDWRITING: S.V. Lipschitz & I.T. Vasilczenko, Central Herbarium USSR 118. 1968 (q.v. also for further details coll.).

EPONYMY: *Potaninia* Maximowicz (1881).

Potier de la Varde, Robert André Léopold (1878-1961), French bryologist and soldier; educated at École militaire de Saint-Cyr; in army until 1918; from then on managing his property Lez Eaux (Dept. de la Manche) and dedicating himself to agriculture, social work and bryology. (*P. de la Varde*).

HERBARIUM and TYPES: PC; duplicates e.g. at BR, MPU. – Letters at G.

BIBLIOGRAPHY and BIOGRAPHY: Barnhart 3: 103; BL 2: 178, 704; IH 1 (ed. 1): 74, (ed. 7): 340; 2: 362; LS suppl. 22140; MW p. 392, suppl. p. 286; Roon p. 89; SBC p. 129; TL-2/see G. Foreau.

BIOFILE: Allorge, V., Rev. bryol. lichenol. 30: 141-144. 1961 (portr., b. 4 Mar 1878, d. 19 Mar 1961); Bull. Soc. bot. France 109: 38-41. 1962 (obit., portr., b. 4 Mar 1878, d. 19 Mar 1961; P's herbarium and the major part of his library were left to PC).
Anon., Rev. bryol. 32: 297. 1963.
Bullock, A.A., Bibl. S. Afr. bot. 91. 1978.
Demaret, F., Rev. bryol. lichénol. 30: 145-153. 1961 (bibl.).
Merrill, E.D., B.P. Bishop Mus. Bull. 144: 152. 1937; Contr. US natl. Herb. 30: 244. 1947.
Sayre, G., Bryologist 80: 515. 1977.
Verdoorn, F., ed., et al., Chron. bot. 2: 131. 1936, 4: 445. 1938.

8204. *Contribution à la flore bryologique de l'Annam* [Extrait de la Revue générale de botanique] 1917. Oct.
Publ.: 1917, p. [1]-16, *pl. 22-25*. *Copies*: G, H. – Reprinted and to be cited from Rev. gén. Bot. 29: 289-304, *pl. 22-25*. 1917. Inf. P. Isoviita.

8205. *Sur trois mousses inédites de la Chine orientale* [Extrait de la Revue générale de Botanique ...] 1918. Oct.
Publ.: 1918, p. [1]-9. *Copy*: H. – Reprinted and to be cited from Rev. gén. Bot. 30: 346-354. 1918. Inf. P. Isoviita.

8206. *Rhachitheciopsis* P. de la V., genre nouveau d'Orthotrichacées de l'Afrique tropicale [Bull. Soc. Bot. France 1926]. Oct.
Publ.: 3 Mai 1926, p. [1-3], *1 pl.* *Copy*: FI. – Reprinted and to be cited from Bull. Soc. bot. France 73: 74-76. 3 Mai 1926.

8207. *Mousses de l'Oubangui* ... Extrait des Archives de Botanique ... 1928. Oct. (*Mouss. Oubangui*).
Publ.: 10 Sep 1928, p. [i]-iv, [1]-152, *pl. 1-4*. *Copies*: B, H, MO(2). – Pages 23-32 ap-

peared separately on 19 Mai 1928. Preprinted from Arch. de Bot. 1, mém. 3, 1927 (publ. 1928).

8208. *Mousses du Gabon* ... Saint-Lo (Imprimerie R. Jacqeline ...) 1936. Oct. (*Mouss. Gabon*).
Publ.: 1936, p. [i], [1]-270, [1, err.]. *Copies*: B, BR, MO, NY. – Issued as Mém. Soc. Sci. nat. Cherbourg, 1936.

8209. *Le genre Fissidens dans la Manche* ... Saint-Lo (Imprimerie René Jacqueline) 1938. Oct. (*Fissidens Manche*).
Publ.: 1938, p. [1]-30. *Copy*: W.R. Buck (NY).

Potonié, Henry (1857-1913), German palaeo– and actuo-botanist of French origin; studied botany in Berlin 1878-1881; assistant to Eichler 1880-1883; Dr. phil. Freiburg i.B. 1884; lecturer on palaeobotany at the royal mining institute 1891; district geologist 1898; professor's title 1900; habil. Berlin 1901, Königl. Landesgeologe 1901, Geheimer Bergrat 1913. (*Potonié*).

HERBARIUM and TYPES: BHU and Herb. Arbeitsstelle für Paläobotanik, Deut. Akad. Wiss. Berlin (Rüffle 1964). – Letters at G.

BIBLIOGRAPHY and BIOGRAPHY: AG 6(2): 913; Andrews p. 314; Barnhart 3: 103; BFM no. 13; BJI 1: 46, 2: 139; BM 4: 1602-1603, 1620, 8: 1024; Bossert p. 316; CSP 11: 52, 12: 585, 17: 981-982; Hegi 4(3): 1518; Kew 4: 348-349; Kleppa p. 330; LS 21148, 38028, suppl. 22141; Moebius p. 131, 143, 379, 382; Morren ed. 10, p. 7; NI 1558, 3: 55; Quenstedt p. 344; Rehder 5: 684; TL-2/6337, see Diels, F.L.E., Müller, K.A.E.; Tucker 1: 565-566; Zep.-Tim. p. 94.

BIOFILE: Andrews, H.N., The fossil hunters 159, 314-315. 1980.
Angersbach, A.L., Naturw. Wochenschr. 28: 776-779. 1913 (P. as a philosopher).
Anon., Ber. deut. bot. Ges. 32: (1). 1915 (d.); Bot. Centralbl. 45: 404. 1891 (Bergakademie), 73: 160. 1898 (Bezirksgeologe), 84: 415. 1900 (prof. title); Bot. Centralbl. 85: 436. 1901 (habil. Berlin), 86: 192. 1901 (Landesgeologe), 123: 128. 1913 (Geh. Bergrat), 125: 16. 1914 (d. 22 Oct 1913); Bot. Jahrb. 14 (Beibl. 10): 8 (Bergakademie), 25 (Beibl. 60): 55. 1898 (Bezirksgeologe, Berlin); Hedwigia 40(1): (22). 1901 (prof. title), 54(2): (125). 1913 (d. 28 Oct 193); Mycol. Centralbl. 3: 270. 193 (d.); Nat. Nov. 23: 66 (prof. title), 228 (habil.) 1901, 24: 681. 1902 (Leopoldina), 35: 553. 1913 (d.); Österr. bot. Z. 48: 159. 1898 (Bezirksgeologe), 51: 38. 1901 (prof. title), 63: 512. 1913 (d.); Science ser. 2. 38: 813. 1913 (d.); Verh. bot. Ver. Prov. Brandenburg 55: 77. 1913, 56: 17. 1915 (d.).
Becker, K. & E. Schumacher, S.B. Ges. naturf. Freunde Berlin ser. 2. 13(2): 146. 1973 (bibl. publ. Gesellsch.).
Branca, W., Naturw. Wochenschr. 28: 753-757. 1913 (obit.; P. founded this journal; "die beiden Todfeinde, Frankreich und Deutschland ... waren in ihm versöhnt"; had french nationality until his appointment at the Geol. Landesanstalt, but was Berlinborn of a French father and a Berlin mother; preferred composer: Wagner; id. fiction character: Peter Schlemihl; original and significant account).
Glass, B. et al., Forerunners of Darwin 324, 327, 329. 1968.
Gothan, W., Ber. deut. bot. Ges. 31: (127)-(136), 1914 (obit., portr., bibl., b. 16 Nov 1857, d. 28 Oct 1913) (see also p. 182, 369); Gesch. Paläobot. Berlin 8-17. 1951.
Kaunhowen, F., Z. deut. geol. Ges. 66: 384-406. 1914 (obit., portr., bibl. of 220 items).
Kerp, J.H.F., Taxon 30: 662. 1981 (on the use of "p-names" by Potonié).
Kneucker, A., Allg. bot. Z. 4: 36. 1898 (lecturer Bergakademie), 7: 20. 1901 (title Professor), 19: 176. 1913 (death).
Poggendorff, J.C., Biogr.-lit. Handw.-Buch 6(3): 2065. 1938.
Potonié, H., Verh. bot. Ver. Prov. Brandenburg 55: 57-61. 1913 (autobiogr. sketches; suppl. notes by R. Potonié; portr.).
Potonié, H. & A. Engler, Verh. bot. Ver. Brandenburg 41: lxvii. 1900 (on the publ. of his well known poster (170-120 cm) *Eine Landschaft der Steinkohlenzeit* (Borntraeger); see also Weisse, Bot. Centralbl. 84: 264-265. 1900, and Potonié's text, Erläuterung zu der Wandtafel, 1899, 40 p.).

Potonié, R., Kosmos 53(11): 554-555. 1957 (100th birthday, portr.), Willdenowia 2(1): 15-22. 1958 (id., on P. as precursor of telome theory and of the "new morphology").
Rüffle, L., Ber. Geol. Ges. D.D.R. 9: 402-404. 1964 (herb.).
Schmid, G., Goethe u.d. Naturwissenschaften 284, 287, 291, 294. 1940.
Taton, R., ed., Science in the 19th century 385. 1965.
Wittrock, V.B., Acta Horti Berg. 3(3): 148, *pl. 130.* 1905 (portr.).

COMPOSITE WORKS: (1) Contributed to EP, *Die natürlichen Pflanzenfamilien*:
Fossile Filicales, 1(4): 473-480. Aug 1900, 481-515. Nov 1900.
Sphenophyllaceae, 1(4): 515-519. Nov 1900.
Fossile Equisetaceae, 1(4): 548-551. Jan 1901.
Calamariaceae, 1(4): 551-558. Jan 1901.
Protocalamariaceae, 1(4): 558-562. Jan 1901.
Fossile Psilotaceae, 1(4): 620-621. Jan 1901.
Fossile Lycopodiaceae, 1(4): 715-716. Nov 1901.
Lepidodendraceae 1(4): 717-739. Nov 1901.
Sigillariaceae, 1(4): 740-753. Nov 1901.
Pleuromoiaceae, 1(4): 754-756. Nov 1901.
Bothrodendraceae, 1(4): 739-740. Nov 1901.
Lepidophyta, 1(4): 779-780. 4 Jan 1902 (Bay), rather Feb 1902.
Cycadofilices, 1(4): 780-798. 4 Jan 1902 (Bay), rather Feb 1902.
Bennettiaceae, Nachtr. 2-4: 14-17 Jul 1897.
(2) Editor: *Naturwissenschaftliche Wochenschrift*, 1-28, 1888-1913.
(3) Contributed "Kultureinflüsse auf Sumpf und Moor" to F. Wahnschaffe et al., *Der Grunewald bei Berlin*, Jena 1907.
(4) Editor: *Palaeobotanische Zeitschrift*, Bd. 1, Heft 1, †, 1912.
(5) With W. Gothan, *Vegetationsbilder der Jetzt– und Vorzeit. pl. 1-8*, 1907-1922.
(6) Collaborator, C.K. Schneider's *Illustriertes Handwörterbuch der Botanik*, ed. 2, 1917. See C.K. Schneider.

EPONYMY: *Potoniea* Zeiller (1899), see ING 3: 1406.

NOTE: Potonié proposed the use of the prefix "p" for names of fossil taxa which are junior homonyms of names already in use for recent genera such as e.g. *Callipteris* and *p-Callipteris* (e.g. Potonié in Naturw. Wochenschrift 15: 313-314. 1900; in EP Nat. Pflanzenfam. 1(4), 1902, and in Abbild. Beschr. foss. Pfl. 1903-1919). This method has not been followed by others and was also abandoned by Potonié himself. The prefix "p" must be treated as a symbol and should be left out of account (see e.g. J.H.F. Kerp 1981).

8210. *Ueber die Zusammensetzung der Leitbündel bei den Gefässkryptogamen.* Der hohen philosophischen Facultät der Universität zu Freiburg i. B. als Dissertation zur Erlangung der Doctorwürde vorgelegt von Henry Potonié aus Berlin. Freiburg in Baden 1884. Oct.
Thesis issue: 1884, p. [1]-46, *1 pl. Copy*: Niedersächs. Staats. und Univ. Bibl. Göttingen, inf. G. Wagenitz.
Commercial issue: 1883, p. [1]-46, *pl. 8. Copies*: G, M. – Reprinted and to be cited from Jahrb. K. bot. Gard. Mus. Berlin 2: 233-278. *pl. 8.* 1883.

8211. *Illustrierte Flora* von Nord– und Mittel-Deutschland mit einer Einführung in die Botanik ... Berlin (Verlag von M. Boas) 1885. Oct. (*Ill. Fl.*).
Ed. 1: Feb-Jun 1885 (p. iv: Apr 1885; im Frühjahr, Potonié in ed. 4), p. [i]-iv, [1]-420. *Copy*: BR. – Published in 10 parts Feb-Jun 1885 (1/2: Feb, 3/4 Mar, 5/6 Apr, 7/8 Mai, 9/10 Jun).
Ed. 2: Feb-Mai 1886 (Bot. Zeit. 26 Feb 1886 and Nat. Nov. Feb(2) 1886, announcements without details; Bot. Centralbl. 6-10 Sep 1886 (rev. J.F. Freyn); Bot. Zeit. rev. 3 Dec 1886; Flora rd. 1-11 Jun 1886), p. [i]-viii, [1]-428. *Copies*: E, M. – Berlin (Verlag von Brachvogel & Boas) 1886. Oct. *Review*: E. Koehne, Bot. Zeit. 44: 819-820. 1886.
Ed. 3: 1-24 Mar 1887 (p. vi: 1 Mar 1887; Nat. Nov. Mar(1) 1887; J. Bot. Apr 1887; Bot. Zeit. 25 Mar 1887; Bot. Centralbl. 25-29 Apr 1887), p. [i]-viii, [1]-511. *Copy*: NY. –

"Dritte wesentlich vermehrte und verbesserte Auflage" Berlin (id.) Oct. – See Vorwort, p. vi, for the numerous collaborators. *review*: E. Koehne, Bot. Zeit. 46: 46-47. 1888.
Ed. 4: Mai 1889 (p. vi: Mai 1889; Nat. Nov. Jun(1) 1889; Bot. Zeit. 14 Jun 1889 ("soeben"), 22 Nov 1889 (review E. Koehne); Flora 20 Jul 1889 (rev.); J. Bot. Sep 1889; Bot. Centralbl. 17 Jun & 3 Oct 1889 (rev.)), p. [i]-viii, [1]-598. *Copies*: G, MO, NY, USDA. – "Vierte wesentlich vermehrte und verbesserte Auflage ..." Berlin (Verlag Julius Springer) 1889. Oct. – See Vorwort p. vi for new collaborators; the collaborators are also mentioned on the cover: P. Ascherson, G. Beck, R. Beyer, R. Caspary, H. Christ, W.O. Focke, J. Freyn, E. Hackel, C. Haussknecht, A. Kerner, M. Kronfeld, E. Loew, G. Leimbach, P. Magnus, C. Müller, F. Pax, A. Peter, Aug. Schulz, P. Taubert, V.B. Wittrock, L. Wittmack, A. Zimmeter, with an appendix on medical plants by W. Lenz.
Ed. 5: Apr 1910 (p. iv: Mar 1910; Nat. Nov. Mai(1) 1910; BR bill dated 3 Jun 1910; publ. announcement Apr 1910; Allg. bot. Z. 15 Mai 1910; ÖbZ 1 Jul 1889; Bot. Centralbl. 26 Jul 1910), in two parts, Jena (Verlag von Gustav Fischer) 1910. Oct. *Copies*: BR, G, NY, USDA.
1 (Text): [i]-vi, [1]-551.
2 (Atlas): [i]-iv, 1-364 (one plate per page).
Ed. 6: 1913 (p. iv: Mar 1913; Nat. Nov. Mai(1) 1913, text Oct (1) 1913; Bot. Jahrb. 9 Dec 1913 (rev. K. Krause); NY Nov 1913; ÖbZ Sep 1913 (text), in two parts, Jena (Verlag von Gustav Fischer) 1913. Oct. *Copies*: B, FAS, G, MO(2), NY, U.
1 (Text): Sep 1913, p. [i]-viii, [1]-562.
2 (Atlas): Mar-Apr 1913, p. [i]-iv, [1]-390.
Ed. 7: see R. Potonié.
Reprints of separate treatments of ed. 3 (publ. Mar-Apr 1887): *Nymphaeaceae* (R. Caspary, p. 211-212), *Rubus* (W.O. Focke, p. 300-315), *Potentilla* (A. Zimmeter, p. 316), *Orobanche* (G. Beck, p. 394), *Hieracium* (A. Peter, p. 449-465).

8212. *Die Pflanzenwelt Norddeutschlands* in den verschiedenen Zeitepochen, besonders seit derEiszeit ... Berlin SW. (Verlag von Carl Habel ...) 1886. Oct. (*Pflanzenw. Nordeutschl.*).
Publ.: 1886 (Bot. Zeit. 31 Dec 1886), p. [1]-32. *Copy*: G. – Reprinted from Samml. Gem.-verst. Wiss. Vortr. N.F. ser. 1, Heft 11, p. 407-436.

8213. *Ueber die Pflanzen-Gattung Tylodendron* ... Berlin (Druck von Mesch & Lichtenfeld, ...) 1888. Oct.
Publ.: 1888 (submitted 11 Nov 1887), p. [i], [114]-126. *Copy*: MO. – Reprinted with special t.p. from Verh. bot. Ver. Brandenburg 29: [114]-126. 1888. Also (extended) with plates in Jahrb. k. preuss. Geol. Landesanst. 1887: 311-331, *pl. 12-13a*, 1888 (Bot. Zeit. 30 Nov 1888).

8214. *Elemente der Botanik* ... Berlin (*Verlag von Julius Springer*) *1888*. Oct. (*Elem. Bot.*).
Ed. 1: Feb-Mar 1888 (Bot. Zeit. 30 Mar 1888; Nat. Nov. Apr(3) 1888; Bot. Centralbl. 21-25 Mai 1888), p. [i-vii], [1]-323. (n.v.).
Ed. 2: Mai-Jun 1889 (p. v: Jan 1889; Bot. Zeit. 14 Jun 1889 (ann.), 28 Jun 1889; Nat. Nov. Jun(3) 1889; Bot. Centralbl. 17 Jun 1889), p. [i-vii], [1]-323. *Copy*: G. – Zweite Ausgabe. Berlin (Verlag von Julius Springer) 1889. Oct.
Ed. 3: Feb-Mar 1894 (p. vi: Feb 1894; Nat. Nov. Apr(1) 1894; Bot. Centralbl. 17 Apr 1894; Bot. Zeit. 1 Mai 1894), p. [i]-vi, [vii, cont.], [1]-343. *Copy*: G. – Dritte, wesentlich verbesserte und vermehrte Auflage. Berlin (Verlag von Julius Springer) 1894. Oct. *Review*: Jost, bot. Zeit. 52: 199-200. 1894.

8215. *Die Flora des Rothliegenden von Thüringen* ... Berlin (In Vertrieb bei der Simon Schropp'schen Hof-Landkartenhandlung (J.H. Neumann)) 1893. Oct. (*Fl. Rothlieg.*).
Publ.: 1893 (p. ix: Jan 1893; Bot. Centralbl. 10 Jan 1894), p. [i*-ii*], [i]-ix, [1]-298, 2 charts, [1, expl. pl.], *pl. 1-34* (uncol.) with text. *Copies*: MO, U, USGS. – Issued as Abh. k. Preuss. geol. Landesanst. N.F. Heft 9, Theil 2.

8216. *Die floristische Gliederung des deutschen Carbon und Perm* ... Berlin (In Vertrieb bei der

Simon Schropp'schen Hof-Landkartenhandlung (J.H. Neumann), ...) 1896. Oct. (*Fl. Glied. Carbon*).
Publ.: Jun-Aug 1896 (p. ii: Jun 1896; Bot. Zeit. 16 Sep 1896), p. [i*, iii*], i-ii, [1]-58, [59-60, err.]. *Copy*: USGS. – Issued as Abh. k. Preuss. geol. Landesanst. N.F. Heft 21.

8217. *Lehrbuch der Pflanzenpalaeontologie* mit besonderer Rücksicht auf die Bedürfnisse des Geologen ... Berlin (Ferd. Dümmlers Verlagsbuchhandlung) [1897-]1899. Oct. (*Lehrb. Pfl.-Palaeont.*).
Ed. 1: in four parts, 1897-1899. *Copies*: BR, FI, G (orig. covers), NY, U(2), USGS.
   1: [i]-vii, 1-112. Feb 1897 (p. vii: Feb 1897; ÖbZ Mar-Apr 1897; Bot. Centralbl. 28 Apr 1897).
   2: 113-208. Aug 1897 (Bot. Centralbl. 29 Sep 1897).
   3: 209-288. Apr 1898 (Bot. Centralbl. 11 Mai 1898; ÖbZ Apr 1898).
   4: [i]-viii, 289-402, *pl. 1-3.* Sep 1899 (p. vi: Sep 1899; ÖbZ Sep 1899).
Ed. 2: in three parts, 1919-1920. "*Lehrbuch der Paläobotanik* von ... H. Potonié ... Zweite Auflage, umgearbeitet von ... W. Gothan [Walther Ulrich Eduard Friedrich Gothan, 1879-1954] mit Beiträgen von ... P. Menzel und ... J. Stoller ..." Berlin (Verlag von Gebrüder Borntraeger ...) 1921. Oct. *Copies*: G, NY, U, USGS.
   1: [1]-160. Oct 1919.
   2: 161-320. Jul 1920.
   3: [i]-vi, [vii], 321-537, [538, err.] late 1920 (t.p. 1921).
See R. Potonié (1958) for the strict editing by Gothan who deleted much that he considered "superfluous". ("überflüssige ... Spekulationen"). It is not certain that Gothan's editing was beneficial.
Ref.: Solms-Laubach, H., Bot. Zeit. 56: 8-9. 1898 (rev. 1(1, 2)).

8218. *Die Metamorphose der Pflanzen* im Lichte palaeontologischer Thatsachen ... Berlin (Ferd. Dümmlers Verlagsbuchhandlung) 1898. Oct. (*Metam. Pfl.*).
Publ.: 1-15 Jan 1898 (p. 4: Dec 1897; Bot. Zeit. 16 Jan 1898; Bot. Centralbl. 2 Feb 1898; Hedwigia 18 Feb 1898), p. [1]-29. *Copies*: FI, MO, U.

8219. *Pflanzen-Vorwesenkunde* im Dienste des Steinkohlen-Bergbaues ... St. Johann (Druck der Saardruckerei) 1899. Oct. (*Pfl.-Vorwesenk.*).
Publ.: 1899 (p. 2: Mar 1899), p. [1]-31. *Copy*: MO. – Reprinted from *Bergmannsfreund* (journal n.v.).

8220. *Eine Landschaft der Steinkohlen-Zeit*. Erläuterung zu der Wandtafel bearbeitet und herausgegeben im Auftrage der Direction der königl. Preuss. geologischen Landesanstalt und Bergakademie zu Berlin ... Leipzig (Verlag von Gebrüder Borntraeger) 1899. Oct. (*Landsch. Steinkohlen-Zeit*).
Publ.: 1899, frontisp. [reduced issue of Wandtafel], p. [1]-40. *Copy*: Rijksmus. Geol. Leiden. – For a brief preliminary note by Potonié see Z. deut. geol. Ges. 1898: 110-111. 1898.

8221. *Die Silur– und die Culm-Flora* des Harzes und des Magdeburgischen. Mit Ausblicken auf die anderen altpaläozoischen Pflanzenfundstellen des Váriscischen Gebirgs-Systems ...Berlin (Im Vertrieb bei der königlichen Geologischen Landesanstalt, ...) (J.H. Neumann)) 1901. Oct. (*Silur– Culm-Fl.*).
Publ.: 1901 (Nat. Nov. Jul(1) 1902), p. [i, iii, v], [1]-183. *Copies*: U, USGS. – Issued as Abh. k. Preuss. Geol. Landesanst. N.F., Heft 36.
Ref.: Solms-Laubach, H., Bot. Zeit. 60(2): 119-120. 16 Apr 1902 (rev.).

8222. *Ein Blick in die Geschichte der botanischen Morphologie* mit besonderer Rücksicht auf die Pericaulom-Theorie ... Jena (Verlag von Gustav Fischer) 1902. Oct.
Publ.: Oct 1902, p. [1]-28. *Copy*: MO. – Reprinted and to be cited from Naturwiss. Wochenschr. N.F. 2: 3-8. 5 Oct, 13-15. 12 Oct, 25-28. 19 Oct 1902. See also *Die Pericaulom-Theorie*, Ber. deut. bot. Ges. 20(8): 502-508. 26 Nov 1902, and Bot. Centralbl. 92: 493-496. 1903. Combined, amended reprint, p. [i, iii], [1]-45. *Copy*: NY. – Continued as *Grundlinien Pfl.-Morphol.* 1912, q.v.
Ref.: Solms-Laubach, H., Bot. Zeit. 61(2): 145-146. 16 Mai 1903 (rev.).

8223. *Abbildungen und Beschreibungen fossiler Pflanzen-Reste* der palaeozoischen und mesozoischen Formationen ... Berlin (Im Vertrieb bei der königlichen geologischen Landesanstalt, ...) 1903-1913, 9 parts. (*Abbild. Beschr. foss. Pfl.*).
1: 1903 (Bot. Centralbl. 12 Apr 1904 (as of "1903")), p. [i]-iv, nos. 1-20 (each number consists of a number of pages with text and illustrations; the number of pages of the numbers varies; each has independent pagination).
2: 1904 (Bot. Zeit. 16 Apr 1905; Bot. Centralbl. 28 Feb 1905 (as of 1904)), p. [i], nos. 21-40.
3: 1905, p. [i], nos. 41-60.
4: 1906, p. [i], nos. 61-80.
5: 1907, p. [i], nos. 81-100, index p. [1]-15.
6: 1909 (Bot. Zeit. rev. 16 Nov 1910), p. [i], nos. 101-120.
7: 1910, p. [i], nos. 121-140.
8: 1912, p. [i], nos. 141-160.
9: 1913, p. [i], nos. 161-180.
Copies: MO, NY, PH, U, USDA.

8224. *Flore dévonienne* de l'étage de H. Barrande ... Leipzig (en commission chez Raimund Gerhard, ancienne maison Wolfgang Gerhard ...) [1904]. Qu. (*Fl. dévon.*).
Co-author: Charles Jean Bernard (1876-1967).
Publ.: Jan-Mar 1904 (fide C.J. Bernard, Bot. Centralbl. 95: 574-576. 31 Mai 1904; Andrews:"1903" see also Bot. Centralbl. 97: 53. 22 Mar 1904), p. [i], [1]-68. *Copies*: PH, USGS. – "Suite de l'ouvrage Système silurien du centre de la Bohème par Joachim Barrande" (1799-1883), see TL-2/1: 125).
Ref.: Solms-Laubach, H., Bot. Zeit. 62(2): 187. 16 Jun 1904 (rev.).

8225. *Die Entstehung der Steinkohle* und der Kaustobiolithe überhaupt (wie des Torfs, der Braunkohle, des Petroleums usw.) ... Fünfte, sehr stark erweiterte Auflage des Heftes "Die Entstehung der Steinkohle und verwandter Bildungen einschliesslich des Petroleums. Berlin (Verlag von Gebrüder Borntraeger ...) 1910. Oct. (*Entst. Steinkohle*).
Ed. 5: 1910 (preface 1910), p. [ii]-x, [xi], [1]-225. *Copies*: BR, LC.
Ed. 6: 1920 (pref. Aug 1920; ÖbZ Jan-Feb 1921), p. [ii]-vii, [viii], [1]-233. *Copies*: LC, U. – Sechste Auflage, durchgesehen von Prof. Dr. W. Gothan.
Earlier editions (n.v.), *ed. 1*: 1905, in "*Tiefbohrwesen*" vol. 3, *ed. 2*: 1905; *ed. 3*: 1905, 53 p.; *ed. 4*: (n.d.).

8226. *Grundlinien der Pflanzen-Morphologie* im Lichte der Palaeontologie ... zweite, stark erweiterte Auflage des Heftes: "Ein Blick in die Geschichte der botanischen Morphologie und die Perikaulom-Theorie". Jena (Verlag von Gustav Fischer) 1912. Oct. (*Grundlin. Pfl.-Morph.*).
Publ.: Jan-Feb 1912 (p. v: Jan 1912; Nat. Nov. Apr(1) 1912; Bot. Centralbl. 30 Apr 1912; BR rd. 21 Aug 1912; NY Mai 1912; ÖbZ Jan-Feb 1912), p. [i]-v, [vii], [1]-259. *Copies*: BR, G, H, MO, NY, U.
Ref.: Diels, L., Bot. Jahrb. 48(Lit.): 35-36. 27 Aug 1912 (rev.).

8227. *Palaeobotanische Zeitschrift* redigiert von H. Potonié. Berlin (Verlag von Gebrüder Borntraeger ...) 1912. †. (*Palaeobot. Z.*).
Vol. 1, part 1, all issued: Nov 1912, p. [1]-84, *pl. 1-3*. *Copy*: PH.

8228. *Paläobotanisches Praktikum* von ... H. Potonié und ... W. Gothan mit je einem Beitrag von ... J. Stoller und A. Franke ... Berlin (Verlag von Gebrüder Borntraeger ...) 1913. Oct. (*Paläobot. Prakt.*).
Co-author: Walther Ulrich Eduard Friedrich Gothan (1879-1954).
Publ.: Jan-Mar 1913 (BR rd. 14 Apr 1913, Bot. Centralbl. 1 Jul 1913), p. [i]-viii, [1]-152. *Copies*: BR, H, U.

Potonié, Robert [Henri Hermann Ernst] (1889-1974), German palaeobotanist, palynologist and coal petrologist; son of Henry Potonié; Dr. phil. Berlin 1920; habil. Berlin

(Techn. Hochsch.) 1922; from 1920-1923 voluntary collaborator, from 1923-1924 assistant with the Preuss. Geol. Landesanst.; 1934-1955 Bezirks-Geologe, Reichsamt Bodenforschung, ultimately at Krefeld. (*R. Potonié*).

HERBARIUM and TYPES: Palaeobotanical collections at the Amt für Bodenforschung, Krefeld. Some pollen preparations in Herb. Arbeitsstelle Paläobotanik, Deut. Akad. Wiss., Berlin.

BIBLIOGRAPHY and BIOGRAPHY: Barnhart 3: 103; BFM no. 2583; BJI 2: 139; BM 8: 1024; Roon p. 89.

BIOFILE: Andrews, H.N., The fossil hunters 316. 1980.
Becker, K. & E. Schumacher, S.B. Ges. naturf. Freunde Berlin N.F. 1(2): 146. 1973 (bibl. publ. Gesellschaft).
Cowan, R.S., Taxon 23: 448. 1974 (death).
Doubinger, J., Pollen et Spores 16(2): 157-160. 1974 (obit., portr.).
Freund, H. & A. Berg, Gesch. Mikroskopie 28, 43. 1963.
Grebe, H., Rev. Palaeobot. Palyn. 17(3/4): 217-220. 1974 (obit., portr, b. 2 Dec 1889, d. 26 Jan 1974).
Kremp, G.O.W., Robert Potonié (1889-1974) mss. (unpubl. obituary).
Poggendorff, J.C., Biogr.-lit. Handw.-Buch 6(3): 2065. 1938, 7a(3): 614-616. 1959. (bibl.).
Potonié, R., Synopsis der Gattungen der *sporae dispersae* vols. 1-6, 1956-1970; see also his Geological sporology and palynology, Palynol. Bull. 1: 7-10. 1965.
Rüffle, L., Ber. Geol. Ges. D.D.R. 9: 404. 1964 (pollen prep.).
Schmid, R. & M.J., *in* J. Arditti, ed., Orchid biol. 1:29, 30, 43, 44. 1977.
Wulff, E.V., Intr. hist. Plant Geogr. 92, 178. 1943.

COMPOSITE WORKS: Editor: *Geologisches Zentralblatt* vols. 69(1)-70(1). 1941-1942.

EPONYMY: *Potonieipollenites* B. Agrali (1965); *Potonieisporites* D.C. Bhardwaj (1955); *Potonieitriradites* D.C. Bharadwaj et V. Sinha (1956).

FESTSCHRIFT: see Glückauf 85, 1949 (p. 945 and foll.), Erdöl u. Kohle 7: 878. 1954.

8229. *Taschenatlas zur Flora von Nord– und Mitteldeutschland* von Henry Potonié überarbeitet von ... Robert Potonié ... 7. Auflage. Jena (Verlag von Gustav Fischer) 1923. Oct. (*Taschenatl. Fl. N.M. Deutschland*).
Author of eds. 1-6 (1885-1913): Henri Potonié, see above.
Publ.: 1923 (Nat. Nov. Aug 1923), p. [i]-vi, 1-409. *Copies*: BR, G, NY, U. The accompanying text of ed. 6 (see Henry Potonié) was not published in a seventh edition.

Pott, Johann Friedrich (1738-1805), German botanist; physician to the Duke of Braunschweig (Brunswick). (*Pott*).

HERBARIUM and TYPES: LE (in herbarium Gorenkianum); some further specimens are at LINN; other material is at H and SBT.

BIBLIOGRAPHY and BIOGRAPHY: Barnhart 3: 103 (d. 13 Apr 1805); Hortus 3: 1202 (J.F. Pott); NI 1: 198; PR 2560, 7281 (ed. 1: 8174); TL-2/1599; Tucker 1: 566; Zander ed. 10, p. 703.

BIOFILE: Herder, F. v., Bot. Centralbl. 55: 260. 1893 (herb. at LE in herb. Gorenkianum).
Jackson, B.D., Notes cat. Linn. herb. 19. 1922 (some plants in LINN). Kukkonen, I., Herb. Christ. Steven 81. 1971.

COMPOSITE WORKS: Published ed. 2 of J.P. Du Roi, *Harbk. Baumz.* (1775-1800), TL-2/1599.

EPONYMY: *Pottia* (Erhardt ex H.G.L. Reichebach) Fürnrohr (1829) and the derived *Pottiella* (Limpricht)Gams (1948) were named for Pott, fide Augier, 1966.

8230. *Index herbarii mei vivi* [Brunsvigae, Typis Viewegianis] 1805. (*Index herb.*). *Editor*: J. L. Hellwig.
Publ.: Jul 1805 (p. [52]), p. [1-52]. *Copy*: NSUB, Inf. G. Wagenitz.

Potter, Michael Cressé (1858-1948), British clergyman and botanist; BA Cambridge 1881; at Cambridge herbarium 1884-1889; professor of botany at Armstrong College, Newcastle, 1889-1925; from 1925-1948 in retirement at New Milton, Hampshire. (*Potter*).

HERBARIUM and TYPES: Unknown.

BIBLIOGRAPHY and BIOGRAPHY: Barnhart 3: 103 (b. 7 Sep 1858); BJI 2: 47; BL 2: 238, 704; BM 4: 1603; CSP 11: 53, 17: 983; Desmond p. 50 (d. 9 Mar 1948); Kelly p. 179; Kew 4: 349; LS 21152-21158a, 38031-38042, suppl. 22153-22154; Morren ed. 10, p. 74; Rehder 5: 684; Stevenson p. 1254; Tucker 1: 566.

BIOFILE: Alexander, N.S., Nature 161: 673-674. 1948 ("... the last of the calorists" opposing first law thermodynamics).
Anon., Bot. Centralbl. 51: 127. 1892 (prof. bot. Newcastle); Bot. Jahrb. 16 (Beibl. 37): 20. 1892 (prof. bot. Newcastle); Proc. Bournemouth nat. Sci. Soc. 38: 73. 1947/1948 (obit., b. 7 Sep 1858; stressed 1900 that bacteria may be pathogenic to plants).
Hawksworth, D.L. & M.R.D. Seaward, Lichenology Brit. Isl. 134. 1977.
Ramsbottom, J., Trans. Brit. mycol. Soc. 30: 3, 4, 13. 1948.
Thomas, M., Nature 161: 590-591. 1948 (d. 9 Mar 1948; hobbies: campanology and dowsing).

COMPOSITE WORKS: (1) Translated E. Warming, *Handbook of systematic botany* (1894).
(2) *Botany*, in The Victoria history of the country of Durham: 35-81. 1905 (fide BL).

NOTE: Published an *Elementary textbook of agricultural botany* (1893, see J. Bot. 31: 378-379. Dec 1873).

Pottier, Jacques Georges (1892-x), French botanist; studied at the École pratique des Hautes Études; Dr. sci. Paris 1920; later at the Faculté des Sciences de Besançon. (*Pottier*).

HERBARIUM and TYPES: Unknown. – Letters at G.

BIBLIOGRAPHY and BIOGRAPHY: Barnhart 3: 103; Hirsch p. 237; Kew 4: 350.

BIOFILE: Anon., Rev. bryol. lichénol. 7: 47, 62, 63. 1920, 53: 30. 1926.
Pottier, J.G., Titres et travaux scientifiques de M. Jacques Pottier, Besançon, ca. 1934, 26 p. (bibl.).

8231. *Recherches sur le développement de la feuille des mousses.* Chartres (Imprimerie Durand ...) 1920. Oct. (*Rech. devel. feuill. mouss.*).
Publ.: 1920 (Rev. bryol. Aug-Sep 1920), p. [i]-viii, [1]-144, *pl. 1-30*, 2 photos. *Copies*: FI, H, NY.

8232. *Nouvelles recherches sur le développement de la feuille des Muscinées* ... Saint-Dizier (Établissements André Brulliard) 1925. Oct.
Publ.: Dec 1925 (bon à tirer 30 Nov 1925, cover of journal issue), p. [1]-60, *pl. 11-44*. *Copies*: G, H. – Reprinted and to be cited from Bull. Soc. bot. France 72: 629-689. *pl. 11-44*. 1925. Inf. P. Isoviita.

8233. *Recherches sur l'anatomie comparée des espèces dans la famille des Elatinacées* et sur le

développement de la tige et de la racine dans le genre Elatine ... Besançon (Imprimerie de l'Est) 1927. Oct. (*Rech. anat. Elatin.*).
Publ.: 1927, p. [1]-157, *pl. 1-19* with text (uncol.). *Copies*: MICH, USDA.

Pottier-Alapetite, Germaine (1894-x), French botanist at the Institut des Hautes Études de Tunis. (*Pottier-Alapetite*).

HERBARIUM and TYPES: TUN.

BIBLIOGRAPHY and BIOGRAPHY: Kew 4: 350; IH 1 (ed. 3): 156, (ed. 4): 177, (ed. 5): 72, 189; Roon p. 89.

BIOFILE: Anon., Rev. bryol. lichénol. 24: 171. 1955.
Verdoorn, F., ed., Chron. bot. 1: 251. 1935.

Potzger, John Ernest (1886-1955), American botanist and ecologist (palynologist); Dr. phil. Butler Univ. 1932; at Dept. of Botany, Butler Univ., Indianapolis, from 1932-1955 (full professor 1948-1955). (*Potzger*).

HERBARIUM and TYPES: BUT and IND; some material at F and US. – Manuscripts SIA 227.

BIBLIOGRAPHY and BIOGRAPHY: Barnhart 3: 104 (b. 31 Jul 1885, d. 18 Sep 1955); Bossert p. 316; IH 1 (ed. 1): 43, (ed. 2): 61; Kew 4: 350.

BIOFILE: Anon., Proc. Indiana Acad. Sci. 39: 15, 328. 1930; Science 122: 637. 1955 (d.).
Billings, W.D., Butler Univ. bot. Stud. 13(1): 3-11. 1956 (obit., bibl., portr.).
Courtemanche, A., Rev. canad. Géogr. 9: 207-210. 1956 (obit., portr.).
Deam, C.C., Fl. Indiana 9, 1018. 1910.
Hamilton, E.S., Bull. Torrey bot. Club 83: 76-78. 1956 (obit.).
Markle, M.S., Proc. Indiana Acad. Sci. 146. 1967.
Verdoorn, F., ed., Chron. bot. 1: 296, 376. 1935.

COMMEMORATIVE VOLUME: Butler University Botanical Studies 13(1), Dec 1956.

Pouchet, Albert Maxime (1880-1965), French amateur mycologist at Lyon; originally a shoemaker; later shop assistant; long time president of the Société linnéenne de Lyon, "homme-champignon" (*A. Pouchet*).

HERBARIUM and TYPES: Unknown.

BIBLIOGRAPHY and BIOGRAPHY: Kelly p. 179; Kew 4: 350; LS suppl. 22155-22160.

BIOFILE: Josseraud, M., Bull. mens. Soc. Linn. Lyon 34: 239-252. 1965 (portr.).

8234. *Monographie des Myxomycètes de France* ... Bourg (Imprimerie nouvelle-Victor Berthod) 1927. Oct. (*Monogr. myxomyc. France*).
Publ.: Jan 1927 (p. 3: 15 Aug 1926; date of Bulletin 30(41); BR rd. journal issue on 15 Mar 1927), p. [1]-71, *pl. 1-3*. *Copies*: G, Stevenson. – Reprinted from Bull. Soc. Natural. Ain 30: 192-262. Jan 1927.

Pouchet, Félix Archimède (1800-1872), French naturalist and physician; Dr. med. Paris 1827; professor of botany at the Jardin des Plantes, and director of the Muséum d'Histoire naturelle, Rouen from 1821-1872. (*F. Pouchet*).

HERBARIUM and TYPES: Unknown.

NAME: Pouchet wrote A.F. Pouchet on his 1829 publication, but F.A. Pouchet on the 1834 and 1835 publications. His later zoological publications are also signed F.A. Pouchet.

BIBLIOGRAPHY and BIOGRAPHY: Barnhart 3: 104 (b. 26 Aug 1800, d. 6 Dec 1872); BM 4: 1604; CSP 4: 996-997, 8: 651; Herder p. 69, 141, 313; Jackson p. 144, 422; Kew 4: 350-351; KR p. 579; Langman p. 593; LS 21163-21182; PR 7282-7285 (err. b. 26 Aug 1810); Rehder 5: 684.

BIOFILE: Anon., J. de Zool. 1: 335-336. 1872.
Beaurain, N. et al., Bull. Soc. Amis Sci. nat. Rouen ser. 2. 13: 175- 229. 1877 (portr.).
Bulloch, W., Hist. bacteriol. 391. 1938.
Carpenter, M.M., Amer. Midl. Natural. 33(1): 81. 1945.
Costantin, J., Ann. Sci. nat., Bot. ser. 10. 16: xiii. 1934 (on generatio spontanea debate; the portrait is possibly wrong).
Crellin, J.K., Dict. sci. biogr. 11: 109-110. 1975 (b. 26 Aug 1800, d. 6 Dec 1872, biogr. refs., notes on generatio spontanea debate with Pasteur and on his recognition of the periodicity of human ovulation).
Freeman, R.B., Brit. nat. hist. books 284. 1980.
Gager, C.S., Brooklyn Bot. Gard. Rec. 27(3): 212. 1938.
Pennetier, G., Un débat scientifique, Pouchet et Pasteur, Rouen 1907 (fide Crellin).

EPONYMY: *Pouchetia* A. Richard ex A.P. de Candolle (1830). *Note*: *Pouchetia* Schütt (1895) honors G. Pouchet (*fl.* 1882-1894), French author on dinoflagellates.

8235. *Histoire naturelle et médicale de la famille des Solanées*, par A.F. Pouchet ... Rouen (F. Baudry, ...) 1829. Oct. (*Hist. nat. Solan.*).
Publ.: 1829, p. [i]-viii, [9]-187. *Copies*: BR, MO, NY.

8236. *Flore ou statistique botanique de la Seine-inférieure*, contenant la description, les propriétés médicales et économiques, et l'histoire abrégée des plantes de ce département; par F.-A. Pouchet, .. Rouen (F. Baudry, ...) 1834, 2 vols. †. Duod. (in sixes). (*Fl. Seine-inf.*).
1: 1834, p. [i*, iii*, v*], [i]-xvi, [1, h.t.], [1]-10.
2: 1834, p. [i, iii], [1]-84.
Copies: B, NY. – Seven "tomes" were anticipated.

8237. *Traité élémentaire de botanique* appliquée, contenant la description de toutes les familles végétales, et celle des genres cultivés ou offrant des plantes remarquables par leurs propriétés ou par leur histoire, par F.A. Pouchet, ... À Rouen (chez E. Legrand, ...) 1835-1836, 2 vols. Oct. (*Traité élém. bot.*).
1: 1835, p. [i*, iii*, v*], [i]-vii, [1]-71, chart, [1]-396.
2: 1896, p. [i, iii], [1]-661.
Copies: NY, PH.

Poulsen, Viggo Albert (1855-1919), Danish botanist; high school teacher 1876-1909; student of and assistant with E. Warming; Dr. phil. Copenhagen 1888; from 1893 lecturer, from 1902 professor of botany at the Copenhagen pharmaceutical school; worked at Bogor, Java 1894-1895. (*Poulsen*).

HERBARIUM and TYPES: C; further material at LCU and P. – Letters at G.

BIBLIOGRAPHY and BIOGRAPHY: Barnhart 3: 104; BJI 1: 46, 2: 139; BM 4: 1604; Bossert p. 316; CSP 11: 56, 17: 989-990; De Toni 1: c; GR p. 679, cat. p. 69; Herder p. 118; Jackson p. 60, 62, 64, 66; Kelly p. 179; Kew 4: 351; KR p. 579; LS 21183-21183a; Morren ed. 10, p. 48; Rehder 5: 685; SK 1: 415, 5: cccxii.

BIOFILE: Anon., Ber. deut. bot. Ges. 37: 471, (160). 1920 (d.); Bot. Centralbl. 54: 352. 1893 (lecturer pharmac. coll.); Bot. Jahrb. 17 (Beibl. 41): 52. 1893 (id.); Bot. Not. 1920: 50 (d.); Österr. bot. Z. 43: 268. 1893 (lecturer), 69: 88. 1920 (d.).
A.S., Farmaceutisk Tidende 29: 661-664. 1919 (obit.).
Christensen, C., Dansk. bot. litt. 1880-1911: 70-73. 1913 (portr., bibl.); Dansk. bot. hist. 1: 674-675, 882, 2: 480-487, 535. 1926 (portr., bibl., b. 31 Mai 1855, d. 16 Oct 1919).
De Toni, G.B., Nuova Notar. 32: 70. 1921 (death).

Kolderup Rosenvinge, L., Bot. Tidsskr. 37: 107-112. 1920 (obit., portr. b. 31 Mai 1855, d. 16 Oct 1919).
Lind, J., Danish fungi; herb. Rostrup 37. 1913.
Mentz, A., Vid. medd. naturh. For. København 71: v-ix. 1920 (obit.).
Paulsen, O., Naturens Verden 1919: 517-519 (obit., portr.).
Petersen, O.G., Dansk biogr. Lex. 13: 260. 1899.
Warming, E., Bot. Tidsskr. 12: 233-235. 1881 (bibl.).

COMPOSITE WORKS: (1) EP, *Nat. Pflanzenfam.*, *Cynocrambaceae*, in 3(1A): 120-124. Aug 1893.
(2) Warming, E., *Symb. bras.*, *Xyridaceae*, 40: 49-120*. 2 Mai 1894.

HANDWRITING: Christensen, C., Dansk. bot. hist. 1: 676. 1926.

EPONYMY: *Poulsenia* Eggers (1898).

8238. *Planternes bygning og liv* en almenfattelig fremstilling frit bearbejdet efter Thomé af V.A. Poulsen. Kjøbenhavn (Andr. Fred. Høst & Sons Forlag) 1878. Oct. (*Pl. bygn. liv.*).
Publ.: 1878 (p. 4: Nov 1877, front.), p. [i*], [1]-249, [250, err.], [i]-xi. *Copies*: H, MO. – Very popular.

8239. *Anatomiske studier over Eriocaulaceerne* ... København (hos brødrene Salmonsen (J. Salmonsen)) 1888. Oct. (*Anat. stud. Eriocaul.*).
Publ.: Dec 1888 (p. [ii]: 24 Nov 1888; preprint dated 1888; Bot. Zeit. 18 Jan 1889), p. [i-v], [1]-166, [167, err.], *pl. 1-7* (uncol.). *Copy*: US. – Thesis Dr. phil. Preprinted or reprinted from Vid. Medd. naturh. For. København 1888: 221-386, *pl. 1-12.* 1888. The plates in the US copy of the reprint may be incomplete.
Ref.: Bay, J.C., Bot. Centralbl. 2(1): 34-36. 1892 (rev.).

Pound, Roscoe (1870-1964), American botanist; Dr. phil. Univ. Nebraska 1897, director Nebraska bot. Survey 1892-1903; Dean College of Law, Nebraska 1903-1907; professor of law Northwestern Univ. 1907-1909; Law School Chicago 1904-1915; LL.D. Mich. Univ. 1913; Harvard Law School 1910-1915; Dean of Harvard Law School 1915-1936, professor of law 1937-1947. (*Pound*).

HERBARIUM and TYPES: NEB; some further material at DNB and E. – Letters and portrait at G.

BIBLIOGRAPHY and BIOGRAPHY: Barnhart 3: 104; BL 1: 195, 196, 313; BM 4: 1605; Bossert p. 316 (b. 27 Oct 1870, d. 1 Jul 1964); CSP 17: 991; GR p. 239; Kelly p. 179; Kew 4: 351; Lenley p. 334; LS 21184-21191, 25027; Pennell p. 616; PH 444; Rehder 5: 685; Stevenson p. 1254; Tucker 1: 566.

BIOFILE: Ewan, J. et al., Short hist. bot. U.S. 125. 1969.
Hedge, I.C. & J.M. Lamond, Index coll. Edinb. herb. 120. 1970.
Pound, R., Amer. Natural. 1895: 1093-1100, 1896: 55-58 (discussion nomenclatural propositions Ascherson/Engler), J. Bot., Morot, 10: 108-112. 1896 (partial transl. with notes by O. Kuntze).
Rogers, D.P., Brief hist. mycol. N. Amer. ed. 2. 30, 45. 1981.
Stieber, M.S., Bull. Hunt Inst. bot. Docum. 2(2): 3 (portr.), 4 (biogr. note). 1980.
Verdoorn, F., ed., Chron. bot. 1: 369. 1936, 2: 328, 4: 81. 1938.

8240. *The phytogeography of Nebraska* I. General survey by Roscoe Pound ... and Frederic E. Clements ... Lincoln, Nebr. (Jacob North & Co., Publishers) 1898. Oct. (*Phytogeogr. Nebraska*).
Co-author: Frederic Edward Clements (1874-1945).
Ed. 1: 1898 (t.p.; copyright 1897; p. iv: Mai 1897; Bot. Gaz. 16 Mai 1898; Nat. Nov. Jul(2-3) 1898; Bot. Centralbl. 28 Dec 1898), p. [i]-xxi, [1]-329, [330], *maps 1-4*. *Copies*: MO, NY. – Constituted the combined thesis of the two authors, accepted by the

University of Nebraska 12 Mai 1897. – The larger part of the stock of the first edition was destroyed by fire. The book was announced by Nat. Nov. Oct(1) 1897.
Ed. 2: 1900 (p. 7: 1 Feb 1900; Bot. Centralbl. 17 Apr 1901; Bot. Gaz. 25 Nov 1901; J. Bot. Mai 1901; Bot. Zeit. 16 Mai 1901; Nat. Nov. Apr(2) 1901), p. [1]-442, [1, expl. maps], 4 maps. *Copies*: BR, FI, G, MICH, MO, NY, US. – Second edition, Lincoln, Neb., U.S.A. (published by the Seminar) 1900. Oct. – University of Nebraska, Botanical Survey of Nebraska conducted by the Botanical Seminar.
Ref.: Cowles, H.C., Bot. Gaz. 25: 370-372, 16 Mai 1898 (ref. ed. 1).
Hollick, A., Science n.s. 13: 981-983. 21 Jun 1901 (rev. ed. 2).

Pourret, [Pourret-Figeac], **Pierre André** (1754-1818), French clergyman and botanist; explorer of the flora of the Pyrenees; originally at Narbonne, in charge of the collections of the brothers Loménie de Brienne; in exile in Spain from 1789 as director of the botanic garden and professor of botany in Barcelona; subsequently at Madrid and Canon of the cathedral at Orense [Galicia], for some time in hiding at Vieïro; ultimately Canon at Santiago. (*Pourr.*).

HERBARIUM and TYPES: MAF and P (rd. 1857); other material at B-W, BM, FI, MAF, MPU, SBT and UPS. – Most of Pourret's collections were lost during the various wars, but the collections made by him for the Brienne family were kept by the family and offered to Pourret in 1812. He refused them and they came into the possession of Barbier. They were recognized by Bonnet in the Barbier herbarium (Joseph-Athanase Barbier, 1767-1846, physician) which had been left to P. in 1846. This Pourret herbarium, dating from before 1789, is now inserted in the general herbarium and in the French herbarium at P. The collectors are, except Pourret (S. France, Pyrenees, Brienne, Paris): W. Aiton(England), C. Allioni (Italy), Asso (Spain), J. Banks (England, Canary Is., Açores, Madeira), Barrera (Roussillon, S. France), P.M.A. Broussonet (Montpellier), A. Cavanilles (Spain), D. Chaix (Dauphiné), R. Desfontaines (Algeria), N.J. Jacquin (Austria), A.L. de Jussieu (*ex herb.*), Ant. de Jussieu (*ex herb.*), P.P. de Lapeyrouse (Pyrenees), M.A.L.C. de Latourette (div.), L.G. Lemonnier (Paris garden), J.B. Leschenault (India), C.L. L'Héritier (Paris, garden plants), C. Linnaeus fil. (Sweden), Mayoral (Spain), L.Née (Spain), C.G. Ortega (Spain), Pech (Narbonne), J.L.M. Poiret (N. Africa), L.C. Richard (Trianon garden), G.A. Scopoli (Italy, Tyrol), J.F. Séguier (Italy, France), D.C. Solander (Cook's ... voyage), J.R. Spielmann (Alsace, Oriet), C.P. Thunberg (Cape), A. Thouin (Paris garden), D. Villars (Dauphiné), and a few specimens by Tournefort (Spain, Orient), Salvador (Spain, Portugal) and J. Ray. The herbarium contains all Pourret types, those of the Cistaceae are kept apart (see Bonnet 1916; Aymonin 1965). A small Pourret herbarium (341 specimens) at P is the *Corona florae Narbonensis et Pyrenaeis* made in 1781, see Guétrot (1931). – Letters and portait at G.

BIBLIOGRAPHY and BIOGRAPHY: AG 2(1): 272, 5(2): 122, 5(4): 19; Backer p. 461; Barnhart 3: 104; BM 4: 1605; Colmeiro 1: cxciv; CSP 12: 586; Dryander 3: 143, 236; Frank 3(Anh.): 78; Herder p. 74; Hortus 3: 1202 (Pourr.); IH 1 (ed. 2): 78, (ed. 3): 98, 99, (ed. 6): 363, 7: 340; ; Kew 4: 351; Lasègue p. 348, 353; Lenley p. 466; PR 7287 (ed. 1: 8182); Rehder 5: 685; RS p. 134; SO 2519; TL-2/see P.J. Bergius; Urban-Berl. p. 384, 415; Zander ed. 10, p. 703, ed. 11, p. 802.

BIOFILE: Anon., Bull. Soc bot. France 24 (bibl.): 47. 1877 (Pourret specimens and brochures for sale).
Aymonin, G., Bull. Soc. bot. France 111: 150-151. 1965. (herb. Pourret).
Bonnet, E., Bull. Mus. Hist. nat. 1916(6): 278-286 (on Brienne, Pourret and Barbier collections).
Candolle, Alph. de, Phytographie 440. 1880 (coll.).
Clos, ,D., Pourret et son histoire des Cistes [Toulouse 1858], 22 p., reprinted from Mém. Acad. Sci. Toulouse. See Bull. Soc. bot. France 5: 291-293. 1858.
Colmeiro, M., Noticia de los trabajos botánicos del abate Pourret en Francia y España. Madrid 1891, 16 p. (copies at BR and G).
Dawson, W.R., Banks correspondence 682-683. 1958 (contents letters to B.).
Granel de Solignac, L. & L. Bertrand, Naturalia monsp. 18: 281. 1967.

Guétrot, Bull. Soc. bot. France 78: 434. 1931 (on herb. *Corona florae narbonensis* at P).
Lapeyrouse, P.P. de, Hist. abr. pl. Pyren. xxviii-xxx. 1813 (on his work on the flora of the Pyrenées; P, in 1813, at Orense [Orenzé]); Suppl. Hist. abr. pl. Pyren. vi. 1818.
Maugeret, A., Bull. Soc. bot. France 9: 594-596. 1864 (on Galibert's biography and Clos' study of the monograph of Cistus).
Pons, S., Bull. Soc. bot. France 40: lxxi-lxxv. 1893 (letter P. to P. de Barréra).
Pourret, P.A., Noticia historica de la familia Salvador de ... Barcelona por Don Pedro Andres Pourret. Barcelona 1796. 32 p. (PR ed. 1, 8183).
Stafleu, F.A., *in* C.L. L'Héritier, Sert. angl. facs. ed. 1963. p. xliii (letter from L.).
Steinberg, C.H., Webbia 32: 35. 1977 (material at FI in H. Desf.?).
Stieber, M.T. et al., Huntia 4(1): 85. 1981 (corr. HU).
Timbal Lagrave, E., Mém. Acad. Sci. Toulouse ser. 7. 4: 438-439. 1872; Bull. Soc. bot. France 21 (bibl.): 239. 1874 (on a Pourret mss. itinéraire pour herboriser dans les Pyrénées; Reliquiae Pourretianae, Bull. Soc. Sci. phys. nat. Toulouse 2: 1-147; reprint Toulouse 1875, iii and 149 (sic) p. Has on p. 5-23 a *Notice biographique sur l'abbé Pourret* extracted from Galibert, La vie et les travaux du botaniste P.A. Pourret, Revue de Toulouse, Juillet (1), 1867, and on p. 104-148 a reprint of the *Chloris narbonensis* (originally publ. in Mém. Acad. Sci. Toulouse ser. 1. 3: 287 ff)).
Willkomm, M., Grundz. Pfl.-Verbr. Iber. Halbinsel 6-7, 1896 (Veg. Erde 1).

EPONYMY: *Pourretia* Ruiz et Pavon (1794); *Pourretia* Willdenow (1800).

Pouzolz [Pouzols], **Pierre Marie Casimir de** (1785-1858), French botanist at Nîmes, "câpitaine en retraite" (1842). (*Pouzolz*).

HERBARIUM and TYPES: MPU; other material at B, E, FI, P. – Letters at G.

BIBLIOGRAPHY and BIOGRAPHY: AG 2(1): 658; Backer p. 462; Barnhart 3: 105 (b. 17 Dec 1785, d. 5 Mar 1858); BL 2: 149, 704; BM 4: 1605; Jackson p. 283; Kew 4: 351; PFC 2(2): xvi; PR 7289-7290 (ed. 1: 8185); Rehder 5: 685; Saccardo 1: 132; Tucker 1: 566; Urban-Berl. p. 384.

BIOFILE: Anon., Bull. Soc. bot. France 10: 176. 1863 (herb. of *Flore du Gard* for sale).
Candolle, Alph. de, Phytographie 440. 1880 (coll. at MPU).
Granel de Solignac, L. & L. Bertrand, Naturalia monsp. 18: 274. 1967 (pl. MPU), 26: 27. 1976 (id. herb. Vichet).
Hedge, I.C. & J.M. Lamond, Index coll. Edinb. herb. 120. 1970.
Lombard-Dumas, A., Bull. Soc. bot. Gard. 47: 526. 1900.
Loret, H., Plantes nouvelles pour le Gard, Montpellier 1880. 10 p. (notes on P.'s herbarium).
Pouzolz, P.C.M., Mém. Soc. Linn. Paris 4(6): 560, cxxxvi. 1826 (cat. pl. Corsica; correct christian names); see also Bull. Linn. Paris, *in* Mém. id. 4, p. 67.
Reynier, A., P.V. Soc. Hort. Bot. Marseille 41: 207-208. 1895 (letter).
Steinberg, C.H., Webbia 32: 35. 1977 (pl. at FI; err. Pouzols).

EPONYMY: *Pouzolsia* Bentham (1849, *orth. var.*). *Note*: *Pouzolzia* Gaudichaud-Beaupré (1830) probably also commemorates him.

8241. *Catalogue des plantes qui croissent naturellement dans le Gard*, pour servir à la formation de la flore de ce département par M. P.M.C. de Pouzolz, ... Nismes [Nîmes] (Imprimerie de Ballivet et Fabre, ...) 1842. Qu. (*Cat. pl. Gard*).
Publ.: 1842, p. [1]-46, [1, err.]. *Copy*: G. – See also H. Loret, *Plantes nouvelles pour le Gard* avec des observations préliminaires sur la flore de Pouzol et sur son herbier départemental, Montpellier, 1880, 10 p., TL-2/5002.

8242. *Flore du département du Gard* ou description des plantes qui croissent naturellement dans ce département ... Montpellier (C. Coulet, ...), Paris (Adrien Delahaye, ...) 1862. Oct. 2 vols. 1862. (*Fl. Gard*).
1: 1862, p. [i], [1]-659, *pl. 1-5* (col. liths.).
2(1): 1862, p. [i], [1]-338.

2(2): 1862(?), p. [339]-644, *pl. 6-7* (col. liths). − "Suite et fin" by Philippe Courcière (x-1885) on p. 505-644.
Copies: B, G, NY, USDA.
Suite et fin: also published separately, p. [i-ii], 503-644, with preface, see Tucker 1: 566. − See also BL 2: 149 for various publications updating the flora, such as B. Martin, Révision de la flore du Gard de Pouzolz, Nîmes 1892, 20 p.

Povah, Albert Hubert William (1889-1975), American mycologist; Dr. phil. Univ. Mich. 1916; at Ann Arbor 1916-1917; lecturer on forest mycology at Syracuse Univ. 1917-1921; at Alabama Polytechnic 1921-1922, at Northwestern Univ., Evanston, Ill. 1922-1925; at Univ. Cincinnati 1929-1930, at Harvard 1929-1930; at Wayne University 1934-1935, ultimately with Detroit high schools. (*Povah*).

HERBARIUM and TYPES: MICH (all orig. coll.); other material at NY (musci).

BIBLIOGRAPHY and BIOGRAPHY: Barnhart 3: 105; LS suppl. 22163-22166; Stevenson p. 1254.

BIOFILE: Crum, H., in lit. to R.S. Cowan 27 Mar 1980 (b. 25 Mar 1889, d. 1 Apr 1975).
Schuster, R.M., Hepat. Anthoc. N. Amer. 1: 88. 1966.
Seaver, F.J., North-Amer. cup-fungi 364. 1942 (bibl.).

Powell, John Wesley (1834-1902), American naturalist, geologist, ethnologist, and administrator; major of artillery, U.S. army, in Civil War 1861-1865; after the war professor of geology in Illinois Wesleyan College and Illinois Normal Univ.; exploring the Rocky Mountains from 1867; from 1871-1879 head of the Survey of the Rocky Mountains; director U.S. Bureau of Ethnology 1879-1902; also Director U.S. Geol. Survey 1881-1894. (*J.W. Powell*).

HERBARIUM and TYPES: US; further material at GH. − Manuscripts Powell Collection, Illinois State Normal University and in Smithsonian Archives.

BIBLIOGRAPHY and BIOGRAPHY: Barnhart 3: 105 (b. 24 Mar 1834, d. 23 Sep 1902); BM 4: 1605; CSP 8: 654, 11: 57, 17: 994; Ewan ed. 1: 355 (index), ed. 2: 177-178 (great detail); Herder p. 18, 36; Lenley p. 334; ME 1: 220-221; Merrill p. 709, 729; PH 1, 1-B, 567; Quenstedt p. 344; PH 1, 1-8, 567; Rehder 5: 685; Tucker 1: 13, 566.

BIOFILE: Anon., Pop. Sci. Monthly 20: 390-397. 1882.
Brewer, N.H., Amer. J. Sci. ser. 4. 14: 377-382. 1902 (obit.).
Bryant, H.C., Cosmos Club Bull. 1(8): 2-5. 1948 (co-founder Cosmos Club, 1878).
Chamberlain, A.F., J. Amer. Folklore 15: 201-203. 1902 (bibl.).
Dall, W.H., Bull. philos. Soc. Washington 14: 300-308. 1905 (biogr. sketch).Darrah, W.C., Powell of the Colorado. Princeton 1951, xiii, 426 p. (main biogr.).
Davis, W.M;, Biogr. Mem. Natl. Acad. Sci. 8: 11-83. 1915 (first main biogr., portr.).
Dellenbaugh, F.S., The romance of the Colorado River. The story of its discovery ... with special reference to the voyages of Powell through the line of the great canyons. ed. 3. New York, London, 1909, xxxvii, 402 p. (ed. 1: 1902).
Gilbert, G.K., Science ser. 2. 16: 561-567. 1902 (obit., portr., b. 24 Mar 1834, d. 23 Sep 1902); lost right arm in battle of Shiloh, civil war); Proc. Washington Acad. Sci. 5: 113-118. 1903; Ann. Rep. Smiths. Inst. 1902: 633-640. 1903 (obit., portr.).
Graham, E.H., Bot. Studies Uinta Basin, Utah, 14-17. 1937 (Ann. Carnegie Mus. 26).
Lincoln, M.D. et al., Open Court, Chicago 16: 705-715. 1902, 17: 14-25, 86-94, 162-174, 228-239, 281-290, 342-351. 1903.
McVaugh, R., Edward Palmer 84. 1956.
Meadows, P., John Wesley Powell, Univ. Nebr. Stud. ser. 2. 10, 106 p. (biogr., bibl., further biogr. refs.).
Merrill, E.D., Bishop Mus. Bull. 144: 152. 1937, Contr. U.S. natl. Herb. 30: 244. 1947 (bibl.).
Merrill, J.W., Amer. Geologist 31(6): 327-332. 1903 (obit., portr.; Survey of the Rocky Mountains in rivalry with Hayden and Wheeler).

Mitten, W., J. Linn. Soc., Bot. 10: 166-195. 1868 (P's mosses described).
Moldenke, H.N., Plant Life 2: 77. 1946 (1948).
Osborn, H., Fragm. entom. Hist. 96. 1937.
Poggendorff, J.C., Biogr.-lit. Handw.-Buch 2: 1189. 1863, 3: 1064. 1898, 5(2): 1001. 1926 (d. 23 Sep 1902).
Powell, J.W., Down the Colorado. Diary of the first trip through the Grand Canyon 1869. Eliot Porter: photographs and Epilogue 1969. New York 1969, 168 p.
Rabbitt, M.C. et al., The Colorado River region and John Wesley Powell., Geol. Surv. prof. Pap. 669, Washington 1969. A collection of papers honoring Powell on the 100th anniversary of his exploration of the Colorado River, 1869-1969. (contains Rabbitt, J.W. Powell, pioneer statesman of federal science, p. 1-21), (frontisp., portr.).
Reveal & Hafen et al., Western America ed. 3. 391, 393. 1970; Reveal, J. in A. Cronquist et al., Intermount. Fl. 1: 56, 57. 1972.
Rodgers, A.D., Amer. bot. 42-43, 49, 57, 137, 237. 1944.
Ross, K., Cosmos Club Bull. 11(10): 3-12. 1958 (P's role in founding the Cosmos Club; id. Natl. Geogr. Soc.; portr., biogr. sketch).
Smith, H.N., Miss. Valley hist. Rev. 34: 37-58. 1947 (establishment U.S. Geol. Survey).
Spitzka, E.A., Amer. Anthropologist ser. 2. 5(4): 585-643. 1903 (study of the brain of J.W.P.).
Stegner, W., beyond the hundredth meridian, John Wesley Powell and the second opening of the West. Boston 1954, xxv, 438 p.
Walcott, C.D. et al., Proc. Washington Acad. Sci. 5: 99-112. 1903 (portr., various papers on P read at commemoration meeting, followed by bibl. by Warman, q.v.).
Ward, L.F., Glimpses of the Cosmos 2: 426-438. 1943 (portr., biogr. sketch).
Warman, P.C., Proc. Washington Acad. Sci. 5: 131-187. 1903 (*main bibl.*, 251 nos.).
Watson, E.S., ed., Professor goes west: Illinois Wesleyan University, Reports on Major John Wesley Powell's expeditions 1867-1934, Bloomington 1954, 138 p., see also Ewan, J., Rhodora 63: 179-180. 1961.
Zittel, K.A. von, Hist. geol. palaeontol. 184, 207, 209, 210, 211, 274. 1901.

8243. *Exploration of the Colorado River* of the West and its tributaries. Explored in 1869, 1870, 1871, and 1872, under the direction of the secretary of the Smithsonian Institution. Washington (Government Printing Office) 1875. Qu. (*Explor. Colorado R.*).
Publ.: 1875 (US copy inscribed 10 Sep 1875), p. [i*], front. [fig. 1], [iii*], [i]-xi, [1]-291, *pl. 2-80. Copy*: US. – Warman no. 11. – For a revised edition see below, *Canyon's of the Colorado*, 1895.

8244. *Canyons of the Colorado* ... Meadville, Pa. (Flood & Vincent, the Chautauqua Century Press) 1895. Oct. (*Canyons Colorado*).
Orig. ed.: 1895, p. [i]-xiv, [15]-400. *Copy*: MO. – A revised and enlarged edition of the *Expl. Colorado Riv.*, Warman 41.
Facsimile ed.: 1964, p. [i*], frontisp., [i]-xiv, [15]-400. *Copy*: US.

Powell, Thomas (1809-1887), British missionary and naturalist; at Upolu, Samoa 1860-1865, also working in the Gilbert and Ellice Islands and other parts of Oceania. (*T. Powell*).

HERBARIUM and TYPES: K (main set); other material at B, BM, E, FH, L, LE, M, MEL, MICH, NA. – *Exsiccatae*: *Mosses of Samoa*, ca. 1867, nos. 1-140, set at FH (see Sayre 1971).

BIBLIOGRAPHY and BIOGRAPHY: Barnhart 3: 105; BB p. 247; CSP 4: 1005, 8: 654, 11: 57; Desmond p. 502 (d. 6 Apr 1887); Urban-Berl. p. 310, 322.

BIOFILE: Bridson, G.D.R. et al., Nat. hist. mss. res. Brit. Isl 255.232. 1980.
Hertel, H., Mitt. Bot. München 16: 419. 1980 (lich. M).
Jackson, B.D., Bull. misc. Inf. Kew 1901: 53 (Samoan pl. at K).
Murray, G., Hist. coll. BM(NH) 1: 174. 1904 (crypt. Samoa).
Sayre, G., Mem. New York Bot. Gard. 19: 239. 1971.

EPONYMY: *Powellia* Mitten (1868). – *Powellia* A.C. Batista et G.E.P. Peres (1964) was named in honor of Dulcie A. Powell, present day Jamaican-American botanist.

Power, Thomas (*fl.* 1845), Irish physician and botanist at the Cork school of medicine. (*Power*).

HERBARIUM and TYPES: Unknown.

BIBLIOGRAPHY and BIOGRAPHY: Barnhart 3: 105; BL 2: 284, 704; BM 4: 1606; Desmond p. 502; Jackson p. 250; Kew 4: 353; PR 7291; Rehder 1: 399; Tucker 1: 567.

BIOFILE: Freeman, R.B., Brit. nat. hist. books 285. 1980.
Knowles, M.C., Proc. R. Irish Acad. 38B(10): 183. 1929 (on his Contributions).
Lett, H.W., Proc. R. Irish Acad. B. 32: 71-72. 1915.
Mitchell, M.E., Bibl. Irish lichenol. 54. 1971.
Praeger, R.L., Proc. R. Irish Acad. 7: cxxxvii. 1901, Bot. Ireland 78. 1934, Some Irish Natural. 145-149. 1949.
Renouf, P.W., Irish Natural. J. 3(11): 238. 1931 (research on fungi in contributions by Denis Murray).

8245. *Contributions towards a fauna and flora of the county of Cork*, ... the Vertebrata by Dr. Harvey. The mollusca, crustacea and echinodermata by J.D. Humphreys. The flora by Dr. Power ...London (John van Voorst, ...), Cork (George Purcell & Co., ...) 1845. Oct. (in fours). (*Contr. fauna fl. Cork*).
Co-authors (*zoologists*): J.R. Harvey, John D. Humphreys.
Publ.: 1845, p. [i*-vi*], [i]-v, [1]-24, [1, 3], [1]-24, *Botanical guide*: [i]-v, [vi], [1]-130. *Copies*: DBN, G, NY, PH.

Pradal, Émile (1795-1874), French botanist and physician; surgeon-dentist at Nantes. (*Pradal*).

HERBARIUM and TYPES: NTM.

BIBLIOGRAPHY and BIOGRAPHY: GR p. 346; IH 2: (in press); Krempelh. 1: 497; LS 21197-21198; PR 7294.

8246. *Catalogue des plantes cryptogames* recueillies dans le département de la Loire-inférieure par Mr. E. Pradal. Nantes (Imprimerie de Mme Ve C. Mellinet) 1858. 18-mo. (*Cat. pl. crypt.*).
Publ.: 1858, p. [1]-254. *Copies*: FH, NY, USDA.

Praeger, Robert Lloyd (1865-1953), Irish botanist; BA Belfast 1885; practicing engineer 1886-1893; assistant librarian National Library of Ireland 1893, chief librarian 1920-1924; in retirement dedicating himself fully to his floristic studies. (*Praeger*).

HERBARIUM and TYPES: DBN (circa 10.000); other material in A, BEL, BM, CGE, E, GH, ILL, K, LIV, NOT. – Letters at G; see also Bridson (1980).

NOTE: We are grateful to Charles Nelson, Dublin, for reading the entry on Praeger and for various valuable additions and suggestions.

BIBLIOGRAPHY and BIOGRAPHY: Backer p. 462; Barnhart 3: 106; BJI 2: 139; BL 2: 704 (many refs.); BM 4: 1569, 1606, 8: 1026; CSP 17: 996-997; Desmond p. 502 (b. 25 Aug 1865, d. 5 Mai 1953); GR p. 755; Hortus 3: 1202 (Praeg.); IH 1 (ed. 6): 363, (ed. 7): 340; 2: (in press); Kew 4: 353-354; Langman p. 594; MW p. 393; LS 22174-22175; Tucker 1: 567; Zander ed. 10, p. 703, ed. 11, p. 802.

BIOFILE: Allen, D.E., The naturalist in Britain 239-240. 1976.
Anon., The New York Times 7 May 1953, p. 31.
Booth, E.M., The flora of County Carlow 6-8. 1979.

Bridson, G.D.R. et al., Nat. hist. mss. res. Brit. Isl., 84.18, 118.9. 1980.
Farrington, A. et al., Irish Natural. J. 11(6): 141-171. 1954 (biogr., portr.; no complete bibl. available yet; by 1901 P had published 121 items).
Fletcher, H.R., Story R. Hort. Soc. 317, 409. 1969.
Freeman, R.B., Brit. nat. hist. books 285. 1980.
Hackney, P., Irish natural. J. 17: 231, 233. 1972 (coll.).
Hawksworth, D.L. & M.R.D. Seaward, Lichenol. Brit. Isl. 198. 1977.
Hedge, I.C. & J.M. Lamond, Index coll. Edinb. herb. 120. 1970.
Kent, D.H., Brit. herb. 73. 1957.
Knowles, M.C., Proc. R. Irish Acad. 38B(10): 190. 1929.
Lousley, J.E., Irish Natural. J. 11: 17-176. 1954 (obit., appreciation).
Mitchell, G.F., Irish Natural. 11(6): 172-175. 1954 (on P's contr. to Quarternary geology).
Mitchell, M.E., Bibl. Irish lichenol. 54-55. 1971.
Murray, G., Hist. coll. BM(NH) 1: 174. 1904.
Nelson, E.C., Glasra 4: 59. 1980.
Piper, R., Introduction to reprint of R.L. Praeger, Nat. hist. Ireland 1972.
Praeger, R.L.,Proc. R. Irish Acad. 7: cxxxvii-cxli. 1901 (bibl. up to 1900); Bot. Ireland 78. 1934; The way that I went ed. 3, 1947 (autobiography); Some Irish naturalists, Tempest 1949, Dundalk 1949, p. 15-37, 146-147.
Robins, J., in E.C. Nelson & A. Brady, eds., Irish gard. hort. 85. 1979 (portr.).
Schollick, H.L., Brit. Pteridol. Soc. Bull. 1(3): 115-118. 1975 (tribute).
T.W.F., Geogr. J. 119(3): 368-369. 1953.
Webb, D.A., Proc. Bot. Soc. Brit. Isl. 1(1): 106-110. 1954 (obit.; personal recollections).
Verdoorn, F., ed., Chron. bot. 2: 29, 197, 214. 1936, 3: 153, 169, 297. 1937, 4: 280, 571. 1938, 5: 123, 302, 512. 1939.

COMPOSITE WORKS: (1) Co-editor, *Irish Naturalist* 1-33, 1892-1924.
(2) Contributed flowering plants, vascular cryptogams and charophytes to the second edition of S.A. Stewart & T.M. Corry, *A flora of the North-East of Ireland*, 1938, lix, 472 p. (rev. A.J. Wilmott, J. Bot. 77: 93-94. Mai 1939. Praeger wrote the chapters on flowering plants, vascular cryptogams and charophytes). – See S.A. Stewart.
(3) *Clare Island Survey* part 10, Phanerogamia and Pteridophyta, p. 1-112, *pl. 1-6*, reprinted with special cover from Proc. R. Irish Acad. 31(10): 1-112. 29 Nov 1911. (Scott. bot. Rev. 1912(1): 56. Jan 1912).

HANDWRITING: Irish Natural. J. 11(6): *pl. 5*. 1954.

MEMORIAL PUBLICATION: The Irish Naturalist's Journal 11(6): 141-176. 1954 (see A. Farrington, J.E. Lousley, G.F. Mitchell).

8247. *The flora of County Armagh* ... Dublin (Webb & Walpole, ...) 1893. Oct.
Publ.: 1893, p. [1]-37, *1 pl.*, 1 map. *Copy*: DBN. – Reprinted and to be cited from Irish Naturalist 2: 11-15 Jan 1893, 34-38. Feb 1893, 59-62. Mar 1893, 91-95. Apr 1893, 127-134. Mai 1893, 155-159. Jun 1893, 182-184. Jul 1893, 212-215. Aug 1893, 228-229. Aug 1893.

8248. *Open-air studies in botany*: sketches of British wild-flowers in their homes ... London (Charles Griffin & Co., ...) 1897. Oct. (*Open-air stud. bot.*).
Ed. 1: Aug 1897 (Bot. Centralbl. 8 Sep 1897; Bot. Zeit. 1 Oct 1897; Nat. Nov. Sep(2) 1897), p. [i*], [i]-xiii, [1]-266, *pl. 1-6* +frontisp. *Copies*: DBN, NY. – Plates by S. Rosamond Praeger.
Ed. 2: 1910, p. [i*], [i]-xiii, [1]-266, *pl. 1-6* + frontisp. *Copies*: BR, G.
Ref.: Britten, J., J. Bot. 35: 158-159. Nov 1897 (rev. ed. 1), 49: 176. Mai 1911 (rev.).

8249. *A tourist's flora of the West of Ireland* ... Dublin (Hodges, Figgis, and Co., Ltd., ...) 1909. Oct. (*Tour. fl. W. Ireland*).
Publ.: Mai 1909 (p. viii: Mar 1909; BR copy inscribed by author 10 Mai 1909; Nat. Nov. Jul(1) 1909), p. [i]-xii, [1]-243, 5 maps, *pl. 1-27*. *Copies*: BR, DBN, MO, NY, USDA.
Ref.: Britten, J., J. Bot. 47: 281. Jul 1909.

8250. *Irish topographical botany* ... Dublin (Published at the Academy House, ...). London, Edinburgh, Oxford 1901. Oct. (*Irish topogr. bot.*).
Publ.: 1 Jul-12 Aug 1901 (p. ix: Mai 1901; published by 13 Aug 1901, see Farrington (1954, p. 159); Nat. Nov. Jan(2) 1902), p. [i]-clxxxviii, [1]-410, *pl. 1-6*. *Copies*: BR, G, MO, USGS. – Issued as Proc. R. Irish Acad. ser. 3. 7, 1901.
Additions: Jan 1904, Irish Natural. 13: 1-15. 1904.
Ref.: Marshall, E.S., J. Bot. 39: 316-318. Sep 1901.

8251. *An account of the genus Sedum* as found in cultivation ... printed for the Royal Horticultural Society by Spottiswoode, Ballantyne & Co. Ltd. ... 1921. Oct. (*Acc. Sedum*).
Publ.: 6 Jun 1921 (journal issue so dated), p. [1]-314, [xi]-xii. *Copies*: NY, U. – Reprinted with special cover (with above title) from J.R. Hort. Soc. 46: 1-314. 1921.
Facsimile ed.: 1967, p. [i-ii], [1]-314. *Copies*: FAS, G, H, MO, NY. – Plant monograph reprints vol. 2, Lehre (J. Cramer) 1967. ISBN 3-7682-0446-4.

8252. *An account of the Sempervivum group* ... London (The Royal Horticultural Society ...) 1932. Oct. (*Acc. Sempervivum*).
Publ.: Aug 1932 (BR rd. 6 Sep 1932; Nat. Nov. Dec 1932), p. [i-iii], [1]-265. *Copies*: BR, DBN, FI, MO, NY, USDA.
Facsimile ed.: 1967, p. [i-v], [1]-265. *Copies*: FAS, H. – Plant monograph reprints vol. 1, Lehre (J. Cramer) 1967, ISBN 3-7682-0445-6.
Ref.: Rendle, A.B., J. Bot. 70: 296-298. Oct 1932.

8253. *The botanist in Ireland* ... Dublin (Hodges, Figgis, & Co. ...) 1934. Oct. (*Bot. Ireland*).
Publ.: 1934 (p. viii: Nov 1934; Nat. Nov. Jul 1935), p. [i]-xii, signatures A-2L = 7 (with paragraphs 1-492 followed by pages 493-587), 6 maps, *pl. 1-44*. *Copies*: FI, G, MO, USDA.
Facsimile ed.: 1974, East Ardsley (n.v.).
Ref.: Willmott, A.J., J. Bot. 73: 57-58. Feb 1935 (rev.).

Praetorius, Ignaz (1836-1908), German (West Prussian) botanist; high school teacher at Conitz; Dr. phil. Breslau 1863; habil. ib. 1863; teacher at the Kön. pedag. Seminar and the Mathiasgymnasium 1863-1864; id. at the Lyceum Hosianum at Braunsberg from 1864-1868, id. at Konitz 1868, with professor's title from 1880; ultimately at Graudenz. (*Praet.*).

HERBARIUM and TYPES: KBG (destroyed).

BIBLIOGRAPHY and BIOGRAPHY: Barnhart 3: 106; BJI 2: 46.

BIOFILE: Abromeit, J., Schr. phys.-ökon. Königsberg 51: 170-171. 1910 (obit., plants in herb. Phys.-ökon. Ges., KBG).
Anon., Bot. Not. 1909: 46 (d. 20 Oct 1908).
Caspary, R., Festschr. 50 j. Best. Preuss. Bot. Ver. 256-258. 1912 (bibl., portr., b. 11 Sep 1836; autobiographical).
Kneucker, A., Allg. bot. Z. 15: 16. 1909 (d. 20 Oct 1908).

EPONYMY: The derivation of *Praetoria* Baillon (1858) was not given.

8254. Königliches Gymnasium zu Konitz. Schuljahr 1888/89. Achtundsechzigster Jahresbericht ... *Zur Flora von Conitz*. Phanerogamen und Gefässkryptogamen ... Conitz (Buchdruckerei von Fr. W. Gebauer) 1889. Qu. (*Fl. Conitz*).
Publ.: Oct 1889 (Bot. Centralbl. 5 Nov 1889; Bot. Zeit. 27 Dec 1889; Nat. Nov. Nov(1) 1889), p. [1]-62, [on p. [63]-73 Schulnachrichten]. *Copy*: MO. – Programm 33, K. Gymnasium Conitz.
Ref.: Reiche, Bot. Jahrb. 12 (Lit.): 23 Dec 1890.

Prahl, Johann Friedrich (*fl.* 1837), German botanist; teacher at the gymnasium of Güstrow (Mecklenburg-Schwerin), later clergyman at Hehen-Horn (Lauenburg). (*J. Prahl*).

HERBARIUM and TYPES: Unknown.

BIBLIOGRAPHY and BIOGRAPHY: Barnhart 3: 106; BM 4: 1608; Herder p. 187; Kew 4: 355; PR 7297 (ed. 1: 8188);Rehder 5: 685; Tucker 1: 567.

BIOFILE: Boll, E.F.A., Fl. Meklenb. 159. 1860.

8255. *Index plantarum, quae circa Gustroviam sponte nascentur,* phanerogamarum ... prostat Gustroviae [Güstrow] apud Opitzium 1837. Qu. (*Index pl. Gustrov.*).
Publ.: 1837, p. [i]-iv, [1]-66. *Copy*: NY.
Nachtrag: 1847, p. [1]-12. *Copy*: NY. – "Nachtrag zu J.F. Prahl's Index plantarum quae circa Gustroviam sponte nascentur, phanerogamarum von J. Drewes." Güstrow (Gedruckt bei Ebert's Erben) 1847. We have no further details on J. Drewes.

Prahl, Peter (1843-1911), German (Schleswig) botanist; born in Danish northern Schleswig; studied medicine at the military hospital of the Friedrich Wilhelms Institut at Berlin; from 1867-1868 assistant physician at the Berlin Charité; military physician at Flensburg 1868-1870, in the Franco-Prussian war of 1870-1871, at Hadersleben 1871-1876 at Flensburg 1876-1879, at Kiel 1879-1888, at Stettin in 1890, at Wandbeck 1890-1892, 1892-1899 at Rostock; on sick leave and later retired living at Lübeck 1899-1911; ardent amateur botanist since his student days. (*Prahl*).

HERBARIUM and TYPES: HBG; further material at B, C, GOET, JE, KIEL, L. – Portrait at G.

BIBLIOGRAPHY and BIOGRAPHY: BFM no. 76; BL 2: 52, 704; BM 4: 1608; Christiansen, see index; CSP 12: 586, 17: 998; DTS 1: 393; GR p. 121; TL-2/3379, see Krause, E.H.L.; Morren ed. 10, p. 14; Tucker 1: 779; Urban-Berl. p. 310, 384.

BIOFILE: Anon., Hedwigia 53: (95). 1912 (d.).; Verh. bot. Ver. Brandenburg 53: (71). 1912 (d).
Ascherson, P., Heimat (Holstein) 22: 196-199. 1912 (obit., portr., b. 24 Mar 1843, d. 23 Oct 1911); Verh. bot. Ver. Brandenburg 53: (48)-(55). 1912 (portr., bibl., b. 24 Mar 1843, d. 23 Oct 1911), see also ib. 29: 132-166. 1888 (on Prahl's criticism of Knuth's *Fl. Schlesw.-Holst.*).
Kneucker, A., Allg. bot. Z. 17: 170. 1911 (death).
Pax, F., Bot. Jahrb. 8 (Lit.): 175. 1887 (Prahl herb. material at KIEL; used by Knuth?).
Prahl, P., Fl. Schlesw.-Holst. 2: 47. 1890.
Schellenberg, G., Beih. Bot. Centralbl. 38(2): 390. 1921 (on P. coll. at Kiel).

NOTE: Preliminary publications for the *Krit. Fl. Schlesw.-Holst.*: Verh. bot. Ver. Brandenburg 14: 101-149. 1872, 18: 1-25. 1876. – Prahl's main bryological publication was his *Laubmoosflora von Schleswig-Holstein*, in Schr. nat. Ver. Schlesw.-Holst. 10: 147-223. 1895.

8256. *Kritische Flora der Provinz Schleswig-Holstein,* des angrenzenden Gebiets der Hansestädte Hamburg und Lübeck und des Fürstenthums Lübeck ... Kiel (Universitäts-Buchhandlung Paul Toeche) 1888-1890, 2 parts. Oct. (*Krit. Fl. Schlesw.-Holst.*).
Collaborators: Rudolf J.D. von Fischer Benzon (1839-1911); Ernst Hans Ludwig Krause (1859-1942).
Ed. 1: two parts. The first part went through five editions; the second part came out only in first edition. *Copies*: B, G, NY.
1. Teil. Schul– und Exkursionsflora, Jun 1888 (p. ix: Mar 1888; G copy sent to Sadebeck by Prahl from publ. house Jun 1888; Bot. Zeit. 27 Jul 1888; J. Bot. Aug 1888; Bot. Centralbl. 9-13 Jul 1888; Nat. Nov. Jul(2) 1888), p. [i]-lxviii, [1]-227.
2. Teil. Kritische Aufzählung ... 1889-1890.
Heft 1: Jun 1889 (Nat. Nov. Jul(1) 1889; Österr. bot. Z. 1 Sep 1889; Bot. Zeit. 26 Jul 1889; Bot. Centralbl. 24 Jul 1889), p. [1]-128.
Heft 2: 1-15 Mai 1890 (p. ix: Jan 1890; Bot. Centralbl. 11 Jun 1890; Nat. Nov. Mai(2) 1890; Bot. Zeit. 27 Jun 1890; ÖbZ 1 Sep 1890), p. 129-345, preface material p. [i]-ix, [1]-63, [64].

Krause did Ranunc.-Saxifr., Labiat.-Chenop.; Fischer-Benzon: Umbell.-Rubiac., Polygon.-Salic. plus historical introduction on p. [1]-64.

Ed. 2 (of *1. Teil* only): Aug 1900 (p. iii: Mai 1900; Bot. Centralbl. 5 Sep 1900; Bot. Zeit. 1 Oct 1900; Nat. Nov. Oct (1) 1900), p. [i*], [i]-vi, [1]-68, [1]-260. *Copies*: G. Zweite vermehrte und verbesserte Auflage ... Kiel (id.) 1900. Oct.

Ed. 3 (id.): 1903, n.v., apparently a simple reprint of ed. 2.

Ed. 4: 1 Mai-15 Jun 1907 (p. iv: Mai 1907; Nat. Nov. Jun(2) 1907), p. [i]-vii, [1]-336. *Copy*: BR. "4. neubearbeitete und verbesserte Auflage ... 4. und 5. Tausend." Kiel (id.) 1907. Oct.

Ed. 5: 1913 (p. vi: Jun 1913; Nat. Nov. Apr(2) 1914), p. [i]-ix, [1]-357. *Copy*: B. – 5. vermehrte Auflage des 1. Teils ... 6. und 7. Tausend. Bearbeitet von P. Junge [Paul Junge 1881-1919, see TL-2/2: 470].

Laubmoosflora (additional to ed. 1, vol. 2): P. Prahl, Schrift. Nat. Ver. Schleswig Holstein 10: 147-223. 1895.

Ref.: Buchenau, F.G.P., Bot. Zeit. 46: 653-655. 1888, 47: 691-692. 1889, 48: 480-482. 1890.
Krause, E.H.L., Bot. Centralbl. 85: 399-400. 1900 (rev.).
Roth, E., Bot. Centralbl. 38: 489-490. 1889, 45: 311-312. 1891 (rev.).

Prahn, Hermann (*fl.* 1887), German botanist. (*Prahn*).

HERBARIUM and TYPES: Unknown.

BIBLIOGRAPHY and BIOGRAPHY: Tucker 1: 567.

8257. *Pflanzennamen.* Erklärung der lateinischen und der deutschen Namen der in Deutschland wildwachsenden und angebauten Pflanzen, der Ziersträucher, der bekanntesten Garten– u. Zimmerpflanzen und der ausländischen Kulturgewächse. (*Pflanzennamen*).
Ed. 1: 1897 (Bot. Zeit. 1 Jun 1897; Nat. Nov. Aug(2) 1897), p. [1]-172. n.v.; see also Allg. bot. Z. 4(4): cover, 15 Apr 1898 (adv.).
Ed. 2: 1909 (Bot. Zeit. 16 Aug 1909; Nat. Nov. Aug(1)1909), p. [i]-iv, [1]-176. *Copies*: B, G, Mägdefrau. – Berlin (Schnetter & Dr. Lindemeyer ...) s.d. Oct.
Ed. 3: 1922 (Nat. Nov. Jun 1922), p. [1]-187. *Copy*: B-S. – Berlin (Verlag Schnetter & Dr. Lindemeyer (Siegfried Cronbach)). 1922.
Ref.: Kienitz-Gerloff, Bot. Zeit. 55(2): 268. 1 Sep 1897.
Peter, A., Bot. Zeit. 67(2): 262. 1 Oct 1909.

Prain, Sir David (1857-1944), British botanist; MB Edinburgh 1882; Indian Medical Service 1884-1887; curator herb. Royal Botanic Garden, Calcutta 1887-1898; superintendent id. 1898-1905; professor of botany Calcutta Medical College 1898-1905; director R.B.G. Kew 1905-1922. (*Prain*).

HERBARIUM and TYPES: CAL and K; other material at A, B, BM, DBN, DD, E, HK, K, L, LE, US, W, WU. – Some letters at G, HU, K and SIA; see also Bridson (1980).

BIBLIOGRAPHY and BIOGRAPHY: Backer p. 462; Barnhart 3: 106; BJI 1: 46, 2: 139; BL 1: 63, 93, 94, 96, 99, 313; Blunt p. 183, 186; BM 4: 1608-1609, 8: 1027-1028; Bossert p. 316; CSP 17: 998; Desmond p. 502-503 (full info.; b. 11 Jul 1857, d. 16 Mar 1944); DNB 1941-1950: 695-696; Hortus 3: 1202 (Prain); IF p. 724; IH 1 (ed. 6): 363, (ed. 7): 340; 2: (in press); Kew 4: 355-358; KR p. 579; Langman p. 594; Lenley p. 334-335; LS 21200, 38054, suppl. 22176; MW p. 393-394; NI 256, 1052, 1559-1560; NW p. 55; PH 209; Rehder 5: 685-686; SK 4: cxix, cxxxiv; ; TL-1/538; TL-2/1290, 2448, 3006, 3218, 4275, 7055. Tucker 1: 567; Urban-Berl. p. 384; Zander ed. 10, p. 703.

BIOFILE: Allan, M., The Hookers of Kew 234. 249-250. 1967.
Anon., Bot. Centralbl. 73: 479. 1898 (appointed superint. Calcutta); Bot. Jahrb. 25 (Beibl. 60): 56. 1898 (superint. Calcutta), 38: (Beibl. 87): 42. 1906 (dir. Kew); Bull. misc. Inf. Kew 1930: 96 (on portr.), 1935: 97; Bull. Soc. bot. France 45: 94. 1898 (dir. bot. gard. Calcutta); Gard. Chron. ser. 3. 115: 144. 1944 (obit.); Gard. Mag. 50: 168.

1907 (portr.), 55: 129. 1912 (portr.); Hedwigia 45: (128). 1906 (app. Kew, 63: (151). 1922 (retirement from Kew); Nat. Nov. 20: 213. 1898 (app. director Cinchona plantations in Bengal), 27: 248. 1905 (FRS), 374. 1905 (superint. R.B.G. Calcutta), 32: 72. 1910 (foreign member Akad. Wiss., München); Österr. bot. Z. 48: 367. 1898 (superint. R.B.G. Calcutta); Österr. bot. Z. 56: 79. 1906 (app. Kew); Proc. Linn. Soc. London 156: 125. 1944; The Times 18 Mar 1944; Tropical Life 1906 (Mai): 72. (portr.); Yearbook Amer. philos. Soc. 1944: 379-383.
Bridson, G.D.R. et al., Nat. hist. mss. res. Brit. Isl. 269.138, 269.288. 1980.
Britten, J., J. Bot. 44: 21-22. 1906 (portr.; appointment).
Bullock, A.A., Bibl. S. Afr. bot. 92. 1978.
Burkill, I.H., Obit. Not. Fellows Roy. Soc. 13(4): 747-770. 1944 (obit., portr., bibl.); Proc. Linn. Soc. London 156: 223-229. 1945 (obit.); Chapt. hist. bot. India 243 [index, many entries]. 1965.
Drewitt, F.D., Apothec. Garden 88, 89. 1922.
Fletcher, H.R., Story R. Hort. Soc. 299, 374. 1969; R. Bot. Gard. Edinburgh 230, 260. 1970 (recomm. for Calcutta).
Gager, C.S., Brooklyn Bot. Gard. Rec. 27(3): 247 (dir. Kew), 261 (dir. Calcutta bot. gard. 1897-1905).
Gagnepain, F., *in* Lecomte, H., Fl. gén. Indoch., tome prél. 46-47, *pl. 13*. 1944 (portr.).
Hedge, I.C. & J.M. Lamond, Index coll. Edinb. herb. 120. 1970.
Hemsley, W.B., J. Kew Guild 2: 289. 1906 (portr.).
Huxley, L., Life letters J.D. Hooker 2: 560. [index]. 1918.
Jackson, B.D., Bull. misc. Inf. Kew 1901: 53.
Kneucker, A., Allg. bot. Z. 12: 36. 1906 (appointm. Kew).
Merrill, E.D., Yearbook Amer. philos. Soc. 1944: 379-383; Chron. bot. 10(3/4): 374-376. 1946 ("the career of Sir David Prain reads more like that of some young American ..."), 14(5/6): 214, 244, 259. 1954 (Bot. Cook's Voy.).
Murray, G., Hist. coll. BM(NH) 1: 174. 1904.
Nelmes and Cuthbertson, Bot. Mag. Dedic. 291-292, portr. 1931.
Prain, D., Memoirs and memoranda,chiefly botanical. Calcutta 1894, vi, 419 p. (mainly reprints from periodicals 1887-1893); Bot. notes and papers, reprints from periodicals 1894-1901, Calcutta 1901 (see Nat. Nov. Jun(1) 1901).
Salisbury, E.J., Nature 153: 426-427. 1944 (obit.).
Smith, W.W., Yearb. R. Soc. Edinburgh 1945: 22-24 (obit.).
Stearn, W.T., Nat. Hist. Mus. S. Kensington 301. 1981.
Taton, R., ed., Science in the 19th century 386. 1965.
Taylor, G., Nature 80: 162-163. 1957 (centenary birth).
Thomson, A.D., New Zealand J. Bot. 18: 409-416. 1980 (letters L. Cockayne to Prain).
Verdoorn, F., ed., Chron. bot. 1: 40, 297. 1935, 2: 185, 187, 196,. 196, 3: 15, 162. 1937, 4: 154, 276. 1938, 5: 262, 281, 299, 314. 1939, 6: 415, 417. 1941, 7: 141. 1943.

COMPOSITE WORKS: (1) Editor, Hooker's *Icones Plantarum* 29 (1906)-30 (1944), *pl. 2801-3000*.
(2) Editor Curtis's *Botanical magazine* 133 (1907)-146 (1920), *pl. nos. 8112-8873*.
(3) *Flora Capensis, Loganiaceae*, with H.S. Cummins, 4, sect. 1(6): 1036-1055. Feb 1909; *Euphorbiaceae*, with N.E. Brown & J. Hutchinson, 5, sect. 2(2): 216-384. Oct 1915.
(4) *Flora tropical Africa, Cycadaceae*, 6, sect. 2(2): 344-354. Nov 1917; edited volumes 6(2) and 9. Co-author *Euphorbiaceae*, 6, Sect. 1(3): 441-576. Oct. 1911.
(5) *Flore générale Indochine, Dioscoreaceae*, with I.H. Burkill, 6(5): 698-720. Feb 1934, 6(6): 721-745. Nov 1934.
(6) *Index kewensis*, editor (in his capacity of director of Kew) of suppl. 3-5, 1908-1921, see B.D. Jackson.

EPONYMY: *Prainea* King ex J.D. Hooker (1888). See ING 3: 1407.

NOTE: Prain readjusted the policy of the *Index kewensis* of providing taxonomic synonymy. In his opinion "uniformity in the delimitation of species as taxonomic units is unattainable" and so "Kew could not dictate nor profitably argue" reductions to taxonomic synonymy. From supplement 4 onward the Index kewensis became the nomenclator it now is: listing names without expressing taxonomic opinions (quotations from Burkill 1944).

8258. *The species of Pedicularis of the Indian Empire* and its frontiers ... Calcutta (Printed at the Bengal Secretariat Press) 1890. Fol.
Publ.: Dec 1890 (p. ii: 1 Jan 1890; p. 196 marked 11 Nov 1890 (registr.)), p. [i*], [i]-ii, [i]-iv, [1, h.t.], [1]-196, chart, *pl. o, o*ª, *1-37*(uncol.; G.C. Das, A.D. Mulla, D. Prain, A.L. Singh). *Copies*: BR, FI, H, NY, U. – Published in and to be cited from Ann. R. Bot. Gard., Calcutta 3: 1-196. Dec 1890. See NI 1560 for artists plates.
Ref.: Franchet, A., Bull. Soc. bot. France 40: 35-37. 1893 (rev.).

8259. *A list of Laccadive plants* ... Calcutta (printed by the Superintendent of Government Printing, India) 1890. Oct.
Publ.: 1890, cover-t.p., p. [1]-23. *Copy*: G. – Reprinted from Sci. Mem. med. Officers Army India, part 5 (journal n.v.). See also his *Botany of the Laccadives, in* J. Bombay nat. hist. Soc. 7: 268-295, 460-486. 1892/93, 8: 57-86, 488. 1893-1894.

8260. *An account of the genus Argemone* [Reprinted from the Journal of Botany 1895]. Oct.
Publ.: Mai-Dec 1895 (in journal), p. [1]-37. *Copy*: FI. – Reprinted and to be cited from J. Bot. 33: 129-135. Mai, 176-178. Jun, 207-209. Jul, 307-312. Oct, 325-333. Nov, 363-371. Dec 1895.

8261. *A revision of the genus Chelidonium* [Extrait du Bulletin de l'Herbier Boissier ... 1895]. Oct.
Publ.: Nov 1895, p. [1]-18. *Copies*: FI, G, U. – Reprinted and to be cited from Bull. Herb. Boissier 3(11): 570-587. 1895.

8261a. *An account of Corydalis persica* Cham. et Schlecht. with remarks on certain allied species of *Corydalis* Vent. ... [Extrait du Bulletin de l'Herbier Boissier ... 1899]. Oct.
Publ.: Mar 1899, p. [1]-16. *Copy*: U. – Reprinted and to be cited from Bull. Herb. Boissier 7(3): 162-177. 1899.

8262. *Bengal plants* a list of the phanerogams, ferns and fern-allies indigenous to, or commonly cultivated in, the lower provinces and Chittagong with definitions of the natural orders and genera, and keys to the genera and species ... Calcutta 1903, 2 vols. Oct. (*Bengal pl.*).
Orig. ed.: 2 volumes. *Copies*: MO, USDA; IDC 7317.
1: 1903 (p. vii: Mar 1903; Nat. Nov. Jan(1) 1904; Bot. Centralbl. 14 Jan 1904), p. [i-ix], [1]-663, map.
2: 1903 (Nat. Nov. Jan(1) 1904; Bot. Centralbl. 14 Jan 1904), p. [i-iii], [663]-1319.
Reprint: 1963, 2 vols. *Copies*: B, G, NY, US. – Botanical Survey of India, Calcutta. – *1*: [i-xi], [1]-489; *2*: [i-iv], [491]-1013.

8263. *The species of Dalbergia of Southeastern Asia* ... Calcutta (Bengal Secretariat Press) 1904. Fol.
Publ.: 1904 (pref.: 5 Apr 1903; Nat. Nov. Oct(2) 1904), p. [i*-iii*], [i]-iii, [i]-iv, [1]-114, *pl. 1-91* (uncol., K.D. Chandra, K.P. Dass, A.L. Singh, A.D. Molla, D.N. Choudhury, auct). *Copies*: B, BR, MO, NY. – Issued as part of, and to be cited from Ann. R. Bot. Gard., Calcutta 10, 1904; see NI 1559 for artists of plates.

8264. *Noviciae indicae*: some additional species of Indian plants ...London (printed by West, Newman & Co., ...) 1905. Oct.
Publ.: 1905, p. [i]-xxviii, [1]-445, *5 pl. Copies*: G, MO. – A reprint of a series of articles previously published in the J. Asiat. Soc. Bengal, vols. 58-73, 1889-1904, listed with precise details on p. vii-viii. The reprint also shows the original pagination.
Reprints journal articles. The journal articles were reprinted with original as well as continuous pagination at the time of publication with the imprint, on the cover of part 1: Calcutta (printed by G.H. House, ...) 1889. *Copy*: NY.
The preface material [i]-xxviii and the index, p. 429-445 were included only in the 1905 reprint.

8265. *Contributions to Indian botany* ... London (Printed by West, Newman & Co., ...) 1906. Oct.

Publ.: 1906, a series of reprints from British and Indian periodicals, 1902-1906, listed in preface. To be cited from original periodicals; p. [i]-viii, 1-432, 1-27 [=433-459]. *Copy*: MO.

8266. *An account of the genus Dioscorea in the East* part I the species which twine to the left ... Alipore, Bengal (Superintendent, Government Printing Bengal Government Press) 1936. Qu.
Co-author: Isaac Henry Burkill (1870-1965).
Publ: 1936, to be cited as Ann. R. Bot. Gard. Calcutta 14(1). *Copies*: BR, NY, PH.
Text: [i*-iii*], [i]-iii, [i]-ii, [1, h.t.], [1]-210, [i]-vi.
Plates: [i*-iii*], [i]-ii, *pl. 1-32, 32-2, 32-3, 32-3, 32-4, 32-5, 33-80, 80-2, 80-3, 81-85* (uncol.). – Merrill (1947) states that p. 211-428 and *pl. 86-150* were printed 1939 but not yet distributed by Aug 1946. However, he had seen p. 427-528 of which a limited distribution had taken place in 1939.
Note: A third volume of reprints from periodicals was *Memoirs and memoranda, chiefly botanical*, Calcutta (Baptist Mission Press) 1894, 406 p. (n.v.) (see C.B. Clarke, J. Bot. 32: 345-347. Nov 1894).

Prantl, Karl Anton Eugen (1849-1893), German (Bavaria-born) botanist; Dr. phil. München 1870; studied with von Naegeli and Radlkofer; 1871-1876 at Würzburg with Sachs; habil. Würzburg 1873; 1876-1889 professor of botany at the forestry college of Aschaffenburg; id. at Breslau 1889-1893; founder, with Adolf Engler, of *Die natürlichen Pflanzenfamilien*. (*Prantl*).

HERBARIUM and TYPES: HBG (esp. Pteridophyta); other material at E; lichens at M and WRSL; the material at B (e.g. Bryophyta) is mainly destroyed, except for the ferns). – Some manuscript material in SIA; letters at G.

BIBLIOGRAPHY and BIOGRAPHY: AG 12(1): 459; Ainsworth p. 273; Backer p. 659; Barnhart 3: 106; BFM no. 2275, 2378; BJI 1: 47, 2: 139; BM 4: 1609, 8: 1028; Bossert p. 317; CSP 8: 655, 11: 59, 12: 586, 17: 999; De Toni 4: xlvi; DTS 1: 229; Frank 3(Anh.): 78; Hegi 6(2): 1242; Herder p. 118, 271; Hortus 3: 1202 (Prantl); IF p. 724; Jackson p. 61, 64, 152, 423; Kew 4: 359; Langman p. 594; Lenley p. 355; LS 21202-21205; Moebius p. 60, 146, 293; Morren ed. 2, p. 10, ed. 10, p. 18; MW p. 394; PR 7286; Rehder 5: 686; Stevenson p. 1254; TL-1/355; TL-2/1710-1711, 6760; Tucker 1: 567; Zander ed. 10, p. 703, ed. 11, p. 802.

BIOFILE: Anon., Amer. J. Pharm. 65: 207-208. 1893; Bot. Centralbl. 34: 32. 1888 (declined call to forestry college Eberswalde), 39: 272. 1889 (appointed Breslau), 54: 96. 1893 (Hedwigia to be edited by Hieronymus/Hennings/Lindau), 54: 127. 1893 (Pax succeeds Prantl at Breslau); Bot. Gaz. 18: 152. 1893; Bot. Zeit. 28: 408. 1870 (asst. München), 31: 607-608. 1873 (habil. Würzberg), 34: 187. 1876 (to Aschaffenburg), 47: 577. 1889 (to Breslau), 51(2): 77 (d.), 138, 208 (library for sale) 1893; Hedwigia 32(1): 1. 1893 (d. 24 Feb 1893), 33 (Repert.): 67. 1894 (library); Leopoldina 29: 59. 1893 (obit.); Nat. Nov. 10: 104 (called to Eberswalde) 132 (not called to Eberswalde). 1888, 11: 256. 1889 (app. prof. bot. Breslau), 16: 47. 1894; Österr. bot. Z. 28: 313. 1878 (appointed Aschaffenburg), 39:384. 1889 (appointed Breslau), 43: 152. 1893 (d.), 44: 120. 1894 (fern herb. for sale).
Bergdolt, E., Karl von Goebel ed. 2: 46, 48, 56, 78, 83. 1941.
Bullock, A.A., Bibl. S. Afr. bot. 92. 1978.
Cohn, F., Jahresber. schles. Ges. vaterl. Cult. 71: 11-14. 1895 (extract Engler 1893).
Engler, A., Bot. Jahrb. 11 (Beibl. 24) 2. 1889 (succeeds E. at Breslau); Ber. deut. bot. Ges. 11: (34)-(39). 1893 (obit., biogr., b. 10 Sep 1849, d. 24 Feb 1893); Bot. Jahrb. 17 (Beibl. 40): 32. 1893 (d. 24 Feb 1893; Nat. Pfl.-Fam. to be continued by Engler alone).
Freund, H. & A. Berg, Gesch. Mikroskopie 294, 297. 1963.
Fritsch, C., Verh. zool.-bot. Ges. Wien 43: 9-12. 1894 ("Carl Prantl als Systematiker" ... "Jünger und Meister der Botanik [haben] an Prantl viel verloren ...").
Gager, C.S., Brooklyn Bot. Gard. Rec. 27(3): 221. 1938 (dir. bot. gard. Breslau 1889-1893).

Harvey-Gibson, R.J., Outlines hist. bot. 163, 170, 259. 1914.
Hedge, I.C. & J.M. Lamond, Index coll. Edinb. herb. 120. 1970 (material via herb. Dörfler).
Hertel, H., Mitt. Bot. München 16: 419. 1980 (lich. at M).
Hellmann, V., Hedwigia 32(2): 45-49. 1893 (obit., portr., bibl.).
Jessen, K.F.W., Bot. Gegenw. Vorz. 44. 1864.
Limpricht, K.G., Abh. Schles. Ges. vaterl. Cultur 71: 11-14. 1893.
Mägdefrau, K., Gesch. Bot. 54, 68, 153, 207. 1973.
Prantl, K., Verh. phys.-med. Ges. Würzburg 9: 84-97. 1876 (important for the early development of his ideas on phylogeny, pteridophytes and spermatophytes).
Pringsheim, E.G., Julius Sachs 186, 209, 210, 271, 273, 274, 288. 1932.
Sayre, G., Bryologist 80: 515. 1977.
Stafleu, F.A., Taxon 21: 503. 1972; Engler und seine Zeit, Bot. Jahrb. 1981 (in press).
Taton, R., ed., Science in the 19th century 381. 1965.
Voigt, A., Bot. Inst. Hamburg 74, 75. 1897.
Weiss, J.E., Ber. bay. bot. Ges. 2: xlix-li. 1893 (portr., bibl.).
Wittrock, V.B., Acta Horti Berg. 3(3): 148. 1905.

COMPOSITE WORKS: with A. Engler, *Die natürlichen Pflanzenfamilien*, co-editor 1887-1893, author of:

Betulaceae, 3(1): 38-46. 28 Nov 1887 (fide Engler in LC cards by Bay).
Fagaceae, 3(1): 47-48. 28 Nov 1887, 49-58. 10 Apr 1888 (id.), Ergänzungsheft 1: 17. 8 Oct 1900.
Magnoliaceae, 3(2): 12-19. 17 Jan 1888 (id.).
Trochodendraceae 3(2): 21-23. 17 Jan 1888 (id.), 273. Mai 1891.
Anonaceae, 3(2): 23-29. 17 Jan 1888 (id.), 273-284. Mai 1891.
Myristicaceae, 3(2): 40-42. 17 Jan 1888 (id.).
Ranunculaceae, 3(2): 43-48. 17 Jan 1888, 49-66. 21 Mai 1888 (id.), 274. Mai 1891.
Lardizabalaceae, 3(2): 67-70. 21 Mai 1888 (id.), 274. Mai 1891.
Berberidaceae, 3(2): 70-77. 21 Mai 1888 (id.), 274. Mai 1891.
Menispermaceae, 3(2): 78-91. 21 Mai 1888 (id.), 275-276. Mai 1891.
Calycanthaceae, 3(2): 92-94. 21 Mai 1888 (id.).
Papaveraceae, 3(2): 130-144. Mar 1889, 145. Mar 1891 (with Jacob Kündig).
Cruciferae, 3(2): 145-192. Mar 1891, 193-206. Mar 1891 (rev. Taubert, P.H.W., Bot. Centralbl. 49: 48-50. 1892).

(2) *Die internationale Polarforschung* 1882-1883. *Die deutschen Expeditionen*, Band II, ed. G. Neumeyer; *Filices* by K. Prantl p. 328, 1890.
(3) Editor *Hedwigia* vols. 26-31. 1887-1893.
(4) Editor: *Arbeiten aus dem k. botanischen Garten zu Breslau*. Band 1, Heft 1, 1892. † see below *Syst. Farne*.
(5) Edited eds. 3 and 4 of Seubert's *Excursionsflora für das Grossherzogthum Baden*, 1880, 1885, Stuttgart (Eugen Ulmer), see Seubert (ed. 3, rev. Bot. Zeit. 38: 676-677. 1 Oct 1880).

HANDWRITING: Candollea 32: 397-398. 1977.

EPONYMY: *Prantleia* Mez (1891) is very likely named for him but Mez does not give the etymology; see ING 3: 1407.

NOTE: Engler (1893) on Prantl: " Da Prantl, wie ich wusste, mit seiner speciellen Kenntniss der Pteridophyten auch sehr umfassende Kenntniss der Kryptogamen überhaupt verband und an stetiges Arbeiten gewöhnt war, zudem auch in der Systematik eine möglichst der Phylogenie entsprechende Anordnung der Formen anstrebte, hielt ich ihn für einen geeigneten Mitarbeiter an dem von mir zuerst projectirten Werk "Die natürlichen Pflanzenfamilien" und bot ihm die Leitung der die Kryptogamen behandelnden Abtheilung an. Leider war es ihm nicht vergönnt, diese Abtheilung über die ersten Anfänge hinaus zu fördern. Dagegen hat er für die Abtheilung der Siphonogamen recht umfassende und werthvolle Bearbeitungen geliefert." and "... wird ein durch werthvollen Arbeiten verewigter Name unter uns und nach uns fortleben!" (Not in the least through Engler und Prantl, F.A.S.).

8267. *Die Inulin*. Ein Beitrag zur Pflanzenphysiologie von der philosophischen Facultät der Universität München gekrönte Preisschrift ... München (Christian Kaiser) 1870. Oct. (*Inulin*).
Publ.: 1870 (preface Apr 1870), p. [i-viii], [1]-72, *1 pl. Copy*: M.

8268. *Lehrbuch der Botanik* für Mittelschulen ... Leipzig (Verlag von Wilhelm Engelmann) 1874. Oct. (*Lehrb. Bot*.).
Ed. 1: Feb-Apr 1874 (p. vi: Jan 1894; Bot. Zeit. 1 Mai 1874), p. [i]-viii, [1]-240. *Copies*: B, MO, NY, US. – "Bearbeitet unter Zugrundelegung des Lehrbuchs der Botanik von Julius Sachs."
Ed. 2: Jul 1876 (p. vii: Jun 1876; Bot. Zeit. 28 Jul 1876), p. [iii]-x, [1]-261. *Copies*: BR, NY.
Ed. 3: 1 Jan-15 Feb 1879 (p. vi: Nov 1878; ÖbZ 1879; Bot. Zeit. 21 Feb 1879; Nat. Nov. Feb(2) 1879), p. [i]-viii, [1]-292. *Copy*: *NY*. – "*Lehrbuch der Botanik* für mittlere und höhere Lehranstalten".
Ed. 4: Jun-Jul 1881 (p. vi: Mai 1881; Bot. Centralbl. 18-21 Jul 1881; Bot. Zeit. 29 Jul 1881; Nat. Nov. Aug(1) 1881), p. [i]-viii, [1]-326. *Copy*: NY.
Ed. 5: Nov-Dec 1883 (p. [vi]: Oct 1883; Bot. Centralbl. 10-14 Dec 1883; Bot. Zeit. 25 Jun 1884; Nat. Nov. Nov(2) 1883, Jan(2) 1884), p. [i]-viii, [1]-335, [336]. *Copies*: LD (inform. O. Almborn), UC. (inf. R. Schmid).
Ed. 6: Mar 1886 (p. vi: Feb 1886; Bot. Centralbl. 29 Mar-2 Apr 1886; Bot. Zeit. 26 Mar 1886; Nat. Nov. Mar(2) 1886), p. [i]-viii, [1]-339. *Copy*: PH.
Ed. 7: 1888 (p. vi: Ostern 1888), p. [i]-viii, [1]-341. *Copies*: NY, PH.
Ed. 8: Oct-Nov 1891 (Bot. Centralbl. 27 Nov 1891; Nat. Nov. Nov(1) 1891; ÖbZ Nov 1891; Allg. bot. Z. Nov 1891), p. [i]-viii, [1]-355. *Copy*: MO.
Ed. 9: 1 Feb-15 Mar 1894 (Nat. Nov. Mar(2) 1894; Bot. Zeit. 1 Mai 1894; Bot. Centralbl. 17 Apr 1894; p. viii: Jan 1894), p. [i]-x, [1]-365. *Copy*: B. – *Prantl's Lehrbuch der Botanik*, herausgegeben und neu bearbeitet von F. Pax [Ferdinand Albin Pax, 1858-1942], Leipzig (id.). Oct.
Ed. 10: Apr 1896 (p. vii: Jan 1896; ÖbZ Apr 1896; Nat. Nov. Mai(1) 1896; Hedwigia Mai-Jun 1896; Bot. Centralbl. 22 Mai 1896; Bot. Zeit. 16 Mai 1896; p. vii: Jan 1896), p. [i]-x, [1]-406. *Copies*: B, G. – Ed.F. Pax.
Ed. 11: 1 Mar-15 Apr 1900 (p. vi: Dec 1899; ÖbZ Apr 1900; Bot. Centralbl. 25 Apr 1900; Bot. Gaz. 19 Jul 1900; Allg. bot. Z. 1900; J. Bot. (Morot) Jun 1900), p. [i]-viii, [1]-455, [456, err.]. *Copies*: B, NY. – Ed.F. Pax.
Ed. 12: 1 Mai-13 Jun 1904 (p. vi: Dec 1903; Bot. Centralbl. 14 Jun 1904; ÖbZ Mar-Apr 1904; Hedwigia 15 Jul 1904; review A.B. Rendle, J. Bot. Jul 1904; Allg. bot. Z. 16 Jul 1904), p. [i]-viii, [1]-478, [1, err.]. *Copies*: B, NY, UC. – Ed. F. Pax.
Ed. 13: 1 Mar-15 Apr 1909 (pref.: Dec 1908; ÖbZ Mar-Apr 1909; Nat.Nov. Apr(2) 1909; Bot. Centralbl. 6 Jul 1909; Bot. Zeit. 16 Jul 1909; Allg. bot. Z. 15 Mai 1909; J. Bot. Jul 1909; pref. Dec 1908), p. [i]-v, [1]-498. *Copies*: B, NY. – Ed.F. Pax.
Ed. 14: Jul-Aug 1916 (Bot. Jahrb. 13 Mar 1917; Nat. Nov. Aug(1, 2) 1916; ÖbZ 1 Sep 1916), p. [i]-vi, [1]-507, [508]. *Copies*: H, M, UC.
English, Ed. 1: Mar-15 Apr 1880 (Bot. Centralbl 12-16 Apr 1880; Nat. Nov. Apr(2) 1880), p. [i]-vii, [1]-332. *Copies*: MICH, NY, PH. – *An elementary textbook of botany* ... the translation revised by S.H. Vines [Sydney Howard Vines, 1849-1934]. Philadelphia (J.B. Lippincott & Co.) 1880. *Other issue*: London (W. Schwan Sonnenschein & Allen, ...) 1880, [i]-vii, [1]-332. *Copy*: UC (inf. G. Schmid).
Engl. ed. 2: Sep 1881 (pref. Apr 1881; Bot. Centralbl. 10-14 Oct 1881; Bot. Zeit. 25 Nov 1881; Nat. Nov. Oct(2) 1881), p. [i]-vii, [1]-344. *Copies*: E, UC. – London (W. Swan Sonnenschein & Allan) 1881.
Engl. ed. 3: Aug 1883 (pref. "1883"; Bot. Zeit. 26 Oct 1883; Bot. Centralbl. 27-31 Aug 1883; Nat. Nov. Sep(1) 1883), as ed. 2, with a few typographical changes.
Engl. ed. 4: 1885 (NY copy inscribed Jul 1885), p. [i]-vii, [1]-344. *Copies*: MO, NY. – London (W. Swan Sonnenschein & Co., ...) 1885. Oct. – As ed. 3.
Engl. ed. 5: 1893 (Bot. Zeit. 1 Oct 1893; Bot. Centralbl. 24 May 1893; Nat. Nov. Mai(2) 1893), p. [i]-vii, [1]-344. *Copies*: E, NY(2). – London (id.) 1892 and 1893. Oct. – Various issues e.g. "twelfth thousand", "fourteenth thousand". – As ed. 3.
Italian: Jan-Mar 1885 (Bot. Zeit. 27 Mar 1885; Bot. Centralbl. 13-17 Apr 1885), p. [i]-

iv, [1]-333. *Copy*: FI. – "Manuale di botanica ...Tradotto sulla quinta edizione originale ..." Transl.: Giuseppe Cuboni (1852-1920). Torino (Ermanno Loescher) 1885. Oct.
Spanish: n.v., translated by de Linares.
Hungarian: n.v., translated by Páter Bela.
Ref.: Bary, H.A. de, Bot. Zeit. 37: 221-223. 1879 (rev.), 39: 546-547. 1881 (rev. ed. 4).
Kienitz-Gerloff, J.H.E.F., Bot. Zeit. 52(2): 362-363. 1894 (rev. ed. 9), 54(2): 269. 1896 (rev. ed. 10).
Goebel, K., Flora 78: 500-501. 4 Jul 1894.
Möbius, M., Bot. Centralbl. 61: 260-261. 1895.
Solms-Laubach, H.M.C.L.F., Bot. Zeit. 58(2): 165-166. 1 Jun 1900, (rev. ed. 11), 62(2): 225-226. 1904 (rev. ed. 12).

8269. *Untersuchungen zur Morphologie der Gefässkryptogamen* ... Leipzig (Verlag von Wilhelm Engelmann) 1875-1882, 2 Hefte. (*Unters. Morph. Gefässkrypt.*).
1: Oct 1875 (p. iv: Jun 1875; Bot. Zeit. 29 Oct 1875 ("neu")), p. [i]-vi, [1]-73, *pl. 1-6* (uncol.).
2: Oct 1881 (p. iv: Jul 1881; Bot. Centralbl. 31 Oct-4 Nov 1881; Nat. Nov. Nov(1) 1881; Bot. Zeit. 30 Dec 1881), p. [i]-vi, [1]-161, *pl. 1-8* (uncol.).
Copies: B, BR, FH, G, MO, NY.
Ref.: Potonié, H., Bot. Centralbl. 10: 351-355. 1882 (extensive review).
G.K., Bot. Zeit. 3: 775. 19 Nov 1875 (rev.).
N.L., Bot. Zeit. 39: 870. 30 Dec 1881 (rev.).
Goebel, K., Bot. Zeit. 40: 152-153. 3 Mar 1882.
Prantl, K., Bot. Zeit. 33: 734-735. 5 Nov 1875 (statement of 19 Sep 1875 on "soeben publicirten Untersuchungen").

8270. *Verzeichniss der im botanischen Garten der königl. Forstlehranstalt Aschaffenburg cultivirten Pflanzen.* Nebst einem Plan zur Orientirung der Studirenden, ... Aschaffenburg (Verlag der C. Krebs'schen Buchhandlung (E. Kriegenherdt)) 1879. Oct. (*Verz. bot. Gart. Aschaffenburg Pfl.*).
Publ.: 1 Mai-15 Jun 1879 (p. 6: Apr 1879; Nat. Nov. Jun(2) 1879; Bot. Zeit. 25 Jul 1879), p. [1]-43, map. *Copy*: MO. – Prantl also published a *Plan des bot. Garten der* ... *Forstlehranstalt Aschaffenburg*, Aschaffenburg 1885 (see Bot. Zeit. 31 Jul 1885).

8271. *Exkursionsflora für das Königreich Bayern.* Eine Anleitung zum Bestimmen der in den bayrischen Gebietsteilen wildwachsenden, verwilderten und häufig kultivirten Gefässpflanzen nebst Angabe ihrer Verbreitung ... Stuttgart (Verlag von Eugen Ulmer) 1884. Oct. (*Exkurs.-Fl. Bayern*).
Publ.: 1-15 Apr 1884 (p. vii: Ostern 1884; Nat. Nov. Apr(2) 1894; Bot. Centralbl. 2-6 Jun 1884; Bot. Zeit. 27 Jun 1884), p. [i]-xvi, [1]-568. *Copies*: B(3), G, USDA.
Re-issue: 1894 (Bot. Centralbl. 17 Apr 1894), p. [i]-xvi, [1]-568. *Copies*: B(2). "Zweite Ausgabe", new t.p. only.
Continuation: H. Poeverlein, Denkschr. k. bot. Ges. Regensburg N.F.7.1, Beilage 1: *Die seit Prantl's "Exkursionsflora* ... *erschienene Literatur* über Bayerns Phanerogamen– und Gefässkryptogamenflora", Regensburg 1899 (see Poeverlein).
Based on Exkurs.-Fl.: Georgii, Adolph, "*Exkursionsflora für die Rheinpfalz.* Eine Anleitung zum Bestimmen ... nach Dr.K. Prantl, Exkursionsflora für das Königreich Bayern frei bearbeitet ..." Stuttgart (Verlag von Eugen Ulmer) 1894. Oct. p. [i]-xx, [1]-215. *Copy*: B.
Ref.: Drude, O., Bot. Zeit. 43: 222-223. 3 Apr 1885.
Engler, A., Bot. Jahrb. 6 (Lit.): 8. 24 Oct 1884.
Freyn, J.F., Bot. Centralbl. 19: 297-299. 1884 (disagrees with P's severe lumping of genera).

8272. *Arbeiten aus dem Königl. Botanischen Garten zu Breslau* ... Erster Band. Erstes Heft ... Breslau (J.V. Kern's Verlag (Max Muller)). Oct. †. (*Arb. bot. Gart. Breslau*).
Publ.: vol. 1(1) (all published): 1892, p. [i-v], [1]-166, *pl. 1*. *Copy*: BR. – Contains K. Prantl, *Das System der Farne* (p. 1-38); Werner Pomrencke, *Vergleichende Untersuchungen über den Bau des Holzes einiger sympetaler Familien* (p. 39-70); Carl Mez, *Spicilegium Lauracearum* (p. 71-166).

Praschil, Wenceslaus Wilhelm (*fl.* 1840), Austrian (Bohemian) botanist and physician at Gleichenberg. (*Praschil*).

HERBARIUM and TYPES: Unknown.

BIBLIOGRAPHY and BIOGRAPHY: PR 7299 (ed. 1: 8189); Rehder 3: 90.

BIOFILE: Maly, J.K., Flora von Steyermark viii. 1868.

8273. Dissertatio inauguralis medico-botanica sistens *plantas venenatas in territiorio vindobonensi sponte crescentes*, quam consensu et auctoritate illustrissimi et magnifici domini praesidis et directoris, perillustris ac spectabilis domini decani, nec non clarissimorum ac celeberrimorum d.d. professorum, pro doctoris medicinae laurea summisque in medicina honoribus ac privilegiis rite et legitime consequendis in antiquissima ac celeberrima Universitate vindobonensi publicae disquisitioni submittit: Wenceslaus Guilelmus Praschil, Bohemus tustensis. In theses adnexas disputabitur in aedibus Universitatis die 1 mensis Augustae 1840. Viennae [Wien] (typis Congregationis Mechitaristicae). [1840]. Oct. (*Pl. venen. vindob.*).
Publ.: 1 Aug 1840, p. [1]-40. *Copy*: WU.

Pratesi, Pietro (*fl.* 1800), Italian botanist; curator of the Pavia botanical garden 1800-1806. (*Pratesi*).

HERBARIUM and TYPES: Unknown. – Letters at G.

BIBLIOGRAPHY and BIOGRAPHY: Kew 4: 359-360; NI 1561; PR 7300 (ed. 1: 8190); Rehder 1: 97; Saccardo 1: 132, 2: 87.

BIOFILE: Gager, C.S., Brooklyn Bot. Gard. Rec. 27(3): 274. 1938 (dir. bot. gard. Pavia).

8274. *Tavole di botanica elementare* disegnate, ed incise da Pietro Pratesi ... [Pavia 1801]. Oct. (*Tav. bot.*).
Publ.: 1801, p. [i], 1-30, [31], *pl. 1-45* (copp., some col.). *Copy*: G.

Pratt (from 1866 Mrs John Pearless), **Anne** (1806-1893), British author of popular botanical books. (*Pratt*).

HERBARIUM and TYPES: Unknown.

BIBLIOGRAPHY and BIOGRAPHY: Barnhart 3: 106; BB p. 248; BL 2: 222, 704; Blunt p. 236, 266; BM 4: 1609; Bossert p. 317; Desmond p. 503 (b. 5 Dec 1806, d. 27 Jul 1893); DNB 46: 284-285; GF p. 71; Jackson p. 20, 40, 41, 214, 236, 237, 239, 240, 245, 486; Kew 4: 360; NI 1562, see also 1: 131; PR 7301-7303; Rehder 5: 686; Tucker 1: 567.

BIOFILE: Allibone, S.A., Crit. dict. Engl. lit. 1660. 1870 (bibl.).
Anon., Bot. Centralbl. 55: 384. 1893, 56: 320. 1893 (death); J. Bot. 31: 288. 1893; Nat. Nov. 15: 385. 1893 (d. 27 Jul 1893).
Britten, J., J. Bot. 32: 205-207. 1894 (b. 5 Dec 1806, d. 27 Jul 1893).
Freeman, R.B., Brit. nat. hist. books 285-288. 1980.
Graham, M., Country Life 161: 1500-1501. 1977 (portr.).

8275. *The flowering plants, grasses, sedges and ferns of Great Britain* and their allies the club mosses, pepperworts, and horsetails ... London (Frederick Warne and Co.) s.d. [? 1855-1866]. Qu. 6 vols. (*Fl. pl. Gr. Brit.*).
Publ.: We list here the copies seen by us. BM 3: 1609 mentions a first edition in 5 vols. (*The flowering plants and ferns of Great Britain*), London [1854-1855] and a third edition, 6 vols. London 1873. Our treatment is therefore obviously incomplete.
USDA copy: 6 vols., s.d. imprint as above.
1: [i]-viii, [1]-288, *pl. 1-47* (col. liths. author).
2: [i]-viii, [1]-355, *pl. 48-51, 51a, 52-70, 71a, 71b, 72-83* (id.).

3: [i]-ix, [1]-410, *pl. 84-110,* [*111-113*], *114-136* (id.).
4: [i]-viii, [1]-328, *pl. 137-189* (id.).
5: [i]-viii, [1]-368, *pl. 190-235, 235a, 236-238* (id.).
6: [i]-x, [1]-319, *pl. 239-313* (id.).
MO and *NY copies:* 6 vols., s.d. imprint: London (Frederick Warne and Co. ...) New York (Scribner, Welford and Co.); composition as in USDA copy except vol. 6: *pl. 238*, 239-313.*
BR copy: 1889, London and New York (Frederick Warne & Co.). 1889. Oct.
1: [i]-xii, [1]-256, *pl. 1-78.*
2: [i]-xii, [1]-247, *pl. 79-157.*
3: [i]-xi, [1]-251, *pl. 158-238.*
4: [i]-x, [xi], [1]-140, *pl. 238*, 239-313.*
New edition: revised by Edward Step (1855-1931), London (Frederick Warne & Co.) and New York 1899-1900. Qu. *Copies:* G, USDA.
1: (in 9 weekly issues) 1899 (Nat. Nov. Oct(1) 1899), p. [i]-xiv, [1]-269, *pl. A-D, 1-77.*
2: (idem) 1899 (Bot. Centralbl. 15 Nov 1899; Nat. Nov. Dec(2) 1899), p. [i]-xii, []-279, *pl. 78-153.*
3: (idem), 1899, p. [i]-xi, [1]-258, *pl. 154-231* (or *232*).
4: (idem), 1900 (Bot. Centralbl. 4 Apr 1900), p. [i]-xi, [1]-215, *pl. 232* (or *231*)-*319.*
Soc. Christ. knowledge ed.: 1886, *The flowering plants of Great Britain,* 3 vols. London (Society ...) s.d. *Copies:* B, G.
1: [i]-x, [1]-256, *pl. 1-78.*
2: [i-ii], [xi]-xviii, [1]-247, *pl.79-257.*
3: [i-ii], [xix]-xxv, [1]-251, 235a, *pl. 158-238.*
The copy at B (incomplete) shows the covers and contents of parts 1 (p. [1]-16, *5 pl.*), 2 (17-32, *6 pl.*) and 15 (p. 225-240, *5 pl.*).

8276. *The ferns of Great Britain,* and their allies the club-mosses, pepperworts and horsetails ... with 63 species printed in colours. London (Fredrick Warne and Co. ...), New York (Scribner, Welford and Armstrong) s.d. Qu. (*Ferns Gr. Brit.*).
Publ.: [date not known], p. [i]-vi, [1]-174, *pl. 273-313* (col. liths., author). *Copies:* G, NY. − Part of vol. 6 of *Fl. pl. Gr. Brit.*
Other issue: [date not known], p. [i]-vi, [1]-174, *pl. 273-313* (id.). *Copy:* BR. − "With 66 species printed in colours." London (Frederick Warne and Co. ...) s.d.
Original issue: 1855, p. [i]-iv, [1]-164, *pl. 273-313* (col. liths). *Copy:* MO. London (printed for the Society for promoting Christian Knowledge, ...) s.d.

8277. *The British grasses and sedges ,...* London (Society for the promotion of Christian Knowledge, ...) s.d. Qu. (*Brit. grass.*).
Publ.: p. [i]-viii, [1]-136, *pl. 238*, 239-272* (col. liths. auct.). *Copy:* USDA. − Part of vol. 6 of *Fl. pl. Gr. Brit.*

8278. *The poisonous, noxious and suspected plants* of our fields and woods ... London (printed for the Society for promoting Christian Kowledge ...) s.d. [1857]. Oct. (*Poison. pl.*).
Publ.: 1857, p. [i]-xii, [1]-208, *44 pl.* (col. liths).

Préaubert, Ernest (1852-1933), French botanist and physicist; teacher at a primary school in Rennes 1872-1873, id. at the Lycée de Rennes for physics 1873-1876; "professeur de sciences" at the Collège de Beauvois 1876-1882; as full professor at the Lycée de Rennes 1882-1902; director of the municipal courses of Rennes 1882-1905; from then on in retirement for health reasons. (*Préaub.*).

HERBARIUM and TYPES: Unknown; if any, at ANG of which P was curator 1929-1934. P's prehistoric collections went to the Musée de Paléontologie, Angers. − Letters and portrait at G.

BIBLIOGRAPHY and BIOGRAPHY: Barnhart 3: 107 (b. 7 Apr 1852); BL 2: 177, 704; CSP 11: 60, 12: 586, 17: 1001-1002; De Toni 1: c; Herder p. 273; Rehder 2: 286; Tucker 1: 567.

BIOFILE: Anon., Bull. Soc. bot. Deux-Sèvres 25: 221, *pl. 3*. 1914 (portr.), 26: *pl. 2* (corrected); Bull. Soc. Étud. sci. Angers 2: 113. 1873, 6/7: 153. 1879.
Surrault, Th., Bull. Soc. Étud. sci. Angers 63: 19-25. 1933 (obit., bibl., portr., b. 7 Jan 1852; d. 18 Aug 1933).
Verdoorn, F., ed., 1: 119. 1935.

8279. *Révision des characées de la flore de Maine-et-Loire* (Imprimerie-Librairie Germain & G. Grassin ...) 1884. Oct.
Publ.: 1884 (in journal, p. 130: Dec 1883; Bot. Zeit. 29 Aug 1884; Nat. Nov. Sep(1) 1884), p. [1]-32. *Copy*: P. – Reprinted and to be cited from Bull. Soc. Étud. sci. Angers 13: 103-130. 1884.

Preble, Edward Alexander (1871-1957), American conservationist, botanist, orand 1 on the staff of the U.S. Biological Survey from 1892, on that of the American Nature Association from 1895.(*Preble*).

HERBARIUM and TYPES: US, – manuscripts and field notes in SIA.

BIBLIOGRAPHY and BIOGRAPHY: Barnhart 3: 106; BL 1: 138, 313; Ewan ed. 2: 178; PH 450.

BIOFILE: Anon., Nature Mag. 41: 35. 1948 (portr.), 50: 455. 1957 (d. 4 Oct 1957).
Hultén, E., Bot. Not. 1920: 324.
McAtee, W.L. Auk 79: 730-742. 1962; – with F. Harper, Univ. Kansas Mus. nat. Hist. misc. Publ. 40: 1-16. 1965 (bibl., portr.).
R.W.W., Nature Mag. 50: 537. 1957 (obit., portr.).
Verdoorn, F., ed., Chron. bot. 2: 306. 1936.
Walter, E. Preble and P.F. Hannah, Nature Mag. 20(2): 71-75. 1932 (biogr., portr., "a modern Thoreau").
Zahniser, H., 34: 307, 350. 1941.

COMPOSITE WORKS: Associate editor *Nature Magazine* 1935-1957.

8280. U.S. Department of Agriculture ... North American Fauna no. 27 ... *A biological investigation of the Athabaska-Mackenzie region* ... Washington (Government Printing Office) 1908. Oct.
Publ.: 26 Oct 1908 (t.p.), p. [1]-574, *pl. 1-25. Copy*: MO. Trees and shrubs on p. 515-534, *pl. 24-25*. – Letters at G.

Preda, Agilulfo (1870-1941), Italian botanist; teacher at the Liceo di Teramo of Pisa; director of the botanical garden of Siena 1916-1920. (*Preda*).

HERBARIUM and TYPES: PI; other material at MOD. – Original herbarium at the Museo Civico, Brescia (fide C. Steinberg).

BIBLIOGRAPHY and BIOGRAPHY: Barnhart 3: 107; BL 2: 344, 362, 401, 704; CSP 17: 1002-1003; De Toni 4: xlvi; Rehder 5: 686; Saccardo 1: 132, 2: 87; TL-2/3302.

BIOFILE: Gager, C.S., Brooklyn Bot. Gard. Rec. 27(3): 288. 1938 (dir. Siena bot. gard.).

EPONYMY: *Predaea* De Toni f. (1936) may be named for this author but no etymology was given by De Toni f.

8281. *Contributo allo studio delle Narcissee italiane* ... Firenze (Stabilimento di Giuseppe Pellas ...) 1896. Oct.
Publ.: Oct 1896 (p. 3: Mai 1895; dates of publ. in journal 15 Apr, 15 Oct), p. [1]-91, [92, index], *pl.* 7 (uncol., auct.). *Copies*: FI, G. – Reprinted and to be cited from: Nuovo Giorn. bot. Ital. ser. 2. 3(2): 214-253, 15 Apr, 3(3): 375-422. 15 Oct 1896. – See also his *Recherches sur le sac embryonnaire de quelques Narcissées*. Bull. Herb. Boissier 5(11): 948-952. Nov 1897.

8281a. *Catalogue des algues de Livourne* ... [Extrait du Bull. Herb. Boissier 1897). Oct. *Publ.*: Nov 1897, p. [1]-36. *Copy*: U. – Reprinted and to be cited from Bull. Herb. Boissier 5(11): 960-965. 1897.

8282. Società botanica italiana. *Flora italica cryptogama* pars II: *Algae* ... Rocca S. Casciano (stabilimento tipografico Cappelli) 1908-1909. Oct. (*Fl. ital. crypt., Alg.*).
1(1): 1 Mar 1909, p. [i*], [i]-lvi, [i]-iii, [1]-462, [1, err.]. Preface by G.B. De Toni (1864-1924); Bibliografia algologica by Preda (first publ. 1905, 41 p.).
1(2): 10 Feb 1908, p. [i], [1]-358. – Florideae.
1(3): 1 Apr 1909, p. [i], 359-462, [1, err.]. – id.
Copies: FH, MICH, NY.

Préfontaine, M. Bruletout de (*fl.* 1763), French colonial administrator in Guyana. (*Préfontaine*).

HERBARIUM and TYPES: Préfontaine is not known to have made collections.

BIBLIOGRAPHY and BIOGRAPHY: Dryander 3: 189; Rehder 3: 99.

BIOFILE: Abonnenc, E. et al., Bibl. Guyane franç. 1: 215-216. 1957.
Lacroix, A., Figures de savants 3: 58, 64, 65, 68. 1938.

8283. *Maison rustique à l'usage des habitants de Cayenne.* Paris 1763. Oct. (n.v.).
Publ.: 1763. – The chapter "Plantes, herbes, arbrisseaux et arbres qui naissent à Cayenne, et dont on y fait usage relativement à divers objets", on p. 135-211 was compiled by de la Sauvage from various other sources (Ewan 1959).
Ref.: Ewan, J., Proc. Amer. philos. Soc. 103(6): 816. 1959.

Preiss, Balthazar (Preis) (1765-1850), German military physician and botanist; Dr. chir. Wien 1792; "Regimentsarzt" at Salzburg 1792-1793; on active duty at various military stations of the "Regensburger Reichstag" 1793-1800; from 1800-1806 again in military and civil practice at Salzburg; from 1806-1830 in Austrian military service at Kuttenberg (Bohemia) and Praha (id.) and at various war stations; from 1830-1833 at Peterwardein, Slavonia; in retirement at Praha 1833-1850. (*B. Preiss*).

HERBARIUM and TYPES: PRC.

BIBLIOGRAPHY and BIOGRAPHY: AG 4: 632; Barnhart 3: 107 (b. 29 Dec 1765, d. 2 Jul 1850, Preiss, not Preis); BM 4: 1610; Frank 3(Anh.): 78; Herder p. 118; Klášterský p. 149; Maiwald p. 101; PR 5176, 7305 (ed. 1: 8194-8195); Saccardo 2: 87-88.

BIOFILE: Anon., Flora 14: 640. 1831 (Staatsarzt und k.k. Rath in Slavonien).
Maiwald, P.V., Jahresber. Stifts-Obergymn. Braunau 1901, p. 15 (bibl., epon.).
Opiz, P.M., Flora 7(2), Beil. 2: 142. 1824.
Weitenweber, W.R., Lotos 2: 171-172. 1852 (b. 29 Dec 1765, d. 2 Jul 1850; obit.).

EPONYMY: *Preissia* Corda (1829); *Preissites* Knowlton (1894); *Pressia* S.O. Lindberg (1877, *orth. var.* of *Preissia* Corda); see ING 3: 1408 and Maiwald (1901).

8284. *Rhizographie, oder Versuch einer Beschreibung und Eintheilung der Wurzeln, Knollen und Zwiebeln der Pflanzen, ihrer verschiedenen Lagen, Formen, Oberflächen, Gränzen und Nebentheile, nebst kurzen Betrachtungen über ihr Entstehen und Fortpflanzen, mit einigen anatomisch-physiologischen Bemerkungen* ...Prag (Zu haben beym Verfasser, ... und bey Anton C. Kronberger, ...) 1823. Oct. (*Rhizographie*).
Publ.: 1823 (Flora 21 Feb 1823: "Exemplare zu haben ..."), p. [i]-xxvi, [27]-256, [257]-270 indexes], [271, err.]. *Copies*: G, GOET, MO.

Preiss, [Johann August] Ludwig (1811-1883), German plant collector; Dr. phil.; in Western Australia 1838-1842; "Privatgelehrter der Botanik und Gutsbesitzer zu Hattorf ... im Harz". (*L. Preiss*).

HERBARIUM and TYPES: All collections sold; the Australian collections are found in many herbaria. Important sets: C, DBN, E, H, HBG, L, LD, LE, M(810), MEL, P, PC, W (see J.G.C. Lehmann, *Pl. Preiss.*, TL-2/4332). It should be noted that Lehmann's herbarium, which contained so many Preiss specimens, was dispersed. For further details see McGillivray (1975), q.v. also for much additional information on *Plantae Preissianae*. – Letters at G and U.

BIBLIOGRAPHY and BIOGRAPHY: AG 4: 631-632; Barnhart 3:107 (b. 21 Nov 1811, d. 21 Mai 1883); BM 4: 1611; Frank 3(Anh.): 78; IH 1 (ed. 6): 363, (ed. 7): 340, 2: (in press); Kew 4: 360; Lasègue p. 283, 308, 506; TL-1/666; TL-2/4332; Urban-Berl. p. 272, 310, 384.

BIOFILE: Anon., Flora 29: 256. 1844 (16 sets of Australian plants for sale).
Beck, G., Bot. Centralbl. 34: 150. 1888 (1104 specimens at W).
Candolle, Alph. de, Phytographie 440: 1880 (coll.).
Carr, D.J. & S.G.M., eds., People and plants in Australia 54-55 (L. Diels), 119-123, 132-138 (S. Ducker). 1981.
Diels, L., Pfl.-W. W. Austral. 47-48, 50, 52. 1906 (Veg. Erde 7).
Gunn, M.D. & L.E. Codd, Bot. expl. S. Afr. 287. 1981 (short visit to Cape).
Hall, N., Botanists of the Eucalypts 104-105. 1978.
Hedge, I.C. & J.M. Lamond, Index coll. Edinb. herb. 120. 1970 (algae in herb. W. Sonder).
Herder, F. v., Bot. Centralbl. 55: 293. 1893 (coll. LE).
Hooker, J.D., Fl. Tasm. cxxvi. 1859.
Jackson, B.D., Bull. misc. Inf. Kew 1901: 53.
Jessen, K.F.W., Bot. Gegenw. Vorz. 468. 1864.
Klatt, F.W., Bonplandia 8: 143. 1860 (on Preiss' herbarium in Lehmann's coll.; was sold and acquired by Agardh, now at LD, see ib. 10: 267. 1862).
Lehmann, J.G.C., Pl. Preiss. 1844-1848, 2 vols. (TL-2/4352).
McGillivray, D.J., Telopea 1(1): 1-18. 1975 (extensive study on P. in W. Australia).
Maiden, J.H., J. W. Austral. nat. Hist. Soc. 6: 22-24. 1909.
Müller, R.H.W. & R. Zaunick, Friedr. Traug. Kützing 244, 291. 1960.
Preiss, J.A.L., Bot. Zeit. 2: 352. 1844 (plants for sale); Flora 25: 539-544. 1842 (note on travels), 33: 160. 1850 (sale pl.).
Seeland, H., Mitt. Herm. Roemer-Mus. Hildesheim 40: 28-29. 1936 (biogr. sketch).
Seemann, B., Bonplandia 6: 50. 1858 (details mentioned in our heading; cogn. Leopoldina: Cunningham).
Souster, J.E.G., W. Austral. Natural. 1(6): 118. 1948.
Stearn, W.T., J. Soc. Bibl. nat. Hist. 1(7): 203-205. 1939.
Steenis-Kruseman, M.J. van, Blumea 25: 38. 1979.
Steinberg, C.H., Webbia 32(1): 35. 1977.

HANDWRITING: Telopea 1(1): 10. 1975.

EPONYMY: *Neopreissia* Ulbrich (1934).

Prenger, Alfred Gerhard (1860-?), German botanist and pharmacist; Dr. phil. Erlangen 1901; student of H. Solederer.(*Prenger*).

HERBARIUM and TYPES: Unknown.

BIBLIOGRAPHY and BIOGRAPHY: Kew 4: 361.

8285. *Systematisch-anatomische Untersuchungen von Blatt und Achse bei den Podalyrieen-Gattungen* der nördlichen Hemisphäre und des Kapgebietes, sowie bei den vier australischen Podalyrieen-Gattungen Brachysema, Oxylobium, Chorizema und Mirbelia. Inaugural-Dissertation zur Erlangung der Doktorwürde der hohen philosophischen Fakultät der Friedrich-Alexanders-Universität Erlangen ... Erlangen (K. b. Hof- und Univ.-Buchdruckerei von Fr. Junge (Junge & Sohn). 1901. Oct. (*Syst.-anat. Unters. Podalyr.*).
Publ.: 1901 (Nat. Nov. Feb(2) 1902), p. [i, iii], [1]-111, [112, vita]. *Copy*: NY. – Born 6 Dec 1890, fide "Vita".

Prentiss, Albert Nelson (1836-1896), American botanist; M.S. Lansing, Mich. 1864; associate principal Kalamazoo, Mich. high school 1862-1863; instructor in botany 1863-1864; professor of botany and horticulture from 1865-1868; id. Cornell Univ. 1868-1896. (*Prentiss*).

HERBARIUM and TYPES: CU (few). Prentiss participated in the Cornell Exploring Expedition to Brazil in 1870 ("Morgan expedition"). R.T. Clausen and R.P. Korf informed us that there are some fern collections at CU but that neither fungi nor angiosperm collections have so far been located. A typescript list of Brazilian ferns collected by Prentiss and others is at CU. Some further material by Prentiss was received by CU in 1900. – Letters at G and PH.

BIBLIOGRAPHY and BIOGRAPHY: Barnhart 3: 108; Bossert p. 317 (b. 22 Mai 1836, d. 14 Aug 1896); CSP 11: 63, 12: 587, 17: 1007; LS 21210-21211, 38060; ME 1: 221; Morren ed. 10, p. 127; PH 132; Rehder 5: 687; Tucker 1: 568.

BIOFILE: Anon., Bot. Centralbl. 66: 206. 1896 (resigned chair for health reasons), 409. 1896 (d.); Bot. Jahrb. 23 (Beibl. 57): 59. 1897 (d. 14 Aug 1896); Bot. Not. 1897: 47; Nat. Nov. 18: 244 (retires), 462. 1896 (d. 14 Aug 1896).
Atkinson, G.F., Bot. Gaz. 21: 283-289, *pl. xix*. 1896. (b. 22 Mai 1836, portr., biogr.), see also 22: 430. 1896 for a correction); Science ser. 2. 4: 523-524. 1896 (obit., d. 15 Aug 1896).
C.E.B., Amer. Natural. 30: 1043-1044. 1896 (obit., b. 22 Mai 1836).
House, H.D., Bibl. bot. N.Y. State 15. 1942.
Kneucker, A., Allg. bot. Z. 2: 204. 1896 (d. 14 Aug 1896).
Rickett, H.W., Index Bull. Torrey Club 80. 1955.
Rodgers, A.D., Liberty Hyde Bailey 501 [index]. 1949.

Prescott, Gerald Webber (1899-x), American algologist; Dr. phil. Iowa State 1928; act. prof. bot. Willamette Univ. 1928-1929; asst. prof. Albion College 1929-1932, assoc., prof. id. 1932-1946; assoc. prof. Michigan State 1946-1949; prof. ib. 1949-1968; in retirement associated with the University of Montana. (*Prescott*).

HERBARIUM and TYPES: MSC and NY, also EMC (fide IH, ed. 6, & 7), other material at CI, DS, F, FH, SI. – Some manuscript material in G and SIA.

BIBLIOGRAPHY and BIOGRAPHY: Barnhart 3: 108; IH 1 (ed. 2): 49, (ed. 3): 59, (ed. 4): 63, (ed. 5): 609 (ed. 6): 363 (ed. 7): 340; NAF ser. 2. 6: 72-73. (bibl.); NW p. 55; Roon p. 90; TL-1/781.

BIOFILE: Acosta-Solis, M., Natural. Viaj. ci. Equador 75. 1968.
Ewan, J. et al., Short hist. bot. U.S. 78, 79. 1969.
Koster, J.Th., Taxon 18: 556. 1969 (algae herb. still private).
Prescott, G.W., List of publications, 2 p.s. s.d. (ca. 1952); The algae, a review. Boston 17 Sep 1968, xi, 436 p.; A contribution to a bibliography of antarctic and subantarctic algae together with a checklist of freshwater taxa reported to 1977. Vaduz 1979 (see Taxon 28: 658. 1979).
Verdoorn, ed., Chron. bot. 3: 295. 1937, 4: 556, 5: 292. 1939.
Wood, R.D., Monogr. Charac. 1: 834. 1965.

COMPOSITE WORKS: *North American Flora, Desmidiales* in ser. 2. 6: 1-84. 17 Aug 1972 (with H.T. Croasdale and W.C. Vinyard).

EPONYMY: *Prescotia* Lindley ex W.J. Hooker (1824) honors John Prescot "of St. Petersburg", which is John D. Prescott (x-1837), correspondent of W. Hooker and J. Lindley.

8286. *Iowa algae* University of Iowa Studies in natural history ... Iowa City (published by the University) 1931. Oct. (*Iowa alg.*).
Publ.: 1 Jul 1931, p. [1]-235. *Copies*: FH, US. – Univ. Iowa Stud. nat. Hist. 13(6).

Presl, Jan Svatopluk (1791-1849), Czech physician and botanist at Praha; professor of botany at the medical faculty. (*J. Presl*).

HERBARIUM and TYPES: PR and PRC. - *Exsiccatae*: *Vegetabilia cryptogamica Boëmiae* collecta a Joanne et Carolo Presl, fasc. 1-2, nos. 1-50, Pragae 1812 (Mar, Mai 1812) see Sayre (1969). Other material at H (herb. Steven).

NAME: Svatopluk: "Heavenly Host" (Barneby 1963).

BIBLIOGRAPHY and BIOGRAPHY: AG 2(2): 279; Barnhart 3: 108; BM 4: 1611-1612; CSP 5: 7-8; Futák-Domin p. 487; Herder p. 472; Hortus 3: 1202 (J. Presl); IH 1 (ed. 2): 200. (ed. 3): 126, (ed. 6): 363, (ed. 7): 240, 2: (in press); Klásterský p. 149 (b. 4 Nov 1791, d. 6 Apr 1849) (many details and biogr. refs.); Kew 4: 361; LS 21213; Maiwald p. 178-179, 282 [index]; PR 631, 7306-7309 (ed. 1: 8198-8200); Quenstedt p. 345; Rehder 5: 687; RS p. 134; SO add. 826a; TL-1/992-993; TL-2/432-433, see Haenke, T.P.X.; Tucker 1: 568; WU 23: 270-275; Zander ed. 10, p. 703; ed. 11, p. 802.

BIOFILE: Anon., Bot. Zeit. 7: 536. 1849 (obit.), 9: 96. 1851; Flora 33: 284. 1850 (d.).
Barneby, R., Garden J. 13: 142. 1963.
Eiselt, J.N., Gesch. Syst. Lit. Insektenk. 92-101. 1836 (J.S. P's entom. system).
Hoffmannová, E., Jan Svatopluk Presl, Karel Bořivoj Presl, Praha, 1973, 299 p. (biogr., portr., extensive discussion of publications).
Kittner, R., Časopis pro mineralogii a geologii 11(4): 491-495. 1966 (portr.).
Kolari, V., Jan Svatopluk Presl und die Tschechische botanische Nomenklatur. Eine lexikalisch-nomenklatorische Studie, Helsinki 1981, xxxii, 422 p. (Ann. Acad. Sci. fenn., diss. human. Litt. 25).
Kops, J., Zdrav. Prac. 25: 558-559. 1975 (n.v.).
Kühnel, J., Thaddaeus Haenke 76, 162. 1960.
Lamson-Scribner, F., Rep. Missouri Bot. Gard. 10: 35-59, *pl. 1-54*. 7 Jun 1899 (Grasses in the Bernhardi herb. descr. by J.S. Presl).
Lindemann, E.v., Bull. Soc. imp. Natural. Moscou 61: 57. 1886.
Matouschek, F., Verh. Zool.-bot. Ges. Wien 50: 276-386. 1900; Bot. Centralbl. 85: 54-55. 1901 (on exsicc.).
Milner, J.D., Cat. portr. Kew 90. 1906.
Němec, B., Čas. nár. Mus., Praha 109: 89-92. 1935.
Poggendorff, J.C., Biogr.-lit. Handw.-Buch 2: 521. 1863 (b. 4 Sep 1791, d. 6 Apr 1849).
Sayre, G., Mem. New York Bot. Gard. 19(1): 41. 1969 (exsicc.).
Slavik, B., Preslia 51: 340. 1979.
Stohandl, F.C., Z. mähr. Landesmus. 1(1/2): [3 p.]. 1901 (lists coll. Franzensmuseum, Brno, now BRNM, with e.g. a set of Veget. crypt. Boëmiae: fide F. Matouschek, Bot. Centralbl. 88: 283-284. 1901; see also Matouschek 1900).
Verdoorn, F., ed., Chron. bot. 4: 188. 1938.
Weitenweber, W.R., Abh. k. böhm. Ges. Wiss. ser. 5. 8(2): 1-27. 1854 (Denkschrift über die Gebrüder Joh. Swat. und Carl. Bor. Presl).

COMPOSITE WORKS: with Friedrich Berchtold (1781-1876):
(1) *O přirozenosti rostlin*, 1820, TL-2/432. Another copy of this rare book is at WU (with minor differences; inf. W. Gutermann): p. [ii, iii, title pages], [1]-322, [1-2, errata]. See also Flora 7: 192. 1824. - A copy was offered by Ronald Meesters (list 24, spring 1980).
(2) Idem, 1823-1835, 3 vols., TL-2/433. Vol. 1 is by both authors; vols. 2 and 3 are by J.S. Presl alone.

EPONYMY: (genera): *Preslaea* C.F.P. von Martius (1827) and *Preslia* Opiz (1824) are dedicated to both J. Presl and his brother K. Presl, q.v.; *Preslea* K.P.J. Sprengel (1827), *orth. var.* of *Preslaea* C.F.P. von Martius); (journal): *Preslia*, Věstnik České Botanické Společnosti. Vol. 1-x, 1914-x. (Also named for K. Presl, q.v.).

8287. *Flora čechica*. Indicatis medicinalibus, oeconomicis technologicisque plantis. Kwětena Česká. Spoznanánjm lekařských, hospodařských a řemeslnických rostlin ... Pragae [Praha] (in comissis apud J.G. Calve) 1819. Oct. †. (*Fl. čech.*).

Co-author: Karel Bořivoj [Boriwag] Presl (1794-1852).
Publ.: 1819 (p. xiv: 2 Jan 1819; rev. Beck Mai-Jun 1820), p. [i]-xiv, [xv-xvi], [1]-224.
Copies: FH, FI, G, MO, NY, Stevenson; IDC 5051.
Other issue: 1820 (t.p.) copy seen in Domín's library at PR by J. Dostál (in lit.).

8288. *Deliciae pragenses*, historiam naturalis spectantes ... volumen primum. Pragae [Praha] (sumptibus Calve) 1832. †. Oct. (*Delic. prag*.).
Co-author: Karel Bořivoj [Boriwag] Presl (1794-1852).
Publ.: Jul 1822 (see Flora 5: 576; sent to booksellers Jul 1822; p. vi: 1 Jul 1821; Beck 15 Sep 1822; Cat. Easter fair, Leipzig 1822), p. [i]-viii, [1]-244. *Copies*: FI, MO, NY; IDC 276.

8289. Jan Swatopluka Presla, ... *Wšobecný rostlinopsis*, čili: popsání rostlin we wšelikém ohledu užitecných a škodliwých ... W Praze [Praha] (W kommissí u Kronbergra a Řiwnáče) 1846. Oct. (*Wšobecný rostl*.).
1: 1846, p. [i*], [i]-xxxii, [1]-1006.
2: 1846, p. [i], 1007-2072, [1-2, err.].
Copies: G, L, NY; IDC 5148. – The NY copy has a second title page facing the regular t.p.: "Nowočeska bibliothéká, ... čislo 7. ..." For a study on the unlisted binomials see Merrill (1950). [approximate English title (Merrill): Universal botany, or descriptions of plants from all parts of the world, especially of useful and harmful species]. Merrill discusses the link with L. Oken, Allgemeine Naturgeschichte, Bot. vols. 1, 2; the binomials of which seem to have been cited almost exclusively by Presl.
Ref.: G.P., Bot. Zeit. 5: 153. 26 Feb 1847.

8290. *Počátkowé rostlinoslowí* ... w Praze [Praha] (w kommissi u Kronbergra a Řiwnáče) 1848. Oct. (*Počatkowé rostl*.).
Publ.: 1848, p. [i*-iii*], [i]-vi, [vii], [1]-564, [1-2, err.]. *Copy*: L.

8291. Třicet a dwa obrazy k *Prwopočátkům rostlinoslowi* .. w Praze [Praha] (Tiskem knížecí arcibiskupské knihtiskárny w Semináři) 1848. Qu. (*Porwopoč. rostl*.).
Publ.: 1848, p. [i], [1]-30, *pl. 1-32* (uncol.). *Copy*: L.

Presl, Karel Bořivoj [Boriwog, Boriwag] (1794-1852), Czech botanist; curator at the Prague national Museum; professor of botany at Prague university 1833-1852. (*K. Presl*).

HERBARIUM and TYPES: PR and PRC (fern herb. and Haenke's coll.); other material at B, BM, GOET, H, HAL and KIEL (destr.) and W. – *Exsiccatae*: see J.S. Presl. – Correspondence with D.F.L. von Schlechtendal at HAL; other letters at G.

NAME: Prince Bořivoj (Boriwog) was by tradition the first Bohemian convert to Christianity (ca. 873) (Barneby 1973).

BIBLIOGRAPHY and BIOGRAPHY: AG 2(2): 279; Backer p. 463; Barnhart 3: 108 (b. 17 Sep 1794); BM 4: 1612; CSP 5: 7-8; DSB 11: 130 (d. 2 Oct 1852); Frank 3 (Anh.): 79; Futák-Domin p. 486-487; GF p. 71; GR p. 672; Hegi 5(4): 2338; Herder p. 472; Hortus 3: 1202 (K. Presl); IH 1 (ed. 7): 340; IF p. 725; Jackson p. 115, 116, 150, 151, 322; Kew 4: 361-362; Klášterský p. 150; Langman p. 595; Lasègue p. 333, 528, 545; LS 21213-21215; Maiwald p. 282 [index]; MD p. 201-203; MW p. 395; NI 1564-1565; PR 1218, 7310- 7324 (ed. 1: 8201-8215); Quenstedt p. 345; Rehder 5: 687; RS p. 134; Saccardo 1: 132, 2: 88, cron. p. xxx; SBC p. 129; SK 1: 416, 4: ccvi; Stevenson p. 1254; TL-1/992-1010, 1270; TL-2/826, see indexes; Tucker 1: 568; WU 23: 275-279; Urban-Berl. p. 322; Zander ed. 10, p. 703, ed. 11, p. 802.

BIOFILE: Alston, A.H.G., J. Bot. 72: 223-226. 1934 (on Haenke's ferns in *Reliq. haenk*.).
Andrews, H.N., The fossil hunters 73. 1980.
Anon., Alman. Akad. Wiss. Wien 2: 187-188. 1852 (bibl.); Bonplandia 2: 267. 1854 (cogn. Leopoldina: Plumier); Bot. Zeit. 1: 831. 1852 (d.), 13: 884. 1855 (herb. for

sale); Flora 36: 32. 1852 (d.); Lotos 6: 15-17. 1816 (description of P's herbarium); Österr. bot. W. 2: 342. 1852 (d. 2 Oct 1852, "organische Wassersucht", dropsy).
Barneby, R.C., J. New York Bot. Gard. 13: 139-140, 142 (*Reliq. haenk.*).
Barnhart, J.H., Bull. Torrey bot. Club 32: 590-591. 1905 (on *Epimel. bot.*).
Beek, G., Bot. Centralbl. 34: 150. 1888 (Pl. sicil. at W).
Candolle, Alph. de, Phytographie 440. 1880 (coll.).
Candolle, Alph. & A. P. de, Prodr. 15(2): 258. 1862 (on *Epimel. bot.*).
Hoffmanová, E., Jan Svatopluk Presl, Karel Borivog Presl, Praha 1973, 299, 2 p. (biogr., portr., disc. publications).
Holttum, R.E., Nov. bot. Inst. bot. Univ. Carol. Prag. 1968; 3-56 (types of P's fern herb.).
Hooker, W.J., J. Bot 4: 286. 1852. (on *Epimel. bot.*).
Jessen, K.F.W., Bot. Gegenw. Vorz. 377, 467. 1864.
Junk, W., Rara 2: 184-185. 1926-1936 (*Reliq. haenk.*).
Kühnel, J., Thad. Haenke 75, 76, 160. 1960.
Kukkonen, I., Herb. Christ. Steven 81. 1909.
Kuntze, O., Rev. gen. pl. 1: cxxviii-cxxxiv. 1891 (Symb. bot.), 3(2): 160. 1898 (*Reliq. haenk.*).
Looser, G., Commun. Museo Concepcion 1: 114-123. 1936 (ferns).
Matouschek, F., Verh. Zool.-bot. Ges. Wien 50: 276-277. 1900.
Merrill, E.D., Enum. Phil. pl. 4: 218-219. 1926; Bishop Mus. Bull. 144: 153. 1937, Contr. U.S. natl. Herb. 30: 245-246. 1947 (bibl).
Milner, J.D., Cat. portr. Kew 90. 1906.
Murray, G., Hist. coll. BM(NH) 1: 175. 1904 (some of the *Reliq. Haenk.*).
Němeč, B., Čas. nár. Mus., Praha 1935: 89-92 (on P's morphology).
Oken, L., Isis 21: 274. 1818, 24: 489. 1831 (on *Symphysia*).
Oliver, F.W., Makers Brit. bot. 145-147. 1913.
Pichi-Sermolli, R.E.G., Webbia 31(1): 255. 1977 (on P's Hymenophyllaceae), 9: 361-366. 1953 (on *Epimel bot.* and Fée's *Gen. fil.*).
Poggendorff, J.C., Biogr.-lit. Handw.-Buch 2: 521. 1863 (b. 17 Feb 1794, d. 2 Oct 1852).
Presl, K.B., Bemerkungen zu einigen Herbarien des F.W. Sieber, Isis 21: 267-276. 1828; Monatschr. Ges. Vaterländ. Mus. Prag 1828: 161-168 (on herb. Haenke).
Schmid, G., Chamisso als Naturforscher 10, 94, 171. 1942.
Seemann, B., Bonplandia 2: 267. 1854 (cogn. Leopoldina *Plumier*), 3: 113. 1855 (on P's *Tent. pterid.*).
Stafleu, F.A., Regn. veg. 71: 339. 1970 (Miquel-Schlecht. corr.), Taxon 22: 679-680. 1973 (re-issue *Reliq. haenk.*).
Stearn, W.T., J. Soc. Bibl. nat. Hist. 1(5): 153-154. 1938 (on *Reliquiae haenkeanae*), J. Soc. Bibl. nat. Hist. 3: 14-16. 1953 (*Prodr. monogr. Lobel., Tent. Pterid.* and other publ.), Taxon 21: 105-111. 1972 (on *Symphysia*), An introduction to K.B. Presl's Reliquiae haenkeanae, Amsterdam 1973, 19 p. in facs., ed. *Reliq. haenk.*
Tempsky, A., Bonplandia 6: 298. 15 Aug 1858 (part 8, adv. dated Jul 1858).
Verdoorn, F., ed., Chron. bot. 4: 188. 1938.
Weitenweber, W.R., Abh. k. böhm. Ges. Wiss. ser. 5. 8(2): 1-27. 1854 (biogr., bibl.).
Wettstein, R. v., Lehrkanzel syst. Bot. deut. Univ. Prag 10. 1899 (fern herb.); Österr. bot. Z. 49(2): 95. 1899 (fern herb. at PRC).

EPONYMY: (genera): *Preslaea* C.F.P. von Martius (1827) and *Preslia* Opiz (1824) are dedicated to both K. Presl and his brother J. Presl, q.v.; *Preslea* K.P.J. Sprengel (1827, orth. var. of *Preslaea* C.F.P. von Martius). (journal): *Preslia*, Věstnik České Botanické Společnosti. Vol. 1-x, 1914-x. (Also named for J.Presl, q.v.).

8292. *Cyperaceae et Gramineae siculae* ... Pragae (apud D. Hartmann) 1820. Oct. (*Cyper. Gramin. sicul.*).
Publ.: 1820 (p. xviii: 2 Dec 1819), p. [i]-xxii, [1]-58. *Copies*: G, NY, WU; IDC 415.

vol.	part	pages	plates	dates
1	1	[i]-xxv, [1]-84	*1-12*	12 Jun-30 Nov 1825
	2	[i], 85-148	*13-25*	Jul-Dec 1827
	3	[i], 149-206	*26-36*	Jul-Dec 1828
	4/5	[i], 207-356, [i*, iii*]	*37-48*	Jan-Jun 1830
2	1	[1]-56	*49-60*	Jan-Jun 1831
	2	57-152	*61-72*	Jan-Jul 1835

8293. *Reliquiae haenkeanae* seu descriptiones et icones plantarum, quas in America meridionali et boreali, in insulis Philippinis et Marianis collegit Thaddeus Haenke, redegit et in ordinem digessit Carolus Bor. Presl, ...cura Musei bohemici... Pragae [Praha] (apud J.G. Calve bibliopolam) 1825-1835, 2 vols. †. Fol. *Reliq. haenk.*).
Collector: Thaddeus Peregrinus Xaverius Haenke (1761-1816) (see TL-2/2: 6-8), vol. 1 has a biography of H. by K. v. Sternberg on p. [iii]-xv.
Publ.: two volumes in seven parts, covers and fascicle t.p.'s dated 1825-1835 (see e.g. copy at G):
1: 1825-1830, p. [i*, iii* h.t. and vol. t.p.], [i]-xv, [1]-356, *pl. 1-48*.
2: 1831-1835, p. [1]-152, *pl. 49-72*.
Copies: BR, G, McVaugh, MICH, MO, NY, U, US; IDC 436.
Plates: Copper engr. of drawings by Franz Both (*1-15, 18, 19*), F.X. Fieber (*16, 17, 20-56, 60-72*), J. Longer (*57-59*). *Authorship*: See below, list of collaborators; K.B. Presl was main author and editor.
Format: folio with every second gathering unsigned. See MD for details (as well as for further bibl. information).
Collaborators (to be cited, for example, as C.A. Agardh *in* K. Presl, etc.):
Agardh, Carl Adolf (1785-1859) – *Algae* 1(1): 8-12.
Bartling, Friedrich Gottlieb (1789-1859) – *Paronychieae, Alsin.* 2(1): 5-20.
Floerke, Heinrich Gustav (1764-1835) – *Lichenes* 1(1): 3-7.
Hornschuch, Christian Friedrich (1793-1850) – *Musci* 1(1): 13.
Meyer, Ernst Heinrich Friedrich (1791-1858) – *Juncaceae* 1(2): 141-146.
Nees von Esenbeck, Theodor Friedrich Ludwig (1787-1837) – *Fungi* 1(1): 1-2.
Opiz, Philipp Maximilian (1787-1858) – *Piperaceae* 1(3): 150-164.
Presl, Jan Svatopluk (1791-1849) – *Taccacaeae* 1(3): 149.
Presl, J.S. & K.B. – *Cyperaceae* 1(3): 165-206.
Presl, J.S. – *Gramineae* 1(4/5): 207-351.
Sternberg, Kaspar Graf von (1761-1838) – Preface [iii]-xv.
Facsimile ed.: 1973, ISBN 90-6123-237-6, p. [i-ii], [1]-19, [introduction by W.T.Stearn], followed by facsimile (lacking the fascicle titles) [i]-xv, [1]-356, [i, repr. cover fasc. 2(1)], [1]-152, [1, h.t. Tabulae], *pl. 1-72. Copies*: B, FAS. "Karl Boriwog Presl *Reliquiae haenkeanae* seu ... Haenke with a map and a bibliographical note by Dr. W.T. Stearn ..." Amsterdam (A. Asher & Co. B.V.) 1973. – The bibliographical "note" by Stearn is a thorough study with ample information on further studies, references and an index to vol. 2 (see also Taxon 22: 679-680. 1973).

8294. *Flora sicula*, exhibens plantas vasculosas in Sicilia aut sponte crescentes aut frequentissime cultas, secundum systema naturale digestas, ... tomus primus. Pragae [Praha] (sumptibus A. Borrosch) 1826. Oct. †. (*Fl. sicul.*).
Publ.: 1826 (p. xii: 31 Nov 1825; Linnaea Oct-Dec 1826; Flora 14 Nov 1826), p. [i]-xlvi, [1]-216, [1, err.]. *Copies*: B, BR, G; IDC 282.
Ref.: Anon., Flora 10: 681-686. 21 Nov 1827 (rev.).

8295. Epistola de *Symphysia*, novo genere plantarum, ad illustrissimum liberum baronem Josephum de Jacquin, ... data a C.B. Presl, ... [Pragae 1827]. Qu. (*Symphysia*).
Dedicatee: Joseph Franz, baron von Jacquin (1766-1839), see TL-2/2: 405-407.
Publ.: Mar 1827 (Stearn 1972; p. ii: 10 Jan 1827; rd. at Regensburg Feb-Mar 1828; fide Flora 11: 192), p. [1-4], *1 pl.* (uncol. copp.). *Copies*: G(2), MO, NY, WU. – See also Oken (1828, 1831) and Stearn (1972).

8296. *Pedilonia*, novum plantarum genus... Pragae [Praha] 1829. Qu. (*Pedilonia*).
Publ.: 15 Mai 1829 (t.p.; rev. Flora 12: 568-569. 28 Sep 1829). p. [1-2]. *Copies*: G, K, PR.

8297. *Didymonema*, novum plantarum genus. Praha 1829. Qu. (*Didymonema*).
Publ.: 30 Mai 1829 (t.p.). *Copy*: K. – Preprinted from *Symb. bot.* 1: 5-6. Oct 1830.

8298. *Thysanachne*, novum plantarum genus. Praha 1829. Qu. (*Thysanachne*).
Publ.: 30 Mai 1829 (t.p.). *Copy*: K. – Preprinted from *Symb. bot.* 1: 11-12. Oct 1830.

8299. *Lepisia*, novum plantarum genus. Praha 1829. Qu. (*Lepisia*).
Publ.: 30 Mai 1829 (t.p.). *Copy*: K. – Preprinted from *Symb. bot.* 1: 9-10. Oct 1830.

8300. *Polpoda*, novum plantarum genus. Praha 1829. Qu. (*Polpoda*).
Publ.: 30 Mai 1829 (t.p.). *Copy*: K. – Preprinted from *Symb. bot.* 1: 1-2. Oct 1830.

8301. *Scyphaea*, novum plantarum genus. Praha 1829. Qu. (*Scyphaea*).
Publ.: 30 Mai 1829 (t.p.). *Copy*: K. – Preprinted from *Symb. bot.* 1: 7-8. Oct 1830.

8302. *Steudelia*, novum plantarum genus. Praha 1829. Qu. (*Steudelia*).
Publ.: 1 Jul 1829 (t.p.). *Copy*: K. – Preprinted from *Symb. bot.* 1: 3-4. Oct 1830.

8303. *Symbolae botanicae* sive descriptiones et icones plantarum novarum aut minus cognitarum ... Pragae [Praha] (sumptibus auctoris. E typographia J. Spurny) 1830-1858. Qu. †. (*Symb. bot.*).
Publ.: In 8 parts (2 vols.) (copy in original covers at G) 1830-1858 (part 8 posthumous, see Bonplandia 6: 298. 15 Aug 1858), p. [i*, iii*, v*], [i]-ii, [1]-76, [1-2, corr., ind.]; [i, iii], [1]-30, [1, index], *pl. 1-80*. *Copies*: G, MICH, MO, NY(2), U; IDC 275.

vol.	part	pages	*plates*	dates
1	1	[i]-ii, [1]-18	*1-10*	Oct 1830
	2	19-32	*11-20*	Jul 1831 (Flora rd 3 Aug 1831)
	3	33-52	*21-30*	Jul 1831 (id.)
	4	49-64	*31-40*	Sep-Dec 1831
	5	65-76, [1-2]	*41-50*	Jan-Feb 1832
		[i*, iii*, v*]	-	1832
2	6	[1]-16	*51-60*	Jul 1834
	7	17-24	*61-70*	Jul 1834
	8	25-30, [1]	*71-80*	Jul 1858
		[i], [iii]		Jul 1858

Imprint covers 1-7: Pragae (apud J.G. Calve); imprint on t.p. of the second volume: Pragae (sumptibus Friderici Tempsky) 1858. The cover of part 4 is dated (by hand) 1831, that of part 5 1832; in the absence of clear indications of later publication these dates have to be accepted. *Plates*: uncol. copper engr. of drawings by F.X. Fieber, Franz Both, J. Skala and the author. *Dates*: see TL-1 for further references.
Ref.: Anon., Ann. Sci. nat., Bot. ser. 2. 1: 357-368. 1834 (rev. *pl. 1-50*).
Schlechtendal, D.F.L. von, Bot. Zeit. 17: 198-199. 1859.

8304. *Repertorium botanicae systematicae*. Excerpta e scriptoribus botanicis, continentia generum et specierum novarum aut melius distinctarum, indicationes iconum generum et specierum jam cognitarum et adnotationes succinctas botanicam systematicam spectantes, sistentia supplementum continuum prodromi systematis naturalis Candollei Systematis vegetabilium Schultesii et Sprengelii ... volumen 1. Pragae [Praha] (Typis filiorum Theophili Haase) 1833-1834. Oct. †. (*Repert. bot. syst.*).
Fasc. 1: Nov-Dec 1833 (p. viii: 30 Oct 1832, Flora 15 Jan 1834), p. [iii]-viii, [1]-184. – P. ii is t.p. fasc. 1, dated 1833.
Fasc. 2: 1834 (Hinrichs 19-25 Dec 1834), p. [i, t.p. vol.; iii, t.p. fasc. 2], 185-385.
Copies: B, BR, FI, G, H, NY(2), U, USDA.

8305. *Bemerkungen über den Bau der Blumen der Balsamineen* ... Prag [Praha] (Druck von Gottlieb Haase Söhne) 1836. Oct. (*Bemerk. Bl. Balsam.*).
Publ.: 1836 (Flora 14 Sep 1836), p. [1]-54, *1 pl.* (uncol.). *Copies*: BR, G, NY. – Reprinted from Abh. k. böhm. Ges. Wiss. ser. 4. 5: 1836.

8305a. *Beschreibung zweier neuen böhmischen Arten der Gattung Asplenium* ... Prag [Praha] (Druck und Papier von Gottlieb Haase Söhne) 1836. Oct. (*Beschr. Art. Asplenium*).
Publ.: 1836 (Flora 14 Sep 1836), p. [1]-11, *pl. 3* (uncol., Corda). *Copy*: BR.

8306. *Prodromus monographiae Lobeliacearum* ... Pragae [Praha] (Typis filiorum Theophili Haase) 1836. Oct. (*Prodr. monogr. Lobel.*).
Publ.: Jul-Aug 1836 (TL-1; Stearn 1954; Flora 14 Sep 1836), p. [1]-52. *Copies*: B, FI, G(2), L, MICH, MO, PH, U, US, USDA; IDC 261. – Also issued, with same pagination and t.p. date (1836), as Abh. k. Böhm. Ges. Wiss. ser. 4. 4: 1-52. 1837. The separate appeared in 1836; it was re-issued with other papers in the 1837 *Abhandlungen*.

8307. *Tentamen pteridographiae, seu genera filicacearum praesertim juxta venarum decursum et distributionem exposita* ... Pragae [Praha] (typis filiorum Theophili Haase) 1836. Oct. (*Tent. pterid.*).
Publ.: shortly before 2 Dec 1836 (Litterar. Zeit. 2 Dec 1836; Hinrichs 4-11 Dec 1836), p. [i], [1]-290, *pl. 1-12* (uncol., auct.). *Copies*: B, BR, FI, G, NY, US; IDC 5207. – Also issued, with same pagination, in Abh. böhm. Ges. Wiss. ser. 4. 5: 1-290. 1837. See also Stearn (1954). the *plates* are uncol. copper engr. (nos. *1-11*) or lith. (12) of drawings by Corda and by the author.
Facsimile ed.: announced, but not yet published by J. Cramer (1974).
Ref.: Pichi-Sermolli, R.E.G., Webbia 25: 290. 1970.

8308. *Hymenophyllaceae. Eine botanische Abhandlung* ... Prag [Praha] (Druck und Papier von Gottlieb Haase Söhne) 1843. Qu. (*Hymenophyllaceae*).
Publ.: late 1843 (Gersdorf 8 Mar 1844; see Stearn 1954, Pichi-Sermolli 1977), p. [1]-70, *pl. 1-12* (uncol., Corda). *Copies*: FI, G, MO, NY(2); US; IDC 7247. – Preprinted from Abh. k. böhm. Ges. Wiss. ser. 5. 3: 93-162. 1845.
Ref.: K.M., Bot. Zeit. 2: 577-581. 16 Aug 1844.
Pichi-Sermolli, R.E.G., Webbia 31: 255. 1977.

8309. *Botanische Bemerkungen* ... Prag (Druck der k.k. Hofbuchdruckerei von Gottlieb Haase Söhne) 1844 [1845, 1846]. Qu.
Publ.: Jul-Dec 1845 (in journal), Jan-Apr 1846 as reprint (TL-1), p. [1]-154. *Copies*: FI, G, MO, NY(2); U; IDC 260. – Originally published and to be cited from Abh. k. Böhm. Ges. Wiss. ser. 5. 3: 431-584. 1845 (when using reprint add 430 to pagination). The reprint carries the date 1844 [sic]. Gersdorf (Leipziger Repertorium 1846(4): 276-277. 13 Nov 1846) writes: "Die Vorrede zu dieser wunderlichen Schrift, wohl nicht ohne Absicht vom 1 April datirt scheint ... eher geschrieben zu sein als der Text." "... noch Niemandem ist es eingefallen solche Studien drucken zu lassen ..." Stearn writes: "... the many new names (now mostly reduced to synonymy) ...".
Ref.: Schlechtendal, D.F.L. von, Bot. Zeit. 4: 470-471. 1846.

8310. *Supplementum tentaminis pteridographiae*, continens genera et species ordinum dictorum Marattiaceae, Ophioglossaceae, Osmundaceae, Schizaeaceae et Lygodiaceae, ... Pragae [Praha] (e Typographia Caes. reg. aulica filiorum Amadei Haase) 1845. Qu. (*Suppl. tent. pterid.*).
Publ.: late 1845 (t.p.; Hinrichs rd. 18-19 Mai 1846; even though it is quite well possible that the preprint came out early 1846, the single Hinrichs reference is not sufficient proof to deviate from the t.p. date), p. [1]-119, [120, err.]. *Copies*: BR, FH, FI, G, MO, NY. – Preprinted from Abh. k. böhm. Ges. Wiss. ser. 5. 4: 261-380. 1847.
Facsimile ed.: announced but not yet published by J. Cramer 1974.
Ref.: Pichi-Sermolli, R.E.G., Webbia 25: 290-291. 1970.
Schlechtendal, D.F.L. von, Bot. Zeit. 4: 487-488. 10 Jul 1846.

8311. *Die Gefässbündel im Stipes der Farrn* ... Prag [Praha] (Druck der k.k. Hofbuchdruckerei von Gottlieb Haase Söhne) 1847. Qu. (*Gefässbündel Farrn*).
Publ.: 1847 (t.p.), p. [i], [1]-48, *pl. 1-7* (uncol., auct.). *Copies*: G, NY(2); IDC 7248. – Preprinted with the addition of a special cover and t.p., from Abh. k. böhm. Ges. Wiss. ser. 5. 5: 307-356, *pl. 1-7*. 1848.
Ref.: Pichi-Sermolli, R.E.G., Webbia 25: 291. 1970.

8312. *Epimeliae botanicae* ... Pragae [Praha] (Typis filiorum Amadei Haase) 1849. Qu. (*Epimel. bot.*).
Publ.: "not much earlier than" Oct 1851(see Barnhart (1905), Stearn (1953), and TL-1), p. [1]-264, *pl. 1-15* (uncol., auct.). *Copies*: FH, FI, G, U, US, USDA; IDC 5335. – Reprinted from Abh. böhm. Ges. Wiss. ser. 5. 6: 361-624, *pl. 1-15*. 1851. Reprint and original must have been published almost simultaneously in October 1851 or somewhat, but not much, earlier. See Pichi-Sermolli (1953) for a discussion of the nomenclature of ferns described by Fée, *Gén. fil.* 3-50. 1850 and Presl's *Epimeliae*.
Facsimile ed.: announced but not yet published by J. Cramer (1974).
Ref.: Hooker, W.J., J. Bot. Kew Gard. Misc. 4: 286-287. 1852.
Schlechtendal, D.F.L. von, Bot. Zeit. 10: 656-657. 1852.

Prestoe, Henry (1842-1923), British gardener; trained at Kew; Government botanist and superintendent Botanic Gardens, Trinidad 1864-1886. (*Prestoe*).

HERBARIUM and TYPES: TRIN; further material at K and US.

BIBLIOGRAPHY and BIOGRAPHY: Barnhart 3: 108; BB p. 248, 341; CSP 11: 63; Desmond p. 503 (b. 6 Jan 1842, d. 24 Sep 1923); Jackson p. 45; Kew 4: 362; Morren ed. 2, p. 34, ed. 10, p. 135; Rehder 5: 688; Tucker 1: 436, 568-569; Urban-Berl. p. 385.

BIOFILE: Anon., Bull. misc. Inf. Kew, add. ser. 1: 56.
Bridson, G.D.R. et al., Nat. hist. mss. res. Brit. Isl. 255.53. 1980.
Gager, C.S., Brooklyn Bot. Gard. Rec. 27(3): 175. 1938 (dir. bot. gard. Trinidad).
Hart, J.H., Ann. Rep. Bot. Gard. Trinidad 1887: 7 (app. Trinidad).
Jackson, B.D., Bull. misc. Inf. Kew 1901: 53 (pl. K).
Urban, I., Symb. ant. 1: 131. 1898, 3: 106. 1902.

EPONYMY: *Prestoea* J.D. Hooker (1883, *nom. cons.*) very likely honors Prestoe but no etymology is given.

8313. Catalogue of plants cultivated in the Royal Botanical Gardens, Trinidad, from 1865-1870. Port-of-Spain ("Chronicle" Printing Office) 1870.
Publ.: 1870 (preface: 1 Mai 1870), p. [i-iv], [1]-105, [i]-xv. *Copies*: DBN(2) (Inf. E.C. Nelson), HH.– Main text published in Proc. Sci. Ass. Trinidad 6: [251]-355. Jun 1869 (*copy*: NY).

Preston, Thomas Arthur (1838-1905), British clergyman and botanist; B.A. Cantab. 1856, master Marlborough College 1858-1885; rector at Thurcaston 1885. (*Preston*).

HERBARIUM and TYPES: Material at BM, CGE, K and LEI.

BIBLIOGRAPHY and BIOGRAPHY: Barnhart 3: 109; BB p. 249; BL 2: 254, 273, 704-705; BM 4: 1612; Bossert p. 317; CSP 8: 658, 11: 64, 17: 1010-1011; Desmond p. 504 (b. 10 Oct 1838, d. 6 Feb 1905); IH 2: (in press); Jackson p. 256; Kew 4: 362; Morren ed. 10, p. 76; PR 7329; Rehder 1: 401, 404; Tucker 1: 569.

BIOFILE: Anon., Bot. Not. 1906: 45 (d.); J. Bot. 43: 104. 1905 (d.); Rep. Marlborough College Nat. Hist. Soc. 53: 102-104. 1905 ("from the pen of an old colleague", reprinted from the Malburian; reminiscences).
Babington, C.C., Mem., Journ. bot. Corr. 474 [index]. 1897.
Britten, J., J. Bot. 43: 362-363. 1905 (obit.).
Freeman, R.B., Brit. nat. hist. books 286. 1980.

Horwood, A.R. & C.W.F. Noel, Fl. Leicestershire ccxxvi. 1933.
Jackson, B.D., Bull. misc. Inf. Kew 1901: 53 (coll. K); Proc. Linn. Soc. 117: 49-51. 1905 (obit., b. 10 Oct 1833, d. 6 Feb 1905).
Kent, D.H., Brit. herb. 72. 1957 (coll.).
Murray, G., Hist. coll. BM(NH) 1: 175. 1904 (coll. BM).
Watson, H.C., Topogr. bot. ed. 2. 553. 1883.

EPONYMY: *Prestonia* R. Brown (1810 & 1811) and the derived *Prestoniopsis* J. Müller-Arg. (1859-1860) honor Charles Preston (1660-1711), British medical doctor and correspondent of Ray and others. *Prestonia* Scopoli (1777) probably also honors Dr. Charles Preston.

8314. *Flora of Marlborough*; with notices of the birds, and a sketch of the geological features of the neighbourhood ... London (John van Voorst, ...) 1863. Duod. (*Fl. Marlborough*).
Ed. 1: 1863 (p. viii: Apr 1863; J. Bot. Aug 1863; BSbF Jan 1864) map, p. [i]-xxiv, [1]-129. *Copies*: G, NY.
Ed. 2: in Report Marlborough College natural History Society, nos. [12, 15] 20, 23 & 24, 1870-1877, n.v., see BM 4: 1612. (only a very few copies extant fide Jackson).
Ed. 3: mentioned as being in preparation, J. Bot. 20: 64. 1882; not published.
Ref.: Anon., J. Bot. 1: 252-253. 1863.

8315. *The flowering plants of Wilts* with sketches of the physical geography and climate of the county ... Published by the Wiltshire Archaeological and Natural History Society, [Devises] 1888. Oct. (*Fl. pl. Wilts*).
Publ.: Nov 1888 (J. Bot. Dec 1888 rev.; Bot. Zeit. 28 Dec 1888; Nat. Nov. Dec(2) 1888), map, [i]-lxix, [1]-436. *Copies*: BR, G, MO, NY, US.
Ref.: Bennett, A., J. Bot. 26: 380-382. 1888.

Preuss, [Carl] Gotllieb Traugott (x-1855), German (Silesian) mycologist and pharmacist at Hoyerswerda from 1834-1855. (*G. Preuss*).

HERBARIUM and TYPES: B (extant, see Hiepko(1978) and Jülich(1974)). Reuss contributed to Rabenhorst's exsiccatae (see also Meyer 1956): *Klotzschii herbarium vivum mycologicum*, cent. 13-20, 1849-1854. – Letters at G.

BIBLIOGRAPHY and BIOGRAPHY: Barnhart 3: 109; BM 4: 1613; Frank 3(Anh.): 79; GR p. 37, cat. p. 69; Hawksworth p. 185; IH 1 (ed. 6): 363, (ed. 7): 340; 2: (in press); Kelly p. 179, 256; LS 21222-21226, 38066-38067; PR 9026; Stevenson p. 1254; TL-1/1275.

BIOFILE: Anon., Bot. Not. 1858: 13 (d.); Bonplandia 3: 343. 1855, 4: 282. 1856; Bot. Zeit. 13: 655. 1855 (d. 11 Jul 1855); Flora 38: 591 (d. 11 Jul 1855, of apoplexy); Österr. bot. W. 5: 334. 1855 (d. 11 Jul 1855).
Ascherson, P., Fl. Prov. Brandenburg 1: 10. 1864.
Jülich, W., Willdenowia 7(2): 261-331. 1974 (inventory herb. Preuss at B; annotated list available taxa).
Kohlmeyer, J., Willdenowia 3(1): 65, 68. 1962.
Meyer, D.E., Willdenowia 1: 573-605. 1956 (discussion of P's publications; published over 300 new species of fungi, list of names with indication of present status; bibl. of Preuss' publ.; biogr., itin.).
Pax, F., Bibl. schles. Bot. 50, 54. 1929.

COMPOSITE WORKS: *Die Pilze Deutschlands*, in J. Sturm, *Deutschlands Flora* Abth. 3, 1848-1862, vol. 6, Hefte:
25/26: 1-48, *pl. 1-24*. 1848 (rev. S-n, Flora 31: 288. 7 Mai 1848).
29/30: 49-96, *pl. 25-48*. 1851 (rev. Fürnrohr, A.E., Flora 34: 348-351. 14 Jun 1851; Schlechtendal, D.F.L. von, Bot. Zeit. 9: 885. 12 Sep 1851).
33/34: 1-48, *pl. 1-12*. 1853 (not mentioned by Meyer, 1956).
35/36: 97-144, *pl. 49-72*. 1862.

HANDWRITING: Willdenowia 7(2): 319-331. 1974.

EPONYMY: *Preussia* Fuckel (1867); *Preussiaster* O. Kuntze (1891)*Preussiella* B.C. Lodha (1978). See ING 3: 1409, Meyer (1956) and Hughes, Canad. J. Bot. 36: 756, 799. 1958.

8316. *Uebersicht untersuchter Pilze*, besonders aus der Umgegend von Hoyerswerda ... [Linnaea vols. 24, 25, 26]. 1851-1855. Oct.
Publ.: A series of papers in *Linnaea*:
24: [99]-128. Mai 1851, 129-153. Jul 1851.
25: [71]-80. Jun 1852, [723]-742. Dec 1853.
26: [705]-742. Sep 1855.
We have seen no repaginated reprints; a reprint with original pagination is at FH. – Published in Linnaea because of the interruption of the publ. of *Die Pilze Deutschlands* in Sturm's *Flora* (see above) after Sturm's death. The diagnoses were meant for the Flora. Their form and contents are heavily influenced by Albertini and Schweinitz, *Consp. fung. lusat.* The latter book deals with the fungi of Niesky in the Upper Lausitz (and not in N. Galicia as erroneously stated by us sub TL-2/75) only 35 km from Hoyerswerda where Preuss worked. For full details see Meyer (1956).

Preuss, Hans (1879-1935), German (Prussian) botanist; high school teacher in Westpreussen and Danzig, later Senator and Stadtschulrat (municipal counsel for education) at Osnabrück. (*H. Preuss*).

HERBARIUM and TYPES: Some material at Berlin.

BIBLIOGRAPHY and BIOGRAPHY: AG 4: 233; Barnhart 3: 109 (b. 3 Aug 1879, d. 25 Apr 1935); BFM no. 180-182, 1280-1282, 1830; BJI 2: 140; CSP 17: 1013; IH 2: (in press); Kew 4: 363; TL-2/2869.

BIOFILE: Christiansen, W. & W., Bot. Schrifttum Schlesw.-Holst. 85, 131, 319. 1936.
Diels, L., Bot. Jahrb. 45 (Lit.): 39-40, 47-48. 1911, 48 (Lit.): 8. 1912.
Mattfeld, J., Verh. bot. Ver. Brand. 76: 115. 1936 (b. 3 Aug 1879, d. 25 Apr 1935).
Szweykowski, J., Prodr. Fl. Hepat. Polon. 1958.

COMPOSITE WORKS: with Höppner, *Fl. Westf. Industriegeb.* 1926, TL-2/2869.

Preuss, Paul Rudolf (1861-?), German (Prussian) botanist and explorer; in Sierra Leone 1886-1888; on the Zintgraff expedition at Kumba 1889-1892; founder and director of the Victoria botanic garden, Cameroon, 1892-1895(-1902); in tropical America 1899-1900; in S.E. Asia 1903-1904. (*P. Preuss*).

HERBARIUM and TYPES: B; other material at A, BM, COI, E, EA, HBG, K, LE, M, MO, P, PRE, S.

BIBLIOGRAPHY and BIOGRAPHY: AG 4: 223; Barnhart 3: 109; BJI 2: 140; CSP 17: 1013; Kew 4: 363; Langman p. 595; LS 22204; Rehder 1: 491, 5: 688; Tucker 1: 569; Urban-Berl. p. 181, 280, 289, 310, 385.

BIOFILE: Anon., Bot. Centralbl. 53: 104 (Leiter Versuchsplantage Kamerun), 271 (id.) 1893; Hedwigia 42: (161). 1903 (prof. title; resigned Victoria position).
Engler, A., Notizbl. k. bot. Gart. Mus. Berlin 3: 1-3. 1900) on Victoria and Buea, Cameroon, as botanical stations).
Gager, C.S., Brooklyn Bot. Gard. Rec. 27(3): 177. 1938 (dir. Victoria bot. gard. 1892-1902).
Hedge, I.C. & J.M. Lamond, Index coll. Edinb. herb. 120. 1970.
Jackson, B.D., Bull. misc. Inf. Kew 1901: 53 (pl. K).
Leeuwenberg, A.J.M., Webbia 19(2): 862. 1965 (dupl. at BM).
Letouzey, R., Fl. Cameroun 7: 15, 53. 1968.
Preuss, P.R., Notizbl. k. bot. Gart. Mus. Berlin 1: 264-265. 1897 (on *Kickxia africana* in Cameroun).

Urban, I., Symb. ant. 5: 10. 1904 (on Exped. C.-Südamerika).
Volkens, G., Notizbl. k. bot. Gart. Mus. Berlin 2: 159-160. 1898 (on Victoria Cameroun botanic garden).

EPONYMY: *Preussiella* Gilg (1897); *Preussiodora* Keay (1958).

8317. Kolonial-wirtschaftliches Komitee. *Expedition nach Central– und Südamerika* ... Berlin (Verlag des Kolonial-wirtschaftlichen Komitees ...) 1901. Oct. (*Exped. C.-Südamer.*).
Publ.: Oct 1901 (p. v: Jul 1901; Bot. Z. 16 Oct 1901; Bot. Jahrb. 1 Nov 1901; Nat. Nov. Nov(2) 1911), p. [i]-xii, [1]-452, *8* unnumbered pl. (gen.), *pl. 1-12* (bot.), chart. *Copies*: B, BR, MO, NY, Wurdack.*Ref.*: Busse, W., Bot. Zeit. 59(2): 347-348. 16 Nov 1901.

8318. *Die Kokospalme und ihre Kultur* ... Berlin (Dietrich Reimer (Ernst Vohsen) ...) [1911]. Oct. (*Kokospalme*).
Publ.: 1911 (pref. Oct 1911), p. [i]-vii, [1]-221, *17 pl. Copy*: L.

Prévost, Jean Louis (*fl.* 1760-1810), French flower and landscape painter. (*Prévost*).

ORIGINAL DRAWINGS: See Lenley p. 335 and NI 1568.

BIBLIOGRAPHY and BIOGRAPHY: DU 229; GF p. 13, 71; Lenley p. 335; NI 1568 and 3: 55; PR 7332 (ed. 1: 8223); Rehder 1: 265, 3: 481.

BIOFILE: Stearn, W.T., Roy. Hort. Soc., Exhib. manuscr. books 1954, p. 35-36.

EPONYMY: *Prevostea* J. Choisy (1825) honors J.L. Prevost, as well as Isaac Benedict Prevost (1755-1819), Swiss professor of plant physiology, and Pierre Prevost, a professor of physics at l'Académie de Genève.

8319. *Collection de fleurs et de fruits*, peints d'après nature, et tirées du portefeuille de J.D. Prévost, avec une description, par F.M.D. Ouvrage composé de 48 planches; destiné à l'usage des artistes, des fabricans, des amateurs, et pouvant servir aux personnes qui se livrent à la peinture. Imprimés en couleur par Langlois. Paris An XIII [1804-1806]. Fol. (*Coll. fl. fruits*) n.v.
Publ.: in 12 fascicles of 4 plates each. The announcement of livr. 1 and 2 in JT (16 pluv. 12, 6 Feb 1804) promises two livraisons each three months. Each plate is accompanied by a text, first written by "F.M.D." later by "M.***" and finally by A.N. Duchesne, who may have been the author of the entire text. Two issues: plain and coloured plates. The announcements in the JT were made on the following dates:

livr.	plates	JT	livr.	plates	JT
1	*1-4*	6 Feb 1804	7	*25-28*	4 Jul 1805
2	*5-8*	6 Feb 1804	8	*29-32*	-
3	*9-12*	13 Jun 1804	9	*33-36*	-
4	*13-16*	7 Oct 1804	10	*37-40*	16 Apr 1806
5	*17-20*	22 Nov 1804	11	*41-44*	8 Mai 1806
6	*21-24*	6 Mar 1805	12	*45-48*	-

For details on the plates (stipple engravings, printed in colour or in black and white) see DU 229.

Reprint: Jean Louis Prévost. *Bouquets*. Eighteen coloured engravings which faithfully reproduce the original watercolour flowers and fruits with captions from the original text and an introduction by Emy Pisčhel. MacDonald London[1969], p. [i-vii], *18 pl.*w.t. Originally published in German, 1959, Schuler Verlagsgesellschaft, Stuttgart (n.v.). – Plates reproduced from a copy in the Coburg library.

Price, Sarah [Sadie] Frances (1849-1903), American naturalist living at Bowling Green, Kentucky. (*S.F. Price*).

HERBARIUM and TYPES: material at A, MO (2912 specimens), MSC, NY. – Portrait at G.

BIBLIOGRAPHY and BIOGRAPHY: Barnhart 3: 110 (d. 3 Jul 1903); BL 1: 183, 313; IH 1 (ed. 7) 340; Lenley p. 335; Pennell p. 616; Rehder 5: 689; Tucker 1: 569.

BIOFILE: Anon., Fern Bull. 11: 85-86. 1903, 12(1): 25. 1904 (obit, portr., d. 3 Jul 1903); J. New York Bot. Gard. 9: 4. 1908.
Clute, W.N. Fern Bull. 12: 25. 1904.

8320. *Flora of Warren County, Kentucky* [New London, Wisconsin 1893]. Oct. (*Fl. Warren Co.*).
Publ.: 1893, p. [1]-31. *Copy*: USDA (with mss annotations by author).

8321. *The fern-collector's handbook and herbarium* an aid in the study and preservation of the ferns of Northern United States, including the district east of the Mississippi and north of North Carolina and Tennessee by Sadie F. Price ... New York (Henry Holt and Company) 1897. Qu. (*Fern-coll. handb.*).
Publ.: Mar 1897 (Nat. Nov. Mai(1) 1897; Bot. Zeit. 16 Mai 1897; Bot. Gaz. 24 Mar 1897; Fern Bull. Jul 1897; Bot. Centralbl. 28 Apr 1897), p. [i-ix], 1-70, [1, index], *pl. 1-72* (uncol.). *Copies*: MO, NY.
Other issue: London 1897, see Bot. Centralbl. 71: 322. 25 Aug 1897, not seen by us.

8322. *Trees and shrubs of Kentucky* [s.l.n.d., 1898]. (*Trees shrubs Kentucky*).
Publ.: 1898 (see TBC 18 Mar 1899 and fide date written on USDA copy by author), p. [1]-6. *Copy*: USDA.

Price, Sarah (*fl.* 1864), British botanical illustrator. (*S. Price*).

HERBARIUM and TYPES: Unknown.

BIBLIOGRAPHY and BIOGRAPHY: BM 4: 1614; Jackson p. 244; Kelly p. 179; Kew 4: 361; NI 1568n.

BIOFILE: Freeman, R.B., Brit. nat. hist. books 287. 1980.
Palmer, J.F., Nova Hedw. 15: 153. 1968.

8323. *Illustrations of the fungi* of our fields and woods drawn from natural specimens by Sarah Price ... [London] (published for the author by Lovell Reeve & Co., ...) 1864. Qu. (*Ill. fung.*).
First series: 1864 (last note dated Feb 1864), p. [i]-xi, *pl. 1-10* w.t. (col., auct.).
Second series: 1865 (p. [iii]: 1 Jun 1865), p. [i*], [i]-xi, *pl. 11-20* (id.).
Copies: BR, NY, Stevenson, USDA.

Prillieux, Édouard Ernest (1829-1915), French agronomist and botanist, especially mycologist and phytopathologist; studied at the Institut agronomique, Versailles, 1849-1855; in various functions at the Sorbonne and the Museum d'Histoire naturelle 1855-1874; lecturer École centr. Arts Manuf. 1874-1876; professor at the Institut agronomique, Versailles 1876; from 1883 inspector general of agricultural teaching, from 1886 director of the Station de pathologie végétale. (*Prill.*).

HERBARIUM and TYPES: Unknown; some material ("ex herb.") at W. – 25 letters to Malinvaud at P; other letters at G.

BIBLIOGRAPHY and BIOGRAPHY: Barnhart 3: 111, 313; BL 1: 106; Bossert p. 318; CSP 5: 20-21, 8: 663, 11: 67-68, 12: 589, 17: 1017; Frank 3(Anh.): 79; Hawksworth p. 185; Herder p. 132, 211; Kelly p. 179-182; Kew 4: 365; LS 21235-21377, 38073-38075, suppl. 22220; Morren ed. 2, p. 21, ed. 10, p. 56; Moebius p. 281; Rehder 5: 689; Stevenson p. 1254; TL-2/4658; Tucker 1: 569-570.

BIOFILE: Anon., Bonplandia 6: 273-274. 1858 (membership Leopoldina); Bot. Centralbl.

79: 78. 1899 (member Acad. Sci., Paris); Bot. Not. 1916: 196 (d. 8 Oct 1915); Bot. Zeit. 34: 766. 1876 (app. Versailles); Hedwigia 58: (88). 1916 (d. 8 Oct 1915), 63:(97) 1922 (d. 7 Oct 1915); J. Bot., Morot 13(6): lvi. 1899 (election Acad. Sci., Paris); Nat. Nov. 38: 97. 1916 (d. 8 Oct 1915); Nature 96: 349. 1915 (d.); Science ser. 2. 42. 866. 1915 (d.); Table art. orig. (1854-1893), Bull. Soc. Bot. France 192-195.
Aymonin, G.G., C.R. Congr. Soc. Sav. 102 (Limoges) (3): 57. 1977 (letters).
Berthault, P., Rev. gén. bot. 28: 193-203. 1916 (obit., b. 11 Jan 1829, d. 7 Oct 1915).
Ciferri, R., Taxon 1: 127. 1952 (coll.).
Costantin, J., Ann. Sci. nat., Bot. ser. 10. 16: lxvi-lxvii. 1934.
Dangeard, P.A., Bull. Soc. bot. France 62: 212-214. *pl. 9*, 1915 (portr.; P was membre fondateur, 1854).
Houard, C., Marcellia 14: xvi. 1915 (obit.).
Marchal, P. & E. Foex, ed., Ann. Service Épiphytes 4: 1-16. 1917 (obit., portr., bibl. by G. Arnaud).
Pinoy, Bull. Soc. mycol. France 32: 1-9. 1916 (b. 11 Jan 1829, d. 7 Oct 1915; portr., bibl.; also reprinted, 9 p.).
Verdoorn, F., ed., Chron. bot. 1: 37. 1935, 2: 128. 1936.

EPONYMY: *Prillieuxia* Saccardo et B. Sydow (1899); *Prillieuxina* Arnaud (1918).

Prince, Arthur Reginald (1900-1969), Canadian botanist; M.A. Harvard 1928; lecturer Nova Scotia Agric. Coll. 1926-1933; high school teacher in the Maritimes and Quebec 1933-1949; from 1949-1966 at Mount Royal College at Calgary (later part of Univ. of Calgary). (*A. Prince*).

HERBARIUM and TYPES: UAC; other material at DS and MICH.

BIBLIOGRAPHY and BIOGRAPHY: Barnhart 3: 111 (b. 28 Oct 1901); Hirsch p. 238; Hortus 3 : 1202 (A. Prince); IH 1 (ed. 6): 363, (ed. 7): 340, 2 : (in press); PH 150.

BIOFILE: Bird, C.D., The blue Jay 28(3): 133-134. 1970 (portr., d. 16 Apr 1969).

EPONYMY: *Princea* Dubard et Dop (1913) is named for Prince, a collector of plants on Madagascar.

Prince, William (1766-1842), American botanist and horticulturalist, proprietor of the "Linnaean botanic Garden" at Flushing, N.Y. (*W. Prince*).

HERBARIUM and TYPES: Material at PH (Schweinitz coll.).

BIBLIOGRAPHY and BIOGRAPHY: Barnhart 3: 111; BM 4: 1615; CSP 5: 22; Lenley p. 335; ME 1: 221, 3: 641; PR ed. 1: 8227; PH 567; Rehder 5: 690; SO 893b; Tucker 1: 570.

BIOFILE: Allibone, S.A., Crit. Dict. Engl. lit. 1690. 1870.
Barnhart, J.H., Mem. Torrey bot. Club 16: 298. 1921, Bartonia 16: 32-33. 1934 (corr. Schweinitz).
Ewan, J. et al., Short hist. bot. U.S. 133. 1969.
Gager, C.S., Brooklyn Bot. Gard. Rec. 27(3): 385. 1938.
Hedrick, H.P., Hist. hort. America 545 [index]. 1950.
Lemmon, K., Golden age plant hunters 145. 1968.
Mears, J.A., Proc. Amer. philos. Soc. 122(3): 169. 1978 (pl. in Schweinitz herb. PH).
Stieber, M.T. et al., Huntia 4(1): 85. 1981 (corr. C.R. Ball).

8324. *Catalogue of fruit and ornamental trees* and plants, bulbous flower roots, green-house plants, &c. &c. Cultivated at the Linnaean botanic garden,... to which is added, a short treatise on their cultivation, &c. Twenty-first edition. New York (printed by T. and J. Swords ...) 1822. Duod. (in sixes). (*Cat. fruit. trees*).
Ed. 21: 1822, p. [i], [1, 9]-140. *Copy*: NY. – See Rehder 1: 38.
Ed. 22: 1823, p. [i]-xii, [13]-143. *Copy*: NY.

Ed. 28: 1832, p. [1]-3, 173-193. *Copy*: FI. – *Annual wholesale catalogue of American trees, shrubs, plants and seeds* ... Providence (W. Marshall and Co. ...) 1832.
Note: We failed to locate other editions.

Pringle, Cyrus Guernsey (1838-1911), American botanist and collector (and Quaker); plant breeder at his farm at East Charlotte, Vt. until 1880; collected extensively especially in the Pacific States and in Mexico between 1880 and 1909; from 1902 at the University of Vermont as keeper of the herbarium. (*Pringle*).

HERBARIUM and TYPES: VT; duplicates in numerous herbaria (e.g. ASU, CM, EMC, FCME, H, MEXU, MO, NTMG, MU, NMC, NY, P, PC, PRC, RM, SMU, TAES, TEX, US, UTD, VT, WRSL, YU, Z), for further details see IH. *Exsiccatae: Mexican fungi*, decas 1, nos. 1-10. 1 Sep 1896 (see Stevenson 1971), e.g. at DAOM and FH. Pringle's plants were distributed under various series names such as *Flora of the Pacific Slope* and *Plantae mexicanae, Musci mexicani, Plantae Arizonae, Plantae Texanae*. A special pamphlet, p. [1-5], copy at USDA, was issued in 1881 giving corrected names of plants "in his several distributions" of *Flora of the Pacific Slope*. – Letters at G, NW, PH.

BIBLIOGRAPHY and BIOGRAPHY: Barnhart 3: 111 (b. 6 Mai 1838, d. 25 Mai 1911); Bossert p. 318; CSP 12: 590, 17: 1019; Ewan ed. 1: 80, 129; IH (ed. 6): 363, (ed. 7): 340; 2: (in press); Kew 4: 366; Lenley p. 335-336; LS 21378-21379; Morren ed. 10, p. 131; NW p. 55 (letters to Piper and Suksdorf); Pennell p. 617; PH 164, 444, 906; Rehder 5: 690; Stevenson p. 1254; Tucker 1: 570; Urban-Berl. p. 310, 385.

BIOFILE: Anon., Bot. Centralbl. 37: 382. 1889 (returned from N. Mexico), 44: 77. 1890 (coll. Mexico); Bot. Not. 1911: 190 (d. 15 Mai 1827, err.); Hedwigia 41: (207). 1902 (keeper VT), 42: (224). 1903 (to Cuba), 52: (83). 1912 (d.); Nat. Nov. 24: 479. 1902 (curator herb. VT); Österr. bot. Z. 40: 354. 1890, 41: 80. 1891; Torreya 2: 94. 1902 (keeper herb. VT), 2: 126. 1902 (fund purchase Pringle herb.), 11: 144. 1911 (b. 15 Mai 1911, err.).
Arthur, J.C., Bot. Gaz. 22: 423. 1896 (on P's Mexican fungi).
Benedict, G.G., Burlington Free Press 2 Jun 1902 (herb. to VT).
Bonner, C.E.B., Mus. Genève 130: 19. 1972 (on Pringle material distributed under the name "Herrera").
Brainerd, E., Rhodora 13: 225-232. 1911 (portr., distr. 500.000 specimens of some 20.000 species; d. 25 Mai 1911).
Burns, G.P., Science ser. 2. 34: 750-751. 1911 (obit.).
Cadbury, H.J., ed., The civil war diary of Cyrus Pringle, Pendle Hill Pamphlet no. 122, Lebanon, Pa., 1962, 39 p.
Cardot, J., Rev. bryol. 36: 67. 1909.
Charette, L.A., Univ. Vermont Alumni Mag. 42: 4-7. 1962 (portr.; account of herbarium and collections).
Davis, H.B., Life and work of Cyrus Guernsey Pringle, Burlington, May 1936, iii, 756 p. (portr., biography; full details on coll., itin., supplement, 14 p., Burlington, Oct 1937).
Day, M., Rhodora 3: 261. 1901 (herb.).
Eggleston, W.W., Bull. Vermont bot. Club 7: 8-11. 1912 (reminiscences).
Evans, P.D., Dict. Amer. biogr. 15: 236-237. 1935 (biogr., b. 6 Mai 1838, d. 25 Mai 1911; further biogr. refs.).
Hedge, I.C. & J.M. Lamond, Index coll. Edinb. herb. 120. 1970.
Hitchcock, A.S. & A. Chase, Man. Grass. U.S. 988. 1951 (epon.).
Jaeger, E.C., Source-book biol. names ed. 3. 319. 1966.
Jones, L.R., Rhodora 4: 171-174. 1902 (on Pringle herb. at VT).
Kneucker, A., Allg. bot. Z. 3: 72. 1897 (returned home from Mexico with 20.000 specimens), 18: 80. 1912 (d.).
Knobloch, I.W., Plant coll. N. Mexico 55. 1979.
McVaugh, R., Edward Palmer 427 [index]. 1956; Contr. Univ. Mich. Herb. 9: 290-292. 1972 (coll. in Jalisco).
Meyer, F.G. & S. Elsasser, Taxon 22: 386. 1973 (pl. at NA).
Moldenke, H.N., List coll. Verbenac. Suppl. 20. 1947 (coll.).

Orcutt, C.R., Science ser. 2. 34: 176. 1911.
Pease, A.S., Fl. N. New Hampshire 34, 36. 1964.
Pringle, C.G., Herbarium 8: 57-58. 1909 (on his Mexican coll.).
Rickett, H.W., Index Bull. Torrey bot. Club 80. 1955.
Rodgers, A.D., Amer. bot. 1873-1892, p. 335 [index]. 1944.
Rugg, H.G., Amer. Fern J. 1(5): 114-115. 1911.
Seymour, F.C., Rhodora 75: 292-303. 1973 (Pringle herbarium VT).
Stevenson, J., Beih. Nova Hedw. 36: 272-273. 1971.
Sutton, S.B., Charles Sprague Sargent 379 [index]. 1970.
Thomas, J.H., Huntia 3: 18. 1969.
Tracy, J.E., New England Mag. ser. 2. 41: 86-91. Sep 1909 (portr.; on P's hybridization of cereals).
Verdoorn, F., ed., Chron. bot. 1: 160. 1935, 3: 321, 374. 1937, 4: 185, 565. 1938, 5: 307. 1939, 6: 238. 1941, 7: 39. 1943.
Wiggins, I.L., Fl. Baja Calif. 42. 1980.
Wittrock, V.B., Acta Horti Berg.3(3): 200. *pl. 149.* 1905 (portr.).

EPONYMY: *Neopringlea* S. Watson (1891); *Pringlella* Cardot (1909); *Pringleochloa* Scribner (1896); *Pringleophytum* A. Gray (1885). *Note: Pringleella* V.F. Brotherus (1924) almost certainly belongs here. *Pringlea* T. Anderson ex J.D. Hooker (1845) was named for Sir John Pringle (1707-1782) who wrote a book on scurvy.

8325. *Musci mexicani first century* ... Leipzig (Theodor Oswald Weigel ...) [1896].
Publ.: 1896 (fide Nat. Nov. Feb(2) 1897), list of names of musci included in the first century for which Weigel had the European distribution rights. Leaflet, 1 p. *Copies*: BR, MO. – Pringle himself also issued 1 p. pamphlets announcing the distribution and contents of the first as well as the second century. Copies of these undated pamphlets are at MO.

8326. C.G. Pringle, *Plantae mexicanae* ... [Charlotte, Vermont] 1887-1906. (*Pl. mexic.*).
Publ.: A series of pamphlets giving the contents of the (almost) annual distribution of Mexican plants. We have seen the following (*copies*: MO):

1887: p. [1]-3.	*1896*: p. [1-4].
1889: p. [1-3].	*1897*: p. [1-3].
1890: p. [1]-3.	*1898*: p. [1-3].
1891: p. [1]-3.	*1899*: p. [1-4].
1892: p. [1]-4.	*1900*: p. [1-3].
1893: p. [1-4].	*1901/1902*: p. [1-4].
1894: p. [1-4].	*1903/1904*: p. [1-4].
1895: p. [1-4].	*1905/1906*: p. [1-4].

Pringsheim, Nathanael [Nathan] (1823-1894), German (Silesian) botanist; Dr. phil. Berlin 1848; Dr. phil. h.c. Jena 1858; habil. Berlin 1850; lecturing in Leipzig and Berlin 1850-1864; professor of botany and director of the botanic garden at Jena 1864-1868; working as Privatgelehrter at Berlin from 1868; one of the founders of the Deutsche botanische Gesellschaft, president 1882-1894; Geheimrath 1888. (*Pringsh.*).

HERBARIUM and TYPES: Unknown. – Pringsheim's library was acquired by B. – Letters to D.F.L. von Schlechtendal at HAL; other letters at G.

BIBLIOGRAPHY and BIOGRAPHY: ADB 53: 120-124; Ainsworth p. 117, 277, 317; Barnhart 3: 111; BM 4: 1616; Bossert p. 318; Cesati p. 318; CSP 5: 22-23, 8: 664, 11: 69, 12: 590, 17: 1020-1021 (further biogr. refs.); De Toni 1: c-ci; DSB 11: 151-155 (bibl.); Frank 3(Anh.): 79; Hawksworth p. 185; Herder p. 118, 132, 253; Jackson p. 76, 79, 95, 99, 107, 149, 157, 158, 493; Kelly p. 182; Kew 4: 367; KR p. 580; LS 21382-21400; Moebius p. 178, 455 [index]; Morren ed. 2, p. 1, ed. 10, p. 6; NAF ser. 2. 1: 72; PR

7334-7340; Rehder 5: 690; Stevenson p. 1254; TL-2/see Kienitz-Gerloff, J.H.E.F., Müller, K.A.E.; Tucker 1: 13, 570.

BIOFILE: Anon., Ber. deut. bot. Ges. 12: 171. 1894 (d.); Bonplandia 1: 224. 1853, 2: 7. 1854, 3: 78, 137. 1855, 4: 99. 1856, 5: 358, 8: 296. 1860; Bot. Centralbl. 54: 319. 1893 (Hung. Acad.), 56: 399. 1893 (70th birthday), 60: 192. 1894 (d.); Bot. Not. 1895: 38 (d.); Bot. Zeit. 19: 48. 1861 (member Akad. Berlin), 22: 164. 1864 (app. Jena), 26: 608. 1868 (to Berlin), 52(2): 318. 1894 (d.); Bull. Soc. bot. France 41: 623. 1894 (1895) (d.); Flora 41: 592. 1858 (dr. phil. h.c. Jena, 43: 767. 1860 (member Akad. Berlin), 42: 283, 284, 413.1864 (app. Berlin, Jena), 50: 173. 1867 (FMLS), 51: 535. 1868 (to Berlin), 53: 43. 1870 (corr. Acad. Paris); Hedwigia 33(1): (32). 1893 (member Acad. St. Petersb.), 33(5): 147. 1894 (d. 6 Oct 1894); Leopoldina 30: 209-210. 1894 (obit.); Monatsber. k. Akad. Wiss., Berlin 7 Jun 1860: 295 (P. elected member Akad.); Nature 50: 580. 1894; Nat. Nov. 16: 460. 1894 (d.), 16: 508. 1894 (Jahrb. to be continued by Pfeffer and Strasburger), 17: 88. 1895 (library given to Deut. bot. Ges. [deposited at B], plus M. 25.000 for its continuation); Österr. bot. Z. 14: 225. 1864 (app. Jena), 20: 123. 1873 (corr. memb. Acad. Sci., Paris), 44: 443. 1893 (d.); Verh. bot. Ver. Brandenburg 36: li, lxxxix. 1895 (d. 6 Oct 1894).
Becker, K. & E. Schumacher, S.B. Ges. naturf. Freunde Berlin ser. 2. 13(2): 146. 1973 (bibl. S.B.).
Bergdolt, E., Karl von Goebel 33, 39, 40, 47, 106, 182. 1941 (important sidelight on mutual relations).
Britten, J., J. Bot. 32: 383-384. 1894 (obit.).
Christiansen, W. & W., Bot. Schrifttum Schlesw.-Holst. 155, 319. 1936.
Cohn, F., Ber. deut. bot. Ges. 13: (10)-(33). 1895 (obit., portr.); Jahrb. wiss. Bot. 28: i-xxxii. 1895 (obit.).
De Toni, G.B., Nuova Notar. 1895: 97-98 (obit.).
De Wildeman, E., La Notarisia 10: 15-16. 1895 (obit.).
Engler, A., Bot. Jahrb. 19 (Beibl. 48): 24. 1894 (death; Jahrb. will be edited by Strasburger and Pfeffer).
Gager, C.S., Brooklyn Bot. Gard. Rec. 27(3): 230. 1938 (dir. bot. gard. Jena 1864-1870 [other sources say 1868]).
Harlay, M., Bull. Soc. mycol. France 11: 142-144. 1895.
Harvey-Gibson, R.J., Outlines hist. bot. 141-144, 161, 170-171, 206-208. 1919.
Hryniewiecki, B., Edward Strasburger 99 (index). 1938.
Hoffmann, P. et al., Wiss. Z. Humb. Univ. Berlin, mat.-nat. 14: 805-806. 1965 (portr.).
Jessen, K.F.W., Bot Gegenw. Vorz. 447. 1864.
Mägdefrau, K., Gesch. Bot. 313 [index]. 1973.
Magnus, P., Hedwigia 34: 14-21. 1895 (bibl.).
O.K., Rev. Belge étrang. 20: 281. 1894 (obit.).
Pringsheim, E.G., Julius Sachs 299 [index]. 1932.
Pringsheim, N., Monatsber. k. Akad. Wiss., Berlin 5 Jun 1860: 397-401. 1860 (Antrittsrede on his developmental studies on lower cryptogams; answer by Ehrenberg on p. 401-403; Biol. Centralbl. 7(5): 129-132. 1887 (polemics).
Reinke, J., Mein Tagewerk 58, 60, 157, 165, 172, 194. 1925 (personal recollections).
Renner, O., Jena Z. Med. Naturw. 78(2): 148. 1947 (P. in Jena).
Reynolds Green, J., A history of botany, 1860-1900, p. 47, 51-52, 227, 237, 291-294, 302, 314, 451.
Sachs, J. v., Gesch. Bot. 218, 225, 229, 344, 401, 478. 1875, Hist. botanique 218, 330, 457. 1892.
Schumann, K., Verh. bot. Ver. Brandenburg 36: xl-xlviii. 1895 (bibl.; b. 30 Nov 1823, d. 6 Oct 1894).
Scott, D.H., Nature 51: 399-402. 1895 (obit.).
Seemann, B., Bonplandia 3: 33. 1855 (Leopoldina cogn. Dutrochet), 6: 50. 1858, 8: 296. 1860 (member K. Preuss. Akad. Wiss., "mosaischen Glaubens").
Stieber, M.T. et al., Huntia 4(1): 85. 1981 (corr. Balfour).
Taton, R., ed., Science 19th cent. 383, 468-469. 1965.
Wittrock, V.B., Acta Hort Berg. 3(2): 139, *pl. 24.* 1903 (portr.), 3(3): 148, xxiii-xxiv, *pl. 80.* 1905 (portr.).
Wunschmann, G., Allg. deut. Biogr. 53: 120-124. 1907 (biogr., b. 30 Nov 1823, d. 6 Oct 1894).

COMPOSITE WORKS: founder and editor of the *Jahrbücher für wissenschaftliche Botanik*, vols. 1-26, 1857-1894. For the "prehistory" of the founding of this journal see Bonplandia 3: 78-79, 125-126, 137-138 (Pringsheim's own reaction), 161-162. 1855, 4: 99. 1856.

EPONYMY: *Pringsheimia* Reinke (1888); *Pringsheimia* Wood (1872); *Pringsheimina* O. Kuntze (1891); *Pseudopringsheimia* Wille (1909); *Pringsheimiella* F. von Höhnel (1920) is a substitute name for *Pringsheimia* Reinke; *Pringsheimia* Schulzer v. Müggenburg (1866) is very probably named for him although the author does not give the derivation of the name.

8327. *Die Entwickelungsgeschichte der Achlya prolifera* ... (Nova Acta Acad. Caes. Leop. Carol. Nat. Cur. vol. xxiii. p. i) [1851]. Qu.
Publ.: 1851 (Flora rd Dec 1851), p. [1]-66 [alternative pagination [395]-460], *pl. 46-50* [*46, 48* partly col. liths. auct.). *Copy*: FH. – Reprinted and to be cited from Nova Acta Leop. 23(1): 395-460, *pl. 46-50*. 1851.
Reprint: Ges. Abh. 2: 1-56. 1895.

8328. *Untersuchungen über den Bau und die Bildung der Pflanzenzelle* ... Erste Abtheilung. Grundlinien einer Theorie der Pflanzenzelle ... Berlin (Verlag von August Hirschwald ...) 1854. Qu. (*Unters. Bau Pfl.-Zell.*).
Publ.: 1854 (p. vi: Mai 1854), p. [i]-vi, [vii], [1]-90, [94, err.], *pl. 1-4* (partly col. liths. auct.). *Copy*: G.

8329. *Über die Befruchtung und Keimung der Algen* und das Wesen des Zeugungsactes ...Berlin (Gedruckt in der Druckerei der königl. Akademie der Wissenschaften) 1855. Oct.
[*Erster Aufssatz*:] Mar-Apr 1855 (in issue for 5 Mar 1855 of Monatsberichte, probably publ. Mar-Apr; rd. Flora Apr 1855), p. [i, iii], [1]-33, *1 pl.* (col. lith. auct.). *Copies*: FH, G; IDC 6395. – Reprinted and to be cited from Monatsber. k. Akad. Wiss., Berlin Mar 1855: 133-165, *1 pl.* (session 5 Mar 1855). – Reprinted *Ges. Abh.* 1: 1-34. 1895.
Zweiter Aufsatz: 1856 (in issue id. 8 Mai 1856, published Mai-Jun; rd. Flora Jul 1856), p. [1]-15, *1 pl.* (col. liths. auct.). *Copy*: FH. – Reprinted from id. Mai 1856: [225]-237, *1 pl.* – Title changed to *Untersuchungen über Befruchtung und Generationswechsel der Algen.* – Reprinted: *Ges. Abh.* 1: 35-48. 1895.
Dritter Aufsatz: 1857 (in issue id. for 11 Jan 1857, publ. Jun-Jul 1857), p. [1]-18. *Copies*: BR, G. – Reprinted from id. Jun 1857: 315-330. – Reprinted: *Ges. Abh.* 1: 49-64. 1895.

8330. *Zur Kritik und Geschichte der Untersuchungen ueber das Algengeschlecht* ... Berlin (Verlag von August Hirschwald ...) 1856 [i.e. 1857]. Oct. (*Krit. Gesch. Algengeschl.*).
First issue: Dec 1856 or Jan 1857 ("Neudamm 5 Dec 1856"; with t.p. dated 1856, cover dated 1857; Bot. Zeit. 30 Jan 1857; ÖbZ 5 Feb 1857; Flora 21 Feb 1857; Bonplandia 15 Feb 1857), p. [i-iv], [1]-75. *Copies*: NY; IDC 6396.
Second issue: 1857 (with t.p. dated 1857), p. [i-iv], [1]-75. *Copy*: G.
Nachtrag: 1860.
Reprint: Ges. Abh. 1: 129-178. 1895, Nachtrag 1: 179-192. 1895.
Ref.: Itzigsohn, H., Bot. Zeit. 15: 22-27. 9 Jan 1857 (reply!).

8331. *Über die Dauerschwärmer des Wassernetzes* und über einige ihnen verwandte Bildungen ... Berlin (Gedruckt in der Druckerei der Königl. Akademie der Wissenschaften) 1861. Oct.
Publ.: Jan 1861 (presented 13 Dec 1860; issue for Dec 1860 of Monatsberichte, prob. publ. Jan 1861; Flora rd. before 28 Apr 1861), p. [1]-21, *1 pl.* (col. lith.). *Copies*: BR, G, NY. – Reprinted and to be cited from Monatsber. k. Akad. Wiss., Berlin 13 Dec 1860: 775-794, *1 pl.* 1861.
French translation: Ann. Sci. nat., Bot. ser. 4. 14: [52]-72. 1860.

8332. *Beiträge zur Morphologie der Meeres-Algen* ... Berlin (Gedruckt in der Druckerei der königl. Akademie der Wissenschaften) 1862. Qu.
Publ.: 1862 (submitted 9 Jan 1862; Bonplandia 1 Nov 1862; Bot. Zeit. 21 Nov 1862), p.

[i-ii], [1]-37, *pl. 1-8* (*pl. 1-2* partly col., *3-8* uncol., liths., auct.). *Copies*: BR, FH, G, NY; IDC 6397. – Reprinted and to be cited from Abh. k. Akad. Wiss., Berlin 9 Jan 1861: [1]-37, *pl. 1-8.* 1862.
Reprint: Ges. Abh. 1: 321-358. 1895.
Ref.: Hermann, Hedwigia 2(13): 98-99. 1863.

8333. *Über die Vorkeime der Charen* ... Berlin (Gedruckt in der Druckerei der königl. Akademie der Wissenschaften) 1862. Oct.
Publ.: Mai 1862 (submitted 28 Apr 1862; printed in Monatsber. for Apr, probably publ. Mai 1862), p. [i], [1]-7. *Copies*: G, NY. – Reprinted and to be cited from Monatsber. k. Akad. Wiss., Berlin 28 Apr 1862: 225-233. 1862.
Reprint: Ges. Abh. 2: 243-252. 1895.
Ref.: Hermann, Bot. Zeit. 20: 220-222. 11 Jul 1862 (rev.).

8334. *Zur Morphologie der Utricularien* ... Berlin (Buchdruckerei der kgl. Akad. der Wissenschaften (G. Vogt), ...) 1869. Oct.
Publ.: Mar 1869 (from Monatsbericht for Feb 1869, probably publ. Mar 1869; BSbF 16 Aug 1869), p. [1]-27, *pl. 1.* *Copy*: G. – Reprinted and to be cited from Monatsber. k. Akad. Wiss., Berlin 15 Feb 1869: 92-116, *pl. 1.* 1869.
Reprint: Ges. Abh. 3: 157-180. 1896.
Ref.: R., Bot. Zeit. 27: 611-613. 10 Sep 1869 (rev.).

8335. *Über Paarung von Schwärmsporen* die morphologische Grundform der Zeugung im Pflanzenreiche ... Berlin (Buchdruckerei der Königlichen Akademie der Wissenschaften (G. Vogt) ...) 1869. Oct.
Publ.: Nov 1869 (from Monatsbericht for Oct 1869), p. [1]- 20. *1 pl.* (col.). *Copy*: U. – Reprinted and to be cited from Monatsber. K. Akad. Wiss., Berlin Oct 1869: 721-738, *1 pl.* 1869. – See also Ann. Sci. nat. 12: 191-208. 1869, Ann. Mag. nat. Hist. 5: 272-278. 1870, Bot. Zeit. 28: 265-272. 1870, Ann. Sci. nat. 12: 211-218. 1869.

8336. *Ueber dem Gang der morphologischen Differenzirung in der Sphacelarien-Reihe* ... Berlin (Buchdruckerei der königl. Akademie der Wissenschaften (G. Vogt)) 1873. Qu.
Publ.: 1873 (subm. 11 Apr 1872, 19 Jun 1873; Bot. Zeit. 26 Dec 1873), p. [i-ii], [137]-191, *pl. 1-11* (partly col. liths. auct. & Vöchting). *Copies*: BR, FH, FI. – Reprinted and to be cited from Abh. k. Akad. Wiss., Berlin 1873: [137]-191, *pl. 1-11.*
Reprint: Ges. Abh. 1: 359. 1895.
Ref.: G.K., Bot. Zeit. 32: 150-154. 6 Mar 1974.

8337. *Über vegetative Sprossung der Moosfrüchte* ... Berlin (Buchdruckerei der Königl. Akademie der Wissenschaften ...) 1876. Oct.
Publ.: Aug-Sep 1876, p. [1]-7, *1 pl.* (col. lith.). *Copy*: G. – Reprinted and to be cited from Monatsber. k. Akad. Wiss., Berlin 10 Jul 1876: 425-429. – Definitive publication (with next paper): Jahrb. Bot. 11: 1-46. 1878.

8338. *Über den Generationswechsel der Thallophyten* und seinen Anschluss an den Generationswechsel der Moose ... Berlin (Buchdruckerei der Königlichen Akademie der Wissenschaften ...) 1877. Oct.
Publ.: Jan-Feb 1877, p. [i], [1]-43 and [869]-911. *Copies*: G, U. – Reprinted, and to be cited from Monatsber. k. Akad. Wiss., Berlin Dec 1876: [869]-911. 1877. – Definitive publication, with previous paper, in Jahrb. Bot. 11: 1-46. 1878.
Ref.: G.K., Bot. Zeit. 35: 357-358. 1 Jun 1877 (rev.).

8339. *Neue Beobachtungen über den Befruchtungsact der Gattungen Achlya und Saprolegnia* ... Berlin (gedruckt in der Reichsdruckerei) 1882. Oct.
Publ.: 26 Oct 1882, see p. 40; Bot. Zeit. 29 Dec 1882; mss submitted 8 Jun 1882), p. [1]-40, *1 pl. Copies*: FI, U. – Reprinted and to be cited from S.B. k. Akad. Wiss., Berlin 1882: 855-890, *1 pl.*

8340. *Gesammelte Abhandlungen von N. Pringsheim* ... Jena (Verlag von Gustav Fischer [Ficher]) 1895-1896, 4 vols. Oct. (*Ges. Abh.*).

1: 1895 (Nat. Nov. Aug(2) 1895; Hedwigia 19 Oct 1895; frontisp. portr. Pringsheim), p. [i]-vi, [vii], [1]-414, *pl. 1-28.*
2: 1 Oct-15 Nov 1895 (Bot. Centralbl. 3 Dec 1895; Nat. Nov. Nov(1) 1895; Bot. Zeit. 16 Nov 1895; ÖbZ Dec 1895), p. [ii]-vi, [1]-410, *pl. 1-32.*
3: Jun 1896 (Nat. Nov. Jul(1) 1896; Bot. Zeit. 1 and 16 Jul 1896; ÖbZ Jul 1896), p. [i]-vi, [1]-389, *pl. 1-13.*
4: Nov-Dec 1896 (Nat. Nov. Dec(2) 1896; Bot. Zeit. 16 Jan 1897), p. [i]-vi, [1]-596, *pl. 1-22.*
Copies: B, BR, G, USDA.

Printz, [Karl] **Henrik** [Oppegaard] (1888-1978), Norwegian botanist; originally at Trondjhem; from 1926 professor of botany and director of the Botanical Institute of the Agricultural College at Aas near Oslo. (*Printz*).

HERBARIUM and TYPES: O; further material at C, H and TRH (p.p. orig.). Archival material at SIA.

BIBLIOGRAPHY and BIOGRAPHY: Barnhart 3: 111; BJI 2: 140; BM 8: 1033; Herder p. 56; Hirsch p. 238; IH 1 (ed. 7): 340, 2: (in press); Kew 4: 367; Kleppa p. 330 [index; bibl.]; MW p. 395-396; Roon p. 90; TL-2/1711, 1834.

BIOFILE: Anon., Österr. bot. Z. 75: 132. 1926 (app. Aas nr. Oslo).
Kneucker, A., Allg. bot. Z. 30/31: 56 (168). 1926 (app. Aas).
Koster, J.Th., Taxon 18: 556. 1969.
Nordhagen, R., Norsk biogr. 11: 181-184. 1952.
Robak, H., Henrik Printz. Minnetale i det Norske Videnskaps-Akademi. Saertrykk av det Norske Videnskaps-Akademis Årbok 1979: 123-127. 1980 repr. 7 p. (obit., portr., b. 15 Sep 1888, 31 Mar 1978).
Verdoorn, F., ed., Chron. bot. 1: 233. 1935, 214. 1937.

COMPOSITE WORKS: EP, *Die natürlichen Pflanzenfamilien*, ed. 2, vol. 3 (*Chlorophyceae* et al.), p. [i]-iv, [1]-463. Jan-Jul 1927 (rev. H. Melchior, Bot. Jahrb. 61 (Lit.): 92. 15 Nov 1927; B. Schussnig, Österr. bot. Z. 76: 326-327. 1927).

EPONYMY: *Printziella* B.V. Skvortzov (1958). *Note: Printzia* Cassini (1825, *nom. cons.*) was dedicated to the memory of Jacob Printz (1740-1779), a pupil of Linnaeus.

8341. *Kristianiatraktens Protococcoideer* ... Kristiania [Oslo] (i kommission hos Jacob Dybwad) 1914. Oct. (*Kristianiatr. Protococc.*).
Publ: 1914 (printed 20 Apr 1914; Nat. Nov. Sep(1, 2) 1916), p. [i]-iv, [1]-123, [124], *pl. 1-7* (uncol., auct.). *Copies*: FH, NY. – Published as Vidensk. Skr. Mat. Naturv. Kl., Oslo, 1913 no. 6.

8342. *Contributiones ad floram Asiae interiores pertinentes* ... Trondjhem (Aktietrykkeriet) 1915 [i.e. 1916] Oct. (*Contr. fl. As.*).
1: 1916 (cover dated 1916; t.p. 1915), p. [i]-xviii, map, [1]-52, *pl. 1-7* w.t. (uncol., auct.). – Issued as Kgl. Norske Vidensk. Selskabs Skr. 1915(4).
2: "1916" i.e. 1919 (printed 6 Mar 1919: p. 13), by B.D.L. Kaalaas (*Einige Bryophyten* ...), p. [1]-13, *pl. 1-2.* – Issued as id. 1918(2).
3: 1931, see below *Veg. Siber.-Mong. front.*
Copies: BR, MICH.

8343. *Some vascular plants from Saghalin* collected by Dr. Ludv. Münsterhjelm in 1914 ... Trondjhem (Aktietrykkeriet) 1917. Oct. (*Vasc. pl. Saghalin*).
Publ.: 1917, p. [1]-16, *pl. 1-4* (uncol.). *Copy*: G. – Issued as Kgl. Norske Vidensk. Selsk. Skr. 1916(3), 1917.

8344. *The vegetation of the Siberian-Mongolian frontiers* (the Sayansk region) ... published by Det Kongelige Norske Videnskabers Selskab [1921]. Qu. (*Veg. Siber.-Mongol. front.*).
Publ.: Mar 1921 (see p. iv; Bergen copy rd. 7 Apr 1921; MO rd. 27 Apr 1921; US rd. 29

Apr 1921; J. Bot. Jun 1921; Nat. Nov. Oct 1921), p. [i-v], 1-458, *pl. 1-16, maps 1-3.*
Copies: B, BR, H, MO, NY, US. – Constitutes part iii of the series *Contr. fl. Asiae.*
Ref.: Krause, K., Bot. Jahrb. 57 (Lit.): 31-32. 1922.

8345. *Die Algenvegetation des Trondjhemsfjordes* ... Oslo (i kommision hos Jacob Dybwad) 1926. Oct. (*Algenveg. Trondjhemsfj.*).
Publ.: 1926 (printed 21 Oct 1926; Nat. Nov. Feb 1927), p. [1]-273, [274], *pl. 1-10,* map. *Copies*: FH, H, NY, USDA. – Skr. Norske Vidensk.-Akad. Oslo I, 1926(5).

Prinz, William Alfred Joseph (1857-1910), Belgian diatomologist and astronomist. (*Prinz*).

HERBARIUM and TYPES: Unknown.

BIBLIOGRAPHY and BIOGRAPHY: Barnhart 3: 111; CSP 17: 10121-10122; De Toni 1: ci, 2: xcvi; Tucker 1: 570.
Prinz, W.A.J.   3094
BIOFILE: Evens, F.M.J.C., Gesch. algol. Belgie 187. 1944 (bibl.).

8346. *Recherches sur la structure de quelques diatomées* contenues dans la "Cementstein" du Jutland ... Bruxelles (A. Manceaux, ...) 1883. Oct.
Publ.: 1883 (Nat. Nov. Oct(1) 1883; Bot. Zeit. 28 Dec 1883; Bot. Centralbl. 22-26 Oct 1883), p. [1]-70, *pl. 1-4. Copy*: FH. – Reprinted and to be cited from Ann. Soc. belge Micr. 8: [5]-74, *pl. 1-4.* 1883.

Prior, Richard Chandler Alexander (formerly Alexander) (1809-1902), British botanist; educated at Charterhouse, Oxford; B.A. 1830; Fellow Royal College of physicians 1840; from 1841 devoting himself to botany, residing at Graz (Austria) 1841-1844, from 1846-1848 in South Africa; 1849-1850 to Canada, U.S.A. and Jamaica; after 1850 living in London and from 1859 on his estate near Taunton. (*Prior*).

HERBARIUM and TYPES: K (30.000); some further material at AK, B, BM, CGE, DBN, E, GOET, OXF, PRE and SAM.

NAME: Until 1859: R.C. Alexander; took name Prior in 1859 when inheriting the estate at Halse near Taunton.

BIBLIOGRAPHY and BIOGRAPHY: Barnhart 3: 112; BB p. 249-250; BL 2: 223, 705; BM 4: 1617; Bossert p. 318; Clokie p. 228-229; CSP 1: 43-44, 6: 564; Desmond p. 506; IH 1 (ed. 6): 363, (ed. 7): 340, 2: (in press); Jackson p. 10, 218; Kanitz 182; Kew 4: 367; Lenley p. 467; Morren ed. 10, p. 77; Pennell p. 617; PR 7342; Rehder 1: 36, 74.

BIOFILE: Anon., Bot. Not. 1903: 34; J. Bot. 41: 32. 1903 (d.); Proc. Linn. Soc. London 115: 35-37. 1903 (obit. based on autobiogr. notes; b. 6 Mar 1809, d. 5 Dec 1902).
Bridson, G.D.R. et al., Nat. hist. mss. res. Brit. Isl. 334.67. 1980.
Britten, J., J. Bot. 41: 108-109. 1903 (b. 6 Mar 1809, d. 8 Dec 1902).
Fawcett, W. & A.B. Rendle, Fl. Jamaica 1: pref. 1910 (in list of "more important collectors").
Freeman, R.B., Brit. nat. hist. books 287. 1980.
Gunn, M.D. and L. Codd, Bot. expl. S. Afr. 287-288. 1981 (coll., eponymy).
Hemsley, W.B., Bull. misc. Inf. Kew 1903: 32 (herb. bequeathed to Kew; 30.000 sheets; 140 volumes from his library came also to Kew), 1909: 318 (Jamaica coll. to be revised by Urban; set of duplicates presented to B).
Jackson, B.D., Bull. misc. Inf. Kew 1901: 54.
Kent, D.H., Brit. herb. 73. 1957.
Murray, G., Hist. coll. BM(NH) 1: 175. 1904 (pl. from Syria, Dalmatia, Italy, S. Afr., presented 1868).
Urban, I., Symb. ant. 1: 57. 1902, 3: 107. 1902 (based on inf. supplied by Prior).

EPONYMY: *Prioria* Grisebach (1860), see ING 3: 1411.

8347. *On the popular names of British plants*, being an explanation of the origin and meaning of the names of our indigenous and most commonly cultivated species, ... London (Williams and Norgate, ...), Edinburgh (id.) 1863. Oct. (*Pop. names Brit. pl.*).
Ed. 1: 1863, p. [i]-xxvi, [1, abbr.], [1]-250, [1-2, add.]. *Copies*: G, MO, NY.
Ed. 2: Nov-Dec 1870 (p. v: 1 Nov 1870; Flora 3 Apr 1871), p. [i]-xxvii, [xxviii], [1]-290. *Copies*: NY, USDA.
Ed. 3: 1879 (J. Bot. Oct 1879; Nat. Nov. Aug(2) 1879), p. [i]-xxvii, [xxviii], [1]-294. *Copy*: B. London (Frederic Norgate, ...) Edinburgh (William & Norgate, ...) 1879. Oct.
Ref.: Anon., J. Bot. 1: 378-384. 1863.
Britten, J., J. Bot. 9: 23-24. 1871, 18: 25. 1880.

Pritchard, Andrew (1804-1882), British microscopist and naturalist; until 1852 running an optical establishment in London, from then on dedicating himself fully to his biological-microscopical interests. (*A. Pritch.*).

HERBARIUM and TYPES: Unknown. – Letters at G.

BIBLIOGRAPHY and BIOGRAPHY: Barnhart 3: 112 (b. 14 Dec 1804, d. 24 Nov 1882); CSP 5: 25; De Toni 1: cii, 2: xcvi; DNB 46: 402-403; Frank 3(Anh.): 79; Jackson p. 158; Kew 4: 368; Nordstedt p. 26; Quenstedt p. 347.

BIOFILE: Allibone, S.A., Crit. dict. Engl. lit. 1693-1694. 1870 (bibl.).
Anon., Amer. Natural. 17(1): 231-232. 1883 (d. 24 Nov 1882); Zool. Anz. 6: 80. 1883 (d.).
Bridson, G.D.R. et al., Nat. hist. mss. res. Brit. Isl. 328.6. 1980.
Freeman, R.B., Brit. nat. hist. books. 287. l980.
Müller, R.H.W. & R. Zaunick, Friedr. Traugott Kützing 291. 1960.
Poggendorff, J.C., Biogr. lit. Handw.-Buch 3: 1071. 1898 (b. 14 Dec 1804, d. 24 Nov 1882, bibl.).

EPONYMY: *Pritchardia* Unger ex Endlicher (1842); *Pritchardia* Rabenhorst (1864) may also be named for him but Rabenhorst does not give the etymology. *Note*: *Pritchardia* B.C. Seemann et H. Wendland ex H. Wendland (1862) was named for William Pritchard, British Consul in the Fiji Islands; *Pritchardiopsis* Beccari (1910) and *Pritchardites* E. Bureau (1896) are related to *Pritchardia* B.C. Seemann et H. Wendland ex H. Wendland; *Pritchardioxylon* Drude (1897) is a substitute name for *Pritchardia* Unger ex Endlicher.

8348. *The natural history of animalcules*: containing descriptions of all the known species of infusoria; with instructions for procuring and viewing them, &c. &c. &c. ... London (Whittaker and Co. ...) 1834. Qu. (*Nat. hist. animalc.*).
Ed. 1: Apr 1834 (plates so dated), frontisp., p. [i], [1]-194, *pl. [1-6]* (uncol. liths. auct. dated). *Copy*: FH.
Ed. 2: 1843 (pref. Jun 1843), p. [i-iii], [1]-83, *pl. 1-12* (id.). *Copy*: FH. – "A general history of animalcules, ... second edition" London (Whittaker and Co., ...), Dublin (Hodges and Smith) 1843. Oct.

8349. *A history of Infusoria*, living and fossil: arranged according to "Die Infusionsthierchen" of C.G. Ehrenberg ... London (Whittaker and Co.) 1842. Oct. (*Hist. Infus.*).
Based on: Christian Gottfried Ehrenberg (1795-1876), *Infusionsthierchen* Jul-Aug 1838, TL-2/1638.
Publ.: 1842 (p. vi: Jun 1841), p. [i]-viii, [1]-439, *pl. 1-12* (col. or uncol. liths. F. Bauer, Unger). *Copies*: FH, LC.
Re-issue: 1845 (note: Apr 1845), p. [i]-viii, [1]-439, *pl. 1-12* (id.). *Copy*: FH.
Ed. 2: 1849, p. [i]-vi, [1]-439, [440], *pl. 1-12* (id.). *Copy*: USDA. – Second edition. London (id.) 1849. Oct.
Ed. [3]: 1852, p. [i]-viii, [1]-704, [1, err.], *pl. 1-24* (uncol. liths.). *Copy*: USDA. – *A history of infusorial animalcules*, living and fossil: ... a new edition, enlarged. London (id.) 1852. Oct.

Ed. 4: 1861 (p. v: 15 Nov 1860), p. [i]-xii, [1]-968, *40 pl.* (mostly uncol liths.). *Copy*: US.
– London (Whittaker and Co., ...) 1861. Oct.

Pritchard, Stephen F. (*fl.* 1836). (*S. Pritch.*).

HERBARIUM and TYPES: Unknown.

BIBLIOGRAPHY and BIOGRAPHY: Barnhart 3: 112; BB p. 250; Jackson p. 353; Kew 4: 368; PR 7343 (ed. 1: 8229); Rehder 1: 498.

8350. *An alphabetical list of indigenous and exotic plants growing on the island of St. Helena*, compiled by Stephen F. Pritchard, Esq. and corrected by Mr. James Bowie, botanist, Ludwigburg Garden, Cape Town, Cape of Good Hope. Printed by G.J. Pike, 11, St. George's Street 1836. Qu. (*Alph. list pl. St. Helena*).
Co-author: James Bowie (c. 1789-1869).
Publ.: 1836, p. [1]-36. *Copy*: G.

Pritzel, Ernst Georg (1875-1946), German botanist; with L. Diels in W. Australia 1900-1901. (*E. Pritz.*).

HERBARIUM and TYPES: B (partly extant); other material at A, AD, BM, BR, DBN, F, GH, L, LAU, M, MEL, MICH, MO, NSW, NY, P, PERTH, PH, PR, S, US, VT, W, WRSL, Z. – P's library is at B (extant). – Letters at G.

BIBLIOGRAPHY and BIOGRAPHY: Barnhart 3: 112; BJI 1: 47, 2: 140; BL 1: 76, 314; BM 4: 1617; Bossert p. 318; CSP 17: 1025; Hortus 3: 1202 (E. Pritz.); IH 1 (ed. 7): 340; Kew 4: 368; Langman p. 596; LS suppl. 22250; PR 1: 121; SK 5: cccxii; TL-2/1711; Urban-Berl. p. 280, 385, 310, 389; Zander ed. 10, p. 704, ed. 11, p. 802 (d. 1946).

BIOFILE: Anon., Hedwigia 39(5): (195). 1900 (journey S. Afr., Austr.), 41(5): (208). 1902 (return from Austr.); Verh. bot. Ver. Brandenburg 76: 115. 1935 (60th birthday).
Diels, L., Pfl.-Welt W. Austral. 6-66, 72. 1906 (Veg. Erde 7); *in* D.J. & S.G.M. Carr, People and plants in Australia 70-74. 1981 (Diels and Pritzel).
Hall, N., Bot. Eucalypts 105. 1978, Suppl. Bot. Eucalypts 15. 1979 (d. 6 Apr 1946).
Hedge, I.C. & J.M. Lamond, Index coll. Edinb. herb. 120. 1970.
Moldenke, H.N., List. coll. Verbenac. suppl. 20. 1947.
Murray, G., Hist. coll. BM(NH) 1: 175. 1904.
Pilger, R., Mitt. Bot. Gart. Mus. Berlin-Dahlem 1(1): 21. 1953 (library at B, extant).
Verdoorn, F., ed., Chron. bot. 6: 302. 1941.

COMPOSITE WORKS: (1) EP, *Die natürlichen Pflanzenfamilien*, (a) *Lycopodiaceae*, 1(4): 563-576. Jan 1901, 577-606. Jan 1901.
(b) *Psilotaceae*, 1(4): 606-619. Jan 1901.
(2) EP, *Die natürlichen Pflanzenfamilien* ed. 2, *Pittosporaceae* in 18a: 265-286. Mai 1930.

EPONYMY: *Pritzeliella* P.C. Hennings (1903).

Pritzel, Georg August (1815-1874), German (Silesian/Prussian) botanist and botanical bibliographer; studied theology and medicine at Breslau (consilium abeundi 1840), continued his medical and botanical studies at Leipzig 1841-1843; from 1844 working at his *Thesaurus*, mainly from Berlin and Leipzig; from 1850 in Berlin, from 1851-1854 employed by the Royal Library as "Hülfsarbeiter"; from 1854-1872 archivist and librarian at the Prussian Academy of Sciences and "custos" at the Royal Library in Berlin; in a psychiatric clinic at Hornheim near Kiel 1872-1874; highly gifted and erudite bibliographer; the man to whose memory "Taxonomic literature" should be dedicated. (*Pritz.*).

HERBARIUM and TYPES: Some material at M. – Letters to Alph. de Candolle and Boissier at G; to Schlechtendal at HAL (17).

NOTE: The Breslau *consilium abeundi* (removal from the University), followed upon a series of incidents such as "whistling in the theatre" and "oral defamation of a police officer" and was founded on the untranslatable offense "wegen unverständiges Tadelns einer obrigkeitlichen Verfügung und beleidigender Schrift ...". Pritzel's sad death was caused by a prolonged *tabes dorsalis*.

BIBLIOGRAPHY and BIOGRAPHY: ADB 26: 612-614; Ainsworth p. 279, 318; Barnhart 3: 112 (b. 2 Sep 1815, d. 14 Jun 1874); BFM no. 3244, 3265; BJI 1: 47; BM 4: 1617, 8: 1033; Bossert p. 318; CSP 5: 25; DTS 1: 230, 320; Frank (Anh.): 80; Herder p. 75, 88, 147; Hortus 3: 1202 (G. Pritz.); HU 2(1): clxxxvi, clxxxviii, cxciii, cc; Jackson p. 3, 15; Kew 4: 368; Langman p. 596; Moebius p. 129, 430, 431; MW p. 396; NI 312 (index); Plesch p. 369; PR 7344-7347 (ed. 1: 8230-8232); Rehder 5: 690; SK 1: cxviii; SO 3662; TL-1/1011-1013; TL-2/see Jessen, K.F.W., Meyer, E.H.F.; Zander ed. 10, p. 703-704, ed. 11, p. 802.

BIOFILE: Anon., Ber. [Monatsber.] Verh. k. Preuss. Akad. Wiss. 28 Jun 1855: 427 (appointed "Archivar" of Akad.); Bonplandia 2: 258. 1854 (elected archivist of K. Akad. Wiss., Berlin), 3: 112, 205. 1855 (Pr. to "Patavia" to consult Moretti's, library), 3: 205. 1855 ("Patavia" should be Pavia; no relation with Batavia=Djakarta), 5: 48. 1857; Bot. Not. 1875: 30; Bot. Zeit. 12: 639. 1854 (app. archivist R. Acad. Berlin), 35: 342-343. 1877 (compl. Thes. ed. 2); Bull. Soc. bot. France 19 (bibl.): 96 (P. said to have disappeared, his house ravaged), 19 (bibl.): 158. 1873 (P. seriously ill), 21 (bibl.): 96. 1874 (d.); Flora 37: 768. 1854 (archives Pruss. Acad.), 57: 303, 503-504. 1874 (d.; inf. on *Thes.*); J. Bot. 65: 294-296. 1927 (on the "new Pritzel", Stapf's *Index lond.*); Österr. bot. Z. 24: 289. 1874 (d. 14 Jun 1874).
Arechevaleta, J., Ann. Mus. nac., Montevideo 4(1): 25-60. 1903 (essentially a translation of O. Kuntze, Rev. gen. pl. 1: cxxii-cxlvi. 1891, see under *Thes. lit. bot.*).
Becker, K. & E. Schumacher, S.B. Ges. naturf. Freunde Berlin ser. 2. 13(2): 146. 1973 (bibl. S.B.).
Britten, J., J. Bot. 12: 288. 1874 (d.).
Candolle, A.P. de, Mém. souvenirs 530. 1862.
Croizat, L., Chron. bot. 12: 134-139. 1951 (on *Specimen bibliographiae botanicae*).
Fontane, Th., Von Zwanzig bis Dreissig, Leipzig 1955, p. 95-96 (on Pritzel's membership of the literary "Herwegh-Klub" in Leipzig; fide Helm 1966.).
Helm, J., Die Kulturpflanze 14: 311-357. 1966 (bio-bibliography with many details!).
Hooker, W.J., London J. Bot. 6: 467-468. 1847 (rev. Thesaurus fasc. 1).
Jackson, B.D., J. Bot. 38: 167-177. 1880 (rev. ed. 2; notes on role Jessen in completing the work); J.R. Hort. Soc. 45(1): 14-21. 1919 (on his Icon. bot. Index).
Jessen, K.F.W., Bot. Zeit. 32: 430-431. l874 (d.), Bot. Gegenw. Vorz. 488 [index]. 1884.
Kuntze, O., Rev. gen. pl. 1: cxxii-cxlvi. 1891 (additions to Thes. lit. bot.).
Pritzel, G.A., Bot. Zeit. 4: 785-790. 1846 (biogr. Johann Wonnecke von Caub).
Seeland, H., Mitt. Herm. Roemer-Mus. Hildesheim 40: 11, 17, 28. 1936 (on A. Schlauter and PR).
Seemann, Bonplandia 6: 50. 1858 (member Leopoldina, cogn. Jonas Dryander stressing association with botanical bibliography).
Simonkai, L., Enum. fl. transsilv. xxvii. 1886.
Stafleu, F.A., Regn. Veg. 71: 309, 316, 317, 328, 339. 1970; Taxon 22: 119-126. 1973 (on the re-publication of his Thesaurus; portr.).
Stapf, O., *Introduction* (on Pritzel) *in* Index lond. 1: ix-xi. 1929.
Urban, I., Symb. ant. 1: 131. 1898.
Wunschmann, G., Allg. deut. Biogr. 26: 612-614. 1888 (b. 2 Sep 1815, d. 14 Jun 1874; biogr.).

COMPOSITE WORKS: (1) in Walpers, G.G., *Repert. bot. syst.* 2, supplementum primum, Ranunculaceae-Cruciferae, p. 737-764. 28-30 Dec 1843 (Helm no. 2, 2a), also published separately with unchanged pagination (n.v.). Pritzel also contributed the *Index ordinum ... et synonymorum* for both vols. 1 and 2 in 2: 995-1029. 28-30 Dec 1843.
(2) Pritzel contributed numerous book-reviews and (brief) biographies to the Botanische Zeitung, signed G.P.

1: 240-242, 262-263, 278-279, 302-303, 731-734, 847-849, 871, 873-876, 882-883. 1843 (figures denote columns).
2: 730-732, 744-746. 1844.
3: 214-215. 1845.
4: 548-549, 701-703. 1846.
5: 77-80, 127-130, 135-136, 153, 156-157, 158-159, 246-248, 303-304, 653-654, 794-796, 810-812, 833-835. 1847 (also 245-246, 518-520, 716-717).
6: 179-180, 246-248, 272, 288-289, 454-455. 1858 (also 70-71, 268-270, 487-488).
10: 386-387. 1852.

HANDWRITING: Taxon 22: 120, 125. 1973; Die Kulturpflanze 14: 318. 1966.

EPONYMY: *Pritzelago* O. Kuntze (1891), *Pritzelia* Walpers (1843), *Pritzelia* Klotzsch (1854-1855), *Pritzelia* F. v. Mueller (1875) very likely is dedicated to G. Pritzel, but v. Mueller does not give the derivation.

8351. *Anemonarum revisio* auctore G.A. Pritzel. Accedunt tabulae sex. Lipsiae [Leipzig] (in commissio apud Leopold Voss) 1842. Oct. (*Anem. revis.*).
Publ.: 24 Jan 1842 (dedication copy for Gustav Kunze, Univ. libr. Leipzig, so signed, fide Helm; New York dedication copy for C.T. Beilschmied dated Feb 1842; Hinrichs 9-12 Feb), p. [i], [1]-142, *pl. 1-6* (uncol. liths. C.H. Reclam). *Copies*: FI, G, H, MO, NY, U. – Preprinted, with independent pagination, from Linnaea 15: [561]-698, *pl. 1-6.* Feb 1842. – The reprint has an *Index* on p. 139-142, not present in the journal. See also Helm (1966) nos. 1 and 1a.
Summary: Z. Ferdinandeum, Innsbruck 8: 191-197. 1842.
Ref.: Anon., Flora 25: Lit.-Ber. 5: 69-79. 14 Sep 1842 (rev.).

8352. *Specimen bibliographiae botanicae*, quod Ernesto Meyer, Med. Dr., botanices professori Regiomontano, nuptias Johannae Isenbartiae cum Dr. Zaddachio, zoologo regio, celebranti gratulurus scripsit G.A. Pritzel, botanicus vagus. Viennae [Wien] (Typis Caroli Gerold) 1845. Oct. (*Spec. bibl. bot.*).
Publ.: 1845, p. [1]-[8]. *Copies*: HH, WU. – See Croizat, L., Chronica botanica 12(4/6): 134-138. 1949 for full details. Only nine copies distributed. See Helm (1966) no. 3 for further details.
Ref.: Anon., [-s], Bot. Zeit. 3: 876-877. 26 Dec 1845. (rev.).

8353. *Thesaurus literaturae botanicae* omnium gentium inde a rerum botanicarum initiis ad nostra usque tempora, quindecim millia operum recensens. Curavit G.A. Pritzel. Lipsiae [Leipzig] (F.A. Brockhaus) [1847-]1851[-1852]. Qu. (*Thes. lit. bot.*; in TL-2: *PR*).
Ed. 1: in six parts, 1847-1852, p. [i]-viii, [1]-547, [548]. *Copies*: B, BR, Ewan, FI, G, H, MO, U, US; IDC 5646.
1: Apr 1847, p. [1]-80.
2: Jul 1847, p. 81-160.
3: Dec 1847, p. 161-240.
4: Apr 1848, p. 241-320.
5: Oct 1850, p. 321-352.
6: Apr 1852, p. 353-547, [i]-viii, (t.p. dated 1851; publisher's notice on completion Bot. Zeit. 10: 304. 23 Apr 1852).

Contains a great many references, especially to repaginated reprints of journal articles and to pre-linnaean literature, which are not incorporated in the more generally used second edition. Pritzel printed a "Vorerinnerung des Verfassers" on the cover of the first part (kindly put at our disposal by G. Buchheim) in which he tells how the book was compiled. An asterisk (*) indicates books consulted by him in Germany, a dagger (†) in "Helvetia, Gallia et Belgio". [In the second edition an asterisk was used for all books seen by Pritzel]. "Ein Repertorium der botanischen Journalliteratur wird unter günstigen Umständen den zweiten Band dieses Werkes bilden, welches durch seinen Reichthum und seine Treue befriedigen und einen dauernden Werth behaupten wird." This repertorium of botanical papers published in journals was never realized. The abbreviations used in the first edition, and in part also in the second edition, were explained on the cover of the second part:

EXPLICATIO SIGLARUM PRAECIPUARUM

D. = Dissertatio.	B.W.T.H.S.H.v.D.Q. = Banks,
Ed. = Editio.	Wilkström, Trautvetter,
s.a. = sine anno.	Haller, Seguier, Henckel,
s.l. = sine loco.	von Donnersmarck,
folio 4.8.12 = forma librorum folio,	Quérard etc.
quarto, octavo, duodecimo dicta.	i.c. xylogr. i.t. = icones
tab. = tabulae.	xylographicae in textu.
col. = colorata.	* Libri in Germania comparati.
praef. = praefatio.	† Libri in Helvetia, Gallia et
ind. = index.	Belgio comparati
pr. = praeside.	Adjutorum, apocryphorum,
fol. = folia.	anonymorum, biographorum,
p. = paginae.	continuatorum,
	commentatorum, collaboratorum,
	collectorum, editorum,
	impressorum,
	peregrinatorum, praesidum,
	praefatorum, pictorum,
	traductorum nomina aptissime
	in locupletissimo indice ad
	calcem operis comparabuntur.

Libraries: consulted for the first edition of the *Thesaurus*: private libraries of H.F. Link, D.F.L. von Schlechtendal, G. Kunze, A.P. de Candolle, A. de Jussieu, B. Delessert, P.B. Webb, J. Gay, J.H. Leveillé, J.P.F.C. Montagne, A.H.R. Grisebach, as well as public libraries in Berlin (Royal library), Wien (Hofnaturalienkabinett; Bot. Gart., Hofbibl., private library emp. Ferdinand), St. Gallen, Paris (Muséum d'Histoire naturelle, municipal library), Bruxelles, Liège, Göttingen, Leipzig, Dresden. For the second edition, for which Pritzel started work as early as 1855; Pritzel was able to work in addition in the libraries of the British Museum and Kew, of the Société botanique de France at Paris; the libraries of San Marco, Venetia and of the Padua Garden, as well as the library of C.F.P. von Martius at München.

Preliminary publication: Einige Berichtungen zu den Materialien zu einem Verzeichniss der jetzt lebenden botanischen Schriftsteller (Linnaea 19(2): 146-192) von Georg Pritzel, *Linnaea* 19(4): 447-464. Dec 1846.

Probeblatt: announced by Bot. Zeit. 4: 555. 7 Aug 1846 (mentions a first proof printed in Paris and a second from Leipzig).

Additamenta [1, by E.A. Zuchold] ad Georgii Augusti Pritzelii Thesaurum Literaurae botanicae collegit et composuit Ernestus Amandus Zuchold. Halis [Halle a. S.] (typis expressum Ploetzianus) 1853 (Flora p. rd. 21 Jul-7 Aug 1853), p. [1]-59, [60]. *Copies*: G, MO. – Reprinted and to be cited from Jahresber. naturw. Ver. Halle 5: 505-582. 1853 (rev. D.F.L. von Schlechtendal, Bot. Zeit. 12: 150. 1854), continued in Abh. naturf. Ges. Halle 2 (Beilage): 67-74. 1855 and Z. ges. Naturwiss. 27: 122-134. 1866.

Additamenta [2, by Ernst Berg (1824-?)]:

1858: Z. Ges. Naturw. 12: 207-244. 1858.

1859: *Additamenta* ad Thesaurum literaturae botanicae. Index librorum botanicorum bibliothecae horti imperialis botanici petropolitani quorum inscriptiones in G.A. Pritzelii Thesauro literaturae botanicae et in *Additamenti* ad Thesaurum illum ab E.A. Zuchold editis desiderantur ... Halis [Halle a. S.] 1859, 40 p., by E. von Berg (St. Petersburg) (reprint with independent pag. from 1858 paper in Z. ges. Naturw.).

1862: *Additamenta* ad Thesaurum literaturae botanicae. Index II librorum ... desiderantur; collegit et composuit [Ernst Berg] ... Petropoli [St. Petersburg] 1862, 21 p. (see BSbF 10: 72. 1863 and Schlechtendal, D.F.L. von, Bot. Zeit. 20: 303. 12 Sep 1862).

1864: *Additamenta* ad Thesaurum literaturae botanicae. Index III. librorum ... Petropoli [St. Petersburg] (Typis Academiae ...) 1864, 69p. (Schlechtendal, D.F.L. von, Bot. Zeit. 22: 92. 1894, Anon., Gartenflora 13: 188-189. 1864).
Omissions: D. Clos, De quelques omissions de la bibliothèque botanique de M. Pritzel, Bull. Soc. bot. France 5: 34-37. 1858.

Ed. 2:1872 (or Dec 1871)-Jan 1882, in seven parts (fasc. 5-7 were edited and partly compiled by Pritzel's friend Karl Friedrich Wilhelm Jessen (1821-1889) (see TL-2/2: 442-443), see also Jackson (1880), Helm (1966) and Stafleu (1973). Helm provides an extensive bibliography of the writings on the Thesaurus, p. [i, iii, v], [1]-576, [1, epilogue by Jessen]. *Copies*: B, BR, FI, G, H, HH, L, MICH, McVaugh, MO, NY, U, US (orig. covers), USDA; IDC 511. – "Editionem novam reformatam curavit G.A. Pritzel. Lipsiae (F.A. Brockhaus) l871-1877. Qu.

1: Dec 1871 or 1-5 Jan 1872 (cover 1872; prospectus announced publ. Oct 1871; Bot. Zeit. 5 Jan 1872 (soeben), 26 Jan 1872; J. Bot. Jan 1872; Flora 21 Jan 1872), p. [1]-80 (nos. 1-2183).
2: Feb-Mar 1872 (cover 1872; J. Bot. Mar 1872; Bot. Zeit. 12 Apr 1872), p. 81-160 (nos. 2184-4552).
3: Apr-Mai 1872 (cover 1872; J. Bot. Jun 1872; Bot. Zeit. 24 Mai 1872), p. 161-240 (nos. 4553-6951).
4: Jul 1872 (cover 1872; Bot. Zeit. 9 Aug 1872), p. 241-320 (nos. 6952-9424).
5-7: Aug 1877 (Bot. Zeit. 25 Mai, 10 Aug 1877; J. Bot. Aug 1877; Flora 1 Sep 1877), p. 321-576 [1, epilogue] (nos. 9425-end).
Facsimile ed. (*1*): 1924, n.v. ("Brandstettersche Obraldruck", see MW p. 396).
Facsimile ed. (*2*): 1950, Milano, Forli, as orig. *Copies*: FAS(2).
Facsimile ed. (*3*): 1972, Koenigstein-Ts./B.R.D. (Otto Koeltz Antiquariat) ISBN 3-87429-035-2, p. [i-vi], [1]-576, [577]. *Copy*: FAS.
Additions and notes: (1) B.D. Jackson, Guide to the literature of botany ... 1881 (TL-2/3217), "6000 titles not given in Pritzel's Thesaurus".
(2) Kuntze, O., Rev. gen. pl. 1: cxxii-cxlvi. 1891, TL-2/4021; reprinted Ann. Mus. nac. Montevideo 4: 25-60. 1903.
Ref.: Bary, A. de, Bot. Zeit. 30: 58-59. 26 Jan 1872 (rev. fasc. 1).
Caspary, R, Bonplandia 3: 75-77. 15 Mar 1815 (extensive review of complete volume).
Grisebach, A.H.R., Gött. gel. Anz. 1848(3) (no. 189): 1886-1891.
Helm, J., Kulturpflanzen 14: 311-357. 1966 (important and detailed study).
Hooker, W.J., J. Bot. 6: 467-468. 1847, J. Bot. Kew Gard. Misc. 4: 319-320. 1852.
Meyer,E., Bot. Zeit. 7: 290-292. 13 Apr 1849 (rev. ed. 1, fasc. 1-4).
Schlechtendal, D.F.L. von, Bot. Zeit. 8: 791-792. 1 Nov 1850.
Trimen, H., J. Bot. 10: 27-28. 1872.

8354. *Iconum botanicarum index* locupletissimus. Die Abbildungen sichtbar blühender Pflanzen und Farnkräuter aus der botanischen und Gartenliteratur des xviii. und xix. Jahrhunderts in alphabetischer Folge zusammengestellt von Dr. G.A. Pritzel. Berolini [Berlin] (in Libraria Friderici Nicolai) [1854-]1855. Qu. (*Icon. bot. index*).
Orig. ed.: in two parts, 1854-1855. *Copies*: B, NY.
1: Oct 1854 (BSbF 8-31 Dec 1854; Bot. Zeit. 3 Nov 1854), columns [i]-iv, [1 p. dedication], col. [v]-xxxii, [1 p. notice 21 Aug], col. [1]-607.
2: Mar 1855 (Bonplandia 15 Mar 1855; Flora 21 Mar 1855; Bot. Zeit. 6 Apr 1855; BSbF Jul-Dec 1855; ÖbZ 18 Oct 1855), col. 608-1184, preface material p. [i*-iii*].
French issue: 1855, with French t.p. and preface. *Copies*: BR, MO, U. – Copies with German t.p. and French preface occur.
English issue: 1855, as orig. except for an English t.p. and preface. *Copy*: PH. – See BM 4: 1617: "*Iconum* ... An alphabetical register ... of phanerogamic plants and ferns ..." Berlin, London 1855. Qu.
Reissue: 1861, "Ed. novis titulis ab auctore repudiata" Berlin 1861 (n.v.). Unchanged, pirated.
Ed. 2: 1866, consisting of a first part which is almost identical with ed. 1, and a second volume bringing the work up to 1865. *Copies*: MO(vol. 2), NY (vols. 1 and 2), PH, U(vol. 2), US; IDC 683.

Pars prima: Aug-Dec 1866 (pref. 1 Aug 1866; Hedwigia Jan 1867), p. [i*-iii*], col. [i]-xxxii, col. [1]-1184. "Zweite bis zu Ende des Jahres 1865 fortgeführte Ausgabe. Erster Theil." Berlin (Nicolaische Verlagsbuchhandlung ...) 1866. Qu.
Pars altera: Aug-Nov 1866 (Vorwort 1 Aug 1866; rev. Flora rd. Dec 1866; ÖbZ 1 Mar 1867; BSbF Jan 1867), col. [i]-xiv, col. [1]-298. Zweiter Theil... Berlin (id.) 1866. Qu.
Modern continuation: See O. Stapf, *Index londinensis*.
Ref.: Bary, A. de, Bot. Zeit. 25: 23. 18 Jan 1867 (rev.).
Fürnrohr, A.E., Flora 39(1): 8-10. 1856 (rev.).
Schlechtendal, D.F.L. von, Bot. Zeit. 12: 778-779. 3 Nov 1854.
*, Flora 50: 11-12, 270, 303. 1867 (rev., ed. 2).

8355. *Die deutschen Volksnamen der Pflanzen.* Neuer Beitrag zum deutschen Sprachschatze. Aus allen Mundarten und Zeiten zusammengestellt ... Hannover (Verlag von Philipp Cohen) 1882 [-1884]. Oct. (*Deut. Volksnam. Pfl.*).
Co-author and editor: Karl Friedrich Wilhelm Jessen (1821-1889).
Orig. ed. in two parts, 1882-1884. *Copies*: BR, G, MICH, U, USDA.
[*1*]: 1 Jan-15 Feb 1882 (p. viii: Jan 1882; Nat. Nov. Feb(2) 1882; ÖbZ 1 Oct 1882; Bot. Centralbl. 13-27 Feb 1882; Bot. Zeit. 31 Mar 1882), p. frontisp. [i]-viii, [1]-448.
[*2*]: 1 Feb-15 Mar 1884 (Bot. Centralbl. 17-21 Mar 1884; Nat. Nov. Mar(2) 1884; J. Bot. Aug 1884; Hedwigia Jul 1884), p. 449-701.
Ed. 2: front., p. [i]-viii, [1]-465. *Copy*: NY. - "2. Ausgabe" Leipzig (Verlag von Otto Lenz (vormals Verlag von Philipp Cohen in Hannover)) s.d. Oct.
Facsimile ed.: 1967, Verlag von P. Schippers N.V., Amsterdam.
Copies: MO, NY.
[*1*]: [i*], [i]-viii, [1]-465.
[*2*]: [2]-241.

Probst, Rudolf (1885-1940), Swiss botanist and physician, studied medicine at Bern 1875-1881; practicing physician at Schaffhausen 1881-1900; from 1900 id. at Solothurn. (*Probst*).

HERBARIUM and TYPES: Städtische Sammlungen Biberach a.d. Riss. Other material at E, GB, L, NY.

BIBLIOGRAPHY and BIOGRAPHY: Barnhart 3: 112; BFM no. 992, 1284, 1285, 1286, 1287, 1288; BL 2: 564, 582, 705; Kelly p. 182; Kew 4: 369; LS suppl. 22257.

BIOFILE: Aellen, P., Verh. Schweiz. naturf. Ges. 1940: 466-469 (bibl.); b. 2 Mai 1855, d. 28 Aug 1940).
Domin, K., Věda přírodní 14(2/3): 88-90. 1933 (obit., portr.).
Hedge, I.C. & J.M. Lamond, Index coll. Edinb. herb. 120. 1970.
Künkele, S. & S. Seybold, Jahresb. Ges. Naturk. Württemberg 125: 153. 1970 (herb. at Biberach a.d. Riss).
Probst, R., Wolladventivflora Mitteleuropas, Solothurn, vii, 193 p. 1949 (portr.) (edited by W. Straub).

Procházka, Jan Svatopluk (1891-1933), Czech botanist at the geological-palaeontological department of the National Museum of Prague; Dr. phil. Praha 1915. (*Proch.*).

HERBARIUM and TYPES: PRC.

BIBLIOGRAPHY and BIOGRAPHY: BM 8: 1034; Futák-Domin p. 489-490; IH 2: (in press); Klášterský p. 150 (b. 13 Mai 1891, d. 30 Jan 1933); Quenstedt p. 347.

BIOFILE: Matouschek, F., Verh. zool.-bot. Ges. Wien 1900: 276-286, see Bauer, Bot. Centralbl. 85: 54-55. 1901 (P. contributed to Fl. crypt. Boëmiae).

Procházka, J.S., Herbarium 58: 88-89. 1922 (on his *herbarium mortuum,* a collection of cyanotypic reproductions of ills. of fossil plants).
VanLandingham, S.L., Cat. Diatom. 6: 3582. 1978, 7: 4218-4219. 1978.

8356. *Katalog českých rozsivek* (Catalogus diatomacearum Bohemiae) ... Praha (V. Komisi Fr. Řivnáče ...) 1923 [1924]. Qu. (*Kat. česk. rozsivek*).
Publ.: Mar 1924 (t.p. 1923; cover dated 1924; cover FH copy inscribed 25 Mar 1924 by author), p. [1]-114. *Copy*: FH. – Archiv pro přírodovědecký výzkum Čech. dil. xvii, čis. 2.

Procopianu-Procopovici, Aurel (*fl.* 1890-1900), Roumanian botanist; inspector at the botanical garden, Bucarest (1896). (*Procopianu*).

HERBARIUM and TYPES: CERN.

BIBLIOGRAPHY and BIOGRAPHY: Barnhart 3: 112; CSP 17: 1027; IH 1 (ed. 3): 46, 2: (in press); Kew 4: 369; Tucker 1: 570.

BIOFILE: Anon., Bot. Centralbl. 66: 144. 1896 (app. Bucarest); Nat. Nov. 18: 216. 1896 (app. Bucarest); Österr. bot. Z. 46: 159. 1896 (app. Bucarest).
Pax, F., Grundz. Pfl.-Verbr. Karp. 55, 63. 1898 (Veg. Erde 2(1)).
Szymkiewicz, D., Bibl. fl. Polsk. 117. 1925.

EPONYMY: *Procopiania* Gusuleac (1928).

8357. *Enumeraţia plantelor vasculare dela Stânca-Ştefănesci* ... Bucuresci (tipografia "Speranţa" ...) 1901. Oct. (*Enum. pl. Stânca-Stef.*).
Publ.: 1901 (p. 7: Jun 1901; Bot. Zeit. 16 Oct 1901; Nat. Nov. Sep(2) 1901), p. [1]-7. *Copy*: USDA. – Reprinted from Publ. Soc. natural. Român., no. 2.

Prodan, Iuliu [Julius] (1875-1959), Roumanian botanist; studied at Cluj (lic. sci. nat. 1899); high school teacher at Gherla, Năsăud, Eger & Zombor 1899-1919; professor of botany at the faculty of agronomy, Cluj 1919-1940. (*Prodan*).

HERBARIUM and TYPES: BUCA and CLA; further material at A, BR, C, E, GH.

BIBLIOGRAPHY and BIOGRAPHY: Barnhart 3: 112; BJI 2: 140; Bossert p. 318; Futák-Domin p. 488; Hirsch p. 238; Hortus 3: 1202 (Prodan); IH 1 (ed. 4): 54, (ed. 5): 48, (ed. 6): 363, (ed. 7): 340, 2: (in press); Kew 4: 370; MW p. 396.

BIOFILE: Anon., Österr. bot. Z. 71: 72. 1922 (app. Cluj).
Degen, A., Magy. bot. Lap. 14: 189. 1915 (bibl.).
Gager, C.S., Brooklyn Bot. Gard. Rec. 27(3): 324. 1938 (dir. bot. gard. Cluj).
Hedge, I.C. and J.M. Lamond, Index coll. Edinb. herb. 120. 1970.
Pax, F., Grundz. Pfl. Verbr. Karp. 2: 280. 1908 (Veg. Erde 10).
Prodan, I., Fl. Roman., 1923 (inside cover vol. 1: list publ.).
Resmeriţă, I., Comun. bot., Rumania [vii], (1969) 1970: 27-37. 1971 (biogr., portr., bibl.).
Sălăgeanu, N., Stud. cerc. Biol., Biol. veg. 11(4): 435-437. 1959 (obit., bibl., portr., d. 27 Jan 1959).
Ştefureac, T.I., Comun. bot. Rumania 12: 11-18. 1971 (on P's bryol. work; bibl.).
Ţopa, E., Univ. Cluj. Grăd. bot., Contr. bot 1960: 9-15 (obit., portr., bibl., b. 1875, d. 27 Feb 1959).
Verdoorn, F., Chron. bot. 1: 246. 1935, 2: 262. 1936, 3: 228. 1937, 6: 233. 1941, 7: 41. 1943.

8358. *Contribuţiune la flora României* ... Bucureşti (Librăriile Socec & Comp. ...) 1914. Qu.
Publ.: 1914 (submitted to Acad. 28 Mar 1914), p. [1]-56, title on cover. *Copy*: G. – Reprinted and to be cited from Anal. Acad. Rom. ser. 2. 36 (Mem. Secţ. ştiinţ.) 10: 1-56. 1914.

8359. *Flora pentru determinarea și descrierea plantelor ce cresc in România* ... Cluj (Institut de Arte grafice, editură și librărie cartea Românească ...) 1923, 2 vols. (*Fl. Român.*).
1: 1923, p. [i]-cxxxiv, [1]-1152, [cxxxv]-cxxxvi.
2: 1923, p. [1]-229, 3 maps, *pl. 1-130*, [*131*], mostly in text.
Copies: B, G, MO, U, USDA. – A second ed., Cluj 1939, was not seen by us.*Ref*.: Hayek, A., Österr. bot. Z. 73: 154. 1924.

8360. *Flora mică ilustrăta a României* ... Cluj (Minerva Institut de Literatura ...) [1928]. Oct. (*Fl. mică ilus. Român.*).
Ed. 1: 1928 (p. [iii]: 24 Jan 1928), p. [i]-lxvi, [1]-518, *pl. 1-26* (uncol.). *Copy*: PH.
Later ed.: 1958, p. [1]-664. *Copy*: UC. – Title as ed. 1961, see below.
Later ed.: 1961 (type set 21 Jun 1961), p. [1]-676. *Copy*: BR. – *Flora mică ilustradă a Repulicii populare Romine*, București (Ministerul agriculturii ...) [1961], 8620 copies, edited by Al. Buia.

8361. *Centaureele României* (Centaureae Romaniae) ... Cluj (Institutul de arte grafice "Ardealul" ...) 1930. Oct. (*Centaur. Român.*).
Publ.: 1930 (Nat. Nov. Apr 1931), p. [i*-ii*], [i]-ii, [1]-256, [1], *pl. 1-32, 32a, 32b, 33, 34-59, 60-63*, 3 maps. *Copies*: G, USDA.

8362. *Conspectul florei Dobrogei* ... Cluj (Tipografia Națională S.A.) 1935. Oct. (*Consp. fl. Dobr.*).
Publ.: 1935, p. [1]-170. *Copy*: G. – Reprinted from Bul. Acad. Stud. agron. Cluj 5(1), 1934 (1935).

Pröll, Alois (*fl.* 1839), Austrian physician and botanist; Dr. med. Wien 1839. (*Pröll*).

HERBARIUM and TYPES: W.

BIBLIOGRAPHY and BIOGRAPHY: IH 2: (in press); PR 7350 (ed. 1: 8236).

BIOFILE: Beck, G., Bot. Centralbl. 34: 150. 1888.
Fritsch, K. et al., Bot. Anst. Wien 75. 1894 (coll.).

8363. Dissertatio inauguralis medico-botanica, sistens tentamen, *Fungos austriacos* esculentos iisque similes virulentos propria investigatione determinandi, quam consensu et auctoritate illustrissimi ac magnifici domini praesidis ac directoris, perillustris ac spectabilis domini decani, nec non clarissimorum ac celeberrimorum d.d. professorum, pro doctoris medicinae laurea rite et legitime obtinenda in antiquissima ac celeberrima Universitate Vindobonensi publicae disquisitioni submittit Aloisius Pröll austriacus. Alumnus C.R. Convictus. In theses adnexas disputabitur in aedibus Universitatis die 9. martii anni mdcccxxxix. Viennae (Ex typographeo viduae Antonii Pichler) 1839. Oct. (*Fung. austriac.*).
Publ.: 9 Mar 1839, p. [i]-vi, [7]-22, [1-2, theses]. *Copies*: FH, WU.

Progel, August (1829-1889), German (Bavarian) physician and botanist; Dr. med. München 1855; practicing in various towns of Oberbayern and Schwaben; ultimately regional physician at Waldmünchen 1876-1889. (*Progel*).

HERBARIUM and TYPES: M (bryoph., lich.; rd. 1925); other material and duplicates at A, B, BR, C, H, L (mainly musci).

BIBLIOGRAPHY and BIOGRAPHY: AG 6(1): 619 (b. 2 Jan 1829, d. 26 Apr 1889); Barnhart 3: 112; BM 4: 1618; CSP 11: 72, 17: 1028; Herder p. 472; IH 2: (in press); Morren ed. 10, p. 20; NI 216, 2248; PR alph.; Rehder 5: 691; SBC p. 129; Urban-Berl. p. 311.

BIOFILE: Anon., Bot. Centralbl. 38: 687. 1889 (d.), Österr. bot. Z. 39: 248. 1889 (d.); Bot. Jahrb. 11 (Beibl. 24): 1. 1889 (d.); Bot. Not. 1890: 45 (d.); Hedwigia 28: 226. 1889 (d. 26 Apr 1889); Leopoldina 25: 113. 1889 (d.).
Familler, I., Denkschr. bay. bot. Ges. Regensb. 11: 8, 13, 15, 20. 1911.

Hertel, H., Mitt. Bot. München 16: 419. 1980 (herb. rd. by M 1925).
Holler, Ber. bot. Ver. Landshut 11: xxxiv-xl. 1889 (obit., bibl., b. 2 Jan 1829, d. 26 Apr 1889).
Mägdefrau, K., Hoppea 37: 145. 1978 (b. 2 Jan 1829, d. 26 Apr 1898).
Sayre, G., Bryologist 80: 515. 1977.
Sudre, H., Bull. Acad. Géogr. bot. 21: 33-68. 1911 (Reliquiae Progelianae, Rubus).
Urban, I., Fl. bras. 1(1): 194. 1906.

COMPOSITE WORKS: (1) *in Flora brasiliensis*:
Gentianaceae, 6(1): 197-246, *pl. 55-66*. 1 Dec 1865. *Loganiaceae*, 6(1): 249-290, *pl. 67-82*. 1 Aug 1868. *Cuscutaceae*, 7: 371-390, *pl. 125-128*. 1 Mar 1871. *Oxalidaceae, Geraniaceae, Vivianiaceae*, 12(2): 473-528, *pl. 102-118*. 1 Dec 1877.
(2) *in* E. Warming, *Symb. fl. Bras. centr.*:
Gentianaceae, 2: 33-38. 11 Oct 1869.
Loganiaceae, 2: 28-33*. 11 Oct 1869.
Cuscutaceae, 9: 316-318. 11 Apr 1871.
Oxalidaceae, 25: 19-24*. 3 Dec 1879.

8364. *Flora des Amtsbezirkes Waldmünchen* ... Landshut (Druck der Jos. Thomann'schen Buchdruckerei) 1882. Oct.
Publ.: 1882, p. [1]-76. *Copies*: B, REG. – Reprinted from Ber. bot. Ver. Landshut 8, 1882.
Ref.: Peter, Bot. Zeit. 41: 367. 1 Jun 1883.

Prokhanov, Yaroslav Ivanovich (1902-1965), Russian botanist; Dr. phil. Leningrad 1924; assistant at the Leningrad botanical garden 1925-1929; on botanical expedition to Mongolia 1926; assistant at the Institute of plant industry 1929. (*Prokh.*).

HERBARIUM and TYPES: LE; important material also in WIR.

BIBLIOGRAPHY and BIOGRAPHY: Barnhart 3: 112; Bossert p. 318; IH 1 (ed. 6): 363, (ed. 7): 340; Kew 4: 370; Langman p. 597; MW p. 396, suppl. p. 287; Roon p. 90; TL-2/3857.

BIOFILE: Vassilczenko, I.T., Bot. Zhurn. 52: 1807-1811. 1967 (obit. portr., bibl., b. 31 Jul 1902, d. 14 Feb 1965).

COMPOSITE WORKS: *in* Komarov, *Fl. URSS*:
Celastraceae, 14: 546-577. 9 Mar 1949.
Malvaceae-Gossypium, 15: 170-184. 12 Dec 1949.
Solanaceae-Lycopersicon, 22: 42-57. 18 Jun 1955.

Prollius, Friedrich (*fl.* 1882), German (Hannoverian) botanist; Dr. phil. Jena 1882. (*Prollius*).

HERBARIUM and TYPES: Unknown.

BIBLIOGRAPHY and BIOGRAPHY: BM 4: 1618; CSP 17: 1029; De Toni 1: cii, 2: xcvii.

8365. *Beobachtungen über die Diatomaceen der Umgebung von Jena*. Inaugural-Dissertation zur Erlangung der philosophischen Doctorwürde der hohen Philosophischen Facultät der Universität Jena vorgelegt von Friedrich Prollius aus Lüneburg. Lüneburg (Druck von Heinrich König) 1882. Oct. (*Beob. Diatom. Jena*).
Publ.: 1882 (Nat. Nov. Apr (2) 1883; Bot. Centralbl. 23-27 Apr 1883; Bot. Zeit. 25 Mai 1883), p. [1]-106, [1-5, cont., index], *pl. 1-2*. *Copies*: FH, US. – The late notices in the journals refer to an issue by Deistung's Verlag, Jena. This may have been a commercial issue, distributed after the dissertation.

Pronville, Auguste de (*fl.* 1818), French rhodologist. (*Pronville*).

HERBARIUM and TYPES: Unknown. – Letters at G.

BIBLIOGRAPHY and BIOGRAPHY: BM 4: 1618; Jackson p. 142; Kew 4: 371; MW p. 396; PR 7351-7352 (ed. 1: 8237-8239); PR 5343; Rehder 5: 691; TL-2/4636; Tucker 1: 571.

COMPOSITE WORKS: translated Lindley's *Rosarum monographia*, see J. Lindley, TL-2/4636: *Monographie du genre Rosier* 1824.

8366. *Nomenclature raisonnée des espèces, variétés et sous-variétés du genre Rosier*, observées au Jardin royal des Plantes, dans ceux de Trianon, de Malmaison, et dans les pépinières des environs de Paris ... à Paris (de l'Imprimerie et dans la librairie de Madame Huzard (née Vallat la Chapelle), ...) 1818. Oct. (*Nomencl. Rosier*).
Publ.: 1818 (Flora 7 Jan 1819: "neu"), p. [1]-119. *Copies*: FI, MO, USDA. – A note on p. 9 indicates that a preliminary publication took place in the Almanach du bon jardinier of 1816, and that the 1818 text was edited by C. Romain Féburier.

Prost, T.C. (x-1848), French postal director and bryologist at Mende (Lozère). (*Prost*).

HERBARIUM and TYPES: in 1862 at Musée de la ville de Mende, dépt. de Lozère (BSbF 9: 246. 1862); other material at AV, CGE, CLF, DBN, G, M (lich.), MANCH, PC. See also H. Loret (1862).

BIBLIOGRAPHY and BIOGRAPHY: Barnhart 3: 113; BM 4: 1618; GR p. 346-347; Kew 4: 371; Krempelh. 1: 457, 612; PR 7353-7354 (ed. 1: 8240-8241).

BIOFILE: Anon., Flora 31: 432. 1848 (d. Apr 1848).
Boulay, J.N., Rev. bryol. 1(2): 21. 1874 (on his *Liste*).
Candolle, Alph. de, Phytographie 441. 1880.
Candolle, A.P. de, Mém. souvenirs 236. 1862.
Fournier, E., Bull. Soc. bot. France 9: 246-248. 1862 (review of Loret, H., L'herbier de la Lozère et M. Prost, Bull. Soc. agr. Lozère, vol. 13, 1862, publ. 26 Dec 1862), (see TL-2/4999).
Loret, H., Bull. Soc. Agr. Industr. Sci. Arts Lozère 13: 81-134. 1862 (see TL-2/4999; L'Herbier de la Lozère et M. Prost).
Malinvaud, É., Bull. Soc. bot. France 40: lxxvi. 1893 (letters Adr. de Jussieu and A.R. Delile to Prost), 46: lxii-lxiii. 1899 (letter A.P. de Candolle to Prost).
Verdoorn, F., ed., Fl. pl. sci. Latin Amer. xxvi. 1945.

EPONYMY: *Prostea* Cambessèdes (1829).

8367. *Liste des mousses, hépatiques et lichens* observés dans le département *de la Lozère, ...* à Mende (Chez J.J.M. Ignon, ...) 1828. Oct. (in fours). (*Liste mousses Lozère*).
Publ.: 1828 (submitted 6 Nov 1827), p. [1]-72. *Copies*: G, NY. – Reprinted from Mém. Soc. Agr. Ville de Mende (1828).

Protić, Georg (Gjorgje) (1864-?), Bosnian cryptogamist (algologist); Dr. phil. Wien 1891. (*Protić*).

HERBARIUM and TYPES: Unknown.

BIBLIOGRAPHY and BIOGRAPHY: Barnhart 3: 113; GR p. 458; Kew 4: 371; LS 21417-21420, suppl. 22271-22272; Nordstedt suppl. p. 13.

BIOFILE: Beck, G., Veg.-Verh. Illyr. Länd. 20, 40. 1901 (Veg. Erde 4).
Dörfler, I., Botaniker-Adressbuch ed. 3. 227. 1909.
Grojonovic, Enum. auct. fl. Dalmat. 8. 1905.
Verdoorn, F., ed., Chron. bot. 1: 443. 1935, 2: 367. 1936.

8368. *Prilozi k poznavanju kremenjašica* (diatomacea) *Bosne i Herzegovine* ... Sarajevo (Zemaljska Štamparija) 1897.
Publ.: 1897, p. [1]-13. *Copy*: G. – Reprinted and to be cited from Glasnika zemaljskog Muzeja u Bosni i Herzegovini 9(2): 313-326. 1897.

8369. *Prilozi k poznavanju flore resina* (alge) *Bosne i Hercegovine* s osobitim obzirom na floru resina okoline Sarajeva, Vareša i Mostarskog blata (isključivši diatomaceje) ... Sarajevo (Zemaljska Štamperiija). 1897.
Publ.: 1897, p. [1]-21. *Copy*: G. – Reprinted and to be cited from Glasnika zemaljskog Muzeja u Bosni Hercegovini 9(4): 539-560. 1897.

8370. *Prilog k oznavanju gljiva Bosne* i Herzegovine ... Sarajevo (Zemalska Štamperija) 1898.
Publ.: 1898, p. [1]-9. *Copy*: G. – Reprinted and to be cited from Glasnika zemaljskog Muzeja u Bosni i Herzegovini 10(1): 93-102. 1898.

Provancher, Léon (1820-1892), Canadian clergyman, botanist, entomologist and conchologist; ord. Québec 1844; minister at various Québec parishes 1844-1854, at St.-Joachim 1854-1862; from 1862-1871 at Portneuf; in retirement at Cap-Rouge nr. Québec from 1871; founder of le Naturaliste canadien. (*Prov.*).

HERBARIUM and TYPES: QPH; further material at QUE. – Some archival material at SIA.

BIBLIOGRAPHY and BIOGRAPHY: Barnhart 3: 113; BL 2: 133, 144, 314; BM 4: 1619; CSP 12: 591, 17: 1033-1034 (further biogr. refs.); IH 1 (ed. 6): 363, (ed. 7): 232, 340, 2: (in press); Jackson p. 366; Kew 4: 372; Morren ed. 2, p. 35, ed. 10, p. 134; PH 211, 567; PR 7355; Rehder 5: 692; Tucker 1: 571.

BIOFILE: Anon., Appleton's Cycl. Amer. biogr. 5: 128. 1888 (b. 10 Mar 1820); Natural. canad. 5: 134. 1873; Nat. Nov. 14: 232. 1892; Ottawa Natural. 6: 44. 1892 (obit.).
Barron, J.R., Natural. canad. 102: 387-591. 1975 (P's colls. of insects; types of Ichneumonidae).
Béique, R., Natural. canad. 95(3): 609-626. 1968 (on P's entomological work).
Chapais, J.C., Natural. canad. 50: 193-195. 1924 (on P. as an educator).
Cinq-Mars, L., Provancheria 1: 5-16. 1967 (biogr., portr.); Natural. canad. 95(1): 7-8. 1968 (hommage, botanical work).
Essig, E.O., Hist. entom. 734-735. 1931 (portr.).
Gray, A., Amer. J. Sci. ser. 2. 35: 445. 1863.
Holland, G.P., L'abbé Provancher 1820-1892, *in* Les pionniers de la science canadienne, Société royale de Canada, Univ. Toronto Press. 1966 (n.v.).
Huard, V.A., Natural. canad. 21(1): 17-21. 1894 (last publ. Provancher); [biography of P., in instalments:] 21: 37-41, 53-56, 85-88, 101-104, 134-137, 149-152, 182-185. 1894; 22: 18-22, 53-57, 117-120, 133-136, 181-185. 1895.; 23: 49-53, 81-84, 113-117, 145-148, 177-180. 1896; 24: 178-182. 1897; 25: 34-37, 52-56, 82-86, 115-118, 133-136, 168-172, 183-187. 1898; 26: 17-21, 41-44, 81-85, 138-142, 150-152, 162-165, 178-182. 1899; La vie et l'oeuvre de l'abbé Provancher, Québec 1926 (biogr, portr., b. 10 Mar 1820, d. 23 Mar 1892), originally published as above in Le Naturaliste canadien, 512 p.
Kucyniak, J., Bryologist 49: 128. 1946 (mosses Fl. canad.).
Maheux, G., Natural. canad. 95(1): 2-6. 1968 (notes on P. on the occasion of the centenary of the journal).
Mallis, A., Amer. entomologists 106-107. 1971 (portr.).
Osborn, H., Fragm. entom. hist. *pl. 9*. 1937 (portr.).
Penhallow, D.P., Proc. Trans. R. Soc. Canada ser. 2. 3: 12, 51. 1897.
Provancher, L., Natural. canad. 5: 134-135. 1873 (autobiogr.).
Robitaille, A., Natural. canad. 57: 254. 1930 (herb. QPH).
Rousseau, J., Contr. Inst. bot. Montréal 44: 39-41. 1922 (A. Gray and the publ. of the *Fl. canad.*).

Rousseau, J. & R. Boivin, Natural. canad. 95(6): 1499-1530. 1968 (the contribution to science of the *Fl. canad.*).
Rumilly, R., Le frère Marie-Victorin et son temps 456 [index]. 1949.

COMPOSITE WORKS: Editor, *Le Naturaliste canadien* vols. 1-20, 1869-1891.

EPONYMY: (genus): *Provencheria* [sic] B. Boivin (1966), see ING 3: 1423 (orthographic error, later corrected by author); (journal): *Provancheria*, Mémoires de l'Herbier Louis-Marie. Faculté d'Agriculture de l'Université Laval. Quebec. Vol. 1-x. 1967-x. — Société Provancher, a society founded in Quebec in 1918 to promote the education and popularization of the natural sciences. Rue Provancher in Cap Rouge, near Quebec.

HANDWRITING: Natural. canad. 102: 404-405. 1975.

8371. *Flore canadienne ou description de toutes les plantes des forêts, champs, jardins et eaux du Canada donnant le nom botanique de chacune, ses noms vulgaires français et anglais, indiquant son parcours géographique, les propriétés qui la distinguent, le mode de culture qui lui convient, etc. Accompagnée d'un vocabulaire des termes techniques et de clefs analytiques permettant de rapporter promptement chaque plante à la famille, au genre et à l'espèce qui la déterminent. Ornée de plus de quatre cents gravures sur bois...* Québec (Joseph Darveau, ...) 1862 [1863], 2 vols. Oct. (in fours). (*Fl. canad.*).
1: Jan 1863 (t.p. 1862 but official registration (p. ii) Jan 1883; Asa Gray acknowledged receipt of a copy on 13 Feb 1883), p. [i*], [i]-xxix, [1]-474.
2: 1863 (t.p. 1862, but see vol. 1), p. [475]-842, [1, err.].
Copies: MO, NY, US, USDA.

Pryor, Alfred Reginald (1839-1881), British botanist; BA Oxon 1862; landed gentleman dedicating himself fully to the flora of Hertfordshire. (*Pryor*).

HERBARIUM and TYPES: formerly at STAL; not traced (came to the Hertfordshire Natural History Society).

BIBLIOGRAPHY and BIOGRAPHY: Barnhart 3: 113 (q.v. for a list of variations of given names); BB p. 250; BL 2: 245 (q.v. for add. refs. *Fl. Herts.*), 705; BM 4: 1620; CSP 11: 75, 12: 591; Desmond p. 506-507 (b. 24 Apr 1839, d. 18 Feb 1884); DNB 46: 437-438; GR p. 410; Kew 4: 372; Lenley p. 336; Rehder 5: 693; TL-2/see Jackson, B.D.; Tucker 1: 572.

BIOFILE: Anon., Bot. Centralbl. 5: 352. 1881 (d.), 8: 93. 1881; Bot. Not. 1882: 30 (d. 10 Feb 1881); J. Bot. 13: 126. 1875 (on proposed Fl. Herts.); 18: 124. 1880 (progress), 19: 96. 1881 (d.), 23: 160. 1885 (difficulties in publ. his Flora posthumously); Nat. Nov. 3: 47. 1881 (d. 18 Feb 1881).
Bridson, G.D.R. et al., Nat. hist. mss. res. Brit. Isl. 186.11, 186,15. 1980.
Britten, J., J. Bot. 19: 276-278. 1881, 26: 58-62. 1888 (rev. *Fl. Herts.*).
Dony, J.G., Fl. Bedfordshire 25, 1953, Fl. Hertfordshire 16. 1967.
Druce, G.C., Fl. Buckinghamshire civ. 1926, Fl. Oxfordshire cxxi. 1927.
Freeman, R.B., Brit. nat. hist. books 287. 1980.
Kent, D.H., Brit. herb. 73. 1957.
Pryor, R.A., J. Bot. 12: 127. 1874 (announces his work on Fl. Herts.).
Verdoorn, F., ed., Chron. bot. 5: 114. 1939.

8372. *A flora of Hertfordshire* by the late Alfred Reginald Pryor, ... edited for the Hertfordshire natural History Society by Benjamin Daydon Jackson, ... with an introduction on the geology, climate, botanical history, etc., of the county by John Hopkinson, ... and the editor. London (Gurney & Jackson, ...) Hertford (Stephen Austin & Sons) 1887. Oct. (*Fl. Hertfordshire*).
Editor: Benjamin Daydon Jackson (1846-1927); *co-author* of introduction: John Hopkinson (1844-1919).
Publ.: Oct-Dec 1887 or Jan 1888 (p. vii: Oct 1887; J. Bot. Feb 1888; Bot. Centralbl. 21-25 Mai 1888), p. [i]-lviii, [1]-588, 3 maps. *Copies*: BR, G, MO, NY, USDA. – Lichens

(p. 520-521) by E.M. Holmes. – The BR copy was signed by the author's mother on 3 Mar 1888.
Preliminary publ.: Notes on a proposed re-issue of the Flora of Hertfordshire [referring to R.H. Webb and W.H. Coleman, *Flora hertfordiensis*, London 1849-1859], Hertford 1875, 14 p., Trans. Watford nat. Hist. Soc. 1: 17-32. 1875 (1878). (n.v.) and *Notes ... Hertfordshire* with supplementary remarks on the botany of the Watford district, published in Trans. Watford nat. Hist. Soc. 1: 63-64, 65-77. 1875 (1878) reprint 16 p. (*copy*: USDA).

Prytz, Lars Johan (1789-1823), Finnish botanist. (*Prytz*).

HERBARIUM and TYPES: H.

BIBLIOGRAPHY and BIOGRAPHY: Barnhart 3: 114; BM 4: 1620; CSP 5: 37; Herder p. 200; IH 2: (in press); PR 7356 (ed. 1: 8243); Rehder 5: 693; Saelan p. 399; TL-2/2610; TR 1093; Tucker 1: 572.

COMPOSITE WORKS: author of parts 6-8 of Hellenius *Diss. hort. abo.*, 1814, TL-2/2610.

8373. *Florae fennicae breviarum*, dissertationibus academicis absolvendum, quarum primam, venia amplissimae Facultatis philosophicae Aboënsis, publice censurae subjiciunt auctor, Laurentius Johannes Prytz, ... et respondens Fredericus Tengström ... in auditorio medico, die v maji mdcccxix, horis ante meridiem consuetis. Aboe (Typis Frenckellianis) [1819]. Qu. (*Fl. fenn. brev.*).
Publ.: 5 Mai 1819, p. [1]-16. *Copies*: H, NY. – This is the first of a series of six dissertations with this title, as follows (*Copies* at NY). Prytz was *author* of all parts.

pars	respondens	pagination	date
1	Fredericus Tengström	[1]-16	5 Mai 1819
2	Carolus Henricus Ringbom	[i], [17]-28	5 Mai 1819
3	Victor Erici Hartwall	[i], 29-44	5 Mai 1819
4	G. A. Wegelius	[i], 45-60	5 Mai 1819
5	Wilhelmus Alex. Nordgren	[i], 61-76	12 Mai 1819
6	Isaacus Reginaldus Eneberg	[i], 77-92	12 Mai 1819

Continuation: by O.E.A. Hjelt *in* Notis. Sällsk. Fauna Fl. fenn. Förh. ser. 2. 7: 245-286. 1869. "Ex schedulis auctoris continuatio".

Przewalski, Nikolai Michailowicz [Przhevalsky, Nikolay Mikhaylovich] (1839-1888), Russian soldier, traveller, geographer and naturalist; from 1864-1867 geography teacher at the Warschau "Junkerschule"; from 1867 on the general staff of the army; explored Manchuria and Central Asia between 1867 and 1885. (*Przew.*).

HERBARIUM and TYPES: LE; other material at B, C, E, FI, K, MO, P. – Collections of plants in first instance worked up by C.J. Maximowicz, q.v.

BIBLIOGRAPHY and BIOGRAPHY: AG 4: 53; Barnhart 3: 114; BM 4: 1620-1621 (lists e.g. his travel reports), 8: 1036; Bossert p. 319; Bret. p. 959-968; CSP 8: 671, 11: 76, 17: 1038-1039 (q.v. for further obituaries); IH 1 (ed. 3): 88, (ed. 6): 363, (ed. 7): 340, 2: (in press); Kew 4: 372; MW p. 396-397 (travel reports), suppl. p. 287-288; Rehder 5: 693; TL-2/5735, 5736; Tucker 1: 572; Urban-Berl. p. 311, 385.

BIOFILE: Anon., Auk 6: 80-81. 1889; Bot. Centralbl. 1/2: 640. 1880 (sources Yellow R.); Mitt. k.k. Geogr. Ges. Wien 31: 583-585. 1888 (obit.).
Embacher, F., Lexikon Reisen 237-239. 1882.
Esakov, V.A., Dict. Sci. biogr. 11: 180-182. 1975 (extensive details, bibl., secondary refs., b. 12 Apr 1839, d. 1 Nov 1888).
Hedge, I.C. & J.M. Lamond, Index coll. Edinb. herb. 120. 1970.
Hedin, S., Ymer, Stockholm 8: 103-105. 1889 (obit.).

Herder, F. v., Bot. Jahrb. 9: 431, 445-446. 1888 (itin.; publ. on plants by e.g. Maximowicz, Diagn. pl. nov. asiat. 1877-1883); Bot. Centralbl. 58: 391. 1894 (coll. at LE).
Jackson, B.D., Bull. misc. Inf. Kew 1901: 54 (pl. K).
Krause, A., Naturw. Wochenschr. 3: 65-67. 1888 (obit.).
Lipskij, V.I., Biogr. bot. sadom. 1913/15: 30-35.
Marthe, F., Verh. Ges. Erd. Berlin 15: 455-462. 1888 (extensive obit.).
Maximowicz, N.M., Acta Horti Petropolitani 10(2): 673-683. 1889 (obit., portr., b. 31 Mar (2 Apr) 1839, d. 20 Oct (1 Nov) 1888).
Nissen, C., Zool. Buchill. 325. 1969 (b. 12 Apr 1839).
P.K., Nature 39: 31-34. 1889 (obit.).
Sayre, G., Mem. New York Bot. Gard. 19(3): 382. 1975 (coll).
Semenova, P.P., Bull. Russ. geogr. Soc. St. Petersburg (Izv. Gosud. rossk. geogr. obsh.) 24: 231-272, 277-280. 1889 (obit., portr.).
Wittrock, V.B., Acta Horti Berg. 3(2): 175. 1903 (b. 11 Apr 1839, d. 1 Nov 1888).

EPONYMY: *Przewalskia* Maximowicz (1882).

Pucci, Angiolo (1851-1935), Italian botanist and horticulturist; professor at the Pomological and Horticultural College at Firenze. (*Pucci*).

HERBARIUM and TYPES: FI.

BIBLIOGRAPHY and BIOGRAPHY: MW p. 397; Rehder 5: 693; Saccardo 1: 133 (b. 2 Jun 1851); Tucker 1: 572.

BIOFILE: Merrill, E.D., Contr. U.S. natl. Herb. 30(1): 246. 1947.
Verdoorn, F., ed., Chron. bot. 1: 195. (d.).

8374. *Les Cypripedium* et genres affines. Histoire, description, synonymie et culture des espèces, variétés et hybrides. Florence (Imprimerie éditrice L. Niccolai ...) 1891. Oct. (*Cypripedium*).
Publ.: 1891, p. [1]-218, [1, 3, 5]. *Copies*: BR, G. – See also Bull. Soc. tosc. Ortic. 15, 1890 (publ. 1891, Bot. Zeit. 29 Mai 1891).

Puccinelli, Benedetto Luigi (1808-1850), Italian botanist; professor of botany at Lucca, director of the botanical garden of the University 1833-1850. (*Puccin.*).

HERBARIUM and TYPES: LUCCA; further material at BM and FI. – Letters at G.

BIBLIOGRAPHY and BIOGRAPHY: AG 2(1): 453; Backer p. 473; Barnhart 3: 114; BL 2: 401, 705 (b. 1805, err.); BM 4: 1621; Bossert p. 319; CSP 5: 37; Herder p. 179; IH 1 (ed. 6): 363, (ed. 7): 340, 2: (in press); Kew 4: 373; PR 7359 (ed. 1: 8244-8245); Rehder 1: 63, 426; Saccardo 1: 133 (b. 11 Feb 1808, d. 25 Mar 1850), 2: 88, cron. p. xxx; Tucker 1: 572.

BIOFILE: Hitchcock, A.S. & A. Chase, Man. Grass. U.S. 988. 1951 (epon.).
Murray, G., Hist. coll. BM(NH) 1: 175. 1904.
Tomei, P.E., Informatore bot. ital. 6: 134-136. 1974 (on herb. at Lucca).

EPONYMY: *Puccinellia* Parlatore (1848, *nom. cons.*), *Puccinellia* Fuckel (1860). See ING 3: 1462.

8375. *Synopsis plantarum in agro lucensi* sponte nascentium ...Lucae [Lucca] (typis Bertinianis) 1841 [-1848]. Oct. †. (*Syn. pl. luc.*).
[*1*]: 1841, p. [1]-326, *9 pl.* (uncol.). *Copy*: FI, G (incomplete). – Original publication in Atti r. Accad. Lucc. Sci. Let. Arti 11: 313-392. *4 pl.* 1841, 12: 1-144, *5 pl.* 1843, 13: 1-96, *2 pl.* 1845, together constituting the first volume (n.v.).
[*2*]: 1848, p. [i], 327-531, i-xxxxix, *5 pl. Copy*: FI. – in id. 14: 91-297, *5 pl.* 1848 (?1852).
Addenda: Parlatore, Giorn. bot. ital. 1(1): 118-123. 1844 (fide id.).
Ref.: Schlechtendal, D.F.L. von, Bot. Zeit. 3: 715-716. 1845, 9: 605. 1851.

8376. *Osservazioni sui funghi dell'agro Lucchese* con tavole a colori ritratte dal vero ed eseguite in litografia da Giuseppe Bertini ... Lucca 1841. (*Osserv. fung. Lucch.*).
Publ.: 1841, p. [1]-18, *4 col. pl. Copy*: FI.

Puel, Timothée (1812-1890), French botanist; practicing physician at Paris; worked mainly on the flora of the dépt. Lot. (*Puel*).

HERBARIUM and TYPES: P; further material at OXF. – Published a series of exsiccatae *Herbier du Lot* (fasc. 1, nos. 1-20, 1860 see BSbF 7: 375, fasc. 2, nos. 21-40, BSbF 9: 399), sets at BR, LE. – Letters at G.

BIBLIOGRAPHY and BIOGRAPHY: AG 2(1): 28; Backer p. 473; Barnhart 3: 114; BL 2: 174, 175, 705; BM 4: 1621; CSP 5: 39, 17: 1041; Hortus 3: 1202 (Puel); IH 1 (ed. 6): 362, (ed. 7): 340; Kew 4: 373; PR 7360-7361,(ed. 1: 8247); Rehder 5: 693; TL-2/see A. Maille; Tucker 1: 572.

BIOFILE: Anon., Bull. Soc. philom. Paris 2: 58-59. 1890 (obit., as F. Puel (sic), d. 28 Feb 1890).
Candolle, Alph. de, Phytographie 441. 1880.

EPONYMY: *Puelia* Franchet (1887).

8377. *Catalogue des plantes vasculaires* qui croissent dans le département *Lot*, ... Cahors (J.-P. Combarieu, imprimeur, ...) [1845-]1852[-1853]. Oct. (*Cat. pl. Lot*).
Publ.: 1845-1853 (t.p. 1852, p. 241: 15 Jul 1852; Puel BSbF 7: 373: "1845-1853"; first vol. presented to session Soc. bot. France 24 Mai 1854, but a second fasc. on 9 Mar 1855?), p. [i], [1]-248. *Copies*: G, NY. – Originally published in parts in the Annuaire statistique du Lot (n.v.) (p. 1-88, 89-120, 121-136, 137-236, suppl. 237-248). Full citation of livraisons by Lucante, Bull. Soc. Ét. Lit. sci. art. Lot 11: 92. 1886 (fide BL), for further details see BL.
Ref.: Schlechtendal, D.F.L. von, Bot. Zeit. 4: 535-536. 1846.

8378. *Catalogue de l'herbier de Syrie* publié par I. Blanche et C. Gaillardot. Paris (Imprimerie de Mme Ve Dondey-Dupré ...) [1854]. Oct. (*Cat. herb. Syrie*).
Authorship: see p. 2, Puel and Alphonse Maille (1813-1865), see TL-2/3: 256. – *Collectors*: Charles Isidiore Blanche (1823-1887), Charles Gaillardot (1814-1883).
Publ.: 1854 (p. 2: 31 Mar 1854), p. [i], [1]-14. *Copy*: HH.

Puerari, Marc Nicolas (1766-1845), Swiss botanist; pupil of Vahl in Copenhagen, teacher in Denmark. (*Puerari*).

HERBARIUM and TYPES: G (presented by him to A.P. de Candolle in 1827). Puerari's library (incl. letters and portrait) is also at G.

BIBLIOGRAPHY and BIOGRAPHY: AG 6(2): 1074; Backer 473-474; Barnhart 3: 114; Hegi 6(2): 1331; Lasègue p. 345; PR (alph.) 10198.

BIOFILE: Baehni, C., Hist. Sci. Genève 44. s.d. (portr.).
Briquet, J., Bull. Soc. bot. Suisse 50a: 389-390. 1940 (b. 19 Nov 1766 based on official records at Genève, d. 11 Jun 1845).
Candolle, Alph. de, Phytographie 441. 1880 (herb. G).
Candolle, A.P., Hist. bot. genev. 48. 1830 (b. 1768, err.) (Mém. Soc. phys. Genève 5: 48. 1830).
Hornemann, J., Nat. Tidskr. 1: 591. 1837.

HANDWRITING: Candollea 32: 399-400. 1977.

EPONYMY: *Pueraria* A.P. de Candolle (1825).

Puerta y Ródenas, Gabriel de la (*fl.* 1891), Spanish botanist, professor of botany at the Universidad Central, Madrid. (*Puerta*).

HERBARIUM and TYPES: Unknown.

BIBLIOGRAPHY and BIOGRAPHY: BL 2: 483, 705; Rehder 5: 693.

8379. *Botánica descriptiva* y determinacion de las plantas indígenas y cultivadas en España de uso medicinal, alimenticio é industrial ... Madrid (administracion de la Revista de Medicina y Cirurgía prácticas ...) 1891. (*Bot. descr.*).
Ed. 1: 1877, n.v., Tratado práctico de determinacion de las plantas indígenas.
Ed. 2: Mar-Jul 1891 (p. 4: Mar 1891; Na. Nov. Aug(2) 1891, p. [i-iv], [1]-669, [1, cont.; 1, err.]. *Copy*: USDA. "2.a edicion corregida ..." Madrid (administracion de la Revista de Medicina y Cirurgía prácticas ...) 1891. Oct.

Pugsley, Herbert William (1868-1948), British botanist; BA London 1889; civil servant in the Admiralty, 1896-1928. (*Pugsl.*).

HERBARIUM and TYPES: BM (30.000), further material at B, C, CGE, K, RTE (Surrey), SLBI, Z. – Letters and portrait at G.

BIBLIOGRAPHY and BIOGRAPHY: Barnhart 3: 114; BL 2: 217, 226, 235, 237, 269, 314, 321, 705; BM 8: 1036; Bossert p. 319; Desmond p. 507 (b. 24 Jan 1868, d. 18 Nov 1947); Hortus 3: 1202 (Pugsl.); MW p 397; NI 1569; Zander ed. 10, p. 704, ed. 11, p. 802.

BIOFILE: Anon., Gard. Chron. 122: 200. 1947 (obit.); The Times 22 Nov 1947 (fide Desmond).
Bridson, G.D.R. et al., Nat. hist. mss. res. Brit. Isl. 84.1, 84.21, 128.1. 1980.
Kent, D.H., Brit. herb. 73. 1957.
Lousley, J.E., Naturalist 824: 13-15. 1948 (obit.).
Ramsbottom, J., Nature 161: 121-122. 1948.
Rendle, A.B., J. Bot. 71: 266-267. Sep 1933 (rev. of his Monograph of Narcissus, subg. Ajax in J. Roy. Hort. Soc. 58: 17-93. 1933, *16 pl.*).
Sandwith, N.Y., Proc. Linn. Soc. London 160: 187-193. 1949 (extensive obit.; "... after Druce's death, Pugsley towered like an Everest above contemporary amateur systematists").
Verdoorn, F., ed., Chron. bot. 1: 171. 1935, 2: 35, 188. 1936, 3: 21, 164. 1937, 4: 86, 533. 1938, 6: 302. 1941.
Wilmott, A.J., Watsonia 1(2): 124-130. 1948 (obit., bibl.).

NOTE: Pugsley's main work: *A prodromus of British Hieracia* appeared shortly after his death in J. Linn. Soc., Bot. 54: 1-356, *pl. 1-17*. 1948.

8380. *The genus Fumaria L. in Britain* [Journal of Botany, Jan-Jul 1912. Supplement]. Oct.
Publ.: Jan-Jul 1912, p. [1]-76, *pl. 519*. *Copy*: U. – Issued as supplement to J. Bot. 50, 1912, p. 1-8. Jan 1912, 9-16. Feb 1912, 17-32. Mar 1912, 33-40. Apr 1912, 41-48. Mai 1912, 49-64. Jun 1912, 65-76. Jul 1912.
Ref.: Anon., The Scottish botanical Review 1(4): 239-240. Oct 1912 (rev.).

8381. *Narcissus poeticus and its allies* ... Issued as a supplement [II] to "Journal of Botany", [August] 1915, ... London (West, Newman & Co., ...) 1915. Oct.
Publ.: Aug 1915, p. [i], frontisp., [1]-44, *1 pl.* (uncol.). *Copies*: G, MO. – To be cited from J. Bot. 53, suppl. 2: 1-44.

Puigdullés, Emilio Maffei (*fl.* 1895), Philippine botanist. (*Puigd.*).

HERBARIUM and TYPES: Unknown.

BIOFILE: Merrill, E.D., Contr. Philipp. bot. 4: 219. 1926.

8382. *Apuntes para el mejor conocimiento*, classificacion y valoracion de las principales *especies*

arboreo-forestales de Filipinas ... Manila (Establecimiento tipográfico de E. Bota ...) 1895. Qu. (*Apuntes conoc. esp. arb.-forest. Filip.*).
Publ.: 1895, p. [i-iii], [1]-151. *Copy*: USDA.

Puiggari, Juan Ignacio (1823-1900/1901), Spanish (Catalan) botanist; Dr. med. Barcelona 1849; in Brazil 1849; practicing at Barcelona 1849-1877; from 1877 in Brazil exploring the cryptogamic flora of São Paulo. (*Puigg.*).

HERBARIUM and TYPES: Material in B, BM, GOET, E, FH, H, H-BR (musci), IAC, K, KIEL, L, M (lich.), NY, P, PC, WU (1400). – Letters at G.

BIBLIOGRAPHY and BIOGRAPHY: Barnhart 3: 114; Colmeiro 1: cxciv; CSP 11: 77, 12: 592; Morren ed. 10, p. 152; SBC p. 129 (d. 1901, fide Geheeb); Urban-Berl. p. 311, 322.

BIOFILE: Anon., Hedwigia 40(6): (204). 1901 (d.).
Geheeb, A., Rev. bryol. 6: 66-67. 1879 (mosses São Paulo); Bot. Centralbl. 8: 161-162. 1881 (review of Noticia cript. Apiahy, Anal. Soc. ci. Argent. 11(4). 1881); Rev. bryol. 27(5). 1900, 28(4): 61. 1901 (mosses São Paulo), 28(4): 83. 1902 (died in April 1901).
Sayre, G., Mem. New York Bot. Gard. 19(3): 382. 1975 (coll.).
Urban, I., Fl. bras. 1(1): 82-83. 1906 (orig. inf.; herb. with P's son; b. 3 Mai 1823, d. 7 Aug 1900; date of death supplied by P's son!).

EPONYMY: *Puiggaria* ("*Puigarria*") Duby (1880); *Puiggariella* Spegazzini (1881); *Puiggariella* V.F. Brotherus (1908); *Puiggarina* Spegazzini (1919).

8383. *Noticia sobre algunas criptógamas nuevas* halladas en Apiahy, Provincia de San Pablo en el Brasil [Anal. Soc. Sci. Argentina 1881]. Oct.
Publ.: 1881 (p. 16: 25 Jan 1881; Bot. Centralbl. 8: 161-162. Oct 1881), p. [1]-16. *Copy*: G. – Reprinted and to be cited from Anal. Soc. Ci. Argent. 11: 201-216. 1881 (first semester).

Puihn, Johann Georg (x-1793), German botanist and physician; regional physician at Kulmbach. (*Puihn*).

HERBARIUM and TYPES: Unknown.

BIBLIOGRAPHY and BIOGRAPHY: BM 4: 1621; Dryander 3: 542; PR 7363 (ed. 1: 8249); Rehder 3: 260.

8384. *Materia venenaria regni vegetabilis* auctore Ioanne Georgio Puihn ... Lipsiae [Leipzig] (apud C.G. Hilscherum) 1785. Oct. (*Mat. venen. regn. veg.*).
Publ.: Oct-Dec 1785 (p. xii: Oct 1785), p. [i]-xii, [1]-196. *Copies*: G, NY, USDA.

Pulle, August Adriaan (1878-1955), Netherlands botanist; Dr. phil. Utrecht 1906; from 1906-1914 lecturer, from 1914-1949 professor of systematic botany at the University of Utrecht; botanical explorer of Suriname; founder of the Utrecht school of systematic botany and long-time editor of the Flora of Suriname. (*Pulle*).

HERBARIUM and TYPES: U; further material at B, BR, BO, G, L, MO, NY, RB. – Letters and portrait at G.

BIBLIOGRAPHY and BIOGRAPHY: Backer p. 474; Barnhart 3: 115 (b. 10 Jan 1878, d. 28 Feb 1955); BFM no. 2289; BJI 2: 141; BL 2: 106, 258, 314; BM 4: 1622, 8: 1036; Bossert p. 319; Hirsch p. 239; IH 1 (ed. 1): 99, (ed. 2): 124; JW 1: 447, 2: 198, 3: 365, 5: 249; Kew 4: 375; Langman p. 598; MW suppl. p. 288; SK 1: 418-419 (itin.), 4: cxix (portr.), cl, cli, 5: cccxii; TL-2/7086, see indexes; Tucker 1: 572.

BIOFILE: Anon., Bot. Centralbl. 120: 80. 1912 (to New Guinea), 123: 128. 1913 (back from New Guinea), 126: 160. 1914 (app. prof. bot. Utrecht); Hedwigia 56(1): (79). 1915 (app. prof. bot. Utrecht); Nat. Nov. 37: 78. 1915 (app. prof. bot. Utrecht); Österr. bot. Z. 64: 464. 1914 (app. prof. bot.).

Gager, C.S., Brooklyn bot. Gard. Rec. 27(3): 305. 1938 (dir. "Cantonspark" 1920 [-1949]).
Kalkman, C. & P. Smit, eds., Blumea 25(1): 16, 19, 66. 1979.
Kneucker, A., Allg. bot. Z. 20: 152. 1915 (app. prof. bot.).
Lam, H.J., Vakbl. Biol. 35: 65-67. 1955.
Lanjouw, J., Meded. Bot. Mus. Herb. Utrecht 55: 1-5. 1939 (portr., list of theses, publ., eponymy); Acta bot. neerl. 4(1): 157-159. 1955 (portr., biogr, b. 10 Jan 1878, d. 28 Feb 1955), West-Indische Gids 36(1): 13-17. 1956; Taxon 4: 54-57. 1955; Meded. Bot. Mus. Utrecht 118a. 1955 (portr.).
Pulle, A.A., Ned. Staatscourant 25 Jan 1907, no. 21, bijvoegsel no. 3, p. 1-2 (Verslag onderzoekingen van ... Pulle ... te Buitenzorg ...), also in Verslag gewone Verg. Wis.– en Nat. Afd. Kon. Ned. Akad. Wet., Amsterdam 29 Dec 1906, p. 467-468; Mensch en natuur, rede ... 26 maart 1930. Utrecht (N.V. A. Oosthoek's Uitgeversmij.) 1930, 24 p.; Afscheidscollege, 26 Jun 1948, Meded. Bot. Mus. Herb. Utrecht 97: 1-36. 1948 (repr. of oil painting; autobiographical).
Sirks, M.J., Ind. natuurond. 295 [index] 1915.
Slooten, D.F. van & C.G.G.J. van Steenis, Bull. Jard. bot. Buitenzorg ser. 3. 16(2): 103-105. 1939 (portr., dedication on 25th anniv. as prof. bot. Utrecht; on P's role in Indonesian botany).
Steenis, C.G.G.J. van, Fl. males. Bull. 4: 85. 1948 (retirement)...
Steenis-Kruseman, M.J. van, in Forbes, R.J. ed., Martinus van Marum 3: 167. 1971.
Verdoorn, F., Chron. bot. 1: 27 (portr.), 28, 32, 213, 219, 383, 413. 1935, 2: 28, 34, 35, 53, 231, 232, 237, 373. 1936, 3: 27, 204, 206. 1937, 4: 84, 92, 109, 248, 444, 561. 1938, 5: 113. 1939, 6: 87, 166, 302. 1941.

FESTSCHRIFT: Mededeelingen van het Botanisch Museum en Herbarium van de Rijksuniversiteit te Utrecht, Prof. Dr. A.A. Pulle Jubileumserie, nos. 55-70, 18 Mai 1939.

DEDICATION: Flora malesiana ser. 1 vol. 5, 1958 was dedicated to the memory of A.A. Pulle with a short tribute by C.G.G.J. van Steenis (with portr.).

COMPOSITE WORKS: (1) *The vegetation*, in L.M.R. Rutten, *Science in the Netherlands East Indies*, Amsterdam 1929, p. 164-191.
(2) Co-editor, *Nova Guinea, Botanique*, vols. 8(3-6), 12, 14, 18, 1911-1936. Pulle himself contributed numerous family treatments to parts 8(2) (publ. 14 Sep 1910), and 8(4).

HANDWRITING: SK 1: cli.

EPONYMY: *Pullea* Schlechter (1914).

NOTE: Pulle was responsible for the revival of professional plant taxonomy in the Netherlands after a period of semi-dormancy following the death of F.A.W. Miquel in 1871. Among the numerous students who took their Dr. phil. degree with him, and other taxonomists closely associated with him, were ("the Utrecht school"):

Amshoff, Gerda Jane Hillegonda (fl. 1939), see TL-2/117.
Bakhuizen van den Brink, Reinier Cornelis Jr. (1911-x), see TL-2/see 274-275.
Brouwer, Jacoba (Dr. phil. 1923).
Donk, Marinus Anton (1908-1972), see TL-2/1502-1503.
Eyma, Pierre Joseph (1903-1945), see TL-2/1735.
Henrard, Jan Theodor (1881-1974) (not a regular student), see TL-2/2647.
Jonker, Fredrik Pieter (1912-x), see TL-2/3405.
Kleinhoonte, Anthonia (1887-1960) (student of F.A.F.C. Went).
Kooper, Willem Johannes Cornelis (Dr. phil. 1927).
Koopmans, Reitze Gerben (Dr. phil. 1928).
Koster, Joséphine Thérèse (1902-x), see TL-2/3885 (promotor H.J. Lam).
Kostermans, André Joseph Guillaume Henri (1907-x), see TL-2/3886.
Lam, Herman Johannes (1892-1977), see TL-2 4133-4134.
Lanjouw, Joseph (1902-x), see TL-2/4202.
Oort, Arend Joan Petrus (1903-x), see TL-2/7085-7086 (student of F.A.F.C. Went).

Ooststroom, Simon Jan van (1906-1982), see TL-2/7087.
Scheygrond, Arie (Dr. phil. 1931).
Slooten, Dirk Fok van (1891-1953).
Stafleu, Frans Antonie (1921-x).
Steenis, Cornelis Gijsbert Gerrit Jan van (1901-x).
Swart, Jan Johannes (1901-1974).
Uittien, Hendrik (1898-1944) (student of F.A.F.C. Went).
Verdoorn, Frans (1906-x).
Vries, Dinand Marius de (Dr. phil. 1929).
Westhoff, Victor (1916-x).

8385. *An enumeration of the vascular plants known from Surinam*, together with their distribution and synonymy. Proefschrift ter verkrijging van den graad van doctor in de plant-en dierkunde, aan de Rijksuniversiteit te Utrecht, ... te verdedigen op Dinsdag 30 Januari 1906 des namiddags te 4 uren ... Leiden (Boekhandel en Drukkerij voorheen E.J. Brill) 1906. Oct. (*Enum. vasc. pl. Surinam*).
Thesis issue: 30 Jan 1906, p. [1]-8, [1]-555, map, [1-4, theses], *pl. 1-17* (uncol.; S. Linschoten). *Copies*: BR, FAS, MO, U, US, USDA.
Commercial issue: Feb 1906 (Bot. Centralbl. 6 Mar 1906; Nat. Nov. Mar(2) 1906; Bot. Zeit. 16 Mai 1906; NY rd. Jun 1906; TBC 13 Jun 1906), p. [3]-8, [1]-555, map, *pl. 1-17* (id.). *Copies*: G, MO, NY.
Ref.: De Wildeman, E., Bot. Centralbl. 104: 458-459. 1907 (rev.).
Krause, K., Bot. Jahrb. 40 (Lit.): 30. 1907 (rev.).

8386. *Neue Beiträge zur Flora Surinams* i[-iv]. [Amsterdam 1907-1925] Oct.
Publ.: Reprints of papers originally published in the Recueil des Travaux botaniques néerlandais, to be cited from the journal.

Beitr. no.	pagination reprint	journal	date
1	[1]-23	4: [119]-141	Aug 1907
2	[1]-43	6: [251]-293	1909
3	-	9(2): [125]-169, *pl. 2-3*	1912
4	-	22: [324]-417	Jan 1926

Copies: FAS, U.

8387. *Lijst van planten* (Vaatkryptogamen en phanerogamen) *die in Suriname voorkomen*, met een geschiedkundig overzicht van het onderzoek naar de flora van Suriname ... Amsterdam (Druk van J.H. de Bussy) 1907. Oct.
Publ.: Dec 1907 (in journal), p. [1]-60. *Copy*: BR. – Reprinted and to be cited from Bulletin van het Koloniaal Museum te Haarlem 38: [33]-92. 1907.

8388. *Mouriria anomala*, eine neue und morphologisch interessante Form der Melastomataceae aus Surinam ... Leide [Leiden] (Librairie et imprimerie ci-devant E.J. Brill) 1909. Oct.
Publ.: 1909, p. [i], 123-140. *Copy*: U. – Reprinted (with special t.p. and original pagination) and to be cited from Ann. Jard. bot. Buitenzorg ser. 2. suppl. 3: 123-130. l909.

8389. *Zakflora voor Suriname*. Deel I. Tabellen tot het determineren van de families en geslachten der wildgroeiende en gekweekte vaatkryptogamen en phanerogamen, die tot nu toe voor Suriname bekend zijn ... [Haarlem] Tweede uitgave van het van Eedenfonds [headline:] Bulletin van het Koloniaal Museum te Haarlem No. 47], [1911] Oct. †. (*Zakfl. Suriname*).
Publ.: Jun 1911 (date on orig. cover; Nat. Nov. Dec 1911), p. [1]-194. *Copies*: FAS, U.

8390. *Problemen der plantengeographie* rede uitgesproken bij de aanvaarding van het ambt van hoogleeraar aan de Rijks-Universiteit te Utrecht op 18 Mei 1914 ... Utrecht (A. Oosthoek) 1914. Oct. (*Probl. pl.-geogr.*).
Publ.: 18 Mai 1914, p. [1]-32. *Copy*: FAS. – Oration accepting professorship at Utrecht.

8391. *Nieuw plantkundig woordenboek voor Nederlandsch Indie* met korte aanwijzingen van het nuttig gebruik der planten en hare beteekenis in het volksleven en met registers der inlandsche en wetenschappelijke benamingen door F.S.A. de Clercq ... na het overlijden van den schrijver voor den druk bewerkt en uitgegeven door Dr. M. Greshoff ... Tweede, herziene en vermeerderde druk bewerkt door A.A. Pulle ... met medewerking voor het taalkundig gedeelte van den heer A.H.J.G. Walbeehm te Leiden. Amsterdam (drukkerij en uitgeverij J.H. de Bussy) 1927. Oct. (*Nieuw plantk. woordenb. Ned.-Indie*).
Original author: F.S.A. de Clercq. *Editor of ed. 1*: Maurits Greshoff (1862-1909).
[*Ed. 1*:]1909, p. [i]-xx, [1]-395, [1-4, ann., request for errata]. *Copy*: U. – 1000 copies issued].
Ed. 2: 1927 (p. x: 23 Apr 1917 for 1927), p. [i]-xxiv, [1]-443. *Copy*: U.

8392. *Mensch en Natuur* rede uitgesproken bij de viering van den 294sten jaardag van de Rijks-Universiteit te Utrecht op 26 Maart 1930 door den rector magnificus Dr. A.A. Pulle ...Utrecht (N.V. A. Oosthoek's Uitgeversmij) [1930]. Oct. (*Mensch en Natuur*).
Publ.: 26 Mar 1930, p. [1]-24. *Copy*: P.A. Florschütz. – Some passages in this oration were criticized by Jac. P. Thysse in the popular nature journal De Levende Natuur 35: 1-5, 49-53, 81-83. 1930. Pulle replied in an open letter, dated 3 Aug 1930, which was distributed in mimeographed form among his colleagues, 4 p. (copy: P.A. Florschütz).

8393. *Flora of Suriname* (Netherlands Guyana) ... Amsterdam ([headline:] Koninklijke Vereeniging Indisch Instituut (voorheen Koloniaal Instituut) [imprint:] Edited [i.e. published] by the Instituut ...) 1932-x. Oct. (*Fl. Suriname*).
1(1): 1932-1943, p. [i]-viii, [1]-524. – For breakdown see table I.
  (*2*): 1953-1968, p. [i]-viii, [1]-441, portr. Pulle.
1(suppl.): 1928, [i-vi], [1]-196, see Posthumus, O., (1928), with special wrapper "Flora of Surinam ... Supplement (p. 1-196). The ferns of Surinam ...".
2(1): 1932-1939, [i]-viii, [1]-500.
  (*2*): 1939-1976, [1]-709.
3(1): 1932-1941, [i]-viii, [1]-456.
  (*2*): 1942-1951, [1]-256.
4(1): 1932-1937, [i]-vii, [1]-513.
  (*2*): 1938-1940, [1]-352.
5(1): 1965, [i-iv], [1]-172.
  (*2*): 1975, [173]-318.
6(1): 1964, [i-viii], [1]-271.
Copies: B, BR, G, MO, NY, U.
Re-issue with orig. type: 1(1): 273-524. Apr 1948.
Re-issues by facsimile: 1(1): 1966 Leiden (E.J. Brill).
  1(2): 1968, Leiden (id.).
  2(1): 1966, Leiden (id.).
  2(2): 1976, Leiden (id.): facsimile of p. 1-383, orig. text of p. 384 left out, p. 385-709 constitute the new additions and corrections of 1976.
  3(1): 1966, Leiden (id.).
  3(2): 1966, Leiden (id.).
Provisional introduction: 1932, 4 p. pamphlet with map of Suriname and a list of main collections and abbreviations. *Copies*: FAS, U.
Editors: 1932-1955: A.A. Pulle, from 1956-1965 still listed (as †)with J. Lanjouw as co-editor.
  1955-1977: Joseph Lanjouw (1902-x), from 1972-1977 with A.L. Stoffers as co-editor.
    1977-x: Antonius Lambertus Stoffers (1926-x), from 1979 with Jan C. Lindeman (1921-x) as co-editor.
Original title: *Flora of Surinam* (Dutch Guyana), on wrappers 1932-1936; from then on *Flora of Suriname* (Netherlands Guyana), until 1947. Now: *Flora of Suriname*. The early volumes have an additional heading on the title page:
  (1) Vereeniging Kolonial Institute te Amsterdam. Mededeeling no. xxx. Afd. Handelsmuseum no. 11. (until 1946).
  (2) Koninklijke Vereeniging Indisch Instituut (voorheen Koloniaal Instituut te Amsterdam) ... 1946-1967.

volume	part	pagination	date	author(s)	content
1	1	1-3	Mai 1932	Markgraf, F.	Gnetaceae
		4-24	Mai 1932	Krause, K.	Loranthaceae
		25-44	Mai 1932	Scheygrond, A.	Amarantaceae
		45-46	Mai 1932	Lanjouw, J.	Balanophoraceae
		47-48	Mai 1932	Ooststroom, S. J. van	Ulmaceae
		49-71	Mar 1934	Eyma, P. J.	Polygonaceae
		72-149	Mar 1934	Uittien, H.	Cyperaceae
		150-153	Mar 1934	Ooststroom, S. J. van	Caryophyllaceae
		154-157	Mar 1934	Mennega, A. M. W.	Proteaceae
		158-161	Mar 1934	Eyma, P. J.	Aizoaceae
		162-170	Jan 1938	Alston, A. H. G.	Selaginellaceae
		171-175	Jan 1938	Alston, A. H. G.	Lycopodiaceae
		176-187	Jan 1938	Jonker, F. P.	Burmanniaceae
		188-190	Jan 1938	Uittien, H.	Thurniaceae
		191-196	Jan 1938	Uittien, H.	Rapateaceae
		197-212	Jan 1938	Uittien, H.	Commelinaceae
		213-224	Jan 1938	Uittien, H. & A. N. J. Heyn	Eriocaulaceae
		225-248	Jan 1938	Lanjouw, J.	Xyridaceae
		249-251	Jan 1938	Lanjouw, J.	Mayacaceae
		252-254	Jan 1938	Lanjouw, J.	Typhaceae
		255-257	Jan 1938	Lanjouw, J.	Haemadoraceae
		258-261	Jan 1938	Vaandrager, G.	Lacistemaceae
		262-272	Jan 1938	Amshoff, G. J. H.	Olacaceae
		273-442	Jan 1943	Amshoff, G. J. H. & Th. Henrard	Gramineae
		443-454	Jan 1943	Boterenbrood, M. J. A.	Amaryllidaceae
		455-460	Jan 1943	Boterenbrood, M. J. A.	Iridaceae
		461-466	Jan 1943	Jonker, F. P.	Triuridaceae
		467-471	Jan 1943	Jonker, F. P.	Hydrocharitaceae
		472-482	Jan 1943	Jonker, F. P.	Alismataceae
		483-485	Jan 1943	Jonker, F. P.	Butomaceae
		486-491	Jan 1943	Ooststroom, S. J. van	Portulacaceae
		492-505	Jan 1943	Uittien, H.	Cyperaceae-Add./Corr.

volume	part	pagination	date	author(s)	content
		506-524	Jan 1943	Jonker, F. P.	Index Vol. 1(1)
1	Suppl.	1-4	1928	Posthumus, O.	Filicales-Gen.
	Suppl.	4-26	1928	Posthumus, O.	Hymenophyllaceae
	Suppl.	26-36	1928	Posthumus, O.	Cyatheaceae
	Suppl.	36-155	1928	Posthumus, O.	Polypodiaceae
	Suppl.	155-156	1928	Posthumus, O.	Parkeriaceae
	Suppl.	156-158	1928	Posthumus, O.	Gleicheniaceae
	Suppl.	158-165	1928	Posthumus, O.	Schizaeaceae
	Suppl.	165	1928	Posthumus, O.	Osmundaceae
	Suppl.	165-167	1928	Posthumus, O.	Salviniaceae
	Suppl.	167	1928	Posthumus, O.	Marsiliaceae
	Suppl.	167-170	1928	Posthumus, O.	Marattiaceae
	Suppl.	170-172	1928	Posthumus, O.	Ophioglossaceae
	Suppl.	173-196	1928	Posthumus, O.	Gen./Index
	2	1-80	1953	Jonker-Verhoef, A. M. E. & F. P. Jonker	Araceae
	(1968 reprint)				
		81-90	1953	Jonker-Verhoef, A. M. E.	Pontederiaceae
		91-92	1953	Jonker, F. P.	Batidaceae
		93	1957	Jonker, F. P.	Batidaceae
		94-148	1957	Smith, L. B.	Bromeliaceae
		149-208	1957	Jonker-Verhoef, A. M. E. & F. P. Jonker	Marantaceae
		209-217	1957	Stoffers, A. L.	Phytolaccaceae
		218-290	1957	Yuncker, T. G.	Piperaceae
		291-292	1957	Lanjouw, J.	Chenopodiaceae
		295-300	1968	Lindeman, J. C. & A. R. A. Görts-van Rijn	Loranthaceae-Corr./Add.
		301-302	1968	Mennega, A. M. W. & A. R. A. Görts-van Rijn	Ulmaceae-Corr./Add.
		303-309	1968	Lindeman, J. C. & A. R. A. Görts-van Rijn	Polygonaceae-Corr./Add.
		310-314	1968	Lindeman, J. C. & A. R. A. Görts-van Rijn	Cyperaceae-Corr./Add.
		315-320	1968	Mennega, A. M. W.	Proteaceae-Corr./Add.
		321-325	1968	Jonker, F. P.	Burmanniaceae

Pages	Date	Author	Family
330-339	1968	Lindeman, J. C. & A. R. A. Görts-van Rijn	Eriocaulaceae-Corr./Add.
340-342	1968	Mennega, A. M. W., J. C. Lindeman & A. R. A. Görts-van Rijn	Xyridaceae-Corr./Add.
343-375	1968	Lindeman, J. C. & A. R. A. Görts-van Rijn	Gramineae-Corr./Add.
376	1968	Maas, P. J. M.	Amaryllidaceae-Corr.
377	1968	Lindeman, J. C.	Iridaceae-Corr./Add.
378-379	1968	Jonker, F. P.	Alismataceae-Corr./Add.
380-412	1968	Jonker-Verhoef, A. M. E. & F. P. Jonker	Araceae-Corr./Add.
413-414	1968	Kramer, K. U. & A. R. A. Görts-van Rijn	Bromeliaceae-Corr./Add.
415-421	1968	Kramer, K. U. & A. R. A. Görts-van Rijn	Piperaceae-Corr./Add.
422-441	1968	Görts-van Rijn, A. R. A.	Index
1-101	Apr 1932	Lanjouw, J.	Euphorbiaceae
102-106	Apr 1932	Lanjouw, J.	Rhamnaceae
107-112	Apr 1932	Petter, H. F. M.	Monimiaceae
113-122	Mar 1934	Ooststroom, S. J. van	Myristicaceae
123-131	Mar 1934	Diels, L.	Menispermaceae
132-144	Mar 1934	Nannenga, E. T.	Anacardiaceae
145	Jan 1936	Nannenga, E. T.	Anacardiaceae
146-243	Jan 1936	Kostermans, A. J. G. H.	Malpighiaceae
244-336	Jan 1936	Kostermans, A. J. G. H.	Lauraceae
337	Sep 1937	Kostermans, A. J. G. H.	Lauraceae
338-344	Sep 1937	Kostermans, A. J. G. H.	Hernandiaceae
345-396	Sep 1937	Uittien, H.	Sapindaceae
397-400	Sep 1937	Went, J. C.	Capparaceae
401-405	Nov 1939	Went, J. C.	Capparaceae
406-425	Nov 1939	Oort, A. J. P.	Polygalaceae
426-456	Nov 1939	Kleinhoonte, A.	Rosaceae
457-470	Nov 1939	Lanjouw, J.	Euphorbiaceae-Corr./Add.

volume	part	pagination	date	author(s)	content
		471-472	Nov 1939	Lanjouw, J.	Monimiaceae-Corr./Add.
		473-475	Nov 1939	Ooststroom, S. J. van	Myristicaceae-Corr./Add.
		476-477	Nov 1939	Jonker, F. P.	Menispermaceae-Corr./Add.
		478-480	Nov 1939	Jonker, F. P.	Malpighiaceae-Corr./Add.
		481-487	Nov 1939	Kostermans, A. J. G. H.	Lauraceae-Corr./Add.
		488-500	Nov 1939	Jonker, F. P.	Index
	2	1-257	Nov 1939	Amshoff, G. J. H.	Papilionaceae
		258-331	Dec 1940	Kleinhoonte, A.	Mimosaceae
		332-340	Dec 1940	Lanjouw, J.	Connaraceae
		341-383	Dec 1940	Fries, R. E.	Annonaceae
		384	Dec 1940	Pulle, A. A.	Droseraceae
		385-709	1976	various authors	additions, corr.
3	1	1-25	Mai 1932	Uittien, H.	Malvaceae
		26-33	Mai 1932	Uittien, H.	Bombacaceae
		34-48	Mai 1932	Uittien, H.	Sterculiaceae
		49-57	Mai 1932	Uittien, H.	Tiliaceae
		58-64	Mai 1932	Uittien, H.	Elaeocarpaceae
		65-118	Mar 1934	Eyma, P. J.	Guttiferae
		119-155	Mar 1934	Eyma, P. J.	Lecythidaceae
		156-157	Mar 1934	Eyma, P. J.	Punicaceae
		158-159	Mar 1934	Eyma, P. J.	Bixaceae
		160	Mar 1934	Muller, F. M.	Araliaceae
		161-163	Jun 1935	Muller, F. M.	Araliaceae
		164-177	Jun 1935	Exell, A. W.	Combretaceae
		178-281	Jun 1935	Gleason, H. A.	Melastomataceae
		282-303	Jun 1935	Sleumer, H. & Uittien, H.	Flacourtiaceae
		304	Jun 1935	Uittien, H.	Canellaceae
		305	Sep 1937	Uittien, H.	Canellaceae
		306-327	Sep 1937	Killip, E. P.	Passifloraceae
		328-336	Sep 1937	Wehlburg, C.	Ochnaceae
		337-341	Feb 1941*	Wehlburg, C.	Ochnaceae
		342-354	Feb 1941	Bremekamp, E.	Turneraceae
		355-365	Feb 1941	Lanjouw, J. &	Quiinaceae

				Author	Family
				Lanjouw, J. & P. F. van Heerdt	Maregraviaceae
		386-408	Feb 1941	Lanjouw, J. & P. F. van Heerdt	Dilleniaceae
		409-411	Feb 1941	Bakhuizen van den Brink, fil., R. C.	Linaceae
		412-421	Feb 1941	Bakhuizen van den Brink, fil., R. C.	Humiriaceae
		422-432	Feb 1941	Jonker, F. P.	Lythraceae
		433-435	Feb 1941	Uittien, H.	Malvaceae-Corr./Add.
		436	Feb 1941	Uittien, H.	Bombacaceae
		437	Feb 1941	Uittien, H.	Sterculiaceae
		438-441	Feb 1941	Uittien, H.	Tiliaceae
		442-443	Feb 1941	Lanjouw, J.	Melastomataceae
		444-456	Feb 1941	Jonker, F. P.	Index
3	2	1-12	Jan 1942	Westhoff, V.	Erythroxylaceae
		13-34	Jan 1942	Jonker, F. P.	Oenotheraceae
		35-43	Jan 1942	Jonker, F. P.	Rhizophoraceae
		44-48	Jan 1942	Jonker, F. P.	Oxalidaceae
		49-55	1951	Jonker, F. P.	Oxalidaceae
		56-158	1951	Amshoff, G. J. H.	Myrtaceae
		159-165	1951	Amshoff, G. J. H.	Aquifoliaceae
		166-172	1951	Stafleu, F. A.	Dichapetalaceae
		173-177	1951	Stafleu, F. A.	Trigoniaceae
		178-199	1951	Stafleu, F. A.	Vochysiaceae
		200-203	1951	Stafleu, F. A.	Zygophyllaceae
		204-251	1951	Swart, J. J.	Burseraceae
		252-256	1951	Stafleu, F. A.	Umbelliferae
4	1	1-65	Mai 1932	Markgraf, F.	Apocynaceae
		66-102	Mai 1932	Ooststroom, S. J. van	Convolvulaceae
		103-110	Mai 1932	Raalte, M. H. van	Loganiaceae

* Inside of front cover: "Printed February 1941". List of publication dates gives January 1941.

volume	part	pagination	date	author(s)	content
		111-112	Mai 1932	Pulle, A.	Pedaliaceae
		113-298	Mai 1932	Bremekamp, C. E. B.	Rubiaceae
		299-301	Mai 1932	Lanjouw, J.	Ericaceae
		302-304	Mai 1932	Lanjouw, J.	Campanulaceae
		305	Jul 1936	Lanjouw, J.	Campanulaceae
		306-333	Jul 1936	Johnston, I. M.	Boraginaceae
		334-353	Jul 1936	Kostermans, A. J. G. H.	Labiatae
		354-399	Jul 1936	Eyma, P. J.	Sapotaceae
		400-427	Jul 1936	Jonker, F. P.	Gentianaceae
		428-430	Jul 1936	Jonker, F. P.	Menyanthaceae
		431-432	Jul 1936	Bottelier, H. P.	Myrsinaceae
		433-442	Sep 1937*	Bottelier, H. P.	Myrsinaceae
		443-467	Sep 1937	Markgraf, F.	Apocynaceae-Corr./Add.
		468-471	Sep 1937	Ooststroom, S. J. van	Convolvulaceae-Corr./Add.
		472-474	Sep 1937	Uittien, H.	Loganiaceae-Corr./Add.
		475-491	Sep 1937	Bremekamp, C. E. B.	Rubiaceae-Corr./Add.
		492-493	Sep 1937	Lanjouw, J.	Ericaceae-Corr./Add.
		494-495	Sep 1937	Lanjouw, J.	Campanulaceae-Corr./Add.
		496-497	Sep 1937	Uittien, H.	Boraginaceae-Corr./Add.
		498	Sep 1937	Kostermans, A.	Labiatae-Corr./Add.
		499-513	Sep 1937	Uittien, H.	Index
		1-86	Jul 1938	Sandwith, N. Y.	Bignoniaceae
	2	87-165	Jul 1938	Koster, J. T.	Compositae
		166-252	Jul 1938	Bremekamp, C. E. B.	Acanthaceae
		253-254	Jul 1938	Lanjouw, J.	Plantaginaceae
		255-256	Jul 1938	Lanjouw, J.	Myoporaceae
		257-321	Jul 1940	Moldenke, H. N.	Verbenaceae
		322-325	Jul 1940	Moldenke, H. N.	Avicenniaceae
		326-352	Jul 1940	Jonker, F. P.	Asclepiadaceae
5	1	1-172	15 Oct 1965	Wessels Boer, J. G.	Palmae
		173-299	1975	Berg, C. C. & P. DeWolf	Moraceae
		300-318	1975	DeRooij, M. J. M.	Urticaceae
		319-330	1979	Jansen-Jacobs, M. J.	Simaroubaceae

pagination	date	author(s)	family	volume	part	pagination	date
356-366	1979	Jansen-Jacobs, M. J. & J. C. Lindeman	Theaceae				
367-369	1979	Lindeman, J. C.	Theophrastaceae				
370-384	1979	Cramer, J.	Nymphaceae				
385-389	1979	Cramer, J.	Cabombaceae				
390-392	1979	Cramer, J.	Nelumbonaceae				
393-411	1979	Maas, P. J. M. & M. J. M. de Rooij	Musaceae				
412-415	1979	Maas, P. J. M.	Cannaceae				
416-441	1979	Maas, P. J. M.	Zingiberaceae				
442-456	1979	Sipman, H.	Liliaceae				
1-271	15 Jun 1964	Florschütz, P. A.	Musci	6	1		

B: *Contents Listed by family* (for Index see 2(2): 689-709, 1976):

content	author(s)	volume	part	pagination	date
Acanthaceae	Bremekamp, C. E. B.	4	2	166-252	Jul 1938
Aizoaceae	Eyma, P. J.	1	1	158-161	Mar 1934
Alismataceae	Jonker, F. P.			472-482	Jan 1943
Alismataceae-Corr./Add.	Jonker, F. P.		2	378-379	1968
Amaranthaceae	Scheygrond, A.		1	25-44	Mai 1932
Amaryllidaceae	Boterenbrood, M. J. A.			443-454	Jan 1943
Amaryllidaceae-Corr./Add.	Maas, P. J. M.			376	1968
Anacardiaceae	Nannenga, E. T.	2	2	132-144	Mar 1934
Anacardiaceae	Nannenga, E. T.		1	145	Jan 1936
Anacardiaceae-Corr./Add.	Jansen-Jacobs, M. J.	2	2	441-443	1976
Annonaceae	Fries, R. E.		2	341-383	Dec 1940
Annonaceae-Corr./Add.	Jansen-Jacobs, M. J.	2	2	658-687	1976
Apocynaceae	Markgraf, F.	4	1	1-65	Mai 1932
Apocynaceae-Corr./Add.	Markgraf, F.			443-467	Sep 1937

* Inside front cover: "Printed September 1937". List of publication dates gives August 1937.

content	author(s)	volume	part	pagination	date
Aquifoliaceae	Amshoff, G. J. H.	3	2	159-165	1951
Araceae	Jonker-Verhoef, A. M. E. & F. P. Jonker	1	2	1-80	1953
Araceae-Corr./Add.	Jonker-Verhoef, A. M. E. & F. P. Jonker			380-412	1968
Araliaceae	Muller, F. M.	3	1	160	Mar 1934
Araliaceae	Muller, F. M.			161-163	Jun 1935
Asclepiadaceae	Jonker, F. P.	4	2	326-352	Jun 1940
Avicenniaceae	Moldenke, H. N.			322-325	Jul 1940
Balanophoraceae	Lanjouw, J.	1	1	45-46	Mai 1932
Batidaceae	Jonker, F. P.	1	2	91-93	1953
Bignoniaceae	Sandwith, N. Y.	4	2	1-86	Jul 1938
Bixaceae	Eyma, P. J.	3	1	158-159	Mar 1934
Bombacaceae	Uittien, H.			26-33	Mai 1932
Bombacaceae-Corr./Add.	Uittien, H.			436	Feb 1941
Boraginaceae	Johnston, I. M.	4	1	306-333	Jul 1936
Boraginaceae-Corr./Add.	Uittien, H.			496-497	Sep 1937
Bromeliaceae	Smith, L. B.	1	2	93-148	1957
Bromeliaceae-Corr./Add.	Kramer, K. U. & A. R. A. Görts-van Rijn			413-414	1968
Burmanniaceae	Jonker, F. P.		1	176-187	Jan 1938
Burmanniaceae	Jonker, F. P.		2	321-325	1968
Burseraceae	Swart, J. J.	3	2	204-251	1951
Butomaceae	Jonker, F. P.	1	1	483-485	Jan 1943
Cabombaceae	Cramer, J.	5	1	385-389	1979
Campanulaceae	Lanjouw, J.	4		302-304	Mai 1932
Campanulaceae	Lanjouw, J.			305	Jul 1936
Campanulaceae-Corr./Add.	Lanjouw, J.			494-495	Sep 1937
Canellaceae	Uittien, H.	3		304	Jun 1935
Canellaceae	Uittien, H.			305	Sep 1937
Cannaceae	Maas, P. J. M.	5	1	412-415	1979
Capparaceae	Went, J. C.	2	1	397-400	Sep 1937
Capparaceae	Went, J. C.			401-405	Nov 1939
Capparaceae-Corr./Add.	Jansen-Jacobs, M. J.	2	2	512-517	1976

Family	Author(s)	Vol.	Pages	Date
Chrysobalanaceae-Corr./Add.	Prance, C. T. & A. R. A. Görts-van Rijn	2	291-292	1957
		2	524-555	1976
Combretaceae	Exell, A. W.	3	164-177	Jun 1935
Commelinaceae	Uittien, H.	1	197-212	Jan 1938
Commelinaceae-Corr./Add.	Lindeman, J. C. & A. R. A. Görts-van Rijn	2	329	1968
Compositae	Koster, J. T.	4	87-165	Jul 1938
Connaraceae	Lanjouw, J.	2	332-340	Dec 1940
Connaraceae-Corr./Add.	Forero, F. & A. R. A. Görts-van Rijn	2	654-657	1976
Convolvulaceae	Ooststroom, S. J. van	4	66-102	Mai 1932
Convolvulaceae-Corr./Add.	Ooststroom, S. J. van		468-471	Sep 1937
Cyatheaceae	Posthumus, O.	Suppl.	26-36	1928
Cyperaceae	Uittien, H.	1	72-149	Mar 1934
Cyperaceae-Corr./Add.	Uittien, H.		492-505	Jan 1943
Cyperaceae-Corr./Add.	Lindeman, J. C. & A. R. A. Görts-van Rijn	2	310-314	1968
Dichapetalaceae	Stafleu, F. A.	3	166-172	1951
Dilleniaceae	Lanjouw, J. & P. F. van Heerdt	1	386-408	Feb 1941
Droseraceae	Pulle, A. A.	2	384	Dec 1940
Elaeocarpaceae	Uittien, H.	3	58-64	Mai 1932
Ericaceae	Lanjouw, J.	4	299-301	Mai 1932
Ericaceae-Corr./Add.	Lanjouw, J.		492-493	Sep 1937
Eriocaulaceae	Uittien, H. & A. N. J. Heyn	1	213-224	Jan 1938
Eriocaulaceae-Corr./Add.	Lindeman, J. C. & A. R. A. Görts-van Rijn	2	330-339	1968
Erythroxylaceae	Westhoff, V.	3	1-12	Jan 1942
Euphorbiaceae	Lanjouw, J.	2	101	Apr 1932
Euphorbiaceae-Corr./Add.	Lanjouw, J.	1	457-470	Nov 1939
Euphorbiaceae-Corr./Add.	Görts-van Rijn, A. R. A.	2	387-424	1976
Filicales-Gen.	Posthumus, O.	Suppl.	1-4	1928
Flacourtiaceae	Sleumer, H. & H. Uittien	3	282-303	Jun 1935

content	author(s)	volume	part	pagination	date
Gentianaceae	Jonker, F. P.	4	1	400-427	Jul 1936
Gleicheniaceae	Posthumus, O.	1	Suppl.	156-158	1928
Gnetaceae	Markgraf, F.		1	1-3	Mai 1932
Gramineae	Amshoff, G. J. H. & J. Th. Henrard			273-442	Jan 1943
Gramineae-Corr./Add.	Lindeman, J. C. & A. R. A. Görts-van Rijn		2	343-375	1968
Guttiferae	Eyma, P. J.	3	1	65-118	Mar 1934
Haemadoraceae	Lanjouw, J.	1	1	255-257	Jan 1938
Hernandiaceae	Kostermans, A. J. G. H.	2	1	338-344	Sep 1937
Hernandiaceae-Corr./Add.	Kubitzki, K. U.	2	2	485-486	1976
Humiriaceae	Bakhuizen van den Brink, fil., R. C.	3		412-421	Feb 1941
Hydrocharitaceae.	Jonker, F. P.	1	1	467-471	Jan 1943
Hymenophyllaceae	Posthumus, O.		Suppl.	4-26	1928
Icacinaceae	Jansen-Jacobs, M. J.	5	1	344-355	1929
Iridaceae	Boterenbrood, M. J. A.		1	455-460	Jan 1943
Iridaceae-Corr./Add.	Lindeman, J. C.		2	377	1968
Labiatae	Kostermans, A. J. G. H.	4	1	334-353	Jul 1936
Labiatae-Corr./Add.	Kostermans, A. J. G. H.	4	1	498	Sep 1937
Lacistemaceae	Vaandrager, G.	1		258-261	Jan 1938
Lauraceae	Kostermans, A. J. G. H.	2		244-336	Jan 1936
Lauraceae	Kostermans, A. J. G. H.			337	Sep 1937
Lauraceae-Corr./Add.	Kostermans, A. J. G. H.			481-487	Nov 1939
Lauraceae-Corr./Add.	Jansen-Jacobs, M. J.	2	2	451-484	1976
Lecythidaceae	Eyma, P. J.	3	1	119-155	Mar 1934
Liliaceae	Sipman, H.	5	1	442-456	1979
Linaceae	Bakhuizen van den Brink, fil., R. C.		1	409-411	Feb 1941
Loganiaceae	Raalte, M. H. van	4		103-110	Mai 1932
Loganiaceae-Corr./Add.	Uittien, H.			472-474	Sep 1937
Loranthaceae	Krause, K.	1		4-24	Mai 1932
Loranthaceae-Corr./Add.	Lindeman, J. C. & A. R. A. Görts-van Rijn		2	295-300	1968
Lycopodiaceae	Alston, A. H. G.		1	171-175	Jan 1938
Lythraceae	Jonker, F. P.	3		422-432	Feb 1941

Family	Author	Vol.	Pages	Date
Malvaceae-Corr./Add.	Uittien, H.	3	1-25	Mai 1932
Malvaceae-Corr./Add.	Uittien, H.	1	433-435	Feb 1941
Marantaceae	Jonker-Verhoef, A. M. E. & F. P. Jonker	2	149-208	1957
Marattiaceae	Posthumus, O.	Suppl.	167-170	1928
Marcgraviaceae	Lanjouw, J. & P. F. van Heerdt	1	373-385	Feb 1941
Marsiliaceae	Posthumus, O.	Suppl.	167	1928
Mayacaceae	Lanjouw, J.	1	249-251	Jan 1938
Melastomataceae	Gleason, H. A.	3	178-281	Jun 1935
Melastomataceae	Lanjouw, J.		442-443	Feb 1941
Menispermaceae	Diels, L.	2	123-131	Mar 1934
Menispermaceae-Corr./Add.	Jonker, F. P.		476-477	Nov 1939
Menispermaceae-Corr./Add.	Jansen-Jacobs, M. J.	2	430-440	1976
Menyanthaceae	Jonker, F. P.	4	428-430	Jul 1936
Mimosaceae	Kleinhoonte, A.	2	258-331	Dec 1940
Mimosaceae-Corr./Add.	Jansen-Jacobs, M. J.	2	611-653	1976
Monimiaceae	Petter, H. F. M.	1	107-112	Apr 1932
Monimiaceae-Corr./Add.	Lanjouw, J.		471-472	Nov 1939
Monimiaceae-Corr./Add.	Jansen-Jacobs, M. J.	2	425-526	1976
Moraceae	Berg, C. C. & G. P. Dewolf	5	173-299	1975
Musaceae	Maas, P. J. M. & M. J. De Rooij	1	393-411	1979
Musci	Florschütz, P. A.	6	1-271	15 Jun 1964
Myoporaceae	Lanjouw, J.	4	255-256	Jul 1938
Myristicaceae	Ooststroom, S. J. van	2	113-122	Mar 1934
Myristicaceae-Corr./Add.	Ooststroom, S. J. van		473-475	Nov 1939
Myristicaceae-Corr./Add.	Jansen-Jacobs, M. J.	2	427-429	1976
Myrsinaceae	Bottelier, H. P.	4	431-432	Jul 1936
Myrsinaceae	Bottelier, H. P.		433-442	Sep 1937
Myrtaceae	Amshoff, G. J. H.	3	56-158	1951
Nelumbonaceae	Cramer, J.	5	390-392	1979
Nymphaeaceae	Cramer, J.	5	370-384	1979
Ochnaceae	Wehlburg, C.		328-336	Sep 1937

content	author(s)	volume	part	pagination	date
Ochnaceae	Wehlburg, C.			337-341	Feb 1941
Oenotheraceae	Jonker, F. P.	3	2	13-34	Jan 1942
Olacaceae	Amshoff, G. J. H.	1		262-272	Jan 1938
Ophioglossaceae	Posthumus, O.		Suppl.	170-172	1928
Osmundaceae	Posthumus, O.			165	1928
Oxalidaceae	Jonker, F. P.	3	2	44-48	Jan 1942
Oxalidaceae	Jonker, F. P.			49-55	1951
Palmae	Wessels de Boer, J. G.			1-172	15 Oct 1965
Papaveraceae	Stoffers, A. L.	5	1	331-334	1979
Papilionaceae	Amshoff, G. J. H.	5	1	1-257	Jul 1939
Papilionaceae-Corr./Add.	Görts-van Rijn, A. R. A., K. U. Kramer & J. C. Lindeman	2	2	556-610	1976
Parkeriaceae	Posthumus, O.	1	Suppl.	155-156	1928
Passifloraceae	Killip, E. P.	3	1	306-327	Sep 1937
Pedaliaceae	Pulle, A.	4	1	111-112	Mai 1932
Phytolaccaceae	Stoffers, A. L.	1		209-217	1957
Piperaceae	Yuncker, T. G.			218-290	1957
Piperaceae-Corr./Add.	Kramer, K. U. & A. R. A. Görts-van Rijn	1	2	415-421	1968
Plantaginaceae	Lanjouw, J.	4	2	253-254	Jul 1938
Polygalaceae	Oort, A. J. P.	2	1	406-425	Nov 1939
Polygalaceae	Görts-van Rijn, A. R. A.	2	2	518-523	1976
Polygonaceae	Eyma, P. J.	1	1	49-71	Mar 1934
Polygonaceae-Corr./Add.	Lindeman, J. C. & A. R. A. Görts-van Rijn		2	303-309	1968
Polypodiaceae	Posthumus, O.		Suppl.	336-155	1928
Pontederiaceae	Jonker-Verhoef, A. M. E.		2	81-90	1953
Portulacaceae	Ooststroom, S. J. van	1		486-491	Jan 1943
Proteaceae	Mennega, A. M. W.			154-	
Rhizophoraceae	Jonker, F. P.	3	2	35-43	Jan 1942
Rosaceae	Kleinhoonte, A.	2	1	426-456	Nov 1939
Rubiaceae	Bremekamp, C. E. B.	4		113-298	Mai 1932
Rubiaceae-Corr./Add.	Bremekamp, C. E. B.			475-491	Sep 1937

Family	Author	Vol.	Pages	Date
Schizaeaceae	..., ...J.	4	354-399	Jul 1936
	Posthumus, O.	1	158-165	1928
Selaginellaceae	Alston, A. H. G.	Suppl.	162-170	Jan 1938
Simaroubaceae	Jansen-Jacobs, M. J.	1	319-330	1979
Sterculiaceae	Uittien, H.	5	34-48	Mai 1932
Sterculiaceae	Uittien, H.	3	437	Feb 1941
Theaceae	Jansen-Jacobs, M. J. & J. C. Lindeman	5	356-366	1979
Theophrastaceae	Lindeman, J. C.	1	367-369	1979
Thurniaceae	Uittien, H.	5	188-190	Jan 1938
Tiliaceae	Uittien, H.	1	49-57	Mai 1932
Tiliaceae	Uittien, H.	3	438-441	Feb 1941
Trigoniaceae	Stafleu, F. A.	2	173-177	1951
Triuridaceae	Jonker, F. P.	1	461-466	Jan 1943
Turneraceae	Bremekamp, E.	3	342-354	Feb 1941
Typhaceae	Lanjouw, J.	1	252-254	Jan 1938
Ulmaceae	Ooststroom, S. J. van	1	47-48	Jan 1938
Ulmaceae-Corr./Add.	Mennega, A. M. W. & A. R. A. Görts-van Rijn	2	301-302	Mai 1932
				1968
Umbelliferae	Stafleu, F. A.	3	252-256	1951
Urticaceae	DeRooij, M. J. M.	5	300-318	1975
Verbenaceae	Moldenke, H. N.	4	257-321	Jul 1940
Vitaceae	Görts-van Rijn, A. R. A.	5	335-343	1979
Vochysiaceae	Stafleu, F. A.	3	178-199	1951
Xyridaceae	Lanjouw, J.	1	225-248	Jan 1938
Xyridaceae-Corr./Add.	Mennega, A. M. W., J. C. Lindeman & A. R. A. Görts-van Rijn	2	340-342	1968
Zingiberaceae	Maas, P. J.	5	416-441	1979
Zygophyllaceae	Stafleu, F. A.	3	200-203	1951

(3) Foundation van Eedenfonds (c/o. Royal Tropical Institute). Amsterdam (1968-x) with imprint Leiden. E.J. Brill.
The imprints show a great variety: until 1950 the Flora was published by the Vereeniging Koloniaal Instituut ("Edited by the Institute"). From 1951-1963 by the "Foundation van Eedenfonds", from 1964 by E.J. Brill (Leiden).

8394. *Remarks on the system of the spermatophytes* by A.A. Pulle ... Utrecht (N.V. A. Oosthoek's Uitg.-Mij) 1937. Oct. (*Rem. syst. spermat.*).
Publ.: Dec 1937 (orig. cover dated 1937; preface dated Dec 1937), p. [1]-17. – The text is a translation of p. 134-139 of the *Compendium* (1938) followed by the review of the system (p. 141-152) as printed in the Compendium. *Copy*: FAS (in orig. cover dated 1937). New family names published in this pamphlet are nomina nuda.
Other issue: Mar 1938 as Meded. bot. Mus. Herb. Utrecht 43: 1-17, chart. 1938.
Ref.: Krause, K., Bot. Jahrb. 69 (Lit.): 33. 23 Dec 1938.

8395. *Compendium van de terminologie, nomenclatuur en systematiek der zaadplanten* ... Utrecht (by N.V. A. Oosthoek's Uitgevers-Maatschappij) 1938. Oct. (*Comp. termin. nom. syst. zaadpl.*).
Ed. 1: Jan-Mar 1938 (J. Bot. Aug 1938), p. [i-iii], [1]-338, chart. *Copies*: FAS(2), U. – Part of the chapter on the system of the Spermatophytes (p. 134-139) was prepublished in English as a separate pamphlet in Dec 1937.
Ed. 2: 1950, p. [i-viii], [1]-370, 1 chart. *Copy*: FAS.
Ed. 3: 1952 (p. [ix]: Sep 1952), p. [i]-viii, [ix], [1]-376, chart. *Copies*: FAS, U.
Continuation: J. Lanjouw et al., *Compendium van de Pteridophyta en Spermatophyta* ... Utrecht (A. Oosthoek ...) 1968, p. [1]-342. *Copies*: FAS, U.
Continuation (2): A.L. Stoffers et al., *Compendium van de Spermatophyta* ... Utrecht/Antwerpen (Bohn, Scheltema & Holkema) 1982, p. [i-xii], 1-310. ISBN 90-3130409-3. *Copies*: FAS, U.
Ref.: Anon., J. Bot. 76: 246. Aug 1938.
Diels, L., Bot. Jahrb. 69 (Lit.): 33. 23 Dec 1938.

8396. *De inventarisatie van het erfdeel der vaderen* rede gehouden bij de opening van het vergrote en verbouwde laboratorium voor bijzondere plantkunde en plantengeographie van de Rijksuniversiteit te Utrecht op 31 Mei l938 ... [Utrecht 1938].
Private issue: with special cover, p. [1]-38, chart. *Copy*: FAS.
Regular issue: as Meded. Bot. Mus. Herb. Utrecht 50: 1-38, chart. 1938.

Pulteney, Richard (1730-1801), British physician, naturalist and historian of science; MD Edinburgh 1764, FRS 1762; surgeon at Leicester and (from 1865) at Blandford. (*Pult.*).

HERBARIUM and TYPES: BM (unlocalised); some lichens at CGE and LINN. – BM possesses the autograph copy of the Catalogue of Plants ... Loughborough (Banksian ms. 90), a mss (no. 26) key to Rheede's *Flora Malabarica*, the autograph copy of the *Catalogue of English plants* and an unpublished *Flora anglica abbreviata* (see BM 4: 1622). Letters at BM, G, LINN and Fitzwilliam Mus. Cambridge.

BIBLIOGRAPHY and BIOGRAPHY: AG 6(2): 202, 208; Barnhart 3: 115; BB p. 250; BM 4: 1622, 8: 1037; Bossert p. 319; CSP 5: 42; Desmond p. 507 (b. 17 Feb 1730, d. 13 Oct 1801); DNB 47: 26-28; Dryander 3: 6, 137, 180, 250, 357, 416, 476, 548, 5: 83; Henrey 2: 736 [index], 3: 104 (nos. 1245-1248); Herder p. 70, 73; IH 2: (in press); Jackson p. 4, 251, 255; Kew 4: 375; Langman p. 598; Lasègue p. 323, 517; LS 21472-21472a; NI 1: 61, 110, 257; Plesch p. 369; PR 7364-7367 (ed. 1: 8252-8254); Rehder 5: 694; SO see index p. 44; Sotheby 617; TL-1/104; TL-2/3058, see W.G. Maton; Tucker 1: 13, 572-573.

BIOFILE: Allen, D.E., The naturalist in Britain 22, 39, 42, 49. 1976.
Allibone, S.A., Crit. dict. Engl. lit. 1705. 1870 (bibl.).
Anon., J. Bot. 9: 31. 1871 (on mss. at LINN); Proc. Linn. Soc. London 1888/1890: 39. 1891 (on portr. by S. Beach).

Baines, J.A., Vict. Natural. 90: 249. 1973 (epon.).
Blunt, W., The compleat naturalist 118, 176, 206. 1971.
Bridson, G.D.R. et al., Nat. hist. mss. res. Brit. Isl. 430 [index]. 1980.
Candolle, Alph. de, Phytographie 441. 1880 (coll.).
Cox, L.R., Proc. malacol. Soc. 24: 121-124. 1940 (on P's Dorset cretac. mollusca).
Dawson, W.R., Smith papers 75-76. 1934 (corr. at LINN); Banks letters 689-690. 1918.
Ewan, J. & N., John Banister 85, 95-96, 104. 1970.
Freeman, R.B., Brit. nat. hist. books 288. 1980.
Hawksworth, D.L. & M.R.D. Seaward, Lichenol. Brit. Isl. 8, 135, 148. 1977 (bibl., mss.).
Horwood, A.R. & C.W.F. Noel, Fl. Leicestershire clxxxix-cxciv. 1933 (portr., "the pioneer of Leicestershire botany", bibl., mss.).
Jeffers, R.H., Proc. Linn. Soc. London 171: 15-26. 1960 (P. and his correspondents; on letters at LINN).
Jessen, K.F.W., Gegenw. Vorz. 187, 361. 1864.
Jourdan, A.J.L., Dict. Sci. méd., Biogr. méd. 6: 512-513. 1824 (bibl.).
Kent, D.H., Brit. herb. 74. 1957 (herb. at BM, unlocalized).
Milner, J.D., Cat. portr. Kew 90-91. 1906.
Murray, G., Hist. coll. BM(NH) 1: 175. 1904 (herb. purchased 1863).
Nissen, C., Zool. Buchill. 326. 1969 (no. 3250).
Rose, H.J., New biogr. dict. 11: 250. 1850 (fide P. Gilbert, Biogr. Lit. Entom. 1977, p. 305).
Sherborn, C.D., Where is the .... collection 111. 1940 (zool. specimens).
Stearn, W.T., Nat. Hist. Mus. S. Kensington 18. 1981.
Williams, F.N., J. Bot. 57: 100. 1919 (on P. and *Fl. lond.*).
Wittrock, V.B., Acta Horti Berg. 3(2): 97. 1903.
Woodward, B.B., Hist. coll. BM(NH) 1: 28, 44. 1904 (mss.).

EPONYMY: *Pultenaea* J.E. Smith (1794).

8397. Dissertatio medica inauguralis, *de Cinchona officinali* Linnaei; sive cortice peruviano: quam, annuente summo numine, ex auctoritate reverendi admodum viri, Gulielmi Robertson, ... Academiae Edinburgenae praefecti; nec non amplissimi senatus academici consensu, et nobilissimae Facultatis medicae decreto; pro gradu doctoratus, summisque in medicina honoribus et privilegiis rite ex legitime consequendis; eruditorum examini subjicit Ricardus Pulteney, R.S.S. Britannus. Ad diem [] Maii, hora locoque solitis ... Edinburgi (Typis Academicis) 1764. Qu. (*Cinch. off.*).
Publ.: Mai 1764, 60 p., *1 pl. Copy*: HU.

8398. *A general view of the writings of Linnaeus* ... London (printed for T. Payne, ... and B. White ...) 1781. Oct. (*Gen. view writ. Linnaeus*).
Ed. [*1*]: 1781, p. [i]-iv, [1]-425, [426]. *Copies*: MO, NY. – SO 33a.
Ed. 2: 1805 (portr. Linn. publ. 16 Mar 1805), frontisp., [i]-xv, *pl.* facing p. [1], [1]-595, *pl.* facing p. 112, 502 and typogr. table facing p. 578. *Copies*: HU, LC, MO, PH, UC, WU; IDC 7726. – SO 33b. – Second edition, with ... Memoirs of the author by W.G. Maton. To which is annexed the diary of Linnaeus, written by himself and now translated into English. [William George Maton (1774-1835)]. Translation of Linnaean diary by Carl Troilius.
French translation: 1789, *1*: [i*-iii*], [i]-vi, [1]-386, *2*: [i-iii], [1]-400, chart. *Copies*: B, BR, FI, MO. SO 33c, 2609. *Revue générale des écrits de Linné*: ... traduit de l'anglais, par L.A. [Louis Aubin] Millin de Grandmaison [1759-1819] avec des notes et des additions du traducteur ... Londres [London] & se trouve à Paris [Chez Buisson, ...) 1789, 2 vols. Oct. (*Rev. gén. écrits Linn.*).

8399. *Historical and biographical sketches of the progress of botany in England* from its origin to the introduction of the Linnaean system ... London (printed from T. Cadell, ...) 1790. Oct. 2 vols. (*Hist. sketches Bot. Engl.*).
Publ.: Mai 1790 (p. v: 28 Feb 1790; sends copies to Smith and others on 13 Mai 1790; letter to Banks 2 Mar 1790; not yet out, 10 Apr 1790 sends advance copy, index not yet printed). *Copies*: MO, NY, UC, USDA.

1: [i]-xvi, [1-4], [1]-360.
2: [i-viii], [1]-352, [1-32, index], [1-2, err.].
German: 1798, D. Richard Pulteney's ... *Geschichte der Botanik* bis auf die neuern Zeiten, mit besonderer Rücksicht auf England. Für Kenner und Dilettanten. Aus dem Englischen mit Anmerkungen versehen von D. Karl Gottlob Kühn, ... Leipzig (in der Weygandschen Buchhandlung) 1798. 2 vols. Oct. (*Gesch. Bot. England*). *Copies*: H, WU.
1: [i]-xii, [1]-260.
2: [261]-566.
French: 1809. *Esquisses historiques et biographiques des progrès de la botanique en Angleterre*, depuis son origine jusqu'à l'adoption du système de Linnée; ... Paris (de l'Imprimerie de Cellot ... Chez Maradan ...) 1809. Oct. 2 vols. (*Esq. hist. bot. Angleterre*). *Copies*: B, BR, FI, G, MO, NY.   *1*: [i]-viii, [1]-374.
2: [i]-vi, [1]-365.

8400. *Catalogues of the birds, shells, and some of the more rare plants of Dorsetshire*. From the new and enlarged edition of Mr. Hutchins's history of that county ... London (printed by J. Nichols, for the use of the compiler and his friends) 1799. Fol. (*Cat. birds, pl. Dorsetshire*).
Ed. 1: 1799, p. [i], [1]-92, portr. (plants on p. 55-92). *Copies*: HU, LINN; IDC 2503.
Ed. 2: Jan-Mai 1813, iv, 110 p., 23 engr. on 12 pl., portr. *Copy*: K. – "Catalogues ... with additions and a brief memoir of the author ... London (printed by and for Nichols, Son, and Bentley, ...) 1813. Fol. – Botany on p. 61-106; published in J. Hutchins, Hist. antiq. County Dorset ed. 2, vol. 3. See Nissen (Zool. Buchill. 3250) and Cox (1960).
Ref.: Cox, L.R., Proc. malac. Soc. 24: 121-124. 1940 (on cretaceous mollusca).

Purchas, William Henry (1823-1903), British botanist and clergyman; B.A. Durham 1857; ord. Church of England 1857; serving at Tickenhall (S. Leicestershire) 1857-1865, Lydney (Gloucestershire) 1865, Gloucester (1866-1870), Alstonfield (N. Staffordshire) 1870-1903; ardent amateur botanist, especially interested in Rubus, Rosa and Hieracium. (*Purchas*).

HERBARIUM and TYPES: Material at BM, CGE, DBY, K, LIV, MANCH, OXF. – Portrait at G.

BIBLIOGRAPHY and BIOGRAPHY: Barnhart 3: 115; BB p. 250; BL 2: 234, 245, 705; BM 4: 1623; Bossert p. 319; Clokie p. 229; CSP 5: 43, 8: 673, 17: 1046; Desmond p. 507; Jackson p. 253; Kew 4: 376; Rehder 5: 694; Tucker 1: 573.

BIOFILE: Babington, C.C., Mem. Journ. bot. corr. 474 [index, many refs.!] 1897.
Edees, E.S., Fl. Staffordshire 20-21. 1972.
Freeman, R.B., Brit. nat. hist. books 288. 1980.
Horwood, A.R. & C.W.F. Noel, Fl. Leicestershire ccxviii. 1933.
Jackson, B.D., Bull. misc. Inf. Kew 1901: 54.
Kent, D.H., Brit. herb. 74. 1957.
Ley, A., J. Bot. 42: 80-82. 1904 (portr., b. 12 Dec 1823, d. 16 Dec 1903).
Riddelsdell, H.J., Fl. Gloucestershire cxxxi. 1948.
Wade, A.E., Fl. Monmouthshire 12, 15. 1970.

8401. *A flora of Herefordshire*. Part i. ... Hereford (printed by William Phillips, High Town, for the Woolhope Naturalist's Field Club) 1867. Oct. †. (*Fl. Herefordshire*).
Publ.: 1867, map, p. [i*-ii*], [1]-25, followed by W.S. Symonds, Flora of Herefordshire, Definition of districts, [1]-25, [26], both issued as part of the Trans. Woolhope Natural. Field Club 1866. *Copy*: PH. – Not continued in this form.
Definitive edition: Apr 1889 (p. xi: Jan 1889; J. Bot. Mai 1889), map, p. [i]-xxxvii, [1]-549, *pl. 1-2* [*3*] (uncol.). *Copies*: HH, NY, UC. – "*A flora of Herefordshire*. Edited by William Henri Purchas, ... and Augustin Ley [1842-1911] ... Hereford (Jakeman and Carver, ...) [1889]. Oct. (in fours). – *Fungi* by W.C. Cooke; *Definition of the botanical districts* ... by W.S. Symonds (re-written).
Ref.: Marshall, E.S., J. Bot. 27: 216-220. 1889 (rev.).

Purdie, William (c. 1817-1857), Scottish gardener and plant collector; originally with Roy. Bot. Gard. Edinburgh; from 1846-1857 superintendent, Botanic Gardens, Trinidad. (*Purdie*).

HERBARIUM and TYPES: Important sets at CGE, E and K; further material at B, COLU, FI, GH, GOET, L, MEDEL, NY, P, TCD, U, US, W.

BIBLIOGRAPHY and BIOGRAPHY: Barnhart 3: 115; BB p. 250-251; CSP 5: 43; Desmond p. 507 (d. 10 Oct 1857); Kew 4: 376; PR (alph.); Rehder 5: 694; Urban-Berl. p. 386.

BIOFILE: Allan, M., The Hookers of Kew 146. 1967.
Anon., Belgique hort. 17: 254, 256. 1857 (bot. expl. Colombia); Bonplandia 5: 352, 364. 1857; Bot. Zeit. 16: 20. 1858 (d.), Flora 41: 111-112. 1858; Gard. Chron. 1857: 792 (d.); Gartenflora 7: 200. 1858 (d.); Österr. bot. Z. 8: 109. 1858 (d.).
Bridson, G.D.R. et al., Nat. hist. mss. res. Brit. Isl. 269.23, 269.290. 1980.
Coats, A.M., The plant hunters 348, 349. 1969.
Ewan, J., Caldasia 5: 90-95. 1948 (rev. Purdieanthus...note on Purdie).
Gager, C.S., Brooklyn Bot. Gard. Rec. 27: 175. 1938 (dir. R. Bot. Gard., Port of Spain).
Hart, J.H., Ann. Rep. R. bot. Gard., Trinidad, 1887, p. [1].
Hedge, I.C. & J.M. Lamond, Index coll. Edinb. herb. 121. 1970.
Hooker, W.J., Bot. Mag. 71: 21. 1847; London J. Bot. 6: 40-41. 1847 (app. Trinidad), J. Bot. Kew Gard. Misc. 9: 374-375. 1857 (d.).
Jackson, B.D., Bull. misc. Inf. Kew 1901: 54 (K).
Linden et Planchon, Troisième voyage de J. Linden 1: lx-lxi. 1863.
Morris, D., Bull. misc. Inf. Kew, add. ser. 1: 55-56. 1898.
Pittier, H., Man. pl. usual. Venezuela 4-5. 1926.
Purdie, W., *in* Hooker, W.J., London J. Bot. 3: 501-533. 1844, 4: 14-27. 1845 (account Jamaican excursions).
Sayre, G., Mem. New York Bot. Gard. 19(3): 382. 1971 (coll.).
Seemann, B., Bonplandia 5: 352. 1857 (d.).
Stearn, W.T., J. Arnold Arb. 46: 268-270. 1965 (itin.).
Urban, I., Symb. ant. 1: 132. 1898 (biogr., itin.), 3: 107-108. 1902 (itin.).

EPONYMY: *Purdiaea* Planchon (1846); Purdieanthus Gilg (1895), see ING 3: 1465.

Purkyně, Emanuel [Ritter von] (1831-1882), Moravian botanist; from 1851-1860 custos at the National Museum, Praha; Dr. phil. Praha 1861; from 1860 professor at the forestry College of Weisswasser (Běla pod Bezděm). (*Purk.*).

HERBARIUM and TYPES: PR (P had also acquired Lehmann's *Potentilla's*). – Letters and portrait at G.

NOTE: not to be confused with the better known physiologist Jan Evangelist Purkyně (1787-1869).

BIBLIOGRAPHY and BIOGRAPHY: AG 12(3): 398; Barnhart 3: 116; CSP 5: 44; Futák-Domin p. 495-496; Hegi 6(2): 1331; Herder p. 293; Hortus 3: 1202 (Purk.); Kanitz 228; Klásterský p. 152 (b. 17 Dec 1831, d. 23 Mai 1882) q.v. for further biogr. refs.; Maiwald p. 91, 112, 138, 236, 237, 243; Morren ed. 2, p. 45; Rehder 5: 694; TL-1 and TL-2/see Opiz, P.M.; Tucker 1: 573; Zander ed. 10, p. 704, ed. 11, p. 803.

BIOFILE: Anon., Bonplandia 8: 361-362. 1860 (acquired remaining stocks Opiz, Tauschanstalt), 9: 91. 1861 (Dr. phil. Praha); Bot. Centralbl. 10: 415. 1882 (d.); Bot. Not. 1883: 30 (d.); Bot. Zeit. 18: 328. 1860 (buys stocks of herb. coll. Opiz), 40: 480. 1882 (d.); Österr. bot. Z. 10: 374. 1860 (has bought stocks Opiz, Tauschanstalt), 11: 131. 1861 (Dr. phil.).; Österr. bot. Z. 32: 211. 1882 (d. 23 Mai 1882).

Purpus, Carl [Karl] **Albert** (1851-1941), German botanical explorer of Mexico and the western United States (1897-1934). (*Purpus*).

HERBARIUM and TYPES: largest sets at A, F, GH, H, HPH, MO, NY, RSDR, UC and US; further material in many herbaria. – Letters at G, NY, and SIA.

NOTE: P's brother Joseph Anton Purpus (1860-1932) was gardener at St. Petersburg (1882-1888) and at Darmstadt (1888-1892). He accompanied his bother on various trips.

BIBLIOGRAPHY and BIOGRAPHY: AG 4: 412, 422; Barnhart 3: 116; CSP 17: 1049; Ewan ed. 1: 285-286, ed. 2: 179; Hortus 3: 1202 (C. Purpus); IH (ed. 6): 363, (ed. 7): 340, 2: (in press); Langman p. 598-601 (bibl.); Lenley p. 337; MW p. 397; Rehder 5: 143, 694-695; Urban-Berl. p. 280, 386; Zander ed. 10, p. 704, ed. 11, p. 803.

BIOFILE: Anon., Darmstädter Tageblatt 9 August 1927 (reprint, 1 p., NY) (on robbery and injury).
Brandegee, T.S., Plantae mexicanae purpusianae, in Univ. Calif. Publ. Bot. 3: 377-396. 1909; 4: 85-95, 177-194, 269-281, 375-388. 1910-1913; 6: 51-77, 177-197, 363-375, 497-504. 1914-1919, 7: 325-331. 1920, 10: 181-188, 403-421. 1922-1924.
Diehl, C., Mitt. deut. dendrol. Ges. 45: xiii, xxiv-xxv. 1933 (obituary of J.A. Purpus with a note on his brother).
Ewan, J., Amer. Midl. Natural. 17: 439. 1936, Madroño 19: 32. 1967.
Hedge, I.C. & J.M. Lamond, Index coll. Edinb. herb. 121. 1970.
Knobloch, J.W., Plant coll. N. Mexico 55. 1979 (b. 26 Feb 1851, d. 17 Jan 1941).
Langman, I., Rev. Soc. mex. Hist. nat. 10: 334-336. 1949 (b. 26 Feb 1853 or 1851; d. 17 Jan 1941).
McVaugh, R., Contr. Univ. Mich. Herb. 9: 292. 1972.
Moran, R., Cact. Succ. J. Amer. 34(1): 8-12. 1962 (portr.).
Purpus, C.A., Mitt. deut. dendrol. Ges. 1899: 144-145 (journey La Sol. Mountains, Utah); Moeller's deut. Gart. Zeit. 29: 293-297, 350, 437-442, 493-494, 509-511, 518-519, 535-537, 541-543. 1914, 30: 24-26, 41-42, 46-47, 55-56. 1915 (account Mexican trip 1912).
Reveal, J., in A. Conquist e.a., Intermountain flora 1: 59. 1972.
Sánchez, M.S., Univ. Calif. Publ. Bot. 51: i-ix, 1-36. 1969 (biogr., portr., detailed study of his Mexican collections and bibl. of studies on P and his colls.; preface by L. Constance; b. 26 Feb 1851).
Standley, P.C., Contr. U.S. natl. Herb. 23: 48. 1920.
Verdoorn, F., ed., Chron. bot. 2: 146. 1936, 6: 331. 1941.
Ziegler, H., Ber. deut. bot. Ges. 81: 564. 1970.

HANDWRITING: Univ. Calif. Publ. Bot. 51: 8. 1969.

EPONYMY: *Purpusia* T.S. Brandegee (1899).

8402. *Die Kakteen der Grand Mesa in West-Colorado.* Von C.A. Purpus in Delta (Verein. Staat. Nord.-Am. Colorado) [Monatschrift für Kakteenkunde 3(4): 49-53. Apr 1893]. Oct.
Publ.: Apr 1893, as cited above; reproduced in facsimile in Kakteen und Orchideen-Rundschau 1977(3): 41-45. 15 Mai 1977.

Pursh, Frederick Traugott [originally Friedrich Traugott Pursch] (1774-1820), German (Saxonian) botanist; studied botany and horticulture at Dresden; to America 1799 as traveller collector, gardener and landscape architect; from 1803-1805[?] in charge of William Hamilton's estate nr Philadelphia (the Woodlands); from 1809-1810 id. of the Elgin Botanic Garden at New York; from 1810-1811 in the West Indies; from 1811 in England, working with A.B. Lambert; ultimately at Montreal. (*Pursh*).

HERBARIUM and TYPES: The Pursh herbarium was acquired by Lambert and dispersed at the Lambert sale. Plants collected by Pursh are at BM, C, K, LIV, MANCH, OXF, PH (e.g. in Barton herbarium). Pursh used not only his own collections but also those of Banks and Sherard. Some of his types are therefore at BM and OXF, others at K and PH.

BIBLIOGRAPHY and BIOGRAPHY: AG 2(1): 375; Backer p. 475-476; Barnhart 3: 116; BB p. 251; BM 4: 1624; Clokie p. 229; DAB 15: 271; Desmond p. 508; DSB 11: 217-219 (by Ewan); Ewan ed. 1: 7, 252; Frank 3 (Anh.): 80; GF p. 71; Henrey 2: 36, 3: 104-105; Herder p. 226; Hortus 3: 1202 (Pursh); IH 1 (ed. 1): 76, (ed. 2): 96, (ed. 3): 122, (ed. 6): 363, (ed. 7): 340; Jackson p. 354, 365, 409, 443; Kew 4: 379; Langman p. 601; Lasègue p. 332, 460, 463, 504; ME 1: 221, 3: 641; MW p. 397; NI 1570; Pennell p. 617; PR 2374, 7369-7371 (ed. 1: 8257-8259); Rehder 5: 695; RS p. 134; SK 4: ccvii; Stevenson p. 1254; TL-1/1015; TL-2/910, 1505, see Hooker, W., Lambert, A.B., Nuttall, T.; Tucker 1: 573; Zander ed. 10, p. 704, ed. 11, p. 803.

BIOFILE: Anon., Appleton's cycl. Amer. biogr. 5: 137. 1888; Bonplandia 1: 42. 1853; Flora 1: 357. 1818 (d.), 3: 542. 1820, 10: 192 (err., said to have been born in Tobolsk, Siberia), 491-496 (correct biogr., see sub C.A. Pursch) 1827; Mém. Soc. Linn. Paris 4: cxiii, 710-712. 1825; Bonplandia 1: 42. 1853 (plants at K).
Barnhart, J.H., Torreya 4: 132-136. 1904 (date Flora); Mem. Torrey bot. Club 16(3): 298. 1921; J. New York Bot. Gard. 24: 109. 1923 (b. 4 Feb 1774); Bartonia 9: 40-41. 1926 (pl. in Barton herb.).
Barnston, J., Canad. Natural. ser. 2. 9: 184-187. 1881.
Blake, S.F., J. Wash. Acad. Sci. 17(13): 351. 1927 (brief sketch).
Boone, W., Hist. bot. W. Virginia 6, 7, 20. 1965.
Bridson, G.D.R. et al., Nat. hist. mss. res. Brit. Isl. 255.234. 1980.
Darlington, ,W., Reliq. Baldw. 1843 (many references, see Ewan in facs. ed. p. xxxviii).
Dawson, J.W., Canad. natural. 2: 78-79. 1857 (monument).
Elkinton, H., Amer.-Germ. Rev. 11(2): 8-11. 1944.
Ewan, J., Proc. Amer. philos. Soc. 96(5): 599-628. 1952 (extensive review of P's activities in N. America and of his botanical associates; bibl. on Pursh; so far main and most critical account of Pursh); Yearb. Amer. philos. Sc. 1955: 160-163; Introduction to F. Pursh, Fl. amer., Sept., p. 5-117. 1979 (biogr. and bibl. detail); Huntia 3(2): 83-86. 1979 (on P's "Prospectus of an expedition to New Mexico and California" [1814?]).
Ewan, J. et al., Short hist. bot. U.S. 170 [index]. 1969.
Ewan, J. & N., John Banister p. 483 [index]. 1970.
Fernald, M.L., Rhodora 40: 354. 1938.
Geiser, S.W., Natural. Frontier 16. 1948.
Graustein, J.E., Thomas Nuttall 476 [index; many refs.]. 1967; Rhodora 56: 275. 1954 (date flora).
Gray, A., Amer. J. Sci. ser. 3. 24: 323-326. 1882; Bot. Gaz. 7: 141-143. 1882.
Harshberger, J.W., Bot. Philadelphia 113-117. 1899.
Hedrick, U.P., Hist. hort. Amer. 404, 405, 424. 1950.
Henckel von Donnersmarck, L.V.F., Flora 3: 542. 1820 (Fredric Pursh is Friedrich T. Pursch).
Hindle, Brook, Pursuit sci. revol. Amer. 305. 1956.
Hooker, W.J., Edinburgh J. Sci. 2: 108-129. 1825 (notes on P on p. 114-119), repr. Amer. J. Sci. 9: 263-284. 1825. (with added footnote on p. 269 by Charles Hooker); J. Bot. Kew Gard. misc. 9: 256. 1857 (erection grave monument).
Humphrey, H.B., Makers N. Amer. bot. 202-204. 1962 (see also Ewan, J., Rhodora 64: 189. 1962).
Jessen, K.F.W., Bot. Gegenw. Vorz. 466. 1864.
Leroy, J.F. et al., Bot. franç. Amér. nord 45, 54, 59, 65, 66, 67, 79, 220. 1957.
McCool, E.W., Bull. Hort. Soc. New York 11: 13. 1961 (incorrect).
McVaugh, R., Bartonia 17: 24-32. 1935 (coll. N.Y., Vermont 1807).
Maxon, W.R., Dict. Amer. biogr. 15: 271. 1935 (b. 4 Feb 1774, d. 11 Jul 1820; biogr.; second refs.; lists traditional errors in Pursh accounts).
Mears, J.A., Proc. Amer. philos. Soc. 122: 166, 167. 1978. (herb. Muhlenberg).
Merrill, E.D. & Hu, Bartonia 25: 27-29. 1949.
Miller, H.S., Taxon 19: 536-537. 1970.
Murrill, A.W., Hist. found. bot. Florida 11. 1945.
Osterhout, G.E., Plant World 10(4): 80-84. 1907. (spring fl. Colorado).
Penhallow, D.P., Trans. R. Soc. Canada ser. 2. 3(4): 3-6, 8, 14, 52. 1897 (d. 11 Jul 1820; on Pursh in Canada; authentic statement on death, burial, reinterment; see also Ewan, 1952, p. 628).

Pennell, F.W., Bartonia 21: 43. 1942, Proc. Amer. philos. Soc. 94: 142-144. 1950.
Provancher, L., Natural. canad. 5: 102. 1873.
Pursch, C.A., Flora 10: 491-496. 1827 (correct biogr. details, given by his brother).
Rafinesque, C.S., Amer. mo. Mag. crit. Rev. 2(3): 170-176, 265-269. 1818 (review *Flora*).
Reichenbach, H.G.L., Flora 10: 491-496. 1827 (corrections of obit. in 10: 192; direct inf. from Carl August Pursch).
Rodgers, A.D., John Torrey 348 [index]. 1942.
Sargent, C.S., Silva 2: 33. 1891, 14: 99. 1902.
Schuyler, A.E., Morris Arb. Bull. 24(2): 28, 31. 1973.
Seemann, B. & W.E.G., Bonplandia 5: 149-150. 1857 (Ein Denkmal für Pursh; d. 11 Jul 1820); J. Bot. 6: 256. 1868 (Pursh's diary found, see Pursh 1869), 8: 63-64. 1870 (quaint account).
Short, C.W. *in* W.J. Hooker, J. Bot. 3: 103, 105, 107-108. 1841.
Spaulding, P., Pop. Sci. Mo. 73: 491, 493, 497. 1908.
Stearn, W.T., Rhodora 45: 415-416, 511-512. 1943 (date flora).
Stieber, F., Huntia 4(1): 85. 1981 (Arch. mat. HU).
Sutton, S.B., Charles Sprague Sargent 281. 1970.
Thomas, J.H., Huntia 3: 13. 1969.
Urban, I., Symb. ant. 3: 108-109. 1902.
Venema, H.J., Jaarb. Ned. dendrol. Ver. 12: 103-106. 1937 (on P. in N. America).

COMPOSITE WORKS: Edited eds. 8-13 of James Donn, *Hortus cantabrig.*, see TL-2no. 1505.

EPONYMY: *Purschia* Post & O. Kuntze (1903, *orth. var.* of *Purshia* A.P. de Candolle ex Poiret); *Purshia* A.P. de Candolle ex Poiret (1816), *Purshia* K.P.J. Sprengel (1817), see ING 3: 1466. *Note*: *Purshia* Rafinesque (1815) is an *orth. var.* of *Burshia* Rafinesque (1803).

8403. II. *Verzeichniss der im Plauischen Grunde und den zunächst angrenzenden Gegenden wildwachsenden Pflanzen* von Friedrich Traugott Pursch, *in* W.G. Becker, ed., Der Plauische Grund bei Dresden, mit Hinsicht auf Naturgeschichte und schöne Gartenkunst ... Nürnberg (in der Frauenholzischen Kunsthandlung) 1799. Qu. (*Verz. Plau. Pfl.*).
Publ.: 1799, plants in part 2(2): [45]-94 ([45-47], 48 [49], 50 ["42"], 51 ["43"], 52-94. *Copies*: HU, PH.

8404. *Flora Americae septentrionalis*; or, a systematic arrangement and description of the plants of North America. Containing, besides what have been described by preceding authors, many new and rare species, collected during twelve years travels and residence in that country, by Frederic Pursh ... London (printed for White, Cochrane, and Co., ...) 1814, 2 vols. Oct. (*Fl. Amer. sept.*).
Publ.: Jan 1814 vel Dec sero 1813. – Asa Gray and Graustein give evidence that the work was published late in 1813. A copy was presented to the Linnean Society on 21 Dec 1813. The book was listed as published in the Monthly Lit. Adv. 105: 2. on 10 Jan 1814 (fide Stearn) and is also listed for that month by Peddie and Waddington. Drawings by William Hooker (1797-1832), the artist of Salisbury's *Paradisus Londinensis*. Copies with coloured plates are rare. *Copies*: BR (col. pl.), FAS (col. pl.), G (uncol.), MICH, MO, NY (uncol.; John Torrey's copy with notes), US (3, uncol.), USDA (uncol.); IDC 263.
1: [i]-xxxvi, [1]-358, *pl. 1-16*.
2: [i-ii], [359]-751, *pl. 17-24*.
Plate *16* should face p. 348, and not 248 as in instr. binders.
Ed. 2 (re-issue): 1816. *Copies*: G (uncol.), MO (crudely coloured pl.), NY (uncol.), USDA (uncol.). – London (printed for James Black and Son, ...) 1816. Oct.
1: [i]-xxxvi, [1]-358, *pl. 1-16*.
2: [i-ii], [359]-751, *pl. 17-24*.
This "second edition" has the same printing errors as ed. 1, e.g. the error of p. 248 instead of p. 348 to place pl. 16 (p. xxvi).
Facsimile ed.: 1979, p. [1]-117, preface and (p. 5-117) introduction by Joseph Ewan,

facsimile: [i]-xxxvi, [1]-358, [i-ii], [359]-751, *pl. 1-24. Copy*: FAS. – "... Edition and introduction by Joseph Ewan. Reprint 1979 ... J. Cramer in der A.R. Gantner Verlag Kommanditgesellschaft FL-9490; ISBN 3-7682-1242-4. – See Taxon 29: 748. 1980, 38: 863. 1981 (separate issue of introduction by J. Ewan).

8405. *Hortus orloviensis*; or, a catalogue of plants cultivated in the Island of Orloff, near St. Petersburgh, by Peter Buck, gardener to his excellency Count Orloff. Arranged by Frederick Pursh. London (Printed by Richard and Arthur Taylor, ...) 1815. Oct. (*Hort. orlov.*).
Publ.: 1815, vii, 72 p. *Copy*: HH. – See Ewan (1952).

8406. *Journal of a botanical excursion* in the Northeastern parts of the states of Pennsylvania and New York, during the year 1807. By Frederick Pursh. Philadelphia (Brinckloe & Marot, ...) 1869. Duod. (*J. bot. excurs.*).
Editor: Thomas Potts James (1803-1882).
Publ.: 1869, p. [i], [1]-87. *Copies*: MICH, MO, USDA.
Reprint: 1923, p. [1]-113, [1-5]. *Copies*: MO, NY. – "... edited by Rv. Wm.M. Beauchamp, ... for the Onondaga historical Association 1923". Reprinted by the Dehler Press, Syracuse, N.Y. [William Martin Beauchamp 1830-1925].
Facsimile reprint of 1923 ed.: 1969, p. [1]-113, [1-5]. *Copies*: MICH, NY. – Ira J. Friedman, Inc. Port Washington, L.I., N.Y., ISBN 87198-073-8.

Purton, Thomas (1768-1833), British botanist and surgeon; practiced in London 1791-1795, later in Alcester. (*Purt.*).

HERBARIUM and TYPES: WOS; fungi at K.

BIBLIOGRAPHY and BIOGRAPHY: Barnhart 3: 116; BB p. 251; BM 4: 1624; CSP 5: 46; Desmond p. 508 (b. 10 Mai 1768, d. 29 Apr 1833); GR p. 410-411; IH 2: (in press); Jackson p. 246; Kelly p. 182; Kew 4: 379; Krempelh. 1: 148, 500; LS 21484; NAF 1(1): 167; NI 1571; PR 7372 (ed. 1: 8260); Tucker 1: 573.

BIOFILE: Bagnall, J.E., Fl. Warwickshire 495-497. 1891.
Bridson, G.D.R. et al., Nat. hist. mss. res. Brit. Isl. 269-292, 302.1. 1980.
Cooper, G., Warwick nat. Hist. Soc. ann. Rep. 20: 12-16. 1974.
Dawson, W.R., Smith papers 76. 1934.
Freeman, R.B., Brit. nat. hist. books 288. 1980.
Hawksworth, D.L. & M.R.D. Seaward, Lichenol. Brit. Isl. 13-14, 136. 1977.
Jackson, B.D., Bull. misc. Inf. Kew 1901: 54 (fungi at K).
Kent, D.H., Brit. herb. 74. 1957.
Lees, E., Bot. Worcestershire lxxxix-xc. 1867.
Petersen, R.H. Mycotaxon 1: 157. 1975.
Wade, A.E., Fl. Monmouthshire 10. 1970.

8407. *A botanical description of British plants*, in the midland counties, particularly of those in the neighbourhood of Alcester; with occasional notes and observations; to which is prefixed a short introduction to the study of botany, and to the knowledge of the principal natural orders ... Stratford-upon-Avon (printed and sold by J. Ward), London (sold also by Longman, Hurst, Rees and Co. ....) 1817-1821, 3 vols. Oct. (in fours). (*Bot. descr. Brit. pl.*).
1: Jul 1817 (p. v: 13 Apr 1817; h.t. dated Jul 1817 in MO copy), p. [i*, iii*], [i]-ix, [1]-361, *pl. 1-3* (col.; by James Sowerby, from English botany). Half title: *A Midland Flora*.
2: Jul 1817 (dated as rd. Jul 1817 in MO copy), p. [i, iii], [363]-795, [796, err.], *pl. 4-8* (id.).
3(1): 1821 (p. viii: 16 Aug 1821), p. [i]-xiv, [xv-xvi, cont.], [1]-333, [334], *pl. 12-14, 17-18, 23-28*. – *An appendix to the Midland flora*; comprising also corrections and additions referring to the two former volumes; and occasional observations tending to elucidate the study of the British fungi. Concluding with a generic and specific index to the whole work, and a general index of synonyms. In two parts ... London (printed for

the author; sold by Longman, Hurst, Rees, Orme, and Brown, ...) 1821. Oct. − Plates may be differently divided over the two parts.
3(2): Nov 1821 (Peterson 1975), p. [i, iii], [335]-575, *pl. 9-11, 15-16, 19-22*.
Copies: MICH, MO, NY, Stevenson; IDC 5090.

Putterlick, Alois [Aloys] (1810-1845), Austrian botanist and physician; custos (adjunct) at the Natural History Museum, Vienna, 1840-1845. (*Putterl.*).

HERBARIUM and TYPES: W; mosses also at MJ (Jihlava). − Letters at G.

BIBLIOGRAPHY and BIOGRAPHY: Backer p. 476; Barnhart 3: 117 (b. 3 Mai 1810, d. 29 Jul 1845); BM 4: 1624; Herder p. 335; Hortus 3: 1202 (Putterl.); IH 2: (in press); Kew 4: 379; Klásterský p. 152 (Alois); MW p. 397; NI 599, 1093, 1441; PR 6665, 7374 (ed. 1: 8262); Rehder 5: 695; SK 4: ccvii; TL-1/919, 1016; TL-2/1679, 4332, 6704; Tucker 1: 573; WU 24: 111-112.

BIOFILE: Beck, G., Bot. Centralbl. 34: 87, 281, 313. 1888 (Custos adjunct W 1840-1845).
Duchartre, P. Rev. bot. 1: 237. 1846.
Fritsch, K. et al., Bot. Anst. Wien 58, 61. 1894.
Neilreich, A., Verh. zool.-bot. Ver. Wien 5: 60. 1855 (Alois).
Oborny, A., Verh. naturf. Ver. Brünn 21(2): 13, 15. 1882 (*Fl. Mähren*).
Pluskal, F., Verh. zool.-bot. Ver. Wien 6: 369. 1856.
Pritzel, S.A., Linnaea 19: 459. 1846 (Aloys, not Alois).
Sayre, G., Bryologist 77: 515. 1977.

COMPOSITE WORKS: see J.G.C. Lehmann, *Pl. Preiss.*, TL-2/4352, and T.F.L. Nees von Esenbeck, *Gen. pl. fl. Germ.* (editor fasc. 22-24, 1843-1845), reviews in Flora by S-n, 26: 717-718. 14 Nov 1843 (fasc. 22), 27: 493. 28 Jul 1844 (23), 29: 255-256. 28 Apr 1846 (24).

EPONYMY: *Putterlickia* Endlicher (1840) very likely commemorates him; *Putterlichia* W.H. Harvey (1868, *orth. var.*).

8408. *Synopsis Pittosporearum* auctore Aloysio Putterlick ... Vindobonae [Wien] (apud Fridericum Beck ...) 1839. Oct. (*Syn. Pittosp.*).
Publ.: Nov-Dec 1839 (SK; Flora 13 Jan 1840), p. [i-vi], [1]-30, [31-32]. *Copies*: FI, G, MO, NY, WU; IDC 5647.

Puydt, Emile de (1810-1891), Belgian horticulturist at Mons. (*Puydt*).

HERBARIUM and TYPES: Unknown.

BIBLIOGRAPHY and BIOGRAPHY: BM 4: 1625; CSP 8: 676; Herder p. 376; Jackson p. 138; Morren ed. 10, p. 46; MW p. 397; NI 1572 (b. 6 Mar 1810, d. 20 Mai 1891); Plesch p. 370; Rehder 5: 605; Sotheby no. 618.

BIOFILE: Duveen, D.I., Amer. Orchid. Soc. Bull. 42(7): 627-629. 1973 (on *Les orchidées*).
Kerckhove de Denterghem, O.C.E.M.G. de, Rev. hort. belg. étrang. 17: 147-151. 1891 (obit., b. 1810, d. 20 Mai 1891, portr.).
Linden, L., Ill. hort. 28: 105-107. 1881 (obit., portr.).

8409. *Les plantes de serre.* Traité théorique & pratique de la culture de toutes les plantes qui demandent un abri sous le climat de la Belgique ... Mons (Hector Manceaux) 1866. 2 vols. 18-mo. (*Pl. serre*).
Publ.: 1866. *1*: [1]-400, *2*: [1]-284. *Copy*: G.
Further editions: reached at any rate a fourth edition, Mons (id.) 1889 (ÖbZ 1 Oct 1889; Nat. Nov. Jun(2) 1889).

8410. *Les Orchidées* histoire iconographique organographie- classification-géographie-collections-commerce-emploi-culture avec une revue descriptive des espèces cultivées en Europe ... Paris (J. Rothschild, ...) 1880. Oct. (*Orchidées*).

Publ.: Jan-Feb 1880 (J. Bot. Mar 1880; Bot. Centralbl. 5-9 Apr 1880; also: 1-9 Jan 1880 (ann.); Bot. Zeit. 28 Mai 1880), p. [i*-ii*], [i]-viii, [1]-348, *pl. 1-50* (chromoliths. by Leroy and Guibert). *Copies*: BR, G. (copy offered by Junk, 177, at Hfl. 1400. =).

Puymaly, André Henri Laurent de (1883-?), French lichenologist; Dr. med. Bordeaux 1912, Dr. sci. nat. 1925; teaching at the University of Bordeaux. (*Puym.*).

HERBARIUM and TYPES: in 1957 still in his private herbarium. – Letters at G.

BIBLIOGRAPHY and BIOGRAPHY: Barnhart 3: 117; GR p. 294, cat. p. 70; Hirsch p. 240; LS suppl. 22323; Roon p. 91; SBC p. 129.

BIOFILE: Anon., Rev. bryol. lichénol. 5: 457. 1931, 9: 169. 1936, 37: 716. 1971.
Puymaly, A.H.L. de, Communication to senior author, 1957 (b. 19 Aug 1883).
Verdoorn, F., ed., Chron. bot. 6: 302. 1941.

8411. Thèses présentées à la Faculté des Sciences de l'université de Paris pour obtenir le grade de docteur ès-sciences naturelles ... 1re thèse. – *Recherches sur les algues vertes aériennes* ... soutenues le [stamped: 30 mars; printed] janvier 1925 ... Bordeaux (Imprimerie L. Delbrel, ...) 1924 [1925]. Oct. (*Rech. alg. vertes aér.*).
Publ.: 30 Mar 1925 (date defense thesis; t.p. 1924), p. [1]-274, [1, thèse], *pl. 1-7* (auct.).
Copies: B, FH.

Quaintance, Altus Lacy (1870-1958), American phytopathologist and economic entomologist; B.S. Florida Agric. Coll. [Univ. Florida] 1893; M.S. Alabama Polytechnic Institute 1894; Dr. Sci. ib. 1915; instructor in biology and professor of entomology at the University of Florida, 1894-1898; biologist at the Georgia Agricultural Experiment Station 1899-1901; State Entomologist of Maryland 1901-1903; in Division of Entomology, U.S.D.A. 1903-1930. (*Quaintance*).

HERBARIUM and TYPES: FLAS, GAES. – Archival material in FH (letters to R. Thaxter) and SIA (134, 7262).

BIBLIOGRAPHY and BIOGRAPHY: Barnhart 3: 117 (b. 19 Dec 1870); BM 4: 1625, 8: 1039; IH 2: (in press); LS 38125-38128, 39098; NW p. 55; PH 59, 150, 939; SIA 134, 7262; Tucker 1: 573-574.

BIOFILE: Anon., Bull. entom. Soc. Amer. 4(3): 114. 1958 (d. 7 Aug 1958); Proc. entom. Soc. Washington 38: 130. 1936 (list of published biogr. refs.).
Cattell, J.M., ed., Amer. men Sci. ed. 2. 382. 1910; ed. 3. 557. 1921; ed. 4. 792. 1927; ed. 5. 905. 1933; ed. 6. 1145. 1938; ed. 7. 1431. 1944.
Gilbert, P., Comp. biogr. lit. entom. 306. 1977.
Mallis, A., American entomologists 487-489. 1971 (portr.).
Porter, B.A., J. econ. Entom. 52(1): 182. 1959 (obit.).

Quandt, Christian (1720-post 1807), Baltic German/Latvian clergyman and missionary; educated with the Moravian brethren; sometime teacher at a boarding-school for boys at Neudietendorf; missionary in Suriname 1768-1780; returned to Europe 1780, ultimately at Herrnhut. (*Quandt*).

HERBARIUM and TYPES: Unknown.

BIBLIOGRAPHY and BIOGRAPHY: Barnhart 3: 118 ("Christlieb").

BIOFILE: Renselaar, H.C. van, Foreword to facs. ed. *Nachr. Suriname* 1968, p. [ii*-v*] (brief biogr. sketch).

8412. *Nachricht von Suriname* und seinen Einwohnern sonderlich den Arawacken, Warauen und Karaiben, von den nützlichsten Gewächsen und Thieren des Landes, den Geschäften der dortigen Missionarien der Brüderunität und der Sprache der Arawacken ... Görlitz (gedruckt bei J.G. Burghart ...) [1807]. Oct. (*Nachr. Suriname*).

Publ.: 1807 (p. 316: 4 Sep 1807), [i]-xiv, [xv-xvi], [1]-316, 1 map, *2 pl.* (copp.). Original edition n.v.
Facsimile ed.: 1968, p. [i], [ii*-v* new text][ii= verso v*], [iii]-xiv, [xv-xvi], [1]-316, 1 map, *2 pl. Copy*: U. - Amsterdam (S. Emmering) 1968, with a foreword by H.C. van Renselaar.

Quartin-Dillon, Richard (x-1841), French botanist, explorer of Abyssinia (1839-1840). (*Quart.-Dill.*).

HERBARIUM and TYPES: P and PC, duplicates e.g. in B, BR, CN, G, K, LE, MO, W.

BIBLIOGRAPHY and BIOGRAPHY: AG 2(2): 466; Backer p. 478; Barnhart 3: 118 (d. 22 Oct 1841); IH 1 (ed. 3): 119, (ed. 6): 363, (ed. 7): 340; Lasègue p. 165-167, 301, 314, 504; PR alph.; Urban-Berl. p. 386.

BIOFILE: Candolle, Alph. de, Phytographie 441. 1880.
Pritzel, G.A., Linnaea 19(3): 460. 1846.
Vallot, J., Bull. Soc. bot. France 29: 187. 1882.
White, A. & B.L. Sloane, The Stapelieae ed. 2: 99, 262, 852. 1937.

COMPOSITE WORKS: (1) A. Richard, *Tentamen florae Abyssinicae* 1847, q.v., is based on the collections made by R. Quartin-Dillon & A. Petit, 1838-1843. This is part 3, tom. 4-5 of *Voyage en Abyssinie*, exécuté pendant ... 1839-1843, 6 vols., atlas, 1845-1851.

EPONYMY: *Quartinia* A. Richard (1840); *Quartinia* Endlicher (1842).

Quehl, Leopold (1849-1922), German (Saxonian) botanist, writer on cacti; grandson of F.L. Jahn; employee in the postal service at Halle; from 1873 interested in cacti, maintaining a large collection of living plants as well as assembling an herbarium. (*Quehl*).

HERBARIUM and TYPES: B (Cactaceae, in part extant).

BIBLIOGRAPHY and BIOGRAPHY: Barnhart 3: 118 (b. 7 Nov 1849, d. 22 Feb 1923); Langman 601-602; SIA 221; Urban-Berl. p. 386; Zander ed. 10, p. 704, ed. 11, p. 803.

BIOFILE: Vaupel, F., Monatsschr. Kakteenk. 32(5): 76-79. 1922 (obit., portr., b. 7 Nov 1849, d. 22 Feb 1922).

Quekett, John Thomas (1815-1861), British surgeon, microscopist and histologist; assistant conservator Hunterian museum 1843, conservator 1852-1861; professor of histology, Royal College of Surgeons 1852; secretary of the Royal microscopical Society, London, 1841-1860. (*Quekett*).

HERBARIUM and TYPES: Unknown. - See J. Quekett microsc. Club 30(3): 57-58. 1965 for manuscripts and collections.

BIBLIOGRAPHY and BIOGRAPHY: Barnhart 3: 118; BB p. 251; BM 4: 1632, 8: 1040, 1041; Bret. p. 701; CSP 5: 53-54; Desmond p. 508 (b. 11 Aug 1815, d. 20 Aug 1861); De Toni 1: cii, 2: xcii; DNB 47: 97-99; Jackson p. 70, 414; Kew 4: 382; Moebius p. 203, 439; PR 7379; Rehder 5: 697.

BIOFILE: Anon., J. Quekett microsc. Club 30(3): 55-59. 1965 (biogr. sketch, portr.).
Bentham, G., Proc. Linn. Soc. London 6: xciii-xciv. 1862 (obit.).
Bridson, G.D.R. et al., Nat. hist. mss. rés. Brit. Isl. 229.527, 271.49. 1980.
Deby, J., Bibl. microsc. microgr. stud. 3(Diat.): 45. 1882.
Freeman, R.B., Brit. nat. hist. books 289. 1980.
Freund, H. & A. Berg, eds., Gesch. Mikroskopie 33, 34. 1963.
Michael, A.D., J.R. microsc. Soc. 1895: 6-7.
Stearn, W.T., Nat. Hist. Mus. S. Kensington 35, 37, 67. 1981.
VanLandingham, S.L., Cat. diat. 4: 2363, 6: 3583, 7: 4221. 1978.

HANDWRITING: J. Quekett microsc. Club 30(3): 56. 1965.

EPONYMY: The Quekett Microscopical Club, London is named for this author. *Note*: *Quekettia* J. Lindley (1839) honors his brother, Edwin J. Quekett (1808-1847), British surgeon and plant anatomist.

8413. *Lectures on histology*, delivered at the Royal College of Surgeons of England in the session 1850-51 ... London (Hippolyte Baillière, ...), New York (id.), Paris (J.B. Baillière, ...), Madrid (Bailly Baillière, ...) 1852-1854. 2 vols. Oct. (*Lect. histol.*).
1: 1852 (p. [iii]: 27 Apr 1852), p. [i]-viii, [1]-215, [216, adv.].
2: 1854, p. [i]-viii, [1]-413.
Copies: BR, NY, PH.

Quélet, Lucien (1832-1899), French mycologist; Dr. med. Strasbourg 1856; practicing physician at Hérimoncourt (Doubs); founder and first president of the Société mycologique de France 1885; prix Desmazières 1878; prix Montagne 1886. (*Quél.*).

HERBARIUM and TYPES: PC; further material (fungi) at UPS. – Contributed to C. Roumeguère, *Fungi exsiccati praecipue gallici*.

BIBLIOGRAPHY and BIOGRAPHY: Ainsworth p. 226, 284, 318; Barnhart 3: 118; BM 1: 379, 4: 1632; Bossert p. 320; CSP 11: 84, 12: 594, 18: 5; DTS 1: 231, 3: xlviii, xxiii; GR p. 294-295, cat. p. 70; Hawksworth p. 185, 197; IH 1 (ed. 6): 363, (ed. 7): 340, 2: (in press); Jackson p. 278, 344; Kelly p. 183-185; Kew 4: 382-383; KR p. 580 (b. 14 Jul 1832, d. 28 Aug 1899); LS 3900, 5848, 21500-21549, 38138-38145; Morren ed. 10, p. 49, 63, 65; NAF 9: 448; NI 1573, 3: 55; Stevenson p. 1255; TL-2/909, 1206, 1896, 4442, 6371.

BIOFILE: Ainsworth, G.C., Dict. fungi ed. 6, 7. 499.
Anon., Bot. Centralbl. 80: 368. 1899; Bot. Not. 1900: 47 (d.); Hedwigia 38: (297). 1899 (d.); Nat. Nov. 21:728. 1899 (d.); Z. Pilzk. 13(5): *pl. 8.* 1934 (portr.).
Boudier, J.L.E., Bull. Soc. mycol. France 15(4): 321-325. 1899 (reprint Lons-le-Saunier, 6 p.), (b. 14 Jul 1832, d. 25 Aug 1899, obit., bibl.); Bull. Soc. bot. France 46: 414-417. 1899 (obit., bibl.), see also p. 128, 353.
Camus, F., Bull. Soc. bot. France 55: 588. 1908 (library and herbarium for sale).
Donk, M.A., Bibliographical note, *in* facsimile ed. of *Champ. Jura Vosges*, Amsterdam 1964, 3 p.
Favre, L., Bull. Soc. neuchat. Sci. nat. 28: 233-238. 1900 (obit., bibl.).
Ferry, R., Rev. mycol. 21: 114-117. 1899 (obit., bibl.).
Fries, E., Commentarius in cel. Quéletii dissertationem: "Sur la classification et la nomenclature des Hyméniés" in "Bulletin de la Société botanique de France 1876" insertam, Uppsala 1876. 10 p.
Galzin, Bull. Ass. vosg. Hist. nat. 1905(8): 124-128. [or 8: 124-128. Jan 1905?].
Gilbert, E.J., Bull. Soc. mycol. France 65(1-2): 5-33. 1949 (une oeuvre, un esprit; bibl.).
Junk, W., Rara 2: 179-181. 1926-1936 (bibl.).
Killermann, S., Z. Pilzk. 11(2): 19-23. 1932 (biogr.).
Magnin, A., Ann. Soc. bot. Lyon 24: 35. 1899 (d.); Arch. fl. jurass. 1: 51-52. 1900 (obit.); Mém. Soc. Émul. Doubs ser. 7. 5: ix. 1900 (obit.).
Müllerot, M., Hopea 34(2): 286. 1976 (list publ. at REG).
Raab, H., Z. Pilzk. 55(12): 179, 180. 1977 (biogr. sketch).
Taton, R., ed., Science in the 19th cent. 383. 1965.
Thiry, G., Bull. Soc. mycol. France 29: 293. 1913 (souvenir de Quélet, portr.).
Wittrock, V.B., Acta Horti Berg. 3(3): 111. *pl. 141.* 1905 (portr.).

COMPOSITE WORKS: (1) With M.C. Cooke, *Clav. syn. hymenomyc. eur.* 1878, see TL-2/1206; IDC 6257.
(2) Collaborator, A. Mougeot, *Fl. Vosges, Champ.* 1887, TL-2/6371.
(3) *L'interprétation des plantes de Bulliard* et leur concordance avec les noms actuels, Toulouse 1895, 39 p., Revue mycologique 1895, see TL-2/909.

EPONYMY: *Queletia* E.M. Fries.

8414. *Les champignons du Jura et des Vosges* ... Paris (J.B. Baillière et fils ...) [1872-1875]. 3 vols. Oct. (*Champ. Jura Vosges*).
Publ.: For a bibliographical analysis of this publication and its supplements see M.A. Donk, Introductory note, to the Asher reprint of 1964. This reprint contains not only the three original parts but also the supplements 4-22. Publication took place as follows:

1: Aug-Dec 1872 (in journal), p. [1]-320, reprinted and to be cited from Mém. Soc. Émul. Montbéliard ser. 2. 5: [43]-332, *pl. 1-23* and [*1*] (partly col. liths.). – The plates were also separately issued with a t.p. as above, marked "planches", p. [i], *pl. 1-23* and [*1*]. – Supplément 1 on p. 317-321.

2: 1873 (after 24 Feb, date of preface), p. [321]-424, reprinted and to be cited from id. ser. 2. 5: [333]-427, *pl. 1-5* (uncol. liths.), [1, expl. pl.]. Plates also separately publ. 1873, p. [i-iv], *pl. 1-5*. – Supplément 2 on p. 338-360.

3: 1875 (BSbF 28 Jan 1876), p. [1]-128, *pl. 1-4* (uncol. liths.), reprinted and to be cited from id. ser. 2. 5: [429]-556. *pl. 1-4*. – Supplément 3 on p. 433-450.

Copies: BR, FH, G, (partly as reprints, partly in journal), MICH, NY, Stevenson.
Supplements: Nos. 1-3 included in main volumes. The other numbers were issued as follows:

Suppl. 4: Bull. Soc. France 23: 324-xl through 332-xlviii. *pl. 2, 3*. 1876, published 1877.
Suppl. 5: Bull. Soc. bot. France 24: 317-xxv through 332-xl. *pl. 5, 6*. 1877, published 1878.
Suppl. 6: Bull. Soc. bot. France 25: [287]-292. *pl. 3*. 1878, published 1879. [Not marked as suppl. 6].
Suppl. 7: Bull. Soc. bot. France 26: 45-54. 1879, published before 16 Feb 1880. [Not marked as suppl. 7].
Suppl. 8: Bull. Soc. bot. France 26: [228-236]. 1879, published Aug 1879-16 Feb 1880 [Not marked as suppl. 8].
Suppl. 1, hors série: Grevillea 8: 37-38. 1879.
Suppl. 2, hors série: Grevillea 8: 115-117, *pl. 131*. 1880.
Suppl. 9: Bull. Soc. Amis Sci. nat. Rouen sér. 2. 15: 151-184, 185-195. *pl. 1-3*. 1879, published 1880. Also issued as a reprint, see below, Quélet & Le Breton 1880.
Suppl. 10: C.R. Ass. franç. Av. Sci. (Reims, 1880) 9: 662-675, *pl. 8-9*. 1881; reprinted p. [1]-15, *pl. 8-9*.
Suppl. 11: C.R. Ass. franç. Av. Sci. (La Rochelle, 1882) 11: 387-412. *pl. 11-12*. 1883; reprinted p. [1]-26, *pl. 11* (only plate) (Bot. Zeit. 26 Oct 1883).
Suppl. 12: C.R. Ass. franç. Av. Sci. (Rouen, 1883) 12: 498-512. *pl. 6-7*. 1884 (Nat. Nov. Jul(2) 1884); reprinted p. [1]-15, *pl. 6-7* (Bot. Zeit. 29 Aug 1884).
Suppl. 13: C.R. Ass. franç. Av. Sci. (Blois, 1884) 13: 277-286. *pl. 8*. 1885; reprinted p. [1]-10. *pl. 8*.
Suppl. 14: C.R. Ass. franç. Av. Sci. (Grenoble, 1885) 14(2): 453. *pl. 12*. 1886; p. [1]-11. *pl. 12*.
Suppl. 15: C.R. Ass. franç. Av. Sci. (Nancy, 1886) 15(2): 484-490. *pl. 9*. 1887; reprinted p.[1]-8. *pl. 9*.
Suppl. 16: C.R. Ass. franç. Av. Sci. (Toulouse, 1887) 16(2): 587-592. *pl. 21*. 1888, reprinted p. [1]-6, *pl. 21* (supplied col. or uncol.).
Suppl. 17: C.R. Ass. franç. Av. Sci. (paris, 1889) 18(2): 508-514. *pl. 15*. 1890; reprinted p. [1]-7, *pl. 15* (p. 7 dated Apr 1890).
Suppl. 18: C.R. Ass. franç. Av. Sci. (Marseille, 1891) 20(2): 464-471, *pl. 2-3*. 1892; reprinted p. [1]-8, *pl. 2-3* (Nat. Nov. Jun 1892, Aug 1892; p. 8: Apr 1892.).
Suppl. 19: C.R. Ass. franç. Av. Sci. (Besançon, 1893) 22(2): 484-490. *pl. 3*. 1894; reprinted p. [1]-7, *pl. 3* (p. 7: Mai 1894).
Suppl. 20: C.R. Ass. franç. Av. Sci. (Bordeaux, 1895) 24(2): 616-622, *pl. 6*. 1896; reprinted p. [1]-8, *pl. 6*.
Suppl. 21: C.R. Ass. franç. Av. Sci. (Saint Étienne, 1897) 26(2): 446-453. *pl. 4*. 1898; reprinted p. [1]-8. *pl. 4*.
Suppl. 22: C.R. Ass. franç. Av. Sci. (Ajaccio, 1901) 30(2): 494-497. *pl. 3*. 1902; reprinted p. [1]-4, *pl. 3*.
Copies reprints: BR, G.

Facsimile ed.: 1964, 701 p., *58 pl.*, preface by M.A. Donk, Amsterdam (A. Asher) 1964, ISBN 90-6123-122-1. *Copy*: PH.

8415. *Catalogue des mousses sphagnes et hépatiques des environs de Montbéliard* ... [Paris (J.B. Baillière et fils 1872]. Oct.
Publ.: Aug-Dec 1872 (in journal), p. [1]-42. *Copies*: FH, G. – Reprinted and to be cited from Mém. Soc. Émul. Montbéliard ser. 2. 5: [1]-42.

8416. *Champignons récemment observés en Normandie*, aux environs de Paris et de La Rochelle, en Alsace, en Suisse et dans les montagnes du Jura et des Vosges par L. Quélet ... Contributions à la flore mycologique de la Seine-inférieure par André Le Breton, ... Rouen (Imprimerie Léon Deshays ...) 1880. Oct. (*Champ. observ. Normandie*).
Co-author: André Le Breton (*fl.* 1880).
Publ.: 1880 (p. 4: Nov 1879), p. [1]-46, [47, expl. pl.], *pl. 1-3*. *Copies*: FH, G, MICH. – Also with pagination [1]-34, [1, expl. pl.] when p. 185-195 have not been added. – Reprinted from Bull. Soc. Amis Sci. nat. Rouen ser. 2. 15: 1879: 151-184, 185-194, *pl. 1-3*; to be regarded as *suppl. 9* to the *Champ. Jura Vosges*. (see above).

8417. *Enchiridion fungorum* in Europa media et praesertim in Gallia vigentium ... Lutetiae [Paris] (Octavi Doin, ...) 1886, 18-mo. (*Enchir. fung.*).
Publ.: Jan-Jun 1886 (p. vi: 25 Jan 1885; t.p. 1886; Nat. Nov. Jul(1) 1886; announced for late Jul 1885 by Bull. Soc. Mycol. France 1: 132. Mai 1885; Bot. Zeit. 30 Jul 1886), p. [i]-vi, [vii-viii], [1]-352). *Copies*: BR, FH, FI, G, H, MICH, NY, PH, Stevenson, USDA.
Ref.: Patouillard, N., Bull. Soc. bot. France 33(bibl): 151-152. Sep-Dec 1886.
Wettstein, R. v., Bot. Centralbl. 30(6): 161-164. 1887 (rev.).

8418. *Flore mycologique de la France* et des pays limitrophes ... Paris (Octave Doin, ...) 1888. 18-mo. (*Fl. mycol. France*).
Publ.: 1888 (p. [xx]: 1 Apr 1888; Bot. Zeit. 27 Jul 1888; Nat. Nov. Jun(3) 1888), p. [i*-iii*], [i]-xviii, [xix-xx], [xxi], 2 charts, [1]-492. *Copies*: BR, FH, G, MICH, NY, Stevenson, USDA.
Facs. ed: 1962, p. [i-viii, incl. 1 chart], [i]-xviii, [xix-xx], [1]-492, chart, Amsterdam (A. Asher & Co.) 1962. ISBN 90-6123-123-x. *Copies*: FAS, MICH.
Essai de table de concordance: L. Magnin & A. Chomette, Essai de table de concordance des principales espèces mycologiques avec la Flore de France & des pays limitrophes de Lucien Quélet, Lons-le-Saunier (Imprimerie et lithographie Lucien Declume) 1906. Oct, p. [i, iii], [1]-100. *Copies*: BR, G, NY, Stevenson.
Facs. ed.: 1963, Amsterdam (A. Asher & Co.), as orig. *Copy*: FAS; ISBN 90-6123-124-8. – *Authors*: Louis-Alexandre Mangin (1852-1937), A. Chomette.

8419. *Flore monographique des Amanites et des Lépiotes* ... Paris (Masson et Cie., ...) 1902. Oct. (*Fl. Amanit. Lépiot.*).
Co-author: Frédéric Bataille (1850-1946).
Publ.: 1902 (p. 88: printing finished 1 Mar 1902; Bot. Centralbl. 20 Aug 1902), p. [1]-88. *Copies*: BR, FH, G, H, MICH, USDA.

Quelle, Ferdinand Friedrich Hermann (1876-1963), German botanist (bryologist) and entomologist; Dr. phil. Göttingen 1902; assistant Botanical Museum Göttingen 1904-1906; from 1908-1939 active as high school teacher (Studienrat). (*Quelle*).

HERBARIUM and TYPES: JE; original herbarium for his thesis: GOET (Bryoph.); further cryptogams at B, L and MANCH.

BIBLIOGRAPHY and BIOGRAPHY: Barnhart 3: 118 (b. 10 Dec 1876); CSP: 18: 5; GR p. 121-122; IH 2: (in press); SBC p. 129; TL-2/4938.

BIOFILE: Anon., Mitt. deut. entom. Ges. 20: 67. 1961 (80th birthday), 22(1): 5. 1963 (d.).
Gilbert, P., Comp. biogr. lit. entom. 307. 1977.

Schuster, R.M., Hepat. Anthoc. N. Amer. 1: 89. 1966.
Wagenitz, G., Letter to authors 1981 (d. 30 Nov 1963); Index coll. princ. herb. Gott. 130. 1982.

HANDWRITING: Wagenitz, G., Index coll. princ. Gott. 208. 1982.

8420. *Göttingens Moosvegetation.* Inaugural-Dissertation zur Erlangung der Doktorwürde der Philosophischen Fakultät der Georg-Augusts-Universität zu Göttingen vorgelegt von Ferdinand Quelle aus Nordhausen. Nordhausen (Druck von Fr. Eberhardt) 1902. Oct. (*Götting. Moosveg.*).
Thesis-issue: Feb-Mai 1902 (oral exam. 4 Feb 1902; ÖbZ Mai 1902; AbZ 15 Jul 1902; Bot. Zeit. 1 Aug 1902; Hedwigia 23 Jun 1902), p. [1]-163, [164, err.], [165, vita]. *Copies*: FH, M, NY.
Commercial issue: Jan-Mai 1902, p. [1]-163, [164]. *Copies*: G, MO-Steere, U-V.
Ref.: Diels, L., Bot. Jahrb. 33 (Lit.): 13-14. 1903.
   Geheeb, A., Bot. Centralbl. 90: 69-70. 22 Jul 1902.
   Wehrhahn, *in* A. Kneucker, Allg. bot. Z. 8: 131-132. 15 Jul 1902.

Quer y Martinez, José (1695-1764), Spanish botanist. (*Quer*).

HERBARIUM and TYPES: MA; other material in G (see Briquet 1919) and S (herb. Montin).

BIBLIOGRAPHY and BIOGRAPHY: AG 5(1): 776; Barnhart 3: 119; BM 4: 1633; Bossert 320; Colmeiro 1: cxciv; Colmeiro penins. p. 163-165; Dryander 3: 26, 47, 145, 481, 500; Herder p. 472 [index]; Hortus 3: 1202; Jackson p. 339; Kew 4: 383; Langman p. 602; Lasègue p. 289, 536; NI 1574, 1: 157; PR 7380; ed. 1: 8276; Rehder 5: 697; RS p. 134; SO 644; TL-1/1017; TL-2/see C.G. Ortega; Tucker 1: 574.

BIOFILE: Briquet, J., Annu. Cons. Jard. bot. Genève 20: 465-478. 1919 (on his botanical coll. at G (through Delessert), b. 26 Jan 1695, d. 19 Mar 1764, portr., biogr., bibl.).
Colmeiro, M., Bot. botanicos Penins. hisp.-lus. 134, 163-165. 1858.
Dandy, J.E., Regn. veg. 51: 18 ff. 1967 (lists generic names).
Fries, Th.M., Bref skrifv. Linné 1(3): 37, 39, 44, 50. 1909, 1(6): 334, 393-401. 1912.
Gomez de Ortega, C., *in* J. Quer, Fl. Españ. 5: p. xi-xxxii. 1784 (biogr.).
Hulth, J.M., Bref Skrifv. Linné 2(1): 406, 410. 1916.
Jessen, K.F.W., Bot. Gegenw. Vorz. 360. 1864.
Lapeyrouse, P., Hist. abr. pl. Pyren. xxvi. 1813.
Willkomm, M., Grundz. Pfl.-Verbr. Iber. Halbins. 5, 26. 1896 (Veg. Erde 1).

EPONYMY: *Quera* Cothenius (1790, *orth. var.*); *Queria* Linnaeus (1753).

8421. *Flora española*, ó historia de las plantas, que se crian en España ... Madrid (por Joachin Ibarra, ...) 1762-1784, 6 vols. (*Fl. españ.*).
Author of vols. 5-6: Casimiro Gomez Ortega (1740-1818).
Publ.: in six volumes, 1762-1784. *Copies*: BR, FI(1-4), HH, G, NY, PH, USDA; IDC 861. – The plates are copper engravings of drawings by Lorenzo Marin menor and Ricarte. Gatherings of 8 p. with vertical chain lines.
   1: 1762 (dedic. 21 Jan 1762), 2 frontisp. p. [i*], [1]-40, map, [1]-402, *pl. 1-4*, *4bis, 5-11*.
   2: 1762 (err. dated 11 Feb 1762), p. [1]-16, [1]-303, *pl. 12-43*.
   3: Dec 1762 or early 1763 (err. dated 4 Dec, authorization 11 Dec 1762), p. [i-xvi], [1]-436, *pl. 1-79*.
   4: Apr-Dec 1764, p. [i, iii], [1]-471, [472], *pl. 1-66* (42 as "24").
   5: 1784, p. [i*, iii*], [i]-xxxii, [1]-538, *pl. 1-7*, [*7bis*], *8-9*, [*9bis*], "Continuacion de la Flora española ...".
   6: 1784, p. [i, iii], [1]-667, *pl. 10-23*.

Quesné, François Alexandre (1742-1820), French author; translator of Linnaeus's *Philosophia botanica*. (*Quesné*).

HERBARIUM and TYPES: Unknown.

BIBLIOGRAPHY and BIOGRAPHY: Barnhart 3: 119 (d. 17 Apr 1820); BM 4: 1634; PR 5426; SO 473; TL-1/729; TL-2/4760.

COMPOSITE WORKS: *Philosophie botanique* de Charles Linné ... 1788, see TL-2/4760 sub no. 18 (other *copy*: PH).

Quételet, Lambert Adolphe Jacques [Adolph Jacob] (1796-1874), Belgian natural scientist; professor of mathematics at Brussels University; director of the Royal Belgian astronomical observatory; perpetual secretary of the Royal Academy of Sciences of Belgium. (*Quételet*).

HERBARIUM and TYPES: Unknown.

BIBLIOGRAPHY and BIOGRAPHY: Backer p. 479; Barnhart 3: 119 (b. 22 Feb 1796, d. 17 Feb 1874); BM 4: 1634; Herder p. 472 (index); Kew 4: 383; PH 364; PR ed. 1: 8279-8280; Rehder 5: 697.

BIOFILE: Anon., Bonplandia 3: 33. 1855, 6: 50. 1858, 9: 12. 1861; Flora 57: 159, 320. 1874 (d.), 62: 404, 718. 1843; Funérailles de Lambert-Adolphe-Jacques Quételet, Bruxelles 1874.
Breure, A.S.H. & J.G. de Bruijn, eds., Levenswerken J.G.S. v. Breda 417. 1979.
Brien, P., *in* Florilège sciences en Belgique 43-68. s.d. (portr., biogr.).
Candolle, Alph. de, Ann. Rep. Smiths. Inst. 1875: 151, 164-165 (obit.).
Collard, A., Ciel et Terre 44(2): 60-74. 1928 (on Q. as mathematician), 44(5): 210-228. 1928 (biogr.), 44(9/10): 310-362. 1928 (biogr. ctd., 32 references to biogr. items, list of portraits), 45(2/3): 89-92. 1929 (on Q. as historian of science), 45(4/5): 127-145. 1929 (ctd), 45(10/11): 302-316. 1929.
Hankins, F.H., Adolphe Quételet as statistician 1968. 135 p.
Huxley, L., Life letters J.D. Hooker 1: 187. 1912.
Iltis, H., Life of Mendel ed. 2, p. 292, 296. 1966.
Jessen, K.F.W., Bot. Gegenw. Vorz. 354. 1884.
Mailly, Éd., Essai sur la vie et les ouvrages de Lambert-Adolphe-Jacques Quételet, *in* Ann. Acad. r. Sci. Belg. 41: 109-297. 1875 (biogr., portr.), also separately. Bruxelles 1875, 191 p.
Pelseneer, P., Ciel et Terre 44(3): 119-129. 1928 (portr., on his work for the Acad. Sci. Belg.).
Poggendorff, J.C., Biogr.-lit. Handw.-Buch 2: 552-553. 1863 (bibl.), 3: 1080. 1898 (bibl.), 6: 2101. 1938 (biogr. refs.).
Schmid, G., Goethe u.d. Naturw. 560 (no. 4239). 1900.
Taton, R., ed., Science in the 19th century 610 [index]. 1965.

EPONYMY: *Queteletia* Blume (1859).

Queva, Charles (*fl.* 1894), French botanist; studied at Lille; from 1902 professor of botany at Dijon. (*Queva*).

HERBARIUM and TYPES: Unknown.

BIBLIOGRAPHY and BIOGRAPHY: Barnhart 3: 119; BM 4: 1634; CSP 12: 594, 18: 7; Rehder 5: 697.

8422. *Recherches sur l'anatomie de l'appareil végétatif des Taccacées & des Dioscorées* ... Lille (Imprimerie L. Danel) 1894. Oct. (*Rech. anat. Taccac. Dioscor.*).
Publ.: Jan-Jun 1894 (Bot. Zeit. 16 Jul 1895; Nat. Nov. Jul(2) 1900), p. [i-iii], [1]-457, *pl. 1-18* (*1-17* uncol. liths., auct.; *18* phot.). *Copies*: BR, G, NY. – Also issued as Mém. Soc. Sci. Lille sér. 4, 20. 1894 (with same pagination).

8423. *Contributions à l'anatomie des monocotylédonées* ... Lille (au siège de l'Université, ...) 1899. Oct. (*Contr. anat. monocot.*).

Publ.: Sep-Dec 1899 (printing finished 1 Sep 1899; Nat. Nov. Jul(2) 1900), p. [i-iii], [1]-162, *pl. 1-11* (uncol. liths.). *Copies*: HH, USDA. – Issued as Trav. Mém. Univ. Lille, tome 7, mém. no. 22.
Continuation: Beih. Bot. Centralbl. 22(1): 30-77. 1 Jun 1907.

Quincy, Charles (*fl.* 1900-1911), French botanist. (*Quincy*).

HERBARIUM and TYPES: AUT.

BIBLIOGRAPHY and BIOGRAPHY: Barnhart 3: 120; BJI 2: 195, 196, 705; IH 2: (in press); Kelly p. 185; LS 21558-21559, 41618.

8424. *Catalogue de la flore Creusotine* comprenant dans un rayon de 20 kilomètres les plantes croissant spontanément, soumises à la grande culture, ou cultivées communément dans les jardins, avec indication de leurs principales propriétés, etc. ... Ouvrage destiné à vulgariser les études élémentaires de botanique locale et à classer les herbiers scolaires ... première partie Creusot (Imprimerie et lithographie G. Martet, ...) 1888.
†. (*Cat. fl. creusot.*).
Publ.: 1888, p. [1]-136. *Copy*: USDA. – Holograph reproduced by lithography; all recto pages 17-127 blank. Preliminary publ.: Feuille jeunes Natural. 11: 107-111. 1881 and Rev. Bot., Bull. mens. Soc. franç. Bot. 3: 294-304. 1885. For later publications see BL 2: 197.

8425. *Florule des alluvions de la Saône aux environs de Chalon* [Proc.-Verb. Soc. Hist. nat. Autun 1898]. Oct.
Publ.: 1898, cover-t.p., p. [1]-15. *Copy*: HH. – Reprinted from Proc.-Verb. Soc. Hist. nat. Autun 1898 (n.v.).

Quintanilha, Aurélio Pereira da Silva (1892-x), Portuguese botanist; Dr. phil. Coimbra 1926; at the Botanical Institute of Coimbra. (*Quintanilha*).

HERBARIUM and TYPES: Some material at LMJ.

BIBLIOGRAPHY and BIOGRAPHY: Ainsworth p. 80, 128, 130, 330; Barnhart 3: 120; IH 1 (ed. 2): 76, (ed. 3): 95, (ed. 4): 104; Kew 4: 385.

BIOFILE: Archer, L.J., Brotéria 44(71): 155-156. 1973 (tribute).
Bullock, A.A., Bibl. S. Afr. bot. 92. 1978.
Exell, A.W. & G.A. Hayes, Kirkia 6(1): 99. 1967.
Fernandes, A., Bol. Soc. Brot. 36: iii-xxx. 1962 (biogr., portr.), Anuar. Soc. Brot. 41: 11-25. 1975.
Nemésio, V., Brotéria 44(71): 175-188. 1975 (biogr.).
Resende, F., Bol. Soc. port. Ci. nat. ser. 2. 4(1): iv-v. 1952 (tribute 60th birthday, portr.), Revista de Biologia 3(2-4): iii-viii. 1963 (id. 70th birthday).
Serra, J.A., Brotéria 44(71): 157-174. 1975 (works, bibl.).
Verdoorn, F., ed., Chron. bot. 1: 30, 31. 1935, 2: 30, 31, 258. 1936, 3: 111, 356. 1937, 4: 85. 1938, 5: 239. 1939.

FESTSCHRIFT: Bol. Soc. Broter. 36, 1962 (portr., biogr. and bibl. by A. Fernandes on p. iii-xxx).

8426. *Contribuição ao estudo dos Synchytrium* ... dissertação para doutortamento ... Coimbra (Imprensa da Universidade) 1926. Oct.
Publ.: 1926, p. [i-vii], [1]-110, *pl. 1-4* (uncol.). *Copy*: FH. – Reprinted and to be cited from Bol. Soc. Broter. ser. 2. vol. 3: 88-195, *pl. 1-4*. 1925 (1926).

8427. *Le problème de la sexualité chez les champignons* recherches sur le genre "Coprinus" ... Coimbra (Imprenta da Universidade) 1933. Oct. (*Probl. sexual. champ.*).
Publ.: 1933 (Nat. Nov. Jan 1934), p. [i-vii], [1]-200. *Copies*: G, NY. – Also issued in Bol. Soc. Broter. ser. 2. 8: 3-99. 1933, with title "Le problème ... chez les Basidiomycètes".

Facsimile ed.: 1968, p. [i*], [i-vii], [1]-100. *Copies*: BR, FAS. Bibliotheca mycologica, Band 13, 3301 Lehre, Verlag von J. Cramer; ISBN 3-7682-0556-8.

Quisumbing, Eduardo (1895-x), Philippine botanist; Dr. phil. Univ. Chicago 1923; in various functions at the College of Agriculture of the University of the Philippines, at the College of Pharmacy of the University of Santo Tomas and with the Division of Botany and National Museum of the Bureau of Science; from 1945-1961 director the National Museum, Manila. (*Quisumb.*).

HERBARIUM and TYPES: PNH; further material in A, AMES, BR, FH, GH, MICH, and UC.

NOTE: Also cited as E. Quisumbing y Argüelles, but not by Q. himself in his curriculum vitae and the information submitted by himself to I.A.P.T. in 1954. (See Philipp. agric. Rev. 12(3): 9). – A major publication, outside our period, is Q's *Medicinal plants of the Philippines*, 1234 p., 1951 (Techn. Bull. Dept. Agric. Philipp. 16; *copies*: HH, USDA).

BIBLIOGRAPHY and BIOGRAPHY: Barnhart 3: 120; BJI 1: 122; Bossert p. 321; IH 1 (ed. 1): 61, (ed. 2): 79, (ed. 3): 100, (ed. 4): 109, (ed. 5): 109, 2: (in press); Kew 4: 385-386; MW p. 381-382, suppl. p. 288; Roon p. 91; SIA 227; SK 1: 420 (portr.), 4: cxxii (portr.), cxxxiii, 85, 5: cccxiii, 8: lxxviii.

BIOFILE: Arditti, J. et al., Orchid biology 1: 143, 154. 1977.
Asis, C.V., Nat. appl. Sci. Bull., Univ. Philipp., 27(1-2): 1-74. 1975 ("Quisumbing and friend", biogr., portr., bibl.).
Jacobs, M., Fl. males. Bull. 29: 2531. 1976, 30: 2740. 1977.
Price, G.R., Kalikasan, Philipp. J. Biol. 5: 2-18. 1975 (biogr. sketch, portr., bibl.), 5: 1-18. 1976.
Quisumbing, E., Curriculum vitae and bibiography, 3 p., mimeographed, Manila [1950] (b. 24 Nov 1895), Philipp. geogr. J. 8: 21-38. 1964 ("Botanical explorations in the Philippines", historical survey 1588-1901, notes on collectors).
Steenis, C.G.G.J. van, Flora males. Bull., 13: 551. 1957, 14: 625. 1959, 15: 709. 1960, 17: 878. 1962 (retirement), 18: 977. 1963.
Verdoorn, F., ed., Chron. bot. 1: 238, 335. 1935, 2: 253, 373. 1936, 4: 274. 1938, 6: 302. 1941.

EPONYMY: X *Quisumbingara* [*Aerides* Loureiro (1790) X *Papilionanthe* Schlechter (1915) X *Vanda* W. Jones ex R. Brown (1820)] L.A. Garay et H.R. Sweet (1974); *Quisumbingia* Merrill (1936).

8428. *Philippine Piperaceae* ... Separate from the Philippine Journal of Science ... volume 43, no. 1. September 1930. Manila (Bureau of Printing).
Publ.: Aug 1930 (USDA copy of journal rd. 26 Aug 1930), cover-t.p., p. [1]-246, *pl. 1-24*. *Copies*: HH, NY, U, US. – To be cited from journal.

Raab, Ludwig (*fl.* 1900), German (Bavarian) botanist and high school teacher at Straubing (Bavaria). (*Raab*).

HERBARIUM and TYPES: M.

BIBLIOGRAPHY and BIOGRAPHY: Tucker 1: 575.

8429. *Die Blütenpflanzen von Straubing und Umgebung* zusammengestellt von Professor Dr. Ludwig Raab. Programm zum Jahresbericht der Kgl. Realschule Straubing vom Schuljahr 1899/1900. Straubing (Cl. Attenkofer'sche Buch– und Accidenzdruckerei) 1900. Oct. (*Bl.-Pfl. Straubing*).
Publ.: Oct-Nov 1900 (Nat. Nov. Dec(2) 1900; Bot. Zeit. 1 Mar 1901), p. [1]-64. *Copy*: M. – Inf. I. Haesler.
Flora straubingensis, Verzeichniss der um Straubing wild wachsenden Gefässpflanzen. Zusammengestelt von Dr. Ludwig Raab, ..." s.l.n.d. p. [i], [73]-112. *Copy*: HH. – Unidentified reprint.

Raatz, Georg Victor (*fl.* 1937), German palaeobotanist; Dr. phil. Berlin 1937. (*Raatz*).

HERBARIUM and TYPES: Unknown.

8430. *Mikrobotanisch-stratigraphische Untersuchung der Braunkohle des Muskauer Bogens* ... Berlin (Im Vertrieb bei der Preussischen Geologischen Landesanstalt ...) 1937. Oct. (*Mikrobot. Unters. Braunk. Muskau*).
Publ.: Dec 1937 (imprimatur 25 Nov 1937, p. 48), p. [1]-48, *pl. 1* with text. *Copy*: Rijksmus. Geol. Leiden. – Issued as Abh. Preuss. Geol. Landesanst. ser. 2, Heft 183. – Dissertation.

Rabanus, Adolf (1890-?), German (Prussian) botanist. (*Rabanus*).

HERBARIUM and TYPES: Unknown.

BIBLIOGRAPHY and BIOGRAPHY: Barnhart 3: 121; BM 8: 1042.

8431. *Beiträge zur Kenntniss der Periodizität und der geographischen Verbreitung der Algen Badens*. Inaugural-Dissertation zur Erlangung der Doktorwürde einer hohen wissenschaftlich-mathematischen Fakultät der Albert-Ludwigs-Universität Freiburg i. Br. ... Naumburg a.d.S. (G. Pätz'sche Buchdruckerei Lippert & Co. ...) 1915. Oct.(*Beitr. Period. Alg. Bad.*).
Publ.: 1915 (ABZ 30 Jun 1915), p. [i-iii], 1-158, *pl. 1-2* (charts). *Copies*: G, M. – Ber. naturf. Ges. Freiburg 21, 1915.
Ref.: Irmscher, E., Bot. Jahrb. 54(Lit.): 24-25. 27 Jun 1916.

Rabenau, [Benno Carl August] Hugo von (1845-1921), German (Prussian) botanist at Görlitz, Oberlausitz; studied in Bern, Leipzig and Halle; participated in the 1870/71 Franco-Prussian war; Dr. phil. Göttingen 1874; high school teacher (girls school) at Görlitz 1875-1885; living in New York and Hoboken 1885-1895 as chemical analyst; curator of the museum of the Naturf. Ges. Görlitz from 1895-1901; perpetual director 1901-1921. (*Rabenau*).

HERBARIUM and TYPES: GLM; further material (400 nos.) from New Jersey and New York at B (for the greater part destroyed).

BIBLIOGRAPHY and BIOGRAPHY: Barnhart 3: 212; BM 4: 1635; IH 2: (in press); Jackson p. 268; Rehder 1: 315, 387, 5: 699; Urban-Berl. p. 386.

BIOFILE: Dunger, W. & G. Vater, Sächs. Heimat-Bl. 26: 251-255. 1980 (Rabenau material in Herbarium lusaticum).
Hartmann, A., Abh. naturf. Ges. Görlitz 29: 6-11. 1922 (obit., portr., based on original sources), also Jahresber. Ver. Schles. Ornith. 7: 115-116. 1922.
Pax, F., Bibl. schles. Bot. 54, 139, 147. 1929.

8432. *Die Gefässkryptogamen, Gymnospermen und monocotyledonischen Angiospermen der* Königl. Preussischen Markgrafschaft *Ober-Lausitz*. Inaugural-Dissertation zur Erlangung der philosophischen Doctorwürde an der Georgia Augusta zu Göttingen ... Halle a.S.

Publ.: 1874 (Bot. Zeit. 19 Jun 1874), p. [1]-100. *Copies*: C (inf. A. Hansen), NSUB (Göttingen; inf. G. Wagenitz).
Commercial issue: 1874, cover-t.p., p. [5]-100. *Copy*: HH. – "*Die Gefässkryptogamen, ... des* Königl. preussischen Markgrafthums Ober-Lausitz ..." Halle a.S. (id.) 1874. Oct.

Rabenhorst, Gottlob Ludwig (1806-1881), German (Brandenburg, Prussian) cryptogamist; founder of Hedwigia; studied pharmacy in Belzig and Berlin 1822-1830; pharmacist at Luckau 1831-1840; Dr. phil. Jena 1841; "Privatgelehrter" dedicating himself entirely to cryptogamy at Dresden 1840-1875, at Meissen 1875-1881; outstanding collector; published numerous extensive series of cryptogamae exsiccatae. (*Rabenh.*).

HERBARIUM and TYPES: Rabenhorst published many series of exsiccatae which are found in numerous herbaria. The specimens distributed in these exsiccatae were collected by himself as well as by numerous collaborators. We list here the main series in limited detail. The collections which had not yet been distributed at the time of his death were in part further distributed in continuations of his series by others. Other material including his personal herbarium, was dispersed by sale. R. Friedländer acquired the remaining stock in 1889; the remaining sets of exsiccatae were first announced for sale by G.A. Kaufmann in Bot. Zeit. 40: 176, 10 Mar 1882. A certain amount of this material reached B which also held (before 1943) the quasi-totality of his published sets. Important collections of sets of Rabenhorst's exsiccatae are found in the world's major herbaria such as B (extant), BM, BPI, C, FH, G, GJO, H, IMI, NY, ODU, PH, S, UPS. – 33 letters to S.O. Lindberg are at Helsinki Univ. Library; letters to D.F.L. von Schlechtendal at HAL.

EXSICCATAE: (1) *Flora lusatiae inferioris exsiccata*, cent. 1-10, see G. Sayre, Mem. New York Bot. Gard. 19(1): 41. 1969. No set known to us.

(2) *Kryptogamen Sammlung*, systematische Uebersicht über das Reich der Kryptogamen in getrockneten Exemplaren. See G. Sayre, Mem. New York Bot. Gard. 19(1): 41-42. 1969. Set at HAL.

(3) *Klotzschii herbarium vivum mycologicum* sistens fungorum per totam Germaniam crescentium collectionem perfectam, Dresden 1842-1855, cent. 3-20.
Note: continuation by Rabenhorst, of J.F. Klotzsch, *Herbarium vivum mycologicum* (cent. 1-2, details see under Klotzsch). For full details on the sets of exsiccatae see J. Kohlmeyer, Index alphabeticus Klotzschii et Rabenhorstii herbarii mycologici, Beih. Nova Hedw. 4, 1962, and J. Stevenson, Taxon 16: 112-119. 1967 and Beih. Nova Hedw. 36: 274-296. 1971, as well as IH 2: 369.
(3a) Idem, *editio nova*, cent. 1-8, Dresden 1855-1858.
(3b) Idem, *editio nova, series secunda*, Dresden 1859-1881. *Note*: for full details see Kohlmeyer (1962), Stevenson (1967, 1971), and IH 2: 369. The centuries are practically all reviewed in Bot. Zeit. (by D.F.L. von Schlechtendal) and Flora (A.E. Fürnrohr), often with the diagnoses of new taxa reprinted.

(4) *Die Bacillarien Sachsens*. Ein Beitrag zur Fauna von Sachsen. Gesammelt und herausgegeben von Dr. L. Rabenhorst. Dresden & Leipzig (Arnold) 1849-1852, fasc. 1-7.
Publ.: Dec 1848-Aug 1852, fasc. 1-7, nos. 1-70, tabulae 1-5 (one tabula issued with each of fasc. 3-7). For full details see G. Sayre, Mem. New York Bot. Gard. 19(1): 95-96. 1969. – Additional information: set 2 was received at Regensburg Jul 1849.
Sets: AWH, B, BM, FH, H, HAL, L, LD, OS, PH, S, W.
Reviews by D.F.L. von Schlechtendal, Bot. Zeit. 7: 87-88. 2 Feb 1849(1), 7: 448. 15 Jun 1849(2), 8: 581 2 Aug 1850(3), 9: 559-560. 1 Aug 1851(4), 9: 685. 19 Sep 1851(5), 10: 354-355. ("254-255"). 14 Mai 1852(6), 10: 725-726. 8 Oct 1852(7); by A.E. Fürnrohr, Flora 32: 446-447. 28 Jul 1849(2), 34: 592. 1851 (3-5), 35: 394. 7 Jul 1852(6), 35: 669-670. 14 Nov 1852(7).
Index: see Rabenhorst (1856), below.

(5) *Die Algen Sachsens*. Gesammelt und herausgegeben von Dr. L. Rabenhorst, Dresden (Leipzig 1848-1860, decades 1-100, nos. 1-1000. – For full details on dates (taken from the title pages, prefaces and indexes) contents, labels, references, etc. see G. Sayre (1969, p. 92-95).
Sets: AWH, B, BM, C, FH, FR, GOET, H, HAL, L, NY, PC, S, STR, W.
Index: see Rabenhorst 1856, below no. 8438. *Continuation*: Die Algen Europas, see below.
Reviews: practically all decades are reviewed (with a listing of the species) by D.F.L. von Schlechtendal in Bot. Zeit. 7 (1849), 18 (1860); many are also reviewed by A.E. Fürnrohr, in *Flora*.

(6) *Algae marinae siccatae* (exsiccatae). fasc. 1-12, nos. 1-600, 1852-1862. See e.g. TL-2/2: 253 (sub R.F. Hohenacker) and G. Sayre (1969, p. 79-80) for further details.
Algae marinae siccatae ... Eine Sammlung europäischer und ausländischer Meeralgen in

getrockneten Exemplaren mit einem kurzen Text versehen von Dr. L. Rabenhorst und G. von Martens [Georg Mathias von Martens (1788-1872), see TL-2/3: 315-317)] ... Herausgegeben von R.F. Hohenacker. Esslingen bei Stuttgart beim Herausgeber und in Commission bei C. Weichardt. – *Other title*: *Algae selectae siccatae*. Eine Auswahl von 50 der in wissenschaftlicher Hinsicht und wegen ihrer Benutzung merkwürdigsten Algen, ...

1: Jan 1852, nos. 1-50 (D.F.L. von Schlechtendal, Bot. Zeit. 10: 117-119. 6 Feb 1852; A.E. Fürnrohr, Flora 35: 648-656. 7 Nov 1882).
2: Mar 1852, nos. 51-100 (id. Bot. Zeit. 10: 269. 9 Apr 1852; id. Flora 35: 648-656. 7 Nov 1882).
3: Oct 1853, nos. 101-150 (1-50) (id. Bot. Zeit. 11: 903. 23 Dec 1853; id. Flora 36: 678-679. 14 Nov 1853).
4: Dec 1854, nos. 151-200 (id. Bot. Zeit. 13: 123-124. 16 Feb 1885; id. Flora 38: 11-13. 7 Jan 1855).
5: 1855, nos. 201-250 (id. Bot. Zeit. 14: 271-272. 11 Apr 1856, as of "1855").
6: Dec 1857, nos. 251-300 (id. Bot. Zeit. 16: 35. 22 Jan 1858; id. Flora 41: 46-47. 21 Jan 1858).
7: 1859, nos. 301-350 (id. Bot. Zeit. 18: 20. 13 Jan 1860).
8: 1860, nos. 351-400 (id. Bot. Zeit. 18: 339-340. 19 Oct 1860).
9: 1860, nos. 401-450 (Bot. Zeit. 19: 304. 11 Oct 1861).
10: 186, nos. 451-500 (Bot. Zeit. 21: 39-40. 30 Jan 1863).
11: 1862, nos. 501-550 (Bot. Zeit. 21: 206-207. 26 Jun 1863).
12: 1862, nos. 551-600 (Bot. Zeit. 23: 28. 20 Jan 1865).
Sets: B, BAS, BM, FH, G, GOET, HAL, KIEL, L, M, NY, STR, UPS, W.
Notes: "Rabenh. Handb." on the labels refers to *Deutschl. Krypt. Fl.* 1844-1848. No. 600 contains the description of *Stenodesmia binervis* Kützing, n. gen. et sp.

(7) *Kryptogamensammlung für Schule und Haus*, Dresden 1854-1855, 500 specimens, see Bot. Zeit. 12: 151-152. 1854; Flora 37: 191-192. 1854; Junk, Rara 1: 19. 1900-1923 and Sayre, G., Mem. New York Bot. Gard. 19(1): 42. 1969. *Text*: *Cursus der Kryptogamensammlung* Dresden 1855. – No set known to us.
Ref.: Rabenhorst, L., Flora 37: 191-192. 28 Mar 1854 (announcement).
Itzigsohn, H., Bot. Zeit. 13: 224-226. 30 Mar 1855.

(8) *Lichenes europaei exsiccati*. Die Flechten Europa's unter Mitwirkung mehrerer nahmhafter Botaniker gesammelt und herausgegeben von Dr. L. Rabenhorst, Neustadt-Dresden (gedruckt bei C. Heinrich) 1855-1879, fasc. 1-36.

Fasc.	pages	numbers	dates
	[i] = t.p.]		
1	[i]-vi	1-25	Aug 1855
2	[i]	26-55	Dec 1855
3	[i]	56-83	Mar 1856
4	[i]	84-112	Sep 1856
5	[i]	113-142	Sep 1856
6	[i]-iv	143-172	Feb 1857
7	[i]	173-200	Mar 1857
8	[i, iii]	201-228	Jul 1857
9	[i]	229-259	Jul 1857
10	[i, iii]	260-310	Dec 1857
11	[i]	311-336	Dec 1857
12	[i]	337-364b	Apr 1858
13	[i]	364-389	Mai 1858
14	[i]	390-416	Sep 1858
15	[i]	417-442	Jun 1859
16	[i]	443-469	Jun 1859
17	[i]	470-490	Dec 1859
18	[i]	497-521	Jun 1860
19	[i]	522-547	Mar 1861
20	[i]	548-573	Mai 1861

Fasc.	pages	numbers	dates
21	[i-iii]	574-598	Oct 1861
22	[i]	599-624	Oct 1861
23	[i, iii]	625-650	Jul 1862
24	[i]	651-675	Nov 1862
25	[i]	676-700	Apr 1863
26	[i]	701-715	Apr 1864
27	[i]	716-750	Jan 1865
28	[i]	751-775	Sep 1866
29	[i]	776-800	Jun 1867
30	[i]	801-825	Aug 1868
31	[i]	826-850	Mai 1869
32	[i]	851-875	Feb 1870
33	[i]	876-900	Dec 1870
34	[i]	901-925	Jan 1872
35	[i]	926-950	Apr 1874
36	[n.v.]	951-974	Mai 1879

Index generum specierum ... fasc. 1-7, n. 1-200, Neustadt, Dresden (Druck von C. Heinrich) [1857], p. [1]-9 (p. 9: 23 Feb 1857).
Dates: the sets were announced and reviewed by Flora, Bot. Zeit. and Hedwigia. D.F.L. von Schlechtendal reviewed sets 1-28 in Bot. Zeit., A. de Bary sets 29-33 in id. The dates given are the earliest months in which they were mentioned in the three periodicals. In addition to these numerous reviews and notes, see G. Sayre, Mem. New York Bot. Gard. 19(1): 156. 1969, and Lynge (1913) p. 113.
Sets: B, BM, BR, DBN, FH, GOET, GJO, H, HAL, IBF, L, M, MICH, MSC, MSTR, NY, PRE, S, SAM, STR, STU, U, US, WRSL.

(9) *Hepaticae europaeae*. Die Lebermoose Europa's unter Mitwirkung mehrerer namhafter Botaniker, dec. 1-66, nos. 1-660, Dresden 1855-1879.

Decas	dates	decas	dates	decas	dates
1-2	Feb-Mar 1855	23-24	Jan-Feb 1863	45-47	Sep-Oct 1869
3-4	Jan-Feb 1856	25-26	Mar-Apr 1863	48-50	Jan-Feb 1871
5-6	Sep-Oct 1856	27-28	Mar-Apr 1863	51-52	Nov-Dec 1871
7-8	Apr-Mai 1857	29-30	Nov-Dec 1863	53-55	Mai-Jun 1872
9-10	Jun-Jul 1859	31-33	Feb-Apr 1863	56-57	Aug-Sep 1873
11-12	Jun-Jul 1859	34-35	Jan-Apr 1864	58-59	Aug-Sep 1873
13-14	Mar-Apr 1860	36-37	Apr-Mai 1864	60-61	Oct-Nov 1874
15-16	Feb-Mar 1861	38-39	1867	62-64	Jan-Feb 1877
17-18	Apr-Mai 1861	40-41	1867	65-66	Apr-Mai 1879
19-20	Jan 1862	42-44	Oct-Nov 1868		
21-22	Apr-Mai 1862				

Sets: (see also IH 2: 233) BM, BR, FH, G, GOET, H, HAL, IBF, K, LAU, LD, LG, PC, STR, W, WRSL.
Index: see below, Rabenhorst 1872 (dec. 1-55).
Co-editor: Carl Moritz Gottsche (1808-1892), see TL-2/1: 974. (Gottsche was co-editor from decas 21 onward).
Dates: taken from the notes and reviews in Bot. Zeit. (by D.F.L. von Schlechtendal, dec. 1-372, and A. de Bary from 42 onward) and from Hedwigia.
Ref.: Junk, W., Rara 1: 113. 1900-1913 (q.v. for numbering; 100 sets published).
Sayre, G., Mem. New York Bot. Gard. 19(2): 242. 1971.

(10) *Die Characeen Europa's* in getrockneten Exemplaren, ... herausgegeben von Dr. A. Braun [in Berlin], Dr. L. Rabenhorst [in Dresden] und Dr. E. Stizenberger [in Constanz]. See under A. Braun, TL-2/1: 307. Additions and corrections to this entry:

Advertisement: Bot. Zeit. 14: 230-232. 1856, Flora 39: 223-224. 1856. – 100 sets will be published. Dated Feb 1856.
1: Oct-Nov 1857, nos. 1-25 (rev. A.E. Fürnrohr, Flora 41: 42-45. 21 Jan 1858).
2: Oct 1857, nos. 26-50 (rev. D.F.L. von Schlechtendal, Bot. Zeit. 17: 383-384. 11 Nov 1859; A.E. Fürnrohr, Flora 42: 669-672. 14 Nov 1859).
3: Sep-Oct 1867, nos. 51-75 (Bot. Zeit. 29 Nov 1867).
4: Mar 1870, nos. 76-100 (rev. A. de Bary, Bot. Zeit. 28: 325. 20 Mai 1870).
5: 1-20 Jun 1878, nos. 101-120 (Bot. Zeit. 21 Jun 1878; Hedwigia Jun 1878).
Additional sets: HAL, MPU.

(11) *Cryptogamae vasculares europaeae*. Die Gefäss-Cryptogamen Europa's. Unter Mitwirkung mehrerer Freunde der Botanik gesammelt und herausgegeben von Dr. Ludwig Rabenhorst. Dresden 1858-1870, fasc. 1-5.

fasc.	nos.	dates	reviews Bot. Zeit.
1	1-25	Mar 1858	16: 95-96. 2 Apr 1858, D. F. L. v. Schlechtendal
2	26-50	Jan 1859	17: 103-104. 18 Mar 1859, id.
3	51-75	Jul 1860	18: 259-260. 20 Jul 1860, id.
4	76-100	Nov 1863	21: 375-376. 27 Nov 1863, id.
5	101-125	Oct 1870	28: 711-712. 4 Nov 1870, A. de Bary

Sets: B, BM, H, IBF, L, LG, MANCH, W.

(12) *Bryotheca europaea*. Die Laubmoose Europa's unter Mitwirkung mehrerer Freunde der Botanik gesammelt und herausgegeben von Dr. L. Rabenhorst. Dresden (Druck von C. Heinrich) 1858-1876, 27 fasc.

fasc.	t.p./cont.	nos.	dates	reviews Bot. Zeit. (Schlechtendal or de Bary)
1	[i, iii]	1-50	Jun 1858	S, 16: 191-192. 25 Jun 1858
2	[i]	51-100	Dec 1858	S, 16: 371-372. 17 Dec 1858
3	[i, iii]	101-50	Dec 1858-Jan 1859	S, 17: 19-20. 14 Jan 1859
4	[i, iii]	151-200	Sep 1859	S, 17: 351-352. 14 Oct 1859
5	[i, iii]	201-250	late 1859 or Jan 1860	
6	[i, iii]	251-300	Feb 1860	S, 18: 56. 10 Feb 1860
7	[i, iii]	301-350	Apr 1861	S, 19: 111-112. 26 Apr 1861
8	[i, iii]	351-400	Mai 1861	S, 19: 127-128. 10 Mai 1861
9	[i, iii]	401-450	Oct 1861	
10	[i, iii]	451-500	Feb 1862	S, 20: 55-56. 14 Feb 1862
11	[i, iii]	501-50	Jul 1862	S, 20: 255. 8 Aug 1862
12	[i, iii]	551-600	Dec 1862	S, 21: 152. 1 Mai 1863
13	[i-iv]	601-650	Sep 1863	S, 21: 320. 16 Oct 1863
14	[i, iii]	651-700	Mar 1864	S, 22: 98-99. 1 Apr 1864
15	[i]	701-750	Feb 1865	S, 23: 68. 24 Feb 1865
16	[i]	751-800	Apr 1865	S, 23: 140. 28 Apr 1865
17	[i]	801-850	Dec 1865 or Jan 1866	
18	[i]	851-900	Jan 1866 (t.p. 1865)	S, 24: 61-62. 23 Feb 1866
19	[i]	901-950	(?) 1867	
20	[i, iii]	951-1000	Dec 1867	
21	[i]	1001-1050	Dec 1868	
22	[i]	1051-1100	Dec 1869	H.S. 28: 15-16
23	[i]	1101-1150	Jul 1871	B, 29: 461. 7 Jul 1871
24	[i]	1151-1200	Mar 1872	

fasc.	t.p./cont.	nos.	dates	reviews Bot. Zeit. (Schlechtendal or de Bary)
25	[i]	1201-1250	Jun 1873	
26	[i]	1251-1300	Feb 1875	
27	[i]	1301-1350	Nov 1875	
28	[i]	1351-1400	Jun 1881	see below
29	[i]	1401-1450	Nov 1884	see below

Index 1-24: see below, Rabenhorst (1872).
Dates: based on notes and reviews in Flora, Bot. Zeit. and Hedwigia.
Further details: Sayre, G., Mem. New York Bot. Gard. 19(2): 243-244. 1971 (mentions also indexes for fasc. 16 and 18).
Continuation: by G. Winter, L. Rabenhorst *Bryotheca europaea et extraeuropaea* ... fasc. 28-29, nos. 1351-1450. Dresden 1881-1884 (fasc. 28: Jun 1881, fasc. 29: Nov 1884). – See Sayre (1971).
Sets: BM, BR, C, E, FH, G, GOET, H, HAL, HBG, IBF, JE, KIEL, L, LAU, LD, MANCH, NY, PC, S, STR, STU, W, WELC.

(13) *Cladoniae europaeae*. Die Cladonien Europa's in getrockneten Exemplaren. Unter Mitwirkung mehrerer Freunde der Botanik gesammelt und herausgegeben von Dr. L. Rabenhorst. Dresden (Druck von C. Heinrich). Fol.
Text: Sep-Nov 1860, p. [i-xii], see below under Rabenhorst 1860, no. 8441.
Exsiccatae: on tabulae i-xxxix, 260 numbers, published Sep-Nov 1860 (Bot. Zeit. rev. 18 Jan 1861; Flora rd. 7 Jan 1861; submitted to Schles. Ges. vaterl. Cultur 29 Nov 1860, fide Flora 44: 79. 1861). (Lynge 1913, p. 113).
Sets: B, FH, FI, H, HAL, KIEL, L, LE, M, MICH, MSTR, PRE, S, U, W, WRSL.
Ref.: Nylander, W., Bot. Zeit. 19: 351-352. 22 Nov 1861 (gives many corrected identifications and other nomenclature).
Sayre, G., Mem. New York Bot. Gard. 19(1): 154. 1969.

(14) *Cladoniae europaeae exsiccati. Supplementum I.* Dresden (Druck von C. Heinrich). Fol.
Text: n.v.
Exsiccatae: on tabulae i-xi, 56 nos. (64 forms), published Jun-Dec 1863 (Hedwigia 2: 82. 1863, suppl. 1: 152. 1863; Bot. Zeit. 22 Jan 1864) (Lynge 1913, p. 113).
Sets: FH, H, S, WRSL.
Ref.: Sayre, G., Mem. New York Bot. Gard. 19(1): 155. 1969.
Schlechtendal, D.F.L. von, Bot. Zeit. 22: 31-32. 1864.

(15) *Die Algen Europa's* (Fortsetzung der Algen Sachsens, resp. Mittel-Europa's) ... Dresden (Druck von C. Heinrich) 1861-1879, decades 1-160, nos. 1101-2600.
Publ.: in 160 decades, Jan 1861(1)-Mar 1879(159), set 160 mentioned by Hedwigia 21: 194. Dec 1882. For full details see G. Sayre, Mem. New York Bot. Gard. 19(1): 93-94. 1969.
Index (nos. 1-2350): see Rabenhorst (1873).
Sets: AWH, B, BM, BR, C, FH, FR, GOET, H, HAL, HBG, IBF, KIEL, L, LD, M, MANCH, MICH, NY, PC, PRE, S, STR, W, WAG, WRSL.
Continuation: *Phycotheca universalis*, by P. Richter & F. Hauck (see e.g. Bot. Monatschr. 3: 125-126. 1885.
Reviews: The earlier decades are usually reviewed in Bot. Zeit. (by D.F.L. von Schlechtendal or A. de Bary).

(16) *Diatomaceae exsiccatae totius terrarum orbis, quas distribuit Dr. L. Rabenhorst.* Semicenturies 1-2, nos. 1-100. Dresden (C. Heinrich) 1871.
Publ.: 1871 (Hedwigia 10: 144. Sep 1871; Bot. Zeit. 29-88. 1 Sep 1871). For further details see G. Sayre, Mem. New York Bot. Gard. 19(1): 96. 1969.
Sets: AWH, BM, FH, L, NY, PH.

BIBLIOGRAPHY and BIOGRAPHY: ADB 27: 89-92 (E. Wunschmann); Ainsworth p. 226, 273, 289, 320, 346; Barnhart 3: 121; BFM no. 3090; Biol.-Dokum. 14: 7271 (partial bibl.); BJI 1: 47; BM 2: 693, 4: 1635; Bossert p. 321; Cesati (Saggio): 319-322; Clokie p. 229; CSP 5: 70, 8: 683-684, 11: 88, 12: 595; De Toni 1: cii-ciii, cxxxix, 2: xcvii-xcviii; DTS 1: 231, 233; Futák-Domin p. 496; GR p. 38, *pl.* [*16*]; GR, cat. p. 70; Hawksworth p. 185, 197; Herder p. 472 [index]; Hortus 3: 1202; IH 1 (ed. 7): 340; Jackson p. 156, 158, 228, 268, 296, 310, 478; Kelly p. 185; Kew 4: 386-387; Krempelh. 2: 174-184; LS 10230, 21567-21618; Maiwald p. 282 [index]; Moebius p. 75, 78, 86, 108; MW p. 398; Morren ed. 2, p. 10; Nordtstedt p. 26, suppl. p. 35; PR 3466, 7381-7387, 8981, 10543, ed. 1: 8283-8286; Rehder 1: 380, 381, 384, 5: 699; Saccardo 1: 134, 2: 88, cron. xxx; SBC 129; SO 858b; Stevenson p. 1255; TL-1/1018-1019; TL-2/see indexes; Tucker 1: 575; Urban-Berl. p. 272, 280, 322.

BIOFILE: Ainsworth, G.C., Dict. fungi ed. 6. 499-500. 1971.
Anon., Bonplandia 3: 33. 1855 (cogn. Leopoldina: Mattuschka), 6: 50. 1858, 7: 69. 1859; Bot. Gaz. 7(5): 52. 1882 (d.); Bot. Not. 1882: 30 (d.); Bot. Zeit. 3: 280. 1845 (Prussian gold medal for science), 22: 32. 1864 (Ritterkreuz des Albrechts-Ordens, Sachsen), 39: 295. 1881 (d.); Bull. Soc. bot. France 28 (bibl.): 60. 1881 (d.); Flora 28: 224. 1845 (Friedrich Wilhelm of Prussia awards the Prussian gold medal for science), 47: 30. 1864 (Albrechts-Orden), 53: 267. 1870 (Prix Desmazières, Acad. Paris), 64: 208. 1881 (d. 24 Apr 1881); Grevillea 9: 137. 1881 (d.); Hedwigia 20(4): 64. 1881 (d.); J. Bot. 19: 286-287. 1881; Leopoldina 17: 102. 1881 (d.); Nat. Nov. 3: 79. 1881 (d.); Österr. bot. Z. 20: 252. 1870 (prix Desmazières), 31: 205. 1881 (d.); Verh. bot. Ver. Brandenb. 23: il. 1881 (d. 24 Apr 1881).
Ascherson, P., Fl. Prov. Brandenburg 1: 10. 1864.
Babington, C.C., Mem. Jour. Bot. Corr. ix, 474. 1897.
Bary, A. de, Bot. Zeit. 39: 435-437. 1881 (obit., b. 22 Mar 1806); excellent evaluation, unusually fair and independent for an obituary).
Bley, C., S.B. Abh. nat. Ges. Isis, Dresden, 1881: 35-38. 1882 (obit.).
Carus, C.G. & C.F.P. von Martius, Eine Altersfreundschaft in Briefen 96, 156. 1939.
Candolle, Alph. de, Phytographie 441-442. 1880.
Deby, J., Bibl. microsc. microgr. Stud. 3, Diat. 45. 1882.
Ewan, J. et al., Short hist. bot. U.S. 77, 79. 1969.
Flotow, J. v., Linnaea 22(3): 353-382. Jun 1849 (Rabenhorst's *Lichenes italici*, coll. by him in 1847).
Frison, Ed., Henri Ferdinand Van Heurck 25, 40. 1959.
Hedge, I.C. & J.M. Lamond, Index coll. Edinb. herb. 121. 1970.
Hein, W.H. & H.D. Schwarz 2: 510-511. 197.
Illig, H., Verh. bot. Ver. Brandenburg 106: 10-11. 1969; Gleditschia 9: 353-360. 1982 (Rabenhorst and the botany of Brandenburg; portr.).
Kolkwitz, R., Ber. deut. bot. Ges. 32: (64). 1914 (P.G. Richter editor of *Krypt. fl.* from 1879).
Lindemann, F. von, Bull. Soc. imp. Natural. Moscou 61(1): 57-58. 1886.
Lynge, B., Nytt. Mag. naturvid. 56: 422-449. 1919 (lists all lichens in exsicc.).
Miège, J. & Wuest, Saussurea 6: 131. 1975 (4700 Rab. exsicc. G.).
Müller, R.H.W. & R. Zaunick, Friedr. Traug. Kützing 20, 292. 1960 ("Kützing hielt wissenschaftlich nicht viel von ihm").
Murray, G., Hist. coll. nat. hist. 1: 175. 1904 (lists Rab. sets of exsicc. at BM).
Pax, F., Bibl. schles. Bot. 6, 7, 53, 54. 1929.
Richter, P., Bot. Centralbl. 7: 379-383. 1881 (obit., bibl., list exsicc., b. 22 Mar 1806, d. 20 Apr 1881), Hedwigia 1881(8): 113-120.
Raab, H., Z. Pilzk. 54(1): 9-11. 1936.
Roumeguère, C., Rev. mycol. 3(11): 25-26. 1881 (obit.).
Sayre, G., Mem. New York Bot. Gard. 19(1): 51. 1969, 19(3): 383. 1975 (further references to Sayre's studies on Rabenhorst exsiccatae are given under the series of exsiccatae themselves); Bryologist 80(3): 515. 1977 (orig. bryoph. destr. at B).
Stafleu, F.A., Regn. veg. 71: 339. 1970.
Steenis-Krusemann, M.J. van, Blumea 25: 43, 45. 1979 (coll. L.).
Stevenson, J., Taxon 16: 112-119. 1967 (on *Fungi europaei exsiccati*).
VanLandingham, S.L., Cat. diat. 6: 3583-3584. 1978, 7: 4221. 1978.
Verdoorn, F., ed., Chron. bot. 2: 80. 1936, 3: 65. 1937, 4: 570. 1938, 5: 122. 1939.
Wagenitz, G., Index coll. princ. herb. Gott. 131. 1982.

Wittrock, V.B., Acta Horti Berg. 3(2): 139. 1903 (b. 22 Mar 1806, d. 24 Apr 1881), 3(3): 148. 1905.
Wood, R.D., Monogr. charac. 1: 834-835. 1965.

COMPOSITE WORKS: (1) Founder of *Hedwigia*, editor of vols. 1-17(10), Mai 1852-Oct 1878 (succeeded by G. Winter).
(2) Preface to J. Nave, *Anleit. Einsamm.* 1864, TL-2/6659.

NOTE: Rabenhorst published a series of reports on his Italian trip to Italy (1847) and a "systematische Uebersicht" of the cryptogams collected in Flora 32: 385-399, 434-444, 1849, 33: 305-313, 322-325, 338-349, 355-363, 372-383, 390-399, 513-525, 529-537, 625-632. 1850.

HANDWRITING: Candollea 32: 401-402. 1977; Gleditschia 9: *pl. xxiii. 1982*.

EPONYMY: *Rabenhorstia* E. Fries (1849); *Rabenhorstia* H.G.L. Reichenbach (1841) probably also honors this author, but the derivation was not given.

8433. *Flora lusatica* oder Verzeichniss und Beschreibung der in der Ober- und Niederlausitz wildwachsenden und häufig cultivirten Pflanzen ... Leipzig (Verlag von Eduard Kummer) 1839-1849, 2 vols. Oct. (*Fl. lusat.*).
1 (Phan.): 14-20 Jul 1839 (TL-1), p. [i]-lxvii, [1-2, err.], [1, h.t.], [1]-336.
2 (Krypt.): 30 Aug-5 Sep 1840 (TL-2), p. [i]-xxii, [1]-507, [508].
Copies: G, HH, MO, NY, USDA.
Preliminary publications: Linnaea 9: 523-565. 1834 (musci), 10: 208-216. 1836 (ferns), 10: 619-640. 1836, 11: 221-247 (Spezielle Übersicht ... Pfl. Niederlausitz); Flora 20: 129-144. 1837 (Chara).
Addenda and corrigenda: L. Rabenhorst, Bot. Centralbl. 1: 189, 236, 325, 341, 365, 381, 404. 1846.
Ref.: Anon., Lit.-Ber. Flora 1839(12): 177-185, 1841(9): 142-147; Linnaea 13 (Litt.): 165-166. 1840.
Kölbing, F.W., Flora 25: 186-192. 1841.

8434. *Deutschlands Kryptogamen-Flora* ... oder Handbuch zur Bestimmung der kryptogamischen Gewächse Deutschlands, der Schweiz, des Lombardisch-Venetianischen Königreichs und Istriens ... Leipzig (Verlag von Eduard Kummer) 1844-1848, 2 vols. Oct. (*Deutschl. Krypt.-Fl.*).
1 (*Pilze*): Aug 1844 (p. iv: Feb 1844; Bot. Zeit. 30 Aug 1844), p. [i]-xxii, [1]-578, [579-582, add.], [583]-613, index, [614, err.].
2(1) (*Lich.*): 1-12 Mar 1845 (p. viii: Dec 1844; Bot. Zeit. 14 Mar 1845; Flora 28 Mar 1845), p. [ii]-xii, [1]-121, [122, err.], [123]-129, [130, index]. – p. [ii]: "*Die Lichenen Deutschlands* ..." alt. t.p.
2(2) (*Algae*): Oct-Nov 1846 (t.p. 1847; Bot. Centralbl. Deutschl. 2 Dec 1846 ("soeben"); Flora 21 Jan 1847), p. [ii]-xix, [xx, err.], [1]-200, [201]-216, index. – p. [ii]: "*Die Algen Deutschlands* ..." alt. t.p.
2(3) (*Leber-, Laubmoose und Farrn*): Lief. *1*: Jul 1848 (Flora 28 Jul 1848), p. [i*], [1]-160.
Lief. *2*: Jan-Feb 1849 (p. iv: Dec 1848), p. [i]-xvi, 161-337, [338]-340, err., [341]-352, index.
Synonymenregister: 1853 (Bonplandia 1 Nov 1853; Flora 28 Oct 1853; Bot. Zeit. 28 Oct 1853), p. [i], [1]-144. – *Synonymenregister zu Deutschlands Kryptogamen-Flora* ... Leipzig (id.) 1853. Oct.
Copies: Almborn, BR, FH, G, H, MO, NY, Stevenson, Teyler, USDA; IDC 5282.
Note: For the *second edition* of Rabenhorst's *Kryptogamen-Flora*, virtually a new work, with a new title and by other authors, see below, TL-2/8450.
Reviews: *1*: Lasch, W.G., Bot. Zeit. 2: 746-748. 25 Oct 1844; –n, Flora 28: 108-110. 21 Feb 1845.
2(1): Flotow, J., Bot. Zeit. 3: 319-325. 9 Mai 1845; Sauter, A.E., Flora 29: 85-90. 14 Feb 1846.
2(2): Schlechtendal, D.F.L. von, Bot. Zeit. 6: 9-12. 7 Jan 1848.
2(3): Müller, K., Bot. Zeit. 7: 369-372. 18 Mai 1849 ("Er hat eine grosse Dreistigkeit,

unbrauchbare botanische Schriften zu liefern ..."); Sauter, A.E., Flora 33: 437-446.
28 Jul 1850; H. Itzigsohn, Bot. Zeit. 7: 441-445. 1849 (rebuttal of Müller's criticism).
Ref.: Isoviita, P., Ann. bot. fenn. 3: 260. 1966 (on *Sphagnum* in 2(3)).

8435. *Botanisches Centralblatt für Deutschland.* Herausgegeben von Dr. L. Rabenhorst. Jahrgang 1846. Leipzig (Verlag von Eduard Kummer) [1846]. Oct. †. (*Bot. Centralbl. Deutschl.*).
Publ.: vol. 1, all published, in 26 dated parts 14 Jan-30 Dec 1846, p. [i]-viii, [1]-552, *pl. 1*. *Copies*: MO, NY, PH.

8436. *Die Süsswasser-Diatomaceen.* (Bacillarien). Für Freunde der Mikroskopie ... Leipzig (Eduard Kummer) 1853. Qu. (*Süssw.-Diatom.*).
Publ.: Mar-Mai 1853 (p. iv: Feb 1853; Bonplandia 1 Aug 1853; Flora 7 Jul 1853; Bot. Zeit. 10 Jun 1853), p. [i]-xii, [1]-72, *pl. 1-10* (with text; uncol. liths. auct.). *Copies*: BR, FH, G, H, M, MICH, NY, PH, USDA; IDC 6398.
Ref.: Anon., Flora 36: 404-408. 7 Jul 1853 (rev.).

8437. *Cursus der Cryptogamenkunde* für Realschulen und höhere Bildungsanstalten, sowie zum Privat-Studium mit Beispielen in natürlichen Exemplaren; oder Text zur Kryptogamensammlung für Schule und Haus ... Dresden (Druck von C. Heinrich) 1855. Oct. (*Cursus Cryptogamenk.*).
Publ.: 1855 (p. vii: Jan 1855; Flora 7 Mar 1855; ÖbZ 17 Mai 1855; Bot. Zeit. 30 Mar 1855), p. [i]-vii, [1]-140, [141]. *Copies*: BR, LC, G. – Text accompanying the series of exsiccatae (q.v. above) *Kryptogamensammlung für Schule und Haus*, Dresden Feb 1854 (Bot. Zeit. 12: 151-152. 3 Mar 1854).
Ref.: Itzigsohn, H., Bot. Zeit. 13: 224-226. 30 Mar 1855.
W.L., Flora 38: 140-142. 7 Mar 1855.

8438. *Alphabetisches Verzeichniss* der Gattungen und Arten welche bis jetzt in Rabenhorst's *Algen und Bacillarien Sachsens* resp. Mitteleuropa's ausgegeben sind. Dresden (Druck von C. Heinrich) 1856. Oct. (*Alph. Verz. Alg. Sachsen*).
Publ.: 1856(p. 4: Mai 1856), p. [1]-17. *Copies*: FI, LC. – Index to the series of exsiccatae *Die Algen Sachsen's* as far as published by 1856, see above under *exsiccatae*. Further index: Stizenberger, Dr. Ludwig Rabenhorst's Algen Sachsens ... Dec 1-100, 1860, 41 p.

8439. *Flora des Königreichs Sachsen.* Nebst Schlüssel zu dem Linné'schen Sexualsysteme und den zum Grunde gelegten natürlichen Systeme ... Dresden (Verlag von C. Heinrich), Leipzig (Karl Friedrich Fleischer) 1859. Oct.
Publ.: Apr 1859 (p. vi: Apr 1859; Flora rd. 28 Apr 1859), p.[i]-lxvi, [lxvii-lxviii, err.], [1]-346], [347 colo.]. *Copy*: HH.
Ref.: Schlechtendal, D.F.L. von, Bot. Zeit. 17: 215-216. 17 Jun 1859.
Fürnrohr, A.E., Flora 42: 582-583. 7 Oct 1859.

Note: Booknumber 8440 has not been used.

8441. *Cladoniae europaeae.* Die Cladonien Europa's in getrockneten Exemplaren. Unter Mitwirkung mehrerer Freunde der Botanik gesammelt und herausgegeben von Dr. L. Rabenhorst. Dresden (Druck von E. Heinrich) 1860. (*Cladon. eur.*).
Publ.: Nov 1860 (Flora late Dec 1860; Schles. Ges. 29 Nov 1860), p. [1]-10 accompanying 39 "tabulae" (i-xxxix) with 260 specimens. *Copies*: FH, FI, L. – A supplement was issued in 1863 (tab. i-xi, 56 nos.), see Bot. Zeit. 22: 31-32. 22 Jan 1864 and Hedwigia 2: 151-152. 1863.
Ref.: Nylander, W., Bot. Zeit. 19: 351-352. 22 Nov 1861.
Schlechtendal, D.F.L. von, Bot. Zeit. 19: 24. 18 Jun 1861, 22: 31-32. 22 Jan 1864 (suppl.).

8442. *Kryptogamen-Flora von Sachsen*, der Ober-Lausitz, Thüringen und Nordböhmen, mit Berücksichtigung der benachbarten Länder ... Leipzig (Verlag von Eduard Kummer) 1863. Oct. †. (*Krypt.-Fl. Sachsen*).

1 (Alg., Bryoph.): 1863 (p. viii: Sep 1862; Bot. Zeit. 7 Apr 1863; BSbB Sep 1863; Flora 28 Feb 1863; J. Bot. Jul 1863), p. [i]-xx, [1]-653.
2(1) (Lich.): Dec 1869 or Jan 1870 (soeben erschienen Hedwigia Dec 1869; ÖbZ 1 Apr 1870; J. Bot. Mar 1870; Bot. Zeit. 12 Nov 1869 (kommt in December)), p. [1]-192.
2(2) (Lich.): Apr 1870 (p. vi: Mar 1870; Bot. Zeit. 13. 28 Mai 1870; BSbF Aug-Dec 1870; Flora 13 Jun 1870; ÖbZ 1 Jun 1870; J. Bot. Oct 1870), p. [i]-xi, 193-400.
Copies: BR, DBN, FH(2), G, H, M, MICH, MO, NY, Stevenson, USDA. – The third part, *Fungi*, was announced for 1871 but did not appear.
Ref.: Arnold, F.C., Flora 53: 91-103. 27 Mar 1870 (2(1)).
Bartsch, Österr. bot. Z. 20: 182-185. 1870.
Bary, A. de, Bot. Zeit. 28: 111-112. 18 Feb 1870.
Hohenbühel-Heufler, L., Österr. bot. Z. 20: 118-120. 1870.
Itzigsohn, H., Bot. Zeit. 21: 148-150. 1 Mai 1863 (rev. vol. 1).
L., Flora 46: 93-96. 28 Feb 1863 (rev.).

8443. *Beiträge zur näheren Kenntniss und Verbreitung der Algen.* Herausgegeben von Dr. L. Rabenhorst ... Leipzig (Verlag von Eduard Kummer) 1863-1865, 2 Hefte. †. (*Beitr. Kenntn. Alg.*).
1: Oct-Nov 1863 (p. 30: 25 Mar 1863; ÖbZ 1 Dec 1863; Hedwigia 1863; Flora rd. Nov 1863), p. [i, iii], [1]-30, *pl. 1-7* (uncol. liths. C. Janisch), p. [1]-16 by Rabenhorst and Carl Janisch (1825-1900), p. 17-22 by Carl August Hantzsch (1825-1880), p. 23-30 by J. Hermann (Characeen).
2: Aug-Sep 1865 (BSbF Jan-Feb 1866; ÖbZ 1 Oct 1865; Flora 27 Sep 1865; Hedwigia Aug-Sep 1865), p. [i, iii], [1]-40, *pl. 1-5* (*1-2* A. Grunow, uncol., *3-5* by F. Cohn, part. col.); p. 1-16 by Albert Grunow (1826-1914), p. 17-32 and 33-40 by Ferdinand Julius Cohn (1828-1898).
Copies: BR, FH, G, M, NY(2), PH; IDC 2784. (no. 1).
Ref.: Groenland, J., Bull. Soc. bot. France 13(bibl.): 1-5. 1866.
H.W.R., Österr. bot. Z. 13: 411-412. 1863, 15: 331-332. 1865.
Schlechtendal, D.F.L. von, Bot. Zeit. 22: 47-48. 1864, 23: 373-374. 1865.

8444. *Flora europaea algarum* aquae dulcis et submarinae ... Lipsiae [Leipzig] (apud Eduardum Kummerum) 1864-1868, 3 vols. (sectiones). Oct. (*Fl. eur. alg.*).
1 (Diat.): Mar-Mai (BSbF Jul 1864; Hedwigia Apr 1864; Flora 10 Jun 1864), p. [i-iii], [1]-359.
2 (Alg. phycochrom.): Jul-Aug 1865 (BSbF Nov 1865; ÖbZ 1 Oct 1865; Flora 19 Aug 1865),p. [i], [1]-319.
3 (Alg. Chlorophyll.) (1): Jan-Feb 1868 (ÖbZ 1 Mar 1868; Hedwigia Dec 1867/Jan 1868; Bot. Zeit. 1 Mai 1868; Flora 11 Apr 1868), p. [1]-320.
3 (id.) (2): Mai-Jun 1868 (see Flora 9 Jun 1868; Hedwigia 7: 80. 1868; ÖbZ 1 Jul 1868; BSbF Dec 1868; BSBB 15 Aug 1868), frontisp., portr., p. [i]-xx, [321]-461. – Main part of preface material may also be bound with vol. 1.
Copies: BR, FH, G, H, HH, M, MO, M, NY, US, USDA; IDC 6399.
Ref.: Anon., Flora 48: 543-544. 1865 (vol. 2).
Bary, A. de, Bot. Zeit. 26: 298-300. 1 Mai 1868 (vol. 3), 26: 496. 1868.
Groenland, J., Bull. Soc. bot. France 15 (bibl.): 193-194. 1869.
Schlechtendal, D.F.L. von, Bot. Zeit. 22: 302-303. 23 Sep 1864 (rev. vol. 1), 23: 392. 22 Dec 1865 (vol. 2).

8445. *Mycologia europaea.* Abbildungen sämmtlicher Pilze Europa's. Gezeichnet und lithographirt von Dr. Gonnermann; mit Text versehen von Dr. L. Rabenhorst ... Neustadt bei Coburg und Dresden (im Selbstverlag der Herausgeber. Druck von C. Heinrich in Dresden) 1869-1882. (*Mycol. eur.*).
Artist: Wilhelm Gonnermann (1806-1884).
1-2 (Amanitae/Agaricini): 1869 (Hedwigia Apr 1869; Flora 10 Sep 1869; Bot. Zeit. 13 Aug 1869), p. [1]-6, *pl. 1-12* (*1-11* col.).
3 (Pezizei): 1869 (Hedwigia Dec 1869; cover: 1870), p. [i, iii], [1]-10, *pl. 1-6* (col.). – Announced as available by Bot. Zeit. on 13 Aug 1869 and Flora 10 Sep 1869, but see preface.
4 (Agarici): Dec 1869 (Hedwigia Dec 1869; cover: 1870), p. [1]-2, *pl. 1-6* (col.).

5-6 (Pyrenomycetes): 1869 (cover 1869), p. [1]-30, *pl. 1-12* (5-8, 11-12 p.p. col.). – The *Synopsis pyrenomycetum europaeorum* is by Berhard Auerswald (1818-1870), the illustrations by Fleischhack.
7 (Polypores, Boletus): Aug-Oct 1870 (Hedwigia Oct 1870; Bot. Zeit. 4 Nov 1870 as publ. "soeben" in Oct 1870), p. [1]-4, *pl. 1-7* (*1-6* col.) (text p. 4 refers to *pl. 8*).
8-9 (Panus et al.): Jan-Feb 1872 (Hedwigia Feb 1872; Flora 53: 128. 11 Mar 1872 lists contents; Bot. Zeit. 1 Mar 1872; letter R. to S.O. Lindberg. 5 Apr 1872: sent), p. [i], [1]-14, *pl.* 7[sic]-*18*.
10 (Agaricus): remained unissued in first instance, *pl. "20-25"* (1-6) distributed in 1882 with a reissue of 7-9 and by R. Friedländer in 1886 (Friedländer bought the remaining stock and distributed some complete sets).
Copies: FH(2), H (in orig. covers, incompl.), ILL (inf. D.P. Rogers), NY, USDA; IDC 883.
Re-issue: Heft 7-9, 1882, see Bot. Centralbl. 10: 44. 1882. Bot. Zeit. 40: 143. 24 Feb 1882.
Commentary: E. Fries, Grevillea 2: 27. 1873 ("many very absurd errors"). See also F.J. Seaver, N. Amer. Cup-fungi 356. 1942; Bail, Flora 52: 446-447. 25 Oct 1869, rev. 1-6; Anon., Flora 52: 471. 20 Nov 1869 (on Rabenhorst's part in the work).
Subscription offer: Bot. Zeit. 26: 807. 20 Nov 1868 (each Heft to consist of 6 plates and text. "Ueber 1000 Tafeln sind fertig und der Text ist im Druck"). See also Bail, Flora 52: 446, 471. 1869 and Bot. Zeit. 28: 155-157. 1870 on the genesis of this work. The plates for Heft 1-6 were ready on 12 Oct 1869. Some of these sets may have been sent out before the accompanying text was ready. This would account for a note by R. on the cover of Heft 4 of the Helsinki copy:"Tafel 1-6 haben Sie erhalten".

8446. *Uebersicht der vom Herrn Professor Dr. Hausknecht im Orient gesammelten Kryptogamen.*,[S.B. Isis 1870]. Oct.
Collector: Heinrich Carl Haussknecht (1838-1903).
Publ.: 1870, p. [1]-16, *pl. 3* (col. lith. Fleischhack). *Copy*: G. – Reprinted and to be cited from S.B. Isis, Dresden 1870(4): 225-241. See also Z. ges. Naturw. Halle 38: 68-69. 1871 and Hedwigia 10: 17-27. 1871.

8447. *Index in Rabenhorst Bryotheca europaeae* fasc. 1-24. No. 1-1200. Dresden (Druck von C. Heinrich) 1872. Qu. (*Index Bryoth. eur.*).
Publ.: Nov 1872 (Hedwigia), p. [1]-17. *Copies*: BR, G, PH. – For exsiccatae see above.

8448. *Index in Gottsche et Rabenhorst Hepaticarum europaearum exsiccatarum* dec. 1-55. Dresden (C. Heinrich) 1872. Qu. (*Index Hepat. eur.*).
Publ.: Sep-Nov 1872 (Hedwigia Nov 1872), p. [i], [1]-8. *Copies*: BR, G.

8449. *Index in* L. Rabenhorst *Algarum europaearum exsiccatarum* (der Algen Europa's mit Berücksichtigung des ganzen Erdballs). Dec. 1-235. No. 1-2350. Dresden (Druck von C. Heinrich) 1873. Qu. (*Index alg. eur. exs.*).
Publ.: 1873, p. [1]-16. *Copies*: BR, NY, PH. – For exsiccatae see above.

8450. Dr. L. *Rabenhorst's Kryptogamen-Flora von Deutschland, Oesterreich und der Schweiz. Zweite Auflage* vollständig neu bearbeitet von [t.p. vol. 1(1):] A. Grunow, Dr. F. Hauck, G. Limpricht, Dr. Ch. Luerssen, P. Richter, Dr. G. Winter u.A. ... Erster Band: Pilze von Dr. Georg Winter. Leipzig (Verlag von Eduard Kummer) [1880-]1884[-1920]. Oct. (*Rabenh. Krypt.-Fl. ed. 2*).
Editorship: No editor is mentioned on the title-pages. However, Rabenhorst's collaborator Paul Gerhard Richter (1837-1913) acted as general editor from 1879 on (fide R. Kolkwitz 1914).
Copies: BR, G, U, USDA; IDC 5399.
Publ.: 14 volumes (Bände) in many various divisions ("Abtheilungen"), published in parts ("Lieferungen"). The above title is that on p. [ii] of vol. 1(1), 1884. *Copies*: BR, G, NY, PH, Stevenson, U, USDA.
Band 1, Abth. 1: 1880-1885 (t.p. 1884), parts 1-13, p. [ii]-viii, [1]-924, col. chart, p. 1-63 [index].
Author: [Heinrich] Georg Winter (1848-1887); index by C. Gustav Oertel (x-1908). –

Alt. t.p.: *Die Pilze Deutschlands, Oesterreichs und der Schweiz* ... I. Abtheilung: Schizomyceten ...
Band 1, Abth. 2: 1884-1887 (t.p. 1887), parts 14-27, p. [ii]-viii, [1]-928, [1]-48 [index].
Author: G. Winter. – *Alt. t.p.*: *Die Pilze* ... II. Abtheilung: Ascomyceten: Gymnoasceen ...
Band Abth. 3: 1887-1896 (t.p. 1896), parts 28-44, 53-56, p. [ii]-viii, [1]-1275, [1]-57 [index].
Author: Heinrich Rehm (1828-1916); index by Franz Otto Pazschke (1843-1922). – *Alt. t.p.*: *Die Pilze* ... III. Abtheilung: Ascomyceten: Hysteriaceen ...
Band 1, Abth. 4: 1892 (t.p. 1892), parts 49-52, p. [ii]-xiv, [1]-505.
Author: Alfred Fischer (1858-1913). – *Alt. t.p.*: *Die Pilze* ... IV. Abtheilung: Phycomycetes.
Band 1, Abth. 5: 1896-1897 (t.p. 1897), parts 57-58, p. [ii-v], [1]-131.
Author: Eduard Fischer (1861-1939) see TL-2/1:834-835. – *Alt. t.p.*: *Die Pilze* ....V. Abtheilung: Ascomyceten: Tuberaceen.
Band 1, Abth. 6: 1898-1903 (t.p. 1901), parts 59-74, p. [ii]-viii, [1]-1016.
Author: Andreas Allescher (1828-1903). *Alt. t.p.*: *Die Pilze* ... VI. Abtheilung: Fungi imperfecti: Hyalin-sporige Sphaerioideen ...
Band 1, Abth. 7: 1901-1903 (t.p. 1903), parts 75-91, p. [ii]-viii], [1]-1072.
Author: A. Allescher; on p. [v]– viii obituary of Allescher by V. Tubeuf.
Band 1, Abt. 8: 1904-1907 (t.p. 1907), parts 92-105, p. [ii]-vi, [vii-viii], [1]-852.
Author: Gustav Lindau (1866-1923); on p. 852 dates of publication 1(1)-1(7). *Alt. t.p.*: *Die Pilze* ... VIII. Abteilung: Fungi imperfecti: Hyphomycetes (erste Hälfte), ... See TL-2/4553.
Band 1, Abt. 9: 1907-1910 (t.p. 1910), parts 106-120, p. [ii]-viii, [1]-983 [984, dates].
Author: G. Lindau. *Alt. t.p.*: *Die Pilze* ... IX. Abteilung: Fungi imperfecti: Hyphomycetes (zweite Hälfte) ... See TL-2/4555.
Band 1, Abt. 10: 1912-1920 (t.p. 1920), parts 121-127, p. [ii]-xi, [1]-472 [1-2, dates; exsicc.].
Author: Hans Schinz (1858-1941). *Alt. t.p.*: *Die Pilze* ... X. Abteilung: Myxogasteres ...". See also under H. Schinz.
Band 2: 1882-1889 (t.p. 1885), parts 1-10, p. [ii]-xxiii, [xxiv], [1]-575, [576], *pl. 1-5* [1, expl. pl].
Author: Ferdinand Hauck (1845-1889). See TL-2/2478. – *Alt. t.p.*: *Die Meeresalgen Deutschlands und Oesterreichs*.
Band 3: 1884-1889 (t.p. 1889), parts 1-14, p. [ii]-xii, [1]-906.
Author: Christian Luerssen (1843-1916), see TL-2/5082. *Alt. t.p.*: *Die Farnpflanzen* oder Gefässbündelkryptogamen (Pteridophyta).
Band 4, Abth. 1: 1885-1889 (t.p. 1890), parts 1-13, p. [ii]-x, [xi, dates], [1]-836.
Author: Karl Gustav Limpricht (1834-1902), see TL-2/4540. – *Alt. t.p.*: *Die Laubmoose Deutschlands, Oesterreichs und der Schweiz* ... I. Abtheilung: Sphagnaceae ... See TL-2/4540 and A. Geheeb, Flora 68: 598-600. 1885.
Band 4, Abth. 2: 1890-1895 (t.p. 1895), parts 14-26, p. [ii-xii], [1]-853.
Author: K.G. Limpricht. – *Alt. t.p.*: *Die Laubmoose* ... II. Abtheilung: Bryinae ... See TL-2/4540.
Band 4, Abth. 3: 1895-1903 (t.p. 1904), parts 27-41, p. [ii-viii], [1]-864, index Abth. 1-3: [1]-79.
Authors: K.G. Limpricht: Hypnaceae and Nachträge pp.; Hans Wolfgang Limpricht (1877-?) end of Nachträge and indexes. – *Alt. t.p.*: *Die Laubmoose* ... III. Abtheilung: Hypnaceae u. Nachträge ... See TL-2/4540. See also W. Mönkemeyer Laubm. Eur. 1927, TL-2/6172.
Band 5: 1895-1897 (t.p.1897), parts 1-12, p. [ii-vii], [1]-765, [ix]-xiii, [1].
Author: Emil Friedrich August Walther Migula (1863-1938), see TL-2/5986. – *Alt. t.p.*: *Die Characeen Deutschlands, Oesterreichs und der Schweiz* ...".
Re-issue: 1900, identical with main edition, except for the date.
Extract: W. Migula, *Syn. Charac. eur.* 1898 (1897), TL-2/5987.
Band 6, Abt. 1: 1905-1911 (t.p. 1906-1911), parts 1-14, p. [ii]-vii, [1]-870 [1, dates].
Author: Karl Müller frib. (1881-1955) see TL-2 /6493. – *Alt. t.p.*: *Die Lebermoose* Deutschlands, Oesterreichs u.d. Schweiz ... I. Abtheilung.

Band 6, Abt. 2: 1912-1916 (t.p. id.), parts 15-28, p. [ii]-vii, [1]-947.
  Author: K. Müller frib. – *Alt. t.p.*: *Die Lebermoose* ... II. Abteilung.
Band 6, Ergänzungsband: 1939-1941, parts 1-2 (3-5 remained unpublished), p. [1]-320. Stocks destroyed during the war.
  Author: K. Müller frib.
Band 6, New edition, 2 vols.., 1954-1957 [1958], see TL-2/6494.
Band 7, 1. Teil: 1930, parts 1-5, p. [ii]-xii, [1]-920.
  Author: Friedrich Hustedt (1886-1968), see TL-2/3162. – *Altern. t.p.*: *Die Kieselalgen Deutschlands, Österreichs und der Schweiz* ... 1. Teil ...
Band 7, 2. Teil: 1931-1959, parts 1-6, p. [ii]-xii, [1]-845.
  Author: F. Hustedt. – *Alt. t.p.*: *Die Kieselalgen* ... 2. Teil ... Leipzig (Akademische Verlagsgesellschaft Geest & Portig K.G.) 1959.
Band 7, 3. Teil: 1961-1966, parts 1-4. †. p. [1]-816.
  Author: F. Hustedt. – *Title on cover*: *Die Kieselalgen* ... 3. Teil ... Leipzig (id.). 1961-1966.
Band 8: 1930, parts 1-2, p. [ii]-xi, [1]-712.
  Author: Karl von Keissler (1872-1965), editor of the volumes on lichens: Alexander Zahlbruckner (1860-1938). – *Alt. t.p.*: *Die Flechtenparasiten* ... Leipzig (Akademische Verlagsgesellschaft m.b.H.) 1930.
Band 9, Abt. 1, Teil 1: 1933-1934, 4 parts, p. [ii]-viii, [1]-695.
  Authors: K. v. Keissler (Moriolaceae), Hermann Zschacke (1867-1937) (Epigloeac., Verrucar., Dermatocarp.). *Editor*: A. Zahlbruckner. – *Alt. t.p.*: *Moriolaceae*, ... *Epigloeaceae*, ... *Dermatocarpaceae*, ... Leipzig (id.) 1934.
Band 9, Abt. 1, Teil 2: 1937-1938, 5 parts, p. [ii]-x, [xi-xii], [1]-846.
  Author: K. v. Keissler. – *Editor*: A. Zahlbruckner. – *Alt. t.p.*: *Pyrenulaceae* bis *Mycoporaceae Coniocarpineae* ... Leipzig (id.) 1938.
Band 9, Abt. 2, Teil 1: 1937-1938, 2 parts, †, [i, abbrev., loose], [1]-404, *pl. 1-2*.
  Author: Karl Martin Redinger (1907-1940). *Editor*: K.M. Redinger. – *T.p.* (on cover; formal t.p. not issued): *Arthoniaceae* ... *Coenogoniaceae* [however, fam. Chiodectonaceae-Coenogoniaceae not issued] ... Leipzig (id.) 1937-1938.
Band 9, Abt. 2, Teil 2: 1940, 2 parts, †, [1]-272, *pl. 1-33*.
  Author: Vilmos Köfaragó-Gyelnik (1906-1945), see TL-2/1: 1027. *Editor*: K. v. Keissler. *T.p.* (on covers; formal t.p. not issued): *Cyanophili* ... Leipzig (id.). 1940. †. (Lichinaceae, Heppiaceae, Pannariaceae), the families Ephebaceae ... Peltigeraceae, announced on cover, were not issued.
Band 9, Abt. 4, I. Hälfte (sic Abt. 3 not issued): 1932-1933, 2 parts, p. [ii]-x, [1]-426, *pl. 1-8*.
  Author: Eduard Frey [-Stauffer] (1888-1974), see TL-2/1: 876-877. *Editor*: Alexand. Zahlbruckner. – *Alt. t.p.*: *Cladoniaceae* ... *Umbilicariaceae* ... Leipzig (id.) 1933.
Band 9, Abt. 4. II. Hälfte: 1931, 2 parts, p. [ii-v], [1]-531, *pl. 1-24*.
  Author: Heinrich Sandstede (1859-1951). *Editor*: A. Zahlbruckner. *Alt. t.p.*: *Die Gattung Cladonia* ... Leipzig (id.) 1931.
Band 9, Abt. 5. Teil 1: 1934-1936, 3 parts, p. [ii-xv], [1]-728.
  Authors: Adolf Hugo Magnusson (1885-1964), (*Acarosporaceae, Thelocarpaceae*) see also TL-2/5243-5252; Christian Friedo Eckhard Erichsen (1867-1945) (*Pertusariaceae*), TL-2/1: 7987. *Editor*: A. Zahlbruckner. *Alt. t.p.*: *Acarosporaceae und Thelocarpaceae* ... *Pertusariaceae*. Leipzig (id.) 1936.
Band 9, Abt. 5, Teil 3: 1936, 2 parts, p. [ii-iv], [1]-309, *pl. 1-2* [1]-10, index.
  Author: Johannes Hillmann (1881-1943). *Editor*: A. Zahlbruckner. –*Alt. t.p.*: *Parmeliaceae* ... Leipzig (id.) 1936.
Band 9, Abt. 5, Teil 4: 1958-1960, 5 parts, p. [ii-xi], [1]-755, *pl. 1-18*.
  Author and editor: K. von Keissler. – *Alt. t.p.*: *Usneaceae* ... Leipzig (Akademische Verlagsgesellschaft Geest & Portig K.G.) 1958-1960. – *Indexes* by J. Huber (Jena). Part 3 was issued with an errata-slip.
Band 9, Abt. 6: 1935, 1 part. †, p. [1]-188, *pl. 1-12*.
  Authors: J. Hillmann (Teloschistaceae), *Bernt Arne Lynge* (1884-1942) (Physciaceae) *Editor*: A. Zahlbruckner. T.p. cover (no formal t.p. issued): Dr. L. Rabenhorst's ... *Teloschistaceae* ... *Physciaceae* ... Leipzig (Akademische Verlagsgesellschaft m.b.H.) 1935.

Band 10, Abt. 2: 1930, 1 part, p. [ii-v], [1]-273.
Authors: Konrad Gemeinhardt. (*Silicoflagellatae*); Jos. Schiller (1877-1960), (*Coccolithineae*). *Editor*: Richard Kolkwitz (1873-1956). *Alt. t.p.*: *Silicoflagellatae ... Coccolithineae* ... Leipzig (id.) 1930.
Band 10, Abt. 3, Teil 1: 1932-1933, 3 parts, p. [ii]-vi, [1]-617.
Authors: Jos. Schiller (*Dinoflagellatae (Peridineae)*). *Editor*: R. Kolkwitz. *Alt. t.p.*: *Dinoflagellatae (Peridineae)* ... 1. Teil ... Leipzig (id.) 1933.
Band 10, Abt. 3, Teil 2: 1935-1937, 4 parts, p. [ii]-vii, [1]-589, [590, err.].
Author: Jos. Schiller. *Editor*: R. Kolkwitz. *Alt. t.p.*: *Dinoflagellatae (Peridineae)* ... Leipzig (id.) 1937.
Band 11: 1937-1939, 6 parts, p. [ii]-x, [i]-ii, [1]-1092.
Author: Adolf Pascher (*1881-1945*) *q.v. Editor*: R. Kolkwitz. – *Alt. t.p.*: *Heterokonten* ... mit einem Beitrag Kultur der Heterokonten von Prof. Dr. W. Vischer [1890-1960] ... Leipzig (id.) 1937-1939.
Band 12, Abt. 4: 1938-1940 (t.p. 1939), 3 parts, p. [ii]-xii, [1]-453.
Author: Konrad Gemeinhardt. *Editor*: R. Kolkwitz. – *Alt. t.p.*: *Oedogoniales* ... Leipzig (id.). 1939.
Band 13, Abt. 1, Teil 1: 1933-1937, 4 parts, p. [ii]-vi, [1]-712, *pl. 1-96*.
Author: Willie Krieger (1886-1954); see TL-2/2: 675. *Editor*: R. Kolkwitz. – *Alt. t.p.*: *Die Desmidiaceen* Europas mit Berücksichtigung der aussereuropäischen Arten ... Leipzig (id.) 1937.
Band 13, Abt. 1, Teil 2: 1939, 1 part †, p. [1]-117, *pl. 97-142*.
Author: W. Krieger. *Editor*: R. Kolkwitz. – T.p. (cover, no formal t.p. issued): *Die Desmidiaceen* ... Leipzig (id.) 1939.
Band 13, Abt. 2: 1941-1944, 4 parts, p. [i, preface], [1]-499.
Authors: R. Kolkwitz and H. Krieger; *Editor*: R. Kolkwitz. T.p. (covers; no formal t.p.'s issued): *Zygnemales* ... Leipzig (Akademische Verlagsgesellschaft Becker & Erler ...) 1941-1944.
Band 14: 1930-1932, 6 parts , p. [ii]-vi, [1]-1196.
Author: Lothar Geitler (1899-x). *Editor*: R. Kolkwitz. – *Alt. t.p.*: *Cyanophyceae* ... Leipzig (Akademische Verlagsgesellschaft m.b.H.) 1932.
Facsimile ed.: 1963, New York (Johnson Reprint), 14 vols. in 42 parts. Available according to commercial announcements , not seen (except for incidental volumes) by us.
TABLES: See p. 474-528.

Rabinowitsch, Lydia (1871-1935), Russian-born mycologist and bacteriologist; Dr. phil. Bern 1894; married the bacteriologist W. Kempner 1898; professor and director of the Bacteriological Institute of the Moabit Hospital, Berlin-Lichtenfeld. (*Rabinowitsch*).

HERBARIUM and TYPES: Unknown.

BIBLIOGRAPHY and BIOGRAPHY: Barnhart 3: 121 (b. 22 Aug 1871); Biol.-Dokum. 14: 7272; BM 4: 1635; LS 21619-21620.

BIOFILE: Bulloch, W., Hist. bacteriol. 391. 1938.

8451. *Beiträge zur Entwickelungsgeschichte der Fruchtkörper einiger Gastromyceten*. Inaugural-Dissertation der hohen Philosophischen Facultät der Universität Bern zur Erlangung der Philosophischen Doctorwürde ... München (Druck von Val. Höfling, ...) 1894. Oct.
Publ.: 29 Oct 1894, (cover dated; p. 2: imprimatur 19 Jul 1894; Nat. Nov. Nov(2) 1894; Bot. Centralbl. 6 Dec 1894), p. [1]-38, *pl. 10-11* (uncol. liths. auct.). *Copy*: G. – Reprinted and to be cited from Flora 79: 385-418. 1894.
Ref.: Büsgen, M., Bot. Zeit. 53(2): 179-180. 1 Jun 1895 (rev.).

Rach, Louis Theodor (1821-1859), German-born botanist and gardener; studied at the horticultural school of Schöneberg 1836-1840; at Berlin-Dahlem 1847-1850, 1852-1856; curator at the St. Petersburg botanical garden, in charge of the carpological and dendrological collections, 1856-1859.(*Rach*).

HERBARIUM and TYPES: Some material (was) at B. – *Continued on p. 529*.

Band	Abt.	Lief.	Pagination	Date	Author	Taxon
1	1		[v]-viii	post Aug 1883	Winter, G.	Foreword
		1	1-32	Dec 1880	Winter, G.	Introduction
			33-67	Dec 1880	Winter, G.	Schizomycetes
			68-72	Dec 1880	Winter, G.	Saccharomycetes
			72-74	Dec 1880	Winter, G.	Basidiomycetes
			74-79	Dec 1880	Winter, G.	Entomophthoreae
			79-80	Dec 1880	Winter, G.	Ustilagineae-Ustilagineae
		2	81-131	Mar 1881	Winter, G.	Ustilagineae-Ustilagineae
			131-144	Mar 1881	Winter, G.	Uredineae-Uredineae
		3	145-224	Mai 1881	Winter, G.	Uredineae-Uredineae
		4	225-270	Aug 1881	Winter, G.	Uredineae-Uredineae
			270-288	Aug 1881	Winter, G.	Tremellineae
		5	289-290	1-15 Nov 1881	Winter, G.	Tremellineae
			290-292	1-15 Nov 1881	Winter, G.	Hymenomycetes
			292-318	1-15 Nov 1881	Winter, G.	Clavariei
			318-352	1-15 Nov 1881	Winter, G.	Thelephorei
		6	353	Jan 1882	Winter, G.	Thelephorei
			354-385	Jan 1882	Winter, G.	Hydnei
			385-416	Jan 1882	Winter, G.	Polyporei
		7	417-480	Apr 1882	Winter, G.	Polyporei
		8	481-560	Jul 1882	Winter, G.	Agaricini
		9	561-624	Oct 1882	Winter, G.	Agaricini
		10	625-688	Dec 1882	Winter, G.	Agaricini
		11	689-752	Jan 1883	Winter, G.	Agaricini
		12	753-832	Mai 1883	Winter, G.	Agaricini
		13	833-864	Sep 1883	Winter, G.	Gasteromycetes
			864-866	Sep 1883	Winter, G.	Phalloidei
1	1	13	866-870	Sep 1883	Winter, G.	Hymenogastrei
			870-886	Sep 1883	Winter, G.	Sclerodermei
			886-891	Sep 1883	Winter, G.	Tulostomei
			892-893	Sep 1883	Winter, G.	Lycoperdinei
			893-915	Sep 1883	Winter, G.	Nidulariei
			915-922	Sep 1883	Winter, G.	Index to Genera
			923-924	Sep 1883	Winter, G.	Corrections
			924	Sep 1883	Winter, G.	

		11-17	Mar 1884	Winter, G.	Gymnoasci
		18-21	Mar 1884	Winter, G.	Pyrenomycetes
		22-42	Mar 1884	Winter, G.	Erysipheae
		43-64	Mar 1884	Winter, G.	Perisporieae
	15	65-82	Sep 1884	Winter, G.	Perisporieae
		82-128	Sep 1884	Winter, G.	Hypocreaceae
	16	129-152	Nov 1884	Winter, G.	Hypocreaceae
		153-161	Nov 1884	Winter, G.	Chaetomieae
		162-187	Nov 1884	Winter, G.	Sordarieae
		187-191	Nov 1884	Winter, G.	Sect. Sphaeriaceae
		191-192	Nov 1884	Winter, G.	Trichosphaerieae
	17	193-219	Apr 1885	Winter, G.	Trichosphaerieae
		220-246	Apr 1885	Winter, G.	Melanommeae
		247-256	Apr 1885	Winter, G.	Ceratostomeae
		257-259	Jul 1885	Winter, G.	Ceratostomeae
	18	259-288	Jul 1885	Winter, G.	Amphisphaerieae
		288-308	Jul 1885	Winter, G.	Lophiostomeae
		308-320	Jul 1885	Winter, G.	Cucurbitarieae
	19	321-333	Jul 1885	Winter, G.	Cucurbitarieae
		334-384	Jul 1885	Winter, G.	Sphaerelloideae
	20	385-405	Oct 1885	Winter, G.	Sphaerelloideae
		405-448	Oct 1885	Winter, G.	Pleosporeae
	21	449-528	Dec 1885	Winter, G.	Pleosporeae
	22	529-533	Mai 1886	Winter, G.	Pleosporeae
		534-554	Mai 1886	Winter, G.	Massarieae
		554-570	Mai 1886	Winter, G.	Clypeosphaerieae
		570-592	Mai 1886	Winter, G.	Gnomonieae
	23	593-594	Aug 1886	Winter, G.	Synopsis of Families
		594-656	Aug 1886	Winter, G.	Valseae
	24	657-736	Aug 1886	Winter, G.	Valseae
	25	737-764	Oct 1886	Winter, G.	Valseae
		764-797	Oct 1886	Winter, G.	Melanconideae
		797-800	Oct 1886	Winter, G.	Melogrammeae
	26	801-810	Dec 1886	Winter, G.	Melogrammeae
		810-842	Dec 1886	Winter, G.	Diatrypeae
		842-864	Dec 1886	Winter, G.	Xylarieae

Band	Abt.	Lief.	Pagination	Date	Author	Taxon
		27	865-880	Mar 1887	Winter, G.	Xylarieae
			881-883	Mar 1887	Winter, G.	Append.-poorly known *Sphaeria* species
			884-893	Mar 1887	Winter, G.	Key to Genera of Sphaerieae
			894-918	Mar 1887	Winter, G.	Dothideaceae
			918-925	Mar 1887	Winter, G.	Laboulbenieae
			925	Mar 1887	Winter, G.	Corrections
			926-928	Mar 1887	Winter, G.	Index to Genera
	3		[iii-v]vi-vii	post Feb 1896	Rehm, H.	Foreword, Contents
		28	[1]-3	Aug 1887	Rehm, H.	Hysteriaceae
1	3	28	3-28	Aug 1887	Rehm, H.	Hysterineae
			28-48	Aug 1887	Rehm, H.	Hypodermieae
			49-52	Aug 1887	Rehm, H.	Dichaenaceae
			52-53	Aug 1887	Rehm, H.	Pseudohysterineae
			53-56	Aug 1887	Rehm, H.	Acrospermaceae
			56-60	Aug 1887	Rehm, H.	Discomycetes
			60-64	Aug 1887	Rehm, H.	Euphacidieae
		29	65-87	Aug 1887	Rehm, H.	Euphacidieae
			87-112	Aug 1887	Rehm, H.	Pseudophacidieae
			112-113	Jan 1888	Rehm, H.	Stictideae
			113-128	Jan 1888	Rehm, H.	Eusticteae
		30	129-185	Jan 1888	Rehm, H.	Eusticteae
			185-190	Aug 1888	Rehm, H.	Ostropeae
			191-198	Aug 1888	Rehm, H.	Tryblidiaceae
			198-208	Aug 1888	Rehm, H.	Heterosphaerieae
		31	209-212	Jan 1889	Rehm, H.	Heterosphaerieae
			213-241	Jan 1889	Rehm, H.	Cenangieae
			241-272	Jan 1889	Rehm, H.	Dermateae
		32	273-277	Apr 1889	Rehm, H.	Dermateae
			277-280	Apr 1889	Rehm, H.	Patellariaceae
			280-291	Apr 1889	Rehm, H.	Patellariaceae-Pseudopatellarieae
			291-336	Apr 1889	Rehm, H.	Patellariaceae-Eupatellarieae
		33	337-382	Jul 1890	Rehm, H.	Patellariaceae-Eupatellarieae
			382-400	Jul 1890	Rehm, H.	Patellariaceae-Calicieae
		34	401-414	Jan 1891	Rehm, H.	Patellariaceae-Calicieae
			414-444	Jan 1891	Rehm, H.	Patellariaceae-Arthonieae
		34	444-445	Jan 1891	Rehm, H.	Bulgariaceae
1	3	34	445-464	Jan 1891	Rehm, H.	Bulgariaceae-Callorieae

36	529-593	Oct 1891	Rehm, H.	Mollisieae-Eumollisieae
	593-608	Oct 1891	Rehm, H.	Mollisieae-Pyrenopezizeae
37	609-647	Jul 1892	Rehm, H.	Mollisieae-Pyrenopezizeae
	647-656	Jul 1892	Rehm, H.	Helotieae
38	657-720	Sep 1892	Rehm, H.	Helotieae
39	721-784	Jan 1893	Rehm, H.	Helotieae
40	785-848	Sep 1893	Rehm, H.	Helotieae
41	849-912	Oct 1893	Rehm, H.	Helotieae
42	913-976	Jun 1894	Rehm, H.	Eupezizeae
43	977-1040	Nov 1894	Rehm, H.	Eupezizeae
	1041-1078	Jan 1895	Rehm, H.	Eupezizeae
44	1078-1104	Jan 1895	Rehm, H.	Ascoboleae

"Lfg. 45-52 sind bereits als IV. Abthlg. erschienen"

53	1105-1134	Jul 1895	Rehm, H.	Ascoboleae
	1134-1142	Jul 1895	Rehm, H.	Rhizineae
	1142-1168	Jul 1895	Rehm, H.	Geoglosseae
54	1169-1171	Nov 1895	Rehm, H.	Geoglosseae
	1172-1232	Nov 1895	Rehm, H.	Helvelleae
55	1233-1245	Apr 1896	Rehm, H.	Helvelleae
55	1246	Apr 1896	Rehm, H.	Hysteriaceae-Nachtr.
	1247-1248	Apr 1896	Rehm, H.	Hypodermieae-Nachtr.
	1248	Apr 1896	Rehm, H.	Acrospermaceae-Nachtr.
	1248-1249	Apr 1896	Rehm, H.	Euphacidieae-Nachtr.
	1249-1251	Apr 1896	Rehm, H.	Pseudophacidieae-Nachtr.
	1251-1253	Apr 1896	Rehm, H.	Eusticteae-Nachtr.
	1253	Apr 1896	Rehm, H.	Tryblidiaceae-Nachtr.
	1254	Apr 1896	Rehm, H.	Heterosphaerieae-Nachtr.
	1255-1256	Apr 1896	Rehm, H.	Cenangieae-Nachtr.
	1256-1259	Apr 1896	Rehm, H.	Dermateae-Nachtr.
	1259-1261	Apr 1896	Rehm, H.	Patellarieae-Nachtr.
	1261-1263	Apr 1896	Rehm, H.	Bulgariaceae-Nachtr.
	1263-1265	Apr 1896	Rehm, H.	Mollisieae-Nachtr.
	1265-1269	Apr 1896	Rehm, H.	Helotieae-Nachtr.
	1269-1270	Apr 1896	Rehm, H.	Eupezizeae-Nachtr.

Band	Abt.	Lief.	Pagination	Date	Author	Taxon
	4	46	[1271]-1272	Apr 1896	Rehm, H.	Corrections
			[1273]-1275	Apr 1896	Anonymous	Generic Index
			[1-3]-57	post Jul 1892	Pazschke, O.	General Index
			[ii-vii]-xiv	Oct 1891	Fischer, A.	Dedication, Foreword, Contents
			[1]3	Oct 1891	Fischer, A.	Phycomycetes
			3-5	Oct 1891	Fischer, A.	Archimycetes
			5-7	Oct 1891	Fischer, A.	Zygomycetes
			7-8	Oct 1891	Fischer, A.	Oomycetes
			8-11	Oct 1891	Fischer, A.	Observations on System
			11-20	Oct 1891	Fischer, A.	Archimycetes
			20-45	Oct 1891	Fischer, A.	Monolpidiaceae
			45-64	Oct 1891	Fischer, A.	Merolpidiaceae
			65-72	Oct 1891	Fischer, A.	Merolpidiaceae
1	4	46	72-85	Oct 1891	Fischer, A.	Holochytriaceae
			85-128	Oct 1891	Fischer, A.	Sporochytriaceae
		47	129-131	Jan 1892	Fischer, A.	Sporochytriaceae
			131-145	Jan 1892	Fischer, A.	Hypochytriaceae
			145-160	Jan 1892	Fischer, A.	Nachträge
			161-177	Jan 1892	Fischer, A.	Zygomycetes
			178-192	Jan 1892	Fischer, A.	Mucoraceae
		48	193-256	Mar 1892	Fischer, A.	Mucoraceae
		49	257-268	Apr 1892	Fischer, A.	Mortierellaceae
			268-283	Apr 1892	Fischer, A.	Chaetocladiaceae
			283-287	Apr 1892	Fischer, A.	Cephalidaceae
			287-309	Apr 1892	Fischer, A.	Oomycetes
			310-320	Apr 1892	Fischer, A.	Oomycetes
		50	321-327	Mai 1892	Fischer, A.	Saprolegniaceae
			327-378	Mai 1892	Fischer, A.	Monoblepharidaceae
			378-383	Mai 1892	Fischer, A.	Peronosporaceae
			383-384	Mai 1892	Fischer, A.	Peronosporaceae
		51	385-448	Jul 1892	Fischer, A.	Peronosporaceae
		52	449-489	Oct 1892	Fischer, A.	Saprolegniaceae-Nachtr.
			490	Oct 1892	Anonymous	General Index
			491-505	Oct 1892	Fischer, A.	Foreword
			[1]2	Jul 1896	Fischer, E.	Tuberaceae
	5	57	3-12	Jul 1896	Fischer, E.	

1	5	58	81-101	Nov 1896	Fischer, E.	Elaphomycetaceen
			101-108	Nov 1896	Fischer, E.	Onygenaceen
			109-110	Nov 1896	Fischer, E.	Hemiasceen
			110-112	Nov 1896	Fischer, E.	Ascoideaceen
			113-118	Nov 1896	Fischer, E.	Protomycetaceen
			118-127	Nov 1896	Fischer, E.	Monascaceen
			[128]	Nov 1896	Anonymous	Generic Index
			[129]-131	Nov 1896	Anonymous	General Index
	6	59	[1]-7	Jun 1898	Allescher, A.	Fungi Imperfecti
			7-64	Jun 1898	Allescher, A.	Sphaerioideae-Hyaline Spored
		60	65-128	Jul 1898	Allescher, A.	Sphaerioideae-Hyaline Spored
		61	129-192	Aug 1898	Allescher, A.	Sphaerioideae-Hyaline Spored
		62	193-256	Oct 1898	Allescher, A.	Sphaerioideae-Hyaline Spored
		63	257-320	Dec 1898	Allescher, A.	Sphaerioideae-Hyaline Spored
		64	321-384	Jan 1899	Allescher, A.	Sphaerioideae-Hyaline Spored
		65	385-448	Jun 1899	Allescher, A.	Sphaerioideae-Hyaline Spored
		66	449-512	Sep 1899	Allescher, A.	Sphaerioideae-Hyaline Spored
		67	513-576	Oct 1899	Allescher, A.	Sphaerioideae-Hyaline Spored
		68	577-640	Dec 1899	Allescher, A.	Sphaerioideae-Hyaline Spored
		69	641-704	Jan 1900	Allescher, A.	Sphaerioideae-Hyaline Spored
		70	705-768	Apr 1900	Allescher, A.	Sphaerioideae-Hyaline Spored
		71	769-832	Jul 1900	Allescher, A.	Sphaerioideae-Hyaline Spored
		72	833-896	Jul 1900	Allescher, A.	Sphaerioideae-Hyaline Spored
		73	897-960	Oct 1900	Allescher, A.	Sphaerioideae-Hyaline Spored
		74	961-992	Dec 1901	Allescher, A.	Generic Index
			[993]-994	Dec 1901	Allescher, A.	Substrate/Host Index
	6		[995]-1016	Dec 1901	Allescher, A.	Corrections
			1016	Dec 1901	Allescher, A.	Index
			[3]-48	post Jun 1903	Tubeuf, F.	Obituary of Allescher
	7	75	[v]-viii	Jan 1901	Allescher, A.	Sphaerioideae-Dark Spored
		76	[1]-64	Apr 1901	Allescher, A.	Sphaerioideae-Dark Spored
		77	65-128	Jun 1901	Allescher, A.	Sphaerioideae-Dark Spored
		78	129-192	Aug 1901	Allescher, A.	Sphaerioideae-Dark Spored
		79	193-256	Dec 1901	Allescher, A.	Sphaerioideae-Dark Spored
			257-294	Dec 1901	Allescher, A.	Nectrioideae
			295-315			

Band	Abt.	Lief.	Pagination	Date	Author	Taxon
		80	316–320	Dec 1901	Allescher, A.	Leptostromaceae
		80	321–384	Dec 1901	Allescher, A.	Leptostromaceae
		81	385–392	Jan 1902	Allescher, A.	Leptostromaceae
		81	393–440	Jan 1902	Allescher, A.	Excipulaceae
			441–443	Jan 1902	Allescher, A.	Sphaeropsideen-Dubious Genera
			444–448	Jan 1902	Allescher, A.	Melanconieae
		82	449–512	Apr 1902	Allescher, A.	Melanconieae
		83	513–576	Mai 1902	Allescher, A.	Melanconieae
		84	577–640	Jul 1902	Allescher, A.	Melanconieae
		85	641–704	Oct 1902	Allescher, A.	Melanconieae
		86	705–768	Jan 1903	Allescher, A.	Melanconieae
		87	769–832	Jan 1903	Allescher, A.	Melanconieae
		88	833–896	Mar 1903	Allescher, A.	Melanconieae
		89	897–957	Apr 1903	Allescher, A.	Melanconieae
			958–961	Apr 1903	Allescher, A.	Generic Index
1	7	90	962–985	Sep 1903	Allescher, A.	Host Index
			986–993	Sep 1903	Anonymous	List of Spp. Illustrated
			995–1024	Sep 1903	Allescher, A.	General Index
		91	1025–1072	Dec 1903	Allescher, A.	General Index
			[v]vi			Foreword
		92	[1]–5	12 Jul 1904	Lindau, G.	Hyphomycetes
		92	[6]–64	16 Mai 1904	Lindau, G.	Mucedinaceae
		93	65–128	16 Mai 1904	Lindau, G.	Mucedinaceae
		94	129–176	30 Jun 1904	Lindau, G.	Mucedinaceae
		95	177–256	15 Jul 1904	Lindau, G.	Mucedinaceae
		96	257–320	3 Apr 1905	Lindau, G.	Mucedinaceae
		97	321–384	10 Mai 1905	Lindau, G.	Mucedinaceae
		98	385–432	20 Jun 1905	Lindau, G.	Mucedinaceae
		99	433–512	15 Jul 1905	Lindau, G.	Mucedinaceae
	8	100	513–546	25 Jul 1906	Lindau, G.	Mucedinaceae
			547–576	30 Aug 1906	Lindau, G.	Dematiaceae
		101	577–640	20 Sep 1906	Lindau, G.	Dematiaceae
		102	641–704	10 Oct 1906	Lindau, G.	Dematiaceae
		103	705–752	15 Nov 1906	Lindau, G.	Dematiaceae
		104	753–833	16 Mai 1907	Lindau, G.	Dematiaceae
		105	834–845	12 Jul 1907	Lindau, G.	Dematiaceae

			[1]–[v]	10 Dec 1910	Lindau, G.	Foreword Dematiaceae
	106		[1]–48	12 Jul 1907	Lindau, G.	Dematiaceae
	107		49–112	20 Dec 1907	Lindau, G.	Dematiaceae
	108		113–176	1 Apr 1908	Lindau, G.	Dematiaceae
9	109		177–240	25 Mai 1908	Lindau, G.	Dematiaceae
			241–285	30 Jul 1908	Lindau, G.	Dematiaceae
	110		[286]–304	30 Jul 1908	Lindau, G.	Stilbaceae
	111		305–368	25 Nov 1908	Lindau, G.	Stilbaceae
1			369–400	7 Apr 1909	Lindau, G.	Stilbaceae
			[401]4-432	7 Apr 1909	Lindau, G.	Tuberculariaceae
	112		433–496	28 Apr 1909	Lindau, G.	Tuberculariaceae
	113		497–560	20 Jun 1909	Lindau, G.	Tuberculariaceae
	114		561–624	25 Jul 1909	Lindau, G.	Tuberculariaceae
	115		625–648	1 Nov 1909	Lindau, G.	Tuberculariaceae
			[649]–688	1 Nov 1909	Lindau, G.	Mycelia Sterilia
	116		689–714	31 Dec 1909	Lindau, G.	Mycelia Sterilia
			[715]–752	31 Dec 1909	Lindau, G.	Mucedinaceae-Nachtr.
	117		753–781	10 Mar 1910	Lindau, G.	Mucedinaceae-Nachtr.
			781–810	10 Mar 1910	Lindau, G.	Dematiaceae-Nachtr.
			810–815	10 Mar 1910	Lindau, G.	Stilbaceae-Nachtr.
			815–816	10 Mar 1910	Lindau, G.	Tuberculariaceae-Nachtr.
	118		817–822	20 Mai 1910	Lindau, G.	Tuberculariaceae-Nachtr.
			822–824	20 Mai 1910	Lindau, G.	Mycelia Sterilia-Nachtr.
			[825]–852	20 Mai 1910	Lindau, G.	Hyphomycetes-key to Genera
			[853]–880	20 Mai 1910	Lindau, G.	Host Index
	119		881–916	20 Jul 1910	Lindau, G.	Host Index
			[917]–922	20 Jul 1910	Lindau, G.	Generic Index, List illustr.
			[923]–944	20 Jul 1910	Lindau, G.	Hyphomycetes-Gen. Index
	120		945–983	10 Dec 1910	Lindau, G.	Hyphomycetes-Gen. Index
			[984]	10 Dec 1910	Lindau, G.	Publication dates
	121	10	[v]–vii	8 Dec 1920	Schinz, H.	Foreword
			viii	8 Dec 1920	Schinz, H.	Corrections
			ix–xi	8 Dec 1920	Anonymous	Contents List
			[1]–64	20 Jul 1912	Schinz, H.	Myxogasteres
1	122	10	65–80	18 Feb 1914	Schinz, H.	Myxogasteres
	122		80	18 Feb 1914	Schinz, H.	Exosporeae
			80–83	18 Feb 1914	Schinz, H.	Ceratiomyxaceae

Band	Abt.	Lief.	Pagination	Date	Author	Taxon
			83-85	18 Feb 1914	Schinz, H.	Endosporeae
			86-128	18 Feb 1914	Schinz, H.	Physaraceae
		123	129-192	17 Mai 1915	Schinz, H.	Physaraceae
		124	193-202	9 Dec 1915	Schinz, H.	Physaraceae
			202-228	9 Dec 1915	Schinz, H.	Didymiaceae
			229-256	9 Dec 1915	Schinz, H.	Stemonitaceae
		125	257-272	28 Aug 1917	Schinz, H.	Stemonitaceae
			272-277	28 Aug 1917	Schinz, H.	Amaurochaetaceae
			278-304	28 Aug 1917	Schinz, H.	Heterodermaceae
			305-310	28 Aug 1917	Schinz, H.	Liceaceae
			310-315	28 Aug 1917	Schinz, H.	Tubulinaceae
		126	315-320	28 Aug 1917	Schinz, H.	Reticulariaceae
			321-323	20 Mar 1918	Schinz, H.	Reticulariaceae
			323-328	20 Mar 1918	Schinz, H.	Lycogalaceae
			329-368	20 Mar 1918	Schinz, H.	Trichiaceae
			368-384	20 Mar 1918	Schinz, H.	Arcyriaceae
		127	385-398	8 Dec 1920	Schinz, H.	Arcyriaceae
			399-409	8 Dec 1920	Schinz, H.	Margaritaceae
			410-412	8 Dec 1920	Schinz, H.	Barbeyella (Incert. Sed.)
			413-446	8 Dec 1920	Schinz, H.	Nachträge
			[447]-470	8 Dec 1920	Anonymous	Index
			[471]-472	8 Dec 1920	Schinz, H.	List of Spp. Illustrated
			[473]	8 Dec 1920	Schinz, H.	Dates of Publication
			[474]	8 Dec 1920	Schinz, H.	List of Exsiccatae
			[v]-viii	Dec 1884	Hauck, F.	Marine Algae-Foreword
			[ix]-xxii	Jan 1885	Hauck, F.	Contents
			[xiii]-xxiii	Jan 1885	Hauck, F.	Bibliography
			[xxiv]	Jan 1885	Hauck, F.	Algae Exsiccatae
2		1	[1]-6	Nov 1882	Hauck, F.	Introduction
			[7]	Nov 1882	Hauck, F.	Rhodophyceae
			[8]-20	Nov 1882	Hauck, F.	Florideae
			21-26	Nov 1882	Hauck, F.	Porphyraceae
			26-37	Nov 1882	Hauck, F.	Squamariaceae
			37-39	Nov 1882	Hauck, F.	Hildenbrandtiaceae
			39-55	Nov 1882	Hauck, F.	Wrangeliaceae
			55-64	Nov 1882	Hauck, F.	Helminthocladiaceae

	110-133	Jan 1883	Hauck, F.	Cryptonemiaceae
	133-149	Jan 1883	Hauck, F.	Gigartinaceae
	149-160	Jan 1883	Hauck, F.	Rhodymeniaceae
4	161-168	Apr 1883	Hauck, F.	Rhodymeniaceae
	169-178	Apr 1883	Hauck, F.	Delesseriaceae
	178-185	Apr 1883	Hauck, F.	Sphaerococcaceae
	186-187	Apr 1883	Hauck, F.	Solieriaceae
	187-189	Apr 1883	Hauck, F.	Hypnaeaceae
	189-197	Apr 1883	Hauck, F.	Gelidiaceae
	197-199	Apr 1883	Hauck, F.	Spongiocarpeae
	199-203	Apr 1883	Hauck, F.	Lomentariaceae
	203-224	Apr 1883	Hauck, F.	Rhodomelaceae
5	225-259	Apr 1883	Hauck, F.	Rhodomelaceae
	259-272	Apr 1883	Hauck, F.	Corallinaceae
6	273-281	Oct 1883	Hauck, F.	Corallinaceae
	282	Oct 1883	Hauck, F.	Phaeophyceae
	282-284	Oct 1883	Hauck, F.	Fucoideae
	285-301	Oct 1883	Hauck, F.	Fucaceae
	[302]-303	Oct 1883	Hauck, F.	Dictyotales
6	304-311	Oct 1883	Hauck, F.	Dictyoteae
	[312]-318	Oct 1883	Hauck, F.	Phaeozoosporeae
	[319]-320	Oct 1883	Hauck, F.	Ectocarpaceae
7	321-350	Jan 1884	Hauck, F.	Ectocarpaceae
	351-369	Jan 1884	Hauck, F.	Mesogloeaceae
	369-380	Jan 1884	Hauck, F.	Punctariaceae
	380-381	Jan 1884	Hauck, F.	Arthrocladiaceae
	382-384	Jan 1884	Hauck, F.	Sporochnaceae
8	383-389	Jul 1884	Hauck, F.	Sporochnaceae
	389-393	Jul 1884	Hauck, F.	Scytosiphonaceae
	394-399	Jul 1884	Hauck, F.	Laminariaceae
	399-402	Jul 1884	Hauck, F.	Ralfsiaceae
	402-403	Jul 1884	Hauck, F.	Lithodermaceae
	403-409	Jul 1884	Hauck, F.	Cuteriaceae
	410	Jul 1884	Hauck, F.	Chlorophyceae
	410-411	Jul 1884	Hauck, F.	Oosporeae
	412-416	Jul 1884	Hauck, F.	Vaucheriaceae
	[417]-421	Jul 1884	Hauck, F.	Chlorozoosporeae

2

Band	Abt.	Lief.	Pagination	Date	Taxon	Author
			[422]-437	Jul 1884	Ulvaceae	Hauck, F.
			437-448	Jul 1884	Confervaceae	Hauck, F.
		9	449-466	Oct 1884	Confervaceae	Hauck, F.
			466-468	Oct 1884	Anadyomenaceae	Hauck, F.
			469-471	Oct 1884	Valoniaceae	Hauck, F.
			471-475	Oct 1884	Bryopsideae	Hauck, F.
			475-477	Oct 1884	Derbesiaceae	Hauck, F.
			477-483	Oct 1884	Codiaceae	Hauck, F.
			483-484	Oct 1884	Dasycladaceae	Hauck, F.
			484-485	Oct 1884	Acetabulariaceae	Hauck, F.
			485-486	Oct 1884	Palmellaceae	Hauck, F.
			[487]-490	Oct 1884	Cyanophyceae-Schizophyceae	Hauck, F.
			[491]-512	Oct 1884	Nostocaceae	Hauck, F.
			513-519	Jan 1885	Chroococcaceae	Hauck, F.
		10	[520]-526	Jan 1885	Nachträge	Hauck, F.
			[527]-547	Jan 1885	Generic key	Hauck, F.
2			[548]-570	Jan 1885	General Index	Anonymous
			[571]-574	Jan 1885	Index of Taxa illustrated	Anonymous
			575	Jan 1885	Corrections	Hauck, F.
			[v]-vii	Jan 1889	Pteridophyta-Foreword	Luerssen, C.
3		1	[1]-5	Mar 1884	Pteridophyta	Luerssen, C.
			6-29	Mar 1884	Filicinae	Luerssen, C.
			29-36	Mar 1884	Hymenophyllaceae	Luerssen, C.
			36-50	Mar 1884	Polypodiaceae	Luerssen, C.
			50-64	Mar 1884	Polypodieae	Luerssen, C.
		2	65-108	Apr 1884	Polypodieae	Luerssen, C.
			108-128	Apr 1884	Aspleniaceae	Luerssen, C.
		3	129-192	Sep 1884	Aspleniaceae	Luerssen, C.
		4	193-256	Mai 1885	Aspleniaceae	Luerssen, C.
		5	257-293	Oct 1885	Aspleniaceae	Luerssen, C.
			293-320	Oct 1885	Aspidiaceae	Luerssen, C.
		6	321-384	Jan 1886	Aspidiaceae	Luerssen, C.
		7	385-448	Jul 1886	Aspidiaceae	Luerssen, C.
		8	449-512	Sep 1886	Aspidiaceae	Luerssen, C.
		9	513-516	Jun 1887	Suborder Osmundaceae	Luerssen, C.
			517-519	Jun 1887		Luerssen, C.

	593-595	Aug 1887	Luerssen, C.	Hydropterides	
	595-606	Aug 1887	Luerssen, C.	Salviniaceae	
	606-622	Aug 1887	Luerssen, C.	Marsiliaceae	
	622-623	Aug 1887	Luerssen, C.	Equisetinae	
	623-640	Aug 1887	Luerssen, C.	Equisetaceae	
11	641-704	Apr 1888	Luerssen, C.	Equisetaceae	
12	705-768	Jan 1889	Luerssen, C.	Equisetaceae	
13	769-781	Apr 1889	Luerssen, C.	Lycopodinae	
14	781-782	Apr 1889	Luerssen, C.	Lycopodiaceae	
	782-832	Apr 1889	Luerssen, C.	Lycopodiaceae	
	833-845	Jun 1889	Luerssen, C.	Isoetaceae	
	845-862	Jun 1889	Luerssen, C.	Selaginellaceae	
	862-877	Jun 1889	Luerssen, C.	Add./Corr.	
	877-888	Jun 1889	Luerssen, C.	Index	
	[889]-906	Nov 1889	Anonymous	Musci-Foreword	
4	[v]-viii	Jul 1885	Limpricht, K. G.	Musci-Introduction	1
	[1]-64	Aug 1885	Limpricht, K. G.	Musci-Introduction	
1	65-128	Dec 1885	Limpricht, K. G.	Musci-Introduction	
2	129-157	Dec 1885	Limpricht, K. G.	Bryineae	
3	157-161	Dec 1885	Limpricht, K. G.	Ephemeraceae	
	161-173	Dec 1885	Limpricht, K. G.	Physcomitrellaceae	
	173-176	Dec 1885	Limpricht, K. G.	Phascaceae	
	176-192	Jul 1886	Limpricht, K. G.	Phascaceae	
4	193-199	Jul 1886	Limpricht, K. G.	Bruchiaceae	
	199-207	Jul 1886	Limpricht, K. G.	Voitiaceae	
	208-211	Jul 1886	Limpricht, K. G.	Acrocarpae	
	212-220	Jul 1886	Limpricht, K. G.	Weisiaceae	
	220-256	Jul 1886	Limpricht, K. G.	Weisiaceae	
5	257-271	Nov 1886	Limpricht, K. G.	Rhabdoweisiaceae	
	271-301	Nov 1886	Limpricht, K. G.	Aongströmiaceae	
	301-304	Nov 1886	Limpricht, K. G.	Dicranaceae	
	304-320	Nov 1886	Limpricht, K. G.	Dicranaceae	
6	321-384	Dec 1886	Limpricht, K. G.	Dicranaceae	
7	385-418	Jul 1887	Limpricht, K. G.	Leucobryaceae	
	418-421	Jul 1887	Limpricht, K. G.	Fissidentaceae	
	422-448	Jul 1887	Limpricht, K. G.	Fissidentaceae	

Band	Abt.	Lief.	Pagination	Date	Author	Taxon
		8	449-459	Oct 1887	Limpricht, K. G.	Fissidentaceae
			459-476	Oct 1887	Limpricht, K. G.	Seligeriaceae
			476-482	Oct 1887	Limpricht, K. G.	Campylosteliaceae
			482-512	Oct 1887	Limpricht, K. G.	Ditrichaceae
			513-517	Mai 1888	Limpricht, K. G.	Ditrichaceae
		9	517-576	Mai 1888	Limpricht, K. G.	Pottiaceae
		10	577-640	Oct 1888	Limpricht, K. G.	Pottiaceae
		11	641-693	Dec 1888	Limpricht, K. G.	Pottiaceae
			693-704	Dec 1888	Limpricht, K. G.	Grimmiaceae
4	1	12	705-768	Oct 1889	Limpricht, K. G.	Grimmiaceae
		13	769-826	Nov 1889	Limpricht, K. G.	Grimmiaceae
			827-834	Nov 1889	Anonymous	Index to Taxa
			[835]-836	Nov 1889	Limpricht, K. G.	Add./Corr.
4	2	14	[1]-64	Jul 1890	Limpricht, K. G.	Orthotrichaceae
		15	65-102	Oct 1890	Limpricht, K. G.	Orthotrichaceae
			102-125	Oct 1890	Limpricht, K. G.	Encalyptaceae
			125-128	Oct 1890	Limpricht, K. G.	Georgiaceae
4	2	16	129-132	Jan 1891	Limpricht, K. G.	Georgiaceae
			132-136	Jan 1891	Limpricht, K. G.	Schistostegaceae
			136-172	Jan 1891	Limpricht, K. G.	Splachnaceae
			172-174	Jan 1891	Limpricht, K. G.	Disceliaceae
			175-192	Jan 1891	Limpricht, K. G.	Funariaceae
		17	193-203	Dec 1891	Limpricht, K. G.	Funariaceae
			203-256	Dec 1891	Limpricht, K. G.	Bryaceae
		18	257-320	Jan 1892	Limpricht, K. G.	Bryaceae
		19	321-384	Nov 1892	Limpricht, K. G.	Bryaceae
		20	385-447	Dec 1892	Limpricht, K. G.	Bryaceae
			447-448	Dec 1892	Limpricht, K. G.	Mniaceae
		21	449-496	Jan 1893	Limpricht, K. G.	Mniaceae
			497-512	Jan 1893	Limpricht, K. G.	Meeseaceae
		22	513-521	Jun 1893	Limpricht, K. G.	Meeseaceae
			521-532	Jun 1893	Limpricht, K. G.	Aulacomniaceae
			532-575	Jun 1893	Limpricht, K. G.	Bartramiaceae
			575-576	Jun 1893	Limpricht, K. G.	Timmiaceae
		23	577-586	Dec 1893	Limpricht, K. G.	Timmiaceae
			586-634	Dec 1893	Limpricht, K. G.	Polytrichaceae

	25		677–691	Aug 1894	Limpricht, K. G.	Cryphaeaceae
			691–704	Aug 1894	Limpricht, K. G.	Neckeraceae
			705–718	Jan 1895	Limpricht, K. G.	Neckeraceae
			718–724	Jan 1895	Limpricht, K. G.	Pterygophyllaceae
			724–745	Jan 1895	Limpricht, K. G.	Fabroniaceae
			745–768	Jan 1895	Limpricht, K. G.	Leskeaceae
	26		769–843	Jun 1895	Limpricht, K. G.	Leskeaceae
			[844]–851	Jun 1895	Anonymous	General Index
			[852]–853	Jun 1895	Limpricht, K. G.	Add./Corr.
		3	[v–vi]	Dec 1903	Limpricht, K. G.	Hypnaceae-Foreword
		3	[viii]	Dec 1903	Limpricht, K. G.	Hypnaceae-publ. dates
	27		[1]–64	Dec 1895	Limpricht, K. G.	Hypnaceae
	28		65–128	Jul 1896	Limpricht, K. G.	Hypnaceae
	29		129–192	Dec 1896	Limpricht, K. G.	Hypnaceae
	30		193–256	Dec 1896	Limpricht, K. G.	Hypnaceae
	31		257–320	Sep 1897	Limpricht, K. G.	Hypnaceae
	32		321–384	Dec 1897	Limpricht, K. G.	Hypnaceae
	33		385–448	Aug 1898	Limpricht, K. G.	Hypnaceae
	34		449–512	Feb 1899	Limpricht, K. G.	Hypnaceae
	35		513–576	Nov 1899	Limpricht, K. G.	Hypnaceae
	36		577–600	Apr 1901	Limpricht, K. G.	Hypnaceae
			601–640	Apr 1901	Limpricht, K. G. & W. Limpricht	Nachträge to Abt. 1
	37		641–704	Dec 1901	Limpricht, K. G. & W. Limpricht	Nachträge to Abt. 1
	38		705–768	Dec 1902	Limpricht, K. G. & W. Limpricht	Nachträge to Abt. 1
	39		769–832	Jun 1903	Limpricht, K. G. & W. Limpricht	Nachträge to Abt. 1
	40		833–836	Dec 1903	Anonymous & W. Limpricht	Indexes & Index
			837–864	Dec 1903	Limpricht, W.	List of Synonyms, Bibliography/Exsiccatae
Suppl.	41		[1–3]–79	Dec 1927	Mönkemeyer, W.	Laubmoose Europas-Foreword
	1		[v]vii	Dec 1927	Mönkemeyer, W.	Laubmoose Europas-Contents
4			[ix]x	Dec 1927	Mönkemeyer, W.	Laubmoose Europas-Introd.
			[1]–67	Apr 1927	Mönkemeyer, W.	

Band	Abt.	Lief.	Pagination	Date	Author	Taxon
			68–123	Apr 1927	Mönkemeyer, W.	Musci-Keys and Synopses
			123–130	Apr 1927	Mönkemeyer, W.	Andreaeaceae
			130–146	Apr 1927	Mönkemeyer, W.	Fissidentaceae
			147–148	Apr 1927	Mönkemeyer, W.	Archidiaceae
			148–161	Apr 1927	Mönkemeyer, W.	Ditrichaceae
			161–162	Apr 1927	Mönkemeyer, W.	Bryoxiphiaceae
			162–171	Apr 1927	Mönkemeyer, W.	Seligeriaceae
			171–229	Apr 1927	Mönkemeyer, W.	Dicranaceae
			229–230	Apr 1927	Mönkemeyer, W.	Leucobryaceae
			231–232	Apr 1927	Mönkemeyer, W.	Calymperaceae
			232–238	Apr 1927	Mönkemeyer, W.	Encalyptaceae
			238–256	Apr 1927	Mönkemeyer, W.	Pottiaceae
			257–336	Jul 1927	Mönkemeyer, W.	Pottiaceae
4	Suppl.	2	336–340	Jul 1927	Mönkemeyer, W.	Cinclidotaceae
			341–380	Jul 1927	Mönkemeyer, W.	Grimmiaceae
			381–382	Jul 1927	Mönkemeyer, W.	Disceliaceae
			382–386	Jul 1927	Mönkemeyer, W.	Ephemeraceae
			386–396	Jul 1927	Mönkemeyer, W.	Funariaceae
			396–397	Jul 1927	Mönkemeyer, W.	Voitiaceae
			397–398	Jul 1927	Mönkemeyer, W.	Oedipodiaceae
			398–409	Jul 1927	Mönkemeyer, W.	Splachnaceae
			409–410	Jul 1927	Mönkemeyer, W.	Schistostegaceae
			410–412	Jul 1927	Mönkemeyer, W.	Georgiaceae
			412–549	Jul 1927	Mönkemeyer, W.	Bryaceae
			549–566	Jul 1927	Mönkemeyer, W.	Mniaceae
			566–569	Jul 1927	Mönkemeyer, W.	Aulacomniaceae
			569–574	Jul 1927	Mönkemeyer, W.	Meeseaceae
			574–575	Jul 1927	Mönkemeyer, W.	Catoscopiaceae
			575–576	Jul 1927	Mönkemeyer, W.	Bartramiaceae
		3	577–588	Dec 1927	Mönkemeyer, W.	Bartramiaceae
			588–591	Dec 1927	Mönkemeyer, W.	Timmiaceae
			592–595	Dec 1927	Mönkemeyer, W.	Ptychomitriaceae
			595–628	Dec 1927	Mönkemeyer, W.	Orthotrichaceae
			629–631	Dec 1927	Mönkemeyer, W.	Hedwigiaceae
			631–633	Dec 1927	Mönkemeyer, W.	Cryphaeaceae
			633–637	Dec 1927	Mönkemeyer, W.	Leucodontaceae

		649-653	Dec 1927	Mönkemeyer, W.	Lembophyllaceae
		653-667	Dec 1927	Mönkemeyer, W.	Fontinalaceae
		668-669	Dec 1927	Mönkemeyer, W.	Climaciaceae
		670-674	Dec 1927	Mönkemeyer, W.	Hookeriaceae
		674-677	Dec 1927	Mönkemeyer, W.	Theliaceae
		677-681	Dec 1927	Mönkemeyer, W.	Fabroniaceae
		681-694	Dec 1927	Mönkemeyer, W.	Leskeaceae
		694-704	Dec 1927	Mönkemeyer, W.	Thuidiaceae
		704-711	Dec 1927	Mönkemeyer, W.	Cratoneuraceae
		711-789	Dec 1927	Mönkemeyer, W.	Amblystegiaceae
		789-844	Dec 1927	Mönkemeyer, W.	Brachytheciaceae
		844-852	Dec 1927	Mönkemeyer, W.	Entodontaceae
		852-867	Dec 1927	Mönkemeyer, W.	Plagiotheciaceae
		867-872	Dec 1927	Mönkemeyer, W.	Sematophyllaceae
		873-893	Dec 1927	Mönkemeyer, W.	Hypnaceae
		893-895	Dec 1927	Mönkemeyer, W.	Rhytidiaceae
		896-901	Dec 1927	Mönkemeyer, W.	Hylocomiaceae
		901-903	Dec 1927	Mönkemeyer, W.	Buxbaumiaceae
		903	Dec 1927	Mönkemeyer, W.	Diphysciaceae
		904-919	Dec 1927	Mönkemeyer, W.	Polytrichaceae
4	Suppl. 3	[921]	Dec 1927	Mönkemeyer, W.	List of European families of Mosses
		[922]-956	Dec 1927	Anonymous	Index
		[957]	Dec 1927	Mönkemeyer, W.	List of Non-European families of Mosses
		[958]-960	Dec 1927	Mönkemeyer, W.	List of Non-European Genera of Mosses
		[v]	Jan 1897	Migula, W.	Characeae-Foreword
5	12	[ix]-xiii	Jan 1897	Migula, W.	Bibliography
	1	[1]-64	Oct 1889	Migula, W.	Characeae
	2	65-128	Dec 1889	Migula, W.	Characeae
	3	129-192	Jan 1890	Migula, W.	Characeae
	4	193-256	Jun 1890	Migula, W.	Characeae
	5	257-320	Jan 1891	Migula, W.	Characeae
	6	321-384	Aug 1891	Migula, W.	Characeae
	7	385-448	Mar 1892	Migula, W.	Characeae
	8	449-512	Jan 1893	Migula, W.	Characeae
	9	513-576	Jul 1894	Migula, W.	Characeae

Band	Abt.	Lief.	Pagination	Date	Author	Taxon
		10	577-640	Sep 1895	Migula, W.	Characeae
		11	641-688	Jan 1896	Migula, W.	Characeae
		12	689-758	Jan 1897	Migula, W.	Characeae
			[759]-765		Anonymous	General Index
6	1		[v]-vii	Apr 1911	Müller, K.	Hepaticae-Foreword
		1	[1]2-64	Dec 1905	Müller, K.	Hepaticae-Introduction
		2	65-128	Mai 1906	Müller, K.	Hepaticae-Introduction
		3	129-132	20 Mar 1907	Müller, K.	Notes for Collectors
6	1	3	133-140	20 Mar 1907	Müller, K.	Systematic Section
			140-192	20 Mar 1907	Müller, K.	Ricciaceae
		4	193-219	20 Jun 1907	Müller, K.	Ricciaceae
			220-256	20 Jun 1907	Müller, K.	Marchantiaceae
		5	257-308	1 Oct 1907	Müller, K.	Marchantiaceae
			309-314	1 Oct 1907	Müller, K.	Jungermanniaceae-Anakrogynae
			314-317	1 Oct 1907	Müller, K.	Sphaerocarpideae
			318-320	1 Oct 1907	Müller, K.	Rielloideae
		6	321-324	20 Mar 1908	Müller, K.	Rielloideae
			324-327	20 Mar 1908	Müller, K.	Elatereae
			327-344	20 Mar 1908	Müller, K.	Aneureae
			345-354	20 Mar 1908	Müller, K.	Metzgerieae
			354-367	20 Mar 1908	Müller, K.	Diplomitrieae (Diplolaeneae, Dilaeneae)
			367-380	20 Mar 1908	Müller, K.	Haplolaeneae
			380-384	20 Mar 1908	Müller, K.	Codonieae
		7	385-395	27 Feb 1909	Müller, K.	Codonieae
			396-400	27 Feb 1909	Müller, K.	Haplomitrieae
			401-404	27 Feb 1909	Müller, K.	Jungermanniaceae-Akrogynae
			404-448	27 Feb 1909	Müller, K.	Epigonantheae
		8	449-512	15 Apr 1909	Müller, K.	Epigonantheae
		9	513-576	15 Sep 1909	Müller, K.	Epigonantheae
		10	577-640	15 Apr 1910	Müller, K.	Epigonantheae
		11	641-704	1 Jun 1910	Müller, K.	Epigonantheae
		12	705-768	Sep 1910	Müller, K.	Epigonantheae
		13	769-832	10 Mar 1911	Müller, K.	Epigonantheae
		14	833-856	10 Apr 1911	Müller, K.	Epigonantheae
			857-870		Anonymous	Indexes

		143-208	20 Jan 1913	Müller, K.	Trigonantheae
	18	209-272	20 Nov 1913	Müller, K.	Trigonantheae
	19	273-300	21 Mar 1914	Müller, K.	Trigonantheae
		300-336	21 Mar 1914	Müller, K.	Ptilidioideae
6	20	337-348	19 Dec 1914	Müller, K.	Ptilidioideae
		349-384	19 Dec 1914	Müller, K.	Scapanioideae
	21	385-464	22 Feb 1915	Müller, K.	Scapanioideae
	22	465-524	1 Mar 1915	Müller, K.	Scapanioideae
		525-528	1 Mar 1915	Müller, K.	Pleurozioideae
	23	529-533	26 Jul 1915	Müller, K.	Pleurozioideae
		533-553	26 Jul 1915	Müller, K.	Raduloideae
		553-592	26 Jul 1915	Müller, K.	Madothecoideae
	24	593-656	7 Oct 1915	Müller, K.	Jubuleae-Lejeunea
	25	657-681	18 Feb 1916	Müller, K.	Lejeunea
		682-709	18 Feb 1916	Müller, K.	Anthocerotales
		710-720	18 Feb 1916	Müller, K.	Nachträge
	26	721-784	28 Apr 1916	Müller, K.	Nachträge
	27	785-818	28 Mai 1916	Müller, K.	Geogr. Distr.
		819-848	28 Mai 1916	Müller, K.	Geogr. Distr.
	28	849-896	27 Aug 1916	Müller, K.	Geogr., Ecol.
		897-947	27 Aug 1916	Anonymous	Indexes
		[I]-VII			
Erg.	1	1-160	1939	Müller, K.	General-Hepaticae
	2	161-320	1940	Müller, K.	General-Hepaticae
6, new edition, 1954-1957[-1958], 2 vols., see TL-2/6494.					
7	1	[v]-ix	Nov 1930	Hustedt, F.	Diatoms-Foreword
		[1]-216	Mar 1927	Hustedt, F.	Diatoms-Introduction
		[217]-219	Mar 1927	Hustedt, F.	Centricae
		219-220	Mar 1927	Hustedt, F.	Discoideae
		220-272	Mar 1927	Hustedt, F.	Coscinodisceae-Melosirinae
	2	273-309	Jan 1928	Hustedt, F.	Coscinodisceae-Melosirinae
		310-333	Jan 1928	Hustedt, F.	Coscinodisceae-Sceletoneminae
		333-464	Jan 1928	Hustedt, F.	Coscinodisceae-Coscinodiscinae
	3	465-467	Jan 1929	Hustedt, F.	Coscinodisceae-Coscinodiscinae
		467-472	Jan 1929	Hustedt, F.	Actinodisceae-Stictodiscinae
		472-483	Jan 1929	Hustedt, F.	Actinodisceae-Actinoptychinae
		483-499	Jan 1929	Hustedt, F.	Actinodisceae-Asterolamprinae

Band	Abt.	Lief.	Pagination	Date	Author	Taxon
7			500-501	Jan 1929	Hustedt, F.	Eupodisceae-Pyrgodiscinae
			501-508	Jan 1929	Hustedt, F.	Eupodisceae-Aulacodiscenae
			508-541	Jan 1929	Hustedt, F.	Eupodisceae-Eupodiscinae
		3	541-542	Jan 1929	Hustedt, F.	Solenioideae
			542-561	Jan 1929	Hustedt, F.	Solenieae-Lauderiinae
			561-608	Jan 1929	Hustedt, F.	Solenieae-Rhizosoleniinae
		4	609	Jan 1930	Hustedt, F.	Biddulphioideae
			609-767	Jan 1930	Hustedt, F.	Chaetocereae
			767-780	Jan 1930	Hustedt, F.	Biddulphieae-Eucampiinae
			781-784	Jan 1930	Hustedt, F.	Biddulphieae-Triceratiinae
		5	785-828	Nov 1930	Hustedt, F.	Biddulphieae-Triceratiinae
			828-864	Nov 1930	Hustedt, F.	Biddulphieae-Biddulphiinae
			864-868	Nov 1930	Hustedt, F.	Biddulphieae-Isthmiinae
			868-890	Nov 1930	Hustedt, F.	Biddulphieae-Hemiaulinae
			891-901	Nov 1930	Hustedt, F.	Anauleae
			901-907	Nov 1930	Hustedt, F.	Euodieae
			908-915	Nov 1930	Anonymous	General Index
			[916]-920	Nov 1930	Anonymous	Index of Species
	2		[v]-ix	Dec 1959	Hustedt, F.	Foreword
		1	[1]-4	Nov 1931	Hustedt, F.	Pennatae
			5-6	Nov 1931	Hustedt, F.	Araphideae
			6-11	Nov 1931	Hustedt, F.	Entopyleae
			11-51	Nov 1931	Hustedt, F.	Tabellarieae-Tabellariinae
			52-91	Nov 1931	Hustedt, F.	Tabellarieae-Licmophorinae
			92-114	Nov 1931	Hustedt, F.	Fragilarieae-Diatominae
			115-176	Nov 1931	Hustedt, F.	Fragilarieae-Fragilariinae
		2	177-260	Jan 1932	Hustedt, F.	Fragilarieae-Fragilariinae
			261-264	Jan 1932	Hustedt, F.	Peronieae
			265-317	Jan 1932	Hustedt, F.	Eunotieae
			317-318	Jan 1932	Hustedt, F.	Monoraphideae
			319-320	Jan 1932	Hustedt, F.	Cocconeideae
		3	321-364	Sep 1933	Hustedt, F.	Cocconeideae
			364-432	Sep 1933	Hustedt, F.	Achnantheae
		4	433	Dec 1933	Hustedt, F.	Achnantheae
			434-576	Dec 1933	Hustedt, F.	Naviculeae
		5	577-736	Mar 1937	Hustedt, F.	Naviculeae

8	3	349-556	Mai 1964	Hustedt, F.	Naviculeae
	4	557-816	Mar 1966	Hustedt, F.	Naviculeae
	1	[v]-viii	Feb 1929	Keissler, K.	Flechtenparasiten
		[1]-26	Jan 1930	Keissler, K.	General Introduction
		[27]28	Jan 1930	Keissler, K.	Myxobacteriales
		28-30	Jan 1930	Keissler, K.	Archangiaceae
		30-33	Jan 1930	Keissler, K.	Polyangiaceae
		33-35	Jan 1930	Keissler, K.	Myxococcaceae
		36	Jan 1930	Keissler, K.	Myxomycetes
		36	Jan 1930	Keissler, K.	Myxogasteres
		37-38	Jan 1930	Keissler, K.	Liceaceae
		38-41	Jan 1930	Keissler, K.	Listerellaceae
		41-42	Jan 1930	Keissler, K.	Didymiaceae
		42-46	Jan 1930	Keissler, K.	Physaraceae
		46	Jan 1930	Keissler, K.	Phycomycetes
		46-47	Jan 1930	Keissler, K.	Mucoraceae
		47-48	Jan 1930	Keissler, K.	Ascomycetes
		48	Jan 1930	Keissler, K.	Protodiscineae
		48-50	Jan 1930	Keissler, K.	Exoascaceae
		50-51	Jan 1930	Keissler, K.	Plectascineae
		51	Jan 1930	Keissler, K.	Aspergillaceae
		52	Jan 1930	Keissler, K.	Discomycetes
		52	Jan 1930	Keissler, K.	Pezizineae
		52-105	Jan 1930	Keissler, K.	Celidiaceae
		105	Jan 1930	Keissler, K.	Helotiaceae
		105-108	Jan 1930	Keissler, K.	Pezizelleae
		108-113	Jan 1930	Keissler, K.	Mollisiaceae
	2	113-118	Jan 1913	Keissler, K.	Pyrenopezizeae, Callorieae
		118-240	Jan 1930	Keissler, K.	Patellariaceae
		241-251	Apr 1930	Keissler, K.	Patellariaceae
		251-261	Apr 1930	Keissler, K.	Stictidaceae
	2	261-264	Apr 1930	Keissler, K.	Excluded Genera-Pezizineae
		264-265	Apr 1930	Keissler, K.	Pyrenomycetes
		265	Apr 1930	Keissler, K.	Perisporiales
8		265-269	Apr 1930	Keissler, K.	Perisporiaceae
		269-274	Apr 1930	Keissler, K.	Microthyriaceae

Band	Abt.	Lief.	Pagination	Date	Author	Taxon
			274-296	Apr 1930	Keissler, K.	Hypocreales
			296-297	Apr 1930	Keissler, K.	Dothideales
			297-304	Apr 1930	Keissler, K.	Phyllachoraceae
			304-325	Apr 1930	Keissler, K.	Pseudosphaeriales
			325-326	Apr 1930	Keissler, K.	Sphaeriales
			326-337	Apr 1930	Keissler, K.	Sphaeriaceae
			337-445	Apr 1930	Keissler, K.	Sphaerellaceae
			445-520	Apr 1930	Keissler, K.	Pleosporaceae
			521	Apr 1930	Keissler, K.	Basidiomycetes
			521	Apr 1930	Keissler, K.	Hymenomycetineae
			521	Apr 1930	Keissler, K.	Thelephoraceae
			521-528	Apr 1930	Keissler, K.	Clavariaceae
			529-531	Apr 1930	Keissler, K.	Fungi Imperfecti
			531-532	Apr 1930	Keissler, K.	Sphaeropsidales
			532-581	Apr 1930	Keissler, K.	Sphaerioidaceae
			581-586	Apr 1930	Keissler, K.	Nectrioidaceae
			586	Apr 1930	Keissler, K.	Melanconiales
			586-589	Apr 1930	Keissler, K.	Melanconiaceae
			589-590	Apr 1930	Keissler, K.	Hyphomycetes
			590-602	Apr 1930	Keissler, K.	Mucedinaceae
			602-623	Apr 1930	Keissler, K.	Dematiaceae
			623-624	Apr 1930	Keissler, K.	Stilbaceae
			624-642	Apr 1930	Keissler, K.	Tuberculariaceae
			643-644	Apr 1930	Keissler, K.	Nachträge
			[645]-678	Apr 1930	Keissler, K.	Host Index
			679-709	Apr 1930	Keissler, K.	General Index
			[710]-712	Apr 1930	Keissler, K.	Add./Corr.
9	1/1	1	[1]-43	Feb 1933	Zschacke, H.	Moriolaceae
			[44]-46	Feb 1933	Zschacke, H.	Epigloeaceae
			46-160	Feb 1933	Zschacke, H.	Verrucariaceae
		2	161-320	Jun 1933	Zschacke, H.	Verrucariaceae
		3	321-480	Aug 1933	Zschacke, H.	Verrucariaceae
9	1/1	4	481-590	Feb 1934	Zschacke, H.	Verrucariaceae
			590-656	Feb 1934	Zschacke, H.	Dermatocarpaceae
			656-668	Feb 1934	Zschacke, H.	Verrucariaceae-Add./Corr.
			[669]-673	Feb 1934	Anonymous	Index of Taxa Illustrated

Vol.	Part	Pages	Date	Author	Subject
9		[xi]		Keissler, K.	Synopsis of Lichens
		[1]-160	post Dec 1937	Keissler, K.	Pyrenulaceae
	2	161-320	Nov 1936	Keissler, K.	Pyrenulaceae
	3	321-421	Jan 1937	Keissler, K.	Pyrenulaceae
		[422]-439	Mar 1937	Keissler, K.	Trypetheliaceae
		[440]-447	Mar 1937	Keissler, K.	Strigulaceae
		[448]-462	Mar 1937	Keissler, K.	Pyrenidiaceae
		[463]-469	Mar 1937	Keissler, K.	Xanthopyreniaceae
	4	481-502	Jul 1937	Keissler, K.	Mycoporaceae
		503-505	Jul 1937	Keissler, K.	Excluded Genera
		[506]	Jul 1937	Keissler, K.	Gymnocarpeae
		[507]-519	Jul 1937	Keissler, K.	Coniocarpineae
		[520]-640	Jul 1937	Keissler, K.	Caliciaceae
	5	641-727	Mar 1938	Keissler, K.	Caliciaceae
		[728]-755	Mar 1938	Keissler, K.	Cypheliaceae
		[756]-773	Mar 1938	Keissler, K.	Sphaerophoraceae
		[774]-784	Mar 1938	Keissler, K.	Add./Corr.
		[785]-786	Mar 1938	Anonymous	Index of Spp. Illustrated
		[787]-793	Mar 1938	Anonymous	Index to Genera
		[794]-841	Mar 1938	Anonymous	General Index
		[842]-846	Mar 1938	Keissler, K.	Add./Corr.
2/1	1	[1]-180	Jan 1937	Redinger, K.	Arthoniaceae
	2	[181]-404	Jan 1939	Redinger, K.	Graphidaceae
2/2	1	[1]-110	1940	Köfaragó-Gyelnik, V.	Lichinaceae
		[111]-134	1940	Köfaragó-Gyelnik, V.	Heppiaceae
	2	[135]-272	1940	Köfaragó-Gyelnik, V.	Pannariaceae
4/1	1	[v-viii]	1933	Frey, E.	Foreword
		[1]-202	Dec 1932	Frey, E.	Cladoniaceae
		[203]-208	Dec 1932	Frey, E.	Umbilicariaceae
	2	209-407	Feb 1933	Frey, E.	Umbilicariaceae
		[408]-411	Feb 1933	Frey, E.	Add./Corr.
		[412]-426	Feb 1933	Anonymous	Index, Expl. pl.
4/2	1	[1]	post Jun 1930	Sandstede, H.	Foreword
		2-240	Jan 1931	Sandstede, H.	Cladonia
	2	241-496	Mar 1931	Sandstede, H.	Cladonia
		[497]-507	Mar 1931	Sandstede, H.	Nachträge

Band	Abt.	Lief.	Pagination	Date	Author	Taxon
9	5/1		[508]-523	Mar 1931	Anonymous	Index of Names
			[524]-53I	Mar 1931	Anonymous	List of Taxa Illustrata
		1	[v]-vii	Mai 1936	Magnusson, A. H.	Foreword
			vii	Mai 1936	Erichsen, C. F. E.	Foreword
			[1]-285	Mar 1935	Magnusson, A. H.	Acarosporaceae
			[286]-318	Mar 1935	Magnusson, A. H.	Thelocarpaceae
		2	[321]-512	Jan 1936	Erichsen, C. F. E.	Pertusariaceae
		3	513-699	Mai 1936	Erichsen, C. F. E.	Pertusariaceae
			[700]701	Mai 1936	Erichsen, C. F. E.	List of Spp. Illustrated
			[702]	Mai 1936	Erichsen, C. F. E.	Additions
			[703]-728	Mai 1936	Anonymous	General Index
			728	Mai 1936	Erichsen, C. F. E.	Corrections
	5/3	1	[1]-3	Mai 1936	Hillman, J.	Foreword
			5-160	Mai 1936	Hillman, J.	Parmeliaceae
		2	161-304	Mai 1936	Hillman, J.	Parmeliaceae
			[305]-309	Mai 1936	Hillman, J.	Additions, Nachträge
			[1]-10	Mai 1936	Anonymous	General Index
	5/4		[v]-viii	Dec 1960	Keissler, K.	Foreword
		1	[3]-160	Oct 1958	Keissler, K.	Usneaceae
		2	161-320	Mar 1959	Keissler, K.	Usneaceae
		3	321-480	Nov 1959	Keissler, K.	Usneaceae
		4	481-640	Mar 1960	Keissler, K.	Usneaceae
		5	641-717	Dec 1960	Keissler, K.	Usneaceae
			718-730	Dec 1960	Keissler, K.	Nachträge
			731	Dec 1960	Keissler, K.	Corrections
			[732]733	Dec 1960	Huber, J.	List of Spp. Illustrated
			[734]-755	Dec 1960	Anonymous	General Index
9	6	1	[1]-36	Dec 1935	Lynge, B.	Teloschistaceae
			[37]-40	Dec 1935	Lynge, B.	Physciaceae-Introd.
			[41]-188	Dec 1935	Lynge, B.	Physciaceae
10	2	1	[1]-26	Nov 1930	Gemeinhardt, K.	Silicoflagellatae-General
			26	Nov 1930	Gemeinhardt, K.	Siphonotestales
			26-77	Nov 1930	Gemeinhardt, K.	Dictyochaceae
			77-79	Nov 1930	Gemeinhardt, K.	Cornuaceae
			79	Nov 1930	Gemeinhardt, K.	Stereotestales
			79-82	Nov 1930	Gemeinhardt, K.	Ebriaceae

10			233-236	Nov 1930	Schiller, J.	Deutschlandiaceae
			237-242	Nov 1930	Schiller, J.	Thoracosphaeraceae
			242-255	Nov 1930	Schiller, J.	Coccolithaceae
			255-257	Nov 1930	Schiller, J.	Coccolithineae-Uncertain Gen.
			258-259	Nov 1930	Schiller, J.	Coccolithineae-Appendix
			260-266	Nov 1930	Anonymous	Bibliography
			[267]	1930	Anonymous	Coccolithineae-Contents
			[268]-273			General Index
	3/1	1	[1]-6	Dec 1931-Jan 1932	Schiller, J.	Dinoflagellata
			6-9	Dec 1931-Jan 1932	Schiller, J.	Desmomonadaceae
			9-10	Dec 1931-Jan 1932	Schiller, J.	Adinimonadaceae
			10-11	Dec 1931-Jan 1932	Schiller, J.	Desmocapsaceae
			12-44	Dec 1931-Jan 1932	Schiller, J.	Prorocentraceae
			45-164	Dec 1931-Jan 1932	Schiller, J.	Dinophysiaceae
			165-192	Dec 1931-Jan 1932	Schiller, J.	Amphisoleniaceae
			192-255	Dec 1931-Jan 1932	Schiller, J.	Ornithocercaceae
			255-256	Dec 1931-Jan 1932	Schiller, J.	Citharistaceae
			255-256		Schiller, J.	Citharistaceae
		2	257-258	Dec 1932	Schiller, J.	Dinophyceae
			258-263	Dec 1932	Schiller, J.	Pronoctilucaceae
			263-272	Dec 1932	Schiller, J.	Gymnodiniaceae
			272-432	Dec 1932	Schiller, J.	Gymnodiniaceae
		3	433-547	Feb 1933	Schiller, J.	Polykrikaceae
			547-553	Feb 1933	Schiller, J.	Warnowiaceae
			554-600	Feb 1933	Schiller, J.	Gymnodiniales-Uncertain Genera
			601-603	Feb 1933	Schiller, J.	General Index
			604-617	Feb 1933	Anonymous	Gymnodiniales-Appendix
10	3/2	1	[1]	Mar 1935	Schiller, J.	Gymnosclerotaceae
			[1]2-8	Mar 1935	Schiller, J.	Blastodiniales
			8-15	Mar 1935	Schiller, J.	Paradiniaceae
			15-19	Mar 1935	Schiller, J.	Blastodiniaceae
			20-53	Mar 1935	Schiller, J.	Syndiniaceae
			53-61	Mar 1935	Schiller, J.	Endodiniaceae
			61-62	Mar 1935	Schiller, J.	Ellobiopsidaceae
			62-71	Mar 1935	Schiller, J.	Peridiniales
			71-74	Mar 1935	Schiller, J.	Ptychodiscaceae
			74-80	Mar 1935	Schiller, J.	

Band	Abt.	Lief.	Pagination	Date	Author	Taxon
			80-92	Mar 1935	Schiller, J.	Glenodiniopsidaceae
			92-123	Mar 1935	Schiller, J.	Glenodiniaceae
			123-160	Mar 1935	Schiller, J.	Peridiniaceae
		2	161-275	Apr 1935	Schiller, J.	Peridiniaceae
			275-320	Apr 1935	Goniaulacaceae	
			320	Apr 1935	Schiller, J.	Congruentidiaceae
		3	321	Mai 1937	Schiller, J.	Congruentidiaceae
			321-327	Mai 1937	Schiller, J.	Protocratiaceae
			327-438	Mai 1937	Schiller, J.	Heterodiniaceae
			438-447	Mai 1937	Schiller, J.	Goniodomaceae
			447-467	Mai 1937	Schiller, J.	Oxytoxaceae
			467-471	Mai 1937	Schiller, J.	Cladopyxiaceae
			471-473	Mai 1937	Schiller, J.	Ostreopsiaceae
			473-480	Mai 1937	Schiller, J.	Podolampaceae
			480	Mai 1937	Schiller, J.	Lissodiniaceae
10	3/2	4	481	Mai 1937	Schiller, J.	Lissodiniaceae
			481-482	Mai 1937	Schiller, J.	Amoebodiniaceae
			482-508	Mai 1937	Schiller, J.	Gloeodiniaceae
			508-510	Mai 1937	Schiller, J.	Dinotrichaceae
			510	Mai 1937	Schiller, J.	Dinocloniaceae
			510-522	Mai 1937	Schiller, J.	Nachträge
			523-570	Mai 1937	Schiller, J.	Bibliography
			[571]-589	Mai 1937	Anonymous	General Index
			[590]	Mai 1937	Schiller, J.	Corrections
			[v]vi	1939	Pascher, A.	Foreword
11		1	[1]-160	Mai 1937	Pascher, A.	Heterokontae-General
		2	161-189	Aug 1937	Pascher, A.	Heterokontae-General
			190-202	Aug 1937	Pascher, A.	Heterokontae-Culture
			[203]-208	Aug 1937	Pascher, A.	Heterokontae-Systematic
			208-236	Aug 1937	Pascher, A.	Heterochloridaceae
			236	Aug 1937	Pascher, A.	Rhizochloridineae
			239-251	Aug 1937	Pascher, A.	Rhizochloridaceae
			251-256	Aug 1937	Pascher, A.	Chlorarachniaceae
			256-274	Aug 1937	Pascher, A.	Myxochloridaceae
			274-277	Aug 1937	Pascher, A.	Heterocapsineae
			277-301	Aug 1937	Pascher, A.	Heterocapsaceae

4		481-632	Feb 1938	Pascher, A.	Pleurochloridaceae	
		632-640	Feb 1938	Pascher, A.	Gloeobotrydaceae	
	5	641-661	Jan 1939	Pascher, A.	Gloeobotrydaceae	
		661-696	Jan 1939	Pascher, A.	Botryochloridaceae	
		696-703	Jan 1939	Pascher, A.	Gloeopodiaceae	
		703-718	Jan 1939	Pascher, A.	Mischococcaceae	
11	5	718-812	Jan 1939	Pascher, A.	Characiopsidaceae	
		812-824	Jan 1939	Pascher, A.	Chloropediaceae	
		825-829	Jan 1939	Pascher, A.	Trypanochloridaceae	
		830-832	Jan 1939	Pascher, A.	Centritractaceae	
	6	833-860	Feb 1939	Pascher, A.	Centritractaceae	
		860-906	Feb 1939	Pascher, A.	Chlorotheciaceae	
		906-912	Feb 1939	Pascher, A.	Heterococcales-Genera Incert. Sed.	
		912-915	Feb 1939	Pascher, A.	Heterotrichineae	
		915-916	Feb 1939	Pascher, A.	Tribonematales	
		916-939	Feb 1939	Pascher, A.	Heterotrichaceae	
		939-991	Feb 1939	Pascher, A.	Tribonemataceae	
		991-992	Feb 1939	Pascher, A.	Heterocloniales	
		992-997	Feb 1939	Pascher, A.	Heterodendraceae	
		997-1023	Feb 1939	Pascher, A.	Heterocloniaceae	
		1023-1024	Feb 1939	Pascher, A.	Botrydiales	
		1024-1055	Feb 1939	Pascher, A.	Botrydiaceae	
		1055-1065	Feb 1939	Pascher, A.	Nachträge	
		1065-1069	Feb 1939	Pascher, A.	Historical Sketch	
		1070-1082	Feb 1939	Pascher, A.	Bibliography	
		[1083]-1092	Feb 1939	Anonomous	Index	
12	1	[v]-ix	Feb 1940	Gemeinhardt, K.	Oedogoniales-Foreword	
		[1]-69	Apr 1938	Gemeinhardt, K.	Oedognoniales-General	
		69-172	Apr 1938	Gemeinhardt, K.	Oedogoniaceae	
	2	173-332	Jun 1939	Gemeinhardt, K.	Oedogoniaceae	
	3	333-444	Feb 1940	Gemeinhardt, K.	Oedogoniaceae	
		[445]-453	Feb 1940	Anonymous	Index	
13	1/1	[1]-3	Jun 1934	Krieger, W.	Desmidiaceae	
		[4]-172	Jun 1934	Krieger, W.	Desmidiaceae-General	
		[173]174	Jun 1934	Krieger, W.	Desmidiales	
		174-223	Jun 1934	Krieger, W.	Mesotaeniaceae	

Band	Abt.	Lief.	Pagination	Date	Author	Taxon
		2	225-375	Dec 1935	Krieger, W.	Desmidiaceae
		3	[377]-536	Aug 1936	Krieger, W.	Desmidiaceae
		4	537-675	Feb 1937	Krieger, W.	Desmidiaceae
			[676]-712, [ii]-vi	1937	Anonymous	Index
13	1/2		1-117	Jun 1939	Krieger, W.	Desmidiaceae
13	2	1	[i]	1941 (foreword 1940)	Kolkwitz, R. & H. Krieger	Zygnemales-Foreword
			[1]-50	1941	Kolkwitz, R.	Zygnemales-General
			[51]-104	1941	Krieger, H.	Zygnemales- Bibliography
			[105]-109	1941	Anonymous	Index
		2	[111]-115	1941	Krieger, H.	Zygnemales-Systematics
			115-196	1941	Krieger, W.	Zygnemaceae
		3	197-294	1941	Krieger, W.	Zygnemaceae
		4	295-482	1944	Krieger, W.	Zygnemaceae
			[483]484	1944	Krieger, W.	Zygnemaceae-Nachträge
			[485]486-499	1944	Anonymous	Index
14		1	1-124	Nov 1930	Geitler, L.	Cyanophyceae
			124	Nov 1930	Geitler, L.	Chroococcales
			124-288	Nov 1930	Geitler, L.	Chroococcaceae
		2	289-292	Jan-Jun 1931	Geitler, L.	Chroococcaceae
			292-312	Jan-Jun 1931	Geitler, L.	Entophysalidaceae
			[313]-315	Jan-Jun 1931	Geitler, L.	Chamaesiphonales
			315-382	Jan-Jun 1931	Geitler, L.	Pleurocapsaceae
			[383]-414	Jan-Jun 1931	Geitler, L.	Dermocarpaceae
			[415]-445	Jan-Jun 1931	Geitler, L.	Chamaesiphonaceae
			445-452	Jan-Jun 1931	Geitler, L.	Siphonemataceae
			452-456	Jan-Jun 1931	Geitler, L.	Enonemataceae
			[457]-459	Jan-Jun 1931	Geitler, L.	Hormogonales
			459-463	Jan-Jun 1931	Geitler, L.	Loriellaceae
			463-464	Jan-Jun 1931	Geitler, L.	Pulvinulariaceae
			464	Jan-Jun 1931	Geitler, L.	Capsosiraceae
		3	465-471	Jul 1931	Geitler, L.	Capsosiraceae
			471-472	Jul 1931	Geitler, L.	Loefgreniaceae
			472-482	Jul 1931	Geitler, L.	Nostochopsidaceae
			482-552	Jul 1931	Geitler, L.	Stigonemataceae

	664-672	Jul 1931	Geitler, L.	Microchaetaceae
4	673-677	Mai 1932	Geitler, L.	Microchaetaceae
	677-801	Mai 1932	Geitler, L.	Scytonemataceae
	801-896	Mai 1932	Geitler, L.	Nostocaceae
5	897-905	Aug 1932	Geitler, L.	Nostocaceae
	906-1056	Aug 1932	Geitler, L.	Oscillatoriaceae
6	1057-1158	Dec 1932	Geitler, L.	Oscillatoriceae
	1158-1160	Dec 1932	Geitler, L.	Hormogonales-Suppl.
	[1161]1165	Dec 1932	Geitler, L.	Chroococcaceae-Add.
	1165-1167	Dec 1932	Geitler, L.	Entophysalidaceae-Add.
	1167-1173	Dec 1932	Geitler, L.	Pleurocapsaceae-Add.
	1173	Dec 1932	Geitler, L.	Stigonemataceae-Add.
	1174-1175	Dec 1932	Geitler, L.	Rivulariaceae-Add.
	[1176]1177	Dec 1932	Geitler, L.	Corrections
	[1178]	Dec 1932	Geitler, L.	Entophysalidaceae-Add.
	[1179]-1196	Dec 1932	Anonymous	General Index

Band	Abt.	Lief.	Date	Band	Abt.	Lief.	Date
I	I	1	Dec 1880			58	Nov 1896
		2	Mar 1881	I	6	59	Jun 1898
		3	Mai 1881			60	Jul 1898
		4	Aug 1881			61	Aug 1898
		5	1-15 Nov 1882			62	Oct 1898
		6	Jan 1882			63	Dec 1898
		7	Apr 1882			64	Jan 1899
		8	Jul 1882			65	Jun 1899
		9	Oct 1882			66	Sep 1899
		10	Dec 1882			67	Oct 1899
		11	Jan 1883	I	6	68	Dec 1899
		12	Mai 1883			69	Jan 1900
		13	Sep 1883			70	Apr 1900
		index	Jan 1885			71	Jul 1900
I	2	14	Mar 1884			72	Jul 1900
		15	Sep 1884			73	Oct 1900
		16	Nov 1884			74	Dec 1901
		17	Apr 1885	I	7	75	Jan 1901
		18	Jul 1885			76	Apr 1901
		19	Jul 1885			77	Jun 1901
		20	Oct 1885			78	Aug 1901
		21	Dec 1885			79	Dec 1901
		22	Mai 1886			80	Dec 1901
		23	Aug 1886			81	Jan 1902
		24	Aug 1886			82	Apr 1902
		25	Oct 1886			83	Mai 1902
		26	Dec 1886			84	Jul 1902
		27	Mar 1887			85	Oct 1902
I	3	28	Aug 1887			86	Jan 1903
		29	Jan 1888			87	Jan 1903
		30	Aug 1888			88	Mar 1903
		31	Jan 1889			89	Apr 1903
		32	Apr 1889			90	Sep 1903
		33	Jul 1890			91	Dec 1903
I	3	34	Jan 1891	I	8	92	16 Mai 1904
		35	Aug 1891			93	30 Jun 1904
		36	Oct 1891			94	15 Jul 1904
		37	Jul 1892			95	3 Apr 1905
		38	Sep 1892			96	10 Mai 1905
		39	Jan 1893			97	20 Jun 1905
		40	Sep 1893			98	15 Jul 1905
		41	Oct 1893			99	25 Jul 1906
		42	Jun 1894			100	30 Aug 1906
		43	Nov 1894			101	20 Sep 1906
		44	Jan 1895			102	10 Oct 1906
I	4	45	Oct 1891			103	15 Nov 1906
		46	Nov 1891			104	16 Mai 1907
		47	Jan 1892			105	12 Jul 1907
		48	Mar 1892	I	9	106	20 Dec 1907
		49	Apr 1892			107	1 Apr 1908
		50	Mai 1892			108	25 Mai 1908
		51	Jul 1892			109	30 Jul 1908
		52	Oct 1892			110	25 Nov 1908
		53	Jul 1895			111	7 Apr 1909
		54	Nov 1895			112	28 Apr 1909
		55	Apr 1896			113	20 Jun 1909
		56	Mai 1896			114	25 Jul 1909
I	5	57	Jul 1896			115	1 Nov 1909

Band	Abt.	Lief.	Date	Band	Abt.	Lief.	Date
		116	31 Dec 1909			23	Dec 1893
		117	10 Mar 1910			24	Aug 1894
		118	20 Mai 1910			25	Jan 1895
		119	20 Jul 1910			26	Jun 1895
		120	10 Dec 1910	4	3	27	Dec 1895
1	10	121	8 Dec 1920			28	Jul 1896
		122	18 Feb 1914			29	Dec 1896
		123	17 Mai 1915	4	3	30	Dec 1896
		124	9 Dec 1915			31	Sep 1897
		125	28 Aug 1917			32	Dec 1897
		126	20 Mar 1918			33	Aug 1898
		127	8 Dec 1920			34	Feb 1899
2		1	Nov 1882			35	Nov 1899
		2	Dec 1882			36	Apr 1901
		3	Jan 1883			37	Dec 1901
		4	Apr 1883			38	Dec 1902
		5	Apr 1883			39	Jun 1903
		6	Oct 1883			40	Dec 1903
		7	Jan 1884			41	Dec 1903
		8	Jul 1884	4	Suppl.	1	Apr 1927
		9	Oct 1884			2	Jul 1927
		10	Jan 1885			3	Dec 1927
3		1	Mar 1884	5		1	Oct 1889
		2	Apr 1884			2	Dec 1889
		3	Sep 1884			3	Jan 1890
		4	Mai 1885			4	Jun 1890
		5	Oct 1885			5	Jan 1891
		6	Jan 1886			6	Aug 1891
3		7	Jul 1886			7	Mar 1892
		8	Sep 1886			8	Jan 1893
		9	Jun 1887			9	Jul 1894
		10	Aug 1887			10	Sep 1895
		11	Apr 1888			11	Apr 1896
		12	Jan 1889			12	Jan 1897
		13	Apr 1889	6	1	1	Dec 1905
		14	Jun 1889			2	Mai 1906
4	1	1	Jul 1885			3	20 Mar 1907
		2	Aug 1885			4	20 Jun 1907
		3	Dec 1885			5	1 Oct 1907
		4	Jul 1886			6	20 Mar 1908
		5	Nov 1886			7	27 Feb 1909
		6	Dec 1886			8	15 Apr 1909
		7	Jul 1887			9	15 Sep 1909
		8	Oct 1887			10	15 Apr 1910
		9	Mai 1888	6	1	11	1 Jun 1910
		10	Oct 1888			12	Sep 1910
		11	Dec 1888			13	10 Mar 1911
		12	Oct 1889			14	10 Apr 1911
		13	Nov 1889	6	2	15	1 Mar 1912
4	2	14	Jul 1890			16	20 Dec 1912
		15	Oct 1890			17	20 Jan 1913
		16	Jun 1891			18	20 Nov 1913
		17	Dec 1891			19	21 Mar 1914
		18	Jan 1892			20	19 Dec 1914
		19	Nov 1892			21	22 Feb 1915
		20	Dec 1892			22	1 Mar 1915
		21	Jan 1893			23	26 Jul 1915
		22	Jun 1893			24	7 Oct 1915

RABENHORST, *Dates Lieferungen*

Band	Abt.	Lief.	Date	Band	Abt.	Lief.	Date
		25	18 Feb 1916		5/1	1	Mar 1935
		26	28 Apr 1916			2	Mai 1936
		27	28 Mai 1916		5/4	1	Oct 1958
		28	27 Aug 1916			2	Mar 1959
	Erg.	1	1939			3	Nov 1959
		2	1940			4	Mar 1960
7	1	1	Mar 1927			5	Dec 1960
		2	Jan 1928		6	1	Dec 1935
		3	Jan 1929	10	2	1	Nov 1930
		4	Jan 1930		3/1	1	Dec 1931/Jan 1932
		5	Nov 1930			2	Dec 1932
7	2	1	Nov 1931			3	Feb 1933
		2	Jan 1932		3/2	1	Mar 1935
		3	Sep 1933			2	Apr 1935
		4	Dec 1933			3	Mai 1937
		5	Mar 1937			4	Mai 1937
		6	Dec 1959	11		1	Mai 1937
7	3	1	Sep 1961			2	Aug 1937
		2	Feb 1962	11		3	Feb 1938
		3	Mai 1964			4	Feb 1938
		4	Mar 1966			5	Jan 1939
8		1	Jan 1930			6	Feb 1939
		2	Apr 1930	12	4	1	Apr 1938
9	1/1	1	Feb 1933			2	Jun 1939
		2	Jun 1933			3	Feb 1940
		3	Aug 1933	13	1/1	1	Jun 1934
		4	Feb 1934			2	Dec 1935
9	1/2	1	Nov 1936			3	Aug 1936
		2	Jan 1937			4	Feb 1937
		3	Mar 1937	13	1/2	1	Jun 1939
		4	Jul 1937		2	1	1941
		5	Mar 1938			2	1941
	2/1	1	Jan 1937			3	1941
		2	Jan 1939			4	1944
	2/2		1940	14		1	Nov 1930
	4/1	1	Dec 1932			2	Jan-Jun 1931
		2	Feb 1933			3	Jul 1931
	4/2	1	Jan 1931			4	Mai 1932
		2	Mar 1931			5	Aug 1932
						6	Dec 1932

RABENHORST'S KRYPTOGAMENFLORA *ed-2, Arrangement by families:* see p. 505-528.

Family	Author			Pages	Figs	Date
Acarosporaceae	Magnusson, A. H.	9	5/1	1	1-59	Mar 1935
Acetabulariaceae	Hauck, F.	2		9	214	Oct 1884
Achnantheae	Hustedt, F.	7	2	3	818-880	Sep 1933
Achnantheae	Hustedt, F.	7	2	4	881	Dec 1933
Acrocarpae	Limpricht, K. G.	4	1	4	212-220	Jul 1886
Acrospermaceae	Rehm, H.	1	3	28	53-56	Aug 1887
Acrospermaceae	Rehm, H.	1	3	55	1248	Apr 1896
Actinodisceae-Actinoptychinae	Hustedt, F.	7	1	3	472-483	Jan 1929
Actinodisceae-Stictodiscinae	Hustedt, F.	7	1	3	467-472	Jan 1929
Adinimonaceae	Schiller, J.	10	3/1	1	9-10	Dec 1931-Jan 1932
Agaricini	Winter, G.	1	1	8	481-560	Jul 1882
Agaricini	Winter, G.	1	1	9	561-624	Oct 1882
Agaricini	Winter, G.	1	1	10	625-688	Dec 1882
Agaricini	Winter, G.	1	1	11	689-752	Jan 1883
Agaricini	Winter, G.	1	1	12	753-832	Mai 1883
Agaricini	Winter, G.	1	1	13	833-864	Sep 1883
Amaurochaetaceae	Schinz, H.	1	10	125	272-277	Aug 1917
Amblystegiaceae	Mönkemeyer, W.	4	Suppl.	3	711-789	Dec 1927
Amoebodiniaceae	Schiller, J.	10	3/2	4	481-482	Mai 1937
Amphisoleniaceae	Schiller, J.	10	3/1	1	165-192	Dec 1931-Jan 1932
Amphisphaerieae	Winter, G.	1	2	18	259-288	Jul 1885
Anadyomenaceae	Hauck, F.	2		9	466-468	Oct 1884
Anauleae	Hustedt, F.	7	1	5	891-901	Nov 1930
Andraeaceae	Mönkemeyer, W.	4	Suppl.	1	123-130	Apr 1927
Aneureae	Müller, K.	6	1	6	327-344	20 Mar 1908
Anthocerotales	Müller, K.	6	2	25	682-709	18 Feb 1916
Aongstroemiaceae	Limpricht, K.	4	1	5	301-304	Nov 1886
Araphideae	Hustedt, F.	7	2		5-6	Nov 1931
Archangiaceae	Keissler, K.	8		1	28-30	Jan 1930
Archidiaceae	Mönkemeyer, W.	4	Suppl.	1	147-148	Apr 1927
Archimycetes	Fischer, A.	1	4	45	3-5	Oct 1891
Archimycetes	Fischer, A.	1	4	45	11-20	Oct 1891
Arcyriaceae	Schinz, H.	1	10	126	368-384	20 Mar 1918
Arcyriaceae	Schinz, H.	1	10	127	385-398	8 Dec 1920
Arthoniaceae	Redinger, K.	9	2/1	1	[1]-180	Jan 1937
Arthrocladiaceae	Hauck, F.	2		7	380-381	Jan 1884

Taxon	Author	Band	Abt.	Lief.	Pag.	Fig.	Date
Ascoboleae	Rehm, H.	1	3	44	1078-1104	*36 figs.*	Jan 1895
Ascoboleae	Rehm, H.	1	3	53	1105-1134	*26 figs.*	Jul 1895
Ascoideaceen	Fischer, E.	1	5	58	110-112		Nov 1896
Ascomycetes	Winter, G.	1	2	14	[1]-3	*4 figs.*	Mar 1884
Ascomycetes	Keissler, K.	8		1	47-48		Jan 1930
Aspergillaceae	Keissler, K.	8		1	51		Jan 1930
Aspidiaceae	Luerssen, C.	3		5	293-320		Oct 1885
Aspidiaceae	Luerssen, C.	3		6	321-384		Jan 1886
Aspidiaceae	Luerssen, C.	3		7	385-448		Jul 1886
Aspidiaceae	Luerssen, C.	3		8	449-512		Sep 1886
Aspidiaceae	Luerssen, C.	3		9	513-516		Jan 1887
Aspleniaceae	Luerssen, C.	3		2	108-128	*84-130*	Apr 1884
Aspleniaceae	Luerssen, C.	3		3	129-192		Sep 1884
Aspleniaceae	Luerssen, C.	3		4	193-256	*131-169*	Mai 1885
Aulacomniaceae	Limpricht, K. G.	4	2	22	521-532	*310A-310B*	Jun 1893
Aulacomniaceae	Mönkemeyer, W.	4	Suppl.	2	566-569	*115*	Jul 1927
Balsamiaceae	Fischer, E.	1	5	57	61-64	*4 figs.*	Jul 1896
Balsamiaceae	Fischer, E.	1	5	58	65		Nov 1896
Barbeyella	Schinz, H.	1	10	127	410-412	*182*	8 Dec 1920
Bartramiaceae	Limpricht, K. G.	4	2	22	532-575	*311-316*	Jun 1893
Bartramiaceae	Mönkemeyer, W.	4	Suppl.	2	575-576		Jul 1927
Bartramiaceae	Mönkemeyer, W.	4	Suppl	3	577-588	*119-123*	Dec 1927
Basidiomycetes	Winter, G.	1	1	1	72-74		Dec 1880
Basidiomycetes	Keissler, K.	8		2	521		Apr 1930
Biddulphieae-Biddulphiinae	Hustedt, F.	7	1	5	828-864	*490-514*	Nov 1930
Biddulphieae-Eucampinae	Hustedt, F.	7	1	4	767-780	*449-455*	Jan 1930
Biddulphieae-Hemiaulinae	Hustedt, F.	7	1	5	868-890	*517-533*	Nov 1930
Buddulphieae-Isthmiinae	Hustedt, F.	7	1	5	868-868	*515-516*	Nov 1930
Biddulphieae-Triceratiinae	Hustedt, F.	7	1	4	781-784	*456*	Jan 1930
Biddulphieae-Triceratiinae	Hustedt, F.	7	1	5	785-828	*457-489*	Nov 1930
Biddulphioideae	Hustedt, F.	7	1	4	609		Jan 1930
Biraphideae	Hustedt, F.	7	2	4	433-434		Dec 1933
Blastodiniaceae	Schiller, J.	10	3/2	1	20-53	*8-42*	Mar 1935
Blastodiniales	Schiller, J.	10	3/2	1	8-15		Mar 1935
Botrydiaceae	Pascher, A.	11		6	1024-1055	*868-899*	Feb 1939
Botrydiales	Pascher, A.	11		6	1023-1024		Feb 1939

Family	Author				Pages	Figs	Date
Bryaceae	Limpricht, K.	4		2	257-320	283-297	Jan 1892
Bryaceae	Limpricht, K.	4		2	321-384	283-297	Nov 1892
Bryaceae	Limpricht, K.	4		2	385-447	298-300	Dec 1892
Bryaceae	Mönkemeyer, K. W.	4		Suppl.	412-549	94-108	Jul 1927
Bryales	Limpricht, K. G.	4		1	157-161	59	Dec 1885
Bryopsideae	Hauck, F.	2			471-475	208	Oct 1884
Bryoxiphiaceae	Mönkemeyer, W.	4		Suppl.	161-162		Apr 1927
Bulgariaceae	Rehm, H.	1		3	444-445		Jan 1891
Bulgariaceae-Callorieae	Rehm, H.	1		3	445-464	27 figs.	Jan 1891
Bulgariaceae Callorieae	Rehm, H.	1		3	465-466		Aug 1891
Bulgariaceae-Bulgarieae	Rehm, H.	1		3	467-501	37 figs.	Aug 1891
Bulgariaceae	Rehm, H.	1		3	1261-1263		Apr 1896
Buxbaumiaceae	Limpricht, K. G.	4		2	634-640	323	Dec 1893
Buxbaumiaceae	Limpricht, K. G.	4		2	641-645	324	Aug 1894
Buxbaumiaceae	Mönkemeyer, W.	4		Suppl.	901-903	222	Dec 1927
Caliciaceae	Keissler, K.	9		1/2	[520]-640	154-181	Jul 1937
Caliciaceae	Keissler, K.	9		1/2	641-727	182-195a	Mar 1938
Calymperaceae	Mönkemeyer, W.	4		Suppl.	231-232	42	Apr 1927
Campylosteliaceae	Limpricht, K. G.	4		1	476-482	150-151	Oct 1887
Capsosiraceae	Geitler, L.	14		2	464		Jan-Jun 1931
Capsosiraceae	Geitler, L.	14		2	465-471	276-282	Jul 1931
Catoscopiaceae	Mönkemeyer, W.	4		Suppl.	574-575	118	Jul 1927
Celidiaceae	Keissler, K.	8			52-105	30-32	Jan 1930
Cenangieae	Rehm, H.	1		3	213-241	32 figs.	Jan 1889
Cenangieae	Rehm, H.	1		3	1255-1256		Apr 1896
Centricae	Hustedt, F.	7		1	[217]-219		Mar 1927
Centritractaceae	Pascher, A.	11			830-832	691-693	Jan 1939
Centritractaceae	Pascher, A.	11			833-860	694-716	Feb 1939
Cephalidaceae	Fischer, A.	1		4	287-299	48-49	Apr 1892
Ceramiaceae	Hauck, F.	2			67-112	23-39	Dec 1882
Ceratiomyxaceae	Schinz, H.	1		10	80-83	38	18 Feb 1914
Ceratostomeae	Winter, G.	1		2	247-256	13 figs.	Apr 1885
Ceratostomeae	Winter, G.	1		2	257-259		Jul 1885
Chaetangiaceae	Hauck, F.	2			65-67	22	Dec 1882
Chaetocereae	Hustedt, F.	7		1	609-767	353-448	Jan 1930
Chaetocladiaceae	Fischer, A.	1		4	283-387	47	Apr 1892

Taxon	Author	Band	Abt.	Lief.	Pag.	Fig.	Date
Chaetomieae	Winter, G.	1	2	16	153-161	*10 figs.*	Nov 1884
Chamaesiphoniaceae	Geitler, L.	14		2	[415]-445	*242-264*	Jan-Jun 1931
Chamaesiphonales	Geitler, L.	14		2	[313]-315		Jan-Jun 1931
Characeae	Migula, W.	5		1	[1]-64	*1-27*	Oct 1889
Characeae	Migula, W.	5		2	65-128		Dec 1889
Characeae	Migula, W.	5		3	129-192	*37-55*	Jan 1890
Characeae	Migula, W.	5		4	193-256	*56-69*	Jun 1890
Characeae	Migula, W.	5		5	257-320	*70-80*	Jan 1891
Characeae	Migula, W.	5		6	321-384	*81-91*	Aug 1891
Characeae	Migula, W.	5		7	385-448	*92-100*	Mar 1892
Characeae	Migula, W.	5		8	449-512	*101-112*	Jan 1893
Characeae	Migula, W.	5		9	513-576	*113-114*	Jul 1894
Characeae	Migula, W.	5		10	577-640	*125-132*	Sep 1895
Characeae	Migula, W.	5		11	641-688	*133-136*	Jan 1896
Characeae	Migula, W.	5		12	689-758	*137-149*	Jan 1897
Characiopsidaceae	Pascher, A.	11		5	718-812	*571-672*	Jan 1939
Chlorarachniaceae	Pascher, A.	11		2	251-256	*162-165*	Aug 1937
Chloropediaceae	Pascher, A.	11		5	812-824	*673-684*	Jan 1939
Chlorophyceae	Hauck, F.	2		8	410		Jul 1884
Chloroteciaceae	Pascher, A.	11		6	860-906	*717-767*	Feb 1939
Chlorozoosporeae	Hauck, F.	2		8	[417]-421	*288-233*	Jul 1884
Chroococcaceae	Geitler, L.	14		1	124-288	*58-141*	Nov 1930
Chroococcaceae	Geitler, L.	14		2	289-292	*142-143*	Jan-Jun 1931
Chroococcaceae	Geitler, L.	14		6	[1161]1165	*767*	Dec 1932
Chroococcaceae	Hauck, F.	2		10	512-519	*228-233*	Jan 1885
Cinclidotaceae	Mönkemeyer, W.	4	Suppl.	2	336-340	*74-75*	Jul 1927
Citharistaceae	Schiller, J.	10	3/1	1	255-256		Dec 1931-Jan 1932
Citharistaceae	Schiller, J.	10	3/1	2	257-258	*252-253*	Dec 1932
Cladonia	Sandstede, H.	9	4/2	1	2-240	*1-8*	Jan 1931
Cladonia	Sandstede, H.	9	4/2	2	241-496	*pls. I-XXXIV*	Mar 1931
Cladoniaceae	Frey, E.	9	4/1	1	[1]-202	*1-32*, *pls. I-III*	Dec 1932
Cladopyxiaceae	Schiller, J.	10	3/2	3	467-471	*537-541*	Mai 1937
Clavariaceae	Keissler, K.	8		2	521-528		Apr 1930
Clavarieï	Winter, G.	1	1	5	292-318	*13 figs.*	1-15 Nov 1881

Taxon	Author	Abt.	Nr.	Seiten	Figs.	Datum
Coccolithineae	Schiller, J.	2	1	242-255	120-134	Nov 1930
Coccolithineae	Schiller, J.	2	1	[89]-170	1-51	Nov 1930
Coccolithineae	Schiller, J.	2	1	255-257	135-137	Nov 1930
Coccolithineae	Schiller, J.	2	1	258-259	fig. A-F	Nov 1930
Coccolithineae	Schiller, J.	3/2	4	510-511	591-592	Mai 1937
Cocconeideae	Hustedt, F.	2	2	319-320	780	Jan 1932
Cocconeideae	Hustedt, F.	2	3	321-364	781-817	Sep 1933
Codiaceae	Hauck, F.		9	477-483	210-212	Oct 1884
Codonieae	Müller, K.	1	6	380-384	224-225	20 Mar 1908
Codonieae	Müller, K.	1	6	385-395	226	27 Feb 1909
Confervaceae	Hauck, F.		7	437-448	192-202	Jul 1884
Confervaceae	Hauck, F.		8	449-466		Oct 1884
Congruentidiaceae	Schiller, J.	3/2	10	320		Apr 1935
Congruentidiaceae	Schiller, J.	3/2	3	321	337	Mai 1937
Coniocarpineae	Keissler, K.	1/2	4	[507]-519	145-153	Jul 1937
Corallinaceae	Hauck, F.		5	259-272	104-116	Apr 1883
Corallinaceae	Hauck, F.		6	273-281		Apr 1883
Cornuaceae	Gemeinhardt, K.	2	10	7-79	65, pls. I-IV	Nov 1930
Coscinodisceae-Coscinodiscinae	Hustedt, F.	1	7	333-334	169-258	Jan 1928
Coscinodisceae-Coscinodiscinae	Hustedt, F.	1	7	465-467	259	Jan 1920
Coscinodisceae-Melosirinae	Hustedt, F.	1	7	220-272	93-114	Mar 1927
Coscinodisceae-Melosirinae	Hustedt, F.	1	7	273-309	115-148	Jan 1928
Coscinodisceae-Sceletoneminae	Hustedt, F.	1	7	310-333	149-168	Jan 1928
Crateroneuraceae	Mönkemeyer, W.	Suppl.	4	704-711	158-159	Dec 1927
Cryphaeaceae	Limpricht, K. G.	2	24	677-691	329-331	Aug 1894
Cryphaeaceae	Mönkemeyer, W.	Suppl.	4	631-633	133	Dec 1927
Cryptonemiaceae	Hauck, F.		3	116-133	42-52	Jan 1883
Cucurbitarieae	Winter, G.	2	3	308-320	15 figs.	Jul 1885
Cucurbitarieae	Winter, G.	2	19	321-333	8 figs.	Jul 1885
Cutleriaceae	Hauck, F.	2	8	403-409	178-181	Jul 1884
Cyanophyceae	Geitler, L.		1	1-124	1-57	Nov 1930
Cyanophyceae-Schizophyceae	Hauck, F.		9	[487]-490		Oct 1884
Cypheliaceae	Keissler, K.	1/2	5	[728]-755	196-203	Mar 1938
Dasycladaceae	Hauck, F.		9	483-484	213	Oct 1884
Delesseriaceae	Lindau, G.		4	169-178	71-75	Apr 1883
Dematiaceae	Lindau, G.	8	100	547-576	2 figs.	30 Aug 1906
Dematiaceae	Lindau, G.	8	101	577-640	25 figs.	20 Sep 1906

Taxon	Author	Band	Abt.	Lief.	Pag.	Fig.	Date
Dematiaceae	Lindau, G.	1	8	102	641-704	*19 figs.*	10 Oct 1906
Dematiaceae	Lindau, G.	1	8	103	705-752	*15 figs.*	15 Nov 1906
Dematiaceae	Lindau, G.	1	8	104	753-833	*23 figs.*	16 Mai 1907
Dematiaceae	Lindau, G.	1	8	105	834-845	*9 figs.*	12 Jul 1907
Dematiaceae	Lindau, G.	1	8	106	[1]-48	*16 figs.*	12 Jul 1907
Dematiaceae	Lindau, G.	1	9	106	49-112	*10 figs.*	20 Dec 1907
Dematiaceae	Lindau, G.	1	9	107	113-176	*20 figs.*	1 Apr 1908
Dematiaceae	Lindau, G.	1	9	108	177-240	*20 figs.*	25 Mai 1908
Dematiaceae	Lindau, G.	1	9	109	241-285	*22 figs.*	30 Jul 1908
Dematiaceae	Lindau, G.	1	9	117	781-810	*9 figs.*	10 Mar 1910
Dematiaceae	Keissler, K.	8		2	602-623	*123-130*	Apr 1930
Derbesiaceae	Hauck, F.	2		9	475-477	209	Oct 1884
Dermateae	Rehm, H.	1	3	31	241-272	*33 figs.*	Jan 1889
Dermateae	Rehm, F.	1	3	32	273-277		Apr 1889
Dermateae	Rehm, F.	1	3	55	1256-1259		Apr 1896
Dermatocarpaceae	Zschacke, H.	9	1/1	4	590-656	*317-342*	Feb 1934
Dermocarpaceae	Geitler, L.	14		2	[383]-414	*212-241*	Jan-Jun 1931
Desmidiaceae	Krieger, W.	13	1/1	1	[1]-3		Jun 1934
Desmidiaceae	Krieger, W.	13	1/1	2	225-375	*9-36*	Dec 1935
Desmidiaceae	Krieger, W.	13	1/1	3	377-536	*37-72*	Aug 1936
Desmidiaceae	Krieger, W.	13	1/1	4	537-675	*73-96*	Feb 1937
Desmidiaceae	Krieger, W.	13	1/2	1	1-117	*98-142*	Jun 1939
Desmidiaceae	Krieger, W.	13	1/1	1	[4]-172	*1-33*	Jun 1934
Desmidiales	Krieger, W.	13	1/1	1	[173]174		Jun 1934
Desmocapsaceae	Schiller, J.	10	3/1	1	10-11		Dec 1931-Jan 1932
Desmomonadaceae	Schiller, J.	10	3/1	1	6-9	*5*	Dec 1931-Jan 1932
Deutschlandiaceae	Schiller, J.	10	2	1	233-236	*1-3*	Nov 1930
Diatoms	Hustedt, F.	7	1	1	[2]-216	*116-118*	Mar 1927
Diatrypeae	Winter, G.	1	2	26	810-842	*17 figs.*	Dec 1886
Dichaenaceae	Rehm, F.	1	3	28	49-52	*5 figs.*	Aug 1887
Dicranaceae	Limpricht, K. G.	4	1	5	304-320	*112-114*	Nov 1886
Dicranaceae	Limpricht, K. G.	4	1	6	321-384	*115-124*	Dec 1886
Dicranaceae	Limpricht, K. G.	4	1	7	385-418	*125-140*	Jul 1887
Dicranaceae	Mönkemeyer, W.	4	Suppl.	1	171-229	*25-40*	Apr 1927
Dichtyochaceae	Gemeinhardt, K.	10	2		26-77	*10-64*	Nov 1930
Dictyotaceae	Hauck, F.	2		6	304-311	*126-130*	Oct 1883

Dinoflagellata	Schiller, J.	10	3/1	4	[1]-6	Dec 1931-Jan 1932
Dinophyceae	Schiller, J.	10	3/1	1	258-263	Dec 1932
Dinophyceae	Schiller, J.	10	3/2	2	512-513	Mai 1937
Dinophysiaceae	Schiller, J.	10	3/1	4	45-164	Dec 1931-Jan 1932
Dinophysiales	Schiller, J.	10	3/2	1	514-517	*503-595*	Mai 1937
Dinotrichaceae	Schiller, J.	10	3/2	4	508-510	*598-603*	Mai 1937
Mönkemeyer, W.	4	Suppl.	3	4	589	Diphysiaceae
Diplomitrieae	Müller, K.	6	1	903	Dec 1927	
Disceliaceae	Limpricht, K. G.	4	2	6	354-367	*212-216*	20 Mar 1908
Disceliaceae	Mönkemeyer, W.	4	Suppl.	16	172-174	*259*	Jan 1891
Discoideae	Hustedt, F.	7	1	2	381-382	*83*	Jul 1927
Discomycetes	Rehm, H.	1	3	28	219-220	Mar 1927
Discomycetes	Keissler, K.	8		1	56-60	Aug 1887
Ditrichaceae	Limpricht, K. G.	4	1	8	52	Jan 1930
Ditrichaceae	Limpricht, K. G.	4	1	9	482-512	*152-157*	Oct 1887
Ditrichaceae	Mönkemeyer, W.	4	Suppl.	1	513-517	*158*	Mai 1888
Dothideaceae	Winter, G.	1	2	27	148-161	*19-23*	Apr 1927
Dothideales	Keissler, K.	8		2	894-918	*41 figs.*	Mar 1887
Ebriaceae	Gemeinhardt, K.	10	2		296-297	*66-69*	Apr 1930
Ectocarpaceae	Hauck, F.	2		6	79-82	Nov 1930
Ectocarpaceae	Hauck, F.	2		7	[319]320	*131-147*	Oct 1883
Elaphomycetaceen	Fischer, E.	1	5	58	321-350	Jan 1884
Elatereae	Müller, K.	6	1	6	81-101	*4 figs.*	Nov 1896
Ellobiopsidaceae	Schiller, J.	10	3/2	1	324-327	20 Mar 1908
Encalyptaceae	Limpricht, K. G.	4	2	15	62-71	*48-59*	Mar 1935
Encalyptaceae	Mönkemeyer, W.	4	Suppl.	1	102-125	*242-249*	Oct 1890
Endodiniaceae	Schiller, J.	10	3/2	1	232-238	*42*	Apr 1927
Endosporeae	Schinz, H.	1	10	122	61-62	Mar 1935
Enonemataceae	Geitler, L.	14	2	2	83-85	18 Feb 1914
Entodontaceae	Mönkemeyer, W.	4	Suppl.	3	452-456	*269-273*	Jan-Jun 1931
Entomothoreae	Winter, G.	1		1	844-852	*301-204*	Dec 1927
Entophysalidaceae	Geitler, L.	14	1		74-79	*1-3*	Dec 1880
Entophysalidaceae	Geitler, L.	14		2	292-312	*144-157*	Jan-Jun 1931
Entophysalidaceae	Geitler, L.	14		6	1165-1167	*768-770*	Dec 1932
Entopyleae	Hustedt, F.	7	2	6	[1178]	Dec 1932
Ephemeraceae	Limpricht, K. G.	4	1	1	6-11	*543-544*	Nov 1931
				3	161-173	*60-72*	Dec 1885

Taxon	Author	Band	Abt.	Lief.	Pag.	Fig.	Date
Ephemeraceae	Mönkemeyer, W.	4	Suppl.	2	382-386	*84*	Jul 1927
Epigloeaceae	Zschacke, H.	9	1/1	1	[44]-46	*16*	Feb 1933
Epigonantheae	Müller, K.	6	1	7	404-448	*228-243*	27 Feb 1909
Epigonantheae	Müller, K.	6	1	8	449-512	*244-266*	15 Apr 1909
Epigonantheae	Müller, K.	6	1	9	513-576	*267-286*	15 Sep 1909
Epigonantheae	Müller, K.	6	1	10	577-640	*287-302*	15 Apr 1910
Epigonantheae	Müller, K.	6	1	11	641-704	*303-321*	1 Jun 1910
Epigonantheae	Müller, K.	6	1	12	705-768	*322-340*	Sep 1910
Epigonantheae	Müller, K.	6	1	13	769-832	*341-357*	10 Mar 1911
Epigonantheae	Müller, K.	6	1	14	833-856	*358-361*	10 Apr 1911
Equisetaceae	Luerssen, C.	3		10	623-640	*193-221*	Aug 1887
Equisetaceae	Luerssen, C.	3		11	641-704		Apr 1888
Equisetaceae	Luerssen, C.	3		12	705-768		Jan 1889
Equisetaceae	Luerssen, C.	3		13	769-781		Apr 1889
Equisetinae	Luerssen, C.	3		10	622-623		Aug 1887
Erysipheae	Winter, G.	1	2	14	22-42	*17 figs.*	Mar 1884
Eunotieae	Hustedt, F.	7	2	2	265-317	*740-779*	Jan 1932
Euodieae	Hustedt, F.	7	1	5	901-907	*542*	Nov 1930
Eupezizeae	Rehm, H.	1	3	42	913-976	*52 figs.*	Jun 1894
Eupezizeae	Rehm, H.	1	3	43	977-1040	*75 figs.*	Nov 1894
Eupezizeae	Rehm, H.	1	3	44	1041-1078		Jan 1895
Eupezizeae	Rehm, H.	1	3	55	1269-1270		Apr 1896
Euphacidieae	Rehm, H.	1	3	28	60-64	*29 figs.*	Aug 1887
Euphacidieae	Rehm, H.	1	3	29	65-87	*6 figs.*	Jan 1888
Euphacidieae	Rehm, H.	1	3	55	1248-1249		Apr 1896
Eupodisceae-Aulacodiscinae	Hustedt, F.	7	1	3	501-508	*281-284*	Jan 1929
Eupodisceae-Eupodiscinae	Hustedt, F.	7	1	3	508-541	*285-309*	Jan 1929
Eupodisceae-Pyrodiscinae	Hustedt, F.	7	1	3	500-501	*280*	Jan 1929
Eusticteae	Rehm, H.	1	3	29	113-128	*114 figs.*	Jan 1888
Eusticteae	Rehm, H.	1	3	30	129-185	*4 figs.*	Aug 1888
Eusticteae	Rehm, H.	1	3	55	1251-1253		Apr 1896
Extuberaceen	Fischer, E.	1	5	57	12-61	*32 figs.*	Jul 1896
Excipulaceae	Allescher, A.	1	7	81	393-440	*19 figs.*	Jan 1902
Exoascaceae	Keissler, K.	8		1	48-50	*24*	Jan 1930
Exoasci	Winter, G.	1	2	14	3-11	*4 figs.*	Mar 1884
Exosporeae	Schinz, H.	1	10	122	80		18 Feb 1914

Fissidentaceae	Limpricht, K. G.	4	1	8	449-459	144-145	Oct 1887
Fissidentaceae	Mönkemeyer, W.	4	Suppl.	1	130-146	15-18	Apr 1927
Florideae	Hauck, F.	2		1	[8]-20		Nov 1882
Fontinalaceae	Limpricht, K. G.	4	2	24	647-677	325-328	Aug 1894
Fontinalaceae	Mönkemeyer, W.	4	Suppl.	3	653-667	143-146	Dec 1927
Fragilarieae-Diatominae	Hustedt, F.	7	2	1	92-114		Nov 1931
Fragilarieae-Fragilariinae	Hustedt, F.	7	2	1	115-176	640-682	Nov 1931
Fragilarieae-Fragilariinae	Hustedt, F.	7	2	2	177-260	683-737	Jan 1932
Fucaceae	Hauck, F.	2		6	285-301	117-125	Oct 1883
Fucoideae	Hauck, F.	2		6	282-284		Oct 1883
Funariaceae	Limpricht, K. G.	4	2	16	175-192	260-262	Jan 1891
Funariaceae	Limpricht, K. G.	4	2	17	193-203	263	Dec 1891
Funariaceae	Mönkemeyer, W.	4	Suppl.	2	386-396	85-87	Jul 1927
Fungi Imperfecti	Allescher, A.	1	6	59	[1]-7		Jun 1898
Fungi Imperfecti	Keissler, K.	8		2	529-531		Apr 1930
Gasteromycetes	Winter, G.	1	1	13	864-866		Sep 1883
Gelidiaceae	Hauck, F.	2		4	189-197	82-85	Apr 1883
Geoglosseae	Rehm, H.	1	3	53	1142-1168		Jul 1895
Geoglosseae	Rehm, H.	1	3	54	1169-1171	39 figs.	Nov 1895
Georgiaceae	Limpricht, K. G.	4	2	15	125-128		Oct 1890
Georgiaceae	Limpricht, K. G.	4	2	16	129-132	250	Jan 1891
Georgiaceae	Mönkemeyer, W.	4	Suppl.	2	410-412	93	Jul 1927
Gigartinaceae	Hauck, F.	2		3	133-149	53-61	Jan 1883
Glenodiniaceae	Schiller, J.	10	3/2	1	92-123	79-119	Mar 1935
Glenodiniopsidaceae	Schiller, J.	11		1	80-92	65-78	Mar 1935
Gloeobotrydaceae	Pascher, A.	11		4	632-640	492-498	Feb 1938
Gloeobtrydaceae	Pascher, A.	11		5	641-661	499-520	Jan 1939
Gloeodiniaceae	Schiller, J.	10	3/2	4	482-508	588	1937
Gloeopodiaceae	Pascher, A.	11		5	696-703	552-556	Jan 1939
Gnomonieae	Winter, G.	1	2	22	570-592	26 figs.	Mai 1886
Goniaulacaceae	Schiller, J.	10	3/2	2	275-320	280-336	Apr 1935
Goniaulacaceae	Schiller, J.	10	3/2	4	522		Mai 1937
Goniodomaceae	Schiller, J.	10	3/2	3	438-447	479-489	Mai 1937
Graphidaceae	Redinger, K.	9	2/1	2	[181]-404	45-107	Jan 1939
Grimmiaceae	Limpricht, K. G.	4	1	11	693-704	190	Dec 1888
Grimmiaceae	Limpricht, K. G.	4	1	12	705-768	191-200	Oct 1889

Taxon	Author	Band	Abt.	Lief.	Pag.	Fig.	Date
Grimmiaceae	Limpricht, K. G.	4	1	13	769-826	*201-211*	Nov 1889
Grimmiaceae	Mönkemeyer, W.	4	Suppl.	2	341-380	*76-82*	Jul 1927
Gymnoasci	Winter, G.	1	2	14	11-17	*19 figs.*	Mar 1884
Gymnocarpeae	Keissler, K.	9	1/2	4	[506]		Jul 1937
Gymnodiniaceae	Schiller, J.	10	3/1	2	272-432	*261-455*	Dec 1932
Gymnodiniaceae	Schiller, J.	10	3/1	3	433-547	*456-576*	Feb 1933
Gymnodiniales	Schiller, J.	10	3/2	1	[1]		Mar 1935
Gymnodiniales	Schiller, J.	10	3/1	3	601-603	*631*	Feb 1933
Gymnosclerotaceae	Schiller, J.	10	3/2	1	[1]-8	*1-5*	Mar 1935
Halopappaceae	Müller, K.	6	2	1	230-233	*114-115*	1930
Haplolaeneae	Müller, K.	6	1	6	367-380	*217-223*	20 Mar 1908
Haplomitrieae	Müller, K.	6	1	7	396-400	*227*	27 Feb 1909
Hedwigiaceae	Mönkemeyer, W.	4	Suppl.	3	629-631	*132*	Dec 1927
Helminthocladiaceae	Hauck, F.	2		1	55-64	*17-21*	Nov 1882
Helotiaceae	Keissler, K.	8		1	105		Jan 1930
Helotieae	Rehm, H.	1	3	37	647-656	*38 figs.*	Jul 1892
Helotieae	Rehm, H.	1	5	38	657-720	*63 figs.*	Sep 1892
Helotieae	Rehm, H.	1	3	39	721-784	*37 figs.*	Jan 1893
Helotieae	Rehm, H.	1	3	40	785-848	*55 figs.*	Sep 1893
Helotieae	Rehm, H.	1	3	41	849-912	*26 figs.*	Oct 1893
Helvelleae	Rehm, H.	1	3	55	1265-1269		Apr 1896
Helvelleae	Rehm, H.	1	3	54	1172-1232	*32 figs.*	Nov 1895
Hemiasceae	Rehm, H.	1	3	55	1233-1245		Apr 1896
Hepaticae	Fischer, E.	1	5	58	109-110		Nov 1896
Hepaticae	Müller, K.	6		1	[1]-64	*1-51*	Dec 1905
Hepaticae	Müller, K.	6		2	65-128	*52-96*	Mai 1906
Hepaticae	Müller, K.	6		3	129-132		20 Mar 1907
Hepaticae	Müller, K.	6		3	133-140		20 Mar 1907
Heppiaceae	Köfaragó-Gyelnik, V.	9	2/2	1	[111]-134	*pls. 15-17*	1940
Heterocapsaceae	Pascher, A.	11		2	277-301	*184-205*	Aug 1937
Heterocapsineae	Pascher, A.	11		2	274-277		Aug 1937
Heterochloridaceae	Pascher, A.	11		2	208-236	*140-154*	Aug 1937
Heterocloniaceae	Pascher, A.	11		6	997-1023	*844-867*	Feb 1939
Heterocloniales	Pascher, A.	11		6	991-992		Feb 1939
Heterococcales	Pascher, A.	11		6	906-912		Feb 1939
Heterococcineae	Pascher, A.	11		2	305-320	*208-211*	Aug 1937

Heterokontae	Pascher, A.	11		1	[1]-160		Mai 1937
Heterokontae	Pascher, A.	11		2	161-189	1-126	Aug 1937
Heterokontae	Pascher, A.	11		2	[203]-208	127-135	Aug 1937
Heterosphaerieae	Rehm, H.	1	3	30	198-208		Aug 1888
Heterosphaerieae	Rehm, H.	1	3	31	209-212	17 figs.	Jan 1889
Heterosphaerieae	Rehm, H.	1	3	55	1254		Apr 1896
Heterotrichaceae	Pascher, A.	11		6	916-939	771-790	Feb 1939
Heterotrichineae	Pascher, A.	11		6	912-915		Feb 1939
Hildenbrandtiaceae	Hauck, F.	2		1	37-39		Nov 1882
Holochytriaceae	Fischer, A.	1	4	46	72-85	9	Oct 1891
Hookeriaceae	Mönkemeyer, W.	4	Suppl.	3	670-674	12-15	Dec 1927
Hormogonales	Geitler, L.	14		2	[457]-459	148-150	Jan-Jun 1931
Hormogonales	Geitler, L.	14		6	1158-1160	766	Dec 1932
Hydnei	Winter, G.	1	1	6	354-385	14 figs.	Jan 1882
Hydropterides	Luerssen, C.	3		10	593-595		Aug 1887
Hylocomiaceae	Mönkemeyer, W.	4	Suppl.	3	896-901	220-221	Dec 1927
Hymenogastrei	Winter, G.	1	1	13	870-886	19 figs.	Sep 1883
Hymenomycetes	Winter, G.	1	1	5	290-292		1-15 Nov 1881
Hymenomycetineae	Keissler, K.	8		2	521		Apr 1930
Hymenophyllaceae	Luerssen, C.	3		1	29-36	38-40	Mar 1884
Hyphomycetes	Lindau, G.	1	8	92	[1]-5		16 Mai 1904
Hyphomycetes	Lindau, G.	1	9	118	[825]-852		20 Mai 1910
Hyphomycetes	Lindau, G.	1	9	119	[923]-944		20 Jul 1910
Hyphomycetes	Lindau, G.	1	9	120	945-983	45-983	10 Dec 1910
Hyphomycetes	Keissler, K.	8		2	589-590		Apr 1930
Hypnaceae	Hauck, F.	2		4	187-189		Apr 1883
Hypnaceae	Limpricht, K. G.	4	3	27	[1]-64		Dec 1895
Hypnaceae	Limpricht, K. G.	4	3	28	65-128		Jul 1896
Hypnaceae	Limpricht, K. G.	4	3	29	129-192		Dec 1896
Hypnaceae	Limpricht, K. G.	4	3	30	193-256	387-392	Dec 1896
Hypnaceae	Limpricht, K. G.	4	3	31	257-320	393-400	Sep 1897
Hypnaceae	Limpricht, K. G.	4	3	32	321-384	401-408	Dec 1897
Hypnaceae	Limpricht, K. G.	4	3	33	385-448		Aug 1898
Hypnaceae	Limpricht, K. G.	4	3	34	449-512	417-425	Feb 1899
Hypnaceae	Limpricht, K. G.	4	3	35	513-576	426-435	Nov 1899
Hypnaceae	Limpricht, K. G.	4	3	36	577-600	436-440	Apr 1901

Taxon	Author	Band	Abt.	Lief.	Pag.	Fig.	Date
Hypnaceae	Limpricht, K. G.	4	2	37	641-704		Dec 1901
Hypnaceae	Limpricht, K. G. & W. Limpricht	4	3	38	705-768		Dec 1902
Hypnaceae	Limpricht, K. G. & W. Limpricht	4	3	39	769-832		Jun 1903
Hypnaceae	Limpricht, K. G. & W. Limpricht	4	3	40	833-864		Dec 1903
Hypnaceae	Mönkemeyer, W.	4	Suppl.	3	873-893	*210-218*	Dec 1927
Hypnaeaceae	Hauck, F.	2		4	187-189	*81*	Apr 1883
Hypochytriaceae	Fischer, A.	1	4	47	131-145	*25-28*	Jan 1892
Hypocreaceae	Winter, G.	1	2	15	82-128	*69 figs.*	Sep 1884
Hypocreaceae	Winter, G.	1	2	16	129-152		Nov 1884
Hypocreales	Keissler, K.	8		2	274-296	*58-61*	Apr 1930
Hypodermieae	Rehm, H.	1	3	28	28-48	*30 figs*	Aug 1887
Hypodermieae	Rehm, H.	1	3	55	1247-1248		Apr 1896
Hysteriaceae	Rehm, H.	1	3	28	[1]-3		Aug 1887
Hysteriaceae	Rehm, H.	1	3		1246		Apr 1896
Hysterineae	Rehm, H.	1	3	28	3-28	*41 figs.*	Aug 1887
Isoetaceae	Luerssen, C.	3		14	845-862	*224*	Jun 1889
Jubuleae-Lejeunea	Müller, K.	6	2	24	593-656		7 Oct 1915
Jungermanniaceae	Müller, K.	6		7	401-404		27 Feb 1909
Jungermanniaceae	Müller, K.	6	1	5	309-314		1 Oct 1907
Laboulbenieae	Winter, G.	1	2	27	918-925	*5 figs.*	Mar 1887
Lejeunea	Mönkemeyer, W.	6	2	27	785-818		28 Mar 1916
Lejeunea	Mönkemeyer, W.	6	2	28	819-947		27 Aug 1916
Lembophyllaceae	Mönkemeyer, W.	4	Suppl.	3	649-653	*142*	D3c 1927
Leptobascseae	Geitler, L.	14		3	659-664	*421-426*	1931
Leptostromaceae	Allescher, A.	1	7	79	316-320	*8 figs.*	Dec 1901
Leptostromaceae	Allescher, A.	1	7	80	321-384	*29 figs.*	Dec 1901
Leptostromaceae	Allescher, A.	1	7	81	385-392	*2 figs.*	Jan 1902
Leskeaceae	Limpricht, K. G.	4	4	25	745-768	*342-343*	Jan 1895
Leskeaceae	Limpricht, K. G.	4	2	26	769-843	*344-352*	Jun 1895
Leskeaceae	Mönkemeyer, W.	4	Suppl.	3	681-694	*153-155*	Dec 1927
Leucobryaceae	Limpricht, K. G.	4	1	7	418-421	*141*	Jul 1887
Leucobryaceae	Mönkemeyer, W.	4	Suppl.	1	229-230	*41*	Apr 1927
Leucodontaceae	Mönkemeyer, W.	4	Suppl.	3	633-637	*134-135*	Dec 1927

Lissodiniaceae	Schiller, J.	10		481	553	Mai 1937
Listerellaceae	Keissler, K.	8		4	23	Jan 1930
Lithodermaceae	Hauck, F.	2		1	177	Jul 1884
Loefgreniaceae	Geitler, L.	14		8	283	Jul 1931
Lomentariaceae	Hauck, F.	2		3	87	Apr 1883
Lophiostomeae	Winter, G.	1	2	4	11 figs.	Jul 1885
Loptobasaceae	Geitler, L.	14		18		Jul 1931
Loriellaceae	Geitler, L.	14		3		Jul 1931
Lycogalaceae	Schinz, H.	1	10	2	274-276	Jan-Jun 1931
Lycoperdiaceae				126	140-142	20 Mar 1918
Lycoperdinei	Luerssen, G.	3			222-223	1889
Lycopodiaceae	Winter, G.	1	1	13	14 figs.	Sep 1883
Lycopodiaceae	Luerssen, C.	3		13		Apr 1889
Lycopodiaceae	Luerssen, C.	3		14		Jun 1889
Lycopodinae	Luerssen, G.	3		13		Apr 1889
Madothecoideae	Müller, K.	6	2	23	160-168	26 Jul 1915
Malleodendraceae	Pascher, A.	11		2	206-207	Aug 1937
Marchantiaceae	Müller, K.	6	1	4	142-156	20 Jun 1907
Marchantiaceae	Müller, K.	6	1	5	157-189	1 Oct 1907
Margaritaceae	Schinz, H.	1	10		172-181	8 Dec 1920
Marsiliaceae	Luerssen, G.	3		10	187-192	Aug 1887
Massarieae	Winter, G.	1	2	22	15 figs.	Mai 1886
Mastigocladaceae	Geitler, L.	14		3	347-354	Jul 1931
Meeseaceae	Limpricht, K. G.	4	2	21	306-307	Jan 1893
Meeseaceae	Limpricht, K. G.	4	2	22	308A-309	Jun 1893
Meeseaceae	Mönkemeyer, W.	4	Suppl.	2	116-117	Jul 1927
Melanconiaceae	Keissler, K.	8		2	113-114	Apr 1930
Melanconiales	Keissler, K.	8		2		Apr 1830
Melanconideae	Winter, G.	1	2	25		Oct 1886
Melanconieae	Allescher, A.	1	7	81	21 figs.	Jan 1902
Melanconieae	Allescher, A.	1	7	82	7 figs.	Apr 1902
Melanconieae	Allescher, A.	1	7	83	33 figs.	Mai 1902
Melanconieae	Allescher, A.	1	7	84	34 figs.	Jul 1902
Melanconieae	Allescher, A.	1	7	85	44 figs.	Oct 1902
Melanconieae	Allescher, A.	1	7	86	26 figs.	Jan 1903
Melanconieae	Allescher, A.	1	7	87	25 figs.	Jan 1903
Melanconieae	Allescher, A.	1	7	88		Mar 1903

Taxon	Author	Band	Abt.	Lief.	Pag.	Fig.	Date
Melanconieae	Allescher, A.	1	7	89	897-957		Apr 1903
Melanommeae	Winter, G.	1	2	17	220-246	31 figs.	Apr 1885
Melogrammeae	Winter, G.	1	2	25	797-800	19 figs.	Oct 1886
Melogrammeae	Winter, G.	1	2	26	801-810		Dec 1886
Merolpidiaceae	Fischer, A.	1	4	45	45-64	8	Oct 1891
Merolpidiaceae	Fischer, A.	1	4	46	65-72	9-11	Oct 1891
Mesogloeaceae	Hauck, F.	2		7	351-369		Jan 1884
Mesotaeniaceae	Krieger, H.	13	1/1	1	174-223	148-157	Jun 1934
Meteoriaceae	Mönkemeyer, W.	4	Suppl.	3	638-639	1-8	Dec 1927
Metzgeriae	Müller, K.	6	1	6	345-354	137	20 Mar 1908
Microchaetaceae	Geitler, L.	14		3	664-672	206-211	Jul 1931
Microchaetaceae	Geitler, L.	14		4	673-677	427-433	Mai 1932
Microthyriaceae	Keissler, K.	8		2	269-274	434-436	Apr 1930
Mischococcaceae	Pascher, A.	11		5	703-718	56-57	Jan 1939
Mniaceae	Limpricht, K. G.	4	2	20	447-448	557-570	Dec 1892
Mniaceae	Limpricht, K. G.	4	2	21	449-497	301-305	Jan 1893
Mniaceae	Mönkemeyer	4	Suppl.	2	549-566	109-114	Jul 1927
Mollisiaceae	Keissler, K.	8		1	108-118		Jan 1930
Mollisieae	Rehm, H.	1	3	35	503		Aug 1891
Mollisieae	Rehm, H.	1	3	55	1263-1265		Apr 1896
Mollisieae-Eumollisieae	Rehm, H.	1	3	35	503-528	62 figs.	Aug 1891
Mollisieae-Eumollisieae	Rehm, H.	1	3	36	529-593	9 figs.	Oct 1891
Mollisieae-Pyrenopezizeae	Rehm, H.	1	3	36	593-608	51 figs.	Oct 1891
Mollisieae-Pyrenopezizeae	Rehm, H.	1	3	37	609-647		Jul 1892
Monascaceen	Fischer, E.	1	5	58	118-127		1897
Monoblepharidaceae	Fischer, A.	1	4	50	378-383	8 figs.	Mai 1892
Monolpidiaceae	Fischer, A.	1	4	45	20-45	62-63	Oct 1891
Monoraphideae	Hustedt, F.	7	2	2	317-318	1-7	Jan 1932
Mortierellaceae	Fischer, A.	1	4	49	268-283	46	Apr 1892
Moriolaceae	Keissler, K.	9	1/1	1	[1]-43	1-15	Feb 1933
Mucedinaceae	Lindau, G.	1	8	92	[6]-64	20 figs.	16 Mai 1904
Mucedinaceae	Lindau, G.	1	8	93	65-128	22 figs.	30 Jun 1904
Mucedinaceae	Lindau, G.	1	8	94	129-176	16 figs.	15 Jul 1904
Mucedinaceae	Lindau, G.	1	8	95	177-256	29 figs.	3 Apr 1905
Mucedinaceae	Lindau, G.	1	8	96	257-320	27 figs.	10 Mai 1905
Mucedinaceae	Lindau, G.	1	8	97	321-384	32 figs.	20 Jun 1905

Mucedinaceae	Lindau, G.	1	9	117	753-781	8 figs.	10 Mar 1910
Mucedinaceae	Keissler, K.	8		2	590-602	115-122	Apr 1930
Mucoraceae	Fischer, A.	1	4	47	178-192	29-33	Jan 1892
Mucoraceae	Fischer, A.	1	4	48	193-256	34-44	Mar 1892
Mucoraceae	Fischer, A.	1	4	49	257-268	45	Apr 1892
Mucoraceae	Keissler, K.	8		1	46-47		Jan 1930
Musci	Limpricht, K. G.	4	1		[v]-viii		Nov 1889
Musci	Limpricht, K. G.	4	1	1	[1]-64	1-36	Jul 1885
Musci	Limpricht, K. G.	4	1	2	65-12	37-50	Aug 1885
Musci	Limpricht, K. G.	4	1	3	129-157	51-58	Dec 1885
Musci	Mönkemeyer, W.	4	Suppl.	1	68-123		Apr 1927
Mycelia Sterilia	Lindau, G.	1	9	115	[649]-688		1 Nov 1909
Mycelia Sterilia	Lindau, G.	1	9	116	689-714		31 Dec 1909
Mycelia Sterilia	Lindau, G.	1	9	118	822-824		20 Mai 1910
Mycoporaceae	Keissler, K.	9	1/2	3	[470]-480	141-143	Mar 1937
Mycoporaceae	Keissler, K.	9	1/2	4	481-502	144	Jul 1937
Myuriaceae	Mönkemeyer, W.	4	Suppl.	3	637-638	136	Dec 1927
Myxobacteriales	Keissler, K.	8		1	[27]-28		Jan 1930
Myxochloridaceae	Pascher, A.	11		2	256-274	166-184	Aug 1937
Myxococcaceae	Keissler, K.	8		1	33-35		Jan 1930
Myxogasteres	Schinz, H.	1		121	[1]-64		20 Jul 1912
Myxogasteres	Schinz, H.	1	10	122	65-80		18 Feb 1914
Myxogasteres	Keissler, K.	8		1	36		Jan 1930
Myxomycetes	Keissler, K.	8		1	36		Jan 1930
Naviculeae	Hustedt, F.	7	2	4	434-576	822-1008	Dec 1933
Naviculeae	Hustedt, F.	7	2	5	577-736	1009-1105	Mar 1937
Naviculeae	Hustedt, F.	7	2	6	737-838	1106-1179	Dec 1959
Naviculeae	Hustedt, F.	7	3	1	[1]-160	1180-1294	Sep 1961
Naviculeae	Hustedt, F.	7	3	2	161-348	1295-1456	Feb 1962
Naviculeae	Hustedt, F.	7	3	3	349-556	1457-1591	Mai 1964
Naviculeae	Hustedt, F.	7	3	4	557-816	1592-1788	Mar 1966
Neckeraceae	Limpricht, K. G.	4	2	24	691-704		Aug 1894
Neckeraceae	Limpricht, K. G.	4	2	25	705-718	332	Jan 1895
Neckeraceae	Mönkemeyer, W.	4	Suppl.	3	640-647	333-335	Dec 1927
Nectrioidaceae	Keissler, K.	8		2	581-586	138-139	Apr 1930
Nectrioideae	Allescher, A.	1	7	79	295-315	112	Dec 1901

519

Taxon	Author	Band	Abt.	Lief.	Pag.	Fig.	Date
Nidularici	Winter, G.	1	1	13	915-922	*8 figs.*	Sep 1883
Nostocaceae	Hauck, F.	2		9	[491]-512	*216-227*	Oct 1884
Nostocaceae	Geitler, L.	14		4	801-896	*513-576*	Mai 1932
Nostocaceae	Geitler, L.	14		5	897-905	*577-583*	Aug 1932
Nostochopsidaceae	Geitler, L.	14		3	472-482	*284-292*	Jul 1931
Oedipodiaceae	Mönkemeyer, W.	4	Suppl.	2	397-398	*89*	Jul 1927
Oedogoniaceae	Gemeinhardt, K.	12	4	1	69-172	*50-181*	Apr 1938
Oedogoniaceae	Gemeinhardt, K.	12	4	2	173-332	*182-401*	Jun 1939
Oedogoniaceae	Gemeinhardt, K.	12	4	3	333-444	*402-539*	Feb 1940
Oedogoniales	Gemeinhardt, K.	12	4	1	[v]-ix		Feb 1940
Oedogoniales	Gemeinhardt, K.	12	4	1	[1]-69		Apr 1938
Onygenaceen	Fischer, E.	1	5	58	101-108	*1-49*	Nov 1896
Oomycetes	Fischer, A.	1	4	45	7-8	*7 figs.*	Oct 1891
Oomycetes	Fischer, A.	1	4	49	310-320		Apr 1892
Oomycetes	Fischer, A.	1	4	50	321-327		Mai 1892
Oosporeae	Hauck, F.	2		8	410-411		Jul 1884
Ophioglossaceae	Luerssen, G.	3		9	534		Jun 1887
Ophioglossaceae	Luerssen, G.	3		9	535-576	*175-183*	Jun 1887
Ophioglosseae	Luerssen, G.	3		10	377-593		Aug 1887
Ornithocercaceae	Schiller, J.	10	3/1	1	192-255	*187-255*	Dec 1931-Jan 1932
Orthotrichaceae	Limpricht, K. G.	4	2	14	[1]-64	*212-227*	Jul 1890
Orthotrichaceae	Limpricht, K. G.	4	2	15	65-102	*228-241*	Oct 1890
Orthotrichaceae	Mönkemeyer, W.	4	Suppl.	3	595-628	*126-131*	Dec 1927
Oscillatoriaceae	Geitler, L.	14		5	906-1056	*584-668*	Aug 1932
Oscillatoriaceae	Geitler, L.	14		6	1057-1158	*669-765*	Dec 1932
Osmundaceae	Luerssen, G.	3		9	517-533	*170-172*	Jun 1887
Ostreopsiaceae	Schiller, J.	10	3/2	3	471-473	*542-543*	Mai 1937
Ostropeae	Rehm, H.	1	3	30	185-190	*14 figs.*	Aug 1888
Oxytoxaceae	Schiller, J.	10	3/2	3	447-467	*490-536*	Mai 1937
Palmellaceae	Hauck, F.	2		9	485-486	*215*	Oct 1884
Pannariaceae	Köfarago-Gyelnik, V.	9	2/2	2	[135]-272	*pls. 18-33*	1940
Paradiniaceae	Schiller, J.	10	3/2	1	15-19	*6-7*	Mar 1935
Parmeliaceae	Hillman, J.	9	5/3	1	5-160	*1-9, pl. 1*	Mai 1936
Parmeliaceae	Hillman, J.	9	5/3	2	161-304	*10-26*	Mai 1936
Patellariaceae	Rehm, H.	1	3	32	277-280		Apr 1889
Patellariaceae	Keissler, K.	8		1	118-240	*34-49*	Jan 1930

Patellariaceae/Calicieae	Rehm, H.	1	3	34	401-414		Jan 1891
Patellariaceae/Eupatellarieae	Rehm, H.	1	3	32	291-336	89 figs.	Apr 1889
Patellariaceae/Eupatellarieae	Rehm, H.	1	3	33	337-382		Jul 1890
Patellariaceae/Pseudopatellariaceae	Rehm, H.	1	3	32	280-291	18 figs.	Apr 1889
Patellarieae	Rehm, H.	1	3	55	1259-1261		Apr 1896
Pennatae	Hustedt, F.	7	2	1	[1]-4		Nov 1931
Pennatae	Schiller, J.	10	3/2	1	123-160	120-157	Mar 1935
Peridiniaceae	Schiller, J.	10	3/2	1	161-275	158-279	Apr 1935
Peridiniales	Schiller, J.	10	3/2	1	71-74		Mar 1935
Peridinium	Schiller, J.	10	3/2	22	517-521	604-610	Mai 1937
Perisporeae	Winter, G.	1	2	14	43-64	56 figs.	Mar 1884
Perisporiaceae	Keissler, K.	8		2	265-269	55	Apr 1930
Perisporiales	Keissler, K.	8		2	265		Apr 1930
Perisporieae	Winter, G.	1	2	15	65-82		Sep 1884
Perisporiaceae	Keissler, K.	8		2	265-269	55	Apr 1930
Peronieae	Hustedt, F.	7	2	2	261-264	738-739	Jan 1932
Peronosporaceae	Fischer, A.	1	4	50	383-384		Mai 1892
Peronosporaceae	Fischer, A.	1	4	51	385-448	64-74	Jul 1892
Peronosporaceae	Fischer, A.	1	4	52	449-489		Oct 1892
Pertusariaceae	Erichsen, C. F. E.	9	5/1	2	[321]-512	1-41	Jan 1936
Pertusariaceae	Erichsen, C. F. E.	9	5/1	3	513-699	42-74	Mai 1936
Pezizeae	Rehm, H.	1	3	35	501-503		Aug 1891
Pezizelleae	Keissler, K.	8		1	105-108	33	Jan 1930
Pezizineae	Keissler, K.	8		1	52		Jan 1930
Phaeophyceae	Hauck, F.	2		6	282		Oct 1883
Phaeozoosporeae	Hauck, F.	2		6	[312]-318		Oct 1883
Phalloïdei	Winter, G.	1		13	866-870	4 figs.	Sep 1883
Phascaceae	Limpricht, K. G.	4	1	3	176-192	65-73	Dec 1885
Phascaceae	Limpricht, K. G.	4	1	4	193-199	74-77	Jul 1886
Phycomycetes	Fischer, A.	1	4	45	[1]-3		Oct 1891
Phycomycetes	Keissler, K.	8		1	46		Jan 1930
Phyllachoraceae	Keissler, K.	8		2	297-304	62-63	Apr 1930
Physaraceae	Schinz, H.	1	10	122	86-128	39-54	18 Feb 1914
Physaraceae	Schinz, H.	1	10	123	129-192	55-69	17 Mai 1915
Physaraceae	Schinz, H.	1	10	124	193-202	73	9 Dec 1915
Physaraceae	Keissler, K.	8		1	42-46		Jan 1930

521

Taxon	Author	Band	Abt.	Lief.	Pag.	Fig.	Date
Physciaceae	Lynge, B.	9	6	1	[37]-40	1-10, pls. 1-12	Dec 1935
Physciaceae	Lynge, B.	9	6	1	[41]-188		Dec 1935
Physcomitrellaceae	Limpricht, K. G.	4	1	3	173-174	63-64	Dec 1885
Plagiotheciaceae	Mönkemeyer, W.	4	Suppl.	3	852-867	205-207	Dec 1927
Plectascineae	Keissler, K.	8		1	50-51		Jan 1930
Pleosporaceae	Keissler, K.	8		2	445-520	87-95	Apr 1930
Pleosporeae	Winter, G.	1	2	20	405-448	15 figs.	Oct 1885
Pleosporeae	Winter, G.	1	2	21	449-528	17 figs.	Dec 1885
Pleosporeae	Winter, G.	1	2	22	529-533		Mai 1886
Pleurocapsaceae	Geitler, L.	14		2	315-382	158-211	Jan-Jun 1931
Pleurocapsaceae	Geitler, L.	14		6	1167-1173	771-779	Dec 1932
Pleurocarpae	Limpricht, K. G.	4		24	645-647		Aug 1894
Pleurochloridaceae	Pascher, A.	11	2	3	333-480	212-335	Feb 1938
Pleurochloridaceae	Pascher, A.	11		4	481-632	336-491	Feb 1938
Pleurozioideae	Müller, K.	6	2	22	525-528		1 Mar 1915
Pleurozioideae	Müller, K.	6	2	23	529-533	152	26 Jul 1915
Podolampaceae	Schiller, J.	10	3/2	3	473-480	544-552	Mai 1937
Polyangiaceae	Keissler, K.	8		1	30-33	22	Jan 1930
Polykrikaceae	Schiller, J.	10	3/1	3	547-553	577-581	Feb 1933
Polypodiaceae	Luerssen, C.	3		1	36-50	41-65	Mar 1884
Polypodieae	Luerssen, C.	3		1	50-64	66-83	Mar 1884
Polypodieae	Luerssen, C.	3		2	65-108		Apr 1884
Polyporei	Winter, G.	1	1	6	385-416	10 figs.	Jan 1882
Polyporei	Winter, G.	1	1	7	417-480	7 figs.	Apr 1882
Polytrichaceae	Limpricht, K. G.	4	2	23	586-634	319-322	Dec 1893
Polytrichaceae	Mönkemeyer, W.	4	Suppl.	3	904-919	223-226	Dec 1927
Porphyraceae	Hauck, F.	2		1	21-26	1-2	Nov 1882
Pottiaceae	Limpricht, K. G.	4	1	9	517-576	159-168	Mai 1888
Pottiaceae	Limpricht, K. G.	4	1	10	577-640	169-180	Oct 1888
Pottiaceae	Limpricht, K. G.	4	1	11	641-693	181-189	Dec 1888
Pottiaceae	Mönkemeyer, W.	4	Suppl.	1	238-256	44-51	Apr 1927
Pottiaceae	Mönkemeyer, W.	4	Suppl.	2	257-336	52-73	Jul 1927
Pronoctilucaceae	Schiller, J.	10	3/1	2	263-272	254-260	Dec 1932
Prorocentraceae	Schiller, J.	10	3/1	1	12-44	6-47	Dec 1931-Jan 1932
Protoceratiaceae	Schiller, J.	10	3/2	3	321-327	338-341	Mai 1937

Family	Author			Pages		Date
Pseudophacidieae	Rehm, H.			1249-1251		Apr 1896
Pseudophaeriales	Keissler, K.	3	55	304-325	64-68b	Apr 1930
Pteridophyta	Luerssen, C.		2	[1]-5		Mar 1884
Pterygophyllaceae	Limpricht, K. G.		1	718-724	336	Jan 1895
Ptilidioideae	Müller, K.	2	25	300-336	92-98	21 Mar 1914
Ptilidioideae	Müller, K.		19	337-348	99-103	19 Dec 1914
Ptychodiscaceae	Schiller, J.		20	74-80	60-64	Mar 1935
Ptychomitriaceae	Mönkemeyer, W.	3/2	1	592-595	125	Dec 1927
Pulvinulariaceae	Geitler, L.	Suppl.	3	463-464	277	Jan-Jun 1931
Punctariaceae	Hauck, F.		2	369-380	158-163	Jan 1884
Pyrenidiaceae	Keissler, K.	1/2	7	[448]-462	135-138	Mar 1937
Pyrenomycetes	Winter, G.	2	3	18-21		Mar 1884
Pyrenomycetes	Keissler, K.		14	264-265		Apr 1930
Pyrenopezizeae	Keissler, K.		2	113-118		Jan 1930
Pyrenulaceae	Keissler, K.	1/2	1	[1]-160	1-64	Nov 1936
Pyrenulaceae	Keissler, K.	1/2	1	161-320	65-102	Jan 1937
Pyrenulaceae	Keissler, K.	1/2	2	321-421	103-127	Mar 1937
Raduloidieae	Müller, K.	2	3	533-553	153-159	26 Jul 1915
Ralfsiaceae	Hauck, F.		23	399-402	176	Jul 1884
Reticulariaceae	Schinz, H.	10	8	315-320	134-137	28 Aug 1917
Reticulariaceae	Schinz, H.	10	125	321-323	138-139	20 Mar 1918
Rhabdoweisiaceae	Limpricht, K. G.	1	126	271-301	102-110	Nov 1886
Rhizineae	Rehm, H.	3	5	1134-1142	10 figs.	Jul 1895
Rhizochloridaceae	Pascher, A.		53	239-251	155-161	Aug 1937
Rhizochloridineae	Pascher, A.		2	236-239		Aug 1937
Rhodomelaceae	Hauck, F.		2	203-224	88-103	Apr 1883
Rhodomelaceae	Hauck, F.		4	225-257		Apr 1883
Rhodophyceae	Hauck, F.		5	[7]		Nov 1882
Rhodymeniaceae	Hauck, F.		1	149-160	[9]	Jan 1883
Rhodomeniaceae	Hauck, F.		3	161-168	62-70	Apr 1883
Rhytidiaceae	Mönkemeyer, W.	Suppl.	4	893-895	219	Dec 1927
Ricciaceae	Müller, K.	1	3	140-192	97-128	20 Mar 1907
Ricciaceae	Müller, K.	1	3	193-219	219-141	20 Jun 1907
Rielloideae	Müller, K.	1	4	318-320	193-194	1 Oct 1907
Rielloideae	Müller, K.	1	5, 6	321-324	195-197	20 Mar 1908
Rivulariaceae	Geitler, L.		3	564-658	356-420	Jul 1931

Taxon	Author	Band	Abt.	Lief.	Pag.	Fig.	Date
Rivulariaceae	Geitler, L.	14		6	1174-1175	*780*	Dec 1932
Saccharomycetes	Winter, O.	1	1	1	68-72	*2 figs.*	Dec 1880
Salviniaceae	Luerssen, C.	3		10	595-606	*184-186*	Aug 1887
Saprolegniaceae	Fischer, A.	1	4	50	327-378	*50-61*	Mai 1892
Saprolegniaceae	Fischer, A.	1	4	52	490		Oct 1892
Scapanioideae	Müller, K.	6	2	20	349-384	*104-111*	19 Dec 1914
Scapanioideae	Müller, K.	6	2	21	385-464	*112-136*	22 Feb 1915
Scapanioideae	Müller, K.	6	2	22	465-524	*137-151*	1 Mar 1915
Schizomycetes	Winter, G.	1	1	1	33-67	*27 figs.*	Dec 1880
Schistostegaceae	Limpricht, K. G.	4	2	16	132-136	*252*	Jan 1891
Schistostegaceae	Mönkemeyer, W.	4	Suppl.	2	409-410	*92*	Jul 1927
Sclerodermei	Winter, G.	1	1	13	886-891	*6 figs.*	Sep 1883
Scytonemataceae	Geitler, L.	14		4	677-801	*437-512*	Mai 1932
Scytosiphonaceae	Hauck, F.	2		8	389-393	*169-171*	Jul 1884
Selaginellaceae	Luerssen, C.	3		14	862-877	*225*	Jun 1889
Seligeriaceae	Limpricht, K. G.	4	1	8	459-476	*146-149*	Oct 1887
Seligeriaceae	Mönkemeyer, W.	4	Suppl.	1	162-171	*24*	Apr 1927
Sematophyllaceae	Mönkemeyer, W.	4	Suppl.	3	867-872	*208-209*	Dec 1927
Silicoflagellatae	Gemeinhardt, K.	10	2		[1]-26	*1-9*	Nov 1930
Silicoflagellatae	Gemeinhardt, K.	10	2		[86], [87]	*pl. 1*	Nov 1930
Siphonemataceae	Geitler, L.	14		2	445-452	*265-268*	Jan-Jun 1931
Siphonotestales	Gemeinhardt, K.	10	2		26		Nov 1930
Sokoloviaceae	Geitler, L.	14	2		563-564	*355*	Jul 1931
Solenieae-Lauderiinae	Hustedt, F.	7	3	3	542-561	*312-321*	Jan 1929
Solenieae-Rhizosoleniinae	Hustedt, F.	7	1	3	561-608	*322-352*	Jan 1929
Solenioideae	Hustedt, F.	7	1	3	541-542		Jan 1929
Solieriaceae	Hauck, F.	2		4	186-187	*80*	Apr 1883
Sordariee	Winter, G.	1	1	16	162-187	*21 figs.*	Nov 1884
Sphaerellaceae	Keissler, K.	8		2	337-445	*73-86*	Apr 1930
Sphaerelloideae	Winter, G.	1	2	19	334-384		Jul 1885
Sphaerelloideae	Winter, G.	1	2	20	385-405	*34 figs.*	Oct 1885
Sphaeriaceae	Winter, G.	1		16	187-191		Nov 1884
Sphaeriaceae	Keissler, K.	8		2	*326-337*	*69-72*	Apr 1930
Sphaeriales	Keissler, K.	8		2	325-326		Apr 1930
Sphaerieae	Winter, G.	1	2	27	*884-893*		Mar 1887
Sphaerioidaceae	Keissler, K.	8	2	2	532-581	*100-111*	Apr 1930

Sphaerioideae-Dark-Spored	Allescher, A.	1	7	79	257-294		Dec 1901
Sphaerioideae-Hyaline-Spored	Allescher, A.	1	6	59	7-64	*1 fig.*	Jun 1898
Sphaerioideae-Hyaline-Spored	Allescher, A.	1	6	60	65-128	*2 figs.*	Jul 1898
Sphaerioideae-Hyaline-Spored	Allescher, A.	1	6	61	129-192	*3 figs.*	Aug 1898
Sphaerioideae-Hyaline-Spored	Allescher, A.	1	6	62	193-256		Oct 1898
Sphaerioideae-Hyaline-Spored	Allescher, A.	1	6	63	257-320	*2 figs.*	Dec 1898
Sphaerioideae-Hyaline-Spored	Allescher, A.	1	6	64	321-384	*12 figs.*	Jan 1899
Sphaerioideae-Hyaline-Spored	Allescher, A.	1	6	65	385-448	*15 figs.*	Jun 1899
Sphaerioideae-Hyaline-Spored	Allescher, A.	1	6	66	449-512	*14 figs.*	Sep 1899
Sphaerioideae-Hyaline-Spored	Allescher, A.	1	6	67	513-576	*23 figs.*	Oct 1899
Sphaerioideae-Hyaline-Spored	Allescher, A.	1	6	68	577-640	*14 figs.*	Dec 1899
Sphaerioideae-Hyaline-Spored	Allescher, A.	1	6	69	641-704		Jun 1900
Sphaerioideae-Hyaline-Spored	Allescher, A.	1	6	70	705-768		Apr 1900
Sphaerioideae-Hyaline-Spored	Allescher, A.	1	6	71	769-832		Jul 1900
Sphaerioideae-Hyaline-Spored	Allescher, A.	1	6	72	833-896		Jul 1900
Sphaerioideae-Hyaline-Spored	Allescher, A.	1	6	73	897-960		Oct 1900
Sphaerioideae-Hyaline-Spored	Allescher, A.	1	6	74	961-992		Dec 1901
Sphaerocarpideae	Müller, K.	6	1	5	314-317		1 Oct 1907
Sphaerococcaceae	Hauck, F.	2		4	178-185		Apr 1883
Sphaerophoraceae	Keissler, K.	9	1/2	5	[756]-773		Mar 1938
Sphaeropsidales	Keissler, K.	8		2	531-532		Apr 1930
Sphaeropsideen	Allescher, A.	1	7	81	441-443		Jan 1902
Splachnaceae	Limpricht, K. G.	4	2	16	136-172		Jan 1891
Splachnaceae	Limpricht, K. G.	4	Suppl.	2	398-409		Jul 1927
Spongiocarpeae	Hauck, F.	2		4	197-199		Apr 1883
Sporochnaceae	Hauck, F.	2		7	382-384		Jan 1884
Sporochnaceae	Hauck, F.	2		8	383-389		Jul 1884
Sporochytriaceae	Fischer, A.	1	4	46	85-128		Oct 1891
Sporochytriaceae	Fischer, A.	1	4	47	129-131		Jan 1892
Spyridiaceae	Hauck, F.	2		3	113-116		Jan 1883
Squamariaceae	Hauck, F.	2		1	26-37		Nov 1882
Stemonitaceae	Schinz, H.	1	10	124	229-256		9 Dec 1915
Stemonitaceae	Schinz, H.	1	10	125	257-272		28 Aug 1917
Stereotestales	Gemeinhardt, K.	10	2	1	79		Nov 1930
Stictidaceae	Keissler, K.	8		2	251-261		Apr 1930
Stictideae	Rehm, K.	1	3	29	112-113		Jan 1888

Taxon	Author	Band	Abt.	Lief.	Pag.	Fig.	Date
Stigonemataceae	Geitler, L.	14		3	482-552		Jul 1931
Stilbaceae	Lindau, G.	1	9	109	[286]-304		30 Jul 1908
Stilbaceae	Lindau, G.	1	9	110	305-368		25 Nov 1908
Stilbaceae	Lindau, G.	1	9	111	369-400		7 Apr 1909
Stilbaceae	Lindau, G.	1	9	117	810-815		10 Mar 1910
Stilbaceae	Lindau, G.	1	9	117	810-815		10 Mar 1910
Stilbaceae	Keissler, K.	8		2	623-624		Apr 1930
Strigulaceae	Keissler, K.	9	1/2	3	[440]-447	43-47	Mar 1937
Syndiniaceae	Schiller, J.	10	3/2	1	53-61		Mar 1935
Syracosphaeraceae	Schiller, J.	10	2		[171]-230		1930
Tabellarieae-Liemophorinae	Hustedt, F.	7	2	1	52-91	51-113	Nov 1931
Tabellarieae-Tabellariinae	Hustedt, F.	7	2	1	11-51	579-626	Nov 1931
Teloschistaceae	Hillman, J.	9	6	1	[1]-36	545-578	Dec 1935
Terfeziaceen	Fischer, E.	1	5	58	66-81	1-4b	Nov 1896
Thamniaceae	Mönkemeyer, W.	4	Suppl.	3	647-649	20 figs.	Dec 1927
Thecatales	Schiller, J.	10	3/2	4	514	140-141	Mai 1937
Thelephoraceae	Keissler, K.	8		2	521	596-597	Apr 1930
Theliaceae	Mönkemeyer, W.	4	Suppl.	3	647-677	96-99	Dec 1927
Thelocarpaceae	Magnusson, A. H.	9	5/1	1	[286]-318	151	Mar 1935
Thelophorei	Winter, G.	1	1	5	318-352	60-64	1-15 Nov 1881
Thelophorei	Winter, G.	1	1	6	353	11 figs.	Jan 1882
Thoracosphaeriaceae	Schiller, J.	10	2	1	237-242	119	Nov 1930
Thuidaceae	Mönkemeyer, W.	4	Suppl.	3	694-704	156-157	Dec 1927
Timmiaceae	Limpricht, K. G.	4	2	22	575-576		Jun 1893
Timmiaceae	Limpricht, K. G.	4	2	23	577-586	317-318	Dec 1893
Timmiaceae	Mönkemeyer, W.	4	Suppl.	3	588-591	124	Dec 1927
Tremellineae	Winter, G.	1	1	4	270-288	23 figs.	Aug 1881
Tremellineae	Winter, G.	1	1	5	289-290		1-5 Nov 1881
Tribonemataceae	Pascher, A.	11		6	939-991	791-839	Feb 1939
Tribonematales	Pascher, A.	11		6	915-916		Feb 1939
Trichiaceae	Schinz, H.	1	10	126	329-368	143-161	20 Mar 1918
Trichosphaerieae	Winter, G.	1	2	16	191-192	7 figs.	Nov 1884
Trichosphaerieae	Winter, G.	1	2	17	193-219	24 figs.	Apr 1885
Trigonantheae	Müller, K.	6	2	15	[1]-80	1-23	1 Mar 1912
Trigonantheae	Müller, K.	6	2	16	81-44	24-40	20 Dec 1912
Trigonantheae	Müller, K.	6	2	17	145-208	41-60	20 Jan 1913

RABENHORST

Trypanochloridaceae	..., A.	1	3	55	1253		Apr 1896
Trypetheliaceae	Pascher, A.	1		5	825-829	685-690	Jan 1939
Tuberaceae	Keissler, K.	9	1/2	3	[422]-439	128-132	Mar 1937
Tuberculariaceae	Fischer, E.	1	5	57	3-12	1 fig.	Jul 1896
Tuberculariaceae	Lindau, G.	1	9	111	[401]-432	14 figs.	7 Apr 1909
Tuberculariaceae	Lindau, G.	1	9	112	433-496	65 figs.	X8 Apr 1909
Tuberculariaceae	Lindau, G.	1	9	113	497-560	41 figs.	20 Jun 1909
Tuberculariaceae	Lindau, G.	1	9	114	561-624	47 figs.	25 Jul 1909
Tuberculariaceae	Lindau, G.	1	9	115	625-648	21 figs.	1 Nov 1909
Tuberculariaceae	Lindau, G.	1	9	117	815-816		10 Mar 1910
Tuberculariaceae	Lindau, G.	1	9	118	817-822	4 figs	20 Mai 1910
Tuberculariaceae	Keissler, K.	8		2	624-642		Apr 1930
Tubulinaceae	Schinz, H.	1	10	125	310-315	131-135	28 Aug 1917
Tulostomei	Winter, G.	1	1	13	892-893	132-133	Sep 1883
Ulvaceae	Hauck, F.	2		8	[422]-437	3 figs.	Jul 1884
Umbilicariaceae	Frey, E.	9	4/1	1	[203]-208	187-191	Dec 1932
Umbilicariaceae	Frey, E.	9	4/1	2	209-407		Feb 1933
Uredineae-Uredineae	Winter, G.	1	1	2	131-144	33-63	Mar 1881
Uredineae-Uredineae	Winter, G.	1	1	3	145-224	21 figs.	Mai 1881
Uredineae-Uredineae	Winter, G.	1	1	4	225-270		Aug 1881
Usneaceae	Keissler, K.	9	5/4	1	[3]-160	1-28	Oct 1958
Usneaceae	Keissler, K.	9	5/4	2	161-320	29-38	Mar 1959
Usneaceae	Keissler, K.	9	5/4	3	321-480	39-54	Nov 1959
Usneaceae	Keissler, K.	9	5/4	4	481-640		Mar 1960
Usneaceae	Keissler, K.	9	5/4	5	641-717	pls. I-XIX	Dec 1960
Ustilagineae-Ustilagineae	Winter, G.	1	1	1	79-80		Dec 1880
Ustilagineae-Ustilagineae	Winter, G.	1	1	2	81-132	25 figs.	Mar 1881
Valoniaceae	Hauck, F.	2		9	469-471	205-207	Oct 1884
Valseae	Winter, G.	1	2	23	594-656	28 figs.	Aug 1886
Valseae	Winter, G.	1	2	24	657-736	2 figs.	Aug 1886
Valseae	Winter, G.	1		25	737-764		Oct 1886
Vaucheriaceae	Hauck, F.	2		8	412-416	182-186	Jul 1884
Verrucariaceae	Zschacke, H.	9	1/1	1	46-160	17-64	Feb 1933
Verrucariaceae	Zschacke, H.	9	1/1	2	161-320	65-154	Jun 1933
Verrucariaceae	Zschacke, H.	9	1/1	3	321-480	155-261	Aug 1933
Verrucariaceae	Zschacke, H.	9	1/1	4	481-590	262-316	Feb 1934
Verrucariaceae	Zschacke, H.	9	1/1	4	656-668	343-344	Feb 1934

Taxon	Author	Band	Abt.	Lief.	Pag.	Fig.	Date
Voitiaceae	Limpricht, K. G.	4	1		208-211	82-84	Jul 1886
Voitiaceae	Mönkemeyer, W.	4	Suppl.	2	396-397		Jul 1927
Weisiaceae	Limpricht, K. G.	4	1	4	220-256	85-96	Jul 1886
Weisiaceae	Limpricht, K. G.	4	1	5	257-271	97-101	Nov 1886
Wernowiaceae	Schiller, J.	10	3/1	3	554-600	582-630	Feb 1933
Wrangeliaceae	Hauck, F.	2		1	39-55	10-16	Nov 1882
Xanthopyreniaceae	Keissler, K.	9	1/2	3	[463]-469	139-140	Mar 1937
Xylarieae	Winter, G.	1	2	26	842-864	20 figs.	Dec 1886
Xylarieae	Winter, G.	1	2	27	865-880		Mar 1887
Zygnemaceae	Krieger, H.	13	2	2	115-196	1-184	1941
Zygnemaceae	Krieger, H.	13	2	3	197-294	185-415	1941
Zygnemaceae	Krieger, H.	13	2	4	295-482	416-779	1944
Zygnemaceae	Krieger, H.	13	2	4	[483]484	4-9	1944
Zygnemales	Kolkwitz, R.	13	2	1	[1]-50	A-Z	1941
Zygnemales	Krieger, H.	13	2	1	[51]-104		1941
Zygnemales	Krieger, H.	13	2	2	[111]-115		1941
Zygomycetes	Fischer, A.	1	4	45	5-7		Oct 1891
Zygomycetes	Fischer, A.	1	4	47	161-177		Jan 1892

Note: For *Lydia Rabinowitsch* and the first part of the entry on *Louis Theodor Rach* see page 473.

BIBLIOGRAPHY and BIOGRAPHY: Barnhart 3: 122 (b. 16 Jan 1821, d. 28 Apr 1859); BM 4: 1636; CSP 5: 71; Herder p. 162, 318; Hortus 3: 1202; MW p. 398, 408; Rehder 2: 657, 5: 699; Urban-Berl. p. 386; Zander ed. 10, p. 704, ed. 11, p. 803; Zep-Tim p. 95.

BIOFILE: Anon., Bonplandia 3: 288. 15 Oct 1855 (rev. of his Ericaceen Thunberg), 4: 240, 282. 1856 (app. St. Petersb.); Bot. Zeit. 14: 624. 1856 (app. St. Petersburg), 17: 328. 1859 (d. 28 Apr 1859); Bull. Soc. bot. France 6: 440. 1859 (obit.); Österr. bot. W. 6: 278. 1856 (app. St. Petersb.).
Ascherson, P., Fl. Prov. Brandenburg 1: 10. 1864.
Cohn, F., Bot. Zeit. 18: 141. 1860.
Lipsky, V.I., Imp. St. Petersb. bot. sada 3: 406-407. 1915 (biogr. sketch; bibl.).
Rach, L., Die Ericaceen der Thunberg'schen Sammlung. Linnaea 26: 767-793. Sep 1855 (repr. rev. Bonplandia 15 Oct 1855).
Regel, E., Gartenflora 8: 192. 1859 (obit. notice).

COMPOSITE WORKS: With E.A. Regel and F. v. Herder, *Verzeichniss der von Herrn Paullowsky und Herrn von Stubendorf in den Jahren 1857/58 zwischen Jakutzk und Ajan gesammelten Pflanzen.* 1859; see under E.A. Regel.

EPONYMY: *Rachia* Klotzsch 1854.

Raciborski, Marjan [Marian, Maryan], (1863-1917), Polish botanist; Dr. phil. München 1894; assistant with K. Goebel 1894-1896; at Buitenzorg (Bogor) Java, with M. Treub, 1896-1897; at Kagok, Tegal, Central Java (Java Sugar Experiment Station), succeeding F.A.F.C. Went, 1897-1898; in charge of tobacco research, Buitenzorg 1898-1900; professor of botany at the agricultural university of Dublany, Lemberg, Poland 1900-1909; extraordinary professor of botany at Lemberg 1903, ord. professor and director of the plant-physiological institute ibid. 1909-1912; director of the botanical institute of the University of Krakau 1912-1917; pioneer of nature protection in Poland. (*Racib.*).

HERBARIUM and TYPES: Collections (incl. exsiccatae) at B, BO, C, FH, GB, K, KRA (impt.), KRAM, L, LD, NSW, P, PC, S, US, W, ZT. – *Exsiccatae*:
1. *Cryptogamae parasiticae in insula Java lectae exsiccatae*, fasc. 1-3, nos. 1-150, 1899-1900. Sets at B, BO, FH, K, W. – See Sayre (1969). – Accompanying text: Paras. Alg. Pilz. Javas (1900), see below, TL-2/8470.
2. *Mycotheca polonica*, fasc 1-4, nos. 1-200, 1909-1910, for accompanying text see below, no. .... Sets at B, C, FH, PC, W.
3. *Phycotheca polonica*, fasc. 1-3, nos. 1-150, 1910-1911. Sets at C, FH, LD, NSW, PC, S, UPS, W. – See Sayre (1969).

BIBLIOGRAPHY and BIOGRAPHY: AG 5(1): 612; Ainsworth p. 123, 322, 346; Andrews p. 315; Backer p. 480-481; Barnhart 3: 122 (b. 16 Sep 1863, d. 27 Mar 1917); BFM no. 1477; Biol.-Dokum. 14: 7277; BJI 1: 47, 2: 142; BM 4: 1636, 8: 1042; Bossert p. 321; CSP 12: 595, 18: 16-17; De Toni 1: ciii, 2: xcviii, 4: xlvi; Futak-Domin p. 496; GR p. 437; GR, cat. p. 70; Hawksworth p. 185; Hortus 3: 1202; IF p. 725-726; IH 1 (ed. 7): 340, 2: in press; Kelly p. 185; Kew 4: 387; LS 21621-21645, 28155-38162; Morren ed. 10, p. 35; NAF 1(1): 168, ser. 2. 6: 73; Nordstedt p. 26, suppl. p. 14 (bibl. Desmid.); Quenstedt p. 350; Rehder 5: 699; SK 1: 422-423 (portr., itin., coll.), 4: 71, 75, 100; cxxiv, cliv, clx, 8: lxxviii; Stevenson p. 1255; Tucker 1: 575; Urban-Berl. p. 262, 322, 386.

BIOFILE: Ainsworth, G.C., Dict. fungi ed. 6. 500. 1971.
Andrews, H.N., The fossil hunters 319-320. 1980.
Anon., Allg. Bot. Z. 2: 140. 1896 (to Buitenzorg), 3: 168. 1897 (to Kagok, Tegal), 9: 196. 1903 (prof. Lemberg), 15: 164. 1909 (ord. prof. Lemberg), 19: 95. 1913 (Flora polonica exsiccata ann.), 23: 32. 1917 (d.); Bot. Centralbl. 71: 384. 1897 (to Kagok, Tegal), 77: 239. 1899 (to Buitenzorg, tobacco), 93: 448. 1903 (prof. Lemberg), 111: 496. 1909 (ord. prof. Lemberg), 120: 64. 1912 (app. Krakau); Bot Jahrb. 24 (Beibl. 58): 22. 1897 (to Kagok, Tegal), 27: (Beibl. 62): 8. 1899 (Buitenzorg, tobacco), 34: (Beibl. 76): 41. 1904 (prof. Lemberg).; Bot. Not. 1917: 141; Bot. Zeit. 57: 32. 1899 (left Kagok Tegal for Buitenzorg), 58: 324. 1900 (app. Lemberg); Hedwigia 35(3):

(96). 1896 (to Buitenzorg), 38(1): (61). 1899 (tobacco research), 39(6): (222). 1900 (to Dublany, Lemberg), 50(2): (89). 1910 (prof. bot.), 52(2): (154). 1912 (dir. bot. gard. Krakau), 59(2): (141). 1917 (d.); Magy. bot. Lap. 15: 312. 1916 (1917) (d.); Mycol. Centralbl. 1(5): 162. 1912 (to Krakau); Nat. Nov. 18: 244, 575. 1896 (to Bot. Gard. Buitenzorg), 19: 331. 1897 (to Kagok, Tegal), 21: 87. 1899 (to Buitenzorg, tobacco), 25: 562. 1903 (to Dublany, Lemberg), 31: 491. 1909 (prof. Lemberg), 34: 387. 1912 (to Krakau); Nuova Notarisia 29: 124. 1918 (d.); Österr. bot. Z. 50: 419. 1900 (to Dublany, Lemberg), 53: 431. 1903 (prof. Lemberg), 59: 367. 1909 (ord. prof. Lemberg), 62: 198. 1912 (to Krakau), 66: 312. 1916 (d.); Verh. zool. bot. Ges. Wien 68: (101). 1918 (d.).

Gager, C.S., Brooklyn Bot. Gard. Rec. 27(3): 316. 1938 (dir. Crakow bot. gard. 1912-1917).

Goebel, K., Ber. deut. bot. Ges. 35: (97)-(107). 1917 (obit., bibl., b. 16 Sep 1863, d. 27 Mar 1917).

Hryniewiecki, B., Edward Strasburger 99 [index]. 1938.

Kulczyński, S., Stud. Mater. Dziejów Nauki Polsk. B 27: 19-46. 1977 (correspondence with W. Kulczyński 1895-1917).

Magnus, P., Hedwigia 30(6): 303-307. 1891 ("Eine Bemerkung gegen Herrn M. Raciborski").

Pax, F., Bibl. schles. Bot. 22, 147. 1929.

Rostafinsky, J., Szafer, W. et al., Kosmos, Lemberg 42: 66-95. 1917 (obit., portr., funeral addresses, Polish with German summary).

Sayre, G., Mem. New York Bot. Gard. 19(1): 42, 96-97. 1969 (Crypt. paras. & Phycoth. polon.).

Siemińska, J., Stud. Mater. Dziejów Nauki Polsk. B 27: 47-62. 1977 (letters B. Eichler to R).

Szafer, W., Concise hist. bot. Cracow 163 [index]. 1969; Ochrona przyrody 29: 9-16. 1963 (the centenary of R's birth; general tribute, Polish with English summary); Regn. veg. 71: 386-387. 1970; Sprawozd. Komis. fizyograf., Krakow 51: xxxvii-xl. 1917 (obit.); Stud. Mater. Dziejów Nauki Polsk. B 27: 5-18. 1977 (on R and K. Goebel).

Szafer, W., ed., The vegetation of Poland 1966 (frontisp. portr. of R.).

Szymkiewicz, D., Bibl. Fl. Polsk. 15-16, 21, 31, 40-41, 55, 117-119. 1925.

Verdoorn, F., ed., Chron. bot. 2: 255. 1936, 6: 263. 1941.

Zittel, K.A. von, Gesch. Geol. Paläont. 789. 1899.

COMPOSITE WORKS: Raciborski is listed as editor and collaborator for M. Raciborski & W. Szafer, *Flora polska*, vol. 1, 1919. See under Szafer.

EPONYMY: *Pseudoraciborskia* Kufferath (1954); *Raciborskia* A.N. Berlese (1888); *Raciborskia* Woloszyńska (1919); *Raciborskia* M. Koczwara (1928); *Raciborskiella* F. von Höhnel (1909); *Raciborskiella* Spegazzini (1919); *Raciborskiella* S. Wislouch (1924); *Raciborskiomyces* Siemaszko (1925).

8452. *Desmidyje okolic Krakowa* ... W Krakowie (W Drukarni Uniwersytetu Jagiellońskiego ...) 1884. Oct. (*Desmid. okol. Krak.*).
Publ: 1884 (p. 4: Dec 1883; Bot. Centralbl. 8-12 Dec 1884; Nat. Nov. Oct(3) 1884), p. [1]-24, *pl. 1. Copy*: FH. – Preprinted from Sprawoz. Komiss. fizyogr., Akad. Umiej. Krakow 19: 3-24. 1885.

8453. Przyczynek do znajomości śluzowców. *Myxomycetum agri cracoviensis genera, species et varietates novae* ... Kraków (w drukarni c.k. Uniwersytetu Jagiellońskiego ...) 1884. Oct.
Publ.: 1884 (Hedwigia Sep 1884; Nat. Nov. Oct(3) 1884), p. [i], [1]-17, *pl. 4. Copy*: FH. – Reprinted and to be cited from Rozpr. Spraw. Wydz. mat.-przyr. Akad. Umiej. (Ber. S.B. Krakau. Akad. Wiss.) 12: [69]-86, *pl. 4*. 1884.

8454. *De nonnullis Desmidiaceis* novis vel minus cognitis, quae in Polonia inventae sunt ... W Krakowie (w Drukarni Uniwersytetu Jagiellońskiego ...) 1885. Qu.
Publ.: Mar-Aug 1885 (p. 2: 15 Lutego (Feb) 1885; Bot. Zeit. 25 Dec 1885; ÖbZ 1 Dec 1885; Nat. Nov. Sep(1) 1885; Hedwigia Sep-Oct 1885), p. [i-ii], [1]-44, *pl. 10-14* (col.

liths. auct.). *Copies*: FH, G, H, M, PH. – Reprinted and to be cited from Pamietnik Wydz. iii Akad. Umiej. w Krakowie 10: 57-100, *pl. 10-14.* 1885.
Ref.: Rothert, K.W., Bot. Centralbl. 29(3): 65-66. 1887.
Richter, P., Hedwigia 24: 267-271. 1885.

8455. *Roślinne pasorzyty karpi* (Saprolegniae). Mit einem deutsch verfassten Resume ... W Krakowie (w Drukarni Uniwersytetu Jagiellónskiego ...) 1885. Oct.
Publ.: 1885 (Bot. Zeit. 24 Sep 1886), p. [i], [1]-20, [1]-3, *pl. 3* (uncol. lith. auct.). *Copy*: FH. – Reprinted and to be cited from Rozpr. Spraw. Wydz. mat.-przyr. Akad. Umiej. 14: [147]-168. 1885(1886).

8456. *De generis Galii* formis, quae in Polonia inventae sunt. Scripsit M. Raciborski [Krakow 1885]. Oct.
Publ.: 1885 (read 20 Apr 1885), p. [1]-10. *Copy*: PH. – Reprinted and to be cited from Rozpr. Spraw. Posiedzeń Wydz. mat.-przyr. Akad. Umiej., Krakow, 14 (Mat.): 263-272. 1885/1886.

8457. *Materyjały do flory glonów polski* ... Kraków (Drukarnia Uniwersytetu Jagiellońskiego ...) 1888. Oct.
Publ : 1888, p. [i], [1]-43. *Copy*: FH. – Reprinted and to be cited from Spraw. Komis. fizyjogr. Akad. Umiej. 22(2): 80-122. 1888. – Materials for a flora of Polish algae.
Ref.: Rothert, K.W., Bot. Centralbl. 38: 702. 21 Mai 1889.

8458. *Desmidyje nowe* (Desmidiaceae novae) ... Kraków (Drukarna Uniwersytetu Jagiellońskiego ...) 1889. Qu. (*Desmid. nov.*).
Publ.: 1889, p. [i-ii], [1]-41, *pl. 5-7* (col. liths. auct.). *Copy*: FH. – Reprinted from Pamietn. Wydz. mat.-przyr. Akad. Umiej. 17, 1889 (journnal issue n.v.).

8459. *Przeglad gatunków rodzaju Pediastrum* ... Krakow (Drukarnia Uniwersytetu Jagiellońskiego ...) 1889. Oct.
Publ.: 1889 (Nat. Nov. Dec(4) 1889, Jan(1) 1890), p. [i], [1]-37, *pl. 2* (col. lith. auct.). *Copy*: FH. – Reprinted and to be cited from Rozpr. Spraw. Wydz. mat.-przyr. Akad. Umiej. 20: [84]-120, *pl. 2.* 1889 (1890). – Also in Bull. Cracov. Acad. Sci. 1889(4): xxxvi-xxxvii. – A monograph of the genus Pediastrum.

8460. *Über die Osmundaceen und Schizaeaceen der Juraformation* [Botanische Jahrbücher, xiii. Bd. 1890]. Oct.
Publ.: Apr 1890 (reprint "Gedruckt im April 1890"), p. [1]-9, *pl. 1.* *Copy*: G. – Preprinted from Bot. Jahrb. 13: 1-9. 15 Jul 1890. The words "Gedruckt im April 1890" appear also in the journal publication. The status of preprint is therefore not entirely clear.

8461. *Desmidya* zebrane przez Dr. E. Ciastonia w podrózy na Około Ziémi ... Kraków (Nakladem Akademii Umiejetności ...) 1892. Oct.
Publ.: 1892 (submitted 7 Mar 1892; Nat. Nov. Aug(2) 1892), p. [i-ii], [1]-32, *pl. 6-7*. *Copies*: FH, M. – Reprinted and to be cited from Rospr. Spraw. Posiedzeń. Wydz. mat.-przyr. Akad. Umiej., Krakow, 2: 361-392, *pl. 6-7.* 1892.
Preliminary publ.: Anz. Akad. Krakau Mar 1892, p. [112]-114. (Desmids coll. by E. Ciaston on circumnavigation by "Saida").

8462. *Cycadeoidea Niedzwiedzkii* nov. sp. ... Kraków (Nakładem Akademii Umiejetności) 1893. Oct.
Publ.: 1893 (read 3 Oct 1892), p. [i-ii], [1]-10, *pl. 7-8*. *Copy*: M (Inf. I. Haesler). – Reprinted and to be cited from Rozpr. Spraw. Posiedzeń. Wydz. mat.-przyr. Akad. Umiej., Kraków 26: 301-310, *pl. 7-8.* 1893. – *Preliminary notice*: Anz. Akad. Wiss. Krakau 1892(8): 355-359. 8 Nov 1892.

8463. *Przyczynek do morfologi jądra komórkowego* nasion kiełkujących ... Kraków (Nakładem Akademii Umiejetności ...) 1893. Oct.
Publ.: 1893 (read 6 Mar 1893), p. [i-ii], [1]-11, *pl. 2*. *Copy*: M. (Inf. I. Haesler). –

Reprinted and to be cited from Rozpr. Spraw. Posiedzeń. Wydz. mat.-przyr. Akad. Umiej., Krakow 26: 362-373, *pl. 9* (sic). 1893. *Preliminary publication*: Anz. Akad. Wis. Krakau 1893(3): 120-123. Mar 1893.

8464. *Nowe gatunki zielenic* ... Kraków (Nakładem Akademii Umiejetności ...) 1893. Oct.
Co-author: Bogumir Eichler (1843-1905).
Publ.: 1893 (Nat. Nov. Dec(1) 1893; mss. submitted 4 Jul 1892), p. [i-ii], [1]-11, *pl. 3* (monochr. lith. auct.). *Copy*: FH. – Reprinted and to be cited from Rozpr. Spraw. Posiedzeń. Wydz. mat.-przyr. Akad. Umiej., Krakow 26: 116-126, *pl. 3*. 1893.

8465. *Die Morphologie der Cabombeen und Nymphaeaceen.* Inaugural-Dissertation zur Erlangung der Doctorwürde bei der hohen philosophischen Facultät der Universität München (ii. Section) eingereicht von Marian Raciborski aus Brzostowa (Polen) ... München (Druck von Val. Höfling, ...) 1894. Oct. (*Morph. Cambomb. Nymphaeac.*).
Publ.: Mar-Mai 1894 (p. 3: 26 Feb 1894; BSbF 13 Jul 1894; Nat. Nov. Nov(1) 1894), p. [1]-38. *Copies*: HH, M. – Preprinted from Flora 79(3): 92-108, *pl. 2*. 29 Oct. 1894.
Commercial reprint: 1894, p. [1]-38. *Copy*: G. – "Die ... Nymphaeaceen ... [Sonderabdruck aus "Flora"]". Preprinted from id.

8466. *Elajoplasty liliowatych* ... Kraków (Nakładem Akademii Umiejetności ...) 1894. Oct.
Publ.: 1894 (mss. submitted 3 Jul 1893), p. [i-ii], [1]-22, *pl. 1* (col.). *Copies*: G, M. – Reprinted or preprinted from Rozpr. Spraw. Posiedzeń Wydz. mat.-przyr. Akad. Umiej., Krakow 27: 1-22, *pl. 1(col.)*, 1895(1894?). *Preliminary publication*: Anz. Akad. Wiss. Krakau 1893(7): 259-271. Jul 1893.

8467. *Flora kopalna ogniotrwałych glinek krakowskich.* (ześc 1. Rodniowce (*Archaegoniatae*). 1894. Oct.
Publ.: 1894 (ÖbZ for Feb 1895, Flora rd. 22 Feb 1895), p. [i-ii], [1]-101, *pl. 6-27*. *Copy*: Fl. – Reprinted and to be cited from Pamietnika mat.-przyr. Akad. Umiej. 18: 143-243. *pl. 6-27*. 1894, see also Bull. Acad. Sci. Cracovie 1890: 31-34.

8468. *Pseudogardneria* nowy rodzaj z rodziny Loganiaceae ... w Krakowie (Nakładem Akademii Umiejetności ...) 1896. Oct.
Publ.: 1896 (read 13 Apr 1896), p. [i-ii], [1]-9. *Copy*: M (Inf. I. Haesler). – Reprinted and to be cited from Rospr. Spraw. Posiedzeń. Wydz. mat.-pzryr. Akad. Umiej., Krakow 32: 313-321. 1896. – Preliminary publication in Bull. int. Acad. Sci. Cracovie 1896(4): 204-208. Apr 1896.

8469. *Die Pteridophyten der Flora von Buitenzorg* ... Leiden (Buchhandlung und Druckerei vormals E.J. Brill) 1898. Oct. (*Pteridoph. Buitenzorg*).
Publ.: Jan-Mar 1898 (ÖbZ 1 Jul 1898; Bot. Centralbl. 25 Mai 1898; Nat. Nov. Apr(2), Mai(2) 1898), p. [ii]-xii, [1]-255. *Copies*: BR, FI, G, HH, NY, PH, USDA. – Second t.p. [ii]: *Flore de Buitenzorg* publiée par le Jardin botanique de l'État. 1re partie. Ptéridophytes ...
Additions: Natuurk. Tijdschr. Ned. Ind. 59: 234. 1900, Bull. Acad. Sci. Cracovie Jan 1902: 54-65.
Ref.: Goebel, K., Flora 85: 319-320. 9 Aug 1898 (rev.).
Karsten, G., Bot. Zeit. 56(2): 197-198. 1 Jul 1898 (rev.).

8470. *Parasitische Algen und Pilze Javas* ... Herausgegeben vom Botanischen Institut zu Buitenzorg. Batavia [Djakarta] (Staatsdruckerei) 1900. 3 parts. Oct. (*Paras. Alg. Pilz. Javas*).
1: Jan-Mar 1900 (p. [3]: 7 Nov 1899; Bot. Zeit. 1 Jun 1900; Bot. Centralbl. 13 Jun 1900; Nat. Nov. Apr(2) 1900), p. [1]-39.
2: Feb-Mai 1900 (Bot. Zeit. 16 Sep 1900; Nat. Nov. Jun(2), Jul(1) 1900), p. [1]-46.
3: Mar-Jul 1900 (ÖbZ Jul 1900; Nat. Nov. Aug(2) 1900; Flora rd. 5 Oct 1900), p. [1]-49.
Copies: BR, FH, H, M, NY, PH, Stevenson, U, USDA.

Facsimile ed.: 1973, p. [i, t.p. Bibl. mycol., ed. J. Cramer], followed by facsimile as above.
Copies: FAS, G, M. – ISBN 3-7682-0855-9.
Additions: Bull. Acad. Cracovie 1909(1): 346-394. 3 Apr 1909.
Ref.: Behrens, W.J., Bot. Centralbl. 84: 48-49, 316-319. 1900.
Hariot, P., Bull. Soc. bot. France 47: 156. 1900 (rev.).

8471. *Choroby tytoniu w Galicyi* ... Lwów [Lemberg] (Wydane Staraniem i Nakładem "Towarzystwa uprawy tytoniu w sniatynie ...) 1902. (*Chorob. tyton.*).
Publ.: 1902, p. [i], [1]-25. *Copies*: MO, U.

8472. *Mycotheca polonica* ... Lwów [Lemberg] (Zwiazkowa drukarnia we Lwowie, ...) 1909-1910. 4 parts (*Mycoth. polon.*).
Publ.: A series of papers in the journal Kosmos accompanying the series of *exsiccatae* with the same title (fasc. 1-4, nos. 1-200):
1: 1909, Kosmos 34: [1166]-1172, reprint with special cover.
2/3: 1910, Kosmos 35: [768]-781, id.
4: 1910, Kosmos 35: [1007]-1012, id.
Copies: BR, G. – ÖbZ mentioned fasc. 1 for Mai 1910, 2/3 on 1 Jan 1911 and 4 on 1 Jan 1912. – Herbarium mentions fasc. 1 (of the exsiccatae themselves) in 1910, nos. 2/3 and 4 in 1911.

Racine, Rudolf (*fl.* 1889), German botanist; Dr. phil. Rostock 1889. (*Racine*).

HERBARIUM and TYPES: Unknown.

BIBLIOGRAPHY and BIOGRAPHY: Barnhart 3: 122.

8473. *Zur Kenntniss der Blütenentwicklung und des Gefässbündelverlaufs der Loasaceen.* Inaugural-Dissertation verfasst und der hohen philosophischen Facultät der Grossherzoglich Mecklenburgischen Landes-Universität Rostock zur Erlangung der Doctorwürde ... Rostock (Carl Boldt'sche Hof-Buchdruckerei) 1889. Oct. (*Kenntn. Blütenentw. Loasac.*).
Publ.: 1889 (Nat. Nov. Nov(2) 1889), p. [1]-46, [1, expl. pl.], *1 pl.* (uncol.). *Copies*: G, NY.

Radais, Maxime Pierre François (1861-?), French botanist; from 1887 connected with the "École de Pharmacie", Paris, ultimately as professor of cryptogamy; Dr. sci. nat. Paris 1894. (*Radais*).

HERBARIUM and TYPES: Unknown.

BIBLIOGRAPHY and BIOGRAPHY: Barnhart 3: 122 (b. 18 Jan 1861); Bossert p. 322; Kelly p. 185; Langman p. 605; LS 7456, 7457, 21646-21649, 38163; Rehder 5: 699; TL-2/2202; Tucker 1: 575.

BIOFILE: Anon., Bot. Centralbl. 119: 160. 1912 (Légion d'honneur).
Wittrock, V.B., Acta Horti Berg. 3(3): 111-112. 1905.

8474. Série A, no 205, no d'ordre 805 Thèses présentées à la Faculté des sciences de Paris pour obtenir le grade de docteur ès sciences naturelles ... 1re thèse. – *Contribution à l'anatomie comparée du fruit des Conifères.* 2me thèse ... soutenues [le 24, mss.] février 1894, ... Paris (E. Bernard et Cie, ...) 1894. Oct. (*Contr. anat. fruit Conif.*).
Publ.: 24 Feb 1894, p. [81]-172, *pl. 1-9* (uncol. liths. auct.). *Copies*: BR, HH.
Ref.: Roth, E., Bot. Centralbl. Beih. 5(3): 181-183. 1895.

8475. École supérieure de Pharmacie de Paris. Thèse présentée au concours d'agrégation du 1er Mai 1894 (section d'histoire naturelle et pharmacie). *La fleur femelle des Conifères* ... Paris (J. Mersch, ...) 1894. Qu. (*Fl. fem. Conif.*).
Publ.: 1 Mai 1894 (thesis for "concours d'agrégation" on that date; Nat. Nov. Jun(1) 1894), p. [1]-103. *Copy*: HH.

Radde, Gustav [Ferdinand Richard] von (1831-1903), German (Danzig)-born botanist and explorer, later naturalized Russian; in the Crimea 1852-1855; on the Russian Baikal expedition 1855-1860; curator at the St. Petersburg Academy of Sciences 1860-1863; from 1863 at Tiflis, exploring the Caucasus; founder and from 1871 onward director of the Caucasian Museum and travelling further in Caucasia, Armenia, Siberia, the Asiatic tropics and Algeria; Dr. phil. h.c. Dorpat. (*Radde*).

HERBARIUM and TYPES: TB; important sets also at LE; other material at B, BP, C, E, FI, H, K, PH and U. – Main set of the 1855-1860 Baikal expedition at LE; main set of voyage in the Tamara (1890-1891) at TB.

BIBLIOGRAPHY and BIOGRAPHY: AG 3: 561; Barnhart 3: 122 (b. 15/27 Nov 1831, d. 2/15 Mar 1903); BJI 1: 47; BM 4: 1636; Bossert p. 322; Bret. p. 617; GR p. 122; Herder p. 57, 200, 235; Hortus 3: 1202; Jackson p. 393; Kelly p. 185; Kew 4: 388; LS 21653; Morren ed. 2, p. 30, ed. 10, p. 100; MW p. 398, suppl. 288; PR 10619; Quenstedt p. 350; Rehder 5: 699; SK 1: 423-424 (itin., coll.), 5: cccxiii; TL-1/1712; TL-2/2671, 3540; TR 1096-1106; Tuckcer 1: 575; Urban-Berl. p. 386; Zander ed. 10, p. 704, ed. 11, p. 803, 804.

BIOFILE: Adelung, N.v., Zool. Zentralbl. 10: 829-831. 1903.
Anon., Auk 20: 458-459. 1903 (obit.); Bilder vom Auslandsdeutschtum 7(10/11): [p. 1]. Oct-Nov 1931 (first page of issue comm. 100th anniversary of R's birth; portr.); Biol.-Dokum. 14: 7286; Bot. Centralbl. 78: 192. 1899 (gold medal), 93: 32. 1903 (d.); Bot. Not. 1904: 46; Brockhaus vollständiges Verzeichniss 1873-1905: 313-314 (bibl. travels on Russian-Persian border);Deut. entom. Z. 1903(1, 2): 7 (vol. 47) (obit.); Flora 50: 255 (to Kasbek), 358-360. 1867 (on the Caucasian Museum at Tiflis), 51: 250-253. 1868 (Taurus Exped.), 54: 201-202. 1871 (note on trip to Lenkoran, E. Caucasus), 379-381. 11 Nov 1871 (on his Ber. biol. geogr. Unters. Kaukasusländ.); Globus 25: 22-24. 1876 (biogr. notes, portr.); Grosse Brockhaus ed. 15. 15: 316. 1933 (d. 16 Mar 1903); Hedwigia 42: (161). 1903 (d.); Magy. bot. Lap. 2: 135-136. 1903; Nat. Nov. 26: 80. 1904 (d.; succeeded by A.N. Kasnakow); Wien entom. Zeit. 22: 108. 1903 (d. 16 Mar 1903).
Asmous, V.C., Chron. bot. 11(2): 107. 1947 (font. hist. bot. ross.).
Blasius, R., J. für Ornithol. 52: 1-49. 1904 (biogr., portr., bibl.).
Borodin, J., Trav. bot. Acad. Sci. St. Petersb. 4: 97-100. 1908 (itin., coll., eponymy).
Candolle, Alph,. de, Phytographie 442. 1880.
Daniel, K., Münchener Koleopterol. Z. 2: 93. 1904 (obit.).
Degen, A. von, Magy. bot. Lap. 2: 135-136. 1903 (obit.).
Drude, O. & O. Taschenberg, Leopoldina 39: 121-128, 135-146. 1903 (biogr., extensive bibl.).
Embacher, F., Lexikon Reisen 241. 1882.
Fedtschenko, B.A., Bot. Zhurn. 30: 40. 1945.
Gilbert, P., Comp. biogr. lit. deceased entom. 307. 1977.
Govorukhina, V.A., Izv. Akad. Nauk. Turkmen. SSR, Biol. 4: 88-91. 1970 (R and his research in Turkmenia).
Hahn, K.F. von, Samml. Kaukas. Mus. 6. 1912 (autobiography of Radde completed by Hahn).
Hantzsch, V., Biogr. Jahrb. deut. Nekrol. 8: 39-45. 1905 (see also col. 90*).
Hedge, I.C. & J.M. Lamond, Index coll. Edinb. herb. 121. 1970.
Hellwald, F. von, Centralasien ed. 2. 289-290. 1880.
Herder, F. von, Bot. Jahrb. 429-456. 1888 (collectors in his herb.); Bot. Centralbl. 58: 390, 391, 392. 1894 (coll. LE).
Jacobi, A., Naturw. Rundschau 18: 309-311. 1903 (obit.).
Kanitz, A., Bot. Zeit. 25: 300-303. 1867 (extensive review of Radde, Berichte über biologisch-geographischen Untersuchungen in den Kaukasusländern, Tiflis 1866.
Kneucker, A., Allg. bot. Z.: 156. 1903 (d.).
Köppen, Th., Journal des Ministeriums der Volksaufklärung, St. Petersburg 1903: 109-128, fide N. v. Adelung, Zool. Centralbl. 10: 829-831. 1903 (extract from Köppen's obit. of Radde).
Kukkonen, I., Herb. Christian Steven 82. 1971.
Lindemann, E. v., Bull. Soc. imp. Natural. Moscou 61(1): 58. 1866.

Lipschitz, S.J., Fl. URSS font. 223 [index]. 1975.
Mutshenko, P., Acta Horti bot. Univ. jurjev. 4: 209-215, portr. opp. p. 157. 1903 (biogr., portr.).
Nissen, C., Zool. Buchill. 327. 1969.
Poggendorff, J.C., Biogr.-lit. Handw.-Buch 3: 1082. 1898, 4(2): 1206. 1904, 7a (suppl.): 520 (bibl.).
Radde, G., Grundz. Pfl.-Verbr. Kaukasusländ. 17-18. 1899 (bibl. publ. on Caucasus); Die Sammlungen des Kaukasischen Museums 2: portr. 1901.
Regel, E., Gartenflora 8: 192. 1859 (will stay another year in Amur region),, 9: 65 (returned to St. Petersburg, 338 (in Crimea), 432 (plans to search for Vogel in Africa). 1860, 12: 207 (accepts position Tiflis), 303 (title of magister from Dorpat). 1863, 14: 386-387. 1865 (travel report Svanetia, Caucasus).
Schüz, E., Die Vogelwelt des südkaspischen Tieflandes 23-24. 1959.
Selsky, H., Bull. Soc. imp. Natural. 30(1): 296-300. 1857.
Winguth, E., Auslanddeutsche 14: 686-688. 1931 (important modern biogr. account based on original documentation).
Wittrock, V.B., Acta Horti Berg. 3(2): 175. 1903, 3(3): 189. 1905.

COMPOSITE WORKS: (1) For *Reisen in den Süden von Ost-Siberien* see F. Herder, TL-2/2671 (for vols. 3 and 4), and E. Regel (for vol. 1); vol. 2 was never issued.
(2) For *Plantae Raddeaneae monopetalae* see F. Herder, TL-2/2671.
(3) For *Nachträge zu den Plantae Raddeaneae* see J.H. Schultes (1880).

EPONYMY: *Raddetes* P.A. Karsten (1887). – The village of *Radde*, Amurskaya region, Siberia, 48.36.N/130.37.E. (Inf. N. Snigirevskaya), originally "Stanitsa-Raddevka".

8476. *Berichte über Reisen im Süden von Ost-Siberien*, im Auftrage der Kaiserlichen Russischen Geographischen Gesellschaft ausgeführt in den Jahren 1855 bis incl. 1859 ... St. Petersburg (Commissionäre der Kaiserlichen Akademie der Wissenschaften ...) 1861. Oct. (*Ber. Reisen Ost-Siberien*).
Publ.: 1861, p. [i*-ii*], [i]-xxii, [xxiii], [1]-719 [720, err.]. *Copies*: NY, U. – Beitr. Kenntn. Russ. Reich. vol. 23.
Atlas.: 1861, 11 col. pl., 2 maps.
Ref.: Regel, E., Gartenflora 11: 130-131. Mar 1862.

8477. *Berichte über die biologisch-geographischen Untersuchungen in den Kaukasusländern*, im Auftrage der Civil-Hauptverwaltung der Kaukasischen Statthalterschaft ausgeführt von Dr. Gustav Radde. Erster Jahrgang ... Tiflis (Buchdruckerei der Civil-Hauptverwaltung) 1866. Qu. (*Ber. Unters. Kaukasus-Länd.*).
Publ.: 1866 (p. x: Feb 1866), p. [i*-vi*], [i]-x, [1]-225, 3 maps (on one pl.), *9 pl. Copies*: LC, NSUB (inf. G. Wagenitz).
Ref.: Regel, E., Gartenflora 17: 156. 1868 (rev.).

8478. *Vier Vorträge über den Kaukasus*. Gotha (Justus Perthes) 1874. Qu.
Publ.: shortly after 30 Mar 1874 (date on cover "geschlossen"), p. [i-iv], [1]-71, 2 maps. *Copy*: BR. – Issued as Mitth. wicht. neue Erf. Geogr. A. Petermann, Ergänzungsheft Nr. 36.

8479. *Die Fauna und Flora des südwestlichen Caspi-Gebietes*. Wissenschaftliche Beiträge zu den Reisen an der Persisch-Russischen Grenze von Dr. Gustav Radde, ... Leipzig (F.A. Brockhaus) 1886. Oct. (*Fauna Fl. südw. Caspi-Geb.*).
Publ.: Mai 1886 (p. viii: Jan 1886; Bot. Zeit. 26 Feb 1886 (ann.), 25 Jun 1886; Bot. Centralbl. 7-11 Jun 1886), p. [i]-viii, [1], [1-425]. *Copies*: BR, M. – Is supplementary to the more general "*Reisen an der persisch-russischen Grenze*", Talysch und seine Bewohner", 1886.
Ref.: G.L., Bot. Monatsschr. 6: 106. 1888 (rev.).

8480. *Karabagh*. Bericht über die im Sommer 1890 im russischen Karabagh von Dr. Gustav Radde und Dr. Jean Valentin ausgeführte Reise. Von Gustav Radde ... Gotha (Justus Perthes) 1890. (*Karabagh*).

Publ.: 9 Dec 1890 (date on cover; Nat. Nov. Jan(1) 1891), p. [i-ii], [1]-56, 1 col. map. *Copies*: GOET, Liverpool City Libr. – Dr. A. Petermanns Mitteilungen aus Justus Perthes' Geographischer Anstalt. Ergänzungsheft Nr. 100.

8481. *Grundzüge der Pflanzenverbreitung in den Kaukasusländern* von der unteren Wolga über den Manytsch-Scheider bis zur Scheitelfläche Hocharmeniens ... Leipzig (Verlag von Wilhelm Engelmann) 1899. Oct. (*Grundz. Pfl.-Verbr. Kaukasusländ.*).
Publ.: Feb-Mar 1899 (Nat. Nov. Mar(2) 1899; ÖbZ Feb-Mar 1899; NY Mar 1899; Bot. Centralbl. 5 Apr 1899; Bot. Zeit. 1 Mai 1899), frontisp., p. [ii]-xii, [1]-500, *6 pl.*, 3 maps. *Copies*: BR, G, H, HH, M, NY, USDA. – See also TL-2/1712. Issued as vol. 3 of A. Engler & O. Drude, *Die Vegetation der Erde*.
Facsimile ed.: 1976, p. [i]-xii, [1]-500, frontisp. and *6 pl.*, 3 maps. *Copy*: FAS. – Vaduz (FL-9490) (A.R. Gantner Verlag K.-G.) 1976.
Ref.: Chodat, R., Bull. Herb. Boissier ser. 1. 7: 493-495. Jun 1899 (rev.).
Dammer, O., Bot. Jahrb. 27: 4-5. 1899.
Kusnezow, N., Bot. Centralbl. 85: 417-430. 1901.
Flahault, C., Bull. Soc. bot. France 46(bibl.): 471-472. 1899.

8482. *Die Sammlungen des Kaukasischen Museums* im Vereine mit Special-Gelehrten bearbeitet, und herausgegeben von Dr. Gustav Radde ... Band II. *Botanik* von Dr. Gustav Radde ... Tiflis (Typographie der Kanzelei des Landeschefs) 1901. Qu. (*Samml. Kaukas. Mus., Bot.*).
Publ.: late 1901 (p. v: Jun 1901, p. xi*: Sep 1901; ÖbZ Feb-Mai 1902; Bot. Zeit. 16 Mai 1902), p. [ii*-vi*], portr., [vii*-viii*], *pl. 1*, [ix*-xi*], [xiii*-xiv*], *pl. 2*, p. i-x, *pl. 3*, [xi], *pl. 4*, [xiii], 1-101, *pl. 5-14* (photos; with text), 1-200, *pl. 15-20* (photos), [11 portr.], 3 maps. *Copies*: FI, H, HH, NSW, US. – Title also in Russian.

Raddi, Giuseppe (1770-1829), Italian botanist at Firenze; on the Austrian Expedition to Brazil (1817-1818), collected also in Madeira (1817) and Egypt (1827-1829); died on the homeward journey from Egypt at Rhodos. (*Raddi*).

HERBARIUM and TYPES: FI and PI. – Other material at B (mainly lost), BP, BR, CGE, E, FH, H, M, PC, PH. – An important part of the original collections is at FI (Brazil, Madeira, Egypt); but PI (according to Savelli, 1918) received another major part through the activity of G. Savi. Flora (13: 286. 1830) reports that the Grand-Duke of Tuscany bought Raddi's private herbarium and donated it, with the Egyptian collections, to PI. – Letters to A.P. de Candolle at G.

BIBLIOGRAPHY and BIOGRAPHY: AG 3: 199; Backer p. 481; Barnhart 3: 122 (b. 9 Jul 1770, d. 6 Sep 1829); BM 4: 1637; Bossert p. 322; CSP 5: 72, 11: 91, 12: 595; Frank 3 (Anh.): 80; GR p. 520, *pl. [28]*; GR, cat. p. 70; Herder p. 472 [index]; Hortus 3: 1202; IF p. 726; IH 1 (ed. 2): 51, (ed. 3): 62, (ed. 6): 363, (ed. 7): 340, 2: (in press); Jackson p. 316, 371, 373; Kew 4: 264, 388; Langman p. 327, 337, 343, 478, 505, 543; LS 21654-21659; NI 1576, also 1: 156, 189; PR 7388-7391, 9033, ed. 1: 8288-8298; Rehder 5: 699; Saccardo 1: 134, 2: 88; Sayre, p. 34, 58, 64, 66; SBC p. 129; TL-1/1020-1022; Tucker 1: 575; Urban-Berl. p. 34; Zander ed. 10, p. 704, ed. 11, p. 803.

BIOFILE: Anon., Ann. Storia nat., Bologna 2: 279-282, 440-441. 1829 (obit., d. 8 Sep 1829, bibl.); Appleton's Cycl. Amer. biogr. 5: 157. 1888; Flora 12: 752. 1829, 13: 286. 1830 (d., herb. to PI).
Bargagli-Petrucci, G. Giuseppe Raddi naturalista e viaggiatore fiorentino. R. Ist. bot. Firenze no. 2. 1922.
Brunner, Flora 10: 585-586, 588-590. 1827 (on material at FI).
Candolle, Alph. de, Phytographie 442. 1880.
Candolle, A.P. de, Mém. souvenirs 546. 1862.
Chiovenda, E., Bull. Soc. bot. ital. 1923: 122-126 (on R. as a mycologist; relations with Persoon).
Francini Corti, E. et al., *in* Giuseppe Raddi, Flora brasiliana, memorie 1819-1828, Roma 1976, iv, 217 p., portr., *7 pl.* (biogr., bibl., reprint of publ.; biogr. note p. 11-25).

Hansen, A., Bocagiana 51: 10. 1980 (visited Madera 1819).
Jessen, K.F.W., Bot. Gegenw. Vorz. 376, 467. 1864.
Licopoli, G., Necrologia Giuseppe Raddi s.l.n.d., 3 p. (ex "Antologia" no. 106).
Marini-Bettalo, G.B., Introduzione, *in* G. Massa et al., Giuseppe Raddi 1976, p. 9-10.
Massa, G. et al., Giuseppe Raddi, Flora brasiliana. Memorie 1819-1828 edite in occasione del primo centenaria dell' emigrazione agricola italiana 1875-1975. Roma 1976 (portr., introd. by G. Massa, G.B. Marini-Bettalo and E. Francini Corti, bibl., reprints of Brasilian publications).
Occhiow, P., Leandra 6/7(7): 166-176. 1977 (biogr., Brazilian work, bibl.).
Parlatore, F., Coll. bot. Florence 6. 1874.
Pritzel, G.A., Linnaea 19(3): 460. 1846 (d. 8 Sep 1829).
Savelli, M., Bull. Soc. bot. ital. 1918: 3-8 (on the sale of the Raddi herb.).
Savi, G. et al., Alta memoria di Giuseppe Raddi, Firenze 1830, 33 p. (bibl., portr., list Egyptian colls.), bibliography reprinted in G. Massa et al. (1976), p. 27-30.
Sayre, G., Mem. New York Bot. Gard. 19(3): 383. 1975 (coll.); Bryologist 80: 516. 1979 (bryoph. types).
Steinberg, C., Webbia 34: 61. 1979.
Stieber, M.T., Montia 3(2): 124. 1979 (mss Chase on Raddi).
Tartini-Salvatici, F., Contin. Atti Accad. Georgof. Firenze, 8(1): 304-309. 1830 (Éloge).
Tassi, Fl., Bull. Lab. Orto Bot. Siena 7: 6, 28. 1905 (d. 8 Sep 1829).
Urban, I., Fl. bras. 1(1): 84-85. 1906 (itin., coll.).
Verdoorn, F., ed., Chron. bot. 4: 223, 224. 1938.
Wittrock, V.B., Acta Horti Berg. 3(2): 165 (d. 8 Sep 1829).

HANDWRITING: *in* G. Massa et al., G. Raddi, Flora brasiliana, memorie 1819-1828, Roma 1976, frontisp.; Candollea 32: 403-404. 1977.

EPONYMY: *Raddia* Bertoloni (1819); *Raddia* Mazziari (1834); *Raddia* A.P. de Candolle ex Miers (1872, *orth. var.* of *Raddisia* P. Leandro de Sacramento); *Raddiella* Swallen (1948); *Raddisia* P. Leandro de Sacramento (1821).

8483. *Delle specie nuove di funghi* ritrovate nei contorni di *Firenze* e non registrate nel Systema naturae di Linnaeo edizione xiii ... Modena (presso la Societa' Tipografica ...) 1806. Qu. (*Sp. nov. fung. Firenze*).
Publ.: 1806, p. [1]-20, *pl. 1-5. Copy*: FI. – Preprinted from Mem. Soc. ital. Modena 13: 345-362. 1807.

8484. *Nova species cryptogamarum* inventa in Florentinis suburbanitatibus, et descriptae in quadam Memoria inserta in volumine Academiae Senensis a Josepho Raddi florentino [Firenze] Anno 1808. Qu. (*Nov. sp. crypt.*).
Publ.: 1808; p. [1-2]. *Copy*: G. – Issued with the reprint of *Alc. sp. nuov. piant. critt.* 1808, q.v. Contains the latin diagnoses of the species described in that publication. See also PR ed. 1: 8288.

8485. *Di alcune specie nuove, e rare di piante crittogame* ritrovate nei contorni di Firenze ... [Siena 1808]. Qu.
Publ.: 1 Jul 1808 (journal issue so dated), p. 1-11, *pl. 1-4. Copies*: FI, G. – Reprinted and to be cited from Atti Accad. Sci. Siena 9: 230-240. *pl. 1-4.* 1808. See also Sayre p. 34.

8486. *Jungermanniografia etrusca* ... Modena (presso la Società tipografica) 1818. Qu. (*Jungermanniogr. etrusca*).
Publ.: 1818 (date t.p. preprint), p. [1]-45, *pl. 1-7* (partly col. copp. engr.). *Copies*: FH, FI, G, H, NY(2). – Preprinted from Mem. Mat. Fis. Soc. ital. Sci. Modena 18: 14-56, *pl. 1-7.* 1820. – The FH copy, originally owned by Charles Lyell, is a special author's copy with the plates on large paper and more coloured details than in the regular issue.
Reprint: 1841 (p. iv: 7 Dec 1840), p. [i]-iv, [1]-28, *pl. 1-7* (uncol. liths.). *Copies*: BR, FH, G(2), H, MO-Steere, NY, U-V, US. – *Editor*: C.G. Nees von Esenbeck (introd. p. [i]-iv); the Raddi text has the original page numbers (journal pagination) in the margin.

8487. *Novarum vel rariorum ex cryptogamia stirpium* in agro florentino collectarum decades due auctore Josepho Raddi Bononiae [Bologna] (typis Annesii de Nobilibus) 1818. Qu.
Publ.: Nov-Dec 1818 or early 1819 (fide Sayre p. 53, 58, q.v.), p. [i], 1-13, *pl. 1-2*. *Copy*: H. – Reprinted or preprinted from Opusc. sci. Bologna 2(6), 349-361, *pl. 15-16* (uncol.). 1818. – A reprint with original pagination is at FH.

8488. *Synopsis filicum brasiliensium* ... Bononiae [Bologna] (ex Typographia Annesii de Nobilibus) 1819. Qu. (*Syn. fil. bras.*).
Publ.: Dec 1819 (1819 on t.p. reprint; imprimatur journal issue 28 Dec 1819), p. [i], [1]-19, *pl. 1-2*. *Copies*: FI, G, M. – Reprinted or preprinted from Opusc. sci., Bologna 3: 279-297, *pl. 11-12* (=*1-2*). 1819. See also Flora 7: 304-317, 324-334, 337-349. 1824, contents reprinted by G. Kunze.

8489. *Di alcune specie nuove di rettili e piante brasiliane memoria* ... Modena (presso la Società Typographica) 1820. Qu. (*Alc. sp. rett. piant. bras.*).
Publ.: 1820 (t.p. reprint), p. [1]-39, *pl. 1-4* (col. copp.). *Copies*: FI, G(2), HH. – Preprinted from Mem. Mat. Fis. Soc. ital. Sci. Modena 18(2): 313-349. 1820 (last page has date 31 Dec 1820, hence fasc. to be dated Jan 1821 (see Sayre p. 66-67). The preprint is clearly dated 1820 and in absence of proof to the contrary this date has to be accepted.
Continuazione: 1822, p. [1]-18. *Copy*: G. – Preprinted from idem 19(1): 58-73. 1823.
Reprint: in G. Massa et al., G. Raddi, Fl. bras., Roma 1976, p. 101-136.

8490. *Quaranta piante nuove del Brasile* ... Modena (presso la Società tipografica) 1820. Qu. (*Quar. piant. nuov. Bras.*).
Publ.: 1820 (t.p. preprint), p. [1]-35, *1 pl.* (no. 5). *Copies*: FI, G, HH, US. – Preprinted from Mem. Mat. Fis. Soc. ital. Sci. Modena 18(2): 382-414. *pl. 5*. 1820 (cf. above, *Alc. sp. rett. piant. bras.*).
Reprint: in G. Massa et al., G. Raddi, Fl. bras. , Roma 1976, p. 69-94.

8491. *Breve osservazione sull' Isola di Madera* fatta nel Tragitto da Livorno a Rio Janeiro [sic] ... Firenze 1821. Oct.
Publ.: 1821, p. [1]-19, *1 pl. Copy*: FI. – Reprinted from "Antologia" fasc. 5 (n.v.).

8492. *Di alcune specie di Pero indiano memoria*... Bologna (Per le Stampe di Annesio Nobili ...) 1821. Qu. (*Alc. sp. Pero*).
Publ.: 1821 (t.p. first printing), p. [1]-7, *pl. 1*. *Copies*: FI, G, HH. – Deals with *Psidium*.
Reprint: in Opusc. sci., Bologna 4: 251-255. *pl. 7*. 1823.
Reprint (2): in G. Massa et al., G. Raddi, Fl. bras., Roma 1976, p. 95-100.

8493. *Crittogame brasiliane* ... Modena (dalla Società tipografica) 1822. Qu. (*Critt. bras.*).
Publ.: 1822 (t.p. preprint), p. [1]-33. *Copies*: G, US. – Preprinted from Mem. Mat. Fis. Soc. ital. Sci. Modena 19(1): 27-57. 1823. – On p. 37: *Frullanoides* Raddi.
Supplement: see Supplemento (1827), below.
Reprint: in G. Massa et al., G. Raddi, Fl. bras. Roma 1976, p. 101-124.

8494. Memoria di Giuseppe Raddi ... *sopra algune piante esculenti del Brasile* e specialmente di una nuova specie di Solano a frutto edule [s.l.n.d.]. Oct.
Publ.: 1822, p. [1]-15, *2 pl. Copy*: FI. – Reprinted from Opusc. Sci. Inghirami 3: 5-17. 1822 (fide CSP).
Reprint: in G. Massa et al., G. Raddi, Fl. bras., Roma 1976, p. 33-40.

8495. *Descrizione di una nuova Orchidea brasiliana* ... Modena (dalla Tipografia Camerale) 1823. Qu.
Publ.: 1823, p. [1]-6, *pl. 1*. *Copies*: G, L. – Reprinted and to be cited from Mem. Mat. Fis. Soc. ital. Sci., Modena 19: 219-222. *pl. 6* [sic] 1823. – Tearsheets of journal publ.: BR, FI.
Reprint: in G. Massa et al., G. Raddi, Fl. bras., Roma 1976, p. 137-140.

8496. *Agrostografia brasiliensis* sive enumeratio plantarum ad familias naturales graminum

et ciperoidarum spectantium, quas in Brasilia collegit et descripsit Josephus Raddius [Lucca (dalla Tipografia ducale ... 1823]. Qu. (*Agrostogr. bras.*).
Publ.: 1823 (fide imprint; Raddi sent a copy to W.J. Hooker on 21 Aug 1824), p. [1]-58, [59, imprint], *pl. 1* (uncol. copp. G. Nerici). *Copies*: FI, G, NY, US. – Reprinted or preprinted from Atti r. Accad. Lucchese Sci. 1823: 331-383, *1 pl.*

8497. *Plantarum brasiliensium nova genera* et species novae vel minus cognitae ... pars i. (Filices). Florentiae [Firenze] (ex Tipographia Aloisii Pezzati) 1825. Fol. †. (*Pl. bras. nov. gen.*).
Publ.: 1825, p. [i-iv], [1]-101, [103, coloph.], *pl. 1-4, 4bis, 5-8, 8bis, 9-22, 22bis, 23-31, 31bis, 32-36, 36bis, 37-54, 54bis, 55-61, 61bis, 62-66, 66bis, 67-68, 68bis, 69, 69bis, 70, 70bis, 71, 71bis, 72, 72bis, 73-84,* (uncol. liths.). *Copies*: BR, FI, G, HH, MICH, MO, NY, PH, U, USDA. – PR (Icon. bot. index p. xxiv, 1855: "97 Schlechte von Dilettanten gezeichnete Lithographien, sign. 1-84, mit 119 Arten"), artists listed by NI 1576.
Facsimile repr.: offered by subscription, but not yet published by A. Asher & Co., Books on botany, 1974.
Ref.: Eschweiler, F.G., Linnaea 2(1): 115-120. 1827 (rev.). Pars ii: never published, see Martini-Bettolo (1976) p. 10. The edition of G. Massa et al. (1976) made up a 1976 title page presenting their compilation of Raddi reprints as part 2 of the *Pl. bras. nova gen.*

8498. *Supplemento alla memoria* di Giuseppe Raddi intitolata *Crittogame brasiliane* inserita nel volumine xix. delle Memorie della Società italiana delle science ... Modena (presso la Tipografia camerale) 1827. Qu. (*Suppl. Critt. bras.*).
Publ.: 1827 (t.p. preprint), p. [1]-14, [1-2, supplemento secondo], *pl. 1-7* (uncol. copp.). *Copies*: G, H. – Preprinted from Mem. Mat. Fis. Soc. ital. Sci., Modena 20: 43-54, *pl. 1-7*. 1827.
Reprint: in G. Massa et al., G. Raddi, Fl. bras., Roma 1976, p. 157-167.

8499. *Melastome brasiliane* ... Modena (presso la Tipografia camerale) 1828. Qu. (*Melast. bras.*).
Publ.: 1828 (t.p. preprint), p. [1]-64, *pl. 1-6*. *Copy*: FI. – Preprinted from Mem. Mat. Fis. Soc. Ital. Sci., Modena 20: 111-172, *pl. 8-13* (= *1-6*). 1829.
Reprint: in G. Massa et al., G. Raddi, F. bras., Roma 1976, p. 169-220.

Raddin, Charles Salisbury (1863-1930), American botanist and mineralogist; MS Northwestern Univ. 1888; vice-president and trustee of the Chicago natural history survey. (*Raddin*).

HERBARIUM and TYPES: Unknown.

BIBLIOGRAPHY and BIOGRAPHY: Barnhart 3: 122 (b. 29 Jan 1863); BL 1: 175, 314; Rehder 5: 699; TL-2/2756.

BIOFILE: Cattell, J.M., ed., American men of science ed. 1: 261. 1906, ed. 2: 383. 1910, ed. 3: 558. 1921, ed. 4: 794. 1927.

COMPOSITE WORKS: With William Kerr Higley (1860-1908), *The flora of Cook County*, 1891, see TL-2/2756, *addition*: publ. Jul-Sep 1891 (Bot. Centralbl. 21 Oct 1891).

8500. *Catalogue of the phaenogamous plants of Evanston and vicinity* for 1883. Evanston (Robt. Vandercook, ...) 1883 [1884]. Oct. (*Cat. phaenog. pl. Evanston*).
Publ.: 1884 (t.p. 1883 but p. [i], preface, dated 1884), p. [i], [1]-26. *Copies*: HH, NY.

Radermacher, Jacobus Cornelius Matthaeus (1741-1783), Dutch botanist; to the Dutch East Indies 1758 as general administrator, ultimately "Raad van Indië" (State Council); founder of the Bataviaasch Genootschap van Kunsten en Wetenschappen. (*Radermacher*).

HERBARIUM and TYPES: L, SBT (Java coll. 1776-1779).

BIBLIOGRAPHY and BIOGRAPHY: Backer 481; Barnhart 3: 122 (b. 30 Mar 1741, d. 24 Dec 1783); CSP 5: 73; Dryander 3: 181; JW 4: 402; Lasègue p. 493, 506; NNBW 2: 1153; PR 7391; Rehder 5: 699; SK 1: 424 (many data!), 4: 21, 38, lxxxix, xcix, 5: cccxiii; SO 102, 104; TL-2/4829.

BIOFILE: Sirks, M.J., Indisch Natuuronderzoek 75-80. 1915.
Steenis, C.G.G.J. & M.J. van, Regn. veg. 71: 371. 1970 (Noroña & R.).

EPONYMY: *Rademachia* Thunberg (1776, orth. var. of *Radermachia* Thunberg); *Radermachera* Zollinger et Moritzi (1855); *Radermachia* Thunberg (1776).

8501. *Naamlijst der planten*, die gevonden worden op het eiland *Java*. Met de beschryving van eenige nieuwe geslagten en soorten ... te Batavia (Gedruckt in d'E: Compagnies Boek-drukkerij, bij Egbert Heemen) 1780-1782, 3 parts. Qu. (*Naaml. pl. Java*).
1: 1780. 1-60.
2: 1781, p. [i], [1]-67, [1]-88 (p. 71 as "59"), 1-40.
3: 1782, p. [i, iii], [1]-102, 1-42, 1-70.
Copies: H (2, 3), HU(1).

Radian, Simion Stefan (1871-1958), Rumanian botanist at the University of Bucarest; director of the Bucarest University Botanical Garden 1937. (*Radian*).

HERBARIUM and TYPES: BUC (destroyed 1944, fide Stefureac 1958).

BIBLIOGRAPHY and BIOGRAPHY: Barnhart 3: 122; Hirsch p. 241; SBC p. 129.

BIOFILE: Gager, C.S., Brooklyn Bot. Gard. Rec. 27(3): 321. 1938.
Sayre, G., Bryologist 80: 516. 1977 (coll. BUC destroyed).
Schuster, R.M., Hep. Anthoc. N. Amer. 1: 89. 1966.
Stefuriac, T.I., Rev. bryol. lichénol. 27: 231-232, *pl. 4*. 1958 (obit., bibl., portr., b. 12 Apr 1871, d. 18 Apr 1858).
Verdoorn, F., ed., Chron. bot. 3: 227. 1937, 5: 509. 1939.

8502. *Sur le Bucegia* nouveau genre d'hépatiques à thalle ... Bucuresci (Institutul de Arte grafice și editură "Minerva" ...) 1903. Oct.
Publ.: Sep 1903 (p. 7: 18 Apr 1903; copy G dedicated to Stephani dated 13 Sep), p. [1]-7. *Copy*: G. – Reprinted and to be cited from Bull. Herb. Inst. Bucarest 3/4: 1-7. 1903.

Radius, Justus Wilhelm Martin (1797-1884), German (Saxonian) physician and botanist; practicing physician and professor of medicine at the University of Leipzig, director of the "medicinische Gesellschaft" in Leipzig. (*Radius*).

HERBARIUM and TYPES: Some material at CGE.

BIBLIOGRAPHY and BIOGRAPHY: Barnhart 3: 123 (b. 14 Nov 1797, d. 7 Mar 1884); BM 4: 1637; Herder p. 318; Hortus 3: 1202; Kew 4: 388; PR 7393, ed. 1: 8300-8301; Rehder 2: 657; Tucker 1: 575.

BIOFILE: Anon., Bonplandia 3: 33 1855, 6: 50. 1858 (cogn. Leopoldina Ludwigius).

EPONYMY: *Radiusa* H.G.L. Reichenbach (1828) was probably named for this author.

8503. *De Pyrola et Chimophila* specimen primum botanicum. Dissertatio quam amplissimi philosophorum ordinis auctoritate ...h.c. die x. mens. mart. a.r.s. mdcccxxi illustris ictorum ordinis venia in auditorio juridico publice defendet Justus Radius ... Lipsiae [Leipzig] 1821. Qu. (*Pyrola & Chimophila*).
Specimen primum, thesis issue, 10 Mar 1821, p. [i-iii], [1]-39, [40, theses], *pl. 1-5* (uncol. liths.). *Copies*: HH, NY, USDA.
Specimen primum, commercial issue: 1821, p. [i], [1]-39, *pl. 1-5* (id.). *Copies*: G(2), HH, M,

NY, WU. – "Dissertatio de Pyrola et Chimophila ..." Lipsiae (apud Hartmannum) 1821. Qu.
Specimen secundum medicum: 11 Jul 1829, p. [i], [1]-33. *Copy*: HH. – "... Dissertatio qua ad audiendam orationem professionis medicae extraordinariae adeunda causa h. et l. const. die xi. mensis Julii a. mdcccxxix habendam observatissime invitat Justus Radius ..." Lipsiae (literis Hirschfeldi).
Specimen secundum, commercial issue: 1829, p. [i], [1]-33. *Copies*: NY, USDA, WU. – Lipsiae (apud Leopoldum Voss) 1829. Qu.
Ref.: Nees von Esenbeck, C.G., Flora 5(2): 12-16. 1822.
Sprengel, K., Neue Entd. 3: 294-296. 1822.

Radlkofer, Ludwig Adolph Timotheus (1829-1927), German (Bavarian) botanist; Dr. med. München 1854; student of M. Schleiden at Jena 1854-1855; Dr. phil. Jena 1855; habil. München 1856; extraord. professor of botany and "Adjunkt" at the Botanical garden and herbarium 1859; regular professor ib. 1863-1913; director of the Botanical museum and state herbarium 1908-1927. (*Radlk.*).

HERBARIUM and TYPES: M (rd 1927).

BIBLIOGRAPHY and BIOGRAPHY: Backer p. 481; Barnhart 3: 123; Biol.-Dokum. 14: 7283-7284; BJI 1: 47, 2: 142; BM 4: 1637, 8: 1043; Bossert p. 322; CSP 5: 73, 6: 746, 8: 686-687, 11: 92, 12: 596, 18: 22; Frank 3 (Anh.): 81; GR p. 122; Herder p. 472 [index]; Hirsch p. 241; Hortus 3: 1202; Jackson p. 76, 98, 100, 104, 144; JW 1: 452; Kew 4: 389-390; Langman p. 605-606; LS 1099, 21661; Moebius p. 189, 209, 346; MW p. 398-399; Morren ed. 2, p. 9, ed. 10, p. 18; Plesch 371; PR 2129-2140, 7394-7397, 8605; Rehder 5: 699-700; Saccardo 1: 134; SO 3486; TL-2/1592, 1713, 2570; see Herzog, T.C.J., Mez, C.C.; Tucker 1: 575-576; Zander ed. 10, p. 704, ed. 11, p. 803.

BIOFILE: Anon., Bonplandia 7: 301. 1859 (succeeds Sendtner at M., in charge of bot. coll.); Bot. Centralbl. 111: 48. 1909 (Geh. Hofrat); Bot. Zeit. 16: 320. 1858 (habil. München), 17: 344. 1859 (in charge of bot. coll. at München), 22: 148 ("137"). 1864 (ordinary prof. bot.), 68: 24. 1910 (80th birthday); Flora 42: 480. 1859 (extraordinary prof. bot. München), 47: 283. 1864 (ordinary prof.); Gartenflora 8: 288. 1859 (app. München), 13: 126. 1864 (regular prof. bot.); Hedwigia 49: (51). 1909 (Geh. Hofrat), 51: (170). 1911 (hon. memb. Pharm. Soc.), 54: (189). 1914 (retirement), 67: (163). 1927 (d.).; Mykol. Centralbl. 3: 205. 1913 (retirement); Nat. Nov. 22: 245. 1900 (member Akad. Wiss. Berlin); Österr. bot. Z. 9: 338. 1859 (extraord. prof. bot. München), 14: 59. 1864 (ord. prof.), 63: 480. 1913 (retirement), 76: 88. 1927 (d.); S.B. k. preuss. Akad. Wiss. 1910(1): 16-17. 1910 "Adresse an Hrn. Ludwig Radlkofer zum 80. Geburtstage am 19. Dezember 1909").
Beck, G., Veg.-Verh. Illyr. Länder 10, 41. 1901 (coll. algae Firenze 1857).
Bergdolt, E., Karl von Goebel ed. 2. 261 [index]. 1941 (e.g. on splitting up bot. inst. between Goebel and Radlkofer; "ich [G] habe wenigstens die besten Räume bekommen ...", some interesting "petites histoires").
Briquet, J., C.R. Soc. Phys. Hist. nat. Genève 47: 5-8. 1930 (obit.; died in the same rooom in which he was born, after 97 yrs; vieil ami de Genève: worked at G almost every year for some time).
Freund, H. and A. Berg, Gesch. Mikroskopie 293, 295. 1963.
Fritsch, F.E., J. Bot. 65: 176-177. 1927 (obit. "vigorous protagonist of the use of anatomical characters in taxonomy; Foreign member Linn. Soc. 1897); Bull. misc. Inf. Kew 1927(5): 220-21(id.).
Hertel, H., Mitt. Bot. München 16: 420. 1980 (Radlkofer herb. rd. by M in 1927).
Herzog, Th., Ber. deut. bot. Ges. 45: (79)-(88). 1927 (obit., portr., signature, b. 19 Dec 1829, d. 11 Feb 1927).
Hoehne, F.L., Jard. bot. São Paulo 165. 1941 (portr.).
Jackson, B.D., Proc. Linn. Soc. 1926/1927: 93-94 (obit.), Nature 119: 432. 1927 (obit.).
Jessen, K.F.W., Bot. Gegenw. Vorz. 446. 1884.
Kneucker, A., Allg. bot. Z. 28/29: 55. 1924, 33: 64. 1928 (d.).
Mägdefrau, K., Gesch. Bot. 119, 146, 158, 252, 270. 1973.

Merrill, E.D., B.P. Bishop Mus. Bull. 144: 153-154. 1937 (Polynes. bibl.); Contr. natl. Herb. U.S. 30(1): 247-248. 1947 (Pacific bibl.).
Pringsheim, E.G., Julius Sachs 125. 1932.
Radlkofer, L., Verh. bot. Ver. Brandenburg 63: 96-97. 1922 (letter to Society as reply to congratulations).
Sachs, J., Gesch. bot. 340, 378, 470. 1875; Hist. bot. 314, 362, 450. 1892.
Ulbrich, E., Verh. bot. Ver. Brandenburg 70: 26-30. 1928 (obit.)
Urban, I., Symb. ant. 1: 132-133. 1898; Fl. bras. 1(1): 194-195. 1906.
Verdoorn, F., ed., Chron. bot. 1: 36. 1935, 5: 464. 1939.
Wittrock, V.B., Acta Horti Berg. 3(3): 148-149, *pl. 128*. 1905 (portr.).

COMPOSITE WORKS: (1) *Sapindaceae*, in E.P., *Nat. Pflanzenfam.* 3(5): 277-320. 2 Apr 1895, 321-366. 7 Mai 1895, 460-462, *Nachtr*. 2-4: 227-229. Oct 1897 (with E. Gilg).
(2) *Sapindaceae*, in Engler, *Pflanzenreich* IV. 165 (Heft 98a-h), p. 1-1539, 8 Dec 1831 – 27 Feb 1934 (for precise dates see TL-2/1: 793); edited (after Radlkofer's death) by Th. Herzog. The former, at the age 97, had not been in a hurry to finish the manuscript, according to Herzog because he was convinced that he would live until 100).
(3) *Sapindaceae*, in Martius, *Fl. bras*. 13(3): 225-658. *pl. 58-123*. 1892-1900 (fasc. 103, 1 Jul 1892; 122: 1 Sep 1897, 124: 1 Apr 1900).
(4) With O. Sendtner (q.v.) and W. Gümbel, *Die Vegetations-Verhältnisse des Bayerischen Waldes*, München 1860.
(5) *Sapindaceae*, in I. Urban, *Symb. ant*. 9(1): 75-76. 1 Jan 1923.
(6) *Sapindaceae*, in E. Warming, *Symbolae* 37: 240-245. 21 Apr 1891.

HANDWRITING: Webbia 32(1): 14. 1977.

THESIS: *Die Kälte als Heilmittel*. Inaugural-Dissertation von Ludwig Radlkofer. München (Druck von Carl Robert Schurich) 1855. Oct., p. [1]-20. *Copy*: M.

EPONYMY: *Radlkofera* Gilg (1897); *Radlkoferotoma* O. Kuntze (1891); *Sinoradlkofera* F.G. Meyer (1977). *Note*: *Radlkoferella* Pierre (1890) is likely also named for this botanist.

8504. *Die Befruchtung der Phanerogamen. Ein Beitrag zur Entscheidung des darüber bestehenden Streites* ... Inaugural-Dissertation Leipzig (Verlag von Wilhelm Engelmann) 1856. Qu. (*Befrucht. Phan.*)
Thesis issue: Jan-Mai 1856 (p. vi: 21 Oct 1855; Flora 21 Mar 1856; BSbF Apr 1856; Bot. Zeit. 13 Jun 1856; Bonplandia 15 Jun 1856), p. [i]-vi, [vii], [1]-36, *pl. 1-3* (uncol. liths auct.). *Copy*: USDA.
Commercial issue: Jan-Mai 1856 (id.), p. [i]-vi, [vii], [1]-36, *pl. 1-3* (partly col. in copy G). Copies: G, HH, M, NY, PH. – Only difference with thesis issue: the word "Inaugural-Dissertation" is absent from the t.p.
French: Ann. Sci. nat., Bot. ser. 4. 5: 220-230. 1856.
Ref.: Hofmeister, W., Bonplandia 4: 180-185. 1856.
Schlechtendal, D.F.L. von, Bot. Zeit. 14: 424-425. 13 Jun 1856 (rev.).
S., Flora 39: 172-173. 21 Mar 1856 (rev.).

8505. *Der Befruchtungsprocess im Pflanzenreiche und sein Verhältniss zu dem im Thierreiche* ... Leipzig (Verlag von Wilhelm Engelmann) 1957. Oct. (*Befrucht.-Proc. Pfl.-Reich.*).
Publ.: Dec 1856 (t.p. 1857; p. viii: 21 Feb 1856; Flora rd. Dec 1856; id. Soc. Sci. nat. Cherbourg; BSbF Oct-Dec 1857), p. [i]-x, [1]-97, [98-102, charts, Nachtr.]. *Copies*: G, H, HH, M(2), NY, PH.
English: Ann. Nat. Hist. 20: 241-262, 344-365, 439-459. 1857.

8406. *Ueber wahre Parthenogenesis bei Pflanzen* ... [Z. wiss. Zool. 8(4) 1857] Oct.
Publ.: 26 Feb 1857 (date on cover journal issue), p. [1]-8. *Copy*: M (Inf. I. Haesler). – Reprinted from Z. wiss. Zool. 8(4): 458-465. 1857. See also Ann. nat. Hist. 20: 204-210. 1857 and Ann. Sci. nat. Bot. ser. 4. 7: 247-252. 1857.

8507. *Ueber das Verhältniss der Parthenogenesis zu den anderen Fortpflanzungsarten. Eine Berichtigung der Einsprüche Prof. A. Braun's gegen meine Anschauungen über die Fort-

pflanzungsverhältnisse der Gewächse ... Leipzig (Verlag von Wilhelm Engelmann) 1858. Oct. (*Verh. Parthenog. Fortpfl.-Arten*).
Publ.: 1858 (p. iv: Jul 1858; BSbF Oct-Dec 1859; Flora rd. Oct 1858), p. [i]-iv, [5]-74, [75, err.]. *Copies*: H, HH, M, PH, USDA.

8508. *Ueber Pausandra*, ein neues Euphorbiaceen-genus von L. Radlkofer. Tab. II. [Flora 1870]. Oct.
Publ.: 22 Mar 1870, p. [1]-15, *pl. 2*. *Copies*: G, M, NY. – Reprinted and to be cited from Flora 53: 81-95. *pl. 2* (uncol. lith. auct.). 1870.

8509. *Conspectus sectionum specierumque generis Serjaniae* ... e monographia generis seorsim editus. Monachii [München] (Typographia Academica F. Straub) mense majo 1874. Qu. (*Consp. sect. sp. Serjan.*).
Publ.: 1-26 Mai 1874 (t.p.; Bot. Zeit. 28 Mai 1875), p. [i], [1]-17 [also: 81-97]. *Copies*: BR, G, HH, M, NY. – Preprinted from *Monogr. Serjania* 81-97. 1875.

8510. *Anhang zur Monographie der Gattung Serjania* ... München (Akademische Buchdruckerei von F. Staub) December 1874. Qu. (*Anh. Monogr. Serjania*).
Publ.: Dec 1874, p. [1]-26. *Copies*: G, H, HH, M, NY. – Preprinted from Monogr. Serjania p. 353-378.

8511. Serjania Sapindacearum genus monographice descriptum. *Monographie* der Sapindaceen-Gattung *Serjania* ... München (Verlag der K.B. Akademie ... Akademische Buchdruckerei von F. Straub) 1875. Qu. (*Monogr. Serjania*).
Publ.: Apr-Jun 1875 (p. xiii: Jul 1874; KNAW Jul-Sep 1875; Bot. Zeit. 28 Mai 1875; Flora 1 Jul 1875; J. Bot. Aug 1875), p. [i]-xviii, [1]-392. *Copies*: BR, FI, G, HH, KNAW, L, M, MICH, NSW, NY, US, USDA.
Ergänzungen: see below (1886); preprints see above (1874).

8512. *Sopra i vari tipi delle anomalie dei tronchi nelle Sapindacee* ... Firenze (Tipografia di M. Ricci ...) 1875. Oct.
Publ.: 1875, p. [1]-8. *Copy*: M (Inf. I. Haesler). – Reprinted and to be cited from Atti Congr. bot. int. Firenze 1874: 60-65. 1875.

8513. *Ueber die Sapindaceen Holländisch-Indiens* ... [Extrait: des "Actes du Congrès international de botanistes etc." tenu à Amsterdam, en 1877], [1879]. Oct.
Publ.: Jan-Feb 1879 (Flora rd. Feb-Apr 1879; Bot. Zeit. 21 Mar 1879), p. [1]-103. *Copies*: BR, G(2), HH(2), M, MICH, NY, U, USDA. – Reprinted and to be cited from Act. Congr. int. Bot., Amsterdam 1877: 70-133, 216-254.

8414. *Ueber Sapindus* und damit in Zusammenhang stehende Pflanzen [Separatabdruck aus den Sitzungsberichten der k. bayer. Akademie der Wissenschaften, Math.-phys. Classe, 1878]. Oct.
Publ.: Jul-Dec 1878 (mss submitted 1 Jun 1878; J. Bot. Apr 1879; Nat. Nov. Mar(1) 1879; Bot. Zeit. 14 Feb 1879), p. [221]-408. *Copies*: FI, G, M, MO, US. – Reprinted with new imprint and title on p. [221] and to be cited from S.B. k. bayer. Akad. Wiss., math.-phys. Cl. 1878. 221-408.

8515. *Ueber Cupania* und damit verwandte Pflanzen [Separatabdruck aus den Sitzungsberichten der k. bayer. Akademie der Wissenschaften, Math.-phys. Classe, 1879]. Oct.
Publ.: Jul-Dec 1879 or Jan-Feb 1880 (mss submitted 5 Jul 1879; BSbF Jan-Mar 1880; Nat. Nov. Mar(2) 1880; Leopoldina rd. 15 Mar-15 Apr 1880), p. [457]-678. *Copies*: H, HH, M, MO, MICH. – Reprinted with new title and imprint on p. [457] and to be cited from S.B. k. bayer. Akd. Wiss., math.-phys. Cl. 1879: 457-678.
Ref.: Dr., Bot. Zeit. 39: 36-38. 1881.

8516. *Ueber die Methoden in der botanischen Systematik, insbesondere die anatomische Methode.* Festrede zur Vorfeier des Allerhöchsten Geburts- und Namenfestes seiner Majestät des Königs Ludwig II. gehalten in der öffentlichen Sitzung der k. Akademie der Wissens-

chaften zu München am 25 Juli 1883 ... München (Im Verlage der k. b. Akademie) 1883. Qu. (*Meth. bot. syst.*).
Publ.: Aug-Oct 1883 (25 Jul 1883: date of lecture; Bot. Zeit. 28 Dec 1883; Bot. Centralbl. 12-16 Nov 1883; Nat. Nov. Dec(1) 1883), p. [1]-64. *Copies*: G, HH, M, PH, US.
Ref.: Pax, F., Bot. Jahrb. 5: 27-28. 31 Dec 1883 (rev.).
Kraus, Bot. Centralbl. 17: 234-236. 1884.

8517. *Ueber die systematischen Werth der Pollenbeschaffenheit bei den Acanthaceen* ... München (Akademische Buchdruckerei von F. Straub) 1883. Oct.
Publ.: Mai-Jun 1883 (read 5 Mai 1883; Bot. Zeit. 7 Sep (journal), 28 Sep 1883; BSbF Sep-Oct 1883; Nat. Nov. Aug(1) 1883), p. [i, on recto of 256], 256-314. *Copies*: FI, G, U. – Reprinted with new t.p. and orig. pagination and to be cited from S.B. k. bayer. Akad. Wiss., math.-phys. Cl. 13(2): 256-314. 1883.

8518. *Ueber einige Capparis-Arten* [Separat-Abdruck aus den Sitzungsberichten der k. b. Akademie der Wissenschaften, Bd. xiv, Heft 1. 1884]. Oct.
Publ.: 1884, p. [101]-182. *Copy*: M (inf. I. Haesler). – Reprinted with new imprint from S.B. k. bayer. Akad. Wiss., Math.-phys. Cl. 14(1): [101]-182. 1884. – *Continuation*: ib. 17(3): [365]-422. 1887 (also reprinted, copy: M).

8519. *Über die Zurückführung von Forchhammeria zur Familie der Capparideen* ... München (Akademische Buchdruckerei von F. Straub) 1884. Oct.
Publ.: 1884 (Nat. Nov.(1) 1884; Bot. Zeit. 25 Jul 1884; Bot. Centralbl. 1-5 Sep 1884; Bot. Jahrb. 24 Oct 1884), p. [i, recto of 58], [58]-100. *Copies*: FI, G. – Reprinted with new t.p. and to be cited from S.B. k. bayer. Akad. Wiss., math.-phys. Cl. 14(1): 58-100. 1884.

8520. *Ueber einige Sapotaceen* [Separat-Abdruck aus den Sitzungsberichten der math.-phys. Classe der k. b. Akad. d. Wiss. Bd. xiv, Heft iii] 1884. Oct.
Publ.: Aug-Oct 1884 (Bot. Zeit. 24 Dec 1884; Nat. Nov. Nov(2) 1884), p. [397]-486. *Copies*: FI, G. – Reprinted with new imprint from S.B. k. bayer. Akad. Wiss., math.-phys. Cl. 14(3): [397]-486.

8521. *Ueber Tetraplacus, eine neue Scrophularineengattung aus Brasilien* ... München (Akademische Buchdruckerei von F. Straub) 1885. Oct.
Publ.: Jun-Oct 1885 (mss. submitted 2 Mai 1885), p. [i, recto of 258], 258-275. *Copy*: FI. – Reprinted with new t.p. from S.B. k. bayer. Akad. Wiss., math.-phys. Cl. 15(2): 258-275. 1885.
Ref.: Koehne, E., Bot. Zeit. 44: 709-710. 15 Oct 1886 (rev.).

8522. Monographiae generis Serjaniae supplementum. *Ergänzungen zur Monographie der Sapindaceen-Gattung Serjania* ... München (Verlag der k. Akademie ...) 1886. Qu. (*Ergänz. Monogr. Serjania*).
Publ.: Aug-Dec 1886 or Jun-Apr 1887 (iv: Jul 1886; Nat. Nov. Mai 1887; G copy rd. Jul 1887; Bot. Centralbl. 9-13 Mai 1887; Bot. Zeit. 29 Apr 1887), p. [i]-x, [1]-195, *pl. 1-9* (*1-8* uncol. liths. Seboth and auct., *9* map by author, col.). *Copies*: BR, FI, G, L, M, MICH. – Also issued as Abh. k. bayer. Akad. Wiss., math.-phys. Cl. 16(1): 1-195. 1886.
Conspectus: Nov 1886, p. [1]-19. *Copy*: M (inf. I. Haesler). – Preprinted or reprinted from *Ergänz. Monogr. Serjania* p. [161]-179. – *Conspectus sectionum specierumque generis Serjaniae auctus* ... Monachii [München] (Typographia academica F. Straub) mense Novembri 1886. Qu. – See also Benecke, Bot. Centralbl. 30: 309-313. 1887.

8523. *Ueber die Arbeit und das Wirken der Pflanze*. Rede an die Studierenden beim Antritte des Rectorates der Ludwig-Maximilians-Universität gehalten am 20. November 1886 ... München (Kgl. Hof– und Universitäts-Buchdruckerei von Dr. C. Wolf & Sohn) 1886. Qu. (*Arb. Wirk. Pfl.*).
Publ.: 20 Nov-31 Dec 1886, p. [1]-24. *Copy*: M (inf. I. Haesler).

8524. *Über die Entwickelung des Pflanzensystems* und den Anteil der Ludwig-Maximilians-Universität an ihr Rede an die Studierenden beim Stiftungsfeste der Ludwig-Maximilians-Universität, gehalten am 25 Juni 1887 ... [Ill. Monatsh. Gartenb. Oct 1887]. Qu.
Publ.: Oct 1887 (in journal), p. [1]-13. *Copy*: M. – Reprinted from Ill. Monatsh. Ges.-Int. Gartenb. 6: 310-322. Oct 1887.

8525. *Ueber die Versetzung der Gattung Dobinea* von den Acerineen zu den Anacardiaceen ... München (Druck der Akademischen Buchdruckerei von F. Straub) 1888. Oct.
Publ.: Dec 1888 (read 8 Nov 1888; reprint dated 1888), p. [i], 385-395. *Copy*: G. – Preprinted or reprinted (with special t.p.) and to be cited from S.B. math.-phys. Cl. k. bayer. Akad. Wiss. 18(3): 385-395. 1888.

8526. *Ueber die Versetzung der Gattung Henoonia* von den Sapotaceen zu den Solanaceen ... München (Druck der Akademischen Buchdruckerei von F. Straub) 1888. Oct.
Publ.: Dec 1888 (read 1 Dec 1888; reprint dated 1888), p. [i], 405-421. *Copy*: G. – Preprinted or reprinted (with special t.p.) and to be cited from S.B. math.-phys. Cl. k. bayer. Akad. Wiss. 18(3): 405-421. 1888.

8527. Th. Durand, *Index generum phanerogamorum.* L. Radlkofer *Sapotaceae.* Bruxellis (ex typis Becquart-Arien ...) 1888. Oct.
Publ.: 1888 (Bot. Centralbl. 27 Nov 1888), p. [1]-6, [1, addenda]. *Copy*: BR. – Reprinted and to be cited from Th. Durand, Index gen. phan. [71]-82, 501. 1888.

8528. *Ueber Nothochilus,* eine neue Scrophularineen-Gattung aus Brasilien, nebst einem Anhange: Ueber zwei neue Touroulia-Arten ... München (Druck der Akademischen Buchdruckerei von F. Straub) 1889. Oct.
Publ.: Jun-Sep 1889 (mss. submitted 1 Jun 1889; Nat. Nov. Oct(2) 1889), p. [i, cover], 213-220. *Copy*: FI. – Reprinted and to be cited from S.B. k. bayer. Akad. Wiss., math.-phys. Cl. 19(2): 213-220. 1889.

8529. *Zur Klärung von Theophrasta* und der Theophrasteen unter Uebertragung dahin gerechneter Pflanzen zu den Sapotaceen und Solanaceen ... München (Druck der Akademischen Buchdruckerei von F. Straub) 1889. Oct.
Publ.: 1889 (Bot. Zeit. 29 Nov 1889; Nat. Nov. Oct(2) 1889), p. [i, cover with text], 221-281. *Copy*: FI. – Reprinted with new t.p. (on cover) and to be cited from S.B. k. bayer. Akad. Wiss., math.-phys. 19(2): 221-281. 1889.

8530. *Conspectus tribuum generumque Sapindacearum* ... Monachii [München] (Typographia Academica F. Straub) m. Junio 1890.
Publ.: Jun 1890, p. [1]-24. *Copies*: BR, HH, USDA. – Reprinted and to be cited from S.B. k. bayer. Akad. Wiss., math.-phys. Cl. 20: 275-296. 1890. This is a reprint from "Ueber die Gliederung ...", see below.

8531. *Ueber die Gliederung der Familie der Sapindaceen* ... München (Druck der Akademischen Buchdruckerei von F. Straub) 1890. Oct.
Publ.: Jun 1890, p. [i, cover with t.p. text], 105-379. *Copies*: BR, FI, G, HH, M. – Reprinted and to be cited from S.B. k. bayer. Akad. Wiss., math.-phys. Cl. 20(1): 105-379. 1890.

8532. *Drei neue Serjania-Arten* ... [Bull. Herb. Boiss. 1, 1893]. Oct.
Publ.: Sep 1893, p. [1]-5 and [463]-468. *Copy*: U. – Issued with double pagination in and to be cited from Bull. herb. Boissier 1: [463]-468. Sep 1893.

8533. Paullinia sapindacearum genus monographice descriptum. *Monographie* der Sapindaceen-Gattung *Paullinia* ... München (Verlag der k. Akademie ...) 1895. Qu. (*Monogr. Paullinia*).
Publ.: Dec 1895-Jun 1896 (p. [3]: Nov 1895; t.p. 1895; Nat. Nov. Aug(1) 1896; Bot. Zeit. 16 Aug 1896; ÖbZ Jul 1886; Bot. Centralbl. 29 Jul 1896), p. [1]-315, *1 pl. C*Copies*: BR, G, HH, M, MO, US. – Preprinted or reprinted from Abh. k. bayer. Akad. Wiss., math.-phys. Cl. 19(1): [67]-381. *1 pl.* 1895 or 1896.

8534. *Conspectus sectionum specierumque generis Paulliniae* ... Monachii (Typographia Academica F. Straub ...) mense decembri 1895. Qu. (*Consp. sect. sp. Paullin.*).
Publ.: Dec 1895, p. [1]-15. *Copies*: BR, M. – Preprinted or reprinted from Abh. k. bayer. Akad. Wiss., math.-phys. Cl. 19(1): [107]-121. 1895 or 1896.

8535. *Eine zweite Valenzuela* ... (Bull. Herb. Boiss. ser. 2. 2. 1902). Oct.
Publ.: 5 Dec 1902, p. [1]-3 and 994-996. *Copy*: U. – Issued with double pagination in Bull. Herb. Boissier ser. 2. 2: 994-996. 5 Dec 1902.

8536. *Guareae species duae costaricenses* ... [Bull. Herb. Boissier ser. 2. 5. 1905]. Oct.
Publ.: 31 Jan 1905, p. [1]-3 and 191-193. *Copy*: U. – Issued with double pagination and to be cited from Bull. Herb. Boissier ser. 2. 5: 191-193. 31 Jan 1905.

8537. *Sapindaceae costaricenses* determinatae novaeque descriptae a L. Radlkofer (locos natales etc. adjecit H. Pittier) [Extrait du Bulletin de l'Herbier Boissier, 2me série. – Tome v (1905) No. 4]. Oct.
Co-author: Henri Pittier. (1857-1950).
Publ.: 31 Mar 1905 (in journal), p. [1]-10. *Copies*: U, US. – Reprinted and to be cited from Bull. Herb. Boiss. ser. 2. 5(4): 319-328. 1905.

8538. *Über die Gattung Allophylus* und die Ordnung ihrer Arten ... München (Verlag der K.B. Akademie der Wissenschaften ...) 1909. Oct.
Publ.: 1909, p. [i, cover with text t.p.], 201-240. *Copy*: BR. – Reprinted and to be cited from S.B. k. bayer. Akad. Wiss., math.-phys. Cl. 38(2): 201-240. 1908.

8539. *New and noteworthy Hawaiian plants* ... Honolulu (Hawaiian Gazette Co., Ltd.) 1911. Oct.
Co-author: Joseph Francis Charles Rock (1881-1962).
Publ.: Sep 1911, p. [1]-14, *pl. 1-6* (photos). *Copies*: BR, FI, G, NY, USDA. – Issued and to be cited as Bot. Bull. Terr. Hawaii, Board Agr. For. 1: 1-14. *pl. 1-6*. 1911.

8540. *New Sapindaceae from Panama and Costa Rica* ... City of Washington (published by the Smithsonian Institution) 1914. Oct. (*New Sapind. Panama*).
Publ.: 9 Feb 1914, p. [i-ii], [1]-8. *Copy*: U.

Radloff, Fredric Wilhelm (1766-1838), Finnish physician and botanist; studied at Uppsala under C.P. Thunberg; Dr. med. Ups. 1788; professor's title 1805; provincial physician in Åland 1790-1799;, id. at Norrtälje 1799-1806; first secretary of the Finnish economic association 1806-1813, med. adjunct and botanical demonstrator at Åbo University 1805-1809; secretary Cabinet Council, Finland 1809-1813; again in Norrtälje 1814-1838. (*Radloff*).

HERBARIUM and TYPES: H and UPS (Thunberg herb. has plants from Åbo collected by R.).

BIBLIOGRAPHY and BIOGRAPHY: Barnhart 3: 123; BM 4: 1638; KR p. 581-582 (b. 19 Sep 1766, d. 18 Apr 1838, bibl.); PR 9278, ed. 1: 10251; Rehder 2: 221; Saelan p. 403-404; TL-2/2610; TR 1107.

BIOFILE: Verdoorn, F., ed., Chron. bot. 3: 24. 1937.

COMPOSITE WORKS: (1) See C.N. Hellenius, TL-2/2610, author of *Diss. hort. abo.* nos. 4-5, 1807.
(2) See C.P. Thunberg, *De Myristica* 1788, of which Radloff was the author.
(3) Radloff was author of *Dissertatio academica, qua nova Ammeos species proponitur*. Praes. Johan Fredric Wallenius. Resp. Josef August Hoeckert. Abo 1810, [i], 13 p., Qu. – See Saelan p. 404-405.

Räsänen, Veli Johannes Paavo Bartholomeus (1888-1953), Finnish botanist; Mag. phil. Helsinki 1914; Dr. phil. ib. 1927; teacher at Kemi 1913-1914, at Lapua 1918-1921;

lecturer at the agricultural college of Kurkijoki 1921-1940; from 1940-1953 id. at the college of animal husbandry at Kuopio; student of E.A. Vainio. (*Räs.*).

HERBARIUM and TYPES: H; other material at BM, BP, H, KUO, LD, M, NY, TUR, UPS.
Exsiccatae: (*1*)*Lichenes Fenniae exsiccati*, fasc. 1-20, nos. 1-1001, Helsinki 1935-1946, sets at BG, BM, BP, C, GB, H, K, KUO, LD, LE, M, NY, O, PC, PRC, TNS, TU, TUR, UC, PS, US, W. – For schedae see below. (2) *Lichenotheca Fennica*, 1946-1960 with (Hakulinen), fasc. 1-32, nos. 1-750 curavit Räsänen; continued by Rainar Hakulinen (1918-x), fasc. 33-52, nos. 751-1300, sets at B, BM, C, COLO, FH, G, GB, GDOR, H, K, KUO, L, LD, LIL, LISU, LWU, M, MICH, MSK, NSW, O, OULU, PC, S, TELA, TMP, TRTC, TU, TUR, UC, UPS, UPSV, US. – Schedae were issued: *Lichenotheca Fennica a Museo Kuopioënsi edita*, Kuopio 1946-1960, with an *Index systematicus lichenum ad fasciculos omnes* (i-lii), Kuopio 1961. For further details see Sayre (1969) and Collander et al. (Kukkonen & Ahti 1973). – 118 letters by Räsänen to K. Linkola are at H-UB.

BIBLIOGRAPHY and BIOGRAPHY: Barnhart 3: 123; Biol.-Dokum. 14: 7288; BL 2: 74, 706; GR 612, *pl.* [*41*], GR, cat. p. 70; Hawksworth p. 185; Hirsch p. 242; IH 1 (ed. 6): 363, (ed. 7): in press, 2: (in press); Kew 4: 386; MW p. 399, suppl. p. 289; SBC p. 129.

BIOFILE: Gyelnik, V. Köfarago, Ann. hist.-nat. Mus. natl. Hung., Bot. 31: 56. 1938.
Fagerström, L., Mem. Soc. Fauna Fl. fenn. 23: 161-163. 1947 (Lich. Fenn. exs. summary).
Häyrén, E., Mem. Soc. Fauna Fl. fenn. 30: 85-86. 1955.
Hakulinen, R., Luonnon Tutkija 57: 84-85. 1953 (obit., portr.); Kuopion Luonnon Yst. Yhd. julk. B 3(2): 1-30. 1956 (bibl., portr., signature); id. 3(4): 1-31. 1961 (list of new taxa and names).
Huuskonen, A.J., Kuopion Luonnon Yst. Yhd. julk. B 2(5): 1-7. 1949 (portr.).
Kujala, V., Luonnon Tutkija 52: 155-156. 1948 (biogr. notes; 60th birthday, portr.).
Kukkonen, I. & T. Ahti, in R. Collander et al., Acta Soc. Fauna Fl. fenn. 81: 645. 1973.
Laundon, J.R., Lichenologist 11: 16. 1979.
Sato, M., Lichenol. Misc. 9: 33. 1953, J. Jap. bot. 29: 32. 1954.
Sayre, G., Mem. New York Bot. Gard. 19(1): 156-157. 1969.
Trass, H.H., Bot. Zhurn. 54(6): 956-957. 1969 (biogr. sketch).
Verdoorn, F., ed., Chron. bot. 2: 118. 1936, 3: 53. 1937.

HANDWRITING: Kuopion Luonnon Yst. Yhd. julk. B 3(2): 2. 1956.

NOTE: (1) See also his *Das System der Flechten*, Acta bot. fenn. 33. 1943, p. [1]-82, *copies*: Almborn, H, and *Suomen jäkäläkasvio*, Kuopio (Kuopion Luonnon Yst. Yhd. julk.) 5, 1951, p. [1]-158, *pl. 1-16*. *Copies*: Almborn, FH, H.
(2) Information on Räsänen was kindly provided by T. Ahti, O. Almborn and P. Isoviita.

EPONYMY: *Raesaeneniolichen* Tomaselli et Ciferri (1952); *Raesaeneniomyces* Ciferri et Tomaselli (1953).

8541. *Die Flechtenflora des Gebiets Ostrobottnia borealis . . .* Helsinki 1926. Oct.
Publ.: 1926, cover t.p., p. [268]-349. *Copies*: Almborn, H, U. – Reprinted and to be cited from Ann. Soc. zool.-bot. fenn. Vanamo 3(8): 268-349. 1926.

8542. *Über Flechtenstandorte und Flechtenvegetation im westlichen Nordfinnland*. Helsinki 1927. Oct.
Publ.: shortly before 12 Mar 1927 (as thesis; inf. P. Isoviita), p. [i-iv], [1]-202. *Copies*: Almborn, BR, FH, G, H, USDA. – Issued and to be cited as Ann. Soc. zool.-bot. fenn. Vanamo 7(1), 1927.

8543. *Die Flechten Estlands* mit einer Bestimmungstabelle der wichtigsten nord– und mitteleuropäischen Flechtenarten und Varietäten I. . . . Helsinki (Suomalainen Tiedeakatemia) 1931. Oct. †.

Publ.: 1931, p. [1]-162, [1, index]. *Copies*: Almborn, FH, FI, H. – Issued and to be cited as Ann. Acad. Sci. Fenn. A. 34(4). 1931.

8544. *Zur Kenntniss der Flechtenflora Feuerlands* sowie der Prov. de Magallanes, Prov. de Chiloë und Prov. de Ñuble in Chile auf Grund des von H. Roivainen gesammelten Materiales dargestellt von Veli Räsänen ... Helsinki 1932. Oct.
Publ.: 19 Oct 1932 (in journal volume; reprints probably published previously, inf. P. Isoviita), p. [i]-vi, [1]-65, [66-68]. *Copy*: H. – Reprinted from Ann. bot. Soc. zool.-bot. Fenn. Vanamo 2(1): [i]-vi, [1]-65, [66-68]. 1932.

8545. *Lichenes Fenniae exsiccati* a Museo botanico Universitatis Helsingiensis editi ... Helsinki 1935-1946.
Schaedae [sic] *ad fasciculos 1-3* (nr. 1-150), quos curavit Veli Räsänen, 1935, p. [i], 1-50, err. slip.
Schedae 4-7, (nr. 151-350): 1936, p. [i], 1-68.
Schedae 8-11, (nr. 351-550): 1939, p. [i], 1-67, [68].
Schedae 12-14, (nr. 551-700): 1940, p. [i], [1]-50.
Schedae 15-17, (nr. 701-850): 1943, p. [i], [1]-50.
Schedae 18-20, (nr. 851-1000) ... Index systematicus lichenum ad fasciculos omnes (i-xx), 1946, p. [1]-63, [64, map].
Copies: Almborn, FH, FI, H, U.

8546. *Die Flechtenflora der nördlichen Küstengegend* am Laatokka-See ... Helsinki 1939. Oct.
Publ.: 1939 (p. viii: Jul 1938), p. [i]-viii, [1]-240. *Copies*: Almborn, G, H, Stevenson. – Issued and to be cited as Ann. bot. Soc. zool.-bot. fenn. Vanamo 12(1), 1939.

Raeuschel [Räuschel], **Ernst Adolf** (*fl.* 1772-1797), German botanist. (*Raeusch.*).

HERBARIUM and TYPES: Unknown.

BIBLIOGRAPHY and BIOGRAPHY: BM 4: 1638; GR p. 38; GR. cat. p. 70; Herder p. 141; Hortus 3: 1202; Jackson p. 14; LS 21663; MW p. 399; PR 7398, ed. 1: 8303; Rehder 1: 76; SO 506; Tucker 1: 522, 576; Zander ed. 10, p. 704, ed. 11, p. 803.

BIOFILE: Howard, R.A., J. Arnold Arb. 56: 240-242. 1975 (on the status of nomina nuda in ed. 3 of the *Nom. bot.*).
Stieber, M.T., Huntia 3(2): 124. 1979 (mss. coll. Agnes Chase).

8547. *Nomenclator botanicus* omnes plantas ab illustr. Carolo a Linné descriptas aliisque botanicis temporis recentioris detectas enumerans. Editio tertia. Curavit Ernestus Adolphus Raeuschel ... Lipsiae [Leizig] (apud Johann Gottlob Feind) 1797. Oct. (*Nomencl. bot.*).
Publ.: 1797 (p. xii: spring 1797), p. [i]-xii, [1]-414. *Copies*: G, HH, MO, PH, U, USDA; IDC 6434. – For editions 1 and 2 see under A.J. Retzius (1772, 1782), see also C. Linnaeus, no. 4817.

Raffles, Sir Thomas Stamford Bingley (1781-1826), British colonial administrator and naturalist; with the East India company in London 1795-1805; at Penang 1805-1811, lieutenant-governor of the Netherlands East Indies 1811-1816; in London 1816-1817; governor of Bencoolen (Bengkulo), Sumatra 1818-1824. (*Raffles*).

HERBARIUM and TYPES: Some material at G and K; collections partly destroyed on his 1824 voyage to England.

BIBLIOGRAPHY and BIOGRAPHY: AG 4: 689; Backer p. 481-483; Barnhart 3: 123; BB p. 251-252; BM 4: 1639; Dawson p. 692; Desmond p. 509 (b. 5 Jul 1781, d. 4 Jul 1826); DNB 47: 161-165; DSB 11: 261-262 (q.v. for further sec. references and bibl.); Frank 3 (Anh.): 81; Lasègue p. 580 [index]; Rehder 3: 360; SK 1: 424-425 (many details), cxxiii, 4: c, clx, 5: ccxxxvii, cccxiii, 8: lxxix; TL-2/see W. Jack.

BIOFILE: Allen, D.E., The naturalist in Britain 104. 1976.
Allibone, S.A., Crit. dict. Engl. lit. 1723-1724. 1878.
Anon., Bull. misc. Inf. Kew 1910: 154 (founded Singapore bot. gard. 1822); Flora 4: 309-319. 1821 (on his History of Java).
Bastin, J.S., The native policies of Sir Stamford Raffles in Java and Sumatra. An economic interpretation. Oxford. 1957, xx, 163 p.; Sir Thomas Stamford Raffles, Liverpool, published by The Ocean Steam Ship Company Limited, 1969, 22 p. and 13 pl. (Raffles-Minto manuscript collection); Straits Times Annual 1971: 58-63. (popular appraisal of R. as a naturalist).
Boulger, D.C., The life of Sir Stamfard Raffles, London, 1899, xv, 403 p., 14 pl. (biogr., portr.); new ed. C. Knight 1973, 207 p.
Bridson, G.D.R. et al., Nat. hist. mss. res. Brit. Isl. 430 [index]. 1980.
Clair, Colin, Sir Thomas Raffles, founder of Singapore, Watsford, Bruce and Gawthorn Ltd., 1963, 104 p.
Collis, M., Raffles, London, 1976, 228 p., 36 pl. (biogr.).
Cook, J.A. Bethune, Sunny Singapore, London, 1907, p. 30-38.
Coupland, R., Raffles 1781-1826, Oxford Univ. Press 1926, 134 p., (biogr., portr.).
Hahn, E., Raffles of Singapore, a biography, Garden City, New York 1946, xv, 587 p., 10 pl.
Jackson, B.D., Bull. misc. Inf. Kew 1901: 54 (material at K).
Jessen, K.F.W., Bot. Gegenw. Vorz. 469. 1864.
Kalkman, C. & P. Smit, eds., Blumea 25(1): 6. 1979.
Miller, H.S., Taxon 19: 537. 1970 (sale Raffles specimens from Lambert herb.).
Poggendorf, J.C., Biogr.-lit. Handw.-Buch 2: 559. 1863.
Raffles, S., Memoir of the life and public services of Sir Thomas Stamford Raffles. London 1830, xvi, 723, 100 p. (portr. biogr.).
Sirks, M.J., Ind. natuurond. 82-84. 1915.
Smith, E., Life Joseph Banks 273, 282, 296. 1911.
Stearn, W.T., Nat. Hist. Mus. S. Kensington 284-285. 1981.
Steenis, C.G. van, Fl. males. 13: 548. 1957, 26: 1996. 1972, 35: 3737. 1982.
Steenis-Kruseman, M.J. van, *in* Forbes R.J. ed., Mart. van Marum 3: 159. 1971.
Swainson, W., Taxidermy 296-300. 1840 (bibl.).
Wurtzburg, C.E., Raffles of the Eastern Isles, London, Hodder and Stoughton 1954, 788 p. (biogr., portr.).

HANDWRITING: Bastin, J., Sir Thomas Stamford Raffles 1969, *pl. 11-12*.

EPONYMY: *Rafflesia* R. Brown (1821).

Rafinesque [-Schmaltz], **Constantin Samuel** (1783-1840), Constantinople-born (French father, Greek mother of German extraction) naturalist and author; in Italy 1795-1802, in the United States 1802-1804(?), in Sicily 1804-1815; settling in the United States in 1815. (*Raf.*).

HERBARIUM and TYPES: P-DU. – Other Rafinesque material in DWC, FI, G, G-DC, LE, NAP, NY, P, PH, PI, W, WIS, WS. – For a recent account of the fate of the Rafinesque herbarium see R.L.S. Stuckey (1971) (see also e.g. Pennell (1945, 1949); the major part of the herbarium was thrown away by Elias Durand, but the P-DU herbarium should always be checked for types where necessary. DWC holds some authentic specimens from the *Autikon botanikon*; another important set is the one at WIS; over fifty types are at PH (Stuckey 1971). – Manuscripts and letters (S.S. Haldeman donation 1849) at PH; other letters by Rafinesque at G and HU; further archival material at SIA.

BIBLIOGRAPHY and BIOGRAPHY: Backer p. 482, 518; Barnhart 3: 123 (b. 22 Oct 1783, d. 18 Sep 1840); BM 4: 1639, 8: 1044-1045; Bossert p. 322; Clokie p. 229; CSP 5: 75-76, 12: 596; DAB 15: 322-324; Dawson p. 692-693; DSB 11: 262-264 (by J. Ewan); Frank 3 (Anh.): 81; Hawksworth p. 185; Herder p. 472 [index]; Hortus 3: 1202; IF suppl. 4: 336-337; IH 1 (ed. 2): 128, (ed. 3): 98, 164, (ed. 6): 363, (ed. 7): 340, 2: in press; Jackson p. 115, 116, 121, 321, 323, 355, 360, 363, 449; Kelly p. 185, 256; Kew 4: 390-392; Langman p. 606; Lasègue p. 313, 346, 462, 505; Lenley p. 467 (Torrey corresp.); LS

RAFINESQUE, *biofile*

21667-21672, 41619; ME 1: 222, 223, 3: 642, 738; Merrill p. 101; Moebius p. 55; MW p. 399, suppl. p. 289; NAF 28B(2): 351-352, ser. 2. 2: 168, ser. 2. 4: 45, ser. 2. 5: 251-252, ser. 2. 7: 44; NI 1577-1581, also 1: 235; Nordstedt p. 26; PR 1995, 7399-7402, ed. 1: 2087, 3850, 8315-8330; Quenstedt p. 350; Rehder 5: 700-701; RS p. 135-136; Saccardo 1: 134-135, 2: 88-89, cron. p. xxx; SBC p. 129; Stevenson p. 1255; TL-1/1023-1036; TL-2/253, 5872, see indexes; Tucker 1: 13 (biographies), 577-580, 634, 662; Zander ed. 10, p. 704, ed. 11, p. 802.

BIOFILE: Allibone, S.A., Crit. dict. Engl. lit. 2: 1724. 1891.
Anon., Appleton's Cycl. Amer. biogr. 5: 159. 1888; Bonplandia 3: 33. 1855 (member Leopoldina, cognomen: Catesbaeus); Evening Sun 9 Oct 1919 (marker placed over grave Raf.); Missouri Bot. Gard. Bull. 15(10): 164-171. 1927 (cites Audubon's description of Rafinesque, 1832); Natl. Cycl. Amer. Biogr. 8: 472-473. 1898; Taxon 31: 388. 1981 (Rafinesque symposium); The New York Sun, 14 Nov 1935 (Audubon's relations to Rafinesque).
Audubon, J.J., The eccentric naturalist. Ornithological biography 1: 455-460. 1832 (on R's visit to Audubon's home); Delineations of American scenery and characters (with an introduction by F.H. Herrick), New York, 1926, p. 97-104 ("The eccentric naturalist"; chapter on Rafinesque).
Baehni, Ch., *in* J.F. Leroy, Bot. franç. Amér. nord 137-147. 1952 (relations Raf. with A.P. de Candolle; material in G-DC).
Barber, L., The heyday of natural history 30, 61-63, 194. 1980.
Barkley, A.H., Ann. med. Hist. 10: 66-76. 1928 (general biogr. notes).
Barnhart, J.H., Torreya 7: 177-182. 1907 (dates *New Flora* and *Sylva telluriana*).; J. New York Bot. Gard. 10: 177-190. 1909, 36: 26-27. 1935 (brief biogr. sketch), 37: 247-248. 1936 (Kentucky friends); Torreya 9: 252. 1909 (brief biogr. sketch); Mem. Torrey Bot. Cl. 298. 1921 (mentioned in Schweinitz-Torrey corr.); Bartonia 9: 41. 1926 (brief biogr. sketch).
Bebb, M.S., Bot. Gaz. 8: 191-192. 1883 (sketch of Raf. from Audubon's diary).
Beck, G.R. von, Bot. Centralbl. 34: 150. 1888 (Raf. herb. mat. in W), in K. Fritsch et al., Bot. Anst. Wiens 75: 1894 (id.).
Berman, A., Bull. Hist. Med. 30(1): 23-31. 1956 ("a striving for scientific respectability"; use of "Medical botany" by medical practitioner).
Betts, E.M., Proc. Amer. philos. Soc. 87: 368-380. 1944 (corr. Raf.-Th. Jefferson).
Boewe, C., Audubon Mag. 59(4): 166-1669. 1957 (general biogr. notes); Proc. Amer. philos. Soc. 102(6): 590-595. 1958 (the manuscripts of C.S. Rafinesque); Books and libraries at the Univ. of Kansas 21: 1-3. 1959 (Rafinesque at Lawrence; on Fitzpatrick coll.).; Filson Club hist. Quart. 35: 28-32. 1961 (Raf. and Dr. Short), 54: 37-49. 1980 (editing Raf. holographs: the case of the Short letters); Names 10(1): 58-60. 1962 (on Mt. Rafinesque, N.Y.); Hist. bot. W. Virginia 6. 1965; ed. Rafinesque: a sketch of his life ... enlarged. 1982 (see e.g. Taxon 31: 800-801. 1982).
Bridson, G.D.R. et al., Nat. his. mss. res. Brit. Isl. 430 [index]. 1980.
Britten, J., J. Bot. 38: 224-225. 1900 (reprint of paper Raf. in *Gard. Mag.* 8: 245-248. 1832 and his *Herbarium Rafinesquianum*).
Brown, L.A., ed., Transylv. Coll. Bull. 15(7): 1-108. 1942 (Raf. memorial papers; each paper separately listed).
Brummitt, R.K., Taxon 21: 303-307. 1972 (Raf. and *Galax*).
Bureau, É., Bull. Soc. bot. France 22: lxxxvii. 1875 (set of Rafinesque plants at Angers, Jard. Bot., now ANGUC).
Call, R.E., The life and writings of Rafinesque, Louisville, Kentucky, 1895, xii, 227 p., reprinted in K.B. Sterling, ed., Rafinesque, autobiography and lives, New York 1978 (biogr., bibl., portr.; the handwriting sample opp. p. 48 reproduces a letter by R. to A.P. de Candolle of 1 Jun 1838 in its entirety); review: Anon., Bot. Gaz. 20: 120-121. 1895.
Camp, W.H., Castanea 6: 80-83. 1941 (identif. of *Adnaria odorata* Raf. and *Arbutus obtusifolius* Raf.); methodology in exegesis Raf.).
Carpenter, M.M., Amer. Midl. Natural. 33(1): 82. 1935 (lists biographies).
Chase, A., Bartonia 17: 40-45. 1936 (no Raf. grass types in P-DU).
Core, E.L., The botanical exploration of the Southern Appalachians p. 25-29. 1971, *in*

P.C. Holt., ed., Distr. hist. biota south. Appalachians, 2, Flora, Blacksburg, Virginia 1971.
Coulter, J.M., Bot. Gaz. 8: 149-152. 1883 (biogr. sketch).
Croizat, L., (under the pseudonym Henricus Quatre), Rafinesque, a concrete case. Forli, Italia 1948, 18 p., (reprinted from Arch. bot. 24(3/4). 1948; proposal to Stockholm Congress to outlaw Rafinesque names not listed in Index kewensis up to suppl. 8. 1930; tries to prove that Raf. was "an arrant lunatic").
Cunningham, G.H., The Gastromycetes of Australia and New Zealand 1944 (treats taxa in R's 1808 *Prospectus*, N.Y. Med. Repos. 2. 5: 355. 1808.
Darlington, W., Reliq. Baldw. 15, 11, 280, 282, 300. 1843; Memorials John Bartram and Humphrey Marshall 25-26. 1849.
Dawson, W.R., The Smith papers 77. 1934 (calendar letters R. to J.E. Smith).
Dayton, W.A., J. Forestry 43(5): 381-382. 1945 (review of *A life of travels*, 1836, repr. Chron. bot. 8: 291-360.
Desvaux, N.A., J. Bot. 4: 176-177, 268. 1814.
Dupre, H., Rafinesque in Lexington 1819-1926, Lexington 1945, xiii, 99 p. (biogr., portr.)..
Dupree, A.H., Asa Gray 1810-1888, p. 100, 279, 400, 437. 1959.
Eifert, V.S., Tall trees and far horizons 177-178. 1965.
Elliott, S., Sketch bot. S. Carol. vi. 1824.
Ewan, J., Rocky Mt. natural. 277. 1950.; *in* J.F. Leroy, ed., Bot. franç. Amér. nord 38. 1957 (lists studies); Rhodora 64: 188. 1962 (criticism of Rafinesque entry in H.B. Humphrey, Makers N. Amer. bot. 1961; J. Arnold Arb. 46(4): 438. 1965 (in letters C.S. Sargent to R.S. Cocks); Editor's introduction to facsimile reprint of C.S. Rafinesque, Florula ludoviciana, New York 1967, p. i-xl; Southw. Louisiana J. 7: 16-17. 1967 (note on *Fl. Ludov.*); *in* C.C.Gillespie, Dict. sci. biogr. 11: 262-264. 1975 (biogr., bibl.); J. Soc. Bibl. nat. Hist. 8(3): 235-243. 1977 (Josiah Hale, Raf.'s pupil).
Ewan, J. et al., Short hist. bot. U.S. 8, 39, 40, 47, 115, 116, 148. 1969 (general).
Fernald, M.L., Rhodora 33: 209-211. 1931, 34: 21-28. 1932, 43: 481-483. 1941, 46: 1-21, 32-57, 310-311, 377-386, 496-497, 48: 5-13. 1946 (misc. studies on Raf. plant names).
Fitzpatrick, F.J., Ann. Iowa 7(3): 196-224. 1905; Iowa Natural. 1(1): 10-21, 38-39. 1905 (add. to Bibl. of Rafinesque); Rafinesque, a sketch of his life with bibliography. Des Moines, Iowa, 1911, 241 p. (main bibliography of 939 items; biogr., portr., Bibliotheca Rafinesquiana (bibl. of secondary lit. on p. 233-239, 134 items); IDC 8314, revised edition by C. Boewe 1982 (see below under Bibliotheca Rafinesquiana).
Fox, W.J., Science ser. 2. 12: 211-215. 1900 (on *Western Minerva*), 14: 498. 1901 (*Fl. lexingt.*, *Amer. florist.*), 14: 617. 1901 (Fl. lexingt. by C.W. Short).
Friesner, R.C., Butler Univ. Bot. Stud. 11: 1-4. 1953 (Raf. on tax. Indiana pl.).
Fryxell, P.A., Taxon 18: 400-413. 1969 (W. Ind. Gossypium sp. of Rafinesque).
Geiser, S.W., Natural. frontier 16, 79, 248-252, 263. 1948.
Genzmer, G.H., Dict. Amer. biogr. 15: 322-324. 1935.
Gerard, W.R., Bull. Torrey bot. Cl. 7(3): 29-30. 1880 (Phalloids) 12: 37-38. (on Reliquiae Rafinesquianae, 29 unpubl. plates by Raf. at NY).
Gilbert, P., Comp. biogr. lit. deceased entom. 308. 1977 (lists biographies).
Gleason, H.A., Rhodora 50: 53-55. 1948 (*Blephilia*).
Goodwin, G.H., Jr., Syst. Zool. 9(1): 1-35. 1960 (unrecorded papers of Rafinesque).
Gordon, Bartonia 22: 8. 1943 (50-60 herb. specimens at DWC of which 20-30 of Autikon bot.).
Graustein, J.E., Thomas Nuttall 476 [index]. 1967.
Gray, A., Amer. J. Sci. 40(2): 221-241. 1841 (notice of the botanical writings of Raf.); reprinted 1841, p. [1]-21. *Copy*: G.
Greene, E.L., Pittonia 2(7): 25-31. 1889 (*Neilliae*), 2(9): 115. 1890 (Fraser's cat.), 2(8): 58-65. 1890 (*Ranunculi*), 2(9): 120-133. 1890 (various genera), 2(10): 173-184. 1891 (misc. nom. notes), 185-195. 1891 (revertible generic names), 2(11): 219-222 (*Spiraea*), 3(16): 186-188. 1897 (nom.), 3(17): 207-212. 1897 (nom.).
Haag, H.B., Transylvania Coll. Bull. 15(7): 91-96. 1942 (extended version in Science 94: 403-406 1941; Raf. and medicinal plants).
Hagen, V.W. von, Natural History 50: 296-303. 1947 ("Rafinesque the unnatural naturalist).

Jaffe, B., Men of Science in America, New York, Simon and Schuster, 1944, p. 104-129 (chapter on Raf., "American science ventures out across new frontiers").
Jessen, K.F.W., Bot. Gegenw. Vorz. 2166. 1864.
Junk, Cat. 214: no. 249. 1979 (descr. copy Medical flora, quotations from DSB).
Kastner, J., A species of eternity, New York, Alfred A. Knopf, 1977 (chapter 11, The gullible genius, p. 240-253, deals with Rafinesque; portr.).
Keep, J., West Amer. Sci. 2: 99-102. 1846 (Audubon anecdote; brief biogr. sketch).
Kuntze, O., Rev. gen. pl. 1: cxxxix-cxli. 1891 (lists 40 bot. titles).
La Roque, A., Sterkiana 16: 1-52. 1964 (a Rafinesque portrait and a translation of Raf., *Monogr. Coquilles bivalves*).
Leroy, J.F. et al., Bot. franç. Amér. Nord 355 [index, 11 entries].
Leonard, W.E., Bull. Minnesota Acad. nat. Sci. 3: 31-35. 1889.
Little, E.L., Jr., Proc. Biol. Soc. Washington 56: 57-65. 1943 (note on unpubl. *Florula columbica*; mss of 1804 not extant).
Lubrecht, H., Early Amer. bot. works (Stechert Hafner n. 353). 40. 1967 (on Florula ludoviciana and its facsimile ed. of 1967).
McVaugh, R., Wrightia 1: 13-52. 1945 (accepts *Triodanis* Raf.).
Mears, J.A., Proc. Amer. philos. Soc. 122(3): 167. 1978 (Raf. specimens in Henry Muhlenberg herbarium).
Meehan, T., Bot. Gaz. 8: 177-178. 1883 (apology for Raf.); Rafinesque. Career of a naturalist of more than ordinary brilliancy. s.l.n.d., 4 p. (*copy* at G; almost certainly identical with FIT 108, text from Philadelphia Public Ledger, February 19, 1891.
Meijer, W., Castanea 38: 261-265. 1973 (contr. Raf. bot. expl. Kentucky).
Merrill, E.D., Science ser. 2. 96: 180-181. 1942 (repr. Autikon); Farlowia 1: 245-262. 1943 (index to Raf. names cellular crypt.); Amer. Fern J. 33: 41-56, 97-105. 1943 (Raf. names ferns); B.P. Bishop Mus. Bull. 144: 154. 1937 (Polynes. bot. bibl.); Proc. Amer. philos. Soc. 86(1): 72-90. 1942 (on a paper by Raf. in Act. Soc. Linn. Bordeaux 6: 261-269. 1834, additions to bibl. of Raf. and to Bibliotheca Rafinesquiana), 87(1): 110-119. 1943 (Raf.'s publ. and world botany); Foreword, in reprint *A life of Travels*, Chron. bot. 8(2): 292-297. 1944; J. Arnold Arb. 29: 202-214. 1948 (nomencl. notes on Raf.); Contr. U.S. natl. Herb. 30(1): 249. 1947 (bibl. Pacific Islands); Harvard Libr. Bull. 2(1): 5-21. 1948 (notes on Raf.'s publ. in Harvard libraries); Index rafinesquianus, Jamaica Plain 1949, ix, 296 p. (see below no. 8550; bibl., analysis, list of all Raf. names).
Mitchill, S.L., Amer. mon. Mag. 2(5): 366. 1818 (rev. Fl. Ludov.).
Murrill, W.A., Hist. found. bot. Florida 11-12. 1945 (brief note).
Nissen, C., Zool. Buchill. 328. 1969 (biogr. info., descr. *Caratteri* (1810) and *Indice d'ittiologia Siciliana* (1810).
Nicolas, J.P., *in* G.H.M. Lawrence, ed., Adanson 1: 228. 1963.
Norton, Mrs. C.F., Transylvania coll. Bull. 15(7): 97-107. 1942 (publ. and mss. in Transylvania library).
Osborn, H., Fragm. entom. Hist. 1: 25-26. *pl. 6.* 1936, 2: 155-157. *pl. 35.* 1946.
Paclt, J., Taxon 9(2): 47-49. 1960 (on Raf.'s "classes" from *Précis* ... 1814).
Peattie, D.C., Rafinesque; madman or genius, reprinted from Nature Magazine, s.d., 2 p., incl. portr. (copy at HU).
Pennell, F.W. & J.H. Barnhart, Monogr. Acad. Sci. Philadelphia 1: 617. 1935 (on herb.); The life and work of Rafinesque, reprinted from Transylvania College Bulletin 15(7): 10-70. 1942 (general evaluation and biogr.); Bartonia 21: 51-52. 1942 (general note), 23: 43-68. 1945 (how Durand acquired Raf.'s herb.), 25: 67-68. 1949 (last sickness), Critical index to *A life of Travels*, Chron. bot. 8(2): 357-360. 1944, Chron. bot. 12: 216-217. 1951 (id.).
Penhallow, D.P., Trans. R. Soc. Can. ser. 2. 3: 9, 10, 22, 32. 1897 (Raf. and Canadian bot.).
Pritzel, G.A., Linnaea 19(3): 460. 1846 (biogr. note); Icon. bot. Index xxiv. 1855 (cites 37 plates from Medical flora).
Quatre, Henricus – see Croizat, L. 1948.
Rafinesque, C.S., [anonymous note evidently inspired by Raf. himself], Flora 8(1) Beil. 4: 144. 1825 (general plans, requests contr. to his museum); A life of travels, Philadelphia 1836, 148 p., reprinted Chron. bot. 8: 293-360. 1944 and in K.B. Sterling, Rafinesque, autobiography and lives, New York 1978; [editorial note evidently based

on inf. given by Raf.], Amer. J. Sci. 34(2): 386-387. 1838 (announcement of *Fl. tellur.* and *New fl.*
Rhoads, S.N., Cassinia 1-12. 1911(1912) (portr.; Raf. as an ornithologist.
Richmond, C.W., Auk 26(1): 37-55. 1909 (repr. Anal. nat. 1815).
Rickett, H.W., Torreya 42: 11-14, 131. 142 (*Cornus*).
Rodgers, A.D., William Starling Sullivant 72, 74, 127. 1940; John Torrey 348 [index, many refs.]. 1942.
Rogers, D.P., Brief hist. mycol. N. Amer. 7. 1978 (note on mycol. taxa); ed. 2: 7, 9. 1981.
Rosen, G. and B. Caspari-Rosen, 400 years of a doctor's life, New York, Henry Schuman, 1947, p. 165-166 (brief sketch on R. as an irregular medical practitioner).
Santesson, R., Sv. bot. Tidskr. 37: 287-303. 1943 (Phalloids).
Sayre, G., Dates publ. musci 34, 55, 1959 (on Prospectus in Med.Repos. 2. 5(4): 350-356. 1808 and on art. in Amer. mon. Mag. 1(6): 429. 1817); Bryologist 80: 516. 1977 (moss coll. perhaps at DWC, PH, WIS).
Schuyler, A.E., Morris Arboretum Bull. 24(2): 28-34. 1973 (general, portr.).
Seaver, F.J., North. Amer. cup-fungi 365. 1942 (Raf.'s Discomycetes).
Short, C.W., in W.J.Hooker, J. Bot. 3: 120. 1841 (from Transylvania J. Med. vol. 35), (brief note).
Silliman, B., Amer. J. Sci. 1(3): 311. 1819, 1(4): 435. 1819, 29(2): 393-394. 1836, 34: 386-387. 1838, 40(2): 237. 1841.
Small, J.K., Bull. Torrey bot. Cl. 24(5):228-236. 1897 (*Tradescantia*).
Smith, C.E., Bartonia 28: 5, 29. 1957.
Sprengel, K., Jahrb. Gewächsk. 1(2): 165-166. 1819. (brief rev. Fl. ludov.); Neue Entd. 1: 142-146. 1820, 2: 206-208. 1821.
Stafleu, F.A., Taxon 17: 296-299. 1968 (on the *Caratteri* and *Fl. ludov.*); in G.S Daniels & F.A. Stafleu, Taxon 22: 228. 1973 (portr. with caption).
Stearn, W.T., Gard. Chron. ser. 3. 114: 60-61. 1943 (accepts Ipheion Raf. 1837).
Stieber, M.T., Huntia 3(2): 124. 1979 (Agnes Chase notes on Raf.).
Sterling, K.B., Rafinesque, autobiography and lives. New York 1978 (Intr. by Sterling p. i-xvi, followed by facs. repr. of *A life of travels* (1836), the biography by R.E. Call (1895) and the biography and bibliography by F.J. Fitzpatrick (1911).
Steudel, E.G. von, Flora 3(1). Beil. 1/2: 33-42. 1820 (on his work and collections, plans "Somiologie", exchanges; taken from Isis).
Strobl, G., Österr. bot. Z. 30: 363-371. 1880 (summary work Raf. on fl. Etna on p. 365).
Stuckey, R.L., Taxon 20(4): 443-459. 1971 (on sale and fate of Raf.'s herb.), abstract Amer. J. Bot. 57: 768. 1970; Brittonia 23(2): 191-208. 1971 (on Raf.'s N. Amer. pl. at PH).
Stuckey, R.L. & M.L. Roberts, Taxon 23(2/3): 365-372. 1974 (additions to the Bibliotheca Rafinesquiana).
Swainson, G., Taxidermy 300-301. 1840 (brief sketch, lists 5 zool. titles).
Townsend, A.C., J. Bibl. nat. Hist. 6(12): 471-476. 1943 (lists Raf. titles at BM-Bloomsbury and BM(NH).
Verdoorn, F., ed., Chron. bot. 1: 336. 1935, 3: 286. 1937, 6: 86, 87, 125, 238. 1941, 7: 236, 237, 342, 343. 1943; [editor and author of texts on Rafinesque portraits] in reprint *A Life of Travels*. Chron. bot. 8(2): 295, *pl. 5-7*. 1944; Rafinesque's proposals to publish a selection of his works, Int. Biohist. comm., Utrecht 1971, 8 p. (contains facs. repr. of the October 1821 proposals).
Vogelenzang, L., [Mink, O. de], Botany, Rare works and herbaria on microfiche, IDZ, Zug, Switzerland 1976, p. 102-108.
Von Hagen, V.W., Nat. Hist. 56: 296-303. 1947 (the unnatural naturalist).
W., V., Flora 5(2): 719. 1822 (quotes story shipwreck).
Weiss, H.B., Rafinesque's Kentucky friends 1-70. 1936 (rev.: J.H. Barnhart, J. New York Bot. Gard. 37: 247-248. 1936).
Whaler, James, Green River, a poem for Rafinesque, 1931.
Wheeler, L.C., Amer. Midl. Natural. 29: 792-795. 1943 (accepts *Paxistima* Raf.), 30: 456-503. 1943 (reduction of Raf. genera of Euphorb.).
Wilbur, R.L., Rhodora 68: 192-208. 1966 (on *Lechea*).
Wilson, J.C. & J. Fiske, Appleton's Cycl. Amer. biogr. 5: 159. 1898.
Wittrock, V.B., Acta Horti Berg. 3(2): 180. 1903 (on portr.).

Wood, R.D., Bull. Torrey bot. Club 75: 282-285. 1948 (*Charac.*), Monogr. Charac. 1: 835. 1965.
Wright, R., Gardener's tribute 193. s.d.
Youmans, W.J., Pioneers of Science in America 182-195. 1896.
Zabel, H., Mitt. deut. dendrol. Ges. 8: 74-77. 1899 (discussion of *Pseva* Raf., *Joxylon* Raf. and *Rosa* sp.).

INDEX:
Index to plant names: Merrill, E.D., *Index rafinesquianus*, 1949, see below no. 8550.

KEYWORDS TO BIOGRAPHICAL and BIBLIOGRAPHICAL REFERENCES:
Autobiography: Rafinesque (all entries), Sterling 1978.
Appalachians, explorations: Core 1971.
Audubon and R.: Anon., 1935; Audubon 1832, 1926; Barnhart 1909, 1926, 1935, 1936; Bebb 1883; Keep 1846.
Baldwin and R.: Darlington 1843.
Bibliography: see below, separate paragraph, and Goodwin 1960, Pritzel 1846, Townsend 1943.
Bibliotheca Rafinesquiana: see below, separate paragraph.
Biography general: Anon., (Appleton) 1888; Barkley 1928; Boewe (all entries); Call (id.); Carpenter 1935; Coulter 1883; Ewan 1950, 1969, 1975; Genzmer 1935; Gilbert 1977; Hagen 1947; Jaffe 1944; Kastner 1977; Meehan 1883; Peattie s.d.; Pennell (all entries).
Candolle, A.P. and R.: Baehni 1952; Call 1895.
Fitzpatrick, coll. Rafinesquiana: Boewe 1959, Fitzpatrick (all items).
Grave: Anon., 1919.
Gray, A. and R.: Dupree 1959, Gray 1841.
Herbarium material: Beck 1888, Bureau 1875; Gordon 1943; Mears 1978; Pennell 1945; Sayre 1977; Stuckey 1971.
Jefferson, Th. and R.: Betts 1944.
Lexington, R. in: Dupree 1945.
Louisiana and R.: Ewan 1967.
Manuscripts: Boewe 1958; Bridson 1980; Little 1943; Norton 1942.
Medicinal pl.: Berman 1956, Haag 1941, 1942; Rosen 1947.
Nomenclature, Rafinesquian: Croizat 1930, 1948; Fernald 1931-1946; Greene (all entries); Kuntze 1891; Merrill (all entries); Paclt 1960.
Nuttall and R.: Graustein 1967.
Plates by R.: Gerard 1880.
Portraits: Fitzpatrick 1905; La Roque 1964; Rhoads 1911; Schuyler 1973; Wittrock 1903.
Schweinitz and R.: Barnhart 1921.
Short and R.: Brewe 1961, 1980; Fox 1901.
Smith, J.E. and R.: Dawson 1934.
Sullivant, W. and R.: Rodgers 1940.
Torrey and R.: Barnhart 1921; Rodgers 1942.
Types: Chase 1936.

CORRESPONDENCE: Charles Boewe, Transylvania University, Lexington, Kentucky, U.S.A., plans to locate holograph letters written by and to Rafinesque and to transcribe, edit, annotate and publish them. For preliminary papers see Boewe (1958, 1980).

EPONYMY: *Schmaltzia* Desvaux ex J.K. Small (1903), *Rafinesquia* Rafinesque (1837, 2 homonyms, see ING 3: 1481), (1838), *Rafinesquia* Nuttall *nom. cons.* (1841). – Mt. Rafinesque, N.Y., U.S.A. was thus named officially 1894 (Boewe 1962); the name was given in 1833.

HANDWRITING: Call, R.E., Life writings Rafinesque, *pl.* opp. p. 48; Proc. Amer. philos. Soc 122(3): 167, fig. 8. 1978 (pre-1805 writing; Bartonia 28: *pl.* 7. 1957.

BIBLIOTHECA RAFINESQUIANA: the literature on Rafinesque continues to grow and the bibliography of his own works still has occasional surprises in store for the Rafinesque connoisseurs. The main publications are the following:

(1) Fitzpatrick, T.J., *Rafinesque*, a sketch of his life with bibliography, Des Moines, Iowa 1911, 241 p.; bibliography of 939 items with the secondary literature *Bibliotheca* on p. 223-239 (134 items).
(2) Merrill, E.D., *A generally overlooked Rafinesque paper*, Proc. Amer. philos. Soc. 86: 72-90. 1942; additions to the Rafinesque bibliography and to the *Bibliotheca* (p. 76-77), 35 further titles.
(3) Merrill, E.D., *Index rafinesquianus*, Jamaica Plain, 1949, 296 p.; additions to the bibliography of Rafinesque on p. 12-14, to the *Bibliotheca* on p. 14-17, 74 further titles.
(4) Stuckey, R.L. and M.L. Roberts, *Additions to the Bibliotheca Rafinesquiana*, Taxon 23(2/3): 365-372. 1974; 127 additional titles to the *Bibliotheca*.
(5) Boewe, C., *Fitzpatrick's Rafinesque*: a sketch of his life with bibliography revised and enlarged, Weston Massachusetts (M. & S. Press) 1982, 327 p.; a reprint of Fitzpatrick with additions to the bibliography on p. 238-260 (items 942-1001 plus reprints and translations and with additions to the Bibliotheca Rafinesquiana (sec. refs.) nos. 135-540. This is now the main bibliographical source for Rafinesque.
(6) *This treatment* of Rafinesque contains a further number of additions, especially to the *Bibliotheca* in our sections BIOGRAPHY and BIBLIOGRAPHY and BIOFILE.
[*Note* on an almost non-botanical item: A second copy of the very rare pamphlet *The Mexicans in 1830*, FIT 579 (one copy seen by FIT in MO) is at G, p. [1]-10 (oct. in size, 6-mo in fold).

8548. Author's edition. Filson Club Publications no. 10. *The life and writings of Rafinesque* ... by Richard Ellsworth Call, ... Louisville, Kentucky (John P. Morton and Company ...) 1895. Qu. (*Life writ. Raf.*).
Author: Richard Ellsworth Call (1856-1917).
Author's edition: Jan 1895 (TBC 26 Feb 1895) front. portr., p. [i]-xii, [1]-227. *Copy*: FAS.
– The first major biography with primary and secondary bibliography. – Author's edition on large paper.
Regular edition: 1895, id. *Copy*: MO.
Facsimile edition: 1978, in K.B. Sterling, Rafinesque, Autobiography and lives, New York (Arno Press ...) 1978. *Copy*: FAS.
Ref.: Anon., Bot. Gaz. 19: 474. 1894 (ann.), 20: 120-121. 1895 (rev.).
Goode, H., Science ser. 2. 1: 384-385. 5 Apr 1894 (rev. biogr. details Raf.).
Hollick, C.A. ["A.H."], Bull. Torrey bot. Club 22(2): 78-82. 1895 (rev.).

8549. *Rafinesque* a sketch of his life with bibliography by T.J. Fitzpatrick, ... Des Moines (the Historical Department of Iowa) 1911. Oct. (*Rafinesque*).
Author: Thomas Jefferson Fitzpatrick (1868-1952).
Publ.: 1911, p. [1]-241, *pl. 1-31* (of which several in text). *Copies*: BR, FAS, G. – Major bibliography of Rafinesque (939 items), with a biographical sketch, a list of portraits and a Bibliotheca Rafinesquiana (bibl. of secondary references) on p. 223-239, 134 items.
Facsimile reprint: 1978, *in* K.B. Sterling, Rafinesque, autobiography and lives, New York (Arno Press). *Copy*: FAS.
Ref.: Geiser, S.W., Amer. Midl. Natural. 2: 150-152. 1911 (rev.).

8550. *Index rafinesquianus*. The plant names published by C.S. Rafinesque with reductions, and a consideration of his methods, objectives, and attainments by Elmer D. Merrill. Jamaica Plain, Massachusetts, U.S.A. (the Arnold Arboretum of Harvard University) 1949. Qu. (*Index Raf.*).
Author: Elmer Drew Merrill (1876-1956).
Publ.: Aug 1949 (see TL-2/5872), p. [i]-ix, [1]-296. *Copies*: G, L, MO, NY, U, US, USDA. – The most important evaluation of Rafinesque's botanical work; analysis of all Rafinesque botanical names; analysis of methods; bibl. of secondary literature additional to Fitzpatrick. – Also treated as TL-2/5872.

8551. *Additions to Michaux's Flora of North America*. In a letter from Mr. Rafinesque, to Dr. [Samuel Latham] Mitchill [1764-1831], dated Palermo, in Sicily, 8th August, 1805. [Med. Repos. second hexade, 3(4): 422-423. 1806]. Oct. (FIT 5).
Publ.: Feb-Apr 1806 (issue for that period of Med. Repos.); to be cited from journal as above (Tearsheet: NY); IDC 8119 (1(8)).

8552. *Pamphysis sicula* sive historia naturalis animalium, vegetabilium, et mineralium quae in Sicilia vel in circuitu ejus inveniuntur opus incoeptum a r.p. Francisco Cupani, tertii Ord. S. Francisci in Pamphyto siculo continuatum suppletumque ab Antonino Bonanno Gervasi, Panormitano et tandum absolutum a Josepho, Stephano et Francisco Paulo Chiarelliis, ... sub auspice C.S. Rafinesque Schmaltz ... ejusque observationibus annotationibusque locupletatum imaginis aereis circiter 700. ab ipsis Cupani schematibus exscriptis et exacte denuo incisis per Salvatorem de Ippolito. Panormi [Palermo] (ex Typographia Jo; Baptistae Giordano ...) 1807. Qu. (*Pamphysis sicul.*). (FIT 7).
Publ.: 1807, p. [1]-8, *1 pl. Copies*: FI, G. – Prospectus for a new version of the *Pamphyton siculum* (1713, see PR 1995). Fitzpatrick quotes this rare pamphlet from Specchio delle Scienze 1(1): 39. 1814. It contains no botanical names; the plate ("no. 4") depicts a *Trifolium* and a *Lathyrus*. See also FIT 23 for a further quote from Specchio delle Scienze.

8553. *Caratteri di alcune nuovi generi* e nuove specie di animali e piante della Sicilia con varie osservazioni sopra i medesimi ... Palermo (Per le Stampe di Sanfilippo ...) 1810. Qu. (*Caratt. nuov. gen.*). (FIT 16, 17).
Publ.: 1810, p. [i-iii], [1]-105, *pl. 1-20* (uncol. copp. auct.). *Copies*: FI, G(2), HH, KU(2), MO, PH, US, USDA; IDC 2136. – See Holthuis & Boeseman (1977) for the description of a copy with part of a previous issue of the first sheet with a t.p. dated 1809. The book was issued as a whole (and not in two parts) in 1810 (after 1 Apr, date on p. [1]), fide Rafinesque's own statement in his *Life of travels*. The first sheet was reprinted in 1810. See Holthuis en Boeseman for full details.
Facsimile ed.: 1967, p. [i-vi], [1]-105, *pl. 1-20. Copy*: FAS. – Amsterdam (A. Asher & Co.) 1967. See Stafleu, F.A., Taxon 17: 296-299. 1968 (the footnote repeats the Fitzpatrick statement on the two parts). The facsimile reprint is combined with a reprint of the *Indice d'ittiologia siciliana* 1810 (FIT 18).

8554. *Statistica generale di Sicilia* de' signori d.d. Giuseppe Emmanuele Ortolani avvocato e mineralogico e Constantino S. Rafinesque Schmalz negoziante e naturalista in due parti nella prima si descrive il fisico della Sicilia, nella seconda il suo morale. Palermo (dalla Reale Stamperia) 1810. Qu. (*Statist. gen. Sicilia*). (FIT 20).
Publ.: 1810, p. [1]-49, 2 maps. *Copies*: G, K.

8555. *An essay on the exotic plants*, mostly European, which have been naturalized, and now grow spontaneously in the Middle States of North America by C.S. Rafinesque Schmaltz. To Samuel L. Mitchill, New York. Palermo, 1st April, 1810. [Med. Repos. third hexade, 2(4): 330-345]. Oct. (FIT 21).
Publ.: Feb-Apr 1811 (issued for that period of Med. Repos.) to be cited from journal as above. (Tearsheet: MO). – IDC 8119 (2(13)).

8556. *Chloris aetnensis* o le quattro florule dell'Etna opusculo del sig. C.S. Rafinesque-Schmaltz Palerme Dicembre 1813. Destinato per essere inserito nella Storia naturale dell'Etna del can. co Recupero dal suo degno nipote il can. co Tes.re D. Agatino Recupero di Catania [Palermo Dec 1813]. Fol. (*Chlor. aetn.*) (FIT 234).
Publ.: Dec 1813 (t.p.) [1815?], p. [1]-15. *Copies*: KU, NY; IDC 5648. – To be inserted in G. Recupero, Storia nat. gen. Etna, Catania 1815. H. Manitz (1982) has seen an original copy of Recupero's work in the University library of Rostock and remarks that since the Recupero book appeared in 1815 this date should be adhered to. We have, however, the two independent items (KY and NY) and they may or may not have been distributed before 1815, and certainly they may have been printed earlier. The date Dec 1813 may well be that of the closing of the manuscript, but the KU and NY copies leave us no choice but to date them from Dec 1813, see ICBN Art. 30 "in the absence of proof establishing some other date, the one appearing on the printed matter must be accepted as correct".
Ref.: Manitz, H., Wiss. Z. Friedr. Schiller Univ. Jena, Math.-nat. R. 31(2): 263-264. 1982.

8557. *Specchio delle scienze* o giornale enciclopedico di Sicilia deposito letterario delle moderne cognizioni, scoperte, ed osservazioni sopra le scienze ed arti e particolarmente

sopra la fisica, la chimica, la storia naturale, la botanica, l'agricoltura, la medicina, il commercio, la legislazione, l'educazione ec. ... Palermo (Dalla Tipografia di Francesco Abate Qm. Domenico) 1814, 2 vols. Qu. (*Specchio sci.*) (FIT 26-229).
1: Jan-Jun 1814, p. [1]-216, *2 pl.*
2: Jul-Dec 1814, p. [1]-196.

number	pages	dates	number	pages	dates
1	1-44	1 Jan 1814	7	1-32	1 Jul 1814
2	45-80	1 Feb 1814	8	33-64	1 Aug 1814
3	81-112, *1 pl.*	1 Mar 1814	9	65-96	1 Sep 1814
4	113-148, *1 pl.*	1 Apr 1814	10	97-128	1 Oct 1814
5	149-180	1 Mai 1814	11	129-160	1 Nov 1814
6	181-216	1 Jun 1814	12	161-196	1 Dec 1814

Copies: FI, G, NY, PH; IDC 5650. – For a complete list of contents see FIT 26-229.

8558. *Précis des découvertes et travaux somiologiques* de Mr C.S. Rafinesque-Schmaltz entre 1800 et 1814 ou choix raisonné de ses principales découvertes en zoologie et en botanique, pour servir d'introduction à ses ouvrages futurs. De Linné le génie il a choisi pour guide. Palerme [Palermo] (Royale Typographie militaire ...) 1814. Oct. (first sign.). Qu. (other sign.) (*Précis découv. somiol.*) (FIT 230).
Publ.: Jun-Dec 1814 (p. 54: 3 Jun 1814; Acad. Paris rd 23 Jan 1815), p. [1]-55, [56, err.]. *Copies*: G(2), KU(2), PH(2); IDC 5650.
Reprint bot. part: in Desvaux, J. Bot. 4: 268-"176" [276]. 1814.
Analysis: Rafinesque, C.S., Specchio sci. 2(8): 43-44. Aug 1844.
Facsimile reprint: 1948, p. [i*-ii*], [1], [1*-2* by E.D. Merrill], 3-55, [56, err.]. *Copies*: KU, NY, PH. – New York (Peter Smith).
Ref.: Stresemann, E., J. für Ornithol. 70: 128-129. 1814 (bird names).

8559. *Principes fondamentaux de somiologie* ou les loix de la nomenclature et de la classification de l'empire organique ou des animaux et des végétaux contenant les règles essentielles de l'art de leur imposer des noms immuables et de les classer méthodiquement par C.S. Rafinesque-Schmaltz. Palerme (de l'Imprimerie de Franc. Abate, aux dépens de l'Auteur) 1814. Qu. (*Princ. fond. somiol.*). (FIT 233).
Publ.: Sep-Dec 1814 (announced by Raf. Specchio sci. 2: 64. Sep 1814; rd Acad. Paris 23 Jan 1815), p. [1]-50, [51, err.], [52, err.]. *Copies*: G, HH, NY, USDA; IDC 5650.
Ref.: Stresemann, E., J. für Ornithol. 70: 128-129. 1814.

8560. *Analyse de la nature* ou tableau de l'univers et des corps organisés (par C.S. Rafinesque ... La nature est mon guide, et Linnéus [sic] mon maître. Palerme (aux dépens de l'auteur) 1815. Qu. (*Anal. nat.*) *(FIT 235)*.
Publ.: Apr-Jul 1815 (Richmond, Iredale, KU copy rd by S.L. Mitchill at New York, 7 Sep 1815), p. [1]-224, frontisp., portr. author. *Copies*: G (has the plates of the *Caratteri* bound in), HH, KU; IDC 8098.
Ref.: Iredale, T., Proc. malac. Soc. 9: 261-263. 1913 (misapplied molluscan generic names).
Richmond, C.W., Auk 26(1): 37-55. 1909.

8561. *Circular address on botany and zoology*; followed by the prospectus of two periodical works; Annals of nature and somiology of North America; By C.S. Rafinesque, ... Chi fa quanto puo, fa quanto deve. Philadelphia (printed for the author, by S. Merritt, ...) 1816. Duod. (in sixes). (*Circ. addr. bot. zool.*). (FIT 236).
Publ.: Mai 1816 (KU copy dedicated to "John Davis, Boston, 20 May 1816"), p. [1]-36. *Copies*: FI, G(2), HH, KU, USDA; IDC 8102.
Announcements: Litterarischer Anzeiger 15: 57-60. 1819, 16: 61-64. 1819 (n.v., fide Fitzpatrick 312-313). ("Dieser Mann kann mehr als Brod essen. So weit hat es Trattinick noch lange nicht gebracht". – A note by the editor pointing at an internal Austrian difficulty).

8562. Art. 4 *Flora Philadelphica prodromus*, or prodromus of the Flora Philadelphica exhibiting a list of all the plants to be described in that work which have as yet been collected. By Dr. William P.C. Barton. Philadelphia 1815. 4to. pp. 100 [review by C.S.R. (Rafinesque) in Amer. mon. Mag. 1(5): 356-359. Sep 1817]. Oct. (FIT 237).
Publ.: Sep 1817, to be cited from journal as indicated. *Tearsheets*: KU, MO. – For Barton, W.P.C., Fl. philadelph. prodr. see TL-2/323.

8563. Art. 4. *A manual of botany* for the Northern States, comprising generic descriptions of all phanerogamous and cryptogamous plants to the north of Virginia ... 12 mo. pp. 164. [Review by C.S.R. (Rafinesque) of A. Eaton, *Man. bot.*, 1817, in Amer. mon. Mag. 1(6): 426-430. Oct 1817]. Oct. (FIT 241).
Publ.: Oct 1817, to be cited from journal as cited. For A. Eaton, *Man. bot.* ed. 1, 18-31 Jul 1817, see TL-2/1614). *Tearsheet* at MO.

8564. *Survey of the progress and actual state of natural sciences in the United States of America*, from the beginning of this century to the present time. [Amer. mon. Mag. 2(2): 81-89. New York, Dec 1817]. Oct. (FIT 251).
Publ.: Dec 1817; to be cited from journal as quoted. *Tearsheet*: MO. – IDC 8119 (3(28)).

8565. [Museum of natural sciences ...] 12. *Description of the Ioxylon pomiferum*, a new genus of North American trees [Amer. mon. Mag. 2(2): 118-119. New York, Dec 1817]. (FIT 252).
Publ.: Dec 1817, to be cited from journal as quoted. *Tearsheet*: MO. – IDC 8119 (3(29)).

8566. [Museum of natural sciences ...] 13. Second decade of undescribed plants [Amer. mon. Mag. 2(2): 119-120. New York, Dec 1917]. (FIT 253).
Publ.: Dec 1817, to be cited from journal as quoted. *Tearsheet*: MO. – The first decade of undescribed plants, FIT 249, ibid. 2(1): 43-44. 1817.

8567. *Florula ludoviciana*; or, a flora of the state of Louisiana. Translated, revised, and improved, from the French of C.C. Robin, by C.S. Rafinesque, ... Quand les matériaux sont imparfaits, l'édifice ne peut pas être complet. New York (published by C. Wiley & Co. ...) 1817. Duod. (in sixes). (*Fl. ludov.*). (FIT 25.).
Publ.: Oct-early Dec 1817, p. [1]-178. *Copies*: BR, FAS, FI, G, H(2), KU, MICH, NY, US, USDA; IDC 5054. – Based on Claude C. Robin, [1750-?], *Voyages dans l'intérieur de la Louisiane*, ... 3: 313-551 (Flore louisianaise). 1807. The cover of the Fl. ludov. has a slightly different title, "*Florula ludoviciana. Flora of Louisiana* by Robin and Rafinesque with a supplement and appendix by C.S. Rafinesque ..." – *Main commentary* and updating of nomenclature: J. Ewan in facs. ed. (below).
Facsimile ed.: 1967, p. [ii*-iv*], i-xl (Editor's introduction, J. Ewan), [i** repr. cover], [1]-178. *Copies*: FAS, PH. – The facsimile ed. was reproduced from the FAS copy of the original. For reviews see J. Ingram, Baileya 16: 147. 1968; A.L. Moldenke, Phytologia 16: 61. 1968; F.A. Stafleu, Taxon 17(3): 296-299. 1968; W.C. Steere, Gard. J. New York Bot. Gard. 19: 27. 1969; R.L. Stuckey, Ohio J. Sci. 68: 128. 1968.
Ref.: Ewan, J., Southw. Louisiana J. 7: 16-17. 1967; see also facs. ed. 1967 above.
Mitchill, S.L., Amer. mon. Mag 2(5): 368. Mar 1818 (rev.).
Sprengel, K.P.J., Jahrb. Gewächsk. 1(2): 165-166. 1819 (rev.).

8568. Art. 2. *Flora Americae septentrionalis*, or a systematic arrangement and description of the plants of North America, &c. By Frederick Pursh ... 1819. [Review by Raf. in Amer. mon. Mag. 2(3): 170-176. Jan 1819, 2(4): 265-269. Feb 1818]. (FIT 258, 264).
Publ.: Jan, Feb 1819, to be cited from journal as quoted above. For Pursh, F., *Fl. Amer. sept.* see TL-2/8404. *Tearsheet copies*: KU, MO.

8569. 18. *Third decade of new species of North-American plants* [in Amer. mon. Mag. 2(3): 206-207. Jan 1818]. Oct. (FIT 262).
Publ.: Jan 1818, to be cited from journal as quoted. *Tearsheet copy*: MO. – IDC 8119 (3(27)).

8570. Art. 3. *Florula bostoniensis*. A collection of plants of Boston ... By Jacob Bigelow,

M.M. Boston, 1814. 8 vo. 800 pp. 280. [review by C.S. Rafinesque in Amer. mon. Mag. 2(5): 342-344. New York, March 1818]. Oct (FIT 265).
Publ.: Mar 1818, to be cited from journal as quoted. *Tearsheet copies*: KU, MO. – For J. Bigelow, *Fl. boston.*, 1814, see TL-2/513.

8571. Art. 3. *A sketch of the botany of South-Carolina and Georgia*. By Stephen Elliott, ... 1817, 5 numbers, 8vo each of 100 pages ... to be continued. [Review by C.R. Rafinesque, Amer. mon. Mag. 3(2): 96-101, New York. Jun 1818]. Oct. (FIT 267).
Publ.: Jun 1818, to be cited from journal as quoted above. *Tearsheet copies*: KU, MO. – For S. Elliott, *Sketch bot. S. Carolina* 1816-1824, see TL-2/1659.

8572. Art. 3. *The genera of North-American plants* and a catalogue of the species to the year 1817. By Thomas Nuttall ... Philadelphia 1818. [Review by C.S. Rafinesque, Amer. mon. Mag. 4(3): 184-196. New York, Jan 1819]. Oct. (FIT 288).
Publ.: Jan 1819, to be cited from journal as quoted. *Tearsheet copies*: G, KU, MO; IDC 8119 (3(21)).– For Th. Nuttall, *Gen. N.-Amer. Pl.* see TL-2/6922.

8573. *Annals of nature* or annual synopsis of new genera and species of animals, plants, &c. discovered in North America by C.S. Rafinesque, ... Exertion unfolds and increases knowledge. First annual number, for 1820 ... [p. 16: printed by Thomas Smith ... Lexington, Ky.]. Qu. †. (*Ann. nat.*). (FIT 370).
Publ.: Mar-Jul 1820 (p. 2: 1 Mar 1820; Torrey copy rd. 31 Jul 1820), p. [1]-16. *Copies*: G, K, KU, MO, NY; IDC 7651. – Plants on p. 11-16.
Facsimile reprint: 15 Jan 1908, p. [i-ii, new t.p.], [1]-16. *Copy*: KU. – Privately reprinted by T.J. Fitzpatrick, Iowa City, Iowa January 15, 1908. (FIT 939).

8574. *Prodrome d'une monographie des Rosiers* de l'Amérique septentrionale, contenant la description de quinze nouvelles espèces et vingt variétés par M. C.S. Rafinesque. Sur le genre *Houstounia* et description de plusieurs espèces nouvelles, etc. Par le même. *Prodrome d'une monographie de turbinolies* fossiles du Kentuki (dans l'Amériq. septentr.). Par MM C.S. Rafinesque et J.D. Clifford. Extraits de la 14me livraison du 5me tome des Annales générales des Sciences physiques. À Bruxelles (de l'imprimerie de Weissenbruch père, ...) [1820]. Duod. (*Prodr. monogr. Rosiers*). FIT 362.
Publ.: 1820, cover p. [i-ii], [1]-20., cover p. [iii-iv]. *Copy*: HH. – Reprinted and to be cited from Ann. gén. Sci. phys., Bruxelles 5(14): 210-220 [Rosiers, FIT 340], 224-227 [*Houstonia* FIT 341), 231-235 [Turbinolies, FIT 343]. 1820. – Third paper with John D. Clifford (1778-1820).

8575. *Western Minerva*, or American annals of knowledge and literature, un peu de tout. food for the mind. First volume for 1821. Lexington, Kentucky. (Published for the editors, by Thomas Smith, ...) 1821. Qu. (*West. Minerva*). (FIT 375-430).
Publ.: 1821 (p. vi: Jan 1821; p. [7]: Jan 1820), p. [i]-vi, [7]-88. *Copies*: PH; IDC 8118. – For further details see the extensive treatment in FIT 375-430, 336, 342, 431, 432 and Merrill, E.D. in facs. repr.
Facsimile repr.: 1949, p [i-iv], [3 p. note by E.D. Merrill], [v]-vi, [7]-88. *Copy*: KU.
Ref.: Anon., Bot. Gaz. 30: 216. 1900 (copy found at PH).
Fox, W.J., Science ser. 2. 12: 211-215. 1900 (descr. PH copy).

8576. *Proposals to publish by subscription a selection of the miscellaneous works and essays of C.S. Rafinesque*, ... in 5 volumes, 12mo. Consisting of a selected collection of tracts, memoirs, dissertations, lectures, reviews, letters, travels, tales, poems, &c. On various subjects ... to be published in Lexington, Ky. [p. 4: Lexington, Ky. October 1821]. (*Prop. publ. works Raf.*). (FIT 435).
Publ.: Oct 1821, p. [1]-4. *Copy*: KU. – See FIT 435 for an extensive description and F. Verdoorn (1971), in facsimile reprint (below).
Facsimile reprint: Jul 1971, 8 p. pamphlet containing the 4 p. facsimile and a commentary, portr., and a note on Rafinesque by F. Verdoorn with a reproduction of the portr. from the *Anal. nat.* (1815). *Copies*: FAS(2). – "*Rafinesque's proposals to publish a selection of his works* (October 1821). A keepsake issued by the International Biohistorical Commision ... (July 1971) [Utrecht, Netherlands].

8577. *First catalogues and circulars of the botanical garden of Transylvania University* at Lexington in Kentucky, for the year 1824. Premiers catalogues et circulaires du jardin botanique de l'université transylvane à Lexington en Kentucky pour l'année 1824. Printed for the Botanical Garden Company, by John M. M'Calla, Lexington, Ky. 1824. Duod. (in sixes). (*First cat. gard. Transylv. Univ.*). (FIT 468).
Publ.: 1824, p. [1]-24. *Copies*: HH, K, KU, NY; IDC 8103. – See FIT for details.

8578. *Neogenyton*, or indication of sixty-six new genera of plants of North America. By C.S. Rafinesque, ... 1825. Dedicated to professor De Candolle of Geneva [Lexington, Ky.]. (*Neogenyton*). (FIT 474).
Publ.: 1825, p. [1]-4. *Copies*: G(2), HH, KU, NY, PH, WU; IDC 5652.
Facsimile reprint: 1914, p. [i-ii], [1]-4. *Copies*: KU, MO, PH. – p. [i]: Neogenyton by C.S. Rafinesque, p. [ii], undated foreword by "The Editor A.M.N." Issued by the Amer. Midl. Natural.

8579. *Medical flora*; or, manual of the medical botany of the United States of North America. Containing a selection of above 100 figures and descriptions of medical plants, with their names, qualities, properties, history, &c. and notes or remarks on nearly 500 equivalent substitutes. In two volumes ... Philadelphia (printed and published by Atkinson & Alexander, ...) 1828. Duod. (in sixes). (*Med. fl.*), (FIT 554, 557).
1: 1828 (deposited 11 Jan 1828), p. [i*-iii*], [i]-xii, [1]-268, *pl. 1-52* (uncol.).
2: 1830 (p. 6: Mai 1830), p. [1]-276, *pl. 53-100* (id.).
Copies: FH, FI, G, HH, KU(2), MO, NY, PH, USDA; IDC 768. – Cover title: "Medical flora and botany of the United States, with one hundred plates ..."
Reissue vol. 1: 1841, p. [i*-iv*], [i]-xii, [1]-259, *pl. 1-52* (id.). (FI 925). *Copies*: HH, KU; IDC 8110. – "*Manual of the medical botany of the United States*; containing a description of fifty-two medical plants, with their names, qualities, properties, history, & c. with remarks on nearly 500 substitutes and fifty-two coloured plates. By Professor Rafinesque, A.M. Philadelphia: 1841."
Reissue vol. 2: see FIT 933. Questionable.
Ref.: Lloyd, J.U. & J.T., J. Amer. pharm. Ass. 20: 918-921. 1931 (acquisition copy *Med. fl.*).
[Schlechtendal, D.F.L. von], Linnaea 9 (litt.): 95-100. 1834 (review).

8580. Eight figures, twenty-five cents. *American manual of the grape vines* and the art of making wine: including an account of 62 species of vines, with nearly 300 varieties. An account of the principal wines, american and foreign. Properties and uses of wines and grapes. Cultivation of vines in America; and the art to make good wines. With 8 figures ... Let every farmer drink his own wine. Philadelphia (printed for the author) 1830. Duod. (in sixes). (*Amer. man. grape vines*) (FIT 558).
Publ.: 1830, p. [1]-64, [1, add.], *2 pl. Copies*: FI, G, KU, NY, USDA; IDC 8093. – A reprint of the article on *Vitis* in Med. fl. 2: 121-180 (= 5-64), *pl. 99-100* (unnumbered in repr.). The reprint is later than the original because it contains at least one correction.

8581. American florist-second series. Eighteen *figures of handsome American and garden flowers–by C.S. Rafinesque*. Philadelphia, 1832. Broadside. (*Fig. Amer. gard. fl.*).
Publ.: 1832, broadside, figures in three rows of six each, nos. 19-36. See FIT 606 and 607 for further details. *Copy*: FH.

8582. *Atlantic journal*, and friend of knowledge. In eight numbers. Containing about 160 original articles and tracts on natural and historical sciences, the description of about 150 new plants, and 100 new animals or fossils. Many vocabularies of languages, historical and geological facts, &c. &c. &c. ... Philadelphia 1832-1833 (two dollars) (*Atl. J.*), (FIT 620-785).
Publ.: 1832-1833, *pl.*, p. [i-iv], [1]-212. *Copies*: KU, NY; IDC 8200.
1(1): 20 Mar 1832, p. [1]-36.
1(2): Summer 1832, p. [37]-82, *pl.*
1 (extra of no. *3*): Sep 1832, p. [83]-90.
1(3): Autumn of 1832, p. 91-122.

1(4): Winter of 1832, p. [123]-154.
1(5): Spring 1833, p. [155]-170.
1(6): Summer 1833, p. [171]-186.
1(7): Autumn 1833, p. [187]-202.
1(8): Winter 1833, p. [203]-212.
Extra of 6: see below *Herbarium Rafinesquianum*.
Facsimile ed.: 1946, [i*, pl., i-iv], [1]-212. *Copies*: BR, FAS, G, MO, NY. – Photolithographed by The Murray Printing Company, Cambridge. Massachusetts for the Arnold Arboretum [Harvard University, Jamaica Plains, Mass. U.S.A.] 1946. (text on p. [ii]).
Extract: J. Bot. 30: 310-311. 1892, "An early evolutionist", reprint of extract letter to Torrey on p. 163-164 "Every variety is a deviation which becomes a Sp. as soon as it is permanent by reproduction.".

8583. Atlantic Journal. – Extra of N. 6. *Herbarium rafinesquianum*. Prodromus. – Pars prima Rarissim. plant. nov. Herbals; or botanical collections of C.S. Rafinesque ... First [second, third] part ... the labor of a whole life! Philadelphia: 1833. Price one dollar. (*Herb. raf.*). (FIT 786-820).
1(1): extra of Atlantic Journal no. 6, summer 1833, p. [1-16.
1(2): extra of id. no. 7, autumn 1833, p. 17-32.
2: extra of id. no. 8, winter 1833, p. 33-48.
3: ? 1833, p. 49-64.
3, continued: ? 1833, p. 65-80.
Copies: G (p. 1-48), HH (p. 1-48), KU (p. 1-32), NY (p. 1-80, only complete copy known); IDC 5653. – Issued separately; free for subscribers to *Atlantic journal*. See also BM 4: 1639.
Ref.: Britten, N.L., J. Bot. 38: 224-225. 1900.
Fernald, M.L., Rhodora 48: 5-13. 1946.

8584. *Bemerkungen zur Encyclopaedia of plants of Loudon*, Lindley and Sowerby von C.S. Rafinesque Professor der Botanik and Naturgeschichte etc. in Philadelphia. Januar 1831 (Aus Loudon's Gardener's Magazine vol. viii. (April 1832). p. 245 [*in* Linnaea 8 (Litt): 67-75. 1834]. Oct.
Publ.: 1834. – A german translation, not mentioned by Fitzpatrick (but listed by Boewe as no. 605) of a paper originally published in The Gardener's Magazine and Register of Rural & Domestic Improvement, conducted by J.C. Loudon, 8(2): 244-248. Apr 1832 ("Art. III. Retrospective criticism"), (FIT 605). J. Britten drew attention to this article with its secondary title "Remarks on the Encyclopedia of plants of Loudon, Lindley and Sowerby" in J. Bot. 224-229. Jun 1900, giving a reprint. (FIT 938). The Linnaea paper is only of historical interest; the 50 notes containing nomenclatural remarks date from Apr 1832. – For J.C. Loudon, *Encycl. Pl.* 1829, see TL-2/5026.

8585. *A life of travels* and researches in North America and South Europe, or outlines of the life, travels and researches of C.S. Rafinesque, ... containing his travels in North America and the South of Europe; the Atlantic Ocean, Mediteranean, Sicily, Azores, &c. from 1802-1835 – with sketches of his scientific and historical researches, &c. ... Philadelphia (printed for the author, ...) 1836. price seventy-five cents. 12-mo (in sixes). (*Life of travels*), (FIT 863).
Publ.: 1836, p. [1]-148. *Copies*: HH, KU, NY; IDC 8109.
Reprint: Chronica botanica 8(2): 298-353. 1944, with an introduction by E.D. Merrill (p. 292-297), a critical index by F.W. Pennell (p. 354-360) and illustrations, provided, with extensive captions, by F. Verdoorn.
Ref.: Brown, R.H., Geogr. Rev. 35: 337-338. 1945 (rev.).
Dayton, W.A., J. Forest. 43: 381-382. 1945 (rev.; misc. info.).
Higinbotham, R., Amer. Midl. Natural. 37: 805-806. 1947 (rev.).
Merriman, D., Amer. J. Sci. 243: 109-110. 1945 (rev.).
Taylor, R.L., William and Mary Quart. ser. 3. 2: 213-221. 1945 (well-informed review).

8586. *New flora and botany of North America* or a supplemental flora, additional to all the

botanical works on North America and the United States, containing 1000 new or revised species ... Philadelphia 1836 [-1838]. Price five dollars – $ 20 for 5 copies (*New Fl.*), (FIT 864-868).
1: Introd. Lexicon, etc.: Dec 1836, p. [1]-100.
2: Neobotanon: Jul-Dec 1837, p. [1]-96.
3: New Sylva: Jan-Mar 1838, p. [1]-96.
4: Neobotanon: late 1838, p. [1]-112. (p. 1 is t.p. for entire work with text as above).
Copies: G, HH, KU, MO, NY, PH, USDA; IDC 5220. – The covers, the first pages of parts 1-3, and the third page of part four have different titles for which see e.g. FIT. For the dates see J.H. Barnhart, Torreya 7: 177-182. 1907.
Facsimile ed: 1946, (cover titles included): [i, cover part i; ii, new imprint and ref. to Barnhart, 1907], [1]-100, [i], [1]-96, [i, cover part 2], [1, t.p. part 4], 3-96 (text part 2), [i], [1]-96, [i, cover part 4], [1, t.p. whole work], [3, t.p. part 2], [4]-122 (text part 4). *Copies*: FAS, G, MO, USDA. – p. [ii]: Photolithographed by the Murray Printing Company Cambridge, Massachusetts, for the Arnold Arboretum [Jamaica Plain, Massachusetts, U.S.A.] 1946.
Ref.: Silliman, B., ed., Amer. J. Si. 34: 386-387. 1839 (notice publ.).

8587. *Flora telluriana* centur. i.-xii. Mantissa synoptica. 200 N. Ord. – N. Gen. – N. Sp. plantarum in orbus tellurianum. Determ. coll. inv. obs. et. descr. ann. 1796-1836. Auctore C.S. Rafinesque, ... Philadelphia 1836[-1838]. Qu. (*Fl. tellur.*), (FIT 869-873).
1: Jan-Mar 1837, p. [1]-101, [103, cont.].
2: Jan-Mar 1837, p. [1]-112.
3: Nov-Dec 1837, p. [1]-100.
4: medio 1838, p. [1]-135 (p. [1] is t.p. whole work with text as above).
Copies: HH, KU, NY, US, USDA; IDC 6158. – The covers and t.p.'s of the four pages are different from the t.p. for the whole work. For full details see e.g. FIT. For dates see J.H. Barnhart, Torreya 7: 177-182. 1907.
Facsimile ed.: 1946 (covers included): [i], [1]-101, [103], [i], [1]-112, [i], [1]-100, [i], [1]-135. *Copies*: BR, FAS, G, US. – Arnold Arboretum, Jamaica Plain, Massachusetts, U.S.A.
Ref.: Silliman, B., ed., Amer. J. Sci. 34: 386-387. 1838 (notice).

8588. *Alsographia americana*, or an American grove of new or revised trees and shrubs of the genera Myrica, Calycanthus, Salix, Quercus, Fraxinus, Populus, Tilia, Sambucus, Viburnum, Cornus, Juglans, Aesculus, &c., with some new genera, monographs, and many new sp. in 330 articles, completing 1405 g. and sp. as a continuation of the Sylva telluriana and North American trees & shrubs, ... Philadelphia 1838. Price one dollar. Qu. (*Alsogr. amer.*), (FIT 883).
Publ.: 1838, p. [1]-76. *Copies*: HH(2), KU, USDA; IDC 8092.

8589. *Sylva telluriana*. Mantis. synopt. New genera and species of trees and shrubs of North America, and other regions of the earth, omitted or mistaken by botanical authors and compilers, or not properly classified, now reduced by their natural affinities to the proper natural orders and tribes ...being a supplement to the Flora telluriana ... Philadelphia (printed for the author and publisher) 1838. Qu. (*Sylva tellur.*), (FIT 885).
Publ.: Oct-Dec 1838 (p. 7: Oct 1838), p. [1]-184. *Copies*: HH, NY, US, USDA; IDC 8117. – See also Sandwith, N.Y., Taxon 4: 70. 1955.
Facsimile ed.: Jan 1943, p. [1]-184. *Copies*: BR, FAS, G, KU. – Arnold Arboretum, Jamaica Plain, Mass.

8590. *American manual of the mulberry trees*. Their history, cultivation, properties, diseases, species and varieties &c., with hints on the production of silk from their barks &c. ... The production of silk by animal agency is one of the miracles of nature. Philadelphia 1839. Duod. (in sixes). (*Amer. man. mulberry trees*), (FIT 894).
Publ.: 1839, p. [1]-96. *Copies*: HH, KU, US, USDA; IDC 8097. – On cover: Publications of the Eleutherium of knowledge, number 5.

8591. *The good book*, and amenities of nature, or annals of historical and natural sciences. Containing selections, of observations, researches and novelties in all the branches of

physical and historical knowledge, with letters of eminent authors – chiefly on zoology, botany, agronomy, geognosy, ethnography ... or organized beings and fossils, nations and languages ... Philadelphia (printed for the Eleutherium of knowledge) 1840. Qu. †. (*Good book*), (FIT 900-921).
Publ.: Jan 1840 (cover of first and only part issued), p. [1]-84. *Copies*: HH, KU, PH, US(2), USDA; IDC 8107. – For a list of contents and further details see e.g. FIT.

8592. *Autikon botanikon.* Icones plantarum select. nov. vel rariorum, plerumque americana, interdum african. europ. asiat. oceanic. &c. centur. xxv. Botanical illustrations by select specimens or self-figures in 25 centuries of 2500 plants, trees, shrubs, vines, lilies, grasses, ferns &c., chiefly new or rare, doubtful or interesting, from North America and some other regions; with accounts of the undescribed, notes, synonyms, localities &c. in 5 parts of 5 centuries each of text with 25 volumes folio of self-figures ... Philadelphia (collected, ascertained and described between 1815 & 1840. Qu. †. (*Autik. bot.*), (FI 897-899).
1, cent. i-v: 1840, [1]-72.
2, cent. vi-x: 1840, [73]-140.
3, cent. xi-xv: 1840, [141]-200.
Copies: HH, MO, NY, USDA; IDC 8101. – Only three out of five scheduled parts published. The work was to have been accompanied by botanical specimens: (t.p.): "the best botanical figures are the objects themselves".
Facsimile ed.: 1942, [i, repr. of cover], ii, note by E.D. Merrill], [1]-200. *Copies*: BR, FAS, G, MO, PH. – Arnold Arboretum, Jamaica Plain, Mass. *Review*: Verdoorn, F., Chron. bot. 7: 236-237. 1942.
Ref.: Merrill, E.D., Science ser. 2-96: 180-184. 1942.
Pennell, F.W., 48: 89-96. 1921.

Rafn, Carl Gottlob (1769-1808), Danish botanist; studied medicine, botany and veterinary sciences at Copenhagen; school teacher and administrator, ultimately in charge of the royal alcohol factory. (*Rafn*).

HERBARIUM and TYPES: C.

BIBLIOGRAPHY and BIOGRAPHY: AG 6(2): 216; Barnhart 3: 123 (b. 31 Jul 1769, d. 17 Mai 1808); BM 4: 1640; Christiansen p. 11, 12; CSP 5: 76; Herder p. 118, 194; Hortus 3: 1202; IH 2: (in press); Jackson p. 333; Kew 4: 392; KR p. 582; LS 21673; PR 7403-7404, ed. 1: 8331-8332; Rehder 1: 352, 5: 143, 701; Tucker 1: 580; Zander ed. 11, p. 803.

BIOFILE: Christensen, C., Dansk. Bot. Hist. 1: 82, 88, 90, 106-108, 110, 177-189, 194, 1924, 2: 118-121. 1924 (biogr., portr).
Fischer-Benzon, R.J.D. von, *in* P. Prahl, Krit. Fl. Schlw.-Holst. 2: 48. 1890.
Fries, E., Bot. Not. 1841: 161-166.
Hansen, A., Bot. Tidsskr. 69: 212. 1974.
Hornemann, J.W., Naturh. Tidsskr. 1: 578. 1837.
Hultén, E., Bot. Not. 1940: 346.
Jessen, K.F.W., Bot. Gegenw. Vorz. 397. 1884.
Nyrop, C. et al., Dansk biogr. lex. 13: 354-358. 1899.
Paulli, R. and C. Christensen, Dansk biogr. leks. 19: 54-57. 1940 (biogr.).
Poggendorff, J.C., Biogr.-Lit. Handw.-Buch 2: 559. 1863.
Warming, E., Bot. Tidsskr. 12: 84-86. 1880 (bibl.).

EPONYMY: *Rafnia* Thunberg (1800).

8593. *Danmarks og Holsteens flora* systematisk, physisk og oekonomisk bearbeydet, et priisskrivt ... Kφbenhavn ["Kiöbenhavn"]1796-1800, 2 vols. Oct. †. (*Danm. Holst. Fl.*).
1: Jan-Jun 1796, (p. x: 24 Jul 1796), engr. t.p., [i*], [i]-x, [1]-722, [723-724, err.].
2: Jun 1800 (p. x: 1 Mai 1800; publ. Jun 1800 fide, Maanedskriftet Minerva 25 Jul 1800), p. [i]-x, [1]-840, [841-842, err.].
Copies: G, HH, LD, MO, NY. – For secondary references and polemics see Christensen 2: 119. 1924. – J. McNeill and H.C. Prentice (1981) have shown that vol. 2 came out

in Jun 1800, is thus later than the competing Gaertner, Meyer and Scherbius, *Oekon. Fl. Wetterau* (TL-2/1926), vol. 2, which was available on the Jubilate-Messe 1800 (Jubilate 1800: 4 Mai according to ALZ 1800(94): 790.
Ref.: McNeill, J. H.C. Prentice, Taxon 30(1): 29. 1981.

Ragot, Jules (*fl.* 1902), French horticulturist; head gardener of the Société d'Horticulture de la Sarthe at Le Mans. (*Ragot*).

HERBARIUM and TYPES: Unknown.

BIBLIOGRAPHY and BIOGRAPHY: Barnhart 3: 123.

BIOFILE: Anon., Bull. Acad. géogr. bot. 9: 16. 1900, 11: 14. 1902, 13: 14. 1904, 19: 13. 1909.

8594. *Notice historique sur la Société d'Horticulture de la Sarthe* suivi du catalogue général des végétaux cultivés dans les jardins de la Société ... Le Mans (Imprimerie de l'Institut de Bibliographie ...) 1902. Oct. (*Not. hist. Soc. Hort. Sarthe*).
Publ.: 1902 (Nat. Nov. Oct(1) 1902), p. [i-iii], [1]-127, [128, cont.]. *Copy*: USDA.

Railonsala, Artturi Nikodemus (orig. A.N. Helenius) (1902-1982), Finnish botanist, teacher and school inspector at Tornio; taraxacologist. (*Railonsala*).

HERBARIUM and TYPES: H, OULU, TUR.

BIBLIOGRAPHY and BIOGRAPHY: Collander p. 427 (bibl.).

Raimann, Rudolf (1863-1896), Austrian botanist; professor of natural history at the "Handelsakademie" in Vienna. (*Raim.*).

HERBARIUM and TYPES: W.

BIBLIOGRAPHY and BIOGRAPHY: Backer p. 659 (err. "German"); Barnhart 3: 124; BJI 1: 47; BM 4: 1640; DTS 1: 233; Hortus 3: 1202; IF suppl. 1: 84; IH 2: (in press); Kew 4: 393; LS 212681-21683; Rehder 5: 701; Tucker 1: 580; Zander ed. 10, p. 705, ed. 11, p. 804.

BIOFILE: Anon., Bot. Centralbl. 69: 192. 1897 (d. 5 Dec 1896).
Kneucker, A., Allg. bot. Z. 3: 20, 56. 1897 (d.).

COMPOSITE WORKS: EP, *Nat. Pflanzenfam.*:(1) Onagraceae, 3(7): 199-208. 24 Oct 1893, 209-223. 21 Nov 1893.
(2) Hydrocaryaceae 7(7): 223-226. 21 Nov 1893.

EPONYMY: *Raimannia* J.N. Rose ex N.L. Britton et A. Brown (1913).

Raimondi, Antonio (1826-1890), Italian-born botanist; studied at Milano; lived in Peru 1850-1890 devoting himself to scientific work, at first as professor of botany in Lima, later as state geologist. (*Raimondi*).

HERBARIUM and TYPES: USM (part of orig. herb.); other material at F and NY.

BIBLIOGRAPHY and BIOGRAPHY: Barnhart 3: 124 (d. 26 Oct 1880); BM 4: 1640; IH 1 (ed. 2): 72, (ed. 3): 90, (ed. 6): 363, (ed. 7): 340, 2: (in press); Morren ed. 10, p. 141; Quenstedt p. 351; Saccardo 2: 89, 126; Zander ed. 10, p. 704-705, ed. 11, p. 803.

BIOFILE: Balta, J., *in* A. Raimondi, El Peru 4: xxvii-xxxvii. 1902 (bibl., 113 nos.).
Embacher, F., Lexik. Reisen 242. 1882.
Macedo, C.M., Bol. Mus. Hist. nat. Javier Prado 4: 431-443. 1940 (portr., biogr., d. 24 Oct 1890).

Poggendorff, J.C., Biogr.-lit. handw.-Buch 3: 1084. 1898 (b. 19 Sep 1826, d. 25 Oct 1890).
Raimondi, A., El Peru, 1: 2-3. 1874, 4: x, xi, xxiv, xxvii-xxxvii. 1902.
Schofield, E.K., Brittonia 30: 404. 1978 (coll. NY).
Weberbauer, A., Pfl.-Welt peruan. Andes 13-14, 35. 1911 (Veg. Erde 12).

EPONYMY: *Raimondia* Safford (1913); *Raimondianthus* Harms (1928).

8595. *El Perú* ... Lima (imprenta del Estado, ...) 1874-1902, 4 vols. (*Perú*).
1: 1874, p. [i*-iii*], [i]-iv, [v, vii], [1]-444.
2: 1876, p. [i]-vii, [ix], [1]-475, *5 pl.*
3: 1879, p. [i]-v, [vii], [1]-610, *4 pl.*
4: 1902, p. [i]-xxxvii, [xxxix], [1]-515, [i]-xiv, [xv].
5: 1913, p. [i], [1]-201.
Copies: John Wurdack (vols. 1-3), LC. – Vol. 4 with a biography and bibliography of Raimondi by J. Balta et al. Imprint vol. 4: Lima (Librería é imprenta Gil ...) 1902; of vol. 5: Lima (imprenta del Estado) 1913.
Notas de viajes para su obra "El Peru", Lima-Peru (imprenta Torres Aguirre) 1942-1948, 4 vols. *Copy*: John Wurdack.
1: 1942, p. [i]-xi, map, [1]-431.
2: 1943, p. [i], map, [1]-288.
3: 1945, p. [i], map, [1]-359.
4: 1948, p. [1], map, [3]-272.

Rainer, Moriz (Moritz) **von und zu Haarbach** (1793-1847), Austrian botanist; for eighteen years banker in Milano; ultimately at Graz. (*Rainer*).

HERBARIUM and TYPES: GJO (rd. 1851); further material in BR (Mart.), E, M, LZ and REG.

BIBLIOGRAPHY and BIOGRAPHY: AG 6(2): 744-745; Barnhart 3: 124; DTS 1: 233, xxiii; IH 2: (in press); Lasègue p. 336, 337; Saccardo 1: 135.

BIOFILE: Anon., Flora 30: 542. 1847 (d.); Österr. bot. W. 1: 12. 1851 (herbarium and library left to GJO).
Hausmann, F. v., Fl. Tirol. 3: 1184. 1854.
Müller, R.H.W. & R. Zaunick, Friedrich Traugott Kützing 136. 1960.

EPONYMY: *Rainera* G. de Notaris (1838); *Campanula raineri* Lena-Perpenti does not commemorate M. Rainer but rather the Archduke Rainer (1783-1853) at the time viceroy of Lombardy and Venice (AG 6(2): 644-645).

Raineri, Rita (1896-x), Italian botanist and paleontologist at Torino. (*Raineri*).

HERBARIUM and TYPES: TO.

BIBLIOGRAPHY and BIOGRAPHY: Andrews p. 316; Barnhart 3: 124; IH 2: (in press); TL-2/see O. Mattirolo.

BIOFILE: Koster, J. Th., Taxon 18: 556. 1969 (coll.).
Verdoorn, F., ed., Chron. bot. 2: 392. 1936, 7: 103. 1943.

8596. *Alghe sifonee fossili della Libia* ... Pavia (Premiato Tipografia successori fratelli Fusi ...) 1922. Oct.
Publ.: Feb 1922 (in journal), p. [i], [1]-15, *1 pl. Copy*: FI. – Reprinted and to be cited from Atti Soc. ital. Sci. nat. 61: [72]-86, *pl. 3.* 1922.

8597. *Alghe fossili mioceniche di Cirenaica* raccolte dall' ing. C. Crema ... Padova (Tipografia Seminario) 1923. Oct. (*Alg. foss. miocen.*).
Publ.: Aug-Dec 1923 (t.p. preprint; p. 23: Jul 1923), p. [1]-23. *Copy*: FI. – Preprinted from Nuova Notarisia 35: 28-46. Jun 1924.

Rainville, Frédéric (x-1779), French-born botanist, private teacher and translator at Rotterdam; student of grasses. (*Rainville*).

HERBARIUM and TYPES: Small herbarium Teyler (Haarlem) (see e.g. van Ooststroom 1958).

BIBLIOGRAPHY and BIOGRAPHY: Dawson p. 693; Dryander 3: 428; IH 2: (in press); JW 4: 402; LS 21684.

BIOFILE: Hazewinkel, H.C., Rotterdams Jaarboekje 1959: 244-248.
Lefebvre, E. & J.G. de Bruyn, Martinus van Marum 6: 38, 40-43, 45, 124. 1976.
Ooststroom, S.J. van & T.J. Reichgelt, Acta bot. neerl. 7: 605-613. 1958 (on Rainville herb. and D. de Gorter, *Fl. vii Prov.*).
Steenis-Kruseman, M.J. van, *in* R.J. Forbes, ed., Martinus van Marum 3: 132-134. 1976 (grass herbarium at Teyler).

Ralfs, John (1807-1890), British botanist; studied medicine in Guy's Hospital, London; med. exam. Apothecaries Hall 1830, College of Surgeons 1832; praticed as a surgeon 1832-1837; settled in Penzance 1837 for health reasons, devoting himself to botany, especially phycology. (*Ralfs*).

HERBARIUM and TYPES: BM (phanerogams at K); other material at CGE, E, FH, H, K, L, LD, MANCH, OXF, PC, SLBI. – *Exsiccatae*: *British algae* (ca. 1830), nos. 1-40, sets at FH, PC. – Manuscript flora of W. Cornwall and Scilly at Penzance library. Letters at BM and K.

BIBLIOGRAPHY and BIOGRAPHY: Barnhart 3: 124 (b. 13 Sep 1807, d. 14 Jul 1890); BB p. 2252; BL 2: 232, 706; BM 4: 1640; Bossert p. 323; Clokie p. 229; CSP 5: 80-81, 12: 597, 18: 36; Desmond p. 509; De Toni 1: ciii, 2: xiii, xciv (bibl.!); DNB 47: 209-210; Frank 3 (Anh.): 81; GR p. 411; Hawksworth p. 24, 137, 198; Herder p. 254; IH 2: (in press); Jackson p. 235, 243; Kew 4: 395; NAF ser.2. 6: 73; NI 1582; Nordstedt p. 26, 35; PR 7406, ed. 1: 8337; Rehder 5: 701; TL-1/1037, TL-2/2455; Tucker 1: 580.

BIOFILE: Allibone, S.A., Crit. dict. Engl. lit. 2: 1730. 1878.
Anon., Bot. Centralbl. 44: 96 1890 (d.); Bot. Not. 1891: 46 (d.); Bull. Soc. bot. France 37 (bibl.): 144. 1890 (d.); J. Bot. 28: 256. 1890 (d.), Morot 4: 316. 1890 (d.); J. Hort. Cottage Gard. 21: 71-72. 1890 (obit.); Nat. Nov. 12: 367. 1890 (d.); Nature 42: 300. 1890; Österr. bot. Z. 40: 394. 1890 (d.); Trans. Penzance nat. Hist. Antiq. Soc. 3: 198. 1890/1891 (d.).
Barone, C., Notarisia 5: 1103-1106. 1890 (obit.).
Bridson, G.D.R. et al., Nat. hist. mss. res. Brit. Isl. 430 [index]. 1980.
Cupp, E.E., Mar. plankton diatoms W. Coast N. Amer. 220. 1943.
Davey, F.H., Fl. Cornwall lvii-lviii. 1909 (portr.).
Debey, J., Bibl. microsc. microgr. Stud. 3. Diat. 1882. p. 46.
Dickinson, C.I., Phycol. Bull. 1: 13. 1952 (Algal herb. at BM but further algae in herb. Hooker at K [algae coll. K now on permanent loan to BM]).
Freeman, R.B., Brit. nat. hist. books 290. 1980.
Fryxell, G.A., Beih. Nova Hedw. 35: 363. 1975.
Groves, H. & J., J. r. microsc. Soc. 1890: 797-799 (obit.); J. Bot. 28: 289-293. 1890 (obit., portr., b. 13 Sep 1807, d. 14 Jul 1890).
Hedge, I.C. & J.M. Lamond, Index coll. Edinb. herb. 121. 1970.
J.D., Nuova Notarisia 1: 241-242. 1890 (obit.).
Kent, D.H., Brit. herb. 74. 1957.
Koster, J.Th., Taxon 18: 556. 1969.
Lewis, A.G., John Ralfs, an old Cornish botanist 1907.
Lousley, E.J., Fl. Isles Scilly 81, 82. 1972.
Marquand, E.D., Trans. Penzance nat. Hist. antiq. Soc. 3: 225-240. 1891 (obit.).
Müller, R.H.W. & R. Zaunick, Friedrich Traugott Kützing 243, 292. 1960.
Murray, G., Hist. coll. BM (NH) 1: 175. 1904 (coll.).

Nordstedt, C.F.O., Bot. Not. 1906: 97-106, J. Bot. 45: 128. 1907 (proposal to accept Ralfs, Brit. Desmid. 1848, as starting point for the nomenclature of Desmids).
Sayre, G., Mem. New York Bot. Gard. 19(1): 97. 1969 (exsiccatae).
Tregelles, G.F., Trans. Penzance nat. Hist. antiq. Soc. 3: 198. 1891 (d.).
VanLandingham, S.L., Cat. Diat. 4: 2364, 6: 3584. 1978, 7: 4222. 1978.
Wittrock, V.B., Acta Horti Berg. 3(2): 97, *pl. 34.* 1903 (portr.).
Wornardt, W.W., Ind. direct. diat. 35. 1968.

COMPOSITE WORKS: joint editor of A. Pritchard, *History of Infusoria*, ed. 4, 1861.

HANDWRITING: J. Bot. 28: pl. opp. p. 289. 1890.

EPONYMY: *Ralfsia* O'Meara (1875); *Ralfsia* Berkeley (1843).

8598. *The British phanerogamous plants & ferns*; arranged on the Linnaean system, and analyzed after the method of Lamarck, with a short comparative analysis of the natural families ... London (Longman, Orme, Brown, Green, and Longmans) 1839. Oct. (*Brit. phaenog. pl.*).
Publ.: 1839 (p. xvi: Jun 1839; J. Bot. (W.J. Hooker) Mai 1840), p. [i]-xvi, [1]-208. *Copies*: E, HH, NY.

8599. *The British Desmidieae* by John Ralfs, ... the drawings by Edward Jenner, ... London (Reeve, Benham, and Reeve, ...) 1848. Oct. (*Brit. Desmid.*).
Publ.: 1 Jan 1848 (conventional date accepted by ICBN 13.1(h)), p. [i]-xxii, [i], [1]-226, *pl. 1-35* (col. liths. Edward Jenner (1803-1872)). *Copies*: BR, E, FH, G, HH, M, NY, PH, US; IDC 6400. – Starting point for the nomenclature of Desmidiaceae, ICBN 13.1(h). See Nordstedt (1906, 1909).
Facsimile ed.: 1962, p. [i*-iv*], [i]-xxii, [i], [1]-236, *c*pl. 1-35 (uncol.). *Copy*: FAS. – Historiae naturalis classica tomus xxviii. Weinheim (J. Cramer) 1962; see Taxon 12: 171. 1963; ISBN 3-7682-0144-9.
Extract (German): s.l.n.d., p. [i], 1-10 (lithogr. mss.), *pl. 1-35* (uncol.). *Copy*: NY. – "Kurzer Auszug aus John Ralfs British Desmidieae mit Zeichnungen. Für einige Freunde".
Ref.: Anon., Gard. Chron. 1848(48): 383 (rev. in issue for 10 Jun 1848).
J.W.B., Amer. J. Sci. ser. 2. 6: 302-303. Nov 1848.
Nordstedt, O., Bot. Not. 1906: 97-118 (proposes *Brit. Desmid.* as starting point work); Motion Congr. int. bot. 1909, TL-2/6870.
Ralfs, J., *in* W.J. Hooker, J. Bot. 5: 12-13. 1846 (announcement, subscription) (see also 7: 392-393. 1848 note on publ.).
Schlechtendal, D.F.L. von, Bot. Zeit. 6: 708-709. 6 Oct 1848 (rev.).

Rallet, Louis (1897-1969), French botanist; Dr. sci. Poitiers 1935; teacher at teacher's college of La Rochelle 1945-1962; president of the Société botanique du Centre-Ouest 1940-1969; prix Coincy 1947. (*Rallet*).

HERBARIUM and TYPES: LR. Some further material MPU.

BIBLIOGRAPHY and BIOGRAPHY: BL 2: 131, 167, 706.

BIOFILE: Barbier, M., Rev. Féd. franç. Soc. Sci. nat. ser. 3. 9: 1-6. 1970 (obit., portr., bibl.).
Contré, E. & J.M. Rouet, Bull. Soc. bot. France 116: 7-29. 1969 (obit., portr., bibl.).
R.D., Ann. Soc. Sci. nat. Charente-marit. 5(2): 37-44. 1970 (obit., portr., bibl., b. 23 Jan 1892, d. 28 Oct 1969).

8600. *Thèses présentées à la Faculté des Sciences de l'Université de Poitiers pour obtenir le grade de docteur ès sciences naturelles par Louis Rallet.* 1re thèse. – *Étude phytogéographique de la Brenne.* 2me thèse. – Propositions données par la Faculté. Soutenues le [] devant la commission d'Examen ... Nantes (Imprimerie Jagueneau frères ...) 1936. Oct.(*Étud. phytogéogr. Brenne*).

Thesis issue: 1936, p. [i-ii], [1]-280, [281], *pl. 1-11*, 2 maps. *Copies*: HH, LC.
Journal issue: Bull. Soc. Sci. nat. Ouest France ser. 5. 5: 1-276. 1935 (1936?).

Ralph, Thomas Shearman (1813-1891), British physician and botanist; emigrated to Australia 1851, to New Zealand 1852-1859; practicing medicine at Melbourne from 1859. (*Ralph*).

HERBARIUM and TYPES: material at B (pterid.), BM, CGE, E, G, K, W.

BIBLIOGRAPHY and BIOGRAPHY: Barnhart 3: 124 (d. 22 Dec 1891); BB p. 292; BM 4: 1640-1641; CSP 5: 81, 8: 689, 11: 96; Desmond p. 509-510; Herder p. 95, 345; HR; Jackson p. 45, 46, 135; Kew 4: 395; LS 21695-21696; NI 1425, 1583; PR 7407; Rehder 5: 701; TL-2/6399; Tucker 1: 580; Urban-Berl. p. 322.

BIOFILE: Allibone, S.A., Crit. dict. Engl. lit. 2: 1731. 1878.
Bridson, G.D.R. et. al., nat. hist. mss. res. Brit. Isl. 275.4. 1980.
Freeman, R.B., Brit. nat. hist. books 290. 1980.
Hedge, I.C. & J.M. Lamond, Index coll. Edinb. herb. 121. 1970.
Hooker, W.J., J. Bot. Kew Gard. misc. 6: 30-31. 1854 (New Zealand plants for sale).
Maiden, J.H., Vict. Natural. 25: 111. 1908 (bibl.).
Murray, G., Hist. coll. BM (NH) 1: 175. 1904.
Sayre, G., Mem. New York Bot. Gard. 19(3): 384. 1975 (coll. New Zealand).

COMPOSITE WORKS: *Opuscula omnia botanica Thomas Johnsoni*, Londini (sumptibus Guiliel. Pamplin) 1847. For further details and on a reissue of the *Descriptio* reprint see J.S.L. Gilmour, Thomas Johnson, Botanical journeys in Kent & Hampstead, Pittsburgh, Pa., 1972 (e.g. p. 1-8). – Review of 1847 publ.: Schlechtendal, D.F.L. von, Bot Zeit. 6: 658-659. 1848.

8601. *Icones carpologicae*; or figures and descriptions of fruits and seeds ... London (William Pamplin, ...) 1849. Qu. †. (*Icon. carpolog.*).
Publ.: Apr-Mai 1849(p. [vii]: Feb 1849; Hooker 1849), p. [i-viii], [1]-48, [49-52, index], *pl. 1-40* (uncol. liths. auct.). *Copies*: E, FI, G, HH, M, NY, PH; IDC 5149. – Only the first part was published; the cover has: "part the first. Leguminosae ... the orders of Rosaceae &c., will appear shortly ...".
Ref.: Hooker, W.J., Bot. Kew Gard Misc. 1: 159-160. Mai 1849.
Schlechtendal, D.F.L. von, Bot. Zeit. 7: 477-478. 29 Jun 1849.

Ramaley, Francis (1870-1942), American botanist and ecologist; M.Sc. Minnesota 1896; Dr. phil. ib. 1899; instructor in botany Univ. Minnesota 1895-1898, ass. prof. biology University of Colorado, Boulder; regular prof. biology and head of the department of biology University of Colorado 1899-1939. (*Ramaley*).

HERBARIUM and TYPES: material at COLO, GH, MIN, NMC, NY, RM, UC, US. – Archival material in SIA.

BIBLIOGRAPHY and BIOGRAPHY: Barnhart 3: 125 (b. 16 Nov 1870, d. 10 Jun 1942); BJI 2: 142; BL 1: 170, 314; BM 4: 1641; CSP 18: 36; Ewan (ed. 1): 287, 355 [index], (ed. 2): 180; IH 2: (in press); Kew 4: 396; NW p. 55; PH 567; Rehder 5: 701; SK 1: 426 (coll. Males.); Tucker 1: 581.

BIOFILE: Alexander, G., Science 96: 102-103. 1942 (obit.).
Allison, E.M., Univ. Colorado Studies 6: 70-71. 1909 (bibl.).
Anon., Bot. Centralbl. 73: 335. 1898 (app. Boulder); Bot. Jahrb. 25 (Beibl. 60): 56. 1898 (app. Boulder), 27: (Beibl. 64): 22. 1900 (prof. biol. ib.); Nat. Nov. 20: 120. 1898 (ass. prof. Univ. Colo.); Nature 150: 262. 1942 (d.); Österr. bot. Z. 48: 159. 1898 (ass. prof. Univ. Colo.), 49: 415. 1899 (prof. biol. ib.).
Cattell, J.M., Amer. men Sci. ed. 1: 261. 1906, ed. 2: 383. 1910, ed. 3: 558. 1921, ed. 4: 795. 1927, ed. 5: 908. 1933, ed. 6: 1150. 1938.
Ewan, J. et al., Short hist. bot. U.S. 16. 1969.

Johnson, E.L., Madroño 6: 260-265., frontisp., portr. 1942 (biogr., portr., bibl.).
Ramaley, F., Plant World 8: 139-150. 1905 (a botanist's trip to Java).
Robbins, W.W., Ecology 23: 385-386. 1942 (obit.).
Verdoorn, F. ed., Chron. bot. 1: 284. 1935, 2: 303. 1936, 3: 271. 1937, 4: 452, 472, 5: 509. 1939, 6: 69. 1941, 7: 233. 1943.

8601a. *Wild flowers and trees of Colorado* ... Boulder, Colorado (A.A. Greenman, University Book Store) 1909. Oct. (*Wild fl. trees Colorado*).
Publ.: Jun-Jul 1909 (p. [v]: Mar 1909; NY rd. Jul 1909; Bot. Zeit. 1 Nov 1909; IDC 14 Aug 1909), p. [i, frontisp.]-vii, [viii], 1-78. *Copies*: HH, MICH, MO, NY, PH, USDA.

8602. *Colorado plant life* ... Boulder, Colorado (Published by the University of Colorado) 1927. (*Colorado pl. life*).
Publ.: Sep 1927 (p. vi), p. [i]-viii, *pl. 1-3* (col), 1-299, *5 uncol. pl. Copies*: HH, MO, NY.
Ref.: Gleason, H.A., Torreya 28(2): 29-30. 23 Apr 1928 (rev.).

Rama Rao, Rao Sahib **Muttada** (1865-?), Indian botanist, conservator of forests, Travancore. (*Rama Rao*).

HERBARIUM and TYPES: FRC.

BIBLIOGRAPHY and BIOGRAPHY: IH 1 (ed. 6): 363, (ed. 7): 340; LS 21698-21700.

8603. *Flowering plants of Travancore* ... Trivandrum (printed at the Government Press) 1914. (*Fl. pl. Travancore*).
Publ.: 1914 (p. viii: 2 Mai 1914), p. [i]-xiv, [1]-448, [449-495]. *Copies*: NY; IDC 6468.
Facsimile reprint: 1976, Dehra Dun (Bishen Singh ...) *Copies*: E, HH, MO.

Ramatuelle, Thomas Albin Joseph d'Audibert de (abbé) (1750-1794), French clergyman and botanist at Aix-en-Provence; died from falling off the roof of a prison during the French revolution. (*Ramat.*).

HERBARIUM and TYPES: Some material in FI(Webb).

BIBLIOGRAPHY and BIOGRAPHY: Barnhart 3: 125 (b. 16 Mai 1750, d. 26 Jun 1914); Bret. p. 132; Dryander 3: 312, 382, 384; Hortus 3: 1202; MW p. 399; Zander ed. 11, p. 804.

BIOFILE: Dawson, W.R., Smith papers 77. 1934.
Duval, H. & A. Reynier, Bull. Soc. bot. France 58: 312-319, 349-358. 1911.
Keefe, Biologist 48: 52. 1966 (epon.).
Steinberg, C., Webbia 32(1): 35. 1977.

EPONYMY: *Ramatuela* Kunth (1825).

Rambert, Eugène (1810-1886), Swiss literary critic and amateur botanist. (*Rambert*).

HERBARIUM and TYPES: LAU.

BIBLIOGRAPHY and BIOGRAPHY: BM 4: 1641; CSP 5: 81, 6: 747, 18: 39; IH 2: (in press); Kew 4: 397; Rehder 1: 435.

BIOFILE: Bonnerot, J., Rev. litt. comp. 28(4): 385-410. 1954 ("une amitié vaudoise", corr. between Rambert and Saint-Beuve).
Chavannes, S., Verh. Schweiz. naturf. Ges. 1886/87: 172-176. (obit.).
Wilczek, E., Bull. Soc. vaud. Sci. nat. 60: 25. 1937 (herb.).

8604. *Les Alpes suisses* ... première [-cinquième] série. Paris (Librairie de la Suisse romande, ...), Genève (Même maison, ...) 1866[- 1875]. Oct. (*Alp. suiss.*).
1: 1866 (p. [vi]: 10 Dec 1865), p. [i-vii], [1]-302.
2: 1866 (p. xiv: 20 Oct 1866), p. [i]-xiv, [xv], [1]-318. - Lausanne (Librairie Delafontaine et Rouge, ...), Paris (Librairie de la Suisse romande, ...).

3: 1869 (p. 334: 26 Jul 1868), p. [i-ii], [1]-334, [335, cont.]. – Bâle et Genève (H. Georg ...).
4: 1871, p. [i-ii], [1]-327, [328, cont.]. – Imprint as vol. 3.
5: 1875, p. [i-ii], [1]-318, [319, cont.]. – Bâle-Genève-Lyon (H. Georg, ...).
Reprint: Lausanne 1888, n.v., 377 p.
Translation of botanical part: *Die Alpenpflanzen* übersetzt aus "Les Alpes Suisses" durch A. Koebisch, Dresden (Huhle in Comm.) 1894, 85 p. (Bot. Centralbl. 4 Jun 1894, Bot. Zeit. 16 Jun 1894), (p. 3-127 of 1888 ed., "Les plantes alpines").

Rambosson, Jean Pierre (1827-1886), French popular author on natural history. (*Rambosson*).

HERBARIUM and TYPES: Unknown.

BIBLIOGRAPHY and BIOGRAPHY: BM 4: 1641; CSP 5: 81, 8: 698, 18: 39; Jackson p. 214; Kew 4: 398; Plesch p. 371; PR 7408; Rehder 1: 95. Tucker 1: 581.

8605. *Histoires et légendes des plantes utiles et curieuses* ... Paris (Librairie de Firmin Didot frères, fils et c^{ie} ...) 1868. Oct. (*Hist. légendes pl.*).
Ed. 1: 1868, p. [i*-ii*], frontisp., [iii*], [i]-v, [1]-371. *Copies*: Edinburgh Univ. Libr., HH, NY, PH.
Ed. 2: 1869, p. [i*-ii*], frontisp., [iii*], [i]-v, [1]-379. *Copies*: FI, G, USDA.
Ed. 3: 1871, p. [i], frontisp., [iii], [1]-420. *Copy*: BR.
Ed. 4: 1881, p. [i-ii], frontisp., [iii], [1]-420. *Copy*: HH.

Ramée, Stanislas Henri de la (1747-1803), fictitious French naturalist, said to have studied in Paris under Buffon, and to have travelled in the West Indies and Central and South America. (*Ramée*).

NOTE: Fictitious botanist appearing in Appleton's: Cyclopaedia of american Biography 5: 164. 1888. His "works" include e.g. a *Nova systema naturae* (2 vols. Paris 1792), and a "*Prodrome des plantes recueillies en Amérique et dans les Indes occidentales* (1789). See Barnhart (1919). Wholly fictitious.

BIOFILE: Barnhart, J.H., J. New York Bot. Gard. 20: 178. 1919.
Conde, J.A., Hist. bot. Cuba 52. 1958 (Conde lists all fictitious botanists in "Appleton", notwithstanding J.H. Barnhart's paper. Conde even contributes to the misinformation by adding "menciona algunas plantas cubanas").
Leon, J.S.S., Mem. Soc. cub. Hist. nat. 3: 182. 1918 (accepts Ramée's existence).

Ramírez, José (1852-1904), Mexican botanist. (*Ramírez*).

HERBARIUM and TYPES: MEXU.

BIBLIOGRAPHY and BIOGRAPHY: Barnhart 3: 125 (b. 12 Nov 1852, d. 11 Apr 1904); BL 1: 149, 150, 314; BM 8: 1046; Bossert p. 323; CSP 12: 598; Kew 4: 398; Langman p. 607-609 (bibl.); Rehder 1: 332, 333, 5: 702; TL-2/5486; Tucker 1: 436, 581.

BIOFILE: Anon., Bot. Not. 1905: 56 (d.).; Torreya 4(5): 77. 1904 (d.).
Castañeda, A.M., La flora del Estado de Jalisco 117-118. 1933.
Leon, N., Bibl. bot.-mexic. 216-217. 1895.
Rzedowski, J., Bol. Soc. Bot. Mexico 40: 3, 5. 1981 (portr.).
Villada, M.M., La Naturaleza ser. 3. 1: iii-x. 1910.

COMPOSITE WORKS: Collaborator for vols. 1-4 of *Datos para la materia médica mexicana*, Mexico, 1894-1907 (vol. 1, 1895: Nat. Nov. Mai(1) 1895; vol. 2, 1898: Mai(2) 1898).

EPONYMY: *Ramirezella* J.N. Rose (1903). *Note*: The source of *Ramirezia* A. Richard (1853) could not be elucidated.

8606. *Sinonimia vulgar y cientifica de las plantas mexicanas* arreglada por el Dr. José Ramírez.... con la colaboracion del Sr. Gabriel V. Alcocer ... Mexico (Oficina tipográfica de la Secretaría de Fomento ...) 1902. Fol. (*Sinon. pl. mexic.*).
Collaborator: Gabriel V. Alcocer (1852-1916).
Publ.: 1902 (p. xii: Mar 1902; NY rd. Dec 1902), p. [i]-xii, [xiii-xiv], [1]-160. *Copies*: G, NY, USDA.

8607. *La vegetación de México.* Recopilación y análysis de las principales clasificaciones propuestas ... México (Oficina tipográfica de la Secretaría de Fomento ...) 1899. Oct. (*Veg. Mexico*).
Publ.: 1899, p. [1]-271, [273, cont.], 9 charts and maps. *Copies*: G, HH, M, MICH, MO, NY, PH, US. – Contains also elements (translated into Spanish) of M. Martens & H.G. Galeotti, Mém. foug. Mexique, 1842, TL-2/5486.

8608. *Indice alfabético* de la obra de Fr. Franciso Jimenez titulada: *Cuatro libres de la naturaleza* y virtudes de las plantas y animales de uso medicinal en la Nuevea España. México (Oficina tipográfica de la Secretaria de Fomento ...) 1900. Qu. (*Ind. cuatro libr. natural.*).
Publ.: 1900, p. [1]-22. *Copy*: FI. – Franciso Ximenez (Jimenez), Quatro libros de la naturaleza, Mexico, 1615, for details see Langman p. 813-814.

8609. *Sinonimia vulgar y cientifica e las plantas mexicanas* ... Mexico (Oficina tipográfica de la secretariá de fomento ...) 1902. Oct. (*Sinon. pl. mexic.*).
Collaborator: Gabriel V. Alcocer (1852-1916).
Publ.: 1902 (p. xii: Mar 1902), p. [i]-xii, [xiii], [1]-160. *Copies*: M, PH.

8610. *Estudios de historia natural* ... Mexico (Imprento de la Secretaría de Fomento) 1904. Oct. (*Estud. hist. nat.*).
Publ.: Oct-Dec 1904 (p. iv: Oct 1904), p. [i]-iv, [1]-311, [1], *pl. 1-34*. *Copies*: BR, HH. – Posthumously published by Gabriel V. Alcocer (1852-1916).

Ramírez Goyena, Miguel (1857-1927), Nicaraguan botanist; studied at the College of Granada (Nic.), connected with Instituto nacional central de Managua for some time, from 1891-1902 in exile in Costa Rica and Honduras; from 1902 again in Nicaragua, working on his Flora nicaraguense. (*Ramírez Goyena*).

HERBARIUM and TYPES: Unknown.

BIBLIOGRAPHY and BIOGRAPHY: Barnhart 3: 125; BL 1: 153, 314, 2: 2; Langman p. 610.

BIOFILE: Montoya, E.F. et al., La prensa literaria Managua, 23 Jul 1977, p. 2-3, 11 (biogr., portr.). (We are grateful to Paul Fryxell who found this newspaper article in a copy used as a herbarium wrapper; the essay is very detailed and is the only major biographical essay of R.G. known to us).

8611. *Flora nicaragüense* por Miguel Ramírez Goyena conteniendo la botanica elemental. Managua-Nicaragua (Compañia Tipográfica Internacional) 1909-1911, 2 vols. Qu. (*Fl. nicarag.*).
1: 1909 (p. v: 27 Aug 1908), p. [i*], [i]-v, [i]-ix, err., [1]-442.
2: 1911, p. [i*], [i]-ix, [other err.], 443-1064, [i]-vi.
Copies: BR, MO, NY, PH, US, USDA.

8612. *Elementos de botanica* conteniendo especies de la *Flora nicaraguense*. Managua (Tipográfia nacional) 1918.(*Elem. bot.*).
Publ.: Nov-Dec 1918 (p. iii: Nov 1918), p. [i, iii], [1]-278. *Copy*: US.

Ramis, Aly Ibrahim (Bey) (1875-1928), Egyptian physician and botanist; Dr. med. München 1901; physician in Cairo 1903-1928; professor of surgery Cairo University. (*Ramis*).

HERBARIUM and TYPES: CAIM.

BIBLIOGRAPHY and BIOGRAPHY: Barnhart 3: 125 (d. 9 Jun 1928); BL 1: 30, 314; IH 1 (ed. 7): in press, 2: (in press); Kew 4: 398.

8613. *Bestimmungstabellen zur Flora von Aegypten* ... Jena (Verlag von Gustav Fischer) 1929. Oct. (*Best.-Tabell. Fl. Aegypt.*).
Publ.: 1929 (p. iv: Mar 1929; NY rd. Aug 1929; PH rd. 12 Sep 1929), p. [i]-iv, [v], [1]-221, [222, colo.]. *Copies*: BR, G(2), HH, M, MO, NY, PH, USDA. – With a preface (p. iii-iv) by Jussub Ibrahim, brother of the author, with some biogr. notes.

Ramis y Ramis, Juan (1746-1819), Spanish (Balearic) botanist at Mahon, Minorca. (*Ramis y Ramis*).

HERBARIUM and TYPES: Unknown.

BIBLIOGRAPHY and BIOGRAPHY: Barnhart 3: 125; BM 8: 1046; Colmeiro 1: cxcv; Colmeiro penins. p. 199; Kew 4: 398; SO 157a.

BIOFILE: Marès, P. & G. Vigineix, Cat. pl. vasc. Baléares p. xi. 1880.

EPONYMY: The source of *Ramisia* Glaziou ex Baillon (1887) could not be determined.

8614. *Specimen animalium, vegetabilium*, et mineralium *in insula Minorica* frequentiorum ad normam Linnaeani sistematis. Exaratum. Accedunt nomina vernacula in quantum fieri potuit. Superioribus annuentibus. Macone Balearium excudebat Petrus Antonius Serra 1814. Qu. (*Specim. anim. veg. Minorica*).
Publ.: 1814, p. [i-viii], [1]-60. *Copy*: NY. – Plants on p. 25-56.

Ramisch, Franz Xavier (1798-1859), Bohemian physician and botanist; Dr. med. Praha 1825; professor of medicine at Prague University; student of parthenogenesis. (*Ramisch*).

HERBARIUM and TYPES: PR.

BIBLIOGRAPHY and BIOGRAPHY: Backer p. 482; Barnhart 3: 125; Frank 3 (Anh.): 81; Futak-Domin p. 497-498; Hegi 5(3): 1574; Jackson p. 100; Kew 4: 398; Klášterský p. 153 (d. 3 Jun 1859); LS 22004-22005; Maiwald p. 205-206, 282.

BIOFILE: Anon., Bonplandia 5: 184. 1857; Flora 20: 729. 1837 (14 Dec 1837 Mercurialis publ.); Lotos 9: 136. 1859 (obit.).
Maiwald, V., Jahresber. Gymn. Braunau 1901: 15, 1902: 158-160.
Opiz, P.M., Flora 7: 142. 1824, 8 (Beil. 4): 51. 1825.
Sachs, J. v., Histoire bot. 437. 1892.
Zuchold, E.A., Jahresber. nat. Ver. Halle 5: 47. 1853.

EPONYMY: *Ramischia* Opiz ex Garcke (1858) is likely named for this author.

8615. *Beobachtungen über die Samenbildung ohne Befruchtung* am Bingelkraute (Mercurialis annua) ... Prag (Gedruckt bei Thomas Thabor, ...) 1837. Oct.
Publ.: 1837 (Flora, distr. at meeting 20 Sep 1837; vorgelegt 6 Nov 1837), p. [1]-26. *Copies*: LD, M, USDA. – Reprinted and to be cited from W.R. Weitenweber, Beitr. ges. Nat.-Heilwiss. 2(3): 426-449. 1837.

Ramond, Louis François Élisabeth de Carbonnières, baron (1753-1829), French botanist; mineralogist, politician and administrator; from 1792-1800 in the Pyrenées; from 1796-1800 teacher at Tarbes; member of the "Corps législatif" 1800-1806; préfet of Puy-de-Dôme 1806-1814; financial administrator from 1815, "phytosociologue avant la lettre". (*Ramond*).

HERBARIUM and TYPES: Plants from the Pyrenées in FI, G, and P.

BIBLIOGRAPHY and BIOGRAPHY: AG 2(2): 17, 5(1): 470, 754; Barnhart 3: 126 (b. 4 Jan 1753, d. 14 Mai 1827); BL 2: 162, 706; BM 4: 1642; Bossert p. 323; Colmeiro 1: cxcv; CSP 5: 88-89; Dryander 5: 69, 88; DSB 11: 272-273; Frank 3 (Anh.): 81; GR p. 295, cat. p. 70; Hawksworth p. 185; Herder p. 57; Hortus 3: 1202; IH 2 : (in press); Jackson p. 278; Kew 4: 399; Lasègue p. 102, 296; LS 21707; MD p. 203-204; Moebius p. 371; PR 7409-7410; Quenstedt p. 31; Rehder 5: 702; SBC p. 129; Tucker 1: 581; Zander ed. 10, p. 705, ed. 11, p. 804.

BIOFILE: Amoureux, Mém. Soc. Linn. Paris 1: 707-708. 1822.
Anon., Grand Larousse encycl. 9: 7. 1964.
Burkhardt, R.W., The spirit of system 37. 1977.
Candolle, A.P. de, Mém. souvenirs 91, 139. 1862.
Cuvier, F., Dict. Sci. nat., portr. 1845; Mém. Acad. r. Sci. Inst. France 9: clxix-cxcv. 1830 (obit., "éloge historique").
Ernst, F., Corona 3(3): 389-405. 1933.
Gerber, C., Bull. Soc. bot. France 75: 653-658. 1928 (biogr.).
Hoefer, L., ed., Nouv. biogr. gén. 41-42: 555-557. 1862 (1968).
Jessen, K.F.W., Bot. Gegenw. Vorz. 335. 1884.
Lacroix, A., Figures de savants 4: 149. 1938.
Lapeyrouse, P., Hist. abr. pl. Pyren. xxx-xxxi. 1813.
Nelson, E.C., Glasra 4: 60. 1980 (plants coll. by "Ramond, Spain" in DBN).
Pée-Laby, E., Bull. soc. Ramond 27: 233-243. 1892 (biogr., Ramond-botaniste).
Poggendorff, J.C., Biogr.-Lit. Handw.-Buch 2: 565. 1863 (bibl.).
Roumeguère, C., Correspondences autographes inédites des anciens botanistes méridionaux, Perpignan 1873, 52 p.
Sayre, G., Bryologist 80: 516. 1977 (musci at PC?).
Schmid, G., Goethe u.d. Naturwiss. 562. 1900.
Steinberg, C.H., Webbia 32(1): 35. 1977 (pl. in Herb. Desfontaines).
Voigt, A., Bot. Inst. Hamburg 15. 1897.
Wittrock, V.B., Acta Horti Berg. 3(2): 117. 1903.
Zittel, K.A. von, Gesch. Geol. Paläont. 146. 1899, Hist. geol. palaeontol. 102, 103. 1901.

EPONYMY: *Bulletin de la Société Ramond*; *Ramonda* L.C. Richard (1805, *nom. cons.*); *Ramondia* Mirbel (1801, *nom. rej.*).

8616. *Observations faites dans les Pyrenées*, pour servir de suite à des observations sur les Alpes, insérées dans une traduction des lettres de W. Coxe, sur la Suisse. À Paris (Chez Belin, ...) 1789. Oct. (*Obs. Pyrenées*).
Publ.: 1789 (p. 452: 6 Apr 1789 approb.), p. [i]-viii, map, [1]-452. *Copy*: G. – "inserted" in a french translation of William Coxe [1747-1828]. *Travels in Switzerland*. In a series of letters, ... 3 vols. 1789.

8617. *Voyages au Mont-Perdu* et dans la partie adjacente des Hautes-Pyrenées, ... à Paris (chez Belin, ...) an ix-1801. Oct. (*Voy. Mont-Perdu*).
Publ.: an ix-1801 (1 Jan-17 Sep 1801), frontisp., p. [i]-iv, [1]-392, *pl. 1-5* (uncol. copp.). *Copies*: G, NY. – Prelim. publ.: J. Mines 7: 35-38. 1797/98; see also Ann. Mus. Hist. nat. 3: 74-84. 1804.
English translation: Journey to the summit of Mont Perdu ..., in J. Pinkerton, a general collection ... Voyages, 4, 1809.

8618. *Mémoire sur l'état de la végétation au sommet du Pic du Midi* de Bagnères; par M.L. Ramond. Lu à l'Académie les 16 Janvier et 13 Mars 1826 [Mém. Mus. Hist. nat. Paris 1826]. Qu.
Publ.: Apr-Jun 1826 (in journal, MD), p. [1]-94. *Copy*: G. – Reprinted and to be cited from Mèm. Mus. Hist. nat. Paris 13(10): 217-282. 1825, publ. Apr-Jun 1826. – See MD 203-204 and BL 2: 162 for further details. MD mentions an article in *Le Globe* with the same title (3: 62. 19 Jan 1826) not seen by us.
Republication: Mém. Acad. roy. Inst. France ser. 2. 6: 81-174. 1827 (with additional

meteorological details); see also Ann. Sci. nat. 8: 96-100. 1826, Nouv. Bull. Soc. philom. 1826: 58-60, Amer. J. Sci. 14: 377-378. 1828.

Ramsbottom, John (1885-1974), British botanist; BA Univ. of Cambridge 1909; at the Botany Department of the British Museum of natural History, from 1910 as mycologist, from 1928 as deputy keeper, from 1930-1951 as keeper; general secretary of British mycological Society 1921-1945; editor of the Transactions 1919-1942; president of the Linnean Society of London 1938-1941. (*Ramsb.*).

HERBARIUM and TYPES: BM. – Letters to W.G. Farlow and R. Thaxter at FH.

BIBLIOGRAPHY and BIOGRAPHY: Ainsworth p. 346 [index]; Barnhart 3: 126; BM 8: 1047; Bossert p. 323; Desmond p. 510 (b. 25 Oct 1885, d. 14 Dec 1974); GR p. 411; IH 1 (ed. 2): 75, (ed. 3): 94, (ed. 4): 103, (ed. 5): 107, (ed. 6): 156, 2: (files); Kelly p. 185-186, 256; Kew 4: 399-400; NI 1: 133, 135, 2: nos. 1584-1585, 3: 50; SO 3722; Stevenson p. 1255; TL-1/1220; TL-2/776, 778, 1290.

BIOFILE: Anon., Bull. Brit. mycol. Soc. 9(1): 58. 1975 (obit. notice); Essex Natural. 28(5): 271-272. 1951 (Victoria Medal of Honour, Roy. Hort. Soc.), 33(2): 110. 1973/74(1975) (obit.); Gard. Chron. ser. 3. 72: 246. 1922 (portr.).; J. Bot. 48: 216. 1910 (app. assistant at BM(NH)), 55: 167. 1917 (temporary protozoologist at Salonica), 66: 32. 1928 (deputy keeper Dept. Bot. BM(NH)), 67: 344. 1929 (keeper of botany BM(NH)); Nature 166: 387-388. 1950 (retirement); Proc. Linn. Soc. 1966: 105-106 (Linnean gold medal); The Times 1 Jun 1938, 17 Dec 1974.
Bergdolt, E., Karl von Goebel ed. 2. 232. 1941.
Bisby, G.R., Trans. brit. mycol. Soc. Fifty-year index 19-20. 1952.
Bridson, G.D.R.et al., Brit. nat. hist. mss. res. Brit. Isl. 269, 293. 1980.
Lloyd, C.G., Mycol. Notes 57: 813, 830. 1919 (portr.).
P.H.G., Trans. brit. mycol. Soc. 65(1): 1-6. 1975 (obit., portr. functions).
Pilát, A., Česká Mykol. 20(3): 202. 1966 (80th birthday).
Ramsbottom, J., Trans. bot. mycol. Soc. 30: 13. 1948 (president of Society 1924, 1946; general secretary 1921-1945, editor 1919-1942).
Stearn, W.T., Nat. Hist. Mus. S. Kensington 301, 303-304, 306-307, 313. 1981.
Verdoorn, F., ed., Chron. bot. 1: 31, 40, 47, 171, 174, 340. 1935, 2: 2, 28, 31, 34, 44, 58, 187, 188, 194, 196. 1936, 3: 2, 26, 27, 32, 164, 165, 169. 1937, 4: 85, 184, 245, 5: 16, 38, 166. 1941.
Wakefield, E.M., Trans. Linn. Soc. 49(1): 1-6. 1966 (tribute, portait).

COMPOSITE WORKS: (1) *Fungi pathogenic to man*, in vol. 8 of *A system of bacteriology in relation to medicine*, London 1931, p. 11-46, 58-70. 1931. (See also J. Bot. 70: 32. Jan 1932). (2) With T.A. Sprague, A.J. Willmott and E.M. Wakefield, *Proposals by the sub-committee on nomenclature*, appointed by the Imperial botanical Conference, London 1924, in T.A. Sprague, ed., International Botanical Congress Cambridge (England), 1930. *Nomenclature proposals by British botanists*. London Aug 1929, p. 7-45, see also J. Bot. 67: 336-339. Dec 1929.

FESTSCHRIFT: Trans. Brit. mycol. Soc. 49(1). 1966.

HANDWRITING: Trans. Brit. mycol. Soc. 65: *pl. 2.* 1975.

EPONYMY: *Ramsbottomia* Buckley (1923).

8619. *A handbook of the larger British fungi* by John Ramsbottom, ... based on the Guide to Sowerby's models of British fungi; in the Department of botany, British Museum (Natural History), by Worthington George Smith [1835-1917] ... London (printed by order of the Trustees of the British Museum ...) 1923. Oct. (*Handb. larg. Brit. fung.*).
Publ.: Apr-Jun 1923 (p. iv: Mar 1923; BR rd. 1 Dec 1923; Nat. Nov. Jul 1923; J. Bot. Aug 1923; NY rd. Aug 1923), p. [i]-iv, [1]-222. *Copies*: BR, DBN, E, FH, G, NY, PH, Stevenson, US, USDA.

Facsimile reprints: Apr 1944, p. [i]-iv, [1]-222; second reprint 1946, third 1948, fourth 1949, fifth 1951, sixth 1961, seventh 1965. *Copy* of seventh reprint: NY.
Ref.: Rea, C., J. Bot. 61: 222-223. Aug 1923 (rev.).
Murrill, W.A., Torreya 23: 107-108. Nov-Dec 1923 (rev.).
Ulbrich, E., Hedwigia 64(2): (83)-(84). 29 Dec 1923.

8620. Presidential address *The taxonomy of fungi* ... Cambridge (at the University Press ...) 1926. Oct.
Publ.: 26 Aug 1926 (in journal), cover with above title, p. [25]-45. *Copy*: FH. – Reprinted and to be cited from Trans. Brit. mycol. Soc. 11(1/2): 25-45. 1926.

8621. Benn's sixpenny library. *Fungi* an introduction to mycology ... London (Ernest Benn Limited ...) [1929]. Oct. (*Fungi*).
Publ.: Mar 1929 (E copy rd. 22 Mar 1929; J. Bot. Mai 1929), p. [1]-80. *Copy*: Natl. Libr. Scotland.

Rand, Edward Lothrop (1859-1924), American botanist and lawyer; BA Harvard College 1881; LL.B and A.M. 1884; admitted to the Massachusetts bar at Boston 1885; one of the founders of the New England Botanical Club and its first corresponding secretary 1895-1921. (*E.L. Rand*).

HERBARIUM and TYPES: GH and NEBC, other material at B (destr.), DPU, PH. – Letters to W.G. Farlow at FH.

BIBLIOGRAPHY and BIOGRAPHY: Barnhart 3: 127 (b. 22 Aug 1859, d. 9 Oct 1924); BL 1: 185, 186, 314; BM 4: 1644; CSP 18: 46; GR p. 239; IH 2: (in press); Kew 4: 400; LS 21709; NW p. 55; PH 444, 450; Rehder 5: 702; Tucker 1: 581; Urban-Berl. p. 311.

BIOFILE: Anon., Hedwigia 66: (119). 1926 (d. 12 Mai 1921, err. for 9 Oct 1924; 12 Mai 1921 is the day on which Mrs. E.L. Rand died); Joint Bull. Vt. bot. bird Clubs 10: 31. 1924 (obit.); Science ser. 2. 60: 427. 1924 (obit.).
Barnhart, J.H. *in* F.W. Pennell, Monogr. Acad. nat. Sci. Philadelphia 1: 617. 1935.
Day, M.A., Rhodora 3: 261-262. 1901 (herb.).
Rickett, H.W., Index Bull. Torrey bot. Cl. 80. 1955.
Robinson, B.L., Rhodora 27: 17-27. 1925 (obit., portr.).

8622. Champlain Society. *A preliminary list of the phaenogams* and vascular cryptogams *of Mt. Desert Island* Maine ... Cambridge, Mass. May 1888. (*Prelim. list phaenog. Mt. Desert Isl.*).
Publ.: Mai 1888, p. [i*], i-ii, 1-43.
Suppl. 1: Mai 1889, p. [i-ii], 1-6.
Suppl. 2: Apr 1890, p. [i-ii], 1-8.
Suppl. 3: Apr 1891, p. [i], err. slip, 1-10.
Suppl. 4: Mai 1892, p. [i], [1]-7.
Copies: HH, NY. – Hectograph printed.

8623. *Flora of Mount Desert Island, Maine*. A preliminary catalogue of the plants growing on Mount Desert and the adjacent islands ... Cambridge (John Wilson and Son ...) 1894. Oct. (*Fl. Mt. Desert Isl.*).
Co-author: John Howard Redfield (1815-1895); geological introduction by William Morris Davis.
Publ.: Aug 1894 (p. 11: 1 Jul 1894; Bot. Centralbl. 7 Sep 1894; Nat. Nov. Nov(1) 1894; Bot. Zeit. 16 Dec 1894; TBC 19 Sep 1894), p. [1]-286. *Copies*: FH, G, HH, MICH, MO, NY, PH, US, USDA. – Request for information: Bull. Torrey bot. Cl. 16(2): 78. 1889.
Ref.: Anon., Bot. Gaz. 19: 385. Sep 1894 (rev.).
Buchenau, F., Bot. Zeit. 52(2): 375-376. 16 Dec 1894 (rev.).
Warnstorf, C.F., Bot. Centralbl. 61: 26-28. 28 Dec 1892 (rev.).

Rand, Eduard Sprague (1834-1897), American horticulturist at Glen Ridge near Boston. (*E.S. Rand*).

HERBARIUM and TYPES: Some Brazilian material (1891-1894) at K.

BIBLIOGRAPHY and BIOGRAPHY: Barnhart 3: 127 (b. 20 Oct 1834); BM 4: 1644; CSP 5: 91; Herder p. 287; Hortus 3: 1202; IH 2: (in press); Jackson p. 138, 448; Langman p. 612; Morren ed. 10, p. 140; Rehder 5: 702; TL-2/2625; Tucker 1: 581.

BIOFILE: Allibone, S.A., Crit. dict. Engl. lit. 2: 1736. 1878.
Anon., Gard. Chron. ser. 3. 22: 315. 1897 (d.).
Jackson, B.D., Bull. misc. Inf. Kew 1901: 54 (pl. K).
Wilson, J.G. & J. Fiske, eds., Appleton's Cycl. Amer. biogr. 5: 169. 1888.

8624. *The Rhododendron* and "American plants". A treatise on the culture, propagation, and species of the Rhododendron; with cultural notes upon other plants which thrive under like treatment, and descriptions of species and varieties; with a chapter upon herbaceous plants requiring similar culture ... Boston (Little, Brown, and Company) 1871. Duod. (*Rhododendron*).
Publ.: 1871 (p. x: Feb 1871), p. [i]-xx, [1]-188. *Copy*: NY.

8625. *Orchids*. A description of the species and varieties grown at Glen Ridge, near Boston, with lists and descriptions of other desirable kinds. Prefaced by chapters on the culture, ... and significance of their names; the whole forming a complete manual of orchid culture ... New York (published by Hurd and Houghton ...) 1876. Oct. (*Orchids*).
Publ.: 1876 (p. xvii: Jan 1876), p. [i]-xxii, [23]-476. *Copy*: NY.

Rand, Isaac (x-1743), British apothecary, gardener and botanist; F.R.S. 1719; curator ("praefectus") of the Chelsea Physick Garden 1724-1743. (*I. Rand*).

HERBARIUM and TYPES: BM; further material at CGE and OXF. – Rand's herbarium was presented to the Chelsea garden by his widow; it was presented to BM by the Apothecaries Company in 1862 together with e.g. the herbarium of J. Ray. The specimens "so far as they were of importance" were incorporated in the general herbarium and the British collection of BM; other specimens are in the BM Sloane herbarium (Dandy 1958).

BIBLIOGRAPHY and BIOGRAPHY: Backer p. 482-483; Barnhart 3: 127; BB p. 252; BM 4: 1644; Clokie p. 229; Desmond p. 510; DNB 47: 268-269; Dryander 3: 97; Frank 3 (Anh.): 81; Henrey 2: 736 [index], 3: 105; Jackson p. 410; Kew 4: 400; Langman p. 612; Lasègue p. 324; PR ed. 1: 8347-8348; Rehder 1: 52; TL-2/see A. Buddle; Tucker 1: 581.

BIOFILE: Allen, D.E., The naturalist in Britain 18-19. 1976.
Allibone, S.A., Crit. dict. Engl. lit. 1736. 1878.
Dandy, J.E., The Sloane herbarium 188-189. 1958.
Drewitt, F.D., Apothec. Garden 56, 65. 1922.
Druce, G.C., Fl. Berkshire cxxxvi. 1897.
Gager, C.S., Brooklyn. Bot. Gard. Rec. 27(3): 241. 1938.
Kent, D.H., Brit. Herb. 74. 1957; Hist. fl. Middlesex 15-16. 1975.
Murray, G., Hist. coll. BM(NH) 1: 176. 1904 (herb.).
Pearsall, W.H., Fl. Surrey 46. 1931.
Richardson, R., Extracts lit. sci. corr. 125. 1835.
Schultes, J.A., Flora 8:36. 1825 (on state of herb. in Apothec. gard.).
Semple, R.H., Mem. bot. gard. Chelsea 41-63. 1878.
Trimen, H. & W.T.T. Dyer, Fl. Middlesex 388-389. 1869.
Wall, C., Hist. worshipf. Soc. Apothec. London 163, 167, 169-171, 174, 176, 178, 221, 271, 282, *pl. 15*. 1968.

EPONYMY: *Randa* Cothenius (1790, *orth. var.*); *Randia* Linnaeus (1753); *Rangia* Grisebach (1866, *orth. var.*).

8626. *Index plantarum officinalium*, quas, ad materiae medicae scientiam promovendam, in

Horto chelseiano, ali ac demonstrari curavit societas pharmaceutica londinensis. Londini [London] (Imprimebat J.W.) 1730. Duod.
Publ.: 1730, p. [i-v], sign. A⁶-H⁶. *Copy*: Houghton library, Cambridge, Mass.

8627. *Horti medici Chelseani*: index compendiarius, exhibens nomina plantarum, quas ad rei herbariae precipue materiae medicae scientiam promovendam, ali curavit Societas pharmacopoeorum londinensium ... Londini (sumptibus auctoris, ...) 1739. Oct. (*Hort. med. chels.*).
Publ.: 1739, p [i-viii], [1]-214, [1, err.]. *Copies*: E, FI. – See Henrey no. 1251, q.v. also, no. 1252, for the 1730 *Index plantarum officinalium*.

Range, Paul Theodor (1879-1952), German (Prussian) geologist and botanist; Dr. phil. Leipzig; geologist with the Preuss. geol. Landesanst. 1904-1906; government geologist for S.W. Africa 1906-1914; on active service in France and the Middle East 1914-1918; lecturer in general geology Univ. Berlin 1921-1932, from 1934 extraord. prof. of geology. (*Range*).

HERBARIUM and TYPES: B (rd. 1951/52, mainly Europe) private S. African herb. at SAM/NBG (rd. 1920); material from S.W. Africa: B (2000 in as far as received before 1943 mainly destroyed), BOL, L, SAM (important coll.); from Egypt: B (id.).

BIBLIOGRAPHY and BIOGRAPHY: Barnhart 3: 128; Biol.-Dokum. 14: 7303; BJI 2: 142; BL 1: 32, 56, 314; BM 8: 1047; Bossert p. 323; Christiansen p. 319; IH 1 (ed. 6): 363, (ed. 7): 340, 2: (in press); Kew 4: 401; Urban-Berl. p. 184, 386.

BIOFILE: Bullock, A.A., Bibl. S. Afr. bot. 93. 1978 (q.v. for bibl. of bot. publ. by R on S.W. Africa).
Gunn, M. & L.E. Codd, Bot. expl. S. Afr. 289-290. 1981 (coll., portr., b. 1 Mai 1879, d. 29 Aug 1952), q.v. for further biographical references).
Poggendorff, J.C., Biogr.-lit. Handw.-Buch 6(3): 2119. 1938 (b. 1 Mai 1879; bibl.).
White, A. & B.L. Sloane, The Stapelieae ed. 2. 28, 340, 1001. 1937.

NOTE: Range published *Die Flora des Namalandes* in 14 parts in Repert. Sp. nov., Fedde 30: 129-158. 1932, 33: 1-22. 1933, 35: 35-42. 1934, 36: 1-19, 97-109, 241-264. 1934, 38: 122-130, 256-280. 1935, 39: 55-60, 283-287. 1936, 43: 251-256. 1938, 44: 101-124. 1938, 45: 320. 1938.

8628. Gesellschaft für Palästina-Forschung 7. Veröffentlichung. *Die Flora der Isthmuswüste*. Auf Grund des vom Verfasser in den Jahren 1915/16 gesammelten, von Herrn Professor Dr. Georg Schweinfurth bestimmten Materials unter Benutzung der einschlägigen Literatur bearbeitet von Dr. Paul Range. Berlin 1921. (*Fl. Isthmus-wüste*).
Author of identifications: Georg August Schweinfurth (1836-1925).
Publ.: 1921 (dated 1921 but reviews seen by us and Nat. Nov. ref. are of 1924), p. [1]-44. *Copy*: USDA. – "Isthmuswüste": area between Suez and Sinai desert. See also Heft 14 of J.L. Wilser, Die Kriegsschauplätze 1914-1918 geologisch dargestellt, 1926.
Ref.: Krause, E., Bot. Jahrb. 58 (lit.): 53. 1923.

Rangel, Eugenio dos Santos (*fl.* 1931), Brazilian agronomist and phytopathologist; head of the serviço de phytopathologia of the Instituto biologico de defessa agricola. (*Rangel*).

HERBARIUM and TYPES: IAC, RB.

BIBLIOGRAPHY and BIOGRAPHY: Barnhart 3: 128; IH 2: (in press); Kelly p. 168; Stevenson p. 1255.

8629. *Contribuçao para o glossario portuguez referente à mycologia o à phytopathologia* ... Rio de Janeiro (Imprensa nacional) 1931. Oct. (*Contr. gloss. mycol. phytopath.*).
Publ.: 1931 (Nat. Nov. Sep 1932; USDA rd. Jul 1932), p. [1]-72. *Copies*: FH, NY, PH, USDA.

Other ed.: 1942, p. [1]-71. *Copy*: USDA. - "Contribuçao ... filopatologià seguido de vocábulos latinos ou alatinados e seus correspondentes em Português ...".

Ranojević, Nikola (*fl.* 1905), Serbian mycologist. (*Ranoj.*).

HERBARIUM and TYPES: Unknown.

BIBLIOGRAPHY and BIOGRAPHY: Kelly p. 186; LS 21713-21715, 38180-38185; Stevenson p. 1255.

BIOFILE: Adamovic, L., Veg.-Verh. Balkanländer 9, 21. 1909.

EPONYMY: *Ranojevicia* Bubák (1910).

NOTE: His three *Beiträge zur Pilzflora Serbiens* were published in Hedwigia 41: 89-103. 1902, and Ann. mycol. 8: 347-402. 1910 (Nat. Nov. Jul 1910), 12(4): 393-421. Aug 1914.

Raoul, Édouard Fiacre Louis (1815-1852), French naval surgeon and botanist; for the Nanto-Bordelaise Company at Akaroa, New Zealand, to serve on l'Allier (1840-1842) and l'Aube (1842-Jan 1843); collected on Banks Peninsula, N.Z., between voyages; later on west coast of Africa; from 1849-1852 teaching medicine at Brest. (*Raoul*).

HERBARIUM and TYPES: P, PC. - Other material at B (pterid. extant), BM, BR, C, FI, G, L, MEL, PH, US.

BIBLIOGRAPHY and BIOGRAPHY: Backer p. 483; Barnhart 3: 129; BM 4:1645; CSP 5: 97; GF 71; Herder p. 229; Hortus 3: 1202; IF p. 726; IH 2: (in press); Jackson p. 398, 403; Lasègue p. 500, 506; LS 21717; NI 1586; PR 7412, ed. 1: 8351; Rehder 1: 506; TL-2/2438; Tucker 1: 581 (first title only); Urban-Berl. p. 311; Zander ed. 10, p. 705, ed. 11, p. 804.

BIOFILE: Cheeseman, T.F., Man. New Zealand fl. ed. 1: xxi-xxii. 1906, ed. 2: xxiii-xxiv. 1925.
Cockayne, L. Veg. N. Zealand 3. 1921.
Godley, E.J., Trans. R. Soc. New Zealand 1(22): 244. 1967 (Raoul arrived at Akaroa, New Zealand, on 17 Aug 1840 on board l'Aube; he was stationed at Akaroa until 11 Jan 1843, being attached to l'Aube 1840-1842, and l'Allier 1842-1843. Akaroa was a settlement of the French Nanto-Bordelaise Company.
Hooker, J.D., Fl. Nov.-Zel. 1: 156. 1852.
Jenkinson, S.H., New Zealanders in science 16, 28, 155. 1940.
Pritzel, G.A., Icon. bot. index xxiv. 1855 (*Choix*: "sehr schöne Zeichnungen").
Sayre, G., Bryologist 80: 516. 1977 (bryoph. PC).
Simpson, M.J.A., New Zeal. J. Bot. 14: 199-202. 1976 (on E.F.L. and E.F.A. Raoul; biogr. notes; b. 23 Jul 1815, d. 30 Mar 1852).

EPONYMY: *Raoulia* J.D. Hooker ex Raoul (1846) and the derived genus *Raouliopsis* S.F. Blake (1938).

8630. *Choix de plantes de la Nouvelle-Zélande* recueillies et décrites par M.E. Raoul ... Paris (Fortin, Masson et Cie. ...) Leipzig (id.) 1846. Qu. (*Choix pl. Nouv.-Zél.*).
Publ.: 1-16 Jan 1846 (BF 17 Jan 1846; Amer. J. Sci. 1 Mai 1846; Flora 21 Mar 1846), p. [i-iii], [1]-53, [55, index pl.], *pl. 1-30* (uncol. copp. Alfred Riocreux; originals at K).
Copies: Edinburgh Univ. Libr., FI, G, HH, M, NY, PH, Teyler, U. - See Godley (1967): majority of species still accepted.
Prelim. publ.:Ann. Sci. nat., Bot. ser. 3. 2: 113-123. Aug 1844.
Ref.: B., Flora 29: 174-176. 21 Mar 1846 (rev.).
Schlechtendal, D.F.L. von, Bot. Zeit. 4: 411-412. 12 Jun 1846 (rev.).

Raoul, Édouard François Armand (1845-1898), French botanist and naval pharmacist; in Formosa 1873 and 1884; on a mission of circumnavigation 1885-1888; member of the High Colonial Council and professor of colonial agriculture at the "École coloniale".(*E.F.A. Raoul*).

HERBARIUM and TYPES: P.

BIBLIOGRAPHY and BIOGRAPHY: Barnhart 3: 129; Hortus 3: 1202; IH 2: (in press); Kew 4: 407; LS 21718-21719; Rehder 3: 91, 727; Tucker 1: 581 (except for first entry), 582.

BIOFILE: Grisard, J., Rev. Sci. nat. appl., Bull. Soc. Acclim. 36: 94-95. 1889 (on his mission of 1885-1888).
Jenkinson, S.H., New Zealanders and science, Christchurch 1950.
Simpson, M.J.A., New Zeal. J. Bot. 14: 199-202. 1976 (biogr. notes, b. 20 Aug 1881; d. 26 Apr 1898).

NOTE: We have not seen a copy of Ch. Hetley & E. Raoul, *Fleurs sauvages et bois précieux de la Nouvelle Zélande*, Paris 1889 (Nat. Nov. Sep (1) 1889), *37 pl.*, see Jenkinson (1950), Northcote-Bode (1971) and Simpson (1976).

8631. *Manuel pratique des cultures tropicales* et des plantations des pays chauds par E. Raoul et P. Sagot. Tome ii. – 1re partie. *Culture du caféier* semis, plantation, taille, cueillette, dépulpation, décorticage, expédition, commerce, espèces et races ... Paris (Augustin Challamel, ...) 1894. Oct. (*Man. cult. trop., Cult. caf.*).
Co-author: Paul Antoine Sagot (1821-1888); collaborator for commercial part: E. Darolles.
Publ.: 1894 (Nat. Nov. Dec(1) 1894; Bot. Zeit. 16 Aug 1895), p. [i-ii], front., [iii], [1]-249. *Copies*: BR, L.
Ed. 2: 1897, p. [i], frontisp., [iii], [1]-251. *Copy*: BR. – By E. Raoul with assistance by E. Darolles.

Rapaics von Rumwerth, Raymund (1885-1954), Hungarian botanist; assistant at the agricultural college of Kolozsvár (Klausenburg, Cluj) 1910, ass. prof. ib. 1912; later at Budapest. (*Rapaics*).

HERBARIUM and TYPES: Material at BP, C and GH.

BIBLIOGRAPHY and BIOGRAPHY: BJI 2: 142; Bossert p. 324 (b. 15 Feb 1885, d. 19 Mar 1954); IH 2: (in press); Kew 4: 407; LS 38188; MW p. 399; Plesch p. 371.

BIOFILE: Anon., Hedwigia 50: (89). 1910 (Kolozsvár), 52: (154). 1912 (asst. prof. ib.), 53:(96). 1912 (called to Debreczen); Österr. bot. Z. 62: 150. 1912 (app. Kolozsvár).
Kneucker, A., Allg. bot. Z. 18: 80. 1912 (app. Kolozsvár).
Verdoorn, F., ed., Chron. bot. 2: 203. 1936, 3: 177, 374. 1937.

Raper, Kenneth Bryan (1908-x), American mycologist; studied at Univ. N. Carolina (AB 1929), George Washington Univ. (AM 1931) and Harvard (Dr. phil. 1936); professor of microbiology Univ. of Wisconsin. (*Raper*).

HERBARIUM and TYPES: Culture collections CBS, ATCC (Washington) and NRRL coll. Peoria Ill.

BIBLIOGRAPHY and BIOGRAPHY: Ainsworth p. 346 (index); Barnhart 3: 129 (b. 11 Jul 1918); Bossert p. 324; Kew 4: 407; Lenley p. 339; Roon p. 92; Stevenson p. 1255.

BIOFILE: Rogers, D.P., Brief hist. mycol. N. Amer. ed. 2: 59, 79. 1981.
Rogerson, C.T., Mycologia index 1027-1028. 1968 (list of publ. in Mycologia).
Verdoorn, F., ed., Chron. bot. 2: 404. 1936, 5: 239. 1939, 7: 340. 1943.

EPONYMY: *Raperia* C.V. Subramanian et C. Rajendran (1976).

NOTE: Main publications outside our period: *Manual of Aspergilli*, 1945 (with Charles Thom; see e.g. J. NYBG 46: 293. Dec 1945), *Manual of the Penicillia*, 1949 (with Charles Thom), reprinted in facsimile 1968, and *The genus Aspergillus* 1965 (with D.I. Fennell).

Rapin, Daniel (1799-1882), Swiss botanist and pharmacist; in pharmacies at Fribourg, Strassbourg, Paris, Genève (pharm. Carouge), at Payerne 1832-1838, at Rolle 1838-1853; in retirement at Yverdon 1853-1857, from 1857 at Plainpalais (Genève). (*Rapin*).

HERBARIUM and TYPES: G; further material at BR, CGE, GOET, MANCH. – The original herbarium was partially destroyed by fire in the University of Geneva. Most of his plants are, however, still at G (Briquet 1940, Burdet, Candollea 32: 405. 1977). The cryptogamic collection of Rapin was acquired by Marc Micheli and presented by the latter's widow to G in 1904. Correspondence: G.

BIBLIOGRAPHY and BIOGRAPHY: AG 4: 296; Barnhart 3: 129; BL 2: 590, 706; BM 4: 1646; Bossert p. 324; CSP 5: 97, 8: 701, 12: 599; IH 2: 572 (with J. Muret) and in press; Jackson p. 345; Kew 4: 407; Lasègue p. 543; Morren ed. 10, p. 110; PR 7413-7414, 10620, ed. 1: 8352-8353; Rehder 1: 434, 2: 286, 287; Tucker 1: 582.

BIOFILE: Anon., Bull. Soc. bot. France 10: 669. 1863.
Briquet, J., Bull. Soc. bot. Suisse 50a: 391-394. 1940 (biogr., bibl., epon., b. 18 Oct 1799, d. 24 apr 1802).
Burdet, H.M., Saussurea 6: 28. 1975 (portr.).
Jaccard, H., Cat. fl. valais. ix. 1895.

HANDWRITING: Candollea 32: 405-406. 1977.

EPONYMY: *Rapinia* Loureiro (1790), *Rapinia* Montrouzier (1860) and the derived *Neorapinia* Moldenke (1955) honor René Rapin (1621-1687), French author on gardening.

8632. *Esquisse de l'histoire naturelle des Plantaginées*, ... Paris (au secrétariat de la Société Linnéenne ...) 1827. Oct.
Publ.: 1827 (p. 4: 1 Aug 1827, corrected to 15 Feb 1827), p. [1]-55. *Copies*: G(2), NY. – Reprinted and to be cited from Ann. (Mém.) Soc. Linn. Paris 6: 437-490. 1827.

8633. *Le guide du botaniste dans le canton de Vaud*. Comprenant la description de toutes les plantes vasculaires qui croissent spontanément dans ce canton, et l'indication de celles qui y sont généralement cultivées pour les usages domestiques ... Lausanne (chez tous les libraires) 1842. Duod. (in sixes). (*Guide bot. Vaud*).
Ed. 1: 1842 (Bot. Zeit. 29 Dec 1843 "published in the year 1843), p. [i*-iii*], [i]-xxiii, [1]-488. *Copies*: BR, G, HH, MO, NY.
Ed. 2: 1862 (p. viii: 31 Mai 1862; BSbF Sep 1863), p. [i]-xxiv, [1]-772. *Copies*: BR, G(2), HH, NY. – "*Le guide* ... *Vaud* Comprenant en outre le bassin de Genève et le cours inférieur du Rhone en Valais ... deuxième edition". Genève et Paris (Joël Cherbuliez, ...) 1862. Duod.

8634. *Flore des plantes vénéneuses* de la Suisse, contenant leur description, l'époque de leur floraison, les lieux ou elles croissent naturellement, l'indication de celles qui y sont employées en médecine, les symptômes qu'elles produisent sur l'économie animale, et les premiers soins à donner dans les empoisonnements. Destinée à l'usage des écoles et des gens de la campagne. Payerne (chez Louis Gueissaz, ...) 1849. Oct. (*Fl. pl. vénén.*).
Publ.: 1849 (but cover NY copy dated 1850), p. [1]-116, [1, cont.], *pl. 1-23* (col. or hand-col. liths.). *Copies*: G (uncol.), NY (hand-col.). – Published anonymously.

8635. *Description de deux nouvelles espèces de Roses* ... Extrait du Bulletin de la Société royale de Botanique de Belgique tome xiv, p. 237 (1875). Oct.
Publ.: 1875, p. [1]-3. *Copy*: G. – Reprinted and to be cited from Bull. Soc. Bot. Belg. 14: 237-239. 1875.

Rapp, Arthur Roman (1854-after 1894), Latvian (Livonian) physician and botanist;

at the State Gymnasium of Dorpat (Jurjew) 1864-1872; studied medicine at Dorpat, Wien and Berlin; practicing physician at Laudohn 1882-1884, Kosenhof, Wainsel; and ultimately in Lemsal 1887-1892; retired to Dorpat 1892 because of ill-health. (*Rapp*).

HERBARIUM and TYPES: Unknown.

BIBLIOGRAPHY and BIOGRAPHY: BM 4: 1646; CSP 18: 54; Rehder 1: 366; Tucker 1: 582.

BIOFILE: Lipschitz, S.J., Florae U.R.S.S. fontes no. 505. 1975.
Klinge, J., *in* A. Rapp, Fl. Umgeb. Lemsal. 1895.
Mühlenbach, V., Apkartraktsts 114: 15. 1978, *also*: personal communication to authors. Apr 1979.
Verdoorn, F., ed., Chron. bot. 3: 280. 1937.

8636. *Flora der Umgebung Lemsals und Laudohns.* Zwei Beiträge zur Flora Livlands ... herausgegeben und mit einer phytogeographischen Einleitung versehen von Dr. botan. J. Klinge. Riga (Druck von W.F. Häcker). 1895. Oct.
Publ.: 5 Apr 1895 (in journal), p. [i]-ix, [1]-94. *Copies*: H, HH, NY. – Reprinted and to be cited from "Festschrift des Naturforscher-Vereins zu Riga in Anlass seines 50-jährigen Bestehens am 27 März (8 April) 1895", in Corr.-Bl. Naturf. Ver. Riga. 1895: 59-160. – Curriculum vitae of Rapp by J. Klinge on p. ix.

Rapp, Severin (1853-1941), German (Württemberg) shoemaker; emigrated to the United States in 1883; settled at Sanford, Florida 1884; from 1907 interested in botany and an active bryological collector who contributed to exsiccatae issued by A.J. Grout and F. Verdoorn; died in an automobile accident. (*S. Rapp*).

HERBARIUM and TYPES: FLAS (15.000); further material at DPU, GOET, H, M, MICH, NY and PH.

BIBLIOGRAPHY and BIOGRAPHY: Barnhart 3: 129 (b. 6 Sep 1853, 19 Oct 1941; original info. q.v.); IH 1 (ed. 7): 340, 2: (in press); Lenley p. 339.

BIOFILE: Grout, A.J., Bryologist 45: 179-180. 1942 (personal recollections).
Hertel, H., Mitt. Bot. München 16: 420. 1980 (lich. M).
Murrill, W.A., Hist. found. bot. Florida 23, 32. 1945.
Sayre, G., Mem. New York Bot. Gard. 19(3): 384. 1975 (mosses distributed in various exsiccatae).
Steere, W.C., Bryologist 45: 28. 1942 (d. 19 Oct 1941); Bot. Rev. 43: 293. 1977.
Verdoorn, F., ed., Chron. bot. 7(7): 353-354. 1943 (d.).
Wagenitz, G., Index coll. princ. herb. Gott. 131. 1982.

Rásky, Klara (1908-1971), Hungarian paleobotanist at the Department of Botany of the Museum of Natural History, Budapest 1938-1971; specialist on fossil of Characeae. (*Rásky*).

HERBARIUM and TYPES: BP.

BIBLIOGRAPHY and BIOGRAPHY: Andrews p. 316-317; Biol.-Dokum. 14: 7311; Roon p. 92.

BIOFILE: Szujkó-Lacza, J., Ann. hist.-nat. Mus. natl. hung. 65: 11-13. 1973 (b. 18 Mar 1908, d. 14 Sep 1971; obit., bibl.).

EPONYMY: *Raskyaechara* H. Horn af Rantzien (1959); *Raskyella* L. Grambast et N. Grambast (1954).

Rasmussen, Rasmus (1871-1962), Danish botanist of the Faroe Islands, born in Midvåg, carpenter, teacher, writer. (*Rasmussen*).

HERBARIUM and TYPES: Natural History Museum Tórshavn, Faroe Isl.; other material at C.

NOTE: Rasmussen used the pseudonym Regin i Lid for his novel *Bábilstornid* (1909). – Information on Rasmussen was kindly provided by Alfred Hansen.

BIOFILE: Hansen, A., Bot. Tidsskr. 58(4): 327-328. 1963.

8637. *Føroya flora* Givid út hevur Føroya málfelag. Tórshavn 1936. Oct. (*Føroya fl.*).
Ed. 1: 1936, p. [i]-xxvi, [1]-160. *Copy*: A. Hansen.
Ed. 2: 1952, p. [i]-xxviii, [1]-232. *Copy*: A. Hansen. – Reprinted by off-set, Tórshavn 1970.

Raspail, François Vincent (1791-1878), French physician and botanist; chemist and popular writer; practiced medicine in Paris, propagated camphor as a medicine (e.g. camphor cigarettes), published on veterinary sciences and meteorology; active political agitator (J. Bot.). (*Raspail*).

HERBARIUM and TYPES: Unknown.

BIBLIOGRAPHY and BIOGRAPHY: AG 2(1): 166; Ainsworth p. 154-155, 346; Barnhart 3: 129 (b. 29 Jan 1791, d. 7 Jan 1878); BM 4: 1646-1647, 8: 1048-1049; Bossert p. 325; CSP 5: 99-102; Herder p. 119, 277; Jackson p. 102, 133; Kew 4: 408; Langman p. 612-613; LS 21726-21728; Moebius p. 200, 203, 438; NI 2: 296, 3: 56; Plesch p. 372; PR 8360-8363, ed. 1: 7417-7418; Quenstedt p. 352; Rehder 5: 703; Sotheby 620; TL-2/905, 912; Zander ed. 10, p. 705, ed. 11, p. 804.

BIOFILE: Anon., Bot. Not. 1879: 31; Bot. Zeit. 6: 390-391. 1848 ("Jakobiner der wüthendsten Sorte, nebenbei praktischer Arzt, der alles mit Kampher heilt, ... von zwei bösen Dämonen, Argwohn und Scheelsucht, gepeinigt ..."); Bot. Zeit. 36: 190. 1878 (d.); J. Bot. 8: 63. 1870, 16: 96. 1878 (d. 6 Jan 1878); Sterbeeckia 3(4): 18-19. 1964.
Blanchard, R., Arch. de Parasitol. 8: 5-87. 1904 (detailed biogr., portr., bibl.).
Freund, H. & A. Berg et al., Gesch. Mikroskopie 315, 320. 1963.
Guédès, M., J. Soc. Bibl. nat. Hist. 5(2): 110-116. 1969.
Jessen, K.F.W., Bot. Gegenw. Vorz. 436-437. 1884.
Sachs, J., Hist. botanique 332. 1892.
P.L., Nouvelle biogr. gén. 41: 671-679. 1918.
Poggendorff, J.C., Biogr.-lit. Handw.-Buch 2: 571. 1863, 3: 1089. 1898 (d. 8 Feb 1878).
Raspail, X., Arch. de Parasitol. 8: 82-87. 1904.
Stieber, M.T., Huntia 3(2): 124. 1979 (notes Agnes Chase).
Wittrock, V.B., Acta Horti Berg. 3(2): 117. 1903, 3(3): 112. 1905.

EPONYMY: *Raspalia* A.T. Brongniart (1826); *Raspallia* Arnott (1841, orth. var.).

8638. *Mémoires sur la famille des Graminées* contenant: 1. la physiologie; 2. la classification des graminées; 3. l'analyse microscopique et le développement de la fécule dans les céréales ... Paris 1825 [-1826]. Oct.
Publ.: A set of independently paged reprints from Ann. Sci. nat. (*Ann.*), preceded by preface material. *Copies*: G, NY, US.
Preface material: dated 1825, p. [i-iii].
  1: Sur la formation de l'embryon dans les Graminées, Mar 1825, p. [1]-48, *pl. 13-14*. – Reprinted from *Ann.* 4: 271-319. Mar 1825.
  2: Essai d'une classification générale des Graminées.
    (*1*): Apr 1825, p. 1-28, chart, *Ann.* 4: 423-451. Apr 1825.
    (*2*): Jul 1825, p. 1-52, *pl. 8-10*, *Ann.* 5: 287-311. Jun 1835, 433-460. Jul 1825. – See also Flora 9: 1-29, 33-40. 1826.
  3: Développement de la fécule ...
    (*1*): Oct 1825, p. [1]-16, *Ann.* 6: 224-239. Oct 1825.
    (*2*): Nov 1825, p. [1]-44, *pl. 16*, *Ann.* 6: 384-427. Nov 1825.
  4: Additions au mémoire sur ... la fécule. Mar 1826, *Ann.* 7: 325-335. Mar 1826.

German transl.: Abhandlung über die Bildung des Embryo in den Gräsern und Versuch einer Klassifikation dieser Familie, translated by B. Trinius, St. Petersburg 1826. Oct., xii, 121 p., 2 pl. (n.v.).

8639. *Nouveau système de physiologie végétale de botanique*, fondé sur méthodes d'observation, qui ont été développées dans le nouveau système de chimie organique, accompagné d'un atlas de 60 planches d'analyses ... Paris (chez J.-B. Baillière, ...), Londres (même maison) 1837, 2 vols. and atlas. Oct. (*Nouv. syst. phys. vég.*).
1: 1837 (p. xxiv: 1 Nov 1836), p. [i]-xxxi, [xxxii, err.], [1]-599.
2: 1837, p. [i]-viii, [1]-658, [659, err.], [660, binder], 4 charts.
Atlas: 1837, p. [1]-91, *pl. 1-60* (uncol. liths.).
Copies: FI, G, NY, PH, US, USDA.
Bruxelles édition: 1837, p. [i]-xii, [1]-450, [i]-xlvii, 2 charts; *atlas*: p. [i-iii], *pl. 1-60* (id.). – "*Nouveau système*... par S.V. Raspail [sic]. Bruxelles (Société typographique belge, ...) 1837. Oct. *Copy*: USDA. – For a detailed description of an 1840 reissue of this edition see Guédès (1969). A re-issue (dated 1838) of the Paris edition has also not been seen by us.

Rassadina, Kseniya Aleksandrovna (1903-x), Russian lichenologist; studied at the University of Leningrad 1922-1926; from 1925 connected with the Komarov Botanical Institute, Leningrad. (*Rass.*).

HERBARIUM and TYPES: LE; some further material at E.

BIBLIOGRAPHY and BIOGRAPHY: GR p. 551 (b. 10 Dec 1903) *pl. 26*; GR, cat. p. 70; MW p. 399, suppl. p. 289-290; SBC p. 130.

BIOFILE: Hedge, I.C. & J.M. Lamond, Index coll. Edinb. herb. 121. 1976.

COMPOSITE WORKS: With Kopaczevskaja, E.G. et al., *Handbook of the lichens of the U.S.S.R.*, vol. 1, 410 p., Leningrad 1971 (in Russian).

NOTE: After our closing date: *Tsetrarija* (*Cetraria*) CCCP (Cetraria in the USSR). Leningrad 1950. Oct., p. [171]-304. – Reprinted and to be cited from Trav. bot. Inst. Komarov ser. 2. 5: 171-304. 1950.

Ráthay, Emerich (1845-1900), Hungarian botanist; from 1894-1900 professor and director of the oenological-pomological institute at Klosterneuburg near Wien. (*Ráthay*).

HERBARIUM and TYPES: Unknown.

BIBLIOGRAPHY and BIOGRAPHY: Barnhart 3: 130 (d. 9 Sep 1900); BJI 1: 47; BM 4: 1647; DTS 1: xxiii, 234-235; Herder p. 119, 132; Jackson p. 89, 105; Kelly p. 186; Kew 4: 409; LS 21732-21766, 38192; Morren ed. 10, p. 31; Rehder 5: 703-704; Stevenson p. 1255; Tucker 1: 582.

BIOFILE: Anon., Bot. Centralbl. 59: 63. 1894 (dir. Klosterneuburg), 84: 208. 1900 (d.), Bot. Jahrb. 18 (Beibl. 46): 15. 1894 (app. Klosterneuburg), 30: (Beibl. 68: 53. 1901 (d.); Bot. Not. 1901: 114 (d.); Hedwigia 33: (147). 1894 (app. Klosterneuburg), 39: (222). 1900 (d.); Österr. bot. Z. 50: 387. 1900 (d. 9 Sep 1900); Kneucker, A., Allg. bot. Z. 6: 212. 1900 (d.).

8640. *Untersuchungen über die Spermogonien der Rostpilze* ... Wien (aus der kaiserlich-königlichen Hof- und Staatsdruckerei ...) 1882. Qu. (*Unters. Spermog. Rostpilze*).
Publ.: Aug-Oct 1882 (read 9 Jun 1882; Nat. Nov. Nov(2) 1882; Bot. Zeit. 22 Dec 1882; Bot. Centralbl. 13-17 Nov 1882), p. [i], [1]-51. *Copy*: G. – Preprinted or reprinted from Denkschr. k. Akad. Wiss., mat.-nat. Cl. 46: 1-51. 1883.
Ref.: Klebs, [E.], Bot. Zeit. 40: 906-908. 22 Dec 1882 (rev.).
Müller, Herm., Bot. Zeit. 40: 908-909. 22 Dec 1882 (rev.).

Rathke, Jens (1769-1855), Norwegian botanist; Mag. theol. Kjøbenhavn 1792; teacher at Christiania (Oslo); professor of zoology at Kjøbenhavn 1810-1813; professor of natural history at Christiania 1813-1845, director of the botanical garden ib. 1816-1843. (*Rathke*).

HERBARIUM and TYPES: C; some further material at B (200 rd. 1839, probably all destroyed) and L.

BIBLIOGRAPHY and BIOGRAPHY: Barnhart 3: 130 (b. 14 Nov 1769, d. 28 Feb 1855); BM 4: 1648; IH 2: (in press); Jackson p. 414; Kew 4: 409; Kleppa p. 9, 18, 298; PR 7423, ed. 1: 8366; Rehder 1: 40.

BIOFILE: Anon., Bonplandia 3: 115. 1855 (d.).
Gager, C.S., Brooklyn Bot. Gard. Rec. 27(3): 313. 1938.
Gilbert, P., Comp. biogr. lit. deceased entom. 309. 1977.
Hansen, A., Bocagiana 51: 10. 1980 (Madeira coll. C).
Hendriksen, K.L., Entom. Meddel. 15(4): 152-153. 1925 (in rev. of Danish entom. hist.).
Hornemann, J., Nat. Tidsskr. 1: 585. 1837 (epon. Rathkea).
Johnson, I.H., Samtiden 63: 551-557. 1954, 64: 213-228. 1955 (biograph. vol.; see also Nordhagen 1954).
Natvig, L.R., Norsk entom. Tidsskr. 7(1/2): 4-5. 1944 (entomology at the royal Frederiks Univ.; biogr. notes, portr.).
Nordhagen, R., Samtiden, Oslo 64: 139-156. 1955 ("Professor Jens Rathke og Henrik Wergelands Blomster").
Rostrup, E., Dansk biogr. Lex. 13: 524-525. 1899.
Wittrock, V.B., Acta Horti Berg. 3(2): 170-171. 1903, 3(3): 184. 1905.

EPONYMY: *Rathkea* H.C.F. Schumacher (1827).

Rattan, Volney (1840-1915), American school teacher and botanist; in California from 1861, settling in Placerville; at Santa Cruz high school 1873-1876; at girl's high school San Francisco 1876-1889; at San José normal school 1889-1906; in retirement living at Berkeley. (*Rattan*).

HERBARIUM and TYPES: DS; further material at GH, JEPS, NY, PH, RSA, US (p.p. orig.), WELC.

BIBLIOGRAPHY and BIOGRAPHY: Barnhart 3: 130 (b. 23 Mai 1840, d. 5 Mar 1915); BM 4: 1649; Bossert p. 325; CSP 12: 599; Hortus 3: 1569; IH 1 (ed. 6): 363, (ed. 7): 240, 2: (in press); Jackson p. 362, 509; Kew 4: 409; ME 1: 38; NW p. 56; Rehder 1: 324; Tucker 1: 582.

BIOFILE: Anon., Torreya 15: 60. 1915 (d.).
Brewer, W.H., *in* S. Watson, Bot. Calif. 2: 558, 559. 1880.
Ewan, J. et al., Short hist. bot. U.S. 44. 1969.
Jepson, W., Madroño 1: 168-170. 1926 (biogr., portr.).
Rodgers, A.D., Amer. bot. 1873-1892, p. 215-216, 253, 276, 280, 300. 1944.
Thomas, J.H., Contr. Dudley Herb. 5(6): 160. 1961 (herb. at DS specimens often lack data).

8641. *Analytical key to West Coast botany*, containing descriptions of sixteen hundred species of flowering plants, growing west of the Sierra Nevada and Cascade Crests, from San Diego to Puget Sound ... San Francisco (A.L. Bancroft and Company) 1887. Oct. (*Anal. key West Coast bot.*).
Ed. 1: Feb-Mar 1887 (p. 3: 22 Jan 1887; TBC Apr 1887; Nat. Nov. Mai 1887; Bot. Centralbl. 18 Mai 1887; Bot. Zeit. 28 Oct 1887), p. [1]-128. *Copies*: HH, NY(3). – Also included in his "*A popular California flora*, ed. 7, 1887.
Ed. [2]: 1888, p. [1]-128. *Copies*: HH, MICH, NY, US. – Idem ed. 8, 1888, 1892, 1894.
Ed. [2a]: 1896, p. [1]-128. *Copy*: NY. – Idem ed. 9, 1896.

Ed. 3: Mar-Apr 1898 (p. 4: 8 Feb 1898; Nat. Nov. Jun(1) 1898), p. [1]-221. *Copies*: G, NY. – "West Coast botany, an analytical key to the flora of the Pacific Coast, in which are described over eighteen hundred species ..." San Francisco (the Whitaker & Ray Co.). 1898.

8642. *A popular California flora*, or manual of botany for beginners. Containing descriptions of exogenous plants growing in Central California, and Westward to the Ocean ... San Francisco (A.L. Bancroft and Company) 1879. Oct. (*Pop. Calif. fl.*).
Ed. 1: Feb-Mar 1879 (p. [3]: Feb 1879; Nat. Nov. Mai(2) 1879), p. [1]-106, [1, err.]. *Copies*: HH, ILL (inf. D.P. Rogers), NY.
Ed. 2: 1880, (p. [iii]: Feb 1880), p. [i]-xviii, [1]-138, [139, err.]. *Copies*: HH, ILL, NY, US, USDA. – Second edition, revised and enlarged. San Francisco (id.) 1880. Oct.
Ed. 3: 1882 (p. iv: Jan 1882), p. [i]-xxviii, [1]-14, 14a-d, [15]-20, 20a-b, 21-38, 38a-b, 39-88, 88a-b, 89-138. *Copy*: NY. – Third edition, revised and enlarged. San Francisco (id.) 1882. Oct.
Ed. 4: 1882 (p. iv: Jan 1882), p. [i]-xxviii, [1]-14, 14a-d, [15]-20, 20a-b, 21-38, 38a-b, 39-88, 88a-b, 89-138. *Copy*: HH.
Ed. 5: 1883, p. [i]-xxviii, [1]-14, 14a-d, 15-20, 20a-b, 21-38, 38a-b, 39-88, 88a-b, 39-138. *Copy*: NY. – Fifth revised edition. San Francisco (id.) 1883. Oct.
Ed. 6: 1885, p. [i]-xxviii, [1]-14, 14a-d, 15-20, 20a-b, 21-38, 38a-b, 39-88, 88a-b, 89-138. *Copy*: USDA. – Sixth revised edition. San Francisco (id.) 1885. Oct.
Ed. 7: 1887, p. [i]-xxviii, [1]-14, 14a-d, 15-20, 20a-b, 21-81, 88a-b, 89-106, [1-12]. *Copies*: HH, NY(2). – Seventh revised edition. San Francisco (The Bancroft Company) 1887. Oct.
Ed. 8(1): Jun 1888 (NY copy rd. Jun 1888; J. Bot. Dec 1888; Bot. Zeit. 28 Dec 1888), p. [i]-xxviii, [1]-14, 14a-d, 15-20, 20a-b, 21-38, 38a-b, 39-88, 88a-b, 89-106, [1-12, glossary]. *Copy*: NY. – Eighth revised edition. San Francisco (id.) 1888.
Ed. 8(2, 3): 1892 and 1894, p. [i]-viii, [1]-14, 14a-d, [15]-20, 20a-b, 21-38, 38a-b, 39-88, 88a-b, 89-106, [1-12]. *Copies*: HH, MICH, NY(2), US. – Eighth revised edition. San Francisco (id.) 1892. Oct. – The second NY copy is dated 1892, rather than 1894, but otherwise identical.
Ed. 9: 1896, p. [i]-xxviii, [1]-14, 14a-d, 15-20, 20a-b, 21-38, 38a-b, 39-88, 88a-b, 89-106, [1-12, glossary]. *Copy*: NY. – Ninth revised edition. San Francisco (The Whitaker & Ray Co.) 1896. Oct.

Rattke, Wilhelm (*fl.* 1884), German botanist. (*Rattke*).

HERBARIUM and TYPES: Unknown.

BIBLIOGRAPHY and BIOGRAPHY: Barnhart 3: 1649; Rehder 1: 388.

8643. *Die Verbreitung der Pflanzen* im allgemeinen und besonders in Bezug auf Deutschland ... Hannover (Helwingsche Verlagsbuchhandlung ...) 1884. Oct. (*Verbr. Pfl.*).
Publ.: Jun-Jul 1884 (p. iv: Feb 1884; Bot. Centralbl. 4-8 Aug 1884; Flora 21 Aug-10 Sep 1884; Bot. Zeit. 25 Jul 1884), p. [i]-vi, [1]-135. *Copies*: G, HH, USDA.
Ref.: J., Österr. bot. Z. 34: 410-411. 1 Nov 1884.
Roth, E., Bot. Centralbl. 19: 332-333. 8-12 Sep 1884.

Rattray, James (*fl.* 1835), Scottish botanist. (*Js. Rattray*).

HERBARIUM and TYPES: Unknown.

BIBLIOGRAPHY and BIOGRAPHY: Barnhart 3: 130; BB p. 253; BM 4: 1649; Jackson p. 234; Kew 4: 410; PR ed. 1: 8367; Roon p. 255; SO 844; TL-2/4348.

BIOFILE: Bridson, G.D.R. et al., Nat. hist. mss. res. Brit. Isl. 229.61. 1980.
Freeman, R.B., Brit. nat. hist. books 290. 1980.
Merrill, E.D., Contr. U.S. natl. Herb. 30(1): 249. 1947 (bibl. Pacific).
Newton, L.M., Phycol. Bull. 1: 18. 1952 (algae at BM).
VanLandingham, S.L., Cat. Diat. 6: 3584. 1978, 7: 4222. 1978.

Rattray, John (1858-1900), Scottish diatomist; MA. Aberdeen 1880. (*Rattray*).

HERBARIUM and TYPES: BM.

BIBLIOGRAPHY and BIOGRAPHY: Barnhart 3: 130; BB p. 253; BM 4: 1649-1650; CSP 18: 60-61; Desmond p. 512 (b. 29 Jun 1858, d. 1900); De Toni 1: ciii-civ, 2: xcix-c, 4: xlvi; LS 21767; MW p. 400.

BIOFILE: Merrill, E.D., Contr. U.S. natl. Herb. 30(1): 249. 1947 (bibl. Pacific). Newton, L.M., Phycol. Bull. 1: 18. 1952 (algae at BM). VanLandingham, S.L., Cat. diat. 4: 2365, 6: 3584, 7: 4222. 1978.

EPONYMY: *Rattrayella* G.B. De Toni (1889). *Note*: *Rattraya* J.B. Phipps (1964) honors Dr. J.M. Rattray, an authority on African grasslands.

NOTE: Not to be confused with J. Rattray (*fl.* 1835-1886), who made a West African journey on board the "Buccaneer" in 1886, collecting in Madeira, Sierra Leone, Conakry, Dakar, Accra, S. Tomé, Principe, St. Paul de Loanda. His first set is at E. – See Oliver, Trans. bot. Soc. Edinb. 16: 475-479. 1886. – For George Rattray (1872-1941), collaborator of J. Hutchinson for the Cycadaceae, Fl. cap. 5 (2 suppl.): 24-44. 1933, see M. Gunn & L.E. Codd, Bot. expl. S. Afr. 290. 1981.

8644. *A revision of the genus Coscinodiscus*, Ehrb., and of some allied genera ... Edinburgh (printed by Neill and Company) 1890. Oct.
Publ.: [Jan-] Jul 1890 (Jul 1890 date on cover NY copy; Nat. Nov. Apr(2) 1890), p. [i]-viii, [1]-244, *pl. 1-2* (uncol. liths auct.). *Copies*: NY, USDA. – Reprinted and to be cited from Proc. R. Soc. Edinburgh 16: 449-692, *pl. 1-2*. 1890.

Ratzeburg, Julius Theodor Christian (1801-1871), German forester, forestry-botanist and zoologist; Dr. med. Berlin 1825; habil. Berlin 1828; professor of natural sciences at the forestry college of Neustadt 1840-1868; in retirement in Berlin 1868-1871; founder of forest entomology. (*Ratzeb.*).

HERBARIUM and TYPES: Unknown. – Letters to D.F.L. von Schlechtendal at HAL.

BIBLIOGRAPHY and BIOGRAPHY: ADB 27: 371-372 (W. Hess); AG 2(1): 319; Barnhart 3: 130 (b. 15 Feb 1801, d. 24 Oct 1871); Biol.-Dokum. 14: 7318, BM 4: 1650, 8: 1050, CSP 5: 106-107, 8: 705-706, 12: 599-600; Frank (Anh.): 81; Herder p. 472 [index]; Jackson p. 102, 104, 166, 167; Kew 4: 410; LS 3742, 21768; Moebius p. 399; MW p. 400; NI 226, 730; PR 1091, 3864, 6143, 7424-7426, ed 1: 1091, 1236, 4227-4228, 8368-8370; Ratzeburg p. 421-429; Rehder 5: 704; TL-1/833; TL-2/708, 2074, 2508, 5895; Tucker 1: 582.

BIOFILE: Anon., Bonplandia 3: 33. 1855, 6: 50. 1858 (cogn. Leopoldina: Gleditsch II); Bot. Zeit. 3: 560. 1845 (Légion d'honneur); Bull. Soc. bot. France 18 (bibl.): 237. 1872 (d.); Flora 34: 48. 1851 (roth. Adler Order, 3rd class, "mit der Schleife"), 55: 16. 1872 (obit.); Jahresber. Schles. Ges. vaterl. Kultur 1871: 352-357. 1872; J. Bot. 10: 64. 1872 (d.); Österr. bot. Z. 22: 103. 1872; Petites nouv. entom. 4: 197-198. 1872; Rev. Mag. de Zool. ser. 2. 23: 157 (d.); Verh. bot. Ver. Prov. Brandenburg 13: [iii]. 1872 (d. 24 Oct 1871).
Ascherson, P., Fl. Provinz Brandenburg 1: 11. 1864; Bot. Zeit. 29: 795-796. 1871 (obit.).
Becker, K. & E. Schumacher, S.B. Ges. naturf. Freunde Berlin ser. 2. 13(2): 146. 1973 (index to publ. in journal).
Danckelmann, B., J. Forst-Jagdwesen 4: 307-323. 1872 (obit. portr., bibl.).
Deyrolle, E., Petites nouvelles entom. 4(49): 197-198. 1872 (d.).
Dohrn, C.A., Entom. Zeit., Stettin 33(1-3): 81. 1872 (obit.).
Gilbert, P., Comp. biogr. lit. deceased entom. 310. 1977.
Guérin-Méneville, F.E., Revue Mag. Zool. ser. 2. 23. 157-158. 1872.
Hess, R., Lebensbilder hervorragender Forstmänner 280-283. 1885 (biogr. sketch, bibl. further biogr. refs.).
Jessen, K.F.W., Bot. Gegenw. Vorz. 382. 1864.

Judeich, J.F. & H. Nitsche, Lehrb. mitteleur. Forstinsektenk. 1: 1-6. 1885 (biogr. sketch, portr.).
Kraatz, G., Berl. entom. Z. 15: viii. 1871 (d.).
Nissen, C., Zool. Buchill. 330-331. 1969.
Phoebus, P., *in* J.T.C. Ratzeburg, Forstwiss. Schriftst.-Lexik. ii-x. 1872 (biographical preface to R's posthumously publ. work).
Poggendorff, J.C., Biogr.-lit. handw.-Buch 2: 573. 1863, 3: 1092. 1898.
Schenkling-Prévôt, Insekten-Börse 13(36): 239-241. 1896 (biogr., portr.).
Wallace, A.R., Trans. entom. Soc. London 1871: lii-liii (obit.), Entomologist 6: 55-56. 1872.
W. Str., Zool. Garten 12: 380. 1871 (obit.).

COMPOSITE WORKS: with J.F. von Brandt, *Abb. Beschr. Deutschl. Giftgew.* 1828-1838, see TL-2/708. *Additional information*: Vol. 1, Heft 2 was available by Apr 1829 (not 1839); Heft 3 came out Jul 1829-Jun 1830; Hefte 8-10 were available by 21 Sep 1834. Vol. 2, preface Jun 1838, also came out in 10 Hefte and was completed by Mar 1839, date of review in Linnaea 12 (Lit.): 242-243. Mar 1838-Mar 1839; on 5 Aug 1839 a complete copy was presented to the Regensb. bot. Ges. (Flora 22: 540. 1839. – A second copy of ed. 2, 1838 is at FI.

EPONYMY: *Ratzeburgia* Kunth (1831).

8645. *Animadversiones quaedam ad peloriarum indolem definiendam spectantes.* Dissertatio inauguralis physiologico-botanica quam con sensu et auctoritate gratiosi medicorum ordinis in Universitate litteraria berolinensi ut summi in medicina et chirurgia honores rite sibi concedantur die xix m. februarii a. mdcccxxv h.l.q.s. publice defendet auctor Julius Theod. Christ. Ratzeburg ... Berolini [Berlin] (formis Brüschckianis) [1825]. Qu. (*Animadv. pelor. spect.*).
Publ.: 19 Feb 1825, p. [1]-28, *pl. 1* (uncol. copp. auct.). *Copy*: NY. – p. 28: extensive "vita".
Commercial issue: 1825, p. [1]-27, *pl. 1* (id.). *Copy*: M. – "*Observationes ad peloriarum indolem definiendam spectantes* ... Berolini" s.d. (*Obs. pelor. spect.*).

8646. *Die Standortsgewächse und Unkräuter Deutschlands* und der Schweiz, in ihren Beziehungen zu Forst-, Garten- und Landwirthschaft und zu anderen Fächern ... Berlin (Nicolaische Verlagsbuchhandlung ...) 1859. Oct. (*Standortsgew. Unkr. Deutschl.*).
Publ.: Apr-Jul 1849 (p. xiii: Mar 1859; ÖbZ 1 Oct 1859; Bot. Zeit. 12 Aug 1859), p. [i]-xxxv, [1]-487, [488,-490, err.], 6 charts, *pl. 1-12* (uncol. liths with text, auct.). *Copies*: G, HH, MICH, MO, NY, USDA.
Ref.: Anon., Österr. bot. Z. 10: 412-413. 1860.
Schlechtendal, D.F.L. von, Bot. Zeit. 17: 279. 12 Aug 1859.

8647. *Die Waldverderbniss* oder dauernder Schade, welcher durch Insektenfrass, Schälen, Schlagen und Verbeissen an lebenden Waldbäumen entsteht. Zugleich ein Ergänzungswerk zu der Abbildung und Beschreibung der Forstinsekten ...Berlin (Nicolaische Verlagsbuchhandlung ...) 1866-1868, 2 vols. Qu. (*Waldverderbniss*).
1: Jan-Apr 1866 (dedic. Nov 1865; VbVB 19 Mai 1866), p. [i]-x, 1-298, *pl. 1-32*, map (*pl. 32*).
2: 1868, p. [i]-xv, 1-464, *pl. 1-5, 34-51*.
Copy: Edinburgh Univ. Libr. (inf. J. Edmonston).
Ref.: Milde, J., Verh. bot. Ver. Brandenburg 7: 223. 19 Mai 1866.

8648. *Forstwissenschaftliches Schriftsteller-Lexikon* ... Berlin (Fr. Nicolaische Verlagsbuchhandlung ...) 1872. Qu. (*Forstwiss. Schriftst.-Lex.*).
Editor: Philipp Phoebus (1804-1880).
Publ.: 1872 (p. x: Mai 1872), p. [i]-x, [xi], [1]-516. *Copies*: HH, NY, USDA. – on p. iii-x a bibliographical preface by the editor.

Rau, Ambrosius (1784-1830), German botanist and mineralogist; professor of natural history, forestry and rural economy at the University of Würzburg. (*A. Rau*).

HERBARIUM and TYPES: Unknown. – The majority of the herbarium collections at Würzburg (WB) were destroyed in World War II.

BIBLIOGRAPHY and BIOGRAPHY: Barnhart 3: 130 (b. 7 Mar 1784, d. 26 Jan 1830); BM 4: 1650; Herder p. 342; Kew 4: 410; PR 7427, ed. 1: 8371; Rehder 2: 284; Tucker 1: 582-583.

BIOFILE: Anon., Flora 2: 524. 1819 (visit Würzburg), 13: 256. 1830 (d. 26 Jan 1830). Poggendorff, J.C., Biogr.-lit. Handw.-Buch 2: 574. 1863 (bibl.).

EPONYMY: *Rauia* C.G.D. Nees et C.F.P. Martius (1823); *Ravia* J.A. Schultes (1824, *orth. var.*).

8649. *Enumeratio rosarum circa Wirceburgum et pagos adjacentes sponte crescentium, cum earum definitionibus, descriptionibus et synonymis, secundum novam methodum disposita et speciebus varietatibusque novis aucta* ... Norimbergae [Nürnberg] (sumptibus Caroli Felssecker) 1816. Oct. (*Enum. ros. Wirceb.*).
Publ.: 1816 (p. 12: 2 Mar 1816), p. [1]-178, [1-2, err.], [1, expl. pl.], *1 pl.* (handcol. copp. I. Sturm). *Copies*: BR (pl. in duplo), G, HH, M, NY, USDA, WU.

Rau, Eugene Abraham (1848-1932), American botanist at Bethlehem, Pa. (*Rau*).

HERBARIUM and TYPES: NY (rd. Jul. 1928); other material at B, BPI, DPU, MICH, PH (major set). – Letters to W.G. Farlow at FH.

BIBLIOGRAPHY and BIOGRAPHY: Barnhart 3: 131 (b. 22 Jul 1848, d. 16 Oct 1932); BL 1: 211, 314; BM 4: 1650; Bossert p. 325; CSP 12: 600; GR p. 239; Jackson p. 359; Kelly p. 186; LS 21771-21771a; Morren ed. 10, p. 129; PH p. 516 (index); SBC p. 130; Stevenson p. 1255; TL-2/see A.B. Hervey; Urban-Berl. p. 280.

BIOFILE: Anon., J. New York Bot. Gard. 23: 260. 1932 (d. 16 Oct 1932); Torreya 34: 23. 1934 (d.).
Britten, N.L., J. New York Bot. Gard. 29: 284-285. 1928 (entire herb. to NY, esp. rich in mosses, but also lichens and algae).
Ewan, J. et al., Short hist. bot. U.S. 91. 1969.
Harshberger, J.W., Bot. Philadelphia 352. 1899 (bibl.).
Rickett, H.W., Index Bull. Torrey bot. Club 80. 1955.
Sayre, G., Bryologist 80: 516. 1977.
Steere, W.C., Bot. Rev. 43: 293. 1977.

EPONYMY: *Rauia* Austin 81880); *Rauiella* H. Reimers (1937).

8650. *Catalogue of North American musci.* Arranged by Eugene A. Rau, and A.B. Hervey, ... Taunton (Printed at Gazette Job Office) 1880. (*Cat. N. Amer. musc.*).
Junior author: Alpheus Baker Hervey (1839-1931).
Publ.: 1880 (TBC Jun 1880; Nat. Nov. Jul(2) 1880; Bot. Zeit. 27 Aug 1880; Bot. Centralbl. 12-16 Jul 1880; BG Jul 1880), p. [1]-52. *Copies*: G, M, NY, US(2); IDC 7610.
Ref.: Geheeb, A., Bot. Centralbl. 5: 363-364. 1881 (rev.).
Hervey, A.B., Bull. Torrey bot. Club 6: 288. 1879 (announcement).

Rauch, Friedrich (1867-?), German botanist; Dr. phil. Erlangen 1895; later active as pharmacist. (*Rauch*).

HERBARIUM and TYPES: Unknown.

BIBLIOGRAPHY and BIOGRAPHY: Barnhart 3: 131; LS 21772, 38195.

BIOFILE: Rauch, F., Beitr. Keim. Ured. Sp. [35]. 1895 (vita).

8651. *Beitrag zur Keimung von Uredineen– und Erysipheen-Sporen* in verschiedenen

Nährmedien. Inaugural-Dissertation zur Erlangung der Doktorwürde der hohen philosophischen Fakultät der Friedrich-Alexanders-Universität Erlangen ... Göttingen (Buchdruckerei von Louis Hofer) 1895. Oct. (*Beitr. Keim. Ured.-Sp.*).
Publ.: Jan-Jun 1895 (Nat. Nov. Jul(2) 1895; Bot. Centralbl. 27 Nov 1895; Bot. Zeit. 1 Apr 1896), p. [1]-34, [35, vita]. *Copy*: USDA.

Rauh, Werner (1913-x), German botanist; Dr. phil. Halle 1937; from 1939 at Heidelberg; from 1947 lecturer, from 1957 extraordinary professor, from 1960 regular professor and director of the Institute of Systematic Botany and of the Botanical Garden; traveller in S. America, Africa and Madagascar; specialist in succulent plants and Bromeliaceae; student of W. Troll. (*Rauh*).

HERBARIUM and TYPES: private and HEID; further material in B, M, PRE, ZSS. – Some archival material at SIA.

BIBLIOGRAPHY and BIOGRAPHY: BFM p. 310-311 (index); Biol.-Dokum. 14: 7324; Bossert p. 325; Hortus 3: 1202; IH 1 (ed. 4: 77, (ed. 5): 76, (ed. 7): 340; Kew 4: 411; MW suppl. 290; NI 3: 56; Urban-Berl. p. 705; Zander ed. 11, p. 804.

BIOFILE: Anon., Kakt. Orchid. Rundschau 2: 64. 1978.
Benson, L., Cact. Succ. J., U.S.A. 47: 57. 1975 (portr.).
Gunn, LM. & L.E. Codd, Bot. expl. S. Afr. 290. 1981 (portr.).
Hegemann, W., Bot. Jahrb. 99: 139-142. 1978 (tribute, portr.).
Hertel, H., Mitt. Bot. München 16: 420. 1980.
Knobloch, I.W., Pl. coll. N. Mexico 56. 1979.
Padilla, V., J. Bromeliad Soc. 23: 83-85. 1973 (biogr. sketch, portr.).
Sayre, G., Bryologist 80: 516. 1977 (herb. private).
Schuder, W., ed., Kürschners deutscher Gelehrten-Kalender ed. 13. 3049. 1980.
Verdoorn, F., ed., Chron. bot. 4: 81. 1938, 5: 72, 106. 1939, 6: 116. 1941.

FESTSCHRIFT: Bot. Jahrb. 99(2/3). 1978.

THESIS: *Beiträge zur Morphologie und Biologie der Holzgewächse*, i, Entwicklungsgeschichte ... arktisch-alpiner Spalierträucher, Nova Acta Leop. ser. 5(30): 289-348. *pl. 47-54.* 1937.

EPONYMY: *Rauhocereus* Backeberg (1956); *Rauhia* Traub (1957); *Rauhiella* G.B.J. Pabst & P.I.S. Braga (1978).

Raulin, Victor Félix (1815-1905), French botanist; Dr. sci. 1848; geological curator at the Museum d'Histoire naturelle, Paris 1838-1846; subsequently professor of geology at the Faculty of Sciences, Bordeaux. (*Raulin*).

HERBARIUM and TYPES: P; some further material at B (no longer extant).

BIBLIOGRAPHY and BIOGRAPHY: AG 4: 699; Barnhart 3: 131 (b. 8 Aug 1915, d. 1 Feb 1905); BM 4: 1651, 8: 1050; Bossert p. 325; GR p. 347; IH 2: (in press); Jackson p. 31; Kew 4: 412; Lasègue p. 528; Quenstedt p. 353; Rehder 1: 411, 452; TL-2/4409; Tucker 1: 583; Urban-Berl. p. 386.

BIOFILE: Poggendorff, J.C., Biogr.-lit. Handw.-Buch 2: 575. 1863 (b. 8 Aug 1815), 3: 1092-1093. 1898 (bibl.), 4(2): 1216. 1904 (bibl.), 5: 1023. 1926 (d. 5 Mar 1905).
Zittel, K.A. v., Gesch. Geol. Paläont.531, 532, 548, 672, 702, 703, 705. 1899.

8652. *Essai d'une division de la France en régions naturelles et botaniques* ... [Bordeaux (Th. Lafargue, ...) 1852]. Oct.
Publ.: 1852, p. [1]-40. *Copy*: LC. – Reprinted and to be cited from Act. Soc. Linn. Bordeaux 18(1): 41-80. 1852.

8653. *Description physique de l'Ile de Crète ... partie botanique* ... Paris (F. Savy, Libraire-éditeur ...) 1869. Qu.

Reprint from journal: 1869, p. [i], 389-642, map, *pl. 1-18. Copy*: FI. – Reprinted and to be cited from Actes Soc. Linn. Bordeaux 24(5/6): 389-642, 1 map. 1869; plates added in reprint.
Reprint from complete work in bookform: 1869, p. [i-ii], [693]-976, 3 maps, *pl. 1-18. Copy*: G.

Raunkiaer, Christen Christiansen (1860-1938), Danish botanist; Mag. Sci. Kjøbenhavn 1885; assistant at the Botanical Garden ib. 1893-1911; lecturer at the University 1909; professor of botany and director of the Botanical Garden ib. 1912-1923, succeeding E. Warming; versatile botanist and ecologist("plant life forms", Taraxacum). (*Raunk.*).

HERBARIUM and TYPES: C; other material at BR, L, MO, NY.

NOTE: Original name Christen Christiansen, the name Raunkiaer was taken from the farm on which he was born.

BIBLIOGRAPHY and BIOGRAPHY: Barnhart 3: 131 (b. 29 Mar 1860, d. 11 Mar 1938); BFM no. 1445, 2609, 2741-2743; BJI 2: 142; BL 1: 47, 2: 49, 54, 56, 706; BM 4: 1651, 8: 1050-1051; Bossert p. 325; Christiansen p. 319 [index] 1936; Collander hist. p. 67; CSP 11: 114, 18: 65; GR p. 685; IH 2: (in press); Kew 4: 412; KR p. 583; Langman p. 613; Lenley p. 340; LS 3264, 21783-21786, 38199, 41622; NI 1586na; Rehder 1: 357, 2: 792; Stevenson p. 1255.

BIOFILE: Anon., Bot. Centralbl. 117: 640. 1911 (app. Copenhagen); Mycol. Centralbl. 1: 33. 1912 (app. prof., dir. Copenhagen); Österr. bot. Z. 62: 39. 1912 (succeeds Warming at Copenhagen).
Christensen, C., Dansk. bot. litt. 1880-1911, p. 105-107. 1913 (bibl., portr.), id. 1912-1939, p. 8-12. 1940 (bibl., portr.). Dansk. bot. hist. 1: 836-845, 882. 1926; Dansk biogr. leks. 19: 265-270. 1940.
Gager, C.S., Brooklyn Bot. Gard. Rec. 27(3): 190. 1938.
Gilbert-Carter, H., Proc. Linn. Soc. 151: 248-251. 1941 (obit.).
Gram, K., Bot. Tidsskr. 44: 255-259. 1938 (obit., portr.).
Hansen, A., Dansk bot. Ark. 21: 70. 1963.
Iltis, H., Life of Mendel 166. 1966.
Lind, J., Danish fungi herb. E. Rostrup 37. 1913.
Paulsen, O., Naturens Verden 22: 241-247. 1938 (obit., portr.); Overs. k. danske vidensk. Selsk. wirksomh. Jun 1937-Mai 1938: 67-89. 1938 (obit., portr.).
Petersen, O.G., Dansk biogr. lex. 13: 528-530. 1899.
Verdoorn, F., ed., Chron. bot. 1: 115. 1935, 2: 112, 193, 196, 239. 1936, 4: 117, 266-268, 271. 1938 (obit.), 5: 134. 1939.

COMPOSITE WORKS: *Sapotaceae*, in E. Warming, *Symb.* 31: 1-10. 8 Mai 1890.

8653a. *Myxomycetes Daniae* eller Danmarks Slimsvampe tilligemed et Forsøg tel en Myxomyceternes Systematik ... Kjøbenhavn (i Kommision hos J. Frimodt ...) 1888. Oct. (*Myxomyc. Dan.*).
Publ.: Nov 1888 (t.p. date preprint; Nat. Nov. Dec(1) 1888), p. [1]-88, *pl. 2-5. Copies*: FH, FI, NY. – Preprinted from Bot. Tidsskr. 17: 20-96, *pl. 2-5*. Nov 188, 97-105. 30 Nov 1889.
Summary (English): 1889, 89-93. *Copy*: FI. – Reprinted from Bot. Tidsskr. 17(3): 106-110. Nov 1889 (Nat. Nov. Mar(1) 1890).
Ref.: Rosenvinge, L.A.K., Bot. Centralbl. 38: 676-678. 1889 (rev.).

8654. *Dansk Exkursions-Flora* eller Nøgle til Bestemmelsen af de danske Blomsterplanter og Karsporeplanter ... Kjøbenhavn (Gyldendalske Bokhandels Forlag (F. Hegel & Son) ...) 1890. Oct. (*Dansk Exkurs.-Fl.*).
Ed. 1: Jan-Jul 1890 (Bot. Centralbl. 3 Sep 1890; Bot. Zeit. 29 Aug 1890; Nat. Nov. Aug(2) 1890), p. [i]-xxxii, [1]-287, [288, err.]. *Copy*: NY.
Ed. 2: Jan-Jun 1906 (Nat. Nov. Jul(2) 1906, Bot. Centralbl. 28 Aug 1906), p. [i]-xxxii,

[1]-286. *Copies*: City Library Aarhus, H. – Anden Udgave. Kjøbenhavn og Kristiania (Gyldendalske Boghandel Nordisk Forlag ...) 1906. Oct. – "There are few additions compared with ed. 1. However, the "life-forms" concept which Raunkiaer had just introduced, is mentioned occasionally. A system of abbreviations for them is given on p. 274". (O. Almborn in litt.).
Ed. 3: Jul-Sep 1914 (Nat. Nov. Oct 1914; Bot. Centralbl. 15 Dec 1904; ÖbZ 15 Mai 1915), p. [i]-xxxvi, [1]-330. *Copy*: LD. – Tredie udgave ved C.H. Ostenfeld [Carl Emil Hansen Ostenfeld, 1873-1931] og C. Raunkiaer, København og Kristiania (id) 1914. Oct. – Contains a more detailed survey of life forms.
Ed. 4: Oct-Dec 1922 (p. vi: Oct 1922), p. [i]-xxxvi, [1]-354. *Copy*: LD. – Fjerde udgave ved C.H. Ostenfeld og C. Raunkiaer. Kjøbenhavn og Kristiania (id.) 1922.
Ed. 5: Mar-Jun 1934 (p. vi: Feb 1934; Nat. Nov. Jul 1934), p. [i]-xxxvii, [1]-363, err. slip. *Copy*: USDA. – Under Medvirkning af K. Wiinstedt og Knud Jessen [1884-1971] ... København (id.) 1934. Oct.
Ed. 6: 1942, p. [i]-xxxi, [1]-370, [1, err.]. *Copies*: BR, G, M. – Dansk Ekskursions-Flora sjette udgave ved K. Wiinstedt. København (id.) 1942. Oct. [Knut Jørgen Frederik Wiinstedt 1878-1964].
Ed. 7: 1950, p. [i]-xxxi, 1-380. *Copy*: USDA. – Syvende udgave ved K. Wiinstedt. København (id.) 1950. Oct.

8655. *De Danske Blomsterplanters Naturhistorie.* Første Bind: Enkimbladede ... Kjøbenhavn (i Kommission hos Gyldendalske Boghandels Forlag ...) 1895-1899. Oct. †. (*Dan. Blomsterpl. naturh.*).
Publ.: 1899 (ÖbZ Mai-Jun 1899; Nat. Nov. Jun(1) 1899; Flora rev. 14 Oct 1899), p. [i]-lxix, [1]-724. *Copies*: FI, G, HH, US, USDA. – 1089 figures in text drawn by Ingeborg Raunkiaer and the author.
Ref.: Rendle, A.B., J. Bot. 37: 333. Jul 1899.
Jost, Bot. Zeit. 57(2): 305-306. 16 Oct 1899.

8656. Académie royale des sciences et des lettres de Danemark extrait du bulletin de l'année 1905. No. 5. *Types biologiques pour la géographie botanique* par C. Raunkiaer [1905] Oct.
Publ.: 1905, cover t.p., p. [347]-437. *Copy*: LD (inf. O. Almborn). – Reprinted and to be cited from Overs. k. danske vidensk. Selsk. Forh. 1905(5): 347-437. – Precursor of *Planterig. livsf.* 1907.
Ref.: Jost, Bot. Zeit. 64: 134-136. 1 Mai 1906 (rev.).

8657. *Planterigets livsformer* og deres betydning for geografien ... Kjøbenhavn og Kristiania [Oslo] (i kommission hos Gyldendalske Boghandel. Nordisk Forlag ...) 1907. Oct. (*Planterig. livsf.*).
Publ.: 1907, frontisp. p. [i-v], [1]-132. *Copies*: GB (inf. B. Peterson), H (inf. P. Isoviita), LD (inf. O. Almborn). – *Preliminary publ.*: Types biologiques ... 1905, see above.

8658. *Livsformernes statistik* som grundlag for biologisk plantegeografi ... København (Bianco Lunas Bogtrykkeri) 1908. Oct.
Publ.: Nov-Dec 1908 (p. 83: 12 Nov 1908), p. [i], [42]-83. *Copy*: HH. – Reprinted from Bot. Tidsskr. 29(1): 42-43. 1908.

8659. *Livsformen hos planter paa ny jord* ... København (Bianco Lunos Bogtrykkeri) 1909. Qu. (*Livsform. pl. ny jord.*).
Publ.: 1909 (Nat. Nov. Nov(2) 1909), p. [1]-70. *Copies*: BR, G, NY, USDA. – Issued as K. Dan. Vid. Selsk. ser. 7. 8(1). 1909.
English transl.: 1934, *in* Life forms pl. 1934, p. 148-200.

8660. *Mosses and lichens collected in the former Danish West Indies* ... Dansk botanisk Arkiv 2(9). 1918. Oct.
Co-author: Frederick Christian Emil Børgesen (1866-1956).
Publ.: 1918, p. [1]-18. *Copy*: U. – Issued as Dansk bot. Ark. 2(9). 1918.

8661. *Über Homodromie und Antidromie* insbesondere bei Gramineen ... København

(Hovedkommisionaer: Andr. Fred. Høst. & Son, ...) 1919. Oct. (*Homodromie & Antidromie*).
Publ.: 1919 (Nat. Nov. Apr 1919), p. [1]-32. *Copy*: FI. – Issued as K. Vidensk. Selsk. Biol. Medd. 1(12). 1919.

8662. *Myxomycetes from* the West Indian Islands *St. Croix, St. Thomas and St. Jan* ... Dansk botanisk Arkiv 5(16). 1928. Oct.
Publ.: 1928, p. [1]-9. *Copy*: U. – Issued as Dansk bot. Ark. 5(16). 1928.

8663. *The life forms of plants* and statistical plant geography being the collected papers of C. Raunkiaer ... Oxford (at the Clarendon Press) 1934. Oct. (*Life forms pl.*).
Publ.: 1934 (p. xvi: Mar 1934; Nature 20 Oct 1934; Nat. Nov. Oct 1934), p.[i-ii], portr., [iii]-xvi, [1]-632. *Copies*: BR, FI, G, HH, MO, NY, USDA. – For plates and figures (one series of numbers) see p. viii-x.
Ref.: Rendle, A.B., J. Bot. 72: 351-353. Dec 1934.
Turrill, W.B., Bull. misc. Inf. Kew 1934: 310. 25 Oct 1934 (rev.).

8664. *Plant life forms* ... Oxford (at the Clarendon Press) 1937. Oct. (*Pl. life forms*).
Translation: Humphrey Gilbert Carter (1884-1969).
Publ.: Jun-Aug 1937 (p. vi: Jun 1937; Nat. Libr. Scotl. rd. 24 Sep 1937), p. [i]-vi, [vii], [1]-104. *Copies*: Natl. Libr. Scotland; MO, PH.
Ref.: Rendle, A.B., J. Bot. 76: 59. 1938.

Raup, Hugh Miller (1901-x), American botanist; Dr. phil. Univ. Pittsburgh 1928; A.M. h.c. Harvard Univ 1945; Dr. sci. h.c. Wittenberg Univ. 1968; instructor in biology Wittenberg College 1923-1925; asst. prof. 1925-1932; research assistant Arnold Arboretum, Harvard University 1932-1934, research associate 1934-1938; asst. prof. plant ecology Harvard 1938-1945; assoc. prof. 1945-1949; prof. bot. 1949-1960; prof. forestry 1960-1967; director Harvard forest 1946-1947; worked extensively in the Mackenzie drainage basin 1926-1939, along the Alaska highway 1943-1948; in N. Greenland 1956-1964. (*Raup*).

HERBARIUM and TYPES: CAN and GH; further material at B, C, DPU, E, F, LD, MICH, MIN, MO, NY, PH, TRT, US. – Archival material at PH and SIA.

BIBLIOGRAPHY and BIOGRAPHY: Barnhart 3: 131; BL 1: 134, 135, 137, 202, 314; Bossert p. 325 (b. 4 Feb 1901); Hortus 3: 1202; PH 209.

BIOFILE: Cattell, J.M. & J., Amer. men Sci. ed. 5. 912. 1933 (b. 4 Feb 1901).
Ewan, J. et al., Short hist. bot. U.S. 122. 1969.
Raup, H.M., [Curriculum vitae and bibliography up to 1981, mimeographed, 5 p. copy submitted to us by author].
Stieber, M.T. et al., Huntia 4(1): 85. 1981 (arch. mat. HU).
Sutton, S.B., Charles Sprague Sargent 66-67. 1970.
Verdoorn, F., ed., Chron. bot. 1: 298, 376, 1935, 2: 330. 1936, 3: 294, 304. 1937, 4: 164. 1938, 5: 295, 308, 501. 1939, 6: 42, 167, 382. 1941, 7: 179, 230. 1943.

8665. *Phytogeographic studies in the Peace and Upper Liard River regions*, Canada with a catalogue of vascular plants ... Jamaica Plain, Mass., U.S.A. (published by the Arnold Arboretum ...) 1934. Oct. (*Phytogeogr. Stud. Peace R.*).
Publ.: 15 Feb 1934 (p. [2]), p. [i], [1]-230, *pl. 1-9*, map. *Copies*: E.G. Voss, USDA. – Issued as Contr. Arnold Arbor. Harv. Univ. 6, 1934.

8666. *Botanical investigations in Wood Buffalo Park* ... Ottawa (J.O. Patenaude, I.S.O. ...) 1935. Oct. (*Bot. invest. Wood Buffalo Park*).
Publ.: Sep 1935 (USDA rd. 19 Oct 1935; PH rd. 17 Oct 1935), p. [i-iv], err. slip, [1]-174, *pl. 1-13*. *Copies*: HH, MO, PH, USDA. – Natl. Mus. Canada Bull. 74, 1935. – *Note*: booknumber 8667 has not been used.

8668. *Botanical studies in the Black Rock Forest* ... Cornwall-on-the-Hudson, New York 1938. Oct. (*Bot. Stud. Black Rock For.*).
Publ.: 1938, p. [i]-vi, 1-161, *1 map. Copies*: HH, USDA. – The Black Rock Forest Bull. 7, 1938.

Rauscher, Robert (1806-1890), Moravian/Austrian administrator and botanist; born and educated at Brünn; studied law at the University of Olmütz; Dr. Jur. 1833; 1829-1841 with the financial administration of Brünn; in Wien as registrar with the "Kammer procurator" 1841-1848; in Linz as "Adjunct bei der k.k. oberösterreichischen Finanzprocurator" (financial registrar) and secretary of the Museum Francisco-Carolinum 1848-1853; in a similar position in Wien 1853-1868; Finanzrath 1864; in retirement in Wien until 1870, subsequently at Linz active in the "Verein für Naturkunde". (*Rauscher*).

HERBARIUM and TYPES: BRNU; other material at C, E, GOET and IBF.

BIBLIOGRAPHY and BIOGRAPHY: Barnhart 3: 131; IH 2: (in press); Klášterský p. 153 (b. 26 Jul 1806, d. 4 Mar 1890); Morren ed. 10, p. 32.

BIOFILE: Anon., Bot. Not. 1891: 46. (d. 4 Mar 1890); Österr. bot. Z. 21: 256. 1871, 40: 182. 1890 (d. 4 Mar 1890).
Elvert, Chr. d', Zur Geschichte der Pflege der Naturwissenschaften in Mähren und Schlesien, Brünn 1868, p. 230.
Iltis, H., Verh. naturf. Ver. Brünn 50: 310. 1912 (herb. to Verein).
Neilreich, A., Verh. zool.-bot. Ver. Wien 5: 64. 1855.
Niessl, G. v., Verh. naturf. Ges. Brünn 28: 38-39. 1890 (herbarium received with relevant correspondence), 29: 21-22. 1891 (obit.).
Schiedermayr, C., Österr. bot. Z. 35: 229-231. 1885 (biogr., portr., bibl.).
Wagenitz, G., Index coll. princ. herb. Gott. 131. 1982.
Wittrock, V.B., Acta Horti Berg. 3(2): 84. 1903.

8669. *Aufzählung der in der Umgebung von Linz wildwachsenden oder im Freien gebauten blüthentragenden Gefäss-pflanzen* ... Linz (Druck von Joseph Wimmer ..) 1871. 2 parts. (*Aufz. Linz Gefäss-Pfl.*).
1: 1871 (ÖbZ 21: 256. 1871), p. [1]-43. (Jahresb. Ver. Naturk., Linz 2, 1871).
2: 1872 (ÖbZ Jul 1872), p. [1]-79, [81-83]. (Ib. 3, 1872).
Copies: LC, M (inf. I.H. Haesler).

Rauth, Franz (1874-?), German botanist; Dr. phil. Erlangen 1901. (*Rauth*).

HERBARIUM and TYPES: Unknown.

BIBLIOGRAPHY and BIOGRAPHY: Barnhart 3: 131 (b. 19 Jul 1874); Kew 4: 412; Tucker 1: 583.

8670. *Beiträge zur vergleichenden Anatomie einiger Genisteen Gattungen.* (Laburnum, Petteria, Spartium, Erinacea, Ulex, Cytisus, Hypocalyptus, Loddigesia). Inaugural-Dissertation zur Erlangung der Doktorwürde der hohen philosophischen Facultät der Kgl. bayer. Friedrich-Alexanders-Universität Erlangen ... Erlangen (Druck der Universitäts-Buchdruckerei von E. Th. Jacob) 1901. Oct. (*Beitr. Anat. Genist. Gatt.*).
Publ.: 1901, p. [1]-58. *Copy*: G.

Rauwenhoff, Nicolaas Willem Pieter (1826-1909), Dutch botanist; Dr. phil.; lecturer at the medical college of Rotterdam and high school teacher ib. 1860-1871; professor of botany and director of the botanical garden at Utrecht (succeeding F.A.W. Miquel) 1871-1896; first Dutch professor of botany to teach plant physiology. (*Rauwenh.*).

HERBARIUM and TYPES: Rauwenhoff is not known to have made a herbarium; R's library is at Utrecht University Library.

BIBLIOGRAPHY and BIOGRAPHY: Backer p. 484; Barnhart 3: 131; Biol.-Dokum. 14: 7332; BM 4: 1651; Bossert p. 325; CSP 5: 109, 8: 707, 11: 115, 12: 600, 18: 66; De Toni 1: civ; Herder p. 282; JW 2: 365, 3: 365, 5: 249; Kew 4: 413; Moebius p. 249, 318; Morren ed. 2, p. 28, ed. 10, p. 90, 91; PR 6744, 6881, 7429; Rehder 5: 704-705; SO 2749, 2777; TL-2/6118; Tucker 1: 583.

BIOFILE: Anon., Bot. Centralbl. 113: 80. 1910 (d.); Bot. Not. 1910: 16 (d.); Bot. Zeit. 29: 239. 1871 (as successor to F.A.W. Miquel at Utrecht), 68: 56. 1910 (d.); Flora 54: 127. 1871 (app. Utrecht); Nat. Nov. 32: 128. 1910 (d.); Nuova Notarisia 21: 114. 1910 (d.); Österr. bot. Z. 21: 148. 1871 (app. Utrecht),60: 87. 1909 (d.).
Gager, C.S., Brooklyn Bot. Gard. Rec. 27: 309. 1938.
Högrell, B., Bot. Not. 1888: 207.
Kneucker, A., Allg. bot. Z. 16: 32. 1910 (d.).
Lefebure, E. & J.G. de Bruyn, Martinus van Marum 6: 418. 1976.
Verdoorn, F., ed., Chron. bot. 1: 35 (secretary Int. bot. Congr. Amsterdam 1865), 37 (president idem 1877).
Visser, R.P.W., Regn. veg. 71: 403-404. 1970 (photograph of R's residence).

EPONYMY: *Rauwenhoffia* Scheffer (1885).

8671. *Inwijdings-Rede over het nut der wetenschap*, zigtbaar in den werkkring der plantkunde en in hare toepassingen. Uitgesproken ter aanvaarding van het lectoraat in de plantkunde, enz. aan de Geneeskundige School te Rotterdam, den 7 Februarij 1860, ... Rotterdam (H.A. Kramers) 1860. Oct. (*Inw.-rede nut wetensch.*).
Publ.: Mar 1860 (rd. by KNAW Mar 1860), p. [1]-32. *Copy*: U.

8672. *Bijdrage tot de kennis van Dracaena draco L.* ... Amsterdam (C.G. van der Post) 1863. Qu. (*Bijdr. Dracaena draco*).
Publ.: 1863 (t.p. of preprint; cover, however, has 1864; 1863 date therefore not completely reliable; Flora 11 Aug 1864), p. [i], [1]-54, *pl. 1-5* (liths. auct., no. 1 col.). *Copies*: G, HH, NY. – Preprinted from Nat. Verh. Kon. Ned. Akad. Wet. 10: [1]-54, *pl. 1-5* (id.).
Ref.: Schlechtendal, D.F.L. von, Bot. Zeit. 22: 395-396. 16 Dec 1864.

8673. *De tegenwoordige richting en beteekenis der planten-physiologie* uit hare geschiedenis toegelicht. Redevoering ter aanvaarding van het gewoon hoogleeraarsambt aan de Hoogeschool te Utrecht den 21sten April 1871 uitgesproken ... Utrecht (J. Greven) 1871. Oct. (*Richt. beteek. pl.-physiol.*).
Publ.: 1871, p. [1]-36. *Copy*: U.

8674. *Sur les premiers phénomènes de la germination des spores des cryptogames* ... [Extrait des Archives Néerlandaises, t. xiv., Amsterdam 1879]. Oct.
Publ: 1879 (BSbF 14 Nov 1879), p. [1]-23, *1 pl.* (col. lith. auct.). *Copy*: G. – Reprinted from Arch. néerl. Sci. exact. nat. 14: 347-369, *pl. 8*. 1879.
Preliminary publ.: Versl. Med. Kon. Ned. Akad. Wet. 14: 320-339, *1 pl.* 1879.
German extract: 11, 18 Jul 1879, p. [1]-8. *Copies*: G, U. – Reprinted from Bot. Zeit. 37: col. 441-448, 457-466. 1879.

8675. *Charles Robert Darwin.* Rede bij de opening der 109[de] algemeene vergadering van het Provinciaal Utrechtsch Genootschap van Kunsten en Wetenschappen te Utrecht den 27 Juni 1882 ... Utrecht (Firma L.E. Bosch & Zoon) 1882. Oct. (*Darwin*).
Publ.: shortly after 27 Jun 1882,(KNAW rd Jul-Sep 1882), p. [1]-29. *Copy*: U. – Also published in the "Verslagen" of the sessions of the Society.

8676. *Over het begrip van leven.* Redevoering uitgesproken op den verjaardag der Universiteit te Utrecht den 26sten Maart 1885, door den Rector magnificus Dr. N.W.P. Rauwenhoff. Utrecht (J.L. Beijers) 1885. Oct. (*Begrip leven*).
Publ.: shortly after 26 Mar 1885 (KNAW rd. Mai 1885), p. [1]-23. *Copy*: U.

8677. *Onderzoekingen over Sphaeroplea annulina* Ag. ... Amsterdam (Johannes Müller) 1887. Qu. (*Onderz. Sphaeroplea annulina*).
Publ.: 1887 (Bot. Zeit. 29 Jul 1887; Nat. Nov. Oct(2) 1887), p. [i], [1]-38, *pl. 1-2* (col. liths. auct.). *Copies*: NY, U. – Preprinted from Nat. Verh. Kon. Ned. Akad. Wet. 26: 1-38, *pl. 1-2*. 1888.
French version: 1888 (BSbF 13 Jan 1888; Nat. Nov. Sep 1888), p. [1]-54, *pl. 1-2* (col. liths. auct.). *Copies*: FH, G. – Reprinted from Arch. néerl. Sci. exact. nat. 22: 91-144. *pl. 3-4*. 1888.
Preliminary publ.: Proc. Kon. Ned. Akad. Wet. 1883/4(1): 4-6, Bot. Centralbl. 15: 398-400. 1883, Bull. Soc. bot. France 31(bibl): 175-176. 1884.

8678. *De geslachtsgeneratie der Gleicheniaceeën* ... [Amsterdam 1890]. Qu.
Publ.: 1890 (p. 2: Oct 1889; Nat. Nov. Aug(2) 1890; Bot. Zeit. 29 Aug 1890; Bot. Centralbl. 28 Mai 1890), p. [1]-54, *pl. 1-7* (uncol. lith. auct.). *Copy*: G. – Issued and to be cited as Nat. Verh. Kon. Ned. Akad. Wet. 28: 1-54, *pl. 1-7*. 1890.
French: 1891, p. [1]-75, *pl. 4-10* (col. *4, 7, 9, 10*) and uncol. liths. auct.). *Copies*: G, U. – *La génération sexuée des Gleicheniacées*, reprinted and to be cited from Arch. néerl. Sci. exact. nat. 24: 157-231, *pl. 4-10*. 1891.
Preliminary publ.: Versl. Afd. Natuurk., Kon. Ned. Akad. Wet. 26 Oct 1889, p– [1]-2. *Copy*: U.
Ref.: Kleinsius, J., Bot. Centralbl. 42: 370-372. 18 Jun 1890 (rev.).

Rauwolff, Leonhart [Leonhard Rauwolf] (1535-1596), German botanist, physician and traveller; M.D. Valence 1562; practiced medicine at Augsburg, Aich and Kempten 1563-1570; city physician Augsburg 1570-1573; travelled in the Near East 1573-1576; again at Augsburg 1576-1588; city physician at Linz 1588-1596; died in Hungary when on a campaign against the Turks. (*Rauwolff*).

HERBARIUM and TYPES: L, four volumes, acquired 1690 with the Isaac Vossius library. Available on IDC microfiche 8303. – The herbarium was taken from Augsburg to Sweden in the thirty years war and given to Queen Christina of Sweden (1626-1689) who gave it to Isaac Vossius. The latter's heirs sold it to the city of Leiden. Vols. 1-3 contain 972 plants from France, Switzerland and Italy (coll. 1560-1563), vol. 4 contains 338 plants from the trip to the Near East.

NOTE: See Frick (1962) for the correct spelling of R's name: Leonhart Rauwolff taken from a letter written by R. to the Augsburg Collegium medicum on 12 Oct 1563.

BIBLIOGRAPHY and BIOGRAPHY: ADB 27: 462-465; Backer p. 484; Barnhart 3: 131; BM 4: 1651; DSB 11: 311-312 (Dannenfeldt; b. 21 Jun 1535, d. 15 Sep 1596); DTS 1: 236, xxiii; Herder p. 57, 211; HU 146; Jackson p. 233; JW 4: 402-403; Kew 4: 413; Lasègue p. 347, 403, 503, 517; Moebius p. 34; NI 1587, 3: 56; Plesch p. 372; PR 3608, 7430, ed. 1: 3920, 8373; Rehder 1: 477, 478; Saccardo 1: 135; SO 635f; Sotheby 621; TL-1/455, 1038; TL-2/2190, 4309; Tucker 1: 583.

BIOFILE: Babinger, F., Arch. Gesch. Naturw. Technik 4: 148-161. 1912.
Blakelock, R.A. & E.R. Guest, Fl. Iraq. 1: 109, 115. 1966.
Blatter, E., Rec. Bot. Surv. India 8(5): 458. 1933.
Candolle, Alph. de, Phytographie 442. 1880.
Cavillier, F., Boissiera 5: 74. 1941.
Coats, A.M., The plant hunters 13-15, 25. 1969.
Dannenfeldt, K.H., Leonhard Rauwolf, sixteenth century physician, botanist, Cambr. Mass. 1968, xiii, 321 p. (rev. R.G.C. Desmond, Kew Bull. 25(2): 374. 1971).
De Toni, G.B., Atti Soc. nat. mat. Modena ser. 4. 7: [7 p.] 1910 (on the third herbarium volume at L).
Frick, K., Südhoffs Arch. gesch. Med. 46(1): 82-83. 1962 (cites a letter by Rauwolff in which he signs himself Leonhart Rauwolff).
Fries, Th.M., Bref Skrifv. Linné ser. 1. 2: 107. 1908, 4: 104. 1910.
Ganzinger, K., Veröff. Int. Ges. Gesch. Pharm. ser. 2. 22: 23-33, 8 figs. 1963 (on Rauwolf and Fuchs).

Hulth, J.M., Bref Skifv. Linné ser. 2. 1: 102, 103. 1916, 2: 146, 147. 1943.
J., Dict. Sci. méd., Biogr. méd. 6: 547. 1824.
Jackson, B.D., J. Bot. 39: 43-45. 1901 (rev. L. Legré, 1900).
Jessen, K.F.W., Gesch. Bot. Gegenw. Vorz. 189. 1884.
Legré, L., J. Bot., Morot 13(5): 160-162. 1899 (on herb. at L); Bull. Soc. bot. France 46: lii-lxi. 1899; La botanique en Provence au xvie siècle, Léonhard Rauwolf, Jacques Raynaudet, Marseille 1900 (TL-2/4309) (reviewed by H. Solms, Bot. Zeit 59: 27. 1901).
Mägdefrau, K., Gesch. Bot. 36. 1973.
Magnin, A., Bull. Soc. bot. Lyon 31: 18. 1906.
Martins, C., Le jardin des plantes de Montpellier 7, 13, 31. 1854.
Murr, J., Neue Übers. Farn-Bl.-Pfl. Vorarlberg xx. 1923.
Poiret, J.L.M., Encycl. méth. Bot. 8: 754. 1808.
Rieppel, F.W., Deut. med. Wochensschr. 80: 653-655. 1955 (contr. to his biogr.; sec. refs.).
Saint-Lager, J.B., Hist. herb. 69-85. 1885.
Saccardo, P.A., Malpighia 17: 269. 1902.
Schinnerl, M., Ber. bayer. bot. Ges. 13: 207. 1912 (herb.).
Stafleu, F.A., Linnaeus and the Linnaeans 162, 165. 1971.
Steenis-Kruseman, M.J. van, Blumea 25: 33. 1979 (see also p. 114).
Verdoorn, F., ed., Chron. bot. 1: 33. 1935.
Wiesner, J., Südhoff's Arch. Gesch. Med. 43: 355-360. 1959 (L. Rauwolff als Altertumsforscher).

COMPOSITE WORKS: J.F. Gronovius, *Fl. orient.* 1755, TL-2/2190, was based on Rauwolff's plants at Leiden.

NOTE: For a detailed description of Rauwolff's travelogue, *Aigentliche Beschreibung der Raiss inn die Morgenländer* see e.g. HU 146 and Dannenfeldt (1963). The fourth part (1583) has forty-two wood cuts of plants.

EPONYMY: *Rauvolfia* Linnaeus (1753); *Rauwolffa* Cothenius (1790, *orth. var.*); *Rauwolfia* Gleditsch (1764, *orth. var.*).

Ravaud, Louis Célestine Mure, abbé (1822-1898), French clergyman and botanist; educated, later teacher, at the Dauphiné seminary of Rondeau; ultimately "curé-archiprêtre" at Villars-de-Lans (Isère). (*Ravaud*).

HERBARIUM and TYPES: GR; further material BR, L, MPU.

BIBLIOGRAPHY and BIOGRAPHY: AG 6(1): 231; Barnhart 3: 131 (b. 17 Oct 1822, d. 10 Apr 1898); BL 2: 144, 706; BM 4: 1651; CSP 5: 109, 8: 707, 18: 66. 1923; GR p. 295; GR cat. p. 70; Hegi 6(2): 1302; Kelly p. 186; LS 21789-21792; Tucker 1: 583.

BIOFILE: Anon., Hedwigia 38: (61). 1899 (d.); Rev. bryol. lich. 26: 24. 1899.
Boullu, Ann. Soc. bot. Lyon 23: 23. 1898.
Husnot, T., Rev. bryol. 26(1): 24. 1899 (d.).
Magnin, A., Bull. Soc. bot. Lyon 32: 42, 133. 1907, 35: 45. 1910.
Pellat, A., Bull. Soc. bot. France 45: 209-210. 1898 (obit.).
Sayre, G., Bryologist 80: 516. 1977.

8679. Abbé Ravaud. Bibliothèque du Touriste en Dauphiné. *Le guide du bryologue et du lichénologue* ainsi que du botaniste à Grenoble et dans ses environs ... Grenoble (Xavier Drevet, ...) 1879. Oct. (*Guide bryol. lichénol. Grenoble*).
Publ.: 1879 (also issued with t.p. dated 1881, n.v.),p. [1]-78. *Copies*: FH, NY. – First volume of the series *Guide bot. Dauphiné*, nos. 1-13; this part also re-issued, s.d. under that title. See below. – A series of publications, also as *Guide du Bryologue et du Lichénologue à Grenoble et dans les environs* also appeared in Revue bryologique 1874-1902: *1*: 17-19. 1874, *2*: 5-8, 44-46. 1875, *3*: 4-6, 35-40. 1876, *4*: 21-23, 54-59, 78-80, 87-92. 1877, *5*: 60-61. 1878, *6*: 37-40, 74-77. 1879, *7*: 106-110. 1880, *8*: 36-40. 1881, *17*: 59-60. 1890, *19*: 59-61. 1892, *22*: 55-60. 1895, *23*: 108-109. 1896, *24*: 40-43, 86-91. 1897, *25*: 85-86, 94-98. 1898, *27*: 9-10. 1900, *29*: 98-103. 1902.

8680. Bibliothèque du Touriste en Dauphiné. *Guide du botaniste dans le Dauphiné.* Excursions bryologiques et lichénologiques suivies pour chacune d'herborisations phanérogamiques ou il est traité des propriétés et des usages des plantes au point de vue de la médecine, de l'industrie et des arts par l'abbé Ravaud ... 1re[-13e] excursion. Grenoble (Xavier Drevet, ...) [1879-1892]. Oct. (*Guide bot. Dauphiné*).
1: 1891 (Nat. Nov. Nov(1) 1891; Bot. Zeit. 26 Feb 1892; Bot. Centralbl. 27 Nov 1891), p. [1]-68. *Copies*: BR (cover: "2; edition"), G. – Second edition of *Guide bryol. lichénol. Grenoble.*
2: 1887 (Bot. Zeit. 28 Jan 1887; Nat. Nov. Jan(1) 1887), p. [1]-32.
3: 1884 (Nat. Nov. Jul 1884; Bot. Centralbl. 4-8 Aug 1884; Bot. Zeit. 29 Aug 1884), p. [1]-40.
4: 1884 (Bot. Zeit. 29 Aug 1884; Nat. Nov. Jul 1884), p. [1]-63.
5/6: 1885 (Bot. Centralbl. 20-24 Apr 1885; Nat. Nov. Mai(1) 1885; Bot. Zeit. 31 Jul 1885), p [1]-35.
5bis: 1886, p. [1]-32. Ed. 2. 1898.
7: 1886 (Bot. Zeit. 28 Jan 1887; Nat. Nov. Jan(1) 1887), p. [1]-62.
8: 1886 (Nat. Nov. Aug(3) 1886), p. [1]-27.
9/10: 1886 (Nat. Nov. Aug(3) 1886; Bot. Zeit. 24 Sep 1886; Bot. Centralbl. 6-10 Sep 1886), p. [1]-48.
11: 1889 (Bot. Centralbl. 4 Sep 1889; Nat. Nov. Sep(1) 1889; Bot. Zeit. 29 Nov 1889), p. [1]-60.
12: 1894 (Bot. Zeit. 16 Jun 1895; Bot. Centralbl. 21 Aug 1894), p. [1]-121.
13: 1894 (Bot. Zeit. 16 Jan 1895; Bot. Centralbl. 21 Aug 1894), p. [1]-67.
Copies: BR, FH, G, NY.

8681. *L'herborisation à la Moucherolle & dans ses alentours* ... Grenoble (Xavier Drevet, ...) 1875. Oct. (*Herb. Moucherolle*).
Publ.: 1875, p. [1]-29. *Copy*: FH. – (Rehder 1: 414: cover dated 1876).

Ravenel, Henry William (1814-1887), American botanist, graduated from South Carolina College 1832; planter at St. John's, S.C. 1832-1853; from 1853-1887 at Aiken, S.C., LL.D. Univ. N. Carolina 1886. (*Ravenel*).

HERBARIUM and TYPES: Cryptogams: BM (14.450 specim.), phanerogams: CHARL. – Other material in B, BPI, DPU, E, FH, FI, GH, K, MANCH, M, MASS, MICH, MO, NY, NYS, PC, QK, US, WELC. Original set of Texas 1869 coll. at BPI. *Exsiccatae*: (1) *Fungi caroliniani*, 5 cent. 1852-1860, sets at BM, BPI, BUF, CUP, F, FH, H, ILL, MASS, NEB, NY, NYS, OS, PC, PH, PUR, STR. (See Stevenson 1971 for extensive details and bibl.). (2) *Fungi americani*, with Mordecai Cubitt Cooke (1825-1914), fasc. 1-8, 1878-1882. Sets at B, BM, BPI, CUP, DBN, E, FH, FI, GH, ILL, K, MANCH, MASS, MO, NEB, NY, WELC. (See Stevenson (1971) for extensive details and bibl.). Letters e.g. at FH and NY. – Other archival material at FH and SIA. For private journal see A.R. Childs (1947).

BIBLIOGRAPHY and BIOGRAPHY: Barnhart 3: 132 (b. 19 Mai 1814, d. 17 Jul 1887); BL 2: 213, 314; Bossert p. 325; Clokie p. 230; CSP 5: 110, 8: 707, 12: 600, 18: 67; DAB 8: 396-397 (by A.R. Childs); GR p. 195; Hawksworth p. 185; IH 1 (ed. 6): 363, (ed. 7): 240; Kew 4: 415; Lenley p. 340, 467; LS 21806-21809a; ME 1: 223, 3: 643; Morren ed. 2, p. 55, ed. 10, p. 130; Rehder 5: 705; SBC p. 130; Stevenson p. 1255; TL-2/465, see J.E. Duby; Tucker 1: 583; Urban-Berl. p. 272.

BIOFILE: Anon., Amer. J. Sci. 35: 263. 1888; Bot. Centralbl. 32: 32. 1887; Bot. Not. 1888: 46 (d.); Bryologist 82(3): 511. 1979 (portr.); Bull. Soc. bot. France 34 (bibl.): 143. 1887 (d.); Bull. Torrey bot. Club 14(9): 197-198. 1887 (d.); Grevillea 16: 22. 1887 (d.); Mycologia 3(3): 161. 1911 (Ravenel herbarium to be revised by E.R. Memminger); Natl. Cycl. Amer. biogr. 10: 47. 1909 (biogr., portr.); Nat. Nov. 9: 232. 1887; Nature 35: 374. 1887 (d.); Recorder, Aiken, S.C. 10 July 1887, repr. J. Mycol. 3(9): 106-107. 1887; The News and Courier (S. Carolina) 20 July 1887 (obit.).
Barnhart, J.H., *in* F.W. Pennell, Monogr. Acad. Nat. Sci. Philadelphia 1: 617. 1935 (coll.).

Candolle, Alph. de, Phytographie 442. 1880.
Childs, A.R., The private journal of Henry William Ravenel 1859-1887. Columbia, S. Carolina 1947, 428 p. (portr.).
Ewan, J. et al., Short hist. bot. U.S. 92. 1969.
Farlow, W.G., Bot. Gaz. 12(8): 194-197. 1887 (obit., bibl.); Ann. Bot. 1: 411. 1888 (bibl.).
Gee, W., Bull. Univ. S. Carolina 72: 38-41. 1918 (biogr., bibl.).
Geiser, S.W., Natural. Frontier 241-242, 280. 1948 (1869 mission); Field & Labor. 4(2): 52. 1936, 27: 180. 1959 (Texas mission 1869).
Hedge, I.C. & J.M. Lamond, Index coll. Edinb. herb. 122. 1970 (material in M.C. Cooke herb. mycol.).
Hesler, L.R., Biogr. sketches N. Amer. mycol., mss. 1975.
Hitchcock, A.S. & A. Chase, Man. grass. U.S. 988. 1951. (*Panicum ravenelii*).
Humphrey, H.B., Makers N. Amer. bot. 207-209. 1961.
Lamson-Scribner, F., Bull. Torrey bot. Club 20: 324-325. 1893.
Murray, G., Hist. coll. BM(NH) 1: 176. 1904 (herb. purchased 1891).
Murrill, W.A., Hist. found bot. Florida 28. 1945.
Rickett, H.W., Index Bull. Torrey bot. Club 80. 1955.
Rodgers, A.D., William Starling Sullivant 177, 184, 188, 224. 1940; John Torrey 228, 263. 1942.
Rogers, D.P., Brief hist. mycol. N. Amer. ed. 1: 12. 1977, ed. 2: 12, 14, 20, 33. 1981.
Sargent, C.S., Silva 8: 160. 1895.
Sayre, G., Mem. New York Bot. Gard. 19(3): 384. 1971 (coll.).
Smith, E., Bartonia 28: 10, 11. 1957.
Steere, W.C., Bot. Rev. 43: 293. 1977.
Stevens, N.E., Sci. Mon. 1919 (Aug): 157-166 (on R. and the civil war); Isis (ed. G. Sarton) 18(1): 133-149. 1932 (mycol. work).
Stevenson, J.A., Taxon 4(8): 181. 1935 (Texas 1869 fungi at BPI).
Sutton, S.B., Charles Sprague Sargent 109. 1970.
Underwood, L.P., J. New York Bot. Gard. 1(3): 40. 1900, 3(27): 57. 1902 (Ravenel material at NY).
Zander, R.H., Bryologist 82: 511. 1979 (portr.).

HANDWRITING: Bartonia 28: *pl. 7.* 1957.

EPONYMY: *Neoravenelia* W.H. Long (1903); *Pleoravenelia* W.H. Long (1903); *Ravenella* M.J. Berkeley (1853); *Ravenelula* Spegazzini (1881).

8682. *Description of a new Baptisia found near Aiken, So. Ca. ... read before the Elliott Society of Natural History, January 18th, 1856.* [Proc. Elliott Soc. 1: 38-39. 1859.] (*Descr. Baptisia*).
Publ.: 1859 (or preprinted?), p. [1]-2, *1 pl. Copies*: HH, LC, PH. – Reprinted or preprinted from Proc. Elliott Soc. nat. Hist. 1: 38-39. 1859.

8683. *Notice of some new and rare phanerogamous plants found in this state* ... [Proc. Elliott Soc. 1: 50-54. 1859].
Publ.: 1859, p. [1]-4. *Copy*: HH. – Reprinted and to be cited from Proc. Elliott Soc. 1: 51-54. 1859.

8684. *A list of the more common native and naturalized plants of South Carolina ... in* South Carolina. Resources and population. Institution and industries, Charleston S.C. (Walker, Evans & Cogswell, ...) 1883. Oct.
Publ.: 1883, p. [312]-359, to be cited from *South Carolina* 1883. *Copies* reprint: HH, NY.

Ravenscroft, Edward James (1816-1890), British printer and publisher at Edinburgh, later associated with his brother in the management of a London Bank. (*Ravenscroft*).

HERBARIUM and TYPES: Unknown.

BIBLIOGRAPHY and BIOGRAPHY: Barnhart 3: 132; BB p. 253; BM 4: 1652; Desmond p. 512 (d. 15 Nov 1890); GF p. 71; Kew 4: 415; NI 1588; Rehder 2: 13, 5: xxv; TL-1/1039; TL-2/see Ch. Lawson, J. Lindley; Tucker 1: 583.

BIOFILE: Anon., Gard. Chron. 1890(2): 605-606 (obit., d. 15 Nov 1890).
Junk, W., Rara 2: 145. 1926-1936; Cat. 214, no. 346. 1979.
Woodward, B.B., Gard. Chron. ser. 3. 36: 36-37. 1904 (see also BM 4: 1652; analysis *Pinet. brit.*).

8685. *The pinetum britannicum* a descriptive account of hardy coniferous trees cultivated in Great Britain ... Edinburgh and London (W. Blackwood & sons, ... and Edward Ravenscroft, ...) [1863]-1884, 3 vols., broadsheet. (*Pinet. brit.*).
Publ.: ... in 52 parts containing one plate each (in one case 2) and illustrated text (643 wood-cuts), between 1863 and 1884. The editor was Charles Lawson (1794-1812), hence "Lawson's Pinetum". Ravenscroft compiled the text (descriptions by J. Lindley, parts 1-3, Andrew Murray (parts 4-37) and Maxwell T. Masters (38-52) and completed the publication of parts 34-52). Woodward (1904) gives a list of the contents of the parts as well as of the dates of receipt in the British Museum. "In paging the work, during its publication in parts, the description of each tree commenced page 1; but each sheet was distinguished by a signature no. in addition to the usual signature". Woodward provides a full collation and proposes a continuous pagination, see our pages 600 and 601.
1: p. [i*-iii*], [i]-ii, [iii-iv], err. slip, [1-110], *pl. 1-15*.
2: p. [111-216], *pl. 16-36*.
3: p. [i], [217-330], *pl.37-53*, addenda.
Copies: NY, UC (inf. D.E. Johnson), USDA; IDC 5655. – The plates are coloured liths. drawn by W. Richardson, J. Black, R.K. Greville and J. Wallace. Five plates were cancelled (and substituted; the NY copy has all five. In all *53 pl.*, 4 photogr., 643 wood engr. in text and one plate containing 3 maps.

Ravenshaw, Thomas Fitzarthur Torin (1829-1882), British clergyman and botanist; Oriel Coll. Oxford MA 1854; curate at Ilfracombe, Devon, 1854-1856; rector of Pewsey Wilts. 1857-1880. (*Ravenshaw*).

HERBARIUM and TYPES: Unknown.

BIBLIOGRAPHY and BIOGRAPHY: Barnhart 3: 132; BB p. 253; BL 2: 304, 706; BM 4: 1652; CSP 5: 110; Desmond p. 512 (d. 26 Sep 1882); Kew 4: 415; Rehder 1: 401, 5: xxi.

BIOFILE: Allibone, S.A., Crit. dict. Engl. lit. 2: 1744. 1878.
Anon., Bot. Rev. 1883: 30; J. Bot. 20: 352. 1882 (d.), 21: 382. 1883.
Boase, F., Modern Engl. biogr. 3: 50. 1965 (d. 26 Sep 1882; Thomas Fitz Arthur Torin R.).
Freeman, R.B., Brit. nat. hist. books 290. 1980.
Hawksworth, D.L. & M.R.D. Seaward, Lichenol. Brit. Isl. 138. 1977.
Martin, W.K. & G.T. Fraser, Fl. Devon 773. 1939.

8686. *Botany of North Devon*, ... Ilfracombe (W. Stewart, ...) [1874]. Oct. (*Bot. N. Devon*).
Publ.: 1874, p. [1]-110, [1-25, ind.]. *Copy*: HH. – "Extracted from Stewart's North Devon Hand Book". 1874. – Other ed. in North Devon Hand Book ed. 4, G. Tugwell, ed., 291-398. 1877.

8687. *A new list of the flowering plants* and ferns growing wild *in the county of Devon*. With their habitats and principal stations ... London (Bosworth & Harrison, ...) 1860. Oct. (*New list. fl. pl. Devon*).
Publ.: 1860, p. [i]-viii, [1]-92, [1, err.]. *Copies*: BR, E; IDC 7249. – "Those who wish to fernize on foot will be aided by Mr. Ravenshaw". London Atheneum 1860(2): 232. – A supplement was published in 1872.

Vol.	publ. in part	Taxon	sign.	orig. pag.	to be re-paged	plate	date BM
1	52	title page, dedication, preface		[1*-iv*], [i]ii			10 Dec 1884
	51	list subscribers, cont., table w/repaging and plate numbers		[iii-iv]			10 Dec 1884
	38	Pinus albicaulis	34	[1]2-4	1-4		10 Dec 1884
	39	Pinus aristata	35	[1]2-4	5-8		10 Dec 1884
	50	Pinus ayacahuite	40	[1]2	9-10	2	10 Dec 1884
	38	Pinus balfouriana	36	[1]2	11-12		10 Dec 1884
	43	Pinus bungeana	37	[1]2-3	13-16		10 Dec 1884
	16	Pinus cembra	16	[1]2-6	17-22	31	30 Jun 1866
	44	Pinus coulteri	38	[1]2-4	23-26		10 Dec 1884
	24	Pinus excelsa	22	[1]2-6	27-32	4	20 Jun 1867
	44	Pinus flexilis	39	[1]2-4	33-36		10 Dec 1884
	19	Pinus insignis	19	[1]2-8	37-44	1, 5[2]	24 Nov 1866
	5	Pinus jeffreyi	8	[1]2	45-46	6	30 Sep 1864
	2	Pinus lambertiana	3	[1]2-8	47-54	7	24 Jul 1863
	51	Pinus laricio	44	[1]2-6	55-60	8[3]	10 Dec 1884
	17	Pinus mandschurica	17	[1]2-4	61-64		30 Jun 1866
	30	Pinus monophylla	27	[1]2-4	65-68	9[4]	16 Sep 1868
	43	Pinus monticola	41	[1]2	69-70		10 Dec 1884
	28/29	Pinus pinaster	26	[1]2-12	71-82	10[5]	2 Mai/16 Sep 1868
	39	Pinus porphyrocarpa	42	[1]2	83-84		10 Dec 1884
	11	Pinus sabiniana	12	[1]2-4	85-88	11	28 Jul 1865
	18	Pinus taeda	18	[1]2-4	89-92	12[6]	24 Nov 1866
	9	Pinus tuberculata	10	[1]2-6	93-98	13	8 Mai 1865
	36/37	Araucaria imbricata	32	[1]2-12	99-110	14[7], 15[8]	21 Feb 1877
2	52	Title page		[1-ii]			10 Dec 1884
	12	Abies albertiana	11	[1]2-4	111-114[a]	16	21 Dec 1865
	32	Abies douglasii	29	[1]2-4	115-118	17	31 Mai 1869
	33	Abies douglasii	29	5-12	119-126	18	9 Oct 1873
	34	Abies douglasii	29	13-20	127-134		9 Oct 1873
	26/27	Abies excelsa	24	[1]2-18	135-156	19, 20[9]	2 May 1868
	4	Abies hookeriana	6	[1]2-4	153-156	21[10]	31 Dec 1863
	4	Abies pattoniana	5	[1]2	157-158	22	31 Dec 1863

3		Picea cephalonica		175-180	27	21 Oct 1863
1	2	Picea nobilis	[1]2-6	181-188	28, 29	24 Jul 1863
	1	Picea pinsapo	[1]2-8	189-190	30	21 Oct 1863
	3	Cupressus lawsoniana	[1]2	191-194	31	7 Mar 1866
	15	Cupressus macrocarpa	[1]2-4	195-198	32, 33[12]	10 Dec 1884
	50	Cupressus nutkäensis	[1]2-4	199-200	34	31 May 1869
	31	Cupressus torulosa	[1]2	201-204	35	2 Dec 1867
	23	Taxodium distichum	[1]2-4	205-214	....	21 Feb 1877
35/36	30	Taxodium montezumae	[1]2-10	215-216	36	21 Feb 1877
36	31	Title page	[1]2			10 Dec 1884
3	52		[i-ii]			
	20	Cedrus atlantica	[1]2-8	217-224	38	9 Apr 1867
	6	Cedrus deodara	[1]2-8	225-232	39	30 Sep 1864
	7	Cedrus deodara	9-16	233-240	40	30 Sep 1864
	8	Cedrus deodara	17-22	241-246	41	8 May 1865
40-42/45-46	33	Cedrus libani	[1]2-52	247-298	42-50[13]	10 Dec 1884
	21	Sequoia wellingtonia	[1]2-8	299-306	37, 51	9 Apr 1867
	22	Sequoia wellingtonia	9-16	307-314	52[14]	20 Jun 1867
	23	Sequoia wellingtonia	17-24	315-322	53	20 Jun 1867
	45	Seed of Coniferae	[1-2]	323-324	....	10 Dec 1884
	45	Prices of Coniferae	[1]2	325-326	....	10 Dec 1884
	46	Index	[1]2-4	327-330	....	10 Dec 1884

Footnotes:

1 Plate 3 published in part 18, 24 Nov 1866
2 Plate 5 published in part 29, 16 Sep 1868
3 Plate 8 published in either part 43 or 39, 10 Dec 1884
4 Plate 9 labeled P. fremontiana
5 Plate 10 published in part 28, 2 May 1868
6 Plate 12 published in part 16, 30 Jun 1866
7 Plate 14 published in part 27, 21 Feb 1877
8 Plate 15 published in part 39 or 43, 10 Dec 1884
9 Plate 20 labeled Abies carpatica, treated as var. of A. excelsa
10 Plate 21 published in part 10, 28 Jul 1865
11 Plate 26 published in part 17, 30 Jun 1866
12 Plate 33 published as Cupressus lambertiana in part 38, 10 Dec 1884
13 Plate 42, 43 published in part 10, 28 Jul 1865 + consists of parts of both Cedrus deodara and C. libani
   Plate 44 published in part 34, 9 Oct 1873
   Plate 48 published in part 47, 10 Dec 1884
   Plate 49 published in part 48, 10 Dec 1884
   Plate 50 published in part 49, 10 Dec 1884
14 Plate 53 published in part 44, 10 Dec 1884
a table on p. 37, Gard. Chron. 16 Jul 1904, cites 28 Dec 1865 for pp. 111-114
b table on p. 37, Gard. Chron. 16 Jul 1904, cites 28 Dec 1865 for pp. 171-174

Ravin, Eugène (*fl.* 1861), French botanist and pharmacist; director of the botanical garden at Auxerre. (*Ravin*).

HERBARIUM and TYPES: Some material at P, PC.

BIBLIOGRAPHY and BIOGRAPHY: BL 2: 213, 706; BM 4: 1652; CSP 8: 707-708, 11: 115, 18: 68; Herder p. 472; Jackson p. 292; Kew 4: 415; PR 10621; Rehder 1: 412, 418; Tucker 1: 583.

8688. *Catalogue raisonné des plantes du Département de l'Yonne* croissant naturellement ou soumises à la grande culture, ... Auxerre (Imprimerie de Perriquet et Rouillé ...) 1861. Oct. (*Cat. rais. pl. Yonne*).
Publ.: 1861 (BSbF Nov 1861), p. [i-iii], [1]-287, [288, err.]. *Copies*: G, NY. – Also issued in Bull. Soc. Sci. Yonne 14: [39]-325. 1861.
Report: Parseval-Grandmaison, J. de, Rapport sur le catalogue des plantes du département de l'Yonne, de M. Eugène Ravin, ... Macon (Imprimerie d'Émile Protat) Nov 1864, p. [1]-13. *Copy*: G. –Reprinted from Ann. Acad. Maçon, Nov. 1864 (journal n.v.).
Eds. 2 and 3: see below, Flore de l'Yonne, 1866, 1883.
Ref.: Fournier, E., Bull. Soc. bot. France 8: 395-396. Nov 1861.

8689. E. Ravin ... *Flore de l'Yonne* catalogue des plantes croissant naturellement ou soumises à la grande culture dans le département. Deuxième edition. Auxerre (chez l'auteur ...) 1866. Oct. (*Fl. Yonne*).
Ed. 2 (for ed. 1, see above *Cat. rais. pl. Yonne*): 1866, p. [i-iii], [1]-334, [335, err.]. *Copies*: G, HH, NY.
Part 2: cryptogames, mousses: 1875 or 1876, p. [i], [21]-136, *pl. 1-76*. *Copies*: G, MO-Steere. – Reprinted and to be cited from Bull. soc. Sci. Yonne, Sci. nat. 29: 21-136, *pl. 1-76*. 1875/1876.
Ed. 3: 1883 (Nat. Nov. Apr(1) 1883 (ann.?); Bot. Centralbl. 29 Oct-2 Nov 1883; Bot. Zeit. 28 Dec 1883), p. [i-iii], [1]-460. *Copies*: BR, USDA.

Ravn, Frederik Kølpin (1873-1920), Danish botanist; Dr. phil. Copenh. 1900; assistant at the Agricultural College of Copenhagen with E. Rostrup 1892-1907; professor of phytopathology ib. 1907-1920. (*Ravn*).

HERBARIUM and TYPES: C, CP.

BIBLIOGRAPHY and BIOGRAPHY: Ainsworth p. 312; Barnhart 3: 132 (b. 10 Mai 1873, d. 24 Mai 1920); BL 2: 55, 706; BM 8: 1051-1052; Bossert p. 325; CSP 18: 68; Kleppa p. 177; KR p. 583; LS 21816-21834, 32065, 38220-38246, 41623; Rehder 2: 173, 180; Stevenson p. 1255.

BIOFILE: Anon., Bot. Centralbl. 107: 80. 1908 (app. Copenhagen); Bot. Not. 1920: 126 (d. 25 Mai 1920, East Orange, New Jersey); Mycologia 12: 335. 1920; Nat. Nov. 30: 173. 1908 (app. Copenhagen); Österr. bot. Z. 58: 175. 1908 (app. Copenhagen), 70: 64. 1921 (d.); Phytopathology 5(5): 283. 1915 (visit to U.S.A.).
Christensen, C., Dansk. Bot. Hist. 1: 854-856, 882. 1926 (biogr., portr.); Dansk. Bot. Litt. 1880-1911, p 170-174 (bibl., portr.), id. 1912-1939, p. 45-48. 1940 (id.); Dansk biogr. leks. 19: 280-282. 1940.
Christensen, H.R., Tidsskr. Landøkon. 1920: 261-264 (portr.).
Dorph-Petersen, K., Naturens Verden 4: 289-301. 1920 (d. 25 Nov 1920 of blood poisoning; b. 10 Mai 1873; obit.).
Helms, J., Dansk Skovfor. Tisskr. 1920: 160-164 (obit.).
Henning, E., K. Landtbruks-Akad. Handl. Tidskr. 59(6): 352-354. 1920 (obit., portr.).
Kneucker, A., Allg. bot. Z. 14: 84. 1908 (app. Copenhagen).
Lind, J., Danish fungi herb. E. Rostrup 37. 1913.
Ostenfeld, C.H., Bot. Tidsskr. 37: 113-120. (obit., portr.).
Ravn, F.K. et al., K. Veterin. Landb. Højsk. Festskrift 521. 1908.
Whetzel, H.H. & H.B. Humphrey, Phytopathology 11: 1-5. 1921.

EPONYMY: *Ravnia* Oersted (1852) was named for P. Ravn (-1839), see below.

NOTE: Peter Ravn (x-1839), Norwegian surgeon went to the Danish West Indies in 1819, stationed at St. Thomas; he collected extensively on St. Thomas, St. Croix, St. Jan and Vieques (Crab Island). Orig. coll. at C (ca. 1700).

8690. F.Kølpin Ravn. Nogle *Helminthosporium-Arter* og de af dem fremdkaldte sygdomme hog byg og havre ... København (i kommision hos Universitetsboghandler G.E.C. Gad ...) 1900. Oct. (*Helminthosp.-Art.*).
Publ.: Sep-Nov 1900 (preface Sep 1900; thesis admitted 20 Nov 1900; date of promotion 19 Dec 1900), slip date prom., p. [i-vii], [1]-220, *pl. 1-2* (col. liths. S. Spies). *Copy*: FH.
– Preprinted from Bot. Tidsskr. 23: 101-321. *pl. 1-2.* Jan 1901.

8691. *Smitsomme Sygdomme hos Landbrugsplanterne* ... København (Hovedkommision: August Bangs Forlag ...) 1914. Oct. (*Smits. Sygd. Landbr.-pl.*).
Publ.: Jul-Sep 1914 (p. vi: Jun 1914; Nat. Nov. Oct 1914), p. [i]-xii, [1]-270. *Copy*: USDA. – Landboskrifter 22, 1914.

Rawitscher, Felix (1890-1957), German botanist; student of F. Oltmanns; Dr. phil. Freiburg 1912; student of Pfeffer 1913-1914; assistant with Oltmanns at Freiburg i.B.; habil. ib. 1921; extraord. prof. forestry botany ib. 1927-1934; professor at the University of São Paulo, Brazil, 1934-1952; Dr. h.c. São Paulo 1955; back in Germany 1952-1957. (*Rawitscher*).

HERBARIUM and TYPES: Unknown.

BIBLIOGRAPHY and BIOGRAPHY: Barnhart 3: 132 (b. 4 Jan 1890); Biol.-Dokum. 14: 7334; BJI 2: 142; Moebius p. 455; Plesch p. 373.

BIOFILE: Anon., Jahrb. Wiss. Bot. 56: 824. 1915; Österr. bot. Z. 76: 88. 1927 (app. Freiburg).
Ferri, M.G., Univ. São Paulo, Fac. filos. Ci. Bol. 224, Bot 15: 7-13. 1958 (obit., b. 4 Jan 1890, d. 18 Dec 1957, portr.).
Verdoorn, F., ed., Chron. bot. 2: 2, 88, 149. 1936, 3: 2, 73, 129, 368. 1937, 4: 189. 1938, 7: 110, 336. 1943.

8692. *Beiträge zur Kenntniss der Ustilagineen.* Inaugural-Dissertation zur Erlangung der Doktorwürde vorgelegt der hohen naturwissenschaftlich-mathematischen Facultät der Albert-Ludwigs-Universität zu Freiburg in Breisgau ... Jena (Verlag von Gustav Fischer) 1912. Oct. (*Beitr. Ustilag.*).
Publ.: 1912 (p. 31: Mai 1912), p. [i], [1]-34, *1 pl. Copy*: FH. – Reprinted or preprinted from Z. Pfl.-Physiol. 4: 673-706. 1912. – Beiträge ... II. in Z. Pfl.-Physiol. 14: 273-296. 1922.

8693. *Die heimische Pflanzenwelt* in ihren Beziehungen zu Landschaft Klima und Boden gemeinverständlich dargestellt ... Freiburg im Breisgau (Herder & Co. ...) 1927. Oct. (*Heim. Pfl.-Welt*).
Ed. 1: 1927 (Nat. Nov. Jul 1927), p. [i]-ix, [1]-238, *pl. 1-11. Copies*: M, USDA.
Ed. 2: 1927 (t.p.; Nat. Nov. Mar 1929 as of "1828"), p. [i]-ix, [1]-238, *pl. 1-11. Copy*: BR.
– "Zweite Auflage", otherwise identical.
Ref.: Engler, A., Bot. Jahrb. 61(Lit.): 84. 1927.
Bartsch, J., Allg. bot. Z. 33: 55 (279)-56(280). 1928.

Rawson, Sir Rawson William (1812-1899), British colonial administrator and pteridologist; clerk Board of Trade 1829; Government secretary Canada 1842; treasurer and paymaster-general Mauritius 1844; colonial secretary Cape of Good Hope (1854-1864); governor Bahamas 1864, id. Jamaica 1865, id. Windward Islands (1869-1875). (*Rawson*).

HERBARIUM and TYPES: BM, other material at B, BP, K, PH and SAM. – Letters at K.

BIBLIOGRAPHY and BIOGRAPHY: Barnhart 3: 132; BB p. 253; BM 4: 1511, 1652, 8: 1052; CSP 8: 708, 11: 116; Desmond p. 513 (b. 8 Sep 1812, d. 20 Nov 1899); IH 2: (in press); Jackson p. 347; PR 6930; Urban-Berl. p. 386.

BIOFILE: Anon., Bot. Not. 1900: 47 (d.); Hedwigia 39: (70). 1900 (d.); J. Bot. 34: 118. 1896, 38: 63. 1900 (obit); R.W., Bull. misc. Inf. Kew 1899: 221-222.
Bayer, A.W., S. Afr. J. Sci. 67: 406. 1971 (John Buchanan used Rawson for his *Revised list of Natal ferns*, 1875).
Fryer, Dict. S. Afr. biogr. 2: 571. 1972.
Gunn, M. & L.E. Codd, Bot. expl. S. Afr. 291. 1981 (coll., portr.).
Harvey, W.H. & O.W. Sonder, Fl. cap. 1: 67. 1859.
Kneucker, A., Allg. bot. Z. 6: 76. 1900 (d.).
Murray, G., Hist coll. BM(NH) 1: 176. 1904.
Urban, I., Symb. ant. 3(1): 109. 1902.

COMPOSITE WORKS: With C.W.L. Pappe (q.v.) *Syn. filic. Afr. austr.* (1858).

EPONYMY: *Rawsonia* W.H. Harvey et Sonder (1860).

Rawton, Olivier de (*fl.* 1882), French economic botanist. (*Rawton*).

HERBARIUM and TYPES: Unknown.

BIBLIOGRAPHY and BIOGRAPHY: BL 2: 105, 706; LS 21823; Plesch p. 373.

8694. Bibliothèque instructive. *Les plantes qui guérissent et les plantes qui tuent* ... Paris (Librairie Furne ...) 1884. Oct. (*Pl. guér. tuent*).
Publ.: 1884 (Bot. Zeit. 23 Jul 1884), p. [i]-viii, [1]-344. *Copy*: USDA. – See Bl 2: 105 for further details.
Spanish transl.: 1887, viii, 348 p., n.v. *Vegetales que curan y vegetales que matan*.

Ray, John (until 1669: Wray) (1627-1705), British naturalist; studied theology at Cambridge, fellow of Trinity College 1649-1662, MA Cambridge 1651, travelled widely in England; in continental Europe (1663-1666) "father" of British botany. F.R.S. 1667. (*Ray*).

HERBARIUM and TYPES: BM (20 vols.); British material also at NOT. The herbarium at BM contains mainly European plants collected by Ray in the years 1663-1665, and is the basis of the *Stirpium Europaearum Sylloge* of 1694. It was given by Ray to Samuel Dale, who in turn bequeathed it to the Apothecaries Company; donated to BM 1862. This "Hortus siccus" consists of 20 books, each containing ca. 30 plants sewn on to the paper. Most of these specimens were collected on Ray's continental tour but some are by others: Matthew Dodsworth, George Wheler, Hans Sloane (see Dandy 1957). For sources, manuscripts et. see Raven pp. xii-xv. For Sherard's contribution in specimens to the *Historia* see Clokie.

BIBLIOGRAPHY and BIOGRAPHY: AG 4: 862; Ainsworth p. 346 [index]; Backer p. 482; Barnhart 3: 132; BB p. 254; BM 4: 1652-1654, 8: 1052; Bossert p. 326; Bret. p. 15-19; Clokie p. 230; Desmond p. 513; De Toni 1: ciii; DNB 47: 339 (G.S. Boulger; "not free from errors and guess-work" Raven); Dryander see index; DSB 11:313-318 (by C. Webster); DTS: 236-238; Frank 3 (Anh.): 81; GR p. 411; Hegi 3: 193; Henry 1: 287 [index], 2: 736-737; Herder p. 472 [index]; Jackson p. 596 [index]; Kelly p. 186-187; Kew 4: 413, 416-417; Lasègue p. 16, 360, 496; LS 21689-21693; Moebius p. 455 [index]; MW p. 400; NI alph. and 1588n, 3: 56; PR 7431-7440, 2041, 3602, 8143, 9425, 9428, ed. 1: 2131, 3913, 7931, 8375-8402, 9092, 10383, 10386; Quenstedt p. 353; Rehder 5: 705; Saccardo 1: 135, 2: 89; SK 4: 14, 15, 21, lxxx, lxxxi, lxxxviii, clx; TL-1/1040-1041; TL-2/1185, 2773, 3496, 6095, 6060, 6075, and indexes; Tucker 1: 584.

BIOFILE: Allen, D.E., The naturalist in Britain 290 [index]. 1976.
Allibone, S.A., Crit. dict. Engl. lit. 2: 1748-1749. 1878 (bibl.).

Andrews, H.N., The fossil hunters 14, 20, 21, 25-27.
Anon., Flora 46: 191. 1863 (on Ray herb. with Apothecaries Cy.); J. Bot. 1: 32. 1863 (herb.), 54: 113. 1916, 55: 167. 1917; Proc. Linn. Soc. London 1888-1890: 40. 1891 (descr. bust).
Arber, A., Isis (ed. G. Sarton) 34(4): 319-324. 1943.
Bagnoll, J.E., Fl. Warwickshire 490-491. 1891.
Baker, J.G., Nat. Hist. Trans. Northumberland 72-74. 1902.
Ballard, Thomas, bookseller, Bibliotheca Rayana: or, a catalogue of the library of Mr John Ray, ... 1708, ii, 30 p. Copy: PH (see also Keynes p. xii).
Berkeley, E. & D.S., Dr. John Mitchell 282 [index]. 1974.
Blunt, W., The compleat naturalist 31, 92, 100-101, 141, 243, 245. 1971.
Boulger, G.S., Trans. Essex Field Cl. 1885: 171-188 (general info.; address).
Bridson, G.D.R. et al., Nat. hist. mss. res. Brit. Isl. 430 [index]. 1980.
Briquet, J., Bull. Soc. bot. Suisse 50a: 394-397. 1940.
Britten, J., J. Bot. 31: 107-108. 1893 (on herb.).
Burdet, H.M., Saussurea 5: 67-100. 1974 (the first Geneva flora; by Ray, *Obs. topogr.* ... 1673).
Cadbury, D.A., Computer mapped Fl. Warwickshire 46-47. 1971.
Candolle, A.P. de, Hist. bot. genev. 8. 1830.
Carpenter, M.M., Amer. midl. Natural. 33: 83. 1945 (refs. to entom. biogr. lit.).
Cooper, G., Ann. Rep. Warwick nat. Hist. Soc. 21: 15-18. 1975.
Crowther, J.G., Founder of British science 94-130. 1960 (biogr., portr.).
Cuvier, G. & A. Du Petit Thouars, Biogr. univ. 37: 155-163. 37, ed. 2. 35: 252-256. 1843.
Dandy, J.E., The Sloane herbarium 189-192, facs. no. 9. 1958 (herb.).
Derham, W., Select remains of the learned John Ray, ... with his life, London 1760, vii, 336 p., portr.; Memorials of John Ray, London 1846 (reprint of Select Remains, 1760); *in* Lancaster, E., ed., The correspondence of John Ray, London 1848, xvi, 502 p. (Ray Society) (reprint of W. Derham, Philosophical letters 1718).
Dony, J.G., Fl. Hertfordshire 11. 1967.
Drewitt, F.D., Apothec. Garden 105 [index]. 1922.
Druce, G.C., Yearb. Pharm. 1910: 401-405; Fl. Buckinghamshire lxxii-lxxiv. 1926; Fl. Oxfordsh. ed. 2. lxxiv-lxxv. 1927.
Edees, E.S., Fl. Staffordshire 14-15. 1972.
Eriksson, G., Bot. hist. Sverige 361 [index]. 1969.
Ewan, J. & N., John Banister 483 [index]. 1970.
Fox, C., Amer. J. Sci. 36: 221-230. 1839.
Freeman, R.B., Brit. nat. hist. books 290-291. 1980.
Fries, Th.M., Bref Skrifv. Linné ser. 1. 2: 146. 1908, 4: 246. 1910, 5: 67, 259, 349. 1911, 7: 141, 142. 1917 (see also J.M. Hulth).
Gilbert, P., Comp. biogr. lit. deceased entom. 310. 1977.
Gilmour, J.S.L., British botanists 16-20. 1944; Watsonia 2: 427-428. 1953 (review of bibl. by Keynes).
Glass, B. et al., Forerunners of Darwin 22, 30, 33-36, 232. 1968.
Green, J. Reynolds, A history of botany in the United Kingdom 67-139. 1914.
Gunawardena, D.C., Proc. Linn. Soc. London 148. 71-73. 1936 (abstract from his unpubl. diss. on Ray, of which a typed copy is at LINN).
Gunn, M. et L.E. Codd, Bot. expl. S. Afr. 31. 1981.
Gunther, R.W.T., Further correspondence of John Ray, London 1928, xxiv, 332 p.
Gunther, R.T., J. Bot. 72: 217-223. 1934 (letters from Ray to P. Courthope).
Haller, A., Hist. stirp. helv. 1: xiv-xv. 1768; Bibl. bot. 1: 500-506. 1771.
Harvey-Gibson, R.J., Outlines hist. bot. 273 [index] 1914.
Hawksworth, D.L. & M.R.D. Seaward, Lichenol. Brit. Isl. 5-6, 138. 1977.
Hornschuch, C.F., Flora 8: 211-212. 1825 (on Ray's work on mosses).
Hulth, J.M., Bref och Skrifv. Linné ser. 1. 7: 141-142. 1917, 8: 28, 33. 1922, ser. 2. 1: 200-202. 1916.
Jenkins, A.C., The naturalists, pioneers of natural history. London 1978 (notice).
Jessen, K.F.W., Bot. Gegenw. Vorz. 488. 1884.
Jourdan, A.J.L., Dict. sci. méd., Biogr. méd. 6: 547-554. 1824 (biogr., bibl.).
Karsten, M.C., The Old Company's garden 83-86, 96. 1951.

Kent, D.H., Brit. herb. 74. 1957; Hist. fl. Middlesex 14. 1975.
Koster, J., Taxon 18: 556. 1969 (types algae).
Lankester, E., ed., Memorials of John Ray, consisting of his life by Dr. [W.] Derham; biographical and critical notices by Sir J.E. Smith and Cuvier and Du Petit Thouars. With his itineraries, etc. London 1846, xii 220 p.; The correspondence of John Ray: consisting of selections of the Philosophical letters published by Dr. [W.] Derham, and original letterss of John Ray, in the collection of the British Museum. London 1848, xvi, 502 p.
Lapeyrouse, P., Hist. abr. pl. Pyren. xxiv-xxv. 1813.
Lett, H.W., Proc. R. Irish Acad. 32: 67. 1916 (Irish mosses).
Lisney, A.A., Bibl. Brit. Lepidopt. 1608-1799, p. 10-24. 1960 (bibl., portr.).
Lütjeharms, W.J., Gesch. Mykol. 261 [index] r936.
Mägdefrau, K., Gesch. Bot. 43-46, 48, 108, 260. 1973.
Magnin, A., Bull. Soc. bot. Lyon 31: 23. 1906, 32: 107. 1907.
Martin, W.K. & G.T. Fraser, Fl. Devon 770. 1939.
Miall, L.L., Early naturalists 99-134. 1912.
Moore, S., Fl. Cheshire lxxxvii. 1899.
Murray, G., Hist. coll. BM(NH) 1: 176. 1904 (herb.).
Nelson, E.C., Works bot. interest... Irish libraries 100-101. 1982 (list Ray holdings in Ireland).
Nicolas, J.P., *in* G.H.M. Lawrence, ed., Adanson 1: 51-153, 161, 164, 167, 170, 175.
Nordenskiöld, E., Hist. biol. 198, 199-202, 208-211, 223, 435, 436. 1949.
Parr, C.B.L. & M. Pollard, Trans. Cambr. bibl. Soc. 3: 335-338. 1962 (bibl. of *Hist pl.*).
Pearsall, W.H., Fl. Surrey 44. 1931.
Perring, F.H. et al., A flora of Cambridgeshire, Cambridge 1964, frontisp. portr. Ray.
Poggendorff, J.C., Biogr.-lit. Handw.-Buch 574-575. 1863.
Pulteney, R. Sketches progr. bot. 1: 189-281. 1790 ("probably still the best account of Ray and his works" Raven 1950), (condensed version, with notes by J.E. Smith, in Rees Cycl. 1819).
Raven, C.E., John Ray, naturalist, his life and works. Cambridge 1942, ed. 2 1910, front. portr. Ray, xx, 506 p. (referred to as Raven in this treatment) (main biogr., portr., discussion of works, b. 29 Nov 1627, d. 17 Jan 1705).
Rendle, A.B., J. Bot. 67: 116-119. 1929 (rev. of R.W.T. Gunther, Further corresp. J. Ray, 1928).
Richardson, R., Extr. lit. sci. corr. 7, 14, 64, 80, 95. 1835.
Riddelsdell, H.J., Fl. Glouc. cx. 1948.
Robinson, J.F., Fl. East Riding Yorks. 10-11. 1902
Rowley, G.D., Cact. Succ. J. Gr. Brit. 28(1): 10-11. 1966 (first key to succ. pl.).
Sachs, J. v., Hist. botanique 7, 63, 72, 397, 486, 552. 1892.
Sawyer, F.C., J. Soc. Bibl. nat. Hist. 4: 97-99. 1963 (on an oil portrait of Ray at BM).
Seward, I.C., John Ray, Cambridge 1937.
Smith, E., Sir Joseph Banks 117. 1911.
Smith, J.E., The life of John Ray, *in* E. Lankester, ed., Memorials of John Ray 1846.
Smit, P. & R.J.C.V. ter Laage et al., Regn. veg. 71: 19, 25-29, 32, 50, 53, 99, 179-182, 185, 188-190, 192, 193. 1971.
Sprengel, K., Gesch. Bot. 2: 40-46. 1818.
Stafleu, F.A., Linnaeus and the Linnaeans 182 [index]. 1971; *in* G.S. Daniels & F.A. Stafleu, Taxon 22: 204. 1973 (portr.).
Stanley, E., Retrospective Review 14: 1-31. 1826 (fide Raven's corr. Ray-Willughby).
Stearn, W.T., Ray, Dillenius, Linnaeus and the Synopsis methodica stirpium britannicarum, intr. to facs. ed. of ed. 3 of Synopsis, 1973; *in* S.M. Walters and C.J. King, Europ. florist. tax. stud. 1-17. 1975.
Stevenson, I.P., J. Hist. Med. 2: 250-261. 1947 (contr. to pl. anim. classif.).
Strohl, J., John Ray 29 November 1627-17 Januar 1705, Zürich, 18 p., reprinted from Neue Zurch. Zeit. 2028, 2035. Nov 1937.
Thompson, R., J. Soc. Bibl. nat. Hist. 7(1): 111-123. 1974 (newly disc. letters of Ray).
Trimen, H., J. Bot. 8: 82-84. 1870 (notes on herb.).
Trimen, H. & W.T. Thistleton Dyer, Fl. Middlesex 372-374. 1869.
Tucker, D.W., J. Soc. Bibl. nat. Hist. 3: 273. 1957 (unrecorded portr; *in* C. Knight, ed., Gallery of Portraits 2: 160-164).

Urness, C., A naturalist in Russia 4, 9, 11, 171, 175. 1967.
Vaughan, J., The wild flowers of Selborne, London, New York 1906 (portr.).
Vines, S.H. *in* Oliver, F.W., Makers Brit. bot. 28-43. 1913.
Walters, S.M., The shaping of Cambridge botany 6-14, 7, 16, 23, 31, 34, 38, 66, 68, 69. 1981.
Ward, B.T., Essex Naturalist 29: 190 1954.
Warren, C.H., Great men of Essex 42-68. 1956 (youth).
Wein, K., Mitt. bayer. bot. Ges. 4(11): 191-196. 1931 (Ray as investigator of the Bavarian flora).
White, J.W., Fl. Bristol 57-60. 1872.
Wilson, A., Fl. Westmorland 70-71. 1938.
Wittrock, V.B., Acta Horti Berg. 3(2): 97, 186. 1903, 3(3): xlvi, 82, *pl. 16.* 1905.
Wolley-Dod, A.H., Fl. Sussex xxxvi-xxxvii. 1937, 1970.
Woodward, B.D., *in* G. Murray, Hist. coll. BM(NH): 46. 1904 (88 autogr. letters; orig. mss. of Derham's Life of Ray).
Zittel, K.A. von, Gesch. Geol. Paläont. 20, 22, 41. 1899; Hist. geol. palaeontol. 17, 19, 31, 410. 1901.

NOTE: *Main bibliography*: G. Keynes, John Ray, a bibliography. London (1951), ("Keynes"), *main biography*: C.E. Raven, John Ray, the naturalist, Cambridge (1942, ed. 2, 1950), ("Raven").

HANDWRITING: Dandy, J.E., The Sloane herbarium, facs. no. 9. 1958.

EPONYMY: The Ray Society, London. England, see e.g. R. Carle, The Ray Society, a bibliographic history, London 1954. – *Janraia* Adanson (1763); *Raia* Cothenius (1790, orth. var.* of *Rajania* Linnaeus); *Raja* J. Burman (1758); *Rajania* Linnaeus (1753); *Rayania* Rafinesque (1840, *orth. var.* of *Rajania* Linnaeus); *Raynia* Rafinesque (1815, *orth. var.* of *Rajania* Linnaeus).

8695. *Catalogus plantarum circa Cantabrigam nascentium*: in qua [sic] exhibentur quotquot hactenus inventae sunt, quae vel sponte proveniunt, vel in agris feruntur; una cum synonymis selectoribus, loci natalibus & observationibus quibusdam oppido raris. Adjiciuntur in gratiam tyronum, index anglo-latinus, index locorum, etymologia nominum, & explicatio quorundam terminorum. Cantabrigiae [Cambridge]: Excudebat Joann. Field, ... 1660. Oct. (*Cat. pl. Cantab. nasc.*).
Publ.: 1660, p. [i-xxx], 1-182, [i], [1]-103, [1]. *Copies*: E, FI, G, USDA. – See Keynes 1, Henrey 24, Raven p. 81-108, Lisney 7. The G has original t.p. "in qua [sic] ...", with the corrected text "in quo" printed on p. [i] of the second part. The USDA copy has both t.p.'s corrected.
Other issue(s): 1660, London (apud Jo. Martin ...). Cancel t.p. only. Henrey 25, Keynes 2, Raven p. 122-123, Lisney 8, 9.
Appendix: 1663, 10 p. (*copy*: E) see Henrey 26, and 27, Keynes 3, Lisney 18.
Appendix: 1685, see Keynes 4, Lisney 11.
Translation: Jan 1975, A.E. Ewen and C.T. Prime, *Ray's Flora of Cambridgeshire* (Catalogus ... nascentium). Wheldon & Wesley Ltd. Hitchin, Herts., p. [i]-vii, [viii], 1-146, *pl. 1-11a, 11b*; ISBN 0-85486-090-8. *Copies*: E, FAS, MO.

8696. *Catalogus plantarum Angliae*, et insularum adjacentium: tum indigenas, tum in agris passim cultus complectens. In quo praeter synonyma necessaria facultates quoque summatim traduntur, una cum observationibus & experimentis novis medicis & physicis ... Londini typis E.C. & A.C. Impensis J. Martyn, ... 1670. Oct. (*Cat. pl. Angl.*).
Ed. 1: 1670, p. [i-xxii], 1-358, [359, err.]. *Copies*: G, MO, PH, USDA. – Henrey 309, Keynes 7, Raven p. 152, 157-159.
Ed. 2: 1677, p. [i-xxviii], 1-311, [313-325, ind.], [326-327, err.], *2 pl*. *Copies*: BR, E, G, NY, USDA. – Henrey 310, Keynes 8.
Additions: 1688. Fasciculus stirpium britannicarum, see below.
Ref.: Greene, E.L., Pittonia 1: 177-183. 1888, 1: 280. 1889.

8697. *Observations* topographical, moral & physiological, *made in a journey through* part of *the Low-Countries*, Germany, Italy, and France: with a catalogue of plants not native of England, found spontaneously growing in those parts, and their virtues. By John Ray, ... whereto is added a brief account of Francis Willughby Esq.; his voyage through a great part of Spain. London (printed for John Martyn, ...) 1673. Oct. (*Observ. journ. Low-Countries*).

Ed. [*1*]: 1673, p. [i-xiv], [xv, err.], 1-499, *4 pl. Copies*: E, G, U. – Appended, but separately paged, *Cat. stirp. ext. region* ..., see below. – Author of Spanish account: Francis Willughby (1635-1672). – Keynes 21.

Ed. 2: 1738, 2 vols. *Travels through the Low-Countries*, Germany, Italy, and France... by the late ... John Ray, ... to which is added an account of travels of Francis Willughby, ... the second edition ... vol. 1; vol. 2: A collection of curious travels and voyages. Containing Dr. Leonhart Rauwolf's Journey into the Eastern Countries ... and also Travels into Greece, ... by the Rev. John Ray ... vol. 2. *Copy*: E. – Keynes 22.

1: p. [i*], [i]-iv, [1]-428.
2: p. [i-x], [1]-489, [i*], 1-44.
For a complete collation see HU 507; se also BM 4: 1652.

Ref.: Burdet, H.M., Saussurea 5: 67-100. 1974.(On first flora of Genève, extracts from Obs. journ. Low-Countries, p. 434-445 of original in facsimile).

8698. *Catalogus stirpium in exteris regionibus* a nobis observatorum, quae vel non omnino vel parce admodum in Anglia sponte proveniunt. Londini (Typis Andreae Clark, ...) 1673. Oct. (*Cat. stirp. ext. region.*).

Publ.: 1673, p. [i-vii], 1-115. *Copies*: E, G, U. – Appended to J. Ray, *Obs. journ. Low-Countries* 1673. See BM 4: 1653, Keynes 22a.

Revised ed.: see below, *Stirp. eur. Syll.* 1694.

8699. *Dictionariolum trilingue*: secundum locos communes, nominibus usitatoribus anglicis, latinis, graecis, .. Londini (typis Andreae Clark, ...) 1675. Oct. (in fours).

Publ.: 1675, p. [i-iii], 1-91. (n.v.).

Facsimile ed.: 1981. The Ray Society, publ. no. 154, ISBN 0-903874-16-4, [i-vi], [1]-23, [p. 1-20 introduction by W.T. Stearn], facsimile, [1-2, general index]. *Copy*: FAS.

8700. *Methodus plantarum nova*, brevitatis & perspicuitatis causa synoptice in tabulis exhibita; cum notis generum tum summorum tum subalternorum characteristicis, observationibus nonnullis de seminibus plantarum & indice copioso ... Londini (impensis Henrici Faitborne [sic] & Joannis Kersey ...) 1682. Oct. (*Meth. pl.*).

Ed. [*1*]: Jun 1682 (Raven p. 195), p.[i-xxii], 1-166, [167-200, ind.]. *Copies*: E, M, USDA; IDC 391. – Henrey 315, Keynes 40, Raven p. 192-200; Adanson, Fam. pl. 1: xix-xxi. 1763.

Ed. [*1*], *Amsterdam issue*: 1682, p. [i-xxii], 1-166, [167-200, ind.], engr. h.t. *Copy*: G. – Henrey 316, Keynes 41 (five extra plates exist facing p. 84, 88, 89, 94, 95. – Issue with cancel title.

Ed. [*1*], *facsimile ed.* (London orig.): 1962, p. [i*-iv*], [i-xxii], 1-166, [167-20, index], 2 charts. *Copies*: FAS, U. – Historiae naturalis classica 26, Weinheim (J. Cramer) 1962.

Ed. [*2*]: 1703, portr., p. [i-xxxiii], 1-202, [203-227]. *Copies*: E, G, NY, U, USDA; IDC 5278. – Henrey 317 (q.v. for notes on history (printed at Leiden, false London imprint), taken from W. Derham (1760) p. 76-77); Keynes 42. – Joannis Raji ... *Methodus plantarum emendata* et aucta in quâ notae maxime characteristicae exhibentur, ... accedit methodus Graminum, Juncorum et Cyperorum specialis eodem auctore. Londini (impensis Samuelis Smith & Benjamini Walford ... et veniunt Amstelaedami apud Janssenio-Waasbergios) 1703. Oct.

Ed. [*2*], *other issue* (cancel t.p.): 1710, portr., p. [i-xxxiii], 1-202, 203-227, ind., 2 charts. *Copies*: BR, NY. – Henrey 43, 318. Keynes. – Joannis Raji ... Methodus ... cyperorum specialis eodem auctore. Prostant Amstelaedami [Amsterdam]. (Rud. & Gerh. Wetstenios HFF) 1710. Oct. – Keynes 44.

Ed. [*2*], *Tübingen reprint*: 1733, p. [i-xxx], 1-196, [197-223]. *Copies*: G, M, NY, PH, USDA. – Joannis Raji ... Methodus ... Cyperorum specialis. Eodem auctore. Londini (apud Christianum Andream Myntsing) 1733. Henrey 329 (the "Londini" is false, the book was printed at Tübingen, fide Séguier, Bibl. bot. 153. 1740), Keynes 45.

8701. *Historia plantarum* species hactenus editas aliasque insuper multas noviter inventas & descriptas complectens. In qua agitur primò de plantis in genere, earúmque partibus, accidentibus & differentiis; deinde genera omnia tum summa tum subalterna ad species usque infimas, notis suis certis & characteristicis definita, methodo naturae vestigiis insistente disponuntur; species singulae accurate describuntur, obscura illustrantur, omissa supplentur, superflua resecantur, synonyma necessaria adjiciuntur; vires denique & usus recepti compendiò traduntur ... Londini (typis Mariae Clark: prostant apud Henricum Faithorne ..) 1686[-1704]. Fol. (*Hist. pl.*).

1: 1686, p. [i-xxiv], [i, impr.], portr., 1-983.
2: 1688, p. [i-viii], 985-1350, [i-ii], 1351-1940 [i.e. 1944], [1-36]. – Tomus secundus: cum duplici indice; ... Londini (typis Mariae Clark ...) 1688. Contains a catalogue of virginian plants by John Banister, discussed extensively by Ewan and Ewan (1970).
3: 1704, p. [i*], [i]-ix, 1-666, [1-40, 43-135, 1-112, 225-255, [257-262], [263, err.]. – Tomus tertius: qui est supplementum duorum praecedentium: ... Londini (apud Sam. Smith & Benj. Walford, ...) 1704. Contains G.J. Kamel (Camellus), *Herbarium aliarumque stirpium in insulâ Luzone ... syllabus*, in App. p. 1-96. (see TL-2/3496). For further bibliographical details of vol. 3 see BM 4: 1653, Henrey 313 and Keynes 48-50; Sotheby 622.
Copies: BR, PH, U, USDA; IDC 390. – Henrey 313, Keynes p. 73-84, nos. 48-50. Raven p. 201-242.
Other issue: 1693-1704; imprint of all three volumes: Impensis Samuelis Smith & Benjamini Walford ...
1: 1693, as original issue but with cancelled t.p.
2: 1693, as original issue but with cancelled t.p.
3: 1704, is original issue.
Copies: G, NY, UC. – Henrey 314, Keynes 51.
Bibliographical details: see C.B.L. Barr and M. Pollard, Trans. Cambr. bibl. Soc. 3: 335-338. 1962 (vols. 1 and 2); see also Henrey, Keynes, Stevenson (1947), Raven p. 201-242, and M. Adanson, Fam. pl. 1: xix-xxi. 1763. – The bibliographical treatment by Barr and Pollard supersedes that of Keynes.

8702. *Fasciculus stirpium britannicarum*, post editum plantarum Angliae catalogum observatorum ... Londini (apud Henricum Faithorne ...) 1888. Oct. (*Fasc. stirp. brit.*).
Publ.: 1688, p. [i-iii], [1]-27. *Copy*: E. – Henrey 312, Keynes 9.

8703. *Synopsis methodica stirpium britannicarum*, in qua tum notae generum charateristicae traduntur, tum species singulae breviter describuntur: ducentae quinquaginta plus minus novae species partim suis locis inseruntur, partim in appendice seorsim exhibentur. Cum indice & virium epitome ... Londini (prostant apud Sam. Smith. ...) 1690. Oct. (*Syn. meth. stirp. brit.*).
Ed. [1]: 1690, p. [ii-xxiv], *2 pl.*, 1-317 [318, ind.; 319, err.]. *Copies*: E, USDA; IDC 393. – Henrey 321; Keynes 54, Raven p. 247-253.
Ed. 2: 1696, p. [ii-xl], 1-346, [347-365, index], [366-368, err.]. *Copies*: BR, USDA. – Henrey 322, Keynes 55, Londini (Impensis S. Smith & B. Walford, ...) 1696. Oct.
Ed. 3: 1724, p. [i-xvi], [1]-288, 281*-288*, 289-482, [483-512], *pl. 1-24*. *Copies*: BR, E(2), PH, Stevenson, U, USDA. – Henrey 323, Keynes 56. Londini (Guilielmi & Joannis Innys ...) 1724. Oct. Henrey 323; HU 460; Keynes 56. – Edited by J.J. Dillenius.
Ed. 3, second issue, 1724, in two volumes. *Copy*: NY. – Henrey 324, Keynes 57.
1: [i-xvi], [1]-288, 281*-288*, *pl. 1-11*.
2: [i], 289-482, [483-512], *pl. 12-24*.
Ed. 3, facsimile ed.: 1973, *in* John Ray, Synopsis methodica stirpium britannicarum editio tertia 1724. Carolus Linnaeus Flora Anglica 1754 & 1759. Facsimiles with an introduction by William T. Stearn, London (The Ray Society) 1973. Stearn's introduction, p. 1-41 provides details on bibliography, importance, collaborators, etc. – Ray Society Publ. no. 149, see also Taxon: 23: 417. 1974. *Copy*: FAS.

8704. *Stirpium europaearum extra britannias nascentium sylloge*. Quas partim observavit ipse, partim è Car. Clusii Historia, ... collegit Joannes Raius. Adjiciuntur catalogi rariorum Alpinarum, & Pyrenaicarum, Baldensium, Hispanicarum grisleii, Graecarum & Orientalium, Creticarum, Aegyptiacarum aliique: ab eodem ... Londini (prostant apud Sam. Smith & Benj. Walford ...) 1694. Oct. (*Stirp. eur. syll.*).

Publ.: 1694, portr., p. [i-xxviii], 45-220, 225-400, [1]-45. *Copies*: BR, E, G, NY, U, USDA. − Revised edition of *Cat. stirp. ext. region.* (1673). − Henrey 320, Keynes 97, Raven p. 282.

8705. Joannis Raii *de variis plantarum methodis dissertatio brevis.* In qua agitur i. de methodi origine & progressu. ii. De notis generum characteristicis. iii. De methodo sua in specie. iv. De notis quas reprobat & rejiciendas censet D. Tournefort. v. De methodo Tournefortiana. Londini (Impensis S. Smith & B. Walford, ...) 1696. Oct. (*De var. pl. meth. diss.*).
Publ.: 1696, p. [i-xiii], 1-48. *Copies*: BR, E, G, PH, U, USDA. − Henrey 311, Keynes 99.

8706. *De methodo plantarum* viri clarissimi d. Augusti Quirini Rivini, physiologiae in Academia Lipsiensi professoris *epistola* ad Joan-Raium, cum ejusdem responsoria: in qua d. Josephi Pitton Tournefort, m.d. Elementa botanica tanguntur. Londini (prostant apud S. Smith, & B. Walford, ...) 1796. Oct. (*De meth. pl. epist.*).
Publ.: 1796, p. [1]-55, [56, err.]. *Copies*: BR, E, G, USDA. − Letter by August Quirinus [Bachmann] Rivinus (1652-1723), with reply by Ray. − see Keynes 55.

Rayner, John Frederick (1854-1947), British florist and naturalist at Swaythling, Southampton. (*Rayner*).

HERBARIUM and TYPES: BMH; further material in OXF.

BIBLIOGRAPHY and BIOGRAPHY: Barnhart 3: 133; BL 2: 242, 243, 244, 706; BM 4: 1657; Clokie p. 230; Desmond p. 514 (d. 19 Jan 1947); GR p. 644; Kelly p. 187; LS 38252-38255, 41630.

BIOFILE: Kent, D.H., Brit. herb. 74. 1957.
Lousley, J.E., Rep. Bot. Soc. Exch. Club Brit. Isl. 13(3): 233- 234. 1948.

COMPOSITE WORKS: *Fungi of the Isle of Wight*, in F. Morey, *A guide to the natural history of the Isle of Wight*, 1909, p. 42-64.

8707. *A standard catalogue of English names of our wild flowers* to which are added the ferns and their allies ... Southampton (A.M. Gilbert & Son, ...) London (Simkin, Marshall, Hamilton, Kent & Co., Ltd., ...) [1927]. (*Stand. cat. Engl. names fl.*).
Publ.: 1927 (MO rd. 1 Jul 1927; J. Bot. Mai 1927), p. [i-iv], [1]-51, [52-56, ind.]. *Copies*: MO, USDA.

8708. *A supplement to* Frederick Townsend's *Flora of Hampshire* and the Isle of Wight ... Swaythling, Southampton (published and sold by the author) [1929]. Oct. (*Suppl. Fl. Hampshire*).
Publ.: 1929 (p. xvi: Mar 1929; J. Bot. Nov 1929), p. [i]-xix, [1]-132. *Copies*: G, USDA. − On p. xvii-xix a memoir, by J. Britten, of Frederick Townsend (1822-1905), taken from J. Bot. Apr 1906. *Fl. Hampshire* ed. 1. 1883, ed. 2: 1904, see under F. Townsend.

Rayss, Tscharna (1890-1965), Bessarabian-born Israeli algologist and mycologist; Dr. phil. Genève 1915, student of R. Chodat 1914-1918; with Savulescu at the department of plant pathology of the Agricultural Research Institute, Bucarest 1918-1934; director of the Laboratory for thallophyta of the Hebrew University Jerusalem 1934-1961. (*Rayss*).

HERBARIUM and TYPES: Israel material: HUJ, also E, PC; Romania: BUCA, C. − Types of algae also at BM, K, PC, UC. − Some archival material in SIA.

BIBLIOGRAPHY and BIOGRAPHY: BJI 2: 143; BM 8: 1053; IH 1 (ed. 2): 64, (ed. 3): 79, (ed. 4): 85, (ed. 5): 85, (ed. 6): 110, (ed. 7): 112, 256; Kew 4: 421; Roon p. 92; Stevenson p. 1255.

BIOFILE: Bodenheimer, F.S., Bull. Res. Counc. Israel 10D: 3-4. 1961 (portr., tribute, followed on p. 5-7 by a bibliography).
Feldmann, J., Bull. Soc. phycol. France 11: 22-23. 1965 (obit.).
Friedmann, I., Rev. algol. ser. 2. 8: 79-81. 1966 (obit., portr., bibl.).
Koster, J.Th., Taxon 18: 556. 1969 (algae types at HUJ).
Moreau, F., Bull. Soc. bot. France 112: 480-487. 1965 (obit., b. 10 Mai 1890, d. 4 Apr 1965, bibl.).
Papenfuss, G.F., Israel J. Bot. 17: 108. 1968 (Red Sea alg.).
Stafleu, F.A., Taxon 14: 206. 1965 (b. 10 Mai 1890, d. 4 Apr 1965).
Timor, B., Israel J. Bot. 14: 1-4. 1965 (obit., portr.).
Verdoorn, F., ed., Chron. bot. 1: 244. 1935, 2: 250, 251, 261, 394. 1936, 3: 216, 228. 1937, 4: 241. 1938, 6: 302. 1941.
Viennot-Bourgin, G., Bull. trim. Soc. mycol. France 81: 113-115. 1965 (obit., portr.).

COMPOSITE WORKS: With T. Săvulescu (q.v.) as senior author: *Materiale pentru flora Besarabiei* 1924, 1926, 1934.

FESTSCHRIFT: Bull. Res. Counc. Israel, 10D(1-4). Jul 1961, 355, v p.

EPONYMY: *Rayssiella* Edelstein et Prescott (1964) likely honors this author, but no etymology was given.

8709. Le *Coelastrum proboscidium* Bohl. Étude de planctologie expérimentale suivie d'une révision de Coelastrum de la Suisse. Thèse présentée à la Faculté des Sciences de l'Université de Genève pour l'obtention du grade de Docteur ès-sciences naturelles par Tscharna Rayss ... Berne (Imprimerie K.J. Wyss) 1915. Thèse no. 566. (*Coelastrum probosc.*).
Thesis issue: shortly after 13 Jul 1915 (date of imprimatur University), p. [i-v], [1]-65, *pl. 1-20* (uncol.). *Copy*: MO.
Regular issue: Jul-Dec 1915 (USDA rd. 12 Sep 1896), p. [ii-vi], [1]-65, [66], *pl. 1-20* (uncol.). *Copies*: BR, ILL, USDA. – Matériaux pour la flore cryptogamique Suisse, vol. 5(2), 1915.
Ref.: Engler, A., Bot. Jahrb. 54(Lit.): 38. 4 Oct. 1916.
G.H., Hedwigia 58: (19)-(21). 8 Jul 1916.
P.H., Bull. Soc. bot. France 66: 258-259. 1919.

Re, Filippo (1763-1817), Italian economic botanist and phytopathologist; professor of agriculture at Bologna; subsequently professor of botany and agriculture and director of the botanical garden at Modena 1814-1817. (*F. Re*).

HERBARIUM and TYPES: SPAL.

BIBLIOGRAPHY and BIOGRAPHY: Ainsworth p. 143, 154; Barnhart 3: 133 (b. 20 Mar 1763, d. 25 Mar 1817); BL 2: 416, 706; BM 4: 1657; Bossert p. 326; CSP 5: 13; Herder p. 364, 393; IH 1 (ed. 3): 131, (ed. 7): 340; Kew 4: 421; LS 21841-21845; PR 7441, ed. 1: 8388-8399; Rehder 5: 765; Saccardo 1: 135-136, 2: 89-90; Tucker 1: 584-585.

BIOFILE: Gager, C.S., Brooklyn bot. Gard. Rec. 27: 268 (1802 briefly dir. bot. gard. Bologna), 275 (1814-1817 id. Modena) 1938.
Poggendorff, J.C., Biogr.-lit. Handw.-Buch 2: 579. 1863 (d. 26 Mar 1817).
De Toni, G.B., Malpighia 20: 273-274. 1906 (at Modena, d. 25 Mar 1817).

8710. *Florae atestinae prodromus*. Mutinae [Modena] (ex Typographia regali) 1816. Oct. (*Fl. atest. prodr.*).
Publ.: 1816, p. [1]-136. *Copies*: FI, HH. – Atesta = Este (prov. Padova).

Re, Giovanni Francesco (1773-1833), Italian botanist; Dr. med. Univ. Torino; practicing physician at Susa; from 1818 professor of botany, chemistry and physics at the Veterinary College at Torino.(*Re*).

HERBARIUM and TYPES: TO (Piemonte), formerly at Orto botanico Sassari, Sardinia (fide Mattirolo 1929).

BIBLIOGRAPHY and BIOGRAPHY: Backer p. 484 (epon. *Reana*); Barnhart 3: 133 (b. 27 Sep 1773, d. 2 Nov 1833); BL 2: 380, 706; BM 4: 1675; CSP 5: 113; De Toni 1: civ, 2: c; GR p. 537; Herder p. 179; Kew 4: 422; LS 21846-21849; MD p. 204-206; PR 7442-7444, ed. 1: 8400-8402; Rehder 5: 705; Saccardo 1: 136, 180, 2: 91, cron. xxx; TL-2/5701.

BIOFILE: Ann. Soc. Linn. Lyon 1847-1849: xiv. 1850 (d. 2 Nov 1833).
Lessona, M., *in* G.F. Re, *La flora segusina* 1881, p. vii-xxv (biogr. Re; b. 27 Sep 1772).
Barraja, E. et al., Il botanico Giovanni Francesco Re, Torino 1909, 127 p. (p. 53-75: O. Mattirolo, discussion of works; bibl.).
Burnat, É., Boissiera 5: 74-75. 1941, Bull. Soc. bot. France 30: cxxvii. 1883.
Caso, B., *in* G.F.Re, Fl. Segus. 1881 (biogr. notice in preface).
Lessone, M., *in* B. Caso, ed., *Fl. segus.*, new ed. 1881, p. [vii]-xxx.
Mattirolo, O., Pubbl. II. cent. fond. Orto bot. Torino, Torino 1929, p. xcvii (herb. Flora di Torino now at TO; had been left to Sassari).
Stieber, M.T. et al., Huntia 4(1): 85. 1980 (arch. mat. HU).

EPONYMY: (journal): *Rea*, Bolletino di informazione del Giardino Sperimentale di acclimatazione per piante alpine ed erbacee perenni de San Bernardino di Trana, Torino, Italy. Vol. 1-x. 1968-x. – *Rea* Bertero ex Decaisne (1833) may honor this author, but no etymology was given.

8711. *Flora segusiensis* sive stirpium in circuitu segusiensi necnon in Montecenisio, aliisque circumeuntibus montibus sponte enascentium enumeratio secundum Linneanum systema auctore J.e Francesco Re ... Taurini [Torino] (ex Typographia Bernardini Barberis, ...) [1805]. Oct. (*in fours*). (*Fl. segus.*).
Publ.: Sep-Dec 1805 (p. 6: 3 Sep 1805), p. [1]-93. [94, err.]. *Copies*: FI, G, NY.
New ed.: Sep-Dec 1881 (p. vi: Sep 1881; Nat. Nov. Feb(2). 1882; Bot. Centralbl. 13-17 Mar 1882; Bot. Zeit. 31 Mar 1882), p. [i]-xxx, [xxxi], 1-405. *Copies*: FI, HH, NY. – "*La flora segusina* di G. Francesco Re riprodotta nel metodo naturale di de Candolle e comentata da Beniamino Caso [1824-1882] ... Torino (Angelo Baglione, ...) 1881. Oct. – For further references see BL 2: 380. – Appendix: 1882, p. [1]-15 (p. [3]: Mar 1882).
Ref.: Mattirolo, O., La flora Segusina dopo gli studi di G.F. Re, Torino 1907, 84 p.(Mem. r. Acad. Torino). (TL-2/5701).

8712. *Ad Floram pedemontanam appendix* doctoris Ioannis Francisci Re ... Taurini [Torino] (ex Typographia Regia) [1821]. Oct.
Appendix: 1821, p. [1]-62, [1]. *Copy*: G.
Appendix altera: 1827, Mem. r. Accad. Sci. Torino 31: 189-224. 1827. *Copies* (reprint with orig. pag.): FI, G, U. – "lecta die 20 junii 1824; volume of Memoria publ. 1827; Flora rd. Jul-Sep 1827.
Appendix tertia: 1829 (lecta 15 Martii 1829), Mem. r. Accad. Sci. Torino 35: 205-222. 1831.

8713. *Flora torinese* del dottore in medicina Giovanni Francesco Re ... Torino (dalla Tipografia Bianco) 1825-1827, 2 vols. in 3. Oct. (*Fl. torin.*).
1: 15 Jan-28 Dec 1825 (MD), p. [1]-372, [1, err.].
2(1): after 11 Feb 1826 (MD), p. [1]-172.
2(2): 1827, p. [173]-340.
Copies: FI, G, HH, M, MO. – For collation and further details see MD and BL.

Rea, Carleton(1861-1946), British barrister and mycologist at Worcester; co-founder, first secretary, and editor of the British mycological Society; educated at Oxford (Magdelen College, grad. 1883), called to the Bar 1881, practicing until 1907, but, thanks to his independent means devoting himself mainly to botany. (*Rea*).

HERBARIUM and TYPES: K (2000); further fungi at CGE, LIV. Drawings of fungi (also by his wife) and mss. at BM; letters to W.G Farlow and R. Thaxter at FH.

BIBLIOGRAPHY and BIOGRAPHY: Ainsworth p. 286, 326, 346; Barnhart 3: 133; BL 2: 274, 706; BM 8: 1053; Bossert p. 326; CSP 18: 79; Desmond p. 514 (b. 7 Mai 1861, d. Jun 1946); Ewan (ed. 2): 181; GR p. 411; Hawksworth p. 185, 197; IH 1 (ed. 3): 93, 2: (in press); Kelly p. 187-188; Kew 4: 422; LS 21850-21852, 38245-38266; Stevenson p. 1255; Tucker 1: 585.

BIOFILE: Ainsworth, G.C., Dict. fungi ed. 6. 502. 1971.
Bisby, G.R., Trans. Brit. mycol. Soc. Fifty-year index 20. 1952 (bibl.).
Blackwell, E.M., Naturalist 877: 59. 1961 (portr.).
Bridson, G.D.R. et al., Nat. hist. mss. res. Brit. Isl. 229.462, 269.296. 1980.
Hesler, L.R., Biogr. sketch N. Amer. mycol. 1975, mss.
Kent, D.H., Brit. herb. 74. 1957 (herb. at BM [now at K]).
Lloyd, C.G., Mycol. Notes 75: 1349. 1925 (portr.).
Palmer, J.T., Chron. cat. lit. Brit. Gasterom. 154. 1968.
Ramsbottom, J., Trans. brit. mycol. Soc. 30: 1-14 (work in Society), 180-185 (obit., portr.) 1948.
Richardson, L. & A.A. Pearson, Trans. Worcestersh. Natural. Club 1945/47: 79-81 (obit. notice, portr.).
Verdoorn, F., Biologia 1(3): 12. 1947 (d.).

COMPOSITE WORKS: Editor, Trans. Worcestershire Natural. Club 1-10(1).

8714. *The botany of Worcestershire* an account of the flowering plants, ferns, mosses, hepatics, lichens fungi, and fresh-water algae, which grow or have grown spontaneously in the county of Worcester with an introduction and a map by John Amphlett, ... and Carleton Rea, ... with the assistance of many friends the mosses and hepatics contributed by J.E. Bagnall, ... with later additions. Birmingham (Cornish Brothers Ltd. ...) 1909. Oct. (*Bot. Worcestershire*).
Co-author: John Amphlett (1845-1918); mosses: James Eustace Bagnall (1830-1918), see TL-2/1: 91.
Publ.: 1909 (Bot. Centralbl. 19 Oct 1909), p. [i]-xxxiii, [1]-654, map (in pocket). *Copies*: BR, E, G, HH, NY, USDA.

8715. *British basidiomycetae* a handbook to the larger British fungi ... Cambridge (at the University Press) 1922. Oct. (*Brit. basidiomyc.*).
Publ.: 1 Aug 1922 (p. [v]: 19 Jun 1922; p. [iv]: publ. 1 Aug 1922, PH rd. 27 Nov 1922), p. [i]-xi, [xii], 1-799. *Copies*: BR, DBN, E, FH, G, H, MO, NY, PH, Stevenson, USDA.
Facsimile ed.: 1968, Bibliotheca mycologia Band 15, Lehre (Verlag von J. Cramer). ISBN 3-7682-0561-4. *Copy*: FAS.
Ref.: Ramsbottom, J., J. Bot. 60: 307-309. Oct 1922 (rev.).

Reade, Oswald Alan (1848-1929), British botanist and pharmaceutical chemist; dispenser in the Royal Navy 1873-1908; stationed at Ascension, Bermuda, Plymouth, Malta and Haslar; collected extensively in Bermuda 1878-1886; in retirement at Lowestoft. (*Reade*).

HERBARIUM and TYPES: K (Bermuda); plants from Malta and mss at BM; some Bermuda ferns at NY.

BIBLIOGRAPHY and BIOGRAPHY: Barnhart 3: 134 (d. 14 Apr 1929); Desmond p. 514; Kew 4: 423; Rehder 1: 336, 3: 717; Tucker 1: 585.

BIOFILE: Anon., Nature 123: 882. 1929 (d.); Pharm. J. Pharmacist 122: 393. 1929 (obit.); J. Bot. 68: 22. 1930 (obit.).
Bridson, G.D.R. et al., Nat. hist. mss. res. Brit. Isl. 229.462. 1980.
Britton, N.L., Fl. Bermuda 548. 1918.
Urban, I., Symb. ant. 1: 134. 1898, 3: 110. 1902.

8716. *Plants of the Bermudas*, or, Somers' Islands ... Bermuda (printed at the "Royal Gazette" Office, Hamilton) 1883. (*Pl. Bermudas*).
Publ.: 1883 (t.p. but cover dated 1885), p. [1]-112, [i]-vii. *Copies*: HH, NY, PH, US, USDA.

Reader, Felix Maximilian (1850-1911), German-born chemist and botanist who emigrated to Australia setting up a chemist's business in Victoria. (*Reader*).

HERBARIUM and TYPES: MEL; further material at A, B, E, K, L, MO, NY. – Issued *Plantae Victoriae Australiae exsiccatae*, Cent. 1-3, 1908 (Herbarium 6: 44. 1908).

BIBLIOGRAPHY and BIOGRAPHY: Barnhart 3: 134; Hortus 3: 1202; HR; IH 2: (in press); LS 21854-21855; Rehder 1: 505; Urban-Berl. 230, 281, 311; Zander ed. 10, p. 386.

BIOFILE: Anon., Vict. Natural. 27: 236. 1911 (extensive collections acquired by MEL).
Maiden, J.H., J. Proc. R. Soc. N.S.W. 55: 165-166. 1922.
Willis, J., Vict. Natural. 66: 105, 125. 1949.

EPONYMY: *Readeriella* Sydow (1908).

Reader, [Henry Charles Lyon] Peter (1850-1929), British clergyman (Dominican priest) and botanist; MA Oxford (Merton College); admitted to the Dominican order 1876, ordained 1878, professed 1881; had various assignations at London, Leicester, Hinckley and Woodchester; from 1926 again at Hinckley. (*P. Reader*).

HERBARIUM and TYPES: BRIST (4000); other material at BM, DBN, E, LIV, LSR, OXF and STO.

BIBLIOGRAPHY and BIOGRAPHY: BL 2: 266, 706; BM 4: 1657; Clokie p. 230; Desmond p. 514-515; IH 1 (ed. 6): 363, 2: (in press); LS 21856.

BIOFILE: Allen, D.E., Newsletter Soc. Bibl. nat. Hist. 15: 10. 1982.
Bridson, G.D.R. et al., Nat. hist. mss. res. Brit. Isl. 84.21. 1980.
Edees, E.S., Fl. Staffordshire 21-22. 1972.
Gumbley, W., Obit. notic. Engl. Dominicans 1555-1952, London 1955, p. 157 (biogr. sketch, b. 12 Aug 1850, d. 11 Mai 1929); "his natural diffidence prevented him leaving any printed record of his [botanical] work").
Hawksworth, D.L. & M.R.D. Seaward, Lichenol. Brit. Isl. 26, 138-139, 198. 1977.
Horwood, A.R. & C.W.F. Noel, Fl. Leicestershire and Rutland ccxxviii, 678. 1933.
Kent, D.H., Brit. herb. 74. 1957.
Mitchell, M.E., Bibl. Irish lichenol. 55. 1971.
Riddelsdell, H.J., Fl. Gloucestershire cxxxviii. 1948.
Wade, A.E., Fl. Monmouthshire 14. 1970.

COMPOSITE WORKS: Contributed *The hepaticae of Gloucestershire* to C.A. Witchell & W.B. Strugnell, *The fauna and flora of Gloucestershire*, 1892.

Réaubourg, Gaston (*fl.* 1906), French pharmacist; Dr. pharm. Paris 1906. (*Réaubourg*).

HERBARIUM and TYPES: Unknown.

BIBLIOGRAPHY and BIOGRAPHY: Barnhart 3: 134; BL 2: 203, 706; Hortus 3: 1202; Kew 4: 424; MW p. 400.

BIOFILE: Fournier, G., Voy. déc. mission. natural. franç. 2: 19. 1932.

8717. Université de Paris. École supérieure de pharmacie année 1905-1906, no. 10. *Étude organographique et anatomique de la famille des Lardizabalées*. Thèse pour l'obtention du diplome de docteur de l'Université de Paris (Pharmacie) présentée et soutenue le 9 juillet 1906 ... Montes-sur-Seine (Am. Beaumont, ...) 1906. Oct. (*Étud. Lardizabal.*).

Publ.: 9 Jul 1906, p. [i-v], [1]-127, [129-131]. *Copy*: NY.
Ref.: Guérin, P., Bull. Soc. bot. France 54: 172-173. 1907 (rev.).

Rebentisch, Johann Friedrich (1772-1810), German (Prussian) botanist, active in the Brandenburger Neumark. (*Rebent.*).

HERBARIUM and TYPES: Unknown.

BIBLIOGRAPHY and BIOGRAPHY: Barnhart 3: 134; BM 4: 1659; Futak-Domin p. 81; GR p. 39; GR, cat. 70; Hawksworth p. 185, 197; Herder p. 187, 244; Jackson p. 302; Kew 4: 424; LS 21859-21861; PR 7447-7448, ed. 1: 8405-8406; Rehder 1: 375, 5: xix; SBC p. 130; Stevenson p. 1255; TL-1/1042; Tucker 1: 585.

BIOFILE: Ascherson, P., Fl. Provinz Brandenb. 1: 11. 1864 (d. 1 Mai 1830, err. [for 1 Mai 1810]); S.B. Ges. naturf. Fr. Berlin 1864 (for 16 Feb): 3 (on manuscript fragment diary of a *Prodromus florae neomarchicae*).

EPONYMY: *Rebentischia* P.A. Karsten (1869) may honor this author, but the etymology could not be determined.

8718. *Prodromus florae neomarchicae* secundum systema proprium conscriptus atque figuris xx coloratis adornatis ... Berolini (impensis Fr. Schüppel) 1804. Oct. (*Prodr. fl. neomarch.*).
Preface (on cryptogams): Carl Ludwig Willdenow (1765-1812).
Publ.: Mai-Jun 1804 (p. xii: 15 Apr 1804; SY p. 20), p. [i]-lxii, [1]-406, *pl. 1-4* (col. copp. Guimpel). *Copies*: G, H, HH, M, NY, USDA; IDC 5246.

8719. *Index plantarum circum berolinum sponte nascentium* adiectis aliquot fungorum descriptionibus ... Berolini (Impensis Fr. Schüppel) 1805. Oct. (*Index pl. berol.*).
Publ.: Sep-Oct 1805 (p. iv: 28 Apr 1805; available Michaelismesse 1805), p. [i]-iv, [5]-46, [1, err.]. *Copies*: FH, G, HH, NY, PH.

Reber, Burkhard (1848-1926), Swiss (Aargau-born) pharmacist and botanist; chief pharmacist at the departmental hospital at Genève 1879-1885; later lecturer for Swiss archeology at the University of Genève. (*Reber*).

HERBARIUM and TYPES: G (includes herb. Henri Feer).

BIBLIOGRAPHY and BIOGRAPHY: Hortus 3: 1202; Kew 4: 424; Rehder 3: 690; TL-2/see H. Feer.

BIOFILE: Briquet, J., Bull. Soc. bot. Suisse 50a: 397-398. 1940 (biogr., bibl., b. 11 Dec 1848, d. 9 Jun 1926).
Miège, J. & J. Wuest, Saussurea 6: 132. 1975 (herb.).
Stieber, M.T. et al., Huntia 4(1): 85. 1981 (arch. mat. HU).

Reboud, Victor Constant (1821-1889), French military physician and botanist; with the French African army 1853-1869, 1871-1880 mainly in Algeria; in retirement at Constantine (Alg.) 1880-1883, and St. Marcellin (Isère, France) 1883-1889. (*Reboud*).

HERBARIUM and TYPES: North African material at AL, CGE, DS, E, FI, MO, W, WRSL (Lich.).

BIBLIOGRAPHY and BIOGRAPHY: AG 12(3): 39; Barnhart 3: 134 (d. 25 Mai 1889); BL 1: 20, 314; CSP 8: 711, 12: 601; GR p. 347; IH 2: (in press); Kelly p. 188; LS 21864; Morren ed. 10, p. 113; Rehder 3: 526, 4: 27; Tucker 1: 585.

BIOFILE: Anon., Bull. Soc. bot. Fr. 36(1): 428. 1889 (d.), Id., Table art. orig. 1854-1893: 199-200.
Cosson, E., Compl. fl. atl. 1: 79-85. 1881 (itin.).

Cosson, E. & Durieu de Maisonneuve, Exp. Sci. Alger. bot. 2: xlii-xliii. 1868.
Magnin, A., Bull. Soc. bot. Lyon 32: 41. 1907.
Maire, R., Proc. conn. bot. Algérie 95, 111-113, *pl. 7.* 1931.
Reboud, V.C., Bull. Soc. bot. France 2: 240-244, 537-539, 785-788. 1855/56, 4: 381-386, 465-473 (letters on Alg. botany; identif. E. Cosson).

EPONYMY: *Reboudia* Cosson et Durieu ex Cosson (1856).

Reboul, Eugène de (1781-1851), French botanist; emigrated to Italy and settled in Florence (calling himself Eugenio). (*Reboul*).

HERBARIUM and TYPES: FI. – Some *Tulipa* types in G-DC; some material also in MW.

BIBLIOGRAPHY and BIOGRAPHY: AG 3: 201; Barnhart 3: 134; BM 4: 1659; Frank 3 (Anh.): 81; IH 2: (in press); Kew 4: 424; PR 7449-7450, ed. 1: 8407-8408; Saccardo 1: 136, 2: 90, cron. xxx; Tucker 1: 585.

BIOFILE: Candolle, Alph. de, Phytographie 442. 1880.
Tassi, Fl., Bull. Lab. Orto bot. 7: 29. 1905.

EPONYMY: *Rebouillia* Raddi (1818, *orth. rej.*); *Reboulea* Kunth (1830); *Reboulia* Raddi (1818), corr. C.G.D. Nees (1846, *nom.* et *orth. cons.*).

8720. *Nonnullarum specierum tuliparum* in agro florentino sponte nascentium propriae *notae* ... [Florentiae (in Archiepiscopali Typographia, ad Crucem Rubram) 1822]. (*Nonnul. sp. tulip. not.*).
Publ.: 1822 (p. 7), p. 1-7. *Copies*: FI, G(3).
Appendix: 12. kal. Maii 1823 (p. [2]), p. [1-2]. *Copies*: FI, G.

8721. *Selecta specierum tuliparum* in agro florentino sponte nascentium synonyma ... [Florentiae (ex Typographia Galilaeiana) 1838]. (*Select. sp. tulip.*).
Publ.: 1838, p. [1]-8. *Copies*: FI, G.

8722. *Sulla divisione del genere Tulipa* in sezioni naturali discorso di Eugenio de Reboul. Firenze (per la Società tipografica) 1847. Oct.
Publ.: 1847, p. [1]-7. *Copy*: FI. – Reprinted and to be cited from Giorn. bot. ital. 2(2), (7-8): 57-61. 1847.

Réchin, Jules (1853-1913), French clergyman and botanist, teacher at Précigni and, from 1876, at the "Collège de Mamers". (*Réchin*).

HERBARIUM and TYPES: ANGUC (bryoph.).

BIBLIOGRAPHY and BIOGRAPHY: AG 4: 528; Barnhart 3: 134 (b. 8 Dec 1853, d. 14 Aug 1913); IH 2: (in press); LS 21866.

BIOFILE: Anon., Bot. Not. 1913: 298 (d.); Bull. Soc. bot. France 60: 473. 1913 (d.);
  Hedwigia 54: (188). 1914 (d.); Monde des Plantes 7: 88. 1898; Rev. bryol. 23: 40. 1896, 25: 55. 1898, 26: 88, 103. 1899, 27: 12, 98. 1900, 31: 31. 1904 (bibl. notices).
Husnot, T., Rev. bryol. 40: 94-95. 1913 (obit., bibl.).
Lamy, D., Occas. pap. Farlow Herb. 16: 126. 1981 (letter to F. Renauld at PC).

Rechinger, Karl (1867-1952), Austrian botanist; student of Richard Wettstein; Dr. phil. Wien 1893; demonstrator and (later) assistant at the University Botanical Garden, Wien 1893-1902; in various functions at the "Naturhistorisches Museum" Wien 1902-1922, from 1918 as curator of the botany department; Regierungsrat 1922. (*Rech.*).

HERBARIUM and TYPES: W; further material at BR, E, G, GB, GJO, GZU, H, LD, M, OXF, US.

BIBLIOGRAPHY and BIOGRAPHY: AG 4: 738, 12(2): 257; Backer p. 484; Barnhart 3: 134; BFM no. 817-818; Biol.-Dokum. 14: 7340; BJI 1: 115-116, 314; BM 4: 1659, 8: 1054-1055; Clokie p. 230; CSP 18: 85-86; DSB; DTS 1: 238, 6(4): 187; Futak-Domin 499; Hegi 6(1): 100; Hirsch p. 243; IF suppl. 1: 84, 2: 38; IH 2: (in press); Kew 4: 425; MW p. 400, suppl. p. 290; Rehder 5: 706; SK 1: 425-429 (itin.); TL-2/387; Tucker 1: 585.

BIOFILE: Anon., Bot. Jahrb. 18 (Beibl. 46): 15. 1894 (provisorischer Assistent WU); Nat. Nov. 24: 559. 1902 (assistant W); Österr. bot. Z. 52: 291. 1903 (assistant W), 55: 210. 1905 (to Samoa), 60: 127. 1910 ("Kustos-Adjunkt"), 67: 184. 1918 ("Kustos", under A. Zahlbruckner as director of the bot. dept. W); Richard Wettstein zum sechzigsten Geburtstag, Wien, s.d. p. [7].
Janchen, E., Richard Wettsten 178. 1933; Cat. fl. austriac. 1: 8, 16, 23, 30, 35, 45. 1956 (bibl.).
Kneucker, A., Allg. bot. Z. 8: 172. 1902 (assistant W), 16: 64. 1910 ("Kustos-Adjunkt"), 23: 47. 1919 ("Kustos").
Merrill, E.D., Contr. U.S. natl. Herb. 30: 249-250. 1947 (Pacific bibl.); B.P. Bishop Mus. Bull. 144: 154. 1957.
Sayre, G., Bryologist 80: 516. 1977.

8723. *Beitrag zur Flora von Persien.* Bearbeitung der von J.A. Knapp im Jahre 1884 in der Provinz Adserbidschan gesammelten Pflanzen. Herausgegeben von Dr. Richard Ritter v. Wettstein II. ... IV. ... [Wien 1889].
II-IV: Late Jun 1889 (Nat. Nov. Jul(1) 1889), p. [1]-10. *Copy*: G. – Reprinted and to be cited from Verh. zool.-bot. Ges. Wien 39: 240-248. *Beitrag I* was by H. Braun, ib.: 213-239.
V-VIII: Late Mar 1894, p. [1]-6. *Copy*: G. – Idem 44: 88-92. 1894.

8724. *Beiträge zur Flora von Oesterreich* [Separat-Abdruck aus der "Oesterr. botan. Zeitschrift" Jahrg. Nr. 10]. Oct.
Publ.: Oct 1891 (in journal), p. [1]-3. *Copy*: G. – Reprinted and to be cited from Österr. bot. Z. 41: 338-340. 1891.

8725. *Beitrag zur Kenntniss der Gattung Rumex* ... [Wien 1891-1892]. Oct.
Publ.: Dec 1891-Feb 1892 (in journal), p. [1]-10. *Copy*: G. – Reprinted and to be cited from Österr. bot. Z. 41: 400-403. 1891, 42: 17-20, 50-53. 1892.

8726. *Botanische und zoologische Ergebnisse einer wissenschaftlichen Forschungsreise nach den Samoainseln,* dem Neuguinea-Archipel und den Salomonsinseln von März bis Dezember 1905 ... Wien (aus der kaiserlich-königlichen Hof- und Staatsdruckerei) 1907-1914. 6 parts. Qu.
1: 1907 (p. 3 = 199: 1 Jul 1907; submitted 11 Jul 1907; ÖbZ Feb 1908; Bot. Zeit. 1 Mai 1909), p. [i], [1]-121, [123], *pl. 1-3* (1-2 col. liths P. Demelius), with text. – Reprinted and to be cited from Denkschr. Akad. Wiss. Wien, math.-nat. Kl. 81: [197]-317, *pl. 1-3* (id.). – Seven papers by T. Reinbold, M. Foslie, F. v. Höhnel, A. Zahlbruckner, F. Stephani, E. Hackel, F. Kohl and G. Mayr.
2: 1908 (p. 2 = 386: 1 Jun 1908; submitted 11 Jun 1908), p. [i], [1]-178, [179], *pl. 1-6* (with text). – Idem 84: [385]-562, *pl. 1-6* (rd). Seven papers by V.F. Brotherus, K. Rechinger, E. Palla, A. Burgerstein, A. Oberwimmer, A. Nalepa, K. Holdhaus.
3: Mai 1910 (p. 3 = 177: 29 Jun 1909; submitted 8 Jul 1909; t.p. 1910; ÖbZ Mai 1910), p. [i], [1]-258, [259], *pl. 1-18* (with text). – Idem 85: [175]-432, *pl. 1-18* (id.). Three papers by J. Bresadola, K. von Keissler, F. Stephani.
4: Nov-Dec 1911 (t.p.; p. 2 = 26 Jun 1911; submitted 6 Jul 1911; ÖbZ Nov-Dec 1911), p. [i], [1]-65, [67], *pl. 1-3* (with text). – Idem 88: [i], [1]-65, *pl. 1-3* (id.). – Four papers by H. & M. Peragallo, A. Zahlbruckner, F. Stephani, O. Pesta.
5: 1913 (t.p.; p. 3 = 3: 16 Oct 1912; submitted 31 Oct 1912), p. [i], [1]-266, [267], *pl. 1-9* (with text). – Idem 89: [i], [443]-708, *pl. 1-9* (id.). Seven papers by G. Hieronymous, U. Martelli, E. Hackel, E. Palla, O. Beccari, F. Gagnepain, R. Schlechter.
6: 1914 (t.p.; p. 2: 8 Feb 1914), p. [i], [1]-75. – Idem 91: [i], [1]-75. Papers by N. Wille and E. Csiki & F. Stephani; and indexes by K. Rechinger.
Copies: HH, M, NY, U (incompl.).

8727. *Streifzüge in Deutsch-Neu-Guinea* und auf den Salomons-Inseln. Eine botanische Forschungsreise von Lili Rechinger und Dr. Karl Rechinger ... Berlin (Dietrich Reimer (Ernst Vohsen)) 1908. Oct. (*Streifz. Deut.-Neu-Guinea*).
Co-author: Lily Rechinger-Favarger (1880-1973).
Publ.: Oct 1908 (ÖbZ Oct 1908; Nat. Nov. Nov(1) 1908; Bot. Zeit. 1 Mar 1909; Bot. Centralbl. 24 Nov 1908; NY rd. Feb 1909), p. [iii]-xii, [1]-106, *pl. 1-27*. *Copies*: NY, PH.

Rechinger, Karl Heinz (1906-x), Austrian botanist and explorer; son of Karl and Lily Rechinger; Dr. phil. Wien 1931; "Demonstrator" at the Botanical Institute, Wien (with Richard Wettstein) 1928-1929; from 1929 in various positions, ultimately as director at the "Naturhistorisches Museum" Wien. (*Rech. fil.*).

HERBARIUM and TYPES: W (complete series from travels in S.W. Asia, in the Flora iranica region (1937-1975 and from Greece 1942); G (private herb.; coll. Iran 1977; complete series of Greek travels 1927-1936; large series of duplicates from Fl. iranica region), B (large series of duplicates from Fl. iranica region), M (id.); other material at AAU, BM, BP, BR, E, G, GB, GOET, H, LD, NY, S, US. – Some archival material in SIA.

BIBLIOGRAPHY and BIOGRAPHY: AG 12(2): 322; Barnhart 3: 135 (b. 16 Oct 1906); BFM p. 311 [index]; Biol.-Dokum. 14: 7340-7342; Bossert p. 326; Futak-Domin p. 499; Hortus 3: 1202; IF (ed. 1): 103, (ed. 2): 128, (ed. 3): 164, (ed. 4): 182, (ed. 5): 198; IH 1 (ed. 6): 295, 363, (ed. 7): 310, 340; Kew 4: 425-429; MW p. 401, suppl. p. 290; NI 3: 44; PH 209; Roon p. 92; SBC p. 130; TL-2/607, 2517; Zander ed. 10, p. 705, ed. 11,p. 804.

BIOFILE: Anon., Österr. bot. Z. 87: 324. 1938 (in charge of bot. dept. W); Taxon 29: 682. 1980 (OPTIMA gold medal).
Bernardi, L., Boissiera 30: frontisp. 1979 (frontisp. portr. K.H. Rechinger; dedication in a limited number of copies).
Blakelock, R.A. & E.R. Guest, Fl. Iraq 1: 110, 116. 1966.
Burgess, R.L. et al., A preliminary bibliography of the natural history of Iran, Shiraz, Iran 1966, Pahlavi Univ. Coll. Arts Sci., Sci. Bull. 1, p. 93-95. (list of publ. on flora of Iran 1939-1964).
Davis, P.H. & J.R. Edmondson, Notes R.B.G. Edinburgh 37(2): 279, 283. 1979 (bibl. Fl. Turkey).
Fedtschenko, B.A., Bot. Zhurn. 30: 40. 1945.
Hedge, I.C. & J.M. Lamond, Index coll. Edinb. herb. 122. 1970.
Rechinger, F. & K.H. Mitt. geogr. Ges. Wien 78: 147-157. 1935 (Reiseskizzen aus dem albanisch-montenegrischen Grenzgebiet).
Rechinger, K.H., Ann. Mus. Goulandris 4: 39-82. 1978 (Meine botanischen Forschungen in Griechenland 1927-1976; itin. 19 journeys, bibl.).
Riedl, H., Ann. naturh. Mus. Wien 75: 1-16. 1971 (1972) (tribute, 65th birthday; bibl.).
Stewart, R.R., Pakistan J. Forestry 17: 354. 1967 (Pakistan coll.).
Verdoorn, F., ed., Chron. bot. 1: 89, 171. 1935, 2: 26, 35, 68, 80, 271. 1936, 3: 12, 65,. 1937, 4: 183, 187, 566, 571. 1938, 5: 123, 288. 1939, 6: 186, 302. 1941.
Wagenitz, G., Index coll. princ. herb. Gott. 131-132. 1982.

COMPOSITE WORKS: (1) Contributed to G. Hegi, *Ill. fl. Mitteleuropa* 3(1) and 3(3).
(2) *Flora iranica*, 1964-x, editor and in part author, see e.g. F.A. Stafleu, Taxon 16: 437-438. 1967, 22: 148-149. 1973, and K.H. Rechinger, ADEVA-Mitt. Graz 20: 7-15. 1969. Rechinger's "magnum opus".
(3) *Flora of lowland Iraq*, by K.H. Rechinger with contributions by P. Aellen, Y.I. Barkoudah, N.L. Bor, B.L. Burtt, C.D.K. Cook, J.E. Dandy, Th.R. Dudley, F. Ehrendorfer, J. Hedge, D. Hillcoat, A. Huber-Morath, A. Patzak, M. Raymond, H. Riedl, H. Schiman-Czeika, S. Snogerup, C.C. Townsend, G. Wagenitz, P. Wendelbo, F. Widder, T.G. Yuncker. Weinheim (Verlag von J. Cramer) publ. 10 Mar 1964, v, 746 p.

FESTSCHRIFT: (1) Ann. Naturhist. Mus. Wien 75, 1971 (with tribute by H. Riedl on p. 1-17.

(2) Ann. Mus. Goulandris 4, 1978 (frontisp., portr., drawing by Niki Goulandris).

EPONYMY: *Rechingerella* J. Fröhlich (1963); *Rechingeria* Servit (1931); *Rechingeriella* Petrak (1940).

8728. *Ergebnisse einer botanischen Reise nach Bulgarien* ... Budapest (Bethlen Gábor Irodalmi és Nyomdai Rt.) 1933. Oct.
Publ.: 1933, cover t.p., p. [5]-58, *pl. 1. Copy*: U. – Reprinted and to be cited from Magy. bot. Lap. 32: 5-58, *pl. 1*. 1933.

8729. *Floristisches aus der Umgebung des Neusiedler Sees* ... Bratislava-Pressburg (Buchdruckerei Karl Angermayer) 1933. Oct.
Publ.: 1933, p. [1]-35. *Copy*: G. Wagenitz. – Reprinted from Verh. heil. naturw.-Ver. Bratislava (Pressburg) ser. 2. 26: 51-83. 1933.

8730. *Zur Kenntniss der Flora der Halbinsel Pelješac* (Sabbioncello) *und einiger Inseln des jugoslavischen Adriagebietes* (Dalmatien) ... Budapest (Bethlen Gábor Irodalmi és Nyomdai RT) 1934.
Publ.: 1934, cover t.p., [i], [23]-42. *Copy*: U. – Reprinted with special t.p. and to be cited from Magy. bot. Lap. 33: [23]-42. 1934.

8731. *Die süd– und zentralamerikanischen Arten der Gattung Rumex.* Vorarbeiten zu einer Monographie der Gattung Rumex. III. [Arkiv för Botanik Band 26 A. N:o 3) [1934]. Oct.
Publ.: 11 Apr 1934, p. [1]-58, *pl. 1-6. Copies*: G, U. – Issued, and to be cited as, Arkiv för Botanik 26A(3): 1-58, *pl. 1-6*. 1934.

8732. *Ein neuer Rumex aus den Nordalbanischen Alpen* ...Budapest (Bethlen Gábor Irodalmi és Nyomdai RT) 1934. Oct. (*Neu. Rumex Nordalb. Alp.*).
Publ.: 10 Jan 1934, p. [1]-7, *pl. 1. Copies*: G, U. – Preprinted from Magy. bot. Lap. 33(1-12): 5-7, *pl. 1*. 1934.

8733. *Dreizehn neue Pflanzenarten aus Griechenland* ... Budapest (Bethlen Gábor Irodalmi és Nyomdai RT) 1934. Oct. (*Dreizehn neue Pfl.-Art. Griechenland*).
Publ.: 10 Jan 1934, p. [i], [7]-22, *pl. 2-7* (uncol., del. Frida Rechinger). *Copies*: G, U. – Preprinted from Magy. bot. Lap. 33(1-12): 8-22. 1934.

8734. *The North-American species of Rumex* ... Chicago, U.S.A. [Field Museum of Natural History ...] 1937. Oct.
Publ.: 24 Jun 1937, p. [1]-151. *Copies*: BR, HH, M, NY. – Issued, and to be cited as Field Mus. Hist., Bot. Ser. 17(1): [1]-151. 1937. – Vorarbeiten zu einer Monographie der Gattung Rumex V.; nos. 1-4 in: (1) Beih. Bot. Centralbl. 49(2). 1932, (2) Repert. Sp. nov. 31; 225-283. 1933, (3) Arkiv Bot. 26A, no. 3, 1933 (see above), 4: Österr. bot. Z. 84: 31-52. 1935. Part (6) Bot. Not. 1939: 485-504, 12: 9-152. 1949, (8): Bot. Not. Suppl. 3: 3. 1954.

8735. *Enumeratio florae constantinopolitanae.* Aufzählung der nach dem Erscheinen von Boissiers Flora orientalis aus der Umgebung von Konstantinopel bekannt gewordenen Farn– und Blüthenpflanzen ... Dahlem bei Berlin [Friedrich Fedde ...] 1938. Oct. (*Enum. fl. constantinop.*).
Publ.: 1 Feb 1938, p. [i], [1]-73, map. *Copies*: BR, G, HH, MICH, U, USDA. – Issued as Beih. Repert. sp. nov. 98. 1938.

8736. *Flora aegaea.* Flora der Inseln und Halbinseln des Ägäischen Meeres ... Wien (In Kommission bei Springer-Verlag Wien ...) 1943. Oct. (*Fl. aegaea*).
Publ.: 1943, p. [i]-xx, 1-924, *pl. 1-25*, [1, expl., maps], 3 maps. *Copies*: BR, HH, M, NY, USDA. – Issued as Denkschr. Akad. Wiss. Wien, math.-nat. Kl. 105(1). 1943.
Facsimile ed.: 1973, as orig. *Copy*: FAS. – Reprint by Otto Koeltz Antiquariat, Koenigstein/Taunus, ISBN 3-87429-044-1. See Taxon 22: 306. 1973.
Supplement: Phyton 1(2-4): 194-228. 1949.

8737. *Neue Beiträge zur Flora von Kreta* (Ergebnisse einer biologischen Forschungsreise nach dem Peloponnès und nach Kreta 1942, im Auftrage des Oberkommandos der Wehrmacht und des Reichsforschungsrates, Nummer 6) ... Wien (in Kommission bei Springer-Verlag, ...) 1943. Oct. (*Neu. Beitr. Fl. Kreta*).
Publ.: 1943, p. [i-iii], [1]-184, 1 map. *Copies*: BR, M. – Issued as Denkschr. Akad. Wiss. Wien, math.-nat. Kl. 105(2, 1). 1943.

Record, Samuel James (1881-1945), American dendrologist and wood technologist; B.A. Wabash 1903, M.A. 1906; M.F. Yale Univ. School of Forestry 1905; Dr. Sci. h.c. 1930; with Division of Forestry, U.S.D.A. 1906-1910; instructor at the Forest School, Yale, 1910; asst. professor of forest products 1911, professor 1917-1939; dean school of forestry and Pinchot Professor of forestry 1939-1945; co-founder (1930) and first secretary-treasurer of the International Association of Wood Anatomists. (*Record*).

HERBARIUM and TYPES: MAD; further material at F, GH, NY, US. – The Samuel James Record Memorial Collection from the School of Forestry at Yale University (Y) is now at MAD. – Some archival material at SIA;

BIBLIOGRAPHY and BIOGRAPHY: Barnhart 3: 135 (b. 10 Mar 1881, d. 3 Feb 1945); BL 2: 314 [index]; Bossert p. 326; Hirsch 243; IH 1 (ed. 6): 165, (ed. 7): 340, 2: (in press); Kew 4: 429; Langman p. 614-615; Lenley p. 340; MW p. 401; Tucker 1: 585.

BIOFILE: Anon., J. New York Bot. Gard. 46:128. 1945; Nat. Cycl. Amer. Biogr. 33: 459-460. 1947 (biogr., portr.).
Cattell, J.M. et al., Amer. men Sci. ed. 3: 563. 1921, ed. 4: 802. 1927, ed. 5: 915. 1933, ed. 6: 1159. 1938, ed. 7: 1449. 1944.
Ferreira de Souza, P., Rodriguesia 3(11): 357-371. 1937 (biogr.).
Garratt, G.A., Chron. bot. 12: 179-180. 1951 (obit.).
Graves, H.S., Trop. Woods 82: 3-9. 1945 (obit., portr.).
Milanez, F.R., Rodriguesia 9(19): 1-7. 1945 (obit.).
Moldenke, H.N., Plant Life 2: 78. 1946 (1948).
Standley, P.C., Trop. Woods 82: 10-13. 1945 (appreciation), followed by a bibliography of 340 nos. on p. 18-37.
Verdoorn, F., ed., Chron. bot. 1: 31, 63, 305. 1935, 2: 31, 51. 1936, 3: 42, 75, 272, 282. 1937, 4:285, 413, 518. 1938, 5: 65, 113, 174, 273, 290, 370, 372, 509. 1939.

COMPOSITE WORKS: (1) Founder, editor and manager of *Tropical Woods* 1925-1945. (2) With Paul C. Standley, q.v., *The forests and flora of British Honduras*, Field Mus. nat. Hist., Bot. ser. 12, 27 Feb 1936.

EPONYMY: *Recordia* Moldenke (1943); *Recordoxylon* Ducke (1943).

8738. *Identification of the economic woods of the United States* including a discussion of the structural and physical properties of wood ... New York (John Wiley & Sons), London (Chapman & Hall, Limited) 1912. Oct. (*Ident. econ. woods U.S.*).
Ed. 1: 1912, frontisp., p. [i]-vii, 1-117, [118], *pl. 1-6* with text. *Copy*: USDA.
Ed. 2: 1919, frontisp., p. [i]-ix, 1-157, [158], *pl. 1-6* with text.
Ref.: W.D., Bull. misc. Inf. Kew 1920: 75-76. 10 Apr 1920 (rev. ed. 2).

8739. *Timbers of tropical America* ... New Haven (Yale University Press), London (Humphrey Milford ...) 1924. Oct. (*Timbers trop. Amer.*).
Junior author: Clayton Dissinger Mell (1875-1945).
Publ.: 1924 (MO rd. 3 Dec 1924, NY rd. Nov-Dec 1924), p. [i]-xviii, [1]-610, frontisp., *pl. 1-50*. *Copies*: BR, HH, MO, NY, PH, US, USDA.
Ref.: T.A.S. [Sprague], Bull. misc. Inf. Kew 1925: 47-48. 16 Feb 1925 (rev.).

8740. *Identification of the timbers of temperate North America* including anatomy and certain physical properties of wood ... New York (John Wiley & sons, Ic.), London (Chapman & Hall, Limited) 1934. Oct. (*Ident. timb. temp. N. Amer.*).
Publ.: 1934 (p. iv: 1 Mar 1934), frontisp., p. [i]-ix, 1-196, [197], *pl. 1-6* with text. *Copies*: U, USDA.

8741. *Timbers of the New World* ... New Haven (Yale University Press), London (Humphrey Milford ...) 1943. Oct. (*Timb. New World*).
Junior author: Robert W. Hess.
Publ.: Mar 1943 (p. vii: 1 Dec 1942; date given in later printings), p. [i]-xv, [1]-640, *pl. 1-58*. *Copies*: BR, G, HH, NY, PH, USDA.
Second printing: Mai 1944, *third*: Oct 1947, *fourth*: Jul 1949.

Redeke, Heinrich Carl (1873-1945), German-born Dutch hydrobiologist, zoologist and botanist; Dr. phil. Amsterdam 1898 (student of Max Weber); at the Dutch marine biological station, Den Helder 1898-1929 (1899-1902 asst. director, from 1902 director); from 1916-1933 also director of the State institute for biological fishery research at Den Helder and (from 1929) at Gouda; from 1934 active at the Utrecht University Library. (*Redeke*).

HERBARIUM and TYPES: L.

BIBLIOGRAPHY and BIOGRAPHY: BL 2: 434, 706; Biol.-Dokum. 14: 7343; BM 4: 1660, 8: 1055-1056; CSP 18: 88; Hirsch p. 244; IH 2: (in press); Nordstedt suppl. p. 13; TL-2/2868; see J.L. Hoek.

BIOFILE: Engel, H., Hydrobiol. Bull. 8: 5-14. 1974 (historical survey of R's work, b. 29 Aug 1873, d. 10 Apr 1945).
Engel, H. et al., Arch. néerl. Zool. 7 (suppl.): i-xx. 1943 (brief biogr. sketch; portr., bibl., tributes).

COMPOSITE WORKS: (1) Editor of *Flora and fauna der Zuiderzee*. Monographie van een brakwatergebied, Den Helder 1922, publ. 1923, p. [i]-viii, [1]-460. Botany, by A.C.J. van Goor, on p. 47-123. *Copies*: BR, HH, MO, PH, USDA. – Supplement 1936, viii, 176, 82 p. *Copies*: id.
(2) With Julie L. Hoek, *Fl. Helder* 1901, TL-2/2868.
(3) With G.M. de Lint and A.C.J. van Goor, *Prodromus eener flora en fauna van het Nederlandsche Zoet- en Brakwaterplankton*, overgedrukt uit: Verhandelingen en rapporten uitgegeven door de Rijksinstituten voor Visscherijonderzoek. Deel 1. Afl. 2: 95-137. 1922 (reprinted with special cover and orig. pag., copy: U).

FESTSCHRIFT: Arch. néerl. Zool. tome vii, suppl., 1943.

Redfield, John Howard (1815-1895), American businessman, botanist, zoologist and palaeontologist; member of a car-wheel manufacturing firm in Philadelphia; after retirement (1885) devoting himself to building up the herbarium of the Academy of Natural Sciences, Philadelphia. (*Redfield*).

HERBARIUM and TYPES: MO(16.447); further material at GH, NA (Martindale herb.), NY, PH, US, WELC. – Archival material at PH and SIA.

BIBLIOGRAPHY and BIOGRAPHY: Barnhart 3: 135 (b. 10 Jul 1815, d. 27 Feb 1895); BL 1: 185, 186, 314; BM 4: 1660; Bossert p. 327; CSP 5: 120, 8: 713, 12: 601, 18: 88; Ewan (ed. 1): 69, 80, 287-288, (ed. 2): 181; Hortus 3: 1202; IH 1 (ed. 6): 363, (ed. 7): 340, 2: (in press); Kew 4: 430; Lenley p. 340, 467; ME 1: 223-224, 3: 643 [index]; Merrill p. 709; Morren ed. 10, p. 130; Nickles p. 864; NW p. 56; PH p. 517 [index]; Quenstedt p. 354; Rehder 5: 707; SIA 7073; Tucker 1: 585.

BIOFILE: Anon., Amer. J. Sci. ser. 3. 49: 485. 1895 (obit., d. 28 Feb 1895); Appleton's Cycl. Amer. Biogr. 5: 205. 1888; Bot. Gaz. 20: 128. 1895 (d.), 21: 181. 1896 (herbarium to be sold by PH to which it had been left to be sold for the benefit of continuing work on the Academy herbarium); Bot. Jahrb. 21 (Beibl. 52): 16. 1895 (d.); Bot. Not. 1896: 44; Nat. Nov. 17: 267. 1895; Recollections of John Howard Redfield, s.l., 1900, ix, 360 p.
Barnhart, J.H., *in* F.W. Pennell, Monogr. Acad. nat. Sci. Philadelphia 1: 617. 1935 (coll.).

Canby, W.M., Bull. Torrey bot. Club 22: 162-171. 1895 (obit., portr., bibl.), see also J. Bot. 33: 313-314. 1895.
Gray, J.L., Letters Asa Gray 836 [index] 1893.
Harshberger, J.W., Bot. Philadelphia 211-239. 1899 (biogr. bibl.).
Kneucker, A., Allg. bot. Z. 1(5): 112. 1895 (d.).
McVaugh, R., Edward Palmer 62, 72, 74, 78. 1956.
Meehan, T., Bot. Gaz. 20(4): 175-176. 1895 (obit.); Fern Bull. 8(2): 25-26. 1900 (obit., portr.); Proc. Acad. nat. Sci. Philadelphia 1895: 292-301. 1896 (obit.; d. 27 Feb 1895); Amer. J. Sci. ser. 3. 49: 485. 1895.
Pennell, F.W., Bartonia 22: 10. 1943.
Redfield, J.H., Recollections, printed for private circulation, 1900, ix, 360 p. (autobiogr.).
Rickett, H.W., Index Bull. Torrey bot. Club 81. 1955.
Rodgers, A.D., Amer. bot. 1873-1892, p. 335 [index]. 1944.
Wittrock, V.B., Acta Horti Berg. 3(3): 200. 1905.

HANDWRITING: Bartonia 28: *pl. 7*. 1956.

COMPOSITE WORKS: With Edward L. Rand (q.v.) *Flora of Mount Desert Island, Maine*, Cambridge 1894.

EPONYMY: *Redfieldia* Vasey (1887).

Redinger, Karl Martin (1907-1940), Austrian lichenologist; Dr. phil. Wien 1931; employed at the "Naturhistorisches Museum" Wien 1931-1937; working on genetics and cytology with F. v. Wettstein at Berlin-Dahlem 1937-1938; cytologist at the German Reichsanstalt für Tabak-forschung nr. Karlsruhe 1938. (*Redinger*).

HERBARIUM and TYPES: Lichens at B (extant), W and WU, some musci at B. *Exsiccatae*: with A. Zahlbruckner, *Lichenes rariores exsiccati* nos. 281-385.

BIBLIOGRAPHY and BIOGRAPHY: Barnhart 3: 135; Biol.-Dokum. 14: 7345; Bossert p. 327; GR p. 438, *pl.* [*23*]; GR, cat. p. 70; Hawksworth p. 185; IH 2: (in press); Kew 4: 430; SBC p. 130.

BIOFILE: Hawksworth, D.L. & M.R.D. Seaward, Lichenol. Brit. Isl. 139. 1977.
Laundon, J.R., Lichenologist 11: 16. 1979.
Mattick, K., Ber. deut. bot. Ges. 58: (70)-(77). 1940 (1941) (obit., portr., bibl.: "begeisterter Nationalsocialist", b. 22 Sep 1907, d. 26 Mar 1940); Willdenowia 1(4): 676. 1956 ("grössere Sendungen" lichens rd.); Herzogia 1: 176. 1969.
Merrill, E.D., Contr. U.S. natl. Herb. 30(1): 250. 1947 (bibl. Pacific).
Mitchell, M.E., Bibl. Irish lichenol. 55. 1971.
Poggendorff, J.C., Biogr.-lit. Handw.-Buch 7a (suppl.): 528. 1971.
Verdoorn, F., ed., Chron. bot. 1: 86-89. 1935, 2: 80. 1936, 3: 65, 295. 1937, 4: 56, 273, 277. 1938, 5: 102. 1939.

8742. *Die Graphidineen der ersten Regnell'schen Expedition* nach Brasilien 1892-1894 ... Stockholm (Almqvist & Wiksells Boktryckeri-A.-B), Berlin (R. Friedländer & Sohn ...), Paris (Librairie C. Klincksieck ...) 1933. 2 parts. Oct.
1 (*Glyphis, Medusulina* and *Sarcographa*): 9 Mar 1934 (in journal; t.p. 1933; p. 20: tryckt 17 Aug 1933; Nat. Nov. Apr 1934), p. [1]-20, *pl. 1*. *Copies*: FI, G, H, U. – Published, and to be cited, as Ark. Bot. 25A(13). 1934.
2 (*Graphina* und *Phaeographina*): Oct-Dec 1933 (t.p. 1933, p. 105: tryckt 14 Sep 1933; Nat. Nov. Dec 1933), p. [1]-105, *pl. 1-7* (uncol.). *Copies*: FI, G, H, U. – Published, and to be cited as Ark. Bot. 26A(1), 1933 (published in volume 11 Apr 1934).
3 (*Graphis* und *Phaeographis*): Mai-Jun 1935 (p. 103: 9 Mai 1935; Nat. Nov. Jul 1935; Rev. bryol. 25 Nov 1936), p. [1]-103, *pl. 1-7*. *Copies*: H, U. – Journal volume (nos. 2-5) issued 22 Oct 1935: Ark. Bot. 27A(3), 1935.
4 (*Opegrapha*): 19 Dec 1940 (in journal), p. [1]-52, *pl. 1-2*. *Copy*: U. – Published and to be cited as Ark. Bot. 7A(3). 1940.

The Swedish expedition to Brasil of 1892-1894 was named after Anders Fredrik Regnell (1807-1884).

8743. *Die Graphidineen der Sunda-Inseln* ... Toulouse (Imprimerie Henri Basuyan & Cie ...) 1936. Oct.
Publ.: 1936, cover t.p., p. [37]-122, *pl. 1-7. Copy*: U. – Reprinted and to be cited from Rev. bryol. lichénol. 9: 37-122, *pl. 1-7.* 1936.

8744. *Thelotremataceae brasilienses* imprimis ex herbario Regnelliano cognitae praetereaque in herbariis Krempelhuberi, Mülleri arg., Nylanderi, Wainionis et Zahlbruckneri asservatae ... Stockholm (Almqvist & Wiksells Boktryckeri-A.-B.), Berlin (R. Friedländer & Sohn ...), London (H.K. Lewis & Co. Ltd. ...), Paris (Librairie C. Klincksieck ...) 1936. Oct.
Publ.: Shortly after 2 Oct 1936 (t.p. 1936, p. 122: "tryckt 2 Oct 1936"; Nat. Nov. Feb 1937), p. [1]-122. *Copies*: G, H, U. – Issued, and to be cited as, Ark. Bot. 28A(8), 1933 (journal issue of nos. 8 and 9 issued 22 Dec 1936).

8745. Dr. L. Rabenhorst's *Kryptogamen-Flora von Deutschland*, Österreich und der Schweiz. IX. Band, 2. Abteilung, 1. Teil. Herausgegeben von Hofrat Dr. A. Zahlbruckner. Wien. Arthoniaceae, Graphidaceae ... Lieferung 1 Arthoniaceae [Lieferung 2 Graphidaceae 1 ...] ... Leipzig (Akademische Verlagsgesellschaft ...) 1937 [-1938]. Oct. (*Krypt. Fl. Deutschl.* 4(2, 1)).
Lief. 1: Jan 1937 (cover dated 1937; rd. at BR 11 Jan 1937), p. [1]-180.
Lief. 2: Jun 1939 (cover dated 1938), p. [181]-404. *pl. 1-2.*
Copies: BR, and see Rabenhorst, *Krypt.-Fl. Deutschl.* ed. 2.
Ref.: Mattick, F., Hedwigia 77(1): (21)-(22). 15 Oct. 1937, 78(2): (101)-(102). 1939.

Redmond, Paul John Dominic (1901-x), American clergyman and botanist; ordained priest 1930; Dr. phil. Cath. Univ. Washington, D.C. 1933. (*Redmond*).

HERBARIUM and TYPES: Material at LAM and LCU.

BIBLIOGRAPHY and BIOGRAPHY: Barnhart 3: 135; BL 1: 186, 314; IH 2: (in press).

8746. *A flora of Worcester county, Maryland* a dissertation submitted to the Faculty of the Graduate School of Arts and Sciences of the Catholic University of America in partial fulfillment of the requirements for the degree of doctor of philosophy ... [Washington, D.C.] (The Catholic University of America ...) 1932. Oct. (*Fl. Worcester Co.*).
Publ.: 1932, p. [1]-104, [105, vita]. *Copies*: HH, NY, USDA. – Contr. Biol. Lab. Cath. Univ. Amer. 11.

Redouté, Pierre Joseph (1759-1840); Luxemburg-born botanical artist who settled in Paris 1782, originally as designer of stage scenery, moving to botanical drawing and painting stimulated by C.L. L'Héritier de Brutelle from 1784, draughtsman and painter to the cabinet of Marie-Antoinette 1787 and associated with G. van Spaendonck; on the staff of the Muséum national d'histoire naturelle, Paris and working with R.L. Desfontaines and A.P. de Candolle after 1795; painter for Joséphine de Beauharnais and Ventenat, later to Marie Louise during the Napoleonic period; during the Restauration working independently as well as for Marie Caroline de Berry. (*Redouté*).

HERBARIUM and TYPES: see under the authors of the texts of Redouté's books. Redouté himself did not publish descriptions and made no herbarium. Some of the plants from the Malmaison garden, used for his paintings, are at P (fide P. Jovet). – For the Redouté originals in the collections of the Muséum d'Histoire naturelle, Paris, see Y. Laissus (ed., 1980); for other details on originals see e.g. Brindle (1963) and the various publications by A. Lawalrée.

BIBLIOGRAPHY and BIOGRAPHY: AG 3: 340; Backer p. 659; Barnhart 3: 135 (b. 18 Aug 1761, d. 18 Jun 1840); Blunt p. 173-182, 302; BM 3: 1108, 4: 661; Bossert p. 327; Dawson p. 695; DSB 11: 343 (by F.A. Stafleu); DU 231-255; Herder p. 282, 342; Hortus

3: 1202; Jackson p. 113, 115, 135, 142; Kew 4: 430-431; Langman p. 616; Lasègue p. 247, 314, 519, 538; Moebius p. 427; MW p. 401; NI 1: 138-143; nos. 1589-1599; see indexes; Plesch p. 374-378; PR 988, 1463, 1465, 2470, 5069, 6191, 7451-7457, 7822, 9731, 9734, ed. 1: 8411-8415 and see p. 539; Rehder 5: 707; RS p. 70-91, 111-121, 132-149, 170-193; Sotheby 625-632; TL-1/see index; TL-2/see indexes; Zander 11, p. 804-805.

BIOFILE: Anderson, F.J., The complete book of 169 Redouté roses. New York 1979, 160 p.
Anon., Flora 23: 543, 736. 1840 (obit.); Bot. Not. 1840: 211 (d.); Erinnerungen an Redouté, Sonderausstellung anlässlig der 10-jährigen Städte-Freundschaft Meudon-Celle, 27 Oktober bis 1 Dezember 1963, 15 pp. Celle. Préfaces par D.J. Leister et par F. Roux-Devillas (n.v.).
Baer, W. & H.W. Lack, Pflanzen auf Porzellan 59, 61, 85. 1979.
Blunt, W., The illustrations of Sertum anglicum, *in* L'Heritier de Brutelle, Sertum anglicum, facsimile edition, Pittsburgh, lxi-lxvi. 1963.
Bonafous, M., Notice historique sur P.-J. Redouté, Turin (Imprimerie Chirio et Mina) 1846, ix p. (extrait de la Biographie Universelle tome lxxviii).
Brindle, J.V., Redouté's drawings, paintings, and prints, *in* G.H.M. Lawrence, A catalogue of Redouteana, Pittsburgh, 73-98. 1963.
Bridson, G.D.R. et al., Nat. hist. mss. res. Brit. Isl. 73.1, 237.15, 275.60. 1980.
Britten, J. and W. Woodward, L'Heritier's botanical works, J. Bot 43: 266-273, 325-329. 1905.
Brockhaus, M., Der Rosenmaler Redouté und seine Rosenbildnisse. Gartenflora 86: 156-158. 1937.
Bultingaire, L., Arch. Mus. natl. Hist. nat. ser. 6. 3: 36. 1928.
Candolle, A.P. de, Mém. souvenirs 60, 62, 478, 512. 1862.
Curmer, P.-J., Redouté *in* Le Muséum d'Histoire naturelle 1: 170-172. 1854.
Dawson, W.R., Smith papers 77. 1934 (letter on (*Liliac.*).
Delcourt, R., A. Lawalrée et al., Pierre Joseph Redouté, 1759-1840 [Catalogue d'une] Exposition [tenue à] Saint-Hubert (Belgique) 7 juin-7 juillet 1964 [s.l.].
Delcourt, R. et A. Lawalrée, Pierre Joseph Redouté, botaniste illustrateur (1759-1840). Lejeunia 13: 5-20. *pl. 1-6.* 1949.
Deleuze, Histoire et description du Muséum d'Histoire naturelle, Paris, 1: 174. 1823.
De Vos, A., Prologue à la mémoire de Pierre-Joseph Redouté 1759-1840, La Belgique horticole 1873: 5-16, reprinted p. [i, t.p.], [5]-16, Gand [Gent] (Imprimerie C. Annoot Braeckman) 1873. Oct.
Duprat, G., Redouté et les vélins du Muséum, *in* Delcourt, Lawalrée, et al. Pierre Joseph Redouté, Saint Hubert, 17-20. 1964.
Fletcher, H.R., Story R. Hort. Soc. 1804-1968, p. 55, 137, 263, 426. 1969.
Forbes, R.J., ed., Martinus van Marum 2: 368, 1970, 3: 27, 153. 1971.
H...n, Redouté, *in* Nouv. Biogr. gén. 41: 830. 1862.
Hardouin-Fugier, E., The pupils of Redouté, 1981, 64 p.
Jessen, K.F.W., Bot. Gegenw. Vorz. 395. 1884.
Jovet, P., Étude botanique de 12 aquarelles et lavis de Malmaison, Bull. Mus. Hist. nat. Paris sér. 2. 23: 416-425; Examen d'une collection de cent "dessins" conservés au Musée de Malmaison, Bull. Mus Hist. nat. Paris ser 2. 23: 426-434. 1951 [description des aquarelles de Redouté, *in*] Billiet, J., Guide-Catalogue officiel: Malmaison, les appartements de Joséphine (nouveaux aménagements), 1951, p. 7-9; Un aspect peu connu de l'oeuvre de P.J. Redouté, Bull. Nat. Plantentuin Belg. 37: 53-60. 1967.
Laissus, Y., ed., Redouté et les vélins du Muséum national d'Histoire naturelle, Paris, 1980 (intr. on Redouté and the "vélins" by editor; 35 reproductions of Redouté originals).
Lawalrée, A., L'Histoire des plantes grasses de De Candolle, Natura mosana 15(4): 53-60. 1962; Letters and documents by Redouté, *in* Lawrence [ed.]. A catalogue of Redouteana, Pittsburgh, 99-117. 1963; Quatre lettres inédites de Redouté à de Candolle. Ardenne et Famenne 1965: 170-175; Natura mosana 18: 27-29. 1965; Manuscrits Redoutéens de la collection Roux-Devillas, Lejeunia ser. 2. 42: 1-11. 1967; Quelques oeuvres des frères Redouté conservées en Belgique ... Ardenne et Famenne 1: 2-9. 1967; Fragments d'une biographie de Redouté. San Miniato,

Auderghem 1969, 29 p.; Les roses de Redouté in folio. Le livre et l'estampe 61-62: 3-5. 1970; *in* M. Wittek et al., Vélins du Muséum, peinture sur vélin de la collection du Muséum national d'Histoire naturelle de Paris. Catalogue. Exposition Bibliothèque Royale Albert Ier. 1974, p. 10-13.
Lawalrée, A., L. Günthart & G. Buchheim, P.J. Redouté. Regensburg & Pittsburgh 1972, vii, 15 p., portr., 17 sketches, 19 col. pl. (facsimile prints of originals at HU).
Lawrence, G.H.M., [ed.], A catalogue of Redouteana exhibited at the Hunt Botanical Library 21 April to 1 August 1963, Pittsburgh [1963]; Huntia 1: 164. 1964 (some letters by R. at HU).
Lefebvre, E. & J.G. de Bruyn, Martinus van Marum 6: 307, 320. 1976.
Lefort, E.L., Bull. Soc.-nat. Luxemb. 54: 62-63. 1950 (on R. and Louis Marchand).
Lefrancq, P., P.-J. Redouté, peintre de fleurs 1759-1840, Musée de Valenciennes Avril-Mai 1954.
Léger, F., Redouté et son temps. Paris 1945. (171 p., 11 pl., portr.).
Leroy, J.F., Les roses de Redouté et de l'Impératrice Joséphine. Sceaux [n.d.] [mimeographed, privately published].
Leroy, J.F. et al., Bot. franç. Amér. nord 355 [index]. 1957.
MacPhail, I., Books illustated by Redouté, *in* Lawrence, A catalogue of Redouteana, Pittsburgh, 33-71, (see also index). 1963 [*Bibliography*, here abbreviated CR].
Madol, R., The life of Redouté, *in* Sitwell and Madol, Album de Redouté, London 1954.
Mägdefrau, K., Gesch. Bot. 64. 1973.
Mannering, , The best of Redoutés Roses. London [1959], 10 p., 24 col. pl.,Mannering, Redoutés fruits and flowers. With the original preface by P.J. Redouté, edited and introduced by Eva Mannering. London [1964].
Nissen, C., Philobiblon 6: 341-342. 1933; Die Botanische Buchillustration, Stuttgart 1: 138-143. 1951 [general discussion; for bibliographical details see part 2 under the appropriate entries].
Redouté, P.J., L'Iconographie appliquée à la botanique en général et aux Roses en particulier. Ann. Encycl. Millin 2: 215-224. 1817.
Ridge, A., The Man who painted roses, the story of Pierre-Joseph Redouté, London (Faber and Faber) 1974, frontip., ii, 404 p.
Rix, M., The art of the botanist 223 [index]. 1981.
Rowley, G., Pierre Joseph Redouté, Raphael of the Succulents, preprinted with separate pagination [Oct. 1956] from Cactus Succulent Soc. Journ. 1956-1957; Redouté redivicus, Rose Annual 107-110. 1975.
Sitwell, S. and R. Madol, Album de Redouté, with twenty-five facsimile colour plates from the edition of 1824 and a new Redouté bibliography. London 1954 [contains an extensive Redouté *bibliography* by Madol, compiled with the assistance of W.T. Stearn], see also J. Soc. Bibl. nat. Hist. 3: 114-115. 1955.
Stafleu, F.A., Redouté, peintre de fleurs, in Lawrence, A catalogue of Redouteana, Pittsburgh, 1-32. 1963; L'Héritier de Brutelle: the man and his work, *in* L'Héritier de Brutelle, Sertum anglicum, facsimile edition, Pittsburgh, xiii-xliii. 1963; Redouté and his circle, *in* Buckman [ed.], Bibliography and natural History 48-65. 1967; Labillardière and the Levant x-xi, *in* Labillardière, Icones plantarum syriae rariorum, facsimile reprint, Lehre [in press]; Linnaeus and the Linnaeans 278, 287. 1971.
Stearn, W.T., Ventenat's ... Jardin de la Malmaison, Journ. Soc. Bibliogr. nat. Hist. 1: 200-201. 1939; Bonpland's Description ... Navarre, J. Arnold Arb. 23: 110-111. 1942; R. Hort., Soc. Exhib. manuscr. books 38-39. 1954; *in* P.J. Redouté, Lilies and related flowers, London 1982, p. 16-22.
Steele, A.R., Flowers for the king 176, 179, 181, 328. 1964.
Stevenson, A., Catalogue of botanical books in the collection of Rachel McMasters Miller Hunt. Pittsburgh 2(2): 440-447. 1961.
Van Schaack, G.B., An exhibit of flowerbooks from the library of the Missouri botanical garden, St. Louis, 1959.
Verdoorn, F., ed., Chron. bot. 2: 339. 1936, 5: 259. 1939.
Wegener, H., Die Rosenabbildungen bis Redouté, Gartenflora 86: 154-156. 1937.
Wittrock, V.B., Iconothecae botanicae, Acta Horti Berg. 3(2): 117. 1903.
Woodward, H., Redoutés works, J. of Bot. 43: 26-30. 1905.
Wright, R., Gardener's tribute 146, 157, 158, 159, 160. 1949.

REDOUTÉ, *Liliac.*

EPONYMY: *Redoutea* Kunth (1822, orth. var.); *Redutea* Ventenat (1800).

8747. *Les Liliacées* ... à Paris (chez l'auteur, au palais national des sciences et arts. De l'Imprimerie de Didot Jeune) an X-1802 [-1816], 8 vols. Fol. (*Liliac.*).
1: Jul 1802-Feb 1804 (see table below), frontisp. portr., p. [i*-v*], [i]-ii, *pl. 1-60*, [i-iii, err. index]. All plates with text.
2: Mar 1804-Apr 1805, p. [i, iii], *pl. 61-120*, [i-iii, err., index]. Id. – Chez l'auteur, Rue de l'Oratoire, Hôtel d'Angivilliers. De l'Imprimerie de Didot jeune an XIII-1805.
3: Jul 1805-Mai 1807, p. [i, iii], *pl. 121-180*, [i-iii, err., index]. Id. – Chez l'auteur, Rue de Seine, Hôtel Mirabeau. De l'Imprimerie de Didot Jeune 1807.
4: Mai 1807-Nov 1808, p. [i-iii], *pl.181-240*, [i-ii, err., index]. Id. – Chez l'auteur, ... Didot jeune 1808.
5: Mar 1809-Sep 1810, p. [i, iii], *pl. 241-300*, [i-ii, index]. Id. – Chez l'auteur ... Didot jeune 1809.
6: 1811-Apr 1812, p. [i-iii], *pl. 301-360*. Id. – Chez l'auteur, ... Didot jeune 1812.
7: 3 Jul 1812-24 Dec 1813, p. [i-iii], *pl. 361-420*, [i-iv, err., ind.]. Id. – Chez l'auteur ... Didot jeune 1813. – *Pl. 372* uncoloured.
8: Mai 1814-Sep 1855, p. [i-iii], *pl. 421-486*, [1]-14. Id. – Chez l'auteur ... Didot jeune 1816. *Copy* GOET: 2 extra plates: *427* (*Narc. interm.*), 427 (id., other drawing), *428* (*N. laetus*), 427 (sic, *N. laetus*), *428* (sic, N. dubius), *430* (*Drimia*).
Copies: BR (1-4), DBN, GOET, HU (1-8), NY (1-7), PH, U (1-8); IDC 7365. – For further details see e.g. CR 10, DU 231, GF p. 71, NI 1597. – Sotheby 626 (by John Collins) describes a special dedication copy in great detail. – For the copy (large folio) at the Royal library, Brussels, see Bot. Zeit. 10: 535. 1852. – Antiquariaat Junk, in its catalogue 225, 1982 (p. 74), offered a copy at Hfl. 150.000.
Authors of text: vols. 1-4 A.-P. de Candolle (*cf.* his Mémoires 497. 1862). Definite proof of de Candolle's share in writing is provided by the Procès-Verbaux Acad. Sci. 3: 440, 491. 1806 where de Candolle is stated to have presented "son ouvrage" with the plates by Redouté. Vols. 5-6 F. Delaroche (fide Pritzel, Woodward). Vols. 7-8 A. Raffeneau Delile (fide Pritzel, Woodward).

vol.	part	*plates*	dates	vol.	part	*plates*	dates
1	1	*1-6*	Jul 1802		28	*163-168*	Mai 1807
	2	*7-12*	Sep 1802		29	*169-174*	Mai 1807*
	3	*13-18*	Nov 1802		30	*175-180*	Mai 1807
	4	*19-24*	[carly 1803]	4	31	*181-186*	Mai 1807
	5	*25-30*	Apr 1803		32	*187-192*	Aug 1807
	6	*31-36*	Jul 1803		33	*193-198*	Nov 1807
	7	*37-42*	Aug 1803		34	*199-204*	Nov 1807
	8	*43-48*	Oct 1803		35	*205-210*	Feb 1808
	9	*49-54*	Jan 1804		36	*211-216*	Mar 1808
	10	*55-60*	Feb 1804		37	*217-222*	Mai 1808
2	11	*61-66*	Mar 1804		38	*223-228*	Jul 1808
	12	*67-72*	Mai 1804		39	*229-234*	Nov 1808
	13	*73-78*	Jul 1804		40	*235-240*	Nov 1808
	14	*79-84*	[Aug?] 1804	5	41	*241-246*	Mar 1809
	15	*85-90*	Sep 1804		42	*247-252*	[1809]
	16	*91-96*	Oct 1804		43	*253-258*	Jun 1809
	17	*97-102*	Nov 1804		44	*259-264*	Jul 1809
	18	*103-108*	Jan 1805		45	*265-270*	[1809]
	19	*109-114*	Mar 1805		46	*271-276*	Nov 1809
	20	*115-120*	Apr 1805		47	*277-282*	Feb 1810
3	21	*121-126*	Jul 1805		48	*283-288*	Apr 1810
	22	*127-132*	Sep 1805		49	*289-294*	[1810]
	23	*133-138*	Nov 1805		50	*295-300*	Sep 1810
	24	*139-144*	Jan 1806	6	51	*301-306*	1811
	25	*145-150*	Feb 1806		52	*307-312*	1811
	26	*151-156*	Apr 1806		53	*313-318*	1811
	27	*157-162*	Jul 1806		54	*319-324*	1811

vol.	part	*plates*	dates	vol.	part	*plates*	dates
	55	*325-330*	1811		68	*403-408*	24 Sep 1813
	56	*331-336*	1811		69	*409-414*	12 Nov 1813
	57	*337-342*	15 Nov 1811		70	*415-420*	24 Dec 1813
	58	*343-348*	28 Dec 1811	8	71	*421-426*	Mai 1814
	59	*349-354*	7 Mar 1812		72	*427-432*	30 Jul 1814
	60	*355-360*	Apr 1812		73	*433-438*	24 Sep 1814
7	61	*361-366*	3 Jul 1812		74	*439-444*	10 Dec 1814
	62	*367-372*	18 Sep 1812		75	*445-450*	11 Feb 1815
	63	*373-378*	6 Nov 1812		76	*451-456*	20 Mai 1815
	64	*379-384*	18 Dec 1812		77	*457-462*	12 Aug 1815
	65	*385-390*	12 Mar 1812		78	*463-468*	12 Aug 1815
	66	*391-...*	Mai 1812		79	*469-474*	2 Mar 1815
	67	*...-402*	Mai 1812		80	*457-486*	Sep 1815

* The 29th part was presented to the Academy as early as 12 Jan 1807. – The dates given here are those provided by Woodward (based on BF and JGLF) with a few additional details derived from the Procès Verbaux of the Académie. The ordinary edition was published in 280 copies. The plates are colour-printed stipple engravings, finished by hand. A large paper (broadsheet) edition (24 × 18 inches) was published in 18 copies between 1807 and 1816. The plates in this edition were finished by hand by Redouté himself. – 468 original drawings were in the library of Eugène de Beauharnais and are now in private hands (USA).

Ref.: CR 10, DU 231, GF p. 71 (Stearn), NI 1597, PR 7453; IDC 5444B.

Redouté, P.J., Catalogue de 486 liliacées et de 186 roses. Paris 1829 (BF 31 Oct 1829).
Woodward, H., J. of Bot. 43: 26-28. 1905.
Stearn, W.T., Herbertia 11: 15-16. 1944 [publ. 1946].
Stafleu, F.A., Catal. Redouteana 20-23. 1963.

8748. *Les Roses*, par P.J. Redouté, peintre de fleurs, ... À Paris (de l'imprimerie de Firmin Didot, ...) 1817-1824, 3 vols. Fol. Qu. (*Roses*).
Author of text: Claude Antoine Thory (1759-1827).
Folio edition: *copy*: GOET, information and analysis G. Wagenitz:
1 (*text*), preceded by "Les Roses par P.J. Redouté. Avis a MM les souscripteurs", p. [1]-8, and "Les Roses par P.J. Redouté. Prospectus", p. [1]-3; text proper: [1-4, h.t., t.p.; 1817], [5]-24 avant propos, 25-138 text for plates, [139]-156 bibliotheca botanica rosarum, [157, table alph.], [158, avis pour la reliure].
1 (*plates*), p. [1-4, h.t., t.p.], portr. Redouté, wreath, plates (unnumbered) to uneven pages 25-137 except to p. 75, in all *56 pl.*
2 (*text*), p. [1-4, h.t., t.p., 1821], 5-122 text for plates, [123, table alph.], [124 relieur].
2 (*plates*), *59 pl.* to uneven pages 5-121.
3 (*text*) 9 p. [1-4, h.t., t.p., 1824], 5-109 text for plates, [110, tableau synoptique], [111]-122 table alphabétique des noms des roses, [123]-125 table alphabétique de quelques articles ..., [126, err.], [127, relieur].
3 (*plates*), plates to uneven pages 5-77, to 78, 79 (twice) and to uneven pages 81-107, in all *54 pl.*
Quarto edition: *copies*: BR (analysis below), NY (all original covers); IDC 5656.
   1: frontisp. portr. Redouté, [1]-156, [157-158], *56 pl.* as in folio ed.
   2: frontisp. wreath, p. [1]-122, [123-124], *59 pl.*, id.
   3: p. [1]-125, [126-127], *54 pl.*, id.
Imprint original covers: Les Roses, par P.J. Redouté, peintre de fleurs; ... 1re [-30me] livraison ... À Paris, chez l'auteur, Rue de Seine, no. 6; et chez Treuttel et Wurtz ... Nicolle, ... À Strasbourg, pour l'Allemagne, chez Treuttel et Wurtz; à Bruxelles, chez de Mat et Lecharlier; À Man[n]heim, chez Dom Artaria; À Londres, chez Bossange et Masson, ... de l'imprimerie de Firmin Didot. 1817 (1-10: 1817, 11-19: 1829, 20-22: 1821, 23: 1819 (was cover part 11, number overprinted), 24-28: 1822, 29-30: 1823).
Dates: The following account is that of Woodward with the addition of some details taken from the announcements in the Bibliographie de la France and the reviews in

the Bulletin Férussac. Every part (except no. 10 which had one, and no. 30 which had none) contained 6 plates, in all 169 plates. The copy at NY is bound with the original covers. It also contains a leaflet issued after the 23rd part stating "cet ouvrage de format grand in-quarto, sur papier vélin, dit Nom-de-Jésus, composé de six planches, et accompagné du texte imprimé chez M. Firmin-Didot est fixé à vingt-cinq francs, par chaque livraison. Il en a été tiré cent exemplaires seulement, sur format grand in-folio, le prix, par chacque livraison, est fixé à cinquante

vol.	part	pages	dates	vol.	part	pages	dates
1	1	1-36	22 Mar 1817	2	16	53-64	8 Jul 1820
	2	37-48	24 Mai 1817		17	65-76	19 Aug 1820
	3	49-60	27 Aug 1817		18	77-92	28 Oct 1820
	4	61-72	1 Nov 1817		19	93-100	12 Jan 1821
	5	73-86	10 Jan 1818		20	101-116	31 Mar 1821
	6	87-98	7 Mar 1818		21	1-4, 117-124	21 Jul 1821
	7	99-110	16 Mai 1818	3	22	5-28	5 Oct 1821
	8	111-122	1 Aug 1818		23	29-40	22 Dec 1821
	9	123-134	17 Oct 1818		24	41-52	16 Mar 1822
	10	135-146	9 Jan 1819		25	53-64	1 Jun 1822
	11	147-158	3 Apr 1819		26	65-76	26 Oct 1823
2	12	5-16	29 Mai 1819		27	77-84	11 Jan 1823
	13	17-28	28 Aug 1819		28	85-96	31 Mai 1823
	14	29-40	4 Dec 1819		29	97-108	22 Nov 1823
	15	41-52	15 Jan 1820		30	1-4, 109-128	6 Mar 1824

Plates: colour printed stipple engravings of watercolors by P.J. Redouté finished by hand.
Special copies: McPhail (Cat. Redout. no. 19) and Lawalrée (1970) mention large paper copies with the engravings in two states, plain and coloured. These (fifteen) copies formed a de luxe issue for which Lawalrée has shown that it contains textual amendments and a portrait of Redouté on which he wears the Légion d'honneur awarded to him on 14 Jan 1825. For further details see. e.g. Sotheby 630-631 and the introduction to the recent facsimile reprint.
Facsimile ed. of first edition, vol. 1: 1974, 2: 1975, 3: 1976 and a fourth volume, 1978, with various studies. *Copy*: A. Lawalrée.
  1: facsimile preceded by a leaf with "Published by de Schutter s.a. Venusstraat 21, B-2000, Antwerp, Belgium", on verso: D/1974/2000/1/Printed in Belgium." Volume ends with a leaf carrying a de Schutter imprint.
  2: idem; front leaf recto as vol. 1, verso: D/1975/2000/1/ISBN 2-8023-0002-4/Printed in Belgium.
  3: idem, front leaf recto as vol. 1, verso: D/1976/2000/1/ISBN 3-8023-003-2/Printed in Belgium.
  4: [i-xi], 14-387, [389] with foreword by George Taylor, introduction by Gordon D. Rowley, a biography and bibliography of Redouté and a biography of Thory by André Lawalrée (with portraits); additional descriptions from ed. 3 by Thory; a commentary on the Roses by Gisèle de la Roche, a bibliography by G.D. Rowley and "some botanists rosarians" by William T. Stearn (inf. A. Lawalrée).
Facsimile ed. of 29 pl.: see Mannering, 1959.
Ref.: Brockhaus, Gartenflora 86: 156-158. 1937.
  Launert, E., Pierre-Joseph Redouté, Les Roses, 170 pl. en couleurs... adapté de l'allemand par André, Lawalrée, Paris-Gembloux, Duculot, 1982 (German original: 1980).
  Lawalrée, A., Le livre et l'estampe 61-62: 3-5. 1970.

8749. *Les Roses* peintes par P.J. Redouté dessinateur ... décrites par C.A. Thorry [sic, for Thory] ... Paris (C.L.F. Panckoucke, éditeur ...) 1824-1826, 2 vols. Oct. (*Roses, ed. 2*).
Publ.: 1824-1826, in 40 parts, each with 4 plates. For dates of publication of parts see Woodward (1905). We saw a complete copy at HU (160 pl.) and two copies of vol. 1 BR (64 pl.) and G (80 pl.).
  1: [i, iii], [1]-4, *pl*. [*1-80*] each with 2[-4] p. text.
  2: [i, iii], *pl*. [*81-160*], id.

Ref.: Woodward, B.B., J. Bot. 43: 28-29. 1905.

8750. *Choix des plus belles fleurs* prises dans différentes familles du règne végétal et de quelques branches des plus beaux fruits groupées quelquefois, et souvent animées par des insectes et des papillons gravées, imprimées en couleur et retouchées au pinceau avec un soin qui doit répondre de leur perfection dédié à ll. aa. rr. les princesses Louise et Marie d'Orléans ... Paris (Chez l'auteur, ...; C.L.F. Pancoucke, ...) 1827. Fol. or Qu. (*Choix plus belles fleurs*).
Publ.: 26 Mai 1827-22 Jun 1833 (see below), p. [i-iv], [1]-17, [18-20], *144 pl.* by Redouté. *Copies*: HU, NY; IDC 5617. – For details see CR 21, DU 235, NI 1591 and Sotheby 628. One volume in 36 parts, 144 plates, stipple-engravings, colour-printed, finished by hand. Each part contained four plates, but since the plates are not numbered the order of issue remains conjectural. Redouté originally planned to issue only 100 plates in 25 parts, but went on until 36. The last part contained 20 pages text by Guillemin. A restricted number of copies was issued in folio (24 frs per part), the majority was in quarto (12 frs per part). Woodward supplies the following list of dates which are the dates upon which the parts were announced in the Bibliographie de la France. These agree very well (taking into account the usual delay) with the dates of announcement in the Revue bibliographique des Pays Bas. [Other editions 1829, and n.d. [1833], with additional text by Guillemin].

part	dates	part	dates	part	dates
1	26 Mai 1827	13	2 Mai 1829	25	5 Feb 1831
2	30 Jun 1827	14	27 Jun 1829	26	16 Apr 1831
3	15 Aug 1827	15	22 Aug 1829	27	11 Jun 1831
4	10 Oct 1827	16	3 Oct 1829	28	13 Aug 1831
5	24 Nov 1827	17	5 Dec 1829	29	5 Nov 1831
6	19 Jan 1828	18	9 Jan 1830	30	31 Dec 1831
7	22 Mar 1828	19	27 Feb 1830	31	17 Mar 1832
9	2 Aug 1828	21	12 Jun 1830	33	21 Jul 1832
10	25 Oct 1828	22	31 Jul 1830	34	29 Sep 1832
11	3 Jan 1829	23	9 Oct 1830	35	22 Dec 1832
12	7 Mar 1829	24	4 Dec 1830	36	22 Jun 1833

8751. *Les Roses*, peintes par P.J. Redouté, chevalier ... décrites et classées selon leur ordre naturel, par C.A. Thory. Troisième édition, publiée sous la direction de M. Pirolle, ... à Paris (chez P. Dufart, ...) 1828-1829, 3 vols. Oct. (*Roses, ed. 3*).
1: 1828, p. [i*-ii*], wreath, [iii*, t.p.], portr. Thory], [i]-vi, biogr. note of Thory by D. Beaumont, [1]-46 and *pl. 1-3* (introductory plates), *45 plates* each with 2 [-4] p. text, [1]-3, [4]. Plates numbered only in index (by groups).
2: 1829, p. [i-ii], Redouté, [1, t.p.], *69 pl.* (id.) with text id., [1]-3, [4].
3: 1829, p. [i-v], *65 pl.* with text id., contents etc. [1]-12.
Copies: BR, MO, NY. – Published in 30 parts (see Woodward, 1905 for dates), in all *181 pl.*, 2 portraits and additional text.
Re-issue: 1835, Paris (chez P. Dufart, ...), St. Pétersbourg (chez J.F. Hauër et cie.), 3 vols., as original issue, *181 pl. Copies*: BR, GOET, NY, USDA.
Ref.: Woodward, B.B., J. Bot. 43: 28-29. 1905.

8752. *Catalogue de 486 Liliacées et de 168 Roses* peintes par P.J. Redouté ... Paris (chez Bossange père, ...) 1829. Oct. (*Cat. Liliac. Roses*).
Publ.: 1829, p. [1]-28. *Copy*: NY.

Redslob, Julius (*fl.* 1863), German botanist. (*Redslob*).

HERBARIUM and TYPES: Unknown.

BIBLIOGRAPHY and BIOGRAPHY: BM 4: 1661; GR p. 39; GR, cat. p. 70; Jackson p. 296; LS 21873; PR 10622.

8753. *Die Moose und Flechten Deutschlands.* Mit besonderer Berücksichtigung auf Nutzen und Nachtheile dieser Gewächse beschrieben von Dr. Julius Redslob ... Leipzig (Wilhelm Baensch Verlagsbuchhandlung) [1862-] 1863 [-1871]. Qu. (*Moose Flecht. Deutschl.*).
Ed. 1: 1862-1871 (published in 8 parts with 4 plates each, part 1 Dec 1862, part 2 in 1862, part 4 in Dec 1871), p. [i-iv], [1]-96, *1-32. Copies*: Cornell Univ., MO.
Ed. 2: 1881 (or 1881-1882), (in 8 parts of *4 pl.* and text, part 1 publ. Oct 1881), p. [i], [1]-88, [89]-96, *pl. 1-32* (col.). *Copy*: O. Almborn. – No date indicated in copy. - Zweite umgearbeitete Auflage. Leipzig (id.). s.d.
Ref.: A.R., Bot. Zeit. 20: 427. 5 Dec 1862 (rev. of Lief. 1).

Redtenbacher, Joseph (1810-1870), Austrian chemist and botanist; Dr. med. Wien 1834; studied with Mitscherlich in Berlin, and with Liebig in Giessen; professor of chemistry Praha 1841-1849, idem at Wien 1849-1870. (*Redtenbacher*).

HERBARIUM and TYPES: Unknown.

BIBLIOGRAPHY and BIOGRAPHY: ADB 27: 542-543 (by Ladenburg); Barnhart 3: 136 (b. 12 Mar 1810, d. 5 Mar 1870); BM 4: 1662, 8: 1056; CSP 5: 122, 8: 713; Kew 4: 431; PH 150; PR 7459, ed. 1: 8417; Rehder 5: 707.

BIOFILE: Anon., [general secretary Akad. Wiss.], Almanach Akad. Wiss. Wien 20: 230-247. 1870.
Konek, F. v., Chemiker-Zeit. 40(82/83): 585-586. 1916 (notes on R. in obit. of A. v. Görgey).
Kohn, M., J. Chem. educ. 24(8): 366-368. 1947 (portr., bibl., brief biogr.).
Ludwig, E., Pharmaz. Post. 43(22): 213-216. 1910 (biogr.).
Neilreich, A., Verh. zool.-bot. Ver. Wien 5: 39. 1855.
Poggendorff, J.C., Biogr.-lit. Handw.-Buch 2: 585. 1863, 3: 1098. 1898, 7a suppl.: 529. 1971.

8754. Dissertatio inauguralis botanica *de caricibus territorii vindobonensis*, quam consensu et auctoritate excellentissimi ac illustratissimi domini praesidis et directoris, perillustris ac spectabilis domini decani, nec non clarissimorum d.d. professorum pro doctoris medicinae laurea rite obtinenda in antiquissima ac celeberrima Universitate vindobonensi, publicae disquisitioni submittit Josephus Redtenbacher, ... In theses adnexas disputabitur in Universitatis aedibus die 3. maji 1834. Vindobonae [Wien] (typis Joannis B. Wallishausser) [1834]. Oct. (*De caric. terr. Vindob.*).
Publ. 3 Mai 1834, p. [1]-40. *Copy*: G. – Only botanical publication; R. turned to chemistry after having written this dissertation under the influence of Mohs, and possibly, J.F. von Jacquin.

Rees, Abraham (1763-1825), British encyclopaedist; student and subsequently tutor at the Hoxton Academy for [protestant] Dissenting Ministers near London; from 1768 pastor of the Presbyterian congregation, Southwark, from 1783 minister in London (congregation of Jewin Street). (*Rees*).

HERBARIUM and TYPES: Unknown.

BIBLIOGRAPHY and BIOGRAPHY: BM 4: 1662; Bret. p. 208; Henrey 2: 737 [index]; Herder p. 88; Kew 4: 433; PR 7460, ed. 1: 9726; Rehder 1: 33; RS p. 136; SK 4: ccviii; SO 286a; TL-1/1046; Tucker 1: 586.

BIOFILE: Allibone, S.A., Crit. dict. Engl. lit. 2: 1761. 1878.
Anon., Gentleman's Magazine, London, 95(2): 181-184. 1825 (obit.).
Collison, R., Encyclopaedias: their history throughout the ages. New York & London, Hafner Publishing Company 1966, ed. 2, p. 109.
Francis, J.W., Old New York; or reminiscences of the past sixty years ... New York, Charles Roe, 1858, p. 158-160 (on American ed. of Cycl.; changes in some theological articles).

Jackson, B.D., J. Bot. 15: 107-108. 1877, 18: 87. 1880, 34: 307-311, 376. 1896 (dating; superseded by Pestana 1979); also, The dates of publication of the various volumes of the Cyclopaedia edited by Rev. A. Rees, London 1880, 2 p., and an attempt to ascertain the actual dates of publication ... Rees's Cyclopaedia, London 1895. (See below).
Nangle, B.C., The Monthy Magazine, first series, indexes 1934, p. 36.
Pestana, H.R., J. Soc. Bibl. nat. Hist. 9(3): 353-361. 1979 (Rees' Cycl. a sourcebook for the history of geology; precise dating).
Poggendorff, J.C., Biogr.-lit. Handw.-Buch 2: 586. 1863 (d. 19 Jun 1825).
Smith, J.E., Mem. 1: 489. 1832 ("5000 copies sell ...").
Walsh, S.P., Anglo-american general encyclopedias. A historical bibliography 1703-1967, New York, London, R.R. Bowker Company 1968, p. 38-39, 104, 174, 229.

EPONYMY: *Reesia* Ewart (1913) honors Bertha Rees, lecturer on botany at the University of Melbourne, Australia.

8755. *The Cyclopaedia*; or, universal dictionary of arts, sciences, and literature by Abraham Rees, ... With the assistance of eminent professional gentlemen. Illustrated with numerous engravings by the most distinguished artists. In thirty-nine volumes ... London (printed for Longman, Hurst, Rees, Orme, & Brown, ...) [1802-] 1819-1820. Qu. (*Cycl.*).
Authors of botanical items: Rev. William Wood (1745-1808), bot. art. as far as *Cyperus*, subsequent articles (botanical as well as bot.-biographical; 3.405 articles): Sir James Edward Smith (1759-1828), some by Rev. William Fitt Drake (1786-1874), medical botany by H. Woodville (fide BM; William Woodville, 1752-1805?). The articles by J.E. Smith are signed "S".
Publ.: in 79 parts, listed below in agreement with the data provided by Pestana (1979). The volumes have no regular pagination; we cite signatures. *Copies*: ILL (inf. D.P. Rogers), M, US, USDA; IDC 495.
1: portr., [i]-viii, sign. B-Z, Aa-Zz, 3A-3Z, 4A-4Z, 5A-5L(6) (note: the letters J and V are not used in the signature alphabet). – The (6) in parentheses indicates the number of pages with text on the last signature. – *Text*: see table below.
2: [i-iii], sign. B-Z, Aa-Zz, 3A-3Z, 4A-4Z, 5A-5L(4).
3: [i-iii], sign. B-Z, Aa-Zz, 3A-3Z, 4A-4Z, 5A-5O(6).
4: [i-iii], sign. B-Z, Aa-Zz, 3A-3Z, 4A-4Z, 5A-5G, 5H(1).
5: [i-iii], sign. B-Z, Aa-Zz, 3A-3Z, 4A-4Z, 5A-5K(7).
6: [i-iii], sign. B-Z, Aa-Zz, 3A-3Z, 4A-4Z, 5A-5H, 5I(3).
7: [i-iii], sign. B-Z, Aa-Zz, 3A-3Z, 4A-4Z, 5A-5M(5).
8: [i-iii], sign. B-Z, Aa-Zz, 3A-3Z, 4A-4Z, 5A-5G(7).
9: [i-iii], sign. B-Z, Aa-Zz, 3A-3Z, 4A-4Z, 5A-5H(7).
10: [i-iii], sign. B-Z, Aa-Zz, 3A-3Z, 4A-4Z, 5A-5C(8).
11: [i-iii], sign. B-Z, Aa-Zz, 3A-3Z, 4A-4Z, 5A-5E(8).
12: [i-iii], sign. B-Z, Aa-Zz, 3A-3Z, 4A-4Z, 5A-5E(6).
13: [i-iii], sign. B-Z, Aa-Zz, 3A-3Z, 4A-4Z, 5A-5H(8).
14: [i-iii], sign. B-Z, Aa-Zz, 3A-3Z, 4A-4Z, 5A-5C(7).
15: [i-iii], sign. B-Z, Aa-Zz, 3A-3Z, 4A-4Z, 5A-5F(4).
16: [i-iii], sign. B-Z, Aa-Zz, 3A-3Z, 4A-4Z, 5A-5E(7).
17: [i-iii], sign. B-Z, Aa-Zz, 3A-3Z, 4A-4Z, 5A-5F(7).
18: [i-iii], sign. B-Z, Aa-Zz, 3A-3Z, 4A-4Z, 5A-5D(3).
19: [i-iii], sign. B-Z, Aa-Zz, 3A-3Z, 4A-4Z, 5A-5E(5).
20: [i-iii], sign. B-Z, Aa-Zz, 3A-3Z, 4A-4Z, 5A-5E(7).
21: [i-iii], sign. B-Z, Aa-Zz, 3A-3Z, 4A-4Z, 5A-5D(8).
22: [i-iii], sign. B-Z, Aa-Zz, 3A-3Z, 4A-4Z, 5A-5G(7).
23: [i-iii], sign. B-Z, Aa-Zz, 3A-3Z, 4A-4Z, 5A-5F(4).
24: [i-iii], sign. B-Z, Aa-Zz, 3A-3Z, 4A-4Z, 5A-5E(8).
25: [i-iii], sign. B-Z, Aa-Zz, 3A-3Z, 4A-4Z, 5A-5E(4).
26: [i-iii], sign. B-Z, Aa-Zz, 3A-3Z, 4A-4Z, 5A-5E(8).
27: [i-iii], sign. B-Z, Aa-Zz, 3A-3Z, 4A-4Z, 5A-5E(8).
28: [i-iii], sign. B-Z, Aa-Zz, 3A-3Z, 4A-4Z, 5A-5E(4).
29: [i-iii], sign. B-Z, Aa-Zz, 3A-3Z, 4A-4Z, 5A-5F(8).

30: [i-iii], sign. B-Z, Aa-Zz, 3A-3Z, 4A-4Z, 5A-5H(5).
31: [i-iii], sign. B-Z, Aa-Zz, 3A-3Z, 4A-4Z, 5A-5D(4).
32: [i-iii], sign. B-Z, Aa-Zz, 3A-3Z, 4A-4Z, 5A-5F(2).
33: [i-iii], sign. B-Z, Aa-Zz, 3A-3Z, 4A-4Z, 5A-5G(4).
34: [i-iii], sign. B-Z, Aa-Zz, 3A-3Z, 4A-4Z, 5A-5B(7).
35: [i-iii], sign. B-Z, Aa-Zz, 3A-3Z, 4A-4Z, 5A-5F(8).
36: [i-iii], sign. B-Z, Aa-Zz, 3A-3Z, 4A-4Z, 5A-5G(2).
37: [i-iii], sign. B-Z, Aa-Zz, 3A-3Z, 4A-4Z, 5A-5C(8).
38: [i-iii], sign. B-Z, Aa-Zz, 3A-3Z, 4A-4Z, 5A-5I(3).
39: [i-iii], sign. B-Z, Aa-Zz, 3A-3Z, 4A-4Z, 5A-5E(6).
Plates *1*: [i-iii], *pl. 1-7, 7*[bis], *8-10, 10*[bis], *11, 11*[bis], *12-13, 13*[bis], *14-21, 22*["xxi"], *23*["Flax"], *24-25, 26*["plate machines Nº. 2"], *27, 28/29, 30-31, [32-36], 37, [38-39], 40*; *1, 1*[bis], *2, 3, 6, 5, 5* ["referred to as Pl. VI"], *7, 8* ["*7*"], *9, 9/10, 11-14; 1-2, 2, 1, 3, 3, 4, 4, 5; 1, 1, 2, 2, 3, 8, 9-16, 16, 18; 1-2; 1-2, 2, 4; 1/2, 9; 1-4; 1-2, 2, 4, 5-10, 11/12, 13-15, 15, 16-18, 18a, 19-23*, "A.B.", *25-30*, A, *32-33, 36, 35, 37, 38-40, 42-47, 58-74, 74*, 75-78; 1-3, 4/5, 1/2; 1-5; 1- 2, 2, 4-11, 11*[bis], *12-13, 14/15, 15-21, 21*, 2 charts; *1-2; 1-2; 1-33*, uncol. liths.
Plates *2*: [i-iii], *pl. 1-4; 1-4; 1-7; 1-2*, [*3*], *5/6, 7/8, 9/10, 11/12; 1-7*; [*1*]; *1-3; 1-2; 1; 1; 1-7, 7, 8-11, 11, 11/12, 13/14, 14, 14, 15-17, 20-21; 1-2; 1-2; 1-6, 7/8, 8/9, 10-12; 1-7, 7, 9, 10-14; 1-*[*2*]; *1-4*[*5*]; *1/2/3, 4, 8-10; 1; 1-2; 1, 1*, 2-5*[*6*]; *7/14, 7-10*[*11*], *12/13, 13, 15; 2, 2-12*[*13*]; *1; 1/2/4, 3, 5, 6/7, 5, 5*, 6, 6/7/8; 1-3, 5; 1-3; 1; 1-10, 10/11, 12-15*[*16*]; *1-11, 11, 12-14; 1-33, 31-47*, uncol. lith.
Plates *3*: [i-iii], *pl. 1-3, 3*, 4-9, 12/13, 13, 14, 14/15, 15; 1, 1, 2, 3/4, 4-5, 4, 3, 8; 1-2*, [2bis]; *1-5, 6/7, 7; 1-2; 1-2, 2-14, 14*, 15, 15, 16-21, 22/23, 23*, [*24*], *25-26, 26*, 27, 27, 28-29, 30/31, 31-34, 34, 35, 35, 37, 40; 1-6; 1-3; 1-3, 3*, 4-8, 9/10, 11-12*, [*12**], *12, 13, 14, 14, 15, 15, 16-22, 23/24, 25*; [*1*], *1-3; 1-4; 1-3; 1*, 2*, 1-25, 27-30, 32-33, 35,* [*36*], *38; 1-15; 1-4, 6, 6, 7-10, 11-13,* [*14*].
Plates *4*: [i-iii], *pl. 1/2, 2/4, 2-3*, [*3*]; *1-19; 1-4; 1-3; 1-2; 1-2; 1, 1, 2-12; 1-13; 1-2,* [*3*]; *1-3, 6, 6, 5, 7-8, 8 Nº. 2, 9-17,* [*18-20*]; *1-4; 1-10*[*11*]; *1; 1; 1-2; 1-4; 1-2; 1-9;* [*1*]; *1-2, 1, 3/4, 5, 3, 7, 8-9, 3; 1/2, 3; 1-2;* [*1*]; [*1*]; [*1*]; *1-5,* [*5**], *6, 8, 8, 9-11; 1/6, 3-6, 6, 7, 7*, 8-9; 3*.
Plates *5* (*Natural history*): [i-iii], *41 pl.* (numbers even more confused; these "Quadrupeds"), 2 of whales; *pl. 1-2, 2, 3-9; 1-2, 5-6, 6, 7, 9; 1-5, 5, 7, 8; 1-9; 2, 80,* [*3*], *4-6, 10, 7,* [*8*], *20; 1-4; 1, 1, 8-9, 6, 2, 5, 4, 3, 7, 6; 1-2;1-2; 1-5, 7-9, 10/11, 11-13; 1-4; 3-5, 5-6; 1-2, 3/4/5, 6-7; 1-12, 12-15; 1-4, 8; 1; 1; 1; 1-3, 11, 13; 1-4; 1-5, 10; 1-2, 3/4/5, 6-7; 1/2/3, 4-6, 4, 7, 3, 2, 5; 1-8, 3, 9-10, 11A, 11B, 12, 13, 14; 1, 16, 17,* [*18-19*]; [*1*], *2-5; 1-2; 1; 1-3, 8; 1, 6/7, 8, 5; 1-2; 1,* [*1**]; *2, 3, 4, 5,* [*6*] [Botany]; *1-2* [fungi]; *1-9* [Botany], *1-2* [Botany]; *1-2; 1/2, 4/5; 1-4;* [*1**]; *1; 1-2, 4-7*.
Plates *6*: [i-iii], 61 maps.

vol.	sect.	part	contents	dates
1	I	1	A-Agoge	2 Jan 1802
	II	2	Agogliastro-Amaranthoides	4 Mai 1802
2	I	3	Amaranthus-Antimony	18 Oct 1802
	II	4	Antimony-Arteriotomy	7 Apr 1803
3	I	5	Artery-Babel-mendeb	22 Sep 1803
	II	6	Babenhausen-Battersea	17 Mar 1804
4	I	7	Battery-Point-Biörnstall	17 Aug 1804
	II	8	Biot-Bookbinding	13 Apr 1805
5	I	9	Bookkeeping-Brunia	1 Jun 1805
	II	10	Brunia-Calvart	26 Dec 1805
6	I	11	Calvary-Cape of Good Hope	18 Feb 1806
	II	12	Cape of Good Hope-Castra	17 Jun 1806
7	I	13	Castramentation-Chalk	1 Oct 1806
	II	14	Chalk-Chronology	9 Feb 1807
8	I	15	Chronometer-Clavaria	18 Mai 1807
	II	16	Clavaria-Colisseum	10 Aug 1807
9	I	17	Collision-Congregation	27 Nov 1807
	II	18	Congregation-Corne	8 Mar 1808
10	I	19	Cornea-Croisade	2 Mai 1808

vol.	sect.	part	contents	dates
	II	20	Croisade-Czyrcassy	2 Jul 1808
11	I	21	D-Deluge	23 Sep 1808
	II	22	Deluge-Dissimilitude	3 Dec 1808
12	I	23	Dissimulation-Dynamics	14 Feb 1809
	II	24	Dynamics-Eloanx	12 Mai 1809
13	I	25	Elocution-Equation	18 Aug 1809
	II	26	Equation-Extremum	25 Nov 1809
14	I	27	Extrinsic-Fibro-cartilage	3 Feb 1810
	II	28	Fibro-cartilage-Food	13 Apr 1810
15	I	29	Food-Froberger	23 Jun 1810
	II	30	Frobisher-Generation	8 Oct 1810
16	I	31	Generation-Gniewe	29 Nov 1810
	II	32	Gnoien-Gretna-Green	25 Jan 1811
17	I	33	Gretry-Hatfield-Regis	8 Mar 1811
	II	34	Hatfield-Regis-Hilbe	22 Apr 1811
18	I	35	Hibiscus-Huysum	3 Jun 1811
	II	36	Huzanka-Increment	20 Aug 1811
19	I	37	Increment-Josephus	14 Sep 1811
	II	38	Josephus-Kilmes	16 Dec 1811
20	I	39	Kiln-Laurenberg	27 Jan 1812
	II	40	Laurenberg-Lights	19 Mar 1812
21	I	41	Lighthouse-Longitude	12 Mai 1812
	II	42	Longitude-Machinery, pl. A	27 Jul 1812
22	I	43	Machinery-Manganese	27 Aug 1812
	II	44	Manganese-Mattheson	4 Nov 1812
23	I	45	Matthew-Metals	11 Dec 1812
	II	46	Metals-Monsoon	9 Feb 1813
24	I	47	Monster-Muscle	30 Mar 1813
	II	48	Muscle-Newton	26 Apr 1813
25	I	49	Newtonian-Oleinae	15 Jul 1813
	II	50	Oleinae-Ozunicze, pl. B	15 Sep 1813
26	I	51	P-Passiflora	27 Nov 1813
	II	52	Passiflora-Pertubation	18 Jan 1814
27	I	53	Pertussis-Picus	22 Mar 1814
	II	54	Picus-Poetics	7 Mai 1814
28	I	55	Poetry-Preaching	14 Jul 1814
	II	56	Preaching-Punjoor	16 Sep 1814
29	I	57	Punishment-Ram, pl. C	14 Dec 1814
	II	58	Ram-Repton	26 Jan 1815
30	I	59	Republic-Rock	21 Mar 1815
	II	60	Rock-Rzemien	1 Jun 1815
31	I	61	S-Sarabanda	11 Jul 1815
	II	62	Sarabanda-Scotium, pl. D	21 Sep 1815
32	I	63	Scotland-Shammy	22 Dec 1815
	II	64	Shammy-Sindy	28 Feb 1816
33	I	65	Sine-Sound	17 Mai 1816
	II	66	Sound-Starboard	27 Jul 1816
34	I	67	Starch-Stuart	26 Oct 1816
	II	68	Stuart-Szydlow	11 Dec 1816
35	I	69	T-Testudo	19 Mar 1817
	II	70	Testudo-Toleration	1 Mai 1817
36	I	71	Tolerium-Tumours	13 Aug 1817
	II	72	Tumours-Vermelho	24 Oct 1817
37	I	73	Vermes-Union	20 Dec 1817
	II	74	Union-Wateeoo	23 Mar 1818
38	I	75	Water-Whitby	29 Mai 1818
	II	76	Whitby-Wren	30 Jul 1818

vol.	sect.	part	contents	dates
39	I	77	Wren-Zyto: Aam-Baldwin, *pl. E*	30 Dec 1818
	II	78	Baldwin-Zolliker, titl., *ol. D*	27 Oct 1819
	III	79	Titles, preface, plates	29 Jul 1820

American ed.: 1806-18.., 41 vols., 83 parts. *Copies*: NY, PH. – "*The Cyclopaedia*"; ... artists. *First American edition*, revised, corrected, enlarged, and adapted to this country, by several literary and scientific charaters ... Philadelphia (published by Samuel F. Bradford, and Murray, Fairman and Co. ...) [1806-18..]. Qu.

1: [i*-ii*], portr., [i]-viii, sign. A-Z, Aa-Zz, 3A-3Z, 4A-4Z, 5A-5N(8) (characters J, V and W not included in signature alphabet); the (8) indicates that 8 p. of the last signature have printed text.
2: [i-iii], sign. A-Z, Aa-Zz, 3A-3Z, 4A-4Z, 5A-5F(6).
3: [i-iii], sign. A-Z, Aa-Zz, 3A-3Z, 4A-4Z, 5A-5H(4).
4: [i-iii], sign. A-Z, Aa-Zz, 3A-3Z, 4A-4Z, 5A-5H(4).
5: [i-iii], sign. A-Z, Aa-Zz, 3A-3Z, 4A-4Z, 5A-5H(4).
6: [i-iii], sign. A-Z, Aa-Zz, 3A-3Z, 4A-4Z, 5A-5H(6).
7: [i-iii], sign. A-Z, Aa-Zz, 3A-3Z, 4A-4Z, 5A-5H(4).
8: [i-iii], sign. A-Z, Aa-Zz, 3-3Z, 4A-4Z, 5A-5H(3).
9: [i-iii], sign. A-Z, Aa-Zz, 3A-3Z, 4A-4Z, 5A-5H(7).
10: [i-iii], sign. A-Z, Aa-Zz, 3A-3Z, 4A-4Z, 5A-5H(6).
11: [i-iii], sign. A-Z, Aa-Zz, 3A-3Z, 4A-4Z, 5A-5K(5).
12: [i-iii], sign. A-Z, Aa-Zz, 3A-3Z, 4A-4Z, 5A-5H(5).
13: [i-iii], sign. A-Z, Aa-Zz, 3A-3Z, 4A-4Z, 5A-5H(5).
14: [i-iii], sign. A-Z, Aa-Zz, 3A-3Z, 4A-4Z, 5A-5I(1).
15: [i-iii], sign. A-Z, Aa-Zz, 3A-3Z, 4A-4Z, 5A-5H(8).
16: [i-iii], sign. A-Z, Aa-Zz, 3A-3Z, 4A-4Z, 5A-5I(1).
17: [i-iii], sign. A-Z, Aa-Zz, 3A-3Z, 4A-4Z, 5A-5H(8).
18: [i-iii], sign. A-Z, Aa-Zz, 3A-3z, 4A-4Z, 5A-5H(6).
19: [i-iii], sign. A-Z, Aa-Zz, 3A-3Z, 4A-4Z, 5A-5F(8).
20: [i-iii], sign. A-Z, Aa-Zz, 3A-3Z, 4A-4Z, 5A-5G(4).
21: [i-iii], sign. A-Z, Aa-Zz, 3A-3Z, 4A-4Z, 5A-5H(6).
22: [i-iii], sign. A-Z, Aa-Zz, 3A-3Z, 4A-4Z, 5A-5H(7).
23: [i-iii], sign. A-Z, Aa-Zz, 3A-3Z, 4A-4Z, 5A-5H(6).
24: [i-iii], sign. A-Z, Aa-Zz, 3A-3Z, 4A-4Z, 5A-5E(2).
25: [i-iii], sign. A-Z, Aa-Zz, 3A-3Z, 4A-4Z, 5A-5H(6).
26: [i-iii], sign. [A-Z, Aa-Zz, 3A-3D], 3E-3Z, 4A-4Z, 5A-5G(7).
27: [i-iii], sign. A-Z, 2A-2Z, 3A-3Z, 4A-4Z, 5A-5G(8).
28: [i-iii], sign. A-Z, 2A-2Z, 3A-3Z, 4A-4Z, 5A-5I(8).
29: [i-iii], sign. A-Z, 2A-2Z, 3A-3Z, 4A-4Z, 5A-5G(8).
30: [i-iii], sign. A-Z, 2A-2Z, 3A-3Z, 4A-4Z, 5A-5G(8).
31: [i-iii], sign. A-Z, 2A-2Z, 3A-3Z, 4A-4Z, 5A-5H(8).
32: [i-iii], sign. A-Z, 2A-2Z, 3A-3Z, 4A-4Z, 5A-5H(3).
33: [i-iii], sign. A-Z, 2A-2Z, 3A-3Z, 4A-4Z, 5A-5G(8).
34: [i-iii], sign. A-Z, 2A-2Z, 3A-3Z, 4A-4Z, 5A-5F(8).
35: [i-iii], sign. A-Z, 2A-2Z, 3A-3Z,4A-4Z, 5A-5H(8).
36: [i-iii], sign. A-Z, 2A-2Z, 3A-3Z, 4A-4Z, 5A-5H(5).
37: [i-iii], sign. A-Z, 2A-2Z, 3A-3Z, 4A-4Z, 5A-5C(8).
38: [i-iii], sign. A-Z, 2A-2Z, 3A-3Z, 4A-4Z, 5A-5F(7).
39: [i-iii], sign. A-Z, 2A-2Z, 3A-3Z, 4A-4L(8).
40: [i-iii], sign. A-Z, 2A-2Z, 3A-3Z, 4A-4Z, 5A-5D(8).
41: [i-iii], sign. A-Z, 2A-2Z, 3A-3Z, 4A-U(5).
Alph. list plates: p. [1]-100.
Plates 1: [i-iii], *236 pl.*
Plates 2: [i-iii], *pl. 1-4, 1-4, 1-7, 1-2, [3], 5, 5, 11, 9, 7, 2-7*, map, *1, 1-3, 1-2, 1, 1-11, [11a], 11/12, 13/14, 14, [14a], 15-17, 20-21,* and *169* other pl.
Plates 3: [i-iii], *193 pl.*
Plates 4: [i-iii], *198 pl.*
Plates 5: [i-iii], *262 pl.* (natural history).

Plates 6: [i-iii], 61 maps.
Copies: PH, US.

8756. *An attempt to ascertain the actual dates of publication of the* various parts of *Rees's Cyclopaedia* by Benjamin Daydon Jackson. London (Pewtress & Co., ...) 1895. Oct.
Publ.: Dec 1895 (p. 4: 4 Dec 1895), p. [1]-7. *Copy*: USDA. – For other papers on Rees' Cyclopaedia by Jackson see above, *biofile*. The dates given by Jackson are now superseded by those given by Pestana (1979).

Reess, Max [Friedrich Timotheus Ferdinand Maria] (1845-1901), German (Baden); Dr. phil. München 1867; bot. assistant München 1867; idem Halle 1867-1868; director botanical garden Halle 1868-1872; habil. Halle 1869; professor of botany and director of the botanical garden at Erlangen 1872-1901. (*Reess*).

HERBARIUM and TYPES: Unknown; some lichens at M. – Letters to W.G. Farlow and R. Thaxter at FH.

NOTE: Name often spelled, erroneously, "Rees".

BIBLIOGRAPHY and BIOGRAPHY: Ainsworth p. 318, 319; Barnhart 3: 137 (b. 10 Jun 1845, d. 14 Sep 1901); BM 2: 574, 4: 1663, 8: 1058; Bossert p. 327; CSP 8: 714, 11: 124-125, 18: 93; Frank 3 (Anh.): 82; GR. p. 122-123, *pl.* [*20*]; Herder p. 70, 157, 244, 245, 260; Jackson p. 161, 169, 173, 426; Kelly p. 188, 256; Kew 4: 433, 434; LS 1900, 21879-21904; Morren ed. 2, p. 9, ed. 10, p. 18; NI 1: 121, 173, 257, 2: 1994, 2079; PR 7461; Rehder 1: 92, 5: 707; Stevenson p. 1255; Tucker 1: 586.

BIOFILE: Anon., Biogr. Jahrb. 6: 84* (further biogr. refs.); Bot. Not. 1902: 94; Bot. Zeit. 25: 176. 1867 (asst. München), 26: 304. 1869 (app. Hallle), 27: 320. 1869 (habil. Halle), 30: 524. 1872 (app. Erlangen); Flora 50: 271. 1867 (app. München), 52: 368. 1869 (habil. Halle), 55: 416. 1872 (called to Erlangen); Hedwigia 40: (59) (retired), (204) (d.). 1901; Leopoldina 37: 90, 95-96. 1901 (obit.); Nat. Nov. 23: 539. 1901 (d.); Österr. bot. Z. 17: 233. 1867 (asst. München), 17: 407. 1867 (asst. Halle), 18: 203. 1868 (dir. bot. gard. Halle), 51: 147 (retired), 467 (d., 56 yrs old). 1901.
Gager, C.S., Brooklyn Bot. Gard. Rec. 27(3): 223. 1938 (dir. bot. gard. Erlangen 1872-1901).
Kneucker, A., Allg. bot. Z. 7: 220. 11901 (d.).
Mägdefrau, K., Gesch. Bot. 146. 1973.
Rikli, M., Biogr. Jahrb. 7: 435-437.
Wittrock, V.B., Acta Horti Berg. 3(3): 149, *pl. 125*. 1905.

EPONYMY: *Reessia* Fisch (1883).

8757. *Zur Entwicklungsgeschichte der Stammspitze von Equisetum*. Inauguraldissertation von Max Reess ... Jena (Druck von Friedrich Frommann) 1867. Oct. (*Entw.-Gesch. Stammsp. Equis.*).
Publ.: 1867, p. [i], [1]-28, *pl. 10-11* (uncol. liths. auct.). *Copies*: FH, NY. – Reprinted or preprinted from Jahrb. wiss. Bot. 6: 209-236, *pl. 10-11*. 1868.

8758. *Dispositio Uredineorum qui in Germaniae Coniferis parasitantur*. Commentatio quam consensu et auctoritate amplissimi philosophorum ordinis in Academia Fridericiana Halensi cum Vitebergensi consociata pro venia legendi rite impetranda die xiii aprilis mdccclxix hora xii in auditorio maximo una cum thesibus publice defendet Maximilianus Reess ... Halis Saxonum [Halle] (typis expressum Gebauerio-Schwetschkianis) [1869]. Oct. (*Disp. Ured. Conif. paras.*).
Publ.: 13 Apr 1869, p. [1]-19, [20, theses]. *Copy*: FI.

8759. *Die Rostpilzformen der deutschen Coniferen* zusammengestellt und beschrieben von Dr. Max Reess, Privatdocenten der Botanik an der Universität Halle ... Halle (Druck und Verlag von H.W. Schmidt) 1869. Qu. (*Rostpilzf. deut. Conif.*).

Publ.: Mai-Jul 1869 (after *Disp. Ured. paras.* because signed "Privatdocent"; BSbF 16 Aug 1869; Flora 10 Sep 1869; Bot. Zeit. 15 Oct 1869), p. [1]-70, *pl. 1-2* (uncol. liths. auct.). *Copies*: FH, H. – Preprinted from Abh. naturf. Ges. Halle 11: 49-118, *pl. 1-2*. Aug-Sep 1869.
Ref.: -n-g, Flora 52, 428-429. 10 Oct 1869 (rev.).

8760. *Über die Entstehung der Flechte Collema glaucescens* Hoffm. ... Berlin (Buchdruckerei der königlichen Akademie der Wissenschaften (G. Vogt) ...) 1871. Oct.
Publ.: Oct-Dec 1871 (in Monatsber. Oct 1871; t.p. reprint; Bot. Zeit. 19 Jan 1872), p. [i], [523]-533, *1 pl.* (partly col. lith. auct.). *Copies*: Almborn, FH, G, NY. – Important work confirming Schwendener's theory of lichen symbiosis.
Ref.: H.S., Bot. Zeit. 30: 270. 19 Apr 1872 (rev.).

8761. *Ueber den Befruchtungsvorgang bei den Basidiomyceten* ... Erlangen (Druck der Universitäts-Buchdruckerei von E.Th. Jacob) 1875. Oct. (*Befr. Vorg. Basidiomyc.*).
Publ.: early Jan 1875 (read 14 Dec 1874; BSbF post 20 Aug 1875; "Neu" Hedwigia Jan 1875; Bot. Zeit. 15 Jan 1875), p. [i, iii], [1]-21. *Copies*: FH, G, NY. – Preprinted or reprinted from S.B. phys.-med. Soc. Erlangen 7: 29-49. Sep-Dec 1875; also published in Jahrb. wiss. Bot. 10: 179-198. 1876. – The FH has a cover: "*Ueber ... Basidiomyceten.* Programm zum Eintritt in die philosophische Facultät und den Senat der k. Friedrich-Alexanders-Universität zu Erlangen".
Ref.: G.K., Bot. Zeit. 33: 78-79. 29 Jan 1875 (rev.).

8762. *Ist der Soorpilz mit dem Kalmpilz wirklich identisch?* [Aus den Sitzungsberichten der physikalisch-medicinalischen Societät zu Erlangen. Sitzung vom 14 Januar 1878]. Oct.
Publ: 14 Jan-14 Feb 1878 (Bot. Zeit. 15 Feb 1878), p. [1]-5. *Copy*: FH. – Reprinted from S.B. phys.-med. Soc. Erlangen 10: 54-58. Sep-Dec 1878. – Preliminary publ. ibid. 9: 190-195. Sep-Dec 1877 (Bot. Zeit. 1877).

8763. *Der botanische Garten zu Erlangen* ... Erlangen (Verlag von Eduard Besold) 1878. Oct. (*Bot. Gart. Erlangen*).
Publ.: Mai-Apr 1878 (p. 3: Easter 1878; Bot. Zeit. 28 Jun 1878), p. [1]-23, 2 maps. *Copies*: H, HH, M.

8764. *Ueber die Natur der Flechten.* Nach einem Vortrag in der Erlanger Philomathie ... Berlin S.W. (Verlag von Carl Habel ...) 1879. Oct. (*Nat. Flechten*).
Publ.: 1879 (Bot. Zeit. 13 Jun 1879; Nat. Nov. Mai(2): 1879), p. [1]-47. *Copies*: Almborn, FH, G, U. – Issued as Heft 320 (p. 1-47) in Samml. gemeinverst. wiss. Vorträge (ed. R. Virchow & Fr. von Holtzendorff) ser. 14: 243-287. 1879.

8765. *Ueber den Parasitismus von Elaphomyces granulatus* [Aus den Sitzungsberichten der physikalisch-medicinischen Societät zu Erlangen. Sitzung vom 10 Mai 1880]. Oct.
Publ.: Jun-Aug 1880 (after 10 Mai 1880; Hedwigia Sep 1880; Bot. Zeit. 24 Sep 1880), p. [1]-5. *Copies*: FH, M. – Preprinted or reprinted from S.B. phys.-med. Soc. Erlangen 12: 103-107. Sep-Dec 1880.

8766. *Ueber die Pflege der Botanik in Franken* in der Mitte des 16. bis zur Mitte des 19. Jahrhunderts nebst einigen Bemerkungen über gegenwärtige Zustände. Rede beim Antritt des Prorectorats der königlich Bayerischen Friedrich-Alexanders-Universität Erlangen am 4. November 1884 gehalten ... Erlangen (Druck der Universitäts-Buchdruckerei von Junge & Sohn) 1884. Qu. (*Pflege Bot.-Franken*).
Publ.: Nov-Dec 1834 (Oration 4 Nov 1884; t.p. 1884; Bot. Centralbl. 12-16 Jan 1885; Bot. Zeit. 30 Jan 1885; Flora 21 Jan-10 Mar 1885; Nat. Nov. Feb(1) 1885), p. [1]-56. *Copies*: G, HH, M, MO, USDA.

8767. *Untersuchungen über Bau und Lebensgeschichte der Hirschtrüffel, Elaphomyces* ... Cassel (Verlag von Theodor Fischer) 1887. Qu.
Co-author: Carl Fisch (1859-?).
Publ.: Jun-Aug 1887 (p. 2: Whitsun 1887; Nat. Nov. Sep(1) 1887; ÖbZ 1 Jan 1888; Bot.

Centralbl. 23-27 Jan 1888), p. [ii-vi], [1]-24, *pl. 1. Copies*: BR, FH. – Issued as Bibl. bot. 2(7), 1887.
Ref.: Büsgen, M., Bot. Zeit. 46: 189-190. 1888.
Burgerstein, A., Österr. bot. Z. 38: 29-30. 1888.
Frank, Bot. Centralbl. 33: 98-101. 1888 (rev.).

8768. *Lehrbuch der Botanik* ... Stuttgart (Verlag von Ferdinand Enke) 1896. Oct. (*Lehrb. Bot.*).
Publ.: Mai-Jun 1896 (p. [v]: Mai 1896; ÖbZ Jul 1896; Bot. Zeit. 16 Jul 1896; Bot. Centralbl. 24 Jun 1896; Flora 4 Jul 1896; Nat. Nov. Jun(2) 1896), p. [i]-x, [1]-453, [454, err.]. *Copies*: H, M, MO.
Ref.: Behrens, J., Bot. Zeit. 55: 101-103. 1897 (rev.).

Regel, [Johann] **Albert von** (1845-1908), Swiss-born Russian physician and botanist at St. Petersburg, explorer of Turkestan and Eastern Asia 1876-1888; oldest son of E.A. von Regel. (*A. Regel*).

HERBARIUM and TYPES: LE; duplicates in B, BM, BR, E, FI, GOET, H, K, MW, NY, PC, W. – Some duplicates distributed as *Iter turkestanicum*.

BIBLIOGRAPHY and BIOGRAPHY: AG 6(1): 25, 30; Barnhart 3: 138; Bret. p. 1036; CSP 8: 714, 11: 125, 18: 95; Herder p. 200, 211, 319; IH 2: (in press); Jackson p. 129; Kew 4: 435-436; Morren ed. 10, p. 148; MW p. 401; Rehder 5: 708; TR 1110-1114; Tucker 1: 586; Urban-Berl. p. 311, 386.

BIOFILE: Coats, A., The plant hunters 59-60. 1969.
Embacher, F., Lex. Reisen 244-245. 1882.
Hedge, I.C. & J.M. Lamond, Index coll. Edinb. herb. 122. 1970.
Herder, F. von, Bot. Centralbl. 55: 262, 268. 1893, 58: 386, 387, 392.; Bot. Jahrb. 9: 446-447. 1888 (notes on travel reports).
Lindemann, E. von, Bull. Soc. Natural. Moscou 61(1): 59. 1886.
Sayre, G., Mem. New York Bot. Gard. 19(3): 384. 1975.
Wagenitz, G., Index coll. princ. herb. Gott. 132. 1982.

EPONYMY: *Albertia* E. Regel et Schmalhausen (1877); *Aregelia* O. Kuntze (1891).

NOTE: For travel reports (Turkestan) see *Gartenflora* 1877: 6-19, 68-70, 103-104, 230-236, 260, 334-341; 1878: 35-40, 106-110, 144-146, 200-202, 227-230, 263-264, 336-338, 363-370; 1879: 35-48, 79-82, 192, 320, 351; 1880: 4-11, 43-50, 68-72, 96, 132-138, 160, 167-177, 197-206, 293-298; 1881: 3-8, 145-150, 206-210, 236-241, 270-274, 337-343; 1882: 355-368, 384; 1883: 15-17, 73-82, 142-145, 176-178, 206-213, 231-238, 268-273; 1884: 4-6, 68-73, 73-89, 111-114, 137-141, 201-204, 256, 259-267, 320; 1885: 261-266, 293-298, 324-330.
Further reports (*Reisebriefe*) in Bull. Soc. Natural. Moscou 51(2): 393-399. 1876, 52(1): 121-127, 350-368. 1877, 52(2): 163-167. 1877, 53(1): 165-205, 397-403. 1878. 54: 124-149. 1879, 56(2): 220-221. 1881, 58(1): 235-241. 1883, 58(2): 220-234, 347-349, 60(1): 167-188. 1885 and in *Geogr. Mitt.* (Petermann): 1877: 36, 75, 359, 391; 1878: 37, 70, 159, 236, 394; 1879: 230, 376, 408, 464; 1880: 70, 116, 205, 315, 399; 1881: 380, 470; 1882: 29, 65, 213, 349, 467; 1883: 68, 231, 461; 1884: 86, 149, 230, 312, 332; 1885: 393-477.

Regel, Constantin [Andreas] **von** (1890-1970), Russian-born botanist; studied at the University of St. Petersburg until 1913; botanical assistant at the Ministry of Agriculture ib. 1913-1917; at the University of Dorpat (Tartu) 1918-1921; Dr. phil. Würzburg 1921; professor of botany at Kaunus University, Lithuania, 1922-1940; in exile in Switzerland 1940-1944; from 1944 at Graz; teaching botany in Bagdad 1952-1955, Istanbul 1956 and Kabul 1958-1959; from 1959-1962 living at Heidelberg; from 1962-1967 at Izmir, later again in Graz and in Zürich. (*C. Regel*).

HERBARIUM and TYPES: Types at G, ISTF, KA; collections at A, B, E, G, GB, ISTF, KA, LE. – *Flora lithuanica exsiccata*, fasc. 1-5, set at H. – Schedae in *Fontes*, see below. – 29 letters to Linkola at Helsinki Univ. Libr.

REGEL, C.

BIBLIOGRAPHY and BIOGRAPHY: Barnhart 3: 138(b. 10 Aug 1890); BFM no. 2943; Biol.-Dokum. 14: 7351-7352; BJI 2: 143; BL 2: 7, 82, 706; Bossert p. 327; IF (ed. 2): 65, (ed. 3): 80, 88; IH 1 (ed. 7): 340, 2: (in press); Kew 4: 434; MW p. 401, suppl. 291; Roon p. 93.

BIOFILE: Asmous, V.C, Chron. bot. 11(2): 167. 1947.
Blakelock, R.A. & E.R. Guest, Fl. Iraq 1: 116. 1966 (coll. Iraq 1953-1954).
Czaja, A.Th., Qualit. Pl. Mat. veg. 20(1/2): 3-5. 1970 (obit.,portr.).
Furrer, E., Vierteljahrsschr. naturf. Ges. Zürich 16(4): 497-501. 1971 (obit., portr.).
Gager, C.S., Brooklyn Bot. Gard. Rec. 27(3): 300. 1938.
Regel, C. v., [bibliography of 209 p., photocopy at ETH, Zürich, fide E. Furrer, 1971].
Šapiraité, S., Lietuvos botanikos bibliografija 1800-1965. Vilnius 1971 (many entries on C. Regel, see index).
Verdoorn, F., ed., Chron. bot. 1: 28, 35, 36, 37, 384. 1935, 2: 7, 28, 29, 53, 226, 228. 1936, 3: 28, 199. 1937, 4: 299, 444, 445. 1938, 5: 123, 526. 1939, 6: 302. 341.
Winkler, E., Geographia helvetica 20(3): 168-169. 1965 (75th birthday).
Wulff, E.V., Intr. hist. pl. geogr. 149, 163. 1943.

EPONYMY: *Neoregelia* L.B. Smith (1934).

8769. Fontes florae Lituaniae-Lietuvos floros šaltiniai ... Kaunas ("Spindulio", B-ves Spaustuvé) 1931-1939. 6 parts.
Publ.: 6 parts, published in and to be cited from Mémoires de la Faculté des sciences de l'Université de Vytautus le Grand and (simultaneously) in Scripta Horti botanici Universitatis Vytauti magni; reprinted with original pagination and cover/reprint title pages. *Copy*: H (inf. P. Isoviita).
1: 1931, cover-t.p., [221]-289. *Mém.* 5: [221]-289. 1931, *Scripta* 1: [221]-289. 1931.
2: 1932, cover-t.p., [1], [3]-69, *3 pl.* – *Mém.* 7: [3]-71, *3 pl.* 1932, *Scripta* 2: [3]-71. *3 pl.* 1932. – Kaunas (Kooperatinés B-vés "Raides" ...); Fontes florae Lituanae [sic].
3: 1935, cover-t.p., [1]-42. – *Mém.* 9: [181]-222. 1935; *Scripta* 3: [87]-128. 1935. – Kaunas (Akc. "Spindulio" ...) Fontes florae Lituanae [sic].
4: 1936, cover-t.p., [47]-81. – *Mém.* 11(2): [47]-81. 1936, *Scripta* 4: [47]-81. 1936. – Kaunas (Kooperatinés B-vés "Raides" ...). – Fontes florea [sic] *Lituanae* [sic].
5: 1937, cover-t.p., [67]-84. – *Mém.* 11(4): [299]-316, 1937, *Scripta* 5: [67]-84. 1937. – *Fontes florae Lituanae* [sic].
6: 1939, cover-t.p., [5]-27. – *Mém.* 13(2): [5]-27. 1939, *Scripta* 6: [5]-27. 1939. – *Fontes florae Lituanae* [sic].

8770. Apie kai kurias Tragopogon rūšis. *Beitrag zur Kenntniss einiger Tragopogon Arten* [Extrait des "Mémoires de la Faculté des Sciences de l'Université Vytautas le Grand 1937, volume ix, fasc. 4]. Oct.
Publ.: 1937 (U received 4 Oct 1938), p. [1]-66, *6 pl. Copy*: U. – Reprinted from V.D.U. Mat. Gamt. F-teto Darbu 11(4) (journal n.v.).
Second part: 1939 (U rd 28 Mai 1940), p. [29]-39, [40], *3 pl. Copy*: U. – Reprinted from id. 13(2) (journal n.v.).

Regel, Eduard [August] von (1815-1892), German (Thuringian) botanist; assistant at the botanical gardens of Göttingen 1833-1837, Bonn 1837-1839, Berlin 1839-1842; head gardener at the University botanical garden Zürich 1842-1855; Dr. phil. Zürich 1855; scientific director of the imperial botanical garden St. Petersburg [1855-]1857-1867, first botanist 1867-1875, director 1875-1892. (*Regel*).

HERBARIUM and TYPES: LE; further material at B (mainly destroyed), G, GOET, H, K, M, MW, P, PH. – Material by Albert Regel was distributed by E.A. von Regel to various herbaria. – Letters by Regel to Miquel at U, to D.F.L. von Schlechtendal at HAL.

BIBLIOGRAPHY and BIOGRAPHY: AG 3: 853, 6(1): 25; Backer p. 485; Barnhart 3: 138 (b. 1/13 Aug 1815, d. 15/27 apr 1892); Biol.-Dokum. 14: 7352; BJI 1: 47; BM 4: 1664; Bossert p. 327; Bret. p. 623, 1062-1064; CSP 5: 127-129, 6: 747, 8: 714-715, 11: 125, 12:

602, 18: 96; DTS 1: 238; Frank 3 (Anh.): 82; Hegi 7: 560; Herder p. 364, 472; Hortus 3: 1202; IF p. 726, suppl. 2: 38; Jackson p. 122, 124, 130, 143, 147, 327, 358, 393, 394, 443, 444; Kew 4: 435-436; Langman p. 617-618; Lenley p. 467; LS 21905-21906; Morren ed. 2, p. 30, ed. 10, p. 95, 96; MW p. 401-408, suppl. p. 291; NI 2079, 2294a, see also 1: 244, 2: 148; PR 7462-7473, 8292, 10623, ed. 1: 8419-8420, 9221; Ratzeburg p. 431-434; Rehder 1: 42, 65, 270, 293, 300, 358, 454-457, 521, 5: 708-710; Stevenson p. 1255; TL-2/999, 1748, 1788, see F. v. Herder, C.A. von Meyer; TR 1115-1139; Tucker 1: 586-587; Urban-Berl. p. 386; Zander ed. 10, p. 705, 706, ed. 11, p. 805.

BIOFILE: Anon., Bonplandia 3: 206. 1855 (leaves Zürich for St. Petersburg; continues Gartenflora), 3: 258. 1855 (Dr. phil. Zürich), 6: 362. 1858 (member Leopoldina, cognomen: Willdenow), 9: 346. 1861 (gold medal); Bot. Gaz. 17(6): 200. 1892 (d.); Bot. Not. 1893: 39 (d.); Bot. Zeit. 13: 532. 1855 (called to St. Petersburg), 13: 728. 1855 (left 10 Sep 1855 for St. Petersburg to become scientific director of the botanical garden; managing director at the time von Kärter), 18: 324. 1860 (descr. of garden activities in Gartenflora), 33: 551. 1875 (general director bot. gard., succeeding Trautvetter), 50: 371 (d.), 500 (succeeded by Batalin) 1892; Flora 38: 528. 1855 (to St. Petersburg), 38: 591. 1855 (Dr. phil. h.c. Zürich), 47: 348. 1864 (Belgian order of Leopold), 51: 73. 1868 (reorganisation garden; R. is now "Oberbotaniker" under v. Trautvetter), 51: 235. 1868 (Prussian Kronenorden III. Kl.), 58: 528. 1875 (succeeds Trautvetter as director of bot. gard. St. Petersburg); J. Bot. 30: 192. 1892 (b. 13 Aug 1815); Leopoldina 28: 107. 1892 (obit.); Nature 46: 60-61. 1892 (obit.); A.J., Neubert's deut. Gart.-Mag. 45: 231-232. 1892 (obit.); Österr. bot. Z. 19: 96. 1969, 26: 75. 1876, 35: 293. 1885, 37: 261. 1887, 42: 223. 1892.
Asmous, V.C., Chron. bot. 7: 199-200. 1942 (portr., biogr., tribute).
Borodin, J., Trav. Mus. bot. Acad. Sci. St. Petersburg 4: 101. 1908.
Britten, N.L. et al., Bull. Torrey bot. Cl. 19: 237. 1892 (memorial).
Buchheim, G., Willdenowia 2(4): 553-558. 1960 (on *Reis. Ostsib.*).
Bullock, A.A., Bibl. S. Afr. bot. 93. 1978.
Candolle, Alph. de, Phytographie 442. 1880 (material from Zürich at G).
Cohn, F., Bot. Zeit. 18: 137-142. 1860 (visit to St. Petersburg bot. gard. under Regel).
Dammer, U., Gartenflora 41: 609. 1892 (protest against Herder's obit. of Regel in Bot. Zeit. 41(11); Herder had spoken of "an unsatisfiable ambition and a ruthless egoism", Dammer speaks of "strictest performance of duty and iron diligence").
Engler, A., Bot. Jahrb. 15 (Beibl. 35): 15. 1892 (d. 27 April 1892.
Fischer von Waldheim, A.A., Imp. S.-Petersb. bot. sad. 3: 128-132. 1913/1915 (biogr., portr.).
Gager, C.S., Brooklyn bot. Gard. Rec. 27: 345. 1938 (director bot. gard. St.Petersb. 1857-1865, 1875-1892).
Govorukhina, V.A., Izvest. Akad. Nauk Turkm. S.S.R., Biol. 1969(6): 69-73 (on R.'s travels in Turkestan 1876-1885).
Herder, F. v., Bot. Centralbl. 50: 191. 1892 (d.), 51: 321-327, 369-374, 401-408. 1892 (brief obit., bibl.), 52: 319-320. 1892 (reply to Schmalhausen), 55: 262. 1893, 58: 386, 387. 1894.
Hertel, H., Mitt. Bot. München 16: 421. 1980 (lich. at M through Nägeli).
Jessen, K.F.W., Bot. Gegenw. Vorz. 394. 1884.
Klotzsch, F., Bonplandia 3: 99-102. 1855 (reply to Regel 1855 on *Aegilops*).
Knapp, J.A., Geheimrath Dr. Eduard August v. Regel, Wien 1892, 46 p., repr. from Verh. zool.-bot. Ges. Wien 42: 260-304. 1892 (obit., bibl. of nearly 1000 titles).
Kuznetzov, N., Acta Hort. bot. Jurj. 2(1): 46-48. 1901 (biogr. note, 50 yrs Gartenflora, portr.).
Lindemann, E. von, Bull. Soc. Natural. Moscou 61: 59. 1886.
Lipsky, V.I., *in* A.A. Fischer v. Waldheim, Imp. S.-Petersb. bot. sad. 3: 133-229, 406-407 (bibl., 2878 journal arts., 233 other titles), 524-525. 1913/1915; major bibl.; portr.).
Morren, É., Belg. hortic. 19: frontisp. portr., Regel, v-xi. 1869 (dedication to E.A. v. Regel with a biogr. sketch and an evaluation).
Nelmes & Cuthbertson, Curtis's Bot. Mag. dedic. 230-232. 1932.
Regel, E.A. v., Bonplandia 2: 206, 219, 220-225, 286-293. 1854, 3: 53-54, 162-172, 322-325. 1855, 4: 243. 1856, 7: 37-40. 1859 (on *Aegilops*). Gartenflora 30: 229-230, pl.

1052. 1881 (portr., with note by Regel himself on other less faithful likenesses; this portrait appeared also in the St. Petersburg illustrated journal Niwa, circulation 60,000, sic); [2 p. printed letter thanking recipients for their contribution to the festivities of his 70th birthday; reminiscenses, dated 3/15 Aug 1855, at G].
Regel, C. v., Mitt. deut. dendrol. Ges. 55: 311-314. 1942 (commmemorative paper, biogr.); Schweizer Garten 1942(11): 301-304, (12): 341-345. (commemorative paper, especially on horticultural and pomological work).
Roth, E., Leopoldina 29: 146-149. 1893 (obit.).
Rübel, E., Gesch. naturf. Ges. Zürich 32. 1947 (R. at Zürich).
Sabersky, M., Regel's 70. Geburtstag, pamphlet 1 p., Berlin 1885 (copy: G; subscription request).
Sargent, C.S., Garden & Forest 5: 252. 1892 (obit.).
Sayre, G., Bryologist 80: 516. 1977.
Schmalhausen, J., Bot. Centralbl. 52: 319. 1892 (criticizes Herder's obituary of R.).
Shetler, S.G., Komarov bot. inst. 29, 35, 41, 190, 198-199. 1967.
Stafleu, F.A., The great Prodromus 30. 1966.
Stapf, O., Bull. misc. Inf. Kew 1913(7): 246-247.
Voit, C. v., S.B. math. phys. Cl. Akad. Wiss. München 23: 113-114. 1894 (obit.).
Wagenitz, G., Index coll. princ. herb. Gott. 132. 1982.
Weiling, F., Decheniana 126: 23-25. 1974 (hybridization exp.).
Wittmack, L., Gartenflora 41: 261-269. 1892 (obit., bibl., portr., list of decorations: highest 1885, Prussian Red Eagle, 2nd class with star, very high for a non-resident, non-establishment Prussian in the diaspora); Gart. Zeit. 4(32): 373-382. 1885 (report on 71st birthday; the subscription resulted in a gift of a 24 person silver service; portr.).
Wittrock, V.B., Acta Horti Berg. 3(2): 175-176, *pl. 21.* 1903, 3(3): 189, lxxx, *pl. 72.* 1905.

COMPOSITE WORKS: (1) Editor of *Gartenflora* 1852-1884, (1852-1930 and index available on IDC), see below.
(2) *Betulaceae*, in DC, Prodromus 16(2): 161-190, 684. med. Jul 1868.
(3) With J.J. Schmitz, q.v., *Flora bonnensis*, Bonn 1841.
(4) F.E.L. von Fischer et al., *Sertum petropolitanum*, see TL-2/1788, decas 3/4, 1869, by E.A. von Regel; *copies*: BR, G, NY.
(5) With E.R. v. Trautvetter, q.v., *Decas plantarum novarum*, 1882.
(6) With O. Heer, Schweizerische Zeitschrift für Land- und Gartenbau, 1843-1847.
(7) *Turkestanskaia Flora*, in A.P. Fedchenko, *Putesh. Turkestan* 3(1): [i-viii], [1]-164, [165], *22 pl. Copy*: M. – TL-2/1748.
(8) *Descriptiones plantarum novarum*, in A.P. Fedchenko, *Putesh. Turkestan* 3(3), p. [i-v], [1]-89. 1882. *Copies*: H, M. – TL-2/1748.

HANDWRITING: Lipschitz, S.V. & I.T. Vasilczenko, Herb. centr. USSR 118, 119. 1968; Chronica botanica 7(5): 199. 1942.

EPONYMY: *Eduardoregelia* Popov ex V.B. Botchantzev (1957); *Euregelia* O. Kuntze (1898); *Regelia* Schauer (1843); *Regelia* C.A.M.Lindman (1890).

NOTE ON BIBLIOGRAPHY: The most complete bibliography of E.A. v. Regel is given by V.I. Lipski (1915): 2878 and 233 nos. Most of the numbers are brief publications in Gartenflora, but the number of papers in other periodicals and of the independent publications is equally impressive. Our choice is of necessity incomplete and somewhat uneven.

8771. *Die Kultur und Aufzählung der* in deutschen und englischen Gärten befindlichen *Eriken* nebst Synonymie und kurzer Charakterisirung und Beschreibung derselben ... Ein Leitfaden für Gärtner und Eriken-Züchter bearbeitet von E. Regel. [Zürich 1843]. (*Kult. Aufz. Eriken*).
Publ.: 1843 (t.p. reprint), p. [1]-189, *3 pl.* (uncol. lith.). *Copies*: HH, M, NY. – "Besonders abgedruckt aus den Verhandlungen des Vereins zur Beförderung des Gartenbaues in den Königlichen Preussischen Staaten [vol. 16], 33ste Lieferung". 16(2): 163-349, *pl. 2, 3, 4.* 1842 (sic). In view of the date on the cover, the journal publication has priority over the reprint.

8772. *Die äusseren Einflüsse auf das Pflanzenleben* in ihren Beziehungen zu den wichtigsten Krankheiten der Kulturgewächse. Ein populärer Vortrag gehalten am 30. März 1847 ... Der Ertrag ist zum Besten der Armen bestimmt. Zürich (Meyer und Zeller) 1847. Oct. (*Äuss. Einfl. Pfl.-Leben*).
Publ.: 1847, p. [1]-32. *Copy*: M.

8773. *Bemerkungen über die Gruppe der Gattung Amaranthus* mit 5-männigen Blüthen. Von C. [sic] Regel in Zürich. (Abdruck aus Flora 1849. Nro. 11). Oct.
Publ.: 21 Mar 1849 (in journal), p. [1]-7. *Copy*: G. – Reprinted and to be cited from Flora 3: 161-167. 21 Mar 1849.

8774. *Einige neue Gattungen der Gesnereen.* Von E. Regel in Zürich. (Abgedruckt aus der Flora Nro. 12. vom 28. März 1849). Oct.
Publ.: 28 Mar 1849, p. [1]-8. *Copy*: G. – Reprinted and to be cited from Flora 32: 177-182. 28 Mar 1839.

8775. *Gartenflora.* Monatschrift für deutsche und schweizerische Garten und Blumenkunde herausgegeben von E. Regel ... [erster Jahrgang] Erlangen (Verlag von Ferdinand Enke). 1852. Oct. (*Gartenflora*).
Publ.: vols. 1-33, 1852-1884, were edited by Regel. *Copies*: NY, US, USDA; IDC 7295.
1: 1852 (monthly issues, dated Jan-Dec), p. [i-iv], [1]-386, [387-388], *pl. 1-34, 35/36* (24 in col.).
2: 1853, p. [i], [1]-386, *pl. 37-72* (24 in col.).
3: 1853, p. [i], [1]-429, *pl. 73-108*.
4: 1855, p. [i], [1]-416, *pl. 109-144* (*115* as "*112*", *116* as "*113*") (24 in col.).
5: 1856, p. [i-ii], [1]-402, *pl. 145-179* (23 in col.).
6: 1857, p. [i], [1]-404, *pl. 180-212* (16 in col.) (no. *193* as "*190*").
7: 1858, p. [i-ii], [1]-404, *pl. 213-244* (21 in col.).
8: 1859, p. [i-ii], [1]-386, *pl. 245-276* (19 in col.).
9: 1860, p. [i-ii], [1]-446, *pl. 277-312* (24 in col.).
10: 1861, p. [i-iv], [1]-450, *pl. 313-348* (24 in col.).
11: 1862, p. [i-ii], [1]-448, *pl. 349-384* (25 in col.), p. [1]-140, suppl.
12: 1863, p. [i-ii], [1]-418, *pl. 385-420* (24 in color).
13: 1864, p. [i-ii], [1]-399, *421-456* (25 in color).
14: 1865, frontisp., p. [i], [1]-411, *pl. 457-492* (24 in color).
15: 1866, p. [i-ii], [1]-402, *pl. 493-528* (24 in color).
16: 1867, p. [i], [1]-411, *pl. 529-564* (24 in color).
17: 1868, p. [i-ii], [1]-397, *pl. 565-600* (24 in color).
18: 1869, p. [i-ii], [1]-394, *pl. 601-636* (24 in color).
19: 1870, p. [i-ii], [1]-394, *pl. 637-672* (24 in color).
20: 1871, p. [i-ii], [1]-396, *pl. 673-708* (24 in color), index p. [i-ii], [1]-120.
21: 1872, p. [i-ii], [1]-395, *pl. 709-744* (24 in color).
22: 1873, p. [i-vi], [1]-396, *pl. 745-780* (24 in color).
23: 1874, p. [i-iv], [1]-396, *pl. 781-816* (24 in color).
24: 1875, p. [i-iv], [1]-395, *pl. 817-851* (24 in color).
25: 1876, p. [i-iii], [1]-416, *pl. 852-887* (24 in color).
26: 1877, p. [i-v], [1]-412,,*pl. 888-923* (24 in color).
27: 1878, p.[i-vi], [1]-408, *pl. 925-959* (24 in color).
28: 1879, p. [i-v], [1]-400, *pl. 960-995* (24 in color).
29: 1880, p. [i]-vi, [1]-404, *pl. 996-1031* (24 in color).
30: 1881, p. [i]-vi, [1]-436, *pl. 1033-1067* (24 in color), index [i], [1]-136.
31: 1882, p. [i-v], [1]-398, *pl. 1068-1103* (24 in color).
32: 1883, p. [i-v], [1]-398, *pl. 1104-1139* (24 in color).
33: 1884, p. [i]-iv, [1]-402, *pl. 1140-1163, 1165-1176* (24 in color).
Gartenflora continued until vol. 87, 1938 and n.s. 1938-1940. The editors succeeding Regel were L. Wittmack, Siegfried Braun, Hugo Fischer, Paul Kache, E. Dageförde and R. Zander.
Imprints: 1-34: Erlangen (Verlag von Ferdinand Enke); 35-42: Berlin (Verlag von Paul Parey), 43-48: Berlin (Selbstverlag des Vereins für beförderung des Gartenbaues ...), 49-57: Berlin (Verlag von Gebrüder Borntraeger), 58-71: Berlin (Kommissionsverlag

von Rudolf Mosse), 73-75: Berlin (Deutsche Gartenbau-Gesellschaft; 76-85: Berlin (Verlag der Gartenflora), 86-87: Berlin/Aachen (Rheinischer Verlag ...).
Circulation: varied, for vol. 14, 1865, the number of copies of loose inserts required was 650; for vol. 17, 1868: 500; for vol. 22: 530; for vols. 24-33: 600.
Issues: with all plates uncolored and with ca. 24 of the ca. 36 plates per volume colored.
Title development: There were many title page-changes, especially because of mutations in the team of co-editors. We list the main mutations:
1: as in heading.
2: *Gartenflora*. Monatschrift für deutsche und schweizerische Garten- und Blumenkunde herausgegeben unter Mitwirkung der tüchtigsten Gärtner, Gartenfreunde und Botaniker von E. Regel, Obergärtner ... Zürich.
3: *Gartenflora*. Monatschrift ... Blumenkunde mit Originalbeiträgen von den Herren Bremi ... [19 collaborators] ... Herausgegeben von E. Regel, Obergärtner ... Zürich
...
4: *Gartenflora*. Monatschrift ... Blumenkunde unter Mitwirkung von Prof. Dr. H.R. Goeppert, ... [10 coll.]. ... Herausgegeben von E. Regel, Obergärtner ... Zürich ...
5: [after Regel's move to St. Petersburg] *Gartenflora*. Monatschrift ... Blumenkunde unter Mitwirkung von Prof. Dr. H.R. Goeppert, ... [7 coll.] ... Herausgegeben von Dr. E. Regel [Regel had received a Dr. phil. h.c. Zürich 1855 when leaving], wissenschaftlichem Direktor ... St. Petersburg, ... Mitherausgeber für Deutschland J. Rinz ... Mitherausgeber für die Schweiz Prof. Dr. O. Heer ... Redactor [sic this office, actually duplicating that of the Herausgeber, Regel, lasted only 2 years: 1856-1857]. Dr. H. Locher, ...
6: *Gartenflora*. Allgemeine Monatschrift ... Blumenkunde unter Mitwirkung vieler Botaniker und Gärtner Deutschlands und der Schweiz herausgegeben von Dr. Eduard Regel, ... Mitherausgeber für Deutschland: H. Jäger, ... Schweiz: E. Ortgies, ... Redactor: Dr.H. Locher, ...
7: *Gartenflora*. Allgemeine Monatschrift für deutsche, russische und schweizerische Garten- und Blumenkunde. Unter Mitwirkung vieler Botaniker und Gärtner Deutschlands, Russlands und der Schweiz, herausgegeben und redigiert von Dr. Eduard Regel, ... für Deutschland: H. Jäger, ... für die Schweiz: E. Ortgies, ... [no Redactor].
8: *Gartenflora*, Allgemeine Monatschrift ... Regel [14 lines of titles, memberships], ... Für Deutschland: H. Jäger, ... Fr. Francke, ... Schweiz: E. Ortgies, ...
9-11: *Gartenflora*, ... Blumenkunde und Organ des Russischen Gartenbau-Vereins in St. Petersburg. Unter ... Regel, ... für Deutschland: H. Jäger, ... Fr. Francke, ... C. Bouché, ... Schweiz: E. Ortgies...
12-15: *Gartenflora*, ... Regel, ... für Deutschland: H. Jäger, ... Fr. Francke, ... C. Bouché, ... Schweiz: E. Ortgies, ... Mitherausgeber für Russland: Dr. F. von Herder.
16: *Gartenflora*, ... Regel, ... für Deutschland: H. Jäger, ... Fr. Francke, ... Paul Sorauer, ... Senoner, ... in Wien [sic]. Mitherausgeber für die Schweiz: E. Ortgies, ... Russland: Dr. F. von Herder, ...
17: *Gartenflora*, ... Regel, ... für Deutschland: H. Jäger, Fr. Francke, Paul Sorauer, ... Senoner, ... E. Mayer, ... Schweiz: E. Ortgies, ... Russland: Dr. F. von Herder, ...
18-19: *Gartenflora*, ... Regel, ...Herder, ... E. Ender, ...
20-21: *Gartenflora*, ... Regel, ... H. Jäger, ... Fr. Francke, ... H. Maurer, ... A. Senoner, ... E. Mayer, ... Schweiz: E. Ortgies, ... Russland: ... Herder, ... E. Ender, ...
22-23: *Gartenflora*, ... Regel, ... H. Jäger, ... Fr. Francke, ... E. Petzold, ... A. Senoner , ... C. Salomon, ... E. Mayer, ...
24: *Gartenflora*, ... Regel, ... H. Jäger, ... Fr. Francke, ... E. Petzold, ... A. Senoner, ... C. Salomon, ... E. Mayer, ... H. Hoffmann, ... Schweiz: ...
25: *Gartenflora*, ... Regel, ...H. Jäger, ... A. Senoner, .. C. Salomon, ... E. Mayer, ... H. Hoffmann, ... Schweiz: ...
26-27: *Gartenflora*, ... Regel, ... H. Jäger, ... E. Mayer, ... A. Senoner, ... L. Beissner, ... H. Hoffmann, ... C. Salomon, ... W. Zeller, ... Schweiz: ...
28: *Gartenflora*, ... Regel, ... H. Jäger, ... E. Mayer, ... A. Senoner, ... L. Beissner, ... H. Hoffmann, ... C. Salomon, ... W. Zeller, ... Prof. Dr. Göppert, ... Schweiz: ...
29: *Gartenflora*, ... Regel, ... [as 28] ... Göppert, ... E. Schmidt, ... Schweiz: ...
30: *Gartenflora*, ... Regel, ... [as 28] ... Göppert, ... M. Kolb, ... E. Schmidt, ... Schweiz": ...

31: *Gartenflora*, ... Regel, ... [as 28], ... Göppert, ... B. Stein, ... H. Zabel, ... E. Schmidt, ... Dr. H.G. Reichenbach, ... Schweiz ...
32: *Gartenflora*, ... Regel, ... [as 28], ... Göppert, ... H. Zabel, ... E. Schmidt, ... Dr. H.G. Reichenbach, ... Schweiz ...
33: *Gartenflora*, ... Regel, ... H. Jäger, ... E. Mayer, ... L. Beissner, ... C. Salomon, ... A. Senoner, ... W. Zeller, ... H. Zabel, ... E. Schmidt, ... D. H.G. Reichenbach, ... Schweiz ...
Continuation beyond E.A. von Regel's editorship:
34: *Gartenflora*. Monatschrift für Garten- und Blumenkunde. Unter Mitwirkung von Dr. Eduard Regel... und Professor Dr. A. Engler, ... herausgegeben von B. Stein, königlicher Garteninspector in Breslau [Berthold Stein, 1847-1899].
35: *Gartenflora* Zeitschrift für Garten- und Blumenkunde ... Unter Mitwirkung von ... [as in 34].
36-39: *Gartenflora* Zeitschrift für Garten- und Blumenkunde. (Begründet von Eduard Regel) ... Unter Mitwirkung von ... L. Beissner ... O. Choné ... F. Cohn ... G. Dieck ... L. Dippel ... O. Drude ... A. Engler ... B. Frank ... H. Gaerdt ... F. Goeschke ... R. Goethe ... L. Graebener ... W. Hampel ... H. Jäger ... F. Jühlke ... L. Kny ... C. Lackner ... H. Mächtig ... C. Mathieu ... I. Möhl ... E. Ortgies ... W. Perring ... F.J. Pfister ... E. von Regel ... H.G. Reichenbach ... C. Salomon ... A. Siebert ... L. Spaeth ... B. Stein ... H. Voechting ... H. Zabel ... herausgegeben von ... L. Wittmack.
40-42: *Gartenflora* Zeitschrift für Garten- und Blumenkunde. (Begründet von Eduard Regel). 40. Jahrgang. Herausgegeben von Dr. L. Wittmack.
43-53: *Gartenflora* ... Jahrgang. Organ des Vereins zur Beförderung des Gartenbaues in den preussischen Staaten. Herausgegeben von Dr. L. Wittmack, ...
54: [as vol. 53 but:] Herausgegeben in der ersten Hälfte von Dr. L. Wittmack, ... in der zweiten Hälfte von Siegfried Braun, ...
55-59: [as vol. 53 but:] Herausgegeben von Siegfried Braun, ...
60-71: *Gartenflora*. Zeitschrift für Garten- und Blumenkunde. Begründet von Eduard Regel 60. Jahrgang 1911. Herausgeber: Deutsche Gartenbau-Gesellschaft ... Schriftleiter: Dr. Hugo Fischer.
73-76: [as 71 but:] Schriftleiter Paul Kache.
77-83: [as 71 but:] Schriftleiter E. Dageförde.
84-86: [as 71 but:] Schriftleiter Dr. Robert Zander, Berlin.
87: Jan-Mar 1938, *Gartenflora*. Monatschrift für Garten- und Blumenkunde vereinigt mit den Veröffentlichungen der Deutschen Gartenbau- Gesellschaft/Schriftleitung Dr. Robert Zander, Berlin.
Ser. 2: Apr-Dec 1938 (no volume number, issued as a whole), *Gartenflora*/Neue Folge/Blätter für Garten- und Blumenkunde/1852 als Monatschrift von Eduard Regel begründet und als solche mit dem 1. Vierteljahr des 87. Jahrgangs 1938 abgeschlossen/April-Dezember 1938/Herausgeber und Selbstverlag Deutsche Gartenbau-Gesellschaft, Berlin. – Editor, as president of the Deutsche Gartenbau-Gesellschaft, Wilhelm Ebert.
Ser. 2: Jan-Dec 1939 (no volume number; issued as a whole).
Ser. 2: Jan-Dec 1940 (id.) id. – With this volume came the end of what was once an outstanding continental European garden journal. The preface ends with "Heil Hitler".

8776. *Die Schmarotzergewächse* und die mit denselben in Verbindung stehenden Pflanzen-Krankheiten. Eine Schilderung der Vegetationsverhältnisse der Epiphyten und Parasiten nebst Anleitung zur Kultur der tropischen Orchideen, Aroideen, Bromeliaceen und Farren und Schilderung der Krankheit des Weines und der Kartoffeln ... Zürich (Verlag von Friedrich Schulthess) 1854. Oct. (*Schmarotzergew.*).
Publ.: Jun-Aug 1854 (ÖbW 31 Aug 1854; Flora 14 Nov-14 Dec 1854; Bot. Zeit. 29 Sep 1854), p. [i-iv], [1]-123, [124], *1 pl*. (uncol. lith.). *Copies*: G, M, NY.
Ref.: Schlechtendal, D.F.L. von, Bot. Zeit. 12: 695-696. 1854.

8777. *Der Obstbau des Kantons Zürich*. Eine Aufzählung und Beschreibung der auf dem Landwirthschaftlichen Feste zu Stäfa im Herbst 1854 ausgestellten Apfelsorten nebst Anleitung zur Kultur der hochstämmigen Obstbäume als Festschrift für das Landwirth-

schaftliche Fest im Oktober 1855 herausgegeben von Verein für Landwirthschaft und Gartenbau im Kanton Zürich. Bearbeitet von Dr. E. Regel ... Zürich (Druck von H. Mahler) 1855. Oct. (*Obstbau Zürich*).
Publ.: 1855, p. [i]-viii, [1]-160. *Copy*:M. – See C. Regel (1942) for notes on the pomology of Regel.

8778. *Die Pflanze und ihr Leben* in ihrer Beziehung zum praktischen Gartenbau ... Zürich (Druck und Verlag von Friedrich Schulthess) 1855. Oct. (*Pfl. u.i. Leben*).
Publ.: 1855 (Bot. Zeit. 25 Nov 1855), p.[i]-xiv, [1]-437. *Copy*: M. – Other t.p.: "*Allgemeines Gartenbuch*. Ein Lehr– und Handbuch für Gärtner und Gartenfreunde ... Erster Band ...".
Ref.: Schlechtendal, D.F.L. von, Bot. Zeit. 13: 825-828. 1855 (rev.).

8779. *Zwei neue Cycadeen*, die im Botanischen Garten zu Petersburg kultivirt werden, nebst Beiträgen zur Kenntniss dieser Familie ... Moskau (In der Buchdruckerei der Kaiserlichen Universität) 1857. Oct.
Publ.: 1857, p. [i-ii], [1]-29, *pl. 3-4*. *Copies*: FI, G, HH, WU. – Reprinted and to be cited from Bull. Soc. Natural. Moscou 30(1): 163-191. 1857. – In his letter to F.A.W. Miquel (at U) of 26 Mar 1856 Regel asks Miquel's opinion on the material and mentions that his publication is "sous presse".

8780. *Florula ajanensis*. Aufzaehlung der in der Umgegend von Ajan wachsenden Phanerogamen und hoeheren Cryptogamen, nebst Beschreibung einiger neuer Arten und Beleuchtung anderer verwandter Pflanzen ... Moskwa (Gedruckt in der Universitaets-Buchdruckerei) 1858. Qu. (*Fl. ajan.*).
Co-author: Heinrich Sylvester Theodor Tiling (1818-1871).
Publ.: Jul-Dec 1858 (p. [ii]: imprimatur 12 Jul 1858 o.s.; ÖbZ 1 Jun 1859), p. [i-v], [1]-128, [i]-ix, [xi]. *Copies*: FI, G, HH.
Other issue: Jul-Dec 1858, p. [i-vi], [1]-128, [i]-ix, [ix, loose]. *Copy*: M. – "*Florula ajanensis* bearbeitet von Dr. E. Regel, ... und Dr. H. Tiling, ... (Aus den Nouveaux Mémoires ...) Moskwa 1858. Preprinted from Nouv. Mém. Soc. Natural. Moscou 11: 1-128. 1859. *Copies* (journal): MICH, MO(2). (p. [i], [1]-128, [i]-ix). – A "*Herbarium der Flora von Ajan*" was offered for sale by the authors in Bot. Zeit. 22:92. 25 Mar 1864.
Ref.: Schlechtendal, D.F.L von, Bot. Zeit. 17: 150-152. 23 Apr 1859 (rev.).
Erman, G.A., Arch. wiss. Kunde Russl. 19: 605-628. 1860 (extensive review).

8781. *Bericht über die erste Blumen– und Pflanzen-Ausstellung* vom 27. April bis Juni 4. Mai 1858 *in St. Petersburg*. St. Petersburg (Buchdruckerei der Kaiserlichen Akademie der Wissenschaften) 1858. Oct.
Publ: 1858, p. [1]-31. *Copy*: M. – Reprinted from St. Petersb. Zeit. 1858 (nos. 100-106) (journal n.v.). – See also Gartenflora 7: 205-216. 1858.

8782. *Vier noch unbeschriebene Peperomeen* des Herbariums des Kaiserlichen botanischen Gartens in St. Petersburg ... Mosquae [Moskva] (Typis Universitatis caesareae) 1859. Oct.
Publ.: Mar-Jul 1859 (censor 26 Mar 1859), p. [1]-4, *1 pl. Copy*: M. – Reprinted and to be cited from Bull. Soc. Natural. Moscou 31(2): 542-545. 1858 (publ. 1859).

8783. *Die Parthenogesis im Pflanzenreiche*. Eine Zusammenstellung der wichtigsten Versuche und Schriften über Samenbildung, ohne Befruchtung nebst Beleuchtung derselben nach eigenen Beobachtungen ... St. Petersburg (Commissionäre der Kaiserlichen Akademie der Wissenschaften: Eggers und Comp. in St. Petersburg, Leopold Voss in Leipzig) 1859. Qu. (*Parthenog. Pfl.-Reich*).
Publ.: Mar-Dec 1859 (mss. submitted 7 Jan 1859; imprimatur Mar 1859), p. [i-ii], [1]-48, *pl. 1-2* (uncol. liths.). *Copies*: G, M. – Issued as Mém. Acad. Sci. St. Petersburg ser. 7, 1(2). 1859.
Ref.: Anon., Bull. Soc. bot. France 6 (bibl.): 815-820. 1860.

8784. *Verzeichniss der* vom Herrn Paullowsky und Herrn von Stubendorf in den Jahren 1857 und 1858 *zwischen Jakutzk und Ajan gesammelten Pflanzen*: Ein Beitrag zur Flora

Ostsiberiens ... Moskau [Moskva] (In der Buchdruckerei der Kaiserlichen Universität) 1859. Oct. (*Verz. Jakutzk. Ajan. Pfl.*).
Co-authors: Louis Theodor Rach (1821-1859), Ferdinand Gottfried Maximilian Theobald von Herder (1828-1896).
Publ.: Jul-Dec 1859 (censor 27 Jun 1859), p. [1]-34. *Copy*: M.

8785. *Beobachtungen ueber Viola epipsila* Ledb. Von E. Regel. Moskau (In der Buchdruckerei der Kaiserlchen Universität) 1860. Oct.
Publ.: 1860 (censor p. 2: 17 Apr 1860), p [1]-4. *Copy*: G. – Reprinted and to be cited from Bull. Soc. imp. Natural. Moscou 1860(2): 535-538. 1860.

8786. *Catalogus plantarum* quae *in horto aksakoviano* coluntur [St. Petersburg ...] 1860. Oct. (*Cat. pl. hort. aksakov.*).
Publ.: 1860,(imprimatur: 3 Oct 1860 o.s.; ÖbZ 1 Feb 1861; Flora 28 Jun-7 Jul 1861), p. [i*-ii*], [i]-vii, [1]-148. *Copies*: BR, FI, G, H, HH, M, NY, US. – See Lipsky (1915) no. 66.

8787. *Tentamen florae ussuriensis* oder Versuch einer Flora des Ussuri-Gebietes nach den von Herrn R. Maack gesammelten Pflanzen bearbeitet ... St. Petersburg (Commissionäre der Kaiserlichen Akademie der Wissenschaften, in St. Petersburg Eggers et Comp., in Riga Samuel Schmidt, in Leipzig Leopold Voss) 1861. Qu. (*Tent. fl.-ussur.*).
Collector: Richard Maack (1825-1886), see TL-2/3: 206.
Publ.: Dec 1861 (t.p.: 1861; imprimatur: Dec 1861; KNAW rd. Jul-Sep 1862), p. [i]-xiii, [1]-228, [1-2], *pl. 1-12* (uncol. lith. J.A. Satory). *Copies*: G, HH, KNAW, M, MO, NY, USDA. – Issued as Mém. Acad. Sci. St. Pétersburg ser. 7, 4(4). 1861.
Russian version: 1862 (imprimatur 24 Jan 1862), p. [i*-ii*], [i]-xvi, [1]-282, [283], *pl. 1-12*. *Copy*: H. "*Opyt flory Usurijskoj strany*, sostavil, po materialam, sobrannym R. Maakom, E. Regel, ... S. Peterburg (v Tipografii v Bezobrazova i komp.) 1862. Qu. – In Russian, descriptions in Latin (Inf. P. Isoviita).

8788. *Monographia Betulacearum* hucusque cognitarum ... Mosquae [Moskva] (typis Universitatis caesareae) 1861. Qu. (*Monogr. Betul.*).
Publ.: Apr-Mai 1861 (imprimatur 6 Apr 1861; Flora 14-21 Aug 1861; Acad. Sci. Stockholm rd. before 11 Dec 1861), p. [i-iii], [1]-129, *pl. 4-17* (uncol. liths.). – *Copy*: BR, FI, G, H, HH, KNAW, L, M, USDA. – Preprinted or reprinted from Nouv. Mém. Soc. Natural. Moscou 13: [59]-187, *pl. 4-17* (id.]. 1861.
Ref.: Schlechtendal, D.F.L. von, Bot. Zeit. 19: 231-232. 9 Aug 1861 (rev.).

8789. *Uebersicht der Arten der Gattung Thalictrum*, welche im Russischen Reiche und den angraenzenden Laendern wachsen ... Moskou (in der Buchdruckerei der Kaiserlichen Universität) 1861. Oct. (*Uebers. Thalictrum*).
Publ: 1861 (imprimatur 27 Jun 1861, o.s.; Flora 28 Mar-31 Mai 1862), p. [i-iii], [1]-50, *pl. 1-3* (uncol. liths.). *Copies*: G, H, HH, KNAW, NY, WU. – Preprinted or reprinted from Bull. Soc. imp. Natural. Moscou 34(1): 14-63, *pl. 1-3*. 1861.

8790. *Reisen in den Süden von Ostsiberien* im Auftrage der Kaiserlichen Russischen geographischen Gesellschaft ausgeführt in den Jahren 1855-1859 durch G. Radde. Botanische Abtheilung. Nachträge zur Flora der Gebiete des Russischen Reichs östlich vom Altai bis Kamtschatka und Sitka, nach den vom G. Radde, Stubendorff, Sensinoff, Rieder und andern gesammelten Pflanzen bearbeitet von E. Regel. Band I. Moskau (in der Buchdruckerei der Kaiserlichen Universität) 1861 [1862]. Oct. (*Reis. Ostsib.*).
Publ.: Volume 1 of the *Reis. Ostsib.* (Bot. Abth.) was published by E.A. Regel. Volume 2 was never published; for vols. 3 and 4 see F.G.M.T. von Herder TL-2/2671.
1(1): Jan 1862 (p. ii: impr. 27 Dec 1861 o.s. [sic]; t.p. 1861 (o.s.); Flora 28 Mar-31 Mai 1862), p. [i]-vii, [1]-211. Reprinted from Bull. Soc. Natural. Moscou 34(2,3): 1-211. 1862 (journal issue approved for publ. 12 Jan 1862).
1(2): Oct 1862 (p. [ii]: impr. 13 Sep 1862 o.s.; Flora 20 Nov 1862-20 Jun 1863; Regel sent a copy to Miquel on 8 Nov 1862 o.s.), p. [i-ii], 213-447, *pl. 1-9* (uncol. liths.). Repr. from id. 34(3): 458-578, 35(1): 214-328. 1862.
Copies: B, BR, G, M, MO, NY, USDA (with general t.p. for vol. 1, dated 1862. – The

publication of 1(1) in the Bull. Soc. natural. Moscou may have been earlier or later; anyhow 27 Dec 1861 o.s. was Jan 1862 n.s., and both the reprint and journal issue came out at the earliest in the course of Jan 1862; see also Buchheim (1960) for further details, especially explaining the occurrence of copies with title pages dated 1861 and 1862. Later complete copies were evidently issued with an 1862 t.p.

8791. *Enumeratio plantarum in regionibus cis– et transiliensibus a cl. Semenovio anno 1857 Collectarum ....* Mosquae [Moskva] (Typis Universitatis caesareae) 1861. Oct.
Co-author: Ferdinand Gottfried Maximilian Theobald von Herder (1828-1896).
Publ.: 1864-1878; a series of papers in Bull. Soc. Natural. Moscou 1864-1878, in part also reprinted with independent pagination. Copies of reprints: H(1-2), HH(2-4), M(1-3).
[1]: 1864 (imprimatur 13 Aug 1864), p. [1]-43, *1 pl.*, from *Bull.* 1864(1): 383-425, *1 pl.*
2: 1866, p. [1]-159, *1 pl.*, from *Bull.* 39(2): 527-571, (3): 1-115. 1866.
3: 1868 (t.p. reprint dated 1868), p. [1]-88, *4 pl.*, from *Bull.* 40(1): 1-22. 1867, 40(3): 124-190. 1868.
4: 1869 (t.p. reprint), p. [1]-177, *1 pl.*, reprint, from *Bull.* 41(1): 59-113. 41(2): 378–459, *1 pl.* 41(4): 269-310. 1869.
Suppl. ["ii"] 1870 (t.p. reprint), p. [i-ii], [1]-47, from *Bull.* 43(2): 237-283. 1870. *Copy*: H.
Further suppl.: *Bull.* 43(3/4): 263-269. 1870, 1871(2): 356-381, 1878(2): 395-396. – TR mentions reprints with independent pagination, not seen by us. The supplements have an alternative title: *Plantae Severzovianae et Borszcovianae* (TR 11 39).

8792. *Sur la valeur de l'espèce* (*Ueber die Idée der Art*). [Amsterdam 1866]. Oct.
Publ.: 1866, p. [1]-39. *Copy*: M. – Reprinted and to be cited from Bull. Congr. int. Bot. Hort. Amsterdam 1865: 159-198. 1866.

8793. *Die Gattung Pleuroplitis* und Andropogon productus erläutert von Dr. E. Regel. [Mélanges biologiques tirés du Bulletin de l'Academie impériale des Sciences de St.-Pétersburg. Tome v.] 1866. Oct.
Publ.: 22 Mar (o.s.; n.s. 3 Apr) 1866, p. [741]-762, *1 pl.* (A. Münster, uncol. lith.). *Copy*: G. – Also published in Bull. Acad. Sci. St. Pétersbourg 10: 364-379. 1866.

8794. *Bemerkungen über die Gattungen Betula und Alnus* nebst Beschreibung einiger neuer Arten ... Moskau (in der Buchdruckerei der kaiserlichen Universität ...) 1866. Oct.
Publ.: Mar 1866 (in journal; journal issue presented to Society on 17 Mar 1866 o.s., see Bull. Soc. Natural. Moscou 39(1), séances, 26. 1866; inf. J.W. Landon; reprint probably published almost simultaneously; Bot. Zeit. rev. 12 Oct 1866; rd. by Acad. Sci. Stockholm before 13 Jun 1866), p. [i-v], [1]-47, *pl. 6-8* (uncol. liths.). *Copies*: G, H, M. – Reprinted from Bull. Soc. Natural. Moscou 38(2): 388-434. *pl. 6-8.* 1866.

8795. *Ruskii dendrologia* ... Sanktpetersburg ... 1870. Oct. (*Rusk. dendrol.*).
1: 1870, p. [1]-32.
2: 1871, p. [i-iv], 33-122.
3: 1873, p. [i-iv], 123-224.
4: 1874, [i-ii], 225-353, [i]-x, index.
5: 1879, p. [i], [354]-473, [i]-xv, index.
6: 1882, p. [i], [475]-542, [i]-iv, index (Bot. Zeit. 27 Oct 1882). *Suppl.* (n.v.): 1, 1883, p. 1-68; 2, 1889, p. 69-194.
Copy: USDA. – See Lipsky 1915, no. 150.

8796. *Animadversiones de plantis vivis* nonnulllis horti botanici imperialis petropolitanae [St. Petersburg 1871]. Oct.
[1]: 1871 (Flora 11 Nov-30 Dec 1871; Bot. Zeit. 9 Feb 1872), p. [1]-12. *Copies*: G, H, M, MO, NY. – Reprinted and to be cited from Acta Horti petrop. 1(1): 89-100. 1871.
[2]: Jan-Apr 1873 (G rd. Mai 1873; Bot. Zeit. 30 Mai 1873; BSbF Dec 1873; Flora 1 Mai-11 Jun 1873), p. [1]-22. *Copies*: G(2), H, M, US. – Reprinted and to be cited from id. 2(2): 305-326. 1873.

8797. *Revisio specierum Crataegorum*, Dracaenarum, Horkeliarum, Laricum et Azalearam [St. Petersburg 1871]. Oct.
Publ.: 1871 (in journal; BSbF Jul 1872; Flora 11 Nov-30 Dec 1871; Bot. Zeit. 9 Feb 1872), p. [1]-64. *Copies*: FI, G(2), H, HH, M, NY. – Reprinted and to be cited from Acta Horti petrop. 1(1): 101-164. 1871.

8798. *Conspectus specierum generis Vitis* regiones Americae borealis, Chinae borealis et Japoniae habitantium ... Petropoli [St. Petersburg] 1873. Oct.
Publ.: 1873 (BSbF after 17 Apr 1874; Flora 11-21 Sep 1873), p. [1]-11. *Copies*: BR, G(2), H, M, NY. – Reprinted and to be cited from Acta Horti petrop. 2: 389-399. 1873.

8799. *Putevoditel po imperatorskomu S.-Petersburgskomu botanicheskomu sadu* ... Sanktpeterburg (Tipografiya V.V. Pratc, ...) 1873. Oct. (*Putev. S. Peterb. bot. sad.*).
Publ.: 1873, p. [1]-147, map. *Copy*: H. – Guide to the imp. St. Petersb. botanical garden. (Inf. P. Isoviita).

8800. *Descriptiones plantarum novarum* in regionibus Turkestanicis a cl. viris Fedjenko, Korolkow, Kuschakewicz et Krause collectis cum adnotationibus ad plantas vivas in horto imperiali botanico petropolitano cultas. Fasciculus i. ... Petropoli 1873 M. Julio [St. Petersburg] 1873. Oct.
1: Jul 1873 (t.p. reprint), p. [1]-57. *Copies*: BR, G(2), H, HH, M, MO(2), USDA. – Reprinted from Acta Horti petrop. 401-457. 1873.
2: Jan-Feb 1874 (Bot. Zeit 2 Apr 1875: J. Bot. Jun 1875; BSbF after 20 Mai 1875; Flora 21 Feb-21 Mar 1875), p. [1]-72. *Copies*: BR, G, HH, M, MO, USDA. – Reprinted from id. 3: 97-168. 1874. – "*Descriptiones ... novarum et minus cognitarum in regionibus ... cultas*".
3: Jan-Jun 1875 (J. Bot. Oct 1875; Bot. Zeit. 30 Jul 1875), p. [1]-17. *Copies*: BR, G(2), HH, M, MO, USDA. – Reprinted from id. 3(2): 281-297. 1875. "*Descriptiones plantarum novarum et minus cognitarum. Fasciculus iii. Auctore E. Regel. Petropolis 1875*".
4: Jan-Sep 1876 (Bot. Zeit. 13 Oct 1876), p. [1]-68. *Copies*: G, HH, M, MO(2). – Reprinted from id. 4(1): 273-340. 1876. Other copies seen have original journal pagination.
5: Jan-Sep 1877 (Bot. Zeit. 19 Oct 1877), p. [1]-56. *Copies*: BR(2), G(2), HH, M, MO, USDA. – Reprinted from id. 5(1): 217-272. 1877.
6: Jan-Oct 1878 (Bot. Zeit. 8 Nov 1878; Not. Nov. Feb(2) 1879), p. [1]-72. *Copies*: BR, G, HH, M, MO(2), USDA. – Reprinted from id. 5(2): 575-646. 1878.
7: 1879 (Nat. Nov. Mar(1) 1880; Bot. Zeit. 3 Dec 1880), p. [1]-263. *Copies*: BR, G, M, US, USDA. – Reprinted from id. 6(2): 287-538. 1879.
Supplementum ad fasciculum vii: 1880 (Flora 1 Jul 1881), p. [1]-8. *Copies*: BR, G, M, MO(2), USDA. – Reprinted from id. 7: 381-388. 1880.
8: 1881 (BSbF Jan-Apr 1882; Bot. Zeit. 31 Mar 1882; Bot. Centralbl. 2-6 Jan 1882; Nat. Nov. Feb(2) 1882), p. [1]-150, 2 charts, 1 map. *Copies*: BR, G, MO, US. – Reprinted from id. 7: 541-690. 1881.
8 (suppl.): Jan-Mai 1883 (Acad. Sci. Stockholm rd. before 6 Jun 1883), p. [1]-11. *Copies*: BR, G, M, MO(2), USDA. – Reprinted from id. 8: 269-279. 1881.
9: late 1884 or Jan-Feb 1885 (Bot. Centralbl. 16-20 Mar 1885; BSbF after Mai 1885); Nat. Nov. Apr(2) 1885; Bot. Zeit. 26 Jun 1885), p. [i], [1]-64, *pl. 1-21* (uncol. liths. U. Dammer). *Copies*: BR, G, M, MO, NY, US, USDA. – Reprinted from id. 8(3): 641-702. 1884.
10: Jan-Aug 1886 (Bot. Zeit. 24 Sept 1886; Nat. Nov. Nov(2) 1886), p. [i-ii], [1]-46, *pl. 10* (uncol. U. Dammer). *Copies*: BR, G(2), M, MO(2), NY, USDA. – Reprinted from id. 9(2) 575-620. 1886. Subtitle: "Loci natales ab Alberto Regelo elaborati sunt".

8801. *Alliorum adhuc cognitorum monographia* ... Petropolis 1875. Oct. (*Allior. monogr.*).
Publ.: Jan-Mai 1875 (Bot. Zeit. 11 Jun 1875; Flora 1 Jun-1 Jul 1875), p. [i], [1]-266. *Copies*: BR, FI, G, H, HH, M, NY(2), PH; IDC 5151. – Also published as Acta Horti petrop. 3(2): 1-266. 1875.
Ref.: G.K., Bot. Zeit. 33: 753. 12 Nov 1875 (rev.).

REGEL, E., 1876.

8802. *Cycadearum generum specierumque revisio* ... St. Petersburg (...) 1876. Oct.
Publ.: 1876 (BSbF after 10 Apr 1877; KNAW rd. Jan 1877), p. [1]-48. *Copies*: FI, G, H, M, NY. – Reprinted and to be cited from Acta Horti petrop. 4: 273-320. 1876.

8803. *Tentamen rosarum monographiae* ... St. Petersburg (...) 1877. Oct. (*Tent. ros. monogr.*).
Publ.: late 1877 (preprint so dated) or perhaps Jan-Feb 1878 (Flora 21 Feb-1 Apr 1878, Bot. Zeit. 22 Mar 1878, BSbF after Aug 1878; J. Bot. Oct 1878; KNAW rd. Mar 1878), p. [1]-114. *Copies*: BR, FI, G(2), H, HH, KNAW, NY, PH, US, USDA; IDC 5152. – Preprinted from Acta Horti petrop. 5(2): 285-398. 1878.
Ref.: Crépin, F., Bull. Soc. Bot. Belg. 16: 21-30. 1877 (1878) (reprint p. [1]-10).

8804. *Descriptiones plantarum novarum rariorumque* a cl. Olga Fedtschenko in Turkestania nec non in Kokania lectarum auctore E. Regel [St. Petersburgh 1882]. Qu.
Publ.: Jan-Mar 1882 (Nat. Nov. Apr(2) 1882; Bot. Zeit. 26 Mai 1882), p. [i-v], [1]-89. *Copies*: BR, FI, G, HH. – Published as Izv. Imp. Obšč. Ljubit. Estesv. Moskovsk. Univ. 34(2). 1882.

8805. *Monographia generis Eremostachys* auctore E. Regel. Loci natales ab Alberto Regel elaborati sunt. Petropoli [St. Petersburg] 1886. Oct. (*Monogr. Eremostachys*).
Publ.: Mai-Jul 1886 (Bot. Centralbl. 23-27 Aug 1886; Nat. Nov. Aug(3) 1886; Bot. Zeit. 30 Jul 1886; Bot. Jahrb. 2 Nov 1886), p. [1]-48, *pl. 1-9* (uncol. liths. U. Dammer). *Copies*: BR, FI, G, H, HH, M, MICH, NY, USDA. – Preprinted or reprinted from Acta Horti petrop. 9(2): 527-574. 1886.

8806. *Descriptiones plantarum* nonnullarum *horti imperialis botanici* in statu vivo examinatarum ... Petropoli [St. Petersburg] 1887. Oct.
Publ.: 1887, p. [1]-15. *Copies*: BR, FI. – Reprinted and to be cited from Acta Horti petrop. 10(1): [363]-377. 1887.

8807. *Allii species Asiae centralis* in Asia media a Turcomania desertisque aralensibus et caspicis usque ad Mongoliam crescentes ... Petropoli [St. Petersburg] 1887. Oct.
Publ: Sep-Dec 1887 (Bot. Zeit. 30 Dec 1887; BSbF after 3 Mar 1888; Nat. Nov. Mar(1) 1888), p. [1]-87, [88], *pl. 1-8* (monochr. liths. H. Bahr). *Copies*: BR, FI, G, H, L, M, MO, NY(2), USDA. – Reprinted and to be cited from Acta Horti petrop. 10: 279-362. 1887.

8808. *Descriptiones et emendationes plantarum* in horto imperiali botanico petropolitano cultarum ... Petropoli [St. Petersburg] 1889. Oct.
Publ.: 1889, p. [1]-14. *Copies*: BR, FI, H, M, NY, US, USDA. – Reprinted and to be cited from Acta Horti petrop. 10: 685-698. 1889.

8809. *Descriptiones plantarum* nonnullarum *horti* imperialis *botanici petropolitani* in statu vivo examinatarum ... [Acta Horti petropolitani, vol. xi, no. 8. 1890]. Oct.
Publ.: 1890, p. [1]-16. *Copy*: NY. – Reprinted and to be cited from Acta Horti petrop. 11(8): 299-314. 1890.

Regnault, Nicolas François (1746-?), French physician and botanist. (*Regnault*).

HERBARIUM and TYPES: Unknown.

BIBLIOGRAPHY and BIOGRAPHY: BM 4: 1666; DU 256; GF p. 62; Kew 4: 436; NI 1600, Nachtr. p. 57; Plesch p. 378; PR 7475, ed. 1: 8421; Rehder 3: 51; Sotheby 634.

8810. *La botanique* mise à la portée de tout le monde ou collection des plantes d'usage dans la médecine, dans les alimens et dans les arts. Avec des notices instructives puisées dans les auteurs les plus célèbres. Contenant la description, le climat, la culture, les propriétés et les vertus propres à chaque plante. Précédé d'une introduction a la botanique. Ou dictionnaire abrégé des principaux termes employés dans cette science ... Paris (chez l'auteur ...) 1774. Fol., 3 vols. (*Botanique*).

Co-author: Geneviève de Nangis Regnault (1746-1892).
1: 1774, p. [i-vi], *pl. 3, pl. 1-295* (col. copp.).
2: 1774, p. [i], [1-295].
3: 1774 [1780], p. [i*], [i]-ii, [1]-12, *pl. 1-174* (id.), p. [1-174].
Copies: BR (incompl.), HH (incompl.), NY (472 pl. as above). – Copies vary in binding (2, 3, 4 vols. text separately or with plates, see BM 4: 1666). The "Table des noms", p. [1]-3, in vol. 3 has a "privilège du Roi" dated 9 Dec 1780. For recent copies offered for sale see Wheldon & Wesley, Cat. 139, no. 65. 1979. (2.400), Sotheby 634, 635 and Junk, Cat. 214, no. 252. 1979 (Hfl. 22.000.-). – See NI Nachtr. p. 57 for a partial publication as "Recueil de plantes". The BM copy has *465 pl.*; Brunel mentions *475 plates*, Blunt (GF) *472* (+ 1 title pl.); Sotheby (634) settles for *472 pl.* (excl. titles) and mentions an additional printed leaf /La botanique ... ordre de la distribution...".

Regnell, Anders Fredrik (1807-1884), Swedish physician and botanist; Dr. med. Uppsala 1837, Dr. phil. h.c. Uppsala 1877; practicing in Stockholm 1836-1839; settled in Brazil 1840 as physician, from 1841-1884 at Caldas, Minas Geraes, collecting from 1841-1874 in Minas Geraes and São Paulo; left a fortune to be spent on science. (*Regnell*).

HERBARIUM and TYPES: S (including 25.000 lichens); duplicates in B, BP, BR, C, E, F, FI, G, GH, GOET, H, HAL, K, LCU, LD, LE, M, NY, O, P, PC, PH, R, U, UPS, US, W, WRSL, WU, Z. – Letters to D.F.L. von Schlechtendal at HAL.

NOTE: (inf. O. Almborn) Regnell's estate was left to subsidize botanical work in tropical South America and was used to finance eleven "Regnellian expeditions" and also to finance the working up of botanical collections from S. America at S and elsewhere. The Regnell herbarium (at S) consisted originally of R.'s own collections and those of Gustaf Anders Lindberg (1832-1900, coll. 1854-1855), Salomon Eberhard Henschen (1845, coll. 1868-1869), Carl Wilhelm Hjalmar Mosén (1841-1887, coll. 1873-1874) and J.F. Widgren (1810-1883). It is now kept as a separate division including all collections at S from "Latin America". See Sparre 1969 and Sayre 1975 for further details and the index to our vol. 3 (e.g.) for references to publications on material collected on Regnellian expeditions. These expeditions were:

1. C.A.M. Lindman and G.O. Malme, 1892-1894, Brazil, Paraguay.
2. G.O. Malme, 1901-1903, Argentina, Paraguay.
3. E. Ekman, 1914-1931, Cuba, Hispaniola.
4. E. Asplund, 1939-1940, Ecuador.
5. F. Fagerlind, 1952-1953, Ecuador.
6. E. Asplund, 1955-1956, Ecuador.
7. G. Harling, 1958-1959, Ecuador.
8. U. Eliasson, 1966-1967, Ecuador.
9. B. Sparre, 1966-1967, Ecuador.
10. U. Eliasson, 1976, Ecuador.
11. R. Santesson et al., 1981, Peru; L. Andersson 1981, Ecuador; B. Sparre, 1981, Ecuador.

BIBLIOGRAPHY and BIOGRAPHY: Barnhart 3: 138; BM 4: 1666; Bossert p. 327; IH 1 (ed. 2): 114, (ed. 3): 146, (ed. 6): 363, (ed. 7): 340, 2: (in press); KR p. 583-584 (b. 8 Jun 1807, d. 12 Sep 1884); Lasègue p. 483, 505; Morren ed. 2, p. 34, ed. 10, p. 139; TL-2/3425; see Henschen, S.E., see index TL-2/3; Urban-Berl. p. 311.

BIOFILE: Anon., General reg. Bot. Not. 186. 1939; Bot. Not. 1884: 168-169 (obit., b. 7 Jun 1807, d. 12 Sep 1884); Bot. Zeit. 1: 222. 1843 (collections from Villa de Caldas, Brazil received).
Candolle, Alph. de, Phytographie 443. 1880.
Dahlgren, K.V.O., Sv. bot. Tidskr. 56(3): 465-470. 1962 (on G.A. Lindberg, Henschen and Regnell in Brazil).
Fries, R.E., Short hist. bot. Sweden 40-41, *pl. 2*. 1950.
Key, A., Lefnadsteckn. K. Sv. Vet.-Akad. Ledam. 3(1): 97-159. 1891 (extensive biogr.).
Lindman, S., Svenska man och kvinnor 6: 227-228. 1949 (portr.).

Rodrigues, J.B., Vellosia ed. 2: 48. 1891 (biogr. sketch).
Sandahl, O.T., Entom. Tidskr. 5: 191-192, 228. 1884 (d.).
Sayre, G., Mem. New York Bot. Gard. 19(3): 384-385. 1975.
Sparre, B., Fauna o. Flora 64: 304-317. 1969 (on the Regnell herb. at S.; 500.000 sheets; 10.000 types; portr.; details on other collectors).
Stieber, M.T., Huntia 3(2): 124. 1979 (mentioned in Chase mss.).
Urban, I., Fl. bras. 1(1): 86. 1906.
Verdoorn, F., ed., Chron. bot. 1: 89, 220. 1935, 3: 310. 1937, 4: 87. 1938.
Wagenitz, G., Index coll. princ. herb. Gott. 132. 1982.
Wittrock, V.B., Acta Horti Berg. 3(2): 182, *pl. 36*. 1903, 3(3): 203. 1905.

COMPOSITE WORKS: *Plantae regnellianae*, Linnaea 2: 511-512. Aug 1849, 513-583. Sep 1849, 23: 443-446. Feb 1850; authors: G. Bentham, F.A.W. Miquel, O.W. Sonder, J.C. Bauer, H.R.F. Grisebach, C.H. Schultz Bip., G. Kunze, E. Hampe.
For further treatments of Regnell material see TL-2/3425, and entries in name-indexes for volumes 3 and 4.

EPONYMY: *Neoregnellia* Urban (1924); *Regnellia* Rodrigues (1877), *Regnellidium* C.A.M. Lindman (1904); see also TL-2/4670.

Reguis, J. Marius F. (1850-?), French physician and mycologist at Allauch nr Marseille, later at Avignon. (*Reguis*).

HERBARIUM and TYPES: Unknown.

BIBLIOGRAPHY and BIOGRAPHY: Barnhart 3: 138; BL 2: 192, 706; BM 4: 1666; CSP 18: 100; Kelly p. 108; Morren ed. 10, p. 65; Plesch p. 379; Rehder 1: 74, 712.

BIOFILE: Cavillier, F.G., Boissiera 5: 75. 1941 (bibl.).

8811. *Nomenclature franco-provençale des plantes* qui croissent dans notre région ... précédée d'un rapport de M. Achintre [Aix (Marius Illy) 1878]. Oct.
Publ.: 1878 (BSbF Nov-Dec 1878 as of "1877"), p. [i], [1]-186. *Copy*: USDA. – Reprinted from Mém. Acad. Sci. Agr. Arts Lettres Aix 11: 1-186. 1878.

8812. *Synonymie provençale des champignons de Vaucluse* ... Marseille (Librairie Bérard, ...) 1886. Oct. (*Syn. provenç. champ. Vaucluse*).
Publ.: 1886 (p. 5: 1 Jan 1886; Nat. Nov. Feb(2) 1887; Bot. Centralbl. 28 Feb 4 Mar 1887; Bot. Zeit. 25 Feb 1887), p. [1]-143. *Copies*: BR, FH.

8813. *Les champignons de la Provence* & du Gard considérés dans leurs rapports avec l'hygiène, la médecine, l'agriculture et l'industrie ... Paris (Librairie J.-B. Baillière et Fils ...) [1895]. Qu. †. (*Champ. Provence*).
Fasc. 1 (all publ.): 1895; p. [i-iii], 1-88. *Copies*: G, Stevenson. – Three fascicles were planned.

Rehder, Alfred (1863-1949), German (Saxony)-born botanist, dendrologist and gardener; at the Zwickau Gymnasium 1876-1881; apprentice to his father, Paul Julius Rehder, director of the park in Waldenburg, Saxony, 1881-1884; studied in Berlin 1884-1886, worked at Frankfurt a.M. and Muskau 1886-1888; head gardener at Darmstadt 1888-1889; id. at Göttingen 1889-1895; associate editor of Möller's Gärtner-Zeitung 1895-1898; to the United States 1898; assistant at the Arnold Arboretum, Jamaica Plain, Mass. 1898-1918; curator of the herbarium ib. 1918-1940; associate professor of dendrology at Harvard University 1934-1940; emeritus professor ib. 1940-1949. (*Rehd.*).

HERBARIUM and TYPES: A; further material at B, C, DPU, GH, NY, US. – Archival material at FH, HH and SIA.

BIBLIOGRAPHY and BIOGRAPHY: AG 4: 358, 5(3): 65; Backer p. 659; Barnhart 3: 138 (b. 4 Sep 1863); BFM 2347, 3012, 3269; BJI 2: 143; BL 1: 4, 121, 314, 2: 707 (index); Biol.-

Dokum. 14: 7356-7357; Bossert p. 328 (d. 21 Jul 1949); CSP 18: 101; Ewan (ed. 1): 288, (ed. 2): 181; GR p. 123; Hirsch p. 244; Hortus 3: 1202; IH 1 (ed. 3): 78; Kew 4: 437-438; Langman p. 618; Lenley p. 341; MW p. 408-410, suppl. p. 291-292; NI 1: 256; NW p. 56; PH 450; Rehder 5: 712-713; SIA 221, 223, 226; SIA-HI 107; SO 3698; Tucker 1: 588; Urban-Berl. p. 386; Zander ed. 10, p. 706, ed. 11, p. 805.

BIOFILE: Anthony, S.W., J. Forestry 48: 338. 1950 (brief obit.).
Boerner, F., Ber. deut. bot. Ges. 68a: 164-166. 1955 (obit.).
Bijhouwer, J.T.P., Jaarb. Ned. Dendrol. Ver. 17: 23-24. 1950 (obit., portr.).
Cattell, J.M., Amer. men Sci. ed. 2: 387. 1910, ed. 3: 565-566. 1921, ed. 4: 806. 1927, ed. 5: 920. 1933, ed. 6: 1165. 1938, ed. 7: 1456. 1944.
Ewan, J. et al., Short hist. bot. U.S. 136. 1969.
Faull, J.H. & C.E. Kobuski, J. Arnold Arb. 21(4): i, portr., 1940 (note on R's retirement on 31 August 1940).
Fletcher, H.R., Story R. Hort. Soc. 344, 348. 1969.
Kobuski, C.E., J. Arnold Arb. 31(1): 1-38. 1950 (obit., eponymy, bibl. by Anneliese Rehder).
Leroy, J.F. et al., Bot. franç. Amér. nord 257, 258, 259. 1957.
Merrill, E.D., Enum. Philipp. pl. 4: 220. 1926.
Pease, A.S., Fl. N. New Hampshire 37. 1964.
Rehder, G., Arnoldia 32(4): 141-156. 1972 (mainly on Jacob Heinrich Rehder, grandfather of A. Rehder and gardener in charge of developing the park in Muskau (Saxony) and on Paul Julius Rehder, Alfred's father and on the latter's youth in Germany; "the making of a botanist").
Rodgers, A.D., Liberty Hyde Bailey 247, 295, 433, 485. 1949.
Rollins, R.C., Taxon 1: 3-4. 1951 (obit., d. 21 Jul 1949).
Schneider, C., Bot. Jahrb. 74: 633. 1949 (obit.).
Shepler, D., The Boston Herald, Sunday 26 Mai 1935 (portr., biogr. sketch).
Sutton, S.B., Charles Sprague Sargent 379 [index] 1970.
Veendorp, H. & L.G.M. Baas-Becking, Hort. acad. lugd.-bot. 216 [index]. 1938.
Venema, J.H. & J.T.P. Bijhouwer, Jaarb. Ned. Dendrol. Ver. 17: 23- 24. 1949 (portr.).
Verdoorn, F., ed., Chron. bot. 1: 164, 167, 298, 314. 1935, 2: 34, 35, 53, 185, 318, 330, 373. 1936, 3: 27, 294. 1937, 4: 37, 557, 561. 1938, 5: 299, 302, 371. 1941, 6: 70, 166. 1941.

COMPOSITE WORKS: (1) Founder of the *Journal of the Arnold Arboretum*, editor of vols. 1-21, 1919-1940.
(2) For contributions to L.H. Bailey, *Cycl. Am. Hort.* and *Stand. Cycl. Hort.* see J. Arnold Arb. 31: 17-32. 1950.
(3) For contributions to C.S. Sargent, *Trees and Shrubs*, see J. Arnold Arb. 31: 24-26. 1950.

EPONYMY: *Rehdera* Moldenke (1935); *Rehderodendron* Hsu-Hu (1932); *Rehderophoenix* Burret (1936).

8814. *Synopsis of the genus Lonicera* ... (from the fourteenth Annual Report of the Missouri Botanical Garden) 1903. Oct.
Publ.: 8 Oct 1903 (date on cover reprint), cover (with above title), p. 27-232, *pl. 1-20* (1-4 uncol. auct.; 5-20 phot.). *Copies*: BR, G, MICH(2), NY, US.
Ref.: Buchenau, F.G.P., Bot. Zeit. 62(2): 59-61. 1904 (rev.).

8815. Publication of the Arnold Arboretum, No. 3. *The Bradley bibliography* a guide to the literature of the woody plants of the world published before the beginning of the twentieth century compiled at the Arnold Arboretum of Harvard University under the direction of Charles Sprague Sargent by Alfred Rehder ... Cambridge (printed at the Riverside Press) 1911-1918, 5 vols. (*Bradley bibl.*) ("Rehder" in TL-2).
1: 1911 (Nat. Nov. Nov(1) 1911), p. [i]-xii, [xiii-xiv], [1]-566.
2: 1912 (Nat. Nov. Jan(1) 1913), p. [i]-vi, [vii-viii], [1]-926.
3: 1915, p. [i]-x, [xi-xii], [1]-806.
4: 1914, p. [i]-xiii, [xv-xvi], [1]-589.

5: 1918, p. [i]-xxxii, [1]-1008.
Copies: BR, FAS, G, NY, USDA.
Facsimile reprint: 1976, as original, Koenigstein-Ts. (Otto Koeltz ...) 1976. ISBN 3-87429-106-5 (-107-3, 108-1, 109-x, 110-3). *Copy*: FAS. – See Taxon 25: 657. 1976, 26: 125. 1977.

8816. Publications of the Arnold Arboretum, no. 9. *A monograph of Azaleas* Rhododendron subgenus Anthodendron ... printed from the income of the William L. Bradley Fund by the University Press Cambridge 1921. Oct. (*Monogr. Azaleas*).
Senior author: Ernest Henry Wilson (1876-1930).
Publ.: 15 Apr 1921 (t.p.), p. [i-ix], 1-219. *Copies*: BR, G, US, USDA.

8817. *Manual of cultivated trees and shrubs* hardy in North America exclusive of the subtropical and warmer temperate regions ... New York (The Macmillan Company) 1927. 16-mo. (*Man. cult. trees*).
Ed. 1: Jan 1927 (p. xvi: Sep 1926; p. ii: Jan 1927; US rd. 2 Mai 1927), p. [i]-xxxvii, 1-930. *Copies*: BR, MICH, MO(2), NY, U, US.
Corrections and emendations: 1935 (p. [ii]: Feb 1935), p. [i-ii], [1]-19. *Copies*: MO, NY, USDA. – Jamaica Plain, Mass. 1935.
Reissue ed. 1: Sep 1934, p. [i]-xxxvii, 1-930. *Copy*: U.
Reprint ed. 1: Oct 1937, p. [i-xxxvii], 1-930. *Copy*: USDA.
Ed. 2 : 1940, frontisp., p. [i]-xxx, 1-996. *Copies*: MO, NY, PH, USDA.
Ed. 2, fifth printing: 1951, id. *Copy*: BR.
Ed. 2, seventh printing: 1956, id. *Copy*: US.
Ed. 2, eleventh printing: 1967, id. *Copy*: USDA.
Ref.: Hastings, G.T., Torreya 40: 170-171. 3 Oct 1940 (rev. ed. 2).
Rendle, A.B., J. Bot 65: 234-235. 1927 (rev. ed. 1).
Taylor, N., Torreya 27: 36-37. 1927 (rev. ed. 1).

8818. *Amendments to the International rules of nomenclature*, ed. 3 ... Arnold Arboretum, Harvard University [Jamaica Plains, Mass., 1934]. (*Amendm. Int. rules nomencl.*).
Publ.: 10 Dec 1934 (p. [4]), p. [1-4]. *Copy*: FAS, U.

8819. *Bibliography of cultivated trees and shrubs* hardy in the cooler temperate of the northern hemisphere ... Jamaica Plain, Massachusetts, U.S.A. (The Arnold Arboretum of Harvard University) 1949. (*Bibl. cult. trees*).
Publ.: 1949, p. [i]-xl, [1]-825. *Copies*: BR, G, MO, NY, PH.
Facsimile ed.: 1978, Koenigstein/Ts. (Otto Koeltz ...), ISBN 3-87429-128-6. *Copy*: FAS. – see J.A. Leussink, Taxon 27: 424. 1978.

Rehm, Heinrich [Simon Ludwig Friedrich Felix] (1828-1916), German (Bavarian) physician, mycologist and lichenologist; studied at Erlangen, München and Heidelberg; Dr. med. 1852; medical studies Praha and Wien 1853-1854; practicing physician at Dietenhofen nr. Ansbach (1854), Sugenheim (1857), Windsheim (1871), "Bezirksgerichtsarzt" (regional medical examiner) Lohr a.M. (1875), id. Regensburg (1878); retired 1898; settled in München-Neufriedenheim 1899; Medizinalrat 1888. (*Rehm*).

HERBARIUM and TYPES: S; parts of herbarium also at M, WU and Z.
Exsiccatae: (1) *Ascomycetes exsiccatae* (1868-1917), 57 fasc., 2175 nos., for schedae see below; fasc. 56 and 57 by F. v. Hoehnel). Sets at: B (extant), BM, BPI, BR, BRU, C, CUP, DAOM, DBN, FH, GOET, H, HAL, IBF, ILL, KIEL, L, M, MICH, MSTR, NCU, NY, OS, PAD, PH, PRE, RSA, S, SOLH, UMO, W, WIS.
(2) *Cladoniae exsiccatae* (1869-1895, 13 fasc., nos. 1-440 of which 151-440 by T.C.G. Arnold; further details: Arnold 1895, Lynge 1919, Sayre 1969). Sets at: B, FH, GOET, H, L, M, MSTR, NY, S, WRSL.
(3) Contributed to P. Sydow, *Mycotheca germanica* and to T. Vestergren, *Micromycetes rariores selecti*.
For an advertisement announcing the sale of the botanical collections left by Rehm upon his death see Herbarium 42: 405. 1916. They included collections of Algae, Characeae, Musci, Hepatics and a phanerogam herbarium of over 5500 species contain-

ing many regular series of exsiccatae issued by other botanists. – Letters to W.G. Farlow and R. Thaxter at FH.

BIBLIOGRAPHY and BIOGRAPHY: Barnhart 3: 139 (b. 20 Oct 1828, d. 1 Apr 1916); BJI 1: 47; BM 4: 1666, 8: 1059; Bossert p. 328; CSP 5: 134, 8: 716, 11: 128, 12: 602, 18: 101; DTS 1: 239-240, 3: xlix-l, Nachtr. 2; GR p. 39, *pl.* [*16*]; Hawksworth p. 185; IH 1 (ed. 6): 363, 2: (in press); Kelly p. 188-189, 256; Kew 4: 438; KR p. 585; Lenley p. 341; LS 1100, 7998, 21918-21962, 38295-38302; Moebius p. 265; Morren, ed. 10, p. 1884; MW p. 411; Quenstedt p. 355; Stevenson p. 1255; Urban-Berl. p. 281, 290.

BIOFILE: Anon., Ber. bayer. bot. Ges. 9: opp. t.p. 1905 (portr.).; Hedwigia 39: (71). 1900 (to München), 42: (162). 1903 (50 yrs. Dr. med.), 58: (88). 1916 (d.); J. Mycol. 2: portr. opp. p. 49. 1905; Mycologia 8: 222. 1916 (d.); Österr. bot. Z. 66: 212. 1916 (d.); Verh. bot. Ver. Brandenburg 58: 215, 677. 1916 (d.).
Arnold, F., Dr. H. Rehm Cladoniae exsiccatae 1869-1895, 1895, 34 p. (suppl. to Ber. Bay. bot. Ges. 4, 1895) (revised nomenclature); Ber. bayer. bot. Ges. 16: 10-13. 1917 (obit., portr., bibl.).
Kneucker, A., Allg. bot. Z. 22: 96. 1916 (d.).
Laundon, J.R., Lichenologist 11(1): 16. 1979 (lich. S).
Lynge, B., Nyt Mag. Naturvid., Oslo 51: 114. 1913 (Ascom. exs.), 56: 449-455. 1919 (Cladoniae exsiccatae, list of all 440 nos. distributed).
Pongratz, L., Acta Albertina ratisb. 25: 114-115. 1963 (biogr. portr.).
Sayre, G., Mem. New York Bot. gard. 19(1): 157-158. 1969 (on Cladoniae exsiccatae).
Wagenitz, G., Index coll. princ. herb. Gott. 133. 1982.

EPONYMY: *Discorehmia* Kirschstein (1936); *Neorehmia* F. von Höhnel (1902); *Rehmia* Krempelhuber (1861); *Rehmiella* Winter (1883); *Rehmiellopsis* Bubák et Kabát ex Bubák (1910); *Rehmiodothis* Theissen et H. Sydow (1914); *Rehmiomycella* E. Müller (1962); *Rehmiomyces* Hennings (1904); *Rehmiomyces* (P.A. Saccardo et P. Sydow) H. Sydow (1904).

8820. *Beiträge zur Flechten-Flora des Allgäu* [Augsburg] 1863. Oct.
Publ.: 1863 (p. 2: Apr 1863), p. [1]-44. *Copies*: Almborn, FH. – Ber. nat. Ver. Augsburg 16: 85-128. 1863.
Weitere Beiträge: 1864 (p. [1]: Mai 1864), p. [1]-9. *Copies*: Almborn, FH. – Ib. 17: [91]-99. 1864.
Dritte Beiträge: 1867 (p. [1]: Feb 1867), p. [1]-5. *Copies*: Almborn, FH. – Ib. 19: [89]-93. 1867.

8821. *Ascomyceten* [fasc. i-xi]. In getrockneten Exemplaren herausgegeben von Dr. med. Rehm. (Separatabdruck aus dem 26. Berichte des Naturhistorischen Vereins in Augsburg) [1881]. Oct.
Publ.: 1881, p. [1]-132. *Copies*: FH, NY. – Reprinted and to be cited from Ber. nat. Ver. Augsburg 26: 1-132. 1881. Contains the text accompanying fasc. 1-11 of the *Ascomycetes exsiccatae* (1868/1879). For detailed treatment of this series and its accompanying texts see J. Stevenson, Beih. Nova Hedw. 36: 316-337. 1971.
Diagnoses and notes: G. Winter, Flora 55: 508-511, 523-527, 542-544. 1872 (account of fasc. 1 and 2).
Texts for fascicles (all by H. Rehm):
12: Hedwigia 20: 33-42, 49-54. 1881, reprint p.1]-16. "Ascomyceten".
13: ib. 21: 65-76, 81-86. 1882, reprint p. [1]-16.
14: ib. 22: 33-41, 52-61. 1883, reprint p. [1]-18.
15: ib. 23: 49-57, 69-77. 1884, reprint p. [1]-17.
16: ib. 24: 7-17, 66-72. 1885.
17: ib. 24: 225-246. 1885, reprint p. [1]-22.
18: ib. 26: 81-98. 1887, reprint p. [1]-18.
19: ib. 27: 163-175. 1888, reprint p. [1]-13.
20: ib. 28: 347-358. 1889.
21: ib. 30(5): 250-262. 1891.
22: ib. 31(6): 299-313. 1892 ("Ascomycetes").

23: ib. 34: (159)-(165). 1895.
24: ib. 35: (145)-(151). 1896.
25: ib. 37: (142)-(144). 1898.
26: ib. 38: (242)-(246). 1899.
27: ib. 39: (192)-(193). 1900.
28: ib. 40: (101)-(106). 1901.
29: ib. 41: (202)-(206). 1902.
30: ib. 42: (289)-293). 1903.
31: ib. 42: (347)-(349). 1903.
32: ib. 43: (31)-(33). 1904.
33: Ann. mycol. 2: 515-521. 1904.
34: ib. 3: 324-331. 1905.
35: ib. 3: 409-417. 1905.
36: ib. 4: 64-71. 1906.
37: ib. 4(5): 404-411. 1906.
38: ib. 5(1): 78-85. 1907.
39: ib. 5(6): 207-213. 1907.
40: ib. 5: 465-473. 1907.
41: ib. 6: 116-124. 1908.
42: ib. 6(5): 485-491. 1908, repr. [1]-7.
43: ib. 7: 134-140. 1909.
44: ib. 7(5): 399-405. 1909.
45: ib. 7(6): 524-530. 1909.
46: ib. 8: 298-304. 1910.
47: ib. 9(1): 1-7. 1911.
48: ib. 9(3): 286-290. 1911.
49: ib. 10: 54-59. 1912.
50: ib. 10: 353-358. 1912.
51: ib. 535-541. 1912.
52: ib. 11: 166-171. 1913.
53: ib. 11: 391-395. 1913.
54: ib. 12: 165-170. 1914.
55: ib. 12: 170-175. 1914.
56/57: ib. 16: 209-224. 1918.
Copies: FH, M, NY.

8822. *Beiträge zur Ascomyceten-Flora der deutschen Alpen* und Voralpen [Hedwigia 1882, nos. 7, 8). Oct.
Publ.: Jul-Aug 1882(in journal; repr.: Bot. Centralbl. 11-15 Sep 1882), p. [1]-17. *Copy*: NY. – Reprinted and to be cited from Hedwigia 21: 97-103. Jul 1882, and 21: 113-123. Aug 1882.

8823. *Ascomycetes lojkani* lecti in Hungaria, Transsylvania et Galicia ... Budapestini (sumptibus auctoris), Berolini (apud R. Friedländer) 1882. Oct. (*Ascomyc. lojk.*).
Collector: Hugo Lojka (1844-1887).
Publ.: Dec 1882 (p. [iv]: 15 Nov 1881; Nat. Nov. Jan(1) 1883; Flora 21 Mar 1883; Hedwigia Mar 1883; Bot. Zeit. 26 Jan 1883; Bot. Centralbl. 8-12 Jan 1883), p. [i-iv], [1]-70. *Copies*: FH, H, NY, Stevenson.
Ref.: Winter, H.G., Bot. Centralbl. 14: 162. 7-22 Mai 1883.

8824. *Revision der Hysterineen im herb. Duby* [Dresden (C. Heinrich) 1886]. Oct.
Publ.: Jul-Oct 1886 (in journal; p. 2: 1 Jun 1886), p. [1]-48. *Copy*: NY. – Reprinted and to be cited from Hedwigia 25: 137-155. Jul-Aug 1886 and 25: 173-202. Sep-Oct 1886.

8825. *Die Pilze Deutschlands*, Oesterreichs, und der Schweiz. III. Abtheilung: *Ascomyceten*: *Hysteriaceen und Discomyceten*, bearbeitet von Dr. H. Rehm,... Leipzig (Verlag von Eduard Kummer) 1896. Oct. (*Pilze Deutschl., Hyster. Discomyc.*).
Publ.: Lieferung 28-44, 53-55 of Band 1 of Dr. L. Rabenhorst's Kryptogamen-Flora von Deutschland, Oesterreich und der Schweiz, ed. 2, 1(3), 1887-1896. For details see

above under Rabenhorst, Krypt.-Fl. ed. 2, TL-2/8450, p. [ii]-viii, [1]-1275, [1]-57.
Copies: M, U, USDA.
Ref.: Magnus, P., Bot. Zeit. 54: 200-201. 1 Jul 1896 (rev.).

8826. *Ascomycetes fuegani* a P. Dusén collecti ... Stockholm (Kungl. Boktryckeriet. P.A. Norstedt & Söner) 1899. Oct. (*Ascomyc. fueg.*).
Collector: Per Karl Hjalmar Dusén (1855-1926).
Publ.: 1899 (pref. Aug 1899; Nat. Nov. Jul(2) 1900), p. [1]-21, [22], *1 pl.*, uncol. lith.
Copies: FH, G, NY, U. – Issued as Bih. k. Sv. Vet.-Akad. Handl. 25, afd. 3, n:o 6.
Other issue: Sv. Exp. Magellansländ. 3(2): 39-58, *pl. 3* (Wiss. Erg. Schwed. Exp. Magellansländ. 1895-1897, vol. 3).

8827. *Beiträge zur Ascomyceten-Flora der Voralpen und Alpen* [Separatabdruck aus der "Oesterreichischen botanischen Zeitschrift", Jahrg. 1903, Nr. 1]. Oct.
1: Jan 1903, p. [1]-6. *Copy*: M. – Reprinted from Öster. bot. Z. 53: 9-14. Jan 1903.
2: Mar 1903, p. [1]-8. *Copy*: M. – Reprinted from id. 54: 81-88. Mar 1903.
3: Aug-Sep 1906, p. [1]-15.*Copy*: M. – Reprinted from id. 56: 291-298. Aug, 341-348. Sep 1906.

8828. *Die Flechten* (Lichenes) *des mittelfränkischen Keupergebietes* ... [Sonderabdruck aus Denkschriften der Kgl. bot. Gesellschaft in Regensburg ... 1905]. Oct.
Publ.: 1905, p. [1]-59. 1 map in text. *Copies*: Almborn, M. – Reprinted and to be cited from Denkschr. bot. Ges. Regensburg 9: [1]-59. 1905.

8829. *Novitates brazilienses* [sic] ... Separat-Abdruck aus Brotéria, Bd. V. 1906 ... Lissabon 1906. Oct.
Co-author: Johann Rick (1869-?).
Publ.: 1906, p. [1]-7. *Copy*: M. – Reprinted and to be cited from Brotéria 5: [223]-228. 1906.

8830. *Ascomycetes philippinenses* collecti a clar. C.F. Baker [Manila] 1913-1916.
Collector: Charles Fuller Baker (1872-1927).
Publ.: a series of 8 papers: *1*: Philip. J. Sci. C. Bot. 8(3): 181-194. Mai 1913.
2: ib. 8(4): 251-263. Jul 1913.
3: ib. 8(5): 391-405. Nov 1913.
4: Leafl. Philip. Bot. 6: 1935-1947. 1 Nov 1913.
5: ib. 6: 2191-2237. 20 Mai 1914.
6: ib. 6: 2257-2281. 27 Jun 1914.
7: ib. 8: 2921-2933. 29 Dec 1915.
8: ib. 8: 2935-2961. 14 Jan 1916.

Rehmann, Anton (1840-1917), Austrian botanist and geographer at Krakau (until 1882); Dr. phil. Krakau 1864; collected and travelled in Southern Russia (1868)and in Natal and Transvaal (1875-1877, 1879-1880); professor of geography at Lemberg (Lwow) from 1882. (*Rehmann*).

HERBARIUM and TYPES: European herbarium: LW; S. African phanerogams: Z; original bryophytes: B (rd. 1892, ca. 10.000; for the greater part destroyed), BR, GRA, Z; important other sets (all taxa): at BM (over 4000), further material at BOL, BP, BR, E, G, GRA, H, JE, K, KRA, L, LW, MANCH, NH, NY, OXF, PC, PRE, S, US, W, WB. – Issued *Flora polonica exsiccata* (900 nos; 1893-1902) with E. Woloszczak; *Musci austroafricani* (1875-1877, nos. 1-424, and 1886, nos. 425-680. 1886; for details see Sayre (1971)); *Hepaticae austro-africanae*, det. F. Stephani (nos. 1-74, for details see Sayre (1971)). For extensive details on S. African collections see Gunn & Codd (1981).

BIBLIOGRAPHY and BIOGRAPHY: AG 2(1): 412, 5(2): 205, 12: 246; Backer p. 486; Barnhart 3: 139 (b. 13 Mai 1840); BJI 1: 48; BM 4: 1667; Bossert p. 328; Clokie p. 230; CSP 5: 234, 8: 716, 11: 128, 12: 602-603, 18: 101; GR p. 730 ; Hegi 6(2): 1225; Herder p. 200; IH 1 (ed. 7): 340; Jackson p. 264, 266, 268, 327; Kew 4: 438; LS 21963; Morren ed. 10, p. 35; Rehder 5: 713; TR 1145-1146; Tucker 1: 588-589; Urban-Berl. p. 124, 311, 387; Zander ed. 10, p. 706, ed. 11, p. 805.

BIOFILE: Anon., Bot. Zeit. 36: 256. 1878 (exsicc.), 37: 48. 1879 (exsicc.), 39: 20. 1881 (offers S. Afr. plants for sale), 40: 537. 1882 (app. Lemberg); Bot. Centralbl. 3/4: 832. 1880 (return second S. Afr. trip); Bull. Soc. bot. France 29 (bibl.): 142. 1882 (app. extraord. prof. Lemberg); Österr. bot. Z. 27: 217-218. 1877 (returned from S. Africa), 32: 240. 1882 (app. Lemberg), 43: 71. 1893 (moss herb. bought by B); Nat. Nov. 4: 144. 1882 (app. Lemberg), 39: 115. 1917 (d.); Hedwigia 59: (141). 1917 (d.); Rev. gen. Bot. 32: 238. 1920 (d.).
Bourdeille de Montresor, Bull. Soc. natural. Moscou ser. 2. 7: 442. 1894.
Bullock, A.A., Bibl. S. Afr. bot. 93. 1978.
Codd, L.E. & M. Gunn, Bothalia 14: 1-14. 1982 (biogr. account, portr., itin.).
Dixon, H.N. & A. Gepp, Bull. misc. Inf. Kew 1923: 193-238. 1923 (on Musci austroafricani).
Doidge, Bothalia 5: 36, 37, 74. 1950.
Geheeb, A., Rev. bryol. 7(6): 111. 1880 (on S. Afr. coll.).
Gunn, M. & L.E. Codd, Bot. expl. S. Afr. 292-294. 1981 (extensive inf. on S. Afr. coll.).
Lindemann, E. v., Bull. Soc. Natural. Moscou 61(1): 59-60. 1886.
Murray, G., Hist. coll. BM(NH) 1: 176. 1904 (2176 phan., 1293 crypt. from S. Afr., 191 pl. from Cherson).
Pax, F., Grundz. Pfl.-Verbr. Karpathen 30, 32-33, 36, 39, 53. 1898 (Veg. Erde 2(1)).
Phillips, E.P., S. Afr. J. Sci. 27: 46. 1930.
Rouppert, K., Sprawozd. komis. fizjograf. 51: xxx-xxxiv. 1917.
Sayre, G., Bryologist 80: 516. 1977 (coll.).
Sim, T.R., Trans. R. Soc. S. Afr. 15: 2. 1926.
Szymkiewicz, D., Bibl. Fl. Polsk. 47, 55-56, 119-120. 1925.
Szafer, W., Concise hist. bot. Cracow 41, 48, 58, 60, 61. 1969 (portr.).
Szyszylowicz, J., Polypetalae thalamiflorae Rehmannianae, Cracoviae 1887, 75 p. (South African pl.).

NAME: Rehmann spelt his name Rehman in Polish publications.

HANDWRITING: Bothalia 14: 7. 1982.

EPONYMY: *Rehmanniella* K.A.F.W. Mueller Hal. (1881).

8831. *Einige Notizen über die Vegetation der nördlichen Gestade des Schwarzen Meeres* [Verh. naturf. Ver. Brünn 10, 1871]. Oct.
Publ.: 1872 (BSbF 19: 241. 1873: publ. 1872; J. Bot. Dec 1872; Bot. Zeit. 8 Nov 1872), p. [1]-85, 2 charts. *Copies*: FI, NY. – Reprinted and to be cited from Verh. naturf. Ver. Brünn, Abh. 10: 1-85, 2 charts. 1871 (publ. 1872).
Ref.: Ascherson, P., Bot. Zeit. 31: 166-174. 1873 (rev.).

Rehsteiner, Hugo (1864-1947), Swiss botanist at St. Gallen. (*Rehsteiner*).

HERBARIUM and TYPES: B (Lack 1980).

BIBLIOGRAPHY and BIOGRAPHY: Barnhart 3: 139; Biol.-Dokum. 14: 7360; BJI 1: 48; LS 21964.

BIOFILE: Ammann, K., pers. comm. in litt. 29 Jan 1981 (d. 16 Nov 1947).
Lack, H.W., Willdenowia 10: 81. 1980 (material in B, extant).
Sayre, G., Mem. New York Bot. Gard. 19(2): 244-245. 1971 (on his *Musci austroafricani*).

8832. *Beiträge zur Entwicklungsgeschichte der Fruchtkörper einiger Gastromyceten*. Inaugural-Dissertation der hohen Philosphischen Facultät der Universität Bern zur Erlangung der philosophischen Doctorwürde vorgelegt an Hugo Rehsteiner ... Leipzig (Druck von Breitkopf & Härtel)] 1892. Oct.
Publ.: 25 Nov-30 Dec 1892 (in journal, Nat. Nov. Jul(1) 1893), p. [i-v], [1]-44, *2 pl.* (uncol. lith. auct.). *Copy*: G. – Reprinted and to be cited from Bot. Zeit. 52: 761-771, 25 Nov, 777-792, 2 Dec, 801-814. 9 Dec, 823-839, 16 Dec, 843-863. 23 Dec, 865-878,

30 Dec 1892. – According to K. Ammann, Bern (in litt.) the dissertation was defended on 15 Jul 1892 almost certainly on the basis of the manuscript (inf. K. Ammann).

Reichard, Johann Jakob (1743-1782), German (Frankfurt a.M.) botanist and physician; studied at Göttingen 1764-1768; Dr. med. ib. 1768; practicing physician Frankfurt a.M. 1768-1773; "Stiftungsarzt" of the Senckenberg Foundation, supervisor of its botanical garden and library, 1773-1782; from 1779 also practicing at the civilian hospital. (*Reichard*).

HERBARIUM and TYPES: FR (in general herbarium; Reichard left his herbarium to the Senckenbergische Stiftung, see Conert 1979).

BIBLIOGRAPHY and BIOGRAPHY: ADB 53: 267-268 (err. d. 27 Jan 1882); AG 7: 67; Backer p. 486; Barnhart 3: 139 (b. 7 Aug 1743, d. 21 Jan 1782); BM 4: 1667; Dryander 3: 84, 91, 117, 156, 352; Frank 3 (Anh.): 82; GR p. 123; Herder p. 22, 187; Hortus 3: 1202; Jackson p. 2, 426; Kew 4: 438; Lasègue p. 23; LS 21969-21971; MW p. 411; PR 5411, 5431, 7478-7480, ed. 1: 5985, 6010, 8428-8426, 11579; Rehder 1: 44, 114, 262, 263, 373; Saccardo 1: 137; SO 11, 320, add. 661a; TL-1/729, 731, 1047-1048; TL-2/4769, 5063; Tucker 1: 589; Zander ed. 10, p. 706, ed. 11, p. 805.

BIOFILE: Baines, J.A., Vict. Natural. 90: 249. 1973.
Blum, J., Ber. Senckenb. naturf. Ges. Wiss., Abh. 1901: 9.
Conert, H.J., Senckenbergiana Biol. 48(C): 6-8, 27. 1967; Jahrb. nass. Ver. Naturk. 104: 6-23. 1979 (J.J. Reichard and his Flora von Frankfurt a.M.; portr., bibl., biogr. sketch; further references).
Egle, K. & G. Rosenstock, Gesch. Bot. Frankfurt a. M. 13-14, 50. 1966.
Faber, J.H., Schr. Berlin Ges. naturf. Fr. 4: 440-447. 1783 (obit., bibl.; "denen Linné und Murray beigestellet zu werden, war für ihn ein würdiger Wunsch, als sein Vermögen auf Kosten des siechen Bürgers zu vermehren.").
Jessen, K.F.W., Bot. Gegenw. Vorz. 385, 394. 1864.
Meusel, Lex. teut. Schriftst. 11: 107. 1811. (bibl.).
Schmid, G., Goethe u.d. Naturwiss. 562. 1940.
Spilger, L., Senckenberg als Botaniker, Abh. senckenberg. naturf. Ges. 458: 1-175. 1941.

EPONYMY: *Reicharda* Cothenius (1790, orth. var. of Reichardia A.W. Roth (1787); *Reichardia* A.W. Roth (1787); *Reichardia* A.W. Roth (1800); *Reichardia* A.W. Roth (1821).

8833. D. Joannis Jacobi Reichard *Flora moeno-francofurtana* enumerans stirpes circa Francofurtum ad Moenum crescentes secundun methodum sexualem dispositus ... Francofurti ad Moenum [Frankfurt am Main] (Typis Henrici Ludovici Broenner) 1772-1778, 2 vols. Oct. (*Fl. moeno-francof.*).

1: 25 Apr-21 Nov 1772 (p. xii: 25 Apr 1772; GGA 21 Nov 1772), p. [i-xii], [1]-112, [113-116, index].

2: Jan-Dec 1778 (GGA 15 Mai 1779; p. viii: 4 Mai 1778), p. [i-viii], [1]-196, [197-206, index], *1 pl.* (uncol. copp.).

Copies: FI, G, H, HH, NY, USDA; IDC 5659. – SO add. 661a. – See also Conert (1979) for the controversy with Senckenberg over this work.

8834. Caroli a Linné ... *Genera plantarum* eorumque characteres naturales secundum numerum, figuram, situm et proportionem omnium fructificationes partium. Editio novissima novis generibus ab ipso perillustri auctore sparsim evulgatis adaucta curante D. Joanne Jacobo Reichard ... Francofurti ad Moenum (apud Varrentrapp filium et Wenner) 1778. Oct. (*Gen. pl.*).

Publ.: 1778 (p. viii: 12 Dec 1777), p. [i]-xxix, [1]-571, [1-44, indexes, err.]. *Copies*: BR, FI, G, MO, NY, PH, TCD, U, US, USDA. – SO 320, STR 7502. Counts as ed. 7 of Linnaeus, *Gen. pl.* (see TL-1/719, TL-2/4714).

Engl. ed.: 1787. *Copies*: MO, USDA. "*The families of plants*, with their natural characters, according to the number, figure, situations, and proportions of all the parts of the fructification. Translated from the last edition, (as published by Dr. Reichard) of the

Genera plantarum, and of the Mantissae plantarum of the elder Linnaeus; and from the Supplementum plantarum of the younger Linnaeus, with all the new families of plants, from Thunberg and L'Héritier. To which is prefix'd an accented catalogue of the names of plants, with the adjectives apply'd to them, and other botanic terms, for the purpose of teaching their right pronunciation ... By a Botanical Society at Lichfield. Lichfield (printed by John Jackson ...) 1787, 2 vols. Oct.
1: [i*, iii*, v*], [i]-lxxx, [1-4, err.], [1]-386.
2: [i, iii], 387-840.
The "Botanical Society at Lichfield", founded by Erasmus Darwin, consisted of Sir Brooke Boothby, (1743-1824), Erasmus Darwin 1731-1802, and the printer John Jackson. See SO 24.

8835. Caroli a Linné ... *Systema plantarum* secundum classes, ordines, genera, species cum characteribus, differentiis, nominibus trivialibus, synonymis selectis, et locis natalibus. Editio novissima novis plantis ac emendationibus ab ipso auctore sparsim evulgatis adaucta curante D. Joanne Jacobo Reichard ... Francofurti ad Moenum [Frankfurt a. Main] (apud Varrentrapp filium et Wenner) 1779-1780, 4 parts. Oct. (*Syst. pl.*).
Publ.: a combination of *Syst. veg.* ed. 13 (Linnaeus, cur. J.A. Murray) (1774), *Species plantarum* (1762-1763) and the *Manatissa* (1767-1771). The 1774 edition of the *Systema vegetabilium* still contained new material supplied by Linnaeus. It is not clear whether any further new material was supplied by Linnaeus but it seems unlikely. See Mikan for a list of errors, which must be attributed to Reichard. – See SO 11. – *Copies*: BR, FI, H, MO, NY, PH, TCD, U, USDA; IDC 7657. – See TL-2/4769, no. 8.
1: Jan-Jul 1779 (London Review Aug 1779), p. [i*-iv*], [i]-xxxii, [1]-778.
2: 1779, p. [i], [1]-674.
3: 1780, p. [i], [1]-972.
4: Nov-Dec 1780 (p. viii: 30 Oct 1780), p. [i-viii], [1]-662, [1-74, indexes], [1, err.].
Ref.: Mikan, J.G., Verzeichniss von nicht weniger als 1700 Schreib– und Druckfehlern und falschen Zitaten, die aus der Reichardschen Ausgabe der Linnéschen Species plantarum in die Willdenowsche übergingen. Mss. at PR or W [?, n.v., cf. PR sub 6218, SO 11a].

8836. Joann. Jacobi Reichard ... *Sylloge opusculorum botanicorum* cum adjectis annotationibus ... pars i. Francofurti ad Moenum [Frankfurt am Main] (apud Varrentrapp filium & Wenner) 1782. Oct. †. (*Syll. opusc. bot.*).
Publ.: 1782, p. [i-iv], [1]-182. *Copies*: BR, B-S, FI, G, H, HH, MO. – Contains reprints of seven smaller papers by C.G. Ludwig (*Diss. sexu pl.*; *Obs. meth. sex. Linn.*), J.E. Stief, A.F. Walther and four titles by C.L. Willich.

8837. Joannis Jacobi Reichard, ... *Enumeratio stirpium horti botanici senckenbergiani*, qui Francofurti ad Moenum est. Francofurti ad Moenum [Frankfurt a. Main] (apud Varrentrapp filium & Wenner) 1782. Oct. (*Enum. pl. hort. senckenb.*).
Publ.: 1782 (p. 5: 15 Jan 1782), p. [1]-68. *Copies*: G, M.

Reichardt, Heinrich Wilhelm (1835-1885), Moravian botanist (of German parents), educated at the gymnasium of Iglau; studied medicine and botany at Wien 1855-1860; student of A. Pokorny, E. Fenzl and F. Unger; Dr. med. Wien 1860; habil. ib. 1860; assistant at the Botanical Garden with Fenzl and lecturer in botany 1860-1866; from 1863 unpaid assistant at the botanical "Hofcabinet" (later Naturh. Hofmus.); from 1866 second "Custos" (curator), first curator 1871, head of the "Cabinet" 1878; id. of the botanical dept. of the Naturh. Hofmuseum from its incorporation in 1884; assistant editor Flora brasiliensis 1870; extraordinary professor of botany at Wien University 1873. (*Reichardt*).

HERBARIUM and TYPES: W (donated 1874); further material at B (extant, in herb. k.k. zool.-bot. Ges.), E, H, PR, WRSL.

BIBLIOGRAPHY and BIOGRAPHY: AG 4: 215, 12(2): 24; Backer p. 486; Barnhart 3: 139 (b. 16 Apr 1835, d. 2 Aug 1885); BJI 1: 48; BL 1: 100, 314, 2: 457, 707; BM 4: 1667; Bossert p. 328; CSP 5: 136, 8: 718-720, 11: 129, 18: 103; De Toni 1: civ, 2: c; DTS 1: 240; Frank

3 (Anh.): 82; Futak-Domin p. 499-500; GR p. 458; Hawksworth p. 185; Hegi 6(2): 1246; Herder p. 473 [index]; IF p. 726; IH 2: (in press); Jackson p. 151, 168, 224, 267, 404; Kelly p. 190; Kew 4: 438; Klášterský p. 153-154; LS 21973-22013; Moebius p. 138; Morren ed. 1, p. 44, ed. 10, p. 29; MW p. 411; PR 7483, 10412; Rehder 5: 713; Saccardo 2: 90; SBC p. 130; Stevenson p. 1255; TL-2/1764, see E. Hackel, F. Hauck; Tucker 1: 589; WU 25: 162-163.

BIOFILE: Anon., Bonplandia 8: 267 (habil.), 325, 326, 348 (ass. Fenzl), 387. 1860, 9: 91. 1861; Bot. Centralbl. 23: 236. 1885 (d.); Bot. Gaz. 10: 371. 1885 (d.); Bot. Jahrb. 7 (Beibl. 13): 2. 1885 (d.); Bot. Not. 1886: 39 (d.); Bot. Zeit. 87: 110-111. 1879 (head of bot. dept. W), 43: 544. 1885 (d.); Bull. Soc. bot. France 25 (bibl.): 237. 1879 (head bot. dept. Nat. Hist. Mus.); Flora 43: 767. 1860 (habil. Wien), 51: 73. 1868 ("custos" at W), 54: 400. 1871 (first "custos"); Leopoldina 21: 162. 1885 (obit.); Österr. bot. Z. 10: 335. 1860 (asst. with Fenzl), 16: 258. 1866 (asst. curator W), 23: 135. 1873 (extraord. prof. bot. Wien), 35: 334. 1885 (d.); Termész. közlön, Budapest 18: 519-520. 1886 (obit.).
Beck, G. v., Ber. deut. Bot. Ges. 3: xvii-xix. 1885 (obit.); Bot. Centralbl. 33: 282, 378, 379, 380. 1888, 34: 87, 158. 1888, in K. Fritsch et al., Bot. Anst. Wiens 63, 64, 75. 1895 (herb. at W); Veg.-Verh. Illyr. Länd. 10, 41. 1901 (Veg. Erde 4); Verh. zool.-bot. Ges. Wien 35: 669-670. 1885 (obit).
P.B., Flora 56: 303, 415. 1873 (extraord. prof. bot. Wien; resigns as secretary zool.-bot. Ges. Wien).
Guglia, O.F., Sonderh. burgenl. Forsch. 5: 139-148. 1973 (biogr.).
Helm, J., Kulturpfl. 14: 353. 1966 (d. by suicide).
Kaemerling, J., Dr. Heinr. Wilh. Reichardt, ein Lebensbild, Mähr. Weisskirchen 1888, 18 p. (obit., written by his cousin Kaemmerling; bibl.; copy at DS (Engler coll.)).
Kanitz, A., Emlékbeszéd Reichardt Henrik Vilmos a m.t. Akadémia kültagja. Budapest 1889, 27 p.
Lack, H.W., Willdenowia 10: 81. 1980 (coll. B).
Lucante, A., Rev. bot. Bull. mens. Soc. franç. bot. 4: 335. 1886 (d.).
Merrill, E.D., B.P. Bishop Mus. Bull. 144: 155. 1937; Contr. U.S. natl. Herb. 30: 251. 1947.
Oborny, A., Verh. naturf. Ver. Brünn 21(2): 15, 22. 1882.
Pluskal, F.S., Verh. zool.-bot. Ver. Wien 6: 369-370. 1856.
Poetsch, J.S. & K.B. Schiedermayr, Syst. Aufz. samenlosen Pfl. xvi. 1872.
Sayre, G., Bryologist 80: 516. 1877 (bryoph. H, W).
Sirks, M.J., Ind. natuurond. 164. 1915.
Szymkiewicz, D., Bibl. Fl. Polsk. 41, 120. 1925.
Urban, I., Fl. bras. 1(1): 195-196. 1906 (biogr. sketch, bibl.).
Wittrock, V.B., Acta Horti Berg. 3(3): 69. 1905.

8838. *Über die Gefässbündel-Vertheilung im Stamme und Stipes der Farne.* Ein Beitrag zur anatomischen und systematischen Kenntniss dieser Familie... Wien (Aus der kaiserlich-königlichen Hof- und Staatsdruckerei...) 1859. Qu.
Publ: 1859 (Flora rd. 7-28 Dec 1859), p. [i], [1]-28, *pl. 1-3* (col. liths.). *Copies*: G(2), U. – Reprinted and to be cited from Denkschr. k. Akad. Wiss., Mat.-Nat. 17: 21-48, *pl. 1-3*. 1859, see also S.B. Akad. Wiss. Wien 25: 513-515. 1857.
Ref.: Anon., Flora 42: 607-608. 14 Oct 1859 (rev.).

8839. *Ueber das Alter der Laubmoose.* Ein Probe-Vortrag gehalten zum Behufe der Habilitation als Privat-Docent für Morphologie und Systematik der Sporenpflanzen... Wien (Druck von Carl Ueberreuter) 1860. Oct.
Publ.: Dec 1860 (mss. submitted 1 Aug 1860), p. [1]-12. *Copies*: G, NY. – Reprinted and to be cited from Verh. zool.-bot. Ges. Wien 10(Abh.): 589-598. Dec 1860.

8840. *Beitrag zur Flora von Niederösterreich* [Aus den Verhandlungen der k.k. zoologisch-botanischen Gesellschaft in Wien... 1861]. Oct.
Publ.: Dec 1861 (read 3 Jul and 7 Aug 1861), p. [1]-8. *Copy*: M. – Reprinted and to be cited from Verh. zool.-bot. Ges. Wien 11 (Abh.): 337-344. Dec 1861.

8841. × *Verbascum Neilreichii* (V. specioso-phlomoides) ein neuer Blendling. [Aus den Verhandlungen der k.k. zoologisch-botanischen Gesellschaft in Wien ... 1861]. Oct.
Publ.: Dec 1861 (read 7 Aug 1861), p. [1]-4. *Copy*: M. – Reprinted and to be cited from Verh. zool.-bot. Ges. Wien 11 (Abh.)(4): [367]-370. Dec 1861.

8842. *Beitrag zur Kenntniss der Cirsien Steiermarks.* [Aus den Verhandlungen der k.k. zoologisch-botanischen Gesellschaft in Wien ... 1861]. Oct.
Publ.: Dec 1861 (read 2 Oct 1861, journal Dec 1861), p. [1]-4. *Copy*: M. – Reprinted and to be cited from Verh. zool.-bot. Ges. Wien 11 (Abh.)(4): [379]-382. 1861.

8843. × *Verbascum pseudo-phoeniceum* (V. Blattaria-phoeniceum) ein neuer Blendling. [Aus den Verhandlungen der k.k. zoologisch-botanischen Gesellschaft in Wien ... 1861]. Oct.
Publ.: Dec 1861 (read 7 Nov 1861, journal Dec 1861), p. [1]-2. *Copy*: M. – Reprinted and to be cited from Verh. zool.-bot. Ver. Wien 11 (Abh.)(4): [403]-404. 1861. (Heft 4 dated 1861, even though it contains also papers read on 4 Dec 1861).

8844. *Ueber zwei neue Arten von Centaurea aus Kurdistan.* [Aus den Verhandlungen der k.k. zoologisch-botanischen Gesellschaft in Wien ... 1863]. Oct.
Publ.: Dec 1863 (read 5 Aug 1863), p. [1]-6. *Copy*: M. – Reprinted and to be cited from Verh. zool.-bot. Ges. Wien 13 (Abh.): [1039]-1044. Dec 1863 or early 1864.

8845. *Reise seiner Majestät Fregatte Novara um die Erde. Botanischer Theil i. Band. Pilze, Leber- und Laubmoose ...* Wien [k.k. Hof- und Staatsdruckerei] 1870. Qu. (*Reise Novara, Pilze, Leber- Laubm.*).
Publ.: 1870 (Flora rd. 29 Jul 1871), p. [i], [131]-196, *pl. 20-36* (J. Seboth, uncol. liths.). *Copies*: BR, H, MICH, MO, Stevenson, U-V. – Reprinted from E. Fenzl, *Reise Novara* 1(3), 1870, see TL-2/1764. – Prepublication of new taxa Verh. zool.-bot. Ges. Wien 16 (Abh.): 373-376. 1866 (fungi), 957-960. 1866 (hepat.) [repr. p. [1]-4. *Copy*: M, publ. Nov-Dec 1866 or early 1867], 18 (Abh.): 193-198. 1868 (musci).
Ref.: X., Flora 55: 58-59. 1872.

8846. *Ueber die Flora der Insel St. Paul* im indischen Ocean. Wien (C. Ueberreuter'sche Buchdruckerei (M. Salzer)). Qu.
Publ.: 1871 (read 4 Jan 1871; Hedwigia Nov 1871), p. [1]-36. *Copies*: FH, G(2). – Reprinted (with special t.p.) and to be cited from Verh. zool.-bot. Ges. Wien 21 (Abh.): 3-36. 1871.

8847. *Carl Clusius' Naturgeschichte der Schwämme Pannoniens ...* Wien (Herausgegeben von der k.k. zoologisch-botanischen Gesellschaft in Wien ...) 1876. Qu.
Publ.: Mar-Apr 1876 (p. 42: 5 Feb 1876; Hedwigia Mai 1876; Bot. Zeit 12 Mai 1876), p. [1]-42. *Copies*: FH, FI, Stevenson. – Reprinted and to be cited from Festschrift k.k. zool.-bot. Ges. Wien 1876: 147-186. 1876. – Identification of 102 fungi described by C. Clusius with species in E. Fries, *Hymenomyc. eur.*

8848. *Beiträge zur Flora der hawaiischen Inseln ...* Wien (Druck der k.k. Hof- und Staatsdruckerei) 1878. Oct.
Publ.: 1878 (submitted to Soc. 11 Mai 1877), p. [i], [1]-30. *Copy*: G. – Reprinted and to be cited from S.B. Akad. Wiss. Wien 75 (Abh. 1)(5): 553-582. Mai 1877, publ. 1878.

8849. *Beitrag zur Phanerogamenflora der hawaiischen Inseln* [Wien 1877]. Oct.
Publ.: 1878, p. [1]-14. *Copy*: G. – Reprinted and to be cited from S.B. Akad. Wiss. Wien 76 (Abh. 1): 721-734. Nov 1877 (publ. 1878).

8850. *Flora der Insel Jan Mayen.* Gesammelt von Dr. F. Fischer, ... bearbeitet unter Mitwirkung von ... Theodor Fries ... Eduard Hackel ... Ferdinand Hauck ... von Dr. H.W. Reichardt ... Wien (aus der kaiserlich-königlichen Hof- und Staatsdruckerei ...) 1886. Qu.
Co-authors: Ferdinand Hauck (1845-1889; algae), Eduard Hackel (1850-1926; grasses), Theodor Magnus Fries (1832-1913; lich.).

Publ.: 1886 (Bot. Monatschr. Nov-Dec 1886; Bot. Centralbl. 14-18 Mar 1887), p. [1]-16.
Copies: FI, G(2), M. – Reprinted and to be cited from *Die internationale Polarforschung 1882-1883*. 3(7): 1-16. 1886.
Ref.: G.L., Bot. Monatschr. 4: 182. 1886.

Reiche, Karl Friedrich (Carlos Federico) (1860-1929), German (Saxonian) botanist; Dr. phil. Leipzig 1885; high school teacher Dresden, 1886-1889; at the same time assistant to O. Drude at the university; to Chile 1889; high school teacher Concepcion 1889-1896; head of the botany dept. National Museum Santiago 1896-1911, concurrently professor of botany at the Medical faculty, Santiago; in Mexico professor of botany and head of the botany dept. Instituto medico nacional 1911-1923; in München 1924-1929 (with trip to Mexico 1926-1927), ultimately at the Botanische Staatssammlungen. (*Reiche*).

HERBARIUM and TYPES: SGO. – Other material at B, CORD, GB, H, M (complete coll. Mexico 1926/1927; rd. 1929; 1069 nos.). – For a list of Reiche's types at SGO see Schick (1971).

BIBLIOGRAPHY and BIOGRAPHY: Barnhart 3: 139 (b. 31 Oct 1860, d. 26 Feb 1929); BFM no. 483; BJI 1: 48, 2: 143-144; Biol.-Dokum. 14: 7364-7365; BL 1: 152, 246, 248, 255, 314; BM 4: 1667, 8: 1060; Bossert p. 328; CSP 18: 103-104; Futak-Domin p. 500; Hortus 3: 1202; Kew 4: 438-439; Langman p. 618-619; LS 22014, 38303; Rehder 5: 713-714; TL-1/1050; TL-21712, see H.A.T. Harms, G. Looser; Tucker 1: 589; Urban-Berl. p. 281, 387.

BIOFILE: Anon., Bot. Jahrb. 23 (Beibl. 57): 62. 1897 (prof. bot. Santiago).
Bergdolt, E., Karl von Goebel ed. 2: 184. 1941.
Hertel, H., Mitt. Bot. München 16: 421. 1980 (Mexico coll. 1926/1927 all in M).
Lohrmann, E., S.B. Abh. naturw. Ges. Isis Dresden 1929: 7-9. 1930 (obit., member Isis 1886; "kein Freund von Artenspalterei").
McVaugh, R., Contr. Univ. Mich. Herb. 9: 293. 1972 (Mexican itin.).
Muñoz, C.P., Indice de Reiche, K., Estudios criticos de la flora chilena. Chile 1942 (IDC 5182), (microfiche publication of the 1942 typescript).
Muñoz Schick, M., Bol. Mus. Nac. Hist. Nat. Chile 32: 377-393. 1971 (Reiche types at SGO).
Percy-Moreau, R.A., Bibl. geobot. Patagonica 31-32, 45, 70-71, 87, 104, 105 (bibl.).
Porter, C.E., Anal. Soc. ci. Argent. 70: 296. 1910 (portr); Rev. chil. Hist. nat. 33: 63-64. 1929/30 (obit., portr.).
Reiche, K.F., Grundz. Pfl.-Verbr. Chile 23-24, 41-43, 372. 1907 (Veg. Erde 8).
Roldan, A., Mem. Soc. Antonio Alzate (Mem. Acad. Nac. Ci. Ant. Alzate) 52: 279-281. 1929 (obit., on Mexican activities).
Ross, H., Ber. deut. bot. Ges. 47: (103)-(110). 1929 (obit., bibl., portr.).
Rzedowski, J., Bol. Soc. bot. Mex. 40: 7-10. 1981.
Schick, M.M., Bol. Mus. nac. Hist. nat. Chile 32: 377-393. 1971 (list of Reiche types at SGO).
Stieber, M.T., Huntia 3(2): 124. 1979 (notes on Reiche in A. Chase mss.).
Verdoorn, F., ed., Chron. bot. 4: 565. 1938.

COMPOSITE WORKS: (1) E.P., *Die natürlichen Pflanzenfamilien*:
(a) *Cistaceae*, 3(6): 299-306. 14 Mai 1895.
(b) *Erythroxylaceae* 3(4): 37-40. Aug 1890.
(c) *Geraniaceae* 3(4): 1-14. Aug 1890.
(d) *Humiriaceae* 3(4): 35-37. Aug 1890.
(e) *Limnanthaceae* 3(5): 136-137. Jan 1892.
(f) *Linaceae* 3(4): 27-35. Aug 1890.
(g) *Oxalidaceae* 3(4): 15-23. Aug 1890, 351-352. Jul 1896.
(h) *Tropaeolaceae* 3(4): 23-27. Aug 1890, 352. Jul 1896.
(i) *Violaceae* 3(6): 322-336. 14 Mai 1895, with Paul Hermann Wilhelm Taubert (1862-1897).
(k) *Plantaginaceae* (exc. Plantago) 4(3B): 363-373. Dec 1895, (Plantago by H. Harms).

THESIS: *Über anatomische Veränderungen, welche in Perianthkreisen der Blüthen während der Entwicklung der Frucht vor sich gehen,* Jahrb. wiss. Bot. 16: 638-687, *pl. 27-28.* 1885. (Bot. Zeit. 29 Jan 1886).

EPONYMY: *Reichea* Kausel (1940); *Reicheella* Pax (1900).

8851. *Sobre el método* que debe seguirse en el estudio comparativo *de la flora de Chile* ... Santiago de Chile (Imprenta Cervantes ...) 1894. Oct.
Publ.: 1894 (p. 23: 21 Mar 1894), p. [1]-23. *Copy*: HH. – Reprinted and to be cited from Anal. Univ. Chile 87: 37-57. 1894.

8852. *Apuntes sobre la vejetacion en la boca del Rio Palena* ... Santiago de Chile (Imprenta Cervantes ..) 1895. Oct.
Publ.: 1895 (p. 35: 12 Oct 1894), p. [1]-35. *Copy*: HH. – Reprinted and to be cited from Anal. Univ. Chile 90: 715-747. 1895 (in "Memoria general de la espedicion esploradora del Rio Palena").

8853. *Flora de Chile* ... Santiago de Chile (Imprenta Cervantes ...) 1896-1911, 6 vols. Oct. (*Fl. Chile*).
Publ.: 1896-1911, in 5 volumes and part of a sixth volume. The text for vols. 1-4 was first published in the Anal. Univ. Chile from Nov 1894 onward as "Estudios criticos sobre la flora de Chile".
1(1): Nov 1894-Jan 1895, p. [i-iii], [1]-127.
1(2): Jan 1895-Oct 1896, p. 129-379.
2(3): 1897 (Bot. Centralbl. 26 Jul 1899), p. [i], [1]-237.
2(4): 1898, p. [i], [239]-397.
3(5): 1899-1900 (t.p. part 5: 1900), p. [1]-209.
3(6): 1901-1902 (t.p. cover: 1902; Bot. Centralbl. 15 Oct 1902; ÖbZ Dec 1902; J. Bot. Feb 1903), p. 211-425, [427, concord.].
4(7): 1902-1903 (t.p. part 7: 1903; Bot. Centralbl. rd. Jan 1905; Bot. Jahrb. 15 Mar 1904), p. [1]-217.
4(8): 1904-1905 (t.p. part 8: 1905; ÖbZ Mar-Mai 1906; Bot. Zeit. 1 Mai 1906), p. [i], [219]-474, "375-382", 483-488, [489].
5(9): 1907 (Bot. Centralbl. 10 Mar 1908), p. [1]-240.
5(10): 1910 (Bot. Centralbl. 11 Apr 1911; ÖbZ Jan 1911), p. [i], 241-463.
6(1)(11): Apr-Jun 1911 (p. [177]: Mar 1911; Bot. Centralbl. 25 Jul 1911), p. [i], [1]-176, [177].
Copies: BR, G, H, HH, M, MICH, MO, NY, PH, US, USDA; IDC 1187.

vol.	pages reprint	vol. Anal.	pages Anal.	dates
1(1)	4-46	88	58-100	Nov 1894
	47-127	90	77-157	Jan 1895
	129-173	90	879-923	Jun 1895
(2)	173-265	91	321-413	Sep 1895
	265-340	93	557-632	Mai 1896
	340-353	93	839-852	Feb 1896
	355-379	94	619-644	Oct 1896
2(3)	1-	97	37-62	1897
		97	289-313	1897
		97	461-489	1897
		97	535-571	1897
	-237	97	725-790	1897
(4)	239-	98	117-175	1897
		98	457-497	1897
		98	695-738	1897
		100	327-371	1898
			531-540	1898
	-397	101	55-76	1898
3(5)	1-46	103	783-828	1899

vol.	pages reprint	vol. Anal.	pages Anal.	dates
	46-125	104	767-847	1899
	126-209	106	965-1048	1900
(6)	211-255	108	767-751	1901
	255-330	109	5-80	1901
	330-381	109	325-376	1901
	381-404	109	565-588	1901
	415-427	-	-	-
4(7)	1-46	111	151-196	1902
	46-128	112	97-179	1903
	123-149	112	269-290	1903
	149-195	112	397-443	1903
	195-217	113	367-389	1903
(8)	219-273	114	147-201	1904
	273-299	114	455-481	1904
	299-302	114	735-738	1904
	302-330	115	91-120	1904
	331-371	115	311-352	1904
	371-391	115	563-583	1904
	391-433	116	169-210	1905
	433-466	116	415-448	1905
	447-489	116	575-606	1905
5(9)	1-	117	189-208	Jul-Aug 1905
		117	451-464	Sep-Oct 1905
		117	481-517	Nov-Dec 1905
		118	147-160	Jan-Feb 1906
		118	323-336	Mar-Apr 1906
		118	503-512	Mai-Jun 1906
		119	141-151	Jul-Aug 1906
		120	187-201	Jan-Feb 1907
		120	809-834	1907
		121	227-250	Jul-Aug 1907
		121	803-835	Nov-Dec 1907
5(10)	241-	122	267-288	Jan-Feb 1908
		123	341-400	1908
			713-722	Sep-Oct 1908
		124	441-463	Jan-Feb 1909
			735-769	Mar-Apr 1909
	-463	125	457-507	Sep-Oct 1909
6(1)	[i], 1-176, [177]			Apr-Dec 1911

8854. *Elementos de la morfolojía i sistemática botanica.* Una introduccion en la flora de Chile ... Santiago de Chile (Imprenta Cervantes ...) 1896. Oct. (*Elem. morf. sist. bot.*).
Publ.: 1896, (p. 2: Sep 1895), p. [i-iii], [1]-113, [1, err.]. *Copy*: HH.

8855. *Los productos vejetales indíjenas de Chile* ... Santiago de Chile (imprenta Cervantes ...) 1901. Qu. (*Prod. vej. Chile*).
Publ.: 1901 (p. 28: 29 Jun 1901), p. [1]-28. *Copy*: HH.

8856. *Kleistogamie und Amphikarpie in der chilenischen Flora* ... Valparaiso (Imprenta del Universo de Gmo. Helfmann ...) 1901. Oct.
Publ.: Oct-Dec 1901 (p. 18: Oct 1901), p. [1]-18. *Copy*: HH. – Reprinted from Verh. Deut. wiss. Ver. Santiago 4, 1901 (journal n.v.).

8857. *La Isla de la Mocha.* Estudios monográficos bajo la cooperacion de F. Germain, M. Machado, F. Philippi y L. Vergara publicados por Carlos Reiche ... Santiago de Chile, 1912. Qu. (*Isla de la Mocha*).
Publ.: 1903 (p. [v]: Dec 1902), p. [i-vii], [1]-104, *pl. 1-12*. *Copy*: HH. – Issued as Anal. Mus. nac. Chile. 1903. Botany on p. 64-101.

8858. *Monotypische Gattungen der chilenischen Flora* ... Santiago de Chile (Sociedad "Imprenta y litografia universo" ...) 1905. Oct.
Publ.: 1905 (p. 16: Apr 1905), p. [1]-16. *Copy*: HH. – Reprinted from Verh. Deut. wiss. Ver. Santiago 5, 1905 (journal n.v.).

8859. *La distribucion geográfica de las compuestas de la flora de Chile* ... Santiago de Chile (Imprenta, Litografia y Encuadernación Barcelona ...) 1905. Qu.
Publ.: 1905 (Bot. Zeit. 24 Jul 1905), p. [1]-44, [45, err.], 2 maps on *1 pl. Copy*: M. – Published as Anal. Mus. nac. Chile, Bot. 17.
Ref.: Buchenau, F., Bot. Zeit. 63: 224-225. 24 Jul 1905 (rev.).

8860. *Grundzüge der Pflanzenverbreitung in Chile* ... Leipzig (Verlag von Wilhelm Engelmann) 1907. Oct. (*Grundz. Pfl.-Verbr. Chile*).
Publ.: 1907 (p. viii: Mar 1906; Nat. Nov. Jan 1908), p. [ii]-xiv, [1]-374, 2 maps, *pl. 1-33*.
Copies: BR, G, HH, NY, PH, USDA; – Issued as vol. 8 of A. Engler & O. Drude, Die Vegetation der Erde, 1907, TL-2/1712.
Facsimile ed.: 1976, A.R. Gantner Verlag K.G., Fl. 9490, Vaduz, ISBN 3-7682-0992-x, as orig. *Copy*: FAS.
Spanish translation: 1934, in 2 parts. *Copies*: G, HH, MICH, NY, U. – *Géografía botánica de Chile* traduccion del Aleman de Gualterio Looser, Imprenta Universitaria, Santiago, 1934-1937 (1938).
1: 1934, p. [3]-423, [424].
2: 1937 (but cover 1938), p. [3]-149, [151].
Ref.: J. Briquet, Bull. Herb. Boissier ser. 2. 8: 1009-1011. 1908 (rev.).
Diels, L., Bot. Zeit. 66(2): 134-137. 1908 (rev.).
Neger, F.W., Bot. Centralbl. 107: 444-446. 28 Apr 1908 (rev.).

8861. *Orchideas chilenses* ensayo de una monografia de las Orquídeas de Chile ... Santiago de Chile (Imprenta, litografia i encuadernacion Barcelona ...) 1910. Qu. (*Orchid. chil.*).
Publ.: 1910, p. [i-iv], 1-88, *2 pl. Copy*: HH. – Issued as Anal. Mus. Nac. Chile, secc. 2 Bot., entrega 18, 1910.

8862. *La vegetación en los alrededores de la capital de México*, ... México (Tipografia económica ...) 1914. Oct. (*Veg. alreded. México*).
Publ.: Jun-Oct 1914 (p. 143: Mai 1914; US rd. 2 Nov 1914), p. [1]-143, [145, err.; 147, cont.), 1 map. *Copies*: HH, MO, US, USDA.
German version: see below (1922).

8863. *Die Vegetationsverhältnisse in der Umgebung der Hauptstadt von Mexico* ... Botanische Jahrbücher, Beiblatt Nr. 129. 1922. Oct.
Publ.: 1 Dec 1922, p. [1]-116, 1 map. *Copy*: M. – Issued as Bot. Jahrb. Beibl. no. 129 (vol. 58(1-3)). – Greatly amended and enlarged German version of *Veg. alreded. México* (1914).

8864. *Flora excursoria en el Valle central de Mexico* claves analiticas y descripciones de las familias y generos fanerogamicos ... Mexico (Talleres graficos de la Nacion) 1926. Oct. (*Fl. excurs. Mexico*).
Publ.: 1926 (p. [3]: Mar 1925), p. [1]-303, [305, bibl.; 307, err.]. *Copies*: G, HH, MICH, MO, NY.
Facsimile ed.: 1963, p. [i], [1]-303, 305, bibl.; 307 err.]. *Copy*: PH. – Mexico, D.F. 1963, Subsecretaria de Eseñanzas Técnica y Superior, Inst. politecn. nac., Comis. Libro de Texto.

8865. *Kreuz und quer durch Mexico*. Aus dem Wanderbuch eines deutschen Gelehrten ... Leipzig (Verlag Deutsche Buchwerkstätten ...) 1930. Oct. (*Kreuz quer Mexico*).
Publ.: 1930, p. [1-2], frontisp., [3]-28, *18 pl*. (photos). *Copy*: LC.

Reichel, Georg Christian (1727-1771), German (Saxonian/Thuringian) physician and botanist; Dr. phil. Leipzig 1759; professor of medicine at the University of Leipzig 1767-1771. (*G. Reichel*).

HERBARIUM and TYPES: Unknown.

BIBLIOGRAPHY and BIOGRAPHY: Backer p. 486; Barnhart 3: 139; BM 4: 1667-1668; CSP 5: 139 (1 no.); Dryander 3: 372; Herder p. 473; PR 4670, 7488 (b. 1727), ed. 1: 5210, 8430; Rehder 1: 133.

BIOFILE: Anon., Cat. gén. Bibl. nat. 148: 273-274. 1938; Brit. Mus. gen. cat. 200: 351. 1963.
Dezeimeris, Dict. hist. méd. 3(2): 792. 1837.
Hartenstein, J.G., Südhoff's Arch. Gesch. Med. 31: 188-200. 1938.
Poggendorff, J.C., Biogr.-lit. Handw.-Buch 7a, suppl. (4): 533. 1971 (baptized 9 Jun 1727, d. 24 Jan 1771, accident).
Schmid, G., Goethe u.d. Naturw. 562. 1940 (R. was Goethe's physician at Leipzig).

EPONYMY: *Reichelia* Schreber (1789) probably honors this author, but no etymology was given; *Reichela* Cothenius (1790, *orth. var.*).

8866. *De vasis plantarum spiralibus* amplissimi philosophorum ordinis in Academia lipsiensi consensu die iii. maii a.o.r. [mdlviii] disserit M. Georg Christian Reichel Mulhusa Thuringus medicinus baccalaureus ... Lipsiae [Leipzig] (ex Officina Breitkopfia). Qu. (*Vasis pl. spiral.*).
Publ.: 3 Mai 1758, p. [i-iii], [1]-44, *1 pl.* (uncol. auct.). *Copy*: USDA.

Reichenau, Wilhelm von (1847-after 1913), German botanist and ornithologist from Dillenburg, Nassau, later director of the Natural History Museum at Mainz. (*Reichenau*).

HERBARIUM and TYPES: Unknown.

BIBLIOGRAPHY and BIOGRAPHY: Barnhart 3: 139; BFM no. 326; BM 4: 1668; CSP 11:: 130, 12: 605, 18: 104; Rehder 3: 211.

BIOFILE: Anon., Bot. Centralbl. 123: 688. 1913 (retirement from directorship Mainz).

8867. *Mainzer Flora.* Beschreibung der wilden und eingebürgerten Blütenpflanzen von Mainz bis Bingen und Oppenheim mit Wiesbaden und dem Rheingau nebst dem Walde von Grossgerau ... Mainz (Verlag von H. Quasthoff). [1900]. (*Mainzer Fl.*).
Publ.: Mai-Jun 1900 (p. vii: Mai 1900; Nat. Nov. Jun(2), Jul(1) 1900; Bot. Centralbl. 4 Jul 1900; Bot. Zeit. 16 Sep 1900; ÖbZ Aug-Sep 1900), p. [i]-xxxvi, [1]-532, *pl. 1-2* (uncol.). *Copies*: BR, G.

Reichenbach, Carl [Karl] **Ludwig** [Freiherr] **von** (1788-1869), German (Württemberg) botanist, chemist and industrialist; studied at Tübingen 1807-1809, specialized in industrial carbonization; from 1822-1839 in charge of the mining industry and metallurgical plants at Blansko (Moravia), studying the chemical products of carbonization as well as the geology of the region; from 1839 residing at his property Reisenberg near Wien as private scientist, developing e.g. his "Odlehre" of physiological magnetism. (*C. Rchb.*).

HERBARIUM and TYPES: AWH (includes herbarium F.W. Sieber (1785-1844); forms nucleus of the Van Heurck herbarium). Correspondence in SIA; letters to D.F.L. von Schlechtendal at HAL; the main part of the Reichenbach papers is in the collection of manuscripts of the Technische Museum, Wien (Hss. Gruppe X, Nr. 192-343, 405-408).

NOTE: R.'s main works on his "Od" concept were "Odisch-magnetische Briefe" (1853) and "Der sensitive Mensch" (1854). The theory ("od" as the carrier of all formative forces in life) is now only of historical interest. His – only – botanical work "Die Pflanzenwelt in ihren Beziehungen zur Sensitivität und zum Ode" Wien 1858 (ed. 2. 1909) deals with this same theme of physiological magnetism. Reichenbach gave the etymology of the word *Od* as follows: "Wodan bezeichnet im Altgermanischen den Begriff des Alldurchdringen, daraus Wuodin, Odon, Odin, die alldurchdringende Kraft.

Od ist somit das Lautzeichen für einer alles in der Natur mit unaufhaltsamer Kraft durchdringenden und durchströmenden Kraft, durchdringenden und durchströmenden Dynamid." (see Bauer, 1911, p. 13). There is no known family relationship with H.G.L. von Reichenbach, who stemmed from a Thuringian family. See R. v. Reichenbach et al. (1869) for further details; and D.F.L. von Schlechtendal, Bot. Zeit. 16: 81-82. 1858 for a review of his 1858 publ.

BIBLIOGRAPHY and BIOGRAPHY: Barnhart 3: 139 (b. 12 Feb 1788, d. 19 Jan 1869); BM 4: 1670, 8: 1060 (Brit. Mus. Gen. cat. 200: 366-368. 1963); CSP 5: 139-14, 8: 720-721; Herder p. 119; IH 1 (ed. 7): 340 (err. sub H.G. Reichenbach); PR 7517; Rehder 3: 347; SIA 7177; TL-2/see C.F. Hochstetter; WU 25: 169-177.

BIOFILE: Anon., Bonplandia 3: 33. 1855, 6: 50. 1858 (cognomen Leopoldina "Orpheus II", more a reference to the descent into Hades (cf. iron industry) than to the loss of Eurydice); Bot. Zeit. 27: 104. 1869 (d.).
Bauer, A., Naturhist.-biogr. Essays, Stuttgart, Verlag von Friedrich Enke, 1911, p. 1-26 ("Erinnerungen an den Freiherrn von Reichenbach"); Karl Ludwig Frhr. von Reichenbach, Wien 1917.
Frison, Ed., Henri Ferdinand van Heurck 1959, p. 19, 38, 39, 40 (hist. herb.); Tijdschr. Stad Antwerpen 6(1): 33-41. 1960 (Reichenbach acquired Sieber herb. in 1837).
Habacher, M., Jahrbuch Geschichte Oberdeutsche Reichsstädte, Esslinger Studien 16: 172-227. 1970 (on C.F. Hochstetter and C.L. v. Reichenbach, extensive study).
Habasker, Frau Dr., Deut. Apoth. Zeit. 103: 704-706. 1963 (corr. with F.W. Sertüner).
Kohn, M., J. Chem. Educ. 32: 188-189. 1955.
Poggendorff, J.C., Biogr.-lit. Handw.-Buch 2: 593-594. 1893, 3: 1101. 1898, 7a: 534-535. 1971 (bibl., sec. refs.).
Reichenbach, R. v. et al., Alman. Akad. Wiss., Wien 19: 326-369. 1869 (biogr.).
Verdoorn, F., ed., Chron. bot. 3: 22. 1937.
Zittel, K.A. v., Gesch. Geol. Paläont. 243-244. 1899; Hist. geol. palaeontol. 165. 1901.

NOTE: (1) For an extensive listing of biographical references on C.L. von Reichenbach see Poggendorff (1971). Most of R's activities were in chemistry.(2) Reichenbach's daughter, Hermine von Reichenbach, is the anonymous author of the paper "*Untersuchungen über zellenartigen Ausfüllungen der Gefässe*, von einem Ungenannten", Bot. Zeit. 3: 225-231, 241-253. 1845 (see Habacher, 1970, p. 200).

Reichenbach, [Heinrich Gottlieb] Ludwig (1793-1879), German (Saxonian) botanist, botanical artist and ornithologist; Dr. phil. Leipzig 1815; Dr. med. ib. 1817; practicing physician ib. 1817-1820; director of the botanical garden at Dresden 1820-1879; director of the natural history cabinet 1820-1874 and professor of natural history at the Königl. chirurg.-med. Academie (college of medicine) 1820-1862; prolific author, popularizer and artist. (*Rchb.*).

HERBARIUM and TYPES: part of R's herbarium was destroyed in the great Zwinger fire of 6 Mai 1849 (see F. Hofmeister et al. 1849, 1850); another part is incorporated in the herbarium of his son, now at W. *Exsiccatae*:
(1) *Lichenes exsiccatae* collecti atque descripti auctoribus L. Reichenbach et C. [Carl] Schubert. *Die Flechten* ... fasc. 1-6, nos. 1-150, 1822-1826 (further details in Sayre 1969). Sets at B, BM, E, H, HAL, K, KIEL, L, LE, LZ , M, NY, O, P, PC, STU, W, WRSL.
− See also Reichenbach & Schubert 1822-1826 (in bibliogr. paragraph below).
(2) *Flora germanica exsiccata*, fasc. 1-26, 1830-1846 (later sets issued with H.G. Reichenbach). For a detailed announcement see H.G. von Reichenbach, Flora 12(1): Beil. 2: 41-46. 1829 and the review in 13(2): 418-424. 1830. Cryptogams (only) in cent. 3, by J.C. Breutel but continued separately. *Sets* of main series: B, BM, BP, BR, BREM, CGE, GFW, GOET, GRO, H, HAL, JE, KIEL, L, LE, LZ (destr.), M, MANCH, MPU, NCY, O, OXF, PRC, PH, S, TU, U, W, WRSL. − Ser. 2. *Cryptog.*, cent. 1-5, 1832-1862, sets at FH, K, M, W. − *Index* to cent. 1-13 and crypt. cent. 1: Index in Herbarium florae germanicae, ... Leipzig s.d., p.[i]-xviii. *Copy*: G.

NOTE: (1) Reichenbach was called *Ludwig* and hence often referred to as L. Reichenbach, also to distinguish him from his son Heinrich Gustav R.
(2) Reichenbach's brother Anton Benedict Reichenbach (1807-1860), high school teacher at Leipzig, was the author of a number of popular books on botany(see e.g. BM 4: 1668, 8: 1060; PR 7181, 7489-7491 (ed. 1: 8051, 8431-8434), Sotheby 636, Tucker 1: 589) such as *Allgemeine Pflanzenkunde* (Leipzig 26 Nov-2 Dec 1837), *Die Pflanzenuhr* (Leipzig 1840), *Botanik für Damen* (Leipzig 1854).
(3) Letters to D.F.L. von Schlechtendal at HAL.

BIOFILE: ADB 27: 667-668 (W. Hess); AG 2(1): 215, 3: 725, 4: 827, 5(2): 186; Andrews p. 317; Barnhart 3: 486; Biol.-Dokum. 14: 7368-7369 (p.p.), 7372; BL 2: 556, 707; Blunt p. 222, 264; BM 4: 1668-1670; Bossert p. 328; Christiansen p. 20, 21; Clokie p. 231; CSP 5: 141-142, 8: 721-722, 11: 130, 12: 595, 605; DTS 1: 240-243; DU 257-260; Frank 3: (Anh.): 82-83; Futak-Domin p. 500; GF p. 72; GR p. 39-40, *pl.* [*16*]; Hegi 5(1): 634; Herder p. 473 [index]; Hortus 3: 1202; IH 2: (in press); Jackson p. 48, 49, 115, 116, 122, 293, 295, 306, 425; Kew 4: 439-440; Langman p. 619; Lasègue p. 314; LS 22019-22023; Maiwald p. 282 [index]; Moebius p. 54; Morren ed. 2, p. 10; MW p. 411; NAF 28B(2): 353, ser. 2. 2: 169, 5: 252. 1965; NI 1601-1608, see also p. 265, 313; Plesch p. 379; PR 7492-7511, 3836, 4049, 4718, 6247, 6318, 9026, ed. 1: 4201, 4453, 5261, 6950, 7011, 8435-8456, 9979; Rehder 5: 714; RS p. 136; Saccardo 2: 90, cron. xxx; SBC p. 130; SIA 7177; SO 835a, 838e, 842a; TL-1/1051-1061, 1275; TL-2/1676, 2454, 2730, 3035, 3722, 6080, 6176, 6177 and see indexes; Tucker 1: 589-590; Urban-Berl. p. 272, 387; Zander ed. 10, p. 705, ed. 11, p. 804.

BIOFILE: Anon., Amer. J. Sci. ser. 3. 19: 77. 1880 (d.); Bonplandia 6: 50. 1858 (cognomen in Leopoldina: Dodonaeus I); Bot. Not. 1880: 28 (d.); Bot. Zeit. 12: 907. 1854 (on éloge of Friedr. Aug. II of Saxony; tribute by R.), 23: 235. 1865 (golden jubilee doctorate), 37: 208. 1879 (d.); Bull. Soc. bot. France 26 (bibl.): 91-92. 1879 (obit.); Flora 3: 256. 1820 (app. Dresden; effective 1 Mai 1820), 21(2): 424. 1838 (Ritterkreuz. k. sächs. Civilverdienstorden), 28: 64. 1845 (app. Director Cabinet of Nat. Hist. Dresden), 33: 47-48. 1850 (on destruction of his coll. in Mai 1879), 48: 237. 1865 (golden jubileee doctorate), 62: 192. 1879 (d.); Ibis, Ser. 4. 3: 384 (d.); J. Bot. 17: 160. 1879 (d.); Leopoldina 17: 19-22, 34-36, 50-54. 1881 (obit., bibl., b. 8 Jan 1793, d. 17 Mar 1879); Nat. Nov. 1: 92. 1879 (d. 27 Mar 1879, err.); Öster. bot. Z. 13: 128. 1863 (adjunct Leopoldina), 15: 269. 1865 (50 yrs. Dr. phil.), 29: 134. 1879 (obit.).
A.R., Ornithol. Centralbl. 4: 56. 1879 (obit.).
Ascherson, P., Fl. Provinz Brandenburg 1: 11. 1864 (on *Fl. saxonica*).
Babington, C.C., Mem., Journ. Bot. corr. 474 [index]. 1897.
Barnhart, J.H., Bartonia 16: 33. 1934 (corr. Schweinitz).
Candolle, A. de, Phytographie 443. 1880. (lists sets of exsicc.).
Carus, V., Zool. Anz. 2: 192. 1879 (d.).
Carus, C.Y. & C.F.P. von Martius, Eine Altersfreundschaft in Briefen 171 [index]. 1939.
Dawson, W.R., Smith letters 77. 1934.
Dolezal, H., Friedrich Welwitsch 35-37, 113. 1974.
Friedrich, S.B. naturw. Ges. Isis, Dresden 1879: 97-104 (obit.).
Frison, Ed., Henri Ferdinand van Heurck 1838-1909, Leiden 1959, p. 39-40 (herb. Reichenbach at AWH is that of Karl von Reichenbach, 1788-1869).
Helm, J., Beitr. Biol. Pfl. 52: 427-428. 1976 (on the engravers of R's iconographies); Festschr. Nissen 351-379. 1973 (on the artists of id.).
Hermann, A.L., Flora 11(1). Erg. Bl. 88-90. 1828 (on R and Friedrich August von Sachsen).
Hofmeister, F., Flora 17(1). Int. Bl. 1: 9-11. 1834 (note on centuries 1-8 of *Fl. germ. exs.*), 17(2), Int. Bl. 2: 23-29. 1834 (note on cent. 8).
Hofmeister, F. et al., Flora 33: 47-48. 1850 (most of R's collections, library, original drawings and correspondence destroyed in the Zwinger fire of Mai 1849 at Dresden; appeal for support to buy books and a new herbarium, as well as for donation of material).
Hofmeister, F., G. Kunze & D.F.L. von Schlechtendal, Aufruf an Botaniker 2 p., s.d.

[1819] (appeal to help rebuilding Reichenbach's collections destroyed in the Zwinger fire. Most of his books, almost his entire herbarium, his carpological collections, letters and manuscripts are all destroyed) (copy: G).
Jessen, K.F.W., Bot. Gegenw. Vorz. 371, 416, Nachtr. 495.
Kanitz, A., Magy. növén. Lap. 3: 63-64. 1879 (obit.).
Kirschleger, F., Fl. Alsace 2: lxxix. 1857.
Lamy, D., Rev. bryol. lich. 42(1): 561-562. 1976 (Rchb. and hepat.).
Lindemann, E. von, Bull. Soc. Natural. Moscou 61(1): 60. 1886.
Mägdefrau, K., Gesch. Bot. 165. 1973.
Müller, K., Natur ser. 2. 5: 230-231. 1879 (obit., portr.).
Müller, R.H.W. & R. Zaunick, Friedr. Traug. Kützing 292 [index]. 1960.
Nees von Esenbeck, C.G., Flora 18(1), Beibl. 1: 1-32. 1835 (extensive discussion of R's botanical works).
Pritzel, G.A., Icon. bot. index 2: xiii, xxiv. 1866.
Reichenbach, H.G.L. & C. Schubert, Flora 5(1): 367-368. 1822 (*Lich. exs.* Heft 1), 6(1): 112. 1823 (id.), 7(1): 369-376. 1824 (Hefte 2-3), 7(1) Beil. 1: 96. 1824 (Heft 4), 9(1) Beil. 2: 97-99. 1826 (Heft 3).
Riedl-Dorn, C., Taxon 30: 727. 1981 (scientific diaries Reichenbach at W).
Rodgers, A.D., William Starling Sullivant 136, 160. 1940.
Schmid, G., Chamisso als Naturforscher 171 [index]. 1942; Goethe u.d. Naturw. 156, 257, 267, 273.
Schmid, G., *in* E.G. Carus and C.F.M. von Martius. Eine Altersfreundschaft in Briefen 136, 160. 1939 (biogr. rep., on his disagreeable character and conflicts with others), see also further refs. listed on p. 171; explains the psychotic habit of Rchb. of calling himself "legitimus praeses" of the Leopoldina after the death of Carus).
Schmid, H., Ber. Senckenb. naturf. Ges. 1878/79: 6. 1887 (obit., member of Senckenb. Ges. from 1822).
Schultes, H.J.A., Flora 5(1): 142, 155-158. 1822 (visit to Rchb.).
Stafleu, F.A., Regn. veg. 71: 308, 309, 335, 339. 1970 (Miquel-Schlechtendal corr.).
Wagenitz, G., Index coll. princ. herb. Gott. 133. 1982.
Wittrock, V.B., Acta Horti Berg. 3(2): 139. 1903, 3(3): xxi, 149, *pl. 49*. 1905.
Zaunick, R., S.B. Abh. naturw. Ges. Isis, Dresden 1935: 153-155. 1936 (on R's significance for the development of the Dresden Isis; extended version of a notice in Dresdner Nachrichten 1934, no. 266, p. 4; Rchb. was an early member of the "Isis" Society and became its spiritual leader; Director of Isis 1836-1866).

COMPOSITE WORKS: (1) *Boraginaceae and Cruciferae* in J. Sturm, Deutschl. Abt. 1. 43: *16 pl.* with text. 1823; *Die Vergissmeinichtarten* (Myosotis) Abt. 1. 42: [i]-xii, table, *16 pl.* with text 1822; *Cruciferae*, Abt. 1, Heft 45: *16 pl.* with text. 1826, *id.* Abt. 1. 48: *16 pl.* with text. 1827.
(2) Editor *Mössler's Handb. Gewächsk.* ed. 2. 1827-1829 (review D.H. Hoppe, Flora 11(2): 529-642. 14 Sep 1818 (vol. 1), 13: 336. 7 Jun 1830 (vols. 2, 3); idem ed. 3. 1833-1834, see TL-2/6176, 6177 (other copy seen of ed. 3: M).
(3) Preface to F.J.F.X. von Miltitz, Handb. bot. Lit., 1829 (TL-2/6080a).

HANDWRITING: Candollea 32: 407-408. 1977.

EPONYMY: *Reichenbachia* K.P.J. Sprengel (1823).

8868. *Florae lipsiensis pharmaceuticae specimen.* Dissertatio inauguralis medica quam gratiosi medicorum ordinis auctoritate pro summis in medicina et chirurgia honoribus rite capessendis die xxiv. mens. januar. [1817] publice defendet auctor Henric. Theophil. Ludov. Reichenbach ... Lipsiae [impressit Joach. Bernh. Hirschfeld) [1817]. Oct. (*Fl. lips. pharm. specim.*).
Publ.: 24 Jan 1817 (date defense thesis), p. [i]-xii, [1]-83. *Copies*: H (2, second incompl.).

8869. *Flora lipsiensis pharmaceutica.* Sistens plantarum agri lipsiensis nunc et olim officinalium venenatarumque diagnoses, descriptiones, synonyma, locos natales, qualitates, vires et usum ... Lipsiae [Leipzig] (apud Carol. Friederic. Franz.) 1817. Oct. (*Fl. lips. pharm.*).

Publ.: After Jan 1817 (p. xii: Jan 1817; thesis Dr. med. defended 24 Jan 1817, see above), p. [i]-xii, [1]-260 (as "248"). *Copies*: MO, NY, USDA.

8870. *Uebersicht der Gattung Aconitum*, Grundzüge einer Monographie derselben, ... Regensburg. 1819. Oct. (*Uebers. Aconitum*).
Publ.: 28 Mar 1819 (see TL-2/2: 307 (sub 3035), issued as an "Extra Beilage" with Flora vol. 2(1), 1819. See e.g. Flora 2: 122-123, 192, 681. 1819, 44: 764. 1861; see also TL-1), p. [1]-84. *Copies*: FI, G, HH, M, NY, U, USDA, WU; IDC 5660. – See also G. Schmid, 1942, p. 95.

8871. *Monographia generis Aconiti* iconibus omnium specierum coloratis illustrata latine et germanice elaborata ... volumen i. Lipsiae [Leipzig] (sumtibus Frid. Christ. Guil. Vogelii) 1820. Fol. †. (*Monogr. Acon.*).
Publ.: 1820-821, p. [i]-iv, [1]-100, *pl. A* (uncol.), *1-18* (7 col. copp., 11 col. liths. by F. Guimpel and auct.). *Copies*: G, H, HH, MICH, MO, NY, USDA; IDC 1179. – See TL-1/1052 and Stearn 1942.
[*1*]: Apr-Mai 1820 (see Flora 28 Mar 1830, for Ostermesse), p. [1]-72, *pl. A, 1-6*.
[*2*]: Jan-Oct 1821, p. 73-100, *pl. 7-18*.
Ref.: Reichenbach, H.G.L., Flora 1: 197-222. 1818, Jahrb. Gewächsk. 1(2): 189-190. Mai 1819.
Sprengel, K., Neue Entd. 3: 285-289. 1822.
Stearn, W.T., J.R. Hort. Soc. 67-297. 1942.

8872. *Amoenitates botanicae dresdenses*. Specimen primum, observationes in Myosotidis genus continens, quas, munus professoris historiae naturalis atque botanices in Academia medico-chirurgica dresdensi ingressus, rei herbariae studiosis offert Henricus Theophilus Ludovicus Reichenbach, ... Dresdae [Dresden] (in Libraria Arnoldiana) 1820. Oct. †. (*Amoen. bot. dresd.*).
Publ.: 1820 (p. 6: Jul 1820; Flora 28 Mai 1821), p. [1]-30, [31, 32]. *Copies*: FI, G, HH, NY.
Ref.: Anon., Flora 4(1): 305-309. 1821.

8873. *Magazin der aesthetischen Botanik* oder Abbildung und Beschreibung der für Gartencultur empfehlungswerthen Gewächse, nebst Angabe ihrer Erziehung, ... Leipzig (Baumgärtnersche Buchhandlung) [1821-] 1822-1824 [-1826]. Qu. †. (*Mag. aesth. Bot.*).
Publ.: 16 Hefte in 2 volumes, 1821-1826. Title as above on covers and on first t.p. Second t.p.: "*Icones et descriptiones plantarum cultarum et colendarum* additi colendi ratione." *Copies*: G, H, NY, USDA (orig. covers); IDC 5661.
1, sect. 1 (fasc. 1-6): 1821-1822, p. [ii]-[iii], *pl. 1-36* with text.
 sect. 2 (fasc. 7-12): 1822-1823, p. [ii]-xviii, *pl. 37-72*.
2, (fasc. 13-16): 1824-1826, cover, *pl. 73-96* with text.

vol.	part	plates	dates	vol.	part	plates	dates
1(1)	1-2	*1-12*	Jan-Jun 1821	2	13-14	*73-84*	1824
	3-4	*13-24*	Jun-Oct 1821		15	*85-90*	1825
	5-6	*25-36*	Jan-Feb 1822		16	*91-96*	Jan-Mar 1826
(2)	7-11	*37-66*	1823				
	12	*67-72*	1823				

Text 200 unnumbered pages; the plates are copper engravings of drawings by H.G.L. von Reichenbach. – The original covers of Hefte 13-16 (vol. 2 †.) (at G) carry a different title: "*Magazin der Garten-Botanik* ..." each Heft with 6 plates and text.
Ref.: GF p. 72-73 (W.T. Stearn).
Anon., Flora 6: 129-137, 319-350, 688, 1823; 8 (Beil. 1): 52-60, 1825, 8(2): 46. 1825.
Schlechtendal, D.F.L. von, Linnaea 1: 281-283. Apr 1826.

8874. *Die Vergissmeinnichtarten* für die Flora Deutschlands des Herrn Jacob Sturm, gezeichnet und beschrieben von H.G. Ludwig Reichenbach ... Myosotis. Echinosper-

mum. Mit xvii. illum. Kupfert. Nürnberg ([b]ei Jacob Sturm) 1822. (*Vergissmeinnichtarten*).
Publ.: 1822 (Flora 21 Aug 1822), p. [i]-xii, *16 pl*. with text, chart. *Copies*: G (without pl.), NY. – Separate issue of same text in J. Sturm, Deutschl. Fl. Abth. 1, Heft 42: [i]-xii, *16 pl*. with text, chart.

8875. *Illustratio specierum Aconiti generis* additis Delphiniis quibusdam, ... Lipsiae [Leipzig] (apud Friedrich Hofmeister) 1823-1827. Fol. (*Ill. sp. Acon. gen.*).
Publ.: 1823-1827 (in 12 Hefte of which the contents and dates are not precisely known), p. [ii-viii, preface material], [1-4, genus descr.], [1-3, analysis], [1-6, synopsis], [1-126, descr. species], [1, binder], *pl. 1-72* (71 copper engr., auct. col.; 1 col. lith.).
Copies: BR, G, H, HH, MO, NY, USDA; IDC 1585. – See binder's instructions for the numbering of the plates. The roman numbers reflect the order of issue. – Alternative title: "*Neue Bearbeitung* der Arten *der Gattung Aconitum* und einiger Delphinien".

Hefte	plates	dates	Hefte	plates	dates
1, 2	*1-12*	Mar-Sep 1823	7, 8, 9	*37-54*	Aug 1825
3, 4	*13-24*	Oct 1823	10, 11, 12	*55-72*	1827
5	*25-30*	Mar 1824			
6	*31-36*	Apr-Dec 1824			

Ref.: Anon., Flora 9: 225-234. 1826 (rev. fasc. 1-5).
Schmid, G., Chamisso als Naturforscher 96. 1942.
Stearn, W.T., J. R. Hort. Soc. 67: 297. 1942.

8876. *Iconographia botanica seu plantae criticae*. Icones plantarum et minus rite cognitarum, indigenarum exoticarumque, iconographia et supplementum, imprimis ad opera Willdenowii, Schkuhrii, Personii [sic], Roemeri et Schultesii, delineatae, et cum commentario succincto editae ... Lipsiae [Leipzig] (apud Fridericum Hofmeister) 1823-1832, 10 vols. Qu. (*Iconogr. bot. pl. crit.*).
Alternative title: "Kupfersammlung kritischer Gewächse oder Abbildungen seltener und weniger genau bekannter Gewächse des In– und Auslandes, als Kupfersammlung und Supplement, vorzüglich zu den Werken von Willdenow, Schkuhr, Persoon, Römer und Schultes, gezeichnet und nebst kurzer Erläuterung herausgegeben von H.G. Ludwig Reichenbach ...".*Publ*.: 10 vols. – An eleventh volume, but with plates numbered *1-110* and with a different alternative title-page (*Agrostographia germanica*) was published in 1834. This volume counts as the first in the series *Icones florae germanicae* (see below). The *plates* (1000) are coloured (USDA)or (other copies seen) uncoloured copper engr. by H.G.L. Reichenbach. The black and white volumes usually came out considerably earlier than their coloured counterparts (see Flora 6: 348. 1823 and letter R. to W.J. Hooker 8 Apr 1831, sending cent. 7 and 8) and also the preface to cent. 1 "auf besonderes Verlangen, doch immer später." It is possible that some of the volumes were published in decades which came out (more or less) monthly. Vol. 3 was published as a whole. – *Copies*: BR, FI, G, H, HH, NY, Teyler, USDA (col.); IDC 5571.
1: 1823 (p. [vii]: Mar 1823), p. [ii-viii], [1]-98, *pl. 1-100* (fig. nos. arabic 1-210).
2: 1824 (p. [vi]: Dec 1824), p. [ii-vii], [1]-96, *pl. 101-200* (211-342).
3: 1825 (p. [vii]: Oct 1825), p. [ii-viii], [1]-92, *pl. 201-300* (343-473).
4: 1826 (p. viii: Sep 1826), p. [ii-viii], [1]-88, *pl. 301-400* (474-586).
5: 1827 (p. viii: 1 Dec 1827), p. [ii-viii], [1]-68, [1]-20 (index), *pl. 401-500* (587-693).
6: 1828 (p. iv: Dec 1828; published in two half-centuries), p. [ii*-v*], [i]-iv, [v-vi], [1]-34, [1]-28, *pl. 501-600* (694-829).
7: 1829 (published in two half-centuries), p. [ii-viii], p. [1]-50, *pl. 601-700* (820-939).
8: 1830, p. [ii-v], [1]-2, [1]-38, *pl. 701-800* (940-1080).
9: 1831 (p. [vii]: 1 Dec 1831), p. [ii-viii], [1]-47, [48, index], *pl. 801-900* (1081-1220).
10: 1832, p. [ii-viii], [1]-32, [1]-42, index, *pl. 900-1000* (1231-1331).
Dates of publication: We have only incomplete information. Most of the volumes came out in decades or sets of decades. We give here information obtained from Flora and Linnaea.

1: *pl. 1-20*: Apr 1823, *pl. 21-60*: Apr-Oct 1823, *pl. 61-100*: Sep-Dec 1823.
2: published as a whole Jan-Mar 1825.
3: *pl. 1-60* published Jan-Jun 1825, *71-100*: Jul-Dec 1825.
4: *pl. 310-330* published by Oct 1826, *331-360* by 1 Jan 1827, *361-400* Jan-Jun 1827.
5: *pl. 401-460* published by Dec 1828, *461-500* late 1827 or early 1828.
6: *pl. 501-550* Jun-Sep 1828, *551-600* Sep 1828-Jun 1829.
7: *pl. 601-700*: t.p. 1829, no further information except that first review known to us was Jul-Dec 1831 in Flora.
8: *pl. 701-750* at any rate in 1830.
9: *pl. 801-900*, in part 1831, later part 1833.
10: *pl. 901-1000*, t.p. 1832, *pl. 951-1000* reviewed by Flora 21 Aug 1834.

8877. *Catechismus der Botanik* als Anleitung zum Selbststudium dieser Wissenschaft, und als botanisches Wörterbuch zu gebrauchen. Gestaltlehre ... Leipzig. 1820. Oct. (*Catech. Bot.*).
Ed. 1: 1820 (preface: 16 Apr 1820), p. [i]-xxxiv, [1]-217, *pl. 1-7*.
Ed. 2: 1824-1826. "Zweite, fast um das Doppelte vermehrte Auflage", Leipzig (in der Baumgärtnerschen Buchhandlung). *Copy*: NY.
  1 (Gestaltlehre): 1825, p. [i]-xxx, [xxxi-xxxii], [1]-252, [253, err.], *pl. 1-7* (col. copp. auct.).
  2 (Physiologie): 1824, p. [ii-xii], [1]-215, [216-217, err.], *pl. 1-3* (uncol. copp. J.C. Zenker). – Alt. t.p. "*Anleitung zur Naturgeschichte des Pflanzenreichs, für die ersten Anfänger ...*".
  3 (Systematik): 1826, p. [i], [1]-306, [307, err.], *pl. 1-5* (uncol. copp.).
English transl. London 1821 (fide PR 7497).

8878. *Iconographia botanica exotica* sive hortus botanicus, imagines plantarum imprimis extra Europam inventarum colligens; cum commentario succincto editus ... Lipsiae [Leipzig] (apud Friedericum Hofmeister) [1824-]1827-1830. Qu. 3 vols. (*Iconogr. bot. exot.*).
Alternative titles (cent. 1-2 and general t.p. 1-3): *Kupfersammlung der* neuesten, oder bisher weniger genau bekannten und verwechselten *ausländischen Gewächse* nebst Angabe ihrer Cultur für Gartenfreunde, ..."; on wrappers of vol. 1: *Decas hortus botanicus*, id. vol. 2, 3: *Iconogr. exot. sive hortus botanicus*.
Publ.: Century 1 was originally published in decades (1, 2 reviewed by Beck 30 Apr 1824). In 1827 Hofmeister took over the publication from the original publisher, and promised one volume of 100 pl. per year. The plates are hand-col. copper engr. by Reichenbach and some by Humm. – The composition of copies varies. We give the composition of the NY copy. – *Copies*: BR, FI, G, HH, NY, Teyler; IDC 5662.
1(1-6): Jan-Jun 1824, p. [i-vi], [1]-32, *pl. 1-60*.
1(6-8): 1825, p. 33-56, *pl. 61-80*.
1(9-10): Jan-Jun 1827 (Flora Jul-Sep 1827; Linnaea Aug-Oct 1827), p. 57-72, [i]-xx, *pl. 81-100*.
2(1-5): 1828, p. [ii-iii], [1]-20, *pl. 101-150*.
2(6-10): 1829, p. 21-36, [1]-4, [i]-xiv, *pl. 151-200*.
3(1-5): 1830, p. [i]-xvi, [1]-18, *pl. 201-250* (copy at G: [i*], [i]-xvi, [xvii-xviii], x-xvi, *pl. 201-250*).

8879. *Uebersicht des Gewaechs-Reichs* in seinen natuerlichen Entwicklungsstufen. Ein Versuch von H.G. Ludwig Reichenbach, ... Erster Theil. Schluessel fuer Herbarien und Gaerten oder Anordnung des Gewaechsreichs nach Classen, Ordnungen, Formationsreihen, Familien, Gruppen, Gattungen und Untergattungen, mit reichhaltigem Register der Gattungen, Untergattungen, Synonymen und franzoesischen Namen. Leipzig (bei Carl Cnobloch) 1828. Oct. (*Uebers. Gew.-Reich.*).
Publ.: Dec 1828 or Jan-Mar 1829 (p. xiv: 1 Dec 1828; Flora 12 (Beil. 2): 48. Mai-Jun 1829, note dated Apr 1829), p. [i]-xiv, [1]-294 [295]. *Copies*: HH, ILL (inf. D.P. Rogers). – Some with latin title: see below, *Consp. regn. veg.*

8880. *Conspectus regni vegetabilis* per gradus naturalis evoluti. Tentamen ... pars prima. Inest clavis herbariorum hortorumque seu dispositio regni vegetabilis secundum classes, ordines, formationes, familias, tribus, genera et subgenera, adiecto indice locupletissimo generum, subgenerum, synonymorum et nominum francogallicorum ... Lipsiae [Leipzig] (apud Carolum Cnobloch) 1828 [1829]. Oct. †. *(Consp. regn. veg.).*

Publ.: Dec 1828 or Jan-Mar 1829 (t.p. 1828; p. xiv pridie cal. Dec. 1828 = 30 Nov 1829; Beck Apr 1829; announcement by Cnobloch, Flora 12(1) Beil. 2: 48. 1829 dated Apr 1829 "erschienen"), p. [i]-xiv, [1]-294, add. 212a-212d, [1, err.]. *Copies*: BR, FH, FI, G, H, M, NY; IDC 5664. – Brizicky(1969) mentions substitute pages 43-44, 145-146.

Other issue: 1828, as orig. but preface and title in German: "*Uebersicht des Gewaechs-Reichs in seinen natuerlichen Entwicklungsstufen. Erster Theil ...*", see above.

Zusätze und Berichtigungen, by H.G.L. Reichenbach *in* Mössler, Handb. Gewächsk. ed. 2. 3: xxxi-xxxiv. 1829.

Ref.: Brizicky, G., Taxon 18: 651-652. 1969 (on subgenera).
Fürnrohr, A.E., Flora 13(1), Erg. Bl. 1-36. 1830.
Schlechtendal, D.F.L. von, Linnaea 5 (Litt.): 2-5. 1830 (rev.).
Schmid, G., Goethe u.d. naturw. 156. 1940 (no. 810) (Sterculiaceae subf. Goetheaceae).

8881. *Flora germanica excursoria ex affinitate regni vegetabilis naturali disposita, sive principia synopseos plantarum in Germania terrisque in Europa media adjacentibus sponte nascentium culturumque frequentius,* ... Lipsiae [Leipzig] (apud Carolum Cnobloch), 1830[-1832]. Duod. (in sixes). *(Fl. germ. excurs.).*

Publ.: 1830-1833 in (at least) seven parts. The copies vary in the order of binding the various parts numbered with roman numerals. Our analysis is still not complete because we could not establish precise dates for the various subdivisions. The dates below are tentative. *Copies*: BR, FH, FI, G, H, HH, M, MO, NY, PH, U, USDA; IDC 283.

part	pages	contents	dates
[1(1)]	[i]-viii	h.t., adv., t.p. dated 1830 [iii] Candido lectore [v]-viii	Mar-Apr 1830
	[ix]-1	consp. linn.; meht. nat., chart	1832
	[1]-136	Acroblastae	Mar-Apr 1830
	137-140, 141^1-140^{20}	Addenda, corrigenda	Jul-Dec 1831 or 1832
[1(2)]	[141]-184	Phylloblastae	Jan-Apr 1831
[1(3)]	[185]-434	Phylloblastae, Synpetalae	Jul-Dec 1831 or 1832
	435-438	Index gen.; 2 maps	Jul-Dec 1832
[2(1)]	[i*-iv*]	h.t., adv., t.p. dated 1830-1832 [iii*], impr.	Jan-Jul 1832
	[i]-iv	dedic. to Hoppe [i], praemonenda [iii-iv; 2 Jun 1832]	Jan-Jul 1832
	v	instr. binder/or/adv. Fl. germ. exs.	
[2(2)]	[435, sic]-647	Calycanthae	1832
	[649]-841	Thalamanthae	1832
	842	nota	1832
	843-873	addenda et corrigenda	1832
	874-878	index generum	1832
[3]	[i]-iv	t.p. and intr. Clavis syn.	Jun 1833
	[1]-140	index generum	Jun 1833

Title pages: 1830: as in heading and "Insunt plantae Acroblastae et Phylloblastae".
1830-1832: id. and "Insunt plantae Acroblastae et Phylloblastae. Accedit I. Conspectus generum ... v mappa orographica ..." .
1833: "*Reichenbachianae florae germanicae clavis synonymica,* simul enumeratio generum, specierum et varietatum, sive index herbariorum ad sublevandum commercium botanophilorum editus. Die Flora von Deutschland nach ihren Gattungen ...". Lipsiae (apud Carolum Cnobloch) 1833. This part was issued separately, p. [i]-iv, [i]-lxxii [sic], [1]-140. *Copies*: HH, M. – See also Linnaea 8 (Litt.): 59-60. 1834.

Associated publication: *Index in Herbarium florae Germanicae,* editum a societate botanicorum

ultra saxaginta, ... herbarii iam centuria xiii. Phanerog. et i. Cryptogamae venales ...". Lipsiae (apud Fr. Hofmeister) s.d., p. [i]-xviii. *Copy*: G. – Index to the first 14 series of exsiccatae with indication of the place of the taxa in the *Fl. germ. excurs*.
Notes: (1) For infrageneric categories in the *Fl. germ. excurs.*, see Brizicky, Taxon 18: 652. 1969.
(2) Cnobloch, (the publisher) announces the book in Flora 13(1). Beil. 1: 32, Jan-Mar 1830: format as Persoon *Synopsis*, but arrangement according to a natural system. "Wird bis zur Ostermesse erscheinen". – "Just received", Flora 14 Mai 1830 ("die ersten Blätter ..." = sect. 1, Acroblastae).
Ref.: Fuchs, H.P., Taxon 7: 51. 1958.
Fürnrohr, A.E., Flora 13(1): 272-285. 14 Mai 1830 (rev. Acroblastae), 16(1). Lit. Ber. 1: 1-13. 28 Jan 1833, 2: 17-23. 7 Mar 1833 (rev. complete work except for Clavis).

8882. *Das Pflanzenreich* in seinem natürlichen Classen und Familien entwickelt und durch mehr als tausend Kupfer gestochene übersichtlich-bildliche Darstellungen für Anfänger und Freunde der Botanik ... Leipzig (Verlag der Expedition des Naturfreundes) 1834[-1835]. Qu. (*Pflanzenreich*).
Publ.: 1834, p. [i]-iv, [1]-62, chart. *Copies*: M, MO, PCS, PH.
Erste Fortsetzung: 1835, p. 63-95, Leipzig (Wagner), n.v.
Other issue: with a cover (only): *Das Universum der Natur*. Zur Unterhaltung und Belehrung über Vor– und Mit-Welt ... Erste Lieferung, Das Pflanzenreich ... Leipzig (id.). 1834, cover around copy as described above, p. [i]-iv, [1]-62, chart. *Copy*: LD.

8883. *Flora exotica*. Die Prachtpflanzen des Auslandes, in naturgetreuen Abbildungen herausgegeben von einer Gesellschaft von Gartenfreunden in Brüssel, mit erläuterndem Text und Anleitung zur Kultur ... Leipzig (bei Friedrich Hofmeister) 1834-1836, 5 vols. Qu. (*Fl. exot.*).
Publ.: 1834-1836, in five volumes with 72 plates each. From Hinrichs' records it looks as if the book was issued (mostly) in sets of 6 Hefte (*36 pl.*) at a time. The plates are hand-col. copp. engr. See note 2, below. *Copy*: NY.
1: 1833-1834, p. [i], [1]-57, 58, *pl. 1-72* (t.p. 1834).
   1(1): *pl. 1-36*: 1830(*1-30*), 1833 (*31-36*).
   1(2): *pl. 37-72*: 26 Jan-1 Feb 1834 (Hinrichs).
2: 1834, p. [i], [1]-48, *pl. 73-144*.
   2(1): *pl. 73-118*: 27 Jul-2 Aug 1834 (Hinrichs).
   2(2): *pl. 119-144*: Aug-Dec 1834.
3: 1835, p. [i], [1]-52, *pl. 145-216*.
   3(1): *pl. 145-176*.
   3(2): *pl. 177-216*.
4: 1835, p. [i], [1]-46, *pl. 217-246, 247/248, 249-288*.
   4(1): *pl. 217-248*.
   4(2): *pl. 249-288*: 18-24 Oct 1835 (Hinrich).
5: 1836, p. [i], [1]-48, [49-50], *pl. 289-326, 328-350, 350a, 351-360*.
   5(1): *pl. 289-324*.
   5(2): *pl. 325-360*: 1-7 Mai 1836 (Hinrichs).
Note: (1) The t.p. of the fifth volume of the NY copy has "sechster Band".
Note (2): An abbreviated version, with German text by Reichenbach of P.C. van Geel, Sertum botanicum, Bruxelles 1827-1832, 4 vols., 594 col. pl. by G. Severeyns. A further edition of van Geel's opus was *Flore de serres* et jardins de Paris ... Paris 1834, 6 vols., 600 col. pl. – *Iconogr. bot. exot.* 3: xvi (translated): The publisher of this work (F. Hofmeister) has bought the plates of the *Sertum botanicum* published in Brussels and publishes them (provided with a German text, serving as a continuation of [the Iconogr. bot. exot.]), in fascicles under the title *Flora exotica*. Die Prachtpflanzen des Auslandes ..." For further notes see J. Helm, Festschr. Nissen 354-355. 1973.
Note (3): D.H. Hoppe, Flora 14(1). Lit. Ber. 2: 17-28. Jan-Feb 1831 reviews "Lieferungen 1-5", *pl. 1-30* as of 1830!

8884. *Agrostographia germanica* sistens icones graminearum et cyperoidearum quas in *Flora germanica* recensuit auctor. Centuria I. Die Gräser und Cyperoideen der deutschen Flora

in getreuen Abbildungen auf Kupfertafeln dargestellt von H.G. Ludwig Reichenbach ... Leipzig (bei Friedrich Hofmeister) 1834. Qu. (*Agrostogr. germ.*).
Orig. ed.: 18-24 Mai 1834 (dec. 1-5), 1836 (dec. 6-10), (t.p. 1834; first part Hinrichs 18-24 Mai 1836; Flora 28 Nov 1836; dec. 6-10: 1836), p. [ii-iii], [1]-50, *pl. 1-110* (hand-col. copp. engr. auct.). *Copies*: BR, G, M, MO, NY, PH, US, USDA.
Note: Was also issued as century 11 of *Iconographia botanica seu plantae criticae* (see above), but counts also as century 1 of *Icones florae germanicae et helveticae* (see below).
Ed. 2: 1850, p. [i], [1]-80, [1, second t.p.], *pl. 72-192* (id.). – *Copies*: BR, G, PH, US.
First t.p.: "*Agrostographia germanica* sistens icones graminearum quas in Flora germanica recensuit auctor Ludovicus Reichenbach. *Editio secunda* emendata et aucta 1850. Die Gräser der deutschen Flora in getreuen Abbildungen auf 121 Kupfertafeln ..." Leipzig (bei Friedrich Hofmeister) 1850. Qu.
Second t.p.: "*Icones florae germanicae et helveticae* ... volumen primum editio secunda emendata et aucta ...".
Ref.: Nees von Esenbeck, C.D., Flora 19(1) Lit. Ber. 2: 29-32. 14 Feb 1836 (rev. of orig. ed.).
Gersdorf, E.G., ed., Repert. ges. deut. Lit. 3: 228-230. 1834 (rev. dec. 1-5), 8: 539. 1836 (rev. dec. 6-10).

8885. *Icones florae germanicae et helveticae*, simul Pedemontanae, Tyrolensis, Istriacae, Dalmaticae, Austriacae, Hungaricae, Transylvanicae, Moravicae, Borussicae, Holsaticae, Belgicae, Hollandicae, ergo Mediae Europae. Iconographia et supplementum ad opera Willdenowii, Schkuhrii, Persoonii, Decandollei, Gaudini, Kochii, aliorumque. Exhibens nuperrimae detectis novitiis additis collectionem compendiosam imaginum characteristicarum omnium generum atque specierum quas in sua Flora germanica excursiora recensuit auctor Ludovicus Reichenbach. Volumen post Agrostographiam editum, secundum. Tetradynamae seu Cruciferae cum Resedeis. Lipsiae [Leipzig] (apud Fridericum Hofmeister) 1837-1838. Qu. (*Icon. fl. germ. helv.*).
Publ.: 25 volumes, of which the first, *Agrostogr. germ.* did not carry this title in the first edition. The first volumes constitute also centuries 11-20 of the *Iconogr. bot. pl. crit.* – The volumes were published, at any rate part of the time, in decuriae, with a varying number of plates. We provide here:
(1) an analysis of each volume.
(2) a table of contents with dates based on notices in Flora, Bot. Zeit., Hinrichs and many other contemporary sources (but data even so incomplete).
(3) a table of contents arranged by family in alphabetical order.
Editors and authors: see below before entry vol. 13.
Issues (a) with coloured plates & Latin text, Qu.
(b) with uncoloured plates & Latin text, Qu.
(c) with partly coloured plates and German text ("wohlfeile halbcolorirte Ausgabe", Oct. entitled *Deutschlands Flora* (1837-1870).
(d) with coloured plates and German text, Qu. Also as *Deutschlands Flora*.
(e) as (d) but with uncoloured plates (probably only from 1896 onward).
Copies: BR (issue b), FI, G (2 copies: issues a and b), H, HH, MO, (issue b), NY (issues a and b), Teyler (issue a), USDA (issue b); IDC 5665.
Artists: vols. 1-12: H.G.L. Reichenbach and C.H. Schnorr, vols. 13-21, Ludwig Reichenbach and H.G. Reichenbach, vols. 22-25 H.G. Reichenbach (in vol. 22), G. Beck v. Mannagetta, F.G. Kohl. The drawings for 19(2) and (3) (1904-1912) were made by J. Poell, the text was by J. Murr, K.H. Zahn and J. Poell. – Originals and proofsheets at W.
Plates: for details see NI 1604 and 1604 Nachtr.
Deutschlands Flora: we have seen only a few volumes of this German language edition of the *Icones* (issue d), namely at BR and G. *Title*: *Deutschlands Flora* mit höchst naturgetreuen, characteristischen Abbildungen aller ihren Pflanzen-Arten in natürlicher Grösse und mit Analysen auf Kupfertafeln, als Beleg für die Flora germanica excursoria und zur Aufnahme und Verbreitung der neuesten Entdeckungen innerhalb Deutschlands und der angrenzenden Länder: Belgien und Holland, Holstein und Schleswig, Ostpreussen, Galizien, Siebenbürgen, Ungarn, Dalmatien, Istrien, Ober-Italien, der Schweiz und Piemont ...". An advertisement included in Flora 31, 1848 mentions the publication of *1185 plates* in 120 *Hefte*. The early volumes came out

concurrently with the original Latin edition; a cheap ("wohlfeile") edition was started before 1843 (see Flora 26(2): 818-819. 28 Dec 1843 (perhaps started 1837).

ANALYSIS ICON. FL. GERM. HELV. (for tables see p. 679-688).

1: 1834-1836 (t.p.; decas 1-5: 1834, decas 6-10: 1836), p. [ii-iii], [1]-50, *pl. 1-110*. "*Agrostographia germanica* sistens icones Graminearum et Cyperoidearum quas in Flora germanica recensuit ...". Although not provided with an *Icon. pl. germ. helv.* t.p. it counts as its first volume. It also counts as cent. 11 of the *Iconogr. bot. pl. crit.*

1 (ed. 2): 1850 (t.p.), p. [i], [1]-80, [, Iconogr. t.p.], *pl. 72-192. Second t.p.: Icones florae Germanicae et helveticae* simul ... volumen primum. Editio secunda emendata et aucta."

2: 1837-1838 (t.p.; dec. 1-2: 26 Mar-1 Apr 1837; 3-5: 19-25 Nov 1837; 6-8: 28 Jan-3 Feb 1838; 9-10: 27 Mai-2 Jun 1838, Flora 21(3) Lit.-Ber. 12: 177. 1838; completed Apr 1838, Flora 28 Apr 1838, 7 Aug 1838; rev. Flora 21(2) Lit.-Ber. 8: 119-126. 7 Sep 1838), p. [ii]-viii, [1]-31, [32, err.], *pl. 1-102. Second t.p.: "Tetradynamiae seu Cruciferae cum Resedeis* in Flora Germanica excursioria recensitae ergo in Germania, Helvetia ... Hollandia provenientes ... tabulae aeneae ciii [sic]. icones cccxxxi ...". First t.p.: in heading.

3: 1838-1839 (t.p.; dec. 1-2: 30 Sep-6 Oct 1838), p. [ii-iii], 1-16, *pl. 1-19, 1-40, 1-16, 16 bis, 17-46. First t.p.: "Icones ...* volumen tertium cui insunt: Papaveraceae, Capparidaceae, Violaceae, Cistineae et Ranunculaceae pro parte." *Second t.p.: "Papaveraceae cum Fumarieis et Berberideis,* Capparideae, Violaceae, Cistineae, Ranunculacearum: Myosurus ... Thalictrum ...".

4: 1840 (t.p.), p. [ii]-viii, 17-28, *pl. 47-128. First t.p.: "Icones ...* volumen quartum cui insunt: Ranunculacearum Anemoneae, Clematideae, Helleboreae, Paeonieae ...". *Second t.p.: Ranunculaceae* [:] Anemoneae, Clematideae, Helleboreae, Paeonicae [sic] ..." – p. [v]-viii; index vol. 3 & 4.

5: Mar 1841-Aug 1842 (t.p. 1841), p. [ii-iii], 1-24, 24b-24c, 25-38, *pl. 129-198, 198b, 199-230. First t.p.: "Icones ...* volumen quintum cui insunt: Rutaceae cum Euphorbiaceis, Sapindaceae, Malvaceae, Oxalidaceae, Caryophyllacearum pars ...". *Second t.p.: Rutaceae cum Euphorbiaceis,* Sapindaceae, Malvaceae, Oxalidaceae, Caryophyllacearum pars ...".

6: 1842-1844 (t.p.: 1844), p. [ii-iii], 39-84, *pl. 231-277, 277b, 278-282, 282b, 283-352. First t.p.: "Icones ...* volumen sextum cui insunt: Caryophyllaceae reliquae, Theaceae, Tiliaceae et cum Lineis Hypericineae ..." . *Second t.p.: "Caryophyllaceae,* Theaceae, Tiliaceae, et cum Lineis Hypericineae ...".

7: Jan-Nov 1845 (t.p.), p. [ii-iii], [1]-40, *pl. 1-51, 51b, 52-82. First t.p.: "Icones ...* volumen septimum Acroblastarum primum cui insunt Isoeteae, Zosteraceae, Aroideae, Potomogetoneae, Alismaceae et Hydrocharideae cum Nymphaeaceis. Accedit supplementum ad Agrostographiam ...". *Second t.p.: Isoeteae, Zosteraceae,* Aroideae, Potamogetoneae, Alismaceae et Hydroclarideae cum Nymphaeaceis ...".

8: 1846 (t.p.), p. [ii-iii], [1]-50, *pl. 193-318. First t.p.: "Icones ...* volumen octavum Acroblastarum secundum cui insunt Cyperoideae: Caricineae, Cyperinae, Scirpinae ...". *Second t.p.: Cyperoideae,* Caricineae, Cyperinae et Scirpineae ...".

9: 1847 (t.p.), p. [ii-iii], [1]-24, *pl. 319, 324, 324b, 325-416, 418. First t.p.: "Icones ...* volumen nonum. Acroblastarum tertium – et cum Agrostographia quartum – cui insunt Typhaceae, Irideae, Narcissineae, Juncaceae ...". *Second t.p.: "Typhaceae,* Irideae, Narcissineae et Juncaceae ...".

10: 1848 (t.p.), p. [ii-iii], [1]-34, *pl. 419-520. First t.p.: "Icones ...* volumen decimum. Acroblastarum tertium – et cum Agrostographia quartum cui insunt Smilaceae et Liliaceae ...". *Second t.p.: Smilaceae et Liliaceae ...".*

11: 1849 (t.p.), p. [ii-iii], [1]-32, *pl. 521-620. First t.p.: Icones ...* volumen decimum primum cui insunt Coniferae, Taxineae, Cytineae, Santalaceae, Elaeagneae, Thymelaeaceae et Amentacearum-Salicineae ...". *Second t.p.: Coniferae, Taxineae, Cytineae, Santalaceae, Elaeagnaceae, Thymelaeaceae et Amentacearum Salicineae ...".*

12: 1850 (t.p.: 1850), p. [ii-iii], [1]-34, *pl. 621-731. First t.p.: Icones ...* volumen decimum secundum cui insunt Amentaceae: Betulineae et Cupuliferae, Urticaceae, Aristolochiaceae, Laurineae et Dipsaceae cum Valerianeis ...". *Second t.p.: Amentaceae: Betulineae et Cupuliferae,* Urticaceae, Aristolochiaceae, Laurineae et Dipsaceae cum Valerianeis ...".

REICHENBACH, H. G. L. (*Icones*)

From volume 13/14 onward the series was written by H.G. Reichenbach (Rchb. fil.), (13-19(1)), 20-22(1); by G.E. Beck von Mannagetta (Beck), (22(2)and 24-25); by J. Murr, K.H. Zahn & J. Poell (19(2); and by F.G. Kohl (23).

13/14: 1850-1851 (t.p.: 1851; p. v.: 28 Dec 1851), p. [ii*-iii*], [ii]-x, [1]-194, *pl. 353-522* (continuing original numbering), *1-170*, (alternative numbering per volume). First t.p.: *Icones ... volumen xiii et xiv auctore H.G. Reichenbach filio ...*". Second t.p.: *Orchideae* in flora germanica recensitae additis Orchideis Europae reliquae, reliqui Rossici imperi, Algerii, ergo tentamen orchidiographiae europaeae ...".

(*1*): 1-32. *pl. 1-60*. 1850.
(*2*): 33-180, *pl. 61-170*. 1851.
(*3*): [i]-x. 28-31 Dec 1851 (or early 1852).
(*4*): 181-194. 28-31 Dec 1851 (or early 1852).

15: 1852-1853 (t.p. 1853)? p. [ii-iii], [1]-106, *pl. 732-891* (cont.), *1-160* (ind.). First t.p.: "*Icones ... recensitarum auctoribus L. Reichenbach et H.G. Reichenbach filio*. Volumen xv. Auctore H.G. Reichenbach filio Lipsiae (sumptibus Ambrosii Abel) ...". Second t.p.: *Cynarocephalae et Calendulaceae* in flora germanica recensitae ...".

15(1): *pl. 1-80*. 1852.
15(2): *pl. 81-160*. 1853.

16: 1853-1854 (t.p.: 1854), p. [ii-iii], [1]-86, *pl. 892-1041* (cont.), *1-150* (ind.). First t.p.: *Icones ... volumen xvi auctore H.G. Reichenbach filio ...*". Second t.p.: *Corymbiferae* in flora germanica recensitae ...". Third t.p. (interim): *Icones ... vol. xvi. Corymbiferae ...* 1853 & 1854.

16(1): *pl. 1-100*. 1853.
16(2): *101-150*. 1854.

17: 1854-1855 (t.p. 1855), p. [ii-iii], [1]-113, *pl. 1042-1116, 1127-1201* (cont.), *pl. 1-150* (ind.). First t.p.: *Icones ... volumen xvii auctore H.G. Reichenbach filio ...*". Second t.p.: *Gentianaceae, Apocynaceae*, Asclepideae, Oleaceae, Styracaceae, Ebenaceae, Aquifoliaceae, Primulaceae, Plumbaginaceae, Bicornes, Caprifoliaceae, Rubiaceae in flora germanica recensitae ...". Third t.p. (interim): *Icones ... vol. xvii. Gentianaceae ...* 1854 & 1855.

18: 1856-1858 (t.p.: 1858), p. [ii-iii], [1]-103, *pl. 1202-1351* (cont.), *pl. 1-150* (ind.). First t.p.: *Icones ... volumen xviii. auctore H.G. Reichenbach filio ...*". Second t.p.: *Labiatae, Verbenaceae*, Heliotropeae, Borragineae, Convolvulaceae superadditis Polygalaceis olim inter Thalamiflores omissis in flora germanica recensitae ...". Pages 1204, 1229, 1283 misnumbered 1205, 1220, 1282.

19(1): 1858-1860 (t.p. 1860), p. [ii-v], [1]-135, *pl. 1352-1440, 1451-1621* (cont.)., *pl. 1-89, 90-260* (ind.). First t.p.: *Icones ... volumen xix. Auctore H.G. Reichenbach filio ...*". Second t.p.: *Cichoriaceae*, Ambrosiaceae, Campanulaceae, Lobeliaceae in flora germanica recensitae. Superadditis Cucurbitaceis ...". third t.p.: (interim): *Icones ... vol. xix. Cichoriaceae ...* 1858 & 1859.

(*1*): *pl. 1-60*. 1858.
(*2*): *pl. 61-150*. 1859.
(*3*): *pl. 151-260*. 1860.

19(2): 1904-1911 (p. v: Oct 1904), p. [ii-v], [1]-341, *pl. 1-308* (by J. Poell). First t.p.: *Icones ... volumen xix(2) auctoribus* Dre J. Murr, K.H. Zahn, J. Poell. Lipsiae et Gerae (Sumptibus Friederici de Zezschwitz) 1906 seq. Second t.p.: *Hieracia* critica vel minus cognita ...".

(*1*): 1-10, *pl. 1-...*. 1904.
(*2*): 11-48, *pl. ...-...* 1905.
(*3*): 49-104, *pl. ...-81*. 1906.
(*4*): 105-152. *pl. 82-...* 1907.
(*5*): 153-184. *pl. ...-...* 1908.
(*6*): 185-240. *pl. ...-174*. 1909.
(*7*): 241-288. *pl. 175-...* 1910.
(*8*): 289-341. *pl. ...-308*. 1911.

For precise dates (each signature came out separately) see p. 341. Second t.p. *part 1*: (pl. 1-81): 1904-1906; *part 2*: (pl. 82-174) 1906-1909, *part 3*: (pl. 175-308): 1909-1912 (sic).

20: 1861-1862 (see p. 125 for dates), p. [ii-iii], [1]-125, *pl. 1622-1841* (cont.), *pl. 1-220* (ind.). First t.p.: *Icones ... volumen xx. Auctore H.G. Reichenbach filio ...*". Second t.p.:

Solanaceae, Personatae, Orobancheae, Acanthaceae, Globulariaceae, Lentibularieae in flora germanica recensitae ...".
- (*1*): [1]-16, *pl. 1-40*. 19 Jun 1861.
- (*2*): 17-32, *pl. 41-80*. 10 Sep 1861.
- (*3*): 33-48, *pl. 81-120*. 21 Dec 1861.
- (*4*): 49-72, *pl. 121-160*. 8 Jul 1862.
- (*5*): 73-88, *pl. 16-200*. 29 Sep 1862.
- (*6*): 89-125, *pl. 201-220*. 22 Dec 1862.

21: 1863-1867 (see p. 100 for dates), p. [ii-iii], [1]-108, *pl. 1842-2051* (cont.), *pl. 1-120* (ind.). First t.p.: *Icones* ... volumen xxi. Auctore H.G. Reichenbach filio ...". Second t.p.: *Umbelliferae* in flora germanica recensitae ...".
- (*1*): *pl. 1-40*. med. Jun 1863.
- (*2*): *pl. 41-70*. late Dec 1863.
- (*3*): *pl. 71-110*. late Dec 1864.
- (*4*): *pl. 111-150*. early Aug 1866.
- (*5*): *pl. 151-190*. med. Jun 1866.
- (*6*): *pl. 191-220*. early Apr 1867.

22: 1867-1903 (t.p. 1903), p. [i-ii], [1]-230, *pl. 2052-2271* (cont.), *1-272* (ind.) and: *148**, *160**, *164**, *169**, *178**, *190**, *191**, *193**, *220**, *220***. First t.p.: *Icones florae germanicae et helveticae simul terrarum adjacentium ergo mediae Europae. Opus auctoribus L. Reichenbach et H.G. Reichenbach fil. conditum, nunc continuatum auctore Dre G. equite Beck de Mannagetta. Volumen xxii auctoribus H.G. Reichenbach filio et G. equite Beck de Mannagetta.* Lipsiae et Gerae [Leipzig, Gera] (sumptibus Friederici de Zezschwitz) 1903. Second t.p.: "*Leguminosae* ...".
- (*1*): 1-16, *pl. 1-40, 46, 47, 57, 79*. 1867. (decas (1-4).
- (*2*): 17-24, *pl. 41-45, 48-56, 58-60*. Jan 1869. (decas 5-6).
- (*3*): 25-32, *pl. 61-78, 80-100*. (decas 7-10).
- (*4*): 33-48, *pl. 101-120*. (decas 11-12).
- (*5*): 49-56, *pl. 121-140*. 1872 (decas 13-14).
- (*6*): 57-64, *pl. 141-160*. (decas 15-16).
- (*7*): 65-80, *pl. 161-180*. (decas 17-18).
- (*8*): 81-88, *pl. 181-200*. (decas 19-20).
- (*9*): 89-96. *pl. 201-220*. Sep 1885 (decas 21-22).
- (*10*): 97-230, *pl. 221-272, 1900-1903, asterisked pl.* 1903; see p. 230 for precise dates of text p. 105-203.

23: 1896-1899 (t.p. 1898-1899), p. [i-iii], [1]-75, [76, err.], *pl. 1-64, 64a, 65-74, 74a, 75, 75[bis]-119, 119a, 120-138, 138a, 138b.* – First t.p.: *Icones* ... [as vol. 21] ... Volumen xxiii. *Auctore* F.G. Kohl [Friedrich Georg Kohl 1855-1910]. Second t.p.: *Onagraceae, Myriophyllaceae, Hippuridaceae, Ceratophyllaceae, Lythraceae, Crassulaceae, Saxifragaceae, Adoxaceae, Grossulariaceae, Araliaceae in flora germanica recensitae* ...". Gerae [Gera] (sumptibus Friederici de Zezschwitz) 1898-1899.
- (*1*): 1-8, *pl. 1-10*. 1896.
- (*2*): 9-16, *pl. 11-20*. 1896.
- (*3/4*): 17-32. *pl. 21-40*. 1896.
- (*5/6*): 33-36. *pl. 41-60*. 1897.
- (*7/8*): 37-44. *pl. 61-80*. 1897.
- (*9/10*): 45-52. *pl. 81-100*. 1898.
- (*11/12*): 53-68. *pl. 101-120*. 1898.
- (*13/14*): 69-83. *pl. 121-138b*. 1899.

24: Sep 1903-Feb 1909 (t.p. 1909; see p. 213 for dates), p. [ii-iii], [1]-213, *pl. 139-151, 151*, 152-301. First t.p.: Icones* ... volumen xxiv. *Auctore* G. Equite Beck de Mannagetta et Lerchenau ...". Second t.p.: *Polygonaceae*, Chenopodiaceae (incl. Amaranthaceis), Lorantaceae, Cornaceae, Spiraeaceae, Philadelphiaceae, Aizoaceae, Thelygonaceae ... Lipsiae et Gerae [Leipzig, Gera] (sumptibus Friederici de Zezschwitz). 1909.
- (*1*): 1-16, *pl. 139-153*. 1903.
- (*2*): 17-48, *pl. 154-188*. 1904.
- (*3*): 49-64, *pl. 189-200*. 1905.
- (*4*): 65-80, *pl. 201-216*. 1906.
- (*5*): 81-112, *pl. 217-250*. 1907.
- (*6*): 113-152, *pl. 251-...* 1908.

(7): 153-160 *pl.* ...-... 1909.
(8): 161-216. *pl.* ...-*301*. 1909.
25(*1*): 1909-1912 (t.p.; p. [1]: Jul 1909, insert p. [v]: Sep 1909), p. [ii-v], [1]-81, *pl. 1-79*.
First t.p.: *Icones* ... volumen xxv, pars 1. *Auctore G. equite Beck de Mannagetta et Lerchenau.* Lipsiae et Gerae [Leipzig, Gera] (sumptibus Friederici de Zezschwitz) 1909-1912. Second t.p.: *Rosaceae.* I. Dryadeae (Potentilleae), Filipenduleae ...".
25(*2*): 1913-1914, 1-40, *pl. 80-119*.
(1): 1-24. *pl. 80-* .1913.
(2): 25-36. *pl. -115*. 1914.
(3): 37-40. *pl. 110-119*. 24 Dec 1914.
Ref.: Hofmeister, F., Flora 21(2) Lit.-Ber. 12: 177-182. 28 Dec 1838 (reply to review in Flora Lit.-Ber. 7 Sep 1838), p. 182-184 (reply reviewer).

8886. *Kupfersammlung zum praktischen deutschen Botanisirbuche* ... Erste Lieferung. Enthält Keimung und Knospung und zweihundert und vier und neunzig Gattungen der deutschen Flora mit ihren Analysen und zwölf netten Kupfertafeln ... Leipzig (Verlag der Wagner'schen Buchhandlung) 1836. †. Oct. (*Kupfersamml. deut. Bot.-Buch.*).
Publ.: 23-29 Oct 1836 (Hinrichs; Flora 28 Nov 1836), p. [i], [1]-16, *12 pl. Copies*: M. (Inf. I. Haesler), NY. – All published.
Ref.: Anon., Flora 20(1) (Lit.-Ber. 5) 72-73. 14 Mai 1837.

8887. *Handbuch des natürlichen Pflanzensystems* nach allen seinen Classen, Ordnungen und Familien, nebst naturgemässer Grüppirung der Gattungen, oder Stamm und Verzweigung des Gewächsreiches, enthaltend eine vollständige Charakteristik und Ausführung der natürlichen Verwandtschaften der Pflanzen in ihrer Richtung aus der Metamorphose und geographischen Verbreitung, wie die fortgebildete Zeit deren Anschauung fordert, ... Dresden und Leipzig (in der Arnoldischen Buchhandlung) 1837. Qu. (*Handb. nat. Pfl.-Syst.*).
Publ.: 1-7 Oct 1837 (Hinrichs; Endlicher sent a copy to Unger 13 Nov 1837), p. [i]-x, [1]-346. *Copies*: FI, H, M, NY, US.
Zweite Ausgabe: Oct 1850 (Bot. Zeit. 24 Oct 1851), p. [i]-x, [1]-346. *Copies*: BR, G. – Differs only in t.p.
Ref.: Anon., Isis (Oken) 1839: 682-685 (rev.).
Schmid, G., Goethe u.d. Naturw. 267, 273, 1900 (p. 19-96 of eds. 1 and 2 contain a chapter on Goethe's "Metamorphosenlehre").

8888. *Der deutsche Botaniker* ... Erster Band. Das *Herbarienbuch*. Erklärung des natürlichen Pflanzensystems, systematische Aufzählung, Synonymik und Register der bis jetzt bekannten Pflanzengattungen zur erleichterten Aufsuchung der Verwandtschaft jeder einzelnen Gattung und Untergattung und zur schnellsten Auffindung derselben im Herbarium, ... Dresden und Leipzig, in der Arnoldischen Buchhandlung) 1841. Oct. (*Deut. Bot. Herb.-Buch*).
Publ.: Jul 1841 (date on p. [ii]), p. [ii]-xcv, [1]-213, [214, err.], synonymorum reductio: [1]-236 [some copies: 237-240, advertisement]. *Copies*: G, HH, L, M, MICH, NY, U; IDC 5665. – Second titlepage: *Repertorium herbarii* sive nomenclator generum plantarum systematicus, synonymicus et alphabeticus, ad usum practicum accomodatus, quo affinitas naturalis et locus cuiusque generis in herbario critissime explorentur, ..." – See O. Kuntze, Rev. gen. pl. 1: cxli, and Brizicky, Taxon 18: 652-653. 1969.
Vol. 2 of der deutsche Botaniker: see *Flora saxonica* 1842.
Note: The copy at M has the *Repertorium* t.p. [ii] and a different t.p. "*Der deutsche Botaniker* herausgegeben von H.G.L. Reichenbach, ... Erster Band. Das Herbarienbuch. Erste Abtheilung: ... unmittelbar folgt: zweite Abtheilung: ... in Herbario ... Juli 1841 [sic: t.p. dated July] preceding p. [v]-x and [1]-213, and p. [xi-xcv], 1-236, [237-240] preceded by the title page as in our heading. This could imply that p. xi-xcv and 1-236 [237-240] came out after Jul 1841.

8889. *Flora saxonica*. Die Flora von Sachsen, ein botanisches Excursionsbuch für das Königreich Sachsen, das Grossherzogthum Sachsen-Weimar-Eisenach, die Herzogthümer Sachsen-Altenburg, Sachsen-Coburg-Gotha und Sachsen-Meiningen, die fürstlich

1	[1]-50	1-110	1834-1836	Reichenbach, L.	Gramineae
2	[1]-31, [32, err.]	1-102	1837-1838	Reichenbach, L.	Cruciferae
3	1-3	1-18	1838-1839	Reichenbach, L.	Papaveraceae
	3	19	1838-1839	Reichenbach, L.	Capparidaceae
	3-6	1-23	1838-1839	Reichenbach, L.	Violaceae
	6-8	24-40	1838-1839	Reichenbach, L.	Cistaceae
	9-16	1-16	1838-1839	Reichenbach, L.	Ranunculaceae
	16 bis, 17-46				
4	17-28	47-128	1840	Reichenbach, L.	Ranunculaceae
5	1-2	129-130	Mar 1841-Aug 1842	Reichenbach, L.	Callitrichaceae
	2-10	131-153	Mar 1841-Aug 1842	Reichenbach, L.	Euphorbiaceae
	11-13	154-160	Mar 1841-Aug 1842	Reichenbach, L.	Rutaceae
	14-16	161-164	Mar 1841-Aug 1842	Reichenbach, L.	Sapindaceae
	17-20	165-182	Mar 1841-Aug 1842	Reichenbach, L.	Malvaceae
	21-24	183-198	Mar 1841-Aug 1842	Reichenbach, L.	Geraniaceae
	24b-24c	198b-199	Mar 1841-Aug 1842	Reichenbach, L.	Oxalidaceae
	25-38	200-230	Mar 1841-Aug 1842	Reichenbach, L.	Caryophyllaceae
	1-56	231-277, 277b,	1842-1844 (vol.)	Reichenbach, L.	Caryophyllaceae
6	278-282, 282b, 283-308				
	56	309-310	1842-1844	Reichenbach, L.	Theaceae
	57-60	311-324	1842-1844	Reichenbach, L.	Tiliaceae
	61-70	325-352	1842-1844	Reichenbach, L.	Hypericaceae
7	[1]-2	1	Jan-Nov 1845 (vol.)	Reichenbach, L.	Isoetaceae
	2-4	2-5	Jan-Nov 1845	Reichenbach, L.	Zosteraceae
	4-7	6-13	Jan-Nov 1845	Reichenbach, L.	Araceae
	7-26	14-50	Jan-Nov 1845	Reichenbach, L.	Potamogetonaceae
	27-30	51, 51b, 52-58	Jan-Nov 1845	Reichenbach, L.	Alismaceae
7	30-34	59-71	Jan-Nov 1845	Reichenbach, L.	Hydrocharitaceae
	35-38	72-82	Jan-Nov 1845	Reichenbach, L.	Gramineae
8	[1]-50	193-318,	1846	Reichenbach, L.	Cyperaceae
9	[1]-3	319-324, 324b,	1847	Reichenbach, L.	Typhaceae
	325-326				
	3-10	327-361	1847	Reichenbach, L.	Iridaceae
	10-14	362-374	1847	Reichenbach, L.	Amaryllidaceae
	14-22	375-416, 418	1847	Reichenbach, L.	Juncaceae

vol.	pag.	plate	date	author(s)	content
10	[1]-3	419-428	1848	Reichenbach, L.	Cyperaceae
	4-6	429-439	1848	Reichenbach, L.	Smilacaceae
	6-32	440-520	1848	Reichenbach, L.	Liliaceae
11	[1]-8	521-539	1849	Reichenbach, L.	Coniferae
	9	540	1849	Reichenbach, L.	Cytineae
	9-11	541-548	1849	Reichenbach, L.	Santalaceae
	12	549	1849	Reichenbach, L.	Elaeagnaceae
	12-15	550-556	1849	Reichenbach, L.	Thymelaeaceae
	15-30	557-619	1849	Reichenbach, L.	Salicaceae
	30	620	1849	Reichenbach, L.	Myricaceae
12	[1]-6	621-638	1850	Reichenbach, L.	Betulaceae
	7-9	639-650	1850	Reichenbach, L.	Fagaceae
	9-14	651-667	1850	Reichenbach, L.	Urticaceae
	14-16	668-672	1850	Reichenbach, L.	Aristolochiaceae
	16	673	1850	Reichenbach, L.	Lauraceae
	16-24	674-707	1850	Reichenbach, L.	Dipsacaceae
	25-31	708-731	1850	Reichenbach, L.	Valerianaceae
13/14	[1]-32	1(353)-60(412)	1850	Reichenbach, H. G.	Orchidaceae
	33-194	61(413)-170(522)	1851	Reichenbach, H. G.	Orchidaceae
15	[1]-106	732-891	1852-1853	Reichenbach, H. G.	Compositae
16	[1]-86	892-1041	1853-1854	Reichenbach, H. G.	Compositae
17	[1]-14	1042-1061, 1199 ["1200"] 1200, 1201	1854-1855	Reichenbach, H. G.	Gentianaceae
	14-16	1062-1065, 1199 ["1200"], 1201	1854-1855	Reichenbach, H. G.	Apocynaceae
	16-18	1066-1071	1854-1855	Reichenbach, H. G.	Asclepiadaceae
	19-22	1072-1077	1854-1855	Reichenbach, H. G.	Oleaceae
	23	1078	1854-1855	Reichenbach, H. G.	Styracaceae
	23-24	1079	1854-1855	Reichenbach, H. G.	Ebenaceae
	24	1080	1854-1855	Reichenbach, H. G.	Aquifoliaceae
	25-51	1080-1127, 1201	1854-1855	Reichenbach, H. G.	Primulaceae
	52-59	1128-1137	1854-1855	Reichenbach, H. G.	Plantaginaceae
	60-68	1138-1151, 1200	1854-1855	Reichenbach, H. G.	Plumbaginaceae
	69	1152	1854-1855	Reichenbach, H. G.	Monotropaceae
	70-71	1153-1156	1854-1855	Reichenbach, H. G.	Pyrolaceae

	52-53	1292-1293	Reichenbach, H. G.	Verbenaceae
	53-79	1294-1333	Reichenbach, H. G.	Borraginaceae
	80	1334	Reichenbach, H. G.	Polemoniaceae
18	80-87	1335-1344	Reichenbach, H. G.	Convolvulaceae
	88-92	1345-1351	Reichenbach, H. G.	Polygalaceae
19(1)	[1]-102	1352-1440, 1451-1577	Reichenbach, H. G.	Compositae
	102-122	1578-1618	Reichenbach, H. G.	Campanulaceae
	122	1618	Reichenbach, H. G.	Lobeliaceae
	122-123	1619-1621	Reichenbach, H. G.	Cucurbitaceae
19(2)	[1]-341	1-308	Murr, J., K. H. Zahn & J. Poell	Compositae
20	[1]-10	1622-1636	Reichenbach, H. G.	Solanaceae
	10-16	1637-1661	Reichenbach, H. G.	Scrophulariaceae
	17-32	1662-1701	Reichenbach, H. G.	Scrophulariaceae
	33-48	1702-1741	Reichenbach, H. G.	Scrophulariaceae
	49-64	1742-1763, 1774-1777	Reichenbach, H. G.	Scrophulariaceae
	65-80	29 Sep 1862	Reichenbach, H. G.	Scrophulariaceae
	81-85, 1826, 1833, 1840, 1841	22 Dec 1862	Reichenbach, H. G.	Scrophulariaceae
	115-117	22 Dec 1862	Reichenbach, H. G.	Orobanchaceae
	85-88	29 Sep 1862	Reichenbach, H. G.	Orobanchaceae
	89-108	22 Dec 1862	Reichenbach, H. G.	Orobanchaceae
	117-118	22 Dec 1862	Reichenbach, H. G.	Orobanchaceae
		1764-1773	8 Jul 1862	Reichenbach, H. G.	Orobanchaceae
		1778-1781	8 Jul 1862	Reichenbach, H. G.	Orobanchaceae
		1782-1810	29 Sep 1862	Reichenbach, H. G.	Orobanchaceae
		1826-1840	22 Dec 1862	Reichenbach, H. G.	Orobanchaceae
	108-109	1811-1815	22 Dec 1862	Reichenbach, H. G.	Acanthaceae
	109-110	29 Sep 1862	Reichenbach, H. G.	Globulariaceae
20		1816-1818	22 Dec 1862	Reichenbach, H. G.	Globulariaceae
	111-114	29 Sep 1862	Reichenbach, H. G.	Lentibulariaceae
		1819-1821	29 Sep 1862	Reichenbach, H. G.	Lentibulariaceae
		1822-1841	22 Dec 1862	Reichenbach, H. G.	Lentibulariaceae
21	[1]-19	1842-1881	mid-Jun 1863	Reichenbach, H. G.	Umbelliferae
	20-35	1882-1911	late Dec 1863	Reichenbach, H. G.	Umbelliferae

REICHENBACH, H. G. L. (*Icones*)

vol.	pag.	plate	date	author(s)	content
	36-57	1912-1951	late Dec 1864	Reichenbach, H. G.	Umbelliferae
	58-73	1952-1991	early Aug 1865	Reichenbach, H. G.	Umbelliferae
	74-96	1992-2031	mid-Jun 1866	Reichenbach, H. G.	Umbelliferae
	97-108	2032-2051	early Apr 1867	Reichenbach, H. G.	Umbelliferae
22	[1]-104	2052-2271	1867	Reichenbach, H. G.	Leguminosae
	105-112		Dec 1900	Beck de Mannagetta, G. E.	Leguminosae
	113-120		Apr 1901	Beck de Mannagetta, G. E.	Leguminosae
	121-128		Jul 1910	Beck de Mannagetta, G. E.	Leguminosae
	129-136		Oct 1901	Beck de Mannagetta, G. E.	Leguminosae
	137-144		Jan 1902	Beck de Mannagetta, G. E.	Leguminosae
	145-152		Apr 1902	Beck de Mannagetta, G. E.	Leguminosae
	153-160		Jul 1902	Beck de Mannagetta, G. E.	Leguminosae
	161-176		Sep 1902	Beck de Mannagetta, G. E.	Leguminosae
	177-230	221-272, 148*, 160*, 169*, 178*, 189*, 190*, 191*, 193*, 195*, 220*, 220**	Apr 1903	Beck de Mannagetta, G. E.	Leguminosae
23	[1]-27	1-29	1896-1899	Kohl, F. G.	Onagraceae
	27	30	1896-1899	Kohl, F. G.	Hippuridaceae
	27-29	31-37	1896-1899	Kohl, F. G.	Ceratophyllaceae
	29-72	39-64, 64a, 65-74, 74a, 75, 75 bis, 76-119, 119a, 120-132	1896-1899	Kohl, F. G.	Crassulaceae
	72-73	133	1896-1899	Kohl, F. G.	Adoxaceae
	73-74	134-138, 138a	1896-1899	Kohl, F. G.	Grossulariaceae
	75	138b	1896-1899	Kohl, F. G.	Araliaceae
24	[1]-4	139-142	Sep 1903	Beck de Mannagetta, G. E. et L. Reichenbach	Loranthaceae
	4-7	143-145	Sep 1903	Beck de Mannagetta, G. E.	Cornaceae
	7-8	146	Sep 1903	Beck de Mannagetta, G. E.	Spiraeaceae
	9-14	147-151, 151*, 152-153	Oct 1903	Beck de Mannagetta, G. E.	Spiraeaceae

			Beck de Mannagetta, G. E.	Aizoaceae
	17-18	157	Dec 1903	Thelygonaceae
	91-96	228-231	Mar 1907	Chenopodiaceae
	97-104	232-242	Jun 1907	Chenopodiaceae
	105-112	243-250	Aug 1907	Chenopodiaceae
24	113-120	251-257	Mar 1908	Chenopodiaceae
	121-136	258-273	Sep 1908	Chenopodiaceae
	137-144	274-281	Nov 1908	Chenopodiaceae
	145-152	282-287, 289	Dec 1908	Chenopodiaceae
	153-160	288, 290-291	Feb 1909	Chenopodiaceae
	161-173	292-295	1909	Chenopodiaceae
	18-24	158-159	Dec 1903	Polygonaceae
	25-32	160-167, 174, 177	Mar 1904	Polygonaceae
	33-40	168-172, 175 181, 184	Mai 1904	Polygonaceae
	41-48	173, 176, 178-180, 182-183, 185	Sep 1904	Polygonaceae
	49-56	186-191, 193-194	Apr 1905	Polygonaceae
	57-64	192, 195-201	Sep 1905	Polygonaceae
	65-72	202-209	Jan 1906	Polygonaceae
	73-80	210-216	Jul 1906	Polygonaceae
	81-88	218-225	Dec 1906	Polygonaceae
	89-90	226-227	Mar 1907	Polygonaceae
24	173-183		1909	Amaranthaceae
		296-301	Feb 1909	Amaranthaceae
25(1)	[1-2]-81	1-79	Sep 1909-Mar 1912	Rosaceae
25(2)	[1]-40	80-119	1913-1914	Rosaceae

Content	author(s)	vol.	pag.	plate	date
Acanthaceae	Reichenbach, H. G.	20	108-109		22 Dec 1862
Acanthaceae	Reichenbach, H. G.				29 Sep 1862
Adoxaceae	Kohl, F. G.	23	72-73	1811-1815	1896-99
Aizoaceae	Beck, G. E. de Mannagetta	24	15-16	133	Oct 1903
Aizoaceae	Beck, G. E. de Mannagetta				Dec 1903
Alismaceae	Reichenbach, L.	7	27-30	155-156	Jan-Nov 1845
Amaranthaceae	Beck, G. E. de Mannagetta	24	173-183	51, 51b, 52-58	1909
Amaranthaceae	Beck, G. E. de Mannagetta				Feb 1909
Amaryllidaceae	Reichenbach, L.	9	10-14	296-301	1847
Apocynaceae	Reichenbach, H. G.	17	14-16	362-374	1854-55
Aquifoliaceae	Reichenbach, H. G.	17	24	1062-1065, 1199 ["1200"], 1201	1854-55
Araceae	Reichenbach, L.	7	4-7	1080	Jan-Nov 1845
Araliaceae	Kohl, F. G.	23	75	6-13	1896-99
Aristolochiaceae	Reichenbach, L.	12	14-16	138b	1850
Asclepiadaceae	Reichenbach, H. G.	17	16-18	668-672	1854-55
Betulaceae	Reichenbach, L.	12	[1]-6	1066-1071	1850
Borraginaceae	Reichenbach, H. G.	18	53-79	621-638	1856-58
Callitrichaceae	Reichenbach, L.	5	1-2	1294-1333	Mar 1841
Campanulaceae	Reichenbach, H. G.	19	102-122	129-130	1858-60
Capparidaceae	Reichenbach, L.	3	3	1578-1618	1838-39
Caprifoliaceae	Reichenbach, H. G.	17	83-87	19	1854-55
Caryophyllaceae	Reichenbach, L.	5	25-38	1170-1175	Mar 1841-Aug 1842
Caryophyllaceae	Reichenbach, L.	6	1-56	200-237	1842-1844
				238-277, 277b, 278-282, 282b, 283-308	
Ceratophyllaceae	Kohl, F. G.	23	27-29	31-37	1896-1898
Chenopodiaceae	Beck de Mannagetta, G. E.	24	91-96	228-231	Mar 1907
Chenopodiaceae	Beck de Mannagetta, G. E.	24	97-104	232-242	Jun 1907
Chenopodiaceae	Beck de Mannagetta, G. E.	24	105-112	243-250	Aug 1907
Chenopodiaceae	Beck de Mannagetta, G. E.	24	113-120	251-257	Mar 1908
Chenopodiaceae	Beck de Mannagetta, G. E.	24	121-136	258-273	Sep 1908
Chenopodiaceae	Beck de Mannagetta, G. E.	24	137-144	274-281	Nov 1908
Chenopodiaceae	Beck de Mannagetta, G. E.	24	145-152	282-287, 289	Dec 1908
Chenopodiaceae	Beck de Mannagetta, G. E.	24	161-173	292-295	Feb 1909

REICHENBACH, H. G. L.

Compositae	Murr, J., K. H. Zahn & J. Poell	19(2)	[1]2-341	1-308	1904-1911
Coniferae	Reichenbach, L.	11	[1]2-8	521-539	1849
Convolvulaceae	Reichenbach, H. G.	18	80-87	1335-1344	1858
Cornaceae	Beck de Mannagetta, G. E.	24	4-7	143-145	Sep 1903
Crassulaceae	Kohl, F. G.	23	29-72	38-64, 64a, 65-74, 74a, 75, 75bis, 76-119, 119a, 120-132	1896-1898
Cruciferae	Reichenbach, L.	2	[1]-31, [32, err.]	1-102	1837-1838
Cucurbitaceae	Reichenbach, H. G.	19(1)	122-123	1619-1621	1858-1860
Cyperaceae	Reichenbach, L.	8	[1]-50	193-318	1846
Cyperaceae	Reichenbach, L.	10	[1]-3	419-428	1848
Cytineae	Reichenbach, L.	11	9	540	1849
Dipsacaceae	Reichenbach, L.	12	16-24	674-707	1850
Ebenaceae	Reichenbach, H. G.	17	23-24	1079	1854-1855
Elaeagnaceae	Reichenbach, L.	11	12	549	1849
Ericaceae	Reichenbach, H. G.	17	72-82	1157-1169	1854-1855
Euphorbiaceae	Reichenbach, L.	5	2-10	131-153	Aug 1842
Fagaceae	Reichenbach, L.	12	7-9	639-650	1850
Gentianaceae	Reichenbach, H. G.	17	[1]-14	1042-1061, 1199 ["1200"], 1200, 1201	1854-1855
Geraniaceae	Reichenbach, L.	5	21-24	183-198	Aug 1842
Globulariaceae	Reichenbach, H. G.	20	109-110		22 Dec 1862
Globulariaceae	Reichenbach, H. G.	20		1816-1818	29 Sep 1862
Gramineae	Reichenbach, L.	1	[1]-50	1-110	1834-1836
Gramineae	Reichenbach, L.	7	35-38	72-82	Jan-Nov 1845
Grossulariaceae	Kohl, F. G.	23	73-74	134-183, 138a	1896-1899
Hippuridaceae	Kohl, F. G.	23	27	30	1896-1899
Hydrocharitaceae	Reichenbach, L.	7	30-34	59-71	Jan-Nov 1845
Hypericaceae	Reichenbach, L.	6	61-70	325-352	1842-1844
Iridaceae	Reichenbach, L.	9	3-10	327-361	1847
Isoetaceae	Reichenbach, L.	7	[1]	1	Jan-Nov 1845
Juncaceae	Reichenbach, L.	9	14-22	375-416, 418	1847
Labiatae	Reichenbach, H. G.	18	[1]-52	1202-1291	1856-1858
Lauraceae	Reichenbach, L.	12	16	673	1850
Leguminosae	Reichenbach, H. G.	22	[1]-104	1-220	1867-1886
Leguminosae	Beck de Mannagetta, G. E.	22	105-112		Dec 1900

REICHENBACH, H. G. L., (*Icones*), *Arr. by families*

Content	author(s)	vol.	pag.	plate	date
Leguminosae	Beck de Mannagetta, G. E.	22	113-120		Apr 1901
Leguminosae	Beck de Mannagetta, G. E.	22	121-128		Jul 1901
Leguminosae	Beck de Mannagetta, G. E.	22	129-136		Oct 1901
Leguminosae	Beck de Mannagetta, G. E.	22	137-144		Jan 1902
Leguminosae	Beck de Mannagetta, G. E.	22	145-152		Apr 1902
Leguminosae	Beck de Mannagetta, G. E.	22	153-160		Jul 1902
Leguminosae	Beck de Mannagetta, G. E.	22	161-176		Sep 1902
Leguminosae	Beck de Mannagetta, G. E.	22	177-230	221-272, 148*, 160*, 169*, 178*, 189*, 190*, 191*, 193*, 195, 220**	Apr 1903
Lentibulariaceae	Reichenbach, H. G.	20	111-114		22 Dec 1862
Lentibulariaceae	Reichenbach, H. G.	20		1819-1821	29 Sep 1862
Lentibulariaceae	Reichenbach, H. G.	20		1822-1841	22 Dec 1862
Liliaceae	Reichenbach, L.	10	6-32	440-520	1848
Lobeliaceae	Reichenbach, H. G.	19	122	1618	1858-1860
Loranthaceae	Beck de Mannagetta, G. E. et L. Reichenbach	24	[1]-4	139-142	Sep 1903
Malvaceae	Reichenbach, L.	5	17-20	165-182	Aug 1842
Monotropaceae	Reichenbach, H. G.	17	69	1152	1854-55
Myricaceae	Reichenbach, L.	11	30	620	1849
Oleaceae	Reichenbach, H. G.	17	19-22	1072-1077	1854-55
Onagraceae	Kohl, F. G.	23	[1]-27	1-29	1896-99
Orchidaceae	Reichenbach, L.	13/14	[1]-194	1(353)-170(522)	1850-51
Orobanchaceae	Reichenbach, H. G.	20	85-88		29 Sep 1862
Orobanchaceae	Reichenbach, H. G.	20	89-108		22 Dec 1862
Orobanchaceae	Reichenbach, H. G.	20	117-118		22 Dec 1862
Orobanchaceae	Reichenbach, H. G.	20		1764-1773	8 Jul 1862
Orobanchaceae	Reichenbach, H. G.	20		1778-1781	8 Jul 1862
Orobanchaceae	Reichenbach, H. G.	20		1782-1810	29 Sep 1862
Orobanchaceae	Reichenbach, H. G.	20		1826-1840	22 Dec 1862
Oxalidaceae	Reichenbach, L.	5	24b-24c	198b-199	Aug 1842
Papaveraceae	Reichenbach, L.	3	1-3	1-18	1838-39
Philadelphiaceae	Beck de Mannagetta, G. E.	24	14-15		Oct 1903

Family	Author				
Pogygalaceae	Reichenbach, H. G.	18	88-92	1334	1856-58
Polygonaceae	Beck de Mannagetta, G. E.	24	18-24	1345-1351	Dec 1903
Polygonaceae	Beck de Mannagetta, G. E.	24	25-32	158-159	Mar 1904
Polygonaceae	Beck de Mannagetta, G. E.	24	33-40	160-167, 174, 177	Mai 1904
Polygonaceae	Beck, G. E. de Mannagetta	24	41-48	168-172, 175, 181, 184	Sep 1904
Polygonaceae	Beck, G. E. de Mannagetta	24	49-56	173, 176, 178-180, 182-183, 185	Apr 1905
Polygonaceae	Beck, G. E. de Mannagetta	24	57-64	186-191, 193-194	Sep 1905
Polygonaceae	Beck, G. E. de Mannagetta	24	65-72	192, 195-201	Jan 1906
Polygonaceae	Beck, G. E. de Mannagetta	24	73-80	202-209	Jul 1906
Polygonaceae	Beck, G. E. de Mannagetta	24	81-88	210-216	Dec 1906
Polygonaceae	Beck, G. E. de Mannagetta	24	89-90	218-225	Mar 1907
Potamogetonaceae	Reichenbach, L.	7	7-26	226-227	Jan-Nov 1845
Primulaceae	Reichenbach, H. G.	17	25-51	14-50	1854-55
Pyrolaceae	Reichenbach, H. G.	17	70-71	1081-1127, 1201	1854-55
Ranunculaceae	Reichenbach, L.	3	9-16	1153-1156	1838-39
Ranunculaceae	Reichenbach, L.	4	17-28	1-16, 16bis, 17-46	1840
Rosaceae	Beck, G. E. de Mannagetta	25	[1-2]-81	47-128	Mar 1912
Rosaceae	Beck, G. E. de Mannagetta	25	[1]-40	1-79	1913-14
Rubiaceae	Reichenbach, H. G.	17	87-103	80-119	1854-55
Rutaceae	Reichenbach, L.	5	11-13	1176-1198, 1201	Aug 1842
Salicaceae	Reichenbach, L.	11	15-30	154-160	1849
Santalaceae	Reichenbach, L.	11	9-11	557-619	1849
Sapindaceae	Reichenbach, L.	5	14-16	541-548	Aug 1842
Scrophulariaceae	Reichenbach, H. G.	20	10-16	161-164	19 Jun 1861
Scrophulariaceae	Reichenbach, H. G.	20	17-32	1637-1661	10 Sep 1861
Scrophulariaceae	Reichenbach, H. G.	20	33-48	1662-1701	21 Dec 1861
Scrophulariaceae	Reichenbach, H. G.	20	49-64	1702-1741	8 Jul 1862
Scrophulariaceae	Reichenbach, H. G.	20	65-80	1742-1763, 1774-1777	29 Sep 1862
Scrophulariaceae	Reichenbach, H. G.	20	81-85, 115-117	1826, 1833, 1840, 1841	22 Dec 1862
Smilacaceae	Reichenbach, L.	10	4-6	429-439	1848
Solanaceae	Reichenbach, H. G.	20	[1]-10	1622-1636	19 Jun 1861
Spiraeaceae	Beck, G. E. de Mannagetta	24	7-8	146	Sep 1903
Spiraeaceae	Beck, G. E. de Mannagetta	24	9-14	147-151, 151*, 152-153	Oct 1903
Styracaceae	Reichenbach, H. G.	17	23	1078	1854-55
Theaceae	Reichenbach, L.	6	56	309-310	1842-44

REICHENBACH, H. G. L., (*Icones*), Arr. by families

Content	author(s)	vol.	pag.	plate	date
Thelygonaceae	Beck, G. E. de Mannagetta	24	17-18	157	Dec 1903
Thymelaeaceae	Reichenbach, L.	11	12-15	550-556	1849
Tiliaceae	Reichenbach, L.	6	57-60	311-324	1842-44
Typhaceae	Reichenbach, L.	9	[1]-3	319-324, 324b, 325-326	1847
Umbelliferae	Reichenbach, H. G.	21	[1]-19	1842-1881	mid-Jun 1863
Umbelliferae	Reichenbach, H. G.	21	20-35	1882-1911	late Dec 1863
Umbelliferae	Reichenbach, H. G.	21	36-57	1912-1951	late Dec 1864
Umbelliferae	Reichenbach, H. G.	21	58-73	1952-1991	early Aug 1865
Umbelliferae	Reichenbach, H. G.	21	74-96	1992-2031	mid-Jun 1866
Umbelliferae	Reichenbach, H. G.	21	97-108	2032-2051	early Apr 1867
Urticaceae	Reichenbach, L.	12	9-14	651-667	1850
Valerianaceae	Reichenbach, L.	12	25-31	708-731	1850
Verbenaceae	Reichenbach, H. G.	18	52-53	1292-1293	1856-58
Violaceae	Reichenbach, L.	3	3-6	1-23	1838-39
Zosteraceae	Reichenbach, L.	7	2-4	2-5	Jan-Nov 1845

Schwarzburgischen und Reussischen Lande, die Herzogthümer Anhalt-Dessen, Anhalt Bernburg und Anhalt-Köthen, die Provinz Sachsen und die Preussische Lausitz. Nebst Schlüssel zum erleichterten Bestimmen der Gattungen nach Linnee's Sexualsystem und deutschem und lateinischem Register ... Des deutschen Botanikers zweiter Band. Dresden und Leipzig (in der Arnoldischen Buchhandlung) 1842. Oct. (*Fl. saxon.*).

Ed. 1: 1842 (Flora 21 Dec 1842, rev.), p. [i]-xlviii, [1]-461, [462-463]. *Copies*: G, H, HH, MICH, NY, PH, USDA; IDC 5665. – *Second volume of Der deutsche Botaniker.* – The PH copy has the 1842 t.p. but the 503 p. of the 1844 ed.

Ed. 2: 1844, p. [i]-xlviii, [1]-503, [504, colo.]. *Copies*: FI, G, HH, NY. – Zweite Ausgabe mit vollständigem Register der deutschen und lateinischen Namen und ihrer Synonymen. Dresden und Leipzig (id.) 1844. Oct.

Register (also separately published): 1844, p. [i-ii], [453]-503. *Copy*: USDA. – "Register zur Flora von Sachsen, enthaltend die deutschen und lateinischen Namen der Gattungen und Arten nebst Synonymen ..." Dresden und Leipzig (id.) 1844. Oct.

Ref.: Anon., Flora 25(2). Lit.-Ber. 8: 120-136. 21 Dec 1842.

8890. Selectus de Seminario hortus botanici dresdensis mdccclxxi. Prolusio de Scleranthis ... Vorläufiger Blick auf Scleranthus. [Dresden 1871]. Qu. (*Prolusio Scleranth.*).

Publ.: 1871, p. [1-2]. *Copies*: G, M.

Reichenbach, Heinrich Gustav (1824-1889), German (Saxonian) botanist; especially orchidologist; Dr. phil. Leipzig 1852; assistant professor for natural history forestry college Tharandt 1848-1853; lecturer University of Leipzig 1852-1855; extra-ordinary professor of botany and curator of the herbarium Leipzig 1855-1863; professor of botany and the "Academic Gymnasium" and director of the botanical garden Hamburg 1863-1889. (*Rchb. fil.*).

HERBARIUM and TYPES: W. – H.G. Reichenbach's herbarium and library were left to W on the condition that "the preserved Orchids and drawings of Orchids" were not to be consulted during the first 25 years after his death. "And so [C.A. Backer, Verkl. Woordenb. 486] not mindful of the lesson in Galat. 5, 26 "Let us not seek vain glory", in order to remain the mourned specialist himself, he made work difficult for others for a quarter of a century after his death." Britten: "It is painful to feel that a career of usefulness and helpfulness should be terminated by an action which, so far as is possible, hinders the development of a branch of science which its perpetrator had spent his life in advancing." – See also Rolfe (1913, 1917). Duplicates at AMES, BM, BR, FI, GB, GRA, H, HBG, K, LZ, MANCH, NY, P, PH, SK, WAG. – For a list of the collectors in the H.G. Reichenbach herbarium see Keissler & Rechinger (1916). Letters by Reichenbach to Miquel at U, to D.F.L. von Schlechtendal at HAL. – The herbarium is available on microfiches: IDC 8800.

BIBLIOGRAPHY and BIOGRAPHY: ADB 53: 272-276 (G. Dilling); Backer p. 486; Barnhart 3: 140; Biol.-Dokum. 14: 7371; BJI1: 48; BM 4: 1670, 1832 (G.W. Schiller); Bossert p. 328; Christiansen p. 319 [index]; CSP 5: 141, 8: 721, 11: 130-131, 12: 605-609, 18: 104-105; DTS 1: 243; Frank 3 (Anh.): 83; Hegi 6(2): 1085; Herder p. 147, 231, 288; Hortus 3: 1202 ("Rchb. f. "); IF p. 726; IH 1 (ed. 6): 363 (citation AWH erroneous), (ed. 7): 340 (id.), 2: (in press); Jackson p. 18, 427, 430; Kew 4: 440; Langman p. 619; Lenley p. 467; Morren ed. 2, p. 12, ed. 10, p. 27; MW p. 411; NI 1086, 1604, 1609, 2118, (b. 3 Jan 1824, d. 6 Mai 1884), see also p. 145, 218; PR 4163, 7512-7516, 4163, 5335, 7081, 8047, 10624, ed. 1: 4610; Rehder 5: 714; RS p. 136; Saccardo 1: 137, 2: 90; SK 4: cxxii; TL-1/1062; TL-2/4622, see Kraenzlin, F.W.L., Murr, J.; Urban-Berl. p. 387; Zander ed. 10, p. 705, ed. 11, p. 804.

BIOFILE: Ames, O.P., Amer. Orch. Soc. Bull. 1(4): 99. 1933 (portr., note on his herbarium), 10(5): 115-116. 1941 (on *Reichenbachia*).
Anon., Ber. Senckenb. naturf. Ges. 1890: v (d.); Bonplandia 3: 33, 176. 1855 (cogn. Leopoldina L.C. Richard; app. prof. Leipzig); Bot. Centralbl. 38: 751. 1889 (d.); Bot. Jahrb. 11 (Beibl. 24) Heft 3. 1889 (d.); Bot. Not. 1890: 39 (on herb.), 46 (d. 6 Mai 1889); Bot. Zeit. 10: 487. 1852 (candidate for Freiburg i.B.), 10: 552. 1852 (habil.

Leipz.), 21: 232. 1863 (app. Hamburg), 22: 156. 1864 (Leopold order), 25: 31. 1867 (identifies coll. Austral. pl. Amalia Dietrich), 47: 355 (d)., 577 (Sadebeck successor). 1889, 48: 285-286. 1890 (no successor for dir. bot. gard.); Bull. Soc. bot. France 36 (bibl.): 95. 1889 (d.); Flora 38: 368. 1855 (e.o. prof. bot. Leipzig), 46: 347. 1863 (prof. acad. Gymnasium, Hamburg), 511. 1863 (Mettenius replaces Rchb. fil. at Leipzig), 47: 348. 1864 (Leopold order), 48: 237. 1865 (foreign member Roy. Hort. Soc.), 48: 431. 1865 (adjunct Leopoldina), 72: 370. 1889 (d.; conditions regarding herb.); Gard. Chron. 1871: 643. (biogr. sketch, portr.), 5: 624-625. 1889 (obit., portr.); Leopoldina 25 (11/12): 14. 1889 (obit.); Mag. növen. Lap. 13: 92-95. 1890 (obit.); Nat. Nov. 1: 135. 1879 (foreign memberLinn. Soc.), 11: 167. 1889 (d.); Nature 40: 83-84. 1889 (obit.); Österr. bot. Z. 13: 262. 1863 (app. Hamburg, 15: 297. 1865 ("Adjunkt" Leopoldina), 39: 280. 1889 (d.), 40: 69. 1890 (herb. and library to W; 62 cases of books, 122 cases of herbarium material; library of 15.000 vols.; herb. 70.000 specimens); 64: 320. 1914 (Orchid herbarium available for consultation from 6 Mai 1914 onward; collection numbers 50.000 specimens and many drawings).

Arditti, J., ed., et al., Orchid biology 1: 254, 291. 1977.
Beck, G.R. v., in K. Fritsch et al., Bot. Anst. Wiens 75. 1894 (herb. at W).
Bridson, G.D.R. et al., Nat. hist. mss. res. Brit. Isl. 269.299, 291.15. 1980.
Britten, J., J. Bot. 27: 197-198. 1889 (herbarium to be kept sealed for 25 yrs), 52: 280. 1914 (sealed cases opened, herb. in good condition).
Burnat, É., Bull. Soc. bot. France 30: cxxvii-cxxviii. 1888.
Carus, C.G. & C.F.P. von Martius, Eine Altersfreundschaft in Briefen 38, 65, 136, 144, 150. 1939 (G. Schmid, ed.).
Cavillier, F., Boissiera 5: 67, 75-77. 1941.
Crépin, F., Bull. Soc. Bot. Belg., C.R. 28(2): 142-143. 1889 (obit.).
Dilling, G., Heinrich Gustav Reichenbach, Hamburg 1890, 20 p., repr. from Jahrb. Hamb. Wiss. Anst. 7: lxxxviii-cviii. 1890 (obit.).
Dolezal, H., Friedrich Welwitsch 245 (index) 1974.
Fischer-Benzon, R.J.D. von, in P. Prahl, Krit. Fl. Schlesw. Holst. 2: 49. 1890.
Frison, Ed., Henri Ferdinand Van Heurck 39. 1959 (Reichenbach herb. at AWH is that of Karl von Reichenbach).
Gager, C.S., Brooklyn Bot. Gard. Rec. 27(3): 229. 1938 (dir. bot. gard. Hamburg 1863-1889).
Godefroy-Lebeuf, A., ed., L'orchidophile 1889 (Jun): 161-165 (obit., herb. left to W on condition that orchids were not to be consulted for 25 yrs. In case of refusal of that condition the order of other recipients was: UPS, GH, P. Parallel with Desfontaines who did the same with his herbarium left to BR (10 yrs.); injustice versus Kew; at P many herbaria remained unopened even without such a provision).
Hoehne, E., Jard. bot. São Paulo 171-172. 1941 (portr.).
Jones, H.G., Taxon 22: 231, 232. 1973 (portr.).
Keissler, K. v. & K. Rechinger, Ann. Naturhist. Hofmus. Wien 30: 13-23. 1916 (list coll. in Rchb. fil. herb.).
Kerchove de Denterghem, O. de, Rev. Hort. belg. étrang. 15: 125-127. 1889 (obit., portr.).
Lengyel, I., Termész. közl. 22: 642-643., 1890 (obit.).
Masters, , & G. Beck, Ann. Bot. 3: 487-488. 1890 (brief curric. vitae; compl. bibl.).
Merrill, E.D., Contr. U.S. natl. Herb. 30(1): 251-252. 1947.
Nelmes & Cuthbertson, Curtis's Bot. Mag. dedic. 1827-1927: 162. 1932.
Rechinger, K., Verh. k.k. zool.-bot. Ges. Wien 30: 431-437. 1916. (On R.'s orchid herb. at W, report on opening R.'s herb. after 25 years; conditions legacy).
Regel, E.A., Gartenflora 4: 180. 1885 (prof. Leipzig), 12: 303. 1863 (app. Hamburg); Gartenflora 38: 315-320. 1889 (obit., portr.).
Reichenbach, H.G., Bonplandia 2: 127-128. 1884 (on continuatuion of Walper's Ann.); in W.J. Hooker, J. Bot. Kew Gard. misc. 1: 221. 1849 (collections on loan to him, and own herbarium not damaged in Zwinger fire. His father's herb. and library partly destroyed).
Reinikka, M.A., A history of the Orchid 215-218. s.d. (portr., brief biogr. sketch).
Riedl-Dorn, C., Taxon 30: 727. 1981 (scientific diaries Reichenbach fil. at W).
Rodigas, É., Iconogr. Orchid. 4(10): 77-78. 1889.
Rolfe, R.A., Orch. Rev. 21: 273-278, 299-301. 1913 (port., on R's herbarium, his at-

titude towards scientific ethics; implied promises with respect to the future of his herbarium, controversy), 25: 219-224. 1917 (biogr. notice; cites a manuscript note by W.B. Hemsley in the Kew copy of the Journal of Botany containing the obituary notice and that of the disposal of his herbarium suggesting indirectly that the ultimate decision was caused by R's objections against Bentham's treatment of the Orchids for the Genera plantarum), repr. Orchid Digest 36: 217-219. 1972.
Sander, F., Reichenbachia ser. 2. 1: i-iii. 1892.
Seemann, B., Bonplandia 2: 123-124. 1854 (reply to Rchb. fil. on continuation Walpers, Ann.), 7: 52-53. 1859 (on Blume's criticism of R.'s orchidol. work).
Stafleu, F.A., Reg. veg. 71: 334-335, 339. 1970 (not successful in obtaining Dutch appointment; corr. Miquel-Schlechtendal).
Stieber, M.T., Huntia 4(1): 85. 1981 (arch. mat. HU).
Swinson, A., Frederick Sander: the Orchid king 97-99. 1970 (on Sanders and R.).
Urban, I., Symb. ant. 1: 134-135. 1898.
Voigt, A., Bot. Inst. Hamburg 38-43, 52, 54. 1897.
Wittrock, V.B., Acta Horti Berg. 3(2): 139-140. 1903, 3(3): 149, *pl. 49.* 1905.

COMPOSITE WORKS: (1) Editor and in part author of H.G.L. Reichenbach, *Icon. fl. germ. helv.*, vols. 13-19(1), 20-22(1). Author of:
Orchidaceae, 13/14: 1-194, *pl. (1)353-170(522)*. 1851.
Compositae, 15: 1-106,*pl. 732-891*. 1853, 16: 1-86, *pl. 892*. 1854, 19: 1-102, *pl. 1352-1577*. 1860.
Gentianaceae through *Rubiaceae*, 17: 1-103, *pl. 1042-1201*. 1854/5.
Labiatae through *Polygalaceae*, 18: 1-92, *pl. 1202-1351*. 1858.
Campan., Lobel. cucurb., 19: 102-123, *pl. 1578-1621*. 1860.
Solanaceae through *Lentibulariaceae* 20: 1-114, *pl. 1622-1826*. 1862.
Umbelliferae, 21: 1-108, *pl. 1842-2051*. 1863-1867.
Leguminosae, 22: 1-104, *pl. 1-220*. 1867-1886.
(2) With E. Warming, *Orchidaceae in Symbolae* 29: 351-368. 24 Nov 1884, 30: 86-99. 10 Mai 1886.
(3) *Orchidaceae, in* B.C. Seemann, Fl. vitiens. 1865-1873.
(4) *Orchidaceae, in* B.C. Seemann, Bot. Herald 214-215. 1854, 417-419. 1857.
(5) *Orchidaceae, in* vols. 1-2 of W.W. Saunders, *Refugium botanicum*, London 1869-1882; see under W.W. Saunders.
(6) *Orchidaceae, in* Walpers, Ann. 6: 1861-1864.
(7) Weekly articles on Orchids in Gard. Chron. 1865-1889. For a list, alphabetical by genera, see CSP 12: 605-609, contr. after 1885: CSP 18: 105.
(8) Co-editor: *Pescatorea*. See J.E. Planchon, TL-2/4622.
(9) For *Reichenbachia* 1888-1894 see under F. Sander.

HANDWRITING: Candollea 32: 409-410. 1977.

EPONYMY: *Reichenbachanthus* Barbosa Rodrigues (1882 "Reichembachanthus"); X *Reichenbachara* L.A. Garay et H.R. Sweet (1966).

8891. Orchideae in flora germanica recensitae additis Orchideis Europae reliquae, reliqui Rossicii imperi, Algerii ergo *Tentamen Orchidographiae Europaeae* iconibus; illustratum ... Lipsiae [Leipzig] (sumptibus Friderici Hofmeister ...) [1850-] 1851. Qu. (*Tent. orchidogr. Eur.*).
Publ.: 1850-1851 (-1852?), p. [ii*-iii*], [ii]-x, [1]-194, *pl. 353-522 (1-170)*. – Issued as vol. 13/14 of H.G.L. Reichenbach, *Icon. fl. germ. helv.* Published in 4 parts; for further details see no. 8885. *Copies*: BR, G, MO, NY, USDA.
Ref.: S., Österr. bot. W. 3: 223-224. 14 Jul 1853 (rev.).

8892. *De pollinis Orchidearum* genesi ac structura et de orchideis in artem ac systema redigendis. Commentatio quam ex auctoritate amplissimi philosophorum ordinis die mensis julii decimo hora decima mdcccli. illustras ictorum ordinis concessu in auditorio juridico pro venia docendi impetranda publice defendet H.G. Reichenbach. Lipsiae [Leipig] (sumptibus F. Hofmeister) 1852. Qu. (*De pollin. Orchid.*).
Publ.: 10 Jul 1852 (date habil.; Bot. Zeit. 6 Aug 1852; Flora 27 Jul-7 Aug 1852), p. [i, iii], [1]-37, [38], *pl. 1-2* (uncol., auct.). *Copies*: BR, FI, H, KNAW, M, NY, PH.

Ref.: Anon., Flora 36: 746-753. 1853.
A.P. Österr. bot. W. 3: 118-119. 14 Apr 1853 (rev.).
Fürnrohr, A.E., Flora 36: 746-752. 1853 (rev.).
Schlechtendal, D.F.L. von, 10: 920-923. 1852 (rev.).

8893. *Catalog der Orchideen-Sammlung von G.W. Schiller zu Ovelgönne an der Elbe.* (Dritte Ausgabe). Hamburg. Gedruckt bei F.H. Nestler & Melle) 1857. Oct. (*Cat. Orch.-Samml. Schiller*).
Ed. 3: 1857 (p. 4: Feb 1857; Gartenflora Feb 1858; Bot. Zeit. 31 Jul 1857), *4 pl.*, p. [1]-80. *Copies*: FI, G, HH, NY. – This edition was the first for which Reichenbach took care of identification and nomenclature.
Ed. 4: 1861 (p. 4: 30 Apr 1861; Bot. Zeit. 8 Nov 1861; Bonplandia 15 Dec 1861; ÖbZ 1 Mar 1862; Gartenflora Jul 1862), p. [1]-76, *3 pl. Copy*: NY. – Hamburg (Druck von Eduard Freyberg).
Ref.: Schlechtendal, D.F.L. von, Bot. Zeit. 15: 535-536. 31 Jul 1857, 19: 335. 8 Nov 1861 (rev.).

8894. *Xenia orchidacea*. Beiträge zur Kenntniss der Orchideen. – Leipzig (F.A. Brockhaus) [1854-]1858-1900, 3 vols. Qu. (*Xenia orchid.*).
Author vol. 3(4-10): Friedrich Wilhelm Ludwig Kraenzlin (1837-1934).
Publ.: 1854-1900 in 30 Hefte (dates on p. 1: 240, 2: 224, 3: 192). The 300 partly col. plates are copper engravings of drawings by H.G. Reichenbach, Wendland, M. Giraud (Gireaud) and F. Kraenzlin. The latter published 3: 65-192, *pl. 231-300*. 1890-1900 after Reichenbach's death. *Copies*: BR, FI, G, HH, MICH, NY, PH, U, USDA; IDC 1261.
1: 1854-1858, p. [i]-x, [1]-246, *pl. 1-100*.
2: 1862-1874, p. [i]-vii, [1]-232, *pl. 101-200*.
3: 1878-1900, p. [i]-vi, [1]-192, *pl. 201-300*.

Vol.	Heft	pages	*plates*	dates
1	1	1-24	*1-10*	1 Apr 1854
	2	25-48	*11-20*	1 Aug 1854
	3	49-72	*21-30*	20 Nov 1854
	4	73-96	*31-40*	10 Sep 1855
	5	97-120	*41-50*	14 Dec 1855
	6	121-144	*51-60*	1 Mar 1856
	7	145-168	*61-70*	1 Mai 1856
	8	169-192	*71-80*	1 Jun 1856
	9	193-216	*81-90*	21 Nov 1856
	10	217-246, i-x	*91-100*	15 Oct 1858
2	1	1-24	*101-110*	30 Apr 1862
	2	25-48	*111-120*	25 Jul 1862
	3	49-72	*121-130*	10 Aug 1863
	4	73-96	*131-140*	28 Mar 1865
	5	97-120	*141-150*	20 Nov 1867
	6	121-144	*151-160*	8 Apr 1868
	7	145-168	*161-170*	17 Feb 1870
	8	169-192	*171-180*	6 Mar 1873
	9	193-208	*181-190*	10 Oct 1874
	10	209-232, i-vii	*191-200*	31 Dec 1874
3	1	1-24	*201-210*	10 Mai 1878
	2	25-48	*211-220*	15 Oct 1881
	3	49-64	*221-230*	1 Mar 1883
	4	65-76	*231-240*	16 Dec 1890
	5	77-92	*241-250*	21 Jan 1892
	6	93-108	*251-260*	25 Nov 1892
	7	109-124	*261-270*	28 Dec 1893
	8	125-140	*271-280*	31 Jul 1894
	9	141-156	*281-290*	12 Aug 1896
	10	157-192, i-vi	*291-300*	27 Feb 1900

Facsimile ed.: announced but not published by A. Asher & Co., Books on botany 1974.
Note: The dates of publication given in the book are in general confirmed by notices in Bonplandia, Flora, J. Bot., Bot. Centralbl., BSbF, Bot. Zeit.
Ref.: Schlechtendal, D.F.L. von, Bot. Zeit. 12: 405. 1854.

8895. *Orchideae quaedam lansbergianae* caracasano e museo splitgerberiano Horti academici Lugduno-Batavo ... [Ned. Kruidk. Arch. 1858]. Oct.
Publ.: 1858, p. [1]-20. *Copies*: BR, G. – (p. 5-20: Orchideae Splitgerberianae surinamenses). Reprinted and to be cited from Ned. Kruidk. Arch. 4: 315-318, 319-336. 1858.

8896. *Beiträge zu einer Orchideenkunde Central-Amerika's.* Hamburg (Druck von Th.G. Meissner) 1866. Qu. (*Beitr. Orchid.-K. C. Amer.*).
Publ.: 1866 (t.p.; BSbF post 14 Aug 1867; Flora 20 Jul 1869). [1]-112, *pl. 1-10* (uncol., Thieme, auct.). *Copies*: BR, G, HH, M, MICH, NY, USDA; IDC 5028. – Published with Verz. Vorles. Hamb. acad. Real-Gymn. 1867-1868, Hamburg 1866.

8897. *Beiträge zur Orchideenkunde* ... Dresden (Druck von E. Blochmann & Sohn) 1869. Qu. (*Beitr. Orchid.-K.*).
Publ.: 1869 (t.p.; mss submitted Sep 1868; but Flora 23 Jun 1871 and Bot. Zeit. 14 Jul 1871), p. [1]-19, *pl. 1-6* (uncol.). *Copies*: BR, FI, G, HH, NY, U, US. – Preprinted from Nova Acta Leop. 35(2): 1870. – The late references in Flora and Bot. Zeit. (1871) might indicate that the t.p. date is not that of publication.

8898. *Beiträge zur systematischen Pflanzenkunde* ... Hamburg (Druck von Th.G. Meissner, ...) 1871. Qu. (*Beitr. syst. Pflanzenk.*).
Publ.: 1871 (Bot. Zeit. 27 Oct 1871), p. [i], [1]-73, [74, index]. *Copies*: BR, G, HH, L, M, PH, USDA.

8899. Henrici G. Reichenbach fil. *Otia botanica hamburgensia* ... Hamburgi. (Typis Theodor. Theophil. Meissneri, ...) 1878-1881. 2 parts. †. (*Otia bot. hamburg.*).
1: 21 Apr 1878 (t.p.: "dieb. fest. paschal. 1878; Nat. Nov. Aug(1) 1879), p. [1]-68. *Copies*: BR, FI, G, HH, M, MICH, NY, U, USDA. – Also published in Verz. Vorl. Hamb. akad. Real-Gymn. 1878-1879, 1878 (contains Jahresber. p. [i]-xviii, Otia [1]-68, Verz. Vorles. p. [69]-71 (copies e.g. G, NY).
2(1): 8 Aug 1881 (vi. ante Idus Augusti ...; Bot. Centralbl. 2-26 Aug 1881; Nat. Nov. Sep(1) 1881; Bot. Zeit. 30 Sep 1881), p. [i], [69]-119. *Copies*: BR, FI, G, M, MICH, NY. – Also published ibid. 1881, p. 37-89 (copy e.g. BR).

8900. *Ueber das System der Orchideen.* [Bull. Congr. int. bot. St.-Pétersbourg 1884].
Publ.: 1884, p. (1)-(20). *Copies*: BR, FI, NY. – Reprinted and to be cited from Bull. Congr. Bot. St.-Pétersbourg 1884, p. [39]-58.

Reichert, Israel (1891-1975), Polish-born Israeli botanist, studied at the University of Berlin 1913-1919, student of A. Engler; in 1920-1921 with Biologische Reichsanstalt, Berlin and the Ist. patologia vegetale, Firenze; Dr. phil. Berlin 1921; from 1921 at the Agricultural Research Station Rehovoth, from 1949-1959 professor at the faculty of agriculture of Hebrew University, Jerusalem; from 1959-1975 in retirement; Israel State Prize 1955; ardent zionist. (*Reichert*).

HERBARIUM and TYPES: TELA; other material at M and PC.

BIBLIOGRAPHY and BIOGRAPHY: Barnhart 3: 140 (b. 5 Aug 1889); Biol.-Dokum. 14: 7380; BJI 2: 144; GR p. 724-725 (b. 5 Aug 1891), *pl.* [*25*], cat. p. 70 (b. 1891); SBC p. 130; Stevenson p. 1255.

BIOFILE: Bizot, M., Rev. bryol. lichénol. 24: 69, 186. 1955 (musci coll. by R.) (hepat. ibid. by S. Jovet-Ast p. 187).
Hertel, H., Mitt. Bot. München 16: 421. 1980 (lich. from TELA at M).
Laundon, J.R., Lichenologist 11(1): 16. 1979 (b. 1891; lichens at TELA).

Oppenheimer, H.R., Israel J. Bot. 15: 83-90. 1966 (tribute on retirement; portr., bibl.).
Verdoorn, F., ed., Chron. bot. 2: 31, 33, 80, 250, 251, 394. 1936, 3: 218, 359. 1937, 4: 274, 568. 1938.
Wulff, E.V., Introd. hist. pl. geogr. 203, 204, 214. 1943.

THESIS: *Die Pilzflora Ägyptens.* Eine mykogeographische Studie, Bot. Jahrb. 36: 598-727, [728], *pl. 2-4.* 29 Nov 1921.

EPONYMY: *Reichertia* H. Karsten (1848) honors Carl Bogislaus Reichert (1811-1883) of Dorpat. Professor of physiology in the mid 19th century.

Reichgelt, Theodorus Johannes (1903-1966), Dutch botanist; school teacher at Wychen and Nijmegen until 1954; at the Rijksherbarium, Leiden working on the Flora neerlandica 1954-1966. (*Reichgelt*).

HERBARIUM and TYPES: L (incl. NBV).

BIBLIOGRAPHY and BIOGRAPHY: BL 2: 423, 425, 426, 707; Bossert p. 328 (b. 12 Feb 1903; d. [5] Jun 1966); IH 1 (ed. 3): 87, (ed. 4): 95, (ed. 5): 94; Kew 4: 440; Roon p. 93; TL-2/see J.H. Kern.

BIOFILE: Kalkman, C. & P. Smit, eds., Blumea 25(1): 19, 48, 117. 1979.
Ooststroom, S.J. van, Gorteria 3(5): 49, 61-63. 1966 (portr., obit.).
Reichgelt, T.J., Curriculum vitae and bibl., mss sent to senior author 1957 (b. 12 Feb 1903).
Rollins, R.C., Taxon 5: 2. 1956 (photogr. in group).

COMPOSITE WORKS: *Flora neerlandica* 1(3), *Cyperaceae* (*Carex*), Amsterdam 1954, 133 p. (with J.H. Kern); *Cyperaceae* excl. *Carex*, Amsterdam 1956, 52 p.

Reid, Clement (1853-1916), British botanist and geologist; with the British Geological Survey 1874-1913; in retirement at Milford-on-Sea; FRS 1899. (*C. Reid*).

HERBARIUM and TYPES: K (in part); other material BM (393 fruits), TCD.

BIBLIOGRAPHY and BIOGRAPHY: Andrews p. 317-318; BB p. 255; BJI 2: 144; BL 2: 223, 707; BM 4: 1671, 8: 1061; CSP 11: 132, 12: 609, 18: 108-109; Desmond p. 516; Kew 4: 441-442; Quenstedt p. 355, (b. 6 Jan 1853, d. 10 Dec 1916); Rehder 1: 404; Tucker 1: 590.

BIOFILE: Andrews, H.W., The fossil hunters 372-376. 1980.
Anon., Nature 98: 312. 1916 (obit.); Science ser. 2. 45: 14. 1917; Who was Who 1916-1928. London, 878. 1929.
Druce, G.C., Bot. Soc. Exch. Club Brit. Isl. 4(5): 465-466. 1917 (obit.).
Groves, J., J. Bot. 55: 145-151. 1917 (obit., portr., b. 6 Jan 1853, d. 10 Dec 1916); Proc. Linn. Soc. 129: 61-64. 1917 (obit.).
G.W.L., Quart. J. Geol. Soc., London 73(1): lxi-lxiv. 1918 (obit.).
J.E.M. and E.T.W., Proc. Roy. Soc. ser. 13. 90: viii-x. 1919 (obit.).
Kent, D.H., Brit. fl. 74. 1957 (coll.).
Nicholson, W.A., Trans. Norfolk Norwich Natural. Soc. 10(3): 292. 1918 (obit.).
Sayre, G., Bryologist 80: 516. 1977.
Woodward, H., ed., Geol. Mag. Dec 6 4(1): 47-48. 1917 (obit.).
Woodward, T.W., Naturalist 1923 (Mar): 118-119.
Zittel, K.A. v., Gesch. Geol. Paläont. 789. 1899.

EPONYMY: *Reidia* R. Wight (1852) is dedicated to Lt. Col. Francis A. Reid (?-1862) the secretary of the Madras Horticultural Society and its garden director.

8901. *The origin of the British flora* ... London (Dulau & Co., ...) 1899. Oct. (*Orig. Brit. fl.*).

Publ.: Mai-Jun 1899 (ÖbZ Mai-Jun 1899; Bot. Zeit. 16 Aug 1899; Nat. Nov. Jun(2) 1899), p. [i]-vi, [vii], [1]-191. *Copies*: E, G, HH, MO, NY, USGS.
Ref.: Groves, H. & J., J. Bot. 37: 441-444. Oct 1899 (rev.).
Weber (Bremen), Bot. Centralbl. 79: 381-383. 14 Sep 1899.

8902. *The fossil flora of Tegelen-sur-Meuse*, near Venloo, in the province of Limburg ... Amsterdam (Johannes Müller) 1907. Oct.
Co-author: Eleanor Mary (May) Reid, née Wynne-Edwards (1860-1953).
Publ.: Sep 1907 (date on cover), p. [1]-26, *pl. 1-3*. *Copy*: USGS. – Issued and to be cited as Verh. Kon. Akad. Wet., Amsterdam, sect. 2. 13(6): 1-26, *pl. 1-3*. 1907.

8903. *Submerged forests* ... Cambridge (at the University Press) 1913. Oct.
Publ.: 1913 (p. [v]: 17 Feb 1913), p. [i-ii], frontisp., [iii-viii], [1]-129. *Copy*: USGS.

8904. *The pliocene floras of the Dutch-Prussian border* ... The Hague (published ... by the Institute for geological exploration of the Netherlands (Rijksopsporing van Delfstoffen)) 1915. Qu. (*Plioc. fl. Dutch-Pruss. border*).
Co-author: Eleanor Mary (May) Reid, née Wynne-Edwards (1860-1953).
Publ.: 1915, p. [ii-vii], 1-178, [179, err.], *pl. 1-20* (photos with text). *Copies*: E, NY, PH, USGS. – Mededelingen van de Rijksopsporing van Delfstoffen no. 6, 1915.
Ref.: Diels, L., Bot. Jahrb. 54 (Lit.): 12. 27 Jun 1916.
Seward, A.C., J. Bot. 54: 146-149. Mai 1916.

Reid, Eleanor Mary (May) (née Wynne-Edwards) (1860-1953), British mathematician and palaeobotanist; married Clement Reid (1897). (*E. Reid*).

HERBARIUM and TYPES: K; other material at TCD.

BIBLIOGRAPHY and BIOGRAPHY: Andrews p. 318; Barnhart 3: 140; BM 4: 1671, 8: 1061-1062; Desmond p. 516 (d. 28 Sep 1953); Kew 4: 442, 443; MW p. 411.

BIOFILE: Andrews, H.N., The fossil hunters 373-381. 1980 (portr.).
Anon., The Times 9 Oct 1953 (fide Desmond).
Edwards, W.N., Nature 173: 190. 1954 (obit.).
Sayre, G., Bryologist 80: 516. 1977.

8905. British Museum (Natural History) Catalogue of Cainozoic plants in the department of geology volume i *The Bembridge flora* ... London (printed by order of the Trustees, ...) 1926. Oct. (*Bembridge fl.*).
Co-author: Marjorie Elizabeth Jane Chandler.
Publ.: 23 Oct 1926 (t.p.), p. [i]-viii, 1-206, *pl. 1-12*. *Copies*: BR, E (copy rd. 6 Dec 1926), NY, PH, USDA, USGS.
Ref.: Rendle, A.B., J. Bot. 65: 231-232. Aug 1927 (rev.).

8906. British Museum (Natural History). *The London clay flora* ... London (printed by order of the Trustees of the British Museum ...) 1933. Oct. (*London clay fl.*).
Co-author: Marjorie Elizabeth Jane Chandler.
Publ.: 25 Nov 1933 (t.p.), p. [i]-viii, 1-561, *pl. 1-33* with text. *Copies*: BR, E, NY, PH, USDA, USGS.
Ref.: Horwood, A.R., Bull. misc. Inf. Kew 1936: 136-137 (rev.).
Rendle, A.B., J. Bot. 73: 53-55. 1935.

Reider, Jakob Ernst von (x-1853), German (Bavarian) horticultural author and judicial magistrate in Bavaria. (*Reider*).

HERBARIUM and TYPES: Unknown.

BIBLIOGRAPHY and BIOGRAPHY: ADB 27: 683; BM 4: 1671; Herder p. 364; NI 1610; PR ed. 1: 8460-8463, 8458; Rehder 1: 268, 5: 714; Tucker 1: 590.

NOTE: Reider also published *"Der schnell unterrichtende Botaniker und Blumist oder vollständiges Verzeichniss aller Blumen- und Zierpflanzen in der Beschreibung der Arten der Blumen, ..."* Nürnberg (i.d. C.A. Zeh'schen Buchhandlung) 1835. Oct. xiv, 696 p. (n.v.), of which Schlechtendal wrote (Linnaea 9 (Litt.): 82-83. 1834): Für den Botaniker ist dies Wirk gar nichts und dem Blumisten wird es nicht viel nützen ...".

8907. *Annalen der Blumisterei* für Gartenbesitzer, Kunstgärtner, Saamenhändler und Blumenfreunde. In Verbindung mit mehreren Blumenfreunden und Kunstgärtnern herausgegeben von J.E. Reider, ... Nürnberg und Leipzig (Verlag von Conrad Heinrich Zeh) 1826-1836, 12 vols. Oct. (*Ann. Blumist.*).
1: 1826, p. [ii-v], portr., editor, [1]-296, [297-300, Intellig.-Bl.], [301-302, err.], 24 pl. (col. copp.). – Alternative t.p.: *"Beschreibung seltner und neuer vorzüglicher Blumen und Ziergewächse* sammt deren Kultur- und Vermehrungsmethoden bewährter Gärtner ...".
2: 1827 (p. 8: Aug 1826), p. [2]-316, *24 pl.* (id.).
3: 1827 (p. 9: Jul 1827), p. [2]-316, *24 pl.* (id.).
4: 1828 (p. 10: Mai 1828), p. [2]-332, *24 pl.* (id.). – "Verlag der C.H. Zeh'schen Buchhandlung".
5: 1829 (p. 12: Apr 1829), p. [2]-300, *24 pl.* (id.).
6: 1830 (p. x: Mar 1830), p. [ii]-x, [11]-334, *24 pl.* (id.).
7: 1831 (p. x: Feb 1831), p. [ii]-x, [11]-316, *24 pl.* (id.).
8: 1832 (p. xi: Jan 1832), p. [ii]-xi, [12]-310, *24 pl.* (id.).
9: 1833 (p. xviii: 21 Feb 1833), p.[ii]-xviii, [1]-310, *24 pl.* (id.).
10: 1834 (p. 16: Feb 1834), p. [2]-324, *24 pl.* (id.).
11: 1835 (p. [8]: 1 Jan 1835), p. [2]-312, *24 pl.* (id.).
12: 1836 (p. 18: Jan 1836), p. [2]-312, *24 pl.* (id.) [*22* in USDA copy].
Copy: USDA. – Also issued without coloured plates.

8908. *Beschreibung aller bekannten Pelargonien* und Anleitung zur Erkennung und Kultur derselben in Verbindung mit mehreren Freunden dieser Blumen ... Nürnberg (Verlag der E.H. Zeh'schen Buchhandlung) 1829. Oct. (*Beschr. Pelargon.*).
Publ.: 1829 (p. viii: 19 Mai 189), p. [i]-viii, [ix-xi], [1]-383, [384, err.]. *Copy*: USDA.

8909. *Bloemkundige beschrijving* van meer dan vijf honderd, 300 nieuwe, uit andere werelddeelen, sedert weinige jaren, in Europa ingevoerde gewassen, als van verscheidenheden van vroeger bekende soorten, de wijze van kweeking en vermeerdering derzelver. Benevens eene verhandeling over den trekbak, ter vervanging van de warme kas; ... Breda (ter Drukkerij van F.P. Sterk) 1830. Duod. (in sixes). (*Bloemk. beschr.*).
Publ.: 1830, p. [i]-vii, [1, err.], [1]-174. *Copy*: NY.

8910. *Die Beschreibung und Kultur der Calceolarien,* Lilien und Rhododendra ... Ulm (In der J. Ebner'schen Buchhandlung) 1834. Oct. (*Beschr. Calceol.*).
Publ.: 1834, p. [i]-xii, [13]-1559, [160, cont.]. *Copy*: HH.

Reimers, Hermann (1893-1961), German (Holstein) botanist, studied at Freiburg and Kiel, in charge of the biology department of the "Deutsche[s] Forschungsinstitut für Textilrohstoffe" at Karlsruhe; Dr. phil. Hamburg 1922; from 1922-1958 at Berlin-Dahlem (1922 assistant; curator 1948); professor at the Humboldt University, Berlin 1947-1951; bryologist and student of the flora of the Mark Brandenburg. (*Reimers*).

HERBARIUM and TYPES: B; other material at NY. – Reimers' private herbarium, library and manuscripts were all destroyed in the night of 1/2 Mar 1943. The material collected and bought after 1943 is extant (50.000 specimens); it forms the nucleus of the new cryptogamic herbarium at B (see e.g. Pilger 1953).

NOTE: Among the manuscripts destroyed in 1943 were an almost completed flora of the Mark Brandenburg and a moss flora of the Balkan peninsula, the fruits of decades of intensive research.

BIBLIOGRAPHY and BIOGRAPHY: Barnhart 3: 141; BFM no. 1348, 3091; Biol.-Dokum. 14:

7389-7391; BJI 2: 144; BM 8: 1062; Bossert p. 328; Christiansen p. 219 [index]; Futak-Domin p. 500; GR p. 40, cat. p. 70; Hirsch p. 245; IF suppl. 3: 217, 4: 337; IH 1 (ed. 1): 10, (ed. 2): 23, (ed. 3): 26, (ed. 4): 27, 2: (in press); MW p. 411-412, suppl. p. 292; SBC p. 160; TL-2/6494; Zep-Tim p. 96 (b. 17 Jun 1893, d. 18 Mai 1961).

BIOFILE: Anon., Fl. Males. Bull. 17: 874. 1962 (d.); Hedwigia 68: (136). 1929 (trip to Kamerun), 73: (199). 1933 (Oberassistent) ; Österr. bot. Z. 83: 79. 1934 ("Oberassistent" at B); Rev. bryol. lichénol. 30: 164. 1961 (d.); Willdenowia 3(1):64. 1962 (d.).
Bischler, H., Rev. bryol. lichénol. 33: 294-295. 1964 (obit., portr; translation of Schultze-Motel 1963).
Eckardt, Th., Willdenowia 4(2): 173. 1966.
Pilger, R., Willdenowia, Mitt. Bot. Gart. Mus. Berlin-Dahlem 1(1): 2, 15-16, 18. 1953 (created a new cryptogamic herbarium at B).
Reimers, H., Rev. bryol. ser. 2. 1: 226. 1928 (brief account of trip to Kamerun).
Sayre, G., Bryologist 80: 516. 1977.
Scholz, H. & I., Willdenowia 3(1): 21-32. 1962 (obit., portr., bibl. of 102 items, b. 17 Jun 1893, d. 1 Mai 1961).
Schultze-Motel, W., Nova Hedw. 6: 1-4. 1963 (obit., portr.).
Schuster, R.M., Hep. Anthoc. N. Amer. 1: 90. 1966.
Verdoorn, F., ed., Chron. bot. 2: 34. 1936, 3: 27. 1937.

COMPOSITE WORKS: (1) Editor of *Hedwigia* 73-80, 1933-Jan 1942.
(2) *Bryophyta*. Moose, in A. Engler's *Syllabus der Pflanzenfamilien* ed. 12. 1: 218-268. 1954.
(3) Corrections to K. Müll. frib., *Leberm. Eur.* 1954-1957, TL-2/6494, p. 1319-1320.

EPONYMY: *Reimersia* Chen (1914).

Rein, Johannes Justus (1835-1918), German (Hessen) geographer and geologist; Dr. phil. 1861 Rostock; high school teacher at Reval 1858-1860, from 1861-1863 private teacher on the Bermuda Islands; travelled in the Canary Islands and Morocco (1872) and Japan (1873-1875); from 1876 professor of geology at Marburg, from 1883 idem at Bonn. (*Rein*).

HERBARIUM and TYPES: B (mainly destroyed), FR, GOET (main set [of algae?] from Bermudas), HBG, K, KIEL, LE, TCD, W, WRSL.

BIBLIOGRAPHY and BIOGRAPHY: Barnhart 3: 141; BJI 1: 48, 2: 144; CSP 8: 723, 11: 133, 12: 609, 18: 112-113; IH 2: (in press); Kew 4: 444; MW p. 412; Rehder 5: 714-715; Tucker 1: 590; Urban-Berl. p. 268, 387.

BIOFILE: Anon., Geogr. Z. 24: 87. 1918 (d.); Bot. Zeit. 34: 544. 1878 (Japanese plants for sale).
Britton, N.L., Fl. Bermuda 547. 1918.
Embacher, F., Lexik. Reisen 245. 1882.
Geyler, H.T., Bot. Zeit. 34: 544. 1876 (offers Japanese plants for sale).
Kerp, J.J., Geogr. Z. 24: 331-342. 1918 (obit., extensive biogr. details, bibl.).
Koponen, T., *in* G.C.S. Clarke, ed., Bryophyte Systematics 163. 1979 (Jap. coll.).
Maximowicz, C.J., Bot. Zeit. 39: 272-277. 29 Apr 1881 (extensive review with botanical notes, of Rein's *Japan nach Reisen und Studien*, vol. 1, 1881).
Oppermann, E., Geogr. Anz. 19: 31-32. 1918 (obit.; R. studied systematic botany at Giessen).
Philippson, A., Geogr. Monatsber. Mar-Apr 1918, *in* Petermanns Mitt. 64: 80. 1918 (obit., b. 27 Jan 1835, d. 23 Jan 1918).
Poggendorff, J.C., Biogr.-Lit. Handw.-Buch 3: 1102. 1898 (b. 27 Jan 1835, bibl.), 4: 1225. 1902 (bibl.), 6(3): 2147. 1937 (d. 23 Jan 1918).
Urban, I., Symb. ant. 1: 135. 1898, 3(1): 110. 1902. (Bermuda pl. at GOET).
Wagenitz, G., Index coll. princ. herb. Gott. 133. 1982 (Bermuda algae; Japanese pl.).
Ziegler, J., Ber. Senckenb. naturf. Ges. 49: 139-142. 1919 (obit., portr.).
Zittel, K.A. von, Gesch. Geol. Paläont. 373, 380. 1899; Hist. geol. palaeontol. 251. 1901.

8910a. *Japan* nach Reisen und Studien im Auftrage der königlich preussischen Regierung dargestellt ... Leipzig (Verlag von Wilhelm Engelmann) 1881-1886, 2 vols. Oct. (*Japan*).
1(Natur und Volk des Mikadoreiches): 1881 (p. viii: Nov 1880), p. [i]-xii, [xiii], [1]-630, *20 pl.*, 2 maps. – *Flora* on p. 153-198.
2(Land und Forstwirthschaft, ...): 1886 (p. vii: Sep 1886), p. [i]-xii, [1]-678, *24 pl.*, 3 maps. – Botany e.g. on p. 266-345.
Copies: GB (inf. B. Petersen), ILL (inf. D.P. Rogers), LC, NSUB (inf. G. Wagenitz).
English translation: 1884, p. [i-ii], front., [iii]-x, [xi], [1]-543, 20 pl., 2 maps. *Copies*: ILL, LC. – "*Japan*: Travels and researches undertaken at the cost of the Prussian government ...". New York (A.C. Armstrong and Son, ...) 1884. Oct. – For translation of vol. 2 see below.
Ed. 2 of vol. 1: 1905 (p. viii: Dec 1904), p. [i]-xiv, [xv], [1]-749, [750-756, list pl.], *pl. 1-26*. *Copies*: NSUB, PH. – Only vol. 1 was republished: "Natur und Volk des Mikadoreiches".
Ref.: Maximowicz, C.J., Bot. Zeit., 39: 272-277. 1881 (rev. of ed. 1 of vol. 1, with additional botanical notes).

8911. *The industries of Japan*. Together with an account of its agriculture, forestry, arts and commerce. From travels and researches undertaken at the cost of the Prussian government ... London (Hodder & Stoughton, ...) 1889. Oct. (*Industr. Japan*).
Publ: 1889, p. [i]-xii, [1]-570, 3 maps. *Copy*: NSUB (inf. G. Wagenitz). – Translation of vol. 2 of *Japan* (1886), see above.

Reinbold, Theodor (x-1918), German botanist (algologist). (*Reinbold*).

HERBARIUM and TYPES: M (marine algae), other material at B, C, FH, HBG, L (types Sargassum, Cladophora), NY, W, WRSL. – Letters to W.G. Farlow at FH.

BIBLIOGRAPHY and BIOGRAPHY: Barnhart 3: 141 (d. 29 Mar 1918); Biol.-Dokum. 14: 7396; BM 4: 1672, 6: 264, 8: 1062; Christiansen p. 158, 159, 160, 319; CSP 18: 113-114; De Toni 4: xlvi; IH 1 (ed. 7): 340; Kew 4: 444; MW p. 412; TL-2/3519; Urban-Berl. p. 268.

BIOFILE: Anon., Bot. Not. 1918: 259 (d.); Hedwigia 60: (178). 1918 (d.); Magy. bot. Lap. 17: 117. 1918 (d.); Nuova Notar. 30: 102. 1919 (d.).
Fischer-Benzon, R.J.D. von, *in* P. Prahl, Krit. Fl. Schlesw. Holst. 2: 49. 1900.
Kneucker, A., Allg. bot. Z. 23: 48. 1919 (d. 29 Mar 1918).
Koster, J.T., Taxon 18: 556. 1969 (types algae).
Merrill, E.D., B.P. Bishop Mus. Bull. 144: 156. 1937.
Papenfuss, G.F., Antarct. Res. Ser. 1: 69. 1964.
Reinke, J., Mein Tagewerk 180. 1925.

COMPOSITE WORKS: (1) Contributed *Phaeophyceae* and *Dictyotales* of East Africa, in A. Engler, *Die Pflanzenwelt Ost-Afrikas* (Deutsch Ost-Afrika vol. 5C), TL-2/1708.
(2) *Die Meeresalgen der Deutschen Südpolar-Expedition 1901-1903, in* K. v. Drygalski, ed., Deutsche Südpolar-Expedition, viii Botanik, p. 177-202. 1908 (Bot. Centralbl. 27 Oct 1908).
(3) *Meeresalgen, in* K. Rechinger, *Bot. zool. Erg. Forschungsreise Samoains.* 1: 200-208. 1907.
(4) *Meeresalgen, in* F. Reinecke, *Die Flora der Samoa-Inseln* 23: 266-275. 15 Sep 1896.

8912. *Die Meeresalgen der deutschen Tiefsee-Expedition 1898-1899* ... [Jena (Verlag von Gustav Fischer) 1907]. Qu. (*Meeresalg. deut. Tiefsee Exped.*).
Publ.: 1907 (mss. rd. 9 Jun 1906; NY rd. Nov. 1907; Bot. Centralbl. 10 Mar 1908; Nat. Nov. Jan(2) 1908), p. [1]-38, *pl. 1-4* (*=55-58*) with text. *Copies*: BR, M, MO, NY. – Reprinted from Chun, C., ed., *Wiss. Erg. Exped. Valdivia* 2(2) (4): [549]-586, *pl. 55-58*. 1907. See G. Karsten, TL-2/3519.
Ref.: Kuckuck, E.H.P., Bot. Zeit. 66(2): 101-102. 16 Mar 1908 (rev.).

Reinecke, Franz (1866-?), German (Silesian) botanist; studied at Breslau; teacher's examination 1888; at Heidelberg 1890-1892 as student of Pfitzer; Dr. phil. Heidelberg 1893; travelled in Samoa 1893-1895; later in Breslau. (*Reinecke*).

HERBARIUM and TYPES: WRSL; other material at B, BISH, E, K, L, STU, US, W.

BIBLIOGRAPHY and BIOGRAPHY: Barnhart 3: 141 (b. 30 Aug 1866); Biol.-Dokum. 14: 7399; BJI 1: 48; BL 1: 127, 314; CSP 18: 114; Kew 4: 445; LS 22028; MW p. 412; Rehder 5: 715; Tucker 1: 590; Urban-Berl. p. 281, 290, 311, 322, 387.

BIOFILE: Anon., Nat. Nov. 18 no. 8976. 1896 (offer of Samoan plants for sale); Österr. bot. Z. 46: 472. 1896 (publishes centuries of Samoan plants, coll. 1893-1895).
Hedge, I.C. & J.M. Lamond, Index coll. Edinb. herb. 122. 1970.
Merrill, E.D., B.P. Bishop Mus. Bull. 144: 156. 1937 (Pacific bibl.); Contr. U.S. natl. Herb. 30(1): 253. 1947 (bibl.).

COMPOSITE WORKS: With assistance of various other botanists *Die Flora der Samoa-Inseln*, Bot. Jahrb. 23: 237-368, *pl. 4-5*. 1896, 25: 578-708. *pl. 8-13*. 1898.

EPONYMY: *Reineckea* Kunth (1844, *nom. cons.*) and *Reineckea* H. Karsten (1858) honor Johann Heinrich Julius Reinecke (1799-1871), German horticulturist.

8913. *Über die Knospenlage der Laubblätter* bei den Compositen, Campanulaceen u. Lobeliaceen. Inaugural-Dissertation zur Erlangung der Doktorwürde der hohen naturwissenschaftlich-mathematischen Fakultät der Ruperto-Carolinischen Universität zu Heidelberg ... Breslau (Druck von Carl Dülfer) 1893. Oct. (*Knospenlage Laubbl.*).
Publ.: Mai-Jun 1893 (ÖbZ Jun 1893; Nat. Nov. Jul(1) 1893; Flora before 14 Aug 1893), p. [i-iii], [1]-63, [64, contents; 65, vita], *1 pl. Copies*: ILL (inf. D.P. Rogers), HH, MO.

8914. *Ueber die Nutzpflanzen Samoas* und ihre Verwendung. (Jahresber. Schles. Ges. vaterl. Cultur 1895). Oct.
Publ.: 1895 (mss. submitted 23 Sep 1895), p. [1]-24. *Copy*: HH. – Reprinted and to be cited from Jahresber. Schles. Ges. vaterl. Cult. 1895 (2c): 22-46. 1895.

8915. *Die Flora der Samoa-Inseln*... [Separat-Abdruck aus Engler's botanischen Jahrbüchern ... Leipzig (Wilhelm Engelmann)]. Oct.
Publ.: in Bot. Jahrb. 23: 337-304. *pl. 3-4*. 15 Sep 1896, 305-368. 24 Nov 1896, 25: 578-708. 23 Dec 1898. – Reprinted with original pagination, but to be cited from journal. With the collaboration of several other botanists. – R. Friedländer & Sohn, Berlin, issued a special reprint "Die Flora der Samoa-Inseln; see Nat. Nov. 24: 218. 1902.
Fresh water algae by W. Schmidt.
Marine algae by Theodor Reinbold (x-1918).
Fungi by Paul Christoph Hennings (1841-1908).
Lichenes by Jean Müller-Argoviensis (1828-1896).
Hepaticae by Franz Stephani (1842-1927).
Musci by Karl August Friedrich Wilhelm Müller-Halensis (1818-1899).
Pteridophyta by Konrad Hermann Heinrich Christ (1833-1933).
Siphonogamen by Otto Warburg (1859-1938).

8916. Sammlung Göschen. *Das Pflanzenreich*. Einteilung des gesammten Pflanzenreiches mit den wichtigsten und bekanntesten Arten ... Leipzig ([in der] G.J. Göschen'sche Verlagshandlung) 1900. Oct. (*Pflanzenreich*).
Co-author: Walter Migula (1863-1938).
Publ.: 1900 (Nat. Nov. Jan(1) 1901; ÖbZ Jan-Feb 1901; Bot. Zeit. 16 Jan 1901), p. [1]-140. *Copies*: G, NY. – Sammlung Göschen no. 122.

8917. *Samoa* von Dr. F. Reinecke. Berlin (Wilhelm Süsserott ...) [1902]. Oct. (*Samoa*).
Publ.: 1902 (p. [vi]: Mar 1902, p. [i-viii], portr., [1]-312, [313, err.], *1 map. Copies*: GOET, LC. – Vegetation on p. 295-301, Nutzpflanzen 302-312. – Süsseroths Kolonialbibliothek 3-4.

Reinecke, Karl L. (1854-1934), German (Saxonian) botanist and high school teacher at Erfurt; teacher's examination 1874; connected with the Erfurt schools until 1917. (*K. Reinecke*).

HERBARIUM and TYPES: Erfurt (Thür. Heimatmuseum); GB, GOET.

BIBLIOGRAPHY and BIOGRAPHY: AG 6(1): 683; Barnhart 3: 141; BFM no. 452, 453, 454, 455; Biol.-Dokum. 14: 7399.

BIOFILE: Wagenitz, G., Index coll.princ. herb. Gott. 134. 1982.
Wand, A., Mitt. thür. bot. Ver. ser. 2. 42: 111. 1935 (b. 30 Mai 1854, d. 1 Dec 1934).

8918. *Flora von Erfurt.* Verzeichniss der im Kreise Erfurt und seiner nächsten Umgebung beobachteten Gefässpflanzen ... Erfurt (Verlag von Karl Villeret ...) 1914. Oct. (*Fl. Erfurt*).
Publ.: 1914 (Nat. Nov. Mar (1, 2) 1915), p. [1]-283. *Copies*: B(2), BR, G. – Reprinted from Jahrb. k. Akad. Wiss. Erfurt, ser. 2. Heft 40: [i*-vi*], [1]-283, [284, colo.]. 1914. – See also his *Neue Beiträge zur Kenntniss der Flora von Thüringen*, Mitt. thür. bot. Ver. 28: 36-43. 1911, 30: 19-22. 1913; further contributions 25: 21-33. 1921, 36: 20-25. 1925 and 38: 43-47. 1929.
Nachtrag: 1919, *in* Jahrb. Akad. Wiss. Erfurt ser. 2. 44/45: [133]-162. 1919; also as reprint with original pagination.
Nachtrag 2: privately published 14 Oct 1923 (see Nachtrag 3: 120. 1928), copy at B.
Nachtrag 3: 1928, Jahrb. Akad. gemeinnütz. Wiss. Erfurt ser. 2. 47: [117]-134. Feb 1928, also as reprint with original pagination.

Reinecke, W. (*fl.* 1886), German botanist at Gernrode a.H. (*W. Reinecke*).

NOTE: We have no information on W. Reinecke except that he published the *Excursionsflora des Harzes*.

8919. *Excursionsflora des Harzes.* Nebst einer Einführung in die Terminologie und einer Anleitung zum Sammeln, Bestimmen und Konservieren der Pflanzen ... Quedlinburg (Verlag von Chr. Friedr. Vieweg's Buchhandlung) [1886]. Oct. (*Excurs.-Fl. Harz.*).
Publ.: Mai-Jun 1886 (p. [iii]: Apr 1886; Nat. Nov. Jul(1) 1886; Bot. Centralbl. 29 Nov-3 Dec 1886; Bot. Zeit. 30 Jul 1886), p. [i-iv], [1]-245. *Copies*: M, USDA. – Rev. in Schr. nat. Ver. Harz. 1: 78-80 (fide G. Wagenitz).
Ref: Freyn, J.F., Bot. Centralbl. 28: 267. 1886 (follows nomenclature of Garcke, *Fl. Deutschl.*).

Reiner, Joseph (1765-1797), Austrian clergyman and botanist at Klagenfurt, court chaplain and town priest. (*Reiner*).

HERBARIUM and TYPES: Unknown. – PI has collection made by a "Reiner".

BIBLIOGRAPHY and BIOGRAPHY: Dryander 3: 154; DTS 1: 244; Herder p. 57, 187, 192; IH 2: (in press); NI p. 186, 187, 265; PR 7522, ed. 1: 8466; Rehder 1: 439.

EPONYMY: *Reineria* Moench (1802).

8920. *Botanische Reisen* nach einigen Oberkärtnerischen und benachbarten Alpen unternommen, und nebst einer ausführlichen Alpenflora und entomologischen Beiträgen als ein Handbuch für reisende Liebhaber herausgegeben von Joseph Reiner ..., und Sigmund von Hohenwarth ... Erste Reise im Jahr 1791 ... Klagenfurt (gedruckt und verlegt bei Carl Friedr. Walliser) 1792. Oct. (*Bot. Reis.*).
Co-author: Sigmund [Freiherr] von Hohenwarth (1745-1825).
Publ.: 1792, p. [i*-ii*], [i]-xi, [1]-270, [1-16], *6 pl.* (*1-5* bot.). Alpenflora on p. 61-254.
Copy: GOET. – A further (imperfect) copy was seen by J.H. Thomas in the Stanford Univ. Libr.
Re-issue: 1793, p. [i*-ii*], [i]-xi, [1]-270, [1-16]. *Copy*: NY. – Ulm (in der Stettinischen Buchhandlung in Kommission) 1793.

Other volume: 1812, 12, 261 p., *10 pl.*, edited by Lorenz Chrysanth von Vest, *Flora alpina* on p. 127-261 (n.v.).

Reinhard, Ludwig (Vasilievič) (1846-1922), Ukrainian algologist; director of the Museum and Garden and extraord. prof. of botany at Odessa University; from 1885/6-1903 director of the botanical garden and morphological department of the Botanical Institute of Charkow Univerity; from 1903 continuing as professor of botany. (*Reinhard*).

HERBARIUM and TYPES: Some material at MW, possibly also in CW.

BIBLIOGRAPHY and BIOGRAPHY: Barnhart 3: 142 (b. 6 Mar 1846); BM 4: 1673; Bossert p. 328 (err. Reinhardt); CSP 18: 115; De Toni 1: civ-cv, 2: c, 4: xlvi; Morren ed. 110, p. 99; Nordstedt suppl. p. 13; Rehder 1: 366, 4: 318; TR 1147-1152 ("Reinhardt").

BIOFILE: Anon., Bot. Jahrb. 7 (Beibl. 14): Heft 3, p. 1. 1886 (app. Charkow), 7: (Beibl. 15), Heft 5, p. 2 (id.); Bot. Not. 1922: 287 (d.); Bot. Zeit. 44: 143. 1886 (app. Charkow); Nat. Nov. 8: 36. 1886 (app. Charkow), 25: 630. 1903 (retired from directorship CW); Österr. bot. Z. 36: 105. 1886 (app. Charkow), 71: 152. 1922 (d.). Lindemann, E. v., Bull. Soc. Natural. Moscou 61(1): 60-61. 1886 ("Reinhardt"). Szymkiewicz, D., Bibl. Fl. Polsk. 120-121. 1925.

Reinhardt, Ludwig (*fl.* 1910), Swiss or German botanist at Basel. (*L. Reinhardt*).

HERBARIUM and TYPES: Unknown.

BIBLIOGRAPHY and BIOGRAPHY: BFM no. 2975; Kew 4: 445; LS 38309; Tucker 1: 590.

8921. *Kulturgeschichte der Nutzpflanzen* ... München (Verlag von Ernst Reinhardt) 1911, 2 vols. Oct. (*Kulturgesch. Nutzpfl.*).
Publ.: 1911, in two volumes. Published as volume 4 of (second t.p.): *Die Erde und die Kultur, die Eroberung und Nutzbarmachung der Erde durch den Menschen. Copies*: BR, G, M, MO, NY, USDA.
1: Nov-Dec 1910 (see preface: Oct 1910 "erschienen"; ÖbZ Nov-Dec 1910; Nat. Nov. Dec(1) 1910), p. [i-vii], [1]-738, *pl. 1-92*.
2: Nov-Dec 1910 (ÖbZ Nov-Dec 1910; Nat. Nov. Dec(1) 1910), p. [i-viii], [1]-756, *pl. 93-168*.
Ref.: Ginzberger, A., Österr. bot. Z. 61: 204. 1911.
Matouschek, F., Bot. Centralbl. 116: 431. 18 Apr 1911 (rev.).

Reinhardt, Otto Wilhelm Hermann (1838-1924), German (Prussian) botanist; student of Alexander Braun at the Friedrich-Wilhelms-Universität, Berlin; Dr. phil. Berlin 1863; high school teacher at Berlin 1865-1887 (Luisenstädtische Gewerbeschule/Oberrealschule); rector of the fourth Berlin "Bürgerschule" 1887-1891; director of the second Realschule 1891-1910; one of the founders of the Bot. Ver. Mark Brandenburg and close associate of P. Ascherson; from 1874 mainly active as malacologist. (*Reinhardt*).

HERBARIUM and TYPES: B (mainly destroyed); other material at GOET and IBF.

NOTE: Not to be confused with [Max] Otto Reinhardt (b. 27 Aug 1854, d. 18 Mar 1935), pupil of Schwendener; Dr. phil. Berlin 1884; plant physiologist at Berlin, whose collection of fungi (107 nos.) is at B (extant). Reinhardt's thesis, *Das leitende Gewebe einiger anomal* [sic] *gebauten Monocotylenwurzeln*, was published 8 Aug 1884 (p. [1]-31, Berlin, Druck von M. Driesner, copy G).

BIBLIOGRAPHY and BIOGRAPHY: Barnhart 3: 142; Biol.-Dokum. 14: 7404; BM 4: 1674; Bossert p. 329; CSP 5: 150, 11: 135,12: 609, 18: 115-116; IH 2: (in press); Kew 4: 445; PR 7523-7524; Saccardo 1: 137; Urban-Berl. p. 311.

BIOFILE: Anon., Bot. Centralbl. 50: 160. 1892 (title professor); Hedwigia 65: (195). 1925 (d.).

Ascherson, P., Fl. Provinz Brandenb. 1: 11. 1864.
Becker, K. & E. Schumacher, Bibl. Schr. Ges. naturf. Fr. Berlin 1834-1942, S.B. Ges. naturf. Fr. Berlin ser. 2. 13(2): 146. 1973 (lists publ. in S.B.).
Stieber, M.T. et al., Huntia 4(1): 85. 1981 (arch. mat. HU).
Ulbrich, E., Verh. bot. Ver. Brandenburg 67(1): 7-11. 1925 (obit., b. 14 Feb 1838, Potsdam; d. 5 Nov 1924, Berlin).
Verdoorn, F., ed., Chron. bot. 2: 139, 140. 1936, 3: 22. 1937, 4: 85. 1938.
Wagenitz, G., Index coll. princ. herb. Gott. 134. 1982.

8922. *Enumeratio muscorum frondosorum* in Marchia brandenburgensis et in ducatu magdeburgensis ad hoc tempus repertorum. Dissertatio inauguralis botanica quam consensu et auctoritate amplissimi philosophorum ordinis in alma litterarum Universitate Friderica Guilelma ad summos in philosophia honores rite capessendos die xi. m. aprilis a. mdccclxiii. h.l.q.s. ... Berolini [Berlin] (typis expressit Hermannus Müller) [1863]. Oct. (*Enum. musc. frond.*).
Publ.: 11 Apr 1863, p. [i-iv], [1]-34, [35, vita], [36, theses]. *Copies*: NY, US; IDC 5193.

8923. *Übersicht der in der Mark Brandenburg bisher beobachteten Laubmoose* ... Berlin (Druck von Hermann Müller) 1863. Oct. (*Übers. Brandenburg Laubm.*).
Publ.: 1863, p. [i], [1]-52. *Copies*: G, NY. – Preprinted from Verh. bot. Ver. Brandenburg 5: [1]-52. 20 Mai 1864 (t.p. 1863).

Reinke, Johannes (1849-1931), German (Mecklenburg) botanist; Dr. phil. 1871; assistant with Bartling in Goettingen 1871-1872; habil. Göttingen 1872; lecturer Bonn 1872-1873; extraordinary professor of botany at Göttingen 1873-1884; regular professor of botany and director of the botanical garden (succeeding A. Engler) at Kiel 1887-1921; in retirement at Preetz (Holstein); Dr. h.c. Kiel 1921; Geh. Regierungsrat 1896; member of the Prussian parliament (Herrenhaus) for Kiel University 1894-1912; philosopher, politician, scientist and vitalist. (*Reinke*).

HERBARIUM and TYPES: KIEL; further material at B, C, FH, L, M, WRSL – Letters to W.G. Farlow at FH.

BIBLIOGRAPHY and BIOGRAPHY: Ainsworth p. 98, 102, 318; Barnhart 3: 142 (b. 3 Feb 1849, d. 25 Feb 1841); Biol.-Dokum. 14: 7409-7411; BJI 1: 48, 2: 144; BM 2: 738, 4: 1674-1675; Bossert p. 329; Cesati p.324-330, Saggio p. 61-62; Christiansen, see index; CSP 8: 724-725, 11: 136, 12: 609-610, 18: 116; De Toni 1: cv-cvi, 4: xlvi-xlvii; Frank 3 (Anh.): 83; GR p. 40, *pl.* [*16*], cat. p. 70; Hawksworth p. 185; Herder p. 473 [index]; Hirsch p. 245; IH 2: (in press); Jackson p. 64, 72, 172, 319, 488, 490; Kelly p. 190; Kew 4: 445; LS 10424; 22033-22043, 38310-38311; Moebius p. 455 [index]; Morren ed. 2, p. 7, ed. 10, p. 12; NI 1611 (see also p. 219); PR 10625; Rehder 5: 715; Saccardo 1: 137; SBC p. 130; Stevenson p. 1255; TL-2/7034, see E.H.P. Kuckuck; Tucker 1: 590; Urban-Berl. p. 268, 311.

BIOFILE: Anon., Bot. Centralbl. 20: 351. 1884 (called to Kiel); Bot. Jahrb. 6 (Beibl. 12): 1. 1885 (called to Kiel); Bot. Zeit. 29: 692. 1871 (asst. GOET), 30: 594. 1872 (habil. ib.), 31: 704. 1873 (e.o. prof. physiol. bot. ib.), 42: 766. 1884 (app. Kiel); Flora 54: 351. 1871 (asst. GOET); Hedwigia 35: (35). 1896 (Geh. Regierungsrath), 72: (114). 1932 (d.); Nat. Nov. 6: 244. 1884 (called to Kiel); Österr. bot. Z. 35: 37. 1885 (called to Kiel), 80: 176. 1931 (d.).
Benecke, W., Ber. deut. bot. Ges. 50: (171)-(202). 1932 (biogr., portr., bibl., d. 25 Feb 1931).
Bergdolt, E., Karl von Goebel. ed. 2. 177, 178, 181, 209-211, 224. 1941 (letters Goebel to Reinke on the latter's general works).
Fischer-Benzon, R.J.D. von, *in* P. Prahl, Krit. Schlesw.- Holst. 2: 49-50. 1890.
Fitting, H., Decheniana 105/106: 8. 1952 (in Bonn).
Gager, C.S., Brooklyn Bot. Gard. Rec. 27(3): 231. 1938 (dir. Kiel bot. gard. 5 Dec 1884-1 Apr 1921).
Harvey-Gibson, R.J., Outlines hist. bot. 172, 205, 208, 209, 241. 1914.
Kluge, M., Johannes Reinke's dynamische Naturphilosophie und Weltanschauung

unter besonderer Berücksichtigung ihrer Herkunft aus der Botanik. Leipzig 1935, vii, 168 p., Stud. Bibl. Gegenwartsphilosophie vol. 17 (main study of Reinke's philosophy and biological ideas; his vitalism and influence on contemporary thinking; extensive bibl.; many secondary refs.; shorter version, 141 p., as thesis. Leipzig 1935).
Koster, J.Th., Taxon 18:556. 1969 (algae KIEL).
Lakowitz, C.W., Ber. westpreuss. bot.-zool. Ver. Danzig 54: 9-11. 1932 (obit.).
Mägdefrau, K., Gesch. Bot. 171, 207. 1973.
Poggendorff, J.C., Biogr.-litt. Handw.-Buch 4(2): 1227. 1904, 6(3): 2149. 1938 (curr. vit.; main works).
Pringsheim, E.G., Julius Sachs 200, 211, 224, 247, 267, 284. 1932.
Reinke, J., Ber. deut. bot. Ges. 43: (19)-(25). 1926 ("Über Botanisieren" approach to diversity seen as problem encountered while in the field); Deutsche Hochschulen und römische Kurie, Leipzig 1911, 59 p., copy: U (shows R.'s interest in politics and religious affairs); Die Welt als Tat. Umrisse einer Weltausicht auf naturwissenschaftlicher Grundlage, ed. 6. Berlin 1915. 505 p., portr. *Copy*: U; Mein Tagewerk, Freiburg i.B. 1925, vii, 495 p. (autobiogr., portr.; main sourrce for facts and ideas; illustrative of Engler era); Naturwissenschaft, Weltanschauung, Religion. Bausteine für eine natürliche Grundlegung des Gottenglaubens, Freiburg i.B. ed. 2/3, 1925, 188 p.
Schellenberg, G., Beih. Bot. Centralbl. 38(2): 390, 391. 1921 (Algae herb. KIEL).
Schmidt, Heinrich (Jena), Die Urzeugung und Professor Reinke, Odenkirchen 1903, 48 p. (copy: B), Der Deutsche Monistenbund im Preussischen Herrenhaus (Reinke contra Haeckel). Eine aktenmässige Darstellung mit Einleitung und Anmerkungen. Brackwede i.W. 1907. 96 p. (copy: B).
Schmidt, O.C., Hedwigia 72: (115)-(116). 1932 (obit., b. 3 Feb 1849, d. 25 Feb 1931).
Schneider, A., Bull. Torrey bot. Club 23: 439-448. 1896 (Reinke's discussions of lichenology).
Wittrock, V.B., Acta Horti Berg. 3(3): 149, *pl. 126*. 1905 (portr.).

PSEUDONYM: Reinke published novels and sketches under the pseudonym Henning von Horst, e.g. *Verbotene Frucht* (1896), *Die Apostelfürsten* (1896) und *Gardensee* (1898, 1926).

COMPOSITE WORKS: A.S. Oersted's *System der Pilze, Lichenen und Algen*. Aus dem Dänischen. Deutsche, vermehrte Ausgabe von A. Grisebach und J. Reinke. Leipzig (Verlag von Wihelm Engelmann) 1873. Oct., p. [i]-viii, [1]-194. *Copies*: BR, USDA. – See TL-2/7034.

EPONYMY: *Reinkella* O.V. Darbishire (1897); *Reinkellomyces* Ciferri et Tomaselli (1953).

8924. *Untersuchungen über Wachsthumsgeschichte und Morphologie der Phanerogamen-Wurzel* ... Bonn (bei Adolph Marcus) 1871. Oct. (*Unters. Wachsth.-Gesch. Phan.-Wurz.*).
Publ.: 1871, p. [ii-iii, v], [1]-50, *pl. 1-2*. *Copies*: BR, M. – Bot. Abh. Morph. Physiol., ed. J. Hanstein, Heft 3.

8925. *Morphologische Abhandlungen* ... Leipzig (Verlag von Wilhelm Engelmann) 1873. Oct. (*Morph. Abh.*).
Publ.: 1873 (p. vi: Mai 1873; Bot. Zeit. 31 Oct 1873), p. [i]-vi, [vii], [1]-122, *pl. 1-7* (uncol. liths. auct.). *Copies*: GOET, LC, LD, MO, U. – The first few sheets came out on 15 Mar 1873 as Festschrift for Johannes Roeper.
Ref.: G.K., Bot. Zeit. 32: 25-28. 9 Jan 1874.

8926. *Entwicklungsgeschichtliche Untersuchungen über die Dictyotaceen des Golfs von Neapel* ... Dresden (Druck von E. Blochmann & Sohn, ...) 1878. Qu. (*Entw.-gesch. Unters. Dictyotac.*).
Publ.: 1878, p. [1]-56, *pl. 1-7* (uncol. liths. auct.). *Copies*: BR, FH, G(2), NY; IDC 669. – Preprinted, or reprinted from Nova Acta Leop. 40(1): 1-56, *pl. 1-7*. 1878.
Ref.: G., Bot. Zeit. 37: 142-144. 1879 (rev.).

8927. *Entwicklungsgeschichtliche Untersuchungen über die Cutleriaceen des Golfs von Neapel* ... Dresden (Druck von E. Blochmann & Sohn, ...) 1878. Qu. (*Entw.-gesch. Unters. Cutleriac.*).

Publ.: 1878 (t.p. reprint; Nat. Nov. Jan(1) 1879), p. [1]-40, *pl. 8-11* (uncol. liths. auct.). *Copies*: FH, FI, G, NY; IDC 669. – Preprinted, or reprinted, from Nova Acta Leop. 40(2): [57]-96, *pl. 8-11.* 1878.
Ref.: G., Bot. Zeit. 37: 142-144. 28 Feb 1879 (rev.).

8928. *Ueber die Entwicklung von Phyllitis, Scytosiphon und Asperococcus* ... [Separat-Abdruck aus Pringsheim's Jahrbüchern für wissenschaftliche Botanik. Band xi. Heft 2. 1878]. Oct.
Publ.: 1878, p. [1]-22, *pl. 11-12. Copy*: FH. – Reprinted and to be cited from Jahrb. wiss. Bot. 11: 262-273, *pl. 11-12.* 1878.

8929. *Untersuchungen ueber die Quellung* einiger vegetabilischer Substanzen ... Bonn (bei Adolph Marcus) 1879. Oct. (*Unters. Quellung*).
Publ.: 1879, p. [i*], [ii-vii], [1]-137, *pl. 1-4* (uncol. liths.). *Copy*: BR. – Bot. Abh. Morph. Physiol. 4(1).

8930. *Algenflora der westlichen Ostsee deutschen Antheils*. Eine systematisch-pflanzengeographische Studie ... Kiel (Druck von Schmidt & Klaunig) 1889. Fol. (*Algenfl. westl. Ostsee*).
Publ.: 1889 (Bot. Zeit. 29 Mar 1889; Hedwigia rd Mar-Apr 1889; Nat. Nov. Jun(1) 1889), p. [i]-xi, [1]-101, *1 map. Copies*: FH, NY, U, USDA. – Published as sechster Bericht der Kommission zur wissenschaftlichen Untersuchung der deutschen Meere, in Kiel für die Jahre 1887 bis 1889. I. Heft. 1889.
Ref.: Bornet, E., Bull. Soc. bot. France 36 (bibl.): 133-135. 1889.
Hauck, F., Hedwigia 28: 198-201. Mai-Jun 1889 (rev.).
Lierau, M., Bot. Centralbl. 38: 821-826. 19 Jun 1889.
Magnus, P., Bot. Zeit. 47: 493-496. 1889 (rev.).
Rosenvinge, K., Bot. Jahrb. 11(lit.): 84-85. 31 Dec 1889.

8931. *Atlas deutscher Meeresalgen*. Im Auftrage des Königlich Preussischen Ministeriums für Landwirthschaft, Domänen und Forsten herausgegeben im Interesse der Fischerei von der Kommision zur wissenschaftlichen Untersuchung der deutschen Meere ... Berlin (Paul Parey) 1889-1892, 2 Hefte. Fol. (*Atlas deut. Meeresalg.*).
Junior authors: Ernst Hermann Paul Kuckuck (1866-1918), Franz Schütt (1859-1921).
1: Aug 1889 (Bot. Centralbl. 10 Sep 1889; Bot. Zeit. 30 Aug 1889; Nat. Nov. Sep(1) 1889; ÖbZ 1 Nov 1889), p. [i-iv], [1]-34, *pl. 1-4, 5/6, 7-11, 12/13, 14-25* (col. liths. F. Schütt, P. Kuckuck).
2(1, 2): Dec 1890 or Jan 1891 (Bot. Centralbl. 11 Feb 1891; Bot. Zeit. 30 Jan 1891; Nat. Nov. Jan(2) 1891), p. 35-54, *pl. 26-35* (P. Kuckuck; some partly col.).
2(3-5): Mar-Apr 1892 (Nat. Nov. Apr(2) 1892; Bot. Jahrb. 2 Aug 1892; Bot. Centralbl. 26 Apr 1892; Bot. Zeit. 29 Apr 1892); J. Bot. Sep 1892), p. [i-iv], 55-70, *pl. 36-50* (as in 2(1, 2)).
Copies: FH, KNAW, MO, NY, U, USDA; IDC 1347.
Ref.: Bornet, E., Bull. Soc. bot. France 36 (bibl.): 133-135. 1889.
Magnus, P., Bot. Jahrb. 15 (Lit.): 101-102. 1892.
Oltmanns, P., Bot. Centralbl. 42: 205. 14 Mai 1890 (rev.).
Rosenvinge, K., Bot. Jahrb. 11 (Lit.): 85-86. 31 Dec 1891 (rev.).
Scott, D.H., J. Bot. 28: 60-63. 1890 (rev.).
H.S., Bot. Zeit. 47: 609-611. 13 Sep 1889 (rev.).

8932. *Beiträge zur vergleichenden Anatomie und Morphologie der Sphacelariaceen* ... Cassel (Verlag von Theodor Fischer) 1891. Qu. (*Beitr. Anat. Morph. Sphacelar.*).
Publ.: 1891 (Bot. Zeit. 3 and 21 Jul 1891; "soeben", Bot. Monatschr. Jun-Jul 1891; Nat. Nov. Jul(2) 1891; ÖbZ Aug 1891; Bot. Centralbl. 12 Aug 1891), p. [ii-iv], [1]-40, *pl. 1-13* (P. Kuckuck; partly col.). *Copies*: BR, FH, G, NY, PH. – Bibliotheca botanica Heft 23, 1891.
Ref.: Magnus, P., Bot. Zeit. 50: 145-147. 4 Mar 1892 (rev.).

8933. *Abhandlungen über Flechten* ... Berlin (Gebrüder Borntraeger ...) 1894. Oct.
1, 2: Aug-Sep 1894 (p. 48: Mar 1894; Nat. Nov Oct(1) 1894), p. [i], [1]-48. – Reprinted and to be cited from Jahrb. wiss. Bot. 26(3): 495-542. Aug-Sep 1894.

3, 4: Mai-Jun 1895 (Nat. Nov. Jul(1) 1895), p. [i], [49]- 160. Id. 28(1): 39-150. Mai-Jun 1895.
4(Schluss): Sep-Oct 1895 (Nat. Nov. Nov(2) 1895), p. [i], [161]-288. Id. 28(3): 359-486. 1895.
5: Mar-Apr 1896 (Nat. Nov. Mai(1) 1896), p. [i], 289-354. Id. 29(2): 171-236. Apr-Mai 1896.
Copies: FH, G, MICH, Stevenson; IDC 5386 (journal).
Ref.: Schneider, A., Bull. Torrey bot. Club 23(11): 439-448. 1896 (review of R's lichenological work), also in Contr. Dept. Bot. Columbia Univ. 5, 1896-1897.

8934. *Untersuchungen über die Assimilationsorgane der Leguminosen* i-iii ... Berlin (Gebrüder Borntraeger) 1896. Oct. (*Unters. Assim.-Org. Legum.*).
Publ.: 1896, p. [i], [1]-156. *Copy*: M. – Reprinted and partly preprinted from Jahrb. wiss. Bot. 30: 1-70, 529-614. 1896.

8935. *Die Assimilationsorgane der Asparageen*. Eine kritische Studie zur Entwickelungslehre ... Berlin (Gebrüder Borntraeger) 1897. Oct.
Publ.: 1897, p. [i], [1]-66. *Copy*: M. – Reprinted and to be cited from Jahrb. wiss. Bot. 31(2): 207-272. 1897.

8936. *Ueber Caulerpa*. Ein Beitrag zur Biologie der Meeres-Organismen. [Kiel und Leipzig (Verlag von Lipsius & Fischer) 1900]. Qu. (*Caulerpa*).
Publ.: Nov-Dec 1899 (Hedwigia 26 Feb 1900 as of 1899; ÖbZ Nov 1899; Nat. Nov Jan(2) 1900), p. [i-iv], [1]-96. *Copy*: G. – Issued as Wiss. Meeeresunters. Kiel, ser. 2. 5(1). 1900.
Ref.: Kuckuck, E.H.P., Bot. Zeit. 58(2): 49-54. 16 Feb 1900 (rev.; also as of 1899).

8937. *Die Entwicklung der Naturwissenschaften* insbesondere der Biologie im neunzehnten Jahrhundert. Rede zur Feier des Jahrhundertwechsels gehalten am 13. Januar 1900 in der Aula der Universität zu Kiel ... Kiel (Universitäts-Buchhandlung. (Paul Toeche)). 1900. Oct. (*Entw. Naturwiss.*).
Publ.: soon after 13 Jan 1900, p. [1]-21. *Copy*: M.

8938. *Studien zur vergleichenden Entwicklungsgeschichte der Laminariaceen* [Kiel (Druck von Schmidt & Klaunig) 1903]. Oct. (*Stud. Entw.-Gesch. Laminariac.*).
Commercial issue: 1-28 Jan 1903 (FH rd Feb 1903; commemorative occasion 18 Jan 1903; Bot. Centralbl. 10 Mar 1903; ÖbZ Jan 1903), p. [1]-67. *Copies*: BR, FH, M.
Festschrift issue (see title): 1-28 Jan 1903, p. [1]-67, [68, colo.]. *Copy*: NY. – Diei natalis serenissimi ac potentissimi principis Guilelmi ii imperatoris regis faustissima sollemnia quorum laetitiae oratio a professore publico ordinario Alfredo Schoene habenda interpretabitur die xxvii mensis januarii mcmiii ... inest: Johannis Reinke dissertatio patrio sermone scripta *de evolutione comparativa Laminariacearum* ... Kiliae [Kiel] (prostat in libraria Lipsii et Fischeri ex officina Schmidtii et Klaunigii) 1903.
Ref.: Kienitz-Gerloff, J.H.E.F., Bot. Centralbl. 92: 273-274. 31 Mar 1903 (rev.).

8939. *Botanisch-geologische Streifzüge* an den Küsten des Herzogthums Schleswig ... Kiel (Druck von Schmidt & Klaunig) 1903. Qu. (*Bot.-geol. Streifz.*).
Publ.: Dec 1903 (Bot. Zeit. 1 Jan 1904; Bot. Centralbl. 16 Feb 1904; Nat. Nov. Jan(2) 1904), p. [ii-iv], [1]-157. *Copies*: BR, FH, G, M, MICH. – Wiss. Meeresunters. Kiel ser. 2. 8 (Ergänzungsheft). – *Related publ.*: Entw.-Gesch. Dünen Westküste Schleswig, S.B. Akad. Wiss. Berlin, Ges.-Sitz. 5. 3, 13: 281-295. 1903.
Ref.: Buchenau, F.G.P., Bot. Zeit. 62(2): 55-57. 16 Feb 1904 (rev.).
Graebner, P., Bot. Jahrb. (Lit.) 33: 48-49. 15 Mar 1909 (rev.).

8940. *Philosophie der Botanik* ... Leipzig (Verlag von Johann Ambrosius Barth) 1905. Oct. (*Phil. Bot.*).
Publ.: Jan-Feb 1905 (p. vi: 16 Nov 1904; Nat. Nov. Feb(2) 1905; Bot. Zeit. 1 Mar 1905; ÖbZ Feb-Mar 1905), p. [i]-vi, [1]-201. *Copies*: BR, FI, G, MO, NY, U. – Natur– und kulturphilosophische Bibliothek Band I.

Ref.: Engler, A., Bot. Jahrb. 37 (Lit.): 1-2. 22 Dec 1905 (rev.).
Günther, K., Bot. Zeit. 63(2): 321-325. (Nov 1905).

8941. *Kritik der Abstammungslehre* ... Leipzig (Verlag von Johann Ambrosius Barth) 1920. Oct. (*Krit. Abst.-Lehre*).
Publ.: 1920 (p. iv: Apr 1920), p. [i]-iv, [v], [1]-133. *Copies*: GOET, U.

8942. *Grundlagen einer Biodynamik* ... Berlin (Verlag von Gebrüder Borntraeger ...) 1922. Oct. (*Grundl. Biodynamik*).
Publ.: 1922 (p. iv: Mar 1922), p. [i]-iv, [v], [1]-160. *Copy*: U.

8943. *Mein Tagewerk* ... Freiburg im Breisgau (Herder & Co.,...) 1925. Oct. (*Tagewerk*).
Publ.: Aug-Dec 1926 (Nat. Nov. Jul 1926), p. [i], portr., Reinke, [iii]-viii, [1]-495. *Copy*: USDA. – Autobiography: important source for Reinke's work, life and thoughts, as well as one of the rare examples of an autobiography of a life scientist of the Engler-era, dealing not only with science but also picturing the social group constituting the "Bildungsbürgertum").

Reinsch, [Edgar] Hugo [Emil] (1809-1884), German (Bavarian) high school teacher and naturalist at Erlangen. (*H. Reinsch*).

HERBARIUM and TYPES: Unknown.

BIBLIOGRAPHY and BIOGRAPHY: BM 4: 1675; CSP 5: 151-152, 8: 725-726, 11: 136, 18: 117; Rehder 1: 383, 532, 3: 408.

BIOFILE: Nissen, C., Zool. Buchill. 336. 1969. – See Poggendorff.

8944. *Taschenbuch der Flora von Deutschland* nach Linnéischem Systeme und Koch'scher Pflanzenbestimmung zum Gebrauche für botanische Excursionen ... Stuttgart (Ad. Becker's Verlag) 1855. Oct. (*Taschenb. Fl. Deutschl.*).
Publ.: 1855, p. [i]-iv, [1]-299, [300, err.]. *Copy*: HH.

Reinsch, Paul Friedrich (1836-1914), German (Bavarian) botanist and palaeontologist; studied at Erlangen and München; high school teacher in Erlangen, Zweibrücken and Baselland; in early retirement at Erlangen. (*Reinsch*).

HERBARIUM and TYPES: M (algae herbarium); other material at BAS, BM, C, E, F, FH, H, K (2237), KIEL, L, MICH, MW, QK, S, US, W.
Exsiccatae: *Herbarium muscorum frondosorum Europae mediae*, 1872-1878, 350-400 nos. (see DTS 1: 244), set at FH; for prospectus see Bot. Zeit. 29: 841-842. 1871. – Contributed to V.B. Wittrock & O. Nordtstedt, *Algae aquae dulcis exsiccatae praecipue scandinavicae* – and to *Flora exsiccata bavarica*. – The phanerogam herbarium (48 fasc.) was offered for sale by Th. Weigel, Leipzig, in Herbarium 49: 471. 1919. – Letters to D.F.L. von Schlechtendal at HAL, to W.G. Farlow at FH.

BIBLIOGRAPHY and BIOGRAPHY: Andrews p. 318; Barnhart 3: 142; Biol.-Dokum. 14: 7414-7415; BJI 1: 48; BM 4: 1675; Bossert p. 329; CSP 5: 152, 6: 748, 8: 726, 11: 136-137, 12: 610, 18: 117-118; De Toni 2: c, 4: xlvii; DTS 1: 244; Frank 3 (Anh.): 83; GR p. 40 (b. 21 Mar 1836, d. 31 Jan 1914), cat. p. 70; Herder p. 28, 254, 433; Jackson p. 71, 77, 80, 106, 156, 163, 164, 220, 294, 506; Kew 4: 446; LS 22046-22053; Morren ed. 10, p. 19; NI 1612-1616, also 3: 23; PH 1; PR 7526; Quenstedt p. 356; Rehder 5: 715; SIA 220; TL-2/6760; Tucker 1: 591; Zander ed. 10, p. 706, ed. 11, p. 805.

BIOFILE: Anon., Ber. deut. bot. Ges. 31: (182). 1913 (1914) (d.); Bot. Not. 1914; 143 (d.); Bot. Zeit. 25: 343-344. 1867 (call for sale), 26: 720. 1868 (id.), 29: 841-842. 1871 (prospectus *Herb. musc. frond. Eur. med.*); Hedwiga 56: (79). 1915 (d.); Österr. bot. Z. 21: 375. 1871 (announcement exsicc. Central-Europ. musci).
Binz, A., Verh. naturf. Ges. Basel 19(3): 141. 1908 (musci BAS).

Candolle, Alph. de, Phytographie 443. 1880 (coll.).
Ewan, J. et al., Short hist. bot. U.S. 108. 1969.
Glück, H., Ber. deut. bot. Ges. 32: (5)-(17). 1913 (obit., bibl., b. 1836, d. 31 Jan 1914).
Kneucker, A., Allg. bot. Z. 20: 48. 1914 (d.).
Koster, J.Th., Taxon 18: 556. 1969 (algae at BM).
Papenfuss, G.F., Antarct. Res. ser. 1: 69. 1964 (bibl. antarctic benthic algae).
Poeverlein, H., Mitt. bayer. bot. Ges. 3: 149-150. 1914 (obit.; "ein ernstes und stilles Forscherdasein").
Prescott, G.W., Contr. bibl. antarct. subantarct. alg. 237, 295. 1979.
Sayre, G., Mem. New York Bot. Gard. 19(3): 385. 1975 (coll.).
VanLandingham, S.L., Cat. Diat. 4: 2365. 1971, 6: 3585. 1978.

EPONYMY: *Reinschia* C.E. Bertrand et B. Renault (1893); *Reinschiella* De Toni (1889); *Reinschospora* J.M. Schopf, L.R. Wilson et R. Bentall (1944).

8945. *Beiträge zur chemischen Kenntniss der weissen Mistel.* (Viscum album Linné) ... Erlangen (Druck von der Ernst Junge'schen Universitätsbuchdruckerei) 1860. Qu. (*Beitr. chem. Kenntn. Mistel*).
Publ.: 1860, p. [1]-26, *1 pl. Copy*: M. – See also Bull. Soc. imp. Natural. Moscou 35: 531-559. 1862.

8946. *Morphologische, anatomische und physiologische Fragmente* Moskau (in der Buchdruckerei der kaiserlichen Universität) 1865.
Publ.: Oct-Dec 1865 (p. ii: 16 Sep 1865), p. [i-ii], [1]-59, *pl. 1-2. Copies*: H, HH, M. – Consolidated reprint of a series of 14 papers published in and to be cited from Bull. Soc. imp. Natural. Moscou 38(2): 1-59. 1865.

8947. *Die Algenflora des mittleren Theiles von Franken* (des Keupergebietes mit den angrenzenden Partien des jurassischen Gebietes) enthaltend die vom Autor bis jetzt in diesen beobachteten Süsswasseralgen und die Diagnosen und Abbildungen von ein und fünfzig vom Autor in diesem Gebiete entdeckten neuen Arten und drei neuen Gattungen ... Nürnberg (Verlag von Wilhelm Schmid) 1867. Oct. (*Algenfl. mittl. Franken*).
Publ.: Jan-Apr 1867 (BSbF Mai 1867; Flora 10 Jul 1867), p. [i]-viii, [1]-238, *pl. 1-13* (uncol. liths. auct.). *Copies*: BR, FH(2), G, H, M, NY(2), USDA; IDC 6402. – (Abh. naturf. Ges. Nürnberg, vol. 3: [xix]-xxiv, [1]-238. *pl. 1-13*. 1866 (1867).

8948. *De speciebus generibusque nonnullis novis ex algarum et fungorum Classe* ... Francofurti ad Moenum [Frankfurt a. Main] (sumptibus Chr. Winter) 1867. Qu.
Publ.: Jan-Mar 1867 (Bot. Zeit. 29 Mar 1867; Flora 10 Jul 1867), p. [1]-36, *pl. 1-6* (1-5 col.; liths. auct.). *Copies*: FH, G, H; IDC 6401. – Reprinted and to be cited from Abh. Senckenb. naturf. Ges. 6: 111-144, *pl. 20-25*. 1866/67 (*Copies* ex journal BR, NY).
Ref.: Bary, A. de, Bot. Zeit. 25: 104. 1867.

8949. *Contributiones ad algologiam et fungologiam* vol. 1 ... Lipsiae [Leipzig] (T.O. Weigel] 1875. Qu. †. (*Contr. algol. fungol.*).
Publ.: 1875 (p. vi: 4 Nov 1876; Grevillea Jun 1875; J. Bot. Aug 1875; ÖbZ Mar 1876; Flora Oct 1876; Bot. Zeit. 16 Apr 1875), p. [i]-xii, [1]-103, [104, err.], *pl.* (Melanophyceae): *1-2, 3, 3a, 4-6, 6a, 7-12, 12a, 13-20, 20a, 21-35, 35a, 36*; (Rhodophyceae):*1-42, 42a, 43-47, 47a, 48-61*; (Chlorophyllophyceae) *1-18*; (fungi:) *1-9* (liths., some in col. ink, auct.) (*131 pl.*). *Copies*: BR, G, LC, MICH, MO, NY, PH; IDC 6403. – The PH and LC copies have a cover dated 1874/1875 "Typis Theodor Haesslein. Norimbergae".

8950. *Beobachtungen über einige neue Saprolegnieae,* über die Parasiten in Desmidienzellen und über die Stachelkugeln in Achlyaschläuchen [Jahrb. wiss. Bot. 11(2). 1878]. Oct.
Publ.: 1878, p. [1]-29, *pl. 14-17. Copy*: M. – Reprinted and to be cited from Jahrb. wiss. Bot. 11(2): 283-311, *pl. 14-17*. 1878.

8951. *Neue Untersuchungen über die Mikrostruktur der Steinkohle* des Carbon, der Dyas und Trias. Beiträge zur Aufhellung des Ursprunges und der Zusammensetzung dieser

Mineralkörper, sowie zur Kenntniss des einfachsten Pflanzenlebens der Vorwelt ... Leipzig (Verlag von T.O. Weigel) 1881. Qu. (*Neue Unters. Mikrostrukt. Steink.*).
Publ.: Mar 1881 (Bot. Jahrb. 5 Apr 1881; Flora 21 Mar-21 Mai 1881), p. [i]-viii, [1]-124, *94 pl.* (uncol. liths. auct.). *Copies*: BR, Delft, NY, PH, USGS. – The book was accompanied by two series of microscopic preparations, ser. 1, 10 specimens; ser. 2, 20 specimens. – Numbering of plates: *1-7, 7a, 7c, 8, 8a, 8b, 9, 9a, 10, 10a, 10b, 10c, 11, 12, 12a, 13, 13a, 14, 14a, 14b, 14c, 15, 15a-15c, 16, 16a, 16b, 17, 17a, 17b, 18, 18a, 19-27, 27a, 28, 28a, 29, 29a, 30-31, 31a, 32, 32a, 33-38, 38a, 39, 39a, 40, 40a, 41-45, 45a, 45b, 46, 46a, 47-49, 49a, 49b, 50-52, 52a, 53, 54, A,* [*B*], [*C*].

8952. *Carbonischer Urwald* Ideallandschaft aus dem Carbonischen Zeitalter der Erde. Verkl. 1: 7. Nach dem Originalgemälde erfunden und gemalt von P.F. Reinsch. Erlangen, 1882. (*Carbon. Urwald*).
Publ.: 1882, p. [1]-4, *3 pl. Copy*: M.

8953. *Mikrophotographien über die Strukturverhältnisse und Zusammensetzung der Steinkohle des Carbon* entnommen von mikroskopischen Durchschnitten der Steinkohle ... [Leipzig (T.O. Weigel) 1883]. Portfolio (*Mikrophot. Strukt.-Verh. Steink.*).
Publ.: 1883 (preface Jul 1882), p. [i], [1]-13, loose photogr. on cardboards nos. *1-12*, [*1*]. *Copy*: BR.

8954. *Micro-Palaeophytologia formationis Carboniferae.* Iconographia et dispositio synoptica plantularum microscopicarum omnium in venis carbonis formationis Carboniferae hucusque cognitarum, eorumque illis proximorum corpusculorum natura vegetabi- lica non incerta, quae inveniuntur et in venis carbonis et in stratis formationum infra supraque Carboniferam sequentium ... Erlangae [Erlangen], Germania (redemptio autoris et apud Theodorum Krische bibliopolam). Londinii (apud Bernardum Quaritch bibliopolam) 1884. Qu. 2 vols. (*Micro-palaeophytol. Carbon.*).
1: 1884 (Bot. Zeit. 26 Sep 1884), p. [i]-vii, [viii], [1]-79, [80], *pl. 1-49, 49a, b, c, 50-66,* photo.
2: 1884 (p. i: 8 Cal Sep 1884), p. [i*], i-ii, [1]-54, [55-56], *pl. 67-95, 72a, b, c, 83a, b, 85a, b, c, d.*
Copies: GOET, GRON, M, PH, USGS.

8955. *Ueber einige neue Desmarestien* [Flora 71, 1888]. Oct.
Publ.: 21 Apr 1888, p. [1]-4. *Copy*: M. – Reprinted and to be cited from Flora 71: 188-192. 21 Apr 1888.

8956. *Familiae Polyedriearum monographia* accedunt species 15 et genera 2 nova ... Venezia (Stab. Tipo-litogr. M. Fontana) 1888. Oct.
Publ.: Nov 1888, p. [1]-27, *5 pl. Copies*: H, M. – Reprinted and to be cited from Notarisia 3(11): 493-516. 1888.

8957. *Die Süsswasseralgenflora von Süd-Georgien* ... [Separat-Abdruck aus dem Werke über die Ergebnisse der deutschen Polar-Expeditionen, Allgemeiner Theil, Band ii, 14, 1890]. Oct.
Publ.: 1890 (Nat. Nov. Jun(2) 1890), p. [i], [1]-37, *pl. 1-4. Copies*: M, NY. – Reprinted and to be cited from Die internazionale Polarforschung 1882-1883, Die deutschen Expeditionen, Allg. Theil, 2(14): 329-365. 1890. – Preliminary publication: Ber. deut. bot. Ges. 6: 144-156. 1888.

8958. *Zur Meeresalgenflora von Süd-Georgien* ... (Separat-Abdruck aus dem Werke über die Ergebnisse der deutschen Polar-Expeditionen Allgemeiner Theil, Band ii, 15, 1890]. Oct.
Publ.: 1890 (Nat. Nov. Jun(2) 1890), p. [1]-85, *pl. 1-19* (on *8*) (partly col. liths. auct.). *Copies*: M, NY(2). – Reprinted and to be cited from die Internationale Polarforschung 1882-1883, Die deutschen Expeditionen, vol. 2, Berlin, p. 366-448. 1890.

Reinwardt, Caspar Georg Carl (1773-1854), German (Prussian)-born Dutch botanist; to Amsterdam in 1787; studied at the Amsterdam Athenaeum; professor of natural history Harderwijk 1800-1808; Dr. h.c. Harderwijk 1800; director of Louis Napoleon's "Jardin du Roi"; 1808-1810; id. of "Cabinet d'histoire naturelle" 1810-1814" and of 's Lands Kabinet van Natuurlijke Historie" 1814-1820; professor of natural history at the Amsterdam Athenaeum 1810-1815; director of agriculture Java 1816-1822, founding (18 Mai 1817) the Buitenzorg (Bogor) botanical garden; from 1823 (appointed 1820)-1845 professor of natural history at Leiden. (*Reinw.*).

HERBARIUM and TYPES: L. – Further material at BR, C, FI, JE, H, LE, NY, PC, U, W, WRSL. The Reinwardt-van Marum correspondence is at the Hollandsche Maatschappij der Wetenschappen, Haarlem. Further documentation on Reinwardt is at the Biohistorical Institute, Utrecht (see e.g. J.A. Koolmees, 1979 and E.J. Tuinstra 1981).

BIBLIOGRAPHY and BIOGRAPHY: Aa 16: 219; ADB 28: 111-113. 1889; AG 3: 443; Backer p. 487 (from 1823 "lost in complete insignificance); Barnhart 3: 142 (b. 3 Jun 1773, d. 6 Mar 1854); BM 4: 1675; Bossert p. 329; Frank 3 (Anh.): 83; Hegi 5(1): 1; Herder p. 68, 70, 167; Hortus 3: 1202; IF p. 726; IH 1 (ed. 2): 70, (ed. 3): 87, (ed. 6)363, (ed. 7): 340, 2: (in press); Jackson p. 395; JW 1: 447, 2: 199, 3: 365-366, 4: 403, 5: 750 (many dutch refs.); Kew 4: 446; Lasègue p. 337, 347, 506; MD 206-208; Moebius p. 421; NNBW 4: 1135; Oudemans 2: 245; PH 567; PR 843, 6652, 7527-7529, ed. 1: 955, 7428, 8449-8471; Rehder 5: 715; RS p. 136; SBC p. 130; SK 1: 429-431, xxxii, itin., portr., coll., 4: 46-48, 51, civ, cv, cxii, clx, 8: i, lxxxix; TL-1/1063, 1375; TL-2/564, 6687; see W.S. Kent, C.G.D. Nees von Esenbeck; Tucker 1: 591; Zander ed. 10, p. 706, ed. 11, p. 805.

BIOFILE: Anon., Alg. Konst.-Letter-Bode 1801(2): 1. (oration Harderwijk), 1811(1): 184-186 (oration Amsterdam), 1815(2): 210. (will go to the Dutch East Indies), 1819(1): 99-101 (first and second consignments of collections lost), 1819(2): 193-194 (third consignment lost); 1821(1): 84-85, 1821(2): 114-115 (two collections arrived safely but a further collection lost in transit), 1823(1): 306-308 (poem by Siegenbeek welcoming R. to Leiden), 1854(10): 73 (d.), 81 (obit.), 1855: 121. (sale library); Bonplandia 2: 105. 1854 (d.), 171 (herb. not to Rijksherb. but to Bot. Gard. Leiden), 267 (cogn. Leopoldina: Rumpf), 3: 115. 1855 (herb. not [yet] at L.); Bot. Not. 1855: 159 (d.); Bot. Zeit. 12: 783-784. 1854 (obit.); Flora 4: 124. 1821 (will occupy Brugman's position at Leiden, 6: 125. 1823 (assumes pro- fessorship Leiden, back from Java), 34: 398-399. 1851 (golden jubilee; medal), 37: 175-176. 1854 (d.); Österr. bot. W. 4: 223. 1854 (d.); Proc. Linn. Soc. London 2: 322-323. 1854 (obit.).
Boelman, H.A.C., Bijdrage tot de geschiedenis der geneeskruidcultuur in Nederlandsch Oost-Indië, Leiden 1936, p. 68-73.
Bom, G.D., Het hooger onderwijs te Amsterdam van 1632 tot onze dagen, Amsterdam 1882, 1: 135-138 (bibl.).
Breure, A.S.H. & J.G. de Bruyn, eds., Leven en werken van J.G.S. van Breda 427 [index]. 1979.
Evers, G.A., Utrecht als Koninklijke residentie 189-207. 1941 (on R.'s role as director of Louis Napoleon's zoo 1808-1810).
Ffolliott & Liversidge, Bull. S. Afr. Libr. 3: 134-142. 1964.
Forbes, R.J., ed., Martinus v. Marum 1: 412 [ed.]. 1969, 3: 15, 17, 24, 136-137, 142-144, 154-160, 171-172, 186-187, 252. 1971 (in part by M.J. van Steenis-Krusemann), see also Lefebvre & de Bruyn 1976).
Fürnrohr, A.E., Flora 37: 175-176. 1854 (obit.).
Gager, C.S., Brooklyn Bot. Gard. Rec. 27(3): 296. (1817-1822 Buitenzorg gard.), 308 (1819 [i.e. 1823]-1845 dir. Leiden bot. gard.).
Geel, van, Handel. Maatsch. Ned. Letterk. Leiden, Levensber. 1854. 87-93 (obit.).
Gunn, M. & L.E. Codd, Bot. Expl. S. Afr. 294. 1981 (collections at the Cape).
Hall, N., Bot. Eucalypts 107-108. 1978.
Hoeven, J. van der, Woorden uitgesproken bij het graf van den hoogleraar C.G.C. Reinwardt, op den 11^{den} Maart 1854, Leiden. 1854. 7p. (obit.).
Holttum, R.E., Taxon 19: 709. 1970.
Hooker, W.J., J. Bot. Kew Gard. Misc. 6: 126. 1854 (d.), 7: 21-23. 1855 (obit.)
Jessen, K.F.W., Bot. Gegenw. Vorz. 469. 1884.
J.G., *in* W.J. Hooker, J. Bot. Kew Gard. Misc. 7: 21-23 1855 (obit., bibliophile, rich library to be dispersed by sale).

Jorissen, W.P., Chemisch Weekblad 8(27): 507-511. 1911 (on R.'s professorship in chemistry and natural history at Amsterdam).
Kalkman, C. & P. Smit, eds., Blumea 25(1): 140 [index]. 1979.
Karsten, M.C., J. bot. Soc. S. Afr. 23: 18-22. 1937 (letters from S. Afr.), S. Afr. biogr.-woordenboek 1: 699-700. 1967 (R. spent six weeks at the Cape in 1816), Aloe, Cact. Succ. Soc. Rhodesia Quart. Newsletter 30: 21-29. 1976.
Kemp, P.H. v.d., Kolon. Tijdschr. 9: 191-207, 630-663. 1920 (on the loss of Reinwardt's collections shipped with the Evertsen, 1849).
Koolmees, J.A., Caspar Georg Carl Reinwardt; unpubl. mss. Biohistorical Institute Utrecht, 1979 (general biography; activities, list of archival material, secondary refs.).
Lefebvre, E. & J.G. de Bruyn, Martinus van Marum 6: 434 [index]. 1976 (letters to van Marum) (see also Forbes, R.J., ed.).
Lindemann, E. von, Bull. Soc. Natural. Moscou 61: 64. 1886.
Poggendorff, J.C., Biogr.-lit. Handw.-Buch 2: 599. 1863.
Pruijs van der Hoeven, C., Album der Natuur 1858: 312-314 (reminescences).
Sayre, G., Bryologist 80: 516. 1977 (coll.).
Sirks, M.J., Ind. Natuuronderz. 301 [index]. 1915.
Stafleu, F.A., Regn. veg. 71: 329, 339. 1970 (Miquel-Schlechtendal corr.); Wentia 16: 20, 26, 39, 1966 (member of the "Instituut" (later KNAW).
Steenis-Kruseman, M.J. van, Blumea 25: 3, 38, 40, 42, 43. 1979.
Treub, M., 's Lands Plantentuin te Buitenzorg 18 Mei 1817-18 Mai 1892, Batavia 1892, p. 1-8 (frontisp. portr., rôle R. as founder).
Tuinstra, E.J., Uit de correspondentie tussen C.G.S. Reinwardt en M. Marum in het tijdperk 1815-1822. Utrecht, Biohistorisch Instituut, mss. 1981 (detailed review corr. with v. Marum).
Veendorp, H. & L.G.M. Baas-Becking, Hortus acad. lugd. batav. 1857-1937, 1938, p. 216 [index].
Vriese, W.H. de, Ons streven naar waren roem. Eenige woorden van dankbare her-innnering, Leiden 1854, 16 p. (obituary); Reis naar het Oostelijk gedeelte van den Indischen Archipel in 1821 door C.G.C. Reinwardt, 1858, p. 1-98, 223-266.
Vrolik, W., Versl. Meded. Kon. Akad. Wet., Amsterdam 2: 214-231. 1854 (obit., bibl.).
Zuchold, E.A., Jahresber. nat. Ver. Halle 5: 48. 1853.

HANDWRITING: Fl. males. ser. 1. 1: cli. 1950.

EPONYMY (genera): *Reinwardtia* Dumortier (1822); *Reinwardtia* Blume ex C.G.D. Nees (1824); *Reinwardtia* K.P.J. Sprengel (1824); *Reinwardtia* Korthals (1841); *Reinwardtiodendron* Koorders (1898); (journal): *Reinwardtia*. Bogor, Indonesia. Vol. 1-x, 1950-x.

8959. Caspari Georgii Caroli Reinwardt *Oratio de ardore, quo historiae naturalis, et in-primis botanices cultores in sua studia feruntur. Publice habita die x. junii, a.* [mdccci] *quum ordinariam chemiae, botanices et historiae naturalis professionem in Academia batava, quae est Hardervici, solemni ritu auspicaretur.* Hardervici [Harderwijk] (apud Everardum Tyhoff, ...) [1801]. Qu. (*Oratio de ardore*).
Publ.: 10 Jun 1801, p. [i-v], [1]-56. *Copies*: G, KNAW, NY, PH, U.

8960. *Rückkunft Reinwardts. Elenchus seminum ...,* Isis 1: 309-319. 1823.
Publ.: See the discussion by v. Steenis in v. Steenis-Kruseman 1960 for the standing of the generic and specific names published in this article. v. Steenis is of the opinion that these names should not be dated from this publication; the great majority were duly published in the *Sylloge plantarum* 2: 1-15, late 1825 or Jan-Feb 1826.
Ref.: Steenis-Kruseman, M.J. van, Fl. males. Bull. 14: 644. 1960.
Steenis, C.G.G.J. van, Fl. males. Bull. 4: 97-99. 1948 (on paper in Sylloge).

8961. Caspari Georgii Caroli Reinwardt *Oratio de augmentis, quae historiae naturali ex Indiae investigatione accesserunt,* habita Leidae a.d. 3. maji mdcccxxiii. Quum in Academia Lugduno-Batava ordinariam chemiae, botanices, et historiae naturalis professionem sol-emniter auspicaretur [Leiden 1823]. Qu. (*Oratio augm. hist. nat. Ind.*).
Publ.: 3 Mai 1823, p. [1]-22. *Copy*: PH.
Dutch translation: 1823, p. [i-v], [1]-72. *Copy*: U. – Redevoering van C.G.C. Reinwardt,

over hetgeen het onderzoek van Indie tot uitbreiding der natuurlijke historie heeft toegebracht, gehouden den 3 Mei 1823, bij het plegtig aanvaarden van het gewoon Hoogleeraarsambt in de Scheikunde, Kruidkunde en Natuurlijke Historie aan de Hoogeschool te Leyden: uit het Latijn vertaald door M. Siegenbeek te Amsterdam, by Johannes van der Hey en Zoon. [1823]. Oct.

8962. *Hepaticae javanicae* editae conjunctis studiis et opera Reinwardti, Dr., Blumii, Dr., et Neesii ab Esenbeck, Dr., A.C.N.C.S.S. Nova Acta Academiae caec. L.C. Naturae Curios. vol. xii, p. i, p. 181-238.
Co-authors: Carl Ludwig Blume (1796-1862); Christian Gottfried Daniel Nees von Esenbeck (1776-1858).
Publ.: Jan-Jun 1825 (announced by Flora 7: 640. 28 Oct 1824; reviewed ib. 8: 385-400. 7 Jul 1825, 401-416. 14 Jul 1828, 417-429. 21 Jul 1825; there is no indication that the mention in Flora 28 Oct 1824 was more than a mere announcement), p. [1]-58 (parallel journal pagination: [181]-238. *Copies*: BR, FH, G, NY, U-V (tear sheets journal). – Reprinted and to be cited from Nova Acta Leop. 181-238. 1825.
Supplement: Jan-Jun 1825 (id.), p. 59-68. ibid. 12(1): [409]-418. 1825. *Copies*: BR, FH, G, NY, U-V (tear sheets journal).

8963. *Observatio de Mangiferae semine polyembryoneo* [Nova Acta Leopoldina 12. 1825]. Qu.
Publ.: Jan-Jun 1825 (see above, Hepaticae javanicae), p. [1]-8. *Copy*: FI. – Reprinted and to be cited from Nova Acta Leop. 12: [339]-346. 1825.

8964. *Musci frondosi javanici*, reddidi conjunctis studiis et opera Reinwardt Dr., et Hornschuchii Dr. A.C.N.C.S.S. [Nova Acta Leopoldina 14, 1828]. Qu.
Co-author: Christian Friedrich Hornschuch (1793-1850).
Publ.: 1828, p. [697]-732, *pl. 39-41* (J. Sturm). *Copies*: FI, G, H, W. – Reprinted and to be cited from Nova Acta Leop. 14: [697]-732. 1828.

8965. *Über den Charakter der Vegetation auf den Inseln des Indischen Archipels.* Vortrag, gehalten in der Versammlung deutscher Naturforscher und Ärzte in Berlin am 20sten September 1828 ... Berlin (gedruckt in der Druckerei der Königl. Akademie der Wissenschaften) 1828. Qu. (*Charak. Veg. Ind. Arch.*).
Publ.: 1828 (Linnaea Jan 1829), p. [1]-18, errata sheet. *Copies*: H, HH, M, NY, U. – See also Hertha 12: 489-503. 1828, and Isis, Oken 22: col. 296-306. 1829 for long extracts. – Note by H.C.D. de Wit (in SK 1(4): civ): "though one of the earliest ... not one of the most successful attempts to ... describe the vegetation of Malaysia in general".

8966. *Enumeratio plantarum in horto Lugduno-Batavo coluntur* [Leiden (Typis J.G. La Lau)] 1831. Oct (in fours). (*Enum. pl. hort. Lugd.-Bat.*).
Publ.: 1831, p. [i-iv], [1]-87. *Copy*: BR.

8967. C.G.C. Reinwardt. *Over het eigenaardige en over de verspreiding der gewassen in de Magellaansche landen.* [Tijdschr. Wis. nat. Wet. 2, 1849]. Oct.
Publ.: 1849, p. [33]-47. *Copies*: REG, U. – Reprinted and to be cited from Tijdschr. wis. nat. Wet. 2: 33-47. 1849. See also Notiz. Nat. Heilk. 11: 193-201. 1849.

Reissek, (Reisseck), **Siegfried** (1819-1871), Austrian-Silesian (born in Teschen) botanist ; from 1838 in Wien, studying medicine and acting as volunteer at the natural history cabinet (the later Naturhist. Museum); from 1845 curator ib. (*Reissek*).

HERBARIUM and TYPES: Some material at M and WRSL (lich.). – Letters to D.F.L. von Schlechtendal at HAL. – A manuscript on the vegetation along the Danube is at W.

BIBLIOGRAPHY and BIOGRAPHY: Ainsworth p. 102, 316, 346; Backer p. 659; Barnhart 3: 143 (b. 11 Apr 1819, d. 9 Nov 1871); Biol.-Dokum. 14: 7423-7424; BM 4: 1676; CSP 5: 155, 8: 727; Frank 3 (Anh.): 83; Futak-Domin p. 500; GR p. 459; Hegi 6(1): 18; Herder p. 119, 224, 230, 335; Hortus 3: 1202; IH 2: (in press); Jackson p. 337, 378; Kanitz no. 286; Kelly p. 190; Kew 4: 446; Klášterský p. 154-155; LS 22063-22071; Moebius p. 200,

281; NI 2248, see also 2: 802; PR 7530-7532; Rehder 5: 716; TL-2/4332; Tucker 1: 591; WU 25: 254-256. Zander ed. 10, p. 706, ed. 11, p. 805.

BIOFILE: Anon., Bonplandia 3: 33. 1855 (cogn. Leopoldina Spallanzani), 4: 122-123, 222. 1856, 6: 50. 1858, 8: 127, 325. 1860; Bot. Not. 1872: 26 (d. 3 Nov 1871); Flora 51: 73. 1868 ("Custos" at botanical Cabinet, Vienna), 54: 400. 1871 (d.);J. Bot. 10: 64. 1872 (d.); Nat. Nov. 1872: 26; Österr. bot. Z. 12: 339. 1862, 21: 375. 1871 (d. 9 Nov 1871).
Beck, G. v., Bot. Anst. Wiens 58. 1894.
Kanitz, A., Bot. Zeit. 29: 812, 854-859. 1871 (obit.); Linnaea 33: 647. 1865 (Hung. publ.).
Neilreich, A., Verh. zool. bot. Ver. Wien 5 (Abh.): 55. 1855.
Oborny, A., Fl. Mähren 13. 1881; Verh. naturf. Ver. Brünn 21(2): 13, 16, 22. 1882 (Reissek worked also on flora of Moravia).
Pluskal, F.S., Verh. zool. bot. Ver. Wien 6 (Abh.): 366. 1856.
Podpera, J., Květena Moravy 1, 1924.
Reichardt, H.W., Almanach Akad. Wiss. Wien, 22: 168-172. 1872 (obit., bibl., b. 11 Apr 1819, d. 9 Nov 1871).
Riedl-Dorn, C., Taxon 30: 727. 1981 (mss. veg. banks Danube nr. Vienna at W).
Urban, I., Fl. bras. 1(1): 196-197. 1906 (biogr. sketch b. 11 Apr 1819, d. 9 Nov 1871).
Zuchold, E.A., Jahresb. nat. Ver. Halle 5: 48. 1853.

COMPOSITE WORKS: (1) Co-author, J.G.C. Lehmann, *Plantae Preissianae*, 1844-1847, TL-2/4332.
(2) *Celastraceae, Ilicineae* et *Rhamneae, Fl. bras.* 11(1): 1-116, *pl. 1-41.* 15 Feb 1861 (fasc. 28).

EPONYMY: *Reissekia* Endlicher (1840).

8968. *Über die selbständige Entwicklung der Pollenzelle zum keimtragenden Pflanze* ... (Acta Acad. caes. Leop. Carol. Nat. Cur. vol. xxi. p. ii) [1845]. Qu.
Publ.: 1845 or Jan-Jul 1846 (Flora rd Sep 1846; Bull. Soc. géol. France rd 7 Dec 1846), p. [1]-26, *pl. 31-34* (col. liths.). *Copy*: HH. – Reprinted and to be cited from Nova Acta Leop. 21(2): 470-492, *pl. 33-34.* 1845.

8969. *Über Endophyten der Pflanzenzelle*, eine gesetzmässige den Samenfaden oder beweglichen Spiralfasern analoge Erscheinung ... Wein (in Commision bei Braumüller und Seidel) 1846. Qu.
Publ.: 1846, p. [i], [1]-16, *pl. 2*. *Copy*: H. – Reprinted and to be cited from Naturw. Abh. ed. W. Haidinger 1: 1-46. 1847 (date of publication of entire volume). – See also Rendic. Lavori Accad. Sci. Napoli 8: 6-70. 1849 and Arch. Sci. phys. nat., Genève 10: 160-164. 1849.

8970. *Die Palmen*. Eine physiognomisch-culturhistorische Skizze ... Wien (Wilhelm Braumüller ...) 1861. Oct. (*Palmen*).
Publ.: 1861, p. [1]-39. *Copies*: HH, NY, USDA. "Aus den populären Vorträgen der k.k. Gartenbau-Gesellschaft am 19. März 1861.
Re-issue: 1870, p. [1]-39. *Copy*: M. – Inserted in F. von Hochstetter et al., Gesammelte naturwissenschaftliche Vorträge, Wien (Wilhelm Braumüller ...) 1870, in which *Die Palmen* is no. iv.

Reitter, Johann Daniel von (1759-1811), German (Württemberg) botanist and forester ("Forstrath"), Württemberg "Hofjäger" 1780-1793 also teaching forestry at Hohenheim; forestry commisioner at Stuttgart 1794-1801; with the Württemberger forestry department 1803-1811. (*Reitter*).

HERBARIUM and TYPES: Unknown.

BIBLIOGRAPHY and BIOGRAPHY: ADB 28: 168-170 (R. Hess); Barnhart 3: 143 (b. 21 Oct 1759, d. 6 Feb 1811); BM 1: 2, 4: 1677; GF p. 73; Herder p. 381; Jackson p. 298; Kew 4:

446-447; Moebius p. 398; NI 1617-1618, see also p. 219; PR 7533, ed. 1: 8473-8474; Rehder 5: 716; Tucker 1: 591.

BIOFILE: Hess, R., Lebensbilder hervorragender Forstmänner 287-288. 1885 (brief biogr. sketch, bibl., sources).
Junk, W., Rara 1: 17. 1900-1913.
Nördlinger, H., ed., Krit. Blätt. Forst-Jagdwiss. 45(1): 170. 1862 (b. 21 Oct 1759, d. 6 Feb 1811).

COMPOSITE WORKS: Editor, *Journal für das Forst- und Jagdwesen* (5 vols., each of 2 Hefte) 1790-1799.

8971. *Abbildung der hundert deutschen wilden Holz-Arten* nach dem Nummern-Verzeichnis im Forst-Handbuch von F.A.L. Burgsdorf. Als eine Beilage zu diesem Werke, heraus gegeben [sic] und Sr. Durchlaucht dem regierenden Herrn Herzog von Wirtemberg unterthänigst zugeeignet von J.D. Reitter, ... und G.F. Abel, Herzogl. Wirtembergischen Hof- und Kupferstecher. Stuttgart (gedrukt [sic] in der Drukerei [sic] der Herzoglichen Hohen Carls-Schule) 1790. Qu. (*Abb. deut. Holz-Art.*).
Artist: Gottlieb Friedrich Abel (1763-?).
Publ.: 1790 (p. iv: 10 Dec 1790), p. [i-viii], [1]-38, *pl. 1-100* (handcol. copp. G.F. Abel). *Copy*: NY.
Re-issue: 1796, p. [i], [1]-38, *pl. 1-100* (id.). – "... Kupferstecher. I. Heft. Stuttgart" (auf Kosten der Herausgeber) 1796. Qu.
Other issue: 1805, fide PR, imprint as 1796.
Note: The reference in the title is to Friedrich August von Burgsdorff (1747-1802), *Anleitung zur sichern Erziehung und zweckmässigen Anpflanzung der einheimischen und fremden Holzarten, welche in Deutschland ... fortkommen.* Berlin 1787.

8972. *Beschreibung und Abbildung der in Deutschland seltener wildwachsenden* und einiger bereits naturalisirten *Holz-Arten* als Fortsezung [sic] von den Abbildungen der Hundert deutschen wilden Holzarten, nach dem Nummern-Verzeichniss im Forsthandbuch von F.A.L. von Burgsdorf. & C. ... I. Heft. Stuttgart (auf Kosten der Herausgeber. Gedruckt in der Cotta'schen Hofbuchdrukcrei [sic] 1803. Qu. † (*Beschr. Abb. Holzart.*).
Artist: Gottlieb Friedrich Abel (1763-?).
Publ.: 1803, (p. [iii]: Apr 1803), p. [i, iii], [1]-28, *pl. 1-25* (hand col. copp. G.F. Abel). *Copies*: MO, NY.

Relhan, Richard (1754-1823), British clergyman and botanist; MA cantab. 1779; rector at Hemingby, Lincoln (Cambridge) 1791; editor of Tacitus; one of the founders of the Linnean Society (London). (*Relhan*).

HERBARIUM and TYPES: LINN; 3 letters at LINN (J.E. Smith corr.).

BIBLIOGRAPHY and BIOGRAPHY: Backer p. 659; Barnhart 3: 143; BB p. 255-256; BM 4: 1678; Dawson p. 697; Desmond p. 516; DNB 48: 6-7; Dryander 3: 24, 136; Frank 3 (Anh.): 83; GR p. 753, cat. p. 70; Henrey 2: 737, 3: 105-106 (nos. 1256-1262); Herder p. 171; IH 2: (in press); Jackson p. 65, 249; Kelly p. 190; Kew 4: 447; LS 22076; NI 1619; Plesch p. 386; PR 7534, ed. 1: 8475-8479; Rehder 1: 395; SBC p. 130; SO add. 700ca-cc; Stevenson p. 1255; SY 13; TL-1/1064; Tucker 1: 591.

BIOFILE: Allibone, S.A., Crit. Dict. Engl. lit. 2: 1771. 1878.
Bridson, G.D.R. et al., Nat. hist. mss. res. Brit. Isl. 431 [index]. 1980.
Dawson, W.R., Smith papers 71. 1934.
Freeman, R.B., Brit. nat. hist. books 293-294. 1980.
Hawksworth, D.L. & M.R.D. Seaward, Lichenol. Brit. Isl. 9, 139. 1977.
Kent, D.H., Brit. herb. 74. 1957 (herb. at LINN?).
Sayre, G., Bryologist 80: 516. 1977 (herb. at LINN).
Wilkinson, H.J., Ann. Rep. Council Yorkshire Philos. Soc. 1906: 45-71. 1907 (material in herb. of Soc.).

EPONYMY: *Relhania* L'Héritier de Brutelle (1789); *Relhanum* S.F. Gray (1821).

8973. Richardi Relhan, A.M. collegii regalis capellani, *Flora cantabrigensis*, exhibens plantas agro cantabrigiensi indigenas, secundum systema sexuale digestas. Cum characteribus genericis, diagnosi specierum, synonymis selectis, nominibus trivialibus, loco natali, tempore inflorescentiae. Cantabrigiae [Cambridge] (Typis academicis excudebat J. Archdeacon, ...) 1785. Oct. (in fours). (*Fl. cantab.*).
Ed. [*1*]: Nov 1785 (see TL-1), p. [i-xxii], [1]-490, 7 *pl.* (uncol. copp. J. Bolton). *Copies*: E, FH, FI, G, HH, KU, MICH, NY, PH, TCD, USDA; IDC 5667.
Suppl. [*1*]: Feb-Mai 1786 (Mon. Rev. London Mai 1786; p. [6]: 23 Jan 1786), p. [1]-39. *Copies*: HH, MICH, NY, PH. – "*Florae cantabrigiensi supplementum ...*" Cantabrigiae (id.) 1786. Oct. (in fours.).
Suppl. alt.: Nov-Dec 1788 (p. [6]: 11 Nov 1788), p. [1]-36. *Copies*: HH, MICH, NY. – Cantabrigiae (id.) 1788. id.
Suppl. 3: 1793 (p. [v]: 15 Mai 1793), p. [i-v], [1]-44. *Copies*: HH, NY. – Cantabrigiae (id.). 1793. id.
Ed. alt.: 23 Apr 1802 (p. [ii], SY p. 13), p. [i]-xii, [xiii, err.], [1]-568. *Copies*: BR, HH, USDA. – Editio altera. Cantabrigiae (Typis academicis excudebat J. Burges, ...). 1802. Oct. (in fours).
Ed. tertia: Apr-Jun 1820 (JGLE Jul 1820; dedic. 30 Mar 1820), p. [i]-xi, [1]-597, 7 *pl.* (uncol. copp. J. Bolton). *Copies*: BR, E, FH, HH, MICH, NY, PH, Stevenson, TCD. – Editio tertia. Cantabrigiae (typis ac sumptibus academicis excudebat J. Smith; ...) 1820. Oct. (in fours).

Remy, Esprit Alexandre (*fl.* 1858), French botanist and physician at Mareuil-le-Port (Marne). (*Remy*).

HERBARIUM and TYPES: Unknown.

NAME: Consistently Remy, not Rémy.

BIBLIOGRAPHY and BIOGRAPHY: Barnhart 3: 143; BL 2: 137, 707; BM 4: 1678; Jackson p. 281; Kew 4: 447; LS 22083; PR 7536-7537; Rehder 1: 411.

8974. *Flore de la Champagne*, description succincte de toutes les plantes cryptogames et phanérogames des départements de la Marne, des Ardennes, de l'Aube et de la Haute-Marne, leurs propriétés médicales, usages économiques, industriels, et intérêt agricole. Manuel d'herborisation. Par M. le docteur Remy père, ... Reims (Imprimerie de E. Luton, ...) 1858. Duod. (*Fl. Champagne*).
Publ.: 1858, p. [i]-xii, [1]-281. *Copies*: HH, NY, USDA. – Blake (BL 2: 137): P.A. Hariot, (Mém. Soc. Acad. Agr., Sci. Arts Dépt. Aube 55: 179. 1891) "states that it is absolutely worthless".

8975. *Essai d'une nouvelle classification de la famille des Graminées ...* Première partie. – Les genres. Paris (Librairie Germer Baillière, éditeur ...). 1861. Oct. †. (*Essai nouv. class. Gramin.*).
Publ.: 1861 (BSbF séance 22 Nov 1861), p. [i]-lx, [1]-308. *Copies*: G, HH, NY, US.
Ref.: Fournier, É., Bull. Soc. bot. France 8: 494-496. Jan 1863 (rev.).

Rémy, [Ezechiel] Jules (1826-1893), French botanist and traveller; high school teacher at Paris 1848-1851; from 1851-ca. 1861 travelling widely around the world; later living at Louvency (Marne). (*J. Rémy*).

HERBARIUM and TYPES: P, PC; duplicates GH and, possibly, M (lich.).

BIBLIOGRAPHY and BIOGRAPHY: Barnhart 3: 143 (b. 2 Sep 1826, d. 2 Dec 1893); BM 4: 1628; Bossert p. 329; CSP 5: 158-159, 18: 130; Ewan (ed. 1): 288-289, (ed. 2): 182; Hortus 3: 1202; IF p. 726; IH 1 (ed. 6): 363, (ed. 7): 340; Jackson p. 371, 374; Kelly p. 190; Kew 4: 447; LS 22084; Morren ed. 10, p. 64; NI 695; PR 7538-7539, ed. 1: 8481; Rehder 5: 716; TL-2/1975; Tucker 1: 591; Urban-Berl. p. 311.

BIOFILE: Acosta-Solis, M., Natural. Viaj. Ci. Ecuador 30. 1968.
Allibone, S.A., Crit. dict. Engl. lit,. 2: 1771. 1878.
Anon., Bull. Soc. bot. France 41: 23. 1894 (d. 1893).
Blatter, E., Fl. Aden 22. 1914 (coll. at Aden).
Brenchley, J.L., Jottings during the cruise of the H.M.S. Curaçoa among the South Sea Islands in 1865, London 1873, p. xii-xvi (n.v.).
Clos, D., Bull. Soc. bot. France 40: 339. 1893 (obit., brief bibl.).
Duchartre, Bull. Soc. bot. France 40: 338. 1893 (d.).
Embacher, F., Lexik. Reisen 246. 1882.
Ewan, J., *in* Cent. Sci. San Francisco 58. 1955; Madroño 19: 31. 1969.
Hertel, H., Mitt. Bot. München 16: 422. 1980 (lich. from R(?) at M).
McCaughey, V., Haw. For. Agr. 16: 26-27, 54. 1919 (collected at Hawaii).
Parry, C.C., Amer. Natural. 9: 15. 1875 (R. in Utah).
Reveal, J.R., *in* A. Cronquist et al. Intermountain fl. 1: 52. 1972.
Sayre, G., Mem. New York Bot. Gard. 19(3): 385. 1975 (coll.).
Schlechtendal, D.F.L. von, Bot. Zeit. 14: 887-891. 1856 (extensive review of R.'s "Description des arbres gigantiques de la Californie, 1856).

COMPOSITE WORKS: C. Gay, *Fl. chil.*, vols. 3-5, 1848-1852, contributed *Saxifragaceae, Calyceraceae, Compositae, Solanaceae, Amarantaceae, Phytolaccaceae, Polygonaceae*.

EPONYMY: *Remya* W. Hillebrand ex Bentham (1873); *Remyella* C. Mueller Hal. (1896). *Note: Remysporites* M.A. Butterworth et R.W. Williams (1958) honors Winfried Remy, German paleobotanist.

8976. *Analecta boliviana*, seu nova genera et species plantarum in Bolivia crescentium ... Pars prima. Parisiis (e typis L. Martinet, ...) 1846. Oct.
Part 1: Dec 1846 (in journal), p. [i, iii], [345]-357, , *pl. 20*. *Copies*: G, HH, NY. – Reprinted and to be cited from Ann. Sci. nat., Bot. ser. 3. 6: 345-357. 1846.
Part 2: Ann. Sci. nat., Bot. ser. 3. 8: 224-240. Oct 1847.

8977. *Monografia de las Compuestas de Chile*, por el señor E. Julio Remy ... sacado de la Flora chilena de Don Claudio Gay ... Paris (en la imprenta de E. Thunot y Ca., ...) 1849. Oct.
Publ.: 1849, p. [i, iii], [257]-482, alt. pagination [5]-317, *15 pl. Copy*: NY. – Reprinted and to be cited from C. Gay, *Fl. chil.* 4: 257-482. 1849.
4610
8978. *Ascension du Pichincha*. Notes d'un voyageur. Lues à la Société d'Agriculture, commerce, sciences et arts du département de la Marne dans la séance du 1er décembre 1857, ... Chalons-sur-Marne (E. Laurent, imprimeur-libraire, ...) 1858. Oct.
Publ.: 1858, p. [i], [1]-30. *Copy*: G. – Possibly a reprint from a publication by the Soc. agr. comm. Sci. arts. dépt. Marne.

8979. *Champignons et truffes* ... Paris (Librairie agricole de la Maison rustique ...) 1861. 18-mo. (*Champ. truffes*).
Publ.: 1861, p. [i-iv], [1, list], *12 pl.* (col. liths.). *Copies*: BR, FH, G(2), MICH, NY, Stevenson.

Renaudet, Georges Benjamin (1852-?), French botanist and phytochemist, sometime "preparateur" at the medical college at Poitiers. (*Renaudet*).

HERBARIUM and TYPES: Unknown.

BIBLIOGRAPHY and BIOGRAPHY: Barnhart 3: 143; LS 22091; PH 567; Plesch p. 380; Rehder 5: 717.

BIOFILE: Souché, B., Fl. Haut Poitou xvii. 1901.

8980. Promenades et excursions. *Les plantes sauvages* dangereuses et utiles ... Paris (Ernest Flammarion, éditeur ...) 1914. Oct. (*Pl. sauv.*).
Publ.: 1914, p. [i-iv], [1]-320. *Copy*: BR.

Renauld, Ferdinand [François Gabriel], (1837-1910), French soldier and botanist (bryologist); from 1856-1887 with the "Spahis" (8 campaigns in Algeria); from 1885-1893 commander of Monaco; Légion d'honneur 1884; in retirement in Vesoul 1893-1897, later in Vence and Nice, from 1908-1910 in Paris. (*Renauld*).

HERBARIUM and TYPES: PC (bryol. herb. and types, rd 1909); other material at B, BM, BR, C, FH, H, L, MICH, NY (isotypes Amer. mat.), US (id.), W.
Exsiccatae: (1) *Musci Americae septentrionalis exsiccati*, 400 nos. 8 fasc. (1892-1908; with J. Cardot). Sets: BM, C, FH, G, H, L, MICH, NY, PC (see e.g. Bull. Herb. Boiss. 4: 1-19, 476-478. 1896).
(2) *Musci mascareno-madagascarienses*, 250 nos. (1892-1895). Sets: B, FH, L, MANCH, W.
(3) *Musci europaei exsiccati*, 350 nos., 7 fasc., 1902?-1908. Sets: BM, FH, PC.
77 letters by Renauld to V.F. Brotherus at Helsinki Univ. Libr. (Koponen 1979), 14 letters to W.F. Farlow at FH. For Renauld's correspondence at PC (826 letters from 119 bryologists, botanists and friends) see Lamy (1981).

BIBLIOGRAPHY and BIOGRAPHY: Barnhart 3: 144 (b. 18 Nov 1837, d. 6 Mai 1910); BL 1: 140, 314, 2: 149, 156, 707; BM 4: 1679, 8: 1065; Bossert p. 329; CSP 12: 611, 18: 133-134; Ewan ed. 1: 233; GR p. 347; Jackson p. 284; Kew 4: 447; Lenley p. 341-342; LS 6806; NAF ser. 2. 3: 41-42; NI 2: 746, also no. 2380; Rehder 5: 717; SBC p. 130 ("Ren."); TL-1/426, 1065; TL-2/2109, 3320, see J. Cardot; Tucker 1: 591; Urban-Berl. p. 311.

BIOFILE: Anon., Bot. Centralbl. 114: 176. 1910 (d.); Bot. Gaz. 21(6): 381. 1896 (isotypes to US); Bot. Not. 176. 1910; Bull. Acad. int. Géogr. bot. 19: 89. 1910 (obit.); Bull. Soc. bot. France 42: 703. 1895 (prix Montagne); Hedwigia 35(2): (62). 1896 (prix Montagne); Rev. bryol. 3(5): 80. 1876 (as lieutenant at Tarbes), 23(1): 16. 1896 (prix Montagne), 33(3): 48. 1906 (prix Desmazières), 38(2): 48. 1911 (herb. bought by Paris).
Britton, E.G., Bull. Herb. Boissier ser. 1. 4: 476-478. 1896 (Criticism Musc. Amer. sept. exs.).
Cavillier, F., Boissiera 5: 77. 1961.
F.C., Bull. Soc. bot. France 57: 624-625. 1910 (obit.).
Holzinger Bryologist 19: 50. 1916 (letter Cardot on publ. of mosses of Madagascar under war conditions).
Kneucker, A., Allg. bot. Z. 16: 144. 1910 (d.).
Koponen, T., *in* G.C.S. Clarke, Bryoph. syst. 164. 1979.
Lamy, D., Occas. Pap. Farlow Herbarium 16: 117-127. 1981 (biogr. sketch, invent. correspondence at PC, list correspondents).
Renauld, F., Rev. bryol. 39(4): 58-61. 1912 (corr. with H. Dupret).
Sayre, G., Bryologist 80: 516. 1977 (coll.); Mem. New York Bot. Gard. 246-247. 1971 (exsiccatae).
Thériot, I., Bryologist 13: 113-116, *pl. 10*, 125-128. 1910 (portr., obit., bibl.); Rev. bryol. 37(5): 106-114. 1910 (obit., bibl., b. 18 Nov 1837, d. 6 Mai 1910), ser. 2. 4(2): 89. 1931 (herb. PC).
Verdoorn, F., ed., Chron. bot. 3: 19. 1937.

COMPOSITE WORKS: (1) With J. Cardot *in* A. Grandidier, *Hist. phys. Madagascar*, vol. 39, Mousses, 1898-1915, see TL-2/2109, (IDC 5936) and below no. 8995.
(2) Co-author with E.M.J. Jeanbernat, *Guide bryol. Pyrénées* 1884-1885, TL-2/3320 (correction: *part 1* incl. Rev. bot. 3: 12-16. 1884, part 2 is a reprint of Rev. bot. 3: 305-338. 1885).
(3) Co-author with T. Husnot of the genus *Hypnum* in *Muscologia gallica* (TL-2/3154).
(4) *Musci* with J. Cardot *in* Th. Durand & H. Pittier, *Primit. fl. costaric.* 1(2): 119-215 (pages Bull.) 1893 and 1(3): 53-80 (pages reprint). 1896.
For further works published with others see nos. 32-57 of the Renauld bibliography by Thériot (1910).

EPONYMY: *Renauldia* C. Mueller Hal. ex Renauld (1891).

8981. *Aperçu phytostatique sur le département de la Haute-Saône* suivi d'un catalogue des plantes vasculaires et des mousses, ... Paris (F. Savy, Libraire ...) 1873. Oct. (*Aperçu phytostat. Haute-Saône*).
Collaborator: D. Laloy (for phanerogams).
Publ.: 1873 (presentation copy BR Mai 1874; Bot. Zeit. 26 Jun 1874; BSbF Dec 1873), p. [1]-398, *1 map*. *Copies*: BR, NY, USDA. – Also issued in Bull. Soc. Agr. Sci. Arts Haute-Saône ser. 3. 4(2): [309]-706.
Other issue: 1873, p. [1]-398, map. *Copies*: G, HH. – "... Catalogue des plantes qui y croissent spontanément". Imprint Vesoul (Lépagnez, libraire) 1873. Oct.

8982. *Recherches sur la distribution géographique des muscinées* dans l'arrondissement *de Forcalquier* et la Chaine de Lure (Basses-Alpes) suivies d'un catalogue des muscinées du bassin principal de la Durance ... Besançon (Imprimerie Dodivers et Cie., ...) 1877. Oct. (*Rech. distr. musc. Forcalquier*).
Publ.: 1876 (p. 87: Jun 1876; BSbF sess. 9 Feb 1877; Bot. Zeit. 16 Mar 1877), p. [i-iv], [1]-87. *Copies*: FH(2), H, NY.
Ref.: Debat, Ann. Soc. bot. Lyon 5: 102-105. 1878.
Husnot, T., Rev. bryol. 4(2): 30. 1877 (rev.).

8983. *Révision de la section Harpidium* du genre Hypnum de la flore française [Extrait des Memoires de la Société d'Émulation du Doubs, séance du 8 novembre 1879].
Publ.: 1880 (Bot. Centralbl. 6-10 Jun 1881; Rev. bryol. Mai-1881 as of 1880; Bot. Zeit. 24 Jun 1881, as of 1880), p. [1]-24. *Copies*: FH, G, H. – Reprinted and to be cited from Mém. Soc. Émul. Doubs 1879, 1880. – See also Rev. bryol. 8: 73-82. 1881, 33: 89-100. 1906, 34: 7-14. 1907 and 39(4): 58. 1912 for a correspondence between Renauld and H. Dupret on this subject. Also F. Renauld, *in* P.T. Husnot, Muscol. gall. 2: 367-395. *pl. 105-113*. Mar 1894 (Warnstorf, Bot. Centralbl. 60: 53-55. 3 Oct 1884 ext. rev. (reprinted with orig. pagination and with date, p. [167]-395, *pl. 105-113*. *Copy*: FH).

8984. *Catalogue raisonné des plantes vasculaires et des mousses* qui croissent spontanément *dans la Haute-Saône* et parties limitrophes du Doubs précédé d'un aperçu sur la géographie botanique ou distribution des plantes dans ce département selon les altitudes et les terrains, avec carte coloriée indiquant les grands divisions géologiques ... Besançon (Marion, Morel et Cie, ...) 1883. Oct. (*Cat. rais. pl. Haute-Saône*).
Publ.: 1883, p. [1]-437. *Copy*: FH. – On p. 399-437 *Supplément* par MM. F. Renauld, C. Flagey, Vendrely et J. Paillot. *Liste des plantes rares ou nouvelles pour ce Département et les parties limitrophes du Doubs* which is reprinted from Mém. Soc. Émul. Doubs, ser. 5. 7: 162-200. 1882. For further literature see BL 2: 156.

8985. *Mousses nouvelles de l'Amérique du Nord* [Extrait du Bulletin de la Société royale de Botanique de Belgique, t. xxvii (1888), première partie]. Oct.
Co-author: Jules Cardot (1860-1934).
Publ.: A series of publications in the Bull. Soc. Bot. Belg. (BSBB) 27-36, 1889-1898 reprinted with new (and sometimes also with journal) pagination; to be cited from journal. *Copies*: G, NY.
1: late 1888 (Nat. Nov. Feb(1) 1889; J. Bot. Mar 1889; Bot. Zeit. 19 Apr 1889), p. [1]-11, *pl. 3-10* (uncol.), BSBB 27(1): [127]-137. *pl. 3-10*. 1888.
2: 7 Feb 1890 (inf. R. Tournay, in copy BR; Bot. Zeit. 21 Mar 1890, J. Bot. Mar 1890), p. [1]-14, *pl. 7-9*, BSBB 28(1): [121]-134, *pl. 7-9*. 1889 (publ. 1890).
3: 1890 or 1891 (Bot. Zeit. 4 Sep 1890; J. Bot. Dec 1891), p. [1]-16, *pl. 2-6*, BSBB 29(1): [145]-160, *pl. 2-6*. 1890 or 1891.
4: 1896 (J. Bot. Dec 1896), p. [1]-7, *pl. 1-2*, BSBB 35(1): [119]-125, *pl. 1-2*. 1896.
5: 1898 (séance 5 Dec 1897), p. [1]-8, *pl. 10-12*, BSBB (C.R.) 36(2): [173]-180.
English translation: in Botanical Gazette (BG) (reprints have journal pagination) (*Copies*: G, NY) (*New Mosses of North America*):
1: Aug 1888, BG 13(8): [197]-203, *pl. 13-20* (uncol.).
2: Apr 1889, BG 14(4): [91]-100, *pl. 12-14*.
3: Feb 1890, BG 15(2): 39-45, *pl. 5-7*. Latin version 3(1).
4: Mar 1890, BG 15(3): 57-62. *pl. 8-9* (also 17: 296. 1892). Latin version: 3(2).
5: Jun 1894, BG 19(6): [237]-240, *pl. 21-22*. Latin version 4.

6: Jul 1896, BG 22(7): [48]-53, *pl. 3-5*, also p. [1]-6, *pl. 3-5*. Latin version 5.

8986. *Flora miquelonensis.* Florule de l'ile Miquelon (Amérique du Nord). Énumération systématique avec notes descriptives des phanérogames, cryptogames, vasculaires, mousses, sphaignes, hépatiques et lichens ... Lyon (Association typographique ...) 1888. Oct. (*Fl. miquelon.*).
Co-author: Ernest Delamare (1833?-1888), Jules Cardot (1860-1934).
Publ.: 1888 (Nat. Nov. Jun(1) 1888; Rev. bryol. Jun 1888; Bot. Centralbl. 27 Apr-4 Mai 1888, BG Jun 1888), p. [1]-78, [79, err.]. *Copy*: USDA.
Ref.: Bescherelle, E., Bull. Soc. bot. France 35 (bibl).: 99-100. post 15 Mai 1888.

8987. *Musci exotici novi vel minus cogniti* ... [Extrait du Bulletin de la Société royale de botanique de Belgique, t. xxix, première partie (1890)]. Oct.
Co-author: Jules Cardot (1860-1934).
Publ.: A series of publications in Bull. Soc. Bot. Belg. (BSBB) 29 (1890)-38 (1903). Reprinted with new and with journal pagination. Reprints' imprint: "Gand, Imp. C. Annoot-Braeckman, ...". To be cited from journal. *Copies reprints*: BR, FH, FI, G, MO, NY.
1: 1890 (NY rd. 26 Nov 1890; Rev. bryol. Jul-Aug 1891; Bot. Zeit. 4 Sep 1891), p. [1]-26; BSBB 29(1):) [161]-186. 1890. – See also F. Renauld, *Prodr. fl. bryol. Madagascar*.
2: 1891 (séance 10 Oct 1891; Rev. bryol. Mar-Apr 1892; Bot. Zeit. 15 Jan 1892), p. [27]-53; BSBB 30(2): [181]-207. 1891.
3: 1892 (séance 12 Jun 1892; Rev. bryol. Sep-Dec 1892; Bot. Zeit. 9 Sep 1892; J. Bot. Sep 1892), p. [55]-78; BSBB 31(2): [100]-123. 1892. – With *Hepaticae* by F. Stephani.
4: 1893 (séance 11 Feb 1893; rd. G 15 Nov 1893; Rev. bryol. Nov-Dec 1893; Nat. Nov. Oct(1) 1893; Bot. Zeit. 16 Sep 1893; J. Bot. Aug 1893; Hedwigia Nov-Dec 1893), p. [79]-111; BSBB 32(2): [8]-40. 1893. – With *Hepaticae* Id.
5: 2 Aug 1894 (rd. G 6 Oct 1894; Bot. Centralbl. 24 Jul 1894; Bot. Zeit. 16 Sep 1894; Hedwigia 25 Oct 1894), p. [113]-133; BSBB 32(1): [101]-121. 1893. – With *Hepaticae* id.
6: 1894 (séance 10 Nov 1894; rd. G 6 Mai 1895; Bot. Zeit. 16 Aug 1895; Hedwigia 19 Oct 1895), p. [135]-163; BSBB 33(2): [109]-137.
7: 1895 (séance 5 Mai 1895; NY rd 27 Jun 1896; Rev. bryol. Nov-Dec 1895; Bot. Zeit. 1 Mai 1896), p. [165]-186; BSBB 34(2): 57-78.
8: 1896 (rd G 5 Jul 1897; Rev. bryol. Jul-Aug 1897; Nat. Nov. Aug(1) 1897; Bot. Zeit. 16 Jun 1897), p. [187]-213; BSBB 35(1): [299]-325.
9: 1899 (Bot. Centralbl. 20 Jun 1900; Rev. bryol. Mai-Jun 1900; Nat. Nov. Oct(2) 1900), p. [215]-256; BSBB 38(1): [7]-48. – *Hepaticae* by F. Stephani.
10: Jan 1905 (Nat. Nov. Jan(1) 1905; Rev. bryol. Mar-Apr 1905), p. [257]-372, preprinted from BSBB 41(1): [7]-122. 8 Mai 1905.

8988. *Musci costariscenses* [sic; parts 2 and 3: costaricenses] ... [Extrait du Bulletin de la Société royale de Botanique de Belgique t. xxxi (1892), première partie, pp. 145-173]. Oct.
Publ.: Three papers published and to be cited from Bull. Soc. Bot. Belg. [BSBB] and reprinted with independent pagination. *Copies*: G, H, NY. – Parts 1 and 2 were also contributions to T.A. Durand & H.F. Pittier, *Primit. fl. costaric.* see TL-2/1592.
1: Oct 1893 (vol. for 1892; rd. G 15 Nov 1893; see TL-2/1592), p. [1]-32; BSBB 31(1): [143]-174. 1892, publ. 1893.
2: 2 Aug 1894 (inf. R. Tournay in copy BR; vol. for 1893; G rd 6 Oct 1894; Hedwigia 25 Oct 1894; Bot. Centralbl. 6 Dec 1894), p. [33]-60; BSBB 32(1): [174]-201. 1893, publ. 1894.
3: 8 Mai 1905 (date on cover journal issue; G reprint rd 9 Jan 1907), p. [61]-86; BSBB 41(1): 123-148. 8 Mai 1905 (for 1902-1903).

8989. *Musci Americae septentrionalis* ex operibus novissimis recensiti et methodice dispositi ... Stenay (Meuse) (J. Cardot), Le Mans (Typographie Edmond Monnoyer ...) 1893. Oct.
Publ.: two papers published in Revue bryologique, vols. 19 and 20, issued as a single reprint (Nat. Nov. Sep(1) 1893; USDA rd 12 Jul 1893) with new pagination, p. [1]-68. *Copies*: FH, G, NY, USDA. – Original publication:

[1]: Rev. bryol. 19(5-6): 65-96. Sep-Dec 1892.
[2]: Rev. bryol. 20(1-2): 1-32. Jan-Apr 1893.

8990. *Mousses nouvelles de l'herbier Boissier* ... [Bull. Herb. Boiss., Janvier 1894]. Oct.
Co-author: Jules Cardot (1860-1934).
1: Jan 1894, p. [1]-2. *Copy*: U. – Issued as and to be cited from Bull. Herb. Bois. ser. 1. 2: 32-33. 1894.
2: Mai 1895, p. [1]-2. *Copy*: U. – Id. ser. 1. 3: 240-241. 1895.

8991. F. Renauld & J. Cardot. *Musci Americae septentrionalis exsiccati. Observations & rectifications* sur les espèces distribuées. Stenay (Typ. et Aut [sic] Cherillard) 1894. Oct.
Co-author: Jules Cardot (1860-1934).
Publ.: Oct 1894 (p. 18), p. [1]-18. *Copies*: G, NY (rd. 13 Mar 1895), lithographed manuscript.

8992. Extrait du Bulletin de l'Herbier Boissier. Tome iv. No. 1. Janvier 1896. *Musci americae septentrionalis exsiccati*. Notes sur quelques espèces distribuées dans cette collection ... [Genève 1896]. Oct.
Publ.: Jan 1896, p. [1]-19. *Copy*: U. – Published and to be cited as Bull.Herb. Boissier 4(1): [1]-19. Jan 1896.

8993. *Prodrome de la flore bryologique de Madagascar* des Mascareignes et des Comores ... Imprimerie de Monaco 1897 [1898]. Qu. (*Prodr. fl. bryol. Madagascar*).
Publ.: 1898 (t.p.1897; back cover 1898; Rev. bryol. Nov-Dec 1898; BG 19 Nov 1898; Bot. Centralbl. 2 Feb 1899; Nat. Nov. Mar(2) 1899), p. [i]-viii, [1]-296, [1, err.], [1-2, index]. *Copies*: BR, FH, FI, G, MICH, MO, MO-Steere, NY, PH, U, USDA.
Supplément: 1909 in *Essai Leucoloma* (q.v.), p. [1]-139, *pl. 1-24* with text.
Ref.: Sydow, F., Hedwigia 38: (104)-(105). 25 Apr 1899 (rev.).
Warnstorf, C., Bot. Centralbl. 77: 201-205. 2 Feb 1899 (rev.).

8994. *Contributions à la flore bryologique de Madagascar* ... Bordeaux (J. Durand, ...) 1898. Oct.
Publ.: 1898 (Rev. bryol. Sep-Oct 1898; Bot. Centralbl. 30 Nov 1898), p. [1]-9, [10, expl. pl.], *pl. 1* (uncol. lith. auct.). *Copies*: FH, G, H. – Reprinted and to be cited from Actes Soc. Linn. Bordeaux 53: 17-23. 1898.

8995. *Histoire physique, naturelle et politique de Madagascar* publiée par Alfred et Guillaume Grandidier. Volumen xxxix Histoire naturelle des plantes *Mousses* par MM F. Renauld et J. Cardot ... Paris (imprimé ... à l'Imprimerie nationale) 1898-1915. Qu. (*Hist. phys. Madagascar, Mousses*).
Authorship: (of series:) Alfred Grandidier (1836-1921) and Guillaume Grandidier (b. 1873); (of *Mousses*:) Renauld with Jules Cardot (1860-1934).
Publ.: For full series see TL-2/2109. The *Mousses* came out as follows:
Texte: 1915 (p. viii: 21 Mai 1914), p. [iii]-viii, [1]-560, [1, cont.], [1, err.].
Atlas: in six fascicles. The plates with A-C numbers, listed separately here, came out with the sixth fascicle and should not have been listed under fasc. 1-5 in TL-2/2109. The BR copy, which has fasc. 1-5 in original state with covers, lacks these A numbers.
1 (fasc. 46 of entire work): 1898, *pl. 1-32*. – "Atlas 1re partie".
2 (fasc. 47): 1899, *pl. 33-64*.
3 (fasc. 48): 1899 (Rev. bryol. Sep-Oct 1900), *pl. 65-106*.
4 (fasc. 51): 1901 (Rev.bryol. Sep-Oct 1901), *pl. 107-143*.
5 (fasc. 55): 1905 (Bot. Centralbl. 14 Nov 1905), *pl. 144-163*.
6 (fasc. 59): 1913, *pl. 3A-B, 7A, 15A, 24A-C, 41A, 42A, 48A, 55A-F, 57A, 62A, 70A, 88A, 89A, 90A, 97A, 111A*.
Copies: BR(1-5), MO, MO-Steere, PH; IDC 5936. – Cardot, in a note in Rev. bryol. 41(4): 76. 1914 mentions the completion of the Atlas and states that 3 copies of the atlas had been distributed by Renauld. These must have been copies of fasc. 1-5, 1898-1905, because Renauld died in 1910. The 24 additional plates were published in 1913. – The 187 lithographs are of drawings by Renauld and Cardot themselves.

The new species were first described by Renauld and Cardot between 1890 and 1905 in the Bulletin de la Société botanique de Belgique. (Musci exotici novi vel minus cogniti, in volumes 29-35, 38). – For the explanations of the plates see Renauld 1909. Plates 1-130 were drawn by Renauld, 131-163 by Cardot. – The text is by Cardot.

Ref.: Cardot, J., Bryologist 19: 75. 1916 (text published. "Only 150 copies are placed on sale, 100 of which are reserved solely for the subscribers to the complete work").
LeRoy Andrews, A. Bryologist 22: 27-28. 1919 (rev.).
Renauld, F., Prodrome de la flore bryologique de Madagascar. Monaco 1897-1909. Qu. (reproduces descriptions and provides, in the supplement of 1909, the explanations of the plates!).

8996. *Musci* [in series: Matériaux pour la flore du Congo] ... [Extrait du Compte-Rendu de la Séance mensuelle du 10 novembre 1900 de la Société royale de Botanique de Belgique ...]. Oct.
Co-author: Jules Cardot (1860-1934).
Publ.: Nov 1900-Feb 1901 (séance 10 Nov 1900; Rev. bryol. Mar-Apr 1901), p. [1]-7. *Copy*: FH. – Reprinted and to be cited from Bull. Soc. Bot. Belg. 39(2): [100]-112. 1900/1901.

8997. Extrait du Bulletin de l'Herbier Boissier. Seconde série (1902). – no. 5. *Mousses des Canaries* récoltées par M.A. Tullgren et coup d'oeil sur la flore bryologique des îles atlantiques ... [Genève 1902]. Oct.
Co-author: Jules Cardot (1860-1934).
Publ.: 30 Apr 1902 (issue dated), p. [1]-21. *Copies*: G, H, U.– Reprinted and to cited from Bull. Herb. Boissier ser. 2. 5: [433]-453. 1902.

8998. *Essai sur les Leucoloma* et supplément au prodrome de la flore bryologique de Madagascar des Mascareignes et des Comores ... Imprimerie de Monaco 1909. Qu. (*Essai Leucoloma*).*Publ.*: 1909 (Rev. bryol. Jul-Aug 1910; Hedwigia 12 Oct 1910; Nat. Nov. Apr(2) 1910), p. [i]-ix, *Leucoloma*: [1]-50; *Supplément Prodrome*: [1]-139, *pl. 1-24* (uncol. liths. auct.) each with 1 p. text, text *pl. 1* on p. [140]. *Copies*: BR, FH, FI, G, MICH, MO-Steere, NY, PH, U. – *Preliminary publ*: Rev. bryol. 28: 66-70, 85-87. 1901.

Renault, Bernard (1836-1904), French botanist and palaeontologist at Autun and Paris; Dr. sci. phys. Paris 1867; Dr. sci. nat. Paris 1879; teacher at a teacher's college at Cluny 1867-1872; from then on at the Muséum d'Histoire naturelle, Paris (préparateur 1872-1876; assistant naturalist 1876-1904); founder and first president of the Société d'Histoire naturelle d'Autun. (*Renault*).

HERBARIUM and TYPES: AUT; fossil preparations at Muséum d'Histoire naturelle, Paris.

BIBLIOGRAPHY and BIOGRAPHY: Andrews 319-320; Barnhart 3: 144 (b. 4 Mar 1836, d. 20 [err.] Oct 1904); BJI 1: 48; BM 2: 705, 4: 1679; Bossert p. 329; CSP 5: 159, 8: 730, 11: 144, 11: 145, 12: 611, 18: 134-135; De Toni 4: xlvii; DSB 11: 372-373 (N. Spjeldnaes); Jackson p. 178, 179, 180, 185, 498; Kelly p. 190; Kew 4: 448; LS 22093-22095, 38324-38330; Moebius p. 378, 381; Morren ed. 10, p. 54; MW p. 412; Quenstedt p. 357; Rehder 3: 707, 4: 491; TL-2/803.

BIOFILE: Alpern, B. et al., Rev. Palaeobot. Palyn. 7: 155-156, 165-166. 1968 (brief biogr. sketch).
Andrews, H.N., The fossil hunters 72, 325-331. 1980.
Anon., Bot. Jahrb. 37 (Beibl. 84): 3. 1906 (d.); Bot. Not. 1905: 56; Bot. Zeit. 62(2): 368. 1904 (d.); Bull. Soc. Bot. Belg. 42(2): 89-90. 1905 (obit.); Hedwigia 35(2): (62). 1896 (prix Trémont), 36(1): (34). 1897 (prix Saintour), 45(1): (37). 1906 (d.); Jenaer Lit. Zeit. 1874 (71): [1-4] (important early international recognition of R's work on silicified Sphenophyllum and Annularia); Nat. Nov. 26: 642. 1904 (d.); Österr. bot. Z. 55: 39. 1905 (d. 16 Oct 1904); Rev. bryol. 28: 83. 1901 (R. Zeiller elected to Acad. Sci., Renault was a good second).

Bornet, É., et al., C.R. Séances Acad. Sci., Paris 137: 1160. 1903 (awarded the Prix Petit d'Ormoy).
Costantin, J., Ann. Sci. nat. Bot. ser. 10. 16: clvi. 1934.
Ewan, J. et al., Short hist. bot. U.S. 104. 1969.
Gaudry, A., Bull. Soc. Sci. nat. Châlon-sur-Saône 30: 156-158. 1904, La Nature 32: 350-351. 1904 (obit.).
Gillot, X., Bull. Soc. bot. France 51: 385, 430-432. 1904 (obit.).
Harvey-Gibson, R.J., Outlines hist. bot. 167, 168. 1914.
Kneucker, A., Allg. bot. Z. 10: 204. 1904 (d.; was president Soc. hist. nat. Autun).
Mägdefrau, K., Gesch. Bot. 248, 298. 1973.
Magnin, A., Bull. Soc. bot. Lyon 32: 138. 1907, 35: 50. 1910.
Renault, B., Bull. Soc. bot. France 42: 287. 1895 (note by R. on F.X. Gillot); Notice sur les travaux scientifiques de Bernard Renault, Autun (Imprimerie Dejussieu père et fils) 1896. Qu., 162 p., *8 pl.* (very detailed bibliography), Supplément ib. 1899, 63 p., *pl. A*, Deuxième Supplément ib. 1901, 84 p.
Rodgers, A.D., Amer. bot. 1873-1892. 174, 187, 195. 1944.
Scott, D.H., J.R. microsc. Soc. ser. 2. 26: 129-145. *pl. 4.* 1906 (biogr., portr., concise bibl., discussion of major works; "few men so distinguished have received such miserably inadequate recognition from their official chiefs ...").
Taton, R., ed., Science in the 19th century 330, 377, 384-385. 1965.
T.R.R.S., Proc. Linn. Soc. London 117: 51-52. 1905 (obit.).
Verdoorn, F., ed., Chron. bot. 1: 12, 23, 37. 1935, 2: 13. 1936.
Zeiller, R., Bot. Centralbl. 96: 495-496. 1904 (obit.); Rev. gén. Sci. 15(23):1057-1058. 1904 (obit.; d. 16 Oct 1904).
Zittel, K.A. von, Gesch. Geol. Paläont. 78, 789. 1899; Hist. geol. palaeontol. 375. 1901.

EPONYMY: *Renaultia* R. Zeiller (1883); *Renaultia* D.R.J. Stur (1883).

8999. *Végétaux silicifiés d'Autun*; observations sur la structure du Dictyoxylon; ... [Paris 1872]. Qu.
Publ.: 1872 (p. 3: 13 Mai 1872), p. [1]-3. *Copy*: USGS. – Reprinted and to be cited from C.R. Acad. Sci., Paris 74: 1295-1298. 1872.

9000. *Recherches sur les végétaux silicifiés d'Autun.* Étude du genre *Myelopteris* ... [Paris (Gauthier-Villars, ...) 1874]. Qu.
Publ.: 1874 (p. 4: 26 Jan 1874). *Copy*:USGS. – Reprinted and to be cited from C.R. Acad. Sci., Paris 78: 257-260. 1874. – See also Ann. Sci. nat., Bot. ser. 5. 20: 154-157. 1874.
Final publ.: 1876, p. [1]-28, *pl. 1-6* (uncol. liths. auct.). *Copy*: G. – Issued as Mém. div. Sav. Acad. Sci. Paris 22(10). 1876.

9001. Contributions à la paleontologie végétale. *Étude sur le Sigillaria spinulosa* et sur le genre Myelopteris, ... Paris (Imprimerie nationale) 1875. Qu.
Publ.: 1875, p. [i-iii], [1]-23. *pl. 1-6* (uncol. liths. auct.). *Copies*: G, USGS. – Issued as Mém. div. Sav. Acad. Sci., Paris 22(9). 1875. – The *Étude sur le genre Myelopteris* occupies id. 22(10). 1876, see above under 1874.

9002. *Recherches sur les végétaux silicifiés d'Autun et de Saint-Étienne.* Étude du genre *Botryopteris* [Paris 1875]. Qu.
Publ.: 1875 (p. 4: 18 Jan 1875), p.[1]-4. *Copy*: USGS. – Reprinted and to be cited from C.R. Acad. Sci., Paris 80: 202-206. 1875. See also Ann. Sci. nat., Bot. ser. 6. 1: 220-240. 1875.

9003. *Sur la fructification de quelques végétaux silicifiés*, provenant des gisements d'Autun et de Saint-Étienne ... [Paris (Gauthier-Villars) 1876]. Qu.
Publ.: 1876 (p. 4: 24 Apr 1876), p. [1]-4. *Copy*: USGS. – Reprinted and to be cited from C.R. Acad. Sci., Paris 82: 992-995. 1876. See also Ann. Sci. nat., Bot. ser. 6. 3: 5-29. 1876.

9004. *Recherches sur les végétaux silicifiés d'Autun et de Saint-Étienne.* Des *Calamodendrées* et de

leurs affinités botaniques probables ... [Paris (Gauthier-Villars, ...)] 1876. Qu.
Publ.: 1876 (p. 4: 4 Sep 1876), p. [1]-4. *Copy*: USGS. – Reprinted and to be cited from C.R. Acad. Sci., Paris 83: 546-549. 1876.

9005. *Recherches sur quelques Calamodendrées* et sur leurs affinités botaniques probables ... [Paris 1876]. Qu.
Publ.: 1876 (p. 3: 11 Sep 1876), p. [1]-3. *Copy*: USGS. – Reprinted and to be cited from C.R. Acad. Sci., Paris 83: 574-576. 1876.

9006. *Fleurs mâles des Cordaïtes* ... [Paris (Gauthier-Villars) 1877]. Qu.
Publ.: 1877 (p. 3: 16 Apr 1877), p. [1]-3. *Copy*: USGS. – Reprinted and to be cited from C.R. Acad. Sci. Paris 84: 782-785. 1877.
Fleurs femelles des Cordaïtes: 1877 (p. (4): 4 Jun 1897), p. [1]-4. *Copy*: USGS. – Id. 84: 1328-1331. 1877.

9007. *Structure des Sigillaires* ... [Paris 1878]. Qu.
Publ.: 1878 (p. 3: 15 Jul 1878), p. [1]-3. *Copy*: USGS. – Reprinted and to be cited from C.R. Acad. Sci., Paris 87: 114-116. 1878.

9008. *Structure comparée des tiges des Lépidodendrons et des Sigillaires* ... [Paris 1878]. Qu.
Publ.: 1878 (p. 3: 9 Sep 1878), p. [1]-3. *Copy*: USGS. – Reprinted and to be cited from C.R. Acad. Sci., Paris 87: 414-416. 1878.

9009. *Structure et affinités botaniques des Cordaïtes* ... [Pais 1878]. Qu.
Publ.: 1878 (p. 4: 7 Oct 1878), p. [1]-4. *Copy*: USGS. – Reprinted and to be cited from C.R. Acad. Sci., Paris 87: 538-541. 1878.

9010. *Recherches sur la structure et les affinités botaniques des végétaux silicifiés* recueillis aux environs d'Autun et de St.-Étienne ... Autun (Imprimerie Dejussieu père et fils) 1878. Oct. (*Rech. vég. silicif.*).
Publ.: 1878, p. [i, iii], [1]-216. *pl. 1-30* (uncol. liths. auct.). *Copies*: BR.
Preliminary publ.: 1876, see above, no. 9004.

9011. *Structure comparée de quelques tiges de la flore Carbonifère* ... Paris (G. Masson, ...) 1879. Qu.
Publ.: 1879, cover t.p., p. [213]-348, *pl. 10-17* (uncol. liths. auct.). *Copy*: USGS. – Reprinted and to be cited from Nouv. Arch. Mus. Hist. nat. 2: 213-348. *pl. 10-17.* 1879. – See also Rev. scientif. 17: 304-306. 1879 and Neu. Jahrb. Mineral. 1881(1): 311-316, (2): 293-297.

9012. *Sur un nouveau groupe de tiges fossiles silicifiées de l'époque houillère* ... [Paris 1879]. Qu.
Publ.: 1879 (p. 3: 6 Jan 1879), p. [1]-3. *Copy*: USGS. – Reprinted and to be cited from C.R. Acad. Sci., Paris 88: 34-36. 1879.

9013. *Étude sur les Stigmaria* rhizomes et racines des Sigillaires ... [Paris 1880].
Publ.: 1880, p. [1]-51, *pl. 1-3* (uncol. lith. auct.). *Copy*: USGS. –Issued as Ann. Soc. géol., Paris 12(1), art. no. 1.

9014. *Sur une nouvelle espèce de Poroxylon* ... [Paris, 1880]. Qu.
Publ.: 1880 (p. 3: 22 Nov 1880), p. [1]-3. *Copy*: USGS. – Reprinted and to be cited from C.R.Acad. Sci., Paris 91: 860-861. 1880.

9015. *Cours de botanique fossile* fait au Muséum d'histoire naturelle ... Paris (G. Masson, éditeur) 1881-1885, 4 parts. Oct. (*Cours bot. foss.*).
1: 1 Dec 1880-23 Jan 1881 (p. [ix] 25 Oct 1880; printer's mark Nov 1880; t.p. 1881; Bot. Centralbl. 24-28 Jan 1881; Bot. Zeit. 28 Jan 1881), frontisp., p. [i-ix], [1]-185, [1, list pl.], *pl. A, 1-20* (uncol. liths., Boirin, with text), in all *22 pl.*
2: Nov-Dec 1881 (p. ii: 25 Oct 1881; t.p. 1882, but rev. BSbF Nov-Dec 1881 with ref. to 1882 t.p. date; Bot. Zeit. 28 Jul 1882), p. [i*-vi*], [i]-ii, [1]-194, [1, list pl.], *pl. 1-24* (id.).

3: Jan-Feb 1883 (p. [vii]: Oct 1882; Bot. Zeit. 30 Mar 1883; BSbF Jan-Feb 1883; printer's mark Dec 1882), p. [i-vii], [1]-241, [1, list pl.], *pl. A, 1-35* (id.).
4: 1885 (Bot. Zeit. 26 Jun 1885), p. [i-iv], [1]-232, [1, list pl.], *pl. A-D, 1-22* (id.).
Copies: BR, E, G, M, PH, U (vol. 3), USGS.
Ref.: Anon., Bull. Soc. bot. France 28 (bibl.): 109-111. Apr-Aug 1881 (rev. part 1), 28 (bibl.): 202-203. Nov-Dec 1881 (rev. part 2), 30 (bibl.): 5-7. Jan-Feb 1883 (rev. part 3).
Steinmann, J.H.C.G.G., Bot. Centralbl. 7: 367-369. 1881.
Félix, Ber. naturf. Ges. Leipzig 1886-1887: 6-16. 1888 (rev. 1-4).

9016. *Sur les Sphenozamites* ... [Paris 1882]. Qu.
Publ.: 1882 (p. 2: 26 Dec 1881), p. [1]-2. *Copy*: USGS. – Reprinted and to be cited from C.R. Acad. Sci., Paris 93: 1165-1166. 1881 (1882).
Note sur les Sphenozamites, 1882, p. [1]-4, *1 pl.* (uncol.). *Copies*: G, USGS. – Reprinted and to be cited from Arch. bot. Nord France, Lille 1(12): 180-184. *1 pl.* Mar 1882 (Bot. Zeit., rev. by Geyler, 40: 769. 3 Nov 1882).

9017. *Sur les Astérophyllites* ... [Paris 1882]. Qu.
Publ.: 1882 (p. 2: 13 Feb 1882), p. [1]-2. *Copy*: USGS. – Reprinted and to be cited from C.R. Acad. Sci., Paris 94: 463-464. 1882.

9018. *Sur les Gnétacées du terrain houiller de Rive-de-Giers* ... [Paris (Gauthier-Villars) 1883]. Qu.
Publ.: 1883 (p. 3: 5 Mar 1883), p. [1]-3. *Copy*: USGS. – Reprinted and to be cited from C.R. Acad. Sci., Paris 96: 660-662. 1883.

9019. *Note pour servir à l'histoire de la formation de la houille*; ... [Paris 1883]. Qu.
[*Note 1*]: 1883 (p. 3: 20 Aug 1883), p. [1]-3. *Copy*: USGS. – Reprinted and to be cited from C.R. Acad. Sci. Paris 97: 511-533. 1883.
Deuxième note: 1883 (p. 3: 5 Nov 1883), p. [1]-3. *Copy*: USGS. – Id. 97: 1019-1021. 1883.
Troisième note: Dec 1883 or 1884 (p. 3: 17 Dec 1883), p. [1]-3. *Copy*: USGS. – Id. 97: 1439-1441. 1883 (1884).
Quatrième note: 1884 (p. 3: 28 Jul 1884), p. [1]-3. *Copy*: USGS. – Id. 99: 200-202. 1884.
See also: Mém. Soc. Sci. Saône-et-Loire 5: 120-124. 1884.

9020. *Sur l'existence du genre Todea dans les terrains jurassiques* ... [Paris, 1883]. Qu.
Publ.: 1883 (p. 3: 8 Jan 1883), p. [1]-3. *Copy*: USGS. – Reprinted and to be cited from C.R. Acad. Sci., Paris 96: 123-129. 1883.

9021. *Sur l'organisation du faisceau foliaire des Sphenophyllum*; ... [Paris, 1883]. Qu.
Publ.: 1883 (p. 3: 10 Sep 1883), p. [1]-3. *Copy*: USGS. – Reprinted and to be cited from C.R. Acad. Sci., Paris 97: 649-651. 1883.

9022. *Considérations sur les rapports des Lépidodendrons, des Sigillaires et des Stigmaria* ... Paris (G. Masson, ...) 1883. Oct.
Publ.: 1883, p. [i-iii], [1]-32. *pl. A* (uncol. lith.). *Copies*: BR, G, USGS. – Reprinted and to be cited from Ann. Sci. nat., Bot. ser. 6. 15: 168-198. 1883. – Reply to Williamson & Hartog, *Les Sigillaires et les Lépidodendrées*, Ann. Sci. nat., Bot. ser. 6. 13: 337-352. 1882. See also Geyler, Bot. Zeit. 42: 139. 29 Feb 1884.

9023. *Sur un nouveau genre de fossiles végétaux*; ... [Paris, 1884]. Qu.
Co-author: [Charles] René Zeiller (1847-1915).
Publ.: 1884 (p. 4: 2 Jun 1884), p. [1]-4. *Copy*: USGS. – Reprinted and to be cited from C.R. Acad. Sci., Paris 98: 1391-1394. 1884.

9024. *Sur un nouveau genre du terrain houiller supérieur*; ... (Paris (Gauthier-Villars) 1884]. Qu.
Co-author: [Charles] René Zeiller (1847-1915).
Publ.: 1884 (p. 3: 7 Jul 1884), p. [1]-3. *Copy*: USGS. – Reprinted and to be cited from C.R. Acad. Sci., Paris 99: 56-58. 1884.

9025. *Sur l'existence d'Astérophyllites phanérogames;* ... [Paris, 1885].
Co-author: [Charles] René Zeiller (1847-1915).
Publ.: 1885 (read 22 Dec 1884; in volume for 1884), p. [1]-3. *Copy*: USGS. – Reprinted and to be cited from C.R. Acad. Sci., Paris 99: 1133-1135. 1884 (1885).

9026. *Sur un Equisetum du terrain houiller supérieur de Commentry* ... [C.R. Acad. Sci. Paris 1885]. Qu.
Co-author: [Charles] René Zeiller (1847-1915).
Publ.: Jan-Mar 1885 (submitted 5 Jan 1885), p. [1]-3. *Copies*: FI, USGS. – Reprinted and to be cited from C.R. Acad. Sci., Paris 100: 71-73. 1885.

9027. *Sur les mousses de l'époque houillère;* ... [Paris 1885]. Qu.
Co-author: [Charles] René Zeiller (1847-1915).
Publ.: 1885 (p. 3: 2 Mar 1885), p. [1]-3. *Copy*: USGS. – Reprinted and to be cited from C.R. Acad. Sci., Paris 100: 660-662. 1885.

9028. *Sur un nouveau type de Cordaïtée;* ... [Paris, 1885]. Qu.
Co-author: [Charles] René Zeiller (1847-1915).
Publ.: 1885 (p. 3: 23 Mar 1885), p. [1]-3. *Copy*: USGS. – Reprinted and to be cited from C.R. Acad. Sci., Paris 100: 867-869. 1885.

9029. *Sur des mousses de l'époque houillère* ... [C.R. Acad. Sci., Paris 1885]. Qu.
Co-author: [Charles] René Zeiller (1847-1915).
Publ.: 2 Mar-8 Mai 1885 (submitted 2 Mar 1885; BSbF rd. séance 8 Mai 1885), p. [1]-3. *Copy*: FI. – Reprinted and to be cited from C.R. Acad. Sci., Paris 100: 660-662. 1885.

9030. *Sur les troncs de Fougères du terrain houiller supérieur;* ... [Paris, 1886]. Qu.
Co-author: [Charles] René Zeiller (1847-1915).
Publ.: 1886 (p. 3: 4 Jan 1886), p. [1]-3. *Copy*: USGS. – Reprinted and to be cited from C.R. Acad. Sci., Paris 102: 64-6. 1886.

9031. *Sur les racines des Calamodendrées;* ... [Paris, 1886]. Qu.
Publ.: 1886 (p. 3: 25 Jan 1886), p. [1]-3. *Copy*: USGS. – Reprinted and to be cited from C.R. Acad. Sci., Paris 102: 227-230. 1886.

9032. *Sur quelques Cycadées houillères* ... [Paris, 1886]. Qu.
Co-author: [Charles] René Zeiller (1847-1915).
Publ.: 1886 (p. 3: 8 Feb 1886), p. [1]-3. *Copy*: USGS. – Reprinted and to be cited from C.R. Acad. Sci., Paris 102: 325-328. 1886.

9033. *Sur l'organisation comparée des feuilles des Sigillaires et des Lépidodendrons;* ... [Paris, 1887]. Qu.
Publ.: Dec 1887 or early 1888 (USGS copy rd 20 Mar 1888; p. 3: 28 Nov 1887), p. [1]-3. *Copy*: USGS. – Reprinted and to be cited from C.R. Acad. Sci., Paris 105: 1087-1089. 1887 (1888).

9034. *Sur les fructifications des Calamodendrons;* ... [Paris, 1886]. Qu.
Publ.: 1886 (p. 3: 15 Mar 1886), p. [1]-3. *Copy*: USGS. – Reprinted and to be cited from C.R. Acad. Sci., Paris 102: 634-637. 1886.

9035. *Sur le genre Bornia F. Roemer;* ... [Paris, 1886]. Qu.
Publ.: 1886 (p. 3: 7 Jun 1886), p. [1]-3. *Copy*: USGS. – Reprinted and to be cited from C.R. Acad. Sci., Paris 102: 1347-1349. 1886.

9036. *Sur les fructifications mâles des Arthriphitus et des Bornia;* ... [Paris 1886]. Qu.
Publ.: 1886 (p. 3: 15 Jun 1886), p. [1]-3. *Copy*: USGS. – Reprinted and to be cited from C.R. Acad. Sci., Paris 102: 1420-1412. 1886.

9037. *Note sur le Clathropodium morieri,* B.R. ... Caen (Imprimerie Henri Delesques ...) 1887. Oct.

Publ.: 1887, p. [1]-11, *pl. 4-5* (uncol. liths.). *Copy*: USGS. –Reprinted and to be cited from Bull. Soc. Linn. Normandie ser. 4. 1: 143-151. 1887.

9038. *Sur les cicatrices des Syringodendron* .. [Paris, 1887]. Qu.
Publ.: Nov-Dec 1887 (p. 3: 24 Oct 1887), p. [1]-3. *Copy*: USGS. – Reprinted and to be cited from C.R. Acad. Sci., Paris 105: 767-769. 1887.

9039. *Sur les Stigmarhizomes*, ... [Paris, 1887]. Qu.
Publ.: Nov-Dec 1887 or early 1888 (p. 3: 7 Nov 1887; copy USGS rd 20 Mar 1888), p. [1]-3. *Copy*: USGS. – Reprinted and to be cited from C.R. Acad. Sci., Paris 105: 890-893. 1887 (or 1888).

9040. *Notice sur les Sigillaires* ... Autun (Imprimerie Dejussieu père et fils) 1888. Oct.
Publ.: 1888, p. [i-ii], [1]-79. *pl. 3-6* (uncol. liths.) with text. *Copy*: USGS. – Reprinted and to be cited from Bull. Soc. Hist. Nat. Autun 1: 121-199, *pl. 3-6*. 1888.

9041. *Études sur le terrain houiller de Commentry*. Livre deuxième. *Flore fossile* par M.M. B. Renault ... et R. Zeiller ... Saint-Étienne (Imprimerie Théolier & Cie ...) 1888-1890, 3 parts in 2 vols. text Oct. and atlas (*Étud. Commentry, fl. foss.*).
Authorship: [Charles] René Zeiller (1847-1915), author of part 1, co-author part 3.
Text (1): 1888 (p. 6: 29 Feb 1888; Bot. Zeit. 29 Mar 1889), p. [1]-366. Bull. Soc. Ind. minér. ser. 3. 3(2). 1888.
(1, 2/3): 1890 (copy rd by Soc. bot. France 12 Dec 1890), p. [369]-746. – Id. ser. 3. 4(2). 1890.
Atlas: *1*: 1888 (Bot. Zeit. 29 Mar 1889), p. [1]-10, *pl. 1-42* (uncol. liths. L. Sohrer). – Extrait de Bulletin de la Société de l'Industrie minérale, ser. 3. 2(2). 1888.
2/3: 1890 (copy rd by Soc. bot. France 12 Dec 1890), p. [i], [1]-14. *pl. 43-75* (uncol. liths. Ch. Thomas, L. Sohrer, Boivin or Boirin). – Atlas de la Société de l'Industrie minérale ser. 3. 2(4).
Copies: BR, L, NY, PH, USGS.
Ref.: Bureau, Éd., Bull. Soc. bot. France 38 (bibl.): 163-165. 1891 (rev.).
4682

9042. *Les plantes fossiles* ... Paris (Librairie J.-B. Baillière et fils ...) 1888. Oct. (*Pl. foss.*).
Publ.: Jul 1888 (p. 7: Mai 1888; Bot. Zeit. 31 Aug 1888; Bot. Centralbl. 5-9 Nov 1888; Soc. bot. Fr. rd 27 Jul 1888), p. [1]-399, [400, cont.]. *Copies*: BR, E (Univ. Libr.), G, NY, PH, U, USGS.
Ref.: Carruthers, W., J. Bot. 27: 25-27 Jan 1889 (rev.).
H.S., Bot. Zeit. 46: 804-806. 14 Dec 1888.

9043. *Sur l'attribution des genres Fayolia et Palaeoxyris*; ... [Paris (Gauthiers-Villars et fils, ...)1888 or 1889]. Qu.
Co-author: [Charles] René Zeiller (1847-1915).
Publ.: Dec 1888 or early 1889 (p. 4: 17 Dec 1888), p. [1]-4. *Copy*: USGS. – Reprinted and to be cited from C.R. Acad. Sci., Paris 107: 1022-1025. 1888 (or 1889).

9044. *Notice sur une Lycopodiacée arborescente* du terrain houiller du Brésil ... [Autun 1890]. Oct.
Publ.: 1890 (p. 16: 1 Jan 1890), p. [1]-16. *pl. 9*. *Copy*: USGS. – Reprinted and to be cited from Bull. Soc. Hist. nat. Autun 3: 109-124. *pl. 9*. 1890. – Also published (abbreviated) in C.R. Acad. Sci., Paris 110: 809-811. Jan-Jun 1890.

9045. *Sur un nouveau genre de tige permo-carbonifère, le G. Retinodendron Rigolloti* ... [Paris, 1892]. Qu.
Publ.: 1892 (p. 3: 16 Aug 1892), p. [1]-3. *Copy*: USGS. – Reprinted and to be cited from C.R. Acad., Paris 115: 339-341. 1892.

9046. *Note sur un nouveau genre de Gymnosperme fossile* du terrain permo-carbonifère d'Autun ... [Autun 1892]. Oct.
Publ.: 1892, p. [1]-6. *pl. 5* with 1 p. text. *Copy*: USGS. – Reprinted and to be cited from Bull. Soc. Hist.-nat. Autun 5: 152-157, *pl. 5*.

9047. *Sur quelques nouveaux parasites des Lépidodendrons* ... Extrait des publications de la Société d'Histoire naturelle d'Autun [1893 or Jan 1894]. Oct.
Publ.: late 1893 or early 1894 (USGS copy rd 26 Feb 1894), p. [1]-12. *Copy*: USGS. – Reprinted from Procès-Verb. Séances 1893: 168-178. 1893 or Jun 1894.

9048. *Notice sur les Calamariées* ... Autun (Imprimerie Dejussieu père et fils ...) 1895[-1898]. Oct.
Publ.: 1895, p. [i-ii], [1]-54, *pl. A, 1-7* (uncol. liths.) with text. – Reprinted and to be cited from Bull. Soc. Hist. nat. Autun 8: 1-54. *pl. A, 1-7.* 1895.
Suite: 1896, p. [i-iv], [1]-50, table, *pl. 1-12* (id.). – Id. 9: 305-354. *pl. 1-12.* 1896.
Suite (3ᵉ partie): 1898, p. [i-iv], [1]-60, *pl. 1-10* (id.). – Id. 11: 377-436. *pl. 1-12.* 1898.
Copy: USGS.

9049. *Sur un mode de déhiscence curieux du pollen de Dolerophyllum*, genre fossile du terrain houiller supérieure ... [Paris 1895]. Qu.
Publ.: 1895 (p. 3: 24 Dec 1894), p. [1]-3. *Copy*: USGS. – Reprinted and to be cited from C.R. Acad. Sci. 119: 1239-1241. 1894 (1895).

9050. Ministère des travaux publics. *Études sur les gîtes minéraux de la France* ... *Bassin Houiller et Permien d'Autun et d'Épinac*, fascicule iv *Flore fossile* deuxième partie par B. Renault ... Paris (Imprimerie nationale) [1893-]1896. Qu. (*Étud. gîtes minér., Bassin Autun Épinac., Fl. foss.*).
Publ. (text): Jun 1896 (Bot. Zeit. 1 Aug 1896; Soc. bot. France rd 26 Jun 1896; Bot. Zeit. 1 Aug 1896; in 2 parts: 1-292, 293-578), p. [i-iv], [1]-578. *Copies*: G, PH, Teyler, U, USGS; IDC 5360.
(Atlas): 1893, *pl. 28-89, A, B* (uncol. liths. various artists) with text.
For part 1, issued as fascicule II, *Flore fossile, première partie*, see R. Zeiller.
Ref.: Zeiller, R., Bull. Soc. bot. France 43(bibl.): 659-665. 1896 (rev.).
Solms-Laubach, H., Bot. Zeit. 54(2): 305-308. 1896 (rev.).

9051. *Notice sur les travaux scientifiques* de M. Bernard Renault ... Autun (Imprimerie Dejussieu père et fils) 1896. Qu. (*Not. trav. Sci.*).
Publ.: 1896, p. [1]-162, *pl. 1-7* with text. *Copies*: BR, USGS.
Suppl.: 1899, p. [1]-63, *pl. A*, [1, expl. pl.]. *Copy*: USGS. – Autun (id.) 1899. Qu.
Suppl. 2: 1901, p. [1]-84. *Copy*: USGS. – Autun (id.) 1901. Qu.

9052. *Notice sur la constitution des lignites* et les organismes qu'ils renferment ... Autun (Imprimerie et librairie Dejussieu) 1898. Oct.
Co-author: Auguste Roche.
Publ.: 1898, p. [i-ii], [1]-39, *pl. 11-13.* w.t. *Copy*: USGS. – Reprinted and to be cited from Bull. Soc. Hist. nat. Autun 12: *pl. 11-13.* 1898. – See also C.R. Acad. Sci., Paris 126: 1828-1831. 1898.

9053. *Sur les marais tourbeux* aux époques primaires ... Paris (Imprimerie nationale) 1900. Oct.
Publ.: 1900 (USGS copy rd 25 Oct 1900; p. 5: Feb 1900), p. [1]-5. *Copy*: USGS. – Reprinted and to be cited from Bull. Mus. Hist. nat., Paris 6(1): 44-46. 1900.

9054. *Sur quelques microorganismes des combustibles fossils* ... Saint-Étienne (Société de l'imprimerie Théolier ...) 1900. Oct.
Publ.: 1899-1900 (reprint rd by L. Ward, U.S.A., 13 Feb 1901), p. [1]-460. *Copy*: USGS. – Reprinted from Bull. Soc. Industrie minérale ser. 3, tome 13(4): 1899, 14(1): 5-159. Jan-Mai 1908, ser. 3. 13(4): 865-1161 (reprint 1-301), 14(1): 5-154 (reprint 302-450), 1163-1169 (158/159) (reprint 451-460). 1900.

9055. *Considérations nouvelles sur les tourbes & les houilles* ... Autun (Imprimerie et librairie Dejussieu) 1900. Oct.
Publ.: 1900, p. [1]-31. *Copy*: USGS. – Reprinted and to be cited from Bull. Soc. Hist. nat. Autun 13: 303-331. 1900.

9056. *Sur un nouveau genre de tige fossile* ... Autun (Imprimerie et librairie Dejussieu) 1901. Oct.
Publ.: 1901, p. [1]-22, *pl. 6-10* with text. *Copy*: U. – Reprinted and to be cited from Bull. Soc. Hist. nat. Autun 13: 405-424, *pl. 6-10*. 1901.

9057. *Sur quelques cryptogames hétérosporées* ... Autun (imprimerie et librairie Dejussieu) 1901. Oct.
Publ.: 1901 (t.p. 1901; Bot. Centralbl. 20 Aug 1902), p. [1]-16. *Copy*: U. – Preprinted or reprinted and to be cited from Bull. Soc. Hist. nat. Autun 14: 339-352. 1901 (or 1902).

9058. *Note sur quelques micro et macrospores fossiles* ... Autun (Imprimerie et librairie Dejussieu) 1902. Oct.
Publ.: 1902 (L.F. Ward, U.S.A., copy rd 21 Jul 1903), p. [i-ii], [1]-22, *pl. 6bis, 7, 7bis, 8-12* (photos, with text). *Copies*: G, USGS. – Reprinted and to be cited from Bull. Soc. Hist. nat. Autun 15: 97-115, *pl. 6bis, 7, 7bis, 8-12*. 1902. – Also in C.R. Soc. sav. 1902: 218-228. 1903, reprinted 1903, p. [1]-15 (*copy*: G).

9059. *Sur quelques micro et macrospores fossiles* ... Paris (Imprimerie nationale) 1903. Oct.
Publ.: 1903, p. [1]-15. *Copy*: G. – Reprinted and to be cited from C.R. Congr. Soc. Sav. 1902 (Sciences): [218]-228.

9060. *Sur une Parkériée fossile* ... [Paris (Gauthier-Villars, ...) 1902]. Qu.
Publ.: 1902 (p. 3: 10 Mar 1902), p. [1]-3. *Copy*: USGS. – Reprinted and to be cited from C.R. Acad. Sci., Paris 134: 618-621. Jan-Jun 1902.

9061. *Sur quelques pollens fossiles.* Prothalles mâles. Tubes polleniques, etc. du terrain houiller; ... [Paris (Gauthier-Villars) 1902]. Qu.
Publ.: 1902 (p. 4: 18 Aug 1902), p. [1]-4. *Copy*: USGS. – Reprinted and to be cited from C.R. Acad. Sci., Paris 135(7): 350-353. 1902.

9062. *Sur quelques algues fossiles* des terrains anciens; ... [Paris (Gauthier-Villars) 1903] Qu.
Publ.: 1903 (p. 4: 2 Jun 1903; USGS copy rd 15 Oct 1903), p. [1]-4. *Copy*: USGS. – Reprinted and to be cited from C.R. Acad. Sci., Paris 136(22): 1340-1343. Jan-Jun 1903.

Renault, Pierre-Antoine (1750-1835), French botanist at Rouen and Alençon; from 1795 professor of natural history at the École centrale de l'Orne at Alençon, establishing a botanical garden and a natural history museum. (*P. Renault*).

HERBARIUM and TYPES: Unknown.

NOTE: Letacq (1907) consistently spells the name Renaut. However, Renault used the spelling Renault on the t.p. of his *Fl. Orne*.

BIBLIOGRAPHY and BIOGRAPHY: BM 4: 1680; Herder p. 473; PR 7541, ed. 1: 8484; Rehder 5: 717; SO add. 785*; Zander ed. 10, p. 706, ed. 11, p. 805-806.

BIOFILE: Letacq, A.L., Bull. Soc. Amis Sci. nat. Rouen ser. 5. 42: 91-94. 1906, also *in* Inv. pl. vasc. crypt. Orne 1906, p. 231-234 (b. 16 Aug 1750, d. 23 Apr 1835).

9063. *Flore du département de l'Orne*, ouvrage élémentaire de botanique, composé de la réunion des systèmes de Tournefort, de Linné et de Jussieu; avec une description exacte des plantes, l'indication des lieux où elles se trouvent, et une notice sur leur usage et leur utilité dans les arts ... à Alençon (De l'imprimerie de Malassis le jeune, ...) an xii [24 Sep 1803-22 Sep 1804]. Oct. (*Fl. Orne*).
Publ.: 24 Sep 1803-22 Sep 1804, p. [i*-vi*], [i]-x, [1]-222. *Copies*: G, HH, NY, USDA.

Rendle, Alfred Barton (1865-1938), British botanist; BA Cantab 1887, M.A. 1891;

B.Sc. London 1887; D.Sci. 1898; connected with the Botany Dept. British Museum, Nat. Hist., from 1888-1906 as assistant, from 1906-1930 as keeper; president Linnean Society, London 1923-1927; F.R.S. 1909; member of the Editorial Committee for the International Rules of botanical Nomenclature 1905-1935; Victoria Medal R.H.S. 1917. (*Rendle*).

HERBARIUM and TYPES: BM.

BIBLIOGRAPHY and BIOGRAPHY: Backer p. 659; Barnhart 3: 144 (b. 19 Jan 1865, d. 11 Jan 1938); BFM no. 2317, 2348, 2944; Biol.-Dokum. 14: 7449; BJI 2: 144; BL 1: 315 [index], 2: 214, 215, 216, 707; BM 4: 1680, 8: 1065; Bossert p. 329;CSP 18: 136; Desmond p. 516; DNB 1931-1940: 730-731; Ewan (ed. 2): 182; Hirsch p. 246; Hortus 3: 1202; Kew 4: 448-449; Lenley p. 342; Moebius p. 202; MW p. 412-413; NI suppl. p. 23; NW p. 56; PH 209; Rehder 5: 717; SO 330, 681a, 910, 2876, 3066a, 3521; TL-1/507, 1220; TL-2/426, 766, 772, 1713, 1744-1745, 1814, 2747, 2808, 7055, see indexes; Tucker 1: 591-592; Zander ed. 10, p. 706, ed. 11, p. 806.

BIOFILE: Anon., Bot. Centralbl. 36: 256. 1888 (assistant BM(NH)), 101: 192. 1906 (keeper bot. BM(NH)), 123: 688. 1913 (general secretary of London Int. bot. Congress 1914 [not held]); Bot. Jahrb. 38 (Beibl. 87): 42. 1906 (keeper bot. BM(NH)); Bull. misc. Inf. Kew 1930: 45-46. 1930 (retirement; brief sketch); Gard. Chron. 70: 256. 1921 (biogr. sketch, obit. 1, 87: 2. 1930; 1938: 65 (obit.); J. Bot. 63: frontisp. portr. 1925. 68: 64. 1930 (retirement), 76: 64. 1938 (d. 11 Jan 1938); Nat. Nov. 28: 229. 1906 (curator BM(NH)); Österr. bot. Z. 87: 160. 1938 (d.).
Arditti, J., ed., Orchid biol. 1: 17, 22 (by R.E. Holttum), 134, 152. 1977.
Bridson, G.D.R. et al., Nat. hist. mss. res. Brit. Isl. 186.19. 1980.
Bullock, A.A., Bibl. S. Afr. bot. 93-94. 1978.
Burkill, I.H., J. Bot. 76: 65-68. 1938 (obit., portr.).
Dolezal, H., Friedrich Welwitsch 192, 194. 1974.
Levyns, M.R.B., Insnar'd with flowers 103. 1977.
Merrill, E.D., B.P. Bishop Mus. Bull. 144: 157. 1937 (Polynes. bibl.); Contr. U.S. natl. U.S. 30(1): 253-254. 1947 (Pacific Isl. bibl.).
Prain, D., Obit. Not. Fellows Roy. Soc. 7(2): 511-517. 1939 (obit., portr.).
Ramsbottom, J., Proc. Linn. Soc. London 150: 327-333. 1938 (obit., portr.; cites J. Briquet on R: "*Rendlius insularis*, species anglica politate et benignitate inter omnes plantas optimes hujus regionis praestans").
Rendle, A.B., Chron. bot. 1: 35-40. 1935 (A short history of the international botanical Congresses).
Sheppard, A.W., J.R. microsc. Soc. ser. 3. 58(1): 86-87. 1938 (obit.).
Sprague, T.A., Nature 141: 400-401. 1938 (obit.), Bull. misc. inf. Kew 1938: 81-82 (obit.).
Stearn, W.T., Nat. Hist. Mus. S. Kensington 291-295, 299-303. 1981.
Stieber, M.T. et al., Huntia 4(1): 85. 1981 (arch. mat. HU).
Verdoorn, F., ed., Chron. bot..1: 9, 28, 31, 34, 35, 39, 40, 47, 63, 171, 172. 1935, 2: 28, 34, 53, 64, 132, 188, 361. 1936, 3: 27, 33, 164, 336. 1937, 4: 72, 83, 268, 269, 297, 480, 533. 1938, 5: 299, 487. 1939.
White, F., C.R. AETFAT 4: 199. 1962 (coll. Rhodesia).

COMPOSITE WORKS: (1) Editor *Journal of Botany*, vols. 63-75, 76 p.p. 1924-1938.
(2) *List of Apetalae and Monocotyledons from East Africa*, in J.W. Gregory, The Great Rift Valley, App. B. 1896.
(3) *Naiadaceae*, in A. Engler, Pflanzenreich iv, 12 (Heft 7): 1-21. 17 Dec 1901 (see also Trans. Linn. Soc. ser. 2. Bot. 5: 379-444. *pl. 39-42*. 1899.
(4) With J. Britten, *List of plants [from Patagonia]*, in H.H. Prichard, *Through the heart of Patagonia*, App. C., London 1902.
(5) With W. Fawcett, *Fl. Jamaica* 1910-1936, 5 vols., see TL-2/1745.
(6) Editor of J. Britten and G.E.S. Boulger, *Biogr. ind. Brit. Irish Bot.* ed. 2, 1931, see TL-2/772 (rev. J. Soc. Bibl. nat. Hist. 1(5): 138. Mai 1938).
(7) Contributions to D. Oliver, *Fl. trop. Afr.*:
(a) With J.G. Baker, *Convolvulaceae*, 4(2) (1): 62-192. Dec 1905.

(b) *Ulmaceae*, 6(2)(1): 1-11. Mar 1916.
(c) *Barbeyaceae*, 6(2)(1): 14-15. Mar 1916.
(d) *Cannabinaceae*,6(2) (1): 16-17. Mar 1916.
(e) *Urticaceae*, 6(2)(2): 240-306. Nov 1917.
(f) With J. Hutchinson, *Moraceae*, 6(2)(1): 17-192. Mar 1916, 2(2): 193-240. Nov 1917.
(8) Botanical editor, *Encyclopaedia brittanica* ed. 11, 1910-1911.
(9) In W.P. Hiern, *Cat. afr. pl.* (TL-2/2747) vol. 2(1), *Monocotyledons and Gymnosperms*, 30 Mai 1899.
(10) *Apetalae, Monocotyledons & Gymnosperms*, in C.W. Andrews et al., *A monograph of Christmas Island*, London 1900.
(11) *Gramineae, tribus Bambuseae*, in C.S. Sargent, *Pl. Wilson.* 1(4). 1914.

EPONYMY: *Rendlia* Chiovenda (1914).

9064. *The classification of flowering plants* ... Cambridge (at the University Press) 1904-1925, 2 vols. Oct. (*Classif. fl. pl.*).
1 (Gymnosp., Monocot.): Mar-Mai 1904, (p. vi: 5 Mar 1904; Soc. bot. France 22 Jul 1904; J. Bot. rev. Aug 1904; TBC 4 Oct 1904; Bot. Centralbl. 5 Jul 1904; Nat. Nov. Jun(2) 1904; Bot. Zeit. 16 Aug 1904 (rev.), p. [i]-xiv, [1]-403.
2 (Dicot.): Oct 1915-Feb 1926 (t.p. 1925; p. [v]: Oct 1925; E copy rd 12 Jul 1926; ÖbZ Feb-Mai 1926; Nat. Nov. Feb 1926), p. [i]-xix, [1]-636.
Copies: BR, E, G, HH, NY, PH, USDA.
Ed. 2 of vol. 1: Jan-Jun 1930 (p. vii: 15 Jun 1929; J. Bot. Aug 1930; Nat. Nov. Aug-Sep 1930), p. [i]-xvi, [1]-412. *Copies*: G, HH, MO, PH, USDA. – Further reprints 1953, 1975/76.
Reprints of vol. 2: 1938 (with cor..), 1952, 1956, 1959, 1963, 1975/76.
Ref.: Fritsch, F.E., Bot. Centralbl. 96: 125. 9 Aug 1904 (rev.).
   Gundersen, A., Torreya 26(4): 70-75. 11 Aug 1926 (rev.).
   Scott-Elliott, G.F., J. Bot. 42: 245-247. Aug 1904 (rev.).

9065. *Amendments to the Paris Code of Botanical Nomenclature* suggested for consideration of the Vienna Congress of 1905 by the botanists of the British Museum and others ... s.l.n.d. [London 1904/1905].
Publ.: possibly 1904 or early 1905, p. [1]-3. *Copy*: G. – The botanists signing the proposals were: W. Carruthers, G. Murray, A. Gepp, E.G. Baker, A.B. Rendle, B.D. Jackson, W.P. Hiern, Spencer le M. Moore, D. Oliver.

9066. *List of British seed-plants and ferns.* Exhibited in the Department of Botany, British Museum (Natural History). London (printed by order of the Trustees of the British Museum) 1907. Oct. (*List Brit. seed.-pl. ferns*).
Co-author: James Britten (1846-1924).
Publ.: Mar 1907 (p. [iii]: Jan 1907; J. Bot. 1907: "will publish immediately", Mar 1907 "just issued"), p. [i-iii], err. slip, 1-43, [44]. *Copies*: NY(3). – For a detailed account of name changes see J. Britten and A.B. Rendle, J. Bot. 45: 99-108. Mar 1907.

9067. *Motion proposing an additional clause to the Rules of 1905 concerning the nomenclature of the algae* suggested for consideration of the Brussels Congress of 1910 by the botanists of the British Museum, and others [London, June 1909]. (*Motion nom. alg.*).
Publ.: Jun 1909 (p. 2), p. [1]-2. *Copy*: FAS. – Motion submitted to the Brussels Congress by A.B. Rendle, J. Britten, E. G. Baker, A. & E.S. Gepp, W.E. St. John Brooks, W. Fawcett, S. Moore, A.L. Smith – Proposal to accept C.A. Agardh, *Syst. alg.* 1824 as the starting point for the nomenclature of all algae.

9068. *A contribution to our knowledge of the flora of Gazaland*: being an account of collections made by C.F.M. Swynnerton ... [London (printed by Taylor and Francis) 1911]. Oct.
Co-author: Edmund Gilbert Baker (1864-1949), Spencer le Marchant Moore (1851-1931), Anthony Gepp (1862-1955); collections and notes by Charles Francis Massey Swynnerton (1877-1938).
Publ.: 21 Sep 1911, p. [1]-245, *pl. 1-7* (uncol.). – Issued and to be cited as J. Linn. Soc., Bot. 40(1) (no. 275). 1911. *Copy*: U.

9069. British Museum, Natural History. *Catalogue of the plants* collected by Mr. & Mrs. P.A. Talbot *in the Oban district* South Nigeria ... London (printed by order of the Trustees of the British Museum ...) 1913. Oct. (*Cat. pl. Oban*).
Co-author: Edmund Gilbert Baker (1864-1949), Spencer le Marchant Moore (1851-1931), Herbert Fuller Wernham (1879-x).
Collectors: Dorothy Amaury Talbot (1871-1916), Percy Amaury Talbot (1877-1945).
Publ.: Jun 1913 (p. [iii]: Apr 1913; J. Bot. Jul 1913; Bot. Centralbl. 1 Jul 1913; ÖbZ Oct-Dec 1913; Nat. Nov. Jun(2) 1913, PH rd 4 Dec 1913), p. [i*], [i]-x, [1]-157, [158-159]. *pl. 1-17* with text. *Copies*: BR, FI, HH, M, MO, NY, PH, USDA; IDC 224.
– On the spine of the original binding: "*Catalogue of Talbot's Nigerian plants*".
Ref.: Green, M.L., Bot. Centralbl. 123: 471-472. 1913 (rev.).
Gagnepain, F., Bull. Soc. bot. France 60: 567-568. 17 Jan 1914 (rev.).
Mildbraed, J., Bot. Jahrb. 51 (Lit.): 14-15. 9 Dec 1913 (rev.).

9070. *Dr. H.O. Forbes's New Guinea plants* [Journal of Botany 1923, supplement]. Oct. (*Forbes's New Guinea pl.*).
Co-authors: E.G. Baker, S. Moore, A. Gepp (see above), Hugh Neville Dixon (1861-1944).
Collector: Henry Ogg Forbes (1851-1932).
Publ.: as a supplement to J. Bot. 61, 1923, p. [1]-64. *Copy*: U. – Published in parts:

part	pages	dates
1	[1]-8	Jun 1923
2	9-16	Jul 1923
3	17-32	Aug 1923
4	33-40	Sep 1923
5	41-48	Oct 1923
6	49-56	Nov 1923
7	57-64	Dec 1923

9071. *Dr. H.O. Forbes's Malayan plants* ... [Journal of Botany 1924-1926, Supplement]. Oct.
Publ.: 1924-1926 as a supplement to J. Bot. vols. 62-64. Texts by E.G. Baker, C. Norman, J.D. Hooker, G. Greves, S. Moore, W.P. Hiern, H.N. Ridley, A. Gepp, H.N. Dixon, R. Paulson, as follows:

vol. (suppl.)	pages	dates	vol. (suppl.)	pages	dates
62	1-16	Jul 1924	63	81-88	Mai 1925
	17-32	Aug 1924		89-96	Jun 1925
	33-40	Nov 1924		97-104	Jul 1925
	41-48	Dec 1924		105-112	Aug 1925
63	49-56	Jun 1925		113-120	Sep 1925
	57-64	Feb 1925		121-128	Oct 1925
	65-72	Mar 1925		129-136	Nov 1925
	73-80	Apr 1925	64	137-149	Jan 1926

9072. *International rules of botanical nomenclature* adopted by the Fifth International Botanical Congress, Cambridge 1930. Supplement to "The Journal of Botany", June 1934. London (printed and published by Taylor and Francis ...) [1934]. Oct. (*Int. rules bot. nomencl.*).
Publ.: 1-21 Jun 1934 (copy FAS sent by Rendle to A.A. Pulle on 21 Jun 1934), p. [1]-29. *Copies*: FAS, U. – Independent publication, issued as a supplement to J. Bot. 72, Jun 1934. – This is an abridged version of the text which was to appear in the official edition of the Rules, *Règles int. nomencl. bot.* ed. 3, Feb-Mar 1935 (TL-2/766), published earlier in order to facilitate the preparation for the Sixth Int. bot. Congress, Amsterdam, 1935. The abridgement consisted "merely in the omission of most of the examples".
Note: Rendle was a member of the Editorial Committee of eds. 1-3 of the Règles. See also ICBN, Edinburgh Code, 1966, p. 400-401.

Reneaulme [Renealmus], **Paul** (1560-1624), French physician and botanist at Blois. (*Reneaulme*).

HERBARIUM and TYPES: Unknown.

BIBLIOGRAPHY and BIOGRAPHY: AG 3: 527; Barnhart 3: 144; Blunt p. 92-93, *pl. 15b*; BM 4: 1680; Dryander 3: 71, 323, 373; Herder p. 84; HU 192; Jackson p. 27; Kew 4: 449; Lasègue p. 518; NI 1621, see also 1: 94-96; Plesch p. 380; PR 7542, ed. 1: 2546, 8488; Rehder 1: 284; Sotheby 639; Tucker 1: 592.

BIOFILE: Bartlett, H.H., 55 rare books no. 23. 1949.
Franchet, A., Fl. Loir-et-Cher v-viii. 1885.
Jessen, K.F.W., Bot. Gegenw. Vorz. 279. 1884.

EPONYMY: *Renealmea* Cothenius (1790, *orth. var.* of *Renealmia* Linnaeus (1753)); *Renealmia* Linnaeus (1753, *nom. rej.*); *Renealmia* Linnaeus f. (1782, *nom. cons.*); *Renealmia* R. Brown (1810). *Note*: The etymology of *Renealmia* Houttuyn (1777) could not be determined.

9073. Pauli Renealmi blaesensis doctoris medici *Specimen historiae plantarum*. Plantae typis aeneis expressae. Parisiis (apud Hadrianum Beys, ...) 1611. Qu. (*Specim. hist. pl.*).
Publ.: 1611, p. [i-vi], [1]-152, [153-154], Jacques-Auguste de Thou (1553-1617): Crambe, Viola, Lilium, Phlogis, Terpsinoe, p. [1]-47; the *Specimen* with 25 etchings of plants. *Copies*: BR, FI, HU, TCD, U; IDC 7331. – For a collation and further details see HU 192.

Renier, Armand-Marie-Vincent-Joseph (1876-1951), Belgian palaeobotanist and geological engineer, educated at the University of Liège; from 1910 lecturer at Liège; with the "Service géologique de Belgique" from 1915 (from 1922 ingénieur en chef; later head of the Geological Service). (*Renier*).

HERBARIUM and TYPES: Unknown.

BIBLIOGRAPHY and BIOGRAPHY: Andrews p. 320; Barnhart 3: 144; Hirsch p. 246.

BIOFILE: Anon., Rev. verviét. Hist. nat. 9(5/6): 60. 1952 (b. 26 Jun 1876, d. 9 Sep 1951).
Grondal, G., Biogr. verviét. 36. 1952; Rev. verviét. Hist. nat. 9(1/2): 60-61. 1952 (b. 26 Jun 1876, d. 9 Sep 1951).

EPONYMY: *Reniera* F. Stockmans (1968).

9074. Musée royal d'histoire naturelle de Belgique. *Flore et faune houillères de la Belgique.* Introduction à l'étude paléontologique du terrain houiller ... Bruxelles (ouvrage édité par le Patrimoine du Musée royal d'histoire naturelle de Belgique ...) 1938. Oct. (*Fl. Faune houill. Belg.*).
Co-authors: François Stockmans, Félix Demanet, Victor van Straelen.
Text: 1938, p. [1]-317.
Plates: 1938, p. [1]-49, *pl. 1-144* (uncol.).
Copies: MO, USDA.

Renner, Otto (1883-1960), German (Bavarian) botanist; Dr. phil. München 1906 (with L. Radlkofer); herbarium assistant with Radlkofer; studied with L. Pfeffer 1907-1908; from 1908-1920 curator of cryptogams with K. Goebel at the plant physiology institute of the University of München; habil. München 1911; extraordinary professor of botany ib. 1913-1920 (from 1917-1919 war service as bacteriologist in Ulm); from 1920-1948 professor of botany and director of the botanical garden at Jena, to the Dutch East Indies 1930-1931; from 1948-1952 professor of botany at München; in retirement continuing his work in München; student of the genetics of *Oenothera*. (*Renner*).

HERBARIUM and TYPES: M.

BIBLIOGRAPHY and BIOGRAPHY: Barnhart 3: 144 (b. 25 Apr 1883); Biol.-Dokum. 14: 7451-7454 (bibl.); BJI 2: 145; Bossert p. 329; Hirsch p. 246; Kew 4: 449; Lenley p. 342; Moebiusp. 257, 265, 355; SBC p. 130.

BIOFILE: Anon., Hedwigia 48: (197). 1909 (curator crypt. M), 52: (84). 1912 (lecturer München), 54: (126). 1913 (prof. München), 62: (147). 1921 (to Jena after Kniep, Würzburg had declined the call), 74: (141). 1934 (members Sächs. Akad. Wiss.); Mycol. Centralbl. 3(4): 205. 1913 (prof. München); Nat. Nov. 35: 444. 1913 (prof. München), 42: 107. 1920 (to Jena succeeding E. Stahl), 42: 153. 1920 (id.), 44: 21. 1922 (call to Kiel); Nature 175: 839. 1955 (brief biogr. note); Österr. bot. Z. 58: 495. 1908, 63: 480. 1913 (extraord. prof. München), 69: 224. 1920 (succeeds E. Stahl at Jena), 84: 240. 1935 (Bayer. Akad. Wiss. corr. member).
Bergdolt, E., Karl von Goebel ed. 2. 184, 207, 227, 230, 231, 243. 1941.
Cleland, R.E., Genetics 53(1): 1-6. 1966 (obit., portr.).
Darlington, C.D., Biogr. Mem. R. Soc. London 7: 207-220. 1961 (biogr., bibl., portr., general evaluation, comparison with Hugo de Vries).
Egle, K. & G. Rosenstock, Gesch. Bot. Frankfurt a.M. 35. 1966 (called to Frankfurt to succeed M. Moebius; declined).
Freund, H. & A. Berg, Gesch. Mikroskopie 336, 343. 1963.
Gager, C.S., Brooklyn Bot. Gard. Rec. 27(3): 230. 1938.
Hertel, H., Mitt. Bot. München 16: 422. 1980 (herb. at M, rd 1956).
Kneucker, A., Allg. bot. Z. 15: 16. 1909 (curator crypt. M), 24/25: 48. 1932 (to Jena).
Künkele, S. & S. Seybold, Jahresh. Ges. Naturk. Württemberg 125: 152. 1970 (coll. Ulm at M).
Mägdefrau, K., Ber. bayer. bot. Ges. 34: 103-113. 1961 (obit., portr., bibl., list of dissertations; d. 7 Jul 1960); Gesch. Bot. 201, 255, 289, 291. 1973.
Oehlkers, F., Ber. deut. bot. Ges. 74(2): (82)-(94). 1961 (obit., portr., bibl.).
Renner, O., Ber. bayer. bot. Ges. 105-112. 1962 (Münchner Antrittsvorlesung 1948, published posthumously; important document for the knowledge of R. and his ideas on his predecessors as well as on botanical issues of the time).
Taton, R., ed., Science in the 19th century 402. 1965.
Verdoorn, F., ed., Chron. bot. 2: 153, 166, 168. 1936, 3: 134. 1937, 5: 208, 209, 463, 512. 1938.

COMPOSITE WORKS: (1) Editor of *Flora* vols. 127-138. 1933-1944.
(2) Editor of *Planta*, vols. 35-45. 1948-1956.
(3) Co-editor (with E. Gäumann), *Fortschritte der Botanik*, vols. 12-17. 1949-1955.

FESTSCHRIFT: Flora ser. 2. 37(1/2). 1943, with a tribute by E. Bünning on p. vii-x, portr.

EPONYMY: *Rennera* Merxmueller (1957).

9075. *Beiträge zur Anatomie und Systematik der Artocarpeen und Concephaleen* insbesondere der Gattung Ficus. Inaugural-Dissertation zur Erlangung der Doktorwürde der hohen philosophischen Fakultät II. Sektion der Kgl. Bayer. Ludwig-Maximilians-Universität München am 4 Mai 1906 vorgelegt von Otto Renner. Leipzig (Wilhelm Engelmann) 1906. Oct. (*Beitr. Anat. Syst. Artocarp.*).
Thesis issue: 4 Mai 1906, p. [i, iii], [319]-448, [1, vita]. *Copies*: G, HH.
Journal issue: Bot. Jahrb. 39(3): [319]-448. 15 Jan 1907.
Ref.: Franz, E., Bot. Centralbl. 111: 440-441. 26 Oct 1906 (rev.).

9076. *Artbastarde bei Pflanzen* ... Berlin (Verlag von Gebrüder Borntraeger ...) 1929. Oct. (*Artbast. Pfl.*).
Publ.: 1929, p. [i]-iv, [1]-161.*Copies*: MO, NY, USDA. – Issued as vol. 2 of *Handbuch der Vererbungswissenschaft* (ed. E. Baur & M. Hartmann) (Lief. 7, 2A).
Ref.: Oehlkers, F., Z. Bot. 22: 538-540. 1930 (rev.).

Rennie, Robert (x-1820), British (Scottish) clergyman and botanist, minister of Kilsyth, Stirlingshire. (*Rennie*).

HERBARIUM and TYPES: Unknown.

BIBLIOGRAPHY and BIOGRAPHY: BM 4: 1681; Dawson p. 702; De Toni 1: cvi; Kew 4: 450; PR 7544, ed. 1: 8492.

BIOFILE: Allibone, S.A. Crit. dict. Engl. lit. 2: 1773. 1878.
Zittel, K.A. von, Hist. geol. palaeontol. 240. 1901.

9077. *Essays on the natural history and origin of peat moss*: the peculiar qualities of that substance; the means of improving it as a soil; the methods of converting it into a manure; and the other economical purposes to which it may be made subservient. By the rev. R. Rennie, Kilsyth, Edinburgh (printed by George Ramsay & Co. for Archibald Constable & Co. Edinburgh; ...) 1807. Oct. (*Essays nat. hist. peat moss*).
Publ.: Oct-Dec 1807 (p. [iii]: 1 oct 1807), p. [i]-vii, [1]-233, [234, err.]. *Copies*: E, NY. – "This is, despite its title, an inspired and far-seeing essay on British vegetational history". J. Edmondson in litt.
Ed. [*2*]: 1810, p. [i]-xvi, [1]-665. *Copies*: E (Univ. Libr.), MO-Steere. – Edinburgh (printed by George Ramsay & Co. for the author; and sold by ...) 1810. Oct.
Abridged version: Massachusetts Agricultural Journal.
Reprinted s.l.n.d. p. [1]-32. *Copy*: PH.

Requien, Esprit (1788-1851), French botanist, malacologist and palaeontologist at Avignon; from 1809 director of the Avignon botanical garden, from 1840 curator of the "Musée d'histoire naturelle"; botanical explorer of the Provence and Corsica. (*Requien*) (*Req.*).

HERBARIUM and TYPES: AV (circa 200.000 sheets, see Granier 1980); other material at B (mainly destr.), BM, CGE, CN, DBN, E, E-GL, FI, G, H, K, LY, M (lich.), MANCH, MPU, MW, OXF, PI, WAG. – See Granier (1980) for a description of the herbarium. Correspondence, autograph collection and historical library at the Musée Calvet, Avignon; the natural history library is at AV. – 59 letters by Requien to J.B. Mougeot are at P (Bibl. Centr.).

BIBLIOGRAPHY and BIOGRAPHY: AG 2(1): 623, 5(2): 46, 399; Barnhart 3: 145 (b. 6 Mai 1788, d. 30 Mai 1851); BM 4: 1635, 1682; Bossert p. 329; Clokie p. 231; CSP 5: 165; CSP 5: 165;Hegi 5(2): 1412; Hortus 3: 1202; IH 1 (ed. 2): 18, (ed. 3): 20, (ed. 6): 363, (ed. 7): 341, 2: (in press);Jackson p. 282; Kew 4: 451; Lasègue p. 102, 296, 322; PFC 1: l, lv, 2(2): xvi-xvii, 3(1): x; PR (alph.): 7046, ed. 1: 7884; Quenstedt p. 328, 488; Rehder 5: 718; Saccardo 1: 137, 2: 90-91, cron. p. xxx; Stevenson p. 1255; Urban-Berl. p. 311, 387; Zander ed. 11, p. 806.

BIOFILE: Anon., Ann. Soc. Linn. Lyon ser. 2. 1: viii, 1850-1852 (d.); Bot. Zeit. 9: 528. 1851 (obit.; d. 29 Mai 1851; herb. left to Avignon); Bull. Soc. géol. France ser. 2. 10: 178. 1853 (d.); Flora 34: 573-574. 1851 (d. 29 Mai 1851 in Corsica; herb. left to Avignon); Österr. bot. W. 1: 315. 1851 (to Corsica for botanical collections; died suddenly 29 Mai 1851).
Candolle, A.P. de, Mém. souvenirs 227. 1862.
Flahault, C., Univ. Montpellier, Inst. bot. 43. 1890.
Granier, J., Candollea 35: 223-229. 1980 (on R. and his herbarium; biogr. note; present state of herb., portr.).
Lacroix, A., Figures de Savants 4: 102. 1938.
Laissus, Y., 103. Congr. natl. Soc. sav. Nancy 5: 237. 1978 (letters to J.B. Mougeot).
Lindemann, E. von, Bull. Soc. Natural. Moscou 61(1): 64. 1886.
Requien, E., [2 letters of Requien to Parlatore on the flora of Corsica and Capraia, 1847]. 15 p. *Copy*: FI (possibly reprint Giorn. bot. ital.).
Reynier, A., Soc. Hort. Bot. Marseille, Rev. mens. 40: 67-74, 123-125, 135-139. 1894 (documents).
Martins, Ch., Jard. pl. Montpellier 54. 1854.
Verdoorn, F., ed., Chron. bot. 2: 122. 1936.

HANDWRITING: Webbia 32(1): 34 (no. 65). 1977; Candollea 32: 411-412. 1977.

EPONYMY: *Requinella* Fabre (1883); *Requienia* A.P. de Candolle (1825).

9078. *Catalogue des végétaux ligneux* qui croissent naturellement *en Corse* ou qui y sont généralement cultivés ... deuxième édition ... Avignon (Imprimerie de Fr. Séguin ainé, ...) 1868. (*Cat. vég. lign. Corse*).
Ed. 1: 1852, "chez G. Marchi, imprimeur-libraire, à Ajaccio" n.v.
Ed. 2: 1868, p. [1]-21. *Copy*: G.

Resvoll, Thekla [Susanne] **Ragnhild** (1871-1948), Norwegian botanist; cand. paedag. Oslo 1899; amanuensis at the botanical laboratory of Christiania (Oslo) University from 1901, lecturer at Oslo Univ. 1903-1936; on Java 1923-1924. (*Resvoll*).

HERBARIUM and TYPES: O; further material at BM and C.

BIBLIOGRAPHY and BIOGRAPHY: Barnhart 3: 145 (b. 22 Mai 1871); Biol.-Dokum. 14: 7464; BL 2: 444, 707; BM 8: 1067; CSP18: 1923; IH 2 (in press); Kew 4: 451; Kleppa p. 330 [index 42 nos.]; SK 1: 432, 5: cccxiv.

BIOFILE: Høeg, O.A., Blyttia 6: 57-61. 1948 (pr. 1949) (obit., portr.); Medd. Norskfarm. Selsk. 10: 133-134. 1948 (obit., b. 22 Mai 1871, d. 14 Jun 1948); Norsk biogr. Leks. 11: 398-399. 1952 (Thekla Susanne Ragnhild R.).
Lynge, B., Veneficus, Oslo, 3(10): 207-208. 1937 (65th birthday), 7(6): 101-102. 1941 (70th birthday; b. 22 Mai 1871).
Nordhagen, R., Norske Vid.-Akad. Oslo Årb. 1949: 29-37. 1950 biogr., portr., bibl.); Blyttia 12: 32. 1954 (memorial).
Wittrock, V.B., Acta Horti Berg. 3(3): 184-185, *pl. 116*. 1905 (portr.).

9079. *Om planter som passer til kort og kold sommer* ... Kristiania [Oslo] (i Kommision hos Morten Johansen) 1917. Oct. (*Pl. pass. kort kold somm.*).
Publ.: 1917 (p. 219: 9 Mar 1917; publication used to obtain Doct. phil. Kristiania; degree awarded 2 Mar 1918), p. [1]-224. *Copy*: H. – Contains on p. [103]-219 a detailed treatment of 56 vascular species (original printing, and a bibliography on p. [221]-224. – Issued as Arch. Math. Naturv. 35(6). 1917.

9080. *Norske fjellplanter* ... Oslo (Den Norske Turistforenning) 1934. Oct. (*Norsk fjellpl.*).
Publ.: 1934 (p. 3: Mai 1934), p. [1]-20, *pl. 1-48*. *Copy*: H.

Resvoll-Holmsen, Hanna [Marie] (1873-1943), Norwegian botanist; sister of T.R. Resvoll; married 1909 Gunnar Holmsen; lecturer plant geography Oslo 1921-1938. (*Resvoll-Holmsen*).

HERBARIUM and TYPES: O and TROM; other material at BM and C.

BIBLIOGRAPHY and BIOGRAPHY: BJI 2: 446, 447, 454, 458, 459, 707; BM 8: 1067; IH 2: (in press); Kew 4: 409, 451; Kleppa 330 (index); Zander ed. 10,p. 706, ed. 11, p. 805.

BIOFILE: Anon., Naturfredn. Norge Årsskr. 1942-1943: 21-22. 1946 (obit., portr.).
Christophersen, E., Blyttia 1: 100-102. 1943 (obit., portr.); Norske Vid.-Akad. Oslo, Årb. 1944: 35-38. 1945 (obit., portr., ref. to bibl. in Bibliografiske Monografier 184-188. 1943 in Norsk bibl. Bibl. 3(5)).
Gleditsch, K.G., Naturfredn. Norge, Årsskr. 1942-1943: 21-22. 1946 (portr.).
Høeg, O.A., Norsk biogr. Leks. 11: 399-400. 1952 (b. 11 Sep 1873, d. 13 Mar 1943).

9081. *Om fjeldvegetationen i det s[t]enfjeldske Norge* ... Kristiania [Oslo] (i Kommission hos Morten Johansen) 1920. Oct.
Publ.: 1920 (preface: 21 Dec 1919), p. [1]-266, [267, err.], 14 pages of plates, map. *Copy*: H. – Issued as Arch. Math. Naturv. 37(1). 1920.

9082. *Svalbards flora* med endel om dens plantevekst i nutid og fortid ... Oslo (J.W. Cappelens Forlag) [1927]. Oct. (*Svalbards fl.*).
Publ.: 1927 (copyright on p. 4), p. [1]-56. *Copies*: GB (inf. B. Peterson), H (inf. P. Isoviita).

9083. *Om planteveksten i grensetrakter* mellem Hallingdal og Valdres ... Oslo (Kommisjon hos Jacob Dybwad) 1932. Oct. (*Plantevekst. grensetr.*).
Publ.: 1932 (mss. presented 1 Mai 1931), p. [1]-50. *pl. 1-10. Copy*: H. – Issued as Skr. Norske Vid.-Akad. Oslo, Mat.-Nat. Kl. 1931 (9).

Retzdorff, Adolf Eduard Willy (1856-1910), German (Prussian) amateur botanist; worked for Caspary in E. Prussia 1876 and with L. Wittmack at the Berlin agricultural museum 1876-1877; trained to be a civil servant 1878-1880; law court clerk 1880-1889; from 1889 provincial tax revenue officer ("Steuersekretär"); from 1890 treasurer of the Bot. Ver. Brandenburg. (*Retzdorff*).

HERBARIUM and TYPES: B (mainly destr.); other material at E, IBF, L, SI. – Retzdorff endowed the Berlin herbarium with a sum to purchase the Warnstorff herbarium.

BIBLIOGRAPHY and BIOGRAPHY: AG 3: 819, 5(2): 28; BFM no. 215; DTS 3: xxxvi, 6(4): 65; IH 2: (in press); Urban-Berl. 387.

BIOFILE: Anon., Nat. Nov. 32: 314. 1910 (d.); Verh. bot. Ver. Brandenburg 52: frontisp. portr., d., ann. and note on p. (83).
Ascherson, P., Verh. bot. Ver. Brandenburg 52: (46)-(50). 1911 (obit., bibl., b. 30 Jul 1856, d. 29 Apr 1910).

Retzius, Anders Jahan (1742-1821), Swedish botanist; studied at Lund; pharmacist's examination Stockholm 1761; Dr. phil. Lund 1766; lecturer chemistry at Lund 1764, id. natural history 1767; "adjunkt" and demonstrator id. 1771; professor's title 1777; regular professor from 1781-1812; later residing in Stockholm; one of the founders of the Royal Physiographical Society at Lund; "a giant of learning" (Krok). (*Retz.*).

HERBARIUM and TYPES: LD (donated 1811; including e.g. herb. E. Acharius); other material at JE, S, SBT. – A large collection of letters to A.J. Retzius and A. Retzius was donated to the library of the Royal Swedish Academy of Sciences, Stockholm, 1883 (Öfv. K. Vet.– Akad. Förh. 40(6): 1).

BIBLIOGRAPHY and BIOGRAPHY: Ainsworth p. 259, 261, 346; Barnhart 3: 146; BL 2: 514, 707; BM 4: 1683-1684, 8: 1068; Bossert p. 329; Christiansen p. 319; CSP 5: 169-171; Dawson p. 702; De Toni 1: cvi; Dryander 3: 82, 129, 170, 232, 233, 296, 304, 312, 521, 645; Frank 3 (Anh.): 84; GR p. 480, *pl.* [*38*]; GR cat. p. 70; Hawksworth p. 185; Herder p. 73, 147, 170, 195; Hortus 3: 1202; IF p. 726; IH 1 (ed. 6): 363, (ed. 7): 341, 2: (in press); Jackson p. 22, 331; Kew 4: 452; KR p. 586-588 (b. 3 Oct 1742, d. 6 Oct 1821; bibl.); Lasègue p. 558; LS 22123-22127; ME 2: 184; MW p. 413; NI 1622, see also 1: 199, 229; PR 7546-7559, ed. 1: 8495-8513; Rehder 5: 719; RS p. 137; SBC p. 130; SK 4: xciii-xciv, clx, ccix (vol. 3 err. 1784); SO see index p. 45; TL-1/ 1066-1069; TL-2/2469-2471 and see indexes; TR 1153; Tucker 1: 592-593; Zander ed. 10, p. 706, ed. 11, p. 806.

BIOFILE: Almborn, O., Bot. Not. 133: 453-454. 1980; (pr. 1981; portr.)
Anon., K. Vet. Akad. Handb. ser. 2. 10: 462-467. 1822; Repert. Pharm. 17(2): 335-336. 1824 (n.v.).
Candolle, Alph. de, Phytographie 443. 1880.
Dawson, W.R., Smith papers 78. 1934.
Eiselt, J.N., Gesch. Syst.-Lit. Insectenk. 46. 1836 (entomol., classes).
Eriksson, G., Bot. hist. Sverige 270, 275, 277, 321. 1969.
Ewan, J., *in* F. Pursh, Fl. Amer. sept. facs. repr. 1979, introd. p. 24.
Fischer-Benzon, R.J.D. von, *in* P. Prahl Krit. Fl. Schlesw.-Holst. 2: 50. 1890.
Fries, Th.M., Bref och Skrifv. Linné 1(2): 266, 316, 350. 1908, 1(5): 170, 177, 186, 194. 1911, & J.M. Hulth 1(7): 98, 114, 115. 1917.

Fries, R.E., Short hist. bot. Sweden 37. 1950.
Fürst, C.M., Kongl. Fysiogr. Sällsk. Handl. ser. 2. 17(10): 1-16. 1907 (portr.; statue at Univ. Lund).
Gertz, O., Kungl. Fysiogr. Sällsk. Förh. 12(20): 1-23. 1942 (bibl.).
Gilbert, P., Comp. biogr. lit. deceased entom. 314. 1977.
Kuntze, O., Rev. gen. pl. 1: cxli. 1891 (on *Obs. bot.*).
Laundon, J.R., Lichenologist 11: 16. 1979.
Lindman, S., Sv. män och kvinn. 6: 242-243. 1949 (biogr., portr.).
Löwegren, Y., Naturaliekabinett Sverige 405 [index]. 1952; Acta Univ. lund. 1(9): 29-38. 1968 (portr., zool. coll.).
Meyer, F.K. & H. Manitz, *in* Reichthümer und Räritäten, Jenaer Reden und Schriften 1974, p. 91 (Retzius material at JE in herb. C.E. Weigel).
Milner, J.D., Cat. portr. Kew 91. 1906.
Percheron, A., Bibl. entom. 2: 9. 1837.
Poggendorff, J.C., Biogr.-lit. Handw.-Buch 2: 610-611 and Erg., 3: 1111.
Rydberg, P.A., Augustana Coll. Libr. Publ. 5: 15. 1907.
Sayre, G., Bryologist 80: 516. 1977.
Scheutz, N.J.W., Bot. Not. 1863: 71.
Stafleu, F.A., Taxon 12: 75-76. 1963 (fasc. 5, 6, *Obs. bot.*), *in* G.S. Daniels & F.A. Stafleu, Taxon 22: 220. 1973 (portr.).
Törje, A., Acta univ. lund. 1(6): 29-35. 1968 (portr.).
Weibull, M. & E. Tegnér, Lunds univ. hist. 1668-1868. 2: 307-310. 1868 (biogr., bibl.).
Wikström, J.E., Consp. 214-221. 1831.
Wittrock, V.B., Acta Horti Berg. 3(2): 67. 1903, 3(3): 49, lxvi, *pl. 32*. 1905 (portr.).

EPONYMY: *Retzia* Thunberg (1776).

9084. *Nomenclator botanicus* enumerans plantas omnes in systematis naturae edit. xii. specier. plantarum edit. ii et mantissis binis a illustr. D. Car. von Linné ... descriptas, Lipsiae [Leipzig] (apud Ioann. Frider. Junium) 1772. Oct. (*Nomencl. bot.*).
Ed. [*1*]: 1772 (NZgS 6 Sep 1773), p. [1/2]-279/280, [1-17, index, 1, err.]. *Copies*: HH, LD, M, MO(2), NY. – See TL-2/4817, 4829.
Ed. nova: 1782, p. [1/2]-279/280, [1-17, index]. *Copies*: HH, M.

9085. D.D. *Fasciculus observationum botanicarum*. Quam, venia amplissimi philosophorum ordinis, praeses Anders Jahan Retzius ... & respondens Magnus Gustaf Sahlstedt ... publico botanophilorum examini submittunt ad diem xxiii martii mdcclxxiv, L.H.S. Lundini [Lund][(Typis Berlingianis) [1774]. Qu. (*Fasc. obs. bot.*).
Publ.: 23 Mar 1774 (t.p.), p. [i-iv], [1]-28. *Copies*: LD, USDA; IDC 5668. – Text occurs again in Obs. bot. 1, 1779.

9086. Andreae Johannis Retzii ... *Observationes botanicae* sex fasciculis comprehensae. Quibus accedunt Joannis Gerhardi Koenig ... Descriptiones monandrarum et epidendrorum in India orientali factae. Lipsiae [Leipzig]. (apud Siegfried Lebrecht Crusium) [1779]-1791. Fol. (*Observ. bot.*).
T.p.: [i] and portr., Jul-Nov 1791.
1: Sep-Oct 1779, p. [1]-38, *pl. 1-2* (uncol. copp.). – "Fasciculus observationum botanicarum primus", Lipsiae [rd.]. 1779. See also above Fasc. obs. bot. 1774.
2: 1781, p. [1]-28, *pl. 1-5* (id.). "Fasciculus ... secundus" id. 1781.
3: 1783, p. [1]-12, 25-76. *pl. 1-3* (id.). "Fasciculus ... tertius ... Joh. Gerh. Koenig *Descriptiones monandrarum* [Johann Gerhard König, 1728-1785, q.v.; Koenig text on p. 45-76]. Id. 1783. – Pages 13-24 were never published.
4: 1786/1787, p. [1]-30, *pl. 1-3* (id.). – "Fasciculus ... quartus" id. 1786.
5: Sep 1788 (t.p. 1789), p. [i-vi], 1-32. *pl. 1-3* (id.). – "Fasciculus ... quintus" id. 1789.
6: Jul-Nov 1791, p. [11]-67 [p. 1-10 were never published]. *pl. 1-3.* – "Fasciculus ... sextus ... et J.G. Koenig *Descriptiones Epidendrorum* [on p. 41-66] id. 1791.
Copies: BR, FI, G, H, LD, M, MO, NY, USDA; IDC 2231. Type material Koenig taxa at C (except for the types of Epidendron, inf. Anne Fox Maule). For Koenig see also C.E.C. Fischer, Bull. misc. Inf. Kew 1932. 49-76. 1933 (study of Lund Koenig material).

Facsimile ed.: Announced but not yet published by Asher & Co. 1974.
Supplementa: Phytogr. Blätter 1: 35-48. 1803.
Plates: copper engravings of drawings by Retzius. Apart from the copy at LD (inf. O. Almborn), we have seen only uncolored copies; KR mentions the existence of coloured copies.

9087. *Florae Scandinaviae prodromus*; enumerans plantas Sveciae, Lapponiae, Finlandiae & Pomeraniae, ac Daniae, Norvegiae, Holsatiae, Islandiae Groenlandiaeque: auctore Andrea Jahanne Retzio ... Holmiae [Stockholm] (Typis Petri Hesselberg) 1779. Oct. (*Fl. Scand. prodr.*).
Ed. [*1*]: Mai-Dec 1779 (p. [xiv]: 1 Mai 1779; GGA 31 Dec 1779), p. [i-xvi], [1]-257, [258-264, ind.], [1, err.]. *Copies*: G, H, HH, LD, USDA; IDC 5669.
Ed. altera: Jan-Jul 1795 (GGA 20 Aug 1795), p. [i]-xvi, [1]-382, [1-2, err.]. *Copies*: BR, FH, FI, G, H, HH, L, LD, MICH, MO, NY, PH, USDA. – Editio altera. Lipsiae [Leipzig] (apud Siegfried Lebrecht Crusius) 1795. Oct.
Suppl. [*1*]: 3 Jun 1805, p. [i], [1]-20. *Copy*: LD (inf. O. Almborn). – Dissertatio sistens supplementum et emendationes in editionem secundam Prodromi Florae Scandinaviae, quam ... sub praesidio D.M. And. J. Retzii, ... pro laurea publico examini subjicit Joh. Chr. Askelöf, ... iii Junii mdcccv. Lundae [Lund] (Literis Berlingianis).
Suppl. 2: 17 Jun 1809, p. [1]-14. *Copy*: LD (inf. O. Almborn). – Dissertatio sistens supplementum secundum et emendationes in editionem secundam Prodromi Florae Scandinaviae, quam ... sub praesidio D.M. And. J. Retzii, ... pro laurea publico examini subjicit Gudmund Sandmark, ... die xvii Junii mdcccix. Lundae [Lund] (Litteris Berlingianis).

9088. Dissertatio botanica sistens *Observationes in methodum tournefortianam* a Cl. Guiart fil. reformatam. Quam, consentiente ampliss. ord. philos. sub praesidio D.M. And. J. Retzii, pro laurea modeste exhibet Carolus Ad. Agardh, Gothoburgensis. In Lyceo carolino die xxix Maji mdcccv. Lundae (Literis Berlingianis) [1805]. (*Obs. meth. tourn.*).
Respondens: Carl Adolph Agardh (1785-1859).
Publ.: 29 Mai 1805 (t.p.), p. [i-ii], [1]-14. *Copies*: LD, NY(2). – KR p. 588 (no. 32, err. as of 1825) rightly attributes this thesis to Retzius.

9089. Försök til en *Flora oeconomica Sveciae* eller Swenska wäxters nytta och skada i hushållningen ... Första [Andra] Delen. Lund (Tryckt hos Professorn, Dokt. Joh. Lundblad, på författarens bekostnad) 1806. Oct. 2 parts. (*Fl. oecon. svec.*).
1: 1806, p. [i*-ii*], [i]-viii, [1]-414.
2: 1806, p. [i-ii], [415]-792, [793-794, err.].
Copies: E, H, LD, NY, USDA. – Retzius obtained a prize of 30 duc. from the Swed. Acad. of medicine for this work.
Appendix: see *Bih. Fl. oecon. Svec.* (1812), below.

9090. Dissertatio botanica *de plantis cibariis romanorum*, quam consentiente amplisssima facultate philos. in Academia carolina, sub praesidio D.M. And. Jah. Retzii, ... pro laurea publicae eruditorum censurae subjicit [part 1:] Joh. Wilh. Zetterstedt, Ostro-Gothus. Die [ ] Aprilis mdcccviii. Lundae [Lund](Litteris Berlingianis) [1808]. Qu. 4 parts. (*Pl. cibar. roman.*).
Publ.: in four parts, each part defended by a different student. *Copies*: G(2), LD, NY, USDA.
[*1*]: Apr 1808, p. [i-ii], [1]-16. Defendens: Johan Wilhelm Zetterstedt (1785-1874).
[*2*]: Apr 1808, p. [i], 17-32. Defendens: Johan Magn. Allgulin, Smolandus.
[*3*]: Apr 1808, p. [i-ii], 33-50. Defendens: C. Sylvan, Scanus.
[*4*]: 28 Apr 1808 ("28" mss. in copy USDA), p. [i-ii], 51-71. Defendens: Gustaf Berggrén, Calmariensis.

9091. *Flora virgiliana* eller försök at utreda de wäxter som anföras uti P. Virgilii Maronis Eclogae, Georgica och Aeneides jämte bihang om romarnes matwäxter ... Lund (tryckt hos Prof. D. Joh. Lundblad, och på dess bekostnad) 1809. Oct. (*Fl. virgil.*).

RETZIUS, A. J.

Publ.: 1809, p. [1]-207. *Copies*: LD, NY, USDA. – The appendix ("Bihang") is a Swedish version of the four dissertations *De plantis cibariis romanorum*, see above.

9092. *Observationum botanicarum pugillus*. Quem speciminis loco cum venia amplissimae facultatis philosophicae sub praesidio D.M. And. J. Retzii, ... publico botanophilorum examini modeste submittit Magnus Christian Retzius, Lundensis, L.H.S. die [14] Novembr. mdcccx. Lundae [Lund] (Litteris Berlingianis) 1810. Qu. (*Obs. bot. pug.*).
Publ.: 14 Nov 1810, p. [1]-23, [24, theses]. *Copies*: HH, LD, USDA; IDC 5483.

9093. *Observationum in Criticam botanicam* C. à Linné, specimum primum [secundum]. Quod consent. ampliss. philosoph. ordine praeside D.M. And. J. Retzio, ... pro laurea exhibet Nicol. Christ. Psilander [2. Jonas Lundqvist], ... die iv [xv] maji mdcccxi. Lundae [Lund] (Litteris Berlingianis) 1811. Qu. (*Obs. Crit. bot.*).
1: 4 Mai 1811, p. [i-ii], [1]-16.
2: 15 Mai 1811, p. [i], 17-27.
Copies: HH, LD, NY.

9094.*Bihang till Flora oeconomica Sveciae*, med Philosophiska facultetens samtycke under inseende af M. Anders Jah. Retzius, för lagerkransen försvaradt af Lars Johan Darin, Götheborgsbo. Den 20 Junii 1812. Lund (Tryckt uti Berlingska Boktryckeriet) [1812]. Qu. (*Bih. Fl. oecon. Svec.*).
Publ.: 20 Jun 1812, p. [1]-20. *Copies*: LD, USDA. – Appendix to *Fl. oecon. Svec.* (1806), see above.

Retzius, [Magnus] **Gustaf** (1842-1919), Swedish physician and zoologist; Dr. med. Stockholm 1871; from 1871 at the Karolinska inst.; professor of histology (later: anatomy) 1877-1890; son of A.A. Retzius. (*G. Retz.*).

HERBARIUM and TYPES: Unknown.

BIBLIOGRAPHY and BIOGRAPHY: Barnhart 3: 146; BM 4: 1684, 8: 1068; Kew 4: 542; KR p. 588 (b. 17 Oct 1842, d. 21 Jul 1919); PH 239; SO 2926, 2958.

BIOFILE: Anon., Bot. Not. 1919: 164 (d.); K. Sv. Vet.-Akad. Årsb. 1922: 239-243, *pl. 1* (obit., photogr. of grave).
Fischer, I., ed., Biogr. Lexik. Ärzte ed. 2/3. 2: 1287. 1962.
Lindman, S. & Nilsonne, U., Sv. män. kvinn. 6: 246-247. 1949 (portr.).
Nissen, C., Zool. Buchill. 337. 1969.
Nordenskiöld, E., Hist. biol. 540, 541, 543. 1949.
Pilger, R., Bot. Jahrb. 40 (Lit.): 19. 1907 (rev. of his "Die Spermien der Fucaceen". Ark. Bot. 5(10). 1906.
Taton, R., ed., Science in the 19th cent. 352, 362. 1965.

EPONYMY: The Hierta-Retzius Foundation was established by Retzius and his wife (b. Hierta) with the Swedish Academy of Sciences to further scientific research and social progress.

9095. *Über die Spermien der Fucaceen* [Arkiv för Botanik Band 5. N:o 10, 1906.]. Oct.
Publ.: 30 Apr 1906 (date of publ. of fasc. 3-4 of vol. 5; Nat. Nov. Mai(1) 1906), p. [1]-9.
Copy: U. – Ark. Bot. 5(10): 1-9. 30 Apr 1906.

Reum, Johann Adam (1780-1839), German (Sachsen-Meiningen) forest botanist and plant physiologist; studied at Jena, Würzburg and Heidelberg; teacher at a forestry college at Zillbach 1805-1811; Dr. phil. Jena 1808; teacher at the forestry college at Tharandt 1811-1816; professor of mathematics and forestry botany at the royal Saxonian forestry academy of Tharandt 1816-1839, also founder and director of the botanical garden of Tharandt from 1816. (*Reum*).

HERBARIUM and TYPES: Unknown.

BIBLIOGRAPHY and BIOGRAPHY: ADB 28: 282-283 (R. Hess); Barnhart 3: 146; BM 4: 1685; CSP 5: 172; Herder p. 119, 381, 393; Jackson p. 207; Kew 4: 452; LS 22131; Moebius p. 397; PR 7561-7564, ed. 1: 8516-8517; Rehder 5: 719; Tucker 1: 593.

BIOFILE: Anon., Bot. Not. 1840: 80. (d. 25 Jul 1839); Flora 22: 560. 1839 (d. 25 Jul 1839).
Gager, C.F., Brooklyn Bot. Gard. Rec. 27(3): 237. 1938.
Gwinner, W.H., Forstl. Mitt., Stuttgart 2: 139-146. 1839-1840 (n.v.).
Hess, R., Lebensbilder hervorragender Forstmänner 290-291. 1885 (biogr. sketch, bibl., sources, b. 16 Mai 1780, d. 26 Jul 1839).
Jahnel, H., Wiss. Z. TU DResden 17: 203-210. 1968 (crit.note; n.v.).
Ratzeburg, J.T.C., Forstwiss. Schrifst.-Lexik. 435-438. 1872 (biogr., bibl., b. 16 Mai 1780, d. 26 Jul 1839).
Tesche, M., Wiss. Z. TU Dresden 29: 911-912. 1980 (brief biogr., critical appraisal; sources).
Verdoorn, F., ed., Chron. bot. 5: 302. 1939.

9096. *Grundriss der deutschen Forstbotanik* ... Dresden (in der Arnoldischen Buchhandlung) 1814. Oct. (*Grundr. deut. Forstbot.*).
Ed. [*1*]: 1814 (p. iv: Aug 1814), p. [i]-vi, [1]-300, [1, err.]. *Copy*: NY.
Ed. 2: 1825 (p. vi: Jul 1825; NY copy inscribed 15 Feb 1826), p. [i]-viii, [1, errr.], [1]-486. *Copies*: NY. – "Forstbotanik. Zweite sehr verbesserte und vermehrte Auflage". Dresden (id.) 1825. Oct. (*Forstbot.*).
Ed. 3: 1837 (p. vi: Jul 1837), p. [i]-viii, [1]-448. *Copy*: NY. – *Forstbotanik* ... Dresden und Leipzig (id.) 1837. Oct. (*Forstbot.*).
Vol. 2: see below, *Deut. Forstkr.*, 1819.

9097. *Die deutschen Forstkräuter.* Ein Versuch, sie kennen, benutzen und vertilgen zu lernen für Forstmänner und Waldeigenthümer ... Dresden (in der Arnoldischen Buchhandlung) 1819. Oct. (*Deut. Forstkr.*).
Publ.: 1819, p. [i]-viii, [1]-111, [112]. *Copy*: GOET (Forstl. Fak.), (inf. G. Wagenitz). – Alternative title: *Grundriss der deutschen Forstbotanik*, ... zweiter Theil".

Reuss, August Emanuel [Rudolf] von (1811-1873), Bohemian physician, botanist, mineralogist and palaeontologist; Dr. med. Praha 1833; physician at Bilin 1834-1849; professor of mineralogy and geology at Praha 1849-1863; professor of mineralogy at the University, Wien 1863-1873. (*A.E. Reuss*).

HERBARIUM and TYPES: B (extant); further material at FI, GOET, IBF, MANCH, PR, W, WU. – The original herbarium at B is incorporated in that of his son August Leopold von Reuss (1841-1924), a collection of circa 30.000 specimens collected by many botanists (inf. H.W. Lack); an important set is also at WU. – For August Leopold von Reuss see e.g. J. Pagel, Biogr. Lex. Ärzte 1369-1370. 1901, ed. 2/3, by I. Fischer, 2: 1287. 1962.

BIBLIOGRAPHY and BIOGRAPHY: ADB 28: 303-305 (Gümbel); Andrews p. 320; Barnhart 3: 146 (b. 8 Jul 1811, d. 26 Nov 1873); BM 4: 1685-1686; CSP 5: 173-175, 6: 748, 8: 735-736, 11: 154, 12: 613; Futak-Domin p. 501; IH 1 (ed. 7): 341, 2: (in press); Kew 4: 452; Klášterský p. 155; Maiwald p. 69, 140, 155, 189; NI 1623; PR 7566, ed. 1: 8519; Quenstedt p. 358; TL-2/1988; WU 25: 350-354. 1873 (bibl.).

BIOFILE: Anon., Bonplandia 6: 51. 1858 (cognomen Leopoldina Saussurea); Lotos 23: 232. 1873 (d.), 24: 51-52. 1874 (obit.); Österr. bot. Z. 13: 337. 1863 (app. Wien), 24: 33, 99. 1874 (d. 26 Nov 1873; herbarium in that of his sons), 60: 46. 1910 (August R. v. Reuss presented the herbarium of his father to WU); Quart. J. Geol. Soc. London 30 (Proc.) xlvii-xlviii. 1874 (obit.); Wiener med. Wochenschr. 61(29): 1885-1886. 1911 (centenary of birth).
Gad, J., ed., S.B. Lotos, Prag 47: 25. 1899 (memorial).
Geinitz, H.B., Leopoldina 9(9/10): 67-72. 1874 (obit., bibl.).
Lack, H.W., Willdenowia 10(1): 81. 1980 (herb.).

Laube, G., Alman. Akad. Wiss. Wien 24: 129-151. 1874 (b. 8 Jul 1811, d. 26 Apr 1873, extensive obit., bibl.; the secretary of the Academy, Schrötter, compares R.'s life and difficulties with that of Agassiz, stating that the latter had better luck in the United States "rich in men ready to take initiatives for the common good, where people are not used to wait until the government steps in ..."); Mitt. Ver. Gesch. Deut. Böhmen 12(5): 193-205. 1874 (obit., bibl.).
Lz. [H. Loretz], Verh. k.k. geol. Reichsanst. 1873(16): 280-282. (obit., d. 26 Nov 1873).
Poggendorff, J.C., Biogr.-Lit. Handw.-Buch 2: 615. 1863 (bibl.), 3: 1113. 1898 (bibl.), 7A, suppl.: 541. 1971 (bibl.).
Wagenitz, G., Index coll. princ. herb. Gott. 134. 1982.
Zittel, K.A. von, Gesch. Geol. Paläont. 862 [index]. 1899; Hist. geol. palaeontol. 555. 1901.

9098. *Die Versteinerungen der Böhmischen Kreideformation*, beschrieben von Dr. August Em. Reuss, ... mit Abbildungen der neuen oder weniger bekannten Arten, gezeichnet von Joseph Rubesch, ... Stuttgart (E. Schweizerbart'sche Verlagsbuchhandlung) 1845-1846, 2 Abt. Qu. (*Verstein. Böhm. Kreideform.*).
1: 1845 (p. iv: 1 Oct 1844), p. [i]-iv, [v], [1]-58. *pl. 1-13* (zool. plates by Reuss and Rubesch).
2: 1846 (p. iv: 1 Jul 1846), p. [i*, iii*], [i]-iv, [1]-148 (botany on p.81-96 by August Joseph Corda), *pl. 14-51* (*46-51* bot; 32 /33 combined; bot. pl. by Corda).
Copies: GOET, GRON, PH, U.
Ref.: Flora 30: 30-32. 1847 (extensive rev.).

Reuss, Christian Friedrich (1745-1813), Danish-born German botanist; Dr. med. 1769; professor of medicine at the University of Tübingen (extraord. 1771, ord. 1796). (*Reuss*).

HERBARIUM and TYPES: Unknown.

BIBLIOGRAPHY and BIOGRAPHY: Barnhart 3: 146 (b. 7 Jul 1745, d. 19 Oct 1813); BM 4: 1686; Dryander 3: 21, 580, 582; Herder p. 393; Jackson p. 10; Kew 4: 452; PR 7567-7569, ed. 1: 8520-8522; Rehder 5: 720; SO 552, 662, 663, 682, 683.

BIOFILE: Lehmann, E., Schwäbische Apotheker 178. 1951.
Poggendorff, J.C., Biogr.-lit. Handw.-Buch 2: 613. 1863, 7a: 542. 1971 (biogr. refs.).

EPONYMY: *Reussia* K.B. Presl (1838) honors F.A.Reuss, early 19th century Bohemian medical doctor and mineralogist. The etymology of *Reussia* Endlicher (1836, *nom. cons.*) could not be determined.

9099. *Compendium botanices* systematis linnaeani conspectum ejusdemque applicationem ad selectiora plantarum Germaniae indigenarum usu medico et oeconomico insignium genera eorumque species continens ... Ulmae [Ulm] (sumtu Stettiniano) 1774. Oct. (*Comp. bot.*).
Ed. [*1*]: 1774 (p. [viii]: 14 Apr 1784), p. [i-viii], [1]-445, [446-467, index], [468, err.]. *pl. 1-10* (uncol. copp.). *Copies*: FI, G, MICH, MO, PH, USDA.
Ed. 2: 1785 (p. [xvi]: Whitsun 1785), p. [i-xvi], [1]-18, chart on "p." 19-22 (single sheet), 23-589. *pl. 1-10* uncol. copp., [1-38, index]. *Copies*: G, HH, MO, NY, PH, USDA. – Editio secunda aucta. Ulmae (id.). 1785. Oct.

9100. *Kenntniss derjenigen Pflanzen die Mahlern und Färbern zum Nutzen* und denen Liebhabern der öconomischen Pflanzenkenntniss zum Vergnügen *gereichen können*. Leipzig (bey Christian Gottlob Hilscher) 1776 (Oct.). (*Kenntn. Pfl. Mahlern*).
Publ.: 1776, p. [i-xviii], [1]-812. *Copy*: NY.

9101. *Dictionarium botanicum* oder botanisches lateinisches und deutsches Handwörterbuch für Ärzte, Cameralisten, Apotheker, Specereyhändler, Kräuterkenner, Bluhmisten, Oekonomen, Gärtner und Fabrikanten nach dem Linneischen System. Leipzig (bey Christian Gottlob Hilscher) 1781. Oct. (*Dict. bot.*).

[*1*]: 1781, p. [i]-xvi, [1]-376.
[*2*]: 1781, p. [1]-485, [486-487, Anhang].
Copy: NY.

9102. D. Christian Friedr. Reuss ... *Primae lineae Encyclopaediae* et methodologiae universe scientiae *medicae* et theoreticae et practicae omniumque eius scientiarum tam praeparantium quam affinium ac subjunctae cui suis historiae litterariae. Tubingae [Tübingen] (in Officina libraria J.G. Cottae) 1783. Oct. (*Prim. lin. encycl. med.*).
Publ.: 1783, p. [i*], [i]-viii, [3]-570, corrig. 570-[572]. *Copy*: E (Univ. Libr.).

Reuss, Georg Christian (*fl.* 1869), German botanist at Ulm. (*G.C. Reuss*).

HERBARIUM and TYPES: Unknown.

BIBLIOGRAPHY and BIOGRAPHY: BM 4: 1687; Herder p. 119; Jackson p. 221; NI 1624, see also 1: 248; PR 7573; Rehder 1: 71.

BIOFILE: Fischer, E., Gutenb. Jahrb. 1933: 211 (no. 84).

9103. *Pflanzenblätter in Naturdruck* mit der botanischen Kunstsprache für die Blattform gesammelt und herausgegeben von Professor Dr. G. Ch. Reuss in Ulm. 42 Foliotafeln mit erläuterndem Text in Oktav. Stuttgart (E. Schweizerbart'sche Verlagshandlung (Eduard Koch)) [1862-1870]. (*Pfl.-Bl. Naturdr.*).
Ed. [*1*]: atlas in 7 Lieferungen 1862-1870, text 1869. *Copy*: NY.
Atlas: 1862-1870 (ÖbZ 1 Jan 1863; issued in Lieferungen), p. [i], *pl. 1-42* (nature prints). T.p. text as above.
Text: 1869 (t.p. so dated; Flora 10 Sep 1869), p. [i]-viii, [1]-176. – "Text zu den 42 Foliotafeln Pflanzenblättern in Naturdruck ..." Stuttgart (id.) 1869. Oct.
Ed. [*2*]: [1869-] 1870 (atlas and text), simple re-issue in 7 Lieferungen. *Copies*: G, NY.
Atlas: p. [i], *pl. 1-42*. T.p. not dated.
Text: p. [i]-viii, [1]-176. T.p. not dated.
Ed. 3: Nov-Dec 1881 (t.p. dated 1882; Nat. Nov. Dec(1) 1881; Bot. Zeit. 30 Dec 1881), simple re-issue. *Copies*: FI, G, USGS.
Atlas: p. [i], *pl. 1-42*.
Text: p. [i]-viii, [1]-176. – Dritte Auflage. Stuttgart (E. Schweizerbart'sche Verlagshandlung (E. Koch)). 1882. Oct.
Ref.: Schlechtendal, D.F.L. von, Bot. Zeit. 20: 414. 28 Nov 1862. (rev. Lief. 1 of atlas; has *pl. 1, 10, 23, 25, 30, 32* and 16 p. text)., 27: 630. 17 Sep 1869 (note on Lief. 5).
J., Gartenflora 12: 109. 1863 (Lief. 1 consists of plates *1, 10, 23, 25, 30, 32* and 16 p. text).

Reuss, Gustáv (1818-1861), Slovak botanist and physician. (*G. Reuss*).

HERBARIUM and TYPES: Unknown.

BIBLIOGRAPHY and BIOGRAPHY: AG 6(1): 176; Barnhart 3:146; BM 4: 1687; Bossert 330; Futak-Domin p. 501; Hortus 3: 1202; Kanitz 227; PR 7574; SO add. 852a; Urban-Berl. p. 706; Zander ed. 11, p. 806.

BIOFILE: Anon., Biologia, Bratislava 27(4): 311-320. 1972 (biogr., b. 4 Jan 1818, d. 12 Jan 1861).
Hendrych, R., Preslia 25: 368-373. 1953 (biogr., portr.).
Pax, F., Grundz. Pfl.-Verbr. Karpathen 20, 33. 1898 (Veg. Erde 2(1)).

9104. *Května slovenska* čili opis všech jevnosnubnýcii na Slovensku divorostancích a mnohých zahradních zrostlin podlé saustavy De Candolle-ovy. S připojeným zrostlinářským názvoslovím, slovoníkem a návodem k určitbě zrostlin podlé saustavy Linné-ovy ... v.b. Štávnici (tiskem Františka Lorbera) 1853. Oct. (*Května slov.*).
Publ.: 1853, p. [i]-lxxv, [1]-496, [1, err.]. *Copy*: G.

Reuss, Leopold (*fl.* 1831), German (Bavarian) botanist and clergyman ("Domvikar") at Passau. (*L. Reuss*).

HERBARIUM and TYPES: Unknown.

BIBLIOGRAPHY and BIOGRAPHY: Jackson p. 301; PR 7575-7576, ed. 1: 8526-8527; Rehder 1: 378, 379; Tucker 1: 593.

9105. *Flora des Unter-Donau-Kreises*, oder Aufzählung und kurze Beschreibung der im Unter-Donau-Kreise wildwachsenden Pflanzen. Mit Angabe des Standorts, der Blühezeit, der ökonomischen, technischen und medizinischen Benützung ... Passau (Druck der Pustet'schen Buchdruckerei) 1831. Oct. (*Fl. Unt.-Donau-Kr.*).
Publ.: 1831 (p. iv: 23 Dec 1830; Flora rd. 8 Feb 1832), p. [i]-iv, [3]-291. *Copies*: NY, REG(2).
Ref.: Anon., Lit. Ber. Flora 2: 25-32. Jan-Feb 1832 (rev.), 2: 236-240. Jul-Dec 1832 (discussion of review).

Reuter, George François (1805-1872), Swiss (French-born) botanist (of Geneva parents), at Genève from 1826, at first active as engraver; from 1835 curator of the Candollean herbarium; with E. Boissier in Spain, 1841 and until 1849 curator of the Boissier herbarium and collaborator of Boissier accompanying him on his travels and collecting trips; from 1849-1872 director of the Genève botanical garden (succeeding Alph. de Candolle). (*Reut.*).

HERBARIUM and TYPES: G (through Barbey to E. Boissier). – Other material at BM, BR, C, CGE, CN, FI, GOET, H, K, KIEL, LAU, M, MANCH, MW, NAP, NCY, OXF, P, PC, P-CO, W, WAG.

BIBLIOGRAPHY and BIOGRAPHY: AG 2(1): 172, 535, 5(1): 165, 795, 6(2): 942, 11 (ed. 2): 167, 12(3): 261; Backer p. 659; Barnhart 3: 147 (b. 30 Nov 1805, d. 23 Mai 1872); BJI 2: 145; BL 2: 571, 707; BM 4: 1687; Bossert p. 330; Clokie p. 231; Colmeiro 1: cxcv; CSP 5: 176-177, 8: 737, 11: 155; DTS 1:395, 6(4): 44; Hegi 5(2): 1196, 6(2): 1348; Herder p. 205; Hortus 3: 1202; IH 1 (ed. 6): 363, (ed. 7): 341, 2: (in press); Jackson p. 447; Kew 4: 452; Lasègue p. 110-111, 297, 503; MW p. 413; PR 943, 946, 5812, 7577-7578, ed. 1: 1060, 6506, 8529-8531; Rehder 5: 720; Saccardo 1: 137; SIA 208; TL-1/114, 115, 116, post 1169; TL-2/606-608, 999, 5405, 5838; Tucker 1: 593; Urban-Berl. p. 387; Zander ed. 10, p. 706, ed. 11, p. 724, 806.

BIOFILE: Anon., Bonplandia 3: 257. 1855 (visit Leipzig); Bot. Not. 1873: 28; Bot. Zeit. 7: 207. 1849 (to Algeria), 8: 72. 1850 (dir. Genève bot. gard.); J. Bot. 10: 224. 1872 (d.); Bull. Soc. bot. France 18 (bibl.): 144. 1872 (d.).
Babington, C.C., Mem., Journ. bot. corr. ix, 474. 1897.
Blakelock, R.A. & E.R. Guest, Fl. Iraq 1: 116. 1966.
Bocquet, G. & M. Mermoud, Arch. Sci., Genève 18(2): 388-397. 1965 (restauration herb. Boissier; herb. Reuter p. 391).
Briquet, J., Ann. Cons. bot. Genève 9: 209-211. 1905 (on history Conservatoire); Bull. Herb. Boissier ser. 2. 7: 48. 1907 (introductory notice to the posthumous publication of Reuter's account of a trip to Norway in Jul-Aug 1861 with E. Boissier); Bull. Soc. bot. Suisse 50a: 398-405. 1940 (biogr., bibl., b. 30 Nov 1805, d. 23 Mai 1872, epon., biogr. sources).
Burdet, H.M. et al., Candollea 36(2): 543-584. 1981 (list of types R. Gymnosp. through Gramin.), 37(2): 381-395. 1982 (id. Iridacées à Potamogetonacées), 37(2): 397-427 1982 (Index to Spanish taxa descr. by Boissier and Greuter), 37(2): 429-438. 1982 (Thesaurus of Boissier and Reuter publ. on Spanish plants; many bibl. details; e.g. corrections to our entry TL2/605.
Burnat, É., Bull. Soc. bot. France 30: cxxviii. 1883 ("peu de botanistes ont su herboriser comme Reuter, auquel rien n'échappait").
Candolle, A.P. de, Mém. souvenirs 469. 1862.
Candolle, Alph. de, Phytographie 443. 1880.
Cavillier, F., Boissiera 5: 77. 1941.

Chodat, R., Bull. Soc. bot. Genève ser. 2. 3: 342-343. 1911 (on herb. Reuter at Inst. bot. Univ. Genève, given by W. Barbey).
Cosson, E., Comp. fl. atl. 1: 86-87. 1881.
Cosson et Durieu, Exp. sci. Algérie Bot. 2: xliii-xliv. 1868.
Fauconnet, Ch., Excursions botaniques dans le Bas-Valais 144-145. 1872.
Gager, C.S., Brooklyn Bot. Gard. Rec. 27(3): 337. 1938.
Gautier, É., Mém. Soc. Phys. Hist. nat. Genève 21(2): 599-602. 1872 (obit.).
Haussknecht, C., Flora 55: 286. 1872 (obit.); Bot. Zeit. 30: 479. 1872; Öster. bot. Z. 22: 237. 1872 (obit. "Brustentzündung").
Lindemann, E., Bull. Soc. Natural. Moscou 61(1): 61. 1886.
Maire, R., Progr. connaiss. bot. Algérie 88. 1931.
Miège & Wuest, Saussurea 6: 132. 1975 (herb.).
Reichenbach, H.G., Bot. Zeit. 30: 590-594. 1872 (see also Briquet 1940, p. 403 some "curieuses erreurs").
Reynier, A., Soc. Hort. Bot. Marseille (Bull.) 40: 202-206. 1894, 41: (487): 27-31. 1895.
Stafleu, F.A., The great prodromus 30. 1966.
Wagenitz, G., Index coll. princ. herb. Gott. 135. 1982.
Weibel, R., Musées de Genève 75: 5-7. 1967.
Willkomm, M., Grundz. Pfl.-Verbr. Iber. Halbinsel 11-12. 1896 (Veg. Erde 1).

COMPOSITE WORKS: (1) *Orobanchaceae*, in A.P. de Candolle, Prodr. 11: 1-45, 717-720. 25 Nov 1847.
(2) With Henri Margot (1807-1894): *Essai fl. Zante*, 1839-1840, see TL-2/5405.
(3) With Pierre Edmond Boissier (1810-1885): *Pugill. pl. Afr. bor. Hispan.* 1852, see TL-2/608; also published in Bibl. univ. Genève. Jan 1852.
(4) With id. *Diagn. pl. nov. hisp.* 1842, see TL-2/606.
(5) With id. *Diagn. pl. orient.* 1842-1859, see TL-2/607.

NOTE: For details on the Boissier and Reuter publications on Spanish plants, see Burdet, H.M. et al. (1981-1982).

HANDWRITING: Candollea 32: 413-414. 1977.

EPONYMY: *Reutera* Boissier (1838).

9106. *Catalogue détaillé des plantes vasculaires* qui croissent naturellement aux environs de Genève, avec l'indication des localités et de l'époque de la floraison... Genève (Librairie A. Cherbuliez, à la Cité) 1832. Duod. (in sixes). (*Cat. pl. vasc. Genève*).
Ed. [1]: 1832, p. [1]-138. *Copies*: BR, G(2), HH, NY.
Supplément: 1841 (p. 6: 31 Mar 1841), p. [1]-51, *1 pl.* (uncol.). *Copies*: G(2), HH, NY. Supplément au catalogue des plantes vasculaires qui croissent naturellement aux environs de Genève, ... Genève (Imprimerie de Ch. Gruaz, ...) 1841. Duod. (id.)
Ed. 2: 1861 (p. viii: 18 Mar 1861; Flora rd. 21 Aug-14 Sep 1861), p. [i]-xvi, [1]-300. *Copies*: BR, FI, G(2), HH, NY. – *Catalogue des plantes vasculaires* qui croissent naturellement aux environs *de Genève* ... deuxième édition entièrement refondue et considérablement augmentée suivie d'une *Monographie des espèces du genre Rubus* des environs de Genève par le Dr E. Mercier[q.v.]. Genève (Librairie allemande de J. Kessmann) 1861. Oct. (*Mercier/Rubus on p. 257-295*).
Additions and annotations: Aug. Schmidely, Bull. Trav. Bull. Soc. bot. Genève 3: 82-155, 160. Oct 1884.
Follow-up publ.: see BL 2: 571-572, and C. Weber, *Catalogue dynamique de la flore de Genève*, Boissiera 12. 1966.

9107. *Essai sur la végétation de la Nouvelle Castille* ... Genève (imprimerie de Jules-Gme Fick, ...) 1843. Qu.
Publ.: 1843, p. [1]-34, *pl. 1* (uncol. lith. Heyland). *Copy*: G. – Reprinted and to be cited from Mém. Soc. phys. Hist. nat. Genève 10: 215-246, *pl. 1*, 1843.

9108. *Quelques notes sur la végétation de l'Algérie* ... Genève (Imprimerie de Ferd. Ramboz et Cie ...) 1852. Oct.

Publ.: Jun 1852 (in journal), p. [1]-27. *Copy*: G. – Reprinted and to be cited from Bibl. univ. Genève 20: 89-113. 1852. See also Belg. hort. 3: 111-119, 147-152. 1853.

9109. *Notulae in species novas vel criticas plantarum horti botanici genevensis* publici juris annis 1852-1868 factae collectae et iterum editae anno 1916 [Annu. Cons. Jard. bot. Genève 18/19, 1916]. Oct. (*Notul. sp. nov. hort. genev.*).
Publ.: 30 Jan 1916 (cover reprint), p. [1]-14. *Copy*: G. – Reprinted and to be cited from Annu. Cons. Jard. bot. Genève 18/19: 239-252. 1916.
Other issue: 30 Jan 1916, p. [1]-16. *Copy*: G. – Reprinted from idem 18/19: 239-254. 1916.
A reprint (edited by J. Briquet) of descriptions of new taxa and critical taxa published by Reuter in the *Catalogues des graines recueillies* en [1852, 1853, 1854, 1855, 1856, 1857, 1861, 1863, 1865, 1867, 1868] et offertes en échange par le Jardin botanique de Genève.

Réveil, [Pierre] Oscar (1821-1865), French botanist and physician at Paris; head pharmacist of the Paris hospitals; professor at the "Faculté de médecine" and at the "École supérieure de pharmacie de Paris", specialist in phytochemistry. (*Réveil*).

HERBARIUM and TYPES: Unknown.

BIBLIOGRAPHY and BIOGRAPHY: Barnhart 3: 147 (b. 20 Mai 1821, d. 7 Jun 1815); BJI 2: 8, 707; NI 568; see also 1: 1; PR 2545, 7579; Rehder 5: 720; Sotheby 641; TL-2/see H.E. Baillon, see F. Hérincq; Tucker 1: 593.

BIOFILE: Anon., Bull. Soc. bot. France, Table art. orig. 200-201, 12: 256. 1865 (d.), 13: 6. 1866 (d.); Cat. gén. livr. impr. Bibl. natl., auteurs 70: 892-897. 1929 (bibl.) (see also 45: 322. 1929 for Règne végétal, sub A. Dupuis).

9110. *Le règne végétal* divisé en traité de botanique, flore médicale, usuelle et industrielle/horticulture théorique et pratique/plantes agricoles et forestières/histoire biographique et bibliographique de la botanique par M.M. O. Réveil ... A. Depuis ... Fr. Gérard ... F. Hérincq ... avec le concours (pour la flore medicale) de M. le docteur Baillon ... et d'après les plus éminents botanistes français et étrangers ... Paris (L. Guérin et Cie, éditeurs ...) [1864-]1870-1871. 17 vols. (9 vols. text, Oct; 8 vols. plates, Qu.). (*Règne vég*.).
Co-authors: Noël Aristide Dupuis (1823-1883), Fr. Gérard, François Hérincq (1820-1891); for *Fl. méd*.: Henri Ernest Baillon (1827-1895).
Publ.: 17 vols., 1864-1871 (precise dates not established) divided over the following series:
Traité de botanique générale par MM. F. Hérincq ... Fr. Gérard ... O. Réveil s.d. [1864-?].
    Copies: BR, NY.
    1: [i*-vii*], [i]-viii, [1]-458.
    2: [i-vii], [1]-427, [i-iii], [1]-79 (dict. étym. termes bot.).
    1 (plates): [i-vii], *pl. 1-52, 29 bis* with text (col. copp.), [1-2].
    2 (plates): [i-vii], *pl. 1-50* id., [i]-ii.
Flore médicale par M.M. A. Dupuis ... O. Réveil ... avec le concours de M. le docteur Baillon ... donnant la description, la culture, la composition chimique (les propriétés curatives ou dangereuses, les usages économiques et industriels des plantes ... 1864-1867. – *Copies*: MO, NY.
    1: [i*-vii*], [i]-xvi, [1]-490.
    2: [i-vii], [1]-488.
    3: [i-vii], [1]-547.
    1 (plates): [i-vii], *pl. 1-50* with text (col. copp.), [1-2].
    2 (plates): [i-vii], *pl. 1-50*, id. [1-2].
    3 (plates): [i-vii], *pl. 1-50*, id. [1-2].
Horticulture. Jardin potager et jardin fruitier par M. F. Hérincq ... Fr. Gérard ... et d'après les plus savants écrits français et étrangers sur la matière ouvrage donnant des notions générales sur la culture du jardin potager et du jardin fruitier et des notions particulières sur chaque plante ... edité par L. Guérin ... 1866. – *Copy*: NY.
    Text: [i-vii], [1]-674.

Plates: [i-vii], *pl. 1-56* with text (col. copp.), [i]-iv.
Horticulture. Végétaux d'ornement par MM. A. Dupuis ... F. Hérincq ... donnant des notions générales sur l'horticulture florale; la culture et la description particulière a chaque plante d'ornement ... s.d. – *Copy*: NY.
Text: [i*-vii*], [i]-xcix, [1]-382.
Plates: [i-vii], *pl. 1-52* with text (col. copp.), [1-3].
Plantes agricoles et forestières par M.A. Dupuis ... Ouvrage donnant la description, la culture et les usages des végetaux dont il traite ... 1867. – *Copy*: NY.
Texte: [i-vii], [1]-566.
Plates: [i-viii], *pl. 1-50* with text (col. copp.), [1-2].
Précis de l'histoire de la botanique pour servir de complément a l'étude du Règne végétal par L. G... suivi d'un appendice de géographie botanique avec cartes par J.A. Barral ... 1869. – Jean Augustin Barral (1809-1884); the identity of L.G. is unknown to us. – *Copies*: MO, NY.
Text (only): [i*-vii*], [i]-ii, [3]-533, [534-542], appendice au Règne végétal, [543, cont.], 4 maps.

Revel, Joseph (1811-1883), French botanist and clergyman; school principal at Villefranche-de-Rouergue (Institution Saint-Joseph). (*Revel*).

HERBARIUM and TYPES: MPU.

BIBLIOGRAPHY and BIOGRAPHY: BL 2: 113, 707; BM 4: 1689; CSP 5: 177, 11: 156-157; Kew 4: 453; Morren ed. 10, p. 69; PR 7580; Rehder 1: 413, 417; Tucker 1: 593-594.

BIOFILE: Anon., Bull. Soc. bot. France 34: 191 (bibl.), 402. 1887 (d.).
Granel de Solignac, L., et al., Naturalia monspeliensia, Bot. 23/24: 148-149. 1973 (herb. at MPU; d. 1883).
Revel, J., Bull. Soc. bot. France 12: 346. 1865 (sends his *Rech. bot. s.o. France*).

9111. *Notice sur les Renoncules batraciennes*, observées dans le département de la Dordogne [Actes Soc. Linn. Bordeaux, 19(2), 1853]. Oct.
Publ.: 1853, p. [1]-11, *1 pl. Copy*: FI. – Reprinted and to be cited from Actes Soc. Linn. Bordeaux 19(2): 114-123. 1853.

9112. *Recherches botaniques faites dans le sud-ouest de la France* ... Bordeaux (chez Coderc, Degréteau et Poujol ...) 1865. Oct.
Publ.: Nov 1865 (author sent a copy to Soc. bot. Fr. 17 Nov 1865), p. [i], [1]-62, *pl. 4*.
Copies: G, HH, NY. – Reprinted, and to be cited from Actes Soc. Linn. Bordeaux 25(5): 353-414. 1865 (last page of fasc. 5 dated 10 Jul 1865).
Ref.: Anon., Bull. Soc. bot. France 12 (C.R.): 345, (bibl.): 258-259, 346. 1866.

9113. *Notes et observations sur quelques plantes rares* litigieuses, nouvelles, ou peu connues *du sud-ouest de la France* ... Rodez (Ve E. Carrère, ...) 1877. Oct.
Publ.: 1877 (t.p. reprint), p. [1]-64. *Copies*: G(2), HH, NY. – Reprinted and to be cited from Congr. sci., France, 40(Rodez, 1, 1874): 221-269. 1877 (journal n.v.).
Ref.: Anon., Bull. Soc. bot. France 25 (bibl.): 77-78. 1878.

9114. *Essai de la flore du sud-ouest de la France* ou recherches botaniques faites dans cette région ... [cover: Paris (chez F. Savy ...)], Villefranche (chez P. Dufour ...) 1885-1889[-1900]. Oct. (*Essai fl. s.o. France*).
1: 19 Mar-15 Mai 1885 (p. [vii]: 19 Mar 1885; J. Bot. Jun 1885; Bot. Centralbl. 27 Jul-7 Aug 1885; Bot. Zeit. 25 Sep 1885; BSbF 26 Jun 1885; Nat. Nov. Mai(1) 1885), p. [i-vii], [1]-431, *1 pl.* (uncol., J. Valadier).
2(1): Aug-Oct 1889 (Bot. Zeit. 31 Jan 1890; BSbF Jul-Dec 1889; Nat. Nov. Oct(2) 1889), p. [i, iii, v], [433]-609. – ..."2me partie des composées". Villefranche (id.) 1889. Oct.
2(2): 1900 (Bot. Zeit. 16 Aug 1901; Bot. Centralbl. 7 Aug 1901; Nat. Nov. Aug(1) 1901), issued together with a re-issue of 2(1): [i-iv], [433]-845. "... continuée et terminée par M. l'abbé Hippolyte Coste [Hippolyte Jacques Coste, 1858-1924] ... 2e partie des

composées aux fougères, inclusivement ". Villefranche (id.), Rodez (imprimerie E. Carrère) 1900. Oct.
Copies: G, HH, NY, US. – Publications de la Société des lettres, sciences et arts de l'Aveyron.
Ref.: Malinvaud, E., Bull. Soc. bot. France 32 (bibl.): 136-137. 1885, 36 (bibl.): 118-119. 1889.

Revelière, Eugène (1822-1892), French botanist and entomologist at Saumur; collected in Corsica 1854 and 1856-1858, later also in Algeria; ultimately living in Corsica. (*Revelière*).

HERBARIUM and TYPES: Some material at GFW, HEID, L, LE, LY, MPU, W. – According to Souché (1901) the original herbarium was with R's brother J. Revelière, Blaise (Loire-Inf.).

BIBLIOGRAPHY and BIOGRAPHY: AG 3: 462; Barnhart 3: 147 (b. 12 Nov 1822, d. 1 Feb 1892).; BL 2: 140, 707; BM 4: 1689; Saccardo 1: 137, 2: 91; TL-2/5465, see P. Mabille.

BIOFILE: Anon., Bull. Soc. bot. Deux-Sèvres 19: *pl. 2*. 1908 (portr.);
Insekten-Börse 9: 9-10. 1892 (d.).
Candolle, A. de, Phytographie 430. 1880 (coll. Mabille & Revelière).
Cosson, E., Comp. fl. atl. 1: 87. 1881.
Cosson et Durieu, Exp. sci. Algérie, Bot. 2: xliv. 1868.
Gilbert, P., Comp. biogr. lit. deceased entom. 315. 1977.
Rey, E., L'Échange, Revue linnéenne 8(87): 31-32. 1892 (obit.).
Souché, B., Fl. Haut Poitou xvii. 1901 (herb., 12 Nov 1822, d. 1 Feb 1892).

COMPOSITE WORKS: with L.J.A. de C. de Marsilly, *Cat. pl. vasc. Corse*, 1872, see TL-2/5465.

Reverchon, Elisée (1835-1914), French plant collector, living at Bollème (Vaucluse) and later in Lyon; collected in S.E. France (1867-1877), Corsica (1878-1880), Sardinia (1881-1882), Crete (1883), Spain (1887-1906) and Algeria 1896-1898. (*Reverchon*).

HERBARIUM and TYPES: Reverchon was a professional plant collector; his plants are found in many herbaria; important sets e.g. at B (extant), G and P/PC. For a note on the reliability of the data on the labels see Burnat (1883).

BIBLIOGRAPHY and BIOGRAPHY: AG 3: 8; Barnhart 3: 147; BL 2: 161, 707; Clokie p. 231; PFC 1: lv; Saccardo 1: 137, 2: 91; TL-2/2702; Urban-Berl. p. 387.

BIOFILE: Anon., Bull. Soc. bot. France 14 (bibl.): 144. 1867, 15 (bibl.): 48. 1868, 17 (bibl.): 47, lxxxiii, 19 (bibl.): 157. 1873, 24 (bibl.): 190. 1877, 25 (c.r.): 255. 1878, 26 (bibl.): 144. 1879, 41 (bibl., suppl.): cccxxxv. 1894 (notes on collecting trips and sale of exsiccatae), 59: 672. 1913 (distr. exsicc.); Flora 30: 224. 1867, 51: 192. 1868 (lives at Briançon; offers plants for sale).
Burnat, É., Bull. Soc. bot. France 30: cxxviii. 1883 (on his collections from S. France; the collections distributed in 1874 contain specimens with erroneous dates: localities are cited which are intentionally erroneous).
Cavillier, F., Boissiera 5: 77-78. 1941.
Hedge, I.C. & J.M. Lamond, Index coll. Edinb. herb. 123. 1970.
Hervier, G.M.J., Excurs. bot. Reverchon, 1905, see TL-2/2702.
Kneucker, A., Allg. bot. Z. 1: 53. 1895, 2: 16. 1896, 8: 19. 1902 (exsicc. Spain, Algeria).
Lack, H.W., Willdenowia 10(1): 73, 81. 1980 (coll. B).
Maire, R., Progr. connaiss. bot. Algérie 152. 1931.
Mattirolo, O., Pubbl. ii. cent. fondaz. Orto bot. Torino 1729-1929, p. cii. 1929 (Sardinian plants in TO).
Wagenitz, G., Index coll. princ. herb. Gott. 135. 1982.
Willkomm, M., Grundz. Pfl.-Verbr. Iber. Halbins. 21-22. 1896 (coll. in Spain 1888-1893).

HANDWRITING: Candollea 32: 425-526. 1977.

Reverchon, Julien (1834-1905), French botanical collector; brother of E. Reverchon settled in Texas in 1856; from 1856-1858 in "La Réunion", from then on living on a farm near Dallas and collecting in various parts of the state. (*J. Reverchon*).

HERBARIUM and TYPES: MO. – Collections from Texas also in A, B, CR, DPU, E, F, FMC, GB, K, LE, MICH, MIN, ND, NY, ORE, P, PH, POM, S, SMU, TAEX, TEX, US, VT, WELC. – Letters to A. Gray and Th. Morong at GH.

BIBLIOGRAPHY and BIOGRAPHY: AG 5(1): 292; Barnhart 3: 147 (b. 3 Aug 1834, d. 30 Dec 1905); Bossert p. 330; CSP 12: 614, 18: 154; IH 1 (ed. 7): 34, 2: (in press); Morren, ed. 11, p. 131; Rehder 5: 721; Urban-Berl. 387.

BIOFILE: Anon., The Dallas Morning News 31 Dec 1905 (obit., portr.; clipping at NY); Torreya 6(1): 18. 1906 (d.).
Geiser, S.W., Southwest Review 14: 331-342. 1929; Naturalists of the frontier 14, 18-19, 77, 215-224, 280. 1948.
Hitchcock, A.S. & A. Chase, Man. grasses U.S. 988. 1951 (epon.).
Rodgers, A.D., Amer. bot. 1873-1892. 123, 125, 208, 216. 1944.
Sargent, C.S., Silva 13: 175-176. 1902 (b. 3 Aug 1837 sic).

EPONYMY: *Reverchonia* A. Gray (1880).

Reverchon, P. (*fl.* 1878-1892), French botanist at Alençon and Angers. (*P. Reverchon*).

HERBARIUM and TYPES: Unknown.

BIBLIOGRAPHY and BIOGRAPHY: Barnhart 3: 147; BL 2: 180, 707; CSP 12: 614, 18: 154; Rehder 1: 417; Tucker 1: 594.

9115. *Flore de la Mayenne.* Catalogue raisonné des plantes vasculaires du Département de la Mayenne ... Angers (Germain & G. Grassin ...) 1892. Oct. (*Fl. Mayenne*).
Publ.: 1891-1892. *Copy*: NY. – Reprinted from Bull. Soc. Étud. Sci. Angers:
 1: late 1891 or Jan-Feb 1892 (Bot. Centralbl. 9 Mar 1892; Bot. Zeit. 24 Jun 1892; Nat. Nov. Mar(1) 1892), p. [1]-101, reprinted from Bull. 20: 139-237. 1891 (for 1890).
 2: Mar-Dec 1892 (Bot. Centralbl. 9 Mai 1893; Nat. Nov. Mai(1) 1893, p. [i], 101-228, reprinted from Bull. 21: 105-224. 1892 (for 1891).

Révoil, Georges (*fl.* 1878-1880), French explorer of Somalia. (*Révoil*).

HERBARIUM and TYPES: Somalian material at P.

BIBLIOGRAPHY and BIOGRAPHY: Barnhart 3:147; BM 4: 1689; CSP 18: 155; IH 2: (in press); Kew 4: 453; Rehder 1: 489; TL-2/1845; Tucker 1: 594.

BIOFILE: Embacher, F., Lexikon Reisen 246. 1882.
Vallot, J., Bull. Soc. bot. France 29: 187. 1882 (in Somalia 1878-1880; plants at P.).

COMPOSITE WORKS: editor: *Faune et flore des Pays Çomalis* (Afrique orientale), 10 parts 1882; of which part ix botanical: A.R. Franchet, *Sert. Somal.* Jun-Jul 1884, see TL-2/1845, p. [1]-70, err. slip, *pl. 1-6* (sic, correction for entry in no. 1845) (uncol. liths. C. Cuisin). *Copies*: BR, HH, NY, PH. – Imprint on p. 70: Paris (Imprimerie Jules Tremblay ...) [1822]. *Review*: E. Roth, Bot. Jahrb. 6 (Lit.): 27-29. 30 Dec 1884.

Revol, J. (*fl.* x-1928), French botanist. (*Revol*).

HERBARIUM and TYPES: GRM.

BIBLIOGRAPHY and BIOGRAPHY: Barnhart 3: 147; BL 2: 120, 707; Kew 4: 453; Tucker 1: 594.

BIOFILE: Anon., Ann. Soc. bot. Lyon 34: 29-316, map. 1910, 42: 51-103. 1922 (suppl.), 43: 75-? 1924 (suppl. 2).

9116. *Catalogue des plantes vasculaires du Département de l'Ardèche* par J. Revol. Introduction de M. le Prof. Flahault. Extrait des Annales de la Société botanique de Lyon t. xxxiv (1909). Paris (Librairie des sciences naturelles Paul Klincksieck ...) 1910. Oct. (*Cat. pl. vasc. Ardèche*).
Introduction: Charles Henri Marie Flahault (1852-1935).
Issue a: Oct-Dec 1910 (p. xxvii: 3 Oct 1910; Nat. Nov. Dec(2) 1910), p. [iii]-xxvii, [29]-316, map. *Copies*: BR, HH, USDA. – Ann. Soc. bot. Lyon 34: [29]-316, map, appendix, xix p. (introd. by Flahault). 1909 (sic). – For follow-up papers see BL 2: 120.
Issue b: Oct-Dec 1910 (id.), p. [i*], [i]-xix, [1]-288. *Copy*: USDA. – *Catalogue* ... Introduction: au sujet de la géographie botanique de l'Ardèche et du Vivarais par Ch. Flahault ... Lyon (A. Rey & Cie, ...) 1910. Oct. – Repaginated reprint of the paper in the Ann. Soc. bot. Lyon.
Supplement: Ann. Soc. bot. Lyon 42: 51-103. 1922.
Supplement [2]: ib. 43: 75. 1924.

Rex, George Abraham (1845-1895), American physician and mycologist in Philadelphia; Dr. med. Univ. Pennsylvania 1868; demonstrator of anatomy at the University of Pennsylvania; conservator of the Microscopical Section of the Academy of natural Sciences 1890-1895. (*Rex*).

HERBARIUM and TYPES: PH; further material at CUP. – Letters at FH.

BIBLIOGRAPHY and BIOGRAPHY: Barnhart 3: 147 (b. 28 Apr 1845; d. 4 Feb 1895); IH 2: (in press); Kelly p. 190; LS 22150-22161a; NAF 1(1): 168; Stevenson p. 1255.

BIOFILE: Anon., Bot. Centralbl. 62: 96. 1895 (d.); Bot. Gaz. 20: 123. 1895 (d.); Bot. Not. 1896: 44; Hedwigia 34: 105. 1895 (d.); Österrr. bot. Z. 45: 248. 1895 (d.).
Harshberger, J.W., Bot. Philadelphia 342-344. 1899 (biogr. sketch; bibl.); *in* H.A. Kelly & W.L. Burrage, Amer. med. biogr. 972-973. 1920.
Kelly, H.A., Cycl. Amer. med. biogr. 2: 316-317. 1912.
Kneucker, A., Allg. bot. Z. 1(5): 112. 1895 (d.).
Lloyd, C.G., Mycol. Notes 56: 798. 1918 (tribute, portr.).
Pennell, F.W., Bartonia 22: 17. 1943 (herb. at PH).
Wilson, E.H., Plant hunting 62. 1927.

Rey-Pailhade, Constantin de (1844-1930), French pteridologist. (*Rey-Pailhade*).

HERBARIUM and TYPES: Some material at MPU (herb. J. de Vichet).

BIBLIOGRAPHY and BIOGRAPHY: Barnhart 3: 148; BL 2: 106, 165, 708; CSP 18: 157; Kew 4: 453.

BIOFILE: Allorge, V., Bull. Soc. bot. France 77: 702. 1930 (d.).
Granel de Solignac, L. et al., Naturalia monspel. Bot. 26: 27. 1976. (pl. MPU).

9117. *Les fougères de France* ... 56 planches intercalées dans le texte et contenant 193 dessins. Paris (Paul Dupont, ...) [1895]. Oct. (*Foug. France*).
Publ.: Jan-Mar 1895 (p. 1 has a note by D. Clos dated 17 Jun 1893; however the first copy mentioned in the press seen by us is that offered to the Société bot. France on 26 Jul 1895, see BSbF 42: 480, 513, 1895. Copy rd by secretary mid Apr 1895, J. Bot. Oct 1895; Nat. Nov. Apr(1) 1895), p. [i-iii], [1]-133, pl. 1-56 (uncol. liths. auct.).
Copies: BR, G(2), HH, NY, USDA; IDC 7250. – Preprinted or reprinted from Soc. Sci. Béziers 16: 37-99, *pl. 1-30*. 1894, 17: 37-106, *pl. 31-56*. 1895.
Ref.: Malinvaud, E., Bull. Soc. bot. France 42: 513-516. 1895.

Reyes y Prosper, Eduardo (1860-1921), Spanish (Valencia-born) botanist; studied at Cordoba, Valencia and Madrid; Dr. Ci. Madrid. (*Reyes y Prosper*).

HERBARIUM and TYPES: Unknown.

BIBLIOGRAPHY and BIOGRAPHY: Barnhart 3: 148 (b. Jan 1860, d. 20 Jun 1921); BJI 2: 141; BL 2: 483, 707; BM 4: 1692; Bossert p. 330; Kew 4: 371.

BIOFILE: Anon., Encicl. univ. ilustr. europ.-amer. 51: 201-202. s.d. (portr., biogr. sketch; d. 20 Jun 1921).

EPONYMY: *Reyesia* C. Gay (1848-1849) honors Ant. Garcia Reyes, 19th century Chilean botanist; *Reyesiella* P.A. Saccardo (1917) was named for S.A. Reyes, a collector of fungi.

9118. *Las Carofitas de España singularmente las que crecen en sus estepas* ... Madrid (Imprenta Artística Española). 1910. Oct. (*Carofit. España*).
Publ.: 1910 (p. 7: 15 Sep 1910), p. [1]-206. *Copy*: G.
Ref.: Reynier, A., Bull. Soc. bot. France 61: 439-440. 1915.

9119. *Las estepas de España y su vegetación* ... Madrid (Est. Tip. "Succesores de Rivadeneyra" ...) 1915. Oct. (*Estepas España*).
Publ.: 1915, p. [1]-304, 28 photos in text. *Copy*: US. – For further details see BL 2: 483.
Ref.: Vaupel, F., Bot. Jahrb. 57 (Lit.): 25-26. 1922.

Reyger, Gottfried (1704-1788), German (Danzig) naturalist. (*Reyger*).

HERBARIUM and TYPES: Unknown.

BIBLIOGRAPHY and BIOGRAPHY: Barnhart 3: 148 (b. 4 Nov 1704, d. 29 Oct 1788); BM 4: 1692, 2289; Herder p. 187; Kew 4: 454; MD p. 208-209; PR 7581-7583, ed. 1: 8533-8535; Rehder 1: 373, 721; SO 647a, 1219.

BIOFILE: Fries, Th.M., Bref Skrifv. Linné 1(6): 55. 1912.
Poggendorff, J.C., Biogr.-lit. Handw.-Buch 2: 616. 1863 (bibl.).

9120. Gottfried Reyger ... *Tentamen florae gedanensis* methodo sexuali adcommodatae ... Dantisci [Danzig] (apud Daniel Ludwig Wedel) 1764-1766, 2 vols. Oct. (*Tent. fl. gedan.*).
[*1*]: 1764 (p. 12: v. kalend. mart. 1763), p. [1]-293, [294-295, err.].
2: 1766, p. [i]-xii, [1]-200. "Tomus ii. Accessit Joannis Philippi Breynii vita Christiani Menzelii Centuria plantarum gedanensium". Dantisci (id.) 1766. Oct. The Menzel reprint on p. [201]-224 has a separate t.p. ([201]): *Centuria plantarum circa nobile Gedanum* sponte nascentium ... Dantisci. Typis Andreae Hünefeldii. 165. Nunc denuo impressa 1766".
Copies: HH, M, NY, USDA; IDC 220.

9121. *Die um Danzig wildwachsende* [sic] *Pflanzen* nach ihren Geschlechtstheilen geordnet, und beschrieben von Gottfried Reyger ... Danzig (bey Daniel Ludewig Wedel) 1768. Oct. (*Danzig wildw. Pfl.*).
Ed [*1*]: 1768, p. [i-xiv], [1]-431. *Copies*: HH, M.
Ed. 2: 1825-1826, 2 vols. *Copies*: "Neue ganz umgearbeitete und vermehrte Auflage. Von Johann Gottfried Weiss ..." Danzig (Im Verlag der S. Anhutschen Buchhandlung ...) 1825-1826. Oct. – Published by Johann Gottfried Weiss (x-1832).
1: 1825 (rev. Flora 21 Sep 1825), p. [i]-cvii), [viii], [1]-541, [542-546], *pl. 1-3*.
2: Jan-Mar 1826, p. [i-iv], [1]-432, [i]-lxviii, [lix-lxviii].
PR: "Ein ganz neues, völlig werthloses Buch. Der zweyte Theil ist aus Schlechtendal's Flora berolinensis fast völlig wörthlich übersetzt;" Schlechtendal (D.F.L. von, in Linnaea 1: 108, 276. Apr 1826): "[part 2] scheint nur eine Uebersetzung von Schlechtendal's Flora Berol. II. Hieraus gibt sich des Buches Werth und Wichtigkeit. Die Kupfertafeln sind schlecht, Kopien aus Weber und Mohr's Taschenbuch. Wozu werden solche Floren geschrieben". See also Flora 8: 553-555. 1825.

Reymond, M.L.C. (*fl.* 1854), French clergyman and botanist at Paris. (*Reymond*).

HERBARIUM and TYPES: Unknown.

BIBLIOGRAPHY and BIOGRAPHY: BM 4: 1692; Jackson p. 276; PR 7584; Rehder 3: 86.

EPONYMY: *Reymondia* H. Karsten (1848) is an *orth. var.* of *Duboisreymondia* H. Karsten (1848) dedicated to E. Dubois-Reymond.

9122. *Flore utile de la France* d'après le système de Linné modifié par Richard, comprenant la description de tous les genres et de toutes les espèces de plantes employées en médecine, dans les arts et dans l'économie domestique, avec un dictionnaire des noms vulgaires, mise à la portée de tous par la suppression des termes scientifiques remplacés par des mots vulgaires, ... Paris, Lyon (Imprimerie et librairie ecclésiastiques de Guyot frères, ...) 1854. Duod. (*Fl. utile France*).
Publ.: 1854 (p. iv: 24 Dec 1853), p. [i*-iv*], [i]-iv, [v], [1]-593. *Copies*: HH, NY.

Reynolds, Gilbert Westacott (1895-1967), Australian-born (of English parents) South-African botanist; specialist in *Aloe*; educated at St. John's College, Johannesburg; served in the 1914-1918 war; practicing as an optician 1917-1960; in retirement in Swaziland 1960-1967. (*Reynolds*).

HERBARIUM and TYPES: PRE; isotypes in EA, K, P, SRGH, TAN; further material at NBG, NY.

BIBLIOGRAPHY and BIOGRAPHY: Bossert p. 330 (d. 7 Apr 1967); Hortus 3: 1202; IH 1 (ed. 7): (in press), 2: (in press); Roon p. 94.

BIOFILE: Anon., The flowering plants of Africa 34(3/4): frontisp., portr. 1961.
Bullock, A.A., Bibl. S. Afr. Bot. 94. 1978 (bibl.).
Exell, A.W. & G.A. Hayes, Kirkia 6(1): 100. 1967 (coll. Fl. Zambesiaca).
Guillarmod, A.J., Fl. Lesotho 54, 64. 1971.
Gunn, M. & L.E. Codd, Bot. expl. S. Afr. 295-297. 1981 (journeys, coll., epon.).
Kimberley, M.J., Excelsa, Salisbury 1: 3-6. 1971 (obit., portr., b. 10 Oct 1895, d. 7 Apr 1967).
L.C.L., Cactus Succ. J. Great Britain 29(3): 41. 1967 (obit.).
Parr, Bull. Afr. Succ. Pl. Soc. 9(2): 69-71. 1974 (biogr. sketch, portr.).
Reynolds, G.W., communication to senior author. 1956.
Stieber, M.T. et al., Huntia 4(1): 85. 1981 (arch. mat. at HU).
Verdoorn, F. et al., Chron. bot. 3: 252. 1937.
White, A. & B.L. Sloane, The Stapelieae ed. 2. 21 [index, many entries]. 1937.

EPONYMY: *Reynoldsia* A. Gray (1854) was named for J.N. Reynolds who collected plants in Chile in the early 19th century.

NOTE: Reynolds' two standard works on African *Aloe* fall outside our period:
(1) *The Aloes of South Africa*, Johannesburg, South Africa, xxiv 520 p., 77 col. pl., published December 1950 (private comm. by author), see also Taxon 1: 30. 1951.
(2) *The Aloes of tropical Africa and Madagascar*, Mbabane, Swaziland, xxii, 537 p., *106 col. pl.* 1966; see Taxon 16: 59, 206. 1967.

Rheede tot Draakestein, Hendrik Adriaan van (1637-1691), Dutch colonial administrator and botanist; employee of the Dutch East India Company; governor of Malabar 1669-1676; in the Netherlands 1677-1684; from 1684 head representative of the East Indian Company in India. (*Rheede*).

HERBARIUM and TYPES: Unknown. – The herbarium of unknown origin acquired by GOET in 1750, called *Planta malabaricae* has no connection with Rheede's *Hortus malabaricus*. It was probably collected by a Dr Hugo, a physician at the Electoral Court of Hanover in the late 17th century (see M.C. Johnston, 1970).

BIBLIOGRAPHY and BIOGRAPHY: Aa 16: 299; Backer p. 490; Barnhart 3: 149 (d. 15 Dec 1691); BM 4: 1692-1693; Bossert p. 330; Dryander 3: 179; Frank 3 (Anh.): 84; Henrey 2: 737, 3: 1271; Herder p. 212; Jackson p. 338; JW 2: 199, 3: 366, 5: 250; Kew 4: 456; Lasègue p. 429, 503, 528; Moebius p. 141, 142; NI 1625, 1: 84-86, 90, 3: 6, 57; PR 2288, 7585-7587, ed. 1: 2638, 8540-8542; Rehder 1: 470; SK 1: lxxxiii, 4: lxxxiii-lxxxiv, clx; SO add. 663*; TL-1/280, 1070, see R. Pulteney; TL-2/566, 1368, 1474, 2472, 2777; Tucker 1: 594.

BIOFILE: Burkill, I.H., Chapt. hist. bot. India 6, 7, 8, 9, 25, 53, 177. 1965.
Busken Huet, C., Het land van Rembrandt ed. 2, 2(2): 69-70.
Candolle, A. de, Phytographie 443. 1880 (on coll. at GOET).
Ewan, J. & N., John Banister 101. 1970.
Fournier, H., in K.S. Manilal et al., Botany and history of Hortus malabaricus. 1980, p. 10.
Fries, Th.M., Bref Skrifv. Linné 1(2): 186. 1908, 1(5): 252. 1911.
Hulth, J.M., Bref Skrifv. Linné 1(8): 18. 1922.
Jessen, K.F.W., Bot. Gegenw. Vorz. 2, 273, 495. 1884.
Johnston, M.C., Taxon 19: 655. 1970 (on coll. at GOET).
Lawrence, G.H.M., ed., Adanson 1: 75, 189, 364. 1963
Veendorp, H. & L.G.M. Baas-Becking, Hortus acad. lugd.-bot. 79, 80-82, 113. 1938.
Kalff, S., Elsevier's geill. Maandschr. 29: 241-257, 312-322. 1905 ("De maecaenas van Malabar"; general biogr. acc.).
King, G., J. Bot. 37: 454-455. 1899.
Mabberley, D.J., Taxon 29: 605. 1980.
Majumdar, N.C. & D.N. Guka Bakshi, Taxon 28: 353-354. 1979 (on Linnaean names typified by ill. in *Hort. malab.*).
Manilal, K.S., ed., Botany and history of Hortus malabaricus, New Delhi 1980 (23 papers on Rheede, the Hort. malab. and modern nomenclature of plants treated by R.).
Merrill, E.D., Chron. bot. 14: 313, 351, 355. 1954.
Poiret, J.L.M., Enc. méth. 8: 754. 1808.
Saccardo, P.A., Malpighia 17: 269. 1902 (lists studies).
Sirks, M.J., Ind. natuurond. 13-24, 42, 44, 60, 62, 185. 1915.
Stafleu, F.A., Linnaeus and the Linneans 133, 166. 1971.
Thothathri, K. & K.K.N. Nair, Taxon 30(1): 43-47. 1981 (Dalbergias in Hort. malab.).
Verdoorn, F., ed., Chron. bot. 1: 381. 1935, 3: 8. 1937, 4: 403. 1938.
Veth, P.J., De Gids 51(3): 423-475, 51(4): 113-161. 1887 (detailed biogr. based on original sources and archives).
Warner, M.F., J. Bot. 58: 291-292. 1920 (on dates of publication).

EPONYMY: *Rhedia* Batsch (1802, *orth. var.*); *Rheedea* Cothenius (1790, *orth. var.*); *Rheedia* Linnaeus (1753); *Rheedja* Linnaeus (1754, *orth. var.*).

9123. *Hortus indicus malabaricus*, continens regni malabarici apud Indos celeberrimi omnis generis plantas rariores, latinis, malabaricis, arabicis, & bramanum characteribus nominibusque expressas, unà cum floribus, fructibus & seminibus, naturali magnitudine à peritissimis pictoribus delineatas, & ad vivum exhibitas. Addita insuper accuratâ earundem descriptione, quâ colores, odores, sapores, facultates, & praecipuae in medicinâ vires exactissimè demonstrantur. Adornatus per Henricum van Rheede, van Draakenstein, ... et Johannem Casearium ... notis adauxit, & commentariis illustravit Arnoldus Syen, ... Amstelodami (sumptibus Joannis van Someren, et Joannis van Dyck) 1678[-1703], 12 vols. Fol. (*Hort. malab.*).

1: 1678, p. [i-xvi], 1-102, [103, index], *pl. 1-57* (uncol. copp. engr.). Engr. t.p. also dated 1678. – Copies occur dated 1686 (see below). Preface material includes engr. t.p. and portrait).

2: 1679, p. [i-viii], 1-110, [111, index], *pl. 1-56*. -Pars secunda de fructicibus ... per Henricum van Reede [sic], tot Draakestein, ... et Johannem Casearium ... notis adauxit, ... Joannis Commelinus [Jan Commelijn, 1629-1692].

3: 1682, engr. t.p. , p. [i-xxiv], p. 1-87 [88, index], *pl. 1-64*. – Pars tertia de arboribus regni malabarici, ... per Henricum van Reede, tot Draakestein, ... et Johannem

Munnicks, ... notis adauxit, ... Johannes Commelinus ... Amstelodami (sumptibus viduae Joannis van Someren, haeredum Joannis van Dyck, & Henrici & viduae Theodori Boom) 1682.

4: 1683, (copies occur with err. date 1673, e.g. BR, NY, U, p. [i-iv], 1-125, [127, index], *pl. 1-61*. – Pars quarta de arboribus regni malabarici, ... [rest as vol. 3] 1783.

5: 1685, p. [i-viii], 1-120, [121, index], *pl. 1-60*. – Pars quinta de arboribus et fruticibus bacciferis regni malabarici, ... [rest as vol. 4] 1785.

6: 1686, p. [i-viii], 1-109, [111, index], *pl. 1-61*. – Pars sexta de varii generis arboribus et fruticibus siliquosis ... per ... Henricum van Rhede tot Draakestein, ... et Theodorum Janson. ab Almeloveen, M.D. Notis adauxit, ... Joannes Commelinus Amstelaedami (id.) 1786.

7: 1688, p. [i-iv], p. 1-111, [113, index], *pl. 1-59*. – Pars septima, de varii generis fruticibus scandentibus; ... per ... Henricum van Rhede tot Draekestein, ... notis adauxit, ... Joannes Commelinus. In ordinem redegit, & latinate donavit Abrahamus à Poot M.D. Amstelaedami (id.) 1688.

8: 1688, p. [i-iv], 1-97 [98, index], *pl. 1-50*. [NY: *51*]. – Pars octava, de varii generis herbis pomiferis & leguminosis; ... per ... Henricum van Rhede tot Drakestein, ... notis adauxit, ... Johannes Commelinus in ordinem redegit, & latinitate donavit Abrahamus à Poot, M,D. Amstelaedami (id.) 1688.

9: 1689, p. [i-viii], 1-170, [171, index], *pl. 1-87*. – Pars nona, herbis et diversis illarum speciebus ... per ... Henricum van Rhede tot Drakestein, ... in ordinem redegit & latinitate donavit Abrahamus A Poot M.D. Notis adauxit, ... Joannes Commelinus. Amstelaedami (id.) 1689.

10: 1690, p. [i-iv], p. 1-187, [189,-190], *pl. 1-94* [NY: *93*]. – Pars decima, de herbis et diversis illarum speciebus ... per ... Henricum van Rhede tot Drakestein, ... in ordinem ... Abrahamus A Poot M.D. Notis adauxit, ... Joannes Commelinus. Amstelaedami (id.) 1690.

11: 1692, p. [i], 1-133, [134, index], *pl. 1-65*. – Pars undecima, de herbis et diversis illarum speciebus ... per ... Henricum van Rhede tot Drakestein, in ordinem redegit ... Abrahamus A Poot M.D. Notis adauxit, ... Joannes Commelinus, ... Amstelaedami (id.) 1692.

12: 1693 (rev. Phil. Trans. 18: 276-280. 1694; t.p. date to be read 1693), p. [i], 1-151, [152-160 indexes], *pl. 1-79*. – Pars duodecima, & ultima, de herbis et diversis illarum speciebus ... per piae memoriae nobilissimum ac generosissimum d.d. Henricum van Rhede tot Drakestein, ... in ordinem redegit ... Abrahamus A Poot M.D. Notes ex parte adauxit, & commentariis illustravit Joannes Commelinus, ... accedit, praeter hujus partis specialem, generalis totius operis index. Amstelaedami (id.) 1693. The book must have been nearly ready by 1692 see Warner (1920) and Fornier (1980). The title-page date should be read [md]cxciii rather than [md]cciii, (D.H. Nicolson).

Copies: BR, G (analysis above based on copy G), NY, PH, TCD, U, USDA; IDC 8304.

Title-page vol. 1: The copies at G and U (text above) have the original t.p., other copies(e.g. NY) may have a title page dated 1686 which essentially repeats that of vol. 6 but with "pars prima" rather than "pars sexta". For further details see Warner (1920).

Editors and authors of notes: Arnoldus Syen (1640-1678).
Johannes Casearius.
Joannes Commelinus (Jan Commelijn) (1629-1692).
Johannes Munnicks.
Theodorus Janson. ab Almeloveen.
Abrahamus à Poot.

Note: Rheede gave names in several languages and scripts. The names on the (794) plates are in Lat(in), Mal(abar), Arab(ic) or Bram. (Brahmi) script. In the text Rheede gives the name used by the Malabar people ("Malabarensis"), the Portuguese ("Lusitanis"), the Dutch ("Belgis"), the Malayans("Malaccis") and the names written in Brahmi ("Brahmanis", "lingua Bramanum", "Bramanum").

Studies on the Hortus Malabaricus:
Commelijn, C., *Fl. malab.* 1769, see TL-2/1184.
Burman, J., *Fl. malab.* 1769, see TL-2/933.
Hill, J., *Hort. malab.* 1774, see TL-2/2777.
Dennstedt, A.W., *Schlüssel Hortus malab.* 1818, see TL-2/1368.

Hamilton, Fr., *A commentary on the Hortus malabaricus*, Trans. Linn. Soc. 13(2): 474-560. 3-19 Dec 1822, 14(2): 171-312. 15 Nov 1825, 15(1): 78-152. 9 Feb 1827, 17: 147-252. 25 Mai 1837; IDC 8308, see also D.J. Mabberley, Taxon 26(5/6): 523-540. 1977.
Dillwyn, L.W., *Rev. Hortus malab.* 2 Apr 1839, see TL-2/1474.
Hasskarl, J.K., *Horti malabarici clavis nova* 1861-1862, see TL-2/2472 – and *Horti malabarici rheedeani clavis locupletissima*, Dresden 1867, Nova Acta Leop. vol. 33, 1867.
Warner, M.F., J. Bot. 58: 291-292. 1920 (dates of publ.; bibl.).
Manitz, H., Taxon 17: 496-501. 1968 (study of Dennstedt's *Schlüssel*).
Johnston, M.C., Taxon 19: 655. 1970 (no herb.; GOET coll.).
Manilal, K.S. et al., Taxon 26: 549-550. 1977 (report on re-investigation of plants described in Hort. malab.).
Manilal, K.S., ed., Botany and history of Hortus malabaricus, New Delhi 1980, 225 p.
Facsimile reprint: announced by Otto Koeltz Antiquariat, Mar 1979. [First volume, publ. Bishen Singh, available 1982].

9124. *Malabaarse Kruidhof*, vervattende het raarste slag van allerlei zoort van planten die in het koninkrijk van Malabaar worden gevonden. Nevens der zelver blommen, vruchten en saden. By een vergaderd door de Ed. Heer Henric van Rheede van Draakestein ... in het Latyn beschreven door Johannes Casearius, bediennar des Goddelyken Woords te Cochin. Met aantekeningen verrykt door Arnoldus Syen, ... vertaalt door Abraham van Poot, M.D. Eerste [tweede] deel handelende van de bomen. In 's Gravenhage (by Rutgert Alberts) 1720. Fol. (*Malab. kruidhof*).
1: 1720, engr. t.p., p. [i], portr., [iii-x], 1-39, [40, index], *pl. 1-57* (uncol. copp.).
2: 1720, p. [i], 1-29, [30], *pl. 1-56* (id.).
Copies: BR, NY.

Rhein, G.F. (1858-?), German botanist; Dr. phil. Kiel 1888. (*Rhein*).

HERBARIUM and TYPES: Unknown.

BIBLIOGRAPHY and BIOGRAPHY: Barnhart 3: 149; Rchdcr 2: 347.

9125. *Beiträge zur Anatomie der Caesalpiniaceen*. Inaugural-Dissertation zur Erlangung der Doctorwürde der philosophischen Fakultät zu Kiel vorgelegt von G.F. Rhein aus Beelitz (Mark). Kiel (Druck von H. Fiencke) 1888. Oct. (*Beitr. Anat. Caesalp.*).
Publ.: Apr-Mai 1888 (Bot. Zeit. 25 Mai 1888; Nat. Nov. Apr(1) 1888), p. [i, iii], [1]-25, [26-28]. *Copies*: FI, G. – Vita on p. [27] has b. 23 Jul 1858.

Rhind, William (*fl.* 1830-1860), British (Scottish) botanist; lecturer in botany, Marischal College, Aberdeen. (*Rhind*).

HERBARIUM and TYPES: Unknown.

BIBLIOGRAPHY and BIOGRAPHY: Barnhart 3: 149; BB p. 256; BM 4: 1693; CSP 5: 183, 8: 741; Desmond p. 517; Jackson 39, 193, 499; Kew 4: 456; NI 1628; PR 7588-7589, ed. 1: 8543-8546; Rehder 1: 86-95.

BIOFILE: Allibone, S.A., Crit. dict. Engl. Lit. 2: 1783. 1878 (bibl.).
Freeman, R.B., Brit. nat. hist. books 295. 1980.

9126. *A history of the vegetable kingdom*; embracing the physiology, classification and culture of plants, with their various uses to man and the lower animals; and their application in the arts, manufactures, and domestic economy Glasgow ... Edinburgh and London (Blackie & Son, ...) [1840-1841]. Qu. (*Hist. veg. kingd.*).
Publ.: 1840-1841 (fide Jackson, p. 499), portr. Linnaeus, engr. t.p., p. [i-iii], ii [=iv]-x, [xi], [1]-711, *pl. 1-12* (uncol., J. Stewart). *Copies*: E (Natl. Libr. Scotland), NY(2).
New edition: 1855, portr. Linnaeus, engr. t.p., p. [i]-xii, [1]-720. *pl. 1-41* (W. Fitch, James Stewart, J.H. Lekeux). *Copies*: E, NY. – Essentially a reissue of the 1840-41 edition

with 29 new plates and an appendix. Plates 19-39 are coloured. – "*A history of the vegetable kingdom*; embracing the physiology of plants with their uses …". Glasgow, Edinburgh and London (id.). 1855. Qu.
Re-issue: 1857, identical with 1855 ed. except for t.p. date *Copies*: MO, NY. – *pl. 19-40* col.
Re-issue: 1860, id. *Copy*: G (*pl.* incompl.). •
Re-issue: 1862, id. *Copy*: USDA.
Revised ed.: 1868, p. [i]-xvi, [1]-744, *pl.1-45*, several col. Supplement on p. 689-712. *Copy*: E (Natl. Libr. Scotland). – "Revised edition, with supplement". London, Glasgow and Edinburgh (id.) 1868.
Reissue of 1868 ed.: 1870, portr. Linn., p. [i]-xvi, [1]-744, *pl. 1-45* (pl. 6, 8, 10, 13-14, 17-19, 27-40 col.). *Copy*: MO.
Re-issue: 1874, p. [i]-xvi, [1]-744, *pl. 1-45*. *Copies*: E (Natl. Libr. Scotland), NY, USDA.
Re-issue: 1877, portr. Linnaeus, engr. t.p., p. [i]-xvi, [1]-744, *pl. 1-45*. *Copies*: BR, NY.

Rhiner, Joseph (1830-1897), Swiss botanist, philologist and private teacher at Schwyz. (*Rhiner*).

HERBARIUM and TYPES: Unknown.

BIBLIOGRAPHY and BIOGRAPHY: Barnhart 3: 149; BFM no. 993; BL 2: 564, 581, 708; BM 4: 1694; DTS 1: 245, xxiii; Jackson p. 345; Kew 4: 456; Rehder 5: 722; Saccardo 2: 91.

BIOFILE: Wartmann, B., Ber. St. Gall. naturw. Ges. 1897/98: 30-32. 1899 (obit., d. 10 Jan 1897; the incidental non-botanical remarks may point at psychological instability, his botany, however, is fully reliable).

9127. *Volksthümliche Pflanzennamen der Waldstätten* nebst Gebrauchs– und Etymologieangaben. Für Landwirthe und Gelehrte zusammengestellt von Jos. Rhiner, Philolog und Botaniker. Schwyz. [Druck von Gebr. Triner] 1866. Qu. (*Volksthüml. Pfl.-Nam. Waldst.*).
Publ.: 1866 (p. vi: 4 Jun 1866), p. [i]-viii, [1]-104. *Copy*: G.
Nachträge: (1)*in* Prodr. Waldst. Gefässpfl. 1870, p. 197-200, see below.
(2) *in* Tabell. Fl. schweiz.-Kant. 1869, p. 200-211.
(3) Jahresber. St. Gallen naturf. Ges. 1893/94: 208-214. 1895.

9128. *Abrisse zur tabellarischen Flora der Schweizer Kantone.* Bulletins de la flore tabellaire des cantons suisses. Jos Rhiner. Schwyz. (Druck und Verlag von (chez) J. Bürgler) 1868 [i.e. 1869]. Qu. (in twos). (*Abrisse tabell. Fl. Schweiz. Kant.*).
1868: publ. 1869 (p. 48 dated 12 Mar 1869), p. [1]-48. *Copies*: BR, G, NY.
1869: publ. 1869, p. [i]-iv, [1]-67. *Copies*: BR, G, HH, NY. – Title on cover: "*Tabellarische Flora der Schweizer-Kantone* sammt standörtlichen Abrissen. Flore tabellaire des cantons Suisses, suivie de bulletins phytostatiques".
Ed. 2: Sep 1897 (ÖbZ for Sep 1897; Nat. Nov. Nov(2) 1897), p. [i]-iv, [1]-64. *Copies*: G, HH.– "*Tabellarische Flora der Schweizer-Kantone.* Zweite Auflage. Flore tabellaire des cantons suisses. Deuxième édition … Schwyz (Im Selbstverlag (chez l'auteur) à Fr. Druck von A. Kälin) 1897. Qu. (*Tabell. Fl. Schweiz.-Kant.*). – See also Ber. St. Gall. naturw. Ges. 1890/91: 118-255. 1892, 1894/95: 173-296. 1896, 1897/98: 283-332. 1899 and below sub *Abrisse* (Esquisse complémentaires) *zur zweiten Tabellarischen Flora*.
Ref.: Roth, E., Bot. Centralbl. 74: 282-283. 18 Mai 1898 (rev.)

9129. *Prodrom der Waldstätter Gefässpflanzen* von Jos. Rhiner. Schwyz (Druck und Verlag von J. Bürgler) 1870. Qu. (*Prodr. Waldst. Gefässpfl.*).
Publ.: 1870 (p. x: 30 Aug 1869), p. [i]-x, [1]-218. *Copies*: BR, G, HH, NY.
Erster Nachtrag: Jan 1872, p. [1]-32. *Copies*: BR,·G, HH, NY.
Ed. 2: see below, *Gefässpfl. Urkantone Zug*, 1870.

9130. *Abrisse* (esquisses complémentaires) *zur zweiten tabellarischen Flora der Schweizerkantone* … St. Gallen (Zollikofer'sche Buchdruckerei) 1892-1899. Oct.
[1]: Jun 1892 (p. iv: Neujahr 1892; Bot. Centralbl. 26 Jul 1892; ÖbZ Jun 1892; Nat. Nov. Sep(2) 1892), p. [i]-iv, [1]-134, [1]. – Reprinted and to be cited from Jahresber. St. Gall. naturw. Ges. 1890/91: 118-255. 1892.

[2]: *Serie 1896*: Aug 1896 (Bot. Centralbl. 9 Sep 1896; Bot. Zeit. 16 Sep 1896; ÖbZ Aug 1896; Nat. Nov. Sep 1896), p. [1]-124. Id. 1894/95: 173-296. 1896. – St. Gallen (Verlag von A. & J. Köppel) 1896.
[3]: *Dritte serie*: 1899 (Nat. Nov. Nov(2) 1899), p. [1]-52. Id. 1897/98: 283-332. 1899. – St. Gallen (Verlag von A. & J. Köppel) 1899.
Copies: BR, G(2), HH (no. 1).
Ref.: Appel, F.C.L.O., Bot. Centralbl. 53: 259. 1893 (rev.).

9131. *Die Gefässpflanzen der Urkantone von Zug* verzeichnet von Jos. Rhiner. Zweite Auflage ... St. Gallen (Verlag von A. & J. Köppel) 1893-1895, 3 Hefte. Oct.
Publ.: 1893-1895, in 3 parts, in the Jahresber. der St. Gallischen naturw. Ges. and reprinted with independent pagination. *Copies*: BR, G, NY. – The "Zweite Auflage" refers to the *Prodr. Waldst. Gefässpfl.* of 1870 as its first ed.
1: Jan-Jun 1893 (Rhiner mailed a copy to Crépin at BR on 12 Jun 1893; Crépin recieved it on 28 Jun 1893; Bot. Centralbl. 28 Jun 1893; Bot. Zeit. 16 Jul 1893; Nat. Nov. Jul(1) 1893), p. [1]-124, [1], repr. from Jahresber. 1891/92: 147-271. 1893.
2: 1894 (Bot. Centralbl. 14 Aug 1894; Bot. Zeit. 16 Aug 1894; Nat. Nov. Aug(1) 1894), p. 125-210, repr. id. 1892/93: 175-260.
3: 1895 (Bot. Zeit. 16 Aug 1895; Nat. Nov. Jul(2) 1895), p. [211]-314, repr. id. 1893/94: 111-207.

Rhode, Johann Gottlieb (1762-1827), German (Prussian Saxony) palaeobotanist; Dr. phil. Breslau 1821; private teacher in Braunschweig, "Hofmeister" in Esthland, later in Berlin, theater director in Breslau, ultimately professor of geography and German language at the military school of Breslau. (*Rhode*).

HERBARIUM and TYPES: Unknown. – Some lichens possibly attributable to Rhode at M (see Hertel 1980).

BIBLIOGRAPHY and BIOGRAPHY: ADB 28: 391; Barnhart 3: 149 (d. 23 Aug 1827); BM 4: 1694; GR p. 124; Herder p. 352; Jackson p. 176; LS 22175; NI 1629; PR 7590, ed. 1: 8545; Rehder 3: 722.

BIOFILE: Andrews, H.N., The fossil hunters 61. 1980.
Hertel, H., Mitt. Bot. München 16: 422. 1980 (lichens from France at M; possibly [?] J.G. Rhode).
Schmid, G., Goethe u.d. Naturw. 564. 1940.
Ward, Ann. Rep. U.S. Geol.-Survey 5: 404. 1885.

EPONYMY: *Rhodea* K.B. Presl (1838); *Rhodeites* Němejc (1937).

9132. *Beiträge zur Pflanzenkunde der Vorwelt.* Nach Abdrücken im Kohlenschiefer und Sandstein aus schlesischen Steinkohlenwerken, ... Breslau (bei Grafs, Barth und Comp.). [1821-1823]. Qu. (*Beitr. Pflanzenk. Vorw.*).
Publ.: in 4 parts, 1821-1823, p. iii-iv, [1]-40, *pl. 1-10* (liths.). *Copies*: BR, PH, Teyler. – No title page issued; title taken from cover of Heft 1 (Teyler).
1: late 1821 (p. iv: Oct 1820; Flora 14 Jan 1822), p. iii-iv,[1]-14, *pl. 1-2*.
2: Jan-Mai 1822 (Flora 7 Jan 1822), p. 15-28, *pl. 3-5*.
3/4: 1823 (Flora 21 Nov 1823), p. 29-40, *pl. 6-10*.
Ref.: Anon., Flora 5(1): 32. 14 Jan 1822 (fasc. 1), 5(1): 330-333. 7 Jun 1822 (fasc. 2), 6(2): 678-687. 21 Nov 1823 (fasc. 3/4) (reviews).
Sprengel, K., Neue Entd. 3: 371-373. 1822.

Rhodes, Philip Grafton Mole (1885-1934), British (Roman-catholic) clergyman, bryologist and mycologist; BA Cambridge, DD Fribourg; curate at Kidderminster, Worcs., professor of Ascott college, Birmingham; parish priest at Evensham. (*Rhodes*).

HERBARIUM and TYPES: mycological herbarium at K (5000); bryological material at BM (moss herb.), CGE, DBN, E, NMW; lichens at BIRA and M; some fungi at W.

BIBLIOGRAPHY and BIOGRAPHY: Barnhart 3: 149; Clokie p. 231; Desmond p. 517 (d. 16 Dec 1934); GR p. 412; IH 1 (ed. 7): 347, 2: (in press); Kew 4: 457; LS 38351, 41637; SBC p. 130.

BIOFILE: Bridson, G.D.R. et al., Nat. hist. mss. Brit. Isl. 84.21. 1980.
Hawksworth, D.L. & M.R.D. Seaward, Lichenol. Brit. Isl. 26, 139, 198. 1977 (lich. BIRA).
Hedge, I.C. & J.M. Lamond, Index coll. Edinb. herb. 123. 1970 (bryo.).
Hertel, H., Mitt. Bot. München 16: 422. 1980.
Jones, D.A., Brit. bryol. Soc., Rep. 1934: 246-247. 1935 (obit.); Proc. Linn. Soc. 147: 187. 1935 (obit.).
Sayre, G., Bryologist 80: 516. 1977.
Verdoorn, F., ed., Chron. bot. 2: 174, 178. 1936.

EPONYMY: *Rhodesia* W.B. Grove (1937).

Ricasoli, Vincenzo, barone (1814-1891), Italian (Tuscany) soldier and amateur horticulturist and botanist at Firenze; fought in the Crimea and played an important role in the Italian "risorgimento"; had a botanical garden on Monte Argentario (Port' Ercole). (*Ricasoli*).

HERBARIUM and TYPES: FI (over 8000); other material at E, M (lich.), TCD.

BIBLIOGRAPHY and BIOGRAPHY: AG 3: 388; Barnhart 3: 150 (b. 13 Feb 1814, d. 20 Jun 1891); Bossert p. 330; Jackson p. 434; Kew 4: 458; Langman p. 624; Morren ed. 10, p. 84; Rehder 5: 723; Saccardo 1: 137-138, 2: 91; Tucker 1: 595.

BIOFILE: Anon., Bot. Jahrb. 14 (Beibl. 32): 69. 1892 (d.).
Arcangeli, G., Bull. Soc. bot. ital. 1892(1): 11-16 (obit., b. 13 Feb 1814, d. 20 Jun 1891).
Fenzi, E.O. et al., Bull. R. Soc. Tosc. Ortic. 6: 193-225. 1881 (commemorative addresses and papers).
Tassi, F., Bull. Lab. Orto bot. Siena 7: 7, 33. 1905 (herb. at FI).
Steinberg, C.H., Webbia 34: 61. 1979 (herb. at FI).

COMPOSITE WORKS: Italian translation of F.G. Baker *Synopsis of Aloineae and Yuccoideae* (J. Linn. Soc. 18: 149-194. 15 Oct 1880, 195-241. 31 Dec 1880) *Rivista delle Yucche* ... (Bull. r. Soc. Tosc. Ortic. 7: 243-247, 270-278, 300-306, 340-343. 1881.

EPONYMY: *Ricasolia* G. De Notaris (1846); *Ricasolia* A. Massalongo (1846); *Ricasoliomyces* Thomas ex Ciferri et Tomaselli (1953).

9133. *Catalogo delle collezioni di piante* coltivate nel giardino del barone Bettine Ricasoli presso il Pellegrino fuori della Barriera di San Gallo. Firenze (Tipografia dei Successori Le Monnier) 1874. Oct. (*Cat. coll. piante Ricasoli*).
Publ.: 1874, p. [i-iii], [1]-98. *Copy*: M (inf. I. Haesler).

9134. *Della utilità dei giardini d'aclimazione* e della naturalizzazione delle piante. Esperimenti nel giardino della Casa bianca presso Port'ercole nel Monte Argentario ... Firenze (Tipografia di Mariano Ricci) 1888. Oct. (*Giard. d'acclimaz.*).
Publ.: 1888 (Bot. Zeit. 31 Mai 1889), p. [i]-xxvii, [1]-87. *Copies*: M, USDA.
Primo supplemento: 1890, p. [i]-viii, [1]-31. *Copy*: M. – "Primo supplemento dal Giugno 1888 al Giugno 1889". Firenze (id.) 1890. Oct.

4935. *Coltivazione* all'aria aperta *di pianti tropicali* e sub-tropicali ... Firenze (Tipografia di Mariano Ricci ...) 1890. Oct.
Publ.: 1890, p. [1]-30. *Copies*: FI, M. – Reprinted and to be cited from Bull. r. Soc. Tosc. Ortic. 15: 102-103(f.f.), 144-151, 171-174, 203-210. 1890.

Ricca, Luigi (x-1879), Italian customs officer and botanist on the French-Italian border in western Liguria. (*Ricca*).

HERBARIUM and TYPES: GE (fide Cavillier 1941); other material at FI.

BIBLIOGRAPHY and BIOGRAPHY: Barnhart 3: 150 (d. 15 Jul 1979 for 1879); BL 2: 363, 708; Kew 4: 458; Moebius p. 455; Roon p. 5: 723; Saccardo 1: 138, 2: 91.

BIOFILE: Burnat, É., Bull. Soc.-bot. France 30: cxxix. 1893 (d. 1881).
Cavillier, F., Boissiera 5: 79. 1941 (d. 15 Jul 1879).

9136. *Catalogo delle piante vascolari* spontanee della zona olearia *nelle due Valli di Diano Marina e di Cervo* [Estratto dagli Atti della Società italiana di scienze naturali. – Vol. xiii, fasc. ii, 1870]. Oct.
Publ.: 1870 (p. [3]: Mar 1870), p. [1]-84. *Copies*: G, HH. – Reprinted and to be cited from Atti Soc. Ital. Sci. nat. Milano 13: 60-143. 1870.

Rich, Obadiah (1783(1782?)-1850), American botanist and bibliographer; living in Boston until 1809, in Georgetown, D.C. and travelling 1809-1816; from 1816-1827 U.S. consul in Valencia, Spain; from 1829 in London as book-seller and antiquary. (*O. Rich*).

HERBARIUM and TYPES: Unknown.

BIBLIOGRAPHY and BIOGRAPHY: Barnhart 3: 150 (b. 25 Nov 1777, d. 20 Jan 1850); BM 4: 1695; LS 22186; ME 1: 224, 3: 374; Rehder 1: 306, 5: xvi; SO add. 802l; Tucker 1: 595.

BIOFILE: Ewan, J. et al., Short hist. bot. U.S. 74. 1969.
Hogan, J.R., The story of the Rich family 6(1, 1): 212-219. 1968; the Rich family association, Kinfolk 13(3-4): 9-14. 1978 (biogr. details); bapt. 13 Jul 1783).
Knepper, A.W., Bibl. Soc. Amer. Papers 49(2): 112-130. 1955 ("Obadiah Rich: bibliopole").
Ricker, P.L., Proc. Wash. Acad. Sci. 8: 491. 1918.
Rogers, D.P., Brief hist. mycol. N. Amer. 7-8. 1979.

NOTE: Sometimes errroneously listed as Oliver Rich, see Barnhart 3: 150. – Obadiah Rich is the author of *Bibliotheca americana nova. A catalogue of books relating to America* ... 2 vols. London 1846, 1: [i-v], [1]-517, 2: [1]-4, [1]-412. *Copies*: GOET (Bound with *Bibliotheca americana vetus*, [i], [1]-16,[1]-8, catalogue [1]-48), LC. – *Facsimile ed.* s.d., Burt Franklin, New York, with new title-page and verso (part 1 with 424 p.). – For the *Bibliotheca americana vetus* see e.g. A.W. Knepper (1955). These two bibliographies were preceded by *A catalogue of books relating principally to America,* ... 1832 (n.v.).

9137. *A synopsis of the genera of American plants,* according to the latest improvements, on the Linnaen [sic] system: with the new genera of Michaux and others. Intended for the use of students in botany ... Georgetown. District of Columbia (Printed by J.M. Carter, ...) 1814. Oct. (in fours). (*Syn. gen. Amer. pl.*).
Publ.: 1814, p. [i]-viii, [1]-167. *Copies*: HU, NY, USDA. – *Note on authorship*: The book is anonymous. The attribution to Rich stems from J.A. Brereton, who, in his *Fl. columb. prodr.* (TL-2/738) credits the synopsis to "O. Rich" on p. 7, see also p. 86. (Note by M.F. Warner, 23 Jun 1915, in USDA copy).

Rich, William (*fl.* 1830-1850), American botanist on the Wilkes Expedition (1838-1842); younger brother of Obadiah Rich. (*W. Rich*).

HERBARIUM and TYPES: US.

BIBLIOGRAPHY and BIOGRAPHY: Barnhart 3: 150; Ewan ed. 1: 168, 337; Lasègue 389, 502; ME 2: 443, 525, 652, 653, 3: 644; SK 1: 434.

BIOFILE: Barnhart, J.H., Bartonia 16: 33. 1934 (bot. corr. Schweinitz).
Ewan, J. et al., Short hist. bot. U.S. 42. 1969.
Graustein, J.E., Thomas Nuttall 332. 1967.
Smith, A.C., Flora vitiensis nova 1: 38. 1979.

Tyler, D.B., The Wilkes Expedition, Philadelphia 1968, p. 41, 71, 124, 157, 184, 200, 263, 267, 276, 297, 311-328, 350, 388-389, 413-414, 433.
Vallot, J., Bull. Soc. bot. France 29: 187. 1882 (pl. from Cape Verde isl.).

Rich, William Penn (1849-1930), American wholesale grocer in Boston 1864-1902; ardent amateur botanist; secretary and librarian of the Massachusetts Horticultural Society 1902-1923; member of the editorial board of Rhodora from 1899. (*W.P. Rich*).

HERBARIUM and TYPES: GH and NEBC.

BIBLIOGRAPHY and BIOGRAPHY: Barnhart 3: 151 (b. 7 Aug 1849, d. 30 Nov 1930); BL 1: 189, 315; CSP 18: 172; IH 2: (in press).

BIOFILE: Anon., Amer. Florist 43: 51. 1914 (portr.); Rev. bryol. 26(3): 56. 1899 (ann. Rhodora; W.P. Rich administrator); Science ser. 2. 72:595. 1930 (d.).
Day, M.A., Rhodora 3: 262. 1901 (on herb.).
Pennell, F.W. & J.H. Barnhart, Monogr. Acad. nat. Sci. Philadelphia 1: 617. 1935.

Richard, Achille (1794-1852), French botanist at Paris; Dr. med. Paris 1820; son of L.C.M. Richard; aide-démonstrateur (1817-1821) and aide-naturaliste (1821-1831) at the Muséum d'Histoire naturelle; professor of botany at the Paris Faculté de médecine in 1831; member Acad. Sci. from 1834. (*A. Rich.*).

HERBARIUM and TYPES: P and PC; a further important set at G and G-DC; further material at B, BR, C, CGE, FI, GH, H, LINN, NEU, W. – The herbarium of L.C.M. and A. Richard (father and son) is incorporated in the herbier général of the British herbarium. It contains many Richard types, but A. Richard in particular worked also on collections that came directly to Paris. Since these are all incorporated in the general herbarium it is usually not difficult to locate the types. The specimens from the Richard herbarium are easily recognizable by the labels. – The herbarium contained the original set of Sagra plants used for A. Richard's *Essai d'une flore de l'Île de Cuba*, as well as the first set of Dillon and Petit plants from Abyssinia. Some of Richard's original drawings and manuscripts are at the Bibliothèque centrale of the *Muséum* in Paris (mss. 566-567).

BIBLIOGRAPHY and BIOGRAPHY: Ainsworth p. 62, 313, 346; Backer p. 494; Barnhart 3: 151 (b. 27 Apr 1794, d. 5 Oct 1852); BL 1: 34, 53, 225, 315; BM 4: 1695; Bossert p. 331; Clokie p. 231-232; Colmeiro 1: cxcv; CSP 5: 185-186, 11: 168, 12: 615; Frank 3 (Anh.): 84; GF p. 73; GR p. 296; GR, cat. p. 70; Herder p. 473 [index]; Hortus 3: 1202; IF p. 726; IH 1 (ed. 6): 363, (ed. 7): 341, 2: (in press); Jackson p. 38, 48, 128, 134, 201, 347; Kew 4: 459; Langman p. 624-625; Lasègue p. 581 [index]; Lenley p. 343; LS 22187; Moebius p. 139; MW p. 413; NI 1630-1632; PR 3647, 7594-7604, 7609, 7610, 7973, ed. 1: 1821, 3164, 8341, 8550-8558, 8564-8565, 10553; Rehder 5: 723; RS p. 137; Saccardo 1: 138, 2: 91; SBC p. 130; SK 4: cvii; SO 2482; TL-1/see p. 524 [index]; TL-2/669, 1556, 1975, 2211, 4962, 7096, and see the indexes; Tucker 1: 595; Zander ed. 10, p. 707, ed. 11, p. 806.

BIOFILE: Anon., Bonplandia 1: 34. 1853 (succession Richard), 2: 267. 1854 (cognomen Leopoldina: Perrault), 4: 282. 1856 (Franqueville, who bought R's herb., has now added that of Steudel); Bot. Zeit. 10: 831. (d.), 14: 151-152, (herb. for sale), 599. (Franqueville, who bought the Richard herbarium for 10.000 francs, has now added that of Steudel to his collections). 1856; Flora 14(2): 639. 1831 (gives up post at Muséum as aide-naturalist; prof. bot. Fac. med.), 36: 32. 1853 (d.); Österr. bot. W. 2: 406. 1852 (d.); Proc. Linn. Soc. 2: 243-244. 1855.
Boiteau, P., C.R. 100. Congr. natl. Soc. Sav. Paris 1975. Sci. 3: 24-25. 1976.
Brongniart, A., Bull. Soc. bot. France 1: 373-386. 1854 (biogr.); Biogr. membr. Soc. Agric. France 1848 à 1853: 399-434. 1865 (biogr., bibl.).
Candolle, A.P. de, Mém. souvenirs 117, 391, 416. 1862.
Cheeseman, T.F., Manual New Zealand flora xxii. 1906.
Fée, A., Nouv. biogr. gén., Didot 42: 187-189. 1866.
Franchet, A., Bull. Soc. bot. France 39: 270. 1892 (herb. Drake del Castillo combines those of Richard and de Franqueville).

Freeman, R.B., Brit. nat. hist. books 295. 1980.
Fries, Th.M. [& J.M. Hulth], Bref skrifv. Linné 1(3): 229. 1909, 1(7): 75-76, 90, 94. 1917.
Guillaumin, A., Les fleurs des Jardins *pl. 16*. 1929 (portr.).
Hainard, R., Trav. Soc. bot. Genève 9: 31. 1968 (portr.).
Hertel, H., Mitt. Bot. München 16: 423. 1980 (coll. Astrolabe (1825-1829) by A.P. Lesson sometimes err. marked as having been collected by Richard).
Hooker, W.J., J. Bot. Kew Gard. Misc. 8: 81-82. 1856 (Richard herbarium up for sale; list of main components).
Huxley, L., Life letters J.D. Hooker 1: 160, 423. 1918.
Jessen, K.F.W., Bot. Gegenw. Vorz. 305, 382, 403, 404, 417, 471-472. 1884.
Kukkonen, I., Herb. Christ. Steven 83. 1971.
Lacroix, A., Figures de savants 3: 95, 96, 4: 175. 1938.
Laundon, J.R., Lichenologist 11(1): 16. 1979.
Maiden, J.H., J. Proc. R. Soc. N.S.W. 44: 145. 1910.
Poggendorff, J.C., Biogr.-lit. Handw.-Buch 2: 629. 1863.
Schmid, G., Goethe u.d. Naturw. 267. 1940.
Stearn, W.T., Taxon 6(7): 186-188. 1952 (on his *Mém. Rubiac.*).
White, A. & B.L. Sloane, Stapelieae ed. 2, p. 96, 268, 851, 852. 1937.
Wittrock, V.B., Acta Horti Berg. 3(3): xl, 112. *pl. 52*. 1905 (portr.).

COMPOSITE WORKS: (1) Contributed to Bory de Saint-Vincent, J.G.B.A., *Dict. class. hist. nat.* 1822-1831, see TL-2/669.
(2) Co-author, J.B.A. Guillemin, *Fl. Seneg. tent.* 1830-1833, see TL-2/2211.
(3) Contributed to A.C.V.D. d'Orbigny, *Dict. univ. hist. nat.* 1839-1849, see TL-2/7096.
(4) Contributed Phanerogamae of Cuba to R. de la Sagra, *Hist. fis. nat. Cuba*, 10, 11, 1845-1850, q.v.
(5) Orchideae of Chile, in C. Gay, *Fl. chil.* vol. 5, 1849-1852, see TL-2/1975.
(6) Dumont d'Urville, J.S.C., *Voy. Astrolabe, Botanique* par MM. A. Lesson et A. Richard [part 2: par A. Richard] 1832-1834, 2 parts, see TL-2/1556. *Copies*: of 2 vols. text and atlas: G, HH, M, MO, NY. – Addition to collation vol. 1: p. [i*-v*], [i]-xvi, [xvii], [1]-376. See also Sotheby 642: some copies have some coloured plates, of which the number seems to vary from copy to copy.
(7) C.A. Lemaire, *Herbier général de l'amateur*, continuation(2), 5 vols. 1841-1850; see A. Richard, co-author, see TL-2/4962 under Loiseleur-Deslongchamps, J.L.A.

HANDWRITING: Webbia 32(1): 34. 1977 (no. 63); Candollea 33: 141-142. 1978.

9138. *Nouveaux élémens de botanique*, appliquée à la médecine à l'usage des élèves qui suivent les cours de la Faculté de médecine et du Jardin du Roi ... Paris (chez Béchet jeune, ...) 1819. Oct. (*Nouv. élém. bot.*).
Ed. [*1*]: 1819, p. [i*-iii*], [i]-xv, [1]-410, *pl. 1-8* with text (uncol.). Copies: G, HH, MO, NY.
Ed. 2: 1822, xviii, 487 p., *pl. 1-8*. (*n.v.*)
Ed. 3: 1825 (Linnaea 1(1), Jan 1826), p. [i]-xxiv, [1]-519, *pl. 1-8*. *Copy*: NY. – Troisième édition, revue, corrigée et augmentée ... Paris (chez Bechet jeune) 1825. Oct.
Ed. 4: 1828 (p. xiv: 1 Feb 1828), p. [i]-xxiv, [1]-593, 386bis (chart), *pl. 1-8*. *Copies*: BR, G, PH. – Quatrième édition, revue, corrigée et augmentée du caractère des familles naturelles du règne végétal; ... Paris (Bechet jeune, ...). Bruxelles (au Dépôt de la Librairie médicale française) 1828. Oct.
Ed. 5: 1833 (p. xiv: 1 Mar 1833), p. [i]-xxiv, [1]-456, [1]-256. *Copies*: G, MO, NY. – Title on p. [1] of second part: De la taxonomie, ou des classifications botaniques en général. – Paris (Béchet jeune, ...) [dépôts Bruxelles, Gand, Liège, Mons]. 1833. Oct.
Ed. 6 (Bruxelles): 1833, p. [i]-xv, [17]-296, *pl. 1-34* (G. Severeyns, uncol. liths.). *Copies*: GOET, NY. – "Nouveaux élémens de botanique et de physiologie végétale; sixième édition revue, corrigée et augmentée des caractères des familles naturelles du règne végétale ... contenant outre les 166 figures qui ornent l'édition originale, plus de 500 sujets propres à faciliter l'étude des familles." Bruxelles (H. Dumont, libraire-éditeur) 1833. Oct. (in fours). – Possibly an unlicensed edition, see also 1837 Bruxelles edition below.

Nouvelle édition par Drapiez: 1837, p. [i]-vii, [9]-343, *pl. 1-46* (some col.). *Copies*: BR, G. – "Nouveaux éléments de botanique et de physiologie végétale, avec le tableau méthodique des familles naturelles, par Ach. Richard, D.M.P., ... nouvelle édition, augmentée d'un précis des propriétés medicamenteuses des végétaux ou de leurs produits, d'âpres les meilleurs traités de matière médicale; et d'un grand nombre de figures pour aider à l'intelligence des caractères des familles du règne végétal; par Drapiez [Pierre Auguste Joseph Drapiez 1778-1856]. Bruxelles (Société typographique belge. Ad. Wahlen et compagnie ...) Londres (Dulau ...) 1837. Oct. (in fours). – Unlicensed edition.

Ed. 6 (Paris; regular): 1838 (p. xii: 1 Jan 1838; Flora 28 Aug 1838; Hinrichs 29 Mai-5 Jun 1838), p. [iii]-xii, [1]-756, *pl. 1-4* (uncol. liths. auct.). *Copies*: FI, NY. – Brongniart (1865) mentions 5 *pl.* – Paris (Béchet jeune, ...) 1838. Oct.

Ed. 7: 1846 (p. vi: 1 Sep 1845; Bot. Zeit. 31 Jul 1846), p. [i*-iii*], [i]-vi, [1]-851. *Copies*: BR, G, MO, NY. – Paris (Béchet jeune, ...) 1846. Oct.

Ed. [8]: 1852 (p. vii: 1 Feb 1852): 1: [i*-iii*], [i]-vii, [1]-332 (première partie, anat. gén.; 3-46; deuxième partie ou organographie végétale 47-327, index 329-332). 2: [i], [1]-301 (troisième partie taxon., classif. 1-27, quatrième partie phytogr., class. 28-295, index 297-301), auteurs: [303-304]).
Copies BR, MO. – The BR copy has the [old] title *Nouveaux éléments* ... Paris (Béchet jeune) [1852]. – The MO copy has a new title page, however, only the one that(classif.). This t.p. (which has the confusing indication "deuxième partie" for the second volume) is bound in front of the two volumes bound in one: *Précis de botanique et de physiologie végétale* ..." Paris (Béchet jeune ...) 1852. Duod. (in sixes). This title was used only for this edition.

Ed. 9: Mar-Mai 1864 (p. iii: Mar 1864; Soc. bot. France rd 13 Mai 1864; Flora 31 Aug 1864), p. [i*-iii*], [i]-vii, [1]-661. *Copies*: BR, FI, G, NY, USDA. – "Neuvième édition augmentée de notes complémentaires par Charles Martins" [Charles Frédéric Martins, 1806-1889] ... Paris (F. Savy, ...) 1864. 18-mo.

Ed. 10: 1870 (p. [v]: Feb 1870; BSbF rd 22 Jul 1870; Bot. Zeit. 1 Nov 1870), p. [i*-iii*], [i]-iv, [v], [1]-663. *Copies*: BR, G, NY. – Dixième édition, ... Charles Martins ... et pour la partie cryptogamique par Jules de Seynes [1833-1912], ... Paris (F. Savy, ...) 1870. Oct. – See also J. de Seynes, Bull. Soc. bot. France 17: 392-394. 1870.

Ed. 11: 1 Apr-12 Mai 1876 (p. xvi: Mar 1876; Soc. bot. France rd 12 Mai 1876), p. [i]-xvi, [1]-710. *Copies*: BR. – "Onzième édition Charles Martins ... Jules de Seynes ..." Paris (Librairie F. Savy ...) 1876. Oct.

Re-issue ed. 11: 1893, fide Bot. Zeit. 52(2): 96. 1894, as of "1893".

Dutch translation: 1831, p. [i], [1]-18, [1]-618, 1-8 (err.), *pl. 1-4* (uncol.). *Copy*: BR. – Achille Richard's ... *Nieuwe beginselen der kruidkunde* en der planten-natuurleer, naar de vierde fransche uitgave vertaald door Hector Livius van Altena, ... met aanteekeningen en bijvoegsels van Claas Mulder [1796-1867], ... te Franeker (by F. Ypma) 1831. Oct. – Based on *Nouv. élém. bot.* ed. 4, 1828.

German translation [1]: 1828, p.[i*], [i]-xxviii, [1]-646, *pl. 1-8* (uncol. liths). *Copy*: M. – "Achilles Richard's *Neuer Grundriss der Botanik* und der Pflanzenphysiologie, nach der vierten mit den Characteren der natürlichen Familien des Gewächsreiches vermehrten u. verbesserten Originalausgabe übersetzt, und mit einigen Zusätzen, Anmerkungen, einem Sach– und Wort-Register versehen von Mart. Balduin Kittel [Martin Baldwin Kittel, 1798-1885, see TL-2/3685-3688], ..." Nürnberg (bei Johann Leonhard Schrag) 1828. Oct. (*Neu. Grundr. Bot.*). – Based on *Nouv. élém. bot.*, ed. 4, 1828.

German translation, ed. 2: 1831, p. [i]-xxxii, [1]-804, *pl. 1-8* (id.). *Copy*: MICH. – "Zweyte, vermehrte und verbesserte Auflage ..." Nürnberg (id.) 1831. Oct. – Flora Jul 1832, review Anon. Flora 16 (Lit.): 235-237. 1831.

German translation, ed. 3: 1840, p. [i]-xxiv, [1]-1111, *16 pl.* (uncol. liths.). *Copies*: GOET, M. – "Achilles Richard's *Grundriss der Botanik* und der Pflanzenphysiologie, nach der sechsten französischen Originalausgabe frei bearbeitet von Martin Balduin Kittel, ... Dritte, vermehrte und verbesserte Auflage ..." Nürnberg (id.) 1840. Oct. (Grundr. Bot. ed. 3). – Based on Nouv. élém. bot. ed. 6 (1838), but "frei bearbeitet", freely revised.

English translation: 1831, by William MacGillivray [1796-1852], "*Elements of botany* and vegetable physiology, including the character of the natural families of plants, ... Edinburgh (William Blackwood), London (T. Cadell, ...) 1831. Oct., p. [i]-xxviii,

[1]-562. *Copies*: E (Univ. Libr.), NY. – We saw no copy of the *New elements of botany*, Dublin 1829 (see Freeman 1980).
Russian translation: Moskva 1835, n.v. fide Brongniart (1865).

9139. *Histoire naturelle et médicale des différentes espèces d'Ipécacuanha du commerce*, ... Paris (Chez Béchet jeune, ...) 1820. Qu. (*Hist. nat. méd. Ipécacuanha*).
Publ.: 1820, p. [i]-vi, [7]-72, *pl. 1-2* (uncol., auct). *Copies*: FI, G. – Thesis for Dr. med.

9140. *Monographie du genre Hydrocotyle de la famille des ombellifères*, ... à Bruxelles (de l'imprimerie de Weissenbruch, ...) 1820. Oct. (*Monogr. Hydrocotyle*).
Publ.: Apr-Jun 1820 (read 7 Feb; Acad. report 28 Feb publ. rd 12 Jun 1820), p. [1]-86, *pl. 52-67*. *Copies*: FI, G, HH, M, US. – Preprinted or reprinted from Ann. gén. Sci. phys. 4: 145-224, *pl. 52-67*. 1820. – The plates are uncol. liths. (and 2 copp. engr.) of drawings by the author and by Bessa & Huet.

9141. *Botanique médicale, ou histoire naturelle et médicale des médicamens, des poisons et des alimens, tirés du règne végétal*; ... à Paris (chez Béchet jeune, libraire, ...) 1823, 2 parts. Oct. (*Bot. méd.*).
1: Dec 1822 (BF 21 Dec 1823, Bull. Férussac Jan 1823), p. [i]-xiv, [1]-448.
2: Jun 1823 (BF 21 Jun 1823), p. [i-iii], [449]-817, 818, err.
Copies: BR, FI, G, NY.
Other editions: incorporated in Richard, A., Élém. hist. nat. méd. 1831 (vol. 1), 1838 (vol. 3) and 1849 (vol. 2).
German translation: 1824-1826, 2 vols. *Copies*: HH, NY. – "A. Richard's *medizinische Botanik*. Aus dem Französischen. Mit Zusätzen und Anmerkungen herausgegeben von Dr. G. Kunze [Gustav Kunze 1793-1851], ... und [vol. 1 only:] Dr. G.F. Kummer [physician at Leipzig]. Berlin 1824-1826, 2 parts. Oct. (*Mediz. Bot.*).
1: 1824 (p. viii: 12 Mai 1824), p. [i]-xiv, [1]-548.
2: 1826 (p. vi: 18 Mar) p. [i]-vi, [549]-1304, 1 chart.

9142. *Monographie de la famille des Élaeagnacées* ... Extrait des Mémoires de la Société d'Histoire naturelle de Paris [1(2) 1824]. Qu. (*Monogr. Élaeagn.*).
Publ.: 1 Mar –10 Apr 1824 (submitted 8 Dec 1823; BF journal issue 10 Apr 1824, see MD p. 180), p. [1]-34, *pl. 24-25* (uncol., auct.). *Copies*: FI, U. – Preprinted or reprinted from Mém. Soc. Hist. nat. Paris ser. 3. 1(2): [375]-408, *pl. 24-25*.

9143. *Monographie des Orchidées des Iles de France et de Bourbon*. Extrait d'un essai d'une flore des Iles de France et de Bourbon ... Paris (Imprimerie de J. Tastu) 1828. Qu. (*Monogr. Orchid. Iles de France Bourbon*).
Publ.: 1 Sep-8 Dec 1828 (in journal; t.p. Sep 1828, BF 8 Dec 1828; "lu, mais retiré" 16 Apr 1827, see PV 8: 519. 1918), p. [i-iii], [1]-83, *pl. 1-11* (uncol., auct.). *Copies*: BR, FI, G, NY. – Preprinted or reprinted from Mém. Soc. Hist. nat. Paris ser. 3. 4: 1-74, *pl. 1-11*. 1828. Type shifted (inf. R. Tournay):

Journal	preprint/reprint	journal	preprint/reprint
1-17	1-17	56-66	64-74
-	18 (blank)	67-71	75-80 (with blanks)
18-21	19-22	72-74	81-83
22-55	23-65 (with blanks)		

9144. *Mémoire sur la famille des Rubiacées contenant les caractères des genres de cette famille et d'un grand nombre d'espèces nouvelles*; ... Paris (imprimerie de J. Tastu, ...) Juillet 1829 [i.e. Dec 1830]. Qu. (*Mém. Rubiac.*).
Publ.: Dec 1830 (t.p preprint Jul 1829 but p. iv dated 1 Dec 1830; further details below), p. [i]-iv, [1]-224, *pl. 1-15* (uncol. copp. auct.). *Copies*: FI, G, HH(2), NY; IDC 5672 (journal-issue). – Preprinted with independent pagination from Mém. Soc. Hist. nat. Paris ser. 3. 5: [81]-304, *pl. 11-25*, publ. Jun 1834 (BF 28 Jun 1834).

The mémoire was read to the Academy of Sciences, Paris, on 7 Jul 1829; printing started soon afterwards. An abstract of this mémoire, as read before the Academy, appeared in Le Globe of 11 July 1829 on p. 438 [n.v.]. On 4 July 1831 the preprint was presented to the Academy (Procès Verbaux 9: 656. 1921) but this cannot have constituted publi- cation for Richard's preface in the completed reprint is dated 1 Dec 1830. Further details on the history of the printing of the book are found in the letters from A. Guillemin to A.P. de Candolle, now at Genève. Guillemin acted as an intermediary between the two authors who were both engaged on the Rubiaceae (de Candolle, Prodr. 4, Sep 1830, Rubiaceae). On 17 Jan 1830 Guillemin reports that 4 sheets had been printed ("généralités") and would be sent to de Candolle. On 8 Jun 1830 Guillemin, who took care of the printing of the fourth volume of the *Prodromus* at Paris, mentions to de Candolle that since the order of treatment of Richard is different, de Candolle would benefit only to a limited extent from the proofs. He sends the uncorrected proofs of the 20-23rd sheets. Richard's preprints were prepared in a very limited number for a few friends. Guillemin inserted the references to Richard in the proofs of the *Prodromus*. On 2 October 1830 Guillemin writes that he has at last received the first 15 copies of the fourth volume of the Prodromus from Wurtz (the printer and publisher).

The 15 copper-engravings are of drawings by A. Richard. Originals in Bibliothèque Centrale, Muséum d'Histoire naturelle, Paris.

Ref.: NI 1630, PR 7600, SK p. ccix.

Kuntze, O., Rev. gen. 3(2): 161. 1898.

Stearn, W.T., Taxon 6: 186-188. 1957; Fl. males. Bull. 14: 645. 1959.

Guillemin, A., Letters to A. de Candolle, Conservatoire Botanique, Genève.

9145. *Plantes nouvelles d'Abyssinie*, recueillies dans la province du Tigré par M. le Dr. Richard Quartin-Dillon, voyageur naturaliste du Muséum d'Histoire naturelle de Paris, ... [Extrait des Annales des Sciences naturelles (Novembre 1840)]. Oct.
Collector: Richard Quartin-Dillon (x-1841).
Publ.: Nov 1840, p. [1]-20. *Copy*: HH. – Reprinted and to be cited from Ann. Sci. nat., Bot. ser. 2. 14: 257-276. Nov 1840.

9146. *Observations sur le genre Quartinia*, ... [Extrait des Annales des Sciences naturelles (Mars 1841)]. Oct.
Publ.: Mar 1841, p. [1]-3, *pl. 14* (uncol. lith. auct.). *Copy*: HH. – Reprinted and to be cited from Ann. Sci. nat., Bot. ser. 2. 15: 179-181. *pl. 14*. Mar 1841.

9147. *Iconographie végétale* ou organisation des végétaux illustrée au moyen de figures analytiques par P.J.F. Turpin ... avec un texte explicatif raisoné et une notice biographique sur M. Turpin par M. A. Richard ... Paris (Imprimerie Panckoucke ...) 1841. Qu. (*Iconogr. vég.*).
Artist: Pierre Jean François Turpin (1775-1840).
Publ.: 1841, p. [i*-iii*], [i]-xii, [1]-144, *pl. 1-2, 2bis, 3-43, 43bis, 44, 44bis, 45-48, 48bis, 49-56, 56bis* (handcol. copp. Turpin). *Copy*: BR. – For a previous edition (*Essai iconogr. élém. philos. vég.* 1820) see under J.L.M. Poiret (*Leçons de Flore*, vol. 3), P.F. Chaumeton (*Fl. méd.* vol. 8) and under P.J.F. Turpin.

9148. *Monographie des Orchidées* recueillis dans la chaine *des Nil-Gherries* (Indes-Orientales) par M. Perrottet, ... décrites par A. Richard, ... Paris (Imprimé chez Paul Renouard ...) 1841. Qu. (*Monogr. Orchid. Nil-Gherries*).
Collector: George Samuel Perrottet (1793-1870).
Publ.: Jan 1841 (t.p. preprint; John Torrey received his copy in New York on 15 Jul 1842), p. [i], [1]-36, *pl. 1-12* (uncol., auct.). *Copies*: BR, FI, G, HH, NY. – Preprinted on large paper from Ann. Sci. nat., Bot. ser. 2. 15: 5-20 Jan 1841, 65-82, *pl. 1-12*. Feb 1841.

9149. *Monographie des Orchidées mexicaines*, précédée de considérations générales sur la végétation du Mexique et sur les diverses stations où croissent les espèces d'Orchidées mexicaines; par M.M. A. Richard et H. Galeotti, ... Extrait des Comptes Rendus des séances de l'Académie des Sciences, tome xviii, séance du 25 mars 1844 [Paris 1844]. Qu.

Co-author: Henri Guillaume Galeotti (1814-1858).
Publ.: 1844 (mss. submitted 25 Mar 1844), p. [1]-16. *Copies*: BR, FI. – Reprinted from C.R. Acad. Sci., Paris, 18: 497-513. 1844.

9150. *Histoire physique, politique et naturelle de l'Ile de Cuba* par M. Ramon de la Sagra ... *Botanique. – Plantes vasculaires*. Par Achille Richard ... Paris (Arthus Bertrand, éditeur, ...) 1845 [1841-1851]. Oct., atlas Fol. (*Hist. phys. Cuba, Pl. vasc.*).
Publ.: 1841-1851 (see below). The text was published also in the Spanish edition (*Hist. fis. Cuba, Fanerogamia*) as volume 10. A second botanical part by Richard came out only in the Spanish edition (vol. 11). We treat here only the French edition and refer to the treatment under Sagra, Ramon de la, for further details.
French ed.: p. [i]-viii, [1]-663.*Copies*: BR, G, NY, MICH; IDC 5707. – For the atlas (whole series) see under Sagra; it does not seem to have been finished before 1851. It remains unclear which edition has priority. Urban (1894) gives priority to the French edition, but Montagne (Ann. Sci. nat. ser. 2. vol. 17: 119. 1842) definitively states that "on ne publie les feuilles de cette édition (i.e. l'édition française) qu'après que celles de l'édition espagnole ont paru". This may refer only to the cryptogamic volume (see Montagne) but may also have applied to the phanerogamic part. Urban's argument is that the Spanish version contains corrections and additions to the French one. A. Richard writes to Bentham on 22 Mai 1843 that his work on the Sagra plants is finished "mais une partie seulement a été publiée." ... "Pour les mousses 21 feuilles ont été imprimées et publiées. 41 ou 42 planches in folio ont été publiées. Lors de mon voyage à Londres il y a quinze mois, j'ai remis à M. M. Brown, Hooker et Lindley le commencement de cette publication."
Brizicky gives the following dates for text and plates:

pages	*plates*	dates
1-336	*1-35*	1841
337-624	*36-44(2)*	1846
625-663, i-viii	*44(3)-89*	1851

A prospectus describing the French edition, and dated Mar 1838, is inserted in the copy at K. The (circa 50) livraisons were to contain four folio plates each accompanied by four sheets of text (Oct.). Each livraison had a printed cover. No copy in its original covers is known to us.
Ref.: BM 4: 1780-1781, NI 1712, 1713, RS p. 139, SK p. ccix.
Urban, I., Bot. Jahrb. 19: 563. 1894; Symb. ant. 1: 144-147. 1898.
Kuntze, O., Rev. gen. 3(2): 162. 1898.
Aguayo, Mem. Soc. Cubana Hist. nat. 18: 153-184. 1946.
Conde, J.A., Hist. bot. Cuba 88-89. 1958.
Steenis-Kruseman, M.J. van, Fl. males. Bull. 13: 741. 1960, 17: 910. 1962.
Brizicky, G., Arnold Arb. 43: 84-86. 1962.

9151. *Tentamen florae abyssinicae* seu enumeratio plantarum hucusque in plerisque Abyssiniae provinciis detectarum et praecipue a beatis doctoribus Richard, Quartin Dillon et Antonio Petit (annis 1838-1843) lectarum auctore Achille Richard ... Parisiis (apud Arthus Bertrand, ...) [1847-1851?]. Oct. (2 vols. text) and Fol. (atlas). (*Tent. fl. abyss.*).
Collectors: Richard Quartin-Dillon (x-1841), Antoine Petit (x-1843).
Publ.: 1847-1850, atlas 1851?, consisting vols. 4 and 5 plus atlas (bot.) of (alternative t.p.'s:) *Voyage en Abyssinie* exécutée pendant les années 1839, 1840, 1841, 1842, 1843 par une commission scientifique composée de MM. Théophile Lefebvre, ... A. Petit et Quartin-Dillon, ... Vignaud, ... publié par ordre du Roi ... troisième partie histoire naturelle. – Botanique par M. A. Richard ... tome quatrième [cinquième and atlas botanique] ..." Paris (id.). – "Partie" 3 of the entire work consists of volumes 4-5 of that same entire work. For a brief account of the *Voyage* as a whole see BM 2: 602.
1(= *Voyage 4*): 1847-1848, p. [i*-vii*], [i]-xi, [1]-472, of which i-xi and [1]-254 on 22 Mai 1847 (BF), p. 255-304 late 1847 or, more likely, Feb 1848, and p. 305-472 and [i*-vii*] on 26 Feb 1848. – Page 272 has a reference to Hooker J. Bot. 7: *pl. 5B*. 1848.
2(= *Voyage 5*): 1850, p. [i-v], [1]-518.

RICHARD, A., *Tent. fl. abyss.*

Atlas: possibly 1851, p. [i-iii], *pl. 1-53, 53bis, 54-102* (uncol., A.C. Vauthier).
Copies: BR, FI, G, HH, M, MO, NY, PH, USDA. – IDC 595.
Facsimile ed.: Uppsala 1982 (ISBN 91-506-0308-6), distr. by Institute of Systematic Botany, Uppsala, with an introduction by W.T. Stearn and an index to the species illustrated in the atlas.
Alph. de Candolle (in a letter to Miquel dated 2 Jan 1848) states that tome 4 was ready by the end of the year 1847. – Richard's treatment is based on material collected by R. Quartin Dillon and Antoine Petit on an expedition to Abyssinia in 1839-1843 sponsored by the Muséum d'Histoire Naturelle. Original set at P. – The 103 copper-plates are from drawings by A.C. Vauthier, the originals of which are also at the Bibliothèque centrale, Muséum d'Histoire naturelle, Paris.
Ref.: BM 2: 602, NI 1633, PR 7604, SK p. ccix.
  Sherborn & Woodward, Ann. Mag. nat. Hist. ser. 7. 8. 162. 1901.
  Steenis-Kruseman, M.J. van, Fl. males. Bull. 15: 741. 1960.
  Baur, Jahrb. Gesch. oberdeut. Reichsst. 16: 238. 1970.

Richard, [Jean Michel] Claude (1784-1868), French botanist and gardener who collected in Sénégal, 1816-1821; founder of the botanic garden Richard-Toll in Sénégal; travelled in Cayenne 1820-1824; from 1830 in charge of the botanical garden of Réunion; collected in Madagascar 1839-1840. (*C. Rich.*).

HERBARIUM and TYPES: plants from Madagascar at G, MO, TCD; from Réunion at B, FI, G, OXF, P; from Sénégal at BM, BR, FI, G, K, M, P, TAN. – The original herbarium which was left to the Muséum de Saint-Denis (Réunion) was destroyed by termites. – For plants attibuted to "Roger", but collected by Claude Richard see J.B. Gillett, Kew Bull. 15(3): 431- 435. 1962.

NOTE: No relation of A. or L.C. Richard.

BIBLIOGRAPHY and BIOGRAPHY: Barnhart 3: 151 (b. 15 Aug 1784); IH 2: (in press); Lasègue p. 83, 191, 302, 315; Tucker 1: 145, 595.

BIOFILE: Boiteau, P., C.R. 100. Congr. natl. Soc. Sav. Paris 1975. Sci. 3: 24-25. 1976.
Sayre, G., Mem. New York Bot. Gard. 19(3): 386. 1975 (coll.).
Vallot, J., Bull. Soc. bot. France 29: 187-188. 1882.

9152. *Catalogue du Jardin de la Réunion* [Saint Denis 1856]. Oct. (*Cat. jard. Réunion*).
Publ.: Nov-Dec 1856 (p. [iii]: Oct 1856), p. [i-v], [1]-113, [115, err.]. *Copy*: USDA. – Title on cover (fide Tucker), *Catalogue des végétaux cultivés au jardin* du gouvernement à l'île *de Réunion*.

Richard, Louis Claude Marie (1754-1821), French botanist, horticulturist and explorer; student of Bernard de Jussieu. Travelled in French Guyana (1781-1785), Brazil (1785) and the Antilles, returned to Paris 1789; professor of botany at the "École de Médecine", Paris 1795-1821; member of the Academy of Sciences. (*Rich.*).

HERBARIUM and TYPES: P and PC; other material at B-Willd., BR, C, CGE, FI, G, GH, H, LINN, NEU, W. – The Richard herbarium containing many of the types of L.C.M. and A. Richard came to P and PC via de Franqueville and the Drake del Castillo collections. It is incorporated in the herbier général but the specimens can easily be recognized by means of their labels. Richard is reported to have made a herbarium from the age of 12 onward. The collection from South America brought back in 1789 contained 3000 specimens. It is not quite clear what part of these 3000 specimens was retained in the Richard herbarium. Many collectors are represented in this herbarium. The plants from Cayenne collected by Leblond, described by L.C. Richard in 1792 (Actes Soc. Hist. nat., Paris, vol. 1) are now at G via the herbarium E.P. Ventenat and, subsequently, B. Delessert.

BIBLIOGRAPHY and BIOGRAPHY: AG 2(2): 61, 4: 569; Ainsworth p. 62, 313; Backer p. 494; Barnhart 3: 151 (b. 4 Sep 1754, d. 7 Jun 1821); BM 4: 1697; Bossert p. 331; CSP 5:

187; Dawson p. 703; Dryander 3: 36, 190, 269, 326; Frank 3 (Anh.):84; Herder p. 473 [index]; Hortus 3: 1202; IF p. 727; Jackson p. 8, 36, 129, 354, 508; Kew 4: 459-460; Langman p. 625; Lasègue p. 581 [index]; ME 3: 364; Moebius p. 85, 139, 141, 431; MW p. 413; NI 1634-1636; PR 1355, 2573, 6674, 7605-7610 [7611 is by A. Michaux], 9256, ed. 1: 1548, 2105, 2863, 5454, 6906, 7454, 8560-8565, 10230; Rehder 1: 338, 5: 723; RS p. 137; SO 741; TL-1/561, 840, 1077-1081; TL-2/809, 910, 1614-1615, 3004, 3142, 4635, 5958, 6745; see A. Michaux; Tucker 1: 596; Urban-Berl. p. 311, 387, 415; Zander ed. 10, p. 707, ed. 11, p. 800.

BIOFILE: Abonnenc, E. et al., Bibl. Guyane franç. 1: 225. 1957.
Anon., Bot. Zeit. 14: 151-152, 599. 1856 (L.C. Richard's herb. incorporated in that of his son A. Richard; all collections offered for sale).
Boiteau, P., C.R. 100. Congr. natl. Soc. Sav. Paris 1975, Sci. 3: 21-24. 1976 (biogr., notes on Richard dynasty).
Candolle, Alph. de, Phytographie 443. 1880 (herb. with de Franqueville).
Candolle, A.P. de, Mém. souvenirs 119, 146, 222. 1862.
Cuvier, F.G., ed., Dict. sci. nat. 61: 211-212. 1845 (biogr. sketch).
Cuvier, G., Mém. Mus. Hist. nat. 12: 349-366. 1825 ("Éloge"), Mem. Acad. Sci., Paris 7: cxciv-ccxii. 1827.
Fée, A., Nouv. biogr. gén., Didot, 42: 184-187. 1866.
Guillaumin, A., Les fleurs des jardins *fl. 16*. 1929 (portr.).
Hooker, W.J., Amer. J. Sci. 9: 268-269. 1825 (on Richard and A. Michaux).
Hooker, W.J., J. Bot. Kew Gard. Misc. 8: 81-82. 1856 (on herb.).
Jackson, B.D., Proc. Linn. Soc. 134 (suppl.): 19. 1922 (R. contributed to L., Mantissa).
Jessen, K.F.W., Bot. Gegenw. Vorz. 305, 402, 404, 426. 1884.
Jourdan, A.J.L., Dict. sci. méd., Biogr. méd. 7: 7-9. 1825.
Kunth, C.S., Biogr. univ. anc. mod. 37: 561-570. 1824; *in* W.J. Hooker, J. Bot. 4: frontisp. portr. 423-433. 1842.
Kukkonen, I., Herb. Christian Steven 83. 1971.
Lacroix, A., Figures de savants 3: 79, 90, 91-96, 107, 157, 216, *pl. 19-20.* 1938.
Milner, J.D., Cat. portr. Kew 91-92. 1906.
Millspaugh, C.F., Publ. Field Colomb. Mus. 1: 449-450. 1902.
Moscoso, R.M., Cat. fl. doming. 1: xxv-xxvi. 1943.
Papavero, N., Essays hist. neotrop. Dipterol. 1: 17-18, 23, 29, 124, 192. 1971.
Prance, G., Acta Amaz. 1(1): 53. 1971.
Sagot, P., Ann. Sci. nat. Bot. ser. 6, 10. 366. 1880.
Sayre, G., Bryologist 80: 516. 1977.
Sherborn, C.D., Ann. Mag. nat. Hist. ser. 8. 13: 367. 1914 (dates Ann. Mus. Hist. nat. 1811).
Stafleu, F.A., *in* C.L. L'Héritier, Sertum angl. facs. ed. xxxi. 1963; *in* Daniels, G.S. & F.A. Stafleu, Taxon 22: 658. 1973.
Stearn, W.T., Regnum vegetabile 71: 343. 1970.
Stieber, M.T., Huntia 3(2): 124. 1979 (notes on R. in A. Chase mss.).
Urban, I., Symb. ant. 3: 111-112. 1902; Fl. bras. 1(1): 87-88. 1906.
Vitt, D.H. & D.G. Horton, Taxon 30: 305-306. 1981 (Bryophytes in Michaux, Fl. bor.-amer. may well have been written by L.C.M. Richard).

COMPOSITE WORKS: (1) For Richard's share in the compilation of A. Michaux, *Flora boreali-americana* 1803, see TL-2/5958. See also Vitt and Horton (1981) for further notes on Richard's rôle in compiling the bryophyte part.
(2) Edited ed. 4 of J.B.F. Bulliard, *Dict. élém. bot.*, 1798, see TL-2/910, see below *Dict. élém. bot.* 1798.
(3) Contributed several descriptions of species of Melastomataceae and Rhexiae to Humboldt, A. & A. Bonpland, Monogr. Melast., TL-2/3142.
(4) Contributed to *Nouv. Dict. Hist. nat.* ed. 2, 1816-1817.
(5) *A botanical dictionary*, being a translation from the French of Louis-Claude Richard, ... with additions from Martyn, Smith, Milne, Wildenow [sic] Acharius, &c. New Haven (printed by Nathan Whiting) 1816. Duod. (in sixes), p. [i]-vi, [7]-14, sign. b_{10}-o_{12}.
Copy: HH. – By Amos Eaton (1776-1842), see TL-2/1615 (in this entry we did not note the existence of this 1816 edition).

RICHARD, L. C. M. 1798

EPONYMY: *Richardia* Kunth (1818). *Note*: The etymology of *Richardella* Pierre (1890) could not be determined.

9153. *Dictionnaire élémentaire de botanique*, par Bulliard, revu et presqu'entièrement refondu par Louis-Claude Richard, ... Ouvrage où toutes les parties des plantes, leurs diverses affections, les termes usités et ceux qu'on peut introduire dans les descriptions botaniques, sont définis et interprétés avec plus de précision qu'ils ne l'ont été jusqu'à ce jour: suivi d'une exposition méthodique de ces mêmes termes, au moyen de laquelle, et à l'aide du Dictionnaire, l'étudiant peut prendre une leçon suivie sur chaque partie des plantes: précédé d'un dictionnaire botanique latin-français ... à Paris (Chez A.J. Dugour et Durand, ...) an vii. Oct. (*Dict. élém. bot.*).
Original author: Jean Baptiste François ["Pierre"] Bulliard (1752-1793). Bulliard's Dictionnaire came out in 1783, see TL-2/910. The Richard edition may count as ed. 4 of Bulliard's Dictionnaire.
Ed. [*1*; ed. 4, Bulliard]: Oct-Nov 1798 (see TL-1/1077, TL-2/910), p. [i*], [i]-lii, [1]-228, *pl. 1-19*, [*20*] with text (uncol. copp. Sellier). *Copies*: NY, US.
Amsterdam ed. (ed. 5, Bulliard): 1800, p. [i*-iv*], [1]-228, [i]-lii, *pl. 1-19*, [*20*], id. *Copies*: BR, MICH, NY. – A truly new printing "À Amsterdam, chez les Libraires associés".
Ed. 2: an viii-1800, p. [i*-iii*], [i]-xlviii, [1]-228, *pl.* (with text) *1-19*, [*20*]. *Copy*: BR, – Seconde édition. À Paris (à la Librairie d'Éducation et des sciences et arts, ...) an viii-1800. Oct. – Not listed sub TL-2/910.
Ed. 2 [*bis*]: an x-1802, (J.T. 13 Jul 1802), p. [i*-iii*], [i]-lxiv, [1]-228, *pl. 1-16, 16bis, 16ter, 17-19* with text, chart. *Copy*: NY. – "Seconde édition, augmentée de l'exposé et du tableau de la méthode de Jussieu". À Paris (chez J.J. Fuchs, ...) an x-1802. Oct.
English translation: by Amos Eaton (1776-1842), see TL-2/1615.
Ed. [*1*]: 1817 (registered 13 Jan 1817), p. [i]-vi, [7]-14, sign. B [1-6]-P[1]. *Copies*:Ewan, MO. – *A botanical dictionary*, being a translation from the French of Louis-Claude Richard, ... with additions from Martyn, Smith, Milne, Wildenow [sic], Acharius, &c." New Haven (published by Hezekiah Howe ...) 1817. Duod. (in sixes) (*Bot. dict.*). – N.B. in our treatment of Bulliard's dictionary in TL-2/1: 407 we stated erroneously that this edition did not exist; it is, however, duly listed sub A. Eaton on p. 719 (TL-2/1615). – Anonymous but by Eaton as is shown by ed. 2.
Ed. 2: 1819, p. [i]-vi, [7]-191. *Copy*: MO. – "*A botanical dictionary*, ... with extensive additions from Martin [sic], Smith, Milne, Willdenow, Acharius, Muhlenberg, Elliott, Nuttall, Pursh, and others. By Amos Eaton, ... second edition, with additions and corrections, and the terms accentuated". New Haven (published by Howe & Spalding ...) 1819. Duod. (in sixes). Preface dated 11 Feb 1819.
Ed. 3: 1828 ("advertisement" dated 29 Aug 1828), p. [1]-53, dictionary sign. 6-11 [1-70]. *Copies*: HU, NY, see also TL-2/1615. "*Botanical grammar and dictionary*; translated from the French, of Bulliard and Richard. By prof. A. Eaton. Third edition, wholly written over, and now including the natural orders of Linneus [sic] and Jessieu [sic]". Albany (printed by Websters and Skinners, ...) 1828. Duod. (in sixes.).
Ed. 4: 1836, p. [1]-125. *Copies*: HU, US. – "Eaton's botanical grammar and dictionary, modernized down to 1876: ..." Albany (published by Oliver Steele ...) 1836. – See TL-2/1615 – ed. 4.

9154. *Démonstrations botaniques*, ou analyse du fruit considéré en général; par M. Louis-Claude Richard, ... publiées par H.A Duval (d'Alençon), ... à Paris (chez Gabon, ...) 1808. Duod. (in sixes). (*Démonstr. bot.*).
Editor: Henri Auguste Duval (1777-1814).
Publ.: Mai 1808 (TL-1), p. [i*-iv*], [i]-xii, [13]-111. *Copies*: BR, G, H-UB, HH, NY(2). – "Opuscule redigé fort à la hâte en sept jours" L.C.M. Richard to W.J. Hooker, 24 Feb 1819 (letter K). See also *Analyse botanique des embryons endorhizes*, below.
German: 1811, p. [i]-xvi, [1]-216, *1 pl.* (uncol. copp.). *Copies*: H-UB, NY. – *Analyse der Frucht und des Saamenkorns von Louis-Claude Richard*, ... nach der Duval'schen Ausgabe übersetzt und mit vielen Zusätzen und Originalzeichnungen vermehrt herausgegeben von F.S. Voigt [Friedrich Siegmund Voigt (1781-1850), translator and author of additions] ... Leipzig (bei Carl Heinrich Reclam) 1811. Oct. (*Anal. Frucht*).

English: 1819, p. [i]-xix, [xx, err.], [1]-99, [100],*pl. 1-6* (uncol. copp. Lindley). *Copies*: G, HH, NY. – *"Observations on the structure of fruits and seeds*; translated from the Analyse du fruit of M. Louis-Claude Richard, comprising the author's latest corrections; and illustrations with plates and original notes by John Lindley [1799-1865], London (John Harding, ...), Norwich (Wilkin and Youngman) 1819., Oct. – TL-2/4635.

9155. *Analyse botanique des embryons endorhizes* ou monocotylédonés, et particulièrement de celui des Graminées: suivie d'un examen critique de quelques mémoires anatomico-physiologico-botaniques de M. Mirbel, par Louis-Claude Richard, ... Paris (de l'Imprimerie de Mme Ve Courcier) 1811. Qu.
Publ.: Jul-Aug 1811 (Sherborn 1914), p. [i-v], [1]-74, *6 pl. Copies*: HH, KU, NY. – Reprinted and to be cited from Ann. Mus. Hist. nat., Paris 17: 223-251. Jul 1811, 442-487, *6 pl.* Aug 1811.

9156. *De Orchideis europaeis annotationes*, praesertim ad genera dilucidanda spectantes ... Parisiis (ex Typographia A. Belin) 1817. Qu. (*De orchid. eur.*).
Publ.: Aug-Sep 1817 (8 Sep 1817 rd by Acad.), p. [i-iv], [1]-39, [1-2, expl. pl.], *1 pl.* (uncol. copp. auct.). *Copies*: FI, M, NY, USDA; IDC 5677. – Preprinted from Mém. Mus. Hist. nat. 4: 23-61, *pl. 5.* 1818.

9157. *Mémoire sur* une famille de plantes dites *les Calycérées*. [ex Mém. Mus. Hist. nat., vol. 6, Paris 1820]. Qu.
Publ.: Nov 1820 (Acad. 4 Dec 1810), p. [1]-55, *3 pl. Copies*: FI, M. – Reprinted (preprinted?) from Mem. Mus. Hist. nat. 6: 28-82, *pl. 10-12.* 1820.

9158. *Mémoire sur* une nouvelle famille de plantes: *les Balanophorées*; ... [ex Mém. Mus. Hist. nat., vol. 8, Paris 1822]. Qu.
Publ.: Nov 1822 (Acad. 18 Nov 1822; Bull. Férussac 1 Mar 1823), p. [1]-32, *3 pl. Copies*: FI, HH.– Reprinted (preprinted?) from Mém. Mus. Hist. nat., Paris, 8: 404-435. *pl. 19-21.* 1822. See also Bull. Soc. philom., Paris, 1822: 54-55. – Edited after the author's death by his son Achille Richard.

9159. L.C. Richard, ... *Commentatio botanica de Conifereis et Cycadeis*, characteres genericos singulorum utriusque familiae, et figuris analyticis eximiè ab auctore ipso ad naturam delineatis ornatos complectens. Opus posthumum ab Achille Richard, filio, ... perfectum et in lucem editum. Stutgardiae [Stuttgart] (sumptibus J.G. Cottae ... ex typographia Pauli Renouard Parisini) 1826. Qu. (*Comm. bot. Conif. Cycad.*).
Editor: Achille Richard (1794-1852).
Publ.: Sep-Nov 1826 (TL-1), p. [i]-xv, [1]-212, *pl. 1-3, 3bis, 4-29* (no. 27 double), (uncol. copp. auct.). *Copies*: FI, G, HH, M, MO, NSW, NY, U; IDC 5019. – Some copies were struck off on wove paper. – Preliminary paper: *Sur les plantes dites Conifères*, Ann. Mus. Hist. nat. 16: 296-299. 1810.
Ref.: Schlechtendal, D.F.L. von, Linnaea 2(3): 496-498. 1827 (rev.).

9160. L.C. Richard, ... *De Musaceis* commentatio botanica sistens characteres hujus-ce familiae generum, ... opus posthumum, ab Achille Richard filio terminato et in lucem editum. Vratislaviae [Breslau] et Bonnae [Bonn] (sumtibus Academiae caes. L.C. naturae curiosorum ...) 1831. Qu. (*De Musac.*).
Editor and, in part, co-author: Achille Richard (1794-1852).
Publ.: 1831, p. [i-iii], [1]-32, *pl. 1-12* (uncol. copp. auct.). *Copies*: FI, HH, NY. – Issued as Nova Acta Leop. 15 (suppl.) 1-32, *pl. 1-12.* 1831.

Richard, Olivier Jules (1836-1896), French lichenologist; studied law in Paris; stationed as magistrate in various towns in Western France; "procureur de la république" at Marennes 1873-1876, at La Roche-sur-Yon 1876-1885, at Poitiers from 1885-1891; outspoken adversary of the Schwendener theory of lichenisation. (*O.J. Rich.*).

HERBARIUM and TYPES: ΛNGUC; other material at CGE, DUKE, H-NYL, KIEL and M (lichens via Arnold and Weddell), 80 letters to W. Nylander at Helsinki Univ. Libr. *Note on name*: Signed himself O.J. Richard on most of his publications, but "Jules Richard" on his *Excurs. bot. Espagne* (1891).

RICHARD, O. J.

BIBLIOGRAPHY and BIOGRAPHY: Barnhart 3: 151; BL 2: 211, 708; BM 4: 1697; CSP 11: 168, 18: 175; GR p. 296; GR cat. p. 70; Herder p. 260; IH 2: (in press); Kelly p. 190-191; Kew 4: 460; LS 22188-22206; Rehder 1: 417, 3: 381; SBC p. 130; TL-2/4301; Tucker 1: 596.

BIOFILE: Anon., Bull. Soc. bot. Deux-Sèvres 24: 257, *pl. 2.* 1913 (portr.); Hedwigia 35: (96). 1896; Rev. mycol. 17: 85. 1895 (1896) (d.), 21: 52. 1899 (bibl.).
Loynes, M. de, Act. Soc. Linn. Bordeaux 50: 257-263. 1896 (obit.).
Malinvaud, E., Bull. Soc. bot. France 43: 25. 1896 (obit.).
Sauzé, J.C. & P.N. Maillard, Cat. pl. phan. Deux-Sèvres 14. 1864.
Souché, B., Fl. Haut Poitou xvii. 1901. (b. 22 Mar 1836, d. 7 Jan 1896).
Verdoorn, F., ed., Chron. bot. 2: 13. 1936.
Wittrock, V.B., Acta Horti Berg. 3(3): 112, *pl. 141.* 1905 (portr.).

9161. *Catalogue des lichens des Deux-Sèvres* ... Niort (L. Clouzot, ...) 1877. Oct. (*Cat. lich. Deux-Sèvres*).
Publ.: Aug-Dec 1877 (p. xvii: Apr 1877; on p. 50 a note on the death of H.A. Weddell on 22 Jul 1877 with brief obituary; Soc. bot. France rd 22 Feb 1878; Acad. Sci. Stockholm rd before 5 Jun 1878), p. [i]-xvii, [xviii], 1-50. *Copies*: Almborn, G(2), H. – Preprinted from Bull. Soc. Stat. Sci. Arts Deux-Sèvres 3: 169-236. 1879.

9162. *Étude sur les substratums des lichens* ... Niort (L. Clouzot, ...) 1883. Oct.
Publ.: Jan-Apr 1883 (Nat. Nov. Mai(2) 1883; Bot. Centralbl. 18-22 Jun 1883; Bot. Zeit. 29 Jun 1883; Soc. bot. France rd 27 Jul 1883), p. [i-iv], [1]-88.*Copies*: Almborn, H, Stevenson. – Reprinted and to be cited from Act. Soc. Linn. Bordeaux 37: 221-308. 1883. – See also his *La synthèse bryo-lichénique*, 1883 (Nat. Nov. Nov(1) 1883), p. [1]-7. (*Copies*: Almborn, BR), published in Le Naturaliste 5: 318-319. 15 Aug 1883 (Bot. Zeit. 28 Dec 1883). – Polemical article on the nature of the gonidia (against E. Bonnier, La Nature 1878(1): 65).

9163. *Instructions pratiques pour la formation et la conservation d'un herbier de lichens* ... Paris (Jacques Lechevalier, ...) [1884]. Oct. (*Instr. format. herb. lich.*).
Publ.: Oct 1884 (Nat. Nov. Nov(1) 1884; rd by Soc. bot. France 12 Dec 1884; Bot. Zeit. 26 Dec 1884), p. [1]-44. *Copies*: Almborn, BR. – Reprinted or preprinted from Bull. Soc. Statist. Sci. Lettr. Arts Deux-Sèvres 1884 (7-9): Jul-Sep.
Ref.: Hue, A.M., Bull. Soc. bot. France 32 (bibl.): 28-30. 1885 (rev.).

9164. *L'autonomie des lichens* ou réfutation du Schwendenérisme. Paris (Jacques Lechevalier ...) 1884. Oct. (*Autonom. lich.*).
Publ.: Aug-Oct 1884 (p. 59: Jul 1884; Bot. Zeit. 31 Oct 1884; Nat. Nov. Oct(1) 1884; Bot. Centralbl. 29 Dec 1884; J. Bot. Jan 1895), p. [1]-59. *Copies*: Almborn, BR, H. – Reprinted or preprinted from Annu. Soc. Émul. Vendée 31, 1884 (n.v.).

9165. *Le procès des lichénologues* ... Extrait du journal Le Naturaliste. – 1884. Paris. Oct.
Publ. 1 Mar 1884 (date journal; p. 7: Feb 1884; Bot. Zeit. 25 Jul 1884; Flora rd Feb-Mar 1884; Nat. Nov. Mai(2) 1884), p. [1]-7. – Reprinted from Le Naturaliste 6: 419-421. 1884. Also published in Rev. mycol. 6: 108-111. 1884. – Polemical article against Léo Errera who had shown himself to be somewhat impatient with the adversaries (such as O.J. Richard) of the Schwendener theory of lichenisation.

9166. *Liste de muscinées* recueillies dans les quatre départements *du Poitou et de la Saintonge* (Vienne, Deux-Sèvres, Vendée, Charente-Inférieure) ... Paris (Nouvelle Librairie médicale et scientifique ...) 1886. Oct. (*Liste musc. Poitou Saintonge*).
Publ.: Aug-Oct 1886 (p. 4: 24 Jun 1886; Rev. bryol. Nov-Dec 1886; Nat. Nov. Nov(1) 1886), p. [1]-26. *Copies*: BR, H. – Reprinted or preprinted from Bull. Soc. Statist. Sci. Lettres Arts Deux-Sèvres 1886 (7-9): 373-396. Jul-Sep 1886.
Ref.: Bescherelle, E., Bull. Soc. bot. France 34 (bibl.): 27-28. 1887 (rev.).

9167. *Florule des clochers et des toitures des églises de Poitiers* (Vienne) ... Paris (Librairie médicale et scientifique ancienne et moderne de Jacques Lechevalier ...) 1888. Oct. (*Fl. clochers toitures Poitiers*).

Publ.: 1888 (Bot. Zeit. 25 Oct 1889; Soc. bot. France rd 9 Nov 1888; Nat. Nov. Feb(1) 1889), p. [1]-50, [51]. *Copies*: HH, NY, USDA.
Ref.: Malinvaud, E., Bull. Soc. bot. France 35 (bibl.): 197-199. 1888 (rev.).

9168. *Excursions botaniques en Espagne* par Jules Richard ancien magistrat. Niort (Imp. & lith. Lemercier & Alliot) 1891. Oct.
Publ.: 1891, p. [1]-29, [31, impr.]. *Copy*: G. – Reprinted or preprinted from Bull. Soc. bot. Deux-Sèvres 1890: 57-83. 1891.

Richards, Herbert Maule (1871-1928), American botanist; Dr. Sci. Harvard 1895; tutor in botany in Barnard College, Columbia Univ., New York 1897, adjunct professor of botany 1903, ord. professor 1906; president of the Torrey botanical Club 1917. (*H. Richards*).

HERBARIUM and TYPES: NY.

BIBLIOGRAPHY and BIOGRAPHY: Barnhart 3: 151 (b. 6 Oct 1871, d. 9 Jan 1928); De Toni 4: xlvii; IH 2: (in press); Lenley p. 343; LS 22207-22210.

BIOFILE: Anon., Jahrb. wiss. Bot. 56: 825. 1915 (student of Pfeffer 1895/6); Nat. Nov. 19: 518. 1897 (tutor bot. Columbia Univ., N.Y.), 25: 242. 1903 (adj. prof. bot. ib.); Torreya 28: 17, 18. 1928 (d.).
Britton, N.L., Fl. Bermuda 548. 1918 (Bermuda algae at NY).
Harper, R.A. et al., J. New York Bot. Gard. 29: 137-138. 1928 (obit.).

EPONYMY: *Richardsiella* J. Elffers et J. Kennedy-O'Byrne (1957) honors Mrs. H.M. Richards, a collector of African plants for Kew.

Richards, Paul Westmacott (1908-x), British botanist, bryologist and forest ecologist; studied at Univ. College London 1925-1927; at Trinity College Cambridge 1927-1933, fellow 1933-1937; Dr. phil. ibid. 1936; Dr. Sci. ib. 1954; demonstrator in botany Cambridge 1938-1945; lecturer 1945-1949; professor of botany University College of North Wales, Bangor 1949-1976; Gold medal Linnean Soc. London 1979. (*P. Richards*).

HERBARIUM and TYPES: BM (on loan from K), NMW and private (musci); other material in many herbaria.

BIBLIOGRAPHY and BIOGRAPHY: BL 1: 244, 315; Bossert p. 331; Clokie p. 232; GR p. 412; IH 1 (ed. 1): 6, (ed. 2): 19, (ed. 3): 22, (ed. 4): 23, (ed. 5): 14, (ed. 7): 341; Kew 4: 461; Langman p. 625; NI 2: 831, 3: 30; Roon p. 94; SBC p. 130; SK 1: 434-435 (portr., bibl.), 4: cxlvi.

BIOFILE: Anon., Nature 163: 354-3455. 1949 (app. Bangor).
Bridson, G.D.R. et al., Nat. hist. mss. res. Brit. Isl. 269.302. 1980.
Hawksworth, D.L. & M.R.D. Seaward, Lichenol. Brit. Isl. 139. 1977.
Jacobs, M., Fl. males. Bull. 33: 3365-3373. 1980 (biogr., bibl., b. 19 Dec 1908).
Kent, D.H., Hist. fl. Middlesex 27. 1975.
Lacey, W.S., Nature, Wales 16(2): 129-131. 1978; Bot. Soc. Brit. Isl. Welsh Reg. Bull. 30: 13-15. 1979.
Prance, G., Acta amaz. 1(1): 53, 56. 1971.
Sayre, G., Bryologist 80: 516. 1977.
Schuster, R.M., Hep. Anthoc. N. Amer. 1: 90. 1966.
Steenis, C.G.G.J. van, Fl. males., Bull. 5: 125. 1949 (app. Bangor), 6: 159. 1950 (id.), 14: 625. 1959.
Verdoorn, F., ed., Chron. bot. 2: 2, 176, 188. 1936, 3: 2, 367, 425, 4: 81, 479, 533. 1938, 5: 260, 317, 318. 1939, 6: 261, 303. 1941.

NOTE: Author of the fundamental treatise *The tropical rainforest*, an ecological study, Cambridge 1952, xviii, 450 p., 15 pl., 2 charts, reprint 1957.

Richardson, Sir John (1787-1865), British (Scottish) arctic explorer and naturalist in the Royal Navy; M.D. Edinburgh 1816; surgeon and naturalist to John Franklin's expeditions of 1819-1822 and 1825-1827; physician to the Royal naval hospital of Haslar 1838; medical inspector of hospitals and fleets 1840-1855; on a search expedition for Franklin 1847-1849; knighted 1846; living in retirement from 1855 at Lancrigg nr Grasmore, Westmoreland. (*Richardson*).

HERBARIUM and TYPES: BM, OXF (moss herb.); further material at CGE, DBN, E, FI, G, GH, H, K, LINN, M, MO, NY, OXF, PH, SFD, WECO. – Letters at K and Scott Polar Research Inst., Cambridge.

BIBLIOGRAPHY and BIOGRAPHY: AG 1: 314; Backer p. 494; Barnhart 3: 152; BB p. 256; BM 4: 1698-1699; Bossert p. 331; Clokie p. 232; CSP 5: 188-189, 6: 748, 8: 743, 744; Desmond p. 518-519; DNB 16: 1119-1121 (by G.S. Boulger), 48: 233-235; Hegi 4(1): 153; Herder p. 226; Hortus 3: 1202; IF p. 727; IH 2: (in press); Jackson p. 223; Kew 4: 462; Lasègue p. 192-193, 581 [index]; LS 22211-22212; ME 3: 402, 423, 437; PR 1214, 4222, 7613, ed. 1: 1384, 4662, 8566; Quenstedt p. 360; Rehder 1: 306, 310; TL-1/534; TL-2/1850, 3003, see T.L. Mitchell; Tucker 1: 596; Zander ed. 10, p. 707, ed. 11, p. 806.

BIOFILE: Allan, M., The Hookers of Kew 83, 97-98, 172. 1967.
Allibone, S.A., Crit. dict. Engl. lit. 2: 1794-1795. 1878 (bibl.).
Anon., Amer. J. Sci. ser. 2. 41: 265. 1866 (obit.); Bot. Zeit. 23: 252. 1865 (b. 5 Nov 1787, d. 5 Jun 1865); Flora 48: 445. 1865 (d.); J. Bot. 3: 231. 1865 (obit.); Le naturaliste Canadien 5: 103. 1873; Österr. bot. Z. 15: 229. 1865 (d.); Proc. Linn. Soc. London 1865/66: lxxxiv-lxxxvi (obit.), 15: xxxvii-xliii. 1867 (obit.).
Babcock, E.B., Rhodora 26: 198-200. 1924 (on citing the *Botanical Appendix*).
Barnhart, J.H., Mem. Torrey bot. Club. 16(3): 299. 1921.
Bridson, G.D.R. et al., Nat. hist. mss. res. Brit. Isl. 431 [index]. 1980.
Currey, M.F. & R.E. Johnson, J. Soc. Bibl. nat. Hist. 5(3): 202-217. 1969 (bibl. printed books).
Embacher, F., Lexikon Reisen 247. 1882.
Geiser, S.W., Naturalists of the Frontier 66, 68, 74. 1948.
Graustein, J.E., Thomas Nuttall 476 [index]. 1967.
Gray, J.L., Letters A. Gray 110, 112, 364. 1893.
Hertel, H., Mitt. Bot. München 16: 423. 1980 (lichens from Canada via Arnold).
Hitchcock, A.S. & A. Chase, Man. grasses U.S. 988. 1951 (epon.).
Hornschuch, C.F., Flora 6(1): 250. 1823, 9(1): 375-376. 1826, 12(1): 185. 1829.
Hooker, W.J., Amer. J. Sci. 9: 281, 283-284. 1825; Bot. Misc. 1: 92-94. 1830 (Hooker prepares Fl. bor. amer., pp. on Richardson material).
Huntley, M.A. et al., J. Soc. Bibl. nat. Hist. 6(2): 98-117. 1972 (bibl. of articles in journals, portraits, eponymy and monuments).
Huxley, L., Life letters J.D. Hooker 2: 561 [index]. 1918.
Johnson, R.E., Sir John Richardson, arctic explorer, ... London 1976, 209 p. (recent biogr., bibl., secondary refs., unpubl. sources, portr.).
Kukkonen, I., Herb. Christ. Steven 84. 1971.
McIlraith, J., Life of Sir John Richardson, London 1868, xi, 280 p. (biogr., portr.).
Miller, H.S., Taxon 19: 537. 1970 (set R.'s plants in Lambert herb.).
Milner, J.D., Cat. portr. Kew 92. 1906.
Murchison, R.I., J. Geogr. Soc. 36: cxxxii-cxxxiv. 1866 (obit.).
Murray, G., *in* Hist. coll. BM(NH) 1: 177. 1904 (119 pl., incl. types W.J. Hooker, rd. from Admiralty 1856).
Nissen, C., Zool. Buchill. 338-339. 1969.
Palmer, T.S., Condor 30: 392.
Penhallow, D.P., Trans. R. Soc. Canada ser. 2. 3 (sect. 4): 8, 11, 14, 52. 1897.
Poggendorff, J.C., Biogr.-lit. handw.-Buch 2: 629-630. 1863, 3: 1119. 1898 (b. 5 Nov 1787, d, 15 Jun 1865).
Polunin, N., Rhodora 38: 412-413. 1936 (plants from Canad. east. Arctic).
Porsild, A.E. & W.J. Cody, Vascular plants continental North Western Territories, Canada 5-7. 1980 (itin., coll.).

Provancher, L., Natural. Canad. 5(3): 103. 1873.
Rodgers, A.D., John Torrey 348 [index]. 1942.
Rolleston, H., J.R. nav. med. Serv. 10(3): 160-172. 1924 (biogr. sketch, portr., memorial).
Sayre, G., Mem. New York Bot. Gard. 19(3): 386. 1975 (coll.).
Schlechtendal, D.F.L. von, Bot. Zeit. 23: 252. 1865 (obit.).
Schultes, J.A., Flora 8(1) Beil. 1: 26, 39-40. 1825.
Short, C.W., *in* W.J. Hooker, J. Bot. 3: 112-113, 114, 116. 1841.
Sprunt, W.H., New Engl. J. Med. 253: 26-27. 1955 (portr., "physician afield").
Stewart, D.A., J.R. nav. med. Serv. 22: 180-187. 1936 (general evaluation, portr.).
Swainson, W., Taxidermy 308-309. 1840.
Voss, E.G., Contr. Univ. Mich. Herb. 13: 41. 1978.

COMPOSITE WORKS: (1) W.J. Hooker, *Fl. bor.-amer.*, 1833-1840, was based on collections made by J. Richardson, T. Drummond and D. Douglas.
(2) For Richardson's numerous zoological publications see BM 4: 1698-1699 and the bibliographies by M.F. Curvey and R.E. Johnson (1969) and by M.A. Huntley et al. (1972).

9169. *Botanical Appendix* to Captain Franklin's narrative of a Journey to the shores of the Polar Sea ... London (printed by W. Clowes, ...) 1823. Qu. (*Bot. App.*).
Author of narrative: Sir John Franklin (1786-1847), see TL-2/1850.
Ed. [1]: Mar 1823 (plates so dated), p. [i*], [1]-40, *pl. 27, 28, 30* (no. 30 col.). *Copies*: E, FH. – Reprinted from J. Franklin, *Narr. journey Polar Sea* 1823, p. [729]-768. *pl. 27, 28, 30*. Bryophytes by Schwaegrichen; lichens and algae by W.J. Hooker.
Ed. [2]: 1823, p. [1]-55, [56], *1 col. pl.* unnumbered, *pl. 27, 28, 29* (uncol.), *30* (col.). Plate 30 dated Mar 1823. *Copies*: FH, G, HH, M, NY, Teyler. – Reprinted from idem p. [729]– [784]. *pl. [26bis], 27-30*, of which p. 769-783, [784] and pl. [26bis] and *29* came out somewhat later than the preceding text and plates. Pages 51-55 contain addenda by Robert Brown (1773-1858).

9170. *Arctic searching expedition*: a journal of a boat-voyage through Rupert's land and the Arctic Sea, in search of the discovery ships under command of Sir John Franklin. With an appendix on the physical geography of North America ... London (Longman, Brown, Green, and Longmans) 1851. Oct. 2 vols. (*Arct. search. exped.*).
1: 1851, p. [i]-viii, fold-out map, [1]-413 [414, colo.], *pl. 1-7, 8* (frontisp.), *9*.
2: Nov-Dec 1851 (p. 426: 20 Oct 1851), p. [i]-vii, [1]-426, *pl. 10*. – Appendix on the physical geography of North America on p. 161-401, no. III: On the geographical distribution of plants north of the 49th parallel of latitude, on p. 264-353; list of *Carices*, p. 344-353, by F. Boott.
Copies: E (Univ. Libr.), HH, L (Univ. Libr.), NY, PH.
American ed.: 1852, p. [i]-xi, [13]-516. *Copies*: HH, PH. -New York (Harper & Brothers, ...) 1852. Duod. – Also with date 1854. *Copy*: HH.
Facsimile reprint: 1969, Greenwood Press, Publishers, New York, as original (imprint on p. ii and iii differs).

9171. *The polar regions* ... Edinburgh (Adam and Charles Black) 1861. Oct. (*Polar regions*).
Publ.: 1861, p. [i]-ix, [1], [1]-400, chart. *Copy*: E (Univ Libr.), (printed by R. & R. Clark who also printed Flora of Turkey vol. 6 in 1978, inf. J. Edmondson).

Richardson, Richard (1663-1741), British (Yorkshire) physician, botanist, antiquary and book-collector, who studied in Oxford (University College, Bachelor of Physics) and Leiden, under Paul Hermann and Boerhaave (M.D. Leiden 1790), F.R.S. 1712; entertained an extensive correspondence with many botanists. (*R. Richardson*).

HERBARIUM and TYPES: BM-Sloane, OXF. – Letters at Royal Society (London) and Bodleian Library (Oxford).

BIBLIOGRAPHY and BIOGRAPHY: Backer p. 494; Barnhart 3: 152 (b. 6 Sep 1663, d. 21 Apr 1741); BB p. 256-257; BM 4: 1699; Bossert p. 331; Clokie p. 232; Desmond p. 519; DNB 16: 1126-1127, 48: 240-241; HU 503 (much info; Index horti bierl.); Jackson p. 4; PR 7614, ed. 1: 8567; Rehder 1: 81; TL-2/see A. Buddle.

BIOFILE: Allen, D.E., The naturalist in Britain 290 [index]. 1976.
Allibone, S.A, Crit. dict. Engl. lit. 2: 1796. 1878.
Anon., J. Bot. 44: 433. 1906 (letters to S. Brewer); North. Gard. 29(1): 20-22. 1975 (letter Philip Miller to R.).
Blackwell, E.M., Naturalist 877: 53. 1961.
Blunt, W., The compleat naturalist 119, 124. 1971.
Berkeley, E. and D.S., John Clayton 58-59, 61, 69, 213. 1963.
Bridson, G.D.R. et al., Nat. hist. mss. res. Brit. Isl. 431 [index]. 1980.
Dandy, J.E., Sloane herb. 194-195. 1958 (herb.).
Ewan, J. & N., John Banister 15. 1970.
Kent, D.H., Brit. herb. 75. 1957.
Moore, S., Fl. Cheshire lxxxvii. 1899.
Nichols, J., Ill. lit. hist. eighttenth cent. 1: 225-252. 1817 (biogr., portr.).
Pulteney, R., Hist. biogr. sketch. progress bot. England 2: 185-188. 1790.
Pearsall, W.H., Fl. Surrey 46. 1931.
Salisbury, R.A., Genera of plants 114. 1866.
Salmon, C.E., Fl. Surrey 46. 1931.
Scott, E.J.L., Index to Sloane manuscr. 452. 1904.
Smith, J.E., Selection corr. Linneus 2: 130-160. 1821.
Wroot, H.E., Naturalist 595: 257-260. 1906 (corr. with Samuel Brewer); Bradford Sci. J. 1912: 234-241 ("an old Bradford botanist", portr.).

HANDWRITING: Dandy, J.E., Sloane herb. facs. 31. 1958.

EPONYMY: *Ricardia* Adanson (1763, *orth. var.* of *Richardia* Linnaeus); *Richarda* Cothenius (1790, *orth. var.* of *Richardia* Linnaeus); *Richardia* Linnaeus (1753).

9172. *Extraits from the literary and scientific correspondence* of Richard Richardson, ... illustrative of the state and progress of botany, and interspersed with·information respecting the study of antiquities and general literature, in Great Britain, during the first half of the eighttenth century ... Yarmouth (printed by Charles Sloman, ...) 1835. Oct. (*Extr. lit. sci. corr.*).
Editor: Dawson Turner (1775-1858).
Publ.: 1835, p. [i*-ii*], portr., [iii*], [i]-lxvi, [1]-451, [453-463, index]. *Copy*: NY. – Important source for the history of British botany.

Richen, Gottfried (1863-?), German-born (Prussian Rheinland) roman catholic clergyman (S.J.) and botanist; from 1889 high shool teacher at Feldkirch (Vorarlberg, Austria). (*Richen*).

HERBARIUM and TYPES: Unknown.

BIBLIOGRAPHY and BIOGRAPHY: AG 12(2): 188; Barnhart 3: 153 (b. 5 Dec 1863); CSP 18: 182; DTS 1: 245, 6(4): 188; Hegi 6(2): 1035, 1262; Rehder 1: 444.

BIOFILE: Janchen, E., Cat. fl. austriac. 1: 47, 49. 1956.
Murr, J., Neue Übers. Farn– Bl.-Pfl. Vorarlberg xx. 1923.

9173. VI. Jahresbericht des öffentlichen Privatgymnasiums an der Stella matutina zu Feldkirch. Veröffentlicht am Schlusse des Schuljahres 1896-97. Inhalt: i. *Die botanische Durchforschung von Vorarlberg und Liechtenstein* von Professor Gottfr. Richen. S.J. ii. Schulnachrichten vom Director. Feldkirch (im Selbstverlage der Anstalt) 1897. Qu. (*Bot. Durchf. Vorarlb. Liechtenst.*).
Publ.: 15 Jul-31 Aug 1897 (p. 39: 15 Jul 1897; Nat. Nov. Sep(2) 1897; Bot. Zeit. 1 Oct 1897; Allg. bot. Z. 15 Oct 1897; Bot. Centralbl. 6 Oct 1897; ÖbZ Jul 1897), p. [1]-90 [Richen], [1]-39 (Schulnachr.). *Copies*: G, M.

Ref.: Kneucker, A. Allg. bot. Z. 3: 200. 15 Dec 1897 (rev.).

9174. *Nachträge zur Flora von Vorarlberg und Liechtenstein* Viertes Stuck. Von Prof. G. Richen S.J. Feldkirch [Bregenz 1907].
Publ.: 1907, p. [1]-12. *Copy*: M. – Reprinted and to be cited from Jahresber. Landesmuseums-Vereins Vorarlberg 44: 49-60. 1907.
Flora von Vorarlberg und Liechtenstein: published in Österr. bot. Z. 47: 78-86. Mar, 137-142. Apr, 179-183. Mai, 213-218. Jun, 245-257. Jul1897.
Nachtrag 1: Österr. bot. Z. 48: 131-134, Apr 171-178. Mai 1898.
Nachtrag 2: 49: 432-436. Dec 1899.
Nachtrag 3: 52: 338-346. Sep 1902.

Richer de Belleval, Pierre (1564-1632), French botanist, founder of the Montpellier botanical garden. (*Richer de Belleval*).

HERBARIUM and TYPES: Unknown.

BIBLIOGRAPHY and BIOGRAPHY: Barnhart 3: 153; BM 1: 130 (sub Belleval); Hegi 4(1): 242; Lasègue p. 518; Moebius p. 419; NI 1: 94-96. PR sub Belleval: 600-603, 9328, ed. 1: (id.): 200, 687-691, 1348, 2697, 3635.

BIOFILE: Anon., Recherches sur la vie et les ouvrages de Pierre Richer de Belleval, fondateur du jardin botanique donné par Henri iv. à la faculté de médecine de Montpellier en 1593 ... Avignon (chez Jean-Albert Joly) 1786, viii, 78 p. (copy at G).
Gérard, Bull. Soc. bot France 40: x. 1893. (Henri IV created bot. gard. Montpellier 8 Dec 1593; R. actual founder).
Legrelle, Bull. Soc. bot. France 40: cclvi-cclx. 1893.
Magnin, A., Bull. Soc. bot. Lyon 31: 24. 1906.
Martins, Ch., Le jardin des plantes de Montpellier, 1854, see index for numerous entries).
Planchon, J.E., Pierre Richer de Belleval fondateur du Jardin des plantes de Montpellier. Montpellier 1869, 72 p. (rev. Bull. Soc. bot. France 17 (bibl.): 125-126. 1870 ; 20: 96-98. 1873.
Ricard, Bull. Soc. bot. France 19: lxxviii-lxxix. 1872 (on unpubl. manuscripts by R.).

EPONYMY: *Richeria* M. Vahl (1797); *Richeriella* Pax et K. Hoffmann (1922).

Richon, Charles Édouard (1820-1893), French botanist and, from 1847, practicing physician at Saint-Amand (Marne). (*Richon*).

HERBARIUM and TYPES: Unknown; the location of the herbarium and iconographies left by him, mentioned by Roze 1894, is not known to us.

BIBLIOGRAPHY and BIOGRAPHY: Ainsworth p. 120, 243, 320; Andrews p. 320; Barnhart 3: 153 (b. 15 Feb 1820, d. 5 Dec 1893); BM 4: 1699; CSP 8: 745, 11: 173, 12: 616, 18: 187; GR p. 296; GR cat. p. 70; Hawksworth p. 197; Jackson p. 172, 496; Kelly p. 191; Kew 4: 463; LS 22216-22239; Morren ed. 10, p. 68; Stevenson p. 1255; Tucker 1: 596.

BIOFILE: Anon., Bull. Soc. bot. France, table articles orig. 185-1893, p. 201-202, 212 (lists BSbF papers), 26: frontisp. portr. 1910 (portr.).
Boudier, E., Bull. Soc. mycol. France 10: 68-71. 1894, J. de Bot., Morot 8: 18-20. 1894 (obit., bibl.).
Duchartre, Bull. Soc. bot. France 40: 338. 1893 (1894) (d. 6 Dec 1893).
Dutertre, E., Mém. Soc. Sci. Arts Vitry-le-François 1896: 33-167 (publication based on a Richon mss., Les stations naturelles des champignons et leurs spores, with 2400 Richon drawings).
Roze, E., Bull. Soc. bot. France 40: 390-393. 1893 (1894) (obit., b. 13 Feb 1820, d. 5 Dec 1893).

EPONYMY: *Richonia* Boudier (1885); *Richoniella* Costantin et Dufour (1916).

9175. *Descriptions et dessins de plantes cryptogames nouvelles* ... Vitry-le-François (Typographie Pessez et Ce, ...) 1879, 2 parts. (*Descr. dess. pl. crypt.*).
1: 1879 (BSbF 22 Apr 1881; Bot. Zeit. 9 Apr 1880), cover (with above text), p. [i], [1]-19. *pl. 1-3* (col. liths. auct.). *Copy*: G.
2: 1879 (Bot. Centralbl. 15-19 Mar 1880; Nat. Nov. Jan(1) 1880; Bot. Zeit. 26 Aug 1881; as of 1879), n.v., 20 p., *3 pl.*

9176. *Atlas des champignons* comestibles et vénéneux de la France et des pays circonvoisins contenant 72 planches en couleur ou sont représentées les figures de 229 types des principales espèces de champignons recherchées pour l'alimentation, et des espèces similaires suspectes ou dangereuses avec lesquelles elles sont confondues dessinées d'après nature avec leurs organes reproducteurs amplifiés par Charles Richon, ... accompagné d'une monographie de ces 229 espèces et d'une histoire générale des champignons comestibles et vénéneux par Ernest Roze ... Paris (Octave Doin, éditeur...), text. Qu. and atlas. Fol. (*Atlas champ.*).
Author of text: Ernest Roze (1833-1900).
Publ.: in fascicles, 1885-1887 (t.p.: 1888). *Copies*: BR, FH, G(2), MICH, MO, NY, PH, Stevenson, USDA; IDC 6306.
Texte: p. [i*-iii*], [i]-xcviii, [1]-265.
Atlas: p. [i]-xii, *pl. 1-72* (col. liths. Ch. Richon).

fasc.	pages	plates	dates
1	i-viii, 1-24	1-8	Dec 1885
2	ix-xvi, 25-48	9-16	Feb 1886
3	xvii-xxiv, 49-72	17-24	Jun 1886
4	xxv-xxxii, 73-96	25-32	Oct 1886
5	xxxiii-xl, 97-120	33-40	Jan 1887
6	xli-xlviii, 121-144	41-48	Apr 1887
7	xlix-lvi, 145-168	49-56	Aug 1887
8	lvii-lxiv, 169-192	57-64	Oct 1887
9	lxv-xcviii, [i]-xii, 193-265	65-72	Nov 1887

The above details stem from notes in the contemporary literature (Nat. Nov., Rev. mycol., Bot. Zeit., Bot. Centralbl., Hedwigia, Bull. Soc. bot. France), in part already brought together 10 Aug 1916 by J.H. Barnhart (mss. in NY copy).

9177. *Catalogue raisonné des champignons qui croissent dans le département de la Marne*, établi d'après les classifications des auteurs modernes, Fries, Quélet, Boudier, Saccardo. Orné de dessins faits d'après nature, lithographiés, représentant les types des principales familles et de deux tableaux de M. E. Roze sur les Agaricinées, avec 4 planches à l'appui de sa classification; suivi d'une table alphabétique d'environ 700 genres accompagnés chacun de la figure et de la dimension des spores qui les caractérisent, et d'une exposition méthodique des champignons vénéneux et comestibles cités dans l'ouvrage, ... Vitry-le-François (Typographie Vve Tavernier et fils) 1889. Oct. (*Cat. champ. Marne*).
Publ.: 1889 (Bot. Zeit. 25 Apr 1890; Soc. bot. France rd 8 Nov 1889; ÖbZ Mar 1890; Nat. Nov. Feb(2) 1890), p. [i*-v*], [i]-xiv, [1]-586, [587, err.], *pl. 1-6*, two charts and *four* unnumbered *pl.* (uncol. liths.). *Copies*: BR, FH, NY, Stevenson, USDA. The cover has a different imprint: Paris (Octave Doin, ...) 1889. – Also issued in Soc. Sci. Arts Vitry-le-François, vol. 15, 1887-1888, 1889.

Richter, [Vincenz] Aladár (1868-1927), Hungarian botanist; school teacher in Transylvania 1891-1895; high school teacher at the first Budapest Staatsgymnasium 1895-1898; habil. Univ. Budapest 1898; head botany dept. Hung. Natl. Mus. 1898-1899; suppl. professor of general botany at Klausenburg (Kolozsvár, Cluj) 1899; director of botanical institute and garden ib. 1900; ordinary professor of general botany ib. ("Ferenze-József" Univ.); also director of institute of syst. bot. ib. succeeding V. Borb-as ib. 1905-1913. (*Al. Richt.*).

HERBARIUM and TYPES: material at B (extant), BP, C, IBF, JE, L, MICH, MPU, P, SI, W.

BIBLIOGRAPHY and BIOGRAPHY: AG 3: 855; Barnhart 3: 154 (b. 5 Jan 1868, d. 11 Jun 1927); Jan 1868, d. 11 Jun 1927); BM 4: 1699; De Toni 4: xlvii; Futak-Domin p. 502-503; GR p. 660; Hortus 3: 1202; Jackson p. 298; LS 22240; Rehder 5: 724; TL-2/see J.C.G. Baumgarten; Urban-Berl. p. 388.

BIOFILE: Anon., Bot. Centralbl. 64: 447. 1895 (from Arad to Budapest as prof. at Staatsgymnasium), 73: 464. 1898 (habil. Budapest, 75: 96. 1898 (head bot. dept. BP), 77: 288. 1899 (suppl. prof. bot. Kolozsvár (Cluj), 83: 32. 1900 (dir. inst. bot. and garden Cluj), 88: 256. 1901 (ord. prof. bot.), 99: 240. 1905 (succeeds V. Borbás), 104: 32. 1906 (id.); Bot. Jahrb. 27 (Beibl. 64): 20. 1900 (app. suppl. prof. bot. Kolozsvár (Cluj)), 31 (Beibl.): 70. 1902 (app. ord. prof. ib.); Hedwigia 38: (114). 1899 (prof. Cluj), 39: (194). 1900 (dir. bot. gard.), 40: (204). 1901 (prof. general bot.), 45: (77). 1906 (dir. inst. plant syst.), 46: (144). 1907 (prof. bot.), 51: (170). 1911 (corr. memb. Akad. Hung.), 54: 189. 1914 (retirement); Nat. Nov. 20: 214 (habil. Budapest), 339 (id.), 420 (head bot. dept. BP), 501 (id.). 1898, 21: 150. 1899 (suppl. prof. Klausenburg (Cluj)), 23: 619. 1901 (ord. prof. Cluj), 27: 505. 1905 (dir. inst. plant systematics, succeeding V. Borbás); Österr. bot. Z. 41: 152, 292 (Banat., Transilvania Versecz). 1891, 46: 79. 1896 (app. Gymnasium Budapest), 48: 367. 1898 (head bot. dept. BP), 50: 307. 1900 (directorship Cluj), 52: 38. 1902 (ord. prof. Cluj), 35: 411. 1905 (succeeds V. Borbás), 76: 330. 1927 (d.).

Borza, Al., Bul. Grăd. bot. Muz. bot. Cluj 71(1-2): 54-57, *pl. 11.* 1927 (obit., portr.).
Gager, C.S., Brooklyn Bot. Gard. Record 27(3): 323. 1938 (dir. bot. gard. Cluj 1901-1903).
Kneucker, A., Allg. bot. Z. 4: 152. 1898 (Privatdoz. Univ. Budapest; head bot. dept. Natl. Mus. [BP], 5: 36. 1899 (to Klausenburg (Cluj)), 7: 220. 1901 (ord. prof. Klausenburg), 12: 208. 1906 (app. Klausenburg), 33: 64 (288). 1928 (d. 11 Jun 1927).
Pax, F., Grundz. Pfl.-Verbr. Karpathen 43-44, 47. 1898 (Veg. Erde 2(1)).
Verseghy, K., Feddes Repert. 68(1): 125. 1963.

9178. *Ueber die Blattstructur der Gattung Cecropia* insbesondere einiger bisher unbekannter Imbauba-Bäume des tropischen Amerika ... Stuttgart (Verlag von Erwin Nägele) 1898. Qu. (*Blattstruct. Cecropia*).
Publ.: 20 Dec 1897 (dedication copy to Radlkofer at M. signed 20 Dec 1897), p. [ii-iv], [1]-25, *pl. 1-8. Copies*: BR, G, GOET. – Issued as Heft 43 (vol. 8) of Bibliotheca botanica.

9179. *Adatok a Marcgraviaceae és az Aroideae* physiologiai-anatomiai és systematikai ismeretéhez ... Beiträge zur physiologisch-anatomischen und systematischen Kenntniss der Marcgra- viaceen und Aroideen ... Budapest 1899. Oct.
Publ.: 1899, cover t.p., p. 27-87, *pl. 2-5. Copies*: BR, M.– Reprinted and to be cited from Termész. Füzetek 22: 27-87. *pl. 2-5.* 1899.

9180. *Növénytani intézete és botanikus kertje.* (1872-1904) ... Kolozsvár [Klausenburg, Cluj] (Ajtai k. Albert könyvsajtója) 1905. Oct.
Publ.: 1905, 331 p. *Copy*: B (biogr.).

9181. *A Marcgraviaceae néhány új alakjáról,* a származás-és az összehasonlító alkattan alapján ... Über einige neue Marcgraviaceen-Arten, auf phylogenetischem und vergleichenden anatomischen Grunde ... [Magy. bot. Lap. 1916, évi 6/12 ...]. Oct.
Publ.: 1916, p. 1-8. *Copy*: M. – Reprinted and to be cited from Magy. bot. lap. 1916: 281-288. – Summary of Richter's publications with this title in Math. Termész. Ertes. 34: 551-586. 1916.

Richter, Berthold (1834-?), German (Silesian) botanist; Dr. med. Breslau 1860. (*B. Richt.*).

HERBARIUM and TYPES: In 1915 in the Upper Silesian Museum in Gleiwitz (Werner, 1915).

BIBLIOGRAPHY and BIOGRAPHY: BM 4: 1699; LS 22245; PR 7619.

BIOFILE: Werner, A. Oberschlesien 14: 208-216. 1915 (on herb.; fide F. Pax, Bibl. schles. Bot. 138. 1929 (herb. at Gleiwitz).

9182. *Commentatio de Favo eiusque fungo.* Dissertatio inauguralis pathologico-botanica quam gratiosi medicorum ordinis in alma literarum universitate viadrina ad summos in medicina et chirurgia honores rite capessendos die xxiii mensis junii a. mdccclx h.l.q.c. publice defendet auctor Bertholdus Richter silesius ... Vratislaviae [Breslau] (typis Brehmeri et Mimuthii) [1860]. Oct. (*Comm. de Favo*).
Publ.: 23 Jun 1860, p. [1]-63, [64], *pl. 1-2* (uncol. lith. auct.). *Copies*: FH, NY.
Ref.: Schlechtendal, D.F.L. von, Bot. Zeit. 18: 290-291. 17 Aug 1860.

Richter, Hermann Eberhard Friedrich (1808-1876), German (Saxonian) botanist and physician; Dr. med. Leipzig 1834. (*H. Richt.*).

HERBARIUM and TYPES: HBG (2500); further material at FI, MW.

BIBLIOGRAPHY and BIOGRAPHY: AG 6(2): 921; Barnhart 3: 153 (b. 14 Mai 1808, d. 24 Mai 1876); Biol.-Dokum. 14: 7503; BM 4: 1700; CSP 5: 194, 8: 745; Herder p. 68, 184; Hortus 3: 1202; IH 1 (ed. 7): (in press); Jackson p. 116; Kew 4: 463; LS 22246-22250; MW p. 414; PR alph., 4718, 5432, ed. 1: 5261, 6011; SO 25, 25a, 26, 845a; TL-1/714, 1082; TL-2/842, 3722; Tucker 1: 596.

BIOFILE: Anon., Bot. Not. 1877: 60(d.).
Laage, R.J.Ch.V. ter, Regn. veg. 71 (Essays in biohistory): 119, 121, 126. 1970.
Lindemann, E. von, Bull. Soc. Natural. Moscou 61: 54-65. 1886.
Reichenbach, H.G., Bot. Zeit. 34: 575. 18786 (obit.).
Sayre, G., Bryologist 80: 516. 1977 (herb. HBG).

COMPOSITE WORKS: With Gustav Theodor Klett (x-1827), *Fl. Leipzig* 1830, TL-2/3722. – Extensive review: Anon., Flora 13(2): 431-438. 21 Jul 1830. – Flora 24, Intell.-Blatt 1: 11. 1841 mentions the existence of a "Prachtausgabe auf Basler Velin papier, in 2 Bänden".

EPONYMY: *Richtera* H.G.L. Reichenbach (1841). *Note*: *Richteria* Karelin et Kirilov (1842) honors Alexander Richter. The derivation of *Richteriella* Lemmermann (1897) was not given.

THESIS: *Problema de via analytica ad certitudinem cognitione medica.* Dissertatio inauguralis medica ... die xv. m. April a.p.ch.n. mdcccxxxiv ... Dresdae (Typis B.G. Teubneri) [1834]. *Copies*: B, REG (fide Hoppea 34: 300. 1976).

9183. Caroli Linnaei systema, genera, species plantarum uno volumine. Editio critica, adstricta, conferta sive *Codex botanicus linnaeanus* textum Linnaeanum integrum ex omnibus systematis, generum, specierum plantarum editionibus, mantissis, additamentis, selectumque ex ceteris ejus botanicis libris digestum, collatum, contractum, cum plena editionum discrepantia exhibens. In usum botanicorum practicum edidit brevique adnotatione explicavit ... Lipsiae [Leizig] (sumptum fecit Otto Wigand) 1835[-1839]. Qu. (*Codex bot. linn.*).
Publ.: in 14 parts between Dec 1835 and Sep 1839 (p. vi: Aug 1835, part 1 Dec 1835, part 14: 15-22 Sep 1839; the contents of the parts not yet known to us), p. [i]-xxxii, [1]-1102. *Copies*: FI, HH, M, NY, U, US, USDA; IDC 7137. – Other t.p.: "*Caroli Linnaei opera.* Editio primo critica, plena, ad editiones veras exacta, textum nullo rei detrimento contractum locosque editionum discrepantes exhibens. Volumen secundum systema vegetabilium libros "diagnostico-botanicos continens". – See SO 25-26, TL-2/4709.
Other issue: Jan-Jun 1840 ("so eben", Jun 1840 adv. publ.), p. [iii]-xxxii, [1]-1102. *Copies*: H-UB, MO, PH, US. – Only difference new t.p. – Lipsiae (sumptum fecit Otto Wigand) 1840. Qu.

Index alphabeticus: by G.L. Petermann (q.v. no. .....), 1840, p. [i]-ix, [1]-202, *Copies*: MO, NY, PH, U. – "*In codicem botanicum linnaeanum index alphabeticus* generum, specierum ac synonymorum omnium completissimus. Composuit atque edidit Dr. Guil. Ludov. Petermann, ... " Lipsiae (sumptum fecit Otto Wigand) 1840. Qu.
Note: Important source for dates of Linnaean works, but also an extremely convenient compilation of Linnaeus's main botanical works, enabling the user to compare at a glance the entries for the same taxon in the various works. The *Supplementum* by C. Linné fil., 1781, is not included in the compilation.
Ref.: Richter, H.B.E., Flora 18(2): 650-655. 7 Nov 1835 (preview).

Richter, Karl (1855-1891), Austrian botanist; Dr. phil. Wien 1877; "Privatgelehrter" in Wien, actively collecting herbarium collections and botanical works; compiler of the first volume of the *Plantae europaeae*. (*K. Richt.*).

HERBARIUM and TYPES: WU (acquired by E. von Halácsy and incorporated in WU with the latter's herbarium; the Greek part of the Halácsy herbarium, which was temporarily on loan at W, is now back at WU (inf. W. Gutermann, 12 Aug 1979, 19 Nov 1980, in lit.)). Other material at C, DBN, GB, MANCH.

BIBLIOGRAPHY and BIOGRAPHY: AG 6(1): 614, 12(3): 228; Barnhart 3: 153 (b. 16 Mai 1855, d. 28 Dec 1891); Biol.-Dokum. 14: 7505; BJI 1: 48; BM 4: 1700; Clokie p. 232; CSP 11: 173-174, 18: 189; DTS 1: 245; Herder p. 119, 433; Hortus 3: 1202; IH 1 (ed. 7): 341, 2: (in press); Jackson p. 77; Kew 4: 463; Morren ed. 10, p. 31; Rehder 5: 724; TL-1/1083; Tucker 1: 596; Zander ed. 10, p. 707, ed. 11, p. 806.

BIOFILE: Anon., Bot. Jahrb. 15 (Beibl. 34): 17. 1892 (d.); Bot. Not. 1892: 47 (d.); J. Bot. 30: 96. 1892 (d.); Nat. Nov. 14: 101. 1892 (d.).
Bennett, A., J. Bot. 29: 75-76. Mar 1891 (Notes on *Potamogeton* as treated by R. in *Pl. eur.* p. 11-16).
Chater, A.O. & R.D. Meikle, Taxon 12: 239. 1963 (infraspecific categories in *Pl. eur.*).
Greuter, W., Candollea 2381): 81-99. 1968 (on subspecific names in *Pl. eur.*).
Sojak, J., Čas. nar. Muz. 15(1)(Bot.): 14. 1982.
Stearn, W.T., J. Bot. 77: 89-91. 1939.
Tournay, R., Bull. Soc. Bot. Belg. 101(2): 323-326. 1968 (on subspecific names in *Pl. eur.*).
Wettstein, R. v., Ber. deut. bot. Ges. 10: (27)-(30). 1892 (obit.), Österr. bot. Z. 42: 72. 1892 (d. 28 Dec 1891; "Richter's Herbarium geht in den Besitz Dr. v. Halásy's über").

COMPOSITE WORKS: Contributed *Convolvulaceae* to O. Stapf, Bot. Erg. Polak. Exp. 1885, p. 22-25.

9184. *Die botanische Systematik* und ihr Verhältniss zur Anatomie und Physiologie der Pflanzen. Eine theoretische Studie ... Wien (Verlag von Georg Paul Faesy ...) 1885. Oct. (*Bot. Syst.*).
Publ.: Aug 1885 (Bot. Centralbl. 7-11 Sep 1885; Nat. Nov. Sep(1) 1885; Bot. Zeit. 25 Sep 1885; J. Bot. Oct 1885), p. [i]-iv, [1]-172, [173, cont.], [174, err.]. *Copies*: G, M.
Ref.: Pax, F., Bot. Jahrb. 7 (Lit.): 86-87. 1886 (rev.).
Koehne, E., Bot. Zeit. 44: 262-271. 16 Apr 1886 (rev.).
Wettstein, R. v., Österr. bot. Z. 36: 136-137. 1886 (rev.).

9185. *Plantae europeae* [vol. 2: *europaeae*]. Enumeratio systematica et synonymica plantarum phanerogamicarum in Europa sponte crescentium vel mere inquilinarum autore Dr. K. Richter [vol. 2: Operis a Dr. K. Richter incepti tomus ii emendavit edditque Dr. M. Gürke] ... Leipzig (Verlag von Wilhelm Engelmann) 1890-1903, 2 vols. *Pl. eur.*).
Authorship: Continued, after Richter's death (1891) by Robert Louis August Max Gürke (1854-1911). Gürke was "Kustos" at Berlin-Dahlem and his types are (were) at B.
1: Oct 1890 (Stearn (1939); Nat. Nov. Nov(1) 1890; J. Bot. Jan 1890; Bot. Zeit. 26 Dec

1890; BSbF Oct-Dec 1890; ÖbZ Nov 1890; Bot. Centralbl. 10 Dec 1890), p. [i]-vi, [vii], [1]-378.
2(1): 1 Jul 1897 (fasc. cover dated), p. [i]-vi, [1]-160.
(2): 31 Jan 1899 (id.), p. 161-320.
(3): 8 Dec 1903 (id.), p. 321-480.
Copies: BR, G, H, HH, MICH, M, MO, NY, PH, US, USDA; IDC 5678. – For contemporary commentaries see A.F. Le Jolis, *Notes Pl. eur.* 1891, TL-2/4363 and G. Rouy, Bull. Soc. bot. France 38: 94-102, 130-142, 223. 1891 (repr. Rouy 1891, see below). See Stearn (1939) for notes on dates and Chater and Meikle (1963) for a discussion of the infrageneric characters used in the *Pl. eur.*
Ref.: Britten, J., J. Bot. 29: 85-88. Mar 1891 (rev.).
Harms, H., Bot. Centralbl. 72: 237-240. 1897 (rev. 2(1), 98: 44-47. 1905 (2(3)).
Jänniche, Bot. Centralbl. 46: 128-129. 1891.
Koehne, E., Bot. Zeit. 49: 723-724. 1891.
Malinvaud, E., Bull. Soc. bot. France 44: 385-386. 1897 (rev. 2(1)).
Rouy, E., Annotations aux Plantae europaeae, fasc. 1. 1891.

Richter, Lajos (Ludwig) (1844-1917), Hungarian clerk and amateur botanist and plant collector in Presburg (Bratislava), later in Budapest. (*L. Richt.*).

HERBARIUM and TYPES: Richter's herbarium was bought by the Rumanian Government (A. Kneucker, Allg. bot. Z. 24/25: 31. 1920); its present location is not known to us. Richter obtained plants from many correspondents all over the world and apparently used several of them for further exchange. Material with his herbarium labels is in many herbaria, e.g. A, AK, B (extant, see Lack 1980), BAF, BP, BUC, BUF, C, CORD, E, F, FI, GB, GH, GOET, L, MANCH, MW, NA (important set), NAP, OXF, PH, US, W, WRSL.

BIBLIOGRAPHY and BIOGRAPHY: AG 4: 866, 6(2): 921; Barnhart 3: 153 (b. 17 Dec 1844, d. 7 Mai 1917); Bossert p. 331; Clokie 232; CSP 18: 190; DTS 1: 396; Futak-Domin p. 503; IH 1 (ed. 6): 363, 2: (in press); Kanitz 317; Morren ed. 10, p. 39; Rehder 2: 106; Urban-Berl. p. 388.

BIOFILE: Anon., Bot. Centralbl. 35: 112. 1888 (has bought herb. Holuby, Steinitz and "the fungi of Kalchbrenner") [sic Kalchbrenner's herbarium is now at BRA! This must be a different part of K.'s collections, see Kotlaba, Taxon 24: 348. 1975.]; Bot. Not. 1917: 206 (d.); Hedwigia 60: (92). 1918 (d.); Österr. bot. Z. 67: 48. 1918 (d. 7 mai 1917).
Degen, A., 16: 182-183. 1917 (obit.).
Goulding, J.H., Rec. Auckland Inst. Mus. 12: 112-114. 1975 (correspondence with Cheeseman; label; handwriting; material at AK).
Kanitz, A., Linnaea 33: 656. 1865.
Lack, H.W., Willdenowia 10: 81. 1980 (coll. in herb. zool.-bot. Ges. Wien at B).
Richter, L., Beiträge zur Flora von Presburg, Corr.-Bl. Ver. Nat. Presburg 2 (4, 5): 997-112. Apr-Mai 1863 (first publication; at age 19).
Wagenitz, G., Index coll. princ. herb. Gott. 135. 1982.

HANDWRITING: Rec. Auckland Inst. Mus. 12: 113. 1975.

Richter, Oswald (1878-1955), Praha-born Austrian botanist; studied at the Deutsche Universität of Praha; habil. ib. 1907; habil. Techn. Hochschule ib. 1909; assistant at the institute for plant physiology of the Deutsche Univ. 1907-1910; id. at the Universität Wien 1910; Adjunkt ib. 1911; extraordinary prof. of botany Wien 1912-1920; ordinary prof. of botany at the Deutsche Techn. Hochschule, Brünn (Brno) 1920. (*O. Richt.*).

HERBARIUM and TYPES: Unknown.

BIBLIOGRAPHY and BIOGRAPHY: Barnhart 3: 154 (b. 1 Jun 1878); Biol.-Dokum. 14: 7505-7506; BJI 2: 145; Kew 4: 463; LS 38362.

BIOFILE: Anon., Ber. deut. bot. Ges. 68: 145. 1955 (d. 8 Apr 1855 at Hannover; was this indeed Oswald Richter, b. 1855?); Bot. Centralbl. 104: 240. 1907 (habil. Deutsche Univ. Prag), 110: 32. 1909 (habil. Techn. Hochschule Prag); Nat. Nov. 29: 207. 1907 (habil. Prag, Deut. Univ., 32: 568. 1910 (assistant pl. physiol. Wien), 35: 76. 1913 (extraordinary professor for plant anatomy and physiology. Univ. Wien Wien), 42: 201. 1930 (professor der Botanik und technischer Mykologie a.d. Deutschen Technischen Hochschule Brünn), Hedwigia 46: (144). 1907 (habil. Prag, Deutsche Univ., anat. and physiol. of plants), 48: (198). 1909 (habil. Prag, Techn. Hochschule, for botany), 50: (143). 1910 (asst. for plant physiology Univ. Wien), 51: 88. 1911 ("Adjunkt" ib.), 53: (235). 1913 (extraord. prof. bot. ib.); Mycol. Centralbl. 2(3): 190. 1913 (extraord. prof. Univ. Wien); Österr. bot. Z. 59: 79. 1909 (habil. Techn. Hochschule, Prag), 60:287. 1910 (assistant pl. physiol. Univ. Wien), 61: 119. 1911 ("Adjunkt"), 62: 495. 1911 (extraord. prof. bot.), 69: 272. 1920 (ord. prof. bot. Deutsche Techn. Hochsch. Brünn).
Kneucker, A., Allg. bot. Z. 16: 144. 1910 (to Vienna), 17: 96. 1911 (Adjunkt Univ. Wien), 19: 48, 176. 1913 (extraord. prof. bot.).
Krumbiegel, I., Gegor Mendel 142. 1967.
VanLandingham, S.L., Cat. diat. 6: 3585. 1978.

9186. *Zur Physiologie der Diatomeeen* (II. Mitteilung). Die Biologie der Nitzschia putrida Benecke ... Wien (aus der kaiserlich-königlichen Hof– und Staatsdruckerei ...) 1909. Qu.
Publ.: 1909, cover, p. [i], [1]-116, *pl. 1-4*, 2 charts. *Copy*: FH. – Reprinted and to be cited from Denkschr. math.-nat. Kl. k. Akad. Wiss., Wien 84: 657-772, *pl. 1-4*, 2 charts. 1909.

9187. *Die Ernährung der Algen* von Oswald Richter (Wien). Leipzig (Verlag von Dr. Werner Klinkhardt) 1911. Qu. (*Ernähr. Alg.*).
Publ.: Jan 1911 (Bot. Centralbl. 17 Jan 1911; Nat. Nov(2) 1910, apparently an announcement), p. [i]-vii, [viii], [1]-192, [193]. *Copies*: FH, USDA. – Monographien und Abhandlungen zur Internationalen Revue der gesammten Hydrobiologie und Hydrographie Band 2.

Richter, Paul Boguslav (1853-1911), German palaeobotanist and high school teacher at Quedlinburg (Saxony). (*P.B. Richt.*).

HERBARIUM and TYPES: US, Fossil collections at S-PA, see Nathorst (1912).

BIBLIOGRAPHY and BIOGRAPHY: Andrews p. 321; Barnhart 3: 154 (b. 12 Dec 1853, d. 9 Oct 1911); BJI 2: 145; BM 4: 1700; CSP 18: 190; Kew 4: 464; Quenstedt p. 360.

BIOFILE: Anon., Geol. Mag. ser. 2. dec. 5. 8: 528. 1911 (brief obit.).
Nathorst, A.G., Palaeobot. Z. 1: 50-51. 1912. (palaeobot. coll. acquired by S-PA).

9188. *Beiträge zur Flora der unteren Kreide Quedlinburgs* ... Leipzig (Verlag von Wilhelm Engelmann) 1906-1909. (*Beitr. Fl. unter. Kreide Quedlinb.*).
1: 1906 (Bot. Centralbl. 17 Jul 1906; Bot. Zeit. 1 Mai 1906; ÖbZ Aug-Sep 1906; Allg. bot. Z. 15 Sep 1906), p. [i*], [i]-iv, [1]-27, *pl. 1-7*.
2: 1909 (Bot. Zeit. 16 Dec 1909; Bot. Centralbl. 22 Mar 1910; Bot. Zeit. 1 Mai 1910; Allg. bot. Z. 15 Jan 1910), p. [1]-12, *pl. 8-13*.
Copy: BR. – Other publications on this subject in Beil. Progr. Gymn. Quedlinburg 1904, p. 3-20, *pl. 1-2* and 1905: 1-19, *pl. 1-4* (n.v.).
Ref.: Salfeld, H., Bot. Zeit. 67: 337. 1909 (rev. 2).
Solms-Laubach, H., Bot. Zeit. 64(2): 167-168. 1906 (rev. 1).

9189. *Beiträge zur Flora der oberen Kreide Quedlinburgs* und seiner Umgebung ... Leipzig (Verlag von Wilhelm Engelmann) 1905-1909. 2 parts. (*Beitr. Fl. ober. Kreide Quedlinb.*).
1: 1905 (Bot. Centralbl. 5 Dec 1905; Bot. Zeit. 1 Dec 1905), p. [i], [1]-18, [19-20], *pl. 1-6* (uncol.).
2: 1909, p. [i], [1]-12, *pl. 8-13*.

Copy: NY. – See also his *Über die Kreidepflanzen der Umgebung Quedlinburgs*, 1, 1904 (20 p., 2 *pl.*), 2, 1905 (19 p., *2 pl.*) in Progr. Kgl. Gymnasium Quedlinburg. 1904, 1905.

Richter, Paul Gerhard (1837-1913), German (Saxonian) algologist; close associate of L. Rabenhorst; high school teacher at Leipzig 1861-1907; editor from 1879 of Rabenhorst's *Krypt.-Fl.* ed. 2. (*P.G. Richt.*).

HERBARIUM and TYPES: Issued *Phykotheka generalis*, Sammlung getrockneter Algen sämmtlicher Ordnungen und aller Gebiete, fasc. 1-15, nos. 1-750. Leipzig 1885-1896; fasc. 1-7 with F. Hauck (q.v. TL-2/2: 100), q.v. for a list of complete sets. The *Phykotheka* is in fact a continuation of L. Rabenhorst's *Die Algen Europas*. Richter contributed also to this latter series.
Richter contributed also to V. Wittrock & O. Nordstedt, *Algae aquae dulcis*. – Correspondence with W.G. Farlow on the Phykotheka at FH.

BIBLIOGRAPHY and BIOGRAPHY: AG 6(2): 922; Barnhart 3: 154 (b. 16 Mai 1937, d. 19 Jul 1913); Clokie p. 232; CSP 5: 745, 4: 174, 12: 616, 18: 190; De Toni 1: cvii, 2: lxiii, ci, cxxv, 4: xlvii; IH 2: (in press); KR p. 589; LS 1343; Morren ed. 10, p. 22; Nordstedt p. 35, suppl. p. 13, 18; TL-2/see F. Hauck; Urban-Berl. p. 268,388.

BIOFILE: Anon., Ber. deut. bot. Ges. 31: (182). 1913 (d.).; Bot. Centralbl. 22: 90. 1885 (announcement *Phykotheka*); Bot. Not. 1914: 31 (d.); Bot. Zeit. 43: 189-190. 1885 (announcement Phykoteka); Hedwigia 54: (125). 1913; Nuova Notarisia 25: 137-138. 1914 (bibl.).
Bary, A. de, Bot. Zeit. 44: 431. 1886 (fasc. 1 of *Phykotheca*).
Dodel-Port, A., Bot. Centralbl. 34: 213-215, 249-250, 283-285. 1888 (rev. fasc. 1-3 *Phykotheca*), 42: 362-367. 1890 (id. fasc. 4-5).
Kolkwitz, R., Ber. deut. bot. Ges. 32: (64)-(67). 1914 (obit., bibl.).
VanLandingham, S.L., Cat. diat. 4: 2366. 1971; 6: 3585. 1978, 7: 4223. 1978.

COMPOSITE WORKS: (1) Fresh water algae (except diat.) in O. Kuntze, *Rev. gen. pl.* 3(2): 385-391. 1898.
(2) Richter signed as "Die Redaktion" (the editors/editor) for L. Rabenhorst's *Krypt. Fl.* ed. 2, in vol. 5, preface to Migula, *Synopsis Characearum*. See under L. Rabenhorst. Kolkwitz (1914) states that Richter was editor for the *Krypt.-Fl.* from 1879.
(3) *Grönländische Süsswasseralgen*, Bibl. bot. Heft 42, 1899, *Botanische Ergebnisse der ... unter Leitung Dr. von Drygalski's ausgesandten Grönlandsexpedition ...*", ed. E. Vanhöffen.
(4) Richter also ran the *Leipziger botanischer Tauschverein*, see e.g. his Doubletten-Verzeichniss, xxviii. Tauschjahr, Jan 1879, 24 p. (copy HH).

Rick, Johann [Johannes] (1869-1946), Austrian mycologist and clergyman (S.J.); high school teacher at the Stella Matutina, Feldkirch, Vorarlberg 1894-1898; from 1899-1902 at Valkenburg, Netherlands, for theological studies; in 1903 at Barro in Portugal studying Portugese; from 1903-1915 teacher at the Jesuit College at S. Leopoldo, Rio Grande do Sul, Brazil; from 1915 doing social work; from 1929 professor of theology in the Seminary of S. Leopoldo; from 1942-1946 at S. Salvador, Rio Grande do Sul. (*Rick*).

HERBARIUM and TYPES: Brazilian fungi in B, BPI, CUP, FH, IAC, IACM, K, MICH, PACA (important set), R, RB, S, SFPA, SI. – Many of these fungi were distributed in the series *Fungi austro-americani*, fasc. 1-15, nos. 1-300, 1904-1911, distributed from Feldkirch by Jos. Rompel; fasc. 16-18 were by Rick and F. Theissen, q.v. For full details, published schedae (in Ann. mycol. 2-9, 1904-1911) see J. Stevenson (1971). The fascicles and their contents were also duly listed by ÖbZ and Bot. Centralbl.

BIBLIOGRAPHY and BIOGRAPHY: Bossert p. 331; CSP 18: 191; DTS 6(4): 67, 1: xxiii, 246, 3: l, 6(4): 67; Hawksworth p. 185; IH 1 (ed. 7): 341, 2: (in press); Kelly p. 191, 256; Kew 4: 464; LS 22254-22270, 38367; Stevenson p. 1255; Urban-Berl. p. 281.

BIOFILE: Burkart, A., Darwiniana 2: 141. 1928 (important set of Brazilian fungi at SI). Fidalgo, P., Rickia 1: 3-11. 1962 (biogr. sketch, portr., bibl.).

Murr, J., Neue Übers. Farn-Bl.-Pfl., Vorarlberg 1/2: xx. 1923.
Rambo, B., Iheringia, Bot. 2: 1-12. 1958 (intr. to Basidiomyc. Rio Grande do Sul, biogr., bibl. portr., b. 19 Jan 1869, d. 6 Mai 1946).
Reitz, R., Anais bot. Herb., Itajai. Sta. Catarina 1: 70-84. 1949 (n.v.).
Rick, J., Iheringia, Bot. 2: 13-14. 1958 (posthumous publ. of his Basidiomyc. in Rio Grande do Sul; portr.).
Stevenson, J., Beih. Nova Hedwigia 36: 338-344. 1971 (on his *Fungi austro-americani*).
Torrend, C., *in* Lloyd, C.G., Mycol. notes 53: [749]-751. 1918 (portr.).

EPONYMY: (genera): *Rickella* Locquin (1952); *Rickia* Cavara (1899); *Rickiella* H. Sydow ex Rick (1904); (journal): *Rickia*, serie criptogamica dos "Arquivos de Botanica do Estado de São Paulo, Brazil. Vol. 1-x, 1962-1963-x.

NOTE: Rick's major study *Basidiomycetes eubasidii in Rio Grande do Sul*-Brasilia, was published by B. Rambo, in Iheringia, Bot. 2-9, 1958-1961 (p. 1-480), with a biographical note on p. 8-12.

9190. *Monographia Sphaerialium astromaticorum Riograndensium* [Broteria, vol. 2, 1933]. Oct.
Publ.: 1933, with double pagination in Broteria 2(3): [133]-145 or [1]-13. 1 Aug 1933 and 2(4): [169]-201 or [14]-46. 1 Nov 1933.

9191. *Monographia das Xylariaceas riograndenses* ... Archivos do Museu nacional vol. xxxvi. Rio de Janeiro [1935]. Qu.
Publ.: 1913, cover, err. slip, p. [i], [41]-71, *6 pl. Copy*: FH. – Issued as Arch. Mus. nac. 36, 1935 (journal issue n.v.).

Ricken, Adalbert (1851-1921), German (Hesse-Nassau, Prussia) clergyman (roman catholic priest); ord. 1873 Fulda; chaplain and parish priest in various locations in the Fulda diocese (1873-1875 Dernbach, 1875-1885 Weimar, 1885-1887 Fritzlar, 1887-1907 Aufenau, from 1907-1921 at Lahrbach; self-taught mycologist; Dr. phil. h.c. Univ. Würzburg. (*Ricken*).

HERBARIUM and TYPES: Unknown; Ricken's original paintings of fungi were destroyed in World War II; a collection of microscopical preparations was at B (now destroyed).

BIBLIOGRAPHY and BIOGRAPHY: Barnhart 3: 154 (b. 30 Mar 1851; d. 1 Mar 1921); Biol.-Dokum. 14: 7515; BM 4: 1701, 8: 1076; Kelly p. 191; Kew 4: 465; LS 38372; NI 1637; Stevenson p. 1255.

BIOFILE: Anon., Nat. Nov. 23: 78. 1921 (d.).; Österr. bot. Z. 70: 152. 1921 (d.).
Kropp, G., Pilz– Kräuterfreund 4(9): 186-187. 1921 (obit., portr.).
Moser, M., Z. Pilzk. 37: 13-18. 1971 (R. and *Cortinarius*).
Pieschel, E., Z. Pilzk. 37: 7-11. 1971 ("Erinnerungen", commemor.; photographs and portr. on cover, *pl. 14* and p. 1-5), Mykol. Mitt.-Blatt 16(1): 30-34. 1972.
Wolfarth, F., Z. Pilzk. 37: 1-6. 1971 (1972).

MEMORIAL PUBLICATION: Der Pilz– und Kräuterfreund 4(9), Mar 1921, Ricken-Gedächtnis-Heft.

EPONYMY: *Rickenella* Raithelhuber (1973) very likely honors this author.

9192. *Die Blätterpilze* (Agaricaceae) *Deutschlands* und der angrenzenden Länder, besonders Oesterreichs und der Schweiz ... Leipzig (Verlag von Theodor Oswald Weigel) [1910-] 1915. Oct. (*Blätterpilze Deutschl.*).
Preface material: Aug-Oct 1915, p. [i*-iii*], [i]-xxiv.
Fasc. 1: Nov-Dec 1910 (p. iv: 1 Aug 1910; ÖbZ Nov-Dec 1910; Nat. Nov. Dec(2) 1910; PH rd 5 Jan 1911), p. [i]-iv, [1]-32, [with *pl. 1-8*].
Fasc. 2: Feb-Mar 1911 (cover dated 1910 but p. ii has references to Jan 1911 reviews; Bot. Centralbl. 6 Jun 1911; Hedwigia 15 Jul 1911; ÖbZ Apr 1911; Nat. Nov. Mar(2) 1911), p. 33-64, [with *pl. 9-16*].

Fasc. 3/4: Jun-Jul 1911 (Bot. Centralbl. 7 Nov 1911; Hedwigia 10 Apr 1912; Nat. Nov. Aug(1) 1911; PH rd 16 Oct 1911), p. 65-128, *pl. 17-32*].
Fasc. 5/6: Mar-Apr 1912 (Hedwigia 14 Dec 1912; ÖbZ Mai 1912; Nat. Nov. Apr(1) 1912; PH rd 4 Jun 1912), p. 129-192, [*pl. 33-48*].
Fasc. 7/8: Sep-Oct 1912 (Hedwigia 14 Dec 1912; ÖbZ Oct-Nov 1912; Nat. Nov. Oct(1) 1912), p. 193-256, [*pl. 49-64*].
Fasc. 9/10: Sep 1913 (PH rd 3 Nov 1913; Mycol. Centralbl. 21 Nov 1913; Bot. Centralbl. 25 Nov 1913; Hedwigia 10 Dec 1913; ÖbZ Sep 1913; Nat. Nov. Oct(1) 1913), p. 257-320, [*pl. 65-80*].
Fasc. 11/12: Oct 1914 (ÖbZ Oct 1914; Nat. Nov. Nov(1, 2) 1914; PH rd 18 Jan 1915), p. 321-384 [*pl. 81-96*].
Fasc. 13/14: Mai-Jun 1915 (Hedwigia 10 Sep 1915; ÖbZ Mai-Jun 1915; Nat. Nov. Jun(1, 2) 1915; PH rd 7 Jul 1915), p. 385-448, [*pl. 97-112*].
Fasc. 15: Aug-Oct 1915 (ÖbZ Aug-Nov 1915; Nat. Nov. Oct(1, 2) 1915; PH rd 27 Dec 1915), p. 449-480.
Plates: 1910-1915, p. [i*-iii*], [i]-vii, [viii], *pl. 1-112* (col.).
Copies: BR, FH, FI, H, M, MICH(2), NY, PH.
Facsimile ed.: announcement for publication in 1980 by Koeltz Scientific Books.
Ref.: Lindau, G., Hedwigia 57: (96)-(97). 1916.

9193. *Vademecum für Pilzfreunde* Taschenbuch zur bequemen Bestimmung aller in Mittel-Europa vorkommenden ansehnlicheren Pilzkörper mit vier Bestimmungstafeln und Zitaten bekannter Bildwerke ... Leipzig (Verlag von Quelle & Meyer) 1918. Oct. (*Vadem. Pilzfr.*).
Publ.: Mai-Jun 1918 (Hedwigia 16 Nov 1918; ÖbZ 1 Sep 1918; Nat. Nov. Jun (1, 2) 1918), p. [ii*], [i]-xx, [1]-334, [1]. *Copies*: FH, G, M.
Ed. 2: Mai-Jun 1920 (ÖBZ, Dec 1920; Nat. Nov. Jun(1, 2). 1920),p. [i]-xxiv, [1]-352. *Copies*: FH, H, MICH, NY. – "Zweite vermehrte und verbesserte Auflage".
Facsimile ed. (of ed. 2): 1969, 3301 Lehre, Verlag von J. Cramer, as original but with special cover. *Copies*: BR, FAS, NY. – ISBN 3-7682-0603-3.

Ricker, Percy Leroy (1878-1973), American botanist; M.S. Univ. Maine 1901; assistant in biology University of Maine until 1901, from then on with USDA. (*Ricker*).

HERBARIUM and TYPES: WIS (10.000 parasitic fungi); collections and types in US; other material in DPU and NY. – Smithsonian Archives SIA 55, 221, S1.B25; letters to W.G. Farlow at FH.

BIBLIOGRAPHY and BIOGRAPHY: Barnhart 3: 154 (b. 27 Mar 1878); BJI 2: 145; BL 1: 172, 315; Bossert p. 331; Hawksworth p. 185; Hortus 3: 1202; IH 1 (ed. 6): 363, (ed. 7): 341, 2: (in press); Kelly p. 192; Lenley p. 343; LS 22265-22270, 38373-38374; MW p. 414; NAF 7: 1085; NW p. 56; Stevenson p. 1255; TL-2/3574.

BIOFILE: Anon., Amer. Mag. 146: 111. 1948 (portr.); Taxon 22: 328. 1973 (d. 2 Feb 1973).
Ewan, J. et al., Short hist. bot. U.S. 14. 1969.
Pennell, F.W. & J.H. Barnhart, Monogr. Acad. Nat. Sci. Philadelphia 1: 617. 1935 (coll.).
Ricker, P.L., Wild Flower 40: 31-32. 1964 (autobiographical).
Verdoorn, F., ed., Chron. bot. 4: 452. 1938.

COMPOSITE WORKS: With W.A. Kellerman, *New gen. fungi*, 1905, see TL-2/3574.

9194. *A preliminary list of Maine fungi* ... Orono, Maine, April, 1902. Oct. (*Prelim. list Maine fung.*).
Publ.: Apr 1902 (t.p.), p. [1]-86, [87, err.]. *Copies*: FH (rd 7 Jun 1902), NY. – The University of Maine Studies no. 3.

Rickett, Harold William (1896-x), English-born American botanist; emigrated to the United States 1910; studied at Harvard Univ. 1913-1915 and Univ. of Wisconsin (Dr.

phil. 1922); in France in World War I 1917-1918; instructor at Univ. Wisconsin 1922-1924; asst. professor of botany at Univ. of Missouri 1924-1928; assoc. professor 1928-1939 ; visiting lecturer Reed College 1937-1938; bibliographer at the New York Botanical Garden 1939-1964 (succeeding J.H. Barnhart in 1942), writing *Wild flowers of the United States* 1964-1973; in retirement at Carmel, California 1973-x. (*Rickett*).

HERBARIUM and TYPES: Material in LIV, MO, NY and UMO. – Professional correspondence at NY; see also Smithsonian Archives SIA 227, 7097.

BIBLIOGRAPHY and BIOGRAPHY: Barnhart 3: 154 (b. 30 Jul 1896); BFM no. 2438, 2466; BL 1: 194, 315; Bossert p. 331; Hirsch p. 247; Hortus 3: 1202; IH 1 (ed. 1): 69, (ed. 2): 88, (ed. 3): 111, (ed. 4): 122, 2: (in press); Kew 4: 465; Langman p. 626; Lenley p. 343, 344, 465; MW suppl. p. 293; NI 1: 159, 3: 11, 80; PH 209; TL-2/781, 5188, see J. Colden; Zander ed. 10, p. 707, ed. 11, p. 806.

BIOFILE: Anon., Taxon 22: 523. 1973 (retired 3 Mai 1973; to Carmel, California).
Rickett, H.W., Index Bull. Torrey bot. Cl. 82. 1955.
Stafleu, F.A., Taxon 23: 387-388. 1974 (brief biogr. sketch).
Stieber, M.T. et al., Huntia 4(1): 85. 1981 (arch. mat. HU).
Verdoorn, F., ed., Chron. bot. 1: 290. 1935.

COMPOSITE WORKS: (1) Editor of *North American flora* 1944-1963 (author of bibliographies of vol. 28B and ser. 2, vols. 2 and 3; treatments of *Cornaceae* 28B(2): 299-311. 28 Dec 1945, *Nyssaceae* (28B(2): 313-316. 28 Dec 1945).
(2) Editor of Bulletin Torrey bot. Club 67(4)-76(3), Apr 1940-Apr 1949.
(3) With W.J. Robbins, q.v., *Botany* a textbook for college and University Students, New York 1929.
(4) Editor, K.K. Mackenzie, N. Amer. Caric. 1940, TL-2/5188.

NOTE: Rickett was the editor and general author of *Wild flowers of the United States*, six volumes (in 14) 1966-1973, see e.g. F.A. Stafleu, Taxon 16: 200-202. 1967 (rev. vol. 1), 22: 143. 1973 (vols. 2-5), and 23: 387-388. 1974. *Index volume*, New York 1975, x, 152 p. – The bibliographical work done by H.W.R. and F.A.S. on the list of *nomina conservanda* in the "Code" (RS; Taxon 10: 70-91, 111-121, 132-149, 170-177. 1961) gave rise to the documentation from which *Taxonomic literature* ed. 1 was compiled.

THESIS: *Fertilization in Sphaerocarpos*, Ann. Bot. 37: 225-259, *pl. 3-4*. Apr 1923. – See also *Regeneration in Sphaerocarpos Donnellii*, Bull. Torrey bot. Club 47: 347-357. 5 Aug 1920.

9195. *Flora of Columbia, Missouri* ... The University of Missouri Studies ... Columbia (University of Missouri) 1931. Oct.
Publ.: 1 Jan 1931, p. [1]-84, map. *Copies*: HH, NY, US, USDA. – Published and to be cited as Univ. Missouri Stud. 6(1): 1-84, map. 1931.

9196. *The green earth* an invitation to botany ... Lancaster, Pennsylvania (The Jacques Cattell Press) [1943]. (*Green earth*).
Publ.: 1943 (MO rd 16 apr 1943), p. [i-iv], [1]-353. *Copies*: HH, MO, NY, PH, USDA.

Rickett, Theresa Cecil, née **Bauchman** (1902-x), American botanist; married 28 Aug 1923, Madison, Wisc., Harold William Rickett; Wisconsin State Teacher's College (Stevens Point) 1919; BS Univ. Wisconsin 1922; high school teacher at Friendship, Wisc. 1919-1920, Columbus, Wisc. 1922-1923. (*T. Rickett*).

HERBARIUM and TYPES: Unknown.

BIBLIOGRAPHY and BIOGRAPHY: Barnhart 3: 154 (b. 9 Apr 1902).

9197. *Wild flowers of Missouri* a guide for beginners ... Illustrated by H.W. Rickett ... Columbia, Missouri (University of Missouri ...) 1937. Oct. (*Wild fl. Missouri*).

Publ.: Mai 1937 (t.p.), p. [1]-144, *40 pl. Copies*: MO(2). – Circular 303, Univ. Missouri Coll. Agric. agr. Extens. Serv.
Ed. 2: 22 Mai 1954 (t.p.), p. [1]-148, *40 pl. Copies*: MO(2). Photographs by H.W. Rickett and E.M. Palmquist. Second edition revised and edited by E.M. Palmquist and C.L. Kucera, Missouri Handbook number 3.
Facsimile reprint: 1968, ISBN 0-8262-0587-9, Univ. Missouri Press, Columbia. *Copy*: MO.

Riddell, John Leonard (1807-1865), American botanist, chemist, physician and inventor; professor of astronomy, geology, botany and chemistry Marietta College 1832; id. chemistry and botany at the Ohio reformed Medical College, Worthington, Ohio 1832-1834; at Cincinnati, Ohio 1834-1836, from 1835-1836 as adjunct professor of botany and chemistry; from 1836 at New Orleans at the Medical College of Louisiana and in other capacities, ultimately as City Post Master. (*Riddell*).

HERBARIUM and TYPES: Riddell material is at BM, DWC, GH, K, LLO, MO, NO, NY, OXF, P ("herbier historique"), PH, US (residue of Ohio herb.), WECO; a small herbarium of Ohio plants is also in the Marietta College Herbarium.

BIBLIOGRAPHY and BIOGRAPHY: Barnhart 3: 155; BL 1: 183, 315; BM 4: 1701; Clokie 232; CSP 5: 197-198, 12: 616; DAB 15: 584-590 (by R. Matas and V. Gray); GR p. 240; Hortus 3: 1202; IH 1 (ed. 7): 341, 2: (in press); Jackson p. 361, 364; Kew 4: 4: 465; Lenley p. 465, 467; LS 22271a; ME 1: 224, 3: 645 [index]; PH 305, 364; PR 7624-7625, ed. 1: 8580-8581; Rehder 5: 724; SIA 7053; Tucker 1: 596; Zander ed. 10, p. 707, ed. 11, p. 806.

BIOFILE: Allibone, S.A., Crit. dict. engl. lit. 2: 1804. 1878.
Anon., Amer. J. Sci., ser. 2. 41: 141-143. 1866.
Armstrong, C., Ohio Natural. 1: 33. 1901 (b. 20 Feb 1807, d. 7 Oct 1863).
Bailey, L.H., Bot. Gaz. 8: 269-271. 1883 (biogr.).
Ewan, J., Southw. Louisiana J. 7: 25. 1967 (bibl.).
Ewan, J. et al., Short hist. bot. U.S. 39, 45, 90. 1969.
Geiser, S.W., Field & Laborat. 4(2): 52. 1936; Natural. Frontier 280. 1948.
Kelly, H.A., Some Amer. med. bot. 154-156. 1929 (portr.).
Pennell, F.W. & J.H. Barnhart, Monogr. Acad. nat. Sci. Philadelphia 1: 617. 1935.
Poggendorff, J.C., Biogr.-lit. Handw.-Buch 3: 1121. 1898.
Riess, K., Tulane Stud. Geol. Paleont. 13(1, 2): 1-110. 1977 (biogr., portr., bibl.; major study).
Rodgers, A.D., William Starling Sullivant 114, 116, 169. 1940.
Rogers, J.G., *in* H.A. Kelly & W.L. Burrage, Amer. med. biogr. 981-982. 1920.
Stuckey, R.L., Trans. Stud. Coll. Phys. Philadelphia 45(5): 270-273. 1978 (portr.).
Wilson, J.G. & J.Fiske, eds., Appleton's Cycl. Amer. biogr. 5:248. 1888.

EPONYMY: *Riddelia* Rafinesque (1838); *Riddellia* Rafinesque (1838, *orth. var.*); *Riddellia* Nuttall (1841).

9198. *A synopsis of the flora of the western states* by John L. Riddell, A.M. Lecturer on chemistry; ... Cincinnati (E. Deming) 1835. Oct. (*Syn. fl. west. states*).
Publ.: Jan-Apr 1835 (in journal), p. [1]-116. *Copies*: G, HH, KU, MICH, MO, NY, PH.
– Reprinted or preprinted from West. J. med. phys. Sci. 7(31): 329-374. Jan 1835; 7(32): 489-556. Apr 1835.
Ref.: Kellerman, W.A., *in* Geology of Ohio plants 59-60. 1895 (q.v. also for preliminary catalogue and a supplement).
Lubrecht, H., Early Amer. bot. works no. 126. 1967.

9199. *A supplementary catalogue of Ohio plants.*Catalogue and descriptions read, and specimens exhibited, before the Western Academy of natural sciences, March 16, 1836 ... Cincinnati (N.S. Johnson, ...) 1836. Qu.
Publ.: Apr 1836 (in journal), p. [1]-28. *Copies*: G, NY, PH. – Reprinted from West. J. med. phys. Sci. 9(36): 567-592. Apr 1836.

9200. *Catalogus florae ludovicianae*, auctore, J.L. Riddell, ... [From the New Orleans Medical and Surgical Journal, vol. viii, May number 1852]. Qu.
Publ.: Mai 1852, p. [743]-764. Separates with original pagination at MO, NY, PH, USDA. – An abridgement of a "Plants of Louisiana", a large manuscript work "communicated to the Smithsonian Institution. Joseph Henry submitted it to Asa Gray for his approval but it was suppressed". (Ewan 1967). A manuscript note by Brother Ephrem, F.S.C. in the USDA copy, dated 5-26-61 (with reference to C. Brown, Fern & Fern allies 1942, p. 4) indicates that the manuscript is no longer at the Smithsonian; it "was cut up and the individual references to the plants collected were pasted on the herbarium sheets of the plants described, and the sheets were deposited in the Gray Herbarium". J.H. Barnhart (mss. NY) also indicates that at least part of the manuscript, (or copies of the descriptions), accompanied by specimens, is at GH. – See also Ewan (1967) no. 82 for further publication in the New Orleans med. surg. J. 9: 609-618. 1853.

Riddelsdell, Harry Joseph (1866-1941), British clergyman and botanist (batologist); subwarden St. Michael's Theological College, Llandoff, Aberdare, 1897-1914; rector, Wigginton, Oxford 1914-1918; Bloxham, Oxford 1918-1936. (*Riddelsdell*).

HERBARIUM and TYPES: BM (incl. *Rubus* types); other material C, CGE, K, LIV, NMW, SWA (200).

BIBLIOGRAPHY and BIOGRAPHY: Barnhart 3: 155; BL 2: 218, 241, 317, 318, 708; Bossert p. 332; Clokie p. 233; Desmond p. 520 (d. 17 Oct 1941); Kew 4: 465-466; Lenley p. 465, 466.

BIOFILE: Anon., Proc. Cotteswold Natural. Field Club 27: 194-196. 1941 (obit., portr.).
Barton, W.C., Nature 149: 376. 1942 (obit.), Proc. Linn. Soc. London 1941/42: 294-295. 1942; Bot. Soc. Exch. Cl. Brit. Is. Rep. 12(5): 462. 1944.
Bridson, G.D.R. et al., Nat. hist. mss. res. Brit. Isl. 84.14, 84.18. 1980.
Ellis, G., Amgueddfa 16: 12-13. 1974.
Kent, D.H., Brit. Herb. 75. 1957 (coll.).
Lewis-Jones, J.L., BCG Newsletter 2(8): 368-369. 1980 (Glamorgan coll., 200, at SW).
Price, W.R., Bot. Soc. Exch. Club Brit. Isl. 12(5): 460-462. 1944 (obit., portr.).
Riddelsdell, H.J. et al., Fl. Gloucestershire clii-clv. 1948.
Verdoorn, F., ed., Chron. bot. 1: 171. 1935, 3: 169. 1937, 7(7): 354. 1943 (d.).
Wade, A.E., Fl. Monmouthshire p. 16.

9201. *A flora of Glamorganshire* [Journal of Botany, London, 1909, supplement]. Oct.
Publ.: Jan-Feb 1909 (Bot. Centralbl. 6 Jul 1909; possibly published in parts but signatures not dated; Nat. Nov. Mar(2). 1909), p. [1]-88. *Copies*: BR. – Issued as a supplement to J. Bot. 45, 1907, see also J. Bot. 47: 397-412. 1909 and Bot. Soc. Exch. Club Brit. Isl. Rep. 10: 666-669. 1934, and BL 2: 318.

9202. *Flora of Gloucestershire* phanerogams, vascular cryptogams, charophyta edited on behalf of the Cotteswold Naturalists' Field Club by the rev. H.J. Riddelsdell, ... G.W. Hedley, ... and W.R. Price ... Cheltenham (the Cotteswold Naturalist's Field Club ...) 1948. Oct. (*Fl. Gloucestershire*).
Publ.: 3 Sep 1948 (on separate slip in most copies), p. [i], frontisp., [iii]-clxxxii, [1]-667, 4 maps. *pl. 1-43*, err. slip. *Copies*: BR, E, G, NY, USDA.

Riddle, Lincoln Ware (1880-1921), American botanist (mycologist, lichenologist), A.B. Harvard 1902; Dr. phil. Harvard 1906; instructor in botany at Wellesley College 1906-1909; professor of botany ib. 1909-1919; assistant professor of cryptogamic botany at Harvard Univ. 1919-1921. (*Riddle*).

HERBARIUM and TYPES: FH (3500 lichens), further material at MICH, NEBC, NY, WELC. – Letters to W.G. Farlow and R. Thaxter at FH; other archival material at NY.

BIBLIOGRAPHY and BIOGRAPHY: Barnhart 3: 155 (b. 17 Oct 1880, d. 16 Jan 1921); GR p. 195, cat. p. 70; Hawksworth p. 185; IH 1 (ed. 6): 363, (ed. 7): 341, 2: (in press); Kelly p. 192; Lenley p. 344; LS 22272-22273, 33375; Nordstedt suppl. p. 13; PH 450; SO p. 130.

BIOFILE: Anon., Bot. Soc. Amer. Publ. 82: 104-105. 1922 (d. 16 Jan 1921); Bryologist 24: 32. 1921 (d.); Hedwigia 64(1): (69). 1921 (d.); Rhodora 23: 28. 1921 (d. 16 Jan 1921).
Cattell, J.M., Amer. men Sci. ed. 2. 394. 1910, ed. 3. 574. 1921.
Ewan, J. et al., Short Hist. bot. U.S. 95. 1969.
Fink, B., Bryologist 24: 33-36. 1921 (obit., portr., bibl.).
Laundon, J.R., Lichenologist 11: 16. 1979.
Rickett, H.W., Index Bull. Torrey bot. Club 82. 1955.
Robinson, B.L., *in* Development of Harvard University p. 345.
Thaxter, R., Rhodora 23: 181-184. 1921 (bibl.).

COMPOSITE WORKS: (1) Contributed the entry on lichens to N.L. Britton, *Fl. Bermuda*, 28 Feb 1918, p. 470-479, TL-2/785.
(2) Id. in N.L. Britton, *Bahama Fl.*, 26 Jun 1920, p. 522-553, TL-2/787.
(3) Member of the board of editors, *Bryologist*, 1911-1921.

EPONYMY: *Riddlea* C. Dodge (1953).

9203. *The North American species of Stereocaulon* ... Chicago (Univ. of Chicago Press) 1910. (*N. Amer. Stereocaulon*).
Publ.: 15 Oct 1910, p. 285-304. *Copy*: Almborn. – Reprinted and to be cited from Bot. Gaz. 50: 285-304. 1910.

9204. *Pyrenothrix nigra*, gen. et sp. nov. ... [Reprinted for private circulation from the Botanical Gazette, vol. lxiv no. 6, December 1917]. Oct.
Publ.: Dec 1917, p. 513-515. *Copy*: Almborn, G. – Reprinted and to be cited from Bot. Gaz. 64(6): 513-515. Dec 1917.

Ridley, Henry Nicholas (1855-1956), British botanist; B.A. Oxford; assistant at the Botany Dept., Brit. Mus. 1880-1888, to Fernando do Noronha 1887; director Botanic Gardens Singapore 1888-1912; in retirement living at Kew, England; Linnean Gold Medal 1950; chief promotor of the introduction of the *Hevea* rubber trees in the Malay Peninsula. (*Ridl.*).

HERBARIUM and TYPES: K; other material at A, AMES, B, BM, (orig. coll. Fernando do Noronha), BO, BR, CAL, DPU, E, H (bryo.), L, MICH, MO, NY, P, PC, SING, US, Z. – Manuscript and drawings at K.

BIBLIOGRAPHY and BIOGRAPHY: Backer p. 494-496; Barnhart 3: 155; BFM no. 2682; BJI 1: 49; BL 1: 90, 94, 104, 121, 129, 130, 315, 2: 322, 708; BM 1: 440, 4: 1702, 8: 1076-1077; Bossert p. 332; CSP 11: 177, 18: 197-198; Desmond p. 520 (b. 10 Dec 1855, d. 24 Oct 1956); DNB 1951-1960: 841-842; Herder p. 279, 288; Hortus 3: 1202; IF suppl. 1: 84, 3: 217, 4: 337; IH 2: (in press); Kew 4:378, 466-468; Langman p. 626; Lenley p. 344, 466-468; LS 22274-22293a, 38382-38394, 41638; Morren ed. 10, p. 71; MW p. 414, suppl. 293; NI 1638; Plesch p. 381; Rehder 5: 725; SK 1: xxix, xxxii, cxxii, 435-437 (extensive details), 4: cxxii, clx (index), 5: cccxiv, 8: lxxx; Tucker 1: 596-597; Urban-Berl. p. 311, 388; Zander ed. 10, p. 707, ed. 11, p. 806.

BIOFILE: Alphonso, A.G., *in* Rubber Centenary Committee, Singapore Rubber Centenary 43-45. 1977 (n.v.).
Anon., Bot. Centralbl. 1-2: 320. 1880 (asst. BM(NH)), 36: 32. 1888 (dir. SING), 51: 256. 1892 (exp. Pahang), 119: 560. 1912 (retirement); Bot. Jahrb. 10 (Beibl. 23): 1. 1889 (app. Singapore); Bot. Zeit. 38: 461. 1880 (app. bot. asst. BM); Bull. misc. Inf. Kew 1928: 158-159 (recipient of Frank N. Meyer medal for distinguished services in plant introduction); Gard. Chron. 1928(1): 330 (biogr. sketch, portr.); J. Bot. 66: 182. 1928; New York Herald Tribune 25 Oct 1956; The Times 12 Dec 1955, 25 Oct 1956 (portr.).

Bridson, G.D.R. et al., Nat. hist. mss. res. Brit. Isl. 431 [index]. 1980.
Burkill, I.H., Proc. Linn. Soc. London 169: 35-38. 1958 (obit.).
Gager, C.S., Brooklyn Bot. Gard. Rec. 27(3): 330. 1938.
Hedge, I.C., & J.M. Lamond, Index coll. Edinb. herb. 123. 1970.
Henderson, M.R & C.G.G.J. van Steenis, Gard. Bull. Straits Settlem. 9(1): 2-30. 1935 (bibl., portr.).
Holttum, R.E., M.A.H.A. Mag. 12(4): 3-6. 1955 (R. and horticulture; portr.); Nature 178: 1092. 1956 (obit.); J. Malay. Branch, Roy. Asiat. Soc. 33(1): 104-109. 1960 (publ. 1964); Taxon 6(1): 1-6. 1957 (personal recoll.; obit.), 19: 710-711. 1970.; *in* J. Arditti, ed., Orch. biol. 1: 17, 19, 20. 1977.
Kent, D.H., Brit. herb. 75. 1957.
Jackson, B.D., Bull. misc. Inf. Kew 4901: 26, 56 (coll.).
Lousley, J.E., Proc. Bot. Soc. Brit. Isl. 2(3): 328-331. 1957 (obit.).
Merrill, E.D., Enum. Philip. pl. 4: 221-222. 1926.
Murray, G., Hist. coll. BM(NH) 177. 1904 (coll.).
Nelmes & Cuthbertson, Curtis's Bot. Mag. dedic. 314-316. 1932 (portr.).
Purseglove, J.W., The Ridley centenary, 10 Dec 1955; Nature 176: 1092-1093. 1955 (100th birthday).
Reinikka, M.A., A history of the Orchid 279-282. 1972 (portr.).
Salisbury, E.J;, Biogr. mem. Roy. Soc. 3: 141-159. 1957 (obit., bibl.).
Sayre, G., Mem. New York Bot. Gard. 19(3): 387. 1975 (coll.).
Slooten, D.F. van, Gard. Bull., Straits Settlem. 9(1): 44-48. 1935 (on R. and the flora of the Neth. Indies).
South, F.W., Gard. Bull., Straits Settlem. 9(1): 31-38. 1935 (R's work on tropical agriculture).
Steenis, C.G.G.J. van, Fl. males. Bull. 7: 191. 1950 (Linn. gold medal), 9: 285. 1952 (96th birthday).
Urban, I., Fl. bras. 1(1): 88-89. 1906.
Verdoorn, F., ed., Chron. bot. 1: 20. 1935, 2: 185, 242, 269. 1936, 3: 162. 1937, 4: 393. 1938, 5: 365, 367. 1939, 5: 417. 1941.
Wulff, E.V., Intr. hist. pl. geogr. 222 [index]. 1943.
Wycherley, P.R., Malay. nat. J. 25: 22-37. 1972 (Ridley and Batu Caves, with checklist of flora).

HANDWRITING: Fl. males. ser. 1. 1: cli. 1950.

FESTSCHRIFT: Gard. Bull. Straits Settlem. 9(1). 1935.

EPONYMY: *Ridleya* (J.D. Hooker) Pfitzer (1900); *Ridleyella* Schlechter (1913); *Ridleyinda* O. Kuntze (1891). *Note*: The derivation of X *Ridleyara* Hort. (1957) could not be determined, but the name very likely honors Ridley.

POSTAGE STAMPS: Christmas Islands 9 c. (1977) yv. 71.

9205. *The Cyperaceae of the West Coast of Africa* in the Welwitsch herbarium... London (Printed for the Linnean Society ...) 1884. Qu.
Collector: Friedrich Martin Josef Welwitsch (1806-1872).
Publ.: Apr 1884 (cover dated), cover t.p., p. 121-172, *pl. 22-23*. *Copy*: U. – Issued and to be cited as Trans. Linn. Soc. London 2(7): 121-172, *pl. 22-23*. Apr 1884.

9206. *The natural history of the island of Fernando de Noronha* based on the collections made by the British Museum expedition in 1887, from the Journal of the Linnean Society, 1890. Oct.
Publ.: 5 Apr 1890 (date in journal; Nat. Nov. Mai(2) 1890), *Botany*: [i* t.p. with above text], [1]-95 [*zoology* (by the same author) 473-570]. *Copy*: E. – Reprinted with special t.p. and original pagination from J. Linn. Soc. 27: 1-95, *pl. 1-4*. 5 Apr 1890, 473-570. 25 Apr 1891.

9207. *The flora of Singapore* [J. Straits Settlem. Roy. Asiat. Soc. 33, 1900]. Oct.

Publ.: 1900 (Nat. Nov. Jul(2) 1900). – Reprinted with original pagination from J. Straits Settl. Roy. Asiat. Soc. 33: 27-196. 1900 *Copies*: FI, G, MO, NY, USDA; IDC 5085.

9208. *Materials for a flora of the Malayan Peninsula* ... Singapore (printed at the Methodist' Publishing House) 1907, 3 parts. Oct. †. (*Mat. fl. Malay. Penins.*).
1: 1907 (J. Bot. Oct 1907), p. [i-iii], [1]-233, [1]-22, index.
2: 1907 (J. Bot. Oct 1907), p. [i-iii], [1]-235, [25]-44.
3: 1907 (Bot. Centralbl. 10 Dec 1907 and 19 Mai 1908 (parts 1-3), p. [i-iii], [1]-197, [49]-75.
General t.p. and indexes: 1908 (date on t.p.), p. [i-iii], [1]-75.
Copies: BR, FI, G, MO, NY, PH, US, USDA. – The original covers and title pages are all dated 1907; a general title page "Monocotyledones vol. 1 ... 1908" was apparently issued in 1908.

9209. *Spices* ... London (Macmillan and Co., Limited ...) 1912. Oct. (*Spices*).
Publ.: 1912 (J. Bot. Mai 1912), frontisp., p. [i]-ix, [1]-449. *Copies*: BR, HH, MO, NY.

9210. *Report on the botany of th Wollaston* expedition to Dutch New Guinea, 1912-13 ...[Trans. Linn. Soc. ser. 2. 9(1). Aug 1916]. Qu.
Publ.: Aug 1916, p. [1]-269, *pl. 1-6* (R.M. Cardew), Trans. Linn. Soc. ser. 2. 9(1): 1-269, *pl. 1-6*. *Copies*: BR, FI, NY, U. – With contributions by "E.G. Baker, S. Moore, H.F. Wernham, C.H. Wright and others. With an introduction by Mr. C.B. Kloss".
Ref.: Anon. [A.B. Rendle], J. Bot. 54: 307-310. Oct. 1916.

9211. *The flora of the Malay Peninsula* ... London (L. Reeve & Co., Ltd. ...) 1922-1925, 5 vols. Oct. (*Fl. Malay. Penins.*).
Illustrations: John Hutchinson (1884-1972).
1: 13-22 Jul 1922, p. [i]-xxxv, [1]-918, [1-36].
2: 25-29 Mai 1923, p. [i]-vi, [1]-672.
3: 30 Apr 1924, p. [i]-vi, [1]-405, [406].
4: 1 Dec 1924, p. [i]-v, [1]-383, [384].
5: 1-6 Aug 1925, p. [i]-v, [1]-470. (BR copy rd 7 Aug 1925; PH copy 21 Aug 1925); J. Bot. 63: 344. Nov 1925 as of "August 1925"; ÖbZ 1 Sep 1926). *Copies*: BR, FI, G, M, MO, NY, PH, US, USDA.
Additions and corrections: E.D. Merrill, Gard. Bull. Straits Settlem. 8(2): 131-133. Jan 1935.
Facsimile ed.: 1967, Amsterdam (A. Asher & Co.), Brook nr. Ashford, Great Britain (L. Reeve & Co.). As original but with additional t.p.'s with Asher/Reeve imprint. – ISBN 90-6123-260-0. *Copy*: FAS.
Ref.: Stafleu, F.A., Taxon 17(2): 227. 1968.
S.T.D., Bull. misc. Inf. Kew 1925: 399-400 (rev.).

9212. *The dispersal of plants throughout the world* ... Ashford, Kent (L. Reeve & Co., Ltd ...) 1930. Oct. (*Dispers. pl.*).
Illustrations: Miss M.B. Moss and author.
Publ.: 1930 (Nat. Nov. Nov 1930; Kew Bull. 20 Jan 1931), p. [i]-xx, [1]-744, *pl. 1-22* (uncol., except *16*: col. frontisp.). *Copies*: BR, E, FI, G, H, MO, NY, U, USDA.
Ref.: Rendle, A.B., J. Bot. 69: 24-26. 1931.
Ulbrich, E., Bot. Jahrb. 64 (Lit.): 47-48. 1931, Hedwigia 70: [75]. 1931.

Rieber, Xaver (1860-1906), German (Hohenzollern) botanist; studied at the Polytechnic and University of Tübingen; from 1881-1884 teacher at the Stuttgart Realgymnasium; teacher at the business school ib. 1886-1893; high school teacher at Ludwigsburg 1893-1896, id. at Ehingen a. Donau 1896-1903, again at Ludwigsburg 1904-1906. (*Rieber*).

HERBARIUM and TYPES: B (extant via G. Lettau); other material at ERZ, M (lich.), STU.

BIBLIOGRAPHY and BIOGRAPHY: GR p. 40-41 (b. 13 Jan 1860, ed. 25 Dec 1906); GR cat.

p. 70; Hawksworth p. 185; IH 2 (in press); Kleppa p. 192; LS 22285-22289; SBC p. 130.

BIOFILE: Anon., Herbarium 51: 8. 1920 (187 lichens from R.'s herb. for sale).

9213. *Zur Flechtenflora der Umgebung von Ehingen* a. D. (Mit einer Tafel). Ein Beitrag zur württembergischen Lichenologie und zur Oberamtsbeschreibung ... Wissenschaftliche Beilage zum Jahresbericht des K. Gymnasiums in Ehingen für das Schuljahr 1900/1901. 1901 Programm Nr. 637. Stuttgart (J.B. Metzlersche Buchdruckerei) 1901. Qu. (*Flechtenfl. Ehingen*).
Publ.: 1901 (t.p.; Nat. Nov. Mai(1). 1902; Bot. Centralbl. 10 Feb 1903), p. [1]-32, *1 pl.*
Copies: Almborn, B, FH.

Riedel, Ludwig (1790-1861), German (Prussian) plant collector and traveller; served in the Prussian army 1813-1815; collected in S. France 1816-1817; from 1820-1836 in Brazil as collector for St. Petersburg; from 1821-1822 in Bahia, 1822-1824 in Rio de Janeiro, 1824-1825 in Minas Geraes; 1825-1829 S. Paulo, Matto Grosso, Alto Amazonas, Para, 1830-1831 back in St. Petersburgh; 1830-1833 Rio de Janeiro; 1833-1835 S. Paulo, Goyaz, Minas Geraes; from 1835 in Rio de Janeiro; from 1836-1838 director of the Rio de Janeiro "Passeio publico" of the national herbarium; director of the botanical section of the Museu nacional 1842-1861. (*Riedel*).

HERBARIUM and TYPES: LE (part of orig. coll.); A, B, BR, C, CN, DPU, F, FI, G, GB, GH, GOET, H, K, KIEL, L, M, MICH, MO, MW, NY, OXF, P, R, RB, S, SI, STU, U, UPS, US, W, Z. – Plants collected by Riedel in France, sent to D.F.L. von Schlechtendal, are at HAL (Werner 1955).

BIBLIOGRAPHY and BIOGRAPHY: Backer p. 495; Barnhart 3: 156; Clokie p. 233; CSP 5: 199, 18: 200; Hortus 3: 1202; IH 1 (ed. 6): 363 (LE, OXF), (ed. 7): 341, 2: (files U); Lasègue p. 331, 338, 346, 478, 505; Moebius p. 411; PR (alph.), ed. 1: 10075; Rehder 1: 342; SK 1: 438; TL-2/see G.H. von Langsdorff; Urban-Berl. p. 388.

BIOFILE: Anon., Flora 5: 112. 1822 (to Brazil, to collect with v. Langsdorff), 13: 680. 1830 (arrived in St. Petersb. from Brazil; 1000 living plants).
Candolle, Alph. de, Phytographie 443. 1880.
Florence, H., Rev. trimens. Inst. hist. geogr. ethn. Brasil. 39(2): 157-182. 1876 (mainly on Langsdorff's travels in Brazil 1825-1829).
Hertel, H., Mitt. Bot. München 16: 423. 1980.
Ihering, H. von, Rev. Mus. Paulista 5: 30-32. 1902.
Komissarov, B.N., The first Russian expedition in Brazil [in Russian], Leningrad 1977 (portr.).
Lindemann, E. v., Bull. Soc. Natural. Moscou 61(1): 62. 1886 (coll. MW).
Nekrassova, V.I., & Proussak, A.V., Bot. Zhurn. 42: 804-813. 1957.
Papavero, N., Essays hist. neotrop. Dipterol. 1: 54, 55, 56, 90. 1972.
Prance, G., Acta amaz. 1(1): 53. 1971.
Sampaio, A.J., Arch. nac. Rio de Janeiro 22: 39, 40, 41, 44, 47. 1919.
Sayre, G., Mem. New York Bot. Gard. 19(3): 387. 1975.
Schmid, G., Chamisso als Naturforscher 12. 1942.
Shetler, S.G., Komarov Bot. Inst. 33, 46, 196, 197. 1967.
Stieber, M.T., Huntia 3(2): 124. 1979 (notes on R. in Agnes Chase mss.).
Trautvetter, R.E., Acta Horti petrop. 2: 239-241. 1873.
Urban, I. Bot. Jahrb. 18 (Beibl. 44): 6-21. 1894 (on Riedel and G.H. von Langsdorff); Fl. bras. 1(1): 89-91. 1906 (itin. Brasill., vita).
Wagenitz, G., Index coll. princ. herb. Gott. 135. 1982.
Warming, E., K. Danske Vid. Selsk. Skr. ser. 6. 6(3): 266-272. 1892 (Lagoa Santa; note on Lund and Riedel's trip of 1833-1835).
Werner, K., Wiss. Z. Martin-Luther-Univ. Halle-Wittenb. 4: 776. 1955 (plants from France in HAL).

EPONYMY: *Riedelia* Chamisso (1832, *nom. rej.*); *Riedelia* C.F. Meisner (1863); *Riedeliella* Harms (1903); *Note*: *Riedelia* D. Oliver (1883, *nom. cons.*); was named for a Mr. Riedel.

Riedelia F. Thiergart et U. Frantz (1963) honors Dipl. Ing. W. Riedel. *Riedelia* A.P. Jousé et V. S. Sheshukova-Poretzkaya (1971) honors Dr. W. Riedel.

Rietsch, Maximilian (*fl.* 1882), French pharmacist and botanist; sometime professor at the school of Pharmacy and Medicine of Marseilles. (*Rietsch*).

HERBARIUM and TYPES: Unknown.

BIBLIOGRAPHY and BIOGRAPHY: BM 4: 1703; CSP 11: 180; Kelly p. 192.

9214. École supérieure de Pharmacie de Paris. Thèse présentée au concours d'agrégation (section des sciences naturelles). *Reproduction des cryptogames* ... Paris (Librairie Germer Baillière ...) 1882. Oct. (*Repr. crypt.*).
Publ.: 1882 (Nat. Nov. Nov(2) 1882; J. Bot. Mai 1883; Bot. Zeit. 29 Dec 1882, 25 Mai 1883), p. [1]-227, [228]. *Copies*: FH, G, NY.

Rietz, Gustaf Einar Du (1895-1967), Swedish botanist, lichenologist and ecologist at the University of Uppsala; Dr. phil. Uppsala 1922, at the Institute for plant biology Uppsala as amanuensis, 1917-1923; at the Botanical Museum ib. as curator 1924-1926; from 1926 again at the Institute for plant biology, from 1931 professor of plant biology; leader of the "Uppsala school of phytosociology". (*Du Rietz*).

HERBARIUM and TYPES: UPSV; other material at A, GB, LD, M, NY, S.

NAME: correctly to be listed as Du Rietz, G.E.

BIBLIOGRAPHY and BIOGRAPHY: GR p. 473-474; Hirsch p. 248; IH 1 (ed. 2): 123, (ed. 3): 158, (ed. 4): 176, (ed. 5): 191, 2: 173; Kelly p. 192; Moebius p. 62; MW p. 414; PH 209; SK 1: 439.

BIOFILE: Anon., Acta phytogeogr. suec. 50: v-viii. 1965 (portr., biogr. note).; Bryologist 70: 282. 1967 (d., portr.); Rev. bryol. lichénol. 35: 406. 1967 (1968) (d. 7 Mar 1967); Taxon 16: 251. 1967.
Lindman, S., Sv. män kvinn. 2: 283. 1944 (portr.).
Rietz, G.E. Du, The fundamental units of biological taxonomy, Sv. bot. Tidskr. 24(3): 333-428. 1930; Uppsala Universitets matrikel 1937-1950, 5 p. (autobiogr. details; complete bibliography up till 1950).
Verdoorn, F., ed., Chron. bot. 1: 257, 258. 1935.

FESTSCHRIFT: Sv. bot. Tidskr. 49(1-2), 1955 (portr., signature); Acta phytogeogr. suec. 50 ("The plant cover of Sweden", dedicated to Du R. by his pupils, x, 314 p., portr.; bibl. of 56 papers by Du Rietz in bibl. of volume by Åke Sjödin.

EPONYMY: *Durietzia* Gyelnik (1935).

THESIS: *Zur methodologischen Grundlage der modernen Pflanzensoziologie*. Akademische Abhandlung welche zur Erlangung der Doktorwürde mit Genehmigung der mathematisch-naturwissenschaftlichen Sektion der weitberühmten Philosophischen Fakultät zu Uppsala am 1. Oktober 1921, 10 Uhr Vormittags, im Hörsaale des Pflanzenbiologischen Instituts öffentlich verteidigt wird von G. Einar Du Rietz ... Wien (Adolf Holzhausen) 1921, p. [1]-270, [271]-272, folded table. Qu. *Copy*: H (inf. P. Isoviita).

9215. *Die europäischen Arten der Gyrophora "anthracina"-Gruppe* ... Stockholm (Almqvist & Wiksells Boktryckeri-A.-B.) 1925. Oct.
Publ.: 2 Oct 1925 (printed 12 Jan 1925, p. 14; issued 2 Oct 1925), cover t.p., [1]-14. *Copies*: H, LD, U. – Issued and to be cited as Ark. Bot. 19(12). 2 Oct 1925.

9216. *Einige von Dr. M. Gusinde gesammelte* [sic] *Flechten aus Patagonien und dem Feuerlande* [Uppsala (Almqvist & Wiksells Boktryckeri-A.-B.) 1926]. Oct.

Publ.: 24 Mar 1926 (printed 3 Feb 1926, issued 24 Mar 1926, fide p. [ii] journal vol.), p. [1]-6. *Copy*: U. – Issued and to be cited as Ark. Bot. 20B, no. 1: 1-6. 1926.

9217. *Vorarbeiten zu einer "Synopsis lichenum"* I. Die Gattungen Alectoria, Oropogon und Cornicularia ... Stockholm (Almqvist & Wiksells Boktryckeri A.-B. ...) 1926. Oct.
Publ.: 9 Dec 1926 (in journal; p. 43: printed 19 Oct 1926; Nat. Nov. Apr 1927), p. [1]-43, *pl. 1-2*. *Copies*: H, LD, NY, U. – Issued, and to be cited as Ark. Bot. 20A(11): 1-43, *pl. 1-2*. 1926.

9218. *The lichens of the Swedish Kamtchatka-Expeditions* ... Stockholm (Almqvist & Wiksells Boktryckeri-A.-B.) 1929. Oct.
Publ.: 19 Nov 1929 (printed 14 Jun 1929; issued 19 Nov fide statement on p. [ii] journal volume), cover t.p., [1]-25, [26], *pl. 1-2*. *Copies*: H, LD, U. – Issued and to be cited as Ark. Bot. 22A(13): [1]-25, [26], *pl. 1-2*. 19 Nov 1929.

9219. *Vegetationsforschung auf soziationsanalytischer Grundlage* ... Berlin und Wien (Verlag von Urban & Schwarzenberg) 1930. Oct.
Publ.: 1930 (p. 480: Abgeschlossen Feb 1929). *Copies*: Almborn, LD. – Issued and to be cited as Handb. biol. Arbeitsmeth. (ed. E. Abderhalden) 11(5): [289]-480. 1930.

9220. *Life forms of terrestrial flowering plants* I. ... Uppsala (Almqvist & Wiksells Boktryckeri-A.-B.) 1931. Oct. (*Life forms terr. fl. pl.*).
1: 1931 (p. [iii]: Jun 1931), p. [i-iii], [1]-95. *Copies*: FAS, LD, U. – Acta phytogeogr. suec. 3.

9221. *Problems of bipolar plant distribution* ... Uppsala (Almqvist & Wiksells Boktryckeri-A.-B.). 1940. Oct.
Publ.: Dec 1940, cover-t.p., p. [215]-282. – Issued and to be cited as Acta phytogeogr. suec. 13.

Rigaud, Antoine (*fl.* 1877], French botanist. (*Rigaud*).

HERBARIUM and TYPES: Unknown.

BIBLIOGRAPHY and BIOGRAPHY: BL 2: 190, 708; Rehder 3: 732.

9222. *Catalogue des plantes vasculaires et des mousses* observées dans les environs *de Boulogne-sur-Mer* ... Boulogne-sur-Mer (Imprimerie [sic, on t.p.: imbrimerie] Camille le Roy, ...) 1877. Oct. (*Cat. pl. Boulogne*).
Publ.: 1877, p. [1]-38. *Copy*: USDA. – See BL 2: 191 for follow-up publ.

Rigg, George Burton (1872-1961), American botanist; B.Sc. Iowa 1896; M.A. Univ. Washington 1909; Dr. phil. Chicago 1914; high school teacher in Iowa 1896-1907; botany and zoology teacher at Lincoln High School, Seattle, Washington 1907; Botany instructor, Univ. of Washington 1909, full professor of botany ib. 1928-1947, head of the botany department 1940-1942; student of the peat resources of the state of Washington. (*Rigg*).

HERBARIUM and TYPES: WTU.

BIBLIOGRAPHY and BIOGRAPHY: Barnhart 3: 157; BL 1: 162, 180, 219, 315; Hortus 3: 1202; Lenley p. 345; NW p. 56; Tucker 1: 597.

BIOFILE: Anon., Ecol. Soc. Bull., Autumn 1962: [2 p.], Resolution of respect. (b. 9 Feb 1872, d. 10 Jul 1961, portr.); The Seattle Times, 10 Jul 1961 (obit., portr., copy in TL-2/files); Who was who in America 6: 344-345 for 1974-1976.
Cattell, J.M., Amer. men sci. ed. 2: 394. 1910, ed. 3: 575. 1921, ed. 4: 820. 1927, ed. 5: 935. 1933.
Hultén, E., Bot. Not. 1940: 325-326. 1940 (Alaska coll.).

Rigg, G.B., List of publications in Univ. Arch., Univ. Washington; dupl. in TL-2/files; Washington historical Quarterly 20(3): 163-173 (hist. bot. State of Washington).

Rikli, Martin Albert (1868-1951), Swiss botanist (plant geographer); studied at Basel, Berlin (with A. Engler and S. Schwendener) and Zürich; Dr. phil. Basel 1895; teacher at a teachers' college in Zürich-Unterstrass 1893-1905; curator of the Botanical Museum of the E.T.H. Zürich 1906-1930; habil. Zürich, E.T.H. 1900; titular professor 1909; travelled widely especially in the mediterranean area. (*Rikli*).

HERBARIUM and TYPES: ZT; other material at C and S.

BIBLIOGRAPHY and BIOGRAPHY: Barnhart 3: 157 (b. 23 Sep 1868); BFM no. 986, 1110, 1690; Biol.-Dokum. 14: 7544; BJI 2:145-146; BL 1: 32, 157, 315, 2: 708 [index]; BM 4: 1704; Bossert p. 332; CSP 18: 209; DTS 6(4): 189; Hirsch p. 248; IH 2: (files); Kew 4: 471-472; KR p. 589; LS 38413; MW p. 414; PFC 1: li, 2: xxvi, 3(2): xvi-xvii; Plesch p. 381; Rehder 5: 726; TL-2/6619; Tucker 1: 597-598; Zander ed. 10, p. 707, ed. 11, p. 807.

BIOFILE: Anon., Bot. Centralbl. 113: 80. 1910 (prof. title); Bot. Zeit. 68: 56. 16 Feb 1910 (prof. title Zürich); Nat. Nov. 32: 128. 1910 (prof. title); Österr. bot. Z. 60: 87. 1910 (prof. title).
Bohny, P., Verh. Schweiz. naturf. Ges. 131: 378-382. 1951 (obit., bibl., portr.).
Keller, C., Neue Zürcher Zeit. 28 Dec 1929 (retirement).
Maire, R., Progr. conn. bot. Algérie 158-159. 1931.
Rübel, E., Gesch. naturf. Ges. Zürich 92-93. 1947 (portr.).
Uehlinger, A., Martin Rikli zum 80. Geburtstag 23. September 1948 [3 p. Schaffhausen, mimeographed].
Verdoorn, F., ed., Chron. bot. 4: 277. 1938, 5: 286. 1939.
Wulff, E., Intr. hist. pl. geogr. 62, 114, 163. 1943.

COMPOSITE WORKS: (1) With C. Schröter (q.v.): *Botanische Exkursionen im Bedretto, Formazza-, und Bosco-Tal*, Zürich 1904.
(2) *Marokko*, in G. Karsten & H. Schenck, *Vegetationsbilder*, 24. Reihe, Heft 4/5, *pl 19-30*, Jena 1934 (also: *Spanien* in 5. Reihe, Heft 6, *6 pl.*Jena 1907, n.v.).
(3) O. Nägeli, *Fl. Zürich*, 1905-1912, TL-2/6619; bibliography and Pteridophyta by Rikli.

EPONYMY: *Rikliella* J. Raynal (1973).

9223. Mitteilungen aus dem botanischen Museum des eidgenössischen Polytechnikums in Zürich. 4. *Die mitteleuropäischen Arten der Gattung Ulex.* Von Dr. M. Rickli [sic, for Rikli] ... Bern (Druck und Verlag von K.J. Wyss) 1898. Oct.
Publ.: 1898. Cover t.p., p. [1]-15. *Copies*: G, M. – Reprinted and to be cited from Ber. schweiz. bot. Ges. 8: 1-15. 1898.

9224. Mitteilungen aus dem botanischen Museum des eidgenössischen Polytechnikums in Zürich. 7. *Die schweizerischen Dorycnien.* Von Dr. M. Rickli [sic, for Rikli] ... Bern (Druck und Verlag von K.J. Wyss) 1900. Oct.
Publ.: 1900, p. [1]-35. *Copy*: M. – Reprinted and to be cited from Ber. schweiz. bot. Ges. 10: 10-44. 1900.

9225. *Botanische Reisestudien* auf einer Frühlingsfahrt durch Korsika ... Zürich (Verlag von Fäsi & Beer) 1903. Oct. (*Bot. Reisestud.*).
Publ.: Oct-Nov 1902 (p. xiii: Sep 1902; NY copy rd Nov 1902; t.p. 1903; ÖbZ Oct-Nov 1902; Allg. bot. Z. 15 Dec 1902; Nat. Nov. Nov(2). 1902), p. [i]-xiii, [1]-140, *pl. 1-16*. *Copies*: G, HH, MO, NY, USDA. – Preprinted from Viertelj.-Schr. naturf. Ges. Zürich 47(3/4): 243-384, *pl. 7-21*. 1902 (1903).
Ref.: Hannig, E., Bot. Zeit. 61: 92. 16 Mar 1903.

9226. Mitteilungen aus dem botanischen Museum des eidgenössischen Polytechnikums

in Zürich. 8. *Beiträge zur Kenntniss der schweizerischen Erigeron-Arten* ... Bern (Buchdruckerei K.J Wyss) 1904, 2 parts. Oct.
1: 1904, p. [1]-16, *2 pl. Copy*: M – Reprinted and to be cited from Ber. schweiz. bot. Ges. 14: 14-29, *pl. 1-2.* 1904.
2: 1904, p. [1]-7. *Copy*: M. – Id. 14: 127-133. 1904.

9227. *Kultur– und Naturbilder von der spanischen Riviera* ... Zürich (Druck von Zürcher & Furrer ...) [1906]. Qu. (*Kult. Naturbild. Span. Riviera*).
Publ.: Oct-Dec 1906 (p. 5: 23 Sep 1906), p. [1]-46, *pl. 1-6. Copies*: BR, M. – Issued as Neujahrsbl. naturf. Ges. Zürich auf das Jahr 1907, 109. Stück.

9228. *Zur Kenntniss der Pflanzenwelt des Kantons Tessin* ... Zürich-Oberstrass (Druck von J.F. Kobold-Lüdi) 1907. Oct.
Publ.: 1907, p. [3]-39. *Copy*: G. – Reprinted and to be cited from Ber. zürch. bot. Ges. 10: 27-63. 1907.

9229. *Botanische Reisestudien von der Spanischen Mittelmeerküste* mit besonderer Berücksichtigung der Litoralsteppe ... Zürich (Verlag von Fäsi & Beer) 1907. Oct. (*Bot. Reisestud. Span. Mittelmeerküste*).
Publ.: 29 Jun 1907 (date on journal cover; p. viii: Mai 1907; ÖbZ Jun-Jul 1907; Allg. bot. Z. 15 Jul 1907; Bot. Centralbl. 10 Sep 1907; Nat. Nov. Jun(2) 1907), p. [i]-viii, [1]-155, *pl. 1-12. Copies*: G, HH, MO. – Reprinted from Viertelj.-Schr. naturf. Ges. Zürich 52(1/2): 1-155, *pl. 1-12.* 1907. – See also: "Zweite naturwissenschaftliche Studienreise nach Spanien, der Küste v. Marokko und den Kanarischen Inseln, Zürich iv (Buchdruckerei J.F. Kobold-Lüdi) Frühjahr 1908, p. [1]-15. *Copy*: G.

9230. *Die Arve in der Schweiz.* Ein Beitrag zur Waldgeschichte und Waldwirtschaft der Schweizer Alpen ... Zürich (Auf Kosten der [Schweizerischen Naturforschenden] Gesellschaft ... gedruckt von Zürcher & Furrer in Zürich ...) 1909. Qu. (*Arve i.d. Schweiz*).
Publ.: 1909 (p. vii: 23 Sep 1908; NYBG Nov 1909; Nat. Nov. Sep(1) 1909). *Copies*: G, HH, NY.
1 (*text*): p. [i]-xxxix, [xl], [1]-455, [456].
2 (*plates and maps*): [i-ii], *pl. 1-9* (photos) each with 2 p. text; 19 maps numbered *10-28* each with 2 p. text, and two unnumbered fold-out maps.
Ref.: Engler, A., Bot. Jahrb.-(Lit.) 44: 22-23. 23 Nov 1909.

9231. *Über die Engelwurz* (Angelica Archangelica L.) [Zürich (Expedition, Druck und Verlag: Art Institut Orell Füssli) 1910]. Oct.
Publ.: Jan-Feb 1910, p. [1]-29. *Copy*: G. – Reprinted and to be cited from Schweiz. Wochenschr. Chemie Pharm. 1910(4-7): 49-56. 22 Jan, 65-71. 29 Jan, 81-88. 5 Feb, 97-105. 12 Feb 1910.

9232. *Vom Mittelmeer zum Nordrand der algerischen Sahara.* Pflanzengeographische Exkursionen ...Zürich (Druck von Zürcher & Furrer) 1912. Oct.
Co-author: Carl Joseph Schröter (1855-1939).
Publ.: 25 Sep 1912 (date on cover reprint; BR rd 20 Dec 1912; Nat. Nov. Nov(2) 1912; journal cover dated 20 Sep 1912), p. [1]-178, *pl. 1-25. Copy*: BR. – Reprinted and to be cited from Viertelj.-Schr. naturf. Ges. Zürich 57: 33-210. 20 Sep 1912 (reprint with orig. pag. at M).
Other issue: s.d. [Sep 1912], p. [i-ii], [1]-178, *pl. 1-25. Copies*: G, NY. – "*Vom Mittelmeer zum Nordrand der Sahara* eine botanische Frühlingsfahrt nach Algerien von Dr. M. Rikli und Dr. C. Schröter ... Mit Beiträgen von C. Hartwich, ... Ed. Rübel, ... L. Rütimeyer, ... und von Herrn und Frau Dr. Schneider-von Orelli ..." Zurich (Verlag: Art. Institut Orell Füssli) s.d.

9233. *Lebensbedingungen und Vegetationsverhältnisse der Mittelmeerländer und der atlantischen Inseln* ... Jena (Verlag von Gustav Fischer) 1912. Oct. (*Lebensbed. Veg.-Verh. Mittelmeerländ.*).

Publ.: 1912 (ÖbZ for Jun-Jul 1912; Bot. Centralbl. 17 Sep 1912, NYBG Nov 1912), p. [i]-xi, [1]-171, *pl. 1-32*. *Copies*: BR, FI, HH, M, NY, PH.

9234. *Natur– und Kulturbilder aus den Kaukasusländern und Hocharmenien* von Teilnehmern der schweizerischen naturwissenschaftlichen Studienreise, Sommer 1912 unter Leitung von Prof. Dr. M. Rikli ... Zürich (Druck und Verlag: Art. Institut Orell Füssli) 1914. Oct. (*Nat.– Kult.-Bilder Kaukasusländ.*).
Publ.: 1914, p. [i]-vii, [1]-317, [1-2], frontisp. and *61 pl. Copy*: GOET. – 15 chapters by various authors: M. Rikli, W.A. Keller, W. Bally, C. Seelig, Bishop Mesrop, E. Rübel, J. Keller, C. Keller (inf. G. Wagenitz).

9235. *Über Flora und Vegetation von Kreta und Griechenland* ... Zürich (Druck von Gebr. Fretz A.G.) 1923.
Co-author: Eduard Rübel (1876-1960).
Publ.: 30 Jun 1923 (date on cover), p. [i-ii, cover, with t.p. text], [105]-227. *Copies*: G, H. – Reprinted or preprinted from Vierteljahrsschr. Naturf. Ges. Zürich 68: 105-337. 1923.

9236. *Von den Pyrenäen zum Nil* Natur– Kulturbilder aus den Mittelmeerländern mit Beiträgen von Pfarrer K. Linder und Dr. H. Weilenmann ... Bern und Leipzig (Verlag Ernst Bicher, ...) [1926]. Oct. (*Pyren. zum Nil*).
Publ.: Jan 1926 (p. ix: 1 Oct 1925; ÖbZ for Jan 1926 in issue for Jun 1926), p. [i]-xi, 1-566, *pl. 1-80* (8 col.). *Copy*: MO.

9237. *Die periodischen Erscheinungen in der Pflanzenwelt der Polarländer* und ein Vergleich mit denjenigen der Alpenflora ... Osterwieck am Harz (A.W. Zickfeldt Verlag) [1926]. Oct.
Publ.: 15 Oct 1926 (date on cover), p. [1]-13. *Copy*: M. – Reprinted from Die pädagogische Warte 33(20), 15 Oct 1926 (journal n.v.).

9238. *Das Pflanzenkleid der Mittelmeerländer* ... Bern (Verlag Hans Huber) [1942-1948]. Oct. (*Pfl.-Kleid. Mittelmeerländ.*).
1: 15 Mai 1942-31 Jul 1943, p. [i]-xvi, 1-436.
2: 25 Jan 1944-14 Jan 1947, p. [i]-xv, 437-1093.
3: 28 Sep 1948, p. [i]-viii, 1095-1418.
Copies: BR, G, H, HH, M, NY, PH, USDA(3). – Issued (first and, for fasc. 1-3 also second edition):

Vol.	fasc.	pages	dates	ed. 2 (unchanged)
1	1	1-128	15 Mai 1942	Mar 1943
	2	129-240	15 Oct 1942	Mar 1943
	3	241-352	15 Mar 1943	Mar 1943
	4	353-436, [i]-xvi	31 Jul 1943	
2	5	437-560	25 Jan 1944	
	6	561-672	26 Jul 1944	
	7	673-784	15 Feb 1945	
	8	785-912	23 Sep 1945	
	9	913-1093, [i]-xv	14 Jan 1947	
3	10	[i]-viii, 1095-1418	23 Sep 1948	

Ref.: Krause, K., Bot. Jahrb. 73 (Lit.): 28-29. 1943.

Riley, John (x-1846), British pteridologist of Papplewick, Nottinghamshire. (*J. Riley*).

HERBARIUM and TYPES: Untraced (see Newman 1847).

BIBLIOGRAPHY and BIOGRAPHY: Barnhart 3: 157; BM 4: 1704; Desmond p. 520; Jackson p. 150; Kew 4: 473; PR 7631, ed. 1: 8587.

BIOFILE: Kent, D.H., Brit. herb. 75. 1957 (herb. not traced).

Newman, E., Phytologist 2: 779-780. 1847 (note on his herbarium offered for sale).

EPONYMY: *Rileya* A. Funk (1979) honors C.G. Riley, Canadian forest pathologist.

9239. *Catalogue of ferns*, after the arrangement of C. Sprengel, with additions from C.B. Presl and references to the authors by whom the species are described: to which is added a synoptical table of C.B. Presl's arrangement of genera ... London (Hamilton, Adams, and Co.; ...) Nottingham (W. Dearden). 1841. Qu. (in twos). (*Cat. ferns*).
Publ.: 1841, p.[i-ii], [1]-29. *Copies*: E, USDA; IDC 7252.

Riley, Laurence Athelstan Molesworth (1888-1928), British (Jersey-born) botanist; volunteer worker at K, on St. George Pacific Expedition 1924-1925; M.A. Oxford 1927. (*L. Riley*).

HERBARIUM and TYPES: K; further material at BM, E, MO.

BIBLIOGRAPHY and BIOGRAPHY: BL 1: 126, 146, 152, 315; Hortus 3: 1202; Langman p. 627.

BIOFILE: Anon., Bull. misc. Inf. Kew 1924: 174 (Pacific exp.).
Kent, D.H., Brit. herb. 75. 1957 (herb. K).
Riley, L.A.M., Bull. misc. Inf. Kew 1925: 26-33, 133-142, 216-231; 1926: 51-56. 1926 (report on botany Pacific exp.).
Sprague, T.A., J. Bot. 66: 118-119. 1928 (obit., d. 13 Mar 1928); Bull. misc. Inf. Kew 1928; 157-158 (obit., d. 13 Mar 1928, bibl.).

Rilstone, Francis (1881-1953), British (Cornish) botanist; headmaster Polperro primary school (batologist); educated at Westminster College, London; teacher at Truro, later at Polperro, headmaster of the County Primary School at Polperro, Cornwall, 1914-1934; in retirement at Perranzabuloe 1934-1953. (*Rilstone*).

HERBARIUM and TYPES: BM (20.000); other material CFN, DBN, LIV, M (lich.), OXF.
– A manuscript *Fungus flora of Cornwall* is in the libraries of the British mycological Society and the Dept. Bot., Univ. Exeter.

BIBLIOGRAPHY and BIOGRAPHY: BL 2: 231, 708; Clokie p. 233; Desmond p. 521; IH 2: (files); Kew 4: 474; SBC p. 130.

BIOFILE: Anon., Rev. bryol. lichénol. 22: 99. 1953 (d. 2 Jan 1953).
Bridson, G.D.R. et al., Nat. hist. mss. res. Brit. Isl. 84.11, 84.18. 1980.
Edees, E.S., Bot. Soc. Brit. Isl. Proc. 1(19: 110-13. 1954 (obit., portr., bibl., b. 5 Nov 1881, d. 22 Jan 1953);herb. 10.000 phan., 10.000 bryo.).
Hertel, H., Mitt. Bot. München 16: 423. 1980 (lich. at M through Berl. bot. Tauschver. [of which Rilstone was a member].
Kent, D.H., Brit. herb. 75. 1957 (coll.).
Palmer, J.T., Nova Hedwigia 15: 155. 1968 (on a mss. *Fungus flora of Cornwall*).
Rilstone, F., A bryophyte flora of Cornwall, i, Musci, Trans. brit. bryol. Soc. 1: 75-100. 1948, ii. Sphagna, ib. 1: 153-155. 1949, iii Hepat., ib. 1: 156-165. 1949; see also J. Bot. 74: 234-236. 1936.
Sayre, G., Bryologist 80: 516. 1977.
Wallace, E.C., Trans. Brit. bryol. Soc. 2(3): 501. 1954 (obit.).

Ringier, Victor Abraham (*fl.* 1823), Swiss (Aargau) physician and botanist; Dr. med. Tübingen. (*Ringier*).

HERBARIUM and TYPES: Unknown.

BIBLIOGRAPHY and BIOGRAPHY: PR 8421 (sub Schübler).

9240. Dissertatio inauguralis botanica *de distributione geographica plantarum Helvetiae*. Quam

consentiente gratiosa facultate medica praeside Gust. Schübler ... pro gradu doctoris medicinae et chirurgiae rite obtinendo publico examini submittit auctor Victor Abraham Ringier helveto-tobiniensis [of Zofingen] mense Julii mdcccxxiii. Tubingiae [Tübingen] (Litteris Schrammianis) [1823]. Oct. (*Distr. geogr. pl. Helv.*)
Authorship: Ringier is clearly indicated as the author on the t.p.; the dissertation is listed under Gustav Schuebler (1787-1834) by PR (8421).
Publ.: Jul 1823, p. [1]-31, [32], charts *1-3*. *Copies*: G, LC.

Ringius, Hans Henric [Henrik] (1808-1874), Swedish clergyman and botanist in Skåne; Dr. phil. Lund 1838; ultimately rector at Silvåkra and Revinge. (*Ringius*).

HERBARIUM and TYPES: Some material at C, GB, and GH. – Issued sets of exsiccatae: *Herbarium normale plantarum rariorum et criticarum Sueciae*, fasc. 1-2, Lund 1835-1836, continued by E.M. Fries; for sets see under Fries.

BIBLIOGRAPHY and BIOGRAPHY: Barnhart 3: 158; BM 4: 1705; KR p. 590 (b. 30 Mar 1808, d. 28 Jun 1874); PR 7633, ed. 1: 8590; TL-2/see E.M. Fries.

BIOFILE: Anon., Flora 20 (Int. Bl. 2): 17-20. 1837 (*Herb. norm.*); Nat. Nov. 1874: 126. Wittrock, V.B., Acta Horti Berg. 3(3): 49, *pl. 103*. 1905 (portr.).

9241. *Herbationes lundenses*, quas, cons. ampliss. philos. ord. lundens., praeside Sv. Nilsson, ... pro laurea p.p. auctor H.H. Ringius, philos. cand., Scanus. In Mus. zool. die xxi Juni mdcccxxxviii. Pars i. Lundae (excudebat C.F. Berling, ...) 1838. Oct. †. (*Herb. lund.*).
Publ.: 21 Jun 1838, p. [i-ii], [1]-16. *Copies*: H (inf. P. Isoviita), LD (inf. O. Almborn).

Rink, Henrik Johannes (1819-1893), Danish (of German parents) geographer, geologist and explorer; Dr. phil. Kiel 1844; participated in the circumnavigation of the *Galathea* 1845-1846 (until Penang), surveying the Nicobars; in Northern Greenland 1848-1855; Royal Inspector (Governor) of South Greenland 1855-1868; Director of the Royal Greenland Board of Trade (Kön. Grönl. Handel) 1871-1882; retired to Christiania (Oslo). (*Rink*).

HERBARIUM and TYPES: C; further material at K, M, MICH. – Rink's collections were among those on which J. Lange based his *Consp. pl. Groenl.* 1880-1894, TL-2/4185. Some correspondence at S; other archival material in the Smithsonian Archives SIA 7073.

BIBLIOGRAPHY and BIOGRAPHY: CSP 5: 209, 8: 752, 11: 186, 12: 620; Herder p. 57; IH 2: (files); Jackson p. 366, 367; Kew 4: 474; PR 7635; Rehder 1: 481. SK 1: 439, 5: cccxiv; TL-2/see J.M.C. Lange.

BIOFILE: Anon., Geogr. Jahrb. 19: 385. 1897 (biogr. details and bibl.); b. 26 Aug 1819, d. 15 Dec 1893); Leopoldina 29: 211. 1893 (obit. not.); Medd. Grønland 16: i-iv. 1896 (obit., portr.); Scott. geogr. Mag. 11: 20. 1895 (obit.).
Brown, R., Geogr. J. 3(1): 65-67. 1894 (obit.).
Christensen, C., Danske bot. hist. 1: 500, 504, 604. 1925, 2: 298-299. 1925 (bibl.; b. 26 Aug 1819, d. 15 Dec 1893).
Embacher, F., Lex. Reisen 248. 1882.
Hertel, H., Mitt. Bot. München 16: 423. 1980 (Greenland lich. at M through Deichmann-Brandt and Arnold).
Jørgensen, C.A. et al., Biol. Skr. Dan. Vid. Selsk. 9(4): 5. 1958.
Kornerup, T., Overs. Medd. Grønland 1876-1912: 105. 1913 (bibl.).
Ostermann, H., Dansk biogr. Leks. 19: 560-564. 1940.
Poggendorff, J.C., Biogr.-Lit. Handw.-Buch 3: 1121. 1898 (b. 26 Aug 1819, d. 15 Dec 1893).
Rydberg, P.A., Augustana Coll. Libr. Publ. 5: 27. 1907.
Steenstrup, K.J.V., Geogr. Tidskr., Kjøbenhavn 12: 162-166. 1894 (obit.).
Warming, E., Bot. Tidsskr. 12: 188-189. 1881 (bibl.).

9242. *Die Nikobarischen Inseln.* Eine geographische Skizze mit specieller Berücksichtigung der Geognosie ... Kopenhagen (Verlag von H.C. Klein ...) 1847. Oct. (*Nikobar. Ins.*).
Publ.: 1847, frontisp., p. [i-viii], [1]-188, 1 map. *Copies*: GOET (inf. G. Wagenitz). Notes on botany on p. 127-139 (see also Flora 31: 398. 1848).

9243. *Grønland geografisk og statistisk beskrevet* ... Kjøbenhavn (I Commission hos Universitetsboghandler Andr. Fred. Høst ...) 1852-1857, 2 vols. (*Grønland*).
1(1): 1852, p.[i-x], [1]-202, [203-206], 1 map, *5 pl*. Botanical notes on p. 154-165; see also W.J. Hooker, J. Bot. 5: 122-124.
1(2): 1855, p. [i-iv], [1]-218, [219-220], *1 pl*.
2: 1857, p.[i]-v, [vi], [1]-416, 3 maps, *3 pl.*, Naturhistorisk Tillaeg [1]-172, in which on p. 106-135, J. Lange, Oversigt over Grønlands planter.
Copies: H (inf. P. Isoviita), GOET (inf. G. Wagenitz).

9244. *Naturhistoriske Bidrag til en Beskrivelse af Grønland,* ... Kjøbenhavn (Louis Kleins Bogtryckkeri) 1857. Oct. (*Naturh. Bidr. Grønland*).
Authorship: J. Reinhardt, J.C. Schiødte, O.M.L. Mørch, C.F. Lütken, J. Lange (botany: p. 106-136) and H.J. Rink.
Publ.: 1857, p. [i], [1]-172, map. *Copy*: USDA. – Constitutes vol. 2 of H.J. Rink, *Grønland geografisk og statistisk beskrevet*, Kjøbenhavn 1857.
English translation: see below, *Danish Greenland*, which has so many alterations that it is virtually a new work.

9245. *Danish Greenland* its people and its products ... edited by Dr Robert Brown, ... London (Henry S. King & Co.) 1877. Oct. (*Dan. Greenland*).
Editor: Robert Brown (1842-1895).
Publ.: Mai-Jun 1877 (Liverpool copy rd 12 Jul 1877; preface dated Mai 1877), p. [i]-xvii, map, [1]-448, *16 pl*. *Copies*: H (inf. P. Isoviita), LC, LIV (city libr., inf. J. Edmondson), p. 409-428, Synopsis of the Greenland flora (409-414 Synopsis by J. Lange; 415-422 Catalogue of the cryptogamous plants by R. Brown, 423-428 general remarks on the Greenland flora by R. Brown).

Riocreux, Alfred (1820-1912), French botanical artist. (*Riocreux*).

HERBARIUM and TYPES: Unknown.

BIBLIOGRAPHY and BIOGRAPHY: Barnhart 3: 159 (b. 8 Jan 1820, d. 15 Mai 1912); Blunt p. 209, 229, 241, 264; Lasègue p. 554; Lenley p. 467; NI see indexes; PR 2126, 2482, 7412, 10023, 10042, ed. 1: 2414, 2802, 8351, 10999; TL-1/113, 151, 256, 272, 273, 652; TL-2/605, 663, 800, 1241, 1245, 1337-1339, 1835, 1975, 2625; TR 3230, 4250-4251, 4361, 4385-4386, 6237, 6653, 6654; Tucker 1: 598.

BIOFILE: J.P., Bull. Soc. bot. France 59: 445. 1912 (obit.).
Bridson, G.D.R. et al., Nat. hist mss. res. Brit. Isl. 269.304. 1980.
Lawalrée, A. et al., Vélins du Muséum 14. 1974.
Rix, M., The art of the botanist 184, 195. 1981.
Rodgers, A.D., John Torrey 227, 244, 254. 1942.
Stafleu, F.A., Taxon 15(8): 324-325. 1966.

EPONYMY: *Riocreuxia* Decaisne (1844).

Rion, Alphonse (1809-1856), Swiss (Valais) clergyman and botanist; teacher at the Lyceum of Sitten; ord. 1832; priest at Zenegyen nr Viège 1834-1835; from 1835 at Sion, ultimately as canon and teacher. (*Rion*).

HERBARIUM and TYPES: Some material at G and RO (via Cesati); Alph. de Candolle 1880 mentions material at the Musée cantonal, Sion, Valais, Switzerland.

BIBLIOGRAPHY and BIOGRAPHY: AG 1: 146, 5(1): 753 (erroneous dates), 5(3): 94, 12(3): 13; Barnhart 3: 159 (b. 12 Jul 1809, d. 8 Nov 1856); BM 4: 1706; Hegi 4(2): 1046; Jackson p. 57, 59, 345; Rehder 1: 436; Tucker 1: 598.

BIOFILE: Anon., Bot. Zeit. 15: 117-118. 1857 (d., left a mss. Fl. Valais [see below, later publ. as *Guide bot. Valais*]); Flora 40: 223. 1857 (d.); Österr. bot. W. 6: 390. 1856 (d.).
Burdet, H.M., Saussurea 6: 29. 1975 (portr.).
Burnat, É., Bull. Murithienne, Soc. Valais Sci. nat., Sion 37: 127-130. 1920 (obit., portr.).
Candolle, Alph. de, Phytographie 444. 1880.
Chiovenda, E,. Nuov. Giorn. bot. ital. 31: 311. 1924.
Jaccard, H., Cat. fl. Valais. x. 1895.

9246. *Guide du botaniste en Valais* par le chanoine Rion publié sous les auspices de la section "Monte-Rosa" du C.A.S. [Swiss alpine club] par R. Ritz et F.-O. Wolf... Sion (Librairie générale de A. Galerini) 1872. Oct. (*Guide bot. Valais*).
Publ.: 1872 (p. xxxii: 14 Jul 1872), p. [i]-xxxii, [1]-252. *Copies*: BR, G, HH, NY.

Ripart, Jean Baptiste Marie Joseph Solange Eugène (1814-1878), French physician and botanist at Bourges. (*Ripart*).

HERBARIUM and TYPES: ANGUC and BR (roses; in herb. O. Dieudonné); further material in COLO (algol., herb.), M (lich.) and PC. – 45 letters to Nylander at H (Univ. libr.).

BIBLIOGRAPHY and BIOGRAPHY: AG 6(1): 311; Barnhart 3: 159; De Toni 1: cvii; GR p. 296; GR cat. p. 70; Hegi 4(2): 1052; IH 2: (files); Kelly p. 192; LS 22327-22329; SBC p. 130; Tucker 1: 598.

BIOFILE: Anon., Bot. Not. 1879: 31 (d.); Bull. Soc. bot. France 25 (bibl.): 143, (C.R.): 230. 1878 (d.), 37 (bibl.): 96. 1890 (herb. for sale; lichens and phanerogams still unsold); J. Bot. 17: 64. 1879 (d.); Rev. Bryol. 6: 16. 1879 (d.).
Genevier, G., Bull. Soc. bot. France 26: 62-64. 1879 (obit., d. 17 Oct 1878).
Hertel, H., Mitt. Bot. München 16: 423. 1980 (French lich., coll. 1862-1872 at M via Arnold).
Huberson, G., Brébissonia 1: 80. 1878 (d. Oct 1878).
Koster, J.Th., Taxon 18: 557. 1969 (algal herb. at COLO).
Sayre, G., Bryologist 80: 516. 1977.
Ripart, J., Bull. Soc. bot. France 13: clxxxv. 1866, 15: xviii, xxxvi. 1868, 23: 158, 210, 258, 307. 1876, 26: 62. 1879.
Wittrock, V.B., Acta Horti Berg. 3(2): 117. 1903, 3(3): 112, *pl. 142.* 1905 (portr.).

EPONYMY: *Ripartia* M. Gandoger (1881); *Ripartitella* Singer (1947); *Ripartites* P.A. Karsten (1879).

Risso, Joseph Antoine (1777-1845), French pharmacist, botanist and high school teacher at Nice; pioneer botanist of the Alpes maritimes (*Risso*).

HERBARIUM and TYPES: NICE (in herb. J.B. Risso); other material at G, G-DC, P (herb. Orangers, rd. 1927; material via herb. Richard), W. – See also Smithsonian Archives SIA 7177.

BIBLIOGRAPHY and BIOGRAPHY: Barnhart 3: 159 (b. 8 Apr 1777, d. 25 Aug 1845); BL 2: 118, 708; BM 4: 1592; Bossert p. 333; CSP 5: 211-212; De Toni 1: cvii; DU 263; Frank 3 (Anh.): 84; GF p. 73; Hegi 5(2): 1116; Herder p. 57, 333; Hortus 3: 1202; IH 1 (ed. 3): 112, (ed. 6): 363, (ed. 7): 341, 2: (file); Jackson p. 128, 288; Kew 4: 475; Lasègue p. 538; LS 22336; MW p. 414; NI 1639-1640; Plesch p. 381-382; PR 7640-7643, ed. 1: 8595-8599; Quenstedt p. 362; Rehder 5: 727; Saccardo 1: 139, 2: 92, cron. p. xxxi; Sotheby 643, 645; TL-1/1084, 1342; Tucker 1: 599; Zander ed. 10, p. 707, ed. 11, p. 807 (b. 7 Aug 1777).

BIOFILE: Anon., Ann. Soc. Linn. Lyon 1849: xiv. 1850 (b. 8 Apr 1777, d. 25 Aug 1845); Bonplandia 3: 33. 1855 (Leopoldina, cogn. Plancius), 6: 338. 1858 (cogn. Leopoldina Plancius).

Bridson, G.D.R. et al., Nat. hist. mss. res. Brit. Isl. 255.130. 1980.
Burnat, É., Bull. Soc. bot. France 30: cxxix. 1883 (bibl.).
Candolle, A.P. de, Mém. souvenirs 482, 483, 527. 1862.
Cavillier, F., Boisierra 5: 79. 1941.
Jessen, K.F.W., Bot. Gegenw. Vorz. 262. 1864.
Nissen, C., Zool. Buchill. 343. 1969.
Wittrock, V.B., Acta Horti Berg. 3(2): 165. 1903, 3(3): 180. 1905.

HANDWRITING: Candollea 33: 143-144. 1978.

EPONYMY: *Rissoa* Arnott (1836). *Note*: The derivation of *Rissoella* J.G. Agardh (1849) was not given.

9247. *Essai sur l'histoire naturelle des orangers*, bigaradiers, limettiers, cedratiers, limoniers ou citronniers, cultivés dans le département des Alpes maritimes; ... À Paris (chez G. Dufour et Cie., ...) 1813. Qu. (*Essai hist.-nat. orangers*).
Publ.: 1813, p. [i], [1]-74, *pl. 3-4*. *Copies*: G(2), HH. – Reprinted and to be cited from Ann. Mus. Hist. nat. Paris 20: 169-212. *pl. 3-4*, 401-431. 1813.

9248. *Histoire naturelle des orangers*, par A. Risso, ... et A. Poiteau ... Paris (Audot, libraire, éditeur de l'Herbier de l'amateur, du Jardin fruitier, ...) 1818-1822. Qu. (in twos). (*Hist. nat. orangers*).
Artist and co-author: Pierre Antoine Poiteau (1766-1854).
Publ.: 1818-1822 (for dates of publication of plates see below), p. [i-iv], [1]-280, *pl. 1-109*. *Copies*: BR(col.), FI, G (uncol.), HH (col.), USDA (col.); IDC 5679. – The 1818-1822 t.p. was evidently publ. 1822; the BR copy has a different t.p., dated 1818, with a different imprint: Paris (Imprimerie de Mme Hérissant le Doux, imprimeur ordinaire du roi et des musées royaux, ...) 1818. – Publication took place in 19 fascicles of 6 plates (except no. 18 with 4 and no. 19 with 3 plates); together 109 color-printed plates by Poiteau. The following dates are those on which the Bibliographie de la France announced the fascicles. The contents of the parts are conjectural. There are no indications that the plates were not issued in numerical order, but there is no definite proof of such regular publication.

part	*plates* (by inference)	dates	part	*plates* (by inference)	dates
1	*1-6*	18 Jul 1818	11	*61-66*	[1819]
2	*7-12*	6 Mar 1819	12	*67-72*	2 Oct 1819
3	*13-18*	6 Mar 1819	13	*73-78*	13 Nov 1819
4	*19-24*	27 Mar 1819	14	*79-84*	1 Jan 1820
5	*25-30*	1 Mai 1819	15	*85-90*	26 Feb 1820
6	*31-36*	8 Mai 1819	16	*91-96*	1 Apr 1820
7	*37-42*	22 Mai 1819	17	*97-102*	13 Mai 1820
8	*43-48*	29 Mai 1819	18	*103-106*	17 Jun 1820
9	*49-54*	26 Jun 1819	19	*107-109*	12 Aug 1820
10	*55-60*	31 Jul 1819			

The BF review of part 14 mentions 10 plates instead of the usual 6, this is probably an error.

The plates are stipple engravings, in the coloured copies printed in colour and finished by hand, all of drawings by Poiteau.
New edition: 1872, p. [i-iii], [1]-228, *pl. 1-29, 29 bis, 30-109* (col. copp.). *Copy*: USDA. – "*Histoire et culture des orangers* par A. Risso et A. Poiteau. Nouvelle edition entièrement revue et augmentée d'un chapitre nouveau sur la culture dans le midi de l'Europe et en Algérie par M.A. du Breuil ... orné de 110 planches en couleur ..." Paris (Henri Plon, ..., G. Masson, ...) 1872. Qu.
English translation: 1853, p. [1]-36. *Copy*: E. – Extracts translated from the Natural history of Orange trees by A. Risso. Third edition [sic, eds. 1 and 2 not seen by us]. Malta (printed at the Malta Mail Office, ...) 1853. Oct. – Translation by Lady Reid, wife

of William Reid, Governor of Malta, who signed the preface 7 Mar 1853. (Inf. J. Edmondson).

9249. *Histoire naturelle des principales productions de l'Europe méridionale* et particulièrement de celles des environs de Nice et des Alpes maritimes; ... À Paris (chez F.-G. Levrault, ...), Strasbourg (id.) 1826, 5 vols. Oct. (*Hist. nat. prod. Eur. mérid.*).
Publ.: in 5 volumes all dated 1826 (but possibly published 1826-1828, see PR). *Copies*: G, HH, NY, PH, USDA; IDC 317. Plates col. or uncol.
1: [i]-xii, [1]-448, 1 chart, 2 maps. − Plants.
2: [i]-vii, [1]-492, *8 pl.* − Plants.
3: [i*-iii*], [i]-xvi, [1]-480, *16 pl.* − Animals.
4: [i*-iii*], [i]-vii, [1]-439, *12 pl.* − Animals.
5: [i]-viii, [1]-403, *10 pl.* − Animals. *Tables vol. 4*: 1829 or 1830 (p. [2]: 25 Dec 1829), p. [1]-28.

9250. *Flore de Nice* & des principales plantes exotiques naturalisées dans ses environs ... Nice (de la Société typographique ...) 1844. Duod. (*Fl. Nice*).
Publ.: 1844, p. [i], [1]-586, *24 pl.* (uncol. P. Geny). *Copies*: BR, FI, G, HH, NY, USDA. − A second copy at HH is a reissue dated 1849 (the 9 in ink)) with a different t.p., lacking the quotation from Amos vii-14, and with "par A. Risso etc. etc." sic.

Ritgen, Ferdinand August Maria Franz von (1787-1837), German (Hessen) physician and botanist; D. med. Giessen 1808; Dr. phil. hon. ib. 1823; ultimately professor of surgery and psychiatry and director of the obstetrical clinic in Giessen. (*Ritgen*).

HERBARIUM and TYPES: Unknown.

BIBLIOGRAPHY and BIOGRAPHY: Biol.-Dokum. 15: 7560-7561; GR p. 125; Herder p. 473; KR p. 590 (b. 11 Oct 1787, d. 14 Apr 1867); PR 10282, ed. 1: 11237; Rehder 1: 260, 5: 727.

BIOFILE: Anon., Bonplandia 3: 33. 1808 (cogn. Leopoldina Röderer), 6: 51. 1858 (full titles); Flora 41: 514. 1858 (order White Falcon, Saxen-Weimar).
Jessen, K.F.W., Bot. Gegenw. Vorz. 351, 416. 1884.
Müllerott, M., Hoppea 34: 301. 1976.
Schmid, G., Goethe u.d. Naturw. 565. 1940.

9251. *Ueber die Aufeinanderfolge des ersten Auftretens der verschiedenen organischen Gestalten* ... Marburg (bei Johann Christian Krieger) 1828. Oct. (*Aufeinanderfolge org. Gest.*).
Preprint: 1828, p. [1]-98, chart. *Copy*: L (Univ. Libr.), issued with independent pagination as preprint from Schr. Ges. Beförd. Ges., Naturw. Marburg vol. 2, 1831. Also issued with this pagination in journal series, completed with journal t.p. dated 1831.
Journal issue: with continuous journal pagination: *Schr.* 2: 41-138. 1831.
Note: Contains on p. 39-98 (79-138) "Andeutungen zu einer natürlichen Gruppirung der Pflanzenwelt".
Ref.: Donk, M.A., Reinwardtia 2: 447-451. 1954, Taxon 12: 118-119. 1963.
Rogers, D.P., Farlowia 4: 33. 1950.

Ritschl, Georg [Adolf] (1816-1866), German (Prussian) philologist and botanist; studied philology in Berlin and Greifswald; teacher's examination 1841; high school teacher at Stettin, Neustettin, and Putbus, from 1844-1866 at the Friedrich-Wilhelms-Gymnasium in Posen; died of cholera. (*Ritschl*).

HERBARIUM and TYPES: Unknown.

BIBLIOGRAPHY and BIOGRAPHY: AG 2(1): 488, 4: 301; Backer p. 496; Barnhart 3: 160 (b. 3 Dec 1816, d. 18 Aug 1866); Biol.-Dokum. 15: 7561; BM 4: 1707; CSP 5: 214; Herder p. 187; PR 7644-7645; Rehder 1: 383; Urban-Berl. p. 388.

BIOFILE: Ascherson, P., Fl. Prov. Brandenburg 1: 11. 1864; Verh. bot. Ver. Brandenburg

8: xviii-xxii. 1866 (obit.); Z. naturw. Ver. Posen Bot. Abt. 1: 3-8. 15 Aug 1894 (id.).Szymkiewicz, D., Bibl. fl. Polsk. 121. 1925 (bibl.).

NOTE: Ritschl's son Julius Ritschl (b. 25 Nov 1850, Posen, d. 13 Jan 1900, Stettin) was also an active botanist and entomologist; see J. Winkelmann, Verh. bot. Ver. Brandenburg 42: xxii-xxiii. 1901, and E. Hering, Stettin. entom. Zeit. 60: 355-356. 1899.

9252. *Flora des Grossherzogthums Posen*, im Auftrage des naturhistorischen Vereins zu Posen herausgegeben ... Berlin (Druck und Verlag von E.S. Mittler und Sohn ...) 1850. Oct. (*Fl. Posen*).
Publ.: 1850 (p. iv: Apr 1850; Bot. Zeit. 28 Nov 1851), p. [i]-xxxii, [1]-291. *Copies*: G, HH, M, NY. – See also Verh. bot. Ver. Brandenburg 2: 105-106. 1860 for additions.

9253. *Zu der öffentlichen Prüfung der Schüler des Königlichen Friedrich-Wilhelms-Gymnasiums zu Posen am 12. April 1851 ladet alle Beschützer, Gönner und Freunde des Schulwesens ehrerbietigst und ergebenst ein A.G. Heydemann, Director und Professor. Inhalt: 1) Beiträge zur Flora des Grossherzogthums Posen.* Vom Gymnasial-Lehrer G. Ritschl. 2) Schulnachrichten ... Posen (h.t. Poznan) (Gedruckt in der Königlichen Hofbuchdruckerei von W. Decker & Comp.) 1851. Qu. (in twos). (*Beiträge Fl. Posen*).
Publ.: 1851, before 12 Apr, p. [i], [1]-25. *Copies*: B-S, E, MO. – Schulnachrichten on p. 27-45.

9254. Zu der öffentlichen Prüfung der Schüler des Königlichen Friedrich-Wilhelms-Gymnasiums zu Posen, welche am Montag dem 6. April 1857 im Saale des Odeums Stattfindet, ladet alle Beschützer, Gönner und Freunde des Schulwesens ehrerbietigst und ergebenst ein Dr. J. Marquardt, Director. Inhalt: 1)*Beitrag zur Flora von Posen*. Von Oberlehrer Ritschl. 2) Schulnachrichten ... Posen [h.t. Poznan] (gedruckt in der Königlichen Hofbuchdruckerei von W. Decker & Comp.). [1857]. Qu. (*Beitrag Fl. Posen*).
Publ.: 1857, before 6 Apr, p. [i], [1]-24, *1 pl.* (*uncol. copp.*). *Copies*: B-S, E, HH. – Schulnachrichten separately paged: [1]-14. See also Verh. bot. Ver. Brandenb. 2: 105-106. 1860 (Neuigkeiten ...).
Ref.: Schlechtendal, D.F.L. von, Bot. Zeit. 19: 350-351. 1861.

9255. *Ueber einige wildwachsende Pflanzenbastarde. Ein Beitrag zur Flora von Posen* ... Posen [h.t. Poznan] (in Commission der E.S. Mittlerschen Buchhandlung ...) 1857. Qu. (*Wildwachs. Pfl.-Bast.*).
Publ.: 1857, p. [i], [1]-24, *1 pl. Copy*: G.
Ref.: Schlechtendal, D.F.L. von, Bot. Zeit. 19: 350-351. 1861 (rev.).

Ritter, Carl of Berlin (1779-1859), German (Prussian) geographer and botanist; private teacher at Frankfurt a.M. 1798-1814, id. at Göttingen 1819; professor of history at the Frankfurt Gymnasium 1819-1820; from 1820 at Berlin, 1820 prof. geogr. Kriegsschule; extraordinary professor of geography at Berlin University 1820, ordinary professor 1825-1859. (*C. Ritter berol.*).

HERBARIUM and TYPES: Unknown. – Letters to D.F.L. von Schlechtendal at HAL.

BIBLIOGRAPHY and BIOGRAPHY: Barnhart 3: 160 (b. 7 Aug 1779, d. 28 Sep 1859); BM 4: 1708; CSP 5: 216; GR p. 125; Herder p. 40, 57; Langman p. 630; PR 7649, ed. 1: 8505-8610; Rehder 5: 727.

BIOFILE: Anon., Bot. Zeit. 11: 512. 1853 (portr. published with motto"Willst du ins Unendliche schreiten, geh nur im Endlichen nach allen Seiten"); Brit. Mus. gen. Cat. 203: 517-521. 1963 (Bibl., 69 nos.); Cat. gén. livr. impr. Bibl. natl. Paris 152: 651-660. 1938 (58 nos. bibl.); Flora 37: 15. 1854 (Maximilians-Orden für Kunst und Wissenschaft), 43: 16. 1860 (d.); Gartenflora 9: 29. 1860 (d.); Österr. bot. Z. 9: 370. 1859 (d. 28 Sep 1859).
Blatter, E., Rec. bot. Surv. India 8(5): 452, 469, 471. 1933.

Jessen, K.F.W., Bot. Gegenw. Vorz. 46, 352. 1884.
Müller, R.H.W. & R. Zaunick, Friedr. Traug. Kützing 234, 292. 1960.
Plott, A., Die Erde 94: 13-36. 1963 (bibl.).
Poggendorff, J.C., Biogr.-lit. Handw.-Buch 2: 654. 1863, 7a: 550-553. 1971 (extensive list of biogr. and other secondary refs. not repeated here).
Richter, O., Der teleologische Zug im Denken Carl Ritters, Leipzig 1905 (n.v.).
Schlechtendal, D.F.L. von, Bot. Zeit. 17: 344. 1859 (obit.).
Schmid, G., Chamisso als Naturforscher 1942, nos. 206, 343, 346.
Zittel, K.A. von, Gesch. Geol. Paläont. 261, 274, 275, 276, 277. 1899; Hist. geol. palaeontol. 178, 181, 183. 1901.

NOTE: (1) Many botanical details in his *Die Erdkunde*, 1817-1859, 19 vols., see BM 4: 1708 and Plott (1963).
(2) Not to be confused with Karl (Carl) Ritter (*fl.* 1836) of Dresden, later Vienna, imperial gardener, ultimately garden director in Hungary, who visited the West Indies and published a *Naturhistorische Reise nach der West-indischen Insel Hayti*, Stuttgart 1836, (206 p., copy: MO), see I. Urban, Symb. ant. 1: 138-139. 1898, 3: 113-114. 1902; plants at W. See below.

COMMEMORATIVE PUBLICATION: Die Erde 1959 (2).

9256. *Über die geographische Verbreitung der Baumwolle* und ihr Verhältniss zur Industrie der Völker alter und neuer Zeit. Erster Abschnitt. Antiquarischer Teil ... Berlin (gedruckt in der Druckerei der Königlichen Akademie der Wissenschaften) 1852. Qu.
Publ.: 1852 (mss. submitted 10 Nov 1851), p. [i], [1]-63. *Copy*: G. – Reprinted and to be cited from Abh. K. Akad. Wiss., Berlin 1851 (Phil.): 297-359. 1852.

Ritter, Christian Wilhelm Jonathan (1775-1821), German (Schleswig) botanist and physician; from 1798-1804 at Altona, 1804-1811 at Crempe, from 1811 at Flensburg. (*Chr. Ritter*).

HERBARIUM and TYPES: Unknown.

BIBLIOGRAPHY and BIOGRAPHY: Barnhart 3: 160(b. 19 Apr 1765); BM 4: 1708; Christiansen p. 12-15, 124; CSP 5: 745; GR p. 125 (b. 19 Apr 1775, d. 10 Sep 1821); LS 22337-22337a; PR 7646-7647, ed. 1: 8600; Quenstedt p. 362; Rehder 1: 376.

BIOFILE: Fischer-Benzon, R.J.D. von, in P. Prahl, Krit. Fl. Schlesw.-Holst. 2: 50. 1890.
Poggendorff, J.C., Biogr.-lit. Handw.-Buch 2: 652. 1863.
Warming, E., Bot. Tidsskr. 12: 106-107. 1880.

9257. *Versuch einer Beschreibung der in den Herzogthümern Schleswig und Holstein*, und auf dem angräntzenden Gebieten der freien Hansestädte Hamburg und Lübeck *wildwachsenden Pflanzen* mit sichtbarer Blüthe. Nebst ihres medicinischen (nach der Erfahrung bewährter Aerzte), oekonomischen und technischen Nutzens. Für deutsche Freunde der Pflanzenkunde nach dem Linnéischen Systeme bearbeitet von Christian Wilhelm Ritter, Doctor der Philosophie. Tondern (Gedruckt auf Kosten des Verfassers) 1816. Oct. (*Vers. Beschr. Schlesw. Holst. Pfl.*).
Publ.: 1816, p. [1]-16, [i]-lxii, [1]-389, [391-394, err.]. *Copy*: NY. Preliminary publ.: *in* D. Hoppe, Bot. Taschenb. 1808: 236-248.

Ritter, Johann Jakob (1714-1784), Swiss botanist; sometime professor at Franeker, Netherlands; ultimately at Gnadenfrei in Silezia. (*J. Ritter*).

HERBARIUM and TYPES: Unknown.

BIBLIOGRAPHY and BIOGRAPHY: Barnhart 3: 160 (b. 15 Jul 1714, d. 23 Nov 1784); Dryander 3: 447; PR 7648, ed. 1: 8603; Rehder 5: 727.

BIOFILE: Spilger, L., Ber. Oberhess. Ges. Nat.-Heilk. Giessen, ser. 2. Naturw. 12: 40-77. 1929 (on his "Flora riedeselia").

EPONYMY: *Rittera* Schreber (1789, also named for Albert Ritter (1681-1755)). *Note*: It is not clear for which Ritter *Rittera* Rafinesque (1840) is named.

9258. Joannis Jacobi Ritteri ... *Tentamen historiae naturalis* ditionis Riedeselioavimontanae in quatuor partes, nempe floram, mineralogiam, faunam et commentatiunculam de aere, aquis et locis etc. divisium. [*in*: Acta phys.-med. Acad. Leop.-Carol. Ephemerides 10, appendix, Norimbergae (Nürnberg) 1754]. Qu.
Publ.: 1754, botany as "Praemonenda circa Floram Riedeselicam" in Act. Leop. Ephem. 10 (App.): 25-114. 1754 (copy of journal in PPAN).
Study: L. Spilger, Ritters Flora riedeselia, Ber. Oberhess. Ges. Nat.-Heilk., Giessen, ser. 2. nat. Abt. 12: 40-77. 1929.

Rittershausen, Paul (*fl.* 1892), German (Hessen-Nassau) botanist; Dr. phil. Erlangen 1892. (*Rittershausen*).

HERBARIUM and TYPES: Unknown.

BIBLIOGRAPHY and BIOGRAPHY: Barnhart 3: 160; BM 4: 1709; Kew 4: 476.

9259. *Anatomisch-systematische Untersuchung von Blatt und Axe der Acalypheen*. InauguralDissertation zur Erlangung der Doctorwürde bei der Hohen philosophischen Facultät der K. Friedrich-Alexander-Universität zu Erlangen ... München (Druck von Val. Höfling, ...) 1892. Oct. (*Anat.-syst. Unters. Acalyph.*).
Publ.: 1892 (t.p. 1892; ÖbZ Dec 1892; Bot. Zeit. 1 Oct 1894, as of 1893; Flora rd 21 Feb 1893; Nat. Nov. Jan(2) 1893), p. [i]-xv, [1]-123, *1 pl.* (uncol.). *Copies*: G(2), M, USDA.

Ritzberger, Engelbert (1868-1923), Austrian pharmacist and botanist at Linz. (*Ritzberger*).

HERBARIUM and TYPES: LI; other material at DPU and E.

BIBLIOGRAPHY and BIOGRAPHY: Barnhart 3: 160 (b. 3 Feb 1868, d. Feb 1923); BFM no. 905.

BIOFILE: Anon., Österr. bot. Z. 72: 124. 1923 (d. Feb 1923).
Janchen, E., Cat. fl. austr. 1: 21. 1956.

NOTE: No. 9260. has not been used.

Ritzema Bos, Jan (1850-1928), Dutch phytopathologist; Dr. phil. Groningen 1874; instructor of zoology Groningen 1874-1876; instructor at Wageningen Agricultural College 1876-1895; director of the Willie Commelin Scholten phytopathological laboratory, Amsterdam 1895-1899; head of the Dutch phytopathological service 1899-1906; director of the Institute for phytopathology, Wageningen 1906-1918; professor of phytopathology at Wageningen Agricultural University 1918-1920. (*Ritz. Bos*).

HERBARIUM and TYPES: Unknown.

BIBLIOGRAPHY and BIOGRAPHY: Biol.-Dokum. 15: 7565; Bossert p. 333 (b. 27 Jul 1850, d. 7 Apr 1928); JW 1: 452, 2: 199, 3: 366-367, 5: 250; LS 22342-22360, 38425-38455, 41649, suppl. 3292-3297; Stevenson p. 1255; Tucker 1: 599.

BIOFILE: Anon., Ardea 9: 20. 1920 (portr.); Bot. Jahrb. 21 (Beibl. 53): 61. 1896 (app. Amsterdam); Hedwigia 35: (35). 1896 (prof. phytopathology Amsterdam); Österr. bot. Z. 45: 488. 1895 (app. Amsterdam); Tijdschr. Planteziekten 34(4): 123-124. 1928 (obit.) (see also 35: 1-2. 1929 on memorial).
Blink, H., Vragen van den Dag 35: 561-570. 1920 (founder of Dutch phytopathology).
Ewan, J. et al., Short hist. bot. U.S. 83. 1969.
Howard, L.O., J. econ. Entom. 21(4): 636-637. 1928 (obit.).

Poeteren, N. van, Anzeiger Schädlingsk. 4(9): 115. 1928 (obit., portr.)
Schoevers, T.A.C., Landbouwk. Tijdschr. 40: 201-204. 1928 (personal recollections).
Waterston, J., Ann. appl. Biol. 16(3): 483-485. 1929 (obit., portr.).

COMPOSITE WORKS: Founder and editor of *Tijdschrift voor Plantenziekten* 1895-1928, vols. 1-33.

Riva, Domenico (ca. 1856-1895), Italian botanist and explorer of Africa with Schweinfurth (1892) and Ruspoli (1892-1893); botanical assistant at Bologna. (*Riva*).

HERBARIUM and TYPES: FT; some further material at BR, RO (fide A. White & B.L. Sloane 1937), and TO.

BIBLIOGRAPHY and BIOGRAPHY: Barnhart 3: 160 (b. 24 Jul 1895); IH 1 (ed. 6): 363, (ed. 7): (in press), 2: (files); Saccardo 1: 139, 2: 92.

BIOFILE: Anon., Bot. Centralbl. 53: 271. 1893 (to East Africa), 63: 384. 1895 (d.); Bot. Not. 1896: 44 (d.); Österr. bot. Z. 45: 446. 1895 (d. 24 Jul 1895, Roma, "durch Selbstmord infolge äusserster Noth").
Gibelli, G., Malpighia 9: 551-552. 1895 (obit.).
White, A. & B.L. Sloane, The Stapelieae 1: 122. 1937 (portr.).

Rivas Goday, Salvador (1905-1981), Spanish botanist; Dr. pharm. Madrid 1926; botanical assistant faculty of pharmacy, Madrid 1929, assoc. prof. mat. farm. veg. ib. 1930-1938; ass. professor of descriptive botany ib. 1939-1942; regular professor of pharmacy Granada 1942-1943; idem Madrid 1943-1975 director of the Inst. bot. Cavanilles 1951-1975. (*Riv.-God.*).

HERBARIUM and TYPES: MAF.

BIBLIOGRAPHY and BIOGRAPHY: BL 2: 709 [index]; Hortus 3: 1202; IH 1 (ed. 1): 60, (ed. 2): 78, 79, (ed. 3): 98, 99, 341; TL-2/see J.C. Mutis; Zander ed. 11, p. 807.

BIOFILE: Anon., Anal. Jard. bot. Madrid 37(2): frontisp. 1980, publ. Mar. 1981 (portrait and caption announcing death on 16 Feb 1981, b. 1 Dec 1905); Lagascalia 10(1): frontisp., [i-ii], 1981 (portr., obit.).
Izco, J., Anal. Inst. bot. Cavanilles 32(2): 9-32. 1975 (biogr. data, portr., bibl.); Lazaroa 3: 5-23. 1981 (portr., bibl.), Anal. Jard. bot. Madrid 38(1): 3-6. 1981 (obit., bibl.).

THESIS: *Révisión de las orquídeas de España*, Anal. Univ. Madrid 1930: 3-36.

FESTSCHRIFT: Anal. Inst. bot. A.J. Cavanilles 32(2). 1975.

EPONYMY: *Rivasgodaya* Esteve Chueca (1973).

Rivas Mateos, Marcelo (1875-1931), Spanish botanist; professor of botany at the Fac. de Farmacia, Barcelona and Madrid. (*Riv. Mat.*).

HERBARIUM and TYPES: MAF.

BIBLIOGRAPHY and BIOGRAPHY: Barnhart 3: 160 (b. 16 Jan 1875); BL 2: 493, 494, 502, 709; BM 4: 1709; CSP 18: 228; GR p. 763-764 (b. 16 Jan 1875, d. 25 Jan 1931); Hirsch p. 249; IH 2(files); Rehder 1: 423.

BIOFILE: Anon., Encicl. univ. illustr. 51: 893. s.d.
Izco, J., Anal.-Inst. Bot. Cavanilles 32(2): 9-32. 1975.

9261. *Botánica criptogámica y en particular de las especies medicinales de la flora española.* Madrid (Librería general de Victoriano Suárez ...) 1925. Oct. (*Bot. cript.*).

Publ.: 1925 (Nat. Nov. Sep 1896), p. [1]-247, *pl. 1-5* (col.). *Copies*: FH, USDA.

9262. *Flora de la provincia de Cáceres* ... Serradilla (Editorial Sánchez Rodrigo) 1931. Oct. (*Fl. Caceres*).
Publ.: 1932 (t.p. 1931; date of printing, p. [310], 25 Jun 1932), p. [1]-307, [310]. *Copy*: USDA. – *Estudios preliminares* para la flora de la provincia de Cáceres, a series of papers, reprinted with original pagination, from Anal. Soc. españ. Hist. nat. 26: 177-215. 1897, 27: 229-256. 1898, 28: 119-252, 413-448. 1899/1900, reprinted p. [1]-208. *Copy*: MO.

Rivière, [Marie] **Auguste** (1821-1877), French horticulturist and botanist; plant introducer, arboriculturist and early orchid hybridizer; gardener at the Faculté de Médecine, Paris 1837-1859; from 1859 director of the Jardin du Luxembourg, from 1868 also of the Jardin du Hamma, Algeria. (*Rivière*).

HERBARIUM and TYPES: Some cultivated orchids at W.

NOTE: Not to be confused with the geologist Alphonse Ennemond Auguste Rivière (1805-1877). Charles Marie Rivière (1845-?) was the botanist's younger brother (fide G. Buchheim, *in* Zander ed. 11).

BIBLIOGRAPHY and BIOGRAPHY: Barnhart 3: 161 (b. 8 Jan 1821, d. 14 Apr 1877); BM 4: 1710; CSP 5: 220-221 [publications by both the geologist and the botanist mixed], 8: 756, 12: 622; Herder p. 277; IH 2: (files); Jackson p. 123, 124, 151, 496; Kelly p. 192-193; Kew 4: 479; Langman p. 631; LS 22362; Morren ed. 2, p. 22, 32; MW p. 414; NI 1641; PR 7650; Rehder 5: 728; Tucker 1: 599; Zander ed. 10, p. 707.

BIOFILE: Anon., Bull. Soc. bot. France 24: 152. 1877 (d.), Table art. orig. 1854-1893, p. 192, 202-203; The Garden 11: 306. 1877 (obit.).
Costantin, J., Ann. Sci. nat. Bot. ser. 10. 16: lxvii-lxviii. 1934.
Duchartre, P., J. Soc. centr. Hort. France ser. 2. 11: 301-310. 1877 (obit.).
Guillaumin, A., Les fleurs des jardins xlix, *pl. 17*. 1929.

9263. *Les fougères* choix des espèces les plus remarquables pour la décoration des serres, parcs, jardins et salons précédé de leur histoire botanique & horticole ... Paris (J. Rothschild, éditeur ...) 1867. Oct. (*Fougères*).
Co-authors: Édouard François André (1840-1911), Ernest Roye (1833-1900).
Publ.: 1867-1869, 2 vols. *Copies*: BR, USDA.
1: 1867 (BSbF after 11 Mar 1867), p.[i-ii], *pl. 1* [is iii-iv], [v]-x, [1]-286, *pl. 1* [sic:]-*75*. (coll. liths.).
2: 1868, p. [1]-242, [243-244], *pl. 1-80* (id.).
Ref.: Anon., Bull. Soc. bot. France 14 (bibl.): 13-15. 1867 (rev.).

9264. *Les bambous* végétation, culture, multiplication en Europe, en Algérie et généralement dans tout le bassin méditerranéen nord de l'Afrique, Maroc, Tunisie, Égypte ... Paris (au Siège de la Société d'Acclimatation ...) 1878. Oct. (*Bambous*).
Co-author: Charles Marie Rivière (1845-?), director of the experimental garden at Hamma, Algeria.
Publ.: 1878 (J. Bot. Aug 1879; Nat. Nov. Jul(1) 1879; Bot. Zeit. 5 Dec 1879), p. [1]-364, [1, err.]. *Copies*: BR, HH. – Originally published in Bull. Soc. Acclim. ser. 3. 5[25]: 221-253, 290-322, 392-421, 460-478, 501-526, 597-645, 666-721, 758-828. 1878.
Ref.: Fenzi, E.O., Bull. Soc. tosc. Ortic. 4: 337-343. 1879, 5: 130-134, 278-281, 303-307, 369-372, 393-402. 1880 (rev.).

Rivière, Charles Marie (1845-?), French horticulturist; brother of Auguste Rivière; director of the experimental garden of Hamma, Algeria (asst. dir. 1868-1877; dir. from 1877). (*C. Rivière*).

HERBARIUM and TYPES: Unknown.

BIBLIOGRAPHY and BIOGRAPHY: Barnhart 3: 161; BM 4: 1710; Hortus 3: 1202; LS 22363, 38460; Morren ed. 2, p. 32, ed. 10, p. 112; MW p. 414; Rehder 5: 728; Tucker 1: 599.

BIOFILE: Cosson, E., Comp. fl. Atl. 1: 87. 1881.

COMPOSITE WORKS: Collaborator, R. de Noter, Orangers 1896, TL-2/6902.

9265. *Le Niaouli* (arbre de la Nouvelle-Calédonie) et le genre Melaleuca en Algérie ... Paris (Au siège de la société [d'Acclimatation] ...). 1883. Oct.
Publ.: 1883, p. [1]-57, *1 pl. Copy*: G. – Reprinted from Bull. Soc. Acclim. [Paris] 1883 (journal n.v.).

Rivinus, August [Augustus] **Quirinus** [Bachmann] (1652-1723), German (Saxonian) botanist and physician; Dr.med. Helmstedt 1676, Dr. med. Leipzig 1677; practicing physician at Leipzig 1677-1688; assessor at the Leipzig medical faculty 1688, professor for physiology and botany 1691; ordinary professor of botany 1701; professor of therapy and botany 1719. (*Riv.*).

HERBARIUM and TYPES: Formerly at Dresden; destroyed by fire 1849.

BIBLIOGRAPHY and BIOGRAPHY: ADB 28: 708 (Pagel); AG 3: 680, 5(1): 383; Barnhart 3: 161 (b. 9 Dec 1652, d. 30 Dec 1723); BM 4: 1710; Bossert p. 333; Dryander 3: 37-38, 43; Frank 3 (Anh.): 84; Hegi 5(1): 635, 5(2): 1177; Henrey 2: 650; Herder p. 84, 141; Jackson p. 1, 29; Kew 4: 479; Lasègue p. 335; Moebius p. 40, 42, 43, 44; NI 1642, see also 1: 103, 161, 162, 164, 196; PR 2284, 7438, 7651-7656, 7657, 7905, 8143, ed. 1: 2632, 6180, 8382, 8611-8616, 8617, 8868, 9092; Rehder 1: 288, 5: 729; RS p. 137; SA 2: 598; TL-1/1085-1092, see C.G. Ludwig; TL-2/1470, 3169, see C.G. Ludwig; Tucker 1: 599-600.

BIOFILE: Britten, J., J. Bot. 29: 310. 1894, 30: 55-56. 1892 (Ordo pl. fl. tetrapet.).
Candolle, Alph. de, Phytographie 444. 1880.
Eriksson, G., Bot. hist. Sverige 71, 143, 153, 164, 196, 235. 1969.
Fries, Th.M., Bref Skrifv. Linné 1(5): 318. 1911.
Greene, E.L., J. Bot. 30: 55. 1892 (with note J. Britten 30: 55-56. 1892) (on Ordo pl. fl. tetrapet.).
Harvey-Gibson, R.J., Outlines hist. bot. 43, 55. 1914.
Heimans, J., Levende Natuur 63(10): 225-231. 1960.
Heller, J.L., Index auctorum et librorum, in C. Linnaeus, Sp. pl., facs. ed. Ray Society 2: 49. 1959.
Hermann, G.S., Vita Rivini, *in Bibl. rivin.* [iii-xvi]. 1727 (biogr., bibl.).
Huth, E., Clavis riviniana, 1891, see TL-2/3169.
Jessen, K.F.V., Bot. Gegenw. Vorz. 266, 267, 281, 460. 1884.
Jourdan, A.J.L., Dict. Sci. méd., Biogr. méd. 7: 31-34. 1825 (bibl.).
Junk, W., Rara 1: 61-62. 1902.
Lischwitz, Oratio panegyrica funebris in obitum Augusti Quirini Rivini, Lipsiae 1724 (fide PR).
Nicolas, J.P., *in* G.H.M. Lawrence, ed., Adanson 1: 154-155. 1963.
Nordenskiöld, E., Hist. biol. 195. 1935.
Poggendorff, J.C., Biogr.-lit. Handw.-Buch 2: 660. 1863.
Rose, H.J., New general biographical dictionary 11: 356. 1850 (n.v.).
Ruhland, J., Über das botanische System des Rivinus, Würzburg 1832, 19 p.
Sachs, J., Gesch. Bot. 611 (index) 1875; Hist. bot. 7, 77, 87, 105. 1892.
Stafleu, F.A, Linnaeus and the Linnaeans 43, 91, 241-242, 271. 1971.
Treviranus, L.C., Linnaea 2: 47-54. 1827 (on tomus supplementorum).
Wittrock, V.B., Acta Horti Berg. 3(2): 140. 1903, 3(3): xix, 149-150, *pl. 19.* 1905 (portr.).
Zaunick, R., *in* facs. ed. *Bibl. rivin.*, Amsterdam 1966, p. [iii*-vii*] (b. 9 Dec 1652, d. 30 Dec 1723; many details).

GENERAL KEY TO RIVINUS' WORKS: E. Huth, *Clavis riviniana*, Frankfurt a.O. 1891 (Nat. Nov. Apr(1) 1891), i, 28 p.; see TL-2/3169.

EPONYMY: *Rivina* Linnaeus (1753); *Rivinia* P. Miller (7154); *Rivinia* Linnaeus (1754, orth. var. of *Rivina* Linnaeus (1753).

9266. D. Augusti Quirini Rivini Lipsiensis, *Introductio generalis in rem herbariam*, Lipsiae [Leipzig] (sumptibus autoris, ... typis Christoph. Güntheri) 1690. Fol. (*Intr. rem. herb.*).
Publ.: 1690, p. [i-viii], 1-39. *Copies*: FI, G, PH, U; IDC 288.
Ed. 2: 1696, p. [i-iv], 1-114, [1]. *Copy*: FI. – "D. Augusti Quirini Rivini lipsiensis, *Introductio generalis in rem herbariam*. Denuo recusa Lipsiae, apud viduam Johannis Heinichii, 1696".
Ed. 3: 1720, n.v.

9267. D.A.Q.R. *Ordo plantarum, quae sunt flore irregulari monopetalo*. Anno m.dc.xc. [1690]. Lipsiae impressus literis Christoph. Fleischeri. Fol. (*Ordo pl. fl. monopetal.*).
Publ.: 1690, p. [i], [1]-22, [23-26, index], [27, h.t. icones], *126 pl.* (uncol. copp.). *Copies*: FI (plates not counted), G (*126 pl.*, has a 2 p. table between p. 5 and 6), PH (*122 pl.*), U (*124 pl.*); IDC 288.

9268. D.A.Q.R. *Ordo plantarum, quae sunt flore irregulari tetrapetalo*. Anno mdcxci [1691]. Lipsiae, typis Christoph. Fleischeri. Fol. (*Ordo pl. fl. tetrapet.*).
Publ.: 1691, p. [i-vi], [1]-20, [21-24 index], *121 pl. Copies*: G, PH (*119 pl.*), U; IDC 288. – PR calls for *119 pl.*

9269. D.A.Q.R. *Ordo plantarum quae sunt florae irregulari pentapetalo*. Anno mdcic [1699]. Lipsiae, typis Joh. Heinrici Richteri. Fol. (*Ordo pl. fl. pentapet.*).
Publ.: 1699, p. [i-vi], [1]-28, [29-32, ind.], [33 h.t. icon. *pl. 1-138, 139a, 139b*. *Copies*: FI (pl. not. counted), G (*154 pl.*), U (*137 pl.*), USDA (*140 pl.*); IDC 288.

9270. *Icones plantarum, quae sunt flore irregulari hexapetalo* [Leipzig s.d.] [publ. 1764]. Fol. (*Icon. pl. fl. hexapet.*).
Publ.: 1764, by C.L. Ludwig, p. [i], *17 pl.* (fide NI & PR). The only copy seen by us, at G, has a t.p. and *84 pl.* See also Junk, Rara (1902); Bibl. rivin. no. 48 cites *23 pl.*

9271. *Officinal-Kräuter*, Leipzig 1717. Fol. (n.v.).
Publ.: 1717, a set of plates of medicinal plants taken from the previous volumes. Junk (1902) mentions 6 p. index and 117 pl.

9272. [*Tomus supplementorum* ad opus botanicum, ... Tabl. aen. 112. Fol. Reg. [1773-1777]. n.v.
Publ.: posthumously between 1773-1777. The title above is taken from *Bibl. rivin.* no. 47. See also Junk (1902). Junk knew of one copy, in the Royal Library Berlin. See also Treviranus (1827).

9273. *Bibliotheca riviniana* sive catalogus librorum philologico-philosophico-historicorum, itinerariorum, in primis autem medicorum, botanicorum et historiae naturalis scriptorum &c. rariorum, quam magno studio et sumptu sibi comparavit D. Aug. Quir. Rivinus, ... vendenda in vaporario collegii rubri a die xxvii. otobr. [1727] more auctionis consueto . Praemissa est vita Rivini descripta per M. Georg Samuel Hermann, ... Lipsiae [Leipzig] (typ. Immanuelis Titii) [1727]. Oct. (*Bibl. rivin.*).
Publ.: 1727, frontisp. portr., p. [i-xiv], [1]-740, index Aaaz-Gggg . (107 p.). (n.v.).
Facsimile ed.: 1966, Amstelodami [Amsterdam] (A. Asher & Co.), facsimile with an extra t.p. and a 5 p. introduction by R. Zaunick. *Copies*: FAS, MO, NY.
Ref.: Stafleu, F.A., Taxon 17: 218-220. 1968.

Robbins, James Watson (1801-1879), American (New England) botanist; Dr. med. Yale 1828; practicing physician at Uxbridge, Mass. 1830-1869, surgeon and physician with a copper-mining company nr Portage Lake, Lake Superior 1839-1863; travelling in Cuba and Texas 1863-1864; from 1864-1879 again in Uxbridge devoting himself to botany. (*J.W. Robbins*).

HERBARIUM and TYPES: Material at GH, MASS, PH, WELC. – See Day (1901).

BIBLIOGRAPHY and BIOGRAPHY: Barnhart 3: 162 (b. 18 Nov 1801, d. 9 Jan 1879); BM 4: 1711; Hortus 3: 1202; Lenley p. 345, 467; ME 1: 225, 2: 443; TL-1/431; TL-2/2124.

BIOFILE: Anon., Amer. J. Sci. ser. 3. 19: 77. 1888 (d.); Bot. Not. 1880. 28 (d.); Bot. Zeit. 37: 260. 1879 (d.); J. Bot. 17: 96. 1879 (d.).
Day, M., Rhodora 3: 262. 1901 (coll.).
Geiser, S.W., Field & Labor. 27: 185. 1959.
Högrell, B., Bot. hist. 1: 255. 1886.
P., Worcester Southern Compendium (Uxbridge, Mass.), 12(3): 18 Jan 1879 (copy: MO).
Pease, A.S., Fl. N. New Hampshire 33, 35. 1964.
Voss, E.G., Contr. Univ. Mich. Herb. 13: 77. 1978.

COMPOSITE WORKS: Contributed *Potamogeton* to A. Gray, Manual, ed. 5, 1867.

Robbins, William Jacob (1890-1978), American botanist; A.B. Lehigh Univ. 1910; Dr. phil. Cornell Univ. 1915; assistant, later instructor for plant physiology at Cornell 1912-1916; professor of botany in the Alabama Polytechnical Institute 1916-1918; on active duty 1918; soil biochemist with USDA 1919; professor of botany and chairman of the Dept. of Botany, Univ. of Missouri, 1919-1937; dean Graduate School 1930-1937; director of the New York Botanical Garden and professor of botany at Columbia Univ., N.Y. 1937-1957; Dr. Sci. h.c. Lehigh 1937, Fordham Univ. 1945. (*W.J. Robbins*).

HERBARIUM and TYPES: Unknown.

BIBLIOGRAPHY and BIOGRAPHY: Ainsworth p. 110, 330; Bossert p. 333; IH 1 (ed. 1): 69, (ed. 2): 88, (ed. 3): 111; Lenley p. 345-349; MW suppl. p. 294; PH 418.

BIOFILE: Anon., Bull. misc. Inf. Kew 1938: 52 (app. NYBG); Current biogr. Yearb. 1956: 515-517 (biogr., portr.); Garden J. 8: 73. 1958 (retirement); J. New York Gard. 38: 249-252. 1937 (biogr. note; portr; b. 22 Feb 1890); Natl. Cycl. Amer. biogr. 1 (1953-1959): 178-179. 1960; New York Botanical Gard., Newsletter 12(8): [2]. 1979 (d. 5 Oct 1978 at New York; obit., portr.).
Cattell, J.M., Amer. men sci. ed. 3: 578. 1921, ed. 4: 824. 1927, ed. 5: 939. 1933, ed. 6: 1190. 1938, ed. 7: 1488. 1944, ed. 8: 2080. 1949.
Ewan, J. et al., Short hist. bot. U.S. 18, 71. 1969.
Hellman, G.T., New Yorker 23: 30-54. 1947 ("Square deal among the fungi").
Kavanagh, F. & A. Hervey, Bull. Torrey bot. Club 108(1): 95-121. 1981 (bigr., port., partial bibl.; complete bibl. on file in archives NY).
Murrill, W.A., Hist. found. Florida 40. 1945.
Rickett, H.W., Index Bull. Torrey bot. Club 83. 1955.
Rogers, D.P., Brief hist. mycol. N. Amer. ed. 2: 60, 63. 1981.
Rogerson, C.T., Mycologia index 1032-1033. 1968 (list of publ. in Mycologia).
Smiley, N., Fairchild Trop. Gard. Bull. 33(4): 28-31. 1978 (portr., biogr. sketch).
Verdoorn, F., ed., Chron. bot. 3: 299, 392. 1937, 4: 81, 165, 184, 233, 285, 287. 1938, 5: 233, 239, 307, 414, 415. 1939, 6: 163, 165, 329, 429. 1941, 7: 38, 138, 357. 1943.

COMPOSITE WORKS: *Botany*: a textbook for college and university students, New York 1929, with H.W. Rickett, q.v.

EPONYMY: W.J. Robbins Plant Science Building, Fairchild Tropical Garden, Miami, Florida (R. was president of the Board of Trustees for the garden 1962-1969).

Robecchi-Bricchetti, Luigi (1855-1926), Italian explorer of Somalia and Ethiopia; studied at the Zürich E.T.H and at Karlsruhe. (*Rob.-Bricch.*).

HERBARIUM and TYPES: FT; further material at B (mostly destroyed), and FI.

BIBLIOGRAPHY and BIOGRAPHY: BM 4: 174; IH 1 (ed. 6): 363, (ed. 7): 341, 2: (files); Urban-Berl. p. 388.

BIOFILE: Cufodontis, C., C.R. AETFAT 4: 246. 1962 (coll.).
Pollacci, G., Atti Ist. bot. Pavia ser. 4. 5: iii-ix. 1934 (biogr., bibl., b. 27 Mai 1855).

9274. *Somalia e Benadir* viaggio di esplorazione nell'Africa orientale prima traversata della Somalia compiuta per incarico della Società geographica italiana ... Milano ("La Poligrafica" ...) [1899]. Oct. (*Somalia & Benadir*).
Publ.: 1899, p. [i]-xix, [1]-726, portr. *Copy*: FI. – In appendix: R. Pirotta, Notizie sulle collezioni botaniche.

Roberg, Lars [Laurentius] (1664-1742), Swedish botanist and physician; Dr. med. Leiden 1689; professor of medicine Uppsala 1697-1740 (predecessor of Linnaeus), director of the Uppsala botanical garden 1740-1742. (*Roberg*).

HERBARIUM and TYPES: Unknown; see Löwegren (1952).

BIBLIOGRAPHY and BIOGRAPHY: Backer p. 496; Barnhart 3: 162; BM 4: 1711, 7: 1081-1082 (list dissertations); Dryander 3: 22, 524; KR p. 590-591 (b. 24 Jan 1664, d. 21 Apr 1742, bibl.); Moebius p. 242; PR 7658-7660, ed. 1: 8618-8626; Quenstedt p. 363; Rheder 5: 729; SO 64, 1333, 1334, 3093, 3597.

BIOFILE: Anon., J. Bot. 61: 159. 1923 (reprint of his *Vegetabilium cum animalibus comparata* of 1737).
Blunt, W., The compleat naturalist 26, 28, 29, 30, 36. 1971.
Eriksson, G., Bot. hist. Sverige 362 [index]. 1969.
Fries, R.E., Short hist. bot. Sweden 18. 1950.
Fries, Th.M., Bref skrifv. Linné 1(2): 5, 7, 216. 1908, 1(6): 444. 1912, 1(7): 58. 1917.
Hulth, J.M., Bref skrifv. Linné 2(1): 337, 368, 372. 1916.
Lillingston, C., Brit. med. J. 4829: 201-202. 1953 (n.v.).
Lindman, S., Sv. män kvinn. 6: 301. 1949 (portr.), ("R. was unkempt and avaricious").
Linnaeus, C., Lars Robergs tal, holne för publique promotioner, vid Upsala Academie, 1747.
Löwegren, Y., Naturaliekabinett i Sverige 168-172, 362, 405. 1952.
Mägdefrau, K., Gesch. Bot. 51. 1973.
Scheutz, N.J.W., Bot. Not. 1863: 71 (epon.).
Sernander, R., Sv. Linné-Sällsk. Årsskr. 11: 101-115. 1925 (portr.).
Stafleu, F.A., Linnaeus and the Linnaeans 6, 17. 1971.
Wittrock, V.B., Acta Horti Berg. 3(3): 50. 1905.

COMPOSITE WORKS: See SO 64, 1333, 1334, 3093, 3597. The *Dissertatio botanica de Planta Sceptrum carolinum*, 1731, defended by J.O. Rudbeck on 19 Jun 1731 under Roberg, is said by Th.M. Fries (Linné 1: 69. 1903) to have been based on a Linnaean manuscript. ("This thesis I compiled in one day at 30 crowns" Linnaeus on the ms.). See also SO 1334, 1335.

NOTE: Roberg's *Vegetabilium cum animalibus comparatio* 1737 was reprinted in facsimile 1923, by B.D. Jackson (privately printed).

EPONYMY: *Roberga* Cothenius (1790, *orth., var.*); *Robergia* Schreber (1789).

Roberge, Michel Robert (x-1864), French (Normandy) mycologist; teacher at several girl schools at Caen. (*Roberge*).

HERBARIUM and TYPES: CN (now at PC?); other material at AHFH, FH, G, H-NYL (lich.), NTM, NYS, PC, PH, UPS. *Exsiccatae: Algues de la Normandie*, fasc. 1-8, nos. 1-200, 1826-1838, with F.J. Chauvin, q.v. See also below for a prospectus.

BIBLIOGRAPHY and BIOGRAPHY: Barnhart 3: 162 (d. 2 Dec 1864, note on his taxa publ. by Desmazières); GR p. 296-297; Hawksworth p. 185; Jackson p. 281; LS 18281; SO 2663; Stevenson p. 1255; TL-2/ see F.J. Chauvin.

BIOFILE: Candolle, Alph. de, Phytographie 444. 1880.
Morière, J., Mém. Acad. Sci. Caen 1866: 494-507 (obit., herb. donated to CN).

EPONYMY: *Robergea* Desmazières (1847).

9274a. *Algues de la Normandie* recueillies et publiées, la partie des Articulées, par M. Roberge, ... et la partie des inarticulées, par J. Chauvin, ... [Caen (Imprimerie de T. Chalopin)] [1826]. Oct. (*Alg. Normandie*).
Publ.: 1826, p. [1]-3. *Copy*: G. – Prospectus of the set of exsiccatae mentioned above. For further details see under F.J. Chauvin.

9275. *Liste des Hypoxylées, Mucédinées et Urédinées* récoltées aux environs *de Caen* ... Caen (Chez F. Le Blanc-Hardel, ...) 1866. Oct.
Editor (after death of author): Pierre Giles Morière (1817-1888).
Publ.: 1866, p. [1]-32. *Copies*: G(2). – Reprinted and to be cited from Bull. Soc. Linn. Normandie 10: 130-157. 1866.

Robert, Gaspard Nicolas (1776-1857), French botanist, gardener and pharmacist; "jardinier"-botaniste du port de Toulon" 1794; Dr. pharm. Paris 1801; pharmacist and botanist and director of the botanical garden "du port de Toulon" 1801-1847; reafforested the "Faron" region nr. Toulon. (*G.N. Robert*).

HERBARIUM and TYPES: 30 specimens from Toulon, sent by Robert are at H (herb. Steven); 1 specimen is mentioned by Lindemann (1886) for MW; some further material at G.

BIBLIOGRAPHY and BIOGRAPHY: AG 3: 784; BL 2: 209, 709; BM 4: 1711; Kew 4: 482; PR 7662, ed. 1: 8627; Tucker 1: 600.

BIOFILE: Kukkonen, I., Herb. Chr. Steven 85. 1971 (30 specimen at H).
Lindeman, E. v., Bull. Soc. natural. Moscou 61: 62. 1886 (1 specimen at MW).
Reynier, Soc. Hort. Bot. Marseille Rev. mens. 1894: 67-74, 125-127, 135-139 (relations with J.L.M. Castagne).
Rix, M., The art of the botanist 223. 1981.
Zaccarie, A., Ann. Soc. Sci. nat. Archéol. Toulon 11: 100-103. 1959 (biogr., b. 23 Dec 1776, d. 10 Jul 1872 based on original documentation).

EPONYMY: *Robertia* Merat (1812).

9276. *Plantes phanérogames qui croissent naturellement aux environs de Toulon*. Brignoles (imprimerie, lithographie et librairie de Perreymond-Dufort) 1838. Qu. (*Pl. phan. Toulon*).
Publ.: 1838 (Flora 28 Aug 1838), p. [1]-116, [117, err.], map. *Copies*: G(2), NY, USDA. For name of author see p. 6.

Robert, Nicolas (1610-1684), French botanical artist for Gaston d'Orléans at Paris. (*N. Robert*).

HERBARIUM and TYPES: Unknown.

NOTE: See NI 1643-1649 for Robert's flower books. A facsimile edition of the 1660 [?] issue of *Variae ac multiformes florum species* was published by Scolar Press, London 1975. – See Laissus (1980) for a major work on Robert's originals.

BIBLIOGRAPHY and BIOGRAPHY: Barnhart 3: 162; Blunt p. 302 [index]; BM 4: 1712; Dryander 3: 66; DU 264; Frank 3 (Anh.): 84; GF p. 73; Herder p. 84; HU 2(1): cxci-cxcii, cxcviii; Jackson p. 29, 110; Lasègue p. 314; NI 1643-1649 (b. 18 Apr 1611, d. 25 Mar 1685) and see indexes; PR 7661, ed. 1: 2648, 8630-8632; Rehder 5: 729; TL-1/577; TL-/876, 3303; Tucker 1: 600.

BIOFILE: Anon., Bull. misc. Inf. Kew 1896: 32 (Kew acquired a set of R.'s engravings).

Blunt, W., The compleat naturalist 18, 116. 1971.
Bridson, G.D.R. et al., Nat. hist. mss. res. Brit. Isl. 431 [index]. 1980.
Brindle, J.V., Bull. Hunt Inst. bot. Docum. 2(1): 3-4. 1980.
Fries, Th.M., Bref skrifv. Linné 1(6): 316. 1912.
Laissus, Y., ed., Nicolas Robert et les vélins du Roy, Paris 1980 (introd. by Yves Laissus; 35 reproductions of Robert originals).
Lawalrée, A. et al., Vélins du Muséum, Bruxelles 1974, p. 3.
Rose, H.J., New gen. biogr. dict. 11: 357. 1850.
Stearn, W.T., Roy. Hort. Soc., Exhib. manuscr. books 1954, p. 23, 24.

EPONYMY: *Robertia* Scopoli (1777); *Robertia* A. Richard ex A.P. de Candolle (1815). *Note*: *Robertia* R. Choubert (1932) honors M. Robert, Belgian geologist; *Robertomyces* Starbäck (1905) honors both Klas Robert Elias Fries, *q.v.* and Oscar Robert Fries (1840-1908), Swedish mycologist.

Robertson, David (1806-1896), British (Scottish) algologist and naturalist at Cumbrae; originally practicing a medical career, then in commerce, from ca. 1860 devoting himself fully to marine zoology and botany; LL.D. h.c. Glasgow 1895; "the naturalist of the Cumbraes". (*D. Robertson*).

HERBARIUM and TYPES: BM (collections Cumbrae, Durham, in herb. W.J. Hooker; orig. at K); also in GLAM (fide Powell and Conway, Brit. Phycol. Soc. 2(4): 267-268. 1963 and Koster (1969)).

BIBLIOGRAPHY and BIOGRAPHY: Barnhart 3: 1643; BB p. 258; BM 4: 1713; CSP 5: 230, 8: 760, 11: 194; Desmond p. 523 (b. 28 Nov 1806, d. 20 Nov 1896); De Toni 4: xlviii; Quenstedt p. 363.

BIOFILE: Anon., Bot. Not. 1897: 47 (d.); J. Bot. 35: 32. 1897, 36: 40. 1898; Proc. Linn. Soc. London 1896/7: 66-67. 1897 (obit.); Quart. J. Geol. Soc. London, Proc. 53: lxiv. 1897 (obit.); Trans. nat. Hist. Soc. Glasgow ser. 2. 5: 18-42. 1900 (obit., b. 28 Nov 1806 old style [sic new style dates from 1751]).
Bridson, G.D.R. et al., Nat. hist. mss. res. Brit. Isl. 158.1. 1980.
Dickinson, C.I., Phycol. Bull. 1: 13. 1952 (algae at K [now BM]).
Kneucker, A., Allg. bot. Z. 3: 56. 1897 (d.).
Koster, J.Th., Taxon 18: 557. 1969.
Newton, L.M., Phycol. Bull. 1: 18. 1952.
Stebbing, Rev. Thom. R.R., The naturalist of Cumbrae, a true story, being the life of David Robertson. London (Kegan Paul, French, Trübner & Co., Ltd.) 1891, x, 398 p. (main biogr., bibl., on p. 385-390; *copy*: Amherst college Library).
T.R.R.S., Geol. mag. ser. 2. dec. 4. 4: 94-96. 1897 (obit.).

EPONYMY: *Robertsonia* A.H. Haworth (1812) honors Benjamin Robertson (c. 1732-1800), English gardener).

Robin, Charles [Philippe] (1821-1885), French biologist and politician; Dr. med. Paris 1846; Dr.Sci. Paris 1847; professor of botany at the Paris "Faculté de médecine" 1847-1885, from 1862 as professor of histology. (*C.P. Robin*).

HERBARIUM and TYPES: Unknown.

BIBLIOGRAPHY and BIOGRAPHY: Ainsworth p. 316; Barnhart 3: 164 (b. 4 Jun 1821, d. 6 Oct 1885); BM 4: 1713; Bossert p. 334; CSP 5: 231-234, 8: 761-762, 11: 194-195; Frank 3 (Anh.): 85; Herder p. 245; Jackson p. 165, 166, 220; Kew 4: 486; Lenley p. 349; LS 22380-22383; PR 7668-7669, ed. 1: 8636; Rehder 1: 36.

BIOFILE: Anon., Bot. Not. 1885: 39 (d.); Bot. Zeit. 43: 790. 1885 (d. 6 Oct 1885); Psyche, Cambridge 5: 36. 1888 (d.).
Bulloch, W., Hist. bacteriol. 392-393. 1938.
Laboulbène, A., Ann. Soc. entom. France ser. 6. 5: 467-472. 1886 (obit.).

Lucante, A., Rev. bot., Bull. mens. Soc. franç. Bot 4: 333. 1886 (b. 4 Jun 1821, d. 6 Oct 1885, obit., brief bibl.).
Magnin, A., Ann. Soc. bot. Lyon 32: 138-139. 1907 (d. 5 Oct 1885), 35: 50. 1910.
Neschniakoff, T., Bull. Soc. natural Moscou 61(1): 205-222. 1886 (obit.).

9277. *Des végétaux qui croissent sur l'homme* et sur les animaux vivants, ... À Paris (chez J.B. Baillière, ...) Londres (chez H. Baillière, ...) 1847. Oct. (*Vég. croiss. l'homme*).
Publ.: 1847 (mentioned by Schlechtendal, Bot. Zeit. 4 Aug 1848), p. [i]-viii, [1]-120, *pl. 1-3* (uncol.). *Copies*: GOET (Inf. G. Wagenitz), H, LC, MO. – For a second edition, virtually a new book, see below.

9278. *Histoire naturelle des végétaux parasites* qui croissent sur l'homme et sur les animaux vivants, ... Paris (chez J.-B. Baillière ...), Londres (chez H. Baillière, ...), New York (id.), Madrid (chez C. Bailly-Baillière, ...) 1853. Oct. (*Hist. nat. vég. paras.*).
Text: 1853 (p. xvi: 20 Mai 1853), p. [i]-xvi, [1]-702, [1-2, err.].
Atlas: 1853, p. [1]-24, *pl. 1-15* (coll. liths. C. Robin and P. Lackerbauer).
Copies: BR, G, H, HH, MO, NY, USDA; IDC 5414.

Robin, Claude C. (1750-?), French clergyman, explorer and botanist; travelled in Louisiana, Florida, Martinique and Ste. Domingo 1802-1806. (*C.C. Robin*).

HERBARIUM and TYPES: Unknown.

BIBLIOGRAPHY and BIOGRAPHY: AG 6(2): 462; Barnhart 3: 164; BM 4: 1713; Herder p. 57, 226; Lasègue p. 462, 505; LS 22384; ME 3: 368, 381; PR 7400, 7667, ed. 1: 8322; Rehder 1: 318; TL-1/1028.

BIOFILE: Anon., Appleton's Cycl. Amer. biogr. 5: 281. 1888.
Ewan, J., Southw. Louisiana J. 7: 12. 1967.
Ewan et al., Short hist. bot. U.S. 38. 1969.
Fournier, P., Voy. déc. sci. mission. natural. franç. 2: 190. 1932.
Lubrecht, H., Early Amer. bot. works 1967, no. 127.
Rogers, D.P., Brief hist. mycol. N. Amer. 7. 1979.
Urban, I., Symb. ant 3(1): 114. 1902.

9279. *Voyages dans l'intérieur de la Louisiane*, de la Floride occidentale, et dans les isles de la Martinique et de Saint-Domingue, pendant les années 1802, 1803, 1804, 1805 et 1806. Contenant de nouvelles observations sur l'histoire naturelle, la géographie, les moeurs, l'agriculture, le commerce, l'industrie et les maladies de ces contrées, particulièrement sur la fièvre jaune, et les moyens de les prévenir. En outre, contenant ce qui s'est passé de plus intéressant, relativement à l'établissement des Anglo-Américains à la Louisiane. Suivis de la Flore louisianaise. Avec une carte nouvelle, ... à Paris (chez F. Buisson, ...) 1807, 3 vols. (*Voy. int. Louisiane*).
1: 1807, p. [i*], port., [iii*], [i]-xii, [1]-346.
2: 1807, p. [i-iii], [1]-511, 1 map.
3: 1807, p. [i]-xii, [1]-551; *Flore louisianaise* on p. 313-551.
Copy: NY. – See also C.S. Rafinesque, *Florula ludoviciana*, "translated, revised and improved" by C.S Rafinesque (see no. ...).

Robin, Jean (1550-1629), French botanist and royal gardener; from 1590 at the Jardin des Plantes, Paris. (*J. Robin*).

HERBARIUM and TYPES: Unknown.

BIBLIOGRAPHY and BIOGRAPHY: AG 6(2): 713; Backer p. 496; Barnhart 3: 164 (d. 25 Apr 1629); BM 4: 1713; Bossert p. 334; Hegi 4(3): 1390; Jackson p. 422; NI alph.; PR 7670-7673, 9671, ed. 1: 6171-6172, 8637-8640, 10643; Rehder 5: 730; Tucker 1: 601.

BIOFILE: Guillaumin, A., Les fleurs des jardins *pl. 16.* 1929 (portr.).
Jessen, K.F.W., Bot. Gegenw. Vorz. 255, 270. 1884.

Lemmon, K., Golden age plant hunters 4, 5. 1968.
Leroy, J.F. et al., Bot. franç. Amér. Nord 19, 66, 142, 149, 151. 1957.
O., Dict. sci. méd., Biogr. méd. 7: 35. 1825.
Warner, M.F., Natl. hort. Mag. 35: 215-220. 1956.

EPONYMY: *Robina* Cothenius (1790, *orth. var.*); *Robinia* Linnaeus (1753) and the derived names *Robincola* Velenovský (1947) and *Robinioxylon* W.R. Müller-Stoll et E. Mädel (1967).

Robinson, Benjamin Lincoln (1864-1935), American botanist; Dr. phil. Strassburg 1889 with H. Solms Laubach; from 1890-1892 assistant to Sereno Watson at the Gray Herbarium, Cambridge, Mass.; curator from 1892-1935; Asa Gray professor of systematic botany 1899-1935. (*B.L. Robinson*).

HERBARIUM and TYPES: GH; duplicates at BM, C, F, K, MO, US, WELC. – Letters e.g. in Smithsonian Archives and at FH.

BIBLIOGRAPHY and BIOGRAPHY: Barnhart 3: 164 (b. 8 Nov 1864, d. 27 Jul 1935); Biol.-Dokum. 15: 7570; BJI 1: 49, 2: 146; BL 1: 104, 109, 140, 158, 160, 161, 315; BM 4: 1714, 7: 1083; Bossert p. 334; CSP 18: 242-243; Hortus 3: 1202; IF suppl. 1: 75, 84; IH 1 (ed. 2): 36, (ed. 3): 41, (ed. 6): 363, (ed. 7): 341, 2: (files); Langman p. 633-635 (Mexican bot. bibl.); Lenley p. 349; LS 22389; MW p. 415; NW p. 56(!); PH p. 520; Rehder 5: 730; SO 493a; Stevenson p. 1256; TL-1/431, 436, TL-2/2124, 2132, 4031; Tucker 1: 601; Urban-Berl. p. 388; Zander ed. 10, p. 707, ed. 11, p. 807.

BIOFILE: Anon., Bot. Centralbl. 47: 223. 1891 (illness), 52: 112, 287. 1892 (curator GH), 81: 319. 1900, 83: 96. 1900 (app. Harvard); Bull. misc. Inf. Kew 1935: 577-578 (obit.); Hedwigia 39: (71), (195). 1900 (app. Harvard), 75: (166). 1936 (d.); J. Bot., Morot 6: 464. 1892 (curator GH, succeeding Sereno Watson); Nat. Nov. 14: 440. 1892 (curator GH), 22: 131. 1900 (app. Harvard); Österr. bot. Z. 42: 428. 1892 (curator GH), 45: 72. 1893 (id.), 50: 108, 347. 1900 (app. Harvard), 84: 320. 1935 (d.); The New York Times 29 Jul 1935; Torreya 35: 116. 1935 (obit.).
Ewan, J. et al., Short hist. bot. U.S. 16, 21, 48, 110. 1969.
Fernald, M.L., Natl. Acad. Sci. biogr. Mem. 17(13): 305-330. 1936 (biogr., portr.; bibl. by L.M. Perry), Proc. Amer. Acad. Arts Sci. 71(10): 539-542. 1937.
Humphrey, H.B., Makers N. Amer. bot. 210-213. 1961.
Merrill, E.D., Science ser. 2. 82: 94, 142-143. 1935 (obit.).
Pennell, F.W. & J.H. Barnhart, Monogr. Acad. nat. Sci. Philadelphia 1: 617. 1935 (coll.).
Rendle, A.B., J. Bot. 73: 300-301. 1935 (obit., F.M.L.S. 1922); Nature 136: 328. 1935 (obit.).
Robinson, B.L., *in* Development of Harvard Univ. 342, 351-352 (Asa Gray prof.).
Rodgers, A.D., Amer. bot. 1873-1892, p. 250, 304, 311, 318. 1944; Liberty Hyde Bailey 502 [index]. 1949.
Sutton, S.B., Charles Sprague Sargent 171, 173, 174, 175. 1970.
Verdoorn, F., ed., Chron. bot. 2: 297, 329, 330. 1936, 6: 238. 1941, 7: 341. 1943.
Wittrock, V.B., Acta Horti Berg. 3(3): 200. *pl. 149.* 1905.

HANDWRITING: Bartonia 28: *pl. 7.* 1957.

COMPOSITE WORKS: (1) Editor-in-chief *Rhodora* 1-30, 1899-1928.
(2) Editor and co-author A. Gray, Man. bot. ed. 7. 18 Sep 1908 (rev. K.K. Mackenzie, Torreya 8: 259-265. 1908, F. Gagnepain, Bull. Soc. bot. France 55: 750-751. 1908), TL-2/2124.
(3) Editor and, in part, author, Asa Gray and Sereno Watson, *Syn. fl. N. Amer.* 1 (1, 2), 1895-1897, TL-2/2132, Reviews by H. Solms-Laubach, Bot. Zeit. 54: 308-311. 1896, and Fr. Buchenau, ib. 55: 262-263. 1897.

EPONYMY: *Robinsonella* J.N. Rose et E.G. Baker (1897). *Note: Robinsona* Cothenius (1790, *orth. var.* of *Robinsonia* Scopoli); *Robinsonetta* G.D. Hanna et A.L. Brigger (1964) honors

J.H. Robinson of Barbados; *Robinsonia* Scopoli (1777, *nom. rej.*) was named for Sir Tancred Robinson (-1748), English botanist and physician; *Robinsonia* A.P. de Candolle (1833, *nom. cons.*) was named for the fictional character, Robinson Crusoe, in the book by Daniel Defoe; *Robinsoniodendron* Merrill (1917) honors Charles Budd Robinson (1871-1913) Canadian-American botanist.

9280. *Beiträge zur Kenntniss der Stammanatomie von Phytocrene macrophylla* Bl. Inaugural-Dissertation der mathematischen und naturwissenschaftlichen Facultät der Kaiser-Wilhelms-Universität Strassburg zur Erlangung der Doctorwürde ... [Leipzig (Breitkopf & Härtel)] 1889. Qu.
Publ.: Oct 1889 (in journal; Nat. Nov. Dec(2) 1890), p. [1]-22, *1 pl. Copies*: G, HH, M. – Reprinted and to be cited from Bot. Zeit. 47: 645-657. 4 Oct 1889, 661-672. 11 Oct 1889, 677-686. 18 Oct 1889, 693-701, *pl. 10.* 25 Oct 1889.

9281. *Descriptions of new plants*, chiefly Gamopetalae, *collected in Mexico* by C.G. Pringle in 1889 and 1890 ... From the Proceedings of the American Academy of Arts and Sciences, vol. xxvi. Issued July 31, 1891. Oct.
Collector: Cyrus Guernsey Pringle (1838-1911).
Publ.: 31 Jul 1891 (cover reprint so dated), cover, p. [i], 164-176. *Copies*: G, NY, USDA. – Reprinted and to be cited from Proc. Amer. Acad. Arts Sci. 26: 164-176. 1891. Contr. Gray Herb. no. 1.
Continuation: 2 Nov 1892 (cover reprint so dated), cover, p. 165-185. *Copy*: USDA. – Id. 27: 165-185. 1892. Contr. Gray Herb. no. 2.

9282. *Descriptions of new plants collected in Mexico* by C.G. Pringle in 1890 and 1891, with notes upon a few other species ... Reprinted from the Proceedings of the American Academy of Arts and Sciences, vol. xxvii. November 2, 1892. Oct.
Collector: Cyrus Guernsey Pringle (1838-1911).
Publ.: 2 Nov 1891 (cover reprint so dated) cover-t.p., p. 165-185. *Copy*: USDA. – Reprinted and to be cited from Proc. Amer. Acad. Arts Sci. 27: 165-186. 1892.

9283. *Additions to the phaenogamic flora of Mexico*, discovered by C.G. Pringle in 1891-92 ... From the Proceedings ... vol. xxviii. [Issued May 13, 1893]. Oct.
Co-author: Henrey Eliason Seaton (1869-1893), *collector*: Cyrus Guernsey Pringle (1838-1911).
Publ.: 13 Mai 1893 (cover-t.p. so dated), cover-t.p., p. 103-115. *Copy*: USDA. – Reprinted and to be cited from Proc. Amer. Acad. Arts Sci. 28: 103-115. 1893. Contr. Gray Herb. ser. 2. no. 3.

9284. *The North American Sileneae and Polycarpeae* ... From the Proceedings ... vol. xxviii. [Issued June 22, 1893]. Oct.
Publ.: 22 Jun 1893 (cover-t.p. so dated), cover-t.p., p. 124-155. *Copy*: USDA. – Reprinted and to be cited from Proc. Amer. Acad. Arts Sci. 28: 124-155. Contr. Gray Herb. ser. 2, no. 5.

9285. *Further new and imperfectly known plants collected in Mexico* by C.G. Pringle in the summer of 1893 ... From the Proceedings ... vol. xxix. [Issued June 29, 1894]. Oct.
Co-author: Jesse More Greenman (1867-1951).
Publ.: 29 Jun 1894 (cover-t.p. so dated), cover-t.p., p. 382-394. *Copy*: USDA. – Reprinted and to be cited from Proc. Amer. Acad. Arts Sci. 29: 382-394. 1894. Contr. Gray herb. ser. 2, no. 7.

9286. *New plants collected* by Messrs. C.V. Hartman and C.E. Lloyd *upon an archeological expedition to northwestern Mexico* under the direction of Dr. Carl. Lumholtz ... From the Proceedings ... vol. xxx. [Issued August 27, 1894]. Oct.
Co-author: Merritt Lyndon Fernald (1873-1950).
Publ.: 27 Aug 1894 (reprint cover so dated), cover-t.p., p. 114-123. *Copy*: USDA. – Reprinted and to be cited from Proc. Amer. Acad. Arts Sci. 30: 114-123. 1894. Contr. Gray herb. ser. 2, no. 8.

9287. *A new genus of Sterculiaceae* and some other noteworthy plants ... Chicago (The University of Chicago Press) 1890. Oct.
Co-author: Jesse More Greenman (1867-1951).
Publ.: Aug 1896, cover t.p., p. 168-170. *Copy*: G. – Reprinted and to be cited from Bot. Gaz. 22(2): 168-170. 1896. *Nephropetalum* Robinson et Greenman (1896).

9288. *Notes upon the flora of Newfoundland* ... Reprinted from the Canadian Record of Science, Jan. and April 1896. Oct.
Co-author: Hermann von Schrenk 81873-1953).
Publ.: Apr 1896, cover, p. [3]-31. *Copies*: BR, M, USDA. – Reprinted with special cover t.p. and original pagination from Canad. Rec. Sci. 7: 3-31. 1896.
Ref.: Buchenau, F., Bot. Zeit. 55(2): 81-83. 16 Mar 1897 (rev.).

9289. *Flora of the Galapagos Islands* ... [Proceedings of the American Academy of Arts and Sciences vol. xxxviii. no. 4. – October 1902]. Oct.
Publ.: 28 Oct 1902 (copies of issue in Contr. Gray Herb. ser. 2. 24 so dated; USDA rd. 30 Oct. 1902), cover, p. [i], [77]-269, [270], *pl. 1-3*. *Copies*: BR, G, M, MO, NY, USDA. – Reprinted and to be cited from Proc. Amer. Acad. Arts Sci. 38(4): [77]-269, [270]. 1902; reprint issued as Contr. Gray Herb. ser. 2, 24.
Preliminary publ.: with J.M. Greenman, Amer. J. Sci. ser. 3. 50: 135-149. 1 Aug 1895.
Ref.: Solms-Laubach, H., Bot. Zeit. 61(2): 39-41. 1 Feb 1903 (rev.).

9290. *Propositions de changements aux lois de la nomenclature botanique de 1867* dont l'adoption est recommandée au Congres International de Nomenclature botanique projeté à Vienne en 1905 par les botanistes attachés à l'Herbier Gray, à l'Herbier cryptogamique et au Musée botanique de l'Université Harvard ... Amendments to the Paris Code of botanical nomenclature ... Cambridge, Massachusetts, U.S.A; 9 June, 1904. Oct. (*Prop. lois nom. bot.*).
Publ.: 9 Jun 1904, p. [1]-32. *Copy*: G. – By B.L. Robinson and other botanists at Harvard.

9291. *Studies in the Eupatorieae* ... Proceedings of the American Academy of Arts and Sciences. Vol. xlii, no. 1. – May, 1906. Oct.
Publ.: 24 May 1906 (reprint issued as Contr. Gray Herb. ser. 2. 32 so dated), cover, p. [1]-48. *Copies*: BR, MO, USDA. – Reprinted and to be cited from Proc. Amer. Acad. Sci. 42(1): 1-48. 1906. Contr. Gray Herb. ser. 2. 32.

9292. *The generic concept in the classification of the flowering plants.* [Reprinted from Science, N.S., vol. xxiii, No. 577, pages 81-92, January 19, 1906]. Oct.
Publ.: 19 Jan 1906, p. [1]-12. *Copy*: M. – Reprinted and to be cited from Science ser. 2. 23: 81-92. 1906.

9293. *A monograph of the genus Brickellia* ... Cambridge, Massachusetts, U.S.A. (Harvard University Press) 1917. Qu. (*Monogr. Brickellia*).
Publ.: 3 Feb 1917 (p. 152 so stamped), p. [1]-151, [152]. *Copies*: BR, HH, NY. – Mem. Gray Herb. no. 1.

9294. Smithsonian Institution United States National Museum. *The woody species of Eupatorium and Ophryosporus occurring in Mexico* ... Washington (Government Printing Office) 1926. Oct.
Publ.: 15 Nov 1926, t.p. on recto of p. 1432, 1432-1470. *Copy*: U. – Reprinted and to be cited from Contr. U.S. natl. Herb. 23: 1432-1470. 1926.

Robinson, Charles Budd, Jr. (1871-1913), Canadian (Nova Scotian) botanist; BA Dalhousie Univ. 1891; teacher at Kentville, N.S. 1892-1893, Pictou, N.S. 1893-1897, 1899-1903, associated with the New York Botanical Garden 1903-1908; Dr. phil. Columbia Univ. 1906; economic botanist with the Bureau of Science Manila 1908-1911, again associated with the New York Botanical Garden 1911-1912, and with the Bureau of Science 1912-1913. (*C.B. Robinson*).

HERBARIUM and TYPES: Material at A, BR, E, F, GH, K, L, MO, NSW, NY, P, PNH, US, WELC. See Sayre (1975) for details on [irregular] exsiccatae such as *Plantae rumphianae amboinenses* and *Reliquiae robinsonianae*. – See also Smithsonian Archives SIA 221, 7275 and archives FH.

BIBLIOGRAPHY and BIOGRAPHY: Backer p. 496; Barnhart 3: 164 (b. 26 Oct 1871, d. 5 Dec 1913); BB 258-259; BJI 2: 146; BL 1: 124, 141, 315; Bossert p. 334; Desmond p. 521; GR p. 740; Hortus 3: 1202; Kew 4: 486-487; PH p. 520 [index]; SK 1: 440-441 (portr.), 4: cxxv, clx; Tucker 1: 601.

BIOFILE: Anon., Bryologist 17: 80. 1914 (obit.); Bull. misc. Inf. Kew 1914: 192 (d.); Hedwigia 55: (71). 1914 (d.); Torreya 6: 131. 1906 (Dr. phil. Columbia on thesis "The Characeae of North America").
Barnhart, J.H., J. New York Bot. Gard. 12: 139. 1911.
Britton, N.L., J. New York Bot. Gard. 15: 106. 1914 (obit.).
Cox, A.J. et al., Obituary Charles Budd Robinson, Jr., Manilla 1914, 1 p. (copy at NY).
Ewan, J. et al., Short hist. bot. U.S. 78. 1969.
Gagnepain, F., Fl. Indochine tome prél. 47. 1944.
Merrill, E.D., Philip. J. Sci., Bot. 9(3): 191-197. 1914 (obit., bibl., full account of murder); Torreya 14: 19, 37. 1914 (obit.), 16: 248-250. 1917 (on Amboina coll. at PNH); Reliquiae robinsonianae, Phil. J. Sci., Bot. 11(5): 243-319. Sep 1916.; Enum. Philip. pl. 4: 222. 1926.
Moldenke, H.N., Plant Life 2: 79. 1946 (1948) (eponymy).
Pennell, F.W. & J.H; Barnhart, Monogr. Acad. nat. Sci. Philadelphia 1: 617. 1935 (Amer. coll.).
Sayre, G., Mem. New York Bot. Gard. 19(3): 388. 1975 (coll.).
Wit, H.C.D. de, *in* J. Arditti, ed., Orchid biol. 1: 297 [index]. 1977.

EPONYMY: *Robinsoniodendron* (Merrill (1917).

9295. *The Characeae of North America* by Charles Budd Robinson submitted in partial fulfilment of the requirements for the degree of doctor of philosophy in the Faculty of pure science of Columbia University. New York June, 1906. Oct.
Publ.: 13 Jun 1906 (NY copy, given by author, so dated), cover, p. [i, reprint t.p.], [244]-308. *Copy*: NY. – Reprinted or preprinted from Bull. New York Bot. Gard. 4(13): 244-308. 1906. – Reprints exist also with a reprint t.p. without the thesis phrase: FH, NY.

9296. Contributions from the New York Botanical Garden – no. 103. *Alabastra philipinensia* ... New York 1908. Oct.
Publ.: 9 Mar 1908, cover t.p., p. 63-75. *Copy*: U. – Reprinted and to be cited from Bull. Torrey bot. Club 35: 63-75. 1908.

9297. Contributions from the New York Botanical Garden – no. 155. *Polycodium* ... New York 1912. Oct.
Publ.: 19 Nov 1912, cover-t.p., p. 549-559. *Copy*: U. – Reprinted and to be cited from Bull. Torrey bot. Club 39: 549-559. 19 Nov 1912.

Robinson, James Fraser (1857-1927), British (English) schoolmaster and naturalist at Hull, Yorkshire. (*J.F. Robinson*).

HERBARIUM and TYPES: HUL (destroyed 1943).

NOTE: Not to be confused with James Frodsham Robinson (1838-1884), writer on the flora of Cheshire and contributor to H.C. Watson, *Topogr. bot.* (see e.g. J. Britten, J. Bot. 42: 300-301. 1904).

BIBLIOGRAPHY and BIOGRAPHY: BL 2: 276, 709; BM 4: 1714; CSP 5: 238, 8: 764; Desmond p. 525; IH 2: (files); TL-2/see J.J. Marshall; Tucker 1: 601.

BIOFILE: Anon., J. Bot. 37: 336. 1899 (not to be confused with James Frodsham Robinson), 38: 500. 1900 (working on *Flora*).
Ellis, G., Amgueddfa 16: 8-10. 1982.
Sheppard, Th., The Naturalist 559: 310. 1903, 843: 126-128. 1927 (obit., portr.).
Woodhead, T.W., The Naturalist 794: 104. 1923 (portr.).

9298. *The flora of the East Riding of Yorkshire*, including a physiographical sketch. By Jas. Fraser Robinson, to which is added a list of the mosses of the Riding by J.J. Marshall. London (A. Brown & Sons, ...) 1902. Oct. (*Fl. East Riding Yorksh.*).
Author of appendix on musci: Joseph Jewison Marshall (1860-1934).
Publ.: 1902 (p. vii: Jul 1902; Bot. Centralbl. 24 Feb 1903; J. Bot. Nov 1902), map, p. [i]-vii, [9]-253. *Copies*: E, HH, MICH, MO, NY. – Also issued as Trans. Hull. Sci. Field natural. Cl. 2, 1902. *Additions*: ib. 3: 98-100, 184-185, 218, 300-302. 1903-1907, 4: 103-104, 105-106. 1909.
Ref.: Rendle, A.B., J. Bot. 40: 393-396. Nov 1902.

Robinson, John (1846-1925), American botanist; educated at Salem, Mass. and Harvard Univ., trustee of Peabody Museum at Salem. (*J. Robinson*).

HERBARIUM and TYPES: Some material at A, L and OXF. – Letter to W.G. Farlow at FH.

BIBLIOGRAPHY and BIOGRAPHY: Barnhart 3: 165 (b. 13 Jul 1846, d. 29 Apr 1925); BL 1: 188, 315; BM 4: 1714; Clokie p. 233; CSP 18: 244, 11: 197; GR p. 240; Herder p. 271; IH 2: (files); Jackson p. 359; Kew 4: 488; PH 520 [index]; Rehder 5: 730; Tucker 1: 601-602.

BIOFILE: Morse, A.P., Rhodora 31: 245-254. 1929 (obit., portr., bibl.).
Pease, A.S., Fl. N. New Hampshire 35, 36, 37, 40. 1964.
Robinson, J., Bull. Essex Inst. 12: 81-94. 1880 (on his predecessors).

9299. *Check list of the ferns of North America* north of Mexico ... Salem, Mass. (The Naturalists' Agency) 1873. Oct.
Ed. (*1*): Apr 1873 (p. [3]: Feb 1873; see BSbF 20 (bibl.): 139. 1873), p. [1-12, printed on one side]. *Copies*: FI, HH, NY, USDA. – Two issues: thick and thin paper. *See also* Ferns of Essex County, Mass., Bull. Essex Inst.7(3): 41-54. Mar 1875, 7: 147-148; 1875, 9: 98. 1877.
Ed. 2: 1876 (p. [3]: 4 Jul 1876; TBC Dec 1876), p. [1-13] thick paper. *Copies*: NY(2), USDA; p. [1-11] thin paper. *Copies*: USDA. – Salem, Mass., (Peabody Academy of Science) 1876).

9300. *Ferns in their homes and ours* ... Salem (S.E. Cassino, ...) 1878. (*Ferns*).
Publ.: 1878 (p. x: 1 Jun 1878), frontisp., p. [i]-xvi, [1]-178, *pl. 1-21*. *Copy*: HH.

9301. *The flora of Essex county*, Massachusetts ... Salem (Essex Institute) 1880. Oct. (*Fl. Essex Co.*).
Publ.: 1880 (TBC Feb 1881; Nat. Nov. Feb(2) 1881; Bot. Zeit. 25 Mar 1881), p. [1]-200.
Copies: BR, FH, HH, NY, PH, US, USDA; IDC 7613.
Preliminary publ.: Bull. Essex Inst. 12: 81-97. 1880.
Ref.: Davenport, G.E., Bot. Gaz. 6(3): 187-188. Mar 1881 (rev.).
Goodale, G.L., Amer. J. Sci. Arts ser. 3. 21: 251. 1881 (rev.).
N.L., Bot. Zeit. 39: 199. 25 Mar 1881.

9302. *The native woods of Essex County*, Massachusetts. An account of the general, distributions and uses, the determinations of the specific gravity, percentage of ash, strength, fuel value, etc., of the woods of the native trees of Essex county, as shown by tests upon specimens furnished by the Peabody Academy of Science ... Salem, Mass. (printed for the Academy) 1885. Oct.
Publ.: 1885, p. [i], 103-137, 2 tables. *Copy*: FH. – Reprinted and to be cited from Ann.

Rep. Trustees Peabody Acad. Sci. 1884: 101-137, 2 tables. Text extracted from Rep. Forests U.S. (C.S. Sargent), Rep. 10. Census US 1880, vol. 9.

9303. *Our trees.* A popular account of the trees in the streets and gardens of Salem, and of the natives trees of Essex County, Massachusetts, with the location of trees, and historical and botanical notes ... Salem (printed by N.A. Horton and Son, office of the Salem Gazette) 1891. Oct. (*Our trees*).
Publ.: 1891 (p. [3]: Oct 1891; USDA rd 3 Feb 1892; J. Bot. Mar 1892, TBC 5 Mar 1892), p. [i-iii], [1]-120. *Copies*: US(2), USDA. – Letters to the Salem Gazette, collected by and published for the Essex Institute.

Robley, Mrs **Augusta J.** (*fl.* 1840), British botanical artist, daughter of Mrs. Jane Wallas Penfold (*Robley*).

HERBARIUM and TYPES: Unknown.

BIBLIOGRAPHY and BIOGRAPHY: Barnhart 3: 166; BB p. 259; BM 4: 1714; Desmond p. 526; GF p. 73; Herder p. 218; Jackson p. 353; Kew 4: 490; NI 1654; Plesch p.383-384; PR 7675, ed. 1: 8642; Rehder 1: 269; Sotheby 648.

BIOFILE: Britten, J., J. Bot. 57: 97-99. 1919.

9304. *A selection of Madeira flowers,* drawn and coloured from nature by Augusta J. Robley, ... London (printed and published by Reeve, Brothers, ...) 1845. Fol. (*Sel. Madeira fl.*).
Publ.: Mai 1845 (Sotheby 648; Bot. Zeit. 1 Aug 1845), t.p., preface, 8 coloured plates with unpaginated text facing. *Copy*: Natl. Libr. Scotland, inf. J. Edmondson. – According to J. Britten (1919) the text is by the Rev. William Lewes Pugh Garnons (1791-1863), Collins (Sotheby 648) suggests that the lithographer R.E.B. is possibly R.E. Branston, or that it may stand for Reeves Brothers.

Robolsky, H. (1796-1849), German botanist and teacher at Neuhaldensleben (Saxony). (*Robolsky*).

HERBARIUM and TYPES: Unknown.

BIBLIOGRAPHY and BIOGRAPHY: Barnhart 3: 166 (b. 5 Nov 1849); PR 7676.

BIOFILE: Anon., Bot. Zeit. 8: 359-360. 1850 (d. 5 Nov 1849).

9305. *Flora der Umgegend von Neuhaldensleben.* Ein Verzeichniss der hier wachsenden Pflanzen, deren Beschreibung, Standort und Blüthezeit ... Neuhaldensleben, 1843. Oct. (*Fl. Neuhaldensleben*).
Publ.: 1843 (p. vi: Mai 1843), p. [i]-vi, [vii-viii], [1]-175, [176, err.], [index 1-20]. *Copy*: NY.
Ed. 2: 1849 (Bot. Zeit. 11 Jan 1850), xxx, 175 p., n.v.

Robson, Edward (1763-1813), British (English) botanist at Darlington; correspondent of W. Withering and J.E. Smith; one of the original fellows of the Linnean Society of London. (*E. Robson*).

HERBARIUM and TYPES: Sunderland Museum (inf. P.S. Davis and G.G. Graham). Other material at LIV.

BIBLIOGRAPHY and BIOGRAPHY: Barnhart 3: 166; BB p. 259; CSP 5: 243; Desmond p. 526 (b. 17 Oct 1763, d. 21 Mai 1813); DNB 49: 62; Dryander 3: 650; Henrey 2: 135, 136; Hortus 3: 1202; Rehder 2: 246; Zander ed. 10, p. 707, ed. 11, p. 807.

BIOFILE: Adamson, R.S., Mem. Proc. lit. philos. Soc. 63: 2. 1920 (friend of John Dalton). Baker, J.G., Nat. Hist. Trans. Northumberland 14(1): 78. 1902 (biogr. sketch).

Blackwell, E.M., Naturalist 877: 54. 1961.
Bolton, J., Hist. fung. Halifax 170. 1792.
Bridson, G.D.R., et al., Nat. hist. mss. res. Brit. Isl. 229.410, 264.8. 1980.
Britten, J., J. Bot. 60: 278. 1922 (on his mss. Suppl. Brit. Fl., 1790, at BM(NH).
Davis, P., Newsletter Soc. Bibl. nat. Hist. 8: 8-9. 1980 (mss.; herb. in Sunderland Museum).
Davis, P.S. & G.G Graham, The authorship of "Plantae rariores agro dunelmensi indigenae", Arch. nat. Hist. 10(2): 335-340. 1981 (biogr. info. on Edward and Stephen Robson and authorship of *Plantae* attributed to Edward).

EPONYMY: *Robsonia* (Berlandier) H.G.L. Reichenbach (1837).

9306. *Plantae rariores agro dunelmensi indigenae* [s.l.n.d., 1798].
Publ.: 1798 (fide P.S. Davis and G.G. Graham in mss.), 3 p. *Copy*: Bodleian Library Oxford (inf. J. Edmondson). Added: 1 p. *Plantae desideratae*. [Mss. addition: E. Robson. Darlington 1/5 mo 98].

Robson, Stephen [1741-1779], British (English) botanist at Darlington (Durham), uncle of Edward Robson; linen manufacturer and grocer. (*S. Robson*).

HERBARIUM and TYPES: YRK.

BIBLIOGRAPHY and BIOGRAPHY: Barnhart 3: 166; BM 4: 1715; Desmond p. 526 (b. 24 Jan 1741, d. 16 Mai 1779; herb. YRK); DNB 49: 62; Dryander 3: 22, 132; Henrey 2: 133-135, 136, 310, 3: 107; Herder p. 95, 171; Jackson p. 232; Kew 4: 491; PR 7677, ed. 1: 8643; Rehder 1: 395; SO 679; Tucker 1: 602.

BIOFILE: Baker, J.G., Nat. Hist. Trans. Northumberland 14(1): 77-78. 1902.
Davis, P., Newsletter Soc. Bibl. nat. Hist. 8: 8-9. 1980 (mss.), Naturalist 106: 67-73. 1981 (on his Hortus Siccus).
Davis, P.S. & G.G. Graham, Arch. nat. Hist. 10(2): 335-340. 1981.
Green, J.J., Friend's Quart. Examiner 1917: 14-31, 265-282 (biogr. essay).
Hawksworth, D.L. & M.R.D, Seaward, Lichenol. Brit. Isl. 8,. 1977.
Kent, D.H., Brit. herb. 75. 1957 (herb. untraced).

9307. *The British flora.* Containing the select names, characters, places of growth, duration, and time of flowering of the plants growing wild in Great Britain. To which are prefixed, the principles of botany ... York (printed by W. Blanchard and Company; ...) 1777. Oct. (in fours). (*Brit. fl.*).
Publ.: 23 Sep 1777 (see J.J. Green, 1917, p. 267), p. [i]-xx, *pl. 1-5* (uncol. copp.), [1]-330, [1-24, indexes]. *Copies*: BR, HH, KU, NY, PH, USDA. – A mss. supplement by E. Robson is at BM.

Robyns, [Frans Hubert Edouard Arthur] Walter (1901-x), Belgian botanist; Dr. sci. Louvain 1923; with the Jardin botanique de l'État (later: ... National) from 1923, assistant 1923-1928, curator 1928-1931; director 1931-1966; in retirement at Louvain and Bruxelles; president of the International Association for Plant Taxonomy 1959-1964. (*Robyns*).

HERBARIUM and TYPES: BR. – Further material e.g. at BM, C, F, G, K (incl. types), MO, NY, P.

BIBLIOGRAPHY and BIOGRAPHY: Barnhart 3: 166 (b. 25 mai 1901); BJI 2: 147; BL 1: 25, 315, 2: 34, 709; BM 7: 1084; Bossert p. 334; Hirsch p. 250; IH 1 (ed. 1): 15, (ed. 2): 30, (ed. 3): 33, 96, (ed. 4): 35, 105, (ed. 5): 28, 109, 2: (files); Kew 4: 491-496; Lenley p. 351; MW p. 415; Roon p. 94.

BIOFILE: Anon., Bull. misc. Inf. Kew 1932: 106 (app. BR); Taxon 1: 36. 1952 (in group portr.), 5: 3. 1956 (in group portr.), 6: 31. 1957 (25 yrs director BR), 15: 287. 1966 (retirement).

Bullock, A.A., Bibl. S. Afr. bot. 96. 1978.
Robyns, W., Taxon 1: 26-27. 1951. (On *Flore du Congo belge*).
Robyns, W., Liste des publications, typescript 1958, files I.A.P.T.; Taxon 13: 301-303. 1964 (presidential address I.A.P.T.: "The unity and the actual needs of plant taxonomy").
Sayre, G., Bryologist 80: 516. 1977.
Stafleu, F.A., Regn. veg. 5: 2. 1954 (in group portr.).
Steenis, C.G.G.J. van, Fl. males. Bull. 13: 551. 1957 (25 yrs director), 21: 1383. 1966 (retirement).
Verdoorn, F., ed., Chron. bot. 91. 1935, 2: 2, 28, 34, 35, 81, 83, 281. 1936, 3: 2, 27, 67, 68, 241, 310. 1937, 4: 183, 186, 245, 470, 561, 565. 1938, 6: 303. 1941, 7: 36, 280. 1943.

EPONYMY: *Robynsia* Hutchinson (1931, *nom. cons.*); *Robynsiella* Suesseguth (1938); *Robynsiochloa* H. Jacques-Félix 81960); *Robynsiophyton* Wilczek (1953). *Note*: *Robynsia* Drapiez (1841, *nom. rej.*) and *Robynsia* Martens et Galeotti (1843) honor "Domini" Robyns 19th century collector and promoter of botany.

FESTSCHRIFT: Bull. Jard. bot. État, Bruxelles, 27, 1957.

9308. *Tentamen monographiae Vangueriae* generumque offinium. Bruxelles (Goemare, Imprimeur du Roi, ...) 1928. Oct. (*Tent. monogr. Vanguerieae*).
Publ.: p. 1-154. Mai 1928, p. 155-159. Aug 1928; cover t.p., [i-iv], [1]-359. *Copies*: BR, G, MICH, NY, PH. – Issued as Bull. Jard. bot. État, Bruxelles 11, fasc. 1 and 2.
Ref.: Krause, K., Bot. Jahrb. 63 (Lit); 62. 1930 (rev.).

9309. *Flore agrostologique du Congo belge et du Ruanda-Urundi* ... Bruxelles (Goemaere, Imprimeur du Roi ...) 1929-1934, 2 parts. Oct. (*Fl. agrost. Congo belge*).
1: Sep-Dec 1929 (p. 8: Sep 1929; J. Bot. Mai 1930; Nat. Nov. Dec 1930), p. [1]-229, *pl. 1-18* (E.L. Thuring) in text, map.
2: Jul-Dec 1934 (p. 8: Jun 1934), p. [1]-386, *pl. 1-54* (Hel. Durand) in text.
Copies: BR, G, HH, MO, PH, USDA.

9310. *Les espèces congolaises du genre Digitaria* Hall ... Bruxelles (Librairie Falk fils, Georges van Campenhout, ...). 1931. Qu. (*Esp. congol. Digitaria*).
Publ.: 1931 (p. [2]: deposit 21 Feb 1931), p. [1]-52, *pl. 1-6* (Hél. Durand) with text, [1].*Copies*: BR, U. – Mém. Inst. Roy. Col. Belge 1(1). 1931.

9311. *L'organisation florale des Solanacées zygomorphes*. Bruxelles (Maurice Lambertin, ...) 1931. Oct. (*Organis. fl. Solanac.*).
Publ.: 1931, p. [1]-82, *pl. 1-6*, with text. *Copy*: U.

9312. *Les espèces congolaises du genre Panicum* L. ... Bruxelles (Hayez, ...) 1932. Qu. (*Esp. congol. Panicum*).
Publ.: 1932 (U rd. 11 Apr 1933), p. [1]-66, [67, cont.], *pl. 1-5* (Hél. Durand) with text.
Copies: BR, U, US. – Issued as Mém. Inst. Roy. colon. Belge 1(6), 1932.

9313. *Essai de révision des espèces africaines du genre Annona* L. ... Gembloux (Imprimerie J. Duculot, éditeur) 1934. Oct.
Co-author: Jean Hector Paul Auguste Ghesquière (1888-x).
Publ.: 1934, cover t.p., [7]-50, *pl. 1-4*. *Copy*: U. – Reprinted and to be cited from Bull. Soc. Bot. Belg. 67(1): 7-50. 1934.

9314. *Flore générale de Belgique* ... Bruxelles (Ministère de l'Agriculture-Jardin botanique de l'État) 1950-x. Oct. (*Fl. gén. Belgique*).
Publ.: in parallel series Pteridophytes, Bryophytes, Spermatophytes. *Copies*: BR, G, H, PH, U.
Ptéridophytes par André Lawalrée (1921-x), 1950 (p. iv: Oct 1950, U rd. 30 Jan 1951), p. [i*, iii*], [i]-iv, (préface W. Robyns), [1]-194, [195], [1, colo.].
Bryophytes:

1(*1*): 1955 (t.p; BR rd. 9 Feb 1956), p. [i*, iii*], [i]-iv, (avant-propos, W. Robyns), [1]-131, [1, colo.]. Author: Constant Vanden Berghen (parts 1-3).
  (*2*): 1956, p. [i, iii], [133]-270, [1, colo.].
  (*3*): 1957 (BR rd. 10 Apr 1958), p. [i, iii], [271]-389. [1, colo.].
2(*1*): 1959 (BR rd. 25 Feb 1960), p. [i*-vii*], [1]-111, [1, colo.]. Authors: Fernand Demaret et Émile Castagne.
  (*2*): 1961 (BR rd. 11 Jan 1962; U rd. 31 Jan 1962), p. [i, iii], [123]-231, [1, colo.]. Authors: id.
  (*3*): 1964 (U rd. 4 Feb 1965), p. [i, iii], [233]-397, [1, colo.]. Authors: id.
3(*1*): 15 Feb 1968, p. [i, iii], [1]-112. Authors: Jean-Louis de Sloover & Fernand Demaret.
Spermatophytes:
1(*1*): 1952 (BR rd. 1 Apr 1952), p. [i, iii*], [1]-170 (preface Walter Robyns)], [1]-170, [1, colo.]. *Author* (1-3): André Lawalrée.
  (*2*): 1953 (BR rd. 23 Mar 1953), p. [i, iii], [171]-349, [1, colo.].
  (*3*): 1954 (BR rd. 18 Mar 1954), p. [i, iii], [351]-505, [1, colo.].
2(*1*): 1955 (BR rd. 12 Apr 1955), p. [i]-v, [1]-120. *Author* (1-3): André Lawalrée.
  (*2*): 1956 (BR rd. 10 Apr 1956), p. [i, iii], [121]-285.
  (*3*): 1957 (BR rd. 27 Mai 1957), p. [i, iii], [287]-490, [1, colo.].
3(*1*): 1958 (BR rd. 12 Mai 1958), p. [i-v], [1]-152. *Authorship*: André Lawalrée (general fasc. 1-3), Joseph Legrain (*Rubus*).
  (*2*): 1959 (BR rd. 29 Mai 1959), p. [i, iii],[153]-306.
  (*3*): 1960 (BR rd. 30 Sep 1960), p. [i, iii], [307]-440.
4(*1*): 1961 (BR rd. 10 Oct 1961), p. [1]-134. *Authorship*: André Lawalrée (in general), James Cullen (*Anthyllis*).
  (*2*): 1963 (BR rd. 19 Mar 1963), p. [i, iii], [135]-237.
  (*3*): 1964 (U rd. 24 Nov 1964), p. [i, iii], [239]-390.
5(*1*): 1966 (BR rd. 19 Aug 1966), p. [i-v], [1]-208, [1, colo.]. Author: André Lawalrée.
Errata: Bryophytes fasc. 1-2, loose, p. [1-2], fasc. 1-2-3, loose, 3 p. one-sided.
Ref.: Robyns, W., Taxon 1: 37-38. 1952.

Roche, Daniel de la (1743-1813), Swiss (Geneva) botanist; Dr. med. Leiden 1766; practicing physician at Genève 1771-1782, at Paris 1782-1813 with a short period (1792-?) at Lausanne. (*D. Roche*).

HERBARIUM and TYPES: G-DC.

NOTE: Treated in TL-2/1 under Daniel Delaroche; we give here a few additions. For publ. see TL-2/1348.

BIBLIOGRAPHY and BIOGRAPHY: AG 3: 539; Barnhart 1: 437; PR 2117-2118, 7453, ed. 1: 2395-2396, 8411; TL-2/5849.

BIOFILE: Briquet, J., Bull. Soc. bot. Suisse 50a: 408-409. 1940 (bibl. epon., sources).
Candolle, A.P. de, Hist. bot. gener. 27, 44. 1830, Mém. souvenirs 72, 123. 1862.
Gautier, L., La médecine à Genève ... fin du xviii. siècle 325, 330, 346, 347, 366, 436, 520. 1906 (fide J. Briquet).
Goldblatt, P., J.S. Afr. Bot. 36(4): 291-318. 1970.
Montet, A. de, Dict. biogr. Genevois Vaudois 2: 385-386. 1878 (fide J. Briquet).

EPONYMY: *Rochea* sect. *Danielia* A.P. de Candolle (1828), *Rochea* A.P. de Candolle (1802), *Larochea* Persoon (1805), *Rochea* Salisbury (1812).

Roche, François de la (1782-1814), Paris-born Swiss botanist; son of D. de la Roche; collaborator of A.P. de Candolle and P.J. Redouté; practicing physician at Paris. (*F. Roche*).

HERBARIUM and TYPES: G-DC, other material at G.

NOTE: For main treatment see TL-2/1: 613 and no. 1349.

BIBLIOGRAPHY and BIOGRAPHY: AG 3: 539; Barnhart 1: 437; BM 1: 437; CSP 2: 217-218; Langman p. 430; PR 2118.

BIOFILE: Candolle, Alph. de, Phytographie 407. 1880.
Candolle, A.P. de, Hist. bot. genev. 27, 44. 1830; Mém. souvenirs 93, 116, 161, 165, 179. 1862.
Desvaux, F., J. Bot. ser. 2. 3: 207-211. 1814.
Montet, A. de, Dict. biogr. Genevois Vaudois 2: 386. 1878 (fide J. Briquet).
Stafleu, F.A., *in* I. MacPhail, Cat. redout. 5-6, 22. 1963. (See also p. 46).

COMPOSITE WORKS: Provided the text for vols. 5, 6, [7?] of P.J. Redouté, *Les Liliacées* (1809-1813), q.v.

EPONYMY: *Rochea* subg. *Franciscea* A.P. de Candolle (1828); *Rochea* A.P. de Candolle (1802) is dedicated to both D. and F. de la Roche.

Rochebrune, Alphonse Trémeau de (1834-1912), French palaeontologist, zoologist and botanist; colonial physician at Saint-Louis, Sénégal; later asst. naturalist at the Paris Muséum d'Histoire naturelle. (*Rochebr.*).

HERBARIUM and TYPES: P; further material at C and FI.

BIBLIOGRAPHY and BIOGRAPHY: Barnhart 3: 166; BL 1: 17, 315, 2: 137, 709; BM 4: 1715; CSP 5: 245, 8: 1112, 11: 199, 12: 623, 18: 247; IH 2: (files); Kew 4: 496; PR 7682; Quenstedt p. 363; Rehder 1: 80, 412, 5: 731; Tucker 1: 602.

BIOFILE: Anon., Bull. Soc. bot. France, Table art. orig. 203. s.d.

9315. *Catalogue raisonné des plantes phanérogames* qui croissent spontanément dans le département *de la Charente* ... À Paris (chez J.B. Baillière, ...) 1860. Oct. (*Cat. pl. phan. Charente*).
Co-author: Alexandre Savatier (1824-1886).
Publ.: Jun-Dec 1860(p. xv: 1 Jun 1860; BSbF rd. 28 Dec 1860, p. [i]-xv, [1]-294. *Copies*: G, HH, NY, USDA.
Ref.: Rochebrune, A.T. de, Bull. Soc. bot. France 7: 504. 1860.

9316. *Toxicologie africaine* étude botanique, historique, ethnographique, chimique, physiologique, thérapeutique, pharmacologique, posologique, etc. sur les végétaux toxiques et suspects propres au continent african et aux iles adjacentes ... Paris (Octave Doin, éditeur ...) [1895-] 1897-1898. Oct. (*Toxicol. afr.*).
1: 1895-1897, in five parts (see below), p. [i*-vii*], [i]-iv, [1]-935. Reprinted and in part preprinted from Bull. Soc. Hist. nat. Autun 8-12. 1895-1899.
2: 1898, one part, remained incomplete, p. [1]-500. Preprinted from Bull. Soc. Hist. nat. Autun 12(2): 1-500. 1899.
Copies: BR, G, HH.

vol.	fasc.	publication in Bulletin	publication as book
1	1	8: 109-300. 1895	1-192. 1896
	2	9: 1-192. 1896	193-384. 1896
	3	10(1): 1-192. 1896	385-576. 1897
	4	11(1): 1-192. 1898	577-768. 1897
	5	12(1): 1-164. 1899	769-932. 1897
2	1	12(2): 1-506. 1899	1-500. 1898

Rochel, Anton (1770-1847), Austrian botanist and surgeon; surgical assistant in the Austrian army 1788-1798; Magister chirurgiae Wien 1792, id. obstetritiae 1798; practicing physician in Moravia, Hungary (Neutra, Rownye) 1798-1820; curator of the Pest botanical garden 1820-1840; travelled to Leningrad and Dresden 1840-1841; settled at Graz in retirement 1841-1847. (*Rochel*).

HERBARIUM and TYPES: DR (destr.); some material at B (extant), BP, BR, E, GJO, H (200), JE, M, W (205, rd. 1846), WU (in herb. Lang). – According to Kanitz (1865) and Ullepitsch (1884), Rochel sold his herbarium to King Johann of Saxony in 1841 against an annuity for life. It went to Dresden where it was destroyed in World War II. A new herbarium, assembled during his stay in Graz, is at GJO. – For details on manuscripts see Kanitz (no. 106).

BIBLIOGRAPHY and BIOGRAPHY: AG 2(1): 362 (herb. at DR fide O. Drude); Backer p. 497; Barnhart 3: 166 (b. 18 Jun 1770, d. 12 Mar 1847); Biol.-Dokum. 15: 7572; BM 4: 1715; CSP 5: 246, 18: 247; Futak-Domin p. 506; Hegi 5(2): 1298; Herder p. 58, 167, 192; Hortus 3: 1202; IH 1 (ed. 7): 341; Jackson p. 264; Kanitz p. 139-145 (no. 106) (biogr. bibl.); Kew 4: 496; Lasègue p. 332, 337; NI 1655; Plesch p. 384; PR 7683-7686, ed. 1: 8646-8649; Rehder 1: 445; Sotheby 651; Tucker 1: 603; Zander ed. 10, p. 707, ed. 11, p. 807.

BIOFILE: Anon., Flora 5: 236. 1832 ("für Botanik geboren"), 334-336. 1822 (Naturh. Misc. vol. 1 "wirklich erschienen"), 14: 30, 31. 1831, 15: 404. 1832 (pl. W), 30: 655-656. 1847.
G.P., Bot. Zeit. 5: 653-654. 1847 (obit.).
Kanitz, A., Bonplandia 10: 363. 1862.
Mecenovic, K., Steyerm. Landesmus. Joanneum, Graz 1976(2) (note on plant collections Rochel recently discovered in GJO; on causes of rarity of R's printed works).
Meyer, F.K. & H. Manitz, in Reichtümer und Raritäten, Jenaer Reden und Schriften 91. 1974.
Pax, F., Grundz. Pfl.-Verbr. Karpathen 1: 11, 30, 33. 1898, 2: 280. 1908.
Simonkai, L., Enum. fl. transsilv. xxviii. 1886.
Ullepitsch, J., Österr. bot. Z. 34: 363-368. 1884 (biogr., b. 18 Jun 1770, d. 12 Mar 1847).

NOTE: Rochel published his *Naturhistorische Miscellen* (1821), *Plantae Banatus rariores* (1828) and *Botanische Reise in das Banat* (1838) at his own expense and sold relatively few copies. When retiring from his position at Pest he burned the stocks, which accounts for their rarity (see Ullepitsch 1884). On the other hand Kanitz (no. 106) notes that copies of this work were still to be obtained from the Verein für Naturkunde, at Presburg (Bratislava).

EPONYMY: *Rochelia* J.J. Roemer et J.A. Schultes (1819, *nom. rej.*); *Rochelia* H.G.L. Reichenbach (1824, *nom. cons.*).

9317. *Pflanzen-Umrisse* aus dem südöstlichen Karpath des Banats von Anton Rochel ... Erste Lieferung mit zweyundachtzig Abbildungen in natürlicher Grösse samt den nöthigen Zergliederungen auf neununddreyssig Tafeln, nach dem Leben gezeichnet, und mit Beschreibungen begleitet ... Wien 1820. (*Pfl.-Umrisse*).
Publ.: 1820. *Copy*: G. – Like Pritzel we have seen only the engraved title page Lieferung 1 with 26 "Umrisse" printed around the text as cited above. A note in Flora (5: 160. 1822) states that the *Pflanzen-Umrisse*, because of the lack of subscriptions, will come out in "Hefte". The editor of Flora had seen a few plates; these may have been the originals of which he speaks in Flora 3: 685. 1820. It is not clear how many plates, if any (except the front page of Lieferung 1, were ever published. The first announcement for subscription ("Pränumerations, Anzeige") was published by Rochel in Flora 3: 601-504. 1820. Kanitz (no. 106) states that this work was never published.

9318. *Naturhistorische Miscellen* über den nordwestlichen Karpath in Ober-Ungarn ... Pesth (Gedruckt bey Johann Thom. von Trattner, auf Kosten des Verfassers). Oct. (*Naturh. Misc.*).
Publ.: 1821, p. [i*-viii*], [i]-xii, [1]-135, [136, err.], map, [1, Nachschr., 10 Nov 1821]. *Copies*: GOET (Inf. G. Wagenitz), HH.

9319. *Plantae Banatus rariores*, iconibus et descriptionibus illustratae. Praemisso tractatu phytogeographico et subnexis additamentis in terminologiam botanicam ... Pestini

[Pest; Budapest] (typis Ludovici Landerer de Füskút, ...) 1828. Fol. (*Pl. Banat. rar.*).
Publ.: 1828 (p. iv: 1 Jan 1828; Flora Jul-Dec 1828), p. [i*-iv*], [i]-iv, 1-84, [85-96, index, prosp.], *pl. 1-40* (uncol., auct.), 2 charts. *Copies*: G, GJO, H, HH, M, MO, NY, REG, USDA.
Ref.: Hoppe, D.H., Flora 12: 609-617, 625-634. 1829 (rev.).

9320. *Botanische Reise in das Banat* im Jahre 1835, nebst Gelegenheits-Bemerkungen und einem Verzeichniss aller bis zur Stunde daselbst vorgefundenen wildwachsenden phanerogamen Pflanzen, sammt topographischen Beiträgen über den südöstlichen Theil des Donau-Stromes im Österreichischen Kaiserthum ... Pesth (bei Gustav Heckenast), Leipzig (bei Otto Wigand) 1838. Oct. (*Bot. Reise Banat*).
Publ.: 1838 (Flora 7 Mar 1839, rd. only 9 Sep 1839), p. [i-x], [1]-90, *1 pl.* (frontisp.). *Copies*: GOET (inf. G. Wagenitz), HH, M, NY. – On p. 32-90 Verzeichniss derjenigen phanerogamen Pflanzen ..."
Ref.: [Schlechtendal, D.F.L. von], Linnaea 12 (Litt): 116-117. Mar 1838-1839.

Rochet d'Héricourt, C.L.X. (*fl.* 1846), French chemist, explorer and geographer; travelled e.g. in Ethiopia 1839-1845. (*Rochet*).

HERBARIUM and TYPES: Ethiopian material, rd. 1845-1850, at P (284) and FI (55).

BIBLIOGRAPHY and BIOGRAPHY: Barnhart 3: 167; BM 4: 1715; CSP 5: 246; Herder p. 58; IH 2: 270 (Héricourt); PR 7687; Rehder 5 : 731.

BIOFILE: Cufodontis, G., C.R. AETFAT 4: 246. 1962 (coll. Ethiopia).
Embacher, F., Lexik. Reisen 249. 1882 ("Doch ist seine Glaubwürdigkeit zweifelhaft").
Lacroix, A., Figures des Savants 4: 177. 1938.
Steinberg, C.H., Webbia 32(1): 35. 1977 (C.E.K. Rochet d'Héricourt; material at FI-Webb).
Vallot, J., Bull. Soc. bot. France 59: 188. 1882.

HANDWRITING: Webbia 32(1): 27. 1977, no. 55 ("C.E.K. Rochet d'Héricourt").

EPONYMY: *Rochetia* Delile (1846).

9321. *Second voyage* sur les deux rives de la mer Rouge *dans le pays des Adels et le Royaume de Choa* ... Paris (Arthur Bertrand, ...) 1846. Oct. (*Sec. voy. Choa*).
Publ.: 1846 [408] dated 5 Oct 1846, p. [i]-xlviii, [1]-406, [407, err.], [408, dated], *pl. 1-16* (in atlas). *Copies*: L, USDA. – Botany on p. [337]-345, [346, note] with e.g. on p. 340 *Tephea* Delile (1846; ING 3: 1732).

Rock, Joseph Francis Charles [until 1933 Joseph Franz Karl] (1884-1962), Austrian-born American botanical explorer, botanist, geographer and ethnologist; to the United States 1905, in Hawaii 1907- 1920; naturalized U.S. citizen 1913; explorer for the Bureau of Plant Industry, USDA, the Arnold Arboretum, the National Geographic Society and other U.S. institutions and agencies, frequently travelling and working in China until 1949; from then on in India, Europe, U.S.A., ultimately on Oahu, Hawaii; "father of Hawaiian botany".(*Rock*).

HERBARIUM and TYPES: A, BISH, BM and E; other material in AMES, B, BISH, BM, BR, C, CAS, F, G, GB, GH, K, L, LU, MO, N, NF, NY, P, PH, S, U, UC, US, W, WELT. – Manuscript copies of diaries and notes and a collection of photographs at E. – Smithsonian Archives SIA 305, 323, 7215, 7270.

BIBLIOGRAPHY and BIOGRAPHY: Barnhart 3: 167 (b. 13 Jan 1884, d. 5 Dec 1962); Biol.-Dokum. 15: 7572; BJI 2: 147; BL 1: 110, 114, 126, 315; BM 7: 1085; Bossert p. 335; Hirsch p. 250; Hortus 3: 1202; IH 1 (ed. 6): 363, (ed. 7): 341, 2: (files); Kew 4: 497; Lenley p. 351; MW p. 415, suppl. p. 294; PH 209; Roon p. 95; SK 1: 440, 8: lxxxi; Tucker 1: 603; Urban-Berl. 388; Zander ed. 10, p. 708, ed. 11, p. 807.

BIOFILE: Anon., Honolulu Star-Bulletin, 16 Dec 1933 (Rock leaves Orient; portr.); Madroño 17: 32. 1963 (d.); Taxon 12: 35. 1963 (d.).
Bridson, G.D.R. et al., Nat. hist. mss. res. Brit. Isl. 138.40. 1980.
Cattell, J.M. et al., Amer. men. sci. ed. 3: 581. 1921, ed. 4 : 829. 1927, ed. 5: 914. 1933, ed. 6: 1196-1197. 1938, ed. 7: 1497. 1944, ed. 8: 2093. 1947.
Chock, A.K., Taxon 12: 89-102. 1963 (obit., bibl., portr.; many details especially on Hawaii period); Amer. hort. Mag. 42(3): 159-167. 1963 (obit., portr.).
Chock, A.K. et al., Newsletter Hawaiian Bot. Soc. 2(1): 1-17. 1963 (Rock memorial edition, biogr. by Chock, bibl. by Chock et al., anec- dote by E.H. Bryan).
Coats, A.M., Plant hunters 117, 126, 139-141. 1969.
Egge, R.G., Principes 24(2): 82-90. 1980 (biogr. sketch).
Ewan, J. et al., Short hist. bot. U.S. 134, 144. 1969.
Fletcher, H.R., Story Roy. Hort. Soc. 297, 299, 344-345. 1969.
Harkness, B., Plants & Gardens 23(3): 65-66. 1968 (portr.).
Hedge, I.C. & J.M. Lamond, Index coll. Edinb. herb. 124. 1970.
Jewett, F.L. & C.L. McCausland, Plant hunters 174-202. 1958.
Lawrence, G.H.M., Huntia 1: 164. 1964 (some letters at HU).
Merrill, E.D., Bull. Bernice P. Bishop Mus. 144: 158-159. 1937 (Polynes. bibl.); Contr. U.S. natl. Herb. 30(1): 255-257. 1947 (bibl. Pacific Isl.).
Rhodes, H.L.J., Baileya 4: 70-80. 1956 (on R's exp. to N.W. China 1924-1927).
Rickett, H.W., Index Bull. Torrey bot. Club 83. 1955.
Sargent, C.S., J. Arnold Arb. 6: 213-216. 1925 (reports on Arnold Arb. exp. N.C. Asia in progress 7: 68-70. 1926 (id.), 7: 245-246. 1926 (id.).
Steenis, C.G.G.J. van, Fl. males. Bull. 7: 191. 1950 (left China), 18: 972. 1963 (d.).
Sutton, S.B., Charles Sprague Sargent 267-275. 342-343, 345. 1970; Arnoldia 32(1): 2-20. 1972; In China's border provinces. The turbulent career of Joseph Rock, botanist, explorer. New York 1974, 334 p., *16 pl.* (impt. biogr.); Bull. Pac. Trop. Bot. Gard. 12(2): 31-36. 1982.
Verdoorn, F., ed., Chron. bot. 1: 264, 287. 1935, 2: 180, 276, 300, 330. 1936, 3: 65, 159, 358. 1937, 4: 184. 1938, 6: 303. 1941.
Walker, E.H., Plant Sci. Bull. 9: 7-8. 1963 (obit.).
Wilson, E.H., J. Arnold Arb. 8: 200-202. 1927 (report on Arnold Arb. N.C. Asia exp. progress) (see also p. 242-243).

EPONYMY: *Rockia* Heimerl (1913).

9322. *Notes upon Hawaiian plants* with descriptions of new species and varieties ... Honolulu (published by the College [of Hawaii]) 1911. Oct. (*Not. Haw. pl.*).
Publ.: Dec 1911, p. [1]-20, *pl. 1-5*. *Copies*: BR, M, PH, US(2), USDA. – Coll. Hawaii Publ., Bull. 1.

9323. *The indigenous trees of the Hawaiian Islands* ... Published under patronage. Honolulu, T.H. 1913. Oct. (*Indig. trees Haw. Isl.*).
Publ.: 26 Jun 1913, p. [i-viii], [1]-518, *pl. 1-215* (in text). *Copies*: BR, FI, G, M, R.S. Cowan, PH, US, USDA.
Reprint: 1974, Kanai, Lawaii, Hawaii (Pacific Tropical Botanical Garden), Rutland, Vermont and Tokyo, Japan (Charles E. Tuttle Company), ISBN 0-8048-1140-7, p. [i]-xx, [1]-548. *Copy*: BR.
Ref.: Engler, A., Bot. Jahrb. 51 (Lit.): 53-545. 1914.
Diels, L., Bot. Jahrb. 51 (Lit.): 29-30. 1913.

9324. *List of Hawaiian names of plants* ... Honolulu (Hawaiian Gazette Co., Ltd.) 1913. (*List Haw. names pl.*).
Publ.: Jun 1913, p. [1]-20. *Copies*: BR, G, H, M, PH, USDA. – Terr. Hawaii, Board Agr. For., Bot. Bull. 2. 1913.

9325. *Palmyra Island* with a description of its flora ... Honolulu (published by the College [of Hawaii]). 1916. Oct. (*Palmyra Isl.*).
Collaborators: O. Beccari, A. Zahlbruckner, U. Martelli, H.L. Lyon, M.A. Howe.

Publ.: 19 Apr 1916, p. [i-ii], [1]-53, *pl. 1-20*, map. *Copies*: FI, M, MO, NY, USDA. – College Hawaii Publ., Bull. 4, 1916.

9326. *The Sandalwoods of Hawaii*. A revision of the Hawaiian species of the genus Santalum ... Honolulu, Territory of Hawaii (Board of Agriculture and Forestry) 1916. Oct. (*Sandalw. Hawaii*).
Publ.: 28 Dec 1916, p. 1-2 cover, [3]-43, *pl. 1-13*. *Copies*: FI, G, NY, US. – Bot. Bull. 3, Terr. Hawaii Board Agr. For.

9327. *The ornamental trees of Hawaii* ... published under patronage. Honolulu, Hawaii, 1917. Oct. (*Orn. trees Hawaii*).
Publ.: Feb 1917, p. [i*-iv*], i-v, [1]-210, *pl. 1-79* and *2 col. pl.* (one as frontisp.). *Copies*: MICH, MO, PH, US, USDA.

9328. *The Ohia Lehua trees of Hawaii*. A revision of the Hawaiian spcies of the genus Metrosideros Banks, with special reference to the varieties and forms of Metrosideros collina (Forster) A. Gray subspecies polymorpha (Gaud.) Rock ... Honolulu Hawaii. 1917. Oct. (*Ohia Lehua trees Hawaii*).
Publ.: 27 Aug 1917, p. [1]-76, *pl. 1-31* (in text). *Copies*: HH, US. – Issued as Terr. Hawaii Board Agr. For., Bot. Bull. 4, 1917.

9329. Publications of the Bernice Pauahi Bishop Museum of Polynesian ethnology and natural history. *A monographic study of the Hawaiian* species of the tribe *Lobelioideae* family Campanulaceae ... Honolulu, H.I. (published by authority of the Trustees) 1919. Oct. (*Monogr. Stud. Haw. Lobelioid.*).
Publ.: 20 Feb 1919, frontisp., p. [i]-xvi, [5]-394, [395], with *pl. 1-217* in text. *Copies*: HH, M, MICH, NY, PH, USDA.
Other issue: 1919 as Mem. Bernice P. Bishop Mus. 7(2). *Copies*: BR, G.
Ref.: N.T., Torreya 19: 228-230. 1919 (rev.).

9330. *The arborescent indigenous legumes of Hawaii* ... Honolulu, Hawaii. 1919. Oct. (*Arboresc. legum. Hawaii*).
Publ.: 9 Jun 1919, p. [2]-53, *pl. 1-18* (in text). *Copy*: US. –Issued as Terr. Hawaii, Board Agr. For., Bot. Bull. 5, 1919.

9331. *The Hawaiian genus Kokia* a relative of the cotton ... Honolulu, Hawaii, 1919. Oct. (*Haw. Kokia*).
Publ.: 9 Jun 1919, p. [2]-22, *pl. 1-7* (in text). *Copy*: US. – Terr. Hawaii, Board Agr. For., Bot. Bull. 6, 1919.

9332. Experiment Station of the Hawaiian Sugar Planters' Association. *The Leguminous plants of Hawaii*. being an account of the native, introduced and naturalized trees, shrubs, vines and herbs, belonging to the family Leguminosae ... Honolulu, Hawaii. July 1920. Oct. (*Legum. pl. Hawaii*).
Publ.: Jul 1920, p. [i]-x, [1]-234, *pl. 1-93* in text. *Copies*: BR, HH, M, MICH, MO, NY, PH, US, USDA.

Rockley, Lady **Alicia Margaret** (formerly Cecil; née Amherst), (1865-1941), British (English) popular botanical author; collected in Mozambique (1899), Rhodesia (1900), Ceylon, Australia, New Zealand, Canada (1927); married (1898) Sir Evelyn Cecil (later Baron Rockley). (*Rockley*).

HERBARIUM and TYPES: K (collected as A.M. Cecil).

BIBLIOGRAPHY and BIOGRAPHY: Barnhart 3: 167; Desmond p. 527 (lists books; d. 14 Sep 1941); IH 2: (in files).

BIOFILE: Anon., J. Bot. 79: 136. 1941 (d.); The Times 15 Sep 1941 (fide Desmond); Who was Who 4 (1941-1950): 992.

Decary, R., Moçambique Docum. trimestr. 21: 114-117. 1940 (Lady Rockley & Madagascar).
Exell, A.W. & G.A. Hayes, Kirkia 6(1): 100. 1967 (coll. Fl. zambes.).
Verdoorn, F., ed., Chron. bot. 7:354. 1943 (d.).

9333. *Wild flowers of the great dominions of the British Empire* ... London (Macmillan and Co., ...) 1935. Oct. (*Wild fl. Brit. emp.*).
Publ.: 1935 (J. Bot. Sep 1935; Nat. Nov. Jul 1935), p. [i]-xii, 1-380. *Copies*: E, NSW.
Ref.: Rendle, A.B., J. Bot. 73: 268-269. Sep 1935 ("Lady Rockley follows the custom of some of the early Victorian botanists who "enriched" their text with quotations from the poets – good, bad, and indifferent).

Rodati, Aloysius (Luigi) (1762-1832), Italian botanist at Bologna; director of the Bologna botanical garden 1792-1802. (*Rodati*).

HERBARIUM and TYPES: Unknown.

BIBLIOGRAPHY and BIOGRAPHY: BM 4: 1716; CSP 5: 249; Jackson p. 432; PR 7691-7692, ed. 1: 8651-8652; Rehder 1: 258; Saccardo cron. p. xxxi; SO 695.

BIOFILE: Hoppe, D.H., Flora 3: 25. 1820.
Gager, C.S., Brooklyn Bot. Gard. Rec. 27: 268. 1938.

EPONYMY: *Rodatia* Rafinesque (1840) may be named for this author.

9334. *Linnaei de plantarum ordine* brevis interpretatio una cum catalogo plantarum quae vel saepius, vel constanter eumdem ordinem eludere visae sunt ... Bononiae [Bologna] (ex typographia Sancti Thomae Aquinatis ...) 1784. Oct. (*Linn. pl. ord.*).
Publ.: 1784, p. [1]-40, *1 pl. Copy*: KSC.

9335. *Index plantarum* quae extant *in horto publico* Bononiae anno mdcccii. Accedunt observationes circa duas species Agaves necnon continuatio historiae horti ejusdem. Bononiae [Bologna] (ex typographia S. Thomae aquinatis) [1802]. Qu. (*Index pl. hort. Bonon.*).
Publ.: 1802, p. [1]-121, *pl. 1-5* (uncol. copp.). *Copies*: G, MO.

Rodegher, Emilio (1856-1922), Italian botanist. (*Rodegher*).

HERBARIUM and TYPES: PAV.

BIBLIOGRAPHY and BIOGRAPHY: Barnhart 3: 167 (b. 12 Sep 1856); BFM no. 1106; BL 2: 369, 709; CSP 18: 249; IH 2: (files); Rehder 1: 430; Saccardo cron. p. xxxi; Tucker 1: 603.

9336. *Prospetto della flora della Provincia di Bergamo* ... Treviglio (stab. tipografico sociale) Novembre 1894 [1895]. Oct. (*Prosp. fl. Bergamo*).
Co-author: Giuseppe Venanzi.
Publ.: Jan 1895 (t.p. dated Nov 1894 but corrected to 1895; Bot. Centralbl. 6 Feb 1895), p. [i]-xviii, [1]-146. *Copies*: G, HH.

9337. *Novissimo prospetto della flora della provincia di Bergamo* [Bergamo 1920]. †.
Co-author: Alcide Rodegher (*fl.* 1920), son of E. Rodegher; sole author for Puntata 3-6.
1: 1920 (p. 4: 8 Feb 1920), p. [1]-50. Atti Ateneo Sci. let. Arti Bergamo 25, comm. 4.
2: 1921, p. [1]-64, Atti 26, comm. 4.
3: Sep 1929, Bergomum n.s. 3(2) (Atti 1927-1929, no. 3), 33-48. 1829. As "Flora della Prov. di Bergamo".
4: Oct 1929, Bergomum n.s. 3(3) (Atti 1927-1929, no. 4): 56-80. 1929.
5: Dec 1929, Bergomum n.s. 3(4) (Atti 1927-1929, no. 5): 81-96. 1929.
6: 1930, Bergomum n.s. 4(1) (Atti 1927-1929, no. 6): 97-112. 1930.
Copy: USDA. – Remained incomplete; *Polypodium-Centaurea*.

Rodet, Henri Jean Antoine (1810-1875), French botanist; teacher at the veterinary college of Lyon; later at Toulouse. (*Rodet*).

HERBARIUM and TYPES: Unknown.

BIBLIOGRAPHY and BIOGRAPHY: Barnhart 3: 167 (b. 2 Oct 1810, d. 24 Oct 1875); BM 4: 1716; Jackson p. 53, 203; PR 7693-7694, ed. 1: 8653; Rehder 5: 732.

BIOFILE: Magnin, Bull. Soc. bot. Lyon 32: 16. 1907 (further refs.).

EPONYMY: *Rodetia* Moquin-Tandon (1849).

9338. *Botanique agricole et médicale* ou étude des plantes qui intéressent principalement les médecins, les vétérinaires et les agriculteurs accompagnée de 160 planches représentant plus de 900 figures intercalées dans le texte ... deuxième édition revue et considérablement augmentée avec la collaboration de C. Baillet .... Paris (P. Asselin, ...) 1872. Oct. (*Bot. agric. méd.*).
Ed. 1: 1857, p. [i]-viii, [1]-856. *Copy*: BR.
Ed. 2: 1872, p. [i]-xix, [xx, err.], [1]-1078. *Copy*: USDA.

Rodgers, Andrew Denny, III (1900-1981), American botanical historian; descendent of William Starling Sullivant; LLB Ohio State University 1925; practiced law at Columbus, Ohio 1926-1933; from 1933 independent author. (*A.D. Rodgers*).

ARCHIVES: Ohio historical Society, Columbus.

NOTE: A.D. Rodgers was the author of several biographies of American botanists and of books dealing with the history of botany often quoted in TL-2: *Noble Fellow: William Starling Sullivant.* (1940). *John Torrey: a story of North American botany* (1942), *American botany 1873-1892: decades of transition* (1944), *John Merle Coulter: missionary in science* (1944); *Liberty Hyde Bailey: a story of American plant sciences* (1949), *Bernard Eduard Fernow: a story of North American forestry* (1951), *Erwin Frink Smith: a story of North American plant pathology* (1952). – All titles, except those on Coulter and Smith were reprinted by Hafner Publishing Company, New York.

Rodigas, Émile (1831-1902), Belgian botanist and zoologist at Gent; director of the local zoological garden, later teacher at and director of the horticultural college. (*Rodigas*).

HERBARIUM and TYPES: BR.

BIBLIOGRAPHY and BIOGRAPHY: Barnhart 3: 167 (b. 2 Apr 1831, d. 14 Nov 1902); BM 4: 1716; CSP 8: 767, 18: 252; Herder p. 474; Hortus 3: 1202; Morren ed. 10, p. 45; MW p. 415; NI alph.; Rehder 5: 732; TL-2/4628, see L. Linden; Tucker 1: 603; Zander ed. 10, p. 708, ed. 11, p. 807.

BIOFILE: André, E., Rev. hortic. 74: 553-554. 1902 (obit., portr.).
Durand, Th., Bull. Soc. bot. Belg. 41(2): 38-39. 1903 (d.).
Nobele, L. de, Fête jubilaire de Émile Rodigas, Compte-Rendu, Gand 1885, 40 p. (portr.).
Oye, P. van, Plantk. Univ. Gent 17. 1960.
Urban, I., Symb. ant. 2: 5. 1900.

COMPOSITE WORKS: (1) Editor (as "secrétaire de la rédaction" later as "rédacteur") of Lucien Linden's *L'illustration horticole* vols. 29-43, 1882-1896.
(2) Co-editor, with Charles Lucien Linden, of *Lindenia*, vols. 1-11, 1885-1895, see TL-2/4628.

NOTE: Émile Rodigas' father, François-Charles-Hubert Rodigas, originally physician and pharmacist, b. 23 Sep 1801, d. 4 Mar 1877; dedicated himself to horticulture from

1847-1859 as professor of horticulture at Lier; returning to his practice as a physician at St. Truiden in 1859. (see e.g. H. Micheels, Biogr. natl. Belgique 19: 600-602. 1907).

Rodin, Hippolyte (1829-1886[?]), French botanist and pharmacist at Beauvais; secretary of the Société d'Agriculture et d'Horticulture de Beauvais (Seine-et-Oise). (*Rodin*).

HERBARIUM and TYPES: Unknown; his herbarium of 4000 species was sold in 1886.

NOTE: E. Malinvaud, in a note in Bull. Soc. bot. France 33 (Bibl.): 48. 1886 announces the sale of Rodin's herbarium and library. From this it might be deduced that Rodin died in 1886; however, further confirmation of this date is still lacking.

BIBLIOGRAPHY and BIOGRAPHY: BL 2: 106, 186, 709; BM 4: 1716; CSP 5: 250, 8: 767, 11: 201; Jackson p. 204; PR 7695; Rehder 1: 413, 3: 86.

BIOFILE: Malinvaud, É., Bull. Soc. bot. France 33 (Bibl.): 48. 1886 (sale herb. libr.).

9339. *Les plantes médicinales et usuelles* de nos champs-jardins-forêts description et usages des plantes comestibles-suspectes-vénéneuses-employées dans la médecine -l'industrie et dans l'économie domestique ... Paris (J. Rothschild, éditeur ...) 1872. Oct. (*Pl. méd. usuel.*).
Ed. 1: 1872, p. [i*-iv*], [i]-iv, [1]-427. *Copy*: G.
Ed. 2: 1875 (fide Bot. Zeit. 28 Mai 1875).
Ed. 3: 1876, xx, 478 p., n.v.
Ed. 5: 1882 (Bot. Zeit. 24 Nov 1882, 27 Apr 1883; Nat. Nov. Mar(1) 1883), p. [i]-xx, [1]-478. *Copy*: BR.
Ed. 10: 1896 (fide Nat. Nov. Apr(1) 1896; Bot. Zeit. 1 Mai 1896 (54(2): 143. 1896)), n.v.
Ed. [..]: 1914, fide Nat. Nov. Mai(2) 1914.
Ref.: [Malinvaud, É.], Bull. Soc. bot. France 19 (bibl.): 239-240. 1873 (rev.).

9340. *Esquisse de la végétation du département de l'Oise* ... Beauvais (Imprimerie d'Achille Desjardins, ...) 1864. Oct.
Première partie: Jun 1864 (NY presentation copy signed 20 Jun 1864 by author; Soc. bot. Fr. rd. 8 Jul 1864), p. [i-iv], [1]-155, [156]. *Copies*: G, NY. – Reprinted and to be cited from Mém. Soc. Acad. Arch. Sci. Arts Oise 5: 333-507. 1863 (1864).
Deuxième partie: Statistique botanique ... par L. Graves révisé, annoté et augmentée par H. Rodin, ib. 5(2): 661-734. 1863 (1864), 6(1): 219-283, 1865, 6(2): 527-574, 772-867. 1867, 7(1): 360-456, 625-715. 1868(-1870), 8(1): 216-255. 1871, 8(2): 713-785. 1873, 9(1): 211-274. 1874, 9(2): 511-543, 771-842. 1876. Original ed. Louis Graves, Cat. pl. Oise, 1857, see TL-2/2117.
Ref.: [Malinvaud, É.], Bull. Soc. bot. France 11(bibl.): 207-210. 1864.

Rodrigues, João Barbosa (1842-1909), Brazilian botanist and traveller, collected widely in Brazil between 1868 and 1897; teacher at the Imperial College Pedro II; associated with F. Freire Allemão 1868-1876; in charge of a chemical factory 1876-1883; director of the Manaos botanical museum 1883-1889; from 1889-1909 director of the botanical garden of Rio de Janeiro. (*Barb. Rodr.*).

HERBARIUM and TYPES: Unknown; most of pollen types destroyed by fire. – Most original drawings came in the hands of C.A. Cogniaux (BR), a set of 550 copies is at K. – Otto Stapf (1909), speaks of an "almost complete absence of specimens" and short diagnoses, concluding that the most important protologue material for R's 573 new species and 25 new genera lies in his drawings, published and unpublished.

NOTE: We treated J.B. Rodrigues under Barbosa Rodrigues in vol. 1 (nos. 299-305) but supply here an amended version. The author is most frequently cited under Barbosa Rodrigues.

BIBLIOGRAPHY and BIOGRAPHY: Barnhart 1: 120 (b. 22 Jun 1842, d. 6 Mar 1909); BL 1: 240, 285; BM 1: 97, 6: 57; Bossert p. 24; Hortus 3: 1188; Jackson p. 120, 374; Kew 4:

500; Langman p. 109-110; Morren ed. 2, p. 35; NI 1656-1660; Rehder 5: 50; TL-2/299-305; Tucker 1: 58; Zander ed. 10, p. 631, ed. 11, p. 717.

BIOFILE: Anon., Bot. Zeit. 48: 835. 1890 (app. Rio de Janeiro).
Cogniaux, A., Bull. Herb. Boissier 1: 425-427. 1893 (bibl. notes on his works).
Grignan, G.T., Rev. hort. 81:300. 1909 (d.).
Hill, J.R., Trans. Proc. Bot. Soc. Edinburgh 24: 48-49. 1912 (obit., R. was foreign corr. member).
Horta, P.P., Rodriguesia 2(5): 181-186. 1936 (on R.'s orchidology).
Ihering, H. von, Rev. Museu Paul. 8: 23-37. 1911 (biogr., bibl.).
Niemeyer de Lavor, J.C., Rodriguésia 31: 278. 1979.
Stapf, O., Bull. misc. Inf. Kew 1909: 225-226 (obit.).
Urban, I., Fl. bras. 1(1): 4-6. 1916 (vita, itin.).
Wittrock, V.B., Acta Horti Berg. 3(2): 181, *pl. 32*. 1903.

EPONYMY: (genera): *Barbosa* Beccari (1887); *Barbosella* Schlechter (1918); *Barbrodria* Luer (l981); *Brodriguesia* Cowan (l981); *Rodrigueziella* [sic] O. Kuntze (l891); (journal): *Rodriguesia*; revista do instituto de biologia vegetatal, jardim botânico e estação biologica do Itatiaya [subtitle varies]. Rio de Janeiro, Brazil. Vol. 1· 1935· × *Barbosaara*, see Orchid Rev. 83: (1979), p. ix. 1975. *Note*: *Rodriguesia* A.T. Brongniart (1843, *orth. var.* of *Rodriguezia* Ruiz et Pavon), *Rodrigueza* Dumortier (1829, *orth. var.* of *Rodriguezia* Ruiz et pavon), *Rodriguezia* Ruiz et Pavon (1794) were named for [Manuel] or Jose Demetrio Rodriguez (1780-1847) according to Bot. Zeit. 5: 357. 1847, and thus the derived names *Rodrigueziopsis* Schlechter (1920) and × *Rodriopsis* [*Ionopsis* Kunth (1816) × *Rodriguezia* Ruiz et Pavon] W. Osment (1969), × *Rodretta* [*Comparettia* Poeppig et Endlicher (1836) × *Rodriquezia* Ruiz et Pavon] Hort. (1958), × *Rodricidium* [*Oncidium* O. Swartz (1800) × *Rodriguezia* Ruiz et Pavon] Hort. (1957) and × *Rodridenia* [*Macradenia* R. Brown (1822) × *Rodriguezia* Ruiz et Pavon] Hort. (1962) as well.

POSTAGE STAMPS: Brazil 40 c. (1943) yv. 413.

9341. *Enumeratio palmarum novarum* quas valle fluminis Amazonum inventas et ad sertum palmarum collectas, descripsit et iconibus illustravit ... Sebastianopolis (apud Brown & Evaristo ...) 1875. Oct. (in fours). (*Enum. palm. nov.*).
Publ.: 1875 (p. 43: "ante diem iii nonas Martius" 1875; Bot. Zeit. 29 Oct 1875; Flora 1 Sep 1876), p. [1]-43, [1, err.]. *Copies*: BR, USDA.
Follow-up: see below *Prot.-App. Enum. palm. nov.* 1879, and *Enum. palm. nov., prot.* 1879.
Ref.: H.W.R., Österr. bot. Z.26: 383. 1 Nov 1876 (rev.).

9342. *Genera et species orchidearum novarum* quas collegit descripsit et iconibus illustravit ... Sebastianopolis 1877-1881 [1882]. Oct. (in fours). (*Gen. sp. Orchid.*).
1: 1877 (p. viii: 20 Jul 1877), p. [i*-v*], frontisp., [i]-vii, [1]-206, chart [1 p.], [i]-x, [xi].
2: 1882 (date on cover; t.p. dated 1882; Nat. Nov. Mai(1) 1883; Soc. bot. Fr. 10 Dec 1882), p. [i*-v*], [i]-vi, [7]-295, chart [1]-2, [1, err.], [i]-xvi. – A note in Bull. Soc. bot. France 29 (bibl.): 183. 1882 suggests that p. 1-136 came out in 1881. Further information required.
Copies: BR, GB, NY, PH, USDA.

9343. *Protesto-appendice ao Enumeratio palmarum novarum* lido no Instituto historico e geographico do Brazil, na sessão de Maio de 1879 ... Rio de Janeiro (Typographia nacional) 1879. Oct. (in fours). (*Prot.-app. Enum. palm. nov.*).
Publ.: 1879 (p. 48: 21 Jul 1879), p. [1]-48, [1, err.], [i]-ii, *pl. 1-2* (uncol. liths. auct.). *Copies*: G(2), M.

9344. *Enumeratio palmarum novarum seguido de un protesto e de novas palmeiras* descriptas ... Rio de Janeiro (Typographia nacional) 1879. (*Enum. palm. nov., prot.*).
Publ.: 1879 (copy at M dedicated to C. Naegeli, dated Rio de Janeiro 15 Sep 1879).
Copy: M. – A re-issue of the *Enum. palm. nov.* of 1875 preceded by a new 1879 t.p. as quoted above, followed by (1) *Enum. palm. nov.* 1875, [1]-43, [1, err.], *followed* by the *Prot.-app. Enum. palm. nov.* 1879, p. [1]-48, [1, err.], [i]-ii of which [1]-38 are the "Protesto" and 39-48 descriptions of new species.

9345. *Les palmiers* observations sur la monographie de cette famille dans la Flora brasiliensis ... Rio de Janeiro (Imprimerie Du-Messager du Brésil ...) 1882. Oct. (in fours). (*Palmiers*).
Publ.: 1882 (US copy rd. 16 Apr 1883; Nat. Nov. Mai(1) 1883; Bot. Centralbl. 28 Mai-1 Jun 1883), frontisp., [i*], [i]-iii, [3]-53, *pl. 1-4* (uncol. liths.). *Copies*: BR, G, U, US, USDA.
Ref.: Drude, O., Bot. Jahrb. 4 (Lit.): 484-486. 1883 (rev.).

9346. *Palmae amazonensis novae* auctore J. Barboza Rodrigues, Direct. Mus. bot. Amaz. 1884-1886. (Extrahido da Vellosia, Contribuçiões do Museu botanico do Amazonas. Anno i. 1887) Manáos (Typographia do "Jornal do Amazonas"). Oct.
Publ.: 1889, p. [i, t.p. iii, errata], [33]-56. *Copy*: C (Inf. A. Hansen). – Reprinted and to be cited from Vellosia 1: 33-56. 1887.

9347. *O Tamakoaré* especies novas da ordem das Ternstroemiaceas ... Manaos (Impresso na typographia do Jornal do Amazonas) 1887. (*Tamakoaré*).
Publ.: 1887 (p. 28: 28 Jul 1887), p. [1]-28, *1 pl.* (uncol. liths.). *Copies*: USDA(2).

9348. *Plantas novas* cultivadas no Jardim botanico do Rio de Janeiro descriptas, classificadas e desentradas ... Rio de Janeiro (Typ. de G. Leuzinger & filhos, ...) 1891. Qu. (*Pl. nov.*).
1: 1891 (p. 27: 1 Aug 1891), p. [i*-iii*], [i]-ii, [1]-37, *pl. 1-9* (uncol. liths. auct.).
2: 1893 (p. 17: 23 Mar 1893; Nat. Nov. Nov(1) 1893), p. [1]-20, *2 pl.* (id.).
3: 1893 (p. 12: 6 Jan 1893; Nat. Nov. Feb(1) 1895), p. [i-v], [1]-12, [13], *pl. 1-2* (id.).
4: 1894 (p. 24: 2 Jan 1894), p. [i-iii], [1]-26, *pl. 1-5.*
5: 1896 (p. 33: 25 Oct 1896; Nat. Nov. Dec(1) 1897), p. [i*-iii*], [i]-ii, [1]-37, *pl. 1-5.*
6: 1898 (p. [v]: 3 Mai 1898), p. [i-v], [1]-31, [i]-ii, *pl. 1-7.*
Copies: R, HH, M, US.

9349. *Vellosia.* Contribuções do Museu botanico do Amazonas. Volume primeiro botanica. 1885-1888 (Segunda edição). Rio de Janeiro (Imprensa nacional) 1891. Qu. (*Vellosia*).
Ed. 1: 1885-1888, n.v. (but see above *Palmas amazonensis novae* 1887); seen xerox at MO: p. [i]-iv, [1]-148, *pl. 1-13.* Manáos (typographia do "Jornal do Amazonas") 1888.
Ed. 2: (volumes 2 and 4 are nonbotanical. Articles by J.B. Rodrigues in vol. *1*: [*1*]-88, [*89*]-112, [*113*].
1: 1891, p. [i]-xix, [1]-133, [1]-6.
3: 1891, p. [i-iii], *pl. 1-13, 1-13, 13a-13q, 14-23, 1-2* (uncol. liths.). Volume terceiro estampas botanica.
Copies: US, USDA.

9350. *Enumeratio plantarum* in horto botanico fluminensi cultarum. Rio de Janeiro (Typ. Jornal do Brazil ...) 1893. Qu. (*Enum. pl. hort. flum.*).
Publ.: 1893 (preface 10 Jan 1893), p. [1]-24. *Copies*: G, USDA.

9351. *Hortus fluminensis* ou breve noticia sobre as plantas cultivadas no Jardim botanico do Rio de Janeiro para servir de guia aos visitantes ... Rio de Janeiro (Typ. Leuzinger ...) 1894 [1895]. Oct. (*Hort. flum.*).
Publ.: 1895 (date on cover; t.p.:1894; Nat. Nov. Dec 1895; Soc. bot. Fr. rd. 12 Aug-31 Dec 1895), p. [i*-v*], [i]-xxxviii, 1-307, [i]-x, [i]-xi, [i]-xvi, [i]-ii, *pl. 1-13* (with captions). *Copies*: ILL (inf. D.P. Rogers), M, USDA; IDC 5313.

9352. *Palmae mattogrossenses* novae vel minus cognitae quas collegit descripsit et iconibus illustravit ... Rio de Janeiro (Typographia Leuzinger) 1898. Qu. (*Palm. mattogr.*).
Publ.: Jan-Apr 1898 (p. xx: 5 Sep 1897; Nat. Nov. Jun(1) 1898; Soc. bot. Fr. rd. Jan-Jun 1898), p. [i]-xx, err. slip, [1]-88, [89, table], [i]-ii, *pl. 1-4, 4a, 5-22, 24-27* (uncol. liths. auct.). *Copies*: BR, ILL, M, U, US.

9353. *Plantae mattogrossenses* ou relação de plantas novas colhidas, classificadas e desenhadas ... Rio de Janeiro (Typographia Leuzinger) 1898. Qu. (*Pl. mattogr.*).

Publ.: 1898 (p. vii: 3 Mar 1898), p. [i]-vii, err. slip, [1]-43, [i]-iii, *pl. 1-13* (uncol. liths. auct.). *Copies*: BR, M, USDA.

9354. *Palmae novae paraguayenses* quas descripsit et iconibus illustravit ... Rio de Janeiro (Typographia Leuzinger) 1899. Qu. (*Palm. paraguay.*).
Publ.: Feb-Mai 1899 (p. ix: 28 Sep 1898, p. 60: 20 Jan 1899; Nat. Nov. Jun(1) 1899; Soc. bot. Fr. rd. Jul-Dec 1899), p. [i]-ix, err. slip, [1]-66, *pl. 1, 2*["*1*"], *3-6* (uncol. liths. auct.). *Copies*: BR, G, M, NY, PH, U, US.

9355. *Palmae hasslerianae novae* ou relação das palmeiras encontradas no Paraguay pelo Dr. Emilio Hassler de 1898-1899 determinadas e desenhades ... Rio de Janeiro (Typographia Leuzinger) 1900. Qu. (*Palm. hassler.*).
Collector: Émile Hassler (1861-1937), see TL-2/2475, 2476.
Publ.: 1900 (p. vii: 11 Jun 1900; p. 16: 15 Aug 1900; Nat. Nov. Dec(1) 1901 as of "1901"; US copy rd. 7 Mar 1902), p. [i]-vii, err. slip, [1]-16, [17, err.]. *Copies*: BR, M, NY, PH, US.

9356. *As Heveas ou Seringueiras*. Informações ... Rio de Janeiro (Imprensa nacional) 1900. Oct. (*Heveas*)
Publ.: 1900, frontisp., [1]-86, *3 pl., 7 charts. Copies*: G(2), M.

9357. *Contributions du Jardin botanique de Rio de Janeiro* par son directeur J. Barbosa Rodrigues i-[iv]. Rio de Janeiro (Typographie L'Étoile du Sud ...) 1901 [-1907]. (*Contr. Jard. bot. Rio de Janeiro*).
1: 1901, p. [1]-22, *pl. 1.*
2: 1901, p. [23]-58, *pl. 2-6.* – "1ᵉʳ volume-n.2." Rio de Janeiro (Estabelecimento Typographico de Brun & Veyssière ...) 1901.
3: 1902, p. [59]-90, *pl. 2-11.* "iii" Rio de Janeiro (Typographie Rua de S. José n. 102). 1902.
4: 1907, p. [89] sic-125, *pl. 12-27.* "iv" Rio de Janeiro (Typographie-Rua do Rosario n. 107). 1907.

9358. *Sertum palmarum brasiliensium* ou relation des palmiers nouveaux du Brésil découverts, décrits et dessinés d'après nature ...Bruxelles (Imprimerie Veuve Monnom ...) 1903. Broadsheet (*Sert. palm. bras.*).
1: 1903 (p. xii: 19 Feb 1903), frontisp. portr. author, p. [i]-xxix, [1]-140, *pl. 1-91* (chromoliths. auct.).
2: 1903 (Nat. Nov. Nov 1903; original price of the two parts together DM 600), p. [i-iii], [1]-114, *pl. 1-83* (id.).
Copies: PH, US.
Ref.: Balis, Hortus belgicus 76. 1962.

9359. *Les noces des palmiers* remarques préliminaires sur la fécondation ... Bruxelles (Imprimerie Ad. Mertens ...) 1903. (*Noces palm.*).
Publ.: 1903 (Nat. Nov. Nov 1903; ÖbZ Sep 1903), p. [i, iii], [1]-90, *pl. 1-7* (partly col. liths.). *Copy*: BR.

9360. Mbaé Kaá Tapyiyetá Enoyndaua ou *a botanica ea nomenclatura indigena memoria* ... Rio de Janeiro (Imprensa nacional) 1905. (*Bot. nomencl. indig.*).
Publ.: 1905, p. [i]-vi, [1]-87, [1, err.]. *Copy*: M.

9361. *A flora brasiliensis de Martius*. Rio de Janeiro (Typ. do Jornal do commercio de Rodrigues & C.) 1907. (*Fl. bras. Mart.*).
Publ.: 1907, p. [1]-32. *Copy*: M.

Rodríguez y Femenías, Juan Joaquín (1839-1905), Spanish (Menorca) botanist, banker and politician and newspaper owner at Mago (Mahon), Menorca. (*J. Rodríguez*).

HERBARIUM and TYPES: FI (orig. herb. or at least major set); algae e.g. at BC, BR, C and L; bryoph. at L; phanerogams e.g. at P. – Letters to W.G. Farlow at FH.

NOTE: Not to be confused with his contemporary Juan José Rodríguez, botanist in Guatemala around 1884.

BIBLIOGRAPHY and BIOGRAPHY: Barnhart 3: 168 (b. 18 Mai, d. 8 Aug 1905); BL 2: 489, 492, 709; BM 4: 1717; Bossert p. 335; Colmeiro 1: cxcvi; CSP 8: 767, 11: 201, 18: 253; De Toni 2: ci, 4: xlviii; Jackson p. 341; Kew 4: 501; Morren ed. 10, p. 52; Rehder 5: 733; Tucker 1: 603; Urban-Berl. p. 272.

BIOFILE: Anon., Bot. Not. 1906: 45 (d.); Encicl. Univ. ilustr. Eur.-Amer. 51: 1319-1320. s.d.; Hedwigia 45: (76). 1906 (d.); Österr. bot. Z. 55: 448. 1905 (d.).
Bornet, E., Bull. Soc. bot. France 52: 490. 1905 (obit.).
Buen, O. de, Bol. Soc. esp. Hist. nat. 6(4): 173-180. 1906 (obit., portr.).
De Toni, G.B., Nuova Notarisia 17: 37. 1906 (obit.).
Gredilla, F., Bol. Soc. esp. Hist. nat. 5(8): 368-373. 1905 (obit.).
Hernandez, F. et al., El Liberal, diario de Union republicana, Mahon 25 (no. 7210), 14 Aug 1905, p. 1-3, (obit., bibl., various tributes).
Kneucker, A., Allg. bot. Z. 11: 208. 1905 (d.).
Sanz, F.M., El Liberal, Mahon 1920, no. 7210, 2 p. (copy BR).
Steinberg, C.H., Webbia 34: 61. 1979 (herb. at FI).
White, J.J., J. Bot. 43: 310. 1905.

EPONYMY: *Rodriguezella* Schmitz (1895).

9362. *Catálogo razonado de las plantas vasculares de Menorca, . . .* Mahon (Tip. de Fábregues Hermanos, ...) 1865-1868. Qu. (*Cat. pl. vasc. Menorca*).
Publ.: Jan-Mai 1869 (we have no indication that the "1865-1868" on the t.p. corresponds with a publication in parts; the book was presented as a whole to the Soc. bot. France on 25 Jun 1869, accompanied by herbarium specimens; p. xxvii is dated Dec 1868), p. [i]-xxx, [xxxi], [1]-116. *Copies*: G, HH, MO, NY.
Suplemento: 1874 ("ObZ 1 Aug 1874), p. [i-ii], [1]-64. *Copies*: G, NY. – Reprinted and to be cited from Anal. Soc. Españ. Hist. nat. 3,(mem.): 5-68. 1874. – Reprint t.p. "suplemento al Catálogo ..." Madrid (imprenta de T. Fortanet ...) 1874. Oct.
Additions: 1878, *in* Bull. Soc. bot. France 25: 238-241. 1878. – See also M. Willkomm, Linnaea 40(1): 1-134. Jan 1876.
Ref.: [Malinvaud, É.], Bull. Soc. bot. France 16(bibl.): 121-122. 1869 (rev.; lists previous Menorcan floras).
Ascherson, P., Bot. Zeit. 31: 220-223. 1873.

9363. *Algas de las Baleares ...* [Anales de la Sociedad Española de Historia natural ... Madrid 1888-1889]. Oct.
Publ.: 1888-1889 (BSbF rd. 13 Dec 1889; Nat. Nov. Mar(2) 1880), p. [1]-96. *Copy*: BR. – Reprinted and to be cited from Anal. Soc. Esp. Hist. nat. 17: [311]-330 [reprint 1-20] 1888, 18: [199]-274 [reprint p. [21]-96]. 1889.

9364. *Datos algológicos ...* Madrid (Establecimiento tipográfico de Fortanet ...) 1889.
1-2: 1889, p. [i-iv], [1]-10, *pl. 3-4*. *Copies*: BR, FH. – Reprinted and to be cited from Anal. Soc. Esp. Hist. nat. 18: [405]-414, *pl. 3-4*. 1889.
3: 1890, p. [i], [97]-100, *pl. 2*. *Copy*: FH. – Id. 19: [97]-100, *pl. 2*. 1890.
4: 1896, cover (with reprint title), p. [155]-160, *pl. 5-6*. *Copy*: FH. – Id. 24: [155]-160, *pl. 5-6*. 1895 (1896).

9365. *Flórula de Menorca ...* Mahón (Imprenta de Francisco Fábregues ...) 1904. Oct. (*Fl. Menorca*).
Publ.: 1904 (p. xv: Oct 1904; Bot. Centralbl. rd. 1906, see 103: 45. 6 Mar 1906), p. [i]-xv, [xvi], [1]-198. *Copies*: BR, FH, G.
Ref.: Flahault, C., Bull. Soc. bot. France 52: 573-576. 1905.

Rodschied, Ernst Karl (x-1796), German (Hanau) physician and botanist; some time practicing in the "Dutch colony Rio Essequibo" (now Guyana). (*Rodschied*).

HERBARIUM and TYPES: GOET (types G.F.W. Meyer, *Prim. fl. esseq.* 1818 (TL-2/5930).

BIBLIOGRAPHY and BIOGRAPHY: Barnhart 3: 168 (b. Jan 1796); Dryander 3: 4; PR 7696, ed. 1: 8655.

BIOFILE: Meyer, G.F.W., Prim. fl. esseq. vii-ix. 1818 (biogr. sketch).
Wagenitz, G., Index coll. princ. herb. Gott. 136. 1982 (herb.).

EPONYMY: *The etyomology of Rodschiedia* P.G. Gaertner, B. Meyer et Scherbius (1800) could not be determined and that of *Rodschiedia* Miquel (1845) was not given.

9366. Dissertatio inauguralis medico-botanica. *De necessitate et utilitate studii botanici.* Quam summi numinis auspicio et gratiosi medicorum ordinis consensu pro gradu doctoris in arte medica et chirurgica rite obtinendo publico eruditorum examini d. [handwriting: xxiv] martii mdccxc subiicit auctor Ernestus Carolus Rodschied hanoviensis. Marburgi (typis novae typograph. academ.) [1790]. (*Necess. util. stud. bot.*).
Publ.: 24 Mar 1790, p. [1]-34. *Copy*: Marburg Univ. Libr. (inf. G. Wagenitz).
Ref.: Hoppe, D.H., Bot. Taschenb. 1792: 189-217.

9367. *Medizinische und chirurgische Bemerkungen über das Klima, die Lebensweise und Krankheiten der Einwohner der hollaendischen Kolonie Rio Essequebo* ... Frankfurt (in der Jaegerschen Buchhandlung) 1796. Oct. (*Med. Bem. Kol. Rio Essequebo*).
Publ.: 1794 (p. xii: 30 Nov 1793), p. [i*], [iii*-iv*], [i]-xii, [1]-320. *Copy*: Koninklijke Bibliotheek, 's Gravenhage (Royal Libr., the Hague). – On p. 21-90 a description of 114 plants; some names antidate those of G.F.W. Meyer (*Prim. fl. Esseq.* q.v. for notes on Rodschied on p. vi-ix. (1818)).

Rodway, Leonard (1853-1936), British (Devon)-born dental surgeon and botanist; in Tasmania 1880-1936; hon. government botanist ib. 1896-1932; lecturer in botany at Univ. Tasmania 1922-1929; director of the herbarium and the botanic garden Hobart 1928-1932. (*Rodway*).

HERBARIUM and TYPES: HO (2500), MEL.– Other material at A, AK, B, CANB, – Library at Royal Society, Tasmania.

BIBLIOGRAPHY and BIOGRAPHY: Barnhart 3: 168 (b. 5 Oct 1853, d. 9 Mar 1936); BJI 2: 147; BL 1: 85, 315; BM 4: 1717; Bossert p. 335; CSP 18: 253; Dawson p. 527; Hortus 3: 1202; IH 1 (ed. 7): 341; Lenley p. 351; LS 22406, suppl. 5237; NI 1661-1662; Rehder 1: 506, 4: 380; SBC p. 130; Stevenson p. 1256; Tucker 1: 603; Urban-Berl. p. 389.

BIOFILE: A.N., Trans. Proc. bot. Soc. Edinb. 32(1): 250. 1936 (obit.).
Anon., Austral. Encycl. 7: 477. 1965; Bull. misc. Inf. Kew 1917(2): 84 (decoration), 1932 (9): 459-560, 1936(4): 287. (obit.); Hedwigia 76: (141). 1936 (d.); Pap. Proc. R. Soc. Tasm. 1936: 94-96. 1937; Who was who 3(1929-1940): 1162; Who's who [Australia] 1933: 268.
A.W.H., Bull. misc. Inf. Kew 1930: 38-41 (report Rodway on Tasmanian Museum herbarium).
Clarke, A.H., Pap. Proc. R. Soc. Tasm. 1928: 163-165. *pl. 30.* 1929 (presented with the first R. Soc. Tasmania medal, tribute, portr.).
Goulding, J.H., Rec. Auckland Inst. Mus. 13: 106. 1976.
Hall, N., Bot. Eucalypts 109-110. 1978.
Moldenke, H.N., Plant Life 2: 79. 1946 (1948) (epon.).
Sayre, G., Bryologist 80: 516. 1977.
Verdoorn, F., ed., Ann. bryol. 9: 153. 1936 (d.); Chron. bot. 3: 58, 60. 1937 (obit., port.).
Wetmore, C.M., Rev. bryol. lichénol. 32: 229. 1963 (coll. lichens Tasm.).

EPONYMY: The source of *Rodwaya* H. Sydow et P. Sydow (1901) was not given.

9368. *The Tasmanian flora* ... with drawings of some typical species ... Tasmania (John Vail, Government Printer, Hobart) 1903. Oct. (*Tasman. fl.*).

Publ.: 1903, p. [i-ii], [1, err.], [iii]-xix, [1]-320, *50 pl. Copies*: MO, NSW, USDA.

9369. Printed for the Tasmanian Field Naturalists' Club. *Ferns of Tasmania* ... Hobart (printed at "The Mercury" Office) [1905]. Oct. (*Ferns Tasmania*).
Publ.: Oct-Dec 1905 (p. [2]: 7 Oct 1905), p. [1]-16. *Copy*: USDA. – Reprint of a series of papers published in "The Tasmanian Mail".

9370. *Trees and shrubs of Tasmanian forests* of the order *Myrtaceae* ... Tasmania (John Vail, Government Printer, Hobart) 1907. (*Trees shrubs Tasman. for., Myrtac.*).
Publ.: 1907 (p. [3]: 19 Jul 1907), p. [1]-30, *20 (22) pl. Copies*: HH, USDA. – Reprinted from "The agricultural Gazette" (orig. publ. n.v.).

9371. *Some wild flowers of Tasmania* ... Hobart (John Vail Government Printer) 1910. Oct. (*Wild fl. Tasmania*).
Publ.: 1910 (preface Feb 1910), p. [i]-viii, [1]-119, *36 pl. Copy*: NSW.
Ed. 2: 1922 (p. [iii]: Apr 1922), p. [i]-ix, [1]-148, *57 pl. Copy*: NSW.

9372. *Tasmanian bryophyta* ... Hobart (The Royal Society of Tasmania) 1914-1916, 2 vols. (*Tasman. bryoph.*).
1 (*Mosses*): Jan-Feb 1914 (Nat. Nov. Apr(1) 1914; publ. in journal in part Apr 1913, see below), p. [i, iii], [1]-163. *Copies*: FH, G, NSW, NY. – Reprinted from the Pap. Proc. R. Soc. Tasm. 1912: 3-24. [reprint also p. 3-24], 87-138 [repr. 25-76], issued Apr 1913; 1913: 177-263 [repr. p. 77-163], issued 14 Feb 1914.
2 (*Hepatices*): 1916, p. [i]-vii, [1]-95. *Copy*: G. – Reprinted from Pap. Proc. R. Soc. Tasm. 1916: 1-95.
Additions: see below, 1916.

9373. *Additions to the bryophyte flora* [preprinted from Pap. Proc. R. Soc. Tasm. 1916]. (*Add. bryoph. fl.*).
Publ.: 11 Jul 1916, 44-47. *Copy*: FH. – Issued separately 11 Jul 1916 as a preprint from Pap. Proc. R. Soc. Tasm. 1916: 44-47. 1917.

9374. *Tasmanian bryophyta* ... [preprinted from Pap. Proc. R. Tasm. Soc. 1916]. Oct. (*Tasman. bryoph.*).
Publ.: 30 Aug 1916, p. 51-143. *Copy*: FH. – Issued separately 30 Aug 1916 as preprint from Pap. Proc. R. Soc. Tasm. 1916: 51-143. 1917.

Roe, John Septimus (1797-1878), English naval officer in the Royal Navy 1813-1827; surveyor general, W. Australia, 1829-1870; explorer and naturalist. (*Roe*).

HERBARIUM and TYPES: Material at G, K and OXF. Manuscripts at the Public Library of Perth (Battye Library).

BIBLIOGRAPHY and BIOGRAPHY: Barnhart 3: 169; BB p. 260; Clokie p. 233; CSP 5: 251; Desmond p. 527 (b. 8 Mai 1797, d. 23 Mai 1878); DNB 49: 88; HR (q.v. for many further biogr. refs.); Lasègue p. 126, 281, 298, 308, 503, 506; SK 8: lxxxi.

BIOFILE: Diels, L., Pfl.-Welt W.-Austr. 54-55, 60, 72. 1906.
Embacher, F., Lexik. Reisen 249. 1882.
Fedtschenko, B.A., Bot. Zhurn. 30: 40. 1945.
Hall, N., Bot. Eucalypts 110. 1978.
Hooker, J.D., Fl. Tasmaniae 1: cxxi-cxxii. 1859.
Lee, I., Early explorers in Australia p. 310, 312, 322, 328, 338, 350- 351, 400, 442, 450, 461, 479, 484, 485. s.d.
Maiden, J.H., J.W. Austral. nat. Hist. Soc. 6: 25-26. 1909 (epon.).
Mercer, F.R., Amazing career; the story of Western Australia's first surveyor general, 1962, 189 p.
Miller, H.S., Taxon 19: 537-538. 1970 (ex Lambert herb. at G).
Norton, J., Proc. Linn. Soc. N.S.W. 25: 777, 778. 1900.

Souster, J.E.S., W. Austral. Naturalist 1(6): 115. 1948 (bot. hist. King George's Sound).
Uren, M., Austral. dict. biogr. 2(1788-1850) (I-Z): 390-392. (extensive biogr.).

EPONYMY: *Roea* Hügel ex Bentham (1837) is likely named for this author but no etymology was gven.

9375. *Report of a journey of discovery into the interior of Western Australia* between 8th September 1848 and 3rd February, 1849 ... [in W.J. Hooker, J. Bot. Kew Gard. Misc. 1854-1855]. Oct.
Publ.: a series of papers in W.J. Hooker, J. Bot. Kew Gard. Misc. as follows: 6: 42-48. Feb 1854, 78-88. Mar 1854, 117-123. Apr 1854, 146-151. Mai 1854, 174-180. Jun 1854, 212-217. Jul 1854, 241-247. Aug 1854, 339-345. Nov 1854, 337-380. Dec 1854, 7: 143-151. Mai 1855.

Roehl, Ernst [Karl Gustav Wilhelm] von (1825-1881), German (Silezian-born Rheinland) palaeobotanist and soldier; in the Prussian army 1843-1869, 1870-1871. (*Roehl*).

HERBARIUM and TYPES: Unknown.

BIBLIOGRAPHY and BIOGRAPHY: ADB 29: 55. 1889 (Gümbel); Andrews p. 321; Barnhart 3: 169 (b. 1 Mai 1825, d. 19 Sep 1881); Biol.-Dokum. 15: 7583; CSP 5: 258, 8: 770; Herder p. 352; NI 1663; PR 7699; Quenstedt p. 370.

BIOFILE: Marck, W. v.d., Verh. nat.-hist. Ver. preuss. Rheinl. Westf. 39: 53-55. 1882 (obit.).
Zittel, K.A. v., Gesch. Geol. Paläont. 512. 1899.

9376. *Fossile Flora der Steinkohlen-Formation Westphalens* einschliesslich Piesberg bei Osnabrueck ... Cassel (Verlag von Theodor Fischer) 1869. Qu. (*Foss. fl. Steinkohlen-Form. Westph.*).
Publ.: Jan-Apr 1869 (p.v.: Sep 1868 in 6 Hefte; Flora 24 Feb 1869 (Lief.); Bot. Zeit. 16 Apr 1869(Lief. 4/5), 7 Mai 1869 (Lief. 6)), p. [i]-v, [1]-191, [192], *pl. 1-32* (uncol. liths.). *Copies*: BR, G, PH, USGS. – Also published as Palaeontographica 18: [1]-191, [192], *pl. 1-32.* 1869.

Röhling, Johann Christoph (1757-1813), German (Hessen) botanist; studied theology at Giessen; private teacher at Frankfurt a.M.; protestant clergyman at Braubach 1792; at Breckenheim 1800, at Massenheim (Hessen) 1802-1813. (*Röhl.*).

HERBARIUM and TYPES: Some musci at STR.

BIBLIOGRAPHY and BIOGRAPHY: ADB 39: 56-57 (F. Otto); Backer p. 660; Barnhart 3: 169 (b. 27 Apr 1757, d. 19 Dec 1813); BM 4: 1717; DTS 1: 246; Frank (Anh.): 85; GR p. 41-42; GR, cat. p. 70; Hawksworth p. 185; Hegi 7: 560; Herder p. 187, 266; Hortus 3: 1202; Jackson p. 293; Kew 4: 502; LS 22410; PR 7700-7701, ed. 1: 8658-8660; Rehder 1: 375, 733; SBC p. 130; Stevenson p. 1256; TL-1/934; TL-2/7090 see W.D.J. Koch and F.K. Mertens; Tucker 1: 604; Zander ed. 10, p. 708, ed. 11, p. 808.

BIOFILE: Sayre, G., Bryologist 80: 516. 1977 (bryoph. at STR).
Seeland, H., Mitt. Herm. Roemer-Mus. Hildesheim 40: 10, 11. 1936.

9377. *Deutschlands Flora* zum bequemen Gebrauche beim Botanisieren. Nebst einer erklärenden Einleitung in die botanische Kunstsprache zum Besten der Anfänger. Ein Taschenbuch ... Bremen (bei Friedrich Wilmans) 1796. Oct. (*Deutschl. Fl.*).
Ed. [1]: 1796, p. [i]-lxiv, [1]-540, [1-20, index]. *Copies*: BM, G, HH(2).
Ed. 2: in three parts, 1812-1813. *Copies*: G, H, HH, M, MO, NY, Stevenson; IDC 5242.
– "*Deutschlands Flora*". Ein botanisches Taschenbuch ... Zweite, ganz umgearbeitete Ausgabe". Frankfurt am Mayn (bei Friedrich Wilmans) 1812-1813. Oct.
Preface material: 1813, p. [i*-iv*]. – General t.p. (1813): "*Deutschlands Flora* oder syste-

matisches Verzeichniss aller in Deutschland entdeckten Gewächsarten; nebst Anleitung zur Kenntniss der äusseren Theile der Pflanzen. Ein Handbuch für Botaniker zum nützlichen Gebrauche beim Unterricht und Selbststudium, auf Exkursionen und in Bibliotheken ... Zweite durchaus umgearbeitete Ausgabe ...". Frankfurt am Mayn (bei Friedrich Wilmans) 1813. Oct.
1 (Kenntniss der äusseren Theile): 1812 (p. xiv: Apr 1811), p. [i]-xxxii, [1]-427, [428, err.], *pl. 1-4.*
2 (Phanerog. Gew.): 1812, p. [i]-xiv, [xv], [1]-586.
3 (Kryptog. Gew.): 21 Nov 1812-Jan 1813 (SY p. 45).
  (1):[i]-x, [1]-210.
  (2) (Flechten): [1]-190.
  (3) (Schwämme): [1]-407.
Anhang: see P.M. Opiz, *Deutschl. crypt. Gew.*, 1817, TL-2/7090.

Ed. 3: 1823-1839, "J.C. Röhlings *Deutschlands Flora*. Nach einem veränderten und erweiterten Plane bearbeitet von Franz Carl Mertens [1764-1831], ... und Wilhelm Daniel Joseph Koch [1771-1849], ... Frankfurt am Main (bei Friedrich Wilmans) 1823-1839, 5 vols. †. *Copies*: BR, G, H, HH, M, MICH, MO, U, USDA, WAG; IDC 5581. – Vols. 1-3 by Mertens and Koch, vols. 4-5 by Koch alone.
1(1): Jan-Mai 1823 (Flora rev. 6 Jul 1823), p. [i]-xxiv, [1-4], [1]-274.
1(2): Jan-Mai 1823 (Flora rev. 14 Jul 1823), p.[i], [275]-891, [892].
2: 1826 (p. iv: Mar-Apr 1826; Linnaea Aug-Oct 1826; Flora 14: Nov 1826), p. [i]-iv, [1]-659, [660].
3: Jun 1831 (Flora 28 Mai 1831 "Druck vollendet", rd. 1 Jun-6 Jul 1831, rev. Mar-Jun 1831), p [i]-viii, [1]-573, [574, err.], [575, impr.].
4: 1833 (Flora rd. 11 Dec 1833) p. [i]-iv, [1]-744, [745, err.], [746, impr.].
5(1): 1839 (Flora rd. 15 Apr 1840), p. [i]-iv, [1]-370.
5(2): not published.
Ref.: Nees von Esenbeck, C.G., Flora 6: 385-396. 7 Jul 1823, 400-413. 14 Jul 1823 (rev. vol. 1).
Anon., Flora 15, Lit.-Ber. 10: 145-160. Mai-Jun 1831 (rev.vol. 3), Lit.-Ber. 9: 131-136. 1840 (rev. vol. 5).

9378. *Moosgeschichte Deutschlands*. Erster Theil. Die Beschreibung aller in Deutschland entdeckten Moosarten enthaltend ... Bremen (bey Friedrich Wilmans) 1800. Oct. (*Moosgesch. Deutschl.*).
Publ.: 1800 (p. xxvi: Aug 1799), p. [ii*], [i]-xli, [1]-436. *Copies*: FH, H, MO-Steere, NY. – Other t.p.: "*Deutschlands Moose*". Nach der neuesten Methode geordnet und beschrieben ...". Bremen (id.) 1800.

Röll, Julius (1846-1928), German (Sachsen-Weimar) bryologist; studied at Jena; Dr. phil. 1873; teacher at Frankfurt a.M. 1874-1876; high school teacher at Darmstadt (Victoriaschule; teachers college) 1876-1911; professor's title 1909; collected in N. America 1888-1889; from 1914-1928 in retirement at Klösterlein bei Aue. (*Röll*).

HERBARIUM and TYPES: WB (fide IH 1 (ed. 7)); other material at B, H-BR (musci), PC, PI, S.

NOTE: For a fundamental criticism of Röll's ideas and practices in the nomenclature of *Sphagnum* see C. Warnstorff, Verh. bot. Ver. Brandenburg 52: 22-38. 1911 (" ... sein oft rücksichtloses Vorgehen gegen andere, die seine Ansichten nicht zu teilen vermögen ..."). See e.g. Röll, *Zur Systematik der Torfmoose*, treated below; and Röll (typescript) Anträge betr. Aendrungen und Zusätze zu den internationalen botanischen Regeln ... in Bezug auf die Nomenclatur der Sphagna (11 p. typescript, n.v.). For a modern evaluation and summary see P. Isoviita (1966): it is proposed to reject the names of Röll's series because they do not apply to species.

BIBLIOGRAPHY and BIOGRAPHY: Barnhart 3: 169 (b. 31 Oct 1846; d. 21 Nov 1928); Biol.-Dokum. 15: 7587-7588; CSP 11: 202, 12: 624, 18: 257-258; DTS 1: 246, 5: lii; Herder p. 266, 433; IH 1 (ed. 7): 341, 2: (files); Kelly p. 193; KR p. 613; Lenley p. 353 (corr. with

E.G. Britton); LS 22415-22416, 38484; Moebius p. 120 (pp. err.); NAF ser. 2. 3: 42; NW p. 56; SBC p. 130; TL-2/see F.F. Meurer; Urban-Berl. 312.

BIOFILE: Anon., Bryologist 32: 68. 1929 (d.); Deut. biogr. Jahrb. 10: 334; Hedwigia 62: (83). 1920 (death of Karl Röll on 6 Sep 1919, Chicago; who enabled his brother Julius to make his trip to N. America 1888/9), 67: (90). 1927 (80th birthday), 68: 351. 1929 (d.; portr.); Österr. bot. z. 78: 96. 1929 (d.); Rev. bryol. ser. 2. 2(1): 62. 1929.
Ewan, J. et al., Short hist. bot. U.S. 96. 1969.
Familler, I., Denkschr. bay. bot. Ges. Regensburg 11: 20. 1911.
Gepp, A., J. Bot. 67: 22. 1929.
Györffy, I., Rev. bryol. ser. 2. 2: 117-125, *pl. 4.* 1929 (obit., bibl., epon.).
Hill, J., Bryologist 13: 105-107. 1910 (on Röll and C.R. Barnes; on his collecting trip to North America 1888/1889).
Isoviita, P., Ann. bot. fenn. 3: 222-225, 260. 1966 (on Röll's taxonomic and nomenclatural opinions; his series of forms "Formenreihen" which he wants to replace species in *Sphagnum*; further literature).
Kneucker, A., Allg. bot. Z. 21: 138. 1916 (Röll offers herb. for sale; details).
Lamy, D., Occ. Pap. Farlow Herb. 16: 126. 1981 (3 letters to F. Renauld at PC).
Sayre, G., Bryologist 80: 516. 1977; Mem. New York Bot. Gard. 19(3): 389. 1975 (coll.).
Schade, A., Naturae lusatica (Bautzen) 5: 31. 1961.
Williams, R.S., Bryologist 39: 39. 1936 (Röll coll. N. Amer. mosses).

EPONYMY: *Roellia* Kindberg (1897). *Note*: *Roella* Linnaeus (1753) and *Roela* Scopoli (1777, *orth. var.*) honor W. Roell, professor of anatomy in Amsterdam in the 18th century).

9379. *Die Thüringer Laubmoose* und ihre geographische Verbreitung ... Frankfurt am Main 1875. Oct.
Publ.: 1875 (Bot. Zeit. 28 Jan 1876), reprint cover (title page), p. 146-299. *Copies*: FH, G, H, MO-Steere. – Reprinted and to be cited from Jahresber. Senckenb. naturf. Ges. 1874-1875: 146-299. 1875.
Continuation: in Deut. bot. Monatschr. as follows:

vol.	pages	dates	vol.	pages	dates
1	81-83	Jun 1883	4	71-74	Mai-Jun 1886
	103-106	Jul 1883		104-107	Jul 1886
	116-118	Aug 1883		134-138	Aug-Sep 1886
	150-152	Oct 1883	5	43-48	Mar 1887
	169bis-171bis	Dec 1883		60-61	Apr 1887
2	73-75	Mai 1884	6	134-138	Aug-Sep 1888
	103-104	Jul 1884	8	155-157	Sep-Oct 1890
	126-127	Aug 1884	9	130-136	Aug-Sep 1891
	147-150	Oct 1884	10	8-14	Jan-Feb 1892
	161-164	Nov 1884			
	189-191	Dec 1884			
3	46-47	Mar 1885			
	57-60	Apr-Mai 1885			
	161-164	Oct-Nov 1885			

Consolidated reprint: s.d., p.[1]-30, [31-68]. *Copy*: G.
Ref.: G.K., Bot. Zeit. 34: 79. 1876 (rev.).

9380. *Beiträge zur Laubmoosflora Deutschlands und der Schweiz* (Separat-Abdruck aus "Flora" 1882 Nr. 11). Oct.
Publ.: 11 Apr 1882 (in journal), p. [1]-14. *Copy*: B. – Reprinted and to be cited from Flora 65: 161-174. 1882.

9381. *Die Torfmoose der thüringischen Flora* ... Sondershausen (Druck von Fr. Aug. Eupel) [1884].

Publ.: 1884, p. [1]-16. *Copy*: H. – Reprinted and to be cited from Abh. Thüring. bot. Ver. Irmischia 4: 1-16. 1884.

9382. *Zur Systematik der Torfmoose* ... Separat-Abdruck aus "Flora", Nr. 32 und 33). Oct.
1 and 2: 11 and 25 Nov 1885 (in journal; repr. Rev. Bryol. Mar 1886), p. [1]-25, *pl. 2*. *Copies*: G, H, U. – Reprinted and to be cited from Flora 68: 569-580. 11 Nov 1885 and 68: 585-595. 21 Nov 1885.
3: 1886, p. [1]-108. *Copies*: G, H. – Id. Flora 69:

33-47. 21 Jan 1886	227-242. 21 Mar 1886
73-80. 11 Feb 1886	328-337. 21 Jul 1886
89-94. 21 Feb 1886	353-370. 11 Aug 1886
105-111. 1 Mar 1886	419-427. 1 Oct 1886
129-137. 21 Mar 1886	467-476. 21 Oct 1886
179-187. 21 Apr 1886	

For a further discussion by Röll of his "Artentypen" und "Formenreihen" in Sphagnum see Bot. Centralbl. 34: 310-314, 338-342, 374-377, 385-389. 1888, and 39: 305-311, 337-344. 1889.
Ref.: Anon., Bot. Gaz. 12: 20-21. Jan 1887 (rev. repr. 3).
Gravet, F., Rev. bryol. 13: 48. 2886, 14: 14-15. 1887 (rev. repr. 1-3).
Warnstorf, C., Bot. Centralbl. 30: 101-102. 1887 (rev.), see also Verh. bot. Ver. Brandenburg 52: 22-38. 1911 for later developments).
Isoviita, P., Ann. bot. fenn. 3: 222-225, 260. 1966 (evaluation).

9383. *"Artentypen" und "Formenreihen" bei den Torfmoosen* [Separat-Abdruck Bot. Centralbl. 36, 1888]. Oct.
Publ.: 1888 (in journal), p. [1]-16. *Copy*: H (Inf. P. Isoviita). – Reprinted and to be cited from Bot. Centralbl. 34: 310-314. 1-15 Nov, 338-342, 16-30 Nov, 374-377, 1-15 Dec, 385-389, 16-31 Dec 1888.

9384. *Unsere essbaren Pilze* in natürlicher Grösse dargestellt und beschrieben mit Angabe ihrer Zubereitung ... Tübingen (Verlag der H. Laupp'sche Buchhandlung) 1889. (*Uns. essb. Pilz.*).
[*Ed. 1*: Die 24 häufigsten essbaren Pilze, 1883].
Ed. 2: 1889 (Bot. Zeit. 27 Sep 1889; Bot. Centralbl. 4 Sep 1889; Nat. Nov. Aug(2) 1889), p. [i]-vi, 1-46, *pl. 1-14*.
Ed. 3: 1891 (Bot. Centralbl. 3 Jun 1891; Bot. Zeit. 26 Jun 1891; Nat. Nov. Jun(1) 1891), p. [i]-viii, 1-48, *pl. 1-14*.
Ed. 4: 1891, p. [i]-viii, [1]-47, [48], *pl. 1-14*. *Copy*: USDA ... Tübingen (Verlag der H. Laupp'schen Buchhandlung) s.d.
Ed. 5: 1895 (p. iv: Apr 1895; Bot. Zeit. 1 Sep 1895; Nat. Nov. Sep(2) 1898; Hedwigia 19 Oct 1895), p. [i]-x, 1-38, *pl. 1-15*.
Ed. 6: Mai 1903 (p. v: Mar 1903; Bot. Monatschr. Mai-Jun 1903; Bot. Zeit. 1 Jun 1903; Bot. Centralbl. 7 Jul 1903; Hedwigia 7 Oct 1903; Nat. Nov. Jun(1) 1903), p. [i]-viii, 1-46, *pl. 1-14*. *Copy*: NY. – Tübingen (id.) 1903. Oct.
Ed. 7: 1908 (Bot. Zeit. 16 Oct 1908; ÖbZ Aug-Sep 1908; Nat. Nov. Aug 1908), p. 1-44, *pl. 1-14*.
Ed. 8: 1918 (Nat. Nov. Aug(1, 2) 1918).

9385. *Vörläufige Mittheilungen über die* von mir im Jahre 1888 *in Nord-Amerika gesammelten* neuen Arten und Varietäten der *Laubmoose*. [Separat-Abdruck Bot. Centralbl. 44, 1890]. Oct.
Publ.: 17 & 24 Dec 1890 (in journal), p. [1]-13, [15]. *Copy*: H (inf. P. Isoviita). – Reprinted and to be cited from Bot. Centralbl. 44: 385-391. 17 Dec, 417-424. 24 Dec 1890.

9386. *Vorläufige Mittheilung über die* von mir im Jahre 1888 *in Nord-Amerika gesammelten* neuen Arten der *Lebermoose* [Separat-Abdruck, Bot. Centralbl. 45, 1891].

Publ.: 18 Feb 1891 (in journal), p. [1]-2. *Copy*: H. (Inf. P. Isoviita). – Reprinted and to be cited from Bot. Centralbl. 45: 203-204. 1891.

9387. *Vorläufige Mittheilungen über die* von mir im Jahre 1888 *in Nord-Amerika gesammelten neuen Varietäten und Formen der Torfmoose* [Separat-Abdruck, Bot. Centralbl. 46, 1891]. Oct.
Publ.: 25 Feb-17 Jun 1891 (in journal), p. [1]-20. *Copy*: H. – Inf. P. Isoviita). – Reprinted and to be cited from Bot. Centralbl. 46: 250-257. 25 Feb 1891, 311-315. 11 Mar 1891, 373-376. 10 Jun 1891, 405-411. 17 Jun 1891.

9388. *Nordamerikanische Laubmoose, Torfmoose und Lebermoose* gesammelt von Dr. Julius Röll in Darmstadt ... [Dresden (Druck von C. Heinrich) 1893].
Publ.: 1893 (see below), p. [181]-321, *pl. 9-10. Copy*: G. – Reprinted and to be cited from Hedwigia 32: 181-203. Jul-Aug, 260-309. Sep-Oct, 334-402. Nov-Dec 1893.
Preliminary publ.: Laubmoose: Bot. Centralbl. 44: 389-391, 417-424. 1890; Lebermoose, ib. 45: 203-204. 1891 (reprint p. [1]-2. *Copy*: G); Torfmoose: ib. 46: 250-257, 311-315, 373-376, 405-411. 1891 (see above).
Nachträge: Hedwigia 35: 58-78. 1896, 36: 41-66. 1897.
Übersicht: 1897, p. [183]-216, *in* Abh. Naturw. Ver. Bremen 14(2): 183-216. Jan 1897 (reprint with special cover and original pagination at H).

9389. *Beiträge zur Laubmoos- und Torfmoosflora von Oesterreich* [Verh. k.k. zool.-bot. Ver. 1897]. Oct.
Publ.: 1898 (Rev. bryol. Jun 1898), p. [1]-13. *Copy*: G. – Reprinted and to be cited from Verh. k.k. zool.-bot. Ver., Wien 47: 659-671. 1898.

9390. *Über die neuesten Torfmoosforschungen* [Separatabdruck aus der "Osterreichischen botanischen Zeitschrift", Jahrg. 1907, Nr. 3 u.f.]. Oct.
Publ.: Mar-Apr 1907 (in journal), p. [1]-14. *Copy*: H (inf. P. Isoviita). – Reprinted and to be cited from Österr. bot. Z. 57: 96-106. Mar, 142-146. Apr 1907. – On G. Roth, *Die europäischen Torfmoose*, 1906.

9391. *Anträge*, betr. Aenderungen und Zusätze zu den internationalen botanischen Regeln von Wien in Bezug auf die *Nomenclatur der Sphagna*. Von Dr. Röll in Darmstadt [1909]. (*Antr. Nomencl. Sphagna*).
Publ.: 1909 (AbZ 20 Jul 1909), p. [1]-10, [1-2, French summary]. *Copy*: FAS. –Mimeographed document submitted to the third International Botanical Congress, Bruxelles 1910. – "Species of *Sphagnum* based on a single form or herbarium specimen, have no value for sphagnology" and several other, taxonomic or methodological rather than nomenclatural proposals.

9392. *Die Thüringer Torfmoose und Laubmoose* und Ihre geographische Verbreitung von Julius Röll. [s.l.n.d., 2 parts, 1914, 1915]. Oct.
1 (Allgemeiner Teil): 1914 (in journal), p. [iii]-xii, [1]-263; originally published and to be cited from Mitt. thür. bot. Ver. ser. 2. 32: 1-263. 1914, preface [ii]-xii and map added to combined reprint of parts 1 and 2 Mar-Jun 1915.
2 (Systematischer Teil): 15 Feb 1915, p. [i-iii], 1-287, map. Issued also as Hedwigia 46: 1-176. 12 Feb 1915, 177-287, map. 18 Feb 1915. Reprint issued later with Allgemeiner Teil.
Copies: BR, FH, MO-Steere, NY.
Ref.: Andrews, A. Leroy, Bryologist 18: 79-80. 1915 (rev.).
Holzinger, J.M., Bryologist 19: 76-78. 1916.
Loeske, L., Bot. Centralbl. 128: 679-680. 15 Jan 1915 (rev.).

9393. *Die Torfmoose und Laubmoose der Umgebung von Erfurt* ... Erfurt (Verlag von Carl Villaret ...) 1915. Oct.
Publ.: 1915, p. [1]-157, [1, err.]. *Copies*: BR, FH. – Reprinted with special cover (with above t.p. text) and to be cited from Jahrb. k. Akad. Wiss. Erfurt ser. 2. 41: 1-157. 1915.
Ref.: Loeske, L., Hedwigia 58(114). 14 Oct 1916 (rev.).

Roemer, [Friedrich] Adolph (1809-1869), German (Hildesheim, Niedersachsen, Hannoverian) palaeobotanist and algologist; studied law at Goettingen and Berlin 1828-1831; law official ("juristischer Beamter") at Hildesheim 1831-1840; id. Bovenden nr. Göttingen 1840-1843; mining official at Clausthal 1843; teacher for geognosy and mineralogy at the mining school Clausthal 1846-1867; brother of [Carl] Ferdinand Roemer. (*A. Roem.*).

HERBARIUM and TYPES: LZ (destroyed); some algae at GOET. – Palaeobotanical collections at Hildesheim (fide Quenstedt).

BIBLIOGRAPHY and BIOGRAPHY: ADB 120-122 (Gümbel); Andrews p. 321; Barnhart 3: 169 (err. d. "1871"); Biol.-Dokum. 15: 7591; BM 4: 1719; CSP 5: 266-267, 8: 775; Frank 3 (Anh.): 85; Herder p. 254; Kew 4: 502; PR 7703-7704, ed. 1: 8663-8664; Quenstedt p. 364; TL-1/1093.

BIOFILE: Böckh, E. et al., Geol. Jahrb., Hannover 76: xxi-xxviii. 1959 (biogr. sketch, portr., primary and secondary bibl.).
Dechen, Verh. naturh. preuss. Rheinl. 27: 23. 1870 (d.).
Kayser, E., Abh. geol. Specialkarte Preussen 2(4): vi-xiv. 1878 (fide E. Böckh et al. 1959).
Koster, J.Th., Taxon 18: 557. 1968 (algae "Römer" at NY).
Nissen, C., Zool. Buchill. 345-346. 1969.
Poggendorff, J.C., Biogr.-lit. Handw.-Buch 2: 674. 1863, 3: 1131. 1898, 7a: (suppl.): 556. 1971 (b. 14 Apr 1809, d. 25 Nov 1869).
Roemer, F., Z. deut. geol. Ges. 22: 96-102. 1870 (obit.; b. 14 Apr 1809, d. 25 Nov 1869).
Seeland, H., Mitt. Herm. Roemer-Mus. Hildesheim 40: 12, 13, 14, 16-18, 20, 22, 29, 30. 1936.
Zittel, K.A. von, Gesch. Geol. Palaeontol. 496-497, 862. 1899; Hist. geol. palaeontol. 387, 390, 448, 502, 519, 520. 1901.

EPONYMY: (journal): *Roemeriana*, eine Schriftenreihe des Geologischen Instituts der Bergakademie Clausthal, Clausthal-Zellerfeld, Germany. vol. 1-8, 1954-1964.

9394. *Die Versteinerungen des norddeutschen Kreidegebirges* ... Hannover (Im Verlag der Hahn'schen Hofbuchhandlung) 1840-1841. Qu. (*Verstein. norddeut. Kreidegeb.*).
1: 1840, p. [i], [1]-48, *pl. 1-7* (uncol. liths. auct.).
2: 1841, p. [i]-iv [dated 14 Apr 1871], 49-145, *pl. 8-20* (id.). *Copies*: PH; Univ. Georgia (Athens).

9395. *Die Versteinerungen des Harzgebirges* ... Hannover (Im Verlage der Hahn'schen Hofbuchhandlung) 1843. Qu. (*Verstein. Harzgeb.*).
Publ.: 1843 (p. [vii*]: 14 Jan 1843), p. [i*-vii*], [i]-xx, [1]-40, *pl. 1-12* (uncol. liths.). *Copies*: GOET (Inf. G. Wagenitz), PH, Univ. Georgia (Athens).

9396. *Die Algen Deutschlands* ... Hannover (Im Verlage der Hahn'schen Hofbuchhandlung) 1845. Qu. (*Alg. Deutschl.*).
Publ.: Apr-Aug 1845 (p. 10: 14 Apr 1895; Hinrichs Sep 1845), p. [i-iv], [1]-72, *pl. 1-11* (uncol. liths). *Copies*: FH(2), G, M, NY.
Ref.: Schlechtendal, D.F.L. von, Bot. Zeit. 3: 869-871. 1845 (rev.).

Römer, Carl (1815-1881), German (Eupen; in 1815 Rheinpreussen) botanist; student of A.L.S. Lejeune; industrial administrator at Namiest (Moravia) ca. 1850-1870; later in Eupen and Quedlinburg. (*C. Röm.*).

HERBARIUM and TYPES: TUB (bryoph. herb.); other material at B (extant), BRNU, GH, L and W.

BIBLIOGRAPHY and BIOGRAPHY: Barnhart 3: 169 (d. 28 Jan 1881); CSP 5: 263, 8: 773, 11: 203; Futak-Domin p. 506; Klášterský p. 158; SK 1: 444; Urban-Berl. p. 272, 312.

BIOFILE: Anon., Bot. Centralbl. 5: 224. 1881 (d.); Bot. Not. 1882: 30 (d.); Österr. bot. Z. 31: 102. 1881 (d.).
Iltis, H., Verh. naturf. Ver. Brünn 50: 311. 1912.
Makowsky, A., Verh. naturf. Ver. Brünn 20: 25-26. 1882 (obit.; died in Halle following an operation).
Oborny, A., Verh. naturf. Ver. Brünn 21(2): 17, 24. 1882.
Römer, C., Österr. bot. Wochenbl. 5: 233-236, 241-243, 249-251, 259-261, 268-269. 1855, 6: 354-355. 1856 (Beitr. Fl. Namiest); Verh. k.k. zool. bot. Ges., Wien 16: 935-942. 1866 (1867) (Laubmoosfl. Namiest; is leaving Namiest).

Roemer, [Carl] **Ferdinand** [von] (1818-1891), German (Hildesheim, Niedersachsen, Hannoverian) geologist and explorer; studied law at Göttingen 1836-1839, natural sciences at Berlin 1840-1842; Dr. phil. Berlin 1842; travelled in Germany 1842-1845, in the United States, especially Texas, 1845-1847; habil. Bonn 1848, "Privatdocent" ib. 1848-1855; professor of mineralogy, geology and paleobotany and director of the mineralogical cabinet (later: museum) at the University of Breslau, 1855-1891; brother of Friedrich Adolph Roemer. (*F. Roem.*).

HERBARIUM and TYPES: Material in the "palaeontological cabinet of Berlin University" (see TL-2/3639). Some marine algae at FR (Conert 1967). Roemer's collections from Texas were studied by G.H.A. Scheele (1808-1864) q.v. – Letters to F.A.W. Miquel in U; Smithsonian Archives SIA 7230; letter to F. Rolle at FR.

BIBLIOGRAPHY and BIOGRAPHY: ADB 53: 451-458 (Hintze, C.); Andrews p. 321; Barnhart 3: 170 (b. 5 Jul 1818 [err.], 14 Dec 1891); Biol.-Dokum. 15: 7591-7592; BM 4: 1718; CSP 5: 264-266, 8: 773-775, 11: 203, 18: 258-259; DAB 16: 91-92; Herder p. 58; Jackson p. 178; Kew 4: 502; ME 3: 646 (index); Merrill p. 487; PH 98, 567, 654; Quenstedt p. 362; Rehder 1: 322; TL-2/806, 3639; Tucker 1: 604.

BIOFILE: Althans, E. et al., Naturw. Wochenschr. 6: 531. 1891 (d.), 7: 49. 1892 (appeal for memorial).
Andrée, K., Z. deut. geol. Ges. 100: 6, 10. 1950 (portr.).
Anon., Amer. J. Sci. ser. 3. 43: 168. 1892 (obit. portr.); Bot. Centralbl. 49: 224. 1892 (d.); Bot. Jahrb. 15 (Beibl. 33): 9. 1892 (d.); Brockhaus' Conversations-Lexikon ed. 13. 13: 804. 1886; Jahres-Ber. schles. Ges. vaterl. Cultur 69: 23-27. 1892 (obit.); Mem. Proc. Manchester Lit. Philos. Soc. ser. 4. 5: 190-191. 1892 (obit.); S.B. Abh. naturw. Ges. Isis, Dresden 1891(2): 33. 1892 (notice).
Boone, W., Hist. bot. W. Virg. 182. 1965.
Breure, A.S.H., & J.G. de Bruijn, eds., Leven werken J.G.S. v. Breda 21, 302, 313, 380, 407, 417. 1979.
Conert, H.J., Senckenbergiana Biol. 48(C): 44. 1967.
Clarke, J.M., James Hall of Albany 172, 445, 472. 1923.
Dames, W., Neu. Jahrb. Mineral. Geol. Palaeont. 1892 (1, Min.): 1-32. 1892 (extensive obituary, bibl.).
Geikie, A., Quart. J. geol. Soc. London 48 (Proc.): 58-60. 1892 (obit.).
Geiser, S.W., Southwest Rev. 17: 421-460. 1932 (portr.); Natural. frontier 17, 19, 109, 116, 137-140, 148-171, 281. 1948 (on his Texas trip).
Gürich, G., *in* Schlesische Lebensbilder 277-279. 1922 (biogr. sketch).
Hintze, C., Chronik Univ. Breslau 6: 106-111. 1892 (obit., b. 5 Jan 1818, d. 14 Dec 1881, "wen die Götter lieben, den nehmen sie mit dem Blitze zu sich", wish fulfilled), also in Jahresb. Schles. Ges. vaterl. Cult. 69: 23-27. 1891.
Hitchcock, A.S. & A. Chase, Man. Grass. U.S. 988. 1951 (epon.).
J.W.G., Geol. Mag. ser. 2. dec. 3. 9: 92-94. 1892 (obit.).
Kumm, P., Schr. naturf. Ges. Danzig, ser. 2. 8(1): 116-145. 1892 ("Leben und Wirken" detailed commemorative address; Roemer was refused a diploma at Göttingen because his brother had refused to acknowledge, under oath, the sovereignty of the new king Ernst August of Hannover).
Lindemann, E. v., Bull. Soc. Natural. Moscou 61(1): 62-63. 1886 (mentions a Roemer herbarium specimen).
Martin, G.P.R., Ber. naturh. Ges. Hannover 105: 11-14. 1961 (letter to Friedrich Rolle, 1827-1887, with facsimile).

Moldenke, H.N., Plant Life 2: 79. 1946 (1948) (epon.).
Nissen, C., Zool. Buchill. 345. 1969.
Pettenkofer, M. v., S.B. mat.-phys. Cl. Akad. Wiss. München 22: 201-202. 1893 (obit.).
Poggendorff, J.C., Biogr.-lit. Handw.-Buch 2: 674-675. 1863, 3: 1131-1132. 1898 (bibl.), 7a(suppl.): 555. 1971.
Rehm, H., Ber. Senckenb. naturf. Ges. 1892: xxiv (obit.).
Seeland, H., Mitt. Herm. Roemer-Mus. Hildesheim 40: 9, 13, 20, 23. 1936.
Simonds, F.W., Amer. Geol. 29(3): 131-140. 1902 (biogr. skech; work on Texas; portr., bibl. of Texas items; "father of the geology of Texas", reprinted Geol. Mag. ser. 2. dec. 4. 9: 412-417. 1902.
Sprengell, Jahresh. naturw. Ver. Fürstent. Lüneburg 12: 40-42. 1893 (obit.).
Struckmann, C., Leopoldina 28: 31-32, 43-46, 63-67. 1892 (obit., partial bibl.).
Taton, R., ed., Science in the 19th century 330. 1965.
Tornier, G., S.B. Ges. naturf. Fr. 1925: 86. 1927.
Trautschold, H., Bull. Soc. Natural. Moscou ser. 2. 6: 132-135. 1893 (obit).
Wilson, C.W., Jr. & J.G. Frank, J. Tennessee Acad. Sci. 19(3): 270-276. 1944, 20(2): 224-227. 1945 (on R's visit to Tennessee in 1847).
Winkler, C.H., Bull. Univ. Texas 18: 8-9. 1915.
Wolkenhauer, W., Geogr. Jahrb. 16: 493. 1893 (b. 5 Jan 1818, d. 14 Dec 1891).
Zittel, K.A. von, Gesch. Geol. Paläont. 862 [index]. 1899; Hist. geol. palaeontol. 147, 364, 393, 394, 395, 408, 448, 449, 453, 463. 1901.

COMPOSITE WORKS: See TL-2/806 for *Lethaea geogn.* (by H.G. Bronn), ed. 3, 1851-1856, F. Roemer co-author. This work was continued by Fritz Frech as *Lethaea geogn.*, 3 Theile, 1876-1908, see e.g. BM 4: 1718, Nissen Zool. Buchill. 345. 1969 (no. 3454).

HANDWRITING: Ber. naturh. Ges. Hannover 105: 12. 1961.

EPONYMY: *Roemeria* F.J.A. N. Unger (1852).

9397. *Texas*. Mit besonderer Rücksicht auf Deutsche Auswanderung und die physischen Verhältnisse des Landes nach eigener Beobachtung geschildert ... Bonn (bei Adolph Marcus) 1849. Oct. (*Texas*).
Publ.: 1849 (preface Aug 1849), p. [i]-xiv, [xv, err.], [1]-464, map. *Copies*: GOET (inf. G. Wagenitz), HH, ILL (inf. O.P. Rogers), MO, PH. – "Verzeichniss der von Dr. Ferdinand Roemer aus Texas mitgebrachten Pflanzen, aufgestellt von Adolph Scheele" on p. 425-449.
English: 1935, p. [i*], [i]-xii, [1]-301. *Copies*: HH, USDA. – *Texas*, with particular reference to German immigration and the physical appearance of the country. Described through personal observation by Dr. Ferdinand Roemer. Translated from the German by Oswald Mueller. San Antonio, Texas (Standard Printing Company) 1935.
Ref.: Schlechtendal, D.F.L. von, Bot. Zeit. 8: 921-926. 27 Dec 1850 (rev.).

9398. *Die Kreidebildungen von Texas* und ihre organischen Einschlüsse ... Mit einem die Beschreibung von Versteinerungen aus paläozoischen und tertiären Schichten enthaltenden Anhange und mit 11 von C. Hohe nach der Natur auf Stein gezeichneten Tafeln. Bonn (bei Adolph Marcus) 1852. Qu. (*Kreidebild. Texas*).
Publ.: 1852 (p. vi: Aug 1852), p. [i]-vi, [1, cont.], [1]-100, *pl. 1-11* (uncol. liths. C. Hohe). *Copy*: PH.

Römer, Heinrich (*fl.* 1843), Swiss botanical artist at Zürich. (*Heinr. Röm.*).

HERBARIUM and TYPES: Unknown. – Letters to D.F.L. von Schlechtendal at HAL.

NOTE: Not to be confused with Hermann Roemer (1816-1891), Niedersachsen (Hannoverian) lawyer, politician, geologist and plant collector, brother of [Friedrich] Adolph and [Carl] Ferdinand [von] Roemer.

BIBLIOGRAPHY and BIOGRAPHY: Rehder 1: 434.

9399. *Flora der Schweiz* oder Abbildungen sämmtlicher Schweizer-Pflanzen zu den Werken von Gaudin, Monnard und Hegetschweiler. Herausgegeben von Heinr. Römer, Maler und Botaniker Zürich (beim Herausgeber in der Enge am Schanzengraben) 1843-1844. (*Fl. Schweiz*).
Publ.: 1843-1844 (later Hefte possibly after 1844), 14 "Hefte" with *8 pl.* each, in all *112 pl.* (handcol. liths.), preceded by four pages introduction. Final t.p. (undated) with imprint Zürich (Schweizerisches Antiquariat). *Copies*: G(2), MO. − References on the plates are to J.F.A.P. Gaudin, *Fl. helv.*, 1828-1833, TL-2/1971, to J. Hegetschweiler-Bodmer, *Fl. Schweiz*, 1840, TL-2/2566, to Jean Piere Monnard, editor of Gaudin, *Syn. fl. Helv.*, 1836, TL-2/1972.

Roemer, Johann Jakob (1763-1819), Swiss (Zürich) botanist; studied medicine and natural history at the University of Zürich 1783-1784, and Göttingen 1784-1786; Dr. phil. 1784; practicing physician at Zürich 1786-1797 and 1804-1819; lectured also at the Zürich medical college; director of the botanical garden of the Naturforschende Gesellschaft Zürich 1797-1819. (*Roem.*).

HERBARIUM and TYPES: In part at Z, another part at BM (via Shuttleworth, see Hoppe 1835); some further material in H-ACH, OXF, PH, UPS (Thunberg). − The large Roemer herbarium at LZ (destroyed) was made by Rudolph Benno von Roemer (x-1871) at Dresden. There are no indications that any J.J. Roemer material was ever at LZ. R.B. von Roemer left his library and herbarium to LZ, see Schenk, Bot. Zeit. 30: 271-272. 1872.

BIBLIOGRAPHY and BIOGRAPHY: ADB 29: 123-124; AG 5(1): 349, 350; Backer p. 497 (epon.); Barnhart 3: 169 (b. 8 Jan 1763, d. 15 Jan 1819); BFM no. 2467; BM 4: 1719-1720; Bossert p. 335; Bret. p. 277; Christiansen, p. 319 [index]; Clokie 233-234; CSP 5: 267, 12: 625, 18: 259; Dawson p. 708-709; Dryander 3: 93, 646; Frank 3 (Anh.): 85; Futak-Domin p. 506; GF p. 74; Hegi 4(1): 19; Herder p. 22, 170; Hortus 3: 1202; HU 745; Jackson p. 5, 115, 226, 447, 478; Kew 4: 502; Langman p. 638; Lasègue p. 347 (herb. with Shuttleworth); LS 22417; MW p. 415-416; NI 1664; PR 1472, 5430, 7706-7715, 84741, ed. 1: 2228, 6009, 8665-8674, 9718; Rehder 5: 733; RS p. 137; SK 4: ccix-ccx; SO 613, 1229, 2466; TL-1/1094-1098, 1241, 1452; TL-2/992, 1442, 1583, and see indexes; Tucker 1: 604; Zander ed. 10, p. 708, ed. 11, p. 807-808.

BIOFILE: Anon., Flora 2: 192. 1819 (Syst. veg. to be continued after Roemer's death with the collaboration of Panzer), 4: 256. 1821 (sale libr.), 5: 46-47. 1827 (sale herb.).
Bridson, G.D.R. et al., Nat. hist. mss. res. Brit. Isl. 229.411. 1980.
Candolle, Alph. de, Phytographie 444. 1880 (coll.; the reference to Leipzig is to the herbarium of baron R.D. von Roemer, of Dresden); Mém. souvenirs 216, 294-295. 1862.
Forbes, R.F., Martinus van Marum 1: 394. 1969, 2: 327-328. 1970.
Gilbert, P., Comp. biogr. lit. deceased entom. 321. 1977.
Hiern, W., J. Bot. 38: 494. 1900 (on *Collectanea*).
Hocquette, M., Bull. Soc. bot. Nord France 19(1): 54-67. 1966 (seven letters by J.J. Roemer to A. Du Petit-Thouars).
Högrell, B., Bot. hist. 1: 223. 1886.
Hoppe, D.H., Flora 2(1): 123-125. 1819 (obit.), 18(2): 573-574. 1835 (R.J. Shuttleworth has bought the Scheuchzer and J.J. Roemer herbaria).
Jessen, K.F.W., Bot. Gegenw. Vorz. 405. 1884.
McVaugh, R.M., Taxon 4(4): 78-86. 1955 (Amer. coll. Humb. & Bonpl. descr. in Roemer & Schultes, Syst. veg.).
Murray, G., *in* Hist. coll. BM(NH) 1: 177. 1904 (herb. Roemer, with Roemer and Schultes types, acquired with Herb. Shuttleworth).
Musgrave, A., Bibl. austral. entom. 1775-1930, 271. 1932.
Nissen, C., Zool. Buchill. 346. 1969.
O., Dict. Sci. méd., Biogr. med. 7: 37-38. 1825 (bibl.).
Pichi-Sermolli, R.E.G., Webbia 8: 439. 1952 (Arch. Bot., date 2(1)), 12: 10. 1955 (id. 2(3)).
Reynolds, G.W., Aloes S. Afr. 94. 1950 (notes on Aloe in Syst. veg. vol. 7).

Römer, J.J., Arch. Bot. 1(3): 172-173. 1798 (letter F.W. Schmidt to Roemer).
Rudio, F., Vierteljahrsschr. naturf. Ges. Zürich 41(1): 209. 1896.
Rübel, E., Gesch. naturf. Ges. Zürich 27, 30. 1947.
Schinz, H.R., Naturw. Anzeiger 2(12): 89-94. 1819 (obit., bibl.).
Steenis-Kruseman, M.J. van, Fl. males. Bull. 12: 490. 1956, 14: 645. 1959 (on Collectanea vol. 1); *in* Forbes, R.J. et al., Mart. van Marum 3: 149. 1971.
Stieber, M.T., Huntia 3(2): 124. 1979 (ref. to notes by A. Chase in her mss. at HU).
Woodward, B.B., *in* Hist. Coll. BM(NH) 1: 46. 1904 (210 original drawings at BM).

COMPOSITE WORKS: (1) *Flora britannica*, auctore Jacobo Edvardo Smith, M.D. ... recudi curavit additis passim adnotatiunculis J.J. Römer, M.D. vol. I [-III]. Turici [Zürich] (typis Henrici Gessneri) 1804 [-1805], 3 vols., see under J.E. Smith (*Fl. brit.*).
(2) J.J. Roemer and P. Usteri published an octavo edition of J. Dickson, *Fasc. pl. Crypt. brit.*, 2 fasc., 1785-1790 in Zürich with a London imprint, see TL-2/1442: Jacobi Dickson *Fasciculus plantarum cryptogamicarum Britanniae* Londini 1785 [and 1790] (prostant venales apud auctorem, in Foro Covent-Garden; & G. Nicol, ...), see p. [ii]. *Copy*: MO.

NOTE: Roemer wrote his name on the title pages of his works as "Römer" when the title was in German, and as "Roemer" when the title was in Latin. In view of the widespread use of "Roemer et Schultes" we keep the transliteration Roemer.

EPONYMY: *Roemera* Trattinick (1802); *Roemeria* Medikus (1792); *Roemeria* Moench (1794); *Roemeria* (Thunberg (1798); *Roemeria* Zea ex J.J. Roemer et J.A. Schultes (1817); *Roemeria* Raddi (1818).

HANDWRITING: Candollea 33: 145-146. 1978.

9400. *Magazin für die Botanik*. Herausgegeben von Joh. Jacob Römer und Paulus Usteri ... Erstes [-Zwölftes] Stück. Zürich ([1-5:]bey Johann Caspar Füessly, [6-12:] bey Ziegler und Söhne) 1787-1791. Oct. (*Mag. Bot.*).
Co-editor: Paulus Usteri (1768-1831).
Publ.: 4 volumes of 3 "Stücke" each. For dates see e.g. ST p. 72-74. *Copies*: BR, FI, G, M, NY, U, USDA; IDC 5685. – The volume title-pages have a different title: "Botanisches Magazin. Band i. No. i.-iii. 1787-1788" etc.

Band	Stück	Pagination	Plates	Dates
1	1	[1]-167, [168]	*1-2*	1 Mar-10 Oct 1787
	2	[1]-164	*1-3*	1 Jan-27 Mar 1788
	3	[1]-158, vol. t.p. [i]		Jan-Apr 1788
2	4	[1]-189, [190-191, err.]	*1-4, 1bis* (col.)	Aug-Sep 1788
	5	[1]-184, 23a-23b		Apr 1789
	6	[1]-191, vol. t.p. [i]	*1-2* (col), *3*	Sep 1789
3	7	[1]-178	*1-4* (col.)	before 10 Jan 1790
	8	[1]-184		1 Feb-17 Apr 1790
	9	[1]-147, vol. t.p. [i]	*pl. 1, [2]*	1-24 Apr 1790
4	10	[1]-200		1 Jul-6 Sep 1790
	11	[1]-192	*1-2* (col.)	Oct-Dec 1790
	12	[1]-205, vol. t.p. [i]	*2 pl.* (col.)	1 Jan-10 Mar 1791

In 1790 the editors each went their own way; J.J. Roemer continued the series as *Neues Magazin für die Botanik* in ihrem ganzem Umfange, Erster Band 1794 (not finished, PR 7707), followed by his *Archiv für die Botanik* (see below). – Usteri started his *Annalen der Botanik* (see sub Usteri). – H. Manitz kindly sent us the text of a paper in press for 1982 giving more precise dates for parts 1, 2, 7, 8, 10 and 12, mainly based on the correspondence between F. Ehrhart and P. Usteri published by Alpers (1905) (n.v.).
Ref.: Manitz, H., Wiss. Z. Friedr. Schiller Univ. Jena, Math.-nat. R. 31(2): 264. 1982.

9401. *Neues Magazin für die Botanik* in ihrem ganzen Umfange. Herausgegeben von J.J. Römer M.D. Erster Band. Zürich 1794. Oct. (*Neu. Mag. Bot.*).

Publ.: Apr-Aug 1794 (p. 329: 21 Mar 1794; publ. before Michaelismesse (see p. 159, 289), p. [i]-viii, 1-336, *pl. 1-4* (uncol. copp.). *Copies*: G, M, NY.

9402. *Archiv für die Botanik.* Herausgegeben von D. Johann Jacob Römer. Erster [-Dritter] Band ... Leipzig (in der Schäferschen Buchhandlung) 1796-1805, 3 vols. (9 "Stücke"). Oct. (*Arch. Bot.*).
Publ.: 1796-1805, three volumes of three "Stücke" each. *Copies*: BR, FI, G, M, NY, USDA; IDC 5681. − For date of 2(1) see Pichi-Sermolli (1952).

Vol.	Stück	pages	plates	dates
1	1	[i]-viii, [ix-x], [1]-134	*1-6, 7* (col.)	1796
	2	[i-viii], [1]-122, table	*3 pl.* (col.),	Mai-Dec 1797
	3	[i]-viii, [1]-186, [187-212]	*1-3* (col.), *4-7*	Jul-Dec 1798
	t.p.	[i*]		Jul-Dec 1798
2	1	[i-viii], [1]-131	*1-3*	(t.p. 1799) Mai-Jul 1800
	2	[i-vi], [133]-318	*1-7, 4, 5*	Mar-Dec 1801
	3	[i]-viii, [319]-490, [491-514]	*1*	Mar-Dec 1801
	t.p.	n.v.		
3	1/2	[i]-vi, [1]-310	*1-3, [4-5]* (p.p. col.)	1803
	3	[i-iv], [311]-464, [465-488]	*1-4*	Apr-Mai 1805
	t.p.	[i*]		Apr-Mai 1805

9403. *Scriptores de plantis hispanicis, lusitanicis, brasiliensibus,* adornavit et recudi curavit I.I. Römer M.D. ... Norimbergae [Nürnberg] (in Officina Raspeana) 1796. Oct. (*Script. pl. hispan.*).
Publ.: 1796 (p. 8: Feb 1796), engr. t.p. and p. [1]-184, *pl. 1-2, 2b, 3-7* (uncol. copp.). *Copies*: FI, G, HH, KU, MO, NY (incompl.), USDA; IDC 5155. −Roemer was editor; contains reprints of Asso, I. de, *Enum. stirp. Arragonia* 1784, D. Vandelli, *Diss. arb. Drac.* 1768, and *Fasc. pl. nov. gen. sp.* 1771 and C. Linnaeus, *Epist. ... ad Dom. Vandelli scriptae, in* Vandelli, *Fl. Lus. Bras. spec.* 1788 (on p. 165-183) (see SO 2465-2466).

9404. *Flora europaea* inchoata a Joh. Jac. Roemer ... Fasciculus i.[-xiv]. Norimbergae [Nürnberg] (ex Officina Raspeana) 1797-1811. Oct. (*Fl. eur.*).
Publ.: 1797-1811, 14 fasc. with in all 112 coloured plates (by Louise Roemer) with descriptive text on (partly) unnumbered pages. *Copies*: FI, G, HH, NY, PH, USDA; IDC 5682. − See also GF p. 74, NI 1664.
1: 1797, p. [1]-30, [1, index], *8 pl.*
2: 1797, p. [33]-62, [1, index], *8 pl.*
3: 1798, p. [i-iii], *8 pl.* w.t.
4: 1799, p. [i-iii], *8 pl.* w.t.
5: 1800, p. [i-iii], *8 pl.* w.t.
6: 1801, p. [i-iii], *8 pl.* w.t.
7: 1801, p. [i-iii], *8 pl.* w.t.
8: 1802, p. [i-iii], *8 pl.* w.t.
9: 1805, p. [i-iii], *8 pl.* w.t.
10: 1805, p. [i-iii], *8 pl.* w.t. (probably publ. Aug-Sep 1805)
11: 1806, p. [i-iii], *8 pl.* w.t. (probably publ. Apr-Mai 1806)
12: 1807, p. [i-iii], *8 pl.* w.t.
13: 1809, p. [i-iii], *8 pl.* w.t.
14: 1811, p. [i-iii], *8 pl.* w.t.
Programma: 1797 (p. [4]: Aug 1797), p. [1-4].

9405. *Catalogus horti botanici* societatis physicae *turicensis.* mdcccii (1802). (*Cat. hort. bot. turic.*).

Publ.: 1802 (p. [12]: Jan 1802), p. [1-66]. *Copies*: M (inf. I. Haesler), NY (photocopy); IDC 7650.

9406. *Collecteana* ad omnem rem botanicam spectantia partim e propriis, partim ex amicorum schedis manuscriptis concinnavit et edidit J.J. Roemer M.D. ... Turici [Zürich] (apud Henricum Gessnerum) [1806-] 1809. Qu. (*Collecteana*).
Publ.: in parts between 1806 and 1810; p. 1-162 (at least) were possibly published in 1806 (see Steenis-Kruseman 1956, 1959), p. [i-x], [1]-314, *pl. 1-4* (uncol. copp. J. Sturm). *Copies*: G, HH, M, MICH, NY (2, of which one on heavier paper) US, USDA; IDC 291. – Contains papers by numerous authors. In the note in the "contenta" on A.Zuccagni's contribution (on p. 113-162) Roemer states that Zuccagni's publication was preprinted; this leaves the possibility open that p. 1-162 were published in Dec 1806 or in 1807 (as usually assumed).

9407. *Versuch eines* möglichst vollständigen *Wörterbuchs der botanischen Terminologie*. Von Dr. Joh. Jacob Römer. Zürich (bey Orell, Füssli und Comp.) 1816. Oct. (*Vers. Wörterb. bot. Termin.*).
Publ.: 1816, p.[i*], [i]-vi, [3]-826. *Copies*: G, HH, NY, USDA.

9408. Caroli a Linné equitis *Systema vegetabilium* secundum classes ordines genera species. Cum characteribus, differentiis et synonymiis, editio nova, speciebus inde ab editione xv. detectis aucta et locupletata. Curantibus Joanne Jacobo Roemer, ... et Jos. Augusto Schultes, ... volumen primum [-septimum]. Stuttgardtiae (sumtibus J.G. Cottae) 1817 [-1830]. 7 vols. Oct. (*Syst. veg.*).
Systema vegetabilium, see TL-2/3, p. 77 (no. 4709), The botanical part of the post-linnaean editions of the *Systema naturae* were frequently published in amended form as *Systema vegetabilium*. The Roemer and Schultes edition refers in its title to ed. 15 (by J.A. Murray, TL-2/6578), counts as ed. 16 and precedes that of K.P.J. Sprengel (1825-1828), which itself is also called ed. 16 but which should count as ed. 17. The numbering, however, is not important; citation should be to Roemer and Schultes, *Syst. veg.*
Authorship: Original author: Carl Linnaeus (1707-1778). *Co-author*, with J.J. Roemer, of this edition: Josef August Schultes (1773-1831), for volumes 1-4. J.A. Schultes continued the book after Roemer's death and is the author of vols. 5 and 6, except for the Umbelliferae in 6: 315-628 which are by K.P.J. Sprengel, and of some special entries attributed to other authors, such as A.W. Roth. Volume 7 was authored by J.A. Schultes and Julius Herman Schultes (1804-1840). These authors also published the *Mantissa* in *volumen primum [secundum, tertium] systematis vegetabilium* ... ex editione Joan. Jac. Roemer ... et Jos. Aug. Schultes, which is treated under J.A. Schultes (Mant. 1 and 2 by J.A. Schultes; Mant. 3 J.H. Schultes co-author).
Note on J.A. Schultes: The indication " Boio" after J.A. Schultes' name on the title pages of the earlier volumes refers to Landishuti bojorum (Landshut in S. Bavaria), where Schultes was professor of botany at the University. This University was still at Ingolstadt when Schultes joined it, from 1800-1826 it was at Landshut; after this it went to München. The indication "bojorum" refers to the Boji, a celtic people in Central Europe and Italy in Roman and pre-Roman time, which had settlements in Bavaria and Bohemia.
Publ.: in 7 vols., 1817-1830 (for origin of dates as given below see TL-1/1097). *Copies*: BR, G, H, HH, MO, NY, PH, U, US(3), USDA; IDC 238. – Roemer and Schultes included various contributions by other authors in their work, such as A.W. Roth, whose new taxa, published in his *Novae plantarum species*, were in part pre-published in "Roemer et Schultes" who had a manuscript copy of Roth's book available to them (see 1: xix; inf. Robert Faden).
Another situation is described by McVaugh (1955): D.F.L. von Schlechtendal made the Willdenow herbarium at Berlin accessible to Roemer and Schultes (see 3: vi), and from vol. 3 on the *Syst. veg.* contains new descriptions based on material from B-Willd., in particular from Humboldt and Bonpland. Such taxa described by Roemer and Schultes in vol. 3 often antidate publication by C.S. Kunth in the *Nova genera* (see McVaugh for further details, and see e.g. Kunth, Linnaea 5: 366-369. 1830 for the reverse situation).
The introductions to the volumes contain information on collaborators and contributors.

1: Jan-Jun 1817 (p. xx: 31 Dec 1816; t.p. 1817), p. [i]-viii, [1]-10, 10a, 10b, 11-642.
2: Nov 1817 (p. viii: 31 Jul 1817; copy sent to A.P. de Candolle on 11 Nov 1817; Flora rev. 30 Jul 1818), p. [i]-viii, [1]-480, [1, h.t.], 481-964, [1, binder, note indicating that part 2 begins at p. 481].
3: Apr-Jul 1818 (p. vi: 20 Mar 1818; copy sent to A.P. de Candolle on 20 Jul 1818, but see Jahrb. Gewächsk. Apr 1818 (1(1): 190. 1818: 'erschienen"), p. [iii]-vi, [vi, h.t.], [1]-584.*4:* Mar-Jun 1819 (rev. by Beck Jul 1819; see Flora 28 Feb 1819, 50 signatures printed), p. [iii]-vi, [vii, h.t.], [i]-lx, [1]-888.
5: Dec 1819 (p. viii: 23 Sep 1819; t.p. 1819; rev. Beck Mai-Jun 1820; Dec 1819, fide McVaugh 1955; thus giving priority to Humboldt and Bonpland, Nov. Gen. Pl. fasc. 12, 21 Nov 1819), p. [i]-viii, [i]-lii, [1]-632, [1-6, err.]. – "Volumen quintum. Inceptum a Joanne Jacobo Roemer, ... post ejus obitum continuatum a Jos. Augusto Schultes, ...".
6: Aug-Dec 1820 (p. viii: Cal. Aug 1820; Beck rev. 21 Feb 1821), p. [i]-viii, portrait Roemer, [i]-lxx, [1]-852, [853-857]. – "Volumen sextum (cum imagine divi J.J. Roemer) concinnatum a Jos. Augusto Schultes, ... Umbelliferas digessit C. Sprengel, ...".
7(1): 1829 (p. vi: Landishuti Bojorum 10 Oct 1828), p. [i]-xliii, [1]-753. – Voluminis septimi pars prima. Curantibus Jos. Augusto Schultes ... et Jul. Herm. Schules, ...".
7(2): Oct-Dec 1830 (p. iv: 30 Sep 1830; t.p. 1830; Flora Sep-Dec 1830), p. [i]-iv, [xlv]-cvii, [755]-1815, [1816, err.]. – Voluminis septimi pars secunda. Curantibus Jos. Augusto Schultes, ... et Jul. Herm. Schultes, ...".
Mantissa in volumen primum [secundum, tertium] Systematis vegetabilium Caroli a Linné ex editione Joann. Jac. Roemer ... et Jos. Aug. Schultes ... curante J.A. Schultes, 3 vols. Stuttgardt 1822, 1824, 1827, *see under:* J.A. Schultes.
Syst. veg.: for other issues see under Linnaeus, 4709.
Ref.: Kunth, C.S., Flora 1: 601-607. 1818 (incl. answer Lehmann), 3(1): 257-271. 1820 (on taxa based on Humboldt and Bonpland material in vol.4); Linnaea 5: 366-369. 1830.
McVaugh, R., Taxon 4: 78-86. 1955 (on taxa based on Humbold and Bonpland material).
Nees von Esenbeck, C.G., Flora 1: 349-362, 365-388. 1818 (extensive review of vols. 1-2 and comparison with A.P. de Candolle, *Regni vegetabilis systema naturale* (vol. 1, 1818)).
-u-, Flora 3(1): 257-271. 1820 (critical note on treatment of the collections made by A. v. Humboldt) (see also McVaugh (1955)).

9409. Caroli a Linné equitis *Systema vegetabilium* secundum classes ordines et *genera* editio nova, generibus inde ab editione xv, detectis aucta et locupletata. Vol. 1. sect. 1 [all published]. Inceptum a Joanne Jacobo Roemer, ... post ejus obitum continuatum a Jos. Augusto Schultes, ... Stuttgardtiae (sumtibus J.G. Cottae) 1820. Oct. †. (*Syst. veg. gen.*).
Publ.: Sep-Dec 1820 (p. iv: Sep 1820), p. [i-iv], [1]-323, [324, err.]. *Copies*: HU, NY. – A reprint of the *Synopsis of the genera* in vols. 1-6 of Roemer and Schultes, *Syst. veg.* See SO 614.

Römer, Julius (1848-1926), Transilvanian botanist from Kronstadt (Hung.: Brasso; after 1920 Braşov, Rumania); studied in Wien, Jena (with E. Haeckel) and Heidelberg; teacher at a girl's college in Kronstadt; Dr. phil. h.c. Breslau 1925. (*Jul. Röm.*).

HERBARIUM and TYPES: Material at E, F, G, GB and LD.

BIBLIOGRAPHY and BIOGRAPHY: AG 6(1): 759; Barnhart 3: 170 (b. 21 Apr 1848, d. 23 Oct 1926); BJI 1: 49, 2: 147; Bossert p. 337; CSP 11: 204, 18: 259-260; Kelly p. 194; Kew 4: 502; Rehder 2: 601, 676.

BIOFILE: Anon., Bul. Grad. bot. Muz. bot. 6(1-2): 72. 1926 (d.); Österr. bot. Z. 74: 216. 1925 (Dr. h.c. Breslau).
Heltmann, H., Ocrot. nat., Bucuresti, 10(1): 59-64. 1966 (portr.; further biogr. references; influence Darwin).
Kneucker, A., Allg. bot. Z. 32: 56(224). 1927 (d.).

Pax, F., Grundz. Pfl.-Verbr. Karpathen 1: 50, 56, 57-58, 60, 63. 1898 (Veg. Erde 2(1)), 2: 280, 281. 1908 (id. 10).
Römer, J., Wanderer nr. 3-4. 1926 (n.v., fide Heltmann; "Mein Weg zur Botanik").
Simonkai, L., Enum. fl. transsilv. xxviii. 1886.
Ungar, K., Verh. Mitt. siebenb. Ver. Naturw. 77: 1-8. 1926/1927 (n.v.); Allg. bot. Z. 32: 2(170)-3(171). 1927 (obit.; d. 23/24 Oct 1927).

9410. *Die Pflanzenwelt der "Zinne" und des Kleinen Hangesteines"*. (Separatabdruck aus der aus Anlass der Wanderversammlung ung. Ärzte und Naturforscher veröffentlichten Festschrift). Kronstadt [Braşov, Brasso in Transylvania] (Buchdruckerei von Johann Gött & Sohn Heinrich) 1892. Oct. (*Pfl.-Welt Zinne*).
Publ.: 1892, p. [1]-69, [i]-xi. *Copy*: G. – We saw no copy of the "Festschrift" from which this paper was reprinted (see title).

9411. *Aus der Pflanzenwelt der Burgenländer Berge* in Siebenburgen ... Wien (Verlag von Carl Graeser) 1898. Oct. (*Pfl.-Welt Burgenländ. Berge*).
Publ.: Aug 1898 (published for Kronstadt session of Aug 1898 of Siebenbürg. Karpathen Verein; p. vi: 31 Mar 1898; the copy at G was sent to E. Haeckel at Jena by the author on 20 Sep 1898; Bot. Zeit. 16 Nov 1898; Bot. Centralbl. 28 Sep 1898; Nat. Nov. Sep(2) 1898), p. [i]-vi, [vii], [1]-119, *pl. 1-30* (col.). *Copies*: G, M.
Ref.: Engler, A., Bot. Jahrb. 26(Lit.): 52. 1899.
Kneucker, A., Allg. bot. Z. 4: 149. 15 Sep 1898 (rev.).
Solms-Laubach, H., Bot. Zeit. 56: 347-348. 16 Nov 1898 (rev.).

Roemer, Max Joseph (1791-1849), German (Bavarian) botanist and "Landrichter" in Aub, later private scientist at Würzburg. (*M. Roem.*).

HERBARIUM and TYPES: Unknown. – Roemer's library was sold by auction after his death.

NOTE: Part of the mystery surrounding this author has been cleared by Müllerott (1975). Even so the biographical details remain scarce. Apparently Roemer was a country judge at Aub who was "relieved" of his duties and who then settled in Würzburg.

BIBLIOGRAPHY and BIOGRAPHY: Barnhart 3: 170; BM 4: 1720; CSP 5: 267; Herder p. 474 (index); Hortus 3: 1202; Jackson p. 39, 121; Kew 4: 502; Langman p. 638; MW p. 416; PR 7716-7718, ed. 1: 8675-8677; Rehder 1: 87, 261, 299; RS p. 137; TL-1/1099; Tucker 1: 604; Zander ed. 10, p. 708, ed. 11, p. 808.

BIOFILE: Müllerot, M., Hoppea 34: 307-308. 1975 (biogr. details; with Nachschrift by M.G. Boar who discovered a death notice, 4 Nov 1849 in the Würzburg municipal archives; R. died at the age of 58, hence 1791 is most probable year of birth).

9412. *Geographie und Geschichte der Pflanzen* von M. Römer, königl. Landrichter in Aub ... München (bei Ernst August Fleischmann) 1841. Oct. (*Geogr. Gesch. Pfl.*).
Publ.: 1841, p. [i], [1]-144. *Copies*: FI, NY. – Reprinted from his Handbuch der allgemeinen Botanik.

9413. *Prospect. Familiarum naturalium regni vegetabilis synopses monographicae* seu enumeratio omnium plantarum, hucusque detectarum secundum ordines naturales, genera et species plantarum, additis diagnosibus, synonymis, novarumque vel minus cognitarum descriptionibus curante M.J. Roemer. Weimar (Druck und Verlag des Landes-Industrie-Comptoirs) 1846. Oct. (*Prospect Fam. nat. syn. monogr.*).
Publ.: Sep 1846 (p. [23]), p. [1]-22, [23]. *Copies*:BR, G, HH, M, USDA. – Text taken from introduction vol. 1. Intended to become a series of monographs of families or parts of families; published irrespective of systematic order; based on existing publications and not on original research. The text on p. [23], dated Sep 1846, mentions the first part as published. The text of this prospectus was not incorporated in the *Fam. nat. syn. monogr.* itself, even though the prospectus as a whole is sometimes bound with it.

9414. *Familiarum naturalium regni vegetabilis synopses monographicae* seu enumeratio omnium plantarum hucusque detectarum secundum ordines naturales, genera et species digestarum, additis diagnosibus, synonymis, novarumque vel minus cognitarum descriptionibus curante M.J. Roemer. Fasc. i [-iv] ... Vimariae [Weimar] (Landes-Industrie-Comptoir) 1846. Oct. † *(Fam. nat. syn. monogr.).*
Publ.: Only four monographs were published. See also above under *Prospect syn. monogr.*
Copies: BR, FI, G, HH, M, MICH, NY, PH, U, USDA; IDC 5686.
1: 14 Sep-15 Oct 1846 (p. viii: 14 Sep 1846; Prospekt: Sep; Hinrichs 15-17 Oct; Flora rd Dec 1846), p. [i]-xii, [1]-151, [152, err.]. – *Hesperides.*
2: Dec 1846 (Hinrichs 9-12 Dec 1846; Flora rd 2 Mar 1847), p. [i]-x, [1]-222. – *Peponiferarum pars prima.*
3: Apr 1847 (Hinrichs 28 Apr-1 Mai 1847; Flora 14 Jul 1847), p. [i]-viii, [ix], [1]-249, [250, err.]. – *Rosiflorae.*
4: Mai-Oct 1847, p. [i]-vi, [vii], [1]-314. – *Ensatae pars prima.*
Ref.: Hooker, W.J., J. Bot. Kew Gard. Misc. 1: 127-128. 1849.
Schlechtendal, D.F.L. von, Bot. Zeit. 4: 896-898. 25 Dec 1846 (fasc. 1), 5: 406-407. 4 Jun 1847 (fasc. 2), 7: 903-905. 1849 (fasc. 3, 4).

Römmp, Herman *(fl.* 1928), German botanist at Tübingen. *(Römmp).*

HERBARIUM and TYPES: Unknown.

BIBLIOGRAPHY and BIOGRAPHY: BFM no. 2809, 2875; Biol.-Dokum. 15: 7594; MW p. 417.

9415. *Die Verwandtschaftsverhältnisse in der Gattung Veronica* Vorarbeiten zu einer Monographie ... Dahlem bei Berlin (Verlag des Repertoriums, ...) 1928. Oct. *(Verw.-Verh. Veronica).*
Publ.: 10 Feb 1928, p. [i, iii], [1]-172. *Copies*: MO, U. – Issued as Rep. sp. nov. regn. veg. Beih. 50.

Rönn, Hans Ludwig Karl (1886-?), German (Hamburg) botanist; Dr. phil. Kiel 1910. *(Rönn).*

HERBARIUM and TYPES: KIEL (Schellenberg 1921).

BIBLIOGRAPHY and BIOGRAPHY: Barnhart 3: 170; Biol.-Dokum. 15: 7594; Christiansen no. E 86.

BIOFILE: Rönn, H.L.K., Myxomyc. n.-östl. Holst., 1911 (vita; b. 28 Mai 1886).
Schellenberg, G., Beih. bot. Centralbl. 38(2): 391. 1921 (myxomyc. herb. at KIEL).

9416. *Die Myxomyceten des nordöstlichen Holsteins.* Floristische und biologische Beiträge. Inaugural-Dissertation zur Erlangung der Doktorwürde der hohen philosophischen Facultät der Königl. Christian-Albrechts-Universität zu Kiel ... Kiel (Druck von Schmidt & Klaunig) 1911. Oct. *(Myxomyc. n.-östl. Holst.).*
Thesis issue: Apr-Jun 1911 (Bot. Centralbl. 4 Jul 1911; Hedwigia 20 Oct 1911; Nat. Nov. Mar(1) 1912), p. [i-iv], 20-76, [1, vita]. *Copy*: FH. – Reprinted or preprinted from Schr. naturw. Ver. Schlesw.-Holst. 15(1): 20-76. 1911.
Ref.: Leeke, G.G.P., Bot. Centralbl. 119: 380-381. 1912.

Roeper, Johannes August Christian (1801-1885), German (Mecklenburg) botanist; studied at the University of Rostock 1817-1819, Berlin 1819-1822, and Göttingen 1822-1824; Dr. med. Göttingen 1823; travelled and studied in Paris 1824-1825 and Genève 1825-1826; extraord. prof. botany Basel 1826-1829; idem ord. prof. 1829-1836; ordinary prof. botany Rostock from 1836, from 1846 also University librarian; Dr. phil. h.c. Basel 1836; Dr. sci. nat. h.c. Tübingen 1873; first author to use flower diagrams. *(Roep.).*

HERBARIUM and TYPES: ROST. – Further material at E-GL and P. – Roeper had acquired the herbarium of J.B.A.P.M. de Lamarck in 1824 during his stay in Paris; it was acquired by the University of Rostock in 1875. Roeper's successor K. von Goebel then sold the Lamarck herbarium to Paris in 1887. A certain number of other collections owned or collected by Roeper seem to have been transferred to P as well. (See Bergdolt) 1940)), but his main herbarium of over 100.000 specimens stayed at Rostock (Oltmans 1894). Roeper's letters to D.F.L. von Schlechtendal are at HAL.

NAME: Roeper spelled his name this way on all his Latin title-pages as well as on some of his German and French language publications; but also Röper on other German publications. We keep "Roeper" because of its common usage.

BIBLIOGRAPHY and BIOGRAPHY: ADB 29: 149-152 (C. Wunschmann); AG 2(2): 482; Backer p. 497 (epon.); Barnhart 3: 170 (b. 25 Apr 1801, d. 17 Mar 1885); BM 4: 1720; Bossert p. 335; CSP 5: 271-272, 18: 260; Frank (Anh.): 85; Herder p. 119, 141, 187, 271, 277, 336, 337; Jackson p. 307; Kew 4: 502; Lasègue p. 335; Moebius p. 38, 153; Morren ed. 2, p. 12, ed. 10, p. 26; NI 815, 1795; PR 1502, 7719-7724, 8618, ed. 1: 2303, 8678-8683, 9574; Rehder 5: 733; RS p. 137; TL-1/1100-1100a; TL-2/712, see Lamarck; Tucker 1: 604.

BIOFILE: Anon., Amer. J. Sci. 31: 22. 1886 (obit.); Biol.-Dokum. 15: 7595; Bonplandia 6(2): 51. 1858 (cogn. Leopoldina Lachenalius); Bot. Centralbl. 3/4: 1213-1214. 1880 (entire Roeper herbarium bought by ROST [N.B. this differs from Oltmann's statement (1894)], 22: 64. 1885 (d.); Bot. Gaz. 10: 297. 1885 (d.); Bot. Not. 1886: 39 (d.); Bot. Zeit. 5: 120. 1847 (first univ. librarian Rostock), 9: 224. 1851 (promotor for W. Hofmeister, awarded a Dr. h.c. at Rostock), 43: 205, 207, 244. 1885 (d.); Bull. Soc. bot. France 32 (bibl.): 48. 1885 (d.); Flora 19: 128. 1836 (call to Rostock, as successor to Floerke, accepted), 30: 147. 1847 (first Univ. librarian Rostock); Österr. bot. Z. 35: 187. 1885 (d.); Verh. bot. Ver. Brand. 27: lvi. 1886 (d.).
Bergdolt, E., Karl von Goebel ed. 2. 29, 31, 34, 36, 47, 57, 210. 1941.
Boll, E., Fl. Mecklenburg 159-160. 1860.
Briquet, J., Bull. Soc. bot. Genève 50a: 411-412. 1940 (biogr.; relations with Genève; studied with A.P. de Candolle; married a Genevoise; significance morphological work; epon., bibl.).
Burckhardt, Fr., Verh. naturf. Ges. Basel 18(1): 108-111. 1905 (on R's Basel period).
Candolle, Alph. de, Phytographie 388. 1880 (herb. at ROST, still private).
Candolle, A.P. de, Mém. souvenirs 331, 392. 1862.
Frison, Ed., Henri Ferdinand van Heurck 21, 32-34. 1959.
Gray, A., Sci. papers 2: 482-483. 1889 (obit.).
Herter, H., Mitt. Bot. München 16: 424. 1980 (lich. M).
Högrell, B., Bot. hist. 1: 261. 1886.
Jessen, K.F.W., Bot. Gegenw. Vorz. 422, 424. 1884.
Lucante, A., Rev. bot., Bull. mens. Soc. franç. Bot. 4: 333. 1885-1886 (d.).
Mägdefrau, K., Gesch. Bot. 165, 274. 1973.
Magnus, P., Leopoldina 21: 170-173. 1885 (obit.); Verh. bot. Ver. Brandenburg 27: xxvii-xxxii. 1886 (obit.).
Martius, C.F.P. von, Flora 20(2) Beibl. 3: 43. 1837 (Roeper will treat the Euphorbiac. for *Fl. bras.*).
Mettenius, C., Alexander Braun's Leben 694-696. 1882.
Müllerot, M., Hoppea 39: 62. 1980 (3 letters Roeper to Irmisch, copies at Sondershausen), Denkschr. Regensb. bot. Ges. 39: 65-66. 1980 (letter Roeper to Thilo Irmisch 11 Jul 1853, informing him that a doctoral h.c. is awarded to him by Rostock; "Männer wie Thilo Irmisch, Wilhelm Hofmeister, Joseph Decaisne zu promovieren ist Hochgenuss!").
Oltmanns, F., Arch. Ver. Fr. Naturgesch. Mecklenburg 47: 109-110. 1894 (Roeper herb. basis of herb. ROST, acquired 1885).
Reinke, J., Mein Tagewerk 493 [index, numerous references] 1925.
Richter, R. et al., Wiss. Z. Univ. Rostock, mat.-naturw. R. 17: 266-267. 1968 (biogr. sketch, portr.).
Roeper, J.A.C., Der Frieden in der Schöpfung kein Frieden in Christo. Ihrer Königlichen Hoheit der Grossherzogin Augusta am 3 Mai 1851 allerunterthänigst gewidmet, ed. 2, 1857, Rostock (Druck von Ludolph Hirsch), 22 p. (copy: G) (ideas).

Sachs, J.S., Gesch. Bot. 155, 400. 1875; J.S., Histoire bot. 150, 383. 1892.
Schlechtendal, D.F.L. von, Bot. Zeit. 1: 16-17. 1843 (herb.).
Schmid, G., Goethe u.d. Naturw. 257-258. 1940.
Stafleu, F.A., Regn. veg. 71: 326, 339. 1970.Struck, *in* Koch, F.E., Arch. Ver. Freunde Naturgesch. Mecklenburg 39: 166-169. 1885 (obit.).
Sydow, P. & C. Mylius, Botaniker-Kalender 1886(2): 4-6 (obit.).
Wittrock, V.B., Acta Horti Berg. 3(3): 150. 1905.

EPONYMY: *Neoroepera* J. Müller-Arg. et F. v. Mueller ex J. Müller-Arg. (1866); *Roepera* A.H.L. Jussieu (1825); *Roeperia* K.P.J. Sprengel (1826); *Roeperia* F. v. Mueller (1857); *Roeperocharis* H.G. Reichenbach (1881).

9417. *Enumeratio Euphorbiarum* quae in Germania et Pannonia gignuntur ... Gottingae [Göttingen] (Typis Caroli Eduardi Rosenbusch) 1824. Qu. (*Enum. Euphorb.*).
Publ.: Dec 1824 (Goethe received a copy on 24 Dec 1824; E. Meyer 12 Dec 1824; Acad. Paris rd. 24 Jan 1825; rev. GGA 21 Feb 1825, id. Flora 28 Mai 1825), p. [i]-viii, [1]-68, *pl. 1-3* (uncol. auct.). *Copies*: BR, E, G, LC, M, PH; IDC 5687.
Variant issue: as above but t.p.: Gottingae (apud Carolum Eduardum Rosenbusch) 1824. *Copies*: HH, NY, USDA.
Ref.: Anon., Flora 8 (Beil. 2): 96. Mar-Apr 1825 ("erschienen").
  Meyer, E., Gött. gel. Anz. 1825: 305-312.
  Schmid, G., Goethe u. d. Naturwiss. 1940, nos. 1686, 1687.
  Nees von Esenbeck, T.F.L., Flora 8: 305-313. 28 Mai 1825.

9418. *Observations sur la nature des fleurs* et des inflorescences, par J. Roeper. [in N.C. Seringe, Mél. bot. 5, 1826]. Oct.
Publ.: 28 Mar 1826, p. [71]-98. *Copies*: HH, USDA. – Issued as no. 5 of N.C. Seringe, Mél. bot. 2: [71]-98. 28 Mar 1826.
Latin version: Linnaea 2: 433-466. Aug-Oct 1826.

9419. *De organis plantarum* scripsit Joannes Roeper ... Basiliae [Basel] (typis Augusti: Wielandi) 1828. Qu. (*Organ. pl.*).
Publ.: 1828 (Goethe received a copy on 22 Oct 1828), p. [i-iii], [1]-23. *Copies*: BR, E, G, NY.
Ref.: Schmid, G., Goethe und die Naturwissenschaften 1940, no. 1690.

9420. *De floribus et affinitatibus Balsaminearum* scripsit Joannes Roeper, ... Basiliae [Basel] (typis J. Georgii Neukirch) 1830. Oct. (*Fl. affin. Balsamin.*).
Publ.: Sep-Dec 1830 (p. ii: 1 Sep 1830), p. [i*], [i]-ii, [3]-70. *Copies*: G(2), H-UB, HH(2), LC, NY, U, USDA, WU. – The NY copy has an 1833 frontispiece presenting flower diagrams of Impatiens and Hydrocera, dated 1833, evidently taken from Linnaea 9: 112-124. *pl. 1.* 1834 (Roeper & G.A. Walker-Arnott, Historia Balsaminearum systematica). For further publications on Balsaminaceae see below.
Criticism by C.A. Agardh: Flora 16: 609-624. 21 Oct 1833.

9421. *Bemerkungen zu Herrn Prof. Agardh's Abhandlung*: Ueber die Deutung der Blumentheile der Balsaminen und die Stelle dieser Pflanzen im Systeme ... [Besonderer Abdruck aus der Flora 1834 ...]. Oct.
Publ.: 14-21 Feb 1834 (in journal), p. [1]-24. *Copies*: G(2). – Reprinted and to be cited from Flora 17: 81-91. 14 Feb, 97-111. 21 Feb 1834. Reply to C.A. Agardh, in Flora 16: 609-624. 21 Oct 1833. Agardh answered in Flora 19: 193-205, 7 Apr, 209-221, *2 pl.* 14 Apr 1836; with a further reply by Roeper in 19: 241-245. 28 Apr 1836.

9422. *Verzeichniss der Gräser Mecklenburgs* von Joh. Röper, ... Rostock (Gedruckt bei Adlers Erben) 1840. Qu. (*Verz. Gräs. Mecklenb.*).
Publ.: 1840, p. [1]-15. *Copies*: LC, NY.
Note: D.F.L. von Schlechtendal wrote four letters to Roeper on his ideas on the morphology of grasses, reproduced in Bot. Zeit. 5: 673-679, 697-703, 1847, 6: 809-816, 841-848. 1848 under the title "*Betrachtungen über die Gräser*, in Briefen an Joh. Roeper".

9423. *Zur Flora Mecklenburgs.* Erster [Zweiter] Theil. Rectorats-Programm von Joh. Roeper, ... Rostock (Gedruckt bei Adler's Erben) 1843 [-1844]. Oct. (*Fl. Mecklenb.*).
1: 1843 (Flora rd. 1-31 Oct 1831; Bot. Zeit. 3 Nov 1843), p. [1]-160, *1 pl.* (uncol., auct.).
– Filices.
2: Oct 1844 (p. 6: Jul 1844; Hinrichs 4-6 Nov 1844, Flora rd. Nov 1844; G copy sent by R. to A.P. de C. in Oct 1844), p. [1]- 296, *1 pl.* (id). – Gramineae.
Copies: G, HH, M, MO, NY, USDA.
Nachtrag: Bot. Zeit. 4: 161-168. 1846.
Ref.: Anon., Lit.-Bih. Bot. Not. 1844(1): 8-10 (rev. 1), 1845(6): 100-102. Jun 1845 (rev. 2).
Schlechtendal, D.F.L. von, Bot. Zeit. 1: 764-765. 1843 (rev.), 2: 860-861. 13 Dec 1844 (id. part 2).
Schleiden, M., Flora 28: 741-743. 1845 (rev.).

9424. *Vorgefasste botanische Meinungen* vertheidigt von Dr. Johannes Roeper ... Rostock (Stiller'sche Hofbuchhandlung (Hermann Schmidt)) 1860. Oct. (*Vorgefasste bot. Meinungen*).
Publ.: 1860 (p. viii: "Ende August 1860"; Flora rd. 1861), p. [i]-viii, 1-74, [1, err.].
Copies: G, NY. – Philosophical commentary. On p. viii Roeper pays a touching tribute to Basel, the town in which he came of age as a morphologist and scientist "Gott, der Herr, segne das theure, das edle Basel".
Ref.: Schlechtendal, D.F.L. von, Bot. Zeit. 19: 176. 1861.

9425. *Botanische Thesen* aufgestellt von Dr. Johannes Roeper ... Rostock (Carl Boldt's Buchdruckerei) 1872. Oct. (*Bot. Thesen*).
Publ.: Aug 1872 (p. vi: 2 Aug 1872; "Neu" Bot. Zeit. 30 Aug 1872), p. [i]-vi, [7]-27.
Copies: G(2), HH, M, NY. – Dedicated to J.C.F. Stempel on the 50th anniversary of his doctorate.

9426. *Der Taumel-Lolch* (Lolium temulentum Linn.) im Bezug auf Ektopie, gewohnheitliche Atrophie und aussergewöhnliche, normanstrebende Hypertrophie festschriftlich betrachtet von Dr. Joh. Röper, ... Hiebei zwei Tafeln. Rostock (Carl Boldt's Buchdruckerei) 1873. Qu. (*Taumel-Lolch*).
Publ.: 1873 (p. vi: 22 Apr 1873), p. [i]-vi, [7]-23. *Copies*: G, M.

Roesch, Charles (*fl.* 1894), French (Alsatian) diatomologist. (*Roesch*).

HERBARIUM and TYPES: Unknown.

BIBLIOGRAPHY and BIOGRAPHY: BM 4: 1720; Nordstedt suppl. p. 13.

9427. *Contribution à l'étude des Diatomées* du territoire *de Belfort* et des environs ... Belfort (Typographie et lithographie Devillers) 1894-1897.
Co-author: Lucien Meyer.
Publ.: four papers, published in Bull. Soc. Belfort. Émul. 13-16, 1894-1897, reprints at FH. *Copy* reprints [1-4]: FH.
[1]: 1894, p. [1]-16, *pl. 1-2* (uncol., auct). Bull. 13: 136-150, *pl. 1-2*. 1894 (journal rd. by LC 12 Jan 1958).
[2]: 1895, p. [1]-10, *pl. 1-2* (id.). Bull. 14: 196-204, *pl. 1-2*. 1895 (rd. LC 4 Mar 1896).
[3]: 1896, p. [1]-22, *pl. 1-6* (col., auct.). Bull. 15: 33-52, *pl. 1-6*. 1896 (LC rd. Oct 1896).
[4]: 1897, p. [1]-12, *pl. 1-2* (uncol., auct) Bull. 16: 310-316. 1897 (LC rd. 20 Apr 1898).

Rösler, Carl August (1770-1858), German (Württemberg) botanist and "Hüttenamts-Buchhalter" in Ludwigsthal (iron-works administrator). (*Rösler*).

HERBARIUM and TYPES: STU; a number of lichens at GZU and M.

BIBLIOGRAPHY and BIOGRAPHY: Barnhart 3: 170; BM 4: 1720; GR p. 125-126.

BIOFILE: Künkele, S., Jh. Ges. Naturk. Württemberg 125: 150, 152. 1970 (herb. material at STU) (Landw. Ver. and Stuttgart-Hohenheim).
Martens, G. von & C.A. Kemmler, Fl. Württemberg. ed. 2. 781. 1865.
Usteri, P., Neue Ann. Bot. 3: 71-73. 1794 (rev. of Beitr. Naturgesch. Wirtemberg Heft 1).

EPONYMY: *Roesleria* Thüman et Passerini (1877) honors the Austrian, Leonard Roesler.

Rössig, Carl Gottlob (1752-1806), German (Saxonian) botanist (rhodologist) and law professor at the University of Leipzig. (*Rössig*).

HERBARIUM and TYPES: Unknown.

BIBLIOGRAPHY and BIOGRAPHY: Barnhart 3: 170 (b. 27 Dec 1752, d. 20 Nov 1806); BM 4: 1720; Bossert p. 335; CSP 11: 206(?); DU 265; GF p. 74; Herder p. 342, 365; Hortus 3: 1203; Jackson p. 142; Kew 4: 502; LS 22426-22427; NI 1665-1666; PR 7725-7727, ed. 1: 8684-8689; Rehder 5: 734; Tucker 1: 604.

BIOFILE: Wittrock, V.B., Acta Horti Berg. 3(2): 140. 1903.

9428. *Oekonomisch-botanische Beschreibung* der verschiedenen und vorzüglichen Arten Ab- und Spielarten *der Rosen* zu näherer Berichtigung derselben für Liebhaber von Lustanlagen und Gärten ... Leipzig (in der von Kleefeldschen Buchhandlung) 1799-1803, 2 vols. Oct. (*Oekon.-bot. Beschr. Rosen*).
1: 1799, p. [i]-xii, [1]-242, [1, err.].
2: 1803, p. [i]-xvi, [1]-247.
Copies: BR (vol. 1), GB (inf. B. Peterson), GOET (inf. G. Wagenitz), NY.

9429. *Die Rosen* nach der Natur gezeichnet und colorirt mit kurzen botanischen Bestimmungen begleitet ... Leipzig (im Industrie-Comptoir) [1802-1820]. (*Rosen*).
Publ.: 12 Hefte, 1802-1820, each consisting of a Heft t.p. and *5 pl.* with unnumbered text pages, *60 pl.* in all, with introductory material (including a French t.p.) and a German and a French t.p. for vol. 2 (Hefte 7-12). – *French title*: "*Les Roses*" désinées et enluminées d'après nature; avec une courte description botanique par M. le D. Roessig. Traduit de l'Allemand par M. de Lahitte". – *Plates*: unsigned (by Luise von Wangeheim, fl. 1800-1820?) "etchings, coloured by hand, some printed in red or green ink ..." (DU 265). A mediocre work, criticised by Redouté.
Preface material: [i-xvii].
1: 6 Hefte, each with 5 pl. and 12 p. text (recto German, verso French).
2: 6 Hefte, each with 5 pl., Heft 7 has 16 p., 9 has 20 p., all others 12 p. text.
Copies: BR, GOET (inf. G. Wagenitz), NY.

9430. *Die Nelken* nach ihren Arten besonders nach der J.C. Etlers in Schneeberg und andern berühmten Sammlungen, in Blättern nach der Natur gezeichnet und ausgemahlt. Mit kurzen Bestimmungen begleitet, nebst einer Einleitung in die verschiedenen Systeme und Vorschläge zu einer einfachern und einer neuen Nomenklatur ... Leipzig (im Industrie-Comptoir), 2 Hefte [1808]. Oct. (*Nelken*).
1: 1808 (p. iv: 6 Apr 1808), p. [i]-iv, [1]-42, *10 pl.* (handcol. copp. engr.).
2: s.d., p. [i], [43]-82, [1, err.], *20 pl.* (id.).
Copies: G, USDA.

Roezl, Benedict (1824-1885), Bohemian horticulturist and seedsman; gardener in various Central European gardens, with VanHoutte at Gent, "chef de culture" at the Belgian school of horticulture; collecting in the United States, Mexico, Cuba and South America between 1854 and 1875; in retirement at Smichov, Praha; one of the most active collectors of living orchids. (*Roezl*).

HERBARIUM and TYPES: Material at K and W (Orch.). – Roezl made relatively few herbarium specimens (but see e.g. Reichenbach 1876); some of his shipments of living orchids went into the ten-thousands: Roezl's objectives had to be mainly commercial and he collected lavishly, apparently not hindered by thoughts about the future of the wild plants in question.

BIBLIOGRAPHY and BIOGRAPHY: AG 2(2): 357; Backer p. 497 (epon.); Barnhart 3: 171 (b. 12 Aug 1824, d. 14 Oct 1885); BM 4: 1721; CSP 11: 206-207, 12: 625; Ewan ed. 1: 291-292, ed. 2: 186; Herder p. 294; Hortus 3: 1202; IH 2: (in files); Langman p. 638-639 (detailed bibl. Mex.); Morren ed. 10, p. 33; Rehder 1: 331, 332, 5: 734; Tucker 1: 604; Zander ed. 10, p. 708, ed. 11, p. 808.

BIOFILE: Acosta-Solis, M., Inst. Ecuat. Ci. nat. Contr. 65: 41. 1968 (in Guyaquil, Ecuador, 1873).
André, E., Rev. hortic. 57: 543-546. 1885 (obit.).
Anon., Bot. Not. 1885: 39 (d.); Bot. Gaz. 11: 47. 1886 (d.); Bot. Jahrb. 7 (Beibl. 13): 2. 1885 (d.); Gard. Chron. ser. 2. 2: 73. 1874 (travels, coll., portr.), 24: 521-522. 1885 (obit., portr.); L'Orchidophile 1885: 140-143, 169-171 (obit., portr.), (Engl. transl.:), Orchid Digest 37: 221-222. 1973; Österr. bot. Z. 35: 445. 1885 (d.); Rev. bot. Bull. mens. Soc. franç. Bot. 4: 335. 1885-1886 (d.); Sieboldia 1: 253-256. 1875 (travels).
Coats, A.M., Plant hunters 60, 341, 348-350, 373, 374-375. 1969.
Eastwood, A., Leafl. west. Bot. 5(6): 103. 1948 (on Lilium Roezlii).
Ewan, J., in Century nat. Sci. San Francisco 58. 1955.
Kanitz, A., Magy. novén. Lap. 10: 9-10. 1886.
Kline, M.C., Amer. Orch. Soc. Bull. 32(8): 611-613 (biogr. sketch).
McVaugh, R., Contr. Univ. Mich. Herb. 9: 294-296. 1972 (detailed info. on R's Mexican collections).
Morren, Éd., Belg. hortic. 30: 5-12. 1880 ("hommage", portr.).
Otto, E., Hamb. Gart.-Bl.-Zeit. 18: 166-168. 1862 (on R's collecting in Mexico), 30: 420-423. 1874 (travel reports).
Poštolka, K., Ziva, Praha 22(6): 211. 1974 (biogr. sketch, memorial).
Parish, S.B., Bot. Gaz. 44(6): 414. 1907 (Palms).
Pynaert, Éd., Revue hort. belge étrang. 11: 246. 1885 (obit.).
Regel, E., Gartenflora 19: 296-297. 1870, 20: 6-8, 36, 37, 301-302. 1871, 21: 369. 1872 (travel reports).
Reichenbach, H.G., Bot. Zeit. 30: 802. 1872 (travel report); Linnaea 41: 1-16. 1846 (based on herb. specimens made by Roezl); Bot. Centralbl. 24: 159. 1885 (obit.; Czech father, German mother); J. Bot. 23: 381. 1885 (obit.).
Reinikka, M.A., A history of the Orchid 219-222. 1972 (portr., biogr.).
Roezl, B., Gartenflora 8: 131-138 (Reiseberichte aus Mexico, ed. E. Ortgies), 276-278. 1859, 9: 120-126, 195. 1890 (Reiseber.), 10: 7-13, 40-46, 119-131. (Reiseber.), 264. 1861, 11: 42, 59-62 (Reiseber.) 1862, 15: 62. 1866, 20: 70, 107. 1871 (Reiseber.); excerpts Reiseber. Hamb. Gart. Blum. Zeit. 18: 166-168. 1862 (in French:) Rev. hortic. 1861: 414-416. 1862, 39-40, 98-99, 118, 179-180 (by W. de Fonvielle), (in English:) Gard. Chron. 1862: 168; Deut. Gaertn.-Zeit. 4: 11-12, 34-36, 58-60, 78-79, 105-108, 129-131, 154-155, 180-181, 227. 1880 (Meine letzte Reise an der Westküste von Mexiko), partial translation by W.B. Hemsley, Gard. Chron. ser. 2. 19: 369, 415-416. 1883 (ascent of the Volcan de Colima), see also Belg. hort. 1880: 5-12 (portr.); La Belgique horticole 32: 68-113. 1882 (Mon dernier voyage à la côte occidentale du Mexique; translated from Deut. Gärtn. Zeit. 1880, "in the beginning of March I finished the packing and sending of 100.000 orchids, among them 22.000 specimens of *Odontoglossum cervantesi*; for which I had paid 5 francs per thousand ...").
Sander, F., Orch. Rev. 60: 130. 1952 (letter F. Sander 17 Feb 1882 to Roezl, on Cattleya aurea).
Schlechtendal, D.F.L. von, Bot. Zeit. 16: 7-8. 1858 (review of Roezl, *Catalogue des graines de Conifères mexicains*, Mexico 1857, 34 p.; we failed to locate a copy of this catalogue; for a latin translation of the taxonomic parts see Schlechtendal, Linnaea 29: 326-356, 699-704. 1858.
Swinson, A., Frederick Sander: the orchid king 22-29. 1970 (portr., relations with Sander; Orchid Rev. 301. 1970.
Verdoorn, F., ed., Chron. bot. 1: 18. 1935.

EPONYMY: *Roezlia* Regel (1871-72); *Roezliella* Schlechter (1918).

9431. *Catalogue des graines de Coniferes mexicains* qui se trouvent chez B. Roezl & C.ie a Napoles pres Mexico. La pluspart découverts, décrits et récoltés par B. Roezl, ancien élève des jardins du comte F. de Thun, à Tetschen en Bohême; du baron Ch. de Hugel, à Vienne; ancien jardinier en chef à Medica, en Gallicie (comte Powlikowsky) et à Telsch, en Moravie (comte Leopold Podstatzky Luhlenstein), ancien chef de culture de l'institut horticole du gouvernement belge, etc. etc. Mexico (Imprimerie de M. Murguia, rue del Aguila de Oro). Iuin, 1857. (*Cat. grain. Conif. mexic.*).
Publ.: Jun 1857 (t.p.), p. [1]-34. *Copy*: C (inf. A. Hansen). – Translation of descriptions into Latin by D.F.L. von Schlechtendal, Linnaea 29: 326-356, 699. 1858.
Ref.: Carrière, E.O., Traité gén. conif. ed. 2. 1867 (reduction of some species).
 Gorden, G., Pinetum suppl. 1862, ed. 2. 1875 (reduction of various Roezl species to synonymy).
 Ortgies, E., Gartenflora 7: 381-382. 1858, 8: 276-278. 1859.
 Parlatore, F., *in* DC., Prodr. 16(2). 1868 (reduces all Roezl species of Pinus to synonymy).
 Shaw, G.R., The pines of Mexico, Publ. Arnold Arb. 1: 3-4. 1909 (reprinted 1961) ("... there is, in the entire list of Roezl's Catalogue, not a single valid species ...").

Roffavier, Georges (1775-1866), French merchant and botanist at Lyon, collaborator of J.B. Balbis and co-founder of the Société Linnéenne de Lyon (1822); student of Gilibert. (*Roffavier*).

HERBARIUM and TYPES: LY (via L. Lortet). – A set of 29 original col. drawings of fungi collected in the dépt. du Rhône, with descriptive mss. text is at BPI (Stevenson), "Fleurieux 1836". Magnin (1907) refers to a further 130 drawings at Lyon.

BIBLIOGRAPHY and BIOGRAPHY: AG 12(3): 543; Barnhart 3: 171 (b. 17 Sep 1775, d. 12 Mar 1866); BL 2: 194, 709; MD 209-210; PR 368 (sub Balbis), ed. 1: 8691; TL-2/281; see L.C. Lortet.

BIOFILE: Candolle, Alph. de, Phytographie 444. 1880 (coll. LY in herb. Lortet).
 Gager, C.S., Brookl. Bot. Gard. Rec. 27(3): 202. 1938 (director of Lyon bot. gard. for three days: 19-21 Aug 1830).
 Magnin, A., Bull. Soc. bot. France 23: clxxxv. 1876 (herb. specimen in herb. Soc. linn. Lyon); Ann. soc. bot. Lyon 19: 14. 1894, 32: 7-8, 62, 103, 125. 1907 (epon., bibl.), 35: 34, 67. 1910; Les Lortet, botanistes lyonnais et le botaniste Roffavier, Lyon 1913, 81 p. (portr., bibl.), also in Ann. Soc. bot. Lyon 37: 29-109. 1912 (1913), 37: 66-72. 1913 (biogr.).

9432. *Supplément à la flore Lyonnaise*, publiée par le docteur J.B. Balbis en 1827 et 1828; ou description des plantes phanérogames et cryptogames découvertes depuis la publication de cet ouvrage; suivi d'un tableau général contenant la nomenclature méthodique des espèces agames décrites dans la Flore lyonnaise, conjointement avec celles qui ont été trouvées depuis la même époque dans les environs de Lyon ... Lyon (imprimerie typographique et lithographique de Louis Perrin, ...) 1835. Oct. (in fours). (*Suppl. fl. Lyonn.*).
Collaborators: Marc Antoine Rollet (1803-1882), Nicolas Charles Seringe (1776-1858).
Publ.: Apr 1835 (see MD p. 210), p. [1]-91, [92]. *Copies*: HU, L. – Published anonymously, distributed by Roffavier in Apr 1835, saying that il représentait le travail de "various botanists of this town ...". Attributed to Roffavier by Magnin (1907, 1910).

Rogers, Charles Coltman [Coltman-Rogers, C.] (1854-1929), British botanist, and country gentleman at Stanage Park, Radnorshire. (*C.C. Rogers*).

HERBARIUM and TYPES: Unknown.

BIBLIOGRAPHY and BIOGRAPHY: Barnhart 3: 171 (b. 12 Mai 1854); BM 7: 1088; Desmond p. 143 (Coltman-Rogers, C.), 528, (d. 20 Mai 1929); MW p. 416.

9433. *Conifers* and their characteristics ... London (John Murray, ...) 1920. Oct. (*Conifers*).

London issue: Jan-Jul 1920 (Nat. Nov. Aug 1920), p. [i]-xiii, [1]-333. *Copy*: LC.
New York issue: 1920, p. [i]-xiii, [1]-333. *Copy*: ILL (inf. D.P. Rogers). – New York (The Macmillan Company) 1920.

Rogers, Charles Gilbert (1864-1937), British (English) forester and botanist; in the Indian Forest Service as assistant conservator, Bengal 1888-1890; instructor Indian forestry school, Dehra Dun 1890-1906; conservator Central Provinces 1906-1910, chief conservator Burma 1913-1919; collected in India, Burma and East Africa. (*C.G. Rogers*).

HERBARIUM and TYPES: Material at BM, DBN, E, K, MO, OXF.

BIBLIOGRAPHY and BIOGRAPHY: Barnhart 3: 171 (d. 25 Nov 1937); BL 1: 95, 315; Desmond p. 528 (d. 18 Nov 1937); IH 2: (in files); Rehder 5: 734.

BIOFILE: Anon., Who was who 3(1929-1940): 1162.
Fischer, C.E.C., Bull. misc. Inf. Kew 1938. 84. (obit.).
Rogers, C.G., A manual of forest engineering for India, 3 vols. Calcutta 1900-1902.
Rogers, L., Proc. Linn. Soc. 150: 333-334. 1938 (obit.).
Verdoorn, F., ed., Chron. bot. 4(3): 262. 1938 (d. 25 Nov 1937).

NOTE: We were unable to locate a copy of his *A preliminary list of the plants of the Andaman Islands*, 1903, ii, 51 p.

Rogers, Donald Philip (1908-x), American botanist (mycologist); B.A. Oberlin Coll. 1929; Dr. phil. Univ. Nebraska 1935, asst. botanist Univ. Nebraska 1929-1930; id. Univ. Iowa 1931-1935 (in Iowa Lakerside Laboratory 1932-1934); instructor botany Oregon State College 1936-1940; idem Brown Univ. 1941-1942; assoc. professor of biology, Amer. International College 1942-1945; asst. professor of botany, Univ. Hawaii 1945-1947; curator of cryptogamic botany, New York Botanical Garden 1947-1957; professor of botany and curator of mycological collections Univ. Illinois 1957-1976; emeritus professor ib. 1976-x; secretary of the Special Committee for Fungi and Lichens of the International Botanical Congresses 1950-1969, taxonomist, nomenclaturist and historian of mycology. (*D.P. Rogers*).

HERBARIUM and TYPES: BISH, CUP, FH, IA, IAC, ILL, NY, OSC, TRTC.

BIBLIOGRAPHY and BIOGRAPHY: Barnhart 3: 171 (b. 5 Feb 1908; orig. info.); Bossert p. 336; IH 1 (ed. 1): 69, (ed. 2): 88, (ed. 3): 111, (ed. 4): 176, (ed. 5): 191; Lenley p. 351-352; PH 209; Roon p. 95; Stevenson p. 1256; TL-2/781.

BIOFILE: Anon., J. New York Bot. Gard. 48: 162-163. 1947 (to NY 1 Sep 1947; portr.).
Rogers D.P., A brief history of mycology in North America. Second Int. mycol. Congr. 2, 1977, 65 p.
Rogerson C.T., Mycologia index 1033-1034. 1968 (list of publ. in Mycologia).
Verdoorn, F., ed., Chron. bot. 2: 404. 1936, 3: 362. 1937, 6: 428. 1941, 7: 88. 1943.

COMPOSITE WORKS: (1) Managing editor *Mycologia* 1948-1957; editor-in-chief 1958-1960.
(2) With A.M. Rogers and E.V. Seeler, Jr., *Index botanicus sistens omnes fungorum species in C.H. Persoonii Mycologia Europeae enumeratas*. Cambridge 1942, 37 p.
(3) Bibliographic note for the reprinted edition, *in* E.M. Fries, *Systema mycologicum*, Facsimile ed., New York 1952, p. i-iv.

EPONYMY: *Rogersella* A.E. Liberta et A.J. Navas (1978); *Rogersia* C.A. Shearer et J.L. Crane (1976); *Rogersiomyces* J.L. Crane et J.D. Schoknecht (1978).

9434. *Notes on the lower Basidiomycetes* ... Iowa City, Iowa (Published by the University [of Iowa]) 1935.
Publ.: 1 Jun 1935 (cover), p. [1]-43, *pl. 1-3*. *Copy*: FH. – Issued as Univ. Iowa Stud. nat. Hist. 17(1): 1-43, *pl. 1-3*. 1935.

Rogers, Frederick Arundel (1876-1944), English-born South African clergyman and botanist; son of W. Moyle Rogers; studied at Oxford; to South Africa for the South African Church Railway Mission 1904-1911, from 1911 as head of this Mission; Archdeacon of Pietersburg, Transvaal 1914; collected plants in S. Africa, Rhodesia, Belgian Congo 1904-1925, Iraq 1929-1930, Syria, Cyprus, Greece and Switzerland 1930. (*F.A. Rogers*).

HERBARIUM and TYPES: A, BAG, BM, BOL, BP, BR, BUL, DPU, EPM (Brit. herb.), FHO, G, GRA, J, K, L, LIV, NH, NY, PRE, S, SAM, SRGH, STE, US.

BIBLIOGRAPHY and BIOGRAPHY: Barnhart 3: 171 (b. 3 Jan 1876); Bossert p 336; Desmond p. 528; IH 1 (ed. 6): 363, (ed. 7): 341, 2: (files); Urban-Berl. p. 389.

BIOFILE: Blakelock, R.A. & E.R. Guest, Fl. Iraq 1: 116. 1966 (coll. Iraq 1929-1930).
Brown, N.E., Bull. misc. Inf. Kew 1921: 289-301 (coll.).
Bullock, F.A., Bibl. S. Afr. bot. 96. 1978.
Exell, A.W. & G.A. Hayes, Kirkia 6(1): 100. 1967 (coll. Fl. zambes.).
Exell et al., Bol. Soc. Broter. ser. 2. 26: 218. 1952 (coll.).
Gomes e Sousa, A. de F., Moçambique, Docum. trimestr. 57-58. 1949 (d. Jun 1944).
Gunn, M. & L.E. Codd, Bot. expl. S. Afr. 298-299. 1981.
White, A. & B.L; Sloane, The Stapelieae 102, 138. 1933, ed. 2. 134, 391. 1937.
Wild, H., C.R. AETFAT 4: 169-170, 185-186. 1962.

NOTE: We were unable to locate a copy of his *Provisional list of flowering plants and ferns found in the district of Albany and Bathurst*, 1909, iv, 21 p.

Rogers, Henry Darwin (1809-1866), American geologist; professor of chemistry and natural philosophy, Carlisle, Pa.; studied in England 1831-1833; M.A. Univ. Pennsylvania 1834; professor of geology ib. 1835-1846; geologist on the Geological Survey of Pennsylvania 1836-1856; from 1846-1857 residing at Boston working on the report of the Geol. Surv. Pa.; from 1857-1866 Regius professor of natural history at Glasgow, Scotland. (*H.D. Rogers*).

HERBARIUM and TYPES: Unknown. – Archival material e.g. in Smithsonian Archives SIA 7230, 7177.

BIBLIOGRAPHY and BIOGRAPHY: BM 4: 1721; CSP 5: 253-254, 8: 769, 18: 268; ME 1: 225, 3: 646-647 (index); Merrill p. 709, 739; PH p. 520 [index]; Quenstedt p. 365.

BIOFILE: Allibone, S.A., Crit. dict. Engl. lit. 2: 1850. 1878 (bibl.).
Anon., Amer. J. Sci. ser. 2. 42: 136-138. 1866 (obit.); Pop. Sci. Mon. 50: 258-264. 1896 (biogr. sketch); Proc. Amer. Acad. Arts Sci. 7: 309-312. 1868 (obit.).
Clarke, J.M., James Hall of Albany 563 [index]. 1923.
Gregory, J.W., Henrey Darwin Rogers, an address. Glasgow 1916, 38 p. (portr., biogr. sketch, bibl. by C.M. Leitch).
Poggendorff, J.C., Biogr.-Lit. Handw.-Buch 3: 1134-1135. 1898 (bibl., d. 29 Mai 1866).
Ruschenberger, W.S.W., Proc. Amer. philos. Soc. 23: 104-146. (obit. of Henry Darwin R.'s brother Robert E., with many notes also on Henry).
Zittel, K.A. von, Gesch. Geol. Paläont. 862 [index]. 1899; Hist. geol.-palaeontol. 296, 304, 309, 357, 442, 443. 1901.

EPONYMY: *Rogersia* Fontaine (1889) was named for William Barton Rogers (1804-1882), American geologist and paleontologist.

9435. *The geology of Pennsylvania* a government survey. With a general view of the geology of the United States, essays on the coal- formation and its fossils, and a description of the coal-fields of North America and Great Britain ... vol. ii. Philadelphia (J.B. Lippincott & Co.) 1858. Qu. (*Geol. Pennsylvania*).
Publ.: in three parts, 1858. *Copies*: NY (Philadelphia imprint), PH (id.), USDA (New York imprint for 2(1): (D. van Nostrand, ...) 1868 sic), USGS (Philadelphia imprint).

1: frontisp., p. [i*-ii*], [i]-xxvii, err. slip, [1]-586, *24 pl.*, *11* geol. sect. (in copy PH *16 pl.*, *8* sect.).
2(1): [i*-ii*], frontisp., [i]-xxiv, [1]-666, *pl. 1-7*, *12 pl.* in text.
  (2): [i], [667]-1045, [1046, err.], *7 pl.* in text, *pl. 1-6*, [7] (sect.), *pl. 1-23* (fossil pl., R.K. Greville, uncol. liths.).
Atlas: 7 folded maps.

Rogers, Julia Ellen (1866-?), American author of popular books on trees; studied at Iowa State University and Cornell Univ., M.S. Cornell 1902; high school teacher in Des Moines 1881-1887, and Cedar Rapids, Iowa, 1892-1900; from 1903 lecturer in civic improvement and nature subjects, residing in New Jersey, later in Long Beach, California. (*J.E. Rogers*).

HERBARIUM and TYPES: Some mosses at NY.

BIBLIOGRAPHY and BIOGRAPHY: Barnhart 3: 171 (b. 21 Jan 1866); PH 98; Tucker 1: 604.

BIOFILE: Anon., Who's who in America 1908-1909: 1613, 15: 1791. 1928, 17: 974. 1932; Women's Who''s who of America 1914-1915: 693. 1976.
Comstock, A.B., Cornell Rural School Leaflet 17(1): 60. 1923 (portr., brief biogr. sketch).

9436. *The tree book* a popular guide to a knowledge of the trees of North America and to their uses and cultivation ... New York (Doubleday, Page & Company) 1905. (*Tree book*).
Publ.: 1905 (NY rd. Feb 1906; TBC 8 Feb 1906; Bot. Centralbl. 8 Mai 1900), p. [i]-xx, [1]-589, *16 col. pl.*, 160 photogr. *Copies*: HH, USDA.
Other issue: 1908, same composition. *Copy*: NY.
Re-issue: 1935 , p. [i]-xviii, [1]-569, *16 col. pl.*, 160 photogr. *Copy*: USDA.

9437. *The book of useful plants* ... Garden City New York (Doubleday, Page & Company) 1913. (*Book usef. pl.*).
Publ.: 1913, p. [i]-xiv, [1]-374. *31 pl.* (incl. frontisp.). *Copies*: NY, USDA.

Rogers, Richard Sanders (1862-1942), Australian physician and botanist (orchidologist) at Adelaide; B.S. Univ. Adelaide 1883; Master of Surgery Edinburgh 1891; practicing physician at Adelaide from 1891; D.Sc. h.c. Adelaide 1936. (*R.S. Rogers*).

HERBARIUM and TYPES: AD; other material at BM, MEL.

BIBLIOGRAPHY and BIOGRAPHY: Barnhart 3: 172 (b. 30 Apr 1942); BJI 2: 147; HR (b. 2 Dec 1861); IH 1 (ed. 6): 363, (ed. 7): 341, 2: (files); TL-2/544.

BIOFILE: Anon., Austral. Encycl. 5: 237, 7: 479-480. 1965; Trans. R. Soc. Australia 66: 2. 1942 (obit., d. 18 Mar 1842); Who's who [Australia] 1941: 576.
Hall, N., Botanists of the Eucalypts 110. 1978.
J.G.W., Austral. J. Sci. 1942 (Apr): 154 (obit).
Osborn, T.G.B., Proc. Linn. Soc. 155: 306-307. 1944 obit.)
Pescott, E.E., Victor. Natural. 49: 196-198. 1932 (portr.), 59(1): 20. 1942 (obit.).
Simmons, J.T., Orchadian 7(3): 65. 1982 (Rogers on Schlechter).

COMPOSITE WORKS: Contributed *Orchidaceae* to J.M. Black, *Fl. S. Austral.* ed. 1, 1822-1829, see TL-2/544.

9438. *Some Australian Orchids*. Reprinted from the "Childrens hour", a paper read in the public schools of South Australia ... Adelaide (R.E.E. Rogers, ...) 1909. (*Austral. Orchid.*).
Publ.: 1909, p. [1]-23. *Copy*: HH (Oakes Ames orchid library). – Original publication in the "Childrens hour" not seen.
Ed. 2: see below.

9439. *An introduction to study of South Australian Orchids* ... second edition, revised and enlarged ... Adelaide (R.E.E.Rogers, Government Printer, ...) 1911. Oct. (*Intr. S. Austral. orchid.*).
Ed. 1: see above, 1909, *Austral. orchid.*
Ed. 2: 1911, p. [i], [1]-52, [53-63, gloss., bibl. index]. *Copies*: NSW, US, USDA.

Rogers, Walter E (1890-1951), American botanist, ecologist and ornithologist; M.S. Iowa 1916; technician St. Louis biol. lab. 1913-1914; asst. bot. Iowa 1914-1917; professor of biology Westminster Coll., Pa., 1917-1919; professor of biology and botany at Lawrence College 1919-1951. (*W.E Rogers*).

HERBARIUM and TYPES: Unknown.

BIBLIOGRAPHY and BIOGRAPHY: Barnhart 3: 172 (b. 24 Feb 1890, d. 5 Oct 1951); Kew 4: 505; PH 209.

BIOFILE: Anon., Auk 69: 350. 1952 (obit.); Science 114: 516. 1951 (d.).
Cattell, J.M. & J. eds., Amer. men Sci ed. 4: 832. 1927, ed. 5: 948. 1933, ed. 6: 1201. 1938, ed. 7: 1502-1503. 1944, ed. 8: 2101. 1949.

9440. *Tree flowers of forest, park and street* by Walter E Rogers ... the drawings from nature by Olga A. Smith ... Appleton, Wisconsin (published by the author) 1935. (*Tree fl.*).
Publ.: 1935 (NY Jan 1936), p. [i-xvi], 1-500, [501, colo.] *122 pl.* (photos). *Copies*: BR, HH, MO, NY, USDA.
Reprint: 1965, Dover Publications, Inc., New York, p. [iii]-xii, 1-499, *122 pl.* in text. *Copies*: FAS, G.

Rogers, William Moyle (1835-1920), British (English) botanist (batologist) and clergyman, vice principal Theological College Cape Town 1860-1862; curate at various locations in Devon, Isle of Wight, Worcestershire and Dorset; ultimately Vicar of Bridgerule Devon 1882-1885; retired to Bournemouth for health reasons. (*W.M. Rogers*).

HERBARIUM and TYPES: HTE (except *Rubus*), BM (*Rubus*). – Further material at BEL, C, K, LIV, MANCH, MO, OXF, SLBI, UCSW.
Exsiccatae: see under E.F. Linton for *British Rubi, in dried specimens* (E.F. & W.R. Linton, R.P. Murray, W.M. Rogers, 1892-1895).

BIBLIOGRAPHY and BIOGRAPHY: AG 6(1): 462; Barnhart 3: 172; BB p. 260; BL 2: 709 [index]; BM 4: 1721; Bossert p. 336; Clokie p. 234; CSP 11: 207-208, 18: 269; Desmond p. 528 (b. 12 Jul 1835, d. 26 Mai 1920); Kew 4: 505; Rehder 5: 734; TL-2/4858; see E.F. and W.R. Linton; Tucker 1: 604; Urban-Berl. p. 389.

BIOFILE: Babington, C.C., Mem. Journ. Bot. Corr. 474. 1897.
Druce, W.M., Bot. Soc. Exch. Cl. Brit. Isl. Rep. 6(1): 105-106. 1921 (obit.); Proc. Linn. Soc. 142: 195-196. 1921 (obit.).
Gunn, M.& L.E. Codd, Bot. expl. S. Afr. 299. 1981.
Horwood, A.R. & C.W.F. Noel, Fl. Leicestershire ccxxvii. 1933.
Kent, D.H., Brit. herb. 76. 1957 (herb.).
Linton, E.F., J. Bot. 58: 161-164. 1920 (obit, portr.).
Martin, W.K. & G.T. Fraser, Fl. Devon 775-776. 1939.
Murray, G., Hist. coll. BM(NH) 1: 177. 1904.
Praeger, R.L., Proc. R. Irish Acad. 7: cxli. 1901; Botanist in Ireland 78. 1934.
White, A. & B.L. Sloane, The Stapelieae ed. 2. 134. 1937.

9441. *An essay at a key to British Rubi* ... London (printed by West, Newman & Co., ...) 1893. Oct.
Publ.: Feb 1893, p. [1]-56. *Copy*: G. – Reprinted and to be cited from J. Bot. 30: 108-114. Apr, 142-145. Mai, 200-205. Jul, 230-235. Aug, 266-272. Sep, 299-305. Oct, 333-341. Nov 1892, 31: 3-10. Jan, 40-49. Feb 1893. – See also "British Rubi again", J. Bot. 32: 374. Dec 1894.

9442. *Rubi notes* [Reprinted from the "Journal of Botany" for February, 1894]. Oct.
Publ.: Feb 1894, p. [1]-10. *Copy*: G. – Reprinted and to be cited from J. Bot. 32: 40-50. Feb 1894.

9443. *On the Rubi list in "London catalogue"*, ed. 9. [Reprinted from the "Journal of Botany" for February, March, and April, 1895]. Oct.
Publ.: Apr 1895, p. [1]-16. *Copy*: G. – Reprinted and to be cited from J. Bot. 33: 45-49. Feb, 77-82. Mar, 100-106. Apr 1895.

9444. *Handbook of British Rubi* ... London (Duckworth and Co. ...) 1900. Oct. (*Handb. Brit. Rubi*).
Publ.: Jul-Aug 1900 (p. ix: Apr 1900; Bot. Zeit. 1 Oct 1900; J. bot. Sep 1900; Bot. Centralbl. 15 Aug 190; NY Oct 1900; Nat. Nov. Sep(1, 2) 1900), p. [i]-xiv, [1]-111. *Copies*: BR, G, HH, NY, USDA.
Ref.: Linton, E.F., J. Bot. 38: 401-403. 1900 (rev.).

Rohde, Michael (1782-1812), German (Bremen) botanist and physician; Dr. med. Göttingen 1804; continued his medical studies at various universities in Germany, Austria and France 1804-1808; practicing physician and active amateur botanist at Bremen 1809-1812; state examination med. Bremen 1809. (*Rohde*).

HERBARIUM and TYPES: LE (via herb. F.K. Mertens) and B (mainly destr., via A.W. Roth); further material at FI.

BIBLIOGRAPHY and BIOGRAPHY: Backer p. 498 (epon.); Barnhart 3: 172 (b. 25 Jul 1782, d. 28 Mai 1812); BM 4: 1732; Herder p. 308; Kew 4: 506; PR 7731, ed. 1: 8693; Rehder 2: 812, 3: 720; Tucker 1: 604.

BIOFILE: Buchenau, Fr., Abh. naturw. Ver. Bremen 1: 237-244. 1866 (1867) (biogr. based on original sources, e.g. curr. vitae submitted by R. 1809 for his med. degree); Bot. Zeit. 25: 216. 1867 (obtained much info. on Rohde from working through the Bremen herb.), 26: 310. 1868 (Rohde material in the herbarium of A.W. Roth, q.v., now mainly destroyed).
Desvaux, N.A., J. de Bot. ser. 2. 1: 93-95. 1813 (appreciation).
Focke, W.O., Abh. naturw. Ver. Bremen 9(3): 329. 1886 (b. 25 Jul 1782, d. 28 Mai 1812).
Steinberg, C.H., Webbia 32(1): 36. 1977 (herb. material at FI).

EPONYMY: *Rhodea* Endlicher (1836, *orth. var.*); *Rohdea* A.W. Roth (1821).

9445. *Monographiae Cinchonae generis specimen* sistens historiam ejus criticam ad introductionem in hoc genus inservientem. Dissertatio inauguralis medica quam illustris medicorum ordinis consensu et auctoritate in Alma Georgia Augusta pro gradu doctoris medicinae et chirurgiae rite consequendo offert auctor Michael Rohde Bremanus ...d. xxx august. mdccciv. Gottingae [Göttingen] (Literis Barmeierianis. Apud J.C. Baier) [1804]. Oct. (*Monogr. Cinch. spec.*).
Publ.: 30 Aug 1804, p. [1]-56. *Copy*: C (inf. A. Hansen). – Completed version: *Monogr. Cinch. tent.*, see below.

9446. *Monographiae Cinchonae generis tentamen*. Fragmentum ex materia medica quod botanice, pharmacognostice, chemice et medice tractavit. Michael Rohde ... Gottingae (impensis VandenHoeck et Ruprecht) 1804. Oct. (*Monogr. Cinch. tent.*).
Publ.: on or shortly after 30 Aug 1804 (date on which degree Dr. med. was awarded), p. [i]-x, [1]-189, [190, err.]. *Copies*: HH, NSUB, NY, USDA. – Preceded by an abbreviated version used as doctor's dissertation *Monogr. Cinch. spec.* (1804) see above.

Rohlena, Joseph (1874-1944), Czech (Bohemian) botanist at Praha; teacher at elementary schools in Lysa nad Labem and Starých Benatky 1893-1895, id. in Praha from 1895-1931; in retirement in Praha; travelled widely in the Balkan peninsula, especially in Montenegro; co-founder of the Czech botanical society in 1912. (*Rohlena*).

HERBARIUM and TYPES: PRC; further material BM, BRNM, C, GB, K, LAU, M, PACA, PR, S, Z. *Exsiccatae: Plantae balcanicae* (montenegrinae) *exsiccatae*. Ser. 1 (nos. 1-150), ser. 2 (nos. 1-100), Praha 1932.

BIBLIOGRAPHY and BIOGRAPHY: AG 5(1): 28, 5(2): 497; Backer p. 660; Barnhart 3: 172 (b. 3 Jan 1874); BFM no. 1505; Biol.-Dokum. 15: 7617; BJI 2: 147; BM 4: 1722; Bossert p. 336; CSP 18: 272; Futak-Domin p. 504-505; Hortus 3: 1202; IH (ed. 2): 100, (ed. 3): 126, (ed. 7): 341; Kew 4: 506; Klášterský p. 157 (d. 26 Jan 1944); Maiwald p. 237, 251, 253.

BIOFILE: Anon., Bot. Centralbl. 93: 48. 1903 (with Bubak on trip to Montenegro); Preslia 9: 95. 1930.
Domin, K., Věda přír. 15: 27-29. 1934 (tribute 60th birthday, portr.).
Morávek, A., Acta Mus. reginaehrad. [Königgrätz], Sci. nat. 15: 177-181. 1974 (biogr., portr., b. 3 Jan 1874, d. 26 Jan 1944).
Novak, F.A., Cas. národn. Mus. 108: 36-43. 1934 (biogr sketch, portr., bibl.), Preslia 24: 415-418. 1944 (obit., bibl.).
Podpera, J., Naše Věda 23: 140. 1944 (obit.).
Verdoorn, F., ed., Chron. bot. 3: 91. 1937, 4: 187, 562. 1938, 5: 309. 1939, 6: 303. 1941.

9447. *Erster Beitrag zur Flora von Montenegro* ... [S.B. böhm. Ges. Wiss. 1902]. Oct.
Publ.: 10 Jul 1902; the first of a series of 12 papers published as follows. *Copies*: M (complete set), partial sets at BR, G, H, HH, LD, MO, U.

no. reprint	pagination	date	publication in journal
1	[1]-26	10 Jul 1902	S.B. böhm. Ges. Wiss. 1902 (32):
		10 Oct 1902	1-26. 1903
2	[1]-37	20 Dec 1903	Id. 1902 (39): 1-37. Jan-Jun 1903
3	[1]-71	25 Jan 1905	Id. 1903 (17): 1-71. Nov-Dec 1903
4	[1]-108		Id. 1904 (38): 1-108. Jan-Jun 1904
5	[1]-143		Id. 1912 (Beitr.) (1): 1-143. 1913
6	[3]-24		Acta bot. Boh. 2: 3-24. 1923
7	1-10		Id. 3: 41-50. 1924
8	[1]-29		Vestn. Král.* 1931 (35): 1-29. 1932
9	[1]-20		Id. 1933 (12): 1-20. 1934
10	[1]-19		Id. 1935 (3): 1-19. 1936
11	[1]-21		Id. 1936 (22): 1-21. 1937
12	[i], [1]-16		Id. 1939 (3): 1-16. 1939

For further papers on the flora of Montenegro see Novák (1952), and the *Consp. fl. montenegr.* (1942), below.
*Vestn. Král. České Spolěcn. Nauk, Tř. mat.-přír. (continuation of S.B. böhm.-Ges. Wiss.).

9448. *Additamenta ad floram dalmaticam* [Acta bot. bohem. 1, 1922]. Oct.
Publ.: 1922, p. [1]-9. *Copy*: M. – Reprinted and to be cited from Acta bot. bohem. 1: 26-34. 1922.

9449. *O vegetačních rozdílech mezi severní a jižní exposicé Čechách.* (Über die Vegetationsunterschiede zwischen der Nord– und Südexposition in Böhmen). [Preslia v. 1927]. Oct.
Publ.: 1927, p. [1]-13. *Copy*: M. – Reprinted and to be cited from Preslia 5: 52-64. 1927.

9450. *Beitrag zur Flora Albaniens* ... Praha (Nákladem Královské České Společnosti Nauk ...) 1937. Oct.
Publ.: 1937, p. [i], [1]-14. *Copy*: M.– Reprinted and to be cited from Věstn. Král. České Společn. Nauk, Tř. mat.- přír. 1937(2): 1-14. 1937.

9451. *Beitrag zur Flora der Hercegovina* ... Praha (Nákladem Královské České Společnosti Nauk ...) 1938. Oct.
Publ.: 1938, p. [i], [1]-19. *Copy*: M. – Reprinted and to be cited from Věstn. Král. České Společn. Nauk, Tř. mat.-přír. 1938(8): 1-19. 1938.

9452. *Beitrag zur Flora des Gebirges Šar planina* ((Šar-Dagh) ... Praha (Nákladem Královské České Společnosti Nauk ...) 1937. Oct.
Publ.: 1937, p. [i], [1]-12. *Copy*: M. – Reprinted and to be cited from Věstn. Král. České Společn. Nauk, Tř. mat.-přír. 1937(16): 1-2. 1937.

9453. *Conspectus Florae montenegrinae* ... Praha (sumptibus Societatis botanicae bohemicae Praha ...) 1942. Oct. (*Consp. Fl. montenegr.*).
Publ.: 1942, p. [1]-506. *Copy*: M. – Issued as Preslia vols. xx-xxi.

Rohlfs, Gerhard [Friedrich] (1831-1896), German (Bremen) traveller and collector in North Africa; studied medicine in Heidelberg, Würzburg and Göttingen; physician in the French colonial army 1855-1860, later exploring Northern Africa. (*Rohlfs*).

HERBARIUM and TYPES: Material at B (mainly destroyed), CORD, K, L, M, P (herb. Cosson).

BIBLIOGRAPHY and BIOGRAPHY: ADB 53: 440-449 (V. Hantzsch); Barnhart 3: 172 (14 Apr 1831, d. 2 Jun 1896); Biol.-Dokum. 15: 7617; BM 4: 1722 (travelogues); Bossert p. 336; Herder p. 58; Rehder 1: 485; TL-2/see A.C.H. Braun, P.F.A. Ascherson; Urban-Berl. p. 389.

BIOFILE: Anon., Bot. Zeit. 31: 704. 1873 (P. Ascherson will be botanist on R.'s Libyan trip); Brockhaus' Conversations-Lexikon ed. 13. 13: 758-759. 1886; Bull. Soc. bot. France 20: 183. 1873, 21 (bibl.): 95. 1874; Flora 50: 367. 1867 (travel rep.), 51: 428. 1868 (Rohlfs and H.W. Schimper), 56:: 431. 1873 (Ascherson to accompany R. to Libya); Polybiblion 77: 80-82. 1896 (biogr. data, bibl.).
Ascherson, P., *in* Durand, E. & G. Barratte, Fl. Libyc. prodr. xxiv-xxv. 1910 (itin.).
Cufodontis, G., C.R. AETFAT 4: 246. 1962 (1867-1868 in Ethiopia).
Embacher, F., Lexikon Reisen 249-251. 1881.
Guenther, K., Gerhard Rohlfs, Lebensbild eines Afrikaforschers, ed. 2, 1912 (main biogr., portr.), viii, 352 p., 44 pl., map.
Letouzey, R., Fl. Cameroun 7: 9. 1968.
Nissen, C., Zool. Buchill. 347. 1969.
Pampanini, R., Prodr. fl. Cirenaica xviii. 1920; Flora Cirenaica xviii. 1930.
Poggendorff, J.C., Biogr.-lit. Handw.-Buch 3: 1136-1137. 1898 (bibl.; brief list of travels; d. 3 Jun 1896), 4(2): 1264. 1904 (d. 2 Jun 1896, add. bibl.).
Reuss, M. & G.W. Hartig, South Atlantic Quart. 74(1): 74-85. 1975 (Bismarck's imperialism amd the Rohlfs mission to East Africa 1884-1885).
Schweinfurth, G., Illustr. deut. Monatschr. (Westermann) 82: 565-578. 1897 (obit., portr.).
Vallot, J., Bull. Soc. bot. France 29: 199. 1882.
Wichmann, A., Peterm. Mitt. J. Perthes geogr. Anst. 42(2): 146-147. 1896 (obit., d. 2 Jun 1896).
Wolkenhauer, W., Deut. geogr. Blätt. 19(4): 165-182. 1896 (obit., bibl.), Globus 70: 31-33. 1896 (obit.).
Zittel, K., Jahresber. geogr. Ges. München 16: 310-313. 1896 (obit.); Gesch. Geol. Paläont. 299, 300, 560. 1899.

EPONYMY: *Rohlfsia* A. Schenk (1880).

9454. *Quer durch Afrika*. Reise vom Mittelmeer nach dem Tschad-See und zum Golf von Guinea ... Leipzig (F.A. Brockhaus) 1874-1875. 2 vols. Oct. (*Quer d. Afrika*).
1: 1874 (p. vi: Jun 1874), p. [i]-x, [1]-352. 1 map.
2: 1875, p. [i]-viii, [1]-298, [299, err.], 1 map.

Copies: GOET (inf. G. Wagenitz), USDA. – *Botanischer Anhang* (p. 276-298) with identifications by P. Ascherson, A. Braun, E. Cosson, A. Garcke, G. Schweinfurth.
Preliminary notice on the botanical results: P. Ascherson, Bot. Zeit. 32: 609-619 ["916"]. 18 Sep, 625-631. 25 Sep, 641-647. 2 Oct 1874. See also G. Rohlfs, Expedition zur Erforschung der Lybischen Wüste, Cassel 1875, xii, 340 p., also with contributions by P. Ascherson, W. Jordan, K. Zittel (P. Ascherson, Bot. Zeit. 34: 334-335. 1876).
Ref.: Ascherson, P., Bot. Zeit. 33: 705-712. 22 Oct 1875 (rev. 1, 2).

9455. *Drei Monate in der Libyschen Wüste* ... Cassel (Verlag von Theodor Fischer) 1875. Oct. (*Drei Monate libysch. Wüste*).
Collaborators: P. Ascherson, W. Jordan, K. Zittel, Ph. Remelé.
Publ.: 1875, p. [i*], [i]-viii, [ix-xii], [1]-340, 16 photos, *11 pl.*, 18 woodcuts, map. *Copies*: BR, GOET (inf. G. Wagenitz), LD (inf. O. Almborn), USDA. – Chapters 9 (p. 229-263) and 10 (p. 264-289) by P. Ascherson.

9456. *Kufra*. Reise von Tripolis nach der Oase Kufra. Ausgeführt im Auftrage der Afrikanischen Gesellschaft in Deutschland von Gerhard Rohlfs. Nebst Beiträgen von P. Ascherson, J. Hann, F. Karsch, W. Peters, A. Stecker ... Leipzig (F.A. Brockhaus) 1881. Oct. (*Kufra*).
Publ.: 1881 (p. [v]: Aug 1881), frontisp., p. [i]-viii, [1, h.t.], [1]-559, [1, h.t.], 21 charts, 3 maps, *11 pl.* (7 h.t.). *Copies*: BR, FI. – Contains in Abth. 2, no. 7: P. Ascherson, *Die aus dem mittlern Nordafrika* ... *bekannt gewordenen Pflanzen*.

Rohr, Julius Philipp Benjamin von (ca. 1737-1793), Danish (German parents) surveyor, agronomist and soldier (first lieutenant of the militia) on the Danish Island of St. Croix 1757-1791; made several trips in the West Indies and adjacent South America between 1784 and 1791 to study cotton culture; died on the high seas when travelling to Guinea. (*Rohr*).

HERBARIUM and TYPES: C (herb. Vahl); other material at B-Willd., BM, BR, PH, UPS (Thunberg). – Rohr sent many of his collections to M. Vahl who described them in his Eclogae. Rohr himself published 8 new (or putatively new) genera in Skr. Nat. Selsk. Kjøbenhavn 2: 205-221. 1792, with species diagnoses of Vahl.

NOTE: Not to be confused with the Saxonian Julius Bernard von Rohr (1688-1742). – The name of the collector of Ethiopian plants J.R. Roth, who was sent to Abessynia in 1841 by the British government, and whose plants are at K, has sometimes erroneously been cited as "Rohr", see Bull. misc. Inf. Kew 1897: 114.

BIBLIOGRAPHY and BIOGRAPHY: Backer p. 660; Barnhart 3: 172; CSP 5: 258; Dawson p. 709; Dryander 3: 36; Herder p. 419; IH 2: (in files); Langman p. 639; Lasègue p. 489, 503; MW suppl. p. 294; PR 8697; Rehder 1: 264; TL-1/1348; see M. Vahl; Zander ed. 10, p. 708, ed. 11, p. 808.

BIOFILE: Christensen, C., Dansk. bot. hist. 1: 116. 1924, 2: 91. 1924.
Gilbert, P., Comp. biogr. lit. entom. 322. 1977.
Hornemann, J., Nat. Tidsk. 1: 578-579. 1837 (epon.).
Jessen, K.F.W., Bot. Gegenw. Vorz. 342. 1884.
Kiaerskou, H., Bot. Tidsskr. 23: 44. 1900 (also Bot. Centralbl. 85: 343. 1901; coll. at C).
Millspaugh, C.F., Publ. Field Mus. 68, Bot. 1: 448. 1902.
Nicolson, D.H., Taxon 30: 489-494. 1981 (see *Note* below; on 1792 Plante-Slaegter).
Papavero, N., Essays hist. neotrop. Dipterol. 1: 20-21, 29. 1971.
Poiret, J.L.M., Enc. méth., Bot. 8: 754-755. 1808.
Rydberg, P.A., Augustana Coll. Libr. Publ. 5: 19. 1907.
Sagot, P., Ann. Sci. nat. Bot. 10: 374. 1880 (in Suriname, 1785 [?]).
Warming, E., Bot. Tidsskr. 12: 82. 1880.

EPONYMY: The etymology of *Rohria* Schreber (1789) was not given and that of *Rohria* M. Vahl (1791) could not be determined; *Rohra* Cothenius (1790, orth. var. of *Rohria* Schreber).

NOTE: See D.H. Nicolson (1981) for a commentary on Plante-Slaegter beskrevne af Hr. Oberst-Lieutenant von Rohr, med tilfoiede Anmaerkninger af Hr. Professor Vahl, Skrivter af Naturhistorie-Selskabet 2(1): 205-221. *pl. 6-10.* 1792. Nicolson provides authorship, nomenclature and current identification of all names.

Rohrbach, Paul (1847-1871), German (Prussian; Berlin) botanist; studied natural sciences in Göttingen with A. Grisebach, in Berlin with A. Braun; Dr. phil. Berlin 1868; curator of the Göttingen University Herbarium. (*Rohrb.*).

HERBARIUM and TYPES: B, GOET. – Rohrbach's work on Silene was based in first instance on the A. Grisebach herbarium (GOET) and furthermore on the collections (made by others) at B, G, LE, M, MO and Z.

BIBLIOGRAPHY and BIOGRAPHY: ADB 29: 62-64 (E. Wunschmann); AG 1: 276; Backer p. 660; Barnhart 3: 172 (b. 9 Jun 1847, d. 6 Jun 1871); Biol.-Dokum. 15: 7619; BM 4: 1723; CSP 8: 770, 12: 626; DTS 1: 247; Herder p. 119, 224, 290, 330, 331, 333, 338; Hortus 3: 1202; Kew 4: 507; Langman p. 639; MW p. 416; PR 7735-7737; Rehder 2: 320; Zander ed. 10, p. 708, ed. 11, p. 808.

BIOFILE: Anon., Bot. Not. 1872: 26 (d.); Bot. Zeit. 29: 475. 1871 (obit.); J. Bot. 9: 288. 1871 (d.); Verh. bot. Ver. Prov. Brandenburg 12: frontisp. death announcement, xlvi: 1870 (d. 6 Jun 1871).
Simonkai, L., Enum. fl. transsilv. xxviii. 1886.
Urban, I., Fl. bras. 1(1): 197. 1906.

COMPOSITE WORKS: (1) *Tropaeolaceae*, in E. Warming, Symb. bot. 27: 145. 1882 (publ. 16 Mai 1883).
(2) *Flora brasiliensis*, 14(2): 221-234, *pl. 53-72*. 1 Feb 1872: *Tropaeolaceae, Molluginaceae, Alsinaceae, Silenaceae, Portulacaceae, Ficoidaceae, Elatinaceae.*

9457. *Über den Blüthenbau und die Befruchtung von Epipogium gmelini* Eine von der Philosophischen Facultät der Georg-August-Universität zu Göttingen gekrönte Preisschrift ... Göttingen (Druck der Universitäts-Buchdruckerei von E.A. Huth) 1866. Qu. (*Bl.-Bau Epipogium gmel.*).
Publ.: 1866 (Flora 27 Nov 21 Dec 1866), p. [1]-28, *2 pl.* (uncol. liths.). *Copies*: G, M.
Ref.: A.W.E., Flora 50: 152-154. 8 Apr 1867.
R. Bot. Zeit. 25: 71-72. 1 Mar 1867 (rev.).

9458. *Morphologie der Gattung Silene*. Botanische Inaugural-Dissertation zur Erlangung der Doctorwürde in der Philosophie vorgelegt der Philosophischen Facultät der Friedrich-Wilhelms Universität zu Berlin und öffentlich zu vertheidigen am 18. Juli 1868 von Paul Rohrbach aus Berlin ... Leipzig (Druck von Breitkopf und Härtel) [1868]. Oct. (*Morph. Silene*).
Publ.: 18 Jul 1868, p. [1]-52. *Copy*: FI. – Identical with first part of his *Monogr. Silene*. – Rohrbach published in addition a *Synopsis der Lychnideen* in Linnaea 36: 170-256. Sep 1869, 257-270. Jan 1870 and *Beiträge zur Systematik der Caryophyllinen*, Linnaea 36: 651-690. Dec 1870 and 37: 183-256. Apr 1872, 257-312. Jul 1872, including *Nachtrag* (by Garcke) 37: 311-312. Jul 1872.

9459. *Monographie der Gattung Silene* ... Leipzig (Verlag von Wilhelm Engelmann) 1868. Oct. (*Monogr. Silene*).
Publ.: 1 Jan 1869 (t.p. 1868; p. viii: Aug 1868; ÖbZ 1 Jan 1869 "erschienen"; Flora 24 Feb 1869), p. [i]-viii, [1]-249, [250, colo.], *pl. 1-2* (uncol. liths auct.). *Copies*: FI, G, HH, M, MICH, NY, PH, US, USDA.
Ref.: Anon., J. Bot. 6: 378-380. Dec 1868 (rev.).
Bary, A. de, Bot. Zeit. 28: 502-504. 1870.

9460. *Ueber die europäischen Arten der Gattung Typha* [Berlin 1870]. Oct.
Publ: 22 Jan 1870, p. 67-104, *1 pl. Copies*: G, HH. – Reprinted and to be cited fron Verh. bot. Ver. Brandenburg 11: 67-104, *1 pl.* 22 Jan 1870.

Ref.: Bary, A. de, Bot. Zeit. 28: 450-455. 15 Jul, 466-468. 22 Jul 1870.

9461. *Beiträge zur Kenntniss einiger Hydrocharideen* nebst Bemerkungen über die Bildung phanerogamer Knospen durch Theilung des Vegetationskegels ... Halle (Druck und Verlag von H.W. Schmidt) 1871. Qu.
Publ.: 1871 (ÖbZ 1 Nov 1871; Bot. Zeit. 28 Apr 1871; Flora 11 Jul 1871), p. [1]-64, *pl. 1-3* (uncol. copp. auct.). *Copies*: FI, G(3), HH, U. – Reprinted and to be cited from Abh. naturf. Ges. Halle 12: 51-114. 1871 [volume: 1873].
Ref.: Anon., Bull. Soc. bot. France 20 (bibl.): 211-212. 1874 (rev.).
X., Flora 55: 312-320. 1872 (rev.).

Rohrer, Rudolph (1805-1839), West-Galician (Krakow-born) printer and botanist, active in Moravia. (*Rohrer*).

HERBARIUM and TYPES: BRNM.

BIBLIOGRAPHY and BIOGRAPHY: AG 6(1): 187; Barnhart 3: 173; BM 4: 1723; Futak-Domin p. 505-506; Herder p. 192; Kew 4: 507; Klášterský p. 158 (b. 1805, d. 27 Dec 1839); PR 7738, ed. 1: 8698; Rehder 1: 439.

BIOFILE: Anon., Preslia 51(2): 160. 1979 (commemorative notice; d. 27 Dec 1839).
Iltis, H., Life of Mendel ed. 2. 50. 1966.
Oborny, A., Verh. naturf. Ver. Brünn 21(2): 11, 12, 13, 15, 16. 1882.
Pax, F., Grundz. Pfl.-Verbr. Karpathen 14, 36. 1898 (Veg. Erde 2(1)); Bibl. Schles. Bot. 20. 1929.
Pluskal, F.S., Verh. Zool. bot. Ges., Wien 6 (Abh.): 365-366. 1856.

9462. *Vorarbeiten zu einer Flora des Mährischen Gouvernements* oder systematisches Verzeichniss aller in Mähren und in dem k.k. östr. Antheile Schlesiens wildwachsenden bis jetzt entdeckten phaenerogamen Pflanzen von Rudolph Rohrer, ... und August Mayer, ... Brünn [Brno] (Gedruckt und im Verlag bei Rudolph Rohrer) 1835. Oct. (*Vorarb. Fl. Mähr. Gouv.*).
Co-author: August Mayer.
Publ.: 1835 (Bot. Ges. Regensb. rd. by 20 Jan 1836), p. [i]-xliv, chart, [1]-217, [218, err.], [NY copy: plus leaf 217 (old text), 218 (new errata), [1-28, index]. *Copies*: E, G, NY. – Cover: "In Commission bei Friedrich Beck, in Wien".
Nachträge: Heinrich Grabowski, Flora 19: 369-375. 1836.
Ref.: [Schlechtendal, D.F.L. von], Linnaea 10 (Litt.): 140-142. 1835-1836 (rev.).

Roig y Mesa, Juan Tomás (1878-?), Cuban botanist; Dr. pharm. Univ. Habana 1910; Dr. Ci. nat. 1912; head of the botany dept. of the Ist. exp. agron. Santiago de las Vegas 1913; teacher natural history at Pinar del Río 1917-1932; again at Santiago de las Vegas as head of the botany dept. from 1932. (*Roig*).

HERBARIUM and TYPES: HAC, HAJB; further material at NY and US.– Barnhart (2: 173) mentions a private herbarium (in 1938) of 9000 sheets.

BIBLIOGRAPHY and BIOGRAPHY: Barnhart 3: 173 (b. 21 Mai 1878, personal comm. Roig to Barnhart); BL 1: 225, 226, 315; IH 1 (ed. 7): 341; Langman p. 639; Lenley p. 352.

BIOFILE: Conde, J.A., Hist. bot. Cuba 54, 97, 111, 302, 303, 315-316, 321 (portr.). 1958 (b. 31 Mai 1877).
Moldenke, H.N., Plant Life 2: 79. 1946 (1948) (epon.).

EPONYMY: *Roigia* Britton (1920).

POSTAGE STAMPS: Cuba 50 c. (1977), yv. bl. 50.

9463. Universidad de la Habana. *Cactáceas de la Flora cubana*. Tesis para el grado de doctor en ciencias naturales ... Habana (Imprenta el Siglo XX de Aurelio Miranda ..) 1912. Oct. (*Cact. fl. cub.*).

Publ.: Mar 1912, p. [1]-58, 6 *pl.* (5 photos & 1 diagr.). *Copy*: US.

9464. *Diccionario botanico de nombres vulgares cubanos* ... Habana (Imprenta y Papeleria de Rambla, Bouza y Ca. ...) 1928, 3 parts. (*Dicc. bot.*).
Publ.: in three parts, 1928, as Boletim numb. 54 of the Estación experimental agronomica, Santiago de las Vegas. *Copies*: HH, MO, NY, US, USDA.
1: Feb 1928 (cover), p. [i*-iii*], [i]-viii, [1]-247, *pl. 1-8.*
2: Feb 1928, p. [i-ii], [249]-598, *pl. 9-38.*
3: Feb 1928, p. [i-iv], [599]-897, [1-6, err.], *pl. 39-48.*
Ed. 2: n.v.
Ed. 3: 1965, 2 parts. *Copy*: NY.
1: [1]-599, *pl. 1-34.*
2: [603]-1142, *pl. 35-61.*

Roivainen, Heikki (1900-x), Finnish botanist (bryologist); collected in Tierra del Fuego, Patagonia and Chile 1928-1929; Dr. phil. Helsinki 1954; senior curator of cryptogams at Helsinki 1956-1968. (*Roivainen*).

HERBARIUM and TYPES: H. – Other material at BA, BP, DBN, E, F, FH, KUO, OULU, S. – 54 letters by R. to K. Linkola at Univ. Libr., Helsinki. – R. continued *Mycotheca fennica* (fasc. 13-18).

BIBLIOGRAPHY and BIOGRAPHY: BL 1: 255, 316, 2: 81, 89, 97, 710; Collander p. 448-450 (bibl.); GR p. 621-622 (b. 30 Dec 1900); IH 1 (ed. 3): 72, (ed. 4): 77, (ed. 5): 77, (ed. 6): 111, (ed. 7): 113; Lenley p. 352; Roon p. 95; SBC p. 130.

BIOFILE: Kotilainen, M.J., Luonnon Tutkija 64: 150-152. 1960 (portr.).
Sayre, G., Bryologist 80: 516. 1977 (coll.).
Verdoorn, F., ed., Chron. bot. 3: 355. 1937, 6: 303. 1941.

FESTSCHRIFT: (dedication): *Karstenia* 17(1). 1977 (portr.).

EPONYMY: *Roivainenia* H. Persson et Grolle (1961).

9465. *Tietoja pihtiputaan ja kinnulan putkilokasvistosta* ... Kuopio (Osakeyhtiö kirjapaino sanan valta) 1927.
Publ.: 1927, cover t.p., [1]-48. *Copy*: H (inf. P. Isoviita). – Issued as Kuopion Luonon Ystävän Yhdistyksen julkaisuja B 1(2).

9466. *Informaciones sobre excursiones botanicas en la Costa oriental de la Patagonia* ... Helsinki 1933.
Publ.: 1933, p. [i, ii], [1]-13, [14]. *Copy*: H (inf. P. Isoviita). – Issued as Ann. bot. Soc. zool.-bot. fenn. Vanamo 4(6).

9467. *Chillania pusilla, eine neue Gattung und Art der Familie Cyperaceae* ... Helsinki 1933.
Publ.: 1933, p. [i], [1]-6. *Copy*: U. – Issued as Ann. bot. Soc. Zool.-bot. fenn. Vanamo 4(7).

9468. *Contribuciones á la flora de Isla Elisabeth, Rio de las Minas y Puerto San Isidor de Prov. de Magallanes, de Puerto Barroso de prov. Chiloë y de los alrededores de Terman de Chillan de Prov. de Ñuble, Chile,* ... Helsinki 1933.
Publ.: 1933, p. [i-ii], [1]-22. *Copies*: H (inf. P. Isoviita), USDA. – Issued as Ann. bot. Soc. Zool.-bot. fenn. Vanamo 4(8).

9469. *Bryological investigations in Tierra del Fuego* by H. Roivainen ... with diagnoses of many new species by Edwin B. Bartram ... 1. Sphagnaceae-Dicranaceae ... Helsinki 1937. †.
Co-author: Edwin Bunting Bartram (1878-1964).

Publ.: 1937 (p. vi: Dec 1936), p. [i]-x, [1]-58. *Copy*: H (inf. P. Isoviita). – Issued as Ann. bot. Soc. Zool.-bot. fenn. Vanamo 9(2). – Further moss taxa described from R.'s material by E.B. Bartram, Farlowia 2: 309-319. 1946. The hepaticae were published by J.J. Engel, Ann. bot. fenn. 13: 132-136. 1976. Notes on *Sphagnum* by P. Isoviita in Ann. bot. fenn. 3 (1966).

Rojas, Teodor (1877-1954), Paraguayan botanist; collected in Paraguay with E. Hassler in 1896 and from then on until 1954; assistant director of the Assuncion botanical garden and curator of the herbaria. (*Rojas*).

HERBARIUM and TYPES: Paraguay, Argentine and Uruguay material in many herbaria e.g. B, BAF, BAI, C, DBN, DPU, E, F, GB, GH, K, L, LIL, M, MO, MVH, MVM, NY, S, SI (impt. set), US. – See under E. Hassler for exsiccatae containing Rojas material.

NOTE: Information on T. Rojas was supplied by J. Angely (in lit. 25 Apr 1962). Rojas was born on 23 Sep 1877 in Asuncion and died in that city on 3 Sep 1954.

BIBLIOGRAPHY and BIOGRAPHY: BJI 2: 147; BL 1: 253, 316; IH 1 (ed. 7): (in press); TL-2/see E. Hassler; Urban-Berl. p. 389.

EPONYMY: *Rojasia* Malme (1905); *Rojasiophyton* Hassler (1910). *Note*: *Rojasianthe* Standley et Steyermark (1940) honors Ulises Rojas, professor of botany in Guatemala.

Rolander, Daniel (1725-1793), Swedish botanist; student of Linnaeus in Uppsala 1744-1754; tutor of Linnaeus fil., 1751-1754; travelled and collected in Suriname 1755-1756 at the invitation of C.G. Dahlberg; returned to Sweden in poor health; died destitute. (*Rolander*).

HERBARIUM and TYPES: C (main collection), H, LINN, SBT. – Rolander's *Diarium surinamense*, quod sub itinere suo conscripsit ...". 2 vols. mss. is at C (see J. Hornemann (1811) and KR p. 593, no. 6).

BIBLIOGRAPHY and BIOGRAPHY: Backer p. 498 (epon.); Barnhart 3: 173 (b. 1725, d. 9 Aug 1793); Dryander 3: 290; IH 2: (files); KR p. 592-593 (b. 1725, d. 9 Aug 1793, bibl.); Lasègue p. 345, 358, 359, 473, 503; PR (alph.); RS p. 137; TL-2/see P.J. Bergius, O. Celsius, C.G. Dahlberg.

BIOFILE: Bernardi, L., Mus. Genève 198: 11. 1979 (epon.).
Blunt, W., The compleat naturalist 183, 190, 201. 1971.
Candolle, Alph. de, Phytographie 444. 1880 (coll. at C).
Christensen, C., Dansk. Bot. Hist. 1: 151. 1924.
Dandy, J.E., Regn. veg. 51: 18. 1967.
Fries, R.E., Short hist. bot. Sweden 31, *pl. 1-2.* 1950.
Fries, Th.M., Bref skrifv. Linné 1(1): 151, 282. 1907, 1(2): 157, 159, 178, 179, 205, 209, 213. 1908, 1(3): 54. 137. 1909, 1(4): 135, 170, 172, 175, 185, 307, 308, 311, 337. 1910, 1(5): 2, 14, 43,52, 320. 1911, 1(6): 273, 427. 1912; Linné 2: 51-55. 1903.
Holthuis, L.B., Zool. Verh. Leiden 44: 19-21. 1959 (biogr. sketch with ref. to Hornemann 1912).
Hornemann, J.W., Om den svenske naturforsker Daniel Rolander og mnscpt. af hans reise til Surinam, Skand. Litt. Selsk. Skr. 7: 457-494. 1811.
Hulth, J.M., Bref skrifv. Linné 2(1): 108. 1916.
Hulth, J.M. and A.H. Uggla, Bref skrifv. Linné 2(2): 84, 87, 92, 197, 198, 224, 225. 288. 1943.
Jackson, B.D., Proc. Linn. Soc. 134 (suppl.):19. 1922 (material LINN).
Löwegren, Y., Naturaliekabinett Sverige 142, 43, 362. 1952 (coll.).
Mägdefrau, K., Gesch. Bot. 54, 55. 1973.
Papavero, N., Essays hist. neotrop. dipterol. 1: 7-9. 1971.
Scheutz, N.J.W., Bot. Not 1863: 71 (epon.).
Smith, J.E., *in* Rees' Cyclopaedia, sub *Rolandra*.

Verdoorn, F., ed., Chron. bot. 1: 382. 1935.
Wilson, J.G. & J. Fiske, Appleton's Cycl. Amer. biogr. 5: 311-312. 1888 (erroneous).

EPONYMY: *Rolandra* Rottbøl (1775).

NOTE: (1) *Doliocarpus*. En ört af nytt genus från America, Sv. Vet.-Akad. Handl. 1756: 256-261, *pl. 9* was translated into German as *Doliocarpus*. Eine neue Gattung Pflanzen aus America, Abh. Schwed.-Akad. Wiss. 18: 246-250. *pl. 9*. 1757. See KR p. 592.
(2) See KR p. 592, nos. 3 and 4 on the descriptions of taxa by Rolander used by C.F. Rottbøl, Descr. plant. quarund. Surinam., *in* Acta lit. Univ. Hafn. 1778: 270, 271, 274, 277, 278, 279, 280, 281 and in Obs. Soc. med. Havn. coll. 2: 248, 251, 258. 1775.

Rolfe, Robert Allen (1855-1921), English botanist, especially orchidologist; brought up as a gardener; came as such to the Royal Botanic Gardens Kew 1879; assistant in the herbarium ib. 1880-1921. (*Rolfe*).

HERBARIUM and TYPES: K. – Some further material at AK.

BIBLIOGRAPHY and BIOGRAPHY: Ainsworth p. 327; Backer p. 498 (epon.); Barnhart 3: 173 (b. 12 Mai 1855, d. 13 Apr 1921); BB p. 260; BJI 1: 49; BL 2: 269, 710; BM 4: 1723; Bret. p. 824; CSP 11: 209-210, 12: 626, 18: 275-276; Herder p. 212, 312, 327; Hortus 3: 1202; Kew 4: 509; Langman p. 641; Morren ed. 10, p. 70; MW p. 416-417; Rehder 5: 734; RS p. 137; SK 4: cxxv, cxxxi; SO add. 872b; TL-2/2448, 4628, 7055; Zander ed. 10, p. 708, ed. 11, p. 808.

BIOFILE: Ames, O., Amer. Orch. Soc. Bull. 2(3): 38-39. 1933 (obit., portr.).
Anon., Bot. Centralbl. 3/4: 832. 1880 (second asst. Kew); Bot. Zeit. 38: 543. 1880 (app. Kew); Bull. misc. Inf. Kew 1912: 305, 1914: 392, 1921: 123-127 (obit.); Hedwigia 63(1): (97). 1922 (d.); Gard. Chron. ser. 3. 69: 74. 1921 (portr., Veitch memorial medal); Gard. Mag. 52: 157. 1909 (portr.).; Garden 85: 207. 1921 (obit., portr.); J. Bot. 59: 152, 182-183. 1921 (obit.); J. Hort. Home Farmer 65: 467. 1912 (portr.); Nature 107: 276-277. 1921 (obit.); Orchid Rev. 29: 5-8. 1921 (obit., portr.), 41: 19-22. 1933 (biogr., portr.).
Arditti, J., ed., Orchid biology 1: 29, 35, 36, 39, 44, 138, 142, 154. 1977.
Bullock, A.A., Bibl. S. Afr. bot. 96. 1978.
Hoehne, C., Jard. bot. São Paulo 175. 1941 (portr.).
J.W., J. Kew Guild 3: 1. 1912 (portr., brief sketch).
Jones, H.G., Taxon 22: 234. 1973 (porr.).
Kent, D.H., Brit. herb. 76. 1957.
Merrill, E.D., Enum. Philipp. fl. pl. 4: 223. 1927; Bishop Mus. Bull. 144: 160. 1937 (bibl. Polyn.); Contr. U.S. natl. Herb. 30(1): 258. 1947 (Pacific bibl.).
Prain, D., Bot. Soc. Exch. Cl. Brit. Isl. Rep. 6(3): 365-367. 1922 (obit.).
Reinikka, M.A., A history of the orchid 275-278. 1972 (biogr., portr.).
Stapf, O., Proc. Linn. Soc. 133: 52-53. 1921 (obit.).
Taylor, P., & P.F. Hunt, Orch. Rev. 84: 311-312. 1976 (biogr. sketch, portr.).
Verdoorn, F., ed., Chron. bot. 6: 415, 416. 1944.
W., Gard. Chron., ser. 3. 69: 204. 1921 (obit., portr.).

EPONYMY: *Allenrolfea* O. Kuntze (1891); *Rolfea* Zahlbruckner (1898); X *Rolfeara* [X *Sophrocattleya* Rolfe (1887); X *Brassocattleya* Hort. (1889] Hort. (1919); *Rolfeella* Schlechter (1924).

COMPOSITE WORKS: (1) Founder and editor of *The Orchid Review* 1893-1920.
(2) Co-editor, with L. Linden, of *Lindenia* vols. 7-9, 1891-1893.
(3) *in* D. Oliver, *Fl. trop. Afr.*:
  Myoporineae 5(2): 262-263. Jun 1900.
  Orchideae 7(1): 12-192. Dec 1897, 7(2): 193-292. Apr 1898.
  Selagineae 5(2): 264-272. Jun 1910.
(4) *in* W.H. Harvey and O.W. Sonder, *Fl. cap.*:
  Myoporineae: 5(1): 92-94. Jun 1901.

Orchideae: 5(3, 1): 3-192. Oct 1912, 5(3, 2): 193-313. Mar 1913.
Selagineae: 5(1, 1): 95-180. Jun 1901.

9470. *The Orchid stud-book*: an enumeration of hybrid orchids of artificial origin, with their parents, raisers, date of first flowering, references to descriptions and figures and synonymy. With an historical introduction and 120 figures and a chapter on hybridising and raising orchids from seed ... Kew (Frank Leslie & Co., ...) 1909. Oct. (*Orchid. stud-book*).
Co-author: Charles Chamberlain Hurst (1870-1949).
Publ.: Jan-Mar 1909 (p. vi: Sep 1908; Nat. Nov. Apr(2) 1909), p. [i*-iv*], i-xlviii, 1-327. *Copies*: BR, MO, NY, USDA.

Rolfs, Peter Henry (1865-1944), American botanist and entomologist at Florida Agricultural Experiment Station at Lake City, Florida 1892-1899; professor of botany at Clemson College and botanist at the Agricultural Experiment Station S. Carolina 1899-1906; director of the Station at Lake City, Florida 1906-1921; dean of the Agricultural College 1915-1921; in charge of the Esc. Sup. Agric. Minaes-Geraes, Brazil 1921-1929; technical consultant Minas Geraes 1929-1933. (*Rolfs*).

HERBARIUM and TYPES: FLAS, MO, NY. – Letters to W.G. Farlow at FH.

BIBLIOGRAPHY and BIOGRAPHY: Barnhart 3: 174 (b. 17 Apr 1865); BJI 2: 147; BM 4: 1724; Bossert p. 336; GR p. 241; IH 2: (files); Kelly p. 193; Lenley p. 353; LS 22438-22445, 38503-38520; Plesch p. 385; Rehder 5: 735; Tucker 1: 605.

BIOFILE: Anon., Bot. Centralbl. 49: 224. 1892 (app. Lake City), 80: 144. 1899 (app. S. Carolina); Bot. Jahrb. 15 (Beibl. 34): 18. 1892 (app. Lake City), 27 (Beibl. 64): 22. 1900 (app. S. Carolina); Florida Entomologist 27: 1-4. 1944 (obit., portr.); Hedwigia 38(6): (298). 1899 (app. S. Carolina); Nat. Nov. 21: 520. 1899 (app. S. Carol.).
Cattell, J.M., ed., Amer. men sci. ed. 1: 273. 1906, ed. 2: 399-400. 1910, ed. 3: 583. 1921, ed. 4: 832. 1927, ed. 5: 948. 1933, ed. 6: 1202-1203. 1938, ed. 7: 1504. 1944.
Murrill, W.A., Hist. found. bot. Florida 37. 1945.
Osborn, H., Fragm. entom. hist. 82, 114, 223-224, 305, 324. *pl. 16.* 1937.
Pennell, F.W. & J.H. Barnhart, Monogr. Acad. nat. Sci. Phila. 1: 618. 1935.
Rodgers, A.D., Liberty Hyde Bailey 148, 153, 452-453, 485-486. 1949.
Verdoorn, F., ed., Chron. bot. 1: 96. 1935, 2: 313, 404. 1936.

Rolland, Eugène (1846-1909), French folklorist who published on French folk names of animals and plants, folk songs, proverbs and superstitions. (*E. Rolland*).

HERBARIUM and TYPES: Unknown.

BIBLIOGRAPHY and BIOGRAPHY: Barnhart 3: 174; BFM no. 312; BL 2: 106, 710; BM 4: 1724; Kew 4: 509; Rehder 1: 74, 518. Tucker 1: 605.

BIOFILE: Anon., Grand Larousse Enc. 9: 328. 1964.

9471. *Flore populaire* ou histoire naturelle des plantes dans leurs rapports avec la linguistique et le folklore ... Paris ([1-3]: Librairie Rolland; [4:] Librairie F. Staude ...; [5-7:] en vente chez les libraires-commissionnaires]) 1896-1904. Oct. †. (*Fl. pop.*).
Authorship vol. 8 (p. 113-218), and 9-11: Henri Gaidoz (1842-1932).
1: 1896 (Bot. Zeit. 1 Mar 1897; J. Bot. Apr 1897; Nat. Nov. Feb(2) 1897), p. [i*-iii*], [i]-iii, [1]-272.
2: 1899 (Bot. Centralbl. 29 Aug 1899; Bot. Zeit. 1 Oct 1899; Nat. Nov. Sep(1) 1899), p. [i-iii], [1]-267, [268, err.].
3: 1900 (Bot. Centralbl. 6 Feb 1901; J. Bot. Apr 1901; Bot. Zeit. 16 Apr 1901; Nat. Nov. Feb(2) 1901), p. [i-iii], [1]-378.
4: 1903 (J. Bot. Jul 1903; Bot. Centralbl. 16 Feb 1904; Bot. Zeit. 1 Dec 1903; Nat. Nov. Sep(1) 1903), p. [i-iii], [1]-263.
5: 1904 (Bot. Centralbl. 11 Apr 1905; Nat. Nov. Feb(1) 1905), p. [i-iii], [1]-415, [416, note].

6: 1906 (Bot. Centralbl. 16 Apr 1907; Nat. Nov. Feb(1) 1907), p. [i-iii], [1]-307.
7: 1908, p. [i-iii], [1]-262, [263, colo.].
8: Jul 1910 (t.p.; Bot. Centralbl. 29 Nov 1910; Nat. Nov. Sep(2) 1910), p. [i-v], [1]-218. – 300 copies printed.
9: Aug 1912 (t.p.; Bot. Centralbl. 4 Feb 1913; Nat. Nov. Dec(1) 1912), p. [i]-viii, [1]-282. – Id.
10: Oct 1913 (t.p.; Bot. Centralbl. 21 Jul 1914; Nat. Nov. Mar(2) 1914), p. [i]-vi, [1]-226. – Id.
11: Sep 1914 (t.p.), p. [i]-vi, [1]-261. – Id.
Copies: BR, NY, U, US, USDA. – A systematic list of plants with their vernacular names in various languages (vols. 1-3) or mostly in French (4-11); The monocots were to have been included in vols. 12, 13.
Facsimile ed.: 1967 (1968) Paris (Éditions G.-P. Maisonneuve et Larose ...). As orig. *Copies*: BR, G.
Preliminary publ. and announcement: Variétés bibliographiques, (organe de la librairie E. Rolland, 1(5): 129-140. Feb-Mar 1889 (treatment Rannunculaceae).
Ref.: Britten, J., J. Bot. 35: 363-365. 1897 (rev. vol. 1), 38: 197-198. 1900 (id. vol. 2).
Blümml, Bot. Centralbl. 86: 281-285. 1901 (rev. 1-3).
Rendle, A.B., J. Bot. 39: 146-148. 1901 (rev. 3).

Rolland, Léon [Louis] (1841-1912), French mycologist; lived at Le Hâvre 1866-1879, later at Paris and Neuilly. (*Rolland*).

HERBARIUM and TYPES: Unknown. – Contributed to C. Roumeguère et al., *Fungi exsiccati praecipue Gallici*.

BIBLIOGRAPHY and BIOGRAPHY: Barnhart 3: 174 (b. 10 Dec 1841, d. 11 Jun 1912); BM 4: 1724; CSP 18: 277; Hawksworth p. 185; Kelly p. 193-194; Kew 4: 509; LS 22446-22495, 38511 (bibl.); NI 1670; PFC 1: li; Saccardo 2: 93; Stevenson p. 1256.

BIOFILE: Anon., Hedwigia 53(1): (95). 1912.
Boudier, Em., Bull. Soc. mycol. France 28: 414-418. 1912 (obit., bibl.); see also Sterbeeckia 3(4): 19. 1964.
Cavillier, F., Boissiera 5: 81. 1941.

EPONYMY: *Rollandina* Patouillard (1905). *Note*: *Rollandia* Gaudichaud-Beaupré (1829) honors a M. Rolland, member of Freycinet's expedition Voyage Autour de Monde.

9472. *Une nouvelle espèce de Stysanus* ... Lons-le-Saunier (Imprimerie et lithographie Lucien Declume ...) 1890. Oct.
Publ.: 10 Apr 1890 (p. 106 journal issue), p. [1]-4. *Copy*: FH. – Reprinted and to be cited from Bull. Soc. mycol. France 6(2): 105-106. 1890.

9473. *Quelques champignons nouveaux du Golfe Juan* ... Lons-le-Saunier (Imprimerie et lithographie Lucien Declume ...) 1891. Oct.
Publ.: 1891, p. [1]-5, *2 pl. Copy*: FH. – Reprinted and to be cited from Bull. Soc. mycol. France 7(4): 211-213. *pl. 14*. 1891. – See also Bull. Soc. mycol. France 17(2) 117-119, *pl. 3, 4* w.t. 13 Mai 1901.

9474. *Excursions mycologiques* dans les Pyrénées et les Alpes maritimes ... Lons-le-Saunier (Imprimerie et Lithographie Lucien Declume ...) 1891. Oct.
Publ.: 30 Jun 1891 (journal issue), p. [1]-16, frontisp. and *pl. 6* (col. lith. auct.). *Copy*: FH. – Reprinted and to be cited from Bull. Soc. mycol. France 7(2): 84-97, *pl. 6*. 1891.

9475. *Espèces nouvelles de la Côte-d'Or* ... [Extrait de la Revue mycologique ... 1894). Oct.
Publ.: 1 Apr 1894 (p. 3; journal issue 1 Apr 1894), p. [1]-3. *pl. 141-144*. *Copy*: FH. – Reprinted and to be cited from Revue mycol. 16: 72-75, *pl. 141-144*. Apr 1894.

9476. *Aliquot fungi novi vel critici Galliae* praecipue meridionales ... Lons-le-Saunier (Imprimerie et Lithographie Lucien Declume) 1896. Oct.
Publ.: 31 Jan 1896 (Nat. Nov. Mai(1) 1896; journal issue 31 Jan 1896), p. [1]-12, *pl. 1* (col. lith. auct.). *Copy*: FH. – Reprinted and to be cited from Bull. Soc. mycol. France 12(1): 1-10, *pl. 1* (id.). 1896.

9477. *De l'instruction populaire sur les champignons* Lons-le-Saunier (Imprimerie et Lithographie Lucien Declume) 1900. Oct.
Publ.: 1900 (Bot. Centralbl. 26 Jun 1901), p. [1]-11. *Copy*: FH. – C.R. Congr. int. Bot. Exp. univ. 1900 (Paris 1-10 Oct) 405-414. 1900.

9478. *Les champignons à l'Exposition de 1900* ... Paris (au Siège de la société [mycologique de France]) 1901. Oct.
Publ.: 28 Feb 1901 (journal issue), p. [1]-13. *Copies*: FH. – Reprinted and to be cited from Bull. Soc. mycol. France 16(4): 211-223. 1900.

9479. *Atlas des champignons de France* Suisse et Belgique ... Paris (Librairie des Sciences naturelles Paul Klincksieck ...) 1906-1910. Oct. (*Atlas champ. France*).
Publ.: The plates were published between Nov 1896 and Nov 1909 in 15 parts each containing 8 numbers. The plates were not published serially but in an irregular order: the fascicles were all accompanied by a "texte provisoire" of 2 p. per livraison in which the numbers are given with short diagnosis of the depicted fungi. Fascicle 15, however, had no such text: it contained *pl. 36, 40, 54, 76, 81, 84, 86 and 113*; and substitute plates *1, 4, 48, 52, 62, 66, 73, 79*. The definitive text appeared after livr. 15, in 1910.
Text (definitive): 1910, p. [i, iii], [1]-126, [127, cont.].
Plates: in 15 livraisons 1906-1909, all dated and provided with contents on the "texte provisoire" (see e.g. copy: FH), in all *pl. 1-120* (col., A. Bassin).
Copies: BR, FH, G, Stevenson, USDA. – 500 copies of the *Atlas* were distributed by the Soc. mycol. France along with its Bulletin.

Rolle, Friedrich (1827-1887), German (Hesse-Nassau) palaeobotanist; Dr. phil. Tübingen 1851; in Austria at Graz for the geognostic survey of Styria 1853-1859; at the Hofmineralien-Cabinett Wien 1859-1860; from 1860 at Homburg as private scientist, studying the geology and palaeontology of the Homburg von der Höhe region. (*Rolle*).

HERBARIUM and TYPES: Unknown.

BIBLIOGRAPHY and BIOGRAPHY: ADB 29: 76-78; Barnhart 3: 174 (b. 16 Mai 1827, d. 10 Feb 1887); BM 4: 1724; Jackson p. 183; KR p. 593; Quenstedt p. 366; WU 26: 299-300 (partial bibl.).

BIOFILE: Jacobi, L., Kleine Presse, Homburg, 239, 10 Oct 1888, fide Verh. k.k. geol. Reichsanst. Wien 1891: 166-167 (brief biogr.; grave).
Poggendorff, J.C., Biogr.-lit. Handw.-Buch. 3(2): 1137-1138. 1898.
Zittel, K.A. von, Gesch. Geol. Paläont. 511, 515, 541, 631, 707. 1899.

9480. *Vergleichende Übersicht der urweltlichen Organismen* besonders nach ihrem inneren Zusammenhange mit denen der jetztlebenden Schöpfung ... Stuttgart (E. Schweizerbart'sche Verlagshandlung und Druckerei) 1851. Duod. (*Vergl. Übers. urweltl. Organ.*).
Publ.: 1851 (p. iv: Nov 1850), p. [i]-vii, [1]-171. *Copy*: G.

Rolli, Ettore (1818-1876), Italian botanist and pharmacist at Rome; professor of botany at Rome University and director of its botanical garden ca. 1851-1870. (*Rolli*).

HERBARIUM and TYPES: RO; other material at FI.

BIBLIOGRAPHY and BIOGRAPHY: AG 3: 113; Barnhart 3: 174 (d. 16 Jan 1876); Bossert p. 327; CSP 5: 261; IH 1 (ed. 7): 341, 2: (in press); Saccardo 1: 140-141, 2: 93.

BIOFILE: Candolle, A. de, Phytographie 444. 1880 (coll. RO).
Caruel, T., Nuovo Giorn. bot. ital. 8(1): 48. 1876 (d.).
Gager, C.S., Brooklyn Bot. Gard. Rec. 27(3): 287. 1938.
Lusina, G., Annali Bot. 25: 169. 1958.

NOTE: a publication *Romanarum plantarum centuria decima octava* was never published; proof sheets of an article meant for the *Giornale arcadico* with pagination 117-144 and 17-32 are in the library of the Ist. bot. Roma (RO) (see Lusina 1958).

9481. *Osservazioni sopra le palme* coltivate nell' Orto botanico, ed in altri luoghi di Roma. Reale Accademia dei Lincei Estratto dalla sessione V, del 10 Aprile 1871. [Roma 1871]. *Publ.*: 1871 (read 10 Apr 1871), p. [1]-9. *Copy*: FI. – Reprinted from Atti r. Accad. Lincei 24: 280-288. 1870 (1871).

Rollins, Reed Clarke (1911-x), American botanist; B.A. Univ. Wyoming 1933; M.S. State College Washington 1936; Dr. phil. Harvard 1941; curator of the Dudley Herbarium of Stanford University 1940-1948; asst. prof. ib. 1940-1941, assoc. prof. ib. 1947-1948; director of the Gray Herbarium of Harvard University 1948-1979, from 1954 Asa Gray professor of systematic botany at Harvard University; specialist on Cruciferae; president of the International Association for Plant Taxonomy 1954-1959; member of the editorial committee for botanical nomenclature 1954-x. (*Rollins*).

HERBARIUM and TYPES: DS (orig. p.p.), GH (most types orig. p.p.) and WYO (orig. p.p.); duplicates at e.g. BR, COLO, DPU, E, GB, MO, NY, RM, TEX, UC, US, USFS, VDB, WS. See Smithsonian Archives SIA 227, 7142.

BIBLIOGRAPHY and BIOGRAPHY: Barnhart 3: 174 (b. 7 Dec 1911); Bossert p. 337; IH 1 (ed. 1): 20, (ed. 2): 36, (ed. 3): 41, (ed.4): 43, (ed. 5): 39, (ed. 6): 50, 363, (ed. 7): 50, 341, 2: (files); Kew 4: 510; Langman p. 641-642; Lenley p. 353; NW p. 56; PH 209; Roon p. 95.

BIOFILE: Anon., Fl. males. Bull. 11: 404. 1955 (Asa Gray prof. syst. bot.); Taxon 3: 94. 1954 (Asa Gray prof. syst. bot.), 5: 2. 1956 (in group photo), 14: 114. 1965 (id.).
Bennett, W.I., Harvard Mag. Nov 1975: 14-17. 1975 (on S.E.M., Lesquerella and Rollins).
Constance, L.C., Taxon 31(3): 401-404. 1982 (biogr., portr.).
Cowan, R.S., Taxon 30: 724. 1981 (Garden Clubs of America's Gold Seal).
Knobloch, I.W., Pl. coll. N. Mexico 59. 1979.
Martin, J., Harvard Mag. Sep-Oct 1977, p. 89-90 (portr.).
Rollins, R.C., Taxon 8: 277-279. 1958 (Taxonomy and the international association ; presidential address at Montreal botanical congress; initiating a widening of the scope of IAPT and Taxon); Occas. Pap. Farlow Herb. 16: 8. 1981 (in group photo).
Shetler, S.G., Fl. N. Amer. Rep. 61, app. D: 3. 1971.
Shetler, S.G. et al., Flora North America, proposal to National Science Foundation, Washington 30 Jun 1971, p. 3 (curr. vit.).
Solbrig, O.T., Taxon 31(3): 405-414. 1982 (Harvard years, bibl.).
Thomas, J., Contr. Dudley Herb. 5(6): 161. 1961.
Verdoorn, F., ed., Chron. bot. 3: 292, 293, 362. 1937, 4: 452. 1938, 5: 516. 1939, 6: 331. 1941, 7: 95. 1943.
Walker, J.W., Rhodora 73: 461. 1971 (*Reedrollinsia* validated).

FESTSCHRIFT: Taxon 31(3). August 1982, with a portrait and bibliography and with biographical articles and tributes by L. Constance, O. Solbrig and R.S. Cowan & F.A. Stafleu.

EPONYMY: *Rollinsia* I.A. Al-Shehbaz (1982).

NOTE: For a synthesis of Rollins long-time work on *Lesquerella*, see R.C. Rollins and E.A. Shaw, *The genus Lesquerella* (Cruciferae) in North America, Cambridge, Mass. 6 Jun 1973, xii, 288 p., rev. B.B. Simpson, Taxon 23: 200-201. 1974. – Companion volume:

Atlas of the trichomes of Lesquerella (Cruciferae), Bussey Institution of Harvard University 1975, 48 p., with U.C. Banerjee. For a complete bibliography see Taxon 31(3): 409-414. 1982.

9482. *The genus Arabis* L. *in the Pacific Northwest* ... (Research Studies of the State College of Washington) 1936. Oct.
Publ.: Mar 1936, p. [1]-52 as Res. Stud. State Coll. Washington 4(1): [1]-52. Mar 1936.
– See also *A monographic study of Arabis in Western America*, Rhodora 43: 289-325. 12 Jul 1941, 348-411. 14 Aug 1911, 425-481. 10 Sep 1941 (also reprinted with orig. pag. and special cover for Contr. Gray Herb. 138, 1941), and *Studies of Arabis of western North America*, Syst. Bot. 6: 55-64. 1981.
Ref.: Constance, L., Madroño 4(1): 37. 6 Jan 1937.

Roloff, Christian Ludwig (1726-1800), German (Prussian) botanist at Berlin. (*Roloff*).

HERBARIUM and TYPES: Unknown.

BIBLIOGRAPHY and BIOGRAPHY: Barnhart 3: 174 (b. 6 Jun 1726, d. 26 Dec 1800); BM 4: 1725; Dryander 3: 122; Jackson p. 424; Kew 4: 510; PR 7745, ed. 1: 8704; Rehder 1: 44; TL-2/see C.L. Krause.

EPONYMY: *Rolofa* Adanson (1763) is likely named for Roloff, but no etymology is given by Adanson.

9483. *Index plantarum* tam peregrinarum quam nostro nascentium coelo quae aluntur Berolini *in horto* celebri *krausiano* ... Berolini [Berlin] (Litteris Christ. Ludov. Kunstii) [1746]. Oct. (*Index pl. hort. kraus.*).
Garden: "in horto celebri krausiano", garden of Christian Ludwig Krause (*fl.* 1753), see TL-2/3915-3916.
Publ.: 1746 (p. [xii]: 12 Nov 1746), p. [i-xvi], [1]-176, [1-16], *4 pl. Copies*: G, MO, NY, USDA.
Note: the proprietor of the garden was Christian Ludwig Krause, the author Christian Ludwig Roloff and the publisher/printer Christian Ludwig Kunst.

Romagnesi, Henri Charles Louis (1912-x), French mycologist, associated with the Laboratoire de Cryptogamie du Muséum d'Histoire naturelle, Paris. (*Romagnesi*).

HERBARIUM and TYPES: Private; will be left to PC.

BIBLIOGRAPHY and BIOGRAPHY: IH 1 (ed. 1): 74, (ed. 2): 95, (ed. 3): 119, (ed. 4): 131, (ed. 5): 140, (ed. 8): 207; Kew 4: 511-512; Roon p. 95.

BIOFILE: Verdoorn, F., ed., Chron. bot. 6: 303. 1941.

Romano, Girolamo (Gerolamo) (1765-1841), Italian botanist and clergyman at Gorgo nr Padua. (*Romano*).

HERBARIUM and TYPES: GE (Saccardo 2: 93); further material at BASSA, G, TO. – We do not know whether the herbarium of G.B. Romano (10.000) at TO, mentioned by O. Mattirolo (Studi veg. Piemonte, Pubbl. 2, 1929, p. xcvi) is that of Girolamo Romano.

BIBLIOGRAPHY and BIOGRAPHY: BL 2: 414, 710; BM 4: 1725; Bossert p. 337; De Toni 1: cviii; Kew 4: 512; LS 22488; PR 7748-7749, ed. 1: 8707-8708; Rehder 1: 425, 426; Saccardo 1: 141 (b. Apr 1765, d. 31 Mai 1841), 2: 93, cron. p. xxxi; Tucker 1: 605.

EPONYMY: *Romana* Vellozo (1825) is "in memoriam D. Romani B. Galli dixi" (see Barnhart 2: 174). The derivation of *Romanoa* Trevisan (1848) was not given.

9484. *Catalogus plantarum italicarum* ... Patavii [Padua] (ex Officina Sociorum titulo Minerva) 1820. Oct. (*Cat. pl. ital.*).

Publ.: 1820, p. [1]-74, [76, colo.]. *Copies*: FI, HH. – List of names without authors.

9485. *Le piante fanerogame euganee* per le nobilissime nozze Cittadella-Malduro ... Padova (Tipografia del Seminario editr.) 1828. Oct. (*Piante fan. eugan.*).
Publ.: 1828, p. [1]-21. *Copy*: FI. – First ed. of 1823 and a third Padova 1831, are mentioned by PR.

Romans, Bernard (ca. 1720-1784), Dutch-born, English-naturalized, American patriot, surveyor and soldier; in Florida ca. 1760-1774, sometime as surveyor for the British government of West Florida; living in New York 1773-1775 as a writer; joined the American patriots in the rebellion against Great Britain 1775; captain in the Pennsylvania artillery 1776-1778; from 1778-1779 living in Wethersfield, Connecticut; working e.g. on a history of the Netherlands; captured by the British 1779, prisoner of war in England 1780-1783; died on the journey back to the United States. (*Romans*).

HERBARIUM and TYPES: Unknown.

BIBLIOGRAPHY and BIOGRAPHY: Barnhart 3: 175 (d. 1784); BB p. 261; CSP 5: 263; Desmond p. 529 (d. 1783); DNB 49: 180-181; ME 2: 10, 11, 3: 350 ("1720-1784"); Rehder 1: 317; Tucker 1: 605.

BIOFILE: Allibone, S.A., Crit. dict. Engl. lit. 2: 1860. 1878 (bibl.).
Ewan, J., Southw. Louisiana J. 7: 9-10. 1967.
Patrick, R.W., Editor's preface and Introduction to Gainsville 1962 facsimile edition of B. Romans, Conc. nat. hist. Florida.
Smith, J.E., Sel. corr. Linnaeus 1: 896. 1821.
Wilson, J.G. & J. Fiske, Appleton's Cycl. Amer. biogr. 5: 313-314. 1888.

9486. *A concise natural history of East and West-Florida*; containing an account of the natural produce of all the Southern part of British America, in the three kingdoms of nature, particularly the animal and vegetable. Likewise, the artificial produce now raised, or possible to be raised, and manufactured there, with some commercial and political observations in that part of the world; and a chorographical account of the same. To which is added, by way of appendix, plain and easy directions to navigators over the bank of Bahama, the coast of the two Floridas, ... By Captain Bernard Romans ... vol. 1. New York (Printed for the author) 1775. (*Conc. nat. hist. Florida*).
Ed. [*1*]: 1775 n.v. A copy of the original issue is at the American Philosophical Society Library; our data stem from the 1962 facsimile, p. [i-iv], "[1]"-"[4]", "[i]"-"[viii]", [1]-342, [5-6], (i)-(lxxxiv), 3 maps, frontisp., 7 *pl.*, table.
Ed. [*2*]: 1776, p. [1], engr. dedic., 3-4, [1]-342, 6 *pl. Copies*: LC, NY, PH.
Facsimile ed.: 1962, Florida facsimile & reprint series, Gainesville (University of Florida Press) 1962, with introduction by Rembert W. Patrick, p. [i]-lii, [new introd.], frontisp., [i-ii], [1]-342, [1], i-lxxxiv, appendix, map. *Copies*: HU, ILL (inf. D.P. Rogers), NY.
Reset edition: 1961, introd. Louise Richardson, Pelican Publishing Company, New Orleans, frontisp., p. [i-xx], 1-291, 3 maps. *Copy*: US.

Rombouts, Johannes Godfried Hendrik (x-1889), Dutch botanist; Dr. med. Leiden 1843. (*Rombouts*).

HERBARIUM and TYPES: Unknown.

BIBLIOGRAPHY and BIOGRAPHY: BL 2: 435, 710; BM 4: 1726; Oudemans 2: 255; PR 7750; Rehder 1: 391; Tucker 1: 605.

9487. *Dissertatio botanico-pharmacologica inauguralis de plantarum radicibus*, quam, annuente summo numine, e auctoritate rectoris magnifici Joannis Matthiae Schrant, ... pro gradu doctoratus, summisque in medicina honoribus et privilegiis, in Academia lugduno-batava, rite ac legitime consequendis, publico ac solemni examini submittit Joannes Gothofredus Henricus Rombouts, amstelodamensis; ad diem xxvii m. aprilis,

anni [mdcccxliii], hora ii. in auditorio majori. Amstelodami (apud Elix & Co., typographos) 1843. Oct. (*De pl. rad.*).
Publ.: 27 Apr 1843, p. [i]-x, [1]-90, [1-4]. *Copies*: NY, U.

9488. *Flora amstelaedamensis*, plantarum quae prope et circa Amstelaedamum sponte nascuntur enumeratio et descriptio. Conscripserunt: J.G.H. Rombouts, ... et J.J.F.H.T. Merkus Doornik, ... Trajecti ad Rhenum [Utrecht] (C. van der Post Jr.), Amstelaedami [Amsterdam] (C.G. van der Post) 1852. Oct. (*Fl. amstelaed.*).
Co-author: J.J.F.H.T. Merkus Doornik (1825-1906).
Publ.: Apr 1852 (p. viii: Nov 1851; KNAW copy inscribed 26 Apr 1852 on cover), p. [i]-lx, [1]-136. *Copies*: HH, KNAW, U, USDA.
Additions: C.M. v.d. Sande Lacoste, Ned. kruidk. Arch. ser. 1. 4: 243-244. 1858.
Facsimile reprint: 1974, Amsterdam.

Romell, Lars (1854-1927), Swedish mycologist; Fil. kand. Uppsala 1855; teacher at Stockholm 1886-1890; patent attorney ibid. 1890-1927; Regnellian curator at the botany department of the Naturh. Riksmuseum ib. 1915-1920; Dr. phil. h.c. Uppsala 1927. (*Romell*).

HERBARIUM and TYPES: S (includes the Bresadola herb.; see also Schäffer 1939). – Published *Fungi exsiccati praesertim scandinavici*, cent. 1-2, Jan 1890, 1895, 40 sets (list of species cent. 1 in Bot. Not. 1890: 151-153), sets at BM, C, CUP, DAO, FH(2), H, HBG, L, LD, M, NY, PC, S, UPS, W, WRSL. – Contributed to T. Vestergren, *Micromycetes rariores selecti praecipue scandinavici*. – Letters to W.G. Farlow at FH.

BIBLIOGRAPHY and BIOGRAPHY: Barnhart 3: 175 (b. 4 Dec 1854, d. 12 Jul 1927); Biol.-Dokum,. 15: 7631; BJI 2: 148; BM 4: 1727; Bossert p. 337; CSP 18: 284; GR p. 480, cat. p. 70; Hawksworth p. 185; IH 1 (ed. 6): 363, (ed. 7): 341, 2: (files); Kelly p. 194; Kew 4: 513; KR 593-594; LS 22490-22495; Stevenson p. 1256; TL-2/4612; Urban-Berl. 281.

BIOFILE: Anon., Bot. Not. 1912: 156, 1916: 64; Hedwigia 68: (48). 1928 (d.); Mycologia 19: 293. 1927 (d.).
Coker, W.C., J. Elisha Mitchell sci. Soc. 43(1/2): 146-151. 1927 (obit., portr., bibl. by L.R. Romell).
Fries, R.E., Short hist. bot. Sweden 74, 118, 126, 154. 1950.
Hartman, I., Industritidningen Norden 1927: [2 p.] (obit., portr., d. 13 Jul 1927).
Hesler, L.R., Biogr. sketch North Amer. mycol. 1975, mss.
Lindman, S., Svenska män och kvinnor 6: 314-315. 1949 (portr.).
Lloyd, C.G., Syn. sect. Ovinus of Polyporus 1911, frontisp. portr. and dedic., Mycol. Notes 16: 160-161. 1904.
Murrill, W.A., Hist. found. bot. Florida 29. 1945.
Romell, L., Bot. Centralbl. 37: [256a] (Inseraten-Beilage) 1889. (announcement *Fungi exs.*); Grevillea 17: 64. 1889 (id.).
Romell, L.G., Sv. bot. Tidskr. 21(3): 370-379. 1927 (obit., portr., bibl.).
Schäffer, J., Z. Pilzk. 12: 20-23, *pl. 1*. 1928 (obit., portr.; description of his archival material and collection); Ark. Bot. 29A(15):1-80. 1940 (Revision der Russula-Sammlung Romells).
Shear, C.L., Mycologia 20(2): 49-51. 1928 (obit., portr.).
Wittrock, V.B., Acta horti berg. 3(2): 68. 1903, 3(3): 50, *pl. 113*. 1905.

COMPOSITE WORKS: (1) Associate editor *Mycologia* vols. 1-19.
(2) *Basidiomycetes from Juan Fernandez*, in C. Skottsberg, *Nat. hist. Juan Fernandez* 2: 465-471. 1926.
(3) *Svampar* (Fungi), in Th.O.B.N. Krok och S. Almquist, *Svensk flora för skolor* ... ii. Kryptogamer, ed. 2 (1898), 3 (1907) and 4 (1917); see TL-2/3955.

NOTE: Romell, Lars-Gunnar (1891-1981), Swedish botanist, son of Lars Romell; Dr. phil. Stockholm 1922, professor h.c. 1948; in various functions at the Forestry Research Institute and at the University of Stockholm 1918-1928 and 1934-1957; professor of forestry and pedology at Cornell University, Ithaca, U.S.A. 1928-1934; published many works on plant ecology, forestry, pedology and physiology.

EPONYMY: *Romellia* Berlese(1900); *Romellia* Murrill (1904); *Romellina* Petrak (1955).

9489. *Hymenomycetes austro-americani* in itinere regnelliano primo collecti I ... Stockholm (Kungl. Boktryckeriet. P.A. Norstedt & Söner) 1901. Oct. (*Hymenomyc. austro-amer.*).
Publ.: 1901 (mss. subm. 14 Nov 1900; Nat. Nov. Apr(2) 1901), p. [1]-61, *pl. 1-3*. *Copies*: BR, G, LD, U. – Issued as Bih. Sv. Vet.-Akad. Handl. 26(3), 16. 1901.

9490. *Hymenomycetes of Lappland* ... Arkiv for Botanik Band 11. N:3. 1912. Oct.
Publ.: 16 Apr 1912 (printed 10 Dec 1911; issued in journal 16 Apr 1912), p. [1]-35, *pl. 1-2*. *Copies*: LD, U. – To be cited as Ark. Bot. 11(3), 1912.

Romieux, Henri (1857-1937), Swiss (Genevese) botanist and administrator; bank employee at Genève 1878-1887; with the federal finance department at Bern 1887-1892; industrial consultant at Genève 1892-1900; member of the "Conseil d'État" (de Genève) from 1900 as head of the public works department; collected extensively in Switzerland as well as in Algeria (1904) and Tunesia (1906). (*Romieux*).

HERBARIUM and TYPES: G (15.000, acquired 1945); further material at A, AK, C, E, L, MO, OXF.

BIBLIOGRAPHY and BIOGRAPHY: AG 12(2): 194, 344; Clokie p. 234; IH 2: (files); Morren ed. 10, p. 110.

BIOFILE: Becherer, A., Actes Soc. helv. Sci. nat. 1938: 463-465 (obit., bibl., b. 20 Apr 1857, d. 21 Nov 1937) (repr. 2 p.).
Miège, J., Musées de Genève 160: 15-17. 1975 (portr.).
Miège & Wuest, Saussurea 6: 134. 1975 (herb. at G).
Thommen, Ed., Bull. Soc. bot. Genève ser. 2. 29: 135-140. 1938 (obit., portr., bibl., b. 20 Apr 1857, d. 21 Nov 1937).

Ronceray, Paul-Louis (1875-1953 (?)), French pharmacist; Dr. pharm. Univ. Paris 1904; Dr. med. ibid. 1911. (*Ronceray*).

HERBARIUM and TYPES: Unknown.

BIBLIOGRAPHY and BIOGRAPHY: GR p. 367 (b. 23 Jun 1875; d. ? 1953); Kelly p. 194; LS 10238, 22510, 38522.

9491. *Contribution à l'étude des lichens à orseille* ... Paris (Maison d'Éditions A. Joanin et Cie. ...) 1904. Oct. (*Contr. lich. orseille*)
Publ.: 1904 (Bot. Centralbl. 28 Dec 1904; Hedwigia 20 Apr 1905), p. [1]-94, [95, cont.], *pl. 1-3*. *Copy*: Stevenson. – Also publ. as Trav. Lab. Mat. méd. École. sup. Pharm., Paris 2, 1905. – "Orseille", orchil, is a red or violet dye obtained from lichens.
Ref.: Hue, A.M., Bot. Centralbl. 102: 67. 1906 (rev.).

Rondelet, Guillaume (1507-1566), French ichthyologist, botanist and physician; MD Montpellier 1537, practicing physician at Maringues and Montpellier; from 1545 regius profesor of medicine at Montpellier; from 1540 also personal physician to François Cardinal Tournon; chancellor of Montpellier 1556-1566. (*Rondelet*).

HERBARIUM and TYPES: Unknown.

BIBLIOGRAPHY and BIOGRAPHY: Ainsworth p. 40; Backer p. 498; Barnhart 3: 175 (b. 27 Sep 1507, d. 30 Jul 1566); BM 4: 1727; Bossert p. 337; DSB 11: 527-528 (secondary literature); Kew 4: 513; PR (alph.), ed. 1: 2429; TL-2/4907; see J. Daléchamps, L. Ghini, M. de L'Obel, G.I. Molina; Tucker 1: 13, 605-606.

BIOFILE: Broussonnet, Notice sur Guillaume Rondelet, 16 p., from Ephém. méd. Montpellier, Jan 1828.
Dulieu, G., Clio medica 1: 89-111. 1966 (biogr., sources, sec. refs.).

Ewan, J. & N., John Banister 13, 334, 339, 354. 1970.
Jeanselme et al., Bull. franç. Hist. Méd. 19: 203-206. 1925.
Joubert, L., Vita Gulielmi Rondeletii 1568 (in his Opera latina 2: 186-193. 1582).
Jourdan, A.J.L., Dict. Sci. méd., Biogr. méd. 7: 48-49. 1825.
Lapeyrouse, P., Hist. abr. pl. Pyren. xx. 1813.
Martins, Ch., Le jardin des plantes de Montpellier 7, 13, 33, 76. 1854.
Nissen, C., Zool. Buchill. 348. 1969.
Oppenheimer, J.M. Bull. inst. Hist. Med. Johns Hopkins Univ. 4: 817-834. 1936 (biogr., b. 25 Sep 1507; analysis of personality, bibl., secondary references).
Planchon, J.E. et G., Rondelet et ses disciples. Montpellier 1886, 2 and 46 p., from Montpellier medical 1866.
Wittrock, V.B., Acta Horti Berg. 3(2): 117. 1903, 3(3): 112. 1905.
Wright, R.L., Gardener's tribute 214-241. 1949 (portr. on p. 128g).

EPONYMY: *Rondeletia* Linnaeus (1753).

Ronniger, Karl (1871-1954), Austrian botanist and "Rechnungsdirector" at the ministry of finances Wien, specialist on *Thymus*. (*Ronniger*).

HERBARIUM and TYPES: material at E, H, L, NY.

BIBLIOGRAPHY and BIOGRAPHY: Barnhart 3: 175 (b. 13 Aug 1871); Biol.-Dokum. 15: 7636; Bossert p. 337; DTS 6(4): 189; Hortus 3: 1202; MW p. 417; NI 2: 94, 244; PFC 2(2): xvii; Zander ed. 10, p. 708, ed. 11, p. 808.

BIOFILE: Anon., Österr. Bot. Z. 72: 376. 1923 (app. "Regierungsrat").
Fukarek, P., Godišnjak, Sarajevo 5(1/2): 460-463 (obit., bibl.).
Janchen, E., Cat. fl. austr. 1: 16, 17, 23, 24, 30, 35, 46. 1956.
Petrak, F., Sydowia 1: 310. 1947 (epon.).
Rechinger, K., Proc. bot. Soc. Brit. Isl. 1(2): 280-281. 1954 (obit.).
Verdoorn, F., ed., Chron. bot. 5: 26. 1939, 6: 303. 1941.

EPONYMY: *Ronnigeria* Petrak (1947).

COMPOSITE WORKS: (1) *Thymus*, in Prodr. fl. penins. Balcan., Rep. Sp. nov. Beih. 30(2): 337-382. 15 Oct 1930 (TL-2/2504); also reprinted with special t.p. and new title: *Die Thymus-Arten der Balkan-Halbinsel* (species balcaniae generis Thymi), Dahlem bei Berlin (Fabeckstr. 49 [address of F. Fedde]) 1930, p. [i], 317-382. *Copy*: H.
(2) *Thymus*, in Grossheim *Fl. kawkasa* 3: 343-347. 1932.

9492. *Bestimmungstabelle für die Thymus-Arten des Deutschen Reiches* ... [Berlin] (Druck und Verlag E.F. Keller's Witwe, ...) 1944. Oct. (*Best.-Tab. Thymus*).
Publ.: 1944, p. [i], cover, [1]-24. *Copy*: J. Jalas. – Veröffentlichungen aus der Zeitschrift "Die deutsche Heilpflanze". Reprinted from Die Deutsche Heilpflanze 10(5) (journal n.v.).

Rooke, Hayman (c. 1722-1806), British (English) botanical artist and soldier, of Whitehaven, Cumberland. (*Rooke*).

HERBARIUM and TYPES: Unknown.

BIBLIOGRAPHY and BIOGRAPHY: Barnhart 3: 176; BB p. 261; BM 4: 1728; Desmond p. 529 (d. 18 Sep 1806); Dryander 3: 322; Henrey 2: 481, 568, 569, 583, 3: 108 (nos. 1277-1279); Kew 4: 514; PR 7755, ed. 1: 8713; Tucker 1: 606.

BIOFILE: Allibone, S.A., Crit. dict. Engl. lit. 2: 1862. 1878.
Freeman, R.B., Brit. nat. hist. books 299. 1980.

9493. *Descriptions and sketches of some remarkable oaks*, in the park at Welbeck, in the County of Nottingham, a seat of his Grace the Duke of Portland. To which are added,

observations on the age and durability of that tree. With remarks on the annual growth of the acorn ... London (printed by J. Nichols for the author, ...) 1790. Qu. (*Descr. sketch. oaks*).
Publ.: 31 Dec 1790 (plates publ. 31 Dec 1790), p. [1]-23, *pl. 1-10* (uncol. copp. auct.).
Copies: HH, MO.

Roper, Freeman Clarke Samuel (1819-1896), British (English) botanist and microscopist; businessman in London, from 1868 resident at Eastbourne. (*Roper*).

HERBARIUM and TYPES: BM (diatoms, 3.580 slides); BTN (herb., 3000); further material at BP and K. – See also under W.B. Hemsley.

BIBLIOGRAPHY and BIOGRAPHY: Barnhart 3: 176; BB p. 261; BL 2: 271, 710; BM 4: 1728; Bossert p. 337; CSP 5: 271, 8: 777, 11: 215, 18: 289; Dawson p. 530 (b. 23 Sep 1819, d. 28 Jul 1896); De Toni 1: cviii, 2: ci-cii; GR p. 413; IH 2: (files); Jackson p. 251; Kelly p. 194; Kew 4: 514; LS 22516-22519; Morren ed. 10, p. 75; Rehder 5: 736; TL-2/see W.B. Hemsley; Tucker 1: 606.

BIOFILE: Anon., J. Bot. 34: 340-341. 1896 (obit., portr.); J.R. microsc. Soc. 1896: 688 (obit., bibl.); Nat. Nov. 18: 462. 1896, 9: 47. 1897; Proc. Linn. Soc. 1896/97: 67. 1897 (obit.).
Bridson, G.D.R. et al., Nat. hist. mss. res. Brit. Isl. 29.412. 1980.
Deby, J., Bibl. microsc. stud. 3 (Diat.): 47. 1882.
Hawksworth, D.L. & M.R.D. Seward, Lichenol. Brit. Isl. 24, 141-142. 1977.
Kneucker, A., Allg. bot. Z. 2: 17. 1896 (d.).
Kent, D.H., Brit. herb. 76. 1957 (coll.).
Koster, J.Th., Taxon 18: 557. 1969 (diat. BM).
Murray, G., Hist. coll. BM(NH) 1: 178. 1904 (Diat. bequeathed 1896; 151 Brit. pl., 538 Amer. pl.).
Roper, F.C.S., Papers Eastbourne natural History Society, suppl. Fauna and Flora, 1873: 5-6. Session 15 Dec 1876 (*Notes on the Eucalyptus globulus* Labillardière).
Stieber, M.T. et al., Huntia 4(1): 85-86. 1981 (arch. mat. HU).
VanLandingham, S.L., Cat. diat. 4: 2366. 1971, 6: 3586. 1978, 7: 4223. 1978.
Watson, H.C., Topogr. bot. ed. 2. 555. 1883.
Wolley-Dod, A.H., Fl. Sussex xlvii. 1970.

EPONYMY: *Roperia* Grunow ex Pelletan (1889).

9494. *Supplement to the fauna and flora of Eastbourne*, together with a list of Eastbourne cretaceous fossils ... For private circulation. December, 1873. Oct. (*Suppl. fauna fl. Eastbourne*).
Publ.: Dec 1873, p. [1]-22, [23, err.]. *Copy*: HH. – Possibly reprinted from the Proc. Eastbourne nat. hist. Soc. 1873 (journal n.v.).

9495. *Flora of Eastbourne*. Being an introduction to the flowering plants, ferns, etc. of the Cuckmere district, East Sussex, with a map ... London (John van Voorst, ...) 1875. Oct. (*Fl. Eastbourne*).
Publ.: 1875 (p. vi: 20 Apr 1875; J. Bot. Jul 1875), p. [i]-vii, map, [ix]-xliii, [1]-165. *Copies*: E, MO, NY, USDA. – See BL 2: 271 for further references.
Ref.: Trimen, H., J. Bot. 13: 246-247. 1875.

Roper, Ida Margaret (1865-1935), British (English) botanist at Bristol; long time secretary and editor of the Bristol Naturalist's Society, president id. 1913-1916. (*I. Roper*).

HERBARIUM and TYPES: LDS (10.000); other material at BRISTM, CGE, K, LIV, NOT, OXF. – Library also at LDS.

BIBLIOGRAPHY and BIOGRAPHY: Barnhart 3: 176 (b. 25 Aug 1865, d. 8 Jun 1935); Clokie p. 234; Desmond p. 530.

BIOFILE: Bridson, G.D.R. et al., Nat. hist. mss. res. Brit. Isl. 52.5, 84.11. 1980.
Burrell, W.H., J. Bot. 74: 78-79. 1936 (herb. to LDS).
H.S.T., Nature 136: 134-135. 1935 (obit.), Bot. Soc. Exch. Cl. Brit. Isl. Rep. 11(1): 19. 1936 (obit.).
Riddelsdell, H.J. et al., Fl. Gloucestershire cxliii. 1948.
Thompson, H.S., J. Bot. 73: 233-234. 1935 (obit.).
Verdoorn, F., ed., Chron. bot. 2: 174. 1936, 3: 162. 1937.

Roques, Joseph (1772-1850), French botanist and physician; connected with the military hospitals and faculty of medicine, Montpellier; later at Paris. (*Roques*).

HERBARIUM and TYPES: Unknown.

BIBLIOGRAPHY and BIOGRAPHY: Barnhart 3: 176 (b. 9 Feb 1772, d. Mai 1850); BM 4: 1728; Herder p. 245, 400, 411; Jackson p. 162, 200, 201, 206, 275; Kelly p. 194; Kew 4: 515; Lasègue p. 581; LS 22519, 38524; NI 1672-1674; Plesch p. 386; PR 7757-7760, ed. 1: 8705-8717; Rehder 3: 55, 56, 57; Sotheby 652-655; Stevenson p. 1256; Tucker 1: 606.

BIOFILE: Wittrock, V.B., Acta Horti Berg. 3(3): 113. 1905.

9496. *Plantes usuelles*, indigènes et exotiques, avec la description de leurs caractères distinctifs et de leurs propriétés médicinales, ... Paris (chez l'auteur, ... et chez Madame veuve Hocquart, ...) 1807-1808. 2 vols. Qu. (*Pl. usuel.*).
1: 1807, p. [i*-v*], [i]-viii, [1]-266, *pl. 1-72* (col. copp. by J. Grasset de Saint-Sauveur), published in two "livraisons".
2: 1808, p. [i-iii], 1-278, [1, err.], *pl. 1-61* (id.).
Copies: FI(BN), G, NY, USDA.
Ed. 2: 1809, 2 vols. *Copy*: HH. – "*Plantes usuelles*, indigènes et exotiques, dessinées et coloriées d'après nature, avec ... seconde édition ..." Paris (id.) 1809.
1: [i*], engr. t.p., [iii*], [i]-viii, [1]-266, pl. 2-72 (id.; pl. 1 = t.p.).
2: [i-iii], 1-278, *pl. 1-61*.

9497. *Phytographie médicale*, ornée de figures coloriées de grandeur naturelle, ou l'on expose l'histoire des poisons tirés du règne végétal, et les moyens de remédier a leurs effets délétères, avec des observations sur les propriétés et les usages des plantes héroïques ... Paris (Chez l'auteur, ... de l'Imprimerie de Didot le Jeune, ...) 1821 [-1824]. Qu. 2 vols. (*Phytogr. méd.*).
Publ.: in 36 "livraisons" of 8-16 p. of text and 5 plates each 1821-1824 (livr. 29 out by Mai 1824). *Copies* (both Qu.): G, NY. – Title pages dated 1821. – A few copies were printed in folio on special paper; a copy is reported by Junk 206, 1977, no. 262 and 214, 1974, no. 259.
1: [i]-xii, [1]-304, *pl. 1-90* (colour printed, touched by hand, by Hocquart.
2: [i-iii], [1]-328, *pl. 91-180* (id.).
Ed. 2: 1835, 3 vols and atlas. *Copy*: GOET (inf. G. Wagenitz). – *Phytographie médicale*, histoire des substances héroïques et des poisons tirés du règne végétal, ... nouvelle édition, entièrement refondue avec un atlas grand in-4 de 150 planches coloriées. Tome premier [-troisième]. Paris (chez B. Cormon et Blanc, ...), Lyon (id.) 1835.
1: [i]-xix, [1]-560.
2: [i-iv], [1]-620.
3: [i-iv], [1]-560.
Atlas: [1]-8, *pl. 1-150* (*id*). *Copy*: also at MICH.
Re-issue of ed. 2, 1845. *Copy*: BR (atlas n.v.). – *Phytographie* ... refondue, et publiée en 3 vol. in 8, beau papier vélin, ... avec un atlas gr. in 4, de 150 planches coloriées ... Paris (Edouard Garnot, libraire) 1845.
1: [i]-xix, [1]-560.
2: [i, iii], [1]-620.
3: [i,iii], [1]-560.
Ref.: Anon., Bull. Linn., Paris 1: 11-12, 19-20. 1824 (ed. 1).

9498. *Histoire des champignons* comestibles et vénéneux, ornée de figures coloriées représ-

entant les principales espèces dans leurs dimensions naturelles; où l'on expose leurs caractères distinctifs, leurs propriétés alimentaires et économiques, leurs effets nuisibles et les moyens de s'en garantir ou d'y remédier, ouvrage utile aux amateurs de champignons, aux médicins, aux naturalistes, aux propriétaires ruraux, aux maires des villes et des campagnes, etc. ... Paris (Hocquart ainé, ...) 1832. Qu. (*Hist. champ.*).
Publ.: 1832 (p. 10: 15 Oct 1831), p. [i-iii], [1]-192, *pl. 1-24* (col. printed, finished by hand, Bordes, Hocquart). *Copies*: FH, G, HH, MICH, Stevenson; IDC 7336.
Ed. 2: 1841, text and atlas, "deuxième édition, revue et augmentée, avec un atlas ...". Paris (Fortin, Masson et Cie., ...) 1841. *Copies*: FH, G, NY, Stevenson, USDA; IDC 6307.
Text: p. [i-iv], [1]-482, [483, er.].
Atlas: p. [1]-4, *pl. 1-24* (as ed. 1).
Reissue of plates with brief notes: 1864, p. [i-iv], [1]-16, *pl. 1-24* (id.). *Copies*: FH, G, NY, Stevenson. – "*Atlas des champignons* ... Extrait de la deuxième édition". Paris (Victor Masson et fils, ...) 1864. Qu.

9499. *Nouveau traité des plantes usuelles*, spécialement appliqué à la médecine domestique, et au régime alimentaire de l'homme sain ou malade; ... Paris (Librairie de P. Dufart, ...) Saint-Pétersbourg (chez J.-F. Haüer et Cie.) 1837. Oct. (*Nouv. traité pl. usuel.*).
1: 1837, p. [i-iii], [1]-574 (in parts).
2: 1837, p. [i-iii], [1]-592 (id.).
3: 1837, p. [i-iii], [1]-560.
4: 1838, p. [i-iii], [1]-599.
Copies: BR, HH, MO, NY.

Rosander, [Karl] **Henrik Andreas** (1873-1950), Swedish botanist; Dr. phil. Uppsala 1906; biology and chemistry teacher at Härnösands grammar school (1911). (*Rosander*).

HERBARIUM and TYPES: Unknown.

BIBLIOGRAPHY and BIOGRAPHY: Barnhart 3: 177; BM 4: 1729; KR p. 596; SO 2895a.

9500. *Studier öfver bladmossornas organisation*. Mössa, vaginula och sporogon. Akademisk afhandling af H.A. Rosander. Disputationsakten äger rum å botaniska lärosalen [Uppsala] den 29 maj 1906 p.v.t. f.m. Oct. (*Stud. bladmoss. org.*).
Publ.: 1-29 Mai 1906, p. [i]-viii, [1]-100. *Copies*: G, MO-Steere, NY (rd. 1 Jan 1907). – p. [ii]: Wretmans Boktryckeri Uppsala, 1906.
Ref.: Arnell, H.W., Bot. Centralbl. 102: 540. 20 Nov 1906 (summary).

Rosanoff, Sergei Matveevič (1840-1870), Russian botanist; librarian of the botanical garden at St. Petersburg; died at sea travelling from Napoli to Palermo. (*Rosanoff*).

HERBARIUM and TYPES: Unknown.

BIBLIOGRAPHY and BIOGRAPHY: Barnhart 3: 177 (d. 3 Dec 1870); BM 4: 1729; CSP 8: 778; De Toni 1: cviii, 4: xlviii; Frank 3 (Anh.): 85; Herder p. 120; Jackson p. 157; Kew 4: 515; LS 22522; Moebius p. 204; PR (alph.); Rehder 5: 736; TR 1171-1172*.

BIOFILE: Anon., Bot. Bot. 1871: 36 (d.); Bot. Zeit. 29: 47. 1871.; Flora 51: 73. 1868 (plant physiol. at bot. gard. St.Petersb.), 54: 63. 1871 (d.); Österr. bot. Z. 16: 232. 1866 (librarian bot. gard. St. Petersburg), 21: 83. 1871 (d. on a sea trip from Napoli to Palermo).
Beccari, Nuova Giorn. bot. ital. 3(1): 105. 1871 (d.).
Petunnikoff, A., Bot. Zeit. 29: 47. 1871 (d.).

EPONYMY: *Rosanovia* Bentham (1876, *orth. var.*); *Rosanowia* Regel (1872).

9501. *Recherches anatomiques sur les Mélobésiées* (Hapalidium, Melobesia, Lithophyllum et Lithothamnium) dédié à M.A. Le Jolis. Par S. Rosanoff cand. sc. nat. ... Cherbourg (Bedelfontaine et Syffert, ...) 1866. Oct.

Publ.: 1866, p. [1]-112, *pl. 1-7* (uncol. liths. auct.). *Copies*: G, NY; IDC 6404. – Reprinted and to be cited from Mém. Soc. Sci. nat. Cherbourg 12: 5-112. *pl. 1-7.* 1866.
Ref.: Schlechtendal, D.F.L. von, Bot. Zeit. 24: 280-282. 7 Sep 1866 (rev.).

Rosbach, Heinrich (*fl.* 1880), German (Prussian) botanist; regional physician at Trier ("Königl. preuss. Kreisphysicus und Sanitätsrath"). (*Rosbach*).

HERBARIUM and TYPES: Unknown.

BIBLIOGRAPHY and BIOGRAPHY: Barnhart 3: 177; BFM no. 327; Biol.-Dokum. 15: 7640; BM 4: 1729; CSP 5: 272, 11: 216, 12: 628; Herder p. 321; Jackson p. 311; Rehder 1: 388.

BIOFILE: Lefort, E.L., Soc. nat. Luxemb. Bull. 54: 118, 135, 137, 140. 1950.

9502. *Saxifraga multifida* nova species und ihre nähere Verwandte: S. cespitosa L., spanhemica Gm., und hypnodes L. ... Gent (Druck von C. Annoot-Braeckman) 1875. Oct.
Publ.: 1875, p. [1]-12. *pl. 1-2. Copy*: M. – Reprinted and to be cited from Bull. Soc. Bot. Belgique 14: 111-120. 1875. See also Corr.-Bl. naturh. Ver. Bonn 1876: 77-81.

9503. *Flora von Trier.* Verzeichniss der im Regierungsbezirke Trier sowie dessen nächster Umgebung wild wachsenden, häufiger angebauten und verwilderten Gefässpflanzen nebst Angabe ihrer Hauptkennzeichen und ihrer Verbreitung ... Trier (Verlag von Eduard Groppe) 1880, 2 parts. (*Fl. Trier*).
Ed. [*1*]: Jan-Mar 1880 (Bot. Centralbl. 15-19 Mar 1880; J. Bot. Jul 1880; Bot. Jahrb. 12 Oct 1880; Bot. Zeit. 16 Apr 1880; Nat. Nov. Mar(1) 1880. *Copy*: BR.
1: [i]-ix, [1]-231.
2: [i]-vi, [1]-197.
Ed. 2 (reissue with new t.p.'s): Jan-Jun 1896 (Bot. Centralbl. 21 Jul 1896; Nat. Nov. Jul(2) 1896), "2. wohlfeile Ausgabe" Trier (Verlag von Heinrich Stephanus) 1896, 1: [i]-ix, [1]-231, 2: [i]-vi, [1]-197. *Copy*: M.
Ref.: Buchenau, F., Bot. Centralbl. 3/4: 1064-1065. 1880 (rev.).
P., Bot. Zeit. 40: 285. 28 Apr 1882 (rev.).

Roscoe, Margaret (née Lace; Mrs Edward Roscoe) (*fl.* 1830), British botanical artist, daughter-in-law of William Roscoe. (*M. Roscoe*).

HERBARIUM and TYPES: Unknown.

BIBLIOGRAPHY and BIOGRAPHY: Barnhart 3: 177; BB p. 261; BM 4: 1729; Desmond p. 530; DU 266; GF p. 74; Jackson p. 407; Kew 4: 515; NI 1676, 1677; Plesch p. 386; PR 7763, ed. 1: 8721; Sotheby 656.

BIOFILE: Dallman, A.A. & M.H. Wood, Trans. Liverpool Bot. Soc. 1: 87. 1909.
Stearn, W.T., Roy. Hort. Soc. Exhib. manuscr. books 42. 1954.

9504. *Floral illustrations of the seasons*, consisting of representations drawn from nature of some of the most beautiful hardy and rare herbaceous plants cultivated in the flower garden, carefully arranged according to their seasons of flowering, with botanical descriptions, directions for culture, &c. By Mrs. Edward Roscoe, ... engraved by R. Havell, Jun. London (published by and for R. Havell, Jun., ... by Baldwin and Cradock, ...) 1829 [-1831]. Qu. (size). (*Fl. ill. seasons*).
Publ.: in 7 fascicles Nov 1829-1831, p. [i]-vi [Nov 1829], [vii], *pl. 1-55* with text (fasc. 1: *pl. 1-8*, 2: *pl. 9-16*, 3: *17-24*, 4: *25-32*, 5: *33-40*, 6: *41-48*, 7: *49-55*). *Copies*: G (fasc. 1, p. [i]-vi, [vii], *pl. 1-8*, with orig. cover), MO, NY. – The G and MO copies have the printed 1829 t.p. as cited above; the NY copy has an engraved 1831 t.p. with abbreviated title. "*Floral, illustrations of the seasons*, consisting of the most beautiful, hardy and rare herbaceous plants, cultivated in the flower garden ..." London (Published by Robt. Havell Jr., ... and Baldwin & Cradock, ...) 1831; it has in addition three half titles preceding *pl. 17, 33* and *49*. – The plates are handcoloured copper engravings by Mrs Roscoe ("line and very fine grained aquatint" DU).

Roscoe, William (1753-1831), British (English) historian, botanist and banker at Liverpool 1774; devoting himself to literary studies 1794; partner in a bank 1799; lost money 1816, bankrupt 1820; founder of the Liverpool botanical garden, 1802. (*Roscoe*).

HERBARIUM and TYPES: LIV (plants as well as letters and original drawings for *Monandr. pl.*); some material also at PH; other letters are at BM, K ad LINN. – Roscoe's library was sold by auction, Liverpool 1816.

BIBLIOGRAPHY and BIOGRAPHY: AG 3: 599; Backer p. 499; Barnhart 3: 177; BB p. 261-262; Blunt p. 214; BM 4: 1729; Bossert p. 338; CSP 5: 274; Dawson p. 711-712; Desmond p. 530 (b. 8 Mar 1753, d. 30 Jun 1831); DNB 49: 222-225 (concise dict. p. 1128); DU 267; Frank 3 (Anh.): 85; GF p. 74; Herder p. 150, 289; Hortus 3: 1202; IH 1 (ed. 7): 341, 2: (files); Jackson p. 121, 413; Kew 4: 515; Langman p. 645; Lasègue 539; Lenley p. 354; MW p. 417-418; NI 1677, see also 1: 129; Plesch p. 386-387; PR 7764-7765, ed. 1: 8720; Rehder 1: 53; RS p. 137; SK 4: ccx; SO 803, add 835b; Sotheby 657-658; TL-/1101; TL-2/937; Zander ed. 10, p. 708, ed. 11, p. 808.

BIOFILE: Allibone, S.A., Crit. dict. Engl. litt. 2: 1865-1868. 1878 (very detailed biogr.).
Baines, E., Hist. Lancastershire 2: 377-379. 1870 (n.v.).
Bartlett, H., 55 rare books no. 55. 1949.
Bioletti, R., Liverpool Libraries Museums & Arts Committee Bulletin 1962/63: 9-13 (acquisition of 231 drawings and 259 sheets of manuscript notes by the Liverpool libraries).
Bridson, G.D.R. et al., Nat. hist. mss. res. Brit. Isl. 432 [index]. 1980.
Chandler, G., William Roscoe of Liverpool, Liverpool 1953, xxxvi, 470 p. (library and de luxe ed.; main modern biography).
Coats, A.M., Gard. Chron. 1964(2): 529, 534 (portr.); Huntia 2: 206. 1965 (portr.).
Cullen, J., Notes R. bot. Gard. Edinburgh 32: 417-721. 1973.
Dallman, A.A. & M.H. Wood, Trans. Liverpool Bot. Soc. 1: 87-88. 1909 (biogr. data, bibl.).
Dawson, W.R., Smith papers 78-83. 1934 (lists extensive corr. Smith/Roscoe).
Graustein, J.E., Thomas Nuttall 11, 56, 57, 74, 75, 86, 89-90, 287. 1967.
Greenwood, E.F., J. Soc. Bibl. nat. Nat. 9(4): 376. 1980.
Milner, J.D., Cat. portr. Kew 92-93. 1906 (portr. at Kew).
Roscoe, Henry, The life of William Roscoe, by his son, Boston, 2 vols. Boston 1833, xii, 370 and x, 374 p. (portr., extensive general biography).
Smith, P., Mem. corr. J.E. Smith 2: 301. 1832.
Stansfield, H., Handb. Guide herb. coll. Liverpool 32, 35, 39, 40, 41, 42, 46, 54, 59. 1935; Liverpool libraries, Museums & Arts Committee Bulletin 5(1,2): 19-61. 1955 (biogr., portr.; Roscoe as a botanist).
Steenis-Krusemann, M.J. van, Fl. males. Bull. 28: 2364. 1974.
Urban, S., Gentleman's Mag. 101: 179-181. 1831 (n.v.).

HANDWRITING: Liverpool Libr. Mus. Arts Comm. Bull. 5(1, 2): fig. 15. 1955.

EPONYMY: *Roscoea* J.E. Smith (1805); *Note*: *Roscoea* Roxburgh (1832) probably also honors W. Roscoe, but no derivation was given.

9505. *Monandrian plants of the order Scitamineae*, chiefly drawn from living specimens in the botanical garden at Liverpool. Arranged according to the system of Linnaeus, with descriptions and observations, by William Roscoe, ... Liverpool (printed by George Smith) [1824-] 1828. Fol. (*Monandr. pl. Scitam.*).
Publ.: in 15 parts 1824-1828, for extensive details see Cullen (1973), who includes an index to plates with name, number, part and date; p. [i-xii], *pl. 1-112* (handcol. liths.) with text. *Copies*: BR, G, HH, NY; IDC 5688. – 150 copies printed.
1/2: Feb-Apr 1824, *pl. 6, 8, 22, 29, 34, 47, 53, 59, 62, 68, 69, 83, 93, 97, 103, 112*.
3/4: distributed before 29 Mai 1824, possibly already in second half of 1824, *pl. 19, 20, 49, 52, 86, 109, 74, 50, 23, 24, 76, 95, 77, 71*.
5/6: Nov-Dec 1825, *pl. 58, 1, 63, 33, 25, 96, 14, 108, 60, 110, 88, 105, 90, 55, 61, 82*.
7/8: 7 Apr-24 Jul 1826, *pl. 12, 13, 36, 80, 17, 18, 30, 48, 111, 73, 56, 94, 104*.

9/10: 7 Oct-8 Dec 1826, *pl. 11, 15, 27, 28, 92, 51, 87, 21, 106, 42, 45, 5, 70, 46, 101, 85, 99*.
11/12: 1827, before 5 Apr, *pl. 9, 10, 43, 44, 57, 54, 35, 38, 84, 91, 16, 102, 76, 64, 89, 39, 98*.
13/14: 23 Oct 1827-29 Feb 1828, *pl. 31, 2, 7, 37, 67, 66, 40, 26, 81, 78, 75,65, 41, 70*.
15: letterpress p. [i-xii], Jun-Aug 1828.
The lithographs of drawings are by several artists, listed in the preface (Thomas Allport; Rebecca Miller, Margaret Roscoe, Mrs J. Dixon, Ellen Yates, Emily Fletcher, Mary Waln) accompanied by letterpress. The numbers assigned to them in the systematical list do not reflect the order of publication. See Sotheby 657 for notes on the lithography.
Preliminary paper: Trans. Linn. Soc. 8: 330-357. 9 Mar 1807 (new arrangement Scitamineae), see also 10: 50-78. 8 Mar 1810 (on artificial and natural systems) and 11: 270-282. 24 Jan 1816 (remarks on Roxburgh's descr. of Monandrous pl.).
Ref.: Cullen, J., Notes R. Bot. Gard. Edinburgh 32: 417-421. 1973.

Rose, Hugh (ca. 1717-1792), British (English) apothecary and botanist at Norwich; teacher of J.E. Smith; associate of W. Hudson. (*H. Rose*).

HERBARIUM and TYPES: LINN (in herb. J.E. Smith).

BIBLIOGRAPHY and BIOGRAPHY: Barnhart 3: 177; BB p. 262; BM 4: 1730; Desmond p. 530 (d. 18 Apr 1792); Dryander 3: 22, 136; Henrey 2: 114-115, 116, 184, 654, 3: 75 (no. 974); IH 2: (files); Jackson p. 33; Kelly p. 194; Kew 4: 516; Lasègue p. 417, 503; PR 7766, ed. 1: 8722; Rehder 1: 83, 250; SO 470; TL-1/729; TL-2/4760.

BIOFILE: Dawson, W.R., Smith papers 83. 1934.
Kent, D.H., Brit. herb. 76. 1957.
Smith, J.E., Trans. Linn. Soc. 7: 297-299. 1884 (biogr. memoirs).
Walter, J.H.F., Trans. Norfolk Norwich Natural. Soc. 9(5): 649, 654-657, 660, 663. 1914 (on Rose and J.E. Smith).

EPONYMY: *Rosea* C.F.P. Martius (1826) and *Rosea* (Klotzsch (1853) honor Valentin Rose, German apothecary of the 19th Century.

9506. *The elements of botany*: containing the history of the science: with accurate descriptions of all the terms of art, exemplified in eleven copper-plates; the theory of vegetables; the scientific arrangement of plants, and names used in botany; rules concerning the general history, virtues, and uses of plants. Being a translation of the Philosophia botanica, and other treatises of the celebrated Linnaeus. To which is added, an appendix, wherein are described some plants lately found in Norfolk and Suffolk, illustrated with three additional copper-plates, all taken from the life. By Hugh Rose, apothecary. London (printed for T. Caddell, ...) 1775. Oct. (*Elem. bot.*).
Publ.: 1775, p. [i]-xii, [1]-472, *pl. 1-11*, *app. 1-3* (uncol. copp.). *Copies*: HH, MICH, MO, NY, PH. – Mainly an English translation of C. Linnaeus, *Philosophia botanica* 1751, see TL-2/4760.

Rose, Joseph Nelson (1862-1928), American botanist; Dr. phil. Wabash, student of J.M. Coulter; assistant botanist USDA, Washington 1888-1896; assistant curator for botany at the Smithsonian Institution 1896-1905; associate curator ib. 1905-1912 and 1923-1928; research associate of the Carnegie Institution of Washington to work on the Cactaceae of the world with N.L. Britton 1912-1923; collected extensively in Mexico, the Central and Southwestern United States and South America; co-author, with N.L. Britton, of the treatments of Leguminosae for the North American Flora. (*J. Rose*).

HERBARIUM and TYPES: US; large sets also at GH and NY; further material at A, BM, C, F, K, LA, MEXU, MO, PH, U, UPS. – Field-books, journals, reports and correspondence (e.g. with J.M. Coulter and N.L. Britton) in Smithsonian Archives.

BIBLIOGRAPHY and BIOGRAPHY: ADB 8: 159-161 (Maxon); Barnhart 3: 177 (b. 11 Jan 1862, d. 4 Mai 1928); BJI 1: 49; BL 1: 132, 150, 152, 316, 2: 148; BM 4: 1730, 8: 1092; Bossert p. 338; CSP 18: 291-292; De Toni 1: cviii; Ewan ed. 2: 187; Hortus 3: 1202; IH (ed. 7): 341; Kew 4: 516; Langman p. 645-650 (impt. bibl); Lenley p. 354; LS 22523-22524; NAF 28B(2): 354; NI (alph.); PH 450, 937; Rehder 5: 737; TL-2/780-781, 786, 1254; Tucker 1: 606; Urban-Berl. p. 281; Zander ed. 10, p. 708, ed. 11, p. 728, 808.

BIOFILE: Acosta-Solis, M., Natural. Viaj. ci. Ecuador 50-51. 1968.
Anon., Hedwigia 38: (249). 1899 (visits Mexico), 40: (204). 1901 (id.), 68: (135). 1929 (d.); J. Bot. 66: 216. 1928 (d.); J. New York Bot. Gard. 29: 139-140. 1928 (d.), 6: 192-193. 1905 (coll. Mexico); J. Washington Acad. Sci. 18(10): 296. 1928 (obit.); Natl. Cycl. Amer. biogr. 27: 449. 1939.; Smithsonian Local Notes 18 Mai 1928, p. 1 (obit.); Torreya 28: 66. 1928 (d.); Who's who in America 1899-1900: 618, 15 (1928-1929): 1799; Who was who in America 1: 1057. 1968.
Barnhart, J.H., J. New York Bot. Gard. 36: 32. 1935.
Benson, L., The cacti of Arizona ed. 2. 9-10. 1950; The cacti of the United States and Canada 10-11. 1982 (portr.).
Berger, A., Z. Sukkulentenk. 3: 281-283. 1928 (obit.; portr.).
Castañeda, A.M., Fl. est. Jalisco 118. 1933.
Castellanos, A., Physis 9: 154-155. 1928 (obit.).
Cattell, J.M., Amer. men. sci. ed. 1: 273. 1906, ed. 2: 400. 1911, ed. 3: 585. 1921, ed. 4: 835. 1927.
Cowan, R.S. & F.A. Stafleu, Brittonia 1981 (relationship J.N. Rose and N.L. Britton; biogr. sketch; work on Cactaceae and Leguminosae).
Deam, C.C., Fl. Indiana 1118. 1910 (Indiana coll. F, US).
Herrera, F.L., Est. fl. Dept. Cuzco 20-21. 1930.
Hunt, D.R., J. Mammilaria Soc. 15(5): 61-62. 1975 (broad species concept in Mammillaria, liberal generic concepts).
Jones, M.E., Contr. West. Bot. 18: 140-141. 1935 (personal recollections; "died disappointed", not a good systematic botanist..." [Jones resented Rose's position regarding genera in the Umbelliferae]).
Junk, W., Rara 2: 215. 1926-1936 (descr. "Cactaceae").
Knobloch, J.W., Pl. coll. N. Mexico 59-60. 1979.
McLeod, M.G., Yearb. Cact. Succ. Soc. America 65-75. 1975.
McVaugh, R., Edward Palmer x, 4, 98, 101, 107, 110, 114, 176, 195, 346, 401. 1956; Contr. Univ. Mich. Herb. 9: 296-298. 1972 (Mexican itineraries).
Mitich, L.W., Cact. Succ. J., U.S.A., 53(6): 299-303. 1981 (portr., bibl.).
N.T., Torreya 19: 200-203. 1919 (rev. vol. 1, Cactaceae, Britton and Rose).
Pennell, F.W. & J.H. Barnhart, Monogr. Acad. nat. Sci. Philadelphia 1: 618. 1935 (coll.).
Rodgers, A.D., Amer. bot. 1873-1892, p 232, 250, 292-293, 296, 299, 310, 315. 1944. Liberty Hyde Bailey 433. 1949, Amer. bot. 1873-1892, 336. 1949 (indexes; many refs.).
Sayre, G., Mem. New York Bot. Gard. 19(3): 389. 1975.
Sutton, S.B., Charles Sprague Sargent 133. 1970.
Trelease, W., Science ser. 2. 67: 598-599. 1928 (obit.).
Verbeek Wolthuys, J.J., Succulenta 10: 125-127. 1928 (obit.).
Wiggins, I.L., Fl. Baja Calif. 43. 1980 (1911 on Albatross exp.).

COMPOSITE WORKS: (1) *Crassulaceae*, in NAF 22: 7-74. 22 Mai 1905 (with N.L. Britton).
(2) *Burseraceae*, in NAF 25(3): 241-261. 6 Mai 1911.
(3) *Mimosaceae*, in NAF 23(1): 1-76. 11 Feb 1928, 23(2): 77-136. 25 Sep 1928, 23(3): 137-194. 20 Dec 1928 (with N.L. Britton).
(4) With N.L. Britton, *New N. Amer. Crassul.* 1903, TL-2/780.
(5) With N.L. Britton, *The Cactaceae*, 1919-1923, 4 vols., TL-2/786.
(6) With J.M. Coulter, *Rev. N. Amer. Umbell.* 1888, TL-2/1254. – Extensive reviews by Humphrey, Bot. Centralbl. 40: 227-230. 1889 and W.M. Canby, Bot. Gaz. 14: 110-111. 1889. – For *Monogr. N. Amer. Umbell.* 1900, see below.

EPONYMY: *Brittonrosea* Spegazzini (1923, also dedicated to N.L. Britton, *q.v.*); *Rosanthus* Small (1910); *Roseanthus* Cogniaux (1896); *Roseocactus* A. Berger (1925); *Roseocereus* Backerberg (1938); *Roseodendron* F. Miranda (9165); *Rhodosciadium* S. Watson (1890). *Note*: The derivation of *Roseia* Fric (1925) could not be determined.

9507. *List of plants collected by Dr Edward Palmer in Lower California and Western Mexico* in 1890. [Contr. U.S. natl. Herb. 1, 1890]. Oct.
Co-author: George Vasey (1822-1893). *Collector*: Edward Palmer (1831-1911).
Publ.: 8 Nov 1890 (in journal), p.[i, iii], 63-90, v-vi. *Copy*: G. – Reprinted and to be cited from Contr. U.S. natl. Herb. 1: 63-90. 8 Nov 1890. – Other publications by Rose and Vasey on United States and Mexican plants collected by Palmer are:
(1) Proc. U.S. natl. Mus. 11: 527-536. 1889 (Lower Calif., coll. 1889).
(2) Contr. U.S. natl. Herb. 1: 1-8. 16 Jun 1890 (S. Calif.).
(3) Ib. 1: 9-28. 16 Jun 1890 (Lower Calif., coll. 1889).
Idem by Rose alone:
(4) Contr. U.S. natl. Herb. 1: 91-116. 30 Jun 1891 (W. Mex., Ariz. 1890).
(5) Ib. 1: 117-127. 30 Jun 1891 (Ariz. 1890).
(6) Ib. 1: 129-134. 25 Sep 1892 (Carmen Isl. 1890).
(7) Ib. 1: 293-392. 31 Jan 1895 (Mexico 1890, 1891).
(8) Ib. 3: 311-319. 14 Dec 1895 (Mexico, U.S., 1891, 1892).

9508. *Studies of Mexican and Central American plants* ... [Contr. U.S. national Herbarium vols. 5-13] Washington (Government Printing Office) 1897-1911, 7 papers. Oct.
Publ.: A series of papers published in the Contr. U.S. natl. Herb. as follows:
1: 27 Aug 1897, *Contr.* 5(3): [i]-vii, 109-144, *pl. 2-17*.
2: 31 Oct 1899, *Contr.* 5(4): [i]viii, 145-200, *pl. 18-25*.
3: 16 Jun 1903, *Contr.* 8(1): [i]-vii, *pl. 1-2* (col.), *3-12*, [1]-55, [i]-iv.
4: 20 Apr 1905, *Contr.* 8(4): [i]-vii, 281-339, ix, *pl. 63-72*.
5: 5 Dec 1906, *Contr.* 10(3): [i]-vii, 79-1322, ix-x, *pl. 16-43*.
6: 12 Apr 1906, *Contr.* 12(7): [i]-vii, 259-302, ix, *pl. 20-27*.
7: 11 Apr 1911, *Contr.* 13(9): [i]-vii, 291-312, ix-x, *pl. 46-67*.
Copies: BR, G, M, MO, NY, US.
Notes on useful plants of Mexico: Contr. U.S. natl. Herb. 5(4): [i]-iv, 209-259, i-vii, *pl. 28-64*. 31 Oct 1899.

9508a. *Agave washingtonensis* and other Agaves ... (From the Ninth Annual Report of the Missouri Botanical Garden). Issued April 20, 1898. Oct.
Publ.: 20 Apr 1898 (reprint), cover t.p., p. 121-126, *pl. 29-31*. *Copies*: U, US. – Reprinted from Ann. Rep. Missouri Bot. Gard. 9: 121-126, *pl. 29-31*. 1898.

9509. *Plants of the Tres Marias Islands* Washington (Government Printing Office) 1899. Oct.
Publ.: 29 Apr 1899, cover-t.p., p. 77-91. *Copy*: US. – Reprinted and to be cited from North American Fauna 14: 77-91. 29 Apr 1899.

9510. *Agave expatriata* and other Agaves ... (Reprinted in advance from the Eleventh Annual Report of the Missouri Botanical Garden). Issued Jun 3, 1899. Oct.
Publ.: 3 Jun 1899 (cover), cover-t.p., p. [1]-5, *pl. 7-10*. *Copy*: U. – Preprinted from Ann. Rep. Missouri Bot. Gard. 11: 79-83. 1900.

9511. *Notes on useful plants of Mexico* ... Reprinted from Contributions from the U.S. National Herbarium vol. v, no. 4, ... 1899. Oct.
Publ.: 31 Oct 1899, cover-t.p., p. 209-259. *Copy*: US. – Reprinted and to be cited from Contr. U.S. natl. Herb. 5(4): 209-259. 31 Oct 1899.

9512. *Monograph of the North American Umbelliferae*, Contributions from the U.S. National Herbarium ... Washington (Government Printing Office) 1900. Oct.
Co-author: John Merle Coulter (1851-1928).
Publ.: 31 Dec 1900, issued as and to be cited from Contr. U.S. natl. Herb. 7(1): 9-256, *fig. 1-65*. *Copy*: US.
Preliminary publ.: See J.M. Coulter & J.N. Rose, Rev. N. Amer. Umbell. 1888, TL-2/1254.
Supplement to the Monograph: Contr. U.S. natl. Herb. 12: 441-451, *pl. 82-83*. 21 Jul 1909.
Ref.: Trelease, W., Bot. Gaz. 31: 130-131. 23 Feb 1901.
Malinvaud, E., Bull. Soc. bot. France 47(bibl.): 528-529. 1901.

9513. *Lenophyllum*, a new genus of Crassulaceae ... City of Washington (published by the Smithsonian Institution) 1904. Oct.
Publ.: 10 Oct 1904 (date on cover-t.p.), cover-t.p., p. 159-162, *pl. 20. Copy*: G. − Reprinted and to be cited from Smiths. misc. Coll. 47: 159-162, *pl. 20*. 1904.

Rose, Lewis Samuel (originally: Rosenbaum) (1893-1973), American (native Californian) plant collector in California; long-time associate of the California Academy of Sciences, from 1930 as a volunteer in its Department of Botany. (*L. Rose*).

HERBARIUM and TYPES: CAS; duplicates in B (extant), BUT, C, DPU, DS, E, F, GB, GH, IAC, JE, K, L, LAM, MIN, NMC, MO, NH, NY, NYS, OKLA, POM, RM, S, U, UC, US.

BIBLIOGRAPHY and BIOGRAPHY: Barnhart 3: 177 (b. 25 Nov 1893); Bossert p. 338; IH 1 (ed. 6): 363, 2: (files); Lenley p. 354.

BIOFILE: Cantelow, E.D. & H.C., Leafl. W. Bot. 8: 98. 1957.
Cowan, R.S., Taxon 23: 448. 1974 (d. 11 Nov 1973).
Ewan, J. et al., Leafl. W. Bot. 7: 78, 91, 94. 1953.
Howell, J.T., Leafl. W. Bot. 7(3): 91, 94. 1953, Madroño 22: 399. 1974 (obit.; Rose's collections constituted a very large portion of the CAS exchange program).
Kearney, T.H., Leafl. W. Bot. 8(12). 24 Nov 1958, fig. 3.
Thomas, J.H., Contr. Dudley Herb. 5(6): 161. 1961; Huntia 3: 28-29. 1969.
Verdoorn, F., ed., Chron. bot. 2: 302, 404. 1936.

Rosén, Eberhard (knighted 1770: Rosenblad)(1714-1796), Swedish physician and botanist; Dr. med. Uppsala 1741 (under N. Rosén); professor of medicine and anatomy at Lund 1744-1784. (*E. Rosén*).

HERBARIUM and TYPES: Some material at LINN and S. − Some plants in UPS (Thunberg) labelled Rosén, may have been collected by E. Rosén.

BIBLIOGRAPHY and BIOGRAPHY: Barnhart 3: 178 (sub Rosenblad); BM 4: 1731; Dryander 3: 169, 475; IH 2: (files); KR p. 596 (b. 16 Nov 1714, d. 21 Mar 1796); PR 7767, ed. 1: 8723-8724; Rehder 1: 351, 3: 719; Tucker 1: 607.

BIOFILE: Anon., Biogr. lexicon öfver namnkunnige svenska män 12: 206-213. 1846.
Eriksson, G., Bot. hist. Sverige 274, 300, 303. 1969.
Fries, Th.M., Bref skrifv. Linné 1(4): 9, 13, 18, 60, 87. 1910.
Jackson, B.D., Proc. Linn. Soc. 134 (suppl.): 19. 1922 (plants from Skåne in LINN).
Lindman, S., Svenska män och kvinnor 6: 352. 1949.
Löwegren, Y., Naturaliekabinett Sverige 121, 122, 124, 130, 149. 1952.Weibull, M. & E. Tegnér, Lunds Univ. Historia 1668-1868, 2: 145-147. 1868 (biogr. data, bibl.).
Wittrock, V.B., Acta Horti Berg. 3(3): 50. 1905.

EPONYMY: *Rosenia* Thunberg 1800 is dedicated to both Eberhard and his brother Nils Rosén (1706-1773).

9514. Eberhardi Rosén ... *Observationes botanicae*, circa plantas qvasdam, Scaniae non ubivis obvias, et partim qvidem, in Svecia hucusque non detectas. Quibus accessit brevis disqvisitio de Strage bovilla, qvotannis in pascuis Christianstadii observata ... Londini Gothorum [Lund] (ex officina directoris, Caroli Gustavi Berling, ...) 1749. Qu. (*Observ. bot.*).
Publ.: Sep-Dec 1749 (see Note on p. [viii]), p. [i-viii], [1]-88, [unnumbered = 89], 89 [is 90, verso of unnumbered 89]. *Copies*: GOET, LD. − See also C. Linnaeus, Skånska resa 285. 1751 in which L. reports on 5 Jul 1749 that the *Observ. bot.* were being printed. The Note on p. [viii] mentioning the date 1 Sep 1749 must have been inserted after the main printing was finished (Inf. O. Almborn). The book is dedicated to A. v. Haller.

Rosen, Felix (1863-1925), German (Leipzig-born) botanist; spent his youth in Jerusalem and Belgrado; studied at Basel (1883) and Strassburg, 1884-1886, with A. de Bary; Dr. phil. Strassburg 1886; assistant at Tübingen 1888-1891; id. with F. Cohn in Breslau 1891; habil. Breslau 1892; first appointment as e.o. prof. of botany at Breslau 1901; travelled in Abyssinia 1904-1905; e.o. professor of botany and director of the Breslau botanical institute 1906; ordinary professor 1920; murdered in Breslau 1925. (*F. Rosen*).

HERBARIUM and TYPES: WRSL.

NOTES (1) After the death of Cohn (1898) Rosen had to leave the Breslau botanical institute from 1901-1906 because of difficulties with J.O. Brefeld. He returned only on 1 Apr 1906 as professor and director of the institute for plant physiology. The press notices of his appointment as extraordinary professor in 1902 are therefore not quite clear.
(2) Friedrich Rosén (baron) (1834-1903) was a Russian (b. St. Petersburg) palaeobotanist.

BIBLIOGRAPHY and BIOGRAPHY: AG 3: 231; Ainsworth p. 122, 321; Barnhart 3: 178 (b. 15 Mar 1863, d. 8 Aug 1925); Biol.-Dokum. 15: 7643-7644; BJI 2: 148; BM 4: 1731; Clokie p. 338; CSP 18: 293; IH 2: (files); Kew 4: 516; LS 5426, 22526-22529, 38539; Moebius p. 56, 189; NI 1: 24, 260.

BIOFILE: Anon., Bot. Centralbl. 52: 318. 1892 (habil. Breslau) 85: 352. 1901 (e.o. prof. bot. Breslau); Bot. Jahrb. 30 (Beil. 68): 53. 1901 (e.o. prof. bot. Breslau), 38 (Beil. 87): 42. 1906 (app. Breslau); Bot. Zeit. 50: 725. 1892 (habil. Breslau), 64: 160. 1906 (ausserord. etatsmäss. Prof., Dir. pfl.-phys. Inst. Breslau); Hedwigia 45: (161). 1906 (app. Breslau), 66: (55). 1926 (murdered 8/9 Aug 1925); Nat. Nov. 23: 65. 1901 (e.o. prof. bot. Breslau), 28: 338. 1986 (e.o. prof. & dir. Breslau); Österr. bot. Z. 56: 247. 1906 (app. Breslau), 74: 288. 1925 (d. 8/9 Aug 1925).
Engler, A., Bot. Jahrb. 33 (Lit.): 5-6. 1903 (review of Rosen, Die Natur in der Kunst, 1903).
Kneucker, F., Allg. bot. Z. 12: 104. 1906 (app.Breslau), 30/31: 56. 1926 (d.).
Oberstein, O., Z. Landwirtschaftskamm. Breslau 29: 1392. 1925 (fide F.A. Pax, 1929).
Pax, F.A., Bibl. Schles. bot. 46, 60, 139, 143, 148. 1929.
Rosen, F., Die Natur in der Kunst – Studien eines Naturforschers zur Geschichte der Malerei, Leipzig 1903, 344 p. (written while banned from his botanical institute).
Schaede, R., Beitr. Biol. Pfl. 14: 261-282. 1926 (main obit., journal issue dedicated to the memory of R.), Ostdeut. Naturwart 1925(9): 450 (n.v.).
Verdoorn, F., ed., Manual of Pteridology 638 [index]. 1938.
Winkler, H., Ber. deut. bot. Ges. 43: (65)-(73). 1926 (obit., bibl.).

COMPOSITE WORKS: (1) Editor of *Beiträge zur Biologie der Pflanzen* 9(2), 1907-14(2) 1925.
(2) *Charakterpflanzen des abyssinischen Hochlandes*, in G.H.H. Karsten & H. Schenk, Vegetationsbilder, ser. 7, Heft 5. Jena 1909, *pl. 25-30*.
(3) F. Rosen contributed to Pauline Cohn's biography of her father Ferdinand Cohn, Breslau 1901, see e.g. p. 107-130.

THESIS: *Ein Beitrag zur Kenntniss der Chytridiaceen*, in F. Cohn, Beitr. Biol. Pfl. 4(3): [253]-266, [267], *pl. 13, 14*. 1887 (rev. Rothert, Bot. Centralbl. 31: 72-74. 18-29 Jul 1887).

9515. *Anleitung zur Beobachtung der Pflanzenwelt* ... Leipzig (Verlag von Quelle & Meyer) 1909. Oct. (*Anleit. Beobacht. Pfl.-Welt*).
Ed. 1: 1909 (p. iii; Mar 1909; ÖbZ Oct/Nov 1909; AbZ 15 Nov 1909; Bot. Centralbl. 18 Jun 1910), p. [i-vi], [1]-155. *Copy*: USDA. – Wissenschaft und Bildung, ed. Paul Herre, no. 42, 1909.
Ed. 2: 1917 (p. [4]: Oct. 1916; ÖbZ 1 Apr 1919), p. [1]-162. *Copy*: BR.

Rosén, Nils (from 1762 knighted Rosén von Rosenstein) (1706-1773), Swedish botanist; studied at Lund; Dr. med. Harderwijk (Netherlands) 1730; professor of medicine (incl. botany) at Uppsala 1740-1742, preceding C. Linnaeus; professor of practical medicine Uppsala 1742; arkiater (royal physician) 1746; librarian Uppsala univ. library 1756-1757; brother of Eberhard Rosén. (*N. Rosén*).

HERBARIUM and TYPES: Unknown.

BIBLIOGRAPHY and BIOGRAPHY: Barnhart 3: 179 (under Rosenstein); Bossert p. 338; KR p. 596 (b. 1 Feb 1706, d. 16 Jul 1773).

BIOFILE: Anon., Biogr. lexicon öfver namnkunnige svenska män 12: 244-252. 1846 (biogr., bibl.).
Blunt, W., The compleat naturalist 28, 38, 40, 65, 76, 80, 135, 149, 223. 1971.
Daniels, G.S., Taxon 25: 20. 1976 (portr. with caption).
Fries, Th.M., Bref skrifv. Linné 1(2): 5, 19, 26, 70, 202. 1908, 1(3): 287. 1909, 1(4): 12, 13, 115, 130, 169, 255, 277. 1910, 1(5): 365 [index], 1(6): 25, 66. 1912, and J.M. Hulth 1(7): 192 [index]. 1917.
Holmgren, G., K. Sv. Vet.-Akad. Levnadst. 8(136): 177-187. *pl. 6*. 1954 (biogr., portr.).
Hulth, J.M., Bref skrifv. Linné 1(8): 7, 8, 16, 21, 31, 42. 1922.
Lindman, S., Svenska män och kvinnor 6: 343. 1949.
Löwegren, I., Naturaliekabinett Sverige 28, 85, 167, 174, 184, 244. 1952.
O., Dict. sci. méd., Biogr. méd. 7: 52-54. 1825.
Scheutz, N.J.W., Bot. Not. 1863: 71.
Strandell, B., Sv. Linnésällsk. Årsskr. 1970-1971: 36-43. 1973 (with English summary).
Wallgren, A., [Linnés ämtetsbroder Rosén von Rosenstein]. Svenska Linnésällsk. Årsskr. 47: 26-42. 1964 (with English summary).

Rosenberg, Caroline Friderike (1810-1902), Hamburg-Altona-born Danish botanist (algologist, bryologist); teacher in Odense ca. 1835; living on the state Hofmansgave on the island of Fünen (Fyn) as foster-daughter of Niels Hofman Bang 1838-1902. (*C. Rosenberg*).

HERBARIUM and TYPES: C; other material at B (destr.), FI, KIEL, MW and S.

BIBLIOGRAPHY and BIOGRAPHY: Barnhart 3: 178 (24 Sep 1810, d. 11 Feb 1902); Urban-Berl. p. 312.

BIOFILE: Anon., Bot. Not. 1903: 34 (d.).
Fischer-Benzon, R.J.D. von, *in* P. Prahl, Krit. Fl. Schlesw.-Holst. 2: 51. 1890.
Rosenvinge, L.K., Bot. Tidsskr. 24: lxvii-lxviii. 1902 (obit.).
Warming, E., Bot. Tidsskr. 12: 180-181. 1881 (bibl.).
Wittrock, V.B., Acta Horti Berg. 3(2): 104, *pl. 18*. 1903 (portr.).

COMPOSITE WORKS: (1) *Flora danica pl. 2618, 2639, 2741, 2747*.
(2) Contributed to T. Jensen, *Bryol. danic.* 1856 (TL-2/3338).
(3) Contributed to J.M.C. Lange, *Haandb. Danske fl.* ed. 1, 1851 (TL-2/4179).
(4) Contributed to P.A.C. Heiberg, *Krit. Overs. Danske Diatom.* 1863, (TL-2/2572).

EPONYMY: *Rosenbergia* Oersted (1856).

Rosenberg, Mary Elizabeth (afterwards: Duffield) (1820-1914), British (English) flower painter of Bath, Somerset; married William Duffield 1850. (*M. Rosenberg*).

HERBARIUM and TYPES: Unknown.

BIBLIOGRAPHY and BIOGRAPHY: Barnhart 3: 178; BB p. 262; BM 4: 1732; Desmond p. 530 (d. 13 Jan 1914); DNB 49: 248; GH p. 74; Jackson p. 41; NI 1678-1679, see also 3: 57; Sotheby 659.

9516. *Corona amaryllidacea* by Miss Rosenberg. Bath (printed and published for the author by C.A. Bartlett, ...) [1839]. Fol. (size). (*Corona amaryll.*).

Publ.: 1839 (p. v: Jun 1839), p. [i-v], *8 pl.* (handcol. liths. auct.) with 1 p. text each. *Copy*: USDA. – According to BM published in four parts.

9517. The *museum of flowers*. London (R. Groombridge & Sons, ...) 1845. (*Mus. fl.*).
Publ.: 1845, unpaginated, t.p., dedication, preface, list of plates, *55 pl.* w.t. *Copy*: E (inf. J. Edmondson; *23 pl.*). According to Sotheby (no. 659) the number of plates varies.

Rosenberg, [Gustaf] **Otto** (1872-1948), Swedish botanist; fil. kand. Uppsala 1895; Dr. phil. Bonn 1899; lecturer in botany at Stockholm University (Stockholms högskola) 1899; profesor of botany ib. (plant anatomy and cytology) 1911-1940; outstanding botanical cytologist, embryologist and precursor of experimental taxonomy. (*O. Rosenberg*).

HERBARIUM and TYPES: Unknown.

BIBLIOGRAPHY and BIOGRAPHY: Barnhart 3: 178 (b. 9 Jun 1872, 11 Nov 1948); Biol.-Dokum. 15: 7647; BJI 2: 148; BM 4: 1731-1732; Bossert p. 338; KR p. 597-599 (bibl.); Moebius p. 350, 354.

BIOFILE: Åberg, B., *in* R.E. Fries, Short hist. bot. Sweden 126. 1950.
Anon., Bot. Centralbl. 120: 112. 1912 (app. prof. bot. Stockholm); Bot. Not. 1900: 48. (app. Stockholms Högskola); Hedwigia 52: (154). 1912 (prof. bot. Stockholm); Mycol. Centalbl. 1: 33. 1912; Nat. Nov. 34: 362. 1912 (prof. bot. Stockholm); Österr. bot. Z. 62: 151, 351. 1912 (prof. bot. Stockholm).
Bergman, B., Sv. Bot. Tidskr. 51: 378-387. 1957 (obit., portr., bibl.).
Dahlgren, K.V.O., *in* R.E. Fries, Short hist. bot. Sweden 95-96. 1950.
Fagerlind, F., Svenska män och kvinnor 6: 349-350. 1949 (portr.).
Fries, R.E., Short hist. bot. Sweden 83, 95, 106, 126. 1950.
Levan, A., K. Fysiogr. Sällk. Forh. Lund 20, 1958, 8 p. (obit.).
Müntzing, A., *in* R.E. Fries, Short hist. bot. Sweden 106. 1930.
Verdoorn, F., ed., Chron. bot. 2: 8, 28, 29, 30, 269, 271. 1936; 3: 25, 27, 49. 1937; 6: 141. 1941.

Wittrock, V.B., Acta Horti Berg. 3(2): 68. 1903, 3(3): 50, *pl. 106.* 1905 (portr.).

FESTSCHRIFT: Sv. bot. Tidskr. 26(1/2). 1932 (portr.).

EPONYMY: *Rosenbergiodendron* Fagerlind (1948).

9518. *Cytologische und morphologische Studien an Drosera longifolia x rotundifolia* ... Uppsala & Stockholm (Almquist & Wiksells Boktryckeri-A.-B.) 1909. Qu.
Publ.: shortly after 15 Mai 1909 (p. 64: mss. submitted 10 Mar 1909; Nat. Nov. Sep(2) 1909),p. [1]-64, [1], *pl. 1-4* (uncol. liths.). *Copies*: US, USDA. – Issued as Sv. Vet.-Akad. Handl. 43(11).

9519. *Chromosomenzahlen und Chromosomendimensionen in der Gattung Crepis* ... Arkiv för Botanik Band 15. N:o 11. [1919]. Oct.
Publ.: shortly after 28 Jun 1918 (printed 28 Jun 1918; issued in journal 28 Oct 1919), p. [1]-16. *Copy*: U. – Issued and to be cited as Ark. Bot. 15(11). 1919.

9520. *Apogamie und Parthenogenesis bei Pflanzen* ... Berlin (Verlag von Gebrüder Borntraeger) 1930. Qu. (*Apogam. Parthenogen. Pfl.*).
Publ.: 1930, p. [i], [1]-66. *Copy*: LD. – Issued as Handbuch der Vererbungswissenschaft (ed. E. Baur & M. Hartmann), Band 2. – Inf. O. Almborn.

Rosendahl, [Carl] **Otto** (1875-1956), American botanist; MS Univ. Minnesota 1902; with the Geol. & Nat. Hist. Survey of Minnesota 1902-1903; studied at the University of Berlin 1903-1905; Dr. phil. Berlin 1905, student of A. Engler; at the Univ. Minnesota from 1901-1903 and 1905-1944 in various functions, as full professor from 1910; from 1944-1956 professor emeritus botany. (*Rosend.*).

HERBARIUM and TYPES: MIN; other material at A, B (destr.), BM, C, CGE, DPU, E, GH, K, MO, MSC, NY, P, SI, US, WTU.

BIBLIOGRAPHY and BIOGRAPHY: Andrews p. 322; Barnhart 3: 178 (b. 24 Oct 1875); Biol.-Dokum. 15: 7650; BJI 2: 148; BL 1: 135, 192, 193,316; Hirsch p. 252; Hortus 3: 1203; IH 1 (ed. 1): 63, (ed. 2): 82, (ed. 3): 103; Kew 4: 517; MW p. 418; NW p. 56; TL-2/4128; Tucker 1: 607; Urban-Berl. p. 389.

BIOFILE: Abbe, E.C., An informal history of the Department of Botany, University of Minnesota 1887-1950, p. 4-5, 7. s.d.Anon., Minneapolis Star 5 Mar 1956; Minnesota Alumnus 4(10). 1944 (portr.); Minnesota Daily 27 Mai 1944; Minnesota Chats 261(11). 19 Mai 1944; St. Paul Dispatch 5 Mar 1956; Taxon 5: 56. 1956 (d. 4 Mar 1956); Univ. Minnesota Senat Docket 4 Jun 1956.
Crowther, G., The Minnesota Alumni Weekly 23 Jan 1943, p. 256-257 (portr., univ. career).
Ownbey, G.B., Publications of C.O. Rosendahl, 1981 (typescript).
Schuster, R.M., Hep. Anthoc. N. Amer. 1: 91. 1966.
Stieber, M.T. et al., Huntia 4(1): 81. 1981 (arch. mat. HU).
Verdoorn, F., ed., chron. bot. 2: 333. 1936, 3: 297. 1937, 4: 314, 370. 1939.

COMPOSITE WORKS: As senior author, with F.K. Butters and Olga Lakela, *A monograph on the genus Heuchera*, 1936, see TL-2/4128 (rev. A.B. Rendle, J. Bot. 75: 237-238. Aug 1937, N.Y. Sandwith, Bull. misc. Inf. Kew 1937: 408).

NOTE: We are grateful to Gerald B. Ownbey for information on Carl Otto Rosendahl.

9521. *Die nordamerikanischen Saxifraginae* und ihre Verwandtschafts-Verhältnisse in Beziehung zu ihrer geographischen Verbreitung. Inaugural-Dissertation zur Erlangung der Doktorwürde genehmigt von der philosophischen Fakultät der Friedrich-Wilhelms-Universität zu Berlin von Carl Otto Rosendahl aus Minnesota, Nordamerika. Tag der Promotion: 12. August 1905. [Leipzig (Druck von Breitkopf & Härbel) 1905]. Oct. (*Nordamer. Saxifrag.*).
Publ.: 12 Aug 1905 (date on t.p. thesis), p. [1]-61, [62, vita]. *Copy*: GOET (inf. G. Wagenitz). – Preprinted (in part) from Bot. Jahrb. 37 (Beil. 83), see below.
Regular publication: in Bot. Jahrb. 37 (Beibl. 83): 1-87, *pl. 4-5*. 22 Dec 1905.

9522. *Guide to the spring flowers of Minnesota* by F.E. Clements, C.O. Rosendahl and F.K. Butters. Minneapolis (University of Minnesota) 1908. Oct. (*Guide spring fl. Minnesota*).
Co-authors: Frederic Edward Clements (1874-1945), Frederic King Butters (1878-1945).
Ed. [*1*]: Apr 1908 (t.p.: Mar 1908; p. [i]: Apr 1908), p. [i*], [i]-iv, [v], [1]-40. *Copy*: USDA. - Issued as Minnesota Plant Studies, I, 1908.
Ed. 2: Apr 1910 (t.p.; p. [i] also Apr 1910; TBC 27 Jul 1911), p. [i*], [i]-iv, [v], [1]-40. *Copies*: BR, USDA.
Ed. 3: Mai 1913 (t.p.; p. [iii]: also Apr 1913), p. [i]-ix, [x], [1]-59. *Copies*: PH, USDA. – "Guide to the spring flowers of Minnesota field and garden third edition.
Ed. 4: n.v.
Ed. 5: 1923 (t.p.), p. [i]-x, [xi], [1]-62. *Copies*: BR, PH. – Authors: C.O. Rosendahl and F.K. Butters.
Ed. 6: Oct-Dec 1931 (p. [iii]:" 1 Oct 1931), p. [i]-xvii, [xviii], 1-89, *pl. 1-22* in text. *Copies*: G, NY, USDA(2). – Authors: C.O. Rosendahl and F.K. Butters. Minneapolis (The University of Minnesota Press) 1931.
Ed. 7: 1937 (preface: 19 Mar 1937), p. [i]-xvii, 1-91, *pl. 1-22* in text. *Copies*: USDA(2).
Ed. 8: 1951 (p. i: Mar 1951; USDA rd. 15 Aug 1951), p. [i*-ii*], i-x, 1-108, *pl. 1-21* in text. *Copies*: BR, USDA. - Minneapolis (Burgess Publishing Co.) [1951].

9523. *Guide to the trees and shrubs of Minnesota* by F.E. Clements, C.O. Rosendahl, and F.K. Butters. Minneapolis (University of Minnesota) 1908. Oct. (*Guide trees shrubs Minnesota*).
Co-authors: Frederick Edward Clements (1874-1945), Frederic King Butters(1878-1945).

Ed. [*1*]: Oct 1908 (t.p.; USDA rd.· Jan 1909; Bot. Centralbl. 6 Apr 1909), p. [i], [1]-28. *Copy*: USDA. – Minnesota Plant Studies II, 1908.
Ed. 2: Sep 1910 (t.p.; p. [iii]: Sep 1910), p. [i]-vii, [1]-30. *Copies*: BR, USDA.

9524. *Guide to the ferns and fern allies of Minnesota* by C.O. Rosendahl and F.K. Butters. University of Minnesota July, 1909. Oct. (*Guide ferns Minnesota*).
Co-author: Frederic King Butters (1878-1945).
Publ.: Jul 1909 (t.p.), p. [i-ii], frontisp., [iii-v], [1]-22, [23, index]. *Copies*: BR, G, NY, PH, USDA. – Minnesota Plant Studies III, 1909.

9525. *Minnesota trees and shrubs* an illustrated manual of the native and cultivated woody plants of the state ... Minneapolis, Minnesota (The University of Minnesota) 1912. Oct. (*Minnesota trees shrubs*).
Co-authors: Frederic Edward Clements (1874-1945), Frederic King Butters (1878-1945).
Publ.: 15 Aug 1912 (p. [ii]: publ. 15 Aug 1912, 3000 copies; PH rd. 6 Jan 1913), frontisp., p. [i]-xxi, 11 [sic]-314. *Copies*: BR, MICH, PH, USDA(2). – Report of the Botanical Survey ix.
Ref.: N.T., Torreya 13: 17-19. 8 Jan 1913 (rev.).

9526. *Guide to the autumn flowers of Minnesota* field and gar9526. .E. Clements, C.O. Rosendahl and F.K. Butters. Minneapolis (University of Minnesota) Jun 1913. Oct. (*Guide autumn fl. Minnesota*).
Co-authors: Frederic Edward Clements (1874-1945), Frederic King Butters (1878-1945).
Publ.: Jun 1913 (t.p., Bot. Centralbl. 21 Apr 1914), p. [i]-xviii, [1]-77. *Copies*: BR, USDA. – Minnesota Plant Studies V, 1913.

9527. *Trees and shrubs of Minnesota* ... Minneapolis, Minnesota (The University of Minnesota Press) 1928. Oct. (*Trees shrubs Minnesota*).
Co-author: Frederick King Butters (1878-1945).
Publ.: 1928 (p. vii: Oct 1928), p. [i-ii], front., [ii]-vii, [ix], 1-385. *Copies*: FI, G, MICH, MO, NY, US, USDA; – "the present volume is not a second edition [of Minnesota trees and shrubs, 1912]".
Trees and shrubs of the Upper Midwest: 1955, University of Minnesota Press, Minneapolis, p. [i-vii], [1]-411. *Copy*: USDA. (rd. 31 Aug 1955). – By Carl Otto Rosendahl; an expanded version of *Trees shrubs Minnesota*; introduction by A. Orville Dahl.
Ref.: Sandwith, N.Y., Kew Bull. 1955: 496 (rev.).

Rosendahl, Friedrich (1881-1942), German (Prussian) lichenologist; studied at Göttingen, München, Münster i.w.; Dr. phil. Münster i.w. 1907; high school teachers exam. Münster 1906; student of W. Zopf and G. Bitter; high school teacher at Münster (1906), Soest (1907), Iserlohn (1907), "studienrat" at the Soest Gymnasium 1908-1926; "Oberstudienrat" Gronau 1926. (*F. Rosend.*).

HERBARIUM and TYPES: Unknown.

BIBLIOGRAPHY and BIOGRAPHY: Barnhart 3: 178 (b. 18 Aug 1881); BM 4: 1732; GR p. 42 (d. 12 Jul 1942), cat. p. 70; LS 38544.

BIOFILE: Rosendahl, F., Lebenslauf, *in* his *Vergl. anat. Unters. Parmel.* 1907.

9528. *Vergleichend anatomische Untersuchungen über die braunen Parmelien*. Inaugural-Dissertation zur Erlangung der Doktorwürde der hohen philosophischen und naturwissenschaftlichen Fakultät der Universität Münster i.W. vorgelegt von F. Rosendahl, Kandidat des hoheren Schulamtes. Münster i.W. (Druck der Aschendorffschen Buchdruckerei) 1907. Oct. (*Vergl. anat. Unters. Parmel.*).
Thesis: Jan-Apr 1907 (Bot. Centralbl. 9 Jul 1907; Bot. Zeit. 16 Mai 1907; Hedwigia 30 Sep 1907), p. [1]-34, [35, bibl.], [36, vita]. *Copy*: NY.
Definitive ed.: late 1907 (mss. submitted 17 Jan 1907; Hedwigia 3 Mar 1908; Nat. Nov. Jan(1) 1908), p. [1]-59, *pl. 1-4* with text (uncol. liths). *Copies*: BR, FH, G, M, NY, USDA. – Published and to be cited as Nova Acta Leop. 87(3): 401-459, *pl. 25-28*

(with double pagination). "Vergleichend-anatomische Untersuchungen über die braunen Parmelien" Halle (Druck von Ehrhardt Karras, Halle a.S.). 1907.
Ref.: Zahlbruckner, A., Bot. Centralbl. 105: 311-314. 24 Sep 1907 (rev. diss.).
Nienburg, Bot. Zeit. 66(2): 233-234. 1 Jul 1908 (rev.).

Rosendahl, Henrik Viktor (1855-1918), Swedish botanist; pharmacist's exam. 1879; med. lic. Karol. Inst. Stockholm 1886; Dr. med. Uppsala and lecturer at Karol. Inst. 1894; professor of pharmacognosy ib. 1902; specialist on pteridophyte taxonomy. (*H. Rosend.*).

HERBARIUM and TYPES: S (important fern collection); some collections from Norway at C; some material (especially ferns) at LD); lichens from Sweden at M. – A series of *Pteridophyta exotica exsiccata* was announced for 1905 by R. in Bot. Zeit. 62: 336. 1904; two centuries of *Filices exoticae exsiccatae*, 1909, were offered for sale in Herbarium 9: 68. 1909.

BIBLIOGRAPHY and BIOGRAPHY: Barnhart 3: 178; Bossert p. 338; IF suppl. 1: 84, 2: 38; IH 2: (files); KR p. 599 (bibl.; b. 12 Dec 1855, d. 11 Aug 1918).

BIOFILE: Birger, S., Sv. bot. Tidskr. 13(2): 228-236. 1919 (portr., obit., bibl.).
Lindman, S., Sv. män och kvinnor 6: 356. 1949 (portr.).
Murbeck, S., Bot. Not. 1916: 257-262, 1917: 81-82 (polemics; reply 1917: 43-46).
Nordstedt, O., Bot. Not. 1918: 215 (obit.).

9529. *Filices novae* ... [Ark. Bot. 14(18) 1916]. Oct.
Publ.: shortly after 24 Mar 1916 (date printing), p. [1]-5, *pl. 1-3. Copies*: LD, U.

9530. *On two collections of ferns made in Madagascar* by Dr. W.A. Kaudern 1911-12, Drs K. Afzelius and B.T. Palm (the Swedish Madagascar Expedition) 1912-13. [Ark. Bot. Band 14(23). 1917]. Oct.
Publ.: shortly after 7 Feb 1917 (date of printing), p. [1]-11, [12]. *Copies*: LD, U.

9531. *De svenska Equisetum-arterna* och deras former. [Ark. Bot., Stockholm, 15(3). 1917]. Oct.
Publ.: shortly after 11 Jun 1917 (date of printing), p. [1]-52. *Copies*: LD, U, USDA.

Rosenstock, Eduard (1856-post 1928), German (Kurhessen) pteridologist; owner of large fern collections; high school teacher at Gotha from ca. 1889. (*Rosenstock*).

HERBARIUM and TYPES: Large collections of exotic ferns at B (extant) and M; other sets at BAF, BM, C, CR, E, GE, L, LAU, MANCH, MICH, NY, OXF, PH, R, SI, STU, UC, US, W. – Rosenstock distributed sets of ferns collected by others as well as by himself under various titles: The following list is given by Urban-Berlin p. 322 (extant at B):
Filices europae, 256 nos., 1901.
Pteridophyta exotica exsiccata (see Bot. Zeit. 62(2): 336. 1904), issued in nine *Lieferungen*:
  1. *Fil. austro-bras.* exs. 145 nos., 1905.
  2. *Fil. austro-bras.* exs., *Fil. Afr. Orient.* exs., *Fil. Javae orient.* 96 nos. 1906.
  3. *Fil. exoticae*, 163 nos. 1907.
  4. *Fil. exs., 100 nos. 1908.*
  5. *Fil. exs., 93 nos. 1909.*
  6. *Fil. novo-guin.* (leg. G. Bamler), Fil. austro-bras. (leg. C. Spannagel, M. Wacket), Fil. Javae orient. (leg. J.P. Mousset) 100 nos. 1910.
  7. *Fil. costaric.* exs. (leg. A. et C. Brade), *Fl. Korean. exs* (leg. T. Taquet, U. Faurie) 110 nos. 1911.
  8. *Fil. boliv.* exs. (leg. O. Buchtien), *mexic.* (leg. Fr. Arsène), sumatr. (leg. J. Winkler), Novae-Caled. (leg. Franc) 100 nos. 1912.
  9. *Fil. Novae Caled. exs.* (leg. Franc), *Fil. mexic.* (leg. Fr. Arsène) 113 nos., 1913.
See Smithsonian Archives SIA 192.

BIBLIOGRAPHY and BIOGRAPHY: Backer p. 499 (epon.); Barnhart 3: 179; Biol.-Dokum.

15: 7653-7654; BM 8: 1093; Clokie p. 234; DTS 1: 2 7, 6(4): 190; GR p. 126; IF p. 727, suppl. 1: 84, 85, 2: 38-39, 3: 217; IH 1 (ed. 6): 363, 2: (files); Kew 4: 518; MW p. 418; Urban-Berl. p. 322.

BIOFILE: Burkhart, A., Darwinia : 142. 1928 (on exsiccatae at SI).
Hedge, I.C. & J.M. Lamond, Index coll. Edinburgh herb. 125. 1970.
Kneucker, A., Allg. bot. Z. 33: 62. 1928 (lists of exsiccatae still available from Rosenstock at Gotha).
Merrill, E.D., B.P. Bishop Mus. Bull. 144: 160. 1937 (bibl. Polynes.).
Rosenstock, E., Österr. bot. Z. 77: 79. 1928 (note by Rosenstock on the continuation of his exsiccatae).

COMPOSITE WORKS: (1) *Contribution à l'étude des Ptéridophytes de Colombie* ... in O. Fuhrman & Eug. Mayor, Voy. expl. sci. Colombie, Neuchâtel 1912, p. [33]-56, *pl. 2-6*. (Mém. Soc. neuchât. Sci. nat. 5).
(2) *Filicales, in* Rosenstock, E. et al., Meded. Rijk's Herb. Leiden 19, 1913 (Th. Herzog's Bolivianische Pflanzen).

EPONYMY: *Rosenstockia* Copeland (1947).

9532. *Aspidium libanoticum* n. sp. ... Mémoires de l'Herbier Boissier 9, 1900. Oct.
Publ.: 2 Mar 1900, p. [1]-2. *Copy*: U. – Issued and to be cited as Mém. Herb. Boissier 2, 1900.

9533. *Beschreibung neuer Hymenophyllaceae aus dem Rijks Herbarium zu Leiden* [Leiden, Firma P.W.M. Trap 1912].
Publ.: 15 Nov 1912, p. [1]-3. *Copy*: HH. – Issued as Meded. Rijks Herbarium 11, 1912.

9534. *Filices palaeotropicae novae herbarii Lugduno-Batavi* [Leiden (Firma P.W.M. Trap) 1917].
Publ.: 10 Mai 1917, p. [1]-8. *Copies*: HH, U. – Issued as Meded. Rijks Herbarium 31, 1917.

Rosenthal, David August (1821-1875), German (Silesian) physician, botanist and theological author; practicing physician as well as municipal physician at Breslau from 1855. (*Rosenthal*).

HERBARIUM and TYPES: Unknown.

BIBLIOGRAPHY and BIOGRAPHY: BL 1: 5, 316, 2: 7, 710; BM 4: 1733; GR p. 174 (b. 16 Apr 1821, d. 29 Mar 1875); Herder p. 120, 411; Jackson p. 202; LS 22535; MW p. 418; PR 7772; Rehder 3: 58, 462; Tucker 1: 607.

9535. *Synopsis plantarum diaphoricarum.* Systematische Uebersicht der Heil-, Nutz– und Giftpflanzen aller Länder ... Erlangen (Verlag von Ferdinand Enke) 1862. Oct. (*Syn. pl. diaph.*).
Publ.: in two parts 1861-1862. *Copies*: H, MICH, MO, USDA.
1: Oct-Nov 1861 (ÖbZ 1 Mar 1862; BSbF Mai 1862; Bonplandia 15 Nov 1861; Bot. Zeit. 15 Nov 1861), p. [1]-480.
2: Aug-Dec 1862 (p. xvii: Jul 1862; J. Bot. Oct 1863), p. [i]-xxvi, 481-1359, [1360-1362].
Ref.: Schlechtendal, D.F.L. von, Bot. Zeit. 19: 344. 15 Nov 1861 (part 1, rev.).

Rosenvinge, (Janus) **Lauritz** (Andreas) **Kolderup** (1858-1939), Danish botanist (algologist); Dr. phil. København 1888; librarian and assistant at the Botanical Garden 1886-1900; from 1895 lecturer at the University of København and from 1900 id. at the Polyteknisk Læreanstalt; professor of botany at the University 1916-1928. (*Rosenvinge*).

HERBARIUM and TYPES: C; other material at E, H, KIEL, LE, LY, M. – Contributed to V.B. Wittrock & O. Nordstedt, *Algae aquae dulcis*. – Letters to W.G. Farlow at FH.

NAME: Often listed (more correctly) as Kolderup Rosenvinge, L.

BIBLIOGRAPHY and BIOGRAPHY: Barnhart 3: 179 (b. 7 Nov 1858, d. 18 Jun 1939); BJI 2: 148; BL 1: 154, 316; BM 8: 581; Bossert p. 318; CSP 10: 436, 12: 401, 16: 399-400; De Toni 1: lxxi-lxxii, 2: cxxii, 4: xlviii; IH 2: 377 (Kolderup); Kelly p. 195; Kew 4: 518-519; Kleppa p. 11, 89, 106; KR p. 601; LS 22536-22538; Morren ed.10, p. 48; Rehder 5: 737; TL-2/4185, 4189, see O. Gallø, C.A. Hesselbo, E.V. strup; Urban-Berl. p. 268, 389.

BIOFILE: Anon., Bot. Jahrb. 29 (Beibl. 66): 31. 1900 (lecturer Polytecnic); Bot. Tidsskr. 40(4): 334. 1929; Nat. Nov. 22: 419. 1900 (lecturer at Polytecnic); Österr. bot. Z. 50: 347. 1900 (app. Polytechnic).
Christensen, C., Dansk biogr. leks. (Engelstoft) 13: 126-127. 1938; Dansk bot. Hist. 1: 882 [index]. 1926; Dansk bot. litt. 1880-1911, p. 82-85. 1913 (portr., biogr. det., bibl.), id. 1912-1939, p. 4-7. 1940 (id.).
Hansen, A., Dansk bot. Ark. 21: 21: 70. 1963 (biogr. refs.).
Koster, J.Th., Taxon 18: 557. 1969.
Lind, J., Danish fungi herb. E. Rostrup 37. 1913.
Paulsen, O., Overs. Vid. Selsk. Virksomhed 1939/40: 103-109. 1940 (obit., portr.); Bot. Tidsskr. 45: 138-140, 258-259. 1940 (on R. and Dansk bot. For.).
Petersen, H.E., Bot. Tidsskr. 45(2): 131-137. 1940 (obit., portr.).
Petersen, J. Boye, Nat. Verden 24: 42-46. 1940 (obit., portr.).
Rosenvinge, L. Kolderup, Metropolitaneren 3(10): 69-72. 1928 (autobiogr. account of his university years; portr.).
Rydberg, L.A., Augustana Coll. Libr. Bull. 5: 31. 1907.
Sayre, G., Bryologist 80: 516. 1977.
Warming, E., Bot. Tidsskr. 12: 236. 1881 (bibl.); Dansk biogr. leks. (C.A. Bricka) 9: 354-355. 1895.
Wittrock, V.B., Acta Horti Berg. 3(2): 104, *pl. 19.* 1903 (portr.).

COMPOSITE WORKS: (1) Editor, *Botanisk Tidsskrift* vol. 19, 1894 – 41, 1931.
(2) Co-author, J.M.C. Lange, *Consp. fl. Groenland.* 3(1), 3(3), 1889, 1891, see TL-2/4185.
(3) Identifications algae in J.M.C. Lange, *Nomencl. Fl. danic.* 1887, see TL-2/4189.

FESTSCHRIFT: Dansk bot. Arkiv 5(6-24), 7 Nov 1928 (portr., dedic.).

EPONYMY: *Rosenvingea* Boergesen (1914); *Rosenvingiella* P.C. Silva (1957).

9536. *Bidrag til Polysiphonia's Morphologi* ... Kjøbenhavn (Hoffensberg & Traps Etabl.) 1884. Oct.
Publ.: 1884, p. [1]-45, *pl. 1-2*, résumé franç. [(1)]-(10). *Copies*: FH, G. – Reprinted and to be cited from Bot. Tidsskr. 14: 11-53, *pl. 1-2*, résumé franç. (1)-(10). 1884. – See also *Sur la disposition des feuilles chez les Polysiphonia,* Bot. Tidsskr. 17: 1-9, *pl.1.* 1888.

9537. *Sur les noyaux des Hyménomycètes* [Extrait des Annales des Sciences naturelles, ... 1886]. Oct.
Publ.: 1886 (Nat. Nov. Mai 1887; Bot. Zeit. 11 Feb 1887), p. [1]-19, *pl. 1. Copy*: FH. – Reprinted and to be cited from Ann. Sci. nat., Bot. ser. 7. 3: 75-93, *pl. 1.* 1886.
Danish version: Bot. Tidsskr. 15: 210-228, *pl. 17.* 1886.
Ref.: Zacharias, E., Bot. Zeit. 45: 94. 11 Feb 1887.

9538. *Undersøgelser over ydre Faktorers Indflydelse paa Organdannelsen hos Planterne* ... Kjøbenhavn (Hos J. Frimodt) 1888. Oct.
Publ.: Apr 1888 (p. [ii]: 28 Mar 1888; Bot. Jahrb. 2 Apr 1889; Nat. Nov. Mai(1, 2) 1888; Hedwigia Oct-Dec 1888), p. [i-iii], [1]-117, *pl. 1-3* (uncol.). *Copies*: FH, USDA. – Reprinted and to be cited from Medd. Dansk. nat. For. Kjøbenhavn 1888: 37-153, *pl. 1-3.* – Thesis for Dr. phil.
French extract: Rev. gén. Bot. 1: 53-62, 123-135, 170-174, 244-255, 304-317. 1889; also reprinted, 1889, p. [1]-51. *Copy*: FH. – "*Influence des agents extérieurs sur l'organisation polaire et dorsiventrale des plantes*".
Ref.: Lagerheim, N.G., Hedwigia 28(1): 60-61. 1899 (rev.).

9539. *Om nogle Vœxtforhold hos Slœgterne Cladophora og Chaetomorpha* [Særtryk af Botanisk Tidsskrift 8. Bind. 1. Hæfte. Kjøbenhavn 1892]. Oct.
Publ.: 1892 (Bot. Zeit. 25 Mai 1892), p. [1]-30, résumé franç. 31-36. *Copy*: FH. – Reprinted and to be cited from Bot. Tidsskr. 18(1): 29-58, résumé franç. 59-64.

9540. *Grønlands Havalger* ... Kjøbenhavn (Bianco Lunos Kgl. Hof-Bogtrykkeri (F. Dreyer). 1893. Oct.
Publ.: 1893 (Hedwigia Jan-Feb 1894; Nat. Nov. Mar(2) 1894), p. [763]-981, *pl. 1-2*. *Copy*: FH (with special cover and t.p.). – Reprinted and to be cited from Medd. Grønland 3: 763-981, *pl. 1-2*. 1893. – List marine algae of Greenland.
French: 1894, *in* Ann. Sci. nat., Bot. ser. 7. 19: 53-164. 1894.
Part 2: Medd. Grønland 20: 1-125. 1898 (Nat. Nov. Nov(1) 1898).
Ref.: Anon., Hedwigia 33: (4)-(6). 1894 (rev.).

9541. *Nye Bidrag til Vest-Grønlands Flora* af L. Kolderup Rosenvinge. 1896. Særtryk af "Meddelelser om Grønland". XV. Kjøbenhavn (Bianco Lunos Hof- og Bogtrykkeri (F. Dreyer)) 1896. Oct. (*Nye Bidr. Vest-Grønl. Fl.*).
Publ.: 1896 (t.p. preprint; Nat. Nov. Nov(1) 1898 in journal), p. [61]-72. *Copy*: FH. – Preprinted from Medd. Grønland 15: 61-72. (résumé franç. 451-452) 1898.

9542. *Det sydligste Grønlands Vegetation* af L. Kolderup Rosenvinge 1896. Særtryk af "Meddelelser om Grønland". XV. Kjøbenhavn (Bianco Lunos Kgl. Hof– Bogtrykkeri (F. Dreyer)). 1896 [corrected to 1897]. Oct. (*Sydl. Grønl. Veg.*).
Publ.: 1897 (author's correction of printed "1896" to "1897" on FH copy; Nat. Nov. Nov(1) 1898 (in journal), p. [73]-249. [250]. *Copy*: FH. – Preprinted from Medd. Grønl. 15: 73-249 (resumé franç. 452-463). 1898.

9543. *Deuxième mémoire sur les algues marines du Groenland* ... Copenhague (Bianco Luno (F. Dreyer), imprimeur de la Cour) 1898. Oct. (*Deux. mém. alg. mar. Groenland*).
Publ.: 1898 (preprint so dated; Bot. Centralbl. 12 Oct 1898), p. [1]-125, [1, expl. pl.], *pl. 1*. *Copy*: FH. – Preprinted from Medd. Gr⁰5nl. 20: 1-125, *pl. 1*. 1899. – For first memoir see above, 1893, *Grønlands Havalger*.

9544. *Om Algevegetationen ved Grønlands Kyster* ... Kjøbenhavn (Bianco Lunos Kgl. Hof– Bogtrykkeri (F. Dreyer)) 1898. Oct. (*Algeveg. Grønl. Kyst.*).
Publ.: 1898 (t.p. preprint so dated; Bot. Zeit. rev. 16 Dec 1898; Nat. Nov. Jul(2) 1899 in journal), p. [129]-242, [243, index]. *Copy*: FH. – Preprinted from Medd. Grønl. 20: 129-243 (resumé franç. 339-346). 1899.
Ref.: Kuckuck, E.H.P., Bot. Zeit. 56(2): 369-374. 1898 (rev.).

9545. *Ueber die Spiralstellungen der Rhodomelaceen* ... Leipzig (Gebrüder Borntraeger) 1902. Oct.
Publ.: Jan-Mai 1902 (Bot. Zeit. rev. 1 Aug 1902; journal issue [37(3)] rd. by USDA 14 Jun 1902), p. [i], [338]-364, *pl. 17*. *Copy*: FH. – Reprinted with special t.p. and to be cited from Jahrb. wiss. Bot. 37(2): 338-364. 1902.
Ref.: Oltmanns, F., Bot. Zeit. 60(2): 232-233. 1 Aug 1902 (rev.).

9546. *The marine algae of Denmark* contributions to their natural history ... København (Bianco Lunos Bogtryckkeri) 1909-1931, 4 parts. (*Mar. alg. Denmark*).
Publ.: in four parts, 1909-1931 as Kgl. Danske Vid. Selsk. Skr. ser. 7, 7(1-4). *Copies*: BR, FH, H, M, NY, PH, USDA.
  1: 1909 (Nat. Nov. Oct(2) 1909), p. [1]-151, 2 maps, *pl. 1-2*. – Introd., Rhodophyceae I. (Bangiales and Nemalionales).
  2: 1917 (ÖbZ 1 Sep 1918; Nat. Nov. Jun 1918, p. [153]-283, [284], *pl. 3-4*. – Rhodophyceae II. (Cryptonemiales).
  3: 1924 (t.p. 1923-1924; USDA rd. Jul 1925), p. [285]-486, [487], *pl. 5-7*. – Rhodophyceae III. (Ceramiales).
  4: 1931 (USDA rd. 2 Mai 1932; Nat. Nov. Sep 1932), p. [489]-627 [628, err.; 629-630 expl. pl.], *pl. 8*. – Rhodophyceae IV. (Gigartinales. Rhodymeniales. Nemastomatales). – See also *Distribution of the Rhodophyceae in the Danish waters*, ib. ser. 9. 6(2), 44 p. 1935.

Ref.: Hariot, P., Bull. soc. bot. France 56: 581-582. 1910 (rev. 1).

9547. *On the marine algae from North-East Greenland* (N. of 76; N. lat.) collected by the "Danmark-Expedition" ... København (Bianco Lunos Bogtrykkeri) 1910. Oct. (*Mar. alg. N.E. Greenland*).
Publ.: 1910 (p. 133: printed 9 Mai 1910; Bot. Centralbl. 14 Mar 1911), p. [91]-133.
Copies: BR, FH, H, G. – Reprinted from Medd. Grønland 43: 91-133. 1910. Also issued as "Arbejd. Bot. Have" København nr. 54 and Danm.-Eksp. Grønl. Nordøstkyst 1906-1908, vol. 3, no. 4 (actually 3).

9548. *The botany of Iceland* edited by L. Kolderup Rosenvinge ... and Eug. Warming ... Copenhagen (J. Frimodt), London (John Wheldon & Co.) 1912-1949, 5 vols. (*Bot. Iceland*).
Editors: vols. 1-2: L. Kolderup Rosenvinge and Johannes Eugenius Bülow Warming (1841-1924); other volumes: Johannes Gröntved (1882-1956), Ove Paulsen (1871-1947), Thorvald Julius Sørensen (1902-1973).
Imprints: vols. 3-5: Copenhagen (Einar Munksgaard ...), London (Humphrey Milford ...).
1(1,1): Sep-Dec 1912 (p. [iii]: Aug 1912; Bot. Centralbl. 3 Dec 1902; ÖbZ 1 Jan 1913; J. Bot. Feb 1913; Nat. Nov. Feb(2) 1913), p. [i-vi], [1]-186. – The marine algae vegetation of Iceland, by Helgi Jónsson. – Rev.: L. Diels, Bot. Jahrb. 119(Lit.): 44-45. Jun 1913.
1(1, 2): 1914 (ÖbZ Oct 1914, J. Bot. Nov 1915), p. [187]-343. – An account of the physical geography of Iceland, by Th. Thoroddson. – Rev.: J. Ramsbottom, J. Bot. 53: 341. 1915.
1(2): 1918 (ÖbZ 1 Jul 1918; J. Bot. Sep 1918; Bot. Jahrb. 13 Dec 1918; Nat. Nov. Aug 1918), p. [i*-iii*], [345]-675, [676, err., 677, cont.]. – Marine Diatoms from the Coasts of Iceland, by E. Østrup, p. 343-394, *pl. 1*, printed 1916; The Bryophyta of Iceland, Aug. Hesselbo, p. 395-675. – Rev. A. Gepp, J. Bot. 56: 277. 1918.
2(1): 1920 (ÖbZ 1 Oct 1920; Nat. Nov. Aug 1920), p. [1]-247, [248], *pl. 1-5*. – Freshwater Diatoms from Iceland, by E. Østrup, 1918, p. [1]-98, *pl. 1-5*; The lichen flora and lichen vegetation of Iceland, p. [101]-247, [248], by O. Gallóe.
2(2)(1): 1923 (J. Bot. Mai 1924), p. [249]-325. – The fresh-water Cyanophyceae of Iceland, 1923, by J. Boye Petersen.
2(2)(2): 1928, p. [325]-447. The aërial algae of Iceland, by J. Boye Petersen.
2(3): Feb-Mar 1932 (p. 607: printing finished Feb 1932; Nat. Nov. Jun 1932), p. [i-iii, vol. t.p.], 449-607, *pl. 6*. – Fungi of Iceland, by Poul Larsen.
3(1): 1930 (J. Bot. Dec 1930; Nat. Nov. Nov 1930), p. [i-iv], [1]-186, [1], *pl. 1-12*. – Studies on the vegetation of Iceland, H. Mølholm Hansen (Thesis).
3(2): 1941, p. [187]-228, *1 pl.*, [1]. – Studies in the larger fungi of Iceland, by M.P. Christiansen.
3(3): 1942, p. [229]-343, *pl. 1-44*. – The Taraxacum-flora of Iceland, by M.P. Christiansen.
3(4): 1945, p. [i-iii], [345]-547. – Studies on the vegetation of the central highland of Iceland, by Steindór Steindórsson.
4(1): 1942, p. [1]-427, [1, err.]. – The Pteridophyta and Spermatophyta of Iceland, by Johs. Gröntved.
Imprint vol. 4(1): as vol. 3.
5(1): 1949 (MO rd. Aug 1949), p. [1]-57, [58-60]. – The flora of Reykjanes Peninsula, Sw.-Iceland, by Emil Hadač.
Copies: BR, PH, USDA.

9549. *Sporeplanterne* (Kryptogamerne) med 513 i teksten trykte figurer eller figurgrupper ... Kjøbenhavn og Kristiania [Oslo] (Gyldendalske Boghandel Nordisk Forlag) 1913. Oct. (*Sporeplanterne*).
Publ.: Mar-Apr 1913 (p. [vi]: Feb 1913; Bot. Centralbl. 13 Mai 1913; Mycol. Centralbl. 4 Jun 1913; Hedwigia 1 Oct 1913; ÖbZ 1 Jun 1913; Nat. Nov. Mai(1) 1913), p. [i-xi], [1]-388. *Copies*: H, USDA.
Ref.: Gepp, A. & E.S., J. Bot. 51: 228-229. 1913 (rev.).

9550. *On the spiral arrangement of the branches in some Callithamnieae* ... København (... Andr. Fred. Høst & Søn, ...) 1920. Oct. (*Spiral arr. branch. Callithamn.*).
Publ.: Jan 1921 (t.p. 1920; printing finished 30 Dec 1920), p. [1]-70. *Copy*: FH. – Issued as Kgl. Danske Vid. Selsk. Biol. Medd. 2(5), 1920 (1921).

9551. *Marine algae collected by Dr. H.G. Simmons during the 2nd Norwegian arctic expedition in 1898-1902* ... Oslo (Kristiania) (A.W. Brøggers Boktrykkeri AS) 1926. Oct. (*Mar. alg. Simmons*).
Collector: Herman Georg Simmons (1866-1943).
Publ.: 1926 (Hedwigia Mai 1927), p. [1]-40. *Copy*: NY. – Rep. Second Norw. arct. exp. Fram 1898-1902, no. 37.

9552. *Phyllophora Brodiaei and Actinococcus subcutaneus* ... København (Bianco Lunos Bogtrykkeri) 1929. Oct. (*Phyllophora brodiaei*).
Publ.: late Aug 1929 (printing finished 24 Aug 1929), p. [1]-40, *1 pl. Copy*: FH. – Issued as Kgl. Danske Vid. Selsk. Biol. Medd. 8(4), and Arb. bot. Have København 115.

9553. *The reproduction of Ahnfeltia plicata* ... København (... Andr. Fred. Høst & Søn, ...) 1931. Oct (*Repr. Ahnfeltia plicata*).
Publ.: early Mar 1931 (printing finished 28 Feb 1931; Hedwigia Apr 1932), p. [1]-29. *Copy*: FH. – Issued as Kgl. Danske Vid. Selsk. Biol. Medd. 10(2), 1931.

9554. *Marine algae from Kangerdlugssuak* ... København (C.A. Reitzels Forlag ...) 1933. (*Mar. alg. Kangerdlugssuak*).
Publ.: 1933 (Nat. Nov. Jun 1934), p. [1]-14. *Copy*: FH. – Medd. Grønl. 104(8). 1933.

9555. *On some Danish Phaeophyceae* ... København (Levin & Munksgaard ...) 1935. Qu. (*Dan. Phaeophyc.*).
Contributions by: Søren Lund.
Publ.: 1935 (Nat. Nov. Feb 1936), p. [1]-40. *Copies*: FH, NY. – Issued as Kgl. Danske Vid. Selsk. Skr., Nat., ser. 9, vol. 6(3).

9556. *Distribution of the Rhodophyceae in the Danish waters* ... København (Levin & Munksgaard) 1935. Qu. (*Distr. Rhodophyc. Dan. waters*).
Publ.: 1935 (Nat. Nov. Feb 1936), p. [1]-43, [44, cont.]. *Copies*: FH, H, NY. – Kgl. Danske Vid. Selsk. Skr., Nat., ser. 9, vol. 6(2).

Roshevitz, Romain U. (Julievich) (1882-1949), Russian botanist (agrostologist). (*Roshevitz*).

HERBARIUM and TYPES: LE; other material e.g. at A, BP, C, DPU, E. – See Smithsonian Archives HI 107.

BIBLIOGRAPHY and BIOGRAPHY: Barnhart 3: 180; BJI 2: 148; Bossert p. 338; IH 2: (files); Kew 4: 519; MW p. 418, suppl. 295; TL-2/3857; Urban-Berl. p. 389.

BIOFILE: Lipschitz, S., Florae URSS fontes nos. 107, 896, 1151, 1156, 1968-1970, 2234, 2235, 2298, 2301. 1975.
Lipsky, V.I., Imp. S.-Pétersb. bot. Sada 3: 412-413. 1915 (portr., bibl., biogr. sketch).
Shetler, S.G., Komarov bot. Inst. 34, 59, 70, 91, 98, 144, 156. 1967.

COMPOSITE WORKS: Contributed the text for many genera of Gramineae and Cyperaceae to *Fl. URSS* vols. 2 (1934) and 3 (1935).

HANDWRITING: S. Lipschitz & I.T. Vasilczenko, Central Herbarium USSR 1968, p. 120.

EPONYMY: *Roshevitzia* N.N. Cvelev (1968).

Rosny, Louis Léon Lucien Prunol de (1837-1916), French orientalist and botanist; founder of the "Société d'ethnographie américaine et orientale", 1858; taught Japanese at the "École des langues orientales", Paris, 1868; assistant director of the École des hautes études, 1886. (*Rosny*).

HERBARIUM and TYPES: Unknown.

BIBLIOGRAPHY and BIOGRAPHY: Barnhart 3: 180 (b. 5 Apr 1837); BM 4: 1733-1734; MW p. 418-419; NI 1018; PR 7774-7775; Rehder 1: 4, 2: 27, 596; Tucker 1: 607.

BIOFILE: Anon., Grand Larousse encycl. 9: 380. 1964.
Sirks, M.J., Ind. natuuronderzoek 13. 1915.

9557. *Notice sur le Thuya de Barbarie* (Callitris quadrivalvis) et sur quelques autres arbres de l'Afrique boréale ... Paris (Just Rouvier, ...) Alger (Dubos frères ...) 1856. Oct. (*Not. Thuya Barbarie*).
Publ.: 1856, p. [i-ii], frontisp., [iii], [1]-19, *pl.* [*2*]. *Copy*: NY. – Reprinted from Bull. Algérie 3: 225-234. 1855 (n.v., see Tucker 1: 607).

Ross, Alexander Milton (1832-1897), Canadian philanthropist and naturalist; studying medicine at NY while a compositor on the Evening Post; Dr. med. 1855; served as surgeon in Nicaragua, the American Civil War, and in the Mexican Army; actively engaged in the anti-slavery struggle 1855-1865; from 1870-1897 back in Canada devoting himself to botany and zoology; died in Detroit, Mich.(*A.M. Ross*).

HERBARIUM and TYPES: Unknown.

BIBLIOGRAPHY and BIOGRAPHY: Barnhart 3: 180 (b. 13 Dec 1832, d. 27 Oct 1897); Jackson p. 366; Kew 4: 519; PR 211; Rehder 1: 309; Tucker 1: 607.

BIOFILE: Anon., Bot. Not. 1898: 46 (d.)
Landon, F., Mich. Hist. Mag. 5(3-4): 364-373. 1921 ("a daring Canadian abolitionist).
Morgan, H.J., Canadian men and women of the time 1898, p. 883-884.
Ross, A.M., Recollections and experiences of an abolitionist; from 1855 to 1865. Toronto, Rowsell and Hutchison 1875, xv, 224 p., *1 pl.*
Wallace, W.S., ed., Dict. Canad. biogr. 2: 573-574. 1945.
Wilson, J.G. & J. Fiske, Appleton's Cycl. Amer. biogr. 5: 327. 1888.

EPONYMY: *Rossioglossum* (Schlechter) Garay et Kennedy (1976) honors John Ross (*fl.* 1830-1840), a collector or orchids in Mexico; *Rossipollis* W. Krutsch (1970) is named for N.E. Ross, palynologist.

9558. *The forest trees of Canada* ... Toronto (Rowsell & Hutchison) 1875.
Publ.: 1875, p. [1]-6. *Copy*: BR. – Names only.

9559. *The flora of Canada* ... Toronto (Rowsell & Hutchison) 1875. (*Fl. Canada*).
Publ.: 1875, p. [1]-29. *Copies*: BR, HH; IDC 7614. – Names only.

Ross, David (ca. 1810-1881), British (Scottish) classics teacher in Edinburgh. (*D. Ross*).

HERBARIUM and TYPES: Unknown.

BIBLIOGRAPHY and BIOGRAPHY: BM 4: 1734; Desmond p. 531 (d. 21 Feb 1881); Jackson p. 43, 280; Kew 4: 519; Rehder 1: 412; Tucker 1: 607.

BIOFILE: Allford, M., (Librarian R.B.G. Edinburgh), letter to J. Ewan of 24 Apr 1961 (d. 21 Feb 1881; death notice in Edinburgh Courant).
Van Tieghem, Ph., Bull. Soc. bot. France 28: 71. 1881 (d.).

9560. *Stray leaves of a naturalist* ... London (Houlston and Wright ...) 1859. 16-mo. (*Stray leaves natural.*).

Publ.: 1859 (p. v: 15 Jun 1859), xl, 205 p. *Copy*: Ewan.

9561. *Some account of a botanical tour in the mountains of Auvergne and Switzerland* ... Edinburgh (R. Grant & Son, ...) London (Simkin, Marshall, & Co.) 1861. Oct. (*Account tour Auvergne & Switzerland*).
Publ.: 1861, p. [1]-60. *Copies*: E (inf. J. Edmondson), Ewan.

9562. *Account of botanical excursions around the environs of Paris*, including Montmorency, Marines en Vexin, Rambouillet, St. Hubert and Dampière, the forest of Fontainebleau, Bois de Vincennes, and the banks of the Marne, Nantes and vicinity, Island of Noirmoutier, etc., in August 1861 ... Edinburgh (R. Grant & Son, ...), London (Simkin, Marshall & Co.) 1862. Oct. (*Acc. excurs. Paris*).
Publ.: 1862, p. [1]-50, err. slip. *Copies*: E, HH, Ewan, NY.

9563. *Account of botanical rambles in the Pyrenees* in August 1862 ... Edinburgh (R. Grant & Son, ...) London (Simkin, Marshall, & Co.) 1863. Oct. (*Acc. rambles Pyrenees*).
Publ.: 1863, p. [1]-67. *Copies*: E, Ewan, HH, NY.

Ross, Hermann (1862-1942), German (Danzig) botanist and cecidiologist; gardener at Greifswald 1880-1882; with Kny and Wittmack in Berlin 1882-1883; worked in Portici 1883-1884; assistant at the Palermo botanical garden 1884-1885; assistant with Wittmack in Berlin 1885-1887; Dr. phil. Freiburg i. Br. 1887; again assistant at Palermo 1887-1890; lecturer ib. 1890-1895; interim dir. Palermo botanical garden 1892-1893; running a milk processing plant at Palermo 1895-1897; "Kustos" (curator) at München with K. Goebel 1897-1902 at the botanical garden; id. with Radlkofer at the botanical museum 1902-1908; "Konservator" 1909, titular professor 1920, "Abteilungsleiter 1925; retired 1927; in retirement working especially on cecidiology. (*H. Ross*).

HERBARIUM and TYPES: M. – Mexican collections (1906) at BAL, BP, C, G, L. Sicilian collections (*Herbarium siculum*, 900 nos., 1898-1916) at B (extant), BP, DBN, DOMO, E, FI, G, GB, GE, GH, GOET, HBG, JE, L, LD, MANCH, NAP, NY, P, PAL, S, WAG, WRSL.

BIBLIOGRAPHY and BIOGRAPHY: Barnhart 3: 180 (b. 8 Mar 1862); BFM no. 3119; Biol.-Dokum. 15: 7663; BJI 1: 49, 2: 148; BL 2: 395, 710; BM 4: 1734; Bossert p. 338; Christiansen p. 319; CSP 12: 629, 18: 303; IF suppl. 1: 85; Kew 4: 519; Langman p. 650; LS 22542a, 38554; NI 1680, 1680n; Rehder 1: 432, 5: 738; Saccardo 1: 142; Tucker 1: 607; Urban-Berl. p. 322, 389.

BIOFILE: Anon., Ber. deut. bot. Ges. 50: 87. 1932; Bot. Centralbl. 69: 400. 1897 (app. München), 41: 368. 1890 ("Privatdocent": Palermo), 50: 191. 1892 (interim director bot. gard. Palermo); Bot. Jahrb. 25 (Beil. 60): 57. 1898 (issues *Herbarium siculum*); Bot. Zeit. 48: 269. 1890 (lecturer Palermo); Hedwigia 48: (197). 1909 (app. München); J. Bot., Morot 3: 392. 1889(Society for exchange of mediterranean plants set up); Nat. Nov. 19: 156. 1897 ("custos" bot. gard. München); Österr. bot. Z. 47: 191. 1897 ("custos" bot. gard. München).
Carus, C.G. & C.F.P. von Martius (G. Schmid, ed.). Eine Altersfreundschaft in Briefen 121. 1939.
G.L., Bot. Monatschrift 16: 69. 1898 (on *Herb. sicul.*).
Hertel, H., Mitt. Bot. München 16: 425. 1980.
Kneucker, A., Allg. bot. Z. 3: 36, 72. 1897 ("Custos" Bot. Gard. München), 4: 67. 1898 (Herb. siculum), 15: 48. 1909 ("Custos" Bot. Mus. München).
McVaugh, R., Contr. Univ. Mich. Herb. 9: 298-299. 1972 (itin. Mexico 1906).
Suessenguth, K., Ber. deut. bot. Ges. 60: 177-185. 1942 (obit., portr., bibl., b. 8 Mar 1862, d. 24 Feb 1942).
Wagenit, G., Index coll. princ. herb. Gott. 137. 1982.

THESIS: *Beiträge zur Kenntniss des Assimilationsgewebes und der Korkentwickelung armlaubiger Pflanzen*. Freiburg 1887, 32 p., *1 pl.* (Bot. Zeit. 25 Nov 1887).

9564. *Anatomia comparata delle foglie delle Iridee* studio anatomico-sistematico ... Estratto dal Giornale Malpighia, vol. vi. 1892. Genova (Tipografia di Angelo Ciminago ...) 1892 [1893]. Oct.
Publ.: 1892-1893 (in journal; p. 99: Jul 1893), p. [1]-100, *pl. 5-8* (col. liths.). *Copy*: U. – Reprinted and to be cited fom Malpighia 6(4): 90-116 (repr. 1-29), 6(5): 179-205 (reprint 29-55), 1892, 7(4): 345-390 (reprint 56-99). 1893.

9565. *Icones et descriptiones plantarum novarum vel rariorum horti botanici panormitani* ... Panormi [Palermo] (ex officina typographica Ignatii Virzì, ex officina cromolographica Cyri Visconti) 1896. Fol. †. (*Icon. pl. hort. bot. panorm.*).
Publ.: 1896 (Bot. Centralbl. 17 Nov 1896; Bot. Zeit. 16 Nov 1896; Nat. Nov. Nov(1) 1896), p. [i], [1]-10, *pl. 1-3* (col. liths.). *Copies*: G, NY, PH, USDA. – Text in part reprinted by H. Ross in Bot. Centralbl. 74: 216-218. 1898.

9566. *Herbarium siculum* herausgegeben von Dr. Hermann Ross ... [München 1898]. (*Herb. sicul.*).
Publ.: 1898 (Bot. Centralbl. 4 Mai 1898), p. [1-3]. *Copy*: G. – Pamphlet introducing this series of exsiccatae and listing the plants in cent. 1.
Erläuterungen und kritische Bemerkungen zum Herbarium siculum I. Centurie, Bull. Herb. Boiss. 7(4): 262-299. Apr 1899 (also separate [1]-38); II. Centurie, ib. ser. 2. 1: 1201-1232. 1901 (also separate [1]-32)
Cent. 1: 1898, nos. 1-100, see above; also listed Bot. Centralbl. 74: 205-206. 4 Mai 1898.
Cent. 2: 1901, nos. 101-200, see above; also listed Bot. Centralbl. 85: 120. 1901.
Cent. 3: 1903, nos. 201-300, (Allg. bot. Z. 9: 76. 15 Apr 1903; ÖbZ 53: 85. 1903.
Cent. 4: 1904, nos. 301-400 (Bot. Centralbl. 25 Oct 1904).
Cent. 5: 30 Mai 1906 (date receipt M), nos. 401-500.
Cent. 6: 1908, nos. 501-600.
Cent. 7: Mar 1909(Bot. Centralbl. 30 Nov 1909; list of contents issued by H. Ross, München 1909, p. [1]-2. *Copy*: G. – Nos. 601-700, ÖbZ 1 Jun 1909.

9567. *Die Pflanzengallen* (Cecidien) Mittel- und Nordeuropas ihre Erreger und Biologie und Bestimmungstabellen ... Jena (Verlag von Gustav Fischer) 1911. Oct. (*Pflanzengallen*).
Ed. [1]: Aug-Sep 1911 (preface p. viii: Aug 1911; ÖbZ Sep 1911; Nat. Nov. Oct(2) 1911), p. [i]-viii, [ix], [1]-350, *pl. 1-10* (with 233 figures by G. Dunzinger). *Copies*: GOET, H, ILL (inf. D.P. Rogers), M, MO, USDA, G. Wagenitz.
Ed. 2: Jan 1927 (p. vi: Nov 1926; Nat. Nov. Feb 1927; NYBG rd. Aug 1927), p. [i]-vi, [vii], [1]-348, *pl. 1-10*. *Copies*: BR, GOET (Pl. phys; Univ. Bibl.) (inf. G. Wagenitz), ILL (inf. D.P. Rogers), LD (inf. O. Almborn), LIV (inf. J. Edmondson), M, USDA. "Zweite vermehrte und verbesserte Auflage, unter Mitwirkung von Dr. H. Hedicke, Berlin [Hans Hedicke]."
Ref.: E.W.S., J. Bot. 49: 372-373. Dec 1911 (rev.).
Fischer, G., Verzeichniss naturw. Werke, Bot. ed. 2 1878-1937, p. 209-210.

9568. *Die Pflanzengallen Bayerns* und der angrenzenden Gebiete von Dr. H. Ross ...mit 325 Abbildungen von Dr. G. Dunzinger ... Jena (Verlag von Gustav Fischer) 1916. Oct. (*Pfl.-Gall. Bayerns*).
Publ.: Aug-Oct 1916 (p. xi: Jul 1916; Nat. Nov. (1, 2) 1916, p. [i]-xi, [xii], [1]-104, figs. in text. *Copies*: BR, GOET (inf. G. Wagenitz), LD (inf. O. Almborn), M (inf. I. Haesler).
Nachtrag 1: 1922, p. 98-141, n.v.

9569. *Praktikum der Gallenkunde* "Cecidologie" Entstehung. Entwicklung. Bau der durch Tiere und Pflanzen hervorgerufenen Gallbildungen sowie Ökologie der Gallenerreger ... Berlin (Verlag von Julius Springer) 1932. Oct. (*Prakt. Gallenk.*).
Publ.: 1932, p. [i]-x, [1]-312. *Copy*: M (inf. I. Haesler). – Biologische Studienbücher, ed. W. Schoenichen, 12.

Ross, [Sir] **James Clark** (1800-1862), British (English) arctic and antarctic explorer; in the Royal Navy, lieutenant 1822; with W.E. Parry in voyages of 1819-1820, 1821-

1823, 1824-1825 and 1827; commander of the antarctic expedition of 1839-1843 (botanist: J.D. Hooker); ultimately (1858) rear-admiral; nephew of Sir John Ross. (*J.C. Ross*).

HERBARIUM and TYPES: K; further material at CGE, E, LD, MO.

NOTES (1): J.C. Ross made collections on several of his journeys; his main claim to fame in botany, however, is that he was in command of the antarctic expedition with the Erebus and Terror, 1839-1843, on which J.D. Hooker took part as a botanist. For Hooker's publications of the botanical results of this expedition see TL-2/2. For the various travelogues see e.g. BM 4: 1734; see also TL-2/2964-2965, 3009.
(2) Sir John Ross (1777-18156), J.C. Ross's uncle, was also an arctic explorer, e.g. in search of the North-West Passage 1818 and 1829-1833; author of TL-2/830, *Voy. explor. Baffin's Bay* an account of the 1819 expedition in which R. Brown published a *List of plants* (see e.g. Barnhart 3: 180, BM 4: 1734, Tucker 607; Allibone, S.A. Crit. dict. Engl. lit. 1874. 1878).

BIBLIOGRAPHY and BIOGRAPHY: AG 5(1): 772; Barnhart 3: 180 (b. 15 Apr 1800, d. 3 Apr 1862); BM 4: 1734; DNB 49: 265; Herder p. 58; PO 2: 698-699; PR 4199, 4228, 8734, ed. 1: 4638-4639, 4667; Rehder 1: 510; TL-2/2964-2965, 3009, see J.D. Hooker, R. McCormick.

BIOFILE: Allan, M., The Hookers of Kew 269 [index]. 1967.
Allibone, S.A., Crit. dict. Engl. lit. 1873.
Anon., Bonplandia 10: 143. 1862 (obit.); Gartenflora 12: 173. 1863.
Bentham, R., Proc. Linn. Soc. London 1862: xciv-xcv.
Bridson, G.D.R. et al., Nat. hist. mss. res. Brit. Isl. 432 [index]. 1980.
Dodge, E.S., The polar Rosses. John and James Clark Ross and their explorations, London. Faber and Faber, 1973, 260 p., *8 pl.*, 3 maps (main modern treatment, q.v. also for earlier literature on R.).
Glenn, R., The botanical explorers of New Zealand 176. 1950.
Huxley, L., Life letters J.D. Hooker 2: 561 [index]. 1918 (many entries!).
Jessen, K.F.W., Bot. Gegenw. Vorzeit 465. 1864.
Jones, A.G.E., Geogr. J. 137: 165-179. 1971.
Nissen, C., Zool. Buchill. 348-349. 1969.
Pennington, P., Great explorers 288-291. 1979.
Polunin, N., Bull. Natl. Mus. Canada 92 (Biol. 26): 21. 1940 (coll. Canada arctic).
Priestley, Raym., Foreword to 1969 ed., 2 vols. of J.C. Ross, Voyage of discovery and research during the years 1839-1843.
Rasky, F., North Pole or bust 7-35. 1977.
Savours, A., Geogr. J. 128: 325-327. 1962 (bibl.).

Ross, Robert (1912-x), British (English) botanist (diatomologist, nomenclaturist), educated at St. Paul's School, London and St. John's College, Cambridge, M.S. Cambridge 1934; on the Cambridge Botanical Expedition to West Africa 1935; with the British Museum (Natural History), Botany Dept., from 1936, in charge of Diatomaceae; deputy keeper Bot. Dept. 1962; keeper 1966; now retired; member of the editorial Committee for the International Code of Botanical Nomenclature 1954-1978. (*R. Ross*).

HERBARIUM and TYPES: BM. – Duplicates e.g. at BR, ENT, FH, LD.

BIBLIOGRAPHY and BIOGRAPHY: BFM see no. 2344; Bossert p. 339; IH 1 (ed. 1): 57, (ed. 2): 74, (ed. 3): 93, 94, (ed. 4): 102, (ed. 5): 107, (ed. 6): 155, (ed. 7): 158; TL-2/66; Zander ed. 11, p. 808 (b. 14 Aug 1912).

BIOFILE: Anon., Taxon 1: 99. 1952 (on Ruwenzori Exp. 1952).; The NIH Record 15(8): 6. 1963 (portr.).
Fryxell, G.A., Nova Hedwigia Beih. 35: 363. 1975 (diatom coll.).
Kent, D.H., Brit. herb. 76. 1957 (British herb., 3000 spec., at BM).
Rollins, R.C., Taxon 5: 2, 3. 1956 (in group portraits).
Stearn, W.T., Nat. Hist. Mus. S. Kensington 311, 313. 1981.

VanLandingham, S.L., Cat. diat. 4: 2366. 1971, 6: 3586. 1978, 7: 4423-4224. 1978.
Verdoorn, F., ed., Chron. bot. 2: 176, 181. 1936, 3: 164, 165. 1937.

EPONYMY: *Rossiella* T.V. Desikachary et C.L. Maheshwari (1958).

Ross-Craig, Stella (Mrs. J.R. Sealy) (1906-x), British botanical artist; contributed to Botanical Magazine, Hooker's Icones plantarum, and Kew Bulletin; artist for the Royal Horticultural Society 1929-1960, and on the staff of the R.B.G. Kew. (*Ross-Craig*).

HERBARIUM and TYPES: Unknown.

BIBLIOGRAPHY and BIOGRAPHY: BFM p. 1428; BL 2: 233, 710; Blunt p. 188, 251-254, 264; Bossert p. 339; MW p. 295; NI 1681, 2341, 2350, Nachtr. 2137, see also 1: 133; TL-1/537; TL-2/3006.

BIOFILE: Daniels, G.S., Artists R.B.G. Kew 52-53. 1974 (portr., b. 19 Mar 1906).
Rix, M., The art of the Botanist 216. 1981.
White, A. & B.L. Sloane, The Stapelieae ed. 2. 21 [index]. 1937.

NOTE: *Drawings of British plants*, parts 1-27, London 1948-1978. Stella Ross-Craig's most significant contribution to British botany, falls outside the period of TL-2 (see e.g. NI 3: 58 and no. 1681 and regular notices in Kew Bulletin).

Rossetti, Corrado (1866-?), Italian botanist; high school teacher of natural history and mathematics at Pisa (1891), in Seravezza (1896) and Querceta (1908). (*Rossetti*).

HERBARIUM and TYPES: PI; further material at B (mainly destr.), BI and GE.

BIBLIOGRAPHY and BIOGRAPHY: Barnhart 3: 181 (b. 7 Feb 1866); BL 2: 402, 710; CSP 18: 306; GR p. 538; LS 22545; Saccardo 1: 142, 2: 93; Urban-Berl. p. 389.

BIOFILE: Dörfler, I., Botaniker-Adressbuch 100. 1896.

9570. *Contribuzione alla Flora della Versilia*... Pisa (Tipografia T. Nistri e C.) 1888. (*Contr. Fl. Versilia*).
Publ.: 1888 (Bot. Zeit. 29 Mai 1889; Nat. Nov. Dec(2) 1888), p. [1]-45. *Copy*: FI. – Reprinted from Atti Soc. Tosc. Sci. nat. 9(1): 384-426. 1888.
Seconda contr.: ib. 12: 120-143. 1892 (Nat. Nov. Nov(1) 1892; Bot. Zeit. 1 Mai 1893); a further note in Proc. Verb. Soc. Tosc. Sci. nat. 1892, n.v.

Rosshirt, Karl (*fl.* 1888), German botanist and high school teacher at Colmar (Elsass; Alsace). (*Rosshirt*).

HERBARIUM and TYPES: Unknown.

BIBLIOGRAPHY and BIOGRAPHY: BL 2: 158, 710.

9571. *Beiträge zur Flora der Umgegend von Colmar* und Ergebnisse von botanischen Ausflügen in die Schweiz ... [Colmar (Buchdruckerei von Wittwe Camill Decker) 1888]. Qu. (*Beitr. Fl. Colmar*).
Publ.: Jul 1888 (for "Schlussfeier" of 4 Aug 1888; Bot. Zeit. 31 Mai 1889; Nat. Nov. Mai(1) 1889), p. [1]-36. *Copy*: MO. – Published in Jahresbericht des Lyceums in Colmar über das Schuljahr 1887-1888, Colmar 1888 (Beiträge p. 1-36; Schulnachrichten p. 37-66).

Rossi, Giovanni Battista (*fl.* 1825), Italian gardener; curator of the botanical garden of Monza, nr. Milano 1825-1843. (*G. Rossi*).

HERBARIUM and TYPES: Unknown.

NOTE: Not to be confused with the Modena botanist and gardener Gaetano (Cajetani) Rossi, 1717-1775.

BIBLIOGRAPHY and BIOGRAPHY: BM 4: 1735; Jackson p. 435; PR 7776, ed. 1: 8736; Rehder 1: 62; Saccardo 1: 142.

BIOFILE: Anon., Flora 10 (Beil.): 91. 1827.

9572. *Catalogus plantarum* existentium *in hortis regiae villae prope Modoetiam.* Modoetiae [Monza] (typis L. Coreetta) 1813. Oct. (*Cat. pl. hort. Modoet.*).
Publ.: 1813 (printed; changed to 1816 in ink in NY copy), p. [1]-75, [77-78, abbr.], 1-16, suppl. [suppl. 1816?]. *Copy*: NY.
1826 ed.: 1826, p. [i]-viii, [1]-83, *3 pl. Copy*: M. − "*Catalogus plantarum horti regii modoetiensis ad annum mdcccxxv*". Mediolani [Milano] (ex Imp. regia Typographia) 1826.

Rossi, Pietro (1738-1804), Italian botanist and entomologist at Pisa. (*P. Rossi*).

HERBARIUM and TYPES: Unknown.

NOTE: for Pietro Rossi (1871-1950) see Atti Soc. ital. Sci. nat. Milano 89(3/4): 242-245. 1950.

BIBLIOGRAPHY and BIOGRAPHY: BM 4: 1735; Dryander 3: 395, 541; Herder p. 411; PR 7777-7778, ed. 1: 8738-8739; Rehder 1: 199, 3: 527; Saccardo 1: 142.

BIOFILE: Baccetti, B., Frustula entomologica 5(3): 1-30. 1962 (on his entomology).
Gilbert, P., Comp. biogr. lit. deceased entom. 324. 1977.
Swainson, W., Taxidermy 311. 1840.

HANDWRITING: Frustula entom. 5(3): 5, 7, 9, 11, 13, 15, 17, 19. 1962.

9573. *De nonnullis plantis* quae pro venenatis habentur observationes et experimenta a Petro Ross Florentiae Instituta. Pisis [Pisa]. (Jo. Paul. Giovanelli cum soc. ...) 1762. Oct. (*Nonnull. pl.*).
Publ.: 1762, p. [i]-lxvi. *Copy*: FI.

9574. *Istoria di ciò che è stato pensato intorno alla fecondazione delle piante*, dalla scoperta del doppio sesso fino a questo tempo. Col aggiunta di nuove sperienze. Del Sig. Pietro Rossi. [t.p. [i]: Memoria di Pietro Rossi estratta dal tomo vii. degl'atti della Società italiana. Verona (per Dionigi Ramanzini) 1794)]. Qu.
Publ.: 1794, p. [i], 1-62. *Copy*: FI. − Reprinted from Mem. Soc. ital. 7: 369-430. 1794.

Rossi, Stefano (1851-1898), Italian botanist; high school teacher at Domodossola. (*S. Rossi*).

HERBARIUM and TYPES: DOMO, FI, TO.

BIBLIOGRAPHY and BIOGRAPHY: BFM no. 1109; BL 2: 380, 710; CSP 18: 308; Rehder 1: 429; Saccardo 1: 142, 2: 94, cron. p. xxxi; Tucker 1: 607.

BIOFILE: Chiovenda, E., Nuov. Giorn. bot. ital. 31: 312. 1924.

EPONYMY: *Stephanorossia* Chiovenda (1911).

9575. *Studi sulla flora ossolana* del D.r Stefano Rossi ... Domodossola (Tipografia Porta) 1883. Oct. (*Stud. fl. ossol.*).
Publ.: 1883 (Bot. Zeit. 25 Jan 1884; Bot. Centralbl. 19-23 Nov 1883; Nat. Nov. Dec(1) 1883), p. [1]-112. *Copies*: FI, G. − See BL 2: 380 for further references and E. Chiovenda, Fl. Alpi Lepont. occ. 1: 73. 1904.
Ref.: Penzig, O., Bot. Centralbl. 20: 77-78. 13-17 Oct 1884 (rev.).

9576. *Flora del Monte Calvario* ... Domodossola (Tipografia Porta) 1883. (*Fl. Monte Calv.*).
Publ.: 1883, p. [1]-15. *Copy*: FI. – "Estratto dalla Cronaca della Fondazione Galletti anno 1883". Unannotated list.
Ref.: Penzig, O., Bot. Centralbl. 20: 77. 13-17 Oct 1884 (rev.).

9577. *Le piante acotiledoni vascolari e le graminacee ossolane* ... Domodossola (Tipografia Porta) 1884. (*Piante acot. vasc. gram. ossol.*).
Publ.: 1884, p. [i-vi], [1]-52. *Copy*: FI. – "Estratto dalla Cronaca della Fondazione Galletti".

9578. *Nuove piante trovate in val d'Ossola* ... Roma (Tipografia della Pace di Filippo Cuggiani ...) 1890.
Publ.: 1890, p. [1]-8, *pl. 1-4* (on 2). *Copy*: FI. – Reprinted and to be cited from Mem. pont. Accad. Nuovi Lincei 6: 63-66, *pl. 1-4* (on 2). 1890.

9579. *Alcune forme vegetali e varietà nuove raccolte nella valle ossolana* ... Roma (Tipografia della Pace di Filippo Cuggiani ...) 1891.
Publ.: 1891, p. [1]-12, *pl. 1-4* (on 2). 1891. *Copy*: FI. – Reprinted and to be cited from Mem. pont. Accad. Nuovi Lincei 7: 81-88, *pl. 1-4* (on 2). 1891.

Rossmässler, Emil Adolph (1806-1867), German (Saxonian) botanist and malacologist at Leipzig; studied theology 1825-1827; popular writer on natural history; professor of natural history (1830-1850) and director of the "Forstbotanischer Garten (1840-1850) at Tharandt, nr. Dresden; member of the 1848 Frankfurter Parliament; released from his functions at Tharandt because of his political activities; from 1850 professional "Volkschriftsteller". (*Rossmässler*).

HERBARIUM and TYPES: a few specimens at REG and MW; 166 plants from Spain at BM (rd. 1856). Issued a series of exsiccatae *Plantae lipsienses, weidanae et tharandtinae* (see Flora 14: 632-638. 1831) in very restricted numbers.

BIBLIOGRAPHY and BIOGRAPHY: Andrews p. 322; Barnhart 3: 181 (b. 3 Mar 1806, d. 8 Apr 1867); BM 1: 72, 4: 1735-1736, 8: 1094; Colmeiro 1: cxcvii; CSP 5: 299, 8: 786; De Toni 2: viii; Herder p. 91, 120, 187, 352, 365, 381; IH 2: (files); Jackson p. 42, 57, 179, 299, 500; KR p. 602; LS 38562; Moebius p. 429 (early use posters); NI 1: 1682, 1682n, see also 1: 249; PR 284, 7779-7783, ed. 1: 8741-8742; Quenstedt p. 367-368; Ratzeburg p. 441-446; Rehder 5: 738; Tucker 1: 607.

BIOFILE: Anon., Bot. Zeit. 5: 783. 1847 (will set up a forestry college in Spain), 6: 471. 1848 (member of Frankfurt parliament); Flora 30: 672. 1847 (intends to go to Madrid to establish a forestry school); Österr. bot. Z. 17: 163. 1867 (d. 7 Apr 1867).
Fischer, E., Gutenberg Jahrb. 1933: 210 (used nature printing in his "Die vier Jahreszeiten").
Friedl, K. & R. Gilsenbach, Das Rossmässlerbüchlein, Berlin 1956, 155 p., 12 pl. (n.v.).
Gager, C.S., Brooklyn Bot. Gard. Rec. 27(3): 237. 1938.
Gilbert, P., Comp. biogr. lit. deceased entom. 324. 1977.
Hess, R., Lebensbilder hervorragender Forstmänner 299-301. 1885 (biogr. sketch).
Lindemann, E. von, Bull. Soc. imp. Natural. Moscou 61(1): 63. 1886.
Mattick, F., Herzogia 1: 174. 1969.
Murray, G., Hist. coll. BM(NH) 1: 178. 1904 (coll.).
Rossmässler, E.A., Aus der Heimath, Leipzig, 1863: 763-768, 769-776, 785-794, 801-810, 815-824 (and an earlier set of pages in same volume not seen by us) (autobiography; Aus der Heimath was a popular journal edited by Rossmässler); Mein Leben und Streben im Verkehr mit der Natur und dem Volke, ed. K. Russ, Hannover 1874, frontisp., ix, 420 p. (autobiogr.).
Schmidt, A., Malakozool. Bl. 14: 183-190. 1866.
St., Zool. Gart., Leipzig 8: 199-200. 1867 (obit. notice).
Ule, O., Natur, Halle, 16: 188-190, 193-195, 217-220. 1867 (R. as a teacher), 19: 220-223. 1870.

Ward, Ann. Rep. U.S. Geol. Survey 5: 413. 1885 (very positive on R.'s *Beiträge zur Versteinerungskunde*, especially with regard to his treatment of dicotyledonous leaves).

FESTSCHRIFT: Hartung, O. et al., *Festschrift zum hundertjährigen Geburtstage Emil Adolf Rossmässlers* am 3. März 1906. Bearbeitet im Auftrage des deutschen Lehrervereins für Naturkunde. Stuttgart (Verlag von K.G. Lutz) 1906, 192 p. (portr., bibl., biogr.).

HANDWRITING: E.A. Rossmässler, Mein Leben und Streben, 1874, p. [ix].

EPONYMY: *Rossmaesslera* H.G.L. Reichenbach (1841) is likely named for this author, but no etymology was given.

9580. *Die Versteinerungen des Braunkohlensandsteins* aus der Gegend von Altsattel in Böhmen (Elnbogener Kreises) lithographiert und beschrieben von E.A. Rossmässler, ... Dresden und Leipzig (in der Arnoldischen Buchhandlung) 1840. Qu. (*Verstein. Braunkohlensandst.*).
Publ.: 1840 (p. vi: Apr 1840), p. [ii]-vi, [1]-42, *pl. 1-12* (uncol. liths. auct.). *Copies*: NY, PH, Zool. Mus. Amsterd. – Second t.p [ii]: *Beiträge zur Versteinerungskunde* ... Erstes Heft. "Die Versteinerungen des Braunkohlensandsteins aus der Gegend von Altsattel in Böhmen". Only "Heft" issued.

9581. *Mikroskopische Blicke* in den innern Bau und das Leben der Gewächse. Populäre Vorlesungen ... Leipzig (Hermann Costenoble) 1852. (*Mikrosk. Blicke*).
Publ.: 1852 (preface: "Wonnemonat" Mai 1852, Bot. Zeit. 10 Dec 1852), p. [i]-viii, [1], [1]-116, *15 pl.* (in pagination). *Copies*: E (Univ. Libr.), H. – Populäre Vorlesungen aus dem Gebiete der Natur vol. 1. – Inf. J. Edmondson.
Ref.: J., Bot. Zeit. 10: 881-883. 10 Dec 1852 (rev.).

9582. *Die Versteinerungen*, deren Beschaffenheit, Entstehungsweise und Bedeutung für die Entwickelungsgeschichte des Erdkörpers, mit Hervorhebung von Repräsentanten der geologischen Epochen. Populaire Vorlesungen ... Leipzig (Hermann Costenoble) 1853. (*Versteinerungen*).
Publ.: 1853, p. [i]-viii, 1-188, *7 pl. Copy*: E (Univ. Libr.). Populäre Vorlesungen aus der Natur 2. – Inf. J. Edmondson.

9583. *Flora im Winterkleide* ... Leipzig (Hermann Costenoble ...) [1854]. Oct. (*Fl. Winterkl.*).
Ed. 1: 1854 (p. viii: 14 Oct 1853), frontisp., p. [i]-viii, [1]-155. n.v.
Ed. 2 (*title issue*): 1856, frontisp., p. [i]-viii, [1]-155. *Copies*: MO, NY.
Ed. 3: 1887 (Bot. Centralbl. 19-23 Dec 1887; Bot. Zeit. 30 Dec 1887), p. [i]-xxiv, [1]-107 (n.v.). "Neu bearbeitet von K.G. Lutz".
Ed. 4: Mar-Jun 1908 (p. vi: 25 Feb 1908; Bot. Centralbl. 25 Aug 1908; Bot. Zeit. 16 Sep 1908; Nat. Nov. Jul(1) 1908), frontisp. portr. author, p. [i]-xxii, [1], [1]-126, *pl. 1-3*. *Copy*: BR. – "Vierte Auflage, bearbeitet von H. Kniep ...mit einer Biographie Rossmässlers ..." Leipzig (Verlag von Dr. Werner Klinkhardt) 1908.
Popular edition: s.d. p. [1]-113. *Flora im Winterkleide*. Mit einer Einleitung und Anmerkungen herausgegeben von R.H. Francé ... Leipzig (Theod. Thomas Verlag ...) s.d. Naturbibliothek 42/43. *Copy*: G.
Ref.: J., Gartenflora 4: 177-179. Mai 1855 (rev.).
Wangerin, W., Bot. Centralbl. 110: 365-366. 1909 (rev. ed. 4).

9584. *Der Wald*. Den Freunden und Pflegern des Waldes geschildert ... Leipzig und Heidelberg (C.F. Winter'sche Verlagshandlung) 1863. Oct. (*Wald*).
Ed. [1]: 1863 (p. x: Oct. 1862; published in parts), p. [i]-xiv, [xv], [1]-628, frontisp., engr. t.p., 2 charts, 17 copper engr. (Ernst Heyn) and 82 woodcuts (A. Thieme). – See C.H. Preller, in Ratzeburg, p. 445 for a critical appraisal. *Copies*: NY, USDA.
Ed. 2: 1870-1871 (p. xi: Sep 1870; in 15 parts), p. [i]-xvi, [1]-671, frontisp., 2 charts, *17 pl. Copy*: G. – "Zweite Auflage, durchgesehen, ergänzt und verbessert von M. Willkomm" [Heinrich Moritz Willkomm 1821-1895]. Leipzig und Heidelberg (id.) 1871. Oct.

Ed. 3: Nov 1880-Jul 1881 (p. xiv: Mai 1880; ÖbZ 1 Jan 1881; in 16 Lieferungen, Heft 1, p. 1-48, Nov 1880; Lief. 14, Jun 1881, 15-16 Jul 1881), p. [i]-xviii, [1]-730, frontisp., 2 charts, *17 pl. Copies*: H, MO, NY. – "Dritte Auflage, ... M. Willkomm, ..." id. 1881.

9585. *Mein Leben und Streben* im Verkehr mit der Natur und dem Volke von E.A. Rossmässler ... nach dem Tode des Verfassers herausgegeben von Karl Russ. Hannover (Carl Rümpler) 1874. Oct. (*Leben u. Streben*).
Publ.: 1874, p. [i]-viii, [1]-420. *Copies*: GOET (inf. G. Wagenitz), GRON.

Rossmann, [Georg Wilhelm] Julius (1831-1866), German (Hessen) botanist; studied medicine at Giessen but devoted himself to botany from 1851 as a student of A. Braun; Dr. phil. Giessen 1853; habil. ib. 1854; extra-ordinary professor of botany ib. 1859-1865; retired to Worms because of illness. (*Rossmann*).

HERBARIUM and TYPES: Unknown. Letters to D.F.L. von Schlechtendal at HAL.

BIBLIOGRAPHY and BIOGRAPHY: Barnhart 3: 181 (d. 21 Jan 1866); BM 4: 1736; CSP 5: 299, 8: 785-786, 12: 632; De Toni 1: cviii; Herder p. 120, 324; Jackson p. 142, 146, 299; Kew 4: 521; LS 22547-22549; PR 7784-7786; Rehder 5: 738; TL-2/2735; Tucker 1: 607.

BIOFILE: Anon., Bot. Zeit. 13: 847. 1855 (habil. Giessen), 21: 376. 1863 (member Naturf. Ges. Halle), 24: 136 (see also H.W., p. 176). 1866 (obit., bibl.); Flora 39: 14. 1856 (habil. Giessen), 39: 175. 1856 (corr. member bot. Ges. Regensb.), 42: 255. 1859 (prof. bot.), 49: 222. 1866 (d.); Bull. Soc. bot. France 13 (bibl.): 191. 1866 (d.); Österr. bot. Z. 9: 201. 1859 (e.o. prof. Giessen), 16: 192. 1866 (d. 21 Jan 1866).
Hess, R., Lebensbilder hervorragender Forstmänner 301-302. 1885 (biogr. sketch, bibl., b. 9 Dec 1831, d. 21 Jan 1866).
Walter, H., Ber. Offenbacher Ver. Naturk. 7: 135-140. 1866 (obit.).
Welcker, H., Bot. Zeit. 24: 176. 1866 (obit. not.).

COMPOSITE WORKS: Co-author and editor of C. Heyer, *Pharm.-Fl. Ober-Hessen* 1863, TL-2/2735.

EPONYMY: *Rossmannia* Klotzsch (1854).

9586. *Beiträge zur Kenntniss der Wasserhahnenfüsse*, Ranunculus sect. Batrachium ... Giessen (J. Ricker'sche Buchhandlung) 1854. Qu. (*Beitr. Kenntn. Wasserhahnenf.*).
Publ.: 1854 (p. vi: Feb 1854; ÖbW 2 Nov 1854; Flora rev. 21 Sep 1854, rd. 28 Apr-28 Jun 1854), p. [i]-vi, [vii], [1]-62, [64, colo.]. *Copies*: G, LC(2), NY.
Ref.: S. [Skofitz], Österr. bot. Wochenbl. 5: 103. 29 Mai 1855 (rev.).
Fürnrohr, A.E., Flora 37: 559-560. 22 Sep 1854.
Schlechtendal, D.F.L. von, Bot. Zeit. 12: 725-727. 13 Oct 1854 (rev.).

9587. *Einladung zu einem Kryptogamen-Tauschvereine* [Giessen (Druck von Wilhelm Keller) 1857].
Publ.: 1857 (p. [3]: Jun 1857), p. [1-4]. *Copy*: G. – Proposal to establish an exchange club for cryptogams.

9588. *Beiträge zur Kenntniss der Phyllomorphose* ... Giessen (J. Ricker'sche Buchhandlung) 1857-1858. Qu. (*Beitr. Kenntn. Phyllomorph.*).
1: 1857 (ÖbW 6 Aug 1857; Flora rd. by 21 Feb 1857), p. [i-viii], [1]-60, *3 pl.*
2: 1858 (ÖbZ 1 Oct 1858; Flora rd. by 28 Mai 1858), p. [i-vi], [1]-26, *8 pl.*
Copy: GOET (Inf. G. Wagenitz).
Ref.: Anon., Flora 41: 174-175. 21 Mar 1858 (rev. 1), 420-422. 14 Jul 1858 (rev. 2).
Schlechtendal, D.F.L. von, Bot. Zeit. 16: 85-86. 26 Mar 1858 (rev. 1), 252-253. 20 Aug 1858 (rev. 2).

9589. *Zur Kenntniss der Wasserhahnenfüsse*, Ranunculus Sect. Batrachium ...[Offenbach a. M. (Druck von Kohler & Teller) 1861]. Oct.

Publ.: 1861, p. [1]-9, *pl. 1-6. Copy*: U. – Reprinted from Ber. Offenbacher Ver. Naturk. 2: 50-58. 1861.

9590. *Beitrag zur Kenntniss der Spreitenformen* in der Familie *der Umbelliferen* ... Halle (Druck und Verlag von H.W. Schmidt) 1864. Qu.
Publ.: 1864, p. [1]-14, *pl. 1-7* (uncol. lith. auct.). *Copies*: M, U. – Reprinted and to be cited from Abh. naturf. Ges. Halle 8: 169-182. 1864.

9591. *Ueber den Bau des Holzes* der in Deutschland wildwachsenden und häufiger cultivirten Bäume und Sträucher. Eine kurze Darlegung der wichtigeren bis jetzt gewonnenen Resultate insbesondere für Forstleute und Techniker ... Frankfurt a.M. (J.D. Sauerländer's Verlag) 1865. Oct. (*Bau Holz.*).
Publ.: 1865 (preface: Nov 1864; ÖbZ 1 Oct 1865; Flora "neu" 22 Jul 1865; Bot. Zeit. rev. 1 Sep 1865), p. [i]-viii, [1]-100, *1 pl. Copies*: GOET (Forstl. Fak.), MO.
Ref.: Anon., Flora 48: 541-542. 15 Nov 1865.
Schlechtendal, D.F.L. von, Bot. Zeit. 23: 271-272. 1 Sep 1865.

Rostafiński, Józef Thomasz (Joseph Thomas) (1850-1928), Polish botanist, born in Warsaw; studied in Jena 1869-1870, Halle 1870-1872 and Strassburg 1872-1873; Dr. phil. Strassburg 1872-1873; Dr. phil. Strassburg (with A. de Bary) 1873; assistant ib. 1873-1876; lecturer at the University of Cracow 1876, e.o. professor of botany 1878, regular professor and director of the botanic garden 1881-1912. (*Rost.*).

HERBARIUM and TYPES: KRA. – Other material at B (mainly destr.), GOET, PC (Some types of Myxomycetes), STR. – Letters to W.G. Farlow at FH.

BIBLIOGRAPHY and BIOGRAPHY: AG 12(1): 194; Ainsworth p. 234, 283, 319; Barnhart 3: 181 (b. 14 Aug 1850, d. 6 Mai 1928); BFM no. 3245; BJI 1: 49; BM 4: 1736, 8: 1094; Bossert p. 339; CSP 8: 786, 11: 225, 12: 632, 18: 310; De Toni 1: cviii-cxix; Frank 3(Anh.): 85; Futak-Domin p. 507; Hawksworth p. 185; Herder p. 200, 245; Jackson p. 156, 157, 171, 330; Kelly p. 195 LS 22550-22551; Moebius p. 74; Morren ed. 2, p. 13, ed. 10, p. 34; MW p. 419; Rehder 5: 738; Stevenson p. 1256; TL-2/1204-1205, 3285-3286; TR 239; Tucker 1: 608; Urban-Berl. p. 389.

BIOFILE: Ainsworth, G.C., Dict. fungi ed. 6. 512. 1971.
Anon., Bot. Zeit. 40: 786. 1882 (regular professor of botany Cracow); Hedwigia 68: (135). 1929 (d.); Mycol. Centralbl. 1: 162. 1912 (succeeded by M. Raciborski as prof. and dir. bot. gard. Cracow); Österr. bot. Z. 32: 310. 1882 (ord. prof. Cracow), 77: 240. 1928 (d. 6 Mai 1928).
Candolle, Alph. de, Phytographie 444. 1880 (types *Prodr. fl. polon.* still in his private herb.).
Dörfler, I., Botaniker-Adressbuch 118. 1896.
Gager, C.S., Brooklyn Bot. Gard. Rec. 27(3): 316. 1938 (dir. bot. gard. Cracow 1876-1912).
Hryniewiecki, B., Précis hist. bot. Pologne 4, 5, 16, 18, 19, 22, 25, 34. 1930; Edward Strasburger 13, 47-49, 52, 55, 59, 94. 1938, Acta Soc. bot. Polon. 20 (suppl.): 47-76. 1951 (biogr., bibl., portr., in Polish; list of biogr. refs.).
Kulczyński, S., Acta Soc. Bot. Polon. 6: 391-395. 1929 (obit.).
Szafer, W., Concise hist. bot. Cracow 163 [index] 1969 (portr.); Regn. Veg. 71: 381, 386. 1970.
Szymkiewicz, D., Bibl. fl. Polsk. 16, 18, 41-42, 122. 1925.
Wittrock, V.B., Acta Horti Berg. 3(2): 84-85, *pl. 23*. 1903 (portr.), 3(3): 69. 1905.

COMPOSITE WORKS: (1) With E. v. G. Janczewsky, *Obs. alg.* 1874, and *Obs. accrois. thalle phéosp.*, 1875, see TL-2/3285-3286.
(2) M.C. Cooke, *Contr. mycol. brit.* 1877, TL-2/1204 is "arranged according to the method of Rostafiński", in part a translation from the Polish (Sluzowce (Mycetozoa) monographia).

EPONYMY: *Rostafinsckia* [sic]Spegazzini (1880); *Rostafinskia* Raciborski (1884).

9592. *Florae Polonicae prodromus*. Uebersicht der bis jetzt im Königreich Polen beobachteten Phanerogamen ... Berlin (R. Friedländer & Sohn) 1873. Oct. (in fours).
Publ.: late 1872-Jan 1873 (rev. Flora 1 Feb 1873; "soeben" Bot. Zeit. Feb 1873), p. [i], [1]-128. *Copies*: FH, G, H, NY, USDA. – Reprinted and to be cited from Verh. zool.-bot. Ges. Wien 22: [81]-208. 1872 [1873?]. The USDA copy has the above text on cover and separate t.p.; the other copies are straight reprints without such a t.p. BR has a reprint with original signatures and pagination.
Ref.: Anon., Bull. Soc. bot. France 21(bibl.): 131-132. 1874.
Kn., Flora 56: 63-64. 1 Feb 1873.
Uechtritz, R.F.C. von, Bot. Zeit. 32: 204-207. 27 Mar 1874, 221-224. 3 Apr 1874 (rev.).

9593. *Versuch eines Systems der Mycetozoen*. Inaugural-Dissertation der Philosophischen Facultät der Universität Strassburg im Elsass für Erlangung der Doctorwürde vorgelegt von Joseph Thomas von Rostafiński aus Warschau. Strassburg (Druck von Friedrich Wolff) 1873. Oct. (*Vers. Syst. Mycetozoen*).
Publ.: 1873 (p. 22: Jan 1873; "neu" Bot. Jahrb. 9 Mai 1873; Hedwigia Mar 1873, Grevillea Apr 1874), p. [i]-iv, [1]-21, [22, curr. vitae; b. 14 Aug 1850]. *Copies*: G, H, NY, Stevenson.
Ref.: Anon., Hedwigia 13: 24-26. Feb 1874 (rev.).
G.K. [Klebs], Bot. Zeit 31: 375-383. 13 Jun 1873 (rev.).

9594. *Quelques mots sur l'Haematococcus lacustris* et sur les bases d'une classification naturelle des algues chlorosporées ... Cherbourg (Imp. Bedelfontaine et Syffert, ...) 1875. Oct.
Publ.: 1875 (Hedwigia Sep 1875; Flora rd. by 1 Sep 1875), p. [i], [137]-154. *Copies*: FH, FI, G. – Reprinted with special t.p. from Mém. Soc. natl. Sci. nat. Cherbourg 19: [137]-154. 1875.
Ref.: G.K. [Klebs], Bot. Zeit. 33: 753-754. 12 Nov 1875.

9595. *Śluzowce* (Mycetozoa) *monografia* przez Dra Józefa Rostafińskiego. Paryz [Paris] (Nakładem Biblioteki Kórnickiej) 1875. Qu. (*Sluzowce monogr.*).
Publ.: 1874-1875 (in journal; reprint 1875; Bot. Zeit. 21 Jan 1876), p. [i-iii], [1]-432, *pl. 1-13* (*1-12* uncol. liths. auct., *13* photo). *Copies*: FH, FI, G, H, NY, PH, Stevenson. – Issued as Pamiet. Towarz. Nauk. Sci. Paryzu 5(4): 1-215. 1874, 6(1): 216-432, *pl. 1-13*. 1875 (the Stevenson copy has the original division).
Suppl.: 1876 (read 3 Feb 1876), p. [1]-43, [1, bibl.]. *Copies*: NY, PH. "Dodatek I do monografi Śluzowców" reprinted from Pamiet. Towarz. Nauk. Sci. Paryzo 8(4): 1-42. *1 pl.* 1876.
English translation (p.p.): 1877, iv, 96 p., *24 pl.* London (n.v.). – *Contributions to Mycologia britannica. The Myxomycetes of Great Britain*. Adaptation by M.C. Cooke.

9596. *Beiträge zur Kenntniss der Tange* ... Heft 1. Ueber das Spitzenwachsthum von Fucus vesiculosus und Himanthalia lorea ... Leipzig (Verlag von Arthur Felix) 1876. †. Oct. (*Beitr. Kenntn. Tange*).
Publ.: 1876 (Bot. Zeit. 14 Apr 1876), p. [i], [1]-18, *pl. 1-3* (uncol. liths. auct.). *Copies*: G, U.

9597. *Über Botrydium granulatum* ... Leipzig (Verlag von Arthur Felix) 1877. Qu.
Co-author: Michael Stepanowitch Woronin (1838-1903).
Publ.: 12, 19 Oct 1877 (in journal), p. [1]-18, *pl. 1-5* (1-4 uncol., 5 col. liths.). *Copies*: BR, FH, FI, G, NY. – Reprinted and to be cited from Bot. Zeit. 35: col. 649-664. 12 Oct. 1877, 665-671, *pl. 7-11*. 19 Oct 1877.

9598. *Hydrurus i jego pokrewieństwo*. Monografija ... Kraków (w drukarni Uniwersytetu Jagiellońskiego, ...) 1882. Oct.
Publ.: Mai 1882 (Bot. Centralbl. 19-23 Jan 1882; Grevillea Sep 1882; J. Bot. Jan 1883; Nat. Nov. Jun(1) 1882; Bot. Zeit. 30 Jun 1882; Hedwigia Jul 1882), p. [1]-34, *1 pl.* *Copies*: G, H, PH. – Page 1: reprint t.p., p. 3-29 reprinted from Rozpr. Akad. umiej.,

Wydz. mat.-przyr. 10: 59-86. *pl. 2.* 1882, p. [31]-34 German resumé: *Hydrurus und seine Verwandtschaft.*
Ref.: Anon., Hedwigia 21: 148-149. Oct 1882.
Klebs, G., Bot. Zeit. 40: 683-687. Oct 1882 (rev.).
N.L., Bot. Zeit. 40: 433. 30 Jun 1882.
Richter, P.G., Bot. Centralbl. 13: 394-395. 1883 (rev. and summary).

9599. *Słownik polskich imign rodzajów oraz wyzszych skupień roślin poprzedzony historyczna rozprawa o źródłach* ... w Krakowie (Nakładem Akademii Umiejetnosci ...) 1900. 'Oct. (*Słown. polsk. imign*).
Publ.: 1900 (p. iv: 7 Jul 1899), p. [i]-iv, [1]-834, [835-836]. *Copy*: NY. – Materiały do historyi jezyka i dyalektologi polskiej. tom. 1.

Roster, Giorgio (x-1968), Italian botanist and horticulturist. (*Roster*).

HERBARIUM and TYPES: Unknown.

NOTE: A different Giorgio Roster, microbiologist and algologist, published *Il pulviscolo atmosferico ed i suoi microrganismi* in 1885; algae collected by him since 1874 are at FI. – See BM 4: 1736.

BIBLIOGRAPHY and BIOGRAPHY: BM 4: 1736; MW p. 419.

9600. *Le palme coltivate o provate in piena aria nei giardini d'Italia* ... Firenze (Tipografia di M. Ricci ...) 1915. (*Palme*).
Publ.: 1913-1915 (in journal; reprint 1915), p. [1]-128, *plates* (see below). *Copy*: FI. – Reprinted and to be cited from Bull. Soc. Tosc. Ortic. 38: 36-40, *pl. 2.* 31 Mar, 82-93. 30 Apr, 107-113, *pl. 3.* 31 Mar, 131-135, *pl. 4.* 30 Jun, 152-158, *pl. 5.* 31 Jul, 178-181. 31 Aug 1913, 194-200. 30 Sep, 218-225, *pl. 7-8.* 31 Oct, 265-269, *pl. 9.* 31 Dec 1913, 39: 13-17, *pl. 1.* 31 Jan, 54-61, *pl. 2.* 31 Mar 1914, 110-113. 15 Mai, 124-129, *pl. 3.* 1 Jun, 150-155. 1 Jul, 169-176. 1 Aug, 191-193. 1 Sep, 213-224, *pl. 4.* 1 Oct, 239-244. 1 Nov 1914, 40: 12-19. 1 Jan, 32-43. 1 Feb 1915.

Rostius, Christopher (1620-1687), German-born Swedish physician and botanist; Dr. med. Leiden 1657; "General-gouvernements-medicus" in the province of Skåne from 1658; first professor of medicine (incl. botany) and "archiater" at Lund University 1667-1676. (*Rostius*).

HERBARIUM and TYPES: Rostius' "Herbarium vivum" (now at LD) is dated 1610 (hence not collected by himself) but was probably brought together in the 1580's or 90's possibly (at least partly) from the Botanical Garden at Leiden. It contains 372 specimens (mainly cultivated plants) glued on to paper sheets bound in one volume and annotated with unitary or binary latin names, sometimes also with vernacular names in German. There are no localities. It was used by Rostius and his successors when teaching botany for the medical students. No doubt it was seen also by Linnaeus.

BIBLIOGRAPHY and BIOGRAPHY: KR p. 602.

BIOFILE: Almborn, O., Bot. Not. 133: 451. 1980.
Gertz, O., Ann. Rep. Lund Grammar School 1917-1918: [i], [1]-41. 1918. (detailed annotated list of Herbarium vivum); Nord. Tidskr. 1918: 565-578. 1918; Österr. bot. Z. 1918: 369-82 (biogr., list).
Lindman, S., Svenska män och kvinnor 6: 370. 1949.
Sjögren, O., Svenskt biogr. lexikon. Ny följd 9: 133-135. 1883.
Weibull, M. & Tegnér, E., Lunds univ. hist. 2: 143. 1868 (biogr., bibl.).

NOTE: The entry on Rostius was submitted by O. Almborn.

Rostkovius, Friedrich Wilhelm Gottlieb [Theophilus] (1770-1848), German (Brandenburg/Prussian) botanist; student of Willdenow; Dr. phil. Halle 1801; physician in Stettin. (*Rostk.*).

HERBARIUM and TYPES: The monograph of Juncus was based on material in B-Willd. − Letters to D.F.L. von Schlechtendal at HAL.

BIBLIOGRAPHY and BIOGRAPHY: Backer p. 499(epon.); Barnhart 3: 181 (d. 17 Aug 1848); BM 4: 1736; Hawksworth p. 185; Hegi 6(1): 92; Herder p. 188, 281; Hortus 3: 1203; Kelly p. 195; LS 22552; PR 7789-7790, 9026, ed. 1: 8745-8746, 9979; Rehder 1: 377; Stevenson p. 1256; TL-1/1275; Tucker 1: 608.

BIOFILE: Ascherson, P., Fl. Prov. Brandenburg 1: 11-12. 1864.
Killermann, S., Z. Pilzk. 6: 129-140. 1927 (biogr. sketch; identification of fungi descr. by Rostkovius; d. 17 Aug 1848).
Kraus, G., Bot. Gart. Univ. Halle 2: 36. 1894.
Raab, H., Schweiz. Z. Pilzk. 49: 154. 1971.

COMPOSITE WORKS: *Pilze* im J. Sturm, *Deutschlands Flora*, Abth. 3.
Vol. 4:
  Heft 5: 1-37, *pl. 1-16*. 1828.
  Heft 10: 37-68, *pl. 17-32*. 1830.
  Heft 16: 69-100, *pl. 33-48*. 1837.
  Heft 17: 101-132, *pl. 49-64*. Sep 1838.
Vol. 5:
  Heft 18: 1-36, *pl. 1-16*. 1839.
  Heft 21-24: 37-132, *pl. 17-48*. 1844.
Vol. 7(1):
  Heft 27-28: *pl. 1-24*. 1848 (A.E. Fürnrohr, Flora 31: 623-624. 14 Oct 1848.
Ref.: Schlechtendal, D.F.L. von, Bot. Zeit. 2: 314-316. 26 Apr 1844 (rev. Heft 21-22). K.M., Bot. Zeit. 3: 363-365. 30 Mai 1845 (rev. Heft 23-24).

EPONYMY: *Rostcovia* K.P.J. Sprengel (1830, *orth. var.*); *Rostkovia* Desvaux (1809). *Note*: *Rostkovites* P.A. Karsten (1881) is probably derived from *Rostkovia* Desvaux.

9601. Dissertatio botanica inauguralis *de Junco* quam consensu facultatis medicae ... ut gradum doctoris medicinae legitime obtineat die iv. april. mdccci. Publice defendet auctor Frid.-Guilielm. Theoph. Rostkovius neomarcho-driseniensis. Halae [Halle] (typis Frid. Aug. Grunerti patr.) [1801]. Oct. (*De Junco*).
Publ.: 4 Apr 1801, p. [i-iv], [1]-58, [59, theses], *pl. 1-2* (uncol. copp. F. Guimpel). *Copies*: HH, NY, PH. − Although defended at Halle, the dissertation was written under the supervision of C.L. Willdenow in Berlin, using his herbarium (now B-Willd.).
Commercial ed.: 1801, p. [i-iv], [1]-58, [59, err.], *pl. 1-2* (id.). *Copies*: G, MICH, NY, USDA(2).

9602. *Flora sedinensis* exhibens plantas phanerogamas spontaneas nec non plantas praecipuas agri Swinemundii ... Sedini [Stettin] (Formis Struckianis) 1824. Duod. (*Fl. Sedin.*).
Co-author: Ewald Luwig Wilhelm Schmidt (1804-1843) (often as W.E.L. Schmidt).
Publ.: 1824 (p. viii: 1 Oct 1824), p. [i]-viii, [1]-411, [1, err.], [1-7, index], *2 pl. Copies*: BR, G, HH, NY, USDA. − Appendix on p. 395. − Sedinum is Stettin (h.t. Szczecin), Swinemundium is Swinemünde (h.t. Swinoujście), cities in Prussian Pommerania.
Ref.: Anon., Flora 8: 555-557. 1925.

Rostock, Michael (1821-1893), German (Saxonian) botanist in the Oberlausitz. (*Rostock*).

HERBARIUM and TYPES: Some bryophytes in MW; further cryptogams at GLM.

BIBLIOGRAPHY and BIOGRAPHY: Barnhart 3: 181; BJI 1: 49; Futak-Domin p. 507; IH 2: (files); Tucker 1: 608.

BIOFILE: Gilbert, P., Comp. biogr. lit. deceased entom. 324. 1977.
Lindemann, E. von, Bull Soc. imp. Natural. Moscou 61: 63. 1886 (coll.).
Mylius, C., Bot. Monatschr. 2: 126. 1884 (on R's *Rubi Sachsens*).

Richter, K., Bautzen. Nachtr. 1926: 161-163. (n.v.).
Schade, A., Natura lusatica, Bautzen 5: 21-22, 36. 1961.

9603. *Phanerogamenflora von Bautzen* und Umgegend, nebst einem Anhange: Verzeichniss Oberlausitzer Kryptogamen [Dresden 1889].
Publ.: 1889, p. [1]-25. – Reprinted and to be cited from S.B. Isis, Dresden 1889 (Abh.): 3-25. 1889.
Ref.: Reiche, K., Bot. Jahrb. 11 (Lit.): 81-82. 31 Dec 1889 (rev.).

Rostovzev, Semen Ivanovich (ca. 1862-1916); Russian botanist. (*Rostovzev*).

HERBARIUM and TYPES: LE, MW; further material at C, H, K and TU (TAM).

BIBLIOGRAPHY and BIOGRAPHY: Barnhart 3: 181; BM 4: 1736; Biol.-Dokum. 15: 7670; Bossert p. 339; CSP 18: 311; Kelly p. 195; LS 22553-22555, 38568-38573; Stevenson p. 1256.

BIOFILE: Anon., Österr. bot. Z. 70: 312. 1921 (d.).
Lipschitz, S., Fl. URSS fontes nos. 902-905, 2332(5). 1975.
Shetler, S.G., Komarov Bot. Inst. facing 59. 1967.

9604. *Beiträge zur Kenntniss der Ophioglosseen*. – 1. Ophioglossum vulgatum L. ... Moskva 1892. Oct. (*Beitr. Kenntn. Ophiogl.*).
Publ.: 1892, p. [1]-108. *Copy*: HH. – Iz "Uchenykh Zapisok" Imp. Moskosk. Univ.

9605. *Isdanie Moskovskago Sel'skokhozayistvennago Instituta. S. Rostovtsev. Posobie k opredeleniyu paraziticheskikh gribov* po rasteniyam-khozyaevam. Moskva (Tipo.-lit. Vysoch. Utv. T-va I.N. Kushnerev i K;) 1896. Oct.
Publ.: 1896, p. [1]-41. *Copy*: M.

Rostrup, [Frederik Georg] Emil (1831-1907), Danish botanist (phytopathologist, mycologist); teacher of science and mathematics at Skaarup (Funen) 1858-1883; "Docent" (1883) lecturer (1889) and professor (1902) Royal Veterinary and Agricultural College, Copenhagen until 1907; Dr. phil. h.c. Copenhagen 1893. (*Rostr.*).

HERBARIUM and TYPES: C and CP; other material at B (mainly destr.), CGE, H, HBG, LD, MICH, M (Lich.), UPS, WRSL. – For a catalogue of the fungi in R's herbarium see J. Lind, *Dan. fung.*, Feb 1913, TL-2/4542. – R. contributed specimens to C. Roumeguère, *Fungi selecti exsiccati*. – Letters to W.G. Farlow at FH.

BIBLIOGRAPHY and BIOGRAPHY: Ainsworth p. 236, 320; Barnhart 3: 181 (b. 28 Jan 1831, d. 16 Jan 1907); BFM no. 1447; BJI 1: 49; BL 2: 49, 55, 57, 710; BM 4: 1736-1737, 8: 1095; Bossert p. 339; Christiansen p. 319; CSP 5: 301, 8: 786, 11: 226, 12: 632, 18: 311-312; GR p. 679, *pl.* [*35*]; GR, cat. p. 70; Hawksworth p. 185; Herder p. 433; IH 1: (ed. 2): 45, (ed. 3): 52, (ed. 6): 363, (ed. 7): 341, 2: (files); Jackson p. 148, 333, 334; Kelly p. p. 195, 256, 257; Kleppa p. 112, 113, 116, 121; KR p. 602; LS 6756, 22556-22946 38563-38566, 41656-41678 (impt. mycol. bibl.); Morren ed. 10, p. 47; NAF 7: 1085; PR 7791-7792, 10548; Rehder 5: 738-739; SBC p. 130; Stevenson p. 1256; TL-2/4185, 4189, 4542; Tucker 1: 608; Urban-Berl. 281, 312.

BIOFILE: Anon., Ber. deut. bot. Ges. 24: (89). 1907 (d.); Bot. Centralbl. 104: 272. 1907 (d.); Bot. Not. 1907: 47 (obit.); Hedwigia 46: (89). 1907; Nat. Nov. 29: 155. 1907 (d.); Österr. bot. Z. 57: 135. 1907; Verh. bot. Ver. Brandenburg 49: lxvii. 1908.
Candolle, A. de, Phytographie 406. 1880.
Christensen, C., Danske bot. litt. 1880-1911, 14-28. 1913 (portr., bibl.); Danske bot. hist. 1: 445-447, 696-711, 882. 1926 (biogr., portr.), 2: 327-356. 1926 (major bibl.; many biogr. refs.).
Häyren, E., Luonnon Ystävä 11(2): 17-18. 1907 (obit.).
Kneucker, A., Allg. bot. Z. 13: 72. 1907 (d.).
Kornerup, T., Overs. Medd. Grønland 1876-1912, p. 97. 1912.

Lind, J., Danish fungi herb. E. Rostrup 1-9. 1913.
Paulsen, O., Bot. Tidsskr. 45: 247. 1940 (portr.).
Raab, H., Z. Pilzk. 57: 67, 70. 1979 (brief biogr. note).
Ravn, F.K. et al., Ber. deut. bot. Ges. 26a: (47)-(55). 1908 (obit, bibl.); K. Veterin. Landbo Højskole, Festskrift 1908, p. 521.
Rosenvinge, L.K., Bot. Tidsskr. 28: 85-198. 1908 (obit., portr.).
Rydberg, P.A., Augustana coll. Libr. Publ. 5: 31: 1907.
Shear, C.L., Phytopathology 12(1): 1-3. 1922.
Warming, E., Bot. Tidsskr. 12: 200-204. 1881 (bibl.).
Whetzel, H.H., Outline hist. phytopath. 81-85. 1918.
Wittrock, V.B., Acta Horti Berg. 3(2): 104, *pl. 19. 1903 (portr.).*

COMPOSITE WORKS: (1) Co-author, J.M.C. Lange, *Consp. fl. Groenland.* 3(1), 3(2) (1889, 1891), see TL-2/4185.
(2) Identified fungi for J.M.C. Lange, *Nomencl. Fl. danic.* 1887, TL-2/4189.

EPONYMY: *Rostrupia* G. Lagerheim (1889).

9606. *Vejledning i den danske Flora.* En populaer Anvisning til at lære at kjende de danske Planter. Kjøbenhavn 1860. (*Dansk fl.*).
Ed. 1: 1860, p. [i]-vii, [1]-247, [248, err.].
Ed. 2: 1864, p. [i]-xviii, [1]-276.
Ed. 3: 1869, p. [i]-xviii, [1]-292.
Ed. 4: 1873, p. [i]-xxi, [1]-374.
Ed. 5: 1878, p. [i]-xix, [1]-446.
Ed. 6: 1882, p. [i]-xviii, [1]-422.
Ed. 7: 1888, p. [i]-xxiii, [1]-424.
Ed. 8: 1896, p. [i]-xxii, [1]-435.
Ed. 9: 1902, p. [1]-xx, [1]-445, 149 fig.
Ed. 10: 1906, p. [i]-xx, [1]-446, 149 fig. – Kjøbenhavn/Kristiania.
Editions by O. Rostrup, after E. Rostrup's death on 16 Jan 1907:
Ed. 11: 1912, p. [i]-xx, 1]-453, 139 fig. – Kjøbenhavn/Kristiania.
Ed. 12: 1917, p. [i]-xx, [1]-456, 139 fig. – id.
Ed. 13: 1922, p. [i]-xx, [1]-474, 143 fig. – id.
Ed. 14: 1925, p. [i]-xx, [1]-475, 143 fig. – Kjøbenhavn.
Editions by C.A. Jørgensen after O. Rostrup's death on 25 Jun 1933:
Ed. 15: 1935, p. [i]-xxii, [1]-477, 142 fig. – København.
Ed. 16: 1943, p. [i]-lxiii, [1]-496, 154 fig. – id.
Ed. 17: 1947, p. [i]-lxiv, [1]-525, 154 fig. – id.
Ed. 18: 1953, p. [i]-lxiv, [1]-527, 154 fig. – id. Four reprints until 1960.
Ed. 19: 1961, p. [1]-561, 141 fig. – København. – Four reprints until 1969.
Editions by A. Hansen , after C.A. Jørgensen's death on 14 Feb 1968:
Ed. 20: 1973, p. [1]-664, 141 fig. – København. – Four reprints until 1979.
Copies: For the above information we are indebted to A. Hansen.

9607. *Afbildning og Beskrivelse af de vigtigste Fodergræsser.* En Vejledning for Landmænd til at lære Græsarterne at kjende og den hensigtsmæstigste Maade at benytte dem paa ... Kjøbenhavn (P.G. Philipsens Forlag. Thieles Bogtrykkeri) 1865. Qu. (*Afb. Fodergraes.*).
Publ.: 1865, p [i-iv], [1]-61, *pl. 1-5* (uncol.), *1-16* (col.). *Copy*: NY.

9608. *Lichenes Daniae* eller Danmarks Laver, ... Kjøbenhavn (G.E.C. Gads Forlag ...) 1869. Oct.
Co-author: Jakob Severin Deichmann Branth (1831-1917).
Publ.: 1869, p. [i], [1]-158, *3-4. Copy*: BR. – Reprinted and to be cited from Bot. Tidsskr. 3: 127-284, *pl. 3-4.* 1869.
Ref.: Stizenberger, E., Bot. Zeit. 28: 309-310. 13 Mai 1870 (rev.).

9609. *Blomsterløse Planter.* Vejledning til Bestemmelse af de i Danmark hyppigst forekommende Svampe, Laver, alger og mosser ... Kjøbenhavn (P.G. Philipsens Forlag ...) 1869. Oct. (*Blomsterl. pl.*).

Publ.: 1869 (p. vi: Apr 1869), p. [i]-vi, [vii], [1]-156. *Copies*: C (inf. A. Hansen), H (inf. P. Isoviita).
Follow-up: see *Vejledning i den Danske Flora* anden del *Blomsterløse Planter* København (Gyldendalske Boghandel ...) 1904, by E. Rostrup with contributions by C. Jensen, L. Kolderup Rosenvinge, Sev. Petersen, p. [i]-xii, [1]-484, [483]. *Copies*: C, H. – ed. 2, 1925 (preface Jul 1925), p. [i*-ii*], [i]-x, [1]-592. *Copies*: C, H, HH. – Reprinted 1967.

9610. *Undersøgelser angaaende Svampeslægten Rhizoctonia* ... Kjøbenhavn (Bianco Lunos Kgl. Hof-Bogtrykkeri (F. Dreyer)) 1886. Oct.
Publ.: shortly after 21 Aug 1886 before 15 Oct 1886 (journal issue 1882(2) printing finished 21 Aug 1886; Bot. Centralbl. 18-22 Oct 1886.), p. [1]-21, *pl. 1-2* (col. liths.). *Copy*: G. – Reprinted and to be cited from Overs. kgl. Danske Vid. Selsk. Forh. 1886(2): 59-77, *pl. 1-2*, résumé: ix-xiv. Résumé also in Rev. mycol. 1887: 6-9.

9611. *Fungi Groenlandiae*. Oversigt over Grønlands Svampe ... Kjøbenhavn (Bianco Lunos Kgl. Hof-Bogtrykkeri (F. Dreyer)). 1888. Oct.
Publ.: 1888 (Bot. Centralbl. 13-17 Aug 1888; Hedwigia rd. Mai-Jun 1888), p. [i], [517]-590. *Copies*: G, NY. – Reprint with a new title and a separate t.p. of a paper entitled "Oversigt over Grønlands Svampe" published in and to be cited from Medd. Grønl. 3: 517-590. 1888. – See also Rev. mycol. 1888: 217-218.
Addenda: 1892 ("1891" on p. [591], journal publ. 1893; Christensen no. 178 dates it 1892), p. [591]-643. *Copy*: NY. – Reprinted and to be cited from Medd. Grønl. 3(3): 591-643. 1893 (publ. 1892?).
Ref.: Zimmermann (Chemnitz), Bot. Centralbl. 36: [3]-7, 1888 (reprint of diagn. new taxa).
Anon., Hedwigia 27: 209-210. Jul-Aug 1888 (rev.).

9612. *Afbildning og Beskrivelse af de farligste Snyltesvampe* i Danmarks Skove ... Kjøbenhavn (P.G. Philipsens Boghandel ...) 1889. Qu. (*Afb. farl. snyltesv.*).
Publ.: Jan-Feb 1889 (p. iv: Jan 1889; Nat. Nov. Mar(2) 1889; Hedwigia rd. Mar-Apr 1889), p. [i-iv], 1]-30, [31, index], *pl. 1-8* (col.). *Copies*: FH, Stevenson, USDA. – Christensen (1926), no. 148 (cites reviews).

9613. *Ustilagineae Daniae* ... Kjøbenhavn (Hoffensberg & Trap's Etabl.). 1890. Oct. (*Ustil. Daniae*).
Publ.: 1890 (Nat. Nov. Mai(2) 1890), p. [1]-54. *Copies*: FH, G, NY. – Reprinted from Bot. Foren. Festskrift p. 117-168. 1890. Christensen (1926) no. 159.
Ref.: Rosenvinge, L.K., Bot. Centralbl. 43: 388. 18 Sep 1890 (rev.).

9614. *Øst-Grønlands Svampe* ... Kjøbenhavn (Bianco Lunos Kgl. Hof-Bogtrykkeri (F. Dreyer)). 1894. Oct. (*Øst-Grønl. Svamp.*).
Publ.: 1894 (t.p. preprint; Bot. Jahrb. 3 Apr 1895), p. [1]-39. *Copies*: H, NY. – Preprinted or reprinted from Medd. Grønland 18: 46-81. 1896 (? 1894). Christensen (1926) no. 202a.

9615. *Hussvampen* en Vejledning for Bygningshaandvaerkere og til Brug i tekniske Skoler ... Kjøbenhavn (Det Nordiske Forlag ...) 1898. Oct. (*Hussvamp.*).
Publ.: 1898 (Nat. Nov. Mai(1) 1898), frontisp., p. [i-iv], [1]-75, [76, cont.]. *Copy*: FH. – See Christensen (1926) no. 235 for further references.

9616. *Fungi from the Faeröes* ... Copenhagen (printed by H.H. Thiele) 1901. (*Fung. Faeröes*).
Publ.: 30 Apr 1901 (reprint so dated; Bot. Centralbl. 24 Jul 1901), cover (with dated t.p. text), p. [304]-316, map. *Copies*: FH, NY. – Preprinted or reprinted from "Botany of the Faröes" 1: 304-316. 1901.
Ref.: Lindau (Berlin), Bot. Centralbl. 88: 338-339. 22 Dec 1901 (rev.).

9617. *Plantepatologi* Haandbog i Læen om Plantesygdomme for Landbrugere Havebrugere og Skovbrugere ... København (Det Nordiske Forlag...) 1902. Oct. (*Plantepatologi*).

Publ.: 1902, p. [i-viii], [1]-640. *Copy*: H (inf. P. Isoviita).

9618. *Norske Ascomyceter* i Christiania Universitetets botaniske Museum ... Christiania [Oslo] (i kommission hos Jacob Dybwad ...) 1904. Oct. (*Norske Ascomyc.*).
Publ.: 1904 (p. 44: printed 21 Apr 1904; Nat. Nov. Aug(1) 1904), p. [1]-44. *Copies*: FH, H, NY. − Issued in Vid.-Selsk. Skr., Mat.-nat. Kl. 1904(4). − See also Christensen (1926) no. 274.

9619. *Norges Hymenomyceter* af Axel Blytt efter Forfatterens død gennemset og afsluttet af E. Rostrup ... Christiania [Oslo] (i kommission hos Jacob Dybwad ...) 1905. Oct. (*Norg. Hymenomyc.*).
Original author: Axel Gudbrand Blytt (1843-1898), mss. completed and edited by E. Rostrup.
Publ.: 1905 (p. 164: printed 3 Mai 1905), p. [1]-164. *Copy*: FH. − Issued in Vid.-Selsk. Skr., Mat.-nat. Kl. 1904(6). 1905. See also Christensen (1926) no. 280.

9620. *Fungi collected by H.G. Simmons on the 2nd Norwegian Polar Expedition, 1898-1902* determined by E. Rostrup ... Kristiania [Oslo] (printed by A.W. Brøgger) 1906. Oct. (*Fung. Simmons*).
Collector: Herman Georg Simmons (1866-1943).
Publ.: 1906 (p. 10: printed Oct 1906; Bot. Centralbl. 12 Feb 1907), p. [1]-10. *Copies*: BR, FH, USDA. − Report of the second Norwegian Arctic Expedition in the "Fram" 1898-1902, no. 9.

Rostrup, Ove Georg Frederik (1864-1933), Danish botanist and phytopathologist; director "Statsanstalten Dansk Frøkontrol" 1891-1902; scientist at the department of plant pathology of the Agricultural college 1902-1933; son of E. Rostrup. (*O. Rostr.*).

HERBARIUM and TYPES: CP; further material at C.

NOTE: Married the entomologist-cecidiologist and collector Sofie Rostrup (1857-1940); Ove Rostrup's sister Asta Rostrup, collected also for her father E. Rostrup.

BIBLIOGRAPHY and BIOGRAPHY: Barnhart 3: 182 (b. 29 Sep 1864, d. 25 Jun 1933); Christiansen p. 319; IH 1 (ed. 2):45, (ed. 3): 52, (ed. 6): 363, (ed. 7): 341, 2: (files); LS 38567, 41679; Rehder 4: 489; Stevenson p. 1256.

BIOFILE: Anon., Hedwigia 74: (139). 1934 (d.).
Christensen, C., Dansk bot. Litt. 1880-1911. 134-135. 1913 (portr., bibl.), 1912-1939. 25. 1940 (bibl., portr.); Dansk bot. Hist. 1: 882 [index]. 1926.
Ferdinandsen, C., Bot. Tidsskr. 42(4): 406-407. 1933 (obit., portr.), Friesia 1(2): 137-138. 1933 (obit., portr.), Dansk bot. Ark. 8(8): 1-4. 1935 (biogr. sketch, portr.).
Lind, J., Danish fungi herb. E. Rostrup 37. 1913.

9621. *Bidrag til Danmarks svampflora* i-ii ... Dansk botanisk Arkiv 2(5), 8(8), 1916, 1935. Oct.
1: Aug 1916, p. [1]-56, *3 pl. Copy*: U. − Issued as Dansk bot. Ark. 2(5). 1916.
2: *1935, p.* [i], [1]-74. *Copy*: U. − Issued as Dansk bot. Ark. 8(8), 1935, with an introductory biographical sketch by C. Ferdinandsen, an English summary, and an alphabetical index to *Bidrag* i and ii by N.F. Buchwald.

Rot von Schreckenstein, Friedrich, Freiherr, zu Immendingen und Bilafingen(1753-1808) "churfürstlich-salzburgischer geheimer Rath" and botanist in Bavaria and Austria. (*Rot von Schr.*).

HERBARIUM and TYPES: Unknown.

BIBLIOGRAPHY and BIOGRAPHY: Barnhart 3: 182 (b. 17 Oct 1753, d. 1808); BM 4: 1738; DTS 1: 247, 248; PR 7793, ed. 1: 8747; Rehder 1: 375.

BIOFILE: Martens, G. von & C.A. Kemmler, Fl. Württemb. ed. 2. 781. 1865.

9622. *Flora der Gegend um den Ursprung der Donau* und des Neckars; dann vom Einfluss der Kinzig in den Rhein. Herausgegeben von den Verfassern der Verzeichnisse der Naturprodukte dieser Gegenden. Donaueschingen (gedruckt und im Verlag bey Aloys Wilibald Hofbuchdrucker) 1804. Oct. (*Fl. Ursprung Donau*).
Co-authors: Josef Meinrad von Engelberg (vols. 2, 3), Johann Nepomuk von Renn (vol. 3).
1: 1804, p. [i-viii], [1]-367, Nachtr. [369]-389, [1-20, index], [1-8, Anleit. Sammler].
2: 1805, p. [1]-645.
3: 1807, p. [1]-536.
4: 1814, p. [1]-567. – "Herausgegeben von Joseph Meinrad von Engelberg ..." Donaueschingen (id.) 1814. Oct.
Copies: GOET (1, 2, inf. G. Wagenitz), NY (1-4).

Rota, Lorenzo (1819-1855), Italian botanist and physician; sometime botanical assistant in the University of Pavia; ultimately practicing physician at Bergamo; died of cholera. (*Rota*).

HERBARIUM and TYPES: In the municipal Museum of Bergamo (A. de Candolle 1880); it is not clear whether this herbarium is still extant. Further material is at FI (inf. C. Steinberg).

BIBLIOGRAPHY and BIOGRAPHY: AG 2(1): 270, 12(1): 288; Barnhart 3: 182 (d. 6 Aug 1855); BL 2: 369, 710; BM 4: 1737; Bossert p. 339; CSP 5: 301; DTS 1: 248, 6(4): 43; Herder p. 179; Hortus 3: 1203; Kew 4: 522; PR 7794-7795, ed. 1: 8748; Rehder 1: 427, 2: 144; Saccardo 1: 142, 2: 94, cron. p. xxxi; SBC p. 130; Tucker 1: 608.

BIOFILE: Anon., Bot. Zeit. 13: 656. 1855 (d.); Bull. Soc. bot. France 2: 496. 1855 (d. of cholera; brief obit.); Flora 38: 592. 1855 (d.).
Candolle, Alph. de, Phytographie 444. 1880.
Sayre, G., Bryologist 80: 516. 1977.

EPONYMY: *Rotaea* Cesati ex Schlechtendal (1851).

9623. *Enumerazione delle piante fanerogame rare della provincia bergamasca* ... Pavia (nella tipografia Fusi e C.) 1843. Oct. (*Enum. piante bergam.*).
Publ.: 1843, p. [1]-38. *Copy*: FI.

9624. *Prospetto della flora della provincia di Bergamo* ... Bergamo (dalla tipografia Mazzoleni) 1853. Oct. (*Prosp. fl. Bergamo*).
Publ.: Apr 1853 (t.p.), p. [1]-104. *Copies*: FI, G, HH, MO, NY.
Ref.: Schlechtendal, D.F.L. von, Bot. Zeit. 12: 383-386. 2 Jun 1854.

Roth, Albrecht Wilhelm (1757-1834), German (Oldenburg) physician and botanist; studied medicine at Halle 1775-1778 and Erlangen 1778; Dr. med. Erlangen 1778; practicing physician at Vegesack nr. Bremen from 1779; regional physician from 1781. (*Roth*).

HERBARIUM and TYPES: B, B-Willd.; other material at BM, BREM (bryo.), DUIS, FH, GOET, HAL, JE, LD, M, PH (algae), S, TCD, W, WELT. – Roth's main herbarium of some 20.000 specimens went originally to the Natural History Museum at Oldenburg (Germany). It was transferred to B in 1925/1926, where it was incorporated in the main herbarium which was for the greater part destroyed in 1943. However, the Roth material in the Willdenow herbarium, the pteridophytes and several types are extant. Roth issued exsiccatae: "*Herbarium vivum plantarum officinalium*, nebst einer Anweisung, Pflanzen zum medicinischen Gebrauche zu sammeln" 8 Hefte, Hannover 1785-1787 (for details see Buchenau 1868). – For his *Herbarium vivum plantarum officinalium* see e.g. Hertel (1982).

NOTE: A.W. Roth's son, C.W. Roth (1810-1881) published on agricultural botany and issued a series of exsiccatae *Landwirtschaftliche Pflanzensammlung*, see GR p. 126.

BIBLIOGRAPHY and BIOGRAPHY: ADB 29: 305 (W.O. Focke); AG 2(1): 118, 12(1): 268; Backer p. 500 (epon.); Barnhart 3: 182 (b. 6 Jan 1757, d. 16 Oct 1834); BM 4: 1737; Bossert p. 339; CSP 5: 301, 8: 786, 12: 633; De Toni 1: cxix, 2: cii; Dryander 3: 4, 13, 47-48, 50, 85, 91, 152, 158, 162, 177, 219, 262, 301, 357, 383, 388, 413; DTS 1: 248; Frank 3(Anh.): 86; GF p. 74; GR p. 42; GR, cat. p. 70; Hawksworth p. 185; Hegi 6(2): 1228; Herder p. 474 [index]; Hortus 3: 1203; IF p. 727; Jackson p. 113, 293; Kelly p. 195; Kew 4: 523; Lasègue p. 427, 503; LS 22950-22954a, 38576; Moebius p. 65, 86, 94, 283; MW p. 419, suppl. p. 295; NI 1: 182, 2: no. 1683; PR 7796-7808, ed. 1: 8749-8756; Rehder 5: 739; RS p. 137-138; SBC p. 130; SK 4: ccx; SO 522, 523, 680, 681; Stevenson p. 1256; SY p. 10, 13, 29, 34; TL-1/1102-1110; TL-2/3120; Tucker 1: 608; Urban-Berl. p. 389, 415; Zander ed. 10, p. 709, ed. 11, p. 808.

BIOFILE: Anon., Flora 11: 704. 1828 (50 yrs Dr. med.), 17: 753-763. 1834 (obit.).
Bridson, G.D.R. et al., Nat. hist. mss. res. Brit. Isl. 255.57. 1981.
Buchenau, F., Bot. Zeit. 26: 305-310. 1868 (Roth's herbarium was sold in 1840 to Oldenburg; in 1868 it was "im grossherzoglichen Naturaliencabinet", 145 fascicles, completely in original condition, circa 20.000 specimens).
Burkill, I.H., Chapt. hist. bot. India 14, 16, 18. 1965.
Diels, L., Notizbl. Bot. Gart. Mus. Berlin-Dahlem 9(88): 888-889. 1926 (herb. Roth bought from Naturh. Mus. Oldenburg).
Focke, W.O., Abh. naturw. Ver. Bremen 19: 280-289. *pl.* 7. 1908 (biogr., portr.).
Heineken, P., Biographische Skizzen verstorbener Bremischer Aerzte und Naturforscher 393-432. 1864 (biogr.).
Hertel, H., Mitt. Bot. München 18: 320. 1982 (on exsicc.).
Jessen, K.F.W., Bot. Gegenw. Vorz. 371, 373-374. 1884.
Koster, J.Th., Taxon 18: 557. 1969.
Mears, J.A., Proc. Amer. philos. Soc. 122: 170. 1978 (some algae labeled A.W. Roth in Muhlenberg herb. PH).
Ménil, A.J. du, Arch. für Pharmacie 117: 345-346. 1851 (evaluation).
Milner, J.D., Cat. portr. Kew 93. 1906.
O., Dict. Sci. méd., Biogr. méd. 7: 56-57. 1825.
Pritzel, G.A., Linnaea 19: 460. 1846.
Sayre, G., Bryologist 80: 516. 1977.
Schmid, G., Goethe u.d. Naturw. 566. 1940.
Stafleu, F.A., Taxon 12: 76. 1963 (on *Tent. fl. Germ.*).
VanLandingham, S.L., Cat. diat. 4: 2366. 1971, 6: 3586. 1978, 7: 4224. 1978.
Wagenitz, G., Index coll. princ. herb. Gott. 138. 1982.
Zuchold, E.A., Jahresb. naturf. Ver. Halle 5: 49. 1853.

COMPOSITE WORKS: See J.J. Roemer and J.A. Schultes, *Syst. veg.* for publication of many of Roth's new taxa of the *Novae plantarum species*; they are identified in the book as having been supplied by Roth; a manuscript copy was available to Roemer and Schultes.

EPONYMY: *Rothia* Lamarck (1792); *Rothia* Persoon (1807, *nom. cons.*). *Note*: the etymology of *Rothia* Schreber (1791, *nom. rej.*) and *Rothia* Borkhausen (1792) could not be determined, but likely honor this author; X *Rothara* Hort. honors the orchid grower Richard Roth.

9625. *Verzeichniss derjenigen Pflanzen*, welche nach der Anzahl und Beschaffenheit ihrer Geschlechtstheile nicht in den gehörigen Klassen und Ordnungen des Linneischen Systems stehen, nebst einer Einleitung in dieses System ... Altenburg (in der Richterischen Buchhandlung) 1781. Oct. (*Verz. Pfl.*).
Publ.: 1781 (p. vi: 14 Jul 1780: third Sunday after Easter), p. [i*], [i]-vi, [1]-216. *Copies*: H (Univ. Libr.) (inf. P. Isoviita), M (inf. I. Haesler); IDC 5689.
Additamentum: see A.W. Roth, Beitr. Bot. 2: 101-124. 1783.

9626. *Beyträge zur Botanik* ... Bremen (bey Georg Ludewig Förster) 1782-1783, 2 vols. Oct. (*Beytr. Bot.*).
1: 1782(preface 22 Aug 1781; GGA 15 Oct 1782; NZgS 22 Sep 1783), p. [i]-viii, [1]-132.
2: 1783 (preface 29 Nov 1782; GGA 15 Nov 1783; NZgS 25 Sep 1783), p. [i]-viii, [1]-190. – Bremen (bey Georg Ludwig [sic] Förster) 1783. *Copies*: G, GOET (inf. G. Wagenitz), M (inf. I. Haesler); IDC 5690.

9627. *Botanische Abhandlungen und Beobachtungen* ... Nürnberg (bei Johann Jacob Winterschmidt) 1787. Qu. (*Bot. Abh. Beobacht.*).
Publ.: 1787, p. [1]-68, *pl. 1-12* (handcol. copp.). *Copies*: G, NY, PH, USDA; IDC 297. – Rev. Allg. Lit. Zeit. 1788: 493. 5 Jun 1788, "das elende Papier und der schlechte Abdruck des Textes ...".

9628. Alberti Guilielmi Rothii, ... *Tentamen florae germanicae* ... Lipsiae [Leipzig] (in Bibliopolio. I.G. Mülleriano) 1788. Oct. 3 vols. (*Tent. fl. Germ.*).
1: Feb-Apr 1788 (p. xvi: 21 Jan 1788; Stafleu 1963), p. [i]-xvi, [1]-560, [561-568]. – Continens enumerationem plantarum in Germania sponte nascentium.
2(1): Apr 1789 (p. ii: 25 Mar 1789; available 26 Apr 1789 at Leipzig Easter Fair), p. [i*-iv*], [i]-ii, [1]-624. – Continens synonyma et adversaria ad illustrationem florae germanicae.
2(2): 1793, p. [i], [1]-593, [594]. – Id.
3(1)(1): Jun-Sep 1799 (p. viii: 14 Sep 1798; TL-1), p. [i], [1]-102. – Continens synonyma et adversaria ad illustrationem florae germanicae. Lipsiae (in Bibliopolo Gleditschiano) 1800.
3(1)(2): Jan-Apr 1800 (Pichi-Sermolli 1952), p. [i]-viii, 103-578, [1-3, err.].
Copies: BR, G, H, HH, MO, NY, US, USDA; IDC 502. – The BR and NY copies have the provisional t.p. for Tomus iii, part i, which accompanied the first 102 p. The remaining part of pars 1, p. 103-578 came out accompanied by the definitive t.p. (dated 1800) for *3(1)*. There was no *3(2)* which was to have contained the lichens and fungi.
Second edition (Linnaean classes 1-13): see below, *Enum. pl. phaen. Germ.*
Ref.: Pichi-Sermolli, R.E.G., Webbia 8: 437-439. 1952.

9629. *Catalecta botanica* quibus plantae novae et minus cognitae describuntur atque illustrantur ... Lipsiae [Leipzig] (in Bibliopolo I.G. Mülleriano [fasc. 2, 3: ... Io.Fr. Gledischiano]). 1797-1806. Oct. 3 parts. (*Catal. bot.*).
1: Jan-Feb 1797 (preface 21 Mar 1796; GGA 11 Feb 1797), p. [i]-viii, [1]-244, [1-2, index pl.] [1-8 index names], *pl. 1-8* (handcol. copp. Jac. Sturm; plates combined 1/2, 3/4, 5/6, 7/8; some pl. by F.G. Hayne).
2: 1800 (p. [viii]: 24 Feb 1799), p. [i-x], [1]-258, [1-2, add.], [1-2 index icon.], [1-5 index, 6-7 err., 8 note], *pl. 1-9* (col. copp. C.F. Mertens, 6/7 and 8/9 comb.).
3: Jan-Jun 1806 (p. [viii]: 12 Sep 1805; mentioned in list Mich. Messe 1805 Sep-Oct), p. [i-viii], [1]-350, [1-2, index pl.], [1-6, index] [1, err.], *pl. 1-12* (handcol. copp. C.F. Mertens).
Copies: BR (photocopy), G, H-UB, MICH, MO, PH, USDA; IDC 296.

9630. *Bemerkungen über das Studium der cryptogamischen Wassergewächse*, ... Hannover (bei den Gebrüdern Hahn) 1797. Oct. (*Bemerk. crypt. Wassergew.*).
Publ.: Feb-Aug 1797 (p. 12: 6 Jan 1797; GGA 2 Sep 1797), p. [1]-109, [1, err.]. *Copies*: G, MO, NY; IDC 5691.

9631. *Neue Beyträge zur Botanik* ... Erster Theil. Frankfurth am Mayn (bei Friedrich Wilmans) 1802. †. (*Neue Beytr. Bot.*).
Publ.: 2 Mai 1802 (p. x: 21 Jan 1802; SY p. 3: Easter Fair 1802), p. [i]-xii, [1]-351.
Copies: G, H-UB, USDA; IDC 5692.
Errata: Sep-Dec 1802, p. [1-2], p.[2] dated 28 Aug 1802. – It is not absolutely certain that copies of the book were indeed available at the 1802 Easter Fair: the listing may have been an announcement. The errata, may also have been issued after the book itself had been issued; the wording on p. [2], however, leaves room for doubt.

9632. *Anweisung Pflanzen* zum Nutzen und Vergnügen *zu sammeln* und nach dem Linneischen Systeme zu bestimmen ... zweite umgearbeitete Auflage. Gotha (in der Ettingerschen Buchhandlung) 1803. Oct. (*Anweis. Pfl. sammeln*).
Ed. 1: 1778 (preface: 6 Jan 1778), 184 p. (copy at H-UB, n.v.).
Ed. 2: 1803 (p. xvi: Dec 1802), p. [i]-xvi, [1-3], [1]-300. *Copy*: H (Univ. Libr.) (Inf. P. Isoviita).

9633. *Botanische Bemerkungen und Berichtigungen* ... Leipzig (in Joachims literarischem Magazin) 1807. Oct. (*Bot. Bemerk. Bericht.*).
Publ.: 1807 (p. vi: Jul 1806; announced for Easter Messe 1806), p. [i]-xiv, [xv, err.], [1]-216, *1 pl.* (handcol. copp.). *Copies*: USDA; IDC 5693.

9634. *Beantwortung* der, von der Botanischen Gesellschaft aufgegebenen *Preissfrage*: Was sind Varietäten im Pflanzenreiche und wie sind sie bestimmt zu Erkennen? Nebst beygefügtem Verzeichnisse der gewöhnlichen in Deutschland vorkommenden Varietäten ... Regensburg (gedruckt bey Johann Baptist Rotermundt) 1811. Oct. (*Beantw. Preissfr.*).
Publ.: 1811, p. [1]-46. *Copy*: GOET (inf. G. Wagenitz).

9635. Alberti Guilielmi Roth, ... *Novae plantarum species* praesertim Indiae orientalis. Ex collectione doct. Benj. Heynii. Cum descriptionibus et observationibus. Halberstadii (sumptibus H. Vogleri) 1821. Oct. (*Nov. pl. sp.*).
Collector: Benjamin Heyne (x-1819).
Publ.: Apr 1821 (p. iv: "ad Portum 1820"; Flora 7 Mai 1821: "so eben erschienen"; Beck 15 Jul 1821), p. [i]-iv, [1]-411, [412, err.]. *Copies*: G, H, HH, M, MO, NY, USDA; IDC 400. – The names were for the greater part prepublished by Roth in Roemer et Schultes' Syst. Veg. vols. 3-5, 1818-1820; see TL-2/9408. This may be of importance in competition with Roxburgh names. A copy of Roth's manuscript was made available by him to R. & S. – Heyne had originally intended to present this set of plants to Willdenow but the latter had died before Heyne's return. Heyne then gave the plants to Roth with the result that the set ended up in the general herbarium of Berlin (B), rather than in the (still extant) Willdenow herbarium at B.
Facsimile ed.: 1975, New York (Oriole Editions), ISBN 0-88211-079-9, p. [i*-iv*, 1975 t.p.'s], [i]-iv, [1]-411, [412]. *Copy*: FAS.
Ref.: Anon., Flora 4: 271-272. 7 Mai (first announcement), 6: 465-477, 481-492. 1823 (extensive review).
Stafleu, F.A., Taxon 24: 685. 1975.

9636. Albert. Guil. Roth, ... *Enumeratio plantarum phaenogamarum in Germania* sponte nascentium. Pars prima. Sectio prior. (classis i-v) [... posterior. (classis vi-xiii)]. Lipsiae [Leipzig] (sumtibus) J.F. Gleditsch) 1827, 2 parts. Oct. †. (*Enum. pl. phaen. Germ.*).
1(1): Oct-Nov 1827 (p. v: 6 Jan 1827; see Flora 10(2), Beibl. 1: 105-106. Dec 1827, but also 10(2): 688. 21 Nov 1827), p. [i]-iv, [1]-1015.
1(2): Oct-Dec 1827, p. [i], [1]-642.
Copies: G, HH, M, NY, PH, USDA; IDC 5398. – Remained incomplete. The manuscript of the second part was ready but the publisher went bankrupt. Roth changed his plan, shortened the descriptions and the synonymy and issued the whole as the *Manuale botanicum*, see below. (Anon., Flora 17(2): 761. 1834).
Ref.: Anon., Flora 13(1), Erg.-Bl. 81-98. 1830.
[Schlechtendal, D.F.L. von], Linnaea 3 (Litt.): 1-2. Jan 1828 (rev.).

9637. *Manuale botanicum* peregrinationibus botanicis accomodatum. Sive prodromus enumerationis plant. phaenogam. in Germania sponte nascentium ... Fasc. i [-iii] ... Lipsiae [Leipzig] (in bibliopolio Hahniano) 1830. 16-mo. (*Man. bot.*).
1 (classis i-viii): Jan-Feb 1830 (p. vi: late Oct 1829; Flora Jan-Mar 1830), p. [i]-vi, [1]-578.
2 (classis ix-xvi): 1830, p. [i], [579]-979.
3 (classis xvii-xxii): Sep-Oct 1830 (Flora Sep-Oct 1830), p. [i], [981]-1467.
Copies: BR, G, HH, M, MO, NY, USDA.

Roth, Ernst [Carl Ferdinand] (1857-1918), German (Berlin/Prussian) botanist and librarian; Dr. phil. Berlin 1883 (student of P. Ascherson); assistant at the Bot. Mus. Berlin 1883-1886; librarian at the Berlin Royal Library (as assistant 1886, later librarian); head librarian of the University Library, Halle 1891; head of the library of the "Leopoldina" 1904. (*E. Roth*).

HERBARIUM and TYPES: GOET (German material).

BIBLIOGRAPHY and BIOGRAPHY: Barnhart 3: 182 (b. 13 Aug 1857, d. 5 Sep 1918); Biol.-Dokum. 15: 7671; BJI 1: 49; BM 4: 1737; Christiansen p. 40, 41, 319; CSP 12: 633, 18: 314; De Toni 2: cii; DTS 1: 248, 397; GR p. 174; Herder p. 433; IH 2: (files); LS 38577; Moebius p. 329; Morren ed. 10, p. 6; Rehder 1: 349, 5: 739; TL-1/932; TL-2/6984; Tucker 1: 608; Zep-Tim. p. 96.

BIOFILE: Anon., Bot. Centralbl. 47: 352. 1891 ("Custos" univ. libr. Halle); Bot. Jahrb. 14 (Beibl. 31): 18. 1891 ("Custos" univ. libr. Halle); Bot. Not. 1919: 152 (d.); Österr. bot. Z. 68: 108. 1919 (d.), 41: 396. 1891 ("Custos" univ. libr. Halle); Deut. biogr. Jahrb. 21(1917-1920): 702. 1928 (further biogr. lit.); Hedwigia 61: (114). 1919 (d.).
Fischer-Benzon, R.J.D. von, *in* P. Prahl, Krit. Fl. Schlesw.-Holst. 2: 51. 1890.
Harms, H., Verh. bot. Ver. Brandenburg 62: 35. 1920 (obit.).
Roth, E.C.F., Pfl. Westküste Eur. [53]. 1883 ("vita", in diss.; studied 1879-1880 in Strassburg with e.g. A. de Bary; 1880-1883 in Berlin with e.g. Ascherson, Eichler and Schwendener).
Urban, I., Bot. Jahrb. 14 (Beibl. 32): 28. 1891 (asst. Berlin 1883-1886).
Wagenitz, G., Index coll. princ. herb. Gott. 7, 138. 1982.

HANDWRITING: Wagenitz, G., Index coll. princ. herb. Gott. 208. 1982.

9638. *Über die Pflanzen, welche den atlantischen Ocean auf der Westküste Europas begleiten.* Eine pflanzengeographische Skizze. Botanische Inaugural-Dissertation zur Erlangung der philosophischen Doctorwürde mit Genehmigung der Philosophischen Facultät der Friedrich-Wilhelms-Universität zu Berlin öffentlich zu verteidigen am 12. November 1883 von Ernst Roth aus Berlin. Hülfsarbeiter am kgl. Bot. Museum ... Berlin 1883. Oct. (*Pfl. Westküste Eur.*).
Thesis issue: 12 Nov 1883, p. [1]-52, [53-54, vita, theses]. *Copies*: G, GOET (inf. G. Wagenitz).
Completed, commercial edition: late 1883 or 1884 (Bot. Zeit. 31 Oct 1884; Nat. Nov. Sep(1) 1884), p. [i], [132]-181. *Copies*: H, MO. – Reprinted and to be cited from Abh. bot. Ver. Brandenburg 25: 132-181. 1883.
Ref.: Pax, F., Bot. Jahrb. 5(Lit.): 68-69. 1884.

9639. *Additamenta ad Conspectum florae europaeae* editum a cl. C.F. Nyman. Beiträge zu C.F. Nyman's Conspectus florae Europaeae ... Berlin (Haude– & Spener'sche Buchhandlung (F. Weidling)). 1886. Oct. (*Add. Consp. fl. eur.*).
Publ.: 1-23 Dec 1885 (t.p. 1886; p. 3: Nov 1885; Nat. Nov. Dec(2) 1885; Bot. Zeit. 25 Dec; Bot. Centralbl. 4-8 Jan 1886; ÖbZ Jan 1886; TL-1), p. [1]-46, [48, colo.]. *Copies*: G(2), H, HH, M, MO, NY, PH, US. – See also under Carl Fredrik Nyman (1820-1893), TL- 2/6984, for *Consp. fl. eur.*
Ref.: Roth, E., Bot. Centralbl. 27: 291. 6-10 Sep 1886.

9640. *Die Verbreitungsmittel der Pflanzen* ... Hamburg (Verlagsanstalt und Druckerei A.G. (vormals J.F. Richter), ...) 1896. Oct. (*Verbreitungsmitt. Pfl.*).
Publ.: 1896, p. [1]-50. *Copy*: M (inf. I. Haesler). – Issued as Samml. gemeinverst. wiss. Vortr. 242; pagination in ser. 2(11), Hefte 241-264: (51)-(98).

Roth, Georg (1842-1915), German (Hessen) forester and botanist (bryologist); studied forestry at Giessen; "Forstassessor" and "Forstrevisor" in Darmstadt, later "Rechnungsrath"; from 1887 in retirement at Laubach (Hessen) dedicating himself entirely to bryology; Forstrat 1898; Dr. h.c. Giessen 1907. (*G. Roth*).

HERBARIUM and TYPES: S; further material (bryoph.) at B, BM, DPU, H, M, PC. –

Roth's herbarium was acquired by S; his library and original drawings (of ca. 8000 species of musci) were offered for sale by Oswald Weigel (s.d.). At least part of these drawings (incl. the plates for the unpublished later volume or volumes of Aussereur. Laubm.) were at MICH at the time of writing this entry (Jan 1981). – 47 letters from R. to Brotherus at H (Univ. Libr.).

BIBLIOGRAPHY and BIOGRAPHY: Barnhart 3: 182 (b. 23 Mar 1842, d. 5 Dec 1915); BJI 2: 149; BM 4: 1737; Bossert p. 339; Christiansen p. 129; CSP 18: 314; IH 2: (files); Kew 4: 523; Lenley p. 355; MW p. 419; Rehder 4: 256; SBC p. 130; Urban-Berl. p. 312.

BIOFILE: Anon., Bot. Not. 1916: 196 (d.); Hedwigia 57: (151). 1916 (d.); Magy. bot. Lap. 15: 116. 1916 (d.); Nat. Nov. 29: 291. 1907 (dr. h.c. Giessen).
Jennings, O.E., Bryologist 19(5): 81. 1916 (d.).
Röll, J., Allg. bot. Z. 21: 132-133. 1915 (obit.); Hedwigia 58: 9-14. 1916 (souvenirs).
Sayre, G., Bryologist 80: 516. 1977 (coll.).
Weigel, O., Herbarium 55: 45. 1921 (herbarium for sale, 105 fasc., mainly mosses); Lager-Katalog ser. 2. 173. s.d. (contains library G. Roth; offer of his "Mikroskopische Zeichnungen von Laubmoosen; drawings of 8000 species on 868 original plates).

9641. *Die europäischen Laubmoose* beschrieben und gezeichnet von Georg Roth, ... Leipzig (Verlag von Wilhelm Engelmann) [1903] 1904-1905. Oct. (*Eur. Laubm.*).
Publ.: in 11 "Lieferungen" and 2 volumes, 1903-1905. *Copies*: BR, FH(2), FI, G, H, M, MICH, MO-Steere, NY, PH, U, USDA.
1: 1903-1904, p. [i]-xiii, [xv-xvi], [1]-598, *pl. 1-52* (uncol.), (final cover "1905").
2: 1904-1905, p. [i]-xvi, [1]-733, *pl. 1-62* (uncol.).

Vol.	Lief.	pages	*plates*	dates
1	1	[i-ii], [1]-128	*1-7, 46-48*	Jul 1903
	2	129-256	*8-16, 49*	Sep 1903
	3	257-368	*17-26*	Oct 1903
	4	369-512	*27-36*	Nov 1903
	5	[i]-xiii, 513-598	*37-45, 50-52*	Apr 1904
2	6	[1]-128	*1-10*	Jun 1904
	7	129-256	*11-20*	Jul 1904
	8	257-384	*21-30*	Sep 1904
	9	385-512	*31-40*	Sep 1904
	10	513-640	*41-50*	Nov 1904
	11	[i]-xvi, 641-733	*51-62*	Mar 1905

Nachtrag: Torfmoose (Sphagnum), see below *Eur. Torfm.*
Ref.: Brotherus, V.F., Bryologist 7: 31-32. 1904 (Lief. 1-3).
  Holzinger, J.M., Bryologist 8: 113-115. 1905.
  Matouschek, F., Bot. Centralbl. 93: 328-329. 1903 (rev. Lief. 1), 93: 632-633. 1903 (Lief. 2), 93: 547-548. 1903 (Lief. 3), 95: 163. 1904 (Lief. 4), 96: 71-72. 1904 (Lief. 5), 96: 147-150. 1904 (Lief. 6), 96: 197. 1904 (Lief. 7), 96: 437-438. 1904, 98: 231-232 (Lief. 10), 98: 500-501. 1905.
  Mildbraed, J., Bot. Jahrb. 34 (Lit.): 15-16. 1904, 36 (Lit.): 28-29. 1905.
  Warnstorf, C.F.E., Bot. Zeit. 62: 11-12, 185, 346-349. 1904, 63(2): 9-11, 168-169. 1905.

9642. *Die europäischen Torfmoose* Nachtragsheft zu den Europäischen Laubmoosen beschrieben und gezeichnet von Georg Roth ... Leipzig (Verlag von Wilhelm Engelmann) 1906. Oct. (*Eur. Torfm.*).
Publ.: Mai 1906 (p. iv: 25 Nov 1905; ÖbZ Mar-Mai 1906; PH d. 23 Jun 1906; Bot. Centralbl. 17 Jul 1906; Bot. Zeit. 1 Jul 1906; Allg. bot. Z. 15 Jul 1906; Hedwigia 11 Jun 1906; Nat. Nov. Jun(1) 1906), p. [i]-viii, [1]-80, *pl. 1-11* (uncol.). *Copies*: BR, FH(2), G, MICH, NY, PH, U, USDA.
Ref.: Andrews, A.L., Bryologist 10(3): 51-52. 1907.
  Brotherus, F., Rev. bryol. 33: 107. 1906.

Matouschek, F., Bot. Centralbl. 104: 50-51. 1907.
Nicolson, W.F., Bryologist 9(6): 102-103. 1906, 10(1): 13. 1907.
Roth, G., Hedwigia 45: (136). 10 Jun 1906.
Warnstorf, C.F.E., Bot. Zeit. 64(2): 227-231. 1 Aug 1906 (rev.).

9643. *Die aussereuropäischen Laubmoose.* Beschrieben und gezeichnet von Dr. Georg Roth, ... Band i, Enthaltend die Andreaeaceae, Archidiaceae, Cleistocarpae, und Trematodonteae ... Dresden (Verlag von C. Heinrich) [1910-] 1911. Oct. †. (*Aussereur. Laubm.*).
Publ.: one volume in four parts, 1910-1911. *Copies*: BR, FH, G, M, MICH, MO-Steere, NY, PH, U; IDC 195.
1: Oct 1910 (ÖbZ Nov-Dec 1910; Bryologist Mar 1911; NY rd. 24 Dec 1910; Bot. Centralbl. 17 Jan 1911; Nat. Nov. Oct(2) 1910), p. [i-ii], [1]-96, *pl. 1-8* (uncol. auct.).
2: Jan 1911 (NY rd. 28 Jan 1911; Rev. bryol. 15 Sep 1911), p. [i-ii], 97-192, *pl. 9-16*.
3: Feb 1911 (Rev. bryol. Mai-Jun 1911; Bot. Centralbl. 6 Jun 1911; Nat. Nov. Mar(1) 1911), p. [i-ii], 193-272, *pl. 17-24*.
4: Mar 1911 (NY rd. 12 Apr 1911; BR rd. 13 Jun 1911; Bot. Centralbl. 6 Jun 1911; ÖbZ 1 Mai 1911; Nat. Nov. Apr(2) 1911), p. [i-ii], 273-331, *pl. 25-33*, preface material [i]-x.
Nachtrag 1: 1913. – Hedwigia 53: 81-98. *pl. 1-2*. 1913, reprinted with book cover, p. [1]-18. *pl. 1-2* (rd. NY. Jan 1914, p. 18: 25 Apr 1912; Rev. bryol. Sep-Oct 1913).
Nachtrag 2: 1914. – Hedwigia 54: 267-274. *pl. 1*. 1914, reprinted p. [1]-8, *pl. 1*.
Nachtrag 3: 1916. – Hedwigia 57: 257-262. *pl.* 1916, reprinted p. [1]-6.
Ref.: Britton, E.G., Bryologist 14: 38-39. 1911, 14: 89. 1911 (rev. parts 1, 2).
Loeske, L., Bot. Centralbl. 122: 437-438. 1913 (rev.).
Schiffner, V., Österr. bot. Z. 61: 452. 1911 (rev.).

Roth, Johannes Rudolph (1814-1858), German zoologist and explorer; professor of zoology München 1843; travelled to Arabia petraea, Palestine, the Libanon and Egypt with M. Erdle and G.H. von Schubert 1836-1837; with W.C. Harris to Schoa (Ethiopia) 1841-1843; in Aden 1847; in Greece and Palestine 1852-1853; ultimately in Palestine, Lebanon and Syria 1856-1858; died of malaria in the Antilibanon. (*J. Roth*).

HERBARIUM and TYPES: M. – Further material at DBN (C. Eur.), FI (Palestine), and K (Syria).

BIBLIOGRAPHY and BIOGRAPHY: ADB 53: 530-533 (dates probably inaccurate; V. Hantzsch); AG 3: 164; Barnhart 3: 182 (b. 4 Sep 1814, d. 26 Jun 1858); BM 4: 1737-1738; Herder p. 218; PR 8153, ed. 1: 9107; Quenstedt p. 368; (b. 14 Sep 1815, d. 25 Jun 1858; dates probably inaccurate).

BIOFILE: Anon., Bonplandia 6: 295-296. 1858 (d. 26 Jun 1858 at Huz-Baba, Antilibanon, of malaria); Bull. misc. Inf. Kew 1897: 114; Bull. Soc. bot. France 5: 312. 1858 (coll. 1856-1858 rd. by M. [N.B. The reference to "Roth" in BSbF 29: 188. 1882 is a mixture of A.W. and J.R. Roth]); Österr. bot. Z. 8: 268. 1858 (plant collections rd. by M), 307. 1858 (d. 28 Jun 1858).
Blatter, E., Fl. Aden 6-7. 1914 (Roth visited Aden 1847); Rec. Bot. Surv. India 8(5): 467. 1937 (Arabia petraea, Palestine, Libanon 1837; Aden 1847).
Bridson, G.D.R. et al., Nat. hist. mss. res. Brit. Isl. 432. 1980.
Embacher, F., Lexik. Reisen 254. 1882.
Hertel, H., Mitt. Bot. München 16: 425. 1980 (coll. M; lich.).
Martius, C.F.P. von, Akad. Denkreden 601-602. 1866.
Roth, J.R., *in* W.J. Hooker, J. Bot. Kew Gard. Misc. 1: 216-219. 1849 (on Aden).
Schlechtendal, D.F.L. von, Bot. Zeit. 16: 240. 1858 (obit.).
Wagner, A., Gelehrte Anz. k. bayer. Akad. Wiss. 48: 25-31, 33-46. 1859 (commemorative address, b. 4 Sep 1815 [sic, err.], bibl.; Roth published many book reviews in Gelehrte Anz.).

COMPOSITE WORKS: See A. von Schenk, *Plantarum species quas in itinere per Aegyptum* ... 1839.

9644. *Schilderung der Naturverhältnisse in Süd-Abyssinien.* Fest-Rede vorgetragen in der öffentlichen Sitzung der k. Akademie der Wissenschaften zu München zur Feier ihres zweiundneunzigsten Stiftungstages am 28 März 1851 ... München (Auf Kosten der Akademie ...) 1851. Qu. (*Schilder. Nat.-Verh. Süd-Abyss.*).
Publ.: 1851 (rd. Geol. Soc. London 1 Jul-31 Oct 1851; address delivered 28 Mar 1851), p. [1]-30. *Copy*: G.

Roth, Wilhelm (1819[?]-1875), German (Silesian) botanist and "Webermeister" [weaver] at Langenbielau in Silezia. (*W. Roth*).

HERBARIUM and TYPES: WRSL; European material collected by a "W. Roth" is at BR.

BIBLIOGRAPHY and BIOGRAPHY: Herder p. 188; IH 2: (files); Jackson p. 310; LS 22955. Pax, F., Bibl. schles. Bot. 109. 1929.

9645. *Laubmoose und Gefäss-Kryptogamen des Eulengebirges*, nebst einer Uebersicht des Floren-Gebiets ... Glatz (Schnellpressendruck von L. Schirmer) 1874. Oct. (*Laubm. Gefäss-Krypt. Eulengeb.*).
Publ.: 1874, p. [1]-30. *Copy*: NY.

9646. *Berichte über das Floren-Gebiet des Eulengebirges*, ... Glatz (Schnelpressendruck von L. Schirmer) 1875. Oct. (*Ber. Fl.-Geb. Eulengeb.*).
Publ.: 22 Jun 1875 or somewhat earlier (copy USDA so stamped; handwritten note "Berlin, 7. v. 75"), p. [1]-32. *Copies*: M, USDA. – "Erste Fortsetzung der 1874 erschienenen Arbeit: "Laubmoose ... Eulengebirges".
Second continuation: 1875, p. [1]-22. *Copy*: M. – "Bericht über das Floren-Gebiet des Eulengebirges vom Webermeister Wilhelm Roth in Langenbielau. Zweite Fortsetzung der 1874 erschienenen Arbeit: "Laubmoose ... Eulengebirges". Breslau (Druck von Robert Nischkowsky). 1875.

Rotheray, Lister (fl. 1900), British author of a flora of Skipton (West Yorkshire). (*Rotheray*).

HERBARIUM and TYPES: SKN; further material at GH and LIV.

BIBLIOGRAPHY and BIOGRAPHY: BL 2: 278, 710.

BIOFILE: Hawksworth, D.L. & M.R.D. Seaward, Lichenol. Brit. Isl. 145. 1977. Kent, D.H., Brit. herb. 76. 1957 (coll.).

9647. *Flora of Skipton* & district, ... Skipton (Edmondson & Co., ...) 1900. Oct. (*Fl. Skipton*).
Publ.: Jun-Sep 1900 (p. vii: Jun 1900; J. Bot. Oct 1900), p. [i]-vii, [1]-133, [134, err.]. *Copies*: NY(2).

Rothert, (Karol) Władisław (Rotert, Vladislav Adol'fovich) (1863-1916), Russian botanist; studied at Dorpat, Dr. phil. 1885; lecturer for plant physiology and anatomy at Kasan 1889; e.o. professor of botany Kasan 1896; professor of botany and head of the plant physiology department at Charkow 1897; corr. member Akad. Krakau 1900; ordinary professor of botany at Odessa 1902; at Riga as Privatgelehrter 1908; to Java 1908-1910; at Warsaw 1910; settled as Privatgelehrter at Krakau 1910; died at St. Petersburg. (*Rothert*).

HERBARIUM and TYPES: Some material at BO (Indonesian), C and KRAM (Sparganium). – Letters to W.G. Farlow at FH.

BIBLIOGRAPHY and BIOGRAPHY: Backer p. 500 (epon.); Barnhart 3: 183; BJI 2: 149; BM 4: 1739; Bossert p. 340; De Toni 2: cii, 4: xlviii; Herder p. 433; IH 2: (files); LS 22959-22967; Moebius p. 311, 319, 334, 337; MW p. 419; Rehder 5: 740; SK 1: 449; Tucker 1: 608.

BIOFILE: Anon., Bot. Centralbl. 40: 96. 1889 (lecturer Kasan), 68: 128. 1896 (e.o. prof. bot. Kasan), 71: 383. 1897 (prof. bot., dir. bot. cabinet Charkow), 81: 79. 1900 (corr. member Akad. Krakow), 90: 368. 1902 (ord. prof. bot. Odessa), 107: 640. 1908 (sent to Java by Russ. Akad.), 108: 272. 1908 (leaves Odessa for Riga), 114: 176, 496. 1910 (Privatgelehrter at Krakau); Bot. Zeit. 47: 721. 1889 (app. Kasan), 60: 336. 1902 (to Odessa), 66: 288. 1908 (to Riga); Jahrb. Wiss. Bot. 56: 825. 1915 (pupil of Pfeffer 1891/92, 1900); Nat. Nov. 18: 531. 1896 (e.o. prof. bot. Kasan), 19: 423. 1896 (e.o. prof. bot. Charkow), 24: 584. 1902 (prof. bot. Odessa), 30: 500. 1908 (resigns post Odessa; to Riga), 32: 521. 1910 (to Krakow, 38: 137. 1916 (d.); Nuova Notarisia 27: 217. 1916 (d.); Österr. bot. Z. 39: 456. 1889 (lecturer Kasan), 46: 473. 1896 (e.o. prof. bot. Kasan), 47: 415. 1897 (prof. bot. Charkow), 52: 467. 1902 (prof. bot. Odessa), 65: 360. 1916 (d.).

Kneucker, A., Allg. bot. Z. 2: 188. 1896 (e.o. prof. bot. Kasan), 3: 168. 1897 (prof. bot. Charkow), 8: 192. 1902 (ord. prof. bot. Odessa), 16: 200. 1910 (at Krakau), 22: 48. 1916 (d.).

Manoilenko, K.V., Vladislav Adol'fovich Rotert 1863-1916. Leningrad, Nauka, 1978, 141 p. (rev. D.V. Lebedev, Bot. Zhurn. 65: 149-150. 1980). (n.v.).

Szafer, W., Concise hist. bot. Cracow 69, 78, 89, 95. 1969.

9648. *Vergleichend-anatomische Untersuchungen über die Differenzen im primären Bau der Stengel und Rhizome krautiger Phanerogamen*, nebst einigen allgemeinen Betrachtungen histologischen Inhalts ... verfasst und behufs Erlangung des Grades eines Magisters der Botanik mit Genehmigung einer Hochverordneten physiko-mathematischen Facultät der Kaiserlichen Universität zu Dorpat zur öffentlichen Vertheidigung bestimmt ... Dorpat (Tartu) (Druck von H. Laakmann's Buch- und Steindruckerei) 1885. Oct. (*Vergl.-anat. Unters. Steng. Rhiz.*).

Publ.: Apr 1885 (imprimatur 15 Mar 1885; Bot. Zeit. 26 Jun 1885; Nat. Nov. Apr(2) 1885; Bot. Zeit. 22 Mai 1885), p. [1]-130, [131, theses]. *Copy*: GOET (inf. G. Wagenitz), ILL (inf. D.P. Rogers), MO (inf. C. Lange).
Ref.: Wieler, Bot. Zeit. 43: 583-588. 11 Sep 1885.

Rothmaler, Werner [Hugo Paul] (1908-1963), German botanist; studied at Jena; from 1933-1940 in Spain and Portugal; at the Berlin Kaiser-Wilhelm Institut (with F. v. Wettstein) 1940-1944; Dr. phil. Univ. Berlin 1943; at the Institut für Kulturpflanzenforschung, Wien, Gatersleben 1943-1950; habil. Univ. Halle, 1947, professor of botany Halle 1950-1953; director of the Institut für Agrobiologie and professor of systematic botany and agricultural biology at the Ernst-Moritz-Arndt-Universität, Greifswald (D.D.R.) 1953-1962. (*Rothm.*).

HERBARIUM and TYPES: JE; other material at B (Flora lusitanica), BC, BCF, BREG[?], G, LISE, MA, MAF, S, W.

NOTE: The Rothmaler herbarium at Bregenz (files Index herbariorum) refers either to a set of his Flora lusitanica, or to a different Rothmaler.

BIBLIOGRAPHY and BIOGRAPHY: Barnhart 3: 183 (b. 20 Aug 1908, d. 13 Apr 1962); Biol.-Dokum. 15: 7683-7685; BFM p. 313 (index); BL 2: 464, 465, 466, 467, 490, 710; Bossert p. 340; Hortus 3: 1203; IF suppl. 4: 337-338; IH 1 (ed. 4): 73, (ed. 7): 341, 2: (files); Kew 4: 525; Langman p. 651; MW suppl. 295; NI 3: 71; Roon p. 96; TL-2/940; Zander ed. 10, p. 709, ed. 11, p. 808.

BIOFILE: Grümmer, G., Taxon 11: 191. 1962 (obit., portr.).
Font Quer, P., Coll. bot. 6: 373-375. 1962 (obit.).
Meyer, F.K. & H. Manitz, *in* Reichthümer und Raritäten Jenaer Reden und Schriften 1974, p. 92, 93 (herb. at JE).
Pinto da Silva, A.R., Agronom. lusit. 24: 253-255. 1962 (obit., portr.).
Rothmaler, W., Kampf ums Dasein, s.l.n.d., p. 132-142, unidentified reprint (copy: FAS).
Scamoni, A., Feddes Repert. Beih. 140: 5. 1963 (obit.).Schwarz, O., Drudea 2: 3-6. 1962 (obit.), 45-54. 1964 (critical notes on Rothmaler's *Kritischer Ergänzungsband* of 1963); Feddes Repert. 68: 1-12. 1963 (obit., bibl.).

Stafleu, F.A., Taxon 11: 177. 1962 (d.).
Steenis, C.G.G.J. van, Fl. males. Bull. 17: 875. 1962 (d.).
Sukopp, H., Willdenowia 2(4): 565. 1960.
Valentine, D.H., Nature 195: 1050. 1962 (obit.).
Verdoorn, F., ed., Chron. bot. 1: 251, 264, 342, 367. 1935, 2: 68, 266, 267. 1936, 3: 2, 55, 225, 233, 370. 1937, 4: 63, 87, 88, 127, 274, 440, 445, 515. 1938, 5: 25, 171, 183, 323, 431, 438. 1939, 6: 186, 303, 331, 332. 1943, 7: 179. 1943.
Wehrli, H., [death notice of R. by the rector of the Ernst-Moritz-Arndt Univ. with brief biogr.], 2 p. 1962.

COMPOSITE WORKS: Editor of *Feddes Repertorium* 1943-1962.

NOTE (1): For a complete bibliography (some 190 numbers) see Schwarz (1963). Rothmaler's *Exkursionsflora* (ed. 1, xxviii, 366 p., Berlin 1952, 30-50. Tausend 1953, xxviii, 366 p., Berlin 1953, re-issue 1956, xxviii, 366 p., Berlin 1956) falls outside our scope. It was succeeded by the *Excursionsflora von Deutschland, Gefässpflanzen* (ed. 1, xlvi, 502 p., Berlin 1958, ed. 2, xlviii, 502 p. Berlin 1961); in 1972 published by Hermann Meusel and Rudolf Schubert, as *Exkursionsflora für die Gebiete der DDR und der BRD*, (ed. 1, 1958, ed. 8, 1976), accompanied by a *Kritischer Ergänzungsband Gefässpflanzen* (1963, other eds. e.g. 1963, 1966, 1976) and by an *Atlas der Gefässpflanzen* (ed. 1, 1959, ed. 4, 1968). The *Taxonomische Monographie der Gattung Antirrhinum*, ii, 124 p., 1956 came out as Beih. 136 of Feddes Repert.

NOTE (2): Rothmaler published a proposal to compile a *Flora europaea* on 15 Jun 1944 in Feddes Repertorium, vol. 53, of which a reprint appeared as "Flora europaea A", p. [1]-18 [20] (*copies*: FAS, J. Jalas, M). The paper mentions that various manuscripts are available and states that the two texts for the first volumes (Pteridophyta through Centrospermae) will be completed by 31 Dec 1944. The list of collaborators includes botanists from Germany and several other continental European and North African countries. The present *Flora europaea* was initiated at the Paris International Botanical Congress, 1954.

NOTE (3): See R.S. Cowan and F.A. Stafleu, Taxon 31(3): 420. 1982 for Rothmaler's proposal to publish a modern botanical bibliography, a plan which contained various elements of the present "TL-2".

EPONYMY: *Rothmaleria* Font Quer (1940).

9649. *Alchemillae columbianae* ... Madrid 1935. Oct. (*Alchem. columb.*).
Publ.: 22 Sep 1935 (date on cover), p. [1]-52, *pl. 1-3*. *Copies*: G, HH, NY. – Trab. Mus. nac. Ci. nat. Jard. bot. Madrid, Bot. 31, 1935.

9650. *Neue Alchemilla-Arten aus dem Stockholmer Naturhistorischen Reichsmuseum* ... (Arkiv för Botanik. Band 28a. N:o 3) [1935].
Publ.: briefly after 17 Dec 1935 (date of printing), p. [1]-7. *Copies*: M, U. – The date of distribution in the journal volume was 10 Oct 1936; however, the fascicles were distributed separately shortly after printing.

9651. *Schedae ad W. Rothmaler Alchemillae exsiccatae* fasciculum primum (1-25). Januario 1938 [Leiden, Netherlands].
Publ.: Jan 1938, p. [1]-10. *Copies*: FAS, FI, H, M, U. Accompanies a set of exsiccatae *Alchemillae exsiccatae*.

9652. *Importância da fitogeografia* nos estudos agronómicos ... Lisboa 1940. Oct.
Publ.: 1940, p. [1]-13, [14], maps 1-13. *Copy*: U. – Reprinted from Palestras agronómicas 2(1), 1939 (1940) (journal n.v.).

Rothman, Georg (Göran) (1739-1778), Swedish physician and botanist; pupil of Linnaeus; Dr. phil. Uppsala 1761; Dr. med. Uppsala 1763; quarantine physician in Stockholm 1770; travelled via Tunis to Tripoli where he stayed 1773-1776; assessor Collegium medicum 1776. (*Rothman*).

HERBARIUM and TYPES: S, SBT; some further material at H and UPS (Thunberg). – The manuscript journal of his trip to Tripoli (1773-1776) is at the Sv. Vet.-Akad. Stockholm.

BIBLIOGRAPHY and BIOGRAPHY: Barnhart 3: 183; BM 4: 1739; KR p. 603 (b. 30 Nov 1739, d. 4 Dec 1778); PR 5500, 9951, ed. 1: 6100, 10935; SO 1307, 2228; TL-1/753; TL-2/4819; see P.J. Bergius.

BIOFILE: Anon., Biogr. lexic. namnk. sv. män 12: 270-292. 1846.
Ascherson, P., *in* E. Durand & C. Baratte, Fl. libycae prodr. xxv. 1910 (on his work in Libya).
Bernardi, L., Mus. Genève 198: 11. 1979 (ill. *Rothmannia*).
Fries, R.E., Short hist. bot. Sweden 30. 1950.
Fries, Th.M., Linné 2: 71. 1903; Bref skrifv. Linné 1(1): 200. 1907, 1(2): 342-347, 352. 1908, 1(5): 203, 230. 1911.
Hedlund, E., Sv. Linné-Sällsk. Årsskr. 20: 4-44. 1937 (detailed biogr. account travels).
Hulth, J.M. & A.H. Uggla, Bref Skrifv. Linné 2(2): 283. 1943.
Lindman, S., Sv. män kvinn. 6: 374. 1949.
Löwegren, Y., Naturaliekabinett i Sverige under 1700-talet 363. 1952 (on nat. hist. cabinet of R.'s father, Linnaeus' teacher, Johan Stensson Rothman, 1684-1763).

COMPOSITE WORKS: (1) Rothman received his Dr. phil. degree in 1761 on a dissertation written by Johan Gottschalk (1709-1785), professor of chemistry, Uppsala, *Dissertatio chemica de origine oleorum in vegetabilius*, 26 Mar 1761, 12 p.
(2) Rothman's medical thesis, under Linnaeus, was written by himself: *De Raphania dissertatio medica*, ... 27 Mai 1763, treated by us under Linnaeus (TL-2/4819). We follow Krok (p. 603) in attributing this thesis to the student and not to the master. – O. Almborn in litt. agrees with this after reading the thesis. "Linnaeus is referred to as "N.D. Praeses" which he would never have written himself. – It is a well-known fact that *De Raphania* launches a most remarkable (not to say peculiar) theory. "*Morbus spasmodicus*" (now known as ergotism) was believed to arise from poisonous seeds of Raphanus raphanistrum mixed with grains of cereals. These ideas were rejected by many contemporary physicians".

EPONYMY: *Rothmannia* Thunberg 1776.

Rothman, Johan Stensson (1684-1763), Swedish physician; high school teacher and botanist; studied at Uppsala; Dr. med. Harderwijk (Netherlands) 1713; regional physician Kronoberg County, S. Sweden from 1714, in addition teacher of logic and physics at Växjö grammar school (with e.g. C. Linnaeus as pupil) 1719-1751; assessor (member of Collegium medicum) 1722. (blind from 1749). (*J. Rothman*).

HERBARIUM and TYPES: Unknown. For his natural history cabinet see Löwegren (1952).

BIBLIOGRAPHY and BIOGRAPHY: Barnhart 3: 183; Dryander 3: 351; KR p. 603 (b. 24 Feb 1684, d. 20 Jul 1763; lists one bot. publ.); LS 22968-22968a; SO 906.

BIOFILE: Anon., Biogr. lexic. namnk. sv. män 12: 269-270. 1846 ("without him Linnaeus would have become a poor priest at Stenbrohult").
Blunt, W., The compleat naturalist 18, 21, 25, 26, 86, 141. 1971 (relationship with Linnaeus).
Fries, Th.M., Linné 1: 15-36. 1903; Bref och skrifv. Linné 1(2): 25-27. 1908, 1(4): 9. 1910.
Hedlund, E., Sv. Linné-Sällsk. Årsskr. 19: 67-200. 1936 (detailed biogr.; handwriting).
Lindman, S., Sv. män kvinn. 6: 373. 1949.
Löwegren, I., Naturaliekabinett Sverige 226, 227, 363. 1952 (non-botanical coll.; biogr. refs.).
Mägdefrau, K., Gesch. Bot. 50, 51. 1973.
Scheutz, N.J.W., Bot. Not. 1863: 71 (epon.).
Stearn, W., *in* C. Linnaeus, Sp. pl., facs. ed. 1957, p. 1: 7-9, 13, 24. 1957.

NOTE: (by O. Almborn): Rothman taught Linnaeus about the system (Tournefort) and the sexuality of plants (Vaillant).

Rothmayr, Julius (*fl.* 1910-1913), Swiss mycologist. (*Rothmayr*).-

HERBARIUM and TYPES: Unknown.

BIBLIOGRAPHY and BIOGRAPHY: Barnhart 3: 183; Kelly p. 195 (bibl.); LS 38580-38605, 41681.

COMPOSITE WORKS: Editor of *Der Pilzfreund*, illustrierte populäre Monatschrift über essbare und giftige Pilze, 1 vol. (Hefte 1-12), Luzern 1910-1911 (all published).

9653. *Essbare und giftige Pilze des Waldes* ... Luzern (Verlag von E. Haag) 1913. 2 vols. Oct. (*Essb. gift. Pilze*).
Original edition of vol. 1: *Essbare und giftige Pilze der Schweiz*, Luzern (E. Haag) 1909, 119 p., *40 pl.* n.v. (Bot. Centralbl. 19 Oct 1909; Nat. Nov. Aug(1) 1909).
Ed. 2 of vol. 1: 1910, 80 p., *40 pl.* n.v. – Essbare und giftige Pilze des Waldes. Luzern (E. Haag) 1910.
Ed. 3 of vol. 1: 1913 (ÖbZ Mar 1914; Bot. Centralbl. 23 Sep 1913; Hedwigia 10 Dec 1913; Nat. Nov. Aug(1) 1913), p. [i]-xvi, [1]-64, [1-2], figs. *1-44* on *42 pl.* w.t. (col.). – Mit 44 Pilzgruppen nach der Natur gemalt von Kunstmaler Georg Troxler, Luzern, Erster Band. 9. bis 12. Tausend. Neue, verbesserte und vermehrte Auflage. *Copy*: BR.
Ed. 1 of vol. 2: 1913 (ÖbZ Mar 1914; Bot. Centralbl. 19 Mai 1914; Nat. Nov. Aug (1) 1913), p. [i]-xvi, [1]-67, [68-70], *pl. 1-44* with text (col.). Zweiter Band. 1. bis 4. Tausend. *Copy*: BR.
Gesammt-Ausgabe des 1. und 2. Bandes: 1914, p. [i]-xvi, [1]-131,[132-34, charts], [i-iv], *80 pl. Copy*: NY.
Reissues: 1916, 1917, 1921. n.v.

Rothpletz, [Friedrich] August (1853-1918), German (Pfalz-born of a Swiss father) palaeontologist; studied at Heidelberg und Zürich; geologist at the Geological Survey of Saxony 1875-1880; travelling in Europe 1880-1882; Dr. phil. Leipzig 1882; in München with K. von Zittel 1882; habil. München 1884; professor of geology and palaeontology at the University of München and director of the Bayerische geologisch-paläontologische Staatssammlung as successor to K. von Zittel 1904-1918. (*Rothpletz*).

HERBARIUM and TYPES: M; Teneriffe plants also at GOET. Palaeontological material in the Paläontologische Staatssammlung, München.

BIBLIOGRAPHY and BIOGRAPHY: Andrews p. 322; Barnhart 3: 183 (b. 25 Apr 1853, d. 27 Jan 1918); BM 4: 1739; CSP 11: 227-228, 12: 634, 18: 319; De Toni 1: cxix, 2: cii, 3: xlviii; DTS 1: 248-249; Ewan ed. 2: 188; Herder p. 433; Jackson p. 189; KR p. 603; Quenstedt p. 368-369.

BIOFILE: Ampferer, O., Verh. k.k. geol. Reichsanst. 1918(3): 59-62. (obit.).
Anon., Nature 101: 109-110. 1918 (obit.).
Broili, F., Jahrb. bayer. Akad. Wiss. 1918: 59-65 (obit.); Neues Jahrb. Mineral. Geol. Paläontol. 1919: i-xx (obit., portr., bibl.); Mitt. Geogr. Ges. München 13: 359-363. 1919 (obit., portr.).
Hertel, H., Mitt. Bot. München 16: 425. 1980 (coll. rd. from R's estate 1928).
Poggendorff, J.C., Biogr.-lit. Handw.-Buch 5: 1072. 1926, 6: 2230. 1938; 7a (suppl.): 563-564. 1971.
Pompeckj, J.F., Z. deut. geol. Ges., Monatsber. 70: 15-35. 1918 (obit.).
Salomon, W., Deut. biogr. Jahrb. 2 (1917-1920): 317-319. 1928.
Sirks, M.J., Indisch natuurond. 243, 252. 1915.
VanLandingham, S.L., Cat. diat. 6: 3586. 1978.
Wagenitz, G., Index coll. princ. herb. Gott. 138. 1982.
Zittel, K.A. von, Gesch. Geol. Paläont. 862 [index]. 1899; Hist. geol. palaeontol. 314, 323, 493. 1901.

EPONYMY: *Rothpletzella* A. Wood (1948).

9654. *Die Flora und Fauna der Culmformation* bei Hainichen in Sachsen ... III. Gratis-Beilage, Botanisches Centralblatt [Cassel (Verlag von Theodor Fischer) 1880, Bot. Centralbl. I. Quartal 1880]. Oct. (*Fl. Fauna Culmform.*).
Publ.: Jan-Mar 1880 (as supplement to Bot. Centralbl. vol. 1, 1880), p. [i], [1]-40, *pl. 1-3* (uncol. liths). *Copies*: BR, GOET.

9655. *Ueber Algen und Hydrozoen im Silur* von Gotland und Oesel ... Uppsala & Stockholm (Almqvist & Wiksells Boktryckeri-A.-B.). 1908. Qu. (*Alg. Hydroz. Silur*).
Publ.: Sep 1908 (p. 25: printed 5 Sep 1908), p. [1]-25, *pl. 1-6*. *Copies*: FH, NY, US, USDA; IDC 452. – K. Sv. Vet.-Akad. Handl. 43(5), 1908.
Ref.: Salfeld, H., Bot. Zeit. 67(2): 73, 261-262. 1909 (criticises Steinmann's review).
Steinmann, Z. indukt. Abst. Vererb.-Lehre 1(4): 405-407. 1909.

9656. *Über die Kalkalgen*, Spongiostromen und einige andere Fossilien aus dem Obersilur Gottlands ... Stockholm (Kungl. Boktryckeriet. P.A. Norstedt & Söner) 1913. Qu. (*Kalkalgen*).
Publ.: 1913, p. [i-iv], [1]-57, *pl. 1-10*. *Copy*: USGS. – Issued as Sv. geol. Undersökn. Afh. quarto ser. C a., no. 10.

Rothrock, Joseph Trimble (1839-1922), American surgeon, botanist, explorer and forester; B.Sc. Harvard 1864; on R. Kennicotts Alaskan exp. 1865-1866; Dr. med. Univ. Pennsylvania 1867; professor of botany in the State Agricultural College of Pennsylvania 1867-1869; practicing physician in Wilkes-Barre; surgeon and botanist to G.N. Wheeler's Exploring Expeditions west of the 100th meridian 1873-1875; professor of botany Univ. Pennsylvania 1877-1891 (-1904); in the West Indies on the "White Cap" 1889-1890; Forestry Commisioner of the State of Pennsylvania 1895-1904; erected a home for open air treatment of tuberculosis at Mont Alto 1902. (*Rothr.*).

HERBARIUM and TYPES: F (private herb.), US (Wheeler exp.); further material at A, B (West Ind., mainly destr.), E, GH, LE, MO, NY, PH. – The plants collected on R. Kennicott's Alaskan and B.C. expedition were lost (coll. F. Pope and J.R. Rothrock) in the Fraser River; the West Indian collections are at PH (formerly PENN) (first set) and F. – Letters to and from S.F. Baird in Smithsonian Archives; letters to W.G. Farlow at FH.

BIBLIOGRAPHY and BIOGRAPHY: Backer p. 660; Barnhart 3: 183; BJI 1: 49; BL 1: 130, 316; BM 4: 1739, 5: 2174 (Wheeler rept.); Bossert p. 340; CSP 5: 303, 8: 787, 11: 228, 12: 634, 18: 319; DAB 16: 188-189; Ewan ed. 1: 165, 293, 294, 304, 340, ed. 2: 188; Herder p. 338; Hortus 3: 1203; IH 2: (files); Jackson p. 357; Kelly p. 195-196; Kew 4: 525; Langman p. 651-652; Lenley p. 355; LS 22969; ME 1: 226, 2: 56, 514, 3: 739; Morren ed. 10, p. 130; NW p. 56; PH p. 521 [index]; PR 7810; Rehder 5: 740-741; TL-2/see E.M. Durand; Tucker 1: 609; Urban-Berl. p. 389.

COMPOSITE WORKS: (1) *Contr. nat. hist. Alaska*, ed. L.M. Turner, p. 61-65. 1886, list of plants by J.T. Rothrock (Arctic ser. publ. Signal Service, U.S. Army no. 2), based on Rothrock's *Sketch of the Flora of Alaska*, Ann. Rep. Smiths. Inst. 1867: 433-463. 1872 (but published by 3 Mar 1870, see Flora 53: 63. 1870).
(2) *List of*, and notes upon, the *lichens* collected by Dr. T.H. Bean *in Alaska* and the adjacent region in 1880, Proc. U.S. natl. Mus. 7(1): 1-9. 3 Jun 1884; identifications by Henrey Willey (1824-1907).

BIOFILE: Allison, E.M., Univ. Colo. Stud. 6: 72-73. 1909.
Anon., Amer. Forestry 28: 414. 1922 (obit.); Daily Local News, West Chester, Penn. 14 Apr 1923 (at HU); Forest leaves 18: 115-116. 1922 (resigns from State Forest Commission, tribute by governor of state); Torreya 9: 269. 1909 (herb. acquired by F.).
Anse, A. de l', Evening Post, New York, 2 Mai 1914.
Brewer, W.H., *in* S. Watson, Bot. Calif. 2: 559. 1880.
Cattell, J.M., Amer. men sci. ed. 1: 274. 1906, ed. 2: 402. 1910, ed. 3: 588. 1921, ed. 4: 839. 1927.

Dudley, S. and D.R. Goddard, Proc. Amer. philos. Soc. 117(1): 37-50. 1973 (R. and forest conservation; secondary refs.).
Graves, H.S. et al., J. Forestry 20(5): 567-568. 1922 (resolution upon death).
Harshberger, J., Bot. Philadelphia 305-313. 1899 (portr.).
Hultén, E., Bot. Not. 1940: 291. (Sketch Fl. Alaska based on Ledebour's Fl. ross.).
Illick, J.S., Joseph Trimble Rothrock, father of Pennsylvania Forestry. 1929, 12 p. (repr. from Penn. German Soc. 34, 1929).
Jaeger, E.C., Source-book biol. names, terms, ed. 3. 320. 1966.
Kelly, H.A., Dict. Amer. med. biogr. 1062-1064. 1928; Some Amer. med. bot. 203-213. 1914, 1929 (autobiogr., portr.; in his autobiogr. R. states that he served at Univ. Penn. only from 1877-1891, afterwards devoting himself almost exclusively to forestry).
Leroy, J.F. et al., Bot. franç. Amér. nord 294, 295, 296. 1957.
McVaugh, R., Edward Palmer 66. 1956.
Millspaugh, C.F., Field. Mus. Publ. Bot. 2(7): 141, 291. 1909 (White Cap trip to West Indies).
Pennell, F.W. & J.H. Barnhart, Monogr. Acad. nat. Sci. Philadelphia 1: 618. 1935.
Pennell, F.W., Bartonia 22: 15. 1943.
Pleasants, H., Jr., Three scientists of Chester County. 1936, p. 37-49.
Reifschneider, O., Biogr. Nevada Botanists 20, 48-50. 1964 (portr.).
Reveal, J., *in* A. Cronquist et al., Intermount. Fl. 1: 56. 1972.
Rodgers, A.D., Amer. bot. 1873-1892, p. 336 [index]. 1944; Bernard Eduard Fernow 620 [index]. 1951.
Sargent, C.S., Silva 8: 92. 1895.
Thomas, J.H., Huntia 3: 17. 1969 (1979).
Urban, I., Symb. ant. 1: 141. 1898, 3: 115. 1902.
Verdoorn, F., ed., Chron. bot. 5: 114, 512. 1939.
Wilson, J.G. & J. Fiske, eds., Appleton's Cycl. Amer. biogr. 5: 334. 1888.
Wirt, G.H., J. Forestry 44: 442-443. 1946 (biogr. sketch, portr.).
Wittrock, V.B., Acta Horti Berg. 3(3): 201. 1905.

EPONYMY: *Rothrockia* A. Gray (1885).

9657. Engineer Department, U.S. Army. Geographical and geological explorations and surveys west of the one hundredth meridian. First Lieut. Geo. M. Wheeler, Corps of Engineers, in charge. *Catalogue of plants* collected in the years 1871, 1872 and 1873, with descriptions of new species. Washington (Government Printing Office) 1874. (*Cat. pl.*).
Authorship: Sereno Watson (1826-1892), author of text p. 5-19; Rothrock is author of p. [21]-62, with the collaboration of Asa Gray, S. Watson, D.C. Eaton, T.P. James, G. Vasey, G. Thurber, and J. Hoopes.
Publ.: Jul-Dec 1874 (p. [2]: 1 Jul 1874), p. [1]-62. *Copies*: US(2), USDA. – Preliminary report; for definitive report see below (1878).

9658. Engineer Department, U.S. Army. *Report* upon *United States geographical surveys* west of the one hundredth meridian, in charge of first lieut. Geo. M. *Wheeler*, ... *vol. vi. – Botany.* Washington (Government Printing Office) 1878. Oct. (*Rep. U.S. geogr. surv., Wheeler*).
Publ.: Jun-Aug 1879 (p. xiii: 10 Mai 1878; t.p. 1878; last part printed Mai 1879; Grevillea Sep 1879; J.Bot. Sep 1879; Nat. Nov. Aug(2) 1879), p. [i]-xx, err. slip, 1-404, *pl. 1-30* (uncol. liths. Sprague). *Copies*: BR, FH, G (p. 53-404), HH, MO, NY, PH, US(2), USDA.
Contents: p. [i]-v, general preface material, p. [vii]-xx, preface material botanical reports; p. 1-14: Notes on Colorado; p. 15-37: Notes on New Mexico; p. 39-52: Notes on economic botany; p. 53-404, *pl. 1-30*: Catalogue of plants by J.T. Rothrock et al.
Alternative t.p. (p. [vii]): *U.S. Geographical surveys* west of the one hundredth meridian ... *Reports upon botanical collections* made in portions of Nevada, Utah, California, Colorado, New Mexico and Arizona, during the years 1871, 1872, 1873, 1874 and 1875. By J.T. Rothrock, ... and the following scientists: Sereno Watson, ... George Engelmann, ... Thos.C. Porter, ... M.S. Bebb, ... William Boott, ... George

Vasey, ... D.C. Eaton, ... Thos.P. James, ... Edward Tuckerman, ... in four chapters and an appendix.
Ref.: Anon., Bull. Soc. bot. France 26 (bibl.): 146. 1880 (rev.).
Engler, A., Bot. Jahrb. 1: 66-69. 1880 (rev.).
W.C.S., Grevillea 8: 16-17. 1879 (rev.).
Dr., Bot. Zeit. 38: 391. 28 Mai 1880 (rev.).

9659. *Catalogue of trees and shrubs native of and introduced in the horticultural gardens* adjacent to Horticultural Hall, *in Fairmount Park*, Philadelphia [Philadelphia 1880]. Oct. (*Cat. trees shrubs gard. Fairmount*).
Publ.: 1880 (p. 3: Mar 1880; the HH copy is dated "June 8th 1880"), p. [1]-99. *Copies*: HH, NY, PH.

9660. *Vacation cruising* in Chesapeake and Delaware Bays ... Philadelphia (J.B. Lippincott & Co.) 1884. Oct. (*Vacation cruis.*).
Publ.: 1884 (Ewan copy inscribed by author 21 Jun 1884), 262 p. *Copy*: Ewan. – "Passing references to vegetation and shore life, a socio-ecological account" (J. Ewan).

Rothschild, Jules (1838-?), French publisher and horticulturist. (*Rothschild*).

HERBARIUM and TYPES: Unknown.

BIBLIOGRAPHY and BIOGRAPHY: BL 2: 106, 710; BM 4: 1739; NI p. 266 (index), no. 1684; TL-2/5042; Tucker 1: 609.

9661. *Les plantes à feuillage coloré* recueil des espèces les plus remarquables servant à la décoration des jardins, des serres et des appartements, par M.M. E.J. Lowe et W. Howard ... traduit de l'anglais par M.J. Rothschild, avec le concours de plusieurs horticulteurs ... Paris (J. Rothschild, éditeur ...) 1865. Oct. (*Pl. feuill. col.*).
Original ed. (English): by Edward Joseph Lowe (1825-1900) and W. Howard, *Beautiful leaved plants*, London 1861, see TL-2/5042.
Ed. 1: 1865, p. [i-viii], [1]-136, *pl. 1-60*. *Copy*: FI.
Ed. 2: 1867, p. [i]-viii, [1]-128, *pl. 1-60*. *Copy*: FI.
Nouvelle serie [vol. 2]: 1867-1870, in parts, p. [1]-8, [1]-124, *pl. 1-60*. *Copies*: BR, FI.
Ed. 3 (of vol. 1): 1874. *Copy*: UC.
Ed. 4 (of vols. 1 and 2): 1880. *Copy*: B.

9662. *Les fougères* choix des espèces les plus remarquables pour la décoration des serres, parcs, jardins et salons précédé de leur histoire botanique & horticole par MM. Aug. Rivière, ... E. André, ... E. Roze, ... publié sous la direction de J. Rothschild ... Paris (J. Rothschild, éditeur ...) 1867-1868, 2 vols. Oct. (*Fougères*).
1: 1867, p. [i]-x, [1]-286, *pl. 1-75* (Riocreux, Faguet, Poiteau).
2: 1868 (BSbF post 7 Feb 1868), p. [1]-242, [1-2, table], *pl. 1-80, 68 [bis]* and [*1*] Selaginella. – "... suivi de l'histoire botanique et horticole par E. Roze ...".
Copies: BR, G.

9663. Botanique populaire illustrée. *Flore pittoresque de la France* anatomie-physiologie-classification-description des plantes indigènes et cultivées au point de vue de l'agriculture, de l'horticulture et de la sylviculture publiée sous la direction de J. Rothschild avec le concours de MM. Gustave Heuzé ... Bouquet de la Grye ... Stanislas Meunier ... J. Pizzetta ... B. Verlot ... à l'usage des lycées ... etc. ... Paris (J. Rothschild, éditeur ...) [1885]. Qu. (*Fl. pittor. France*).
Ed. 1: 1885, n.v.
Ed. 2: 1885(?), p. [i]-xvi, [1]-473, *pl. 1-82*, map. *Copies*: HH, USDA. – possible simply a reissue with new t.p. – HH has also a "quatrième édition" with same collation.
Ref.: Malinvaud, E., Bull. Soc. bot. France 32 (bibl.).: 233-234. 1885 (post 15 Nov) or 1886 (rev. ed. 1).

Rottbøll, Christen Friis (1727-1797), Danish botanist; Dr. med. Kjøbenhavn 1755;

professor of medicine ("designatus") ib. 1756; studied botany with Linnaeus 1756-1757; travelled in Holland and France; director of the Copenhagen botanical garden 1770-1797; ord. professor of medicine 1776-1797. (*Rottb.*).

HERBARIUM and TYPES: C; some further material in LINN ("styled "Friis", Jackson, 1922), LIV, NY and PH. – The Rottbøll herbarium at C is available on microfiche IDC 2202.

BIBLIOGRAPHY and BIOGRAPHY: AG 2(1): 543; Backer p. 500; Barnhart 3: 183 (b. 3 Apr 1727, d. 15 Jun 1797); BM 4: 1740; Bossert p. 340; Dawson p. 713; Dryander 3: 4, 6, 33, 83, 123, 167, 178, 189, 213, 216, 252; Herder p. 147, 163, 226, 227, 411; Hortus 3: 1203; IH 2: (files); Jackson p. 21, 111, 145, 375, 444; Kew 4: 525-526; Lasègue p. 345, 473; MW p. 419; NI 1: 201, 227, 2: no. 1685; PR 4798, 7811-7815, ed. 1: 5340, 8764-8770; Rehder 5: 741; RS p. 138; SO 2430, 2438; Tucker 1: 609; Zander ed. 10, p. 709, ed. 11, p. 809.

BIOFILE: Candolle, Alph. de, Phytographie 444. 1880 (herb. at C).
Christensen, C., Dansk. Bot. Hist. 1: 69, 78-79, 84, 134, 150-152. 1924 (portr., biogr.), 2: 53-55. 1924 (bibl.; biogr. data).
Fries, Th.M., Bref Skrifv. Linné 1(6): 224. 1912.
Gager, C.S., Brooklyn Bot. Gard. Rec. 27(3): 190. 1938 (dir. bot. gard. Copenhagen 1778-1797).
Hornemann, J., Nat. Tidskr. 1: 566-567. 1837.
Hulth, J.M., Bref Skrifv. Linné 2(1): 418, 419. 1916, with A.H. Uggla, 2(2): 197, 198. 1943.
Jackson, B.D., Proc. Linn. Soc. 134 (suppl.): 19. 1922 (material LINN).
Rydberg, P.A., Augustana Coll. Libr. Publ. 5: 15. 1907.
Stafleu, F.A., *in* G.S. Daniels & F.A. Stafleu, Taxon 23: 44. 1974 (portr.).
Warming, E., Bot. Tidsskr. 12: 68-69. 1880 (bibl.).
Wittrock, V.B., Acta Horti Berg. 3(2): 104-105. 1903.

EPONYMY: *Rotbolla* Zumaglini (1849, *orth. var.* of *Rottboellia* Linnaeus f. (1780)); *Rotbollia* T. Nuttall (1818, *orth. var.* of *Rottboellia* Linnaeus f. (1780)); *Rottboelia* Scopoli (1777); *Rottboelia* Dumortier (1829, *orth. var.* of Rottboellia Linnaeus f. (1780)); *Rottboella* Linnaeus f. (1782, *orth. var.* of *Rottboellia* Linnaeus f. (1780)); *Rottboellia* Linnaeus f. (1780, *nom. cons.*); *Rottbolla* Lamarck (1792, *orth. var.* of *Rottboellia* Linnaeus f. (1780)); *Rottbollia* A.L. Jussieu (1789, *orth. var.* of *Rottboellia* Linnaeus f. (1780)).

9664. *Botanikens udstrakte Nytte* ... Kiøbenhavn (Findes tilkiøbs i den Mummiske Boghandling, ...) 1771. Oct. (*Bot. udstr. Nytte*).
Publ.: 1771, p. [1]-63, *1 pl.* (uncol. copp.). *Copies*: G, MO, NY.

9665. *Descriptiones plantarum rariorum* iconibus illustrandas, cum earum, quae primo proximeque prodituro fasciculo continebuntur, elencho, programmate, quo lectiones in horto botanico ao. 1772. Auspicatur, indicit Christianus Friis Rottbøll, ... Havniae [Kjøbenhavn] (litteris N. Mölleri, ...) 1772. Oct. (*Descr. pl. rar.*).
Publ.: 1772 (GGA 3 Apr 1773), p. [1]-32. *Copies*: HH, NY; IDC 5694. – The designations *Schoenoides*, *Scirpoides*, and *Cyperoides* are token words, not to be considered as generic names, see ICBN Art. 20.4 and 43. Note 1. – This octavo publication was preliminary to the *Descr. icon. rar. pl.* of 1773 in which Rottbøll published the formally correct names for the above token designations. *Chondropetalum* is a new generic name.
Ref.: Dandy, J.E., Regn. veg. 51: 18. 1967.

9666. *Descriptionum et iconum rariores* et pro maxima parte novas *plantas* illustrantium liber primus. Conscriptus a Christiano Friis Rottböll, ... Hafniae [Kjøbenhavn] (sumptibus Societatis typographicae ...) 1773. Fol. (*Descr. icon. rar. pl.*).
Publ.: Jan-Jul 1773 (NZgS 23 Aug 1773), p. [i-viii], [1]-71, [72, err.], *pl. 1-21* (uncol. copp. G. Haas, M. Haas and N. Abilgaard). *Copies*: M, MO, NY, PH, USDA; IDC 2259.
Editio nova: 1786, p. [i-viii], [1]-71, [72, err.], *pl. 1-21* (id.). *Copies*: FI, G(2), MO, NY,

Teyler, U. – "Descriptiones et icones rariorum et pro maxima parte novarum plantarum ... editio nova ..." Havniae [Kjøbenhavn] (Impensis Gyldendalii, ...) 1786. Fol. – Title-page only changed; actually a re-issue.

9667. *Descriptiones rariorum plantarum*, nec non materiae medicae atque oeconomicae *e terra surinamensi* fragmentum placido ampliss. professorum examini, pro loco in consistorio rite tenendo disputaturus, subjicit Christianus Friis Rottbøll, ... Spartam defendentis ornante ... ad d. [] maji a. [1776], Havniae (Typis N.C. Höpffneri, ...) [1776]. Qu. (*Descr. rar. pl. surin.*).
Publ.: 1776, p. [i-vi], [1]-34. *Copy*: HH. – Also published in Acta litt. Univ. Hafn. 1778: [267]-304, *pl. 1-5.* Plants collected by Rolander.
Ed. 2: 1798, p. [1]-22, *pl. 1-5* (uncol. copp.). *Copies*: E, FI, G, NY, U; IDC 2260. – "Descriptiones plantarum quarundam surinamensium, cum fragmento materiae medicae et oeconomicae surinamensis ... Editio secunda. Emendatior. ... Hafniae et Lipsiae (apud Johannem Henricum Schubothe) 1798. Fol.

Rottenbach, Heinrich (1835-1917), German (Thuringian) botanist; high school teacher at Meiningen 1871-1895 (from 1877 with professor's title); in retirement in Berlin 1895-1915; ultimately at Einhausen nr Meiningen. (*Rottenbach*).

HERBARIUM and TYPES: Unknown.

BIBLIOGRAPHY and BIOGRAPHY: Barnhart 3: 183 (b. 28 Mar 1835, d. 4 Mai 1917); BFM no. 313; BL 2: 405, 710; Biol.-Dokum. 15: 7688; Christiansen A588; DTS 1: 249, 6(4): 190; Jackson p. 311, 507; Morren ed. 10, p. 26; Rehder 1: 388.

BIOFILE: Anon., Bot. Not. 1917: 206. (d.); Hedwigia 59: (141). 1917 (d.).
Harms, H., Verh. bot. Ver. Brandenburg 59: 41-46. 1918 (obit., bibl.).

9668. Programm zur öffentlichen Prüfung, welche mit den Zöglingen der Realschule in Meiningen Donnerstag, den 21. März 1872, ... *Thüringens polypetale Dicotyledonen* mit hypogynischer Insertion; ... Meiningen (Druck der Keyssner'schen Hofbuchdruckerei) 1872. Qu. (*Fl. Thüring. 1*).
Publ.: 1872, p. [1]-29 (followed by "Schulnachrichten"). – Title on p. [3]: *Zur Flora Thüringens*, insbesondere des Meininger Landes. Erster Beitrag: Polypetale Dicotyledonen mit hypogynischer Insertion. (*Fl. Thüring. 1*). *Copy*: GOET (inf. G. Wagenitz).
– For 2.-8. Beitrag see below.

9669. Programm zur öffentlichen Prüfung der Zöglinge der Realschule in Meiningen, ... *Zur Flora Thüringens*. *Zweiter Beitrag*; ... Meiningen (Druck der Keyssner'schen Hofbuchdruckerei) 1877. Progr. nr. 570. Qu. (*Fl. Thüring.*, 2).
Erster Beitrag: see above, 29 p., Programm Realschule Meiningen 1872; *Zur Flora Thüringens*, insbesondere des Meininger Landes. *1. Beitrag*: Polypetale Dicotyledonen mit hypogynischer Insertion. (*Fl. Thüring. 1*).
Zweiter Beitrag: 1877, p. [1]-32, *Programm* 1877, no. 570. *2. Beitrag*: Polypetale Dicotyledonen mit perigynischer und epigynischer Insertion. (*Fl. Thüring. 2*). *Copy*: GOET (inf. G. Wagenitz).
Dritter Beitrag: Jan-Mar 1880 (*Programm* no. 594, for 18-19 Mar 1880; Bot. Centralbl. Sep 1880; Nat. Nov. Nov(2) 1880; Bot. Zeit. 25 Feb 1881), p. [1]-22 (entire *Programm* p. [1]-39). *Copy*: MO. (*Fl. Thüring. 3*).
Vierter Beitrag: Jan-Mar 1882 (*Programm* no. 623, for 30-31 Mar 1882; Bot. Centralbl. 31 Jul-4 Aug 1882), p. [1]-11 (entire *Programm* p. [1]-25). *Copy*: MO. (*Fl. Thüring. 4*).
Fünfter Beitrag: 1 Jan-18 Feb 1883 (*Programm* no. 634, for 15-16 Mar 1883; Bot. Centralbl. 19-23 Feb 1883; Nat. Nov. Jun(1) 1883), p. [1]-17 (entire *Programm* p. [1]-31). *Copy*: MO. (*Fl. Thüring. 5*).
Sechster Beitrag: Jan-Mar 1884 (*Programm* no. 640, for 3-4 Apr 1884; Bot. Zeit. 27 Jun 1884; Nat. Nov. Mai(1) 1884), p. [1]-20 (entire *Programm* p. [1]-35). *Copy*: MO. (*Fl. Thüring. 6*).
Siebenter Beitrag: Jan-Mar 1885 (*Programm* no. 638 for 26-27 Mar 1885; Bot. Monatschr. Apr-Mai 1885; Nat. Nov. Jun(1, 2) 1885), p. [1]-16. *Copy*: MO. (*Fl. Thüring. 7*).

Achter Beitrag: Jan-Apr 1889 (*Programm* no. 696 for 12 Apr 1889; Bot. Centralbl. 3 Dec 1889; Bot. Zeit. 27 Dec 1889; Nat. Nov. Nov(1) 1889), p. [1]-18 (entire *Programm* p. [1]-40), "Zur Flora Thüringens, insbesondere des Meininger Landes ..." *Copy:* MO. (*Fl. Thüring. 8*).

See also: Notizen zur Flora Thüringens, Verh. bot. Ver. Brandenburg 20: 101-102. 1878.

Rottler, Johan Peter (1749-1836), Strassburg (Alsatia) born Danish missionary, traveller, orientalist and botanist; Dr. phil. Wien 1795; at Tranquebar (South of Madras on Coromandel Coast) 1776-1806; subsequently with English missionaries at Madras, travelling in the Ganges region and Ceylon (1788, 1795). (*Rottler*).

HERBARIUM and TYPES: K and LIV; other material at C, ER, FI, H, H-Ach(Lich.), JE, LE, M (important set), MH, MO, MW, NY. – Manuscript catalogue at K; letters to J.E. Smith at LINN.

BIBLIOGRAPHY and BIOGRAPHY: AG 7: 412; Backer p. 500 (epon.); Barnhart 3: 183 (d. 27 Jan 1836); BB p. 262; CSP 5: 304; Desmond p. 532; Frank 3(Anh.): 86; Hegi 7: 560; Hortus 3: 1203; IH 1 (ed. 6): 363, 2: (files); Lasègue p. 131, 137, 138, 331, 337, 430, 503, 558; PR (*alph.*); Rehder 1: 471; SK 4: ci; Urban-Berl. p. 389, 415; Zander ed. 10, p. 709, ed. 11, p. 809.

BIOFILE: Anon., Bull. misc. Inf. Kew 1926: 221-222; J. Bot. 10: 128. 1872 (R.'s herbarium transferred to K).
Bridson, G.D.R. et al., Nat. hist. mss. res. Brit. Isl. 269.308. 1980.
Burkill, I.H., Chapt. hist. bot. India 17, 42, 49, 55, 71, 116. 1965.
Christensen, Dansk. Bot. Hist. 1: 119. 1924, 2: 98-100. 1924 (bibl.).
Cleghorn, H., Bot. Gaz., London 3: 55. 1851 (brief biogr. sketch).
Dawson, W.R., Smith papers 83. 1934.
Foulkes, T., Madras J. Lit. Sci. 22: 1-17. 1861 (biogr. memoir).
Gunn, M. & L.E. Codd, Bot. explor. S. Afr. 301. 1981.
Jackson, B.D., Bull. misc. Inf. Kew 1901: 57.
Jessen, K.F.W., Bot. Gegenw. Vorz. 469. 1884.
Lindemann, E. v., Bull. Soc. imp. Natural. Moscou 61(1): 63. 1886.
Martius, C.P.F. von, Flora 17(1): 3. 1834 (Rottler material in Wallich herb.); 34(1): 3, 5. 1851 (Rottler plants at M).
Müller, D., Bot. Tidsskr. 52: 56-57. 1955 (letters R. to Vahl at C).
Rottler, J.B., Neu. Schr. Ges. naturf. Fr. Berlin 4: 180-224. *pl. 3-5*. 1803 (Bot. Bem. Reise Frankenbar nach Madras).
Smith, J.E., *in* Rees' Cycl. sub Rottlera.
Stansfield, H., Bull. Liverpool Mus. 1957: 27-42 (on Rottler and G.F. Klein's coll. at LIV).
Verdoorn, F., ed., Chron. bot. 2: 11. 1936.
Wight, R., Bot. Misc. 2: 95. 1831.

HANDWRITING: Bull. Liverpool Mus. 1957: 20, 32.

EPONYMY: *Roettlera* M. Vahl (1805); *Rottlera* Willdenow (1797); *Rottlera* J.J. Roemer et J.A. Schultes (1817, *orth. var.* of *Roettlera* M. Vahl); *Rottleria* S.E. Bridel (1826).

Roucel, François Antoine (1736-1831), Belgian (Baden-born) botanist; town physician of Aalst (Flanders, Belgium) from 1777. (*Roucel*).

HERBARIUM and TYPES: GENT.

BIBLIOGRAPHY and BIOGRAPHY: Barnhart 3: 183 (d. 6 Oct 1831); BL 2: 38, 711; BM 4: 1740-1741; Herder p. 172, 175; IH 2: (files); Jackson p. 277; Kew 4: 526; PR 7817-7818, ed. 1: 7213, 8772, 8773; Rehder 1: 392, 406; SBC p. 130; SO add. 782a; SY p. 16; TL-1/1113; TL-2/4349; Tucker 1: 609.

BIOFILE: Candolle, Alph. de, Phytographie 444. 1880.

Crépin, F., Guide du botaniste en Belgique 225-226, 427-428. 1878.
Lejeune, A.L.S., Fl. Spa 1: 4, 5. 1811.
Micheels, H., Biogr. natl. Belgique 20: 198-202. 1908-1910 (b. 1736).
Pritzel, G.A., Linnaea 19: 460. 1846.
Sayre, G., Bryologist 80: 516. 1977.

EPONYMY: *Roucela* Dumortier (1822).

9670. *Traité des plantes les moins fréquentes, qui croissent naturellement dans les environs des villes de Gand, d'Alost, de Termonde & Bruxelles, rapportées sous les dénominations des modernes & des anciens, & arrangées suivant le systême de Linnaeus: avec une explication des termes de la nomenclature botanique, les noms françois & flamands de chaque plante, les lieux positifs ou elles croissent, & des observations sur leurs usages dans la médecine, dans les alimens, dans les arts & métiers. Par Mr. Roucel.* Paris (chez MM. Bossange, & Compagnie), Bruxelles (chez Lemaire, ...) 1792. Oct. (*Traité pl. Gand*).
Publ.: 1792, p. [i*-iii*], [i]-xxix, [1]-118, [1, err.]. *Copies*: G, NY.
Other issue: 1792, p. [i*-iii*], [i]-xxix, [1]-118, [1, err.]. *Copies*: BR, FAS, HH, MO. – "Traité ...Par Mr Roucel. A Bruxelles (Chez Lemaire, ...) et à Paris (Chez MM. Bossange, & Compagnie) 1792. Oct.

9671. *Flore du Nord de la France, ou description des plantes indigènes et de celles cultivées dans les Départemens de la Lys, de l'Escaut, de la Dyle et des Deux-Nèthes, y compris les plantes qui naissent dans les pays limitrophes de ces départemens; ouvrage de près de trente ans de soins et de recherches, dans lequel les plantes sont arrangées suivant le systême de Linné, et décrites par genres et espèces, avec des observations de l'auteur. On y a joint les lieux positifs où elles naissent, et leurs propriétés reconnues dans la médecine, dans les alimens et dans les arts...* Paris (Chez Me. Veuve Richard, Libr., ...) an xi (1803). Oct. 2 vols. (*Fl. Nord France*).
Publ.: 28 Jan 1803 (JT). *Copies*: BR, G, HH, KNAW, NY, PH, USDA; IDC 5695.
1: [i]-xxvi, [1]-465.
2: [i, iii], [1]-548.
Follow-up: Lejeune, A.L.S., Fl. Spa 1813, TL-2/4349.

Roumeguère, Casimir (1828-1892), French cryptogamist and conchologist; founder of the Revue mycologique. (*Roum.*).

HERBARIUM and TYPES: BR, PC; other material at AUT, CLF, G, H, IASI, MPU, MI, NY, PC, W.
Exsiccatae: (1) for *Algues de France*, see under A. Mougeot. For century xiv see Rev. mycol. 1893: 81. For *Reliquiae brébisonianae* see Sayre, G., Mem. New York Bot. Gard. 19(1): 98. 1969.
(2) for *Stirpes cryptogamae vogeso-rhenanae* (fasc. 16 by Roumeguère) see under J.B. Mougeot.
(3) *Bryologie de l'Aude*, also as *Mousses du département de l'Aude*, 11 fasc., 1870, nos. 1-276. See Sayre, G., Mem. New York Bot. Gard. 19(2): 240-248. 1971. Set at PC.
(4) *Fungi selecti gallici exsiccati*, cent. 1-74, 1879-1898. For details, including a list of published references see J. Stevenson, Beih. Nova Hedw. 36: 347-367. 1971 (dec. 1-41 as *Fungi gallici exsiccati*). *Sets* at: AUT, B, BPI, BR, CLF, CUP, DAOM, FH, H, I,ILL, K, L, LG, M, MPU, NHES, NY, NYS, PI, RUTPP, U, UMO, VC.
A selection of 400 specimens from the *Fungi gallici exsiccati* was published 1884-1887, as *Champignons qui envahissent les végétaux cultivés* by the "Bureaux de la Revue mycologique, ... Toulouse". Set at BR. – A further selection was published 1895 as *Genera fungorum exsiccata* (200 specimens).
(5) *Lichenes gallici exsiccati*, cent. 1-6, 1880-1884:
  1: 1879, notes Rev. mycol. 2: 31. 1880; Bot. Centralbl. 1/2: 157-158. 1880.
  2: 1880, Rev. mycol. 2: 159, 197. 1880; Bot. Centralbl. 3/4: 1407-1408. 1880.
  3: 1881, Rev. mycol. 3: 32-33. 1881; Bot. Centralbl. 7: 249-250. 1881.
  4: 1882, Rev. mycol. 4: 105. 1882; Bot. Centralbl. 11: 215-216. 1882.
  5: 1882, Rev. mycol. 5: 1882; Bot. Zeit. 40: 674. 29 Sep 1882.
  6: 1884.

Sets: BM, DUKE, G, L, LG, M, MPU, MSC, NY, OC, PC, S, U, UPS.
Ref.: Sayre, G., Mem. New York Bot. Gard. 19(1): 160. 1969.
(6): *Genera lichenum europaeaorum exsiccata*, 100 lichens appartenant à 100 genres ou sousgenres distincts, préparés pour l'étude et distribués systématiquement. Toulouse 1895. – Mainly taken from Roumeguère, *Lichenes gallici exsiccati*.
Ref.: Zahlbruckner, A., Bot. Centralbl. 66: 373-374. 1896.
Sayre, G., Mem. New York Bot. Gard. 19(1): 159-160. 1969.
Letters e.g. at G and FH.

BIBLIOGRAPHY and BIOGRAPHY: Ainsworth p. 277, 319, 320; Barnhart 3: 184 (b. 15 Aug 1828, d. 29 Feb 1892); BM 4: 1742; Bossert p. 340; CSP 5: 308, 8: 789, 11: 230, 12: 635-636, 18: 326; De Toni 1: cxxxix, 2: lxxxii; GR p. 297; GR, cat. 70; Hawksworth p. 185; Herder p. 170; IH 1 (ed. 6): 363, 2: (files); Jackson p. 10, 160, 164, 227, 229, 278, 280, 284, 291; Kelly 196-199; Kew 4: 527; Lenley p. 355; LS 2791, 3883, 4136a, 22978-23110, 23440-23441, 31963; Morren ed. 10, p. 69; MW p. 419; PR 7819-7820; Rehder 5: 742; SBC p. 130; Stevenson p. 1256; TL-1/see C.G. Nestler; TL-2/see A.N. Berlese, G. Bresadola, J.A. Mougeot, V. Mouton and 5052, 6372, 6455; Tucker 1:609; Urban-Berl. p. 281.

BIOFILE: Anon., Bot. Centralbl. 58: 191. 1892 (d.); Bot. Not. 1893: 39 (d.); Bot. Gaz. 17: 200. 1892 (d.); Bull. Soc. bot. France 26 (bibl.): 95. 1879 (gold medal Soc. arts sci. Carcassonne), table art. orig,. (1854-1903), p. 204-205; Flora 53: 43. 1870 (gold medal Carcassonne for his *Bryol. Aude*); Grevillea 20: 113. 1892 ("he has done some good work in his time..." sic); Österr. bot. Z.: 20: 123. 1870 (gold medal), 42: 223. 1892 (d.); Rev. mycol. 14: 57. 1892.
Astre, G., La vie de Benjamin Balansa 1947, p. 102, 153; Bull. Soc. bot. Hist. nat. Toulouse 101: 183, 186-187. 1966.
Boudier, M., Bull. Soc. mycol. France 8: 70. 1892 (obit.).
Clos, D., Rapport sur l'ouvrage de M. Casimir Roumeguère, ... qui a obtenu le grand prix de l'année 1857, Toulouse 1857, 8 p., reprinted from Mém. Acad. Sci. Toulouse (on *Descr. fig. Mousses Lichens Bordeaux*).
Debeaux, O., Rev. bot. 10: 664-65. 1892 (obit.).
Henriques, J., Bol. Soc. Brot. 10: 256. 1892 (obit.).
Debeaux, O., Rev. bot. 10: 664-665. 1892 (obit.).
Laundon, J.R., Lichenologist 11: 16. 1979.
Levi-Morenos, D., Notarisia 7: 1395. 1892 (obit.).
Roumeguère, C., Correspondances autographes inédites des anciens botanistes méridionaux. I. Pierre Barrera. II. Ramond & Picot de Lapeyrouse. Perpignan 1873, 52 p. (repr. from Bull. Soc. agr. sci. litt. Pyrén.-orient 20, 1873); Stat. bot. Haute-Garonne 93, 94, 95-97, 98-101. 1876 (bibl.).
Sayre, G., Bryologist 80: 516. 1977.
Wittrock, V.B., Acta Horti Berg. 3(3): 113, *pl. 142*. 1905 (portr.).

COMPOSITE WORKS: (1) founder and editor of the first mycological journal *Revue mycologique*, 1(1879)-13(1891).
(2) see A. Mougeot, *Algues* 1887, TL-2/6372.

HANDWRITING: Candollea 33: 147-148. 1978.

EPONYMY: *Roumegueria* (P.A. Saccardo) P.C. Hennings (1908); *Roumegueriella* Spegazzini (1880); *Roumeguerites* P.A. Karsten (1879).

9672. *Des lichens* et en particulier des lichens des environs de Toulouse pouvant être utilisés dans l'économie domestique, la médecine et les arts industriels ... Extrait du Journal d'Agriculture pratique et d'économie rurale, pour le Midi de la France, Septembre 1860.
Publ.: Sep 1860, p. [1]-9. *Copy*: G. – Reprinted from J. Agr. prat. Midi France ser. 3. 11: 385-393. Sep 1860. – Part of introduction to a manuscript *Monographie des lichens du bassin de la Garonne*, awarded a gold medal by the Toulouse Academy. See D. Clos (1857).

9673. *Cryptogamie illustrée* ou histoire des familles naturelles des plantes acotyledones d'Europe coordonnée suivant les dernières classifications et complétée par les recherches scientifiques les plus récentes. *Famille des Lichens* contenant 927 figures, représentant, pour chaque genre, la plante de grandeur naturelle et l'anatomie de ses différents organes de végétation et de reproduction, dessinés au microscope composé ... Paris (J.-B. Baillière ...), Toulouse (F. Gimet) 1868. Qu. *C(Crypt. ill., Lichens).

Publ.: 1868 (p. [iii]: 15 Apr 1868; BSbF Feb-Mai 1868; Flora 20 Mai 1868), p. [i-iii], [1]-73, *21 pl.* (uncol..). *Copies*: BR, FI, G, MICH, NY, PH.
Ref.: Hoffmann, Bot. Zeit. 27: 431. 25 Jun 1869 (rev.).

9674. *Cryptogamie illustrée* ou histoire des familles naturelles des plantes acotylédones d'Europe coordonnée suivant les dernières classifications et complétée par les recherches scientifiques les plus récentes. Famille des *Champignons* contenant 1,700 figures représentant, à ses différents âges, la plante de grandeur naturelle et l'anatomie de ses organes de végétation et de reproduction, dessinés au microscope composé ... Paris (J.-B. Baillière & fils ... F. Savy ...), Londres (H. Baillière ...), Madrid (C. Bailly-Baillère ...), New York (Baillière-Brothers ...) 1870. Qu. (*Crypt. ill., Champ.*).

Publ.: 1 Sep 1869-Jan 1870 (Bot. Zeit. 27: 631. 17 Sep 1869, publ. 1 Sep 1869; Flora 27 Dec 1869; BSbF rd. before 22 Jul 1870; mentioned as published, at least p. 2, on 28 Jan 1870 by de Seynes, (1870; rev. Apr-Mai 1870; see refs.), p. [i-ii], [1]-164, *23 pl.* (monochr. liths.). *Copies*: BR, FH, FI, G, Stevenson, USDA.
Index synonymique de la famille des Champignons. Complément du tome ii de la Cryptogamie illustrée. 1873, cover, p. 1-20, lithographed manuscript. (BSbF rev. post 17 Apr 1874; mentioned in 19(bibl.): 98. 1872 [sic] as published).
Second printing ("2ᵉ tirage") announced as still available by R. in Bull. Soc. bot. France 22(bibl.): 4. 1875.
Ref.: Anon., Bull. Soc. bot. France 17(bibl.): 56-57. Apr-Mai 1870 (but on p. 48, Feb-Mar 1870; published, review to come).
Fabre, A., Gaz. chir.-méd., Toulouse, 10 Jan 1870 (published, reviewed).
Fournier, E., Bull. Soc. bot. France 16 (bibl.): 192. 1892 (will be publ. Dec 1859).
Seynes, J. de, Bull. Soc. bot. France 17: 59-61. 1870 (séance 28 Jan 1870, "le second volume paru ...").

9675. *Bryologie du département de l'Aude.* Mémoire envoyé au concours de la médaille d'or de 200 francs, ouvert par la Société des Arts et Sciences de Carcassonne sur une question scientifique concernant le département de l'Aude et couronné par cette Société ... Carcassonne (Typographie de Louis Pomiés, ...) 1870. Oct. (*Bryol. Aude*).

Publ.: 1870, p. [1]-100, *1 pl.* (uncol. lith,.). *Copies*: FI, G, H, MO, MO-Steere, NY. – Reprinted from Mém. Soc. Sci. Arts Carcassonne 3: 1870. On p. [4] of the cover Roumeguère announces the publication of an accompanying set of exsiccata, 11 fascicles of 25 numbers each, of the mosses of the Dépt. de l'Aude.
Ref.: Anon., Bull. Soc. bot. France 16(bibl.): 143. Aug 1869.
Roumeguère, C., Bull. Soc. bot. France 16: 310-311. 1869 (séance 12 Nov 1869; gold medal awarded 1 Jul 1869; printing ordered by 1 Oct 1869; brief digest; see also 16: 435-448. 1869 for a *Catalogue des mousses du département de l'Aude*; a preliminary publication with restricted synonymy and less bibl. refs.

9676. *Une confusion dans les fleurs poétiques*, que distribue l'Académie des Jeux floraux ... Toulouse (Imprim. Louis et Jean-Matthieu Douladoure, ...) 1874. (*Confus. fl. poet.*).

Publ.: 1874, p. [1]-34. *Copy*: G. – Reprinted from l'Echo de la Province 19 & 20 Nov 1874 (journal n.v.); rev. Bull. Soc. bot. France 21(bibl.): 198. 1875.

9677. *Glossaire mycologique.* Etymologie et concordance des noms vulgaires ou patois avec les noms français ou latins des principaux champignons alimentaires et vénéneux du Midi de la France ... Perpignan (Imprimerie Charles Latrobe ...) 1875. Oct. (*Gloss. mycol.*).

Publ.: 1875, p. [i, iii], [1]-43. *Copies*: BR, FH, G(2). – Reprinted from Bull. Soc. agr. sci. litt. Pyrén.-orient 21: 217-259. 1874 (journal volume dated 1874).
Ref.: Anon., Bull. Soc. bot. France 22 (bibl.): 134-135. 1875.

9678. *Statistique botanique du Département de la Haute-Garonne* ... Paris (J.-B. Baillère [sic] et fils, ...) 1876. Oct. (*Stat. bot. Haute-Garonne*).
Publ.: 5 Apr 1876 (in journal; BSbF séance 28 Jul 1876; Bot. Zeit. 22 Sep 1876), p. [1]-101, *1 pl.* (uncol. lith.). *Copies*: BR, FH, G, H, HH(2), NY, U. – Reprinted from l'Echo de la Province of 5 Apr 1876 (journal n.v.).

9679. *Nouveaux documents sur l'histoire des plantes cryptogames et phanérogames des Pyrénées. Correspondances scientifiques inédites* échangées par Picot de Lapeyrouse, Pyrame de Candolle, Léon Dufour, C. Montagne, Auguste de St. Hilaire et Endress avec P. de Barrera, Coder & Xatart mises en lumière et annotées ... Paris (J.-B. Baillière et fils, ...) 1876. Oct. (*Corr. sci. inéd.*).
Publ.: 1876 (BSbF rd. 10 Nov 1876; Bot. Zeit. 20 Apr 1877, frontisp.), p.[1]-164, facsimile repr. handwriting Lapeyrouse. *Copies*: BR, FH, HH(2). – Reprinted from Bull. Soc. agr. Sci. litt. dépt. Pyrén.-orient. 22, 1876.

9680. *Les lichens neo-granadins et ecuadoriens* récoltés par M. Ed. André. [Extrait de la Revue mycologique numéro 4 Octobre 1879, page 160-171]. Oct.
Collector: Edouard François André (1840-1911).
Publ.: 4 Oct 1879, p. [1]-15 (on p. 15: Toulouse, imprimerie Douladoure). *Copy*: G. – Reprinted and to be cited from Rev. mycol. 1: 160-171. Oct 1879.

9681. *La mycologie des environs de Collioure* ou catalogue des funginées de cette localité par M.C. Roumeguère ... Bellac 1879. Oct. (*Mycol. Collioure*).
Publ.: 1879 (BSbF rd. 14 Nov 1879; Grevillea Dec 1879; Bot. Zeit. 5 Nov 1880), p. [i], [121]-144, [1-3, err.]. *Copy*: NY. – Reprinted from Sériziat, *Études sur Collioure et ses environs*, Bellac 1879.

9682. *Flore mycologique du département de Tarn-et-Garonne.* Agaricinées ... Étude qui a obtenu la médaille d'or de 400 fr. au concours de 1877, ... Montauban (imprimerie et lithographie Forestié, ...) 1879 [1880]. Oct. (*Fl. mycol. Tarn-et-Garonne*).
Publ.: 1880 (date on cover; t.p. dated 1879; BSbF issue for Nov-Dec 1880; Bot. Centralbl. 31 Jan-4 Feb 1881; BSbF copy presented 11 Feb 1881; J. Bot. Feb 1881; Bot. Zeit. 24 Jun 1881), p. [1]-278, *pl. 1-8* (uncol. liths.). *Copies*: BR, FH, G, NY, Stevenson; IDC 6308.
Ref.: Zimmermann (Chemnitz), Bot. Centralbl. 7: 194-196. 1881 (rev.).
N.L., Bot. Zeit. 39:407, 711. 1881.

9683. *Reliquiae mycologicae libertianae* series altera reviserunt C. Roumeguère & P.A. Saccardo ... Toulouse (Typographie Henri Montaubin, ...) 1881. Oct.
Co-author: Pier Andrea Saccardo (1845-1920).
Collector: Marie-Anne Libert (1782-1865) see TL-2/4496-4497. Roumeguère received the Libert collections of fungi on loan from BR and published the following papers on them together with P.A. Saccardo (the latter author of parts 3, 4).
1: Rev. mycol. 2(8): 15-24. 1 Jan 1880, following his general notes on the collections and letters on p. 7-14. We have seen no independently paged reprint of this publication.
2 (Series altera, see above): 1 Jul 1881 (BSbF rd. 11 Nov 1881; Nat. Nov. Oct(1) 1881; Bot. Centralbl. 3-7 Oct 1881; Hedwigia Oct 1881), p. [1]-21, *pl. 19-20. Copy*: FH. – Reprinted and to be cited from Rev. mycol. 3(11): 39-59, *pl. 19-20*. 1881. Saccardo junior author.
3 (Series tertia): 1 Oct 1883 (Hedwigia Oct 1883, Nat. Nov. Dec(2) 1883), p. [1]-7, *pl. 39-41. Copies*: FH, FI. – Reprinted and to be cited from Rev. mycol. 5: 233-239. *pl. 39-41*. 1883. Saccardo senior author.
4 (Series iv): Jan 1884 (Hedwigia Dec 1883, Feb 1884; Bot. Zeit. 28 Mar 1884; Nat. Nov. Feb(2) 1884), p. [1]-15, *pl. 42-46. Copies*: FH, G. – Reprinted and to be cited from Rev. mycol. 6: 26-39. *pl. 42-46*. 1884. Saccardo senior author.
Ref.: Winter (Leipzig), Bot. Centralbl. 8: 290-291. 1881 (rev. no. 2).

Roupell, Arabella Elizabeth (née Pigott), (1817-1914), British (English) botanical artist; married Thomas Boone Roupell of the East India Company; travelled at the Cape (S. Africa) 1843-1845; back in England 1858. (*Roupell*).

HERBARIUM and TYPES: Unknown; paintings at Cape Town University.

BIBLIOGRAPHY and BIOGRAPHY: Backer p. 500 (epon.); Barnhart 3: 184; BB p. 262; BM 4: 1742; Desmond p. 532 (b. 23 Mar 1817, d. 31 Jul 1914); GF p. 74; Jackson p. 347; Kew 4: 528; NI 1687, 1687n, see also 1: 122; Plesch p. 387-388; PR 7821; Rehder 1: 494.

BIOFILE: Bullock, A.A., Bibl. S. Afr. bot. 97. 1978.
Bird, Allen, The paintings of A.R., *in* More Cape flowers by a Lady, Johannesburg 1964; Arabella Roupell: pioneer artist of Cape flowers, S. Afr. nat. Hist. Publ. Co. 1975, 5 p., 18 pl., portr.
Gunn, M. & L.E. Codd, Bot. expl. S. Afr. 301-302. 1981 (further refs.).
Parry, V.T.H., Kew Bull. 32: 260. 1979 (rev. A. Bird 1975).
Richings, F.G., Quart. Bull. S. Afr. Libr. 27(1): 4-14. 1972 (biogr.).
Stearn, W.T., Roy. Hort. Soc. Exhib. manuscr. books 46. 1954 (no. 107).
Tyrell-Glynn, W. & M.L. Levyns, Fl. afr. 57. 1963.

HANDWRITING: Quart. Bull. S. Afr. Libr. 27: 9. 1972.

EPONYMY: *Roupellia* Wallich et W.J. Hooker ex Bentham (1849) honors the Roupell family, Charles Roupell of Charleston, South Carolina, his grandson Dr. Roupell of Welbeck Street, London, another grandson Thomas Boone Roupell in the East India Company, as well as the latter's wife, Arabella Elizabeth; *Roupallia* Hasskarl (1857, *orth. var.*); *Roupellina* (Baillon) Pichon (1950).

9684. *Specimens of the flora of South Africa* by a Lady [London 1850]. Fol. (*Spec. fl. S. Afr.*).
Publ.: Mar 1850 (Hooker's J. Apr 1850; binding finished by 18 Mar 1850, letters to W.J. Hooker, Richings 1972), title-page (coloured plate), dedic. [i], preface [iii, dated Sep 1849], list of 103 subscribers, *9 plates* (nos. 1-8, *pl. 9*, unnumbered, on last page of letterpress), letterpress accompanying plates [1-17], in all *10 pl.* and 23 p. *Copies*: BR, USDA. – The book is anonymous, but N. Wallich and W.J. Hooker, *in* W.J. Hooker Mag. 4466. 1 Sep 1849, and W.J. Hooker, J. Bot. Kew. Gard. Misc. 2: 127-128. Apr 1850 (rev.) disclose the identity of the author. The botanical text is by William Henry Harvey (1811-1866). See also W. Junk, Rara 2: 638. 1926-1936.
More Cape flowers by a Lady. The paintings of Arabella Roupell. Text by Allan Bird. The South African Natural History Publication Company, Johannesburg, 1964, p. [i-xv], *12 pl.* (col.), [1-2, bibl.]. *Copies*: MO, NY.

Rouppert, Kazimierz (Stefan) (1885-1963), Polish botanist; Dr. phil. Krakow 1909; professor of botany at Krakow until 1939; at the Treub Laboratory Buitenzorg (Bogor), Java 1926; from 1939 at the University of Budapest. (*Rouppert*).

HERBARIUM and TYPES: KRA; further material at DPU.

BIBLIOGRAPHY and BIOGRAPHY: Barnhart 3: 184; Futak-Domin p. 509; Hirsch p. 254; IH 2: (files); Kelly p. 199; Kew 4: 528; LS 38615-38617, 41682; SK 1: 449-450.

BIOFILE: Szymkiewicz, D., Bibl. Fl. Polsk. 16, 31, 32, 42, 56, 122-123. 1925.
Szafer, W., Concise hist. bot. Cracow 61, 77, 78, 80, 84, 86. 1969.

9685. *Wrażenia ogrodnicze z jawy* ... Kraków (Nakładem Krakowskiego Towarzystwa Ogrodniczego) 1927. Oct.
Publ.: 1927, p. [1]-30, *13 pl. Copies*: G, H.

Rousseau, Jean Jacques (1712-1778); French writer and philosopher. (*Rousseau*).

HERBARIUM and TYPES: following the eighteenth century tradition Rousseau accompanied his interest in botany by making herbarium specimens, and even by the acquisition of plants collected by others. The following herbaria are known to have been made by Rousseau (enumeration Briquet 1940):
1. A folio sized herbarium of 1500-2000 plants, sold by him in 1775-1776, of which the catalogue of 1770 was at B. (See Urban-Berl. p. 416-418).

2. A quarto-sized herbarium (10 volumes) which belonged originally to Mlle de Girardin, was kept at B (see Urban-Berl. p. 416-618).
3. The quarto-sized herbarium originally containing 1500 specimens made in Paris and Ermenonville towards the end of Rousseau's life, given by Thérèse Levasseur to Le Bègue de Presle is now at P (available on IDC 6213). For its history see Lanjouw and Uittien 1940. This herbarium has great scientific value because it contains part of the collections made by Fusée Aublet. – Available on microfiche IDC 6213.
4. A herbarium made at Scoraille (Auvergne) of which the location is now unknown.
5. A herbarium of 200 plants made for Marguérite-Madeleine Delessert (see Jaccard 1893).
6. A small herbarium made for Mlle de Girardin (see Roux).
7. A small herbarium made for Mlle Julie Boy de la Tour (seee Dufour).
8. A small herbarium sent to Malesherbes in 1773.
9. A herbarium sent to the Duchess of Portland in 1773.

NOTE: Rousseau's interest in botany was awakened in 1763 or 1764, shortly after he had gone to Switzerland, folllowing the outlawing of his Émile (see Jansen 1885, de Beer 1954). He entered into relations with various botanists like Abraham Gagnebin, Bernard de Jussieu, Michel Adanson and, later in life, Guettard and Fusée Aublet. The *Lettres élémentaires sur la Botanique* (1781) were written at the request of Madeleine Catherine [Julie] Boy de la Tour, (Mme Étienne Delessert, b. 1747) to the benefit of the instruction of her eldest daughter Marguérite-Madeleine Delessert (sister of Benjamin Delessert). The *Lettres* greatly influenced both Marguérite-Madeleine and Benjamin and may well have been the initial stimulus for the latter's interest in botany. On the one hand the *Lettres* did much to popularize botany and may have been the initial cause for the setting up of one of the world's greatest herbaria (Benjamin Delessert, now at G); on the other hand it provided Pierre-Joseph Redouté with an opportunity to publish a famous illustrated edition containing some of his best work.

BIBLIOGRAPHY and BIOGRAPHY: BM 8: 1099; Bossert p. 340; Dryander 3: 23; GR p. 644; Hegi 2: 55-56, 171, 654, 3: nos. 1281-1288; Herder p. 95; IH 1 (ed. 6): 363, (ed. 7): 343, 2: (files); Jackson p. 7, 35; Kelly p. 199; Kew 4: 529; Lasègue 43-45, 54, 61, 551, 552; Moebius p. 407, 424, 433, 434; NI 1688, see also 1: 147; PH 247; Plesch p. 388, 389, 390; PR 5927, 6062, 7454, 7822-7824, ed. 1: 2552, 6627, 6751, 8412, 8775, 11606; Rehder 1: 93, 5: 742; SO 701-709, add. 700d, 700e, 2471, 2475, 2680; Sotheby 661-662; TL-1/810; TL-2/5568, see G.L.L. de Buffon; Tucker 1: 610; Urban-Berl. p. 389, 416-418.

BIOFILE: Allen, D.E., The naturalist in Britain 40, 49, 53, 54, 74. 1976.
Amoureux, Mém. Soc. Linn. Paris 1: 704-706. 1822 (letters to A. Gouan).
Anon., Cat. gén. livr. impr. Bibl. natl., Paris, Auteurs 157: 410-535. 1939 (bibl.).
Baehni, C., Aesculape 28(4): 62, 112-119. 1938 (on *Lettres*; portr. Marguérite-Madeleine Delessert).
Beer, G. de, Ann. Sci. 10(3): 189-223. 1954 (free translation, with notes and quotations, of Jansen 1885).
Biers, P.M., Rev. bryol. ser. 2. 1(1): 49. 1928 (on two bryologists: corr. between Rousseau and Lamoignon de Malesherbes).
Blunt, W., The compleat naturalist 100, 214. 1971.
Bord, B., Aesculape Mai 1938: 6 (portr.) (expos. botanistes genevois).
Bridson, G.d.R. et al., Nat. hist. mss. res. Brit. Isl. 275.65. 1980.
Briquet, J., Bull. Inst. natl. genevois 41: 131-137. 1914 ("Jean-Jacques Rousseau botaniste"); Bull. Soc. bot. Genève 50a: 417-420. 1940 (bibl.; list herb.).
Burkhardt, R.W., The spirit of system 14, 15, 17, 34, 145, 215. 1977.
Candolle, Alph. de, Phytographie 444. 1880.
Candolle, A.P. de, Mém. Soc. phys. Genève 5: 42. 1830; Mém. souvenirs 2, 86, 300, 302. 1862.
Cohn, F., Deutsche Rundschau 47: 364-385. 1886 (J.J.R. als Botaniker).
Cummings, B., J. Bot. 54: 80-84. 1916 ("Rousseau as botanist").
Dufour, T., Pages inédites de J.J.R. sér. 2: 110-117. 1907 (herb. Julie Boy de la Tour).
Earnest, E., John and William Bartram 2, 87, 133, 141, 146, 149. 1940.

Eichler, A.W., Jahresber. Schles. Ges. vaterl. Cultur 1886: 153-154. 1887 (on the Rousseau herbarium at B).
Fagin, N.B., William Bartram 42, 56, 57. 1933.
Flahault, C., Univ. Montpellier, Inst. bot. 36. 1890 (R. at Montpellier).
Foggitt, G., Rep. Bot. Soc. Exch. Club Brit. Isl., Annals 1932: 283. (occasional mention of "a large herbarium of French plants supposed to have belonged to Jean Jacques Rousseau").
Fries, Th.M., Bref Skrifv. Linné 1(3): 240. 1909, 1(5): 322. 1911, 1(6): 223. 1912.
Gagnebin, B., ed., Lettres sur la botanique par Jean-Jacques Rousseau. Club des Libraires de France. 1962, xxxv, 305 p., 8 col. photos (contains texts of letters on botany to various recipients and a fragment of Rousseau's botanical dictionnary).
Glass, B. et al., Forerunners of Darwin 468 [index]. 1968.
Godet, P. & M. Boy de la Tour, eds., Lettres inédites de J.J. Rousseau, Paris 1911, p. 65.
Gonod d'Artemare, E., Bull. Acad. int. Géogr. bot. 8(114): 145-152. 1899 (early history of herb. no. 3/Ermenonville).
Heine, H.H., Jahrb., Ver. Schutz. Alpenpfl. 27: [4 p.]. 1962.
Jaccard, P., Bull. Soc. vaud. Sci. nat. 30: 85-88. 1893 (on a 200 pl. herb. for Marguérite-Madeleine Delessert).
Jansen, A., Jean-Jacques Rousseau als Botaniker. Berlin (Georg Reimer) 1885, vi, 308 p. (see Moebius, M., 1885 for review).
Jessen, K.F.W., Bot. Gegenw. Vorz. 337-339, 378, 396. 1884.
Hocquette, M., Bull. Soc. bot. Nord France 16(1): 17-20. 1963 (on herbaria and *Lettres*).
Lacroix, A., Figures de savants 3: 84, 85, 4: 74, 212. 1938.
Lanjouw, J. & H. Uittien, Med. Bot. Mus. Utrecht 75: 133-170. 1940 (on herb. no. 3).
Lefébvre, Mém. Soc. Linn. Paris 1: 41-48. 1822.
MacPhail, I., Cat. Redout. 52-53, 66. 1963.
Magnin, A., Bull. Soc. bot. Lyon 31: 48-49, 70. 1906 (on botanical field trips around Lyon and Grenoble; bibl.), 32: 115. 1907, 35: 26, 32, 67-68. 1910.
Moebius, M., Bot. Centralbl. 24: 194-199. 1885 (rev. Jansen 1885); Gartenwelt 16: 374-378, 384-389. 1912 (J.J. Rousseau als Botaniker).
Nicolas, J.P. and Stafleu, F.A., *in* G.H.M. Lawrence, Adanson 1: 388 [index]. 1963 (on relations with Adanson).
Ottevanger, K., (transl.), J.J. Rousseau, Botany, a study of pure curiosity, London 1979, 154 p.
Roux, C., Ann. Soc. linn. Lyon 60: 101-120. 1913 (on R.'s botanical excursions at Grande Chartreuse in 1768 and Mont Pilat in 1769).
Smith, J.E., Sel. corr. Linnaeus 2: 552. 1821 (letter R. to Linnaeus).
Stafleu, F.A., Taxon 3: 247. 1954 (herb. no. 3, acquired by P.); *in* L'Héritier, Sert. angl. facs. ed. 1963, p. xvi, xx; *in* J. MacPhail, Cat. Redout. 10, 23. 1963; Linnaeus and the Linnaeans, 281, 282, 296, 298. 1971.
Starobinski, J., Gesnerus 22: 83-94. 1964 ("Rousseau et Buffon").
Wittrock, V.B., Acta Horti Berg. 3(2): 117. 1903, 3(3): 113, 210-211. 1905.

EPONYMY: *Roussaea* A.P. de Candolle (1839, *orth. var.* of *Roussea* J.E. Smith); *Roussea* J.E. Smith (1789); *Rousseaua* Post et O. Kuntze (1903, *orth. var.* of *Roussea* J.E. Smith); *Rousseauvia* Bojer (1837, *orth. var.* of *Roussea* J.E. Smith); *Roussoa* J.J. Roemer et J.A. Schultes (1818, *orth. var.* of *Roussea* J.E. Smith); *Russea* J.F. Gmelin (1791, *orth. var.* of *Roussea* J.E. Smith). *Note*: *Roussoella* P.A. Saccardo (1888) and *Roussoellopsis* I. Hino et K. Katumoto (1965) honor Maria Rousseau, Belgian mycologist.

POSTAGE STAMPS: France 15 f. (1956) yv. 1084; 1,00 + 0,20 f. (1978) yv. 1990; Switzerland 5 + 5 c. (1962) yv. 693.

9686. *Lettres élémentaires sur la botanique* [Collection complète des oeuvres de J.J. Rousseau, citoyen de Genève. Tome quatorzième. contenant le iv. volume des Mélanges. À Genève 1782]. Oct. (*Lettr. élém. bot.*).
Publ.: 1782, p. [i-ii], [1]-535. *Copy*: G. – Previous editions not seen; this was possibly the first publication. – Oeuvres ed. Peyrou et Moultou.
Other ed.: 1789, in Oeuvres complètes de J.J. Rousseau, nouvelle édition, tomes 5, 6, 1789 [Paris]. *Copies*: BR, NY.

1: engr. t.p., p. [1]-393.
2: engr. t.p., p. [1]-507.
Plates: 1789, p. [i], *pl. 1-38, 1-6* (col. copp., J. Aubry). *Copies*: E, G, HH, MO. – "*Recueil de plantes coloriées*, pour servir à l'intelligence des lettres élémentaires sur la botanique de J.J. Rousseau ... à Paris (Chez Poinçot, ...) 1789. – The NY copy of the text has these plates added. Plates *1-6* illustrate the fragment of a botanical dictionary. – Vol. 38bis of the Poinçot ed. of J.J. Rousseau: Oeuvres compl.
English ed.: *Letters on the elements of botany*, by Thomas Martyn, ed. 1, 1785, see TL-2/5568.
Ed. 8: 1815, p. [i]-xx, [1]-434. *Copies*: NY, PH. – London (printed for White, Cochrane and co., ...) 1815. Oct. (in fours).
German ed.: J.J. Rousseau's Botanik für Frauenzimmer in Briefen an die Frau von L.** Frankfurt und Leipzig 1781, 126 p. Oct. *Copy*: MO. – Modern German ed. *Zehn botanische Lehrbriefe für eine Freundin*, Insel Taschenbuch 366. 1979; Botanik für artige Frauenzimmer, 65 ills. P.J. Redouté, Hanau 1980.
Russian ed.: Moskwa 1810, by Wladimir Ismailow (n.v.).
Recent reprint of original texts: see B. Gagnebin (1962).

9687. *La botanique de J.J. Rousseau*, contenant tout ce qu'il a écrit sur cette science; l'exposition de la méthode botanique de M. de Jussieu; la manière de former les herbiers par M. Haüy. À Paris (chez F. Louis, ...) an x-1802. Duod. (*Bot. Rousseau, ed. duod.*).
Ed. 1: 1802, p. [i]-xxiv, [1]-322. *Copies*: BR, G, H-UB, NY.
Ed. 2: 1823, p. [i*-ii*], [i]-viii, [3]-340, *pl. 1-8* with text (uncol.). *Copy*: G. "... sur cette science, augmentée de l'exposition de la méthode de Tournefort, de celle du système de Linné, d'un nouveau dictionnaire de botanique, et de notes historiques, etc. Par M.A. Deville, médecin. Seconde édition ... " à Paris (chez François Louis, ...) 1823. Duod.

9688. *La botanique de J.J. Rousseau*, ornée de soixante-cinq planches, d'après les peintures de P.J. Redouté. Imprimerie de H. Perronneau. Paris (Delachaussée, ... Garnery, ...) xiv = 1805. Fol. (*Bot. Rousseau ed. fol.*).
Orig. ed.: 1805 (published in 7 parts), p. [i*, iii*], lettres [1]-81, fragm. dict. [83]-122, *pl. 1-65* (stipple engr., printed in colour and finished by hand after paintings by P.J. Redouté). *Copies*: BR, PH.
Other issue: 1805, p. [i]-viii, [ix-x], [1]-124, *pl. 1-65* (id.). *Copy*: NY. – *La botanique* ... P.J. Redouté. Paris (Delachaussée, ...Garnery, ...) xiv = 1805. Fol.
Quarto ed.: 1805 (n.v.), see MacPhail no. 16. Quarto, p. [2], [v]-xi, [1], [1]-159, [1].
Quarto ed.: 1821, Paris Baudoin frères, 1821. n.v.
Quarto ed.: 1822, p. [i]-xi, [1]-153, [154], *pl. 1-65* (id.). *Copy*: G. – Paris, Baudoin frères, ... 1822. Qu.
Octavo ed.: 1824, p. [1]-468, [1]-7, *pl. 1-65* (id. but coloured). *Copies*: BR, NY. – Oeuvres complètes de J.J. Rousseau, mises dans un nouvel ordre, ... par V.D. Musset-Pathay. Philosophie. *Lettres sur la botanique*, suivies d'une introduction à l'étude de cette science, et de fragments pour un dictionnaire des termes d'usage en botanique. Paris (chez P. Dupont, ...) 1824. Oct. (*Lettres bot. 1824*).
N.B. We have made no attempt to treat all editions and issues of this book.

9689. *Le botaniste sans maître*, ou manière d'apprendre seul la botanique au moyen de l'instruction commencée par J.J. Rousseau. Continuée et complettée [sic] dans la même forme par M. de C. Paris (chez Levrault, Schoell et Comp., ...) Winterthour (chez Steiner-Ziegler, ...) 1805. Duod. (*Bot. sans maître*).
Author: Joseph Philippe de Clairville (1742-1830).
Publ.: 1805, p. [i*], [i]-xxiv, [1]-297, *pl. 1-6*. (Other edition of the original *Lettres*; 300 copies printed. *Copy*: BR.

Rousseau, [Joseph Jules Jean] **Jacques** (1905-1970), Canadian botanist; student of Marie-Victorin; Dr. phil. Univ. Montréal 1934; demonstrator of botany Univ. Montreal 1926-1928; assistant, later associate professor of botany ib. 1928-1944; director of the Montreal botanical garden 1944-1957; director of human history branch, Natl. Mus. Canada Ottawa 1957-1959; professor at the Musée de l'Homme, Paris 1959-1962; professor at the Centre d'Études nordiques, Université Laval, Québec City 1962-1970;

member of the editorial committee for botanical nomenclature 1954-1964. (*J. Rousseau*).

HERBARIUM and TYPES: MT and MTJB; other material at A, C, CAN, DAO, DPU, E, GB, GH, H, K, NY, QUE, QMP, US, WELC.

BIBLIOGRAPHY and BIOGRAPHY: Ainsworth p. 299; Barnhart 3: 184 (b. 5 Oct 1905); BL 1: 141, 145, 316; BM 8: 1099; Bossert p. 340; GR p. 740; IF (ed. 2): 84, (ed. 3): 105; IH 1(ed. 7): 341; PH 209; Roon p. 96.

BIOFILE: Caron, F., Arctic 24(2): 151-152. 1971 (obit.).
Cooke, A., Polar Record 15: 556-557. 1971 (obit.).
Ewan, J. et al., Short hist. bot. U.S. 119. 1969.
Kucyniak, J., Bryologist 49: 137. 1946.
Löve, A., Taxon 20(1): 153-156. 1971 (obit., portr.).
Morisset, P., Natural. canad. 97: 497-498. 1970 (obit., portr.).
Pomerleau, R., Natural. canad. 98: 215-224. 1971 (portr.).
Rollins, R.C., Taxon 5: 2, 3. 1956 (in group photos).
Rousseau, J., Curriculum vitae et bibliographie, 60 p., mimeogr. s.d.n.l. (Montréal 1955) (copy: FAS).
Rumilly, R., Le frère Marie-Victorin et son temps 457 [index]. 1949.
Stafleu, F.A., Taxon 3: 219. 1954 (portr.).
Smit, P. & R.J.C.V. ter Laage, eds., Essays in biohistory 21, 50, 53, 219, 220. 1970 (Regn. veg. 71).
Verdoorn, F., ed., Chron. bot. 3: 78, 81, 310. 1937, 4: 436. 1938, 6: 464. 1941.

MEMORIAL VOLUME: Le Naturaliste canadien 98(3), 1971 (portr., biogr. by R. Pomerleau p. 215-224).

9690. *Études floristiques sur la région de Matapédia* (Québec). Notes sur la flore de Saint-Urbain, Comté de Charlevoix (Quebec) ... Ottawa (F.A. Acland ...) 1931. Oct.
Publ.: 1931, p. [i]-v, 1-36, *pl. 1-2. Copy*: MICH. – Musée national du Canada, Bulletin no. 66, série biol. no. 17.

Rousseau, Marietta, née Hannon (1850-1926), Belgian self-taught mycologist; from 1908 connected with the Jardin Botanique de l'État, Bruxelles; married E.J. Rousseau 1871. (*M. Rousseau*).

HERBARIUM and TYPES: BR; other material at CUP.

BIBLIOGRAPHY and BIOGRAPHY: Barnhart 3: 184; BM 4: 1742; CSP 13: 667-668; GR p. 691; Hawksworth p. 185; IH 1 (ed. 7): 341; Kelly p. 199; Stevenson p. 1256; TL-2/627-629, 1592.

BIOFILE: Beeli, M., Natural. Belg. 7: 18-20. 1926 (obit., bibl.).
Dwyer, J.D., Taxon 22: 563. 1973 (coll. in Panama 1914).
Raab, H., Z. Pilzk. 57. 61-62, 70. 1979.

COMPOSITE WORKS: (1) with E.C. Bommer, *Fungi, in* T.A. Durand & H. Pittier, *Prim. fl. costaric.* 1(3): 81-96. 1896.
(2) with E.C. Bommer, *Cat. champ. Bruxelles.* 1879, TL-2/627.
(3) with E.C. Bommer, *Fl. mycol. Bruxelles* 1884. TL-2/628. (rev.: H.G. Winter, Bot. Centralbl. 24: 2-3. 1885).

EPONYMY: *Roussoella* P.A. Saccardo (1888), *Roussoellopsis* I. Hino et K. Katumoto (1965).

9691. Expédition antarctique belge. *Résultats du Voyage du S.Y. Belgica* en 1897-1898-1899 sous le commandement de A. de Gerlache de Gomery. Rapports scientifiques ... *Botanique Champignons* par Mmes E. Bommer et M. Rousseau. Anvers [Antwerpen] (Imprimerie J.E. Buschmann ...) 1905 Qu. (*Résult. Voy. Belgica, Champ.*).
Co-author: Elisa Caroline Bommer, née Destrée, wife of the pteridologist Joseph Édouard Bommer. (TL-2/630).

Publ.: shortly after 1 Apr 1905 (p. 2; Bot. Centralbl. 13 Feb 1906; Nat. Nov. Dec 1905), p. [1]-15, *pl. 1-5* (uncol. H. Durand). *Copies*: G, MO, Stevenson, USDA. – See TL-2/629.
Preliminary publ.: Bull. Acad. r. Belg., Sci. 1900(8): 640-646.
Ref.: Jonge, A.E. de, Bot. Centralbl. 104: 90. 1907 (rev.).
Lindau, G., Hedwigia 45: (88)-(89). 1906.

Roussel, Alexandre Victor (1795-1874), French military pharmacist and botanist from Melun, Seine-et-Marne; stationed in various garrisons in France (e.g. Toulon, Metz) 1820-1835; stationed in Algeria as principal pharmacist for the military hospitals 1835-1838; ultimately principal pharmacist at the Paris Hôpital du Val-de-Grace. (*A. Roussel*).

HERBARIUM and TYPES: P & PC; other material at BR, DBN, FI, G, M, MPU (herb. J. Vichet). – Letters to J.B. Mougeot at P (Bibl. centr. 137 nos., 1838-1858). Library dispersed by sale (5 Apr 1875).

BIBLIOGRAPHY and BIOGRAPHY: AG 6(1): 163; Bossert p. 340 (b. 28 Jul 1795, d. 17 Dec 1874); CSP 8: 790, 11: 230; GR p. 348-349; Jackson p. 369 (err. sub E. Roussel); LS 23117, 23118 (err. sub E. Roussel); TL-2/see P.T. Husnot (in index err. as E. Roussel); TR 6236.

BIOFILE: Anon., Bull. Soc. bot. France 22: 109. 1875 (herb. bought by P.).
Cosson, E., Comp. fl. atl. 1: 87. 1881.
Cosson, E. & Durieu, Exp. sci. Alger. Bot. 2: xliv. 1868.
Hertel, H., Mitt. Bot. München 16: 425. 1980.
Husnot, T., Rev. Bryol. 2: 16 (d.), 48 (sale library), 104 (herb. bought by P). 1875.
Maire, R., Progr. conn. bot. Algérie 23, 180, 188. 1931.
Roumeguère, C., Bull. Soc. bot. France 22: 6-9. 1875 (obit., b. 28 Jul 1795, d. 17 Dec 1874; "Que dans ce mouvement des hommes et des choses nous restions attachés aux études qui ont si bien rempli notre temps ..." [Roussel]).
Seynes, J. de, Bull. Soc. bot. France 21: 381. 1874 (1875) (d.).
Urban, I., Symb. ant. 1: 141. 1898.

EPONYMY: *Rousselia* Gaudichaud-Beaupré (1830), fide Roumeguère 1875.

HANDWRITING: Webbia 32(1): 34. 1977 (no. 66).

Roussel, Ernest (*fl.* 1860), French pharmacist and mycologist at Évreux; studied at the École de médecine et de pharmacie de Rouen (thesis 1860). (*E. Roussel*).

HERBARIUM and TYPES: Unknown.

BIBLIOGRAPHY and BIOGRAPHY: BM 4: 1742; GR p. 349 (err.); Kelly p. 199; LS 23115; PR 7825.

9692. *Des champignons comestibles et vénéneux* qui croisssent dans les environs *de Paris*. Travail présenté comme sujet de thèse, devant l'École de médecine et de pharmacie de Rouen, par Ernest Roussel (d'Évreux) pharmacien ... Paris (Libraire Victor Masson et Fils ...) 1860. Oct. (*Champ. comest. vénén. Paris*).
Publ.: 1860, p. [1]-68. *Copy*: NY.
Ref.: Fournier, E., Bull. Soc. bot. France 8: 112. Apr 1861.

Roussel, Henri François Anne de (1748-1812), French (Normandy) botanist; Dr. med. Caen 1771 or 1772; professor of medicine at the University of Caen 1773; professor of medical botany ib. 1776; from 1801 professor of botany at the Faculté des Sciences, Caen (preceding J.V.F. Lamouroux); director of the Jardin botanique de la ville de Caen 1786-1797 and 1801-1812. (*Roussel*).

HERBARIUM and TYPES: CN; further material at DBN.

ROUSSEL, H. F. A. de

BIBLIOGRAPHY and BIOGRAPHY: Barnhart 3: 185 (b. 11 Jul 1748, d. 12 Feb 1812); BM 4: 1742-1743; CSP 5: 309; De Toni 1: cxix; Hawksworth p. 185; Lasègue p. 442, 504; PR 7826-7827, ed. 1: 8778-8779; Rehder 1: 406, 3: 55; SO add. 759a; Stevenson p. 1256; TL-2/see J.V.F. Lamouroux.

BIOFILE: Desvaux, N.A., J. de Bot. ser. 2. 1: 141-144. 1813 (b. 1747; obit.).
Gager, C.S., Brooklyn Bot. Gard. Rec. 27(3): 196. 1938.
Hardouin, L. et al., Cat. pl. vasc. Calvados 58. 1848.
Lange, M., Notice sur H.-F.-A. Deroussel, Caen (P. Chalopin) 1812, 24 p. (obit. read before the Acad. Sci. Caen 17 Jul 1812; provides further details on his career and position at Caen which were more involved than shown above because of various reorganisations).
Letacq, A.L., Bull. Soc. Amis Sci. nat. Rouen: ser. 5. 42: 90-91. 1907 (biogr.).
Lignier, O., Essai sur l'histoire du Jardin des Plantes de Caen 84-90. 1904 (various data on R.'s career and functions [this and other accounts do not always agree on various dates and appointments; further research needed]), see also Mém. Acad. Sci. Arts Caen 1891(2): 89-91.
Poggendorff, J.C., Biogr.-Lit. Handw.-Buch 3: 1148. 1898.

9693. *Tableau des plantes usuelles* rangées par ordre, suivant les rapports de leurs principes & de leurs propriétés ... à Caen (de l'imprimerie de L.J. Poisson, ...) 1792. Oct. (*Tabl. pl. usuel.*).
Publ.: 1792, p. [1]-175. *Copy*: BR.

9694. *Flore du Calvados* et terreins adjacents, composée suivant la méthode de Jussieu ... à Caen (De l'imprimerie de Louis-Jean Poisson, ...) Quatrième année républicaine [23 Sep 1795-16 Sep 1796]. Oct. (in fours). (*Fl. Calvados*).
Ed. [*1*]: 23 Sep 1795-16 Sep 1796 (is 4th year of the republic), p. [1]-268. *Copies*: G, HH, USDA.
Ed. 2: 1806, p. [1]-371, [372]. *Copies*: FH, NY, PH, Stevenson; IDC 5346. – "*Flore du Calvados* ... Jussieu, comparée avec celle de Tournefort et de Linné. IIe. Edition, dans laquelle les cryptogames sont distribuées par séries, où l'on a réuni quelques genres nouveaux ..." à Caen (De l'imprimerie de F. Poisson, ...) 1806. Oct. (in fours).

Rouville, Paul Gervais de (1823-1907), French geologist, palaeontologist and botanist; Dr. sci. Montpellier 1853; professor of geology and mineralogy at the University of Monpellier 1862-1894. (*Rouville*).

HERBARIUM and TYPES: Unknown.

BIBLIOGRAPHY and BIOGRAPHY: Barnhart 3: 185; CSP 5: 311, 8: 791, 11: 231-232, 18: 331-332; Kew 4: 529; PR 7829; Quenstedt p. 369 (b. 25 Mai 1823, d. 29 Nov 1907).

BIOFILE: Delage, A., Bull. Soc. géol. France ser. 4. 8: 211-222. 1908 (obit., bibl., b. 25 Mai 1823, d. 29 Nov 1907).
Poggendorff, J.C., Biogr.-lit. Handw.-Buch 4(2): 1278. 1904 (bibl., b. 25 Mai 1823).
Zittel, K.A. von, Gesch. Geol. Paläont. 532, 604, 672. 1899; Hist. geol. palaeobot. 450. 1901.

COMPOSITE WORKS: Co-editor (with C.H.M. Flahault) *Revue des Sciences naturelles* ser. 3, vol. 1-4, 1881-1885.

9695. Thèse de botanique. *Monographie du genre Lolium* ... Montpellier (Boehm, imprimeur de l'Académie, ...) 1853. Qu. (*Monogr. Lolium*).
Publ.: 1853, p. [i]-viii, [9]-57, *pl. 1-3* (uncol. liths., S.A. Node). *Copies*: G, LC, MICH, NY.

Roux, Honoré (1812-1892), French stevedore and naturalist at Marseille. (*Roux*).

HERBARIUM and TYPES: 200 specimens at G; Roux offered his herbarium for sale in 1881 (5000 specimens).

BIBLIOGRAPHY and BIOGRAPHY: Barnhart 3: 185 (b. 19 Nov 1812, d. 3 Dec 1892); BL 2: 192, 710; GR p. 349; Morren ed. 10, p. 65; PR ed. 1: 4351; Rehder 1: 416; TL-2/3125; Tucker 1: 610.

BIOFILE: Burnat, E., Bull. Soc. bot. France 30: cxxix-cxxx. 1883.
Cavillier, F., Boissiera 5: 82-83. 1941.
Fournier, E., Bull. Soc. bot. France 28(bibl.): 192. 1881 (herb. of 5000 specimens for sale).

COMPOSITE WORKS: Collaborator, E. Huet du Pavillon, *Cat. pl. Provence* 1889; TL-2/3125.

9696. *Catalogue des plantes de Provence* spontanées ou généralement cultivées ... Marseille (Typographie et lithographie Marius Olive ...) 1881[-1891]. Oct. (*Cat. pl. Provence*).
Publ.: in 5 fascicles between 1881 and 1892 (p. 1-104: 1881, possibly 1879-1881 in journal; p. 105-170. 1881-1883, in journal; complete work Jul 1892), p. [i*-iii*], [i]-vii, [viii], [1]-654, [655, dated 1 Sep 1891]. *Copies*: G, HH, USDA. – Originally published (in parts) in Bull. Soc. bot. hort. Provence).
Additions: L. Legré, Bull. Soc. bot France 38: 393-402. 1892.
Supplément: 1893, p. [657]-696. *Copies*: HH, USDA. – BSbF séance 28 Jul 1893, Marseille 1893 (p. 696: "manuscrit posthume, mis au courant par une commission de la Société d'Horticulture et de Botanique de Marseille". BSbF séance 8 Jul 1892, copy rd. BSbF 39(bibl.): 48, Jun-Dec 1892 "vient de paraître".
Ref.: Malinvaud, E., Bull. Soc. bot. France 39(bibl.): 109-111. Jun-Dec 1892 (rev.) (as of 1891, 655 p.), 40(bibl.): 119. 1893 (suppl. Marseille 1893, 40 p.).

Roux, Nisius (1854-1923), French botanist at Lyon; employee at various sericultural enterprises from 1906 in retirement dedicating himself almost entirely to botany; general secretary of the Société linnéenne de Lyon. (*N. Roux*).

HERBARIUM and TYPES: LY; further material at DBN, E, MPU (herb. J. de Vichet).

BIBLIOGRAPHY and BIOGRAPHY: Barnhart 3: 185 (d. 12 Mar 1923); BL 2: 114, 711; CSP 18: 334; Morren ed. 10, p. 65; PFC 1: li, 2(2): xxvii, 3(2): xii; Saccardo 1: 142.

BIOFILE: Anon., Bot. Not. 1923: 384(d.); Bull. Soc. bot. France 60: 584. 1914 (coll. in Corsica).
Beauverie, J., Soixante-huitième Congrès des Sociétés savantes 1935, p. 209 (herb. Nisius Roux at LY).
Meyran, O., Ann. Soc. bot. Lyon 43: 77-82. 1924 (obit., portr., bibl.).
Molliard, M., Bull. Soc. bot. France 70: 248. 1923 (d.).
Thiébaut, J., Monde des Pl. 24 (26-141): 1. 1923 (obit.).

NOTE (homonymy): (1) Auguste Roux (1855-1882), algologist and curator of the Cosson herbarium; later lecturer at the École supérieure des Sciences, Alger, 1880-1882; travelled in Algeria and Tunisia (see e.g. F. Cosson, Comp. fl. atl. 1: 87-89. 1881 and R. Maire, Progr. connaiss. bot.bot. Algérie 1931, p. 114-115).
(2) Claudius (Jean Antoine) Roux (1872-1961), botanist and geologist at Lyon, see GR p. 349.
(3) Charles Constant François Marie le Roux (1885-1947), Dutch physician in the Dutch East-Indian Army, see SK 1: 450.
(4) Eugène Roux (1853-1933), mycologist.
(5) Jacques Roux (1773-1822), Geneva botanist; collected in Spain, S. France and Switzerland (J. Briquet, Bull. Soc. bot. Suisse 50a: 420-421. 1940).
(6) Jean Roux (1876-1938), Swiss zoologist, with F. Sarasin author of *Nova Caledonia* with botanical contributions by Schinz and Guillaumin.
(7) Pierre-Paul-Émile Roux (1853-1933), French bacteriologist.
(8) Jean Louis Florent Polydore Roux (1792-1833), French naturalist.
(9) Wilhelm Roux (1850-1924), German plant anatomist and palaeobotanist at Breslau, Innsbruck and Halle.

EPONYMY: *Rouxia* Husnot (1899). *Note: Rouxia* Brun et Héribaud (1893) honors Maxime Roux (1891-?), diatomologist.

9697. *Herborisation au Mont Seneppé* et à La Salette ... Lyon (Association typographique ...) 1890. Oct.
Publ.: 1890 (Bot. Zeit. 29 Aug 1890), p. [1]-20. *Copy*: G. – Reprinted and to be cited from Bull. Soc. bot. Lyon 1889 (1890).

9698. *Herborisation au Col de Chavière* et au Mont Thabor ... Lyon (Association Typographique ...) 1891. Oct.
Publ.: 1891 (Bot. Centralbl. 14 Oct 1891), p. [1]-15. *Copy*: G. – Reprinted from Ann. Soc. bot. Lyon 17: 169-181. 1891.

Rouy, Georges [C.Ch.] (1851-1924), French secretary of the press syndicate and amateur floristic botanist in Paris. (*Rouy*).

HERBARIUM and TYPES: LY (via Herb. R. Bonaparte); other material at FI, G, GB, MPU, P, PH. – The herbarium of Roland Bonaparte, which included the Rouy herbarium (acquired 1906), was offered by Bonaparte's daughter to P; this institution, however, accepted only the vascular cryptogams and the remaining part of the Bonaparte herbarium (including that of Rouy) went to LY. See Malinvaud (1889) for a description of the herbarium by 1889, Anon. (1900) for the situation in 1900. – Letters from Rouy to Malinvaud (100) are at P.

BIBLIOGRAPHY and BIOGRAPHY: AG 3: 765, 5(2): 465; Backer p. 660; Barnhart 3: 185 (b. 2 Dec 1851, d. 25 Dec 1924); BFM no. 1411, 1412; BJI 1: 49; BL 2: 107, 505, 508, 711; BM 2: 599, 4: 1743-1744, 8: 1100; Bossert p. 340; Colmeiro 1: cxcvi-cxcvii; CSP 11: 232-233, 12: 636, 18: 335-336; DTS 1: 249, 6(4): 190; Hortus 3: 1203; IF p. 727, suppl. 2: 39; IH 1 (ed. 2): 77, (ed. 3): 97, (ed. 6): 363, (ed. 7): 341, 2: (files); Kew 4: 530; Morren ed. 10, p. 57; PFC 1: lii, 2(2): xvii, xxvii; Rehder 5: 742; Saccardo 2: 94, cron. xxxi; TL-1/1114; TL-2/see J.N. Boulay, E.G. Camus; Tucker 1: 610; Zander ed. 10, p. 709, ed. 11, p. 809.

BIOFILE: Anon., Bull. Soc. bot. France, Table articles originaux 1854-1893, p. 206-208 (lists the great number of contr. to BSbF); Hedwigia 45: (205). 1906 (Roland Bonaparte acquired the Rouy herbarium); J. Bot., Morot 17(1) (bibl.): xii. 1903 (ann. *Rev. bot. syst.*); Monde des Pl. 3: 241-243. 1894 (bibl.); Rapport sur l'herbier de M. Georges Rouy, Lons-le-Saunier 1900, 8 p. (repr. from Act. Congr. int. Bot. Paris 1900).
Aymonin, G.G., C.R. Congr. Soc. Sav. (Limoges) 102(3): 57. 1977 (100 letters to Malinvaud at P).
Beauverd, G., Bull. Herb. Boiss. ser. 2. 5: 551-556. 1905 (on the treatment of some of R.'s names in the Index botanique universel).
Cavillier, F., Boissiera 5: 83. 1941.
Dolezal, H., Friedrich Welwitsch 125, 136. 1974.
Heywood, V., Taxon 7(4): 89-93. 1958.
Lecomte, H., *in* G. Rouy, Consp. Fl. France v-x. 1927 (biogr.).
Malinvaud, E., Bull. Soc. bot. France 36: cclxxx-cclxxxviii. 1889 (1890) (report on the Rouy herbarium; list of collections), 40: (bibl. 1 Jul 1893): 63-64. 64. 1893 (announcement of *Fl. France*).
Morot, L., J. de Bot. 4: 348. 1890 (announcement request for material of rare and critical species).
Saint-Lager, J.B., Les nouvelles flores de France, Paris 1894.
Wittrock, V.B., Acta Horti Berg. 3(3): 113. 1905.

COMPOSITE WORKS: Founder and editor of *Revue de botanique systématique et de géographie botanique*, tome 1 & 2, 1903-1904, all published.

HANDWRITING: Candollea 33: 149-150. 1978.

EPONYMY: *Rouya* Coincy (1901).

9699. Matériaux pour servir à la révision de la flore portugaise. Par M.G. Rouy. I. – *Sur quelques Graminées du Portugal* [Bull. Soc. bot. France 28, 1881]. Oct.
Publ.: Apr-Mai 1881 (Séance 28 Jan 1881; KNAW journal rd Mai 1881; Geol. Soc. London id. 25 Apr-20 Jun 1881), p. [1]-7. *Copies*: M, MO(2). – Reprinted and to be cited from Bull. Soc. bot. France 28(1): 36-42. 1881.

9700. *Excursions botaniques en Espagne* par M.G. Rouy. [Extrait du Bulletin de la Société botanique de France. Tome xxviii, séance 27 Mai 1881, publ. 1882].
Publ.: Jul-Aug 1882 (Bot. Centralbl. rev. 13-17 Nov 1882; Bot. Zeit. 27 Oct 1882; Nat. Nov. Aug(2) 1882; journal issue Jul-Aug 1882), p. [1]-19. *Copies*: G, HH, MO(2). – Reprinted and to be cited from Bull. Soc. bot. France 28(3): 153-171. Jul-Aug 1881.
Continuation (1) Oct-Nov 1882 (Nat. Nov. Feb(2) 1883; journal issue Oct-Nov 1882), p. [1]-21. *Copies*: G, HH, MO(2). – Reprinted and to be cited from Bull. Soc. bot. France 29(1): 40-47. Jul-Aug 1881, 29(2): 108-114, 29: 120-127. Oct-Nov 1882).
Continuation (2) in Bull. Soc. bot. France 31: 33-41, 52-56. Feb-Apr, 71-75, Mai-Jun, 269-279, Jun-Jul 1884, 33(6): 524-529. Feb-Mar 1887, 35: 115-124. Apr-Jun 1888. – Reprint 46 p. n.v. (1888), (Nat. Nov. Apr(2) 1889).

9701. *Étude des Diplotaxis européens* de la section Brassicaria (Godr. et Gren., Flore de France, i, p. 78) …[Revue des Sciences naturelles Juin 1882]. Oct.
Publ.: Jun 1882, p. 423-436. *Copy*: FI. – Reprinted and to be cited from Rev. Sci. nat. ser. 3. 1: 423-436. Jun 1882. – Bot. Zeit, 27 Oct 1882; Nat. Nov. Aug(2) 1882).
Ref.: Freyn, J.F., Bot. Centralbl. 12: 266-267. 13-17 Nov 1882 (rev.).
Anon., Bull. Soc. bot. France 29(bibl.): 94. 1882 (bibl. for Mai-Jun 1882).

9702. *Matériaux pour servir à la revision de la flore portugaise* accompagnés de notes sur certaines espèces ou variétés critiques de plantes européennes. [Extrait du Journal Le Naturaliste 1882-1884, Paris]. Oct.
Publ.: in small installments in *Le Naturaliste*, 1882-1884, see below for dates. Reprinted (but to be cited from the journal) in two sets:
[*1*]: 1 Jan –1 Sep 1882 (Bot. Zeit. 30 Mai 1883; Nat. Nov. Feb(1) 1883), p. [1]-52. *Copies*: BR, G, HH, M. – Deals with Labiatae.
[*2*]: 1 Oct 1882-1 Jun 1884, p. [1]-23, [25]-70. *Copies*: BR, G, HH. – Deals with Scrophulariaceae. – Bot. Zeit. 31 Oct 1884; Nat. Nov. Sep(1) 1884.
Publication in Le naturaliste, 4(-6), 1882-1884 as follows:

	pages		pages	
[1]	2-4	1 Jan 1882	156-158	15 Oct 1882
	14-15	15 Jan 1882	180-182	1 Dec 1882
	19-21	1 Feb 1882	189-191	15 Dec 1882
	30-31	15 Feb 1882	244-246	1 Apr 1883
	36-38	1 Mar 1882	284-286	15 Jun 1883
	43-44	15 Mar 1882	307-309	1 Aug 1883
	52-54	1 Apr 1882	341-342	1 Oct 1883
	60-62	15 Apr 1882	349-351	15 Oct 1883
	91-93	15 Jun 1882	365-367	15 Nov 1883
	115-116	1 Aug 1882	423-424	1 Mar 1884
	124-126	15 Aug 1882	428-429	15 Mar 1884
[2]	131	1 Sep 1882	452-454	1 Mar 1884
	147-149	1 Oct 1882	469-470	1 Jun 1884

Ref.: Malinvaud, E., Bull. Soc. bot. France 30 (bibl.): 31-32. 1883 (for Jan-Feb 1883; rev. of [1]), 31(bibl.): 107-108. 1884 (rev. of [2]).

9703. *Excursions botaniques en Espagne en 1881 et 1882* par G. Rouy. Orihuela, Murcia, Velez-Rubio, Hellin, Madrid, Irun. Montpellier (Typographie et lithographie Boehm et fils …) 1883. Oct.

Publ.: 1883 (Bot. Zeit. 30 Mai 1884; Nat. Nov. Apr(2) 1884), p. [3]-86, [87, colo.].
Copies: G(2), HH, MO, NY. – Reprinted and to be cited from Rev. Sci. nat. ser. 3. 2: 228-256, 557-564. 1883, 3: 58-81, 229-250. 1884.
Diagnoses d'espèces nouvelles pour la flore de la Péninsule ibérique: Le Naturaliste 5: 372-373. 1 Dec 1883, 6: 405-406. 1 Feb 1884, 6: 557-558. 15 Nov 1884, 9: 178. 15 Oct 1887, 9: 199. 15 Nov 1887, 13: 248. 15 Oct 1891.
Excursions (Mai-Jun 1883): 1884, p. [1]-46. *Copies*: G, HH. – Reprinted from Bull. Soc. bot. France 31: 33-41, 52-56, 71-75, 269-279. 1884, continued in 33: 524-529. 1886, 35: 115-124. 1888. -"Excursions en Espagne [Mai-Juin 1883] par G. Rouy ... Denia-Madrid-Aranjuez. Paris (Société anonyme des imprimeries réunies, à ...) 1884. Oct.

9704. *Suites à la Flore de France* de Grenier et Godron diagnoses des plantes signalées en France et en Corse depuis 1855 par G. Rouy, ... fascicule i. Extrait du Naturaliste ... Paris (Émile Deyrole ...) 1887. Oct.
Publ.: 1887 (J. Bot. Morot 1 Feb 1888; Nat. Nov. Mar(1) 1888), p. [1]-194. *Copies*: BR, G, HH, MICH, NY. – Reprinted and to be cited from *Naturaliste* 1882-1887. This reprint contains only the first part of the texts in the *Naturaliste*; the series continued until 1892.

Année	pages	dates	Année	pages	dates
6	538	15 Oct 1884		43-44	15 Feb 1888
	547-548	1 Nov 1884		68-70	15 Mar 1888
	563-564	1 Dec 1884		84-86	1 Apr 1888
7	19-21	1 Feb 1885		131-132	1 Jun 1888
	27-28	15 Feb 1885		195-196	15 Aug 1888
	37-39	1 Mar 1885		213-215	15 Sep 1888
	43-44	15 Mar 1885		272-273	1 Dec 1888
	52-53	1 Apr 1885		285-286	15 Dec 1888
	59-61	15 Apr 1885	11	10-11	1 Jan 1889
	76-77	15 Mai 1885		55-56	1 Mar 1889
	84-85	1 Jun 1885		82-84	1 Apr 1889
	124-125	15 Aug 1885		95-96	15 Apr 1889
	133-135	1 Sep 1885		122-123	15 Mar 1889
	140-141	15 Sep 1885		131-132	1 Jun 1889
	154-156	15 Oct 1885		217-219	15 Sep 1889
	163-164	1 Nov 1885		235-237	1 Oct 1889
	171-174	15 Nov 1885		256-257	1 Nov 1889
8	197-198	1 Jan 1886		271-272	15 Nov 1889
	203-204	15 Jan 1886		280-282	1 Dec 1889
	215	1 Feb 1886		293-294	15 Dec 1889
	222-223	15 Feb 1886	12	7-8	1 Jan 1890
	229-230	1 Mar 1886		18-19	15 Jan 1890
	316	15 Aug 1886		38	1 Feb 1890
	325-326	1 Sep 1886		68	15 Mar 1890
	333-334	15 Sep 1886		84-85	1 Apr 1890
	339-340	1 Oct 1886		108	1 Mai 1890
	349-351	15 Oct 1886		119	15 Mai 1890
	346-357	1 Nov 1886		178-179	1 Aug 1890
	373-375	1 Dec 1886		205-207	1 Sep 1890
	380-382	15 Dec 1886		238-239	15 Oct 1890
	387-389	1 Jan 1887		263-264	15 Nov 1890
	395-396	15 Jan 1887	13	11-12	1 Jan 1891
	406-408	1 Feb 1887		21-22	15 Jan 1891
	413-415	15 Feb 1887		94	15 Apr 1891
9	8-11	15 Mar 1887	14	81-83	1 Apr 1892
	55-56	15 Mai 1887		92-93	15 Apr 1892
	115-116	1 Aug 1887		129-130	1 Jun 1892
	153-155	15 Sep 1887		138-140	15 Jun 1892
10	30-32	1 Feb 1888		165-166	15 Jul 1892

Année	pages	dates	Année	pages	dates
14	197-198	15 Aug 1892	14	229-230	1 Oct 1892
	207-208	1 Sep 1892		244-246	15 Oct 1892
	219-220	15 Sep 1892			

Ref.: Malinvaud, E., Bull. Soc. bot. France 34(bibl.): 183-185. 1887.
Morot, L., J. Bot. 2: 20-21. 1888.

9705. *Annotations aux Plantae europaeae* de M. Karl Richter ... fascicule i ... Paris (Librairies-imprimeries réunies, ...) 1891. Oct.
Publ.: 1891, p. [1]-21. *Copies*: BR, G, HH. – Reprinted and to be cited from Bull. Soc. bot. France 38(2): 94-102, Mai 1891, 38(3): 130-142. Jun 1891, 38(4): 223 (footnote). Oct 1891.

9706. *Flore de France* ou description des plantes qui croissent spontanément en France, en Corse et en Alsace-Lorraine par G. Rouy ...et J. Foucaud ... Asnières (chez G. Rouy ...), Rochefort (chez J. Foucaud) ... 1893 [-1913], vol. 1 [-14]. Oct. (*Fl. France*).
Co-authors: vols. 1-3 (3 up to Caryoph.): Julien Foucaud (1847-1904); vols. 6-7 Edmond Gustave Camus (1852-1915); vols. 4-5, 8-14, by G. Rouy alone. The genus *Rubus* was by J.N. Boulay (6: 30-149. 1900). – "La collaboration a parfois de graves inconvénients ..." (Rouy, vol. 8, p. vi).

1: Nov 1893 (p. xv: 5 Sep 1893; Bot. Zeit. 1 Jun 1894; rev. BSbF Mar 1894; J. Bot. Jan 1894; Bot. Centralbl. 14 Mar 1894; Nat. Nov Dec(1) 1893; *however*: "fascicule 1. Paris 1890 ... 194 pg." Oct(1) 1890), p. [i]-lxvi, [lxvii], [1]-264.

2: Apr 1895 (copy presented to Soc. bot. France 26 Apr 1895; Bot. Centralbl. 6 Jun 1895; Nat. Nov. Jul(1) 1895), p. [i]-xi, [1]-349.

3: Jul-Aug 1896 (Bot. Centralbl. 29 Sep 1896; ÖbZ for Sep 1896; BSbF nouvelles 15 Aug 1896, just publ.; copy presented to Soc. bot. Fr. 13 Nov 1896; Bot. Zeit. 1 Dec 1896; Nat. Nov. Aug (2) 1896), p. [i-iii], [1]-382.

4: 1-12 Nov 1897 (Bot. Centralbl. 30 Dec 1897; BSbF nouvelles 25 Nov 1987, just publ.; Rouy presents copy to Soc. bot. France 12 Nov 1897; ÖbZ for Nov 1897; Nat. Nov. Nov(2) 1897), p. [i-iii], [1]-313. – Continuée par G. Rouy. – Imprint: On souscrit chez G. Rouy (... Asnières ...), Les fils d'Émile Deyrolle (...Paris) ... 1897.

5: Jan 1899 (t.p.; p. iv: 21 Dec 1898; Bot. Centralbl. 12 Apr 1899; presented to Soc. bot. France 27 Jul 1900; Bot. Zeit. 1 Mai 1899; J. de Bot. Mar 1899; Nat. Nov. Mar(1) 1899, p. [i-iv], [1]-344. – Imprint: as vol. 4.

6: Jun 1900 (t.p.; p. [v]: 27 Apr 1900; rd. by 1 Oct 1900 BSbF; Bot. Centralbl. 12 Sep 1900; ÖbZ for Jul 1900; Nat. Nov. Aug(2) 1900), p. [i-v], [1]-489. – Continuée par G. Rouy ... et E.G. Camus. – On souscrit chez G. Rouy (... Asnières ...), E.G. Camus (.. Paris), Les fils d'Émile Deyrolle (Paris).

7: Nov 1901 (t.p.; J. de Bot. Dec 1901; BSbF 24 Jan 1902; ÖbZ for Feb-Mar 1902; Bot. Centralbl. 3 Mar 1902; Nat. Nov. Mar(1) 1902), p. [i-v], [1]-440. – As vol. 6.

8: Apr 1903 (t.p.; p. vi: 20 Apr 1903; J. de Bot. Jun-Jul 1903; BSbF 15 Jul 1903; ÖbZ for Jun 1903; Bot. Zeit. 16 Aug 1903; Bot. Centralbl. 16 Feb 1904; Nat. Nov. Aug(2) 1903), p. [i]-vi, [1]-406. – As vol. 4.

9: Mar 1905 (t.p.; Bot. Centralbl. 3 Oct 1905; Bot. Centralbl. 20 Feb 1906; Nat. Nov. Apr(2) 1905; see BSbF 53: 160), p. [i]-vi, [1]-490. – As vol. 4.

10: Feb 1908 (t.p.; Bot. Centralbl. 14 Jul 1908; ÖbZ for Mar 1908; Nat. Nov. Mai(1) 1908; BSbF 8 Mai 1908), p. [i-iv], [1]-404. – As vol. 4.

11: Jul 1909 (t.p.; Bot. Centralbl. 30 Nov 1909; ÖbZ for Aug-Sep 1909; Bot. Zeit. 16 Aug 1909; Nat. Nov. Oct(1) 1909), p. [i-iv], [1]-429. – "*Flore de France* ... par G. Rouy ... en vente chez Les fils d'Émile Deyrolle ... Paris.

12: Nov 1910 (t.p.; Bot. Centralbl. 11 Apr 1911; ÖbZ for Jan 1911; Nat. Nov. Apr(1) 1911), p. [i-ii], [1]-505. – As vol. 11.

13: Mai 1912 (t.p.; Bot. Centralbl. 6 Aug 1912; ÖbZ for Mai 1912; Nat. Nov. Jul(1) 1912; BSbF 24 Mai 1912), p. [i]-viii, [1]-548. – As vol. 11.

14: Apr 1913 (t.p.; p. viii: 2 Dec 1912; Hedwigia 1 Oct 1913; Bot. Centralbl. 1 Jul 1913; ÖbZ for Apr 1913; BSbF 22 Jul 1913), p. [i]-viii, [1]-562, err. slip. – As vol. 11.

Copies: BR, FI, G, HH, MICH, MO, NY, PH, USDA; – The volumes were issued in the Annales de la Société des Sciences naturelles de la Charente-Inférieure. The original covers had as text: Académie de La Rochelle. Société des Sciences naturelles de la Charente-Inférieure. Annales de [1893-1912] Flore de France par ... La Rochelle [1893-1913]. From vol. 5 onward the years for which the Annales were issued were those preceding the year of publication.

Supplement: see below, *Consp. Fl. France*, 1927.

Subspecies: Rouy and Foucaud's names for subspecies should be interpreted as normal trinomial subspecific combinations made by the authors, notwithstanding their somewhat unorthodox form (Heywood 1958).

Reviews by É. Malinvaud in Bull. Soc. bot. France 41: 155-159. 1894 (vol. 1), 42: 203-207. 1895 (vol. 2), 43: 755-756. 1896 (vol. 3); 44: 497-500. 1897 (vol. 4), 46: 339-341. 1899 (vol. 5), 47: 287-288, 385-390. 1900 (vol. 6), 49: 213-219. 1902 (vol. 7). (vol. 8 by F. Gagnepain 52: 50-51. 1905), 53: 160-162. 1906 (vol. 9). (Vol. 10 by J. Poisson 55: 415-516. 1908).

Polemics by G. Rouy: s.d. "Flore de France par G. Rouy & J. Foucaud", 2 p. s.d. (*copy*: G); 1894, 2 p. printed letter (*copy*: G) in reply to a paper by Malinvaud in J. Bot., Morot, 12 Jul 1894; 4 p. printed letter "Asnières (Seine) 12 juillet 1894" by Rouy; 3p. *Réponse au nouvel article de M. Rouy*, Asnières 26 Jul 1894; see also below G. Rouy, *Contes fantast.* 1905, 12 p.

Announcement: On 5 Nov 1891 Rouy published a 3 p. announcement of the work as a whole, giving details on the proposed publication and requesting material of some species which had not yet been seen.

Dictionnaire inventoriel (by H. Léveillé) (with only one level of infraspecific taxa (subsp. + proles = "races") and *Conspectus*, see below G. Rouy (1927).

Ref.: Saint-Lager, J.B., Ann. Soc. bot. Lyon 19: 81-109. 1894 ("Les nouvelles flores de France. Étude bibliographique").

9707. *Illustrationes plantarum Europae rariorum* auctore G. Rouy. Diagnoses de plantes rares ou rarissimes de la flore européenne accompagnées de planches représentant toutes les espèces décrites. Reproduction photographique des exemplaires existant dans les grandes collections botaniques et notamment dans l'herbier Rouy. Fascicule i-xx. Paris (chez les fils d'Émile Deyrolle, ...) 1895-1905. (*Ill. pl. Eur.*).

1: 1895 (Nat. Nov. Mai(1) 1895), p. [i], [1]-8, preface, [1]-8 descr. text, *pl. 1-25* (uncol. photogr. in portfolio).
2: 1895 (BSbF Dec 1895), p. [i], 9-16, *pl. 26-50*.
3: 1895 (Nat. Nov. Mar(2) 1896), p. [i], 17-24, *pl. 51-75*.
4: 1895, p. [i], 25-32, *pl. 76-100*.
5: 1896 (BSbF by 27 Nov 1896), p. [i], 33-40, *pl. 101-125*.
6: 1896 (BSbF 25 Jun 1897), p. [i], 42-50, *pl. 126-150*.
7: 1896, p. [i], 51-58, *pl. 151-175*.
8: 1897 (Nat. Nov. Dec(2) 1897), p. [i], 59-66, *pl. 176-200*.
9: 1898, p. [i], 67-74, *pl. 201-225*.
10: 1898, p. [i], 75-82, *pl. 226-250*.
11: 1899 (ÖbZ for Sep 1899; Bot. Zeit. 1 Mar 1900; Nat. Nov. Dec(1) 1899), p. [i], 83-90, *pl. 251-275*.
12: 1899 (ÖbZ for Dec 1899), p. [i], 91-98, *pl. 276-300*.
13: 1900, p. [i], 99-107, *pl.301-325*.
14: 1900 (ÖbZ for Jan-Feb 1901), p. [i], 108-115, *pl. 326-350*.
15: 1901 (ÖbZ for Jul-Aug 1901), p. [i], 116-123, *pl. 351-375*.
16: 1901 (BSbF Feb 1902; ÖbZ for Jan 1902), p. [i], 124-131, *pl. 376-400*.
17: 1902 (NY rd. by Dec 1902; Nat. Nov. Aug(1) 1902), p. [i], 132-139, *pl. 401-425*.
18: 1903 (Bot. Centralbl. 16 Feb 1904; ÖbZ for Aug 1903; Bot. Zeit. 1 Dec 1903; NY rd. by Oct 1903), p. [i], 140-147, *pl. 426-450*.
19: 1904 (J. Bot. Oct 1904; NY rd. by Nov 1904; Nat. Nov. Oct(1) 1904), p. [i], 148-155, *pl. 451-475*.
20: 1905 (Bot. Centralbl. 13 Jan 1905; J. Bot. Oct 1905; NY rd. Nov 1905; Nat. Nov. Apr(2) 1905), p. [i], 156-168, *pl. 476-500*.

Copies: G, H, MO, NY.

Prospectus: Dec 1894, Asnières, p. [1-4]. *Copy*: G. – Illustrationes plantarum Europae rariorum ... Georges Roux ... Asnières, Décembre l894. – Number of copies: 150.

Ref.: Malinvaud, É., Bull. Soc. bot. France 42: 208, 697. 1895 (ann.; rev. fasc. 1), 43(bibl.): 643-646. 1896 (rev. fasc. 1-5), 45 (bibl.): 327-331. 1898 (fasc. 6-9).

9708. *Conspectus des espèces françaises du genre Spergularia* Pers. [Bull. Herb. Boiss. ser. 1. 3. 1895] Oct.
Publ.: Mai 1895, p. [1]-3. *Copy*: U. – Issued and to be cited as Bull. Herb. Boissier ser. 1. 3: 222-224. Mai 1895.

9708a. *Sur l'application rigoureuse de la règle d'antériorité de la dénomination binaire* dans la nomenclature. [Bull. Herb. Boiss. ser. 1. 5. 1897]. Oct.
Publ.: Jan 1897, p. [1]-6. *Copy*: U. – Issued and to be cited as Bull. Herb. Boiss. ser. 1. 5: 60-65. Jan 1887.
Continuation: Apr 1897, p. [1]-6. *Copy*: U. – *Questions de nomenclature*, ib. ser. 1. 5: 273-278. Apr 1897 (reprint: p. 1-6).

9709. *Icones plantarum Galliae rariorum.* Atlas iconographique des plantes rares de France et de Corse... Fascicule i planches 1-50. On souscrit chez G. Roux... Asnières (Seine). Les fils d'Émile Deyrolle... Paris... Octobre 1897. Portfolio. †. (*Icon. pl. rar. France*).
Publ.: Oct 1897 (t.p.; BSbF nouvelles 15 Nov 1897; copy presented 12 Nov 1897; Nat. Nov. Nov(2) 1897), p. [i], *pl. 1-50* (uncol. photos), [i, index]. *Copies*: G, HH, MO, NY.

9710. *Lettres sur quelques plantes de la flore Française* [Bull. Herb. Boissier (2me sér.) 1905]. Oct.
Publ.: 31 Mai 1905, p. [1]-7 (with note on p. 551). *Copy*: U. – To be cited from Bull. Herb. Boissier ser. 2. 5: 544-551; followed by a reply by G. Beauverd on p. 551-556, repr. [1]-6. Discussion of some entries in *Index botanique universel*.

9711. *Les contes fantastiques de M. Malinvaud* réfutation de ses inexactitudes... Un papier anonyme. [Asnières 1905]. (*Contes Malinvaud*).
Publ.: Oct-Dec 1905 (p. 10: Oct 1905), p. [1]-12. *Copy*: G.

9712. *Conspectus de la Flore de France* ou catalogue général des espèces, sous-espèces, races, variétés, sous-variétés et formes hybrides contenues dans la "Flore de France", de Georges Rouy... Paris (Paul Lechevalier...) 1927. Oct. (*Consp. Fl. France*).
Publ.: 1927 (p. [320]: 15 Aug 1927, printer's date; Nat. Nov. Feb 1928), p. [i-iii], portr., [v]-xv, [xvi], [1]-319, [320]. *Copies*: BR, FI, G, HH, M, NY, PH, USDA. – Other t.p.: Flore de France ou description... Alsace-Lorraine par Georges Rouy... Supplément Conspectus de la Flore de France...".
Precursor: H. Léveillé, *Dictionnaire inventoriel de la Flore française*... espèces et races, Le Mans (chez l'auteur), 40 double pages (1 printed/1 blank). – M. Kerguélen informs us that this is a (badly) compiled list taken from Rouy, *Fl. France* with Rouy's subspecies and proles, all treated as "races".

Rovirosa, José N. (1849-1901), Mexican botanist. (*Rovirosa*).

HERBARIUM and TYPES: MEXU; other material at B (mainly destr.), COLU, GH, K, NY, PH, UC, US. – A.R. Smith (Fl. Chiapas 1981) notes that the most complete set may be at PH. (see also Mears 1981).

BIBLIOGRAPHY and BIOGRAPHY: Barnhart 3: 186 (b. 9 Apr 1849, d. 23 Dec 1901); BM 4: 1744; Bossert p. 340; CSP 18: 337; IF p. 727, suppl. 1: 85; IH 1 (ed. 6): 363, 2: (files); Kew 4: 530; Langman p. 653-654; PH p. 521 [index]; Rehder 5: 742; Urban-Berl. p. 389.

BIOFILE: Breedlove, D.F., Fl. Chiapas, Intr. 25. 1981.
Leon, N., Bibl. bot.-mexic. 225, 292, 360-366. 1895.
Mears, J.A., Proc. Acad. nat. Sci. Philadelphia 133: 161. 1981 (primary set herb. mat. PH).
Rzedowski, J., Bol. Soc. bot. Mexico 40: 3, 6. 1981.

9712a. *Pteridografía del Sur de México* ó sea clasificación y descripción de los helechos de esta región, precedida de un bosquejo de la flora general, ... México (Imprenta de Ignacio escalante ...) 1909. Qu. (*Pteridogr. Sur México*).
Publ.: 1910 (p. iv: Nov 1908; cover dated 1910; t.p. 1909; Hedwigia 15 Jul 1911; Nat. Nov. Jan(2) 1911), portr. author, p. [i]-iv, [1]-298, *pl. 1-7, 7a, 8-14, 14bis, 15-28, 28bis, 29-38, 38a, 39-70* (uncol. liths. auct.). *Copies*: BR, G, MO, NY, USDA.
Ref.: G.H., Hedwigia, Beibl. 58(2): (115). 14 Oct 1916 (rev.).

Rowlee, Willard Winfield (1861-1923), American botanist; D.L. Cornell 1888; Dr. sci. 1893; instructor of botany at Cornell University 1889-1893; assistant professor of botany 1896-1906; ordinary professor of botany 1906-1923. (*Rowlee*).

HERBARIUM and TYPES: CU; further material GH, K, MO, NY, US.

BIBLIOGRAPHY and BIOGRAPHY: Barnhart 3: 186 (b. 15 Dec 1861, d. 8 Aug 1923); Bossert p. 340; CSP 18: 338; GR p. 241; Hortus 3: 1203; Langman p. 653; LS 23128; NW p. 56; Rehder 5: 743; SIA 222, 224; Tucker 1: 610-611; Zander ed. 10, p. 709, ed. 11, p. 809.

BIOFILE: Anon., Allg. bot. Z. 12: 168. 1906 (ord. prof. bot. Cornell Univ.); Bot. Centralbl. 66: 80. 1896 (asst. prof. bot. Cornell Univ.); Bot. Jahrb. 21(Beibl. 54): 31. 1896 (asst. prof. bot. Cornell Univ.), 38 (Beibl. 38): 42. 1906 (ord. prof.); Bot. Soc. Amer. 86: 92-93. 1924 (obit. resolution); Bot. Zeit. 54: 160. 1896 (prof. bot. Cornell); Cornell Alumni News Ithaca N.Y. 25: 506. 1923 (d. 8 Aug 1923, obit.); J. New York Bot. Gard. 24: 189. 1923 (obit.); Nat. Nov. 18: 244. 1896 (asst. prof. bot. Cornell Univ.); Who's who in America 1906-1907: 1535.
Cattell, J.M., Amer. men Sci. ed. 1: 275. 1906, ed. 2: 403. 1910, ed. 3: 589. 1921.
Rickett, H.W., Index Bull. Torrey bot. Club 84. 1955.
Urban, I., Symb. ant. 5: 11. 1904.

Rowntree, Lester (Gertrud Ellen Lester) (1879-1979), British-born (Cumberland) naturalist, horticulturist and writer; emigrated to America 1889; married to Bernard Rowntree 1908-1930; settled in California where she ran a wild flower firm. (*Rowntree*).

HERBARIUM and TYPES: CAS; other material at A, COLO, LAM, NY, SD, TEX. – Main Mexican herbarium destroyed by fire (McVaugh 1972).

BIBLIOGRAPHY and BIOGRAPHY: Barnhart 3: 186 (b. 16 Feb 1879); IH 2: (files); KR p. 4: 533; Langman p. 653.

BIOFILE: Anon., California Native Plant Society Bull. Mai-Jun 1979, p. 7 (d.); Fremontia 7(1): 23. 1979 (d. 21 Feb 1978 (for 1979), obit.); Gard. Chron. Amer. 51: 243. 1947 (portr.).
Barker, P.A., Amer. hort. Mag. 44: 32-35. 1963 (a visit with Lester Rowntree).
Cantelow, E.D. & H.C., Leafl. W. Bot. 8: 98. 1957.
Hamann, S., Calif. Hort. Soc. J. 35: 73-76. 1974 ("Mountain mystic"), Fremontia Jan 1976: 3-8 ("The wildflower lover at ninety-seven").
Howell, J.T., Leafl. W. Bot. 7(3): 94. 1953 (impt. set Calif. pl. at CAS).
Lenz, L.W., Rancho Santa Ana Botanic Garden 1927-1977, p. 41, 84. 1977.
McVaugh, R., Contr. Univ. Mich. Herb. 9: 299. 1972.
O'Connor, N.G., Baileya 11(2): 49-53. 1963 (biogr. sketch, portr.).
Pearce, F.O., Pacific horticulture 40(2): 52. 1979 (obit., portr.).
Rowntree, L., Fremontia Jan 1976: 9-16. Jan 1977: 15-19, Jul 1977: 13-18. ("Sierra wildflowers", has autobiographical elements, portr.).
Woolfenden, J., Calif. hort. J. 29(4): 98-102, 126. 1968 (portr., biogr. sketch).

9713. *Hardy Californians* ... New York (The Macmillan Company) 1936. Oct. (*Hardy Calif.*).
Publ.: 1936, p. [i]-xiv, [xv], 1-255, map Calif. on endpapers. *Copy*: E (inf. J. Edmondson).
Ref.: Robinson, G.W., Bull. misc. Inf. Kew 1937: 279-280 (rev.).

9714. *Flowering shrubs of California* and their value to the gardener ... Stanford (Stanford University Press ...) [1939]. Oct. (*Fl. shrubs Calif.*).
Ed. [*1*]: Oct 1939, p. [i]-xii, 1-317. *Copies*: HH, NY, USDA.
Second printing: Feb 1948. *Copy*: G.

Roxas Clemente y Rubio, Simon de (1777-1827), Spanish botanist; librarian of the Jardin botánico de Madrid. (*Roxas*).

HERBARIUM and TYPES: MA; further material BM.

BIBLIOGRAPHY and BIOGRAPHY: AG 2(1): 176, 5(1): 203; Barnhart 1: 357; BL 2: 507, 711; BM 4: 1745; Bossert p. 77, 336; Colmeiro 1: clxviii-clxix; Colmeiro penins. p. 195-197; GR p. 759; GR, cat. p. 70; Herder p. 474; Jackson p. 340; Kew 1: 569; LS 5223-5224; PR 7739-7742, ed. 1: 4411, 8783-8784; Rehder 1: 421; TL-2/see Lagasca y Segura, M.; Tucker 1: 611.

BIOFILE: Anon., Flora 10: 736. 1827 (d.).
Murray, G., Hist. coll. BM(NH) 1: 178. 1904.
Pardo, Simon de Rojas Clemente y Rubio y el primer centenario de su muerto. Valencia 1927, 27 p.

NOTE: We treated this author very briefly in vol. 2 (no. 1141) under Clemente y Rubio, Simon de Rojas. We give more details here under the anyhow preferable "Roxas Clemente y Rubio", following the spelling of his name on the title-pages of his publications.

EPONYMY: *Clementea* Cavanilles (1803); *Clementea* Cavanilles (1804).

9715. *Ensayo sobre las variedades de la vid comun* que vegetan en Andalucía, con un índice etimológico y tres listas de plantas en que se caracterizan varias especies nuevas, por Don Simon de Roxas Clemente y Rubio ... Madrid (En la imprenta de Villalpando) 1807. Qu. (*Ens. var. vid. com.*).
Publ.: 1807 (p. viii: 1 Mar 1807), p. [i*-ii*], [i]-xviii, [1]-324, [1, err.], *1 pl.* (col.).
Copies: FI, G, HH, M, NY, USDA (the HH and USDA copies have an additional frontisp. with 1 p. expl.).
Illustrated ed.: 1879 (Bot. Zeit. 28 Mai 1880; Nat. Nov. Dec(2) 1879), p. [i-ii], engr. t.p., [iii]-xxv, [i]-ix, [1]-149, 2 pl., *pl. 1-38* (col.), [4 p. mss. reproduced]. *Copies*: BR, HH. – "*Ensayo* ... Andalucía por Don Simon de Rojas Clemente y Rubio edicion ilustrada hecha de real órden, en honra del autor y en memoria de la primera exposicion vinícola nacional celebrada en España, ... Madrid (Imprenta estereotipía Perojo, ...) 1879. Qu.
French: 1814, p. [i]-xvi, [1]-418, [419-420, index]. *Copy*: USDA. – *Essai sur les variétés de la vigne* qui végètent en Andalousie ... traduit par M. le. M.is de Caumels, ... Paris (de l'imprimerie de Poulet ...) 1814. Oct. (*Essai var. vigne*).
German: 1821, p. [i]-xii, [1]-308, 4 (6?) charts, *1 pl. Copy*: NY. – *Versuch über die Varietäten des Weinstocks* in Andalusien ... Aus dem Französischen ... ins Deutsche übersetzt durch Anton Albert Freyherrn von Mascon ... Grätz (bey Franz Ferst, ...) 1821. (*Vers. Var. Weinst.*).

9716. *Memoria sobre el cultivo y cosecha del algodon* en general y con aplicacion a España, particularmente a Motril, ... Madrid (en la Imprenta Real) 1818. Oct. (*Mem. cult. algodon*).
Publ.: 1818, p. [1]-43. *Copy*: FI.

9717. *Tentativa sobre la liquenologia geografica de Andalucia* por D. Simon de Rojas Clemente. Trabajo ordenado conforme á los manuscritos del autor, por D. Miguel Colmeiro ... Madrid (por Aguado, ...) 1863. Oct.
Editor: Miguel Colmeiro y Penido (1816-1901).
Publ.: 1863 (BSbF rd. 11 Mar 1864), p. [1]-22. *Copy*: G. – Reprinted and to be cited from Revista de los Progresos de las Ciencias 14(1): 39-58. 1863.
Ref.: Anon., Bull. Soc. bot. France 11(bibl.): 34-35. Jul 1864.

9718. *Plantas que viven espontaneamente en el termino de Titaguas*, pueblo de Valencia, enumeradas en forma de índice alfabético ... Madrid (por Aguado, ...) 1864. Oct.
Publ.: 1864 (BSbF rd. 21 Apr 1865), p. [1]-72. *Copies*: G, MO, NY. – Reprinted and to be cited from Revista de los Progresos de las Ciencias 14: 429-445, 544-567, 568-576. 1864.

Roxburgh, William (1751-1815), British (Scottish) botanist and physician; MD Edinburgh 1876; with the East India Company in the Madras Medical Service 1776-1780; superintendent Samalkot (Samul Cattah) botanic garden 1781-1793; superintendant of the Calcutta botanic garden 1793-1813; in London 1806-1813; travelled to the Cape of Good Hope (1798, 1799, 1813-1814) and St. Helena 1814; died at Edinburgh. (*Roxb.*).

HERBARIUM and TYPES: Main collection at K (mainly in the Wallich herbarium, originally at LINN). "Roxburgh's practice seems to have been to distribute specimens he collected without keeping any significant personal herbarium. It thus frequently happened that specimens were later misidentified by him with species that he had described at an earlier date, he having kept no material for comparison" (Frances M. Jarret, personal communication). Considerable sets also BM, BR, E, G and LIV; small sets at A, B (Willd.), C, DBN, E, FI, NY, OXF, P, PH, UPS (Thunb.), drawings at CAL, BM and K. Manuscript of *Flora indica* with authograph notes by Roxburgh and Robert Brown at BM. For the Roxburgh *Flora indica* drawings at K, which are often of great importance for the typification of Roxburgh species, see Sealy (1957, 1975). For drawings at the India Office Library, see M. Archer (1962). – Letters at BM and LINN.

BIBLIOGRAPHY and BIOGRAPHY: AG 3: 119; Backer p. 500 (epon.); Barnhart 3: 187 (b. 3 Jun 1751, d. 18 Feb 1815); BB p. 263; Blunt p. 166, 193, 264; BM 4: 1745; Bossert p. 341; Bret. p. 237-246, 1090; CSP 5: 314; Dawson p. 713-720, 954; Desmond p. 533; De Toni 2: cii; DNB 49: 368-370; Dryander 3: 130, 303, 497, 534, 625, 649, 652; DU 269; Frank 3 (Anh.): 86; GF p. 74; Henrey 2: 738 [index], 3: no. 1289; Herder p. 164, 212; Hortus 3: 1203; IF p. 727; IH 1 (ed. 2): 35, (ed. 3): 33, (ed. 6): 363, (ed. 7): 341; Jackson p. 353, 383, 384, 388, 451; Kew 4: 533; Lasègue p. 144-145, 581 [index] (Roxburgh Sr. did not collect in Malay Arch.); LS 23129; Moebius p. 421; MW p. 419; NI 1689; Plesch p. 390; PR 1247, 7831-7838, ed. 1: 618, 1383, 8785-8792; Rehder 5: 743; RS p. 138; SK 1: 450-451, xxxii, cxlvii, 4: clx, clxxi, cxxv, 5: ccx, ccxlv, ccxxxvii; SO 805a-c; Sotheby 665; TL-1/1115-1118; TL-2/378, 566, 3583; see W. Carey; Tucker 1: 611; Urban-Berl. p. 415; Zander ed. 10, p. 709, ed. 11, p. 809.

BIOFILE: Allibone, S.A., Crit. dict. Engl. lit. 2: 1885. 1878.
Archer, M., Natural history drawings in the India office library, London 1962, p. 20-23, 65-66, 102.
Bridson, G.D.R. et al., Nat. hist. mss. res. Brit. Isl. 432 [index]. 1982.
Britten, J., J. Linn. Soc. Bot. 45: 47-48. 1920 (Cape coll.).
Burkill, I.H., Chapt. hist. bot. India 243 [index]. 1965.
Candolle, Alph. de, Phytographie 444-445. 1880 (coll.).
Coats, A.M., Gard. Chron. 160(20): 14. 1966 (portr.; on Henna and Frankincense); Quest for plants 148-149. 1969; Plant hunters 148-149, 150, *pl. 15*. 1969.
Dawson, W.R., Smith papers 83-84. 1935.
Gager, C.S., Brooklyn Bot. Gard. Rec. 27(3): 260. 1938.
Gunn, M. & L.E. Codd, Bot. Expl. S. Afr. 303. 1981.
Hall, N., Bot. Eucalypts 111. 1978.
Hara, H., Fl. Eastern Himalaya 674. 1966 (on Pl. Coromandel).
Holttum, R., Taxon 19: 708. 1970.
Huxley, L., Life letters J.D. Hooker 1: 473, 2: 181, 183. 1918.
Jessen, K.F.W., Bot. Gegenw. Vorz. 2, 469. 1864.
Karegeannes, C., Begonian 46: 261, 280. 1979 (R. as chronicler of Indian Begonias).
Ker, J.B., *Crinum* 1817, 16 p. (several nomina nuda in *Hort. bengal.* 1814 validated).
King, G., J. Bot. 37: 457-458. 1899; Ann. r. bot. Gard. Calcutta 5: 1-9. 1895 (portr., memoir, bibl.).
Kuntze, O., Rev. gen. pl. 1: cxli. 1891.

Lemmon, K., Golden age plant hunters 74, 118. 1968.
Lyte, C., Sir Joseph Banks 176. 1980.
Mabberley, D.J., Taxon 26: 523-524. 1977 (Fl. ind. ed. 1832, vol. 1, publ. 14 Jan 1832), 31: 65-66. 1982 (*Swietenia febrifuga*).
Merrill, E.D., Taxon 1: 124-125. 1952 (on "Hamou" and "Romoa" in Fl. indica errors for Honimoa); Chron Bot. 14(56): 171, 198, 220, 237, 294, 348. 1954 (botany Cook's Voy.).
Miller, H.S., Taxon 19: 538. 1970 (Roxburgh specimens from Lambert sale at G).
Morton, C.V. (D.B. Lellinger ed.), Contr. U.S. Nat. Herb. 38: 283-396. 1974 (W. Roxburgh's fern types).
Murray, G., Hist. coll. BM(NH) 1: 178. 1904.
Pillery, G., Investigations on Cetacea 9: 11-21. 1981 (on W. Roxburgh and M.J. Lebeck on Cetaceae).
Poggendorff, J.C., Biogr.-lit. Handw.-Buch 2: 708. 1863.
R.A.R., Gard. Chron. 1896(1): 781-782.
Rix, M., The art of the botanist 97, 107, 108, 113, 182. 1981.
Sealy, J.R., Kew Bull. 11: 297-399. 1957, Endeavour 34: 84-89. 1975 (on R's collection of drawings of Indian plants at Kew; Kew Bull. paper has list of holdings).
Sen, J., Nature 207: 1234-1235. 1965.
Smith, E., Life Joseph Banks xi, 114-116, 118. 1911.
Stafleu, F.A., in G.S. Daniels & F.A. Stafleu, Taxon 23(1): 52. 1974 (portr.).
Stansfield, H., Handb. guide herb. coll. Liverpool 41, 44, 46, 47. 1935 (portr., coll.).
Stearn, W.T., Roy. Hort. Soc., Exhib. manuscr. books 33-34. 1954 (*Pl. Coromandel*).
Steenis-Krusemann, M.J. von, Fl. males. Bull. 24: 1801. 1970, 26: 2018. 1972 (Pl. Coromandel).
Verdoorn, F., ed., Chron. bot. 1: 206. 1935, 4: 154. 1938, 5: 116. 1939.
White, A. & B.L. Sloane, The Stapelieae ed. 1: 16, 20, 48, 52, 53. 1933, ed. 2: 21 [index]. 1937.
Wight, R., in W.J. Hooker, Bot. Misc. 2: 90-91, 95. 1831.
Thomson, T., in W.J. Hooker J. Bot. Kew Gard. Misc. 9: 11-12. 1837.
Wood, D., Notes R. Bot. Gard. Edinburgh 29: 211-212. 1969 (dates parts 1-4 of vol. 3, Pl. Coromandel).
Woodward, B.B., Hist. coll. BM(NH) 1: 46. 1904 (mss. copy Flora indica; index to his botanical mss., 36 water colour drawings).

COMPOSITE WORKS: *An alphabetical list of plants*, seen by Dr. Roxburgh growing in the Island of St. Helena, in 1813-14, *in* A. Beatson, *Tracts St. Helena*. Jan 1816, p. 293ff. See TL-2/378, *copies*: GOET (inf. G. Wagenitz), LC.

HANDWRITING: Webbia 32(1): 14. 1977; Fl. males. 1: cli.

NOTE: James Roxburgh (1802-1884) and his brother Bruce R.(1797-1861), sons of William Roxburgh were responsable for the publication of the *Flora indica*. The preface to the 1832 edition is signed by James Roxburgh. A third son, John Roxburgh (*fl.* 1777-1824) collected at the Cape (1798-1804), see Desmond p. 533, M. Gunn & L.E. Codd, Bot. Expl. S. Afr. 302 (1981); J. Britten, J. Bot. 56: 202-203. 1918 (denying John''s existence), D. Prain, J. Bot. 57: 28-34. 1919 (showing that John Roxburgh did exist) and Dawson 713 (two letters by John R. to Joseph Banks). William Roxburgh Jr. (*fl.* 1780-1806) collected in India and the Malay Archipelago.

EPONYMY: *Roxburghia* W. Jones ex Roxburgh (1795).

9719. *A botanical description of a new species of Swietenia* (Mahogony), *with experiments and observations on the bark thereof, in order to determine and compare its powers with those of Peruvian bark, for which it is proposed as a substitute. Addressed to the honourable Court of directors of the United East-India Company, by their most obedient, humble servant, Wm. Roxburgh.* s.l.n.d. [1793].
Publ.: 1793 (in journal), p. [i], [1]-24. *Copy*: Natl. Libr. Scotland . – Reprinted from Medical facts and observations 6: 127-153. 1793 (journal at LC).

9720. *Plants of the Coast of Coromandel*; selected from drawings and descriptions presented to the hon. court of directors of the East India Company, by William Roxburgh, M.D. Published by their order, under the direction of Sir Joseph Banks, Bart. P.R.S. ... London (printed by W. Bulmer and Co. ...) 3 vols. 1795-1820; broadsheet. (*Pl. Coromandel*).

1: 1795-1798, p. [i*, t.p.], [i]-vi, colums [1]-68, indexes p. [1-2], *pl. 1-100* (on *pl. 1*: Mackenzie omnes fecit).
2: 1799 (t.p.: 1798)-1805, p. [i-iv], columns [1]-56, *pl. 101-200*.
3: 1811-1820(t.p. 1819), p. [i-iv], colums [1]-98, *pl. 201-300*.
Copies: BM, BR, NY, Teyler, U, USDA; IDC 5228. – The plates are line and stipple copper engravings, coloured by hand, engraved (nos. 1-250) by D. Mackenzie and (251-300), by J. Girlin, E.S. Weddell or R.B. Peake, after drawings made by various Indian artists. For original drawings see above under "Herbarium and types" and see e.g. M. Archer (1962), copies at K. – For dates see e.g. v. Steenis-Kruseman (SK), D. Wood, RS, and TL-1; for a collation see Sotheby 665.

vol. & part	pages	plates	dates
1(1)	1-28	*1-25*	Mai 1795
(2)	29-40	*26-50*	Nov 1895
(3)	41-56	*51-75*	Aug 1796
(4)	57-68	*76-100*	Jan-Mar 1798
2(1)(5)	1-16	*101-125*	Mai 1799
(2)(6)	17-28	*126-150*	Mai 1800
(3)(7)	29-40	*151-175*	Apr 1802
(4)(8)	41-56	*176-200*	Mai 1805
3(1)(9)	1-20	*201-225*	Jul 1811
(2)(10)	21-44	*226-250*	Mai 1815
(3)(11)	45-72	*251-275*	18 Feb 1820
(4)(12)	73-98	*276-300*	18 Feb 1820

Ref.: Anon., Flora 6(2): 465-477, 481-492. 1823 (rev. vol. 3).
Desmond, R., Hortulus aliquando 2: 22-41. 1977.
Wood, D., Notes R. Bot. Gard. Edinburgh 29(2): 211-212. 1969.
Facsimile ed.: 1982, Bishen Singh Mahendra Pal Singh, Dehra Dun, 1982 (vol. 2 seen).

9721. *Hortus bengalensis*, or a catalogue of the plants growing in the honourable East India Company's Botanic Garden at Calcutta. Serampore (printed at the Mission Press) 1814, sign. of 4 p. (*Hort. bengal.*).
Publ.: Jun-Dec 1814 (p. xii: 4 Jun 1814), p.[i]-v, [i]-xii, [1]-104, [105]. *Copies*: G, HH, NY, PH; IDC 5696.
Facsimile ed.: 1980, ISBN 90.70153.15.7, Boerhaave Press (p.o. Box 1051, Leiden 2302 BB, Holland), p. [i*-ii*, new t.p. and colo.], followed by facsimile. *Copy*: FAS.
Notes: on p. 77-105 "A catalogue of plants ..." dated 1813. The actual date 1814 applies to the entire book. See Robinson for a list of names that are validly published in this work. Most names are nomina nuda, but several specific epithets are validly published by means of references to other publications. Several of the nomina nuda were validated by description in G. Don's *General system* (see Sprague (1925) for a list).
Index: A holograph *index* to *Hort. bengal.*, written by Bernard M'Mahon, is at PH (fide J. Ewan et J. Soc. Bibl. nat. Hist. 3: *pl. 3*. 1960.
Ref.: SK p. ccx (v. Steenis).
Alston, I.A.H.G., Ann. Roy. Bot. Garden Peradeniya 11: 299-300. 1930.
Kuntze, O., Rev. gen. cxli. 1891.
Robinson, C.B., Philip. Journ. Sci. & C. Bot. 7: 411-419. 1912.
Sprague, T.A., Bull. misc. Inf. Kew 1925: 312-314.

9722. *Observations of the late Dr. William Roxburgh*, botanical superintendant of the honourable East India Company's garden at Calcutta, *on the various specimens of fibrous vegetables*, the produce of India, which may prove valuable substitutes for Hemp and Flax, on some future day, in Europe. Edited by a friend, to whom he transmitted them from St. Helena, during his illness; and now published at the expence of the East India

Company, for the information of the residents, and the benefit that may arise therefrom throughout the settlements in India. London (printed by J. Darling, ...) 1815. (*Observ. fibrous veg.*).
Publ.: 1815, p. [1]-78. *Copy*: FI.

9723. Appendix. *An alphabetical list of plants*, seen by Dr. Roxburgh growing on the Island *of St. Helena*, in 1813-1814. [*in* Alexander Beatson, Tracts relative to the Island of St. Helena; ... London (printed by W. Bulmer and Co., ...) 1816, p. 293-326]. Qu.
Publ.: 1816, p. [293]-326, as above. *Copy*: LC. – We have seen no independent reprint.
Amended version: in F. Antommarchi, Derniers moments de Napoléon, 2: [255]-425. 1825 [*copy* at BR], *Esquisse de la flore de Sainte-Hélène*; see also Alexander Watson, *Flora Sta. Helenica* 1825.

9724. *Flora indica*; or descriptions of Indian plants, by the late William Roxburgh, ... edited by William Carey, D.D. to which are added descriptions of plants more recently discovered by Nathaniel Wallich, ... Serampore (printed at the Mission Press) 1820-1824, 2 vols. Oct. (in fours). †. (*Fl. ind.*).
Annotations and additions: Nathaniel Wallich (1786-1854); editor: William Carey (1761-1834).
1: 1820 (prob. Jan-Jun), p. [1]-7, [1]-493.
2: 1824 (prob. Mar-Jun, p. v: Mar 1824), p. [i]-v, [1]-588.
Copies: G, HH, MO, NY, PH, USDA; IDC 300.
Facsimile ed.: 1975, New York (Oriole Editions), 2 vols., ISBN 0-882-1177-2, with an introduction by D.H. Nicolson. *1*: [i]-ix, and facsimile, *2*: [i*-iv*], and facsimile. *Copies*: FAS, MO.
Note: This first edition contains only part of the material in the original manuscript (published in 1832, see below), but was in some places quite heavily annotated by N. Wallich who included many new taxa based on material collected by himself. The new Roxburgh taxa are marked "R", those of Wallich "Wall.". The latter's notes are signed N.W. The first edition ("Carey/Wallich ed.") remained unfinished (ends with *Posoqueria*, Pent. Monog.); the same coverage (excluding Wallich's notes and taxa) forms vol. 1 of ed. 2, 1832. (Inf. D.H. Nicolson).
Ref.: Kuntze, O., Rev. gen. pl. 1: cxli.1891.
Stafleu, F.A., Taxon 24: 685-686. 1975.

9725. *Flora indica*; or, descriptions of Indian plants. By the late William Roxburgh, ... Serampore (printed for W. Thacker and Co. Calcutta and Parbury, Allen and Co. London) 1832, 3 vols. Oct. †. (*Fl. ind. ed. 1832*).
Editor: William Carey (1761-1834).
1: 14 Jan 1832, (Review in J. Asiat. Soc. Bengal 1(4): 131-139. Apr 1832; see Mabberley 1977), p. [i-vii], [1]-741.
2: 1832, p. [i]-vi, [1]-691.
3: Oct-Dec 1832 (p. vi: 7 Sep 1832), p. [i]-viii, [1]-875, err. slip.
Copies: G, HH, M, MO, NY, PH, U, USDA; IDC 5697.
4: Jan 1844 (cryptogams), edited by William Griffith (1810-1845), Calcutta, s.d., p. [i], [1]-58, [i]-ii. *Copies*: E, HH. –Reprinted with independent pagination from Calcutta J. nat. Hist. 4: 463-520. Jan 1844.
Note: "The present edition ["Carey ed."] of the Flora, to be completed in four [sic] volumes, will ... consist of the Mss. left with me by the late Dr. Roxburgh, without any addition".
The additions in ed. 1 (see above) were the "invaluable notes and additions of Dr Wallich which Roxburgh's heirs "did not consider themselves at liberty to make use of without the permission of that eminent botanist". [Carey, in advertisement preceding vol. 1]. This edition is complete but excludes the Wallich notes and new taxa of ed. 1. All Roxburgh taxa in vol. 1 were previously published in ed. 1, 1820-1824.
Reprint ("literatim"): 1874, p. [i*-iii*], [i]-vi, [vii], [1]-763, [i]-lxiv. *Copies*: BR, FI, MICH, MO, USDA. – Preface by Charles Baron Clarke 81832-1906) on p. [i]-vi. – The reprint includes the Roxburgh cryptogams as published by Griffith in Calcutta J. nat. Hist. 4: 463-520. 1844. – The MO copy has a cutting from a Madras Newspaper

of 21 Mar 1896, reviewing G. King's tribute to Roxburgh of 1895, signed "Englishman".
Reprint: 1971 (n.v.) New Delhi.
Ref.: Anon., J. Asiat. Soc., Bengal. 1(4): 131-139. Apr 1832 (review of vols. 1-3; however, vol. 3 has a preface dated 7 Sep 1832).

Roy, John (1826-1893), Scottish botanist (desmidiologist); educated in the Normal College of the Church of Scotland, Edinburgh; school teacher at Brackmuirhill and at Old Bridge of Don nr. Aberdeen; from 1863 in charge of the Skene Square Public School at Aberdeen; LL.D.h.c. Aberdeen 1889. (*Roy*).

HERBARIUM and TYPES: in herb. J. Keith (1825-1905), and OXF. Contributed to V. Wittrock & O. Nordstedt, *Algae aquae dulcis exsiccatae praecipue Scandinavicae*.

BIBLIOGRAPHY and BIOGRAPHY: Barnhart 3: 187 (b. 24 Feb 1826, d. 18 Dec 1893); BB p. 263; BL 2: 294, 711; BM 4: 1745; Clokie p. 235; Desmond p. 533; De Toni 1: cxix, 4: xlviii; MW p. 419.

BIOFILE: Anon., Bot. Centralbl. 57: 320. 1894 (d.).; Bot. Not. 1894: 47(d.); J. Bot. 32: 64. 1894 (d.).
J.W.H.T., Ann. Scott. nat. Hist. 3: 72-75. 1894 (obit., portr., bibl.).
Kent, D.H., Brit. herb. 76. 1957.

EPONYMY: *Roya* W. West et G. S. West (1896).

Royen, Adriaan van (1704-1779), Dutch botanist; Dr. med. Leiden 1728; lecturer in botany and medicine Leiden 1729; director of the botanic garden 1730-1754; regular professor of botany and medicine 1732-1754 (for botany) 1732-1775 (for medicine); associated with C. Linnaeus during the latter's years in Holland; correspondent of A. v. Haller. (*Royen*).

HERBARIUM and TYPES: L; further material in BM, G, LINN. – The van Royen herbarium contains a sizeable number of Linnaean types (see e.g. Wijnands, in press).

BIBLIOGRAPHY and BIOGRAPHY: Aa 16: 531-532 (b. 17 Nov 1704; AG 5(2): 50; Backer p. 501 (epon.); Barnhart 3: 187 (b. 11 Sep 1704, d. 28 Feb 1779); BM 4: 1758; Clokie p. 235, 279; Dryander 3: 2, 103, 192, 363; GR p. 709; Hegi 5(2): 1060; Hortus 3: 1203; HU 515; Jackson p. 441; JW 1: 447, 2: 199, 3: 367 (further biogr. refs.); Kew 4: 539; KR p. 603 (b. 17 Nov 1704); Lasègue p. 66, 356; LS 23131; NI 3: 25, 28; NNBW 10: 846 (b. 11 Nov 1704, d. 28 Feb 1779); PR 3393, 7840-7844, ed. 1: 3713, 8795-8798; Rehder 5: 743; SK 1: 451; SO 3611, 3612; TL-1/1151; see M. Houttuyn; TL-2/2048, 6047; see J. Burman, J. Gaertner, M. Houttuyn; Tucker 1: 611-612; Zander ed. 10, p. 709.

BIOFILE: Anker, J., Centaurus 2 11: 247-250. 1951 (Oeder's botanische Reise; visit to Leiden).
Blunt, W., The compleat naturalist 99, 123, 124. 1971.
Bonnet, E., Lettres de Linné à David van Royen, Bull. Herb. Boiss. 3: 16. 1895.
Bridson, G.D.R. et al., Nat. hist. mss. res. Brit. Isl. 255.12. 1980.
Candolle, Alph. de, Phytographie 445. 1880.
Ewan, J., Regn. veg. 71: 23, 29-31, 42, 50. 1970 (plant collectors in America, backgrounds for Linnaeus).
Fries, Th.M., Bref Skrifv. Linné 1(3): 184. 1909, 1(4): 3, 104. 1910, 1(5): 245. 1911, 1(6): 32. 1912.
Gager, C.S., Brooklyn Bot. Gard. Rec. 27(3): 308. 1938.
Hulth, J.M., Bref Skrifv. Linné 1(8): 66, 67. 1922, 2(1): 427 [index] 1916, with A.H. Uggla 2(2): 296 [index]. 1943.
Jackson, B.D., Proc. Linn. Soc. 134 (suppl.): 19. 1922 (coll. LINN).
Jessen, K.F.W., Bot. Gegenw. Vorz. 289, 297, 312. 1884.
Kalkman, C. & P. Smit, eds., Blumea 25(1): 140. 1979.
Kobus, J.C. & W. de Rivecourt, Bekn. biogr. handw.-boek Nederland 2: 725-726. 1857.

Koster, J.Th., Taxon 18: 557. 1969.
Morton, A.G., Hist. bot. Sci. 269. 1981.
Murray, G., Hist. coll. BM(NH) 1: 178. 1904 (plants from East Indies in BM).
Richardson, R., Extr. lit. sci. corr. 338. 1835.
Stafleu, F.A., *in* G.H.M. Lawrence, ed., Adanson 1: 157-159, 164, 172, 177. 1963; Linnaeus and the Linnaeans 45, 87, 109, 129, 158, 159-161, 194. 1971.
Steenis-Kruseman, M.C. van, Blumea 25: 34. 1979.
Veendorp, H. & L.G.M. Baas Becking, Hortus acad. lugd.-bot. 104, 110, 113, 114, 115-129, 131, 168. 1938.
Wijnands, O., The botany of the Commelins, in press (on the van R. herbarium).
Winkler Prins, A., Geill. encycl. ed. 2. 13: 120. 1887 (b. 17 Nov 1704, d. 28 Feb 1779).
Wittrock, V.B., Acta Horti Berg. 3(3): 168. 1905.

COMPOSITE WORKS: (1) *Chionanthus,* Linnaeus, Gen. pl. 335. 1737; *Adenanthera, Acalypha, Phaca, in* Linnaeus Coroll. gen. 7, 13, 19. 1737; *Achyronia, in* Linnaeus Gen. pl. ed. 2. 346. 1742; *Randia, Cyanella,* ib. ed. 5. 74, 149. 1754.
(2) A set of 40 drawings and accompanying etchings of mostly African Erica species is referred to as *Ericetum africanum* of A. van Royen. The original drawings are at L, format 26.7-28 x 21-21.8 cm, numbered *1-10, 12-29, 29.1, 30-40,* unsigned. The etchings are marked "P: Cattell: del." and "J: v: d: Spyk fecit", the set of etchings at L has no. 11 but lacks 16. Corrections in the handwriting of A. van Royen show that this was a series of proofs. The original herbarium specimens in the herb. van Royen(L) of circa 30 plates have been traced. The drawings follow the specimens closely. [Information P.W. Leenhouts, Leiden]. - Another set of etchings is reported from BM(NH).
(3) Preface to P. Miller, *Groot kruidk. woordenb.* 1745, TL-2/6047.

HANDWRITING: Candollea 33: 151-152; Clokie p. 279.

EPONYMY: *Royena* Linnaeus (1753); *Royenia* K.P.J. Sprengel (1830, *orth. var.*).

9726. Disser. botanico-medica inauguralis, *de anatome & oeconomia plantarum.* Quam, annuente deo ter opt. max. ex auctoritate magnifici rectoris, ... nec non amplissimi senatus academici consensu, & nobilissimae facultatis medicae decreto, per gradu doctoratus summisque in medicina honoribus & privilegiis ritè ac legitimè consequendis, publico ac solenni examini submittit Adrianus van Royen, lugd. bot. Ad diem 23. Februarii 1728. hora locoque solitis ... Lugduni batavorum [Leiden] (apud S. Luchtmans) [1728]. Qu. (*Anat. oecon. pl.*).
Publ.: 23 Feb 1728, 46 p. *Copies:* AMD (Univ. Libr.), L (Univ. Libr.), MO.

9727. Adriani van Royen, *Oratio* qua jucunda, utilis, ac necessaria, medicinae cultoribus commendatur doctrina botanica. Habita ix. maji mdccxxix. Cum publicum institutiones botanicas praelegendi munus in Academia lugduno-batava inchoaret. Lugduni batavorum [Leiden] (apud Samuelem Luchtmans) 1729. Qu. (*Oratio*).
Publ.: 9 Mai 1729, p. [1]-26, poemata: [27-44], by Joh. Burman, F. v. Oudendorp, H. Snakenburg, D. van Royen]. *Copy:* U (Univ. Libr.).

Book number 9728 has not been used.

9729. Adriani van Royen, *Carmen elegiacum de amoribus et connubis plantarum* quum ordinariam medicinae & botanices professionem in Batava, quae est Leidae, Academia auspicaretur, dictum xix junii mdccxxxii ... Lugduni Batavorum (apud Samuelem Luchtmans, ...) 1732. Qu. (*Carmen amor. connub. pl.*).
Publ.: 1732, p. [i-ii], [1]-34, [35-40, poemata by J.O. Schacht, D. van Royen]. *Copies:* M, NY, U (Univ. Libr.).
Ref.: Richardson, R., Extracts lit. sci. corr. 366. 1835.

9730. Adriani van Royen, M.D. ... *Florae leydensis prodromus,* exhibens plantas quae in horto academico lugduno-batavo aluntur. Lugduni batavorum (apud Samuelem Luchtmans ...) 1740. Oct. (*Fl. leyd. prodr.*).
Publ.: 1740 (p. iv: 26 Sep 1739), p. [i-iv], map garden, clavis classium fold-out, [v-lxxii],

1-538, [1-30, indices]. *Copies*: BR, G, H-UB, HH, HU, MO, NY, PH, U, USDA; IDC 299. – The fold-out plate with a key to the classes is sometimes attributed to Linnaeus; Linnaeus's influence on Royen's *Clavis classium* and *Methodus naturalis praeludium* is evident, although Linnaeus himself did not give the 20 classes.
Ref.: Stafleu, F.A., Linnaeus and the Linnaeans p. 159-161. l971.

9731. *Elegia* Adriani van Royen, quum botanices professionem poneret, publice dicta ix. kal. junii [1754]. Lugduni batavorum [Leiden] (apud Samuelem Luchtmans et filios) 1754. Qu. (*Elegia*).
Publ.: 9 Jun 1754, p. [1]-22, [23-24]. *Copy*: U (Univ. Libr.). – Poems read when resigning professorship in botany.

Royen, David van (1727-1799), Dutch botanist and physician; succeeded his uncle Adriaan van Royen as professor of botany at the University of Leiden and director of its botanical garden 1754 (official appointment 1756)-1786. (*D. Royen*).

HERBARIUM and TYPES: L; some further material at LINN.

BIBLIOGRAPHY and BIOGRAPHY: Aa 16: 532-533; Barnhart 3: 187; BM 4: 1758, 8: 1107; Dawson p. 721-722; Dryander 3: 94; Frank 3(Anh.): 86; JW 1: 447, 2: 199 (b. 30 Dec 1727, d. 19 Apr 1799), 3: 367, 5: 251; NNBW 10: 847; PR 7845-7846, 8541, ed. 1: 8799, 9490; SO 306, 2541, 2544, 2550, 2582; TL-2/4483; Zander ed. 10, p. 709, ed. 11: 809.

BIOFILE: Berkeley, E. and D.S., John Clayton 70, 82, 145, 154. 1963; Dr. Alexander Garden 53, 73, 91. 1969.
Bonnet, E., Lettres de Linné à David van Royen, Genève 14 p., 1895; issued as Bull. Herb. Boissier 3: 13-26. 1895.
Bridson, G.D.R. et al., Nat. hist. mss. res. Brit. Isl. 433 [index]. 1980.
Ewan, J., Regn. veg. 41: 30, 31, 41, 50. 1970 (plant coll. in America, backgrounds for Linnaeus).
Hulth, J.M., Bref Skrifv. Linne 2(1): 97-99. 1916, 2(2)(with A.H. Uggla): 218, 289, 290, 291. 1943.
Jackson, B.D., Proc. Linn. Soc. 134 (suppl.): 19. 1922 (coll. LINN).
Kalkman, C. & P. Smit, eds., Blumea 25(1): 120. 1979.
Kobus, J.C. & W. de Rivecourt, Bekn. biogr. handw.-boek Nederland 2: 726. 1857.
Stafleu, F.A., Regn. veg. 71: 336. 1970; Linnaeus and the Linnaeans 18, 109, 173. 1971.
Steenis-Kruseman, M.J., Blumea 25: 34. 1979 (coll. L).
Stieber, M.T. et al., Huntia 4(1): 86. 1981 (corr. Allioni HU).
Veendorp, H. & L.G.M. Baas-Becking, Hortus acad. lugd.-bat. 117, 118, 129. 1938.

COMPOSITE WORKS: *Cunila, Pedalium, in* C. Linnaeus' Syst. nat. ed. 10, p. 1359, 1375. 1759; *Schwenkia, in* C. Linnaeus, Gen. pl. ed. 6, p. 577 ("567") 1764 (see also below 1766); *Leea, in* C. Linnaeus, Mant. pl. 1: 17. 1767; *Codon, in* C. Linnaeus , Syst. nat. ed. 12. 2: 291. 1767.

9731a. Davidis van Royen *Oratio de hortis publicis*, praestantissimis scientiae botanicae adminiculis, habitu xiv. junii mdccliv quum ordinarium botanicis professionem in Batava, quae Leidae est, Academia auspicaretur. Lugduni batavorum [Leiden] (apud Samuelem Luchtmans & filios) 1754. Qu. (*Oratio hort. publ.*).
Publ.: 14 Jun 1754, p. [i-vi], 1-23, [24-27]. *Copies*: U (Univ. Libr.) (2).

9732. *Novae plantae Schwenckia*, dictae a celeb. C. Linnaeo, in Gener. plant. editione vi. p. 567. Ex celeb. D. van Rooyen [sic] charact. mss. 1761. communicata brevis descriptio et delineatio cum notis characteristicis Hagae comitum ['s Gravenhage; Den Haag] (apud Jacobus van Karnebeek) 1766. Oct. (*Nov. pl. Schwenckia*).
Publ.: 1766, p. [i-viii], *1 pl.* (col. copp.). *Copies*: B-S, BR, G, Kon.-Bibl. 's Gravenhage. – Issued as an appendix to Martin Wilhelm Schwencke (1707-1785), *Kruidkundige beschrijving der in– en uit-landsche gewassen*, welke heedendaagsch meest in gebruik zijn, 's Gravenhage, 1766, q.v. – The only independent copy of this pamphlet known to us is that at G, pagination [ii-viii]. For an extensive discussion see Heine (1963). – Van Royen sent a copy of this appendix to Linnaeus on 21 Feb 1767 (see Bonnet 1895, p.

22). In his correspondence Linnaeus continues his use of the spelling *Schwenkia*.
Ref.: Heine, H., Kew Bull. 16(3): 465-469. 1963.

Royer, Charles Louis Alexis (1831-1883), French botanist; landowner at Quincy, Côte-d'-Or. (*Royer*).

HERBARIUM and TYPES: AUT.

NOTE: Not to be confused with the Namur (Belgian) pomologist Auguste Philippe Antoine Royer (1796-1867) (see e.g. Ed. Morren, Bull. Féd. Soc. Hort. Belgique 1868: 207-219).

BIBLIOGRAPHY and BIOGRAPHY: Barnhart 3: 187 (d. 18 Dec 1883); BL 2: 141, 711; BM 4: 1758; CSP 5: 315, 8: 793, 11: 235, 18: 340, 12: 637; Jackson p. 83; Kew 4: 540; Moebius p. 327; Rehder 5: 743; Tucker 1: 612.

BIOFILE: Anon., Bull. Soc. bot. France, Table art. orig. 208-210; C.R. Soc. roy. Bot. Belg. 23(2): 6. 1884 (d.); Rev. bot. 2: 337. 1884, 4: frontisp. portr., 1886.
Malinvaud, E., Bull. Soc. bot. France 30: 314-315. 1883 (d. 18 Dec 1853).
Magnin, A., Bull. Soc. bot. Lyon 35: 50. 1910.

9733. *Flore de la Côte-d'Or* avec déterminations par les parties souterraines... Paris (Librairie F. Savy...) [1876-] 1881-1883. 2 vols. Oct. (*Fl. Côte-d'Or*).
1: 1876-1881 (Nat. Nov. Nov(1) 1881; Bot. Zeit. 30 Dec 1881; Bot. Centralbl. 7-11 Nov 1881; first fascicle, p. xxvii and 57 is reviewed by BSbF 23: 25-26. Jul-Dec 1876, rd. 19 Jun 1876 by Society; J. Bot. Dec 1881), p. [1]-xxvii, [1]-346.
2: Mai 1883 (Bot. Centralbl. 11-15 Jun 1883; BSbF séance 8 Jun 1883; Bot. Zeit. 27 Jul 1883; rev. BSbF for books publ. Mar-Mai 1883), p. [i, iii], [347]-693, [1, cont.].
Copies: BR, G, HH, M, MO, NY, USDA. – The flora has a key based solely on subterranean organs.
Ref.: Malinvaud, E., Bull. Soc. bot. France 28 (bibl.): 146-147 1881 (rev. vol. 1), 30: 53-55. 1883 (rev. vol. 2).
N.L., Bot. Zeit. 40: 223. 1882.

Royle, John Forbes (1800-1858), British botanist and physician, educated at Edinburgh; MD München 1833; surgeon East India Company, Bengal 1819; curator of the Saharunpur garden 1833; to England 1831; professor of materia medica King's College, London 1837-1856; secretary Horticultural Society London 1851-1858. (*Royle*).

HERBARIUM and TYPES: LIV (12.000 Indian pl.), BM (orig. bryoph.), other material at B (mainly destr.), BR, BSD, CAL, CGE, DD, E, G, K, LE, PH (major coll.), TCD. – Letters at Kew and the Royal Society of Arts (London). – According to Hitchcock (2: 22) Royle's types are at K; this is based on Alph. de Candolle's erroneous statement (Phytographie 445. 1880) that the original Royle herbarium is at K.

BIBLIOGRAPHY and BIOGRAPHY: AG 6(2): 10; Backer p. 501 (epon.); Barnhart 3: 187 (d. 2 Jan 1858); BB p. 263-264; Blunt p. 166; BM 4: 1758, 8: 1107; CSP 5: 316, 12: 637; Desmond p. 533; DNB 49: 375-376; Frank 3 (Anh.): 86; GF p. 74; GR p. 413; Hortus 3: 1203; IH 1 (ed. 2): 74, (ed. 3): 92, (ed. 7): 341, 2: (files); Jackson p. 20, 385, 388; Kew 4: 450-451; Langman p. 654; Lasègue p. 128, 149, 154, 346, 432, 503, 527; LS 23132; MW p. 419; NI 1690; see also 1: 129, 130, 240, 3: 58; Plesch p. 390; PR 617, 7849-7854, ed. 1: 717, 8802-8809; Rehder 5: 743; SK 5: ccx-ccxi; Sotheby 666; TL-1/1119; TL-2/422; Tucker 1: 612; Urban-Berl. p. 389; Zander ed. 10, p. 709, ed. 11, p. 809.

BIOFILE: Allan, M., The Hookers of Kew 205. 1967.
Allibone, S.A., Crit. dict. Engl. lit. 2: 1886. 1878.
Anon., Bonplandia 3: 33. 1855 (member Leopoldina, cogn. Heyne), 6: 34 (d.), 338. 1858 (Leopoldina cogn. Heyne); Bot. Zeit. 7: 206-207. 1849 (distr. duplicates), 16: 68, 127-128. 1858 (d.), 18: 311. 1860; Bull. Soc. bot. France 4: 1070. 1858; Cottage Gardener 19: 225, 249-250. 1857 (obit.); Flora 41: 47-48. 1858 (obit.); Gard. Chron. 1858: 20-21 (obit.); Österr. bot. Z. 8: 75, 139. 1858 (d.); Proc. Linn. Soc. 1858: xxxi-

xxxvii (obit.); Proc. R. soc. London 9: 547-548. 1859 (obit.); Taxon 2: 183. 1953 (rediscovery Royle herb. in LIVU).
Bor, N.L., Kew Bull. 1954(3): 453-460. (Indian species of *Agrostis* collected by R.).
Bridson, G.D.R. et al., Nat. hist. mss. res. Brit. Isl. 433 [index]. 1980.
Burkill, I.H., Chapt. hist. bot. India 243 [index, 24 entries]. 1965.
Candolle, A. de, Phytographie 445. 1880 (the Kew set mentioned by A.D.C. was not the orig. herb.).
Coats, A.M., The plant hunters 150, 156. 1969; The quest for plants 150-151. 1969.
Crawford, D.G., A history of the Indian medical service 1600-1913, 2: 149-151, 171. 1914.
E.O., Gartenflora 7: 198. 1858 (d.).
Fletcher, H.R., Story R. Hort. Soc. 157, 162, 170, 275.
Gager, C.S., Brooklyn Bot. Gard. Rec. 27: 264. 1938 (dir. Saharunpur bot. Gard. 1823-1831).
Gray, J.L., Letters A. Gray 112, 151. 1893.
Greenwood, E.F., J. Soc. Bibl. nat. Hist. 9(4): 379. 1980.
Harrison, S., Taxon 27: 21-33. 1978 (recent work on R.'s herb.; revised list of type specimens in herb. at LIV).
Heywood, V.H., Taxon 5: 11. 1956 (herb. presented to Liverpool Royal Institution 1859; to Univ. Liverpool 1948, later to City Museum (LIV)).
Huxley, L., Life letters J.D. Hooker 1: 44, 468, 473, 2: 286. 1918.
Jessen, K.F.W., Bot. Gegenw. Vorz. 2, 469. 1884.
King, G., J. Bot. 37: 462. 1899.
Kuntze, O., Rev. gen. pl. 1: cxli. 1891.
Nissen, C., Zool. Buchill. 350. 1969.
Poggendorff, J.C., Biogr.-lit. Handw.-Buch 2: 708. 1863.
Rix, M., The art of the botanist 183, 184, 186. 1981.
Sayre, G., Bryologist 80: 516. 1977 (bryo. at BM, CGE, NY-Mitten).
Sch., Bonplandia 6: 221-222. 1858 (obit.).
Schlechtendal, D.F.L. von, Bot. Zeit. 16: 127-128. 1858.
Soejarto, D.D. et al., Taxon 30: 652-656. 1981 (on Podophyllum and the Royle herb.).
Stansfield, H., North West Naturalist 24: 250-265. 1953 (rediscovery of Royle's herbarium; now at LIV), Museums J. 52(12): 292-295, *pl. 43-44*. 1953 (id.), Pharmac. J. 170: 74-75, 78-79. 1953 (id.); Liverpool Libr. Mus. Arts Comm. Bull. 3(3): 5-38. 1954 (description Royle herb.); Hist. Congr. int. Bot., Rapp. Comm. avant le Congrès, sect. 21-27, p. 173-175. 1954 (on type herb., *Ill. bot. Himal. Mts.* at LIV).
Stearn, W.T., J. Arnold Arb. 24: 484-487. 1943 (on Ill. bot. Himal. Mts.); Roy. Hort. Soc. Exhib. manuscr. books 1954, no. 103.
Stewart, R.R., Pakistan J. Forestry 17: 355. 1967 (Royle had native collector in Kashmir who travelled with shawl merchants); Taxon 28: 8. 1979 (R's work in Kashmir pl.).
Subba Reddy, D.V., Bull. Inst. Hist. Med. Hyderabad 3(2): 79-87. 1973 (J.F.R., "botanist-medical historian-teacher and benefactor of British Empire").

EPONYMY: *Roylea* Wallich ex Bentham (1829).

9734. *Illustrations of the botany* and other branches of the natural history *of the Himalayan Mountains* and of the flora of Cashmere ... London (Wm. H. Alland and Co., ...) [1833] 1839[-1840]. 2 vols., Qu. (*Ill. bot. Himal. Mts.*).
1 (text): 1833-1840, frontisp., p. [[i*]-viii*, [sign. b, table: ix*-xiv*], sign. b, introd. [sic]: [v]-xxxvi, [text runs on:] xxix*-xxxvi*, [xxxvii]-lxxviii, [lxxix-lxxx, [1]-472. (analysis copy: G).
2 (plates): 1833-1840, frontisp., p. [i]-iv, *pl. 1-63*, "*63a or 79*", "*75*" [= 76], *75a, 77, 78*, "*99 or 78a*", *80-82*, "*83 or 100*", *83[bis]*, "*84a or 99*", *84-97* (analysis copy: G).
Copies: BR, G, M, MICH, MO, Teyler, U, USDA, IDC 695. – Plates (col. or uncol. liths.) by Vishnupersaud, with some by Capt. Cantley, C.M. Curtis, Miss Drake, J.T. Hart, W. Saunders, Luchmun Sing, J.D.C. Sowerby, J.O. Westwood. See also NI 1690; NI mentions *pl. 92* in 3 states. The col. frontisp. of vol. 1 is by R.Smith (*pl. 66*). The plate "84a or 99" seems to occur also as "84a or 98". – For the dates see TL-1, Sprague (1933), Stearn (1943) and below. Part 11 was donated to the Geological

Society of London on 20 Apr 1840 (see Trans. Geol. Soc. London ser. 2. 6, suppl.: 12. 1842).

part	pages	plates	dates
1	1-40	*4, 11-18, 22*	Sep 1833
2	v-xii, 41-72	*1, 19-21, 23-28*	Mar 1834
3	xiii-xx, 73-104	*2, 5, 29, 31-35, 37, 38*	Jun 1834
4	105-136	*30, 39, 40, 42, 44-46, 64 (Rhododendron) 76 as 75 (Primula), 78*	Sep 1834
5	137-176	*3, 41, 48-51, 57, 62, 63, 64*	Jan 1835
6	177-216	*7, 36, 43, 55, 56, 58, 60, 61, 75 as 75a (Phlomis, Salvia) & frontispiece to vol. 1*	Apr 1835
7	217-248	*8, 9, 47, 52, 59, 64 (Olea), 67-69, 71, 77*	24 Aug 1835
8	249-288	*53, 54, "63a or 79", 66, 70, 72, 73, 80, 87, 88*	
9	289-336	*10, 81, 82, 83 (Procris), 84-86, 90, 100, ("83 or 100 Putranjiva")*	Mai 1836 (cover: "Apr 1835")
10	337-384	*"78a or 99", "84a or 99", 89, 91-96, & frontispiece to vol. 2*	Feb 1839
11	xxi-lxxx, 385-472	t.p.'s, dedic., preface, lists etc. *6, 97*	Mar-Apr 1840

The new species fall into two categories (Sprague 1933):
   a) those with figures accompanied by analyses which date from the publication of the plate where this is earlier than the corresponding text.
   b) those without analyses, which date from the publication of the description (if any) in the text.
Sprague and Stearn (1943) give lists of the new species with their precise dates. – The contents of the parts were given on the last page of the original covers (copy at U).
Ref.: Sprague, T.A., Bull. misc. Publ. Kew 1933: 378-390. 1933 (dates).
   Stearn, W.T., J. Arnold Arb. 24: 484-487. 1943 (dates).

9735. *Essay on the productive resources of India* ... London (Wm. H. Allen and Co.) 1840. Oct. (*Essay prod. resourc. India*).
Publ.: 1840, p. [i]-x, [1]-451. *Copy*: G.

9736. *On the culture and commerce of cotton in India*, and elsewhere; with an account of the experiments made by the hon. East India Company up to the present time. Appendix: papers relating to the great industrial exhibition ... London (Smith, Elder & Co., ...) 1851. Oct. (*Cult. comm. cotton India*).
Publ.: 1851, p. [i]-xvi, 1-607. *Copy*: E (Univ. Libr.).

9737. *Fibrous plants of India* fitted for cordage, clothing, and paper. With an account of the cultivation and preparation of flax, hemp, and their substitutes ... London (Smith, Elder, and Co., ...) 1855. Oct. (*Fibr. pl. India*).
Publ.: Mar 1855 (p. x: 24 Feb 1855; ÖbZ 17 Mai 1855; Bonplandia 15 Apr 1855), p. [i]-xiv, 1-403, [404, err.]. *Copies*: BR, E, L, E(Univ. Libr.), FI, NY, USDA; IDC 5156.

Roze, Ernest (1833-1900), French administrator at the ministry of finances Paris; self-taught amateur botanist, mycologist, bryologist and botanical historian. (*Roze*).

HERBARIUM and TYPES: P, PC. – *Exsiccatae*: *Muscinées des environs de Paris* (with É. Bescherelle), fasc. 1-10, nos. 1-250, 1861-1866; see also A. Camus (1903). Sets at BM, FH, H (has a complete set plus an unpublished fascicle), LG, PC. – Fasc. noticed in Bull. Soc. bot. France 8: 328. Aug 1861(announcement), 9: 209. 1862 (1 & 2 rd. 25 Apr 1862), 9: 447. 1863 (3 & 4 rd. 28 Nov 1862), 10: 105. Oct 1863 (5 rd. 27 Feb 1863), 10:

538. Jun 1864 (7 & 8 rd. 11 Dec 1863), 11: 310. 1864 (9 rd. 9 Dec 1864), 13 (bibl.): 47. 1888 (10 rd. 1866). – For further details see G. Sayre, Mem. New York Bot. Gard. 19(2): 248. 1971. – Contributed to C. Roumeguère, *Fungi exsiccati praecipue Gallici.* Thirty letters to E. Malinvaud at P (Aymonin 1977).

BIBLIOGRAPHY and BIOGRAPHY: Ainsworth p. 120, 243, 297, 308; Barnhart 3: 187 (b. 17 Mai 1833, d. 25 Mai 1900); BM 2: 386, 4: 1758; Bossert p. 341; CSP 5: 317, 8: 795, 11: 236, 12: 637, 18: 341-342; De Toni 1: cxix, 4: xlix; Frank 3(Anh.): 86; GR p. 349; Hawksworth p. 185; Herder p. 247; IH 2: (files); Kelly p. 199-200; Kew 4: 541; LS 23134-23210a; Morren ed. 2, p. 21, ed. 10, p. 57; MW p. 414, 420; NI 1691, see also 1: 69, 150; Plesch p. 391; PR 7650; Rehder 5:744; SBC p. 130; Stevenson p. 1256; TL-2/602; Tucker 1: 612; Urban-Berl. p. 312.

BIOFILE: Anon., Bot. Not. 1901: 114 (d.); Bull. Soc. bot. France, Table art. orig. 1854-1893, p. 210-213; Hedwigia 36: (34). 1897 (president Soc. mycol. France 1897), 39: (194). 1900 (d.); Nat. Nov. 22: 458. 1900 (d.); Österr. bot. Z. 50: 307. 1900 (d.).
Aymonin, G.G., C.R. Congr. Soc. Sav. (Limoges) 102(3): 57. 1977 (letters to E. Malinvaud).
Camus, E., Bull. Soc. bot. France 50: 227-239. 1903 (in biogr. Bescherelle).
Candolle, Alph. de, Phytographie 445. 1880.
Cornu, M., Bull. Soc. bot. France 47: 177, 179-185. *pl. 7.* 1900 (portr., obit.).
Kneucker, A., Allg. bot. Z. 6: 196. 1900 (d.).
Loynes, de, Actes Soc. Linn. Bordeaux 55: cx. 1900 (d.).
Seynes, J. de, Bull. Soc. mycol. France 16(3): 164-174. 31 Jul 1900 (obit., bibl., portr.).
Wittrock, V.B., Acta Horti Berg. 3(3): 113. 1905.

COMPOSITE WORKS: (1) Contributed a chapter on the history of botany to J. Rothschild, ed., *Les fougères,* vol. 1, 1867, and wrote the "histoire botanique et horticole des Selaginelles" for vol. 2, 1868; see under J. Rothschild.
(2) See C. Richon for Richon & Roze, *Atlas des champignons.* 1885 (NI 1691).

EPONYMY: *Rozea* Bescherelle (1872); *Rozella* Cornu (1872); *Rozellopsis* Karling ex Cejp (1959); *Rozites* P.A. Karsten (1879).

9738. *Recherches biologiques sur l'Azolla filiculoides,* Lamarck; ... [Paris (Gauthier-Villars et fils, ... 1888]. Oct.
Publ.: 1888, p. [1]-13. *pl. 19* (uncol. lith. C. Rolet). *Copy*: FH. – Reprinted and to be cited from Mém. Soc. philom. centenaire 216-227. *pl. 19.* 1888.

9739. *La flore d'Étampes* en 1747 d'après Descurain et Guettard [extrait du Journal de Botanique. N⁰ˢ des 1ᵉʳ et 16 Avril 1889). Oct.
Publ.: 1 and 16 Apr 1889 (in journal), p. [1]-12. *Copy*: HH. – Reprinted and to be cited from J. de Bot. 3: 124-128. 1 Apr, 141-148. 16 Apr 1889. – Notes on Jean Étienne Guettard (1715-1786), *Observ. pl.* 1747, TL-2/2207.

9740. *Histoire de la pomme de terre* traitée aux point de vue historique, biologique, pathologique, culturel et utilitaire ... Paris (J. Rothschild, éditeur ...) 1898. Oct. (*Hist. pomme de terre*).
Publ.: 1898 (Bot. Zeit. 1 Feb 1899; Nat. Nov. Jan(1) 1899), p. [i*-iii*], [i]-xii, *1 pl.* (col.), [1]-464. *Copies*: MO, NY, USDA.
Ref.: Bornet, Ed., Bull. Soc. bot. France 46: 79-80. 1899.
Jackson, B.D., J. Bot. 37: 232-235. Mai 1899 (rev.).
Saint-Lager, J.B., Ann. Soc. bot. Lyon 23: 49-51. 1898 (rev.).

9741. *Recherches biologiques sur l'Azolla filiculoides,* ... [ex Mém. Soc. Philom. centenaire 1788-1888, 1888].
Publ.: 1888, p. [1]-13, *pl. 19. Copy*: BR. – Reprinted and to be cited from Mém. Soc. philom. à l'occasion de sa fondation 1788-1888, Paris 1888, p. 215*-229*.

Rozier, François (Jean-François), (1734-1793), French botanist, agronomist and clergyman; educated at the Lyon seminary of Saint-Irenée; director of the École vétérinaire de Lyon and in charge of its botanical garden 1765-1766; professor at the "Académie royale" Lyon; associated with J.J. Rousseau ca. 1767-1768 in Paris (ca. 1770-1780) where he founded the *Journal de Physique* (1771); travelled in Corsica and the Netherlands (1777); at Béziers, 1780, shortly afterwards back in Lyon, working on his *Cours compl. agric.*; during the French revolution "curé constitutionnel" of the Lyon parishes; died through war action 1793. (*Rozier*).

HERBARIUM and TYPES: In the library of the Lyon Palais des Arts (fide Magnin 1906; no further details known to us); some material in the Giseke herbarium at E.

BIBLIOGRAPHY and BIOGRAPHY: Backer p. 660; Barnhart 3: 187 (b. 23 Jan 1734, d. 29 Sep 1793); BM 4: 1758, 1759; Bossert p. 341; Dryander 3: 20, 459, 568, 640; Frank 3(Anh.): 86; GR p. 349-350; Herder p. 393; Kew 4: 542; KR p. 603; Lasègue p. 344; LS 23211; NI (alph.), see also 3: 95; Plesch p. 391, 392; PR 5084-5085, 7855, ed. 1: 1803, 5577-5578, 7781, 8807-8809, 11678; Rehder 5: 744; SO 282, 716, 717, 2631; Sotheby 667; TL-2/4229.

BIOFILE: Anon., Grand Larousse encycl. 9: 426. 1964.
Dandy, J.E., Regn. veg. 51: 18. 1967.
Dugour, A.J., *in* F. Rozier, *Cours compl. agric.* 10: [i]-xvi. 1800 (biogr.).
Faivre, J.J.A.E., Bull. Soc. bot France 23: viii. 1876.
Gager, C.S., Brooklyn Bot. Gard. Rec. 27(3): 202. 1938 (dir. bot. gard. École vét. Lyon 1765-1766).
Guillaumin, A., Les fleurs des jardins xliii, *pl. 15.* 1929 (portr.).
Hedge, I.C. & J.M. Lamond, Index coll. Edinb. herb. 125. 1970.
Latourrette, M.A.L.C. de, Démonstr. élém. bot. ed. 4. 1796, ed. J.E. Gilibert, TL-2/4229, contains a Notice sur la vie et les écrits de M. l'abbé Rozier by J.E. Gilibert, see 1: lix-lxviii, 3: 269.
Magnin, A., Bull. Soc. bot. Lyon 31: 3, 39 (further biogr. refs.), 47 (dir. bot. gard.) 1906, 32: 113. 1907 (further refs.), 35: 22 (id.), 66, 67. 1910.
Margadant, W.D., *in* G.H.M. Lawrence, ed., Adanson 1: 336. 1963 (see also J.P. Nicolas ib. p. 74).
Stieber, M.T. et al., Huntia 4(1): 86. 1981 (corr. Adanson, HU).
Wittrock, V.B., Acta Horti Berg. 3(3): 113. 1905.

COMPOSITE WORKS: (1) collaborated with M.A.L.C. de Latourette in the *Démonst. élém. bot.* (ed. 1, 1766 through ed. 4, 1796), see TL-2/4229.
(2) founder and editor of the *Journal de Physique.* 1771-1780.

9741a. *Cours complet d'agriculture* théorique, pratique, économique, et de médecine rurale et vétérinaire; suivi d'une méthode pour étudier l'agriculture par principes: ou dictionnaire universel d'agriculture; par une société d'agriculteurs, & rédigé par l'abbé Rozier, prieur commandataire de Nanteuil-le-Haudoin, ... Paris (Rue et Hôtel Serpente) 1781-1805, 12 vols. Qu. (*Cours compl. agric.*).

1: 1781, p. [i]-vi, [vii-viii], [1]-702, table, *pl. 1-21* (uncol. copp. Sellier). *Re-issue*: 1791 (copy: BR), title-page issue; same reprint); other *re-issue* 1797, à Paris (chez Delalain, ...) (copy: NY).
2: 1782, p. [i-viii], [1]-683, table, *pl. 1-27, 4bis*. Re-issue 1785, à Paris (chez Delalain ...) (copy: NY).
3: 1783, p. [i-vi], [1]-685, *pl. 1-20*. – Re-issue 1785, id.
4: 1783, p. [i, iii], [1]-694, *pl. 1-11*. – Re-issue 1786, id.
5: 1785, p. [i-vi], [1]-747, *pl. 1-29, 2bis, 15bis*.
6: 1785, p. [i, iii], [1]-735, [736, err.], *pl. 1-24*. Re-issue 1786 (copy: BR), title-page issue.
7: 1786, p. [i*, iii*], [i]-iv, [1]-760, *pl. 1-22*.
8: 1789, p. [i, iii], [1]-8, [1]-709, [710], *pl. 1-40*.
9: 1796, p. [i-vi], [1]-674, *pl. 1-18*.
Note: vols. 10-12 were continued by others after Rozier's death. Vol. 9, which came out posthumously was still by Rozier and carried the same imprint as vol. 1; vols. 10-12 have different title-pages.
10: An viii-1800, p. [i*-ii*], frontisp., [iii*-iv*], [i]-xvi, [1]-499, [500 adv.] table, *pl. 1- 29*. "*Cours complet d'agriculture* ... rédigé par Rozier. Tome dixième rédigé par les

citoyens Chaptal, ... Dussieux, Lasteyrie et Cadet-de-Vaux, ... Parmentier, Gilbert, Rougier-Labergerie, et Chambon ... à Paris" (à la Librairie d'Éducation et des Sciences et Arts, ...) Contains a biography of Rozier by A.J. Dugour on p. [i]-xvi. – Botanical authors among the above are Jean Antonin Claude (comte) Chaptal de Chanteloup ("citoyen Chaptal" during the revolution) (1756-1832), Antoine Alexis François Cadet de Vaux (1743-1829), Charles Philippe (comte) de Lasteyrie-Dusaillant (1759-1849), Antoine Auguste Parmentier (1737-1813) and Jean Baptiste (baron) Rouzer de la Bergerie (1762-1836).

11: 1805, p. [i*-iv*], [i]-lvi, chart 1-3, [1]-492, *pl. 1-15, 2bis.* – *Cours complet d'agriculture*, rédigé par MM. Rozier, Chaptal, Parmentier, Delalause, Mongez, Lasteyrie, Dussieux, Gilbert, Rouzier de la Bergerie, etc. etc. Tome onzième, rédigé par MM. A. Thouin, Parmentier, Biot, de Chassiron, Chabert, Lasteyrie, de Perthuis, Cotte, Sonnini, Fromage, Chaumontel, Tollard aîné, Bosc, Curaudau. Précédé d'un discours sur l'exposition et la division méthodique de l'économie rurale, ... accompagné de tableaux synoptiques, ... destinés à server de tables au douze volumes du Cours complet. Par M.A. Thouin, ... à Paris (chez Marchant, ... Drevet, ... Crapart, Caille et Ravier, ...) 1805. Qu. – New botanists among the editors: André Thouin (1747-1824), Jean Baptiste Biot (1774-1862), Louis Augustin Guillaume Bosc (1759-1828), Charles Nicolas Sigisbert Sonnini de Manoncour (1751-1812).

12: 1805, p. [i-iii], [1]-668, table, *pl. 1-10.* – "Tome douzième, rédigé par MM. A. Thouin, ... Perthuis, Roard, Cotte, ... Curaudau. Formant le complément de cet ouvrage, et contenant les découvertes et améliorations faites en agriculture, art vétérinaire et économie rurale, depuis vingt ans ...".

Copies: BR (1-12), G(1-7), NY.

Second edition: 1809, 7 vols. *Copy*: BR (vols. 1-6). – Cours complet d'agriculture pratique d'économie rurale et domestique, et de médecine vétérinaire, par l'abbé Rozier; rédigé par ordre alphabétique: ouvrage dont on a écarté toute théorie superflue, et dans lequel on a conservé les procédés confirmés par l'expérience et recommandés par Rozier, par M. Parmentier et les autres collaborateurs que Rozier s'était choisis. On y a ajouté les connaissances pratiques acquises, depuis la publication de son ouvrage, sur toutes les branches de l'agriculture et l'économie rurale et domestique. Par Messieurs: Sonnini, Tollard aîné, Lamarck, Chabert, Lafosse, Fromage Defeugré, Cadet-de-Vaux. Lamerville, Cossigny, Curaudau, Chevalier, Lombard, Cadet-Gassicourt, Poiret, Chaumontel, Louis Dubois, V. Demusset, Demusset de Cogners ... Paris (chez F. Buisson, ... Léopold Collin, ... D. Colas, ...) 1809. 7 vols. Oct. (*Cours compl. agr. prat.*).

1: [i-ii], frontisp. portr. of Rozier, [iii]-vi, [vii-viii], [1]-598, (details on authors on p. v-vi).
2: n.v.
3: [i-iv], [1]-603, [1, err.], *pl. 1-2.*
4: [i-ii], 1]-655, [656], *pl. 1-6.*
5: [i-iii], [1]-600, *pl. 1-3.*
6: [i*-iii*], [i]-xxxii, [1]-510.
7: n.v.

Rozin, A., (*fl.* 1791), Belgian (Liège) botanist and physician. (*Rozin*).

HERBARIUM and TYPES: Unknown.

BIBLIOGRAPHY and BIOGRAPHY: Barnhart 3: 187; BM 4: 1759; PR 7856, ed. 1: 8810; Rehder 1: 406.

BIOFILE: Crépin, F., Guide bot. belg. 225, 428. 1878.

9742. *Herbier portatif* des plantes qui se trouvent dans les environs de Liège, avec leur description & classification selon le système de Linné. Précédé d'un discours sur la botanique. Par A. Rozin, Méd. Premier cahier. 1791. Oct. (*Herb. portat.*).
Publ.: 1791, p. [i]-viii, [1]-72, [1, err.], 37 dried specimens. *Copies*: BR, also BM (see 4: 1759).

Rubel, Franz (*fl.* 1778), Moravian physician and botanist at Sternberg; Dr. med. Wien 1778. (*Rubel*).

HERBARIUM and TYPES: Unknown.

BIOFILE: Barnhart 3: 188; BM 4:1759; Dryander 3: 535; Kelly p. 201; LS 23212; PR 7857, ed. 1: 8811.

9743. Dissertatio inauguralis botanico-medica *de Agarico officinali* quam annuente inclyta facultate medica in antiquissima ac celeberrima Universitate vindobonensi publicae disquisitioni submittit Franciscus Rubel, Moravus sternbergensis disputabitur in Universitatis palatio die [ ] mensis februarii anno m.dcc.lxxviii. Vindobonae [Wien] (apud Josephum Gerold) [1778]. Oct. (*Agaric. off.*).
Publ.: Feb 1778, p. [1]-42, [1-4 theses], *1 pl. Copy*: NY. – Also published in N.J. Jacquin, Misc. austr. 1(6): 164-203, *1 pl.* 1778 (or 1779, see TL-2/3248). – See also DTS 1: 249-250.

Ruchinger, Giuseppe [baptised: Josef] (1761-1847), German (Bavarian) gardener and botanist; from 1798-1815 in Triest; head gardener of the Botanical Garden of S. Giobbe in Venezia. (*Ruchinger*).

HERBARIUM and TYPES: "Ruchinger" collections at W (algae), further algae at HBG. – Possibly partly collected by G. Ruchinger's son Giuseppe Maria Ruchinger 1809-1879. A large collection of plants from Venezia collected by "G. Ruchinger" was at B (now mainly destroyed).

NOTE: Ruchinger had three sons: Giuseppe Maria Ruchinger (1809-1879) who succeeded his father as head of the Venetia botanical garden; Giuseppe Ruchinger, Jr. (1802-1856) professor of pathology at Praha, and Francesco R., physician at Verona. Ruchinger Sr. collected algae and was in correspondence with e.g. C.A. Agardh and F.T. Kützing (see Müller & Zaunick 1960). – G.M. Ruchinger published a *Cenni storici dell' I.R. Orto botanico in Venezia*, 1847. Oct. (PR 10628), see below. – For information on G. Ruchinger Jr. (b. 1802, d. 9 Mar 1856, not 1855), from 1834 physician in Praha, see GR p. 538.

BIBLIOGRAPHY and BIOGRAPHY: BM 4: 1759; Cesati p. 337; De Toni 1: cxix; Herder p. 474; IH 2: (files); Jackson p. 323; LS 23209 (err. G.M. Ruchinger); PR 7858 (d. 1855, err.), ed. 1: 8816; Rehder 1: 425; Saccardo 1: 142-143, 2: 94, (b. 17 Mar 1761, d. 18 Mar 1847), cron. p. xxxi; Tucker 1: 612.

BIOFILE: Hooker, J.D., *in* W.J. Hooker, London J. Bot. 6: 51-52. 1847.
Müller, R.H.W. & R. Zaunick, Friedrich Traugott Kützing 148, 149, 152, 154, 155, 293. 1960.
Saccardo, P.A., Stor. lett. Fl. venet. 100, 106. 1869.

9744. *Flora dei Lidi veneti* ... in Venezia (presso Gio. Giacomo Fuchs ...) 1818. Oct. (*Fl. Lidi veneti*).
Publ.: 1818 (p. [x]: 10 Jan 1818), p. [i-xii], 1-304. *Copies*: BR, E, FI, G, HH, MO, NY, USDA. – Some copies have only four preliminary pages: [i-iv].
Ref.: Anon., Flora 2(2): 390-393. 7 Jul 1819 (rev.).

Ruchinger, Giuseppe Maria (1809-1879), Italian gardener and botanist at the botanical garden of Venetia; succeeding his father Giuseppe [Josef] Ruchinger (1761-1847) as director of this garden. (*G.M. Ruchinger*).

HERBARIUM and TYPES: See under G. Ruchinger.

BIBLIOGRAPHY and BIOGRAPHY: Barnhart 3: 188 (d. 26 Sep 1879); BM 4: 1759; Bossert p. 341; Jackson p. 438; Kew 4: 542; PR 7859, 10628, ed. 1: 8817; Rehder 1: 63, 3: 529; Saccardo 1: 143 (b. 13 Dec 1809, d. 26 Dec 1879), 2: 94.

BIOFILE: Anon., Bot. Centralbl. 1/2: 96. 1880 (d. 26 Dec 1879); Bot. Zeit. 38: 437. 1880 (d.); Gartenflora 12: 336. 1863 (botanic garden closed as such, Ruchinger continues it as a commercial garden).
Kanitz, A., Magy. növen. Lap. 4: 77-78. 1880 (d.).

9745. *Cenni storici dell'* imp. regio *orto botanico in Venezia* e catalogo delle piante in esso coltivate compilato per cura del giardiniere Giuseppe M. Ruchinger ... Venezia (Nell' i.r. priv. stabilimento Antonelli) 1847. Oct. (*Cennia ort. bot. Venezia*).
Publ.: 1847 (p. xiii: Jul 1847), p. [iii]-xiii, [xiv], 1/2-149/150. *Copies*: FI, NY, USDA.

Rudbeck, Johan Olof (1711-1790), Swedish natural scientist; "assessor" at the Swedish mining college 1753; ultimately president (1778) of this college; son of Olaus Rudbeck. (*J.O. Rudbeck*).

HERBARIUM and TYPES: Some material at H.

BIBLIOGRAPHY and BIOGRAPHY: Barnhart 3: 188; BM 4: 1759; IH 2: (files); KR p. 604 (b. 25 Feb 1711, d. 27 Oct 1790); PR 7659, ed. 1: 8622; SO 1334.

BIOFILE: Fries, R.E., Short hist. bot. Sweden 15. 1950.
Fries, Th.M., Linné 1: 69. 1903 (on Sceptrum carol.).
Hulth, J.M., Bref Skrifv. Linné 2(1): 337. 1916.
Swederus, M.B., Bot. trädgård. Upsala 1655-1807. Falun 1877.

NOTE: For information on the Rudbecks we are specially indebted to O. Almborn.

9746. D.D. Dissertatio botanica de planta *Sceptrum carolinum* dicta, quam, consensu facultatis medicae praeside Laurentio Robergio ... ventilandam sistit auctor Johannes Olavus Rudbeck Ol. fil. In Audit. Gust. maj. ad diem xix. jun. anni mdccxxxi horis solitis. Upsalis, Literis Wernerianis. [1731]. Qu. (*Sceptr. carol.*).
Publ.: 19 Jun 1731, p. [i-viii], [1]-17 ["71"], [18-23]. *Copies*: LD, MO. – According to Th.M. Fries (1903; see also SO 1334), this was mainly the work of Linnaeus of whom an original manuscript "Caroli Linnaei Dissertatio de Planta Sceptro Carolino" is kept at the Univ. Libr., Uppsala with a note on the first leaf "I composed this dissertation in one day for 30 daler kopparmynt. For this another has acquired the glory" (transl. SO). – The plant is now called *Pedicularis Sceptrum-carolinum* L. (Sp. Pl. 608. 1753). The text is also included in Skrifter af Carl v. Linné 4(1): 243-256 [-259]. 1908.

Rudbeck, Olaus [Olof] **Johannis** [Sr.] (1630-1702), Swedish physician, polyhistorian and botanist, "adjunkt" (1655) and assistant professor of anatomy and botany at the University of Uppsala 1658, full professor of medicine ib. 1660-1691; founder and director of the Uppsala University botanical garden (now Hortus linnaeanus) 1655-1691. (*O.J. Rudbeck*; *Rudbeck Sr.*).

HERBARIUM and TYPES: Library, manuscripts, other collections and practically all finished woodblocks for his *Campus Elysius* were also destroyed in the Uppsala fire, 16 Mai 1702. See Krok and Löwegren (1952). Some Rudbeck material may still be in BM-Sloane (see Dandy).

BIBLIOGRAPHY and BIOGRAPHY: Barnhart 3: 188; BM 4: 1759, 8: 1108; Bossert p. 341; Dryander 3: 67, 124-125, 605; DSB 11: 586-588 (by S. Lindroth); Hegi 6(1): 503; Henrey 2: 186, 3: no. 1290; Herder p. 84, 147; Jackson p. 30,32, 445; Kew 4: 543; KR p. 604-605 (b. 13 Sep 1630, d. 17 Sep 1702, bibl.); Lasègue p. 391, 502; Moebius p. 419; NI 1692-1694; PR 7860, ed. 1: 4631, 8819-8826, 10959; Rehder 1: 292, 5: 744; TL-1/1120-1121, 1235; TL-2/892; Tucker 1: 612

BIOFILE: Amoreux, P.J., Mém. Soc. Linn. Paris 4: 118-131. 1826 (on Campi Elysii ...).
Annerstedt, C., Upsala univ. hist. 2. 1908-1909.
Blunt, W., The compleat naturalist 26-27, 149. 1971.

Dandy, J.E., The Sloane herbarium 196. 1958 (coll. and letters).
Fries, R.E., Short hist. bot. Sweden 15-16, 18, 114. 1950.
Gager, C.S., Brooklyn Bot. Gard. Rec. 27: 333. 1938.
Hofsten, N. von, *in* S. Lindroth, Swedish men of science 1650-1950, 33-41. 1952.
Hulth, J.M., Bref skrifv. Linné 1(8): 105. 1922.
J., Dict. sci. méd., Biogr. méd. 7: 67-68. 1825 (bibl.).
Lindroth, S., Sv. män kvinn. 6: 382-385. 1949 (portr. in colour); J. Hist. Med. 12(2): 209-219. 1957 (on Harvey, Descartes and young Rudbeck).
Löwegren, Y., Naturaliekabinett i Sverige 62-66, 363, 405 [index]. 1952.
Palmblad, W.F., Biogr. lex. namnk. sv. män. 12: 314-334. 1846.
Percheron, A., Bibl. entom. 2: 20. 1837.
Rudbeck, J.R.G., Campus Elysii eller Glyssis wald af Olof Rudbeck far och sort. Några biogr. anteckn. Samlaren 1911.
Scheutz, N.J.W., Bot. Not. 1863: 71 (epon.).
Sjögren, O., ed., Sv. biogr. lex. Ny följd 9: 488-489. (add. to Palmgren).
Swederus, M.B., Bot. trädgård. Upsala 1655-1807. Falun 1877; Bot. Zeit. 37: 25-27. 1879 (on the 10 unpubl. volumes of the Campus Elysius).
Wikström, J.E., Consp. lit. bot. Suec. 224-232. 1831.
Wittrock, V.B., Acta Horti Berg. 3(2): 68-69. 1903, 3(3): liv, 50-51, *pl. 17*. 1905 (portr.).

COMPOSITE WORKS: (1) Rudbeck started the publication of a work, *Campus Elysius*, which would depict the known species of plants. He planned for it to contain some 11.000 pictures. The work was not completed. Together with his son, O.O. Rudbeck (see below) he published two volumes, vol. 2 in 1701, and vol. 1 in 1702 (sic). However, the drawings, the blocks (3200 were ready) and the quasi-totality of stock of the two volumes were destroyed in the great Uppsala fire of 16 Mai 1702. Two copies of vol. 1 and nineteen of vol. 2 were saved. For extensive details see Swederus (1879), KR (p. 604-605) and Lindroth (DSB 11: 587-588). A facsimile edition of vol. 1 was published in 1863. Some of the manuscripts were saved. - A number of the plates in Linnaeus's library were published by J.E. Smith(see below). The titles are: *Campi Elysii liber secundus*, opera Olai Rudbeckii, patris et filii, editus. Upsalae. 1701. Fol. ([4], 239 p.); IDC 6188. *Copies*: LD, SBT, UPS. *Campi Elysii liber primus*, opera Olai Rudbeckii, patris et filii, editus. Upsalae. Fol. 1702 [224 p.]. *Copies*: UPS (compl.), Royal libr. Stockh. (incompl.). - Photolith. reprint, ed. G.E. Klemming, Stockholm (P.H. Mandel) 1863. *Copy*: LD. - See also *Reliquiae Rudbeckianae*, below.
(2) Rudbeck published *Hortus botanicus* variis exoticis indigenisque plantis instructus, Uppsala, 1685 (*copy*: LD-Univ.); this was reprinted Uppsala 1930 (iv, 124 p.).

HANDWRITING: Dandy, J.E., The Sloane herbarium, facsimile no. 76 (see also p. 246). 1958.

9747. *Reliquiae rudbeckianae*, sive, Camporum elysiorum libri primi, olim ab Olao Rudbeckio patre et filio, Upsaliae anno 1702 editi, quae supersunt, adjectis nominibus linnaeanis. Accedunt aliae quaedam icones caeteris voluminibus rudbeckianis aut destinatae, aut certe haud omnino alienae, hactenus ineditae. Cura Jacobi Edvardi Smith. Londini (impensis editoris) 1789. Fol. (*Reliq. rudb.*).
Editor: James Edward Smith (1759-1828).
Publ.: Apr-Mai 1789 (TL-1/1235; ST p. 78-79), p. [i-vii], 1-35, *pl. 1-35* (uncol. copp.).
Copies: BR, G, LD-Univ., USDA; IDC 5742. - SO 732.

Rudbeck, Olaus [Olof] **Olai** (1660-1740), Swedish physician and botanist; Dr. med. Utrecht 1690; professor of anatomy Uppsala 1690, and botany 1691, succeeding his father, O.J. Rudbeck; full professor 1702; director of the botanical garden 1691-1740; "arkiater" 1739; travelled in Lappland 1695; teacher of Linnaeus. (*O.O. Rudbeck*; *Rudbeck Jr.*).

HERBARIUM and TYPES: Some phanerogams at BM and H. - All pre-1702 material at Uppsala destroyed. See O.O. Rudbeck and Dandy (1958).

BIBLIOGRAPHY and BIOGRAPHY: AG 4: 133; Backer p. 502 (epon.); Barnhart 3: 188; BM

4: 1759-1760, 8: 1108; Bossert p. 341; Dryander 3: 12, 170, 196, 197, 252, 288, 478, 606; DSB 11: 588; Henrey 3: 1290; Herder p. 84, 147, 254, 365, 419; IH 2: (files); KR pp. 605-607 (b. 15 Mar 1660, d. 23 Mar 1740); Lasègue p. 97, 391, 502; NI 1692-1694; PR 4189, 7861, 7860-7874, ed. 1: 8826-8835, 10959, 11420; Rehder 1: 78, 5: 744; SO 732, 2521, 3071; Tucker 1: 612.

BIOFILE: Blunt, W., The compleat naturalist 21, 26-28, 29, 30, 35, 36, 38, 40, 135, 150. 1971.
Bridson, G.D.R. et al., Nat. hist. mss. res. Brit. Isl. 255.4, 255. 58. 1980.
Dandy, J.E., The Sloane herbarium 196. 1958.
Dörfler, I., Botaniker-Porträts 3/4: 21. 1907 (portr.).
Eriksson, G., Bot. hist. Sverige 361 [index]. 1969.
Fée, A.L.A., Mém. Soc. Sci. Lille 1832(1): 85-87. 1832 (corr. with Linnaeus).
Fries, R.E., Short hist. bot. Sweden 16. 1950.
Fries, Th.M., Bref Skrifv. Linné 1(1): 311. 1907, 1(3): 94, 229. 1909.
Gager, C.S., Brooklyn Bot. Gard. Rec. 27(3): 333. 1938.
Hulth, J.M., Bref Skrifv. Linné 1(8): 23, 105, 1922. 2(1): 337. 1916.
Hulth, J.M. & A.H. Uggla, Bref Skrifv. Linné 2(2): 131, 133, 282, 283, 284. 1943.
J., Dict. sci. med., Dict. méd. 7: 68-69. 1825.
Lindroth, S., Sv. män kvinn. 6: 385. 1949 (portr.).
Löwegren, Y., Naturaliekabinett Sverige 405 [index]. 1952.
Mägdefrau, K., Gesch. Bot. 51. 1973.
Palmblad, W.F., Biogr. lex. namnk. sv. män 12: 334-340. 1846.
Stafleu, F.A., Linnaeus and the Linnaeans 6. 1971.
Swederus, M.B., Bot. trädgård. Upsala 1655-1807. Falun 1877.
Wittrock, V.B., Acta Horti Berg. 3(2): 69. 1903, 3(3): 51. 1905.

COMPOSITE WORKS: With his father O.J. Rudbeck, q.v., *Campus Elysius*.

EPONYMY: *Rudbeckia* Linnaeus (1753).

HANDWRITING: J.E. Dandy, The Sloane herbarium, facs. no. 76. 1958.

9748. *Disputatio medica inauguralis, de fundamentali plantarum notitia rite acquirenda. Quam, auspice deo opt. max. auctoritate magnifici rectoris, ... nec non amplissimi Senatus Academici consensu, nobilissimaeque Facultatis medicae decreto pro gradu doctoratus summisque in medicina honoribus & privilegiis rite & legitime consequendis, publico examini subjicit Olavus Rudbeck, ... ad diem 29 Aug. horis locoque solitis*. Trajecti ad Rhenum [Utrecht] (ex officina Francisci Halma, ...) 1790. Qu. (*Fund. pl. not.*).
Orig. ed.: 29 Aug 1690, p. [1]-25, [26-28]. *Copies*: LD, U. – The poem on p. [28] is dated 5 Sep 1690 (sic). – Written by Linnaeus's teacher, the work may be regarded as foreshadowing the Linnaean reform in plant systematics.
Reprinted: 1691, Augustae vindelicorum [Augsburg] (apud L. Kronigerum & haeredes Th. Gaebelii, ...) 1691. Duod., 57 p.

9749. C.G. *Propagatio plantarum botanico-physica, quam experientia & rationibus stabilitam, figuris aeneis exornatam, et huic nostro climati adcommodatum evulgat Olavus Rudbeck Ol. fil. anno m.dc.lxxxvi*. Upsalae (Excudit Henricus Curio S.R.M. & Acad. Upsal. Bibliop.). [1686]. (*Propag. pl. bot.-phys.*).
Publ.: 1686, p. [i-xvi], 1-142, *pl. 1-5*, 2 unnumbered pl., 15 vignettes. *Copies*: FI (inf. C. Steinberg), LD-Univ. (inf. O. Almborn). For further details see KR p. 606, no. 1 (coll.: tit., 142 p., [4], 16 vign., 5 pl.).

Rudberg, August (1842-1912), Swedish clergyman and botanist; ordained 1872; parish priest in various locations; dean from 1901. (*Rudberg*).

HERBARIUM and TYPES: Unknown.

BIBLIOGRAPHY and BIOGRAPHY: Barnhart 3: 188; BL 2: 551, 552, 553, 711; Bossert p. 341; KR p. 607 (b. 1 Jan 1842, d. 18 Jan 1912); Rehder 2: 117.

BIOFILE: Anon., Bot. Not. 1912: 44 (d.); Sv. bot. Tidskr. 5: 454. 1912 (d.).
Kilander, S., Västgöta litteratur 1972: 11-25 (portr.; mss. at Skara).
Månson, T., Sv. män. kvin. 6: 391. 1949.
Sylvén, N., Sv. bot. Tidskr. 6: 333-334. 1912 (obit., portr.).
Wittrock, V.B., Acta Horti Berg. 3(3): 51. 1905.

9750. *Förteckning öfver Västergötlands fanerogamer* och kärlkryptogamer med uppgift om växeställen och frekvens efter kollega Ernst Linnarssons och lektor Bror Forssells med fleres anteckningar ... Mariestad (Länstidningens tryckeri. Abr. Berg. & C:o) 1902. Oct. (*Förteckn. Västergötl. fan.*).
Collaborators: Ernst Josef Emanuel Linnarsson (1837-1897), Karl Bror Jakob Forssell (1856-1898).
Publ.: 1902 (p. vi, Summer 1902; Nat. Nov. Jan(3) 1903; reviewed Bot. Not. 1902: 191. 15 Sep 1902), p. [i]-xiii, [xiv-xv], [1]-129. *Copies*: G, GB, LD, USDA. – For further details and follow-up see BL 2: 551-552.
Additions: Witte, H., Bot. Not. 1902: 271-282. 18 Dec 1902, "Tillägg till Rudberg: Växtförteckning öfver Västergötland".

Rudge, Edward (1763-1846), British (English) botanist and antiquary, living on the Abbey estate at Evesham. (*Rudge*).

HERBARIUM and TYPES: BM. – The Guyana plants colleced by Joseph Martin are in the Rudge herbarium. They were captured from the French in May 1803 and acquired by Rudge and Lambert. Circa 400 Martin specimens were bought by BM at the Lambert sale; another 772 were included in Rudge's herbarium. Stearn and Williams (1957) give a list of the plants described by Rudge with indication of types and synonymy. Rudge's own herbarium, consisting of 4318 specimens (in addition to the Martin specimens) was presented to BM in 1847. It contained e.g. Dickson's exsiccatae, Fuci from Dawson Turner, Ericae from R.A. Salisbury. Some further specimens from Rudge's herbarium (not collected by him) are at FI.

BIBLIOGRAPHY and BIOGRAPHY: Backer p. 502 (epon.); Barnhart 3: 188; BB p. 264; BM 4: 1760; CSP 5: 322, 6: 751; Desmond p. 534 (b. 27 Jun 1763, d. 3 Sep 1846); DNB 49: 383-384; GF p. 74; Herder p. 226; Hortus 3: 1203; IF p. 727; Jackson p. 375; Kew 4: 543; NI 1695-1696, 1695n; PR 7875, ed. 1: 8836; Rehder 1: 338, 500, 2: 356; TL-1/1122; Tucker 1: 612; Zander ed. 10, p. 709, ed. 11, p. 809.

BIOFILE: Allibone, S.A., Crit. dict. Engl. lit. 2: 1888. 1878.
Anon., Bot. Zeit. 5: 135-136. 1847 (d.); Flora 30: 147-148. 1847 (d.).; Proc. Linn. Soc. 1: 337-338. 1849 (obit.);.
Beer, G. de, The sciences were never at war 128. 1960.
Britten, J., J. Bot. 55: 344-345. 1917 (on herb. at BM).
Dawson, W.R., Smith papers 84. 1934 (2 letters to J.E. Smith).
G.P., Bot. Zeit. 5: 135. 1847 (obit.).
Meyer, G.F.W., Prim. fl. esseq. 198-201. 1818, repr. J. Bot. 50: 63-64. 1912 (tribute to Anne Rudge, wife of E. Rudge, who illustrated his works).
Murray, G., Hist. coll. BM(NH) 1: 178. 1904.
Sagot, P., Ann. Sci. nat., Bot. ser. 6. 10: 366-367, 375. 1880.
Sims, J., Curtis's Bot. Mag. 23: *pl. 935.* 1806, 51: *pl. 2465.* 1824.
Stearn, W.T., & H.J. Williams, Bull. Jard. bot. Bruxelles 27: 243-265. 1957 (on J. Martin's plants and the Pl. Guian. rar. icon.).
Steenis-Krusemann, M.J. van, Fl. males. Bull. 15: 741. 1960.
Steinberg, C.H., Webbia 32(1): 36. 1977.
Wolley-Dod, A.H., Fl. Sussex xlviii. 1937.

EPONYMY: *Rudgea* R.A. Salisbury (*1807*).

9751. *Plantarum Guianae rariorum icones et descriptiones* hactenus ineditae: auctore Edvardo Rudge, ... volumen i. Londini (Sumptibus auctoris, typis Richardi Taylor et Soc. ...) 1805[-1806]. Fol. †. (*Pl. Guian.*).

Publ.: 1805-1806, in four (out of eight planned) fascicles, p. [1]-32, *pl. 1-50* (uncol. copp., Anne Rudge, née Nouaille). *Copies*: G, NY, Teyler; IDC 5698. – The contents listed below are conjectural.
1: Apr 1805, p. [1]-12, *pl. 1-12*.
2: Jun 1805, p. 13-20, *pl. 13-24*.
3: Jun 1805, p. 21-28, *pl. 25-36*.
4: Apr-Mai 1806, p. 29-32, *pl. 37-50*.
A facsimile reprint, 1974 (announced by J. Cramer) remained unpublished.

Rudio, Franz (1811-1877), German (Hessen) botanist and pharmacist; Dr. phil. Giessen; emigrated to Brazil 1858, collected plants near Rio de Janeiro in 1859; settled in the colony of Porto do Cachoeiro, Espirito Santo. (*Rudio*).

HERBARIUM and TYPES: Plants from Brazil at B (mainly destr.), GOET, L. – Herbarium specimens in "Museum Wiesbaden" by 1936.

BIBLIOGRAPHY and BIOGRAPHY: Barnhart 3: 188; Biol.-Dokum. 15: 7711; BM 4: 1760; IH 2: (files); Kew 4: 543; Rehder 1: 383; Urban-Berl. p. 290, 312, 389.

BIOFILE: Jessen, K.F.W., Bot. Gegenw. Vorz. 231. 1884.
Urban, I., Fl. bras. 181): 92. 1906.
Wagenitz, G., Index coll. princ. herb. Goett. 138. 1982.

9752. *Uebersicht der Phanerogamen und Gefässcryptogamen von Nassau.* Im Auftrage der botanischen Section zusammengestellt von Franz Rudio zu Weilburg [Wiesbaden (Druck von W.G. Riedel) 1851]. Oct.
Publ.: 1851(p. vi: Apr 1851), p. [i*, t.p. journal], [i]-vi, [1]-136, [i]-vi, *pl. 1*. *Copies*: HH, MO. – Published and to be cited as Jahrb. Ver. Nat. Herzogth. Nassau 7(1): [i]-vi, [1]-136, [i]-vi, *pl. 1*. 1851.
Nachtrag: Ib. 8(2): 166-199. 1852 (gives locations).

Rudolph, Johann Heinrich (1744-1809), German (Thuringian) botanist; Dr. med. Jena 1781, from 1783 at St. Petersburg as "professor of botany"; from 1804 member of the St. Petersburg Academy of Sciences. (*J.H. Rudolph*).

HERBARIUM and TYPES: LE; further material at C, H, MW.

BIBLIOGRAPHY and BIOGRAPHY: Barnhart 3: 189; BM 4: 1760; CSP 5: 322; Dryander 3: 161; Herder p. 188; Hortus 3: 1203; IH 2: (files); Kew 4: 544; MW p. 420; PR 7876, ed. 1: 8837; TR 1177-1180; Zander ed. 10, p. 709, ed. 11, p. 809.

BIOFILE: Herder, F. v., Bot. Centralbl. 55: 260. 1893 (coll. LE).
Lindemann, E. v., Bull. Soc. Natural. Moscou 61(1): 63-64. 1886 (108 specimens in L's herb. at MW).
Lipsky, V.I., Imp. S.-Petersb. bot. sada 3: 422-424. 1915 (biogr., bibl.).

9753. Dissertatio inauguralis botanica sistens *Florae jenensis plantas* ad polyandriam monogyniam Linnaei pertinentes. Quam rectore academiae magnificentissimo serenissimo duce ac domino Carolo Augusto ... auctoritate gratiosi medicorum ordinis a.d. xxi februarii mdcclxxxi pro licentia summos in utraque medicina honores consequendi publice defendet auctor Jo. Henricus Rudolph jenensis. Jenae (ex officina Fickelscherriana) [1781]. Qu. (*Fl. jen.*).
Publ.: 21 Feb 1781, p. [1]-26. *Copies*: BM, E (photocopy of BM copy), NY.

Rudolph, Karl (1881-1937), Bohemian botanist of German (Saxonian) origin; studied botany, palaeontology and geology in Wien 1900-1905; student of R. v. Wettstein; Dr. phil. Wien 1905; assistant with F. Czapeck in Czernowicz 1906-1910 and Praha 1910-1913; with G. Beck in Praha 1913-1914; on active duty in the Austrian army 1914-1919; habil. Praha 1919, extraordinary professor of plant systematics at the German University, Praha 1924, "wirklicher" (paid) extraordinary professor ib. 1931; developed quaternary palaeobotany with F. Firbas in Central Europe. (*K. Rudolph*).

RUDOLPH, L.

HERBARIUM and TYPES: PRC; further material at CERN.

BIBLIOGRAPHY and BIOGRAPHY: Barnhart 3: 189; BJI 2: 149; Biol.-Dokum. 15: 7713; Bossert p. 342 (b. 11 Apr 1881, d. 2 Mar 1937); Futak-Domin p. 511; IH 2: (files); Klášterský p. 159-160; Quenstedt p. 371; Tucker 1: 612.

BIOFILE: Anon., Hedwigia 50: (193). 1910 (to Praha), 77: (109). 1938 (d.); Österr. bot. Z. 73: 160. 1924 (professorship), 82: 186. 1933, 86: 240. 1937 (d.); Preslia 49: 65. 1977 (brief memo); Richard Wettstein zum sechzigsten Geburtstag 7. 1923.
Domin, K., Věda přír. 18: 123. 1937 (obit.).
Firbas, F., Forschungen u. Fortschritte 13: 215-216. 1937 (obit.).
Firbas, F. & A. Pascher, Ber. deut. bot. Ges. 55: (277)-(292). 1937 (obit., portr., bibl.).
Grohmann, E., Teplitz-Schönauer Anzeiger 8. 1937 (n.v.).
Janchen, E., Richard Wettstein 186. 1933.
Kka., Čas. národn. Musea 111: 77. 1937.
Kneucker, A., Allg. bot. Z. 16: 164. 1910 (assistant Praha).
Pax, F., Bibl. schles. Bot. 20, 101, 182. 1929.
Pohl, F., Natur und Heimat 8: 3-8. 1937 (obit., bibl.).
Pop, E., Bul. Grad. bot. Muz. bot. Cluj 17(3-4): 165-168. 1937 (obit.).
Rudolph, K., Beih. Bot. Centralbl. 45(2, 1) 1928, 180 p. (Die bisherigen Ergebnisse der botanischen Mooruntersuchungen in Böhmen; see also F. Vierhapper, Österr. bot. Z. 78: 286-288. 1929); Natur und Heimat 8: 8-9. 1937 (autobiogr., sketch).
Sigmond, J., Sudetendeut. Forst– und Jagdzeit. 1937 (n.v.).
Szymkiewicz, D., Bibl. fl. Polsk. 123. 1925.
Verdoorn, F., Chron. bot. 4(2): 175, 177-178. 1938 (obit., b. 11 Apr 1881, d. 2 Mar 1937, portr.).

EPONYMY: *Rudolphisporis* W. Krutzsch (1963). *Note*: *Rudolfiella* Hoehne (1944) honors Friedrich Richard Rudolf Schlechter, *q.v.*; *Rudolphia* Medikus (1787) honors Elias Rudolf Camerarius (1641-1695).

Rudolph, Ludwig (ca. 1813-1896), German (Prussian) high school teacher and botanist in Berlin. (*L. Rudolph*).

HERBARIUM and TYPES: Unknown.

BIBLIOGRAPHY and BIOGRAPHY: BM 4: 1760; Herder p. 167; Jackson p. 223; Kew 4: 544; MW p. 420; PR 7877-7879; Rehder 1: 249, 299, 2: 619; Tucker 1: 612.

BIOFILE: Anon., Flora 36: 31. 1852 (gold medal for science and arts, Prussia); Nat. Nov. 18: 482. 1896 (d. 26 Sep 1896, 83 yrs old).

9754. *Atlas der Pflanzengeographie* über alle Theile der Erde für Freunde und Lehrer der Botanik und Geographie nach den neuesten und besten Quellen entworfen und gezeichnet von Ludwig Rudolph ordentlichem Lehrer an der städtischen höheren Töchterschule zu Berlin. Berlin (Verlag der Nicolaischen Buchhandlung) 1852. Fol. (*Atlas Pfl.-Geogr.*).
Publ.: 1852 (p. [vi]: Mai 1852; Bot. Zeit. 1 Dec 1852 "ist erschienen"; Bonplandia 1 Mar 1853 "soeben"; id. Flora 28 Mar 1853), engr. t.p., p. [i-vi], *9 maps* each with text. *Copies*: BR, FI, G, NY, USDA.
Ed. 2: Jan-Feb 1864 (p. [vi]: Oct 1863; rev. Bot. Zeit. 18 Mar 1864), cover, p. [i-vii], *9 maps* each with text. *Copy*: NY. – Zweite Auflage. Berlin (Nicolaische Verlagsbuchhandlung)(G. Parthey) 1864. Fol.
Ref.: Schlechtendal, D.F.L. von, Bot. Zeit. 22: 81-82. 18 Mar 1854 (rev.).

9755. *Die Pflanzendecke der Erde*. Populäre Darstellung der Pflanzengeographie für Freunde und Lehrer der Botanik und Geographie. Nach den neusten und besten Quellen zusammengestellt und bearbeitet von Ludwig Rudolph, ... Berlin (Verlag der Nicolai'schen Buchhandlung) 1853. Oct. (*Pfl.-Decke Erde*).
Ed. [*1*]: 1853 (p. viii: 30 Sep 1852; ÖbZ 3 Feb 1853; Bonplandia 1 Mar 1853; Flora 28

Mar, 21 Apr 1853; Bot. Zeit. 29 Apr 1853), p. [i]-xiv, [xv], [1]-416. *Copies*: BR, FI, G, M.
Ed. 2: 1859 (p. viii: Jan 1859), frontisp. (col.), p. [i]-xiv, [xv], [1]-450, *pl. 1-12* (uncol.). *Copies*: E (Univ. Libr.), G, NY. – Zweite vermehrte Ausgabe ... Berlin (Nicolaische Verlagsbuchhandlung (G. Parthey) 1859. Oct.
Supplementheft: 1859, Berlin (Nicolai) Oct., 34 p. (n.v.).
Russian translation: 1861, Moscow, see Rehder 1: 399.
Ref.: K.I., Bot. Zeit. 19: 32. 23 Jan 1861 (rev. ed. 2).

Rudolphi, Friedrich [Karl Ludwig] (1801-1849), German (Schleswig-Holstein, then Danish) botanist and physician; studied pharmacy in Neustrelitz, Stralsund and Hamburg; from 1826 medicine at the University of Greifswald; Dr. phil. Rostock 1829; Dr. med. Greifswald 1830; from 1830 practicing at Ratzeburg; from 1837 regional physician of the Lauenburg district. (*F. Rudolphi*).

HERBARIUM and TYPES: HBG (algae); other algae at MANCH; some further material at MW. – Letters to D.F.L. von Schlechtendal at HAL.

BIBLIOGRAPHY and BIOGRAPHY: Barnhart 3: 189 (b. 18 Sep 1801, d. 27 Apr 1849); BM 4: 1761; CSP 5: 323, 12: 638; DTS 1: 250, 6(4): 190; Frank 3 (Anh.): 86; Herder p. 142; Hortus 3: 1203; LS 23219; PR 7880, ed. 1: 8838; Rehder 1: 260; Saccardo 1: 143; Stevenson p. 1256.

BIOFILE: Anon., Archives Pharm. 110: 221-222. 1849 (obit., b. 20 Sep 1801); Bot. Zeit. 7: 607 (d.), 728 (note) 1849, 8: 31. 1850; Flora 9: 588. 1826 (in S. Germany), 10(1): 123-124. 1827 (stay in Triest), 33: 284. 14 Mai 1850 (d.).
Fischer-Benzon, R.J.D. von, *in* P. Prahl, Krit. Fl. Schleswig-Holstein 2: 51. 1890.
Hausmann, F. v., Fl. Tirol 1185. 1854.
Saccardo, P.A. et al., Michelia 2: 217-218. 1881 (bibl. Ital. mycol., one paper).
Schmid, G., Chamisso als Naturforscher 256. 1942.

9756. *Systema orbis vegetabilium* quod gratiosi medicorum ordinis consensu et auctoritate dissertatione inaugurali ad summos in medicina et chirurgia honores rite impetrandos aut d. [ ] cal. [ ] a. mdcccxxx in auditorio majori h.a.m.s. publice defendet auctor Frid. Carol. Lud. Rudolphi ratzeburgensis, ... Gryphiae [Greifswald] (typis F. Guil. Kunike, ...) [1830]. Oct. (*Syst. orb. veg.*).
Pul.: 1830 (Flora Feb-Mai 1831), p. [i-iii], [1]-80. *Copy*: G. – Rudolphi also published *Plantarum vel novarum vel minus cognitarum descriptiones*, Linnaea 4: 114-120. Jan 1829, 387-395. Jul 1829, 509-515. Nov-Dec 1829.
Ref.: [Schlechtendal, D.F.L. von], Linnaea 5 (Litt.): 69-71. 1830 (rev.).

Rudolphi, [Israel] Karl Asmund [Asmus] (1771-1832), Swedish-born German (Jewish) naturalist, physician, and numismatologist; studied in Greifswald; Dr. phil. 1793; Dr. med. 1794; lecturer ("Privatdozent") Greifswald 1793-1798; prosector ib. 1798; professor of medicine 1808; professor of medicine and anatomy at the University of Berlin 1810; Geh. Medicinalrath 1817. (*Rudolphi*).

HERBARIUM and TYPES: MW (fide Karavaev, 1975); material at H, L and UPS (Thunberg). The material cited by Urban at B (140 species mainly from Haiti), came from P.A. Poiteau and is extant in herb. Willdenow).

BIBLIOGRAPHY and BIOGRAPHY: ADB 29: 577-579; Barnhart 3: 189; Biol.-Dokum. 15: 7714; BM 4: 1760-1761; Bossert p. 342; CSP 5: 322-323; Hegi 4(2): 581; Herder p. 58, 120; Hortus 3: 1203; Kew 4: 544; KR p. 607-608 (b. 14 Jul 1771, d. 28 Nov 1832); Moebius p. 165; PR 7881-7882, 9795, ed. 1: 7421, 8840-8841; Ratzeburg p. 446-448; Rehder 5: 744; SO 2661, 3384-3386; Stevenson p. 1256; TL-2/4281, 6680; Urban-Berl. p. 415; Zander ed. 10, p. 709, ed. 11, p. 809.

BIOFILE: Dittrich, M., Wiss. Z. Univ. Greifswald 16 (math.-nat. R. 3): 249-277. 1967 (general importance for natural sciences, biogr. notes; portr.; further biogr. refs); Forschung-Praxis-Fortbildung 18: 356-360. 1967.

Freund, H. & A. Berg, Gesch. Mikroskopie 81, 211. 1963.
Gilbert, P., Comp. biogr. lit. entom. 327. 1977.
Hoffmann, P. et al., Wiss. Z. Humb. Univ. Berlin, Math. Nat. 14: 802. 1965.
Karavaev, M.N., Byull. Mosk. Obsch. Ispyt. Prir., Biol. 80(3): 146-153. 1975 (herbarium Rudolphi discovered in MW).
Link, H.F., Med. Zeit. 1833(4): 17-20 (obit.).
Lühe, M., Arch. parasit. 3(4): 549-577. 1901 (biogr., bibl., portr., "father of helminthology").
Mägdefrau, K., Gesch. Bot. 140, 276. 1973.
Müller, J., Abh. k. Akad. Wiss. Berlin 1835: xvii-xxxviii. 1837 (obit., bibl.).
Nordenskiöld, E., Hist. biol. xii [index]. 1949.
Nuttall, G.H.F., Parasitology 13(4): 402-403. 1921.
Sachs, J., Gesch. Bot. 227, 276, 288, 289-291. 1875; Hist. botanique 219, 277. 1892.
Scheutz, N.J.W., Bot. Not. 1863: 71 (epon.).
Schmid, G., Goethe u.d. Naturw. 566. 1940; Chamisso als Naturforscher p. 17, nos. 13, 119, 184, 202, 346, 357. 1942.
Taton, R., ed., Science 19th cent. 366. 1965.
Wittrock, V.B., Acta Horti Berg. 3(2): 140. 1903, 3(3): 150. 1905.

EPONYMY: *Neorudolphia* N.L. Britton (1924); *Rudolphia* Willdenow (1801).

9757. *Anatomie der Pflanzen* von Karl Asmund Rudolphi ... Eine von der Königl. Societät der Wissenschaften in Göttingen gekrönte Preisschrift ... Berlin (In der Myliussischen Buchhandlung) 1807. Oct (*Anat. Pfl.*).
Publ.: 1807, p. [i]-xvi, [1]-286, [287, err.], *pl. 1-6* (uncol. copp. Besemann). *Copies*: E (Univ. Libr.), G, H-UB, NY, USDA.

Rübel, (Rübel-Blass) **Eduard August** (1876-1960), Swiss (German-American father and Swiss mother) botanist; Dr. phil. E.T.H. Zürich 1901; engaged in botanical research from 1904, especially in geobotany; founder (with his sisters) of the Geobotanische Forschungsinstitut Rübel 1918; "Privatdocent" E.T.H. Zürich 1917-1934. (*Rübel*).

HERBARIUM and TYPES: RUEB; other material e.g. at C, M (lich.).

BIBLIOGRAPHY and BIOGRAPHY: AG 12(2): 63; Barnhart 3: 189 (b. 18 Jul 1876, d. 24 Jun 1960); BFM no. 1028, 1049, 1660; Biol.-Dokum. 15: 7723; BJI 2: 149-150; BL 1: 32, 316, 2: 573, 575, 711; BM 8: 1109; Bossert p. 341; Hegi 6(1): 1260; Kew 4: 544-545; KR p.4; Langman p. 654-655; Moebius p. 374; PFC 2(2): xxvii; TL-2/720.

BIOFILE: Anon., Bot. Centralbl. 126: 192. 1914 (foundation for plant-geographical research).
Furrer, E., Viertelj.-Schr. naturf. Ges. Zürich 105: 331-333. 1960.
Landolt, E., Neue Zürcher Zeit. 8 Oct 1980, Fernausgabe nr. 233. p. 33 (on Rübel and Braun-Blanquet).
Lüdi, W., Verh. Schweiz. naturf. Ges. 140: 237-240. 1960 (obit., portr.), Eduard August Rübel 1876-1960, 24 p., s.d. [1961] (souvenirs, photographs), also in Ber. Geobot. Inst. E.T.H. 32, 1961; Ber. geol. Inst. E.T.H., Stiftung Rübel 32: [5]-24. 1961.,
Rübel, E., Bot. Gaz. 84: 428-439. (ecology, plant geogr., geobotany, history and aim); Eine allgemeine schweizerische Akademie und die bestehende Naturforscher-Akademie, Sonderabdruck a.d. Neuen Schweizer Rundschau, Feb 1934, 15 p.; Gesch. naturf. Ges. Zürich 1947, 2 p. (portr.).
Rübel-Kolb, Eduard, Eduard Rübel-Blass 1876-1960. Zürich 44 p. (reprinted from Neujahrsbl. 1970 zum Besten des Waisenhauses Zürich, no. 133; biogr., portr.
Schröter, C., Die Naturwissenschaften 24(41): 641-642. 1936 (biogr. sketch; 60th birthday).
Verdoorn, F., ed., Chron. bot. 1: 30, 39, 40, 260, 264. 1935, 2: 28, 30, 32, 57, 278, 299. 1936, 3: 45, 109, 243. 1937, 4: 91, 276. 1938, 5: 136, 137, 283, 286, 526. 1939.
Weber, C., Pl. Sci. Bull. 7(1): 6. 1961 (obit.).

EPONYMY: *Rubelia* Nieuwland (1916).

9758. *Pflanzengeographische Monographie des Berninagebietes* ... Leipzig (Verlag von Wilhelm Engelmann) [1911-]1912. Oct.
Publ.: 4 Jul 1911-20 Feb 1912 (in journal; reprint Aug-Sep 1911; Bot. Centralbl. 30 Apr 1912; A.B.Z. 15 Oct 1912; Nat. Nov. Apr(2) 1912), frontisp., p.[i]-x, [1]-615, *pl. 1-36*, err. slip, 1 map. *Copies*: NY, USDA. – Reprinted ((part 1) and preprinted (part 2)) and to be cited from Bot. Jahrb. 47: i-vi, 1-296. 4 Jul 1911, 297-616. 20 Feb 1912, *pl. 1-36*.
Ref.: Boulger, G.S., J. Bot. 50: 351-352. Nov 1912 (rev.).
Kneucker, A., Allg. bot. Z. 17: 159. 15 Oct 1912 (review of complete reprint of 615 p.).

9759. *Geobotanische Untersuchungsmethoden* ... Berlin (Verlag von Gebrüder Borntraeger ...) 1922. Oct. (*Geobot. Unters.-Meth.*).
Publ.: Mar-Mai 1922 (p. v: 28 Feb 1921; BR rd. 1 Dec 1922; ÖbZ 1 Dec 1922; Nat. Nov. Jun 1922), p. [i]-xii, [1]-290, *1 pl*. *Copies*: BR, G, M, NY, PH.
Ref.: Markgraf, F., Bot. Jahrb. 58 (Lit.): 44-45. 1 Feb 1923 (rev.).
Vierhapper, F., Österr. bot. Z. 71: 282. 1922.

9760. *Pflanzengesellschaften der Erde* ... Bern-Berlin (Verlag von Hans Huber) [1930]. Oct. (*Pfl.-Ges. Erde*).
Publ.: 1930 (p. 12: 28 Sep 1928; Nat. Nov. Aug 1930), p. [i*-ii*], [i]-viii, 3-464, map (10 col.) by H. Brockmann-Jerosch. *Copies*: G, M, MO, NY, USDA(2).
Ref.: Diels, L., Bot. Jahrb. 63: 142-143. 10 Oct 1930.
Ginzberger, A., Österr. bot. Z. 79: 376-377. 1930.

Rückert, Ernst Ferdinand (1794-1843), German (Saxonian) botanist and practicing physician at Königsbrück. (*Rückert*).

HERBARIUM and TYPES: Unknown.

BIBLIOGRAPHY and BIOGRAPHY: Barnhart 3: 189 (d. 21 Jul 1843); BM 4: 1761; Herder p. 188; Jackson p. 310; PR 7883, ed. 1: 8843-8844; Rehder 1: 380.

BIOFILE: Anon., Bot. Zeit. 1: 856. 1843 (d. 21 Jul 1843); Flora 27: 175. 1844 (d. 21 Jul 1843).
Carus, C.G. & C.F.P. von Martius, Eine Altersfreundschaft in Briefen 50, 142. 1939 (ed. G. Schmid).

9761. *Flora von Sachsen*, ein practischer und bequemer Wegweiser auf heimathlichen botanischen Excursionen durch die Pflanzenwelt des Königreichs Sachsen, der sächsischen Herzogthümer und sächsischen Grenzprovinzen, für unstudierte Freunde vaterländischer Pflanzenkunde, d.i. einfache und deutliche Beschreibung sämmtlicher, im Königreiche Sachsen und dessen anliegenden Provinzen wildwachsenden Pflanzen, mit genauer Angabe ihrer Standorte, wie ihres technischen und offizinellen Gebrauchs, zum Handgebrauch und Selbstunterricht beim Botanisiren, für Apotheker, Land– und Forstwirthe, Schullehrer und sonstige Freunde vaterländischer Gewächskunde ... Grimma & Leipzig (Verlags-Comptoir) [1844]. 2 vols. Oct. (*Fl. Sachsen*).
Publ.: 2 vols., 1844 (fide PR), according to Pritzel an illegitimate reprint (or re-issue) of the 1840 *Beschreibung* (see below) with just a new t.p. *Copies*: G, NY.
1: [i]-viii, [ix], [1]-306, [1, err.]. (p. viii: 2 Feb 1840).
2: [i], [1]-302.
Other issue: s.d., Wurzen (Verlags-Comptoir), 2 vols., as above. *Copies*: HH, M.
Original issue: Leipzig 1840, fide Pritzel. – We have not been able to find a copy of this *Beschreibuung der am häufigsten wildwachsenden und kultivirten phanerogamen Gewächse*, Farnkräuter, so wie einiger officinellen Moose und Schwämme Sachsens und der angränzenden preussischen Provinzen, mit Angabe ihrer nützlichen und schädlichen Eigenschaften. Leipzig 1840, viii, 306 p. and 302 p. [see e.g. Linnaea 14(Litt.): 104-105. 1840/41].

Rüggeberg, Hermann Karl August (1886-?), German lichenologist; studied natural sciences at Göttingen (1906-1908) and München 1908; Dr. phil. Göttingen 1910; assistant at the Bromberg Kaiser-Wilhelm-Institut für Landwirthschaft 1910-1911; id. at Botanical Garden Göttingen 1911-1912; teacher at Hameln (1912), Nienburg (1913) and from 1914 at Celle; in forced retirement 1939-1945; retired 1952. (*Rüggeberg*).

HERBARIUM and TYPES: GOET (lich.).

BIBLIOGRAPHY and BIOGRAPHY: GR p. 126-127; LS 38635.

9762. *Die Lichenen des östlichen Weserberglandes*. Inaugural-Dissertation zur Erlangung der Doktorwürde der philosophischen Facultät der Georg-August-Universität zu Göttingen eingereicht von Hermann Rüggeberg aus Hardegsen ... Göttingen (Druck von Wilh. Riemschneider in Hannover) 1910. Oct. (*Lich. östl. Weser.*).
Publ.: 1910 (p. [ii]: 6 Jul 1910; Bot. Centralbl. 25 Jul 1911), p. [i-iii], [1]-82, [83, vita].
  Copies: FH(2), USDA. – Preprinted from Jahresber. Niedersächs. bot. Ver. 3: 1-82. 1911.
Ref.: Leeke (Neubabelsberg) Bot. Centralbl. 119: 172. 13 Feb 1912 (rev.).

Rueling, Johann Philipp (1741-?), German (Niedersachsen) physician in Einbeck and Nordheim; studied medicine at Göttingen. (*Ruel.*).

HERBARIUM and TYPES: Unknown.

BIBLIOGRAPHY and BIOGRAPHY: AG 5(1): 428; Backer p. 502; Barnhart 3: 190; BM 4: 1762; Dryander 3: 33; Herder p. 142; PR 7886-7889, ed. 1: 8846-8849; RS p. 138; SO 663a, 663b.

BIOFILE: Baines, J.A., Vict. Natural. 90: 249. 1973 (epon.).
Lütjeharms, W.J., Gesch. Mykol. 249. 1936.

COMPOSITE WORKS: *Verzeichniss der an und auf dem Harz wildwachsenden Bäume, Gesträuche und Kräuter in*: Christoph Wilhelm Jakob Gatterer, *Anleitung den Harz zu bereisen*, Göttingen, 2: 186-247. 1786.

EPONYMY: *Ruelinga* Cothenius (1790, *orth. var.* of *Ruelingia* Ehrhart (1788)); *Ruelingia* Ehrhart (1788, *nom. rej.*); *Rulingia* Haworth (1819, *orth. var.* of *Ruelingia* Ehrhart (1788); *Rulingia* R. Brown (1820, *nom. cons.*).

9763. *Commentatio botanica de ordinibus naturalibus plantarum*. Quam speciminis inauguralis loco pro summis, in arte salutari honoribus d. xvii. Septembr. a. mdcclxvi. In se collatis gratioso medicorum ordini offert Joannes Philippus Rüling ... Goettingae (Litteris Frieder. Andr. Rosenbusch) [1766]. Qu. (*Comm. bot.*).
Publ.: 17 Sep 1766, p. [i], [1]-36, chart. *Copies*: M (inf. I. Haesler), NSUB (inf. G. Wagenitz).
Reprint: in P. Usteri, Delect. opusc. bot. 2: 431-462. 1793.

9764. *Ordines naturales plantarum* commentatio botanica ... Goettingae (sumtibus Vid. Abrah. Vandenhoeck) 1774. Oct. (*Ord. nat. pl.*).
Publ.: 1774 (GGA 14 Feb 1775), p. [i-v], [1]-112, [113, err.], 1 chart (identical with the 1766 chart). *Copies*: M (inf. I. Haesler), NSUB (inf. G. Wagenitz); IDC 863.

Rümpler, Theodor (1817-1891), German botanist, specialist on succulent plants, horticulturist and popular author on gardening. (*Rümpler*).

HERBARIUM and TYPES: Unknown.

BIBLIOGRAPHY and BIOGRAPHY: BM 4: 1762; Herder p. 7, 88, 358, 365, 368; Hortus 3: 1203; Kew 4: 546; Langman p. 655; NI 3: 67; Rehder 5: 745; TL-2/1812; Tucker 1: 612-613; Zander ed. 10, p. 709-710, ed. 11, p. 810.

BIOFILE: Anon., Monatschr. Kakteenk. 1(3): 46. 1891 (d.); Anon., Österr. bot. Z. 41: 260. 1891 (d. 23 Mai 1891).

COMPOSITE WORKS: (1) Published the second edition of Carl Friedrich Förster, *Handbuch der Cacteenkunde* 1884-1886, see TL-2/1812, p. [i]-xv, [1]-1029. *Copies*: MO, NSW, USDA(2).
(2) Editor, with Groenland, of *Vilmorins illustrierte Blumengärtnerei* 1872-1875, see under Vilmorin, ed. 2, edited by Rümpler, 1879.
(3) Editor, *Illustriertes Gartenbau-Lexikon*, eds. 1 and 2, 1882, 1890 (ed. 3, 1901 edited by L. Wittmack).

9765. *Die Sukkulenten* (Fettpflanzen und Kakteen) Beschreibung, Abbildung und Kultur derselben von Theodor Rümpler. Nach dem Tode des Verfassers herausgegeben von Prof. Dr. K. Schumann ... Berlin (Verlag von Paul Parey ...) 1892. Oct. (*Sukkulenten*).
Editor: Karl Moritz Schumann (1851-1904).
Publ.: 1892 (p. iv*: Sep 1892; ÖbZ for Nov 1892; Nat. Nov. Nov(1) 1902), p. [i*-iv*], [i]-ii, [1]-263. *Copies*: HH, NY, USDA.

Rüppell [Rüppel], **[Wilhelm Peter] Eduard [Simon]** (1794-1884), German (Frankfurt) zoologist, mineralogist and explorer; studied at Pavia 1817-1821; Dr. med. h.c. Giessen 1827; travelled in Egypt and Nubia 1817, 1822-1827, and Arabia and Ethiopia 1831-1834, Italy 1816-1817, 1844 and Egypt 1850; donated his archeological, mineralogical and natural history collections to the Senckenbergische Gesellschaft at Frankfurt; curator of these collections 1841-1843, 1846-1847, 1854-1855 and 1858-1859; later especially active as numismatologist. (*Rüppell*).

HERBARIUM and TYPES: FR; some orchids in W.

NOTE: For the rich biographical and other literature on Rüppell we refer to Poggendorff (7a, 1971). – We follow the main biographer, R. Mertens (1949), in accepting the generally used spelling "Rüppell"; however, Poggendorff insists "Rüppel, nicht Rüppell".

BIBLIOGRAPHY and BIOGRAPHY: ADB 29: 707-714 (W. Stricker); AG 4: 566; Backer p. 503 (epon.); Barnhart 3: 190 (b. 20 Nov 1794, d. 10 Dec 1884), 192 (err. as Ruppel); BM 4: 1762-1763; CSP 5: 337-339; Herder p. 58; IH 1 (ed. 6): 363, (ed. 7): 341, 2: (files); Lasègue p. 333, 437, 504; PR ed. 1: 8851; Rehder 1: 487.

BIOFILE: Anon., Bonplandia 6: 51. 1858 (Leopoldina cogn. Bruce); Flora 12: 93. 1829 (goes to Ethiopia), 15: 36. 1832 (colls. rd.), 17(1): 46. 1834 (news from Ethiopia), 192. 28 Mar 1834 (back in Italy), 18(2): 661. 1835; Ibis ser. 5. 3: 238, 336-338. 1885 (obit.); Proc. Linn. Soc. 1883/1886: 106. 1886 (obit.).
Blatter, E., Rec. Bot. Surv. India 8(5): 464. 1933.
Blum, J., Ber. Senckenb. naturf. Ges., Abh. 1901: 18.
Boissier, E., Fl. Orient. 1: xxi. 1867.
Bridson, G.D.R. et al., Nat. hist. mss. res. Brit. Isl. 227.15, 229.286, 274.42. 1980.
Candolle, Alph. de, Phytographie 445. 1880 ("Ruppell").
Conert, H.J., Senckenbergiana, Biol. 48(c): 16. 1967 (Red sea algae FR as well as his Abyssinian, Arabian and Sinai coll.).
Embacher, F., Lexikon Reisen 254-255. 1882.
Jessen, K.F.W., Bot. Gegenw. Vorz. 471. 1884.
Knoblauch, A., Ber. Senckenb. naturf. Ges. Frankfurt 48: 54-57. 1919 (brief biogr., coll. FR).
Kobelt, W., Ber. Senckenb. naturf. Ges. 1895: 3-18 (commemoration).
Mertens, R., Eduard Rüppell, Leben und Werke eines Forschungsreisenden, Frankfurt a.M., Verlag Waldemar Kramer, iv, 388 p., *44 pl.* 1949 (main biogr., bibl., portr.); Natur und Volk 77: 46-50. 1947 (biogr. sketch, portr.).
Nissen, C., Zool. Buchill. 351. 1916 (also Ill. Vögelbücher 799-801).
Poggendorff J.C., Biogr.-lit. Handw.-Buch 2: 715. 1863, 3: 1153. 1898, 7a, Suppl.: 566. 1971 (further biogr. refs., Rüppel, not Rüppell).

Schmidt, H., Ber. Senckenb. naturf. Ges. 1885: 95-160. (Éloge, portr.).
Vallot, J., Bull. Soc. bot. France 29: 188-189, 203. 1882 ("Ruppel").
Zittel, K.A. von, Gesch. Geol. Paläont. 1899 (see index); Hist. geol. palaeontol. 405. 1901.

COMPOSITE WORKS: For publications on Rüppell's botanical collections see J.B.G.W. Fresenius, TL-2/1857-1858, *Beitr. Fl. Aegypt.*; J.G. Agardh, Museum Senckenbergianum 2: 169-174. 1837 (Red sea algae), and C.H. Schultz Bip., Flora 25: 417-431, 433-442. 1842 and Mus. Senckenb. 3: 45-60. 1845 (Compositae). Rüppell's own account of his travels (without botany) is his *Reisen in Nubien, Kordofan und dem peträischen Arabien vorzüglich in geographisch-statistischer Hinsicht*, Frankfurt am Main (bei Friedrich Wilmans) 1829. Oct., p. [i]-xxiv, [1]-388, [1-2, err., colo.], *pl. 1-8*, 4 maps. *Copy*: Teyler.

EPONYMY: *Rueppelia* A. Richard (1847).

Ruge, Georg (*fl.* 1893), German (Hannoverian) botanist; Dr. phil. München 1893. (*Ruge*).

HERBARIUM and TYPES: Unknown.

BIBLIOGRAPHY and BIOGRAPHY: Barnhart 3: 190.

9766. *Beiträge zur Kenntniss der Vegetationsorgane der Lebermoose*. Inaugural-Dissertation zur Erlangung der Doctorwürde bei der hohen philosophischen Facultät (ii. Section) der kgl. bayer. Ludwig-Maximilians-Universität zu München eingereicht von Georg Ruge aus Neuhaus a.d. Oste (Hannover) ... München (Druck on Val. Höfling, ...) 1893. Oct. (*Beitr. Veg.-Org. Leberm.*).
Publ.: 14 Aug 1893 (in journal; Nat. Nov. Feb(2) 1894), p. [1]-38, *pl. 4*. *Copy*: G. – Reprinted or preprinted from Flora 77: 279-312, *pl. 4*. 1893.
Ref.: Roth, E., Bot. Centralbl. 60: 229-230. 7 Nov 1894.
    Kienitz-Gerloff, J.H.E.F., Bot. Zeit. 52: 57-58. 1894.

Rugel, Ferdinand (Ignatius Xavier) (1806-1878), German (Württemberg, Oberschwaben) born American botanical explorer, pharmacist and surgeon; practicing in Bern as pharmacist 1827-1840 and collecting in Switzerland, France, Spain and Sicily; to the United States, 1840, for R.J. Shuttleworth, collecting widely in S.W. states 1840-1848 and Cuba 1849; severing his ties with Shuttleworth 1849, after 1850 collecting occasionally in Tennessee and Texas; settled in Knoxville, Tennessee 1849, working for a wholesale drug firm; after the Civil War living in Jefferson County, Tennessee. (*Rugel*).

HERBARIUM and TYPES: NA (via J.C. Martindale and PHIL); BM (first set of 1840-1849 American plants via Shuttleworth); further material in B (extant), BAS, BERN, BR, DBN, DPU, E, F, G, GB, GH, GOET, K, KIEL, L, LE, LIVU, LY, LZ, MANCH, MICH, MIN, MO, NY, OXF, PH (major set), US (1100), W.

BIBLIOGRAPHY and BIOGRAPHY: AG 4: 454; Barnhart 3: 190 (b. 24 Dec 1806; d. 31 Dec 1878); Bossert p. 235; IH 1 (ed. 6): 363, 2: (files); Lenley p. 468; Urban-Berl. p. 390.

BIOFILE: Anon., Bot. Zeit. 3: 218-224. 1845 (list coll. Florida, Alabama deposited with R.J. Shuttleworth); Bull. Torrey bot. Club 6: 311-312; 1879 (brief obit. , "left a large collection of both shells and plants" which is for sale); Flora 21(1) Intell.-Bl. 29-30. 1838, 23(2) Intell.-bl. 1-2. 1840, 27: 175. 1844 (visited Flora), 57: 463-464, 496, 527. 1874 (sale herb.).
Corgan, J.X., J. Tennessee Acad. Sci. 53(1): 5-7. 1978 (brief sketch).
Cullen, J., Taxon 21: 730. 1972 (pl. LIVU).
Ewan, J., *in* R.W. Long & O. Lakela, Fl. trop. Florida 4, 5. 1971.
Geiser, S.W., Field and Laboratory 16: 113-119. 1948 (portr.), 18: 112. 1950 (b. 17 Dec 1806, Wolfegg, Württemberg), 27: 190. 1959 (coll. Texas 1878); see also Pharmaz. Zeit. 85: 346-347. 1949.
Hollick, A., Bull. Torrey bot. Club 6: 311. 1879 (d.; herbarium to be sold).

Lehmann, E., Schwäbische Apotheker 113-115, 205. 1951 (b. Altorf (= Weingarten), Oberschwaben 17 Dec 1806, sic date from Weingärtner Bürgerbuch, biogr. sketch; d. 31 Jan 1879).
Lindsley, P., *in* L.B. Hill, ed., A history of Greater Dallas and vicinity 2: 184-185. 1909 (brief biogr. note of F. Rugel and his son Joseph).
M., Bot. Zeit. 2: 110-111. 1844 (Tennessee plants rd. by Shuttleworth).
Mears, J.A., Proc. Acad. nat. Sci. Philadelphia 133: 161. 1981.
Meyer, F.G. & S. Elsasser, Taxon 22: 383-385, 391. 1973 (herb. NA; d. 1879).
Murray, G., Hist. coll. BM(NH) 1: 178. 1904 (coll. via Shuttleworth).
Sargent, C.S., Silva 9: 110. 1896.
Shuttleworth, R.J., Bot. Zeit. 3: 218-224. 1845 (list of plants coll. 1843 in Florida and Alabama, distr. by R.J. Shuttleworth).
Stearn, W.T., J. Arnold Arb. 46: 270. 1965.
Urban, I., Symb. ant. 3(1): 115-116. 1902.
Wagenitz, G., Index coll. princ. herb. Gött. 139. 1982.

HANDWRITING: Taxon 22: 385, 391. 1973.

EPONYMY: *Rugelia* Shuttleworth ex Chapman (1860).

Ruhland, Wilhelm [Willy] **Otto Eugen** (1878-1960), German (Schleswig) botanist; student of A. Engler at Berlin; Dr. phil. 1899; collaborator at the Botanical Museum ib. 1899-1903; Privatdocent Univ. Berlin 1903-1911; assistant at the Biologische Reichsanstalt Berlin-Dahlem 1905-1911; extraordinary professor of botany at Halle 1911-1918; full professor of botany at Tübingen Univ. 1918-1922; professor of botany and director of the botanical garden and institute of the University of Leipzig 1922-1945; honorary professor at Erlangen 1948-1956; Dr. h.c. Erlangen 1949. (*Ruhland*).

HERBARIUM and TYPES: Types at B (partly destroyed).

BIBLIOGRAPHY and BIOGRAPHY: Barnhart 3: 191 (b. 7 Aug 1878, d. 5 Jun 1960); BFM p. 313 (index); Biol.-Dokum. 15: 7747; BJI 1: 49-50, 2: 150; BM 4: 1764, 8: 1110; Bossert p. 342; Kelly p. 201; LS 23225-23235, 38640-38648; Moebius p. 105, 121, 247, 248, 252, 254; MW p. 420; Stevenson p. 1256; TL-2/1711, 1713; Urban-Berl. p. 281, 312, 390; Zander ed. 11, p. 809; Zep-Tim p. 97.

BIOFILE: Anon., Bot. Centralbl. 99: 176. 1905 ("Hilfsarbeiter" Kais. Biol.-Anst. Berlin-Dahlem), 117: 560. 1911 (to Halle), 119: 160. 1912 (id.); Bot. Zeit. 41: 47. (travels in the Cyrenaica, collects for Ascherson), 672 (d. 23 Aug 1883) 1883.; Hedwigia 45: (38). 1906 (asst. Kais. Biol. Anst. Berlin-Dahlem), 51: (244). 1911 (app. Regierungsrat and member of the Kais. Biol. Anstalt Berlin-Dahlem) 1911, 60: (178). 1918 (app. Tübingen), 64(1): (70). 1923 (to Leipzig), 78: (75). 1938 (corr. member Akad. Berlin); Nat. Nov. 25: 197. 1903 (habil. Berlin), 33: 551. 1911 (to Halle, "Otto Ruhland"), 40: 372. 1918 (app. Tübingen), 43: 175. 1921 (call to Heidelberg); Österr. bot. Z. 61: 495. 1911 (app. Halle, succeeding Fitting), 71: 152. 1922 (app. Leipzig).
Bullock, A.A., Bibl. S. Afr. bot. 97. 1978.
Gager, C.S., Brooklyn Bot. Gard. Rec. 27(3): 233, 237. 1938.
Mägdefrau, K., Gesch. Bot. 212, 214, 291. 1973.
Poggendorff, J.C., Biogr.-lit. Handw.-Buch 6: 2241-2242. 1938, 7a(3): 849. 1959 (biogr. detail, bibl., secondary refs.).
Renner, O., Z. Naturforsch. 3b: 141-144. 1948, Planta 36: [portr., facs]. 1948.
Schumacher, W., Planta 52: 1-2. 1958 (tribute, portr.), 54(5): portr. 1960.
Verdoorn, F., ed., Chron. bot. 1: 154, 168. 1935, 3: 122, 126, 137, 147. 1937, 4: 85. 1938, 5: 119. 1939.

COMPOSITE WORKS: (1) *Eriocaulaceae*, *in* A. Engler, Pflanzenreich 13(4, 30): 1-294. 27 Mar 1903 (rev. K. Schumann, Bot. Centralbl. 93, 120-122. 1903).
(2) *Eriocaulaceae*, *in* J. Perkins, Fragm. Fl. Philip. 139. 1904.
(3) *Eriocaulaceae*, *in* I. Urban, Symb. ant. 1(3): 472-494. 15 Jan 1900.

(4) *Eriocaulaceae, in* EP, Nat. Pflanzenfam. ed. 2. 15a: 39-57. 1930.
(5) *Sphagnales, Andreales, Bryales* (gen.) in EP, Nat. Pflanzenfam. 10: 101-105, 126-128, 132-142. 1924.
(6) *Ergebnisse der Biologie*, eds., K. v. Frisch ... W. Ruhland, vol. 1, 1926.
(7) Founder, with Hans Winkler, of *Planta*; editor vols. 1-51. 1925-1958.

HANDWRITING: Planta 54: 445. 1960.

EPONYMY: *Ruhlandiella* Hennings (1903).

Ruhmer, Gustav [Gustaf] **Ferdinand** (1853-1883), German (Oberschlesien) botanist; "Hilfsarbeiter" at the Botanical Museum Berlin-Dahlem 1877-1883; collected in Lybia 1882-1883; studied the flora of Brandenburg and Thüringen. (*Ruhmer*).

HERBARIUM and TYPES: Material at B (mainly destr.), BP, BR, C, CR, E, FI, GOET, JE, L, MANCH, S, W.

BIBLIOGRAPHY and BIOGRAPHY: AG 2(2): 451; Backer p. 503; Barnhart 3: 191 (b. 21 Mai 1853, d. 23 Aug 1883); LS 23236; Rehder 1: 271, 388; Urban-Berl. p. 390; Zep-Tim. p. 97-98.

BIOFILE: Anon., Bot. Centralbl. 15: 326. 1883; Bot. Not. 1884: 39 (d.); Bot. Zeit. 41: 47 (coll. Lybia, distr. by Ascherson), 672 (d.). 1883; Bull. Soc. bot. France 29 (bibl.): 144. 1882 (coll. Cyrenaica); Verh. bot. Ver. Brandenburg 24: xlvi. 1883 (d.).
Ascherson, P., Bot. Jahrb. 4, Beil. 2: 1. 1883 (coll. Cyrenaica); Bot. Zeit. 41: 672. 1883 (d.; plants will be distributed by Ascherson); *in* Durand, E. & G. Barratte, Plantes de Tripolitaine, Florae Lybicae prodr. xxv. 1910.
Kanitz, A., Magy. növen. Lap. 7: 122-123. 1883 (d.).
Müllerott, M., Hoppea 39: 62. 1980 (letter to Irmisch).
Pampanini, R., Fl. Cyrenaica xviii. 1930.
Ruhmer, G., Jahrb. Bot. Gart. Berlin 1: 222-259. 1881 (Die in Thüringen bisher wild beobachteten und wichtigeren cultivirten Pflanzenbastarde).
Wagenitz, G., Index coll. princ. Gött. 139. 1982.

Ruijs (Ruys), **Johannes Marinus** (*fl.* 1884), Dutch botanist and zoologist; participated in the Dutch Polar Expedition of 1882-1883; Dr. phil. Amsterdam 1884. (*Ruijs*).

HERBARIUM and TYPES: L, NBV.

BIBLIOGRAPHY and BIOGRAPHY: BM 4: 1764; CSP 18: 358; IH 2: (files); JW 1: 452, 3: 367 (Ruys); Kelly p. 201-202; LS 38649; Rehder 1: 241, 349.

9767. *De verspreiding der phanerogamen van arktisch Europa.* (Een bijdrage tot de plantengeographie). Academisch proefschrift ter verkrijging van den graad van doctor in de plant– en dierkunde, aan de Universiteit van Amsterdam, op gezag van den Rector magnificus Dr. J.G. de Hoop Scheffer, ... te verdedigen op Vrijdag 28 November 1884, des namiddags te 3 ure, door Johannes Marinus Ruijs ... Kampen (Stoomdrukkerij, Laurens van Hulst) 1884. Oct. (*Verspr. phan. arkt. Eur.*).
Publ.: 28 Nov 1884, p. [i, iii, v], [1]-155. *Copies*: HH, KNAW, MO, NY.

Ruiz Lopez, Hipólito (1754-1815), Spanish botanist and explorer; studied natural sciences at Madrid under C.G. Ortega and A. Palau; leader of the Spanish *Expedición botánica* of 1777-1788 to the vice-kingdom of Peru (now Peru & Chile). (*Ruiz*).

HERBARIUM and TYPES: MA. – Some of the collections were perhaps dispersed in later years by Pavon without retaining the originals at MA. The main part of the important set sold to Lambert is now at BM. Other significant sets are at B-Willd., B (see Lack 1979), FI, G, OXF and US. Other material at AMES, CGE, F, FW, G, H, HAL, K, M, MO, MPU, NY, W. For details on Lambert's set of Ruiz materials see H.S. Miller (1970). Pavon also sold to Lambert some of the manuscripts of the Flora, a copy of the itinerary and manuscripts on Cinchona. These manuscripts are also at BM, but other manuscripts are still at MA.

BIBLIOGRAPHY and BIOGRAPHY: AG 4: 865; Backer p. 503 (epon.); Barnhart 3: 191; BM 4: 1765, 8: 1110; Bossert p. 342; Clokie p. 235; Colmeiro 1: clxxx; Colmeiro penins. p. 135; CSP 5: 326; De Toni 1: cxix; Dryander 3: 190; DSB 11: 605-606; Frank 3 (Anh.): 86; GF p. 74; Henrey 2: 35, 36, 3: 109; Herder p. 226; Hortus 3: 1203; IF p. 727; IH 1 (ed. 2): 51, 78, 92, (ed. 3): 62, 98, 117, (ed. 6): 363, (ed. 7): 341, 2: (files); Jackson p. 127, 377; Kew 4: 546-547; Langman p. 656; Lasègue p. 244-247, see also p. 581 (index); NI 1698-1699, 3: 58-59; PR 7891-7903, 5011, 6859, ed. 1: 5508, 7660, 8854-8866; Rehder 1: 340, 346, 5: 745; RS p. 138; TL-1/1123-1125, see J.F. Jacquin, A.B. Lambert; TL-2/3143, 4147, see J. Dombey, J.F. Jacquin, A.B. Lambert; Tucker 1: 613; Urban-Berl. p. 268, 390, 415; Zander ed. 10, p. 709, ed. 11, p. 809.

BIOFILE: Alvarez Lopez, E, Anal.-Inst. bot. Cavanilles 11(2): 547-557. 1953 (comm. on Ruiz in account journey Hernández), 12: 5-112. 1954.
Anon., Bot. Zeit. 16: 572. 1852 (Ruiz material in Fielding herb. left to OXF); Flora 4: 78. 1821, 13: 693. 1830, 15: 542. 1832, 24: 569, 572. 1841.
Arango, J.J., Rev. Acad. Colomb. Ci. 9: 30, portr. 1953.
Barreiro, A.J., Investigación y Progreso 5(7-8): 107-108. 1931.
Bellot, F., Lagascalia 7: 51-53. 1977 (departure from Cadiz).
Bridson, G.D.R. et al., Nat. hist. mss. res. Brit. Isl. 229.45. 198.
Candolle, Phytographie 445. 1880 (662 specimens at KBG; 640 at GFW).
Coats, A.M., The plant hunters 338, 363-367, 373. 1969.
Croizat, L., Lilloa 18: 295-329. 1949 (biogr.).
Dahlgren, B.E., Field Mus. Bot. 13(1, 1): 5-8. 1936 (in preface J.F. Macbride, Fl. Peru).
Hatcher, (transl.), An historical eulogium on don Hippolito Ruiz Lopez, first botanist, chief of the Expedition to Peru and Chile, Salisbury 1831, 55 p.; french version in Guillemin, Arch. Bot. 1: 377-381. 1833 (with facs. handwriting).
Herrera, F.L., Est. Fl. Dept. Cuzco 8. 1930.
Hiepko, P., Willdenowia 8: 389-400. 1978.
König, Ch. & J. Sims, Ann. Bot. 1(1): 12-14. 1804.
Kühnel, J., Thad. Haenke 275 [index]. 1960.
Lack, H.W., Willdenowia 9: 177-198. 1979 (the S. Amer. coll. of H. Ruiz et al. at B; lists of specimens, illustrations; detailed study of extant material at B).
Langman, I., Econ. Bot. 19: 441-443. 1965 (rev. Steele).
Muñoz Medina, D.J., Hist. desarr. bot. España 15. 1968 (Ruiz was a student of Gómez Ortega).
Murray, G., Hist. Coll. BM(NH): 1: 178. 1904 (1500-1700 sp., Cinchona barks; mss travels all purchased at Lambert sale).
Miller, H.S., Taxon 19: 538-540, 547-549. 1970 (on Lambert's Ruiz material).
Pichi-Sermolli, R.E.G., Nuov. Giorn. bot. ser. 2. 56(4): 699. 1959.
Reiche, K., Grundz. Pfl.-Verbr. Chile 5-6. 1907 (Veg. Erde 8).
Sandwith, N.Y., Kew Bull. 1953: 186-189.
Stafleu, F.A., in C.L. L'Héritier, Sert. angl. facs. ed. 1963, p. xx-xxi; in G.S. Daniels and F.A. Stafleu, HI-IAPT portraits of botanists, Taxon 23(1): : 108. 1974 (portr.).
Stearn, W.T., Nat. Hist. Mus. S. Kensington 285. 1981 (coll.).
Steele, A.J., Flowers for the King 1964 (extensive study of the Expedición botánica; many refs. to original sources) (rev. N.Y. Sandwith, Kew Bull. 19: 499-500. 1965).
Schultes, J.A., Flora 8(1), Beil. 1: 5, 6, 34. 1825.
Schultes, R.E., Bot. Mus. Leafl. 28(1): 87-122. 1980 (on R. as an ethnopharmacologist in Peru and Chile; portr.).
Turrill, W.B., Bull. misc. Inf. Kew 1920: 58.
Urban, I., Symb. ant. 3: 116. 1902.
Weberbauer, O., Pfl.-Welt peruan. Anden 2-4. 1911 (Veg. Erde 12).
Woodward, B.B., in G. Murray, Hist. Coll. BM(BH) 1: 47. 1904 (mss. descr. of plants 1777-1788, autograph of Relacion, and manuscripts on Cinchona at BM).

EPONYMY: *Ruiza* Cothenius (1790, *orth. var.*); *Ruizia* Cavanilles (1786); *Ruizia* Pavon (1794); *Ruizodendron* R.E. Fries (1936).

HANDWRITING: Candollea 32: 377-418. 1977; Arch. Bot. Guillemin 1: pl. facing p. 380.

9768. *Quinologia*, o tratado del árbol de la quina ó cascarilla, con su descripcion y la de otros especies de quinos nuevamente descubiertas en el Perú; del modo de beneficiarla, de su eleccion, comercio, virtudes, y extracto elaborado con cortezas recientes, y de la eficacia de este, comprobada con observaciones; á que se añaden algunos experimentos chímicos, y noticias acerca del analisis de todas ellas. Por Don Hipólito Ruiz, ... Madrid (en la oficina de la viuda é hijo de Marin) 1792. Qu. (*Quinologia*).

Publ.: Jan-Aug 1792 (p. [iv]: 15 Aug 1791; Italian transl. preface 10 Sep 1792), p. [i-xvi], 1-103, [104-107, contents]. *Copies*: FI, G, HH(2), MO, USDA. – For a supplement see below, *Supl. Quinologia*, 1801. For details on genesis and contents of the book see Steele (1964). – See also A.B. Lambert, *Ill. Cinchona* 1821, TL-2/4147, with extracts from *Quinologia*.

Italian translation: Oct-Dec 1792 (p. xxii: 10 Sep 1792), p. [i]-xxxii, 1-139. *Copies*: FI, HH, NY. – *Della China* e delle altre sue specie nuovamente scoperte e descritte da D. Ippolito Ruiz, ... prima traduzione dall'originale Spagnuolo stampato in Madrid 1792. In Roma (dalla Stamperia Giunchiana ...) 1792. Oct. (*Della China*).

German translation: 1794, p. [i-viii], [1]-106, [1-4, contents]. *Copies*: G, MO. – *Von dem officinellen Fieberrindenbaum* und den andern Arten desselben, die neuerlich Hippolitus Ruiz, ... entdeckte und beschrieb zuerst aus dem Spanischen ins Italienische, und aus diesem ins Deutsche übersetzt. Göttingen (im Vandenhoeck-Ruprechtischen Verlage). Oct. (*Officin. Fieberrindenbaum*).

Ref.: Steele, A.R., Flowers for the King 192-197, 203, 207. 1964.

9769. *Flora peruvianae, et chilensis prodromus*, sive novorum generum plantarum peruvianarum, et chilensium descriptiones, et icones. Descripciones y láminas de los nuevos géneros de plantas de la flora del Perú y Chile por Don Hipólito Ruiz y Don Joseph Pavon, botánicos de la Expedicion del Perú, y de la real Academia médica de Madrid. De órden del Rey. Madrid (en la imprenta de Sancha) 1794. Fol. (*Fl. peruv. prodr.*).

Co-author: José Antonio Pavon (1754-1844).

Publ.: early Oct 1794, p. [i*-iii*], i-xxii, 1-153, [154, err.], *pl. 1-37* (uncol. copp. Isidorus Galvez del.). *Copies*: FI, G, MO, NY, PH, Teyler, USDA; IDC 1293.

The best testimony on the publication of Ruiz and Pavon's *Prodromus* is that by A.J. Cavanilles, who wrote to James Smith on 14 October 1794 that the book was ready and that it had been presented to the King. This agrees with Steele's statement (1964, p. 232) that the book was ready for distribution in September and that a few copies found their way shortly into other hands. It is at any rate certain that Cavanilles saw a copy in October 1794, and obtained one later. Cavanilles refers to the *Prodromus* in the preface to his *Icones* vol. 3, dated lo January 1795. Furthermore a copy arrived in Rome early in 1795 (Steele 1964). Although the "distribution" was extremely limited, it must be assumed that the book was available as at present required by the *International Code of Botanical Nomenclature* (Art. 29.1, ed. 1978) from the date of the presentation to the King onward (early October 1794). Shortly afterwards, however, the release of the book was stopped, probably because of various intrigues caused by unpleasant local botanical competition. A few copies were sent abroad by the Spanish government in 1795, but the bulk (567 copies) was released only on 13 June 1796 (Steele p. 233, footnote 26).

The copy that had reached Italy early 1795, came into the hands of Gaspar Xuárez, a botanist residing in Rome, but a native of Tucumán (Argentina). With the permission of Ruiz, Xuárez started the preparation of a new edition of the *Prodromus* in order to rescue it from oblivion. The Xuárez edition, which is in Latin only, contains several corrections and many taxonomic and historical notes. It was published in 1797 and is now extremely rare. Sir Joseph Banks's copy (British Museum, Bloomsbury) and the copy at the New York Botanical Garden are among the very few copies known outside Italy. An extensive review of this edition was published by Schrader (1799). Reviews of the original edition were few, if any, owing to the exceptional circumstances of its distribution.

Pritzel (Thesaurus no. 7894) mentions the presence, in the Delessert copy (now at Genève), of an additional plate with description of the genus *Beauharnoisia*. This plate seems to be unique. The genus *Beauharnoisia* Ruiz and Pavon is usually dated from its publication in the *Annales du Muséum d'Histoire naturelle* 11: 71-73. *t. 9.* 1808. For criticism by Cavanilles of the *Prodromus* and the answer by Ruiz see the *Respuesta* 1796 (see below).

Facsimile ed.: 1965 (i.e. 1966) Historiae naturalis classica, tomus xliii, Lehre (J. Cramer), ISBN 3-7682-0282-8, p. [ii*-v**], new t.p. and note on publication by F.A. Stafleu of which the main contents are reproduced above), [i*, iii*], i-xxii, 1-153, *pl. 1-37* (id.) – Followed by a facs. of the *Fl. peruv. chil.* vols. 1-3. *Copies*: BR, FAS. – See Taxon 15: 237. 1966.

Ed. 2: 1797, p. [i*-vi*], i-xxvi, 1-152, *pl. 1-37*. *Copies*: FI, NY; also BM-Bloomsbury (n.v.). – Editio secunda auctior, et emendatior. Romae (in typographio Paleariniano) 1797. Fol. – See note above under original edition.

Ref.: Cavanilles, A.L., Letter to James Edward Smith, The Linnean Society of London, Smith mss. 3. 98.

Schrader, H.A., J. Bot. 1799(1): 150-181.

Stafleu, F.A., Intr. note in Fl. peruv. chil. prodr., facs. ed. p. [v**]. 1965.

Steele, A.R., Flowers for the king, Durham p. 227-228, 232-233, 246-247. 1964.

Xuárez, G., Lecturis [Introduction to] Florae peruvianae et chilensis prodromus ... editio secunda, auctior et emendatior, Romae 1797, p. i-viii.

9770. *Respuesta* para desengaño del público *à la impugnacion* que ha divulgado prematuramente el presbítero don Josef Antonio Cavanilles *contra el pródromo* de la flora del Perú, é insinuacion de algunos de los reparos que ofrecen sus obras botánicas, por Don Hipólito Ruiz, ... en Madrid (en la imprenta de la viuda é hijo de Marin) 1796. Qu. (*Respuesta*).

Publ.: 1796 (p. 101: 11 Sep 1795), p. [1]-100. *Copies*: G, HH(2). – For details on the Ruiz-Cavanilles battle see Steele (1964, p. 235-240). The *Respuesta* was written in answer to Cavanilles' criticism of the Prodromus in his own *Icones* (vol. 3, 1795, preface) (TL-2/1061).

9771. *Flora peruviana, et chilensis*, sive descripciones, et icones plantarum peruvianarum, et chilensium, secundum systema linnaeanum digestae, cum characteribus plurium generum evulgatorum reformatis. Auctoribus Hippolyto Ruiz, et Josepho Pavon, ... superiorum permissu. [Madrid] (typis Gabrielis de Sancha) 1798-1802, 3 vols. Fol. †. (*Fl. peruv.*).

Co-author: José Antonio Pavon (1754-1844).

1: medio 1798 (probably Jul 1798), p. [i*-ii*], i-vi, 1-78, *pl. 1-106* (see below).

2: Sep 1799, p. [i*], i-ii, 1-76, *pl. 107-222*.

3: Aug 1802, p. [i*], i-xxiv, 1-95, *pl. 223-325*.

Copies: FI, G, MO, NY, USDA; IDC 9. – The MO copy has the text of the three volumes as above, but the plates are in two volumes with special t.p. 5: *pl. 1-152*, and *pl. 153-325*.

Volume 4 (pl. 326-425) was not issued by Ruiz and Pavon. Copies of the plates were acquired by O. Rich and given to several botanists (sets at BM(NH), LINN, G, Inst. de France). Technically this can be regarded as effective publication, but the date is not to be ascertained. Final publication of *plates and text*: *Anal. Inst. bot. Cavanilles* 12(1): 113-195 (1954), 13: 5-70 (1955), 14: 717-784 (1956), 15: 115-241 (1957); issued separately as one volume, Madrid 1957 (new names listed as Ruiz et Pavon apud Gomez in Index kewensis, suppl. 13. – For competition between names quoted from the plates distributed by O. Rich and at least one name (*Weinmannia*) used by C.S. Kunth in Humboldt, Bonpland et Kunth 6: 49-58. 1823, see our note in TL-2/2: 371 (sub no. 3143). The first references to the plates of volume 4 date from 1830.

Volume 5 (pl. 426-495) was published in *Anal. Inst. bot. Cavanilles* 16: 353-462. (1958), 17: 377-495 (1959) (new names listed in Index kewensis suppl. 13).

The *Flora* must always be consulted together with the *Prodromus* and the *Systema*. The latter book was published after vol. 1, but before vol. 2 of the *Flora*. The *Prodromus* contains the valid publication of many generic names later used in the *Flora*. The artists who provided the drawings for the published plates were José Brunete, Isodorus Galvez, Francisco Pulgar, J. Rivera and José Rubio. The original edition numbered 300 copies on regular paper and 150 copies on special paper("fino").

Facsimile ed. (of vols. 1-3 and plates): 1965 (i.e. 1966), bound in one volume with the facsimile edition of the *Prodromus*, see above under *Fl. peruv. chil. prodr.* Lehre (J.

Cramer), part of Historiae naturalis classica, tomus xliii. *Copies*: BR, FAS. – See Taxon 15: 237. 1966. Preceded by a one page bibliographical note by F.A. Stafleu, the main contents of which are given above. ISBN 3-7682-0283-6.

Ref.: Alvarez Lopez, E., Anal. Inst. bot. Cavanilles 12: 5-112. 1954.

Sandwith, N.Y., Kew Bull. 1953: 186-189.

Stafleu, F.A., The publication of the Flora peruviana, 1 p. preceding facsimile in Hist. nat. class. xliii, Lehre 1965.

Steele, A.R., Flowers for the King 225, 249, 250, 253, 254, 259, 261, 262, 325. 1964.

9772. *Systema vegetabilium florae peruvianae et chilensis*, characteres prodromi genericos differentiales, specierum omnium differentias, durationem, loca natalia, tempus florendi, nomina vernacula, vires et usus nonnullis illustrationibus interspersis complectens. Auctoribus Hippolyto Ruiz, et Josepho Pavon, ... tomus primus. Superiorum permissu [Madrid] (typis Gabrielis de Sancha) 1798. Oct. †. (*Syst. veg. fl. peruv. chil.*).

Co-author: José Antonio Pavon (1754-1844).

Publ.: late Dec 1798 (after vol. 1 but before vol. 2 of the *Fl. peruv. chil.*), p. [i*], [i]-vi, 1-455, [466]. *Copies*: FI, G, HH, M, NY; IDC 1288. – For dates see RS p. 138, TL-1, Steele (1904) and Stafleu (1965).

Ref.: Stafleu, F.A., The publication of the *Flora peruviana*, 1 p. Hist. nat. class. xliii, Lehre 1965 (facs. of *Prodromus* and of *Flora*).

Steele, A.D., Flowers for the king 254-256, 262, 301. 1964.

9773. Hippolyti Ruiz, Regiae Academiae med. Matrit. Socii, et peruvianae expedit. botanici primarii, *de vera fuci natantis fructificatione*. Commentarius. Superiorem permissu. Matriti [Madrid] (apud viduam e filium Petri Marin) 1798. Qu. (*Vera fuci fruct.*).

Publ.: 1798, p. [1]-38, *1 pl.* (uncol. copp. Isid. Galvez). *Copies*: B-S, HH, NY.

9774. *Disertacion sobre la raiz de la ratánhia*, específico singular contra los fluxos de sangre, que se insertó en el primer tomo de las memorias de la real Académia médica de Madrid, y se ilustra nuevamente con notas y con la estampa de la planta, por su autor Don Hipólito Ruiz, ... Madrid (en la Imprenta de la viuda é hijo de Marin) 1799. Qu. (*Dissert. raiz ratánhia*).

Publ.: *1799, p.* [*i-xvi*], *1-47, 1 pl.* (uncol. copp. Isid. Galvez). *Copy*: HH.

9775. *Suplemento á la Quinologia*, en el qual se aumentan las especies de Quina nuevamente descubiertas en el Perú por Don Juan Tafalla, y la Quina naranjada de Santa Fé con su estampa. Añadese la repuesta á la memoria de las Quinas de Santa Fé, que insertó Don Francisco Zea en los Anales de Historia natural, y la satisfaction á los reparos ó dudas ciudadano Jussieu sobre los géneros del pródromo de la Flora del Perú y Chile. Por Don Hipólito Ruiz y Don Josef Pavon, ... Madrid (en la imprenta de la viuda e hijo de Marin) 1801. Qu. (*Supl. Quinologia*).

Co-author: José Antonio Pavon (1754-1844).

Publ.: 1801, p. [i*], [i-viii], [1]-154. *Copies*: FI, G, HH. – See also A.B. Lambert, *Ill. Cinchona* 1821, TL-2/4147.

Ref.: Steele, A.R., Flowers for the King 203-207, 239. 1964 ("the invective supplement" q.v. for many details on the conflict with F.A. Zea).

Zea, F.A., Anal. Hist. nat. 2: 196-235. 1800.

9776. Hippolyti Ruiz *ad* clar. vir. *A.L. Jussieum*, botan. professorem parisiensem, *epistola*, in qua ejus dubiis circa nova plantar. genera in Flora peruviana, et in D. Cavanilles operibus constituta respondetur. Matriti [Madrid] (ex typographia Mariniana) 1801. Qu.

Publ.: 1801, p. [i], 121-154. *Copies*: FI, HH. – Reprinted and to be cited from *Supl. Quinologia* 1801, see above, p. 121-154 (text also in vol. 3 of Fl. peruv. vol. 3).

Note: Booknumber 9777 has not been used.

9778. *Memoria sobre la legitima Calaguala* y otras dos raices que con el mismo nombre nos vienen de la America meridional. Por Don Hipolito Ruiz, .. Madrid (en la imprenta de D. José del Collado) 1805. Qu. (*Mem. Calaguala*).

Publ.: 1805, p. [i-iv], [1]-60, *4 pl.* (uncol. copp. Is. Galvez).*Copies*: FI (no. pl.), G (*4 pl.*) HH *1 pl.*). – PR 7903 cites one plate (Polypodium Calaguala).

9779. *Memoria de las virtudes y usos de la raiz de la planta* llamada *Yallhoy* en el Perú. Por Don Hipolito Ruiz, ... Madrid (en la imprenta de D. José del Collado) 1805. Qu. (*Mem. Yallhoy*).
Publ.: 1805 , p. [i-iv], [1]-35, *1 pl. Copies*: G, HH. – *Monnina polystachya* Ruiz et Pavon, *Syst. Fl. peruv.* 169. 1798.

9780. *Memoria sobre las virtudes y usos de la planta* llamada en el Perú *Bejuco e la Estrella*. Por Don Hipolíto Ruiz, ... Madrid (en la imprenta de D. José del Collado) 1805. Qu. (*Mem. Bejuco*).
Publ.: 1805, p. [i-iv], [1]-52, *1 pl. Copies*: G, HH. – Aristolochia fragrantissima.

9781. *Memoria sobre* las virtudes y usos de *la raiz de Purhampuy* ó China peruana, por Don Hipólito Ruiz, ... ilustrada y aumentada por el licenciado don Antonio Ruiz, ... Madrid (Imprenta de don José del Collado) 1821. Qu. (*Mem. raiz Purhampuy*).
Publ.: 1821, p. [1]-96. *Copy*: HH.

9782. *Relación del viaje* hecho a los reynos del Perú y Chile por los botánicos y dibuxantes enviados para aquella expedición, extractado de los diarios por el orden que llevó en estos su autor Don Hipólito Ruiz. Publicado por primera vez ... revisada y anotada por el ... R.P.A.J. Barreiro, O.S.A. Madrid (Est. Tipografico Huelves y Compañia ...) 1931. Oct. (*Relac. viaje*).
Publ.: 1931, p. [1]-558, [1]. *Copies*: BR, PH, US.
English translation: 28 Mar 1940, Field Mus. Bot. 21: 1-372. 1940, by B.E. Dahlgren. – "Travels of Ruiz, Pavón, and Dombey in Peru and Chile (1777-1788) ..."
Ed. 2: 1952, 2 vols., [i]-xliv, [1]-526 and [i-xi], [1]-244, [245], *pl. 1-20. Copy*: J. Wurdack.
– Edited and annotated by Jaime Jaramillo-Arango. – Rev.: N.Y. Sandwith, Kew Bull. 1953: 186-189 (mss. of *Relación* probably sold to A.B. Lambert by Pavon after Ruiz's death).

Ruiz, Sebastian Joseph Lopez, (*fl.* 1802), Spanish botanist. (*S. Ruiz*).

HERBARIUM and TYPES: Unknown.

BIBLIOGRAPHY and BIOGRAPHY: PR 7904, ed. 1: 8867.

9783. *Defensa y demostracion* del verdadero descubridor de las Quinas del Reyno de Santa Fé, con varias noticias utiles de este especifico, en contestacion á la memoria de Don Francisco Antonio Zea. Su autor el mismu descubridor D. Sebastian Josef Lopez Ruiz, ... en Madrid (en la emprenta de la viuda é hijo de Marin) 1802. Qu. (*Def. demostr.*).
Publ.: 1802, p. [1]-24. *Copies*: G, HH.

Rumphius, (baptized Rumpf), **Georg Eberhard** (1628-1702), German-born Dutch naturalist; enlisted with the Dutch West Indian Company for Brazil 1846 (but taken prisoner by the Portuguese); enlisted with the Dutch East Indian Company 1652, in Java 1652, to the Moluccas in the military service of the Company; commissioned officer 1655; in civil service as merchant to Ambon 1657 (Larike), from 1660 at Hitoe; first merchant at Amboina 1666 (appointment not ratified); blind from 1670; staying in Ambon mainly dedicating himself to the writing of his publications. (*Rumph.*).

HERBARIUM and TYPES: Almost entirely lost. Most of Rumphius' books, manuscripts and collections were destroyed by fire on 11 Jan 1687. Rumphius sold part of his collections (herbarium specimens as well as shells) to Cosimo II de Medici in 1682. Some of these collections are still at FI (Martelli 1903) although difficult to recognize because of the absence of the old labels. A few specimens may have found their way to Vienna (W) and to Gaertner (now TUB). The manuscript of the *Herbarium amboinense* (partly in second copy, becuase the first was lost at sea) is at the University Library of Leiden. The original copy for Appendix 2 and some drawings are at the University Library of Utrecht. See Greshoff (1902) for further details on the various manuscripts.

NOTE(1): Rumphius was (probably) born in Hanau (Hessen) Germany, a city founded by Dutch refugees; his mother had Dutch first names. It is therefore quite well possible that he was of Dutch origin, which would explain his perfect usage of Dutch (see H.C.D. de Wit, 1952).

NOTE(2): Main *bibliography*: G.P. Rouffaer & W.C. Mullcr, *in* Rumphius Gedenkboek, 1702-1902, p. 165-221. Contains also a list of manuscripts and an extensive critical list of secondary literature.

BIBLIOGRAPHY and BIOGRAPHY: ADB 29: 663-667. 1889 (many errors); Backer p. 503 (epon.); Barnhart 3:191 (d. 15 Jun 1702); Blunt p. 121, 138, 263; BM 4: 1766, 8: 1111; Bossert p. 342 (Rumpf); Bret. p. 27; Dryander 3: 181, 286; Frank 3 (Anh.): 87; GR p. 127; HA 1: 616-617; Henrey 2: 191; Herder p. 212; HU 518 (extensive treatment Herb. amb.), see also 2(1): clvii, clxx, clxxxv; Jackson p. 378; JW 2: 199-200, 3: 367, 4: 404, 5: 251 (many Dutch references) Kew 4: 547; Lasègue p. 76, 493, 506, 528, 535; LS 23244; Moebius p. 141, 142, 367, 368; MW p. 420; NI 1: 79, 86, 87, 90, 184, 2: nos. 1700-1701, 3: 59; NNBW 3: 1104 (M.J. Sirks); PR 846, 1393, 3849, 3974, 4663, 7908-7909, ed. 1: 958, 1583, 4370, 8874-8875, 10632; Quenstedt p. 371; Rehder 1: 464; RS p. 138; SA 1: 154, 2: 600; SK 1: xxx, 452-453 (many data), 2: lxxxvi, 4: 13, 17, 21-23, 26, 36, 60, 104, lxxxv-lxxxix, 5: cccxvii, 8: lxxxiii; TL-1/1126; TL-2/566, 930-932, 2473, 2654, 4777, 5457, 5860, 6111; see J. Burman; Tucker 1: 613.

BIOFILE: Anon., Fl. males. Bull. 1(5): 129. 1949 (monument); Grote Winkler Prins Encycl. 16: 642. 1973; Koloniaal Museum te Haarlem, Verslag der Rumphiusherdenking Amsterdam 1903, 11 p. (suppl. to Verslag Kol.-Mus. 1902, Bull. no. 28) (on Rumphius medal).

Baillon, H., Dict. bot. 3: 753. 1891.
Ballintijn, G., Rumphius, de blinde ziener van Ambon, Utrecht 1944, viii, 191 p.
Bernardi, L., Musées de Genève 123: 11-12. 1972 (portr.).
Bickmore, A.S., Travels in the East Indian Archipelago, London 1868, p. 250-252 (visit to R.'s grave).
Bloys van Treslong Prins, P.C., Tijdschr. Ind. Taal-, Land– en Volkenkunde 69: 426-435. 1929 (transcription of various official documents regarding R.; signs himself geogr. Everhardus Rumphius).
Blume, C.L., Rumphia 1835-1848, 4 vols. (see TL-2/566; frontisp. portr. in vol. 1).
Boedijn, K.B., *in* H.C. de Wit, Rumphius memorial volume 289-294. 1959 (on R's fungi).
Bridson, G.D.R. et al., Nat. hist. mss. res. Brit. Isl. 224.210. 1980.
Britten, J., J. Bot. 56: 362-365. 1918 (reviews of Merrill (1917) and description of two rare printed indexes to the Herb. amb.).
Burman, J., Index alter in omnes tomos herbarii amboinensis. Leiden 1769, 32 p.
Busken Huet, Ed., Het land van Rembrand 2(2): 91-99. 1884.
Candolle, Alph. de, Phytographie 445. 1880 (no herb. known).
Dandy, J.E., Regn. veg. 51: 7-8. 1967 (no new generic names in Auctuarium).
Du Petit-Thouars, L.M.A.A., *in* Biogr. univ. anc. mod., ed. L.G. Michaud, 39: 317-322. 1825.
Engel, H., Bijdr. Dierk. 27: 310-311. 1939 (sec. refs.; on zool. coll.).
Fosberg, F.R., Taxon 30: 218-227. 1981 (on *Casuarina litorea*, a valid Rumphian name and on names in Stickman's Herb. amboin. 1754; proposal to invalidate names in left hand column).
Fries, Th.M., Bref Skrifv. Linné 1(3): 19. 1909, 1(4): 213. 1910, 1(5): 143, 259. 1911.
Gaertner, J., Hanauisches Magazin vom Jahr 1785, 8: 337-344. 1785 (first critical biography; see Rouffaer & Muller p. 177).
Goebel, K., *in* Greshoff et al., Rumphius Gedenkboek 59-62. 1902 ("Rumphius als botanischer Naturforscher").
Greshoff, M., Encycl. Ned.-Ind. 3: 464-468. 1902.
Haller, A. v., Bibl. bot. 1: 615-617. 1771.
Hamilton, F., Mem. Wern. nat. Hist. Soc. 5(2): 307-383. 1826 (commentary on *Herb. amb.*).
Harting, P., Album der natuur 1885: 1-15.

Hartwich, C., *in* M. Greshoff et al., Rumphius Gedenkboek 1902. p. 79-88 (on American plants in Herb. amboin.).
Hasskarl, J.K., Neuer Schlüssel zu Rumph's Herbarium amboinense. Halle 1866, 247 p. (see TL-2/2473).
Heeres, J.E., *in* M. Greshoff et al., Rumphius Gedenkboek 1-16. 1902 (biography based on Leupe 1871).
Henschel, A.W.E., Vita G.E. Rumphii, 1833, xiv, 216 p. (see TL-2/2654; for review see W.H. de Vriese, Tijdschr. nat. Gesch. Physiol. 1 (boek beschr.): 208-214. 1834); also issued as Clavis Rumphiana botanica et zoologica, Breslau 1833.
Hickson, S.J., Mem. Proc. Manchester lit. philos. Soc. 70: 17-28. 1926 (biogr. sketch).
Hoëvell, W.R. van, Tijdschr. Neerl. Ind. 2(2): 25-30. 1839 (mentions a doctorate h.c. Hanau for Rumphius. This is erroneous; Hanau never had a university).
Hulth, J.M., Bref Skrifv. Linné 1(8): 18, 1922, 2(1): 118. 1916.
Hulth, J.M. & A.H. Uggla, Bref Skrifv. Linné 2(2): 52, 59, 60, 63, 64, 65, 83, 96, 146. 1943.
Kalkman, C. & P. Smit, eds., Blumea 25: 5, 57. 1979.
Karsten, M.C., The old company's garden 57, 58, 79, 125. 1951.
King, G., J. Bot. 37: 455. 1849.
Krikorian, A.D., Principes 26(3): 107-121. 1982 (R. on coconut "stones").
Kuntze, O., Rev. gen. pl. 1: cxlii. 1892.
Leupe, P.A., De Navorscher 16: 207. 1866 (some preliminary remarks); Nat. Verh. Kon. Akad. Wet. Amsterdam 13: 1-36. 1871 (biogr., for a detailed review of this first true biography of Rumphius see Rouffaer & Muller, p. 183-185).
Lotsy, J.P., *in* M. Greshoff et al., Rumphius Gedenkboek 1902, p. 46-58 (on the Rumphius manuscripts in the Netherlands).
Martelli, U., Bull. Soc. bot. ital. 1902: 90 (coll. FI); Le collezioni di G.E. Rumpf. Firenze 1903, 213 p. see TL-2/5475.
Meeuse, A.D.J., Biologist 47: 42-54. 1965 ("Straddling two worlds" biogr. sketch).
Merrill, E.D., An interpretation of Rumphius' Herbarium amboinense, Manila 1917 (authoritative detailed study); Chron. bot. 14: 376 [index]. 1954 (botany Cook's Voyage).
Milner, J.D., Cat. portr. Kew. 93. 1906.
Nissen, C., Zool. Buchill. 352-353. 1969.
O., Dict. sci. méd., Biogr. méd. 7: 73-74. 1825.
Papageno, [S. Kalff], Java-Bode Feb 1896: nos. 30, 33, 36, 38, 41, 44 ("de Indische Plinius").
Perry, L.M., *in* Honig & Verdoorn, Sci. scient. Neth. Indies 295-308. 1945 (based on M.J. Sirks 1915).
Poggendorff, J.C., Biogr.-lit. Handw.-Buch 2: 270-721. 1863, 7a (suppl.): 566-568. 1971 (extensive secondary bibl.).
Poiret, J.L.M., Enc. méth. 8: 755. 1808.
Robyns, A.G., Bull. Miss. Bot. Gard. 56(3): 4-5. 1968 (on the Kapok tree and R's herbal).
Rouffaer, G.P. & W.C. Muller, *in* M. Greshoff et al., Rumphius Gedenkboek 165-221. 1902 (extensive bibliography).
Sarton, G., Isis 27: 242-257. 1937 (Rumphius, Plinius indicus); Sarton on the history of science 189-196. 1962.
Schöppel, F.A., Deut. ges. Nat.-Völkerk. Ostasiens, Nachtr. 52: 14-18. 1939.
Siebert, K., Encycl. Nederl. Indie 3: 640-645. 1919.
Sirks, M.J., Indisch natuurond. 24-62, 301 [index]. 1915 (biogr.).
Slooten, D.F. van, *in* H.C.D. de Wit et al., Rumphius memorial volume 295-338. 1959 (on R. as an economic botanist).
Smith, J.J., Philip. J. Sci. C. Bot. 12: 249-262. 1917 (Amboina orch. coll. by C.B. Robinson with ref. to Rumphius).
Stafleu, F.A., Linnaeus and the Linnaeans 109, 166, 168, 169. 1971.
Stearn, W.T., Roy. Hort. Soc., Exhib. manuscr. books 29. 1954.
Targioni-Tozzetti, G., Le collezioni di Giorgio Everardo Rumpf acquisitate dal Granduca Cosimo iii de' Medici, Firenze 1903, 213 p. (edited by Ugulino Martelli).
Turner, R.D., Johnsonia 3: 326-327. 1958 (on zool. work).
Verdoorn, F., ed., Chron. bot. 1: 381. 1935, 3: 56. 1937.

Warburg, O., *in* M. Greshoff et al., Rumphius Gedenkboek 1902, p. 63-78 (botanical exploration Moluccas after R.).
Wit, H.C.D. de, Indones. J. nat. Sci. 108: 161-172. 1952; Taxon 1: 101-110. 1952; Rumphius memorial volume, Amsterdam 1959, 462 p., 27 pl. (on p. 339-460 a checklist to Rumphius's Herbarium amboinense by H.C. D. de Wit and on p. 1-26 a biogr. of R. also by de Wit); Belmontia 4(4). 1960, Georgius Everhardus Rumphius, 26 p. (portr.); *in* J. Arditti, ed., Orchid biol. 47-94. 1977 (Orchids in Rumphius' Herbarium amboinense).
Wittrock, V.B., Acta Horti Berg. 3(2): 155. 1903, 3(3): 173. 1905.
Zanefeld, J.S., *in* H.C.D. de Wit et al., Rumphius memorial volume 277-288. 1959 (identification of R's algae; general contr. to phycology by Rumphius).

EPONYMY (books): C.L. Blume, *Rumphia* 1836-1849, see TL-2/566; (taxa): *Rumpfa* Cothenius (1790, *orth. var.*); *Rumpfia* Linnaeus (1754, *orth. var.*); *Rumphia* Linnaeus (1753).

9784. Georgii Everhardi Rumphii, ... *Herbarium amboinense*, plurimas complectens arbores, frutices, herbas, plantas terrestres & aquaticas, quae in Amboina, et adjacentibus reperiuntur insulis, adcuratissime descriptas juxta earum formas, cum diversis denominationibus, cultura, usu, ac virtutibus. Quod & insuper exhibet varia insectorum animaliumque genera, plurima cum naturalibus eorum figuris depicta. Omnia magno labore ac studio multos per annos conlegit, & duodecim libris Belgicis conscripsit ... Nunc primum in lucem edidit & in Latinum sermonem vertit Joannes Burmannus, ... pars prima. Amsteleadami [Amsterdam] (apud Franciscum Changuion, Joannem Catuffe, Hermannum Uytwerf) Hagae comitis ['s Gravenhage) (apud Petrum Gosse, ...) Ultrajecti [Utrecht] (apud Stephanum Neaulme) 1741 (in sets of two). Fol. (*Herb. amboin.*).
Editor: Johannes Burman (1707-1779), see TL-2/927-933.
Copies: BR, HU, NY, U, USDA; IDC 292.
Collation: see HU 518 and Rouffaer & Muller p. 168-170. – The pagination cited below is that of the copy at U. The HU copy has the plate *123* in vol. 5; a plate *84[bis]* in vol. 3. – In all there are *696* plates.
1: 1741, p. [i-xxxvi], frontisp. and 2 portr., 1-200, *pl. 1-82*. – *Preface contents*: [ii], Lat. t.p.; frontisp. allegor.; [iii], t.p. Dutch; portr. J. Burman; [v-vii, dedic.]; [viii-xiii, preface], [xvi, author's note]; [xv, reply to text p. xiv]; [xvi epigramma], [xvii-xviii, poem A. v. Ommering], [xix-xxii, Burman's introd.]; [xxii-xxxvi, poems by P. Burman, J.O. Schacht, J. Balde]; portr. Rumphius; 1-200 scientific text, *pl. 1-82* uncol. copp. – *Imprint*: as in heading.
Dutch title: "Het amboinsche kruid-boek. Dat is, beschryving van de meest bekende boomen, heesters, kruiden, land– en water-planten, die men in Amboina, en de omleggende eylanden vind, na haare gedaante, verscheide benamingen, aanqueking, en gebruik: mitsgaders van eenige insecten en gediertens, voor 't meeste deel met de figuren daar toe behoorende, allen met veel moeite en vleit in veel jaaren vergadert, en beschreven in twaalf boeken, door Georgius Everhardus Rumphius, ... nagezien uitgegeven door Joannes Burmannus, ... die daar verscheide benamingen en zyne aanmerkingen heeft bygevoegt. Eerste deel."
2: 1741, p. [ii-iii, Lat. & Dutch t.p.], 1-270, *pl. 1-87. Imprint*: as vol. 1.
3: 1743, p. [ii-iii, Lat. & Dutch t.p.], 1-128, 131-218, *pl. 1-84, 84[bis], 85-107, 107[bis], 108-127, 127[bis], 128-141. Imprint*: as vol. 1. The gap in pagination (between 128 and 131) occurs in all copies.
4: 1743, p. [ii-iii, Lat. & Dutch t.p.], 1-154, *pl. 1-82, Imprint*: as vol. 1.
5: 1747, p. [ii, Lat. t.p.; iii, Dutch t.p.; v-vi, author's pref.; vii-viii, dedic.], 1-492, *pl. 1-75, 75[bis], 76-184. Imprint*: Amsteleadami (apud Franciscum Chaguion, Hermannum Uytwerf), Hagae comitis (apud Petrum Gosse, ... Antonium van Dole). – *Pl. 123* lacking in U copy. For a variant issue see Rouffaer & Muller, p. 169, note 3.
6: 1750, p. [ii, Lat. t.p., iii, Dutch t.p., v-viii poem iter in Bras.], 1-256, *pl. 1-90. Imprint*: Amsteleadami (apud Franciscum Changuion, Hermannum Uytwerf), Hagae comitis (apud Petrum Gosse, ... Antonium van Dole).
7: see below, *Auctuarium*, 1755.
Variant issues: see Rouffaer & Muller p. 168-170.

RUMPHIUS, *Herb. amboin.*

Re-issue of vols. 1-6, 1750, Amsterdam (Meinard Uytwerf).
Other indexes: (1) J. Burman, *Index alter in omnes tomos herbarii amboinensis*, Leiden 1769, TL-2/932.
(2) A.W.E.T. Henschel, *Vita Rumphii*, Breslau 1833, TL-2/2654; for a detailed critical discussion see Rouffaer & Muller p. 178-182.
(3) J.K. Hasskarl, *Neuer Schlüssel zu Rumph's Herbarium amboinense*, Halle 1866, TL-2/2473.
(4) E.D. Merrill, an interpretation of Rumphius' Herbarium amboinense, Manila 1917, TL-2/5860, see also J.H., Bull. misc. Inf. Kew 1918: 244-246.
(5) H.C.D. de Wit, *A checklist to Rumphius's Herbarium amboinense, in* Rumphius memorial volume, Amsterdam 1959, p. 339-460.
Manuscripts: see Lotsy (1902, p. 46), Rouffaer & Muller (1902, p. 165), Targioni-Tozzetti (1903), NI 1701.
Linnaean study: Herbarium amboinense, see TL-2/4777 and F.R. Fosberg, Taxon 30: 218-227. 1981.

9785. Georgi Everhardi Rumphii, ... *Herbarii amboinensis auctuarium*, reliquas complectens arbores, frutices, ac plantas, quae in Amboina, et adjacentibus demum repertae sunt insulis, omnes accuratissime descriptae, & delineatae juxta earum formas, cum diversis indicis denominationibus, cultura, usu, ac viribus; nunc primum in lucem editum, & in Latinum sermonem versum, cura & studio Joannis Burmanni, ... qui varia adjecit synonyma, suasque observationes. Amstelaedami [Amsterdam] (apud Mynardum Uytwerf, & viduam ac filium S. Schouten) 1705. Fol. (*Herb. amboin. auctuar.*).
Co-auhor: Johannes Burman (1707-1779) see TL-2/927-933.
Publ.: on or around 22 Sep 1755 (fide letter by Burman to Linnaeus of 20 Sep 1755; postface 20 Jun 1755; GGA 6 Sep 1756), p. [ii, iii, Lat. & Dutch t.p.], [1]-74, *pl. 1-29*, *Index universalis*: [75-94], postface [95] by J. Burman. [p. 74 is sign. T¹, p. [75] is T²]. *Copies*: BR, NY, U, USDA; IDC 292. The *Index universalis* in sex tomos et Auctuarium Herbarii amboinensis Cl. Georgii Everhardi Rumphii, by J. Burman is in general based on C. Linnaeus, *Herb. amb.* 11 Mai 1754 (TL-2/4777) but contains in addition new binary names given by "B". (J. Burman). This index is the first place in literature in which an author other than Linnaeus or his pupils uses Linnaean binomials.
Dutch t.p.: Het auctuarium, ofte vermeerdering, op het amboinsch kruyd-boek. Dat is, beschryving van de overige boomen, heesters, en planten, die men in Amboina, en de omleggende eilanden vind, allen zeer accuraat beschreven en afgebeeldt na der zelver gedaantes, met de verscheide Indische benamingen, aanqueking, en gebruik ...".

Rupin, Ernest (Jean Baptiste) (1845-1909), French botanist, archeologist and artist; in the French civil service until 1873; founder and first curator of the Musée E. Rupin, Brive (Corrèze). (*Rupin*).

HERBARIUM and TYPES: Musée Rupin, Brive (Corrèze); further material at PC.

BIBLIOGRAPHY and BIOGRAPHY: BL 2: 139, 711; BM 4: 1767; CSP 18: 361; GR p. 350 (b. 6 Mai 1845, d. 24 Oct 1909); IH 2: (files); LS 23247.

EPONYMY: *Rupinia* Spegazzini et Roumeguère (1879). *Note*: The etymology of *Rupinia* Linnaeus f. (1782) and *Rupinia* Corda (1829) could not be established.

9786. *Catalogue des plantes vasculaires du Département de la Corrèze* ... Brive (Marcel Roche, ...) 1884. Oct. (*Cat. pl. Corrèze*).
Publ.: 1879-1884 (in journal; repr. Bot. Zeit. 28 Nov 1884; BSbF séance 27 Jun 1884; Nat. Nov. Oct(3) 1884), p. [1]-277 [as "377"]. *Copies*: G, H. – Originally published in Bull. Soc. sci. hist. archéol. Corrèze 1: 687-693. 1879, 2: 247-256, 433-446, 607-621, 817-827. 1879-1880, 3: 181-196, 355-375, 525-544, 725-747. 1881, 4: 243-252, 419-433, 587-609. 1882, 5: 325-363. 1883.
Ref.: Malinvaud, E., Bull. Soc. bot. France 31 (bibl.): 10-14. 1884 (rev.).

9787. *Catalogue des mousses, hépatiques et lichens de la Corrèze* ... Limoges (Imprimerie-

Librairie Limousine Ve H. Ducourtieux ...) 1895. Qu. (*Cat. mouss. hepat. lich. Corrèze*).
Publ.: 1893-1895 (in journal; reprint 1895; Bot. Centralbl. 20 Aug 1895; BSbF séance 26 Jul 1895; Nat. Nov. Jul(2) 1895), p. [1]-92. *Copies*: BR, G, MO(2), PH. – Issued as supplement to issues of la Revue scientifique du Limousin (copy journal p.p. in G):
  p. [1-4], in 1(25), 15 Jan 1895.
  p. [5]-12, in 1(4), Apr 1893.
  p. 13-20, in 1(6), Jun 1893.
  p. 21-28, in 1(8), Aug 1893.
  p. 29-36, in 1(10), Oct 1893.
  p. 37-44, in 1(12), 15 Dec 1893.
  p. 45-52, in 2(15), 15 Mar 1894.
  p. 53-60, in 2(18), 15 Jun 1894.
  p. 61-68, in 2(21), 15 Sep 1894.
  p. 69-76, in 2(22), 15 Oct 1894.
  p. 77-84, in 2(24), 15 Dec 1894.
  p. 85-92, in 3(25), 15 Jan 1895.
Ref.: Bescherelle, É., Bull. Soc. bot. France 41 (bibl.): 647. 1894 (1895).
  Hue, A., Bull. Soc. bot. France 43 (bibl.): 75-76. 1896.

Rupp, Herman Montague Rucker (1872-1956), Australian botanist; B.A. Univ. Melbourne 1897; ord. minister church of England 1899; serving in various parishes in Victoria, Tasmania and New South Wales; amateur orchidologist; Clarke Medal, Royal Society NSW 1949. (*Rupp*).

HERBARIUM and TYPES: MEL and NSW; other material in A, AK, G, K, W.

BIBLIOGRAPHY and BIOGRAPHY: Bossert p. 342; HR (b. 24 Dec 1872, d. 2 Sep 1956); IH 1 (ed. 2): 116, (ed. 3): 148, (ed. 4): 165, (ed. 5): 178; Kew 4: 548; Lenley p. 356; NI 17011; Zander ed. 10, p. 710, ed. 11, p. 810.

BIOFILE: Anon., Austral. encycl.7: 519. 1965; Proc. Linn. Soc. N.S.W. 82: 4. 1957 (obit.).
Blaxell, D.F., Orchadian 6: 42. 1978 (Orch. N.S.W. publ. late Jan 1944; at any rate between 13 Jan and 27 Feb).
Rupp, H.M.R., Austral. Orch. Rev. 6(2): 41-42, 64-65. 1941 (memories of an orchid lover).
Willis, J.H., Vict. Natural. 66: 127-128. 1949, 73: 105-110. 1956 (obit., portr.).

9788. *Guide to the orchids of New South Wales* ... Australia (Angus & Robertson Limited ...) 1930. Oct. (*Guide orchid. N.S.W.*).
Publ.: 1930 (Nat. Nov. 1931), p. [i]-viii, [1]-157. *Copies*: G, NY, USDA.

9789. *The orchids of New South Wales* ... Issued from the National Herbarium, Sydney, as a part of the Flora of New South Wales. December 1943. Oct. (*Orchid. N.S.W.*).
Publ.: late Jan 1944 (see Blaxell (1978); t.p. dated 1943), frontisp., p. [i]-xv, [1]-152, *pl. 1-23* (in text). *Copies*: FI, G, HH, NY, PH, USDA.
Facsimile ed.: 1969, issued by the National Herbarium of New South Wales, Sydney, as *Flora of New South Wales*, no. 48 Orchidaceae. 1969; with a suplplement by D.J. McGillivray, p. [i*-iv*], [i]-xv, [1]-176, [177]. *Copies*: BR, FI, HH, NY, PH. – Portrait of Rupp on p. [iv*]. – Taxon 19: 644. 1970.
Ref.: Woodward, C.H., J. New York Bot. Gard. 46: 70. 1945.

Ruppius, (Rupp), **Heinrich Bernard** (1688-1719), German (Thuringian) botanist. (*Ruppius*).

HERBARIUM and TYPES: Unknown.

BIBLIOGRAPHY and BIOGRAPHY: AG 1: 355; Backer p. 504 (epon.); Barnhart 3: 192 (d. 7 Mar 1719); BM 4: 1767; Bret. p. 79; Dryander 3: 29, 160; Frank 3 (Anh.): 87; GR p. 127 (b. 27 Aug 1688, from church registers Giessen); Hegi 1: 140, 5(1): 621; Herder p. 188; Hortus 3: 1203; Jackson p. 305; Kew 4: 549; LS 23248; Moebius p. 48, 57; PR

7913, ed. 1: 8879; Rehder 1: 372-373; TL-2/2307; Tucker 1: 613; Zander ed. 10, p. 710, ed. 11, p. 810.

BIOFILE: Baines, J.E., Vict. Natural. 90: 250. 1973 (epon.).
Drude, O., Hercyn. Florenbez. 5-6. 1902 (Veg. Erde 6).
Fries, Th.M., Bref Skrifv. Linné 1(4): 43. 1910.
Hulth, J.M., Bref Skrifv. Linné 2(1). 138, 139. 1916, and A.H. Uggla 2(2): 25, 26. 1943.
Jessen, K.F.W., Bot. Gegenw. Vorz. 268, 276, 277. 1864.
Oligschläger, F.W., Flora 16: 339. 1833.
Stafleu, F.A., Linnaeus and the Linnaeans 246. 1971.
Wein, K., Mitt. thüring. bot. Ver. ser. 2. 40: 42-58. 1931 (on R as a bryologist); updates nom. bryoph. *Fl. jen.*).
White, A. & B.L. Sloane, The Stapeliae 14. 1933, ed. 2. 78. 1937.

EPONYMY: *Ruppa* Cothenius (1790, *orth. var.*); *Ruppia* Linnaeus (1753).

9790. Henr. Bernh. Ruppii *Flora jenensis* sive enumeratio plantarum, tam sponte circa Jenam, & in locis vicinis nascentium, quam in hortis obviarum, methodo conveniente in classes distributa, figurisque rariorum aeneis ornata: in usum botanophilorum jenensium edita à Jo. Henr. Schutteo, ... cui accedit supplementum. Francofurti & Lipsiae (apud Ernestum Claud. Bailliar.) 1718. Oct. (*Fl. jen.*).
Ed. 1: 1718 (p. xii; 6 Jan 1718), p. [i-xii], table, [1]-376, [377-472 indexes], *pl. 1-3* (uncol. copp.). *Copies*: M, NY.
Ed. 2: 1726, p. [i-viii], table, [1]-311, [312-436]. *Copies*: G, H, MICH, NY, Teyler, USDA. "... botanophilorum jenensium edita multisque in locis correcta et aucta". Francofurti & Lipsiae (apud Ernestum Claud. Bailliar.) 1726. Oct.
Ed. 3: 1745 (p. [xvi], [sic not vi]: 2 Sep 1744), p. [i-xvi], chart, [1]-416, [1-22, index; 1, err.], *pl. 1-6* (uncol. copp. C.J. Rollinus). *Copies*: BR, G, HH, M, NY, Stevenson, USDA. – See also TL-2/2307, edited by A. von Haller (1708-1777), " Alberti Haller ... Flora jenensis Henrici Bernhardi Ruppii ex posthumis auctoris schedis et propriis observationibus aucta et emendata accesserunt plantarum rariorum novae icones. Jenae (sumptibus Christ. Henr. Cunonis) 1745. Oct.

Rupprecht, Johann Baptist (1776-1846), Austrian botanist and gardener in Vienna. (*Rupprecht*).

HERBARIUM and TYPES: Unknown (some "Rupprecht" material from Germany in BR). Kukkonen (1971) mentions two *Pisum* specimens in H (herb. Steven) "which are probably from this Ruprecht [sic]".

NOTE: For an obituary of the German micromycologist Heinrich Rupprecht (1890-1969) see H.J. Conert & D. Mollenhauer, Natur und Museum 99(5): 215-219. 1969 (coll. of fungi at FR; b. 21 Nov 1870, d. 1 Jan 1969).

BIBLIOGRAPHY and BIOGRAPHY: Barnhart 3: 192; BM 4: 1767, 8: 1112; Herder p. 305; Kew 4: 549; MW p. 420; PR 7928 (d. 15 Oct 1846), ed. 1: 8880.

BIOFILE: Anon., Bot. Zeit. 5: 120. 1847 (d.); Flora 19(1): 127-128. 1836 (garden exhibit of Chrys. ind. and potatoes), 20(2): 734. 1837 (list of Vitis, potato and Chrys. ind. cultivars publ. Wien 1837), 30(1): 114-115. 1847 (d. 14 Sep 1846).
Kukkonen, I., Herb. Christian Steven 87. 1971.

9791. Ueber das *Chrysanthemum indicum*, seine Geschichte, Bestimmung und Pflege. Ein botanisch-praktischer Versuch ... Wien (Gedruckt bey A. Strauss's sel. Witwe) 1833. Oct. (*Chrysanth. ind.*).
Publ.: 1833 (p. 8: Dec 1832; Hinrichs Verz. 30 Aug-5 Sep 1835 [sic]), p. [i], [1]-211. *Copies*: G, M, NY, REG (fide Hoppea 34: 308).
Ref.: Anon., Flora 17: 235. 21 Apr 1834 (presented at meeting 12 Mar 1834; brief review).

Ruprecht, Franz Josef (1814-1870) Austro-Bohemian botanist; studied medicine in Praha 1830-1836 (Carl-Ferdinand University); Dr. med. Praha 1838; practicing physician Praha 1838; curator of the botanical collections of the Academy of Sciences at St. Petersburg 1839 at the invitation of Trinius; from 1851-1855 connected with the St. Petersburg botanical garden; from 1855 director of the Botanical Museum of the Academy. (*Rupr.*).

HERBARIUM and TYPES: LE (coll. after 1839); other material at B(mainly destr.), BM, FH, FI, GOET, H, K, KAZ(?), KIEL, L, LY, MW, P, PR (Bohemian material coll. 1835-1838; duplicates from Russian coll.), S, TCD. – According to a note in Lotos (1852: 190) R. left his Bohemian herbarium to the University of Kasan (KAZ); this herbarium was brought together between 1830 and 1839. – Letters to D.F.L. von Schlechtendal at HAL.

NOTE: (1) According to Maximowicz (1871) in his biography of Ruprecht, R.'s father was a supply officer in the Austrian army, stationed at Praha, who moved around with the army during the Napoleonic years. Ruprecht was born in Freiburg in Breisgau 1 Nov 1814 and moved around with the family until the end of the war. Ruprecht Sr. then moved back to Praha and Franz Josef grew up and went to school in that city. – Ruprecht married the Baltic-German (Riga) Caroline Meinshausen (1847), sister of C.A. Meyer's wife. Ruprecht acquired Russian nationality after his marriage. Klášterský et al. (1970) state that Ruprecht was born at Praha.
(2) In some of his publications Ruprecht calls himself "Frans Joseph", in Russian he is also called "Franz Joseph Ivanovich".
(3) While still in Praha Ruprecht collected for Opiz and H.L. Reichenbach and worked on *Tentamen agrostographia universalis* (1838). Through Trinius he was appointed curator of the botanical collections of the St. Petersburg academy in 1839. In the following years Ruprecht travelled widely all over the territory of what is now the Soviet Union and published critical regional floras, e.g. the *Flora ingrica*. Ruprecht became an associate member of the St. Petersburg Academy in 1848, extraordinary member in 1853 and "ordinarius" in 1857. From 1851-1855 he was assistant director of the Imperial Botanic Garden, in 1855 he succeeded C.A. Meyer as director of the Botanical Museum of the Academy, a position which he held until his death. Ruprecht's scientific achievements cover a broad field: regional floras, phycology, phytogeography, geography and soil science (has made an extensive study for instance of the Tschornosjom).

BIBLIOGRAPHY and BIOGRAPHY: ADB 29: 748-753; AG 2(1): 263; Andrews p. 322; Backer p. 504 (epon.); Barnhart 3: 192; BM 4: 1602, 1767-1768; Bossert p. 342; Bret. p. 623; CSP 5: 339, 8: 799-800, 18: 362; De Toni 1: cxix; Frank 3(Anh.): 87; GR p. 438-439, cat. p. 70; Herder p. 473 [index]; Hortus 3: 1203; IF p. 728, suppl. 1: 85; IH 1(ed. 6): 363, (ed. 7): 341, 2: (files); Jackson p. 48, 328, 330, 392; Kew 4: 549; Klášterský p. 160; KR p. 608-609; Lasègue p. 340, 398, 503; LS 23249-23251; Maiwald p. 91, 121-123, 188; Moebius p. 88, 91; MW p. 392, 420-421; NI 1557, 1702, 1702n, 1704; PR 6178, 6209, 6869, 7280, 7914-7927, 9522, ed. 1: 6888, 8172, 8881-8885, 10485; Ratzeburg p. 448-449; Rehder 5: 746; RS p. 138; SBC p. 130; TL-1/1127-1128; TL-2/5909, 5978, 7128, see E. Fenzl; TR 1181-1215 (p. 219, 240-247); Tucker 1: 613-614; Urban-Berl. p. 390; Zander ed. 10, p. 710, ed. 11, p. 810.

BIOFILE: Anon., Bonplandia 9: 14. 1861 (visits Caucasus); Bot. Not. 1871: 36 (d.); Bot. Zeit. 5: 120. 1847 (d.), 28: 616. 23 Sep 1870 (d.); Flora 53: 490. 1871 (d.); J. Bot. 8: 366. 1870 (C.J. Maximowicz succeeds F.J. Ruprecht); Österr. bot. Z. 21: 62. 1871 (d.); Russk. biogr. slov. 17: 610-614. 1908.
Asmous, C.V., Chron. bot. 11(2): 108. 1947 (historical publ. on Bot. Mus. Acad. Nauk).
Candolle, Alph. de, Phytographie 445. 1880.
Cohn, F., Bot. Zeit. 18: 141, 142, 143. 1860.
Dolezal, H., Friedrich Welwitsch 36, 77, 214. 1974.
Fischer, F.E.L. von, *in* W.J. Hooker, Lond. J. Bot. 3: 271. 1844.
Harms, H., *in* I. Urban & P. Graebner, Festschrift Paul Ascherson 302-326. 1904 (generic nomenclature *Flora ingrica*).
Hooker, J.D., *in* W.J. Hooker, Lond. J. Bot. 5: 531-532. 1846.

Jessen, K.F.W., Bot. Gegenw. Vorzeit 375, 470. 1884.
Koster, J.Th., Taxon 18: 557. 1969 (algal coll.).
Lindemann, E. v., Bull. soc. imp. Natural. Moscou 61(1): 64. 1886.
Lipsky, V.I., Imp. S.-Petersb. bot. sada 3: 426-432. 1915 (biogr., bibl., portr.).
Maximowicz, C.J., Bull. Acad. Sci. St.-Pétersbourg 16, suppl. 1- 1-21. 1871 (obit., portr., bibl. by Trautvetter).
Opiz, P.M., Lotos 1852: 191 (has sold his plant collections to the University of Kasan).
Radde, G., Grundz. Pfl.-Verbr. Kaukasusländ. 5-6, 19. 1899 (Veg. Erde 3).
Regel, E.A., Gartenflora 9: 371-372. 1860 (arrived in Tiflis), 11: 44-45. 1862 (returned from trip Caucasus), 11: 234-235. 1862 (travel report Caucasus), 14: 220. 1865 (Caucasian primulas), 15: 88-89. 1866 (origin Tschornosjom/black earth).
Ricker, P.L., Proc. biol. Soc. Washington 21: 16. 1908 (dates Mémoires ... St. Petersb.).
Sayre, G., Bryologist 80: 516. 1977.
Shetler, S.G., Komarov bot. Inst. 235 [index]. 1967.
Stieber, M.T., Huntia 3(2): 124. 1979 (mss. notes on R. in Agnes Chase corr.).
Wagenitz, G., Index coll. princ. herb. Gött. 139. 1982.
Wittrock, V.B., Acta Horti Berg. 3(2): 176. 1903, 3(3): 189. *pl. 122.* 1905.
Wulff, E., Intr. hist. pl. geogr. 2, 3, 9, 23. 1943.

COMPOSITE WORKS: (1) A. Postels & F.J. Ruprecht, *Illustrationes algarum*, Petropoli 1840, see under Postels.
(2) with F. Osten-Saken, *Sert. tianschan.* 1869, TL-2/7128.

HANDWRITING: S. Lipschitz & I.T. Vasilczenko, Central Herbarium U.S.S.R. 121. 1968.

EPONYMY: *Ruprechtia* C.A. Meyer (1840); *Ruprechtiella* K. Yendo (1913); *Ruprechtiella* Kylin (1924).

9792. *Tentamen agrostographiae universalis* exhibens characteres ordinum, generumque dispositionem naturalem cum distributione geographica, adjectis tabulis analyticis. Auctore Dre. F.J. Ruprecht. Pragae [Praha] (typis filiorum Theophili Haase) 1838. Oct. (*Tent. agrostogr.*).
Publ.: Aug 1838 or somewhat later (p. [4]: Jul 1838; Ruprecht received the doctor's title on 1 Aug 1838), p. [i-iii], [1]-48. *Copy*: G.
Thesis issue: 1 Aug 1838 (t.p.) (n.v., but see Trautvetter 1871, p. 43, no. 3). Has thesis t.p. but lacks preface dedication and ordo iii, Saccharineae (n.v.).

9793. *Bambuseae* auctore F.J. Ruprecht. Ex Actis Acad. caes. petrop. ... Petropoli [St. Petersburg] (typis Academiae caesareae scientiarum) 1839. Qu. (*Bambuseae*).
Publ.: Sep-Dec 1839 (read 6 Sep 1839; t.p. preprint 1839; Flora 28 Dec 1840), p. [i], [1]-74, [75], *pl. 1-18* (uncol. liths., 1-17 by auct.) *Copies*: FI, NY, US. – Preprinted from Méd. Acad. Sci. St. Pétersbourg ser. 6. 5(2): 91-166, *pl. 1-18*. Feb 1840 (Ricker 1908). – *pl. 18* is a vegetation picture by A. Postels. – Title in journal publ.: Bambuseas monographice exponit ... accedunt tabulae 17 cum analysibus specierum".

9794. *Species graminum stipaceorum*. Auctoribus C.B. Trinius et F.J. Ruprecht. Ex Actis Academ. Imp. scient. Petrop. ... Petropoli [St. Petersburg] (typis Academiae imperialis scientiarum) 1842. Qu. (*Sp. gram. stipac.*).
Senior author: Carl Bernhard von Trinius (1778-1844).
Publ.: 1842 (submitted 10 Jan 1842; t.p. 1842), p. [i-ii], [1]-189. *Copies*: FI, G, MICH. – Preprinted from Mém. Acad. sci. St. Pétersbourg 7(2): 1-189. Mar 1843 (for date journal see Ricker, Proc. biol. Soc. Washington 21: 17. 1908. – Title in journal: "Gramina agrostidea, iii. Callus obconibus (Stipacea)".

9795. *Neue Beobachtung über Oscillaria*; von F.J. Ruprecht (Lu le 9 février 1844). [St. Pétersbourg 1844].
Publ.: 1844 or early 1845, p. [1]-2. *Copy*: FH. – Reprinted or preprinted from Bull. Acad. Sci. St. Pétersbourg, phys. math. 3(2): 29. 1845.

9796. *Flores samejedorum cisuralensium* [Beiträge zur Pflanzenkunde des Russischen Reiches

... Zweite Lieferung. St. Petersburg (Buchdruckerei der Kaiserlichen Academie der Wissenschaften) 1845]. Oct. (*Fl. samojed. cisural.*).
Publ.: Jun 1845 (date on p. [1], actually date of printing), p. [i-vii], [1]-67, *pl. 1-6* (uncol. liths. W. Pape, T.A. Satory. *Copies*: BR, FI, G, H, MO, NY, U, USDA. – See also TL-2/5909, under TR 1182.
Reprint: 1846 in *Symb. pl. Ross.* q.v. below.
Ref.: Gray, A., Amer. J. Sci. ser. 2. 3: 309. 1847 (rev.).

9797. *Distributio cryptogamarum vascularium in Imperio Rossico* auctore F.J. Ruprecht [Beiträge zur Pflanzenkunde des Russischen Reiches ... Dritte Lieferung. St. Petersburg (Buchdruckerei der Kaiserlichen Akademie der Wissenschaften) 1845]. Oct. (*Distr. crypt. vasc. Ross.*).
Publ.: Jul 1845 (p. 1; actually date of printing), p. [ii-iii], [1]-56. *Copies*: FI, H, MO, NY, U, USDA. – See also TL-2/5909; TR 1183.
Reprint: 1846, in *Symb. pl. Ross.* q.v. below.

9798. *In historiam stirpium florae petropolitanae diatribae* auctore F.J. Ruprecht. [Beiträge zur Pflanzenkunde des Russischen Reiches ... Vierte Leiferung, St. Petersburg (Buchdruckerei der Kaiserlichen Akademie der Wissenschaften) 1845]. Oct. (*Hist. stirp. fl. petrop.*).
Publ.: Nov 1845 (p. 1; actually date of printing), p. [ii-iii], [1]-93. *Copies*: FI, H, MO, NY, U, USDA. – See also TL-2/5909; TR 1184.
Reprint: 1846, in *Symb. pl. Ross.* q.v. below.

9799. *Symbolae ad historiam et geographiam plantarum Rossicarum* auctore F.J. Ruprecht. Petropoli [St. Petersburg], (typis et impensis Academiae imperialis scientiarum) 1846. Oct. (*Symb. pl. Ross.*).
Publ.: Nov-Dec 1846 (imprimatur Nov 1846), p. [i-vii], [1]-242, [1], *pl. 1-6. Copies*: G, H, HH, MO, NY, USDA. – A reprint of the three preceding items: *Fl. samajed. cisural.* p. [v-vii], 1-67; *Distr. crypt. vasc. Ross.* on p. 69-124; *Hist. stirp. fl. petrop.* on p. 125-217; added: *Curae posteriores* on p. 219-234, *index*, p. 235-242, and contents p. [1]. The first three items to be cited from the original publications, Jun, Jul and Oct 1845.

9800. *Bemerkungen über den Bau und das Wachsthum einiger grossen Algen-Stämme*, und über die Mittel, das Alter derselben zu bestimmen ... St. Petersburg (Typographie der Kaiserl. Akademie der Wissenschaften) 1848. Qu. (*Bemerk. Algen-Stämme*).
Publ.: Mar 1848 (as preprint; so dated), p. [1]-14. *pl. 6. Copy*: FH. – Preprinted from Mém. Acad. Sci. St. Pétersbourg 8(2): [59]-70. Mar 1849 (fide P.L. Ricker, Proc. Biol. Soc. Washington 21: 17. 1908).
Ref.: Schlechtendal, D.F.L. von, Bot. Zeit. 8: 261-263. 1850 (rev.).

9801. *Die Vegetation des rothen Meeres* und ihre Beziehung zu den allgemeinen Sätzen der Pflanzen-Geographie ... Aus den Mémoires ... Tome vi [sic for ser. 6, vol. 8], besonders abgedruckt. St. Petersburg (gedruckt bei der Kaiserlichen Akademie der Wissenschaften) 1849. Qu. (*Veg. roth. Meer.*).
Publ.: Mar 1849 (in journal, fide P.L. Ricker, Proc. biol. Soc. Washington 21: 17. 1908), p. [i], [1]-14. *Copies*: E (Univ. Libr.), FH, FI, G. – Reprinted or preprinted from Mém. Acad. Sci., St. Petersb. 8(2): 71-[84]. Mar 1849.
Ref.: Fürnrohr, A.E., Flora 32(2): 766-768. 28 Dec 1849 (rev.).

9802. *Ueber die Verbreitung der Pflanzen im nördlichen Ural*. Nach den Ergebnissen der geographischen Expedition im Jahre 1847 und 1848 [Beiträge zur Pflanzenkunde des Russischen Reiches. Siebente Lieferung. St. Petersburg (Buchdruckerei der Kaiserlichen Akademie der Wissenschaften) 1850 ...]. Oct. (*Verbr. Pfl. Ural*).
Publ.: Aug 1850 (imprimatur on p. 2 of cover), p. [ii-iii], [1]-84. *Copies*: FI, G, H, MO, NY, U, USDA. – Also published in part, in Bull. Acad. Sci. St. Pétersbourg, phys.-math. 8: 273-297. 1850 and Mélanges biol. 1: 74-107. 1853; see also TR 1188, Maximowicz-Trautvetter (1871) nos. 22 and 29. See also TL-2/5909.
Reprint with minor differences: see below, *Flora boreali-uralensis* 1854.
Ref.: Schlechtendal, D.F.L. von, Bot. Zeit. 8: 824-828. 15 Nov 1850 (rev.).

9803. *Tange des Ochotskischen Meeres.* Bearbeitet von Dr. F.J. Ruprecht [*in* A.T. von Middendorff, Reise Sibir. 1(2), Lief. 2. 1851]. Qu. (*Tange Ochotsk. Meer.*).
Author of main work: Alexander Theodor von Middendorff (1815-1894), see TL-2/5978 for his *Reise Sibir.*).
Publ.: Jan-Mai 1851 (rev. Bot. Zeit. 13 Jun 1851; Easter fair, Leipzig, fide Bot. Zeit. 9: 422), p. [191]-435, *pl. 9-18* (liths. printed in a single color; auct. del.). *Copies*: BR, G, NY. – Reprinted from A.T. von Middendorff, *Reise Sibir.* 1(2), Lief. 2: [191]-435, *pl. 9-18*. 1851. – TL-1/1127, TR 1189, TL-2/5978.
Facsimile repr.: 1978, Vaduz, Bibliotheca phycologica Band 39, ISBN 3-7682-1184-3, p. [i-iv], [193]-435, *pl. 9-18*. *Copies*: FAS, G. – *Phycologia ochotiensis*, "Tange des Ochotskischen Meeres", Vaduz (J. Cramer ...) 1978. – See Taxon 27: 424. 1978.
Note: Ruprecht presents a revised nomenclature of the marine algae treated by him based on extensive historical research. For a critique see A. Le Jolis, Mém. Soc. Sci. nat. Cherbourg 4: 65-84. 1856 [TL-2/4358] "Quelques remarques sur la nomenclature générique des algues".
Ref.: Schlechtendal, D.F.L. von, Bot. Zeit. 9: 444-447. 13 Jun 1851 (rev.; cites the reprint as of 1850).

9804. *Über das System der Rhodophyceae* von F.J. Ruprecht ... St. Petersburg (Buchdruckerei der Kaiserlichen Akademie der Wissenschaften) 1851. Qu. (*Syst. Rhodophyc.*).
Publ.: 1851 (t.p. preprint; read 14 Feb 1851), p. [1]-30, *1* lith. chart. *Copies*: FH, FI, G. – Preprinted from Mém. Acad. St. Petersb. ser. 6. 9 (Mém. Sci. nat. 7): [25]-54. Nov 1855 (date fide Ricker 1908).

9805. *Neue oder unvollständig bekannte Pflanzen aus dem nördlichen Theile des Stillen Oceans* ... St. Petersburg (Buchdruckerei der Kaiserlichen Akademie der Wissenschaften) 1852. Qu. (*Neu. Pfl. Still. Ocean*).
Publ.: 1852 (t.p. 1852: read 30 Jan 1852), p. [i], [1]-26, *pl. 1-8* (uncol. liths. Prüss, auct.). *Copies*: FH, G, NY. – Preprinted from Mém. Acad. Sci. St. Pétersbourg 9(2) (Mém. Sci. nat. 7): [55]-82, *pl. 1-8*. Nov 1855. (date fide Ricker (1908)).

9806. *Flora boreali-uralensis.* Ueber die Verbreitung der Pflanzen im nördlichen Ural. Nach den Ergebnissen der Ural-Expedition in den Jahren 1847-1848 ... St. Petersburg (Buchdruckerei der Kaiserlichen Akademie der Wissenschaften) 1854. Qu. (*Fl. bor.-ural.*).
Publ.: Mai 1854 (imprimatur 27 Mai 1854), p. [1]-49, [50], *3 pl.* (uncol. liths.). *Copies*: G, H, MO, PH. – A reprint with minor differences of *Verbr. Pfl. Ural* 1850 (see above). Preprinted from E. [von] Hofmann, *Der nördliche Ural* 2, 1856. See also TR 1188 and Maczimowicz-Trautvetter 1871, no. 29. The book by Hofmann was also published in Russian, ca. 1857. – Also Flora severnogo Urala 1854, 1856 (n.v.).

9807. *Revision der Umbelliferen aus Kamtschatka.* Von F.J. Ruprecht. [Beiträge zur Pflanzenkunde des Russischen Reiches ... Eilfte und letzte Lieferung ... St. Petersburg (Commissionäre der Kaiserlichen Akademie der Wissenschaften ...) 1859]. Oct. (*Revis. Umbell. Kamtschatka*).
Publ.: Nov 1859 (imprimatur Nov 1859), p. [ii-iv], [1]-30. *Copies*: MO, NY, U, USDA. – See also TL-2/5909, TR 1199. – Summary in Bull. phys. math. Acad. Sci. St. Pétersbourg 17: 106-108. 1859.

9808. *Bemerkungen über einige Arten der Gatttung Botrychium.* von F.J. Ruprecht [Beiträge zur Pflanzenkunde des Russischen Reiches ... Eilfte und letzte Lieferung ... St. Petersburg (Commissionäre der Kaiserlichen Akademie der Wissenschaften ...) 1859]. Oct. (*Bemerk. Botrychium*).
Publ.: Nov 1859 (imprimatur Nov 1859), p. [31]-43. *Copies*: MO, NY. – Reprinted from Beitr. Pfl.-K. Russ. Reiches 11: 31-43. 1859. – Summary in Bull. phys. math. Acad. St. Petersburg 17: 47-48. 1859. See also TL-2/5909; TR 1200.

9809. *Flora ingrica* sive historia plantarum gubernii petropolitani auctore F.J. Ruprecht, ... Vol. 1 ... Petropoli [St. Petersburg] (apud Eggers et soc.), Rigae [Riga] (apud Sam. Schmidt), Lipsiae [Leipzig] (apud Leopoldum Voss) 1860. Oct. †. (*Fl. ingr.*).

Publ.: Mai 1860 (imprimatur Mai 1860), p. [i]-xxvi, [1]-670. *Copies*: BR, G, H, HH, NY, USDA. – *Note*: the signatures are dated: 1: Jan 1853, 42: Mai 1856; the preface was printed Mai 1860. It is not clear whether the signatures were distributed on any scale individually or in groups before Mai 1860. The main chapters: Polyp. Thalam., p.1-240, ready Aug 1843; Polyp. Calyc., p. 241-476, by late 1854; Gamop. germ. inf., p. 477-670, by Mai 1856, were distributed among some friends and correspondents possibly shortly after printing according to Maximowicz (1871), p. 9. At the time this must have constituted "effective publication" (... "nachdem er einzelnen Freunden und Correspondenten auch schon früher die einzelnen Lieferungen mitgetheilt hatte ...). See also TR 1202 and Maximowicz-Trautvetter (1871) no. 49. For an extensive commentary on the generic nomenclature see Harms (1904); see also the Flora ingrica itself, ratio operis, p. i-xxvi: "Justitia nomenclaturae fundamentum; aliud non existit" goes back to pre-linnaean generic names.

Ref.: Harms, H.A.T., *in* I. Urban & P. Graebner, Festschrift Ascherson 302-326. 1904.

9810. *Decas plantarum amurensium* sive tabulae botanicae x ex itinerario D. Maack seorsum editae a F.J. Ruprecht. Petropoli [St. Petersburg] (Typis Academiae caesareae scientiarum) 1859. Fol. (*Dec. pl. amur.*).
Publ.: 19 Oct-31 Dec 1859 (p. [ii]: 19 Oct 1859), p. [i-iii], *10 pl.* (double, uncol. liths.). *Copy*: G. – Preprinted or reprinted from Richard Maack [1825-1886, see vol. 3, p. 206], Puteshestvie na Amur ... v 1855 ghodu [Journey to Amur ... in 1855], St. Petersburg 1859. – See also Maximowicz-Trautvetter (1871) no. 46.
Ref.: Anon., Bull. Soc. bot. France 7: 527-528. 1861 (summary; lists, plates).

9811. *Flora caucasi*. Pars i. Auctore F.J. Ruprecht ... St.-Pétersbourg (Commissionaires de l'Académie impériale des sciences ...) 1869. Qu. †. (*Fl. caucasi*).
Publ.: Dec 1869-Apr 1870 (imprimatur Dec 1869; t.p. 1869; présenté le 12 Décembre 1867 [sic]; Flora 20 Mai 1870; Bot. Zeit. 3 Jun 1870), p. [i-iv], [1]-302., *pl. 1-6* (uncol. liths. W. Pape). *Copies*: FI, G, H, M, MO, NSW, NY, U. – Issued as Mém. Acad. Sci. St.-Pétersbourg ser. 7. 15(2). 1869.
Ref.: Anon., Bull. Soc. bot. France 18 (bibl.): 181-182. Jul 1872 (rev.).
Kanitz, A., Bot. Zeit. 28: 514-518. 12 Aug 1870 (rev.).

Rusby, Henry Hurd (1855-1940), American botanist, physician and plant explorer; collected in New Mexico and Arizona before going to medical school; Dr. med. New York 1884; to South America for Parke-Davis Co. in search of cocaine in Bolivia and Peru and subsequently crossing the Amazon region from west to east 1884-1886; professor of botany and materia medica, New York College of Pharmacy (the later Dept. Pharmacy, Columbia Univ.) 1888-1905; dean of the faculty ib. 1905-1933; dean emeritus 1933-1940; one of the incorporators of the New York Botanical Garden and associated with it from its organization in 1898; curator of economic collections 1900-1911; honorary curator id. 1912-1940; member of the board of scientific directors 1900-1933; in later years travelling in Venezuela, Columbia, Mexico and again in Brazil, as well as throughout the United States; intrepid collector, "always ... an amazing story to tell, ... always in trouble". (*Rusby*).

HERBARIUM and TYPES: Rusby's pre-1884 herbarium (esp. Arizona and New Mexico) was sold to Parke-Davis Co., Detroit and is now at MICH; the trip to South America 1884-1886 was also for Parke-Davis and the first set is also at MICH, but 6000 specimens went to NY; a special collection of economic plants is at NY and so are the first sets of Rusby's later trips, including that of the Mulford Biological Expedition to the Amazon Basin (1931-1932), incl. note books. Further material at: A, AMES, B (mainly destr.), BM, BKL, BOL, BUF, C, CM, CU, CGE, CORD, DPU, E, ECON, F, G, GH, K, LA, LE, M, MANCH, MICH, MIN, MO, MSC, NY, PH, S, TRIN, U, US, W, WELC, WRSL, Z and other herbaria. – Archival material in FH, NY, PH and SIA.

BIBLIOGRAPHY and BIOGRAPHY: Backer p. 660; Barnhart 3: 192; BJI 2: 150; BL 1: 238, 316; BM 4: 1768; Bossert p. 342 (b. 26 Apr 1855, d. 18 Nov 1940); CSP 12: 640, 18: 362; DAB suppl. 2: 590-592 (F. Hart); De Toni 4: xlix; Ewan ed. 1: 294, ed. 2: 189-190; GR p. 241; Hortus 3: 1203; IH 1 (ed. 6): 363, (ed. 7): 341, 2: (files); Kelly p. 202; Kew 4:

550; Langman p. 657-658; Lenley p. 356; LS 23252, 23253, 38654; NW p. 56; PH p. 522 (index); Rehder 5: 746; RS p. 138; TL-2/773, 781, see M. Bang, P.D. Knieskern; Tucker 1: 614; Urban-Berl. p. 312, 390; Zander ed. 10, p. 710, ed. 11, p. 810.

BIOFILE: Anon., Bot. Centralbl. 39: 367. 1889 (app. New York Coll. Pharm.); Bot. Jahrb. 11 (Beibl. 25): 9. 1889 (app. N.Y. Coll. Pharm.); J. Amer. pharm. Ass. 4(3):265-266. 1915 (portr. and caption); J. New York Bot. Gard. 41: 277. 1940; New York Herald Tribune 6 Sep 1925 (portr., sketch of life and work); New York Sun 1 May 1929 (proposed for Hanbury Medal), 9 May 1929 (rd. Hanbury Gold Award); New York Telegram 23 May 1929 (Rusby brings editor of "Time" magazine to court on Russian and Spanish ergot).
Ballard, C.W., Science ser. 2. 93: 53-54. 1941 (obit.).
Bender, G.A., Great moments in pharmacy 158-164. 1966 ("wresting the jungle's secrets"); Pharm. Hist. 23: 71-85. 1981.
Bonisteel, Wm. J., Torreya 41: 14-15. 1941 (obit.).
Cattell, J.M. & J., Amer. men Sci. 275. 1906, ed. 2: 404. 1910, ed. 5: 961-962. 1933, ed. 6: 1220. 1938.
Ewan, J. et al., Short hist. bot. U.S. 12, 17, 99, 150. 1969.
Hastings, G.T., Torreya 34: 17. 1934 (rev. Jungle memories).
Herzog, Th., Pflanzenwelt Boliv. Anden 2, 19. 1923 (Veg. Erde 15).
Kibby, A.L., Afield with plant lovers and collectors 387-390. 1953 (corr. H.N. Patterson) (fide Ewan).
Leonard, J.W., ed., Who's who in America 1899-1900: 623, 19: 2119. 1934, 21: 2242. 1940, 22: 1902. 1942.
Lipp, F., Garden J. 24: 70-75. 1974 (portr., "one of the giants").
MacCreagh, G., White waters and black, New York, Grosset & Dunlop, The Century Company [1926], xiv, 404 p., 62 pl., map, reprinted 1961, New York, Doubleday, xvii, 335 p., 8 pl. (semi-scurrilous tongue-in-cheek description of an expedition led by Rusby although neither he nor any member of the party is mentioned by name. Half-serious. The "eminent botanist" was O.E. White; the "eminent leader" H.H. Rusby).
Moldenke H.N., J. New York Bot. Gard. 42: 38. 1941 (6000 Bolivian coll. at NY); Plant Life 2: 80. 1946 (1948); Chron. bot. 7: 230-231. 1943 (extensive obit.).
Murrill, W.A., Hist. found. bot. Florida 19-20. 1945 ("He always had an amazing story to tell and was always in trouble").
Pennell, F.W. & J.H. Barnhart, Monogr. Acad. nat. Sci. Philadelphia 1: 618. 1935.
Prance, G., Acta amaz. 1(1): 57. 1971.
Rickett, H.W., Index Bull. Torrey bot. Cl. 84-85. 1955.
Rodgers, A.D., Amer. bot. 1873-1892, p. 336 [index]. 1944.
Rusby, H.H., Jungle memories, New York and London, Whittlesey House, 1933, xiii, 388 p., *16 pl.* (autobiogr.).
Sayre, G., Mem. New York Bot. Gard. 19(3): 390. 1975 (colls.).
Verdoorn, F., ed., Chron. bot. 1: 280. 1935, 3: 311. 1937, 4: 278, 565. 1938, 6: 234, 237. 1941, 7: 226. 1943.
Woodward, C.H., J. New York Bot. Gard. 42: 43-45. 1941 (obit., portr.).

COMPOSITE WORKS: (1) With Thomas Morong et al., *List Pter. Sperm. N.E. North America* 1893-1894, see TL2-6342.
(2) *Phyllonomaceae, in* NAF 22(2): 191. 18 Dec 1905.

NOTE: (1) William Sanford Rusby (b. 1852), brother of H.H. Rusby, collected widely in the United States from 1873 until 1883 or later. His main collections went to Parke, Davis & Co., the herbarium of which is now at MICH (inf. H.H. Rusby to J.H. Barnhart, see Barnhart 2: 192).
(2) Henry H. Rusby collected for Parke, Davis & Co., Detroit in South America (search for cocaine), in Bolivia and Para and Terr. Rondonia (Brazil) 1885-1886. The first set of these collections went to Parke, Davis & Co., and is now at MICH. A second visit to Bolivia and Brazil took place on 1921-1922 (on the Mulford Biological Expedition); the first set of this trip is at NY. – Rusby travelled in the Orinoco valley and other parts of Venezuela in 1893 and 1896; in Mexico 1908-1910, Columbia 1917.
(3) See under P.D. Knieskern for Rusby's role in securing the Knieskern herbarium.

EPONYMY: *Rusbya* N.L. Britton (1893); *Rusbyanthus* Gilg (1895); *Rusbyella* Rolfe ex H.H. Rusby (1896).

9812. *Plants of New Mexico and Arizona* ... Franklin, N.Y. [1881]. (*Pl. New Mexico Arizona*).
Publ.: 1881, p. [1]-4. *Copy*: NY.

9813. *An enumeration of the plants collected by Dr. H.H. Rusby in South America*, 1885-'86. I. General features of the region traversed. Enumeration of the Thallophyta. (Reprinted from the Bulletin of the Torrey Botanical Club, vol. xv, no. 7. [1888]. Oct.
Authorship: nos. 1, 24-32 by Rusby, no. 2 by Elizabeth Gertrude Britton (1858-1934) et al. (see below), no. 3 by E.G. Britton; the remainder by Nathaniel Lord Britton (1859-1934).
Publ.: A series of papers, by various authors, and reprinted as Contributions from the Herbarium of Columbia College no. 6.

number	Bull. Torrey bot. Club		reprint	author
1	15: 177-184.	2 Jul 1888	[1]-8	H. H. Rusby
2	23: 471-499.	28 Dec 1896	[9]-28, 28a-i	E. G. Britton et al.
3	15: 247-253.	3 Oct 1888	29-35	E. G. Britton
4	16: 13-20.	12 Jan 1889	37-44	N. L. Britton
5	16: 61-64.	8 Mar 1889	45-48	N. L. Britton
6	16: 163-160.	8 Jun 1889	49-56	N. L. Britton
7	16: 189-192.	6 Jul 1889	57-60	N. L. Britton
8	16: 259-262.	5 Oct 1889	61-64	N. L. Britton
9	16: 324-327.	10 Dec 1889	65-68	N. L. Britton
10	17: 9-12.	15 Jan 1890	69-72	N. L. Britton
11	17: 53-60.	10 Mar 1890	73-80	N. L. Britton
12	17: 91-94.	10 Apr 1890	81-84	N. L. Britton
13	17: 211-214.	12 Aug 1890	85-88	N. L. Britton
14	17: 281-284.	9 Nov 1890	89-92	N. L. Britton
15	18: 35-38.	12 Feb 1891	93-96	N. L. Britton
16	18: 107-110.	4 Apr 1891	97-100	N. L. Britton
17	18: 261-264.	21 Sep 1891	101-104	N. L. Britton
18	18: 331-334.	15 Nov 1891	105-108	N. L. Britton
19	19: 1-4.	15 Jun 1892	109-112	N. L. Britton
20	19: 148-151.	5 Mai 1892	113-116	N. L. Britton
21	19: 263-266.	10 Sep 1892	117-120	N. L. Britton
22	19: 371-374	15 Dec 1892	121-124	N. L. Britton
23	20: 117-140	10 Apr 1893	125-128	N. L. Britton
24	25: 495-500.	10 Sep 1898	129-134	H. H. Rusby
25	25: 542-545.	15 Oct 1898	135-138	H. H. Rusby
26	26: 145-158.	18 Mar 1899	139-152	H. H. Rusby
27	26: 189-200.	10 Apr 1899	153-164	H. H. Rusby
28	27: 22-31.	22 Jan 1900	165-174	H. H. Rusby
29	27: 69-84.	17 Feb 1900	175-190	H. H. Rusby
30	27: 124-137.	24 Mar 1900	191-204	H. H. Rusby
31	28: 301-313.	21 Mai 1900	205-217	H. H. Rusby
32	29: 694-704.	30 Dec 1902	219-229	H. H. Rusby

Numbers 1-23 re-issued together in a single volume with a special cover: Contributions from the Herbarium of Columbia College. – No. 6. An Enumeration ... i-xxiii. By N.L. Britton. An account of the region traversed by Dr. Rusby; the algae and fungi determined by Prof. W.G. Farlow, the lichens by Dr. J.W. Eckfeldt, the musci and pteridophyta by Elizabeth G. Britton, the hepaticae by Dr. Richard Spruce. (Reprinted from the Bulletin of the Torrey botanical Club 1886-1896), p. [1]-128. *Copy*: US.

9814. *Revised names of plants of New Jersey*, extracted from Britton's State Catalogue [reprinted from the Druggists' Bulletin, July, 1890.

Publ.: Jul 1890, p. [1]-23. *Copy*: USDA. – Reprinted from Drugg. Bull., Detroit, 4: 219-229. Jul 1890.

9815. *Botanical collecting in the tropical Andes.* [Reprinted from Bulletin of Pharmacy, April 1891].
Publ.: Apr 1891, p. [1]-20. *Copies*: NY, U. – Reprinted and to be cited from Bull. Pharm. 5(4): 157-163. Apr 1891.

9816. *The botanical names of the U.S. pharmacopoeia.* [Reprinted from the Bulletin of Pharmacy, July 1892].
Publ.: Jul 1892, p. [1]-8. *Copy*: G. – Reprinted and to be cited from Bull. Pharm. 6(7): 305-312. Jul 1892. – Also in Proc. Amer. Pharm. Ass. 40: 204-217. 1892.

9817. *Botany at the Rochester meeting* of the American Association for the Advancement of Science. [Reprinted from the Bulletin of Pharmacy, December 1892].
Publ.: Dec 1892, p. [1]-23. *Copy*: G. Reprinted from Bull. Pharm. 6(12): 652-658. Dec 1892. – Contains a report by Rusby on the basic proposals put forward by American botanists on 19 Aug 1892 at the Rochester meeting of the A.A.A.S. in reply to O. Kuntze's Rev. gen. pl. of 1891 and the proposals made in 1892 by P. Ascherson, A. Engler, K. Schumann and I. Urban. The German proposals were:
1. starting point generic names 1752/1753.
2. nomina nuda and semi-nuda to be rejected.
3. similar names differing in the last syllable to be retained.
4. proposal to set up a list of nomina conservanda.

The Rochester answer (proposed by H.H. Rusby, N.L. Britton, J.M. Coulter, F.W. Coville, L.M. Underwood, L.F. Ward, and W.A. Kellerman) accepted points 1-3 in its own rules affirming the 1867 Paris code with the following restrictions:
1. priority is fundamental.
2. starting point 1753, Linnaeus, *Species plantarum*.
3. specific epithets to be retained on transfer (negation of the Kew Rule).
4. absolute homonym rule; later homonym always to be rejected.
5. publication of generic names by (1) distr. printed matter of description; (2) by citation of the name with citation of included species with or without diagnosis.
6: publication of specific names by (1) distr. as in 5, (2) publishing a binomial with reference to a previously publ. species.
7. similar generic names do not differ necessarily when ending differs; they are to be rejected if they differ slightly in the spelling.
8. citation in parenthesis of author of epithet in original combination.

9818. *Parke, Davis & Co.*, manufacturing chemists. Detroit, Mich., U.S.A. *Herbarium department* ... H.H. Rusby Curator [Detroit 1880-1881?]. (*Parke, Davis herb. dept.*).
Publ.: 1880 or 1881, p. [1]-7. *Copies*: HH, NY, USDA. – Updated list of New Mexico and Arizona material for exchange.

9819. *An enumeration of the plants collected in Bolivia by Miguel Bang*, with descriptions of new genera and species. By Henry H. Rusby ... [Mem. Torrey bot. Club 3(3). 28 Apr 1893].
Collector: Miguel Bang. (*fl.* 1883).
[*1*]: 28 Apr 1893,p [1]-67. Mem. Torrey bot. Cl. 3(3): [1]-67. 1893. – "On the collections of Mr. Miguel Bang in Bolivia".
[*2*]: 27 Apr 1895, p. [203]-274. Mem. Torrey bot. Cl. 4(3): 203-274. 1895.
[*3*]: 17 Nov 1896, p. [1]-130. Mem. Torrey bot. Cl. 6(1): 1-130. 1896.
[*4*]: 5 Sep 1907, p. 309-470. Bull. New York Bot. Gard. 4(14): 309-470. 1907.
Copies: BR, G, HH, M, MICH, NY.

9820. *New genera of plants from Bolivia* ... (Reprinted from the Bulletin of the Torrey Botanical Club, ... 1893]. Oct.
Publ.: Nov 1893 (in journal), cover-t.p., p. 429-434, *pl. 167-170*. *Copy*: US. – Contr. Herb. Columbia Coll. 40, reprinted and to be cited from Bull. Torrey bot. Club 20(11): 429-434, *pl. 167-170*. Nov 1893.

9821. *Essentials of vegetable pharmacognosy* a treatise on structural botany. – Designed especially for pharmaceutical and medical students, pharmacists and physicians ... New York (D.O. Haynes & Co.) 1895. Oct. (*Essent. veg. pharmacogn.*).
Co-author: (part 2, minute structure of plants): Smith Ely Jelliffe (1866-1945).
Publ.: 1895, p. [i-vi], [1]-149, [1, err.]. *Copies*: USDA(2).

9822. *Morphology and histology of plants* designed especially as a guide to plant-analysis and classification, and as an introduction to pharmacognosy and vegetable physiology ... New York (published by the authors) 1899. (*Morph. histol. pl.*).
Co-author: (histology, p. 245-355): Smith Ely Jelliffe (1866-1945).
Publ.: 1899 (p. iv: Sep 1899; Bot. Zeit. 16 Aug 1900), p. [i]-xi, [1]-378. *Copies*: NY, USDA.

9823. *A manual of structural botany* an introductory text-book for students of science and pharmacy ... Philadelphia and New York (Lea & Febiger) 1911. Oct. (*Man. struct. bot.*).
Publ.: 1911, p. [iii]-viii, 17 [sic]-248. *Copy*: MO. – Also with an English imprint, London (J. & A. Churchill).
Ref.: Boulger, G.S., J. Bot. 50: 202-203. Jan 1912 (rev.).

9824. *Descriptions of three hundred new species of South American plants*, with an index to previously published South American species by the same author. By H.H. Rusby, M.D. Published by the author ... New York, N.Y. ... 1920. Oct. (*Descr. S. Amer. pl.*).
Publ.: 20 Dec 1920 (PH rd. 17 Feb 1921), p. [1]-170. *Copies*: HH, MICH, NY, PH, US, USDA.

9825. *Jungle memories* ... New York and London (Whittlesey House McGraw-Hill Book Company, Inc.) 1933. Oct. (*Jungle mem.*).
Publ.: 1933 (p. viii: Jun 1933), p. [i]-xiii, [1]-388, *pl. 1-16*. *Copies*: MICH, MO.

Russ, Georg Philipp (*fl.* 1868), German botanist; high school teacher at Hanau. (*Russ*).

HERBARIUM and TYPES: Unknown.

9826. *Flora der Gefäss-Pflanzen der Wetterau.* Zum Gebrauch auf botanischen Excursionen ... 1. Lieferung [Hanau (Druck der Waisenhaus-Buchdruckerei) 1868]. Oct.
Publ.: 1868, p. [i], [1]-120. *Copies*: LC, M. – Issued as Ber. Wetterau. Ges. Nat. Hanau 1863-1867, Abh. [6]. 1868.

Russegger, Joseph von (1802-1863), Austrian traveller; trained as a mining engineer in Schemnitz (now Banska Stiavnica); in Austrian government service 1825; general manager at Böckstein nr Gastein 1827-1835; travelled in the Near East and Europe 1835-1841; mining director at Wieliczka 1841; at Hull 1843; director of the mining academy and of the lower Austrian mining industry at Schemnitz 1850-1863; ennobled (Ritter) 1853. (*Russegger*).

HERBARIUM and TYPES: The plants of Russegger's trips to the Near East between 1835 and 1841 were collected by T. Kotschy and are at W. See TL-2/1765.

BIBLIOGRAPHY and BIOGRAPHY: BM 4: 1768-1769; CSP 5: 342-343, 8: 800; Herder p. 58; Lasègue p. 383, 502; NI (alph.); PR 6869; Quenstedt p. 372; Rehder 1: 480; TL-2/1765, 3887, 3895.

BIOFILE: Anon., Alman. k. Akad. Wiss., Wien 2: 196-201. 1852 (bibl., list of functions and honors) see also P. Wolf 1864; Bot. Zeit. 1: 590. 1843 (app. Hull).
Embacher, F., Lexik. Reisen 255. 1882.
Jessen, K.F.W., Bot. Gegenw. Vorz. 471. 1884.
Nissen, C., Zool. Buchill. 354. 1969.
Poggendorff, J.C., Biogr.-lit. Handw.-Buch 2: 724-725. 1863 (biogr. sketch, bibl., b. 18 Nov 1802, Salzburg).

Wolf, P., Alman. k. Akad. Wiss., Wien 14: 158-163. 1864 (bibl.).
Zittel, K.A. von, Gesch. Geol. Paläont. 421, 558, 559. 1899.

COMPOSITE WORKS: See E. Fenzl and C.G.F. Kotschy, *Abb. Thiere Pfl. Syr.* 1843, TL-2/1765 and 3887 as well as *Reliq. Kotschy*. 1868, TL-2/3895 and sub G.A. Schweinfurth.

EPONYMY: *Russeggera* Endlicher et Fenzl (1839).

9827. *Reisen in Europa, Asien und Afrika*, mit besonderer Rücksicht auf die naturwissenschaftlichen Verhältnisse der betreffenden Länder, unternommen in den Jahren 1835 bis 1841 ... Stuttgart (E. Schweizerbart'sche Verlagshandlung) 1841-1849, 4 vols. text. Oct. and atlas Fol. (*Reisen*).

1(1): 1841 (p. 6: 20 Aug 1841), p. [i], [1]-469, [470]. – Alt. title: "*Reise in Griechenland, Unteregypten, im nördlichen Syrien und südöstlichen Kleinasien* ...
1(2): 1843, p. [471]-1102. *Botanik* by E. Fenzl, on p. 881-970. For a reprint of the botanical part see TL-2/1765 and 3887. – Alt. t.p.: as in 1(1), dated 1843.
2(1): 1843, p. [i, iii], [1]-635, [636].
2(2): 1844 (p. 6: 12 Mar 1843), p. [1]-778.
2(3): 1849, p. [i, iii], [1]-360, [1-6, index, also to plants in 1(2)].
3: 1847 (p. 8: 12 Jun 1846), p. [1]-291, [292].
4: 1848(-1849), p. [1]-758, [1-2 note by E. Fenzl dated 1 Apr 1849].
Atlas: 73 pl., 19 maps.
Copies: G, PH, Teyler.

Russell, Alexander (ca. 1715-1768), British (Scottish) physician and naturalist, MD Glasgow; physician to the English factory at Aleppo, Syria, ca. 1740-1753; to St. Thomas's Hospital London 1759-1768. (*Al. Russell*).

HERBARIUM and TYPES: BM; some further material at MO, S. See also V. Allorge (1968). – The material at BM contains Russell's Aleppo plants (originally in herb. Banks), and the types of Banks and Solander of their text in Nat. hist. Aleppo ed. 2. 2: 237-271. 1794 (Murray 1904). Some Aleppo material was collected by Alexander Russell's half-brother Patrick [1727-1805].

BIBLIOGRAPHY and BIOGRAPHY: Backer p. 504 (epon.); Barnhart 3: 193; BB p. 264; BM 4: 1769; Desmond pp. 534 (d. 28 Nov 1768); DNB 49: 426-427; Dryander 3: 472; Frank 3 (Anh.): 87; Hegi 5(2): 1032; Herder p. 58; Jackson p. 510; Kew 4: 551; Lasègue p. 131, 137; NI 583, see also 1: 110; PR, ed. 1: 886; Rehder 1: 479; Tucker 1: 614.

BIOFILE: Allorge, V., Rev. bryol. lichénol. 35: 406. 1968 (in bryol. herb. Raymond Benoist, at PC, historical collections by "Alexander Russell" sic. It is not clear who is meant; this may have been e.g. Alexander Russell 1814-1878, Scottish naturalist).
Anon., Gentleman's Mag. 1771 (March): 109-111 (obit.).
Boissier, E., Fl. Orient 1: xxiii. 1867.
Fothergill, J., *in* J.C. Lettsom, Memoir of John Fothergill 1786, p. 241-259 (obit.).
Fox, R.H., Dr, John Fothergill and his friends 118-120. 1919.
Fries, Th.M., Bref Skrifv. Linné 1(5): 40. 1911.
Hulth, J.M. & A.H. Uggla, Bref Skrifv. Linné 2(2): 168. 1943.
Munk, W., The roll of the Royal College of Physicians of London ed. 2. 2: 230-231. 1878 (brief sketch).
Murray, G., Hist. coll. BM(NH) 1: 179. 1904 (coll. BM).
Nissen, C., Zool. Buchill. 355. 1969.
Poggendorff, J.C., Biogr.-Lit. Handw.-Buch 2: 725. 1863.
Smith, E., Life of Joseph Banks 116. 1911.

EPONYMY: *Russela* Cothenius (1790, *orth. var.* of *Russelia* Linnaeus f. (1782)); *Russelia* [sic] Linnaeus f. (1782); *Note*: *Russelia* N.J. Jacquin (1760) probably also honors Alexander Russell.

9828. *The natural history of Aleppo*, and parts adjacent. Containing a description of the city, and the principal natural productions in its neighbourhood; together with an account of the climate, inhabitants and diseases; particularly of the plague, with the methods used by the europeans for their preservation. By Alex. Russell, M.D. London (printed for A. Millar, in the Strand) mdccclvi [sic, should be 1756]. Qu. (*Nat. hist. Aleppo*).

Ed. [*1*]: 1756 (NZgS 3 Feb 1757), p. [i]-viii, [1]-266, [1-9, index; 10 err., *pl. 1-16* (uncol. copp. G.D.E. Ehret, J.S. Miller). *Copies*: FI, G, Teyler. – Plants on p. 30-46.

Dutch translation: 1762, p. i-vi, 1-294, *pl. 1-8. Naauwkeurige en natuurlijke beschrijving van de stad Aleppo* ... uyt het Engels vert. door L.T. Gronovius. Leiden (Corn. de Pecker & Corn. van Hoogeveen jr.). 1762. Qu. (*Naauwk. beschr. Aleppo*).

Ed. 2: 1794, in 2 vols. *The natural history of Aleppo.* Containing a description of the city, and the principal natural productions in its neighbourhood. Together with an account of the climate, inhabitants, and diseases; particularly of the plague ... The second edition. Revised, enlarged, and illustrated with notes. By Pat. Russel [Patrick Russell (1727-1805)] 2 vols. London (Printed for G.G. and J. Robinson, ...) 1794. Qu. *Copies*: G, PH, USDA.

1: frontisp., p. [iii]-xxiv, [1]-446, [i]-xxiii, [xxiv, err.], *pl. 1-4*.
2: p. [i]-vii, [1]-430, i-xxxiv, [1-25, index; 26, err.], *pl. 1/2, 3-16* (uncol. copp.,id., nos. *9-16* plants). Text on plants (J. Banks and D.C. Solander) on p. 237-271.

German translation of ed. 2: 1797-1798, 2 vols. *Naturgeschichte von Aleppo*, enthaltend eine Beschreibung der Stadt, und der vornehmsten Naturerzeugnisse in ihrer Nachbarschaft, zugleich mit einer Nachricht von dem Himmelsstriche, den Einwohnern, und ihren Krankheiten, insbesondere der Pest; von Alexander Russell ... Zwote Ausgabe. Durchgesehen vermehrt und mit Anmerkungen erläutert von Patrick Russell, ... Uebersezt [sic] mit einigen Anmerkungen von Johann Friedrich Gmelin [1748-1804], ... Göttingen (bei Johann Georg Rosenbusch) 1797-1798, 2 vols. Oct. (*Naturgesch. Aleppo*). *Copy*: USDA.

1: 1797 (p. xxviii: 17 Apr 1797), p. [i]-xxxii, [1]-440, [1]-176, map.
2: 1798, *Beschreibung der Thiere und Gewächse* p. [i-iii], [1]-280. *Nachricht von dem Zustande der Gelehrsamkeit*, p. [iii]-vi, [1]-242. – Botany on p. 139-197, *pl.9-16* of *Beschreibung*.

Note: see A.B. Rendle (1900) and A. Eig (1937), cited under Patrick Russell, below, for notes on ed. 2.

Russell, Anna (née Worsley) (1807-1876), British (English) mycologist at Kenilworth; married Frederick Russell. (*An. Russell*).

HERBARIUM and TYPES: Some material at OXF. – 733 original water-colour drawings of higher fungi at BM.

BIBLIOGRAPHY and BIOGRAPHY: Barnhart 3: 193; BB p. 264-265; BL 2: 229, 711; BM 4: 1769; Clokie p. 236; CSP 8: 800; Desmond p. 534 (b. Nov 1807, d. 11 Nov 1876); Jackson p. 257; LS 23255; NI 2: 267; Rehder 1: 399, 5: xx; Tucker 1: 614.

BIOFILE: Anon., J. Bot. 15: 32. 1877 (obit.).
Bagnall, J.E., Fl. Warwickshire 505. 1891.
Bridson, G.D.R. et al., Nat. hist. mss. res. Brit. Isl. 229.527. 1980.
Druce, G.C., Fl. Berkshire clxvii-clxviii. 1897.
Pearsall, W.H., Bot. Soc. Exch. Club Brit. Isl., 10(1) (Rep. 1932): 287. 1933.
Riddelsdell, H.J., Fl. Gloucestershire cxxi. 1948.
Watson, H.C., Topogr. Bot. ed. 2. 555. 1883.
White, J.W., Fl. Bristol 78-79. 1912.
Woodward, B.B., in G. Murray, Hist. Coll. BM(NH) 1: 47. 1904 (730 drawings bequeathed 1876; 5 presented 1886).

COMPOSITE WORKS: *A catalogue of the plants in the neighbourhood of Newbury*, in E.W. Gray, ed., The history and antiquities of Newbury ... Speenhamland 1839, p. 310-340. Reprinted 31 p., 1839. See also An. Russell, Phytologist 3: 716. 1849.

Russell, John sixth duke of Bedford (1766-1839), British nobleman and maecenas residing at Woburn Abbey, Bedfordshire; succeeded to the dukedom 1802; lord-lieutenant of Ireland 1806-1807; patron of George Gardner (1812-1849); published the Woburn Abbey series of botanical works. (*J. Russell*).

HERBARIUM and TYPES: Unknown. – Letters at K, LINN.

BIBLIOGRAPHY and BIOGRAPHY: AG 4: 72; Barnhart 3: 194; BB p. 265; BM 4: 1769-1770; Bossert p. 342; Dawson p. 723-724; Desmond p. 535 (b. 6 Jul 1766, d. 20 Oct 1839); DNB 49: 454; NI (alph.); PR 2959-2962, alph., ed. 1: alph., 3265-3268.

BIOFILE: Allan, M., The Hookers of Kew 105, 109. 1967.
Bridson, G.D.R. et al., Nat. hist. mss. res. Brit. Isl. 255.62, 277.5. 1980.
Burdet, H.M., Musées de Genève 153: 9-14. 1975 (portr.).
Dawson, W.J., Smith papers 84. 1934 (letters to J.E. Smith).
Hooker, W.J., Copy of a letter addressed to Dawson Turner, ... on the occasion of the death of the late Duke of Bedford: particularly in reference to the services rendered by his Grace to botany and horticulture. Glasgow (printed by George Richardson ...) 1840. Qu., 25 p., plate Bedfordia salicina (*copies*: HH, LC).
Nelmes & Cuthbertson, Curtis's Bot. Mag. Dedic. 1827-1829, p. 19-20 (portr.).
Pritzel, G.A., Linnaea 19: 460-461. 1846.

COMPOSITE WORKS: Published and wrote introductions to the volumes of the "Woburn Abbey series", see James Forbes, TL-2/1815 (*Salict. woburn.*), 1816 (*Hort. woburn.*), 1817 (*Pinet. woburn.*), and under G. Sinclair for *Hortus gramineus woburnensis* and *Hortus ericaeus*. For notes on the series see e.g. H.M. Burdet (1975).

EPONYMY: *Bedfordia* A.P. de Candolle (1833). *Note: Russellia* J.B. Risatti (1973) honors E.E. Russell, American geologist; *Russellites* S.H. Mamay (1968) was named for Mr. and Mrs. Mart Russell of Weymour, Texas, U.S.A.; *Russellodendron* N.L. Britton and J.N. Rose (1930) was probably named for Paul G. Russell (1889-1963), American botanist who collected with Rose.

Russell, John Lewis (1808-1873), American botanist; Unitarian clergyman and botanist at Salem, Mass.; A.B. Harvard 1828, A.M. id. 1836; Div. school id. 1831; vicar at Chelmsford, Hingham, Brattleboro, Kennebunk (Mass.); from 1853 again at Salem devoting himself to cryptogamic botany and writing of popular articles on natural history in addition to occasional preaching. (*J.L. Russell*).

HERBARIUM and TYPES: B (mainly destr), BUF, FH, NEBC (lich.), VT.

BIBLIOGRAPHY and BIOGRAPHY: Bossert p. 342 (b. 2 Dec 1808, d. 7 Jun 1873); GR p. 196; GR, cat. p. 70; Herder p. 18; Lenley p. 468; ME 1: 226, 3: 648 [index]; SBC p. 130; Urban-Berl. p. 290.

BIOFILE: Anon., Bryologist 82: 511. 1979 (portr.); Bull. Essex Inst. 5: 51, 103-104, 108-109. 1874; Meehan's Monthly 3: 78-79. 1903; Proc. Amer. Acad. Arts Sci. 9: 321. 1874 (obit.).
Ewan, J. et al., Short hist. bot. U.S. 92. 1969.
Graustein, J.E., Thomas Nuttall 224, 225, 226, 249, 334. 1967.
Robinson, J., Bull. Essex Inst. 12: 92. 1880.
Sayre, G., Bryologist 80: 516. 1977 (bryol. coll. at BUF, FH, VT).
Schuster, R.M., Hep. Anthoc. N. Amer. 1: 91. 1966.

9829. *Hydrothyria venosa*, a new genus and species of the Collemaceae. Read before the Essex Institute, June 8, 1853, by John Lewis Russell, ...[Proc. Essex Inst. 1, 1856]. Oct. *Publ.*: 1856, p. [1]-12. *Copy*: G. – Reprinted and to be cited from Proc. Essex Inst. 1: 188-191. 1856.

Russell, Patrick (1727-1805), British (Scottish) physician and naturalist; MD Edinburgh; physician at the English factory at Aleppo 1753-1772; succeeding his half-brother Alexander; botanist to the East India Company at Madras 1785-1789; in London 1772-1785, 1789-1805. (*P. Russell*).

HERBARIUM and TYPES: BM (Aleppo and Indian material), E, K.

BIOFILE: AG 5(2): 92; Backer p. 504 (epon.); Barnhart 3: 194; BB p. 265; BM 4: 1770; Bossert p. 342; CSP 5: 345; Dawson p. 724-726; Desmond p. 535 (b. 6 Feb 1727, d. 2 Jul 1805); DNB 49: 469-470; Dryander 3: 437; IH 2: (files); PR ed. 1: 8887; Rehder 3: 315.

BIOFILE: Bridson, G.D.R. et al., Nat. hist. mss. res. Brit. Isl. 248.32, 278.4. 1980.
Eig, A., J. Bot. 185-186. 1937 (on Nat. Hist. Aleppo, ed. 2).
Martius, C.F.P., Flora 17: 3. 1834 (on Russell coll. in Wallich herb.) Miller, H.S., Taxon 19: 540. 1970 (material Lambert herb. from India not traced).
Nissen, C., Zool. Buchill. 355. 1969.
Rendle, A.B., J. Bot. 38: 81-82. 1900 (on Nat. Hist. Aleppo, ed. 2).
Roxburgh, W., Pl. Coromandel 1795 (see preface).
Smith, E., Life Joseph Banks 115, 116-120. 1911 (quotes letters to Banks).
Wight, R., in W.J. Hooker, Bot. misc. 2: 95. 1831.

COMPOSITE WORKS: (1) Editor, W. Roxburgh, *Pl. Coromandel* 1795-1819 (see preface). (2) Editor, A. Russell, *Nat. Hist. Aleppo*, ed. 2, 1794. q.v.

Russell, Paul George (1889-1963), American botanist; with the U.S. Department of Agriculture at Beltsville, MD. (*P.G. Russell*).

HERBARIUM and TYPES: NY and US. – Collected with J.N. Rose and P.C. Standley (q.v.) in Mexico 1910; with M.J. Souviron in 1930; alone 1927.

BIBLIOGRAPHY and BIOGRAPHY: Barnhart 3: 194 (b. 24 Apr 1889); Bossert p. 343; Hortus 3: 1203; IH 2: (files); Lenley p. 356; MW p. 421; Roon p. 98; Zander ed. 10, p. 810.

BIOFILE: Knobloch, I.W., Plant Coll. N. Mexico 60. 1979 (d. 3 Apr 1963).
McVaugh, R., Contr. Univ. Mich. Herb. 9: 299. 1972.

Russell, Thomas Hawkes (1851-1913), British (English) solicitor and botanist. (*T.H. Russell*).

HERBARIUM and TYPES: Russell had a private bryological herbarium.

BIBLIOGRAPHY and BIOGRAPHY: Barnhart 3:194; BB p 265; BM 4: 1770; CSP 18: 367; Desmond p. 535 (b. 30 Mar 1851, d. 31 Jul 1913); Kew 4: 553.

BIOFILE: Jackson, B.D., Proc. Linn. Soc. 1913-14: 61 (obit.).

9830. *Mosses and liverworts* an introduction to their study, with hints as to their collection and preservation ... London (Sampson Low, Marston & Company, L.) 1908. Oct. (*Moss. liverw.*).
Ed. [*1*]: May 1908 (p. ix: 13 Jan 1908; J. Bot. Aug 1908; Bot. Centralbl. 6 Oct 1908; J. Bot. 49: 104; Nat. Nov. Jul(2) 1908), frontisp., p. [i]-xiii, [1]-200, *pl. 1-10* (by author). *Copies*: G, NSW.
Ed. 2: 1910 (p. viii: 25 Jul 1910, J. Bot. Mar 1911; Bot. Centralbl. 23 Mai 1911), frontisp., p. [i]-xvi, [1]-211, *pl. 1-4, 4b, 5, 5a, 6-10* (id.). *Copy*: MO. – "New and revised edition". London (id.) 1910. Oct.
Ref.: Gepp, A., J. Bot. 46: 270-271. 1908 (rev.); Bot. Centralbl. 108: 387. 1908 (rev. ed. 1); J. Bot. 49: 103-104. 1911 (gives date ed. 1); Bot. Centralbl. 13 Jun 1911.

Russow, Edmund [August Friedrich] (1841-1897), Esthonian botanist (of Baltic German origin); studied at Dorpat (Tartu) University 1860-1865; Dr. phil. 1871; habil. Dorpat 1865; lecturer plant anatomy and physiology ib. 1867; succeeding M. Willkomm as professor of botany and director of the botanical garden at Dorpat 1874-1895; professor emeritus 1895-1897; specialist on *Sphagnum*, the Esthonian flora and on the plant morphology of vascular cryptogams. (*Russow*).

HERBARIUM and TYPES: TU; other material at B (mainly destr.), H and S.

BIBLIOGRAPHY and BIOGRAPHY: AG 3: 730; Barnhart 3: 194 (b. 24 Feb/8 Mar 1841, d. 11/23 Apr 1897); BM 4: 1772; Bossert p. 343; CSP 8: 1879, 11: 244, 12: 640, 18: 370; DTS 1: 250; Frank 3 (Anh.): 87; Herder p. 120, 121, 201, 266, 433; Jackson p. 82; Kew 4: 553; Langman p. 658; Moebius p. 121, 125, 182, 190, 212, 219; Morren ed. 2, p. 29, ed. 10, p. 97; PR 7930-7932; Rehder 5: 747; SBC p. 130; TR 1215-1218; Urban-Berl. p. 312, 390.

BIOFILE: Anon., Bot. Centralbl. 70: 303. 1897 (d.); Bot. Zeit. 32: 271. 1874 (app. prof. bot. Dorpat), 55: 160. 1897 (d. 11 Apr 1897); Hedwigia 36: (100). 1897 (d.); Leopoldina 33: 92. 1897 (obit.); Nat. Nov. 19: 264. 1879 (d.); Termész. Közlöny 30(352): 654. 1898 (d.); Österr. bot. Z. 47: 271. 1897 (d.).
Blumenthal, E., Ber. Senckenb. naturf. Ges. 1898: v-vi (obit.).
Bureau, E. & F. Camus, Rev. bryol. 24: 95-96. 1897 (obit.).
Dehio, Sankt.-Petersburger medizinische Zeitschrift (Wochenschrift) 22: 157-158. 1897 (obit.).
Gager, C.S., Brooklyn Bot. Gard. Rec. 27(3): 192. 1938.
Harvey-Gibson, R.J., Outlines hist. bot. 170, 171, 175. 1914.
Isoviita, P., Ann. bot. fenn. 3: 260. 1966 (Sphagnum publ.).
Kennel, J. von, S.B. Naturf. Ges. Univ. Jurjew (Dorpat) 11: 245-258. 1896 (obit.).
Kneucker, A., Allg. bot. Z. 3: 104. 1897 (d.).
Kupffer, K.R., Korrespondenzbl. Naturf.-Ver. Riga 40: 43-51. 1898 (obit.).
Kusnezow, N.J., Bot. Centralbl. 68: 257. 1896 (coll. TU), 71: 265-269. 1897 (obit., bibl.).
Martenson, I., Pharm. Z. Russland 36: 236-237. 1897 (obit.).
Morot, L., J. de Bot. 11: 206. 1897 (d.).
Russow, Emma, Bot. Gaz. 24: 307-308. 1897 (R. left two collections to be sold: 3750 micr. preparations and a Sphagnum collection of 314 fascicles, 3-4000 micr. preparations, 300 slides).
Sayre, G., Bryologist 80: 516. 1977 (bryo. at TU, H-SOL, S).
Winkler, C., Ber. deut. bot. Ges. 15: (46)-(55). 1897 (obit.).

EPONYMY: *Russowia* C. Winkler (1890).

9831. *Flora der Umgebung Reval's* von Edmund Russow, stud. botan. ... Dorpat [Tartu] (Druck von Heinrich Laakmann) 1862. Oct. (*Fl. Reval*).
Publ.: Mar-Apr 1862 (imprimatur censor 22 Mar 1862 on p. [ii]), p. [i-ii], [1]-122. *Copies*: G, HH. − Preprinted or reprinted from Arch. Naturk. Liv-, Est− Kurl. ser. 2. 3: 1-120. 1860 (p. 1 of Archiv is p. 3 of reprint, p. 122 Archiv is p. 120 repr.). Flora of Reval (Tallinn), Russow's native city. TR 1216.
Ref.: Schlechtendal, D.F.L. von, Bot. Zeit. 23: 218. 7 Jul 1865.

9832. *Beiträge zur Kenntniss der Torfmoose* (mit 5 lithographirten Tafeln). Eine zur Erlangung der Magisterwürde Verfasste und mit Genehmigung einer hochverordneten physiko-mathematischen Facultät der kaiserlichen Universität zu Dorpat zur öffentlichen Vertheidigung bestimmte Abhandlung von Edmund Russow, Gehilfen des Directors des botanischen Gartens. Dorpat [Tartu] (Druck von Heinrich Laakmann) 1865. Oct. (*Beitr. Torfm.*).
Thesis-issue: Dec 1865 (imprimatur Univ. Dorpat 1 Dec 1865), p. [1]-82, [83 ("161")], [84, ("162")], theses), *pl. 1-5* (uncol. liths., auct.). *Copies*: FH, H, NY. − Thesis for "M.A."; for Russow's doctor's dissertation see *Histiol. Entw.-Gesch. Marsilia* 1871. − Preprinted from Archiv naturk. Liv-, Est-, Kurlands ser. 2. 7(1): 83-162. Mai-Dec 1867 (censor 30 Apr 1867).
Commercial edition: Dec 1865, p. [1]-82, *pl. 1-5* (id.). *Copies*: BM (inf. S.W. Greene), BR, FH, H, USDA. − TR 1217. Pages 4-80 correspond with p. 84-160 in the Archiv (see above).
Ref.: Milde, J., Österr. bot. Z. 16: 396-398. 1 Dec 1866 (rev.); Hedwigia 5: 171-172. Nov 1866; Bot. Zeit. 24: 361-362. 16 Nov 1866.

9833. *Histiologie und Entwickelungsgeschichte der Sporenfrucht von Marsilia*. Eine zur Erlangung der Würde eines Doctor's der Botanik, mit Genehmigung einer hochverordneten physiko-mathematischen Facultät der kaiserl. Universität zu Dorpat, öffentlich zu vertheidigende Abhandlung von mag. Edmund Russow, Docenten der Botanik an der Universität Dorpat. Dorpat [Tartu] (Druck von Heinrich Laakmann) 1871. Oct. (*Histiol. Entw.-Gesch. Marsilia*).
Publ.: 29 Apr-13 Mai 1871 (imprimatur University 29 Apr 1871; doctor's promotion 13 Mai 1871; KNAW rd. Dec 1871; Bot. Zeit. 28 Jul 1871), p. [1]-80, [1-2, theses]. *Copies*: FH, FI. – TR 1215.

9834. *Vergleichende Untersuchungen* betreffend die Histiologie [Histiographie und Histiogenie) der vegetativen und sporenbildenden Organe und die Entwickelung der Sporen *der Leitbündel-Kryptogamen*, mit Berücksichtigung der Histiologie der Phanerogamen, ausgehend von der Betrachtung der Marsiliaceen von Dr. Edmund Russow, ... St. Pétersbourg (Commissionaires de l'Académie impériale des sciences ...) 1872. Qu. (*Vergl. Unters. Leitb.-Krypt.*).
Publ.: Sep-Oct 1872 (imprimatur p. [ii]: Sep 1872; Bot. Zeit. 13 Dec 1872), p. [i]-ix, [x, err.], [1]-207, chart, *pl. 1-11* (liths. auct., 1-9 uncol., 10-11 col.). *Copies*: G, H, NY. – Issued as Mém. Acad. imp. Sci. St.-Pétersbourg, ser. 7. 19(1). 1872. – TR 1216.
Ref.: Anon., Bull. Soc. bot. France 20 (bibl.): 91-94. 1873 (rev.).

9835. *Betrachtungen über das Leitbündel und Grundgewebe* aus vergleichend morphologischem und phylogenetischem Gesichtspunkt, ... Dorpat (Druck von Schnakenburg's litho- und typogr. Anstalt) 1875. Qu. (*Betracht. Leitb. Grundgewebe*).
Publ.: 24 Nov 1875 (presented to Alexander von Bunge, as Festschrift on the occasion of the 50th anniversary of his doctorship (Winkler 1897, p. (51)); printed 31 Oct 1871 (p. ii); Bot. Zeit. 7 Jan 1876), p. [i-iii], [1]-78. *Copy*: G.
Ref.: Loew, E., Bot. Jahresber., Just., 3: 375, 396. 1876.

9836. *Ueber die Boden- und Vegetationsverhältnisse zweier Ortschaften an der Nordküste Estlands* ... Dorpat [Tartu] (Druck von C. Mattiesen) 1886. Oct.
Publ.: Nov-Dec 1886 (p. ii: imprimatur 4 Nov 1886; Bot. Centralbl. 7-11 Mar 1887; Bot. Jahrb. 24 Jun 1887), p. [i-ii], [1]-49. *Copy*: USDA. – Preprinted or reprinted from S.B. Naturf.-Ges. Dorpat 8(1): 93-142. 1886.
Ref.: Herder, F. v., Bot. Centralbl. 31: 303-308. 1887 (rev.).

9837. *Zur Anatomie* resp. physiologischen und vergleichenden Anatomie *der Torfmoose* ... Dorpat [Tartu] (Druck von C. Mattiesen) 1887. Qu. (*Anat. Torfm.*).
Publ.: 27 Dec 1887 (Festschrift for Alexander Graf Keyserling 27 Dec 1887; censor 23 Dec 1887; Bot. Zeit. 27 Apr 1888; Rev. bryol. Jul-Aug 1888; Nat. Nov. Aug(1) 1888), p. [i-iii], [1]-35, *pl. 1-5*. *Copies*: FH, H. – Issued as Schr. Naturf. Ges. Dorpat vol. 3, 1887.
Ref.: Warnstorf, C.F.E., Bot. Centralbl. 5: 354-362. 1888 ("wahrhaft epochemachend in der Sphagnologie").
Kienitz-Gerloff, J.H.E.F., Bot. Zeit. 46: 335-336. 25 Mai 1888.

9838. *Zur Kenntniss der Subsecundum- und Cymbifoliumgruppe europäischer Torfmoose*, nebst einem Anhang, enthaltend eine Aufzählung der bisher im Ostbalticum beobachteten Sphagnum-Arten und einen Schlüssel zur Bestimmung dieser Arten ... [Arch. Nat. Liv-, Est-, Kurlands ser. 2. 10(11). 1894, Dorpat (Druck von C. Mattiesen)]. Oct. (*Kenntn. eur. Torfm.*).
Publ.: Feb-Mar 1894 (imprimatur journal 5 Feb 1894; Bot. Zeit. 1 Jun 1894; Rev. bryol. Jul-Aug 1894; Bot. Centralbl. 3 Oct 1894; Nat. Nov. Sep(2) 1894), p. [1]-167. *Copies*: BR, H. – Preprinted or reprinted from Arch. Nat. Liv-, Est-, Kurlands ser. 2. 10(4): [361]-527. 1894. – For further journal publications on *Sphagnum* see E.J. Winkler (1897) and CSP 18: 370.
Ref.: Warnstorf, C.F.E., Bot. Centralbl. Beih. 4: 211-216. 1894 (rev.).

Rutenberg, Diedrich Christian (1851-1878), German (Bremen) plant collector and traveller; studied natural sciences at Jena with E. Häckel; travelled to South Africa 1877

(Cape, Orange Free State, Transvaal, and Madagascar 1877-1878; murdered by his Sakalava carriers while exploring the Meningaza river. (*Rutenberg*).

HERBARIUM and TYPES: BREM; further material at B (mainly destr.), BRNU(?), M (Madagascar lichens ex BREM), P (pteridoph. Madagascar).

BIBLIOGRAPHY and BIOGRAPHY: Barnhart 3: 194 (b. 11 Jun 1851, d. 25 Aug 1878); BM 8: 1117; Urban-Berl. p. 312.

BIOFILE: Abh. naturw. Ver. Bremen 9: 333. 1886 (d.); Anon., J. Bot. 20: 157. 1882 (on Reliquiae Rutenbergianae).
Buchenau, F., Abh. Naturw. Ver. Bremen 7: 1-4. 1880, 13: 87-90. 1896 ("Christian Rutenberg's Ende"); Bot. Zeit. 38: 189. 1880 (Buchenau has received R.'s collections and requests help in working them up).
Christensen, C., Dansk. bot. Arkiv 7: viii. 1932 (pteridoph. P).
Embacher, F., Lexik. Reisen 255-256. 1882.
Hertel, H., Mitt. Bot. München 16: 426. 1980 (lich. M).
Humbert, H., C.R. AETFAT 4: 133. 1962 (J.M. Hildebrandt retrieved R.'s coll.).
Koehne, E., Bot. Jahrb. 4(4): 411. 1883.

COMPOSITE WORKS: The Madagascar plants collected by Rutenberg were retrieved in various ways and came in the hands of Franz Georg Philipp Buchenau (1831-1906) at Bremen who published a series of papers *Reliquiae rutenbergianae* in which he and many other specialist authors described the collections (605 species, 5 new genera, 168 new species and varieties). The series was published in the *Abh. naturw. Ver. Bremen* (see also Buchenau 1896):
1, in 7(1): 1-54, *2 pl.* Nov 1880 (on p. 4 sec. refs. on Rutenberg and his travels).
2, in 7(2): 177-197, *1 pl.* Apr 1881.
3, in 7(2): 198-214, *1 pl.* Apr 1881.
4, in 7(3): 239-264, *1 pl.* Apr 1882.
5, in 7(3): 335-365, *1 pl.* Apr 1882.
6, in 9(2): 115-138. Feb 1885.
7: in 9(4): 401-403. Jan 1887.
8: in 10(3): 369-396, *pl.* index Jan 1889.
Authors: O. Böckeler, F. Buchenau, R. Caspary, C.B. Clarke, A. Cogniaux, O. Drude, A. Engler, W.O. Focke, J. Freyn, A. Garcke, A. Geheeb, C.M. Gottsche, C. Hausknecht, O. Hoffmann, F. Körnicke, E. Koehne, F. Kränzlin, A. v. Krempelhuber, Chr. Luerssen, K. Müller, J. Müller arg., R.A. Rolfe, K. Schumann, H.M.C.L.F. Solms-Laubach, I. Urban, W. Vatke.

EPONYMY: *Neorutenbergia* M. Bizot et T. Pocs (1974); *Rutenbergia* Geheeb et E. Hampe ex E. Bescherelle (1880).

Ruthe, Johannes Friedrich (1788-1859), German (Hildesheim, Nieder-Sachsen) botanist and zoologist; studied medicine at the University of Berlin 1811; teacher at the Plamann College 1813, high school teacher at Frankfurt a. Oder 1823, at the Kölnische Gymnasium Berlin 1825, at the Berliner Städtische Gewerbeschule 1829-1842; in retirement at Berlin 1842-1859; student of the flora of the Mark Brandenburg, in his later years of the entomology of that region; natural sciences teacher of Theodor Fontane. (*Ruthe*).

HERBARIUM and TYPES: B (mainly destr.). – Issued a *Flora der Mittelmark in getrockneten Exemplaren*, 1 cent. (issued in 10 decades) 1820-1822 (see Anon., Flora 3: 749-750. 1820, 4: 175-176. 1821 (dec. 1-4) and 5: 655-656. 1822 (dec. 10). – Ruthe left a rich herbarium which was put up for sale (Ascherson 1860); this was incorporated in his son Rudolf's herbarium (at B, see below), now almost entirely destroyed (Mar 1943). – Letters to D.F.L. von Schlechtendal at HAL.

BIBLIOGRAPHY and BIOGRAPHY: AG 3: 749; Barnhart 3: 195; BM 4: 1773; CSP 5: 346-347; GR p. 128 (b. 16 Apr 1788, d. 24 Aug 1859); Herder p. 188; Hortus 3: 1203; Kew

4: 554; LS 29270; PR 7933, ed. 1: 8888; Rehder 1: 377; SBC p. 130; Tucker 1: 615.

BIOFILE: Anon., Bot. Zeit. 2: 144. 1844 (retirement).
Ascherson, P., Verh. bot. Ver. Brandenburg 2: 211-216. 1860 (obit.; rich herbarium still for sale); Fl. Provinz Brandenburg 1: 12. 1864.
Fontane, Th., Sämtl. Werke, ed. Hauser, 4: 282 (natural sciences teacher of Fontane; inf. G. Wagenitz).
Kraatz, G., Berliner entom. Z. 4: 101-102. 1860 (obit., portr., b. 16 Apr 1788, d. 24 Aug 1859).
Ruthe, J.F., Leben, Leiden und Widerwärtigkeiten eines Niedersachsen, Berlin 1841. (autobiogr.).
Schmid, G., Chamisso als Naturforscher 18, 172. 1942.

EPONYMY: *Ruthea* Opatowski (1836); *Ruthea* Bolle (1862). *Note*: *Ruthiella* C.G.G.J. van Steenis (1965) was dedicated to Miss Ruth van Crevel, illustrator of Flora Malesiana.

9839. *Flora der Mark Brandenburg* und der Niederlausitz ... Erste Abtheilung: Phanerogamen ... Berlin (verlegt bei Heinr. Adolph Wilhelm Logier) 1827. Oct. †. (*Fl. Mark Brandenburg*).
Publ.: 1827 (p. x: 16 Mar 1827; Linnaea Aug-Oct 1827), p. [ii]-xxiv, [1]-491, [492-493], [494 adv.]. *Copies*: G, HH, NY. – Other t.p.: *Versuch einer Naturgeschichte der Mark Brandenburg* und der Niederlausitz ... Pflanzen". No further volumes issued.
Ed. 2: 27 Apr-3 Mai 1834 (p. x: Apr 1834; Hinrichs; Flora 21 Jun 1834), p. [ii]-xxvi, [1]-687, [688-692], *pl. 1-2* (uncol. liths.). *Copies*: G, HH, NY, PH.
Ref.: [Schlechtendal, D.F.L. von], Linnaea 2(3): 482-483. Aug-Oct. 1827 (rev. ed. 1), 9 (Litt.): 25-26. 1834 (rev.ed. 2).

Ruthe, [Johann Gustav] Rudolf (1823-1905), German (Prussian) veterinary surgeon; son of Johann Friedrich Ruthe; amateur bryologist and coleopterologist; studied at the Berlin veterinary college; practicing veterinarian at Bärwalde (Brandenburg) 1849-1882; regional veterinary surgeon ("Kreistierarzt") Usedom-Wollin in Swinemünde (Pomerania) 1882-1904. (*R. Ruthe*).

HERBARIUM and TYPES: B (mainly destroyed; contained also his father's herbarium as well); other material at BM, BP, DBN, E, GB, GOET, H, L, MW, S.

BIBLIOGRAPHY and BIOGRAPHY: AG 3: 749; Barnhart 3: 195 (b. 1 Nov 1823, d. 12 Nov 1905), CSP 0. 002, 11: 245; Biol. Dokum. 15: 7766; DTS 1: 250; IH 2: (files); Morren ed. 10, p. 8; Urban-Berl. p. 312, 390.

BIOFILE: Anon., Hedwigia 45: (76). 1906 (d. 12 Nov 1905, Swinemünde); Verh. bot. Ver. Brandenburg 47: lxix. 1906 (d.).
Ascherson, P., Verh. bot. Ver. Brandenburg 2: 216. 1860, 47: li-lvi. 1906 (obit., bibl.); Bot. Zeit. 19: 167-168. 1861.
Kneucker, A., Allg. bot. Z. 11: 208. 1905 (d.); Rev. bryol. 33: 43-44. 1906.
Maas, W., Monatschr. Kakteenk. 16(4): 62-63. 1906 (obit., portr.).
Sayre, G., Bryologist 80: 516. 1977 (coll.).
Wagenitz, G., Ind. coll. princ. Herb. Gott. 140. 1982.

Rutherford, Daniel (1749-1819), British (Scottish) botanist, chemist and physician; MD Edinburgh 1772; practicing physician at Edinburgh 1775-1786; professor of botany and regius keeper, Royal Botanic Garden, Edinburgh 1786-1819 (*Rutherford*).

HERBARIUM and TYPES: Unknown.

BIBLIOGRAPHY and BIOGRAPHY: Barnhart 3: 195; BB p. 265; BM 4: 1773; Bossert p. 343; CSP 5: 347; Dawson p. 726; Desmond p. 535 (b. 3 Nov 1749, d. 15 Dec 1819); DNB 50: 5-6; Henrey 2: 201, 202; Jackson p. 16; PR (alph.); Rehder 1: 258; SO 581.

BIOFILE: Allibone, S.A. Crit. dict. engl. lit. 2: 1903. 1878.

Bridson, G.D.R. et al., Nat. hist. mss. res. Brit. Isl. 433 [index]. 1980.
Dawson, W.R., Smith papers 84. 1934.
Fletcher, H.R. & W.H. Brown, Royal Bot. Gard. Edinburgh 1670-1970; 68-80. 1970 (on R. and his gardeners at E).
Gager, C.S., Brooklyn Bot. Gard. Record 27(3): 242. 1938.
Hedge, I.C. & J.M. Lamond, Index coll. Edinburgh herb. 125. 1970 (no specimens from his herb. found in E).
Huntress, E.H., Proc. Amer. Acad. Arts Sci. 77(2): 50-51. 1949.
Oliver, F.W., Makers Brit. bot. 1913: 290-291.
Poggendorff, J.C., Biogr.-lit. Handw.-Buch 2: 726. 1823.
Wittrock, V.B., Acta Horti Berg. 3(2): 97. 1903.

9840. *Characteres generum plantarum*; ex systemate vegetabilium Linnaei, et Horto kewensi; praecipue excerpti; quibus accedit series ordinum naturalium, olim a Linnaeo ipso, et nuper a cl. Jussiaeo proposita. In usus academicos. Edinburgi 1793. Oct. (in fours). (*Char. gen. pl.*).
Publ.: 1793, chart, p. [i-v], [1]-2, err., [1]-138. *Copies*: E, NY, USDA (lacks t.p.). – SO 581. – Authorship attributed to D. Rutherford e.g. by J.M. Hulth, Bibl. Linn. 1(1): 145. 1907.

Rutström, Carl Birger (1758-1826), Swedish botanist; Dr. phil. Uppsala 1772; Dr. med. Harderwijk 1793; botanical demonstrator in Åbo 1794-1798; in various administrative positions at Stockholm from 1799. (*Rutstr.*).

HERBARIUM and TYPES: Lapland (trip 1782) material at H and S; Rutström's herbarium went to Linköping high school fide KR p. 609 and N.C. Kindberg (Bot. Not. 1863: 160). B. Peterson informs us that the herbarium is still at Linköping, with some material deposited at S.

BIBLIOGRAPHY and BIOGRAPHY: Barnhart 3: 195; GR p. 480; GR, cat. p. 70; KR p. 609 (b. 22 Nov 1758, d. 13 Apr 1826); LS 23272; PR 7934-7935, ed. 1: 8889-8890; Saelan p. 430-431; SBC 131; TL-2/see C. Alströmer.

BIOFILE: Eriksson, G., Bot. Hist. Sverige 284, 296, 325. 1969.
Hässler, A., Bot. Not. 1933: 465. 1933 (herbarium at Linköping high school).
Löwegren, Y., Naturaliekabinett Sverige 337, 363. 1952; Naturaliesaml. och naturhist. underv. vid läroverken 141, 147, 148. 1974 (herbarium at Linköping high school).

EPONYMY: *Rutstroemia* P.A. Karsten (1871).

9841. *Positiones physiologici medici et botanici argumenti*, quas adnuente summo numine, ex auctoritate rectoris magnifici ... amplissimique senatus academici consensu et nobilissimae facultatis medicae decreto, pro gradu doctoratus, summisque in medicina honoribus et privilegiis rite ac legitime consequendis, eruditorum examini submittit Carolus Birgerus Rutström, ... a.d. xxi. maji mdccxciii. H.L.Q.S. Hardervici [Harderwijk], (apud Joannem Moojen, ...) [1793]. Qu. (*Posit. phys. med. bot. argum.*).
Publ.: 21 Mai 1793, p. [1]-8. *Copy*: AMD (Univ. Libr.).

9842. *Spicilegium plantarum cryptogamarum Sueciae*, pro argumento publicae disputationis, in regiae Academiae aboënsis, aud. maj. venia experientissimae Facultatis medicae, d. xviii Dec. h.a. mdccxciv, horis solitis agendae, propositum, vindicimus auctore Carolo Birgero Rutström, ... et Johanne Gustavo Haartman, ... Aboae (Typis Frenckellianis) [1794]. Qu. (*Spic. pl. crypt. Suec.*).
Publ.: 3 Dec 1794 (see end of dedication; thesis defended 18 Dec 1794), p. [i-iv], [1]-20. *Copies*: C (inf. A. Hansen), GB (inf. B. Peterson), H (inf. P. Isoviita), HU and private copy O. Almborn. Dr. Almborn provided us with the following note: "As stated by Krok, Rutström was the author of the thesis defended by Haartman. 20 species (1 fern, 14 lichens, 5 fungi) were recorded mainly from prov. Halland (S.W. Sweden). Most of the species described as new have been sunk into synonymy. Only one (*Lichen demissus*) seems to have survived. It is now *Lepidoma demissum* (Rutstr.) Choisy, syn.

Lecidea demissa (Rutstr.) Ach., *Psora demissa* (Rutstr.) Hepp.-cf. Krempelhuber 1: 82-83. 1867".

Ruttner, Franz (1882-1961), Sudeten-German (Bohemia-born) limnologist, hydrobiologist and botanist; studied with H. Molisch at Praha (Karls-Universität); Dr. phil. ib. 1906; assistant at the Biological Station of Lunz am See, Austria, 1906-1908; director ib. 1908– 1957; habil. University of Vienna 1924; professor of hydrobiology 1927; with A. Thienemann on the Deutsche limnologische Sunda-Expedition 1928-1929; Einar Naumann medal 1942. (*Ruttner*).

HERBARIUM and TYPES: Indonesian material at L; lichens from Java at W.

BIBLIOGRAPHY and BIOGRAPHY: Backer p. 504 (epon.); BM 4: 1774; Biol.-Dokum. 15: 7769-7770; Christiansen p. 148; SK 1: 455-456; SK 8: lxxxiii.

BIOFILE: Anon., Österr. bot. Z. 88: 79. 1939 (member Leopoldina).
Braun, R., Schweiz. Z. Hydrol. 14(2): 483-484. 1952 (rev. of ed. 2 of his Grundriss der Limnologie, 1952).
Brehm, V., Hydrobiologia 20: 193-201. 1962 (obit., portr., bibl.).
Findenegg, Arch. Hydrobiol. 58(2): 244-251.1962 (obit., portr., bibl.).
Jaag, O., Schweiz. Z. Hydrol. 14: 443-445. 1952.
Poggendorff, J.C., Biogr.-lit. Handw.-Buch 7a: 861-863. 1959 (extensive bibl. and secondary refs.).
Schimitschek, E., Anzeiger Schädl.-Kunde 34: 171. 1961 (obit., b. 12 Mai 1882, d. 17 Mai 1961).
Steenis, C.G.G.J. von, Fl. males. Bull. 18: 972. 1963.

NOTE: R.'s main work was *Grundriss der Limnologie*, Berlin 1940, ed. 2, Berlin 1952, ed. 3, Berlin 1963; translation: *Fundamentals of Limnology*, Toronto 1953.

FESTSCHRIFT: Int. Rev. ges. Hydrobiol. Hydrogr. 42(4-6), 1942, 43. 1943.

9843. *Die Mikroflora der Prager Wasserleitung* ... Prag (Kommissionsverlag von Fr. Rivnáč ...) 1906. Oct. (*Mikrofl. Prag. Wasserleit.*).
Publ.: Jan-Jul (Nat. Nov. Aug(1) 1906), p. [i]-iv, [1]-47. *Copies*: MO, USDA. – Issued as Arch. naturw. Landesdurchf. Böhmen 13(4), 1906. Inaugural dissertation.
Ref.: Wille, N., Bot. Zeit. 65(2): 227-229. 1907.

Rutty, John (1697-1775), British (English) physician; MD Leiden 1723; physician at Dublin 1724-1775. (*Rutty*).

HERBARIUM and TYPES: Unknown.

BIBLIOGRAPHY and BIOGRAPHY: Barnhart 3: 195; BB p. 265; BL 2: 285, 711; BM 4: 1774; Desmond p. 536 (b. 25 Dec 1697, d. 26 Apr 1775); DNB 50: 31-32; Dryander 3: 550; Rehder 3: 253.

BIOFILE: Anon., Notes and Queries. ser. 2. 7: 147-148, 264 (ref. to biogr. sketch in Dublin Quart. J. med. Sci., Mai 1847 by J. Osborne), 324, 423. 1859.
Colgan, N., Fl. county Dublin xxii-xxiii. 1904 (on R.'s work on this flora).
Davies, G.L.H., Irish Geogr. 12: 92-98. 1979, Western Natural. 7: 92. 1978 (on R.'s geological work).
Fox, R.H., Dr. John Fothergill 41, 128. 1919.
Gilbert, P., Comp. biogr. lit. entom. 328. 1977.
Knowles, M.C., Proc. R. Irish Acad. 38 (B, 10): 180. 1929 (lichens).
Mitchell, M.E., Bibl. Irish lichenol. 56. 1971.
Nelson, E.C., J. Soc. Bibl. nat. Hist. 9(3): 291. 1979 (note on R.'s annotated copy of Threlkeld's Syn. stirp. Hibern. in Royal Irish Academy, Dublin); Huntia 4(2):1 133. 1982; Occ. Pap. Natl. Bot. Gard. Glasnevin 1982: 106 (list of copies of *Essay* in Ireland).

Praeger, R.L., Some Irish naturalists 150. 1949.
Rutty, J., A spiritual diary and soliloquies, Dublin 1776 (posthumously publ. autobiogr.; mainly religious).
Widdess, J.D.H., A history of the Royal College of Physicians of Ireland 1654-1963, London 1963,p. 73-75.

EPONYMY: *Ruttya* W.H. Harvey (1842); X *Ruttyruspolia* [*Ruttya* W.H. Harvey X *Ruspolia* Lindau (1896)] Meeuse et Wet (1961).

NOTE: We are grateful to E.C. Nelson, Dublin, who supplied most of the information for this entry on J. Rutty. Rutty's botanical work is recorded in mss. in the Royal Irish Academy, Dublin (minute books of the Dublin physico-historical Society, his own annotated copy of Threlkeld) as well as in his book on Dublin.

9844. *An essay towards a natural history of the county of Dublin*, accomodated to the noble designs of the Dublin Society; affording a summary view. i. Of its vegetables, with their mechanical and oeconomical uses, and as food for men and cattle; a catalogue of our vegetable poisons; and a botanical kalendar, exhibiting the respective months in which most of the simples in use are found in flower. ii. Of its animals. iii. Of its soil, ... iv. Of the nature of the climate, ... Dublin (printed by W. Sleater, ... for the author) 1772. 2 vols. Oct. (*Essay nat. hist. Dublin*).
Publ.: 2 vols., 1772. *Copies*: DBN, GOET, HU, ILL, KU, PH, TCD.
1: [i]-xiv, [1]-4, [1]-392, *pl. 1-5* (*pl. 1* is à fold-out frontisp. map). Plants on p. 33-261. See also BL 2: 285. – Errata sheet for both volumes attached to 1: xiv in copy ILL.
2: [i]-v, [vi], [1]-488, tables 1-7, fold-out tables 1-4, [1].

Ruys, Johannes (1856-1933), Dutch mycologist. (*J. Ruys*).

HERBARIUM and TYPES: Unknown.

BIBLIOGRAPHY and BIOGRAPHY: Barnhart 3: 195 (b. 24 Mai 1856); JW 1: 447, 2: 200.

BIOFILE: Anon., Med. Ned. mycol. Ver. 21: frontisp. portr. 1973.

9845. *De paddenstoelen van Nederland* naar verschillende bronnen bewerkt door Joh. Ruys ... 's Gravenhage (Martinus Nijhoff) 1909. Oct. (*Paddenst. Nederland*).
Publ.: 1909 (p. vi: Aug 1909; Hedwigia 26 Apr 1910; Nat. Nov. Nov(1) 1909), p. [i]-vi, [vii, cont.], [1]-461, *1 pl. Copies*: BR, L, MICH, NY, Stevenson.
Ref.: Weevers, Th., Bot. Centralbl. 117: 514. 1911 (rev.).

Ruys, Jan Daniel (1897-1954), Dutch botanist; Dr. phil. Utrecht 1925; director of the Moerheim nursery 1926-1954. (*J.D. Ruys*).

HERBARIUM and TYPES: Some material from Algeria at U.

BIBLIOGRAPHY and BIOGRAPHY: Barnhart 3: 195 (b. 27 Apr 1897); BFM no. 2988; BJI 2: 150; BL 2: 421, 711; Hortus 3: 1203; JW 2: 200.

BIOFILE: Anon., Gard. Chron. ser. 3. 91: 472. 1932 (obit., portr.).
V., Jaarb. Nederl. dendrol. Ver. 20: 30-31. 1955 (obit., portr., b. 27 Apr 1897, d. 31 Jul 1954).
Verdoorn, F., ed., Chron. bot. 1: 44, 5, 53. 1936, 3: 44. 1937, 4: 197, 200, 297. 1938, 5: 64, 241. 1939, 6: 303. 1941.

9846. *Contribution à l'histoire du développement des Mélastomatacées.* Énumération des plantes phanérogames angiospermes examinées au point de vue de la karyologie. Proefschrift ter verkrijging van den graad van doctor in de wis– en natuurkunde aan de Rijksuniversiteit te Utrecht, ... te verdedigen op Maandag 26 Januari 1925, des namiddags 4 uur, door Jan Daniel Ruys, ... Leiden (Boekhandel en drukkerij voorheen E.J. Brill) 1925. Oct. (*Contr. hist. dével. Mélastom.*).

Publ.: 26 Jan 1925, p. [i**, iii**, theses., loose], [i*], [i]-vi, [vii], [1]-123, *pl. 1-3. Copy*: U.
Ref.: Engler, A., Bot. Jahrb. 60 (Lit.): 26. 15 Dec 1925.

Ruysch [Ruijsch], **Frederik** (1638-1731), Dutch physician and anatomist; M.D. Leiden 1664; practicing physician in the Hague and Amsterdam; praelector of anatomy for the Amsterdam surgeon's guild 1666-1731; in addition professor of botany at the Amsterdam Athenaeum illustre and superintendent of its botanical garden 1685. (*Ruysch*).

HERBARIUM and TYPES: Ruysch possessed a large natural history cabinet which was predominantly anatomical and zoological, but which contained also an herbarium. The entire cabinet was sold to Czar Peter of Russia, 1717 and the herbarium (mostly of garden plants) is still at LE. Another herbarium collected in Holland about 1657, was acquired by Sir Hans Sloane, now at BM (Hortus siccus 111, 112). Some other Ruysch specimens are also at BM.

NOTE: Ruysch was mainly an human anatomist. He is often cited, however, because of his herbaria at BM and LE, and because he published Jan Commelijn's *Horti med. amstelod.* 1697-1701 (TL-2/1187). For a biographical summary, a bibliography and secondary refs. see G.A. Lindeboom (DSB 12: 39-42. 1975).

BIBLIOGRAPHY and BIOGRAPHY: Backer p. 504; Barnhart 3: 504; BM 4: 1774; Bossert p. 343; Clokie p. 236; DSB 12: 39-42 (G.A. Lindeboom); Frank 3 (Anh.): 87; Hegi 5(4): 2363; Henrey 2: 79, 426, 429; JW 2: 200, 3: 367, 5: 251; NI (alph.); NNBW 3: 1108-1109 (b. 28 Mar 1638); PR (alph.) and 1833, ed. 1: 1946; TL-2/1187, see H. Boerhaave.

BIOFILE: Appleby, J.H., Arch. nat. hist. 10(3): 385-386. 1982 (on R. and his cabinet).
Bridson, G.D.R. et al., Nat. hist. mss. res. Brit. Isl. 224.103. 1980.
Dandy, J.E., The Sloane herbarium 197. 1958.
Engel, H., Bijdr. Dierk. 27: 310. 1939 (further biogr. refs.).
Fries, Th.M., Bref Skrifv. Linné 1(4): 306. 1910, 1(6): 37. 1912.
Herder, F. v., Bot. Centralbl. 55: 259. 1893 (on herb. Ruysch at LE: "almost worthless", described by Steller, Musei imp. petrop. 1(2), 1745).
Löwegren, Y., Naturaliekabinett Sverige 58. 1952.
Nordenskiöld, E., Hist. biol. 170, 171, 185. 1949.
Murray, G., Hist. coll. BM(NH) 1: 179. 1904 (herb.).
Nissen, C., Zool. Buchill. 355. 1969.
Scheltema, P., Het leven van Frederik Ruijsch, Sliedrecht 1886, 115 p. (biogr., bibl.).
Stafleu, F.A., Linnaeus and the Linnaeans 165. 1971.
Talbott, J.H., Biogr. hist. med. 1970, p. 148-149.
Verdoorn, F., ed., Chron. bot. 3: 23. 1937, 4: 85. 1938.

EPONYMY: *Ruyscha* Cothenus (1790, *orth. var.*); *Ruyschia* N.J. Jacquin (1760); *Ruyschiana* P. Miller (1754); *Ruyschioxylon* H. Hoffmann (1884).

HANDWRITING: Dandy, J.E., The Sloane herbarium 1958, facsimile no. 49.

Ryan, Elling (1849-1905), Norwegian botanist and pharmacist; pharmac. exam. 1871; later manager of a chemical factory at Graesvik nr Frederikstad; amateur bryologist. (*Ryan*).

HERBARIUM and TYPES: O (original bryophyte herb.), TRH (phan.); other material at GB, H, L.

BIBLIOGRAPHY and BIOGRAPHY: Barnhart 3: 195 (b. 24 Oct 1849, d. 25 Apr 1905); BM 4: 1775; KR p. 609; Lenley p. 357; SBC p. 131; TL-2/2246, 2249, 2251.

BIOFILE: Anon., Hedwigia 45: (38). 1906; Rev. bryol. 32(4): 84. 1905 (obit. not.).

Hagen, I., Fra E. Ryans mosherbarium Trondhjem 1907, 36 p., K. Norske Vid. Selsk. Skr. 1907, no. 1 (biogr., portr., description of herbarium).
Kneucker, A., Allg. bot. Z. 11: 144. 1905 (d.).
Sayre, G., Bryologist 80: 516. 1977 (coll.).
Wittrock, V.B., Acta Horti Berg. 3(3): 185. 1905.

COMPOSITE WORKS: With I. Hagen, *Iagttag. mosern. udbr.* 1896, see TL-2/2246. See also I. Hagen, *Musci Norvegiae borealis* 1899-1904, TL-2/2249.

EPONYMY: *Ryanaea* A.P. de Candolle (1824, *orth. var.*) and *Ryania* M. Vahl (1796, *nom. cons.*) are both named for J. Ryan (*fl.* 1770-1790), plant collector.

Rydberg, Pehr Axel (1860-1931), Swedish botanist; studied at Skara high school; emigrated to the United States 1882; teacher at Luther Academy, Wahon, Nebraska 1884-1890; studied at Lincoln Univ., Nebraska 1890-1895; student of C.E. Bessey; professor of natural sciences and mathematics Upsala College, Kenilworth, New Jersey 1895-1899; Dr. phil. Columbia Univ. 1898; assistant curator New York Botanical Garden 1899-1905; curator ib. 1906-1931. (*Rydb.*).

HERBARIUM and TYPES: NY (35.000); duplicates in many herbaria, e.g.; A, B (mainly destr.), BUF, C, CS, E, GH, HBG, K, KANU, KSC, L, LE, MIN, MO, MONT, MONTU, NEB, NMC, NY, NYS, RM, US. Field notes books Montana (1896), Utah and New Jersey (1905-1908), Kansas (1929) and Western United States at NY. – Smithsonian Archives SIA 220, 221, 222, 224, 226, 227, 239, HI 107.

BIBLIOGRAPHY and BIOGRAPHY: Barnhart 3: 196 (b. 6 Jul 1860, d. 25 Jul 1931); BFM no. 1601, 1602; BJI 2: 50, 150; BL 1: 145, 162, 170, 195, 196, 214, 316; BM 4: 1775, 1940, 8: 1119; Bossert p. 343; CSP 18: 377; DAB 16: 269-270 (by J.H. Barnhart); Ewan ed. 1: 355 [index], ed. 2: 191; Hawksworth p. 185; Hirsch p. 256; Hortus 3: 1203; IH 1 (ed. 2): 88, (ed. 3): 111, (ed. 6): 363, (ed. 7): 341, 2: (files); Kew 4: 556; KR p. 610-612; Langman p. 659-660; Lenley p. 357; MW p. 421; NAF 28B(2): 355-356; NI 1706; NW p. 56; PH 209, 450; Rehder 5: 747-748; RS p. 138; SO 2913-2914, 2918, 2957; Stevenson p. 1256; TL-1/1129; TL-2/779, 781, see T.D.A. Cockerell; Tucker 1: 615; Urban-Berl. p. 390; Zander ed. 10, p. 710, ed. 11, p. 810.

BIOFILE: Allison, E.M., Univ. Colo. Studies 6: 73, 74. 1909 (coll. Colorado).
Anon., Columbia Alumni News 23: 14. 1931 (d.).; The Grace abounding 20(7), August 1931 (obit., role in Lutheran church); J. New York Bot. Gard. 1: 35. 1900 (accession Rydberg herb.); Torreya 31: 128. 1931 (d.).
Barnhart, J.H., J. New York Bot. Gard. 32: 229-233. 1931 (obit., portr.).
Becker, H.F., Gard. J. 10: 68-69. 1960 (letter T.D.A. Cockerell to P.A. Rydberg).
Blankinship, J.W., Montana Agric. Col. Stud. Bot. 1: 13-14, 23. 1905 (bibl. Montana; expl. Montana).
Boone, W., Hist. bot. W. Virginia 175, 178. 1965.
Cattell, J.M., Amer. men sci. 276-277. 1906, ed. 2: 405. 1910.
Ewan, J. et al., Short hist. bot. U.S. 19. 1969.
Jones, M.E., Contr. West. Bot. 18: 141-142. 1935 (personal recollections; subordinated his judgement to that of Britton "for the sake of financial support"; "silly Astragaloid genera").
Murrill, W.A., Hist. found. bot. Florida 20. 1945.
Pennell, F.W. & J.H. Barnhart, Monogr. Acad. Nat. Sci. Philadelphia 1: 618. 1935.
Reveal, J., *in* A. Cronquist et al., Intermountain fl. 1: 63, 64. 1972.
Rickett, H.W., Index Bull. Torrey bot. Cl. 85-87. 1955 (bibl. of his numerous contributions).
Rydberg, P.A., Scandinavians who have contributed to the knowledge of the flora of North America, Augustana Library Publ. 1907, Contr. New York Bot. Gard. 100, 1907; p. 42-43 on Rydberg himself.
Sargent, C.S., Silva 14: 69-70. 1902 (biogr. note).
Small, J.K., Fl. S.E. United States 1903, see preface.

Verdoorn, F., ed., Chron. bot. 6: 235. 1941.

COMPOSITE WORKS: Contributions to NAF:
Sparganiaceae, 17(1): 5-10. 30 Jun 1909.
Elodeaceae, 17(1): 67-71. 30 Jun 1909.
Hydrocharitaceae, 17(1): 71-74. 30 Jun 1909.
Portulacaceae, 21(4): 279-328. 29 Dec 1932.
Penthoraceae, 22(1): 75. 22 Mai 1905, 22(6): 548. 30 Dec 1918.
Parnassiaceae, 22(1): 77-80. 22 Mai 1905, 22(6): 548-549. 1918.
Hydrangeaceae, 22(2): 159-178. 18 Dec 1905, 22(6): 555-557. 30 Dec 1928.
Phyllonomaceae, 22(2): 191. 18 Dec 1905.
Rosaceae, 22(3): 239-292. 12 Jun 1908, 22(4): 293-388. 20 Nov 1908, 22(5): 389-480. 23 Dec 1913, 22(6): 481-533. 30 Dec 1918.
Saxifragaceae add., with J.K. Small, 22(6): 549-555. 30 Dec 1918.
Fabaceae, 24(1): 1-34. 25 Apr 1919, 24(2): 65-136. 22 Jan 1920, 24(3): 137-200. 16 Jul 1923, 24(4): 201-250. 4 Oct 1924, 24(5): 251-314. 19 Feb 1929, 24(6): 315-378. 15 Mai 1929, 24(7): 379-462. 26 Jul 1929.
Balsaminaceae, 25(2): 93-96. 3 Jun 1910.
Limnanthaceae, 25(2): 97-100. 3 Jun 1910.
Zygophyllaceae, with A.M. Vail, 25(2): 103-116. 3 Jan 1910.
Lennoaceae, 29(1): 19-20. 31 Aug 1914.
Pyrolaceae, 29(1): 21-32. 31 Aug 1914.
Ericaceae, 29(1): 33-102. 31 Aug 1914.
Carduales, 33(1): 1. 15 Sep 1922.
Ambrosiaceae, 33(1): 3-44. 15 Sep 1922.
Carduaceae, 33(1): 45-46. 15 Sep 1922, 34(1): 1-75. 31 Dec 1914, 34(2): 83-146. 28 Jul 1915, 34(2): 147-180. 28 Jul 1915, 34(3): 181-216. 29 Dec 1916, 34(3): 217-288. 29 Dec 1916, 34(4): 289-301. 22 Jun 1927, 34(4): 303-308. 22 Jun 1927, 34(4): 309-360. 22 Jun 1927.

EPONYMY: *Rydbergia* E.L. Greene (1898); *Rydbergiella* F. Fedde et P. Sydow. (1906).

9847. University of Nebraska. *Flora of Nebraska*. Published by the Botanical Seminar. xvi. Part 21. *Rosales*. Lincoln, 1895. (issued December 30, 1895). (*Fl. Nebr., Rosales*).
Publ.: 30 Dec 1895, p. [1]-82. *Copies*: BR, G, HH. – Title on cover: "*Flora of Nebraska*" Edited by the members of the Botanical Seminar of the University of Nebraska. Part 21. *Rosales*. Lincoln, Nebraska, U.S.A. Published by the Seminar 1895". Editorial committee: C.E. Bessey, Roscoe Pound, F.E. Clements.
Preliminary publication: "A revision of the nomenclature of the Nebraska Polypetalae" in Bot. Surv. Nebraska 3: 20-39. 18 Jun 1894.
Ref.: Small, J.K., Bull. Torrey bot. Club 23(5): 217. 1896.

9848. U.S. Department of Agriculture, Division of Botany ... *Flora of the Sand Hills of Nebraska* ... Washington (Government Printing Office) 1895. Oct.
Publ.: 14 Sep 1895, p. [i]-v, 133-203, *pl. 1-2*. *Copies*: G, HH, MO, NY, PH, U, US, USDA. – Issued and to be cited as Contr. U.S. natl. Herb. 3(3): i-v, 133-203, *pl. 1-2*. 1895. See BL 1: 196.

9849. U.S. Department of Agriculture. Division of botany ... *Flora of the Black Hills of South Dakota* ... Washington (Government Printing Office) 1896. Oct.
Publ.: 13 Jun 1896, p. [i]-v, 463-536, *pl. 18-20*. *Copies*: BR, HH, MO, NY, PH, U, US, USDA. – Issued and to be cited as Contr. U.S. natl. Herb. 3(8): [i]-v, i-iv, 463-536. *pl. 18-30*.
Additions: A.C. McIntosh, Black Hills Engineer 16: 160-167. 1928 (fide BL).

9850. Contributions from the Herbarium of Columbia College ... *Notes on Potentilla*. i-iv, New York 1896-1897. Oct.
Publ.: A series papers published in Bull. Torrey bot. Club 23-24, 1896-1897 ("Bull.") and published as reprints in the series Contr. Herb. Columbia Coll. ("Contr.").
1: Jun 1896, Contr. no. 96; Bull. 23: 244-248. Jun 1896.

2: Jul 1896, Contr. no. 96 (combined with no. 1); Bull. 23: 259-265. Jul 1896.
3: Aug 1896, Contr. no. 99; Bull. 23: 301-306. Aug 1896.
4: Oct 1896, Contr. no. 103; Bull. 23: 394-399, *pl. 274-275*. Oct 1896.
5: Nov 1896, Contr. no. 105; Bull. 23: 411-435, *pl. 276-277*. Nov 1896.
6: 28 Jan 1897, Contr. no. 110; Bull. 24: 1-13, *pl. 287-288*. 28 Jan 1877.

9851. *The North American species of Physalis and related genera* ... Memoirs of the Torrey Botanical Club vol. iv. no. 5. 1896. Oct.
Publ.: 15 Sep 1896, cover-t.p., p. [297]-394. *Copy*: HH. – Issued as and to be cited from Mem. Torrey bot. Club 4(5). 1896.

9852. A report upon the *grasses and forage plants of the Rocky Mountain region* ... Washington (Government Printing Office) 1897. Oct. (*Grass. forage pl. Rocky Mt.*).
Publ: 1897 (p. 2: 20 Nov 1896), p. [1]-48. *Copy*: USDA. – Bull. 5, U.S. Dept. Agr. Div. Agrost.

9853. *A monograph of the North American Potentilleae.* [Memoirs from the Department of Botany of Columbia University. Vol. ii.] issued November 25th, 1898. Qu. (*Monogr. N. Amer. Potent.*).
Publ.: 25 Nov 1898 (t.p.), p. [i-iv], [1]-223, [224], *pl. 1-112* (drawings by F. Emil and author). *Copies*: G, NY, PH, US [imprint: "New York, 1898"], USDA. – See also his *Notes on Potentilla* 1-6, Bull. Torrey bot. Club 23 and 24, 1896-1897 (Contr. Herb. Columbia College). Further studies on Potentilla, ib. 28: 173-183. 1901 and *Notes on Rosaceae* 4, ib. 37: 487-502. 1900.

9854. Contributions from the New York Botanical Garden – no. 2. *New species from the western United States* ... New York 1899. Oct.
Publ.: 16 Oct 1899, cover t.p., p. 541-546. *Copy*: U. – Reprinted and to be cited from Bull. Torrey bot. Club 26: 541-546. 16 Oct 1899.

9855. Contributions from the New York Botanical Garden – no. 4. *Delphinium carolinianum and related species* ... New York 1899. Oct.
Publ.: 15 Nov 1899, cover-t.p., p. 582-587. *Copy*: U. – Reprinted and to be cited from Bull. Torrey bot. Club 26: 582-587. 15 Nov 1899.

9856. Contributions from the New York Botanical Garden ... *Studies on the Rocky Mountain flora* i-xxix, New York 1900-1913. Oct.
Publ.: A series of 29 papers published in Bull. Torrey bot. Club 27-40. 1900-1913 ("Bull.") and published as reprints in the series Contr. N.Y. Bot. Gard. ("Contr.").

1: 21 Apr 1900, Contr. no. 5; Bull. 27: 169-189. 21 Apr 1900.
2: 26 Oct 1900, Contr. no. 9, Bull. 27: 528-538. 26 Oct 1900 .
3: 29 Dec 1900, Contr. no. 10, Bull. 27: 614-636. 29 Dec 1900.
4: 31 Jan 1901, Contr. no. 12, Bull. 28 20-38. 31 Jan 1901.
5: 21 Mai 1901, Contr. no. 14, Bull. 28: 266-284. 266-284. 21 Mai 1901.
6: 30 Sep 1901, Contr. no. 15, Bull. 28: 499-513. 30 Sep 1901.
7: 24 Mar 1902, Contr. no. 22, Bull. 29: 145-160. 24 Mar 1902.
8: 24 Apr 1902, Contr. no. 23, Bull. 29: 232-246. 24 Apr 1902.
9: 30 Dec 1902, Contr. no. 30, Bull. 29: 480-693. Dec 1902.
10: 24 Apr 1903, Contr. no. 30, Bull. 30: 247-262. Apr 1903.
11: 29 Jul 1904, Contr. no. 55, Bull. 399-410. Jul 1904.
12: 1 Nov 1904, Contr. no. 59, Bull. 31: 555-575. Oct 1904.
13: 9 Jan 1905, Contr. no. 62, Bull. 31: 631-655. Dec 1904.
14: 19 Apr 1905. Contr. no. 66, Bull. 32: 123-168. 1905.
15: 6 Dec 1905, Contr. no. 73, Bull. 32: 597-610. 1905.
16: 7 Apr 1906, Contr. no. 82, Bull. 33: 137-161. 1906.
17: 27 Feb 1907, Contr. no. 87, Bull. 34: 35-50. 1907.
18: 10 Oct 1907, Contr. no 96, Bull. 34: 417-437. 1907.
19: 1 Oct 1909, Contr. no. 128, Bull. 36: 511-541. 1 Oct 1907.
20: 28 Dec 1909, Contr. no. 129. Bull. 36: 675-698. 28 Dec 1909.

21: 31 Mar 1910, Contr. no. 131, Bull. 37: 127-148. 31 Mar 1910.
22: 21 Jul 1910, Contr. no. 134, Bull. 37: 313-335. 21 Jul 1910.
23: 5 Oct 1910, Contr. no. 137, Bull. 37: 443-471. 5 Oct 1910.
24: 30 Nov 1910, Contr. no. 140, Bull. 37: 541-557. 30 Nov 1910.
25: 15 Feb 1911, Contr. no. 141, Bull. 38: 11-23. 15 Feb 1911.
26: 13 Apr 1912, Contr. no. 151, Bull. 39: 99-111. 13 Apr 1912.
27: 23 Jul 1912, Contr. no. 153, Bull. 39: 301-328. 23 Jul 1912.
28: 18 Mar 1913, Contr. no. 156, Bull. 40: 43-74. 18 Mar 1913.
29: 12 Sep 1913, Contr. no. 160, Bull. 40: 461-485. 12 Sep 1913.

9857. *Catalogue of the flora of Montana* and the Yellowstone National Park ... New York 1900. Oct. (*Cat. fl. Montana*).
Publ.: 15 Feb 1900, p. [i]-xi, [1]-492. *Copies*: BR, HH, MO, NY, PH, US, USDA; IDC 7615. – Issued as Mem. New York Bot. Gard. vol. 1. See also his *Rarities from Montana*, Bull. Torrey bot. Club 24: 188-192, 243-253, 292-299, *pl. 304-307*. 1897, and his *Composition of the Rocky Mountain Flora*, Science, ser. 2. 12: 870-873. 1900.
Supplement: J.W. Blankinship, Montana Agric. Coll. Sci. Stud. 1: 33-109. 1905.
Ref.: Nelson, A., Bot. Gaz. 30: 61-64. 1900.

9858. Contributions from the New York Botanical Garden – no. 19. *The American species of Limnorchis and Piperia*, north of Mexico ... New York 1901. Oct.
Publ.: 25 Nov 1901, cover-t.p., p. 605-643. *Copy*: U. – Reprinted and to be cited from Bull. Torrey bot. Club 28: 605-643. 25 Nov 1901.

9859. Contributions from the New York Botanical Garden – n. 37. *Some generic segregations* ... New York 1903. Oct.
Publ.: 16 Mai 1903, cover-t.p., p. 271-281. *Copy*: U. – Reprinted and to be cited from Bull. Torrey bot. Club 30: 271-281. Mai 1903.

9860. Contributions from the New York Botanical Garden – no. 76, *Astragalus and its segregates* as represented in Colorado ... New York 1906. Oct.
Publ.: 22 Jan 1906, cover-t.p., p. 657-669. *Copy*: U. – Reprinted and to be cited from Bull. Torrey bot. Club 32: 657-658. 1906.

9861. *Flora of Colorado* ... Fort Collins, Colorado (published by the Experiment Station) 1906. Oct. (*Fl. Colorado*).
Publ.: Aug 1906 (see J. New York Bot. Gard. 7: 217. Sep 1906 and Bull. Torrey bot. Club 33: 511. 5 Oct 1906, inf. J.W. Landon), p. [i]-xxii, [1]-447, [448, err.]. *Copies*: FH, G, HH, MICH, MO, NY, PH, US, USDA. – Published as Bulletin 100, Agric. Exp. Station, Colorado Agric. Coll.
Ref.: Nelson, Aven, Torreya 6: 237-239. 26 Nov 1906 (rev.).

9862. Contributions from the New York Botanical Garden – no. 112. *Notes on Philotria* Raf. ... New York 1908. Oct.
Publ.: 29 Sep 1908, cover-t.p., p. 457-465. *Copy*: U. – Reprinted and to be cited from Bull. Torrey bot. Club 35: 457-465. 29 Sep 1908.

9863. *Notes on Rosaceae* i-xiv. ... New York 1908-1923. Oct.
Publ.: A series of papers published in *Bull. Torrey bot. Club* ("Bull.") and reprinted in *Contr. N.Y. Bot. Gard.* ("Contr."). *Copy*: U.
1: 30 Nov 1908, Bull. 35: 535-542. No 1908, Contr. 46.
2: 14 Aug 1909, Bull. 36: 397-407. 14 Aug 1909, Contr. 125.
3: 29 Jul 1910, Bull. 37: 375-386. 29 Jul 1910, Contr. 136.
4: 28 Oct 1910, Bull. 37: 487-502. 28 Oct 1910, Contr. 138.
5: 7 Mar 1911, Bull. 38: 78-89. 7 Mar 1911, Contr. 142.
6: 21 Aug 1911, Bull. 38: 351-368. 21 Aug 1911, Contr. 145.
7: 22 Jul 1914, Bull. 41: 399-332. 2 Jul 1914. Contr. 170.
8: 29 Oct 1914, Bull. 41: 483-503. 29 Oct 1914, Contr. 173.
9: 17 Mar 1915, Bull. 42: 117-160. 17 Mar 1915, Contr. 178.
10: 22 Sep 1915, Bull. 42: 463-479, *pl. 26-27*. 22 Sep 1915, Contr. 182.

11: 27 Feb 1915 (on cover Contr.; on p. 65 reprint: 26 Feb 1917), Bull. 44: 65-84. 27 Feb 1917, Contr. 196.
12: 10 Mar 1920 (on cover Contr.; on p. 45 reprint: 13 Mar 1920), Bull. 47: 45-66. 10 Mar 1920, Contr. 220.
13: 1 Aug 1921 (on cover Contr.; on p. 159 reprint: 5 Aug 1921), Bull. 48: 159-172. 1 Aug 1921.
14: 23 Feb 1923, Bull. 50: 61-71. 1923, Contr. 241.

9864. Contributions from the New York Botanical Garden – no. 148. *List of plants collected on the Peary Arctic Expedition* of 1905-1906 by Dr. L.J. Wolf, and of 1908-09 by Dr. J.W. Goodsell. New York 1912. Oct.
Publ.: 24 Jan 1912, p. [1]-11. *Copies*: BR, U. – Reprinted and to be cited from Torreya 12: 1-11. 1912.

9865. *Flora of the Rocky Mountains* and adjacent plains. Colorado, Utah, Wyoming, Idaho, Montana, Saskatchewan, Alberta, and neighboring parts of Nebraska, South Dakota, North Dakota, and British Columbia ... New York (published by the author) 1917. 16-mo. (*Fl. Rocky Mts.*).
Publ.: 31 Dec 1917 (see J. New York Bot. Gard. 19: 66. 1918; p. iii: Nov 1917, US rd. 25 Mar 1916, GH copy rd. 5 Mar 1918; see also Fernald 1918 who prefers"early in 1918"), p. [i]-xii, 1-1110. *Copies*: HH, MICH, MO, NY, PH, US.
Preliminary studies: Studies on the Rocky Mountain flora 1-29, *in* Bull. Torrey bot. Club 27-40, 1900-1913.
Ed. 2: 21 Jan 1923 (p. iv: Nov 1922; see Fernald (1936); "recently" J. NYBG Mar 1923; PH rd. 29 Mai 1923), p. [i]-xii, 1-1143, [1144]. – Additions on p. 1111-1144. *Copies*: G, HH, MICH, PH, USDA. – Second edition, New York (published by the author) 1922. 16-mo.
Facsimile ed. vol. 2: 1954, p. [i]-xii, [1]-1143, [1144]. *Copies*: BR, FI. – Hafner Publishing Co. New York 1954. *Second printing*: 1969, idem. *Copy*: FAS.
Ref.: Ewan, J., Rocky Mt. natural. 295. 1950, ed. 2. 191. 1981.
Fernald, M.L., Rhodora 38: 329-331. 1936.
Stafleu, F.A., Taxon 19: 135. 1970.
Standley, P.C., Torreya 18: 91-94. 4 Jun 1918 (rev. ed. 1).

9866. *Key to the Rocky Mountain flora*. Colorado, Utah, Wyoming, Idaho, Montana, Saskatchewan, Alberta, and parts of Nebraska, South Dakota, North Dakota, and British Columbia ... New York (published by the author) 1919. (*Key Rocky Mt. fl.*).
Publ.: Mar 1919 (p. [3]: Sep 1918, US rd. 18 Apr 1919), p. [1]-305. *Copies*: US, USDA. – A reprint of the keys of Rydberg's *Fl. Rocky Mts.*, see above.

9867. *Notes on Fabaceae* – i-xiii ... New York 1923-1931. Oct.
Publ.: A series of 13 papers published in *Bull. Torrey bot. Club* ("Bull.") and reprinted as Contr. N.Y. Bot. Gard. ("Contr."). *Copy*: U.
1: 25 Mai 1923, Bull. 50: 177-187. 25 Mai 1923, Contr. 244.
2: 17 Aug 1923, Bull. 50: 261-272. 17 Aug 1923, Contr. 247.
3: Jan 1924, Bull. 51: 13-23. 8 Feb 1924, Contr. 253.
4: 18 Mai 1925, Bull. 52: 143-156. 18 Mai 1925, Contr. 270.
5: 9 Jun 1925, Bull 52: 229-235. 9 Jun 1925, Contr. 272.
6: 20 Nov 1925, Bull. 52: 365-372. 20 Nov 1925, Contr. 275.
7: 1 Apr 1926, Bull. 53: 161-169. 1 Apr 1926, Contr. 280.
8: 21 Feb 1927, Bull. 54: 13-23. 21 Feb 1927, Contr. 287.
9: 16 Mai 1927, Bull. 54: 321-336. 16 Mai 1927, Contr. 289.
10: 15 Mar 1928, Bull. 55: 119-134. 15 Mar 1928, Contr. 297.
11: 7 Apr 1928, Bull. 55: 155-164. 7 Apr 1928, Contr. 300.
12: 17 Feb 1930, Bull. 56: 539-554. 17 Feb 1930, Contr. 314.
13: 20 Jun 1931, Bull. 57: 397-407. 20 Jun 1931, Contr. 320.

9868. *Genera of North American Fabaceae* i-vii, 1923-1930. Oct.
Publ. 1923-1929; a series of 7 papers especially published in Amer. J. Bot. ("AJB") and reprinted as Contr. N.Y. Bot. Gard. ("Contr."). *Copy*: U.

1: Nov 1923, AJB 10: 485-498. Nov 1923, Contr. 250.
2: Aug 1923, AJB 11: 470-482, *pl. 33-36*. Aug 1924, Contr. 257.
3: Mar 1928, AJB 15: 195-203. Mar 1928, Contr. 296. – *Tribe Psoraleae.*
4: Jul 1928, AJB 15: 425-432. Jul 1928, Contr. 301. – *Tribe Psoraleae* (continued).
5: 14 Jan 1929, AJB 15: 584-594. Dec 1928, publ. 14 Jan 1929, Contr. 303. – *Astralagus and related genera.*
6: Apr 1929, AJB 16: 177-206. Apr 1929, Contr. 309. – *Astralagus and related genera* (continued).
7: Mar 1930, AJB 17: 231-238. Mar 1930, Contr. 316. – *Astralagus and related genera* (continued).

9869. Contributions from the New York Botanical Garden – no. 255. *The section Tuberarium of the genus Solanum in Mexico and Central America* ... New York 1924. Oct.
Publ.: 17 Mai and 13 Jun 1924 (in journal), cover-t.p., p. 145-154, 167-176. *Copy*: U. – Reprinted and to be cited from Bull. Torrey bot. Club 51: 145-154. 17 Mai 1924, 167-176. 13 Jun 1924.

9870. Contributions from the New York Botanical Garden – no. 260. *Some senecioid genera.* New York 1924. Oct.
Publ.: 18 Sep & 24 Oct 1924, cover-t.p., p. 369-378, 409-428. *Copy*: U. – Reprinted and to be cited from Bull. Torrey bot. Club 51: 369-378. 18 Sep 1924, 409-420. 24 Oct 1924.

9871. Contributions from the New York Botanical Garden – no. 294. *New species from the Blue Ridge.* New York 1927. Oct.
Publ.: 4 Nov 1927, cover t.p., p. 84-90, *pl. 3-5*. *Copy*: U. – Reprinted and to be cited from Torreya 27: 84-90, *pl. 3-5*. 4 Nov 1927.

9872. Contributions from the New York Botanical Garden – 311. *Scylla or Charybdis* ... New York 1929. Oct.
Publ.: 1929, cover t.p., p. 1539-1551. *Copy*: U. – Reprinted and to be cited from Proc. Int. Congr. Pl. Sci., Ithaca, 2: 1539-1551. 1929. – Reflections on nomenclature and on splitting or not splitting genera. Pragmatical approach; enlightening for understanding Rydberg's taxonomy of small units.

9873. *Flora of the prairies and plains of Central North America* ... New York (published by The New York Botanical Garden) 1932. Oct. (*Fl. Plains N. Amer.*).
Publ.: 14 Apr 1932 (see J. New York Bot. Gard. 33: 138, 201. 1932), p. [i]-vii, 1-969. *Copies*: BR, G, HH, MICH, MO, NY, PH, US(2), USDA. – New taxa previously published in Brittonia 1: 79-104. Oct 1931.
Facsimile ed.(*1*): 1965, [i]-vii, [1]-969. *Copy*: FAS. – Published for the New York Botanical Garden by Hafner Publishing Company, New York and London 1965.
Facsimile ed.(*2*): 1971, 2 vols., [i-iv], 1-503 and [1-2], 503-969. *Copy*: FI. – ISBN 0-486-22585-2.
Ref.: Diels, L., Bot. Jahrb. 66 (Lit.): 25. 1934.
Gleason, H.A., Torreya 32(5): 130-133. 18 Oct 1932.
Rendle, A.R., J. Bot. 71: 53-54. Feb 1933.
Rosendahl, C.O., J. New York Bot. Gard. 33: 201-204. Sep 1932.
Stafleu, F.A., Taxon 15: 329. 1966.

Rylands, Thomas Glazebrook (1818-1900), British (English) wire manufacturer and diatomist at Warrington (Lancashire). (*Rylands*).

HERBARIUM and TYPES: BM (orig. diat.; incl. 6000 slides); other material at AWH, BP and GL (Brit. herb.). Rylands correspondence is also at BM (presented 1907 by Miss Martha G. Rylands). The Diatom herbarium includes also that of Christopher Johnston (of Lancaster); other slides are by Walker Arnott, R.K. Greville, Gregory and G. Norman. – Dallman and Wood 1909 mention a herbarium at Warrington Museum (WRN)(Rylands was one of the founders of the Warrington Natural History Society).

BIBLIOGRAPHY and BIOGRAPHY: Ainsworth p. 102, 314, 347; Barnhart 3: 196; BB p. 265; BM 4: 1776; CSP 5: 349; Desmond p. 536 (b. 24 Mai 1818, d. 14 Feb 1900); De Toni 1: cxix, 2: ciii; Hortus 3: 1203; IH 2: (files); Zander ed. 11, p. 810.

BIOFILE: Anon., J. Bot. 45: 455-456. 1907 (diatom herb. to BM; list letters in Rylands collection of letters from and to diatomologists).
Bridson, G.D.R. et al., Nat. hist. mss. res. Brit. Isl. 433 [index]. 1980.
Dallman, A.A. & M.H. Wood, Trans. Liverpool bot. Soc. 1: 88. 1909.
Fryxell, G.A., Nova Hedwigia Beih. 35: 363. 1975 (coll.).
Kent, D.H., Brit. herb. 76. 1957 (Brit. herb. at GL).
Newton, L.M., Phycol. Bull. 1: 18. 1952.
Radcliffe, R.D., Proc. Linn. Soc., London 114: 41-42. 1902 (obit.).

EPONYMY: *Rylandsia* Greville et Ralfs ex Greville (1861).

Rytz, August Rudolf Walther (Rytz-Miller) (1882-1966), Swiss botanist; Dr. phil. Bern 1907; habil. Bern 1911; curator of botanical collections Bot. Inst. Bern, 1915-1920; extraordinary professor of botany Univ. Bern 1920-1952; moving force in setting up the alpine garden Schynige Platte, nr. Interlaken; vice-director University Botanical Garden Bern 1952-1966; co-founder of the Bernische Botanische Gesellschaft (1918); outstanding worker on the flora of the Bernese Alps and historian of botany. (*Rytz*).

HERBARIUM and TYPES: BERN (rd 1920).

BIBLIOGRAPHY and BIOGRAPHY: Barnhart 3: 196 (b. 13 Jan 1882); BFM no. 957, 975; BJI 2: 151; BL 2: 565, 569, 570, 711; Hirsch p. 256; IH 2: (files); Kew 4: 556; LS 23274, 38662; MW suppl. p. 296; NI 1709, 1708n; PFC 3(2): xvii; Plesch p. 393; TL-2/1793.

BIOFILE: Anon., Hedwigia 57(1): (76). 1915 (curator Bern).
Rytz, W., Mitt. naturf. Ges. Bern 5: 26. 1923 (herb.); Verh. naturf. Ges. Basel 44(1): 1-222. 1933 (Das Herbarium Felix Platters. Ein Beitrag zur Geschichte der Botanik der xvi. Jahrhunderts).
Welten, M., Mitt. naturf. Ges. Bern 24: 103-113. 1967 (obit., portr., bibl.).
Verdoorn, F., ed., Chron. bot. 1: 30, 132, 275. 1936, 3: 240. 1937.

NOTE: Rytz published a *Schweizerische Schulflora*, ed. 1 (1923), ed. 4 (1954).

9874. *Geschichte der Flora des bernischen Hügellandes* zwischen Alpen und Jura ... Bern (Buchdruckerei K.J.Wyss) 1912. Oct. (*Gesch. Fl. bern. Hügell.*).
Publ.: 1912 (t.p. preprint; Nat. Nov. Dec(2)) 1912), p. [i]-ix, [1]-169, chart. *Copy*: G. – Preprinted from Mitt. naturf. Ges. Bern 1912: 53-221. 1913.
Ref.: Baumann, E., Bot. Centralbl. 122: 570-572. 1913 (rev.).

9875. *Pflanzenaquarelle des Hans Weiditz* aus dem Jahre 1529 (Die Originale zu den Holzschnitten im Brunfels'schen Kräuterbuch ... Bern (Verlag Paul Haupt, ...) 1936. (*Pfl.-Aquar. Weiditz*).
Publ.: 1936 (p. 4: Oct 1934), p. [1]-44, *pl. 1-15* (col.). *Copy*: U. – Contains reproductions of some of the originals for the woodcuts in Otto Brunfels, *Herb. vivae eicon.* 1530, see TL-2/852. These originals were included in the herbarium of Félix Platter at BERN.
– For further details, see Rytz, *Das Herbarium Felix Platters* (Rytz 1933).

Saage, Martin Joseph (1803-?), German (Bohemian-born) botanist and high school teacher at the gymnasium of Braunsberg (Eastern Prussia). (*Saage*).

HERBARIUM and TYPES: Unknown.

BIBLIOGRAPHY and BIOGRAPHY: Barnhart 3: 1776; Herder p. 70; PR 7938, ed. 1: 8900; Rehder 1: 381.

9876. *Catalogus plantarum phanerogamarum circa Brunsbergam* sponte crescentium cum clave

Linneana et systemate naturali in usum discipulorum conscriptus a M.J. Saage. Brunsbergae [Braunsberg; h.t. Braniewo] (impressit C.A. Heyne), [1846]. Oct. (*Cat. phan. Brunsb.*).
Publ.: 1846 (see Flora 5: 635. 1847 for date and review), p. [1]-88. *Copy*: NY.
Ref.: J., Bot. Zeit. 5: 635. 3 Sep 1847 (rev.).

Sabbati, Liberato (ca. 1714-?), Italian (Umbrian) botanist and gardener; curator of a botanical garden and surgeon in Roma. (*Sabbati*).

HERBARIUM and TYPES: Sabbati made a number of herbaria mainly for teaching purposes, but also to sell or to present to his benefactors. Two of these herbaria are at BM, (Britten 1904, Ardagh 1942), six others in the Bibliotheca Casanatense et Alessandrina in Roma of which one is in 19 volumes, compiled between 1756 and 1776, called *Theatrum botanicum romanum*. Two other herbaria (*Hortulus practico-botanicus* and *Theatrum botanicum*) are in the Corsiniana library in Rome (Pirotta & Chiovenda 1902, Pirotta 1903, Béguinot 1899, 1900, and Saccardo 2: 94-95.

BIOFILE: Barnhart 3 : 197; BM 4: 1777; Dryander 3: 112, 149; Frank 3 (Anh.): 87; Herder p. 179; Kew 4: 558; Plesch p. 393; PR 7939, ed. 1: 8902; Rehder 1: 424; Saccardo 1: 143, 2: 94-95, cron. p. xxxi; TL-1/122; TL-2/634; see G. Bonelli, N. Martelli; Tucker 1: 615.

BIOFILE: Ardagh, J., J. Bot. 80: 120. 1942 (a further Sabbati herbarium purchased by BM).
Béguinot, A., Boll. Soc. geogr. ital. ser. 4. 1(3): 198-207. Mar 1900 (on a Sabbati collection in the Visconti Lyceum in Rome); Bull. Soc. bot. ital. 1899 (9-10): 305-306 (on a Sabbati herbarium in the Bibl. Corsiniana, "Hortulus practico-botanico").
Britten, J., J. Bot. 42: 148-151. 1904 (on two Sabbati herbaria in BM).
Murray, G., in Hist. coll. BM(NH) 1: 179. 1904.
Pirotta, R., Ann. di Bot. 1: 59-61. 1903 (on a Sabbati herbarium in the Corsiniana library).
Pirotta, R. & E. Chiovenda, Malpighia 16: 49-157. 1902 (on the Sabbati herbaria in Rome).
Saccardo, P.A., Malpighia 17: 270-271. 1902 (notes on studies on Sabbati).

COMPOSITE WORKS: See G. Bonelli, *Hort. rom.* 1772-1793, TL-2/634. L. Sabbati probably supplied most of the data for this book: "... adumbrationem dirigente Liberato Sabbati ..." The species descriptions in volumes 6-8 are attributed to Constantino Sabbati, see e.g. Dryander 3: 113. We have no further information on C. Sabbati except that he too was a gardener in Roma, somewhat later than Liberato Sabbati (Saccardo 1: 143.).

EPONYMY: *Sabbata* Vellozo (1825); *Sabbatia* Moench (1794). *Note: Sabatia* Adanson (1763) is very likely also named for Sabbati but no etymology is given (see Adanson, Fam. Pl. 1: 27. 1763).

9877. *Synopsis plantarum* quae *in solo Romano* luxuriantur studio, et labore Liberati Sabbati ... breviori forma, & facilitate descriptarum juxta methodum tournefortianum liber primus ... Ferrariae [Ferrara] (apud Josephum Barbieri ...) 1745. Qu. (*Syn. pl. solo Rom.*).
Publ.: 1745 (imprimatur 19 Mar 1745), p. [i-ix], 1-50, [51-54; charts, err.], *2 pl. Copies*: FI, G.
Other issue: 1754, p. [i]-viii, [ix], 1-50, [51-54, charts, err.], *2 pl. Copy*: NY. – Preface material different, main text same. "*Collectio plantarum* quae *in solo Romano* luxuriantur opera, et studio Liberati Sabbati ... breviori forma, & facilitate descriptarum editio primo romana cum figuris aeneis et indice denominationum, virtutum, atque facultatum locupletissimo. Romae", 1754. Qu. To be taken into account for botanical nomenclature.

Sabine, [Sir] **Edward** (1788-1883), British (Dublin-born) astronomer, geophysicist, zoologist and explorer; brother of Joseph Sabine; astronomer to arctic expeditions led by

SABINE, E.

J. Ross (1818) and W.E. Parry 1819-1820; collected plants in Melville Island and Greenland; president Royal Society 1861-1871. (*E. Sabine*).

HERBARIUM and TYPES: BM; further material at C, CGE, FI. – Letters at BM and K.

BIBLIOGRAPHY and BIOGRAPHY: AG 2(1): 438; Backer p. 266; Barnhart 3: 197; BM 4: 1777; Bret. p. 251-252; CSP 5: 351-354, 8: 805-806, 11: 251; Dawson p. 727; Desmond p. 536 (b. 14 Oct 1788, d. 26 Jun 1883); DNB 50: 74-78; IH 2: (files); Lasègue p. 324, 326, 329, 373, 396, 457; PR 4214, ed. 1: 4652, 4787; TL-2/2994.

BIOFILE: Allan, M., The Hookers of Kew 123. 1967.
Allibone, S.A., Crit. dict. Engl. lit. 2: 1909. 1878.
Anon., Proc. R. Soc. London 51: xliii-li. 1892 (obit.).
Bridson, G.D.R. et al., Nat. hist. mss. res. Brit. Isl. 75.49, 143.21, 278.37. 1980.
Embacher, F., Lexik. Reisen 256-257. 1882.
Fletcher, H.R., The story of the Royal Hort. Soc. 1804-1968, 67, 95-96, 100. 1969.
Huxley, L., Life letters J.D. Hooker 1: 15, 42, 49, 145, 2: 75, 127, 133, 149. 1918.
Lemmon, K., Golden age plant hunters 125, 176. 1968.
Murray, G., Hist. coll. BM(NH) 1: 179. 1904.
Poggendorff, J.C., Biogr.-lit. Handw.-Buch 2: 728-730. 1863, 3: 1158. 1898 (bibl.).

COMPOSITE WORKS: (1) W.J. Hooker, *Some account of a collection of arctic plants formed by Edward Sabine* ... Trans. Linn. Soc. 14: 360-394, *pl. 13*. 1824. See TL-2/2994 and ... (W.E. Parry 1824).
(2) See also W.E. Parry, J. Voy. North-West Passage, suppl. to appendix p. cclxi-cccx. 1824 and R. Brown, *Chlor. melvill.* 1823. The plants from Melville Island were collected by E. Sabine. See TL-2/831.

Sabine, Joseph (1770-1837), British (English) barrister and horticulturist; inspector-general of assessed taxes 1808-1835; secretary of the Horticultural Society of London 1816-1830; elder brother of Edward Sabine, founder of the Transactions of the [later Royal] Horticultural Society. (*Sabine*).

HERBARIUM and TYPES: Unknown. – Letters at K and LINN; mss. at LINN.

BIBLIOGRAPHY and BIOGRAPHY: AG 1: 210; Backer p. 505; BB p. 266; BL 1: 53, 316; BM 4: 1777; Bossert.p. 344; Bret. p. 251; CSP 5: 354-355, 12: 642; Dawson p. 728; Desmond p. 536 (b. 6 Jun 1770, d. 24 Jan 1837); DNB 50: 79; Frank 3 (Anh.): 87; Herder p. 305; Hortus 3: 1203; MW p. 421-422; PR 7940, ed. 1: 8904; Rehder 5: 749; Tucker 1: 615-616; Zander ed. 10, p. 710, ed. 11, p. 810.

BIOFILE: Allibone, S.A., Crit. dict. Engl. lit. 2: 1909. 1878.
Anon., Flora 20: 288. 1837 (rev.).
Bridson, G.D.R. et al., Nat. hist. mss. res. Brit. Isl. 433 [index]. 1980.
Charlesworth, E., Mag. nat. Hist. 1837: 390-392 (obit.).
Coats, A.M., Plant hunters 304, 307, 312. 1969.
Dony, J.G., Fl. Hertfordshire 12. 1967.
Druce, G.C., Gard. Chron. ser. 3. 75: 106. 1924 (letter Sabine to William Baxter).
Fletcher, H.R., Story Roy. Hort. Soc. 1804-1968, p. 555. 1969 (index; many entries; info. on his part in the activities of the RHS).
Kernan, J., *in* J. Harrison, Floricult. cabinet 5: 94-95. 1837 (note on death and horticultural work of S.).
Lemmon, K., Golden age plant hunters 228. 1968 [index].
Nelmes & Cuthbertson, Curtis's Bot. Mag. dedic. 1827-1927, p. 10-12. 1932.
Palmer, T.S., The Condor 30(5): 293-294. 1928 (epon.).
Pryor, R.A., Fl. Hertfordshire xli-xlii. 1887.
Texnier Le, Le Jardin 24 (suppl. 20 Dec 1910): 30-31. 1910 (assessment of his horticultural work).

EPONYMY: *Sabinea* A.P. de Candolle (1825).

Sabransky, Heinrich (1864-1915), Austrian botanist (batologist) and physician; born in Presburg (Bratislava); studied medicine in Wien; practicing physician at Tramin 1892; at Mayrhofen 1896; regional physician at Söchau (Steiermark), 1896-1915. (*Sabransky*).

HERBARIUM and TYPES: W (17.558); further material at B (extant), BP, C, CERN (500 Rubus), E, GB, H, WRSL.

BIBLIOGRAPHY and BIOGRAPHY: AG 6(1): 165, 12(1): 374; Barnhart 3: 197 (b. 28 Apr 1864, d. 23 Dec 1915); BFM no. 901; Biol.-Dokum. 15: 7783; DTS 1: 250, 6(4): 191; Rehder 5: 749; SBC 131; Tucker 1: 616.

BIOFILE: Anon., Bot. Not. 1917: 141 (err. d. 28 Dec 1916); Hedwigia 59: (141). 1917 (d. 28 Dec 1916, err.); Magy. bot. lap. 15: 312. 1916 (d. 29 Dec 1916, err.); Österr. bot. Z. 66: 312. (d. 24 Dec 1916, err.), 408 (herb. to W). 1916.
Gáyer, G., Westungarischer Grenzbote 46, 1917, nr. 15389, 4 old. (n.v. fide Hedwigia 60: (58). 1918).
Hayek, A. v., Verh. zool.-bot. Ges. Wien 67: 216-217. 1917 (obit., portr., bibl.).
Janchen, E., Cat. fl. austr. 1: 31, 46. 1956.
Kneucker, A., Allg. bot. Z. 23: 32. 1917 (d. 24 Dec 1916, err.), 25: 31. 1920 (herb. to W).
Lack, H.W., Willdenowia 10: 81. 1980 (coll. at B extant).
Pax, F., Grundz. Pfl.-Verbr. Karpathen 46. 1898 (Veg. Erde 2(1)); Bibl. schles. Bot. 20. 1929.
Szymkiewicz, D., Bibl. Fl. Polsk. 123. 1925.

Saccardo, Domenico (1872-1952), Italian botanist; lecturer (assistant) in botany at the University of Bologna 1899; assistant plant pathologist at the viticultural college Conegliano 1900; assistant at the phytopathological station of Roma 1902; free lecturer at the University of Roma; collected in Ethiopia between 1936 and 1939; son of P.A. Saccardo. (*D. Sacc.*).

HERBARIUM and TYPES: PAD; *Exsiccatae*: *Mycotheca italica*, nos. 1-1750, and 19 suppl. nos. 1897-1913. – Sets at BM, DAO, FH, H, L, M, MO, NY, PC, S.

BIBLIOGRAPHY and BIOGRAPHY: Barnhart 3: 197 (b. 14 Nov 1872); BL 2: 337, 416, 712; BM 4: 1777; Bossert p. 344; CSP 18: 391; De Toni 4: xlix; GR p. 521 (d. 16 Aug 1952), *pl.* [*28*]; GR, cat. p. 70; Kelly p. 202; LS 23304-23311, 31067, 38669; Saccardo 2: 95; Stevenson p. 1256; TL-2/543; Tucker 1: 616.

BIOFILE: Anon., Bot. Centralbl. 89: 175. 1902 (app. Roma) 77: 176. 1899 (app. Univ. Bologna), 82: 399. 1900 (app. Conegliano), 89: 175. 1902 (app. Roma); Hedwigia 38: (114). 1899 (app. "professor" at Bologna, actually regular assistant), 39: (195). 1900 (app. Conegliano). Nat. Nov. 24: 187. 1902 (asst. phytopath. station Roma); Österr. bot. Z. 49: 311. 1899 (asst. Bologna).
Österr. bot. Z. 52: 171. 1902 (asst. phytopath. station Roma).
Cufodontis, G., C.R. AETFAT 4: 246. 1962 (coll. Ethiopia 1936-1939).
Kneucker, A., Allg. bot. Z. 5: 36. 1899 (app. Bologna), 6: 172. 190 (app. Conegliano), 8: 72. 1902 (app. Roma).
Laundon, J.R., Lichenologist 11: 16. 1979 (herb. PAD).

COMPOSITE WORKS: (1) For contributions to P.A. Saccardo, *Sylloge fungorum* (vols. 17, 18) see under that author.
(2) *Supplemento micologico* alla "Flora veneta crittogamica" [by G. Bizzozero, see TL-2/543], 1900.

9878. *Contribuzione alla micologia veneta e modenese* [Genova (Tipografia Ciminago) 1898]. Oct.
Publ.: 1898 (p. [1]: Jul 1898), p. [1]-27, *pl. 7-8*. *Copy*: G. – Reprinted and to be cited from Malpighia 12: 201-228. 1898.

Saccardo, Francesco (1869-1896), Italian botanist; studied at the University of Padova; Mag. sci. nat. 1892; assistant at the school of ecology and viticulture at Avellino 1895-1896; nephew of P.A. Saccardo. (*F. Sacc.*).

HERBARIUM and TYPES: Lichens at PAD.

BIBLIOGRAPHY and BIOGRAPHY: Barnhart 3: 197(b. 7 Jul 1869, d. 6 Oct 1896); BL 2: 416, 712; Bossert p. 344; CSP 18: 391; De Toni 4: xviii; GR p. 538; Kew 4: 558-599; LS 23312-23316; Saccardo 1: 144, 2: 95; Stevenson p. 1256.

BIOFILE: Anon., Bot. Centralbl. 62: 192. 1895 (app. Avellino), 64: 432. 1895 (d.); Bot. Jahrb. 21 (Beibl. 53): 61. 1896 (app. Avellino), 21 (Beibl. 57): 59. 1897 (d. 6 Oct 1896).; Bot. Not. 1897: 47 (d.); Nat. Nov. 18: 70 (app. Avellino), 332 (d.). 1896; Österr. bot. Z. 46: 39 (app. Avellino), 473 (d.). 1896.
De Toni, G.B., Nuovo Notarisia 7: 154-156. 1896 (obit., bibl.).

9879. *Saggio di una flora analitica dei licheni del Veneto* aggiuntavi l'enumerazione sistematica delle altre specie italiane. Padova (Stab. Prosperini) 1894. Oct. (*Sagg. lich. veneto*).
Publ.: 1894 (list of new taxa dated Aug 1894; t.p. date 1894; Bot. Centralbl. 23 Jan 1895; Nat. Nov. Jan(1, 2), Jun(1, 2) 1895), p. [i-iv], [1]-164, *pl. 1-13* (p.p. col. liths. auct.). *Copies*:: FH, H. – Preprinted from Atti Soc. veneto trent. Sci. nat. Padova 2(1): 83-241, *pl. 1-13*. 1895. – See also F. Saccardo & A. Fiori, Contribuzione alla lichenologia del Modenese e Reggiano, Atti Soc. Natural. Modena ser. 3. 13: [170]-207. Jul-Dec 1895 (p. 207: Jul 1895; t.p. journal 1894).

9880. *Florula del Montello* (Provincia di Treviso). Padova (R. Stabilimento, Prosperini) 1895. Oct. (*Flor. Montello*).
Publ.: Mai 1895 (p. 5: Jan 1895; journal issue for Mai 1895; Bot. Zeit. 1 Dec 1895; Nat. Nov. Jun(1,2) 1895), p. [1]-15, err. slip. *Copy*: G. – Preprinted or reprinted from Bull. Soc. veneto trent. Sci. nat. 6(1): 5-18. Mai 1895.
Ed. 2, riveduta e aumentata da P.A. Saccardo e A. Trotter, Treviso 1920, 26 p., (n.v., fide BL).

Saccardo, Pier Andrea (1845-1920), Italian botanist (mycologist); Dr. phil. Padua 1867; assistant to R. de Visiani at Padua University 1866-1872; professor of natural sciences at the Padua Technical Institute 1869-1879; professor of botany and director of the botanical garden of the University of Padua 1879-1915. (*Sacc.*).

HERBARIUM and TYPES: PAD (69.000 fungi and further material); for a description and a catalogue of the Saccardo herbarium see G. Gola (1930). Saccardo's mycological library and manuscripts are also at PAD. Some other types are at Portici and Napoli (see Ciferri, 1952). – *Exsiccatae*: *Mycotheca veneta* 1874-1881, nos. 1-1600: B (extant), BM, BUCM, DBN, E, FH, GE, HBG, K, L, M, PRE, S, SIENA, STR, TLA, TO, W, WRSL. (Index to cent. 1-9 *in* Michelia 1: 101-116. 1 Jun 1877, also by A. Vido in 1: 553-619. 1 Sep 1879. – *Bryotheca tarvisiana centuria*, Padova/Treviso 1864, nos. 1-100 at W. – For an index see below, 1872, *Musci tarvisini*; see also Sayre (1971). – 44 letters to W.G. Farlow at FH.

BIBLIOGRAPHY and BIOGRAPHY: Ainsworth p. 320, 347 [index]; Backer p. 505; Barnhart 3: 197 (b. 23 Apr 1845, d. 11 Feb 1920); BFM no. 1064, 3034, 3035, 3092; BL 2: 337, 338, 411, 416, 712; BM 4: 1777-1778, 8: 1120; Bossert p. 344; CSP 5: 357, 8: 806-807, 11: 252-253, 12: 642, 18: 391-392; De Toni 1: cxix-cxx; DTS 1: 252, 3: l, 6(4): 191; Frank 3 (Anh.): 87; GR p. 521, *pl.* [*28*], cat. p. 70; Hawksworth p. 185, 197; Herder p. 245, 266, 433, 434; Hortus 3: 1203; IH 1 (ed. 6): 363, (ed. 7): 341; Jackson p. 65, 92, 316, 322, 511; Kelly p. 202-204; Kew 4: 270; KR p. 614; Langman p. 661; Lenley p. 357; LS 2783, 2791, 2793a, 3552, 3886, 4877, 7995a, 23107-23445, 24435, 38670-38679, 41690, suppl. 23845-23876; Moebius p. 107; Morren ed. 2, p. 26, ed. 10, p. 85; MW p. 422, suppl. p. 296; NAF 7: 1086, 9: 450; NI 1710-1711, 2: 152-155, 244, 3: 12, 59; PFC 1: lii; PH 906; PR 7941-7944; Rehder 5: 750; Saccardo 1: 143, 2: 95, cron. xxxi; SBC p. 131; SK 1: cxviii; Stevenson p. 1256; TL-1/1130-1131; TL-2/1174, 1215, 5698, 6588, see indexes; Tucker 1: 616.

BIOFILE: Ainsworth, G.C., Dict. fungi ed. 6. 514-515. 1971.
Anon., Atti Ist. venet. Sci. 50: ix-x. 1892 (titles and memberships); Bot. Centralbl. 47: 32. 1891 (price Lincei), 58: 415. 1894 (member Soc. natural. Moscou), 96: 256. 1904 (app. corr. memb. Lincei); Bot. Zeit. 38: 190. 1880 (app. Padua); Bull. Soc. bot. France 67: 86. 1920 (d.); Flora 63: 96. 1880 (app. Padua); Hedwigia 33: (112). 1894 (member Soc. Natural. Moscou), 62(1): 83. 1920 (d.); Nat. Nov. 22: 81. 1920 (d.); Nuova Notarisia 32: 70-71. 1921 (d.); Österr. bot. Z. 30: 139. 1880 (app. Padua), 69: 88. 1920 (d.).
Béguinot, A., Atti Accad. sci. Venet.-trent.-istr. ser. 3. 11: xvii-xviii. 1921 (obit.).
Cappelletti, C., Informatore bot. ital. 3(1): 8-10. 1970.
Ciferri, R.A., Taxon 1: 126. 1952 (coll., types).
De Toni, G.B., Atti r. Ist. venet. Sci. 79(1): 1-36. 1920, also as Commemorazione del prof. Pier Adrea [sic] Saccardo. Venezia 1920, 36 p., (obit., portr., bibl.).
Gager, C.S., Brooklyn Bot. Gard. Rec. 27(3): 277. 1938.
Gola, G., L'erbario micologico di P.A. Saccardo. Catalogo. Padova. Tipografia Editrice Antoniana, 1930, xvi, 328 p., 2 p. (portr. Saccardo, list of collectors; catalogue). Issued as Supplemento i, Atti Accad. sci. Venet.-trent.-istr., vol. 21.
Jackson, B.D., Proc. Linn. Soc. 133: 53-54. 1921 (obit., F.M.L.S. 1916).
Junk, W., Rara 1: 40, 110-111. 1900-1913 (descr. Syll. fung. 1-22).
Laundon, J.R., Lichenologist 11: 16. 1979.
Lloyd, C.G., Mycol. notes 29: 365-366. 1908, 52: 734. 1917 (portr.), 70: 1220. 1920, 83: 946. 1921 (d.).
Mattirolo, O., Bull. Soc. bot. ital. 1920(1-3): 2-3. 1920 (obit.); Reale Accad. naz. Lincei ser. 5. 30(5): 149-160. 1920 (obit.).
Murrill, W.A., Mycology 12: 164. 1920 (d.).
Saccardo, D., Nuovo Giorn. bot. ital. ser. 2. 27: 58-74. 1920 (bibl.).
Sayre, G.A., Mem. New York Bot. Gard. 19(2): 249. 1971 (*Bryotheca tarvisiana*); Bryologist 80: 516. 1977.
Traverso, G.B., Nuovo Giorn. bot. ital. ser. 2. 27: 39-74. 1920 (obit., portr., bibl. by D. Saccardo).
Trotter, A. et al., *in* P.A. Saccardo, Syll. fung. 23: ix-xxxii. 1925 (biogr., list of obituaries; bibl.).
Wittrock, V.B., Acta Horti Berg. 3(2): 165. 1903, 3(3): 180, *pl. 136*. 1905 (portr.).

COMPOSITE WORKS: (1) With C. Roumeguère, q.v., *Reliquiae mycologicae libertianae* series 2-4, 1881-1884, see TL-2/9683.
(2) With Albert Julius Otto Penzig (1856-1929), *Diagnoses fungorum novorum*, ser. 1-3, 1897-1902, see under O. Penzig, TL-2/7648.
(3) With William Trelease (1857-1945) and Charles Horton Peck (1833-1917), *The fungi of Alaska*, *in* W. Trelease, ed., Harriman Alaska Expedition, *Alaska* volume 5, *Cryptogamic botany*, New York 1904 (reissued as Harriman Alaska Series volume 5, Washington 1910), p. 11-64, *pl. 2-7* (rev. Bot. Zeit. 16 Jun 1904).
(4) With A.J.O. Penzig, *Icon. fung. jav.*, 2 vols., 1903, see under O. Penzig, TL-2/7655.
(5) Editor, *Flora italica cryptogamica*.

HANDWRITING: Candollea 33: 153-154. 1978.

EPONYMY: (genera): *Chaetosaccardinula* Batista (1962); *Neosaccardia* Mattirolo (1921); *Pasaccardoa* O. Kuntze (1891); *Phaeosaccardinula* P.C. Hennings (1905); *Saccardaea* Cavara (1894); *Saccardia* M.C. Cooke (1878); *Saccardinula* Spegazzini (1885); *Saccardoella* Spegazzini (1879); *Saccardomyces* Hennings (1904); *Saccardophytum* Spegazzini (1902); (journal): *Saccardoa*: Monographiae mycologicae, ed. R. Ciferri, 2 vols. Pavia, Italy. 1960-1963. *Note*: *Saccardoa* Trevisan (1869) is also likely named for this author.

NOTE: *Main bibliography* in G.B. de Toni, Commemorazione del Prof. Pier Andrea Saccardo, Venezia 1920, p. 22-36, but see also D. Saccardo in G.B. Traverso (1920) and LS.

9881. *Prospetto della flora trevigiana* ossia enumerazione sistematica delle piante finora osservata spontanee o naturalizzate nella provincia di Treviso aggiuntevi le denomi-

nazioni vernacole e varie osservazioni ... Venezia (dal priv. stab. di G. Antonelli edit.) 1864. Oct.
Publ: 1863-1864 (in journal), p. [i], [1]-156. *Copies*: FI, G. – Reprinted and to be cited from Atti Soc. venet.-trent.-istr. Sci. nat. ser. 3. 8(10): 1087-1132. 1863 (session 19 Jul), 9(3): 427-445 (session 27 Dec 1863), 9(4): 481-498 (session 27 Jan 1864), 9(5): 605-638 (session 21 Feb 1864), 9(5): 837-877 (session 20 Apr 1864). 1864.
Ref.: Schlechtendal, D.F.L. von, Bot. Zeit. 23: 219. 1865 (rev.).

9882. *Sulla flora fossile della formazione oolitica* del barone Achille de Zigno ... (Estratta dagli Atti del R. Istituto veneto di Scienze, Lettere ed Arti) [1868]. Oct.
Publ.: 1868, p. [1]-23. *Copy*: FI. – Reprinted and to be cited from Atti Ist. venet. Sci. nat. ser. 3. 13: 1562-1584. 1868.

9883. *Breve illustrazione delle crittogame vascolari trivigianae* aggiuntavi l'enumerazione di quelle fino ad oggi note nella flora veneta ... Venezia (Tipografia del Commercio di Marco Visentini) 1868. Qu.
Publ.: Jul 1867-Jul 1868 (in journal; reprint BSbF post 15 Oct 1868; Flora 24 Feb 1869), p. [i-ii], [1]-69. *Copies*: FI, G, NY. – Reprinted and to be cited from A.P. Ninni e P.A. Saccardo, eds., Comment. fauna fl. gea Veneto Trentino 1(1): 24-40. Jul 1867, 1(3): 150-163. Jan 1868, 1(4): 191-200. Apr 1868, 1(5): 225-251. Jul 1868.

9884. *Catalogo delle piante vascolari del Veneto* e di quelle più estesamente coltivate compilato da R. de Visiani ... P.A. Saccardo ... Venezia (Stabilimento priv. di Giuseppe Antonelli) 1869. Qu.
Co-author: Roberto de Visiani (1800-1878).
Publ.: 1868-1869, p. [i-ii], [1]-292. *Copies*: FI, USDA. – Reprinted and to be cited from Atti Ist. venet. Sci. ser. 3. 14(1): 71-111 (for Nov 1868), 14(2): 303-349 (session 22 Nov 1868), 14(3): 479-519 (session Dec 1868), 14(3): 703-737 (session 30 Jan 1869), 14(6): 1091-1139 (session 21 Mar 1869), 14(8): 1503-1545 (session 23 Mai 1869), 14(9): 1735-1776 (session 20 Jan 1869). 1869.

9885. *Della storia e letteratura della flora veneta* sommario ... Milano (Valentino e Mues libraj-editori) 1869. Oct. (*Stor. lett. fl. venet.*).
Publ.: 1869 (p. x: Jan 1869; Flora 10 Sep 1869; Bot. Zeit. 24 Sep 1869), p. [i]-x, [xi], [1]-208. *Copies*: FI, USDA.

9886. *Le piante dell'agro veneto* esposte in quadri dicotomici nella forma a grappe. Saggio letto all' Accademia di Scienze, Lettere ed Arti di Padova nella Tornata i. Maggio 1870 ... Padova (Tipografia G.B. Brandi) 1870.
Publ.: 1870 (read 1 Mai 1870), p. [1]-11, 2 charts. *Copy*: FH. – Also published in Rev. Accad. Sci. Lett. Padova 19: 53-59. 2 charts. 1870.

9887. *Musci tarvisini* enumerati tabulisque dichotomicis strictim et comparate descripti ... [Padova (Stab. Prosperini) 1872]. Oct.
Publ.: Jun 1872 (Bot. Zeit. 31 Jan 1873; published in journal issue for Jun 1872), p. [5]-47, [48, colo.]. *Copies*: FH, G, NY. – Reprinted from Atti Soc. venet.-trent. Sci. nat. Padova 1(1): 21-63. Jun 1872.

9888. *Mycologiae venetae specimen* ... Patavii [Padova] (Typ. P. Prosperini) 1873. Oct. (*Mycol. venet. spec.*).
Publ.: Jul, Dec 1873 (in journal, see below; if preprinted Jul[?]; J. Bot. Nov 1874; Bot. Zeit. 13 Nov 1874), p. [1]-215, [216], *pl. 4-17* (col. liths. auct., with each 1 p. expl.). *Copies*: FH, FI, G, NY, G, NY, PH, Stevenson. – Preprinted or reprinted from Atti Soc. venet.-trent. Sci. nat. 2(1): 53-96. Jul 1873, 97-264. Dec 1873.
Ref.: Thümen, F.K.A.E.J. von, Hedwigia 13: 171-172. Nov 1874 (rev.).

9889. *Di alcune nuove ruggini* (funghi uredinei) osservati nell'agro veneto. Padova (Tipografia G.B. Randi) 1874. Oct. (*Alc. nuov. ruggini*).
Publ.: 1874 (read Accad. Pad. 11 Apr 1874; Hedwigia Mai 1875), p. [1]-14, [1, expl.

pl.], *1 pl. Copies*: FH(2). – Reprinted from Riv. period. R. Accad. Padova 24: 199-212. 1874 (journal n.v.).

9890. *Fungi veneti novi vel critici*, a P.A. Saccardo observati [Nuovo Giorn. bot. ital. 5: 269-298. Pisa 1873].
Publ.: a series of 5 publications in various journals (*copies*: FH, FI, NY):
1: 20 Oct 1873, published *in* Nuovo Giorn. bot. ital. 5: 269-298. 20 Oct 1873.
2: Oct 1875, *in* Nuovo Giorn. bot. ital. 7(4): 299-329. Oct 1875 (but mentioned by Hedwigia Sep 1875 sic).
3: Mai 1875, p. [1]-8. *Copy*: NY. – Reprinted and to be cited from Hedwigia 14: 68-76. 1875; *editio secunda in* Michelia 2(4): 446-452. 15 Nov 1879.
4: Jul 1875, p. [1]-41. *Copies*: FH, NY. – Reprinted and to be cited from Atti Soc. venet.-trent. Sci. nat. 4(1): 101-141. Jul 1875 (Hedwigia Sep 1875).
5: Apr 1876, *in* Nuovo Giorn. bot. ital. 8(2): 162-211. Apr 1876.
6: 1 Jun 1877, *in* Michelia 1: 1-72.
7: 15 Jan 1878, *in* Michelia 1: 133-221.
8: 15 Jun 1878, *in* Michelia 1: 239-275.
8 (app.): 1 Jul 1878, *in* Michelia 1: 351-355.
9: 15 Nov 1878, *in* Michelia 1: 361-434, 435-445.
10: 1 Sep 1879, *in* Michelia 1: 539-546.
11: 25 Apr 1880, *in* Michelia 2: 154-176.
12: 5 Mar 1881, *in* Michelia 2: 241-301.
12 (app.): 5 Mar 1881, *in* Michelia 2: 377-383.
13: 1 Dec 1882, *in* Michelia 2: 528-563.

9891. *Sommario di un corso di botanica* già tenuto nella r. Università di Padova ... ii. edizione. Padova (Prem. Tipografia Ed.F. Sacchetto) 1874. Oct. (*Somm. corso bot.*).
Ed. 1: 1871, autograph, fide A. Trotter (1925), p. xxx, no. 195.
Ed. 2: 1874, p. [1]-223, *pl. 1-2* (uncol. liths). *Copy*: MO.
Ed. 3: 1880, 313 p., *3 pl.* Padova (F. Sacchetto), n.v. (Nat. Nov. Jan(1) 1882; Bot. Zeit. 27 Jan 1882, as of "1881").
Ed. 4: 1898, 343 p., *3 pl.* Padova (Seminario), n.v.

9892. *Nova ascomycetum genera* [Grevillea 4: 21-22. Sep 1875]. Oct.
Publ.: Sep 1875, in Grevillea 4: 21-22. Sep 1875. Reprint with original pagination at FH.

9893. *Conspectus generum pyrenomycetum italicorum* systemate carpologico dispositorum auctore P.A. Saccardo. [Atti Soc. venet.-trent. Sci. nat. 4(1), Padova (Stab. Prosperini) 1875]. Qu. (*Consp. gen. pyrenomyc.*).
Publ.: Jan-Jun 1875 (Grevillea Sep 1875; Hedwigia Jul 1875; Bot. Zeit. 10 Dec 1875), p. [1]-24, table. *Copies*: FH, FI. – Preprinted from Atti Soc. venet.-trent. Sci. nat. 4(1): 77-100. table. Oct 1875.
Corrigenda: Hedwigia 15: 6. Jan 1876.
Summary: Nuovo Giorn. bot. ital. 8(1): 11-15. Jan 1876.
Ref.: Anon., Grevillea 4: 30-32. Sep 1875 (strongly critical anonymous review).

9894. *Fungi italici* autographice delineati (additis nonnullis extra-italicis, asterisco notatis) ... Patavii [Padova] (Sumpt. auctoris (Lithogr. P. Fracanzani)) 1877-1886. (*Fung. ital.*).
Publ.: Mai 1877-Mai 1886, in 38 fascicles, p. [1]-14. *pl. 1-1500* (four illustrations ("plates") per leaf; partly col. liths.). *Copies*: BR, FI, FH, G, MICH, NY(2), PH, Stevenson, USDA; IDC 5704. – The plates are dated individually; they were issued, however, in sets of fascicles as follows (dates confirmed by Nat. Nov., Bot. Zeit. and Hedwigia):
fasc. 1-4: *pl. 1-160*. Mai 1877.
C*5-8: pl. 161-320. Feb 1878.
C*9-12: pl. 321-480. Nov 1878.
13-16: *pl. 481-640*. Jun 1879 (on t.p. of cover, Mai 1879; on t.p. of completed work).
17-28: *pl. 641-1120*. Jul 1881.

29-32: *pl. 1121-1280*. Aug 1882.
**33-36*: *pl. 1281-1440*. Jun 1883.
37-38: [1]-14, *pl. 1441-1500*. Mai 1886.

Index alphabeticus fungorum italicorum autographice delineatorum ... indicem ordinavit alumnus Augustus Berlese ... N. 1-1280. [Aug 1882], p. [1]-19. *Copy*: NY. – Issued with fasc. 29-32. Also published in Michelia 2: 509-527. 1 Dec 1882.

Index iconum fungorum ductu et consilio P.A. Saccardo, by J.B. Traverso, in Saccardo, Syll. fung. 19/20, 1910-1911.

Note: The way of issuing the plates, in sets and the dates on the covers of the sets and on the general t.p., are confirmed by contemporary notices in e.g. Grevillea, Bull. Soc. bot. France, Bot. Centralbl., J. Bot. and Bot. Zeit.

Commentarium (by P.A. Saccardo): in *Michelia*: to *pl. 1-160* in 1: 73-100. 1 Jun 1877; to *pl. 161-320* in 1: 326-350. 1 Jul 1878; to *pl. 321-640*, in 1: 488-500. 1 Sep 1879.

9895. *Michelia* commentarium mycologicum fungos in primis italicos illustrans curante P.A. Saccardo ... Patavii [Padova] (Typis seminarii) [1877-]1879-1882, 2 vols. Oct. (*Michelia*).
Publ.: 8 fascicles in 2 volumes, see below. *Copies*: BR, FH, G, H, MO, NY, PH, Stevenson, USDA.

vol.	number	pages	dates	vol.	number	pages	dates
1	1	1-116	1 Jun 1877	2	6	1-176	25 Apr 1880
	2	117-276	15 Jan 1878		7	177-384	5 Mar 1881
	3	277-356	1 Jul 1878		8	385-682	1 Dec 1882
	4	357-452	15 Nov 1878			[i-iii]	1 Dec 1882
	5	453-619, [i, iii]	15 Sep 1879 (cover)	[date 1(5): on p. [1]: 1 Sep 1879]			

Authorship: most articles by P.A. Saccardo; others by C. Spegazzini, A. Vido, R. Cobelli, O. Penzig, A. Berlese, R. Pirotta.
Facsimile ed.: 1969, Amsterdam (A. Asher & Co.), 2 vols. (n.v.).

9896. *Spegazzinia* novum hyphomycetum genus auctore P.A. Saccardo ... Patavii 15 julio 1879. (*Spegazzinia*).
Publ.: 15 Jul 1879, 1 p. text with *1 pl.*, both lithographed. *Copy*: FI. – Constitutes valid publication of *Spegazzinia* Saccardo 1879.

9897. *Aggiunte alla flora trevigiana* ... [Venezia (Tip. Antonelli) 1880]. Qu.
Co-author: Giacomo Bizzozero (1852-1885).
Publ.: 1880 (ÖbZ 1 Dec 1880; Bot. Zeit. 25 Mar 1881), p. [1]-39, [40]. *Copies*: FI, G. – Reprinted from Atti Ist. venet. Sci. nat. ser. 5. 6(7-8): 681-719. 1880 (session 25 Apr 1880).

9898. *Sylloge fungorum* omnium hucusque cognitorum [vol. 1] digessit P.A. Saccardo. Patavii [Padua] (sumptibus auctoris typis seminarii) 1882. Oct. (*Syll. fung.*).
1: 13 Jun 1882 (p. viii: 13 Jun 1882, t.p.: idem; Bot. Centralbl. 24-28 Jul 1882; BSbF séance 28 Jul 1882; J. Bot. Aug 1882; Hedwigia 1882; Nat. Nov. Jul(3) 1882; Bot. Zeit. 29 Sep 1882), p. [i]-xix, [1]-767. – Other t.p.: *Sylloge pyrenomycetum* omnium hucusque cognitorum ... vol. i ...".
2: 13 Jun 1883 (t.p.: BSbF 27 Jul 1883; Bot. Centralbl. 30 Jul-3 Aug 1883; TBC Aug 1883; Nat. Nov. Aug(2) 1883; Hedwigia Sep 1883), p. [i, iii], [1]-815 [815=i], [ii]-lxix, addenda, [1]-77, index, "vol. ii. Pyrenomycologiae universae continuatio et finis". Other t.p.: *Sylloge pyrenomycetum* ... vol. ii ..." – See also P.A. Saccardo, *Genera pyrenomycetum* Nov 1883, below, TL-2/9900.
3: 15 Dec 1884 (t.p.; Bot. Centralbl. 9-13 Feb 1885; BSbF 27 Feb 1885; Bot. Zeit. 27 Mar 1885; Grevillea Mar 1885; J. Bot. Mar 1855; Nat. Nov. Feb(2) 1885; Hedwigia Jan-Feb 1885), p. [i, iii], [1]-860. – Other t.p.: "*Sylloge sphaeropsidearum et melanconiearum* omnium hucusqua cognitarum ...".
4: 10 Apr 1886 (t.p.; Bot. Centralbl. 10-14 Mai 1886; BSbF 28 Mai 1886; Bot. Zeit. 25

Jun 1886; Grevillea Jun 1886; J. Bot. Jun 1886; Bot. Gaz. Jul 1886; Hedwigia Jul-Aug 1886; Nat. Nov. Mai(2) 1886), p. [i, iii], [1]-807, [808, err.], [809-810]. – Other t.p.: "*Sylloge hyphomycetum* omnium hucusque cognitorum ...". For *additamenta ad vol. 1-4* see below; list of names omitted from vol. 4: Grevillea 17: 19-21. 1888.

5: 28 Mai 1887 (t.p.; BSbF 28 Jul 1887; Bot. Centralbl. 8-12 Aug 1887; Bot. Zeit. 26 Aug 1887; Grevillea Sep 1887; Bot. Gaz. Sep 1887; Nat. Nov. Jul(1) 1887; Bot. Zeit. 8 Jul 1887), p. [i, iii-iv], [1]-1146. – Other t.p.: *Sylloge hymenomycetum* omnium hucusque cognitorum. Digessit P.A. Saccardo collaborantibus J. Cuboni [Giuseppe Cuboni, 1852-1920] et V. Mancini [Vincenzo Mancini, 1853-?]. Vol. i. *Agaricineae*.

6: 1 Aug 1888 (t.p.; Bot. Centralbl. 27 Nov 1888; Bot. Zeit. 28 Dec 1888; Grevillea Dec 1888; BSbF 11 Jan 1889; Nat. Nov. Nov(2) 1888; Bot. Zeit. 30 Nov 1888), p. [i, iii-iv], [1]-928. – Other t.p.: *Sylloge hymenomycetum* omnium hucusque cognitorum digessit P.A. Saccardo collaborantibus J. Cuboni et V. Mancini. Vol. ii. *Polyporeae*, Hydneae, Thelephoreae, Clavarieae, Tremellineae.

7(1): 15 Mar 1888 (t.p.; Bot. Centralbl. 7-11 Mai 1888; BSbF 11 Mai 1888; Grevillea Jun 1888; Bot. Gaz. Jun 1888; Bot. Zeit. 27 Jul 1888; Nat. Nov. Mai(1, 2) 1888; Flora 21 Apr 1888), p. [i, iii], [1]-498, [i]-xxx. – "Vol. vii. (pars i) *Gasteromycetae Phalloideae* auctore Ed. Fischer [Eduard Fischer 1861-1939, see TL-2/1: 834-835] *Nidulariaceae*, Lycoperdaceae et Hymenogastraceae auctore Doct. J.B. De-Toni [Giovanni Batista De Toni, 1864-1924, see TL-2/1421-1424], *Phycomyceteae* Mucoraceae, Peronosporaceae, Saprolegniaceae, Entomophthoraceae, Chytridiaceae, Protomycetaceae auctoribus Doct. A.N. Berlese [Augusto Napoleone Berlese, 1864-1903, see TL-2/472-474] et J.B. De Toni[.] *Mycomyceteae*, Eumyxomyceteae et Monodineae auctore Doct. A.N. Berlese.

7(2): 28 Oct 1888 (t.p.; Grevillea Dec 1888; BSbF 11 Jan 1889; Nat. Nov. Dec(1) 1888; Bot. Zeit. 30 Nov 1888), p. [i, iii], [449]-882, xxxi-lix. – General title page for vol. 7: *Sylloge fungorum* ... vol. vii. Gasteromyceteae, Phycomyceteae, Myxomyceteae, Ustilagineae et Uredineae auctoribus A.N. Berlese, J.B. De-Toni et Ed. Fischer ..." Other t.p., for 7(2) only: "*Sylloge Ustilaginearum et Uredinearum* ..." on cover: "Ustilagineae et Uredineae auctore Doct. J.B. De-Toni". The page numbers 449-498 are used for the second time in 7(2); 7(1) also runs to p. 498.

8: 20 Dec 1889 (t.p.; ÖbZ Feb 1890; Bot. Centralbl. 5 Mar 1890; BSbF 14 Mar 1890; Bot. Zeit. 14 Mar 1890, 25 Apr 1890; Nat. Nov. Mar(1) 1890), p. [iii]-xvi, [1]-1143. – *Sylloge fungorum* ... vol. viii. *Discomyceteae* et *Phymatosphaeriaceae* auctore P.A. Saccardo [.] *Tuberaceae, Elaphomycetaceae, Onygenaceae* auctore Doct. J. Paoletti [Giulio Paoletti, 1865-1936] *Laboulbeniaceae* auctore prof. A.N. Berlese[.] *Saccharomycetaceae* auctore doct. J.B. De-Toni[.] *Schizomycetaceae* auctoribus doct. J.B. De-Toni et Com. V. Trevisan [Vittore Benedetto Antonio Trevisan de Saint-Léon, 1818-1897]. – Other t.p.: *Sylloge discomycetum et Phymatosphaeriacearum* accedunt *Tuberaceae* ... Trevisan ..."

9: 15 Sep 1891 (t.p.; Bot. Centralbl. 7 Oct 1891; Bot. Zeit. 16, 30 Oct 1891; ÖbZ Oct 1891; Grevillea Dec 1891; Nat. Nov. Oct(1) 1891), p. [i], 1-1141. – *Sylloge fungorum* ... vol. ix. *Supplementum universale* sistens genera et species nuperius edita nec non ea in Sylloges additamentis praecedentibus jam evulgata nunc una systematice disposita. Pars i. *Agaricaceae – Laboulbeniaceae* ...

10: 30 Jun 1892 (t.p.; Bot. Centralbl. 26 Jul 1892; ÖbZ Jul 1892; Bot. Zeit. 5 Aug, 2 Sep 1892; Grevillea Sep 1892; Nat. Nov. Jul(3) 1892), p. [iii]-xxx, 1-964. – "*Sylloge fungorum* ... vol. x. *Supplementum universale* ... disposita Pars ii ... *Discomyceteae-Hyphomyceteae* Additi sunt [p. 741-808] *Fungi fossiles* auctore doct. A. Meschinelli [Aloysius (Luigi) Meschinelli 1865-?] ..." For Meschinelli, *Fung. foss. iconogr.* 1898, see TL-2/5875. Large parts of the 1898 publication are identical with parts of Meschinelli's continuation to the Sylloge (1892). Most "*-ites*" names proposed by Meschinelli (1892) are not valid because they lack a description; some are valid because accompanied by a brief diagnosis; others are taken from other authors. See also Andrews who often notes: genus ... *ites*: Meschinelli erroneously attributed this genus to [author X] who was the author of the living generic name from which the *-ites-* name was derived. [Note by Gea Zijlstra].

11: Jul 1895 (t.p.; Bot. Centralbl. 4 Sep 1895; Hedwigia Sep-Oct 1895; ÖbZ Sep 1895; Hedwigia 1895; Nat. Nov. Aug(2) 1895), p. [i-vii], [1]-753. – *Sylloge fungorum* ... vol. xi. *Supplementum universale* pars iii. Adjectus est index operis universalis ..."

12(1)(fasc. 1): late 1896 or Jan 1897 (t.p.; Bot. Centralbl. 14 Jan 1897; Bot. Zeit. 16 Jan

1897; ÖbZ for Jan 1897; Nat. Nov. Nov(2) 1896), p. [iii]-viii, [ix-xii], [1]-639. – *Sylloge fungorum* vol. xii. – pars 1. *Index universalis et locupletissimus generum*, specierum, subspecierum et varietatum [hospitumque] in toto opere (vol. i-xi) expositorum auctore P. Sydow [Paul Sydow, 1851-1925]. Berolini [Berlin] (fratres Borntraeger ...) 1897. – Preface material probably issued with *12(1)(fasc. 2)*.

12(1)(fasc. 2): Jan-Jun 1897 (J. Bot. Morot 16 Jul 1897; Bot. Centralbl. 21 Jul 1897; ÖbZ Jul 1897; Bot. Zeit. 16 Aug 1897; Nat. Nov. Jun(1) 1897), p. 641-1053. – *Sylloge fungorum* ... vol. xii. – pars i. – fasc. 2. *Index universalis* ... [P-Z]. ... Berolini (Fratres Borntraeger ...) 1897. – When issued the fascicle contained the preface material to vol. 12, a general volume t.p., a t.p. for 12 pars 1; pages 873-1053 are in theory a third fascicle (see W. Junk 1900-1913).

12(2)(fasc. 1): [= *vol. 13, pars 1*]: Dec 1897 (t.p.: 1898; BSbF 8 Jul 1878; Bot. Zeit. 16 Nov 1898; Bot. Centralbl. 23 Nov 1898; Nt. Nov. Dec(2) 1897), p. [i], [1]-624. – *Sylloge fungorum* ... vol. xii. – pars ii. – fasc. 1. [could also be counted as vol. 13, fasc. 1]. *Index universalis* et locupletissmus *nominum plantarum hospitum* specierumque omnium fungorum has incolentum quae usque ad finem 1897 excepsit P. Sydow. Berolini (Fratres Borntraeger ...) 1898. [A.K.] – Sometimes also bound with t.p. vol. 13.

13: 1898 (p. vi: Aug 1898; Hedwigia 31 Dec 1898; Nat. Nov. Nov(1) 1898), p. [iii]-vi, [vii], 625-1340. – *Sylloge fungorum* ... vol. xiii. (vol. xii. – pars ii. – fasc. 2) [sic for fasc. 1 see *12(2) fasc. 1*]. *Index universalis* et locupletissimus *nominum plantarum hospitum* ... anni 1897 innotuerunt concinnavit P. Sydow. Berolini(Fratres Borntraeger ...) 1898.

Other issue: 1898, vol. 13 here containing the text of both 12(2) fasc. i and of vol. 13, Berlin issue: [i]-vi, [1]-1340, [1341, abbr.]. "*Sylloge fungorum* ... vol. xiii. Index universalis. (Fratres Borntraeger ...) 1898. *Copy*: NY.

Other issue: 1898, p. [iii]-vi, [vii], [1]-340. "*Index universalis et locupletissimus nominum plantarum hospitum* ..." Berolini (Fratres Borntraeger ...) 1898. Independent publication of Sydow's index. *Copy*: FI.

14: 20 Aug 1899 (t.p.; Bot. Centralbl. 5 Oct 1899; ÖbZ for Sep 1899; Bot. Zeit. 16 Nov 1899; Nat. Nov Sep(2) 1899), p. [i-vi], [1]-1316. – *Sylloge fungorum* ... vol. xiv. *Supplementum universale* pars iv. [for parts 1-3 see *Sylloge* vols. 9-11, above] auctoribus P.A. Saccardo et P. Sydow. Adjectus est index totius operis. Patavii [Padova] (sumptibus P.A. Saccardo Typis Seminarii] die xx Augusti mdcccic. [USDA copy bound in two parts, p. [i]-vi, [1]-724 and [i], [725]-1316, each provided with the above 20 August 1899 t.p.]. – On p. [1]-62 *Tab. comp. gen. fung. omn.*

15: 1901 (p. viii: Oct 1900; Bot. Zeit. 1 Jun 1901; Nat. Nov. Mar(1) 1901), p. [iii]-viii, [1]-455. – *Sylloge fungorum* ... vol. xv. *Synonymia* generum, specierum subspecierumque in vol. i-xiv descriptorum auctore E. Mussat [Émile Victor Mussat, 1833-1902] Parisiis [Paris] (Octave Doin edidit ...) 1901.

16: 1 Feb 1902 (t.p.; J. Bot. Apr 1902; Bot. Centralbl. 23 Sep 1902; Bot. Centralbl. 1 Jul 1902; Hedwigia 21 Apr 1902; ÖbZ for Mar 1902; Nat. Nov. Apr(2) 1902), p. [i-viii], [1]-694, index generum: [695]-1291. – *Sylloge fungorum* ... vol. xvi. *Supplementum universale* pars v. [for parts 1-3 see vols. 9-11 and 14 above] auctoribus P.A. Saccardo et P. Sydow. Adjectus est index totius operis ... Patavii [Padova] (Sumptibus P.A. Saccardo Typis Seminarii) die 1. Februarii mcmii.

17: 25 Mai 1905 (t.p.; Bot. Centralbl. 22 Aug 1905; Bot. Zeit. 7 Oct 1905; Nat. Nov. Jul(2) 1905), p. [i]-cvii, [1]-470, 471-991. – *Sylloge fungorum* ... vol. xvii. *Supplementum universale* pars vi. [parts 1-5: vols. 9-11, 13, 16, see above] *Hymenomyceteae – Laboulbeniomycetae* auctoribus P.A. Saccardo et D. Saccardo fil. Adjecta est bibliotheca mycologica auctore J.B. Traverso ... Patavii [Padova] (sumptibus P.A. Saccardo Typis Seminaris) die 25 maji mcmv. [Domenico Saccardo fil. 1872-1952; Giovanni Batista Traverso 1878-1955].

18: 30 Jan 1906 (t.p.; Bot. Zeit. 16 Apr 1906; ÖbZ Mai-Jun 1906; BSbF 23 Mar 1906; Nat. Nov. Mar(2) 1906), p. [i]-vii, [1]-838, [839, err.]. – "*Sylloge fungorum* ... vol. xviii. *Supplementum universale* pars vii. *Discomycetae – Deuteromycetae* auctoribus P.A. Saccardo et D. Saccardo fil. Adjectus est index universalis una cum singulorum generum familia et anno institutionis ..." Patavii [Padova] (id.) die xxx januarii mcmvi. – Contains on p. 741-838 an *Index universalis* ... in toto opere (vol. i-xviii) expositorum; also reprinted separately (99 p.).

19: 31 Mar 1910 (t.p.; ÖbZ for Apr 1910; J. Bot. Jul 1910; Nat. Nov. Mai(2) 1910), p. [i]-xi, [1]-1158. *Sylloge fungorum* ... vol. xix. *Index iconum fungorum* enumerans eorundem figuras omnes hucusque editas ab auctoribus sive antiquis sive recentioribus. Ductu et consilio P.A. Saccardo congessit J.B. Traverso. A-L. Patavii (id.) die xxxi Martii mcmx.

20: 25 Mai 1911 (t.p.; ÖbZ for Mai-Jun 1911; J. Bot. Jul 1911; Bot. Centralbl. 15 Aug 1911; Nat. Nov. Jun(1) 1911), p. [i], [1]-1310. – "*Sylloge fungorum* ... vol. xx. *Index iconum fungorum* enumerans ... recentioribus ductu et consilio P.A. Saccardo congessit J.B. Traverso. M-Z addito supplemento indicis totius". Patavii (id.) die xxv Maji mcmxi.

21: 15 Mar 1912 (t.p.; ÖbZ for Mar 1912; Bot. Centralbl. 18 Jun 1912; J. Bot. Jun 1912; Nat. Nov. Mai(1) 1912), p. [iii]-xv, [1]-928. – "*Sylloge fungorum* ... vol. xxi. *Supplementum universale* pars viii [parts 1-7: vols. 9-11, 13, 16, 17, 18, see above]. *Hymenomycetae-Phycomycetae* auctoribus P.A. Saccardo et Alex. Trotter [Alessandro Trotter 1874-x] ... Patavii (id.) die xv Martii mcmxii. – Signatures dated (not dates of publication). This holds also for vols. 22-25.
Anastatic reprint: 1914 (Bot. Centralbl. 10 Nov 1914), by R. Friedländer & Sohn, Berlin 1914.

22: 20 Aug 1913 (t.p.; ÖbZ for Sep 1913; Bot. Centralbl. 25 Nov 1913; Mycol. Centralbl. 17 Dec 1913; Nat. Nov. Sep(1) 1913). *Sectio i*: p. [i-vi], [1]-822, *sectio ii*: [i-ii], 823-1612. – "*Sylloge fungorum* ... vol. xxii. *Supplementum universale* pars ix. *Ascomycetae-Deuteromycetae* (editae usque ad finem anni mcmx. Auctoribus P.A. Saccardo et Alex. Trotter ... Patavii (id.) die xx Augusti mcmxiii. – T.p., Sectio i: "*Sylloge ... supplementum universale* pars ix. – Sect. i. Pag. 1 ad pag. 822. *Ascomycetae* ..." – *T.p.* Sectio ii: "*Sylloge* ... *supplementum universale* pars ix. – Sect. ii. pag. 823 ad finem Deuteromycetae *Supplementum universale*, parts 1-8, vols. 9-11, 13, 16, 17, 18, 21.

23: 15 Apr 1925 (t.p.; ÖbZ for Oct-Nov 1925), portr. Saccardo, p. [i]-xxxii, [1]-1026. – *Sylloge fungorum* ... vol. xxiii. *Supplementum universale* pars x. *Basidiomycetae* curante Alex. Trotter (Collab. P.A. et Dominicus Saccardo, G.B. Traverso, A. Trotter) ..." Abellini [Avellino] (sumptibus coheredum Saccardo Typis Pergola) die xv Aprilis mcmxxv.
Other issue: ÖbZ 75: 130. 1926 mentions a Berlin (R. Friedländer & Sohn) issue, 1925.

24(1): 15 Jul 1926 (t.p.; Nat. Nov. Oct 1926), p. [i-ii], [1]-703. – *Sylloge fungorum* ... vol. xxiv sectio i. *Supplementum universale* pars x *Phycomycetae, Laboulbeniomycetae Pyrenomycetae* p.p. curante Alex. Trotter (Collab. P.A. et Dom. Saccardo, G.B. Traverso, A. Trotter) ..." Abellini [Avellino] (id.) die 15 Julii mcmxxvi.

24(2): 25 Apr 1928 (t.p.), p. [i-ii], 705-1438. – *Sylloge fungorum* ... vol. xxiv. Sectio ii. *Supplementum universale* part x *Pyrenomycetae* p.p., Discomycetae, Appendix (addenda ad vol. xxiii-xxiv) Curante Alex Trotter (Collab. P.A. et Dom. Saccardo, G.B. Traverso, A. Trotter) ..." Abellini[Avellino] (id.) die 25 Aprilis mcmxxviii.

25: 25 Jun 1931 (t.p.; Nat. Nov. Nov 1931), p. [i-ii], [1]-1093. – *Sylloge fungorum* ... vol. xxv. *Supplementum universale* pars x *Myxomycetae, Myxobacteriaceae, Deuteromycetae, Mycelia sterilia* curante Alex. Trotter (Collab. P.A. et Dom. Saccardo, G.B. Traverso, A. Trotter) ..." Abellini [Avellino] (id.) die 25 Juni mcmxxxi.

26: 1972, p. [iii]-ix, [xi], 1-1563. – "*Sylloge fungorum* ... vol. xxvi. *Supplementum universale* pars xi. Auctore Alex. Trotter. *Mycobacteriales, Myxomycetes, Phycomycetes, Ascomycetes, Basidiomycetes, Deuteromycetes, Fungi fossiles* Descriptiones recensuit et in ordinem systematicum disposuit Edith K. Cash ..." New York, London (Johnson reprint corporation) 1972. With portr. of Trotter. First published 1972. [Edith Katherine Cash, 1890-x].

Copies: BR, FI, FH, G, H, NY, PH, US, USDA; IDC 5395.
Reprint Friedländer: vols. 1-11, 14, 16-18, 21 may have in addition the imprint "Iterum impressum apud R. Friedländer & Sohn Berlin" in copies: NY(2-11, 16), USDA (2-11). This is apparently a later facsimile reprint (vols. 1-11 advertised to have been reprinted in Bot. Zeit. 16 Jul 1902). Junk (1900-1913) states that vols. 1-11, 14, 16 were reprinted this way and that the original edition of the Sylloge numbered 275-350 copies. – Most sets are mixed.
Additamenta ad volumina i-iv: 31 Dec 1886 (t.p.); Bot. Zeit. 25 Mar 1887; Grevillea Mar 1887; Bot. Zeit. 11 Feb 1887; Hedwigia Jan-Feb 1887; Nat. Nov. Feb(1) 1887), p. [i]-iv, [1]-484. – *Sylloge fungorum* ... *Additamenta* ad volumina i.-iv. Curantibus doct. A.N.

Berlese et P. Voglino ... Patavii [Padua] (sumptibus auctorum Typis Seminarii) xxxi decembris mdccclxxxvi. *Copy*: BR. – Pietro Voglino (1864-1933).
Appendix: Feb 1896, p. [i*], i-l. *Copies*: FI(2), U. – "*Sylloge fungorum* ... Appendix sistens Elenchum fungorum novorum qui post editum vol. xi Sylloges usque ad finem Decembris mdcccxcv innotuerunt ..." Dresdae [Dresden] (ex Hedwigia vol. xxxv n.7, febr. 1896) separately paged appendix.
Key to Saccardos Sylloge fungorum: F.E. Clements, A key to Saccardos Sylloge fungorum, includes all the genera in vol. 9 to 18 ... Minneapolis 1912. Oct. (n.v.).
Reprint Edwards Brothers: 1944 (or later), Ann. Arbor Mich., n.v. – H. Lubrecht informs us that the reprint "was done under Governments Alien Property Custodian permission shortly after World War II". See also Wheldon & Wesley, Cat. 153, no. 3992.
Reprint Johnson Reprint Corporation: 1967, 25 vols. (in 28).
Authorship: apart from P.A. Saccardo himself, the following authors took care of part or the entire text of various volumes:

Berlese, Augusto Napoleone (1864-1903), *7(1)*, *8*, add. *1-4*.
Cash, Edith Katherine (1890-x) editor vol. *26*.
Cuboni, Giuseppe (1852-1920), *5*, *6*.
De Toni, Giovanni Batista (1864-1924), *7(1, 2)*, *8*.
Fischer, Eduard (1861-1939), *7(1)*.
Mancini, Vincenzo (1853-?), *5*, *6*.
Meschinelli, Aloysius [Luigi] (1865-?), *10*.
Mussat, Émile Victor (1833-1902), *15*.
Paoletti, Giulio (1865-1936), *8*.
Saccardo, Domenico (1872-1952), *17*, *18*, *23*, *24(1, 2)*, *25*.
Sydow, Paul (1851-1925), *12(1, 2)*, *13*, *14*, *16*.
Traverso, Giovanni Batista (1878-1955), *17*, *19*, *20*, *23*, *24(1, 2)*, *25*.
Trevisan de Saint-Léon, Vittore Benedetto Antonio (1818-1897), *8*.
Trotter, Alessandro (1874-?), *21*, *23*, *24(1, 2)*, *25*, *26*.
Voglino, Pietro (1864-1933), add. *1-4*.
Reviews vol. 1: Anon., Bull. Soc. bot. France 29(bibl.): 124-125. 1882; Cooke, M.C., Grevillea 11: 34-35. 1882; J. Bot. 20: 286. 1882; Kohl (Strassbourg), Bot. Centralbl. 13: 396. 1883. – See also Saccardo's reply to M.C. Cooke's reviews in Grevillea 11: 66-67. 1882. Cooke reproaches S. to propose a system equivalent to the old Linnean system, but S. thinks that Cooke's own system is closest to that of Tournefort.
Vol. 2: Cooke, M.C., Grevillea 12: 34-35. 1883; Kohl (Marburg), Bot. Centralbl. 16: 131-132. 1883.
Vol. 3: Cooke, M.C, J. Bot. 23: 124-127. 1885; Patouillard, N., Bull. Soc. bot. France 32(bibl.): 113-114. 1885; Penzig, O., Bot. Centralbl. 29: 129-132. 1887 (vols. 3 & 4).
Vol. 4: Cooke, M.C., Grevillea 14: 130-131. 1886; Grove, W.B., J. Bot. 24: 247-248. 1886 ("... as great a triumph as any of the three that have preceded it"); Penzig, O., Bot. Centralbl. 29(5): 129-132. 1887; Patouillard, N., Bull. Soc. bot. France 33: 101-102. 1886.
Addit. 1-4: Anon., Hedwigia 26: 63-64. 1887; Patouillard, N., Bull. Soc. bot. France 34(bibl.): 79-80. 1887.
Vol. 5: Patouillard, N., Bull. Soc. bot. France 34(bibl.): 106-107. 1887; Peck, C.H., Bot. Gaz. 12: 278-279. 1887; De Toni, J.B., Bot. Centralbl. 34: 322-324. 1888.
Vol. 6: Cooke, M.C., Grevillea 17: 60-61. 1889; Patouillard, N., Bull. Soc. bot. France 36(bibl.): 72-73. 1889; Anon., Bot. Gaz. 14: 85-86. 1889.
Vol. 7: Cooke, M.C., Grevillea 16: 102-104. 1888; Anon., Bot. Gaz. 14: 20-21. 1889.
Vol. 7(2): Anon., Bot. Gaz. 14: 85-86. 1889.
Vol. 8: De Toni, G.B., Bot. Centralbl. 44: 216-219. 1890.
Vol. 9: Cooke, M.C., Grevillea 20: 39. 1891; Patouillard, N., Bull. Soc. bot. France 38: 168. 1891; Anon., Bot. Gaz. 16: 316-317. 1891; Möbius, M., Bot. Centralbl. 50: 326. 1892.
Vol. 10: Möbius, M., Bot. Centralbl. 54: 228-229. 1893; Patouillard, N., Bull. Soc. bot. France 39(bibl.): 71. 1892; Anon., Bot. Gaz. 17: 331. 1892.
Vol. 11: Anon., Bot. Gaz. 20: 507. 1895.
Vol. 12(1, 1): Arthur, J.C., Bot. Gaz. 23: 381-382. 1897; Anon., Hedwigia 36: (18). 1897. (1,2): Arthur, J.C., Bot. Gaz. 24: 379. 1897; Lindau, G., Bot. Centralbl. 75: 386-387. 1898, Lindau, G., 79: 323. 1899, Hedwigia 37: (18)-(19). 1898.

Vol. 13: Lindau, G., Hedwigia 37: (231). 1898; Patouillard, N., Bull. Soc. bot. France 46(bibl.): 505-506. 1899; Arthur, J.C., Bot. Gaz. 27: 399-400. 1899.
Vol. 14: Lindau, G., Bot. Centralbl. Beih. 9(5): 335. 1900.
Vol. 15: Lindau, G., Bot. Centralbl. 2 Mai 1901.
Vol. 16: Perrot, E., Bull. Soc. bot. France 48(bibl.): 445-446. 1901.
Vol. 17: Patouillard, N., Bull. Soc. bot. France 52: 476. 1905.

9899. *Flora briologica della Venezia* ... Venezia (Stabilimento di G. Antonelli) 1883. Qu. (*Fl. briol. Venezia*).
Co-author: Giacomo Bizzozero (1852-1885).
Publ.: Aug-Nov 1883 (p. 7: 12 Jul 1883; Bot. Zeit. 3-7 Dec 1883; BSbF séance 14 Dec 1883; Bot. Zeit. 25 Jan 1884; J. Bot. Jan 1884; Hedwigia Dec 1883), p. [1]-111. *Copies:* BR, FH, FI, H, NY. – Preprinted or reprinted from Atti Soc. venet. Sci. nat. ser. 6. 1(9): 1283-1314 (repr. p. 5-36; session 16 Jul 1883), 1(10): 1319-1324 (repr. p. 37-42, type shifted), 1325-1393 (reprint p. 43-111; session 14 Aug 1883). 1883.

9900. *Genera pyrenomycetum* schematice delineata ... Illustratio adcommodata ad usum Sylloges pyrenomycetum ejusdem auctoris ... Patavii [Padova] (Lithogr. P. Fracanzani) Nov. 1883. (*Gen. pyrenomyc.*).
Publ.: 1-15 Nov 1883 (cover: Nov; Bot. Centralbl. 19-23 Nov 1883; Nat. Nov. Dec(1) 1883; Hedwigia Dec 1883; Bot. Zeit. 28 Dec 1883), cover-t.p., p. [1]-6 (expl. pl.), *pl. 1-14* (uncol. liths.). *Copies:* FH, G, NY, USDA.
Ref.: Kohl (Marburg), Bot. Centralbl. 17: 1-2. 31 Dec 1883-4 Jan 1884.

9901. *Conspectus generum discomycetum* hucusque cognitorum ... [Separat-Abdruck aus "Botanisches Centralblatt". Bd. xviii. 1884, Cassel (Verlag von Theodor Fischer).].
Publ.: Mai 1884 (Bot. Centralbl. 25 Jul 1884; Grevillea Jun 1884; BSbF séance 9 Mai 1884; Nat. Nov. Mai(2) 1884; Hedwigia Mai 1884), p. [1]-16. *Copies:* FH, G. – Reprinted from Bot. Centralbl. 18: 213-220. [12-16] Mai, 247-256 [19-23] Mai 1884.

9902. *Catalogo dei funghi italiani* ... Varese (Tipografia Ferri di Maj e Malnati ...) 1884. Oct. (*Cat. fung. ital.*).
Co-author: Augusto Napoleone Berlese (1864-1903).
Publ.: Apr-Mai 1884 (t.p. preprint; p. 261: 10 Mar 1884; Bot. Zeit. 28 Aug 1885; Bot. Centralbl. 2: 18-22 Mai 1889), p. [1]-108. *Copies:* FH, FI, NY, USDA. – Preprinted or reprinted from Atti Soc. critt. ital. 4(4): 261-368. 1885.

9903. *Micromycetes sclavonici novi* ... [Toulouse (Imp. Fond.-gén. Sud-Ouest) 1884]. Oct.
Co-author: Stephan Schulzer von Müggenburg (1802-1892).
Publ.: Mar-Aug 1884 (in journal), p. [1]-12. *Copies:* FH, FI. – Reprinted from Revue mycol. 6: 68-80. 1884. – See also Hedwigia 23: 41-44. Mar; 77-80. Mai; 89-91. Jun; 107-112. Jul; 125-128. Aug 1884.

9904. *Miscellanea mycologia* ... [Venezia (Tip. Antonelli) 1884]. Qu.
Ser. [i] (nos. 1-5): 1884 (Bot. Centralbl. 9-13 Mar 1885; Hedwigia Mar 1884), p. [1]-29, [30]. *Copies:* FH, FI, G. – Reprinted from Atti Ist venet. Sci. nat. ser. 6. 2(3): 435-463. (Session 27 Jan) 1884.
Ser. ii: 1885 (Bot. Zeit. 29 Mai 1885; Grevillea Jun 1885; Bot. Centralbl. 27 Apr-1 Mai 1885), p. [1]-32, [1, caption pl., 2, coloph.], *pl. 8-11. Copies:* FH, G. – Reprinted from Atti Ist. venet. nat. ser. 7. 3: 711-742. 1885.

9905. *Funghi delle Ardenne* contenuti nelle Cryptogamae Ard6863. e della signora M.A. Libert. – Riveduti da P.A. Saccardo. [Estrattto dalla Malpighia ... 1887, 1888]. Oct.
Collector: Marie-Anne Libert (1782-1865), see TL-2/3: 6-7.
Publ.: 1887-1888 (in journal), p. [1]-27. *Copy:* FH. – Reprinted and to be cited from Malpighia 1(5): 211-219. 1887, 1( ): 454-459. 1887, 2(1): 18-25. 1888; 2( ): 234-241. 1888.

9906. *Mycetes malacenses.* Funghi della penisola di Malacca raccolti nel 1885 dell'ab. Benedetto Scortechini ... [Venezia (Tip. Antonelli) 1888]. Qu.

Collector: Benedetto Scortechini (1845-1886).
Publ.: Feb-Apr 1888 (p. 4: Jan 1888; Bot. Centralbl. 25 Mai 1888; Grevillea Jun 1888; Bot. Centralbl. 9-20 Apr 1888; Hedwigia Mai-Jun 1888; Nat. Nov. Apr(3) 1888), p. [1]-42, *pl. 5-7* (col. liths. A.N. Berlese). *Copies*: FH, G. – Reprinted from Atti Ist. Sci. nat. ser. 6. 6: 387-428, *pl. 5-7*. 1888.
Ref.: Patouillard, N., Bull. Soc. bot. France 35(bibl.): 87-88. 1888 (rev.).

9907. *Mycetes siberici* ... [Bull. Soc. Bot. Belg. 28. 1889]. Oct.
Co-authors: Augusto Napoleone Berlese (1864-1903), Giovanni Batista De Toni (1864-1924), Giulio Paoletti (1865-1935), Francesco Saccardo (1869-1896).
Publ.: Jul-Aug 1889 (p. 2: 15 Mai 1889; Grevillea Dec 1889; Nat. Nov. Sep(2) 1889), p. [1]-44, *pl. 4-6*. *Copy*: FH. – Reprinted and to be cited from Bull. Soc. Bot. Belg. 28: 77-120, *pl. 4-6*. 1889.
Pugillus alter, auctore P.A. Saccardo, Bull. Soc. bot. ital. (report session 12 Mar 1893) 1893(4): 213-221. – *Copies*, reprints with original pagination: FI, NY.
Pugillus tertius, P.A. Saccardo, 1896 (p. 258: 15 Mar 1896 p. [1]-24, *2 pl. Copies*: FI, NY. – Reprinted and to be cited from Malpighia 10: 258-280, *2 pl.* 1896.

9908. *Notes mycologiques* ... Lons-le-Saunier (Imprimerie et Lithographie Lucien Declume ...) 1890. Oct.
Publ.: Oct 1890 (reprint so dated; Nat. Nov. Jan(1) 1891), p. [1]-11, *pl. 14* (col. lith. auct.). *Copy*: FH. – Reprinted and to be cited from Bull. Soc. mycol. France 5(4): 115-123. 1890.
Series ii: 1896, Lons-le-Saunier (id.), p. [1]-20, *pl. 5-7*. *Copy*: FI. – Reprinted and to be cited from Bull. Soc. mycol. France 12: 64-81, *pl. 5-7*. 1896.
Continued as *Notae mycologicae* series 3-25, 29. 1903-1920. See below.

9909. *Fungi aliquot mycologiae romanae addendi* ... Roma (tipografia della R. Accademia dei Lincei) 1890. Qu.
Publ.: 1890 (Nat. Nov. Jan(1) 1891), p. [1]-10, *pl. 20* (col. lith. auct.). *Copies*: FH, G, NY. – Reprinted from Annu. r. Ist. bot. Roma 4(1): 192-199. *pl. 20*. 1890.

9910. *Pugillus mycetum australiensium* auctoribus J. Bresadola et P.A. Saccardo ... [Malpighia 4, 1890]. Oct.
Co-author: Don Giacomo Bresadola (1847-1929), see TL-2/739-745.
Publ.: 1890 (p. (1): Sep 1890, p. [1]-15, *pl. 9* (uncol. lith. auct.). *Copy*: FH. – Reprinted and to be cited from Malpighia (7): 259-301, *pl. 9*. 1890.

9911. *Fungi abyssinici* a cl. O. Penzig collecti ... [Estratto dal Giornale Malpighia, Anno v, fasc. vi. Genova (Tip. di Angelo Ciminago, ...) 1891]. Oct.
Collector: Albert Julius Otto Penzig (1856-1929).
Publ.: 1891, p. [1]-14, [15, colo.], *pl. 20*. *Copies*: FH, FI, G, NY. – Reprinted and to be cited from Malpighia 5: 274-287. *pl. 20*. 1891.

9912. *Chromotaxia* seu nomenclator colorum polyglottus additis speciminibus coloratis ad usum botanicorum et zoologorum ... Patavii [Padova] (Typis Seminarii) 1891. (*Chromotaxia*).
Publ.: 5 Mar 1891 (date of distr. copies, fide notice of distribution by author on slip in copy BR; p. 4: 25 Feb 1891; J. Bot. Apr 1891; ÖbZ Mai 1891; Bot. Zeit. 24 Apr 1891; Hedwigia 8 Apr 1895; Nat. Nov. Mar(2): 1891), p. [1]-22. *pl. 1-2* (color charts). *Copies*: BR, FH, FI, G, H, HH, MO, NY, PH, Stevenson.
Editio altera: 1894 (Bot. Zeit. 16 Feb 1895, 1 Jul 1895, Bot. Centralbl. 27 Feb 1895), p. [1]-22, *pl. 1-2*. id. *Copies*: HH, Stevenson, U, USDA.
Editio terzia: 1912, p. [1]-22, *pl. 1-2* (id.). *Copy*: M.
Ref.: Möbius (Heidelberg), Bot. Centralbl. 47: 361-362. 1891 (rev.).

9913. *I funghi mangerecci* più comuni e più sicuri della regione veneta e di gran parte d'Italia ... Padova, Verona (Fratelli Drucker ...) 1891. Oct. (*Fung. manger.*).
Co-author: N.D'Ancona.
Publ.: early Jan 1891 (p. 8: Oct 1890; Bot. Zeit. 27 Mai 1891; Nat. Nov. Jan(2) 1891), p.

[3]-12, *pl. 1-15* (col. liths. A. Berlese, each with 1 p. expl.). *Copies*: FH, G, Stevenson. – Pubblicazioni della Società d'Igiene per la città e provincia di Padova.

9914. *L'Azolla caroliniana in Europa* nota. Venezia (Tip. Antonelli). 1892.
Publ.: 1892 (p. 6: 18 Jan 1892; Bot. Zeit. 1 Feb 1893; Nat. Nov. Jun(2) 1893), p. [1]-6, [7, colo.]. *Copies*: FI, G. -Reprinted and to be cited from Atti Ist. venet. Sci. Ser. 7. 3: 831-836. 1892.

9915. *Fungilli aliquot herbarii regii bruxellensis* observante P.A. Saccardo [Extrait du Compte-rendu ... Société royale de Botanique de Belgique ... 1893]. Oct.
Publ.: 1893 (C.R. séance 4 Dec 1892; G copy rd. 27 Mar 1895), p. [1]-15. *Copies*: FI, G, NY. – Reprinted and to be cited from Bull. Soc. Bot. Belg. 31(2): 224-239. 1893.

9916. *Il numero delle piante* ... Genova (Tip. Sordo-Muti) [1893]. Oct.
Publ.: 1893 (Nat. Nov. Jun(2) 1893), p. [1]-9. *Copies*: M, NY. – Reprinted and to be cited from Atti Congr. bot. int. 1892: 434-439. 1893.

9917. *I nomi generici dei funghi* e la riforma del D.r O. Kuntze ... Genova – Tip. Sordo-muti [1893]. Oct.
Publ.: 1893 (Bot. Centralbl. 16 Aug 1893; Nat. Nov. Aug(1) 1893), p. [1]-6. *Copies*: FH, FI, G. – Reprinted and to be cited from Atti Congr. bot. int. 1892: 434-439. 1892. – For Carl Ernst Otto Kuntze and his nomenclatural reforms see TL-2/2: 698-702.

9918. *Florula mycologica lusitanica* sistens contributionem decimam ad eamdem floram nec non conspectum fungorum omnium in Lusitania hucusque observatorum ... Coimbra (Imprensa da Universidade) 1893. Oct.
Publ.: 1893 (p. 9: 1 Mai 1893; Nat. Nov. Apr(1) 1894), p. [1]-63. *Copies*: FH, FI, G, NY. – Reprinted and to be cited from Bol. Soc. brot. 11: 9-81. 1893. – Tenth contribution in a series by various authors.
Contributio duodecima: 1902, Bol. Soc. brot. 19: 156-171. 1902, also reprinted s.l.n.d. p. [1]-16. *Copy*: FI. – "Florula mycologicae lusitanicae".

9919. Fungi. *Fungilli novi europaei et asiatici* [Reprinted from "Grevillea" 1893]. Oct.
Publ.: Mar 1893 (in journal), p. [1]-5, pl. 184 (uncol. lith. auct.). *Copy*: G. – Reprinted and to be cited from Grevillea 21: 65-69. Mar 1893.

9920. *Il primato degli italiani nella botanica*. Discorso letto el 5 Novembre 1893 nell'aula magna della r. Università di Padova per l'inaugurazione dell'anno accademico ... Padova. (Tipografia Gio. Batt. Randi) 1893. Oct. (*Primato ital. bot.*).
Publ.: 5 Nov-31 Dec 1893 (Nat. Nov. Feb(1) 1894), p. [3]-82, [1, cont.]. *Copies*: FH, G.
Summary: 1894, p. [1]-5, [6]. *Copies*: BR, G. – Reprinted and to be cited from Malpighia 7: 483-487. 1894. Chronological account of Italian "firsts" in botany from Empedocles through Gasparrini.
Ref.: Solla (Vallombrosa), Bot. Centralbl. 58: 158-160. 1894.

9921. *La botanica in Italia*. Materiali per la storia di questa scienza ... Venezia (Tipografia Carlo Ferrari) 1895, 1901, 2 vols. (*Bot. Italia*).
1: Oct-Nov 1895 (p. 236, printing finished Oct 1895; J. Bot. Apr 1896; ÖbZ Dec 1895; Nat. Nov. Dec(1, 2) 1895, p. [1]-236. Also issued as Mem. ist. venet. Sci. 25(4). 1895.
2: 1901 (p. 172: 1 Jul 1901, printing finished; J. Bot. Nov 1901; Nat. Nov. Sep 1901), p. [i]-xv, [1]-172. Also issed as Mem. Ist. venet. Sci. 26(6). 1901.
Copies: BR, FI, G, NY, Teyler, USDA. – 100 copies printed of the book issue.
Facsimile reprint: "Available", see Koeltz, Katal. 235, no. 750. 1974 (n.v.).
Note: major source of information on the history of Italian botany: biogr. details, refs to bibl. and biogr., lists of major works, accounts of botanical gardens, chronology of Italian botanical achievements, location herbaria, portraits, busts.
Summary of botanical gardens and list of botanists: Malpighia 8: 476-539. 1895.

9922. *I prevedibili funghi futuri* seconda la legge d'analogia saggio di P.A. Saccardo. Venezia (Tip. Ferrari) 1896.

Publ.: Dec 1896 (G rd. 29 Dec 1896), p. [1]-7, 4 charts. *Copies*: FH, FI, G, M, NY. – Reprinted and to be cited from Atti Ist. venet. Sci. ser. 7. 8: 45-51. 1896.

9923. *Fungi aliquot brasilienses phyllogeni* [Extrait du Bulletin de la Société royale de botanique de Belgique t. xxxv (1896), première partie; Bruxelles]. Oct.
Publ.: 1896 (p. [1]: ult. Mar 1896; Nat. Nov. Jan(2) 1897), p. [1]-6, *pl. 3-4* (uncol.). *Copies*: FH, FI, G. – Reprinted and to be cited from Bull. Soc. Belg. 35(1): [127]-132, *pl. 3-4.* 1896.

9924. *Enumerazione dei funghi della Valsesia* raccolti dal Ch. Ab. Antonio Carestia ... Genova (Tipografia di Angelo Ciminago ...) 1897. Oct.
Co-author: Giacopo [abate] Bresadola (1847-1929). *Collector*: Antonio (abate) Carestia (1825-1908).
Publ.: 1897 (p. 5: Aug 1897), p. [1]-87. *Copy*: G. – Reprinted and to be cited from Malpighia 11: 241-325. 1897.
Continuation: 1900 (Bot. Centralbl. Aug 1900), p. [1]-28. *Copies*: FI, G. – Reprinted and to be cited from Malpighia 13: 425-452. 1900.

9925. *Tabulae comparativae generum fungorum omnium* Patavii [Padova] (Typis Seminarii) 30 Aug 1898. Oct. (*Tab. compar. gen. fung.*).
Publ.: 30 Aug 1898, p. [i], [1]-62. *Copy*: NY. – Reprinted from P.A. Saccardo, Syll. fung. 14: 1-62. 1898.

9926. *La iconoteca dei botanici* nel r. istituto botanico di Padova. [Malpighia 13, 1899]. Oct.
Publ.: Aug 1899, p. [1]-35. *Copy*: FH. – Reprinted and to be cited from Malpighia 13: 89-123. 1899. – List of portraits of botanists at PAD.
Supplemento: Malpighia 15: 414-437. 1901.

9927. *Funghi dell'Isola del Giglio* (Estratto da: S.Sommier, L'isola del Giglio e la sua flora. Torino, C. Clausen, 1900). Oct.
Publ.: 1900, p. [1]-8. *Copies*: FH, FI, NY. – Reprinted from S. Sommier, *L'isola del Giglio e la sua flora* 138-143. 1900.

9928. *Nouvelles espèces de champignons de la Cote-d'Or* ... Lons-le-Saunier (Imprimerie et Lithographie Lucien Declume) 1900. Oct.
Publ.: 1900 (Bot. Centralbl. 11 Jul 1900), p. [1]-7, *pl. 2*. *Copy*: FI. – Reprinted and to be cited from Bull. Soc. mycol. France 16(1): 21-25. *pl. 2.* 1900.

9929. *Manipolo di micromiceti nuovi* [Rendic. Congr. bot. Palermo ... 1902]. Oct.
Publ.: Sep-Dec 1902 (p. 15: 31 Aug 1902; Bot. Centralbl. 24 Mar 1903; Hedwigia 15 Feb 1903), p. [1]-15, *1 pl.* (col. lith.). *Copies*: FI, NY. – Reprinted from Rendiconti del Congresso botanico di Palermo, Maggio 1902 (Rendiconti n.v.).

9930. *Progetto di un lessico dell'antica nomenclatura botanica* comparata alla linneana ed elenco bibliografico delle fonti relative ... [Malpighia 17, 1902]. Oct.
Publ.: 1903, p. [1]-39. *Copy*: FI. – Reprinted and to be cited from Malpighia 17: 241-278. – An attempt to start a bibliography of the secondary literature on pre-linnean authors and their main publications.

9931. *Notes mycologiques* ... [Extrait des "Annales mycologici" (vol. 1, no. 3, 1903)]. Oct.
Senior author: René Charles Joseph Ernest Maire (1878-1949).
Publ.: Mai 1903 (in journal), p. [1]-5. *Copy*: FH. – Reprinted and to be cited from Ann. mycol. 1(3): 220-224. Mai 1903.

9932. *Notae mycologicae* [series 3-25, 29 in various journals, 1903-1920].
Notes mycologiques: 1, Oct 1890, see above no. TL-2/9908; but see also Maire et Saccardo 1903, below.
Notes mycologiques: 2, 1896, see above under no. TL-2/9908.
Notae mycologicae: series 3-19, published in *Annales* mycologici; sometimes reprinted with

independent pagination; most reprints with cover (with title) but original pagination (not cited here); all papers to be cited from journal:

Series	Ann. mycol.	pages	dates	reprint
3	1(1)	24-29	30 Jan 1903	
4	2(1)	12-19, *pl. 3*	15 Feb 1904	[1]-8
5	3(2)	165-171	10 Mai 1905	
6	3(6)	505-516	10 Feb 1906	
7	4(3)	273-278	5 Jan 1906	
8	4(1)	490-494, *pl. 10*	15 Jan 1907	
9	5(2)	177-179	15 Mai 1907	
10	6(6)	553-569, *pl. 24*	31 Dec 1908	
11	7(5)	432-437	10 Nov 1909	
12	8(3)	333-347	15 Jun 1910	
13	9(3)	249-257	1 Jun 1911	
14	10(3)	310-322	10 Jun 1912	
15	11(1)	14-21	15 Mar 1913	
16	11(4)	312-325	15 Sep 1913	
17	11(6)	546-568	31 Dec 1913	
18	12(3)	282-314	30 Jun 1914	
19	13(2)	115-138	10 Jun 1915	

Ser. 20: Nuov. Giorn. bot. ital. 23(2): 184-234. 1 Mai 1916; preprint 15 Apr 1916, p.[1]-52. *Copy*: FH.
Ser. 21: in id. 24(1): 31-43, Jan 1917; reprint p. [1]-15. *Copy*: FH. – Pugillo di funghi della Val d'Aosta.
Ser. 22: in Atti Mem. Accad. Sci. Padova 23 (or 33): 157-195. 1917, reprint p. [i-ii], [1]-39. Padova (Tipografia Batt. Randi) 1917. Oct. – "*Manipolo di funghi nuovi o più notevoli di Spagna, Francia, Calabria, America, Giappone Eritrea e della Repubblica di S. Marino*". *Copy*: FH.
Ser. 23: in Atti Accad. sci. venet.-trent.-istr. ser. 3. 10: 57-94. 1917 (vol. publ. 1919), reprint p. [1]-40, Padova (Tipografia all' Università ...) 1917. Oct. – *Fungi philippinenses*. *Copies*: FH, FI.
Ser. 24: in Bull. Orto bot. Univ. Napoli 6: 39-73. 1921, preprint p. [1]-35 (rd. by Farlow 11 Nov 1918; signed by author Avellino 1 Sep 1918), 1. *Fungi singaporenses bakeriani*, 2. *Fungi abellinenses novi*. Finito stampare il dì 24 Agosto 1918.
Ser. 25: in Madonna Verona 1918: 1-24 (n.v.), reprint p. [i-ii], [1]-24, Verona (Tipografia cooperativa) 1918. *Fungi veronenses*. *Copy*: FH (rd. 11 Nov 1918).
Ser. 26-28: not published.
Ser. 29: in Mycologia 12(4): 199-205. 1920 . Reprint with original pagination at NY.

9933. *Adjonctions au Code de Paris de 1867* proposées par quelques botanistes italiens. Florence, Juin 1904. (*Adjonct. Code Paris*).
Authorship: Giovanni Arcangeli (1840-1921), Antonio Bottini (1850-1931), Émile Levier (1839-1911), Caro Benigno Massalongo (1852-1928), Pier Andrea Saccardo, Carlo Pietro Stefano (1848-1922) and Jules Cardot (1860-1934).
Publ.: 23 Jun 1904 (p. 12), p. [i-ii], [1]-12. *Copies*: FAS, G. – Tries to introduce the rejection of *nomina semi-nuda* to protect well-established generic names; rejection of orthographic variants only if taxa are closely related. Text in French, Italian and German.
English translation: by T.N. Williams, J. Bot. 42: 233-236. 1904.

9934. *Notions supplémentaires* présentées au Congrès international de botanique de Vienne par P.A. Saccardo ... Avellino, juin 1904. (*Not. suppl.*).
Publ.: Jun 1904, p. [1]. *Copies*: FAS, FI. – Proposal to identify and precisely date separates and to preserve the journal pagination; to give the derivation of new specific and generic names and to cite generic and specific names in systematic and floristic works with their date of valid publication ("date de leur création").

SACCARDO, P. A. 1904

9935. *Le reliquie dell'erbario micologico di P.A. Micheli* ... [Estratto dal Bulletino della Società botanica italiana ... 1904]. Oct.
Publ.: 20 Jun 1904 (date of journal issue), p. [1]-10. *Copy*: FI. – Reprinted and to be cited from Bull. Soc. bot. ital. 1904(5): 221-230. 20 Jun 1904. – Deals with fungi; dealt with (mostly) by P.A. Micheli (q.v.) in his *Nova plantarum genera* 1729 (TL-2/5974), paper not quoted by us under Micheli.

9936. *Micromiceti italiani* nuovi o interessanti ... [Estratto dal Bulletino della Società botanica italiana ... 1904]. Oct.
Co-author: Giovanni Battista Traverso (1878-1955).
Publ.: 20 Jun 1904 (read 8 Mai 1904; journal issued 20 Jun 1904), p. [1]-15. *Copy*: FI. – Reprinted and to be cited from Bull. Soc. ital. bot. 1904(5): 207-221. 20 Jun 1904.

9937. *I codici botanici figurati* e gli erbari di Gian Girolamo Zannichelli, Bartolomeo Martini e Giuseppe Agosti esistenti nell'Istituto botanico di Padova (con un'appendice sull'erbario di L. Pedoni). Studio storico e sinonimico ... Venezia (Officine grafiche di C. Ferrari) 1904. Oct. (*Cod. bot. fig.*).
Publ.: Jun-Sep 1904 (p. 122: imprimatur 10 Jun 1904; Hedwigia 29 Oct 1904; Nat. Nov. Oct(1) 1904), p. [1]-122, *pl. 6. Copies*: FI, G, M. – Also issued as Atti Ist. venet. sci. ser. 8, vol. 6 (vol. 63) 1-122. 1904. – On herbaria and codices of Giuseppe Agosti (1715-1786), see TL-2/66 (*De re botanica tractatus* 1770), Bartolomeo de Martinis (Martini) (x-1720), Lorenzo Pedoni; Gian Girolamo Zannichelli (1662-1729) all at PAD. – On the Giuseppe Agosti herbarium at PAD see also P.A. Saccardo, *Sul rinvenimento di un antico erbario* dell'abate Conte Giuseppe Agosti botanico bellunese ...Padova (R. Stab. P. Prosperini) 1905, p. 5-13, reprinted from Atti Accad. Sci. venet. ser. 2. 1: 5-13. 1904.

9938. *Des diagnoses et de la nomenclature mycologiques*. Propositions par P.A. Saccardo ...]. Estratto dal Bulletino della Società botanica italiana ... 1904]. Oct.
Publ.: 31 Jul 1904 (in journal), p. [1]-6. *Copies*: FI, G, U. – Reprinted and to be cited from Bull. Soc. bot. ital. 1904(6): 281-286, 31 Jul 1904. – Attempts to regulate contents and form of diagnoses of new taxa of fungi; introduction of nomina anamorph. for fungi imperfecti.
Italian: Ann. mycol. 2(2): 195-198. 1904.

9939. *La flora delle vette di Feltre*. Saggio. Venzia (Officine grafiche di C. Ferrari) 1905. Qu.
Co-author: Giovanni Battista Traverso (1878-1955).
Publ.: 1905 (p. 76: printing approved 27 Feb 1905), p. [1]-76. *Copies*: FI, M. – Reprinted and to be cited from Atti Ist. venet. Sci. 64(2): 823-902. 1905.

9940. *Chi ha creato il nome "fanerogame"*. Estratto dal Bulletino della società botanica italiana (Adunanza della Sede di Firenze del 14 Gennaio 1906). Oct.
Publ.: 20 Mar 1906 (date journal), p. [1]-3. *Copy*: FI. – Reprinted and to be cited from Bull. Soc. bot. ital. 1906(1-2): 25-27. 1906. – The term "phanerogams" is shown to have been used first by G.F.B. de Saint-Amans, J. Sci. utiles 1791(17/18): 283, 285, 291.

9941. *Fungi aliquot africani* lecti a Cl. A. Moller, Is. Newton et A. Sarmento ... [Extr. do Bol. da Soc. Brot., vol. xxi, 1904-1905]. Oct.
Collectors: Adolpho Frederico Moller (1842-1920; orig. herb. WI), Isaac Newton (*fl.* 1900; Portuguese), Alberto Arthur Sarmento (*fl.* 1900).
Publ.: Mar-Jun 1906 (journal issue rd. USDA 21 Jul 1906; p. 217: Mar 1906), p. [1]-9. *Copy*: FI. – Reprinted from Bol. Soc. Broter. 21: 209-217. 1906.

9942. *Un manipolo della flora del Monte Cavallo* desunto dalle iconografie inedite di G.G. Zannichelli. Nota. Venezia (Premiate officine grafiche di C. Ferrari) 1907. Qu.
Publ.: 1907 (mss. submitted 19 Mai 1907), p. [1]-18. *Copy*: FI. – Reprinted and to be cited from Atti r. ist. Veneto Sci. 66(2): 625-642. 1907.

9943. *Di un operetta sulla flora della Corsica* di un autore pseudonimo e plagiario. Nota. Venezia (Officine grafiche di C. Ferrari) 1908. Qu.
Publ.: 1908 (p. 5: printing authorized 8 Mai 1908), p. [1]-5. *Copy*: FI. – Reprinted and to be cited from Atti Ist. venet. Sci. 67(2): 717-721. 1908. – The pseudonymous work *Storia naturale dell'isola di Corsica*, Firenze 1774, of which the dedication is signed by Stataneo Gresalvi, is shown to be attributable to Salvatore Ginestra, of Bastia (anagram). In his introduction Ginestra gives much of the credit to G. Maratti, but the book is actually a plagiary of works by Paolo Boccone and Luigi Amando Jaussin.

9944. *La cronologia della flora italiana. Notizia preliminare*. Padova (R. Stab. P. Prosperini) 1908.
Publ.: 1908, p. [1]-7. *Copies*: BR, FI, MO. – Reprinted and to be cited from Atti Accad. venet.-trent. Padova ser. 2. 5: 1-5. 1908.

9945. *Cronologia della flora italiana* ossia repertorio sistematico delle più antiche date ed autori del rinvenimento delle piante (fanerogame e pteridofite) indigene, naturalizzate e avventizie d'Italia e della introduzione di quelle esotiche più comunemente coltivate fra noi ... Padova (Tipografia del Seminario) Marzo 1909. A spese dell'autore (*Cron. fl. ital.*).
Publ.: Mar 1909 (t.p.; Bot. Zeit. 16 Aug 1909; Bot. Centralbl. 17 Aug 1909; ÖbZ Mai 1909; Nat. Nov. Apr(1) 1909), p. [ii*], [i]-xxxvii, [1]-390. *Copies*: BR, FI, G, HH, NY, U, USDA. – Other t.p. [ii*]: "*Flora analytica d'Italia* dei professori Adriano Fiori, Giulio Paoletti e Augusto Béguinot volume v ...". For the first four volumes of this flora see A. Fiori, TL-2/1782. – The motto of Saccardo's *Cronologia*, "*Ego plantavi*: lascio ad altri la palma di migliorare e perfezionare l'opera mia" is also attributable to, among many other works, TL-2. – The cronologia brings a review of the increase in numbers of known species, statistics, a bibliography of works on the Italian flora, dates of early records; lists of species known at certain periods and many data on plant introduction. In the Roman (classical) period, for instance, 408 species were known; at the beginning of the 19th century 8000, "A monument of labour, learning, and scholarship ... so far no other country has produced anything to be compared with this *Cronologia* ..." (J. Britten 1910).
Preliminary note: Atti Accad. sci. venet.-trent.-istr. ser. 2. 5: 1-5. 1908, see above.,
Facsimile ed.: 1971, Edagricole [Edizioni Agricole], Bologna, with an introduction by Carlo Cappelletti and a portrait of Saccardo, p. [ii-vi], [iii]-xxxvii, [1]-390. *Copy*: BR.
Ref.: Britten, J., J. Bot. 48: 80-82. 1910 (rev.).
Malinvaud, E., Bull. Soc. bot. France 57(2): 158-160. 1910.
Peter, Bot. Zeit. 67: 234-235. 1909.

9946. *Da quale anno debba cominciare la validità della nomenclatura scientifica delle crittogame* ... [Estratto dagli "Annales mycologici" (vol. vii, no. 4, 1909]. Oct.
Publ.: 1909, p. [1]-6. *Copies*: FAS, FH, FI, NY. – Reprinted and to be cited from Ann. mycol. 7(4): 339-342. 1909. – Proposal to accept Linnaeus, Sp. pl. 1753 as starting-point book for nomenclature of all cryptogams. The good long term reasons provided by Saccardo for this proposal were not honored by the Bruxelles Congress of 1910 where the botanists preferred short term ad hoc solutions. Even so works not using binary nomenclature (e.g. Gleditsch, 1753 and Haller, 1768) and Adanson (Fam. pl. 1763) would have to be left out of account in botanical nomenclature. – The Sydney botanical Congress (1981) followed part of Saccardo's advice for the fungi, even though with some restrictions.

9947. *La flora trevigiana*. Notizie storiche e bibliografiche di P.A. Saccardo. Treviso (Stab. Tipogr. Ditta Luiga Zoppelli) 1910. (*Fl. trevig.*).
Publ.: 1910, p. [1]-28. *Copy*: FI. – Also published in Atti Ateneo Treviso 1900 (n.v.).

9948. *Fungi ex insula Melita* (Malta) *lecti* a doct. Alf. Caruana Gatto et Doct. Giov. Borg ... [Estratto dal Bull. della Soc. bot. ital. ...1912 (1913)]. Oct.
Collectors: John Borg (1873-?), Alfredo Caruana Gatto (*fl.* 1910).
Publ.: Dec 1912 (Nat. Nov. Feb(1) 1914; journal issue for Dec 1912), p. [1]-13. *Copy*: G. – Reprinted and to be cited from Bull. Soc. bot. ital. 1912: 314-327. Dec 1912.

Continuation (1): Jan 1914, p. [1]-19. *Copy*: FI. – Id. Nuovo Giorn. bot. Ital. 21(1): 110-126. 1914.
Continuation (2): Jan 1915, p. [1]-55. *Copy*: FI. – Id. 22(1): 24-76. 1915.
See also: P.A. Saccardo, *Fungi, in* S. Sommier & A. Caruana Gatto, Fl. melit. nov. 388-435. 1915.

9949. *Fungi tripolitani* a R. Pampanini anno 1913 lecti ... [Estratto dal Bull. della Soc. bot. ital. ... 1913]. Oct.
Collector: Renato Pampanini (1875-1949).
Publ.: Oct-Nov 1913 (read 24 Sep 1913; Nat. Nov. Dec(1) 1913; journal issue for Oct-Nov 1913), p. [1]-7. *Copy*: FI. – Reprinted and to be cited from Bull. Soc. ital. 1913: 150-156. – See also A. Trotter, Ann. mycol. 10: 508-514. 1912.
Continuation: With A. Trotter, Ann. mycol. 11: 409-420. 1913.

9950. Società botanica italiana. *Flora italica cryptogama Hymeniales* seu Hymenomycetes digessit P.A. Saccardo adjuvante Ab. Hier. Dalla Costa qui praeterea Conspectus synopticos concinnavit tabulae xi schemata generum referentes. Rocca S. Casciano (Stabilimento Tipografico L. Cappelli) Gennaio 1915-Marzo 1916. Oct. (*Fl. ital. crypt., Hymeniales*).
Co-author: Girolamo (Hieronymo) Dalla Costa.
1: Jan 1915, p. [i-iii], [1]-576, *pl. 1-6*. Fasc. 14 of *Flora*.
2: Mar 1916, p. [i-iii], 577-1386, err. slip., *pl. 7-11*. – Fasc. 15 of *Flora*.
Copies: BR, FH, FI, G, MICH, NY, Stevenson, USDA. – For full series see under G.B. Traverso.
Note: on nomenclature followed in *Fl. ital. crypt.*: P.A. Saccardo & G.B. Traverso, Bull. Soc. bot. ital. 1907: 22-28.(reprinted p. [1]-7, *copy*: FH).

9951. *Flora tarvisiana renovata* enumerazione critica delle piante vascolari finora note nella provincia di Treviso. Aggiuntevi le specie più comunemente coltivate e i nome dialettali ... Venezia (Premiate officine grafiche C. Ferrari) 1917. Oct. (*Fl. tarvis. renov.*).
Publ.: Sep-Dec 1917 (FI copy signed by author Apr 1918; printing authorized 5 Sep 1917), p. [1]-309. *Copies*: FI, M. – Also published in Atti Ist. venet. Sci. 76: 1237-1545. 1917.

Sachs, [Ferdinand Gustav] Julius von (1832-1897), German (Silesian) botanist; outstanding plant physiologist; draftsman with J.E. Purkyné at Praha 1849-1851; studied at Praha University 1851-1856; Dr. phil. ib. 1856; habil. ib. 1857; teaching plant physiology at Praha 1857-1859; assistant at the Forestry College of Tharandt (nr. Dresden) 1859; id. at Poppelsdorf (nr. Bonn) 1859-1867; professor of botany Freiburg i.B. 1867-1868; full professor of botany at Würzburg 1868-1897; Geheimrat 1877; most influential nineteenth century botanist in promoting experimental botany. (*Sachs*).

HERBARIUM and TYPES: WB (destroyed); some material at BR (in herb. O. Dieudonné). – Sachs collected a Silesian herbarium in his youth at Breslau; this was stolen from him; the herbarium at WB was only for teaching purposes. – Letters to D.F.L. von Schlechtendal at HAL. – Letters to W.G. Farlow at FH.

BIBLIOGRAPHY and BIOGRAPHY: Ainsworth p. 198, 283, 306, 308, 321; Biol.-Dokum. 15: 7788-7789; BJI 1: 50; BM 4: 1778-1779; Bossert p. 344; Collander hist. p. 42, 44, 45, 49, 94; CSP 5: 357-358, 8: 807-808, 11: 253, 18: 395; De Toni 1: cx, 2: cii; DSB 12: 58-60 (M. Bopp); Frank 3(Anh.): p. 87-88; GF p. 128; Herder p. 70, 95, 120, 271; IH 2: (files); Jackson p. 3, 54, 71, 72, 74, 90, 465; Kew 4: 560; Lenley p. 357; LS 23449-23454; Maiwald p. 69, 206, 230, 246; Moebius p. 455 [index]; Morren ed. 2, p. 10, ed. 10, p. 20; NI 1: 14, 204, 257, 3: 59; Plesch p. 394; PR 7948-7952, 10629; Rehder 5: 750; SO 866, 867; TL-2/see indexes; Tucker 1: 616.

BIOFILE: Allen, C.E., Bull. Torrey bot. Club 60: 341-346. 1933 (Sachs, the last of the botanical epitomists).
Anon., Ber. Senckenb. naturf. Ges. Frankf. a.M. 1897: xv-xvi. (obit.); Bonplandia 9: 159. 1861; Bot. Centralbl. 36: 223. 1888 (declined call to München), 70: 336. 1896

(d.); Bot. Zeit. 19: 136. 1861 (call to Bonn), 20: 456. 1862 (professor's title), 25: 39. 1867 (app. Freiburg), 26: 304. 1868 (call to Würzburg), 55: 192. 1897 (d.), 55(2): 192. 1897; Flora 50: 29-30. 1867 (called to Freiburg i.B.), 51: 236. 1868 (called to Würzburg), 60: 560. 1877 (order Bayer. Krone); Hedwigia 36: (101). 1897 (d.); J. Bot. Morot 11: 206. 1896 (d.); Leopoldina 33: 89, 94-95. 1897 (obit.); Nat. Nov. 2: 142. 1880 (corr. memb. Akad. München), 19: 311. 1897 (d.), 21: 230. 1899 (G. Kraus succeeds Sachs); Österr. bot. Z. 11: 237. 1861 (app. Poppelsdorf), 13: 56. 1863 (professor's title), 17: 90. 1867 (app. Freiburg), 18: 203. 1868 (app. Würzburg), 47: 231. 1897 (d.).

Bachmann, Fr., Planta 17: i-xviii. 1932 (biogr., portr., evaluation).
Bergdolt, E., Karl von Goebel ed. 2. 261 [index]. 1941.
Bessey, C.E., Amer. Naturalist 31: 713-714. 1897 (brief obit. not.).
Burgeff, H. et al., Bericht ü.d. Festsitzung anl. d. hundertst. Wiederkehr d. Geburtst. v. J. von Sachs, Würzburg 1932, 23 p., reprinted from Verh. phys.-med. Ges. Würzburg ser. 2, 57. 1932.
Campbell, D.H., Bull. Torrey bot. Club 60: 331-333. 1933 (influence of Sachs' textbook on American botany).
Conert, H.J., Senckenbergiana, Biol. 48(C): 34. 1967 (on Sömmerring Preis 1881).
Darwin, F., Nature 56: 201-202. 1897 (obit.).
Fletcher, H.R., Roy. Bot. Gard. Edinburgh 174-176, 198. 1970.
Freund, H. & A. Berg et al., Gesch. Mikroskopie 373 [index]. 1963.
Glass, B. et al., Forerunners of Darwin 115, 162, 329. 1968.
Goebel, K., Flora 84: 101-130. 1897 (portr., biogr., bibl.); Science 7: 662-668, 695-702. 1898 (translation of Flora obit. by E.D. Shipley).
Harvey, R.B., Plant Physiology 4: 155-157. *pl. 3* (portr.). 1929.
Harvey Gibson, R.J., Outlines hist. bot. 274 [index]. 1914.
Hauptfleisch, P., München. med. Wochenschr. 44(26): 709-711. 1897 (obit.); Professor Julius von Sachs. Gedächtnisrede, Würzburg 1897, 41 p. also published in Verh. phys.-med. Ges. Würzburg 31: 425-465. 1898 (obit., portr., bibl.).
Henriques, J., Bot. Soc. Broter. 15: 3-5. 1898 (obit.).
Hottes, C.F., Ann. Missouri bot. Gard. 19: 15-30. 1932 (contr. to botany by Sachs).
Hryniewiecki, B., Edward Strasburger 10, 47, 48, 88. 1938.
Hygen, G., Blyttia 5: 33-40. 1947 (evaluation 50 years after his death; portr.).
James, W.O., Endeavour 28: 60-64. 1969 (Sachs and the 19th century renaissance of botany).
Kralik, D. von, Forschungen und Fortschritte, Nachr.-Bl. deut. Wiss. Techn. 8: 363-364. 1932 (100th birthday).
Lütjeharms, W.J., Gesch. Mykol. 10, 16, 23, 33, 231. 1936.
Mägdefrau, K., Gesch. Bot. 313 [index]. 1973.
Matsumura, J., Bot. Mag. Tokyo 11: 353-357. 1897 (obit., Japanese).
Milner, J.D., Cat. portr. Kew 94. 1906.
Morton, A.G., Hist. bot. Sci. 149, 420, 424-428, 463. 1981.
Noll, F., Julius von Sachs, ein Nachruf, Braunschweig 1897, 21 p. (obit., many original details), reprinted from Naturw. Rundschau 12(36/37): 495 f. 1897; Bot. Gaz. 25: 1-2. 1898 (abridged version of Noll 1897, portr., biogr. sketch).
Nordenskiöld, E., Hist. biol. 551, 575-578, 580, 587, 605. 1949.
Oliver, F.W., Makers Brit. bot. 331 [index]. 1913.
Overbeck, F., Decheniana 105/106: 19-23, *pl. 1.* 1952 (on Sachs at Poppelsdorf 1860-1867).
Pringsheim, E.G., Julius Sachs, der Begründer der neueren Pflanzenphysiologie 1832 bis 1897, Jena 1932, xii, 302 p., *13 pl.* (rev. L. Diels, Bot. Jahrb. 66(Lit.): 8. 1934; Lotos 80: 1-4. 1932); Regnum vegetabile 71: 163-168. 1970 (additions to the main biogr.).
Reinke, J., Mein Tagewerk 493 [index]. 1925.
Roberts, H.F., Plant hybrid. before Mendel 81. 1965 (Sachs' opinion of Sprengel).
Rodgers, A.D., Amer. bot. 1873-1892. p. 366 [index]. 1944.
Schmid, G., Goethe u.d. Naturw. 1900, nos. 1825, 1868, 1887.
Skinner, H.M., The Student, Valparaiso, Indiana, U.S.A. 1892: 257-261.
Taton, R., ed., Science in the 19th century 612 [index]. 1965.
True, R.H., Bull. Torrey bot. Club 60: 335-340. 1933 (Sachs the man and the teacher).
Veer, P.H.W.A.M. de, Leven en werk van Hugo de Vries 58-61. 1969 (2 portr.).

Vines, S.H., Proc. R. Soc. London 62: xxiv-xxix. 1898 (obit.).
Voit, G., S.B. Akad. Wiss. München, mat.-phys. Cl. 28: 478-487. 1899 (obit.).
Wittrock, V.B., Acta Horti Berg. 3(2): 140-141, *pl. 24.* 1903, 3(3): xxvi-xxvii, 150, *pl. 88.* 1905 (portr.).

THESIS: *Physiologische Untersuchungen über die Keimung der Schminkbohne* (Phaseolus multiflorus), Wien 1859, 65 p., S.B. Akad. Wiss. Wien 37: 57-119. 1859.

MAIN BIOGRAPHY and BIBLIOGRAPHY: E.G. Pringsheim, *Julius Sachs,* der Begründer des neueren Pflanzenphysiologie 1832-1897, Jena (Gustav Fischer) 1932, xii, 303 p. (biogr., portr, handwriting), part of correspondence, secondary references, bibl.). *Nachtrag* in Essays in Biohistory, Regn. reg. 71: 163-168. 1970.

HANDWRITING: E.G. Pringsheim, Julius Sachs, pl. between p. 240-241.

NOTE: (1): Julius Sachs, b. Baltimore Md. 6 Jul 1849, d. 2 Feb 1934, was an American educator who wrote on science education in *Torreya.*
(2) We have no information on Franz Jakob Sachs, who published a dissertation *De Ulmo* in Strassburg 1738 [Dissertatio inauguralis medica *de Ulmo* quam consensu gratiosi medicorum ordinis pro honoribus atque privilegiis doctoralibus obtinendis ad diem xi. Aprilis mdccxxxviii ... Argentorati, Literis heredum Johannes Pastorii, p. [i-iv], [1]-36. *Copy*: USDA].
(3) During his years at Praha, Sachs published in Czech in the Bohemian journal Živa (edited by J.E. Purkyně) under the name *Julia Saxa.* The papers in Živa are listed e.g. by Goebel (1897) (titles in German translation) and Pringsheim (1932).
(4) Carl (Karl) Sachs (1853-1878), German traveller and zoologist, wrote e.g. *Aus den Llanos,* Schilderungen einer naturwissenschaftlichen Reise nach Venezuela, Leipzig 1879, vii, ii, 369 p., *1 pl.*; travelled in Venezuela 1876-1877.

COMPOSITE WORKS: (1) Hofmeister, W.F.B., ed., *Handbuch der physiologischen Botanik* ... vol. 4, 1865 by J. Sachs, *Handbuch der Experimental-Physiologie der Pflanzen.*
(2) *Arbeiten des botanischen Instituts ... Würzburg,* vols. 1-3, Leipzig, 1871-1888.
(3) *Gesammelte Abhandlungen über Pflanzen-Physiologie,* 2 vols., Leipzig 1892-1893, x, 1293 p., *10 pl.* – See e.g. F. Noll, Bot. Jahrb. 16 (Lit.): 42-43. 27 Jun 1893.

EPONYMY: *Sachsia* Grisebach (1866); *Sachsia* Bay (1894).

9952. *Über einige neue mikroskopisch-chemische Reactionsmethoden* ... Wien (aus der k.k. Hof– und Staatsdruckerei ...) 1859. Oct.
Publ.: 1859, p. [1]-24, *2 col. pl. Copy*: M. – Reprinted and to be cited from S.B. kais. Akad. Wiss., Wien, math.-nat. Cl. 36: 5-36. 1859.

9953. *Handbuch der Experimental-Physiologie der Pflanzen.* Untersuchungen über die allgemeinesten Lebensbedingungen der Pflanzen und die Functionen ihrer Organe ... Leipzig (Verlag von Wilhelm Engelmann) 1865. Oct. (*Handb. Exp.-Physiol. Pfl.*).
Publ.: 1865 (p. vi: 8 Oct 1865), p. [ii]-ix, [1]-514. *Copy*: U. – Handb. physiol. Bot., ed. W. Hofmeister, vol. 4, 1865.

9954. *Lehrbuch der Botanik* nach dem gegenwärtigen Stand der Wissenschaft ... Leipzig (Verlag von Wilhelm Engelmann) 1868. Oct. (*Lehrb. Bot.*).
Ed.[1]: 1868 (p. iv: 27 Jun 1868; Flora 30 Sep 1868), p. [i]-xii, [1]-632. *Copies*: BR, G, HH, NY, U.
Ed. 2: 1870 (p. iv: 24 Mai 1870; rev. A. de Bary, Bot. Zeit. 11 Nov 1870), p. [i]-xii, [1]-688. *Copies*: FH, FI, G, HH, H, M, NY. – "Zweite vermehrte und theilweise umgearbeitete Auflage ..." Leipzig (id.) 1870. Oct.
Ed. 3: Nov-Dec 1872 (p. vi: 5 Nov 1872; Litter. Centralbl. 19 Jul 1873; published in two parts: 1-336. Nov 1872, 317-end. Dec 1872, see Bot. Zeit. 30: 820, 883. 1872), p. [i]-xvi, [1]-848. *Copies*: BR, FI, G, H, HH, NY. – Dritte abermals vermehrte und stellenweise neubearbeitete Auflage ..." Leipzig (Verlag von Wilhelm Engelmann) 1873. Oct. – Part 3 issued separately as Grundzüge der Pflanzen-Physiologie, 1873, see below.

Ed. 4: 1874 (p. vi: 2 Mai 1874; Flora review dated 6 Oct 1874; Bot. Zeit. 7 Aug 1874), p. [i]-xvi, [1]-928. *Copies*: G, H, HH, M, MO, NY, U(2), USDA. – Vierte umgearbeitete Auflage ...".
Continuation: see below, 1882, *Vorles. Pfl.-Physiol.*
French: 1874 (BSbF Jul 1874: completed; Dec 1873: fasc. 1-3 ready, four to follow; J. Bot. Nov 1873: fasc. 1), p. [i]-xliii, [1]-1120. *Copies*: FI, G, HH, NY, USDA. – *Traité de botanique* conforme à l'état présent de la science par J. Sachs ... traduit de l'allemand sur la 3ᵉ édition et annoté par Ph. Van Tieghem [Philippe Édouard Léon] Van Tieghem (1839-1914)] ... Paris (Librairie F. Savy ...) 1874. Oct. (*Traité bot.*). – Based on *Lehrb. bot.* ed. 3, 1873.
English: 1875 (p. [vii]: Feb 1875; J. Bot. Mai 1875), p. [i]-xii, [1]-858. *Copies*: E, MICH, MO, NY, PH, USDA. – *Text-book of botany* morphological and physiological ... translated and annotated by Alfred W. Bennett [1833-1902], ... assisted by W.T. Thiselton Dyer [1843-1928]", ... Oxford (at the Clarendon Press) 1875 (*Text-book bot.*). – Based on *Lehrb. Bot.* ed. 3, 1873, with material from ed. 4, 1874 in footnotes. – *Reviews*: H. Trimen, J. Bot. 13: 213-216. 1875.
English, ed. 2: 1882, (p. vii: Aug 1882; Nat. Nov. Oct(1) 1882; Bot. Zeit. 24 Nov 1882), p. [i]-xii, [1]-980. *Copies*: G, HH, MICH, NY, PH, USDA. "*Text-book of botany* morphological and physiological ... edited, with an appendix, by Sydney H. Vines [1849-1934]. Review: A.W.B., J. Bot. 20: 348-349. 1882.

9955. *Ueber den gegenwärtigen Zustand der Botanik in Deutschland*. Rede zur Feier des 290. Stiftungstages der Julius-Maximilians-Universität gehalten von Professor Dr. Julius Sachs, zeitigem Rector, am 2. Januar 1872. Würzburg (Druck der F.E. Thein'schen Buchdruckerei) [1872]. Qu. (*Zustand Bot. Deutschl.*).
Publ.: 1872 (shortly after 2 Jan 1872, date of address; Flora 11 Jun 1872), p. [1]-28. *Copies*: G, M, U.

9956. *Grundzüge der Pflanzen-Physiologie.* Separatabdruck des dritten Buchs der dritten Auflage des Lehrbuchs der Botanik (1873) ... Leipzig (Verlag von Wilhelm Engelmann) 1873. Oct. (*Grundz. Pfl.-Physiol.*).
Publ.: 1873, (Bot. Zeit. 31 Oct 1873), p. [i]-viii, [1]-270. *Copy*: U. – Reprinted from Lehrb. Bot. ed. 3, 1873.

9957. *Geschichte der Botanik* vom 16. Jahrhundert bis 1860 ... Auf Veranlassung und mit Unterstützung seiner Majestät des Königs von Bayern Maximilian ii, herausgegeben durch die historische Commission bei der königl. Academie der Wissenschaften. München (Druck und Verlag von R. Oldenbourg) 1875. Oct. (*Gesch. Bot.*).
Publ.: 1875 (p. x: 22 Jul 1875; J. Bot. Dec 1875; Bot. Zeit. 29 Oct 1875), p. [iv]-xii, [1]-612. *Copies*: FI, G, HH, MICH, MO, NY, PH, U(2).– p. [iv]: "Geschichte der Wissenschaften in Deutschland, Neuere Zeit. Fünfzehnter Band ..." – Some copies have an advertisement for this series on p. [i-ii].
Facsimile ed.: 1966, p. [iii-vi, new preface material], [vii]-xii, [1]-608. *Copies*: FAS, NY. – New York (Johnson Reprint Corporation), Hildesheim (Georg Olms Verlagsbuchhandlung) 1966. Oct. – The index on p. 609-612 of the original is not reproduced in his reprint.
English(1): 1890 (p. xii: 24 Mar 1889; ÖbZ for Mai 1890; Bot. Zeit. 28 Mar 1890; Bot. Centralbl. 19 Mar 1890; Flora 5 Jul 1890; Nat. Nov. Mar(1). 1890), p. [i]-xv, [xvi, err.], [1]-568. *Copies*: E, NY, PH. – "*History of botany* (1530-1860) by Julius von Sachs ... authorised translation by Henry E.F. Garnsey [1826-1903], ...revised by Isaac Bayley Balfour [1853-1922], ... "Oxford (at the Clarendon Press) 1890. Oct. (*Hist. bot.*). – This English translation contains a new introduction by Sachs giving the general outlines of his views on the development of botany. He also notes to have overestimated the ideas of C. Darwin and C. v. Naegeli.
English(2): 1906 (Bot. Centralbl. 7 Mai 1907; Nat. Nov. Feb(1) 1907), p. [i]-xv, [1]-568. *Copies*: MICH, U, USDA. – "Second impression" Oxford (id.) 1906. Oct.
Reissue: 1967, New York (Russell & Russel). *Copy*: NY.
Continuation: J. Reynolds Green, *A history of botany* 1860-1900 being a continuation of Sachs "History of botany, 1530-1860" ... Oxford (at the Clarendon Press) 1909.
French: 1892 (p. xiii: 22 Jul 1891; Bot. Zeit. 1 Apr 1893; Nat Nov. Oct(2) 1892), p. [iii]-

xvi, [1]-584. *Copies*: BR, FAS, FI, G, NY, U. – "*Histoire de la botanique* du xvi siècle a 1860 par le Dr Julius von Sachs ... traduction française par Henry de Varigny [1855- ?] ..." Paris (C. Reinwald & Cie., ...) 1892. Oct. (*Hist. de bot.*). – Contains an introduction by Sachs similar to that to the English translation.

9958. *Vorlesungen über Pflanzen-Physiologie* ... Leipzig (Verlag von Wilhelm Engelmann) 1882. Oct. (*Vorles. Pfl.-Physiol.*).
Ed. [*1*]: Jul, Oct 1882 (p. v: 27 Jun 1882; published in 2 parts: p. 1-432. Jul 1882 (Bot. Zeit. 25 Aug 1882; Nat. Nov. Aug(1) 1882), part 2: p. 433-991. Oct 1882 (Nat. Nov. Nov(2) 1882; Bot. Zeit. 29 Dec 1882), p. [i]-xi, [xii], [1]-991. *Copy*: U. –Copy BM lists viii extra pages.
Ed. 2: 1887 (Bot. Zeit. 30 Dec 1887; Nat. Nov. Dec(2) 1887), p. [i]-xii, [1]-884. *Copies*: C, G, GB, H, ILL, LD, NSUB, U, US.
English: 1887 (Nat. Nov. Aug(1) 1887), p. [i]-xiv, [xv], [1]-836. *Copies*: ILL, MO, USDA. – "Lectures on the physiology of plants ... translated by H.M. Ward [Henry Marshall Ward, 1854-1906] ..." Oxford (at the Clarendon Press) 1887. Oct.
Ref.: Noll, F., Bot. Zeit. 46: 319-321. 18 Mai 1888 (rev.).

9959. *Gesammelte Abhandlungen über Pflanzen-Physiologie* ... Leipzig (Verlag von Wilhelm Engelmann) 1892-1893, 2 vols. Oct. (*Ges. Abh. Pfl.-Physiol.*).
1: 1892 (p. vii: 31 Jul 1892; Bot. Zeit. rev. 23 Dec 1892), p. [i]-vii, [ix-x], [1]-674.
2: 1893 (Bot. Zeit. rev. 1 Mai 1893, p. [i-iii], [675]-1243, *pl. 1-10*. *Copies*: U(2).

Sachse, Carl Traugott (1815-1863), German (Saxonian) naturalist and high school teacher at Dresden. (*Sachse*).

HERBARIUM and TYPES: Unknown.

BIBLIOGRAPHY and BIOGRAPHY: BM 4: 1779; CSP 5: 359, 8: 808; Herder p. 22; PR 7954; Rehder 1: 224, 383.

BIOFILE: Poggendorff, J.C., Biogr.-lit. Handw.-Buch 3: 1159. 1898 (b. 18 Dec 1815, d. 19 Nov 1863, bibl.).
Reichenbach, H.G.L., S.B. Isis, Dresden 1864(1-3): 1-6. (obit.; accounts by his wife and friends).

COMPOSITE WORKS: *Allgemeine deutsche naturhistorische Zeitung*, ed. C.T. Sachse, vols. 1-2, Dresden 1846-1847.

9960. *Zur Pflanzengeographie des Erzgebirges* ... [*in* Programm des Gymnasiums zu Dresden, womit zu dem Valedictions-Actus am 2. April ergebenst einladet das Lehrer-Collegium. 1 Zur Pflanzengeographie vom Gymnasiallehrer Carl Tr. Sachse ... Dresden (Druck von E. Blochmann & Sohn) 1855]. Oct. (*Pfl.-Geogr. Erzgeb.*).
Publ.: shortly before 2 Apr 1855, in Programm p. [i], [1]-41, (*Programm* as a whole: [1]-75). *Copies*: G, NY. – Bot. Zeit. 13: 406. 8 Jun 1855.
Note: Sachse also published "Beobachtungen über die Witterungs– und Vegetationsverhältnisse des Dresdner Elbthales während der Jahre 1847 bis 1852, ... Jahresber. Ges. Nat.-Heilk. Dresden 1853, 24 p., see A.E. Fürnrohr, Flora 37(1): 8-14. 1854.

Sacleux, Charles (1856-1943), French missionary, linguist and plant collector in Zanzibar and adjoining East Africa 1879-1898; from then on teacher at a missionary college at Chevilly and working at the Museum d'Histoire naturelle, Paris; ordained 1878. (*Sacleux*).

HERBARIUM and TYPES: P, PC.

BIBLIOGRAPHY and BIOGRAPHY: Barnhart 3: 198; BL 1: 63, 316; IH 1 (ed. 6): 363, (ed. 7): 341, 2: (files); Rehder 1: 536; Tucker 1: 616.

BIOFILE: Fournier, P., Voy. déc. Sci. mission. natural. franç. 2: 168, 174-176. 179, 180, 237. 1932 (bibl.; b. 1856; still alive 1932).

Le Gallo, Naturaliste canadien 77(3-4): 96-111. 1950 (obit., portr., bibl.).
Pellegrin, F., Bull. Soc. bot. France 77: 450-452. 1930 (on his collections and publications, portr., prix Gandoger).

9961. *Essai de catalogue des plantes de Zanzibar*, Pemba, Mombassa, Amou et de la Grande terre en face de ces îles jusqu'à la ligne de partage des eaux entre la côte et les grands lacs ... Extrait du Dictionnaire Français-Swahili par le R.P.Ch. Sacleux s.l.n.d. Oct. *Publ*: 1891, p. [i]-xxxvi. *Copy*: BR. – A reprint of the "Appendice" in Ch. Sacleux, *Dictionnaire français-swahili*, Zanzibar (mission des P.P. du St. Esprit), Paris (30, Rue Lhomond) 1891, p. [i]-xix, [1]-989, appendice: [i]-xxxvi, [a]-d err. *Copy*: LC. – The imprint on p. ii gives "1888-1891" which might indicate that the book came out in parts.

Sadebeck, Richard Emil Benjamin (1839-1905), German (Silesian) botanist; studied at Breslau with Goeppert, Cohn and Koerber; Dr. phil. ib. 1864; high school teacher in Berlin 1865-1876; teacher at the Johanneum high school at Hamburg 1876; subsequently director of the Botanical Museum (1883-1901) and the Botanical Garden ib.; in retirement from 1901. (*Sadebeck*).

HERBARIUM and TYPES: M (fungi herbarium; Exoascus), HBG; other material in GOET, M (lich.), WRSL. – Letters to W.G. Farlow at FH.

BIBLIOGRAPHY and BIOGRAPHY: Backer p. 660; Barnhart 3: 198 (b. 20 Mai 1839, d. 11 Feb 1905); Biol.-Dokum. 15: 7779-7780; BJI 1: 50, 2: 151; BL 1: 57, 316; BM 4: 1779; Christiansen 319 [index]; CSP 8: 808, 11: 255, 12: 643, 18: 397-398; DTS 1: 252-253; Frank 3 (Anh. p. 88; Futak-Domin p. 516; Herder p. 434; IF p. 728; IH 1 (ed. 6): 363, (ed. 7): 341, 2: (files); Jackson p. 95, 151, 152; Kew 4: 561; KR p. 614; LS 23456-23483; Moebius p. 125, 137; Morren ed. 10, p. 27; MW p. 422; PR 7955; Rehder 5: 750-751; Stevenson p. 1256; TL-2/see G.H.E.W. Hieronymus; Tucker 1: 616-617; Urban-Berl. p. 290.

BIOFILE: Anon., Biogr. Jahrb. 10: 239*. 1907 (further biogr. refs.).; Bonplandia 1: 217. 1853; Bot. Centralbl. 88: 360. 1901 (retirement), 98: 208. 1905 (d.); Bot. Not. 1906: 45 (d.); Bot. Monatschr. 15: 160. 1897 (decoration); Bot. Zeit. 34: 352. 1876 (app. Johanneum, Hamburg), 39: 403. 1881 (president of a new bot. soc. at Hamburg), 47: 577. 1889 (in charge of Hamburg bot. gard.), 60: 96. 1892 (retirement), 63: 80. 1905 (d.); Hedwigia 44: (137). 1905 (d. 11 Feb 1905), 40: (175). 1901 (retires from HBG); Nat. Nov. 3: 71. 1881 (S. sets up a bot. society at Hamburg), 11: 291. 1889 (assumes directorship Bot. Gard. Hamburg), 24: 70. 1902 (retirement); Österr. bot. Z. 55: 123. 1905 (d.); Verh. bot. Ver. Brandenb. 46: lxxxv. 1905.
Esdorn, I., Deut. Apoth.-Zeit. 117(1): 23. 1977 (brief biogr.).
Fischer-Benzon, R.J.D. von, *in* P. Prahl, Krit. Fl. Schlesw.-Holst. 2:51-52.
Hertel, H., Mitt. Bot. München 16: 426. 1980.
Pax, F., Schlesiens Pflanzenwelt 12, 14. 1915 (associated with v. Uechtritz), Bibl. schles. Bot. 90, 109, 110, 129, 148. 1929.
Voigt, A., Bot. Inst. Hamburg 47, 63, 64, 68, 74. 1897.
Wagenitz, G., Index coll. princ. herb. Gott. 140. 1982.

COMPOSITE WORKS: (1) in Engler and Prantl, *Nat. Pflanzenfam.*:
Pteridophyta, Einleitung, 1(4): 1-91. 1898.
Hymenophyllaceae, 1(4): 91-113. 21 Jul 1899.
Marsiliaceae, 1(4): 403-422. 17 Apr 1900, 799, 4 Jan 1902.
Salviniaceae, 1(4): 383-384. 12 Mai 1900, 385-402. 17 Dec 1900.
Equisetaceae, 1(4): 520-548. 1900.
Hydropteridineae, 1(4): 381-383. 12 Mai 1900.
Isoetaceae, 1(4): 756-768. 17 Oct 1901, 768-779. 4 Jan 1902.
Selaginellaceae (collaborating with G.H.E.W. Hieronymus, 1(4): 621-624. Jan 1901, 625-672. Sep 1901, 673-715. Nov 1901.
(2) Collaborator, A. Schenk et al., *Handbuch der Botanik* 1: 1-326. 1879-1882 (*Die Gefässkryptogamen*); other contributions in vols. 3(1). 1884, 3(2). 1887, 4. 1890.

9962. *De montium inter Vistritium et Nissam fluvios sitorum flora.* Dissertatio inauguralis botanica quam scripsit et gratiosi philosophorum ordinis consensu et auctoritate in alma litterarum Universitate viadrina ad summos in philosophia honores rite capessendos die xii. m. augusti a. mdccclxiv hora xi. publice defendet auctor Richard Sadebeck, ... Vratislaviae [Breslau] (Typis officinae A. Neumanni) [1864]. Qu. (*Mont. int. Vistrit. Niss. fluv.*).
Publ : 12 Aug 1864, p. [i-iv], [1]-42, [43-44, vita]. *Copies*: G, MO, NY.
Ref.: Anon., Flora 17: 592. 30 Nov 1864 (rd. Nov), 48: 172. 1865 (discussed at meeting Schles. Ges. vaterl. Cult. 17 Nov 1864).

9963. *Ueber die Entwickelung des Farnblattes* (Eine morphologische Studie) ... Berlin (Buchdruckerei von Gustav Lange (Paul Lange) ... 1874. Qu. (*Entw. Farnblatt.*).
Publ.: Dec 1873 (t.p. reprint: 1874; rev. Bot. Zeit. 9 Jan 1874), p. [1]-16, *1 pl.* (uncol. lith. auct.). *Copy*: G. – See also his *Zur Wachsthumsgeschichte des Farnwedels*. Ver. Bot. Ver. Brandenb. 15: 116-132. 1873 ; summary Bot. Zeit. 32: 28-30. 1874.
Ref.: G.K., Bot. Zeit. 32: 28-30. 9 Jan 1874.

9964. *Beobachtungen und Untersuchungen über die Pilzvegetation in der Umgegend von Hamburg* von Profesor Dr. R. Sadebeck Vorsitzender der Gesellschaft für Botanik zu Hamburg. Überreicht vom Verfasser [Hamburg (Druck von Ferdinand Schlotke) 1881]. (*Beobacht. Pilzveg. Hamburg*).
Publ.: on or just before 8 Aug 1881 (see p. [ii]: p. 21: 20 Jul 1881; Bot. Centralbl. 5-9 Dec 1881; Bot. Zeit. 27 Jan 1882), p. [ii-iii], [1]-21, [22]. *Copies*: G, USDA. – Preprinted from Verh. Ges. Bot. Hamburg vol. 1, 1881.
Ref.: N.L., Bot. Zeit. 40: 64. 27 Jan 1882.

9965. *Untersuchungen über die Pilzgattung Exoascus* und die durch dieselbe um Hamburg hervorgerufenen Baumkrankheiten ... Hamburg (gedruckt bei Th.G. Meissner, ...) 1884.
Publ.: 1884 (Hedwigia Sep 1884), p. [91]-124, *pl. 1-4* (uncol. liths. auct.). *Copies*: BR, FH, G, U. – Reprinted from Jahresb. wiss. Anst. Hamburg 1: 91-124, *pl. 1-4.* 1884.
Preliminary notice: Tagebl. deut. Naturf., Eisenach 1882: 193-194. (n.v.).
Ref.: Büsgen, M., Bot. Zeit. 42: 655-656. 10 Oct 1884.

9966. *Kritische Untersuchungen über die durch Taphrina-Arten hervorgebrachten Baumkrankheiten* ... Hamburg (Gedruckt bei Lütcke & Wulff, ...) [1890]. Oct. (*Krit. Unters. Taphrina Baumkrankh.*).
Publ.: 1890 (Bot. Zeit. 28 Nov 1890; Hedwigia Nov-Dec 1890; Nat. Nov. Nov(1) 1891, sic), p. [1]-37, *pl. 1-5* (liths. auct.). *Copies*: FH, FI, Stevenson, U. – Preprinted or reprinted from Jahrb. Hamb. wiss. Anst. 8: 59-95, *pl. 1-5.* 1890 (1891).
Ref.: Niedenzu, F., Bot. Jahrb. 13 (Lit.): 18-19. 20 Mar 1891 (rev.).
Kienitz zu Gerloff, J.H.E.F., Bot. Zeit. 49: 108-109. 13 Feb 1891 (rev.).

9967. *Die tropischen Nutzpflanzen Ostafrikas* ihre Anzucht und ihr ev. Plantagenbetrieb ... Eine orientirende Mittheilung über einige Aufgaben und Arbeiten des Hamburgischen Botanischen Museums und Laboratoriums für Waarenkunde ... Hamburg (Commissions-Verlag von Lucas Gräfe & Sillem). 1891. Oct.
Publ.: 1891(p. 26: Apr 1891; Bot. Zeit. 27 Nov 1891; ÖbZ for Dec 1891; Nat. Nov Sep(2) 1891), p. [1]-26. *Copies*: FH, HH, M, MO, U. – Reprinted and to be cited from Jahrb. Hamb. wiss. Anst. 9(1): 203-228. 1891.
Ref.: Brick (Hamburg), Bot. Centralbl. 51: 247-249. 1892 (rev.).

9968. *Die parasitischen Exoasceen.* Eine Monographie ... Hamburg (Gedruckt bei Lütcke & Wulff, ...) 1893. Oct. (*Paras. Exoasc.*).
Publ.: 1893 (p. 110: 8 Jul 1893; Nat. Nov. Aug(2) 1893), p. [1]-110, *pl. 1-3* (uncol. liths. auct.). *Copies*: BR, FH, FI, G, M, NY, Stevenson, U.
Ref.: Kienitz-Gerloff, J.H.E.F., Bot. Zeit. 51(2): 325-329. 1 Nov 1893.

9969. *Filices Camerunianae Dinklageanae* ... Hamburg (Commissions-Verlag von Lucas Gräfe & Sillem) 1897. Oct.

Publ.: 1897 (Nat. Nov. Jan(2) 1898, p. [1]-18, *1 pl.* (uncol. lith.). *Copies*: BR, G(2), U. – Reprinted and to be cited from Jahrb. Hamb. wiss. Anst., 14 (Beih. 3): 1-18. 1896.

9970. *Die wichtigeren Nutzpflanzen* und deren Erzeugnisse *aus den deutschen Colonien*. Ein mit Erläuterungen versehenes Verzeichniss der Colonial-Abtheilung des Hamburgischen Botanischen Museums ... Hamburg (Commissions-Verlag von Lucas Gräfe & Sillem). Oct.
Publ.: 1897 (ÖbZ for Dec 1897; Nat. Nov. Jan(1) 1898), p. [1]-138, [139, err.]. *Copies*: BR, G(2), USDA. – Reprinted and to be cited from Jahrb. Hamb. wiss. Anst. 14 (Beiheft 3): 19-156. 1896.

9971. *Die Kulturgewächse der deutschen Kolonien* und ihre Erzeugnisse. Für Studierende und Lehrer der Naturwissenschaften, Plantagenbesitzer, Kaufleute und alle Freunde kolonialer Bestrebungen nach dem gegenwärtigen Stande unserer Kenntnisse ... Jena (Verlag von Gustav Fischer) 1899. Oct. (*Kulturgew. deut. Kolon.*).
Publ : Dec 1898 (t.p. 1899; p. vi: Aug 1898; Bot. Zeit. 16 Jan 1899; ÖbZ for Nov 1898; Fischer pamphlet Nov 1898:"soeben ist erschienen" Nat. Nov. Dec(2) 1898), p. [i]-xiii, [1]-366. *Copies*: BR, FI, H-UB, M, NY, PH, USDA.
Ref.: Gilg, E., Bot. Jahrb. 26 (Lit.): 68-69. 18 Apr 1899 (rev.)
Fischer, Gustav (Verlag), Advertising pamphlet Jena, Nov 1898, 8 p. (extensive description).
Solms-Laubach, H., Bot. Zeit. 57(2): 81-83. 1899.
Solederer, H., Flora 86: 110. 28 Jan 1899.

9972. *Die Raphiabast* ... Hamburg (Commissions-Verlag von Lucas Gräfe & Sillem) 1901. Oct.
Publ.: 1901 (Nat. Nov. Jan(1) 1902), p. [1]-42, *pl. 1-2* (uncol. lith. auct.). *Copy*: BR. – Reprinted and to be cited from Jahrb. Hamb. wiss. Anst. 18 (Beiheft 3): [1]-42, *pl. 1-2*. 1901.
Ref.: Warburg, O., Bot. Zeit. 60(2): 217-219. 1902.

Sadler, John (1837-1882), Scottish botanist at Edinburgh; assistant to J.H. Balfour 1854; acting secretary Botanical Society of Edinburgh 1858-1879; curator Royal Botanic Garden 1879-1882. (*John Sadler*).

HERBARIUM and TYPES: E; mosses at BM; further material at CGE. – Letters at K.

BIBLIOGRAPHY and BIOGRAPHY: Barnhart 3: 198 (b. 3 Feb 1837, d. 9 Dec 1882); BB p. 266; BL 2: 301, 306, 309, 712; BM 4: 1779-1780; Bossert p. 344; Clokie p. 236; CSP 5: 360, 8: 808-809, 11: 256, 12: 643, 18: 398; Desmond p. 537; Herder p. 235; Jackson p. 246, 254; LS 7061, 23484-23487; TL-2/287; Tucker 1: 617.

BIOFILE: Anon., Bot. Centralbl. 12: 424. 1882 (d.); Gard. Chron. 15: 76, 81, 1879 (app. curator Bot. Gard. Edinb.; portr.), 18: 793. 1882 (obit.); J. Bot. 21: 81-82, 382. 1883 (obit.); Proc. Perthshire Soc. nat. Sci. 2(3): 87-88. 1883 (obit.).
Balfour, J.B., Trans. Bot. Soc. Edinburgh 16: 11-15. 1886 (obit.).
Bridson, G.D.R. et al., Nat. Hist. mss. res. Brit. Isl. 138.43, 240.90, 1980.
Craig, W., [Hist.] Trans. Berwickshire Natural. Club 10. 72-83. 1882, repr. 8 p. (copy at E).
Fletcher, H.R. and W.H. Brown, Roy. bot. Gard. Edinburgh 123, 146, 182, 187, 189-141. 1970.
Freeman, R.B., Brit. nat. hist. books 303. 1980.
Hawksworth, D.L. & M.R.D. Seaward, Lichenology Brit. Isl. 20, 145, 198. 1977.
Hedge, I.C. & J.M. Lamond, Index coll. Edinb. herb. 125. 1970.
Murray, G., Hist. coll. BM(NH) 1: 179. 1904 (360 mosses, 150 lichens).
Sayre, G., Bryologist 80: 516. 1977.
Stuart, C., J. Hort. Cottage Gard. ser. 3. 6: 55-56. 1883 (a botanical ramble with J.S.; from North British Advertiser).
Trail, J.W.H., ed., Scott. Natural. 7: 43-44. 1884 (obit.).

SADLER, JOHN

COMPOSITE WORKS: (1) J.H. Balfour (1808-1884), *Fl. Edinburgh* 1863; mosses, hepatics and lichens by J. Sadler; see TL-2/287; id. ed. 2, Edinburgh (Adam and Charles Black) 1871, p. [i]-vi, [1]-183, map. *Copy*: E (not mentioned under no. 287).
(2) *Mr. John Sadler's list of arctic cryptogamic* and other plants, collected by Robert Brown, Esq., during the summer of 1861, on the islands of Greenland, in Baffin's Bay and Davis' Strait, and presented to the Herbarium of the Botanical Society, in Trans. bot. Soc. Edinburgh 7: 374-375. 1862, and in R. Jones, Manual of the natural history, geology, and physics of Greenland and the neighbouring regions ... London 1875, p. 253-254.

9973. *Narrative of a ramble among the wild flowers on the Moffat Hills* in August 1857; with a list of plants to be found in the district ... Moffat (William Muir) 1857. Oct. (*Narr. ramble Moffat Hills*).
Publ.: 1857 (preface Oct 1857), p. [1]-64, map and 4 plates of dried specimens (see BM 4: 1779). *Copy*: E.

9974. *Notice of Salix sadleri* (Syme) *and Carex frigida* (Allioni) *both recently discovered in the Highlands of Scotland* ... Edinburgh (printed by Neill and Company) 1874. Oct.
Publ: 1874 (plants collected on 7 Aug 1874), p. [t.p. and 3 p. unnumbered)]. *Copy*: E. – Reprinted and to be cited from Trans. bot. Soc. Edinburgh 12: 209-211. 1874. Inf. J. Edmondson.

Sadler, Joseph (1791-1849), Hungarian botanist; assistant for chemistry and botany at the University of Budapest 1815-1819; Dr. med. Univ. Budapest 1820; subsequently curator at the National Hungarian Museum for zoology and mineralogy; ultimately professor of botany and director of the University Botanic Garden, Budapest 1834-1849. (*Jos. Sadler*).

HERBARIUM and TYPES: BP (ca. 8000); other material at BR, FI, GOET, HBG, JE, KIEL, LZ (destr.), PRC. *Exsiccatae*: *Agrostotheca hungarica*, 75 nos., see Sauter, Flora 21: 516-517. 1838; Anon., Linnaea 10 (Litt.): 125, 191-192. 1836, 12 (Litt.): 22-23. 1839 and Priszter (1976). Part of the herbarium of Jos. Sadler's brother Michael Sadler is at GJO. For other exsiccatae, published between 1821 and 1841 see Priszter (1976/1977). – (Fasc. 1-14 of Hungarian plants reviewed by Flora 4: 141-144. 1821, 5: 64. 1822, 6(1) Beil. 2: 48. 1823, 8(2) Beil. 2: 46-47. 1825, as A' Magyar plánták' száritott Gyüjteménye, 1823-1830. – Letters to D.F.L. von Schlechtendal at HAL.

BIBLIOGRAPHY and BIOGRAPHY: AG 2(1): 320; Backer p. 506; Barnhart 3: 198 (b. 6 Mai 1791); BM 4: 1780; Bossert p. 344; Futak-Domin p. 516-517; Hegi 5(2): 1116; Herder p. 193, 271; IF p. 728; IH 1 (ed. 7): in press, 2: (files); Jackson p. 268; Kanitz p. 81-84, no. 124; Kew 4: 561; PR 7956-7961, p. 574, ed. 1: 8911-8915; Rehder 1: 445.

BIOFILE: Anon., Bonplandia 3: 34. 1855 (cogn. Leopoldina Kitaibel), 6: 339. 1858; Flora 5(1): 64, 176, 236-237, 287. 1822, 6(1): 31. 1823, 8(2): 46-47, 573. 1825.
Kanitz, A., Linnaea 33: 555-562. 1865.
Priszter, S., Bot. Közlem. 63(3): 217-229. 1976 (1977) (on his exsiccatae, 1823-1841).
Simonkai, L., Enum. fl. transsilv. xxviii-xxix. 1886.
Wagenitz, G., Index coll. princ. herb. Gott. 140. 1982.

EPONYMY: *Sadleria* Kaulfuss (1824).

9975. *Verzeichniss der um Pesth und Ofen wildwachsenden phanerogamischen Gewächse* mit Angabe ihrer Standorte und Blüthezeit ... Pesth (Bei Konrad Adolf Hartleben) 1818. Duod. (*Verz. Pesth wildw. Gew.*).
Publ: 1818 (p. vi: 1 Jan 1818), p. [i]-vi, [7]-79. *Copies*: G, NY.

9976. *Flora comitatus pestiensis* ... Pestini [Pest, Budapest] (typis nobilis Matthiae Trattner de Petróza) 1825-1826. 2 parts. Oct. (*Fl. comit. pest.*).
Ed. 1, in 2 vols. *Copies*: G, M, MO, NY, US, USDA; IDC 5336.
1: 1825 (p. vi: 10 Mar 1825), p. [i-vi], [1]-335, [336, err.].
2: 1826, p. [1]-398, [399, err.].

Ed. 2: 1840 (p. [iv]: 25 Nov 1839; Flora 21 Dec 1840), p. [i-iv], [1]-499. *Copies*: FI, G, HH, KNAW, NY. The copy at G has two frontisp. [Oudinot, Talleyrand] not issued with the original ed. – *Flora comitatus pestiensis* in uno volumine comprehensa ... editio secunda. Pesthini (apud Kilian et Comp.) 1840. Oct.
Ref.: Anon., Flora 14(2) (Lit. 19): 283-291. 1831 (rev.).

9977. *De filicibus veris hungariae*, transylvaniae, croatiae et litoralis hungarici solennia instaurationis semisecularis regiae Universitatis Hungaricae die xxviii junii anni mdcccxxx recolens disserit Josephus Sadler ... decanus. Budae [Buda; Budapest] (typis regiae Universitatis Hungaricae) [1830]. Oct. (*Fil. ver. hung.*).
Publ.: 28 Jun 1830, p. [1]-69, [70]. *Copies*: E, FI, G, HH, MO; IDC 7254.

Sadler, Michael (*fl.* 1831), Hungarian botanist; brother of Joseph Sadler. (*M. Sadler*).

HERBARIUM and TYPES: Some of the material listed under Joseph Sadler may have been collected by Michael Sadler.

BIBLIOGRAPHY and BIOGRAPHY: Barnhart 3: 198; Futak-Domin p. 517; IH 1 (ed. 7): 341, 2: (files); Kanitz p. 104; PR 7962, ed. 1: 8916; Rehder 2: 102; Tucker 1: 617.

BIOFILE: Simonkai, L., Enum. fl. transsilv. ix. 1886.

9978. Specimen inaugurale sistens *synopsin Salicum hungariae*, quam annuentibus magnifico domino praeside et directore, spectabili domino decano ac clarissimis dominis professoribus pro doctoris medicinae laurea rite consequenda in alma ac celeberrima Universitate regia hungarica pesthinensi publicae eruditorum disquisitioni submittit Michael Sadler ... theses adnexae publice defendentur in aedibus Facultatibus die [ ] 1831. Pesthini [Pest, Budapest] (typis Trattner-Karolyanis) [1831]. Oct. (*Syn. Salic. hung.*).
Publ.: 1831, p. [1]-32. *Copy*: E. – Inf. J. Edmondson.

Saelan, [Anders] Thiodolf (1834-1921), Finnish physician and botanist; Dr. med. Helsinki 1865; amanuensis at the Botanical Museum ib. 1859-1866; practicing physician 1861-1865; physician (1865) and head-physician at Lappvik hospital 1868-1904; professor's title 1877; Dr. phil. h.c. 1907. (*Saelan*).

HERBARIUM and TYPES: H. – Manuscripts and some correspondence at H; 124 letters from Saelan to Nylander and 289 from Nylander to Saelan at H-UB.

BIBLIOGRAPHY and BIOGRAPHY: AG 12(1): 156; Barnhart 3: 198 (b. 20 Nov 1834, d. 24 Jun 1921); BJI 1: 56, 2: 151; BL 2: 61, 63, 67, 70, 85, 92; BM 4: 1780, 8: 1121; Bossert p. 344; Collander p. 464-465 (bibl.); Collander hist. p. 157 [index]; CSP 11: 256, 18: 399; GR p. 622; Herder p. 201, 294, 305, 331, 332, 337; IH 2: (files); Kew 4: 561; KR p. 614; LS 23488, 38685; Morren ed. 10, p. 97; PR 5570, 6781, 7963, 7964; Rehder 5: 751; Saelan p. 435-445 (bibl.); SBC p. 131; TL-2/4927, 6935, 6926, see A.H. Hjelt, A. Kihlmann, S.S. Murbeck; TR 1223-1226; Tucker 1: 617.

BIOFILE: Anon., Bot. Not. 1921: 173 (d.); Medd. Soc. Fauna Fl. fenn. 48: 1. 1925 (portr.).
Carpelan, T. & L.O. Th. Tudeer, Helsingin yliopisto 2: 830-835. 1924 (fide Collander).
Elfving, F., Acta Soc. Fauna Fl. fenn. 50: 108, 225. 1921 (portr.); Soc. Sci. fenn., Årsbok 1(B, 8): 1-16. 1923 (biogr., portr., bibl.).
Fagerström, L., Acta Soc. Fauna Fl. fenn. 71(2): 1-20. 1954 (on p. 8-12 account of S.'s excursion to eastern Nyland, 1856).
Häyrén, E., Finska Trädgårdsodl. 15: 162-163. 1921 (fide Collander).
Hagelstam, J. Finska läkaresällsk. Handl. 63: 503-511. 1921 (fide Collander).
Sayre, G., Bryologist 80: 516. 1977 (coll.).
Ulvinen, A., Kymenlaakson luonto 12: 18-20. 1971 (a pioneer of the floristic investigation of Kymenlaakso; portr.); Kymenlaakson Luonnon Ystävien julkaisuja 2: 1-27. 1974 (on an 1856 diary of his trip to Kymenlaakso; portr.).

COMPOSITE WORKS: Co-author, with E. Lönnrot, of *Flora fennica,* ed. 2, 1866, see TL-2/4927. Herbarium Lönnrot: TUR.

EPONYMY: *Saelania* S.O. Lindberg (1878).

9979. *Herbarium musei fennici.* Förteckning öfver Finska Musei växtsamling, utgifven af Sällskapet pro fauna et flora fennica och uppgjord af W. Nylander och Th. Saelan ... Helsingfors (Finska Litteratur-Sällskapets Tryckeri) 1859, Oct. (in fours). (*Herb. mus. fenn.*).
Co-author: William Nylander (1822-1899), see TL-2/6946. We give here a slightly amended treatment of eds. 1 and 2.
Ed. 1: Jun 1859 (11 Apr 1859 preface etc. accepted; publ. Jun 1859, fide Elfving (1921), p. 108; inf. P. Isoviita; Flora rd. 7 Jul 1859), p. [1]-118, map. *Copies:* BR, FI, G, H.
Ed. 2: Mar-13 Mai 1889 (inf. P. Isoviita; Bot. Centralbl. 9 Jul 1889; Bot. Zeit. 28 Jun 1889; Flora rd. 1 Nov 1889; Nat. Nov. Apr(1) 1890), p. [i]-xix, [1]-156, 2 maps. *Copies:* BR, FI, G, H, NY, PH, US, USDA. – *Herbarium musei fennici* enumeratio plantarum Musei fennici quam edidit Societas pro fauna et flora fennica editio secunda. I. *Plantae vasculares* curantibus Th. Saelan, A. Osw. Kihlman, Hj. Hjelt ... Helsingforsiae [Helsinki] (ex officina typographica heredum J. Simelii) 1889. Oct. (in fours). *Note:* The second part, II. *Musci,* by J.O. Bomansson & V.F. Brotherus, p. [i]-vii, [viii], [1]-77, [i, index], 1 map, came out Nov-Dec 1894.
Ref.: Anon., Bull. Soc. bot. France 6: 430-431. 1859 (rev.).
Bennett, A., J. Bot. 27: 220-222. Jul 1889 (rev. ed. 2).
Drude, O., Bot. Jahrb. 11 (Lit.): 82. 31 Dec 1889 (id.).
Brotherus, V.F., Bot. Centralbl. 40: 377-379. 1889 (id.).
Schlechtendal, D.F.L. von, Bot. Zeit. 17: 268-269. 5 Aug 1859 (rev.).

Note: Booknumber 9980 has not been used.

9981. *Musci lapponiae kolaënsis* auctoribus V.F. Brotherus et Th. Saelan ... Helsingforsiae (ex officina typographica heredum J. Simelii) 1890. Oct. (*Musc. lapp. kolaens.*).
Senior author: Viktor Ferdinand Brotherus (1849-1929).
Publ.: 1890 (read 1 Feb 1890; Rev. bryol. Aug-Sep 1891; Bot. Gaz. Sep 1891), p. [1]-100, map. *Copy:* FI, H. – Published as Acta Soc. Fauna Fl. fenn. 6(4), 1890; also included in *Wiss. Erg. Finn. Exp. Kola* 1887-1892, see Saelan p. 54.

9982. *Finlands botaniska litteratur* till och med år 1900 .... Helsingfors 1916. Oct. (*Finl. bot. litt.*), ("Saelan" in TL-2).
Publ.: Dec 1916 (p. vi: Dec 1916), p. [i]-xi, [1]-633. *Copies:* BR, FAS, G, H, NY, USDA. – Acta Soc. Fauna Fl. fenn. 43(1). 1916. The journal issue has p. [i*-iv*] journal preface material. – The major Finnish botanical bibliography up to and including 1900. For a continuation see R. Collander, V. Erkamo, P. Lehtonen, *Bibliographia botanica fenniae 1901-1905,* Helsinki Sep 1973, Acta Soc. Fauna Fl. fenn. 81, 1973. ("Collander" in TL-2).

Säve, Carl [Fredrik] (1812-1876), Swedish linguist and botanist; Dr. phil. Uppsala 1848; lecturer (1849) and professor of scandinavian languages at Uppsala 1859-1876. (*Säve*).

HERBARIUM and TYPES: High school at Visby.

NOTE: Per (Magnus) Arvid Säve (1811-1887), brother of C.F. Säve, also published on the Gothland flora, see KR p. 693.

BIBLIOGRAPHY and BIOGRAPHY: BM 4: 1815; Herder p. 195; Kew 4: 562; KR p. 693 (b. 22 Oct 1812, d. 28 Mar 1876); PR 7965, Rehder 1: 353.

BIOFILE: Löwegren, Y., Årsb. sv. underv. hist. 132: 230. 1974 (herb. at Visby).
Sjögren, G., Sv. män kvin. 7: 412-413. 1954 (biogr., portr; mainly on his linguistic work [not really a good linguist], no mention of his bot. publ.).

9983. *Synopsis florae Gothlandicae.* Quam venia experientiss. Facult. med. Upsal. praeside doct. Georgio Wahlenberg ... p.p. auctor Carolus Säve ... in Audit. Linnaean. die xx. maji mdcccxxxvii h.a.m.s. P. I. [Upsaliae [Uppsala] (Excudebant Regiae Academiae Typographi) [1837]. Oct. (*Syn. fl. Gothl.*).
1: 1-20 Mai 1837 (defended 20 Mai 1837; Hinrichs rd. 29 Sep 1837), p. [i-iv, dedications by Säve], [1]-16.
2: 1-20 Mai 1837 (id.), p. [i-ii], 17-34. − p. [i]: "Synopsis ... Wahlenberg ... p.p. Arvidus Sundberg ... P. II. ..."
Copies: C, GB, H, HH, LD, NY,B. Peterson. − The Helsinki copy is dated for receipt 27 Mai 1837 by K.F. Thedenius. − Dissertation defended under G. Wahlenberg, q.v.; Säve was respondens for part 1; Krok (p. 745) and Almborn (in litt.) note that Säve was the author of the entire text; part 2 was defended by Arvid Sundberg.

Safford, William Edwin ("Ned") (1859-1926), American botanist and conchologist; United States Naval Academy grad. 1880; from 1880-1902 in the U.S. Navy, 1899-1900 as lieutenant governor of Guam; with the Office of Economic and Systematic Botany of the Bureau of Plant Industry (USDA) of the United States 1902-1926. (*Saff.*).

HERBARIUM and TYPES: US; mss. and corr. also at US(SIA).

BIBLIOGRAPHY and BIOGRAPHY: Barnhart 3: 199 (b. 14 Dec 1859, d. 10 Jan 1926); BJI 2: 151; BL 1: 117, 124, 316; Bossert p. 345; CSP 18: 400; Hirsch p. 257; Hortus 3: 1203; IH 2: (files); Kew 4: 562; Langman p. 662-663; Lenley p. 357; MW p. 422; NW p. 56; PH 150; Plesch p. 394; SIA 189, 208, 221, 7073, 7183, 7275; Zander ed.10, p. 710, ed. 11, p. 810.

BIOFILE: Anon., Amer. Fern J. 16: 129. 1927 (obit.); The Evening Star, Washington D.C. 11 Jan 1926 (obit.).
Barnes, W.C., Science ser. 2. 63: 418. 1926 (obit.).
Ewan, J., *in* R.W. Long & O. Lakela, Fl. trop. Florida 6. 1971 (portr.).
Humphrey, H.B., Makers of North American botany 213-215. 1961.
Kearney, T.H. & E.J. Smith, J. Heredity 17(10): 365-367. 1926 (obit., portr.).
Kellerman, W.A., *in* Geol. Ohio pl. 69. 1895.
Knobloch, I.W., Pl. coll. N. Mexico 61. 1979.
McVaugh, R., Edward Palmer 428. 1956; Contr. Univ. Mich. Herb. 9: 302. 1972 (Mexican trip of 1907).
Merrill, E.D., B.P. Bishop Mus. Bull. 144: 161. 1937 (Polyn. bot. bibl.).
Nelsen, F.J., U.S. Naval Inst. Proc. 78(8): 851-861. 1952 (on S. as lieutenant governor of Guam; portr.).
Peattie, D.C., Amer. botanist 32(2): 63-66. Apr 1926 (obit.).
Rickett, H.W., Index Torrey bot. Club 87. 1955.
Rogers, D.P., Brief hist. mycol. N. Amer. ed. 2. 1. 1981.
Stieber, M.T. et al., Huntia 4(1): 86. 1981 (arch. mat. HU).

EPONYMY: *Saffordia* Maxon (1913); *Saffordiella* Merrill (1914).

9984. *The useful plants of the Island of Guam* ... Washington (Government Printing Office) 1905. Oct. (*Usef. pl. Guam*).
Publ.: 8 Apr 1905, p. [i-ii], [1]-416, *pl. 1-69*, map (= *pl. 70*). *Copies*: BR, MICH, MO, NY, PH, U, US, USDA. − Issued as Contr. U.S. natl. Herb. 9, 1905.
Ref.: Diels, L., Bot. Jahrb. 36 (Lit.): 21-22. 1905 (rev.).
Gagnepain, F., Bull. Soc. bot. France 52: 482-484. 1905 (rev.).
Solms-Laubach, H., Bot. Zeit. 63(2): 266. 1 Sep 1905 (rev.).

9985. *Cactaceae of Northeastern and Central Mexico* together with a synopsis of the principal Mexican genera ... Washington (Government Printing Office) 1909.
Publ.: 1909 (Nat. Nov. Mar(2) 1910; p. iii of journal issue dated 12 Jun 1909), p. [i, reprint t.p.], 525-563, *pl. 1-15*. *Copies*: G, USDA. − Reprinted from Smithson. Rep. 1908: 525-563, *pl. 1-15*. 1909.

9986. *Raimondia*, a new genus of Annonaceae from Colombia ... Washington (Government Printing Office) 1913. Oct.
Publ.: 11 Feb 1913, p. [i, repr. t.p.], 217-219, *pl. 52-53*. *Copy*: G. – Reprinted and to be cited from Contr. U.S. natl. Herb. 16: 217-219. *Pl. 52-53*. 11 Feb 1913.

9987. *Classification of the genus Annona* with descriptions of new and imperfectly known species ... Washington (Government Printing Office) 1914. Oct.
Publ.: 17 Jun 1914 (Nat. Nov. Jul(1) 1914), p. [i]-ix [journal pref.], [1]-68, *pl. 1-41*, xi-xii[index]. *Copies*: BR, G, H, HH, MO, NY, U, US. – Reprinted and to be cited from Contr. U.S. natl. Herb. 18(1): 1-68, *pl. 1-41*, xi-xii. 1914.

Sageret, Augustin (1763-1851), French land-owner, agriculturist, plant hybridizer and botanist. (*Sageret*).

HERBARIUM and TYPES: Unknown.

BIBLIOGRAPHY and BIOGRAPHY: Backer p. 506; Barnhart 3: 199; BM 4: 1780; CSP 5: 363, 12: 643; Herder p. 120; Kew 4: 563; PR 7967-7970, ed. 1: 8919-8921; Rehder 5: 751; Tucker 1: 617.

BIOFILE: Berg, E. de, Add. Thes. lit. bot. 31. 1859.
Iltis, H., Life of Mendel ed. 2. 119. 1966.
Jussieu, Adr. de, Mém. Agric. Écon. rur., Soc. Agric., Paris. 1852(2): 443-464. 1853 (obit.; b. 27 Jul 1763, d. 23 Mar 1851; bibl. in footnotes).
Roberts, A.F., Plant hybridizers before Mendel 120-123. 1965 (judgement Kölreuter).

EPONYMY: *Sageretia* A.T. Brogniart (1827).

9988. *Mémoire sur les Cucurbitacées*, principalement sur le melon, avec des considérations sur la production des hybrides, des variétés, etc.; ... à Paris (de l'Imprimerie de madame Huzard, ...) 1826. Oct.
Publ.: 1826, p. [1]-60. *Copy*: NY. – Reprinted and to be cited from Mém. Agric. Écon. rur., Soc. Agric., Paris 30: 435-492. 1825 (1826).
Deuxième mémoire: 1827, p. [1]-118. *Copy*: NY. – Id. 32: 1-116. 1827. "*Deuxième mémoire sur les Cucurbitacées, ...*" Paris (id.) 1827. Oct.

Sagorski, Ernst (Adolf) (1847-1929), German (Saarbrücken/Prussian) botanist; studied at Bonn 1865-1868; high school teacher at the Königl. Landesschule, Pforta near Naumburg, Thüringen 1870-1905; Dr. phil. h.c. Halle 1905; in retirement at Almrich. (*Sagorski*).

HERBARIUM and TYPES: JE (in herb. Haussknecht); further material in A, B (extant), BP, C, E, GB, GOET, HAL, K, MANCH, STU, W, WRSL. – Sagorski was also in charge of the "Thüringischer [later: "europäischer] botanischer Tauschverein", the numerous extra specimens of this society also came to JE after Sagorski's death.

BIBLIOGRAPHY and BIOGRAPHY: AG 4: 754, 6(1): 116, 12(1): 449; Backer p. 506; Barnhart 3: 199 (b. 26 Mai 1847, d. 8 Feb 1929); Biol.-Dokum. 15: 7803; BJI 1: 50; BM 4: 1780; CSP 18: 402; DTS 6(4): 191; Futak-Domin p. 517; Hegi 7: 560; Hortus 3: 1203; IF p. 728; IH 1 (ed. 6): 363, (ed. 7): 341; PFC 1: lii, 2(2): xvii; Rehder 5: 751; Tucker 1: 617; Urban-Berl. p. 390.

BIOFILE: Anon., Bot. Centralbl. 65: 175. 1896 (Dr. phil. h.c. Halle).
Bornmüller, J.F.N., Mitt. Thür. bot. Ver. ser. 2. 39: x-xiii. 1930 (herb. Sagorski to Herb. Haussknecht).
Kneucker, A., Allg. bot. Z. 1(1): 24. 1895 (Thür. bot. Tauschverein exchange list), 3: 18, 206. 1897 (id.), 8: 208. 1902 (id.).
Pax, F., Grundz. Pfl.-Verbr. Karpathen 32, 40, 44, 58. 1898 (Veg. Erde 2(1)); Bibl. schles. Bot. 7. 1929.
Poggendorff, J.C., Biogr.-lit. Handw.-Buch 4(2): 1299. 1904, 6: 2268. 1940.

Sagorski, E., Bot. Monatschr. 15: 329. 1897 (on his trip to Montenegro).
Szymkiewicz, D., Bibl. fl. Polsk. 123. 1925.
Wagenitz, G., Index coll. princ. herb. Gott. 141. 1982.
Wein, K., Mitt. Thür. bot. Ver. ser. 2. 39: xvii-xx. 1930 (obit., portr., bibl.)

COMPOSITE WORKS: Collaborator for W.D.J. Koch, *Syn. deut. schweiz. Fl.* ed. 3, by E. Hallier, 1890-1907, TL-2/3804. For Lieferung 1 of this work, Jan 1890, see extensive review by C. Haussknecht, Bot. Centralbl. 45: 185-190. 1891.

NOTE: Sagorski published *Plantae criticae Thuringae* i-v, in Bot. Monatschr. 6: 145-146. Oct 1888, 7: 6-7 Jan 18889, 38-42. Mar 1889, 77-79. Jul, 132-133. Sep-Oct 1889.

9989. *Die Rosen der Flora von Naumburg* a/S. nebst den in Thüringen bisher beobachteten Formen ... (Beilage zum Jahresbericht der Königl. Landesschule Pforta. 1885) Naumberg a/S. (Druck von H. Sieling) Qu. (*Ros. Fl. Naumberg*).
Publ.: Mar-Jun 1885 (Bot. Zeit. 26 Jun 1885; Nat. Nov. Jun(1, 2). 1885; Bot. Monatschr. Jun 1885), p. [i], [1]-48, *pl. 1-4* (uncol. liths). *Copies*: BR, H, HH, M, MO. – Issued as an addendum to the Jahresber. k. Landesschule, Pforta 1885, Programm Nr. 222. ii. Abhandlung.

9990. *Flora der Centralkarpathen* mit specieller Berücksichtigung der in der Hohen Tatra vorkommenden Phanerogamen und Gefäss-Cryptogamen nach eigenen und fremden Beobachtungen ... Leipzig (Verlag von Eduard Kummer) 1891, 2 vols. Oct. (*Fl. Centralkarpath.*).
Co-author: Gustav Schneider (1834-1900).
1 (Einleitung, Flora der Hohen Tatra nach Standorten): Nov 1890 (p. xii: Oct 1890; publ. Nov 1890, see vol. 2, p. 590; t.p.1891; J. Bot. Jun 1891; Bot. Centralbl. 26 Nov 1890; Bot. Zeit. 12. 26 Dec 1890; ÖbZ for Nov 1890; Nat. Nov. Nov(1) 1890), p. [ii]-xvi, [1]-209, [210, err.]. Second t.p. (p. [ii]): "*Flora carpatorum centralium* phanerogamarum et cryptogamarum vascularium praecipue in Tatrae Magnae montibus regionibusque adjacentibus sponte crescentium enumerationem et descriptionem continens ..." Leipzig (id.) 1891.
2 (Systematische Uebersicht und Beschreibung ...): Dec 1890-Jan 1891 (p. 591: early Dec 1890; Bot. Zeit. 12 Dec 1890, 30 Jan 1891; ÖbZ for Jan 1891), p. [ii]-vii, [viii], [1]-591, [592], i-lvi [index], *pl. 1-2* (uncol.). – Second t.p.: as in vol. 1.
Copies: BR, G, H, HH, M, MO, NY, PH, USDA.
Ref.: Pax, F., Bot. Jahrb. 13: 7-8. 20 Mar 1891 (rev.).
Taubert, P.H.W., Bot. Centralbl. 46: 273-274, 274-275. 20 Mai 1891 (rev.); Bot. Zeit. 50: 147-148. 4 Mar 1892.

9991. *Ueber den Formenkreis der Anthyllis vulneraria L. sensu amplissimo* ... Naumburg a. S. (Druck von H. Sieling) [1908-1909]. Oct.
Publ.: Mar l908-Feb 1909 (in journal), p. [i], [1]-50. *Copies*: B, G, M. – Reprinted and to be cited from Allg. bot. Z. 14: 40-43. Mar 1908, 55-58. Apr 1908, 89-93. Jun 1908, 124-134. Jul-Aug 1908, 154-157. Sep 1908, 172-175. Oct 1908, 184-189 Nov 1908, 204-205. Dec 1908, 15, 7-11. Jan 1909, 19-23. Feb 1909.

Sagot, Paul Antoine (1821-1888), French botanist and plant collector; Dr. med. Paris 1848; practicing physician at Coulanges-sur-Yonne 1848-1853; naval surgeon in French Guyana and briefly on the French Antilles 1854-1859; in Ténériffe 1864-1865; professor of natural sciences at the École normale spéciale de Cluny; 1865-1877; from 1877-1881 at Dijon; from 1881-1888 at Melun nr. Paris, associated with the Paris Muséum d'Histoire naturelle. (*Sagot*).

HERBARIUM and TYPES: P, PC, P-CO. – Other material in: B, BM, BR, DBN, E, F, FI, G, GH, GOET, K, L, LY, MANCH, MO, MPU, NY, S, U, W.– Sagot left his Guyana herbarium and a manuscript Flore de la Guyane to P/PC.

BIBLIOGRAPHY and BIOGRAPHY: AG 12(3): 466; Barnhart 3: 199 (b. 14 Jun 1821, d. 8 Sep 1838); BJI 1: 50; BL 1: 251, 316; BM 4: 1547, 1780; Bossert p. 345; CSP 5: 363, 8:

809-810, 11: 257, 12: 643, 18: 402; Herder p. 434; Hortus 3: 1203; IH 1 (ed. 6): 363, (ed. 7): 341, 2: (files); Jackson p. 274, 376; Lenley p. 468; Morren, ed. 2, p. 18, ed. 10, p. 65; MW p. 422; PR 7971-7972; Rehder 5: 751-752; SBC p. 131; Tucker 1: 617; Urban-Berl. p. 312, 390.

BIOFILE: Abonnenc, E. et al., Bibl. Guyane franc. 1: 231-232. 1957 (bibl.).
Anon., Bot. Not. 1890: 46 (d.); Bot. Zeit. 24: 404. 1866 (app. Cluny), 25: 344. 1867 (app. Cluny), 47: 322. 1889 (d.); Bull. Soc. bot. France 21: 99. 1874 (society receives a volume of Sagot reprints; for contents see Abonnenc et al. 1957), 35 (bibl.): 207. 1888, 37 (bibl): 216. 1890 (herbarium of 3875 species to P/PC), Table gén. art. orig. 1854-1893, p. 200, 213-214. 1900; Flora 50: 285. 1867 (app. Cluny); Österr. bot. Z. 17: 334. 1867 (app. Cluny).
Bureau, E. & J. Poisson, Bull. Soc. bot. France 36: 372-378. 1889 (obit., bibl.).
Candolle, Alph. de, Phytographie 445-446. 1880 (coll.).
Duchartre, P.E.S., Bull. Soc. bot. France 35: 371-372. 1888 (obit.).
Gaudin, A.J., Bot. 2: 428. 1889 (bibl.).
Magnin, A., Bull. Soc. bot. Lyon 32: 36, 131. 1907.
Sayre, G., Bryologist 80: 516. 1977.
Wagenitz, G., Index coll. princ. Gott. 141. 1982.

HANDWRITING: Wagenitz, G., Index coll. princ. Gott. 209. 1982.

EPONYMY: *Sagotanthus* Van Tieghem (1897); *Sagotia* Duchassaing et Walpers (1851, *nom. rej.*); *Sagotia* Baillon (1860, *nom. cons.*).

NOTE: Backer (p. 506) refers to a J.L.A. Sagot (b. 1805, d.?) who served as surgeon with the French navy from 1824-1854, stationed in the French Antilles 1827-1829 and 1836-1837; in French Guyana 1829-1830. This information was obtained from the French Naval Department. Backer states that nothing shows that Paul Sagot ever collected in Guiana or Guadeloupe. Backer has not been followed in this by later French bibliographers and is in contradiction with P. Sagot's own statement in his *Cat. pl. Guyane franç.* "Guyane française, pays que j'ai habité cinq ans, et ou j'ai herborisé avec ardeur, ...". The Paul Sagot biography by E. Bureau and J. Poisson (1889) is also clear about Paul Antoine Sagot being the Sagot who collected in Guiana and described its flora. The botanical eponyms all commemorate P.A. Sagot.

9992. *Études sur la végétation des plantes potagères d'Europe à la Guyane française*... (Extrait du Journal de la Société impériale et centrale d'Horticulture ... 1860 ...). Oct.
Publ.: Feb 1860 (in journal), p. [3]-24. *Copy*: G. – Reprinted and to be cited from J. Soc. imp. centr. Hort. 6: 113-134. 1860.
German translation: Flora 48: 105-110, 122-125. 1865.

9993. *Principes généraux de géographie agricole*... Paris (au bureau de la Revue du Monde colonial ...) 1862. Oct.
Publ.: 1862 (BSbF Feb 1863), p. [1]-47. *Copy*: MO. – Reprinted and to be cited from Revue du Monde colonial ser. 2. 7: 89-100, 5: 205-219, 378-392. 1862.

9994. *Quelques souvenirs d'herborisations* à propos de la relation qui lie la végétation à la nature du sol. [Mém. Soc. acad. Maine-et-Loire 1871]. Oct.
Publ.: 1871, p. [1]-15. *Copy*: G. – Reprinted and to be cited from Mém. Soc. Acad. Maine-et-Loire 26 (Sci): 21-35. 1871.

9995. *Catalogue des plantes* phanérogames et cryptogames vasculaires de la Guyane française par M. le Dr P. Sagot, ancien chirurgien de marine, etc. [series of papers in Ann. Sci. nat., Bot. 1880-1885]. Oct.
Publ.: in Ann. Sci. nat., Bot. as follows:
[1]: *in* ser. 6. 10: 361-382. Mar 1880.
[2]: *in* ser. 6. 11: 134-180. Jun 1881.
[3]: *in* ser. 6. 12: 177-211. Apr 1882.
[4]: *in* ser. 6. 13: 283-336. Jul 1882.

[5]: *in* ser. 6. 15: 303-336. Jun 1883.
[6]: *in* ser. 6. 20: 181-216. Jan-Apr 1885.
P. 379 of first article dated "Dijon, mars 1881".

9996. *Remarques sur les Mélastomacées de la Guyane française* [Extrait Bull. Soc. Bot. Belg. 22(2). 1883]. Oct.
Publ.: 6 Mai-11 Jun 1883 (read 6 Mai 1883; Bot. Centralbl. 11-15 Jun 1883), p. [1]-7. *Copies*: G, U. – Reprinted and to be cited from Bull. Soc. Bot. Belg. 22(2): 71-77. 1883. Introductory note on P. Sagot by Alfred Cogniaux.

9997. *Les différentes espèces dans le genre Musa* (Bananier), leur groupement naturel. Courtes indications sur les caractères distinctifs de chacune et sur l'intérêt alimentaire ou ornemental de plusieurs, ... Extrait du Journal de la Société nationale d'Horticulture de France ... 1887. Oct.
Publ.: Apr-Mai (or Mai-Jun) 1887, p. [1]-34. *Copy*: G. – Reprinted and to be cited from J. Soc. natl. Hort. France ser. 3. 9: 238-249. Apr 1887, 285-305. Mai 1887 (cahiers for Apr and Mai 1887 contain reports of séances" in those months and were possibly published shortly afterwards.

Sagra, Ramón de la (1798-1871), Spanish economist, agriculturist and botanist; director of a botanical garden at Havana, Cuba and professor of botany at the University 1822-1835; working on his collections and his *Hist. fis. Cuba* at Paris from 1836. (*Sagra*).

HERBARIUM and TYPES: Sagra's first collections, made mainly by himself, went to A.P. de Candolle and are at G. Duplicates are in FI and P (incl. P-JU). The main Cuban herbarium, however, collected by Sagra and his collaborators, was brought to Paris by him for the preparation of the *Historia fisica*. The types of the phanerogams are in the Richard herbarium, which is now at P. On A. Richard's death (1854) the Sagra Cuban collections had not yet been distributed. The next owner, however, the Comte de Franqueville, distributed duplicates. Drake del Castillo bought the Franqueville-Richard herbarium and presented it (for the greater part) to P. Most of the types of the phanerogams will therefore be found in the herbier général of P. The cryptogams went to Montagne and are at PC. – Other Sagra material (including the duplicates distributed by de Franqueville) is at B (mainly destr.), BR, F, FI, G, K, LE, LUB, NY, PH, W and Z.

BIBLIOGRAPHY and BIOGRAPHY: BL 1: 225, 316; BM 4: 1780-1781; Bossert p. 345; Colmeiro penins. p. 202-203; CSP 3: 857, 6: 709, 8: 167-168, 10: 518; GF p. 74; Herder p. 57, 226, 363; Jackson p. 369, 370, 449; KR p. 4: 563 (and sub La Sagra); Langman p. 663; Lasègue p. 265; Lenley p. 357; LS 23491; MW p. 422; NI 1712-1714; PH 567; PR 7973, ed. 1: 8340-8342; Quenstedt p. 373; Rehder 5: 751; RS p. 139; TL-1/884, 1074, 1075, 1132-1134; TL-2/6241, 6242; Tucker 1: 617; Urban-Berl. 312, 390.

BIOFILE: Aguayo, J., Mem. Soc. cub. Hist. nat. 18: 153-184. 1946 (on Hist. fis. Cuba).
Anon., Bot. Zeit. 14: 151. 1856 (Cuban herbarium included in herb. Ach. Richard), 29: 604. 1871 (d.); Bull. Soc. bot. France 18: 189. 1871 (d.); Flora 54: 303. 1871 (d. at Cortaillod, Neuchatel).
Candolle, A.P. de, Mém. souv. 146, 463, 464. 1862.
Candolle, Alph. de, Ann. Rep. Smiths. Inst. 1875. 162-163; Phytographie 446. 1880.
Conde, J.A., Hist. bot. Cuba 74-79, 278-280. 1908.
Jessen, K.F.W., Bot Gegenw. Vorz. 467. 1884.
Leon, J.S.S., Mem. Soc. cub. Hist.nat. 3: 190-191.1918.
Nissen, C., Zool. Buchill. 356. 1969.
Papavero, N., Ess. hist. neotrop. dipterol. 1: 179-180, 192. 1971.
Poggendorff, J.C., Biogr.-lit. Handw.-Buch 2: 735. 1863, 3: 1161. 1898.
Sagra, R. de la, Bull. Soc. bot. France 3: 229-230. 1856 (presents a complete copy of the *Flora cubana* at the session of 25 Apr 1856); Catálogo cronológico de las obras publicadas por Don Ramon de la Sagra en los años comprendidos desde 1823 à 1845, s.l.n.d. p. [1]-19. *Copy*: G (bibl.).
Sayre, G., Bryologist 80: 517. 1977.

Stahl, A., Estud. Fl. Puerto Rico ed. 2. 1: 40-41. 1936.
Trelles, C.M., Bibl. ci. cuban., Mantanzas 183-186. 1918 (bibl. of Hist. fis. Cuba).
Urban, I., Symb. ant. 1: 109, 114, 141-147. 1898, 3: 117-118. 1902.
Wilson, J.G. & J. Fiske, Appleton's Cycl. Amer. biogr. 5: 367-368. 1888.

COMPOSITE WORKS: Editor, *Anales de Ciencias, Agricultura, Comércio, y Artes*, Habana 1-3, 1827-1829 (rev. D.F.L. v. Schlechtendal, Linnaea 3 (Litt.): 158. 1878, 7 (Litt.): 53-57. 1832.

HANDWRITING: Candollea 33: 155-156. 1978.

EPONYMY: *Sagraea* A.P. de Candolle (1828).

9998. *Principios fundamentales* para servir de introduccion a la Escuela botanica-agricola del Jardin botanico de la Habana. Dispuestos para la catedra del establecimiento ... Habana (por don Tiburcio Campe, ...) 1824. Qu. (*Princ. fond.*).
Publ.: 1824 (p. [iii]: 20 Oct 1824), p. [i-viii], [1]-151. *Copy*: HH.

9999. *Historia economico-politica y estadistica de la Isla de Cuba* ó sea de sus progresos en la poblacion, la agricultura, el comercio y las rentas ... Habana (Imprenta de las viudas de Arazoza y Soler, ...) 1831. Qu. (*Hist. econ.-pol. Cuba*).
Publ.: 1831, p. [i*, iii*], [i]-xiii, [1-3], [1]-386, [1], chart. *Copies*: G, NY, PH, USDA.

10000. *Historia fisica politica y natural de la Isla de Cuba* ... Segunda parte. Historia natural. Tome ix. *Botanica*. Paris (en la Libreria de Arthus Bertrand ...) Madrid (Establecimiento tipographico de Don Francisco de P. Mellado, ..) 1845. Fol. (*Hist. fis. Cuba, Bot.*).
Tomo ix: 1845, p. [i-v], [1]-64, [1]-316, 321-328, *pl. 1-20* (coll. copp. Riocreux). *Copies*: MO, NY. – Pages [1]-64 contain a general introduction by Sagra; the remainder of the volume is occupied by J.P.F.C. Montagne, *Plantas celulares o criptogamia* (see also TL-2/6242 and Aguayo (1946)). Pages 317-320 are missing; 321-325 have an alphabetical index, 327-328 a list of the plates. These plates are usually bound with those for tomes x and xi in the atlas volume. The Sagra introduction is not included in the french edition.
Tomo x: 1845, p. [i-v], [1]-319. *Copies*: G, MO, NY. – For plates see Atlas. – Contains A. Richard, Fanerogamia o plantas vasculares. – "Tomo x. Botanica" ... Paris, Madrid (id.) 1845. Fol.
Tomo xi: 1850, p. [1]-339, [1-2]. *Copies*: G, MO, NY. For plates see Atlas. – Contains second part of A. Richard, Fanerogamia. – "Tomo xi. Botanica". Paris (Imprenta de Maulde y Renou) 1850. Fol. This part of the phanerogams is not included in the french edition. – Also as *Flora cubana* ó descripcion botánica usos y aplicaciones de las plantas reunidas en la isla de Cuba ... Tomo iii, Fanerogamia. Paris (id.) 1853. Fol., p. [1]-339, [1-2]. *Copy*: HH.
Tomo xii: 1855 (Sagra presented a complete copy of the "Flora cubana" to the Soc. bot. France on 25 Apr 1856), p. [i-iv], *pl. 1-20* [already listed above, belong to vol. 9] and 102 plates illustrating vols. 10 and 11 numbered *1-12, 12bis, 13-28, 28bis, 29-36, 36bis, 37-38, 38bis, 39-40, 40bis, 41-44, 44[1]-44[3], 45-47, 47bis, 48-49, 49bis, 50-54, 54bis, 54ter, 55-59, 59bis, 60-77, 78[bis, sic], 79-89*. – *Copy*: USDA; plates NY copy bound with vols. 10 and 11. – "Tomo xii. Atlas de Botanica". Imprint,as in vol. 11, dated 1855.
Ref.: Aguayo, J., Mem. Soc. cub. Hist. nat. 18: 153-184. 1946 (extensive analysis, q.v. for further details).
Urban, I., Symb. ant. 1: 143-147. 1898 (provides e.g. a list of plates with names of taxa with references to the text).
Sagra, R. de la, Sucinta noticia del origen, objeto y estado presente de la Historia politíca &, de Cuba, Madrid 16 Feb 857 (n.v.; fide C.M. Trelles).

10.001. *Histoire physique, politique et naturelle de l'Ile de Cuba* par M. Ramon de la Sagra, ... Botanique. – *Plantes cellulaires*, par Camille Montagne. Paris (Arthur Bertrand, ...) 1838-1842. Oct. (*Hist. phys. Cuba, Bot. Pl. cell.*).
Author: Jean Pierre François Camille Montagne (1784-1866).

Publ.: late 1842, at any rate after 1841 (see TL-2/6242), p. [i*-v*], [i]-x, [1]-549. *Copies*: G, HH, MO, NY, Stevenson, Teyler; IDC 5707. – Accompanied by *pl. 1-20 crypt.* in the atlas (see below). – For further details see TL-2/6242.
Preprint of text on fungi: see J.P.F.C. Montagne, *Esq. org. phys. champ.* 1841, TL-2/6241.

10.002. *Histoire physique, politique et naturelle de l'Ile de Cuba* par M. Ramon de la Sagra, ... *Botanique. – Plantes vasculaires.* Par Achille Richard. Paris (Arthus Bertrand, ... [1840-]1845[-1851]. Oct. (*Hist. phys. Cuba, Bot. Pl. vasc.*).
Author: Achille Richard (1794-1852).
Publ.: 1840-1851 (in an unknown number of parts; see below), p. [i]-viii, [1]-663. *Copies*: BR, G, MICH, MO, NY; IDC 5707. – For plates see atlas below.

pages	plates	dates
1-336	*1-35*	1841
337-624	*36-44(2)*	1846
625-663, i-viii	*44(3)-89*	1851

The above dates and contents are conjectural, see the detailed discussion under A. Richard (1845) (TL-2/9150). The second part of the *Plantes vasculaires* appeared only in the Spanish edition, see above. The first part is the french version of vol. 10 of the Spanish version.

10.003. *Histoire physique, politique et naturelle de l'Ile de Cuba* par M. Ramon de la Sagra, ... *Atlas* ... Paris (Arthus Bertrand ...) [1840-1851]. Fol. (*Hist. phys. Cuba, Atl.*).
Publ.: 1840-1851, see above under *Pl. phan.*, issued in an unknown number of fascicles (in principle each with four plates). – Contains the plates for the *Hist. phys. Cuba, Pl. cell.* (*1-20*, Crypt.) as well as for the *Pl. vasc.* (102 nos. *Pl. vasc.*). The atlas of the Teyler copy is constituted as follows: p. [i, t.p.], *Pl. vasc. 1-12, 12bis, 13-36, 36bis, 37-38, 38bis, 39-40, 40bis, 41-44, 44[1]-44[3], 45-47, 47bis, 48-49, 49bis, 50-54, 54bis [54ter], 56, 56[bis], 57-59, 59bis, 60-77, 78[bis; the proper no. 78 was not issued], 79-89*, [1 list *pl. crypt.*], *pl. 1-20*. The cryptogamic plates are colored copper engr. by A. Riocreux; the vasc. pl. plates are uncoloured copper engr. by J. Gontier and Vautier. – Zoological plates not listed.
Copies: G, NY, PH, Stevenson, Teyler. – See also under Spanish edition for notes on the *Atlas* (vol. 12) and for secondary references.

10.004. *Énumeration des espèces zoologiques et botaniques de l'île de Cuba* utiles a acclimater dans d'autres régions analogues du globe ... Paris (Imprimerie de L. Martinet, ...) 1859. Oct.
Publ.: Mai, Jun-Dec 1859 (in journal), cover-t.p., p. [1]-31. *Copy*: G. – Reprinted and to be cited from Bull. Soc. imp. zool. Acclim. 6: 169-184. Mai 1859, 237-251. Jun-Dec 1859.

10.005. *Icones plantarum in Flora cubana* descriptarum ex Historia physica, politica et naturali a Ramon de la Sagra ... edita excerptae. Introductio in Flora cubana Ramon de la Sagra. Cryptogamia (20 tabulae) a claro Camillo Montagne elaboratae. Phanerogamia (102 tabulae) a claro A. Richard elaboratae. Parisiis. (J.B. Baillière et filiis, ...). Londini (Hipp. Baillière). New York (Baillière fratres). Madrid (C. Bailly-Baillière) 1863. Fol. (*Icon. pl. Fl. cub.*).
Publ.: 1863, p. [i-vii], Sagra, Introd.: [1]-64, plates Montagne *Pl. cell.: 1-20* (col. copp. Riocreux), plates Richard *Pl. vasc.: 1-12, 12bis, 13-28, 28bis, 29-36, 36bis, 37-38, 38bis, 39, 40, 40bis, 41-44, 44[1]-44[3], 45-47, 47bis, 48-49, 49bis, 50-54, 54bis, 54ter, 55-59, 59bis, 60-89* (uncol. copp. Vauthier, and J. Gonthier. *Copies*: G, NY, PH. – Publication of the Sagra introduction (64 p.) of the Spanish version vol. 9(1), followed by the plates from the atlas (vol. 12).
Ref.: L.C.T., Bot. Zeit. 22: 6-8. 1 Jan 1864.

Sahlberg, Carl Reinhold (1779-1860), Finnish botanist; med. lic. Turku (Åbo) 1810; adjunct for medicine and demonstrator for botany at Turku 1810-1813; adjunct for natural history and museum inspector ib. 1813-1818; professor of natural history 1818-

1828; moved with the university after the great Turku fire of 1827 to Helsinki in the same function until 1840, also in charge of the establishment (1833) and development of a university botanical garden (*C. Sahlberg*).

HERBARIUM and TYPES: H; other material in P, S and UPS-Thunberg (plants from Turku bot. gard.). – Correspondence at H-UB.

BIBLIOGRAPHY and BIOGRAPHY: Barnhart 3: 199 (d. 18 Oct 1860); BM 4: 1781, 8: 1121; Collander hist. p. 14-17, 20, 59, 102; IH 1 (ed. 6): 363, ed. 7: 341; LS 23493; PR 7974; Saelan 445; TR 1227-1228.

BIOFILE: Anon., Flora 54: 303. 1871 (d.).
Elfving, F., Acta Soc. Fauna Fl. fenn. 50: 225-226. 1921 (portr.).
Gager, C.S., Brookl. Bot. Gard. Rec. 27(3): 193. 1938.
Kukkonen, I., Herb. Christ. Steven 87. 1971.
Kukkonen, I. & K. Viljamaa, Ann. bot. fenn. 10: 312, 333. 1973.
Löwegren, Y., Naturaliekabinett i Sverige under 1700-talet 363. 1952.
Saalas, U., Luonnon ystävä 38: 174-180. 1934; Carl Reinhold Sahlberg, Luonnontutkija, yliopisto-ja maatalousmies, Helsinki 1956 (1957), 480 p. , published as Historiallisia tutkimuksia 47, 1957 (biogr., bibl., portr., German summary on p. 452-479), 20 p. index to personal names published separately; Acta entom. fenn. 14: 1-255. 1958 (biogr. of R.F. Sahlberg, son of C.R.S.).
Törnroth, L.H., Acta Soc. Sci. fenn. 6: i, 1-7. 1861 (obit.).

HANDWRITING: Ann. bot. fenn. 10: 333. 1973.

NOTE: Information on Sahlberg was kindly supplied by P. Isoviita.

10.006. Dissertatio academica, *de progressu cognitionis plantarum cryptogamicarum*, cujus particulam primam, cons. ampl. Fac. fil. Aboënsi, publico examini submittunt Carolus Regin. Sahlberg, ... & Gustavus Guilielm. Rönnbäck, ... in Audit. Maj. die 6 Junii 1804. h.a.m.s. Aboae (typis Frenckellianis) [1804]. Qu. (*Progr. cogn. pl. crypt.*).
Publ.: 6 Jun 1804, p. [i], [1]-16. *Copies*: H, LD.

Sahlén, Anders Johan (1822-1891), Swedish botanist; Dr. phil. Lund 1850; teacher at Vänersborg 1850-1854; id. at Lidköping 1855; adjunct (1856), later lecturer (1861) at the State high school of Skara. (*Sahlén*).

HERBARIUM and TYPES: Some material in Skara High School.

BIBLIOGRAPHY and BIOGRAPHY: Barnhart 3: 200 (b. 22 Mar 1822, d. 23 Jul 1891); BL 2: 553, 712; BM 4: 1782; CSP 8: 810, 18: 403; KR p. 614 (b. 2 Mar 1822, d. 23 Jul 1891); Morren ed. 10, p. 106; PR 7975; Rehder 1: 355.

BIOFILE: Anon. [=Nordstedt, O.], Bot. Not. 1851: 181 (rev. Wenersborgs fl.), 1891: 172 (2 Mar 1822, d. 24 Jul 1891).
Kilander, S., Skaraborgsnatur 18: 63-72. 1981 (portr., S. as floristic author); Västgötalitteratur 1972: 11-25. (portr.; on S.' ms. on the flora of Västergötland; records from school boys not always reliable).
Wittrock, V.B., Acta Horti Berg. 3(2): 69. 1903 (b. 22 Mar 1822), 3(3): 51. *pl. 103*. 1905.
Zakariasson, C., Förord (preface) to reprint of Wenersborgs flora: [iii*-v**]. 1980.

10.007. *Wenersborgs flora* eller kort beskrifning på de växter, som förekomma närmast omkring Wenersborg samt på Halle– och Hunneberg, till den studerande ungdomens tjenst ... Mariestad (tryckt hos A.A. Berg) 1854. Duod. (*Wenersborgs fl.*).
Publ.: 1854 (p. [i]: 26 Mai 1854; Bot. Not. Nov-Dec 1854), p. [i*], [i]-viii, [1]-190, err. [191]-192. *Copies*: LD, NY, B. Peterson.
Additions: Kindberg, N.C., Tillägg till Sahléns Wenersborgsflora, Bot. Not. 1863 (1-2): 12-14. 1863; Sahlén, A.J., Nya växtlokaler i Vestergötland, Bot. Not. 1863: 60-62.

Facsimile ed.: 1980, ed. C. Zakariasson, 1980. – Has reproductions of two maps of the area (from 1781, 1808), not present in the original issue.

Sahni, Birbal (1891-1949), Indian palaeobotanist; MA Cantab. 1918; D.Sc. Univ. London 1929; at Emmanuel College Cambridge 1911-1914; with A.C. Seward at the Botany School ib. 1914-1919; professor of botany at Lucknow University 1921-1949; D.Sc. Cambr. 1929; FRS 1936. (*Sahni*).

HERBARIUM and TYPES: Palaeobotanical collections at the Birbal Sahni Institute, Lucknow. – Some recent Indian plants at K.

BIBLIOGRAPHY and BIOGRAPHY: Andrews p. 323, 324; Barnhart 3: 200 (b. 14 Nov 1891); BJI 2: 151; BM 8: 1122; Bossert p. 345.

BIOFILE: Andrews, H.N., The fossil hunters 167-177. 1980.
Anon., Science 110: 56. 1949.
Bhatnagar, A.K., Botanica, Delhi 22(4): 132-135. 1972 (portr.).
Boureau, É., Bull. Soc. bot. France 100: 207-213. 1953 (obit.; bibl.).
Evans, P. & J. Coates, Palaeobotanist 1: 44-45 (palaeobot. Assam Tert.).
Gupta, S.M., Birbal Sahni, 1978, 87 p.
Halle, T.G., Palaeobotanist 1: 22-41 1952 (palaeobot. work).
Hoeg, D.A., Nature 172: 104-105. 1953 (on B. Sahni Institute, Lucknow).
Hsü, B. et al., Palaeobotanist 1: 56-60. 1952 (bibl.).
Maheswari, B., Palaeobotanist 1: 17-21. 1952 (contr. living plants).
Narayana Rao, S.R., Palaeobotanist 1: 46-48. 1952 (Indian geology).
Rao, A.R., Palaeobotanist 1: 9-16. 1952 (28 yrs at Univ. Lucknow).
Sahni, M.R., Palaeobotanist 1: 1-8. 1952 (biogr.); J. palaeont. Soc. India 3: 1-14. 1958 (biogr., bibl., portr.).
Sitholey, R.V., Palynol. Bull. 1: 5-6. 1965 (Sahni's contr. palynology).
Stewart, R.R., Pakistan J. For. 17: 355. 1967 (coll. K).
Thomas, H.H., Nature 164: 645. 1949; Obit. not. R. Soc. London 19: 265-277. 1950.
Walkom, A.B., Palaeobotanist 1: 42-43 (Australian palaeobot.).

COMPOSITE WORKS: With A.C. Seward, q.v., *Indian Gondwana plants*: a revision, Mem. Geol. Surv. India ser. 2. 7(1): [i-iii], [1]-54, *pl. 1-7*. 1920.

EPONYMY: *Sahnia* Vishnu-Mittre (1953); *Sahnianthus* V.B. Shukla (1944); *Sahnioxylon* M.N. Bose et S.C.D. Sah (1954); *Sahnipushpam* V.B. Shukla (1948 & 1950). *Note*: *Sahniocarpon* S.D. Chitaley et G.V. Patil (1973) and *Sahnisporites* D.C. Bhardwaj (1955) probably also honor Sahni, as does the Birbal Sahni Institute of Paleobotany, Lucknow, India.

MEMORIAL VOLUMES: The Palaeobotanist vol. 1, 1952, Lucknow (portr., biogr., bibl., tributes); Journal of the Palaeontological Society of India, vol. 3, 1958 (biogr.).

10.008. *Revisions of Indian fossil plants*: part i. – *Coniferales* (a. impressions and incrustations) ... Calcutta (Government of India Central Publication Branch) 1928. Qu. (*Revis. Ind. foss. pl., Conif.*).
1: 1928 (expl. pl. dated 28 Apr 1928), p. [i-v], [1]-49, [1, expl. pl.], *pl. 1-6* with text. – Mem. Geol. Surv. India ser. 2. 11(1).
2: 1931 (p. 124: 25 Feb 1931), p. [i-v], [51]-124, *pl. 7-15* with text. – Part ii. *Coniferales* (b. Petrifactions), Mem. Geol. Surv. India ser. 2. 11(2).
Copy: USGS.

10.009. *Homoxylon rajmahalense*, gen. et sp. nov., a fossil angiospermous wood, devoid of vessels, from the Rajmahal Hills, Behar ... Calcutta (Government of India Central Publication branch) 1932. Qu. (*Homoxylon*).
Publ.: 1932 (p. 19: 24 Mai 1932), p. [i-v], [1]-19, *pl. 1-2* with text. *Copy*: USGS. – Mem. Geol. Surv. India ser. 2. 20(2). 1932.

10.010. *A petrified Williamsonia* (W. Sewardiana, sp. nov.) from the Rajmahal Hills, India ... Calcutta (Government of India Central Publication Branch) 1932. 1932. Qu. (*Williamsonia*).
Publ.: 1932 (caption pl. 3: 15 Aug 1932), p. [i-v], [1]-19, *pl. 1-3* with text. *Copy*: USGS. - Mem. Geol. Surv. India 20(3), 1932.

Sahut, Félix (1835-1904), French horticulturist, viticulturist and botanist at the Établissement d'Horticulture Claude Sahut, Montpellier. (*Sahut*).

HERBARIUM and TYPES: Unknown.

BIBLIOGRAPHY and BIOGRAPHY: Barnhart 3:200 (b. 29 Mai 1835, d. 5 Mai 1904); BM 4: 1782; CSP 18: 403-404; Kew 4: 564; LS 23494-23495; MW p. 422; Rehder 5: 752; Saccardo 1: 144; TL-2/6902; Tucker 1: 617.

BIOFILE: André, Ed., Rev. horticole 76: 254-255. 1904 (obit., portr.).
Anon., Bull. Acad. int. Géogr. bot. 13: 217. 1904 (d.); Félix Sahut 1835-1904, notes bibliographiques et biographiques, Montpellier, s.d., 7 p. (titles bibl.).
Aymard, J., Ann. Soc. Hort. Hist. nat. Hérault 44: 84-87. 1904 (obit.).
Mandon, L., Ann. Soc. Hort. Hist. nat. Hérault 50: 206-207. 1910 (portr., pres. soc. 1889).
Sahut, F., Mélanges agricoles, horticoles, viticoles, botaniques, climatologiues, etc. (1888-1897). Montpellier 1893 (assemblage of 60 papers, with bibl.).

COMPOSITE WORKS: Collaborator, R. de Noter, *Orangers* 1896, TL-2/6902.

10.011. *Le Lac Majeur* et les Iles Borromée leur climat caractérisé par leur végétation ... Montpellier (Imprimerie centrale du Midi – Hamelin frères ...) 1883. (*Lac Majeur*).
Publ.: 1883 (Bot. Zeit. 25 Apr 1884), p. [1]-67, [68]. *Copies*: FI, G.

10.012. *Les vignes américaines* leur greffage et leur taille. Étude raisonnée de la possibilité de reconstituer les vignobles et des moyens de défense pour les conserver ... Montpellier (Camille Coulet, ...), Paris (A. Delahaye et E. Lecrosnier, ...) 1887. Oct. (*Vign. amér.*).
Publ.: 1887 (p. viii: 15 Jan 1887), p. [i]-viii, [9]-782. *Copy*: HH.
German transl.: 1891, p. [i]-viii, [1]-411. *Copy*: HH. – Die amerikanischen Reben, ihr Schnitt und ihre Veredlung. Studie über die Möglichkeit der Wiederherstellung der durch die Reblaus zerstörten Weingärten und die zu ihrer Erhaltung dienenden Vertheidigungsmittel ... Mit Genehmigung des Verfassers ins Deutsche übertragen und bearbeitet vom Nikolaus Freiherrn von Thümen. Hannover (Verlag von Philipp Cohen) 1891. Oct.

10.013. *Les Eucalyptus* aire géographique de leur indigénat et de leur culture historique de leur découverte description de leurs propriétés forestières, industrielles, assainissantes, médicinales, etc. Guide théorique et pratique de leur culture. Avec figures intercalées dans le texte et une carte de la Tasmanie ... Montpellier (Camille Coulet, ...) Paris (A. Delahaye & E. Lecrosnier, ...) 1888. Oct. (*Eucalyptus*).
Publ.: 1888 (p. vii: 5 Jan 1888; Bot. Zeit. 27 Apr 1888; Nat. Nov. Feb(2)), p. [i]-vii, [3]-212, 1 map. *Copy*: USDA. – Reprinted from Bull. Soc. Languedoc Géogr., Montpellier 8: 340-374, 552-571. 1885, 9: 106-135, 291-308, 428-449. 1886, 10: 66-84, 176-187, 305-321, 434-455. 1887.

Sailer, Franz Seraphin (1792-1847), Austrian botanist and clergyman; ordained 1816; chaplain at Gallneukirchen 1816-1821; curate at Altenberg and ultimately parson at Pöstingberg until 1835; "Konsistorialrath" at Linz 1835-1847. (*Sailer*).

HERBARIUM and TYPES: Unknown.

BIBLIOGRAPHY and BIOGRAPHY: BM 4: 1782; Herder p. 193; PR 7976-7977, ed. 1: 8922-8923; Rehder 1: 440.

BIOFILE: Poetsch, J.S. & K.B. Schiedermayr, Syst. Aufz. samenlosen Pfl. vii-viii. 1872. Janchen, E., Cat. fl. austr. 1: 20. 1956.

10.014. *Die Flora Oberöstreichs.* Beschrieben von Franz Seraph. Sailer, ... Linz (In Commission bey Quirin Haslinger, ...) 1841. 2 vols. Oct. (*Fl. Oberöstr.*).
1: 1841, p. [i]-lii, [1]-348.
2: 1841, p. [i]-xliv, [1]-361.
Copy: G.

10.015. *Flora der Linzergegend* und des obern und untern Mühlviertels in Oberösterreich, oder Aufzählung der allda wildwachsenden Pflanzen mit kenntlichen Blüthen, mittelst Angabe ihrer deutschen, lateinischen und vulgaren Namen ... (*Fl. Linzergegend*).
Publ.: 1844 (Int. Bl. Bot. Zeit. 4 Apr 1845, adv.), p. [i]-vi, [1]-54. Extract from his *Fl. Oberöstr.*– n.v.
Ref.: Schlechtendal, D.F.L. von, Bot. Zeit. 3: 305-306. 2 Mai 1845 (rev.).

Sainsbury, George Osborne King (1880-1957), New Zealand barrister, solicitor and bryologist; practiced at Gisborne, New Zealand 1903-1911, farming 1911-1917, practicing sollicitor 1917-1946. (*Sainsbury*).

HERBARIUM and TYPES: WELT (18.350); further bryophytes (incl. exsicc.) at AK, BR, CHR, COL, FH, H (Bryoph.), LD, MEL, MICH, NY, PH, WELC. *Exsiccatae: Musci exsiccati Novae-Zelandiae* fasc. 1, 1931 (nos. 1-50), see Sayre (1971).

BIBLIOGRAPHY and BIOGRAPHY: Barnhart 3: 200 (b. 1 Jun 1880, d. 22 Jul 1957); Bossert p. 345; HR; IH 1 (ed. 7): 341; SBC p. 131.

BIOFILE: Anon., Rev. bryol. ser. 2. 3: 147, 209. 1931, 4: 203, 218. 1932, Rev. bryol. lichenol. 5: 233. 1933, 7: 123. 1934, 9: 161. 1936, 11: 130. 1938, 18: 90-93. 1949, 19: 247. 1950, 21: 294, 303. 1952, 24: 389. 1955 (bibl. notes), 26: 269. 1957 (d.).
Carr, D.J. & S.G.M., eds., People and plants in Australia 177. 1781.
Hodgson, E.A., Rev. bryol. lichénol. 27(1-2): 104-106. 1958 (obit., bibl.).
Martin, W., Bryologist 60: 363-367. 1957 (obit., bibl., portr.).
Sayre, G., Mem. New York Bot. 19(2): 249. 1971 (exc.), 19(3): 390-391. 1975 (coll.); Bryologist 80: 517. 1977.
Willis, J.H., Victorian Naturalist 74: 118-120. 1957 (obit., portr.).

EPONYMY: *Sainsburia* Dixon (1941).

Saint-Amans, Jean Florimond Boudon de (1748-1831), French botanist and soldier; from 1768-1773 with a French expeditionary force in the French Antilles; from 1773 private scientist at Agen, sometime professor of natural history at the École centrale of Lot-et-Garonne; from 1800-1831 president of the conseil-général of the département Lot-et-Garonne, also devoting himself to archeology and numismatics. (*Saint-Amans*).

HERBARIUM and TYPES: Unknown.

BIBLIOGRAPHY and BIOGRAPHY: AG 6(1): 809; Barnhart 3: 200 (b. 24 Jun 1748, d. 28 Oct 1831); BM 4: 1782; Bossert p. 345; Colmeiro 1: cxcvii; CSP 5: 364; Dawson p. 728-729; Dryander 3: 142; GF p. 298; GR, cat. p. 71; Hawksworth p. 185, 197; Herder p. 175; Hortus 3: 1204; Kelly p. 205; Kew 4: 564; PR 1164, 7978-7980, ed. 1: 2167, 8925-8929; Rehder 1: 406, 407, 3: 212; SO 2621, 2657; TL-1/1135; TL-2/795; see L.A. Chaubard; Tucker 1: 618; Zander ed. 10, p. 720, ed. 11, p. 822.

BIOFILE: Anon., Act. Soc. linn. Bordeaux 6(33): (24). 20 Oct 1833 (d.); Ann. Soc. Linn. Lyon 1847-1849: xiv. 1850 (d. 23 Oct 1845); La France littéraire 8: 316-319. [date not noted] (bibl.).
Clos, D., Bull. Soc. bot. France 40: 243-250. 1893 (on Chaubard and the *Fl. agen.*).
Debeaux, D., Rev. de Bot. 13: 11-13. 1895 (in his Révision de la Flore Agenaise); Révis. fl. agen. 11-13. 1898. Lapeyrouse, P., Hist. abr. pl. Pyren. xxx. 1813.

Milner, J.D., Cat. portr. Kew 94. 1906.
Percheron, A., Bibl. entom. 2: 25. 1837.

EPONYMY: *Amansia* J.V.F. Lamouroux(1809) and *Amansites* (A.T. Brongniart) A.T. Brongniart (1849) may be named for this author.

NOTE: P.A. Saccardo (Bull. Soc. bot. ital. 1906: 25-27. 1906) points out that the term "phanerogams" stems from Saint-Amans, J. Sci. util., Bertholon, 1791 (17/18): 283, 285, 291.

10.016. *Fragmens d'un voyage* sentimental & pittoresque *dans les Pyrénées*, ou lettre écrite de ces montagnes ... à Metz (chez Devilly, ...) 1789. Oct. (*Fragm. voy. Pyrén.*).
Publ.: 1789, p. [i*, iii*, v*], [i]-iv, [1]-259. *Copy*: G. – Botany ("*Le bouquet des Pyrénées*, ou catalogue des plantes observées dans ces montagnes, pendant les mois de juillet et août de l'année 1788") on p. 189-259.

10.017. *Flore agenaise* ou description méthodique des plantes observées dans le Département de Lot-et-Garonne et dans quelques parties des départemens voisins ... Agen (Prosper Noubel, ...) 1821. Oct. (*Fl. agen.*).
Publ.: 20-28 Apr 1821 (BF 28 Apr 1821; Donk 1957; Petersen 1975), p. [1]-61, [i], [1]-632. *Copies*: BR, FH, G, HH, MICH, MO, NY, USDA; IDC 5082; for *plates* see *Bouquet* (1821) below.
Cryptogams: see also L. de Brondeau, *Rec. pl. crypt. agen.* 1828-1830; TL-2/795; *Révision*": see J.O. Debeaux, *Rév. pl. phan. agen.* 1895, TL-2/1331.
Notes: (*1*) See Saccardo, P.A., in Bull. Soc. bot. ital. 1906: 225-227 on Saint-Amans' statement on the origin of the word Phanerogamae; Saccardo states that the term stems from Saint-Amans.
(2) For L.A. Chaubard's part in writing the *Fl. agen.* see Clos (1893).
Ref.: Clos, D., Bull. Soc. bot. France 40: 243-250. 1893.
Donk, M.A., Taxon 6: 253. 1957.
Petersen, R.H., Mycotaxon 1: 156. 1975.

10.018. *Le bouquet du département de Lot-et-Garonne*, ou fascicule de quelques plantes de ce département, nouvelles, rares, point ou mal figurées dans les ouvrages de botanique, et décrites dans la Flore agenaise ... Agen (Prosper Noubel, ...) 1821. (*Bouquet Lot-et-Garonne*).
Publ.: 1821, p. [i, iii], *pl. 1-12* (uncol. liths.). *Copies*: BR, FH, G, HH, MICH, MO, NY, USDA. – Plates accompanying the *Fl. agen.* (1821); usually bound with the publication.

St. Brody, Gustavus A. Ornano (1828-1901), French-born botanist and teacher; Dr. phil. Göttingen, settled in England as a young man; as French teacher at the Crypt Grammar School, Gloucester. (*St. Brody*).

HERBARIUM and TYPES: GLR (1100).

BIBLIOGRAPHY and BIOGRAPHY: Barnhart 3: 200; BB p. 266; BL 2: 264, 712; BM 4: 1782; CSP 8: 810; Desmond p. 537 (d. 22 Nov 1901); Jackson p. 262; Kew 4: 564; PR 7981; Rehder 1: 400.

BIOFILE: Anon., J. Bot 40: 127-128. 1902 (obit.).
Kent, D.H., Brit. herb. 76. 1957.
Riddelsdell, H.J. et al., Fl. Gloucestershire cxxix-cxxx. 1948.
White, J.W., Fl. Bristol 93-95. 1872.

10.019. *The flora of Weston*, and its immediate neighbourhood, including the habitats of Brean-Down, Uphill, Hutton Wood, Worle or Weston Hill, Sand-Point, and Birnbeck Island: with critical remarks on our doubtful species, and botanical memoranda of Steep Holm, Chapwick Moor, and Cheddar Cliffs. Compiled for the use of the young ... Weston-Super-Mare: may be had of all booksellers 1856. Oct. (in 4-s). (*Fl. Weston*).

Publ.: 1856 (p. viii: Sep 1856), p. [i]-viii, 1-174. *Copies*: E, MO, USDA. – See also his New Gloucestershire plants, J. Bot. 4: 121-123. 1866, 7: 147-148. 1869.

Saint-Cyr, Dominique Napoléon (1826-1899), Canadian (Quebec) botanist and entomologist; teacher at Lennoxville 1846-1850, at Ste-Anne de la Pérade 1850-1876; provincial administrator 1875-1881; from 1881 dedicating himself to the setting up of a public museum; curator of the Musée de l'Instruction publique (Provincial Government Museum) 1886-1899. (*Saint-Cyr*).

HERBARIUM and TYPES: QMP (in QUE).

BIBLIOGRAPHY and BIOGRAPHY: Barnhart 3: 200 (b. 4 Aug 1826, d. 5 Mar 1899); BL 1: 146, 316; Tucker 1: 618.

BIOFILE: Caron, O., Natural. canad. 57: 183, 253. 1930 (on herbaria).
Huard, V.A., ed., Natural. canad. 26(3): 45-47, 59-63. 1899 (obit.).
Kucyniak, J., Bryologist 49: 129. 1946 (bryological work).
Penhallow, D.P., Trans. R. Soc. Can. ser. 3(4): 22, 53. 1897.
Provancher, L., Natural. canad. 5(7): 225-229. 1873.
Steere, W.C., Bot. Rev. 43: 293. 1977.
Treffry, J.E., Canad. entom. 31: 102. 1899 (obit. not.).

Saint-Gal, Marie Joseph (1841-1932); French botanist and assistant professor of botany and forestry at the Agricultural College at Grand-Jouan, Loire-Inférieure 1864-1869; id. at Grignon 1869-1870; on active service with the Armée de la Loire 1870; regular professor again at Grand-Jouan 1870-1895; transferred with the college to Rennes 1895-1901; in retirement at Fougères (Rennes). (*Saint-Gal*).

HERBARIUM and TYPES: Unknown.

BIBLIOGRAPHY and BIOGRAPHY: BL 2: 166, 173, 712; BM 4: 1783; CSP 12: 644; Jackson p. 81, 284; Kelly p. 205; Morren ed. 10, p. 63; Tucker 1: 618.

BIOFILE: Corbineau, R., Bull. Soc. Sci. nat. Ouest France 73: 20-21. 1973 (biogr. sketch).
Daniel, L., Rev. bretonne bot. pure appl. 1931: 25-32. 1932 (obit., bibl.).

10.020. *Flore des environs de Grand-Jouan* contenant la description des végétaux vasculaires qui poussent spontanément dans un rayon de 12 à 16 kilomètres autour de l'école d'agriculture de Grand-Jouan et celle des végétaux le plus ordinairement cultivés par l'agriculteur, le forestier et le maraîcher par St-Gal, M.-J., ... Nantes (MM. Douillard frères, ...) 1874. 18-mo. (*Fl. Grand-Jouan*).
Publ.: 1874, p. [i]-xlvi, [1]-521, [1, cont.]. *Copies*: G, HH, MO, NY, USDA.
Supplément: 1885 (Bot. Zeit. 27 Feb 1885; Bot. Centralbl. 30 Mar-3 Apr 1885; Nat. Nov. Mar(1) 1885), p. [i], [1]-29. *Copy*: G. – *Supplément à la Flore des environs de Grand-Jouan* par Marie-Joseph St-Gal, ... Nantes (Mme Ve Camille Mellinet, ...) 1885 (*Suppl. Fl. Grand-Jouan*).
Ref.: Malinvaud, E., Bull. Soc. bot. France 22 (bibl.): 188-189. 1875 (rev.), 32 (bibl.): 85. post 15 Mai 1885.

10.021. *Liste des plantes qui croissent spontanément dans le département de la Loire-Inférieure* et qui ne sont pas décrites dans la Flore des environs de Grand-Jouan ni dans le supplément publié en 1885 par Marie Joseph St.-Gal ... Nantes (Mme Ve Camille Mellinet, ...) 1885. Duod. (*Liste pl. Loire-Infér.*). *Publ.*: Jan-Mai 1885 (Bot. Centralbl. 8-12 Jun 1885; Bot. Zeit. 26 Jun 1885; Nat. Nov. Jan (1,2) 1885), p. [1]-48. *Copies*: B, MO.

10.022. *Catalogue raisonné des végétaux* spontanés ou cultivés *en Ille-et-Vilaine* récoltés de 1895 à 1900 ... Rennes (Fr. Simon, ...) 1900. Oct. (*Cat. Vég. Ille-et-Vilaine*).
Junior author: E. Demarquet.
Publ.: 1900 (p. xi: 15 Jan 1900; Bot. Centralbl. 11 Apr 1900; Nat. Nov. Jun(2) Jul(1) 1900), p. [i]-xvi, [1, err.], [1]-175. *Copy*: G.

Saint-Germain, J.J. de (*fl.* 1784), French botanist. (*Saint-Germain*).

HERBARIUM and TYPES: Unknown.

BIBLIOGRAPHY and BIOGRAPHY: Dryander 3: 42, 211; Kew 4: 565; Langman p. 664; MW p. 423; PR 7982, ed. 1: 8932-8933; Rehder 1: 258; SO add. 695*.

EPONYMY: *Germainia* Post et O. Kuntze (1903, *orth. var.* of *Germanea* Lamarck); *Germanea* Lamarck (1788). *Note*: *Germainia* Balansa et Poitrasson (1873) was named for Rodolphe Germain, plant collector.

10.023. *Manuel des végétaux*; ou catalogue Latin et François, de toutes les plantes, arbres & arbrisseaux connus sur le globe de la terre jusqu'à ce jour, rangés selon le système de Linné, par classes, ordres, genres & espèces, avec les endroits où ils croissent; les plantes des environs de Paris y sont spécialement indiquées, avec une table françoise. Ouvrage très-utile aux Botanistes, pour leur faciliter l'arrangement d'un herbier, d'un grainier & d'un jardin. Par M.J.J. de St. Germain. À Paris (Chez P.M. Delaguette, ...) 1784. Oct. (in fours) (*Man. vég.*).
Publ.: 1784 (privilège 11 Jun 1784), p. [i]-xi, [1]-378, [1-4]. *Copies*: G, HH, NY, PH, USDA. – The HH copy has preface material p. [i]-xxxix, [xl].
Suite du Manuel des végétaux, ou les présens de Pomone. Paris 1786, Duod. 191 p. n.v.

Saint-Hilaire, Auguste François César Prouvençal de (1779-1853), French explorer, botanist and entomologist; self-taught naturalist of independent means; in Hamburg during the revolution; in Orléans in the early years of the century; later at Paris associated with A.L. de Jussieu; in Brazil and Uruguay 1816-1822; back in France dedicating himself to the publication of the results of his Brazilian journeys; member of the Institut de France (Académie des Sciences from 1830). (*St. Hil.*).

HERBARIUM and TYPES: P, PC. – Other material at B, F, FI, G, H-NYL (Lich.), K, MO, MPU, NA, NY, ORM, US. – Saint Hilaire left his Brazilian herbarium, including the types of the new taxa, published in the books listed here, to P. It is now incorporated in the general herbarium. The plants from Auvergne, collected by Dutour de Salvert, were left to the Muséum d'Histoire naturelle in Clermont-Ferrand (now probably at CLF); the French and Swiss herbarium went to the city of Orléans (now in ORM). The original drawings by Turpin for the *Flora Brasiliae meridionalis* were left to the Orléans art museum, the library went to Montpellier. Saint-Hilaire's correspondence with Moquin-Tandon is at P (114 letters); letters to D.F.L. von Schlechtendal are at HAL, fide Bonnet 1913 (not located by Dwyer 1955); ten lists of plants collected in Brazil are at P (Dwyer 1955).

NOTE: Dutour de Salvert was St. Hilaire's brother in law; the latter dedicated the genus *Salvertia* (Vochysiaceae) to him.

BIBLIOGRAPHY and BIOGRAPHY: Backer p. 662; Barnhart 3: 200 (b. 4 Oct 1779, d. 30 Sep 1853); BM 1: 497, 4: 1783; Bossert p. 345; CSP 5: 366-369, 6: 752; DSB 12: 72; GF p. 74; GR p. 298; GR, cat. p. 71; Herder p. 58, 95, 175, 226, 279, 317, 411; Hortus 3: 1205; IF p. 328; IH 1 (ed. 6): 363, (ed. 7): 341, 2: (files); Jackson p. 114, 371, 373, 374; Kew 4: 565; Lasègue p. 582 [index]; Lenley p. 358; LS 23502; NI 1715-1717; Plesch p. 394; PR 7983-7993, ed. 1: 7081, 8934-8950, 10756; Rehder 5: 753; RS p. 139; SK 4: ccxi; Sotheby 668; TL-1/267, 561, 1136-1140; TL-2/110, 3142, 6238, see indexes; Tucker 1: 618-619; Urban-Berl. p. 390; Zander ed. 10, p. 721, ed. 11, p. 823.

BIOFILE: Angely, J., Bol. InPaBo 8: 3-39. 1958 (general information).
Anon., Almanak agricola brasileiro 18: 186-192. 1929 (general account, picture of bust in Museo nacional); Bonplandia 1: 187 (illness), 216 (death). 1853, 2: 267. 1854 (cogn. Leopoldina Jacquin); Bot. Zeit. 11: 767-768. 1853 (obit., not.); Cat., gén. livr. impr. Bibl. natl., Paris 160: 767-771. 1941 (bibl.); Flora 36: 600. 1853, 37: 277. 1854 (d.); London J. Bot. 1: 204. 1842; Proc. Linn. Soc. London 2: 323-325. 1855 (obit.).

Arechavaleta, J., Fl. Uruguaya 2: xxxix. 1902.
Bennett, J.J., Proc. Linn. Soc. London 2: 323-325. 1854, repr. Chron. bot. 10: 5-6. 1946.
Bonnet, Ed., Bull. Soc. bot. France 60: lxxxviii-ci. 1913 (letters Aug. St.-Hilaire to Moquin-Tandon).
Candolle, A.P. de, Mém. souvenirs 217, 223, 411, 414. 1862.
Candolle, Alph. de, Phytographie 446. 1880.
Chevalier, A., Bull. Soc. bot. France 76: 3-10. 1929 (work on dynamic phytogeography).
Deleuze, M., Hist. descr. Mus. r. Hist. nat. 1: 179, 2: 712-713. 1823 (notes on collections).
Dreuzy, R. de, *in* A. de Saint-Hilaire, Voy. Rio Grande do Sul, Orleans 1887, introduction and portrait (reprinted Chron. bot. 10: 10-11. 1946).
Dwyer, J.D., Ann. Missouri Bot. Gard. 42: 153-170. 1955 (catalogues of plants collected by St. Hil., 10 books at P.).
Embacher, F., Lexikon Reisen 257. 1882.
Flahault, C., Univ. Montpellier, Inst. Bot. 45: 1890 (library to Montpellier).
Geoffroy Saint-Hilaire et al., Rapport voy. Aug. St. Hil. Brésil, Paris 1823, 8 p. (see below TL-2/10.029).
Gray, J.L., Letters A. Gray 24, 167, 168. 1893.
Guillaumin, A., Les fleurs des jardins xxxix, *pl. 14*. 1929.
Herter, W., Bot. Jahrb. 74(1): 119-149, map. 1945 (Auf den Spuren der Naturforscher Sellow and Saint-Hilaire; collection localities).
Hoehne, C., O Jardim botânico de São Paulo 180-181. 1941 (portr.).
Jenkins, A.E., Chron. bot. 10: 5-21. 1946 (biogr. sketch, preceding a reprint of him "Esquisse de mes voyages au Brésil et Paraguay ..." portr., biogr., bibl. details, chronology itineraires); Regnum vegetabile 71: 111-113. 1970 (tributes to St. Hil. by A. v. Humboldt; *Rhexia hilariana*).
Jessen, K.F.W., Bot. Gegenw. Vorz. 404, 431, 467. 1884.
Lacroix, A., Figures de savants 4: 176, 178. 1938.
Léotard, S., ed., Le bibliophile du Bas-Languedoc 1: 1-16,49-64, 79-144. 1879, 2: 49-96, 97-102, 145-146, 193-195. 1879, 4: 1-5. 1881; 9: 1-6. 1886 (Letters inédites de Moquin-Tandon à Auguste de Saint Hilaire, copy at G), republication in book form: Lettres inédites de Moquin-Tandon à Auguste de Saint-Hilaire, Clermont-L'Hérault, 1893, 311 p., n.v.).
Letacq, A.L., Bull. Soc. Amis Sci. nat. Rouen ser. 5. 42: 101-102. 1907.
Martius, C.F.P. von, Flora 20 (Beibl. 1-4): 31-32. 1837.
Moquin-Tandon, C.H.B.A., Bonplandia 2: 27-28. 1854; Le bibliophile du Bas-Languedoc 1878 (1): 1-8, (2): 49-54, (3): 97-103, (5): 193-195, 1881 (1): 1-5, 1886 (2): 1-6. (lettres inédites de M.-T. à de St. Hil., copy at G); *in* Michaud, Biogr. univ. ed. 2. 37: 327-329.
Niemeyer, J.C. de Lavõr, Rodriguésia 32(54): 378, 400. 1980.
Papavero, N., Essays hist. neotrop. dipterol. 1: 115-123. 1971.
Planchon, J.E., Revue hort. ser. 4. 3: 176-180. 1854 (obit.).
Rizzini, C.T., Rodriguesia 32(52): 253. 1980 (bicentenary).
Roumeguère, C., Nouveaux documents sur l'histoire des plantes ... des Pyrénées. 1876 (letters).
Sachs, J., Hist. de bot. 157. 1892.
Saint-Hilaire A. de, Esquisse de mes voyages au Brésil et Paraguay, Chron. bot. 10(1): 1-61. 1966.
Sampaio, A.J. de, Bol. Mus. nac. Rio de Janeiro 4(4): 1-31. 1928 ("bibliography" actually a general account of his publ. on the Brazilian flora).
Santa Helena, H., Natureza em Revista 6: 4-8. 1979 (general account; Brazilian secondary refs.).
Sayre, G., Mem. New York Bot. Gard. 19(3): 391. 1975 (coll.).
Stieber, M.T., Huntia 3(2): 124. 1979 (notes in Chase mss.), 4(1): 86. 1981 (arch. mat. HU).
Urban, I., Fl. bras. 1(1): 92-98. 1906 (extensive itin.).
Verdoorn, F., ed., Pl. pl. Sci. Latin Amer. xxvi, xxvii. 1945 (travelogues).
Wilson, J.G. & J. Fiske, Appleton's Cycl. Amer. biogr. 5: 370-371. 1888.

HANDWRITING: Candollea 33: 157-158. 1978.

EPONYMY: *Augusta* P. Leandro de Sacramento (1821, *nom. rej.*); *Hilairanthus* Van Tieghem (1898); *Hilairella* Van Tieghem (1904); *Hilaria* Kunth (1816); *Hilariophyton* Pichon (1946); *Sanhilaria* Baillion (1888).

POSTAGE STAMPS: Brazil 1.20 cr. (1953) yv. 547.

10.024. *Notice sur soixante-dix espèces et quelques variétés de plantes phanérogames trouvées dans le Département du Loiret*, depuis la publication de la Flore orléanaise de M. l'Abbé Dubois; ... [Orléans (de l'Imprimerie de Huet-Perdoux) 1810]. Oct.
Publ.: 1810, p. [1]-47. *Copies*: B, FI, G, HH. – Reprinted and to be cited from Bull. Sci. phys. méd. Agr. Orléans 1: 97-103, 134-143, 210-219, 264-285. 1810.

10.025. *Observations sur le genre Hyacinthus*; par Aug. de S.-Hilaire. [Orléans (de l'Imprimerie de Huet-Perdoux) 1810]. Oct.
Publ.: 1810, p. [1]-6. *Copies*: FI, G. – Reprinted and to be cited from Bull. Sci. phys. med. Agric. Orléans 2: 200-205. 1810.

10.026. *Réponse aux reproches* que les gens du monde font à l'étude de *la botanique* ... A Orléans (de l'Imprimerie de Huet-Perdoux) 1811. Oct. (*Réponse reproches bot.*).
Publ.: 1811 (read 26 Nov 1810), p. [3]-30. *Copies*: G(2), NY.

10.027. *Mémoire sur les plantes auxquelles on attribue un placenta central libre* et sur la nouvelle famille des Paronychiées, par Auguste de Saint-Hilaire; suivi d'une note sur la même famille par M. A.L. de Jussieu, et d'observations sur le genre Glaux. Paris (de l'imprimerie de A. Belin, ...) 1816. Qu. (*Mém. pl. plac. centr. libr.*).
Publ.: 1815-1816 (in journal), p. [i], [1]-109. *pl. 4. Copies*: G(2). – Reprinted and to be cited from Mém. Mus. Hist. nat. 2: 40-61, *pl. 4*. 1815, 120-126. 1815, 195-208. 1815, 261-291, 1815, 377-382. 1816. See also Bull. Soc. philom. Paris 1815: 16-24. Feb 1815, 37-42. Mar 1815 for a preliminary notice.
Second Mémoire: ib. 4: 381-394. 1818.
Troisième Mémoire: see below *Monogr. Primul.* 1838.
The *Observations sur le genre Glaux*, also included in the first *Mémoire* is by Dutour de Salvert and A. de Saint-Hilaire, originally published in Mém. Mus. Hist. nat. 2: 393-395. 1815/1816.

10.028. *Mémoire sur les Cucurbitacées*, Les Passiflorées et le nouveau groupe des Nandhirobées ... Paris (Imprimerie de A. Belin, ...) 1823. Qu.
Publ.: 1823 (as reprint; in journal 1819 and Mar 1823), p. [i], [1]-47, *pl. 24-25*, seconde partie [1]-32. *Copies*: BR, G, HH, NY. – Reprinted and to be cited from Mém. Mus. Hist. nat. 5: 304-350, *pl. 24-25*. 1819 (publ. before Mar 1820), and 9: 190- 221. Mar 1823 ("1822").

10.029. *Rapport sur le voyage de M. Auguste de Saint Hilaire dans le Brésil* et les missions du Paraguay ... Paris (de l'imprimerie de J. Smith) 1823. Qu. (*Rapp. voy. Brésil*).
Authors of report: E. Geoffroy Saint Hilaire, R.L. Defontaines, P.A. Latreille, A.T. Brongniart, A.L. de Jussieu.
Publ.: 1823, p. [1]-8. *Copy*: G.

10.030. *Du Pelletiera*, nouveau genre de plantes de la famille des Primulacées (Triandrie monogynie. Lin.), recueilli dans le Brésil méridional et la province Cisplatine; ... [Orléans 1823]. Oct.
Publ.: 1823, p. [1-2]. *Copy*: HH. – Reprinted and to be cited from Ann. Soc. roy. Sci. Belles-Lettres Arts Orléans 5: 141-142. 1823. Previously published in Mém. Mus. Hist. nat. 9: 365. 1822.

10.031. *Aperçu d'un voyage dans l'intérieur du Brésil*, la province cisplatine et les missions dites du Paraguay ... Paris (imprimerie de A. Belin, ...) 1823. Qu.
Publ.: Jul 1823 (in journal, see MD p. 179), p. [i, iii], [1]-73.*Copies*: FI, G(2). – Reprinted and to be cited from Mém. Mus. Hist. nat. 9: 307-380. 1822 (publ. Jul 1823).

10.032. *Plantes usuelles des brasiliens, ...* Paris (Grimbert, libraire, ...) 1824[-1828]. Qu. (*Pl. usuel. bras.*).
Co-authors (parts 9-14): Jacques Cambessèdes (1799-1863), Adrien Henri Laurent de Jussieu (1797-1853).
Publ.: in 14 parts 1824-1828, see below, p. [i-vi], [1]-5, *pl. 1*, [1-5], *pl. 2*, [1-5], *pl. 3*, [1]-2, *pl. 4*, [1]-4, *pl. 5*, [1]-6, *pl. 6*, [1]-3, *pl. 7*, [1]-4, *pl. 8*, [1]-3, *pl. 9*,[1]-5, *pl. 10*, [1]-6, *pl. 11*, [1]-3, *pl. 12*, [1]-3, *pl. 13*, [1]-4, *pl. 14*, [1]-4, *pl. 15*, [1]-3, *pl. 16*, [1]-5, *pl. 17*, [1]-4, *pl. 18*, [1]-4, *pl. 19*, [1]-4, *pl. 20*, [1]-3, *pl. 21*, [1]-6, *pl. 22*, R*[1]-3, *pl. 23*, [1]4, *pl. 24*, [1]-6, *pl. 25*, [1]-8, *pl. 26-28*, [1]-4, *pl. 29*, [1]-6, *pl. 30*, [1]-4, *pl. 31*, [1]-3, *pl. 32*, [1]-7, *pl. 34*, [1]-4, *pl. 35*, [1]-4, *pl. 36*, [1]-4, *pl. 37*, [1]-4, *pl. 38*, [1]-4, *pl. 39*, [1]-4, *pl. 40*, [1]-4, *pl. 41*, [1]-3, *pl.42*, [1]-4, *pl. 43*, [1]- 2, *pl. 44*, [1]-2, *pl. 45*, [1]-6, *pl. 46*, [1]-7, *pl. 47-48*, [1]-3, *pl. 49*, [1]-3, *pl. 50*, [1]-3, *pl. 51*, [1]-3, *pl. 52*, [1]-3, *pl.53*, [1]-7, *pl. 54, 55*, [1]-4, *pl. 56*, [1]-4, *pl. 57*, [1]-3, *pl. 58*, [1]-4, *pl. 59*, [1]-2, *pl. 60*, [1]-3, *pl. 61*, [1]- 2, *pl. 62*, [1]-5, *pl. 63*, [1]-5,*pl. 64*, [1]-4, *pl. 65*, [1]-4, *pl. 66*, [1]-4, 67, [1]-3, *pl. 68*, [1]-3, *pl. 69*, [1]-3, *pl. 70*. *Copies*: BR, G, HH, M, NY, PH, US, USDA; IDC 5708.

part	*plates*	dates	part	*plates*	dates
1	*1-5*	31 Mar 1824	9	*41-45*	3 Mar 1827
2	*6-10*	12 Jun 1824	10	*46-50*	21 Apr 1827
3	*11-15*	14 Aug 1824	11	*51-55*	16 Jun 1827
4	*16-20*	6 Nov 1824	12	*56-60*	15 Sep 1827
5	*21-25*	3 Jan 1825	13	*61-65*	31 Mar 1828
6	*26-30*	30 Apr 1825	14	*66-70*	4 Aug 1828
7	*31-35*	22 Jun 1825	1 (substitute)	*1-5*	13 Aug 1827
8	*36-40*	21 Dec 1825			

The text for each plate has its own pagination. The above dates are based on BF, Revue bibliographique des Pays-Bas, Bulletin Férussac, and on the dates of receipt by the Académie. – The prospectus (BF, Aug. 1823) promised 50 fascicles! Flora (7: 111, 135) mentions the first fascicle already on 21 Feb 1824 and reviews it on 7 Mar 1824.
Ref.: Junk, W., Rara 15, 1900.
M., Flora 7: 135-137. 1824 (rev. fasc. 1).
Steenis-Kruseman, M.J.W., Fl. males. Bull. 19: 1148. 1964.
Woodward, B.B., J. Bot. 42: 86-87. 1904.

10.033. *Histoire des plantes les plus remarquables du Brésil* et du Paraguay; comprenant leur description, et des dissertations sur leurs rapports, leurs usages, etc., avec des planches, en partie coloriées ... tome premier. À Paris (chez A. Belin, ...) 1824[-1826]. Qu. †. (*Hist. pl. remarq. Brésil*).
Publ.: 1824-1826, in 6 parts, p. [i*-viii*], [i]-lxvii, [1]-355, *pl. 1-30* (col. and (mostly) uncol. liths. E. Blanchard). *Copies*: BR, G, HH, MO, NY; IDC 5157. – Contains as an introduction an "Esquisse des voyages de l'auteur, considérés principalement sous le rapport de la botanique" which was reprinted by A.E. Jenkins, Chron. bot. 10(1): 24-61. 1946. Only one volume (355 pp., 30 plates) was published. The plans called for one or two volumes of 10 fascicles each. The book was offered for sale with coloured and with uncoloured plates, the former also on "grand raisin vélin satiné". The contents of the parts are unknown. Contemporary reviews (Bull. Férussac, Isis) and the information on the receipt of the parts by the Académie are in part contradictory. The history of the book merits further investigation. Parts 1 and 2 together (p. 1-80, *t. 1-8*) were issued Jun 1824; parts 3 and 4 Nov 1825 and 5 and 6 possibly Jul 1826.
Ref.: Kuntze, O., Rev. gen. pl. 1: cxliii. 1931.
Schäfer, W., Flora 9: 192. 28 Mar 1826 (fasc. 1-4).
Schrader, H.A., Goett. gel. Anz. 1827: 409-413.
Steenis-Kruseman, M.J. v., Fl. males. Bull. 19: 1147-1148. 1964.
Zuccarini, J.G., Flora 8: 33-47. 21 Jan 1824 (rev. p. i-lxvii and 1-80, *pl. 1-8* (rev.).

10.034. *Flora Brasiliae meridionalis, ...* Parisiis (apud A. Belin, ...) 1824[-1833]. Qu. (*Fl. Bras. merid.*).

SAINT-HILAIRE, *Fl. Bras. merid.* 1824

Co-author for parts 5-22 Jacques Cambessèdes (1799-1863), and Adrien Henri Laurent de Jussieu (1797-1853). Family treatments signed by respective authors.
Publ.: in 24 parts, 1825-1833, see below. *Copies*: BR (Qu.), FI(Qu.), G(Qu.), MO(Fol.), NY(Qu.); IDC 309.
1: [i*-iii*], [i]-iv, [1]-167, [i, h.t.], [169, t.p. 1827]-395, [397-index], [399-400, index], *pl. 1-63, 63bis, 64-67, 67bis, 68-82* (plates 81-82 came out in first fasc. of vol. 2). The t.p. for part 1 has Saint-Hilaire as the sole author, that for part 2 [169] Saint-Hilaire, Adr. Jussieu and J. Cambessèdes.
2: [1]-381, [382-384], *pl. 83-159* [there is no plate *160*, neither is there any text for this number; never issued].
3(1): [1]-"86"[=84], *pl. 161-176*. – Title page mentions St. Saint-Hilaire, Adr. Jussieu and J. Cambessèdes as authors.
3(2): [i, iii], 85-160, *pl. 177-192*. – Title page mentions Saint-Hilaire as sole author. – There is no text for plate 192 (*Callisthene fasciculata*).

vol. & part	pages	*plates*	dates	vol. & part	pages	*plates*	dates
1(1)	1-40	*1-8*	23 Feb 1824	2(13)	81-116	*97-104*	10 Oct 1829
(2)	41-80	*9-16*	26 Mar 1825	(14)	117-156	*105-112*	8 Mai 1830
(3)	81-120	*17-24*	11 Jun 1825	(15)	157-196	*113-120*	24 Jul 1830
(4)	121-168	*25-32*	16 Nov 1825	(16)	197-236	*121-128*	Oct-Nov 1830
(5)	169-200	*33-40*	23 Apr 1827	(17)	237-276	*129-136*	4 Dec 1830
(6)	201-240	*41-48*	18 Jul 1827	(18)	277-316	*137-144*	25 Feb 1832
(7)	241-280	*49-56*	5 Apr 1828	(19)	317-356	*145-152*	27 Oct 1832
(8)	281-320	*57-64*	28 Jun 1828	(20)	357-384	*153-159*	3 Aug 1833
(9)	321-360	*65-72*	29 Sep 1828	3(21)	1-40	*161-168*	4 Feb 1833
(10)	361-400	*73-80*	6 Dec 1828	(22)	41-84 ("86")	*169-176*	4 Mai 1833
2(11)	1-40	*81-88*	9 Mai 1829				
(12)	41-80	*89-96*	12 Jul 1829	(23)	85-120	*177-184*	28 Oct 1833
				(24)	121-160	*185-192*	9 Dec 1833

Folio ed.: 3 vols., copy at MO. – The following information on the contents of this copy was provided by Mrs. C. Lange. Text on t.p.'s as in Quarto ed.
1: t.p. date 1825, p. [i*-iii*], [i]-iv, [1]-132, [second t.p. dated 1827], [133]-305, [i-iii], col. *pl. 1-67, 67bis, 68-82*.
2: t.p. date 1829, p. [i-ii], [1]-275, [i-iii], col. *pl. 83-128*, [129 missing from this copy but was originally present], 130-159. [There is no *pl. 160*].
3: t.p. date 1832, p. [1]-65, [second t.p. dated 1833, i*], [i-iii], 73-93, 96, 93, 96-128, col. *pl. 161-192*.
The dates above for the quarto edition are those of the Bibliographie de la France. Parts 1, 20, 21, 23 and 24 were received earlier by the Académie des Sciences than by BF; the Acad. dates are given for these parts (see M.J. van Steenis-Kruseman 1972). The date of part 16 was taken from Rev. bibl. Pays Bas.
Most copies are of the quarto edition (with uncol. pl.); the folio edition, with coloured plates, was issued in a small number of copies; the copy at MO is described by N.C. Horner (1953). Another copy is at the Bibliothèque nationale, Paris.
The plates are from drawings by Eulalia Delile and P.J.F. Turpin, numbered 1-191; two plates bear the number 63, another 67bis, but no. 160 is missing. The plate number 63 labeled *Larnottea tomentosa* should have been cancelled and replaced by 67bis (*Caryocar brasiliense*). See also GF p. 74 and Sotheby 668.
Ref.: Horner, N.C., Missouri Bot. Gard. Bull. 41: 57-59. 1953.
Steenis-Kruseman, van, Fl. males. Bull. 11: 453. 1955, 26: 2018-2019. 1972.

10.035. *Mémoire sur le genre Tozzia* ... [Mém. Mus. Hist. Nat. 14, 1827]. Qu.
Publ.: Mar-Mai 1827 (in journal), p. [11]sic-16. *Copies*: G, HH. – Reprinted and to be cited from Mém. Mus. Hist. nat. 14: 94-99. Mar-Mai 1827; see also Nouv. Bull. Soc. Philom., Paris 1826: 191-192. Dec 1826 for a preliminary publ.

10.036. *Premier mémoire sur la famille des Polygalées*, contenant des recherches sur la sym-

étrie de leurs organes, ... Extrait des Mémoires du Muséum d'Histoire naturelle. [Paris (imprimerie de A. Belin) 1828]. Qu.
Co-author: Christian Bénédict Alfred Moquin-Tandon (1804-1863).
Publ.: 1828, p. [1]-63, *pl. 27-31*. *Copies*: FI, G, HH, NY. – Reprinted and to be cited from Mém. Mus. Hist. nat. 17: 314-375, *pl. 27-31*. 1828.
Deuxième mémoire: 1832, p. [1]-35. *Copies*: FI, G, NY, WU. – Reprinted and to be cited from id. 19: 305-339. 1830. – *Deuxième mémoire sur la famille des Polygalées*, contenant principalement l'examen de leurs rapports et la comparaison de leurs déviations du type symétrique avec celles que présentent quelques autres familles ... [Paris (id.) 1832]. Qu.
Second mémoire: 1832, p. [1]-31. *Copy*: HH. – Reprinted from Ann. Soc. roy. Sci. Belles-Lettres Arts Orléans 12: 6-54[?]. 1832 (journal n.v.). – Other edition of "deuxième memoire".

10.037. *Conspectus Polygalearum Brasiliae meridionalis*; ... Orléans (Imprimerie de Danicourt-Huet ...) 1828. Oct.
Publ.: 17 Nov 1828 (Acad. Paris), p. [1]-18. *Copies*: FI, G, WU. – Reprinted and to be cited from Ann. Soc. roy. Sci. Belles-Lettres Arts Orléans 9: 44-59. 1828.

10.038. *Mémoire sur la symétrie des Capparidées* et des familles qui ont le plus de rapports avec elles; ... (extrait des Annales des Sciences naturelles, juillet 1830) [Paris (Imprimerie de Ve Thuau, ...) 1830]. Oct.
Publ.: Jul 1830 (in journal; reprint rd. Acad. 8 Nov 1830), p. [1]-8. *Copies*: FI, G(2), HH. – Reprinted and to be cited from Ann. Sci. nat. 20: 318-326. Jul 1830.

10.039. *Voyage dans les provinces de Rio de Janeiro et de Minas Geraes*; ... Paris (Grimbert et Dorez, ...) 1830. 2 vols. Oct. (*Voy. Rio de Janeiro*).
1: 1830 (p. xiv: 21 Mar 1830), p. [i], front., [iii]-xvi, [1]-458.
2: 1830, p. [i], front., [iii]-vi, [1]-478, [1, err.].
Copies: G, H (vol. 2), HH, NY. – First volume of the "Voyages dans l'Intérieur du Brésil".

10.040. *Observations sur le genre Anacardium* et les nouvelles espèces qu'on doit y faire entrer ... [Paris (Crochard, ... Imprimerie de Ve Thuau, ...) 1831. Oct.
Publ.: Jul 1831 (in journal), p. [1]-8. *Copies*: FI, G. – Reprinted and to be cited from Ann. Sci. nat. 23: 268-274. Jul 1831.

10.041. *Tableau de la végétation primitive* dans la province *de Minas Geraes*; ... [Ann. Sci. nat. 24, 1831]. Oct.
Original publ.: Sep 1831, in Ann. Sci. nat. 24: 64-102. Sep 1831.
Reprint, new type, quarto format, but text identified on p. 30 as "Extrait des Annales sciences naturelles ... Cahier de Septembre 1831, p. [1]-30. *Copies*: FI, HH, NY(2).
Ed. 1837: 1837 (prob. Sep-Dec), p. [i], [1]-49. *Copies*: BR, U. – *Tableau géographique de la végétation primitive* dans la province *de Minas Geraes* ... Paris (A. Pihan de la Forest ...) 1837. Oct. – Also issued in Nouv. Ann. Voyages 1837(3): 170-218. 1837 (Paris (Librairie de Gide, ...) 1837.
Reissue: 1837, as ed. 1837 but with a label stuck on the title-page "seconde édition, revue et corrigée". This refers probably to the original publication in the Ann. Sci. nat. *Copies*: G, HH, NY; IDC 5562.

10.042. *Voyage dans le district des diamans* et sur le littoral du Brésil, suivi de note sur quelques plantes caractéristiques et d'un précis de l'histoire des révolutions de l'empire brésilien, depuis le commencement du règne de Jean vi jusqu'à l'abdication de D. Pedro ... Paris (Librairie-Gide ...) 1833, 2 vols. Oct. (*Voy. distr. diam.*).
1: 1833, p. [i]-xx, [1]-402, [1, err.].
2: 1833, p. [i-iii], [1]-456.
Copies: G, HH, NY, J. Wurdack. – Vol. 2 of the "Voyages dans l'intérieur du Brésil". – See also F. Dunal's review of the *Voyage*, p. [1]-13, s.d., *copy*: G (journal unidentified).
Ref.: Guillemin, A., Arch. Bot. 2: 444. 25 Nov 1833 (rev.).

10.043. *Observations sur la famille des Amaranthacées*; ... Extrait du premier volume des Archives de Botanique [1, 1833]. Oct.
Publ.: 13 Mai 1833 (in journal), p. [1]-11. *Copy*: HH. – Reprinted and to be cited from Arch. Bot., Guillemin 1: 402-412. 1833.

10.044. *Observations sur le genre Escallonia* [Extrait des Archives de Botanique 2, 1833]. Oct.
Publ.: 16 Sep 1833 (in journal), p. [1]-7. *Copy*: HH. – Reprinted and to be cited from Arch. Bot., Guillemin 2: 225-231. 16 Sep 1833.

10.045. *Observations sur le genre Cuphea* [Extrait des Archives de Botanique 2, 1833]. Oct.
Publ.: 25 Nov 1833, p. [1]-10. *Copy*: HH. – Reprinted and to be cited from Arch. Bot., Guillemin 2: 385-393. 25 Nov 1833.

10.046. *Première mémoire sur la structure et les anomalies de la fleur des Résédacées*, ... [Orléans (Imprimerie de Danicourt-Huet) 1837]. Qu.
Orig. publ.: 1837[?], p. [1]-20. *Copy*: HH. – Reprinted from Ann. Soc. r. Sci. Belles-Lettres, Arts Orléans 13, 1837(?) (journal publ. n.v.; reprint not dateed).
Reprint: Aug 1837, p. [1]-30. *Copies*: FI, G, HH, MO, NY. – Reprinted and to be cited from Mém. Acad. Sci. Paris 15: 1-30. 1837 (see below under *Deuxième mémoire*).

10.047. *Deuxième mémoire sur les Résédacées*; corrigé et augmenté, ... Montpellier (Jean Martel Ainé, ...) 1837. Qu. (*Deux. mém. Réséd.*).
Publ.: Dec 1837 (t.p. of preprint; BF Jan 1838; mss. presented at Orléans 17 Mar 1837), p. [1]-42. *Copies*: FI, G, HH(2), MO, NY. – Originally published in, or reprinted from Mém. Soc. roy. Sci. Bell.-Lettr. Arts Orléans 1: 93-132. 1837.
Original publication of text in Mém. Acad. Sci. 15: 313-354. Aug 1837. The first memoir was (see above) reprinted together with the original publication of the second one in the Mém. Acad. Sci. Paris 15: 1-30, 313-354. 1838 (probably issued already in 1837, because the Academy received them on 7 Aug 1837). In his preface to the second edition the author states (of the first) "je dois la considérer comme non avenue" because of the many mistakes. This separate edition appeared perhaps very late in 1837, but most probably in early Jan 1838 (B.F. 1838: 15. 17 Jan 1838) and was reprinted or preprinted, with independent pagination, from the Mém. Soc. Roy. Sci. Belles-Lettres et Arts d'Orléans, vol. 1. This means also that the part of the Mém. Acad. Sci. Paris 15, containing the two memoirs must have been published in 1837. Since the first publication of the second memoir was evidently unauthorized, it seems advisable to follow the separate edition. The original publication of e.g. *Caylusea* is therefore in Mém. Acad. Sci. Paris 15: 342. 1837 ("1838"), but not accepted by its author.
Preliminary publ.: Ann. Sci. nat. Bot. ser. 2. 7: 371-376. Jun 1837, C.R. Acad. Sci., Paris 2: 31-34. 1836.
Ref.: Schmid, G., Goethe und die Naturwissenschaften 267. 1900 (no. 1750).

10.048. *Histoire de l'indigo*, depuis l'origine des temps historiques jusqu'a l'année 1833; ... Orléans (Imprimerie de Danicourt-Huet) 1837. Oct.
Publ.: 1837, p. [1]-13. *Copy*: HH. – Reprinted from Mém. Soc. r. Sci. Belles-Lettres Arts Orléans 1, 1837 (journal n.v.).
Reprint (or earlier publication): Feb 1837, in Ann. Sci. nat., Bot. ser. 2. 7: 110-119. 1837.

10.049. *Mémoire sur les Myrsinées*, les Sapotées et les embryons parallèles au plan de l'ombilic, ... [Mém. Acad. Sci. Paris, 1838]. Qu.
Orig. issue (as preprint or reprint from journal): 1838, p. [1]-50, table. *Copies*: HH, WU. – Preprinted or reprinted from Mém. Acad. Sci. Paris 16: 117-167. 1838.
Second issue (with label "Seconde édition, revue et corrigée") stuck on p. [1]: 1838, p. [1]-50, table. *Copy*: G. – The label may refer to the preliminary publ. (see below) as the original.
Preliminary publ.: Ann. Sci. nat., Bot. ser. 2. 5: 193-225. Apr 1836.

10.050. *Monographie des Primulacées* et des Lentibulariées du Brésil méridional; ...

(Extrait des Comptes rendus des séances de l'Académie des Sciences, séance du 19 novembre 1838). Qu.
Co-author: Frédéric de Girard.
Publ.: 19 Nov-31 Dec 1838 (in the absence of information of publication at a later date), p. [1]-4. *Copies*: FI, G, HH. − Reprinted or preprinted from C.R. Séances Acad. Sci., Paris 1838 (séance 19 Nov 1838): 868-870. 1838. − *Note*: together with a note in C.R. 1837 (séance 6 Nov 1837): 645-650, this text constitutes the third *Mémoire sur le placenta central*, see above. This is a preliminary and abbreviated treatment.
Definitive text: Feb, Mar 1839 (publication in *Annales*), p. [1]-36, *2 pl. Copies*: FI, HH. − Reprinted and to be cited from Ann. Sci. nat. Bot. ser. 2. 11: 85-99, *pl. 4*. Feb 1839, 149-169, *pl. 5*. Mar 1838, 382. Jun 1839. − Title "*Monographie* ... méridional et de la République Argentine".
Later edition: 1840 (t.p. reprint), p. [i], [1]-48, *2 pl. Copies*: B, G, NY. − Reprinted from Mém. Soc. roy. Sci. Belles-Lettr. Arts Orléans 2: 201-248, *2 pl.* 1838 (date on t.p. of journal vol. 2). − Reprint t.p.: *Monographie des Primulacées* ...méridional et de la République Argentine ...". Orléans (Imprimerie de Danicourt-Huet) 1840. Oct. − The Genève copy has a handwritten note by the author: "2ª édition corrigée" referring to the publication in the *Annales* as the original.

10.051. *Leçons de botanique* comprenant principalement la morphologie végétale, la terminologie, la botanique comparée, l'examen de la valeur des caractères dans les diverses familles naturelles, etc., ... Paris (P.J. Loss, ...) 1840. Oct. (*Leçons bot.*).
Publ.: 1840 (published in two parts, 1840-1841 fide Amer. J. Sci. 41: 371. 1 Oct 1841; Flora (see below) 21 Dec 1841), p. [i*-vii*], [i]-viii, [1]-930, *pl. 1-24* (uncol. liths. S.A. Node). *Copies*: G, H, MO, NY.
Re-issue: 1841, p. [i*-vii*], [i]-viii, [1]-930, *pl. 1-24* (id.). *Copy*: HH. − "Ouvrage adopté par le conseil royal de l'instruction publique". Paris (P.-J. Loss, ...) 1841. Oct.
Re-issue: 1847, p. [i*-vii*], [i]-viii, [1]-930, *pl. 1-24* (id.). *Copy*: BR. − "Ouvrage adopté par le Conseil royal de l'Instruction publique". Paris (Librairie agricole de Dusacq ...) 1847. Oct.
Ref.: Kirschleger, F., Flora 24(2), Lit. Ber. 10: 148-160. 21 Dec, 11: 161-176. 28 Dec, 12: 177-182. 28 Dec 1841 (extensive review).

10.052. *Revue de la Flore du Brésil méridional* ... [Extrait des Annales des Sciences naturelles (Mars 1842)]. Oct.
Co-author: [Edmond] Louis René Tulasne (1815-1885).
Publ.: Mar 1842 (in journal), p. [1]-15, *pl. 6-7*. *Copies*: B, FI. − Reprinted and to be cited from Ann. Sci. nat., Bot. ser. 2. 17: 129-143, *pl. 6-7*. Mar 1842.

10.053. *Voyage aux sources du Rio de S. Francisco* et dans la province de Goyaz ... Paris (Arthus Bertrand, ...) 1847-1848, 2 vols. Oct. (*Voy. Rio S. Francisco*).
1: 1847, p. [i-iii], [1]-380.
2: 1848 (p. xv: 10 Jan 1848), p. [i]-xv, [1]-349, [1, err.].
Copy: NY. − Constitutes the third volume of the "Voyages dans l'intérieur du Brésil".
Ref.: Schlechtendal, D.F.L. von, Bot. Zeit. 8: 609-613. 16 Aug 1850, 8: 625-629. 23 Aug 1850 (rev.).

10.054. *Voyage dans l'intérieur du Brésil*, ... Ixelles lez Bruxelles (Delevigne et Callewaert, ...) 1850, 2 vols. Duod. (*Voy. intér. Brésil*).
1: 1850, p. [i-ii], frontisp., [iii], [1]-212.
2: 1850, p. [i-ii], frontisp., [iii], [1]-208.
Copy: HH.

10.055. *Voyage dans les provinces de Saint-Paul* et de Sainte-Catherine; ... Paris (Arthus Bertrand, ...) 1851, 2 vols. Oct. (*Voy. Saint-Paul*).
1: 1851, p. [i]-vi, [1]-464.
2: 1851, p. [i-iii], [1]-423, [424, err.].
Copy: NY. − Volume four of the "Voyages dans l'intérieur du Brésil". See Jenkins (1946) for a review of all of Saint-Hilaire's travelogues.

St. John, Harold (1892-x), American botanist; Dr. phil. Harvard 1920; assistant, later associate professor of botany at the State College of Washington 1920-1929; professor of botany at the University of Hawaii 1929-1958; Whitney visiting professor at Chatham College, Pittsburgh, Pa. 1958-1959; teaching at Saigon 1959-1960; continuing botanical research in retirement in Hawaii. (*St. John*).

HERBARIUM and TYPES: BISH; further material at A, DS, GH, M, MO, NY, NYS, PH, WELT, WS, WTU. – St. John donated his professional correspondence relative to taxonomic studies of the vegetation of the American Northwest to WS.

BIBLIOGRAPHY and BIOGRAPHY: Barnhart 3: 201 (b. 25 Jul 1892); BL 1: 110, 111, 126, 138, 141, 145, 186, 220, 316; BM 8: 1123; Bossert p. 345; Ewan ed. 1: 240, 296, 332, ed. 2: 192; GF p. 249; Hortus 3: 1205; IF suppl. 3: 217, 4: 339-340, IH 1 (ed. 1): 41,(ed. 2): 60, (ed. 3): 74, (ed. 4): 79, (ed. 5): 79, (ed. 6): 115, 363, (ed. 7): 116, 343, 2: (files); Kew 4: 566-570; Langman p. 664; Lenley p. 358; MW p. 423; MW suppl. p. 297; NW p. 35-37, 56; PH 209; Rehder 5: 753; Roon p. 108; Zander ed. 10, p. 721, ed. 11, p. 822.

BIOFILE: Anon., Bull. Amer. Ass. bot. Gard. Arbor. 10(1): 22. 1976 (award of merit); Pac. Sci. 33(4): 435-447. 1979 (1980) (career synopsis; bibl.).
Bakhuizen van den Brink, R.C. & C.G.G.J. van Steenis, Fl. males. Bull. 26: 2042. 1972 (additions to St. John's evaluation of J.R. & G. Forster, *Char. gen.* of 1971).
Cattell, J.M. & J., eds., Amer. men Sci. ed. 5: 967. 1933, ed. 6: 1227. 1938, ed. 7: 1536. 1944, ed. 8: 2150. 1949.
Merrill, E.D., B.P. Bishop Mus. Bull. 144: 161-162. 1937 (Polynes. bot. bibl.); Contr. U.S. natl. Herb. 30: 261-263. 1947 (Bot. bibl. Pacific); Chron. bot. 14: 183, 260, 264, 351, 353. 1954 (Botany Cook's Voyage).
Pennell, F.W. & J.H. Barnhart, Monogr. Acad. nat. Sci. Philadelphia 1: 618. 1935 (coll.).
Rickett, H.W., Index Bull. Torrey bot. club 87. 1955.
Rigg, G.B., Washington hist. Quarterly 20: 166-167. 1929.
Smith, A.C., Flora vitiens. nova 1: 76. 1979.
Stafleu, F.A., ed., Taxon 7: 180. 1958 (retirement).
Steenis, C.G.G.J. van, Fl. males. Bull. 7: 191. 1950 (at Leiden), 14: 626. 1959 (to Italy), 15: 708. 1960 (at Saigon), 22: 1528. 1968 (back from Italy).
Theobald, W.L., Bull. Pac. Trop. Bot. Gard. 12(3): 60-63. 1982 (award).
Thomas, J.H., Huntia 3: 20. 1969 (1979).

NOTE: See also H. St. John, *Nomenclature of Plants*. A text for the application by the case method of the International Code of botanical Nomenclature, New York 1958, 157 p.

FESTSCHRIFT: Pacific Science 33(4). 1979.

10.056. *Sable Island*, with a catalogue of its vascular plants ... Boston (Printed for the [Boston] Society [of Natural History]). March 1921. Oct. (*Sable Isl.*).
Publ.: Mar 1921 (t.p.), p. [1]-103. *pl. 1-2. Copies*: BR, MO, NY. – Issued as Proc. Boston Soc. nat. Hist. 36(1): 1-103, *pl. 1-2*, and as Contr. Gray Herb. ser. 2. no. 62.

Booknumber 10.057 has not been used.

10.058. *A botanical exploration of the North Shore of the Gulf of St. Lawrence* including an annotated list of the species of vascular plants ... Ottawa (F.A. Acland, ...) 1922. Oct. (*Bot. explor. St. Lawrence*).
Publ.: 25 Mar 1922 (date on cover and t.p.), p. [i*], [i]-iii, [1]-130, 2 maps, *pl. 1* (frontisp.), 2-6 (in text). *Copies*: BR, FI, MICH, MO, NY, PH, USDA. – Issued as Memoir 126, Victoria Memorial Museum (Canada Dept. Mines).

10.059. *An annotated catalogue of the vacular plants of Benton County*, Washington ... Contribution from the Department of Botany State College of Washington, no. 9. 1928. Oct. *Co-author*: George Neville Jones (1904-1970).

Publ.: 21 Oct 1928, p. 73-93. *Copies*: FI, G, MICH, U, E.G. Voss. – Reprinted with special t.p.-cover from Northwest Science 2: 73-93. 21 Oct 1928.

10.060. *Plants of the headwaters of the St. John river*, Maine ... Pullman, Washington, June 28, 1929. Oct.
Publ.: 28 Jun 1929, (cover-t.p), p. 28-58. *Copies*: FI, MICH, U. – Reprinted with special cover and to be cited from Res. Stud. State Coll. Washington 1(1): 28-58. 28 Jun 1929.

10.061. *Flora of Mt. Baker* ... Reprinted without change from Mazama ... December 1929. Oct.
Co-author: Edith Hardin [later: English] (1897-1979).
Publ.: Dec 1929, p. [i], [52]-102, [1, index]. *Copies*: R.S.Cowan, FI, MICH, U. – Reprinted with a special t.p. and to be cited from Mazama 11: 52-102. Dec 1929.

10.062. *Weeds of the pineapple fields of the Hawaiian Islands* ... Honolulu (published by the University of Hawaii) 1932. Oct. (*Weeds pineapple fields*).
Co-author: Edward Yataro Hosaka (1907-1961).
Publ.: Sep 1932, p. [1]-196. *80 pl.* (uncol.). *Copies*: G, MO, NY, PH, US, USDA. – University of Hawaii Research Publication no. 6.

10.063. *The plants of Mt. Rainier National Park* Washington ... Reprinted from "The American Midland naturalist" vol. 18, no. 6, p. 952-985, November, 1937 Notre Dame, Ind. (The University Press) 1937. Oct.
Co-author: Fred Adelbert Warren (1902-ca. 1940).
Publ.: Nov 1937, p. [i], 952-985. *Copies*: FI, MICH. – Reprinted with special t.p. and to be cited from Amer. Midl. Natural. 18(6): 952-985. Nov 1937.

10.064. *Flora of Southeastern Washington* and of adjacent Idaho ... Pullman, Wash. (Students Book Corporation) 1937. Oct. (*Fl. S.-e. Washington*).
Collaborators: Charles Piper Smith (1877-1955) for *Lupinus*; John Hendley Barnhart (1871-1949) bibl.
Ed. [*1*]: 19 Jun 1937, frontisp., p. [i]-xxv, 1-529. *Copies*: BR, MICH, MO, NY, PH, US, USDA.
Ed. 2: 1 Mar 1956, p. [i]-iii, map, v-xxv, 1-531. *Copy*: G. "Revised edition 1956".
Ed. 3: Nov 1963, p. [i]-xxix, 1-583. *Copies*: MO, US, USDA. – See Taxon 13: 147. 1964.
Ref.: Constance, L., Madroño 4(4): 132-134. 4 Oct 1937 (rev. ed.).
  Daubenmire, R.F., Leaf. W. Bot. 2: 199-200. 1939 (range extensions).
  Fassett, N.C., Torreya 37: 127-129. 29 Dec 1937.
  Rendle, A.B., J. Bot. 76: 28. Jan 1938 (rev.).
  Sprague, T.A., Bull. misc. Inf. Kew 1937: 519-520. 30 Dec 1937 (rev.).

10.065. *Identification of Hawaii plants*: a key to the families of dicotyledons of the Hawaiian Island, descriptions of the families, and list of the genera ... University of Hawaii Occasional papers, no. 36. [1938]. (*Ident. Hawaii pl.*).
Co-author: Francis Raymond Fosberg (1908-x).
[*1*]: 28 Jun 1938 (copies so stamped), p. [1]-53. – Univ. Hawaii Occ. Pap. 36.
2: 20 Feb 1940 (id.), p. [1]-47. – Univ. Hawaii Occ. Pap. 41.
Copies: R.S. Cowan, FI, MICH, MO, E.G. Voss.
Note: See also H. St. John, List and summary of the flowering plants in the Hawaiian Islands. Hawaii, 30 Aug 1973, 519 p., Mém. Pacif. Bot. Gard. no. 1, see also Taxon 24: 165. 1975.

Saint-Lager, Jean Baptiste (1825-1912), French physician, botanist, botanical historian and bibliographer at Lyon; laureat École de Médecine Lyon 1847; Dr. med. Paris 1850; practicing physician at Lyon 1850-1862; from 1862 private scientist, travelling for botanical and medical studies 1862-1890; one of the founders of the Société botanique de Lyon; librarian of the Société linnéenne de Lyon and the Société botanique de Lyon until 1911 and of the Palais des Arts. (*St. Lag.*)

HERBARIUM and TYPES: G; other material at C and W. – Fifty letters to C. Malinvaud are at P.

BIBLIOGRAPHY and BIOGRAPHY: Barnhart 3: 201 (b. 4 Dec 1825, d. 29 Dec 1912); BJI 1: 50; BL 2: 109, 113, 115, 194, 712; BM 4: 1784, 8: 1123; Bossert p 345; CSP 11: 261, 12: 644, 18: 406-407; Herder p. 89; Hortus 3: 1205; Jackson p. 121; Kew 4: 570; LS 23503; Morren ed. 10, p. 64; NI 1: 244; PFC 1: lii; Plesch p. 395; Rehder 5: 753; Saccardo cron. p. xxxi; TL-1/1141; TL-2/1024, 5790; Tucker 1: 619; Zander ed. 10, p. 721, ed. 11, p. 731, 823.

BIOFILE: Anon., Bull. Soc. bot. France, Table gén. art. orig. 214. 1899; Journal de Genève, 20 Sep 1933 (herbarium left to G by his son).
Aymonin, G.G., C.R. Congr. Soc. Sav. (Limoges) 102(3): 57. 1977 (letters to Malinvaud at P);
Burnat, E., Bull. Soc. bot. France 30: cxxx. 1883 (coll. Alpes maritimes).
Candolle, Alph. de, Nouv. rem. nom. bot. 19, 20, 35. 1883 (rejects S.'s proposals for nomenclatural reform)Cavillier, F., Boissiera 5: 84. 1941.
Durand, Th., Bull. Soc. bot. Belg. 21(2): 7-15. 1882.
Lamy, D., Occ. pap. Farlow Herb. 16: 125, 126. 1981.
Malinvaud, E., Bull. Soc. bot. France 41 (bibl.): 320. 1894 (app. "officier de l'Instruction publique").
M.C., La Tribune de Genève 12 Oct 1933 (herb. to G).
Miège & Wuest, Saussurea 6: 133. 1975 (herb. G).
Roux, C. & O. Meyran, La vie et les travaux du docteur J.-B. Saint Lager, Lyon 1913, 39 p., also in Ann. Soc. bot. Lyon 38: 1-39. 1913 (biogr., bibl.).

COMPOSITE WORKS: With M.A.A. Méhu and L.A. Cusin, *Herborisations dans les montagnes d'Hauteville*, 1876, see TL-2/5790.

10.066. *Catalogue des plantes vasculaires du bassin du Rhone* ... Lyon-Genève-Bale (Librairie H. Georg). [1873-]1883. Oct. (*Cat. fl. bass. Rhone*).
Publ.: 1873-1883 in seven parts and preface material; main part, issued as supplements to the Annales de la Société botanique de Lyon. *Copies*: G, H, HH, MO, NY, USDA (orig. covers). – Information on publ. (in copy G) by L. Garcin, 31 Mai 1894.
Preface: 1883 (with consolidated volume; Bot. Zeit. 27 Apr 1883), p. [i*-iv*].
Advertisement: 1875, p. [i]-viii.
Preface (orig.): 1873, p. [1]-10.
1(orig..): 1873, p. [1]-32, with Ann. 1.
1(ed. 2): 1880, p. [1]-32f.
2: 1874, p. 33-114, with Ann. 2(1).
3: 1875, p. 115-186, with Ann. 3(1).
4: 1876, p. 187-334, with Ann. 4(2).
5: 1877, p. 335-495, with Ann. 6(1).
6: 1881, p. 495-638, with Ann. 9(1).
7: 1882, p. 689-886, with Ann. 10(1).
Ref.: Durand, Th., Bull. Soc. bot. Belg. 21(2): 7-15. 1882 (crit. notes).

10.067. *Réforme de la nomenclature botanique* ... Lyon (Association typographique ...) 1880. Oct. (*Réform. nom. bot.*).
Publ.: 1880 (J. Bot. Aug 1880; BSbF for Jan-Mar 1880; Nat. Nov. Jun(1) 1880; Bot. Zeit. 28 Mai 1880; Flora 1 Jun 1880), p. [i, iii], [1]-154, [1]. *Copies*: BR, G, H, HH, NY. – Also issued in Ann. Soc. bot. Lyon, Mém. 7: 1-154. 1880. – "The work of a classical purist who would unhesitatingly alter every botanical name, generic and specific alike, which does not square with the canons laid down by the author" and "The airy manner in which Dr. Saint-Lager proposes to constitute new generic names, as *Glechonion* [for Glechoma], and *Meladendron* [for Melaleuca], as well as to set aside names like *Anteuphorbium* and *Aphaca*, which were perfectly well known as single names long before Linnaeus came into existence, shows his eminent unfitness for universal censorship" ... "it is fortunately powerless to set aside accepted nomenclature ; what Salisbury could not do, seventy years ago, will not be accomplished

now by Dr. Saint-Lager" (B.D. Jackson 1880). Most of the names suggested by Saint-Lager in this work are later orthographic variants which can be rejected. The main changes proposed by Saint-Lager concern:
1. Changing specific names not in accordance with grammar, including differences of opinion on grammatical gender.
2. Replacing two-word epithets by new ones rather than hyphenate them.
3. Re-instating greek suffixes for greek words (Dianthos).
4. The rejection of "barbaric" epithets, eponyms or with a "ridiculous" connotation.

Ref.: Jackson, B.J., J. Bot. 18: 278-281. 1880 (rev.).
Fournier, E., Bull. Soc. bot. France 27 (bibl. for Jan-Mar 1880), 21-23. 1880 (rev.).
Knapp, J.A., Österr. bot. Z. 30: 204-205. 1880.
P., Bot. Zeit. 39: 355-356. 3 Jun 1881 (rev.).

10.068. *Nouvelles remarques sur la nomenclature botanique* ... Paris (J.B. Baillière, ...) 1881. Oct.

Publ.: 1881 (KNAW Jul-Sep 1881; Flora 21 Mar 1881; 1 Jun 1881; ÖbZ 1 Mai 1881; BSbF rd. 22 Apr 1881; Bot. Zeit. 27 Mai 1881; Nat. Nov. Mar(2) 1881), p. [i], [1]-55. *Copies*: BR, G, H. – Reprinted and to be cited from Ann. Soc. bot. Lyon, Mém. 8: 149-204. 1880 (1881), with the addition of one page (55) of errata to the 1880 *Remarques* with the motto "medice, cura te ipsum". The publication brings further precisions of correct grammatical coining of generic and specific names especially with respect to those of greek origin. Eponyms have to be based on the latinised version of the names of the respective botanists; tautonymous specific names in meaning (not simply in form), that is, where the epithet repeats the idea of the generic name, have to be abandoned (*Sagittaria sagittifolia*).

Ref.: Koehne, E., Bot. Centralbl. 6: 41-43. 1881 (rev.).
–, Bull. Soc. bot. France 27(bibl.): 229-230. 1881 (rev. of books for Nov-Dec 1880).

10.069. *Des origines des sciences naturelles* ... Lyon (Association typographique ...) 1882. Oct.

Publ.: 1882 (read 11 Jun 1882), p. [i], [1]-34. *Copy*: G. – Reprinted and to be cited from Mém. Acad. Sci. Bell.-Lettr. Arts Lyon 26 (cl. sci.): 27-161. 1882. – Inaugural address as member of the Académie des Sciences, ... Lyon, read on 11 Jul 1882. Saint-Lager develops here the same guidelines for zoological nomenclature as set out by him for botany in 1880 and 1881.

10.070. *Quel est l'inventeur de la nomenclature binaire* Remarques historiques. – ... Paris (Librairie J.B. Baillière & fils ...) 1883. Oct.

Publ.: 1883 (t.p. reprint; Bot. Zeit. 31 Aug 1883; BSbF lit. for Mar-Mai 1883), p. [i, iii], [1]-16. *Copies*: BR, FI, G, MICH. – Reprinted and to be cited from Ann. Soc. Linn. Lyon 29: 367-382. 1882 (1883). – Tournefort is said to have clearly stated that the names of species should be binary (Tournefort, Inst. rei herb. 1: 63 "Nomina plantarum sunt quaedam veluti definitiones, quorum prima vox genus plantae, cetera differentiam exprimunt". However, binary names for species are found in many early botanical publications such as *Cassia fistularis* with Albert the Great and many others were in use already in classical times: *Salix viminalis* with Virgilius (Georg. 2: 446).

Ref.: [Malinvaud, E.?], Bull. Soc. bot. France 30(bibl.): 66-67. 1883.

10.071. *Recherches historiques sur les mots plantes mâles et plantes femelles* ... Paris (J.B. Baillière ...) 1884. Oct.

Publ.: 1884 (t.p. reprint; Bot. Zeit. 28 Nov 1884; BSbF rd. 14 Nov 1884; Nat. Nov. Jan(2) 1885), frontisp., p. [1]-48. *Copies*: FI, NY. – Reprinted and to be cited from Ann. Soc. bot. Lyon 11 (Mém.): 1-48. 1883 (1884). Classical botanists, e.g. Theophrastus, were acquainted with the phenomenon of sexuality in plants, e.g. in Palms. The words "male" and "female" were later used for various connotations: less versus more fertile, larger versus smaller, for morphological analogy and metaphorically ("Orchis"), as an expedient to distinguish closely related species. Furthermore the epithets "mas" and "foemina" should be banned from nomenclature: *Cornus mas* should become *C. erythrocarpa*, etc.

Ref.: Malinvaud, E., Bull. Soc. bot. France 31(bibl.): 200-201. 1884 (1885).
Koehne, E., Bot. Zeit. 43: 820-822. 18 Dec 1885.

10.072. *Histoire des herbiers* ... Paris (J.B. Baillière et fils, ...) 1885. Oct.
Publ.: late 1885, or, more probably, Jan-Mai 1886 (t.p. reprint 1885; TBC Jul 1886 "just received; Bot. Zeit. 30 Jul 1886; BSbF séance 28 Mai 1886; the author sent a copy to É. Burnat in 1886(NY); journal issue dated 1886; Nat. Nov. Jun(2) 1886; KNAW rd. Oct 1886), p. [i-iii], [1]-120. *Copies*: BR, G, HH, MICH, MO, NY. – Reprinted and to be cited from Ann. Soc. bot. Lyon 13 (Mém.): 1-120. 1886. Describes herbaria from Jean Girault (1558) through C. Bauhin.
Preliminary publication: *in* Bull. trimestr. Soc. bot. Lyon, Proc. Verb. 1885: 61-64. Apr-Jun 1885.
Ref.: Anon., Bull. Torrey bot. Club 12: 129-131. Dec 1885 (review taken from preliminary notice).
Flahault, C., Bull. Soc. bot. France 33(bibl.): 154-155. 1886.
Schumann, Bot. Zeit., 44: 629-632. 10 Sep 1886.

10.073. *Recherches sur les anciens herbaria* ... Paris (J.B. Baillière et fils, ...) 1886. Oct.
Publ.: 1886 (t.p. reprint journal issue so dated; Bot. Zeit. 25 Mar 1887; BSbF séance 11 Mar 1887; Nat. Nov. Mar(2) 1887), p. [1]-45. *Copies*: BR, G, H, NY. – Reprinted and to be cited from Ann. Soc. bot. Lyon 13: 237-281. 1886.

10.074. *Le procès de la nomenclature botanique* et zoologique ... Paris (Librairie J.B. Baillière et fils ...) 1886. Oct.
Publ.: 1886 (t.p. reprint; BSbF séance 11 Mar 1887; Bot. Zeit. 26 Nov 1886), p. [i-iii], [1]-54. *Copies*: BR, FI, G, HH(2), NY. – Reprinted and to be cited from Ann. Soc. Linn. Bot. 32: 265-318. 1885 (1886). – Saint-Lager holds (1) that grammatical precision should be strictly observed in technical language and (2) that the "right" of priority does not imply acceptance of grammatically defective nomenclature. Some 18.000 botanical names (19% of the number of then current names of plants) should have to be corrected. He replies to Candolle (1883) who had declined to accept the Saint-Lager reforms.
Ref.: Malinvaud, E., Bull. Soc. bot. France 34(bibl.): 94-95. 1887.

10.075. *Vicissitudes onomastiques de la Globulaire vulgaire* ... Paris (Librairie J.B. Baillière et fils ...) 1889. Oct.
Publ.: 1889 (date cover-t.p. reprint; BSbF séance 12 Jul 1889; ÖbZ 1 Aug 1889; Bot. Centralbl. 6 Aug 1889; Bot. Zeit. 29 Sep 1889), p. [i, cover-t.p.], [1]-24, followed by *Note sur quelques plantes de la Haute-Maurienne*, p. [1]-12. *Copies*: FI, U (Biohist. Inst.). – *Globularia* paper reprinted and to be cited from Ann. Soc. bot. Lyon 16 (Mém): 233-256. 1889. The *Note* is also reprinted from the Ann. Soc. bot. Lyon.
Ref.: Malinvaud, E., Bull. Soc. bot. France 36(bibl.): 124-125. 1889.
Möbius, M., Bot. Centralbl. 41: 109-110. 1890 (rev. Globularia), 42: 26. 2 Apr 1890 (review of *Note*).

10.076. *La priorité des noms de plantes* ... Paris (Librairie J.-B. Baillière et fils ...) 1890. Oct.
Publ.: 1890 (t.p. reprint; copy sent by author rd. at Genève Mai 1891; Bot. Zeit. 29 Aug 1890), p. [1]-31. *Copies*: BR, G(2), H, HH, NY. – Reprinted and to be cited from Ann. Soc. bot. Lyon 16 (Mém.):257-285. 1889 (1890). – Even though Saint-Lager rejects the idea of stability of plant names as a prime consideration, he rejects the idea of unlimited priority. He too wants to start with Linnaeus, at least for the phanerogams. However, no bryologist would like to start again with Linnaeus: a different starting point will be necessary.
Ref.: Malinvaud, E., Bull. Soc. bot. France 37(bibl.): 94-95. 1890 (rev.).
Schiffner, V.F., Bot. Centralbl. 52: 219-220. 1892 (rev.).

10.077. *Considérations sur le polymorphisme de quelques espèces du genre Bupleurum* ... Paris (Librairie J.B. Baillière et fils ...) 1891. Oct.
Publ.: 1891 (date reprint; Bot. Centralbl. 15 Apr 1891; ÖbZ for Mar 1891; Bot. Zeit. 29 Mai 1891; BSbF séance 13 Mar 1891), p. [i], [1]-24. *Copy*: FI. – Reprinted and to be cited from Ann. Soc. bot. Lyon 17: 51-74. 1891.
Ref.: Schiffner, V.F., Bot. Centralbl. 52: 273-274. 1892.

10.078. *La guerre des Nymphes* suivie de la nouvelle incarnation de Buda ... Paris (Librairie J.B. Baillière et fils ...) 1891. Oct.
Publ.: 1891 (t.p. reprint; J. Bot. Sep 1891; BSbF séance 13 Nov 1891; Nat. Nov. Oct(1) 1893), p. [1]-39. *Copies*: BR, G(2), H, HH. – Reprinted and to be cited from Ann. Soc. bot. Lyon 17(Mém.): 183-219. 1891. – On *Nymphaea* vs. *Castalia*, and on the use of *Buda* Adanson and *Tisso* Adanson. – "Those who have read the author's former lucubrations on nomenclature will not expect to find here any practical solution of difficulties, and will therefore not be disappointed". (Britten 1891).
Ref.: Britten, J., J. Bot. 29: 288. 1891.

10.079. *Aire géographique de l'Arabis arenosa* et du Cirsium oleraceum ... Paris (Librairie J.B. Baillière et fils ...) 1892. Oct.
Publ.: 1892 (date reprint; Bot. Centralbl. 26 Jul 1892; Bot. Zeit. 1 Aug 1893; BSbF séance 12 Feb 1892; BSbF séance 8 Jul 1892), p. [i], [1]-15. *Copy*: FI. – Reprinted and to be cited from Ann. Soc. bot. Lyon 18: 29-43. 1892.
Ref.: Malinvaud, E., Bull. Soc. bot. France 39(bibl.): 111. 1892.
Schiffner, V.F., Bot. Centralbl. 53: 194. 1 Feb 1893.

10.080. *Note sur le Carex tenax* ... Paris (Librairie J.B. Baillière et fils ...) 1892. Oct.
Publ.: 1892 (t.p. reprint; Bot. Centralbl. 10 Aug 1892; ÖbZ for Jul 1892), p. [1]-12. *Copy*: FI. – Reprinted and to be cited from Ann. Soc. bot. Lyon 18: 45-54. 1892.
Ref.: Malinvaud, E., Bull. Soc. bot. France 39(bibl.): 112. 1892.
Schiffner, V.F., Bot. Centralbl. Beih. 3(7): 507. 1892.

10.081. *Un chapitre de grammaire* à l'usage des botanistes ... Paris (Librairie J.B. Baillière et fils ...) 1892. Oct.
Publ.: 1892 (t.p. reprint; ÖbZ for Sep 1892; BSbF séance 11 Nov 1892; Bot. Zeit. 16 Feb 1893), p. [1]-23. *Copies*: BR, FI, G(2). – On composite adjectives and connecting vowels. Reprinted and to be cited from Ann. Soc. bot. Lyon 18: 75-95. 1892.
Ref.: Malinvaud, E., Bull. Soc. bot. France 39(bibl.): 192-193. 1892 (rev.).

10.082. *Onothera ou Oenothera*. Les anes et le vin ... Paris (Librairie J.B. Baillière et fils ...) 1893. Oct.
Publ.: 1893 (t.p. reprint; BSbF séance 23 Jun 1893; BF 1 Jul 1893), p. [1]-22. *Copies*: BR, FI. – Reprinted and to be cited from Ann. Soc. bot. Lyon 18: 143-162. 1893. – On Onothera (Onothes: wild donkeys) and Oenothera (oinos: wine and thes: wild beast).

10.083. *Les nouvelles flores de France*. Étude bibliographique ... Paris (Librairie J.B. Baillière et fils ...) 1894. Oct.
Publ.: 1894 (t.p. reprint; read 24 Jul 1894; ÖbZ Nov-Dec 1894), p. [1]-31. *Copies*: BR, G, H, HH, M, NY. – Reprinted and to be cited from Ann. Soc. bot. Lyon 19(Proc. Verb.): 81-109. 1894.
Ref.: Malinvaud, E., Bull. Soc. bot. France 42(bibl.): 201-202. 1895.

10.084. *Les Gentianella du groupe Grandiflora* ... [Lyon (Association typographique) 1895]. Oct.
Publ.: 1895 (in journal; Bot. Centralbl. 20 Aug 1895; ÖbZ for Jun-Jul 1895; BSbF séance 24 Mai 1895), p. [1]-13. *Copies*: FI, G, MO, US. – Reprinted and to be cited from Ann. Soc. bot. Lyon 20(Mém.): 1-13. 1895. – In reprint bound with "L'appétence chimique des plantes et la concurrence vitale" on p. 15-32.
Ref.: Höck (Luckenwalde), Bot. Centralbl. 67: 83-85. 1896 (rev.).
Malinvaud, E., Bull. Soc. bot. France 42(bibl.): 373-374. 1895.

10.085. *La vigne du Mont Ida et le Vaccinium* ... Paris (Librairie J.B. Baillière et fils ...) 1896. Oct.
Publ.: Apr 1896 (t.p. reprint; ÖbZ for Apr 1896; Bot. Centralbl. 22 Mai 1896; Bot. Zeit. 16 Aug 1896), p. [1]-37. *Copies*: BR, FI. – Reprinted and to be cited from Ann. Soc. bot. Lyon 20(Mém.): 73-107. 1895 (1896). – *Vaccinium* in classical usage probably referred to Hyacinthus; V. vitis-idaea (red berries) did not grow on Mount Ida.
Ref.: Harms, H.A.T., Bot. Centralbl. 69: 24. 1896.
Malinvaud, E., Bull. Soc. bot. France 43(bibl.): 424-426. 1896.

10.086. *Genre grammatical des noms génériques.* Grandeur et décadence du Nard ... Paris (Librairie J.B. Baillière et fils ...) 1897. Oct.
Publ.: 1897 (BSbF séance 8 Jul 1898; ÖbZ for Jan-Feb 1898), p. [1]-28. *Copies*: BR. – Reprinted and to be cited from Ann. Soc. bot. Lyon 22 (Mém.): 35-60. – History of the name *Nardus*.
Ref.: Malinvaud, E., Bull. Soc. bot. France 44(bibl.): 599-600. 1897.

10.087. *Histoire de l'Abrotonum.* Signification de la desinence ex de quelques noms de plantes ... Paris (J.B. Baillière ...) 1900. Oct.
Publ.: 1900 (t.p.reprint; Bot. Centralbl. 17 Jul 1901), cover-t.p., p. [1]-48. *Copy*: BR. – Reprinted and to be cited from Ann. Soc. bot. Lyon 24 (Mém.): 131-147. 1899, 25 (Mém.): 1-6, 21-42. 1900. – *Abrotonum* instead of *Abrotanum*; ex in Ilex, Ulex et al. is probably the same as "ac", something sharp or pointed.
Ref.: Jackson, B.D., J. Bot. 39: 284. Aug 1901 (rev.).
Malinvaud, E., Bull. Soc. bot. France 47(bibl.): 394-396. 1900 (rev.).

10.088. *Le perfidie des synonymes* dévoilée à propos d'un Astragale ... Lyon (Association typographique ...) 1901. Oct.
Publ.: 1901 (date reprint; BSbF séance 27 Dec 1901), p. [i], [1]-15. *Copies*: FI, U (Bioh. inst.). – Reprinted and to be cited from Ann. Soc. bot. Linn. 26(Mém.): 113-127. 1901.
Ref.: Malinvaud, E., Bull. Soc. bot. France 48(bibl.): 453-455. 1901.

10.089. *La perfidie des homonymes* Aloès purgatif et bois d'Aloès aromatique ... [Lyon (Imprimerie A. Rey, ...) 1903]. Oct. (*Perfid. homon.*).
Publ.: 1903 (fide Roux & Meyran, 1914, p. 37), p. [1]-12. *Copies*: BR, M, NY. – Reprinted and to be cited from Ann. Soc. Linn. Lyon 49: 83-95. 1902 (1903).

Saint-Moulin, Vincentius Josephus de (1804-1837), Belgian (Hainaut) botanist; studied at the University of Gent; Dr. med. Univ. Liège 1831; practicing physician at Houdeng-Aimeries 1831-1837. (*Saint-Moulin*).

HERBARIUM and TYPES: Unknown.

BIBLIOGRAPHY and BIOGRAPHY: BM 4: 1785; CSP 5: 369-370; Herder p. 386; PR 7994-7995, ed. 1: 8951; Rehder 5: 754; Tucker 1: 619.

BIOFILE: Matthieu, E., Biogr. natl. Belgique 21: 116-117. 1911-1913.

10.090. *Commentatio botanico-oeconomica, de quibusdam arboribus in Belgio cultis,* ... Trajecti ad Rhenum [Utrecht] (apud Joh. Altheer ...) 1827. Oct. (*Comm. arb. Belg. cult.*).
Publ.: 1827 (answer to prize question, prize rd. 26 Mar 1827), p. [1]-116. *Copies*: BR, NY, U. – Also issued with same pagination in Ann. Acad. Rheno-Traiect. 1828.

10.091. Vincentii Josephi de Saint-Moulin, ex Houdeng-Aimeries, in Hannonia; ... *Commentatio, de monographia Quercus roboris,* in certamine literario civium Academiarum belgicarum, die viii mensis Februarii a. mdcccxxviii, ex sententia ordinis disciplinarum mathematicarum et physicarum in Academia lugduno-batava praemio ornata. Lugduni Batavorum [Leiden] (apud S. et J. Luchtmans, ...) 1828. Qu. (*Comm. monogr. Quercus rob.*).
Publ.: 1828 (prize awarded 8 Feb 1828), p. [i], [1]-73, [74, expl. pl.], *pl. 1-2* (uncol. copp.). *Copy*: NY.

Saint-Simon, Maximilien Henri, Marquis **de** (1720-1799), French-Dutch horticulturist; engaged in the French-Italian wars as aide-de-camp to the Prince de Conti; in 1758 he retired to the Netherlands, living at the estate of Nieuw Amelisweerd near Utrecht. (*Saint-Simon*).

HERBARIUM and TYPES: de Saint-Simon is not known to have made a herbarium. In his work he lists over two thousand named varieties of Hyacinths, a considerable number for the period. Many of these varieties were grown by George Voorhelm at Haarlem; some of them also at de Saint-Simon's private estate near Utrecht (Nieuw Amelisweerd), where a few are still found (naturalized) at present. The authors of TL-2 enjoy[ed] seeing these historical flowers each year during their early-spring joint work at Utrecht and Brussels.

BIBLIOGRAPHY and BIOGRAPHY: BM 4: 1790; HU 602, 2(1): cc; Jackson p. 134; Kew 4: 571; NI 1718; Plesch p. 395; PR 7996, ed. 1: 8955; Sotheby 669; TL-1/1142.

BIOFILE: Fries, Th.M., Bref Skrifv. Linné 1(3): 261. 1909.
Hulth, J.M. & A.H. Uggla, Bref Skrifv. Linné 2(2): 246. 1943.
Roberts, J. Roy. Hort. Soc. 60: 203. 1935.

10.092. *Des Jacintes*, de leur anatomie, reproduction et culture ... sic parvis componere magne solebam. À Amsterdam [de l'imprimerie de Claas Eel] 1768. Qu. (*Jacintes*).
Publ.: 1 Mai 1768 (p. [168]: printing finished 28 Mar 1768; the author sent a copy to C. Linnaeus on 1 Mai 1768), p. [i]-iv, [5]-164, [165-168], *pl. 1-10* (uncol. copp.), *catalogues des jacintes* connues en 1767: [1]-15. *Copies*: BR, MO, NY, U; IDC 5709. – The book was published anonymously. The 10 plates are by Jacob van der Schley (1715-1779); the title-engraving by J. van Hiltrop. – See HU 602 for an analytical collation.
Note(1): The anonymous *Traité sur la connaissance et la culture des Jacintes*, ... à Avignon (Chez Louis Chambeau, ...) 1759. Duod. (*Copy*: BR), p. [1]-156, was written by Jean Paul Rome d'Ardene, see PR ed. 1, no. 264.
Note(2): The Utrecht copy has bound in, between p. [i] and [iii], a "Discours adressée à l'Académie Royale des Sciences de Berlin, avec un livre intitulé: Des Jacinthes, de leur anatomie, réproduction et culture".

Saint-Yves, Alfred [Marie Augustine] (1855-1933), French botanist (agrostologist) and soldier; at the École polytechnique 1875-1879; following a military career 1879-1905, ultimately in the rank of "Commandant"; in retirement at Nice; in army service 1914-1918. (*Saint-Yves*).

HERBARIUM and TYPES: LAU, but genera *Festuca* (except those of the Alpes maritimes), *Avena* sect. *Avenastrum*, *Spartina* and *Brachypodium* at G (containing types in papers published in *Annuaire du Conservatoire* and *Candollea*). Some further material at C.

BIBLIOGRAPHY and BIOGRAPHY: AG 12(3): 33; Backer p. 662; Barnhart 3: 201 (b. 7 Mai 1855, d. 8 Oct 1933); Bossert p. 345, 379; IH 2: (files); Kew 4: 571-572; Langman p. 664-65; MW p. 423; Zander ed. 10, p. 722, ed. 11, p. 823.

BIOFILE: Cavillier, F., Candollea 6: 25-43. 1935 (obit., bibl.); Boissiera 5: 84. 1941. Litardière, R. de, Bull. Soc. bot. France 81: 46-53. *pl. 1.* 1934 (biogr., bibl.).
Wilczek, E., Bull. Soc. vaud. Sci. nat. 60: 21, 22, 26. 1937.

HANDWRITING: Candollea 33: 159-160. 1978.

EPONYMY: *Yvesia* A. Camus (1927).

10.093. *Les Festuca* de la section Eu-Festuca et leurs variations dans les Alpes maritimes ... Genève, Bale, Lyon (Georg & Co. ...) 1913. Oct.
Publ.: 1 Sep 1913 (p. 218), p. [i-iii], [1]-218, *pl. 1-7* with text. *Copies*: BR, G, MICH, NY, US. – Reprinted and to be cited from Annu. Cons. Jard. bot. Genève 17: 1-218. 1913.

10.094. *Contribution à l'étude des Festuca* (subgen. Eu-Festuca) *de l'Amérique du Nord et du Mexique* ... [Extrait de Candollea, ... ii, p. 229-316. 1925]. Oct.
Publ.: Jul 1925, p. [1]-88. *Copies* G, HH. – Reprinted and to be cited from Candollea 2: 229-316. 1925.

10.095. *Contribution à l'étude des Festuca* (subgen. Eu-Festuca) *de l'Amérique du Sud* ... [Extrait de Candollea, ... vol. iii, p. 151-315. 1927]. Oct.

Publ.: Aug 1927 (J. Briquet informed A.S. Hitchcock [fide copy US] that although dated "Mars 1927" the publication came out only in Aug 1927), p. [1]-165. *Copies*: MICH, US. – Reprinted and to be cited from Candollea 3: 151-315. 1927.

10.096. *Tentamen. Claves analyticae festucarum* veteris orbis (subgen. Eu-Festucarum) ad subspecies, multas varietates et nonnullas subvarietates usque ducentes ... Rennes (Imprimeries Oberthur) 1927. Oct. (*Tent. clav. anal. festuc.*).
Publ.: Sep-Dec 1927 (p. 4: Sep 1287), p. [i-iii], [1]-124. *Copies*: G, H, HH, M, MICH, NY. – Also issued in Revue bretonne de Botanique 2: 1-124. 1927.

10.097. *Contribution à l'étude des Festuca*. (subgen. Eu-Festuca) de *l'Orient* Asie et région méditerranéenne voisine ... [Extrait de Candollea, ... vol. iii, p. 321-466. 1928]. Oct.
Publ.: 31 Aug 1928 or shortly afterwards (date printing finished), p. [1]-146. *Copies*: G, HH, MICH. – Reprinted and to be cited from Candollea 3: 321-466. 1928.

10.098. *Contribution à l'étude des Festuca* (subgen. Eu-Festuca) de *l'Afrique australe* et de l'Océanie ... [Extrait de Candollea, ... vol. iv, p. 65-129. 1929]. Oct.
Publ.: 30 Nov 1929 or somewhat later (date printing finished), p. [1]-65. *Copies*: G, MICH. – Reprinted and to be cited from Candollea 4: 65-129. 1929.

10.099. *Festuca Font-Queri* St. Yv., sp. nova ... Barcinone [Barcelona] 1930. Oct. (*Festuca Font-Queri*).
Publ.: 10 Mar 1930, cover-t.p., p. [5]-7. *Copy*: G. – Preprinted from Cavanillisia 3(1-5): 10 Jul 1930.

10.100. *Festuca de la Nouvelle Zélande* (Herbier du professeur Wall) ... Candollea ... vol. iv, p. 293-307. Tiré à part. Genève Février 1931. Oct.
Publ.: Feb 1931, p. [1]-15. *Copy*: G. – Reprinted and to be cited from Candollea 4: 293-307. 1931.

10.101. *Contribution à l'étude des Avena sect. Avenastrum* (Eurasie et région méditerranéenne) ... Genève (Imprimerie Jent. S.A. ...) 1931. Oct.
Publ.: Jul 1931, cover-t.p., p. [1]-152, *pl. 4*. *Copies*: MICH, US. – Reprinted and to be cited from Candollea 4: 353-504, *pl. 4*. 1931.

10.102. *Festucae novae* et loci novi Festucarum jam cognitarum (subgen. Eu-Festuca) ... [Candollea, ... v, p. 101-141. Tiré à part. 1932]. Oct.
Publ.: Dec 1932, p. [1]-41. *Copy*: G. – Reprinted and to be cited from Candollea 5: 101-141. 1932.

10.103. *Monographia Spartinarum* ... Candollea ... vol. v, p. 19-100. Tiré à part Genève Décmbre 1932. Oct.
Publ.: Dec. 1932, cover-t.p., p. [1]-82, *pl. 1-10*. *Copies*: BR, G, H, HH, MICH, US.– Reprinted and to be cited from Candollea 5: 19-100, *pl. 1-10*. 1932.

10.104. *Contribution à l'étude des Brachypodium* (Europe et région méditerranéenne). Oeuvre posthume ... Candollea ... vol. v, p. 427-493. Genève Mai 1934. Oct.
Publ.: Mai 1934, cover-tp., p. [1]-67, *pl. 13-17*. *Copies*: G, US. – Reprinted and to be cited from Candollea 427-493, *pl. 13-17*. 1934.

Sakurai, Kyuichi (1889-1963), Japanese bryologist. (*Sakurai*).

HERBARIUM and TYPES: MAK (300.000); further material B (ferns, extant) and H (Bryo.). – 32 letters to V. Brotherus in H-UB.

BIBLIOGRAPHY and BIOGRAPHY: Barnhart 3: 202; IH 1 (ed. 6): 363, (ed. 7): 341, 2: (files); MW p. 424; MW suppl. p. 298-299; SBC p. 131.

BIOFILE: Anon., Rev. bryol. lichénol. 6: 211. 1933, 8: 118, 231. 1935, 9: 154. 1936, 21: 294-295. 1952, 22: 111. 1953.

Koponen, T., *in* G.C.S. Clarke, ed., Bryophyte systematics 166, 167. 1979.
Mizushima, U., Misc. bryol. lichenol. 3(4): 57-59. 1963 (obit. (japanese), portr., bibl.).
Sayre, G., Bryologist 80: 517. 1977 (coll.).

EPONYMY: *Sakuraia* V.F. Brotherus (1925).

Index to Titles

Each book treated in the foregoing text has been assigned a number printed in bold face; these are given in this index at the end of each entry. Because both short titles (the first significant words of the full title) and abbreviations of short titles are given, this index includes both. The initial definite and indefinite articles are omitted from titles. The unabbreviated short titles are alphabetized word by word, but the abbreviated ones are treated as if they are single words with the omission of punctuation and spaces. If a title is mentioned under two book numbers, both are given in this index.

Since the same title, at least in its short form, may have been used by two or more authors, each entry includes the name of the author in the form recommended at the end of the first paragraph of the section treating him and his books.

Titles of exsiccatae are included, printed in italics, and titles of journals taken from the surname of botanists are printed in capital letters. Both kinds of entry include, in parentheses, the surname of the person concerned preceded by the initial of his first name with the word "see" in italics. In addition there are two, less frequently encountered kinds of references which should be identified and explained: (1) It sometimes happens that in the treatment of one book by author "A", the title of a book by author "B" is mentioned; this situation is indicated in the title-index by parenthetically instructing the user to "see" the number assigned to "A's" book; (2) Similarly a book by author "B" may be mentioned under author "A" in paragraphs other than those describing his books; this is indicated by the index-notation "see author A'.

We have attempted to eliminate inconsistencies in the abbreviations used in the first volume; so, if there are differences between abbreviations in volume one and four, those used herein may be considered the standard.

Abb. Beschr. Deutschl. Giftgew. (Phoebus) 7875
Abb. Beschr. Deutschl. Giftgew. *(see* J. Ratzeburg)
Abb. deut. Holz-Art. (Reitter) 8971
Abbild. Beschr. Cact. (Pfeiff.) 7817
Abbild. Beschr. foss. Pfl. (Potonié) 8223
Abbild. Schwämme *(see* C. Persoon)
Abbildung der hundert deutschen wilden Holz-Arten (Reitter) 8971
Abbildungen und Beschreibung blühender Cacteen (Pfeiff.) 7817
Abbildungen und Beschreibungen fossiler Pflanzen-Reste (Potonié) 8223
Abbildung und Beschreibung seltener Gewächse *(see* G. Panzer)
Abb. Thiere Pfl. Syr. *(see* J. Russegger)
Abhandlung über die Bildung des Embryo in den Gräsern und Versuch einer Klassifikation dieser Familie (Raspail) 8638
Abhandlung über die essbaren Schwämme (Pers.) 7735
Abhandlung über Flechten (Reinke) 8933
Abh. essb. Schwämme (Pers.) 7734, 7735
Abh. Gebrauch Bäum. (H. Koch) 3787
About mushrooms (J. A. Palmer) 7238
Abrégé des plantes médicinales (Patout) 7496
Abrisse tabell. Fl. Schweiz. Kant. (Rhiner) 9128
Abrisse zur tabellarischen Flora der Schweizer Kantone (Rhiner) 9128
Abrisse zur zweiten tabellarischen Flora der Schweizerkantone (Rhiner) 9130
Abr. pl. méd. (Patout) 7496
Acc. excurs. Paris (D. Ross) 9562

Account of botanical excursions around the environs of Paris (D. Ross) 9562
Account of botanical rambles in the Pyrenees (D. Ross) 9563
Account of Corydalis persica (Prain) 8261a
Account of the genus Argemone (Prain) 8260
Account of the genus Dioscorea in the East (Prain) 8266
Account of the genus Sedum (Praeger) 8251
Account of the physical geography of Iceland (Rosenvinge) 9548
Account of the Sempervivum group (Praeger) 8252
Acc. rambles Pyrenees (D. Ross) 9563
Acc. Sedum (Praeger) 8251
Acc. Sempervivum (Praeger) 8252
Acc. tour Auvergne & Switzerland (D. Ross) 9561
Actual dates of publication of Rees's Cyclopaedia (Rees) 8756
Adatok a Marcgraviaceae és az Aroideae (Al. Richt.) 9179
Ad bryophytorum cisuralensium cognitionem additamentum (Podp.) 8078
Add. bryoph. fl. (Rodway) 9373
Add. Consp. fl. eur. (E. Roth) 9639
Additamenta ad Conspectum florae europaeae (E. Roth) 9639
Additamenta ad floram agri Nyssani (Petrović) 7791
Additamenta ad floram dalmaticam (Rohlena) 9448
Additions à la flore de la Meuse (Pierrot) 7925
Additions au catalogue des champignons de la Tunisie (Pat.) 7476
Additions to the bryophyte flora (Rodway) 9373
Additions to the phaenogamic flora of Mexico (B. L. Robinson) 9283
Addition to Michaux's Flora of North America (Raf.) 8551
Adjonct. Code Paris (Sacc.) 9933
Adjonctions au Code de Paris de 1867 (Sacc.) 9933
Adnotationes ad floram et faunam Hercegovinae (Pant.) 7290
Adnot. fl. faun. Herceg. (Pant.) 7290
Aërial algae, Botany of Iceland (see J. Petersen)
Aërial algae of Iceland (Rosenvinge) 9548
Äuss. Einfl. Pfl.-Leben (Regel) 8772
Äusseren Einflüsse auf das Pflanzenleben (Regel) 8772
Afb. farl. Snyltesv. (Rostr.) 9612
Afb. Fodergræs. (Rostr.) 9607
Afbildning og Beskrivelse af de farligste Snyltesvampe (Rostr.) 9612
Afbildning og Beskrivelse af de vigtigste Fodergræsser (Rostr.) 9607
Agaric. Michigan (see L. Pennington)
Agaric. off. (Rubel) 9743
Agaricum campestrem veneno in patria infamem Acta (Picco) 7896
Agave expatriata (J. Rose) 9510
Agave washingtonensis (J. Rose) 9508
Aggiunte alla flora trevigiana (Sacc.) 9897
Aggiunte e correzioni al Prodromo della flora cirenaica (Pamp.) 7273
A. Gray, Manual (see J. W. Robbins)
Agricultural varieties of the Cowpea (Piper) 7948
Agric. var. Cowpea (Piper) 7948
Agrostografia brasiliensis (Raddi) 8496
Agrostographia germanica (Rchb.) 8876, 8884
Agrostogr. bras. (Raddi) 8496
Agrostogr. germ. (Rchb.) 8884
Agrostotheca hungarica (see Jos. Sadler)
Aire géographique de l'Arabis arenosa (St. Lag.) 10079
Alabastra philipinensia (C. B. Robinson) 9296
Alaska Cryptogamic botany (see P. Saccardo)
Alchem. columb. (Rothm.) 9649
Alchemillae columbianae (Rothm.) 9649
Alc. nuov. ruggini (Sacc.) 9889
Alc. sp. Pero (Raddi) 8492
Alc. sp. rett. piant. bras. (Raddi) 8489
Alcune forme vegetali e varietà nuove raccolte nella valle Ossolana (S. Rossi) 9579
Alcune notizie sul Lago d'Arquà-Petrarca (see G. Paoletti)
Alcune nuove ruggini (Sacc.) 9889
Alcune specie di alghe del Mar di Sargasso (Picc.) 7908
Algae aquae dulcis exsiccatae praecipue scandinavicae (see P. Reinsch)
Algae aquae dulcis (see L. Rosenvinge)
Algae aquae dulcis (see P. G. Richter)
Algae marinae siccatae (see G. Rabenhorst)
Algae of Marion County (C. Palmer) 7231
Algae selectae siccatae (see G. Rabenhorst)
Algas de las Baleares (J. Rodríguez) 9363
Alg. Deutschl. (A. Roem.) 9396
Algen Deutschlands (A. Roem.) 9396
Algen Deutschlands (Rabenh.) 8434
Algen Europa's (see G. Rabenhorst)
Algen Europas (see P. G. Richter)
Algenfl. mittl. Franken (Reinsch) 8947
Algenflora der westlichen Ostsee deutschen Antheils (Reinke) 8930
Algenflora des mittleren Theiles von Franken (Reinsch) 8947
Algenfl. westl. Ostsee (Reinke) 8930

Algen Sachsen's (Rabenh.) 8438
Algen Sachsens (see G. Rabenhorst)
Algenvegetation des Trondjhemsfjordes (Printz) 8345
Algenveg. Trondjhemsfj. (Printz) 8345
Algeveg. Grønl. Kyst. (Rosenvinge) 9544
Alghe dell'isola del Giglio *(see* A. Piccone)
Alghe del viaggio di circumnavigazione della Vettor Pisani (Picc.) 7903
Alghe fossili mioceniche di Cirenaica (Raineri) 8597
Alghe sifonee fossili della Libia (Raineri) 8596
Alg. Hydroz. Silur (Rothpletz) 9655
Alg. Normandie (Roberge) 9274a
Algues de France (see C. Roumeguère)
Algues de la Normandie (Roberge) 9274a
Algues de la Normandie (see M. Roberge)
Algues marines (Perrot) 7714
Algues *(see* C. Roumeguère)
Alg. viagg. Vettor Pisani (Picc.) 7903
Alien plants and apophytes of Greenland (M. Porsild) 8177
Alien pl. Greenland (M. Porsild) 8177
Aliquot fungi novi vel critici Galliae (Rolland) 9476
A.L. Jussieum, epistola (Ruiz) 9776
Allgemeine deutsche naturhistorische Zeitung *(see* C. Sachse)
Allgemeine Morphologie der Pflanzen (Pax) 7546
Allg. Morph. Pfl. (Pax) 7546
Allii species Asiae centralis (Regel) 8807
Allior. monogr. (Regel) 8801
Alliorum adhuc cognitorum monographia (Regel) 8801
Almagesti botanici mantissa (Pluk.) 8064
Almagestum botanicum (Pluk.) 8064
Aloes of South Africa *(see* G. Reynolds)
Aloes of tropical Africa and Madagascar, Mbabane, Swaziland *(see* G. Reynolds)
Alpenpflanzen (Rambert) 8604
Alpes suisses (Rambert) 8604
Alphabetical list of indigenous and exotic plants growing on the island of St. Helena (S. Pritch.) 8350
Alphabetical list of plants in the Royal Botanic Gardens. Peradeniya (T. Parsons) 7417
Alphabetical list of plants of St. Helena (Roxb.) 9723
Alphabetical list of plants *(see* W. Roxburgh)
Alphabetical register...of phanerogamic plants and ferns (Pritz.) 8354
Alphabetisches Verzeichniss Algen und Bacillarien Sachsens (Rabenh.) 8438
Alph. list pl. Peradeniya (T. Parsons) 7417
Alph. list pl. St. Helena (S. Pritch.) 8350
Alph. Verz. Alg. Sachsen (Rabenh.) 8438

Alpine vegetation of the southern Rockies (Penland) 7633
Alp. suiss. (Rambert) 8604
Alsogr. amer. (Raf.) 8588
Alsographia americana (Raf.) 8588
Amallospora (Penz.) 7651
Amaltheum botanicum (Pluk.) 8064
Amboinsche kruid-boek (Rumph.) 9784
Amendments to the International rules of nomenclature (Rehd.) 8818
Amendments to the Paris Code of Botanical Nomenclature (Rendle) 9065
Amendm. Int. rules nomencl. (Rehd.) 8818
American manual of the grape vines (Raf.) 8580
American manual of the mulberry trees (Raf.) 8590
American Naturalist (see A. Packard, D. Penhallow)
American species of Canavalia and Wenderothia (Piper) 7953
American species of Limnorchis and Piperia (Rydb.) 9858
Amerikanischen Reben (Sahut) 10012
Amer. man. grape vines (Raf.) 8580
Amer. man. mulberry trees (Raf.) 8590
Amoen. bot. dresd. (Rchb.) 8872
Amoenitates botanicae dresdenses (Rchb.) 8872
Analecta boliviana (J. Rémy) 8976
Anales de Ciencias, Agricultura, Comércio *(see* R. Sagra)
Anales del Museo nacional *(see* R. Philippi)
Anal. fl. Bruxell. (Piré) 7960
Anal. key West Coast bot. (Rattan) 8641
Anal. nat. (Raf.) 8560
Anal. Pfl.-Schlüss. (Peterm.) 7756
Analyse botanique des embryons endorhizes (Rich.) 9155
Analyse de la nature (Raf.) 8560
Analyse der Frucht (Rich.) 9154
Analyse des familles et des genres de la flore Bruxelloise (Piré) 7960
Analytical key to West Coast botany (Rattan) 8641
Analytischer Pflanzenschlüssel (Peterm.) 7756
Anat. comp. Gentian. (Perrot) 7713
Anat. oecon. pl. (Royen) 9726
Anatomia comparata delle foglie delle Iridee (H. Ross) 9564
Anatomia fisiologica delle Zygophyllaceae (E. Pantan.) 7287
Anatomical characters of the seeds of Leguminosae *(see* L. Pammel)
Anatomie comparée des Gentianacées (Perrot) 7713
Anatomie der Pflanzen (Rudolphi) 9757
Anatomisch-systematische Untersuchung

1085

des Blattes der Melastomaceen (Paléz.) 7205
Anatomisch systematische Untersuchung des Blattes der Melastomaceen (Pfl.) 7834
Anatomisch-systematische Untersuchung von Blatt und Axe der Acalypheen (Rittershausen) 9259
Anatomiske studier over Eriocaulaceerne (Poulsen) 8239
Anat. Pfl. (Rudolphi) 9757
Anat. stud. Eriocaul. (Poulsen) 8239
Anat.-syst. Unters. Acalyph. (J. Ritter) 9259
Anat. syst. Unters. Blatt. Melast. (Paléz.) 7205
Anat.-syst. Unters. Blatt. Melast. (Pfl.) 7834
Anat. Torfm. (Russow) 9837
Anemonarum revisio (Pritz.) 8351
Anem. revis. (Pritz.) 8351
Anfangsgründe der botanischen Terminologie (Plenck) 8058
Anhang zur Monographie der Gattung Serjania (Radlk.) 8510
Anh. Monogr. Serjania (Radlk.) 8510
Animadversiones de plantis vivis (Regel) 8796
Animadversiones et dilucidationes circa varias fungorum species (Pers.) 7724
Animadversiones in structuram ac figuram foliorum in plantis (J. E. Pohl) 8105
Animadversiones quaedam ad peloriarum indolem definiendam spectantes (Ratzeb.) 8645
Animadv. pelor. spect. (Ratzeb.) 8645
Animadv. struct. fig. fol. (J. E. Pohl) 8105
Anleit. Beobacht. Pfl.-Welt. (F. Rosen) 9515
Anleit. Best. Gräs. (Polscher) 8146
Anleit. Einsamm. *(see* G. Rabenhorst)
Anleitung den Harz zu bereisen *(see* J. Rueling)
Anleitung zur Beobachtung der Pflanzenwelt (F. Rosen) 9515
Anleitung zur Bestimmung der Gräser (Polscher) 8146
Anleitung zur Naturgeschichte des Pflanzenreichs (Rchb.) 8877
Annales de Chimie et de Physique *(see* A. Parmentier)
Annali di Botanica *(see* P. Pirotta)
Annals of nature (Raf.) 8573
Annals of the Bolus Herbarium *(see* H. Pearson)
Annals r. Bot. Gard. Peradeniya *(see* T. Petch)
Ann. Blumist. (Reider) 8907
Ann. nat. (Raf.) 8573
Annotated catalogue of the flowering plants of Missouri (E. J. Palmer) 7233
Annotated catalogue of the vascular plants of Benton County (St. John) 10059
Annotated flora of the Chicago area (Pepoon) 7660
Annotations aux Plantae europaeae (Rouy) 9705
Annot. fl. Chicago (Pepoon) 7660
Ann. Rep. State Bot. (Peck) 7595
Annual report, Report of the Botanist (Peck) 7595
Annual wholesale catalogue of American trees, shrubs, plants and seeds (W. Prince) 8324
Antillarum insularum natur. icones *(see* C. Plumier)
Antotrofia (A. Piccioli) 7884
Anträge Nomenclatur der Sphagna (Röll) 9391
Antr. Nomencl. Sphagna (Röll) 9391
Anweis. Pfl. sammeln (Roth) 9632
Anweisung Pflanzen zu sammeln (Roth) 9632
Aperçu d'un voyage dans l'intérieur du Brésil (St. Hil.) 10031
Aperçu phytostat. Haute-Saône (Renauld) 8981
Aperçu phytostatique sur le département de la Haute-Saône (Renauld) 8981
Aperçu sur la distribution des espèces végétales dans les Alpes de la Savoie (E. Perrier) 7698
Apogamie Parthenogenesis bei Pflanzen (O. Rosenberg) 9520
Appendix to Captain Parry's Journal of a second voyage (W. Parry) 7412
Appendix to the Midland flora (Purt.) 8407
App. Parry J. sec. voy. (W. Parry) 7412
Appunti fl. micol. Mt. Generoso (Penz.) 7644
Appunti sulla flora micologia del Monte Generoso (Penz.) 7644
Apuntes conoc. esp. arb.-forest. Filip. (Puigd.) 8382
Apuntes para el mejor conocimiento especies arboreo- forestales de Filipinas (Puigd.) 8382
Apuntes sobre la familia de las Nictagíneas (D. Parodi) 7392
Apuntes sobre la vejatacion en la boca del Rio Palena (Reiche) 8852
Arabis (Rollins) 9482
Arb. arbust. Venez. (Pitt.) 7984
Arb. bot. Gart. Breslau (Prantl) 8272
Arbeiten aus dem k. botanischen Garten zu Breslau *(see* K. Prantl)
Arbeiten aus dem Königl. Botanischen Garten zu Breslau (Prantl) 8272
Arbeiten des botanischen Institute *(see* J. Sachs)

Arb. legum. (Pitt.) 7988
Arboles y arbustos del orden de las leguminosas (Pitt.) 7988
Arboles y arbustos de Venezuela (Pitt.) 7984
Arbor. Barres (Pardé) 7323
Arborescent indigenous legumes of Hawaii (Rock) 9330
Arboresc. legum. Hawaii (Rock) 9330
Arboretum muscaviense (Petzold) 7801
Arboretum national des Barres (Pardé) 7323
Arbor. muscav. (Petzold) 7801
Arbor. Vallombrosa (L. Piccioli) 7894
Arb. Wirk. Pfl. (Radlk.) 8523
Arch. Bot. (Roem.) 9402
Archives de botanique *(see* H. Perrier)
Archiv für die Botanik (Roem.) 9402
Archiv für Protistenkunde *(see* A. Pascher)
Arctic searching expedition (Richardson) 9170
Arctostaphylos (Parry) 7405
Arct. search. exped. (Richardson) 9170
Aristoloches (L. Planch.) 8027
Aroideae maximilianae (Peyr.) 7808
Aroid. maximil. (Peyr.) 7808
Artbastarde bei Pflanzen (Renner) 9076
Artbast. Pfl. (Renner) 9076
Art de l'indigotier (Perrottet) 7719
Artentypen und Formenreihen bei den Torfmoosen (Röll) 9383
Artenzahl als pflanzengeographischer Charakter (Palmgr.) 7245
Art indigotier (Perrottet) 7719
Arve in der Schweiz (Rikli) 9230
Arve Schweiz (Rikli) 9230
Ascension du Pichincha (J. Rémy) 8978
Ascomyceten (Rehm) 8821
Ascomycetes exsiccatae (see H. Rehm)
Ascomycetes fuegani (Rehm) 8826
Ascomycetes lojkani (Rehm) 8823
Ascomycetes philippenses (Rehm) 8830
Ascomyc. fueg. (Rehm) 8826
Ascomyc. lojk. (Rehm) 8823
As Heveas ou Seringueiras (Barb. Rodr.) 9356
Aspidium libanoticum (Rosenstock) 9532
Assimilationsorgane der Asparageen (Reinke) 8935
Asterodon (Pat.) 7469
Astragalus and its segregates (Rydb.) 9860
Astralagus alopecuroides (Pamp.) 7261
Atlante de botanica popolare *(see* G. Pasquale)
Atlante di piante medicinali (F. Pasq.) 7425
Atlante piante med. (F. Pasq.) 7425
Atlantic journal (Raf.) 8582
Atlas champ. (Richon) 9176
Atlas champ. Eur. (Pilát) 7935

Atlas champ. France (Rolland) 9479
Atlas de champignons de l'Europe (Pilát) 7935
Atlas der Alpenflora *(see* E. Palla)
Atlas der Pflanzengeographie (L. Rudolph) 9754
Atlas des champignons (Richon) 9176
Atlas des champignons de France (Rolland) 9479
Atlas des champignons *(see* E. Roze)
Atlas deut. Meeresalg. (Reinke) 8931
Atlas deutscher Meeresalgen (Reinke) 8931
Atlas of the trichomes of Lesquerella *(see* R. Rollins)
Atlas Pfl.-Geogr. (L. Rudolph) 9754
Atlas zur Beschreibung der Reise. (Pohl) 8104
Atl. j. (Raf.) 8582
Auctuarium ad catalogum plantarum horti botanici Florentini (G. Piccioli) 7888
Aufeinanderfolge org. Gest. (Ritgen) 9251
Aufg. region. Moorforsch. (L. Post) 8189b
Aufzählung der in der Umgebung von Linz wildwachsenden (Rauscher) 8669
Aufzählung der von G. Woronoff gesammelten Cirsien (Petr.) 7780
Aufz. Linz Gefäfass-Pfl. (Rauscher) 8669
Aus den Llanos *(see* J. Sachs)
Aus der Pflanzenwelt der Burgenländer Berge (Jul. Röm) 9411
Aussereur. Laubm. (G. Roth) 9643
Aussereuropäischen Laubmoose (G. Roth) 9643
Australian freshwater phytoplankton. Protococcoideae (Playf.) 8045
Austral. Orchid. (R. S. Rogers) 9438
Autik. bot. (Raf.) 8592
Autikon botanikon (Raf.) 8592
Autonom. lich. (O. J. Rich.) 9164
Azolla caroliniana (Sacc.) 9914

Bacillariák vagy Kovamoszatok (Pant.) 7294
Bacillariales *(see* J. Pavillard)
Bacillarien des Balatonsees (Pant.) 7292
Bacillarien Sachsens (see G. Rabenhorst)
Bacill. Balatonsees (Pant.) 7292
Bacill. vagy Kovamoszatok (Pant.) 7294
Bahama Fl. *(see* L. Riddle)
Balano (Petzh.) 7798
Balano et Calamosyringe (Petzh.) 7798
Balaton kovamoszatai (Pant.) 7292
Bambous (Rivière) 9264
Bambuseae (Rupr.) 9793
Bambuseas monographice exponit (Rupr.) 9793
Bartonia *(see* F. Pennell)
Basidiomycetes eubasidii in Rio Grande do Sul *(see* J. Rick)

Basidiomycetes from Juan Fernandez *(see* L. Romell)
Bau Holz. (Rossmann) 9591
Bayerischen Arten, Formen und Bastarde der Gattung Potentilla (Poeverl.) 8096
Bayer. Potentilla (Poeverl.) 8096
Beantwortung Preissfrage (Roth) 9634
Beantw. Preissfr. (Roth) 9634
Beauties Flora *(see* C. Pope)
Bedeut. Griech. Pfl.-Namen (W. Petzold) 7803
Bedeutung des Griechischen für das Verständnis der Pflanzennamen (W. Petzold) 7803
Befrucht. Phan. (Radlk.) 8504
Befrucht.-Proc. Pfl.-Reich. (Radlk.) 8505
Befruchtung der Phanerogamen (Radlk.) 8504
Befruchtungsprocess im Pflanzenreiche (Radlk.) 8505
Befr. Vorg. Basidiomyc. (Reess) 8761
Begrip leven (Rauwenh.) 8676
Beiträge zu einer Orchideenkunde Central-Amerika's (Rchb. fil.) 8896
Beiträge zur Anatomie der Caesalpiniaceen (Rhein) 9125
Beiträge zur Anatomie und Systematik der Artocarpeen und Concephaleen (Renner) 9075
Beiträge zur Ascomyceten-Flora der deutschen Alpen (Rehm) 8822
Beiträge zur Ascomyceten-Flora der Voralpen und Alpen (Rehm) 8827
Beiträge zur Biologie der Laubmoosrhizoiden (H. Paul) 7517
Beiträge zur Biologie der Pflanzen *(see* F. Rosen)
Beiträge zur chemischen Kenntniss der weissen Mistel (Reinsch) 8945
Beiträge zur Entwickelungsgeschichte der Fruchtkörper einiger Gastromyceten (Rabinowitsch) 8451
Beiträge zur Entwicklungsgeschichte der Fruchtkörper einiger Gastromyceten (Rehsteiner) 8832
Beiträge zur Flechten-Flora des Allgäu (Rehm) 8820
Beiträge zur Flora der hawaiischen Inseln (Reichardt) 8848
Beiträge zur Flora der oberen Kreide Quedlinburgs (P. B. Richt.) 9189
Beiträge zur Flora der Umgegend von Colmar (Rosshirt) 9571
Beiträge zur Flora der unteren Kreide Quedlinburgs (P. B. Richt.) 9188
Beiträge zur Flora des Grossherzogthums Posen (Ritschl) 9253
Beiträge zur Flora von Bolivia (Perkins) 7686

Beiträge zur Flora von Croatien und Dalmatien (Posch.) 8186
Beiträge zur Flora von Oesterreich (Rech.) 8724
Beiträge zur Gebirgskunde Brasiliens (Pohl) 8104
Beiträge zur Kenntniss der fossilen Bacillarien Ungarns (Pant.) 7291
Beiträge zur Kenntniss der Laboulbenien (Peyr.) 7806
Beiträge zur Kenntniss der Ophioglosseen (Rostovzev) 9604
Beiträge zur Kenntniss der Periodizität und der geographischen Verbreitung der Algen Badens (Rabanus) 8431
Beiträge zur Kenntniss der Phyllomorphose (Rossmann) 9588
Beiträge zur Kenntniss der schweizerischen Erigeron-Arten (Rikli) 9226
Beiträge zur Kenntniss der Stammanatomie von Phytocrene macrophylla (B. L. Robinson) 9280
Beiträge zur Kenntniss der Tange (Rost.) 9596
Beiträge zur Kenntniss der Torfmoose (Russow) 9832
Beiträge zur Kenntniss der Ustilagineen (Rawitscher) 8692
Beiträge zur Kenntniss der Vegetationsorgane der Lebermoose (Ruge) 9766
Beiträge zur Kenntniss der Vegetationsverhältnisse Krains (Paulin) 7524
Beiträge zur Kenntniss der Wasserhahnenfüsse (Rossmann) 9586
Beiträge zur Kenntniss einiger Hydrocharideen (Rohrb.) 9461
Beiträge zur Kenntniss von Vegetation und Flora der kanarischen Inseln (Pit.) 7972
Beiträge zur Laubmoosflora Deutschlands und der Schweiz (Röll) 9380
Beiträge zur Laubmoos- und Torfmoosflora von Oesterreich (Röll) 9389
Beiträge zur Morphologie der Meeres-Algen (Pringsh.) 8332
Beiträge zur Morphologie und Biologie der Holzgewächse *(see* W. Rauh)
Beiträge zur Morphologie und Systematik der Cyperaceen (Pax) 7544
Beiträge zur näheren Kenntniss und Verbreitung der Algen (Rabenh.) 8443
Beiträge zur Orchideenkunde (Rchb. fil.) 8897
Beiträge zur Pflanzenkunde der Vorwelt (Rhode) 9132
Beiträge zur Pilzflora Serbiens *(see* N. Ranojević)
Beiträge zur Pilzflora von Mähren (Petr.) 7782

Beiträge zur Systematik der Caryophyllinen (Rohrb.) 9458
Beiträge zur Systematik der Orchideen (Pfitz.) 7832
Beiträge zur systematischen Pflanzenkunde (Rchb. fil.) 8898
Beiträge zur vergleichenden Anatomie einiger Genisteen Gattungen (Rauth) 8670
Beiträge zur vergleichenden Anatomie und Morphologie der Sphacelariaceen (Reinke) 8932
Beiträge zur Versteinerungskunde (Rossmässler) 9580
Beitrag zur Flechtenkunde Niederösterreichs (Poetsch) 8092
Beitrag zur Flora Albaniens (Rohlena) 9450
Beitrag zur Flora der Hercegovina (Rohlena) 9451
Beitrag zur Flora des Gebirges Šar planina (Rohlena) 9452
Beitrag zur Flora des ungarischen Tieflandes (Pokorny) 8127
Beitrag zur Flora Ehstlands (Pahnsch) 7186
Beitrag zur Flora von Alsen (H. Petersen) 7760
Beitrag zur Flora von Niederösterreich (Reichardt) 8840
Beitrag zur Flora von Persien (Rech.) 8723
Beitrag zur Flora von Posen (Ritschl) 9254
Beitrag zur Keimung von Uredineen- und Erysipheen-Sporen (Rauch) 8651
Beitrag zur Kenntnis des Ovulums (Pax) 7542
Beitrag zur Kenntniss der Cirsiens Steiermarks (Reichardt) 8842
Beitrag zur Kenntniss der Gattung Rumex (Rech.) 8725
Beitrag zur Kenntniss der Laubmoose und Flechten (Poetsch) 8093
Beitrag zur Kenntniss der Spreitenformen (Rossmann) 9590
Beitrag zur Kenntniss einiger Tragopogon Arten (C. Regel) 8770
Beitrag zur Phanerogamenflora der hawaiischen Inseln (Reichardt) 8849
Beitr. Anat. Caesalp. (Rhein) 9125
Beitr. Anat. Genist. Gatt. (Rauth) 8670
Beitr. Anat. Morph. Sphacelar. (Reinke) 8932
Beitr. Anat. Syst. Artocarp. (Renner) 9075
Beitr. Biol. Laubm.-Rhiz. (H. Paul) 7517
Beitr. chem. Kenntn. Mistel (Reinsch) 8945
Beitr. Fl. Aegypt. (see E. Rüppell)
Beitr. Fl. Alsen (H. Petersen) 7760
Beitr. Fl. Bolivia (Perkins) 7686
Beitr. Fl. Colmar (Rosshirt) 9571
Beitr. Fl. Croat. (Posch.) 8186
Beitr. Fl. Ehstlands (Pahnsch) 7186
Beitr. Fl. ober. Kreide Quedlinb. (P. B. Richt.) 9189
Beitr. Fl. Posen (Ritschl) 9253, 9254
Beitr. Fl. unter. Kreide Quedlinb. (P. B. Richt.) 9188
Beitr. foss. Bacill. Ung. (Pant.) 7291
Beitr. Keim. Ured.-Sp. (Rauch) 8651
Beitr. Kenntn. Alg. (Rabenh.) 8443
Beitr. Kenntn. Ophiogl. (Rostovzev) 9604
Beitr. Kenntn. Ovul. (Pax) 7542
Beitr. Kenntn. Phyllomorph. (Rossmann) 9588
Beitr. Kenntn. Tange (Rost.) 9596
Beitr. Kenntn. Wasserhahnenf. (Rossmann) 9586
Beitr. Morph. Syst. Cyper. (Pax) 7544
Beitr. Orchid.-K. (Rchb. fil.) 8897
Beitr. Orchid.-K. C. Amer. (Rchb. fil.) 8896
Beitr. Period. Alg. Bad. (Rabanus) 8431
Beitr. Pflanzenk. Vorw. (Rhode) 9132
Beitr. syst. Pflanzenk. (Rchb. fil.) 8898
Beitr. Torfm. (Russow) 9832
Beitr. Ustilag. (Rawitscher) 8692
Beitr. Veg.-Org. Leberm. (Ruge) 9766
Beitr. Veg.-Verh. Krains (Paulin) 7524
Belladone (Pauquy) 7532
Bem. Algen-Stämme (Rupr.) 9800
Bembridge fl. (E. Reid) 8905
Bembridge flora (E. Reid) 8905
Bemerk. Bl. Balsam. (K. Presl) 8305
Bemerk. Botrychium (Rupr.) 9808
Bemerk. crypt. Wassergew. (Roth) 9630
Bemerkungen auf einer Reise in die südlichen Statthalterschaften des russischen Reichs (Pall.) 7228
Bemerkungen über das Studium der cryptogamischen Wassergewächse (Roth) 9630
Bemerkungen über den Bau der Blumen der Balsamineen (K. Presl) 8305
Bemerkungen über den Bau und das Wachsthum einiger grossen Algen-Stämme (Rupr.) 9800
Bemerkungen über die Gattungen Betula und Alnus (Regel) 8794
Bemerkungen über die Gruppe der Gattung Amaranthus (Regel) 8773
Bemerkungen über einige Arten der Gattung Botrychium (Rupr.) 9808
Bemerkungen zu Herrn Prof. Agardh's Abhandlung (Roep.) 9421
Bemerkungen zur Encyclopaedia of plants of Loudon (Raf.) 8584
Bengal pl. (Prain) 8262
Bengal plants (Prain) 8262

Beobacht. Pilzveg. Hamburg (Sadebeck) 9964
Beobachtungen über Bau und Entwicklung epiphytischer Orchideen (Pfitz.) 7827
Beobachtungen über die Diatomaceen der Umgebung von Jena (Prollius) 8365
Beobachtungen über die Samenbildung ohne Befruchtung (Ramisch) 8615
Beobachtungen über die Witterungs- und Vegetationsverhältnisse (Sachse) 9960
Beobachtungen über einige neue Saprolegnieae (Reinsch) 8950
Beobachtungen ueber Viola epipsila (Regel) 8785
Beobachtungen und Untersuchungen über die Pilzvegetation in der Umgegend von Hamburg (Sadebeck) 9964
Beob. Diatom. Jena (Prollius) 8365
Ber. Fl.-Geb. Eulengeb. (W. Roth) 9646
Berichte über das Floren-Gebiet des Eulengebirges (W. Roth) 9646
Berichte über die biologisch-geographischen Untersuchungen in den Kaukasusländern (Radde) 8477
Berichte über Reisen im Süden von Ost-Sibirien (Radde) 8476
Bericht über die erste Blumen- und Pflanzen-Ausstellung (Regel) 8781
Ber. Reisen Ost-Siberien (Radde) 8476
Ber. Unters. Kaukasusländ. (Radde) 8476
Beschr. Abb. Holz-Art. (Reitter) 8972
Beschr. Art. Asplenium (K. Presl) 8305a
Beschr. Calceol. (Reider) 8910
Beschreibung aller bekannten Pelargonien (Reider) 8908
Beschreibung neuer Bacillarien (Pant.) 7293
Beschreibung neuer Hymenophyllaceae (Rosenstock) 9533
Beschreibung seltner und neuer vorzüglicher Blumen und Ziergewächse (Reider) 8907
Beschreibung und Abbildung der in Deutschland seltener wildwachsenden Holz-Arten (Reitter) 8972
Beschreibung und Kultur der Calceolarien (Reider) 8910
Beschreibung und Synonymik der in deutschen Gärten lebend vorkommenden Cacteen (Pfeiff.) 7816
Beschreibung zweier neuen böhmischen Arten der Gattung Asplenium (K. Presl) 8305a
Beschreibuung der am häufigsten wildwachsenden und kultivirten phanerogamen Gewächse (Rückert) 9761
Beschr. neu. Bacill. (Pant.) 7293
Beschr. Pelargon. (Reider) 8908
Beschr. Synon. Cact. (Pfeiff.) 7816

Bestimmungstabelle für die Thymus-Arten des Deutschen Reiches (Ronniger) 9492
Bestimmungstabellen zur Flora von Aegypten (Ramis) 8613
Best.-Tabell. Fl. Aegypt. (Ramis) 8613
Best.-Tab. Thymus (Ronniger) 9492
Betracht. Leitb. Grundgewebe (Russow) 9835
Betrachtungen über das Leitbündel und Grundgewebe (Russow) 9835
Beytrag zur Geschichte des ostindischen Brodbaums (Panz.) 7301
Beytr. Bot. (Roth) 9626
Beytr. Gesch. Brodbaums (Panz.) 7301
Beyträge zur Botanik (Roth) 9626
Bibl. cult. trees (Rehd.) 8819
Bibliographia botanica fenniae (Saelan) 9982
Bibliographie des diatoméees (see J. Pelletan)
Bibliography of cultivated trees and shrubs (Rehd.) 8819
Bibliotheca americana nova (see O. Rich)
Bibliotheca americana vetus (see O. Rich)
Bibliotheca riviniana (Riv.) 9273
Bibl. rivin. (Riv.) 9273
Bidrag til Danmarks svampflora (O. Rostr.) 9621
Bidrag till kännedom om Ålands vegetation och flora (Palmgr.) 7239
Bidrag til Polysiphonia's Morphologi (Rosenvinge) 9536
Bignoniacées de la région malgache (H. Perrier) 7708
Bignon. malg. (H. Perrier) 7708
Bihang till Flora oeconomica Sveciae (Retz.) 9094
Bih. Fl. oecon. Svec. (Retz.) 9094
Bijdrage tot de kennis van Dracaena draco L. (Rauwenh.) 8672
Bijdr. Dracaena draco (Rauwenh.) 8672
Bilder ur Nordens Flora (Palmstr.) 7247
Billeder of Nordens Flora (Palmstr.) 7247
Biogéographie des plantes de Madagascar (H. Perrier) 7707
Biogéogr. Madag. (H. Perrier) 7707
Biogr. ind. Brit. Irish Bot. (see A. Rendle)
Biological investigation of the Athabaska-Mackenzie region (Preble) 8280
Bladmossfloran i sydvästra Jämtland (Perss.) 7738
Blätterpilze Deutschl. (Ricken) 9192
Blätterpilze Deutschlands (Ricken) 9192
Blattstruct. Cecropia (Al. Richt.) 9178
Bl.-Bau Epipogium gmel. (Rohrb.) 9457
Blick in die Geschichte der botanischen Morphologie (Potonié) 8222
Bloemk. beschr. (Reider) 8909
Bloemkundige beschrijving (Reider) 8909

Bloemnam. Zuidnederl. dial. (Pauwels) 7534
Blomsterløse Planter (Rostr.) 9609
Blomsterl. Pl. (Rostr.) 9609
Bl.-Pfl. Straubing (Raab) 8429
Blütenpflanzen von Straubing und Umgebung (Raab) 8429
Boleti of the United States (Peck) 7598
Book of useful plants (J. E. Rogers) 9437
Book usef. pl. (J. E. Rogers) 9437
Boscia (Pestal.) 7743
Bot. Abh. Beobacht. (Roth) 9627
Bot. agric. méd. (Rodet) 9338
Botanica (Parl.) 7352
Botánica criptogámica (Riv. Mat.) 9261
Botánica descriptiva (Puerta) 8379
Botanica ea nomenclatura indigena memoria (Barb. Rodr.) 9360
Botanica farmaceutica (Pollacci) 8135
Botanica in Italia (Sacc.) 9921
Botanical Appendix (Richardson) 9169
Botanical collecting in the tropical Andes (Rusby) 9815
Botanical collector's guide (Penh.) 7630
Botanical description of a new species of Swietenia (Roxb.) 9719
Botanical description of British plants (Purt.) 8407
Botanical dictionary (Rich.) 9153
Botanical Dictionary (see L. Richard)
Botanical exploration of the North Shore of the Gulf of St. Lawrence (St. John) 10058
Botanical grammar and dictionary (Rich.) 9153
Botanical investigations in Wood Buffalo Park (Raup) 8665
Botanical magazine (see D. Prain)
Botanical names of the U.S. pharmacopoeia (Rusby) 9816
Botanical observations in Western Wyoming (Parry) 7404
Botanical studies in the Black Rock Forest (Raup) 8668
Botanical tour in the highlands of Perthshire (Pamplin) 7275
Botanicon americanum (see C. Plumier)
Botanikens udstrakte Nytee (Rottb.) 9664
Botanik, Wiss. Erg. deut. Z.-Afr. Exped. (see J. Perkins)
Botanique (Regnault) 8810
Botanique agricole et médicale (Rodet) 9338
Botanique cryptogamique (Payer) 7560
Botanique de J.J. Rousseau (Rousseau) 9687, 9688
Botanique médicale (A. Rich.) 9141 9688
Botanique médicale (A. Rich.) 9141
Botanische Abhandlungen und Beobachtungen (Roth) 9627

Botanische Bemerkungen (K. Presl) 8309
Botanische Bemerkungen und Berichtigungen (Roth) 9633
Botanische Durchforschung von Vorarlberg und Liechtenstein (Richen) 9173
Botanische Excursion in das Araukanerland (Phil.) 7858
Botanische Excursion in die Provinz Aconcagua (see R. Philippi)
Botanische Exkursionen im Bedretto, Formazza-, und Bosco-Tal (see M. Rikli)
Botanische Garten der Universität Heidelberg (Pfitz.) 7828
Botanische Garten zu Erlangen (Reess) 8763
Botanische Mitteilung über Hydrastis canadensis (Jul. Pohl) 8107
Botanische Reise in das Banat (Rochel) 9320
Botanische Reisen (Reiner) 8920
Botanische Reise nach der Provinz Valdivia (see R. Philippi)
Botanische Reisestudien (Rikli) 9225
Botanische Reisestudien von der Spanischen Mittelmeerküste (Rikli) 9229
Botanischer Wegweiser in der Gegend von Spalato (Petter) 7793
Botanisches Centralblatt für Deutschland (Rabenh.) 8435
Botanisches Centralblatt (see D. Penhallow)
Botanische Systematik (K. Richt.) 9184
Botanische Terminologie (Pehersd.) 7611
Botanische Thesen (Roep.) 9425
Botanische und zoologische Ergebnisse einer wissenschaftlichen Forschungsreise nach den Samoainseln (Rech.) 8726
Botanische Zeitung (see A. Peter)
Botanisch-geologische Streifzüge (Reinke) 8939
Botaniska anteckningar från dyröya och några angränsande öar (Petterson) 7795
Botanisk Tidsskrift (see L. Rosenvinge)
Botaniste sans maître (Rousseau) 9689
Botanist in Ireland (Praeger) 8253
Botanographia americana (see C. Plumier)
Botany at the Rochester meeting (Rusby) 9817
Botany of Iceland (Rosenvinge) 9548
Botany of North Devon (Ravenshaw) 8686
Botany of the Canadian Eastern Arctic (Polunin) 8147
Botany of the Laccadives (Prain) 8259
Botany of Worcestershire (Rea) 8714
Botany (see H. Rickett)
Botany (see W. J. Robbins)
Bot. App. (Richardson) 9169
Bot. Bemerk. Bericht. (Roth) 9633
Bot. Canad. E. Arctic (Polunin) 8147
Bot. Centralbl. Deutschl. (Rabenh.) 8435

Bot. Centralbl. *(see* A. Pascher)
Bot. coll. guide (Penh.) 7630
Bot. cript. (Riv. Mat.) 9261
Bot. crypt. (Payer) 7560
Bot. descr. (Puerta) 8379
Bot. descr. Brit. pl. (Purt.) 8407
Bot. Durchf. Vorarlb. Liechtenst. (Richen) 9173
Bot. explor. St. Lawrence (St. John) 10058
Bot. farm. (Pollacci) 8135
Bot. Gart. Erlangen (Reess) 8763
Bot. Gart. Heidelberg (Pfitz.) 7828
Bot.-geol. Streifz. (Reinke) 8939
Bothalia *(see* I. Pole-Evans)
Bot. Herald *(see* H. G. Reichenbach)
Bot. Ireland (Praeger) 8253
Bot. Italia (Sacc.) 9921
Bot. méd. (A. Rich.) 9141
Bot. Mitt. Hydrastis (Jul. Pohl) 8107
Bot. N. Devon (Ravenshaw) 8686
Bot. nomencl. indig. (Barb. Rodr.) 9360
Bot. Reis. (Reiner) 8920
Bot. Reise Banat (Rochel) 9320
Bot. Reisestud. (Rikli) 9225
Bot. Reisestud. Span. Mittelmeerküste (Rikli) 9229
Bot. Rousseau, ed. duod. (Rousseau) 9687
Bot. Rousseau, ed. fol. (Rousseau) 9688
Bot. sans maître (Rousseau) 9689
Bot. Soc. Exch. Club Brit. Isles, Report *(see* W. Pearsall)
Bot. Stud. Black Rock For. (Raup) 8668
Bot. Syst. (K. Richt.) 9184
Bot. Termin. (Pehersd.) 7611
Bot. Thesen (Roep.) 9425
Bot. udstr. Nytte (Rottb.) 9664
Bot. Wegweis. Spalato (Petter) 7793
Bot. Wood Buffalo Park (Raup) 8665
Bot. Worcestershire (Rea) 8714
Bot. zool. Erg. Forschungsreise Samoains *(see* T. Reinbold)
Bouquet des Pyrénées (Saint-Amans) 10016
Bouquet du département de Lot-et-Garonne (Saint-Amans) 10018
Bouquet Lot-et-Garonne (Saint-Amans) 10018
Bouquets (Prévost) 8319
Bradley bibl. (Rehd.) 8815
Bradley bibliography (Rehd.) 8815
Breconshire border *(see* R. W. Phillips)
Breve illustrazione delle crittogame vascolari trivigianae (Sacc.) 9883
Breve osservazione sull' Isola di Madera (Raddi) 8491
Brief sketch of the development of botanical science in South Africa (E. Phillips) 7862
Brit. basidiomyc. (Rea) 8715
Brit. Desmid. (Ralfs) 8599

Brit. ferns (Plues) 8062
Brit. fl. (S. Robson) 9307
Brit. fl. pl. (Perrin) 7709
Brit. grass. (Plues) 8063
Brit. grass. (Pratt) 8277
British algae (see J. Ralfs)
British basidiomycetae (Rea) 8715
British Desmidieae (Ralfs) 8599
British ferns (Plues) 8062
British flora (S. Robson) 9307
British flowering plants (Perrin) 7709
British grasses (Plues) 8063
British grasses and sedges (Pratt) 8277
British lichens (W. Phillips) 7872
British phanerogamous plants & ferns (Ralfs) 8598
Brit. phaenog. pl. (Ralfs) 8598
Bromatologia (Plenck) 8056
Bryi generis sectionis Erythrocarpa species europaeae (Podp.) 8080
Bryogeographische Studien (Pfeff.) 7810
Bryogeogr. Stud. (Pfeff.) 7810
Bryol. Aude (Roum.) 9675
Bryol. danic. *(see* C. Rosenberg)
Bryological investigations in Tierra del Fuego (Roivainen) 9469
Bryologie de l'Aude (see C. Roumeguère)
Bryologie du département de l'Aude (Roum.) 9675
Bryologische Beiträge aus Südböhmen (Podp.) 8074
Bryologist (see L. Riddle)
Bryophyta exsiccata reipublicae čechosloveniae (see J. Podpěra)
Bryophyta of Iceland (Rosenvinge) 9548
Bryotheca belgica (see L. Piré)
Bryotheca čechoslovenica (see Z. Pilous)
Bryotheca europaea et extraeuropaea (see G. Rabenhorst)
Bryotheca europaea (see G. Rabenhorst)
Bryotheca regni Hungariae exsiccatae (see M. Péterfi)
Bryotheca romanica (see Z. Pilous)
Bryotheca tarvisiana centuria (see P. Saccardo)
Bryum generis monographiae prodromus *(see* J. Podpěra)
Bulletino bibliografico della Botanica italiana (see R. Pichi-Sermolli)
Bulletin Torrey bot. Club (see H. Rickett)

Cactaceae *(see* J. Rose)
Cactáceas de la Flora cubana (Roig) 9463
Cact. fl. cub. (Roig) 9463
Calendarium botanicum (W. Phelps) 7839
Calend. bot. (W. Phelps) 7839
Campi Elysii liber primus *(see* O. J. Rudbeck)
Campi Elysii liber secundus *(see* O. J. Rudbeck)
Campus Elysius *(see* O.J. Rudbeck, O.O. Rudbeck)

Canadian hepaticae (see W. Pearson)
Canadian Record of Science *(see* D. Penhallow)
Canyons Colorado (J. W. Powell) 8244
Canyons of the Colorado (J. W. Powell) 8243, 8244
Caratteri di alcune nuovi generi (Raf.) 8553
Carbonischer Urwald (Reinsch) 8952
Carbon. Urwald (Reinsch) 8952
Carices fulvellae (*see* A. Palmgren)
Carl Clusius' Naturgeschichte der Schwämme Pannoniens (Reichardt) 8847
Carmen amor. connub. pl. (Royen) 9729
Carmen elegiacum de amoribus et connubis plantarum (Royen) 9729
Carofitas de España (Reyes y Prosper) 9118
Carofit. España (Reyes y Prosper) 9118
Cataceae of Northeastern and Central Mexico (Saff.) 9985
Cat. afr. pl. *(see* A. Rendle)
Catal. bot. (Roth) 9629
Catalecta botanica (Roth) 9629
Cat. alg. Violante (Picc.) 7898
Catalog der Orchideen-Sammlung von G.W. Schiller (Rchb. fil.) 8893
Catalogo dei funghi italiani (Sacc.) 9902
Catalogo de la flora venezolana *(see* H. Pittier)
Catalogo delle alghe raccolte durante le crociere del cutter Violante (Picc.) 7898
Catalogo delle collezioni di piante Ricasoli (Ricasoli) 9133
Catalogo delle piante del giardino botanico di Firenze (A. Piccioli) 7885
Catalogo delle piante dell'orto botanico veronese (Pollini) 8139
Catalogo delle piante vascolari del Veneto (Sacc.) 9884
Catalogo delle piante vascolari nelle due Valli di Diano Marina e di Cervo (Ricca) 9136
Catalogo del real orto botanico di Napoli (Pasq.) 7428
Catálogo de plantas de Toledo (Pomata) 8148
Catálogo razonado de las plantas vasculares de Menorca (J. Rodríguez) 9362
Catalogue de la flore Creusotine (Quincy) 8424
Catalogue de l'herbier de Syrie (Puel) 8377a
Catalogue de 486 Liliacées et de 168 Roses (Redouté) 8752
Catalogue des algues de Livourne (Preda) 8281a
Catalogue des arbres et plantes, cultivés dans les jardins de M.r Joseph Parmentier (J. Parm.) 7381

Catalogue des Diatomées de l'Ile Campbell (P. Petit) 7769
Catalogue des diatomées provenant de Madagascar (P. Petit) 7774
Catalogue des fougères, prêles et lycopodiacées des environs du Mont-Blanc (Payot) 7568
Catalogue des graines de Coniferes mexicains (Roezl) 9431
Catalogue des lichens des Deux-Sèvres (O. J. Rich.) 9161
Catalogue des mousses, hépatiques et lichens de la Corrèze (Rupin) 9787
Catalogue des mousses sphagnes et hépatiques des environs de Montbéliard (Quél.) 8415
Catalogue des plantes (Sagot) 9995
Catalogue des plantes cryptogames (Pradal) 8246
Catalogue des plantes cultivées àl'école de botanique du Port de Brest (T. Pichon) 7911
Catalogue des plantes cultivées dans le jardin de M.r Joseph Parmentier (J. Parm.) 7380
Catalogue des plantes cultivées dans les jardins de M.r Joseph Parmentier (J. Parm.) 7379
Catalogue des plantes de Madagascar (H. Perrier) 7704
Catalogue des plantes de Provence (Roux) 9696
Catalogue des plantes du jardin botanique et d'acclimatation du gouvernement à Pondichéry (Perrottet) 7720
Catalogue des plantes phanérogames d'Avignon (Palun) 7249
Catalogue des plantes plus ou moins rares de Turnhout (Pâque) 7318
Catalogue des plantes qui croissent naturellement dans le Gard (Pouzolz) 8241
Catalogue des plantes vasculaires de Genève (Reut.) 9106
Catalogue des plantes vasculaires de Montmédy (Pierrot) 7927
Catalogue des plantes vasculaires du bassin du Rhone (St. Lag.) 10066
Catalogue des plantes vasculaires du Département de la Corrèze (Rupin) 9786
Catalogue des plantes vasculaires du Département de l'Ardèche (Revol) 9116
Catalogue des plantes vasculaires et des mousses de Boulogne-sur-Mer (Rigaud) 9222
Catalogue des plantes vasculaires et spontanées du Département de la Vendée *(see* N. Pontarlier)
Catalogue des plantes vasculaires Lot (Puel) 8377
Catalogue des plantes vasculaires sponta-

1093

nées du département d'Ille-et-Vilaine (Picq.) 7914
Catalogue des végétaux cultivés au jardin de Réunion (C. Rich.) 9152
Catalogue des végétaux ligneux (Philippar) 7845
Catalogue des végétaux ligneux en Corse (Requien) 9078
Catalogue des vignes américaines (Planch.) 8024
Catalogue détaillé des plantes vasculaires de Genève (Reut.) 9106
Catalogue du Jardin de la Réunion (C. Rich.) 9152
Catalogue dynamique de la flore de Genève (Reut.) 9106
Catalogue général des Diatomées (M. Perag.) 7665
Catalogue méthodologique des végétaux cultivés jardin des plantes Versailles (Philippar) 7846
Catalogue of American and foreign plants (W. Peck) 7605
Catalogue of books relating principally to America (see O. Rich)
Catalogue of English plants (see R. Pulteney)
Catalogue of ferns (J. Riley) 9239
Catalogue of fruit and ornamental trees (W. Prince) 8324
Catalogue of Hepaticae (Anacrogynae) in the Manchester Museum (Pearson) 7591
Catalogue of mosses (Roum.) 9675
Catalogue of North American musci (Rau) 8650
Catalogue of phaenogamous and acrogenous plants (Elm. Palmer) 7232
Catalogue of plants (Rothr.) 9657
Catalogue of plants collected in Nevada (see T. Porter)
Catalogue of plants collected in the Salton Sink (Parish) 7334
Catalogue of plants cultivated in the Royal Botanical Gardens, Trinidad (Prestoe) 8313
Catalogue of plants found in Oneida County (Paine) 7192
Catalogue of plants (see H. Patterson)
Catalogue of the bryophyta and pteridophyta found in Pennsylvania (Porter) 8184
Catalogue of the flora of Montana (Rydb.) 9857
Catalogue of the flora of Vermont (G. Perkins) 7682
Catalogue of the flowering plants of Schenectady County (Paige) 7188
Catalogue of the flowering plants of Vermont (G. Perkins) 7682

Catalogue of the fossil remains, described as fern stems and petioles (Posth.) 8202
Catalogue of the herbarium of the late Dr. Charles C. Parry (Parry) 7408
Catalogue of the phaenogamous and vascular cryptogamous plants of Illinois (H. Patt.) 7503
Catalogue of the phaenogamous plants of Evanston and vicinity (Raddin) 8500
Catalogue of the plants in the neighbourhood of Newbury (see An. Russell)
Catalogue of the plants in the Oban district (Rendle) 9069
Catalogue of trees and shrubs native of and introduced in the horticultural gardens in Fairmount Park (Rothr.) 9659
Catalogue phytostatique (Payot) 7567
Catalogue raisonné des champignons qui croissent dans le département de la Marne (Richon) 9177
Catalogue raisonné des plantes cellulaires le la Tunisie (Pat.) 7475
Catalogue raisonné des plantes de Montluçon (Pérard) 7667
Catalogue raisonné des plantes du Département de l'Yonne (Ravin) 8688
Catalogue raisonné des plantes introduites dans les colonies françaises de Bourbon et de Cayenne (Perrottet) 7718
Catalogue raisonné des plantes phanérogames de la Charente (Rochebr.) 9315
Catalogue raisonné des plantes vasculaires de Savoie (E. Perrier) 7701
Catalogue raisonné des plantes vasculaires et des mousses (Renauld) 8984
Catalogue raisonné des végétaux en Ille-et-Vilaine (Saint-Gal) 10022
Catalogues des graines recueillies (Reut.) 9109
Catalogues of the birds, shells, and some of the more rare plants of Dorsetshire (Pult.) 8400
Catalogus diatomacearum Bohemiae (Proch.) 8356
Catalogus florae ludovicianae (Riddell) 9200
Catalogus horti botanici turicensis (Roem.) 9405
Catalogus plantarum americanarum (Plum.) 8067
Catalogus plantarum Angliae (Ray) 8696
Catalogus plantarum circa Cantabrigam nascentium (Ray) 8695
Catalogus plantarum horti botanici florentini (A. Piccioli) 7883
Catalogus plantarum horti botanici florentini (G. Piccioli) 7888
Catalogus plantarum in hortis regiae villae prope Modoetiam (G. Rossi) 9572
Catalogus plantarum in horto aksakoviano (Regel) 8786

Catalogus plantarum italicarum (Romano) 9484
Catalogus Plantarum madagascariensium (Palacký) 7195
Catalogus plantarum phanerogamarum circa Brunsbergam (Saage) 9876
Catalogus plantarum vascularium chilensium (F. Philippi) 7848
Catalogus praevius plantarum in itinere ad Tarapaca a Friderico Philippi lectárum (Phil.) 7856
Catalogus stirpium in exteris regionibus (Ray) 8698
Cat. Amer. pl. (W. Peck) 7605
Cat. arbr. Parm. (J. Parm.) 7381
Cat. birds, pl. Dorsetshire (Pult.) 8400
Cat. bryoph. pterid. Pennsylvania (Porter) 8184
Cat. champ. Bruxelles (see M. Rousseau)
Cat. champ. Marne (Richon) 9177
Cat. coll. piante Ricasoli (Ricasoli) 9133
Cat. diatomées (see J. Pelletan)
Cat. Diatom. Ile Campbell (P. Petit) 7769
Catech. Bot. (Rchb.) 8877
Catechismus der Botanik (Rchb.) 8877
Cat. ferns (J. Riley) 9239
Cat. fl. bass. Rhone (St. Lag.) 10066
Cat. fl. Creusot. (Quincy) 8424
Cat. fl. Montana (Rydb.) 9857
Cat. fl. pl. Schenectady Co. (Paige) 7188
Cat. fl. Vaud. (see H. Pittier)
Cat. fl. Vermont (G. Perkins) 7682
Cat. foss. fern stems (Posth.) 8202
Cat. foug. Mont-Blanc (Payot) 7568
Cat. fruit trees (W. Prince) 8324
Cat. fung. ital. (Sacc.) 9902
Cat. gén. Diatom. (M. Perag.) 7665
Cat. giard. bot. Firenze (A. Piccioli) 7885
Cat. grain. Conif. mexic. (Roezl) 9431
Cat. Hepat. Manchester Mus. (Pearson) 7591
Cat. herb. Parry (Parry) 7408
Cat. herb. Syrie (Puel) 8377a
Cat. hort. bot. turic. (Roem.) 9405
Cat. jard. bot. Pondichéry (Perrottet) 7720
Cat. jard. Réunion (C. Rich.) 9152
Cat. lich. Deux-Sèvres (O. J. Rich.) 9161
Cat. mouss. hepat. lich. Corrèze (Rupin) 9787
Cat. N. Amer. musc. (Rau) 8650
Cat. Orch.-Samml. Schiller (Rchb. fil.) 8893
Cat. ort. bot. Napoli (Pasq.) 7428
Cat. Peale's Mus. (P. Beauv.) 7210
Cat. phaenog. pl. (Elm. Palmer) 7232
Cat. phaenog. pl. Evanston (Raddin) 8500
Cat. phan. Brunsb. (Saage) 9876
Cat. phytostat. (Payot) 7567
Cat. piante orto veron. (Pollini) 8139
Cat. pl. (Rothr.) 9657

Cat. pl. Angl. (Ray) 8696
Cat. pl. Boulogne (Rigaud) 9222
Cat. pl. Cantab. nasc. (Ray) 8695
Cat. pl. cell. Tunisie (Pat.) 7475
Cat. pl. Corrèze (Rupin) 9786
Cat. pl. crypt. (Pradal) 8246
Cat. pl. école bot. Brest (T. Pichon) 7911
Cat. pl. Gard (Pouzolz) 8241
Cat. pl. hort. aksakov. (Regel) 8786
Cat. pl. hort. florent. (A. Piccioli) 7883
Cat. pl. hort. florent. (G. Piccioli) 7888
Cat. pl. hort. Modoet. (G. Rossi) 9572
Cat. pl. Illinois (H. Patt.) 7503
Cat. pl. intr. colon. (Perrottet) 7718
Cat. pl. ital. (Romano) 9484
Cat. pl. Lot (Puel) 8377
Cat. pl. Madag. (H. Perrier) 7704
Cat. pl. madagasc. (Palacký) 7195
Cat. pl. Oban (Rendle) 9069
Cat. pl. Oneida Co. (Paine) 7192
Cat. pl. Parm. (J. Parm.) 7379
Cat. pl. Parm. (J. Parm.) 7380
Cat. pl. phan. Avignon (Palun) 7249
Cat. pl. phan. Charente (Rochebr.) 9315
Cat. pl. Provence (Roux) 9696
Cat. pl. Provence (see H. Roux)
Cat. pl. Salton Sink (Parish) 7334
Cat. pl. vasc. Ardéche (Revol) 9116
Cat. pl. vasc. chil. (F. Philippi) 7848
Cat. pl. vasc. Corse (see E. Relevière)
Cat. pl. vasc. Genève (Reut.) 9106
Cat. pl. vasc. Menorca (J. Rodríguez) 9362
Cat. pl. vasc. Montmédy (Pierrot) 7927
Cat. rais. pl. Haute-Saône (Renauld) 8984
Cat. rais. pl. Savoie (E. Perrier) 7701
Cat. rais. pl. Yonne (Ravin) 8688
Cat. stirp. ext. region (Ray) 8697, 8698
Cat. trees shrubs gard. Fairmount (Rothr.) 9659
Cat. Vég. Ille-et-Vilaine (Saint-Gal) 10022
Cat. vég. jard. Versailles (Philippar) 7846
Cat. vég. lign. (Philippar) 7845
Cat. vég. lign. Corse (Requien) 9078
Caulerpa (Reinke) 8936
Ceanothus (Parry) 7407
Cennia ort. bot. Venezia (G. M. Ruchinger) 9745
Cenni fl. Assab (Pasq.) 7433
Cenni storici dell' orto botanico in Venezia (G. M. Ruchinger) 9745
Cenni sulla flora di Assab (Pasq.) 7433
Census Acacia (Pescott) 7740
Census catalogue of Irish fungi (see G. Pethybridge)
Census NSW fresh-wat. alg. (Playf.) 8046
Census of New South Wales fresh-water algae (Playf.) 8046
Census of the genus Acacia in Australia (Pescott) 7740

Centaureele României (Procopianu) 8361
Centaur. Român. (Procopianu) 8361
Centralamer. Lobeliac. *(see* J. Planchon)
Centuria plantarum circa nobile Gedanum (Reyger) 9120
Ceratium macroceros (Penard) 7627
Česká Mykologie *(see* A. Pilát)
Cetraria in the U.S.S.R. *(see* K. Rassadina)
Champ. comest. vénén. Paris (E. Roussel) 9692
Champignons à l'Exposition de 1900 (Rolland) 9478
Champignons Algéro-Tunisiens (Pat.) 7485
Champignons comestibles et vénéneux de Paris (E. Roussel) 9692
Champignons de la Guadeloupe (Pat.) 7481
Champignons de la Guadeloupe (Pat.) 7481, 7484
Champignons de la Guadeloupe, recueillis par le R.P. Duss (Pat.) 7486
Champignons de la Nouvelle Calédonie (Pat.) 7444
Champignons de la Provence (Reguis) 8813
Champignons de l'Équateur (Pat.) 7455
Champignons de Montpellier (L. Planch.) 8025
Champignons du Jura et des Vosges (Quél.) 8414
Champignons du Nord de l'Afrique (Pat.) 7480
Champignons du Vénézuela (Pat.) 7445
Champignons et truffes (J. Rémy) 8979
Champignons figurés et déséchés (see N. Patouillard)
Champignons ...Haut-Orénoque (Pat.) 7445
Champignons nouveaux (Pat.) 7478
Champignons qui envahissent les végétaux cultivés (see C. Roumeguère)
Champignons récemment observés en Normandie (Quél.) 8416
Champignons recueillis par M. Seurat (Pat.) 7490
Champ. Jura Vosges (Quél.) 8414, 8416
Champ. Montpellier (L. Planch.) 8025
Champ. observ. Normandie (Quél.) 8416
Champ. Provence (Reguis) 8813
Champ. truffes (J. Rémy) 8979
Chapitre de grammaire (St. Lag.) 10081
Chapters on fossil botany (Pattison) 7508
Chapt. foss. bot. (Pattison) 7508
Characeae of North America (C. B. Robinson) 9295
Characeen Europa's (see G. Rabenhorst)
Characteres generum plantarum (Rutherford) 9840

Charakteristik der Thierpflanzen (Pall.) 7222
Charakterpflanzen des abyssinischen Hochlandes *(see* F. Rosen)
Charak. Veg. Ind. Arch. (Reinw.) 8965
Char. gen. pl. (Rutherford) 9840
Charles Robert Darwin (Rauwenh.) 8675
Check list N. Amer. Polypet. (H. Patt.) 7505
Check-list of North American Gamopetalae (H. Patt.) 7504
Check-list of North American plants (H. Patt.) 7506
Check list of North American Polypetalae (H. Patt.) 7505
Check-list of plants of Grand Canyon National Park (Patraw) 7497
Check list of the ferns of North America (J. Robinson) 9299
Checklist to Rumphius's Herbarium amboinense (Rumph.) 9784
Checkl. N. Amer. Gamopet. (H. Patt.) 7504
Checkl. N. Amer. pl. (H. Patt.) 7506
Checkl. pl. Grand Canyon (Patraw) 7497
Chi ha creato il nome fanerogame (Sacc.) 9940
Chillania pusilla (Roivainen) 9467
Chlor. aetn. (Raf.) 8556
Chlorideas (Parodi) 7394
Chloris aetnensis (Raf.) 8556
Chloris melvilliana (W. Parry) 7410
Chlor. melvill. *(see* E. Sabine)
Choix de plantes de la Nouvelle-Zélande (Raoul) 8630
Choix des plus belles fleurs (Redouté) 8750
Choix pl. Nouv.-Zél. (Raoul) 8630
Choix plus belles fl. (Redouté) 8750
Chorizanthe (Parry) 7406
Chorob. tyton. (Racib.) 8471
Choroby tytoniu w Galicyi (Racib.) 8471
Chromosomenzahlen und Chromosomendimensionen in der Gattung Crepis (O. Rosenberg) 9519
Chromotaxia (Sacc.) 9912
Chron. hist. pl. (Pickering) 7913
Chronological history of plants (Pickering) 7913
Chronological observations on introduced animals and plants (Pickering) 7912
Chrysanthemum indicum (Rupprecht) 9791
Chrysanth. ind. (Rupprecht) 9791
Chrysomonaden aus dem Hirschberger Grossteiche (Pasch.) 7421
Chrysomon. Hirschb. Grossteich. (Pasch.) 7421
Chrysosplenio (J. Pall.) 7221
Cinch. off. (Pult.) 8397
Circ. addr. bot. zool. (Raf.) 8561

Circular address on botany and zoology (Raf.) 8561
Cirsiotheca universa (see F. Petrak)
Cladon. eur. (Rabenh.) 8440, 8441
Cladoniae austriaceae (see I. Poetsch)
Cladoniae europaeae (Rabenh.) 8440, 8441
Cladoniae europaeae exsiccati. Supplementum I (see G. Rabenhorst)
Cladoniae europaeae (see G. Rabenhorst)
Cladoniae exsiccatae (see H. Rehm)
Cladonies de la flore de France (Parrique) 7401
Clare Island Survey *(see* R. Praeger)
Classif. fl. pl. (Rendle) 9064
Classificacíon natural de las plantas (Pitt.) 7989
Classification du genre Mentha (Pérard) 7668
Classification of flowering plants (Rendle) 9064
Classification of the genus Annona (Saff.) 9987
Classifications et des méthodes en histoire naturelle (Payer) 7558
Classif. méth. hist. nat. (Payer) 7558
Classif. nat. pl. (Pitt.) 7989
Clav. class. (Perleb) 7692
Clave analítica de las familias de plantas fanerógamas de Venezuela (Pitt.) 7981
Clave analítica de las familias de plantas superiores de la América tropical (Pitt.) 7986
Clave para la determinacion de los generos de Gramineas...de Buenos Aires (Parodi) 7395
Clav. fam. pl. Amér. trop. (Pitt.) 7986
Clav. fam. pl. Venez. (Pitt.) 7981
Clavis classium (Perleb) 7692
Clavis riviniana *(see* A. Rivinus)
Clav. syn. hymenomyc. eur. *(see* L. Quélet)
Cod. bot. fig. (Sacc.) 9937
Codex botanicus linnaeanus (H. Richt.) 9183
Codex botanicus Linnaeanus (Peterm.) 7753
Codex bot. linn. (H. Richt.) 9183
Codicem botanicum linnaeanum index alphabeticus (Peterm.) 7753
Cod. linn. index (Peterm.) 7753
Coelastrum proboscidium (Rayss) 8709
Collatio nominum brotherianorum (Parfitt) 7331
Coll. bot. Florence (Parl.) 7373
Coll. econ. fung. (F. Patterson) 7499
Collecteana (Rocm.) 9406
Collected mycological papers (Pat.) 7495
Collection de fleurs et de fruits (Prévost) 8319
Collection of economic and other fungi (F. Patterson) 7499

Collections botaniques Florence (Parl.) 7373
Collectio plantarum in solo Romano (Sabbati) 9877
Coll. fl. Fruits (Prévost) 8319
Coll. mycol. pap. (Pat.) 7495
Coll. nom. broth. (Parfitt) 7331
Colorado plant life (Ramaley) 8602
Colorado pl. life (Ramaley) 8602
Colpo d'occhio sulla vegetatione d'Italia (Parl.) 7362
Coltivazione di pianti tropicali (Ricasoli) 9135
Colt. Salici (L. Piccioli) 7892
Coltura dei Salici (L. Piccioli) 7892
Comm. arb. Belg. cult. (Saint-Moulin) 10090
Comm. bot. (Ruel.) 9763
Comm. bot. Conif. Cycad. (Rich.) 9159
Comm. de Favo (B. Richt.) 9182
Commentarius Schaefferi icones pictas (Pers.) 7729
Commentatio botanica de Conifereis et Cycadeis (Rich.) 9159
Commentatio botanica de ordinibus naturalibus plantarum (Ruel.) 9763
Commentatio botanico-oeconomica, de quibusdam arboribus in Belgio cultis (Saint-Moulin) 10090
Commentatio de Favo eiusque fungo (B. Richt.) 9182
Commentatio de fungis clavaeformibus (Pers.) 7725
Commentatio de fungis clavaeformibus in Coryphaei clavarias Ramariasque *(see* C. Persoon)
Commentatio, de monographia Quercus roboris (Saint-Moulin) 10091
Comm. fung. clav. (Pers.) 7725
Comm. monogr. Quercus rob. (Saint-Moulin) 10091
Common Indian trees (R. Parker) 7340
Common Ind. trees (R. Parker) 7340
Comm. Schaeff. icon. pict. (Pers.) 7729
Comp. bot. (Reuss) 9099
Compendio della flora italiana *(see* G. Passerini)
Compendio di botanica *(see* G. Pasquale)
Compendium botanices (Reuss) 9099
Compendium van de Pteridophyta en Spermatophyta (Pulle) 8395
Compendium van de Spermatophyta (Pulle) 8395
Compendium van de terminologie, nomenclatuur en systematiek der zaadplanten (Pulle) 8395
Comp. termin. nom. syst. zaadpl. (Pulle) 8395
Concise natural history of East and West-Florida (Romans) 9486

Conc. nat. hist. Florida (Romans) 9486
Confus. fl. poét. (Roum.) 9676
Coniferas novas nullas (Parl.) 7368
Conifères (Pardé) 7325
Conifers (C. C. Rogers) 9433
Conif. nov. (Parl.) 7368
Considérations générales sur les anomalies des Orchidées (Penz.) 7647
Considérations nouvelles sur les tourbes & les houilles (Renault) 9055
Considérations sur la méthode naturelle en botanique (Parl.) 7369
Considérations sur le polymorphisme de quelques espèces du genre Bupleurum (St. Lag.) 10077
Considérations sur les rapports de Lépidodendrons, des Sigillaires et des Stigmaria (Renault) 9022
Consid. méth. nat. (Parl.) 7369
Conspectul florei Dobrogei (Procopianu) 8362
Conspectus de la Flore de France (Rouy) 9712
Conspectus des espèces françaises du genre Spergularia (Rouy) 9708
Conspectus florae Koreae (Palib.) 7207
Conspectus Florae montenegrinae (Rohlena) 9453
Conspectus generum discomycetum (Sacc.) 9901
Conspectus generum pyrenomycetum italicorum (Sacc.) 9893
Conspectus methodi plantarum naturalis (Perleb) 7689
Conspectus muscorum europaeorum *(see* J. Podpěra)
Conspectus Polygalaearum Brasiliae meridionalis (St. Hil.) 10037
Conspectus regni vegetabilis (Rchb.) 8880
Conspectus sectionum specierumque generis Paulliniae (Radlk.) 8534
Conspectus sectionum specierumque generis Serjaniae (Radlk.) 8509
Conspectus sectionum specierumque generis Serjaniae auctus (Radlk.) 8522
Conspectus specierum generis Vitis (Regel) 8798
Conspectus tribuum generumque Sapindacearum (Radlk.) 8530
Consp. fl. Dobr. (Procopianu) 8362
Consp. fl. France (Rouy) 9712
Consp. fl. Groenland. *(see* E. Rostrup)
Consp. fl. Groenland *(see* L. Rosenvinge)
Consp. fl. Koreae (Palib.) 7207
Consp. fl. montenegr. (Rohlena) 9447
Consp. Fl. montenegr. (Rohlena) 9453
Consp. gen. pyrenomyc. (Sacc.) 9893
Consp. meth. pl. nat. (Perleb) 7689
Consp. regn. veg. (Rchb.) 8880
Consp. sect. sp. Paullin. (Radlk.) 8534

Consp. sect. sp. Serjan. (Radlk.) 8509
Contes fantastiques de m. Malinvaud (Rouy) 9711
Contes Malinvaud (Rouy) 9711
Contr. algol. fungol. (Reinsch) 8949
Contr. anat. foglia, Oleac. (Pirotta) 7967
Contr. anat. fruit Conif. (Radais) 8474
Contr. anat. monocot. (Queva) 8423
Contr. champ. Madagascar (Pat.) 7494
Contr. classif. crucif. (Pomel) 8151
Contr. étud. Pulmonaria (P. Parm.) 7382
Contr. fauna fl. Cork (Power) 8245
Contr. fl. As. (Printz) 8342
Contr. fl. Asiae (Printz) 8344
Contr. fl. Bacino (Paol.) 7305
Contr. fl. Buceg. (Panţu) 7296
Contr. fl. Bucureşt. (Panţu) 7297
Contr. fl. Ceahlau. (Panţu) 7295
Contr. fl. crypt. Roum. (Popovici) 8162
Contr. Fl. Derbyshire (Painter) 7193
Contr. fl. Greenland (M. Porsild) 8176
Contr. fl. Maroc (Pit.) 7975
Contr. fl. mycol. Mt. Ciahlău (Popovici) 8164
Contr. fl. mycol. Roum. (Popovici) 8163
Contr. fl. Paraguay (D. Parodi) 7391
Contr. fl. Paraguay, Amarant. (D. Parodi) 7393
Contr. fl. Transcaucas. occid. (Palib.) 7209
Contr. Fl. Versilia (Rossetti) 9570
Contr. gloss. mycol. phytopath. (Rangel) 8629
Contr. hist. dével. Mélastom. (J. D. Ruys) 9846
Contr. histol. Laurac. (Perrot) 7712
Contribuçao para o glossario portuguez referente à mycologia o à phytopathologia (Rangel) 8629
Contribución al estudio de las gramíneas del género Paspalum (Parodi) 7399
Contribuciones à la flora de Isla Elisabeth (Roivainen) 9468
Contribuciones à la flora del Paraguay (D. Parodi) 7391
Contribuciones à la flora del Paraguay Amarantáceas (D. Parodi) 7393
Contribuição ao estudo dos Synchytrium (Quintanilha) 8426
Contribution à la flore bryologique de l'Annam (P. de la Varde) 8204
Contribution à la flore cryptogamique de la Roumanie (Popovici) 8162
Contribution à la flore mycologique de la Roumanie (Popovici) 8163
Contribution à l'anatomie comparée du fruit des Conifères (Radais) 8474
Contribution à l'étude de la flore du Maroc (Pit.) 7975
Contribution à l'étude de la flore mycolgique du Mont Ciahlău (Popovici) 8164

Contribution à l'étude des Avena sect. Avenastrum (Saint- Yves) 10101
Contribution à l'étude des Brachypodium (Saint-Yves) 10104
Contribution à l'étude des champignons de Madagascar (Pat.) 7494
Contribution à l'étude des Diatomées (Roesch) 9427
Contribution à l'étude des Festuca de l'Afrique australe (Saint-Yves) 10098
Contribution à l'étude des Festuca de l'Amérique du Nord (Saint-Yves) 10094
Contribution à l'étude des Festuca de l'Amérique du Sud (Saint-Yves) 10095
Contribution à l'étude des Festuca de l'Orient (Saint-Yves) 10097
Contribution à l'étude des Laboulbéniacées d'Europe (F. Picard) 7881
Contribution à l'étude des lichens à orseille (Ronceray) 9491
Contribution à l'étude des lichens des îles Canaries (Pit.) 7973
Contribution à l'étude des Muscinées des îles Canaries (Pit.) 7971
Contribution à l'étude des Ptéridophytes de Colombie *(see* E. Rosenstock)
Contribution à l'étude du genre Pulmonaria (P. Parm.) 7382
Contribution à l'étude histologique des Lauracées (Perrot) 7712
Contribution à l'histoire du développement des Mélastomatacées (J. D. Ruys) 9846
Contributiones ad algologiam et fungologiam (Reinsch) 8949
Contributiones ad floram Asiae interiores pertinentes (Printz) 8342
Contributions à la classification méthodique des crucifères (Pomel) 8151
Contributions à la flore bryologique de Madagascar (Renauld) 8994
Contributions à l'anatomie des monocotylédonées (Queva) 8423
Contributions à l'étude des champignons extra-européens (Pat.) 7443
Contributions à l'étude des Primulacées sino-japonaises (Petitm.) 7776
Contributions à l'histoire de la flore de la Transcaucasie occidentale (Palib.) 7209
Contributions du Jardin botanique de Rio de Janeiro (Barb. Rodr.) 9357
Contributions to a knowledge of the biology of the Richmond River (Playf.) 8041
Contributions to Indian botany (Prain) 8265
Contributions to Mycologia britannica (Rost.) 9595
Contributions to the botany of the State of New York (Peck) 7597
Contributions towards a fauna and flora of the county of Cork (Power) 8245

Contribution to knowledge of the Mesogloiaceae (Parke) 7338
Contribution to our knowledge of the flora of Gazaland (Rendle) 9068
Contribution to the Flora of Derbyshire (Painter) 7193
Contribution to the flora of the Leribe Plateau (E. Phillips) 7859
Contribuțiune la flora Bucegilor (Panțu) 7296
Contribuțiune la flora României (Procopianu) 8358
Contribuțiuni la flora Bucureștilor (Panțu) 7297
Contribuțiuni la flora Ceahlaului (Panțu) 7295
Contribuțiuni nouă flora ceahlăului (Panțu) 7298
Contributo alla conoscenza della flora del Cadore (Pamp.) 7274
Contributo allo studio delle Narcissee italiane (Preda) 8281
Contribuzione alla flora del Bacino di Primiero (Paol.) 7305
Contribuzione alla Flora della Versilia (Rossetti) 9570
Contribuzione alla micologia veneta e modenese (D. Sacc.) 9878
Contribuzione all'anatomia comparata della foglia I. Oleaceae (Pirotta) 7967
Contr. Jard. bot. Rio de Janeiro (Barb. Rodr.) 9357
Contr. lich. Canaries (Pit.) 7973
Contr. lich. orseille (Ronceray) 9491
Contr. Mesogloiac. (Parke) 7338
Contr. Musc. Canaries (Pit.) 7971
Contr. mycol. brit. *(see* J. Rostafinský)
Contr. nat. hist. Alaska *(see* J. Rothrock)
Corcir. fl., cent. 1-3 (Pieri) 7921
Coroll. gen. *(see* A. Royen)
Corona amaryll. (M. Rosenberg) 9516
Corona amaryllidacea (M. Rosenberg) 9516
Corona florae Narbonensis et Pyrenaeis (see P. Pourret)
Correspondances scientifiques inédites (Roum.) 9679
Corr. sci. inéd. (Roum.) 9679
Cours bot. foss. (Renault) 9015
Cours compl. agric. (Rozier) 9741a
Cours complet d'agriculture (Rozier) 9741a
Cours de botanique fossile (Renault) 9015
Critical review of the flora of Moscow *(see* A. Petunnikov)
Critt. bras. (Raddi) 8493
Crittogame brasiliane (Raddi) 8493
Crociera Corsaro, Alg. (Picc.) 7901
Crociera del Corsaro Alghe (Picc.) 7901
Cron. fl. ital. (Sacc.) 9945

Cronologia della flora italiana (Sacc.) 9944, 9945
Crypt.-Fl. (G. Pabst) 7175
Crypt. ill., Champ. (Roum.) 9674
Crypt. ill., Lichens (Roum.) 9673
Cryptogamae exsiccatae *(see* F. Petrak)
Cryptogamae parasiticae in insula Java lectae exsiccatae *(see* M. Raciborski)
Cryptogamae pyrenaicae *(see* X. Philippe)
Cryptogamae vasculares europaeae *(see* G. Rabenhorst)
Cryptogamen-Flora (G. Pabst) 7175
Cryptogamie illustrée Champignons (Roum.) 9674
Cryptogamie illustrée Famille des Lichens (Roum.) 9673
Cult. comm. cotton India (Royle) 9736
Culture des cactées (F. Palmer) 7234
Curso elemental de botánica *(see* A. Paláu)
Cursus Cryptogamenk. (Rabenh.) 8437
Cursus der Cryptogamenkunde (Rabenh.) 8437
Cursus der Kryptogamensammlung *(see* G. Rabenhorst)
Cycadearum generum specierumque revisio (Regel) 8802
Cycadeoidea Niedzwiedzkii (Racib.) 8462
Cycl. (Rees) 8755
Cycl. Am. Hort. *(see* A. Rehder)
Cyclopaedia (Rees) 8755
Cyclostomella (Pat.) 7474
Cyperaceae et Gramineae siculae (K. Presl) 8292
Cyperaceae of the West Coast of Africa (Ridl.) 9205
Cyper. Gramin. sicul. (K. Presl) 8292
Cypripedium (Pucci) 8374
Cytologische und morphologische Studien an Drosera longifolia x rotundifolia (O. Rosenberg) 9518

Dan. Agaric. (S. Petersen) 7767
Dan. Blomsterpl. naturh. (Raunk.) 8655
Dan. Greenland (Rink) 9245
Danish Greenland (Rink) 9244, 9245
Danmarks og Holsteens flora (Rafn) 8593
Danm. Holst. fl. (Rafn) 8593
Dan. Phaeophyc. (Rosenvinge) 9555
Danske Agaricaceer (S. Petersen) 7767
Danske arter af slaegten Ceramium (H. E. Petersen) 7761
Dansk Exkurs.-Fl. (Raunk.) 8654
Dansk Exkursions-Flora (Raunk.) 8654
Dansk fl. (Rostr.) 9606
Dansk oeconomisk Urte-Bog (Paulli) 7528
Dansk oecon. Urte-Bog (Paulli) 7528
Danzig wildwachsende Pflanzen (Reyger) 9121
Danzig wildw. Pfl. (Reyger) 9121
Da quale anno debba cominciare la validità della nomenclatura scientifica delle crittogame (Sacc.) 9946
Darwin (Rauwenh.) 8675
Datos algológicos (J. Rodríguez) 9364
Datos para la materia médica mexicana *(see* J. Ramírez)
De Agarico officinali (Rubel) 9743
De anatome & oeconomia plantarum (Royen) 9726
De caricibus terrritorii vindobonensis (Redtenbacher) 8754
De caric. terr. Vindob. (Redtenbacher) 8754
Decas hortus botanicus (Rchb.) 8878
Decas plantarum amurensium (Rupr.) 9810
Decas plantarum novarum *(see* E. Regel)
De Chrysospl. (J. Pall.) 7221
De Cinchona officinali (Pult.) 8397
Dec. pl. amur. (Rupr.) 9810
De Danske Blomsterplanters Naturhistorie (Raunk.) 8655
De distributione geographica plantarum Helvetiae (Ringier) 9240
De embryologie generum Asteris et Soladaginis (Palm) 7230a
Def. demostr. (S. Ruiz) 9783
Defensa y demostracion (S. Ruiz) 9783
Defensa y demostracion del verdadero descubridor de las Quinas (Ruiz) 9777
Defens. descubr. Quinas (Ruiz) 9777
De filicibus veris hungariae (Jos. Sadler) 9977
De floribus et affinitatibus Balsaminearum (Roep.) 9420
De fundamentali plantarum notitia (O. O. Rudbeck) 9748
De generis Galii (Racib.) 8456
De geslachtsgeneratie der Gleicheniaceeën (Rauwenh.) 8678
De Houtsoorten van Suriname (J. Pfeiff.) 7814
De inventarisatie van het erfdeel der vaderen (Pulle) 8396
De Junco (Rostk.) 9601
De la Belladone (Pauquy) 7532
Deliciae pragenses (J. Presl) 8288
Delic. prag. (J. Presl) 8288
De l'instruction populaire sur les champignons (Rolland) 9477
Della Corcirese flora. Centuria prima (Pieri) 7921
Della storia e letteratura della flora veneta (Sacc.) 9885
Della utilità dei giardini d'aclimazione (Ricasoli) 9134
Delle specie nuove di funghi Firenze (Raddi) 8483
Delondre-Bouchardatschen China-Rinden (Phoebus) 7876

Delondre-Bouch. China-Rind. (Phoebus) 7876
Delphinium carolinianum (Rydb.) 9855
De methodo plantarum epistola (Ray) 8706
Démonst. élém. bot. *(see* F. Rozier)
Démonstrations botaniques (Rich.) 9154
Démonstr. bot. (Rich.) 9154
De montium inter Vistritium et Nissam fluvios sitorum flora (Sadebeck) 9962
De Musac. (Rich.) 9160
De Musaceis (Rich.) 9160
De Myristica *(see* F. Radloff)
De necessitate et utilitate studii botanici (Rodschied) 9366
Denkwürdigkeiten einer Reise nach dem russischen Amerika *(see* A. Postels)
De nonnullis Desmidiaceis (Racib.) 8454
De nonnullis plantis (P. Rossi) 9573
De Orchideis europaeis annotationes (Rich.) 9156
De orchid. eur. (Rich.) 9156
De organis plantarum (Roep.) 9419
De paddenstoelen van Nederland (J. Ruys) 9845
De Physica de fungorum generatione (Picco) 7896
De plantarum plethora (Plaz) 8049
De plantarum radicibus (Rombouts) 9487
De plantarum sub diverso coelo nascentium cultura (Plaz) 8051
De plantis cibariis romanorum (Retz.) 9090
De pl. cult. (Plaz) 8051
De pollinis Orchidearum (Rchb. fil.) 8892
De pollin. Orchid. (Rchb. fil.) 8892
De progressu cognitionis plantarum cryptogamicarum (C. Sahlberg) 10006
De Pyrola et Chimophila (Radius) 8503
De Raphania *(see* G. Rothman)
De re botanica tractatus (Sacc.) 9937
De Saccharo (Plaz) 8050
Descr. Baptisia (Ravenel) 8682
Descr. champ. nouv. *(see* C. Persoon)
Descr. dess. pl. crypt. (Richon) 9175
Descr. icon. rar. pl. (Rottb.) 9666
Descripción de las nuevas plantas (Phil.) 7852
Description de deux genres nouveaux de la famille des Euphorbiacées (Planch.) 8005
Description de deux nouvelles espèces de Roses (Rapin) 8635
Description des plantes de l'Amérique (Plum.) 8066
Description d'un genre nouveau du groupe des Thismiées (Planch.) 8010
Description d'un genre nouveau, voisin du Cliftonia (Planch.) 8007
Descriptiones Epidendrorum (Retz.) 9086

Descriptiones et emendationes plantarum (Regel) 8808
Descriptiones et icones rariorum et pro maxima parte novarum plantarum (Rottb.) 9666
Descriptiones monandrarum (Retz.) 9086
Descriptiones plantarum botanici petropolitani (Regel) 8809
Descriptiones plantarum horti imperialis botanici (Regel) 8806
Descriptiones plantarum novarum (Regel) 8800
Descriptiones plantarum novarum rariorumque (Regel) 8804
Descriptiones plantarum novarum *(see* E. Regel)
Descriptiones plantarum quarundam surinamensium (Rottb.) 9667
Descriptiones plantarum rariorum (Rottb.) 9665
Descriptiones rariorum plantarum e terra surinamensi (Rottb.) 9667
Description of a new Baptisia (Ravenel) 8682
Description of the Ioxylon pomiferum (Raf.) 8565
Description physique de l'Ile de Crète ... partie botanique (Raulin) 8653
Descriptions and sketches of some remarkable oaks (Rooke) 9493
Descriptions et dessins de plantes cryptogames nouvelles (Richon) 9175
Descriptions of new plants (B. L. Robinson) 9281
Descriptions of new plants collected in Mexico (B. L. Robinson) 9282
Descriptions of three hundred new species of South American plants (Rusby) 9824
Descriptionum et iconum rariores plantas (Rottb.) 9666
Descrizione di due nove specie di piante orientali (Parl.) 7359
Descrizione di una nuova Orchidea brasiliana (Raddi) 8495
Descr. nuev. pl. (Phil.) 7852
Descr. pl. Amér. (Plum.) 8066
Descr. pl. rar. (Rottb.) 9665
Descr. rar. pl. surin. (Rottb.) 9667
Descr. S. Amer. pl. (Rusby) 9824
Descr. sketch. oaks (Rooke) 9493
Des diagnoses et de la nomenclature mycologiques (Sacc.) 9938
De serie vegetabili (Pontén) 8154
Desmid. now. (Racib.) 8458
Desmid. okol. Krak. (Racib.) 8452
Desmidya (Racib.) 8461
Desmidyje nowe (Racib.) 8458
Desmidyje okolic Krakowa (Racib.) 8452
De soli diff. (J. E. Pohl) 8106

De soli differentia in cultura plantarum (J. E. Pohl) 8106
De speciebus generibusque nonnullis novis ex algarum et fungorum Classe (Reinsch) 8948
De svenska Equisetum-arterna (H. Rosend.) 9531
De tegenwoordige richting en beteekenis der planten- physiologie (Rauwenh.) 8673
Det sydligste Grønlands Vegetation (Rosevinge) 9542
De Ulmo (see J. Sachs)
Deut. Bot. Herb.-Buch) (Rchb.) 8888
Deut. Fl. (Pilling) 7940
Deut. Forstkr. (Reum) 9097
Deut. Holzzucht (Pfeil) 7824
Deutsche Botaniker Herbarienbuch (Rchb.) 8888
Deutsche Flora (Pilling) 7940
Deutsche Holzzucht (Pfeil) 7824
Deutschen Foäuter (Reum) 9097
Deutschen Volksnamen der Pflanzen (Pritz.) 8355
Deutsche Schul-Flora (Pilling) 7940
Deutsche Südpolar-Expedition Botanik (see T. Reinbold)
Deutschlands Flora (Peterm.) 7757
Deutschlands Flora (Rchb.) 8885
Deutschlands Flora (Röhl.) 9377
Deutschlands Flora (see F. Rostkovius)
Deutschlands Kryptogamen-Flora (Rabenh.) 8434
Deutschlands kryptogamische Giftgewächse (Phoebus) 7875
Deutschlands Moose (Röhl.) 9378
Deutschl. Fl. (Peterm.) 7757
Deutschl. Fl. (Röhl.) 9377
Deutschl. Krypt.-Fl. (Rabenh.) 8434
Deutschl. krypt. Giftgew. (Phoebus) 7875
Deut. Volksnam. Pfl. (Pritz.) 8355
Deux. expéd. antarct. franç., Diatom. (M. Perag.) 7666
Deuxième expédition antarctique française Diatomées (M. Perag.) 7666
Deuxième mémoire sur la famille des Polygalées (St. Hil.) 10036
Deuxième mémoire sur les algues marines du Groenland (Rosenvinge) 9543
Deuxième mémoire sur les Cucurbitacées (Sageret) 9988
Deuxième mémoire sur les Résédacées (St. Hil.) 10047
Deux. mém. alg. mar. Groenland (Rosenvinge) 9543
Deux. mém. Réséd. (St. Hil.) 10047
De variis plantarum methodis dissertatio brevis (Ray) 8705
De var. pl. meth. diss. (Ray) 8705

De vasis plantarum spiralibus (G. Reichel) 8866
Développement de la fécule (Raspail) 8638
De vera fuci natantis fructificatione (Ruiz) 9773
De verspreiding der phanerogamen van arktisch Europa (Ruijs) 9767
De Viola (Pio) 7943
De Volcamero (Panz.) 7302
Diagnoses fungorum novorum (Penz.) 7648
Diagnoses fungorum novorum (see P. Saccardo)
Diagn. pl. nov. hisp. (see G. Reuter)
Diagn. pl. orient. (see G. Reuter)
Di alcune specie di Pero indiano memoria (Raddi) 8492
Di alcune specie nuove, di piante crittogame (Raddi) 8485
Di alcune specie nuove di rettili e piante brasiliane memoria (Raddi) 8489
Diatomaceae exsiccatae (see G. Rabenhorst)
Diatomaceae of the Hull district (see R. Philip)
Diatom. Baie Villefranche (H. Perag.) 7662
Diatomées (Pelletan) 7620
Diatomées collection J. Tempère & H. Peragallo (see H. Peragallo)
Diatomées de France (see H. Peragallo, P. Petit)
Diatomées de la Baie de Villefranche (H. Perag.) 7662
Diatomées du midi de la France (H. Perag.) 7661
Diatomées du monde entier (see H. Peragallo)
Diatomées du monde entier (see H. Peragallo)
Diatomées marines de France (H. Perag.) 7664
Diatomées planctoniques (see J. Pavillard)
Diatomées rares des cotes françaises (P. Petit) 7773
Diatomées recoltées en Cochinchine (P. Petit) 7775
Diatom. mar. France (H. Perag.) 7664
Diatom. midi France (H. Perag.) 7661
Diatoms of the United States (see R. Patrick)
Dicc. bot. (Roig) 9464
Dicc. bot. bras. (J. de A. Pinto) 7942
Diccionario botanico (Roig) 9464
Diccionario das plantas uteis (H. Pereira) 7673
Diccionario de botanica brasileira (J. de A. Pinto) 7942
Dicc. pl. uteis (H. Pereira) 7673
Dict. abr. bot. (J. Philib.) 7844
Dict. bot. (Reuss) 9101
Dict. bot. (see J. Poisson)

Dict. class. hist. nat. *(see* A. Ricd)
Dict. élém. bot. (Rich.) 9153
Dict. élém. bot. *(see* L. Richard)
Dictionariolum trilingue (Ray) 8699
Dictionarium botanicum (Reuss) 9101
Dictionnaire abrégé de botanique (J. Philib.) 7844
Dictionnaire des sciences naturelles *(see* P. Beauv.)
Dictionnaire élémentaire de botanique (Rich.) 9153
Dictionnaire français-swahili (Sacleux) 9961
Dictionnaire universel de botanique (J. Philib.) 7843
Dict. Sci. nat. *(see* P. Petit-Radel)
Dict. univ. bot. (J. Philib.) 7843
Dict. univ. hist. nat. *(see* A. Richard)
Didymonema (K. Presl) 8297
Différentes espèces dans le genre Musa (Sagot) 9997
Disertacion botanica (Pav.) 7539
Disertacion sobre la raiz de la ratánhia (Ruiz) 9774
Disert. raiz ratánhia (Ruiz) 9774
Dispersal of plants throughout the world (Ridl.) 9212
Dispers. pl. (Ridl.) 9212
Dispositio methodica fungorum (Pers.) 7727
Dispositio Uredineorum qui in Germaniae Coniferis parasitantur (Reess) 8758
Disp. Ured. Conif. paras. (Reess) 8758
Dissertação Solanaceas brasileiras (Pizarro) 7993
Dissertatio academica, qua nova Ammeos species proponitur *(see* F. Radloff)
Dissertatio botanica de Hydrocotyle *(see* J. Pontén)
Dissertatio botanica de Planta Sceptrum carolinum *(see* L. Roberg)
Dissertatio chemica de origine oleorum in vegetabilius *(see* G. Rothman)
Dissertatio de Pyrola et Chimophila (Radius) 8503
Dissertatio inauguralis de plantarum exhalationibus (J. L. Palmer) 7236
Dissertation physico-médicale, sur les truffes (Pennier) 7640
Diss. hort. abo. *(see* F. Radloff)
Diss. pl. exhal. (J. L. Palmer) 7236
Diss. Solan. bras. (Pizarro) 7993
Diss. truffes (Pennier) 7640
Distr. crypt. vasc. Ross. (Rupr.) 9797, 9799
Distr. geogr. pl. Helv. (Ringier) 9240
Distribucion geográfica de las compuestas de la flora de Chile (Reiche) 8859
Distributio cryptogamarum vascularium in Imperio Rossico (Rupr.) 9797

Distribution of the Rhodophyceae (Rosenvinge) 9556
Distribution of the Rhodophyceae in the Danish waters (Rosenvinge) 9546
Distr. Rhodophyc. Dan. waters (Rosenvinge) 9556
Di un operatta sulla flora della Corsica (Sacc.) 9943
Di un viaggio botanico al Gargano (Pasq.) 7431
D. Joanne Georgio Volcamero (Panz.) 7302
Doliocarpus. Eine neue Gattung Pflanzen aus America *(see* D. Rolander)
Doliocarpus. En ört af nytt genus från America *(see* D. Rolander)
Doplňky ku květeně moravské (Podp.) 8077
Drabae asiaticae (Pohle) 8110
Drab. asiat. (Pohle) 8110
Drei Monate in der Libyschen Wüste (Rohlfs) 9455
Drei Monate libysch. Wüste (Rohlfs) 9455
Drei neue Serjania-Arten (Radlk.) 8532
Dreizehn neue Pflanzenarten aus Griechenland (Rech. fil.) 8733
Dreizehn neue Pfl.-Art. Griechenland (Rech. fil.) 8733
Dr. H.O. Forbes's Malayan plants (Rendle) 9071
Dr. H.O. Forbes's New Guinea plants (Rendle) 9070
Dritter Beitrag zur Kryptogamenkunde Oberösterreichs (Poetsch) 8094
Drogues simples d'origine végétale (G. Planch.) 8003
Drogues simpl. vég. (G. Planch.) 8003
Due nuov. gen. monocot. (Parl.) 7367
Due nuovi generi di piante monoctiledoni (Parl.) 7367
Dussiella (Pat.) 7452

Eine zweite Valenzuela (Radlk.) 8535
Einige Aufgaben der regionalen Moorforschung (L. Post) 8189b
Einige neue Gattungen der Gesnereen (Regel) 8774
Einige Notizen über die Vegetation der nördlichen Gestade des Schwarzen Meeres (Rehmann) 8831
Einiges über die Gymnospermen (Plitzka) 8059
Einige von Dr. M. Gusinde gesammelte [sic]Flechten aus Patogonien und dem Feuerlande (Du Rietz) 9216
Einige Worte über die subalpine Flora des Meissners (Pfeiff.) 7818
Einladung zu einem Kryptogamen-Tauschvereine (Rossmann) 9587

Einwanderungswege der Flora (Palmgr.) 7246
Elajoplasty liliowatych (Racib.) 8466
Elegia (Royen) 9731
Elem. bot. (E. Perkins) 7680
Elem. bot. (H. Rose) 9506
Élém. bot. (Payer) 7563
Elem. bot. (Pollini) 8138
Elem. bot. (Ramírez Goyena) 8612
Elem. Bot. (Potonié) 8214
Élém. bot. indoch. *(see* A. Pételot)
Elementa ad floram pricipatus Bulgariae (Pančic) 7282
Elementary textbook of agricultural botany *(see* M. Potter)
Elementary textbook of botany (Prantl) 8268
Elementa terminologiae botanicae (Plenck) 8058
Elemente der Botanik (Potonié) 8214
Elementi di botanica (Pollini) 8138
Elementos de botanica (Ramírez Goyena) 8612
Elementos de la morfolojía i sistemática botanica (Reiche) 8854
Elementos de la nomenclatura botánica (Plenck) 8058
Éléments de botanique (Payer) 7563
Elements of botany (A. Rich.) 9138
Elements of botany (E. Perkins) 7680
Elements of botany (H. Rose) 9506
Elem. fl. Bulg. (Pančic) 7282
Elem. morf. sist. bot. (Reiche) 8854
Elem. termin. bot. (Plenck) 8058
Elenc. alg. Corsaro (Picc.) 7906
Elench. pl. vasc. (Pančic) 7280
Elenchus plantarum vascularium (Pančic) 7280
Elenchus zoophytorum (Pall.) 7222
Elench. zooph. (Pall.) 7222
Elenco dei muschi di Liguria (Picc.) 7897
Elenco delle alghe crociera del Corsaro alle Baleari (Picc.) 7906
El Perú (Raimondi) 8595
Elvellacei britannici (see C. Plowright, W. Phillips)
Embryol. Aster. et Solidag. (Palm) 7230a
Enchir. fung. (Quél.) 8417
Enchiridion fungorum (Quél.) 8417
Encore la Chlamydomyxa (Penard) 7628a
Encycl. *(see* J. Poiret)
English index to the plants of India (Piddington) 7918
Enkele bloemnamen in de Zuidnederlandsche dialecten (Pauwels) 7534
Ensayo de una flora fanerogámica gallega (Planellas) 8029
Ensayo fitogeográfico sobre el partido de Pergamino (Parodi) 7398
Ensayo fl. gallega (Planellas) 8029

Ensayo pl. usual. Costa Rica (Pitt.) 7979
Ensayo sobre las plantas usuales de Costa Rica (Pitt.) 7979
Ensayo sobre las variedades de la vid comun (Roxas) 9715
Ens. var. vid. com. (Roxas) 9715
Entfernung als pflanzengeographischer Faktor (Palmgr.) 7242
Entstehung der Steinkohle (Potonié) 8225
Entst. Steinkohle (Potonié) 8225
Entw. Farnblatt. (Sadebeck) 9963
Entw.-Gesch. Stammsp. Equis. (Reess) 8757
Entw.-gesch. Unters. Cutleriac. (Reinke) 8927
Entw.-gesch. Unters. Dictyotac. (Reinke) 8926
Entwickelung des Keimes der Gattung Selaginella (Pfeff.) 7811
Entwickelungsgeschichte der Achlya prolifera (Pringsh.) 8327
Entwicklung der Naturwissenschaften (Reinke) 8937
Entwicklungsgeschichtliche Untersuchungen über die Cutleriaceen (Reinke) 8927
Entwicklungsgeschichtliche Untersuchungen über die Dictyotaceen (Reinke) 8926
Entw. Keim. Selaginella (Pfeff.) 7811
Entw. Naturwiss. (Reinke) 8937
Entwurf Anordn. Orch. (Pfitz.) 7831
Entwurf einer natürlichen Anordnung der Orchideen (Pfitz.) 7831
Énum. champ. Guadeloupe *(see* N. Patouillard)
Énum. champ. Tunisie (Pat.) 7460
Enum. diagn. Cact. (Pfeiff.) 7815
Enumeraţia plantelor vasculare dela Stânca-Ştefănesci (Procopianu) 8357
Enumeratio diagnostica Cactearum (Pfeiff.) 7815
Enumeratio Euphorbiarum (Roep.) 9417
Enumeratio florae constantinopolitanae (Rech. fil.) 8735
Enumeratio muscorum frondosorum (Reinhardt) 8922
Énumération des champignons observés en Tunisie (Pat.) 7460
Énumération des champignons récoltés par les rr. pp. Farges et Soulié (Pat.) 7472
Énumération des espèces zoologiques et botaniques de l'ile̦ de Cuba (Sagra) 10004
Énumération des mousses du Mont-Blanc (Payot) 7569
Enumerationis plantarum phaenogamarum lipsiensium specimen (Pappe) 7313
Enumeration of the plants collected by Dr. H.H. Rusby (Rusby) 9813
Enumeration of the plants collected in Bolivia by Miguel Bang (Rusby) 9819

Enumeration of the Pteridophytes and Spermatophytes of the San Bernardino Mountains (Parish) 7335
Enumeration of the vascular plants known from Surinam (Pulle) 8375
Énumération succincte des espèces de la famille des Nymphéacées (Planch.) 8011
Enumeratio palmarum novarum (Barb. Rodr.) 9341
Enumeratio palmarum novarum seguido de un protesto e de novas palmeiras (Barb. Rodr.) 9344
Enumeratio plantarum florae palatinatu sponte crescentium (Petif) 7768
Enumeratio plantarum horti à Demidof (Pall.) 7226
Enumeratio plantarum hucusque cognitarum insulae Cypri (Poech) 8085
Enumeratio plantarum in Dalmatia (Portenschl.) 8179
Enumeratio plantarum in horto botanico fluminensi cultarum (Barb. Rodr.) 9350
Enumeratio plantarum in horto Lugduno-Batavo (Reinw.) 8966
Enumeratio plantarum in regionibus cis- et transiliensibus a cl. Semenovio anno 1857 Collectarum (Regel) 8791
Enumeratio plantarum phaenogamarum in Germania (Roth) 9636
Enumeratio plantarum phanerogamicarum districtus quondam naszódiensis (Porcius) 8169
Enumeratio plantarum vascularium florae Serbiae (Pančić) 7284
Enumeratio rosarum circa Wirceburgum (A. Rau) 8649
Enumeratio seminum (Parl.) 7370
Enumeratio stirpium horti botanici senckenbergiani (Reichard) 8837
Enumeratio systematica specierum plantarum medicinalium (Pers.) 7737
Enumerazione dei funghi della Valsesia (Sacc.) 9924
Enumerazione delle piante fanerogame rare della provincia bergamasca (Rota) 9623
Enum. Euphorb. (Roep.) 9417
Enum. fl. constantinop. (Rech. fil.) 8735
Enum. hort. Demidof (Pall.) 7226
Enum. musc. frond. (Reinhardt) 8922
Enum. palm. nov. (Barb. Rodr.) 9341
Enum. palm. nov., prot. (Barb. Rodr.) 9341, 9344
Enum. piante bergam. (Rota) 9623
Enum. pl. Cypr. (Poech) 8085
Enum. pl. Dalmatia (Portenschl.) 8179
Enum. pl. fl. palat. (Petif) 7768
Enum. pl. hort. flum. (Barb. Rodr.) 9350
Enum. pl. hort. Lugd.-Bat. (Reinw.) 8966
Enum. pl. hort. senckenb. (Reichard) 8837
Enum. pl. lips. spec. (Pappe) 7313
Enum. pl. med. (Pers.) 7737
Enum. pl. phaen. Germ. (Roth) 9636
Enum. pl. phan. naszód. (Porcius) 8169
Enum. pl. Stânca-Stef. (Procopianu) 8357
Enum. ros. Wirceb. (A. Rau) 8649
Enum. semin. (Parl.) 7370
Enum. vasc. pl. Surinam (8375)
Epimel. bot. (K. Presl) 8312
Epimeliae botanicae (K. Presl) 8312
Erbario crittogamico italiano (see A. Piccone, G. Passerini)
Erdkunde *(see* C. Ritter berol.)
Ergänz. Monogr. Serjania (Radlk.) 8522
Ergänzungen zur Monographie der Sapindaceen-Gattung Serjania (Radlk.) 8522
Ergebnisse der Biologie *(see* W. Ruhland)
Ergebnisse einer botanischen Reise nach Bulgarien (Rech. fil.) 8728
Ericetum africanum *(see* A. Royen)
Ernähr. Alg. (O. Richt.) 9187
Ernährung der Algen (O. Richt.) 9187
Erster Beitrag zur Flora von Montenegro (Rohlena) 9447
Erysiphaceae of Nebraska (Pool) 8155
Esboz. form. veg. Venez. (Pitt.) 7983
Esbozo da las formaciones vegetales de Venezuela (Pitt.) 7983
Esp. congol. Digitaria (Robyns) 9310
Esp. congol. Panicum (Robyns) 9312
Espèces congolaises du genre Digitaria (Robyns) 9310
Espèces congolaises du genre Panicum (Robyns) 9312
Espèces critiques d'hyménomycètes (Pat.) 7468
Espèces européennes du genre Orthotrichum (E. Piccioli) 7886
Espèces nouvelles de la Côte-d'Or (Rolland) 9475
Esp. eur. Orthotrichum (E. Piccioli) 7886
Esplorazione botanica del Dodecaneso (Pamp.) 7269
Esplorazione botanica dell' isola di Rodi (Pamp.) 7267
Esquisse de la flore de Sainte-Hélène (Roxb.) 9723
Esquisse de la végétation du département de l'Oise (Rodin) 9340
Esquisse de l'histoire naturelle des Plantaginées (Rapin) 8632
Esquisse des voyages de l'auteur le rapport de la botanique (St. Hil.) 10033
Esquisse organographique et physiologique sur la classe des champignons *(see* J. Pfund)
Esquisses historiques et biographiques des progrès de la botanique en Angleterre (Pult.) 8399
Ess. Agrostogr. (P. Beauv.) 7215

Essai de catalogue des plantes de Zanzibar (Sacleux) 9961
Essai de la flore du sud-ouest de la France (Revel) 9114
Essai de révision des espèces africaines du genre Annona (Robyns) 9313
Essai d'une classification générale des Graminées (Raspail) 8638
Essai d'une division de la France en régions naturelles et botaniques (Raulin) 8652
Essai d'une iconographie des végétaux (Poir.) 8115
Essai d'une monographie des fougères françaises (Palouz.) 7248
Essai d'une nouvelle agrostographie (P. Beauv.) 7215
Essai d'une nouvelle classification de la famille des Graminées (Remy) 8975
Essai fl. Guinée franç. (Pobég.) 8072
Essai fl. s.o. France (Revel) 9114
Essai fl. Zante (see G. Reuter)
Essai géogr. bot. Alpes (Pamp.) 7259
Essai hist.-nat. orangers (Risso) 9247
Essai iconogr. vég. (Poir.) 8115
Essai Leucoloma (Renauld) 8998
Essai monogr. foug. franç. (Palouz.) 7248
Essai nerv. feuill. dicot. (Payer) 7557
Essai nouv. class. Gramin. (Remy) 8975
Essai propr. méd. pl. (see K. Perleb)
Essai sur la flore de la Guinée française (Pobég.) 8072
Essai sur la géographie botanique de Alpes (Pamp.) 7259
Essai sur la nervation des feuilles dans les plantes dicotylées (Payer) 7557
Essai sur la végétation de la Nouvelle Castille (Reut.) 9107
Essai sur les Leucoloma (Renauld) 8998
Essai sur les variétés de la vigne (Roxas) 9715
Essai sur l'histoire naturelle des orangers (Risso) 9247
Essai tax. Hyménomyc. (Pat.) 7483
Essai taxonomique sur les familles et les genres des Hyménomycètes (Pat.) 7483
Essai var. vigne (Roxas) 9715
Essay at a key to British Rubi (W. M. Rogers) 9441
Essay nat. hist. Dublin (Rutty) 9844
Essay on the exotic plants (Raf.) 8555
Essay on the productive resources of India (Royle) 9735
Essay prod. resourc. India (Royle) 9735
Essays nat. hist. peat moss (Rennie) 9077
Essays on the natural history and origin of peat moss (Rennie) 9077
Essay towards a natural history of the county of Dublin (Rutty) 9844

Essbare und giftige Pilze der Schweiz (Rothmayr) 9653
Essbare und giftige Pilze des Waldes (Rothmayr) 9653
Essb. gift. Pilze (Rothmayr) 9653
Essentials of vegetable pharmacognosy (Rusby) 9821
Essent. veg. pharmacogn. (Rusby) 9821
Estepas de España y su vegetación (Reyes y Prosper) 9119
Estepas España (Reyes y Prosper) 9119
Estud. hist. nat. (Ramírez) 8610
Estudio crítico de Agropyron (Parodi) 7400
Estudios criticos sobre la flora de Chile (Reiche) 8853
Estudios de historia natural (Ramírez) 8610
Estudios preliminares para la flora de la provincia de Cáceres (Riv. Mat.) 9262
Étud. bot. (Palustre) 7250
Étude de la famille des Pipéracées (Plusz.) 8071
Étude des Diplotaxis européens (Rouy) 9701
Étude des Lardizabalées (Réaubourg) 8717
Étude des tufs de Montpellier (G. Planch.) 7997
Étude phytogéographique de la Brenne (Rallet) 8600
Étude Sapot. (L. Planch.) 8026
Études de botanique (Palustre) 7250
Études floristiques sur la région de Matapédia (J. Rousseau) 9690
Études sur Collioure et ses environs (Roum.) 9681
Études sur la géographie botanique de l'Italie (Parl.) 7375
Études sur la végétation des plantes potagères d'Europe à la Guyane française (Sagot) 9992
Études sur les collections botaniques des frères Crouan (Picq.) 7917
Études sur les gîtes minéraux de la France ... Bassin Houiller et Permien d'Autun et d'Épinac (Renault) 9050
Études sur les Nymphéacées (Planch.) 8012
Études sur le terrain houiller de Commentry. Flore fossile (Renault) 9041
Étude sur le nouveau genre Hennecartia (J. Poiss.) 8119
Étude sur les Géraniées (Picard) 7880
Étude sur le Sigillaria spinulosa (Renault) 9001
Étude sur les produits de la famille des Sapotacées (G. Planch.) 8002
Étude sur les produits de la famille des Sapotées (L. Planch.) 8026

Étude sur les Stigmaria (Renault) 9013
Étude sur les substratums des lichens (O. J. Rich.) 9162
Étud. géogr. bot. Italie (Parl.) 7375
Étud. gîtes minér., Bassin Autun Épinac., Fl. foss. (Renault) 9050
Étud. Lardizabal. (Réaubourg) 8717
Étud. phytogéogr. Brenne (Rallet) 8600
Étud. Piper. (Plusz.) 8071
Étud. pl. vasc. Grèce *(see* M. Petitmengin)
Étud. prod. Sapotac. (G. Planch.) 8002
Étud. tufs Montpellier (G. Planch.) 7997
Eucalyptus (Sahut) 10013
Eucalyptus globulus (Planch.) 8022
Eur. Laubm. (G. Roth) 9641
Europäischen Arten der Gyrophora anthracina-Gruppe (Du Rietz) 9215
Europäischen Laubmoose (G. Roth) 9641
Europäischen Torfmoose (G. Roth) 9642
Eur. Torfm. (G. Roth) 9642
Evol. ci. nat. Venez. (Pitt.) 7982
Evolución de las ciencias naturales en Venezuela (Pitt.) 7982
Evolutione comparativa Laminariacearum (Reinke) 8938
Examen chymique des pommes de terre (A. Parm.) 7377
Examen d'un ouvrage (Paul.) 7522
Exam. ouvr. (Paul.) 7522
Exam. pommes de terre (A. Parm.) 7377
Excurs.-Fl. Harz. (W. Reinecke) 8919
Excursion botanique dans les montagnes du Bourbonnais *(see* A. Pérard)
Excursion nach dem Ranco-See in der Provinz Valdivia *(see* R. Philippi)
Excursions botaniques en Espagne (O. J. Rich.) 9168
Excursions botaniques en Espagne (Rouy) 9700
Excursions botaniques en Espagne en 1881 et 1882 (Rouy) 9703
Excursionsflora des Harzes (W. Reinecke) 8919
Excursionsflora für das Grossherzogthum Baden *(see* K. Prantl)
Excursions mycologiques (Rolland) 9474
Exerc. bot. (J. Philib.) 7842
Exercices de botanique (J. Philib.) 7842
Exkurs.-Fl. Bayern (Prantl) 8271
Exkursionsflora ...erschienene Literatur (Prantl) 8271
Exkursionsflora für das Königreich Bayern (Prantl) 8271
Exkursionsflora für die Rheinpfalz (Prantl) 8271
Exped. antarct., bot. *(see* P. Petit)
Exped. C.-Südamer. (P. Preuss) 8317
Expedition nach Central- und südamerika (P. Preuss) 8317
Expl. filos. fund. bot. (Paláu) 7201

Explicacion de la filosofia, y fundamentos botanicos (Paláu) 7201
Explicatio partis IV Phytographiae Leonardi Pluc'neti (Phelsum) 7840
Explic. Phytogr. Pluc'n. (Phelsum) 7840
Exploraciones botanicas y otras en la Cuenca de Maracaibo (Pitt.) 7985
Exploration of the Colorado River (J. W. Powell) 8243
Exploration scientifique du Maroc. Botanique (Pitt.) 7974
Explor. Colorado R. (J. W. Powell) 8243
Explor. Cuenca de Maracaibo (Pitt.) 7985
Explor. sci. Maroc, Bot. (Pit.) 7974
Explor. sci. Tunisie, Ill. champ. (Pat.) 7461
Exposé prod. Enghien (J. Parm.) 7382
Exposé succinct de produits d'Enghien (J. Parm.) 7382
Extraits from the literary and scientific correspondence (R. Richardson) 9172
Extr. lit. sci. corr. (R. Richardson) 9172

Fam. fl. pl. (Pollard) 8136
Familiae Polyedriearum monographia (Reinsch) 8956
Familiar lectures on botany (A. Phelps) 7838
Familiarum naturalium regni vegetabilis synopses monographicae (M. Roem.) 9414
Families of flowering plants (Pollard) 8136
Families of plants (Reichard) 8834
Famil. lect. bot. (A. Phelps) 7838
Famille des Malvacées (Payer) 7561
Familles naturelles des plantes, Algues et Champignons (Payer) 7559
Familles naturelles des plantes *(see* J. Payer)
Fam. Malv. (Payer) 7561
Fam. nat. syn. monogr. (M. Roem.) 9414
Fam. pl., alg. champ. (Payer) 7559
Farne Krains (Paulin) 7525
Fasciculus observationum botanicarum (Retz.) 9085
Fasciculus plantarum cryptogamicarum Britanniae *(see* J. J. Roemer)
Fasciculus stirpium bitannicarum (Ray) 8702
Fasc. obs. bot. (Retz.) 9085
Fasc. stir. brit. (Ray) 8702
Fauna and flora of Gloucestershire *(see* P. Reader)
Fauna Fl. südw. Caspi-Geb. (Radde) 8479
Fauna und Flora des südwestlichen Caspi-Gebietes (Radde) 8479
Faune et flore des Pays Çomalis *(see* G. Révoil)
Feddes Repertorium *(see* W. Rothmaler)

Fern-collector's handbook and herbarium (S. F. Price) 8321
Fern-coll. handb. (S. F. Price) 8321
Ferns (J. Robinson) 9300
Ferns Gr. Brit. (Pratt) 8276
Ferns of Bawean (Posth.) 8200
Ferns of Great Britain (Pratt) 8276
Ferns of Surinam (Posth.) 8199
Ferns of Tasmania (Rodway) 9369
Ferns Surinam (Posth.) 8199
Ferns Tasmania (Rodway) 9369
Fertilization in Sphaerocarpos *(see* H. Rickett)
Festuca (Saint-Yves) 10093
Festuca de la Nouvelle Zélande (Saint-Yves) 10100
Festucae novae (Saint-Yves) 10102
Festuca Font-Queri (Saint-Yves) 10099
Fibrous plants of India (Royle) 9737
Fibr. pl. India (Royle) 9737
Fig. Amer. gard. fl. (Raf.) 8581
Fig. foss. Cornwall (J. Phillips) 7871
Figures and descriptions of the Palaeozoic fossils of Cornwall, Devon, and West Somerset (J. Phillips) 7871
Figures of handsome American and garden flowers (Raf.) 8581
Fil. Afr. Orient. exs. (see E. Rosenstock)
Fil. austro-bras. exs. (see E. Rosenstock)
Fil. austro-bras. (see E. Rosenstock)
Fil. austro-bras. (see E. Rosenstock)
Fil. boliv. exs. (see E. Rosenstock)
Fil. costaric. exs. (see E. Rosenstock)
Fil. exoticae (see E. Rosenstock)
Fil. exs. (see E. Rosenstock)
Filices Camerunianae Dinklageanae (Sadebeck) 9969
Filices europae (see E. Rosenstock)
Filices exoticae (see H. Rosendahl)
Filices madagascarienses (Palacký) 7196
Filices novae (H. Rosend.) 9529
Filices palaeotropicae (Rosenstock) 9534
Fil. Javae orient. (see E. Rosenstock)
Fil. Korean. exs. (see E. Rosenstock)
Fil. madagasc. (Palacký) 7196
Fil. mexic. (see E. Rosenstock)
Fil. Novae Caled. (see E. Rosenstock)
Fil. novo-guin. (see E. Rosenstock)
Fil. ver. hung. (Jos. Sadler) 9977
Finlands botaniska litteratur (Saelan) 9982
Finl. bot. litt. (Saelan) 9982
First catalogues and circulars of the botanical garden of Transylvania University (Raf.) 8577
First cat. gard. Transylv. Univ. (Raf.) 8577
Fl. aegaea (Rech. fil.) 8736
Fl. affin. Balsamin. (Roep.) 9420
Flagellaten und Rhizopoden (Pasch.) 7422

Fl. agen. (Saint-Amans) 10017
Fl. agri belgr. (Pančic) 7277
Fl. agr. nyss. (Petrovič) 7791
Fl. agrost. Congo belge (Robyns) 9309
Fl. ajan. (Regel) 8780
Fl. Algérie *(see* P. Petit)
Fl. Amanit. Lépiot. (Quél.) 8419
Fl. Amer. sept. (Pursh) 8404
Fl. amstelaed. (Rombouts) 9488
Fl. aquat. alg. Macéd. (Petkoff) 7778
Fl. atacam. (Phil.) 7851
Fl. atest. prodr. (F. Re) 8710
Fl. Bari (Palanza) 7197
Fl. Bermuda *(see* L. Riddle)
Fl. Bienitz (Peterm.) 7754
Fl. bor.-amer. *(see* J. Richardson)
Fl. bor.-ural. (Rupr.) 9806
Fl. Bourb. (Pérard) 7670
Fl. bras. Mart. (Barb. Rodr.) 9361
Fl. Bras. merid. (St. Hil.) 10034
Fl. bras. *(see* J. Peyritsch, L. Radlkofer), S. Reissek)
Fl. briol. Venezia (Sacc.) 9899
Fl. brux. (Passy) 7438
Fl. bryol. Guinée (Parfitt) 7330
Fl. bryol. Mont-Blanc (Payot) 7573
Fl. Caceres (Riv. Mat.) 9262
Fl. Calvados (Roussel) 9694
Fl. canad. (Prov.) 8371
Fl. cantab. (Reitter) 8973
Fl. cap. med. prodr. (Pappe) 7315
Fl. cap. *(see* H. Pearson, R. Rolfe)
Fl. Caracorùm (Pamp.) 7272
Fl. caucasi (Rupr.) 9811
Fl. čech. (J. Presl) 8287
Fl. Centralkarpath. (Sagorski) 9990
Fl. centr. Belgique (Piré) 7956
Fl. centr. Kazakh. (Pavlov) 7538
Fl. Champagne (Remy) 8974
Fl. Chile (Reiche) 8853
Fl. chil. *(see* A. Richard, J. Rémy)
Fl. Cidlina (Pospichal) 8187
Fl. clochers toitures Poitiers (O. J. Rich.) 9167
Fl. Colorado (Rydb.) 9861
Fl. comit. pest. (Jos. Sadler) 9976
Fl. Conitz (Praet.) 8254
Fl. corcir. (Pieri) 7922
Fl. Costa Rica (Polak.) 8131
Fl. Côte-d'Or (Royer) 9733
Fl. cruc. (R. Peña) 7626
Fl. crypt. cell. Toulouse (Pée-Laby) 7610
Fl. Deutschl. *(see* J. Sturm)
Fl. Deutschl. Wäld. (Pernitzsch) 7694
Fl. dévon. (Potonié) 8224
Fl. Disko Isl. (M. Porsild) 8175
Fl. Eastbourne (Roper) 9495
Fl. East Riding Yorksh. (J. F. Robinson) 9298
Flechten des Bezirkes Steyr (Pehersd.) 7612

Flechten des mittelfränkischen Keupergebietes (Rehm) 8828
Flechten Estlands (Räs.) 8543
Flechtenfl. Ehingen (Rieber) 9213
Flechtenflora der nördlichen Küstengegend (Räs.) 8546
Flechtenflora des Gebiets Ostrobottnia borealis (Räs.) 8541
Flecht. Steyr (Pehersd.) 7612
Fl. Edinburgh (see John Sadler)
Fl. emblems (H. Phillips) 7869
Fl. Erfurt (K. Reinecke) 8918
Fl. Eritrea (Pirotta) 7969
Fl. españ. (Quer) 8421
Fl. Essex Co. (J. Robinson) 9301
Fl. est. Mt. d'Oropa (Pellanda) 7616
Fl. eur. (Roem.) 9404
Fl. eur. alg. (Rabenh.) 8444
Fleur femelle des Conifères (Radais) 8475
Fleurs mâles de Cordaïtes (Renault) 9006
Fleurs sauvages et bois précieux de la Nouvelle Zélande (see E. F. A. Raoul)
Fl. excurs. Mexico (Reiche) 8864
Fl. exot. (Rchb.) 8883
Fl. exs. rhen. (Poeverl.) 8100
Fl. Fauna Culmform. (Rothpletz) 9654
Fl. Faune houill. Belg. (Renier) 9074
Fl. faune Virgile (Paul.) 7523
Fl. fem. Conif. (Radais) 8475
Fl. fenn. brev. (Prytz) 8373
Fl. forest. Cochinch. (Pierre) 7923
Fl. forojul. syll. (Pirona) 7964
Fl. foss. braid. (Peola) 7657
Fl. foss. sused. (Pilar) 7933
Fl. fotogr. Sanremo (Panizzi) 7285
Fl. France (Rouy) 9706
Fl. Gangetic plain (see R. Parker)
Fl. Gard (Pouzolz) 8242
Fl. gén. Belgique (Robyns) 9314
Fl.-gén. France (see P. Poiteau)
Fl. gén. Indochine (see C. Pitard, F. Pellegrin)
Fl. germ. excurs. (Rchb.) 8881
Fl. Glied. Carbon (Potonié) 8216
Fl. Gloucestershire (Riddelsdell) 9202
Fl. gramin. (Peterm.) 7749
Fl. Grand-Jouan (Saint-Gal) 10020
Fl. Helder (see H. Redeke)
Fl. Herefordshire (Purchas) 8401
Fl. Hertfordshire (Pryor) 8372
Fl. hist. (H. Phillips) 7868
Fl. Iakut. (Petrov) 7790
Fl. ill. seasons (M. Roscoe) 9504
Fl. ind. (Roxb.) 9724
Fl. ind. ed. 1832 (Roxb.) 9725
Fl. Indiana dunes (Peattie) 7593
Fl. ingr. (Rupr.) 9809
Fl. Isthmus-wüste (Range) 8628
Fl. ital. (Parl.) 7361
Fl. ital. crypt., Alg. (Preda) 8282
Fl. ital. crypt., Gasterales (Petri) 7788
Fl. ital. crypt., Hymeniales (Sacc.) 9950
Fl. Italia sett. (Perini) 7679
Fl. Ital. sup. (Pass.) 7434
Fl. Jamaica (see A. Rendle)
Fl. jen. (J. H. Rudolph) 9753
Fl. jen. (Ruppius) 9790
Fl. Kärnten (Pacher) 7178
Fl. kawkasa (see K. Ronniger)
Fl. Leipzig (see H. Richter)
Fl. leyd. prodr. (Royen) 9730
Fl. Lidi veneti (Ruchinger) 9744
Fl. Linzergegend (Sailer) 10015
Fl. lips. excurs. (Peterm.) 7751
Fl. lips. pharm. (Rchb.) 8869
Fl. lips. pharm. specim. (Rchb.) 8868
Fl. litt. médit. (Penz.) 7654
Fl. Lorraine (Petitm.) 7777
Fl. ludov. (Raf.) 8567
Fl. Lübeck (K. Petersen) 7764
Fl. lusat. (Rabenh.) 8433
Fl. Madagascar (see H. Perrier)
Fl. Malay. Penins. (Ridl.) 9211
Fl. marchig. (Paolucci) 7308
Fl. Mark Brandenburg (Ruthe) 9839
Fl. Marlborough (Preston) 8314
Fl. Mayenne (P. Reverchon) 9115
Fl. Mayombe (Pellegr.) 7618
Fl. Mecklenb. (Roep.) 9423
Fl. med. Napoli (Pasq.) 7426
Fl. Menorca (J. Rodríguez) 9365
Fl. mică ilus. Român. (Procopianu) 8360
Fl. miquelon. (Renauld) 8986
Fl. Moden. (Pirotta) 7966
Fl. moeno-francof. (Reichard) 8833
Fl. Mont-Blanc, Phan. (Payot) 7572
Fl. Monte Calv. (S. Rossi) 9576
Fl. Montello (F. Sacc.) 9880
Fl. Mt. Desert Isl. (E. L. Rand) 8623
Fl. mycol. Bruxelles (see M. Rousseau)
Fl. mycol. France (Quél.) 8418
Fl. mycol. Tarn-et-Garonne (Roum.) 9682
Fl. N. Amer. (see L. Pennington)
Fl. Namur Luxemb. (Pâque) 7321
Fl. Năseud. (Porcius) 8170
Fl. Nebraska (N. Petersen) 7765
Fl. Nebr., Rosales (Rydb.) 9847
Fl. Neuhaldensleben (Robolsky) 9305
Fl. nicarag. (Ramírez Goyena) 8611
Fl. Nice (Risso) 9250
Fl. Niederhessen (Pfeiff.) 7819
Fl. Nord France (Roucel) 9671
Fl. nouv. chaine jurass. (P. Parm.) 7383a
Fl. n.w. coast (Piper) 7950
Fl. Oberöstr. (Sailer) 10014
Fl. oecon. Svec. (Retz.) 9089
Fl. oesterr. Küstenl. (Pospichal) 8187a
Flora aegaea (Rech. fil.) 8736
Flora agri belgradensis (Pančic) 7277
Flora agri nyssani (Petrovič) 7791

Flora alpina (Reiner) 8920
Flora Americae septentrionalis (Pursh) 8404
Flora Americae septentrionalis (Raf.) 8568
Flora amstelaedamensis (Rombouts) 9488
Flora analitica d'Italia *(see* G. Paoletti)
Flora and fauna der Zuiderzee *(see* H. Redeke)
Flora anglica abbreviata *(see* R. Pulteney)
Flora Bohemiae et Moraviae exsiccata (see F. Petrak)
Flora Bohemiae et Moraviae exsiccata (see R. Picbauer)
Flora Bohemiae et Moraviae exsiccata (see R. Picbauer, J. Podpěra)
Flora bonnensis *(see* E. Regel)
Flora boreali-americana *(see* L. Richard)
Flora boreali-uralensis (Rupr.) 9802, 9806
Flora Brasiliae meridionalis (St. Hil.) 10034
Flora brasiliensis de Martius (Barb. Rodr.) 9361
Flora brasiliensis *(see* A. Progel, O. Petersen, P. Rohrbach)
Flora briologica della Venezia (Sacc.) 9899
Flora britannica *(see* J. J. Roemer)
Flora cantabrigensis (Reitter) 8973
Flora capensis *(see* E. Phillips, D. Prain)
Flora carpatorum centralium (Sagorski) 9990
Flora caucasi (Rupr.) 9811
Flora caucasica critica *(see* I. Palibin)
Flora čechica (J. Presl) 8287
Flora central'nogo Kazakhstana (Pavlov) 7538
Flora comitatus pestiensis (Jos. Sadler) 9976
Flora corcirensis (Pieri) 7922
Flora corcirese *(see* M. Pieri)
Flora cruceña (R. Peña) 7626
Flora cryptogama Sequaniae exsiccata (see J. Paillot)
Flora ČSR *(see* A. Pilát)
Flora cubana (Sagra) 10000
Flora Dalmatica (see W. Pamplin)
Flora danica *(see* C. Rosenberg)
Flora de Chile (Reiche) 8853
Flora de Costa Rica (Polak.) 8131
Flora dei contorni di Parma (Pass.) 7435
Flora dei Lidi veneti (Ruchinger) 9744
Flora de la provincia de Caceres (Riv. Mat.) 9262
Flora de la republica Argentina y Paraguay (D. Parodi) 7390
Flora del Cadore *(see* R. Pampanini)
Flora del Caracorùm (Pamp.) 7272
Flora della colonia Eritrea (Pirotta) 7969
Flora del Langhiano torinese (Peola) 7658
Flora della provincia di Roma *(see* P. Pirotta)

Flora della Repubblica di San Marino (Pamp.) 7271
Flora della terra di Bari (Palanza) 7197
Flora delle vette di Feltre (Sacc.) 9939
Flora dell'Italia settentrionale (Perini) 7679
Flora del Modenese e del Reggiano (Pirotta) 7966
Flora del Monte Calvario (S. Rossi) 9576
Flora der Centralkarpathen (Sagorski) 9990
Flora der Gefäss-Pflanzen der Wetterau (Russ) 9826
Flora der Gegend um den Ursprung der Donau (Rot von Schr.) 9622
Flora der Insel Jan Mayen (Reichardt) 8850
Flora der Isthmuswüste (Range) 8628
Flora der Linzergegend (Sailer) 10015
Flora der Mark Brandenburg (Ruthe) 9839
Flora der Mittelmark in getrockneten Exemplaren (see J. Ruthe)
Flora der Provinz Preussen (Patze) 7509
Flora der Samoa-Inseln (Reinecke) 8915
Flora der Samoa-Inseln *(see* F. Reinecke)
Flora der Samoa-Inseln *(see* T. Reinbold)
Flora der Schweiz (Heinr. Röm.) 9399
Flora der Umgebung Lemsals und Laudohns (Rapp) 8636
Flora der Umgebung Reval's (Russow) 9831
Flora der Umgegend von Neuhaldensleben (Robolsky) 9305
Flora der Umgegend von Schweidnitz (F. Peck) 7603
Flora des Amtsbezirkes Waldmünchen (Progel) 8364
Flora des Bienitz (Peterm.) 7754
Flora des Flussgebietes der Cidlina und Mrdlina (Pospichal) 8187
Flora des Grossherzogthums Posen (Ritschl) 9252
Flora des Königreichs Sachsen (Rabenh.) 8439
Flora des Namalandes *(see* P. Range)
Flora des oesterreichischen Küstenlandes (Pospichal) 8187a
Flora des Rothliegenden von Thüringen (Potonié) 8215
Flora des Unter-Donau-Kreises (L. Reuss) 9105
Flora din fostulŭ districtŭ Romanescŭ alŭ Năsĕuduluĭ (Porcius) 8170
Florae atestinae prodromus (F. Re) 8710
Florae cantabrigiensi supplementum (Reitter) 8973
Florae capensis medicae prodromus (Pappe) 7315
Florae fennicae breviarum (Prytz) 8373

Florae forojuliensis syllabus (Pirona) 7964
Florae jenensis plantas (J. H. Rudolph) 9753
Florae leydensis prodromus (Royen) 9730
Florae ligusticae synopsis (Penz.) 7649
Florae lipsiensis pharmaceuticae specimen (Rchb.) 8868
Florae Polonicae prodromus (Rost.) 9592
Florae Scandinaviae prodromus (Retz.) 9087
Florae senegambiae tentamen *(see* G. Perrottet)
Flora española (Quer) 8421
Flora europaea (Roem.) 9404
Flora europaea algarum (Rabenh.) 8444
Flora excursoria en el Valle central de Mexico (Reiche) 8864
Flora exotica (Rchb.) 8883
Flora exsiccata austro-hungarica (see G. Pernhoffer)
Flora exsiccata bavarica (see H. Poeverlein, P. Reinsch)
Flora exsiccata Carniolica (see A. Paulin)
Flora exsiccata reipublicae Bohemicae Slovenicae (see J. Podpěra)
Flora exsiccata rhenana (Poeverl.) 8100
Flora exsiccata rhenana (see H. Poeverlein)
Flora fennica (Saelan) 9980
Flora fossile braidese (Peola) 7657
Flora fossilis susedana (Pilar) 7933
Flora fotografata Sanremo (Panizzi) 7285
Flora germanica excursoria (Rchb.) 8881
Flora germanica exsiccata (see L. Reichenbach)
Flora historica (H. Phillips) 7868
Flora Iakutiae (Petrov) 7790
Flora im Winterkleide (Rossmässler) 9583
Flora indica (Roxb.) 9724, 9725
Flora ingrica (Rupr.) 9809
Flora ingrica (see F. Ruprecht)
Flora iranica (see R. H. Rechinger)
Flora Italiae superioris (Pass.) 7434
Flora italiana (Parl.) 7361
Flora italica cryptogama, Algae (Preda) 8282
Flora italica cryptogama Gasterales (Petri) 7788
Flora italica cryptogama Hymeniales (Sacc.) 9950
Flora italica cryptogamica (see P. Saccardo)
Flora italica exsiccata (see R. Pampanini)
Flora jenensis (Ruppius) 9790
Flora kopalna ogniotrwałych glinek krakowskich (Racib.) 8467
Floral emblems (H. Phillips) 7869
Floral illustrations of the seasons (M. Roscoe) 9504
Flora lipsiensis excursoria (Peterm.) 7751
Flora lipsiensis pharmaceutica (Rchb.) 8869

Flora lithuanica exsiccata (see C. Regel)
Flora lusatiae inferioris exsiccata (see G. Rabenhorst)
Flora lusatica (Rabenh.) 8433
Flora marchigiana (Paolucci) 7308
Flora medica Provincia di Napoli (Pasq.) 7426
Flora mică ilustradă a Repulicii populare Romine (Procopianu) 8360
Flora mică ilustrăta a României (Procopianu) 8360
Flora miquelonensis (Renauld) 8986
Flora moeno-francofurtana (Reichard) 8833
Floram pedemontanam appendix (Re) 8712
Flora neerlandica (see T. Reichgelt)
Flora nicaragüense (Ramírez Goyena) 8611
Flora Oberöstreichs (Sailer) 10014
Flora oeconomica Sveciae (Retz.) 9089
Flora of Canada (A. M. Ross) 9559
Flora of Colorado (Rydb.) 9861
Flora of Colorado (see H. Patterson)
Flora of Columbia, Missouri (Rickett) 9195
Flora of Cook County *(see* C. Raddin)
Flora of County Armagh (Praeger) 8247
Flora of Disko Island (M. Porsild) 8175
Flora of Eastbourne (Roper) 9495
Flora of Essex county (J. Robinson) 9301
Flora of Glamorganshire (Riddelsdell) 9201
Flora of Gloucestershire (Riddelsdell) 9202
Flora of Herefordshire (Purchas) 8401
Flora of Hertfordshire (Pryor) 8372
Flora of lowland Iraq *(see* R. H. Rechinger)
Flora of Marlborough (Preston) 8314
Flora of Mount Desert Island, Maine *(see* T. Porter, J. Redfield)
Flora of Mount Rainier (Piper) 7945
Flora of Mt. Baker (St. John) 10061
Flora of Nebraska (N. Petersen) 7765
Flora of Nebraska, Rosales (Rydb.) 9847
Flora of Pennsylvania (Porter) 8183
Flora of Reykjanes Peninsula (Rosenvinge) 9548
Flora of Singapore (Ridl.) 9207
Flora of Skipton (Rotheray) 9647
Flora of Southeastern Washington (Piper) 7949
Flora of Southeastern Washington (St. John) 10064
Flora of Suriname *(see* O. Posthumus)
Flora of Syria and Egypt (Post) 8191
Flora of Syria, Palestine, and Sinai (Post) 8191
Flora of the Black Hills of South Dakota (Rydb.) 9849

Flora of the Conowingo barrens (Pennell) 7635
Flora of the East Riding of Yorkshire (J. F. Robinson) 9298
Flora of the Galapagos Islands (B. L. Robinson) 9289
Flora of the Indiana dunes (Peattie) 7593
Flora of the Jaroslaw district *(see* A. Petrowsky)
Flora of the Malay Peninsula (Ridl.) 9211
Flora of the North-East of Ireland *(see* R. Praeger)
Flora of the northwest coast (Piper) 7950
Flora of the Pacific Slope (see C. Pringle)
Flora of the Palouse region (Piper) 7944
Flora of the Pays d'Enhaut (Pitt.) 7976
Flora of the prairies and plains of Central North America (Rydb.) 9873
Flora of the Rocky Mountains (Rydb.) 9865
Flora of the Sand Hills of Nebraska (Rydb.) 9848
Flora of the State of Washington (Piper) 7946
Flora of Warren County, Kentucky (S. F. Price) 8320
Flora of Weston (St. Brody) 10019
Flora of Worcester County, Maryland (Redmond) 8746
Flora okoline Niša (Petrovič) 7791
Flora palermitana (Parl.) 7360
Flora panormitana (Parl.) 7347
Flora parisiensis (Poit.) 8122
Flora pentru determinarea (Procopianu) 8359
Flora peruvianae, et chilensis prodromus (Ruiz) 9769
Flora peruviana, et chilensis (Ruiz) 9771
Flora Philadelphica prodromus (Raf.) 8562
Flora pliocenică dela Borsec (Pop) 8158
Flora polonica exsiccata (see A. Rehmann)
Flora polska (see M. Raciborski)
Flora popolare italiana (Penz.) 7656
Flora popolare ligure (Penz.) 7650
Flora principatus Serbiae (Pančic) 7279
Flora pyrenaea *(see* A. Penzig)
Flora republicii socialiste România *(see* E. Pop)
Flora romana (Pirotta) 7968
Flora rossica (Pall.) 7227
Flora saxonica (Rchb.) 8889
Flora sedinensis (Rostk.) 9602
Flora *(see* O. Renner)
Flora segusiensis (Re) 8711
Flora Sequaniae exsiccata (Paill.) 7191
Flora Sequaniae exsiccata ou herbier de la Flore du Franche-Comté (see J. Paillot)
Flora sicula (K. Presl) 8294
Flora srednej Sibiri *(see* M. Popov)

Flora Sta. Helenica (Roxb.) 9723
Flora Straubingensis (Raab) 8429
Flora Suriname (Pulle) 8393
Flora tarvisiana renovata (Sacc.) 9951
Flora telluriana (Raf.) 8587
Flora torinese (Re) 8713
Flora tremesnensis (Pampuch) 7276
Flora trevigiana (Sacc.) 9947
Flora tridentina exsiccata (see C. Perini)
Flora tropical Africa *(see* D. Prain)
Flora und Fauna der Culmformation (Rothpletz) 9654
Flora URSS *(see* I. Palibin)
Flora veneta crittogamica *(see* D. Saccardo)
Flora veronensis (Pollini) 8145
Flora vesuviana (Pasq.) 7430
Flora virgiliana (Retz.) 9091
Flora von Deutschlands Wäldern (Pernitzsch) 7694
Flora von Deutsch-Ostafrika (Peter) 7748
Flora von Erfurt (K. Reinecke) 8918
Flora von Kärnten (Pacher) 7178
Flora von Lübeck un Umgebung (K. Petersen) 7764
Flora von Niederhessen und Münden (Pfeiff.) 7819
Flora von Sachsen (Rückert) 9761
Flora von Scheyern (Popp) 8165
Flora von Südhannover (Peter) 7746
Flora von Trier (Rosbach) 9503
Flore agenaise (Saint-Amans) 10017
Flore agrostologique du Congo belge (Robyns) 9309
Flore analytique de poche de la Lorraine (Petitm.) 7777
Flore analytique du centre de la Belgique (Piré) 7956
Flore analytique et descriptive des cryptogames cellulaires des environs de Toulouse (Pée-Laby) 7610
Flore aquatique et algologique de la Macédoine (Petkoff) 7778
Flore canadienne (Prov.) 8371
Flore de Chamounix (Payot) 7567
Flore de Chile *(see* F. Philippi)
Flore de France (Rouy) 9706
Flore de la Champagne (Remy) 8974
Flore de la Côte-d'Or (Royer) 9733
Flore de la Mayenne (P. Reverchon) 9115
Flore de l'Yonne (Ravin) 8689
Flore de Namur et Luxembourg (Pâque) 7321
Flore de Nice (Risso) 9250
Flore des environs de Grand-Jouan (Saint-Gal) 10020
Flore de serres (Rchb.) 8883
Flore des plantes vénéneuses (Rapin) 8634
Flore des Pyrénées (Philippe) 7847
Flore des Rives de la Touque (Poplu) 8161

Flore des Serres *(see* J. Planchon)
Flore d'Étampes (Roze) 9739
Flore dévonienne (Potonié) 8224
Flore d'Oware (P. Beauv.) 7212
Flore du Bourbonnais (Pérard) 7670
Flore du Calvados (Roussel) 9694
Flore du département de l'Orne (P. Renault) 9063
Flore du département du Gard (Pouzolz) 8242
Flore du littoral méditerranéen (Penz.) 7654
Flore du Mayombe (Pellegr.) 7618
Flore du Nord de la Françe (Roucel) 9671
Flore estiva dei Monti d'Oropa (Pellanda) 7616
Flore et faune de Virgile (Paul.) 7523
Flore et faune houillères de la Belgique (Renier) 9074
Flore forestière de la Cochinchine (Pierre) 7923
Flore générale de Belgique (Robyns) 9314
Flore générale Indochine *(see* D. Prain)
Flore gramineo (Peterm.) 7749
Flore médicale (Poir.) 8115
Flore médicale (Réveil) 9110
Flore monographique des Amanites et des Lépiotes (Quél.) 8419
Flore mycologique de la France (Quél.) 8418
Flore mycologique du département de Tarn-et-Garonne (Roum.) 9682
Flore nouvelle de la chaine jurassique (P. Parm.) 7383a
Flore ou statistique botanique de la Seine-inférieure (F. Pouchet) 8236
Flore pittoresque de la France (Rothschild) 9663
Flore populaire (E. Rolland) 9471
Flores samejedorum cisuralensium (Rupr.) 9796
Flore utile de la France (Reymond) 9122
Fl. orient *(see* L. Rauwolff)
Floristische Gliederung des deutschen Carbon und Perm (Pontonié) 8216
Floristisches aus der Umgebung des Neusiedler Sees (Rech. fil.) 8729
Fl. Orne (P. Renault) 9063
Florula aethiopico-aegyptica *(see* F. Parlatore)
Florula ajanensis (Regel) 8780
Florula atacamensis (Phil.) 7850, 7851
Florula bostoniensis (Raf.) 8570
Florula bruxellensis (Passy) 7438
Florula del Montello (F. Sacc.) 9880
Flórula de Menorca (J. Rodríguez) 9365
Florula gaditana (Pérez Lara) 7678
Florula ludoviciana (Raf.) 8567
Florula ludoviciana. Flora of Louisiana (Raf.) 8567

Florula mycologica lusitanica (Sacc.) 9918
Florule bryologique de la Guinée française (Parfitt) 7330
Florule bryologique Mont-Blanc (Payot) 7573
Florule de Saint-Domingue *(see* P. Poiteau)
Florule des alluvions de la Saône aux environs de Chalon (Quincy) 8425
Florule des clochers et des toitures des églises de Potiers (O. J. Rich.) 9167
Florule du Mont-Blanc, Phanérogames (Payot) 7572
Fl. Oware (P. Beauv.) 7212
Flowering plants and ferns of Great Britain (Pratt) 8275
Flowering plants, grasses, sedges and ferns of Great Britain (Pratt) 8275
Flowering plants of Great Britain (Pratt) 8275
Flowering plants of South Africa (Pole-Evans) 8132
Flowering plants of South Africa *(see* E. Phillips)
Flowering plants of Travancore (Rama Rao) 8603
Flowering plants of Wilts (Preston) 8315
Flowering shrubs of California (Rowntree) 9714
Flowers and flowering plants (Pool) 8157
Fl. palerm. (Parl.) 7360
Fl. Palouse reg. (Piper) 7944
Fl. panorm. (Parl.) 7347
Fl. paris. (Poit.) 8122
Fl. Parma (Pass.) 7435
Fl. Pays d'Enhaut (Pitt.) 7976
Fl. Pennsylvania (Porter) 8183
Fl. peruv. (Ruiz) 9771
Fl. peruv. prodr. (Ruiz) 9769
Fl. pittor. France (Rothschild) 9663
Fl. pl. (Pool) 8157
Fl. Plains N. Amer. (Rydb.) 9873
Fl. plioc. Borsec (Pop) 8158
Fl. pl. Gr. Brit. (Pratt) 8275
Fl. pl. S. Afr. (Pole-Evans) 8132
Fl. pl. Travancore (Rama Rao) 8603
Fl. pl. vénén. (Rapin) 8634
Fl. pl. Wilts (Preston) 8315
Fl. pop. (E. Rolland) 9471
Fl. pop. ital. (Penz.) 7656
Fl. pop. ligure (Penz.) 7650
Fl. Posen (Ritschl) 9252
Fl. Preuss. (Patze) 7509
Fl. Pyren. (Philippe) 7847
Fl. Reval. (Russow) 9831
Fl. Rives Touque (Poplu) 8161
Fl. Rocky Mts. (Rydb.) 9865
Fl. Rom. (Pirotta) 7968
Fl. Român. (Procopianu) 8359
Fl. ross. (Pall.) 7227
Fl. Rothlieg. (Potonié) 8215

Fl. Sachsen (Rückert) 9761
Fl. samojed. cisural. (Rupr.) 9796
Fl. San Marino (Pamp.) 7271
Fl. São Paulo (see E. Palla)
Fl. S. Austral. (see R. S. Rogers)
Fl. saxon. (Rchb.) 8889
Fl. Scand. prodr. (Retz.) 9087
Fl. Scheyern (Popp) 8165
Fl. Schweiz. (Heinr. Röm.) 9399
Fl. Sedin. (Rostk.) 9602
Fl. segus. (Re) 8711
Fl. segusina (Re) 8711
Fl. Seine-inf. (F. Pouchet) 8236
Fl. Seneg. tent. (see A. Richard)
Fl. Serbiae (Pančic) 7279
Fl. s.e. Washington (Piper) 7949
Fl. S.-e. Washington (St. John) 10064
Fl. shrubs Calif. (Rowntree) 9714
Fl. sicul. (K. Presl) 8294
Fl. Skipton (Rotheray) 9647
Fl. Südhannover (Peter) 7746
Fl. Suriname (Pulle) 8393
Fl. Surrey (see W. Pearsall)
Fl. Syria (Post) 8191
Fl. Tambov (see A. Petunnikov)
Fl. tarvis. renov. (Sacc.) 9951
Fl. tellur. (Raf.) 8587
Fl. Thüring. 1 (Rottenbach) 9668
Fl. Thüring. 2-8 (Rottenbach) 9669
Fl. torin. (Re) 8713
Fl. tremesn. (Pampuch) 7276
Fl. trevig. (Sacc.) 9947
Fl. Trier (Rosbach) 9503
Fl. trop. Afr. (see A. Rendle, R. Rolfe)
Fl. Unt.-Donau-Kr. (L. Reuss) 9105
Fl. Ursprung Donau (Rot von Schr.) 9622
Fl. URSS (see N. Pavlov, E. Pobedimova, M. Popov, Y. Prokhanov, R. Roshevitz)
Fl. utile France (Reymond) 9122
Fl. veron. (Pollini) 8145
Fl. vesuv. (Pasq.) 7430
Fl. virgil. (Retz.) 9091
Fl. vitiens. (see H. G. Reichenbach)
Fl. Vosges, Champ. (see L. Quélet)
Fl. Warren Co. (S. F. Price) 8320
Fl. Washington (Piper) 7946
Fl. Westf. Industriegeb. (see H. Preuss)
Fl. Westmorland (see J. Pickard)
Fl. Weston (St. Brody) 10019
Fl. Winterkl. (Rossmässler) 9583
Fl. Worcester Co. (Redmond) 8746
Fl. Yonne (Ravin) 8689
Føroya fl. (Rasmussen) 8637
Føroya flora (Rasmussen) 8637
Försök till en systematisk uppställning af vextställena (H. v. Post) 8192
Förs. uppställn. vextställena (H. v. Post) 8192
Förteckn. Göteb. fanerog. (J. E. Palmér) 7235

Förteckning öfver Vasstergötlands fanero-gamer (Rudberg) 9750
Förteckning över Göteborgs och Bohus Läns fanerogamer (J. E. Palmér) 7235
Förteckn. Västergötl. fan. (Rudberg) 9750
Fonds de la Mer (P. Petit) 7769
Fontes fl. Lituaniae (C. Regel) 8769
Fontes florae Lituaniae (C. Regel) 8769
Forbes's New Guinea pl. (Rendle) 9070
Forest fl. Andaman Isl. (C. Parkinson) 7342
Forest flora for the Punjab (R. Parker) 7339
Forest flora of the Andaman Islands (C. Parkinson) 7342
Forest fl. Punjab (R. Parker) 7339
Forests and flora of British Honduras (see S. Record)
Forest trees of Canada (A. M. Ross) 9558
Formation et aspect du relief actuel de Cé-vennes (Pantel) 7288
Formenkr. Cirsium eriophorum (Petr.) 7779
Formenkreis des Cirsium eriophorum (Petr.) 7779
Form. relief Cévennes (Pantel) 7288
Forstbot. (Reum) 9096
Forstbotanik (Reum) 9096
Forstwissenschaftliches Schriftsteller-Lexikon (Ratzeb.) 8648
Forstwiss. Schriftst.-Lex. (Ratzeb.) 8648
Fortschritte der Botanik (see O. Renner)
Forty trees Common in India (R. Parker) 7340
Foss. fl. Steinkohlen-Form. Westph. (Roehl) 9376
Fossile Flora der Steinkohlen-Formation Westphalens (Roehl) 9376
Fossil flora of Tegelen-sur-Meuse (C. Reid) 8902
Fossilium, catalogus (see O. Posthumus)
Fougères (Rivière) 9263
Fougères (Rothschild) 9662
Fougères de France (Rey-Pailhade) 9117
Fougères (see E. Roze)
Foug. France (Rey-Pailhade) 9117
Fragmens d'un voyage dans les Pryénées (Saint-Amans) 10016
Fragmenta florae Philippinae (Perkins) 7684
Fragments mycologiques (Pat.) 7447
Fragmentum synopseos plantarum phane-rogamum (Poepp.) 8089
Fragm. fl. Philipp. (Perkins) 7684
Fragm. fl. Philipp. (see R. Pilger, W. Ruhland)
Fragm. Syn. Pl. (Poepp.) 8089
Fragm. voy. Pryén. (Saint-Amans) 10016
Freiherrn von Hochberg botanischer Garten zu Hlubosch (Pohl) 8102

Freshwater algae of the Lismore district (Playf.) 8043
Fresh-water Cyanophyceae, Botany of Iceland *(see* J. Petersen)
Fresh-water Cyanophyceae of Iceland (Rosenvinge) 9548
Fresh-water Diatoms from Iceland (Rosenvinge) 9548
Frullaniae madagascarienses (Pearson) 7587
Frull. madag. (Pearson) 7587
Führ. Bot. Gart. (K. Peters) 7758
Führ. bot. Gart. Breslau (Pax) 7547
Führer durch den königlichen botanischen Garten Breslau (Pax) 7547
Führer in die Pflanzenwelt (Postel) 8195
Führer zu einem Rundgang durch die Freiland-Anlagen des Botanischen Gartens (K. Peters) 7758
Führ. Pflanzenw. (Postel) 8195
Fund. pl. not. (O. O. Rudbeck) 9748
Fung. agrum. (Penz.) 7642
Fung. Alaska (Peck) 7602
Fung. austriac. (Pröll) 8363
Fung. Bav. Palat. nasc. (Pers.) 7729
Fung. Bav. Palat. nasc. *(see* C. Persoon)
Fung. Faeröes (Rostr.) 9616
Funghi agrumicoli (Penz.) 7642
Funghi della Mortola (Penz.) 7643
Funghi delle Ardenne (Sacc.) 9905
Funghi dell'Isola del Giglio (Sacc.) 9927
Funghi mangerecci della regione Marchigiana (Paolucci) 7310
Funghi Mortola (Penz.) 7643
Funghi parassiti dei vitigni (Pirotta) 7965
Fungi (Ramsb.) 8621
Fungi abellinenses novi (Sacc.) 9932
Fungi abyssinici (Sacc.) 9911
Fungi albanici et bosniae exs. (see F. Petrak)
Fungi aliquot africani (Sacc.) 9941
Fungi aliquot brasilienses phyllogeni (Sacc.) 9923
Fungi aliquot mycologiae romanae addendi (Sacc.) 9909
Fungi americani (see H. Ravenel)
Fungi austro-americani (see J. Rick)
Fungi britannici (see C. Plowright, W. Phillips)
Fungi caroliniani (see H. Ravenel)
Fungi carpatici lignicoli (see A. Pilát)
Fungi collected by H.G. Simmons (Rostr.) 9620
Fungi eichleriani (see F. Petrak)
Fungi ex insula Melita (Sacc.) 9948
Fungi exsiccati praecipue gallici (see L. Quélet, E. Roze, L. Rolland)
Fungi exsiccati praesertim scandinavici (see L. Romell)
Fungi from the Faeröes (Rostr.) 9616
Fungi gallici exsiccati (see C. Roumeguère)

Fungi Groenlandiae (Rostr.) 9611
Fungi italici (Sacc.) 9894
Fungilli aliquot herbarii regii bruxellensis (Sacc.) 9915
Fungilli novi europaei et asiatici (Sacc.) 9919
Fungi longobardiae exsiccatae (see G. Pollaccii)
Fungi of Alaska (Peck) 7602
Fungi of Alaska *(see* P. Saccardo)
Fungi of Ceylon *(see* T. Petch)
Fungi of Iceland (Rosenvinge) 9548
Fungi of the Isle of Wight *(see* J. Rayner)
Fungi pathogenic to man *(see* J. Ramsbottom)
Fungi philippinenses (Sacc.) 9932
Fungi polonici exs. (see F. Petrak)
Fungi selecti exsiccati (see C. Plowright, G. Passerini)
Fungi selecti gallici exsiccati (see C. Roumeguère)
Fungi singaporenses bakeriani (Sacc.) 9932
Fung. ital. (Sacc.) 9894
Fungi tripolitani (Sacc.) 9949
Fungi veneti novi vel critici (Sacc.) 9890
Fungi veronenses (Sacc.) 9932
Fung. mang. (Paolucci) 7310
Fung. manger. (Sacc.) 9913
Fungorum novorum decas prima (Pat.) 7489
Fungos austriacos (Pröll) 8363
Fung. Simmons (Rostr.) 9620
Fungus flora of Cornwall *(see* F. Rilstone)
Further new and imperfectly known plants collected in Mexico (B. L. Robinson) 9285

Ganoderma (Pat.) 7450
Gartenbast. Hieracium (Peter) 7745
Gartenflora (Regel) 8775
Gatt. Pfl. (Planer) 8030
Gatt. Pyrenomyz. (Petr.) 7785
Gattung Boscia (Pestal.) 7743
Gattungen der mitteleuropäischen Scirpoideen (Palla) 7220
Gattungen der Pflanzen (Planer) 8030
Gattungen der Pyrenomyzeten, Sphaeropsideen, und Melanconieen (Petr.) 7785
Gattung Plantago in Zentral- und Südamerika (Pilg.) 7938
Gattung Pleuroplitis (Regel) 8793
Gazapos bot. (Pau) 7512
Gazapos botáicos (Pau) 7512
Gefässbündel Farrn (K. Presl) 8311
Gefässbündel im Stipes der Farrn (K. Presl) 8311
Gefässkrypt., Gymn. monocot. Angiosp. Ober-Lausitz (Rabenau) 8432
Gefässkryptogamen, Gymnospermen und moncotyledonischen Angiospermen der

Ober-Lausitz (Rabenau) 8432
Gefässpflanzen der Urkantone von Zug (Rhiner) 9131
Gen. Armeriae (F. Petri) 7787
Gen. cat. fl. Vermont (G. Perkins) 7681
Genera et species orchidearum novarum (Barb. Rodr.) 9342
Genera fungorum exsiccata (see C. Roumeguère)
General catalogue of the flora of Vermont (G. Perkins) 7681
Genera lichenum europaeaorum exsiccata (see C. Roumeguère)
General view of the writings of Linnaeus (Pult.) 8398
Genera of North American Fabaceae (Rydb.) 9868
Genera of North-American plants (Raf.) 8572
Genera of South African flowering plants (E. Phillips) 7861
Genera plantarum (Reichard) 8834
Genera plantarum venezuelensium (Pitt.) 7992
Genera pryenomycetum (Sacc.) 9900
Genere Armeriae (F. Petri) 7787
Generic concept in the classification of the flowering plants (B. L. Robinson) 9292
Gen. pl. (Reichard) 8834
Gen. pl. ed. 2 *(see* A. Royen)
Gen. pl. ed. 6 *(see* D. Royen)
Gen. pl. fl. Germ. *(see* A. Putterlick)
Gen. pl. *(see* A. Royen)
Gen. pl. venez. (Pitt.) 7992
Gen. pyrenomyc. (Sacc.) 9900
Genre grammatical des noms génériques (St. Lag.) 10086
Genres nouveaux de Sapotacées (Pierre) 7923a
Gen. S. Afr. fl. pl. (E. Phillips) 7861
Gen. sp. Orchid. (Barb. Rodr.) 9342
Gentianella du groupe Grandiflora (St. Lag.) 10084
Genus Aspergillus *(see* K. Raper)
Genus Fumaria (Pugsl.) 8380
Genus Lesquerella *(see* R. Rollins)
Genus Trachelomonas (Playf.) 8042
Gen. view writ. Linnaeus (Pult.) 8398
Geobotanische Untersuchungsmethoden (Rübel) 9759
Geobot. Unters.-Meth. (Rübel) 9759
Géografia botánica de Chile (Reiche) 8860
Geographie und Geschichte der Pflanzen (M. Roem.) 9412
Geogr. Gesch. Pfl. (M. Roem.) 9412
Geol. mineral. *(see* J. Phillips)
Geologisches Zentralblatt *(see* R. Potonié)
Geology of Pennsylvania (H. D. Rogers) 9435
Geol. Pennsylvania (H. D. Rogers) 9435

Ges. Abh. Pfl.-Physiol. (Sachs) 9959
Gesammelte Abhandlungen über Pflanzen-Physiologie (Sachs) 9959
Gesammelte Abhandlungen über Pflanzen-Physiologie *(see* J. Sachs)
Gesammelte Abhandlungen von N. Pringsheim (Pringsh.) 8340
Gesch. Bot. (Sachs) 9957
Gesch. Fl. bern. Hügell. (Rytz) 9874
Geschichte der Botanik (Pult.) 8399
Geschichte der Botanik (Sachs) 9957
Geschichte der Flora des bernischen Hügellandes (Rytz) 9874
Geschichte der Wissenschaften in Deutschland (Sachs) 9957
Giard. d'acclimaz. (Ricasoli) 9134
Giornale botanico italiano (Parl.) 7355
Gli arboreti sperimentali di Vallombrosa (L. Piccioli) 7894
Globulaires (G. Planch.) 7994
Glossaire mycologique (Roum.) 9677
Gloss. bot. (F. Plée) 8055
Gloss. mycol. (Roum.) 9677
Glossologie botanique (F. Plée) 8055
Gnetales (H. Pearson) 7583
Göttingens Moosvegetation (Quelle) 8420
Götting. Moosveg. (Quelle) 8420
Gommes résines (Pellet.) 7621
Gomm. résin. (Pellet.) 7621
Good book (Raf.) 8591
Gram. bonaer. (Parodi) 7395
Gramina agrostidea (Rupr.) 9794
Gramina of the subantarctic islands of New Zealand *(see* D. Petrie)
Gramíneas bonaerenses (Parodi) 7395
Graphidineen der ersten Regnell'schen Expedition (Redinger) 8742
Graphidineen der Sunda-Inseln (Redinger) 8743
Grass. Britain (Parn.) 7389
Grasses and forage plants of the Rocky Mountain region (Rydb.) 9852
Grasses of Britain (Parn.) 7389
Grasses of Iowa (Pammel) 7255
Grasses of Scotland (Parn.) 7388
Grass. forage pl. Rocky Mt. (Rydb.) 9852
Grass. Iowa (Pammel) 7255
Grass. Scotland (Parn.) 7388
Gray, Man. bot. *(see* B. L. Robinson)
Gray's Manual of Botany *(see* E. J. Palmer)
Great Rift Valley *(see* A. Rendle)
Green earth (Rickett) 9196
Grönländische Süsswasseralgen *(see* P. G. Richter)
Grønland (Rink) 9243
Grønland geografisk og statistisk beskrevet (Rink) 9244
Grønlands Havalger (Rosenvinge) 9540
Groot kruidk. woordenb. *(see* A. Royen)
Growth, development and life-history in the Desmidiaceae (Playf.) 8039

Grundlagen einer Biodynamik (Reinke) 8942
Grundl. Biodynamik (Reinke) 8942
Grundlinien der Pflanzen-Morphologie (Potonié) 8226
Grundlinien Pfl.-Morphol. (Potonié) 8222
Grundlin. Pfl.-Morph. (Potonié) 8226
Grundr. deut. Forstbot. (Reum) 9096
Grundriss (Reum) der deutschen Forstbotanik (Reum) 9097
Grundriss der deutschen Forstbotanik (Reum) 9096
Grundz. Entw. (Pacz.) 7181
Grundz. Morph. Orchid. (Pfitz.) 7829
Grundz. Pfl.-Physiol. (Sachs) 9956
Grundz. Pfl.-Verbr. Chile (Reiche) 8860
Grundz. Pfl.-Verbr. Karpath. (Pax) 7549
Grundz. Pfl.-Verbr. Kaukasusländ. (Radde) 8481
Grundzüge der Entwickelung der Flora in Südwest-Russland (Pacz.) 7181
Grundzüge der Pflanzen-Physiologie (Sachs) 9956
Grundzüge der Pflanzenverbreitung in Chile (Reiche) 8860
Grundzüge der Pflanzenverbreitung in den Karpathen (Pax) 7549
Grundzüge der Pflanzenverbreitung in den Kaukasusländern (Radde) 8481
Guareae species duae costaricenses (Radlk.) 8536
Guerre des Nymphes (St. Lag.) 10078
Guida alle escursioni botaniche (L. Piccioli) 7889
Guida della Provincia di Roma (see P. Pirotta)
Guida escurs. bot. (L. Piccioli) 7889
Guide autumn fl. Minnesota (Rosend.) 9526
Guide bot. (E. Perrier) 7699
Guide bot. Dauphiné (Ravaud) 8680
Guide bot. Suisse franç. (Payot) 7570
Guide bot. Valais (Rion) 9246
Guide bot. Vaud (Rapin) 8633
Guide bryol. lichénol. Grenoble (Ravaud) 8679
Guide bryol. Pyrénées (see F. Renauld)
Guide de l'herborisateur en Belgique (Pâque) 7320
Guide du botaniste (E. Perrier) 7699
Guide du botaniste au jardin de la Mer de Glace (Payot) 7566
Guide du botaniste dans le canton de Vaud (Rapin) 8633
Guide du botaniste dans le Dauphiné (Ravaud) 8680
Guide du botaniste en Valais (Rion) 9246
Guide du botaniste Suisse française (Payot) 7570
Guide du bryologue et du lichénologue Grenoble (Ravaud) 8679

Guide ferns Minnesota (Rosend.) 9524
Guide herbor. Belgique (Pâque) 7320
Guide Mer de Glace (Payot) 7566
Guide orchid. N.S.W. (Rupp) 9788
Guide Pyrenees (Packe) 7180
Guide spring fl. Minnesota (Rosend.) 9522
Guide to the autumn flowers of Minnesota (Rosend.) 9526
Guide to the ferns and fern allies of Minnesota (Rosend.) 9524
Guide to the natural history of the Isle of Wight (see J. Rayner)
Guide to the orchids of New South Wales (Rupp) 9788
Guide to the Pyrenees (Packe) 7180
Guide to the spring flowers of Minnesota (Rosend.) 9522
Guide to the trees and shrubs of Minnesota (Rosend.) 9523
Guide trees shrubs Minnesota (Rosend.) 9523
Gundzüge einer vergleichenden Morphologie der Orchideen (Pfitz.) 7829
Gymnospermen (Plitzka) 8059

Haandb. Danske fl. (see C. Rosenberg)
Handb. Brit. Rubi (W. M. Rogers) 9444
Handb. Exp.-Physiol. Pfl. (Sachs) 9953
Handb. Gewächsk. (Peterm.) 7750
Handb. larg. Brit. fung. (Ramsb.) 8619
Handb. nat. Pfl.-Syst. (Rchb.) 8887
Handbook of British Rubi (W. M. Rogers) 9444
Handbook of Nebraska trees (Pool) 8156
Handbook of systematic botany (see M. Potter)
Handbook of the larger British fungi (Ramsb.) 8619
Handbook of the lichens of the U.S.S.R. (see K. Rassadina)
Handbuch der Botanik (see E. Pfitzer, R. Sadebeck)
Handbuch der Cacteenkunde (see T. Rümpler)
Handbuch der Experimental-Physiologie der Pflanzen (Sachs) 9953
Handbuch der Experimental-Physiologie der Pflanzen (see J. Sachs)
Handbuch der Gewächskunde (Peterm.) 7750
Handbuch der Pflanzengeographie (see G. Poirault)
Handbuch der physiologischen Botanik (see J. Sachs)
Handbuch der Vererbungswissenschaft (Renner) 9076
Handbuch des natürlichen Pflanzensystems (Rchb.) 8887
Handwörterbuch der Botanik (see O. Porsch)

Handwörterbuch Naturwissenschaften (see R. Pilger)
Harbk. Baumz. (see J. Pott)
Hardy Calif. (Rowntree) 9713
Hardy Californians (Rowntree) 9713
Hawaiian genus Kokia (Rock) 9331
Haw. Kokia (Rock) 9331
Hedwigia (see K. Prantl, G. Rabenhorst, H. Reimers)
Heimische Pflanzenwelt (Rawitscher) 8693
Heim. Pfl.-Welt (Rawitscher) 8693
Helminthosp.-Art. (Ravn) 8690
Helminthosporium-Arter (Ravn) 8690
Hennecartia (J. Poiss.) 8119
Hepat. Brit. Isl. (Pearson) 7590
Hepaticae austro-africanae (see A. Rehmann)
Hepaticae britannicae exsiccatae (see W. Pearson)
Hepaticae europaeae (see G. Rabenhorst)
Hepaticae javanicae (Reinw.) 8962
Hepaticae knysnanae (Pearson) 7585
Hepaticae madagascarienses (Pearson) 7589
Hepaticae natalenses (Pearson) 7584
Hepaticae of Gloucestershire (see P. Reader)
Hepaticae of the British Isles (Pearson) 7590
Hepat. knysn. (Pearson) 7585
Hepat. madag. (Pearson) 7589
Hepat. natal. (Pearson) 7584
Herb. amboin. (Rumph.) 9784
Herb. amboin. auctuar. (Rumph.) 9785
Herbarii amboinensis auctuarium (Rumph.) 9785
Herbarium aliarumque stirpium in insulâ Luzone (Ray) 8701
Herbarium amboinense (Rumph.) 9784
Herbarium cecidiologicum (see F. Pax)
Herbarium der Flora von Jaroslaw (see A. Petrowsky)
Herbarium florae Rossicae (see A. Petunnikov)
Herbarium muscorum frondosorum Europae mediae (see P. Reinsch)
Herbarium musei fennici (Saelan) 9979
Herbarium normale plantarum rariorum et criticarum Sueciae (see H. Ringius)
Herbarium rafinesquianum (Raf.) 8583
Herbarium siculum (H. Ross) 9566
Herbarium vivum mycologicum (see G. Rabenhorst)
Herbarium vivum plantarum officinalium (see A. Roth)
Herbationes lundenses (Ringius) 9241
Herb. Chirac (Planch.) 8023
Herbier des Alpes (see V. Payot)
Herbier du Lot (see T. Puel)
Herbier général de l'amateur *(see* A. Richard)
Herbier portatif (Rozin) 9742
Herb. lund. (Ringius) 9241
Herb. Moucherolle (Ravaud) 8681
Herb. mus. fenn. (Saelan) 9979
Herbor. artif. (A. Plée) 8052
Herborisation au Col de Chavière (N. Roux) 9698
Herborisation au Mont Seneppé (N. Roux) 9697
Herborisations artificielles (A. Plée) 8052
Herborisations dans les montagnes d'Hauteville *(see* J. Saint-Lager)
Herborizaciones por Valldigna, Jálisa y Sierra Mariola (Pau) 7513
Herb. portat. (Rozin) 9742
Herb. raf. (Raf.) 8583
Herb. sicul. (H. Ross) 9566
Heterokonten (Pasch.) 7424
Heveas (Barb. Rodr.) 9356
Hieracia Naegelianae exsiccatae (see A. Peter)
Hieracia seckauensia exsiccata (see G. Pernhoffer)
Hieracien der Umgebung von Seckau in Ober-Steiermark (Pernh.) 7693
Hieracien Mittel-Europas *(see* A. Peter)
Hippophaës rhamnoides (Palmgr.) 7240
Hist. bot. (Sachs) 9957
Hist. champ. (Roques) 9448
Hist. Cuba *(see* J. Pfund)
Hist. cult. veg. (H. Phillips) 7866
Hist. de bot. (Sachs) 9957
Hist. econ.-pol. Cuba (Sagra) 9999
Hist. fis. Cuba, Bot. (Sagra) 10000
Hist. fis. Cuba, Fanerogamia (A. Rich.) 9150
Hist. fis. nat. Cuba *(see* A. Richard)
Hist. Infus. (A. Pritch.) 8349
Histiol. Entw.-Gesch. Marsilia (Russow) 9832, 9833
Histiologie und Entwickelungsgeschichte der Sporenfrucht von Marsilia (Russow) 9833
Hist. légendes pl. (Rambosson) 8605
Hist. nat. méd. Ipécacuanha (A. Rich.) 9139
Hist. nat. orangers (Risso) 9248
Hist. nat. Orangers *(see* P. Poiteau)
Hist. nat. prod. Eur. mérid. (Risso) 9249
Hist. nat. Solan. (F. Pouchet) 8235
Hist. nat. vég. paras. (C. P. Robin) 9278
Histoire botanique et horticole des Selaginelles *(see* E. Roze)
Histoire de la botanique (Sachs) 9957
Histoire de la botanique *(see* J. Payer)
Histoire de l'Abrotonum (St. Lag.) 10087
Histoire de la pomme de terre (Roze) 9740
Histoire de l'indigo (St. Hil.) 10048
Histoire des champignons (Roques) 9448
Histoire des plantes les plus remarquables du Brésil (St. Hil.) 10033

Histoire et culture des orangers (Risso) 9248
Histoire naturelle des orangers (Risso) 9248
Histoire naturelle des principales productions de l'Europe méridionale (Risso) 9249
Histoire naturelle des végétaux parasites (C. P. Robin) 9278
Histoire naturelle et médicale de la famille des Solanées (F. Pouchet) 8235
Histoire naturelle et médicale l'Ipécacuanha (A. Rich.) 9139
Histoire philosophique, littéraire, économique des plantes de l'Europe (Poir.) 8116
Histoire physique, naturelle et politique de Madagascar, Mousses (Renauld) 8995
Histoire physique, politique et naturelle de l'Ile de Cuba, Botanique. - Plantes cellulaires (Sagra) 10001
Histoire physique, politique et naturelle de l'Ile de Cuba Botanique. - Plantes vasculaires (A. Rich.) 9150
Histoire physique, politique et naturelle de l'Ile de Cuba, Botanique. - Plantes vasculaires (Sagra) 10002
Histoires et légendes des plantes utiles et curieuses (Rambosson) 8605
Histol. Ébén. (P. Parm.) 7383
Histologie comparée des Ebénacées (P. Parm.) 7383
Historia das plantas alimentares e de gozo do Brazil (Peckolt) 7606
Historia economico-politica y estadistica de la Isla de Cuba (Sagra) 9999
Historia fisica politica y natural de la Isla de Cuba, Botanica (Sagra) 10000
Historia plantarum (Ray) 8701
Historia plantarum in Palatinatu electorali (Pollich) 8137
Historical and biographical sketches of the progress of botany in England (Pult.) 8399
Historie des herbiers (St. Lag.) 10072
Historie physique, politique et naturelle de l'Ile de Cuba, Atlas (Sagra) 10003
History of botany (Sachs) 9957
History of cultivated vegetables (H. Phillips) 7866
History of Infusoria (A. Pritch.) 8349
History of infusorial animalcules (A. Pritch.) 8349
History of Infusoria (see J. Ralfs)
History of the vegetable kingdom (Rhind) 9126
Hist. philos. pl. Eur. (Poir.) 8116
Hist. phys. Cuba, Atlas (Sagra) 10003
Hist. phys. Cuba, Bot. Pl. cell. (Sagra) 10001

Hist. phys. Cuba, Bot. Pl. vasc. (Sagra) 10002
Hist. phys. Cuba, Pl. vasc. (A. Rich.) 9150
Hist. phys. Madagascar, Mousses (Renauld) 8995
Hist. phys. Madagascar (see F. Renauld)
Hist. pl. (Ray) 8701
Hist. pl. alim. Brazil (Peckolt) 7606
Hist. pl. Palat. (Pollich) 8137
Hist. pl. remarq. Brésil (St. Hil.) 10033
Hist. pomme de terre (Roze) 9740
Hist. sketches bot. Engl. (Pult.) 8399
Hist. stirp. fl. petrop. (Rupr.) 9798, 9799
Hist. veg. kingd. (Rhind) 9126
Hochberg bot. Gart. (Pohl) 8102
Homodromie & Antidromie (Raunk.) 8661
Homoxylon (Sahni) 10009
Homoxylon rajmahalense (Sahni) 10009
Honey plants of Iowa (Pammel) 7258
Honey pl. Iowa (Pammel) 7258
Hort. bengal. (Roxb.) 9721
Hort. bot. frib. (Perleb) 7691
Hort. donat. (Planch.) 8013
Hort. flum. (Barb. Rodr.) 9351
Horticultural register and general magazine (Paxt.) 7553
Horticulture. Jardin potager (Réveil) 9110
Horticulture. Végétaux d'ornement (Réveil) 9110
Horti et provinciae veronensis plantae (Pollini) 8141
Horti med. amstelod. (see F. Ruysch)
Horti medici Chelseani (I. Rand) 8627
Hort. malab. (Rheede) 9123
Hort. med. Chels. (I. Rand) 8627
Horto botanico friburgensi (Perleb) 7691
Hort. orlov. (Pursh) 8405
Hort. panciat. (G. Piccioli) 7887
Hort. reg. (Paxt.) 7553
Hort. rom. (see L. Sabbati)
Hort. thuret. (G. Poirault) 8112
Hortulus practico-botanicus (see L. Sabbati)
Hortus bengalensis (Roxb.) 9721
Hortus botanicus (see O. J. Rudbeck)
Hortus cantabrig. (see F. Pursh)
Hortus donatensis (Planch.) 8013
Hortus ericaeus (see J. Russell)
Hortus fluminensis (Barb. Rodr.) 9351
Hortus gramineus woburnensis (see J. Russell)
Hortus indicus malabaricus (Rheede) 9123
Hortus orloviensis (Pursh) 8405
Hortus panciaticus (G. Piccioli) 7887
Hortus thuretianus antipolitanus (G. Poirault) 8112
Hort. veron. pl. (Pollini) 8141
Houtsoort. Suriname (J. Pfeiff.) 7814
How to know ferns (F. Parsons) 7415

How to know the ferns (F. Parsons) 7415
Hussvamp. (Rostr.) 9615
Hussvampen (Rostr.) 9615
Hydrothyria venosa (J. L. Russell) 9829
Hydrurus i jego pokrewienβtwo (Rost.) 9598
Hydrurus und seine Verwandtschaft (Rost.) 9598
Hymenomyc. austro-amer. (Romell) 9489
Hyménomycètes (Pat.) 7441
Hymenomycetes austro-americani (Romell) 9489
Hyménomycètes d'Europe (Pat.) 7442
Hymenomycetes of Lappland (Romell) 9490
Hyménomyc. Eur. (Pat.) 7442
Hymenophyllaceae (K. Presl) 8308

Iagttag. mosern. udbr. *(see* E. Ryan)
Ic. fl. Germ. Helv. *(see* J. Pöll)
I codici botanici figurati (Sacc.) 9937
Icon. bot. index (Pritz.) 8354
Icon. carpolog. (Ralph) 8601
Icon. descr. fung. (Pers.) 7728
Icones carpologicae (Ralph) 8601
Icones et descriptiones fungorum (Pers.) 7728
Icones et descriptiones plantarum novarum vel rariorum horti botanici panormitani (H. Ross) 9565
Icones florae germanicae (Rchb.) 8876
Icones florae germanicae et helveticae (Rchb.) 8884, 8885
Icones fungorum javanicorum (Penz.) 7655
Icones pictae specierum rariorum fungorum (Pers.) 7731
Icones plantarum Galliae rariorum (Rouy) 9709
Icones plantarum in Flora cubana (Sagra) 10005
Icones plantarum medicinalium (Plenck) 8057
Icones plantarum, quae sunt flore irregulari hexapetalo (Riv.) 9270
Icones Plantarum *(see* D. Prain)
Icon. fl. germ. helv. (Rchb.) 8885
Icon. fl. germ. helv. (Rchb. fil.) 8891
Icon. fl. germ. helv. *(see* H. G. Reichenbach)
Icon. fung. jav. (Penz.) 7655
Icon. fung. jav. *(see* P. Saccardo)
Iconographia botanica exotica (Rchb.) 8878
Iconographia botanica seu plantae criticae (Rchb.) 8876, 8884
Iconographia florae italicae *(see* G. Paoletti)
Iconographie des champignons (Paul.) 7519

Iconographie des champignons de Paulet (Paul.) 7519
Iconographie des conifères (Pardé) 7324
Iconographie végétale (A. Rich.) 9147
Iconogr. bot. exot. (Rchb.) 8878
Iconogr. bot. pl. crit. (Rchb.) 8876, 8885
Iconogr. conif. (Pardé) 7324
Iconogr. exot. sive hortus botanicus (Rchb.) 8878
Iconogr. vég. (A. Rich.) 9147
Iconoteca dei botanici (Sacc.) 9926
Icon. pict. sp. fung. (Pers.) 7731
Icon. pl. Fl. cub. (Sagra) 10005
Icon. pl. fl. hexapet. (Riv.) 9270
Icon. pl. hort. bot. panorm. (H. Ross) 9565
Icon. pl. med. (Plenck) 8057
Icon. pl. rar. France (Rouy) 9709
Iconum botanicarum index (Pritz.) 8354
Ideen Revis. Gräs. (Panz.) 7303
Ideen zu einer künftigen Revision der Gattungen der Gräser (Panz.) 7303
Ident. econ. woods U.S. (Record) 8738
Ident. Hawaii pl. (St. John) 10065
Identification of Hawaii plants (St. John) 10065
Identification of the economic woods of the United States (Record) 8738
Identification of the timbers of temperate North America (Record) 8740
Ident. timb. temp. N. Amer. (Record) 8740
I funghi mangerecci (Sacc.) 9913
Iles Canaries (Pit.) 7972
Ill. alg. (Postels) 8197
Ill. bot. Himal. Mts. (Royle) 9734
Ill. champ. Tunisie (Pat.) 7461
Ill. Cinchona (Ruiz) 9768
Ill. Fl. (Potonié) 8211
Ill. fl. Mitteleuropa *(see* R. H. Rechinger)
Ill. fl. Mitteleur. *(see* F. Petrak)
Ill. Fl. N. Mitt. Deutschl. *(see* A. Peter)
Ill. fung. (S. F. Price) 8323
Ill. geol. Yorkshire (J. Phillips) 7870
Ill. pl. (Pall.) 7230
Ill. pl. Eur. (Rouy) 9707
Ill. sp. Acon. gen. (Rchb.) 8875
Illustration des espèces nouvelles rares ou critiques de champignons de la Tunisie (Pat.) 7461
Illustrationes algarum (Postels) 8197
Illustrationes algarum *(see* F. Ruprecht)
Illustrationes plantarum (Pall.) 7230
Illustrationes plantarum Europae rariorum (Rouy) 9707
Illustrations de la partie botanique (Pat.) 7461
Illustrations of the botany of the Himalayan Mountains (Royle) 9734
Illustrations of the fungi (S. F. Price) 8323
Illustrations of the geology of Yorkshire (J. Phillips) 7870

Illustrations of the Nueva Quinologia of Pavon (Pav.) 7540
Illustratio specierum Aconiti generis (Rchb.) 8875
Illustrierte Flora (Potonié) 8211
Illustrierte Flora des Bismarck-Archipels (Peekel) 7609
Illustrierte Naturgeschichte der drei Reiche *(see* A. Pokorny)
Illustrierte Naturgeschichte des Pflanzenreiches *(see* A. Pokorny)
Illustriertes Handwörterbuch der Botanik *(see* H. Potonié)
Illustrirtes Gartenbau-Lexikon *(see* T. Rümpler)
Il Ruwenzori *(see* P. Pirotta)
Immigrant plants of Southern California (Parish) 7336
Importância da fitogeografia (Rothm.) 9652
Im Schatten der Cordillera (Poepp.) 8090
In codicem botanicum linnaeanum index alphabeticus (H. Richt.) 9183
Ind. bot. Pers. Syn. meth. fung. *(see* C. Persoon)
Ind. cuatro libr. natural. (Ramírez) 8608
Index Algarum europaearum exsiccatarum (Rabenh.) 8449
Index alg. eur. exs. (Rabenh.) 8449
Index botanicus Persoonii Synopsi methodica fungorum *(see* C. Persoon)
Index botanicus sistens omnes fungorum species in C.H. Persoonii Mycologia Europeae enumeratas *(see* D. P. Rogers)
Index Bryoth. eur. (Rabenh.) 8447
Index filicum, supplementum quartum *(see* R. Pichi- Sermolli)
Index generum phanerogamorum. Sapotaceae (Radlk.) 8527
Index Hepat. eur. (Rabenh.) 8448
Index herb. (Pott) 8230
Index herbarii mei vivi (Pott) 8230
Index Hort. bengal. (Roxb.) 9721
Index iconum fungorum (Sacc.) 9894
Index in Gottsche et Rabenhorst Hepaticarum europaearum exsiccatarum (Rabenh.) 8448
Index in Herbarium florae Germanicae (Rchb.) 8881
Index in Rabenhorst Bryotheca europaeae (Rabenh.) 8447
Index kewensis *(see* D. Prain)
Index linnaeanus in Leonhardi Plukenetii (Pluk.) 8065
Index linnaeanus in Leonhardi Plukenetii,...opera botanica *(see* L. Plukenet)
Index linn. Pluk. (Pluk.) 8065
Index of fungi (Petr.) 7786
Index of Fungi, a supplement to Petrak's lists (Petr.) 7786

Index pl. agr. erfurt. (Planer) 8032
Index plantarum circum berolinum sponte nascentium (Rebent.) 8719
Index plantarum horti imperatoriae medico-chirurgicae Academiae (Pers.) 7732
Index plantarum in horto krausiano (Roloff) 9483
Index plantarum in horto publico Bononiae (Rodati) 9335
Index plantarum officinalium (I. Rand) 8626
Index plantarum, quae circa Gustroviam sponte nascentur (J. Prahl) 8255
Index plantarum, quas in agro Erfurtensi sponte provenientes (Planer) 8032
Index pl. berol. (Rebent.) 8719
Index pl. Gustrov. (J. Prahl) 8255
Index pl. hort. Bonon. (Rodati) 9335
Index pl. hort. kraus. (Roloff) 9483
Index pl. India (Piddington) 7918
Index Raf. (Raf.) 8550
Index Rafinesquianus (Raf.) 8550
Indian Gondwana plants *(see* B. Sahni)
Indication de quelques plantes nouvelles (E. Perrier) 7696
Indice alfabético Cuatro libres de la naturaleza (Ramírez) 8608
Indici plantarum erffurtensium (Planer) 8033
Indici pl. erffurt. (Planer) 8034
Indic. pl. nouv. (E. Perrier) 7696
Indigenous trees of the Hawaiian Islands (Rock) 9323
Indig. pl. Lanarkshire (W. Patrick) 7498
Indig. trees Haw. Isl. (Rock) 9323
Industries of Japan (Rein) 8911
Industr. Japan (Rein) 8911
Influence des agents extérieurs sur l'organisation polaire et dorsiventrale des plantes (Rosenvinge) 9538
Informaciones sobre excursiones botanicas en la Costa oriental de la Patagonia (Roivainen) 9466
In historiam stirpium florae petropolitanae diatribae (Rupr.) 9798
Inledende studier over polymorphien hos Anthriscus silvestris (H. E. Petersen) 7761a
In nonnullas Filaginis Evacisque observationes (Parl.) 7349
I nomi generici dei funghi (Sacc.) 9917
Inst. bot. (Petagna) 7743a
Institutiones botanicae (Petagna) 7743a
Instr. format. herb. lich. (O. J. Rich.) 9163
Instructions pratiques pour la formation et la conservation d'un herbier de lichens (O. J. Rich.) 9163
Internationale Polarforschung *(see* K. Prantl)

International rules of botanical nomenclature (Rendle) 9072
Intr. bot. (J. Philib.) 7841
Introductio generalis in rem herbariam (Riv.) 9266
Introduction à l'étude de la botanique (J. Philib.) 7841
Introduction to study of South Australian Orchids (R. S. Rogers) 9439
Introduction to the study of the South African grasses (E. Phillips) 7863
Intr. rem. herb. (Riv.) 9266
Intr. S. Afr. grass. (E. Phillips) 7863
Intr. S. Austral. orchid. (R. S. Rogers) 9439
Int. rules bot. nomencl. (Rendle) 9072
Inulin (Prantl) 8267
Inwijdings-Rede over het nut der wetenschap (Rauwenh.) 8671
Inw.-Rede nut wetensch. (Rauwenh.) 8671
Iowa alg. (Prescott) 8286
Iowa algae (Prescott) 8286
I prevedibili funghi futuri (Sacc.) 9922
Irish Naturalist (see R. Praeger)
Irish topographical botany (Praeger) 8250
Irish topogr. bot. (Praeger) 8250
Isla de la Mocha (Reiche) 8857
Ist der Soorpilz mit dem Kalmpilz wirklich identisch? (Reess) 8762
Istoria di ciò che è stato pensato intorno alla fecondzione delle piante (P. Rossi) 9574
Iter per Poseganam Sclavoniae (Piller) 7939
Iter Poseg. Sclavon. (Piller) 7939

Jacintes (Saint-Simon) 10092
Jahrbücher für wissenschaftliche Botanik (see N. Pringsheim)
Jahrb. wiss. Bot. (see W. Pfeffer)
Japan (Rein) 8910a
Jard. bot. École méd. (Poit.) 8123
Jardin botanique de l'École de médecine de Paris (Poit.) 8123
J. bot. excurs. (Pursh) 8406
J. Bot. (see C. Persoon)
Jeune bot. (A. Plée) 8053
Jeune botaniste (A. Plée) 8053
Journal de Micrographie (see J. Pelletan)
Journal de Physique (see F. Rozier)
Journal für das Forst- und Jagdwesen (see J. Reitter)
Journal of a botanical excursion (Pursh) 8406
Journal of a second voyage for the discovery of a North-West passage (W. Parry) 7411
Journal of a third voyage for the discovery of a North-West Passage (W. Parry) 7413
Journal of a voyage for the discovery of a North-West passage (W. Parry) 7409
Journal of Botany (see A. Rendle)
Journal of Pomology and horticultural Science (see G. Pethybridge)
Journal of the Arnold Arboretum (see A. Rehder)
Journey to Ararat (Parrot) 7402
Journey to the summit of Mont Perdu (Ramond) 8617
J. sec. voy. (W. Parry) 7411
J. third voy. (W. Parry) 7413
Jungermanniografia etrusca (Raddi) 8486
Jungermanniogr. etrusca (Raddi) 8486
Jungle mem. (Rusby) 9825
Jungle memories (Rusby) 9825
J. voy. N.-W. pass. (W. Parry) 7409

Kälte als Heilmittel (see L. Radlkofer)
Kakteen der Grand Mesa in West-Colorado (Purpus) 8402
Kalkalgen (Rothpletz) 9656
Karabagh (Radde) 8480
Katalog českých rozsivek (Proch.) 8356
Kat. česk. rozsivek (Proch.) 8356
Kenntn. Blütenentw. Loasac. (Racine) 8473
Kenntn. eur. Torfm. (Russow) 9838
Kenntniss der enigen Pflanzen die Mahlern und Färbern zum Nutzen gereichen können (Reuss) 9100
Kenntn. kleinst. Lebensf. (Perty) 7739
Kenntn. Pfl. Mahlern (Reuss) 9100
Key Rocky Mt. fl. (Rydb.) 9866
Key spring fl. Riley (Paull) 7527
Key to the Rocky Mountain flora (Rydb.) 9866
Key to the Spring flora of Riley (Paull) 7527
Kirschlorbeer (Pančic) 7283
Kl. Ausl. Steyr (Pehersd.) 7613
Kleine Auslese der interessantesten Pflanzen aus der Flora von Steyr (Pehersd.) 7613
Kleistogamie und Amphikarpie (Reiche) 8856
Klíč k úplné květeně republiky Čeckoslovenské (Podp.) 8083
Klíč květ. Českoslov. (Podp.) 8083
Klima und Pflanzenleben auf Ostgrönland (see A. Pansch)
Klotzschii herbarium vivum mycologicum (see G. Preuss, G. Rabenhorst)
Knospenlage Laubbl. (Reinecke) 8913
Kokospalme (P. Preuss) 8318
Kokospalme und ihre Kultur (P. Preuss) 8318
Kreidebild. Texas (F. Roem.) 9398

Kreidebildungen von Texas (F. Roem.) 9398
Kreuz und quer durch Mexico (Reiche) 8865
Kristianiatraktens Protococcoideer (Printz) 8341
Kristianiatr. Protococc. (Printz) 8341
Krit. Abst.-Lehre (Reinke) 8941
Krit. Fl. Schlesw.-Holst. (Prahl) 8256
Krit. Gesch. Algengeschl. (Pringsh.) 8330
Kritik der Abstammungslehre (Reinke) 8941
Kritische Blätter für Forst- und Jagdwissenschaft (see F. Pfeil)
Kritische Flora der Provinz Schleswig-Holstein (Prahl) 8256
Kritische Untersuchungen über die durch Taphrina-Arten hervorgebrachten Baumkrankheiten (Sadebeck) 9966
Kritisch-systematische Original-Untersuchungen über Pyrenomyzeten (Petr.) 7785
Krit. Overs. Danske Diatom. (see C. Rosenberg)
Krit. Unters. Taphrina Baumkrankh. (Sadebeck) 9966
Kruidkundige beschrijving der in- en uitlandsche gewassen (D. Royen) 9732
Krypt. Fl. Deutschl. (Redinger) 8745
Krypt.-Fl. Deutschland (see A. Pascher)
Krypt.-Fl. Sachsen (Rabenh.) 8442
Krypt.-Fl. (see P. G. Richter)
Kryptogamenflora für Anfänger (see R. Pilger)
Kryptogamen-Flora von Deutschland (Redinger) 8745
Kryptogamen-Flora von Sachsen (Rabenh.) 8442
Kryptogamensammlung für Schule und Haus (Rabenh.) 8437
Kryptogamensammlung für Schule und Haus (see G. Rabenhorst)
Kryptogamen Sammlung (see G. Rabenhorst)
Kufra (Rohlfs) 9456
Kult. Aufz. Eriken (Regel) 8771
Kult. Naturbild. Span. Riviera (Rikli) 9227
Kulturgeschichte der Nutzpflanzen (L. Reinhardt) 8921
Kulturgewächse der deutschen Kolonien (Sadebeck) 9971
Kulturgew. deut. Kolon. (Sadebeck) 9971
Kultur und Aufzählung der Eriken (Regel) 8771
Kultur- und Naturbilder von der spanischen Riviera (Rikli) 9227
Kupfersamml. deut. Bot.-Buch. (Rchb.) 8886
Kupfersammlung der ausländischen Gewächse (Rchb.) 8878

Kupfersammlung kritischer Gewächse oder Abbildungen seltener und weniger genau bekannter Gewächse (Rchb.) 8876
Kupfersammlung zum praktischen deutschen Botanisirbuche (Rchb.) 8886
Květena Moravy (Podp.) 8081
Května slov. (G. Reuss) 9104
Května slovenska (G. Reuss) 9104

Labrador Coast (Packard) 7179
Lac Majeur (Sahut) 10011
Landschaft der Steinkohlen-Zeit (Potonié) 8220
Landsch. Steinkohlen-Zeit (Potonié) 8220
Landwirtschaftliche Pflanzensammlung (see A. Roth)
L'antotrofia (A. Piccioli) 7884
Lasy Białowiczy (Pacz.) 7182
Laubm. Gefäss-Krypt. Eulengeb. (W. Roth) 9645
Laubmoose und Gefäss-Kryptogamen des Eulengebirges (W. Roth) 9645
L'autonomie des lichens (O. J. Rich.) 9164
Leben der Pflanze (see R. Pilger)
Lebensbedingungen und Vegetationsverhältnisse der Mittelmeerländer und der atlantischen Inseln (Rikli) 9233
Lebensbed. Veg.-Verh. Mittelmeerländ. (Rikli) 9233
Leben u. Streben (Rossmässler) 9585
Leberm. Eur. (see H. Reimers)
Lebermoose (G. Pabst) 7176
Leçons bot. (P. Parm.) 7387
Leçons bot. (St. Hil.) 10051
Leçons de botanique (P. Parm.) 7387
Leçons de botanique (St. Hil.) 10051
Leçons de flore (Poir.) 8115
Leçons fam. nat. (Payer) 7564
Leçons fl. (Poir.) 8115
Leçons sur les familles naturelles des plantes (Payer) 7564
Lect. histol. (Quekett) 8413
Lectures on histology (Quekett) 8413
Lectures on the physiology of plants (Sachs) 9958
Leguminosae of Porto Rico (Perkins) 7685
Leguminous plants of Hawaii (Rock) 9332
Legum. pl. Hawaii (Rock) 9332
Lehrb. Bot. (Prantl) 8268
Lehrb. Bot. (Reess) 8768
Lehrb. Bot. (Sachs) 9954
Lehrb. Naturgesch. Pflanzenr. (Perleb) 7690
Lehrb. Pfl.-Palaeont. (Potonié) 8217
Lehrbuch der Botanik (Prantl) 8268
Lehrbuch der Botanik (Reess) 8768
Lehrbuch der Botanik (Sachs) 9954
Lehrbuch der Botanik (see F. Pax)

Lehrbuch der Naturgeschichte des Pflanzenreichs (Perleb) 7690
Lehrbuch der Paläobotanik (Potonié) 8217
Lehrbuch der Pflanzenpalaeontologie (Potonié) 8217
Leipziger botanischer Tauschverein *(see* P. G. Richter)
Leitende Gewebe einiger anomal gebauten Monocotylenwurzeln *(see* O. Reinhardt)
Lejeuneae madagascarienses (Pearson) 7588
Lejeun. madag. (Pearson) 7588
Lenophyllum (J. Rose) 9513
Lepisia (K. Presl) 8299
Lespedeza (Pieters) 7930
Lethaea geogn. *(see* F. Roemer)
Letters on the elements of botany (Rousseau) 9686
Lettre de M. Palisot (P. Beauv.) 7216
Lettr. élém. bot. (Rousseau) 9686
Lettres bot. 1824 (Rousseau) 9688
Lettres élémentaires sur la botanique (Rousseau) 9686
Lettres sur la botanique (Rousseau) 9688
Lettres sur quelques plantes de la flore Française (Rouy) 9710
Lex. gen. phan. (T. v. Post) 8193
Lex. gen. phan. (T. v. Post) 8194
Lexicon generum phanerogamarum (T. v. Post) 8193, 8194
Lezioni bot. comp. (Parl.) 7354
Lezioni di botanica comparata (Parl.) 7354
L'Herbier de Chirac (Planch.) 8023
L'herborisation à la Moucherolle & dans ses alentours (Ravaud) 8681
Lichenen des östlichen Weserberglandes (Rüggeberg) 9762
Lichenen Deutschlands (Rabenh.) 8434
Lichenes Daniae (Rostr.) 9608
Lichenes europaei exsiccati (see G. Rabenhorst)
Lichenes exsiccatae (see L. Reichenbach)
Lichenes exsiccati ex herb. Dr. H.E. Hasse relicti (see C. Plitt)
Lichenes Fenniae exsiccatae (see V. Räsänen)
Lichenes Fennieae exsiccati (Räs.) 8545
Lichenes gallici exsiccati (see C. Roumeguère)
Lichenes rariores exsiccati (see K. Redinger)
Lichenotheca Fennica (see V. Räsänen)
Lichens (Roum.) 9672
Lichens du Finistère (Picq.) 7916
Lichens du Finistère (see C. Picquenard)
Lichens neo-granadins et ecuadoriens (Roum.) 9680
Lichens of Ireland (see L. Porter)
Lichens of the Swedish Kamtchatka-Expeditions (Du Rietz) 9218
Lichens recueillis sur le Massif du Mont-Blanc (Payot) 7574

Lichen xanthostigma *(see* C. Persoon)
Lich. Finistère (Picq.) 7916
Lich. östl. Weserbergl. (Rüggeberg) 9762
Life and work of Cyrus Guernsey Pringle *(see* F. Pennell)
Life and writings of Rafinesque (Raf.) 8548
Life forms of plants (Raunk.) 8663
Life forms of terrestrial flowering plants (Du Rietz) 9220
Life forms pl. (Raunk.) 8663
Life forms terr. fl. pl. (Du Rietz) 9220
Life of travels (Raf.) 8585
Life writ. Raf. (Raf.) 8548
Lijst van planten (Pulle) 8387
Liliac. (Redouté) 8747
Liliacées (Redouté) 8747
Liliacées *(see* F. Roche)
L'illustration horticole *(see* É. Rodigas)
Limites nat. fl. (Planch.) 8018
Limites naturelles des flores (Planch.) 8018
Lindenia *(see* É. Rodigas)
Linnaei de plantarum ordine (Rodati) 9334
Linn. pl. ord. (Rodati) 9334
L'interprétation des plantes de Bulliard *(see* L. Quélet)
Liostephania (Payne) 7565
Lista de algunas plantas medicinales (Paccard) 7177
Lista pl. med. (Paccard) 7177
Lista provisional de las gramíneas señaladas en Venezuela (Pitt.) 7991
List Brit. seed-pl. ferns (Rendle) 9066
List Canad. Hepat. (Pearson) 7586
Liste de muscinées du Poitou et de la Saintonge (O. J. Rich.) 9166
Liste des champignons récoltés en Basse-Californie (Pat.) 7473
Liste des champignons recueillis à San Thomé (Pat.) 7493
Liste des Diatomées et des Desmidiées (P. Petit) 7770
Liste des Hypoxylées, Mucédinées et Urédinées de Caen (Roberge) 9275
Liste des mousses et hépatiques (Parfitt) 7332
Liste des mousses, hépatiques et lichens de la Lozère (Prost) 8367
Liste des plantes qui croissent spontanément dans le département de la Loire-Inférieure (Saint-Gal) 10021
Liste des plantes rares ou nouvelles pour ce Département et les parties limitrophes du Doubs (Renauld) 8984
Liste des plantes vasculaires observés dans l'arrondissement de Montmédy (Pierrot) 7926
Liste Diatom. Desmid. (P. Petit) 7770
Liste mouss. hépat. (Parfitt) 7332

Liste mouss. Lozère (Prost) 8367
Liste musc. Poitou Saintonge (O. J. Rich.) 9166
Liste pl. Loire-Infér. (Saint-Gal) 10021
Liste pl. vasc. Montmédy (Pierrot) 7926
List fung. Epping Forest (A. Pearson) 7582
List fungi, Ustil., Ured. (F. Patterson) 7501
List Haw. names pl. (Rock) 9324
List of British seed-plants and ferns (Rendle) 9066
List of Canadian Hepaticae (Pearson) 7586
List of fungi (Ustilaginales and Uredinales) (F. Patterson) 7501
List of Hawaiian names of plants (Rock) 9324
List of Laccadive plants (Prain) 8259
List of lichens in Alaska *(see* J. Rothrock)
List of mosses of the State of New York (Peck) 7594
List of plants collected by Dr. Edward Palmer in Lower California and Western Mexico (J. Rose) 9507
List of plants collected in the vicinity of Oquawka (H. Patt.) 7502
List of plants collected on the Peary Arctic Expediton (Rydb.) 9864
List of succulent plants (Peacock) 7580
List of the Carices of Pennsylvania (Porter) 8181
List of the flowering plants indigenous to Otago (Petrie) 7789
List of the fungi of Epping Forest (A. Pearson) 7582
List of the more common native and naturalized plants of South Carolina (Ravenel) 8684
List of vascular plants of the Nugsuaq peninsula (M. Porsild) 8174
List of wild flowering plants (Paley) 7204
List pl. Nugsuaq penins. (M. Porsild) 8174
List pl. Oquawka (H. Patt.) 7502
List poison. pl. S. Afr. (E. Phillips) 7860
List Pter. Sperm. N.E. North America *(see* H. Rusby)
List succ. pl. (Peacock) 7580
List wild fl. pl. (Paley) 7204
Lit. Bayer. Phan.-Fl. (Poeverl.) 8097
Literatur über Bayerns Phanerogamen- und Gefässkryptogamenflora (Poeverl.) 8097
Little book of Lespedeza (Pieters) 7930
Livsformen hos planter paa ny jord (Raunk.) 8659
Livsformernes Statistik (Raunk.) 8658
Livsform. pl. ny jord. (Raunk.) 8659
Lomatophyllum et les Aloe de Madagascar (H. Perrier) 7703
London clay fl. (E. Reid) 8906

London clay flora (E. Reid) 8906
Lopharia (Pat.) 7470
L'organisation florale des Solanacées zygomorphes (Robyns) 9311
Lyst der plant-dieren (Pall.) 7222

Madeira fl. (Penfold) 7629
Madeira flowers, fruits, and ferns (Penfold) 7629
Mag. aesth. Bot. (Rchb.) 8873
Magazin der aesthetischen Botanik (Rchb.) 8873
Magazin der Garten-Botanik (Rchb.) 8873
Magazin für die Botanik (Roem.) 9400
Mag. Bot. (Roem.) 9400
Magnolie (Pamp.) 7264
Mainzer Fl. (Reichenau) 8867
Mainzer Flora (Reichenau) 8867
Maison rustique à l'usage des habitants de Cayenne (Préfontaine) 8283
Malabaarse Kruidhof (Rheede) 9124
Malab. Kruidhof (Rheede) 9124
Malayan fern studies (Posth.) 8203
Malay. fern stud. (Posth.) 8203
Malpighia *(see* P. Pirotta)
Man. arboriste belg. (Poederlé) 8086
Man. bot. (Raf.) 8563
Man. bot. (Roth) 9637
Man. Brit. Discomyc. (W. Phillips) 7873
Man. cult. trees (Rehd.) 8817
Man. cult. trop., Cult. caf. (E. F. A. Raoul) 8631
Manipolo della flora del Monte Cavallo (Sacc.) 9942
Manipolo di alghe del Mar Rossa (Picc.) 7907
Manipolo di funghi nuovi (Sacc.) 9932
Manipolo di micromiceti nuovi (Sacc.) 9929
Manipolo di piante nuove (Pamp.) 7260
Man. N. Amer. Gymnosp. (Penh.) 7631
Man. pl. Oregon (M. Peck) 7604
Man. pl. usual. Venez. (Pitt.) 7987
Man. poison. pl. (Pammel) 7256
Man. struct. bot. (Rusby) 9823
Mant. pl. *(see* D. Royen)
Man. trop. fruits (Popenoe) 8160
Manual de las plantas usuales de Venezuela (Pitt.) 7987
Manuale botanicum (Roth) 9636, 9637
Manual of Aspergilli *(see* K. Raper)
Manual of botany (Raf.) 8563
Manual of botany (Raf.) 8563
Manual of cultivated trees and shrubs (Rehd.) 8817
Manual of poisonous plants (Pammel) 7256
Manual of structural botany (Rusby) 9823
Manual of the British Discomycetes (W. Phillips) 7873

Manual of the higher plants of Oregon (M. Peck) 7604
Manual of the medical botany of the United States (Raf.) 8579
Manual of the natural history, geology, and physics of Greenland *(see* John Sadler)
Manual of the North American Gymnosperms (Penh.) 7631
Manual of the Penicillia *(see* K. Raper)
Manual of tropical and subtropical fruits (Popenoe) 8160
Manual of wayside plants of Hawaii (Pope) 8159
Manuel de l'arboriste et du forestier belgiques (Poederlé) 8086
Manuel des végétaux (Saint-Germain) 10023
Manuel géographie botanique *(see* G. Poirault)
Manuel pratique des cultures tropicales, Culture du Caféier (E. F. A. Raoul) 8631
Man. vég. (Saint-Germain) 10023
Man. wayside pl. Hawaii (Pope) 8159
Mar. alg. Denmark (Rosenvinge) 9546
Mar. alg. Kangerdlugssuak (Rosenvinge) 9554
Mar. alg. N.E. Greenland (Rosenvinge) 9547
Mar. alg. Simmons (Rosenvinge) 9551
Marcgraviaceae néhány új alakjáról (Al. Richt.) 9181
Maria Antonia (Parl.) 7358
Marine algae collected by Dr. H.G. Simmons (Rosenvinge) 9551
Marine algae from Kangerdlugssuak (Rosenvinge) 9554
Marine algae of Denmark (Rosenvinge) 9546
Marine algae vegetation of Iceland (Rosenvinge) 9548
Marine Cyanophyceae from Easter Island, in Nat. Hist. Juan Fernandez *(see* J. Petersen)
Marine Diatoms from the Coast of Iceland (Rosenvinge) 9548
Materiale pentru flora Besarabiei *(see* T. Rayss)
Materials for a flora of the Malayan Peninsula (Ridl.) 9208
Materialy dlya poznaniya rastitel'nosti severnoy Rossii (Pohle) 8109
Materia medica de fungis (Picco) 7896
Matériaux pour la flore atlantique (Pomel) 8149
Matériaux pour la flore médicale de Montpellier (G. Planch.) 7999
Matériaux pour servir à la revision de la flore portugaise (Rouy) 9702

Materia vegetabilis (N. Pereb.) 7672
Materia venenaria regni vegetabilis (Puihn) 8384
Materyjały do flory glonów polski (Racib.) 8457
Mat. fl. atl. (Pomel) 8149
Mat. fl. Malay. Penins. (Ridl.) 9208
Mat. fl. méd. Montpellier (G. Planch.) 7999
Matières premières d'origine végétale (Perrot) 7715
Mat. orig. vég. (Perrot) 7715
Mat. veg. (N. Pereb.) 7672
Mat. venen. regn. veg. (Puihn) 8384
Mazzetto di fiori (Pass.) 7436
Mazz. fior. (Pass.) 7436
Med. Bem. Kol. Rio Essequebo (Rodschied) 9367
Med.-bot. cat. St. John's (Porcher) 8166
Med. fl. (Raf.) 8579
Medical flora (Raf.) 8579
Medical flora and botany of the United States (Raf.) 8579
Medicinal plants of the Philippines (Pardo) 7327
Medicinal plants of the Philippines *(see* E. Quisumbling)
Medicinal properties of the cryptogamic plants of the United States (Porcher) 8167
Medico-botanical catalogue of St. John's (Porcher) 8166
Medizinal-Pflanzen *(see* G. Pabst)
Medizinische Botanik (A. Rich.) 9141
Medizinische und chirurgische Bemerkungen über das Klima, die Lebensweise und Krankheiten der Einwohner der hollaendischen Kolonie Rio Essequebo (Rodschied) 9367
Med. pl. Philipp. (Pardo) 7327
Meeresalg. deut. Tiefsee Exped. (Reinbold) 8912
Meeresalgen der deutschen Tiefsee-Expedition (Reinbold) 8912
Meeresalgenflora von Süd-Georgien (Reinsch) 8958
Mein Leben und Streben (Rossmässler) 9585
Mein Tagewerk (Reinke) 8943
Melast. bras. (Raddi) 8499
Mélast. Madag. (H. Perrier) 7705
Mélastomacées de Madagascar (H. Perrier) 7705
Melastome brasiliane (Raddi) 8499
Meleth. bot. (Picco) 7896
Melethemata inauguralia (Picco) 7896
Mém. arilles (Planch.) 8004
Mem. Bejuco (Ruiz) 9780
Mem. Calaguala (Ruiz) 9778
Mem. cult. algodon (Roxas) 9716

Mémoires sur la famille des Graminées (Raspail) 8638
Mémoire sur la famille des Guttifères (Planch.) 8014
Mémoire sur la famille des Rubiacées (A. Rich.) 9144
Mémoire sur la symétrie des Capparidées (St. Hil.) 10038
Mémoire sur le genre Tozzia (St. Hil.) 10035
Mémoire sur le Papyrus des anciens (Parl.) 7364
Mémoire sur le placenta central (St. Hil.) 10050
Mémoire sur les arilles (Planch.) 8004
Mémoire sur les Balanophorées (Rich.) 9158
Mémoire sur les Calycérées (Rich.) 9157
Mémoire sur les Cucurbitacées (Sageret) 9988
Mémoire sur les Cucurbitacées (St. Hil.) 10028
Mémoire sur les Lemna (P. Beauv.) 7217
Mémoire sur les Myrsinées (St. Hil.) 10049
Mémoire sur les plantes auxquelles on attribue un placenta central libre (St. Hil.) 10027
Mémoire sur l'état de la végétation au sommet du Pic du Midi (Ramond) 8618
Memoria de la planta Bejuco e la Estrella (Ruiz) 9780
Memoria de la planta Yallhoy (Ruiz) 9779
Memoria i catalogo de las plantas cultivadas en el Jardin botanico (F. Philippi) 7849
Memoria sobre el cultivo y cosecha del algodon (Roxas) 9716
Memoria sobre la legitima Calaguala (Ruiz) 9778
Memoria sobre la raiz de Purhampuy (Ruiz) 9781
Mém. Papyrus (Parl.) 7364
Mém. pl. cult. (F. Philippi) 7849
Mém. pl. plac. (St. Hil.) 10027
Mem. raiz Purhampuy (Ruiz) 9781
Mém. Rubiac. (A. Rich.) 9144
Mem. Yallhoy (Ruiz) 9779
Mensch en Natuur (Pulle) 8392
Menthotheca universa (see F. Petrak)
Metamorphose der Pflanzen (Potonié) 8218
Metam. Pfl. (Potonié) 8218
Méth. anal.-comp. bot. (Peyre) 7804
Meth. bot. syst. (Radlk.) 8516
Méthode analytique-comparative de botanique (Peyre) 7804
Methodik der Blütenbiologie *(see* O. Porsch)
Methodus plantarum nova (Ray) 8700
Meth. pł. (Ray) 8700

Meth. pl. epist. (Ray) 8706
Mexikanischen und zentralamerikanischen Arten der Gattung Cirsium (Petr.) 7783
Miceti patogeni (see G. Pollacci)
Michelia (Sacc.) 9895
Micologia ligustica (Pollacci) 8134
Micromiceti italiani (Sacc.) 9936
Micromycetes rariores selecti praecipue scandinavici (see L. Romell)
Micromycetes rariores selecti (see N. Patouillard)
Micromycetes sclavonici novi (Sacc.) 9903
Micro-Palaeophytol. Carbon. (Reinsch) 8954
Micro-Palaeophytologia formationis Carboniferae (Reinsch) 8954
Microscope (Pelletan) 7619
Mikrobotanisch-stratigraphische Untersuchung der Braunkohle des Muskauer Bogens (Raatz) 8430
Mikrobot. Unters. Braunk. Muskau (Raatz) 8430
Mikroflora der Prager Wasserleitung (Ruttner) 9843
Mikrofl. Prag. Wasserleit. (Ruttner) 9843
Mikrophotographien über die Strukturverhältnisse und Zusammensetzung der Steinkohle des Carbon (Reinsch) 8953
Mikrophot. Strukt.-Verh. Steink. (Reinsch) 8953
Mikrosk. Blicke (Rossmässler) 9581
Mikroskopische Blicke (Rossmässler) 9581
Minnesota trees and shrubs (Rosend.) 9525
Minnesota trees shrubs (Rosend.) 9525
Miscellanea mycologia (Sacc.) 9904
Miscellanea zoologica (Pall.) 7223
Misc. zool. (Pall.) 7223
Miss. sci. Cap Horn, Bot. *(see P. Petit)*
Mitteleuropäischen Arten der Gattung Ulex (Rikli) 9223
Mittelmeer zum Nordrand der algerischen Sahara (Rikli) 9232
Mittelmeer zum Nordrand der Sahara (Rikli) 9232
Modif. fl. Montpellier (G. Planch.) 7996
Modifications de la flore de Montpellier (G. Planch.) 7996
Mössler's Handb. Gewächsk. *(see L. Reichenbach)*
Monandrian plants of the order Scitamineae (Roscoe) 9505
Monandr. pl. Scitam. (Roscoe) 9505
Mongolischen Völkerschaften (Pall.) 7225
Mongol. Völkersch. (Pall.) 7225
Monogr. Acon. (Rchb.) 8871
Monografia de las Compuestas de Chile (J. Rémy) 8977
Monografia del Carpino (L. Piccioli) 7895

Monografia del Castagno (L. Piccioli) 7893
Monografia delle Fumariée (Parl.) 7356
Monografické studie o českých druzích rodu Bryum (Podp.) 8075
Monographia Betulacearum (Regel) 8788
Monographia Cyphellacearum Cechosloveniae (Pilát) 7934
Monographia das Xylariaceas riograndenses (Rick) 9191
Monographia do milho e da mandioca (Peckolt) 7606
Monographiae Cinchonae generis specimen (Rohde) 9445
Monographiae Cinchonae generis tentamen (Rohde) 9446
Monographiae generis Verbasci prodromus (Pfund) 7835
Monographiae phanerogamarum *(see* J. Planchon)
Monographia generis Aconiti (Rchb.) 8871
Monographia generis Eremostachys (Regel) 8805
Monographia Spartinarum (Saint-Yves) 10103
Monographia Sphaerialium astromaticorum Riograndensium (Rick) 9190
Monographic study of the Hawaiian Lobelioideae (Rock) 9329
Monographic study of Thelypodium (Payson) 7578
Monographie de la famille des Élaeagnacées (A. Rich.) 9142
Monographie der asiatischen Arten der Gattung Melica (Papp) 7312
Monographie der europäischen Arten der Gattung Melica (Papp) 7311
Monographie der Gattung Acer (Pax) 7543
Monographie der Gattung Mollinedia (Perkins) 7683
Monographie der Gattung Silene (Rohrb.) 9459
Monographie des lichens du bassin de la Garonne (Roum.) 9672
Monographie des Myxomycètes de France (A. Pouchet) 8234
Monographie des Orchidées des Iles de France et de Bourbon (A. Rich.) 9143
Monographie des Orchidées des Nil-Gherries (A. Rich.) 9148
Monographie des Orchidées mexicaines (A. Rich.) 9149
Monographie des Orchidées recueillies dans les Nil-Gherries *(see* G. Perrottet)
Monographie des Primulacées (St. Hil.) 10050
Monographie du genre Hydrocotyle (A. Rich.) 9140
Monographie du genre Hyptis (Poit.) 8121
Monographie du genre Lolium (Rouville) 9695
Monographie du genre Pleurosigma (H. Perag.) 7663
Monographie du genre Rosier *(see* A. Pronville)
Monographie Paullinia (Radlk.) 8533
Monographie Serjania (Radlk.) 8511
Monographische Übersicht über die Arten der Gattung Primula (Pax) 7545
Monograph of Azaleas (Rehd.) 8816
Monograph of Christmas Island *(see* A. Rendle)
Monograph of the British Uredineae and Ustilagineae (Plowr.) 8060
Monograph of the genus Brickellia (B. L. Robinson) 9293
Monograph of the genus Lesquerella (Payson) 7577
Monograph of the Isoetaceae (N. Pfeiff.) 7823
Monograph of the North American Potentilleae (Rydb.) 9853
Monograph of the North American Umbelliferae (J. Rose) 9512
Monograph of the section Oreocarya of Cryptantha (Payson) 7579
Monograph on the genus Heuchera *(see* C. Rosendahl)
Monogr. Azaleas (Rehd.) 8816
Monogr. Betul. (Regel) 8788
Monogr. Brickellia (B. L. Robinson) 9293
Monogr. Brit. Ured. (Plowr.) 8060
Monogr. Bryum (Podp.) 8075
Monogr. Camellia *(see* C. Pope)
Monogr. Carpino (L. Piccioli) 7895
Monogr. Castagno (L. Piccioli) 7893
Monogr. Cinch. spec. (Rohde) 9445
Monogr. Cinch. tent. (Rohde) 9446
Monogr. Cyphell. Čech. (Pilát) 7934
Monogr. Élaeagn. (A. Rich.) 9142
Monogr. Eremostachys (Regel) 8805
Monogr. Fumar. (Parl.) 7356
Monogr. Hydrocotyle (A. Rich.) 9140
Monogr. Lolium (Rouville) 9695
Monogr. Melast. *(see* L. Richard)
Monogr. Mollinedia (Perkins) 7683
Monogr. myxomyc. France (A. Pouchet) 8234
Monogr. N. Amer. Potent. (Rydb.) 9853
Monogr. N. Amer. Umbell. *(see* J. Rose)
Monogr. Orchid. Iles de France Bourbon (A. Rich.) 9143
Monogr. Orchid. Nil-Gherries (A. Rich.) 9148
Monogr. Paullinia (Radlk.) 8533
Monogr. Pleurosigma (H. Perag.) 7663
Monogr. Primula (Pax) 7545
Monogr. Serjania (Radlk.) 8509

Monogr. Serjania (Radlk.) 8511
Monogr. Silene (Rohrb.) 9458, 9459
Monogr. Stud. Haw. Lobelioid. (Rock) 9329
Monogr. Verbasci prodr. (Pfund) 7835
Monotypische Gattungen der chilenischen Flora (Reiche) 8858
Monte generoso (Penz.) 7641
Mont. int. Vistrit. Niss. fluv. (Sadebeck) 9962
Moose Flecht. Deutschl. (Redslob) 8753
Moose und Flechten Deutschlands (Redslob) 8753
Moosgesch. Deutschl. (Röhl.) 9378
Moosgeschichte Deutschlands (Röhl.) 9378
Morcellement de l'espèce en botanique et le jordanisme (Planch.) 8020
Morph. Abh. (Reinke) 8925
Morph. Cambomb. Nymphaeac. (Racib.) 8465
Morph. histol. pl. (Rusby) 9822
Morphologie der Cabombeen und Nymphaeaceen (Racib.) 8465
Morphologie der Gattung Silene (Rohrb.) 9458
Morphologische Abhandlungen (Reinke) 8925
Morphologische, anatomische und physiologische Fragmente (Reinsch) 8946
Morphologische Studien über die Orchideenblüthe (Pfitz.) 7830
Morphology and histology of plants (Rusby) 9822
Morphology of Thismia americana (N. Pfeiff.) 7822
Morph. Silene (Rohrb.) 9458
Morph. Stud. Orchideenbl. (Pfitz.) 7830
Mosses and lichens collected in the former Danish West Indies (Raunk.) 8660
Mosses and liverworts (T. H. Russell) 9830
Mosses of Samoa (see T. Powell)
Moss. liverw. (T. H. Russell) 9830
Motion nom. alg. (Rendle) 9067
Motion proposing an additional clause to the Rules of 1905 concerning the nomenclature of the algae (Rendle) 9067
Mouriria anomala (Pulle) 8388
Mousses de la Belgique (see L. Piré)
Mousses de l'Oubangui (P. de la Varde) 8207
Mousses des Canaries (Renauld) 8997
Mousses du département de l'Aude (see C. Roumeguère)
Mousses du Gabon (P. de la Varde) 8208
Mousses nouvelles de l'Amérique du Nord (Renauld) 8985
Mousses nouvelles de l'herbier Boissier (Renauld) 8990

Mouss. Gabon (P. de la Varde) 8208
Mouss. Oubangui (P. de la Varde) 8207
Musci (Renauld) 8996
Musci Americae septentrionalis (Renauld) 8989
Musci americae septentrionalis exsiccati. Notes (Renauld) 8992
Musci Americae septentrionalis exsiccati. Observations & rectifications (Renauld) 8991
Musci Americae septentrionalis exsiccati (see F. Renauld)
Musci austro-africani (see A. Rehmann)
Musci bohemici (see J. Poech)
Musci čechoslovenici exsiccati (see Z. Pilous)
Musci costaricenses (Renauld) 8988
Musci europaei exsiccati (see F. Renauld)
Musci exotici novi vel minus cogniti (Renauld) 8987
Musci exsiccati Novae-Zelandiae (see G. Sainsbury)
Musci frondosi javanici (Reinw.) 8964
Musci insulae rossicae prope Vladivostok (Podp.) 8084
Musci lapponiae kolaënsis (Saelan) 9981
Musci marcareno-madagascarienses (see F. Renauld)
Musci mexicani first century (Pringle) 8325
Musci mexicani (see C. Pringle)
Muscinées des environs de Spa (Piré) 7961
Musci Norvegiae borealis (see E. Ryan)
Musc. ins. ross. Vladiv. (Podp.) 8084
Musci occidentali-americani (see C. Piper)
Musci tarvisini (Sacc.) 9887
Musci tarvisini (see P. Saccardo)
Musc. lapp. kolaens. (Saelan) 9981
Muscol. (P. Beauv.) 7218
Muscologia gallica (see F. Renauld)
Muscologie ou traité sur les mousses (P. Beauv.) 7218
Museum of flowers (M. Rosenberg) 9517
Mus. fl. (M. Rosenberg) 9517
Musgos de Venezuela (Pitt.) 7990
Musg. Venez. (Pitt.) 7990
Mushrooms (F. Patterson) 7500
Mushrooms (J. A. Palmer) 7238
Mushrooms (Peck) 7599
Mushrooms Amer. (J. A. Palmer) 7237
Mushrooms and other common fungi (F. Patterson) 7500
Mushrooms and their use (Peck) 7599
Mushrooms of America (J. A. Palmer) 7237
Mycetes malacenses (Sacc.) 9906
Mycetes malacenses (see G. Paoletti)
Mycetes siberici (Sacc.) 9907
Mycétologie (Paul.) 7520
Mycol. Collioure (Roum.) 9681
Mycol. eur. (Pers.) 7734, 7736

Mycol. eur. (Rabenh.) 8445
Mycologia europaea (Pers.) 7736
Mycologia europaea (Rabenh.) 8445
Mycologiae venetae specimen (Sacc.) 9888
Mycologia *(see* D. P. Rogers, L. Romell)
Mycologie des environs de Collioure (Roum.) 9681
Mycol. ven. spec. (Sacc.) 9888
Mycotheca carpathica (see F. Petrak)
Mycotheca eichleriana (see F. Petrak)
Mycotheca generalis (see F. Petrak)
Mycotheca germanica (see H. Poeverlein, H. Rehm)
Mycotheca italica (see D. Saccardo)
Mycotheca polonica (Racib.) 8472
Mycotheca polonica (see M. Raciborski)
Mycotheca veneta (see P. Saccardo)
Mycoth. polon. (Racib.) 8472
Mykologische Notizen (Petr.) 7784
Myxomyc. Dan. (Raunk.) 8653a
Myxomyceten der Flora von Buitenzorg (Penz.) 7652
Myxomyceten des nordöstlichen Holsteins (Rönn) 9416
Myxomycetes Daniae (Raunk.) 8653a
Myxomycètes des environs de Montpellier (Jul. Pavillard) 7535
Myxomycetes from St. Croix, St. Thomas and St. John (Raunk.) 8662
Myxomycetum agri cracoviensis genera, species et varietates novae (Racib.) 8453
Myxomyc. Fl. Buitenzorg (Penz.) 7652
Myxomyc. n.-östl. Holst. (Rönn) 9416

Naamlijst der planten Java (Radermacher) 8501
Naaml. pl. Java (Radermacher) 8501
Naauwk. beschr. Aleppo (Al. Russell) 9828
Naauwkeurige en natuurlijke beschrijving van de stad Aleppo (Al. Russell) 9828
Nachricht von Suriname (Quandt) 8412
Nachr. Suriname (Quandt) 8412
Nachträge zu den Plantae Raddeaneae *(see* G. Radde)
Nachträge zur Flora von Vorarlberg und Liechtenstein (Richen) 9174
N. Amer. Caric. *(see* H. Rickett)
N. Amer. Stereocaulon (Riddle) 9203
Napoléone (P. Beauv.) 7211
Napoléone impériale (P. Beauv.) 7211
Narcissus poeticus and its allies (Pugsl.) 8381
Narrative of an attempt to reach the North Pole (W.Parry) 7414
Narrative of a ramble among the wild flowers on the Moffat Hills (John Sadler) 9973
Narrative of four journeys (W. Paterson) 7439

Narr. attempt North Pole (W. Parry) 7414
Narr. journey Polar Sea (Richardson) 9169
Narr. journ. Hottent. (W. Paterson) 7439
Narr. ramble Moffat Hills (John Sadler) 9973
Nat. Flechten (Reess) 8764
Nat. Gruppe Davall. (Pérez Arb.) 7674
Nat. hist. Aleppo (Al. Russell) 9828
Nat. Hist. Aleppo *(see* P. Russell)
Nat. hist. animalc. (A. Pritch.) 8348
Native woods of Essex County (J. Robinson) 9302
Nat.- Kult.-Bilder Kaukasusländ. (Rikli) 9234
Nat. Pflanzenfam. ed. 2 *(see* W. Ruhland)
Nat. Pflanzenfam. *(see* V. Poulsen, L. Radlkofer, R. Raimann, W. Ruhland, R. Sadebeck)
Natürliche Gruppe der Davalliaceen (Pérez Arb.) 7674
Natürliche Pflanzenfamilien, ed. 2 *(see* F. Pax, R. Pilger)
Natürliche Pflanzenfamilien *(see* H. Paul, A. Peter, O. Petersen, E. Pfitzer, T. v. Post, H. Potonié, K. Prantl, H. Printz, E. Pritzel, C. Reiche)
Natural history of Aleppo (Al. Russell) 9828
Natural history of animalcules (A. Pritch.) 8348
Natural history of Orange trees (Risso) 9248
Natural history of the island of Fernando de Noronha (Ridl.) 9206
Naturaliste canadien *(see* L. Provancher)
Nature Magazine *(see* E. Preble)
Naturgeschichte von Aleppo (Al. Russell) 9828
Naturh. Bidr. Grønland (Rink) 9244
Naturhistorische Miscellen (Rochel) 9318
Naturhistoriske Bidrag til en Beskrivelse af Grønland (Rink) 9244
Naturh. Misc. (Rochel) 9318
Natur- und Kulturbilder aus den kaukasusländern und Hocharmenien (Rikli) 9234
Natur und Volk des Mikadoreiches (Rein) 8910a
Naturwissenschaftliche Reise nach Mossambique (W. Peters) 7759
Naturwissenschaftliche Wochenschrift *(see* H. Potonié)
Naturw. Reise Mossambique (W. Peters) 7759
Natuurlijke historie der plant-dieren (Pall.) 7222
Názorná Květ. (Polívka) 8133
Názorná květena (Polívka) 8133
Necess. util. stud. bot. (Rodschied) 9366
Nelken (Rössig) 9430

Nella regione dei Laghi Equatoriali (Piscicelli) 7970
Neobotanon (Raf.) 8586
Neogenyton (Raf.) 8578
Nervation der Pflanzenblätter (Pokorny), 8126
Nerv. Pfl.-Blätt. (Pokorny) 8126
Neu. Beitr. Fl. Kreta (Rech. fil.) 8737
Neu. Conif. Alp. (Pančic) 7281
Neue Alchemilla-Arten (Rothm.) 9650
Neue Bearbeitung der Gattung Aconitum (Rchb.) 8875
Neue Beiträge zur Flora Surinams (Pulle) 8386
Neue Beiträge zur Flora von Kreta (Rech. fil.) 8737
Neue Beiträge zur Kenntniss der Cirsien der Kaukasus (Petr.) 7781
Neue Beiträge zur Kenntniss der Flora von Thüringen (K. Reinecke) 8918
Neue Beobachtungen über den Befruchtungsact der Gattungen Achlya und Saprolegnia (Pringsh.) 8339
Neue Beobachtung über Oscillaria (Rupr.) 9795
Neue Beyträge zur Botanik (Roth) 9631
Neue Beytr. Bot. (Roth) 9631
Neue Conifere in den östlichen Alpen (Pančic) 7281
Neue Methode die Pflanzen zu trocknen (Pluskal) 8070
Neue oder unvollständig bekannte Pflanzen aus dem nördlichen Theile des Stillen Oceans (Rupr.) 9805
Neuer Grundriss der Botanik (A. Rich.) 9138
Neuer Rumex aus den Nordalbanischen Alpen (Rech. fil.) 8732
Neuer Schlüssel zu Rumph's Herbarium amboinense (Rumph.) 9784
Neuer Versuch einer systematischen Eintheilung der Schwämme (Pers.) 7727
Neues Magazin für die Botanik (Roem.) 9401
Neue Unters. Mikrostrukt. Steink. (Reinsch) 8951
Neue Untersuchungen über die Mikrostruktur der Steinkohle (Reinsch) 8951
Neu. Mag. Bot. (Roem.) 9401
Neu. Meth. Pfl. trockn. (Pluskal) 8070
Neu. Pfl. Still. Ocean (Rupr.) 9805
Neu. Rumex Nordalb. Alp. (Rech. fil.) 8732
New and noteworthy Hawaiian plants (Radlk.) 8539
New and rare freshwater algae (Playf.) 8047
New Britton and Brown (see E. J. Palmer)
New fl. (Raf.) 8586

New flora and botany of North America (Raf.) 8586
New genera of plants from Bolivia (Rusby) 9820
New gen. fungi (see P. Ricker)
New genus of Sterculiaceae (B. L. Robinson) 9287
New list fl. pl. Devon (Ravenshaw) 8687
New list of the flowering plants in the county of Devon (Ravenshaw) 8687
New Mosses of North America (Renauld) 8985
New N. Amer Crassul. (see J. Rose)
New or noteworthy plants from Colombia and Central America (Pitt.) 7980
New plants collected upon an archaeological expedition to Northwestern Mexico (B. L. Robinson) 9286
New species from the Blue Ridge (Rydb.) 9871
New Species from the western United States (Rydb.) 9854
New Sylva (Raf.) 8586
Niaouli (C. Rivière) 9265
Nieuwe beginselen der kruidkunde (A. Rich.) 9138
Nieuw plantkundig woordenboek voor Nederlandsch Indie (Pulle) 8391
Nieuw plantk. woordenb. Ned.-Indie (Pulle) 8391
Nikobar. ins. (Rink) 9242
Nikobarischen Inseln (Rink) 9242
Noces des palmiers (Barb. Rodr.) 9359
Noces palm. (Barb. Rodr.) 9359
Növénytani intézete és botanikus kertje (Al. Richt.) 9180
Nomenclator botanicus (Pfeiff.) 7821
Nomenclator botanicus (Raeusch.) 8547
Nomenclator botanicus (Retz.) 9084
Nomenclature franco-provençale des plantes (Reguis) 8811
Nomenclature of Plants (see H. St. John)
Nomenclature proposals by British botanists (see J. Ramsbottom)
Nomenclature raisonnée des espèces, variétés et sous- variétés du genre Rosier (Pronville) 8366
Nomencl. bot. (Pfeiff.) 7820, 7821
Nomencl. bot. (Raeusch.) 8547
Nomencl. bot. (Retz.) 9084
Nomencl. Fl. danic. (see E. Rostrup, L. Rosenvinge)
Nomencl. Rosier (Pronville) 8366
Nonnullarum specierum tuliparum notae (Reboul) 8720
Nonnull. pl. (P. Rossi) 9573
Nonnul. sp. tulip. not. (Reboul) 8720
Nordamerikanische Laubmoose, Torfmoose und Lebermoose (Röll) 9388

Nordamerikanischen Arten der Gattung Cirsium (Petr.) 7783
Nordamerikanischen Saxifraginae (Rosend.) 9521
Nordamer. Saxifrag. (Rosend.) 9521
Nordens Flora (Palmstr.) 7247
Norges Hymenomyceter (Rostr.) 9619
Norg. Hymenomyc. (Rostr.) 9619
Norrländiska torfmosse studier (L. Post) 8188
Norrl. torfmossestud. (L. Post) 8188
Norske Ascomyc. (Rostr.) 9618
Norske Ascomyceter (Rostr.) 9618
Norske fjellpl. (Resvoll) 9080
North American flora (see F. Pennell, C. Pollard, G. Prescott, H. Rickett, J. Rose, H. Rusby, P. Rydberg)
North American Sileneae (B. L. Robinson) 9284
North American species of Aquilegia (Payson) 7576
North American species of Festuca (Piper) 7947
North American species of Physalis (Rydb.) 9851
North-American species of Rumex (Rech. fil.) 8734
North-American species of Stereocaulon (Riddle) 9203
Notae mycologicae (Sacc.) 9932
Notas bótanicas à la flora española (Pau) 7511
Notas sobre algunas plantas usuales del Paraguay (D. Parodi) 7390
Not. bot. (Pasq.) 7432
Not. bot. (Pierre) 7923a
Not. bót. fl. espan. (Pau) 7511
Note micologiche (Penz.) 7642
Note microlitologiche sopra i calcari (D. Pantan.) 7286
Note pour servir à l'histoire de la formation de la houille (Renault) 9019
Notes botaniques (Pierre) 7923a
Notes et observations sur quelques plantes rares de sud-ouest de la France (Revel) 9113
Notes forage pl. Iowa (Pammel) 7253
Notes in entomogenous fungi (see T. Petch)
Notes mycologiques (Sacc.) 9908, 9931
Notes on Fabaceae (Rydb.) 9867
Notes on Philotria (Rydb.) 9862
Notes on Potentilla (Rydb.) 9850
Notes on Rosaceae (Rydb.) 9863
Notes on the flora of Vermont (G. Perkins) 7682
Notes on the grasses and forage plants of Iowa (Pammel) 7253
Notes on the lower Basidiomycetes (D. P. Rogers) 9434

Notes on the medicinal Cinchona barks of New Granada (see E. Poeppig)
Notes on useful plants of Mexico (J. Rose) 9511
Notes sur des plantes nouvelles ou peu connues de la Savoie (E. Perrier) 7697
Note sul genere Mycosyrinx (Penz.) 7653
Notes upon Hawaiian plants (Rock) 9322
Notes upon the flora of Newfoundland (B. L. Robinson) 9288
Note sur le Carex tenax (St. Lag.) 10080
Note sur le Clathropodium morieri (Renault) 9037
Note sur les Sphenozamites (Renault) 9016
Note sur quelques micro et macrospores fossiles (Renault) 9058
Note sur quelques plantes de la Haute-Maurienne (St. Lag.) 10075
Note sur trois champignons des Antilles (Pat.) 7487
Note sur trois espèces d'Hydnangium (Pat.) 7492
Note sur trois espèces mal connues d'hyménomycètes (Pat.) 7449
Note sur un nouveau genre de Gymnosperme fossile (Renault) 9046
Not. fl. Belfort (Parisot) 7337
Not. Haw. pl. (Rock) 9322
Not. hist. Soc. Hort. Sarthe (Ragot) 8594
Notice historique sur la Société d'Horticulture de la Sarthe (Ragot) 8594
Notice of Salix sadleri (John Sadler) 9974
Notice of some new and rare phaenogamous plants found in this state (Ravenen) 8683
Notice sur la constitution des lignites (Renault) 9052
Notice sur la flore des environs de Belfort (Parisot) 7337
Notice sur les Calamariées (Renault) 9048
Notice sur les plantes vénéneuses (Pierrot) 7924
Notice sur les Renoncules batraciennes (Revel) 9111
Notice sur les Sigillaires (Renault) 9040
Notice sur les travaux scientifiques (Renault) 9051
Notice sur le Thuya de Barbarie (Rosny) 9557
Notice sur soixante-dix espèces et quelques variétés de plantes phanérogames Loiret (St. Hil.) 10024
Notice sur une Lycopodiacée arborescente (Renault) 9044
Noticia sobre algunas criptógamas nuevas (Puigg.) 8383
Notions supplémentaires (Sacc.) 9934
Notizia sulla Pachira alba (Parl.) 7353

Notizie botaniche (Pasq.) 7432
Not. microlit. calc. (D. Pantan.) 7286
Not. Pachira alba (Parl.) 7353
Not. pl. vénén. (Pierrot) 7924
Not. suppl. (Sacc.) 9934
Not. Thuya Barbarie (Rosny) 9557
Not. trav. Sci. (Renault) 9051
Notulae in species novas vel criticas plantarum horti botanici genevensis (Reut.) 9109
Notul. sp. nov. hort. genev. (Reut.) 9109
Nouv. Contr. Crassulac. malgach. *(see* H. Perrier)
Nouv. dict. hist. nat. *(see* A. Parmentier)
Nouv. Dict. Hist. nat. *(see* L. Richard)
Nouveau système de physiologie végétale de botanique (Raspail) 8639
Nouveau traité des plantes usuelles (Roques) 9499
Nouveaux élémens de botanique (A. Rich.) 9138
Nouveaux matériaux pour la flore atlantique (Pomel) 8150
Nouv. élém. bot. (A. Rich.) 9138
Nouvelles espèces de champignons de la Cote-d'Or (Sacc.) 9928
Nouvelles flores de France (St. Lag.) 10083
Nouvelles observations sur la fructification des mousses et des lycopodes (P. Beauv.) 7214
Nouvelles observations sur les Tulipes de la Savoie (E. Perrier) 7700
Nouvelles recherches bryologiques (Piré) 7958
Nouvelles recherches sur le développement de la feuille des Muscinées (Pottier) 8232
Nouvelles remarques sur la nomenclature botanique (St. Lag.) 10068
Nouv. mat. fl. atl. (Pomel) 8150
Nouv. observ. mousses (P. Beauv.) 7214
Nouv. syst. phys. vég. (Raspail) 8639
Nouv. traité pl. usuel. (Roques) 9499
Nova ascomycetum genera (Sacc.) 9892
Nova elementa ad floram principatus Bulgariae (Pančic) 7282
Novae plantae Schwenckia (D. Royen) 9732
Novae plantarum species (Roth) 9635
Novae plantarum species *(see* A. Roth)
Nova genera ac species plantarum (Poepp.) 8091
Nova Guinea, Botanique *(see* A. Pulle)
Nova plantarum americanarum genera (Plum.) 8067
Nova plantarum genera (Sacc.) 9935
Novarum vel rariorum ex cryptogamia stirpium (Raddi) 8487
Nova species cryptogamatum (Raddi) 8484

Nova systema naturae *(see* S. Ramée)
Nov. gen. sp. pl. (Poepp.) 8091
Noviciae indicae (Prain) 8264
Novissimo prospetto della flora della provincia di Bergamo (Rodegher) 9337
Novitates brazilienses (Rehm) 8829
Nov. pl. amer. (Plum.) 8067
Nov. pl. Schwenckia (D. Royen) 9732
Nov. pl. sp. (Roth) 9635
Nov. sp. crypt. (Raddi) 8484
Nowe gatunki zielenic (Racib.) 8464
Nowočeska bibliothéká (J. Presl) 8289
Nueva contr. fl. Granada (Pau) 7515
Nueva contribución al estudio de la flora de Granada (Pau) 7515
Numb. checkl. N. Amer. pl. (H. Patt.) 7507
Numenkl. Revis. höher. Pfl.-Gr. *(see* T. v. Post)
Numero delle piante (Sacc.) 9916
Nuove alghe del viaggio di circumnavigazione della Vettor Pisani (Picc.) 7905
Nuove contribuzioni alla flora marina del Mar Rosso (Picc.) 7909
Nuove piante trovate in val d'Ossola (S. Rossi) 9578
Nuov. gen. sp. monocot. (Parl.) 7366
Nuovi generi e nuove specie di piante monocotiledoni (Parl.) 7366
Nye Bidrag til Vest-Grønlands Flora (Rosenvinge) 9541
Nye Bidr. Vest-Grønl. Fl. (Rosenvinge) 9541

Obs. accrois. thalle phéosp. *(see* J. Rostafiṅsky)
Obs. alg. *(see* J. Rostafiṅsky)
Obs. bot. pug. (Retz.) 9092
Obs. Crit. bot. (Retz.) 9093
Observatio de Mangiferae semine polyembryoneo (Reinw.) 8963
Observationes ad peloriarum indolem definiendam spectantes (Ratzeb.) 8645
Observationes botanicae (E. Rosén) 9514
Observationes botanicae (Retz.) 9086
Observationes in methodum tournefortianam (Retz.) 9088
Observationes mycologicae (Pers.) 7724
Observations botaniques sur le genre Sonchus (Picard) 7879
Observations faites dans les Pyrenées (Ramond) 8616
Observations l'histoire naturelle de la Vallée d'Aspe (Palassou) 7200
Observations made in journey through the Low-Countries (Ray) 8697
Observations of the late Dr. William Roxburgh on the various specimens of fibrous vegetables (Roxb.) 9722

Observations on the structure of fruits and seeds (Rich.) 9154
Observations sur la famille des Amaranthacées (St. Hil.) 10043
Observations sur la nature des fleurs (Roep.) 9418
Observations sur le genre Anacardium (St. Hil.) 10040
Observations sur le genre Cuphea (St. Hil.) 10045
Observations sur le genre Escallonia (St. Hil.) 10044
Observations sur le genre Hyacinthus (St. Hil.) 10025
Observations sur le genre Quartinia (A. Rich.) 9146
Observations sur quelques espèces critiques du genre Hieracium (Paiche) 7187
Observations sur quelques plantes d'Italie (Parl.) 7348
Observationum botanicarum pugillus (Retz.) 9092
Observationum botanicarum specimen (Panz.) 7300
Observationum in Criticam botanicam (Retz.) 9093
Observ. bot. (E. Rosén) 9514
Observ. bot. (Retz.) 9086
Observ. bot. spec. (Panz.) 7300
Observ. fibrous veg. (Roxb.) 9722
Observ. hist. nat. Vallée d'Aspe (Palassou) 7200
Observ. journ. Low-Countries (Ray) 8697
Observ. mycol. (Pers.) 7724
Obs. meth. tourn. (Retz.) 9088
Obs. Pyrenées (Ramond) 8616
Obstbau des Kantons Zürich (Regel) 8777
Obstbau Zurrich (Regel) 8777
Oekon.-bot. Beschr. Rosen (Rössig) 9428
Oekonomisch-botanische Beschreibung der Rosen (Rössig) 9428
Oelkörper der Lebermoose (Pfeff.) 7812
Österreichischen Galeopsisarten (Porsch) 8171
Österreichs Holzpflanzen (Pokorny) 8129
Österr. Galeopsis. (Porsch) 8171
Österr. Holzpfl. (Pokorny) 8129
Øst-Grønlands Svampe (Rostr.) 9614
Øst-Grønl. Svamp. (Rostr.) 9614
Oeuvres complètes de P. Poivre *(see* P. Poivre*)*
Officinal-Kräuter (Riv.) 9271
Ohia Lehua trees Hawaii (Rock) 9328
Ohia Lehua trees of Hawaii (Rock) 9328
Om Algevegetation ved Grønlands Kyster (Rosenvinge) 9544
Om fjeldvegetationen i det Østenfjeldske Norge (Resvoll- Holmsen) 9082
Om nogle Væxtforhold hos Slægterne Cladophora og Chaetomorpha (Rosenvinge) 9539
Om planter som passer til kort og kold sommer (Resvoll) 9079
Om planteveksten i grensetrakter (Resvoll-Holmsen) 9083
Om Vegetationen paa de dansk-vestindiske Øer (Paulsen) 7529
Onderzoekingen over Sphaeroplea annulina (Rauwenh.) 8677
Onderz. Sphaeroplea annulina (Rauwenh.) 8677
Onothera ou Oenothera (St. Lag.) 10082
On some Danish Phaeophyceae (Rosenvinge) 9555
On some principles of stelar morphology (Posth.) 8198
On the attachment organs of the common corticulous Ramalinae *(see* L. Porter*)*
On the collections of dried plants obtained in South-West Africa by the Percy Sladen Memorial Expeditions *(see* H. Pearson*)*
On the culture and commerce of cotton in India (Royle) 9736
On the ferns of Sumba (Posth.) 8201
On the marine algae from North-East Greenland (Rosenvinge) 9547
On the popular names of British plants (Prior) 8347
On the Rubi list in London catalogue (W. M. Rogers) 9443
On the spiral arrangement of the branches in some Callithamnieae (Rosenvinge) 9550
On two collections of ferns made in Madagascar (H. Rosend.) 9530
Oocystis and Eremosphaera (Playf.) 8044
Open-air stud. bot. (Praeger) 8248
Open-air studies in botany (Praeger) 8248
Opera historiam naturalem spectantia *(see* J. Petiver*)*
O přirozenosti rostlin *(see* J. Presl*)*
Opuscula omnia botanica Thomas Johnsoni *(see* T. Ralph*)*
Opyt flory Usurijskoj strany (Regel) 8787
Orangers *(see* F. Sahut*)*
Oratio (Royen) 9727
Oratio augm. hist. nat. Ind. (Reinw.) 8961
Oratio de ardore (Reinw.) 8959
Oratio de augmentis, quae historiae naturali ex Indiae investigatione accesserunt (Reinw.) 8961
Oratio de hortis publicis (D. Royen) 9731
Oratio hort. publ. (D. Royen) 9731
Orchidaceae in Symbolae *(see* H. G. Reichenbach*)*
Orchidaceele din România (Panțu) 7299
Orchid. chil. (Reiche) 8861

Orchideae quaedam lansbergianae (Rchb. fil.) 8895
Orchideae Splitgerberianae surinamenses (Rchb. fil.) 8895
Orchideas chilenses (Reiche) 8861
Orchidées (Puydt) 8410
Orchid N.S.W. (Rupp) 9789
Orchid Review *(see* R. Rolfe) Lindenia *(see* R. Rolfe)
Orchid. Român. (Panțu) 7299
Orchids (E. S. Rand) 8625
Orchids from Sikkim (Pantl.) 7289
Orchids of New South Wales (Rupp) 9789
Orchids of the Sikkim Himalaya *(see* R. Pantling)
Orchids of Victoria (Pescott) 7742
Orchid stud-book (Rolfe) 9470
Orchid. Victoria (Pescott) 7742
Ordines naturales plantarum (Ruel.) 9764
Ord. nat. pl. (Ruel.) 9764
Ordo plantarum quae sunt florae irregulari pentapetalo (Riv.) 9269
Ordo plantarum, quae sunt flore irregulari monopetalo (Riv.) 9267
Ordo plantarum, quae sunt flore irregulari tetrapetalo (Riv.) 9268
Ordo pl. fl. monopetal. (Riv.) 9267
Ordo pl. fl. pentapet. (Riv.) 9269
Ordo pl. fl. tetrapet. (Riv.) 9268
Organic remains of a former world (Js. Parkinson) 7343
Organis. fl. Solanac. (Robyns) 9311
Organ. pl. (Roep.) 9419
Organ. remains (Js. Parkinson) 7343
Orig. Brit. fl. (C. Reid) 8901
Origines des sciences naturelles (St. Lag.) 10069
Origin of the British flora (C. Reid) 8901
Ornamental trees of Hawaii (Rock) 9327
Orn. trees Hawaii (Rock) 9327
Osservazioni intorno al Viaggio al Lago di Garda e al Monte Baldo del dottor Ciro Pollini (Pollini) 8142
Osservazioni sopra le palme (Rolli) 9481
Osservazioni sui funghi dell'agro Lucchese (Puccin.) 8376
Osserv. fung. Lucch. (Puccin.) 8376
Otia botanica hamburgensia (Rchb. fil.) 8899
Otia bot. hamburg. (Rchb. fil.) 8899
Oulaisten pitäjän kasvisto (T. Parsons) 7418
Our trees (J. Robinson) 9303
Ouvrage économique sur les pommes de terre, le froment et le riz (A. Parm.) 7377
O vegetačních rozdílech mezi severní a jižní exposicé Čechách (Rohlena) 9449
Over het begrip van leven (Rauwenh.) 8676
Over het eigenaardige en over de versprei-ding der gewassen in de Magellaansche landen (Reinw.) 8967

Paddenst. Nederland (J. Ruys) 9845
Paläobotanisches Praktikum (Potonié) 8228
Palaeobotanische Zeitschrift (Potonié) 8227
Palaeobotanische Zeitschrift *(see* H. Potonié)
Paläobot. Prakt. (Potonié) 8228
Palaeobot. Z. (Potonié) 8227
Paléontologie végétale (Pelourde) 7624
Paléontol. vég. (Pelourde) 7624
PALLASIA *(see* P. Pallas)
Palmae amazonensis novae (Barb. Rodr.) 9346
Palmae hasslerianae novae (Barb. Rodr.) 9355
Palmae mattogrossenses (Barb. Rodr.) 9352
Palmae novae paraguayenses (Barb. Rodr.) 9354
Palme (Roster) 9600
Palmen (Reissek) 8970
Palm. hassler. (Barb. Rodr.) 9355
Palmiers (Barb. Rodr.) 9345
Palm. mattogr. (Barb. Rodr.) 9352
Palm. paraguay. (Barb. Rodr.) 9354
Palmyra Isl. (Rock) 9325
Palmyra Island (Rock) 9325
Pampayaco (Poepp.) 8090
Pamphysis sicul. (Raf.) 8552
Pamphysis sicula (Raf.) 8552
Paradisi in sole parodisus terrestris *(see* Jn. Parkinson)
Paras. Alg. Pilz. Javas (Racib.) 8470
Paras. Exoasc. (Sadebeck) 9968
Parasitische Algen und Pilze Javas (Racib.) 8470
Parasitischen Exoasceen (Sadebeck) 9968
Parke, Davis & Co. Herbarium department (Rusby) 9818
Parke, Davis herb. dept. (Rusby) 9818
Parte práct. bot. (Paláu) 7202
Parte práctica de botánica (Paláu) 7202
Parthenogesis im Pflanzenreiche (Regel) 8783
Parthenog. Pfl.-Reich (Regel) 8783
Patterson's numbered check-list of North American plants (H. Patt.) 7507
Paxton's botanical dictionary (Paxt.) 7555
Paxton's Fl. Gard. (Paxt.) 7556
Paxton's flower garden (Paxt.) 7556
Paxton's Magazine of botany (Paxt.) 7554
Paxton's Mag. Bot. (Paxt.) 7554
Pedilonia (K. Presl) 8296
Pelletiera (St. Hil.) 10030
Perfid. homon. (St. Lag.) 10089
Perfidie des homonymes (St. Lag.) 10089

Perfidie des synonymes (St. Lag.) 10088
Pericaulom-Theorie (Potonié) 8222
Peridineae of New South Wales (Playf.) 8048
Péridiniacées du Léman (Penard) 7628
Péridin. Léman (Penard) 7628
Periodischen Erscheinungen in der Pflanzenwelt der Polarländer (Rikli) 9237
PERSOONIA (see C. Persoon)
Perú (Raimondi) 8595
Pescatorea (see J. Planchon, H. G. Reichenbach)
Petrak's lists (Petr.) 7786
Petrified Williamsonia (Sahni) 10010
Pflanzen als Gesteinsbildner (Pia) 7878
Pflanzenaquarelle des Hans Weiditz (Rytz) 9875
Pflanzenblätter in Naturdruck (G. C. Reuss) 9103
Pflanzendecke der Erde (L. Rudolph) 9755
Pflanzengallen (H. Ross) 9567
Pflanzengallen Bayerns (H. Ross) 9568
Pflanzengarten (see F. Pfuhl)
Pflanzengeografische Studien (Palacký) 7194
Pflanzengeographie von Polen (Pax) 7551
Pflanzengeographie von Rumänien (Pax) 7552
Pflanzengeographische Monographie des Berninagebietes (Rübel) 9758
Pflanzengeographische Studien über die Halbinsel Kanin (Pohle) 8108
Pflanzengesellschaften der Erde (Rübel) 9760
Pflanzenkleid der Mittelmeerländer (Rikli) 9238
Pflanzennamen (Prahn) 8257
Pflanzen Palästinas (Post) 8191
Pflanzen-physiognomische Studien auf Torfmoosen (L. Post) 8189
Pflanzenreich (Peterm.) 7752
Pflanzenreich (Rchb.) 8882
Pflanzenreich (see F. Pax, J. Perkins, E. Pfitzer, R. Pilger, L. Radlkofer, A. Rendle, W. Ruhland)
Pflanzenreich (Reinecke) 8916
Pflanzen-Teratologie (Penz.) 7646
Pflanzen-Umrisse (Rochel) 9317
Pflanzen-Vorwesenkunde (Potonié) 8219
Pflanzenw. Afrik. (see R. Pilger)
Pflanzenwelt der Zinne und des Kleinen Hangesteines (Jul. Röm.) 9410
Pflanzenwelt Norddeutschlands (Potonié) 8212
Pflanzenwelt Ost-Afrikas (see T. Reinbold)
Pflanzenwelt von Costa-Rica (Polak.) 8131
Pflanzenw. Norddeutschl. (Potonié) 8212
Pflanze und ihr Leben (Regel) 8778
Pfl.-Aquar. Weiditz (Rytz) 9875
Pfl.-Bl. Naturdr. (G. C. Reuss) 9103

Pfl.-Decke Erde (L. Rudolph) 9755
Pflege Bot.-Franken (Reess) 8766
Pfl.-Gall. Bayerns (H. Ross) 9568
Pfl.-Geogr. Erzgeb. (Sachse) 9960
Pfl.-Geogr. Polen (Pax) 7551
Pfl.-geogr. Stud. (Palacký) 7194
Pfl.-geogr. Stud. Kanin (Pohle) 8108
Pfl.-Ges. Erde (Rübel) 9760
Pfl. Gesteinsbild. (Pia) 7878
Pfl.-Kleid. Mittelmeerländ. (Rikli) 9238
Pfl.-physiogn. Stud. Torfm. (L. Post) 8189
Pfl.-Teratol. (Penz.) 7646
Pfl. u. i. Leben (Regel) 8778
Pfl.-Umrisse (Rochel) 9317
Pfl.-Vorwesenk. (Potonié) 8219
Pfl.-Welt Burgenländ. Berge (Jul. Röm.) 9411
Pfl.-Welt Zinne (Jul. Röm.) 9410
Pfl. Westküste Eur. (E. Roth) 9638
Phanerogamenflora von Bautzen (Rostock) 9603
Pharm.-Fl. Ober-Hessen (see J. Rossmann)
Phil. Bot. (Reinke) 8940
PHILIPPIA (see R. Philippi)
Philosophie botanique (see F. Quesné)
Philosophie der Botanik (Reinke) 8940
Phlebophora (Pat.) 7467
Phlyctospora maculata (Pat.) 7462
Phycologia ochotiensis (Rupr.) 9803
Phycotheca polonica (see M. Raciborski)
Phykotheka (see P. G. Richter)
Phyllophora Brodiaei (Rosenvinge) 9552
Physiographical sketch (Parry) 7403
Physiologische Untersuchungen über die Keimung der Schminkbohne (see J. Sachs)
Physiotyp. pl. austr. (see A. Pokorny)
Phytogeographic studies in the Peace and Upper Liard River regions (Raup) 8666
Phytogeography of Nebraska (Pound) 8240
Phytogeogr. Nebraska (Pound) 8240
Phytogeogr. Peace R. (Raup) 8666
Phytographia (Pluk.) 8064
Phytographie médicale (Roques) 9497
Phytogr. méd. (Roques) 9497
Phytologist (see W. Pamplin)
Phytoplankton of the British Lakes (see W. Pearsall)
Piante acotiledoni vascolari e le graminacee ossolane (S. Rossi) 9577
Piante acot. vasc. gram. ossol. (S. Rossi) 9577
Piante dell'agro veneto (Sacc.) 9886
Piante di Bengasi e del suo territorio (Pamp.) 7265
Piante fanerogame euganee (Romano) 9485
Piante fan. eugan. (Romano) 9485

Piante fossili terziarie (Paolucci) 7309
Piante foss. terz. (Paolucci) 7309
Piante legn. ital. (L. Piccioli) 7890
Piante legnose italiane (L. Piccioli) 7890
Piante nuove della Republicca di San Marino (Pamp.) 7266
Piante San Marino (Pamp.) 7266
Piante vascolari raccolte Hu-peh (Pamp.) 7262
Pierre Richer de Belleval (Planch.) 8017
Pilze Deutschlands Ascomyceten: Hysteriaceen und Discomyceten (Rehm) 8825
Pilze Deutschlands, Deutschlands Flora *(see* G. Preuss)
Pilze Deutschl., Hyster. Discomyc. (Rehm) 8825
Pilzflora Ägyptens *(see* I. Reichert)
Pilzfreund *(see* J. Rothmayr)
Pilz- und Kräuterfreund *(see* A. Ricken)
Pinet. brit. (Ravenscroft) 8685
Pinetum britannicum (Ravenscroft) 8685
Pinet. woburn. *(see* J. Russell)
PITTIERA *(see* H. Pittier)
PITTONIA *(see* J. Pittoni)
Pl. Alméria (Pau) 7516
Pl. amer. (Plum.) 8069
Plan des bot. Garten der...Forstlehranstalt Aschaffenburg (Prantl) 8270
Plankton of the Sydney water-supply (Playf.) 8040
Plantae Arizonae (see C. Pringle)
Plantae balcanicae (see J. Rohlena)
Plantae Banatus rariores (Rochel) 9319
Plantae canarienses (see C. Pitard)
Plantae chilenses (see R. Philippi)
Plantae costaricenses exsiccatae (see H. Pittier)
Plantae criticae Thuringae *(see* E. Sagorski)
Plantae Dalmaticae (see F. Petter)
Plantae europeae (K. Richt.) 9185
Plantae lipsienses, weidanae et tharandinae (see E. Rossmässler)
Plantae mattogrossenses (Barb. Rodr.) 9353
Plantae mexicanae (Pringle) 8326
Plantae mexicanae (see C. Pringle)
Plantae moravicae novae vel minus cognitae (Podp.) 8079
Plantae novae (Parl.) 7351
Plantae Pondicerianae (see G. Perrottet)
Plantae postianae (Post) 8190
Plantae Preissianae *(see* S. Reissek)
Plantae Raddeaneae monopetalae (see G. Radde)
Plantae rariores agro dunelmensi indigenae (E. Robson) 9306
Plantae regnellianae *(see* A. Regnell)
Plantae rumphianae amboinenses (see C. B. Robinson)
Plantae Senegambiae (see G. Perrottet)
Plantae Severzovianae et Borszcovianae (Regel) 8791
Plantae Texanae (see C. Pringle)
Plantae tinneanae *(see* J. Peyritsch)
Plantae tripolitanae (Pamp.) 7263
Plantae tunetanae (see C. Pitard)
Plantae varvicenses selectae (W. G. Perry) 7723a
Plantae Victoriae Australiae exsiccatae (see F. Reader)
Plantarum americanarum fasciculus primus [-decimus] (Plum.) 8069
Plantarum Brasiliae icones et descriptiones (Pohl) 8103
Plantarum brasiliensium nova genera (Raddi) 8497
Plantarum Guianae rariorum icones et descriptiones (Rudge) 9751
Plantarum species quas in itinere per Aegyptum *(see* J. Roth)
Plantas de Alméria (Pau) 7516
Plantas de Persia y de Mesopotamia (Pau) 7514
Planta *(see* O. Renner, W. Ruhland)
Plantas medicinales de Filipinas (Pardo) 7326
Plantas medicinales más usadas en Bogotá (Pérez Arb.) 7676
Plantas medicinales y venenosas de Colombia (Pérez Arb.) 7677
Plantas novas (Barb. Rodr.) 9348
Plantas nuevas Chilenas (Phil.) 7857
Plantas que viven espontaneamente en el termino de Titaguas (Roxas) 9718
Plantas útiles de Colombia (Pérez Arb.) 7675
Plantas útiles de Colombia ensayo de botánica colombiana (Pérez Arb.) 7675
Plantas venenatas in territorio vindobonensi sponte crescentes (Praschil) 8273
Plant cover of Sweden *(see* G. Du Rietz)
Plant ecology and floristics of Salton Sink (Parish) 7334
Plantepatologi (Rostr.) 9617
Planterigets livsformer (Raunk.) 8657
Planterig. livsf. (Raunk.) 8657
Planternes bygning og liv (Poulsen) 8238
Plantes à feuillage coloré (Rothschild) 9661
Plantes agricoles et forestières (Réveil) 9110
Plantes alimentaires chez tous les peuples (Pailleux) 7190
Plantes cryptogames Florule du Mont-Blanc (Payot) 7571
Plantes de serre (Puydt) 8409
Plantes du Maroc (see C. Pitard)
Plantes fossiles (Renault) 9042
Plantes introduites à Madagascar (H. Perrier) 7706

Plantes médicinales de France (Perrot) 7716
Plantes médicinales de la Guinée (Pobég.) 8073
Plantes médicinales et usuelles (Rodin) 9339
Plantes nouvelles d'Abyssinie (A. Rich.) 9145
Plantes nouvelles pour le Gard (Pouzolz) 8241
Plantes phanérogames de Fréjus (Perreymond) 7695
Plantes phanérogames qui croissent naturellement aux environs de Toulon (G. N. Robert) 9276
Plantes principales de la région de Kisantu (Pâque) 7322
Plantes qui guérissent et les plantes qui tuent (Rawton) 8694
Plantes sauvages (Renaudet) 8980
Plantes usuelles (Roques) 9496
Plantes usuelles des brasiliens (St. Hil.) 10032
Plantes vasculaires du département de la Vienne (Poirault) 8113
Plantevekst. grensetr. (Resvoll-Holmsen) 9083
Plant life forms (Raunk.) 8664
Plants of Lake St. Clair (Pieters) 7928
Plants of Mt. Rainier National Park (St. John) 10063
Plants of New Mexico and Arizona (Rusby) 9812
Plants of North Elba (Peck) 7600
Plants of Rock Creek Lake Basin (Peirson) 7615
Plants of Southern California (Parish) 7333
Plants of the Bermudas (Reade) 8716
Plants of the Coast of Coromandel (Roxb.) 9720
Plants of the headwaters of the St. John river (St. John) 10060
Plants of the Tres Marias Islands (J. Rose) 9509
Plants of Western Lake Erie (Pieters) 7929
Pl. Banat. rar. (Rochel) 9319
Pl. Bermudas (Reade) 8716
Pl. bras. Icon. descr. (Pohl) 8103
Pl. bras. nov. gen. (Raddi) 8497
Pl. bygn. liv (Poulsen) 8238
Pl. cibar. roman. (Retz.) 9090
Pl. columb. *(see* J. Planchon)
Pl. Coromandel (Roxb.) 9720
Pl. Coromandel *(see* P. Russell)
Pl. crypt., Fl. Mont-Blanc (Payot) 7571
Pl. eur. (K. Richt.) 9185
Pl. feuill. col. (Rothschild) 9661
Pl. foss. (Renault) 9042
Pl. guér. tuent (Rawton) 8694

Pl. Guian. (Rudge) 9751
Pl. introd. Madag. (H. Perrier) 7706
Pliocene floras of the Dutch-Prussian border (C. Reid) 8904
Plioc. fl. Dutch-Pruss. border (C. Reid) 8904
Pl. Kisantu (Pâque) 7322
Pl. Lake St. Clair (Pieters) 7928
Pl. life forms (Raunk.) 8664
Pl. mattogr. (Barb. Rodr.) 9353
Pl. med. (Pérez Arb.) 7676
Pl. med. Filip. (Pardo) 7326
Pl. méd. France (Perrot) 7716
Pl. méd. Guinée (Pobég.) 8073
Pl. méd. usuel. (Rodin) 9339
Pl. med. venen. Colombia (Pérez Arb.) 7677
Pl. mexic. (Pringle) 8326
Pl. morav. (Podp.) 8079
Pl. New Mexico Arizona (Rusby) 9812
Pl. nov. (Barb. Rodr.) 9348
Pl. nov. (Parl.) 7351
Pl. pass. kort kold somNorske fjellplanter (Resvoll) 9080
Pl. Persia Mesop. (Pau) 7514
Pl. phan. Fréjus (Perreymond) 7695
Pl. phan. Toulon (G. N. Robert) 9276
Pl. pleth. (Plaz) 8049
Pl. post. (Post) 8190
Pl. Preiss. *(see* A. Putterlick)
Pl. rad. (Rombouts) 9487
Pl. Rock Creek Lake Basin (Peirson) 7615
Pl. sauv. (Renaudet) 8980
Pl. S. Calif. (Parish) 7333
Pl. serre (Puydt) 8409
Pl. tripol. (Pamp.) 7263
Pl. usuel. (Roques) 9496
Pl. usuel. bras. (St. Hil.) 10032
Pl. útil. Colombia (Pérez Arb.) 7675
Pl. varvic. sel. (W. G. Perry) 7723a
Pl. vasc. Vienne (Poirault) 8113
Pl. venen. vindob. (Praschil) 8273
Pl. Wilson *(see* A. Rendle)
Pl. W. Lake Erie (Pieters) 7929
Počátkowé rostl. (J. Presl) 8290
Počátkowé rostlinoslowí (J. Presl) 8290
Pochi stud. diatom. (Pedicino) 7608
Pochi studi sulle diatomee (Pedicino) 7608
Pocket botanical dictionary (Paxt.) 7555
Pocket bot. dict. (Paxt.) 7555
Podaxon (Pat.) 7453
Podaxon squamosus (Pat.) 7458
Poisonous, noxious and suspected plants (Pratt) 8278
Poison. pl. (Pratt) 8278
Polar regions (Richardson) 9171
Pollenstatistika perspectiv på jordens klimathistoria (L. Post) 8189c
POLLICHIA *(see* J. Pollich)
Polpoda (K. Presl) 8300

Polycodium (C. B. Robinson) 9297
Polymorphism and life history in the Desmidiaceae (Playf.) 8038
Polyporus bambusinus (Pat.) 7457
Pomarium britannicum (H. Phillips) 7865
Pom. brit. (H. Phillips) 7865
Pomona toscana (A. Piccioli) 7882
Pom. tosc. (A. Piccioli) 7882
Pop. Calif. fl. (Rattan) 8642
Pop. names Brit. pl. (Prior) 8347
Popular California flora (Rattan) 8641, 8642
Popular description of the indigenous plants of Lanarkshire (W. Patrick) 7498
Positiones physiologici medici et botanici argumenti (Rutstr.) 9841
Posobie k opredeleniyu paraziticheskikh gribov (Rostovzev) 9605
POSTELSIA (see A. Postels)
Post. phys. med. bot. argum. (Rutstr.) 9841
Potag. cur. (Pailleux) 7190
Potager d'un curieux (Pailleux) 7190
Potentillen Centralrusslands (Petunn.) 7796
Praemonenda circa Floram Riedeselicam (J. Ritter) 9258
Prakt. Gallenk. (H. Ross) 9569
Praktikum der Gallenkunde (H. Ross) 9569
Précis de botanique et de physiologie végétale (A. Rich.) 9138
Précis de l'histoire de la botanique (Réveil) 9110
Précis des découvertes et travaux somiologiques (Raf.) 8558
Prelim. Cat. (see J. Poggenburg)
Preliminary list of Maine fungi (Ricker) 9194
Preliminary list of the known poisonous plants found in South Africa (E. Phillips) 7860
Preliminary list of the phaenogams of Mt. Desert Island (E. L. Rand) 8622
Prelim. list Maine fung. (Ricker) 9194
Prelim. list phaenog. Mt. Desert Isl. (E. L. Rand) 8622
Prelud. fl. columb. (see J. Planchon)
Première mémoire sur la structure et les anomalies de la fleur des Résédacées (St. Hil.) 10046
Premier mémoire sur la famille des Polygalées (St. Hil.) 10036
PRESLIA (see J. Presl)
Prilog k oznavanju gljiva Bosne (Protíco) 8370
Prilozi k poznavanju flore resina (alge) Bosne i Hercegovine (Protíc) 8369
Prilozi k poznavanju kremenjaşica (diatomacea) Bosne i Herzegovine (Protíc) 8368

Primae lineae Encyclopaediae (Reuss) 9102
Primato degli italiani nella botanica (Sacc.) 9920
Primato ital. bot. (Sacc.) 9920
Prim. elenc. fung. San Martino (Peyronel) 7809
Prime linee per una geografica algologica marina (Picc.) 7900
Prim. fl. costaric. (Pitt.) 7978
Prim fl. costaric. (see M. Rousseau)
Primit. fl. costaric. (Renauld) 8988
Primit. fl. costaric. (see F. Pax, F. Renauld)
Primitiae florae costaricensis (Pitt.) 7978
Prim. linee geogr. algol. mar. (Picc.) 7900
Prim. lin. encycl. med. (Reuss) 9102
Primo elenco delle piante dei Monti Sibillini (Paolucci) 7307
Primo elenco di funghi di Val San Martino (Peyronel) 7809
Primo elenco di funghi parmensi (Pass.) 7437
Primo elenc. piante Monti Sibill. (Paolucci) 7307
Primule italiane (Paol.) 7306
Princ. fond. somiol. (Raf.) 8559
Princ. fund. (Sagra) 9998
Principes de la méthode naturelle (G. Planch.) 7995
Principes fondamentaux de somiologie (Raf.) 8559
Principes généraux de géographie agricole (Sagot) 9993
Principios fundamentales (Sagra) 9998
Princ. méth. nat. (G. Planch.) 7995
Princ. stelar morph. (Posth.) 8198
Priorité des noms de plantes (St. Lag.) 10076
Problema de via analytica ad certitudinem cognitione medica (see H. Richter)
Problème chez les Basidiomycètes (Quintanilha) 8427
Problème de la sexualité chez les champignons (Quintanilha) 8427
Problemen der plantengeographie (Pulle) 8390
Problems of bipolar plant distribution (Du Rietz) 9221
Probl. pl.-geogr. (Pulle) 8390
Probl. sexual. champ. (Quintanilha) 8427
Procès de la nomenclature botanique (St. Lag.) 10074
Procès des lichénologues (O. J. Rich.) 9165
Prod. fl. bryol. Madagascar (Renauld) 8993
Prodr. aethéogam. (P. Beauv.) 7213
Prodr. Apoc. (L. Planch.) 8028
Prodr. fl. ciren. (Pamp.) 7270
Prodr. fl. neomarch. (Rebent.) 8718
Prodr. fl. penins. Balcan. (see K. Ronniger)

Prodr. monogr. Lobel. (K. Presl) 8306
Prodr. monogr. Rosiers (Raf.) 8574
Prodrom der Waldstätter Gefässpflanzen (Rhiner) 9129
Prodrome de la flore bryologique de Madagascar (Renauld) 8993
Prodrome des cinquième et sixième familles de l'Aethéogamie (P. Beauv.) 7213
Prodrome des plantes recueillies en Amérique *(see* S. Ramée)
Prodrome d'une monographie des Rosiers (Raf.) 8574
Prodromo della flora Cirenaica (Pamp.) 7270
Prodromus (Page) 7183
Prodromus eener flora en fauna van het Nederlandsche Zoet- en Brakwaterplankton *(see* H. Redeke)
Prodromus florae neomarchicae (Rebent.) 8718
Prodromus florae nova-granatensis *(see* J. Planchon)
Prodromus monographiae Lobeliacearum (K. Presl) 8306
Prodromus of British Hieracia *(see* H. Pugsley)
Prodromus *(see* F. Parlatore, J. Planchon, E. Regel, G. Reuter)
Prodr. Waldst. Gefässpfl. (Rhiner) 9129
Productos vejetales indijenas de Chile (Reiche) 8855
Produits des Apocynées (L. Planch.) 8028
Prod. vég. Afr. occid. (Perrot) 7717
Prod. vej. Chile (Reiche) 8855
Progetto di un lessico dell'antica nomenclatura botanica (Sacc.) 9930
Progr. cogn. pl. crypt. (C. Sahlberg) 10006
Prolusio de Scleranthis (Rchb.) 8890
Prolusio Scleranth. (Rchb.) 8890
Propagatio plantarum botanico-physica (O. O. Rudbeck) 9749
Propag. pl. (O. O. Rudbeck) 9749
Prop. lois nom. bot. (B. L. Robinson) 9290
Proposals by the sub-committee on nomenclature *(see* J. Ramsbottom)
Proposals to publish by subscription a selection of the miscellaneous works and essays of C.S. Rafinesque (Raf.) 8576
Propositions de changements aux lois de la nomenclature botanique de 1867 (B. L. Robinson) 9290
Proposta d'un nuovo genere di Leguminosi (Pasq.) 7429
Prop. publ. works Raf. (Raf.) 8576
Prospect. Familiarum naturalium regni vegetabilis synopses monographicae (M. Roem.) 9413
Prospect Fam. nat. syn. monogr. (M. Roem.) 9413
Prospect for pollen analysis (L. Post) 8189c

Prospectus du traité des champignons (Paul.) 7521
Prospetto della flora della provincia di Bergamo (Rota) 9624
Prospetto della flora della Provincia di Bergamo (Rodegher) 9336
Prospetto della flora trevigiana (Sacc.) 9881
Prosp. fl. Bergamo (Rodegher) 9336
Prosp. fl. Bergamo (Rota) 9624
Prosp. Traité champ. (Paul.) 7521
Prot.-app. Enum. palm nov. (Barb. Rodr.) 9343
Protesto-appendice ao Enumeratio palmarum novarum (Barb. Rodr.) 9343
PROVANCHERIA *(see* L. Provancher)
Provisional list of flowering plants and ferns *(see* F. A. Rogers)
Prwocoč. rostl. (J. Presl) 8291
Prwopočátkům rostlinoslowi (J. Presl) 8291
Przeglad gatunków rodzaju Pediastrum (Racib.) 8459
Przyczynek do morfologi jądra komórkowego (Racib.) 8463
Pseudogardneria (Racib.) 8468
Pterid. Buitenzorg (Racib.) 8469
Pteridografia del Sur de México (Rouy) 9712a
Pteridogr. Sur México (Rouy) 9712a
Pteridophyta and Spermatophyta of Iceland (Rosenvinge) 9548
Pteridophyta exotica exsiccata (see E. Rosenstock, H. Rosendahl)
Pteridophyten der Flora von Buitenzorg (Racib.) 8469
Pteridophyten der Flora von Buitenzorg (Racib.) 8469
Pugill. pl. Afr. Bor. Hispan. *(see* G. Reuter)
Pugillus mycetum australiensium (Sacc.) 9910
Putesh. Turkestan *(see* E. Regel)
Putevoditel po imperatorskomu S.-Petersburgskomu botanicheskomu sadu (Regel) 8799
Putev. S. Peterb. bot. sad. (Regel) 8799
Pyrenothrix nigra (Riddle) 9204
Pyren. zum Nil (Rikli) 9236
Pyrola et Chimophila (Radius) 8503

Quaranta piante nuove del Brasile (Raddi) 8490
Quar. piant. nuov. Bras. (Raddi) 8490
Quarterly Review of Biology and human Biology *(see* R. Pearl)
Quel est l'inventeur de la nomenclature binaire (St. Lag.) 10070
Quelques champignons asiatiques nouveaux ou peu connus (Pat.) 7464
Quelques champignons de Java (Pat.) 7479

Quelques champignons de la Chine (Pat.) 7454
Quelques champignons du Thibet (Pat.) 7466
Quelques champignons nouveaux du Golfe Juan (Rolland) 9473
Quelques champignons nouveaux récoltés au Mexique par Paul Méry (Pat.) 7477
Quelques espèces nouvelles de champignons africains (Pat.) 7471
Quelques espèces nouvelles de champignons extra-européens (Pat.) 7459
Quelques mots sur l'Haematococcus lacustris (Rost.) 9594
Quelques notes sur la végétation de l'Algérie (Reut.) 9108
Quelques souvenirs d'herborisations (Sagot) 9994
Quer Afrika (Rohlfs) 9454
Quer durch Afrika (Rohlfs) 9454
Quinologia (Ruiz) 9768
Quinquinas (G. Planch.) 7998

Rabenh. Krypt.-Fl. ed. 2 (Rabenh.) 8450
Rabenhorst's Kryptogamen-Flora Zweite Auflage (Rabenh.) 8450
Raccolta di vocaboli botanici e forestali italiani e tedeschi (L. Piccioli) 7891
Racc. vocab. bot. (L. Piccioli) 7891
Rafinesque (Raf.) 8549
Rafinesque's proposals to publish a selection of his works (Raf.) 8576
Raimondia (Saff.) 9986
Raphiabast (Sadebeck) 9972
Rapport sur les spécimens botaniques au Kordofan (Pfund) 7836
Rapport sur le voyage de M. Auguste de Saint Hilaire dans le Brésil (St. Hil.) 10029
Rapp. spec. bot. Kordofan (Pfund) 7836
Rapp. voy. Brésil (St. Hil.) 10029
Rare plants of Southeastern Pennsylvania (Porter) 8182
Rare pl. s.e. Pennsylvania (Porter) 8182
Rariorum plantarum in Sicilia sponte provenientium (Parl.) 7346
Rar. pl. Sicilia (Parl.) 7346
Ratia (Pat.) 7491
Ray's Flora of Cambridgeshire (Ray) 8695
REA (see G. Re)
Rech. alg. vertes aér. (Puym.) 8411
Rech. anat. Elatin. (Pottier) 8233
Rech. anat. Genêts (Pellegr.) 7617
Rech. anat. Taccac. Dioscor. (Queva) 8422
Rech. Casuarina (J. Poiss.) 8118
Rech. devel. feuill. mouss. (Pottier) 8231
Rech. distr. musc. Forcalquier (Renauld) 8982
Recherches anantomiques sur les Mélobésiées (Rosanoff) 9501

Recherches anatomiques et taxinomiques sur les Onothéracées et les Haloragacées (P. Parm.) 7385
Recherches anatomiques et taxinomiques sur les Rosiers (P. Parm.) 7386
Recherches anatomiques sur la classification des Genêts et des Cytises (Pellegr.) 7617
Recherches biologiques sur l'Azolla filiculiodes (Roze) 9741
Recherches botaniques faites dans le sud-ouest de la France (Revel) 9112
Recherches bryologiques (Piré) 7958
Recherches d'histoire végétale (G. Poirault) 8111
Recherches historiques sur les mots plantes mâles et plantes femelles (St. Lag.) 10071
Recherches sur la distribution géographique des muscinées de Forcalquier (Renauld) 8982
Recherches sur la flore méridionale de Madagascar (Poiss.) 8117
Recherches sur la flore pélagique de l'Étang de Thau (Jul. Pavillard) 7536
Recherches sur l'anatomie comparée des espèces dans la famille des Elatinacées (Pottier) 8233
Recherches sur l'anatomie de l'appareil végétatif des Taccacées & des Dioscorées (Queva) 8422
Recherches sur la structure de quelques diatomées (Prinz) 8346
Recherches sur la structure et les affinités botaniques des végétaux silicifiés (Renault) 9010
Recherches sur le développement de la feuille des mousses (Pottier) 8231
Recherches sur le sac embryonnaire de quelques Narcissées (Preda) 8281
Recherches sur les algues vertes aériennes (Puym.) 8411
Recherches sur les anciens herbaria (St. Lag.) 10073
Recherches sur les Casuarina (J. Poiss.) 8118
Recherches sur les Épilobes de France (P. Parm.) 7384
Recherches sur les Strophanthus (Payrau) 7575
Recherches sur les végétaux silicifiés d'Autun et de Saint-Étienne Botryopteris (Renault) 9002
Recherches sur les végétaux silicifiés d'Autun et de Saint-Étienne. Calamodendrées (Renault) 9004
Recherches sur les végétaux silicifiés d'Autun Myelopteris (Renault) 9000
Recherches sur l'évolution et la valeur anatomique et taxonomique du péricycle

des Angiospermes *(see* C. Pitard)
Recherches sur quelques Calamodendrées (Renault) 9005
Rech. fl. mérid. Madagascar (Poiss.) 8117
Rech. fl. pélag. Étang Thau (Jul. Pavillard) 7536
Rech. Onothér. (P. Parm.) 7385
Rech. Strophanthus (Payrau) 7575
Rech. vég. silicif. (Renault) 9010
Redevvoering van C.G.C. Reinwardt, over hetgeen het onderzoek van Indie tot uitbreiding der natuurlijke historie (Reinw.) 8961
Réforme de la nomenclature botanique (St. Lag.) 10067
Réform. nom. bot. (St. Lag.) 10067
Refugium botanicum *(see* H. G. Reichenbach)
Regeneration in Sphearocarpos Donnellii *(see* H. Rickett)
Reg. hort. bot. belgr. (Pančic) 7284
Register zur Flora von Sachsen (Rchb.) 8889
Regius hortus botanicus belgradensis (Pančic) 7284
Reg. Laghi equat. (Piscicelli) 7970
Reg. mar. *(see* E. Petit)
Règne vég. (Réveil) 9110
Règne végétal (Réveil) 9110
Reichenbachianae florae germanicae clavis synonymica (Rchb.) 8881
Reichenbachia *(see* H. G. Reichenbach)
Reise Ararat (Parrot) 7402
Reise Atacama (Phil.) 7850
Reise Brasil. (Pohl) 8104
Reise Chile (Poepp.) 8090
Reise durch die Wueste Atacama (Phil.) 7850
Reise durch Russland *(see* P. Pallas)
Reise durch verschiedene Provinzen des russischen Reichs (Pall.) 7224
Reise im Innern von Brasilien (Pohl) 8104
Reise in Chile, Peru (Poepp.) 8090
Reise in die Barbarey (Poir.) 8114
Reise in die Krym und den Kaukasus *(see* J. Parrot)
Reise in Griechenland (Russegger) 9827
Reisen (Russegger) 9827
Reisen an der persisch-russischen Grenze (Radde) 8479
Reisen in das Land der Hottentotten (W. Paterson) 7439
Reisen in den Süden von Ostsiberien (Regel) 8790
Reisen in den Süden von Ost-Siberien *(see* G. Radde)
Reisen in Europa, Asien und Afrika (Russegger) 9827
Reisen in Nubien, Kordofan *(see* E. Rüppell)

Reise Novara, Pilze, Leber- Laubm. (Reichardt) 8845
Reise Novara Pilze, Leber- und Laubmoose (Reichardt) 8845
Reise russ. Reich. (Pall.) 7224
Reise südl. Statthaltersch. russ. Reich. (Pall.) 7228
Reise zum Ararat (Parrot) 7402
Reis. Land Hottent. (W. Paterson) 7439
Reis. Ostsib. (Regel) 8790
Reis ter ontdekking van eene noordwestelijke door vaart (W. Parry) 7409
Relación del viaje (Ruiz) 9782
Relazione sulle piante raccolte nel Karakoram *(see* P. Pirotta)
Relazione sullo stato della prima Calabria ulteriore (Pasq.) 7427
Reliq. haenk. (K. Presl) 8293
Reliq. Kotschy *(see* J. Russegger)
Reliq. rudb. (O. J. Rudbeck) 9747
Reliquiae brébisonianae (see C. Roumeguère)
Reliquiae haenkeanae (K. Presl) 8293
Reliquiae mycologicae libertianae (Roum.) 9683
Reliquiae mycologicae libertianae *(see* P. Saccardo)
Reliquiae robinsonianae (see C. B. Robinson)
Reliquiae rudbeckianae (O. J. Rudbeck) 9747
Reliquiae Rudbeckianae *(see* O. J. Rudbeck)
Reliquiae rutenbergianae *(see* D. Rutenberg)
Reliquie dell'erbario micologico di P.A. Micheli (Sacc.) 9935
Rel. viaje (Ruiz) 9782
Remarks on the system of the spermatophytes (Pulle) 8394
Remarques sur la nomenclature phytogéographique (Jul. Pavillard) 7537
Remarques sur les Mélastomacées de la Guyane française (Sagot) 9996
Remarques sur l'organisation de quelques champignons exotiques (Pat.) 7456
Rem. nomencl. phytogéogr. (Jul. Pavillard) 7537
Rem. syst. spermat. (Pulle) 8394
Repert. bot. syst. (K. Presl) 8304
Repert. bot. syst. *(see* G. Pritzel)
Repertorium botanicae systematicae (K. Presl) 8304
Repertorium herbarii (Rchb.) 8888
Réponse aux reproches la botanique (St. Hil.) 10026
Réponse reproches bot. (St. Hil.) 10026
Report of a journey of discovery into the interior of Western Australia (Roe) 9375
Report of the State Botanist on edible fungi of New York (Peck) 7601
Report on the botany of the Wollaston (Ridl.) 9210

Report on the Tertiary plants of British Columbia (Penh.) 7632
Report on the United States and Mexican Boundary Survey, Botany of the Boundary *(see* C. Parry)
Report: The Committee on Agriculture (Perrine) 7711
Report United States geographical surveys, Wheeler, Botany (Rothr.) 9658
Repr. Ahnfeltia plicata (Rosenvinge) 9553
Repr. crypt. (Rietsch) 9214
Reproduction des cryptogames (Rietsch) 9214
Reproduction of Ahnfeltia plicata (Rosenvinge) 9553
Rep. U.S. geogr. surv., Wheeler (Rothr.) 9658
Resources of the Southern fields (Porcher) 8168
Resour. S. fields (Porcher) 8168
Respuesta (Ruiz) 9770
Respuesta à la impugnacion contra el pródromo (Ruiz) 9770
Result. algol. Violante (Picc.) 7899
Resultati algologici delle crociere del Violante (Picc.) 7899
Résultats botaniques du voyage à l'Océan glacial sur le bateau brise-glace Ermak (Palib.) 7208
Résultats du Voyage du S.Y. Belgica, Botanique Champignons (M. Rousseau) 9691
Résult. bot. voy. Ermak (Palib.) 7208
Résult. Voy. Belgica, Champ. (M. Rousseau) 9691
Rev. gen. pl. *(see* P. G. Richter)
Revised names of plants of New Jersey (Rusby) 9814
Revis. Ficinia (H. Pfeiff.) 7813
Revis. Ind. foss. pl., Conif. (Sahni) 10008
Revisio generis Enkianthus (Palib.) 7206
Révision de la section Harpidium (Renauld) 8983
Revisión de las gramíneas argentinas del género Diplachne (Parodi) 7396
Revisión de las gramíneas argentinas del género Sporobolus (Parodi) 7397
Révisión de las orquídeas de España *(see* S. Rivas Goday)
Revision der Gattung Ficinia (H. Pfeiff.) 7813
Revision der Hysterineen im herb. Duby (Rehm) 8824
Revision der Umbelliferen aus Kamtschatka (Rupr.) 9807
Révision des characées de la flore de Maine-et-Loire (Préaub.) 8279
Révision des diatomées de la Guadeloupe (P. Petit) 7772
Revision of the genus Chelidonium (Prain) 8261

Revision of the genus Coscinodiscus (Rattray) 8644
Revision of the nomenclature of the Nebraska Polypetalae (Rydb.) 9847
Revision of the North American species of Verbena (L. M. Perry) 7721
Revisions of Indian fossil plants, Coniferales (Sahni) 10008
Revisio specierum Crataegorum (Regel) 8797
Revista Argentina de Agronomia *(see* L. Parodi)
Revista Chilena de Historia natural *(see* C. E. Porter)
Revista de la Sociedad Cubana de Botánica *(see* A. Ponce de Leon)
Revis. Umbell. Kamtschatka (Rupr.) 9807
Rev. monogr. Mentha (Pérard) 7669
Rev. N. Amer. Umbell. *(see* J. Rose)
Rev. N. Amer. Verbena (L. M. Perry) 7721
Revue bryologique *(see* H. Philibert)
Revue de botanique systématique et de géographie botanique *(see* G. Rouy)
Revue de la famille des Simaroubées (Planch.) 8008
Revue de la Flore du Brésil méridional (St. Hil.) 10052
Revue des mousses acrocarpes de la flore belge (Piré) 7958
Revue des Sciences naturelles *(see* P. Rouville)
Revue horticole *(see* P. Poiteau)
Revue monographique du genre Mentha (Pérard) 7669
Revue mycologique *(see* C. Roumeguère)
Rhachitheciopsis (P. de la Varde) 8206
Rhinantheen Niederbayerns (Poeverl.) 8098
Rhizographie (B. Preiss) 8284
Rhododendron (E. S. Rand) 8624
Rhodora *(see* B. L. Robinson)
Ric. bot. Lago Tana (Pic.-Ser.) 7910
Ricerche botaniche nella regione del Lago Tana (Pic.-Ser.) 7910
Richt. beteek. pl.-physiol. (Rauwenh.) 8673
RICKIA *(see* J. Rick)
Risposto di Eleuterio Benacense (Pollini) 8142
Rivista delle Yucche *(see* V. Ricasoli)
RODRIGUESIA *(see* J. Rodrigues)
ROEMERIANA *(see* A. Roemer)
Rollandina (Pat.) 7488
Romanarum plantarum centuria decima octava *(see* E. Rolli)
Rosliny polskie *(see* B. Pawlowski)
Rosarum monographia *(see* A. Pronville)
Rosen (Rössig) 9429
Rosen der Flora von Naumburg (Sagorski) 9989

Roses (Redouté) 8748
Roses, ed. 2 (Redouté) 8749
Roses, ed. 3 (Redouté) 8751
Ros. Fl. Naumberg (Sagorski) 9989
Rostpilzf. deut. Conif. (Reess) 8759
Rostpilzformen der deutschen Coniferen (Reess) 8759
Rousseau's Botanik für Frauenzimmer in Briefen an die Frau (Rousseau) 9686
Rubi notes (W. M. Rogers) 9442
Rückkunft Reinwardts. Elenchus seminum (Reinw.) 8960
Rusk. dendrol. (Regel) 8795
Ruskii dendrologia (Regel) 8795

Słuzowce monografia (Rost.) 9595
Słuzowce monogr. (Rost.) 9595
Sable Isl. (St. John) 10056
Sable Island (St. John) 10056
SACCARDOA (see P. Saccardo)
Saggio di osservazioni alberi (Pollini) 8140
Saggio di studi intorno alla distribuzione geografica delle alghe d'acqua dolce e terrestri (Picc.) 7904
Saggio di studi naturali sul territorio mantovano (Paglia) 7185
Saggio di una flora analitica dei licheni del Veneto (F. Sacc.) 9879
Saggio di una monografia del genere Eutypa (Paol.) 7304
Sagg. lich. Veneto (F. Sacc.) 9879
Sagg. osserv. alb. (Pollini) 8140
Sagg. stud. nat. mantov. (Paglia) 7185
Salict. woburn. (see J. Russell)
Samml. Kaukas. Mus., Bot. (Radde) 8482
Sammlungen des Kaukasischen Museums (Radde) 8482
Samoa (Reinecke) 8917
Sandalw. Hawaii (Rock) 9326
Sandalwoods of Hawaii (Rock) 9326
Sapindaceae costaricenses (Radlk.) 8537
Saprolegniae (Racib.) 8455
Saxifraga multifida (Rosbach) 9502
Sceptr. carol. (J. O. Rudbeck) 9746
Sceptrum carolinum (J. O. Rudbeck) 9746
Schedae ad floram exsiccatam reipublicae bohemicae slovenicae (Podp.) 8082
Schedae ad W. Rothmaler Alchemillae exsiccatae (Rothm.) 9651
Sched. fl. exs. bohem. (Podp.) 8082
Schilder. Nat.-Verh. Süd-Abyss. (J. Roth) 9644
Schilderung der Naturverhältnisse in Süd-Abyssinien (J. Roth) 9644
Schlesiens Pflanzenwelt (Pax) 7550
Schles. Pflanzenw. (Pax) 7550
Schmarotzergew. (Regel) 8776
Schmarotzergewächse (Regel) 8776
Schnell unterrichtende Boder) 8907
Schutzscheide deut. Equiset. (Pfitz.) 7825

Schweizerischen Dorycnien (Rikli) 9224
Science in the Netherlands East Indies (see A. Pulle)
Scientific and descriptive catalogue of Peale's Museum (P. Beauv.) 7210
Scriptores de plantis hispanicis, lusitanicis, brasiliensibus (Roem.) 9403
Script. pl. hispan. (Roem.) 9403
Scroph. e. N. Amer. (Pennell) 7638
Scrophulariaceae of eastern temperate North America (Pennell) 7638
Scrophulariaceae of the Central Rocky Mountain States (Pennell) 7637
Scrophulariaceae of the southeastern United States (Pennell) 7636
Scrophulariaceae of the western Himalayas (Pennell) 7639
Scroph. w. Himal. (Pennell) 7639
Scylla or Charybdis (Rydb.) 9872
Scyphaea (K. Presl) 8301
Second decade of undescribed plants (Raf.) 8566
Second voyage dans le pays des Adels et le Royaume de Choa (Rochet) 9321
Second voyage de Pallas (Pall.) 7228
Section Tuberarium of the genus Solanum (Rydb.) 9869
Sec. voy. Choa (Rochet) 9321
Selecta specierum tuliparum (Reboul) 8721
Select. fung. Gr. Brit. (Plues) 8061
Selection of Madeira flowers (Robley) 9304
Selection of the eatable funguses of Great Britain (Plues) 8061
Select list of plants found in Warwickshire (W. G. Perry) 7723
Select list pl. Warwicks. (W. G. Perry) 7723
Select plants of Southern California (see S. Parish)
Select. sp. tulip. (Reboul) 8721
Sel. Madeira fl. (Robley) 9304
Septobasidium (Pat.) 7463
Septobasidium langloisii (Pat.) 7482
Serta Florea Svecana (Palmberg) 7230d
Serta fl. svec. (Palmberg) 7230d
Sert. palm. bras. (Barb. Rodr.) 9358
Sert. Somal. (see G. Révoil)
Sert. tianschan (see F. Ruprecht)
Sertum benguelense (see J. Peyritsch)
Sertum botanicum (Rchb.) 8883
Sertum palmarum brasiliensium (Barb. Rodr.) 9358
Sertum petropolitanum (see E. Regel)
Ser. veg. (Pontén) 8154
Silur- Culm-Fl. (Potonié) 8221
Silur und die Culm-Flora (Potonié) 8221
Silva capensis (Pappe) 7316
Sinonimia vulgar y cientifica de las plantas mexicanas (Ramírez) 8606, 8609

Sinon. pl. mexic. (Ramírez) 8606
Sinon. pl. mexic. (Ramírez) 8609
Siphonieae verticillatae (Pia) 7877
Siphon. vertic. (Pia) 7877
Sirobasidium *(see* N. Patouillard)
Sistema de los vegetables (Paláu) 7203
Sist. Veg. (Paláu) 7203
Skepperia (Pat.) 7465
Sketch nat. hist. Yarmouth (Paget) 7184
Sketch of the botany of South-Carolina and Georgia (Raf.) 8571
Sketch of the Flora of Alaska *(see* J. Rothrock)
Sketch of the natural history of Yarmouth (Paget) 7184
Skizzen zur Organographie und Physiologie der Classe der Schwämme *(see* J. Pfund)
Skogsträdpollen i sydsvenska torvmosselagerföljder (L. Post) 8189a
Słownik polskich imign (Rost.) 9599
Słownik wyrazów botanicznych (Plawski) 8035
Słown. polsk. imign (Rost.) 9599
Słw. wyraz. bot. (Plawski) 8035
Smitsomme Sygdomme hos Landbrugsplanterne (Ravn) 8691
Smits. Sygd. Landbr.-pl. (Ravn) 8691
Sobre el método de la flora de Chile (Reiche) 8851
Sobre las especies chilenas del jenero Polyachyrus (Phil.) 7853
Somalia & Benadir (Rob.-Bricch.) 9274
Somalia e Benadir (Rob.-Bricch.) 9274
Some account of a botanical tour in the mountains of Auvergne and Switzerland (D. Ross) 9561
Some account of a collection of arctic plants formed by Edward Sabine *(see* E. Sabine)
Some Australian Orchids (R. S. Rogers) 9438
Some generic segregations (Rydb.) 9859
Some new or less known Desmids (Playf.) 8036
Some senecioid genera (Rydb.) 9870
Some Sydney Desmids (Playf.) 8037
Some vascular plants from Saghalin (Printz) 8343
Some wild flowers of Tasmania (Rodway) 9371
Sommario di un corso di botanica (Sacc.) 9891
Somm. corso bot. (Sacc.) 9891
Sopra algune piante esculenti del Brasile (Raddi) 8494
Sopra i vari tipi delle anomalie dei tronchi nelle Sapindacee (Radlk.) 8512
Sopra la teoria della riproduzione vegetale (Pollini) 8144

Soya (Pailleux) 7189
Soy bean (Piper) 7952
Spaltöffnungsapparat im Lichte der Phylogenie (Porsch) 8172
Spaltöffnungsapp. Phylog. (Porsch) 8172
Sp. astragal. (Pall.) 7229
Sp. Cotoni (Parl.) 7372
Spec. bibl. bot. (Pritz.) 8352
Specchio delle scienze (Raf.) 8557
Specchio sci. (Raf.) 8557
Spec. fl. S. Afr. (Roupell) 9684
Specie dei Cotoni (Parl.) 7372
Species astragalorum (Pall.) 7229
Species graminum stipaceorum (Rupr.) 9794
Species of Dalbergia of Southeastern Asia (Prain) 8263
Species of Pedicularis of the Indian Empire (Prain) 8258
Species plantarum (Pers.) 7733
Specim. anim. veg. Minorica (Ramis y Ramis) 8614
Specimen animalium, vegetabilium in insula Minorica (Ramis y Ramis) 8614
Specimen bibliographiae botanicae (Pritz.) 8352
Specimen historiae plantarum (Reneaulme) 9073
Specimens of the flora of South Africa (Roupell) 9684
Specim. hist. pl. (Reneaulme) 9073
Spegazzinia (Sacc.) 9896
Sp. gram. stipac. (Rupr.) 9794
Sphaeriacei britannici (see W. Phillips, C. Plowright)
Sphagna čechoslovenica exsiccata (see Z. Pilous)
Sphaignes de la flore de Belgique (Piré) 7957
Spices (Ridl.) 9209
Spicilège de la flore bryologique des environs de Montreux- Clarens (Piré) 7963
Spicilegium Lauracearum (Prantl) 8272
Spicilegium plantarum cryptogamarum Sueciae (Rutstr.) 9842
Spic. pl. crypt. Suec. (Rutstr.) 9842
Spigolature per la ficologia ligustica (Picc.) 7902
Spiral arr. branch. Callithamn. (Rosenvinge) 9550
Spirito sci. nat. (Parl.) 7357
Spirogyra des environs de Paris (P. Petit) 7771
Spirogyra Paris (P. Petit) 7771
Sp. nov. fung. Firenze (Raddi) 8483
Sporeplanterne (Rosenvinge) 9549
Sp. pl. (Pers.) 7733
Stämme des Pflanzenreiches (Pilg.) 7937
Stämme Pflanzenr. (Pilg.) 7937
Standard catalogue of English names of our wild flowers (Rayner) 8707

Stand. cat. Engl. names fl. (Rayner) 8707
Stand. Cycl. Hort. *(see* A. Rehder)
Standortsgewächse und Unkräuter Deutschlands (Ratzeb.) 8646
Standortsgew. Unkr. Deutschl. (Ratzeb.) 8646
Stat. bot. Haute-Garonne (Roum.) 9678
Statist. bot. Somme (Pauquy) 7533
Statist. gen. Sicilia (Raf.) 8554
Statistica generale di Sicilia (Raf.) 8554
Statistique botanique de Département de la Haute-Garonne (Roum.) 9678
Statistique botanique ou flore du département de la Somme (Pauquy) 7533
Stato conosc. veg. Italia *(see* R. Pampanini)
Steudelia (K. Presl) 8302
Stirp. advers. nov. (Pena) 7625
Stirpes cryptogamae vogeso-rhenanae (see C. Roumeguère)
Stirp. eur. syll. (Ray) 8704
Stirpium adversaria nova (Pena) 7625
Stirpium europaearum extra britannias nascentium sylloge (Ray) 8704
Stoich. bot. (Ponerop.) 8153
Stoicheia botanikes (Ponerop.) 8153
Storia illustrata del regno vegetabile *(see* A. Pokorny)
Storia naturale dell'isola di Corsica (Sacc.) 9943
Stor. lett. fl. ven. (Sacc.) 9885
Stray contributions to the flora of Greenland I-V (M. Porsild) 8176
Stray leaves natural. (D. Ross) 9560
Stray leaves of a naturalist (D. Ross) 9560
Streifz. Deut.-Neu-Guinea (Rech.) 8727
Streifzüge in Deutsch-Neu-Guinea (Rech.) 8727
Structure comparée de quelques tiges de la flore Carbonifère (Renault) 9011
Structure comparée des tiges des Lépidodendrons et des Sigillaires (Renault) 9008
Structure des Sigillaires (Renault) 9007
Structure et affinités botaniques des Cordätes (Renault) 9009
Stud. bladmoss. org. (Rosander) 9500
Stud. bot. agrum. (Penz.) 7645
Stud. Embryosack. Angiosperm. (Palm) 7230b
Stud. Entw.-Gesch. Laminariac. (Reinke) 8938
Stud. fl. ossol. (S. Rossi) 9575
Studi botanici sugli agrumi (Penz.) 7645
Studien über die Schwärmer einiger Süsswasseralgen (Pasch.) 7420
Studien über Konstruktionstypen des Embryosackes der Angiospermen (Palm) 7230b
Studien über mitteleuropäische Delphinien (Pawl.) 7541

Studien zur vergleichenden Entwicklungsgeschichte der Laminariaceen (Reinke) 8938
Studier öfver bladmossornas organisation (Rosander) 9500
Studier öfver löfängsområdena på Aland (Palmgr.) 7241
Studier over danske aërofile alger (J. B. Petersen) 7762
Studies in entomogenous fungi *(see* T. Petch)
Studies in the American Phaseolineae (Piper) 7954
Studies in the Eupatorieae (B. L. Robinson) 9291
Studies in the larger fungi of Iceland (Rosenvinge) 9548
Studies in the vegetation of Pamir (Paulsen) 7531
Studies of Mexican and Central American plants (J. Rose) 9508
Studies of plant life (Pepoon) 7659
Studies on the biology and taxonomy of soil algae (J. B. Petersen) 7763
Studies on the Rocky Mountain flora (Rydb.) 9856, 9865
Studies on the vegetation of Iceland (Rosenvinge) 9548
Studies on the vegetation of the central highland of Iceland (Rosenvinge) 9548
Studies on the vegetation of the Transcaspian lowlands (Paulsen) 7530
Studi organografici sui fiori e sui frutti delle Conifere (Parl.) 7371
Studi sulla flora ossolana (S. Rossi) 9575
Stud. organogr. conif. (Parl.) 7371
Stud. pl. life (Pepoon) 7659
Stud. Schwärm. Süssw.-Alg. (Pasch.) 7420
Stud. soil alg. (J. B. Petersen) 7763
Stud. veg. Pamir (Paulsen) 7531
Study of Allocarya (Piper) 7951
Subalp. Fl. Meissn. (Pfeiff.) 7818
Submerged forests (C. Reid) 8903
Sud. dansk. aërof. alg. (J. B. Petersen) 7762
Süd- und zentralamerikanischen Arten der Gattung Rumex (Rech. fil.) 8731
Süsswasseralgenflora von Süd-Georgien (Reinsch) 8957
Süsswasser-Diatomaceen (Rabenh.) 8436
Süsswasserflora (Pasch.) 7423
Süsswasser-Flora Deutschlands, Österreichs und der Schweiz (Pasch.) 7423
Süsswasserflora von Mitteleuropa (Pasch.) 7423
Süssw.-Diatom. (Rabenh.) 8436
Süssw.-Fl. Mitteleur. *(see* H. Paul)
Suite mém. hist. nat. Pyrénées (Palassou) 7198
Suite mémoires l'histoire naturelle Pyrénées (Palassou) 7198

Suites à la Flore de France (Rouy) 9704
Sukkulenten (Rümpler) 9765
Sulla botanica (Parl.) 7350
Sulla divisione del genere Tulipa (Reboul) 8722
Sulla flora fossile della formazione oolitica (Sacc.) 9882
Sulle alg. (Pollini) 8143
Sulle alghe (Pollini) 8143
Sullo spirito delle scienze naturali (Parl.) 7357
Sul rinvenimento di un antico erbario (Sacc.) 9937
Suma agrostológica *(see* L. Parodi)
Šumsko drvece i Šiblje u Srbiji (Pančic) 7278
Suplemento á la Quinologia (Ruiz) 9775
Supl. Quinologia (Ruiz) 9775
Suppl. App. Parry's Voy. (W. Parry) 7410
Suppl. Critt. bras. (Raddi) 8498
Supplément à la flore Lyonnaise (Roffavier) 9432
Supplementary catalogue of Ohio plants (Riddell) 9199
Supplément aux mémoires pour servir à l'histoire naturelle des Pyrénées (Palassou) 7199
Supplément du catalogue raisonné des plantes de l'arrondissement de Montluçon (Pérard) 7667
Supplemento alla memoria Crittogame brasiliane (Raddi) 8498
Supplemento micologico *(see* D. Saccardo)
Supplement to Flora of Hampshire (Rayner) 8708
Supplement to the Appendix of Captain Parry's Voyage (W. Parry) 7410
Supplement to the fauna and flora of Eastbourne (Roper) 9494
Supplementum tentaminis pteridographiae (K. Presl) 8310
Suppl. fauna fl. Eastbourne (Roper) 9494
Suppl. Fl. Hampshire (Rayner) 8708
Suppl. fl. Lyon. (Roffavier) 9432
Suppl. mém. hist. nat. Pyrénées (Palassou) 7199
Suppl. tent. pterid. (K. Presl) 8310
Sur des mousses de l'époque houillère (Renault) 9029
Sur la famille Guttifères (Planch.) 8015
Sur la formation de l'embryon dans les Graminées (Raspail) 8638
Sur la fructification de quelques végétaux silicifiés (Renault) 9003
Sur la place du genre Favolus (Pat.) 7451
Sur l'application rigoureuse de la règle d'antériorité (Rouy) 9708a
Sur l'attribution des genres Fayolia et Palaeoxyris (Renault) 9043
Sur le Bucegia (Radian) 8502

Sur le genre Bornia (Renault) 9035
Sur les affinités des genres Henslowia (Planch.) 8006
Sur les Astérophyllites (Renault) 9017
Sur les charactères et l'origine botanique du Jaborandie (G. Planch.) 8001
Sur les cicatrices des Syringodendron (Renault) 9038
Sur les fructifications des Calamodendrons (Renault) 9034
Sur les fructifications mâles des Arthriphitus et des Bornia (Renault) 9036
Sur les Gnétacées (Renault) 9018
Sur les marais tourbeux (Renault) 9053
Sur les mousses de l'époque houillère (Renault) 9027
Sur les noyaux des Hyménomycètes (Rosenvinge) 9537
Sur les premiers phénomènes de la germination des spores des cryptogames (Rauwenh.) 8674
Sur les productions végétales indigènes ou cultivées de l'Afrique occidentale française (Perrot) 7717
Sur les racines des Calamodendrées (Renault) 9031
Sur les Sphenozamites (Renault) 9016
Sur les Stigmarhizomes (Renault) 9039
Sur les troncs de Fougères du terrain houiller supérieur (Renault) 9030
Sur l'existence d'Astérophyllites phanérogames (Renault) 9025
Sur l'existence du genre Todea (Renault) 9020
Sur l'organisation comparée des feuilles des Sigillaires et des Lépidodendrons (Renault) 9033
Sur l'organisation du faisceau foliaire des Sphenophyllum (Renault) 9021
Sur quelques algues fossiles (Renault) 9062
Sur quelques champignons extra-européens (Pat.) 7448
Sur quelques cryptogames hétérosporées (Renault) 9057
Sur quelques Cycadées houillères (Renault) 9032
Sur quelques espèces de Meliola (Pat.) 7446
Sur quelques Graminées du Portugal (Rouy) 9699
Sur quelques micro et macrospores fossiles (Renault) 9059
Sur quelques microorganismes des combustibles fossils (Renault) 9054
Sur quelques nouveaux parasites des Lépidodendrons (Renault) 9047
Sur quelques pollens fossiles (Renault) 9061

Sur trois espèces cactiformes d'Euphorbes (J. Poiss.) 8120
Sur trois mousses inédites de la Chine orientale (P. de la Varde) 8205
Sur une nouvelle espèce de Poroxylon (Renault) 9014
Sur une Parkériée fossile (Renault) 9060
Sur un Equisetum du terrain houiller supérieur de Commentry (Renault) 9026
Sur un mode de déhiscence curieux du pollen de Dolerophyllum (Renault) 9049
Sur un nouveau genre de fossiles végétaux (Renault) 9023
Sur un nouveau genre de tige fossile (Renault) 9056
Sur un nouveau genre de tige permo-carbonifère, le G. Retinodendron Rigolloti (Renault) 9045
Sur un nouveau genre du terrain houiller supérieur (Renault) 9024
Sur un nouveau groupe de tiges fossiles silicifiées de l'époque houillère (Renault) 9012
Sur un nouveau type de Cordaïtée (Renault) 9028
Survey of the progress and actual state of natural sciences in the United States of America (Raf.) 8564
Svalbards fl. (Resvoll-Holmsen) 9082
Svalbards flora (Resvoll-Holmsen) 9082
Sv. bot. (Palmstr.) 7247
Svenska Taphrinaarter (Palm) 7230c
Svenska Taphrinaarter Uppsala (Palm) 7230c
Svensk botanik (Palmstr.) 7247
Svensk flora för skolor *(see* L. Romell)
Svensk Skolbotanik (Palmstr.) 7247
Svod botanicheskikh terminov (Petunn.) 7797
Svod bot. termin. (Petunn.) 7797
Sydl. Grønl. Veg. (Rosenvinge) 9542
Sydowia *(see* F. Petrak)
Syllabus der Pflanzenfamilien *(see* H. Reimers)
Syll. fung. (Sacc.) 9898
Sylloge fungorum (Sacc.) 9898
Sylloge fungorum ... Additamenta (Sacc.) 9898
Sylloge fungorum Elenchum fungorum novorum (Sacc.) 9898
Sylloge fungorum Index iconum fungorum (Sacc.) 9898
Sylloge fungorum *(see* D. Saccardo)
Sylloge fungorum *(see* G. Paoletti)
Sylloge fungorum Supplementum universale (Sacc.) 9898
Sylloge fungorum Synonymia (Sacc.) 9898
Sylloge hyphomycetum (Sacc.) 9898
Sylloge opusculorum botanicorum (Reichard) 8836
Sylloge pyrenomycetum (Sacc.) 9898
Sylloge sphaeropsidearum et melanconiearum (Sacc.) 9898
Syll. opusc. bot. (Reichard) 8836
Syll. Pfl. Fam. *(see* R. Pilger)
Sylva fl. (H. Phillips) 7867
Sylva florifera (H. Phillips) 7867
Sylva tellur. (Raf.) 8589
Sylva telluriana (Raf.) 8589
Symb. antill. *(see* J. Pierre)
Symb. ant.*(see* R. Pilger, L. Radlkofer)
Symb. bot. (K. Presl) 8303
Symb. bot. *(see* P. Rohrbach)
Symb. bras. *(see* J. Peyritsch, V. Poulsen)
Symbolae ad historiam et geographiam plantarum Rossicarum (Rupr.) 9799
Symbolae botanicae (K. Presl) 8303
Symbolae *(see* O. Petersen, L. Radlkofer)
Symb. pl. Ross. (Rupr.) 9799
Symphysia (K. Presl) 8295
Syn. deut. schweiz. Fl. *(see* E. Sagorski)
Syn. fil. Afr. austr. (Pappe) 7317
Syn. fil. bras. (Raddi) 8488
Syn. fl. Colorado (Porter) 8180
Syn. fl. deut. schweiz. Fl. *(see* E. Palla)
Syn. fl. Gothl. (Säve) 9983
Syn. fl. N. Amer. *(see* B. L. Robinson)
Syn. fl. west. states (Riddell) 9198
Syn. fung. Carol. sup. (Pers.) 7734
Syn. gen. Amer. pl. (O. Rich) 9137
Syn. meth. fung. (Pers.) 7730
Syn. meth. stirp. brit. (Ray) 8703
Synonymenregister zu Deutschlands Kryptogamen-Flora (Rabenh.) 8434)
Synonymia botanica (Pfeiff.) 7820
Synonymiae botanicae supplementum primum (Pfeiff.) 7820
Synonymie provençale des champignons de Vaucluse (Reguis) 8812
Synonymik des botanischen Klassen-, Familien-, Gattungs- und Sektionsnamen (Pfeiff.) 7820
Synopsin Salicum hungariae (M. Sadler) 9978
Synopsis der Lychnideen (Rohrb.) 9458
Synopsis filicum Africae australis (Pappe) 7317
Synopsis filicum brasiliensium (Raddi) 8488
Synopsis florae Gothlandicae (Säve) 9983
Synopsis methodica fungorum (Pers.) 7730
Synopsis methodica stirpium britannicarum (Ray) 8703
Synopsis of Aloineae and Yuccoideae *(see* V. Ricasoli)
Synopsis of North American Crataegi *(see* E. J. Palmer)
Synopsis of the flora of Colorado (Porter) 8180

Synopsis of the flora of the western states (Riddell) 9198
Synopsis of the genera (Roem.) 9409
Synopsis of the genera of American plants (O. Rich) 9137
Synopsis of the genus Lonicera (Rehd.) 8814
Synopsis of the medical flora of the state of Mississippi (Phares) 7837
Synopsis Pittosporearum (Putterl.) 8408
Synopsis plantarum (Pers.) 7732
Synopsis plantarum diaphoricarum (Rosenthal) 9535
Synopsis plantarum in agro lucensi (Puccin.) 8375
Synopsis plantarum in solo Romano (Sabbati) 9877
Synopsis plantarum phaenogamarum agro lipsiensi indigenarum (Pappe) 7314
Syn. Pittosp. (Putterl.) 8408
Syn. pl. (Pers.) 7732
Syn. pl. agr. lips. (Pappe) 7314
Syn. pl. diaph. (Rosenthal) 9535
Syn. pl. luc. (Puccin.) 8375
Syn. pl. solo Rom. (Sabbati) 9877
Syn. provenç. champ. Vaucluse (Reguis) 8812
Syn. Salic. hung. (M. Sadler) 9978
Syst.-anat. Unters. Podalyr. (Prenger) 8285
Syst. Aufz. Krypt. (Poetsch) 8095
Syst. Bl.-Pfl. (Pilg.) 7936
Syst. charact. pl. (C. Pereb.) 7671
Systema characterum plantarum (C. Pereb.) 7671
Systema mycologicum (see D. P. Rogers)
Systema orbis vegetabilium (F. Rudolphi) 9756
Systema plantarum (Reichard) 8835
Systematisch-anatomische Untersuchungen von Blatt und Achse bei den Podalyrieen-Gattungen (Prenger) 8285
Systematische Aufzählung Kärnten wildwachsenden Gefässpflanzen (Pacher) 7178
Systematische Aufzählung samenlosen Pflanzen (Kryptogamen) (Poetsch) 8095
Systematische Übersicht der Phanerogamen (G. Piep.) 7920
Systema vegetabilium (Roem.) 9408
Systema vegetabilium Editio decima quinta (Pers.) 7726
Systema vegetabilium florae peruvianae et chilensis (Ruiz) 9772
Systema vegetabilium genera (Roem.) 9409
System der Blütenpflanzen (Pilg.) 7936
System der Farne (Prantl) 8272
System der Gramineae (see R. Pilger)

System der Pilze, Lichenen und Algen (see J. Reinke)
Syst. nat. ed. 10 (see D. Royen)
Syst. nat. ed. 12 (see D. Royen)
Syst. orb. veg. (F. Rudolphi) 9756
Syst. pl. (Reichard) 8835
Syst. Rhodophyc. (Rupr.) 9804
Syst. Übers. Phan. (G. Piep.) 7920
Syst. veg. (Roem.) 9408
Syst. veg. ed. 15 (Pers.) 7726
Syst. veg. fl. peruv. chil. (Ruiz) 9772
Syst. veg. gen. (Roem.) 9409
Syst. veg. (see A. Roth)

Tab. anal. fung. (Pat.) 7440
Tab. compar. gen. fung. (Sacc.) 9925
Tabellarische Flora der Schweizer-Kantone (Rhiner) 9128
Tabell. Fl. Schweiz.-Kant. (Rhiner) 9128
Tab. gen. char. (Piddington) 7919
Tableau de la végétation primitive de Minas Geraes (St. Hil.) 10041
Tableau des familles végétales (Piré) 7959
Tableau des plantes usuelles (Roussel) 9693
Tableau géographique de la végétation primitive de Minas Geraes (St. Hil.) 10041
Tabl. encycl. (see J. Poiret)
Tabl. pl. usuel. (Roussel) 9693
Tabulae analyticae fungorum (Pat.) 7440
Tabulae comparativae generum fungorum omnium (Sacc.) 9925
Tabular view of the generic characters in Roxburgh's Flora indica (Piddington) 7919
Tagewerk (Reinke) 8943
Tamakoaré (Barb. Rodr.) 9347
Tange des Ochotskischen Meeres (Rupr.) 9803
Tange Ochotsk. Meer. (Rupr.) 9803
Taraxacum flora of Iceland (Rosenvinge) 9548
Taschenatlas zur Flora von Nord- und Mitteldeutschland (R. Potonié) 8229
Taschenatl. Fl. N.M. Deutschland (R. Potonié) 8229
Taschenb. Bot. (Peterm.) 7755
Taschenb. Fl. deutschl. (H. Reinsch) 8944
Taschenbuch der Botanik (Peterm.) 7755
Taschenbuch der Flora von Deutschland (H. Reinsch) 8944
Tasman. bryoph. (Rodway) 9372
Tasman. bryophyta (Rodway) 9374
Tasman. fl. (Rodway) 9368
Tasmanian bryophyta (Rodway) 9372, 9374
Tasmanian flora (Rodway) 9368
Taumel-Lolch (Roep.) 9426
Tav. Anat. piante aquat. (Parl.) 7376

Tav. bot. (Pratesi) 8274
Tavole di botanica elementare (Pratesi) 8274
Tavole per una Anatomia delle piante aquatiche'opera (Parl.) 7376
Taxonomic and distributional study of some diatoms from Siam and the Federated Malay States *(see* R. Patrick)
Taxonomy of fungi (Ramsb.) 8620
Tent. agrostogr. (Rupr.) 9792
Tentamen agrostographiae universalis (Rupr.) 9792
Tentamen agrostographia universalis *(see* F. Ruprecht)
Tentamen. Claves analyticae festucarum (Saint-Yves) 10096
Tentamen dispositionis methodicae fungorum (Pers.) 7727
Tentamen florae abyssinicae (A. Rich.) 9151
Tentamen florae Abyssinicae *(see* R. Quartin-Dillon)
Tentamen florae bohemicae (Pohl) 8101
Tentamen florae gedanensis (Reyger) 9120
Tentamen florae germanicae (Roth) 9628
Tentamen florae ussuriensis (Regel) 8787
Tentamen historiae naturalis (J. Ritter) 9258
Tentamen monographiae Vangueriae (Robyns) 9308
Tentamen Orchidographiae Europaeae (Rchb. fil.) 8891
Tentamen pteridographiae (K. Presl) 8307
Tentamen rosarum monographiae (Regel) 8803
Tentativa sobre la liquenologia geografica de Andalucia (Roxas) 9717
Tentative revision of Alchemilla Sect. Lachemilla (L. M. Perry) 7722
Tent. clav. anal. festuc. (Saint-Yves) 10096
Tent. disp. meth. fung. (Pers.) 7727
Tent. fl. abyss. (A. Rich.) 9151
Tent. fl. bohem. (Pohl) 8101
Tent. fl. gedan. (Reyger) 9120
Tent. fl. Germ. (Roth) 9628
Tent. fl.-ussur. (Regel) 8787
Tent. monogr. Vanguerieae (Robyns) 9308
Tent. Orchidogr. Eur. (Rchb. fil.) 8891
Tent. pterid. (K. Presl) 8307
Tent. rev. Lachemilla (L. M. Perry) 7722
Tent. ros. monogr. (Regel) 8803
Teor. riprod. veg. (Pollini) 8144
Term. bot. (Planer) 8031
Termini botanici (Planer) 8031
Tertiären und quartären Versteinerungen Chiles (Phil.) 7854
Tert. pl. Brit. Columb. (Penh.) 7632
Tert. quart. Verstein. Chil. (Phil.) 7854

Testigo en lu alborada de Chile (Poepp.) 8090
Texas (F. Roem.) 9397
Text-book bot. (Sachs) 9954
Text-book of botany (Sachs) 9954
Thalamifl. Wien (Patzelt) 7510
Theatrum botanicum romanum (see L. Sabbati)
Theatrum botanicum (see Jn. Parkinson)
Theatrum botanicum (see L. Sabbati)
Thelotremataceae brasilienses (Redinger) 8744
Theoria de symptomatibus quae fungorum venenatorum esum consequi solent (Picco) 7896
Thesaurus literaturae botanicae (Pritz.) 8353
Thes. lit. bot. (Pritz.) 8353
Third decade of new species of North-American plants (Raf.) 8569
Thistles Iowa (Pammel) 7254
Thistles of Iowa (Pammel) 7254
Through the heart of Patagonia *(see* A. Rendle)
Thüringens polypetale Dicotyledonen (Rottenbach) 9668
Thüringer Laubmoose (Röll) 9379
Thüringer Torfmoose und Laubmoose (Röll) 9392
Thymus-Arten der Balkan-Halbinsel *(see* K. Ronniger)
Thysanachne (K. Presl) 8298
Tietoja pihtiputaan ja kinnulan putkilokasvistosta (Roivainen) 9465
Tijdschrift voor Plantenziekten *(see* J. Ritzema Bos)
Timbers of the New World (Record) 8741
Timbers of tropical America (Record) 8739
Timbers trop. Amer. (Record) 8739
Timber trees and forests of North Carolina (Pinchot) 7941
Timb. New World (Record) 8741
Timb. trees N. Carolina (Pinchot) 7941
Todaroa (Parl.) 7374
Tomus supplementorum ad opus botanicum (Riv.) 9272
Torfmoose der thüringischen Flora (Röll) 9381
Torfmoose und Laubmoose der Umgebung von Erfurt (Röll) 9393
Tour. fl. W. Ireland (Praeger) 8249
Tourist's flora of the West of Ireland (Praeger) 8249
Toxicol. afr. (Rochebr.) 9316
Toxicologie africaine (Rochebr.) 9316
Tracts St. Helena *(see* W. Roxburgh)
Traek af Vegetationen i Transkaspiens Lavland (Paulsen) 7530
Traek veg. Transkasp. (Paulsen) 7530
Træer buske (O. Petersen) 7766

Træer og buske (O. Petersen) 7766
Traité arbr. fruit. *(see* P. Poiteau)
Traité bot. (Sachs) 9954
Traité champ. (Paul.) 7519
Traité champ. comest. (Pers.) 7734
Traité chataigne (A. Parm.) 7378
Traité de botanique (Sachs) 9954
Traité de la chataigne (A. Parm.) 7378
Traité des champignons (Paul.) 7519
Traité des fougères de l'Amerique (Plum.) 8068
Traité des plantes de Gand (Roucel) 9670
Traité d'organogénie comparée de la fleur (Payer) 7562
Traité drog. simpl. (G. Planch.) 8000
Traité élém. bot. (F. Pouchet) 8237
Traité élémentaire de botanique (F. Pouchet) 8237
Traité foug. Amér. (Plum.) 8068
Traité organogén. fl. (Payer) 7562
Traité pl. Gand (Roucel) 9670
Traité pratique de la détermination des drogues simples (G. Planch.) 8000
Traité sur la connaissance et la culture des Jacintes (Saint-Simon) 10092
Traité sur les champignons comestibles (Pers.) 7734
Tratado práctico de determinacion de las plantas indígenas (Puerta) 8379
Travels in Switzerland (Ramond) 8616
Travels of Ruiz, Pavón, and Dombey in Peru and Chile (Ruiz) 9782
Travels through the Low-Countries (Ray) 8697
Travels through the southern provinces of the Russian empire (Pall.) 7228
Trav. s. prov. Russ. emp. (Pall.) 7228
Tree book (J. E. Rogers) 9436
Tree fl. (W. E. Rogers) 9440
Tree flowers of forest, park and street (W. E. Rogers) 9440
Trees Amer. (R. Piper) 7955
Trees and shrubs of Kentucky (S. F. Price) 8322
Trees and shrubs of Tasmanian forests, Myrtaceae (Rodway) 9370
Trees and Shrubs *(see* A. Rehder)
Trees, n.-e. U.S. (Parkhurst) 7341
Trees of America (R. Piper) 7955
Trees, shrubs and vines of the northeastern United States (Parkhurst) 7341
Trees shrubs Kentucky (S. F. Price) 8322
Trees shrubs Minnesota (Rosend.) 9527
Trees shrubs of Minnesota (Rosend.) 9527
Trees shrubs Tasman. for., Myrtac. (Rodway) 9370
Tropical plants (Perrine) 7710
Tropical rainforest *see* P. Richards)
Tropical Woods *(see* S. Record)
Tropischen Nutzpflanzen Ostafrikas (Sadebeck) 9967

Tsetrarija (Cetraria) CCCP *(see* K. Rassadina)
Turkestanskaia Flora *(see* E. Regel)
Types biologiques pour la géographie botanique (Raunk.) 8656
Types de chaque famille (F. Plée) 8054
Types fam. (F. Plée) 8054

Ueber Algen und Hydrozoen im Silur (Rothpletz) 9655
Über anatomische Veränderungen, welche in Perianthkreisen der Blüthen während der Entwicklung der Frucht vor sich gehen *(see* C.Reiche)
Über Artenzahl und Areal (Palmgr.) 7243
Über Botrydium granulatum (Rost.) 9597
Ueber Calamiten und Steinkohlenbildung (Petzh.) 7799
Ueber Caulerpa (Reinke) 8936
Ueber Cupania (Radlk.) 8515
Ueber das Alter der Laubmoose (Reichardt) 8839
Ueber das System der Orchideen (Rchb. fil.) 8900
Über das System der Rhodophyceae (Rupr.) 9804
Ueber das Verhältniss der Parthenogenesis zu den anderen Fortpflanzungsarten (Radlk.) 8507
Ueber das Wandern der Pflanzen (Pokorny) 8128
Ueber dem Gang der morphlogischen Differenzirung in der Sphacelarien-Reihe (Pringsh.) 8336
Ueber den Bau des Holzes (Rossmann) 9591
Ueber den Befruchtungsvorgang bei den Basidiomyceten (Reess) 8761
Über den Blüthenbau und die Befruchtung von Epipogium gmelini (Rohrb.) 9457
Über den Charakter der Vegetation auf den Inseln des indischen Archipels (Reinw.) 8965
Ueber den Formenkreis der Anthyllis vulneraria L. sensu amplissimo (Sagorski) 9991
Ueber den gegenwärtigen Zustand der Botanik in Deutschland (Sachs) 9955
Über den Generationswechsel der Thallophyten (Pringsh.) 8338
Ueber den Parasitismus von Elaphomyces granulatus (Reess) 8765
Ueber den Ursprung der Alpenpflanzen (Pokorny) 8130
Über die Anden zum Amazonas (Poepp.) 8090
Ueber die Arbeit und das Wirken der Pflanze (Radlk.) 8523
Ueber die Aufeinanderfolge des ersten

Auftretens der verschiedenen organischen Gestalten (Ritgen) 9251
Über die Befruchtung und Keimung der Algen (Pringsh.) 8329
Ueber die Blattstructur der Gattung Cecropia (Al. Richt.) 9178
Ueber die Boden- und Vegetationsverhältnisse zweier Ortschaften an der Nordküste Estlands (Russow) 9836
Über die Dauerschwärmer des Wassernetzes (Pringsh.) 8331
Über die Engelwurz (Rikli) 9231
Über die Entstehung der Flechte Collema glaucescens (Reess) 8760
Ueber die Entwickelung des Farnblattes (Sadebeck) 9963
Über die Entwickelung des Pflanzensystems (Radlk.) 8524
Ueber die Entwicklung von Phyllitis, Scytosiphon und Asperococcus (Reinke) 8928
Ueber die europäischen Arten der Gattung Typha (Rohrb.) 9460
Ueber die Flora der Insel St. Paul (Reichardt) 8846
Über die Gattung Allophylus (Radlk.) 8538
Über die Gefässbündel-Vertheilung im Stamme und Stipes der Farne (Reichardt) 8838
Über die geographische Verbreitung der Baumwolle (C. Ritter berol.) 9256
Ueber die Gliederung der Familie der Sapindaceen (Radlk.) 8531
Über die Gliederung der Karpathenflora (Pax) 7548
Ueber die Idée der Art (Regel) 8792
Ueber die Kalkagen (Rothpletz) 9656
Über die Knospenlage der Laubblätter (Reinecke) 8913
Ueber die Methoden in der botanischen Systematik, insbesondere die anatomische Methode (Radlk.) 8516
Ueber die Natur der Flechten (Reess) 8764
Über die neuesten Torfmoosforschungen (Röll) 9390
Ueber die Nutzpflanzen Samoas (Reinecke) 8914
Über die Osmundaceen und Schizaeaceen der Juraformation (Racib.) 8460
Ueber die Pflanzen-Gattung Tylodendron (Potonié) 8213
Über die Pflanzen, welche den atlantischen Ocean auf der Westküste Europas begleiten (E. Roth) 9638
Ueber die Pflege der Botanik in Franken (Reess) 8766
Über die Polyembryonie bei Eugenia (v. d. Pijl) 7932

Ueber die Sapindaceen Holländisch-Indiens (Radlk.) 8513
Ueber die Schutzscheide der deutschen Equisetaceen (Pfitz.) 7825
Über die selbständige Entwicklung der Pollenzelle (Reissek) 8968
Über die Spermien der Fucaceen (M. Retz.) 9095
Ueber die Spiralstellungen der Rhodomelaceen (Rosenvinge) 9545
Ueber die systematischen Werth der Pollenbeschaffenheit bei den Acanthaceen (Radlk.) 8517
Ueber die Verbreitung der Pflanzen im nördlichen Ural (Rupr.) 9802
Ueber die Versetzung der Gattung Dobinea (Radlk.) 8525
Ueber die Versetzung der Gattung Henoonia (Radlk.) 8526
Ueber die Vertheilung des Gerbstoffes in den Zweigen und Blättern unserer Holzgewächse *(see* W. Petzold)
Über die Vorkeime der Charen (Pringsh.) 8333
Über die Zurückführung von Forchhammeria zur Familie der Capparideen (Radlk.) 8519
Ueber die Zusammensetzung der Leitbündel bei den Gefässkryptogamen (Potonié) 8210
Ueber einige Capparis-Arten (Radlk.) 8518
Ueber einige Cirsium aus dem Kaukasus (Petr.) 7781
Ueber einige neue Desmarestien (Reinsch) 8955
Über einige neue mikroskopisch-chemische Reactionsmethoden (Sachs) 9952
Über einige Pilze aus der Familie der Laboulbenien (Peyr.) 7805
Ueber einige Sapotaceen (Radlk.) 8520
Ueber einige wildwachsende Pflanzenbastarde (Ritschl) 9255
Über Endophyten der Pflanzenzelle (Reissek) 8969
Ueber Endophyten in der Pflanzenzelle *(see* S. Reissek)
Über Flechtenstandorte und Flechtenvegetation im westlichen Nordfinnland (Räs.) 8542
Über Flora und Vegetation von Kieta und Griechenland (Rikli) 9235
Über Homodromie und Antidromie (Raunk.) 8661
Ueber Hypnum turgescens (H. Paul) 7518
Ueber Nothochilus (Radlk.) 8528
Über Paarung von Schwarmsporen (Pringsh.) 8335
Ueber Pausandra (Radlk.) 8508
Uebers. Aconitum (Rchb.) 8870

Ueber Sapindus (Radlk.) 8514
Übers. Brandenburg Laubm. (Reinhardt) 8923
Übers. Gatt. Monim. (Perkins) 7687
Übers. Gatt. Styrac. (Perkins) 7688
Uebers. Gew.-Reich. (Rchb.) 8879
Uebersicht der Arten der Gattung Thalictrum (Regel) 8789
Uebersicht der Gattung Aconitum (Rchb.) 8870
Übersicht der in der Mark Brandenburg bisher beobachteten Laubmoose (Reinhardt) 8923
Übersicht der in Krain bisher nachgewiesenen Formen aus der Gattung Alchemilla (Paulin) 7526
Uebersicht der Phanerogamen und Gefässcryptogamen von Nassau (Rudio) 9752
Uebersicht der vom Herrn Professor Dr. Haussknecht im Orient gesammelten Kryptogamen (Rabenh.) 8446
Uebersicht des Gewaechs-Reichs (Rchb.) 8879
Uebersicht des Gewaechs-Reichs (Rchb.) 8880
Übersicht des natürlichen Systems der Pflanzen (Pfitz.) 7833
Uebersicht über die Arten der Gattung Gagea (Pasch.) 7419
Übersicht über die Gattungen der Monimiaceae (Perkins) 7687
Übersicht über die Gattungen der Styracaceae (Perkins) 7688
Uebersicht untersuchter (G. Preuss) 8316
Übers. Krain Achemilla (Paulin) 7526
Uebers. Kurhessen Pfl. *(see* L. Pfeiffer)
Ueber spontane und künstliche Gartenbastarde der Gattung Hieracium sect. Piloselloidea (Peter) 7745
Übers. Syst. Pfl. (Pfitz.) 7833
Uebers. Thalictrum (Regel) 8789
Ueber Tetraplacus (Radlk.) 8521
Über vegetative Sprossung der Moosfrüchte (Pringsh.) 8337
Ueber Vorkommen und Biologie von Laboulbeniaceen (Peyr.) 7806
Ueber wahre Parthenogenesis bei Pflanzen (Radlk.) 8506
Ueber zwei neue Arten von Centaurea aus Kurdistan (Reichardt) 8844
Undersøgelser angaaende Svampeslægten Rhizoctonia (Rostr.) 9610
Undersøgelser over ydre Faktorers Indflydelse paa Organdannelsen hos Planterne (Rosenvinge) 9538
Une confusion dans les fleurs poétiques (Roum.) 9676
Une nouvelle espèce de Stysanus (Rolland) 9472

United States Exploring Expedition. The geographical distribution of animals and plants (Pickering) 7912
United States species of Lycoperdon (Peck) 7596
Universum der Natur (Rchb.) 8882
Unsere essbaren Pilze (Röll) 9384
Uns. essb. Pilz. (Röll) 9384
Unters. Assim.-Org. Legum. (Reinke) 8934
Unters. Bacill. (Pfitz.) 7826
Unters. Bau Pfl.-Zell. (Pringsh.) 8328
Unters. foss. Hölz. (Platen) 8034
Unters. Morph. Gefässkrypt. (Prantl) 8269
Unters. Quellung (Reinke) 8929
Unters. Spermog. Rostpilze (Ráthay) 8640
Untersuchungen fossiler Hölzer (Platen) 8034
Untersuchungen über Bau und Entwicklung der Bacillariaceen (Pfitz.) 7826
Untersuchungen über Bau und Lebensgeschichte der Hirschtrüffel Elaphomyces (Reess) 8767
Untersuchungen über Befruchtung und Generationswechsel der Algen (Pringsh.) 8329
Untersuchungen über den Bau und die Bildung der Pflanzenzelle (Pringsh.) 8328
Untersuchungen über die Assimilationsorgane der Leguminosen (Reinke) 8934
Untersuchungen über die Pilzgattung Exoascus (Sadebeck) 9965
Untersuchungen ueber die Quellung (Reinke) 8929
Untersuchungen über die Spermogonien der Rostpilze (Ráthay) 8640
Untersuchungen über Drosophyllum lusitanicum *(see* A. Penzig)
Untersuchungen über Wachsthumsgeschichte und Morphologie der Phanerogamen-Wurzel (Reinke) 8924
Untersuchungen über zellenartigen Ausfüllungen der Gefässe *(see* C. Reichenbach)
Untersuchungen zur Morphologie der Gefässkryptogamen (Prantl) 8269
Unters. Wachsth.-Gesch. Phan.-Wurz. (Reinke) 8924
Urban, Symb. ant. *(see* W. Ruhland)
Uredineen Süddeutschlands (see H. Poeverlein)
Urspr. Alpenpfl. (Pokorny) 8130
Usef. pl. Guam (Saff.) 9984
Useful plants of the Island of Guam (Saff.) 9984
U.S. Expl. Exped., Geogr. distr. pl. (Pickering) 7912
U.S. Geographical surveys, Reports upon botanical collection (Rothr.) 9658
U.S. Lycoperdon (Peck) 7596

Ustilagneae Daniae (Rostr.) 9613
Ustil. Daniae (Rostr.) 9613

Vacation cruising (Rothr.) 9660
Vademecum für Pilzfreunde (Ricken) 9193
Vadem. Pflanzenw. (Postel) 8196
Vadem. Pilzfr. (Ricken) 9193
Valdemecum für Freunde der Pflanzenwelt (Postel) 8196
Var. diff. Ceratophyllum (Pearl) 7581
Varenfl. Java *(see* O. Posthumus)
Variae ac multiformes florum species *(see* N. Robert)
Variation and differentiation in Ceratophyllum (Pearl) 7581
Vasc. pl. Saghalin (Printz) 8343
Vascular flora of Coös county (Pease) 7592
Vasis pl. spiral. (G. Reichel) 8866
Veg. alreded. México (Reiche) 8862
Vég. Bretagne (Picq.) 7915
Vég. croiss. l'homme (C. P. Robin) 9277
Veg. dansk-vestind. Øer (Paulsen) 7529
Vegetabilia cryptogamica Boëmiae (see J. Presl)
Vegetabilia in itinere iberico austro-meridionali lecta (Porta) 8178
Vegetabilium cum animalibus comparatio *(see* L. Roberg)
Vegetación de México (Ramírez) 8607
Vegetación en los alrededores de la capital de México (Reiche) 8862
Végétation de la Bretagne (Picq.) 7915
Vegetation des rothen Meeres (Rupr.) 9801
Végétation malgache (H. Perrier) 7702
Vegetation of the District lying South of Dublin *(see* G. Pethybridge)
Vegetation of the Siberian-Mongolian frontiers (Printz) 8344
Vegetationsbilder aus den Südkarpathen *(see* F. Pax)
Vegetationsbilder aus Nordrussland *(see* R. Pohle)
Vegetationsbilder der Jetzt- und Vorzeit. *(see* H. Potonié)
Vegetationsbilder *(see* R. Pohle, M. Rikli)
Vegetations forschung auf soziationsanalytischer Grundlage (Du Rietz) 9219
Vegetations-Verhältnisse des Bayerischen Waldes *(see* L. Radlkofer)
Vegetationsverhältnisse des Kyffhäuser Gebirges (Petry) 7792
Vegetationsverhältnisse in der Umgebung der Hauptstadt von Mexico (Reiche) 8863
Vegetationsverhältnisse von Gera (Pietsch) 7931
Vegetationsverhältnisse von Iglau (Pokorny) 8124
Végétaux inférieurs (Piré) 7962

Végétaux qui croissent sur l'homme (C. P. Robin) 9277
Végétaux silicifiés d'Autun (Renault) 8999
Vég. inf. (Piré) 7962
Veg. itin. iber. (Porta) 8178
Vég. malg. (H. Perrier) 7702
Veg. México (Ramirez) 8607
Veg. roth. Meer. (Rupr.) 9801
Veg. Siber.-Mongol. front. (Printz) 8344
Veg.-Verh. Gera (Pietsch) 7931
Veg.-Verh. Iglau (Pokorny) 8124
Veg.-Verh. Kyffhäus. Geb. (Petry) 7792
Vejledning i den danske Flora (Rostr.) 9606
Vellosia (Barb. Rodr.) 9349
Vera fuci fruct. (Ruiz) 9773
Verbascum Nielreichii (Reichardt) 8841
Verbascum pseudo-phoeniceum (Reichardt) 8843
Verbenaceae of China (P'ei) 7614
Verben. China (P'ei) 7614
Verbreitung der Pflanzen (Rattke) 8643
Verbreitungsmittel der Pflanzen (E. Roth) 9640
Verbreitungsmitt. Pfl. (E. Roth) 9640
Verbr. Pfl. (Rattke) 8643
Verbr. Pfl. Ural (Rupr.) 9802
Verbr. Pfl. Ural (Rupr.) 9806
Vergissmeinichtarten *(see* L. Reichenbach)
Vergissmeinnichtarten (Rchb.) 8874
Vergl. anat. Unters. Parmel. (F. Rosend.) 9528
Vergl.-anat. Unters. Steng. Rhiz. (Rothert) 9648
Vergleichend anatomische Untersuchungen über die braunen Parmelien (F. Rosend.) 9528
Vergleichend-anatomische Untersuchungen über die Differenzen im primären Bau der Stengel und Rhizome krautiger Phanerogamen (Rothert) 9648
Vergleichende Übersicht der urweltlichen Organismen (Rolle) 9480
Vergleichende Untersuchungen der Leitbündel-Kryptogamen (Russow) 9834
Vergleichende Untersuchungen über den Bau des Holzes einiger sympetaler Familien (Prantl) 8272
Vergl. Übers. urveltl. Organ. (Rolle) 9480
Verg. Unters. Leitb.-Krypt. (Russow) 9834
Verh. Parthenog. Fortpfl.-Arten (Radlk.) 8507
Vers. Arzneikr. Pfl. *(see* K. Perleb)
Vers. Beschr. Schlesw. Holst. Pfl. (Chr. Ritter) 9257
Vers. Gesch. amer. Agave (Petters) 7794
Verspr. phan. arkt. Eur. (Ruijs) 9767
Vers. Syst. Mycetozoen (Rost.) 9593

Verstein. Böhm. Kreideform. (A. E. Reuss) 9098
Verstein. Braunkohlensandst. (Rossmässler) 9580
Versteinerungen (Rossmässler) 9582
Versteinerungen der Böhmischen Kreideformation (A. E. Reuss) 9098
Versteinerungen des Braunkohlensandsteins (Rossmässler) 9580
Versteinerungen des Harzgebirges (A. Roem.) 9395
Versteinerungen des norddeutschen Kreidegebirges (A. Roem.) 9394
Verstein. Harzgeb. (A. Roem.) 9395
Verstein. norddeut. Kreidegeb. (A. Roem.) 9394
Versuch einer Beschreibung der in den Herzogthümern Schleswig und Holstein wildwachsenden Pflanzen (Chr. Ritter) 9257
Versuch einer Geschichte der amerikanischen Agave (Petters) 7794
Versuch einer Monographie der Gattung Portulaca (Poelln.) 8087
Versuch einer Naturgeschichte der Mark Brandenburg (Ruthe) 9839
Versuch einer phylogenetischen Erklarung des Embryosackes und der doppelten Befruchtung der Angiospermen (Porsch) 8173
Versuch eines Systems der Mycetozoen (Rost.) 9593
Versuch eines Wörterbuchs der botanischen Terminologie (Roem.) 9407
Versuch über die Varietäten des Weinstocks (Roxas) 9715
Vers. Var. Weinst. (Roxas) 9715
Vers. Wörterb. bot. Termin. (Roem.) 9407
Verwandtschaftsverhältnisse in der Gattung Veronica (Römmp) 9415
Verw.-Verh. Veronica (Römmp) 9415
Verz. Antofagasta Pfl. (Phil.) 7856
Verz. bot. Gart. Aschaffenburg Pfl. (Prantl) 8270
Verzeichniss der an und auf dem Harz wildwachsenden Bäume, Gesträuche und Kräuter (see J. Rueling)
Verzeichniss der Gräser Mecklenburgs (Roep.) 9422
Verzeichniss der im botanischen Garten der königl. Forstlehranstalt Aschaffenburg cultivirten Pflanzen (Prantl) 8270
Verzeichniss der im Plauischen Grunde (Pursh) 8403
Verzeichniss der in der Umgebung von Seckau wachsenden Phanerogamen (Pernh.) 7693
Verzeichniss der in der Umgegend von Weissenburg Gefässpflanzen (W. Petzold) 7802

Verzeichniss derjenigen Pflanzen (Roth) 9625
Verzeichniss der neuen Arten (Petr.) 7786
Verzeichniss der um Pesth und Ofen wildwachsenden phanerogamischen Gewächse (Jos. Sadler) 9975
Verzeichniss der von Friedrich Philippi auf der Hochebene der Provinzen Antofagasta und Tarapacá gesammelten Pflanzen (Phil.) 7856
Verzeichniss der von Herrn Paullowsky und Herrn von Stubendorf Jakutzk und Ajan gesammelten Pflanzen (see L. Rach)
Verzeichniss der von Vidal Gormaz gesammelten Gefässpflanzen (Phil.) 7855
Verzeichniss der zwischen Jakutzk und Ajan gesammelten Pflanzen (Regel) 8784
Verz. Gräs. Mecklenb. (Roep.) 9422
Verz. Jakutzk Ajan Pfl. (Regel) 8784
Verz. Pesth wildw. Gew. (Jos. Sadler) 9975
Verz. Pfl. (Roth) 9625
Verz. Plau. Pfl. (Pursh) 8403
Verz. Weissenburg Gefässpfl. (W. Petzold) 7802
Viage al desierto de Atacama (Phil.) 7850
Viagem no interior do Brasil (Pohl) 8104
Viagg. bot. Gargano (Pasq.) 7431
Viaggio alla catena del Monte Bianco (Parl.) 7363
Viaggio al Lago di Garda (Pollini) 8142
Viaggio del botanico fiorentino Pier Antonio Micheli a Verona ed al Monte Baldo (Pamp.) 7268
Viaggio per le parti settentrionali di Europa (Parl.) 7365
Viagg. Lago di Garda (Pollini) 8142
Viagg. Micheli (Pamp.) 7268
Viagg. Monte Bianco (Parl.) 7363
Viagg. parti sett. Eur. (Parl.) 7365
Viaje de exploracion al Valle del Rio Grande de Térraba (Pitt.) 7977
Viaje Rio Grande Térraba (Pitt.) 7977
Vicissitudes onomastiques de la Globulaire vulgaire (St. Lag.) 10075
Victoria history of the country of Durham (see M. Potter)
Victoria regia (Planch.) 8009
Vier noch unbeschriebene Peperomeen (Regel) 8782
Vier und zwanzig Vegetation-Ansichten (see A. Postels)
Vier Vorträge über den Kaukasus (Radde) 8478
View Malay. Isl. (Pennant) 7634
View of the Malayan Isles (Pennant) 7634
Vign. amér. (Sahut) 10012
Vigne du Mont Ida et le Vaccinium (St. Lag.) 10085

Vignes amér. (Planch.) 8021
Vignes américaines (Planch.) 8021
Vignes américaines (Sahut) 10012
Vignes des tropiques du genre Ampelocissus (Planch.) 8024
Vignes sauvages (Planch.) 8019
Vilmorins illustrierte Blumengärtnerei *(see* T. Rümpler)
Viola (Pio) 7943
Vlaamsche volksnamen der planten (Pâque) 7319
Vlaamsche volksnam. pl. (Pâque) 7319
Vocabulario botanico friulano *(see* G. Pirona)
Vörläufige Mittheilungen über die in Nord-Amerika gesammelten Laubmoose (Röll) 9385
Volksbenenn. bras. Pfl. (Peckolt) 7607
Volksbenennungen der brasilianischen Pflanzen (Peckolt) 7607
Volksthümliche Pflanzennamen der Waldstätten (Rhiner) 9127
Volksthüml. Pfl.-Nam. Waldst. (Rhiner) 9127
Vollständige Synonymik (Pfeiff.) 7820
Vollst. Pflanzensyst. *(see* G. Panzer)
Vollst. Synon. (Pfeiff.) 7820
Von dem officinellen Fieberrindenbaum (Ruiz) 9768
Von den Pyrenäen zum Nil (Rikli) 9236
Vorarbeiten zu einer Flora Bayerns (Poeverl.) 8099
Vorarbeiten zu einer Flora des Mährischen Gouvernements (Rohrer) 9462
Vorarbeiten zu einer Synopsis lichenum (Du Rietz) 9217
Vorarbeiten zur Kryptogamenflora von Unter-Oesterreich (Pokorny) 8125
Vorarb. Fl. Mähr. Gouv. (Rohrer) 9462
Vorarb. Krypt.-Fl. U.-Oesterr. (Pokorny) 8125
Vorgefasste botanische Meinungen (Roep.) 9424
Vorgefasste bot. Meinungen (Roep.) 9424
Vorläufige Mittheilungen in Nord-Amerika gesammelten Torfmoose (Röll) 9387
Vorläufige Mittheilung über die in Nord-Amerika gesammelten Lebermoose (Röll) 9386
Vorläuf. Nachr. Geschl. Pfl. *(see* W. Pfeffer)
Vorles. Pfl.-Physiol. (Sachs) 9958
Vorlesungen über Pflanzen-Physiologie (Sachs) 9958
Vorzüglich lästige Insekten (Pohl) 8104
Voyage autour du monde *(see* A. Postels)
Voyage aux sources du Rio de S. Francisco (St. Hil.) 10053
Voyage dans le district des diamans (St. Hil.) 10042

Voyage dans les provinces de Rio de Janeiro et de Minas Geraes (St. Hil.) 10039
Voyage dans les provinces de Saint-Paul (St. Hil.) 10055
Voyage dans l'intérieur du Brésil (St. Hil.) 10054
Voyage en Abyssinie (A. Rich.) 9151
Voyage en Abyssinie *(see* R. Quartin-Dillon)
Voyage en Barbarie (Poir.) 8114
Voyages au Mont-Perdu (Ramond) 8617
Voyages dans l'intérieur de la Louisiane (C. C. Robin) 9279
Voyages dans l'intérieur de la Louisiane (Raf.) 8567
Voyages de M.P.S. Pallas (Pall.) 7224
Voyages du professeur Pallas (Pall.) 7224
Voyage to the North Pole (Phipps) 7874
Voy. Astrolabe, Botanique *(see* A. Richard)
Voy. Barbarie (Poir.) 8114
Voy. distr. diam. (St. Hil.) 10042
Voy. intér. Brésil (St. Hil.) 10054
Voy. int. Louisiane (C. C. Robin) 9279
Voy. Mont-Perdu (Ramond) 8617
Voy. North Pole (Phipps) 7874
Voy. Rio de Janeiro (St. Hil.) 10039
Voy. Rio S. Francisco (St. Hil.) 10053
Voy. Saint-Paul (St. Hil.) 10055
Voy. Uranie *(see* C. Persoon)
Vraie nature de la fleur des Euphorbes (Planch.) 8016
Výsl. bryol. výzk. Moravy (Podp.) 8076
Výsledky bryologického výzkuma Moravy (Podp.) 8076

Wald (Rossmässler) 9584
Waldverderbniss (Ratzeb.) 8647
Walpers, Ann. *(see* H. G. Reichenbach)
Wandern Pfl. (Pokorny) 8128
Wasserpflanzen und Sumpfgewächse in Deutsch-Ostafrika (Peter) 7747
Wasserpfl. Deut.-Ostafr. (Peter) 7747
W. Austral. Orchid. (Pelloe) 7623
Wayside plants and weeds of Shanghai (Porterf.) 8185
Wayside pl. Shanghai (Porterf.) 8185
Weed fl. Iowa (Pammel) 7257
Weed flora of Iowa (Pammel) 7257
Weeds of South Africa (E. Phillips) 7864
Weeds of southwestern Wisconsin (Pammel) 7251
Weeds of the pineapple fields of the Hawaiian Islands (St. John) 10062
Weeds pineapple fields (St. John) 10062
Weeds S. Afr. (E. Phillips) 7864
Weeds Wisc. Minn. (Pammel) 7251
Wenersborg fl. (Sahlén) 10007
Wenersborg flora (Sahlén) 10007
West Australian Orchids (Pelloe) 7623

West Coast botany, and analytical key to the flora of the Pacific Coast (Rattan) 8641
Western Minerva (Raf.) 8575
West. Minerva (Raf.) 8575
Wichtigeren Nutzpflanzen (Sadebeck) 9970
Wild fl. Brit. emp. (Rockley) 9333
Wild fl. Calif. (M. Parsons) 7416
Wild fl. Missouri (T. Rickett) 9197
Wild flowers and trees of Colorado (Ramaley) 8601a
Wild flowers of California (M. Parsons) 7416
Wild flowers of Missouri (T. Rickett) 9197
Wild flowers of the great dominions of the British Empire (Rockley) 9333
Wild flowers of the United States *(see* H. Rickett)
Wild flowers of Western Australia (Pelloe) 7622
Wild fl. Tasmania (Rodway) 9371
Wild fl. trees Colorado (Ramaley) 8601a
Wild fl. W. Australia (Pelloe) 7622
Wildwachsende Thalamifloren der Umgebungen Wien's (Patzelt) 7510
Wildwachs. Pfl.-Bast. (Ritschl) 9255
Williamsonia (Sahni) 10010
Wiss. Erg. deut. Zentr.-Afr. Exped. *(see* R. Pilger)
Wiss. Erg. Deut. Zentral-Afrika-Exp. Botanik *(see* F. Pax)
Woody plants of Western Wisconsin (Pammel) 7252
Woody species of Eupatorium and Ophryosporus occurring in Mexico (B. L. Robinson) 9294
Wrażenia ogrodnicze z jawy (Roupperet) 9685
Všobecný rostl. (J. Presl) 8289
Všobecný rostlinopsis (J. Presl) 8289

Xenia orchid. (Rchb. fil.) 8894
Xenia orchidacea (Rchb. fil.) 8894

Zakflora voor Suriname (Pulle) 8389
Zakfl. Suriname (Pulle) 8389
Zur Anatomie der Torfmoose (Russow) 9837
Zur Entwicklungsgeschichte der Stammspitze von Equisetum (Reess) 8757
Zur Flechtenflora der Umgebung von Ehingen (Rieber) 9213
Zur Flora Mecklenburgs (Roep.) 9423
Zur Flora Thüringens. Zweiter Beitrag (Rottenbach) 9669
Zur Flora von Conitz (Praet.) 8254
Zur Kenntniss des Florencharakters des Nadelwaldes (Palmgr.) 7244
Zur Kenntniss der Blütenentwicklung und des Gefässbündelverlaufs der Loasaceen (Racine) 8473
Zur Kenntniss der Flechtenflora Feuerlands (Räs.) 8544
Zur Kenntniss der Flora der Halbinsel Pelješac (Rech. fil.) 8730
Zur Kenntniss der Gattung Echeveria (Poelln.) 8088
Zur Kenntniss der Pflanzenwelt des Kantons Tessin (Rikli) 9228
Zur Kenntniss der Subsecundum- und Cymbifoliumgruppe europäischer Torfmoose (Russow) 9838
Zur Kenntniss der Wasserhahnenfüsse (Rossmann) 9589
Zur Kenntniss kleinster Lebensformen (Perty) 7739
Zur Klärung von Theophrasta (Radlk.) 8529
Zur Kritik und Geschichte der Untersuchungen ueber das Algengeschlecht (Pringsh.) 8330
Zur methodologischen Grundlage der modernen Pflanzensoziologie *(see* G. Du Rietz)
Zur Morphologie der Utricularien (Pringsh.) 8334
Zur Naturgeschichte der Torfmoore (Petzh.) 7800
Zur Pflanzengeographie des Erzgebirges (Sachse) 9960
Zur Physiologie der Diatomeen (O. Richt.) 9186
Zur Synonymie einiger Hipocratea-Arten (Peyr.) 7807
Zur Systematik der Gattung Eriophorum (Palla) 7219
Zur Systematik der Torfmoose (Röll) 9382
Zustand Bot. Deutschl. (Sachs) 9955
Zwei neue Cycadeen (Regel) 8779
Zweite naturwissenschaftliche Studienreise nach Spanien, der Küste v. Marokko (Rikli) 9229

Index to Names

This index contains the names of authors, collectors, botanical explorers, patrons of botanists, botanical artists, and other collaborators, as well as personal surnames that have been immortalized in names of genera. Also included are abbreviations listed for author names. The names of ships that carried early exploring expeditions to remote parts of the world are included in this index too, because the user may only remember that much of the subsequent reports of the trip. There are, however, a number of explanations and cautions: (1) the abbreviations of author names are included in the index when they are really abbreviations, i.e. some letters have been dropped. Such abbreviations are alphabetized as if they were complete words, including one or more initials, with the omission of punctuation and spaces. In this index the initial follows the abbreviation even though the suggested abbreviation in the text is preceded by the initial; (2) diacritical marks have been deleted in this index except for the German Umlaut, whereas the ø is alphabetized as oe; (3) names beginning with "M", "Mc", or "Mac" are treated as if spelled "Mac" in combination with the rest of the name as one word (Macdonald); (4) similarly, double names whether hyphenated or not, names beginning with "Da", "De", "Des", "Do", "Du", "La", "Le" etc. are alphabetized as if they were one word under the first letter of the first word ("Candolle" for "de Candolle" is one of several exceptions where established usage does not conform this convention). If there is chance of confusion in locating such names, the index lists also the last part of the names under the appropriate letter (the user can find "de Candolle" in both the "d" and "c" part of the index). Hyphenated double names are not, however, cross-referenced under the last part; (5) for purposes of the index, "van", "von", "y", "zu" are ignored; (6) names of authors of publications cited in the biography/bibliography sections are omitted from the index; (7) "Aiton, W. 72" refers to that part of the text just preceding the treatment of book number 72, as well as to the book itself. "Aiton, W. 83" means there is a reference to Aiton, W. in the information concerning book 83, which is by an author other than Aiton, W. "Aiton, W. *see* Banks" indicates mention of Aiton, W. in the author-paragraphs under "Banks"; (8) an author's name with the notation "*alph.*" indicates a person, who is included in the text for one or more reasons, but who did not publish items treated in this guide (Abich, H., *alph.*); (9) the notation "*v.*" is used to connect the abbreviation of a personal name with the name in full to explain the meaning of the suggested abbreviation. It is also used as a means of cross-referencing double names, maiden names of female botanists who have published under both names, and for anyone, who has used more than one name for his publications; (10) references to Linnaeus occur so frequently that they are omitted from this index.

Several entries which were omitted from the index to names in the third volume are included now. Some mistakes in the indexes of volumes 1-3 have been entered here correctly. Finally, because an index to all the volumes will be published after the last volume has been issued, the present

index (as well as the index to titles) has to be viewed as a temporary but, hopefully, a useful guide to the contents of this volume.
For authors between Pabst and Sakurai see also the indexes in vols. 1-3. References in previous volumes are not repeated here.

Abbate, E. see Pirotta, P.R.
Abderhalden, E. 9219; see Porsch, O.
Abel, G.F. 8971-8972
Abilgaard, N. 9666
Acharius, E. 7247, 9153; see Persoon, C.H., Retzius, A.J.
Achintre, J.F. 8811; see Perreymond, J.H.
Adams, C.D. 1745, 5170
Adams, J. see Pethybridge, G.H.
Adams, W.H.D. 5345
Adanson, A. see Payer, J.-B.
Adanson, M. 7559, 9946; see Payer, J.-B., Plumier, C., Rousseau, J.J., Rozier, F.
Adlerskron, M.B. von v. Behaghel von Adlerskron, M.
Adriaensen, J. 7318
Åkerman, A. 4783
Aellen, P. see Rechinger, K.H.
Afzelius, K. 9530
Agard, A.E. see Pocock, M.A.
Agardh, C.A. 8293, 9088, 9421; see Pocock, M.A., Ruchinger, G., Sr.
Agardh, J.G. see Nordstedt, C.F.O., Pocock, M.A., Preiss, L., Rüppell, E.
Agassiz, J.L.R. see Packard, A.S., Reuss, A.E. von
Agosti, G. 9937
Agrelius, A.J. 7247
Aimé, A. Ponce de Leon y v. Ponce de leon y Aimé, A.
Aiton, W. see Pourret, P.A.
Albertia see Regel, A. von
Albertini, J.B. von 8316
Albertis, E. d' v. d'Albertis, E.
Albert the Great 10.070
Alcocer, G.V. 8606, 8609-8610
Alexander, R.C. v. Prior, R.C.A.
Alfredus 5927
Alken, S. 5670
Allemão e Cysneiro, F.F. see Barbosa Rodrigues, J.
Allenrolfea see Rolfe, R.A.
Allescher, A. 8450
Allgaier 6851
Allgulin, J.G. 9090
ALLIER see Raoul, É.F.L.
Allioni, C. see Pourret, P.A.
Allport, T. 9505
Alluard, G. 6784
Alm, J. 4848
Almagia, R. 7969

Almeida Pinto, J. de v. Pinto, J. de Almeida
Almeida Pinto, Z. d' v. d'Almeida Pinto, Z.
Almeloveen, T.J. van 9123
Almquist, S.O.I. see Romell, L.
Alston, A.H.G. 8393
Altena, H.L. van 9138
Amans, J.F.B. de Saint v. Saint-Amans, J.F.B. de
Amansia see Saint-Amans, J.F.B. de
Amansites see Saint-Amans, J.F.B. de
Amati, P. 7943
Ames, O. 9438
Amherst, A.M. v. Rockley, A.M.
Amphlett, J. 8714
Amshoff, G.J.H. 8393; see Pulle, A.A.
Ancona, N. D' v. D'Ancona, N.
Anderson, N.J. 7247
Andersson, L. see Regnell, A.F.
Andersson, N.J. 7759
André, E.A. 9680
André, E.F. 5377, 6784, 9263, 9662
Andrews, C.W. see Rendle, A.B.
Anguillara, L. see Micheli, P.A.
Antommarchi, F. 9723
Aosta v. Hélène Duchess of Aosta
Appel, P. 8071
Arago, D.F.J. 7096
Arbeláez, E. Pérez v. Pérez Arbeláez, E.
Arcangeli, G. 7361, 9933
Archer, W.H. see McCoy, F.
Ardene, J.P.R. d' v. d'Ardene, J.P.R.
Aregelia see Regel, A. von
Argüelles, E. Quisumbing y v. Quisumbing, E.
Aristoteles 5927; see Nees von Esenbeck, C.G.D.
Armari, B. 7969
Arnold, F.C.G. 6941, 6945, 6948, 6960; see Perktold, J.A., Richard, O.J., Richardson, J., Rink, H.J., Ripart, J.B.M.J.S.E.
Arnold. T.C.G. see Rehm, H.S.L.F.F.
Arnott, G.A.W. 9420; see Norman, G., Rylands, T.G.
Arruda da Camara, M. 7942
Arsène, Frère G. see Rosenstock, E.
Artis, E.T. 5368
Ascherson, P.F.A. 8211, 9454-9456, 9817; see Pound, R., Reinhardt, O.W.H., Rohlfs, G.F., Roth, E.C.F., Ruhland, W.O.E., Ruhmer, G.F.

Ashe, W.W. 7941
Askelöf, J.C. 9087
Asplund, E. *see* Regnell, A.F.
Asso, I. de 9403
Asso e del Rio, I.J. de *see* Pourret, P.A.
ASTROLABE *see* Richard, A.
AUBE *see* Raoul, É.F.L.
Aubert *see* Oldham, T.
Aublet, J.B.C.F. *see* Monti, G.L., Rousseau, J.J.
Aubreginea see Pellegrin, F.
Aubréville, A. *see* Pellegrin, F.
Aubriet, C. *see* Plumier, C.
Aubry, J. 9686
Audibert de Ramatuelle, T.A.J. d' *v.* Ramatuelle, T.A.J.d'Audibert de
Audubon, J.J.L. *see* Meisel, M., Rafinesque-Schmaltz, C.S.
Auerswald, B. 8445
Augusta see Saint-Hilaire, A.F.C.P. de
Austin, C.F. 7586
Austroplenckia see Plenck, J.J. von
Autran, E.J.B. 8190
Avetta, C. 7435
Avillino, G. 7426

Babbage, B.H. 6398
Babington, C.C. *see* Pallas, P.S.
Bachmann, A.Q. *v.* Rivinus, A.Q.
Backer, C.A. *see* Posthumus, O.
Badé, W.F. 6539
Badillo, V. *see* Pittier, H.F.
Bagnall, J.E.8714
Bahi, J.F. 8058
Bahr, H. 8807
Baikie, W.B. *see* Mann, G.
Bailey, L.H. *see* Mendel, G.J., Rehder, A., Rodgers, A.D.
Baillet, C. 9338
Baillon, H.E. 7560, 7564, 9110; *see* Mueller Argoviensis, J.
Baines, T. 7556
Baird, S.F. *see* Rothrock, J.T.
Baker, C.F. 8830, 9932
Baker, E.G. 9065, 9067-9071, 9210
Baker, F.G. *see* Ricasoli, V.
Baker, J.G. *see* Porter, L.E.
Baker, L.E. *v.* Porter, L.E.
Bakhuizen van den Brink, R.C., Jr. 8393; *see* Pulle, A.A.
Balansa, B. *see* Müller, Karl (Halle)
Balbis, J.B. 9432; *see* Roffavier, G.
Balde, J. 9784
Baldwin, W. *see* Rafinesque-Schmaltz, C.S.
Balfour, I.B. 9957
Balfour, J.H. 6565; *see* Pringsheim, N., Sadler, John
Ball, C.R. 7255
Bally, W. 9234

Balta, J. 8595
Bamler, G. *see* Rosenstock, E.
Banerjee, U.C. *see* Rollins, R.C.
Bang, M. 9819
Banister, J. 8701
Banks, J. 5038, 6078, 7344, 7728, 7874, 8399, 9720, 9769, 9828; *see* Loureiro, J. de, Marsden, W., Menzies, A., Miller, P., Mitchell, J. (1711-1768), Pallas, P.S., Parkinson, Sydney, Paterson, W., Phipps, C.J., Plumier, C., Pourret, P.A., Pursh, F.T., Roxburgh, W., Russell, Alexander (*ca.* 1715-1768), Russell, P.
Baratte, J.F.G. 7475
Barbey, W. *see* Reuter, G.F.
Barbier, J.-A. *see* Pourret, P.A.
Barbosa see Barbosa Rodrigues, J.
×*Barbosaara see* Barbosa Rodrigues, J.
Barbosa Rodrigues, J. *9341-9361*
Barbosella see Barbosa Rodrigues, J.
Barb. Rodr. v. Barbosa Rodrigues, J.
Barbrodria see Barbosa Rodrigues, J.
Barkoudha, Y.I. *see* Rechinger, K.H.
Barnes, C.R. *see* Röll, J.
Barnhart, J.H. 5802, 10.064; *see* Nash, G.V., Rickett, H.W.
Bärnkron, G. Pernhoffer von *v.* Pernhoffer von Bärnkron, G.
Barral, J.A. 9110
Barrande, J. 8224
Barratte, J.F.G. 7475
Barreiro, A.J. 9782
Barrera, P. de 9679; *see* Pourret, P.A.
Barrington, R.M. *see* More, A.G.
Barrotia see Parrot, J.J.F.W.
Bartling, F.G. 8293; *see* Reinke, J.
Barton, B.S. *see* Pursh, F.T.
Barton, W.P.C. 8562; *see* Pursh, F.T.
Bartram, E.B. 9469
Bary, H.A. de *see* Murray, C.R.M., Oltmanns, F., Pfeffer, W.F.P., Rabenhorst, G.L., Rosen, Felix, Rostafiński, J.T., Roth, E.C.F.
Bataille, F. 8419
Batalin, A.F. *see* Regel, E.A. von
Bâthie, Perrier de la *v.* Perrier de la Bâthie
Bathiea see Perrier de la Bâthie, H.
Bathiorhamnus see Perrier de la Bâthie, H.
Battaleni, F. 6305
Bauchman, T.C. *v.* Rickett, T.C.
Bauer, E. 4564
Bauer, F.A. 7410, 8349
Bauer, J.C. *see* Regnell, A.F.
Bauhin, C. 5230, 10.072
Baur, E. 9076, 9520
Bavoux, E.V. 518; *see* Billot, P.C., Paillot, J.
Baxter, W. *see* Sabine, J.

INDEX TO NAMES

Bay, J.C. see Prantl, K.A.E.
Bazille, L. 8024
Beach, S. see Pulteney, R.
Bean, T.H. see Rothrock, J.T.
Beatson, A. 9723; see Roxburgh, W.
Beattie, R.K. 7944, 7949, 7950
Beauv., P. v. Palisot de Beauvois, A.M.F.J.
Beauchamp, W.M. 8406
Beauharnais, E. de 8747
Beauharnais, J. de see Redouté, P.J.
Beaumont, D. 8751
Beauverd, G. 9710
Beauvois, A.M.F.J. Palisot de *v.* Palisot de Beauvois, A.M.F.J.
Bebb, M.S. 7507, 9658
Beccari, O. 7684, 8726, 9325
Beck, G. 7423, 8211, 8885; see Pantocsek, J., Pascher, A., Rudolph, K.
Becker, W.G. 8403
Beckmann, J. 4829
Bedford, Duke of *v.* Russell, J.
Bedfordia see Russell, J.
Bègue de Presle, Le *v.* Le Bègue de Presle
Béguinot, A. 9945
Behaghel von Adlerskron, M. 7402
Beilschmied, C.T. 8351
Beissner, L. 8775
Bela, Pater 8268
BELGICA 9691
Bellar, G. 6784
Belleval, P. Richer de *v.* Richer de Belleval, P.
Bellynck, A.A.A.A. see Morren, C.J.É.
Belvisia see Palisot de Beauvois, A.M.F.J.
Benacense, E. 8142
Bennett, A.W. 9954
Benoist, R. 7704; see Russell, Alexander (*ca.* 1715-1768)
Bentham, G. 6242, 6739, 7194, 9150; see Nees von Esenbeck, C.G.D., Regnell, A.F., Reichenbach, H.G.
Benzon, R.J.D. von Fischer *v.* Fischer Benzon, R.J.D. von
Berchtold, F. 7835; see Presl, J.S.
Berens, E. 5728
Berg, C.C. 8393
Berg, E. 8353
Bergerie, J.B. Rouzier de la *v.* Rouzier de la Bergerie, J.B.
Berggrén, G. 9090
Berggren, S. 6865
Berghen, C. Vanden *v.* Vanden Berghen, C.
Bergsma, C.A. see Miquel, F.A.W.
Berkeley, M.J. see Massee, G.E.
Berlese, A.N. 9894-9895, 9898, 9902, 9906-9907, 9913
Berlin, A. 6859

Bernard, C.J. 8224
Bernhardi, J.J. see Presl, J.S.
Bernoulli, C.G. 6516
Berry, M.C. de see Redouté, P.J.
Bertelsen, H. 7584
Berthelot, S. 7374
Bertini, G. 8376
Bertoni, M. see Pittier, H.F.
Bescherelle, É. 7475; see Müller, Karl (Halle), Roze, E.
Besemann, C.A. 7724, 7728, 9757
Bessa, P. 4960, 5967, 9140
Bessey, C.E. 8947; see Pool, R.J., Rydberg, P.A.
Betche, E. 8046
Beyer, R. 8211
Bianchi, C. 4794
Bigelow, J. 8570
Bijlaart, J.J. 7223
Bilafingen, F. Rot von S. zu I. und *v.* Rot von Schreckenstein, F.
Bilberg, G.J. 7247
Billot, P.C. 7697; see Paillot, J.
Biot, J.B. 9741a
Bird, A. 9684
Bisby, G.R. see Petch, T.
Bishop, B.P. *v.* Pauahi, B.
Bismarck see Rohlfs, G.F.
Bitter, F.A.G. see Rosendahl, F.
Bizzozero, G. 9897, 9899
Black, J. 8685
Black, J.M. see Rogers, R.S.
Blanchard, E. 10.033
Blanche, C.I. 8378
Blanco, F.M. 4881
Blankinship, J.W. 9857
Blume, C.L. 6685, 6687, 8962; see Reichenbach, H.G.
Blunt, W. 4492
Blytt, A.G. 9619
Bobart, J. 6334
Boccone, P. 9943
Boddaert, P. 7222
Boeckeler, J.O. 7759; see Rutenberg, D.C.
Børgesen, F.C.E. 7529, 8660
Boerhaave, H. 7302; see Plumier, C., Richardson, R.
Boirin 9015, 9041
Bois, D.G.J.M. 7190, 7775
Boissier, P.E. 8191, 8735, 8990; see Paiche, P., Poeppig, E.F., Pritzel, G.A., Reuter, G.F.
Boivin 9041
Bolle, C.A. 7759
Bolton, J. 8973
Bolus, H. 8132
Bomansson, J.O. 9979
Bommer, E.C. 9691; see Rousseau, M.
Bommer, J.É. 7978, 9691

Bonaparte, R.N. Prince see Rouy, G.C.C.
Bondt, J. de see Piso, W.
Bonelli, G. see Sabbati, L.
Bonnier, E. 9162
Bonpland, A.J.A. 9408, 9771; see Peck, W.D., Roemer, J.J.
Booth, T.J. see Nuttall, T.
Boothby, B. 8834
Boott, W. 9658
Bor, N.L. see Rechinger, K.H.
Borbás, V. von see Porcius, F., Richter, Aladár
Borchgrevink 7589
Bordes, J. 9498
Borg, J. 9948
Borge, O.F.A. 7423
Borgen, M.G. 7587, 7589
Bornet, J.-B.É. see Mazé, H.P.
Bornmüller, J.F.N. 7284
Borszczow, E.G. 8791
Bos, J. Ritzema v. Ritzema Bos, J.
Bosc, L.A.G. 9741a
Boselli, E. 7969
Bosio, J.C.B. Mutis y v. Mutis y Bosio, J.C.B.
Bossu 4921
Boterenbrood, M.J.A. 8393
Both, F. 8293, 8303
Bottelier, H.P. 8393
Bottini, A. 9933
Bouché, C. 8775
Boudier, J.L.É. 9177
Boulay, J.N. 9706
Boulaye, de la 7228
Boulger, G.E.S. 7709
Bouly de Lesdain, M. see Parrique, G., Mouret, M.
Bouquet de la Grye 9663
Bowie, J. 8350
Boy de la Tour, J. v. Boy de la Tour, M.C.
Boy de la Tour, M.C. see Rousseau, J.J.
Bracelin, Mrs. H.P. see Mexia, Y.E.J.
Brackenridge, W.D. see Pickering, C.
Brade, A.C. see Rosenstock, E.
Brade, Alfr. see Rosenstock, E.
Bradley, R. see Pfeiffer, L.K.G.
Bradley, W.L. 8815-8816
Brand, A. 7684
Brandt, J.F. von 7875; see Ratzeburg, J.T.C.
Branstorr, R.E. 9304
Braun, A.C.H. 7759, 7787, 8507, 9454; see Peters, W.C.H., Petri, F., Rabenhorst, G.L., Reinhardt, O.W.H., Rohrbach, P., Rossmann, J.
Braun, H. 8723
Braun, S. 8775
Braun-Blanquet, J. see Rübel, E.A.
Braunschweig, Duke of see Pott, J.F.

Brébisson, L.A. de see Roumeguère, C.
Brefeld, J.O. see Rosen, Felix
Bremekamp, C.E.B. 8393
Bremekamp, E. 8393
Bremi-Wolf, J.J. 8775
Brereton, J.A. 9137
Bresadola, G. 8726, 9910, 9924; see Romell, L.
Breuil, M.A. du 9248
Breutel, J.C. see Reichenbach, H.G.L.
Brewer, S. see Richardson, R.
Breyne, J.P. 9120
Bridel, S.-E. de see Lindberg, S.O., Necker, N.J. de, Nees von Esenbeck, C.G.D.
Brieger, F.G. see Pax, F.A.
Brienne, Loménie de v. Loménie de Brienne
Briquet, J.I. 9109, 10.095; see Peragallo, M., Pitard, C.-J.M.
Britten, J. 9066-9067; see Oliver, D., Rendle, A.B.
Britton, E.G. 9813; see Mitten, W., Röll, J.
Britton, N.L. 5170, 6772, 7944, 9813-9814, 9817; see Palmer, E.J., Riddle, L.W., Rose, J.N., Rydberg, P.A.
Brittonrosea see Rose, J.N.
Brizi, U. see Pirotta, P.R.
Broca 7911
Brockhaus, F.A. see Morren, C.J.É.
Brockmann-Jerosch, H. 9760
Brodriguesia see Barbosa Rodrigues, J.
Brody, G.A.O. St. v. Saint Brody, G.A.O.
Brondeau, L. de 10.017
Brongniart, A.T. 5500, 10.029
Bronn, H.G. see Roemer, F. von
Brooks, W.E. St. John v. St. John Brooks, W.E.
Brotherus, V.F. 7331, 8726, 9979, 9981; see Lindberg, S.O., Loeske, L., Paris, É.G., Renauld, F.F.G., Roth, G., Sakurai, K.
Broussonet, P.M.A. see Pourret, P.A.
Brouwer, J. see Pulle, A.A.
Brown, A. see Palmer, E.J.
Brown, N.E. see Oliver, D.
Brown, R. 7410, 9150, 9169; see McNab, W.R., Mitchell, T.L., Pallas, P.S., Persoon, C.H., Pohl, J.B.E., Ross, J.C., Roxburgh, W., Sabine, E.
Brown, R. (1842-1895) 9245; see Sadler, John
Brown, W.L. 7125
Bruce, J. see Rüppell, E.
Brugmans, S.J. see Reinwardt, C.G.C.
Bruletout de Préfontaine, M. v. Préfontaine, M. Bruletout de
Brunete, J. 9771

1163

INDEX TO NAMES

Brunfels, O. 9875
Brunnthaler, J. 7423
Bruyn, J.G. de 5573
Bryoporteria see Porter, C.E.
Bubák, F. *see* Rohlena, J.
Bubani, P. *see* Penzig, A.J.O.
BUCCANEER *see* Rattray, John
Buchanan, J. *see* Rawson, R.W.
Buchenau, F.G.P. *see* Rutenberg, D.C.
Büchner, E.W.G. 4931
Buchoz, P.J. 4828
Buchtien, O. *see* Rosenstock, E.
Buchwald, N.F. 9621
Buck, M.W. 7416
Buck, P. 8405
Buffon, G.L.L. de 7096; *see* Ramée, S.H. de la, Rousseau, J.J.
Buia, A. 8360
Bulliard, J.B.F. 5052, 9153; *see* Quélet, L.
Bullo, G.S. *see* Paoletti, G.
Bunge, A.A. von 9835
Burgerstein, A. 8726
Burgsdorff, F.A.L. von 8971-8972
Burkill, I.H. 8266
Burman, J. 5787, 8069, 9727, 9784-9785; *see* Plumier, C.
Burman, N.L. *see* Linnaeus, C.
Burman, P. 9784
Burnat, E. 5085, 5375, 6393, 8169, 10.072; *see* Michalet, E.
Burshia see Pursh, F.T.
Burtt, B.L. *see* Rechinger, K.H.
Buscalioni, L. 7970
Buschler, R. 7970
Bush 8024
Butcher, F.D. 7257
Bute, J.S. Earl of *see* Plumier, C.
Butters, F.K. 9522-9527; *see* Rosendahl, O.
Butze, H. 8090

Cadet de Gassicourt, L.C. 9741a
Cadet de Vaux, A.A.F. 9741a
Caldesi, L. 7361
Call, R.E. 8548
Callisthenes *see* Martius, C.F.P. von
Camara, M. Arruda da *v.* Arruda da Camara, M.
Cambessèdes, J. 10.032, 10.034
Camerarius, E.R. *see* Rudolph, K.
Campbell, W.S. 5258
Camus, E.G. 9706
Camus, P.X. *v.* Philippe, X.
Canal von Hochberg, J.M. de 8102; *see* Pohl, J.B.E.
Candolle, A.C.P. de 7684, 7978; *see* Micheli, M., Planchon, J.É.
Candolle, A.L.P.P. de 4934, 7372, 9151, 10.074; *see* Macreight, D.C., Malin-vaud, L.J.E., Meisner, C.F., Micheli, M., Miquel, F.A.W., Moquin-Tandon,C.H.B.A., Mueller Argoviensis, J., Nees von Esenbeck, C.G.D., Planchon, J.É., Pritzel, G.A., Reuter, G.F.
Candolle, A.P. de 5024, 5205, 5497, 6225, 6739, 7250, 8304, 8353, 8578, 8711, 8747, 8885, 9104, 9144, 9408, 9423, 9679; *see* Mueller Argoviensis, J., Prost, T.C., Puerari, M.N., Raddi, G., Rafinesque-Schmaltz, C.S., Redouté, P.J., Regel, E.A. von, Reuter, G.F., Roche, F. de la, Roeper, J.A.C., Sagra, R. de la
Cantley, Capt. 9734
Canu, J.D.E. 7215
Cappelletti, C. 9945
Capua, E. di 7969
Carano, E. *see* Pirotta, P.R.
Carbonnières Ramond, L.F.É. de *v.* Ramond, L.F.É. deCarbonnières
CARCASS *see* Phipps, C.J.
Cardew, R.M. 9210
Cardot, J. 7926-7927, 7961, 8985-8987, 8990-8991, 8995-8997, 9933; *see* Renauld, F.F.G.
Carestia, A. 9924
Carey, J. *see* Nicholson, G.
Carey, W. 9724-9725
Cario, O.R. 6516
Carolath, Princess of *see* Pfeil, F.W.L.
Carruthers, W. 9065
Carter, H.G. 8664
Caruel, T. 7361, 7376; *see* Micheli, P.A., Pokorny, A.
Carus, C.G. *see* Reichenbach, H.G.L.
Casearius, J. 9123-9124
Cash, E.K. 7501, 9898
Caso, B. 5701, 8711
Casoni, V. 7435
Caspar, S.J. 7423
Caspary, J.X.R. 7030, 8211; *see* Luerssen, C., Retzdorff, A.E.W., Rutenberg, D.C.
Cassebeer, J.H. *see* Pfeiffer, L.K.G.
Castagne, É.E.E.J. 9314
Castagne, J.L.M. *see* Robert, G.N.
Catesby, M. *see* Panzer, G.W.F., Rafinesque-Schmaltz, C.S.
Catherine II, the Great, of Russia 7227; *see* Pallas, P.S.
Cattell, P. *see* Royen, A. van
Caub, J.W. von *see* Pritzel, G.A.
Caumels, de 9715
Cavanilles, A.J. 9769-9770, 9776; *see* Pourret, P.A.
Cecil, A.M. *v.* Rockley, A.M.
Cecil, E. *v.* Rockley, E.
Cejp, K. 7935
Cervi, J. 6085a

Cesati, V. de 5696; *see* Pasquale, G.A., Passerini, G., Rion, A.
Chabert 9741a
Chaetosaccardinula see Saccardo, P.A.
Chaix, D. *see* Pourret, P.A.
Chambon 9741a
Chamisso, L.A. von *see* Mertens, K.H.
Chandler, M.E.J. 8905-8906
Chandra, K.D. 8263
Chanteloup, J.A.C. Chantal de *v.* Chaptal de Chanteloup, J.A.C.
Chapeckia see Peck, C.H.
Chaptal 5959
Chaptal de Chanteloup, J.A.C. 9741a
Charcot, J. 7666
Charles, V.K. 7500
Charpentier de Cossigny, J.F. *v.* Cossigny, J.F. Charpentier de
Charpin, A. *see* Mouterde, P.
Chase, M.A. *see* Palisot de Beauvois, A.M.F.J., Parodi, L.R., Perkins, J.R., Philippi, R.A., Pilger, R.K.F.,Pittier, H.F., Poeppig, E.F., Pohl, J.B.E., Poiret, J.L.M., Poiteau, P.A., Raddi, G., Raeuschel, E.A., Rafinesque-Schmaltz, C.S., Raspail, F.V., Regnell, A.F., Reiche, K.F., Richard, L.C.M., Riedel, L., Roemer, J.J., Ruprecht, F.J., Saint-Hilaire, A.F.C.P. de
Chassiron, de 9741a
Chaubard, L.A. 10.017; *see* Saint-Amans, J.F.B. de
Chaumeton, F.P. 8115, 9147
Chaumontel 9741a
Chauvin, F.J. 9274a; *see* Roberge, M.R.
Cheeseman, T.F. *see* Maiden, J.H., Richter, L.
Chermezon, H. 7704; *see* Perrier de la Bâthie, H.
Chevalier 9741a
Chiarelli, F.P. 8552
Chiarelli, J. 8552
Chiarelli, S. 8552
Chilton, C. *see* Petrie, D.
Chiovenda, E. 7968-7969, 9575
Chirac, L. 8023
Chirac, P. *see* Magnol, P.
Chodat, R.H. 7259; *see* Pampanini, R., Rayss, T.
Chomette, A, 8418
Choné, O. 8775
Choudhury, D.N. 8263
Choux, P. 7704
Christ, K.H.H. 7978, 8211, 8915
Christensen, C.F.A 7704
Christiansen, C. *v.* Raunkiaer, C.C.
Christiansen, M.P. 9548
Christina, Queen of Sweden *see* Rauwolff, L.

Christmann, G.F. 7301; *see* Panzer, G.W.F.
Chun, C. 8912
Ciaston, E. 8461
Cittadella-Malduro 9485
Clairville, J.P. de 9689
Clark, C. 7258
Clarke, C.B. 9725; *see* Rutenberg, D.C.
Clarke, W.R. 4776
Clementea see Roxas Clemente y Rubio, S. de
Clemente y Rubio, S. de Roxas *v.* Roxas Clemente y Rubio, S. de
Clements, F.E. 8240, 9522-9523, 9525-9526, 9847, 9898
Clercq, F.S.A. de 8391
Clifford, G. *see* Murray, J.A.
Clifford, J.D. 8574
Clos, D. 8353, 9117
Clusius, C. 8704, 8847; *see* Neilreich, A.
Cobelli, R. 9895
Cockayne, L. *see* Prain, D.
Cockerell, T.D.A. *see* Rydberg, P.A.
Cocks, R.S. *see* Rafinesque-Schmaltz, C.S.
Codd, L.E.W. 8132
Coder 9679
Cogners, V. Demusset de *v.* Demusset de Cogners, V.
Cogniaux, C.A. 7958, 9996; *see* Barbosa Rodrigues, J., Rutenberg, D.C.
Cohn, F.J. 8443, 8775; *see* Pax, F.A., Rosen, Felix, Sadebeck, R.E.B.
Cohn, P. *see* Rosen, Felix
Cohrs, A. 8187a
Colaneri, M. 7608
Cole, G.A.J. *see* McArdle, D.
Coleman, W.H. 8372
Collan, K. *v.* Linkola, K.
Collander, R. 9982
Collin, E.B. 8003
Collins, J. 8064
Colmeiro y Penido, M. 7512, 9717
Coltman-Rogers, C. *v.* Rogers, C.C.
Commelijn, J. 9123; *see* Ruysch, F.
Constans, L. 7556
Cook, C.D.K. *see* Rechinger, K.H.
Cook, J. *see* Nelson, D., Pallas, P.S., Parkinson, Sydney, Pourret, P.A., Roxburgh, W., Rumphius, G.E., St. John, H.
Cook, W.S. 7258
Cooke, M.C. 5656, 8401, 9595, 9898; *see* Persoon, C.H., Phillips, W., Plowright, C.B., Quélet, L., Ravenel, H.W., Rostafiński, J.T.
Cooley, W.D. 7402
Copeland, E.B. 7684
Corbière, L. 7971-7972
Corda, A.K.J. 8307-8308, 9098

INDEX TO NAMES

Corrêa da Serra, J.F. 4644
Corry, T.M. see Praeger, R.L.
Cortesi, F. 7969; see Pirotta, P.R.
Cosimo II de Medici see Rumphius, G.E.
Cosimo III de Medici 5475
Cossigny, J.F. Charpentier de 9741a
Cosson, E.St.-C. 9454; see Michalet, E., Moquin-Tandon, C.H.B.A., Poiret, J.L.M., Pomel, A.N., Reboud, V.C., Rohlfs, G.F., Roux, N.
Costa, G. Dalla v. Dalla Costa, G.
Coste, H.J. 9114
Cotta, J.F. see Nees von Esenbeck, C.G.D.
Cotte 9741a
Coulter, J.M. 8180, 9512, 9817; see Rodgers, A.D., Rose, J.N.
Courcière, P. 8242
Courthope, P. see Ray, J.
Coville, F.V. 6729, 9817
Coxe, W. see Ramond, L.F.É. de Carbonnières
Cramer, C.E. see Nägeli, C.W. von
Cramer, J. 8393
Cratty, R.I. 7258
Crépin, F. 9131
Crevel, R. van see Ruthe, J.F.
Croasdale, H.T. see Prescott, G.W.
Crouan, H.M. see Picquenard, C.-A.
Crouan, P.L. see Picquenard, C.-A.
Csiki, E. 8726
Cuboni, G. 5335, 8268, 9898
Cuisin, C.E. 5279, 7654; see Révoil, G.
Cullen, J. 9314
Cunningham, A. see Mudie, R., Preiss, L.
Cunningham, J.C. 7257
Cupani, F. 8552
Curaudau 9741a
Curtis, C.M. 9734
Cyr, D.N. Saint v. Saint-Cyr, D.N.
Czapek, F.J.F. see Rudolph, K.
Czurda, V. 7423

Da Camara, M. Arruda v. Arruda da Camara, M.
Dageförde 8775
Dahl, A.O. 9527
Dahl, E. 5131
Dahlberg, C.G. see Rolander, D.
Dahle 7589
Dahlgren, B.E. 9782
Dainelli, G. 7272
d'Albertis, E. 7899, 7901
Dale, S. see Ray, J.
Dalla Costa, G. 9950
Dalla Costa, H. v. Dalla Costa, G.
Dalla Torre, K.W. von see Perktold, J.A.
Dalman, G. 8191
d'Almeida Pinto, Z. 7942

Dalton, J. see Robson, E.
Dammer, C.L.U. 8800, 8805; see Regel, E.A. von
Dana, Mrs. W.S. v. Parsons, F.T.
D'Ancona, N. 9913
Dandy, J.E. see Rechinger, K.H.
Danielia see Roche, D. de la
Dannenfeldt, J.C. Pittoni von v. Pittoni, J.C.
Dardana, J.A. 7896
d'Ardene, J.P.R. 10.092
Darin, L.J. 9094
Darolles, E. 8631
Darwin, C.R. 6337, 8675, 9957; see Mac Leod, J., Moe, N.G., Müller, H.L.H.,Nägeli, C.W. von, Naudin, C.V., Römer, J.
Darwin, E. 4799, 8834
Darwin, F. see Müller, H.L.H.
Das, G.C. 8258
Da Serra, J.F. Corrêa v. Corrêa da Serra, J.F.
Da Silva, A.R. Pinto v. Pinto da Silva, A.R.
Da Silva, Quintanilha, A.P. v. Quintanilha, A.P. da Silva
Dass, K.P. 8263
Dassen, M. see Miquel, F.A.W.
Dassier, A. 6908
Daubrée, L. 7323
d'Audibert de Ramatuelle, T.A.J. v. Ramatuelle, T.A.J.d'Audibert de
Davenport, G.E. 7507
Davies, T. 8064
Davis see Nash, G.V.
Davis, B. 8189a
Davis, H.B. see Pennell, F.W.
Davis, J. 8561
Davis, W.M. 8623
De Almeida Pinto, J. v. Pinto, J. de Almeida
De Asso e del Rio, I.J. v. Asso e del Rio, I.J. de
De Barrera, P. v. Barrera, P. de
De Beauharnais v. Beauharnais, de
De Beauvois, A.M.F.J. Palisot v. Palisot de Beauvois, A.M.F.J.
Debeaux, J.O. 7511, 10.017
De Belleval, P. Richer v. Richer de Belleval, P.
De Berry, M.C. v. Berry, M.C. de
De Bondt, J. v. Bondt, J. de
De Brienne, Loménie v. Loménie de Brienne
Deby, J.M. 7620
Decaisne, J. see Mirbel, C.F.B. de, Roeper, J.A.C.
De Canal von Hochberg, J.M. v. Canal von Hochberg, J.M. de

INDEX TO NAMES

De Carbonnières Ramond, L.F.É. *v.* Ramond, L.F.É. deCarbonnières
De Caumels *v.* Caumels, de
De Chanteloup, J.A.C. Chaptal *v.* Chaptal de Chanteloup, J.A.C.
De Clairville, J.P. *v.* Clairville, J.P. de
De Clercq, F.S.A. *v.* Clercq, F.S.A. de
De Cogners, V. Demusset *v.* Demusset de Cogners, V.
De Cossigny, J.F. Charpentier *v.* Cossigny, J.F. Charpentier de
De Dieudonné, O.F.C.M.J. *v.* Dieudonné, O.F.C.M.J. de
De Fabrega, H.F. Pittier *v.* Pittier, H.F.
De Febres, Z. Luces *v.* Luces de Febres, Z.
Defeugré 9741a
De Filippe, F. *v.* Filippi, F. de
Defoe, D. *see* Robinson, B.L.
De Gassicourt, L.C. Cadet *v.* Cadet de Gassicourt, L.C.
Degen, A. von *see* Porcius, F.
De Girard, F. *v.* Girard, F. de
De Girardin, Mlle *v.* Girardin, Mlle de
De Grandmaison, L.A.M. *v.* Grandmaison, L.A.M. de
De Hoop Scheffer, J.G. *v.* Hoop Scheffer, J.G. de
De Hugel, C. *v.* Hugel, C. de
Deichmann-Branth, J.S. 9608; *see* Rink, H.J.
De Ippolito, S. *v.* Ippolito, S. de
Dekin, A. 7438
De la Bâthie, Perrier *v.* Perrier de la Bâthie
De la Bergerie, J.B. Rouzier *v.* Rouzier de la Bergerie, J.B.
De Laborde, B. *v.* Laborde, B. de
De la Boulaye *v.* Boulaye de la
De la Lachenal, W. v. Lachenal, W. de
De la Escalera, F. Martínez *v.* Martínez de la Escalera, F.
De la Gautrois, J.F.C.L.C. Petit *v.* Petit, J.F.C.L.C.
De la Grye, Bouquet *v.* Bouquet de la Grye
De Lahitte 9429
Delalause 9741a
Delamare, E. 8986
De Lamoignon de Malesherbes, C.G. *v.* Malesherbes, C.G. deLamoignon
De la Peyronie, Gauthier *v.* Gauthier de la Peyronie
De la Puerta y Rodénas, G. *v.* Puerta y Rodénas, G. de la
De la Ramée, S.H. *v.* Ramée, S.H. de la
Delaroche *v.* Roche, de la
De la Roche *v.* Roche, de la
De la Sagra, R. *v.* Sagra, R. de la
De la Sauvage *v.* Sauvage, de la

De Lasteyrie-Dusaillant, C.P. *v.* Lasteyrie-Dusaillant, C.P. de
De la Tour, M.C. Boy *v.* Boy de la Tour, M.C.
De la Varde, R.A.L. Potier *v.* Potier de la Varde, R.A.L.
Delavay, P.J.M. 7454
De Leon, J. Ponce *v.* Ponce de Leon, J.
De Leon y Aimé, A. Ponce *v.* Ponce de Leon y Aimé, A.
De Lesdain, M. Bouly *v.* Bouly de Lesdain, M.
Delessert, J.P.B. 4717, 8353, 9769; *see* Noroña, F., Palisot de Beauvois, A.M.F.J., Quer y Martinez, J., Richard, L.C.M., Rousseau, J.J.
Delessert, M.-M. *see* Rousseau, J.J.
Delessert, Mme É. *v.* Boy de la Tour, M.C.
Delile, A.R. 6196, 8747; *see* Michaux, A., Prost, T.C.
Delile, E. 10.034
De Linares *v.* Linares, de
Delitsch, O. *see* Petermann, W.L.
De Long, G.W. 6539
De Longchamp, P.B. Pennier *v.* Pennier de Longchamp, P.B.
Delpy, E. 7923, 7923a
Del Rio, I.J. de Asso e *v.* Asso e del Rio, I.J. de
De Luigné, S. *v.* Luigné, S. de
De Malesherbes, C.G. de Lamoignon *v.* Malesherbes, C.G. de Lamoignon
De Malzine, O. de P.A.H.R. *v.* Malzine, O. de P.A.H.R. de
Demanet, F. 9074
Demaret, F.M.H. 9314
Demarquet, E. 10.022
De Martinis, B. *v.* Martinis, B. de
Demelius, P. 8726
De Menezes, C.A. *v.* Menezes, C.A. de
Demidof, P.A. 7226
Démidoff, A. de 8013
Demusset de Cogners, V. 9741a
De Nangis Regnault, G. *v.* Nangis Regnault, G. de
De Nemours, P.S. Dupont *v.* Dupont de Nemours, P.S.
De Notaris, G. 5598; *see* Pedicino, N.A.
De Oliveira, P. Erichsen *v.* Erichsen de Oliveira, P.
De Otero, L. Pagán *v.* Pagán de Otero, L.
De Palézieux, P. *v.* Palézieux, P. de
De Parseval-Grandmaison, J. *v.* Parseval-Grandmaison, J. de
De Perrier de la Bâthie, E.P. *v.* Perrier de la Bâthie, E.P. de
De Poederlé, E.J.C.G.H. d'Olmen *v.* Poederlé, E.J.C.G.H. d'Olmen

INDEX TO NAMES

De Pouzolz, P.C.M. *v.* Pouzolz, P.C.M. de
De Préfontaine, M. Bruletout *v.* Préfontaine, M. Bruletout de
De Presle, Le Bègue *v.* Le Bègue de Presle
De Pronville, A. *v.* Pronville, A. de
De Puydt, E. *v.* Puydt, E. de
De Puymaly, A.H.L. *v.* Puymaly, A.H.L.
De Rawton, O. *v.* Rawton, O. de
De Reboul, E. *v.* Reboul, E. de
De Reygadas, Mrs. A.A. *v.* Mexia, Y.E.J.
De Rochebrune, A.T. *v.* Rochebrune, A.T. de
De Rooij, M.J.M. 8393
De Rosny, L.L.L. Prunol *v.* Rosny L.L.L. Prunol de
De Rouville, P.G. *v.* Rouville, P.G. de
De Roxas Clemente y Rubio, S. *v.* Roxas Clemente y Rubio, S. de
De Saint-Amans, J.F.B. *v.* Saint-Amans, J.F.B. de
De Saint-Germain, J.J. *v.* Saint-Germain, J.J. de
De Saint Hilaire, A.F.C.P. *v.* Saint-Hilaire, A.F.C.P. de
De Saint-Léon, V.B.A. Trevisan *v.* Trevisan de Saint-Léon, V.B.A.
De Saint-Moulin, V.J. *v.* Saint-Moulin, V.J. de
De Saint-Sauveur, J. Grasset *v.* Grasset de Saint-Sauveur, J.
De Saint-Simon, M.H. *v.* Saint-Simon, M.H. de
De Salvert, Dutour *v.* Dutour de Salvert
De Saussure, H.B. *v.* Saussure, H.B. de
Descartes, R. *see* Rudbeck, O.J.
Descurain 9739
Desfontaines, R.L. 4829, 7539, 10.029; *see* Meneghini, G.G.A., Poiret, J.L.M., Pourret, P.A., Ramond, L.F.É. de Carbonnières, Redouté, P.J., Reichenbach, H.G.
De Sloover, J.-L. 9314
Desmazières, J.B.H.J. *see* Roberge, M.R.
De Tavera, T.H.J. Pardo *v.* Pardo de Tavera, T.H.J.
Déterville *see* Parmentier, A.A.
De Thou, J.-A. *v.* Thou, J.-A. de
De Thun, F. *v.* Thun, F. de
De Toni, G.B. 8282, 9898, 9907; *see* Paoletti, G., Piccone, A.
De Varigny, H. *v.* Varigny, H. de
De Vaux, A.A.F. Cadet *v.* Cadet de Vaux, A.A.F.
De Vichet, J. *v.* Vichet, J. de
Deville, A. 9687
De Vilmorin, P.L.F.L. *v.* Vilmorin, P.L.F.L. de

De Wildeman, É.A.J. 7778
DeWolf, P. 8393
De Zigno, A. *v.* Zigno, A. de
d'Héricourt, C.L.X. Rochet *v.* Rochet d'Héricourt, C.L.X.
Di Capua, E. *v.* Capua, E. di
Dichl, W.W. 7501
Dickson, J. *see* Roemer, J.J., Rudge, E.
Dickson, O. 6859
Dieck, G. 8775
Dieffenbach, C.E. *see* Mueller, J.B.
Diels, F.L.E. 7704, 8393; *see* Pritzel, E.G.
Dierbach, J.H. 7735
Dieterich, C.F. 4829
Dietrich, A. *see* Reichenbach, H.G.
Dietrich, D.N.F. 7820; *see* Poselger, H.
Dieudonné, O.F.C.M.J. de *see* Ripart, J.B.M.J.S.E., Sachs, J. von
Dilenius, J.J. 8065
Dillon, R.Q. *v.* Quartin-Dillon, R.
Dinklage, M.J. 9969
Dinsmore, J.E. 8191
Dioscorides 7625
Dippel, L. 8775
Di Savoia, L.A. *v.* Savoia, L.A. di
Discorehmia see Rehm, H.S.L.F.F.
Dixon, H.N. 9070-9071
Dixon, Mrs. J. 9505
Doassans, J.E. *see* Patouillard, N.T.
Dodart, D. *see* Mirbel, C.F.B. de
Dodge, C.W. *see* Llano, G.A.
Dodoens, R. *see* Reichenbach, H.G.L.
Dodson, C.H. *see* Pijl, L. van der
Dodsworth, M. *see* Ray, J.
d'Olmen de Poederlé, E.J.C.G.H. *v.* Poederlé, E.J.C.G.H. d'Olmen
Dolneus, M. *see* Petiver, J.
Dombey, J. 9782
Domin, K. 8083, 8287
Don, D. 4637
Don, G., Jr., 5026-5027, 9721; *see* Loddiges, C.L.
Donk, M.A. 7483, 8414; *see* Pulle, A.A.
Doornik, J.J.F.H.T. Merkus *v.* Merkus Doornik, J.J.F.H.T.
Dörfler, I. *see* Moggridge, J.T., Moris, G.G., Prantl, K.A.E.
d'Orléans, G. *see* Robert, N.
Dos Santos Rangel, E. *v.* Rangel, E. dos Santos
Douglas, D. *see* Richardson, J.
Drake, S.A. 9734
Drake, W.F. 8755
Drake del Castillo, E. *see* Richard, A., Richard, L.C.M., Sagra, R. de la
Drapiez, P.A.J. 9138
Drascher, W. 8090
Drewes, J. 8255
Druce, G.C. *see* Pamplin, W., Piquet, J., Pugsley, H.W.

Drude, C.G.O. 8481, 8860, 8775; see Neger, F.W., Poirault, G., Reiche, K.F., Rutenberg, D.C.
Drummond, T. see Richardson, J.
Dryander, J. see Pritzel, G.A.
Drygalski, E. von see Reinbold, T., Richter, P.G.
Dubois, F.N.A. 10.024
Dubois, L. 9741a
Dubois-Reymond, E. see Reymond, M.L.A.
Duboisreymondia see Reymond, M.L.A.
Du Breuil, M.A. v. Breuil, M.A. du
Duby, J.É. 8824
Duchesne, A.N. 8319
Dudley, T.R. see Rechinger, K.H.
Duffield, M.E. v. Rosenberg, M.E.
Duffield, W. see Rosenberg, M.E.
Dufour, L.M. 9679
Dugdale, W. 7723
Dugès, A.A.D. see Oliva, L.
Duhamel, P.M. see Perrier, A.
Duhamel du Monceau, H.L. see Loiseleur-Deslongchamps, J.L.A., Michel, É.
Dulau see Pamplin, W.
Duménil, P.C.R.C. 7736
Dumont d'Urville, J.S.C. see Richard, A.
Dunal, M.F. 10.042; see Planchon, J.É.
Dunker, W.B.R.H. 6535
Dunzinger, G. 9567-9568
Du Petit-Thouars, L.-M.A.A. see Roemer, J.J.
Dupont de Nemours, P.S. see Poivre, P.
Dupret, H. 8983; see Renauld, F.F.G.
Dupuis, N.A. 9110
Durand, E.M. see Rafinesque-Schmaltz, C.S.
Durand, H. 9309-9310, 9312, 9691
Durand, T.A. 6465, 7978-7979, 8527, 8988; see Pax, F.A., Pittier, H.F., Renauld, F.F.G., Rousseau, M.
Du Rietz, G.E. *9215-9221*
Durietzia see Du Rietz, G.E.
Dusén, P.K.H. 8826
Duss, A. 7481, 7484, 7486
Dussieux 9741a
Dutour de Salvert 10.027; see Saint-Hilaire, A.F.C.P. de
Dutrochet, R.J.H. see Pringsheim, N.
Duval, H.A. 9154
Dyer, C.B. 6062a
Dyer, R.A. 8132

Eastwood, A. see Palmer, Edward
Eaton, A. 8563, 9153; see Richard, L.C.M.
Eaton, D.C. 9657-9658
Ebert, W. 8775
Eckeberg, C.G. v. Ekeberg, C.G.
Eckfeldt, J.W. 9813
Eduardoregelia see Regel, E.A. von
Ehrenberg, C.G. 7739, 8349; see Pringsheim, N.
Ehrenberg, G.R. v. Ehrenberg, C.G.
Ehrendorfer, F. see Rechinger, K.H.
Ehret, G.D. 4767, 9828
Ehrhart, J.B. 7528
Ehrhart, J.F. 9400; see Meyer, E.H.F., Murray, J.A.
Eichhorn, E. see Poeverlein, H.
Eichler, A.W. 5170, 5538, 6444; see Miquel, F.A.W., Potonié, H., Roth, E.C.F.
Eichler, B. 8464; see Raciborski, M.
Eichler, C.A. see Petrak, F.
Eichwald, K.E.I. von see Nordmann, A.D. von
Ekeberg, C.G. 7122-7123
Ekman, E. see Regnell, A.F.
Elfving, F. 4590
Eliasson, U. see Regnell, A.F.
Elkan, L. 7509
Elliott, G.F.S. see M'Andrew, J.
Elliott, S. 8571, 9153
Ellis, J.B. see Morgan, A.P.
Emil, F. 9853
Empedocles 9920
ENDEAVOUR 7344; see Parkinson, Sydney
Ender, E. 8775
Endler, T. 8104
Endlicher, S.L. 7820, 8091, 8887
Endress, P.A.C. 9679
Eneberg, I.R. 8373
Engel, J.J. 9469
Engelberg, J.M. von 9622
Engelhardt, M. von see Parrot, J.J.F.W.
Engelmann, G. 9658; see Parry, C.C.
Engler, H.G.A. 6358, 7758, 7978, 8481, 8775, 8860, 8943, 9817; see Lindau, G., Magnus, P.W., Migula,E.F.A.W., Müller, E.G.O., Mueller Argoviensis, J., Müller, K.A.E.,Nel, G.C., Niedenzu, F.J., Nordtstedt, C.F.O., Paul, H.K.G., Pax, F.A., Perkins, J.R., Peter, A., Petersen, O.G., Pfeffer, W.F.P., Pfitzer, E.H.H., Pilger, R.K.F., Pohl, Julius, Post, T.E. von, Potonié, H., Poulsen, V.A., Pound, R., Prantl, K.A.E., Printz, H., Pritzel, E.G., Radlkofer, L.A.T., Raimann, R., Reiche, K.F., Reichert, I., Reimers, H., Reinbold, T., Reinke, J., Rendle, A.B., Rikli, M.A., Rosendahl, O., Ruhland, W.O.E., Rutenberg, D.C., Sadebeck, R.E.B.
Engler, V. see Pax, F.A.
English, E. v. Hardin, E.
Erdle, M. see Roth, J.R.
EREBUS see Ross, J.C.

INDEX TO NAMES

Erichsen, C.F.E. 8450
Erichsen de Oliveira, P. see Paul, H.K.G.
Erkamo, V. 9982
ERMAK 7208
Ernst August, King of Hannover 5935
Errera, L.A. 9165
Escalera, F. Martínez de la v. Martínez de la Escalera, F.
Eschweiler, F.G. 6454
Español, C. Pau y v. Pau y Español, C.
Etheridge, E. 7870
Etler, J.C. 9430
Ettingshausen, C. von see Pokorny, A.
Ettl, H. 7423
Euganco, C. 8142
Euregelia see Regel, E.A. von
Evans, A.W. 6782
Evelyn, J. 6130
EVERTSEN see Reinwardt, C.G.C.
Ewan, J.A. 5958, 8404, 8567
Ewen, A.E. 8695
Exell, A.W. 8393
Eyma, P.J. 8393; see Pulle, A.A.

Fabrega, H.F. Pittier de v. Pittier, H.F.
Faegri, K. 8189a; see Pijl, L. van der
Fagerlind, F. see Regnell, A.F.
Faguet, A. 5381, 5383, 5386, 7562, 8118
Fairchild, D. see Meijer, F.N.
Famintzin, A.S. see Poirault, G.
Farges, P.G. 7472.
Farlow, W.G. 6348, 7159, 9813, 9932; see Pammel, L.H., Ramsbottom, J., Rand, E.L., Rau, E.A., Rea, C., Reess, M.F.T.F.M., Rehm, H.S.L.F.F., Reinbold, T., Reinke, J., Reinsch, P.F., Renauld, F.F.G., Richter, P.G., Ricken, J., Riddle, L.W., Robinson, John, Rodríguez y Femenías, J.J., Rolfs, P.H., Romell, L., Rosenvinge, L. Kolderup, Rostafiński, J.T., Rostrup, E., Rothert, W., Rothrock, J.T., Saccardo, P.A., Sadebeck, R.E.B., Sachs, J. von
Farr, E.R. 4778
Faurie, U.J. 7329; see Rosenstock, E.
Fawcett, B. see Lowe, E.J.
Fawcett, W. 5170, 9067; see Moore, S. Le Marchant
Febres, Z. Luces de v. Luces de Febres, Z.
Féburier, C.R. 8366
Fedchenko, A.P. 8800; see Regel, E.A. von
Fedchenko, O.A. 8804
Fedde, F.K.G. see Pax, F.A., Ronniger, K.
Fedorov, W. 7402
Fée, A.L.A. 8312; see Presl, K.B.
Fée, P. see Paul, H.K.G.

Feer, H. see Reber, B.
Felix, J.P. see Platen, P.L.
Fellmann, N.I. see Nylander, F.
Femenías, J.J. Rodríguez y v. Rodríguez y Femenías, J.J.
Fennell, D.I. see Raper, K.B.
Fenzl, E. 5538, 7808, 8845, 9827; see Reichardt, H.W.
Ferdinand, of Austria 8353
Ferdinandsen, C. 9621
Fernald, M.L. 9286; see Palmer, E.J.
Fernow, B.E. see Rodgers, A.D.
Fieber, F.X. 8293, 8303
Fielding, H.B. see Ruiz Lopez, H.
Filhol, M. 6963
Filippi, F. de 7272; see Pirotta, P.R.
Fiori, A. 9879, 9945
Firbas, F. see Rudolph, K.
Fisch, C. 8767
Fischer, A. 8450
Fischer, E. 8450, 9898
Fischer, F.E.L. von 8850; see Panzer, G.W.F., Regel, E.A. von
Fischer, H. 8775
Fischer Benzon, R.J.D. von 8256
Fisher, A. see Parry, W.E.
Fitch, J.N. 7829
Fitch, W.H. 6274, 7540, 7829, 8062-8063, 9126
Fitting, J.T.G.E. see Ruhland, W.O.E.
Fitzpatrick, H.M. 7731
Fitzpatrick, T.J. 8549; see Rafinesque-Schmaltz, C.S.
Flagey, C. 7190, 8984
Flahault, C.H.M. 9116; see Pavillard, Jules, Rouville, P.G. de
Flamini, G. Scarabelli Gommi v. Scarabelli Gommi Flamini, G.
Fleischhack 8445-8446
Flemin, C.J. 7554
Fletcher, E. 9505
Floerke, H.G. 8293; see Roeper, J.A.C.
Florschütz, P.A. 8393
Focke, W.O. 8211; see Rutenberg, D.C.
Förster, C.F. see Rümpler, T.
Fontane, T. see Ruthe, J.F.
Font y Quer, P. 7678
Forbes, H.O. 9070-9071
Forbes, J. see Russell, J.
Forbes, R.J. 5573
Forero, F. 8393
Forni, A. 5648, 7435
Forrest, J. 6428
Forssell, K.B.J. 9750
Forster, J.G.A. see St. John, H.
Forster, J.R. 7439; see St. John, H.
Forster, J.W. see Oliver, D.
Forster, T.I.M. 7730
Fosberg, F.R. 10.065
Foslie, M.H. 8726

Fossier, L. 7519
Fothergill, J. 7344
Foucaud, J. 9706
Franc, I. see Rosenstock, E.
Francé, R.H. 9583; see Pilger, R.K.F.
Franchet, A.R. see Révoil, G.
Franciscea see Roche, F. de la
Francke, F. 8775
Frank, B. 8775
Franke, A. 8228
Franklin, J. 9169-9170; see Richardson, J.
Franqueville, A. de see Moritzi, A., Richard, A., Richard, L.C.M., Sagra, R. de la
Franz I, Emperor of Austria 6006, 8103-8104
Franzpetrakia see Petrak, F.
Fraser, J. see Nuttall, T.
Frech, F. see Roemer, F. von
Freddi, V. 8141
Frederik Willem Karel, Prince of Oranje 7801
French, J.T. 6930
Frère Gasilien *v.* Parrique, G.
Fresenius, J.B.G.W. see Rüppell, E.
Frey, E. 8450
Freycinet, H.L.C. de Saulces de see Rolland, L.L.
Freyn, J.F. 8211; see Rutenberg, D.C.
Friedländer, R. 8445; see Rabenhorst, G.L.
Friedrich August von Sachsen see Reichenbach, H.G.L.
Friedrich Wilhelm III, of Prussia 6686
Friedrich Wilhelm IV, of Prussia 6694, 7759
Fries, E.M. 7247, 7736, 8847, 9177; see Mougeot, J.B., Ringius, H.H., Rogers, D.P.
Fries, H. 7235
Fries, M. 7247
Fries, R.E. 8393
Fries, T.M. 5132, 8850
Friis v. Rottbøll, C.F.
Frisch, K. von see Ruhland, W.O.E.
Fritsch, K., Jr. 5009; see Pokorny, A.
Fromage 9741a
Fuchs, L. see Rauwolff, L.
Fuckel, K.W.G.L. 6827
Fürnrohr, A.E. see Rabenhorst, G.L.
Fuhrmann, O. 4562; see Rosenstock, E.
Fukuyama, N. 5576
Funck, H.C. 6689; see Nees von Esenbeck, C.G.D.
FURY 7411-7413

Gaerdt, H. 8775
Gaertner, J. see Rumphius, G.E.
Gaertner, P.G. 8593
Gäumann, E.A. see Renner, O.

Gagnebin de la Ferrière, A. *see* Rousseau, J.J.
Gagnepain, F. 8726
Gaidoz, H. 9471
Gaillard, A. 7445
Gaillardot, C. 8378
Gal, M.J. Saint *v.* Saint-Gal, M.J.
GALATHEA *see* Rink, H.J.
Galeotti, H.G. 8607, 9149
Gallenkamp, W. 4510
Gallesio, G. 8144
Gallóe, O. 9548
Gallway, B.P. 7159
Galvez, I. 7114, 9769, 9771, 9773-9774, 9778
Gambel, W. *see* Nuttall, T.
Gamsenegg und Möderndorf, M. Jabornegg von *v.* Jabornegg von Gamsenegg und Möderndorf, M.
Gandoger, M. *see* Porcius, F.
Gane, J. 4765
Garcia Reyes, A. *v.* Reyes, A.G.
Garcin, L. 10.066
Garcke, C.A.F. 7759, 9454, 9458; *see* Rutenberg, D.C.
Gardner, G. *see* Russell, J.
Garnery 4494
Garnons, W.L.P. 7629, 9304
Garnsey, H.F. 9957
Garovaglio, S. *see Penzig, A.J.O.*
Gasilien, Frère v. Parrique, G.
Gasparrini, G. 9920
Gassicourt, L.C. Cadet de *v.* Cadet de Gassicourt, L.C.
Gatterer, C.W.J. *see* Rueling, J.P.
Gatto, A.C. 9948
Gaudin, J.F.A.P. 8885, 9399
Gauthier de la Peyronie 7224
Gautrois, J.F.C.L.C. Petif de la *v.* Petif, J.F.C.L.C.
Gay, C. 6253, 7857, 8977; *see* Rémy, J.
Gay, J. 8353
Geel, P.C. van 8883
Geheeb, A. *see* Newton, I., Rutenberg, D.C.
Geissler, C.G.H. 7229-7230; *see* Pallas, P.S.
Geitler, L. 7423, 8450
Gemeinhardt, K. 8450
Genevier, G. *see* Maillard, P.N.
Geny, P. 9250
Geoffroy Saint Hilaire, E. 10.029
Georgii, A. 8271
Gepp, A. 9065, 9067-9068, 9070-9071
Gepp, E.S. 9067
Gérard, F. 9110
Gerlache de Gomery, A. de 9691
Gerloff, J. 7423
Germain, F. 8857

Germain, J.J. de Saint *v.* Saint-Germain, J.J. de
Germain, R. see Saint-Germain, J.J. de
Germainia see Saint-Germain, J.J. de
Germanea see Saint-Germain, J.J. de
Gervasi, A.B. 8552
Gessner, J. *see* Linnaeus, C.
Gestro, R. 7901
Geus, A. 5552
Ghesquière, J.H.P.A. 9313
Gibelli, G. 7966; see Passerini, G.
Gilg, E.F. *see* Perkins, J.R.
Gilibert, J.E. 6574
Gillet, J. 7322
Gillot, F.X. *see* Renault, B.
Gilman, J.C. 7257
Gilmour, J.S.L. 4492
Ginestra, S. 9943
Gioberti, V. *see* Parlatore, F.
Giovanelli, J. von *see* Portenschlag-Ledermayer, F. von
Giralt, J. Planellas *v.* Planellas Giralt, J.
Girard, F. de 10.050
Girardin, Mlle de *see* Rousseau, J.J.
Giraud, M. 8894
Girault, J. 10.072
Gireaud, M. *v.* Giraud, M.
Girieud, J. 6847
Girlin, J. 9720
Gisbert, E. Pomata y *v.* Pomata y Gisbert, E.
Giseke, P.D. 8031, 8065; see Montin, L.J., Murray, J.A., Mygind, F. von, Osbeck, P., Plukenet, L., Rozier, F.
Gjöwell, C.C. 4848
Glaziou, A.F.M. 5485
Gleason, H.A. 8393; see Otis, C.H., Palmer, E.J.
Gleditsch, J.G. 5073, 9946; see Ratzeburg, J.T.C.
Glijm, C. 7148
Gloner, P. 4626
Glück, C.M.H. 7423
Gmelin, F.G. 7236
Gmelin, J.F. 4709, 7911, 9828
Gmelin, S.G. 7224; see Pallas, P.S.
Goday, S. Rivas *v.* Rivas Goday, S.
Godfrin, J. 7777
Godron, D.A. 9704
Goebel, K.I.E. 7831; see Neger, F.W., Oltmans, F., Raciborski, M., Radlkofer, L.A.T., Reinke, J., Renner, O., Roeper, J.A.C., Ross, H.
Goeppert, H.R. 8775; see Sadebeck, R.E.B.
Görgey, A. von *see* Redtenbacher, J.
Görts-van Rijn, A.R.A. 8393
Goeschke, F. 8775
Goethe, J.W. von 6676, 6679, 9417, 9419; see Reichel, G.C.

Goethe, R. 8775
Goldman, E.A. *see* Nelson, E.W.
Gommi Flamini, G. Scarabelli *v.* Scarabelli Gommi Flamini, G.
Gonnermann, W. 8445
Gonthier, J. *v.* Gontier, J.
Gontier, J. 10.003, 10.005
Goodsell, J.W. 9864
Goor, A.C.J. van see Redeke, H.C.
Gormaz, F.V. 7855
Gormaz, F. Vidal *v.* Gormaz, F.V.
Gorter, D. de 4799; see Rainville, F.
Gothan, W.U.E.F. 8217, 8225, 8228; see Potonié, H.
Gottschalk, J. *see* Rothman, G.
Gottsche, C.M. 8448; see Rabenhorst, G.L., Rutenberg, D.C.
Gouan, A. *see* Rousseau, J.J.
Goyena, M. Ramírez *v.* Ramírez Goyena, M.
Grabowski, H.E. 9462
Graebener, L. 8775
Graebner, K.O.R.P.P. 7684, 7758
Graff, R. 6433-6434
Grandidier, A. 8995; see Renauld, F.F.G.
Grandidier, G. 8995
Grandmaison, L.A.M. de 8398
Grasset de Saint-Sauveur, J. 9496
Graves, L. 9340
Gray, A. 7504, 7504, 7913, 8371, 9200, 9657; see Palmer, E.J., Pammel, L.H., Parry, C.C., Provancher, L., Rafinesque-Schmaltz, C.S., Reverchon, J., Robinson, B.L.
Gray, E.W. *see* Russell, Anna
Green, J.R. 9957
Green, M.L. 4778
Greene, E.L. *see* Nieuwland, J.A., Parry, C.C.
Greenman, J.M. 9285, 9287, 9289
Gregory, J.W. *see* Rendle, A.B.
Gregory, W. *see* Rylands, T.G.
Grenier, J.C.M. 9704; see Philippe, X.
Gresalvi, S. *v.* Ginestra, S.
Greshoff, M. 8391
Greuter, W. *see* Mouterde, P.
Greves, G. 9071
Greville, R.K. 8685, 9435; see Lunan, J., Neill, P., Rylands, T.G.
Griffith, W. 9725; see M'Clelland, J.
GRIPER 7409
Grisebach, A.H.R. 5170, 7290, 7375, 8353; see Örstedt, A.S., Reinke, J., Rohrbach, P., Regnell, A.F.
Groddeck, E. *see* Nachtigal, G.H.
Grönblad, R. 7423
Groenewald, E.W. *see* Persoon, C.H.
Groenewald, J. *see* Persoon, C.H.
Groenewegen, J.B. 7148
Groenland, J. *see* Rümpler, T.

Gröntved, J. 9548
Gronovius, J.F. see Rauwolff, L.
Gronovius, L.T. 9828
Grootaers, L. 7534
Grosser, W. see Pax, F.A.
Grossheim, A.A. see Ronniger, K.
Grout, A.J. see Rapp, S.
Grunow, A. 8443, 8450
Grye, Bouquet de la v. Bouquet de la Grye
Gualteri, G. 7112
Gueinzius, W. see Poeppig, E.F.
Gümbel, W.T. see Radlkofer, L.A.T.
Günther, A. 4740
Guérin, L. 9110
Gürke, R.L.A.M. 9185
Guettard, J.É. 9739; see Rousseau, J.J.
Guiart, L.D. 9088
Guibert 8410
Guichard, A. see Paillot, J.
Guichard, P. see Paillot, J.
Guidi, P. 7285
Guillaumin, A. see Roux, N.
Guillemin, J.B.A. 8750, 9144
Guillo-Kastner, Mrs. 7324
Guimpel, F. 8718, 8871, 9601
Gusinde, M. 9216
Gyelnik, V.K. 8450
Györfy, I. see Péterfi, M.

Haarbach, M. Rainer von und zu v. Rainer von und zu Haarbach, M.
Haartman, J.G. 9842
Haas, C. van Overeem de v. Overeem, C. van
Haas, M. 9666
Haberlandt, G.J.F. see Porsch, O.
Habirshaw, F. see Pelletan, J.
Hackel, E. 8211, 8726, 8850; see Pax, F.A., Porcius, F.
Hadač, E. 9548
Haeck, P. 7318
Haeckel, E.H.P.A. 9411; see Müller, H.L.H., Ortman, A.E., Römer, J., Rutenberg, D.C.
Haeckel, I. 7748
Haenke, T.P.X. 8293; see Presl, K.B.
Häusler, J. 7423
Hagen, K.G. 4872
Hagle, F.J. 7581
Hakulinen, R. see Räsänen, V.J.P.B.
Halácsy, E. von see Richter, K.
Haldeman, S.S. see Rafinesque-Schmaltz, C.S.
Hale, J. see Rafinesque-Schmaltz, C.S.
Hall, H.C. van 6088
Hallenberg, G. 4839
Haller, A. von 9514, 9790, 9946; see Mueller Argoviensis, J., Royen, A. van
Hallier, E. see Sagorski, E.A.

Hamberg, A. 6823
Hamilton, W. see Pursh, F.T.
Hampe, G.E.L. see Liebmann, F.M., Müller, Karl (Halle), Perktold, J.A., Regnell, A.F.
Hampel, W. 8775
Hann, J. 9456
Hannon, M. v. Rousseau, M.
Hansen, A. 9606
Hansen, H.M. 9548
Hantzsch, C.A. 8443
Hao, K.-S. 4862
Hardin, E. 10.061
Hariot, P.A. 7473, 7489; see Petit, P.C.M.
Harling, G. see Regnell, A.F.
Harmand, J.H.A.J. 7574, 7973
Harms, H.A.T. see Müller, E.G.O., Reiche, K.F.
Harnier, R. 7818
Harriman, E.H. 7602; see Saccardo, P.A.
Harris, T.W. see Pickering, C.
Harris, W.C. see Roth, J.R.
Harrison, J. 7553
Hart, A. v. Phelps, A.
Hart, J.T. 9734
Hartlaub, G. see Pansch, A.
Hartman, C.V. 9286
Hartman, R.W. see Lindberg, S.O.
Hartmann, C.J. 8090; see Poeppig, E.F.
Hartmann, M. 9076, 9520
Hartog, M.M. 9022
Hartwall, V.E. 8373
Hartwich, C. 9232
Harvey, J.R. 8245
Harvey, W. see Rudbeck, O.J.
Harvey, W.H. 9684; see Pike, N.
Hasskarl, J.K. 7759, 9784
Hassler, E. 9355; see Rojas, T.
Hatting, M. see Persoon, C.H.
Hauck, F. 8450, 8850; see Rabenhorst, G.L., Richter, P.G.
Haüy 9687
Hauman, L. see Parodi, L.R.
Haussknecht, H.C. 8211, 8446; see Rutenberg, D.C., Sagorski, E.A.
Havell, R., Jr. 9504
Hayden, A. 7257 7258
Hayden, F.V. 8180; see Porter, T.C., Powell, J.W.
Hayne, F.G. 9629
Hazen, T.E. see Pennell, F.W.
Heart, A. v. Phelps, A.
Heart, L. see Phelps, A.
HECLA 7409, 7411-7414
Hedelin, C. 6643
Hedge, J. see Rechinger, K.H.
Hedicke, H. 9567
Hedley, G.W. 9202

INDEX TO NAMES

Hedwig, J. *see* Lindberg, S.O., Mohr, D.M.H.
Heeg, M. 8095
Heer, O. von 8775; *see* Moritzi, A., Regel, E.A. von
Heerdt, P.F. van 8393
Heering, W.C.A. 7423
Hegetschweiler-Bodmer, J. 9399
Hegi, G. *see* Rechinger, K.H.
Hehn, J. 7402
Heiberg, P.A.C. *see* Rosenberg, C.F.
Heldreich, T. von 7356
Hélène, Duchess of Aosta 7970; *see* Piscicelli, M.
Helenius, A.N. *v.* Railonsala, A.N.
Hellwig, J.L. 8230
Helmert, W.O. 6659
Helwing, G.A. 4933
Hemsley, W.B. *see* Oliver, D., Reichenbach, H.G.
Hennecart, J. 8119
Hennings, P.C. 8915; *see* Prantl, K.A.E.
Henrard, J.T. 8393; *see* Pulle, A.A.
Henri, C. 7344
Henri III, King of France *see* Pena, P.
Henri IV, King of France *see* Richer de Belleval, P.
Henry, Aug. *see* Oliver, D.
Henry, J. 9200
Henschel, A.W.E.T. 9784
Henschen, S.E. *see* Regnell, A.F.
Herbert, W. *see* Phillips, H.
Herbst, J.F.W. 7222
Hercules *see* Oken, L.
Herder, F.G.M.T. von 8775, 8784, 8791; *see* Radde, G.F.R. von, Regel, E.A. von
Héricourt, C.L.X. Rochet d' *v.* Rochet d'Héricourt, C.L.X.
Hérincq, B. 7654, 8028
Hérincq, F. 9110
Hermann, F.J. *see* Porter, T.C.
Hermann, G.S. 9273
Hermann, J. 8443
Hermann, P. *see* Richardson, R.
Hernández, F. *see* Maximilian, A.P. zu Wied-Neuwied, Poeppig, E.F., Ruiz Lopez, H.
Herre, P. 9515
Herrera *see* Pringle, C.G.
Hervey, A.B. 8650
Herzog, T.C.J. *see* Paul, H.K.G., Radlkofer, L.A.T., Rosenstock, E.
Hess, R.W. 8741
Hesselbo, A. 9548
Hesselman, H. 5248
Heteman, S. 7555
Hetley, C. *see* Raoul, É.F.A.
Heufler zu Rasen und Perdonegg, L.S.J.D.A. *see* Perktold, J.A.

Heurck, H.F. Van *v.* Van Heurck, H.F.
Heuzé, L.G. 9663
Heydemann, A.G. 9253
Heyland, J.C. 9107
Heyn, A.N.J. 8393
Heyn, E. 9584
Heyne, B. 9635; *see* Royle, J.F.
Heynig, H. 7423
Hiern, W.P. 9065, 9071
Hieronymus, G.H.E.W. 7390, 8726; *see* Parodi, D., Pax, F.A., Prantl, K.A.E., Sadebeck, R.E.B.
Higley, W.K. *see* Raddin, C.S.
Hiitonen, H.I.A. 4566
Hilairanthus see Saint-Hilaire, A.F.C.P. de
Hilaire, A.F.C.P. de Saint *v.* Saint-Hilaire, A.F.C.P. de
Hilairella see Saint-Hilaire, A.F.C.P. de
Hilaria see Saint-Hilaire, A.F.C.P. de
Hilariophyton see Saint-Hilaire, A.F.C.P. de
Hildebrandt, J.M. 6505; *see* Rutenberg, D.C.
Hill, J. *see* Meyer, E.H.F.
Hillcoat, D. *see* Rechinger, K.H.
Hillmann, J. 8450
Hiltrop, J. van 10.092
Hirn, K.E. 6870
Hitchcock, A.S. 4778, 6893, 8194, 10.095
Hitler, A. 8775
Hjelt, H. 9979
Hjelt, O.E.A. 8373
Hochberg, J.M. de Canal von *v.* Canal von Hochberg, J.M. de
Hochstetter, C.F. *see* Reichenbach, C.L. von
Hochstetter, C.G.F. von 8970
Hochstetter, W.C. *see* Loescher, E.
Hocquart, E., 9497-9498
Hodge, C. *see* Patrick, R.M.
Hoeckert, J.A. *see* Radloff, F.W.
Höhnel, F.X.R. von 7785, 8726; *see* Rehm, H.S.L.F.F.
Hoffman, C.A.F. 7258
Hoffmann, H. 8775
Hoffmann, J.B. *see* Persoon, C.H.
Hoffmann, O. *see* Rutenberg, D.C.
Hofman Bang, N. *see* Rosenberg, C.F.
Hofmann, E. von 9806
Hofmeister, F. 8878, 8883
Hofmeister, W.F.B. 9953; *see* Micheli, M., Roeper, J.A.C., Sachs, J. von
Hogg, R. 8061
Hohe, C. 9398
Hohenacker, R.F. *see* Noë, F.W., Orphanides, T.G., Philippi, R.A., Rabenhorst, G.L.
Hohenwarth, S. von 8920
Holandre 6052
Holden, S. 7554

Holdhaus, K. 8726
Holmes, E.M. 8372
Holmsen, G. *see* Resvoll-Holmsen, H.M.
Holmskjold, T. 7725; *see* Persoon, C.H.
Holuby, J.L. 6739, 7194, 9071; *see* Richter, L., Ross, J.C.
Hooker, W. 8404
Hooker, W.J. 6102, 6237, 6243, 7412-7414, 8496, 8876, 9150, 9154, 9169, 9684; *see* Mann, G., Mohr, D.M.H., Planchon, J.É., Prescott, G.W., Ralfs, J., Richardson, J., Robertson, D., Sabine, E.
Hooper, W.H. *see* Parry, W.E.
Hoopes, J. 9657
Hoop Scheffer, J.G. de 9767
Hopkinson, J. 8372
Hoppe, D.H. 5756
Hornschuch, C.F. 8293, 8964
Horst, H. von *v.* Reinke, J.
Hosaka, E.Y. 10.062
Hosokawa, T., 5576
House, H.D. 7595
Houtte, L.B. Van *v.* Van Houtte, L.B.
Houttuyn, M. 4794
Howard, J.E. 7540
Howard, W. 9661
Howe, M.A. 9325
Howell, J.T. 7416
Hubbard, C.E. 7055
Huber, J. 8450
Huber-Morath, A. *see* Rechinger, K.H.
Hudson, W. *see* Rose, H.
Hue, A.M. 7475
Huet, N., Jr. 9140
Hugel, C. de 9431
Hughes, D.K. *v.* Popenoe, D.K.
Hugo *see* Rheede tot Draakestein, H.A. van
Hulting, J. 7032
Humbert, H. *see* Perrier de la Bâthie, H.
Humboldt, F.W.H.A. von 9408, 9771; *see* Nees von Esenbeck, C.G.D., Persoon, C.H., Roemer, J.J., Saint-Hilaire, A.F.C.P. de
Humm 8878
Humphreys, J.D. 8245
Hunneman *see* Pamplin, W.
Hurst, C.C. 9470
Husnot, P.T. 8983; *see* Renauld, F.F.G.
Hustedt, F. 7423, 8450
Hutchins, J. 8400
Hutchinson, J. 9211; *see* Rattray, John
Huter, R. *see* Porta, P.
Huth, E. *see* Rivinus,, A.Q.

Immendingen und Bilafingen, F. Rot von S. zu *v.* Rot vonSchreckenstein, F.
Ippolito, S. de 8552

Irmisch, J.F.T. *see* Roeper, J.A.C., Ruhmer, G.F.
Irvine, A. 7275; *see* Pamplin, W.
Isenbart, J. 8352
Ismailow, W. 9686
Isoviita, P. 9469
Iversen, H. 7585
Iversen, J. 8189a
Ives, J.C. *see* Newberry, J.S.

Jabornegg von Gamsenegg und Möderndorf, M. 7178
Jackson, B.D. 8353, 8372, 8756, 9065; *see* Mueller, F.J.H. von, Roberg, L.
Jackson, J. 8834
Jacquin, J.F. von 8295, 8754
Jacquin, N.J. von 4852, 9743; *see* Pourret, P.A., Saint-Hilaire, A.F.C.P. de
Jäger, H. 8775
Jahn, F.L. *see* Quehl, L.
Jakšsic, S. *see* Pančic, J.
James, T.P. 8406, 9657, 9658
James I of England *see* L'Obel, M. de
Janchen, E. *see* Pilger, R.K.F.
Jañez, J.I. Otero *v.* Otero Jañez, J.I.
Janisch, C. 8443
Janraia see Ray, J.
Jansen-Jacobs, M.J. 8393
Jaramillo-Arango, J. 9782
Jatta, A. 7197
Jaussin, L.A. 9943
Jeanbernat, E.M.J. *see* Renauld, F.F.G.
JEANNETTE 6539
Jean VI, of Brazil 10.042
Jefferson, T. *see* Rafinesque-Schmaltz, C.S.
Jelliffe, S.E. 9821-9822
Jenkins, A.E. 10.033
Jenner, E. 8599
Jensen, C. 9609
Jensen, T. *see* Rosenberg, C.F.
Jessen, K. 8654
Jessen, K.F.W. 8353, 8355; *see* Pritzel, G.A.
Jesup, H.G. *see* Moore, G.T.
Jewitt, O. 7554
Jimenez, F. *v.* Ximenez, F.
Jiménez, J.A. Pavon y *v.* Pavon y Jiménez, J.A.
Jobin 7654
Jørgensen, C.A. 9606
Jörgensen, E.H. 4506
Johan Maurits van Nassau *see* Marcgrave, G., Piso, W.
Johann, King of Saxony *see* Rochel, A.
John, H. St. *v.* St. John, H.
John Brooks, W.E. St. *v.* St. John Brooks, W.E.
Johnson, G.W. 8061
Johnson, T. *see* Ralph, T.S.

Johnston, C. *see* Rylands, T.G.
Johnston, I.M. 8393
Jolyclerc, N.-M.-T. 7726
Jones, G.N. 10.059; *see* Mead, S.B.
Jones, M.E. *see* Rose, J.N.
Jones, R. *see* Sadler, John
Jones, W.A. 7404
Jongmans, W.J. *see* Posthumus, O.
Jonker, F.P. 8393; *see* Pulle, A.A.
Jonker-Verhoef, A.M.E. 8393
Jónsson, H. 9548
Jordan, A. 8020
Jordan, W. 9454-9455
Joséphine de Beauharnais *v.* Beauharnais, J. de
Jovet-Ast, S. *see* Reichert, I.
JUBILEE 7131
Jühlke, F. 8775
Jürgens, G.H.B. *see* Möhring, P.H.G.
Julius Caesar 6839
Jumelle, H.L. 7704
Junge, P. 8256
Junghuhn, F.W. 6107; *see* Persoon, C.H.
Jurišic, Ž.J. 7277, 7284
Jussieu, A. de 4712, 8353; *see* Pourret, P.A.
Jussieu, A.H.L. de 10.032, 10.034; *see* Palisot de Beauvois, A.M.F.J., Prost, T.C.
Jussieu, A.L. de 5026-5027, 6196, 7250, 7732, 7911, 9063, 9153, 9687, 9694, 9775-9776, 9840, 10.027, 10.029; *see* Palisot de Beauvois, A.M.F.J., Plumier, C., Pourret, P.A., Saint-Hilaire, A.F.C.P. de
Jussieu, B. de *see* Linnaeus, C., Richard, L.C.M., Rousseau, J.J.

Kaalaas, B.D.L. 8342
Kache, P. 8775
Kaemerling, J. *see* Reichardt, H.W.
Kajanus, B. *v.* Nilson, B.
Kalchbrenner, K. *see* Richter, L.
Kalm, P. *see* Linnaeus, C., Pennant, T.
Kalmus, J. *see* Niessl von Mayendorf, G.
Kamel, G.J. 8701
Kamikôti, S. 5576
Kanitz, A. 6723, 8169; *see* Porcius, F.
Karcher, A. 5771
Karsch, F. 9456
Karsten, G.H.H. *see* Pax, F.A., Pohle, R.R., Rikli, M.A., Rosen, Felix
Karsten, H. *see* Poeppig, E.F.
Kasnakow, A.N. *see* Radde, G.F.R. von
Kaudern, W.A. 9530
Kauffman, C.H. *see* Pennington, L.H.
Kaufmann, G.A. *see* Rabenhorst, G.L.
Kavina, K. 7935
Kearny, T.H. 7255
Keissler, K. von 8450, 8726

Keith, J. *see* Roy, J.
Keller, C. 8090, 9234
Keller, J. 9234
Keller, W.A. 9234
Kellerman, W.A. 9817
Kellogg, H.S. 7257
Kelly, H.A. *see* Peck, C.H.
Kempner, W. *see* Rabinowitsch, L.
Kennedy, E.B. *see* MacGillivray, J.
Kennedy, J. 7183; *see* Page, W.B.
Kennicott, R. *see* Rothrock, J.T.
Kenoyer, L.A. 7258
Kent, E. *see* Phillips, H.
Kern, J.H. *see* Reichgelt, T.J.
Kerndl, J.L. 8057
Kerner von Marilaun, A.J. 8211; *see* Pernhoffer von Bärnkron, G., Peyritsch, J.
Keyserling, A. 9837
Kiaer, F.C. 7585
Kibler, J. 7939
Kihlman, A.O. 9979
Killick, D.J.B. 8132
Killip, E.P. 7987, 8393; *see* Pennell, F.W.
Kindberg, N.C. 10.007
King, C.J. 4492
King, C.M. 7257, 7258
King, G. 7289, 9725; *see* Pantling, R.
Kirchner, E.O.O. *see* Pax, F.A.
Kirchner, G. 7801
Kitaibel, P. *see* Sadler, Joseph
Kittel, M.B. 9138
Kittlitz, F.H. von *see* Postels, A.P.
Klatt, F.W. 7759
Klein, G.F. *see* Rottler, J.P.
Kleinhoonte, A. 8393; *see* Pulle, A.A.
Klemming, G.E. *see* Rudbeck, O.J.
Klett, G.T. *see* Richter, H.E.F.
Klinge, J. 8636
Kloss, C.B. 9210
Klotzsch, J.F. 7759; *see* Preuss, G.T., Rabenhorst, G.L.
Knapp, J.A. 8723
Knappe, K.F. 7227
Kneucker, J.A. *see* Murr, J.
Kniep, H. 9583
Kniep, K.J. *see* Renner, O.
Knight, C. 6466
Knorr, G.W. *see* Noeggerath, J.J.
Knowles, M.C. *see* Muir, J., Porter, L.E.
Knuth, P.E.O.W. *see* Prahl, P.
Knuth, R.G.P. *see* Pax, F.A.
Kny, C.I.L. 8775; *see* Pantanelli, E.F., Perkins, J.R., Ross, H.
Koch, C.L. *see* Philippi, R.A.
Koch, G.F. 7768
Koch, K.H.E. *see* Meyer, C.A. von
Koch, W.D.J. 8885, 8944, 9377; *see* Palla, E., Sagorski, E.A.
Koebisch, A. 8604

Koehne, B.A.E. *see* Rutenberg, D.C.
Kölreuter, J.G. *see* Sageret, A.
König, J.G. 9086
Koenig, K.D.E. *see* M'Clelland, J.
Koerber, G.W. *see* Massalongo, A.B., Sadebeck, R.E.B.
Körnicke, F.A. *see* Rutenberg, D.C.
Kohl, F.G. 8726, 8885
Kolb, M. 8775
Kolderup Rosenvinge, L. *v.* Rosenvinge, L. Kolderup
Koldewey, K. *see* Pansch, A.
Kolenati, F.A.R. 5916
Kolkwitz, R. 7423, 8450
Komarov, V.L. 7790
Kooper, W.J.C. *see* Pulle, A.A.
Koopmans, R.G. *see* Pulle, A.A.
Kopaczevskaja, E.G. see Rassadina, K.A.
Korolkow, N.I. 8800
Korshinsky, S.I. *see* Petunnikov, A.N.
Koster, J.T. 8393; *see* Pulle, A.A.
Kostermans, A.J.G.H. 8393; *see* Pulle, A.A.
Kotschy, C.G.T. 7808; *see* Russegger, J. von
Kraenzlin, F.W.L. 8894; *see* Pfitzer, E.H.H., Rutenberg, D.C.
Kräusel, R. *see* Pax, F.A.
Kramer, K.U. 8393
Kraus, G.K.M. *see* Sachs, J. von
Krausch, H.K. 7423
Krause, C.L. 9483
Krause, E.H.L. 8256
Krause, E.L. *see* Müller, H.L.H.
Krause, H. 8800
Krause, K. 8393; *see* Pilger, R.K.F.
Krelage, J.H. 7148
Krempelhuber, A. von 8744; *see* Rutenberg, D.C.
Krets, H. *see* Mutel, A.
Krieger, W. 8450
Kriel, S. *v.* Phillips, S.
Krok, T.O.B.N. *see* Romell, L.
Kronfeld, E.M. 8211
Kubitzki, K.U. 8393
Kucera, C.L. 9197
Kuckuck, E.H.P. 8931-8932
Kühn, K.G. 8399
Kükenthal, G. 7748
Kündig, J. *see* Prantl, K.A.E.
Kützing, F.T. *see* Nees von Esenbeck, C.G.D., Rabenhorst, G.L., Ruchinger, G., Sr.
Kulczyński S. *see* Pawlowski, B., Raciborski, M.
Kummer, G.F. 9141
Kung, H.-W. 4862
Kunst, C.L. 9483
Kunth, C.S. 5958, 7759, 9408, 9771; *see* Lucae, A.F.T., Miers, J., Née, I.C.S.

Kuntze, C.E.O. 8193-8194, 8353, 9817, 9917; *see* Malinvaud, L.J.E., Post, T.E. von, Pound, R., Pritzel, G.A., Richter, P.G.
Kunze, G. 6102, 6104, 8351, 8353, 8488, 9141; Petermann, W.L., Poeppig, E.F., Regnell, A.F.
Kuschakewicz 8800
Kusnetzov, N.I. *see* Palibin, I.V.
Kylin, J.H. *see* Norén, C.O.G.

La Bâthie, Perrier de *v.* Perrier de la Bâthie
La Bergerie, J.B. Rouzier de *v.* Rouzier de la Bergerie, J.B.
Laborde, B. de 7439
La Boulaye, de *v.* Boulaye de la
Lace, E. *v.* Roscoe, M.
Lachenal, W. de *see* Roeper, J.A.C.
Lackerbauer, P. 9278
Lackner, C. 8775
Lacoste, C.M. van der Sande *v.* Sande Lacoste, C.M. van der
Laerstadt, L.L. *v.* Laestadius, L.L.
La Escalera, F. Martínez de *v.* Martínez de la Escalera, F.
Laestadius, L.L. 7247
Lafosse 9741a
Lagarde, J.J. 7535
La Gautrois, J.F.C.L.C. Petif de *v.* Petif, J.F.C.L.C.
Lager, J.B. Saint *v.* Saint-Lager, J.B.
Lagerheim, N.G. von 7455
La Grye, Bouquet de *v.* Bouquet de la Grye
Lahitte, de *v.* De Lahitte
Laissus 7699
Lakela, O. *see* Rosendahl, O.
L'allier *v.* Allier
Laloy, D. 8981
Lam, H.J. *see* Pulle, A.A.
Lamarck, J.B.A.P.M. de 7224, 8598, 9741a; *see* Pallas, P.S., Roeper, J.A.C.
Lambe, L.M. 7632
Lambert, A.B. 8103, 9768, 9782; *see* Martius, C.F.P. von, Michaux, A., Pallas, P.S., Paterson, W., Pavon y Jiménez, J.A., Pursh, F.T., Raffles, T.S.B., Richardson, J., Roe, J.S., Roxburgh, W., Rudge, E., Ruiz Lopez, H., , Russell, P.
Lamerville 9741a
Lamoignon de Malesherbes, C.G. de *v.* Malesherbes, C.G. de Lamoignon
Lamouroux, J.V.F. *see* Roussel, H.F.A. de
Lamson-Scribner, F. 7255
Landerer von Walddorf, H. 6186
Láng, A.F. *see* Rochel, A.

Lange, J.M.C. 7008, 9243-9245; see Petersen, O.G., Rink, H.J., Rosenberg, C.F., Rosenvinge, L. Kolderup
Langlès, L.M. 7224
Langsdorff, G.H. von see Riedel, L.
Lanjouw, J. 8393, 8395; see Pulle, A.
Lansberge, J.G. van 8895
La Peyronie, Gauthier de v. Gauthier de la Peyronie
Lapeyrouse, P.P. de 9679; see Pourret, P.A.
La Puerta y Rodénas, G. de v. Puerta y Rodénas, G. de la
Lara, J.M. Pérez v. Pérez Lara, J.M.
La Ramée, S.H. de v. Ramée, S.H. de la
Larios, J. Pérez v. Pérez Larios, J.
La Roche, de v. Roche, de la
Larochea see Roche, D. de la
Larsen, P. 9548
La Sagra, R. de v. Sagra, R. de la
La Sauvage, de v. Sauvage, de la
Lasser, T. see Pittier, H.F.
Lasteyrie-Dusaillant, C.P. de 9741a
La Tour, M.C. Boy de v. Boy de la Tour, M.C.
Latourette, M.A.L. Claret de see Pourret, P.A., Rozier, F.
Latreille, P.A. 10.029; see Newman, E.
L'AUBE see AUBE
La Varde, R.A.L. Potier De v. Potier de la Varde, R.A.L.
Lawalrée, A.G.C. 8748, 9314
Lawson, C. 8685
Leandri, J.D. 7704
Lebeck, M.J. see Roxburgh, W.
Le Bègue de Presle see Rousseau, J.J.
Leblond, J.B. see Richard, L.C.M.
Le Breton, A. 8416
Leche, J. see Lindblom, A.E.
Lecomte, P.H. 7704
Ledebour, C.F. von see Rothrock, J.T.
Ledsebe, R. 7794
Lee, A. see Parkinson, Sydney
Lefebvre, E. 5573
Lefebvre, T. 9151
Legendre v. Philibert, J.C.
Legendre, F. 6128
Legrain, J. 9314
Legré, L. 9696
Legros, G. 6784
Lehmann, J.G.C. see Preiss, L., Purkyně, E. von
Lehtonen, P. 9982
Leigh, C.M. v. Pope, C.M.
Leimbach, G. 8211
Leitgeb, H. see Peyritsch, J.J.
Lejeune, A.L.S. 5292, 9671; see Römer, C.
Le Jolis, A.F. 9501, 9803
Lekeux, J.H. 9126

Lemmermann, E.J. 7423
Lemmon, J.G. see Plummer, S.A.
Lemonnier, L.G. see Pourret, P.A.
Lenz, W. 8211
Leon, J. Ponce de v. Ponce de Leon, J.
Leon y Aimé, A. Ponce de v. Ponce de Leon y Aimé, A.
Le Rat, A.-J. 7491
Le Ratia 7491
Le Roux, C.C.F.M. v. Roux, C.C.F.M. le
Leroy, 8410
Leschenault de la Tour, J.B.L.C.T. see Mueller, F.J.H. von, Pourret, P.A.
Leske, N.G. 8105
Lesson, A.P. see Richard, A.
Lestiboudois, T.G. 7215
Le Testu, G.M.P.C. 7618
Lettau, G. see Rieber, X.
Leussink, J.A. 4778
Levasseur, T. see Rousseau, J.J.
Léveillé, A.A.H. 9706, 9712
Léveillé, J.H. 8353
Levier, E. 9933; see Müller, Karl (Halle)
Ley, A. 8401
L'Héritier de Brutelle, C.L. 8834; see Morren, C.F.A., Pourret, P.A., Redouté, P.J.
Libert, M.-A. 9683, 9905
Lichtenstein, A.G.G. 7732
Licopoli, G. 7431
Lid, Regin i v. Rasmussen, R.
Liebig, J. von see Redtenbacher, J.
Liepoldt, W. 7806, 7808
Limminghe, A.M.A. de see Morren, C.J.É.
Limpricht, H.W. 8450; see Pax, F.A.
Limpricht, K.G. 8450
Linares, de 8268
Lincoln, A. v. Phelps, A.
Lincoln, S. see Phelps, A.
Lind, J.W.A. see Rostrup, E.
Lindau, G. 6979, 7978, 8450; see Pilger, R.K.F., Prantl, K.A.E.
Lindberg, G.A. see Regnell, A.F.
Lindberg, H. see Norrlin, J.P.
Lindberg, S.O. 6961, 8445; see Nyman, C.F., Rabenhorst, G.L.
Lindeman, J.C. 8393
Lindeman, M. see Pansch, A.
Lindemann, E.E. von see Pax, F.A., Rudolph, J.H.
Linden, C.L. see Rodigas, É.,
Linden, J.J. see Pancher, J.A.I.
Linden, L. see Rodigas, É., Rolfe, R.A.
Lindenberg, J.B.W. see Mohr, D.M.H.
Linder, K. 9236
Lindinger, L. 7972
Lindley, J. 6102, 6688, 7555-7556, 8584, 8685, 9150, 9154; see Nees von Esenbeck, C.G.D., Prescott, G.W.

INDEX TO NAMES

Lindley, Mrs. 4642
Lindman, C.A.M. 7247; see Regnell, A.F.
Lingelsheim, A. von see Pax, F.A.
Linhart, G. see Porcius, F.
Link, J.H.F. 8353; see Persoon, C.H.
Linkola, K. see Räsänen, V.J.P.B., Regel, C.A. von, Roivainen, H.
Linnaeus, S. see Nordstedt, C.F.O.
Linnaeus, U.C. see Nordstedt, C.F.O.
Linnarsson, E.J.E. 9750
Linné, Carl von, filius 4851, 6620, 8834, 9183; see Naezén, D.E., Pourret, P.A., Rolander, D.
Linné, C. von v. Linnaeus, C.
Linschoten, S. 8385
Lint, G.M. de see Redeke, H.C.
Lipman, T. v. Lippmaa, T.
Lippert, X.J. 4799
Lipsky, W.H. v. Lipsky, V.I.
Liro, J.I. v. Lindroth, J.I.
Lisa, M. 6327
Lister, A. see Lister, G.
Litke, F.P. v. Lütke, F.P.
Litwinow, D.I. v. Litvinov, D.I.
Lloyd, C.E. 9286
Lloyd, C.G. see Patterson, F.
L'Obel, M. de 7625, 7999; see Morren, C.F.A., Pena, P.,
Locher, H. 8775
Lönberg, E.G. 4529
Lönnrot, E. see Saelan, T.
Loeske, L. 7518
Loew, E. 8211
Lohwag, H. see Litschauer, V.
Lojka, H. 8823
Lomanto 7744
Lombard 9741a
Loménie de Brienne see Pourret, P.A.
Long, G.W. De v. De Long, G.W.
Longchamp, P.B. Pennier de v. Pennier de Longchamp, P.B.
Longer, J. 8293
Looser, G. 8860
Lopez, H. Ruiz v. Ruiz Lopez, H.
Lopez Ruiz, S.J. v. Ruiz, S.J. Lopez
Lorentz, P.G. see Müller, H.L.H., Müller, Karl (Halle)
Lorenzana see Lorenz, A.
Lorenzochloa see Parodi, L.R.
Lortet, L.-C. see Roffavier, G.
Lort-Phillips, E. see Phillips, W.
Loudon, J.C. 8584
Louis, L. 6372
Louise d'Orléans 8750
Louise von Preussen 7801
Louis Napoleon see Reinwardt, C.G.C.
Louis Philippe, King of France see Philippi, R.A.
Lounsberry, C.C. 7258
Lowe, E.J. 9661

Luces de Febres, Z. see Pittier, H.F.
Ludwig, C.F. see Radius, J.W.M.
Ludwig, C.G. 8836
Ludwig, C.L. 9270
Lühnemann, G.H. 7730
Lünemann, G.H. v. Lühnemann,G.H.
Luerssen, C. 8450; see Mez, C.C., Rutenberg, D.C.
Lütkemüller, J. 7423
Lütken, C.F. 9244
Luetke, F.P. 8197; see Postels, A.P.
Luhlenstein, L.P. v. Podstatzky Luhlenstein, L.
Luigné, S. de 7212
Lumholtz, C. 9286
Lund, P.W. see Riedel, L.
Lund, S. 9555
Lundqvist, J. 9093
Lusina, G. 7968
Lutz, K.G. 9583
Luxford, G. see Newman, E.
Lyell, C. 8486
Lynge, B.A. 8450
Lyon, H.L. 9325
Lyons, I. see Phipps, C.J.

Maack, R. 8787, 9810
Maas, P.J.M. 8393
Mabille, P. see Revelière, P.
McCormick, R. see Parry, W.E.
Macedo, M.A. de v. Macedo, A. de
McGillivray, D.J. 9789
MacGillivray, W. 9138
Machado, M. 8857
Mackenzie, D. 9720
Mac Kerc, M.J. v. M'Ken, M.J.
M'Mahon, B. 9721
Macoun, J. 7179, 7586; see Pearson, W.H.
Mächtig, H. 8775
Maehara, K. v. Mayebara, K.
Maffei, S.L. 8135
Magnin, A. see Paillot, J.
Magnol, A. see Magnol, P.
Magnol, P. 8023
Magnus, P.W. 8211
Magnusson, A.H. 8450
Maiden, J.H. 8046
Maille, A. 8378
Main, J. 7553
Maire, R.C.J.E. 9931; see Petitmengin, M.G.C.
Makielski, L. see Newcombe, F.C.
Malaspina, A. see Née, L.
Malbranche, Mrs. see Malbranche, A.F.
Malesherbes, C.G. de Lamoignon de see Rousseau, J.J.
Malinvaud, L.J.E. 9711; see Poisson, H.L., Prillieux, É.E., Rouy, G.C.C., Roze, É., Saint-Lager, J.B.

1179

INDEX TO NAMES

Malmberg, A.J. v. Mela, A.J.
Malme, G.O. see Regnell, A.F.
Malperia see Palmer, Edward
Malte, M.O. 5197
Maly, J.C. v. Maly, J.K.
Mancini, V. 9898
Mangin, L.-A. 8418
Mangin, L.N. 7495
Mansel, J.C. v. Mansel-Pleydell, J.C.
Mappus, M., Sr. see Mappus, M.
Maratti, G. 9943
Marcgrave, G. 5546; see Piso, W.
Marchand, L. see Redouté, P.J.
Marggraf, G. v. Marcgrave, G.
Marichal, H.N. see Pontarlier, N.C.
Marie-Antoinette see Redouté, P.J.
Marie d'Orléans 8750
Marie Louise see Redouté, P.J.
Marie-Victorin, Frère see Rousseau, J.J.J.J.
Marin, L., Jr. 8421
Markgraf, F. 8393
Marloth, H.W.R. 8132
Marquardt, J. 9254
Marshall, J.J. 9298
Martelli, U. 8726, 9325
Martens, É. see Martens, M.
Martens, G.M. von see Martens, E.K. von, Rabenhorst, G.L.
Martens, M. 8607
Martin, B. 8242
Martin, J. see Rudge, E.
Martin, J.N. 7257, 7258
Martin, W. 9153
Martindale, I.C. see Planchon, J.É., Redfield, J.H., Rugel, F.
Martinez, J. Quer y v. Quer y Martinez, J.
Martínez de la Escalera, F. 7514
Martini, B. v. Martinis, B. de
Martinis, B. de 9937
Martins, C.F. 9138
Martius, C.F.P. von 7390, 8353, 9361; see Marchal, É., Masters, M.T., Maximilian A.P. zu Wied-Neuwied, Meisner, C.P., Mez, C.C., Micheli, M., Mueller Argoviensis, J., Müller, K.A.E., Nees von Esenbeck, C.G.D., Panzer, G.W.F., Parodi, D., Petersen, O.G., Peyritsch, J.J., Pohl, J.B.E., Progel, A., Radlkofer, L.A.T., Reissek, S., Rohrbach, P.
Martyn, T. 9686
Marum, M. van see Persoon, C.H., Reinwardt, C.G.C.
Mary II, Queen of England see Plukenet, L.
Mascon, A.A. von 9715
Massalongo, A.B. see Parlatore, F., Pollini, C.

Massalongo, C.B. 9933
Masters, M.T. 8685
Mathieu, C. 8775
Maton, W.G. 8398
Mattioli, P.A.G. see Moretti, G.
Mattirolo, O. see Passerini, G.
Mattolini, A. 7374, 7376
Mattuschka, H.G. von see Rabenhorst, G.L.
Maugini, A. 7270
Maurer, H. 8775
Maury, P.J.B. 7475, 7477
Maximilian A.P. zu Wied Neuwied 6686
Maximilian I, Emperor of Austria 7808
Maximilian Joseph I 5519, 5522-5523, 5526, 5528
Maximowicz, C.J. see Przewalski, N.M., Ruprecht, F.J.
Maxon, W.R. 5090; see Parsons, M.E.
Maxwell, F.B. 7659
Mayer, A. 9462
Mayer, E. 8775
Mayor, E. 4562; see Rosenstock, E.
Mayoral see Pourret, P.A.
Mayr, G. 8726
Mazé, H.P. 7772
Mazziari, A.D. see Pieri, M.T.
Medikus, L.W. see Medikus, F.K.
Meinshausen, C. see Ruprecht, F.J.
Meisner, C.F. 6098, 6120
Meisner, C.F.A. see Meisner, C.F.
Meissner 8024
Melchior, H. see Pilger, R.K.F.
Mell, C.D. 8739
Melliss, Mrs. J.C. 5815
Memminger, E.R. see Ravenel, H.W.
Mendel, G.J. 5287; see Makowsky, A., Peter, A.
Meneghini, G.G.A. see Pantanelli, D.
Meneses, C.A. de C*v. Menezes, C.A. de
Menici, A. 7372
Mennega, A.M.W. 8393
Menzel, C. 9120
Menzel, P.J. 8217
Mercier de Coppet, E. 9106
Merck, J.H. see Mueller, J.S.
Merkus Doornik, J.J.F.H.T. 9488
Merrill, E.D. 5678, 7255, 8550, 8558, 8575, 8585, 8592, 9211, 9784
Mertens, C.F. 9629
Mertens, F.C. v. Mertens, F.K.
Mertens, F.K. 9377; see Rohde, M.
Mertens, K.H. see Persoon, C.H., Postels, A.P.
Merwe, F.Z. van der 8132
Mesa, J.T. Roig y v. Roig y Mesa, J.T.
Meschinelli, A.L. 5874-5875, 9898
Mesrop, Bishop 9234
Messala, Count see Margot, H.

Messeri, A. *see* Pichi-Sermolli, R.E.G.
Méthérie, M. de la 7216
Mettenius, C. *v.* Braun, C.
Mettenius, G.H. *see* Reichenbach, H.G.
Metzger, B. 6662
Metzler, J.A. *see* Milde, C.A.J.
Meunier, E.S. 9663
Meusel, H. *see* Rothmaler, W.H.P.
Mexia, J.A. *see* Mexia, Y.E.J.
Meyen, F.J.F. *see* Münter, J.A.H.A.J.
Meyer, B. 8593
Meyer, C.A. von *see* Petunnikov, A.N., Ruprecht, F.J.
Meyer, E.H.F. 7509, 8293, 8352
Meyer, G.F.W. 9367; *see* Rodschied, E.K.
Meyer, K. *see* Pax, F.A.
Meyer, L. 9427
Meyer, Mrs. C.A. von *see* Ruprecht, F.J.
Mez, C.C. 7684, 8272; *see* Nicholson, W.E.
Michaux, A. 5961, 8551, 9137; *see* Richard, L.C.M.
Michaux, F.A. 5958; *see* Michaux, A.
Micheletti 5605
Micheli, M. *see* Rapin, D.
Micheli, P.A. 7268, 9935
Middendorff, A.T. von 9803
Migula, E.F.A.W. 7423, 8450, 8916; *see* Richter, P.G.
Mikan, J.C. 8101
Milan, King of Serbia *see* Petrovič, S.
Mildbraed, G.W.J. *see* Pax, F.A., Perkins, J.R., Pilger, R.K.F.
Milde, C.A.J. 7571
Miller, J. *v.* Mueller, J.S.
Miller, P. *see* Richardson, R.
Miller, R. 9505
Millspaugh, C.F. 7507; *see* Patterson, H.N.
Milne, C. 9153
Minden, M.D. von 7423
Minto *see* Raffles, T.S.B.
Miquel, F.A.W. 5895, 6694, 6703, 8779, 8790, 9151; *see* Mueller, Karl (Halle), Nees von Esenbeck, C.G.D.Persoon, C.H., Petermann, W.L., Pfeiffer, L.K.G., Pulle, A.A., Rauwenhoff, N.W.P., Regel, E.A. von, Regnell, A.F., Reichenbach, H.G., Reichenbach, H.G.L., Reinwardt, C.G.C., Roemer, F. von
Mirbel, C.F.B. de 7212, 9155
Mitch. v. Mitchell, J. (1711-1768)
Mitchell, J. (1711-1768) *6129*
Mitchell, J. (1762-?) *6130*
Mitchell, W.R. 7659
Mitchella see Mitchell, J. (1711-1768)
Mitchill, S.L. 8551, 8555, 8560
Mitch. Stanst., J. v. Mitchell, J. (1762-?)

Mitscherlich, E.A. *see* Redtenbacher, J.
Mitten, W. *see* Mann, G., Mason, F., Menzies, A., Royle, J.F.
Mittenburg, L. Mitterpacher von *v.* Mitterpacher von Mittenburg, L.
Mitterpacher von Mittenburg, L. 7939
Moçiño, J.M. *see* McVaugh, R., Menzies, A., Pavon y Jiménez, J.A.
Moebius, M.A.J. *see* Pfitzer, E.H.H., Renner, O.
Möderndorf, M. Jabornegg von Gamsenegg und *v.* Jabornegg von Gamsenegg und Möderndorf, M.
Möhl, I. 8775
Möller, A. *v.* Möller, F.A.G.J.
Mönkemeyer, W. 7423, 8450
Mørch, O.M.L. 9244
Mössler, J.C. 8880; *see* Reichenbach, H.G.L.
Moggridge, M. *see* Moggridge, J.T.
Mohr, D.M.H. 9121
Mohs, F. 8754
Moldenhawer, J.J.P. *see* Oudemans, C.A.J.A.
Moldenke, H.N. 8393; *see* Moldenke, C.E.
Molina, J.I. *v.* Molina, G.I.
Molisch, H. *see* Oppenheimer, H.R., Ruttner, F.
Molla, A.D. 8258, 8263
Moller, A.F. 9941
MOLLER *see* Postels, A.P.
Moloney, A. *v.* Moloney, C.A.
Mongez 9741a
Monnard, J.P. 9399
Monnet, R. 6226
Montagne, J.P.F.C. 8353, 9679, 10.000-10.001, 10.005; *see* Pfund, J.D.C., Sagra, R. de la
Monti, G. 6261
Monti, G.L. see Moris, G.G.
Monti, L. *see* Monti, G.L.
Montin, L.J. *see* Quer y Martinez, J.
Moore *see* Plukenet, L.
Moore, S. le Marchant 9065, 9067-9071, 9210
Moquin-Tandon, C.H.B.A. 10.036; *see* Poiret, J.L.M., Saint-Hilaire, A.F.C.P. de
Moreau, Mrs. F. *v.* Jeanneau, V.
Moretti, G. *see* Pritzel, G.A.
Morey, F. *see* Rayner, J.F.
Morgagni *see* Phoebus, P.
Morgan, C.L. *see* Prentiss, A.N.
Morgan, D.J. *see* Phillips, R.W.
Morgenbesser, M. 5059
Mori, A. 7361
Mori, K. 5576
Moricand, Jr. *see* Moricand, M.É.
Moricand, S. *v.* Moricand, M.É.

1181

INDEX TO NAMES

Morière, P.G. 9275
Moriot, J. see Peragallo, M.
Moris, J.H. v. Moris, G.G.
Morison, J.H. 7913
Morong, T. see Reverchon, J.
Morren, C.J.É. see Morren, C.F.A.
Morris, L. see Pennant, T.
Morris, R. see Pennant, T.
Morris, W. see Pennant, T.
Morse, W.J. 7952
Mosén, C.W.H. see Regnell, A.F.
Moss, M.B. 9212
Mougeot, A. v. Mougeot, J.A.
Mougeot, J.A. de see Quélet, L., Roumeguère, C.
Mougeot, J.B. 6942-6945; see Persoon, C.H., Requien, E., Roussel, A.V.
Moulin, V.J. de Saint v. Saint-Moulin, V.J. de
Moultou 9686
Mousset, J.P. see Rosenstock, E.
Moziño Suarez de Figueroa, J.M. v. Moçiño, J.M.
Müggenburg, S. Schulzer von v. Schulzer von Müggenburg, S.
Müller, Fritz see Möller, F.A.G.J.
Mueller, J.S. 9828
Müller, K.A.E. 8211
Müller, Karl (Freiburg) 5643, 8450
Müller, Karl (Halle) 7759, 8915; see Mohr, D.M.H., Peters, W.C.H., Polakowsky, H., Rutenberg, D.C.
Mueller, O. 9397
Müller, W.O. 7175-7176, 7940
Mueller Argoviensis, J. 8744, 8915; see Rutenberg, D.C.
Münster, A. 8793
Münsterhjelm, L. 8343
Mütter, O. 7544
Muhlenberg, G.H.E. 9153; see Palisot de Beauvois, A.M.F.J., Persoon, C.H., Pursh, F.T., Ott, J., Rafinesque-Schmaltz, C.S., Roth, A.W.
Muir, D. v. Moore, D.
Muir, J. (1838-1914) *6539*
Muir, J. (1874-1914) see Muir, J. (1838-1914)
Mulder, C. v. Mulder, N.
Mulder, G.J. 7737
Mulder, N. *6540*, 9138
Mulford see Rusby, H.H.
Mulgrave, C.J. Phipps v. Phipps, C.J.
Muller, F. 7956
Muller, F.M. 8393
Muller, P.J. v. Müller, P.J.
Munck af Rosenschöld, E. see Parodi, D.
Munnicks, J. 9123
Murphy and Co. see Mason, F.A.
Murr, J. 8885; see Pöll, J.
Murray, A. 8685

Murray, G.R.M. 9065
Murray, J.A. 4709, 6845a, 7726, 8835, 9408; see Reichard, J.J.
Mussat, É.V. 9898
Musset-Pathay, V.D. 9688
Mutis y Boscio, J.C.B. 5428

Nägeli, C.W. von 9344; 9957; see Moritzi, A., Peter, A., Porcius, F., Prantl, K.A.E., Regel, E.A. von
Nakai, T. 6141
Nalepa, A. 8726
Nangis Regnault, G. de 8810
Nannenga, E.T. 8393
Napoleon Bonaparte see Nectoux, H., Redouté, P.J.
Nassau, Johan Maurits van v. Johan Maurits van Nassau
Naudin, C.V. 6406
Née, L. see Née, I.C.S., Pavon y Jiménez, J.A., Pourret, P.A.
Nees von Esenbeck, C.G.D. 6097, 8486, 8962; see Nees von Esenbeck, T.F.L.
Nees von Esenbeck, T.F.L. 6678, 8293; see Miquel, F.A.W., Putterlick, A.
Negri, G. 7971-7972; see Pampanini, R.
Nemours, P.S. Dupont de v. Dupont de Nemours, P.S.
Neobathiea see Perrier de la Bâthie, H.
Neopallasia see Pallas, P.S.
Neoparodia see Parodi, D.
Neoparrya see Parry, C.C.
Neopatersonia see Paterson, W.
Neopaxia see Pax, F.A.
Neopeckia see Peck, C.H.
Neoporteria see Porter, C.E.
Neopreissia see Preiss, L.
Neopringlea see Pringle, C.G.
Neorapinia see Rapin, D.
Neoravenelia see Ravenel, H.W.
Neoregelia see Regel, C.A. von
Neoregnellia see Regnell, A.F.
Neorehmia see Rehm, H.S.L.F.F.
Neoroepera see Roeper, J.A.C.
Neorudolphia see Rudolphi, K.A.
Neorutenbergia see Rutenberg, D.C.
Neosaccardia see Saccardo, P.A.
Nerici, G. 8496
Neumeyer, G. see Prantl, K.A.E.
Newton, I. 9941
Newton, W.C.F. see Newton, L.
Nicolai I, Czar of Russia 8197
Nicolson, D.H. 9724
Nilsson, S. 9241
Nilsson-Degelius, G.B.F. v. Nilsson, G.B.F.
Ninni, A.P. 9883
Nissolle, G. v. Nisolle, G.
Niven, D.J. v. Niven, J.
Nobbe, F. see Neger, F.W.

Node, S.A. 9695, 10.051
Noezens see Pallas, P.S.
Nonne, J.P. 8032
Nordgren, W.A. 8373
Nordstedt, C.F.O. see Reinsch, P.F., Richter, P.G., Rosenvinge, L. Kolderup, Roy, J.
Nordwall, E. 4848
Norman, C. 9071
Norman, G. see Philip, R.H., Rylands, T.G.
Noroña, F. see Radermacher, J.C.M.
Norris 7913
Northrop, J.I. 6892
Northumberland, Percy H. of v. Percy, H., of Northumberland
Nouaille, A. v. Rudge, A.
NOVARA 8845
NOWELL, W. see Nowell, J.
Nuttall, L.W. 6073
Nuttall, T. 8572, 9153; see Rafinesque-Schmaltz, C.S.
Nylander, W. 5290, 6743, 8744, 9979; Massalongo, A.B., Mudd, W.A., Picquenard, C.-A., Ripart, J.B.M.J.S.E., Saelan, T.
Nyman, C.F. 9639

Oberwimmer, A. 8726
Oeder, G.C. see Rosenberg, C.F., Royen, A. van
Oersted, A.S. see Reinke, J.
Oertel, C.G. 8450
Østrup, E. 9548
Offner, J. 7701
Oken, L. 8289
Oldenburg, G.C. Oeder von v. Oeder, G.C. von Oldenburg
Oliveira, P. Erichsen de v. Erichsen de Oliveira, P.
Oliver, D. 9065
Olivier, C.A. see Olivier, E.
Olmen de Poederlé, E.J.C.G.H. d' v. Poederlé, E.J.C.G.H. d'Olmen
Olsson, P.H v. Olsson-Seffer, P.H.
Oltmans, F. see Rawitscher, F.
Olufsen, O. 7530-7531
Ommering, A. van 9784
Oort, A.J.P. 8393; see Pulle, A.A.
Ooststroom, S.J. van 8393; see Pulle, A.A.
Ophioparodia see Parodi, D.
Opiz, P.M. 8293, 9377; see Petters, F., Poech, J.A., Pohl, J.B.E., Purkyně, E, von, Ruprecht, F.J.
Orbigny, C.V.D. d' 6248-6251
Orloff, Count 8405
Orpheus see Reichenbach, C.L. von
Ortega, C.G. de 8421; see Paláu y Verdéra, A., Pourret, P.A., Ruiz Lopez, H.

Ortgies, E. 8775
Ortolani, G.E. 8554
Ostenfeld, C.E.H. 6157, 8654
Otero, L. Pagán de v. Pagán de Otero, L.
Otto, C.F. 7817
Otto, Ed. *7140-7141*
Otto, Ernst *7142*
Otto, F. see Otto, Ed.
Oudemans, C.A.J.A. 6107
Oudendorp, F. van 9727
Oudinot 9976

Pabst, C. *alph.*
Pabst, G. *7175-7176*
Pabst, G.F.J. see Pabst, G.
Pabstia see Pabst, G.
Paccard, E. *7177*
Pacher, D. *7178*
Packard, A.S., Sr. see Packard, A.S.
Packard, A.S. *7179*
Packe, C. *7180*
Pacz. v. Paczoski, J.C.
Paczoski J.C. *7181-7182*
Pagán de Otero, L. *7135*
Page, M.M. see Page, W.B.
Page, W.B. *7183*
Pagella see Page, W.B.
Paget, C.J. *7184*
Paget, J. *7184*
Pagetia see Paget, J.
Paglia, E. *7185*
Pahnsch, G. *7186*
Paiche, P. *7187*
Paige, E.W. *7188*
Paill. v. Paillot, J.
Pailleux, A. *7189-7190*
Pailleux, N.A. v. Pailleux, A.
Paillot, J. *7191*, 8984
Paine, J.A. *7192*
Painter, W.H. *7193*
Palacký, J.B. *7194-7196*
Palackya see Palacký, J.B.
Palanza, A. *7197*
Palasso, P.B. v. Palassou, P.B.
Palassou, P.B. *7198-7200*
Palaua see Paláu y Verdéra, A.
Paláu y Verdéra, A. *7201-7203;* see Ruiz Lopez, H.
Palava see Paláu y Verdéra, A.
Palavia see Paláu y Verdéra, A.
Paley, F.A. *7204*
Paley, W. see Paley, F.A.
Paleya see Paley, F.A.
Paléz. v. Palézieux, P.
Palézieux, P. de *7205*
Palhinha, R.T. *alph.*
Palhinhaea see Palhinha, R.T.
Palib. v. Palibin, I.V.
Palibin, I.V. *7206-7209*

INDEX TO NAMES

Palibinia see Palibin, I.V.
Palibiniopteris see Palibin, I.V.
Palisota see Palisot de Beauvois, A.M.F.J.
Palisot de Beauvois, A.M.F.J. *7210-7218*
Pall., J. v. Pallas, J.D.
Pall. v. Pallas, P.S.
Palla, E. *7219-7220*, 8726
Pallas, J.D. *7221*
Pallas, P.S. *7222-7230*
Pallasa see Pallas, P.S.
Pallasia see Pallas, P.S.
Pallassia see Pallas, P.S.
Pallisat de Beauvois, A.M.F.J. *v.* Palisot de Beauvois, A.M.F.J.
Palm, B.T. *7230a-7230c*, 9530
Palmberg, J.O. *7230d*
Palmer, C.M. *7231*
Palmer, Edward *alph.*, 9507
Palmer, E.J. *7233*
Palmer, Elmore *7232*
Palmer, F.T. *7234*
Palmer, J.A. *7237-7238*
Palmér, J.E. *7235*
Palmer, J.F. *see* Palmer, J.A.
Palmer, Johann Ludwig *7236*
Palmer, John Linton *see* Palmer, J.A.
Palmerella see Palmer, Edward
Palmeria see Palmer, J.A.
Palmeriamonas see Palmer, C.M.
Palmerocassia see Palmer, Edward
Palmgr. v. Palmgren, A.
Palmgren, A. *7239-7246*
Palmomyces see Palm, B.T.
Palmquist, E.M. 9197
Palmstr. v. Palmstruch, J.W.
Palmstruch, J.W. *7247*
Palmstruckia see Palmstruch, J.W.
Palouz. v. Palouzier, É.
Palouzier, É. *7248*
Palun, M. *7249*
Palustre *7250*
Pammel, E.C. *7258*
Pammel, L.H. *7251-7258*
Pamp. v. Pampanini, R.
Pampanini, R. *7259-7274*, 9949
Pamplin, W. *7275*
Pampuch, A. *7276*
Pančic, J. *7277-7284*
Pancher, J.A.I. *alph.*
Pancheria see Pancher, J.A.I.
Panchezia see Pancher, J.A.I.
Panchic, J. *v.* Pančic, J.
Panciatichi, N. 7887
Pancio, G. *v.* Pančic, J.
Pandé, S.K. *alph.*
Panizzi-Savio, F. *7285*
Pansch, A. *alph.*
Pant. v. Pantocsek, J.
Pantan., D. v. Pantanelli, D.
Pantan., E. v. Pantanelli, E.F.

Pantanelli, D. *7286; see* Pantanelli, E.F.
Pantanelli, E.F. *7287*
Pantel, C. *7288*
Panțu, Z.C. *7295-7299*
Pantl. v. Pantling, R.
Pantling, R. *7289*
Pantlingia see Pantling, R.
Pantocsek, J. *7290-7294; see* Murbeck, S.S.
Pantocsekia see Pantocsek, J.
Pantzu, Z.V. *v.* Panțu, Z.V.
Panz. v. Panzer, G.W.F.
Panzer, G.W. *7302; see* Panzer, G.W.F.
Panzer, G.W.F. *7300-7303; see* Roemer, J.J.
Panzera see Panzer, G.W.F.
Panzeria see Panzer, G.W.F.
Paol. v. Paoletti, G.
Paoletti, G. *7304-7306*, 9898, 9907, 9945
Paolucci, L. *7307-7310*
Pape, G.K. von *alph.*
Pape, W. 9796, 9811
Papenfuss, G.F. *alph.*
Papenfussia see Papenfuss, G.F.
Papenfussiella see Papenfuss, G.F.
Papenfussiomonas see Papenfuss, G.F.
Papp, C. *7311-7312*
Pappe, C.W.L. *v.* Pappe, K.W.L.
Pappe, K.W.L. *7313-7317*
Pappea see Pappe, K.W.L.
Pâque, É. *7318-7322*
Pâques, É. *v.* Pâque, É.
Par. v. Paris, É.G.
Pardé, L.G.C. *7323-7325*
Pardo de Tavera, T.H.J. *7326-7327*
Parey, P. *see* Maatsch, R.F.T.
Parfitt, E. *alph.*
Paris, É.G. *7328-7332*
Paris, J.É.G.N. *v.* Paris, É.G.
Parish, C.S.P. *see* Parish, S.B.
Parish, S.B. *7333-7336*
Parish, W.F. *7333; see* Parish, S.B.
Parishella see Parish, S.B.
Parishia see Parish, S.B.
Parisia see Paris, É.G.
Parisot, C.L. *7337*
Park, M. see Parke, M.
Park, O.W. *7258*
Parka see Parke, M.
Parke, H.C. *see* Nash, G.V.
Parke, M. *7338*
Parke-Davis & Co. 9818; *see* Rusby, H.H.
Parker, A.K. *see* Parker, R.N.
Parker, C.S. *see* Parker, R.N.
Parker, R.N. *7339-7340*
Parkerella see Parker, R.N.
Parkeria see Parker, R.N.
Parkeroidea see Parker, R.N.
Parkhurst, H.E. *7341*

Parkia see Parke, M.
Parkinson, C.E. *7342*
Parkinson, James *7343*, 5368
Parkinson, John *alph.;* 4909, 6779
Parkinson, Stanfield *7344*
Parkinson, Sydney *7344*
Parkinsona see Parkinson, John
Parkinsonia see Parkinson, John
Parks, H.B. *7345*
Parks, H.E. *alph.*
Parksia see Parks, H.E.
Parlatore, F. *7346-7376*, 6795, 8375
Parlatorea see Parlatore, F.
Parlatoria see Parlatore, F.
Parm., A. v. Parmentier, A.A.
Parm., J. v. Parmentier, J.J.G.
Parm., P. v. Parmentier, P.E.
Parmentaria see Parmentier, A.A.
Parmentariomyces see Parmentier, A.A.
Parmentier, A.A. *7377-7378*, 9741a; *see* Molina, G.I.
Parmentier, J.J.G. *7380-7381a*
Parmentier, P.E. *7382-7387*
Parmentiera see Parmentier, A.A.
Parmentieria see Parmentier, A.A.
Parn. v. Parnell, R.
Parnell, R. *7388-7389*
Parodi v. Parodi, L.R.
Parodi, D. *7390-7393; see* Parodi, L.R.
Parodi, L.R. *7394-7400*
Parodia see Parodi, D.
Parodianthus see Parodi, L.R.
Parodiella see Parodi, D., Parodi, L.R.
Parodiellina see Parodi, D.
Parodiellinopsis see Parodi, D.
Parodiodendron see Parodi, L.R.
Parodiodia see Parodi, D.
Parodiodoxa see Parodi, L.R.
Parodiopsis see Parodi, D.
Parrique, G. *7401*
Parrot, J.J.F.W. *7402*
Parrotia see Parrot, J.J.F.W.
Parrotiopsis see Parrot, J.J.F.W.
Parry v. Parry, C.C.
Parry, C.C. *7403-7408*
Parry, Mrs. E.R. 7408; *see* Parry, C.C.,
Parry, W.E. *7409-7414; see* Ross, J.C., Sabine, E.
Parrya see Parry, W.E.
Parryella see Parry, C.C.
Parryodes see Parry, W.E.
Parryopsis see Parry, W.E.
Parseval-Grandmaison, J. de 8688
Parsons, F.T. *7415*
Parsons, J. *see* Parsons, F.T.
Parsons, M.E. *7416*
Parsons, T.H. *7417*
Parsonsia see Parsons, F.T.
Parvela, A.A. *7418*
Pasaccardoa see Saccardo, P.A.

Pascal, D.B. *alph.*
Pascalia see Pascal, D.B.
Pasch. v. Pascher, A.
Pascher, A. *7419-7424*, 8450; *see* Migula, E.F.A.W., Paul, H.K.G.
Pascherella see Pascher, A.
Pascheriella see Pascher, A.
Pascherina see Pascher, A.
Pascherinema see Pascher, A.
Pasq. v. Pasquale, G.A.
Pasq., F. v. Pasquale, F.
Pasquale, F. *7425*
Pasquale, G.A. *7426-7433*, 7425; *see* Pasquale, F.
Pass. v. Passerini, G.
Passerini, G. *7434-7437*
Passerini, N. *see* Passerini, G.
Passeriniella see Passerini, G.
Passerinula see Passerini, G.
Passy, A.F. *7438*
Pasteur, L. *see* Pouchet, F.A.
Pat. v. Patouillard, N.T.
Patarolo, F.R. 8143-8144
Paterson, T.V. *see* Paterson, W.
Paterson, W. *7439*
Patersonia see Paterson, W., Patterson, F.
Pathersonia see Paterson, W., Patterson, F.
Patouillard, N.T. *7440-7495*
Patouillardea see Patouillard, N.T.
Patouillardiella see Patouillard, N.T.
Patouillardina see Patouillard, N.T.
Patoullardiella see Patouillard, N.T.
Patout, M.R. *7496*
Patraw, P.M. *7497*
Patrick, R.M. *alph.*
Patrick, R.W. 9486
Patrick, W. *7498*
Patris, J.B. *see* L'Héritier de Brutelle, C.-L.
Patschosky, J.C. *v.* Paczoski, J.C.
Patschotsky, J.C. *v.* Paczoski, J.C.
Patt., H. v. Patterson, H.N.
Patt., P. v. Patterson, P.M.
Patterson, E. *see* Patterson, F.
Patterson, E.H.N. *see* Patterson, H.N.
Patterson, F. *7499-7501*
Patterson, H.N. *7502-7507; see* Rusby, H.H.
Patterson, P.M. *alph.*
Pattersonia see Paterson, W., Patterson, F.
Pattison, S.R. *7508*
Patzak, A. *see* Rechinger, K.H.
Patze, C.A. *7509*
Patzea see Patze, C.A.
Patzelt, J.E. *7510*
Pau y Español, C. *7511-7516*
Paua see Pau y Español, C.
Pauahi, B. *see* Pau y Español, C.
Pauahia see Pau y Español, C.
Pauella see Pau y Español, C.

Pauia see Pau y Español, C.
Paul. v. Paulet, J.J.
Paul, H.K.G. *7517-7518*, *7423*
Paul, J.T. *see* Paul, H.K.G.
Paulet, J.J. *7519-7523*
Pauletia see Paulet, J.J.
Pauli, J. *v.* Paulli, J.
Paulia see Paul, H.K.G.
Paulin, A. *7524-7526*
Paulina see Paul, H.K.G.
Paull, C.L.F. *7527*
Paulli, J. *7528*
Paulli, S. *see* Paulli, J.
Paullia see Paulli, J.
Paullinia see Paulli, J.
Paullowsky *8784*
Paulomyces see Paul, H.K.G.
Paulophyton see Paul, H.K.G.
Paulsen, O.V. *7529-7531*, *9548*; see Ostenfeld, C.E.H.
Paulsenella see Paulsen, O.V.
Paulseniella see Paulsen, O.V.
Paulson, R. *alph.*, *9071*
Pauquy, C.L.C. *7532-7533*
Pauwels, J.L.H. *7534*
Pav. v. Pavon y Jiménez, J.A.
Pavillard, Jean *see* Pavillard, Jules
Pavillard, Jules *7535-7537*
Pavillardia see Pavillard, Jules
Pavillardinium see Pavillard, Jules
Pavlov, N.V. *7538*
Pavlova see Pavlov, N.V.
Pavona see Pavon y Jiménez, J.A.
Pavonia see Pavon y Jiménez, J.A.
Pavon y Jiménez, J.A. *7539-7540*, *9769*, *9771-9772*, *9775*, *9779*, *9782*; *see* Ruiz Lopez, H.
Pawl. v. Pawlowski, B.
Pawlowski, B. *7541*
Pax, F.A. *7542-7552*, *7978*, *8211*, *8268*; *see* Müller, E.G.O., Porcius, F., Prantl, K.A.E.
Paxia see Pax, F.A.
Paxina see Pax, F.A.
Paxiodendron see Pax, F.A.
Paxiuscula see Pax, F.A.
Paxt. v. Paxton, J.
Paxton, J. *7553-7556*
Paxtonia see Paxton, J.
Payer, J.-B. *7557-7564*
Payera see Payer, J.-B.
Payeria see Payer, J.-B.
Payne, F.W. *7565*
Payot, V. *7566-7574*
Payrau, V. *7575*
Payson, E.B. *7576-7579*
Pazschke, F.O. *8450*
Peacock, J.T. *7580*
Peake, R.B. *9720*
Peale, C.W. *7210*

Pearl, R. *7581*
Pearless, Mrs. J. *v.* Pratt, A.
Pearsall, W. Harold *see* Pearsall, W. Harrison
Pearsall, W. Harrison *alph.*
Pearson v. Pearson, W.H.
Pearson, A.A. *7582*
Pearson, H.H.W. *7583*
Pearson, W.H. *7584-7591*
Pearsonia see Pearson, H.H.W.
Pearsoniella see Pearson, H.H.W.
Peary, R.E. *9864*
Pease, A.S. *7592*
Peattie, D.C. *7593*
Pech, J. *see* Pourret, P.A.
Peck *v.* Peck, C.H.
Peck, C.H. *7594-7602*; *see* Saccardo, P.A.
Peck, D. *see* Peck, W.D.
Peck, F.G.M. *7603*
Peck, M.E. *7604*
Peck, R.E. *see* Peck, C.H.
Peck, W.D. *7605*
Peckia see Peck, C.H., Peck, W.D.
Peckichara see Peck, C.H.
Peckiella see Peck, C.H.
Peckifungus see Peck, C.H.
Peckisphaera see Peck, C.H.
Peckolt, T. *7606-7607*
Peckoltia see Peckolt, T.
Pedersen, M. *v.* Porsild, M.P.
Pedicino, N.A. *7608*
Pedoni, L. *9937*
Pedro, of Brazil *10.042*
Peekel, G. *7609*
Peekelia see Peekel, G.
Peekeliodendron see Peekel, G.
Peekeliopanax see Peekel, G.
Pée-Laby, E. *7610*
Pehersd. v. Pehersdorfer, A.
Pehersdorfer, A. *7611-7613*
P'ei Chien *7614*
Peirce, G.J. *alph.*
Peirson, F.W. *7615*
Pellanda, G. *7616*
Pellegr. v. Pellegrin, F.
Pellegrin, F. *7617-7618*
Pellegrinia see Pellegrin, F.
Pellegriniodendron see Pellegrin, F.
Pellet. v. Pelletier, J.
Pelletan, J. *7619-7620*
Pelleteria see Pelletier, J.
Pelletier, J. *7621*
Pelletier, P.J. *v.* Pelletier, J.
Pelletiera see Pelletier, J.
Pelletieria see Pelletier, J.
Pelloe, E.H. *7622-7623*
Pelloe, Mrs. T. *v.* Pelloe, E.H.
Pelourde, F. *7624*
Pelourdea see Pelourde, F.
Pena *v.* Pena, P.

INDEX TO NAMES

Pena, P. *7625*
Peña, R. *7626*
Penada, J. *see* Nardo, G.D.
Penaea see Pena, P.
Penantia see Pennant, T.
Penard, E. *7627-7628a*
Penardia see Penard, E.
Penfold, J.W. *7629; see* Robley, Mrs. A.J.
Penh. v. Penhallow, D.P.
Penhallow, D.P. *7630-7632*
Penhallowia see Penhallow, D.P.
Penland, C.W.T. *7633*
Pennant, T. *7634*
Pennanta see Pennant, T.
Pennell, F.W. *7635-7639,* 8585
Pennellia see Pennell, F.W.
Pennellianthus see Pennell, F.W.
Pennier de Longchamp, P.B. *7640*
Pennington, L.H. *alph.*
Penz. v. Penzig, A.J.O.
Penzig, A.G.O. *v.* Penzig, A.J.O.
Penzig, A.J.O. *7641-7656,* 9895, 9911
Penzigia see Penzig, A.J.O.
Penzigiella see Penzig, A.J.O.
Penzigina see Penzig, A.J.O.
Peola, P. *7657-7658*
Pepoon, H.S. *7659-7660*
Pepper, O.M. 7581
Perag., H. v. Peragallo, H.
Perag., M. v. Peragallo, M.
Peragallia see Peragallo, H.
Peragallo, H. *7661-7664,* 8726; *see* Peragallo, M.
Peragallo, M. *7665-7666,* 7620, 7664, 8726
Peralta, M.C. 8131
Pérard, A.J.C. *7667-7670*
Percy, H., of Northumberland *see* Pierce, N.B.
Pereb., C. v. Pereboom, C.
Pereb., N. v. Pereboom, N.E.
Pereboom, C. *7671; see* Pereboom, N.E.
Pereboom, N.E. *7672*
Pereira, H. *7673*
Pereira, J. *see* Pereira, H.
Pereiria see Pereira, H.
Pérez Arb. v. Pérez Arbeláez, E.
Perez, L. *see* Pérez Lara, J.M.
Pérez Arbeláez, E. *7674-7677*
Perezia see Pérez Lara, J.M.
Pereziopsis see Pérez Lara, J.M.
Pérez Lara, J.M. *7678*
Perezlaria see Pérez Lara, J.M.
Pérez Larios, J. *see* Pérez Lara, J.M.
Périer, L. 7769
Perini, A. *7679; see* Perini, C.
Perini, C. *7679*
Perkins v. Perkins, J.R.
Perkins, G.H. *7681-7682*

Perkins, J.R. *7683-7688; see* Ruhland, W.O.E.
Perkins, Mrs. E.E. *7680*
Perkolt, T. *v.* Peckolt, T.
Perktold, J.A. *alph.*
Perleb, C.J. *v.* Perleb, K.J.
Perleb, K.J. *7689-7692*
Perlebia see Perleb, K.J.
Pernh. v. Pernhoffer von Bärnkron, G.
Pernhoffer von Bärnkron, G. *7693*
Pernitzsch, H. *7694*
Perrault *see* Richard, A.
Perreymond, J.H. *7695*
Perreymondia see Perreymond, J.H.
Perrier, A. *alph.*
Perrier, E. v. Perrier de la Bâthie, E.P. de
Perrier, H. v. Perrier de la Bâthie, H.
Perriera see Perrier de la Bâthie, H.
Perrieranthus see Perrier de la Bâthie, H.
Perrierastrum see Perrier de la Bâthie, H.
Perrierbambus see Perrier de la Bâthie, H.
Perrier de la Bâthie, E.P. de *7696-7701; see* Perrier de la Bâthie, H.
Perrier de la Bâthie, H. *7702-7708*
Perrier de la Bâthie, J.M.H.A. *v. Perrier de la Bâthie, H.*
Perrieriella see Perrier de la Bâthie, H.
Perrierodendron see Perrier de la Bâthie, H.
Perrierophytum see Perrier de la Bâthie, H.
Perrin, I.S. *7709*
Perrin, Mrs. H. *v.* Perrin, I.S.
Perrine, H. *7710-7711*
Perring, W. 8775
Perrot, A. *see* Perrot, É.C.
Perrot, É.C. *7712-7717*
Perrotia see Perrot, É.C.
Perrotiella see Perrot, É.C.
Perrottet, G.G.S. *v.* Perrottet, G.S.
Perrottet, G.S. *7718-7720,* 9148
Perrottetia see Perrottet, G.S.
Perroud, L.F. *alph.*
Perry, Captain *see* Perry, L.M.
Perry, L.M. *7721-7722*
Perry, N.F. *v.* Bracelin, Mrs. H.P.
Perry, W.G. *7723-7723a*
Perrya see Perry, L.M.
Pers. v. Persoon, C.H.
Persohn, C.D. *v.* Persoon, C.D.
Persoon, C.D. *see* Persoon, C.H.
Persoon, C.H. *7724-7737,* 8876, 8881, 8885; *see* Mougeot, J.B., Raddi, G., Rogers, D.P.
Persoon, J.M. *v.* Storm, J.M.
Persoonia see Persoon, C.H.
Persooniana see Persoon, C.H.
Persooniella see Persoon, C.H.
Perss. v. Persson, N.P.H.
Persson, N.P.H. *7738*
Perssonia see Persson, N.P.H.
Perssoniella see Persson, N.P.H.

1187

Perthes, J. *see* Petermann, W.L.
Perthuis, de 9741a
Perty, J.A.M. *7739*
Pertya see Perty, J.A.M.
Pescott, E.E. *7740-7742*
Pestal. v. Pestalozzi, A.
Pestallozia see Pestalozzi, A.
Pestalopezia see Pestalozzi, A.
Pestalosphaeria see Pestalozzi, A.
Pestalotia see Pestalozzi, A.
Pestalotiopsis see Pestalozzi, A.
Pestalozza, F. *see* Pestalozzi, A.
Pestalozzi, A. *7743*
Pestalozzi, J.A. *see* Pestalozzi, A.
Pestalozzia see Pestalozzi, A.
Pestalozziella see Pestalozzi, A.
Pestalozzina see Pestalozzi, A.
Pestalozzites see Pestalozzi, A.
Pesta, O. 8726
Petagna v. Petagna, V.
Petagna, L. *alph.*
Petagna, V. *7744.*
Petagnaea see Petagna, V.
Petagnana see Petagna, V.
Petagnia see Petagna, V.
Petch, T. *alph.*
Petchia see Petch, T.
Pételot, A. *alph.*
Pételot, P.A. *v.* Pételot, A.
Petelotia see Pételot, A.
Petelotiella see Pételot, A.
Peter, A. *7745-7748,* 8211; *see* Porcius, F.
Peter, Czar of Russia *see* Ruysch, F.
Peter, G.A. *v.* Peter, A.
Peter, R. *see* Peter, A.
Péterfi, M. *alph.*
Péterfi, S. *see* Péterfi, M.
Peterfiella see Péterfi, M.
Peteria see Peter, A.
Peterm. v. Petermann, W.L.
Petermann, A.H. *see* Petermann, W.L.
Petermann, W.L. *7749-7757,* 9183
Petermannia see Petermann, W.L.
Peterodendron see Peter, A.
Peters, C. *v.* Peters, K.
Peters, K. *7758*
Peters, T.M. 6193
Peters, W.C.H. *7759,* 9456
Petersen, H. *7760,* 8090
Petersen, H.E. *7761-7761a*
Petersen, J.B. *7762-7763,* 9548
Petersen, K. *7765*
Petersen, O.G. *7766*
Petersen, R.H. 7595
Petersen, S. *7767,* 9609
Petersenia see Petersen, H.E.
Petersia see Peters, W.C.H.
Petersianthus see Peters, W.C.H.
Petesia see Peter, A.
Pethybr. v. Pethybridge, G.H.

Pethybridge, G.H. *alph.*
Petif, E. *see* Petif, J.F.C.L.C.
Petif, J.F.C.L.C. *7768*
Petif de la Gautrois, J.F.C.L.C. *v.* Petif, J.F.C.L.C.
Petimenginia see Petitmengin, M.G.C.
Petit, A. 9151; *see* Quartin-Dillon, R., Richard, A.
Petit, E.C.N. *alph.*
Petit, Félix *see* Petit, P.C.M.
Petit, François *see* Petit, P.C.M.
Petit, N.E.C. *v.* Petit, E.C.N.
Petit, P.C.M. *7769-7775,* 7620
Petita see Petit, P.C.M.
Petitia see Petit, P.C.M.
Petitm. v. Petitmengin, M.G.C.
Petitmengin, M.G.C. *7776-7777*
Petitmenginia see Petitmengin, M.G.C.
Petit-Radel, P. *alph.*
Petiver, J. *alph.; see* Plukenet, L.
Petivera see Petiver, J.
Petiveria see Petiver, J.
Petkoff, S.P. *7778*
Petkov, S.P. *v.* Petkoff, S.P.
Petkovia see Petkoff, S.P.
Petr. v. Petrak, F.
Petrak, F. *7779-7786; see* Petch, T., Picbauer, R., Podpěra, J.
Petrakia see Petrak, F.
Petrakiella see Petrak, F.
Petrakina see Petrak, F.
Petrakiopeltis see Petrak, F.
Petrakiopsis see Petrak, F.
Petrakomyces see Petrak, F.
Petrescu, C.C. *alph.*
Petri v. Petri, L.
Petri, F. *7787*
Petri, L. *7788*
Petri, R.J. *see* Petri, F.
Petrie, D. *7789*
Petriella see Petri, F., Petrie, D.
Petriellidium see Petri, F.
Petrov, V.A. *7790*
Petrová, J. *see* Petrov, V.A.
Petrovanella see Petrov, V.A.
Petrovič, S. *7791*
Petrovsky, A.S. *v.* Petrowsky, A.S.
Petrow, J. *7732*
Petrowski, A.S. *v.* Petrowsky, A.S.
Petrowsky, A.S. *alph.*
Petry, A. *7792*
Petry, L.C. 6531
Petter, F. *7793; see* Petters, F.
Petter, H.F.M. 8393
Pettera see Petter, F.
Petteria see Petter, F.
Petters, F. *7794; see* Petter, F.
Petterson, B.J. *7795*
Petunn. v. Petunnikov, A.N.
Petunnikov, A.N. *7796-7797*

Petzh. v. Petzholdt, G.P.A.
Petzholdt, G.P.A. *7798-7800*
Petzholdtia see Petzholdt, G.P.A.
Petzi, F. von S. *alph.*
Petzold v. Petzold, E.A.
Petzold, C.E.A. *v.* Petzold, E.A.
Petzold, E.A. *7801,* 8775
Petzold, K.F.A. *v.* Petzold, E.A.
Petzold, K.W. *v.* Petzold, W.
Petzold, W. *7802-7803*
Peyl, J. *alph.*
Peylia see Peyl, J.
Peyr. v. Peyritsch, J.J.
Peyre, B.L. *7804*
Peyritsch, J.J. *7805-7808*
Peyritschia see Peyritsch, J.J
Peyritschiella see Peyritsch, J.J.
Peyronel, B. *7809*
Peyronelia see Peyronel, B.
Peyronelina see Peyronel, B.
Peyronellaea see Peyronel, B.
Peyronellula see Peyronel, B.
Peyronie, Gauthier de la *v.* Gauthier de la Peyronie
Peyrou 9686
Pfeff. v. Pfeffer, W.F.P.
Pfeffer, W.F.P. *7810-7812; see* Lindroth, J.I., Newcombe, F.C., Pantanelli, E.F., Peirce, G.J., Popovici, A.P., Pringsheim, N., Rawitscher, F., Renner, O., Richards, H.M., Rothert, W.
Pfeiff. v. Pfeiffer, L.K.G.
Pfeiff., H. v. Pfeiffer, H.H.
Pfeiff., J. v. Pfeiffer, J.P.
Pfeiff., N. v. Pfeiffer, N.E.
Pfeiffer, C.G.L. *v.* Pfeiffer, L.K.G.
Pfeiffer, H.H. *7813*
Pfeiffer, J.P. *7814*
Pfeiffer, K.G.L. *v.* Pfeiffer, L.K.G.
Pfeiffer, L.C.G. *v.* Pfeiffer, L.K.G.
Pfeiffer, L.G.K. *v.* Pfeiffer, L.K.G.
Pfeiffer, L.K.G. *7815-7821*
Pfeiffer, N.E. *7822-7823*
Pfeiffera see Pfeiffer, L.K.G.
Pfeifferago see Pfeiffer, L.K.G.
Pfeil, F.W.L. *7824*
Pfeil Jr., 7824
Pfister, F.J. 8775
Pfitz. v. Pfitzer, E.H.H.
Pfitzer, E.H.H. *7825-7833; see* Reinecke, F.
Pflaum, F. *7834*
Pfuhl, F.C.A. *alph.*
Pfund, J.D.C. *7835-7836*
Phaeosaccardinula see Saccardo, P.A.
Phares, D.L. *7837*
Phelps, A. *7838*
Phelps, J. *see* Phelps, A.
Phelps, K.D. *see* Phelps, A.
Phelps, W. *7839*

Phelps, W.H. *see* Phelps, A.
Phelps, W.H., Jr. *see* Phelps, A.
Phelpsiella see Phelps, A.
Phelsum, M. van *7840; see* Plukenet, L.
Phil. v. Philippi, R.A.
Philagrius *see* Nocca, A.
Philib., H. v. Philibert, H.
Philib., J. v. Philibert, J.C.
Philibert, H. *alph.*
Philibert, J.C. *7841-7844*
Philibertella see Philibert, H.
Philibertia see Philibert, H.
Philibertiella see Philibert, H.
Philip, R.H. *alph.*
Philippar, F.A. *v.* Philippar, F.H.
Philippar, F.H. *7845-7846*
Philippe, X. *7847*
Philippe (Saint-Mandrier) *see* Philippe, X.
Philippi, F. *7848-7849,* 7856, 8857
Philippi, F.H.E. *v.* Philippi, F.
Philippi, R.A. *7850-7858; see* Philippi, F.
Philippia see Philippi, R.A.
Philippiamra see Philippi, R.A.
Philippicereus see Philippi, R.A.
Philippiella see Philippi, R.A.
Philippimalva see Philippi, R.A.
Philippinaea see Philippi, R.A.
Philippiregis see Philippi, R.A.
Philippo, R. *see* Philippi, R.A.
Philippodendron see Philippi, R.A.
Philippodendrum see Philippi, R.A.
Phillips, E.P. *7859-7864,* 8132
Phillips, F.W. *see* Phillips, W.
Phillips, H. *7865-7869*
Phillips, J. *7870-7871*
Phillips, R.A. *alph.*
Phillips, R.W. *alph.*
Phillips, S. *see* Phillips, E.P.
Phillips, T.R. *see* Phillips, R.W.
Phillips, W. *7872-7873; see* Plowright, C.B.
Phillips, W.S. *alph.*
Phillipsia see Phillips, J., Phillips, W.
Phillipsiella see Phillips, W.
Phinney, A.J. *alph.*
Phipps, C.J. *7874*
Phippsia see Phipps, C.J.
Phoebus, P. *7875-7876,* 8648
Pia, J. von *7877-7878*
Piaea see Pia, J. von
Piaella see Pia, J. von
Pianella see Pia, J. von
Picard v. Picard, C.
Picard, C. *7879-7880*
Picard, F. *7881*
Picarda, L. *see* Picard, C.
Picardaea see Picard, C.
Picardenia see Picard, C.
Picb. v. Picbauer, R.

INDEX TO NAMES

Picbauer, R. *alph.*
Picc. v. Piccone, A.
Piccioli, A. *7882-7885*
Piccioli, E. *7886*
Piccioli, G. *7887-7888*
Piccioli, L. *7889-7895*
Picciuoli, G. *v.* Piccioli, G.
Picco, V. *7896*
Piccone, A. *7897-7909*
Picconi, J.B. *see* Piccone, A.
Picconia see Piccone, A.
Picconiella see Piccone, A.
Pichi-Sermolli, R.E.G. *7910*
Pichon, T. *7911*
Pichonia see Pichon, T.
Pickard, J.F. *alph.*
Pickel, B.J. *alph.*
Pickering, C. *7912-7913*
Pickering, Mrs. 7913
Pickeringia see Pickering, C.
Pickett, F.L. *alph.*
Picq. v. Picquenard, C.-A.
Picquenard, C.-A. *7914-1917*
Picradenia see Picard, C.
Pic.-Ser. v. Pichi-Sermolli, R.E.G.
Picus, V. *v.* Picco, V.
Piddington, H. *7918-7919*
Piddingtonia see Piddington, H.
Piep. v. Pieper, P.A.
Piep., G. v. Pieper, G.R.
Piepenbr. v. Piepenbring, G.H.
Piepenbring, G.H. *alph.*
Pieper, G.R. *7920*
Pieper, P.A. *alph.*
Pierce, N.B. *alph.*
Piercea see Pierce, N.B.
Piercy, H. *v.* Percy H., of Northumberland
Pieri, M.T. *7921-7922*
Pierranthus see Pierre, J.B.L.
Pierre, J.B.L. *7923-7923a*
Pierrea see Pierre, J.B.L.
Pierreodendron see Pierre, J.B.L.
Pierrina see Pierre, J.B.L.
Pierrot, P. *7924-7927*
Pieters, A.J. *7928-7930*
Pietsch, F.M. *7931*
Pigott, A.E. *v.* Roupell, A.E.
Pijl, L. van der *7932*
Pike, N. *alph.*
Pike, Z.M. *see* Pike, N.
Pikea see Pike, N.
Pilar, G. *7933*
Pilát, A. *7934-7935*
Pilatia see Pilát, A.
Pilg. v. Pilger, R.K.F.
Pilger, R.K.F. *7936-7938*, 7684
Pilgeria see Pilger, R.K.F.
Pilgeriella see Pilger, R.K.F.
Pilgerochloa see Pilger, R.K.F.

Pilgerodendron see Pilger, R.K.F.
Pillans, N.S. *alph.*
Pillansia see Pillans, N.S.
Piller, M. *7939*
Pilling, F.O. *7940*
Pilous, Z. *alph.; see* Podpěra, J.
Pinatzis, L. *alph.*
Pinchot, G. *7941*
Pinkerton, J. 7874, 8617
Pinoy, P.E. *alph.*
Pinoyella see Pinoy, P.E.
Pinto, A. *see* Pinto, J. de Almeida
Pinto, J. de Almeida *7942*
Pinto, Z. d'Almeida *v.* d'Almeida Pinto, Z.
Pintoa see Pinto, J. de Almeida
Pinto da Silva, A.R. *alph.*
Pio, G.B. *7943*
Piper v. Piper, C.V.
Piper, C.V. *7944-7954*
Piper, R.U. *7955*
Piperia see Piper, C.V.
Piquet, J. *alph.*
Piquetia see Piquet, J.
Piré, L.A.H.J. *7956-7963*
Pirea see Piré, L.A.H.J.
Pireella see Piré, L.A.H.J.
Pirolle 8751
Pirona, G.A. *7964*
Pirona, G.J. *see* Pirona, G.A.
Pirotta, P.R. *7965-7969*, 9274, 9895
Pirottaea see Pirotta, P.R.
Pirottantha see Pirotta, P.R.
PISANI, VETTOR *v.* VETTOR PISANI
Pischel, E. 8319
Piscicelli, M. *7970*
Piso, W. *alph.*, 5546
Pisoa see Piso, W.
Pisonia see Piso, W.
Pisoniella see Piso, W.
Pit. v. Pitard, C.-J.M.
Pitard, C.-J.M. *7971-7975*
Pitardia see Pitard, C.-J.M.
Pitcher, Z. *alph.*
Pitcheria see Pitcher, Z.
Pitt. v. Pittier, H.F.
Pittier, H.F. *7976-7992*, 6465, 8131, 8537, 8988; *see* Pax, F.A., Petersen, N.F., Renauld, F.F.G., Rousseau, M.
Pittiera see Pittier, H.F.
Pittier de Fabrega, H.F. *v.* Pittier, H.F.
Pittierella see Pittier, H.F.
Pittierothamnus see Pittier, H.F.
Pittoni, J.C. *alph.*
Pittonia see Pittoni, J.C.
Pittoniotis see Pittonia, J.C.
Pittoni von Dannenfeldt, J.C. *v.* Pittoni, J.C.
Pius, VI, Pope 5469
Pizarro, J.J. *7993*

Pizzetta, J. 9663
Planch. v. Planchon, J.É.
Planch., G. v. Planchon, G.
Planch., L. v. Planchon, L.D.
Planchon, F.G. *v.* Planchon, G.
Planchon, G. *7994-8003; see* Perrot, É.C.
Planchon, J.É. *8004-8024; see* Planchon, G., Planchon, L.D.
Planchon, L.D. *8025-8028*
Planchonella see Planchon, J.É.
Planchonia see Planchon, J.É.
Plancius see Risso, J.A.
Planellas Giralt, J. *8029*
Planer, J.J. *8030-8033*
Planera see Planer, J.J.
Platen, P.L. *8034*
Platter, F. 9875; *see* Rytz, A.R.W.
Plawski, A. *8035*
Playf. v. Playfair, G.I.
Playfair, G.I. *8036-8048*
Plaz, A.W. *8049-8051*
Plaza, J. *see* Plaz, A.W.
Plazia see Plaz, A.W.
Plée, A. *8052-8053*
Plée, F. *8054-8055*, 8052
Pleijel, C.G.W. *alph.*
Plenck, J.J. von *8056-8058*
Plenckia see Plenck, J.J. von
Plenk, J.J. von *v.* Plenck, J.J. von
Pleoravenelia see Ravenel, H.W.
Plinius *see* Rumphius, G.E.
Plitt, C.C. *alph.*
Plitzka, A. *8059*
Plowr. v. Plowright, C.B.
Plowright, C.B. *8060; see* Petch, T., Phillips, W.
Plowrightia see Plowright, C.B.
Plowrightiella see Plowright, C.B.
Plues, M. *alph.*
Pluesia see Plues, M.
Pluk. v. Plukenet, L.
Plukenet, L. *8064-8065*, 7840
Plukeneta see Plukenet, L.
Plukenetia see Plukenet, L.
Plum. v. Plumier, C.
Plumeria see Plumier, C.
Plumeriopsis see Plumier, C.
Plumier, C. *8066-8069; see* Presl, K.B.
Plumiera see Plumier, C.
Plumieria see Plumier, C.
Plummer, S.A. *alph.*
Plummera see Plummer, S.A.
Pluskal, F.S. *8070*
Pluska-Moravičanský, F.S. *v.* Pluskal, F.S.
Plusz. v. Pluszczewski, E.
Pluszczewski, E. *8071*
Pobed. v. Pobedimova, E.G.
Pobedimova, E.G. *alph.*
Pobég. v. Pobéguin, C.H.O.

Pobéguin, C.H.O. *8072-8073*
Pobeguinea see Pobéguin, C.H.O.
Pocock, M.A. *alph.*
Pocock, S.A.J. *see* Pocock, M.A.
Pococke, R. *see* Pocock, M.A.
Pocockia see Pocock, M.A.
Pocockiella see Pocock, M.A.
Pocockipites see Pocock, M.A.
Podp. v. Podpěra, J.
Podpěra, J. *8074-8084*
Podstatzky Luhlenstein, L. 9431
Poech, J.A. *8085*
Poechia see Poech, J.A.
Poederlé, E.J.C.G.H. d'Olmen de *8086*
Pöll, J. *alph.*, 8885
Poelln. v. Poellnitz, K. von
×*Poellneria see* Poellnitz, K. von
Poellnitz, K. von *8087-8088*
Poellnitzia see Poellnitz, K. von
Poepp. v. Poeppig, E.F.
Poeppig, E.F. *8089-8091*
Poeppigia see Poeppig, E.F.
Poetsch, I.S. *8092-8095*
Poetschia see Poetsch, I.S.
Poeverl. v. Poeverlein, H.
Poeverlein, H. *8096-8100*, 8271
Poggenb. v. Poggenburg, J.F.
Poggenburg, J.F. *alph.*
Pohl v. Pohl, J.B.E.
Pohl, J.B.E. *8101-8104*
Pohl, J.E. *8105-8106*
Pohl, Joseph 7683; *see* Pohl, J.B.E., Pohl, Julius
Pohl, Julius *8107*
Pohlana see Pohl, J.B.E.
Pohle, R.R. *8108-8110*
Pohlia see Pohl, J.B.E.
Pohliella see Pohl, J.B.E., Pohl, Julius
Poinçot 9686
Poir. v. Poiret, J.L.M.
Poirault v. Poirault, J.P.F.
Poirault, G. *8111-8112*
Poirault, J.P.F. *8113; see* Poirault, G.
Poirault, M.H.G. *v.* Poirault, G.
Poiret, A. 8116
Poiret, J.L.M. *8114-8116*, 9147, 9741a; *see* Pourret, P.A.
Poiretia see Poiret, J.L.M.
Poiss. v. Poisson, H.-L.
Poiss., J. v. Poisson, J.
Poisson, H.L. *8117*
Poisson, J. *8118-8120*
Poissonia see Poisson, J.
Poissoniella see Poisson, J.
Poit. v. Poiteau, P.A.
Poiteau, A. 7736; *see* Poiteau, P.A.
Poiteau, P.A. *8121-8123*, 9248; *see* Rudolphi, K.A.
Poivre, P. *alph.*
Poivrea see Poivre, P.

Pokorny, A. *8124-8130;* see Reichardt, H.W.
Pokorny, F. *see* Pokorny, A.
Pokornya see Pokorny, A.
Polak. v. Polakowsky, H.
Polakowskia see Polakowsky, H.
Polakowsky, H. *8131*
Pole-Evans, I.B. *8132*
Polevansia see Pole-Evans, I.B.
Polg. v. Polgár, S.
Polgár, S. *alph.*
Polichia see Pollich, J.A.
Polívka, F. *8133,* 8083
Polívka, J. *see* Polívka, F.
Poll. v. Pollich, J.A.
Pollacci, G. *8134-8135*
Pollaccia see Pollacci, G.
Pollard, C.L. *8136*
Pollard, H.M. *see* Pollard, C.L.
Pollexf. v. Pollexfen, J.H.
Pollexfen, J.H. *alph.*
Pollexfenia see Pollexfen, J.H.
Pollich, J.A. *8137,* 7768
Pollicha see Pollich, J.A.
Pollichia see Pollich, J.A.
Pollini, C. *8138-8145,* 5586, 6301; *see* Massalongo, A.B.
Pollinia see Pollini, C.
Pollinidium see Pollini, C.
Polliniopsis see Pollini, C.
Polscher, W. *8146*
Polunin, N.V. *8147*
Pomata y Gisbert, E. *8148*
Pomel see Martius, T.W.C.
Pomel, A.N. *8149-8151*
Pomelia see Pomel, A.N.
Pomrencke, W. 8272
Ponce de Leon v. Ponce de Leon, J.
Ponce de Leon, J. *8152*
Ponce de Leon y Aimé, A. *alph.*
Ponerop. v. Poneropoulos, E.
Poneropoulos, E. *8153*
Pontarl. v. Pontarlier, N.C.
Pontarlier, N.C. *alph.*
Pontén, J.P. *8154*
Pool, R.J. *8155-8157*
Poot, A. van 9123-9124
Pop, E. *8158*
Pope v. Pope, W.T.
Pope, A. *see* Pope, C.M.
Pope, C.M. *alph.*
Pope, F. *see* Rothrock, J.T.
Pope, W.T. *8159*
Popenoe v. Popenoe, F.W.
Popenoe, D.K. *alph.*
Popenoe, F.W. *8160; see* Popenoe, D.K.
Popenoe, W. *v.* Popenoe, F.W.
Poplu, M.C. *8161*
Popov, M.G. *alph.*
Popovici, A.P. *8162-8164*

Popoviocodonia see Popov, M.G.
Popp, B. *8165*
Porcher, F.P. *8166-8168*
Porcius, F. *8169-8170*
Porsch, O. *8171-8173*
Porsild v. Porsild, A.E.
Porsild, A.E. *alph.,* 8175
Porsild, M.P. *8174-8177; see* Porsild, A.E.
Porsildia see Porsild, A.E.
Porta, P. *8178*
Portaea see Porta, P.
Portales, D. *see* Porta, P.
Portalesia see Porta, P.
Portea see Porta, P.
Portenschl. v. Portenschlag-Ledermayer, F. von
Portenschlagia see Portenschlag-Ledermayer,F. von
Portenschlagiella see Portenschlag-Ledermayer, F. von
Portenschlag-Ledermayer, F. von *8179*
Porter v. Porter, T.C.
Porter, C.E. *alph.*
Porter, C.L. *alph.*
Porter, D.R. 7257
Porter, G. *see* Porter, C.E.
Porter, H.C. *see* Porter, C.E.
Porter, L.E. *alph.*
Porter, R.K. *see* Porter, C.E.
Porter, T.C. *8180-8184,* 9658
Porterandia see Porter, C.E.
Porteranthus see Porter, T.C
Porterella see Porter, T.C.
Portères, R. *see* Porter, C.E.
Porteresia see Porter, C.E.
Porterf. v. Porterfield, W.M., Jr.
Porterfield, W.M., Jr. *8185*
Porteria see Porter, C.E.
Porterinema see Porter, C.E.
Porterula see Porter, C.E.
Portland, Duchess of *see* Rousseau, J.J.
Portland, Duke of 9493
Portphillipia see Philippi, R.A.
Posch. v. Poscharsky, G.A.
Poscharsky, G.A. *8186*
Poscharsky, O. *see* Poscharsky, G.A.
Poselg. v. Poselger, H.
Poselger, H. *alph.*
Pospichal, E. *8187-8187a*
Pospíšil, V. *alph.*
Post v. Post, G.E.
Post, L. *v.* Post, E.J.L. von
Post, E.J.L. von *8188-8189c*
Post, G.E. *8190-8191*
Post, G.H. 7020
Post, H.A. von *8192; see* Post, T.E. von
Post, T.E. von *8193-8194*
Postel, E.A.W. *8195-8196*
Postels, A.P. *8197,* 9793
Postelsia see Postels, A.P.

INDEX TO NAMES

Posth. v. Posthumus, O.
Posthumus, O. *8198-8203,* 8393
Postia see Post, G.E., Post, H.A. von
Potanin, G.N. *alph.*
Potaninia see Potanin, G.N.
Potier de la Varde, R.A.L. *8204-8209*
Potonié v. Potonié, H.
Potonié, H. *8210-8228,* 6337, 8229; *see* Peter, A., Potonié, R.H.H.E.
Potonié, R.H.H.E. *8229*
Potoniea see Potonié, H.
Potonieipollenites see Potonié, R.H.H.E.
Potonieisporites see Potonié, R.H.H.E.
Potonieitriradites see Potonié, R.H.H.E.
Pott, J.F. *8230*
Potter, M.C. *alph.*
Pottia see Pott, J.F.
Pottiella see Pott, J.F.
Pottier, J.G. *8231-8233*
Pottier-Alapetite, G. *alph.*
Potzger, J.E. *alph.*
Potztal, E. *see* Pilger, R.K.F.
Pouchet, A.F. v. Pouchet, F.A.
Pouchet, A.M. *8234*
Pouchet, F.A. *8235-8237*
Pouchet, G. *see* Pouchet, F.A.
Pouchetia see Pouchet, F.A.
Poulsen, V.A. *8238-8239*
Poulsenia see Poulsen, V.A.
Pound, R. *8240,* 9847
Pourr. v. Pourret, P.A.
Pourret, P.A. *alph.*
Pourret-Figeac, P.A. *v.* Pourret, P.A.
Pourretia see Pourret, P.A.
Pouzols, P.M.C. de *v.* Pouzolz, P.M.C. de
Pouzolsia see Pouzolz, P.M.C. de
Pouzolz, P.M.C. de *8241-8242,* 5002
Pouzolzia see Pouzolz, P.M.C. de
Povah, A.H.W. *alph.*
Powell, D.A. *see* Powell, T.
Powell, J.W. *8243-8244*
Powell, T. *alph.*
Powellia see Powell, T.
Power, T. *8245*
Powlikowsky 9431
Pradal, É. *8246*
Praeger, R.L. *8247-8253; see* McArdle, D., Pethybridge, G.H.
Praeger, S.R. 8248
Praet. v. Praetorius, I.
Praetoria see Praetorius, I.
Praetorius, I. *8254*
Prahl v. Prahl, P.
Prahl, J.F. *8255*
Prahl, P. *8256*
Prahn, H. *8257*
Prain, D. *8258-8266*
Prainea see Prain, D.
Prance, C.T. 8393

Prantl, K.A.E. *8267-8272,* 8097; *see* Pax, F.A.
Prantleia see Prantl, K.A.E.
Praschil, W.W. *8273*
Pratesi, P. *8274*
Pratt, A. *8275-8278*
Préaub. v. Préaubert, E.
Préaubert, E. *8279*
Preble, E.A. *8280*
Preda, A. *8281-8282*
Predaea see Preda, A.
Préfontaine, M. Bruletout de *8283*
Preiss, B. *8284*
Preiss, J.A.L. *v.* Preiss, L.
Preiss, L. *alph.,* 5807
Preissia see Preiss, B.
Preissites see Preiss, B.
Prenger, A.G. *8285*
Prentiss, A.N. *alph.*
Prescot, J. *v.* Prescott, J.D.
Prescotia see Prescott, G.W.
Prescott, G.W. *8286*
Prescott, J.D. *see* Prescott, G.W.
Presl, J.S. *8287-8291,* 8293
Presl, K.B. *8292-8312,* 8287-8288, 9239; *see* Presl, J.S.
Preslaea see Presl, J.S.
Presle, Le Bègue de *v.* Le Bègue de Presle
Preslea see Presl, J.S.
Preslia see Presl, J.S.
Presnall, C. *see* Patraw, P.M.
Pressia see Preiss, B.
Prestele, F.J.U. 7817
Prestoe, H. *8313*
Prestoea see Prestoe, H.
Preston, C. *see* Preston, T.A.
Preston, T.A. *8314-8315*
Prestonia see Preston, T.A.
Prestoniopsis see Preston, T.A.
Prêtre, J.G. 7212, 7215, 7217-7218
Preuss, C.G.T. *v.* Preuss, G.T.
Preuss, G.T. *8316*
Preuss, H. *alph.*
Preuss, P.R. *8317-8318*
Preussia see Preuss, G.T.
Preussiaster see Preuss, G.T.
Preussiella see Preuss, G.T., Preuss, P.R.
Preussiodora see Preuss, P.R.
Prevost, I.B. *see* Prévost, J.L.
Prévost, J.L. *8319*
Prevost, P. *see* Prévost, J.L.
Prevostea see Prévost, J.L.
Price, S. *8323*
Price, S.F. *8320-8322*
Price, W.R. 9202
Prichard, H.H. *see* Rendle, A.B.
Prill. v. Prillieux, É.E.
Prillieux, É.E. *alph.,* 4658
Prillieuxia see Prillieux, É.E.

INDEX TO NAMES

Prillieuxina see Prillieux, É.E.
Prime, C.T. 8695
Prince *see* Prince, A.R.
Prince, A.R. *alph.*
Prince, W. *8324*
Princea see Prince, A.R.
Pringle, C.G. *8325-8326*, 9281-9283, 9285; *see* Pennell, F.W.
Pringle, J. *see* Pringle, C.G.
Pringlea see Pringle, C.G.
Pringleella see Pringle, C.G.
Pringlella see Pringle, C.G.
Pringleochloa see Pringle, C.G.
Pringleophyllum see Pringle, C.G.
Pringsh. v. Pringsheim, N.
Pringsheim, N. *8327-8340; see* Pfeffer, W.F.P.
Pringsheimia see Pringsheim, N.
Pringsheimiella see Pringsheim, N.
Pringsheimina see Pringsheim, N.
PRINS CARL *see* Osbeck, P.
Printz, H. *8341-8345*
Printz, J. *see* Printz, H.
Printz, K.H.O. *v.* Printz, H.
Printzia see Printz, H.
Printziella see Printz, H.
Prinz, W.A.J. *8346*
Prior, R.C.A. *8347*
Prioria see Prior, R.C.A.
Pritch., A. v. Pritchard, R.C.A.
Pritch., S. v. Pritchard, S.F.
Pritchard, A. *8348-8349; see* Ralfs, J.
Pritchard, S.F. *8350*
Pritchard, W. *see* Pritchard, A.
Pritchardia see Pritchard, A.
Pritchardiopsis see Pritchard, A.
Pritchardioxylon see Pritchard, A.
Pritchardites see Pritchard, A.
Pritz. v. Pritzel, G.A.
Pritz., E. v. Pritzel, E.G.
Pritzel, E.G. *alph.*
Pritzel, G.A. *8351-8355*, 5924, 5958; *see* Poeppig, E.F.
Pritzelago see Pritzel, G.A.
Pritzelia see Pritzel, G.A.
Pritzeliella see Pritzel, E.G.
Probst, R. *alph.*
Proch. v. Procházka, J.S.
Procházka, J.S. *8356*
Procopiania see Procopianu-Procopovici, A.
Procopianu-Procopovici, A. *8357*, 7295
Proctor, G.R. 1745, 5170
Prodan, I. *8358-8362*
Prodan, J. *v.* Prodan, I.
Pröll, A. *8363*
Progel, A. *8364*
Prokh. v. Prokhanov, Y.I.
Prokhanov, Y.I. *alph.*
Prollius, F. *8365*

Pronville, A. de *8366*
Prosper, E. Reyes y *v.* Reyes y Prosper, E.
Prost, T.C. *8367*, 4999
Prostea see Prost, T.C.
Protić, G. *8368-8370*
Proust, L. 7972
Prov. v. Provancher, L.
Provancher, L. *8371*
Provencheria see Provancher, L.
Prüss 9805
Prunol de Rosny, L.L.L. *v.* Rosny, L.L.L. Prunol de
Pryor, A.R. *8372*
Prytz, L.J. *8373*
Przew. v. Przewalski, N.M.
Przewalski, N.M. *alph.*
Przewalskia see Przewalski, N.M.
Przhevalsky, N.M. *v.* Przewalski, N.M.
Pseudoparodia see Parodi, D.
Pseudoparodiella see Parodi, D.
Pseudopavonia see Pavon y Jiménez, J.A.
Pseudopetrakia see Petrak, F.
Pseudopiaea see Pia, J. von
Pseudopohlia see Pohl, J.B.E.
Pseudopringsheimia see Pringsheim, N.
Pseudoraciborskia see Raciborski, M.
Psilander, N.C. 9093
Pucci, A. *8374*
Puccin. v. Puccinelli, B.L.
Puccinelli, B.L. *8375-8376*
Puccinellia see Puccinelli, B.L.
Puel, F. *v.* Puel, T.
Puel, T. *8377-8378*
Puelia see Puel, T.
Puerari, M.N. *alph.*
Pueraria see Puerari, M.N.
Puerta y Ródenas, G. de la *8379*
Pugsl. v. Pugsley, H.W.
Pugsley, H.W. *8380-8381*
Puigarria see Puiggari, J.I.
Puigd. v. Puigdullés, E.M.
Puigdullés, E.M. *8382*
Puigg. v. Puiggari, J.I.
Puiggari, J.I. *8383*
Puiggaria see Puiggari, J.I.
Puiggariella see Puiggari, J.I.
Puiggarina see Puiggari, J.I.
Puihn, J.G. *8384*
Pulgar, F. 9771
Pulina see Paul, H.K.G.
Pulle, A.A. *8385-8396*, 8199, 9072; *see* Posthumus, O.
Pullea see Pulle, A.A.
Pult. v. Pulteney, R.
Pultenaea see Pulteney, R.
Pulteney, R. *8397-8400*
Purchas, W.H. *8401*
Purdiaea see Purdie, W.
Purdie, W. *alph.*

Purdieanthus see Purdie, W.
Purfoot, T. *7625*
Purk. v. Purkyně, E. von
Purkyně, E. von *alph.*
Purkyně, J.E. *see* Purkyně, E. von, Sachs, J. von
Purpus, C.A. *8402*
Purpus, J.A. *see* Purpus, C.A.
Purpus, K.A. *v.* Purpus, C.A.
Purpusia see Purpus, C.A.
Pursch, C.A. *see* Pursh, F.T.
Pursch, F.T. *v.* Pursh, F.T.
Purschia see Pursh, F.T.
Pursh, F.T. *8403-8406*, 8568, 9153; *see* Peck, W.D.
Purshia see Pursh, F.T.
Purt. v. Purton, T.
Purton, T. *8407*
Putterl. v. Putterlick, A.
Putterlichia see Putterlick, A.
Putterlick, A. *8408*
Putterlickia see Putterlick, A.
Puydt, E. de *8409-8410*
Puym. v. Puymaly, A.H.L.
Puymaly, A.H.L. de *8411*

Quaintance, A.L. *alph.*
Quandt, C. *8412*
Quart.-Dill. v. Quartin-Dillon, R.
Quartin-Dillon, R. *alph.*, 9145, 9151; *see* Richard, A.
Quartinia see Quartin-Dillon, R.
Quehl, L. *alph.*
Quekett, E.J. *see* Quekett, J.T.
Quekett, J.T. *8413*
Quekettia see Quekett, J.T.
Quél. v. Quélet, L.
Quélet, L. *8414-8419*, 9177
Queletia see Quélet, L.
Quelle, F.F.H. *8420*
Quensel, C. 7247
Quera see Quer y Martinez, J.
Queria see Quer y Martinez, J.
Quer y Martinez, J. *8421*
Quesné, F.A. *alph.*
Quételet, A.J. *v.* Quételet, L.A.J.
Quételet, L.A.J. *alph.*
Queteletia see Quételet, L.A.J.
Queva, C. *8422-8423*
Quincy, C. *8424-8425*
Quintanilha, A.P. da Silva *8426-8427*
Quisumb. v. Quisumbing, E.
Quisumbing, E. *8428*
× *Quisumbingara* see Quisumbing, E.
Quisumbingia see Quisumbing, E.

Raab, L. *8429*
Raalte, M.H. van 8393
Raatz, G.V. *8430*
Rabanus, A. *8431*

Rabenau, B.C.A.H. von *v.* Rabenau, H. von
Rabenau, H. von *8432*
Rabenh. v. Rabenhorst, G.L.
Rabenhorst, G.L. *8433-8450*, 7424, 8745, 8825; *see* Lindberg, S.O., Lynge, B.A., Magnusson,A.K., Martens, G.M. von, Pascher, A., Preuss, G.T., Richter, P.G.
Rabenhorstia see Rabenhorst, G.L.
Rabinowitsch, L. *8451*
RACECOURSE *see* Phipps, C.J.
Rach, L.T. *alph.*, 8784
Rachia see Rach, L.T.
Racib. v. Raciborski, M.
Raciborski, M. *8452-8472*; *see* Rostafińsky, J.T.
Raciborskia see Racoborski, M.
Raciborskiella see Raciborski, M.
Raciborskiomyces see Raciborski, M.
Racine, R. *8473*
Radais, M.P.F. *8474-8475*
Radde, G.F.R. von *8476-8482*, 8790
Raddetes see Radde, G.F.R. von
Raddi, G. *8483-8499*; *see* Persoon, C.H.
Raddia see Raddi, G.
Raddiella see Raddi, G.
Raddin, C.S. *8500*
Raddisia see Raddi, G.
Rademachia see Radermacher, J.C.M.
Radermacher, J.C.M. *8501*, 4829
Radermachera see Radermacher, J.C.M.
Radermachia see Radermacher, J.C.M.
Radian, S.S. *8502*
Radius, J.W.M. *8503*
Radiusa see Radius, J.W.M.
Radlk. v. Radlkofer, L.A.T.
Radlkofer, L.A.T. *8504-8540*, 7684, 7828, 9178; *see* Palézieux, P. de, Peyritsch, J.J., Prantl, K.A.E., Renner, O., Ross, H.
Radlkofera see Radlkofer, L.A.T.
Radlkoferella see Radlkofer, L.A.T.
Radlkoferotoma see Radlkofer, L.A.T.
Radloff, F.W. *alph.*
Räs. v. Räsänen, J.P.B.
Räsänen, V.J.P.B. *8541-8546*
Raesaeneniolichen see Räsänen, V.J.P.B.
Raesaeneniomyces see Räsänen, V.J.P.B.
Raeusch. v. Raeuschel, E.A.
Raeuschel, E.A. *8547*, 4817
Raf. v. Rafinesque-Schmaltz, C.S.
Raffles, T.S.B. *alph.*
Rafflesia see Raffles, T.S.B.
Rafinesque-Schmaltz, C.S. *8548-8592*, 9279
Rafinesquia see Rafinesque-Schmaltz, C.S.
Rafn, C.G. *8593*
Rafnia see Rafn, C.G.
Ragot, J. *8594*

INDEX TO NAMES

Raia see Ray, J.
Railonsala, A.N. *alph.*
Raim. v. Raimann, R.
Raimann, R. *alph.*
Raimannia see Raimann, R.
Raimondi, A. *8595*
Raimondia see Raimondi, A.
Raimondianthus see Raimondi, A.
Rainer, Archduke see Rainer von und zu Haarbach, M.
Rainera see Rainer von und zu Haarbach, M.
Raineri, R. *8596-8597*
Rainer von und zu Haarbach, M. *alph.*
Rainville, F. *alph.*
Raja see Ray, J.
Rajania see Ray, J.
Ralfs, J. *8598-8599*, 6870
Ralfsia see Ralfs, J.
Rallet, L. *8600*
Ralph, T.S. *8601*
Ramaley, F. *8601a-8602*
Rama Rao, M. *8603*
Ramat. v. Ramatuelle, T.A.J. d'Audibert de
Ramatuela see Ramatuelle, T.A.J. d'Audibert de
Ramatuelle, T.A.J. d'Audibert de *alph.*
Rambert, E. *8604*
Rambo, B. see Rick, J.
Rambosson, J.P. *8605*
Ramée, S.H. de la *alph.*
Ramírez, J. *8606-8610*
Ramirezella see Ramírez, J.
Ramírez Goyena, M. *8611-8612*
Ramirezia see Ramírez, J.
Ramis, A.I. *8613*
Ramis, J.I. 8613
Ramis y Ramis, J. *8614*
Ramisch, F.X. *8615*
Ramischia see Ramis y Ramis, J.
Ramisia see Ramis y Ramis, J.
Ramond, L.F.É. de Carbonnières *8616-8618*
Ramonda see Ramond, L.F.É. de Carbonnières
Ramondia see Ramond, L.F.É. de Carbonnières
Ramsb. v. Ramsbottom, J.
Ramsbottom, J. *8619-8621*
Ramsbottomia see Ramsbottom, J.
Rand, E.L. *8622-8623;* see Porter, T.C.
Rand, E.S. *8624-8625*
Rand, I. *8626-8627*
Rand, Mrs. E.L. see Rand, E.L.
Randa see Rand, I.
Randia see Rand, I.
Range, P.T. *8628*
Rangel, E. dos Santos *8629*
Rangia see Rand, I.

Ranoj. v. Ranojević, N.
Ranojević, N. *alph.*
Ranojevicia see Ranojević, N.
Rao, M. Rama *v.* Rama Rao, M.
Raoul v. Raoul, É.F.L.
Raoul, É.F.A. *8631;* see Raoul, É.F.L.
Raoul, É.F.L. *8630*
Raoulia see Raoul, É.F.L.
Raouliopsis see Raoul, É.F.L.
Raoult 6371
Rapaics von Rumwerth, R. *alph.*
Raper, K.B. *alph.*
Raperia see Raper, K.B.
Rapin, D. *8632-8635*
Rapin, R. see Rapin, D.
Rapinia see Rapin, D.
Rapp v. Rapp, A.R.
Rapp, A.R. *8636*
Rapp, S. *alph.*
Rásky, K. *alph.*
Raskyaechara see Rásky, K.
Raskyella see Rásky, K.
Rasmussen, R. *8637*
Raspail, F.V. *8638-8639*
Raspail, S.V. *8639*
Raspalia see Raspail, F.V.
Raspallia see Raspail, F.V.
Rass. v. Rassadina, K.A.
Rassadina, K.A. *alph.*
Rat, A.-J. Le *v.* Le Rat, A.-J.
Ráthay, E. *8640*
Rathke, J. *alph.*
Rathkea see Rathke, J.
Rattan, V. *8641-8642*
Rattke, W. *8643*
Rattray v. Rattray, John
Rattray, G. see Rattray, John
Rattray, James *alph.*
Rattray, J. (fl. 1835-1886) see Rattray, John
Rattray, J.M. see Rattray, John
Rattray, John *8644*
Rattraya see Rattray, John
Rattrayella see Rattray, John
Ratzeb. v. Ratzeburg, J.T.C.
Ratzeburg, J.T.C. *8645-8648*, 5895
Ratzeburgia see Ratzeburg, J.T.C.
Rau v. Rau, E.A.
Rau, A. *8649*
Rau, E.A. *8650*
Rauch, F. *8651*
Rauh, W. *alph.*
Rauhia see Rauh, W.
Rauhiella see Rauh, W.
Rauhocereus see Rauh, W.
Rauia see Rau, A., Rau, E.A.
Rauiella see Rau, E.A.
Raulin, V.F. *8652-8653*
Raunk. v. Raunkiaer, C.C.

Raunkiaer, C.C. *8653a-8664;* see Ostenfeld, C.E.H.
Raunkiaer, I. 8655
Raup, H.M. *8665-8668*
Rauscher, R. *8669*
Rauth, F. *8670*
Rauvolfia see Rauwolff, L.
Rauwenh. v. Rauwenhoff, N.W.P.
Rauwenhoff, N.W.P. *8671-8678*
Rauwenhoffia see Rauwenhoff, N.W.P.
Rauwolf, L. *v.* Rauwolff, L.
Rauwolff, L. *alph.*, 8697
Rauwolffa see Rauwolff, L.
Rauwolfia see Rauwolff, L.
Ravaud, L.C.M. *8679-8681*
Ravenel, H.W. *8682-8684*
Ravenella see Ravenel, H.W.
Ravenelula see Ravenel, H.W.
Ravenscroft, E.J. *8685*
Ravenshaw, T.F.A.T. *v.* Ravenshaw, T.F.T.
Ravenshaw, T.F.T. *8686-8687*
Ravia see Rau, A.
Ravin, E. *8688-8689*
Ravn, F.K. *8690-8691*
Ravn, P. see Ravn, F.K.
Ravnia see Ravn, F.K.
Rawitscher, F. *8692-8693*
Rawson,R.W. *alph.*, 7317
Rawsonia see Rawson, R.W.
Rawton, O. de *8694*
Ray, J. *8695-8706;* 5137, 5562; Plukenet, L., Pourret, P.A., Preston, T.A., Rand, I.
Rayania see Ray, J.
Raymond, M. *see* Rechinger, K.H.
Rayner, J.F. *8707-8708*
Raynia see Ray, J.
Rayss, T. *8709*
Rayssiella see Rayss, T.
Rchb. v. Reichenbach, H.G.L.
Rchb., C. v. Reichenbach, C.L. von
Rchb. fil. v. Reichenbach, H.G.
Re v. Re, G.F.
Re, F. *8710*
Re, G.F. *8711-8713,* 5701
Rea, C. *8714-8715*
Rea, Mrs. C. *see* Rea, C.
Rea see Re, G.F.
Reade, O.A. *8716*
Reader v. Reader, F.M.
Reader, F.M. *alph.*
Reader, H.C.L.P. *v.* Reader, P.
Reader, P. *alph.*
Readeriella see Reader, F.M.
Réaubourg, G. *8717*
Rebent. v. Rebentisch, J.F.
Rebentisch, J.F. *8718-8719*
Rebentischia see Rebentisch, J.F.
Reber, B. *alph.*

Reboud, V.C. *alph.*
Reboudia see Reboud, V.C.
Rebouillia see Reboul, E. de
Reboul, E. de *8720-8722*
Reboulea see Reboul, E. de
Reboulia see Reboul, E. de
Rech. v. Rechinger, K.
Rech. fil. v. Rechinger, K.H.
Réchin, J. *alph.*
Rechinger, F. *8733*
Rechinger, K. *8723-8727; see* Rechinger, K.H., Reinbold, T.
Rechinger, K.H. *8728-8737*
Rechingerella see Rechinger, K.H.
Rechinger-Favarger, L. 8727; see Rechinger, K.H.
Rechingeria see Rechinger, K.H.
Rechingeriella see Rechinger, K.H.
Reclam, C.H. 8351
Record, S.J. *8738-8741*
Recordia see Record, S.J.
Recordoxylon see Record, S.J.
Recq de Malzine, O. de P.A.H. *v.* Malzine, O. de P.A.H.R. de
Recupero A. 8556
Redeke, H.C. *alph.*
Redfield, J.H. *alph.*, 8623; see Porter, T.C.
Redfieldia see Redfield, J.H.
Redinger, K.M. *8742-8745,* 8450
Redmond, P.J.D. *8746*
Redouté, P.J. *8747-8752,* 9429, 9686, 9688; see Roche, F. de la, Rousseau, J.J.
Redoutea see Redouté, P.J.
Redslob, J. *8753*
Redtenbacher, J. *8754*
Redutea see Redouté, P.J.
Reed, C.T. see Plitt, C.C.
Reede tot Draakestein, H.A. van *v.* Rheede tot Draakestein, H.A. van
Reedrollinsia see Rollins, R.C.
Rees, A. *8755-8756*
Rees, B. *see* Rees, A.
Rees, M. *v.* Reess, M.F.T.F.M.
Reesia see Rees, A.
Reess, M.F.T.F.M. *8757-8768*
Reessia see Reess, M.F.T.F.M.
Regel v. Regel, E.A. von
Regel, A. von *alph.*, 8800, 8805; *see* Regel, E.A. von
Regel, C.A. von *8769-8770*
Regel, E.A. von *8771-8809; see* Maack, R., Regel, A. von
Regel, J.A. von *v.* Regel, A. von
Regelia see Regel, E.A. von
Regin i Lid *v.* Rasmussen, R.
Regnault, G. de Nangis *v.* Nangis Regnault, G. de
Regnault, N.F. *8810*

INDEX TO NAMES

Regnell, A.F. *alph.*, 4664, 4669, 8742, 8744, 9489; *see* Malme, G.O.A., Romell, L.
Regnellia see Regnell, A.F.
Regnellidium see Regnell, A.F.
Reguis, J.M. *8811-8813*
Rehd. v. Rehder, A.
Rehder, A. *8814-8819*
Rehder, J.H. *see* Rehder, A.
Rehder, P.J. *see* Rehder, A.
Rehdera see Rehder, A.
Rehderodendron see Rehder, A.
Rehderophoenix see Rehder, A.
Rehm, H.S.L.F.F. *8820-8830*, 8450
Rehman, A. *v.* Rehmann, A.
Rehmann, A. *8831*
Rehmanniella see Rehmann, A.
Rehmia see Rehm, H.S.L.F.F.
Rehmiella see Rehm, H.S.L.F.F.
Rehmiellopsis see Rehm, H.S.L.F.F.
Rehmiodothis see Rehm, H.S.L.F.F.
Rehmiomycella see Rehm, H.S.L.F.F.
Rehmiomyces see Rehm, H.S.L.F.F.
Rehsteiner, H. *8832*
Reichard, J.J. *8833-8837*, 5063, 7202
Reicharda see Reichard, J.J.
Reichardia see Reichard, J.J.
Reichardt, H.W. *8838-8850*
Reiche, C.F. *v.* Reiche, K.F.
Reiche, K.F. *8851-8865*
Reichea see Reiche, K.F.
Reicheella see Reiche, K.F.
Reichel, G.C. *8866*
Reichela see Reichel, G.C.
Reichelia see Reichel, G.C.
Reichembachanthus see Reichenbach, H.G.
Reichenau, W. von *8867*
Reichenbach, A.B. *see* Reichenbach, H.G.L.
Reichenbach, C.L. von *alph.; see* Reichenbach, H.G.L., Reichenbach, H.G.
Reichenbach, H. von *see* Reichenbach, C.L. von
Reichenbach, H.G. *8891-8900*, 7759, 8775, 8885; *see* Pöll, J., Reichenbach, H.G.L.
Reichenbach, H.G.L. *8868-8890*, 8891; *see* Pöll, J., Reichenbach, C.L. von, Reichenbach, H.G., Ruprecht, F.J.
Reichenbach, K.L. von *v.* Reichenbach, C.L. von
Reichenbach, L. *v.* Reichenbach, H.G.L.
Reichenbachanthus see Reichenbach, H.G.
×*Reichenbachara see* Reichenbach, H.G.
Reichenbachia see Reichenbach, H.G.L.
Reichert, C.B. *see* Reichert, I.
Reichert, I. *alph.*
Reichertia see Reichert, I.
Reichgelt, T.J. *alph.*
Reid, C. *8901-8904; see* Reid, E.M.

Reid, E.M. *8905-8906*, 8902, 8904
Reid, F.A. *see* Reid, C.
Reid, Lady 9248
Reid, M. *v.* Reid, E.M.
Reid, W. 9248
Reider, J.E. von *8907-8910*
Reidia see Reid, C.
Reighard, J.E. 7928
Reimer, C.W. *see* Patrick, R.M.
Reimers, H. *alph.*
Reimersia see Reimers, H.
Rein, J.J. *8910a-8911*
Reinbold, T. *8912*, 8726, 8915
Reinecke v. Reinecke, F.
Reinecke, F. *8913-8917; see* Reinbold, T.
Reinecke, J.H.J. *see* Reinecke, F.
Reinecke, K.L. *8918*
Reinecke, W. *8919*
Reineckea see Reinecke, F.
Reiner, J. *8920*
Reineria see Reiner, J.
Reinhard, L.V. *alph.*
Reinhardt *v.* Reinhardt, O.W.H.
Reinhardt, J. 9244
Reinhardt, L. *8921*
Reinhardt, L.V. *v.* Reinhard, L.V.
Reinhardt, M.O. *see* Reinhardt, O.W.H.
Reinhardt, O. *v.* Reinhardt, M.O.
Reinhardt, O.W.H. *8922-8923*
Reinke, J. *8924-8943*
Reinkella see Reinke, J.
Reinkellomyces see Reinke, J.
Reinsch v. Reinsch, P.F.
Reinsch, E.H.E. *v.* Reinsch, H.
Reinsch, H. *8944*
Reinsch, P.F. *8945-8958*
Reinschia see Reinsch, P.F.
Reinschiella see Reinsch, P.F.
Reinschospora see Reinsch, P.F.
Reinw. v. Reinwardt, C.G.C.
Reinwardt, C.G.C. *8959-8967*, 6687
Reinwardtia see Reinwardt, C.G.C.
Reinwardtiodendron see Reinwardt, C.G.C.
Reisseck, S. *v.* Reissek, S.
Reissek, S. *8968-8970*, 7808
Reissekia see Reissek, S.
Reitter, J.D. von *8971-8972*
Reitter, K. *see* Pax, F.A.
Relhan, R. *8973*
Relhania see Relhan, R.
Relhanum see Relhan, R.
Remelé, P. 9455
Remy v. Remy, E.A.
Remy, E.A. *8974-8975*
Rémy, E.J. v. Rémy, J.
Rémy, J. *8976-8979*
Remy, W. *see* Rémy, J.
Remya see Rémy, J.
Remyella see Rémy, J.
Remysporites see Rémy, J.

Renaudet, G.B. *8980*
Renauld, F.F.G. *8981-8998,* 7191; see Paillot, J., Paris, É.G., Philibert, H., Réchin, J., Röll, J.
Renauldia see Renauld, F.F.G.
Renault v. Renault, B.
Renault, B. *8999-9062*
Renault, P.-A. *9063*
Renaultia see Renault, B.
Renaut, P.-A. *v.* Renault, P.-A.
Rendle, A.B. *9064-9072,* 5170
Rendlia see Rendle, A.B.
Renealmea see Reneaulme, P.
Renealmia see Reneaulme, P.
Renealmus, P. *v.* Reneaulme, P.
Reneaulme, P. *9073*
Renier, A.-M.-V.-J. *9074*
Reniera see Renier, A.-M.-V.-J.
Renn, J.N. von 9622
Renner, O. *9075-9076*
Rennera see Renner, O.
Rennie, R. *9077*
Renselaar, H.C. van 8412
Req. v. Requien, E.
Requien, E. *9078*
Requienia see Requien, E.
Requinella see Requien, E.
Resvoll, T.R. *9079-9080; see* Resvoll-Holmsen, H.M.
Resvoll, T.S.R. *v.* Resvoll, T.R.
Resvoll-Holmsen, H.M. *9081-9083*
Retz. v. Retzius, A.J.
Retz., G. v. Retzius, G.
Retzdorff, A.E.W. *alph.*
Retzia see Retzius, A.J.
Retzius, A.A. *see* Retzius, A.J., Retzius, G.
Retzius, A.J. *9084-9094,* 4817
Retzius, G. *9095*
Retzius, M.C. 9092
Retzius, M.G. *v.* Retzius, G.
Retzius-Hierta, Mrs. *see* Retzius, G.
Reum, J.A. *9096-9097*
Reuss v. Reuss, C.F.
Reuss, A.E.R. von *9098*
Reuss, A.L. von *see* Reuss, A.E. von
Reuss, C.F. *9099-9102*
Reuss, F.A. *see* Reuss, C.F.
Reuss, G. *9104*
Reuss, G.C. *9103*
Reuss, L. *9105*
Reussia see Reuss, C.F.
Reut. v. Reuter, G.F.
Reuter, G.F. *9106-9109; see* Macreight, D.C.
Reutera see Reuter, G.F.
Réveil, O. *9110*
Réveil, P.O. *v.* Réveil, O.
Revel, J. *9111-9114*
Revelière, E. *alph.*

Revelière, J. *see* Revelière, E.
Reverchon v. Reverchon, E.
Reverchon, E. *alph.; see* Reverchon, J.
Reverchon, J. *alph.*
Reverchon, P. *9115*
Reverchonia see Reverchon, J.
Révoil, G. *alph.*
Revol, J. *9116*
Rex, G.A. *alph.*
Reyes, A.G. *see* Reyes y Prosper, E.
Reyes, S.A. *see* Reyes y Prosper, E.
Reyes y Prosper, E. *9118-9119*
Reyesia see Reyes y Prosper, E.
Reyesiella see Reyes y Prosper, E.
Reygadas, Mrs. A.A. de *v.* Mexia, Y.E.J.
Reyger, G. *9120-9121*
Reymond, M.L.C. *9122*
Reymondia see Reymond, M.L.C.
Reynolds, G.W. *alph.*
Reynolds, J.N. *see* Reynolds, G.W.
Reynoldsia see Reynolds, G.W.
Rey-Pailhade, C. de *9117*
Rhedia see Rheede tot Draakestein, H.A. van
Rheedea see Rheede tot Draakestein, H.A. van
Rheede tot Draakestein, H.A. van *9123-9124; see* Pulteney, R.
Rheedia see Rheede tot Draakestein, H.A. van
Rheedja see Rheede tot Draakestein, H.A. van
Rhein, G.F. *9125*
Rhind, W. *9126*
Rhiner, J. *9127-9131*
Rhode, J.G. *9132*
Rhodea see Rhode, J.G.
Rhodeites see Rhode, J.G.
Rhodes, P.G.M. *alph.*
Rhodesia see Rhodes, P.G.M.
Rhodosciadium see Rose, J.N.
Ricardia see Richardson, R.
Ricarte 8421
Ricasoli, V. *9133-9135*
Ricasolia see Ricasoli, V.
Ricasoliomyces see Ricasoli, V.
Ricca, L. *9136*
Rich. v. Richard, L.C.M.
Rich., A. v. Richard, A.
Rich., C. v. Richard, C.
Rich., O.J. v. Richard, O.J.
Rich, O. 9771
Rich, Obadiah *9137; see* Rich, W.
Rich, Oliver *v.* Rich, Obadiah
Rich, W. *alph.*
Rich, W.P. *alph.*
Richard, A. *9138-9151,* 6242, 9122, 9158-9160, 10.000, 10.002, 10.005; *see* MacGillivray, W., Quartin-Dillon, R., Perrottet, G.S., Richard, C., Richard,

INDEX TO NAMES

L.C.M., Risso, J.A., Sagra, R. de la
Richard, C. *9152*
Richard, J.M.C. *v.* Richard, C.
Richard, L.C.M. *9153-9160, 7732; see*
  Nestler, C.G., Pourret, P.A., Reichenbach, H.G., Richard, A., Richard, C.
Richard, O.J. *9161-9168*
Richarda see Richardson, R.
Richardella see Richard, L.C.M.
Richardia see Richard, L.C.M., Richardson, R.
Richards, H.M. *alph.*
Richards, Mrs. H.M. *see* Richards, H.M.
Richards, P.W. *alph.*
Richardsiella see Richards, H.M.
Richardson v. Richardson, J.
Richardson, J. *9169-9171*
Richardson, L. 9486; *see* Phillips, J.
Richardson, R. *9172; see* Miller, P.
Richardson, W. 8685
Richen, G. *9173-9174*
Richer de Belleval, P. *alph.;* 8017
Richeria see Richer de Belleval, P.
Richeriella see Richer de Belleval, P.
Richon, C.É. *9175-9177; see* Roze, É.
Richonia see Richon, C.É.
Richoniella see Richon, C.É.
Richt., Al. v. Richter, Aladár
Richt., B. v. Richter, B.
Richt., H. v. Richter, H.E.F.
Richt., K. v. Richter, K.
Richt., L. v. Richter, L.
Richt., O. v. Richter, O.
Richt., P.B. v. Richter, P.B.
Richt., P.G. v. Richter, P.G.
Richter, Aladár *9178-9181*
Richter, Alexander *see* Richter, H.E.F.
Richter, B. *9182*
Richter, H.E.F. *9183,* 7753
Richter, K. *9184-9185*
Richter, L. *alph.*
Richter, O. *9186-9187*
Richter, P.B. *9188-9189*
Richter, P.G. *alph.,* 8450; *see* Rabenhorst, G.L.
Richter, V.A. *v.* Richter, Aladár
Richtera see Richter, H.E.F.
Richteria see Richter, H.E.F.
Richteriella see Richter, H.E.F.
Rick, J. *9190-9191,* 8829
Rickella see Rick, J.
Ricken, A. *9192-9193*
Rickenella see Ricken, J.
Ricker, P.L. *9194*
Rickett v. Rickett, H.W.
Rickett, H.W. *9195-9196,* 9197; *see* Rickett, T.C.
Rickett, T.C. *9197*
Rickia see Rick, J.
Rickiella see Rick, J.

Rickli, M. *v.* Rikli, M.A.
Riddel *see* Loureiro, J. de
Riddelia see Riddell, J.L.
Riddell, J.L. *9198-9200*
Riddellia see Riddell, J.L.
Riddelsdell, H.J. *9201-9202*
Riddle, L.W. *9203-9204*
Riddlea see Riddle, L.W.
Ridl. v. Ridley, H.N.
Ridley, H.N. *9205-9212,* 9071
Ridleya see Ridley, H.N.
×*Ridleyara see* Ridley, H.N.
Ridleyella see Ridley, H.N.
Ridleyinda see Ridley, H.N.
Rieben 5779
Rieber, X. *9213*
Riedel, L. *alph.*
Riedel, Mr. *see* Riedel, L.
Riedel, W. *see* Riedel, L.
Riedelia see Riedel, L.
Riedeliella see Riedel, L.
Rieder, J.G. von 8790
Riedl, H. *see* Rechinger, K.H.
Rieth, A. 7423
Rietsch, M. *9214*
Rietz, G.E. Du *v.* Du Rietz, G.E.
Rigaud, A. *9222*
Rigg, G.B. alph.
Rigo, G. *see* Porta, P.
Rikli, M.A. *9223-9238*
Rikliella see Rikli, M.A.
Riley, C.G. *see* Riley, J.
Riley, J. *9239*
Riley, L.A.M. *alph.*
Rileya see Riley, J.
Rilstone, F. *alph.*
Ringbom, C.H. 8373
Ringier, V.A. *9240*
Ringius, H.H. *9241*
Rink, H.J. *9242-9245*
Rinz, J. 8775
Rio, I.J. de Asso e del *v.* Asso e del Rio, I.J. de
Riocreux, A. *alph.,* 6237, 6653-6654, 8630, 10.000, 10.003, 10.005
Riocreuxia see Riocreux, A.
Rion, A. *9246*
Ripart, J.B.M.J.S.E. *alph.*
Ripartia see Ripart, J.B.M.J.S.E.
Ripartitella see Ripart, J.B.M.J.S.E.
Ripartites see Ripart, J.B.M.J.S.E.
Risso, J.A. *9247-9250; see* Panizzi-Savio, F.
Risso, J.B. *see* Risso, J.A.
Rissoa see Risso, J.A.
Rissoella see Risso, J.A.
Ritgen, F.A.M.F. von *9251*
Ritschl v. Ritschl, G.A.
Ritschl, G.A. *9252-9255*
Ritschl, J. *see* Ritschl, G.A.

INDEX TO NAMES

Ritter, A. *see* Ritter, J.J.
Ritter, C. (Berlin) *9256*
Ritter, C. (Dresden) *v.* Ritter, K.
Ritter, C.W.J. *9257*
Ritter, J.J. *9258*
Ritter, K. *see* Ritter, C. (Berlin)
Rittera see Ritter, J.J.
Ritter berol., C. v. Ritter, C. (Berlin)
Rittershausen, P. *9259*
Ritz, R. 9246
Ritzberger, E. *alph.*
Ritz. Bos v. Ritzema Bos, J.
Ritzema Bos, J. *alph.*
Riv. v. Rivinus, A.G.
Riva, D. *alph.*
Rivas Goday, S. *alph.*
Rivasgodaya see Rivas Goday, S.
Rivas Mateos, M. *9261-9262*
Rivera, J. 9771
Riv.-God. v. Rivas Goday, S.
Rivière, A. *9263-9264*, 9662; *see* Rivière, C.M.
Rivière, A.E.A. *see* Rivière, A.
Rivière, C.M. *9265*, 9264; *see* Rivière, A.
Rivière, M.A. *v.* Rivière, A.
Rivina see Rivinus, A.Q.
Rivinia see Rivinus, A.Q.
Rivinus, A.Q. *9266-9273*, 8706
Riv. Mat. v. Rivas Mateos, M.
Roard 9741a
Robbins, J.W. *alph.*
Robbins, W.J. *alph.*; *see* Rickett, H.W.
Rob.-Bricch. v. Robecchi-Bricchetti, L.
Robecchi-Bricchetti, L. *9274*
Roberg, L. *alph.*, 9746
Roberga see Roberg, L.
Roberge, M.R. *9274a-9275*
Robergea see Roberge, M.R.
Robergia see Roberg, L.
Robert, G.N. *9276*
Robert, M. *see* Robert, N.
Robert, N. *alph.*
Robertia see Robert, G.N., Robert, N.
Robertomyces see Robert, N.
Robertson, B. *see* Robertson, D.
Robertson, D. *alph.*
Robertson, W. 8397; *see* Pallas, P.S.
Robertsonia see Robertson, D.
Robin, C.C. *9279*, 8567
Robin, C.P. *9277-9278*
Robin, J. *alph.*
Robina see Robin, J.
Robincola see Robin, J.
Robinia see Robin, J.
Robinioxylon see Robin, J.
Robins, I.S. *v.* Perrin, I.S.
Robinson, B.L. *9280-9294*
Robinson, C.B., Jr. *9295-9297*; *see* Rumphius, G.E.
Robinson, James Fraser *9298*

Robinson, James Frodham *see* Robinson, James Fraser
Robinson, J.H. *see* Robinson, B.L.
Robinson, John *9299-9303*
Robinson, T. *see* Robinson, B.L.
Robinsona see Robinson, B.J.
Robinson Crusoe *see* Robinson, B.L.
Robinsonella see Robinson, B.J.
Robinsonetta see Robinson, B.J.
Robinsonia see Robinson, B.L.
Robinsoniodendron see Robinson, C.B., Jr.
Robley, Mrs A.J. *9304*, 7629
Robolsky, H. *9305*
Robson, E. *9306*, 9307; *see* Robson, S.
Robson, S. *9307*; *see* Robson, E.
Robsonia see Robson, E.
Robyns, D. *see* Robyns, W.
Robyns, F.H.E.A.W. *v.* Robyns, W.
Robyns, W. *9308-9314*
Robynsia see Robyns, W.
Robynsiella see Robyns, W.
Robynsiochloa see Robyns, W.
Robynsiophyton see Robyns, W.
Roca, J.A. 4997
Roche, A. 9052
Roche, D. de la *alph.*; *see* Roche, F. de la
Roche, F. de la *alph.*, 8747
Roche, G. de la 8748
Rochea see Roche, D. de la, Roche, F. de la
Rochebr. v. Rochebrune, A.T. de
Rochebrune, A.T. de *9315-9316*
Rochel, A. *9317-9320*, 8179
Rochelia see Rochel, A.
Rochet d'Héricourt, C.E.K. *v.* Rochet d'Héricourt, C.L.X.
Rochet d'Héricourt, C.L.X. *9321*
Rochetia see Rochet d'Héricourt, C.L.X.
Rock, J.F.C. *9322-9332*, 8539
Rock, J.F.K. *v.* Rock, J.F.C.
Rockia see Rock, J.F.C.
Rockley, A.M. *9333*
Rockley, E. *see* Rockley, A.M.
Rodati, A.L. *9334-9335*
Rodatia see Rodati, A.L.
Rodegher, A. 9337
Rodegher, E. *9336-9337*
Rodénas, G. de la Puerta y *v.* Puerta y Rodénas, G. de la
Rodet, H.J.A. *9338*
Rodetia see Rodet, H.J.A.
Rodgers, A.D. *alph.*
Rodhea see Rhode, M.
Rodigas, É. *alph.*
Rodigas, F.-C.-H. *see* Rodigas, É.
Rodin, H. *9339-9340*
×*Rodretta see* Barbosa Rodrigues, J.
×*Rodricidium see* Barbosa Rodrigues, J.
×*Rodridenia see* Barbosa Rodrigues, J.

Rodrigues, J. Barbosa v. Barbosa Rodrigues, J.
Rodriguesia see Barbosa Rodrigues, J.
Rodriguez, J.D. *see* Barbosa Rodrigues, J.
Rodríguez, Juan José *see* Rodríguez y Femenías, J.J.
Rodriguez, M. *v.* Rodriguez, J.D.
Rodríguez y Femenías, J.J. *9362-9365*
Rodriguezella see Rodríguez y Femenías, J.J.
Rodrigueziella see Barbosa Rodrigues, J.
Rodriguezopsis see Barbosa Rodrigues, J.
×*Rodriopsis see* Barbosa Rodrigues, J.
Rodriqueza see Barbosa Rodrigues, J.
Rodriquezia see Barbosa Rodrigues, J.
Rodschied, E.K. *9366-9367*
Rodschiedia see Rodschied, E.K.
Rodway, L. *9368-9374*
Rodwaya see Rodway, L.
Roe, J.S. *9375*
Röderer see Ritgen, F.A.M.F. von
Roea see Roe, J.S.
Röhl. v. Röhling, J.C.
Roehl, E.K.G.W. von *9376*
Röhling, J.C. *9377-9378*
Roela see Röll, J.
Röll, J. *9379-9393*
Röll, K. *see* Röll, J.
Roell, W. *see* Röll, J.
Roella see Röll, J.
Roellia see Röll, J.
Roem. v. Roemer, J.J.
Roem., A. v. Roemer, A.
Röm., C. v. Römer, C.
Roem., F. v. Roemer, F. von
Röm., Jul. v. Römer, J.
Roem., M. v. Roemer, M.J.
Roemer, A. *9394-9396; see* Roemer, F. von, Römer, Heinrich
Römer, C. *alph.*
Roemer, C.F. von *v.* Roemer, F. von
Roemer, F.A. *v.* Roemer, A.
Roemer, F. von *9397-9398; see* Roemer, A., Römer, Heinrich
Römer, Heinrich *9399*
Roemer, Hermann *see* Römer, Heinrich
Römer, J. *9410-9411*
Roemer, J.J. *9400-9409*, 8876, 9635; *see* Meisner, C.F., Roth, A.W.
Roemer, K.F. 6109
Roemer, L. 9404
Roemer, M.J. *9412-9414*
Roemer, R.B. von *see* Roemer, J.J.
Roemera see Roemer, J.J.
Roemeria see Roemer, F. von, Roemer, J.J.
Römmp, H. *9415*
Rönn, H.L.K. *9416*
Rönnbäck, G.G. 10.006
Roep. v. Roeper, J.A.C.

Roeper, J.A.C. *9417-9426*, 8925
Roepera see Roeper, J.A.C.
Roeperia see Roeper, J.A.C.
Roeperocharis see Roeper, J.A.C.
Roesch, C. *9427*
Rösler, C.A. *alph.*
Roesler, L. *see* Rösler, C.A.
Roesleria see Rösler, C.A.
Rössig, C.G. *9428-9430*
Roettlera see Rottler, J.P.
Roezl, B. *9431*
Roezlia see Roezl, B.
Roezliella see Roezl, B.
Roffavier, G. *9432*
Roger, J.F. *see* Richard, C.
Rogers, A.M. *see* Rogers, D.P.
Rogers, C.C. *9433*
Rogers, C.G. *alph.*
Rogers, D.P. *9434*
Rogers, F.A. *alph.*
Rogers, H.D. *9435*
Rogers, J.E. *9436-9437*
Rogers, R.E. *see* Rogers, H.D.
Rogers, R.S. *9438-9439*
Rogers, W.B. *see* Rogers, H.D.
Rogers, W.E *9440*
Rogers, W.M. *9441-9444; see* Rogers, F.A.
Rogersella see Rogers, D.P.
Rogersia see Rogers, D.P., Rogers, H.D.
Rogersiomyces see Rogers, D.P.
Rohde, M. *9445-9446*
Rohlena, J. *9447-9453*
Rohlfs, G.F. *9454-9456; see* Nachtigal, G.H.
Rohlfsia see Rohlfs, G.F.
Rohr, J.B. von *see* Rohr, J.P.B. von
Rohr, J.P.B. von *alph.*
Rohra see Rohr, J.P.B. von
Rohrb. v. Rohrbach, P.
Rohrbach, P. *9457-9461*
Rohrer, R. *9462*
Rohria see Rohr, J.P.B. von
Roigia see Roig y Mesa, J.T.
Roig y Mesa, J.T. *9463-9464*
Roivainen, H. *9465-9469*, 8544
Roivainenia see Roivainen, H.
Rojas, T. *alph.*
Rojas, U. *see* Rojas, T.
Rojas Clemente y Rubio, S. de *v.* Roxas Clemente y Rubio, S. de
Rojasia see Rojas, T.
Rojasianthe see Rojas, T.
Rojasiophyton see Rojas, T.
Rolander, D. *alph.*, 9667
Rolandra see Rolander, D.
Rolet, C. 9738
Rolfe, R.A. *9470; see* Rutenberg, D.C.
Rolfea see Rolfe, R.A.
×*Rolfeara see* Rolfe, R.A.
Rolfeella see Rolfe, R.A.

Rolfs, P.H. *alph.*
Rolland v. Rolland, L.L.
Rolland *see* Rolland, L.L.
Rolland, E. *9471*
Rolland, L.L. *9472-9479*
Rollandia see Rolland, L.L.
Rollandina see Rolland, L.L.
Rolle, F. *9480;* see Roemer, F. von
Rollet, M.A. 9432
Rolli, E. *9481*
Rollinius, C.J. 9790
Rollins, R.C. *9482*
Rollinsia see Rollins, R.C.
Rolofa see Roloff, C.L.
Roloff, C.L. *9483*
Romagnesi, H.C.L. *alph.*
Roman *see* Romano, G.
Romana *see* Romano, G.
Romano, G. *9484-9485*
Romano, G.B. see Romano, G.
Romanoa see Romano, G.
Romans, B. *9486*
Rombouts, J.G.H. *9487-9488*
Romell, L. *9489-9490*
Romell, L.-G. *see* Romell, L.
Romellia see Romell, L.
Romellina see Romell, L.
Romieux, H. *alph.*
Rompel, J. *see* Rick, J.
Ronceray, P.-L. *9491*
Rondelet, G. *alph.*
Rondeletia see Rondelet, G.
Ronniger, K. *9492*
Ronnigeria see Ronniger, K.
Ronzani 8145
Rooke, H. *9493*
Roper v. Roper, F.C.S.
Roper, F.C.S. *9494-9495*
Roper, I.M. *alph.*
Roperia see Roper, F.C.S.
Roques, J. *9496-9499*
Rosander, H.A. *9500*
Rosander, K.H.A. *v.* Rosander, H.A.
Rosanoff, S.M. *9501*
Rosanovia see Rosanoff, S.M.
Rosanowia see Rosanoff, S.M.
Rosanthus see Rose, J.N.
Rosbach, H. *9502-9503*
Roscoe v. Roscoe, W.
Roscoe, M. *9504,* 9505
Roscoe, Mrs. E. *v.* Roscoe, M.
Roscoe, W. *9505;* see Roscoe, M.
Roscoea see Roscoe, W.
Rose, H. *9506*
Rose, J.N. *9507-9513;* see Russell, J., Russell, P.G.
Rose, L.S. *alph.*
Rose, V. *see* Rose, H.
Rosea see Rose, H.
Roseanthus see Rose, J.N.

Roseia see Rose, J.N.
Rosén, E. *9514;* see Rosén, N.
Rosen, Felix *9515*
Rosén, Friedrich *see* Rosen, Felix
Rosén, N. *alph.; see* Rosén, E.
Rosenbaum, L.S. *v.* Rose, L.S.
Rosenberg, C.F. *alph.*
Rosenberg, G.O. *v.* Rosenberg, O.
Rosenberg, M.E. *9516-9517*
Rosenberg, O. *9518-9520*
Rosenbergia see Rosenberg, C.F.
Rosenbergiodendron see Rosenberg, O.
Rosenblad, E. *v.* Rosén, E.
Rosend. *v.* Rosendahl, O.
Rosend., F. *v.* Rosendahl, F.
Rosend., H. *v.* Rosendahl, F.H.V.
Rosendahl, C.O. *v.* Rosendahl, O.
Rosendahl, F. *9528*
Rosendahl, H.V. *9529-9531*
Rosendahl, O. *9521-9527.*
Rosenia see Rosén, E.
Rosenschöld, E. Munck af *v.* Munck af Rosenschöld, E.
Rosenstein, N. Rosén von *v.* Rosén, N.
Rosenstock, E. *9532-9534*
Rosenstockia see Rosenstock, E.
Rosenthal, D.A. *9535*
Rosenvinge, J.L.A. Kolderup *v.* Rosenvinge, L. Kolderup
Rosenvinge, L. Kolderup *9536-9556,* 9609
Rosenvingea see Rosenvinge, L. Kolderup
Rosenvingiella see Rosenvinge, L. Kolderup
Rosén von Rosenstein, N. *v.* Rosén, N.
Roseocactus see Rose, J.N.
Roseocereus see Rose, J.N.
Roseodendron see Rose, J.N.
Roshevitz, R.U.J. *alph.*
Roshevitzia see Roshevitz, R.U.J.
Rosny, L.L.L. Prunol de *9557*
Ross, A.M. *9558-9559*
Ross, D. *9560-9563*
Ross, H. *9564-9569*
Ross, J. (1777-1856) *see* Ross, J.C.
Ross, J. (fl. 1830-1840) *see* Ross, A.M.
Ross, J.C. *alph.; see* Lyall, D., Sabine, E.
Ross, N.E. *see* Ross, A.M.
Ross, R. *alph.*
Ross-Craig, S. *alph.*
Rossetti, C. *9570*
Rosshirt, K. *9571*
Rossi, C. *v.* Rossi, G.
Rossi, G. *see* Rossi, G.B.
Rossi, G.B. *9572*
Rossi, P. (1738-1804) *9573-9574*
Rossi, P. (1871-1950) *see* Rossi, P. (1738-1804)
Rossi, S. *9575-9579*
Rossiella see Ross, R.
Rossioglossum see Ross, A.M.

Rossipollis see Ross, A.M.
Rossmässler, E.A. *9580-9585*
Rossmaesslera see Rossmässler, E.A.
Rossmann, G.W.J. *v.* Rossmann, J.
Rossmann, J. *9586-9591*
Rossmannia see Rossmann, J.
Rost. v. Rostafiński, J.T.
Rostafinsckia see Rostafiński, J.T.
Rostafinskia see Rostafiński, J.T.
Rostafiński, J.T. *9592-9599*
Rostcovia see Rostkovius, F.W.G.
Roster, G. (*fl.* 1874-1885) see Roster, G. (x-1968)
Roster, G. (x-1968) *9600*
Rostius, C. *alph.*
Rostk. v. Rostkovius, F.W.G.
Rostkovia see Rostkovius, F.W.G.
Rostkovites see Rostkovius, F.W.G.
Rostkovius, F.W.G. *9601-9602*
Rostkovius, F.W.T. *v.* Rostkovius, F.W.G.
Rostock, M. *9603*
Rostovzev, S.I. *9604-9605*
Rostr. v. Rostrup, E.
Rostr., O. v. Rostrup, O.G.F.
Rostrup, A. see Rostrup, O.G.F.
Rostrup, E. *9606-9620;* see Ravn, F.K., Rostrup, O.G.F.
Rostrup, F.G.E. *v.* Rostrup, E.
Rostrup, O.G.F. *9621,* 9606
Rostrup, S. see Rostrup, O.G.F.
Rostrupia see Rostrup, E.
Rot von Schr. v. Rot von Schreckenstein, F.
Rot von Schreckenstein, F. *9622*
Rota, L. *9623-9624*
Rotaea see Rota, L.
Rotbolla see Rottbøll, C.F.
Rotbollia see Rottbøll, C.F.
Rotert, V.A. *v.* Rothert, W.
Roth v. Roth, A.W.
Roth, A.W. *9625-9637,* 9408; see Rohde, M., Roth, J.R.
Roth, C.W. see Roth, A.W.
Roth, E.C.F. *9638-9640*
Roth, G. *9641-9643,* 9390
Roth, J.R. *9644;* see Rohr, J.P.B. von
Roth, R. see Roth, A.W.
Roth, W. *9645-9646*
Rothacker, R.R. *7257*
×*Rothara* see Roth, A.W.
Rotheray, L. *9647*
Rothert, K.W. *v.* Rothert, W.
Rothert, W. *9648*
Rothia see Roth, A.W.
Rothm. v. Rothmaler, W.H.P.
Rothmaler, W.H.P. *9649-9652*
Rothmaleria see Rothmaler, W.H.P.
Rothman v. Rothman, G.
Rothman, G. *alph.*

Rothman, J.S. *alph.;* see Rothman, G.
Rothmannia see Rothman, G.
Rothmayr, J. *9653*
Rothpletz, A. *9654-9656*
Rothpletz, F.A. *v.* Rothpletz, A.
Rothpletzella see Rothpletz, A.
Rothr. v. Rothrock, J.T.
Rothrock, J.T. *9657-9660;* see Porter, T.C.
Rothrockia see Rothrock, J.T.
Rothschild, J. *9661-9663;* see Roze, É.
Rottb. v. Rottbøll, C.F.
Rottboelia see Rottbøll, C.F.
Rottbøll, C.F. *9664-9667;* see Rolander, D.
Rottboella see Rottbøll, C.F.
Rottboellia see Rottbøll, C.F.
Rottbolla see Rottbøll, C.F.
Rottbollia see Rottbøll, C.F.
Rottenbach, H. *9668-9669*
Rottler, J.P. *alph.*
Rottlera see Rottler, J.P.
Rottleria see Rottler, J.P.
Roucel, F.A. *9670-9671*
Roucela see Roucel, F.A.
Roum. v. Roumeguère, C.
Roumeguère, C. *9672-9683,* 6455; see Malbranche, A.F., Manoury, C.A., Mougeot, J.B., Nestler, C.G., Niel, P.E., Passerini, C., Plowright, C.B., Quélet, L., Rolland, L.L., Rostrup, E., Roze, É.
Roumegueria see Roumeguère, C.
Roumegueriella see Roumeguère, C.
Roumeguerites see Roumeguère, C.
Roupallia see Roupell, A.E.
Roupell, A.E. *9684*
Roupell, C. see Roupell, A.E.
Roupell, Dr. see Roupell, A.E.
Roupell, T.B. see Roupell, A.E.
Roupellia see Roupell, A.E.
Roupellina see Rouypell, A.E.
Rouppert, K. *9685*
Roussaea see Rousseau, J.J.
Roussea see Rousseau, J.J.
Rousseau v. Rousseau, J.J.
Rousseau, E.J. see Rousseau, M.
Rousseau, J.J. *9686-9689;* see Rozier, F.
Rousseau, J.J.J. *9690*
Rousseau, M. *9691*
Rousseaua see Rousseau, J.J.
Rousseauvia see Rousseau, J.J.
Roussel v. Roussel, H.F.A. de
Roussel, A.V. *alph.*
Roussel, E. *9692*
Roussel, H.F.A. de *9693-9694*
Rousselia see Roussel, A.V.
Roussoa see Rousseau, J.J.
Roussoella see Rousseau, M.
Roussoellopsis see Rousseau, M.
Rouville, P.G. de *9695*

Roux v. Roux, H.
Roux, A. *see* Roux, N.
Roux, C.C.F.M. le *see* Roux, N.
Roux, C.J.A. *see* Roux, N.
Roux, E. *see* Roux, N.
Roux, H. *9696*
Roux, Jacques *see* Roux, N.
Roux, Jean *see* Roux, N.
Roux, J.L.F.P.*see* Roux, N.
Roux, M. *see* Roux, N.
Roux, N. *9697-9698*
Roux, P.-P.-É. *see* Roux, N.
Roux, W. *Roux, N.*
Rouxia see Roux, N.
Rouy, G.C.C. *9699-9712*
Rouya see Rouy, G.C.C.
Rouzier de la Bergerie, J.B. 9741a
Rovirosa, J.N. *9712a*
Rowlee, W.W. *alph.*
Rowley, G.D. 8748
Rowntree, B. *see* Rowntree, L.
Rowntree, G.E.L.*v.* Rowntree, L.
Rowntree, L. *9713-9714*
Roxas Clemente y Rubio, S. de *9715-9718*
Roxb. v. Roxburgh, W.
Roxburgh, B. *see* Roxburgh, W.
Roxburgh, James *see* Roxburgh, W.
Roxburgh, John *see* Roxburgh, W.
Roxburgh, W., Jr. *see* Roxburgh, W.
Roxburgh, W. *9719-9725*, 7919, 9505, 9635; *see* Russell, P.
Roxburghia see Roxburgh, W.
Roy, J. *alph.*
Roya see Roy, J.
Roye, E. 9263
Royen v. Royen, A. van
Royen, A. van *9726-9731; see* Royen, D. van
Royen, D. van *9731a-9732*, 9727, 9729
Royena see Royen, A. van
Royenia see Royen, A. van
Royer, A.P.A. *see* Royer, C.L.A.
Royer, C.L.A. *9733*
Royle, J.F. *9734-9737*
Roylea see Royle, J.F.
Roze, E. *9738-9741*, 9176-9177, 9662
Rozea see Roze, É.
Rozella see Roze, É.
Rozellopsis see Roze, É.
Rozier, F. *9741a*
Rozier, J.-F. *v.* Rozier, F.
Rozin, A. *9742*
Rozites see Roze, É.
Rubel, F. *9743*
Rubelia see Rübel, E.A.
Rubesch, J. 9098
Rubio, J. 9771
Rubio, S. de Roxas Clemente y *v.* Roxas Clemente y Rubio, S. de
Ruchinger v. Ruchinger, G., Sr.

Ruchinger, F. *see* Ruchinger, G., Sr.
Ruchinger, G., Jr. *see* Ruchinger, G., Sr.
Ruchinger, G., Sr. *9744; see* Ruchinger, G.M.
Ruchinger, G.M. *9745; see* Ruchinger, G., Sr.
Ruchinger, J. *v.* Ruchinger, G., Sr.
Rudbeck Jr. v. Rudbeck, O.O.
Rudbeck Sr. v. Rudbeck, O.J.
Rudbeck, J.O. *9746; see* Roberg, L.
Rudbeck, O.J. *9747; see* Rudbeck, O.O.
Rudbeck, O.O. *9748-9749*, 9747; *see* Rudbeck, J.O., Rudbeck, O.J.
Rudbeckia see Rudbeck, O.O.
Rudberg, A. *9750*
Rudge, A. 9751
Rudge, E. *9751*
Rudgea see Rudge, E.
Rudio, F. *9752*
Rudolfiella see Rudolph, K.
Rudolph, J.H. *9753*
Rudolph, K. *alph.*
Rudolph, L. *9754-9755*
Rudolphi v. Rudolphi, K.A.
Rudolphi, F.K.L. *9756*
Rudolphi, I.K.A. *v.* Rudolphi, K.A.
Rudolphi, K.A. *9757; see* Nordmann, A.D. von
Rudolphia see Rudolph, K., Rudolphi, K.A.
Rudolphisporis see Rudolph, K.
Rübel, E.A. *9758-9760*, 9232, 9234-9235
Rübel-Blass, E.A. *v.* Rübel, E.A.
Rückert, E.F. *9761*
Rüggeberg, H.K.A. *9762*
Ruel. v. Rueling, J.P.
Rueling, J.P. *9763-9764*
Ruelinga see Rueling, J.P.
Ruelingia see Rueling, J.P.
Rümpler, T. *9765*
Rüppel, E. *v.* Rüppell, E.
Rüppell, W.P.E.S. *v.* Rüppell, E.
Rueppelia see Rüppell, E.
Rüppell, E. *alph.*
Rütimeyer, L. 9232
Ruge, G. *9766*
Rugel, F.I.X. *alph.*
Rugel, J. *see* Rugel, F.
Rugelia see Rugel, F.
Ruhland, W.O.E. *alph.*, 7684
Ruhlandiella see Ruhland, W.O.E.
Ruhmer, G.F. *alph.*
Ruijs, J.M. *9767*
Ruijsch, F. *v.* Ruysch, F.
Ruiz, A. 9781
Ruiz, S.J. Lopez *9783*
Ruiza see Ruiz Lopez, H.
Ruizia see Ruiz Lopez, H.
Ruiz Lopez, H. *9768-9782; see* Pavon y Jiménez, J.A.

INDEX TO NAMES

Ruizodendron see Ruiz Lopez,H.
Ruland, K. 6818
Rulingia see Rueling, J.P.
Rumpf, G.E. *v.* Rumphius, G.E.
Rumpfa see Rumphius, G.E.
Rumpfia see Rumphius, G.E.
Rumph. v. Rumphius, G.E.
Rumphia see Rumphius, G.E.
Rumphius, G.E. *9784-9785,* 4829; *see* Petiver, J., Reinwardt, C.G.C., Robinson, C.B., Jr.
Rumwerth, R. Rapaics von *v.* Rapaics von Rumwerth, R.
Rupin, E.J.P. *9786-9787*
Rupinia see Rupin, E.J.P.
Rupp v. Rupp, H.M.R.
Rupp, H.B. *v.* Ruppius, H.B.
Rupp, H.M.R. *9788-9789*
Ruppa see Ruppius, H.B.
Ruppia see Ruppius, H.B.
Ruppius, H.B. *9790*
Rupprecht, H. see Rupprecht, J.B.
Rupprecht, J.B. *9791*
Rupr. v. Ruprecht, F.J.
Ruprecht, F.J. *9792-9811,* 8197
Ruprecht, F.J.I. *v.* Ruprecht, F.J.
Ruprecht Sr. *see* Ruprecht, F.J.
Ruprechtia see Ruprecht, F.J.
Ruprechtiella see Ruprecht, F.J.
Rusby, H.H. *9812-9825,* 6342
Rusby, W.S. *see* Rusby, H.H.
Rusbya see Rusby, H.H.
Rusbyanthus see Rusby, H.H.
Rusbyella see Rusby, H.H.
Ruschenberger, W.W. 7913
Ruspoli, E. see Riva, D.
Russ, G.P. *9826*
Russ, K. *9585*
Russea see Rousseau, J.J.
Russegger, J. von *9827*
Russeggera see Russegger, J. von
Russela see Russell, Alexander (*ca.* 1715-1768)
Russelia see Russell, Alexander (*ca.* 1715-1768)
Russell, Alexander (1814-1878) *see* Russell, Alexander (*ca.* 1715-1768)
Russell, Alexander (*ca.* 1715-1768) *9828; see* Russell, P.
Russell, Anna *alph.*
Russell, E.E. *see* Russell, J.
Russell, F. *see* Russell, Anna
Russell, J. *alph.*
Russell, J.L. *9829*
Russell, M. *see* Russell, J.
Russell, Mrs. M. *see* Russell, J.
Russell, P. *alph.,* 9828; *see* Russell, Alexander (*ca.* 1715-1768)
Russell, P.G. *alph.; see* Russell, J.
Russell, T.H. *9830*

Russellia see Russell, J.
Russellites see Russell, J.
Russelodendron see Russell, J.
Russow, E.A.F. *9831-9838*
Russowia see Russow, E.A.F.
Rutenberg, D.C. *alph.*
Rutenbergia see Rutenberg, D.C.
Ruthe v. Ruthe, J.F.
Ruthe, J.F. *9839; see* Ruthe, R.
Ruthe, J.G.R. *v.* Ruthe, R.
Ruthe, R. *see* Ruthe, J.F.
Ruthe, R. *alph.*
Ruthea see Ruthe, J.F.
Rutherford, D. *9840*
Ruthiella see Ruthe, J.F.
Rutstr. v. Rutström, C.B.
Rutström, C.B. *9841-9842*
Rutstroemia see Rutström, C.B.
Rutten, L.M.R. *see* Pulle, A.A.
Ruttner, F. *9843*
Rutty, J. *9844*
Ruttya see Rutty, J.
×*Ruttyruspolia see* Rutty, J.
Ruys, J. *9845*
Ruys, J.D. *9846*
Ruys, J.M. *v.* Ruijs, J.M.
Ruysch, F. *alph.*
Ruyscha see Ruysch, F.
Ruyschia see Ruysch, F.
Ruyschiana see Ruysch, F.
Ruyschioxylon see Ruysch, F.
Ryan, E. *alph.*
Ryan, J. *see* Ryan, E.
Ryanaea see Ryan, E.
Ryania see Ryan, E.
Rydb. v. Rydberg, P.A.
Rydberg, P.A. *9847-9873*
Rydbergia see Rydberg, P.A.
Rydbergiella see Rydberg, P.A.
Rylands, M.G. *see* Rylands, T.G.
Rylands, T.G. *alph.*
Rylandsia see Rylands, T.G.
Rytz, A.R.W. *9874-9875*
Rytz-Miller, A.R.W. *v.* Rytz, A.R.W.

Saage, M.J. *9876*
Sabatia see Sabbati, L.
Sabbata see Sabbati, L.
Sabbati, C. *see* Sabbati, L.
Sabbati, L. *9877*
Sabbatia see Sabbati, L.
Sabine v. Sabine, J.
Sabine, E. *alph.; see* Sabine, J.
Sabine, J. *alph.; see* Sabine, E.
Sabinea see Sabine, J.
Sabransky, H. *alph.*
Sacc. v. Saccardo, P.A.
Sacc., D. v. Saccardo, D.
Sacc., F. v. Saccardo, F.
Saccardaea see Saccardo, P.A.

Saccardia see Saccardo, P.A.
Saccardinula see Saccardo, P.A.
Saccardo, D. *9878*, 9898
Saccardo, F. *9879-9880*, 9907
Saccardo, P.A. *9881-9951*, 7442, 7602, 7642, 7648, 7655, 7785-7786, 9177, 9683, 9880, 10.017; *see* Paoletti, G., Penzig, A.J.O., Saccardo, D., Saccardo, F., Saint-Amans, J.F.B. de
Saccardoa see Saccardo, P.A.
Saccardoella see Saccardo, P.A.
Saccardomyces see Saccardo, P.A.
Saccardophytum see Saccardo, P.A.
Sachs, C. *see* Sachs, J. von
Sachs, F.G.J. von *v.* Sachs, J. von
Sachs, F.J. *see* Sachs, J. von
Sachs, J. 8268
Sachs, Julius (1849-1934) *see* Sachs, J. von
Sachs, J. von *9952-9959*, 7831, 8268; *see* Micheli, M., Pfeffer, W.F.P., Prantl, K.A.E.
Sachs, K. *v.* Sachs, C.
Sachse, C.T. *9960*
Sachsia see Sachs, J. von
Sacleux, C. *9961*
Sadebeck, R.E.B. *9962-9972*, 8256; *see* Reichenbach, H.G.
Sadler, John *9973-9974*
Sadler, Joseph *9975-9977; see* Sadler, M.
Sadler, M. *9978; see* Sadler, Joseph
Sadleria see Sadler, Joseph
Saelan, A.T. *v.* Saelan, T.
Saelan, T. *9979-9982*
Saelania see Saelan, T.
Säve, C.F. *9983*
Säve, P.M.A. *see* Säve, C.F.
Saff. v. Safford, W.E.
Safford, W.E. *9984-9987*
Saffordia see Safford, W.E.
Saffordiella v. Safford, W.E.
Sageret, A. *9988*
Sageretia see Sageret, A.
Sagorski, E.A. *9989-9991*
Sagot, J.L.A. *see* Sagot, P.A.
Sagot, P.A. *9992-9997*, 8631
Sagotanthus see Sagot, P.A.
Sagotia see Sagot, P.A.
Sagra, R. de la *9998-10.005*, 9150; *see* Pfund, J.D.C., Richard, A.
Sagraea see Sagra, R. de la
Sahlberg, C.R. *10.006*
Sahlberg, R.F. *see* Sahlberg, C.R.
Sahlén, A.J. *10.007*.
Sahlstedt, M.G. 9085
Sahni, B. *10.008-10.010*
Sahnia see Sahni, B.
Sahnianthus see Sahni, B.
Sahniocarpon see Sahni, B.
Sahnioxylon see Sahni, B.

Sahnipushpam see Sahni, B.
Sahnisporites see Sahni, B.
Sahut, C. *see* Sahut, F.
Sahut, F. *10.011-10.013*
SAIDA 8461
Sailer, F.S. *10.014-10.015*
Sainsburia see Sainsbury, G.O.K.
Sainsbury, G.O.K. *alph.*
Saint-Amans, J.F.B. de *10.016-10.018*, 9940
Saint-Beuve *see* Rambert, E.
Saint Brody, G.A.O. *v.* St. Brody, G.A.O.
Saint-Cyr, D.N. *alph.*
Saint-Gal, M.J. *10.020-10.022*
Saint-Germain, J.J. de *10.023*
Saint-Hilaire, A.F.C.P. de *10.024-10.055*, 9679
Saint John, H. *v.* St. John, H.
Saint-Lager, J.B. *10.066-10.089*.
Saint-Léon, V.B.A. Trevisan de *v.* Trevisan de Saint-Léon, V.B.A.
Saint-Moulin, V.J. de *10.090-10.091*
Saint-Sauveur, J. Grasset de *v.* Grasset de Saint-Sauveur, J.
Saint-Simon, M.H. de *10.092*
Saint-Yves, A.M.A. *10.093-10.104*
Sakurai, K. *alph.*
Sakuraia see Sakurai, K.
Salisbury, R.A. 8404, 10.067; *see* Rudge, E.
Salmon see Pearsall, W. Harrison
Salomon, C. 8775
Salter, J.W. *see* Phillips, J.
Salvador *see* Pourret, P.A.
Salvert, Dutour de *v.* Dutour de Salvert
Salvertia see Saint-Hilaire, A.F.C.P. de
Sande Lacoste, C.M. van der 6119, 9488
Sander, F. *see* Reichenbach, H.G., Roezl, B.
Sandler, W. 8103
Sandmark, G. 9087
Sandstede, H. 8450
Sandwith, N.Y. 8393
Sanhilaria see Saint-Hilaire, A.F.C.P. de
Santapau, H. *see* Pau y Español, C.
Santesson, R. *see* Regnell, A.F.
Santos Rangel, E. dos *v.* Rangel, E. dos Santos
Sarasin, F. *see* Roux, N.
Sargent, C.S. 8815, 9302; *see* Palmer, E.J., Rafinesque-Schmaltz, C.S., Rehder, A., Rendle, A.B.
Sarmento, A.A. 9941
Sasaki, S. 5576
Satory, J.A. 5909, 8787, 9796
Satterlee, M. 7415
Saunders, W.W. 9734; *see* Reichenbach, H.G.
Saussure, H.B. de *see* Reuss, A.E. von

Sauvage, de la 8283
Sauvageau, C.F. 7475
Savatier, A. 9315
Savès, T. 6455; see Müller, Karl (Halle)
Savi, G. see Raddi, G.
Savoia, L.A. di see Pirotta, P.R.
Savulescu, T. see Rayss, T.
Saxa, J. see Sachs, J. von
Say, T. see Pickering, C.
Scarabelli Gommi Flamini, G. 5619
Scarpellini, F. 7968
Schacht, J.O. 9729, 9784
Schaeffer, J.C. 7729; see Persoon, C.H.
Schaerer, L.E. 5595, 5599
Schaldemose, F. 8090
Scheele, G.H.A. 9397; see Roemer, F. von
Scheffer, J.G. de Hoop v. Hoop Scheffer, J.G. de
Schenck, H. see Pax, F.A., Rikli, M.A., Rosen, Felix
Schenk, A. see Sadebeck, R.E.B.
Schenk, A. von see Roth, J.R.
Schenk, H. see Pohle, R.R.
Schenk, J.A. von see Pfeffer, W.F.P., Pfitzer, E.H.H.
Scherbius, J. 8593
Scheuchzer, J.J. see Roemer, J.J.
Scheygrond, A. 8393; see Pulle, A.A.
Schiedermayr, K.B. 8095; see Poetsch, I.S.
Schiemann, K. 7402
Schiffner, V.F. 7423
Schiller, G.W. 8893
Schiller, J. 7423, 8450
Schilling, A.J. 7423
Schiman-Czeika, H. see Rechinger, K.H.
Schimper, H.W. see Rohlfs, G.F.
Schimper, W.P. 7328; see Neuweiler, E.
Schinz, H. 8450; see Roux, N.
Schiødte, J.C. 9244
Schkur, C. 8876, 8885
Schlauter, A. see Peter, A., Pritzel, G.A.
Schlechtendal, D.F.L. von 6102, 6112, 8353, 9121, 9408, 9422, 9431; see Miquel, F.A.W., Müller, Karl (Halle), Pamplin, W., Passerini, G., Petzold, E.A., Peyritsch, J.J., Pfeifer, L.K.G., Philippi, R.A., Phoebus, P., Planchon, J.É., Poeppig, E.F., Pohl, J.B.E., Pokorny, A., Presl, K.B., Pringsheim, N., Pritzel, G.A., Rabenhorst, G.L., Ratzeburg, J.T.C., Regel, E.A. von, Regnell, A.F., Reichenbach, C.L. von, Reichenbach, H.G.L., Reichenbach, H.G., Reinsch, P.F., Reinwardt, C.G.C., Reissek, S., Riedel, L., Ritter, C. (Berlin), Römer, Heinrich, Roeper, J.A.C., Rossmann, J., Rostkovius, F.W.G., Rudolphi, F.K.L., Ruprecht,

J.B., Ruthe, J.F., Sachs, J. von, Sadler, Joseph, Saint-Hilaire, A.F.C.P. de
Schlechter, F.R.R. 7684, 8726; see Maatsch, R.F.T., Rogers, R.S., Rudolph, K.
Schleiden, M.J. see Nägeli, C.W. von, Radlkofer, L.A.T.
Schley, J. van der 10.092
Schlosser, J.C. 6725
Schmalfuss, A. 5952
Schmalhausen, J.T. see Regel, E.A. von
Schmaltzia see Rafinesque-Schmaltz, C.S.
Schmidt, C.F. 6183
Schmidt, E. 8775
Schmidt, E.L.W. 9602
Schmidt, F.W. see Roemer, J.J.
Schmidt, W. 8915
Schmidt, W.E.L. v. Schmidt, E.L.W.
Schmitz, J.J. see Regel, E.A. von
Schnee, L. see Pittier, H.F.
Schneider, C.K. see Porsch, O., Potonié, H.
Schneider, G. 9990
Schneider-Orelli, Mrs. 9232
Schneider-Orelli, O. 9232
Schnorr, C.H. 8885
Schoene, A. 8938
Schönfeld, H. von 7423
Schoenichen, W. 9569
Schönland, S. 8132
Scholander, P.F. 5131
Schott, H.W. 7808
Schott, K. see Pax, F.A.
Schrader, H.A. 9769
Schramm, A. 7772
Schrant, J.M. 9487
Schreckenstein, F. Rot von v. Rot von Schreckenstein, F.
Schrenk, H. von 9288
Schröter, C.J. 9232; see Rikli, M.A.
Schröter, J.S. 7222
Schube, T. 7544
Schubert, C. see Reichenbach, H.G.L.
Schubert, G.H. von see Roth, J.R.
Schubert, R. see Rothmaler, W.H.P.
Schuebler, G. 9240
Schütt, F. 8931
Schultes, J.A. 8304, 8876, 9408-9409, 9635; see Roemer, J.J., Roth, A.W.
Schultes, J.H. 9408; see Radde, G.F.R. von
Schultz Bipontinus, C.H. see Loescher, E., Regnell, A.F., Rüppell, E.
Schulz, A.A.H. 8211
Schulz, R. see Pax, F.A.
Schulze, E. see Pantanelli, E.F.
Schulzer von Müggenburg, S. 9903
Schumann, K.M. 9765, 9817; see Rutenberg, D.C.
Schutte, J.H. 9790

Schwaegrichen, C.F. 9169; see Motelay, L.
Schwan, J.P. 7139
Schweinfurth, G.A. 8628, 9454; see Riva, D., Russegger, J. von
Schweinitz, L.D. de 7734, 8316; see Muhlenberg, G.H.E., Prince, W., Rafinesque-Schmaltz, C.S., Reichenbach, H.G.L., Rich, W.
Schwencke, W. 9732
Schwendener, S. 6083, 8760, 9164-9165; see Pfeffer, W.F.P., Reinhardt, O.W.H., Richard, O.J., Rikli, M.A., Roth, E.C.F.
Scopoli, G.A. see Martens, G.M. von, Pourret, P.A.
Scortechini, B. 9906
Scott, W. see Nuttall, T.
Sealy, Mrs. J.R. v. Ross-Craig, S.
Seaton, H.E. 9283
Seboth, J. 8522, 8845
Seeler, E.V., Jr. see Rogers, D.P.
Seelig, C. 9234
Seemann, B.C. see Reichenbach, H.G.
Seemen, K.O. von 7684
Séguier, J.F. see Pourret, P.A.
Selleny, J. 7808
Sellier 9153, 9741a
Sellow, F. see Saint-Hilaire, A.F.C.P. de
Semenov, P. 8791
Senckenberg, J.C. 8833
Sendtner, O. see Perktold, J.A., Radlkofer, L.A.T.
SENIAVIN 8197; see Postels, A.P.
Senoner, A. 8775
Sensinow, M.S. 8790
Seringe, N.C. 9418, 9432; see Otth, A.
Sériziat, C.V.E. 9681
Sernander, J.R. 5248, 8189
Sertüner, F.W. see Reichenbach, C.L. von
Sessé, M. see McVaugh, M., Menzies, A., Pavon y Jiménez, J.A.
Setchell, W.A. see Parks, H.E.
Seubert, M.A. see Prantl, K.A.E.
Seurat, L.G. 7490
Severeyns, G. 8883, 9138
Severzoff, N.A. 8791
Seward, A.C. 7583; see Pocock, M.A., Sahni, B.
Seynes, J. de 7997, 9138
Shaffer, R.L. 5178
Shaw, E.A. see Rollins, R.C.
Shear, C.L. 7255
Sherard, J. see Plumier, C., Pursh, F.T.
Sherard, W. see Ray, J.
Shibata, K. 5283
Shimada, H. 5576
Short, C.W. see Rafinesque-Schmaltz, C.S.

Shuttleworth, R.J. see Roemer, J.J., Rugel, F.
Sieber, F.W. see Reichenbach, C.L. von
Siebert, A. 8775
Siegenbeek, M. 8961; see Reinwardt, C.G.C.
Siegle 6851
Sillén, N.J. 5106
Silva, A.R. Pinto da v. Pinto da Silva, A.R.
Silva Quintanilha, A.P. da v. Quintanilha, A.P. da Silva
Silvestri, C. 7262
Simmons, H.G. 9551, 9620
Simon, M.H. de Saint v. Saint-Simon, M.H. de
Simonkai, L. see Porcius, F.
Simpson, N.D. 6590
Sinclair, G. see Russell, J.
Singh, A.L. 8258, 8263
Singh, B. 9123
Singh, L. 9734
Sinoradlkofera see Radlkofer, L.A.T.
Sipe, F.P. 7258
Sipman, H. 8393
Skala, J. 8303
Skottsberg, C.J.F. see Petersen, J.B., Romell, L.
Sleumer, H. 8393
Sloane, H. see Petiver, J., Plukenet, L., Rand, I., Ray, J., Richardson, R., Rudbeck, O.J., Ruysch, F.
Slooten, D.F. van see Pulle, A.A.
Sloover, J.-L. de v. De Sloover, J.-L.
Small, J.K. 8183-8184
Smith, A.J. 7415
Smith, A.L. 9067
Smith, C.A. 7861
Smith, C.P. 10.064
Smith, E.F. see Rodgers, A.D.
Smith, F.T. v. Parsons, F.T.
Smith, F.W. 7554
Smith, J. 7255
Smith, J.D. 1592, 7978; see Micheli, M.
Smith, J.D. 7978
Smith, J.E. 4910, 7724, 7730, 8399, 8755, 9153, 9747, 9769; see Link, J.H.F., MacGillivray, W., Martyn, T., Maton, W.G., Menzies, A., Noehden, H.A., Peck, W.D., Persoon, C.H., Rafinesque-Schmaltz, C.S., Relhan, R., Robson, E., Roemer, J.J., Roscoe, W., Rose, H., Rottler, J.P., Rudbeck, O.J., Rudge, E., Russell, J.
Smith, L.B. 8393
Smith, O.A. 9440
Smith, R. 9734
Smith, W.G. 8061, 8619
Snakenburg, H. 9727
Sniktau v. Patterson, E.H.N.

INDEX TO NAMES

Snogerup, S. *see* Rechinger, K.H.
Soderstrom, T.R. *see* McClure, F.A.
Sørensen, T.J. 9548
Sohrer, L. 9041
Solander, D.C. 7344, 9828; *see* Parkinson, Sydney, Phipps, C.J., Pourret, P.A., Russell, Alexander (*ca.* 1715-1768)
Solereder, H. *see* Prenger, A.G.
Solla, R.F. 7889
Solms-Laubach, H.M.C.L.F. *see* Peter, A., Robinson, B.L., Rutenberg, D.C.
Sommier, S. 9927, 9948; *see* Piccone, A.
Sonder, O.W. *see* Preiss, L., Regnell, A.F.
Songeon, A. 7696-7698
Sonnini de Manoncour, C.N.S. 9741a
Sorauer, P. 8775
Soto, M. 5334
Soulié, J.A. 7472
Souviron, M.J. *see* Russell, P.G.
Sowerby, J. 7724, 7730, 8407, 8619; *see* Noehden, H.A., Persoon, C.H.
Sowerby, J.C.D. 8584
Sowerby, J. de C. 9734
Spaendonck, G. van *see* Redouté, P.J.
Spaeth, L. 8775
Spallanzani, L. *see* Reissek, S.
Spannagel, C. *see* Rosenstock, E.
Sparre, B. *see* Regnell, A.F.
Spegazzini, C. 5647, 9895
Spencer, H. *see* Naudin, C.V.
Speranzini, F. 7895
Spielmann, J.R. *see* Pourret, P.A.
Spies, S. 8690
Spix, J.B. von *see* Martius, C.F.P von, Pohl, J.B.E.
Splitgerber, F.L. 8895
Sprague, I. 9658
Sprague, T.A. 7055; *see* Ramsbottom, J.
Sprengel, C.K. *see* Müller, H.L.H., Sachs, J. von
Sprengel, K.P.J. 6845a, 8304, 9239, 9408
Spruce, R. 9813
Spyk, J. van der *see* Royen, A. van
Stackhouse, J. 7522
Stafleu, F.A. 4492, 4778, 8393, 9769, 9771; *see* Pulle, A.A., Rickett, H.W.
Stahl, C.E. *see* Renner, O.
Standley, P.C. *see* Record, S.J., Russell, P.G.
Stapf, O. 8354; *see* Otto, F., Popenoe, D.K., Pritzel, G.A., Richter, K.
St. Brody, G.A.O. *10.019*
Stearn, W.T. 4776, 5170, 8293, 8699, 8703, 8748, 9151; *see* Linnaeus, C.
Stecker, A. 9456
Steenis, C.G.G.J. van *see* Pulle, A.A.
Steetz, J. 7759
Stefano, C.P. 9933

Stein, B. 8775
Steindórsson, S. 9548
Steinitz, W. *see* Richter, L.
Steinmejer, J.B. 7230d
Steller, G.W. *see* Ruysch, F.
Stempel, J.C.F. 9425
Step, E. 8275
Stephani, F. 7591, 8502, 8726, 8915, 8987; *see* Rehmann, A.
Stephanorossia see Rossi, S.
Sterler, A. *see* Mayrhoffer, J.N.
Sternberg, K.M. von 8293
Steudel, E.G. 7820; *see* Richard, A.
Steven, C. von *see* Pott, J.F., Robert, G.N., Rupprecht, J.B.
Stewart, J. 9126
Stewart, S.A. *see* Oliver, D., Praeger, R.L.
Stewart, W. 8686
Steyermark, J.A. 7233
St. Hil. v. Saint-Hilaire, A.F.C.P. de
Stickman, O. *see* Rumphius, G.E.
Stief, J.E. 8836
Stieglitz, F. *see* Poetsch, I.S.
Stizenberger, E. 6948, 6971, 8438; *see* Rabenhorst, G.L.
St. John, H. *10.056-10.065*
St. John Brooks, W.E. 9067
St. Lag. v. Saint-Lager, J.B.
Stockmans, F. 9074
Stockmayer, S. 8095
Stoffers, A.L. 8393, 8395
Stoller, J. 8217, 8228
Stone Pacha 7836
Storm, J.M. *see* Persoon, C.H.
Strabo, 5928
Straelen, V. Van *v.* Van Straelen, V.
Strasburger, E.A. *see* Oltmanns, F., Pfeffer, W.F.P., Porter, C.E., Pringsheim, N.
Straub, W. *see* Probst, R.
Strugnell, W.B. *see* Reader, P.
Stubendorff, J. von 8784, 8790
Sturm, J. 7736, 8316, 8649, 8874, 9406, 9629; *see* Nees von Esenbeck, T.F.L., Preuss, C.G., Reichenbach, H.G.L., Rostkovius, F.W.G.
Sullivant, W.S. *see* Rafinesque-Schmaltz, C.S., Rodgers, A.D.
Sundberg, A. 9983
Sundercombe, E.H. *v.* Pelloe, E.H.
Surian, J.D.*see* Plumier, C.
Suringar, W.F.R. *see* Molkenboer, J.H.
Susanna see Phillips, E.P.
Suzuki, S. 5576
Suzuki, T. 5576
Swart, J.J. 8393; *see* Pulle, A.A.
Swartz, O.P. 7247; *see* Mohr, D.M.H.
Swynnerton, C.F.M. 9068
Sydow, H. 7785; *see* Petch, T.

INDEX TO NAMES

Sydow, P. 9898; see Poeverlein, H., Rehm, H.S.L.F.F.
Syen, A. 9123-9124
Sylvan, C. 9090
Symonds, W.S. 8401
Szabó, Z. von see Pax, F.A.
Szafer, W. see Pawlowski, B., Raciborski, M.

Tacitus see Relhan, R.
Tafalla, J. 9775
Tafulla see Pavon y Jiménez, J.A.
Talbot, D.A. 9069
Talbot, P.A. 9069
TALISMAN see Poirault, G.
Talleyrand 9976
Tanfani, E. 7361
Taquet, T. see Rosenstock, E.
Taubert, P.H.W. 8211; see Reiche, K.F.
Taurenhayn, J. see Molisch, H.
Tausch, F.I. 7137
Tavera, T.H.J. Pardo de v. Pardo de Tavera, T.H.J.
Taylor, G. 8748
Taylor, T. see Menzies, A.
Tchihatcheff, P. de 7375
Tela, P. 7943
Tempère, J.A. 7664; see Peragallo, H., Petit, P.C.M.
Tengström, F. 8373
Tenore, V. 7425; see Pasquale, G.A.
Terracciano, A. 7419; see Pirotta, P.R.
TERROR see Ross, J.C.
Testu, G.M.P.C. le v. Le Testu, G.M.P.C.
Thaxter, R. see Quaintance, A.L., Ramsbottom, J., Rea, C., Reess, M.F.T.F.M., Rehm, H.S.L.F.F., Riddle, L.W.
Thedenius, K.F. 7247, 9983
Theissen, F. see Rick, J.
Theophrastos, E. 7522, 10.071
Thieme, A. 8896, 9584
Thienemann, A. see Ruttner, F.
Thiselton Dyer, W.T. 9954
Thoday, M.G. 7583
Thom, C. see Raper, K.B.
Thomas, C. 9041
Thomas, J.B., Jr. 7327
Thomé, O.W. 8238
Thompson, D'Arcy, W. see Müller, H.L.H.
Thornam, C. 7016, 7024, 7031
Thoroddson, T. 9548
Thorry, C.A. v. Thory, C.A.
Thory, C.A. 8748-8749, 8751
Thou, J.-A. de 9073
Thouin, A. 9741a; see Pourret, P.A.
Threlkeld, C. see Rutty, J.
Thümen, N. von 10.012

Thun, F. de 9431
Thunberg, C.P. 8834; see Pappe, K.W.L., Peck, W.D., Poiret, C.P., Pontén, J.P., Pourret, P.A., Radloff, F.W., Roemer, J.J., Rohr, J.P.B. von, Rosén, E., Rothman, G., Roxburgh, W., Rudolphi, K.A., Sahlberg, C.R.
Thurber, G. 9657
Thuret, G.A. see Michaux, A.
Thuring, E.L. 9309
Thysse, J.P. 8392
Tiling, H.S.T. 8780
Tollard 9741a
Tonnelier 7228
Torén, O. 7122-7123
Toro, R.A. 7135; see Overholts, L.O.
Torrey, J. 7019, 8404, 8573, 8582, 9148; see Mead, S.B., Muhlenberg, G.H.E., Parry, C.C., Rafinesque-Schmaltz, C.S., Rodgers, A.
Torssell, G. 5894
Tour, M.C. Boy de la v. Boy de la Tour, M.C.
Tournefort, J.P. de 4798, 5974, 7110, 7201, 7302, 8705-8706, 9063, 9088, 9687, 9694, 9877, 9898, 10.070; see Pittoni, J.C., Plukenet, L., Pourret, P.A., Rothman, J.S.
Tournon, F. see Rondelet, G.
Townsend, C.C. see Rechinger, K.H.
Townsend, F. 8708
Trattinnick, L. 8179, 8561; see Portenschlag-Ledermayer, F. von
Trautvetter, E.R. von see Regel, E.A. von
Traverso, J.B. 9894, 9898, 9936, 9939, 9950
Trelease, W. 4546, 5946, 7507, 7602; see Saccardo, P.A.
Treub, M. see Raciborski, M.
Trevisan de Saint-Léon, V.B.A. 9898
Trew, C.J. see Morren, C.J.É.
Triana, J. 8014-8015; see Planchon, J.É.
Trinius, C.B. von 8638, 9794; see Meyer, C.A. von, Ruprecht, F.J.
Troilius, C. 8398
Troll, W. see Rauh, W.
Trotter, A. 9880, 9898, 9949
Troxler, G. 9653
Tschermak, E. von see Mendel, G.J.
Tubeuf, C. von 8450
Tuckerman, E. 9658
Türckheim, H. von 6516
Tugwell, G. 8686
Tulasne, E.L.R. 10.052
Tullgren, H.A. 8997
Turner see Mohr, D.M.H.
Turner, D. 9172; see Rudge, E.
Turner, L.M. see Rothrock, J.T.

INDEX TO NAMES

Turpin, P.J.F. 8115, 8122, 9147, 10.034; see Saint-Hilaire, A.F.C.P. de
Turpin, R. 7842
Tuscany, Grand-Duke of see Raddi, G.

Uechtritz, R.F.C. von see Sadebeck, R.E.B.
Uhlworm, O. see Pascher, A.
Uittien, H. 8393; see Pulle, A.A.
Underwood, L.M. 5192, 8060, 9817
Unger, F.J.A.N. 8349, 8887; see Mohl, H. von, Reichardt, H.W.
Urban, I. 5170, 6461; see Pierre, J.B.L., Pilger, R.K.F., Plumier, C., Prior, R.C.A., Radlkofer, L.A.T., Ruhland, W.O.E., Rutenberg, D.C.
Urban, S. 7612
Usteri, A. see Palla, E.
Usteri, P. 4688, 9400; see Roemer, J.J.

Vaandrager, G. 8393
Vahl, M. see Puerari, M.N., Rohr, J.P.B. von, Rottler, J.P.
Vaillant, S. see Montagne, J.P.F.C., Plukenet, L., Plumier, C., Rothman, J.S.
Vainio, E.A. see Räsänen, V.J.P.B.
Valadier, J. 9114
VALDIVIA 8912
Valentin, J. 8480
Vandelli, D. 9403
Vanden Berghen, C. 9314
Van Heurck, H.F. see Piquet, J., Reichenbach, C.L. von
Vanhöffen, E. see Richter, P.G.
Van Houtte, L.B. 8009; see Pabst, C., Roezl, B.
Van Straelen, V. 9074
Van Tieghem, P.É.L. 1303, 9954; see Mangin, L.A.
Varde, R.A.L. Potier de la v. Potier de la Varde, R.A.L.
Varigny, H. de 9957
Vasey, G. 7507, 9507, 9657, 9658; see Neally, G.C.
Vatke, W. see Rutenberg, D.C.
Vaucher, J.P.É. see Milde, C.A.J.
Vauthier, A.C. 9151, 10.003, 10.005
Vaux, A.A.F. Cadet de v. Cadet de Vaux, A.A.F.
Vecchi, O. 6658
Velenovský, J. see Pilát, A.
Venanzi, G. 9336
Vendrely, X. 7191, 8984
Ventenat, É.P. see Redouté, P.J., Richard, L.C.M.
Venus, C.W. 7247
Verdéra, A. Paláu y v. Paláu y Verdéra, A.
Verdoorn, F. 8576, 8585; see Pulle, A.A., Rapp, S.

Vergara, L. 8857
Verhelst, E. 8137
Ver Huell, Q.M.R. 6091, 6102
Verlot, P.B.L. 9663
Vernadsky, V.I. see Němec, B.Ř.
Verschaffelt, E. see MacLeod, J.
Veselý, R. 7935
Vest, L.C. von 8920
Vestergren, T. see Patouillard, N.T., Rehm, H.S.L.F.F., Romell, L.
VETTOR PISANI 7903, 7905
Vichet, J. de see Petitmengin, M.G.C., Planchon, L.D., Pouzolz, P.M.C, de, Rey-Pailhade, C. de, Roussel, A.V., Roux, N.
Vido, A. 9895; see Saccardo, P.A.
Vieillard, E. see Pancher, J.A.I.
Vignaud 9151
Viguier, R. see Perrier de la Bâthie, H.
Villars, D. see Mougeot, J.B., Nestler, C.G., Pourret, P.A.
Vilmorin, P.L.F.L. de see Rümpler, T.
Vilmorin-Andrieux 6784
Vinciguerra, D. 7272
Vines, S.H. 8268, 9954
Vinyard, W.C. see Prescott, G.W.
Virgilius 7523, 9091, 10.070
Vischer, W. 7424, 8450
Vishnupersaud 9734
Visiani, R. de 5618, 9884; see Pančic, J., Saccardo, P.A.
Voechting, H. 8336, 8775
Vogel, E. see Radde, G.F.R. von
Vogelenzang, L. 7495, 7595
Voglino, P. 9898
Voigt, F.S. 9154
Voigtlaender-Tetzner, W. 8100
Volckamer, J.G. 7302
Voorhelm, G. see Saint-Simon, M.H. de
Vossius, I. see Rauwolff, L.
Vouaux, L. 7973
Vries, D.M. de see Pulle, A.A.
Vries, H. de see MacLeod, J., Mendel, G.J., Renner, O.
Vuillaume, A. 7927
Vukotinovíc, L.F. 6725

Wacket, M. see Rosenstock, E.
Wagenitz, G. see Rechinger, K.H.
Wagner, F. 7759
Wahlberg, P.F. 7247
Wahlenberg, C. 7247, 9983; see Mohr, D.M.H.
Wahnschaffe, F. see Potonié, H.
Wainio, E.A. 8744
Wakefield, E.M. see Ramsbottom, J.
Walbeehm, A.H.J.G. 8391
Walddorf, H., Landerer von see Landerer von Walddorf, H.
Wall, A. 10.100

Wallace, J. 8685
Wallenius, J.F. *see* Radloff, F.W.
Wallich, N. 9684, 9724-9725; *see* Rottler, J.P., Roxburgh, W., Russell, P.
Waln, M. 9505
Walpers, G.G. *see* Pritzel, G.A., Reichenbach, H.G.
Walther, A.F. 8836
Wambaugh, F. *v.* Patterson, F.
Wangeheim, L. von 9429
Warburg, O. 7684, 8915
Ward, H.M. 9958
Ward, L.F. 9054, 9058, 9817
Warming, J.E.B. 7423, 9548; *see* Meisner, C.F., Petersen, O.G., Peyritsch, J.J., Potter, M.C., Poulsen, V.A., Progel, A., Radlkofer, L.A.T., Raunkiaer, C.C., Reichenbach, H.G., Rohrbach, P.
Warnstorf, C.F.E. *see* Retzdorff, A.E.W., Röll, J.
Warren, F.A. 10.063
Washington, G. *see* Michaux, A.
Watson, A. 9723
Watson, H.C. *see* Robinson, James Fraser
Watson, S. 7505, 7507, 8180, 9657-9658; *see* Parry, C.C., Patterson, H.N., Robinson, B.L.
Wawra von Fernsee, H. 7808; *see* Peyritsch, J.J.
Webb, P.B. 7374, 7539, 8353; *see* Mercier, M.P., Moquin-Tandon, C.H.B.A., Parlatore, F., Pavon y Jiménez, J.A.
Webb, R.H. 8372
Weber, C. 9106
Weber, F. 9121; *see* Lindenberg, J.B.W.
Weber, M.C.W. *see* Redeke, H.C.
Weddell, E.S. 9720
Weddell, H.A. 9161; *see* Richard, O.J.
Weems, J.B. 7255
Wegelius, G.A. 8373
Wehlburg, C. 8393
Weiditz, H. 9875
Weigel, C.E. von *see* Martin, A., Retzius, A.J.
Weigel, T. *see* Reinsch, P.F.
Weilenmann, H. 9236
Weiss, J.G. 9121
Weissenborn, J.F. 6655
Weitenweber, W.R. 8615
Weitz, C. 6689-6690
Welwitsch, F.M.J. 9205; *see* Lister, A.
Wendelbo, P. *see* Rechinger, K.H.
Wendland 8894
Went, F.A.F.C. *see* Pulle, A.A., Raciborski, M.
Went, J.C. 8393
Wercklé, C. *see* Pittier, H.F.

Wergeland, H. *see* Rathke, J.
Wernham, H.F. 9069, 9210
Wessels Boer, J.G. 8393
West, G. 6930
Westhoff, C.F. *see* Mohr, D.M.H.
Westhoff, V. 8393; *see* Pulle, A.A.
Westphal, A. 4796
Westwood, J.O. 9734
Wettstein, R. von 8723; *see* Petrak, F., Porsch, O., Rechinger, K., Rechinger, K.H., Redinger, K.M., Rothmaler, W.H.P., Rudolph, K.
Wheatley, F. *see* Pope, C.M.
Wheeler, G.M. 9657-9658
Wheeler, G.N. *see* Rothrock, J.T.
Wheeler, W.M. *see* Powell, J.W.
Wheler, G. *see* Ray, J.
White, C. *see* Miller, P.
White, G. *see* Pennant, T.
White, O.E. *see* Rusby, H.H.
Whitfield 6924
Widder, F. *see* Rechinger, K.H.
Widgren, J.F. *see* Regnell, A.F.
Wiesner, J. von *see* Molisch, H.
Wiinstedt, K.J.F. 8654
Wikström, J.E. 4854
Wild, T. 6683, 6700, 6702
Wilhelm II, Emperor of Germany 8938
Wilhelm IV, King of Hannover 5935
Wilkens, C.F. 7222
Wilkes, C. 7912; *see* Pickering, C., Rich, W.
Willdenow, C.L. 7732, 8718, 8876, 8885, 9153, 9408, 9601, 9635; *see* Pallas, P.S., Panzer, G.W.F., Pavon y Jiménez, J.A., Poiteau, P.A., Regel, E.A. von, Richard, L.C.M., Rohr, J.P.B. von, Rostkovius, F.W.G., Roth, A.W., Roxburgh, W., Rudolphi, K.A., Ruiz Lopez, H.
Wille, F.N. 7423
Wille, N. 8726
Willem I, King of the Netherlands 7801
Willem III, King of England *see* Plukenet, L.
Willey, H. *see* Rothrock, J.T.
Williams, L.H.J. 4492
Williamson, W.C. 9022
Willich, C.L. 8836
Willkomm, H.M. 9584; *see* Russow, E.A.F.
Willkomm, W. 9362
Willmott, A.J. *see* Ramsbottom, J.
Willughby, F. 8697; *see* Ray, J.
Wilser, J.L. 8628
Wilson, A. *see* Pickard, J.F.
Wilson, E.H. 8816; *see* Oliver, D.
Wilson, M. *see* Pethybridge, G.H.
Winck, N.J. *see* Mackay, J.T.
Winkler, H. *see* Pax, F.A., Ruhland, W.O.E.

INDEX TO NAMES

Winkler, J. *see* Rosenstock, E.
Winter, H.G. 8450, 8821; *see* Rabenhorst, G.L.
Winter, H.G. 8450
Winterschmidt, A.W. 7011
Wit, H.C.D. de 9784
Witchell, C.A. *see* Reader, P.
Withering, W. *see* Robson, E.
Witte, H. 7148, 9750
Wittmack, M.C.L. 8211, 8775; *see* Retzdorff, A.E.W., Ross, H., Rümpler, T.
Wittrock, V.B. 8211; *see* Reinsch, P.F., Richter, P.G., Rosenvinge, L. Kolderup, Roy, J.
Wolf, F.-O. 9246
Wolf, L.J. 9864
Wolf, L.L. 6174
Wolff, C. von *see* Mohl, H. von
Wollaston, A.F.R. 9210
Woloszczak, E. *see* Rehmann, A.
Wood, W. 8755
Woodville, W. 8755
Worley, G. 6930
Woronin, M.S. 9597
Woronow, G.J.N. 7780
Wray, J. *v.* Ray, J.
Wright, C.H. 9210
Wright, P.E. *see* Müller, H.L.H.
Wynne-Edwards, E.M. *v.* Reid, E.M.

Xatart, B.J.P. 9679
Ximenez, F. 8608
Xuárez, G. 9769

Yates, E. 9505
Yocum, L.E. 7258

Yuncker, T.G. 8393; *see* Rechinger, K.H.
Yves, A.M.A. Saint *v.* Saint-Yves, A.M.A.
Yvesia see Saint-Yves, A.M.A.

Zabel, H. 8775
Zaddach, E.G. 8352
Zahlbruckner, A., 7423, 8450, 8726, 8744-8745, 9325; *see* Rechinger, K., Redinger, K.M.
Zahn, K.H. 8885; *see* Pöll, J.
Zakariasson, C. 10.007
Zander, R. 8775
Zangheri, P. 7435; *see* Pampanini, R.
Zannichelli, G.G. 9937, 9942
Zanon, V. 7265
Zarb, J.H. 7836
Zaunick, R. 9273
Zea, F.A. 9775, 9783
Zehner, J. 8091
Zeiller, C.R. 7624, 9023-9030, 9032, 9041, 9043; *see* Renault, B.
Zeller, W. 8775
Zenker, J.C. 6456, 8877
Zetterstedt, J.W. 9090
Zigno, A. de 9882
Zimmermann, F. 8100
Zimmeter, A. 8211; *see* Porcius, F.
Zintgraff, E. *see* Preuss, P.R.
Zittel, K.A. von 9454-9455; *see* Rothpletz, A.
Ziz, J.B. 7768
Zopf, W. *see* Rosendahl, F.
Zschacke, H. 8450
Zuccagni, A. 9406
Zuchold, E.A. 8353
Zwanziger, G.A. 7178

vol.	pages	plates	date
1	[i]-xii, [1]-68	*1-44* engr. t.p.	Aug 1795 p. xii: 14 Mar 1795
2	[i]-xvi, [1]-72	*45-92*	Jun 1797 p. vi: 14 Dec 1796
3	[i]-xiv, [1]-80	*93-138*	Aug 1799 p. iv: 8 Feb 1799
4	[i*], [i]-clxxx, [i], [1]-80, [38, ind.], [1, corr.]	*139-182*	Dec 1820

The above analysis is based on a copy with hand coloured plates in the Stevenson library. Copies vary in binding. Other *copies*: B, BR, NY. – Data based on documentation BH.
Vol. 1, engraved t.p., as above
Vol. 2, printed t.p. "... mit 48 ..."
Vol. 3, printed t.p. "... mit 46 ..."
Vol. 4, has three t.p.'s:
   a: *printed*. "Jacob ... Willdenow. IV Theil, Anhänge und Nachträge. Mit 44 illuminierten Kupfern. Fortgesetzt und mit einer Einleitung und einer erklärenden Übersicht sämmtlicher Tafeln versehen von Dr. Ch. G. Nees von Esenbeck und Dr. Th. Fr. Ludw. Nees von Esenbeck. Berlin Bey G. Reimer 1820."
   b: *printed*, new t.p. for entire work, issued 1820, bound in various places: "Beschreibung der um Halifax wachsende Pilze, enthaltend 241 Pilzarten in 900 Figuren auf 182 Kupfertafeln, alle von dem Verfasser ... gegründet von James Bolton ... Aus dem Englischen mit Anmerkungen von Carl Ludwig Willdenow. Fortgesetzt ... von ... Esenbeck. Berlin ... 1820."
   c: *engraved*, "Historia fungorum circa Halifax sponte nascentium tomi iv, A. Agaricus ... L. Mucor."
Ref.: BM 1: 192; LS 3294; NI 196; PR 962.
Barnhart, NAF 9(6): 430. 1916.

Bolus, Harry (1834-1911), British born South African banker and botanist. (*H. Bolus*).

HERBARIUM and TYPES: BOL. Other coll. e.g. BUL, GRA, NH, NU, PRE, SAM, STE and outside S. Africa. Exsiccatae: *Herbarium austro-africanum*, with Mac Owan, q v.
Ref.: IH 2: 84.
Tölken, Index herb. austro-afr. 58. 1971.

BIBLIOGRAPHY and BIOGRAPHY: AG 5(1): 394; Barnhart 1: 214; BB p. 37; BL 1: 28, 54; BM 1: 192-193, 6: 106; Bossert p. 44; CSP 9: 288, 12: 174, 13: 665; Kew 1: 291.
Pearson, S. Afr. J. Science 8: 59-79. 1911 (portr., bibl., itineraries)
Anon., Bull. misc. Inf. Kew 1911: 275-277, 319-322.
Pearson, Rep. S. Afr. Ass. Adv. Sci. 8: 59-79. 1912 (bibl., portr.)
Stapf, Proc. Linn. Soc. London 1911/12: 42-44.
Harloth, Fl. S. Afr. 1: x. 1913 (portr.)
Verduyn de Boer, Botanists at the Cape. 1929 (portr.)
Nelmes and Cuthbertson, Bot. Mag. Dedic. 1827-1927: 270-272. 1932 (portr.)
White, Stapeliae 1: 105. 1937.
Hutchinson, A botanist in Southern Africa 645. 1946.
Lütjeharms, The life and work of Harry Bolus, Cape Town. 1965 (mimeogr.)
Tyrrell-Glynn and Levyns, Flora africana 7-8. 1968.
Jessop, S. Afr. biogr. woordenb. 1: 92-95. 1968.

EPONYMY: *Bolusafra* O. Kuntze (1891); *Bolusanthus* Harms (1906); *Bolusia* Bentham (1873); *Bolusiella* Schlechter (1918); *Neobolusia* Schlechter (1895).
Note: *Bolusanthemum* Schwantes (1928) is dedicated to his daughter-in-law Harriet Margaret Louisa Bolus née Kensit (q.v.)

625. *Icones orchidearum austro-africanarum extra-tropicarum*; or, figures with descriptions, of